HANDBOOK OF
AIR CONDITIONING
HEATING AND
VENTILATING

HANDBOOK OF AIR CONDITIONING HEATING AND VENTILATING

EUGENE STAMPER, Editor

Professor of MECHANICAL ENGINEERING AND ASSISTANT DEAN,
New Jersey Institute of Technology

RICHARD L. KORAL, Consulting Editor

Editor, BUILDING SYSTEMS DESIGN

THIRD EDITION

Industrial Press Inc., 200 Madison Ave., New York 10016

Library of Congress Catalog Card Number 78-71559

ISBN 0-8311-1124-0

Printed in The United States of America

THIRD EDITION
Sixth Printing

GENERAL CONTENTS

A more detailed list of contents will be found at the begin-ning of each section; cross index is in the back of the book.

CONTENTS (Concluded)

PREFACE

The purpose of this Handbook is to provide necessary information to engineers, architects, contractors, plant engineers and others who need data on air conditioning, heating, refrigeration, and ventilation systems in the most convenient, time-saving manner possible. With this objective in mind, emphasis is placed wherever possible on data in the form of tables, formulas, graphs, maps, and short, well illustrated text that are immediately useful in solving problems of design, installation, and operation.

In this Third Edition, much of the second edition has been re-written and material on plumbing and drainage, available elsewhere, eliminated to make room for new topics without increasing the physical size of the volume. It remains, in the literal sense, a handbook. Also eliminated are mathematical tables which the electronic hand calculator has made obsolete.

In preparing this edition, the editors were fortunate in being able to call upon respected engineers in our industry for guidance in updating each section. The names of these consulting editors appear on the frontispieces of the sections for which they were responsible. The section on heating was extensively revised, as was the subject of noise in air conditioning systems. Climatic data are updated; there is much new material on modern air conditioning systems and changes in almost every section too numerous to enumerate here.

Many of the pages appearing here were extracted from the bimonthly periodical, *Building Systems Design* and there remain many valuable pages of data from its predecessor, *Air Conditioning, Heating and Ventilating*. In addition, much of the material herein was written by specialists whose contributions are credited on the frontispieces of each section of this Handbook. Inclusion of their material is an indication of the high regard the editors hold for their work.

An important additional source of material used in this and prior editions is various copyrighted material from professional societies and technical committees of industry associations who were most gracious in granting permission to reproduce tables, charts, illustrations, and text. Governmental agencies, too, have been an important source of data. Each of these are credited in the pages on which they appear.

In planning and gathering this material, coordinating it, and putting it into final form, every effort has been made to prevent errors but, in a work of this scope, they can creep in. The editors and publisher will be grateful to readers who call any such errors to their attention.

The editors gratefully acknowledge their debt to the editor of the first edition of this handbook, the former, long-time editor of *Air Conditioning, Heating, and Ventilating*, Clifford Strock. He conceived the idea of satisfying the need for a practical handbook of the arts and sciences encompassed herein, and whose genius and style are indelibly impressed on this work, even in its third revision.

Eugene Stamper and Richard L. Koral

Section 1

CLIMATIC DATA

Contents

(Contents concluded on next page)

Contents *(Concluded)*

Authorities

Section Editors: *Eugene Stamper* and *Richard L. Koral*

Donal W. Boyd, Canadian Climatic Data
Sigmund Fritz, Daily Total Solar Radiation
Irving F. Hand, Solar Energy Received on Horizontal Surfaces, Hourly Insolation on Cloudless Days

C.O. Mackey and *L.T. Wright, Jr.*, Heat Gains through Walls and Roofs
Clifford Strock, Climatic Maps, The Degree-Day

AIR CONDITIONING AND HEATING DATA

The following data for heating design and air conditioning design and criteria have been selected from tables in the Air Force, Army, and Navy Manual, "Engineering Weather Data."

The data listed are:

Heating Design Data: Dry-bulb temperatures (F) that are equalled or exceeded 99 and 97½ percent of the time, on the average, during the coldest 3 consecutive months. ·For United States sites (including Alaska and Hawaii), data for the months of December, January, and February are used. For sites outside the United States, the coldest 3 consecutive months are determined from the monthly mean dry-bulb temperature.

Air Conditioning Design Data: Dry-bulb and wet-bulb temperature (°F) that are equalled or exceeded 1, 2½, 5, and 10 percent of the time, on the average, during the warmest 4 consecutive months. For United States sites (including Alaska and Hawaii), data for the months of June through September are used. For sites outside the United States, the warmest 4 consecutive months are determined from the monthly mean wet-bulb temperature.

Air Conditioning Criteria Data: The number of hours, on the average, that the dry-bulb temperatures of 93°F and 80°F and the wet bulb temperatures of 73°F and 67°F are equalled or

exceeded during the warmest 6 consecutive months. For United States sites (including Alaska and Hawaii), data for. the months of May through October are used. For sites outside the United States, the warmest 6 consecutive months are determined from the monthly mean wet-bulb temperature.

Beginning below, the first table covers states of the United States. Following, on page 1-12, is a table which includes territories of the United States as well as foreign countries.

Listings are given in alphabetical order of geographical area; city or military station. The location of military sites must be obtained from the appropriate service. The data given were compiled by the USAF ETAC, at the direction of the U. S. Department of Defense.

Defense contractors may obtain data (other than specific locations and elevations of military stations) by written request to the USAF ETAC (MAC), Building 159, Navy Yard Annex, Washington, D.C. 20333. Requests for nonmilitary site data should be forwarded to the National Oceanic and Atmospheric Administration (NOAA), Washington Science Center, Rockville, Md. 20852.

Private interests should consult private meteorologists. A reference list of names and addresses may be obtained from the American Meteorological Society, 45 Beacon Street, Boston, Mass. 02108.

AIR CONDITIONING AND HEATING DESIGN CLIMATIC DATA

State and Station	Location N Lat	Location W Long	Location Elev.	Heating Design Data Percent 99	Heating Design Data Percent 97½	Air Conditioning Criteria Data, Percent 1	2½	5	10	1	2½	5	10	Air Conditioning Design Data Dry Bulb, F 93	80	Wet Bulb, F 73	67
	Deg, Min.		Ft.	Dry Bulb, F		Dry Bulb, F				Wet Bulb, F				Hours			
ALABAMA:																	
Anniston	33 35	85 51	602	18	22	96	94	93	90	79	78	77	75	119	1372	839	2699
Bates Field, Mobile	30 41	88 14	217	26	30	95	93	91	88	78	77	77	76	80	1269	1500	3555
Birmingham	33 34	86 45	630	18	22	97	94	93	90	79	78	77	76	119	1372	839	2699
Fairhope	30 33	87 45	25	28	31	95	93	91	88	81	80	79	79	73	1818	2505	3598
Gadsden	34 01	86 00	554	18	22	97	94	93	90	80	79	78	77	119	1372	839	2699
Huntsville	34 43	86 35	632	19	22	97	94	92	89	79	78	78	77	105	1240	1124	2526
Maxwell AFB				24	28	98	96	94	92	80	79	78	77	256	1743	1755	3207
McClellan, Fort				18	22	97	94	93	90	79	78	77	76	119	1372	839	2699
Mobile	30 42	88 02	71	28	31	96	94	92	89	81	80	79	79	106	1844	2505	3598
Montgomery	32 18	86 24	201	23	26	98	96	94	92	80	79	78	77	184	1611	1286	3111
Selma	32 25	87 00	79	23	27	98	96	94	91	81	80	79	77	225	1689	1835	3218
Thomasville	31 55	87 45	385	27	30	98	96	93	89	80	79	78	77	137	1591	1561	3213
Tuscaloosa	33 14	87 37	170	22	26	98	96	94	90	80	79	78	76	180	1533	1171	2872

AIR CONDITIONING AND HEATING DESIGN CLIMATIC DATA

State and Station	N Lat (Deg)	Lat (Min)	W Long (Deg)	Long (Min)	Elev. (Ft.)	Heating 99 (Dry Bulb, F)	Heating 97½ (Dry Bulb, F)	AC DB 1	AC DB 2½	AC DB 5	AC DB 10	AC WB 1	AC WB 2½	AC WB 5	AC WB 10	Hours 93 DB	Hours 80 DB	Hours 73 WB	Hours 67 WB
ALASKA:																			
Anchorage	61	10	149	59	105	-25	-20	74	71	68	64	63	61	59	57	0	6	1	6
Attu	52	48	173	10E	92	20	23	54	53	52	50	52	51	50	49	0	0	0	0
Barrow	71	18	156	47	31	-45	-42	58	54	50	46	54	51	48	44	0	0	0	0
Fairbanks	64	49	147	52	440	-53	-50	82	78	75	71	64	63	61	59	1	53	1	9
Juneau	58	22	134	35	20	-7	-4	75	71	68	63	66	64	62	58	0	6	0	12
Whittier	60	47	148	41	31	0	4	70	67	64	61	60	59	57	56	0	0	0	0
ARIZONA:																			
Flagstaff	35	08	111	40	6993	0	5	84	82	80	78	61	60	59	58	0	176	0	0
Kingman AFS						16	20	103	100	97	94	70	70	69	67	324	1662	1	343
Phoenix	33	27	112	04	1083	31	34	109	107	105	102	80	79	78	77	1170	2771	1025	2277
Tucson	32	07	110	56	2584	30	33	105	102	101	98	76	74	73	71	795	2445	173	1422
Yuma	32	40	114	36	206	37	40	111	109	107	104	79	78	77	76	1422	3148	947	1874
ARKANSAS:																			
Camden	33	35	92	51	155	19	26	100	97	95	92	80	79	79	78	239	1618	1824	3128
Fayetteville	36	00	94	10	1259	13	17	97	95	92	88	78	77	76	75	108	1207	716	1925
Hot Springs	34	31	93	03	630	18	23	100	96	94	91	79	78	77	76	191	1559	1190	2494
Little Rock	34	44	92	14	265	15	20	99	96	94	90	80	79	78	77	163	2014	1324	2740
Pine Bluff Arsenal						18	23	100	97	94	91	81	80	80	78	176	1641	1613	2802
Texarkana	33	27	94	00	368	19	26	99	97	95	92	80	79	79	78	259	1730	1824	3128
CALIFORNIA:																			
Anaheim	33	51	117	55	100	31	34	93	90	88	84	72	71	70	68	32	728	32	700
Bakersfield	35	25	119	03	494	27	30	107	104	100	96	72	71	70	68	501	1648	19	520
Barstow	34	54	117	02	2105	19	22	107	104	101	97	73	72	71	69	545	1763	35	502
Berkeley	37	52	122	17	200	37	39	79	77	75	72	66	64	63	61	0	13	0	0
Burbank	34	12	118	22	725	31	34	96	93	90	86	72	70	69	68	96	775	6	346
Chula Vista NAAS						37	40	88	81	77	72	72	71	70	68	13	207	18	526
Compton	33	53	118	13	65	32	35	84	81	79	76	72	70	69	67	3	152	19	378
Coronado	32	38	117	08	10	33	35	78	75	73	71	71	69	68	67	1	24	2	317
Edwards AFB						21	24	104	102	100	96	71	70	68	66	533	1556	15	284
Fresno	36	43	119	49	282	27	30	102	100	97	93	73	72	71	69	370	1364	36	539
Inglewood	33	54	118	21	99	41	43	84	81	78	75	71	69	68	67	9	125	5	400
Livermore	37	41	121	46	478	26	29	101	97	93	89	69	67	66	64	168	873	1	90
Long Beach	33	49	118	09	43	36	38	87	84	81	78	72	70	69	67	10	250	19	378
Los Angeles Intl Aprt	33	56	118	23	122	41	43	84	81	78	75	71	69	68	67	9	125	5	400
Los Angeles Post Office	34	03	118	14	312	35	38	92	88	85	82	72	70	69	68	23	576	6	346
Merced	37	20	120	31	152	30	32	102	99	96	93	73	72	70	69	330	1299	45	611
Moffett Field NAS						34	36	85	80	76	73	68	67	65	64	3	107	1	84
Monterey, Presidio of	36	36	121	54	100	34	37	74	71	68	66	64	63	61	60	0	14	0	8
Oakland	37	44	122	12	18	35	38	85	79	74	70	66	64	63	61	2	89	0	10
Palo Alto	37	27	122	10	54	34	36	85	80	76	73	68	67	65	64	3	107	1	84
Pasadena	34	09	118	09	864	31	34	97	93	90	86	72	70	69	68	81	924	6	346
Pomona	34	03	117	45	934	32	34	102	99	97	94	75	74	72	70	348	1213	91	684
Richmond	37	55	122	21	20	35	38	85	79	74	70	66	64	63	61	2	89	0	10
Riverside	33	58	117	22	900	34	36	101	98	96	93	73	71	70	68	326	1209	35	518
Sacramento	38	31	121	30	23	29	31	100	96	93	88	72	70	69	67	188	1006	13	351
San Bernardino	34	07	117	19	1125	31	33	101	98	96	93	75	73	71	69	318	1188	84	692
San Diego	32	44	117	10	37	43	45	84	81	78	76	71	69	68	67	4	130	2	317
San Francisco	37	37	122	23	90	36	38	81	77	74	70	65	63	62	60	2	50	0	12
San Jose	37	20	121	53	95	36	38	87	83	79	76	68	67	65	64	6	157	1	84
San Luis Obispo	35	18	120	40	300	32	34	89	85	82	79	65	64	63	61	11	329	0	7
San Rafael	37	58	122	33	31	33	35	89	85	81	77	73	70	68	66	14	228	38	288
Santa Barbara	34	26	119	50	13	33	35	79	77	74	72	67	66	65	64	2	44	0	38
San Ysidro	32	33	117	03	350	35	42	86	81	77	74	72	70	69	67	8	168	10	421
Stockton	37	54	121	15	27	31	33	101	98	95	91	71	70	69	67	244	1017	9	330
Sunnyvale	37	23	122	02	30	34	36	85	80	76	73	68	67	65	64	3	107	1	84
Torrance	33	50	118	19	80	32	35	84	81	79	76	72	70	69	67	3	152	19	378
Van Nuys	34	13	118	30	794	31	34	96	93	90	86	72	70	69	68	96	775	6	346

AIR CONDITIONING AND HEATING DESIGN CLIMATIC DATA

State and Station	N Lat	W Long	Elev.	Heating Design Data Percent 99	97½	Air Conditioning Criteria Data, Percent — Dry Bulb, F 1	2½	5	10	Wet Bulb, F 1	2½	5	10	AC Design Dry Bulb, F 93	80	Wet Bulb, F 73	67
	Deg, Min.		Ft.	Dry Bulb, F		Dry Bulb, F				Wet Bulb, F				Hours			
COLORADO:																	
Colorado Springs, Peterson Field	38 49	104 42	6170	-1	4	90	88	86	83	63	62	61	59	9	508	0	0
Denver	39 46	104 53	5332	-3	2	92	90	88	85	65	64	63	61	26	647	0	3
Pueblo	38 17	104 31	4639	-1	4	98	96	93	89	68	67	66	64	147	909	0	50
Trinidad	37 16	104 20	5743	1	5	93	91	89	86	66	65	64	63	41	743	1	18
CONNECTICUT:																	
Bridgeport	41 10	73 08	7	8	12	88	86	83	80	77	76	75	73	6	302	301	1150
Groton	41 20	72 03	14	8	11	89	86	83	80	77	75	74	72	7	292	227	1050
Hartford	41 44	72 39	20	3	6	91	89	86	83	77	76	74	73	18	476	253	1106
New Haven	41 16	72 53	6	8	11	87	84	81	78	77	76	75	73	3	193	301	1150
New London NAVSTA				8	11	89	86	83	80	77	75	74	72	7	292	227	1050
Stamford	41 03	73 32	10	8	12	88	86	83	80	76	75	73	71	8	345	196	894
Waterbury	41 33	73 02	843	1	4	89	86	84	80	76	74	73	71	4	344	163	889
DELAWARE:																	
Bethany Beach	38 32	75 03	20	17	20	89	87	84	82	79	78	77	75	7	461	680	1760
Dover AFB				13	15	93	90	88	84	79	78	77	75	31	683	680	1760
Lewes	38 46	75 09	10	16	19	90	87	85	82	79	78	77	75	8	456	634	1743
Wilmington	39 40	75 36	78	12	16	93	90	88	84	79	77	76	74	29	623	351	1191
DISTRICT OF COLUMBIA:																	
Andrews AFB				11	15	93	91	88	85	78	77	76	74	34	761	543	1682
Washington National Aprt	38 50	77 02	14	14	17	95	93	90	87	78	77	76	75	72	1002	709	1881
FLORIDA:																	
Bartow	27 57	81 47	130	37	40	96	94	92	90	80	79	78	78	147	1730	2807	4018
Cocoa Beach	28 14	80 36	9	40	43	90	89	87	86	81	80	79	79	11	2113	3242	4174
Daytona Beach	29 11	81 03	61	36	39	93	92	90	88	81	80	79	79	37	1597	2800	4165
Fort Lauderdale NOL				41	43	91	90	89	88	81	80	79	79	1	2052	2315	3254
Jacksonville	30 25	81 39		29	32	95	93	92	90	81	80	79	78	198	1792	1835	3646
Key West NAVSTA				55	58	90	89	88	87	80	79	79	78	4	3131	3840	4381
Miami	25 47	80 17	9	44	47	91	90	89	88	79	79	78	78	12	2388	3293	4248
Orlando	28 33	81 20	106	35	39	93	92	90	88	79	78	78	77	51	1609	2613	3958
Pensacola NAS				29	32	92	90	89	87	81	81	80	79	14	1884	2594	3629
St. Augustine	29 54	81 19	10	36	40	94	92	90	87	81	80	79	78	55	1777	2800	4165
St. Petersburg	27 47	82 38	35	39	42	93	92	91	89	81	80	79	78	55	2154	3032	4082
Tallahassee	30 26	84 20	68	30	33	95	93	91	89	80	79	79	78	84	1538	2021	3567
Tampa	27 58	82 32	36	35	39	93	92	90	88	80	79	78	77	56	1786	2014	4013
Venice	27 05	82 26	21	40	44	92	91	89	87	80	80	79	78	16	1589	3039	4085
GEORGIA:																	
Albany	31 35	84 10	225	26	30	98	96	94	92	80	79	78	77	289	1747	1847	3323
Athens	33 57	83 19	798	21	25	97	94	92	89	78	77	76	75	127	1398	724	2446
Atlanta	33 39	84 25	976	19	24	95	93	91	88	78	77	76	75	73	1004	740	2397
Augusta	33 22	81 58	182	20	23	97	95	93	90	80	79	78	77	174	1431	1445	2912
Columbus	32 31	84 56	385	23	26	98	96	94	91	80	79	78	77	218	1511	1518	3047
Macon	32 42	83 39	356	24	28	98	96	94	92	80	79	78	77	206	1549	1520	3069
Marietta NAS				17	21	95	93	91	88	78	77	76	75	82	1154	740	2397
Savannah	32 04	81 05	48	24	27	95	93	91	88	82	81	80	79	91	1442	1931	3405
Valdosta	30 50	83 17	265	28	31	96	94	92	90	80	79	78	77	139	1539	1898	3464
HAWAII:																	
Hickam AFB				62	64	86	85	84	83	75	74	73	72	0	1342	312	3999
Kahoku	21 43	157 59	11	57	59	87	85	84	83	75	75	74	74	0	1113	571	4197
Pearl Harbor				62	64	86	85	84	83	75	74	73	72	0	1342	312	3999
Wheeler AFB				57	58	84	83	82	80	74	73	72	72	0	576	162	3322

AIR CONDITIONING AND HEATING DESIGN CLIMATIC DATA

State and Station	Location N Lat	W Long	Elev.	Heating Design Data Percent 99	97½	Air Conditioning Criteria Data, Percent Dry Bulb 1	2½	5	10	Wet Bulb 1	2½	5	10	Air Conditioning Design Data Dry Bulb, F 93	Wet Bulb, F 80	73	67
	Deg, Min.		Ft.	Dry Bulb, F		Dry Bulb, F				Wet Bulb, F				Hours			
IDAHO:																	
Arco	43 38	113 19	5320	-25	-20	97	95	92	87	62	61	59	57	115	606	0	0
Boise	43 34	116 13	2857	5	11	97	94	91	87	68	66	64	62	96	679	2	62
Idaho Falls	43 31	112 04	4744	-10	-6	90	87	84	81	65	63	62	60	6	356	0	7
Lewiston	46 23	117 02	1419	4	12	98	96	93	89	67	66	65	63	153	878	1	56
Pocatello	42 55	112 36	4449	-8	-2	94	91	88	85	65	63	62	60	56	601	0	3
Twin Falls	42 34	114 28	3770	-5	0	100	97	94	88	66	65	63	61	171	841	0	26
ILLINOIS:																	
Arlington Heights	42 02	87 58	680	-6	-1	92	89	86	83	78	76	74	72	23	525	290	1029
Champaign	40 08	88 16	743	-1	3	94	92	89	86	78	77	76	74	55	788	445	1392
Chicago	41 47	87 45	614	-4	1	94	92	89	85	78	76	75	73	59	727	298	1109
Cicero	41 51	87 46	608	-2	3	93	90	87	84	78	76	75	73	30	622	312	1339
Decatur	39 51	88 58	670	0	3	94	92	90	87	79	77	76	75	61	935	582	1645
Evanston	42 00	87 44	600	-5	0	92	89	86	82	78	76	74	72	28	424	290	1029
Joliet	41 32	89 05	590	-3	2	94	91	88	85	79	77	75	73	49	676	324	1150
Kankakee	41 07	87 52	631	-4	1	94	91	88	85	78	76	75	73	39	659	298	1109
Moline	41 27	90 31	594	-7	-3	93	91	88	85	78	77	75	73	43	703	391	1259
O'Hare International Aprt	41 59	87 54	667	-6	-1	92	89	86	83	78	76	74	72	23	525	290	1029
Peoria	40 40	89 41	662	-3	2	94	91	88	85	78	77	76	74	44	612	410	1227
Quincy	39 56	91 11	762	-2	4	95	92	90	87	80	79	77	75	66	794	596	1584
Rock Island Arsenal				-6	-2	94	91	88	84	79	77	76	74	37	633	388	1236
Springfield	39 50	89 40	602	-1	3	94	92	90	87	79	77	76	75	61	935	582	1645
INDIANA:																	
Bloomington	39 10	86 32	796	3	7	95	92	90	86	79	78	76	75	63	848	564	1656
Columbus	39 13	85 55	632	3	7	95	92	90	86	79	78	76	75	63	848	564	1656
Evansville	38 03	87 32	400	6	11	96	94	91	88	81	79	78	76	97	1090	619	1753
Fort Wayne	41 00	85 12	828	-2	3	93	90	87	84	77	76	75	73	29	581	264	1072
Gary	41 34	87 21	600	-2	3	92	89	86	83	78	76	74	72	23	525	290	1029
Hammond	41 37	87 33	590	-2	3	92	89	86	83	78	76	74	72	23	525	290	1029
Indianapolis	39 44	86 16	793	2	6	94	92	89	86	78	77	76	74	45	763	417	1462
Lafayette	40 25	86 56	637	-1	3	94	92	89	86	78	77	76	74	55	788	445	1392
Michigan City	41 42	86 50	650	-2	3	92	89	87	83	77	76	74	72	19	563	288	1116
Muncie	40 12	85 23	950	-2	4	94	91	88	85	78	77	76	74	42	655	417	1462
South Bend	41 42	86 19	773	-2	3	92	89	87	83	77	76	74	72	19	563	288	1116
Terre Haute	39 27	87 18	581	3	7	94	91	88	85	79	78	77	75	50	834	501	1596
IOWA:																	
Cedar Rapids	41 53	91 42	863	-7	-2	93	90	87	83	80	78	76	74	32	540	403	1387
Des Moines	41 32	93 39	963	-8	-4	95	92	89	85	78	77	76	74	56	738	429	1306
Dubuque	42 24	90 42	1080	-7	-4	89	86	84	81	76	74	73	71	8	381	168	694
Iowa City	41 38	91 33	653	-6	-1	94	91	88	84	80	78	76	74	43	622	403	1387
Sioux City	42 24	96 23	1113	-9	-5	96	93	90	86	78	77	75	73	79	796	362	1171
Waterloo	42 33	92 24	878	-13	-11	92	90	87	84	79	77	76	74	8	425	350	1157
KANSAS:																	
Dodge City	37 46	99 58	2592	3	7	99	97	95	91	74	73	72	70	235	1135	85	1098
Kansas City	39 08	94 38	800	4	8	99	96	94	91	78	77	76	75	200	1092	669	1852
Riley, Fort				-1	4	101	98	95	91	80	78	77	75	217	1235	765	1921
Salina	38 49	97 34	1271	3	7	104	101	98	94	80	78	76	75	346	1390	665	1888
Topeka	39 03	95 41	885	3	6	99	96	93	90	78	77	76	75	176	1154	656	1820
Wichita	37 40	97 20	1392	3	8	101	98	95	91	77	76	75	74	248	1290	502	1949
KENTUCKY:																	
Ashland	38 28	82 38	550	14	18	95	93	90	87	78	77	76	75	67	979	526	1891
Covington	39 04	84 40	888	6	10	94	92	89	86	77	76	75	73	45	764	342	1435
Lexington	38 02	84 36	989	6	10	95	92	90	87	78	77	76	74	62	954	448	1687
Louisville	38 11	85 44	488	6	11	95	93	91	87	78	77	76	75	86	1041	700	1938
Owensboro	37 46	87 09	420	8	12	98	96	93	90	79	78	77	76	171	1195	772	1959

AIR CONDITIONING AND HEATING DESIGN CLIMATIC DATA

State and Station	N Lat (Deg, Min.)	W Long (Deg, Min.)	Elev. (Ft.)	Heating 99 (Dry Bulb, F)	Heating 97½ (Dry Bulb, F)	AC Dry Bulb, F 1	AC Dry Bulb, F 2½	AC Dry Bulb, F 5	AC Dry Bulb, F 10	AC Wet Bulb, F 1	AC Wet Bulb, F 2½	AC Wet Bulb, F 5	AC Wet Bulb, F 10	Design DB 93 (Hours)	Design DB 80 (Hours)	Design WB 73 (Hours)	Design WB 67 (Hours)
LOUISIANA:																	
Baton Rouge	30 32	91 09	67	27	30	95	94	93	90	81	80	79	78	133	1634	2032	3478
Lake Charles	30 13	93 09	32	29	33	95	93	92	90	80	79	79	78	103	1807	2512	3602
New Orleans	29 59	90 15	20	31	35	94	92	91	89	81	80	79	78	66	1733	2609	3665
Shreveport	32 28	93 49	251	21	27	99	97	95	93	80	79	78	78	311	1846	2008	3198
MAINE:																	
Augusta	44 19	69 48	350	-10	-5	89	86	83	80	74	72	70	68	9	298	63	448
Bangor	44 48	68 46	61	-8	-4	88	85	81	78	75	73	71	68	5	220	87	448
Bar Harbor	44 27	68 21	67	-7	-2	85	82	79	76	73	71	69	67	0	137	30	264
Lewiston	44 02	70 15	199	-10	-5	89	86	83	80	74	72	70	68	9	298	63	448
Portland	43 39	70 19	61	-6	0	89	86	82	78	75	73	71	69	9	236	78	494
Winter Harbor	44 24	68 01	11	-7	-2	83	80	77	74	72	70	68	66	0	103	23	231
MARYLAND:																	
Aberdeen Proving Ground				12	14	89	87	85	82	79	77	76	75	1	500	559	1482
Annapolis USNA				15	19	91	88	86	83	79	77	76	76	12	651	626	1946
Baltimore, Friendship Aprt	39 11	76 40	197	12	15	94	91	89	86	79	78	77	75	53	833	573	1617
Bethesda NATNAVMEDCEN				14	17	95	92	90	87	79	78	77	75	60	913	610	1739
Cumberland	39 39	78 45	945	8	11	92	90	87	84	78	76	75	73	25	627	390	1360
Hagerstown	39 39	77 45	660	0	6	93	90	87	83	78	77	76	74	26	560	399	1446
Ocean City	38 20	75 05	11	17	20	89	87	84	82	79	78	77	75	7	461	680	1760
MASSACHUSETTS:																	
Boston Army Base	CASTLE			6	10	91	88	85	82	76	74	73	71	20	420	133	815
Cambridge	42 23	71 05	30	6	10	91	88	85	82	76	74	73	71	20	420	133	815
Fall River	41 43	71 08	190	7	11	85	82	80	77	75	73	72	70	1	164	121	877
Lawrence	42 43	71 07	155	-5	-1	92	89	86	83	76	74	73	70	29	506	157	785
Lynn	42 28	70 55	50	6	10	91	88	85	82	76	74	73	71	20	420	133	815
Nantucket	41 15	70 04	12	13	16	78	76	74	72	73	71	70	69	0	15	20	603
New Bedford	41 39	70 55	90	7	11	85	82	80	77	75	73	72	70	1	164	121	877
North Truro AFS				8	12	85	82	80	77	75	73	72	70	1	164	121	877
Pittsfield	42 26	73 13	1169	-2	1	86	83	81	77	73	72	70	67	2	199	47	352
Quincy	42 14	71 00	20	8	12	90	86	83	79	75	74	72	70	16	294	118	525
Salem	42 32	70 52	21	0	4	91	88	85	82	76	75	73	71	15	408	162	764
Springfield	42 07	72 35	190	-4	1	90	88	85	81	76	74	73	71	11	426	165	825
Watertown Arsenal				6	10	91	88	85	82	76	74	73	71	26	420	133	815
Worcester	42 16	71 52	986	-4	2	89	86	83	80	75	73	72	70	7	297	95	675
MICHIGAN:																	
Ann Arbor	42 16	83 44	926	-4	0	90	87	85	82	76	74	73	71	8	418	120	762
Battle Creek	42 18	85 14	939	1	5	92	89	86	83	76	74	73	71	25	511	169	833
Bay City	43 36	83 52	593	-3	2	91	88	85	82	77	75	73	71	11	422	159	686
Benton Harbor	42 08	86 26	635	-1	3	90	87	84	81	76	74	73	71	7	359	122	693
Dearborn	42 18	83 12	650	4	8	92	89	86	82	76	75	74	72	21	495	185	908
Detroit	42 24	83 00	626	4	8	92	89	86	82	76	75	74	72	21	495	185	908
Flint	42 58	83 44	766	-4	-1	91	88	86	83	77	75	74	72	15	509	191	787
Grand Rapids	42 54	85 40	681	3	7	91	88	85	82	76	75	73	71	12	420	160	774
Houghton	47 10	88 30	1079	-8	-4	84	80	77	73	72	70	68	66	0	97	28	237
Kalamazoo	42 17	85 36	955	1	5	92	89	86	83	76	74	73	71	25	511	169	833
Lansing	42 47	84 36	874	-7	-1	90	86	84	80	77	75	73	71	7	323	163	749
Marquette	46 34	87 24	734	-9	-5	84	81	78	74	72	70	68	66	1	114	27	228
Muskegon	43 10	86 14	627	0	4	88	85	82	79	76	74	73	71	4	247	124	712
Pontiac	42 38	83 16	935	-4	0	90	88	85	82	76	75	73	71	9	414	146	765
Saginaw	43 26	83 52	601	1	3	91	88	85	81	77	75	73	71	15	392	160	718
Willow Run	42 14	83 32	777	1	5	91	88	85	82	76	75	73	71	12	410	146	825
Ypsilanti	42 14	83 39	715	1	5	91	88	85	82	76	75	73	71	12	410	146	825

AIR CONDITIONING AND HEATING DESIGN CLIMATIC DATA

State and Station	Location			Heating Design Data Percent		Air Conditioning Criteria Data, Percent								Air Conditioning Design Data			
	N Lat	W Long	Elev.	99	97½	Dry Bulb, F				Wet Bulb, F				Dry Bulb, F		Wet Bulb, F	
						1	2½	5	10	1	2½	5	10	93	80	73	67
	Deg, Min.		Ft.	Dry Bulb, F										Hours			
MINNESOTA:																	
Bemidji	47 30	94 55	1394	-36	-30	90	87	83	79	73	71	69	67	15	258	38	331
Duluth	46 50	92 11	1417	-19	-15	85	82	79	75	73	70	68	66	0	132	29	244
Hastings	44 46	92 50	695	-14	-9	92	89	86	82	77	75	74	71	19	496	195	794
International Falls	48 36	93 24	1126	-29	-24	86	82	79	75	72	69	68	65	3	152	18	199
Little Falls	45 58	94 23	1135	-26	-20	91	88	85	82	76	74	72	70	15	402	101	547
Minneapolis	44 53	93 15	838	-14	-9	92	89	86	82	77	75	74	71	19	496	195	794
Minneapolis-St. Paul International Airport	44 53	93 12	859	-14	-9	92	89	86	82	77	75	74	71	19	496	195	794
Rochester	44 00	92 29	1021	-19	-13	90	87	84	80	77	75	74	72	12	336	223	856
Worthington	43 37	95 36	1593	-12	-7	92	88	85	81	76	74	73	70	25	470	138	681
MISSISSIPPI:																	
Biloxi	30 24	88 54	18	30	32	93	92	90	89	82	81	80	79	48	2052	2599	3675
Greenville	33 23	91 03	132	21	24	98	96	94	92	81	80	79	78	247	1686	1734	2962
Gulfport	30 22	89 03	42	30	32	93	92	90	89	82	81	80	79	48	2052	2599	3675
Jackson	32 20	90 13	332	23	26	98	96	94	91	79	78	78	77	212	1611	1592	3050
Natchez	31 32	91 22	168	27	29	98	96	93	90	80	80	79	78	155	1674	2117	3302
Vicksburg	32 21	90 53	295	23	26	98	96	94	91	79	78	78	77	212	1611	1592	3050
MISSOURI:																	
Columbia	38 58	92 22	785	4	8	97	94	92	89	79	77	76	75	116	1066	596	1730
Hannibal	39 43	91 22	712	-1	4	96	93	91	87	80	79	77	75	76	867	596	1584
Jefferson City	38 34	92 11	557	4	9	98	94	92	89	79	77	76	75	134	1080	638	1813
Joplin	37 10	94 30	985	5	11	98	95	92	89	78	77	76	75	141	1058	696	2019
Kansas City	39 07	94 35	750	4	8	99	96	94	91	78	77	76	75	200	1092	669	1852
St. Joseph	39 46	94 55	817	-2	2	96	93	91	88	79	78	77	75	94	1125	589	1636
St. Louis	38 45	90 23	564	5	10	98	95	93	89	79	78	77	75	149	1151	676	1866
Springfield	37 14	93 23	1270	5	11	97	94	91	88	78	76	75	74	103	964	550	1862
Wood, Fort Leonard				5	11	97	94	91	88	78	76	75	74	103	964	550	1862
MONTANA:																	
Billings	45 48	108 32	3583	-16	-11	94	91	88	83	68	66	65	63	54	515	0	42
Butte	45 58	112 30	5529	-24	-16	86	83	80	76	59	58	57	55	1	175	0	0
Great Falls	47 29	111 21	3664	-21	-17	91	87	84	79	64	63	61	59	15	286	0	6
Helena	46 36	112 00	3898	-23	-19	90	87	84	79	65	63	61	60	8	242	0	6
Missoula	46 55	114 05	3200	-7	-2	91	88	85	80	65	63	62	60	16	303	0	4
NEBRASKA:																	
Grand Island	40 58	98 19	1856	-6	-2	98	95	92	88	76	75	74	72	127	864	204	1069
Kearney	40 42	99 05	2146	-10	-5	97	94	90	87	77	75	74	72	84	896	209	1006
Lincoln	40 49	96 42	1189	-4	0	100	96	93	89	77	76	74	74	163	1000	454	1414
North Platte	41 08	100 42	2787	-6	-2	97	94	90	86	74	73	71	70	95	724	77	745
Omaha	41 18	95 54	982	-4	-1	97	94	91	87	79	78	76	74	106	901	496	1443
NEVADA:																	
Carson City	39 10	119 46	4675	6	10	92	90	88	85	62	61	60	58	23	574	0	0
Elko	40 50	115 47	5079	-13	-6	94	92	90	86	64	62	61	59	56	661	0	3
Ely	39 17	114 51	6262	-7	-2	90	88	86	83	60	59	58	56	4	519	0	1
Las Vegas	36 04	115 10	2180	23	25	108	106	104	101	72	71	70	68	943	2360	6	380
Reno	39 30	119 47	4400	7	11	93	91	89	86	63	61	60	58	38	647	0	0
NEW HAMPSHIRE:																	
Concord	43 12	71 30	354	-8	-2	91	88	85	82	75	74	72	70	6	398	93	644
Manchester	43 00	71 28	250	-5	1	92	89	86	83	76	74	73	70	29	506	157	785
New Castle	43 04	70 43	14	-2	3	88	85	83	79	75	73	71	69	4	293	100	610
Pease AFB				-2	3	88	85	83	79	75	73	71	69	4	293	100	610
Portsmouth NAVBASE				-2	3	88	85	83	79	75	73	71	69	4	293	100	610

AIR CONDITIONING AND HEATING DESIGN CLIMATIC DATA

State and Station	N Lat	W Long	Elev.	Heating Design Data Percent 99	97½	Air Conditioning Criteria Data, Percent Dry Bulb, F 1	2½	5	10	Wet Bulb, F 1	2½	5	10	Air Conditioning Design Data Dry Bulb, F 93	80	Wet Bulb, F 73	67
	Deg, Min.		Ft.	Dry Bulb, F		Dry Bulb, F				Wet Bulb, F				Hours			
NEW JERSEY:																	
Atlantic City	39 27	74 35	67	11	15	91	88	85	82	78	77	76	75	17	473	576	1558
Bayonne NSC				11	15	94	91	88	84	77	76	75	73	40	592	344	1290
Camden	39 55	75 04	20	7	10	94	91	88	85	79	77	76	75	38	679	498	1464
Clifton	40 52	74 10	175	10	14	93	90	87	83	77	76	75	73	32	533	344	1290
Dix, Fort				10	14	93	90	87	84	78	77	76	74	32	616	438	1400
Dover	40 55	74 35	570	2	6	91	89	86	83	78	76	75	73	15	626	280	1045
Elizabeth	40 40	74 12	33	11	15	94	91	88	84	77	76	75	73	40	592	344	1290
Jersey City	40 43	74 04	135	11	15	94	91	88	84	77	76	75	73	40	592	344	1290
McGuire AFB				10	14	93	90	87	84	78	77	76	74	32	616	438	1400
Monmouth, Fort				8	12	93	90	88	84	78	77	76	74	34	599	416	1416
Newark	40 43	74 10	10	11	15	94	91	88	84	77	76	75	73	40	592	344	1290
Patterson	40 55	74 09	100	11	15	94	91	88	84	77	76	75	73	40	592	344	1290
Perth Amboy	40 31	74 17	20	11	15	92	89	86	83	78	77	76	74	25	494	367	1348
Picatinny Arsenal				2	6	91	89	86	83	78	76	75	73	15	626	280	1045
Trenton	40 16	74 49	197	3	9	91	88	85	82	78	77	76	74	15	448	438	1400
NEW MEXICO:																	
Alamogordo	32 54	105 58	4300	16	20	99	97	95	92	69	68	67	66	283	1519	1	269
Albuquerque	35 03	106 37	5314	14	17	96	94	92	89	66	65	64	63	120	1130	0	20
Carlsbad	32 21	104 15	3276	17	20	102	100	98	95	72	72	71	70	463	1779	7	988
Clovis	34 24	103 12	4280	14	17	97	95	93	90	70	69	68	67	171	1199	7	321
Holloman AFB				18	22	100	98	96	93	70	69	68	67	369	1718	2	365
Roswell	33 24	104 32	3612	16	19	101	99	97	94	71	70	69	68	416	1617	9	681
Sandia Base				14	17	96	94	92	89	66	65	64	63	120	1130	0	20
Santa Fe	35 36	106 05	6308	-4	-1	92	90	88	85	65	64	63	62	15	686	0	2
White Sands	32 17	106 45	3909	18	22	100	98	96	93	70	69	68	67	369	1718	2	365
NEW YORK:																	
Albany	42 45	73 48	277	-1	2	91	88	85	81	76	75	73	71	20	420	140	759
Binghamton	42 13	75 59	1638	-1	3	88	85	82	78	73	72	71	69	4	255	53	548
Brooklyn NAVYSHIPYD				11	15	94	91	88	84	77	76	75	73	40	592	344	1290
Buffalo	42 56	78 44	715	3	7	88	86	83	80	75	74	72	70	4	347	107	731
Corning	42 08	77 03	930	1	5	92	88	85	82	75	73	72	70	19	420	73	604
Drum, Camp				-17	-11	88	85	82	79	75	73	71	70	1	206	73	585
Elmira	42 10	76 54	954	1	5	92	88	85	82	75	73	72	70	19	420	73	604
Geneva	42 53	77 00	615	0	2	91	87	85	81	75	74	72	70	14	394	106	742
Hempstead	40 43	73 38	80	10	14	91	88	85	82	77	76	75	73	20	456	328	1271
Kennedy Airport	40 39	73 47	16	12	16	90	87	84	81	77	76	75	73	15	389	317	1279
Johnstown	43 01	74 23	688	-9	-4	90	87	85	81	75	73	72	70	9	383	92	639
Mitchell AFB				10	14	91	88	85	82	77	76	75	73	20	456	328	1271
Montauk AFS				9	13	87	84	81	78	76	75	74	72	4	225	246	1153
New Rochelle	40 50	73 47	70	12	15	93	90	87	84	77	76	75	73	33	648	323	1288
New York	40 42	74 01	10	10	14	93	90	87	84	77	76	75	73	33	648	323	1288
Niagara Falls	43 06	78 56	597	4	7	88	86	83	80	75	74	72	70	3	335	137	767
Olean	42 05	78 27	1420	-8	-3	91	88	85	81	72	71	70	68	12	376	29	466
Poughkeepsie	41 38	73 53	140	3	7	93	90	87	84	78	77	75	73	38	623	332	1247
Rochester	43 07	77 40	543	3	6	91	88	85	81	75	74	72	70	19	419	129	724
Rome	43 14	75 28	445	-8	-4	89	86	83	80	75	73	72	70	4	328	102	667
Syracuse	43 04	76 16	408	0	3	91	88	85	82	76	74	73	71	16	433	122	797
U. S. Military Academy (West Point)				2	6	92	89	86	83	78	76	74	72	24	522	267	1059
Yonkers	40 56	73 53	50	11	15	94	91	88	84	77	76	75	73	40	592	344	1290
NORTH CAROLINA:																	
Asheville	35 26	82 29	2096	13	17	91	88	86	83	75	74	73	72	14	610	212	1309
Cape Hatteras	35 16	75 33	13	24	28	87	86	85	83	81	80	80	79	2	934	1712	2720
Charlotte	35 14	80 56	769	18	22	95	93	91	88	77	76	76	75	94	1138	736	2366
Fayetteville	35 03	78 51	95	17	20	97	94	92	89	80	79	78	77	138	1260	1243	2614
Greensboro	36 05	79 57	891	16	19	94	91	89	86	77	76	75	74	50	916	525	1972
Raleigh-Durham Airport	35 52	78 47	444	19	23	96	93	91	88	79	78	77	76	99	1031	829	2305
Rocky Mount	35 58	77 48	81	19	21	97	93	90	88	79	78	77	76	67	1169	1214	2512
Wilmington	34 14	77 57	46	24	27	93	91	89	87	82	81	80	79	26	1246	1684	3031
Winston-Salem	36 07	80 12	967	14	18	92	90	88	85	77	76	75	74	20	806	414	1800

AIR CONDITIONING AND HEATING DESIGN CLIMATIC DATA

State and Station	Location N Lat	Location W Long	Location Elev.	Heating Design Data Percent 99	Heating Design Data Percent 97½	Air Conditioning Criteria Data, Percent Dry Bulb, F 1	2½	5	10	Wet Bulb, F 1	2½	5	10	Air Conditioning Design Data Dry Bulb, F 93	809	Wet Bulb, F 73	67
	Deg, Min.		Ft.	Dry Bulb, F		Dry Bulb, F				Wet Bulb, F				Hours			
NORTH DAKOTA:																	
Bismarck	46 46	100 45	1660	-25	-20	94	90	87	82	74	72	70	67	43	471	56	371
Fargo	46 54	96 48	899	-24	-20	92	88	85	81	76	74	72	70	26	409	100	516
Grand Forks AFB				-25	-22	92	88	85	81	74	72	70	68	21	372	56	408
Minot	48 15	101 17	1714	-24	-20	91	88	84	79	72	70	68	66	20	310	23	228
OHIO:																	
Akron-Canton Airport	40 55	81 26	1236	1	6	89	87	84	81	75	73	72	70	7	416	124	919
Canton	40 48	81 23	1054	1	6	89	87	84	81	75	73	72	70	7	416	124	919
Cincinnati	39 06	84 26	483	4	8	93	91	88	85	79	78	77	75	35	802	703	1786
Cleveland	41 24	81 51	805	2	7	92	89	86	82	76	75	74	72	21	523	199	987
Dayton	39 54	84 12	1003	1	7	93	90	88	84	77	75	74	73	29	679	255	1241
Columbus	40 00	82 53	833	-1	4	92	90	88	85	76	75	74	72	9	602	243	1234
Marietta	39 25	81 26	627	1	7	96	93	90	86	78	77	75	74	78	813	399	1491
Springfield	39 55	83 49	1020	4	8	93	90	88	85	77	76	75	73	35	793	360	1346
Toledo	41 36	83 48	692	0	6	93	90	87	83	77	75	74	72	36	590	242	1023
Wright-Patterson AFB				3	8	93	90	88	85	77	76	75	73	35	739	360	1346
Youngstown	41 16	80 40	1196	3	8	90	87	84	81	75	73	72	70	12	388	115	783
OKLAHOMA:																	
Bartlesville AFS				4	10	99	95	93	90	78	77	76	75	153	1364	842	1991
Enid	36 24	97 53	1250	10	14	103	100	98	94	78	77	76	75	390	1543	797	2207
Norman	35 14	97 25	1175	11	15	101	98	96	93	78	77	76	75	307	1579	805	2367
Oklahoma City	35 24	97 36	1311	10	14	100	97	95	92	78	77	76	75	240	1439	762	2300
Stillwater	36 08	97 05	910	7	12	99	96	93	90	80	79	78	77	155	1444	1140	2468
Tulsa	36 11	95 54	674	12	16	102	99	96	93	79	78	78	76	301	1621	1083	2372
OREGON:																	
Corvallis	44 38	123 12	205	20	25	93	89	86	82	69	68	66	64	28	391	1	100
Eugene	44 07	123 13	361	20	23	94	91	87	83	69	67	65	63	42	441	2	70
Hermiston	45 49	119 17	624	3	10	101	96	93	89	68	67	65	64	159	938	0	56
Klamath Falls	42 09	121 43	4091	1	5	89	87	84	80	63	62	61	60	6	339	0	1
Portland	45 36	122 36	24	22	25	89	85	81	77	69	67	66	63	13	208	4	95
Salem	44 55	123 00	209	18	23	92	88	84	80	69	67	66	63	26	296	11	123
PENNSYLVANIA:																	
Allentown	40 39	75 26	379	8	12	92	89	86	83	77	75	74	72	24	509	254	1123
Altoona	40 18	78 19	1468	-5	0	89	87	84	81	75	74	72	70	6	351	103	703
Erie	42 05	80 12	732	4	6	87	85	82	79	76	74	73	71	0	269	139	782
Harrisburg	40 13	76 51	347	10	14	95	92	88	85	77	76	75	73	55	744	435	1437
Johnstown	40 20	78 55	1210	3	6	89	87	85	82	75	74	72	71	6	467	100	866
Philadelphia Army Depot	PIER 98 S			12	16	93	91	88	84	79	77	76	75	38	702	495	1507
Philipsburg	40 53	78 05	1923	-7	-2	88	85	82	79	74	73	71	69	1	264	79	615
Pittsburgh	40 30	80 13	1151	4	9	90	87	85	82	75	74	72	71	9	471	136	991
Reading	40 20	75 58	266	11	14	93	90	87	85	77	76	75	73	38	788	435	1437
Wilkes-Barre-Scranton Airport	41 20	75 44	940	0	4	89	87	84	81	75	74	73	71	5	400	118	875
York	39 56	76 43	460	11	14	95	92	88	85	77	76	75	73	55	744	435	1437
RHODE ISLAND:																	
Kingston	41 29	71 32	100	5	10	87	84	82	79	77	75	74	72	3	230	221	1053
Pawtucket	41 52	71 22	97	5	10	90	86	83	80	76	75	74	72	6	316	180	915
Providence	41 44	71 26	55	5	10	90	86	83	80	76	75	74	72	6	316	180	915
Woonsocket	42 00	71 31	400	5	10	90	86	83	80	76	75	74	72	6	316	180	915
SOUTH CAROLINA:																	
Charleston AFB				24	27	94	91	89	87	81	80	79	78	56	1252	1760	3184
Columbia	33 57	81 07	222	23	26	98	95	93	90	79	79	78	77	172	1359	1285	2807
Greenville	34 50	82 24	1039	18	23	95	93	90	87	77	76	75	74	83	1083	637	2256
Myrtle Beach AFB				22	25	92	89	88	86	81	80	79	78	25	1204	1763	3084
Parris Island				26	29	96	93	91	89	81	80	79	78	115	1515	1994	3436
Spartanburg	34 58	81 57	816	18	23	95	93	90	87	77	76	75	74	83	1083	637	2256
Sumter	33 56	80 19	169	24	27	96	94	92	89	80	79	78	77	132	1372	1371	2841

AIR CONDITIONING AND HEATING DESIGN CLIMATIC DATA

State and Station	N Lat Deg	Min	W Long Deg	Min	Elev. Ft.	Heating 99 Dry Bulb F	Heating 97½ Dry Bulb F	AC 1 Dry	AC 2½ Dry	AC 5 Dry	AC 10 Dry	AC 1 Wet	AC 2½ Wet	AC 5 Wet	AC 10 Wet	93 Hrs	80 Hrs	73 Hrs	67 Hrs
SOUTH DAKOTA:																			
Black Hills Army Depot						-11	-6	95	92	88	84	71	69	68	66	57	569	11	242
Huron	44	23	98	13	1289	-17	-13	96	93	89	85	76	75	73	71	77	644	189	776
Mitchell	43	42	98	00	1295	-17	-12	98	94	90	86	77	76	74	72	93	760	207	832
Rapid City	44	02	103	03	3168	-11	-6	95	92	88	84	71	69	68	66	57	569	11	242
Sioux Falls	43	34	96	44	1422	-15	-11	92	90	87	83	76	75	73	71	26	498	181	794
TENNESSEE:																			
Chattanooga	35	02	85	12	688	17	21	98	96	93	90	78	78	77	76	170	1250	835	2371
Johnson City	36	19	82	23	1730	14	18	93	91	88	85	77	76	75	73	40	850	301	1636
Kingsport	36	31	82	30	1284	12	17	93	91	89	87	77	75	74	73	41	1126	389	1847
Knoxville	35	49	83	59	974	10	16	93	91	89	86	77	76	75	73	33	1009	427	2113
Memphis	35	03	89	59	282	17	21	98	96	94	91	80	79	78	77	207	1509	1332	2631
Nashville	36	07	86	41	606	12	17	98	95	93	90	79	78	77	76	162	1295	913	2307
Oak Ridge	36	02	84	14	914	14	21	94	92	90	87	77	76	75	74	46	1039	553	2158
TEXAS:																			
Abilene	32	26	99	41	1759	17	21	101	99	97	95	75	74	74	72	482	2005	298	2360
Amarillo	35	14	101	46	3700	8	13	98	96	93	90	71	70	69	68	189	1176	8	710
Austin	30	18	97	42	615	25	29	101	100	98	95	79	78	78	77	547	2243	2201	3457
Beaumont	30	05	94	15	34	29	33	95	93	92	90	80	79	79	78	103	1807	2512	3602
Brownsville	25	54	97	26	16	37	40	95	93	92	91	80	80	79	79	119	2295	3299	3994
Corpus Christi NAS						29	35	95	94	93	91	81	80	80	79	154	2531	3090	3872
Dallas NAS						19	25	102	100	98	95	79	78	78	77	539	2304	1609	3001
El Paso	31	48	106	24	3920	20	24	100	98	96	93	70	69	68	67	356	1860	6	505
Fort Worth	32	45	97	20	701	20	24	101	99	99	96	79	78	77	76	571	2207	1457	2983
Galveston	29	16	94	52	32	31	35	90	89	88	87	82	81	81	80	30	2639	2982	3790
Houston	29	39	95	17	51	29	33	96	94	92	90	80	80	79	78	139	1894	2675	3695
Laredo AFB						33	37	103	101	100	98	79	78	78	77	841	2756	2510	3784
Lubbock	33	39	101	50	3243	12	17	98	96	94	91	73	72	70	69	229	1341	39	968
Paris	33	38	95	27	530	16	22	100	98	96	93	78	77	76	75	342	1838	1167	2822
San Antonio	29	32	98	28	796	28	32	99	97	96	94	78	77	76	75	421	2004	1927	3463
Tyler	32	21	95	24	515	21	27	101	98	96	93	80	79	78	78	305	1855	1967	3193
Waco	31	37	97	13	500	21	26	101	100	98	96	79	78	78	76	536	2194	1873	3215
Wichita Falls	33	59	98	31	1039	15	19	103	100	98	96	77	76	75	74	520	2047	787	2570
UTAH:																			
Logan	41	44	111	49	4778	-7	0	94	92	89	85	66	65	63	62	53	603	0	14
Ogden	41	12	112	01	4440	6	11	95	93	90	87	67	65	64	62	70	791	0	18
Provo	40	13	111	43	4448	2	6	96	93	91	88	67	66	65	63	92	877	0	27
Salt Lake City	40	47	111	58	4224	2	6	96	93	91	88	67	66	65	63	92	877	0	27
VERMONT:																			
Burlington	44	28	73	09	331	-12	-7	88	85	83	79	74	73	71	69	5	284	73	579
North Concord AFS						-21	-15	78	74	71	68	69	68	66	64	0	9	0	103
St. Johnsbury	44	25	72	01	699	-13	-8	91	88	85	81	74	73	71	69	12	335	73	579
Winooski	44	29	73	10	190	-12	-7	88	85	83	79	74	73	71	69	5	284	73	579
VIRGINIA:																			
Belvoir, Fort						14	17	93	91	88	86	79	78	77	75	34	839	709	1869
Cape Henry	36	56	76	01	16	21	24	94	92	89	86	79	78	78	76	54	974	982	2298
Charlottesville	38	02	78	31	870	11	15	93	90	88	85	79	77	76	74	28	826	462	1576
Dulles International Airport	38	57	77	27	291	8	13	94	91	89	85	78	77	76	74	54	841	549	1654
Hampton	37	01	76	21	20	19	22	94	92	89	86	80	79	78	76	58	952	1105	2374
Lynchburg	37	20	79	12	955	14	18	92	89	87	84	77	76	75	73	24	661	406	1660
Petersburg	37	13	77	25	15	15	18	97	94	92	88	79	78	77	76	120	1006	795	1991
Richmond	37	30	77	20	180	15	18	97	94	92	88	79	78	77	76	120	1006	795	1991
Roanoke	37	19	79	58	1174	15	18	94	91	89	86	76	74	73	72	45	799	262	1508
Staunton	38	09	79	05	1480	12	15	93	90	87	84	76	74	73	72	31	692	262	1508
Virginia Beach	36	51	75	59	15	21	24	94	92	89	86	79	78	78	76	54	974	982	2298
Yorktown	37	14	76	31	25	19	22	94	92	89	86	80	79	78	76	58	952	1105	2374

AIR CONDITIONING AND HEATING DESIGN CLIMATIC DATA

State and Station	Location N Lat	W Long	Elev.	Heating Design Data Percent 99	97½	Air Conditioning Criteria Data, Percent 1	2½	5	10	1	2½	5	10	Air Conditioning Design Data Dry Bulb, F 93	80	Wet Bulb, F 73	67
	Deg, Min.		Ft.	Dry Bulb, F		Dry Bulb, F				Wet Bulb, F				Hours			
WASHINGTON:																	
Aberdeen	46 59	123 49	12	25	29	76	71	69	66	65	63	62	61	1	17	0	13
Richland	46 20	119 20	396	-4	2	98	94	91	87	71	69	68	66	103	689	8	192
Seattle-Tacoma Airport	47 27	122 18	451	21	25	85	81	77	72	66	64	63	61	4	91	0	17
Spokane	47 37	117 31	2357	0	5	91	88	85	81	64	63	61	60	16	363	0	6
Tacoma	47 15	122 30	100	18	23	85	81	78	74	68	66	64	62	4	117	2	59
Walla Walla	46 06	118 17	1206	7	1	95	92	88	85	69	68	66	64	51	601	2	104
Yakima	46 34	120 32	1062	7	12	95	92	89	84	68	67	65	63	60	551	0	67
WEST VIRGINIA:																	
Charleston	38 22	81 36	989	9	14	92	90	88	85	76	75	74	73	29	779	349	1563
Elkins	38 53	79 51	1973	1	5	87	84	82	80	74	73	72	70	1	343	85	820
Huntington	38 25	82 27	565	14	18	95	93	90	87	78	77	76	75	67	979	526	1891
Morgantown	39 38	79 55	1248	7	11	90	88	85	82	77	76	74	73	9	482	258	1248
Parkersburg	39 16	81 34	840	5	11	93	90	87	84	78	77	75	74	30	632	399	1491
Wheeling	40 11	80 39	1190	4	9	90	87	85	82	75	74	72	71	9	471	136	991
WISCONSIN:																	
Beloit	42 30	89 02	780	-9	-4	93	90	87	83	78	76	75	73	33	583	308	1031
Green Bay	44 29	88 08	699	-12	-7	88	85	82	79	75	73	72	69	4	264	100	574
La Crosse	43 56	91 17	672	-11	-7	90	88	85	82	77	75	74	71	13	476	266	944
Madison	43 08	89 20	866	-9	-5	92	89	86	82	77	75	73	71	19	485	200	888
Milwaukee	42 57	87 54	704	-6	-2	90	87	84	80	77	75	73	71	13	358	155	809
Oshkosh	44 03	88 32	760	-13	-7	92	88	85	82	75	73	72	69	19	388	100	574
WYOMING:																	
Casper	42 55	106 28	5321	-11	-5	92	90	87	84	63	62	60	59	24	546	0	1
Cheyenne	41 09	104 49	6144	-2	2	89	86	84	81	63	62	60	59	3	370	0	0
Sheridan	44 46	106 58	3946	-12	-7	95	92	89	85	66	64	63	61	60	589	0	16
Sundance AFS Site 2				-26	-19	84	80	77	72	64	62	60	59	0	100	0	0
AFRICA:																	
Algeria:																	
Algiers			194	39	42	95	92	89	85	77	76	75	74	63	752	426	1715
Ethiopia:																	
Addis Ababa			8038	34	36	85	82	79	75	66	65	64	62	0	188	0	15
Ghana:																	
Accra			88	66	67	91	90	89	88	83	82	81	81	13	2796	4040	4368
Ivory Coast:																	
Abidjan			65	66	67	93	92	90	88	83	82	81	80	39	2839	3931	4368
Libya:																	
Nalut			2100	29	31	101	98	95	91	69	68	68	67	250	1521	0	377
Morocco:																	
Rabat			204	38	40	92	87	85	82	75	74	73	72	37	390	148	1678
Tangier			239	38	40	89	86	83	80	72	71	70	69	6	278	23	688
Nigeria:																	
Lagos			10	71	72	94	93	91	89	84	83	82	81	64	2922	4023	4362
Somalia:																	
Kismayo			33	65	66	93	92	81	89	83	82	82	81	35	2244	4127	4344

AIR CONDITIONING AND HEATING DESIGN CLIMATIC DATA

Station	Location		Elev. Ft.	Heating Design Data Percent — Dry Bulb, F		Air Conditioning Criteria Data, Percent — Dry Bulb, F				Wet Bulb, F				Air Conditioning Design Data — Dry Bulb, F		Wet Bulb, F	
				99	97½	1	2½	5	10	1	2½	5	10	93	80	73	67
														Hours			
South Africa:																	
Pretoria			4491	28	31	90	87	85	82	70	69	68	67	5	570	4	446
Sudan:																	
Khartoum			1279	53	56	107	105	102	99	79	78	77	76	1476	3851	980	2600
Tunisia:																	
Tunis			217	35	36	99	95	92	88	77	76	74	73	114	1184	284	1760
United Arab Republic:																	
Cairo			233	45	46	100	98	96	94	76	75	74	73	449	2090	327	2524
Zaire:																	
Kinshasa			1066	59	60	93	92	90	88	81	81	80	79	38	2071	2456	4145
ANTARCTICA:																	
McMurdo Sound			80	-40	-36	37	35	34	32	33	32	31	29	0	0	0	0
ASIA:																	
Afghanistan:																	
Kabul			5955	6	9	98	96	93	90	66	65	64	62	147	1037	0	14
Burma:																	
Rangoon			18	60	62	100	98	95	93	85	84	83	81	352	2985	4129	4416
Cambodia:																	
Phnom Penh			39	65	67	98	97	95	94	83	82	81	80	408	3531	3975	4416
China:																	
Hong Kong			109	48	50	93	91	89	87	81	80	80	79	24	2613	4398	4416
India:																	
Burdwan			106	46	48	96	92	90	88	86	85	84	83	235	2055	4195	4416
Calcutta			21	52	54	94	92	90	88	86	85	84	83	115	3356	4241	4385
Krishnagar			48	41	42	96	93	91	89	86	85	84	83	166	3707	4100	4364
New Delhi			703	39	41	110	107	105	102	83	82	82	81	1151	3464	2565	3429
Indonesia:																	
Djakarta			26	71	71	89	87	86	85	80	79	79	78	0	1523	3232	4368
Iran:																	
Meshed			3104	6	12	99	96	93	90	71	70	68	66	182	1155	16	268
Teheran			4002	14	18	103	100	97	94	75	74	73	71	356	1798	105	1464
Iraq:																	
Baghdad			111	29	32	113	111	108	105	75	73	72	70	1374	3198	95	980
Israel:																	
Tel Aviv			128	38	40	96	93	91	88	74	73	72	71	111	1368	83	1474
Japan:																	
Nagasaki			87	28	30	88	86	84	82	79	78	77	76	1	477	791	2062
Osaka			26	30	31	92	91	89	87	83	81	80	79	1	1066	1390	2323
Tokyo			9	29	31	90	88	86	84	81	80	79	78	8	850	1456	2424
Yokohama			10	29	31	90	88	86	84	81	80	79	78	8	850	1456	2424

AIR CONDITIONING AND HEATING DESIGN CLIMATIC DATA

Station	Elev. Ft.	Heating Design Data Percent — Dry Bulb, F		Air Conditioning Criteria Data, Percent — Dry Bulb, F				Wet Bulb, F				Air Conditioning Design Data — Dry Bulb, F		Wet Bulb, F		Hours
		99	97½	1	2½	5	10	1	2½	5	10	93	80	73	67	
Jordan:																
Amman	2548	33	34	97	94	91	88	68	67	66	65	100	1129	0	57	
Korea:																
Inchon	50	5	7	91	89	87	84	81	79	78	77	16	653	910	1849	
Laos:																
Vientiane	531	50	53	95	93	91	89	83	82	82	81	77	2063	3333	4400	
Lebanon:																
Beirut	111	44	46	93	91	90	87	78	77	76	74	40	1595	604	2882	
Malaysia:																
Kuala Lumpur	127	70	71	93	91	90	88	82	82	81	80	44	2694	3754	4376	
Singapore	33	70	70	95	93	92	90	82	81	80	80	124	2735	4041	4416	
Nepal:																
Katmandu	4388	31	32	89	87	86	84	78	77	76	75	12	1111	924	3038	
Pakistan:																
Karachi	70	47	49	97	95	93	91	81	81	80	80	209	3560	4077	4391	
Lahore	702	30	33	107	104	102	98	83	82	81	80	1088	3360	2709	3471	
Saudi Arabia:																
Dhahran AB		45	48	111	110	108	106	86	85	84	83	1811	3956	2219	3783	
Sri Lanka (Ceylon):																
Colombo	24	66	68	88	87	86	85	81	80	80	79	0	2870	4195	4416	
Taiwan:																
Shu Lin Kou	450	41	45	91	89	87	85	81	80	79	78	9	1273	2185	3766	
Tainan	75	46	49	90	89	88	87	84	83	82	81	15	2370	3726	4324	
Taipei	26	44	47	93	91	89	87	83	82	81	80	30	1950	2854	3985	
Thailand:																
Bangkok	39	61	63	97	95	93	91	82	82	82	81	217	2960	4376	4392	
Turkey:																
Ankara	2825	7	12	94	92	89	85	66	65	64	63	53	626	0	15	
Istanbul	130	28	30	91	88	86	83	75	74	73	71	10	492	117	1039	
Vietnam:																
Saigon (Ho Chi Minh City)	30	65	67	93	91	89	87	85	84	83	81	57	1744	3421	4359	
Yemen (P.D.R.):																
Aden	22	66	67	102	100	98	97	85	84	83	83	798	3988	4363	4416	
ATLANTIC OCEAN:																
Ascension Island:																
Ascension Island AAFB		68	69	86	86	85	84	77	76	75	75	0	1489	1381	4283	
Azores:																
Lajes Field, Terceira		46	49	80	78	77	75	73	72	71	70	0	39	32	1213	
Villa do Porto	330	51	52	80	78	76	74	70	69	68	67	0	45	0	456	

AIR CONDITIONING AND HEATING DESIGN CLIMATIC DATA

Station	Location		Elev. Ft.	Heating Design Data Percent		Air Conditioning Criteria Data, Percent								Air Conditioning Design Data			
						Dry Bulb, F				Wet Bulb, F				Dry Bulb, F		Wet Bulb, F	
				99	97½	1	2½	5	10	1	2½	5	10	93	80	73	67
				Dry Bulb, F										Hours			
Bermuda:																	
Bermuda NAVSTA				53	55	87	86	85	84	79	78	78	77	0	1369	2358	3799
Iceland:																	
Keflavik			164	14	17	59	58	56	55	54	53	53	52	0	0	0	0
AUSTRALIA:																	
Canberra			1837	24	26	91	87	84	80	69	68	66	65	15	281	1	113
Perth			210	38	40	100	96	93	88	76	74	73	71	46	834	4112	4332
CARIBBEAN SEA:																	
Bahamas:																	
Grand Bahama AAFB				47	49	89	88	88	87	80	80	79	79	0	2495	3410	4215
San Salvador AAFB				62	64	90	89	88	87	80	80	79	78	1	2822	3869	4376
Barbados:																	
Christs Church			195	68	70	89	88	87	86	81	80	80	79	0	3300	4300	4392
Cuba:																	
Guantanamo Bay NAS				64	66	95	94	92	91	82	81	80	79	145	2817	3958	4413
Havana			80	59	60	91	90	89	87	81	81	80	79	6	1853	2985	3974
Dominican Republic:																	
Sabana de la Mar			36	61	63	88	87	86	85	81	80	79	78	0	2074	3412	4412
Santo Domingo			57	63	64	92	90	88	86	81	80	80	79	15	2132	3517	4412
Haiti:																	
Port-au-Prince			121	62	63	97	95	93	91	82	81	80	79	187	2455	2575	4406
Jamaica Island:																	
Vernam AFB				62	64	92	91	90	89	80	79	79	78	16	2083	3199	4398
Leeward Islands:																	
Anguilla			213	66	69	88	87	86	86	80	79	78	78	0	2950	4260	4392
Barbuda			205	68	70	88	87	86	86	81	80	79	79	0	3447	4287	4392
Puerto Rico:																	
Martin Pena				66	67	89	88	88	87	79	79	78	78	0	2122	3497	4342
San Juan NAS				65	66	90	89	88	86	80	79	79	78	2	2132	3880	4416
Trinidad:																	
Waller AFB				65	67	89	88	87	86	79	78	78	78	1	1719	3924	4415
Virgin Islands:																	
Alexander Hamilton Fld, St Croix				69	70	88	87	86	86	80	79	78	78	0	2634	4097	4391
Truman Field, St Thomas			15	69	71	89	88	87	87	79	79	78	78	0	2952	4041	4391
CENTRAL AMERICA:																	
Canal Zone:																	
Balboa			15	72	73	91	90	89	87	81	81	80	79	10	1904	4268	4416
Costa Rica:																	
San Jose			3760	53	54	84	83	82	79	72	71	71	70	0	375	2	1720

AIR CONDITIONING AND HEATING DESIGN CLIMATIC DATA

Station	Location		Elev. Ft.	Heating Design Data Percent		Air Conditioning Criteria Data, Percent								Air Conditioning Design Data			
						Dry Bulb, F				Wet Bulb, F				Dry Bulb, F		Wet Bulb, F	
				99	97½	1	2½	5	10	1	2½	5	10	93	80	73	67
				Dry Bulb, F		Dry Bulb, F				Wet Bulb, F				Hours			
El Salvador:																	
San Salvador			2238	54	56	95	93	91	88	77	76	75	74	125	1765	1059	3532
Guatemala:																	
Guatemala City			4873	48	51	81	80	78	77	68	67	66	66	0	114	1	179
Honduras:																	
Tegucigalpa			3250	47	50	89	87	85	82	73	72	71	70	30	857	26	1261
Nicaragua:																	
Managua			208	66	67	91	90	89	88	81	80	79	78	45	2085	3508	4411
EUROPE:																	
Austria:																	
Innsbruck			1909	1	6	87	84	81	77	67	65	64	62	5	134	0	38
Vienna			600	6	11	88	86	83	79	72	70	69	67	6	269	9	294
Belgium:																	
Brussels			328	15	19	83	79	76	73	70	68	67	65	0	62	7	117
Iseghem			49	15	19	81	78	75	72	71	69	68	66	1	53	6	234
Mons			148	16	19	82	78	76	73	69	67	66	64	0	58	0	88
Denmark:																	
Copenhagen			43	19	22	79	76	74	71	68	66	64	62	0	23	2	36
Finland:																	
Helsinki			30	-16	-9	77	74	72	69	66	65	63	61	0	8	0	14
France:																	
Bordeaux			157	25	27	88	85	82	79	74	73	71	70	5	258	63	651
Fontainebleau			250	18	23	86	83	80	76	71	69	67	66	5	161	9	205
Limoges			935	19	23	87	84	82	79	71	69	68	66	5	251	16	214
Marseille			246	25	28	90	87	84	81	72	71	69	68	12	370	5	445
Nancy			764	9	13	83	80	77	74	69	67	65	63	1	86	2	92
Nantes			121	26	28	86	83	80	77	70	69	68	66	3	164	0	190
Orleans			390	15	19	89	85	81	77	71	69	67	65	10	206	12	202
Paris			315	22	25	84	81	78	74	70	68	67	65	3	106	8	186
Rheims			272	14	18	85	82	79	76	70	68	66	64	2	139	10	143
Tours			350	22	26	88	85	81	78	71	69	67	65	9	211	12	212
Verdun			1017	3	8	83	79	75	72	69	67	65	63	3	67	5	97
Germany:																	
Aachen			669	20	23	81	78	75	72	69	67	65	63	0	44	3	67
Bayreuth			1600	3	9	82	79	75	72	67	65	63	61	1	54	2	32
Berlin			180	6	12	84	81	78	74	68	67	66	64	3	102	0	90
Bonn			197	14	18	81	78	76	73	69	67	66	64	0	38	3	82
Bremen			30	16	20	79	75	73	70	68	66	64	62	0	18	2	45
Bremerhaven			20	15	18	78	75	72	69	66	65	65	63	1	15	0	6
Frankfurt			425	8	13	86	82	79	74	70	68	66	64	6	142	8	135
Heidelberg			359	8	14	88	85	81	77	73	70	68	66	11	207	31	262
Mannheim			359	8	14	88	85	81	77	73	70	68	66	11	207	31	262
Munich			1740	5	9	86	83	80	76	68	66	64	62	2	148	2	41
Mainz			450	11	16	85	82	78	74	69	68	66	64	4	124	3	127
Nuremberg			1050	1		83	79	77	74	67	65	64	62	1	68	1	30
Stuttgart			880	8	13	85	81	78	74	70	68	66	64	2	121	6	126
Wiesbaden AB				11	16	85	82	78	74	69	68	66	64	4	124	3	127

AIR CONDITIONING AND HEATING DESIGN CLIMATIC DATA

Station	Elev. Ft.	Heating Design Data Percent		Air Conditioning Criteria Data, Percent								Air Conditioning Design Data			
		99	97½	1	2½	5	10	1	2½	5	10	93	80	73	67
		Dry Bulb, F		Dry Bulb, F				Wet Bulb, F				Dry Bulb, F		Wet Bulb, F	
												Hours			
Greece:															
Athens	351	33	36	94	91	88	86	71	70	70	69	35	891	3	732
Larisa	246	23	26	99	95	93	89	76	74	73	71	143	1171	150	1351
Patras	15	35	38	95	92	90	87	70	69	69	68	53	1004	1	505
Hungary:															
Budapest	394	10	14	90	86	84	80	72	71	70	68	10	334	23	498
Ireland:															
Dublin	155	24	27	74	72	70	67	65	64	62	60	0	1	0	3
Shannon Aprt	8	25	28	76	73	71	68	65	64	63	61	0	4	0	13
Italy:															
Cagliari	3	35	37	94	91	88	84	77	75	74	72	41	539	205	1146
Milan	341	18	22	89	87	84	81	76	75	74	73	3	393	250	1122
Naples	220	34	36	91	88	86	83	74	73	72	71	15	570	59	1318
Rome	377	30	33	94	92	89	86	73	72	70	69	47	782	29	603
Verona	239	14	18	92	88	86	83	73	72	71	69	15	590	36	727
Netherlands:															
Amsterdam	5	20	23	79	76	73	70	63	62	61	60	0	2	0	0
The Hague	10	20	23	79	76	73	70	63	62	61	60	0	2	0	0
Norway:															
Oslo	308	3	7	81	77	74	70	65	64	62	61	0	35	0	5
Poland:															
Warsaw	394	3	8	84	81	78	75	71	70	63	66	1	124	15	251
Portugal:															
Lisbon	313	37	39	89	86	83	80	69	68	67	66	6	313	0	149
Spain:															
Barcelona	312	33	36	86	84	82	81	77	75	74	72	0	333	234	1119
Cadiz	79	39	41	91	89	87	84	74	73	72	71	14	617	74	1470
Madrid	2188	28	30	94	92	90	87	71	69	68	66	69	879	16	244
San Sebastian	846	30	32	78	76	74	72	69	68	67	66	0	10	1	138
Sevilla	89	32	33	102	100	98	94	79	77	76	74	366	1383	487	1557
Sweden:															
Goteborg	55	8	12	77	74	71	68	63	61	60	58	0	10	0	1
Stockholm	146	-3	1	78	74	72	69	64	62	60	59	0	14	0	0
Switzerland:															
Bern	1877	11	15	82	79	76	73	68	66	64	63	0	47	0	41
United Kingdom:															
Aberdeen (Scotland)	79	25	28	70	68	66	63	62	61	59	58	0	0	0	0
Edinburgh (Scotland)	441	26	28	74	71	68	66	62	61	59	58	0	1	0	0
Greenwich	149	24	27	82	79	76	72	66	65	64	62	1	53	0	9
London	82	24	26	81	78	74	71	68	66	65	63	0	48	2	66
Marlborough	424	24	26	80	77	74	71	68	66	64	62	0	29	1	47
Woburn	291	24	26	80	77	75	71	67	66	64	62	0	29	0	39
Union of Soviet Socialist Republics:															
Moscow (Russia)	505	-11	-6	84	81	78	74	69	67	65	63	1	102	2	88
Yugoslavia:															
Belgrade	453	9	13	92	89	86	82	74	73	72	70	23	436	95	819

AIR CONDITIONING AND HEATING DESIGN CLIMATIC DATA

Station	Location Elev. Ft.	Heating Design Data Percent Dry Bulb, F		Air Conditioning Criteria Data, Percent Dry Bulb, F				Wet Bulb, F				Air Conditioning Design Data Dry Bulb, F		Wet Bulb, F	
		99	97½	1	2½	5	10	1	2½	5	10	93	80	73	67
												Hours			
MEDITERRANEAN:															
Cyprus:															
Nicosia	716	33	35	102	99	96	93	76	75	74	72	292	1745	260	1776
Malta:															
Valetta	233	43	46	91	88	86	84	77	76	75	74	19	787	394	2420
NORTH AMERICA:															
Canada:															
Beausejour AS, Man.		-34	-27	88	85	81	77	74	72	69	67	7	223	48	327
Churchill, Man.	115	-41	-39	78	73	69	63	67	64	61	57	0	2	0	3
Cut Throat Island, Lab.	30	-20	-18	78	74	70	65	64	62	60	57	0	22	0	8
Frobisher Bay	68	-40	-38	62	58	55	51	56	53	50	47	0	0	0	0
Gander, Nfld.	482	-4	0	80	77	74	70	68	66	65	62	1	36	2	85
Goose Bay, Nfld.	144	-26	-23	83	77	73	69	67	64	62	60	2	51	2	30
Halifax, N.S.	136	2	8	78	75	73	70	69	68	66	64	0	21	0	157
Ottawa, Ont.	339	-18	-11	89	85	82	79	75	73	71	69	9	292	108	636
Prince George, B.C.	2218	-41	-30	81	77	73	69	65	63	61	59	0	52	0	19
Saskatoon, Sask.	1645	-33	-29	87	83	80	75	68	66	64	62	3	148	3	59
Seven Islands, Que.	190	-19	-16	74	71	69	65	65	63	61	59	0	3	0	13
Vancouver, B.C.	16	7	13	76	74	72	69	67	65	64	62	0	9	0	35
Winnepeg, Man.	786	-34	-27	88	85	81	77	74	72	69	67	7	223	48	327
Yarmouth, N.S.	136	2	9	76	73	71	68	68	67	65	63	0	10	0	75
Yellowknife, N.W.T.	682	-46	-44	77	74	71	68	64	62	61	59	0	8	0	4
Greenland:															
Thule AB		-34	-33	55	52	50	47	47	45	44	42	0	0	0	0
Mexico:															
Mexico City	7575	34	36	83	81	79	76	61	60	59	58	0	171	0	0
PACIFIC OCEAN:															
Johnston Island:															
Johnston Island AFB		71	72	86	86	85	84	79	78	78	77	0	2456	3576	4416
Mariana Islands:															
Andersen AFB, Guam		73	74	87	86	85	84	81	80	79	79	0	2292	4364	4416
Marshall Islands:															
Eniwetok		77	78	90	90	88	87	83	82	80	80	3	3815	4411	4416
Kwajalein		77	78	90	90	88	87	83	82	80	80	3	3815	4411	4416
Midway Island:															
Midway NAVSTA		58	59	84	83	82	81	77	76	76	75	0	876	1765	3814
New Zealand:															
Wellington	415	35	37	76	74	72	70	67	66	64	63	0	2	0	26
Okinawa:															
Onna Point	180	50	51	90	89	88	87	83	82	81	81	1	2323	3327	4108
Philippines															
Bataan Ocean Petroleum Depot		73	74	94	92	91	90	81	80	80	79	62	3541	4362	4392
Manila	109	73	74	94	92	91	90	81	80	80	79	62	3541	4362	4392
Quezon City	50	73	74	94	92	91	90	81	80	80	79	62	3541	4362	4392

AIR CONDITIONING AND HEATING DESIGN CLIMATIC DATA

Station	Location	Heating Design Data Percent		Air Conditioning Criteria Data, Percent								Air Conditioning Design Data			
	Elev.	99	97½	1	2½	5	10	1	2½	5	10	Dry Bulb, F 93	80	Wet Bulb, F 73	67
	Ft.	Dry Bulb, F		Dry Bulb, F				Wet Bulb, F				Hours			
Wake Island:															
Wake Island	11	71	72	88	87	87	86	80	80	80	79	0	3459	4229	4413
SOUTH AMERICA:															
Argentina:															
Buenos Aires	89	32	34	91	89	86	83	74	73	72	71	15	616	102	1144
Bolivia:															
La Paz	12001	29	31	73	71	70	67	57	56	54	53	0	0	0	0
Brazil:															
Belem	33	71	72	90	89	88	87	80	79	78	77	1	1590	3707	4368
Rio de Janeiro	61	58	59	86	85	84	83	78	77	76	75	0	750	651	2769
Chile:															
Santiago	1706	30	31	94	92	89	85	72	70	69	67	51	683	10	380
Colombia:															
Bogota	8399	45	46	73	72	71	69	62	61	59	58	0	0	0	0
Ecuador:															
Quito	9350	37	38	79	78	77	74	66	65	63	61	0	3	0	26
French Guiana:															
Cayenne	20	68	69	93	91	90	88	83	83	82	81	44	2723	4221	4392
Guyana:															
Georgetown	6	70	70	89	88	87	86	82	81	80	80	0	2767	4150	4392
Paraguay:															
Asuncion	456	43	46	100	97	95	92	81	81	80	79	279	2109	2156	3728
Peru:															
Lima	394	51	52	88	87	85	83	76	75	74	73	0	688	429	2847
Surinam:															
Zanderij	70	66	67	93	92	90	88	83	83	82	81	70	2723	4172	4389
Uruguay:															
Montevideo	72	33	35	90	88	85	82	73	72	71	70	11	449	44	796
Venezuela:															
Caracas	3420	50	52	84	83	81	78	70	69	69	68	0	288	0	729

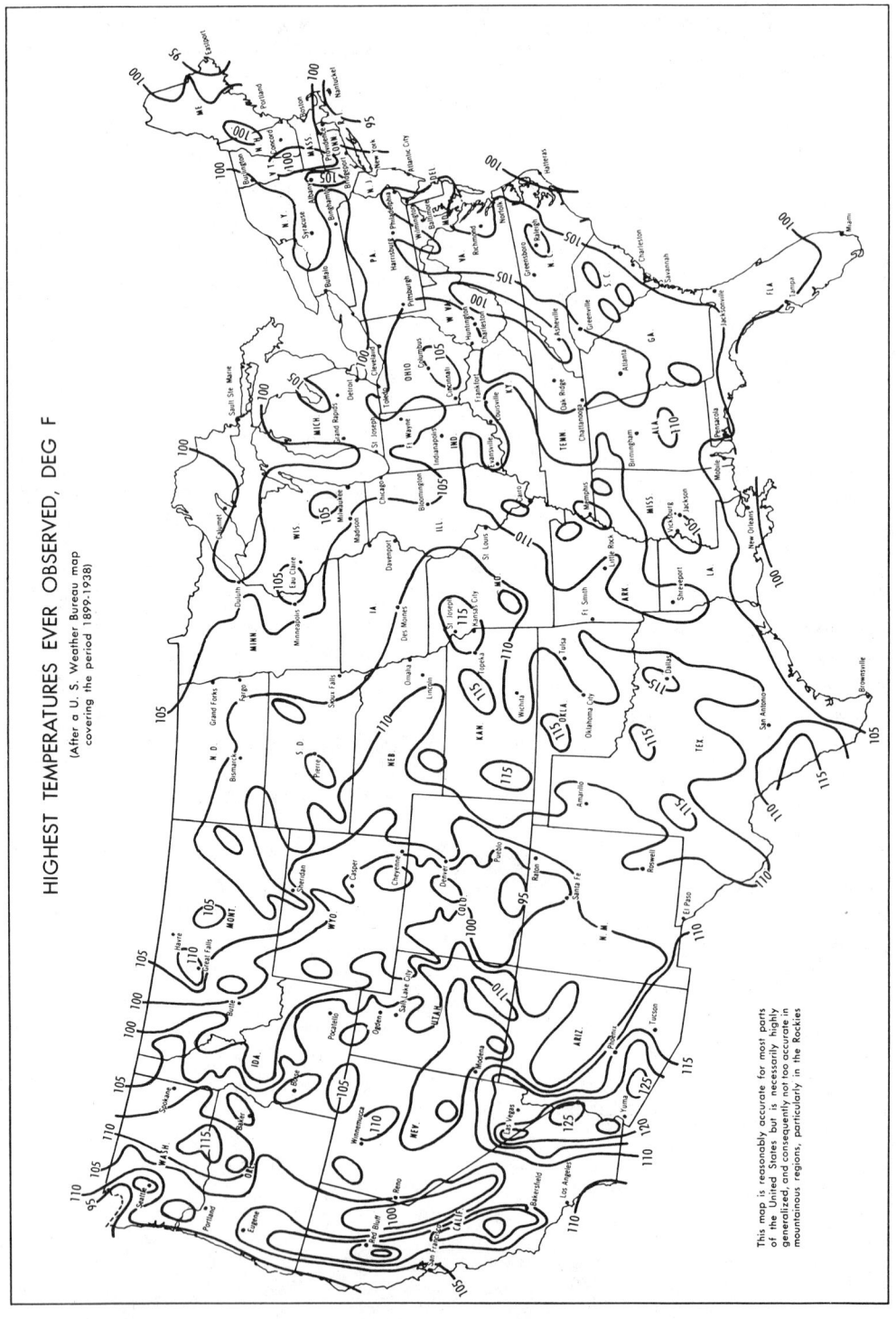

HIGHEST TEMPERATURES EVER OBSERVED, DEG F

(After a U. S. Weather Bureau map
covering the period 1899-1938)

This map is reasonably accurate for most parts
of the United States but is necessarily highly
generalized, and consequently not too accurate in
mountainous regions, particularly in the Rockies

SUMMER OUTSIDE WET BULB DESIGN TEMPERATURE, DEG F
(Based on standards of Air-conditioning and Refrigeration Institute)

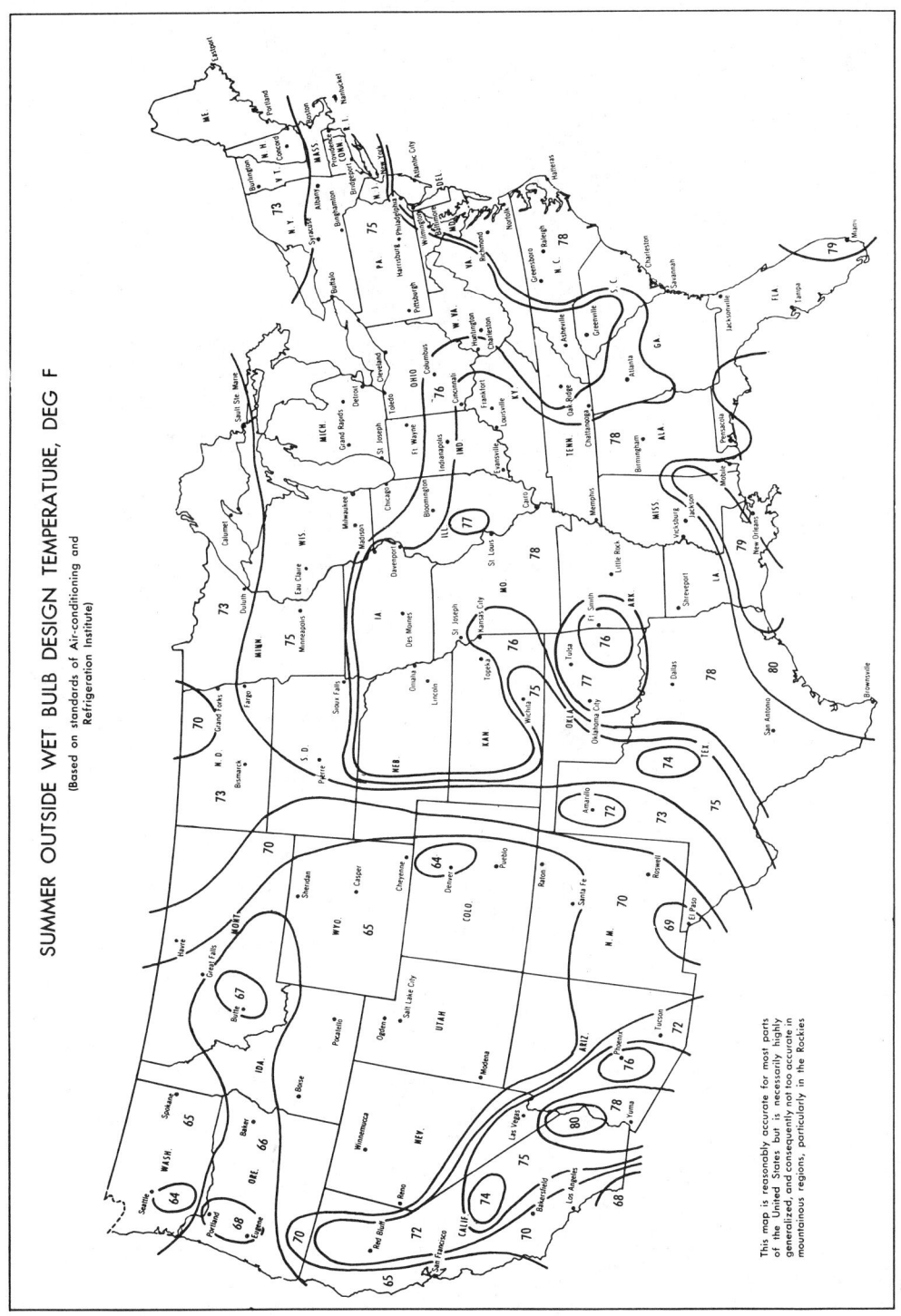

This map is reasonably accurate for most parts of the United States but is necessarily highly generalized and consequently not too accurate in mountainous regions, particularly in the Rockies.

AVERAGE JULY DRY BULB TEMPERATURE, DEG F

(After a U. S. Weather Bureau map based on reports from 200 first order Weather Bureau stations covering the period 1899-1938)

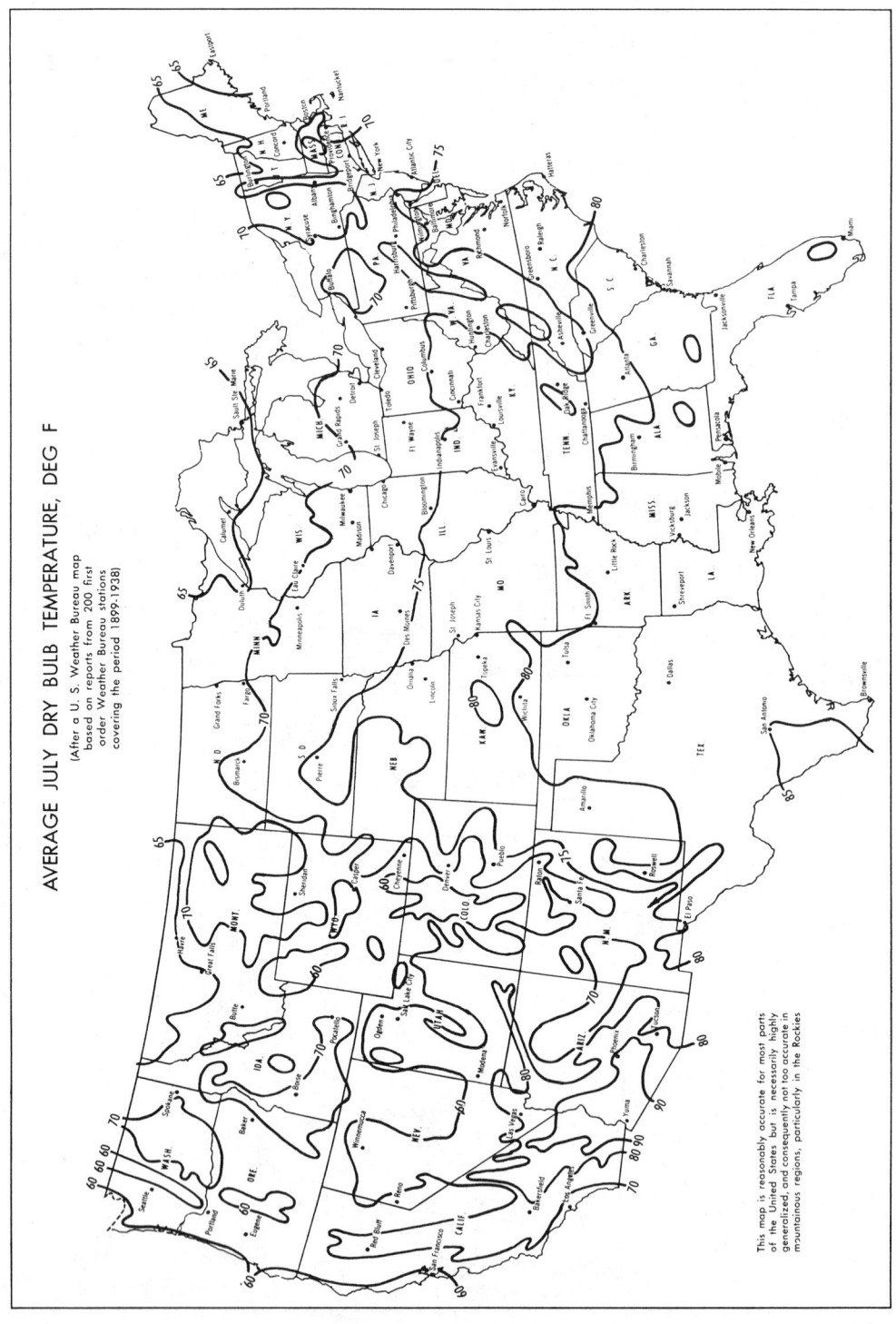

This map is reasonably accurate for most parts of the United States but is necessarily highly generalized, and consequently not too accurate in mountainous regions, particularly in the Rockies

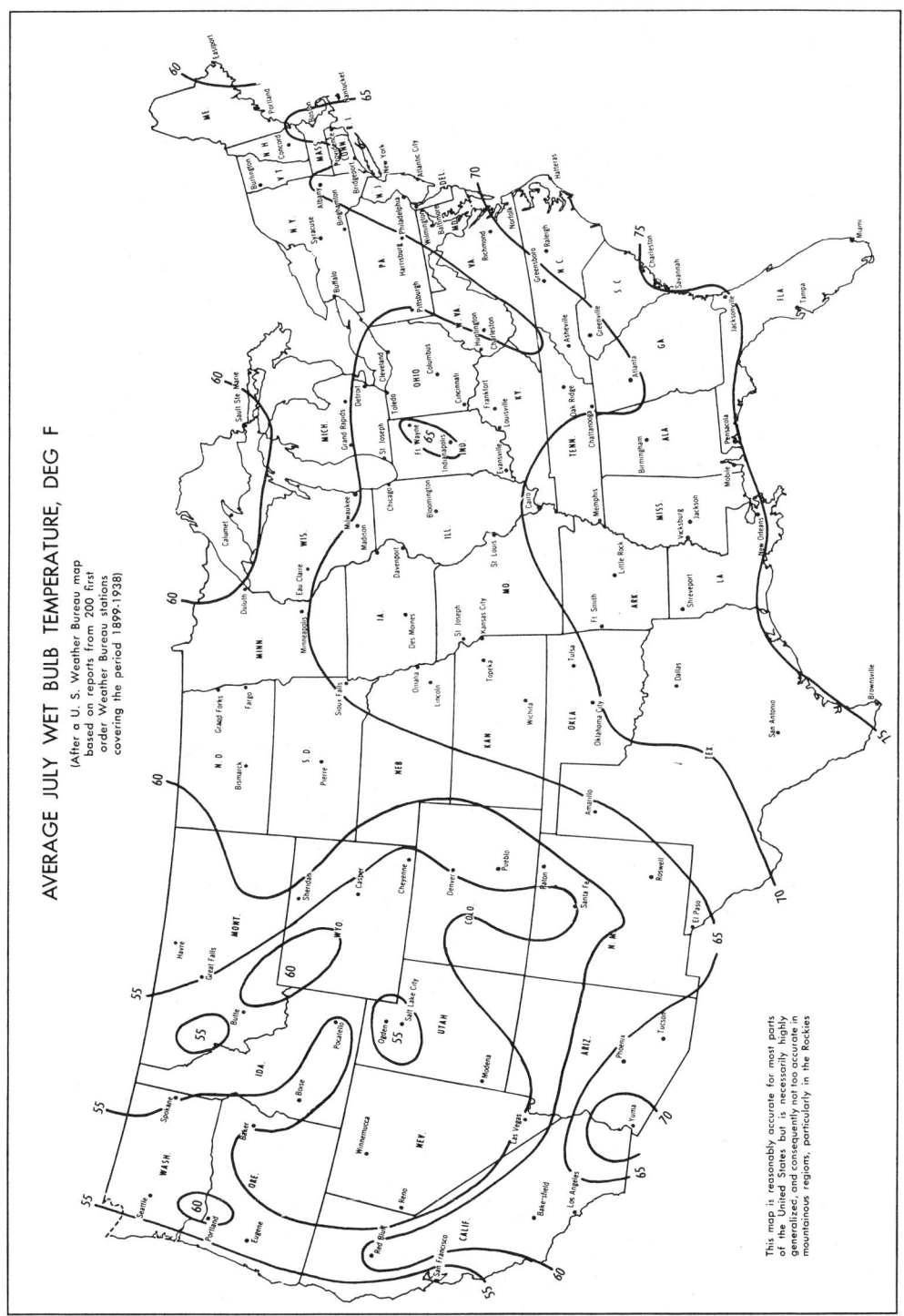

AVERAGE JULY WET BULB TEMPERATURE, DEG F

(After a U. S. Weather Bureau map based on reports from 200 first order Weather Bureau stations covering the period 1899-1938)

This map is reasonably accurate for most parts of the United States but is necessarily highly generalized, and consequently not too accurate in mountainous regions, particularly in the Rockies

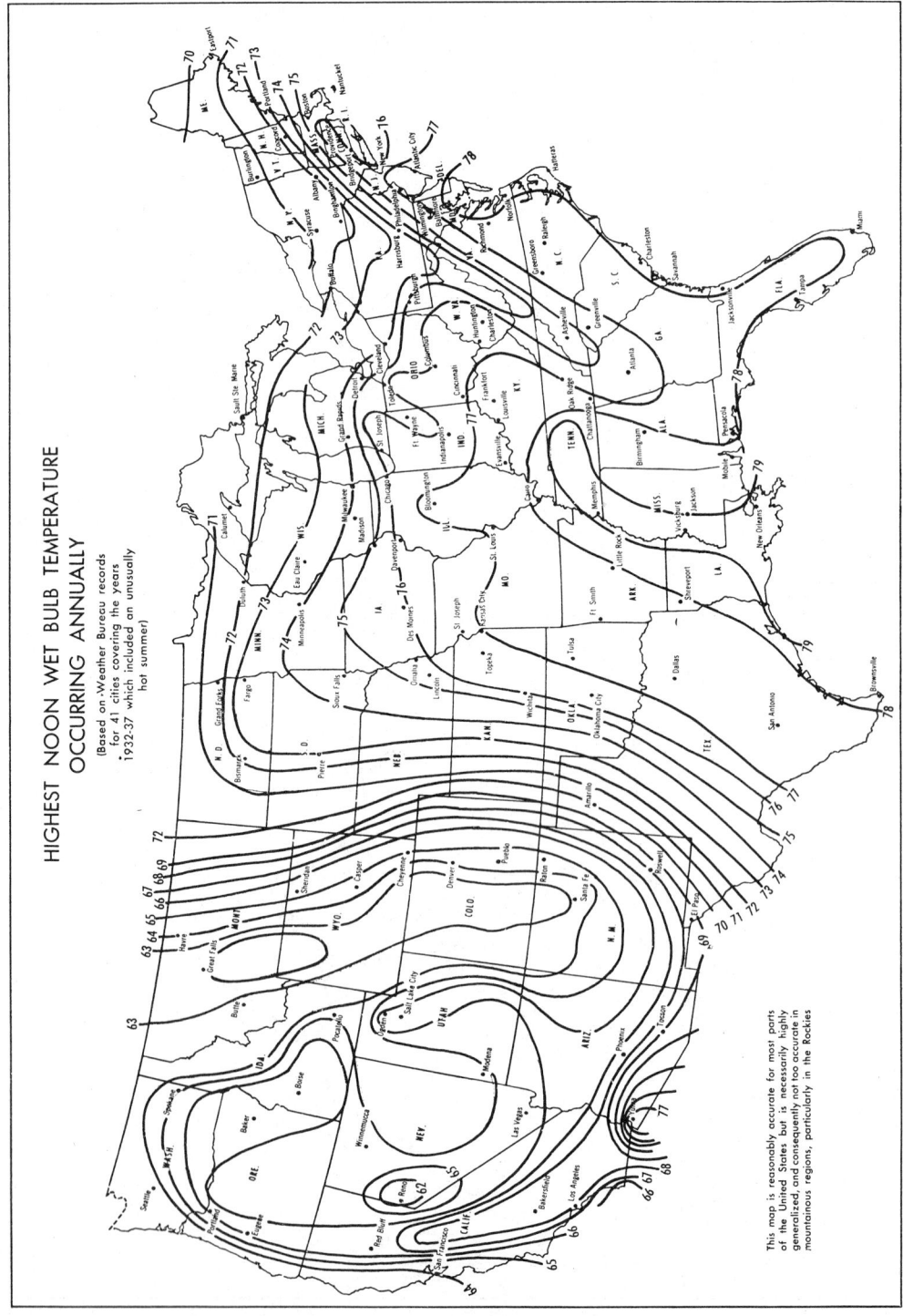

HIGHEST NOON WET BULB TEMPERATURE OCCURRING ANNUALLY

(Based on Weather Bureau records for 41 cities covering the years 1932-37 which included an unusually hot summer)

This map is reasonably accurate for most parts of the United States but is necessarily highly generalized, and consequently not too accurate in mountainous regions, particularly in the Rockies

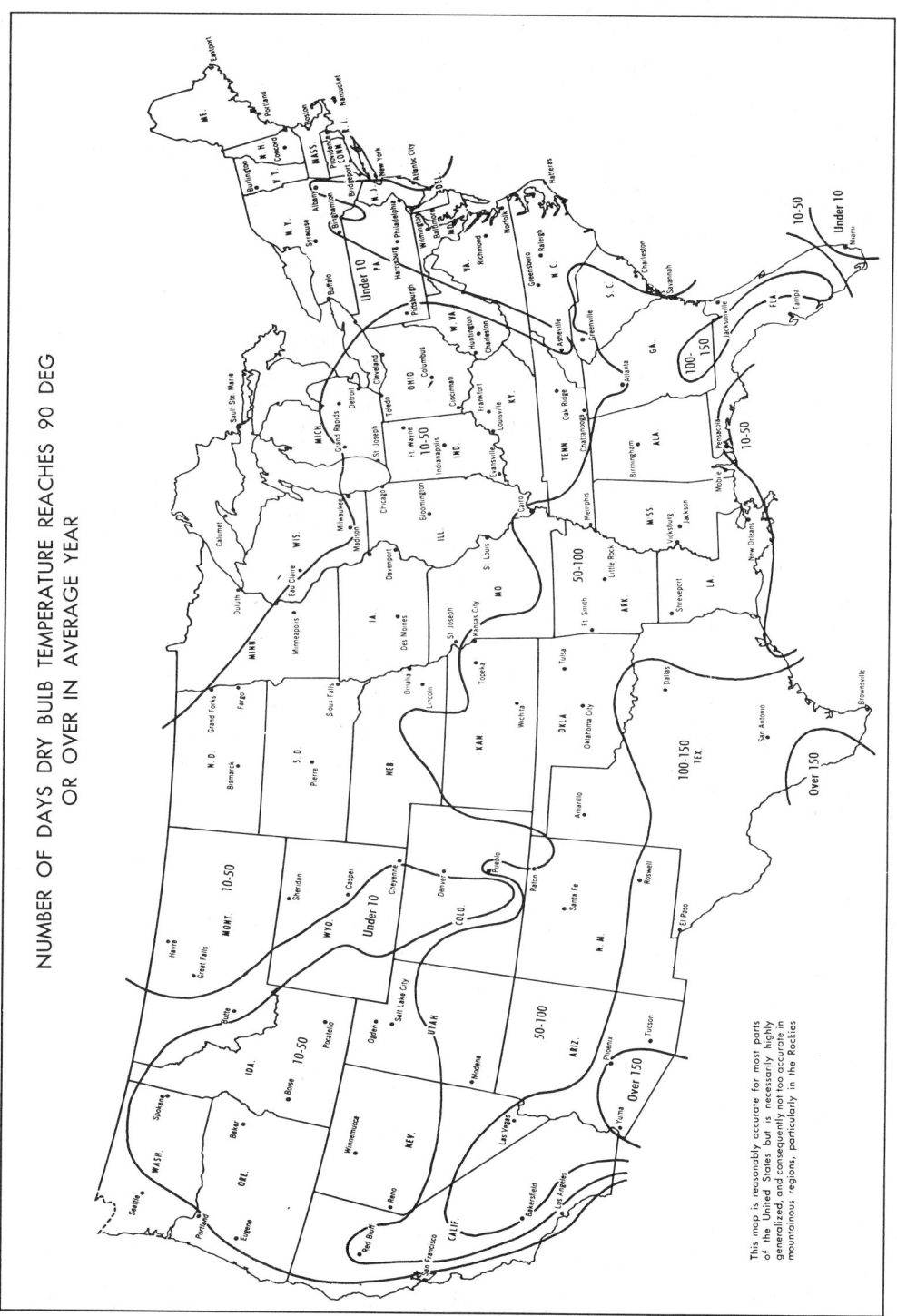

NUMBER OF DAYS DRY BULB TEMPERATURE REACHES 90 DEG OR OVER IN AVERAGE YEAR

This map is reasonably accurate for most parts of the United States but is necessarily highly generalized, and consequently not too accurate in mountainous regions, particularly in the Rockies

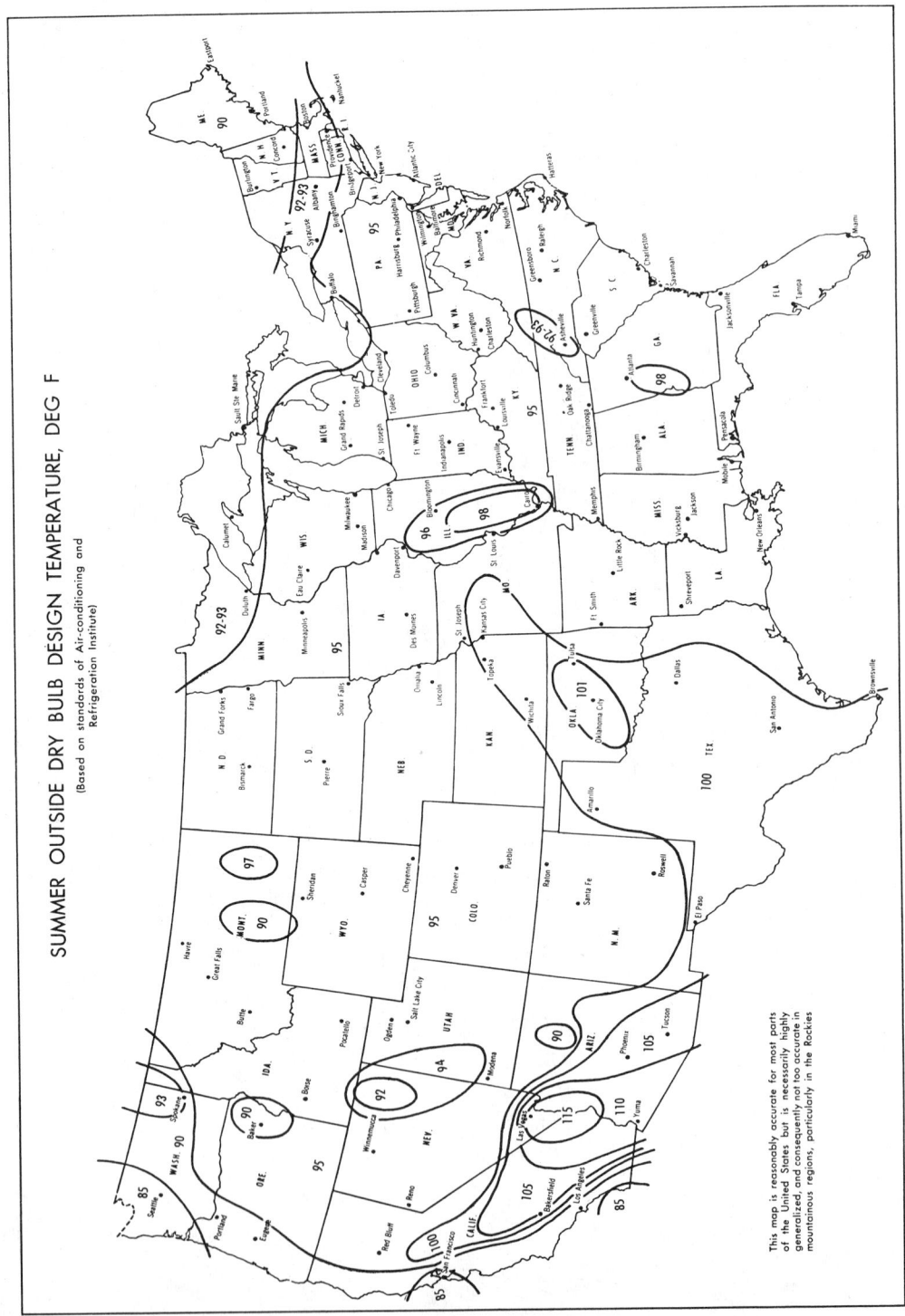

SUMMER OUTSIDE DRY BULB DESIGN TEMPERATURE, DEG F

(Based on standards of Air-conditioning and Refrigeration Institute)

This map is reasonably accurate for most parts of the United States but is necessarily highly generalized, and consequently not too accurate in mountainous regions, particularly in the Rockies

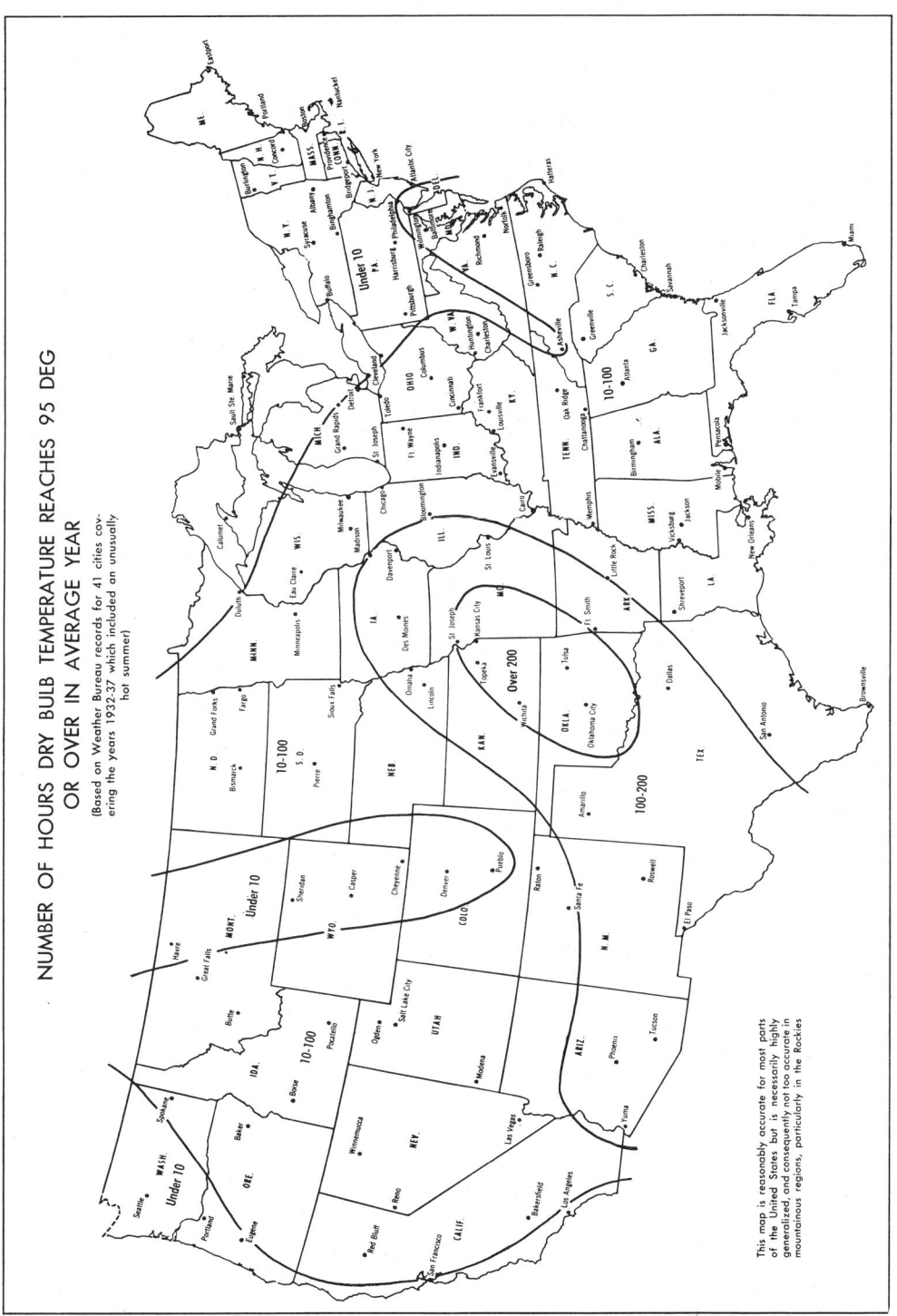

NUMBER OF HOURS DRY BULB TEMPERATURE REACHES 95 DEG OR OVER IN AVERAGE YEAR

(Based on Weather Bureau records for 41 cities covering the years 1932-37 which included an unusually hot summer)

This map is reasonably accurate for most parts of the United States but is necessarily highly generalized, and consequently not too accurate in mountainous regions, particularly in the Rockies

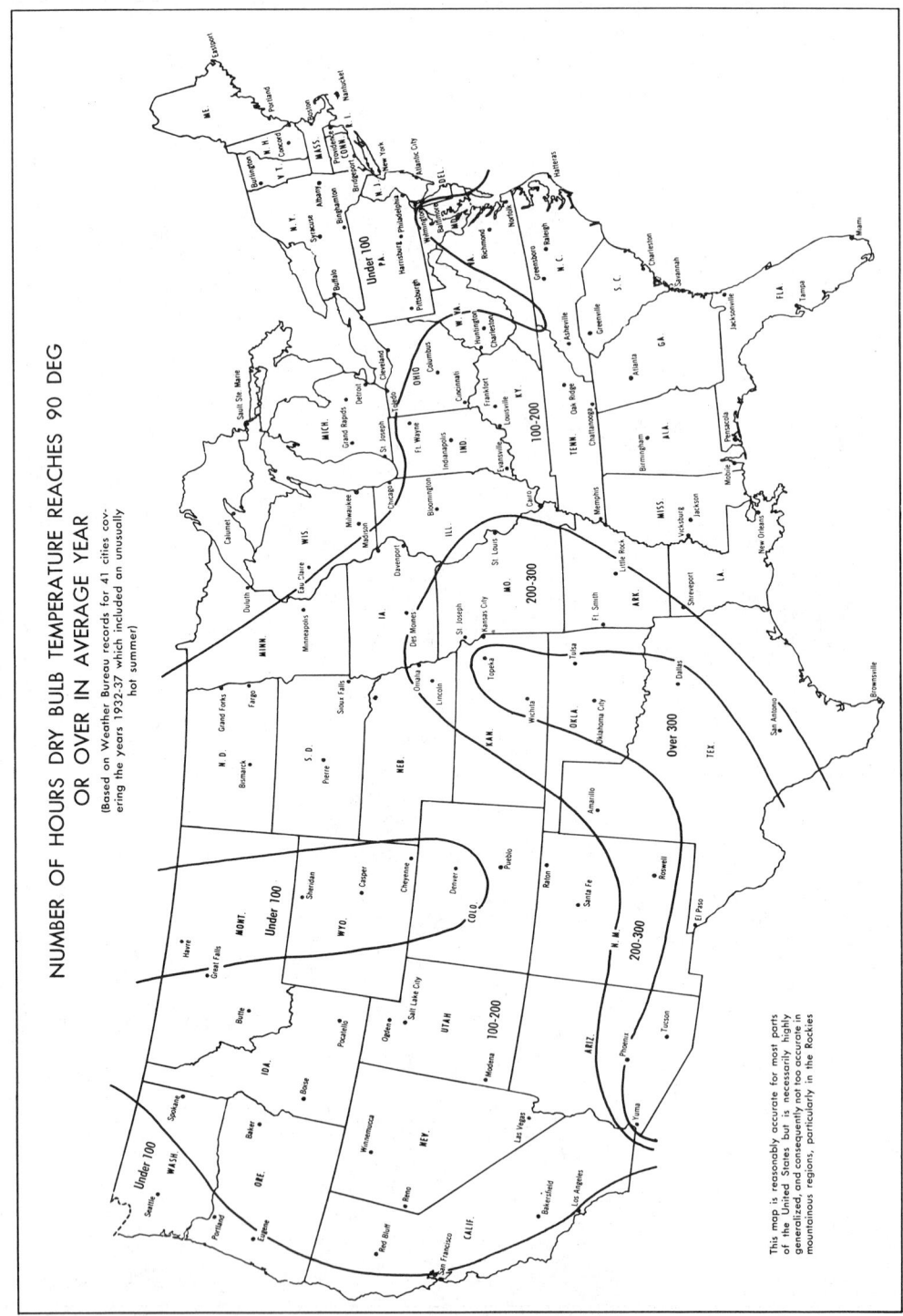

NUMBER OF HOURS DRY BULB TEMPERATURE REACHES 90 DEG
OR OVER IN AVERAGE YEAR

(Based on Weather Bureau records for 41 cities covering the years 1932-37 which included an unusually hot summer)

This map is reasonably accurate for most parts of the United States but is necessarily highly generalized, and consequently not too accurate in mountainous regions, particularly in the Rockies

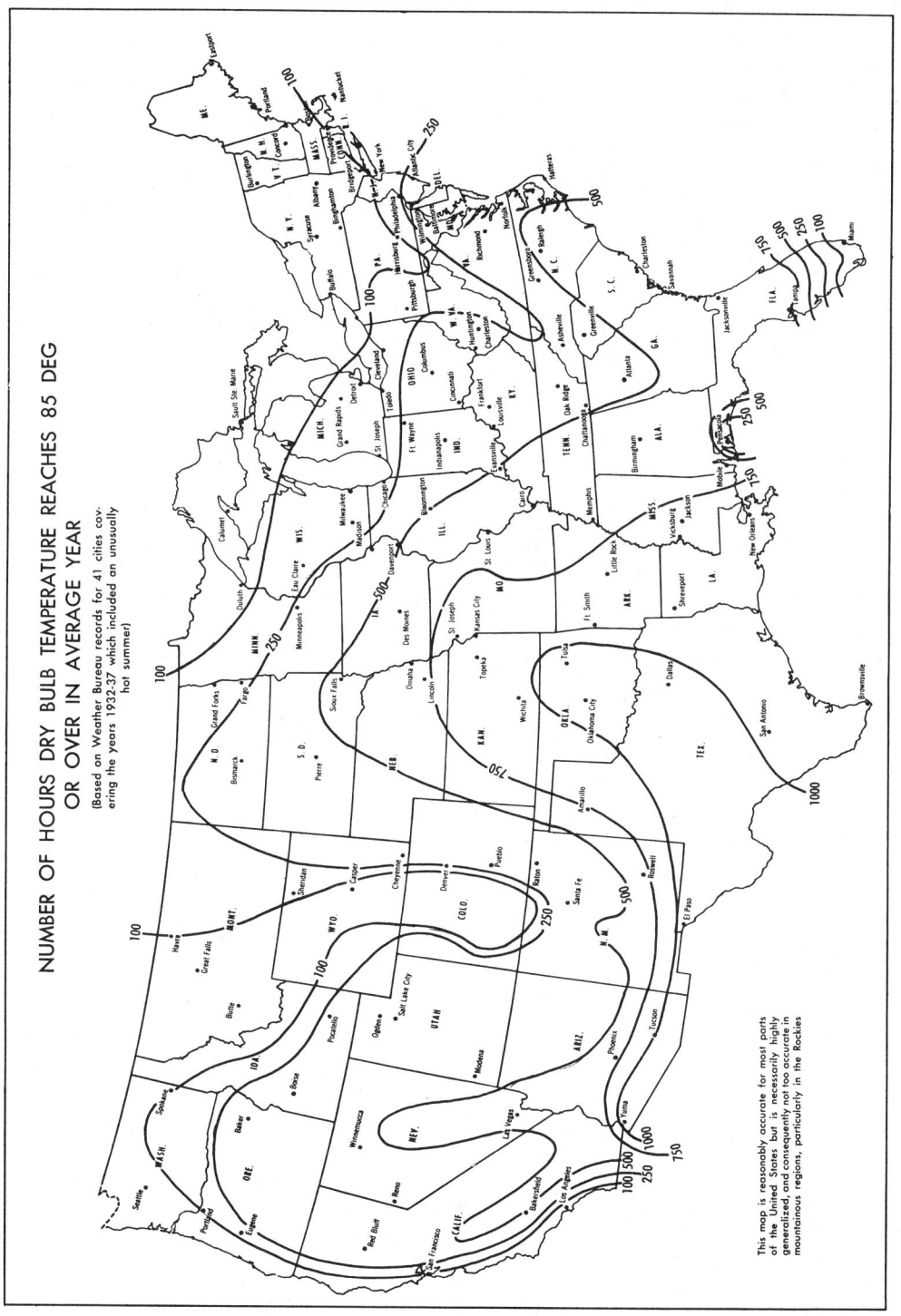

NUMBER OF HOURS DRY BULB TEMPERATURE REACHES 85 DEG
OR OVER IN AVERAGE YEAR

(Based on Weather Bureau records for 41 cities covering the years 1932-37 which included an unusually
hot summer)

This map is reasonably accurate for most parts
of the United States but is necessarily highly
generalized, and consequently not too accurate in
mountainous regions, particularly in the Rockies

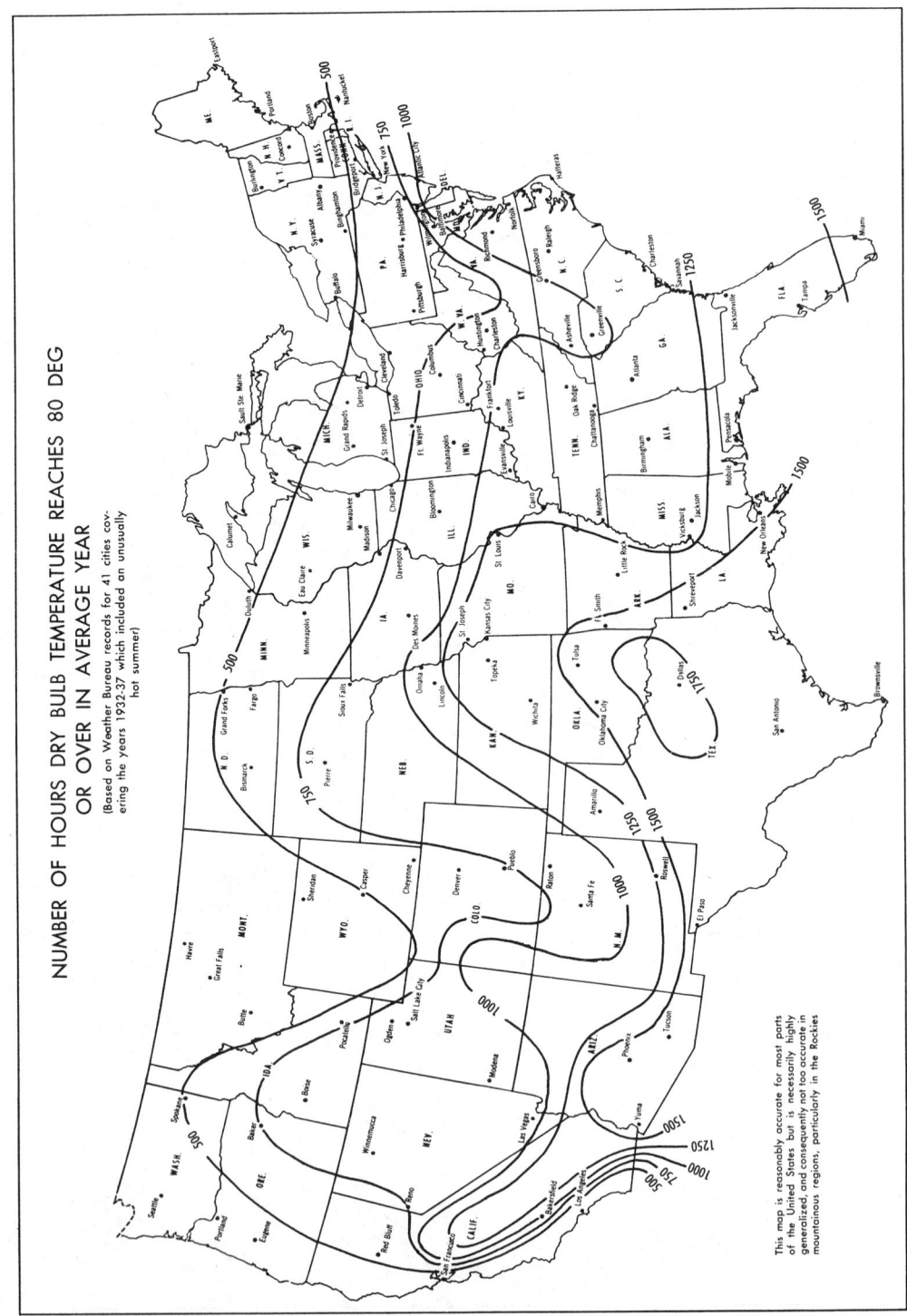

NUMBER OF HOURS DRY BULB TEMPERATURE REACHES 80 DEG OR OVER IN AVERAGE YEAR

(Based on Weather Bureau records for 41 cities covering the years 1932-37 which included an unusually hot summer)

This map is reasonably accurate for most parts of the United States but is necessarily highly generalized, and consequently not too accurate in mountainous regions, particularly in the Rockies

COOLING DEGREE-DAYS *

The problem of developing a cooling degree-day unit for air conditioning work similar to the degree-day unit for heating has been complicated by the fact that the energy consumption for summer cooling is affected not only by the dry bulb temperature but also by the relative humidity. What follows is a brief discussion of a cooling degree-day unit as developed by U. S. Weather Bureau personnel.

The first step was the development of an equation to approximate the ASHRAE Effective Temperature. This equation, derived by J. F. Bosen, Office of Climatology, U. S. Weather Bureau, is:

$$DI = 0.4\ (t_d + t_w) + 15 \qquad (1)$$

which can also be expressed as

$$DI = .55\ t_d + .2t_{dp} + 17.5 \qquad (2)$$

where DI = Discomfort Index

$\quad t_d$ = dry bulb temperature,

$\quad t_w$ = wet bulb temperature,

$\quad t_{dp}$ = dew point temperature,

and where the temperatures are simultaneous readings. The second equation is intended for future use when dew point temperatures will be more readily available; at the present time equation (1) is used.

The DI is *not* a temperature reading but an index in that the higher the value the greater the degree of discomfort.

The second step was the application of this index so that the resulting figure would be proportional to the energy consumed for air conditioning. This was done by Earl C. Thom, also of the office of Climatology, U. S. Weather Bureau, who has proposed

(a) for a given day, adding the four values of the DI for six hour intervals, (b) dividing by 4, and (c) subtracting this average from a base of 60. This yields a number proportional to the total of sensible and latent heat loads imposed by outside air. The daily cooling degree-day values can be accumulated into monthly and annual totals.

The accompanying map shows the *average* number of cooling degree-days for 1953 to 1957.

The material presented should be useful for reference and study but there are a number of refinements which may also need to be given consideration. The Discomfort Index value should prove basically satisfactory for a cooling degree-day system but more work needs to be done to ascertain whether or not 60 is the best possible base. Then there is the consideration of the effects of radiation and of surface wind speed on human comfort. It is believed that air-conditioning installations of office buildings, hotels, and hospitals are of major interest. Since the variations in radiation will have less effect on the comfort of the occupants of these buildings, such structures usually being well insulated, no doubt the variations of temperature and moisture content of the lower layers of the atmosphere, as expressed by these cooling degree-day values, will be found to correlate sufficiently well with the energy needed to operate air-conditioning equipment without considering the effects of radiation.

Table 1 shows values of the Discomfort Index when the Index reached maximum values.

Table 1. Selected Examples of Maximum Discomfort Index, 1953-57

Station	Date	Maximum Dry Bulb Temperature, Deg F	Discomfort Index at Time of Maximum Temperature	Cooling Degree-Day Value for This Date
Yuma, Ariz.	July 31, 1957	119	92	25
Yuma, Ariz.	June 24, 1957	120	91	20
Kansas City, Mo.	July 13, 1954	112	89	23
Charleston, S. C.	Aug. 17, 1954	102	87	20
Bakersfield, Calif.	Aug. 3, 1955	110	87	18
Miami, Fla.	Aug. 6, 1954	98	86	20
Oklahoma City, Okla.	July 12, 1954	107	86	19
New Orleans, La.	July 16, 1954	96	85	19
North Platte, Neb.	July 29, 1955	103	85	17
Chicago, Ill.	July 27, 1955	100	84	15
Los Angeles, Calif.	Aug. 31, 1955	101	84	9
Raleigh, N. C.	July 3, 1955	98	83	15
Las Vegas, Nev.	Aug. 30, 1955	108	83	13

* While the current use of Cooling Degree-Days is not firmly established, this material is included for historical reference.

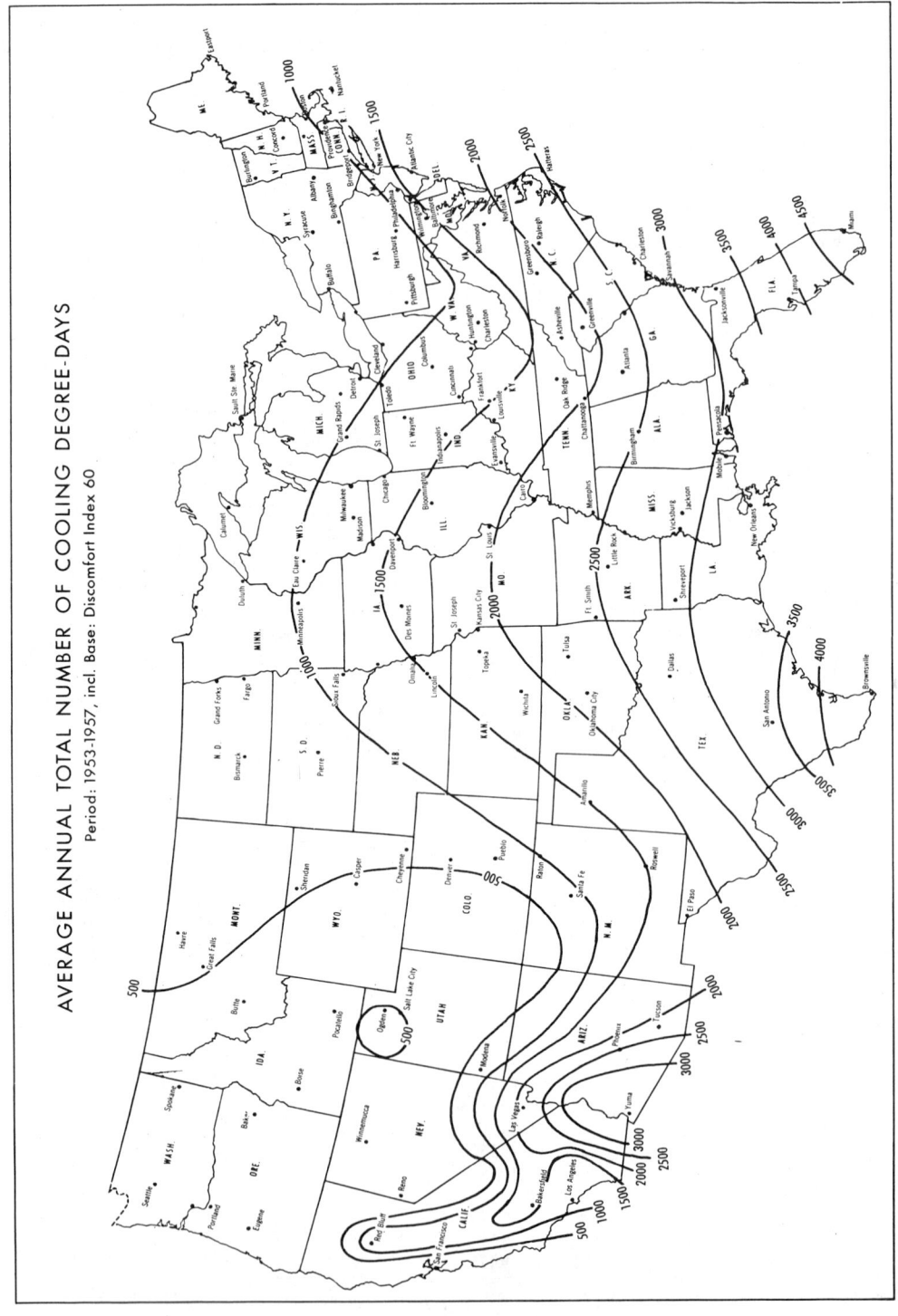

AVERAGE ANNUAL TOTAL NUMBER OF COOLING DEGREE-DAYS

Period: 1953-1957, incl. Base: Discomfort Index 60

HEAT GAINS THROUGH WALLS AND ROOFS

Sol-Air Method

The tables on the four pages which follow show the rate of heat gain through walls at various hours of a design day in mid-July. They are based on the sol-air temperature concept of Mackey and Wright (see credits), and take into account:

> Temperature difference between inside and outside air,
> Solar radiation,
> Time lag in heat transfer,
> Heat storage,
> Surface color effects,
> Summer film coefficients, and
> Wall orientation

The tables are calculated on an indoor dry bulb temperature of 80 deg F. If another indoor temperature, t, is used, add $U(80-t)$ to the tabulated values. Outside air temperatures of the design day reach a high of 95 deg at 3 p.m. and a pre-dawn low of 74 deg (see Fig. 1).

Solar radiation calculations are for a North latitude of 40 degrees but as U. S. Weather Bureau maps elsewhere in this section show, the effect of latitude on daily total solar radiation in summer is slight. It is believed that the effect of latitude for anywhere in the United States is probably less than 3%.

Summer film coefficients of 4 (7.5 mph wind) for outside walls and 1.2 for inside ceilings were used in basic calculations. Hence, application of tabulated U values is conservative for most areas in the United States. It is suggested that reference be made to the Daily Total Solar Radiation maps elsewhere in this section which indicate that in certain areas, such as near the Great Lakes, engineering judgment may safely dictate reductions of tabulated values by 5 or 10%.

Sol-Air Temperature

Sol-air temperature is the hypothetical temperature of outdoor air which, by convection and conduction only, would result in the same rate of heat transfer to a building surface as is accomplished by the combined effects of the actual air temperature and solar radiation. Thus, for example, although peak air temperature on a design day may be 95 deg at 3 p.m., the peak sol-air temperature for that day would be 144 deg and would occur at 1 p.m. on a horizontal roof. The diurnal march of air and sol-air temperatures is shown for variously oriented surfaces in Fig. 1. Note that these temperatures are an index of instantaneous rates of heat entry into

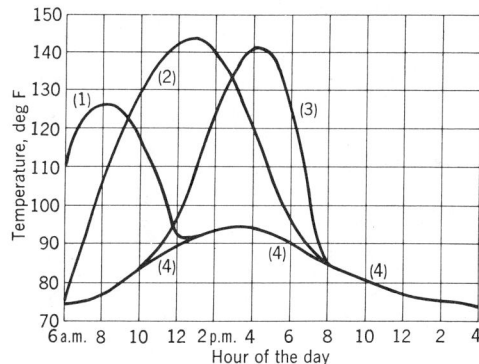

Fig. 1. Diurnal march of air and sol-air temperatures on a design day. Curves represent sol-air temperatures on (1) east wall, (2) horizontal roof, and (3) west wall. Curve (4) is outside air dry bulb temperature. Sol-air temperature on a south wall would follow a curve of expected shape with peak of 115 at 1 p.m. Sol-air temperature on a north wall (or shaded wall) is the same as air dry bulb.

surfaces, not into rooms. Heat gains within rooms are further modified by lag and other factors mentioned above.

Example: On a typical sunny July day, what is the heat gain due to sun heat and air temperature difference at 4 p.m. and at 6 p.m. through uninsulated walls of a wood frame building with wood sheathing, 25/32-inch wood siding, and plaster on studding? The siding is painted light color. Net wall areas are: North and south, 20,000 sq ft each; east and west, 10,000 sq ft each.

Solution: The table for wood frame walls of this construction shows the following heat gains in Btu per (hour) (square foot):

	4 p.m.	6 p.m.
North	3.6	3.4
South	5.3	4.6
East	5.3	4.6
West	8.4	7.4

Multiplying these values by respective wall areas and adding products for each hour gives total loads as follows:

	4 p.m.	6 p.m.
North	72,000	68,000
South	106,000	92,000
East	53,000	46,000
West	84,000	74,000
Total	315,000	280,000

Walls of One Material				Hours of the Day								
Construction	Thick-ness. Inches	U Value	Facing	8	10	Noon	2	4	6	8	10	12
				Heat Gain, Btu per (Hr) (Sq Ft)								
Plain Brick	4	.63	N*	1.2	2.5	4.4	6.3	7.5	8.2	8.1	6.9	5.6
			NE	10.7	12.6	12.6	12.6	12.6	11.3	10.0	8.1	6.3
			E	12.6	18.2	18.2	17.0	15.7	14.5	11.9	9.4	6.9
			SE	5.6	11.3	14.4	14.4	13.8	13.2	11.3	9.4	6.9
			S	1.8	5.0	10.6	13.8	15.1	14.4	12.6	10.8	8.2
			SW	1.2	3.1	6.9	15.1	20.1	19.5	17.0	13.8	10.7
			W	2.5	3.1	6.3	12.6	20.7	24.6	21.4	17.6	13.8
			NW	6.3	5.6	6.9	9.4	12.6	17.6	18.3	15.8	12.6
No Plaster on Walls	8	.41	N*	2.0	1.6	1.6	2.0	2.8	3.6	4.1	4.5	4.1
			NE	3.6	4.5	4.9	5.3	5.3	5.7	5.7	5.7	5.3
			E	3.2	4.5	6.6	7.8	7.7	7.3	6.9	6.8	6.1
			SE	2.4	2.4	3.6	5.3	6.1	6.1	6.1	6.1	5.7
			S	2.9	2.4	2.8	4.1	6.1	6.9	6.9	6.9	6.1
			SW	2.8	2.4	2.0	2.8	5.3	7.3	8.6	8.2	6.9
			W	3.6	2.8	2.4	2.8	4.5	7.3	9.4	9.8	8.6
			NW	3.2	2.8	2.4	2.8	3.6	5.3	7.3	7.7	7.3
	12	.31	N*	2.1	2.4	2.7	2.7	2.7	2.7	2.4	2.1	1.8
			NE	3.1	3.4	3.7	4.0	4.3	4.3	4.0	3.7	3.1
			E	3.4	3.4	4.0	4.9	5.5	5.5	4.9	4.3	4.0
			SE	3.1	3.4	3.7	4.0	4.3	4.6	4.3	3.7	3.4
			S	3.1	3.1	3.4	3.7	3.7	4.0	4.3	4.6	3.4
			SW	3.4	3.4	3.4	3.7	3.7	3.7	4.3	4.6	4.3
			W	4.0	4.0	4.0	4.0	4.0	4.0	4.3	4.6	4.9
			NW	3.4	3.7	3.7	4.0	4.0	3.7	3.7	3.7	3.7
Stone	8	.67	N*	2.6	2.6	2.6	4.0	5.3	6.7	7.3	7.3	7.3
			NE	5.3	6.7	7.3	8.0	8.0	8.7	9.3	8.7	8.0
			E	6.0	8.7	11.3	11.3	11.3	11.3	10.7	10.0	8.7
			SE	3.3	4.0	6.0	8.0	9.3	10.0	10.0	9.3	8.0
			S	3.3	3.3	4.6	7.3	10.0	10.7	10.7	10.0	9.3
			SW	3.3	2.6	3.3	5.3	9.4	12.7	12.7	12.0	10.2
			W	4.6	4.0	3.3	5.3	8.7	12.7	15.4	14.0	12.7
			NW	4.0	3.3	3.3	4.6	6.7	10.0	12.0	12.0	10.7
No Plaster on Walls	12	.55	N*	3.8	3.3	2.7	2.7	3.3	3.8	4.4	4.9	5.5
			NE	4.9	4.9	4.9	5.5	5.5	5.5	6.0	6.6	6.6
			E	4.9	4.9	5.5	6.6	7.1	7.7	7.7	7.1	7.2
			SE	4.9	4.4	3.8	4.4	5.5	6.0	6.6	7.1	7.2
			S	4.9	4.4	4.4	3.8	4.9	6.0	7.1	7.1	7.7
			SW	4.9	4.4	3.8	3.3	3.8	5.5	7.7	8.2	7.7
			W	6.0	4.9	4.4	3.8	4.4	5.5	7.7	9.3	9.3
			NW	5.5	4.4	3.8	3.8	3.8	4.9	6.6	7.7	7.7
	16	.47	N*	3.7	3.2	2.3	1.8	1.8	2.3	3.7	4.7	5.1
			NE	4.2	3.7	2.8	2.8	2.3	2.8	3.2	3.7	4.7
			E	4.2	3.7	3.2	3.7	4.2	4.7	6.5	6.1	5.6
			SE	4.2	4.2	3.7	3.2	3.2	3.7	4.2	5.1	5.6
			S	4.2	4.2	3.7	3.2	3.7	4.2	5.1	5.6	4.7
			SW	4.2	3.7	3.2	3.2	2.8	3.2	3.2	5.2	6.1
			W	4.7	3.7	3.2	3.2	3.2	3.7	5.1	6.5	7.0
			NW	4.2	3.7	3.2	2.8	2.8	3.2	3.7	4.7	5.1

*Use also for shade.

Walls of One Material				Hours of the Day								
Construction	Thickness, Inches	U Value	Facing	8	10	Noon	2	4	6	8	10	12
				Heat Gain, Btu per (Hr) (Sq Ft)								
Concrete	2	.99	N*	4.0	8.9	11.9	14.8	14.8	13.8	10.9	7.9	4.0
			NE	22.7	22.7	22.7	22.7	20.8	16.8	12.7	7.9	4.0
			E	27.6	32.6	31.8	29.8	29.8	20.8	15.8	9.9	4.9
			SE	20.8	26.7	26.7	25.7	23.7	19.8	14.9	9.9	4.9
			S	2.8	17.8	23.8	25.7	23.7	19.8	15.8	10.9	5.9
			SW	5.0	11.9	23.7	32.6	33.7	28.7	22.8	15.9	9.9
			W	4.0	9.9	20.8	34.6	39.6	37.6	27.6	20.8	13.9
			NW	4.0	8.9	15.9	26.7	32.6	29.6	23.7	17.9	11.9
Walls 140-lb Density	4	.85	N*	2.6	5.6	7.5	11.9	11.9	11.9	10.2	8.5	6.0
			NE	15.3	16.2	17.0	18.7	17.0	15.3	12.8	9.4	6.0
			E	19.6	22.1	22.1	23.0	19.6	17.0	13.6	10.2	6.8
			SE	11.0	17.8	19.5	21.3	18.8	17.0	13.6	10.2	6.8
			S	3.4	9.4	16.2	20.4	19.5	17.0	14.5	11.0	7.65
			SW	2.6	6.8	14.4	24.7	26.4	23.8	19.5	15.3	11.0
			W	3.4	5.9	11.9	23.8	29.8	29.8	24.7	19.6	13.6
			NW	2.6	5.1	10.2	18.7	23.8	24.6	21.3	16.3	11.9
No Plaster on Walls	8	.67	N*	1.3	2.0	4.0	5.4	7.4	8.0	8.0	8.0	6.7
			NE	7.4	9.4	9.4	10.0	10.7	10.7	10.0	8.7	7.4
			E	7.4	12.0	13.4	14.1	13.4	12.7	11.4	10.0	7.4
			SE	3.4	6.7	10.0	11.4	12.0	12.0	10.7	9.4	7.4
			S	2.0	3.4	7.4	10.7	12.7	12.7	12.0	10.0	8.7
			SW	2.0	2.7	5.4	10.7	14.7	16.1	14.7	12.7	10.0
			W	3.4	3.4	4.7	9.4	15.4	18.7	18.1	15.4	12.7
			NW	8.0	3.4	4.0	7.4	12.0	15.4	15.4	13.4	10.7
	12	.55	N*	2.8	2.2	2.2	2.8	3.9	5.0	6.1	6.1	5.5
			NE	4.4	5.0	6.0	6.0	6.6	7.2	7.2	7.2	6.6
			E	3.9	5.5	7.7	8.8	8.8	8.8	8.8	8.3	7.2
			SE	2.8	3.3	5.0	6.6	7.2	7.7	7.2	7.7	7.2
			S	3.3	2.8	3.3	5.5	7.7	8.8	8.8	8.3	7.7
			SW	3.3	2.2	2.2	3.9	7.2	9.9	10.4	9.9	8.8
			W	3.9	3.3	2.8	3.9	6.7	9.9	12.1	11.6	10.5
			NW	3.3	2.8	2.8	3.3	5.0	7.7	9.3	9.9	8.8
	16	.46	N*	3.2	2.8	2.3	2.3	2.8	3.2	4.2	4.6	4.6
			NE	4.2	4.2	4.2	4.6	4.6	5.0	5.5	5.5	5.5
			E	4.2	3.7	4.3	6.0	6.4	6.4	6.4	6.4	6.4
			SE	3.7	3.2	2.8	2.8	3.8	5.0	6.9	6.9	6.9
			S	3.7	3.2	3.2	3.7	4.6	5.5	6.0	6.4	6.4
			SW	4.2	3.2	2.8	2.8	3.7	5.0	6.9	6.9	6.9
			W	4.6	4.2	4.2	3.2	3.7	5.1	6.9	8.3	7.8
			NW	4.2	3.7	4.2	3.2	3.2	4.2	5.5	6.5	6.9
	20	.41	N*	3.3	2.9	2.5	2.5	2.1	2.5	2.9	3.3	3.3
			NE	4.1	3.7	3.7	3.7	3.7	3.7	4.1	4.5	4.5
			E	4.5	4.1	4.1	4.9	4.9	5.3	5.3	5.3	5.3
			SE	4.1	3.7	3.3	3.3	3.7	4.1	4.5	4.5	4.9
			S	4.1	3.7	3.3	3.3	3.3	3.7	4.1	4.9	4.9
			SW	4.5	3.7	3.3	2.9	2.9	3.3	4.5	5.3	5.3
			W	5.3	4.9	4.5	4.1	3.7	4.1	4.5	5.7	6.6
			NW	4.5	4.1	3.7	3.3	3.3	3.3	3.7	4.5	4.9

*Use also for shade.

Wood Frame Walls				Hours of the Day								
Construction	U Value	Facing	Light or Dark Color	8	10	Noon	2	4	6	8	10	12
				Heat Gain, Btu per (Hr) (Sq Ft)								
Wood Siding 25/32-inch Wood Sheathing Plaster on Studding no Insulation	.24	N*	LorD	.96	1.9	2.9	3.4	3.6	3.4	2.6	1.9	.96
		NE	L	4.3	4.8	5.0	5.0	4.6	4.1	3.1	1.9	.96
		NE	D	7.0	7.0	6.7	6.2	5.5	4.6	3.4	2.2	.96
		E	L	5.8	6.2	6.2	6.0	5.3	4.6	3.4	2.2	.96
		E	D	9.4	9.4	8.6	7.9	6.7	5.3	3.8	2.4	.96
		SE	L	4.3	5.5	5.8	5.8	5.3	4.6	3.4	2.4	1.2
		SE	D	6.7	8.4	8.2	7.5	6.5	5.3	4.1	2.6	1.2
		S	L	1.7	3.6	5.0	5.5	5.3	4.6	3.6	2.4	1.4
		S	D	2.4	4.8	7.0	7.2	6.5	5.5	4.3	2.9	1.7
		SW	L	.96	2.6	5.0	7.0	7.2	6.7	4.8	3.6	2.2
		SW	D	1.2	3.1	6.7	9.8	10.0	9.1	6.5	4.8	2.9
		W	L	.96	2.2	4.3	7.2	8.4	7.4	6.0	4.3	2.9
		W	D	.96	2.4	5.8	10.0	12.0	10.1	8.4	6.2	4.1
		NW	L	.96	2.2	3.6	5.8	7.0	6.5	5.3	3.8	2.4
		NW	D	.96	2.2	4.3	7.7	9.6	8.9	7.2	5.3	3.6
Same as above but with Rock Wool Fill	.06	N*	LorD	0	.24	.48	.66	.78	.78	.72	.60	.42
		NE	L	.66	.90	.96	1.0	1.0	1.0	.84	.66	.42
		NE	D	1.2	1.3	1.4	1.3	1.3	1.1	.96	.72	.42
		E	L	.90	1.2	1.3	1.3	1.3	1.1	.96	.72	.48
		E	D	1.5	1.9	1.8	1.7	1.6	1.4	1.1	.84	.48
		SE	L	.42	.90	1.1	1.2	1.2	1.1	.96	.72	.48
		SE	D	.72	1.4	1.6	1.6	1.4	1.3	1.1	.84	.54
		S	L	.06	.48	.90	1.1	1.2	1.1	.96	.78	.54
		S	D	.12	.60	1.2	1.5	1.5	1.4	1.1	.90	.60
		SW	L	.06	.30	.98	1.3	1.6	1.5	1.2	1.0	.72
		SW	D	.06	.36	1.0	1.8	2.2	2.0	1.7	1.3	.90
		W	L	.06	.30	.66	1.2	1.7	1.7	1.5	1.2	1.0
		W	D	.12	.30	.78	1.7	2.3	2.4	2.1	1.7	1.2
		NW	L	.06	.24	.54	.96	1.4	1.5	1.3	1.1	.78
		NW	D	.06	.24	.60	1.2	1.9	2.0	1.7	1.4	1.0
Wood Siding 25/32-inch Rigid Insulation for Sheathing Plaster on Studding	.22	N*	LorD	.66	1.5	2.4	3.1	3.3	3.1	2.4	2.0	1.1
		NE	L	3.7	4.0	4.2	4.4	4.2	3.7	2.9	2.0	1.1
		NE	D	5.7	6.0	5.7	5.5	5.1	4.2	3.1	2.2	1.1
		E	L	4.8	5.3	5.5	5.3	4.8	4.2	3.3	2.2	1.1
		E	D	7.7	8.1	7.7	7.1	6.2	5.1	3.7	2.4	1.1
		SE	L	3.5	4.6	5.1	5.1	4.8	4.2	3.3	2.2	1.3
		SE	D	5.5	7.1	7.1	6.6	6.0	4.8	3.7	2.6	1.3
		S	L	1.3	3.1	4.4	4.8	4.8	4.2	3.3	2.4	1.3
		S	D	1.8	4.2	6.0	6.4	6.0	5.1	4.0	2.9	1.5
		SW	L	.66	2.2	4.2	6.2	6.4	5.7	4.6	3.3	2.2
		SW	D	.88	2.6	5.7	8.6	8.8	7.7	6.0	4.4	2.9
		W	L	.66	1.8	3.7	6.2	7.5	6.8	5.5	4.2	2.9
		W	D	.88	2.0	4.6	8.6	10.5	9.7	7.9	5.9	4.0
		NW	L	.66	1.5	2.8	5.1	6.2	5.9	4.8	3.5	2.4
		NW	D	.66	1.8	3.5	6.4	8.4	8.1	6.6	4.8	3.3

*Use also for shade.

Wood Frame Walls			Hours of the Day								
Construction	U Value	Facing	8	10	Noon	2	4	6	8	10	12
			Heat Gain, Btu per (Hr) (Sq Ft)								
Brick Veneer 25/32 Wood Sheathing Plaster on Studding	.29	N*	.87	.87	1.5	2.3	2.9	3.5	3.5	3.5	2.9
		NE	3.5	4.6	4.9	4.9	5.2	4.9	4.6	4.1	3.5
		E	4.1	6.7	7.5	7.3	7.0	6.4	5.5	4.6	3.5
		SE	1.7	3.5	5.2	5.8	5.8	5.5	5.2	4.3	3.5
		S	.87	1.7	3.5	5.5	6.4	6.1	5.8	4.9	4.1
		SW	.87	1.2	2.3	4.9	7.5	8.1	7.5	6.4	5.2
		W	1.7	1.4	2.0	4.3	7.5	9.9	9.6	8.1	6.7
		NW	1.2	1.2	1.7	3.2	5.5	7.5	7.8	6.7	5.5
Brick Veneer 25/32 Wood Sheathing Plaster on Studding Rock Wool Fill	.069	N*	.28	.28	.28	.41	.55	.69	.76	.76	.76
		NE	.62	.83	.96	.96	1.0	1.0	1.0	.97	.90
		E	.55	.96	1.4	1.4	1.4	1.3	1.2	1.1	.97
		SE.	.35	.48	.83	1.1	1.2	1.2	1.1	1.0	.90
		S	.35	.35	.48	.83	1.2	1.3	1.3	1.2	1.0
		SW	.41	.28	.34	.69	1.2	1.6	1.6	1.4	1.2
		W	.55	.48	.41	.62	1.1	1.7	1.9	1.8	1.5
		NW	.41	.35	.35	.48	.76	1.2	1.4	1.4	1.2
Brick Veneer 25/32 Rigid Insulation for Sheathing Plaster on Studding	.22	N*	.66	.66	1.1	1.5	2.2	2.4	2.6	2.6	2.2
		NE	2.6	3.5	3.5	3.7	3.7	3.7	3.5	3.3	2.6
		E	2.6	4.6	5.3	5.3	5.1	4.6	4.2	3.7	2.8
		SE	1.1	2.2	3.5	4.2	4.2	4.2	3.7	3.5	2.9
		S	.88	1.1	2.2	3.7	4.6	4.6	4.4	4.0	3.1
		SW	.88	.66	1.5	3.3	5.3	5.9	5.5	5.1	4.0
		W	1.3	1.1	1.5	2.9	5.1	6.8	6.8	6.4	5.1
		NW	1.1	.88	1.1	2.0	3.5	5.3	5.5	5.1	4.2

Flat Roofs					Hours of the Day							
Construction		Deck Thickness, Inches	U Value	Aluminum or Black	8	10	Noon	2	4	6	8	10	12
					Heat Gain, Btu per (Hr) (Sq Ft)							
Concrete Roofs	No Ceiling No Insulation	2	.78	A	12.5	21.0	26.5	27.3	24.2	19.5	14.8	9.3	5.5
				B	30.4	50.0	59.2	57.6	48.3	36.6	25.7	15.5	7.8
		4	.70	A	7.7	14.7	20.1	24.5	23.1	19.6	16.1	12.6	8.4
				B	16.1	32.2	44.8	50.4	45.5	36.4	27.3	18.9	11.9
		6	.63	A	4.4	8.2	13.8	17.6	18.2	17.0	14.5	11.3	8.2
				B	9.4	20.1	32.8	40.3	39.6	34.0	27.1	20.1	13.8
	No Ceiling 2" Insulation	4	.11	A	.7	1.3	2.2	3.0	3.2	3.0	2.5	2.1	1.5
				B	1.5	3.2	5.3	6.7	6.6	6.0	4.9	3.9	2.6
		6	.11	A	.7	1.2	2.0	2.7	3.1	3.0	2.5	2.1	1.6
				B	1.3	2.7	4.7	6.3	6.6	6.0	5.0	3.8	2.7
	Plaster Ceiling 2" Insulation	4	.10	A	.6	1.2	2.0	2.6	2.8	2.7	2.3	1.9	1.4
				B	1.3	2.8	4.6	5.9	6.2	5.5	4.4	3.4	2.4
		6	.10	A	.6	1.0	1.8	2.4	2.7	2.7	2.3	1.9	1.5
				E	1.2	2.4	4.1	5.5	6.0	5.5	4.6	3.5	2.6
Wood Roofs	No Ceiling No Insulation	1½	.43	A	3.9	8.2	11.6	13.3	13.3	11.6	9.0	6.9	4.3
				B	9.9	19.3	27.0	30.0	27.5	22.4	16.7	11.6	7.3
	No Ceiling 2" Insulation	1½	.10	A	.6	.9	1.6	2.3	2.7	2.6	2.4	2.0	1.6
				B	1.1	2.2	3.9	5.3	5.9	5.5	4.6	3.6	2.7
	Plaster Ceiling 2" Insulation	1½	.09	A	.5	.8	1.4	2.1	2.4	2.4	2.2	1.8	1.4
				B	1.0	1.9	3.4	4.8	5.2	5.0	4.1	3.3	2.2

*Use also for shade.

DIRECT HEAT GAINS FROM THE SUN

On this and subsequent pages are data for applying a direct method of estimating the sun heat which passes through a wall, roof, or glass. This will be an advantage where special shading orientation, or fenestration conditions exist.

Walls and Roofs. Only a percentage of the radiant heat from the sun is absorbed by the wall or roof, and of this heat only a part reaches the interior of the building because of the resistance of the wall or roof construction. This statement is represented by the following formula:

$$H_s = A\ I\ b\ g \qquad (1)$$

H_s = sun heat reaching the room, Btu per hr,
A = area affected by sun, sq ft,
I = intensity of solar radiation striking the area, Btu per (hr) (sq ft),
b = decimal part of I absorbed by the surface, and
g = decimal part of b transmitted to the inside.

Values for Solar Intensity (I). From measurements of the amount of solar radiation received on a surface normal to the sun in midsummer, Table 4 on intensity (I) has been calculated. Values for sloped roofs may be approximated somewhere between the amount received on a flat roof and that received on a vertical wall and will depend on the slope of the roof in the relation, $I = I_n \cos i$, where I_n is intensity normal to the sun and i is the angle of incidence. Values of I_n can be obtained from the charts of Insolation on Cloudless Days elsewhere in this book (see Index), but June 21 values should be reduced by about 20% for most midsummer calculations in average atmospheres.

Absorption of Solar Energy (b). The factor b in equation (1) is the fraction of solar heat which is absorbed. Table 1 lists values of solar absorption collected from various sources for a large variety of building materials.

Amount of Solar Energy Conducted (g). The heat absorbed at the outer surface of a wall or roof meets with further resistance before it reaches the interior. This resistance is expressed in formula (1) by the letter g. The heat resistivity and the heat lag of the building material as well as the intermittent character of the solar radiation tend to restrict the flow of heat through the wall. (See Table 2.) The estimated sun heat is added to the cooling load at a time later than the hour of calculation by the amount of lag shown in the following table.

Time Lag in Periodic Heat Flow through Walls

Construction	Lag, Hours
Walls:	
Brick, 4-inch	2.3
8-inch	5.5
12-inch	8.0
Concrete, solid or block	
2-inch	1.0
4-inch	2.6
6-inch	3.8
8-inch	5.1
10-inch	6.4
12-inch	7.6
Glass	
window	0.0
block[1]	2.0
Stone	
8-inch	5.4
12-inch	8.0
Frame	
wood, plaster,	
no insulation	0.8
insulated	3.0
Brick veneer, plaster	
no insulation	3.0
insulated	5.5
Roofs:[2]	
Light construction	0.7 to 1.3
Medium construction	1.4 to 2.4
Heavy construction	2.5 to 5.0

[1] Note: Glass block transmits, in addition to delayed conducted load, an instantaneous sun load which for smooth-faced block is about 0.45 times that of plain window glass, and for diffusing block about 25% of plain glass transmission.

[2] Note: Where applicable, the lags for wall materials can be used for roofs. Lags of materials added to a built-up structures are additive.

Area Affected by Sun (A). The values of A are obtained by measurement. Special attention should be given to those areas on which the sun shines at the time of the peak load.

Glass. In the case of windows,

$$H_g = ASI_g \qquad (2)$$

H_g = sun heat transmitted by exposed window glass, Btu per hr,
A = net area of glass actually exposed to direct rays of sun, sq ft,
S = shading factor, expressed as decimal of heat transmitted by glass, and
I_g = intensity of solar heat transmitted by clean window glass exposed to sun, Btu per (hr) (sq ft).

Table 1. — Solar Radiation (b) Absorbed by Various Substances

(Figures are expressed as decimals of the intensity of solar radiation striking the surface)

Brick		Roofing Materials	
Glazed, white................	0.26	Asbestos-cement, white........	0.42
Glazed, ivory to cream........	0.35	Asbestos-cement, 6 months' exposure....................	0.61
Common, light red...........	0.55	Asbestos-cement, 12 months' exposure....................	0.71
Common, red.................	0.68	Asbestos-cement, 6 years' exposure, very dirty............	0.83
Wire-cut, red................	0.52	Asbestos-cement, red..........	0.69
Mottled purple..............	0.77		
Blue........................	0.89	Asphalt, new.................	0.91
Limestone		Asphalt, weathered...........	0.82
Light.......................	0.35	Bitumen-covered roofing sheet, brown....................	0.87
Dark.......................	0.50	Bitumen-covered roofing sheet, green....................	0.86
Sandstone			
Light fawn..................	0.54	Bituminous felt..............	0.88
Light grey..................	0.62	Bituminous felt, with aluminized surface....................	0.40
Red........................	0.73	Slate, silver grey.............	0.79
Marble		Slate, blue grey..............	0.87
White.....	0.44	Slate, greenish grey, rough......	0.88
Dark.......................	0.66	Slate, dark grey, smooth.......	0.89
Granite		Slate, dark grey, rough........	0.90
Reddish....................	0.55	Tile, clay, machine made, red....	0.64
Metals		Tile, clay, machine made, dark purple.....................	0.81
Steel, vitreous enameled, white..	0.45	Tile, clay, hand made, red.......	0.60
Steel, vitreous enameled, green..	0.76	Tile, clay, hand made, reddish brown...................	0.69
Steel, vitreous enameled, dark red........................	0.81	Tile, concrete, uncolored.......	0.65
Steel, vitreous enameled, blue....	0.80	Tile, concrete, brown.........	0.85
Galvanized iron, new.........	0.64	Tile, concrete, black..........	0.91
Very dirty.............	0.92		
White, washed.............	0.22		
Copper, polished.............	0.18		
Tarnished..............	0.64		
Lead sheeting, old...........	0.79		
Paints			
Aluminum...................	0.54		
Cellulose, white.............	0.18		
Cellulose, yellow.............	0.33		
Cellulose, orange.............	0.41		
Cellulose, signal red..........	0.44		
Cellulose, dark red...........	0.57		
Cellulose, brown.............	0.79		
Cellulose, bright green........	0.79		
Cellulose, dark green..........	0.88		
Cellulose, dark blue...........	0.91		
Cellulose, black.............	0.94		

Where specific material is not mentioned above, an approximate value may be assigned by use of the following rough color guide.

1.—For white, smooth surfaces, use	0.25 to 0.40
2.—For grey to dark grey, use	0.40 to 0.50
3.—For green, red and brown, use..	0.50 to 0.70
4.—For dark brown to blue, use....	0.70 to 0.80
5.—For dark blue to black, use	0.80 to 0.90

Table 2. — Values of g for Walls and Roofs

Coefficient U for Wall or Roof	Decimal Part of Absorbed Solar Radiation Transmitted to Inside
0.10	0.02
0.15	0.03
0.20	0.04
0.25	0.05
0.30	0.062
0.35	0.075
0.40	0.085
0.45	0.095
0.50	0.105
0.60	0.13
0.70	0.15

Table 3. — Shading Factor (S) for Glass Windows and Skylights

(In Decimals of Heat Transmitted Without Shade)

Total shading, regardless of source	00
Tree shade, heavy.............	0.20 to 0.25
Tree shade, light.............	0.50 to 0.60
Awning, first floor............	0.25 to 0.30
Awning, upper floors..........	0.15 to 0.20
Inside window shade, light colored, fully drawn........	0.45 to 0.55
Inside Venetian blind, light colored....................	0.45 to 0.55
Inside draperies, light colored...	0.80 to 0.90

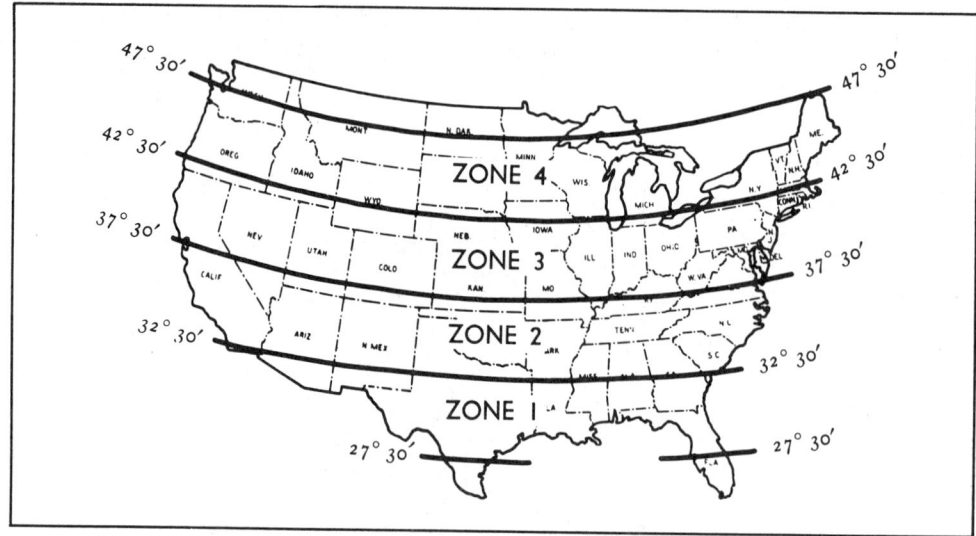

Fig. 1 U.S. zones of solar intensity with reference to Tables 4 to 7.

Sun Heat Through Glass (I_g). Table 5 gives values of I_g for four different zones of the United States shown in Fig. 1. The values in the table are for a clear day in August.

Shading Factors (S). Table 3 gives values of the shading factor S of equation (2). Note these factors are applied to the area A, not always total glass area.

Glass Area (A). It should be kept in mind that the area, A, is the net area of glass actually exposed to the sun's rays. This can be calculated as shown in the section on sun angles.

Screens or Special Glass

Tables 6 and 7 show solar heat transmitted through single glass windows in the summer when ordinary glass is used with shading screens and when heat absorbing glass is used. The figures are entirely independent of the amount of heat entering the room due to the temperature differential between inside and outside air.

Table 6 shows sun heat gain in Btu per (hr) (sq ft) through a plain glass window equipped with a shading screen of the Koolshade type.

Figures in Table 7 apply to ordinary thicknesses of heat-absorbing glass. The figures are conservative; that is, commercial heat-absorbing glass is available for which claims are made for somewhat smaller transmission values than those given here.

To use the tables, determine the zone of the city where the installation is located by referring to the

map, Fig. 1. to the proper table, and opposite the time of day and under the direction faced by the window find the solar heat entering the room per square foot during the hour in question.

Example: What will be the summer heat gain at 3 p.m. through a window 7 ft × 5 ft, ordinary single glass, in the SW wall of a Washington, D. C., building, and how much heat would be saved by using heat-absorbing glass?

Solution: Fig. 1 shows that Washington lies in Zone 3. Opposite 3 p.m. and under SW in Table 5, Zone 3, find 131 for ordinary glass and in Table 7, Zone 3, find 101 for heat-absorbing glass. These figures are per square foot, so that the transmission of the plain glass window during the hour from 2:30 to 3:30 p.m. is estimated at (7 × 5) × 131, or 4585 Btu.

The heat-absorbing glass transmission would be (7 × 5) × 101 = 3535 Btu for the same hour. The saving would be 4585 — 3535 or 1050 Btu for one hour.

In the case of skylights in this building, the gain would be 172 and 133 Btu per square foot for the hour in question in Washington for plain and heat-absorbing glass, respectively.

The foregoing information on plain or heat-absorbing glass deals with unshaded windows. If there are shade trees outside the window the heat gain from the sun is sharply cut down, sometimes to the point where it becomes of little moment. The efficacy of various types of shading on windows is summarized in Table 3.

Table 4.—Intensity (I) of Solar Radiation in Summer Striking Walls and Roofs

30° N: Zone 1, Fig. 1

Sun Time	Vertical Walls Facing								Horizontal Surface
	E	SE	S	SW	W	NW	N	NE	
6 AM	113	55					35	105	20
7	180	104					35	153	75
8	204	133					15	155	145
9	180	134						120	200
10	136	120	32					73	254
11	72	85	47					18	283
Noon		37	52	37					296
1			46	82	70	18			277
2			31	115	132	71			245
3			10	130	173	116			195
4				128	195	148	15		138
5				105	182	153	35		75
6				55	113	105	35		20

35° N: Zone 2, Fig. 1

Sun Time	E	SE	S	SW	W	NW	N	NE	Horizontal Surface
6 AM	113	50					33	103	24
7	180	102					28	148	79
8	204	138	8				10	146	145
9	180	147	27					108	200
10	136	140	54					58	250
11	72	112	71	10				12	277
Noon		71	77	55					290
1		11	70	99	70	12			277
2			52	129	132	56			242
3			26	140	173	104			192
4			7	136	195	140	10		138
5				109	181	148	28		60
6				56	113	103	33		24

40° N: Zone 3, Fig. 1

Sun Time	E	SE	S	SW	W	NW	N	NE	Horizontal Surface
6 AM	113	45					31	102	26
7	180	100					21	143	81
8	204	143	15				5	137	145
9	180	160	44					96	197
10	136	162	75					43	244
11	72	141	94	22				6	270
Noon		105	102	72					282
1		21	92	115	70	6			265
2			73	144	134	42			236
3			43	152	172	92			190
4			14	145	195	132	5		138
5				114	180	143	21		81
6				58	113	102	31		26

45° N: Zone 4, Fig. 1

Sun Time	E	SE	S	SW	W	NW	N	NE	Horizontal Surface
6 AM	113	42					29	100	30
7	180	99					14	138	83
8	204	148	22					128	143
9	180	173	62					83	192
10	136	183	97					28	247
11	72	167	118	33					260
Noon		136	130	90					270
1		32	116	132	70				257
2			93	158	132	27			229
3			59	164	172	80			185
4			22	154	195	123			138
5				118	182	138	14		83
6				59	113	100	29		29

Table 5.—Intensity (I_g) of Solar Radiation in Summer Transmitted by Clean Glass Windows

30° N: Zone 1, Fig. 1

Sun Time	Vertical Plain Glass Facing								Horizontal Skylights
	E	SE	S	SW	W	NW	N	NE	
6 AM	104	43					23	96	8
7	167	86					14	149	56
8	185	110						135	123
9	160	110						95	182
10	107	90	9					42	234
11	39	52	17						260
Noon		10	21	10					272
1			16	51	39				256
2			8	87	104	40			226
3				106	154	91			175
4				107	178	130			119
5				86	167	150	14		56
6				43	104	96	23		8

35° N: Zone 2, Fig. 1

Sun Time	E	SE	S	SW	W	NW	N	NE	Horizontal Skylights
6 AM	104	39					20	95	10
7	167	84					9	134	59
8	185	117						125	125
9	160	124	8					81	181
10	107	114	20						230
11	39	83	38						255
Noon		38	44	23					267
1			38	68	39				251
2			22	103	104	26			221
3			7	120	155	78			174
4				116	178	120			118
5				91	167	134	9		59
6				45	104	95	20		10

40° N: Zone 3, Fig. 1

Sun Time	E	SE	S	SW	W	NW	N	NE	Horizontal Skylights
6 AM	104	34					18	93	14
7	167	82					5	130	62
8	185	123						115	125
9	160	138	17					68	178
10	107	137	44					15	224
11	39	114	62						248
Noon		74	71	39					260
1			61	86	39				244
2			42	119	104	14			216
3			16	131	155	65			172
4				125	178	110			118
5				97	167	130	5		62
6				47	104	93	18		14

45° N: Zone 4, Fig. 1

Sun Time	E	SE	S	SW	W	NW	N	NE	Horizontal Skylights
6 AM	104	29					16	92	16
7	167	86						124	63
8	185	128	5					106	123
9	160	150	32					54	173
10	107	160	66					6	218
11	39	139	89	9					239
Noon		108	98	57					250
1		9	87	105	39				235
2			64	135	104	6			210
3			31	145	155	52			166
4			5	135	178	100			117
5				102	167	124			63
6				48	104	92	16		16

Table 6.—Intensity (I_g) of Solar Radiation in Summer Transmitted by Plain Glass Windows Covered by Shading Screen

30° N: Zone 1, Fig. 1

Sun Time	Windows with Shading Screens Facing								Horizontal Sky-lights
	E	SE	S	SW	W	NW	N	NE	
6 AM	27	9					6	24	—
7	57	17					6	42	—
8	42	16						21	—
9	21	12						10	—
10	12	9	1					4	—
11	4	4	2					1	—
Noon		1	2	1					—
1			2	4	4	1			—
2			1	9	12	4			—
3				12	21	10			—
4				16	42	21			—
5				17	57	42			—
6				9	27	24			—

35° N: Zone 2, Fig. 1

Sun Time	E	SE	S	SW	W	NW	N	NE	Hori-
6 AM	36	12					5	33	—
7	59	20					2	41	—
8	42	18						20	—
9	21	14	1					10	—
10	12	11	2					3	—
11	4	6	3						—
Noon		2	4	2					—
1			3	6	4				—
2			2	11	12	3			—
3			1	14	21	10			—
4				18	42	20			—
5				20	59	41			—
6				12	36	33			—

40° N: Zone 3, Fig. 1

Sun Time	E	SE	S	SW	W	NW	N	NE	Hori-
6 AM	39	13					5	33	—
7	61	22					1	39	—
8	43	21						17	—
9	22	17	2					8	—
10	12	13	4					2	—
11	4	9	6						—
Noon		4	6	4					—
1			6	9	4				—
2			4	13	12	2			—
3			2	17	22	8			—
4				21	43	17			—
5				22	61	39			—
6				13	39	33			—

45° N: Zone 4, Fig. 1

Sun Time	E	SE	S	SW	W	NW	N	NE	Hori-
6 AM	46	16					4	39	—
7	64	25						37	—
8	45	25	1					16	—
9	23	20	4					7	—
10	12	16	7					1	—
11	4	11	9						—
Noon		5	9	5					—
1			9	11	4				—
2			7	16	12	1			—
3			4	20	23	7			—
4			1	25	45	16			—
5				25	64	37			—
6				16	46	39			—

Table 7.—Intensity (I_g) of Solar Radiation in Summer Transmitted by Windows Made of Single Heat Absorbing Glass

30° N: Zone 1, Fig. 1

Sun Time	Heat Absorbing Glass Windows Facing								Horizontal Sky-lights
	E	SE	S	SW	W	NW	N	NE	
6 AM	80	33					18	74	6
7	129	66					11	115	43
8	143	85						104	95
9	123	85						73	140
10	82	69	7					32	180
11	30	40	13						200
Noon		8	16	8					210
1			12	39	30				197
2			6	67	80	31			174
3				81	119	70			135
4				82	137	100			92
5				60	129	116			43
6				33	80	74			6

35° N: Zone 2, Fig. 1

Sun Time	E	SE	S	SW	W	NW	N	NE	Hori-
6 AM	80	30					15	73	8
7	129	65					7	103	45
8	143	90						96	96
9	123	95	6					62	139
10	82	88	15						177
11	30	64	29						196
Noon		29	34	18					206
1			29	52	30				193
2			17	79	80	20			170
3			5	93	120	60			134
4				90	137	93			91
5				70	129	103			45
6				35	80	73			8

40° N: Zone 3, Fig. 1

Sun Time	E	SE	S	SW	W	NW	N	NE	Hori-
6 AM	80	26					14	72	11
7	129	63					4	100	48
8	143	95						89	96
9	123	107	13					52	137
10	82	106	34					12	173
11	30	88	48						191
Noon		57	55	30					200
1			47	66	30				188
2			32	92	80	11			167
3			12	101	120	50			133
4				96	137	85			91
5				75	129	100	4		48
6				36	80	72	14		11

45° N: Zone 4, Fig. 1

Sun Time	E	SE	S	SW	W	NW	N	NE	Hori-
6 AM	80	22					12	71	12
7	129	62						96	48
8	143	99	4					81	95
9	123	116	25					42	133
10	82	123	51					5	168
11	30	107	69	7					184
Noon		83	75	44					193
1		7	67	81	30				181
2			49	104	80	5			182
3			24	112	120	40			128
4			4	104	137	77			90
5				79	129	96			49
6				37	80	71	12		12

SUMMER SOL-AIR TEMPERATURES FOR EIGHT CITIES

The sol-air temperature at each hour, for materials which do not directly transmit solar radiation, may be found from the equation:

$$t_s = t_a + bI/h \qquad (1)$$

t_s = sol-air temperature, F;

t_a = dry-bulb temperature of outdoor air in the shade, F;

I = intensity of total solar and sky radiation incident upon the surface, Btu per hour per sq ft;

b = absorptivity of surface for incident total radiation (dimensionless);

h = film coefficient of heat transfer between surface and air, Btuh per sq ft per deg F. *(Actually, this coefficient depends upon wind velocity; for simplicity, a constant value of 4 is used.)*

Between sunset and sunrise, the night sol-air temperature is:

$$t_s = t_a - eI'/4 \qquad (2)$$

e = surface emissivity;

I' = intensity of nocturnal radiation. Strictly, a value of the net rate of heat transfer by radiation from a black surface to the cold sky, Btu per hour per square foot should be used in equation (2), but, for simplicity, a constant value of 40 is used, and half of this constant value, or 20, assumed for vertical surface of various orientations.

The following relationship has been established for the general case:

$$K^m = I_{es}/I_c \qquad (3)$$

I_c = solar constant;

I_{es} = solar radiation at the earth's surface normal to the sun;

m = air mass; and

K = transmission coefficient for air mass.

It will be seen that the intensity of solar radiation at the earth's surface normal to the sun can be calculated from this relation for any air mass and solar altitude, provided atmospheric depletion coefficients for the air mass number are available. Atmospheric transmission coefficients can be taken from the Manual of Meteorology, and other sources.

Since weather data during the period of July 16 through August 15 are taken for analysis, therefore, August 1 (18° declination, north) is chosen for calculating radiation data.

Table 1 gives outdoor air temperatures for various cities, the temperature listed as equalled or exceeded by only 5% of the total hours in July and August.

Table 2 summarizes and gives the average daily design values for the different cities.

Table 3 gives calculated design sol-air temperatures for a horizontal surface and four various orientations of solar absorptivities (emissivities) of 1.0, 0.7 or 0.4. For example, white asbestos cement has a solar absorptivity of 0.50, asphalt waterproofing course 0.95, white brick tile 0.30, and blue brick tile 0.88.

Example 1. Find sol-air temperature in Austin, Tex., at 2:00 p.m. for a horizontal surface whose solar absorptivity, b, is 0.5.

In Table 3 for Austin find, in the absence of data for $b = 0.5$, $t_s = 170.8$ at 2:00 p.m. for $b = 1.0$.

From Table 1, $t_a = 101$ for Austin at 2 p.m.

Eq. (1) can be written (h taken as 4, b is 1)

$$b\,I/4 = t_s - t_a$$
$$I = 4(t_s - t_a)/b =$$
$$4(170.8 - 101)/1 = 4 \times 69.8$$

Where $b = 0.5$,
$$t_s = t_a \times b\,I/4 = 101 + (0.5)$$
$$(4 \times 69.8)/4 = 135.9 \text{ F}$$

Note: In this, as in many cases, the same result will be obtained more readily by interpolating directly between values of t_s from Table 3 for values of b above and below. In this case,

b	t_s
0.7	149.9 F
0.5	x
0.4	128.9 F

$$x = 135.9 \text{ F}$$

Example 2. Estimate, for Austin, Tex., the rate of heat flow into a thick wall facing West with overall heat transmission coefficient, $U = 0.20$, where solar absorptivity, $b = 0.4$.

In the case of a thick wall with a large heat storage capacity, solar heat flow into the room may be practically constant over the entire 24-hr period. If we assume an inside design temperature, $t_i = 78$ F, then from Table 2, for Austin, for a vertical wall facing West, and for $b = 0.4$, *average* $t_s = 94.0$ F. Then,

$$Q/A = U \ (average \ t_s - t_i) =$$
$$0.20(94\text{-}78) = 3.2 \text{ Btuh/sq ft}$$

Table 1. Outdoor Air Temperature

(Hourly outdoor dry-bulb air temperature equalled or exceeded on not more than five per cent of days in the period from 1951 through 1960)

Time of Day	Austin, Tex.	Kansas City, Mo.	Los Angeles, Calif.	Miami, Fla.	Minneapolis, Minn.	New Orleans, La.	New York, N. Y.	Seattle, Wash.
0	85	88	72	84	80	80	80	70
1 am	83	87	71	83	79	79	79	68
2	82	86	71	83	77	79	78	66
3	80	85	71	82	75	79	77	64
4	79	84	71	82	75	79	76	62
5	78	83	71	82	75	79	76	62
6	78	83	71	82	76	80	76	65
7	80	85	73	84	78	83	77	69
8	83	88	76	86	81	85	81	72
9	88	91	80	88	83	88	84	76
10	91	94	83	90	87	90	87	78
11	95	97	84	91	89	92	89	81
12	97	98	85	91	90	93	91	83
1 pm	100	100	85	91	91	93	92	86
2	101	101	84	91	92	94	93	88
3	102	102	83	91	94	94	93	88
4	102	102	82	90	93	94	93	88
5	101	101	80	89	93	92	91	88
6	100	100	77	88	91	89	89	86
7	97	98	75	86	88	86	87	84
8	93	95	74	84	86	84	85	81
9	90	93	73	84	84	83	84	77
10	88	91	73	84	82	82	82	75
11	87	90	73	84	81	81	81	72

Table 2. Comparison of Average Daily Design Values for Different Cities

Surface Orientation	Solar Absorptivity, b	Seattle, Wash.	Los Angeles, Calif.	Minneapolis, Minn.	Kansas City, Mo.	Austin, Tex.	New Orleans, La.	New York, N. Y.	Miami, Fla.
		Average Daily Sol-Air Temperatures, F							
Outdoor air in shade	0	76.2	76.5	84.2	92.6	90.0	85.8	84.2	86.3
Horizontal	1.0	98.7	99.6	107.0	116.2	112.9	109.1	101.2	109.8
	0.7	91.9	92.7	100.2	109.1	106.2	102.1	95.5	102.8
	0.4	85.2	85.8	93.3	102.0	99.4	95.0	90.3	95.7
Vertical, facing NORTH	1.0	78.6	78.6	86.5	94.8	92.1	87.9	87.1	88.6
	0.7	77.9	78.0	85.8	94.2	91.5	87.3	86.4	87.9
	0.4	77.2	77.4	85.1	93.5	90.9	86.6	85.8	87.2
Vertical, facing EAST	1.0	89.4	88.0	97.0	104.8	101.3	97.2	91.0	97.5
	0.7	85.4	84.6	93.2	101.1	98.0	93.8	88.9	94.1
	0.4	81.5	81.2	89.3	97.5	94.5	90.3	87.4	90.8
Vertical, facing SOUTH	1.0	86.4	81.9	93.7	100.2	94.1	89.9	89.4	89.1
	0.7	83.4	80.3	90.8	97.9	92.8	88.7	88.0	88.2
	0.4	80.3	78.7	88.0	95.6	91.8	87.4	86.6	87.4
Vertical, facing WEST	1.0	87.2	86.4	95.0	103.0	103.1	95.6	95.1	96.0
	0.7	83.9	83.4	91.7	99.9	96.9	92.7	92.2	93.1
	0.4	80.6	80.5	88.5	96.8	94.0	89.7	89.0	90.1

Table 3. Summer Sol-Air Temperatures for Various Cities

Austin, Tex.

Time of Day	Horizontal			North			East			South			West		
	Solar Absorptivity, b														
	1.0	0.7	0.4	1.0	0.7	0.4	1.0	0.7	0.4	1.0	0.7	0.4	1.0	0.7	0.4
	Sol-Air Temperature, t_s, F (July 16-Aug 15)														
0	75.0	77.0	80.0	80.0	81.5	83.0	80.0	81.5	83.0	80.0	81.5	83.0	80.0	81.5	83.0
1 am	73.0	75.0	78.0	78.0	79.5	81.0	78.0	79.5	81.0	78.0	79.5	81.0	78.0	79.5	81.0
2	72.0	74.0	77.0	77.0	78.5	80.0	77.0	78.5	80.0	77.0	78.5	80.0	77.0	78.5	80.0
3	70.0	72.0	75.0	75.0	76.5	78.0	75.0	76.5	78.0	75.0	76.5	78.0	75.0	76.5	78.0
4	74.0	75.0	76.5	76.5	77.3	78.0	76.0	77.3	78.0	76.0	77.3	78.0	76.0	77.3	78.0
5	78.0	78.0	78.0	78.0	78.0	78.0	78.0	78.0	78.0	78.0	78.0	78.0	78.0	78.0	78.0
6	85.8	83.5	81.1	88.3	85.2	82.1	110.4	100.6	90.9	79.8	79.3	78.7	79.8	79.3	78.7
7	104.6	97.2	89.8	91.1	87.8	84.4	134.9	118.5	102.0	83.5	82.5	81.5	83.0	82.1	81.2
8	125.5	112.8	100.0	89.1	87.3	85.4	142.9	125.0	107.0	87.8	86.4	85.0	87.0	85.8	84.6
9	135.5	128.3	111.0	92.8	91.4	89.9	141.7	125.6	109.5	98.6	95.5	92.3	92.5	91.2	89.9
10	160.8	139.9	118.9	96.3	94.7	93.1	131.6	119.5	107.3	107.9	102.8	97.8	96.3	94.7	93.1
11	172.7	149.4	126.1	100.5	98.9	97.2	118.6	111.6	104.5	116.2	109.8	103.4	100.8	99.1	97.4
12	176.7	152.8	128.9	102.5	100.9	99.2	102.8	101.1	99.3	119.4	112.0	105.6	102.8	101.1	99.4
1 pm	177.7	154.4	131.1	105.5	103.9	102.2	105.8	104.1	102.3	121.2	114.8	108.4	123.6	116.6	109.5
2	170.8	149.9	128.9	106.3	104.7	103.1	108.0	105.9	103.8	117.9	112.8	107.8	139.9	128.3	116.6
3	159.5	142.3	125.0	106.8	105.4	104.0	110.5	108.0	105.5	112.6	109.5	106.3	151.6	136.8	121.9
4	144.5	131.8	119.0	108.1	106.3	104.4	111.0	108.3	105.6	106.8	105.4	104.0	156.9	140.5	124.0
5	125.6	118.2	110.8	112.1	108.8	105.4	109.3	106.8	104.3	104.5	103.4	102.5	150.6	135.8	120.9
6	107.8	105.5	103.1	110.3	107.2	104.1	104.5	103.2	101.8	101.8	101.3	100.7	129.6	120.8	111.9
7	97.0	97.0	97.0	97.0	97.0	97.0	97.0	97.0	97.0	97.0	97.0	97.0	97.0	97.0	97.0
8	88.0	89.0	90.5	90.5	91.3	92.0	90.5	91.3	92.0	90.5	90.5	92.0	90.5	90.5	92.0
9	80.0	82.0	85.0	85.0	86.5	88.0	85.0	86.5	88.0	85.0	85.0	88.0	85.0	85.0	88.0
10	78.0	80.0	83.0	83.2	84.5	86.0	83.2	84.5	86.0	83.2	83.2	86.0	83.2	83.2	86.0
11	77.0	79.0	82.0	82.0	83.5	85.0	82.0	83.5	85.0	82.0	82.0	85.0	82.0	82.0	85.0

Kansas City, Mo.

Time of Day	Horizontal			North			East			South			West		
0	78.0	81.0	84.0	83.0	84.5	86.0	83.0	84.5	86.0	83.0	84.5	86.0	83.0	84.5	86.0
1 am	77.0	80.0	83.0	82.0	83.5	85.0	82.0	83.5	85.0	82.0	83.5	85.0	82.0	83.5	85.0
2	76.0	79.0	82.0	81.0	82.5	84.0	81.0	82.5	84.0	81.0	82.5	84.0	81.0	82.5	84.0
3	80.0	81.5	83.0	82.5	83.5	84.0	82.5	83.5	84.0	82.5	83.5	84.0	82.5	83.5	84.0
4	84.0	84.0	84.0	84.0	84.0	84.0	84.0	84.0	84.0	84.0	84.0	84.0	84.0	84.0	84.0
5	85.0	85.0	85.0	85.0	85.0	85.0	85.0	85.0	85.0	85.0	85.0	85.0	85.0	85.0	85.0
6	93.4	90.3	87.1	94.0	90.7	87.4	120.3	109.1	97.9	85.2	84.5	83.9	85.0	84.4	83.8
7	111.5	103.6	95.6	93.6	91.0	88.4	141.4	124.5	107.6	88.5	87.5	86.4	88.1	87.2	86.3
8	130.4	117.7	105.0	92.3	91.0	89.7	147.9	130.0	112.0	97.0	94.3	91.6	92.0	90.8	89.6
9	146.8	130.0	113.3	95.8	94.3	92.9	144.7	128.6	112.5	109.0	103.6	98.2	95.6	94.2	92.8
10	160.6	140.6	120.6	99.1	97.6	96.0	134.6	122.4	110.2	119.9	112.1	104.4	99.1	97.6	96.0
11	170.6	148.5	126.5	102.4	100.8	99.2	121.0	113.8	106.6	128.1	118.8	109.5	102.5	100.9	99.2
12	174.0	151.2	128.4	103.5	101.8	100.2	103.9	102.1	100.4	131.0	121.1	111.2	103.7	102.0	100.3
1 pm	173.6	151.5	129.5	105.4	103.8	102.2	106.4	104.5	102.5	131.1	121.8	112.5	123.1	116.2	109.3
2	167.6	147.6	127.6	106.1	104.6	103.0	108.5	106.2	104.0	126.9	119.1	111.4	139.3	127.8	116.3
3	157.8	141.0	124.3	106.8	105.3	103.9	110.7	108.1	105.5	120.0	114.6	109.2	151.5	136.6	121.8
4	144.4	131.7	119.0	106.3	105.0	103.7	111.0	108.3	105.6	111.0	108.3	105.6	156.7	140.3	123.9
5	127.5	119.6	111.6	109.6	107.0	103.4	109.4	106.9	104.4	104.5	103.5	102.4	152.1	136.8	121.5
6	110.4	107.3	104.1	111.0	107.6	104.4	105.5	103.8	102.2	102.2	101.5	100.9	133.8	123.6	113.5
7	98.0	98.0	98.0	98.0	98.0	98.0	98.0	98.0	98.0	98.0	98.0	98.0	98.0	98.0	98.0
8	95.0	95.0	95.0	95.0	95.0	95.0	95.0	95.0	95.0	95.0	95.0	95.0	95.0	95.0	95.0
9	88.0	89.5	91.0	90.5	91.3	92.0	90.5	91.3	92.0	90.5	91.3	92.0	90.5	91.3	92.0
10	81.0	84.0	87.0	86.0	87.5	89.0	86.0	87.5	89.0	86.0	87.5	89.0	86.0	87.5	89.0
11	80.0	83.0	86.0	85.0	86.5	88.0	85.0	86.5	88.0	85.0	86.5	88.0	85.0	86.5	88.0

Table 3. Summer Sol-Air Temperatures for Various Cities (Continued)

Los Angeles, Calif.

Time of Day	Horizontal			North			East			South			West		
	Solar Absorptivity, b														
	1.0	0.7	0.4	1.0	0.7	0.4	1.0	0.7	0.4	1.0	0.7	0.4	1.0	0.7	0.4
	Sol-Air Temperature, t_s, F (July 16-Aug 15)														
0	62.0	65.0	68.0	67.0	68.5	70.0	67.0	68.5	70.0	67.0	68.5	70.0	67.0	68.5	70.0
1 am	61.0	64.0	67.0	66.0	67.5	69.0	66.0	67.5	69.0	66.0	67.5	69.0	66.0	67.5	69.0
2	61.0	64.0	67.0	66.0	67.5	69.0	66.0	67.5	69.0	66.0	67.5	69.0	66.0	67.5	69.0
3	61.0	64.0	67.0	66.0	67.5	69.0	66.0	67.5	69.0	66.0	67.5	69.0	66.0	67.5	69.0
4	66.0	67.5	69.0	68.5	69.3	70.0	68.5	69.3	70.0	68.5	69.3	70.0	68.5	69.3	70.0
5	71.0	71.0	71.0	71.0	71.0	71.0	71.0	71.0	71.0	71.0	71.0	71.0	71.0	71.0	71.0
6	79.8	77.2	74.5	81.6	78.4	75.2	105.3	95.0	84.7	73.0	72.4	71.8	72.8	72.2	71.7
7	98.4	90.8	83.2	82.9	80.0	77.0	129.2	112.4	95.5	76.5	75.4	74.4	76.1	75.2	74.2
8	118.4	105.7	92.9	82.1	80.3	78.4	135.8	117.8	99.9	80.7	79.3	77.9	80.0	78.8	77.6
9	136.9	119.8	102.8	84.8	83.3	81.9	130.7	115.4	100.3	93.9	89.7	85.6	84.6	83.2	81.8
10	151.3	130.8	110.3	88.2	86.6	85.1	123.7	111.5	99.3	103.9	97.6	91.4	88.2	86.6	85.1
11	159.9	137.1	114.4	89.5	87.8	86.2	107.6	100.5	93.5	109.7	102.0	94.3	89.5	87.8	86.2
12	163.0	139.6	116.2	90.3	88.7	87.1	90.8	89.0	87.3	112.2	104.0	95.9	90.8	89.0	87.3
1 pm	160.9	138.1	115.4	90.5	88.8	87.2	90.7	89.1	87.4	110.7	103.0	95.3	108.2	101.2	94.3
2	152.3	131.8	111.3	89.2	87.6	86.1	91.2	89.0	86.9	104.9	98.6	92.4	122.7	111.1	99.5
3	139.8	122.8	105.8	87.8	86.3	84.9	91.6	89.0	86.4	96.9	92.7	88.6	129.7	115.7	101.7
4	124.4	111.7	98.9	88.1	86.3	84.4	91.0	88.3	85.6	86.7	85.3	83.9	136.7	120.3	103.9
5	105.4	97.8	90.2	89.9	87.0	84.0	88.3	85.8	83.3	83.5	82.4	81.4	131.1	115.8	100.4
6	85.8	83.2	80.5	87.6	84.4	81.2	82.0	80.5	79.0	79.0	78.4	77.8	108.0	98.7	89.4
7	75.0	75.0	75.0	75.0	75.0	75.0	75.0	75.0	75.0	75.0	75.0	75.0	75.0	75.0	75.0
8	69.0	70.5	72.0	71.5	72.3	73.0	71.5	72.3	73.0	71.5	72.3	73.0	71.5	72.3	73.0
9	63.0	66.0	69.0	68.0	69.5	71.0	68.0	69.5	71.0	68.0	69.5	71.0	68.0	69.5	71.0
10	63.0	66.0	69.0	68.0	69.5	71.0	68.0	69.5	71.0	68.0	69.5	71.0	68.0	69.5	71.0
11	63.0	66.0	69.0	68.0	69.5	71.0	68.0	69.5	71.0	68.0	69.5	71.0	68.0	69.5	71.0

Miami, Fla.

Time of Day	Horizontal			North			East			South			West		
0	74.0	77.0	80.0	79.0	80.5	82.0	79.0	80.5	82.0	79.0	80.5	82.0	79.0	80.5	82.0
1 am	73.0	76.0	79.0	78.0	79.5	81.0	78.0	79.5	81.0	78.0	79.5	81.0	78.0	79.5	81.0
2	73.0	76.0	79.0	78.0	79.5	81.0	78.0	79.5	81.0	78.0	79.5	81.0	78.0	79.5	81.0
3	72.0	75.0	78.0	77.0	78.5	80.0	77.0	78.5	80.0	77.0	78.5	80.0	77.0	78.5	80.0
4	77.0	78.5	80.0	79.5	80.3	81.0	79.5	80.3	81.0	79.5	80.3	81.0	79.5	80.3	81.0
5	82.0	82.0	82.0	82.0	82.0	82.0	82.0	82.0	82.0	82.0	82.0	82.0	82.0	82.0	82.0
6	88.9	86.9	84.8	91.8	88.9	85.9	111.8	102.9	93.9	83.7	83.2	82.7	83.5	83.1	82.6
7	107.9	100.7	93.6	96.4	92.7	88.9	138.4	122.1	105.8	87.4	86.4	85.4	87.1	86.1	85.2
8	129.1	116.2	103.2	93.4	91.2	88.9	146.0	128.0	110.0	90.8	89.3	87.9	90.0	88.8	87.6
9	146.4	128.8	111.4	92.8	91.3	89.9	141.8	125.7	109.5	94.1	92.2	90.4	92.7	91.3	89.9
10	160.9	139.6	118.4	95.3	93.7	92.1	130.6	118.4	106.2	103.1	99.1	95.2	95.3	93.7	92.1
11	170.2	146.5	122.7	96.5	94.9	93.2	114.6	107.6	100.5	107.6	102.6	97.6	96.8	95.0	93.3
12	172.3	147.9	123.5	96.5	94.9	93.2	96.8	95.0	93.3	108.5	103.2	98.0	96.8	95.0	93.3
1 pm	170.2	146.5	122.7	96.5	94.9	93.2	96.8	95.0	93.3	107.6	102.6	97.6	114.6	107.5	100.5
2	161.9	140.6	119.4	96.3	94.7	93.1	97.8	95.8	93.7	104.1	100.1	96.2	130.1	118.4	106.7
3	149.4	131.8	114.4	95.8	94.3	92.9	99.5	96.9	94.4	97.1	95.2	93.4	141.0	126.0	111.0
4	133.1	120.2	107.2	97.4	95.2	92.9	99.0	96.3	93.6	94.8	93.3	91.9	145.0	128.5	112.0
5	112.9	105.7	98.6	101.4	97.7	93.9	97.2	94.7	92.3	92.4	91.4	90.4	138.3	123.5	108.7
6	94.9	92.9	90.8	97.8	94.9	91.9	92.3	91.0	89.7	89.7	89.2	88.7	115.0	106.9	98.8
7	86.0	86.0	86.0	86.0	86.0	86.0	86.0	86.0	86.0	86.0	86.0	86.0	86.0	86.0	86.0
8	79.0	80.5	82.0	81.5	82.3	83.0	81.5	82.3	83.0	81.5	82.3	83.0	81.5	82.3	83.0
9	74.0	77.0	80.0	79.0	80.5	82.0	79.0	80.5	82.0	79.0	80.5	82.0	79.0	80.5	82.0
10	74.0	77.0	80.0	79.0	80.5	82.0	79.0	80.5	82.0	79.0	80.5	82.0	79.0	80.5	82.0
11	74.0	77.0	80.0	79.0	80.5	82.0	79.0	80.5	82.0	79.0	80.5	82.0	79.0	80.5	82.0

Table 3. Summer Sol-Air Temperatures for Various Cities (Continued)

Minneapolis, Minn.

Time of Day	Horizontal			North			East			South			West		
	Solar Absorptivity, b														
	1.0	0.7	0.4	1.0	0.7	0.4	1.0	0.7	0.4	1.0	0.7	0.4	1.0	0.7	0.4
	Sol-Air Temperature, t_s, F (July 16-Aug 15)														
0	70.0	73.0	76.0	75.0	76.5	78.0	75.0	76.5	78.0	75.0	76.5	78.0	75.0	76.5	78.0
1 am	69.0	72.0	75.0	74.0	75.5	77.0	74.0	75.5	77.0	74.0	75.5	77.0	74.0	75.5	77.0
2	67.0	70.0	73.0	72.0	73.5	75.0	72.0	73.5	75.0	72.0	73.5	75.0	72.0	73.5	75.0
3	70.0	71.5	73.0	72.5	73.3	74.0	72.5	73.3	74.0	72.5	73.3	74.0	72.5	73.3	74.0
4	75.0	75.0	75.0	75.0	75.0	75.0	75.0	75.0	75.0	75.0	75.0	75.0	75.0	75.0	75.0
5	76.2	75.9	75.5	79.1	77.8	76.6	84.1	81.3	78.6	75.5	75.4	75.2	75.5	75.4	75.2
6	87.9	84.4	80.8	87.0	83.7	80.4	116.3	104.2	92.1	78.4	77.7	77.0	78.1	77.5	76.9
7	105.1	96.9	88.8	84.3	82.2	80.5	134.8	117.8	100.7	81.6	80.5	79.4	81.2	80.2	79.3
8	122.4	110.0	97.6	85.3	84.0	82.7	140.4	122.5	104.8	93.1	89.5	85.8	85.0	83.8	82.6
9	136.6	120.6	104.4	87.7	86.3	84.9	136.2	120.3	104.3	105.5	98.7	92.0	87.5	86.1	84.8
10	150.3	131.3	112.3	91.9	90.4	89.0	127.6	115.5	103.3	118.4	109.0	99.6	91.9	90.4	89.0
11	158.8	137.9	116.9	94.3	92.7	91.1	113.5	106.1	98.8	126.0	114.9	103.8	94.3	92.7	91.1
12	161.8	140.2	118.7	95.3	93.7	92.1	96.6	94.6	92.7	129.2	117.4	105.7	95.4	93.8	92.2
1 pm	160.8	139.9	118.9	96.3	94.7	93.1	98.0	95.9	93.8	128.0	116.9	105.8	113.7	106.9	100.1
2	155.3	136.3	117.3	96.9	95.4	94.0	99.9	97.5	95.2	123.4	114.0	104.6	129.7	118.4	107.1
3	147.6	131.6	115.4	98.7	97.3	95.9	102.8	100.2	97.5	116.5	109.7	103.0	142.9	128.2	113.6
4	134.4	122.0	109.6	97.3	96.0	94.7	102.0	99.3	96.6	105.1	101.5	97.8	147.5	131.1	114.8
5	120.1	112.0	103.8	99.3	97.4	95.5	101.5	99.0	96.4	96.6	95.5	94.4	144.5	129.0	113.6
6	102.9	99.4	95.8	102.0	98.7	95.4	96.9	95.1	93.4	93.4	92.7	92.0	127.6	116.6	105.7
7	89.2	88.9	88.5	92.1	90.8	89.6	89.4	89.0	88.6	88.5	88.4	88.2	96.2	93.7	91.3
8	86.0	86.0	86.0	86.0	86.0	86.0	86.0	86.0	86.0	86.0	86.0	86.0	86.0	86.0	86.0
9	79.0	80.5	82.0	81.5	82.3	83.0	81.5	82.3	83.0	81.5	82.3	83.0	81.5	82.3	83.0
10	72.0	75.0	78.0	77.0	78.5	80.0	77.0	78.5	80.0	77.0	78.5	80.0	77.0	78.5	80.0
11	71.0	74.0	77.0	76.0	77.5	79.0	76.0	77.5	79.0	76.0	77.5	79.0	76.0	77.5	79.0

New Orleans, La.

Time of Day	Horizontal			North			East			South			West		
0	70.0	73.0	76.0	75.0	76.5	78.0	75.0	76.5	78.0	75.0	76.5	78.0	75.0	76.5	78.0
1 am	69.0	72.0	75.0	74.0	75.5	77.0	74.0	75.5	77.0	74.0	75.5	77.0	74.0	75.5	77.0
2	69.0	72.0	75.0	74.0	75.5	77.0	74.0	75.5	77.0	74.0	75.5	77.0	74.0	75.5	77.0
3	69.0	72.0	75.0	74.0	75.5	77.0	74.0	75.5	77.0	74.0	75.5	77.0	74.0	75.5	77.0
4	74.0	75.5	77.0	76.5	77.2	78.0	76.5	77.2	78.0	76.5	77.2	78.0	76.5	77.2	78.0
5	79.0	79.0	79.0	79.0	79.0	79.0	79.0	79.0	79.0	79.0	79.0	79.0	79.0	79.0	79.0
6	87.8	85.5	83.1	90.3	87.2	84.1	112.4	102.6	92.9	79.8	79.3	78.7	81.8	81.3	80.7
7	107.6	100.2	92.8	94.1	90.8	87.4	137.9	121.5	105.0	81.8	81.3	80.7	86.0	85.1	84.2
8	127.5	114.8	102.1	91.1	89.3	87.4	144.9	127.0	109.0	86.5	85.5	84.5	89.0	87.8	86.6
9	145.5	128.3	111.0	92.8	91.4	89.9	141.7	125.6	109.5	89.8	88.4	87.0	92.5	91.2	89.9
10	159.8	138.9	117.9	95.3	93.7	92.1	130.6	118.5	106.3	98.6	95.2	92.3	95.3	93.7	92.1
11	169.7	146.4	123.1	97.5	95.9	94.2	115.6	108.6	101.5	106.9	101.8	96.8	97.8	96.1	94.4
12	172.7	148.8	124.9	98.5	96.9	95.2	98.8	97.1	95.3	113.2	106.8	100.4	98.8	97.1	95.4
1 pm	170.7	147.4	124.1	98.5	96.9	95.2	98.8	97.1	95.3	115.4	108.0	101.6	116.6	109.6	102.5
2	163.8	142.9	121.9	99.5	97.7	96.1	101.0	98.9	96.8	114.2	107.8	101.4	132.9	121.3	109.6
3	151.5	134.3	117.0	98.8	97.4	95.9	102.5	100.0	97.4	110.9	105.8	100.8	143.6	128.8	113.9
4	136.5	123.8	111.0	100.1	98.3	96.4	103.0	100.3	97.6	104.9	101.5	98.3	148.9	132.5	116.0
5	116.6	109.2	101.8	103.1	99.8	96.4	100.3	97.8	95.3	98.8	97.4	96.0	141.6	126.8	111.9
6	96.8	94.5	92.1	99.3	96.2	93.1	93.5	92.2	90.8	95.5	94.5	93.5	118.6	109.8	100.9
7	86.0	86.0	86.0	86.0	86.0	86.0	86.0	86.0	86.0	86.0	86.0	86.0	86.0	86.0	86.0
8	79.0	80.5	82.0	81.5	82.2	83.0	81.5	82.2	83.0	81.5	82.2	83.0	81.5	82.2	83.0
9	73.0	76.0	79.0	78.0	79.5	81.0	78.0	79.5	81.0	78.0	79.5	81.0	78.0	79.5	81.0
10	72.0	75.0	78.0	77.0	78.5	80.0	77.0	78.5	80.0	77.0	78.5	80.0	77.0	78.5	80.0
11	71.0	74.0	77.0	76.0	77.5	79.0	76.0	77.5	79.0	76.0	77.5	79.0	76.0	77.5	79.0

Table 3. Summer Sol-Air Temperatures for Various Cities (Continued)

New York, N. Y.

Time of Day	Horizontal			North			East			South			West		
	Solar Absorptivity, b														
	1.0	0.7	0.4	1.0	0.7	0.4	1.0	0.7	0.4	1.0	0.7	0.4	1.0	0.7	0.4
	Sol-Air Temperature, t_s, F (July 16-Aug 15)														
0	70.0	73.0	76.0	75.0	77.0	78.0	75.0	77.0	78.0	75.0	77.0	78.0	75.0	77.0	78.0
1 am	70.0	73.0	76.0	74.0	76.0	77.0	74.0	76.0	77.0	74.0	76.0	77.0	74.0	76.0	77.0
2	78.0	71.0	74.0	72.0	74.0	75.0	72.0	74.0	75.0	72.0	74.0	75.0	72.0	74.0	75.0
3	72.0	73.0	75.0	72.0	73.0	74.0	72.0	73.0	74.0	72.0	73.0	74.0	72.0	73.0	74.0
4	72.0	73.0	75.0	74.0	75.0	75.0	74.0	75.0	75.0	74.0	75.0	75.0	74.0	75.0	75.0
5	76.0	76.0	76.0	76.0	76.0	76.0	76.0	76.0	76.0	76.0	76.0	76.0	76.0	76.0	76.0
6	78.1	77.3	77.0	80.1	79.0	77.1	89.1	85.0	81.1	77.5	77.4	76.2	77.5	77.4	76.2
7	88.9	85.9	81.7	85.0	83.1	82.1	106.3	98.2	91.1	82.4	82.7	81.0	82.4	82.7	81.0
8	106.2	98.2	90.0	85.3	84.3	83.2	113.8	105.0	95.0	86.6	85.5	83.4	85.2	84.2	83.3
9	122.6	111.5	97.4	90.3	89.3	88.2	120.4	109.3	99.3	97.1	94.5	90.8	90.0	89.1	88.2
10	131.3	117.4	103.3	91.7	91.6	90.7	114.2	106.1	98.0	104.5	98.7	94.0	92.5	91.1	90.0
11	140.7	124.6	108.7	93.9	93.0	92.0	106.6	101.4	97.4	111.0	104.9	98.0	94.3	92.7	92.1
12	148.2	129.1	113.1	97.3	95.0	93.0	97.5	95.5	94.5	115.2	108.4	101.7	98.4	96.3	94.2
1 pm	149.9	132.1	114.0	98.3	96.0	95.0	98.3	96.0	95.0	117.4	110.0	103.6	108.5	104.4	99.1
2	145.9	130.0	113.0	99.2	97.2	96.2	99.2	97.2	96.2	112.5	107.3	101.0	125.9	116.2	107.4
3	140.2	125.3	110.2	99.2	97.2	96.2	99.2	97.2	96.2	102.1	99.3	97.2	136.2	123.1	110.9
4	129.3	118.3	106.3	99.2	97.2	96.2	99.2	97.2	96.2	99.2	98.4	96.3	145.4	130.3	114.1
5	115.7	108.0	100.0	103.0	100.0	97.0	97.5	96.0	95.0	97.5	96.0	95.0	140.3	126.2	112.0
6	101.4	97.4	91.3	106.1	101.0	96.0	92.9	92.0	91.0	92.9	92.0	91.0	134.7	120.5	107.4
7	88.0	87.0	86.0	102.3	98.2	94.0	90.0	89.0	89.0	90.0	89.0	89.0	115.3	107.2	99.0
8	85.0	85.0	85.0	85.0	85.0	85.0	85.0	85.0	85.0	85.0	85.0	85.0	85.0	85.0	85.0
9	79.0	81.0	82.0	81.0	82.0	83.0	81.0	82.0	83.0	81.0	82.0	83.0	81.0	82.0	83.0
10	72.0	75.0	78.0	77.0	78.0	80.0	77.0	78.0	80.0	77.0	78.0	80.0	77.0	78.0	80.0
11	71.0	74.0	77.0	76.0	77.0	79.0	76.0	77.0	79.0	76.0	77.0	79.0	76.0	77.0	79.0

Seattle, Wash.

Time of Day	Horizontal			North			East			South			West		
0	60.0	63.0	66.0	65.0	66.5	68.0	65.0	66.5	68.0	65.0	66.5	68.0	65.0	66.5	68.0
1 am	58.0	61.0	64.0	63.0	64.5	66.0	63.0	64.5	66.0	63.0	64.5	66.0	63.0	64.5	66.0
2	56.0	59.0	62.0	61.0	62.5	64.0	61.0	62.5	64.0	61.0	62.5	64.0	61.0	62.5	64.0
3	59.0	60.5	62.0	61.5	62.3	63.0	61.5	62.3	63.0	61.5	62.3	63.0	61.5	62.3	63.0
4	62.0	62.0	62.0	62.0	62.0	62.0	62.0	62.0	62.0	62.0	62.0	62.0	62.0	62.0	62.0
5	64.0	63.4	62.8	68.0	66.2	64.4	75.5	71.4	67.4	62.8	62.5	62.3	62.8	62.5	62.3
6	77.6	73.8	70.0	76.0	72.7	69.4	106.4	94.0	81.6	67.5	66.7	66.0	67.2	66.5	65.9
7	96.2	88.0	79.9	74.5	72.8	71.2	126.0	108.8	91.8	72.6	71.5	70.4	72.2	71.2	70.3
8	112.9	100.6	88.4	76.3	75.0	73.7	131.2	113.5	95.7	85.1	81.2	77.2	75.9	74.8	73.6
9	128.4	112.6	97.0	80.7	79.3	77.9	129.1	113.2	97.2	100.3	93.0	85.7	80.4	79.1	77.8
10	139.8	121.3	102.7	82.4	81.4	79.9	118.7	106.5	94.3	111.6	101.6	91.5	82.8	81.4	79.9
11	149.0	128.5	108.2	86.2	84.6	83.1	105.6	98.2	90.9	120.2	108.4	96.7	86.2	84.6	83.1
12	153.0	132.0	111.0	88.3	86.7	85.1	90.0	87.9	85.8	124.5	112.0	99.6	88.3	86.7	85.1
1 pm	154.0	133.5	113.2	91.2	89.6	88.1	93.3	91.1	88.9	125.2	113.4	101.7	108.5	101.7	95.0
2	149.8	131.3	112.7	92.8	91.4	89.9	96.1	93.7	91.2	121.6	111.6	101.5	125.5	114.2	103.0
3	140.4	124.6	109.0	92.7	91.3	89.9	96.9	94.2	91.5	112.3	105.0	97.7	136.6	122.0	107.4
4	128.9	116.6	104.4	92.3	91.0	89.7	97.0	94.3	91.6	101.1	97.2	93.2	142.2	125.9	109.7
5	115.2	107.0	98.9	93.5	91.8	90.3	96.5	94.0	91.4	91.6	90.5	89.4	139.6	124.2	108.6
6	98.6	94.8	91.0	97.0	93.7	90.4	92.1	90.2	88.4	88.5	87.7	87.0	123.6	112.3	101.0
7	86.0	85.4	84.8	90.0	88.2	86.4	86.0	85.4	84.8	84.8	84.5	84.3	96.2	92.6	88.9
8	81.0	81.0	81.0	81.0	81.0	81.0	81.0	81.0	81.0	81.0	81.0	81.0	81.0	81.0	81.0
9	72.0	73.5	75.0	74.5	75.3	76.0	74.5	75.3	76.0	74.5	75.3	76.0	74.5	75.3	76.0
10	65.0	68.0	71.0	70.0	71.5	73.0	70.0	71.5	73.0	70.0	71.5	73.0	70.0	71.5	73.0
11	62.0	65.0	68.0	67.0	68.5	70.0	67.0	68.5	70.0	67.0	68.5	70.0	67.0	68.5	70.0

YEAR-ROUND SUN ANGLES FOR THE U. S.

Heat gains from the sun are among the most important in estimating the air conditioning load, and data for calculating them appear on subsequent pages. The angle between the sun's rays and the building surfaces affects heat gain, and in many cases it is advisable to consider shading due to adjacent buildings, window recesses, or inner courts of multi-wing structures when determining the individual room load and simultaneous maximum load.

The table of sun angles shows azimuth and altitude in simple form covering twelve months so that shading can be analyzed and the incident angle on any exposure can be determined.

Altitude angle is the angle in a vertical plane between the earth's surface and the sun's rays. *Azimuth angle* is that between the vertical plane of the sun's rays and due North. *Incident angle* is that between the sun's rays and a perpendicular to the surface on which the rays are impinging.

For engineering purposes, the angles presented here for four zones of the United States are well within the margins of error in heating calculations.

The incident angle for vertical surfaces can be calculated by:

(1) From Table 1, determine the wall azimuth, i.e., the angle Y between a perpendicular to the wall and due North.

(2) Determine the solar azimuth for the hour and month in question from the table of solar azimuth and altitude angles. (Note that solar azimuth is rotation eastward from North in the morning and westward from North in the afternoon.)

(3) Find the angle X between wall and solar azimuths. When the angle Y (wall azimuth) is eastward and the solar azimuth is westward (or vice versa) the angle X between them is $360 - (Y +$ sun azimuth). When both Y and solar azimuths are east (or west), simply subtract the larger from the smaller to find X.

(4) Find the incident angle by solving the formula:

cos incident angle = cos altitude angle $\times$ cos X.

For horizontal surfaces, the incident angle is 90° minus the altitude angle.

Table 2 is a condensed table of cosines included for convenience. Complete tables are in math section.

Example: What is the incident angle in latitude 35°N (a) on a Southeast wall on August 24 at 11 a.m. and (b) on a flat roof at the same time?

Solution: (1) From Table 1 for a Southeast wall $Y = 135°$; (2) From the main table the azimuth for August 24, 11 am., 35°N, is 146°; (3) Subtracting (1) from (2), $146° - 135° = 11°$; (4) From the main table, the altitude angle is 62°; (5) The cosine of 11° is 0.982 and of 62° is 0.469 so that

Table 1. — Angle Between Due North and Wall Perpendicular
(Wall Azimuth)

Wall Facing	Angle Y Deg.
North	0
Northeast	45
East	90
Southeast	135
South	180
Southwest	135
West	90
Northwest	45

Table 2. — Cosines of Angles

Angle, Deg.	Cos	Angle, Deg.	Cos	Angle, Deg.	Cos	Angle, Deg.	Cos	Angle, Deg.	Cos
0	1.000	35	.819	75	.259	110	.342	145	.819
5	.996	40	.766	80	.174	115	.423	150	.866
10	.985	45	.707	85	.087	120	.500	155	.906
15	.966	50	.643	90	.000	125	.574	160	.940
20	.940	55	.574	95	.087	130	.643	165	.966
25	.960	60	.500	100	.174	135	.707	170	.985
30	.866	65	.423	105	.259	140	.766	175	.996
		70	.342					180	1.000

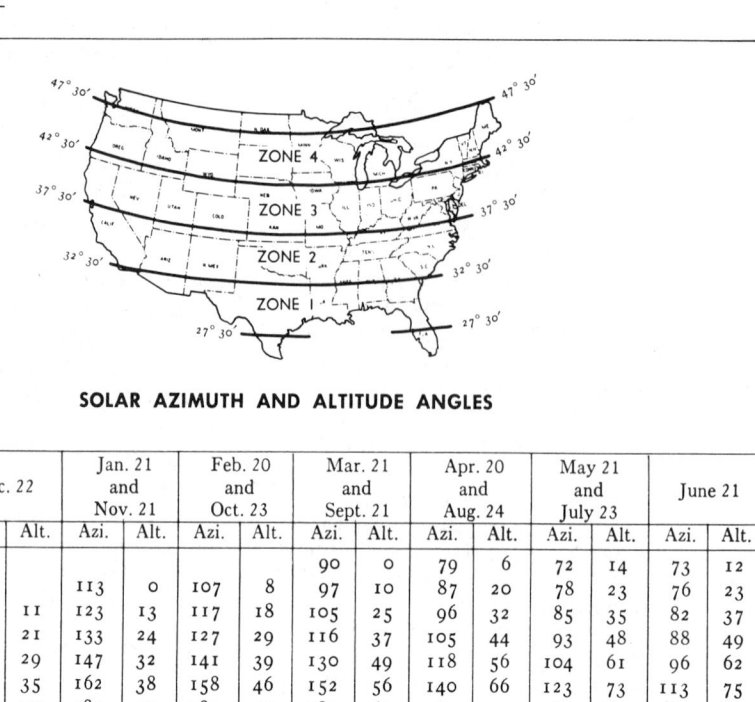

SOLAR AZIMUTH AND ALTITUDE ANGLES

Zone 1

Sun Time AM	PM	Dec. 22 Azi.	Alt.	Jan. 21 and Nov. 21 Azi.	Alt.	Feb. 20 and Oct. 23 Azi.	Alt.	Mar. 21 and Sept. 21 Azi.	Alt.	Apr. 20 and Aug. 24 Azi.	Alt.	May 21 and July 23 Azi.	Alt.	June 21 Azi.	Alt.
6	6							90	0	79	6	72	14	73	12
7	5			113	0	107	8	97	10	87	20	78	23	76	23
8	4	125	11	123	13	117	18	105	25	96	32	85	35	82	37
9	3	136	21	133	24	127	29	116	37	105	44	93	48	88	49
10	2	148	29	147	32	141	39	130	49	118	56	104	61	96	62
11	1	163	35	162	38	158	46	152	56	140	66	123	73	113	75
NOON		180	37	180	40	180	49	180	60	180	71	180	80	180	83

Zone 2

AM	PM	Azi.	Alt.	Azi.	Alt.	Azi.	Alt.	Azi.	Alt.	Azi.	Alt.	Azi.	Alt.	Azi.	Alt.
6	6							90	0	80	7	73	15	75	11
7	5					108	8	98	10	90	20	81	24	78	24
8	4	126	10	124	11	118	17	108	24	98	31	90	35	85	37
9	3	137	17	135	21	128	27	120	35	108	42	99	48	94	48
10	2	149	24	148	28	143	36	134	45	122	54	112	60	106	61
11	1	164	29	163	32	160	41	155	52	146	62	133	69	127	73
NOON		180	31	180	35	180	44	180	56	180	66	180	75	180	77

Zone 3

AM	PM	Azi.	Alt.	Azi.	Alt.	Azi.	Alt.	Azi.	Alt.	Azi.	Alt.	Azi.	Alt.	Azi.	Alt.
6	6							90	0	81	8	74	16	75	13
7	5					108	8	98	9	91	19	84	25	80	25
8	4	126	9	124	9	118	15	110	22	101	30	93	36	88	37
9	3	138	14	136	16	131	24	123	32	112	41	104	47	100	48
10	2	151	20	149	24	145	32	137	41	128	51	120	58	114	60
11	1	165	24	163	28	162	37	157	47	152	59	143	66	138	69
NOON		180	25	180	29	180	39	180	50	180	61	180	70	180	73

Zone 4

AM	PM	Azi.	Alt.	Azi.	Alt.	Azi.	Alt.	Azi.	Alt.	Azi.	Alt.	Azi.	Alt.	Azi.	Alt.
6	6							90	0	82	8	75	16	77	15
7	5							101	8	93	18	87	26	83	26
8	4	127	9	125	8	119	13	111	20	103	30	96	35	92	37
9	3	139	8	137	13	132	22	124	29	117	39	109	46	105	47
10	2	152	13	150	17	146	27	140	38	133	47	125	55	121	56
11	1	165	18	164	22	163	32	160	44	154	54	148	63	146	64
NOON		180	22	180	24	180	34	180	45	180	56	180	65	180	69

cos incident angle = cos altitude angle
$$\times \cos X = .47 \times .98 = .46$$

Since the cosine of the incident angle = .46, refer to a table of cosines and find that the angle whose cosine is .46 is 62° 25′.

For the roof, the incident angle is 90° − 62° = 28° for which the cosine is .88.

Example: What is the incident angle in latitude 35°N on a Southwest wall on August 24 at 11 a.m.?

Solution: (1) From Table 1, $Y = 135°$ westward. (2) From the main table, solar azimuth is 146° eastward. (2) $X = 360 − (135 + 146) = 79$. (4) From the main table, the altitude angle is 62°, and

cos incident angle = cos altitude angle
$$\times \cos X = 0.469 \times 0.191 = 0.090$$

A table of cosines (see Trigonometric Functions in Section 9) shows the nearest angle whose cosine is 0.090 to be 84°50′.

Shadow Forecast and Eave Overhang

At height H the roof projection in the direction of solar azimuth to shade a window with lower sill at height W is (see Fig. 1):
$$G = (H − W) \times \text{cot altitude angle}$$

The eave overhang E at height H perpendicular to a wall oriented at angle Y from due North is
$$E = G \times \cos X$$
where $X =$ wall-solar azimuth as previously defined.

In terms of wall and sill heights,
$$E = (H − W) \times \text{cot altitude angle} \times \cos X$$

Projection P of eaves to right or left of the window will be

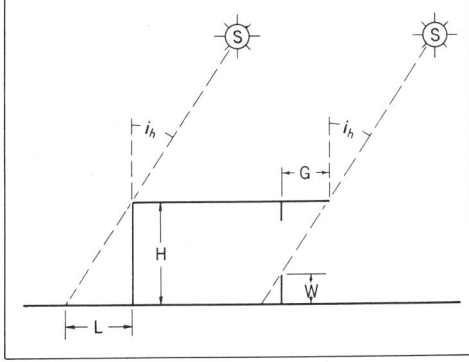

Fig. 1. Geometry of sun angles:
$$L = H \tan i_h$$
$$G = (H − W) \tan i_h$$
$$(H − W) = G \tan \text{altitude}$$
$$i_h = 90° − \text{altitude}$$

$$P = E \times \tan X = G \times \sin X$$

The horizontal length L of shadow cast by an object of height H is
$$L = H \times \tan (90 − \text{altitude angle})$$
or $L = H \times \text{cot altitude angle}$

Depth of shadow D perpendicular to a wall oriented at angle Y' from due North is
$$D = L \times \sin X' = H \times \text{cot altitude angle}$$
$$\times \sin (Y' − \text{azimuth})$$

Where $X' = 360 − (Y' + \text{solar azimuth})$. Note that Y' is always on the opposite side of the North-South axis from the solar azimuth and X' is always more than 90°—otherwise there would be no shadow.

Example: What are the minimum required dimensions of a horizontal visor symmetrically placed 10 ft above the lower sill of a window 3 ft wide in a wall facing southwest at latitude 35°N if complete shading is required at 2 p.m. on May 21?

Solution: Solar azimuth, 2 p.m. May 21, latitude 35°N (Zone 2) = 112° solar altitude = 60° (from main table)

$H − W = 10$ ft (given)
$Y = 135$ (from Table 1)
$X = 135 − 112 = 23°$ (Y and azimuth both westward)
$E = G \times \cos X$
$\quad = (H − W) \times \text{cot altitude} \times \cos X$
$\quad = 10 \times 0.577 \times 0.920 = 5.3$ ft
$P = E \times \tan X$
$\quad = 5.3 \times 0.424 = 2.25$ ft

Visor will overhang 5.3 ft and its width will be window width plus $2P$, i.e., $3 + 4.5 = 7.5$ ft.

Example: What would be dimensions of the visor in the preceding example if shading were required from 1 p.m. to 3 p.m.?

Solution: Since the cotangent of the altitude angle increases with decreasing solar altitude, maximum overhang will be required at minimum altitude. Similarly, since the tangent of X increases as (Y − azimuth) increases, maximum projection P will be required at minimum azimuth. Minimum altitude and azimuth for the example hours occur at 3 p.m., and procedure for sizing the visor would be as before but using the angle values for 3 p.m.

Notice that in shadow forecast a diagram such as Fig. 2 is helpful in assuring that wall orientation and solar angles used in calculations are realistic. Otherwise rather cumbersome attention must be given to the positive and negative values of trigonometric functions in various quadrants of orientation.

Example: What fraction of a Southwest facing window 4 ft high with sill 6 ft below a 4-ft overhang

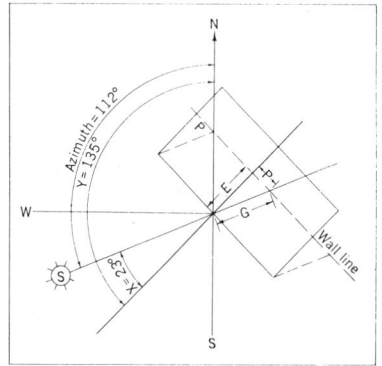

Fig. 2. Geometry of overhang, latitude 35°N.,
2 p.m., May 21.
$E = G \cos X = (H - W) \cot$ altitude $\times \cos X$
$P = E \tan X = G \sin X$

would be shaded in latitude 35N on August 24 at 3 p.m.?

Solution: With both Y and azimuth west,
$X = 135 - 108 = 27$

Altitude, from main table, $= 42$

A sketch of the problem situation shows that the formula for overhang applies:

$E = (H - W') \times \cot$ altitude $\times \cos X$

where $(H - W')$ is the extent of shadow below the eaves.

Then

$$(H - W') = \frac{E}{\cot \text{alt} \times \cos X} = \frac{4}{1.11 \times .891}$$
$$= 4 \text{ (approximately)}$$

Thus, the shadow will fall on 2 ft of wall and on $(4 - 2)$ ft of window. This is 2/4 or 50% of a 4-ft high window.

Example: If the window in the previous example is inset one foot from the outside face of the wall, what fraction of the window will be shaded at 3 p.m. August 24?

Solution: One-foot inset gives 4-ft eave the effect of 5-ft overhang. Hence, depth of shadow will be

$$(H - W') = \frac{5}{1.11 \cdot .891}$$
$$= 5 \text{ (approximately)}$$

Thus $\frac{5 - 2}{4} = 75\%$ of the glass will be shaded.

Note that this neglects the fact that the vertical side of the setback will also shade the glass in an area $(4 - 3) \times P = 1 \times 1 \times \tan X = .510$ sq ft.

Example: What will be the fraction of glass shaded in an east facing window 3 ft wide by 5 ft high with a setback of 6 inches, no roof overhang,

August 1, 40°N, at 9 a.m.?

Solution: From the table of solar angles, zone 3:
Azimuth July 23 = 104
Azimuth August 24 = 112
Interpolating, azimuth August 1 = 106
Altitude July 23 = 47
Altitude August 24 = 41
Interpolating, altitude August 1 = 45

From Fig. 1, vertical cast of shadow $S = (H - W')$
$= G \tan$ altitude.

From Table 1, wall-solar azimuth, X, $= 106 - 90$
$= 16$

From Fig. 2, horizontal cast of shadow $P = E$
$\tan X$, where $E = .5$ ft and $G = \dfrac{E}{\cos X}$

Area of glass unshaded will be:
$$A_U = (5 - S) \times (3 - P) = 12.8 \text{ sq ft}$$
Fraction shaded will be:
$$\frac{15 - 12.8}{15} = 15\%$$

Example: A southwest facing wall in latitude 35°N (zone 2) has a 3-ft overhang 6 ft above the lower sill of a window 4 ft high by 3 ft wide. At what hour will the peak instantaneous heat gain through the window glass occur on July 23?

Solution: Peak gain through glass occurs when the combined effect of shading and incident angle are at a minimum; i.e., when the function, $A_U I \cos i$, is maximum, where A_U is unshaded area of glass, I is solar intensity normal to the sun's rays and i is the incident angle.

Since shading obviously will be minimum when altitude is minimum, a trial calculation at 6 p.m. shows, letting $S =$ vertical cast of shadow $= E/(\cot$ altitude $\times \cos X)$

$$S = \frac{3}{\cot 15 \cos 62} = 1.71 \text{ ft}$$

Thus the whole window is in sunshine and
$$A_U I \cos i = 12 I \cos 15 \cos 62 = 5.44 I$$

Similarly, at 5 p.m., when altitude $= 24$ and $X = 54$

$$S = \frac{3}{\cot 24 \cos 54} = 2.28$$
$$A_U = 3 \times 3.72 = 11.16$$
$$\cos i = \cos 54 \cos 24 = .537$$
$$\text{and } A_U I \cos i = 5.99 I$$

Thus, in spite of some shading, the heat gain would be higher at 5 p.m. than it would at 6 p.m. with no shading. A calculation for 4 p.m., altitude $= 35$, $X = 45$, shows $S = 2.97$; and $A_U I \cos i = 5.27 I$.

This confirms peak gain at 5 p.m. Note that without roof overhang the peak value would coincide with maximum value of $I \cos i$ at 4 p.m.

DAILY TOTAL SOLAR RADIATION
Cloudless Days

Summations of the average clear-day solar energy received on horizontal surfaces for the 15th of each month are shown on accompanying U. S. maps as isolines of langleys per day. One langley (ly) equals one gram-calorie per square centimeter. To convert to Btu per square foot, multiply ly units by 3.68.

Values were computed from U. S. Weather Bureau records of solar radiation measurements available at several stations and from estimates of the amount of atmospheric water vapor on clear days at those same stations and numerous additional ones.

Obviously it would be impractical to compute the solar radiation for each point in the mountainous terrain, but the isolines have been drawn to account for the major changes of elevation. However, where the albedo[1] is markedly different from the places where solar radiation is observed as in places where snow cover persists for long periods of time or where the dust content is higher as in larger cities, the radiation would be different from the values shown. For those places it would be necessary to make additional correction.

With regard to the radiation isolines themselves, it is at once obvious that the latitudinal change of the amount of radiation which falls on the United States during a cloudless day is very marked in winter and very small in midsummer. In December, for example, central Texas receives about 350 ly per day while only about 135 ly per day, or less than half the Texas amount, falls along the northern boundary of the country. The longitudinal variation is affected somewhat by the height of the terrain, but is generally quite small. The predominance of the latitudinal change is, of course, due to the lower solar altitude and smaller number of hours of daylight in the north as compared to the south.

In the early summer months, the distribution is considerably changed. During these months, the cloudless day radiation is not much different in the Lake region from that in the Gulf States, a belt of maximum radiation extends east-west through the center of the country, and the elevation of the ground determines the location of the absolute maxima of radiation. This elevation effect should be anticipated from the fact that the greater number of hours of daylight in the north counterbalances the lower solar altitude there. Furthermore, when the amount of precipitable water vapor is large, changes

[1] The albedo of a body (such as the ground) is the ratio of solar energy reflected by the body to the solar energy incident upon it. For example, a freshly fallen snow cover will reflect about 90% of the solar energy which falls on it; the albedo of the snow cover is 0.90. Albedo of bare ground is generally about 0.10.

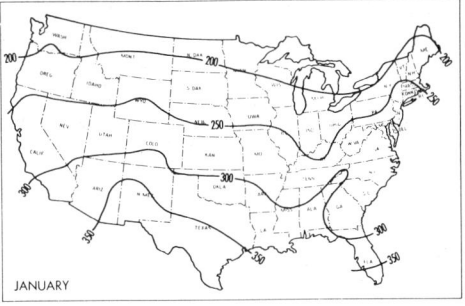

JANUARY

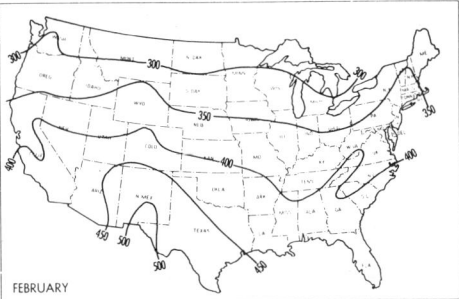

FEBRUARY

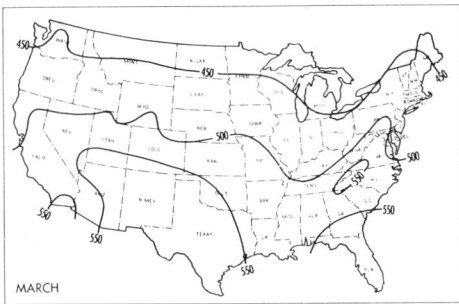

MARCH

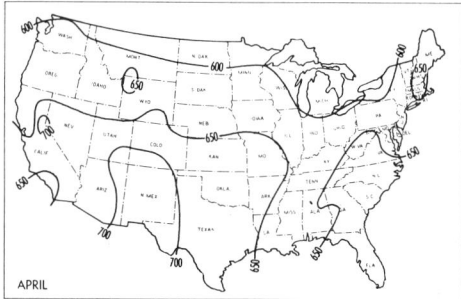

APRIL

in vapor amount affect the radiation relatively little. Therefore, although the south has more water vapor than the north, in summer when large amounts of vapor are present everywhere the difference does not

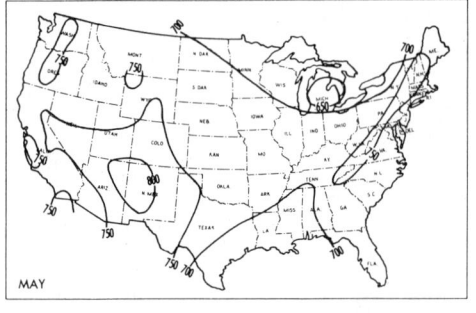

MAY

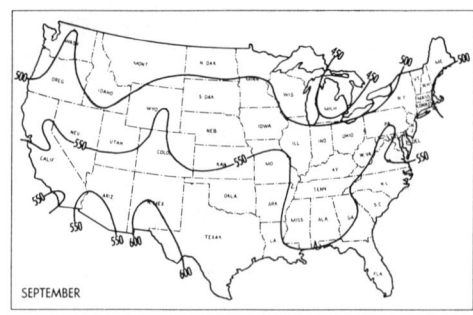

SEPTEMBER

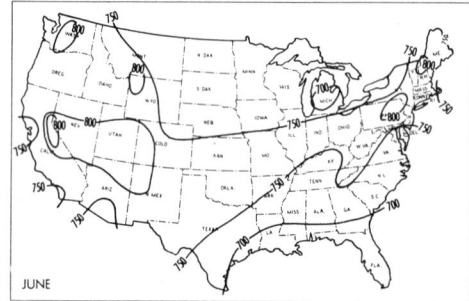

JUNE

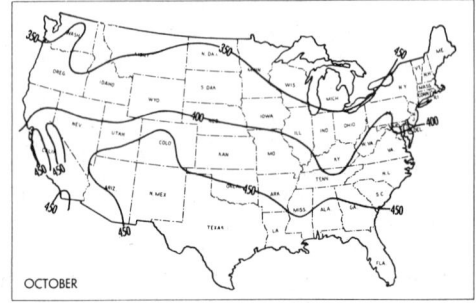

OCTOBER

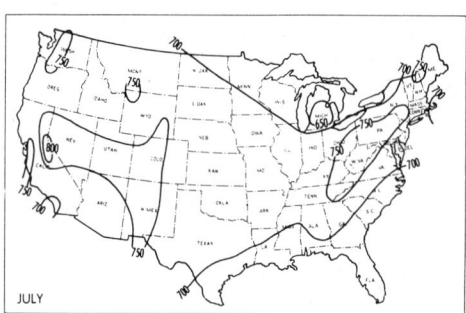

JULY

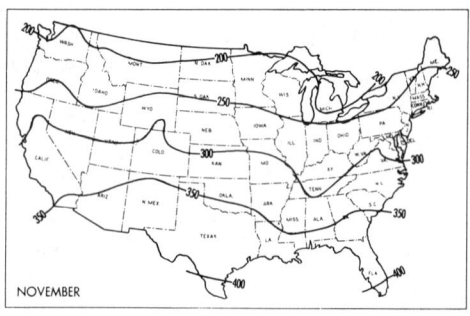

NOVEMBER

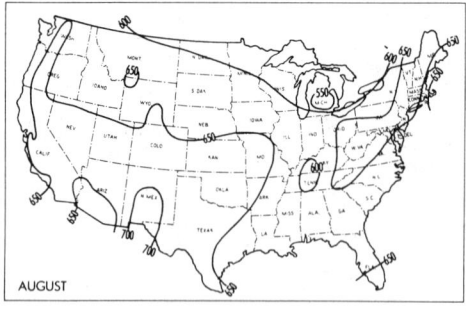

AUGUST

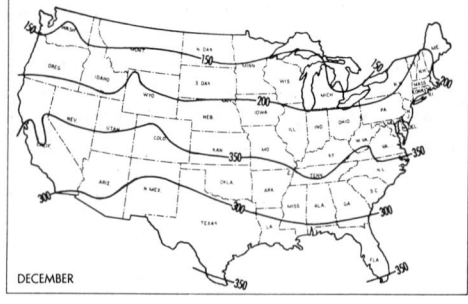

DECEMBER

affect the radiation distribution markedly.

A comparison between winter and summer radiation shows that, starting roughly with the same value everywhere in summer, namely about 750 ly per day, the radiation during cloudless days decreases in December to about 20% of the summer value in the north, and to about 50% of the summer value in the south.

DAILY TOTAL SOLAR RADIATION

Average Days

The accompanying U. S. maps show average daily totals of solar radiation received on a horizontal surface during average days for each month. Isolines are of langleys per day. One langley (ly) equals one gram-calorie per square centimeter. To convert to Btu per square foot, multiply ly units by 3.68.

The basic solar radiation data were obtained from continuously recording pyrheliometers at eleven U. S. Weather Bureau stations. Radiation was calculated for many other stations from records of sunshine hours and previously estimated values of radiation on cloudless days by the equation

$$Q/Q_0 = .35 + .61\ S$$

where $Q =$ average radiation for all days during the month, ly per day

$Q_0 =$ cloudless day radiation, ly per day

$S =$ ratio of actual recorded sunshine hours to the number of hours of possible sunshine.

Discussion of the Charts

The radiation charts show much the same broad features as the cloudless day. The radiation values are low during winter (centering on December) when the solar trajectories are low, and high in the months of high solar trajectory, centering on June. In December there is a relatively great change from south to north, while in June there is relatively small latitudinal variation in Q with a weak maximum in the neighborhood of 40 degrees north. The most obvious features of the longitudinal distribution in both winter and summer are relatively low values in the Great Lakes region due to a combination of relatively high cloudiness and important cloudless day radiation losses, and maximum values in the Southwest, except on the immediate coast, due to a combination of relatively sparse cloud cover and slight cloudless day depletions. In winter the depleting influence of clouds is greater than is the case during summer, since on the average more clouds are present nearly everywhere in winter.

The average error in the basic data (that is, the difference between Q as computed through use of the equation and Q as measured) amounts to about 4.5% of the measured Q. Since the computations for the "sunshine stations" were corrected for this difference, the expected error in them is of the order of 5% or less. In using the charts at points other than those at which Q is actually computed or measured, there is an additional source of possible

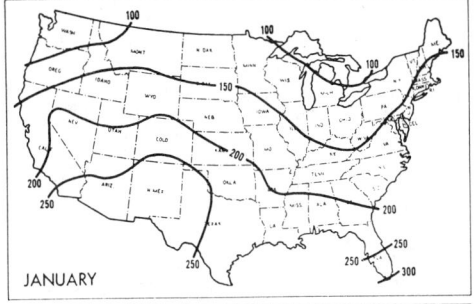

JANUARY

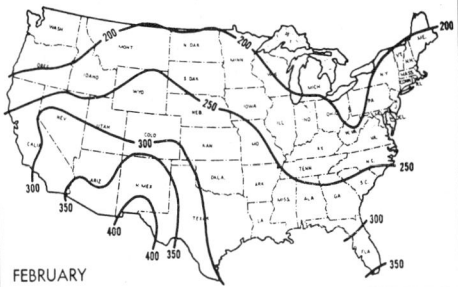

FEBRUARY

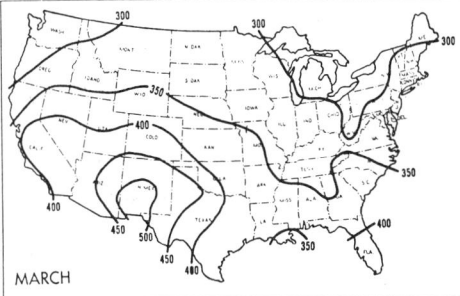

MARCH

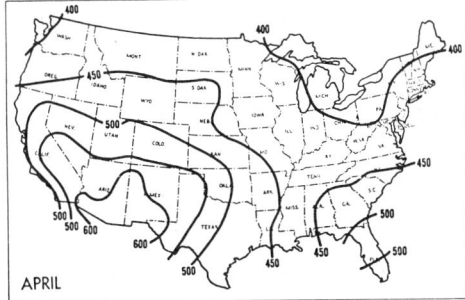

APRIL

error, important in some areas, unimportant in others. This arises from the fact that in some areas there are important local variations in factors influencing Q_0 and S, and hence similar important

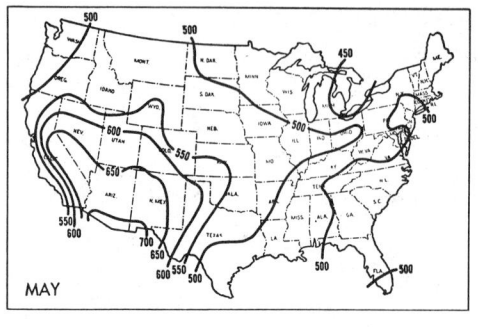

MAY

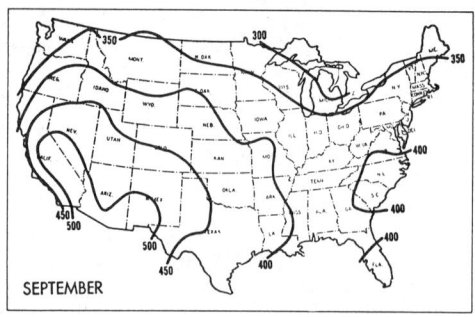

SEPTEMBER

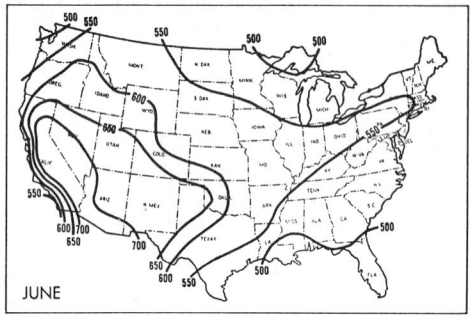

JUNE

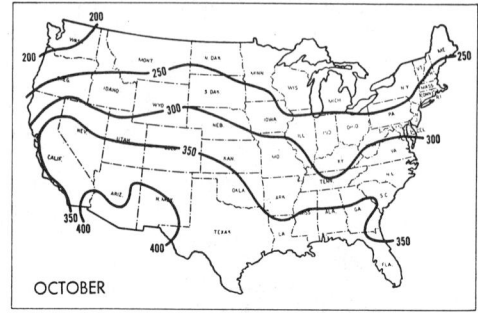

OCTOBER

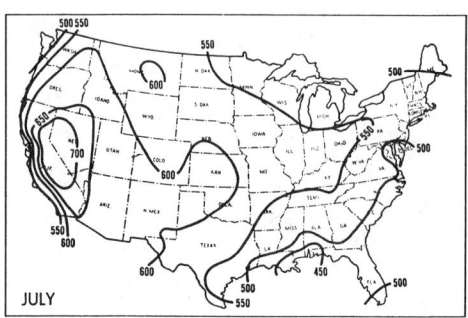

JULY

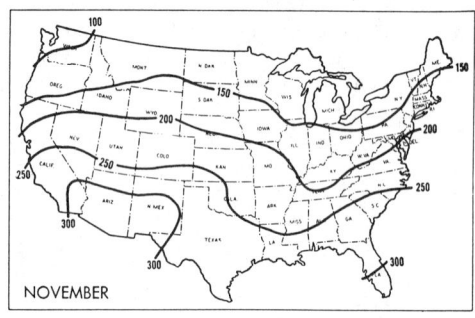

NOVEMBER

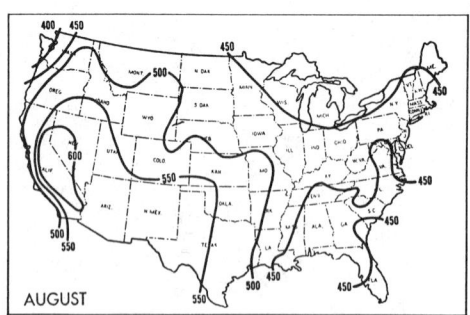

AUGUST

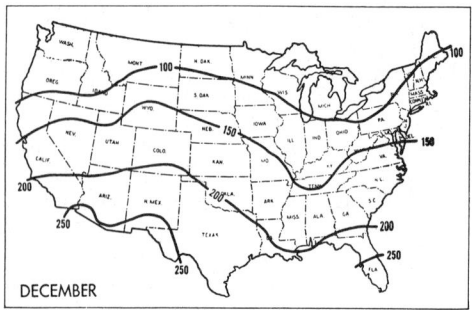

DECEMBER

local variations in Q, while in other areas local variations are negligible. The most important factors affecting local variations are atmospheric pollution, elevation, ground reflection, and degree of cloudiness. Thus, Q will have large local variations in large cities, mountainous areas, in regions having abrupt changes in average snow cover, and on the shores of large bodies of water.

SOLAR ENERGY RECEIVED ON A HORIZONTAL SURFACE

The accompanying data, prepared by I. F. Hand of the U. S. Weather Bureau, Blue Hill Observatory, Milton, Mass., show the amount of energy received by a horizontal surface in various locations in the U. S. and two stations in Canada. The figures are in Btu per square foot per average day for the various months as indicated.

Table 1 shows the relationship between daily average receipt of solar energy and (1) number of hours of sunshine per year; (2) average cloudiness; (3) yearly number of clear days, and; (4) percentage of daylight hours the sun shines. Table 2 covers the period October through March, inclusive.

Table 4 lists normals of the daily receipt of solar energy by months for the year, in terms of Btu per square foot per day.

Table 1. — Sunshine Data for Whole Year

Solar Energy, Btu per Sq Ft. per Day	Hours of Sunshine	Average Cloudiness: Scale, 1–10	Clear Days, 0–3 Cloudiness	Percent. of Daylight Hours Sun Shines
Under 1000	2177	6.6	86	48
1001–1200	2339	6.2	93	51
1201–1400	2643	5.5	123	58
1401–1600	2826	5.1	134	63
1601–1800	3055	4.6	166	68
1801–2000	3145	4.4	179	71
Over 2000	3776	2.6	246	84

Table 2. — Sunshine Data for Winter Months
(October through March, incl.)

Solar Energy, Btu per Sq Ft. per Day	Hours of Sunshine	Average Cloudiness: Scale, 1–10	Clear Days, 0–3 Cloudiness	Percent. of Daylight Hours Sun Shines
201–400	82	7.6	4.1	29
401–600	118	6.8	6.6	41
601–800	150	5.8	9.3	49
801–1000	171	5.3	11.1	55
1001–1200	197	5.1	11.7	59
1201–1400	211	4.8	12.1	64
1401–1600	242	4.7	13.3	68
1601–1800	245	4.4	14.4	72
1801–2000	271	4.2	16.0	73

Table 3. — Receipt of Solar Energy As It Varies with Latitude

City	North Latitude	Yearly Aver., Btu per Sq Ft per Day	City	North Latitude	Yearly Aver., Btu per Sq Ft per Day
Miami, Fla.	25° 49	1497	Salt Lake City, Utah	40° 46	1442
Brownsville, Tex.	25 55	1840	New York, N. Y.	40 46	1054
Gainesville, Tex.	29 39	1471	Sayville, N. Y.	40 46	1291
Apalachicolas, Fla.	29 45	1675	State College, Pa.	40 48	1154
New Orleans, La.	30 02	1250	Lincoln, Neb.	40 52	1354
Lake Charles, La.	30 13	1568	Upton, N. Y.	40 52	1269
El Paso, Tex.	31 48	2037	Cleveland, Ohio	41 24	1312
Fort Worth, Tex.	32 49	1699	Newport, R. I.	41 30	1258
La Jolla, Calif.	32 52	1526	Put-in-Bay, Ohio	41 39	1242
Charleston, S. C.	32 54	1575	East Wareham, Mass.	41 46	1204
Griffin, Ga.	33 14	1578	Blue Hill, Mass.	42 13	1243
Riverside, Calif.	33 32	1558	Boston, Mass.	42 21	1110
Santa Maria, Calif.	34 56	1817	Medford, Ore.	42 22	1575
Albuquerque, N. M.	35 03	1892	Ithaca, N. Y.	42 27	1119
Hatteras, N. C.	35 15	1594	Twin Falls, Ida.	42 33	1438
Oak Ridge, Tenn.	35 55	1307	East Lansing, Mich.	42 42	998
Las Vegas, Nev.	36 05	1822	Madison, Wis.	43 05	1218
Nashville, Tenn.	36 07	1253	Toronto, Ont., Can.	43 40	1078
Stillwater, Okla.	36 08	1467	St. Cloud, Minn.	45 35	1352
Fresno, Calif.	36 46	1670	Caribou, Me.	46 52	1227
Davis, Calif.	38 32	1633	Spokane, Wash.	47 37	1374
Washington, D. C.	38 56	1234	Seattle, Wash.	47 39	1160
Columbia, Mo.	38 57	1388	Glasgow, Mont.	48 11	1455
Seabrook, N. J.	39 30	1316	Winnipeg, Man., Can.	49 53	1216
Grand Lake, Col.	40 15	1573			

Table 4. — Daily Averages, Btu Per Sq. Ft., of Solar Energy Received on a

Horizontal Surface, By Months

(Yearly averages are given in Table 3)

City	Jan	Feb	Mar	Apr	May	Jun	Jul	Aug	Sep	Oct	Nov	Dec
Miami, Fla.	1100	1284	1535	1756	1852	1771	1749	1716	1528	1351	1232	1085
Brownsville, Tex.	1240	1465	1782	1919	2144	2590	2472	2336	1934	1786	1221	1192
Gainesville, Fla.	1011	1255	1587	1937	2066	1904	1823	1624	1446	1181	974	841
Apalachicola, Fla.	1192	1542	1616	1941	2074	2192	2103	2007	1646	1513	1317	959
New Orleans, La.	756	915	1207	1487	1590	1697	1491	1439	1402	1258	952	804
Lake Charles, La.	1041	1114	1531	1771	2118	2151	1926	2052	1594	1439	1262	819
El Paso, Tex.	1328	1546	2125	2524	2716	2731	2531	2435	2103	1786	1428	1207
Fort Worth, Tex.	1015	1218	1734	1875	2140	2332	2303	2258	1779	1587	1277	867
La Jolla, Calif.	930	1181	1550	1882	2007	2066	2015	1830	1572	1273	1107	904
Charleston, S. C.	923	1232	1664	2059	2288	2166	1989	1945	1509	1203	1137	786
Griffin, Ga.	1063	1107	1181	2103	2288	2273	2170	2066	1546	1358	1044	738
Riverside, Calif.	974	1151	1506	1823	2007	2207	2184	2011	1712	1321	1018	782
Santa Maria, Calif.	1070	1380	1882	2251	2506	2399	2428	2369	1945	1594	1114	867
Albuquerque, N. M.	1133	1354	1834	2236	2494	2749	2502	2299	2018	1712	1284	1085
Hatteras, N. C.	941	1063	1550	2103	2229	2266	2229	2125	1587	1269	1015	756
Oak Ridge, Tenn.	642	852	1055	1483	2103	2000	1838	1708	1517	1129	753	598
Las Vegas, Nev.	963	1292	1956	2111	2362	2771	2539	2332	2044	1483	1166	845
Nashville, Tenn.	524	753	1089	1557	1838	1934	1867	1668	1439	1125	779	465
Stillwater, Okla.	923	1004	1520	1801	1838	2196	1889	1937	1565	1255	900	775
Fresno, Calif.	664	1037	1539	2096	2406	2642	2576	2288	1889	1380	915	605
Davis, Calif.	738	1026	1417	1982	2387	2642	2605	2303	1834	1336	745	576
Washington, D. C.	568	738	1225	1513	1716	1867	1808	1631	1373	1085	745	539
Columbia, Mo.	598	930	1229	1631	1801	2077	2221	1856	1624	1299	731	664
Seabrook, N. J.	686	908	1321	1668	1897	2007	1838	1771	1336	1052	771	535
Grand Lake, Colo.	790	1144	1624	2030	2177	2362	2236	1989	1720	1328	863	613
Salt Lake City, Utah	572	908	1424	1882	2015	2192	2303	2247	1631	1026	661	443
New York, N. Y.	450	705	956	1339	1572	1646	1620	1351	1166	897	546	395
Sayville, N. Y.	635	904	1236	1631	1970	2066	1812	1694	1292	1070	686	498
State College, Pa.	506	642	1015	1428	1572	1845	1889	1683	1321	915	605	424
Lincoln, Neb.	686	930	1247	1576	1852	2052	2122	1775	1509	1114	771	613
Upton, N. Y.	568	790	1181	1550	2030	2081	1779	1683	1328	1111	642	487
Cleveland, Ohio	373	675	1030	1550	2140	2214	2192	1934	1705	1041	487	406
Newport, R. I.	583	845	1196	1535	1786	1963	1860	1683	1358	1092	668	524
Put-in-Bay, Ohio	509	756	1137	1476	1734	2092	2000	1830	1387	1026	557	399
East Wareham, Mass.	557	790	1085	1498	1720	1897	1768	1668	1299	996	627	542
Blue Hill, Mass.	601	923	1196	1465	1756	1911	1838	1708	1343	1077	601	498
Boston, Mass.	454	745	1085	1328	1661	1823	1690	1502	1184	937	502	406
Medford, Ore.	391	768	1232	2107	2790	2590	2804	2494	1753	1063	550	362
Ithaca, N. Y.	435	760	926	1173	1624	1867	1845	1697	1299	940	476	391
Twin Falls, Idaho	613	827	1269	1705	2184	2303	2280	1985	1646	1255	738	450
E. Lansing, Mich.	384	649	945	1284	1395	1638	1653	1432	1048	819	380	343
Madison, Wis.	539	797	1166	1498	1727	1904	1993	1690	1321	959	557	443
Toronto, Ont.	351	605	1004	1317	1668	1926	1756	1627	1144	797	399	347
St. Cloud, Minn.	627	878	1461	1734	2070	2066	2118	1667	1343	1048	646	561
Caribou, Me.	531	745	1192	1697	1771	1963	2015	1690	1343	937	450	391
Spokane, Wash.	443	731	1240	2111	1782	2269	2487	2144	1587	923	491	280
Seattle, Wash.	229	328	1033	1823	1867	2280	2170	1753	1255	627	325	229
Glasgow, Mont.	576	900	1446	1852	2362	2494	2435	2011	1395	900	642	450
Winnipeg, Man.	524	745	1225	1461	1945	2048	2052	1808	1166	841	417	365

HOURLY INSOLATION ON CLOUDLESS DAYS
at the Time of Solstices and Equinoxes

A graphic presentation of the diurnal march of insolation received on several surfaces at the time of the solstices, June 21 and December 21, and the equinoxes, March 21 and September 21, is given for six latitudes in Figures 1 to 18, inclusive.

In the illustrations, the numbers on the curves refer to the surface on which energy is received as follows:

(1) Normal incidence.
(2) Horizontal surface.
(3) Vertical surface facing South.
(4) Vertical surface facing North.
(5) Vertical surface facing East during the morning and West during the afternoon. Dashed lines are for East and West walls during afternoon and morning respectively.
(6) Diffuse on horizontal surface.

The calculations were made for average clear sky conditions, and application of the data for local conditions must take into account the degree of atmospheric contamination, humidity, and elevation above sea level. For example, values of total insolation in large industrial cities might be 20% lower than here

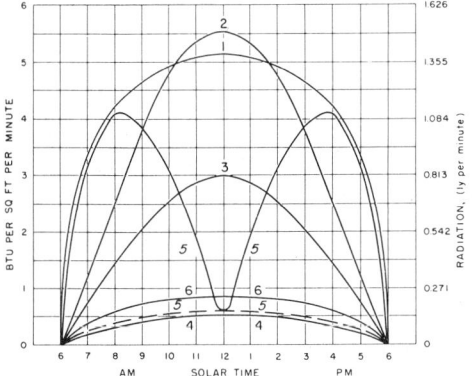

Fig. 1. March 21 and Sept. 21, latitude 25°N.

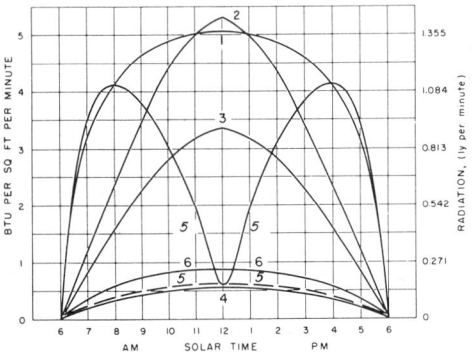

Fig. 2. March 21 and Sept. 21, latitude 30°N.

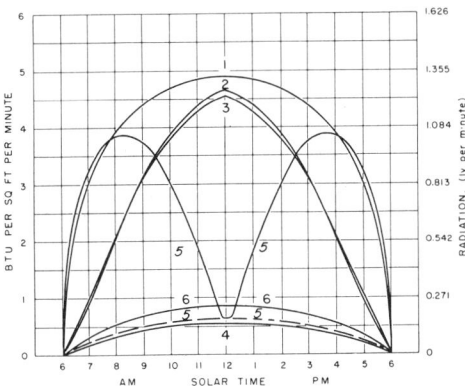

Fig. 4. March 21 and Sept. 21, latitude 40°N.

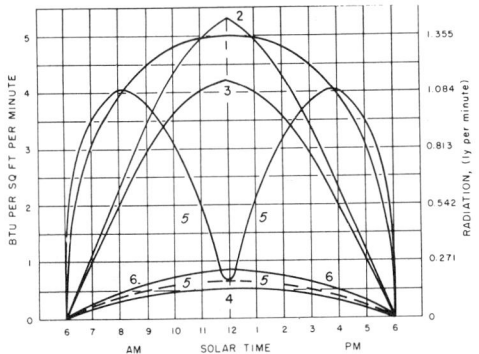

Fig. 3. March 21 and Sept. 21, latitude 35°N.

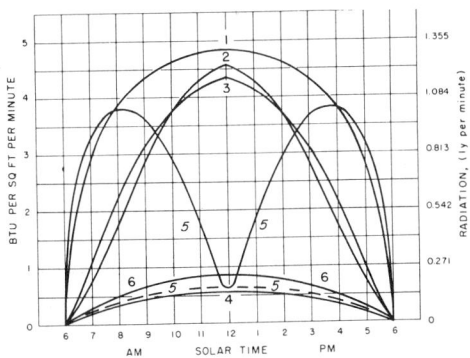

Fig. 5. March 21 and Sept. 21, latitude 45°N.

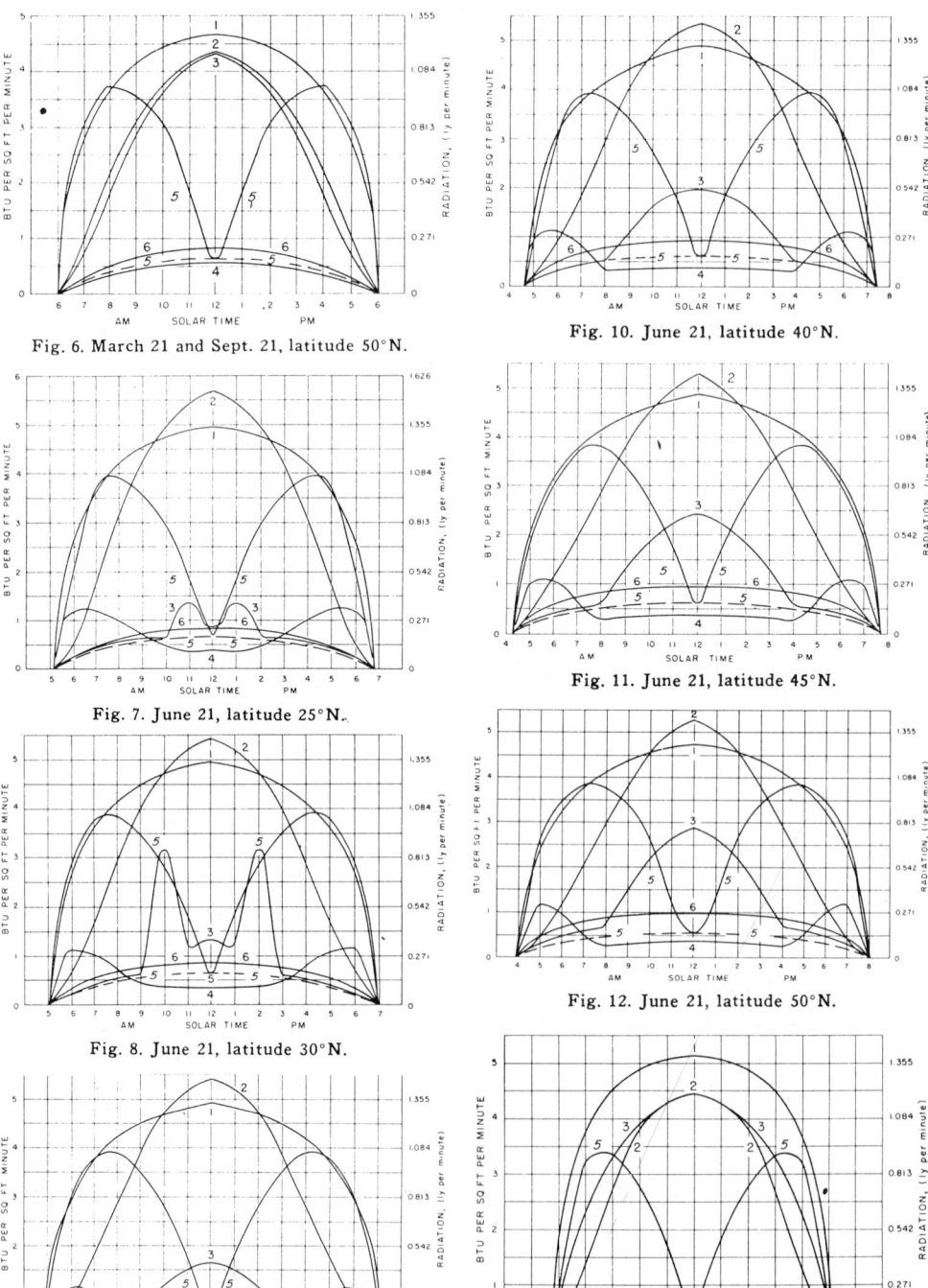

Fig. 6. March 21 and Sept. 21, latitude 50°N.

Fig. 7. June 21, latitude 25°N.

Fig. 8. June 21, latitude 30°N.

Fig. 9. June 21, latitude 35°N.

Fig. 10. June 21, latitude 40°N.

Fig. 11. June 21, latitude 45°N.

Fig. 12. June 21, latitude 50°N.

Fig. 13. Dec. 21, latitude 25°N.

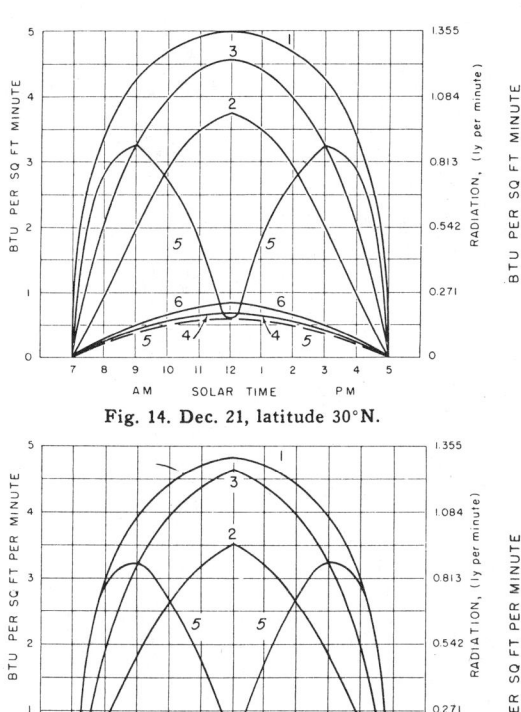

Fig. 14. Dec. 21, latitude 30°N.

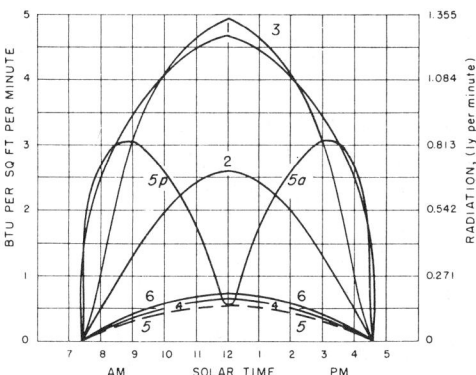

Fig. 15. Dec. 21, latitude 35°N.

Fig. 16. Dec. 21, latitude 40°N.

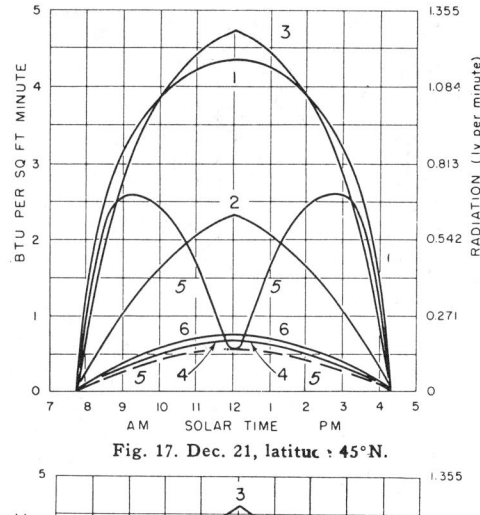

Fig. 17. Dec. 21, latitude 45°N.

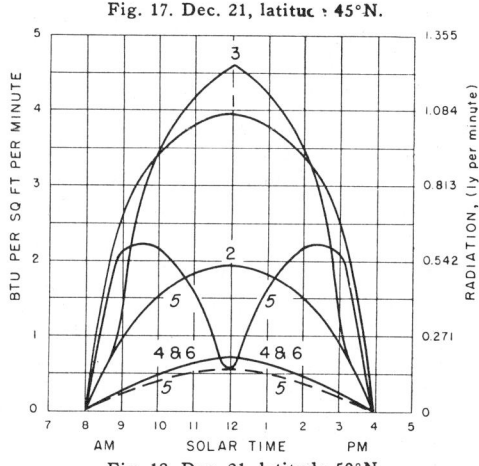

Fig. 18. Dec. 21, latitude 50°N.

At Mount Evans and Pike's Peak, Colorado, the diffuse radiation on cloudless days at noon might not exceed 4% of the total, whereas in cities where soft coal is burned in quantity without benefit of smoke-eliminating devices, the diffuse component might comprise 40% of the total, or even more in extreme cases.

As a basis for making the calculations, I. F. Hand and the staff at the U. S. Weather Bureau Blue Hill Observatory in Milton, Mass., first obtained the mean of normal incidence received at various stations throughout the United States and, by simple trigonometrical formulas computed the amount to be expected on various surfaces facing the cardinal points. Values of total solar and sky energy received on a horizontal surface at a large number of stations were plotted against latitude; and the amounts to be expected at the latitudes listed were obtained through simple interpolation.

given, while we would expect considerably higher values of total solar energy at high elevations, as for example, Estes Park and Grand Lake, Colorado.

These variations would be reversed in the case of the diffuse component, for an increase in atmospheric contamination increases the indirect energy.

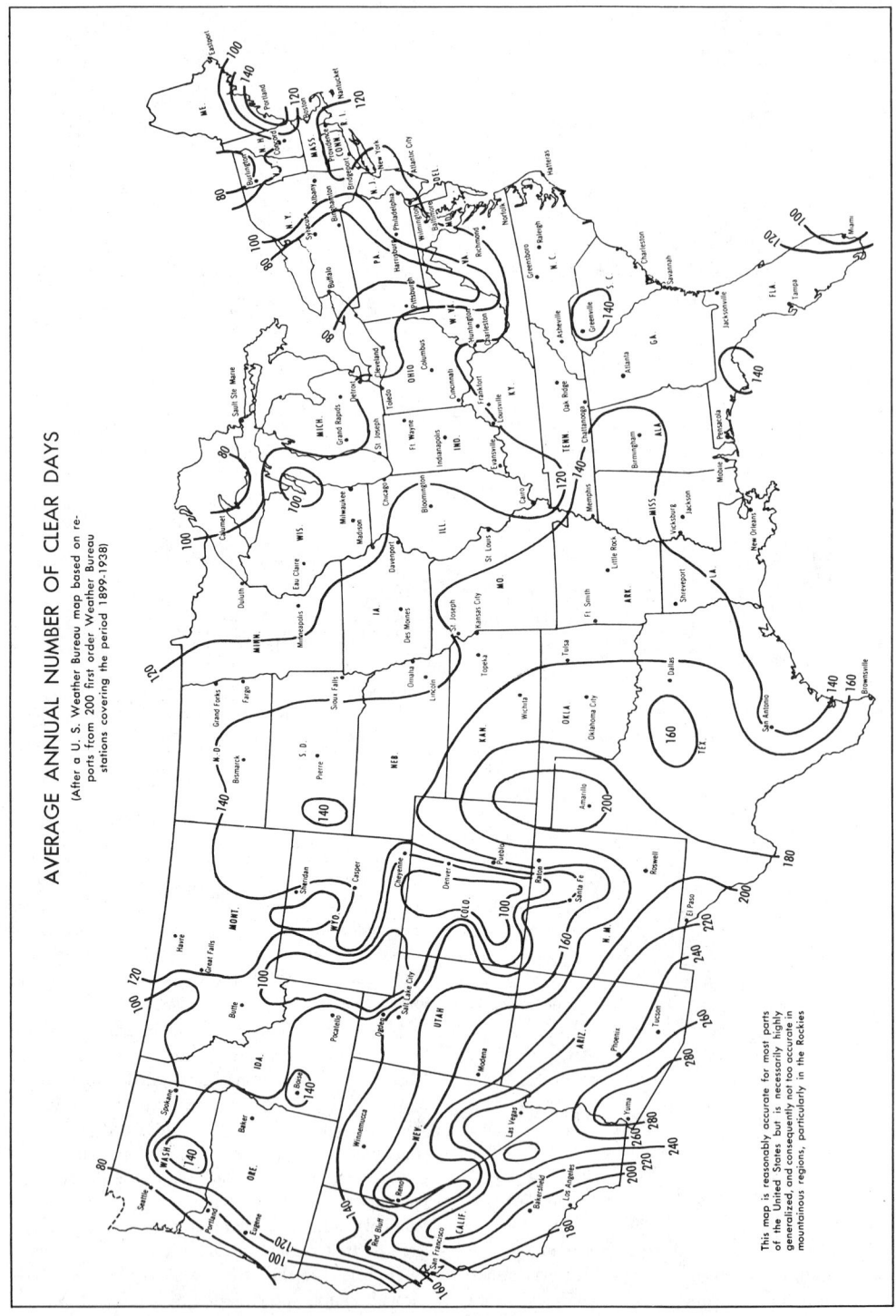

AVERAGE ANNUAL NUMBER OF CLEAR DAYS

(After a U. S. Weather Bureau map based on reports from 200 first order Weather Bureau stations covering the period 1899-1938)

This map is reasonably accurate for most parts of the United States but is necessarily highly generalized, and consequently not too accurate in mountainous regions, particularly in the Rockies

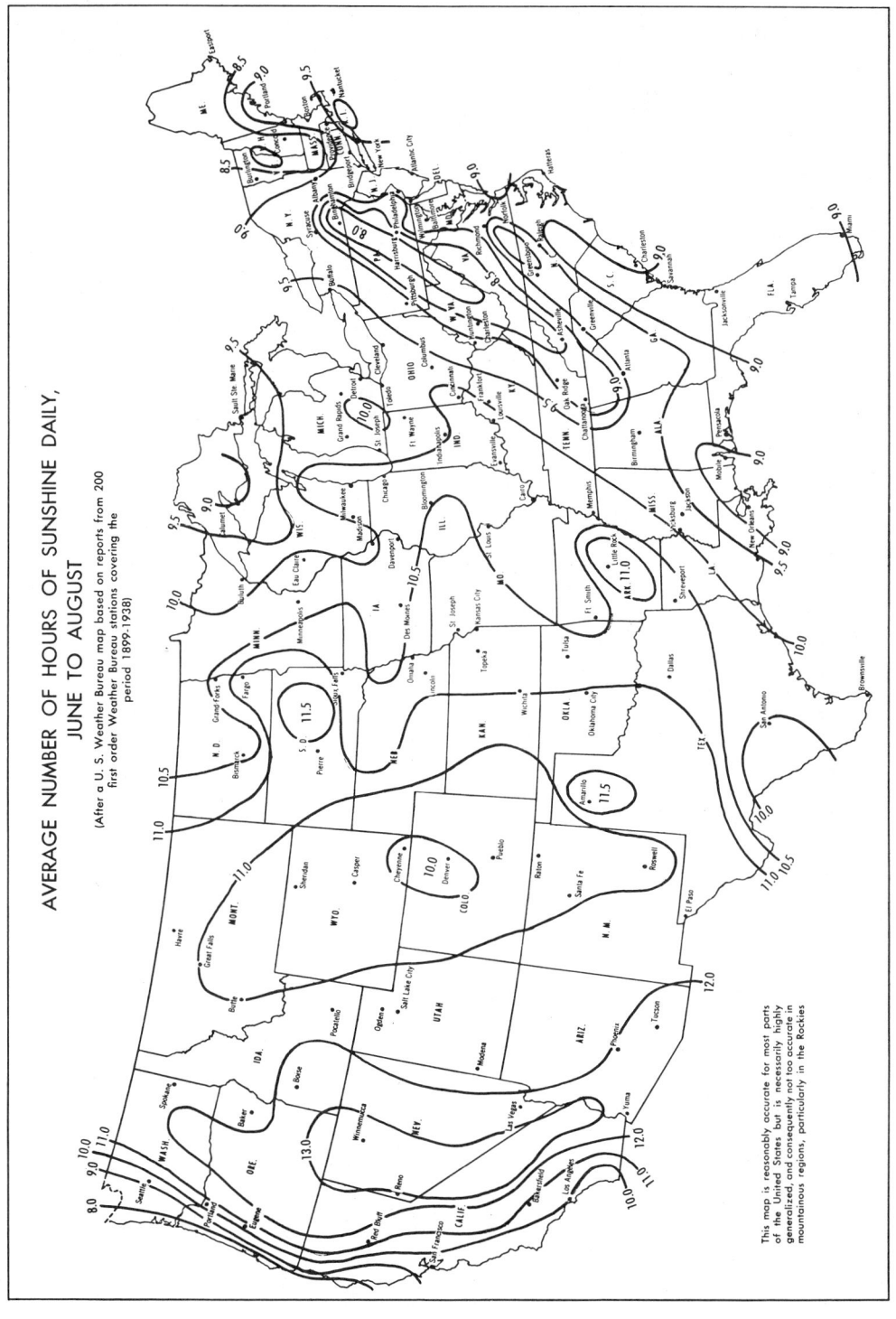

AVERAGE NUMBER OF HOURS OF SUNSHINE DAILY,
JUNE TO AUGUST

(After a U. S. Weather Bureau map based on reports from 200
first order Weather Bureau stations covering the
period 1899-1938)

This map is reasonably accurate for most parts
of the United States but is necessarily highly
generalized, and consequently not too accurate in
mountainous regions, particularly in the Rockies

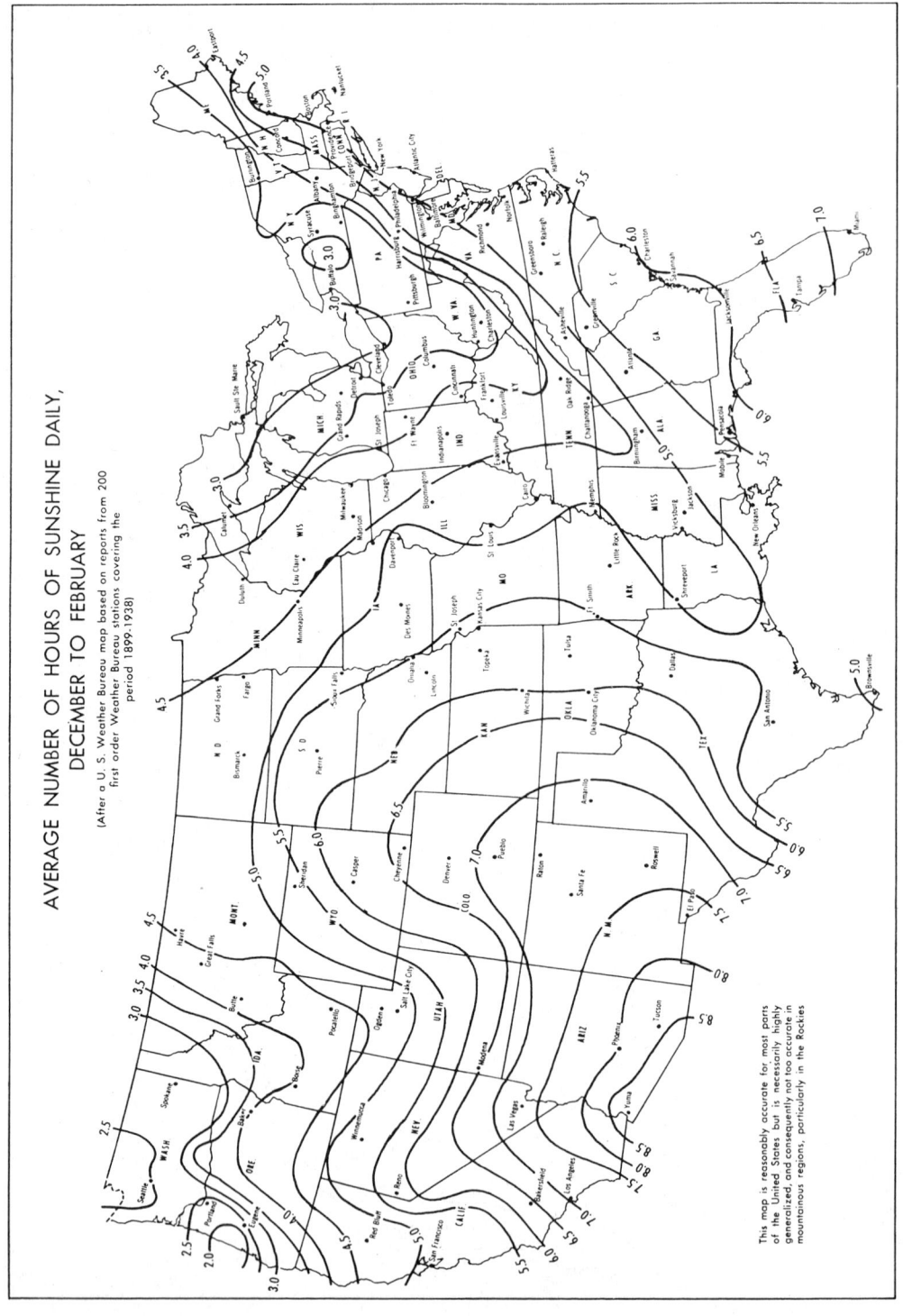

AVERAGE NUMBER OF HOURS OF SUNSHINE DAILY,
DECEMBER TO FEBRUARY

(After a U. S. Weather Bureau map based on reports from 200
first order Weather Bureau stations covering the
period 1899-1938)

This map is reasonably accurate for most parts
of the United States, but is necessarily highly
generalized, and consequently not too accurate in
mountainous regions, particularly in the Rockies

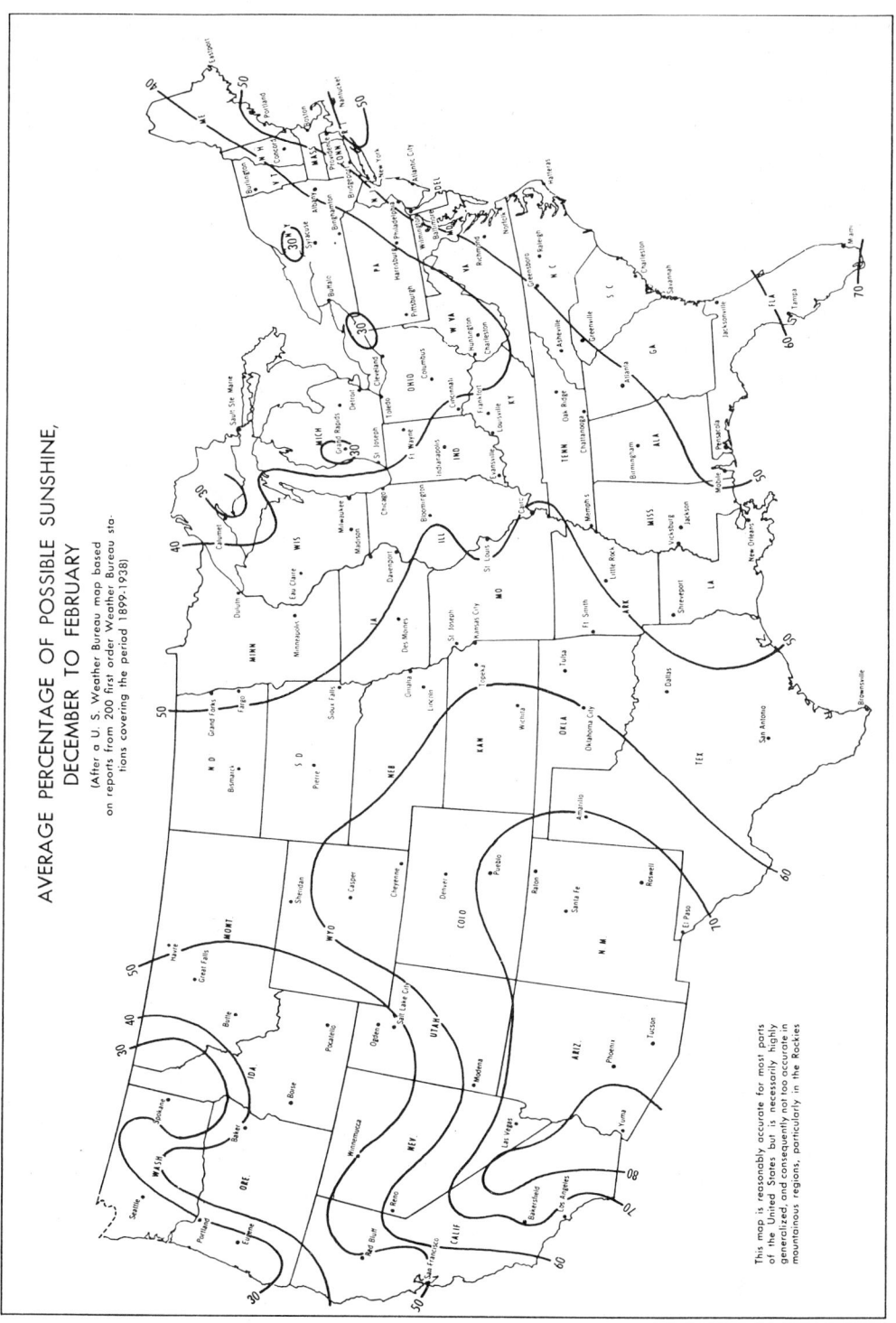

AVERAGE PERCENTAGE OF POSSIBLE SUNSHINE,
DECEMBER TO FEBRUARY

(After a U. S. Weather Bureau map based
on reports from 200 first order Weather Bureau stations covering the period 1899-1938)

This map is reasonably accurate for most parts
of the United States but is necessarily highly
generalized, and consequently not too accurate in
mountainous regions, particularly in the Rockies

CANADIAN WINTER AND SUMMER DESIGN TEMPERATURES *

PROVINCE Station	Latitude °	'	Longitude °	'	Elev, Feet	WINTER Design Dry Bulb 99%	97½%	SUMMER Design Dry Bulb and Mean Coincident Wet Bulb 1%		2½%		5%		Mean Daily Range	Design Wet Bulb 1%	2½%	5%
ALBERTA																	
Calgary	51	6	114	1	3540	-27	-23	84	63	81	61	79	60	25	65	63	62
Edmonton	53	34	113	31	2219	-29	-25	85	66	82	65	79	63	23	68	66	65
Grande Prairie	55	11	118	53	2190	-39	-33	83	64	80	63	78	61	23	66	64	62
Jasper	52	53	118	4	3480	-31	-26	83	64	80	62	77	61	28	66	64	63
Lethbridge	49	38	112	48	3018	-27	-22	90	65	87	64	84	63	28	68	66	65
McMurray	56	39	111	13	1216	-41	-38	86	67	82	65	79	64	26	69	67	65
Medicine Hat	50	1	110	43	2365	-29	-24	93	66	90	65	87	64	28	70	68	66
Red Deer	52	11	113	54	2965	-31	-26	84	65	81	64	78	62	25	67	66	64
BRITISH COLUMBIA																	
Dawson Creek	55	44	120	11	2164	-37	-33	82	64	79	63	76	61	26	66	64	62
Fort Nelson	58	50	122	35	1230	-43	-40	84	64	81	63	78	62	23	67	65	64
Kamloops	50	43	120	25	1133	-21	-15	94	66	91	65	88	64	29	68	66	65
Nanaimo	49	11	123	58	230	16	20	83	67	80	65	77	64	21	68	66	65
New Westminster	49	13	122	54	50	14	18	84	68	81	67	78	66	19	69	68	66
Penticton	49	28	119	36	1121	0	4	92	68	89	67	87	66	31	70	68	67
Prince George	53	53	122	41	2218	-33	-28	84	64	80	62	77	61	26	66	64	62
Prince Rupert	54	17	130	23	170	-2	2	64	59	63	57	61	56	12	60	58	57
Trail	49	8	117	44	1400	-5	0	92	66	89	65	86	64	33	68	67	65
Vancouver	49	11	123	10	16	15	19	79	67	77	66	74	65	17	68	67	66
Victoria	48	25	123	19	228	20	23	77	64	73	62	70	60	16	64	62	60
MANITOBA																	
Brandon	49	52	99	59	1200	-30	-27	89	72	86	70	83	68	25	74	72	70
Churchill	58	45	94	4	115	-41	-39	81	66	77	64	74	62	18	67	65	63
Dauphin	51	6	100	3	999	-31	-28	87	71	84	70	81	68	23	74	72	70
Flinflon	54	46	101	51	1098	-41	-37	84	68	81	66	79	65	19	70	68	67
Portage laPrairie	49	54	98	16	867	-28	-24	88	73	86	72	83	70	22	76	74	71
The Pas	53	58	101	6	894	-37	-33	85	68	82	67	79	66	20	71	69	68
Winnipeg	49	54	97	14	786	-30	-27	89	73	86	71	84	70	22	75	73	71
NEW BRUNSWICK																	
Campbellton	48	0	66	40	25	-18	-14	85	68	82	67	79	66	21	72	70	68
Chatham	47	1	65	27	112	-15	-10	89	69	85	68	82	67	22	72	71	69
Edmundston	47	22	68	20	500	-21	-16	87	70	83	68	80	67	21	73	71	69
Fredericton	45	52	66	32	74	-16	-11	89	71	85	69	82	68	23	73	71	70
Moncton	46	7	64	41	248	-12	-8	85	70	82	69	79	67	23	72	71	69
Saint John	45	19	65	53	352	-12	-8	80	67	77	65	75	64	19	70	68	66
NEWFOUNDLAND																	
Corner Brook	48	58	57	57	15	-5	0	76	64	73	63	71	62	17	67	66	65
Gander	48	57	54	34	482	-5	-1	82	66	79	65	77	64	19	69	67	66
Goose Bay	53	19	60	25	144	-27	-24	85	66	81	64	77	63	19	68	66	64
St. Johns	47	37	52	45	463	3	7	77	66	75	65	73	64	18	69	67	66
Stephenville	48	32	58	33	44	-3	4	76	65	74	64	71	63	14	67	66	65
NORTHWEST TERRITORY																	
Fort Smith	60	1	111	58	665	-49	-45	85	66	81	64	78	63	24	68	66	65
Frobisher	63	45	68	33	68	-43	-41	66	53	63	51	59	50	14	54	52	51
Inuvik	68	18	133	29	200	-56	-53	79	62	77	60	75	59	21	64	62	61
Resolute	74	43	94	59	209	-50	-47	57	48	54	46	51	45	10	50	48	46
Yellowknife	62	28	114	27	682	-49	-46	79	62	77	61	74	60	16	64	63	62
NOVA SCOTIA																	
Amherst	45	49	64	13	65	-11	-6	84	69	81	68	79	67	21	72	70	68
Halifax	44	39	63	34	83	1	5	79	66	76	65	74	64	16	69	67	66
Kentville	45	3	64	36	40	-3	1	85	69	83	68	80	67	22	72	71	69
New Glasgow	45	37	62	37	317	-9	5	81	69	79	68	77	67	20	72	70	69
Sydney	46	10	60	3	197	-1	3	82	69	80	68	77	66	19	71	70	68
Truro	45	22	63	16	131	-8	-5	82	70	80	69	78	68	22	73	71	70
Yarmouth	43	50	66	5	136	5	9	74	65	72	64	70	63	15	68	66	65

*Data supplied by Donald W. Boyd, Meteorologist, Division of Building Research, National Research Council of Canada. Reprinted with permission.

PROVINCE Station	Latitude ° '	Longitude ° '	Elev, Feet	WINTER Design Dry Bulb 99%	WINTER Design Dry Bulb 97½%	SUMMER Design Dry Bulb and Mean Coincident Wet Bulb 1%		SUMMER Design Dry Bulb and Mean Coincident Wet Bulb 2½%		SUMMER Design Dry Bulb and Mean Coincident Wet Bulb 5%		Mean Daily Range	Design Wet Bulb 1%	Design Wet Bulb 2½%	Design Wet Bulb 5%
ONTARIO															
Belleville	44 9	77 24	250	-11	- 7	86	73	84	72	82	71	20	75	74	73
Chatham	42 24	82 12	600	0	3	89	74	87	73	85	72	19	76	75	74
Cornwall	45 1	74 45	210	-13	- 9	89	73	87	72	84	71	21	75	74	72
Hamilton	43 16	79 54	303	- 3	1	88	73	86	72	83	71	21	76	74	73
Kapuskasing	49 25	82 28	752	-31	-28	86	70	83	69	80	67	23	72	70	69
Kenora	49 48	94 22	1345	-32	-28	84	70	82	69	80	68	19	73	71	70
Kingston	44 16	76 30	300	-11	- 7	87	73	84	72	82	71	20	75	74	73
Kitchener	43 26	80 30	1125	- 6	- 2	88	73	85	72	83	71	23	75	74	72
London	43 2	81 9	912	- 4	0	87	74	85	73	83	72	21	76	74	73
North Bay	46 22	79 25	1210	-22	-18	84	68	81	67	79	66	20	71	70	68
Oshawa	43 54	78 52	370	- 6	- 3	88	73	86	72	84	71	20	75	74	73
Ottawa	45 19	75 40	413	-17	-13	90	72	87	71	84	70	21	75	73	72
Owen Sound	44 34	80 55	597	- 6	- 2	84	71	82	70	80	69	21	73	72	70
Peterborough	44 17	78 19	635	-13	- 9	87	72	85	71	83	70	21	75	73	72
St. Catharines	43 11	79 14	325	- 1	3	87	73	85	72	83	71	20	76	74	73
Sarnia	42 58	82 22	625	0	3	88	73	86	72	84	71	19	76	74	73
Sault Ste. Marie	46 32	84 30	675	-17	-13	85	71	82	69	79	68	22	73	71	70
Sudbury	46 37	80 48	1121	-22	-19	86	69	83	67	81	66	22	72	70	68
Thunder Bay	48 22	89 19	644	-27	-24	85	70	83	68	80	67	24	72	70	68
Timmins	48 34	81 22	965	-33	-29	87	69	84	68	81	66	25	72	70	68
Toronto	43 41	79 38	578	- 5	- 1	90	73	87	72	85	71	20	75	74	73
Windsor	42 16	82 58	637	0	4	90	74	88	73	86	72	20	77	75	74
PRINCE EDWARD ISLAND															
Charlottetown	46 17	63 8	186	- 7	- 4	80	69	78	68	76	67	16	71	70	68
Summerside	46 26	63 50	78	- 8	- 4	81	69	79	68	77	67	16	72	70	68
QUEBEC															
Bagotville	48 20	71 0	536	-28	-23	87	70	83	68	80	67	21	72	70	68
Chicoutimi	48 25	71 5	150	-26	-22	86	70	83	68	80	67	20	72	70	68
Drummondville	45 53	72 29	270	-18	-14	88	72	85	71	82	69	21	75	73	71
Granby	45 23	72 42	550	-19	-14	88	72	85	71	83	70	21	75	73	72
Hull	45 26	75 44	200	-18	-14	90	72	87	71	84	70	21	75	73	72
Megantic	45 35	70 52	1362	-20	-16	86	71	83	70	81	69	20	74	72	71
Montreal	45 28	73 45	98	-16	-10	88	73	85	72	83	71	17	75	74	72
Quebec	46 48	71 23	245	-19	-14	87	72	84	70	81	68	20	74	72	70
Rimouski	48 27	68 32	117	-16	-12	83	68	79	66	76	65	18	71	69	67
St. Jean	45 18	73 16	129	-15	-11	88	73	86	72	84	71	20	75	74	72
St. Jerome	45 48	74 1	556	-17	-13	88	72	86	71	83	70	23	75	73	72
Septilles	50 13	66 16	190	-26	-21	76	63	73	61	70	60	17	67	65	63
Shawinigan	46 34	72 43	306	-18	-14	87	72	84	70	82	69	21	74	72	71
Sherbrooke	45 24	71 54	595	-25	-21	86	72	84	71	81	69	20	74	73	71
Thetford Mines	46 4	71 19	1020	-19	-14	87	71	84	70	81	69	21	74	72	71
Trois Rivieres	46 21	72 35	50	-17	-13	88	72	85	70	82	69	23	74	72	71
Val d'Or	48 3	77 47	1108	-32	-27	85	70	83	68	80	67	22	72	70	68
Valleyfield	45 16	74 6	150	-14	-10	89	73	86	72	84	71	20	75	74	72
SASKATCHEWAN															
Estevan	49 4	103 0	1884	-30	-25	92	70	89	68	86	67	26	72	70	69
Moose Jaw	50 20	105 33	1857	-29	-25	93	69	89	67	86	66	27	71	69	68
North Battleford	52 46	108 15	1796	-33	-30	88	67	85	66	82	65	23	69	68	66
Prince Albert	53 13	105 41	1414	-42	-35	87	67	84	66	81	65	25	70	68	67
Regina	50 26	104 40	1884	-33	-29	91	69	88	68	84	67	26	72	70	68
Saskatoon	52 10	106 41	1645	-35	-31	89	68	86	66	83	65	26	70	68	67
Swift Current	50 17	107 41	2677	-28	-25	93	68	90	66	87	65	25	70	69	67
Yorkton	51 16	102 28	1653	-35	-30	87	69	84	68	80	66	23	72	70	68
YUKON TERRITORY															
Whitehorse	60 43	135 4	2289	-46	-43	80	59	77	58	74	56	22	61	59	58

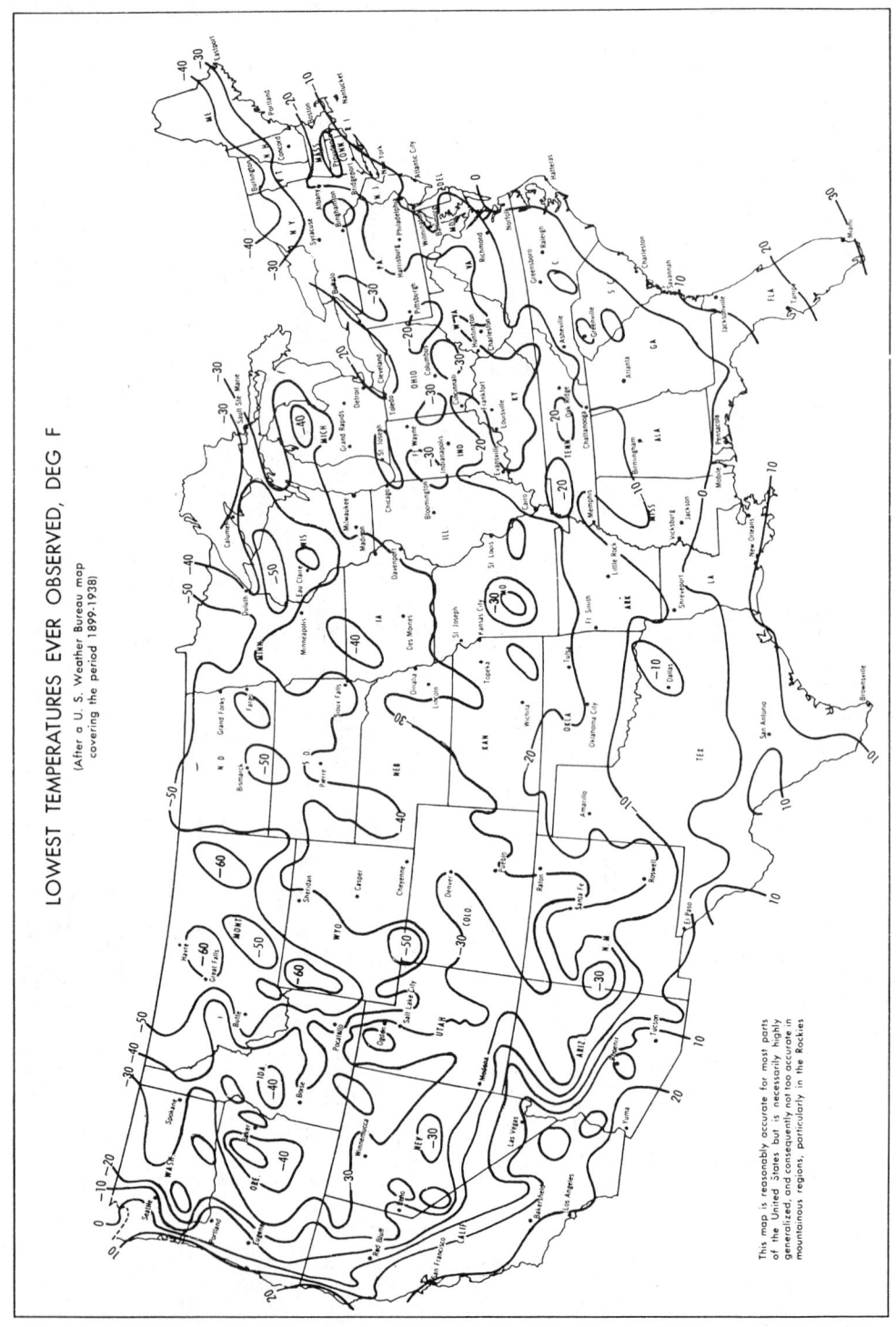

LOWEST TEMPERATURES EVER OBSERVED, DEG F

(After a U. S. Weather Bureau map
covering the period 1899-1938)

This map is reasonably accurate for most parts
of the United States but is necessarily highly
generalized, and consequently not too accurate in
mountainous regions, particularly in the Rockies

FREQUENCY OF LOW TEMPERATURES

In order to provide a basis for selecting heating design temperatures, an analysis of weather in 120 cities was made by Clark M. Humphreys, at that time on the staff of Carnegie Institute of Technology. In making his study, he selected 23 years as the average number of years' data on which to base his study. The study included analysis of both daily minimum and daily mean temperatures, as shown in the table which follows.

This table shows the *average number of times per year*, during the period studied, *that the daily minimum and daily mean temperatures were as low as or lower than the temperature indicated in the body of the table*. Frequency of recurrence ranging from once in three years to five times in one year is included.

As an example of the use of this table it is shown that in Birmingham, Ala., a daily minimum temperature of 8F or lower occurred on an average of once every two years. A daily mean temperature of 18F or lower occurred with the same frequency. A daily mean temperature of 17F occurred on an average of three times per year and a daily mean temperature of 26F or lower occurred on an average of three times per year.

Conclusions from the study were that:

(1) The daily mean temperature is a better indication of heating load than the daily minimum temperature. It is therefore a more logical basis for the determination of design temperature;

(2) The design temperature should be based upon the probable frequency of recurrence of low temperatures, rather than upon the lowest daily mean or daily minimum temperature on record;

(3) There is no one frequency of recurrence which should be used for the selection of all design temperatures. It will depend upon the type of building or the purpose for which it is to be used, upon the severity of the weather in the district in which the building will be built, and upon local weather peculiarities, and

(4) Any attempt to predict weather conditions in one city from weather data for another city should be made with caution.

Average Frequency of Occurrence of Low Temperatures

(See text for exact definition of data)

State and City	Once in 3 Years		Once in 2 Years		Once per Year		Twice per Year		3 Times per Year		4 Times per Year		5 Times per Year	
	Minimum	Mean	Minimum	Mean	Minimum	Mean	Minimum	Mean	Minimum	Mean	Minimum	Mean	Minimum	Mean
ALABAMA														
Birmingham....	7	16	8	18	11	21	15	25	17	26	19	28	20	29
Mobile........	18	27	19	28	22	30	24	33	27	35	28	36	29	38
ARIZONA														
Flagstaff.......	−18	2	−17	3	−14	6	−11	8	−8	10	−6	12	−5	13
Phoenix........	24	38	25	39	26	40	27	41	28	42	29	43	29	43
Yuma.........	29	39	29	40	30	42	31	44	32	45	33	46	33	46
ARKANSAS														
Fort Smith.....	−3	9	1	12	5	15	9	18	12	21	14	22	15	23
Little Rock.....	5	13	6	15	9	18	12	20	14	22	16	23	18	25
CALIFORNIA														
Eureka........	25	33	27	34	28	36	29	37	31	38	31	39	32	39
Fresno.........	24	32	25	33	27	35	28	36	29	37	29	38	30	38
Los Angeles....	35	42	36	42	37	43	38	45	39	46	40	47	40	47
San Francisco ..	33	38	34	39	35	40	36	41	37	42	38	43	39	44
COLORADO														
Denver........	−16	−7	−14	−6	−11	−3	−8	1	−5	5	−3	7	−1	9
Durango.......	−19	0	−18	1	−15	3	−12	6	−10	8	−8	9	−7	10
Grand Junction.	−17	−1	−16	0	−12	2	−9	4	−6	7	−4	9	−2	10
Trinidad.......	−18	−2	−16	1	−9	4	−5	9	−3	11	−1	13	0	15
CONNECTICUT														
New Haven....	−8	1	−6	3	−3	7	1	10	3	12	5	13	6	14

Average Frequency of Occurrence of Low Temperatures — Continued

State and City	Once in 3 Years		Once in 2 Years		Once per Year		Twice per Year		3 Times per Year		4 Times per Year		5 Times per Year	
	Minimum	Mean	Minimum	Mean	Minimum	Mean	Minimum	Mean	Minimum	Mean	Minimum	Mean	Minimum	Mean
FLORIDA														
Jacksonville....	21	30	22	32	25	35	28	37	30	39	31	40	32	41
Tampa........	28	40	30	40	33	43	35	45	36	46	37	47	38	48
GEORGIA														
Atlanta........	5	15	6	16	11	20	14	23	16	26	18	27	19	27
Savannah......	18	26	20	28	22	32	25	35	27	36	28	37	29	38
IDAHO														
Lewiston.......	−15	−4	−13	−3	−10	0	−5	4	0	8	3	11	5	12
Pocatello.......	−20	−9	−18	−8	−16	−6	−11	−1	−8	3	−6	5	−4	7
ILLINOIS														
Cairo..........	−4	4	−2	5	1	9	4	12	7	14	9	16	10	18
Chicago........	−14	−6	−13	−5	−9	−1	−6	1	−5	3	−3	5	−2	6
Springfield.....	−15	−5	−13	−3	−8	0	−5	3	−3	5	−2	6	0	8
INDIANA														
Evansville......	−7	1	−5	3	−1	6	2	10	4	12	6	14	7	15
Fort Wayne.....	−14	−2	−12	−1	−8	1	−6	3	−5	5	−3	6	−1	7
Indianapolis....	−11	−2	−10	−1	−8	2	−4	5	−3	7	−1	8	0	10
IOWA														
Dubuque.......	−24	−11	−22	−10	−18	−8	−15	−5	−13	−3	−10	−2	−9	−1
Keokuk........	−18	−8	−17	−7	−13	−3	−9	0	−7	2	−5	4	−4	5
Sioux City......	−24	−16	−23	−15	−20	−11	−16	−8	−14	−6	−12	−3	−11	−3
KANSAS														
Concordia......	−15	−6	−14	−5	−10	−2	−8	2	−5	3	−3	5	−2	6
Dodge City.....	−12	−3	−11	−2	−9	1	−6	5	−4	7	−1	9	0	10
Goodland......	−17	−9	−16	−7	−14	−3	−10	1	−7	4	−6	6	−5	8
Wichita........	−10	−1	−8	0	−6	3	−2	6	1	9	2	11	3	12
KENTUCKY														
Lexington......	−8	1	−6	2	−4	5	0	8	2	10	4	13	6	15
LOUISIANA														
New Orleans....	22	28	23	30	26	33	29	36	31	37	32	39	33	40
Shreveport.....	11	21	13	22	16	25	19	26	21	28	23	30	24	31
MAINE														
Eastport.......	−17	−10	−16	−7	−12	−4	−10	0	−7	1	−6	3	−5	4
Portland.......	−14	−5	−12	−3	−9	1	−6	3	−4	5	−2	6	−1	8
Van Buren......	−42	−23	−41	−22	−37	−19	−34	−15	−32	−13	−30	−11	−28	−10
MARYLAND														
Baltimore......	2	10	4	11	6	14	9	16	10	18	12	19	13	20
Cumberland....	−6	6	−5	7	−2	10	1	12	4	14	5	15	6	17
MASSACHUSETTS														
Boston.........	−10	0	−7	2	−3	5	0	9	2	11	3	12	4	14
MICHIGAN														
Alpena........	−18	−8	−17	−6	−13	−4	−10	−1	−9	1	−8	2	−7	3
Detroit........	−10	−1	−9	0	−5	2	−3	.5	−1	6	0	8	1	9
Grand Rapids...	−11	−1	−10	0	−7	2	−2	5	−1	7	0	8	1	9
Marquette......	−18	−10	−17	−9	−14	−6	−12	−4	−10	−2	−9	−1	−8	0
MINNESOTA														
Duluth.........	−33	−23	−32	−22	−29	−20	−26	−16	−25	−14	−23	−13	−22	−12
Minneapolis....	−27	−19	−25	−18	−23	−15	−20	−12	−18	−10	−17	−8	−16	−7
Moorhead......	−35	−25	−32	−24	−31	−20	−28	−18	−26	−16	−24	−15	−23	−14

Average Frequency of Occurrence of Low Temperatures — Continued

State and City	Once in 3 Years		Once in 2 Years		Once per Year		Twice per Year		3 Times per Year		4 Times per Year		5 Times per Year	
	Minimum	Mean	Minimum	Mean	Minimum	Mean	Minimum	Mean	Minimum	Mean	Minimum	Mean	Minimum	Mean
MISSISSIPPI														
Vicksburg......	11	20	13	22	16	24	20	27	21	29	23	31	24	32
MISSOURI														
St. Joseph......	−16	−7	−15	−6	−12	−3	−9	0	−6	2	−4	4	−2	6
St. Louis.......	−10	−1	−8	1	−3	4	0	8	2	10	3	11	5	13
Springfield.....	−11	−1	−9	0	−5	4	−1	8	1	11	3	12	5	13
MONTANA														
Billings........	−36	−23	−34	−22	−31	−19	−26	−14	−22	−11	−20	−9	−18	−7
Hayre..........	−42	−31	−39	−29	−36	−26	−33	−23	−30	−20	−28	−17	−26	−15
Helena.........	−32	−25	−31	−24	−29	−21	−25	−17	−22	−13	−20	−11	−17	−8
Kalispell.......	−26	−16	−25	−16	−22	−13	−18	−10	−16	−7	−13	−5	−11	−3
NEBRASKA														
Lincoln........	−20	−10	−18	−9	−16	−6	−13	−3	−10	−1	−9	0	−7	1
North Platte....	−22	−12	−21	−10	−18	−8	−15	−4	−12	−2	−11	0	−10	1
NEVADA														
Logandale......	14	28	14	29	16	31	19	32	20	34	21	35	21	36
Reno..........	−10	5	−7	7	−3	9	0	13	2	15	5	17	7	18
Searchlight.....	14	22	15	23	17	25	20	28	22	30	23	31	24	32
Tonopah.......	−5	2	−4	4	2	9	4	11	7	13	8	14	9	15
Winnemucca....	−25	−8	−24	−6	−19	−1	−13	4	−8	8	−6	10	−5	11
NEW HAMPSHIRE														
Concord........	−20	−8	−18	−6	−16	−1	−12	1	−10	3	−8	5	−7	6
NEW JERSEY														
Atlantic City...	0	9	1	10	4	13	7	16	9	18	11	19	12	20
NEW MEXICO														
Roswell........	−4	9	−2	10	2	14	6	18	9	20	10	22	12	23
Santa Fe.......	−3	7	−2	9	0	11	2	13	4	15	5	16	6	17
NEW YORK														
Albany........	−18	−9	−16	−7	−12	−2	−8	2	−6	4	−4	6	−3	7
Buffalo........	−8	−3	−7	−1	−5	3	−2	5	−1	7	0	8	2	9
New York......	−6	4	−3	6	0	10	4	12	6	14	7	15	8	17
NORTH CAROLINA														
Asheville.......	−2	9	−1	10	3	11	7	17	9	19	11	21	12	22
Raleigh........	8	16	8	18	11	20	14	23	15	24	17	25	18	26
Wilmington.....	12	21	13	23	16	25	19	28	21	30	23	31	24	32
NORTH DAKOTA														
Bismarck.......	−39	−27	−37	−25	−33	−22	−29	−20	−27	−17	−26	−16	−25	−14
Devil's Lake....	−40	−30	−38	−29	−35	−26	−32	−23	−30	−21	−28	−20	−27	−19
Williston.......	−39	−29	−37	−27	−34	−24	−31	−22	−29	−20	−27	−19	−26	−18
OHIO														
Cincinnati......	−10	0	−9	1	−5	3	−2	6	0	9	2	11	3	12
Cleveland......	−7	1	−6	2	−4	4	−1	7	0	8	2	9	3	10
Columbus......	−9	0	−9	0	−7	3	−2	6	0	8	1	10	3	11
OKLAHOMA														
Oklahoma City..	−4	4	−2	5	1	8	4	12	7	15	9	16	10	18
OREGON														
Baker..........	−21	−8	−19	−7	−15	−3	−11	1	−8	4	−5	6	−4	7
Medford.......	3	15	6	16	9	20	13	24	15	26	17	27	18	28
Portland.......	11	15	12	16	14	19	16	21	18	22	19	24	21	25

Average Frequency of Recurrence

Temperatures, F

Average Frequency of Occurrence of Low Temperatures — Concluded

State and City	Once in 3 Years Minimum	Mean	Once in 2 Years Minimum	Mean	Once per Year Minimum	Mean	Twice per Year Minimum	Mean	3 Times per Year Minimum	Mean	4 Times per Year Minimum	Mean	5 Times per Year Minimum	Mean
PENNSYLVANIA														
Philadelphia....	1	9	2	10	4	12	8	15	10	17	11	18	12	19
Pittsburgh......	−7	1	−6	2	−4	4	−1	7	1	9	3	11	5	12
Scranton........	−9	0	−7	3	−5	5	−1	8	1	9	2	10	3	11
Somerset........	−20	−4	−17	−3	−14	0	−10	3	−8	6	−6	8	−4	9
RHODE ISLAND														
Providence.....	−8	0	−6	2	−2	6	0	9	3	11	4	12	5	13
SOUTH CAROLINA														
Charleston......	17	23	18	25	22	29	25	33	27	34	28	35	29	36
Columbia.......	12	21	13	22	16	26	19	28	21	30	22	31	24	32
SOUTH DAKOTA														
Huron.........	−31	−21	−29	−19	−26	−16	−23	−14	−22	−12	−20	−10	−18	−9
Rapid City.....	−28	−19	−27	−18	−24	−15	−21	−12	−18	−9	−17	−7	−15	−6
TENNESSEE														
Knoxville.......	2	9	3	11	6	14	9	19	12	21	14	23	15	24
Memphis.......	5	11	6	12	8	15	11	18	14	20	15	22	16	23
TEXAS														
Amarillo.......	−5	4	−4	5	−2	7	1	13	4	15	6	17	8	18
Brownsville.....	25	32	26	35	30	37	31	39	33	40	34	42	35	43
El Paso........	14	25	16	25	18	27	19	30	21	32	22	33	23	34
Fort Worth.....	7	16	10	18	12	21	15	23	17	26	18	27	19	28
San Antonio....	17	22	18	25	21	29	24	32	25	34	26	35	27	36
UTAH														
Modena........	−23	−9	−22	−6	−17	−2	−13	2	−11	5	−9	6	−7	9
Salt Lake City..	−8	1	−6	2	−2	7	3	12	5	14	7	15	8	16
VERMONT														
Burlington......	−25	−17	−23	−13	−19	−8	−16	−6	−14	−3	−12	−2	−11	−1
VIRGINIA														
Lynchburg.....	0	13	2	14	4	15	8	18	9	20	11	22	13	23
Norfolk........	9	16	10	17	13	20	16	22	17	24	19	26	20	27
Wytheville	−5	5	−4	6	−1	9	2	12	5	14	6	16	8	17
WASHINGTON														
Seattle.........	14	19	15	20	17	23	19	24	21	26	23	28	24	29
Spokane........	−15	−3	−13	−2	−9	0	−7	2	−4	5	−1	7	1	9
Yakima........	−14	−1	−11	1	−8	3	−4	5	0	7	1	10	3	11
WEST VIRGINIA														
Elkins.........	−14	0	−11	1	−8	3	−5	6	−3	8	−1	10	1	12
Parkersburg....	−9	3	−7	4	−3	6	0	9	2	12	4	13	5	15
WISCONSIN														
Green Bay......	−24	−13	−23	−12	−20	−10	−17	−8	−15	−6	−14	−4	−12	−3
LaCrosse.......	−27	−16	−26	−15	−24	−12	−21	−9	−18	−7	−16	−6	−15	−5
Milwaukee.....	−18	−10	−16	−9	−14	−5	−11	−3	−9	−1	−7	0	−6	2
WYOMING														
Encampment...	−30	−12	−26	−11	−22	−6	−17	−1	−14	1	−12	4	−10	5
Lander.........	−36	−22	−34	−20	−27	−16	−23	−12	−21	−9	−19	−7	−18	−5
Sheridan.......	−35	−22	−33	−21	−29	−19	−25	−14	−23	−11	−21	−9	−19	−7
Yellowstone Park	−34	−24	−33	−22	−29	−18	−25	−13	−20	−9	−18	−6	−16	−4

Average Frequency of Recurrence — Temperatures, F

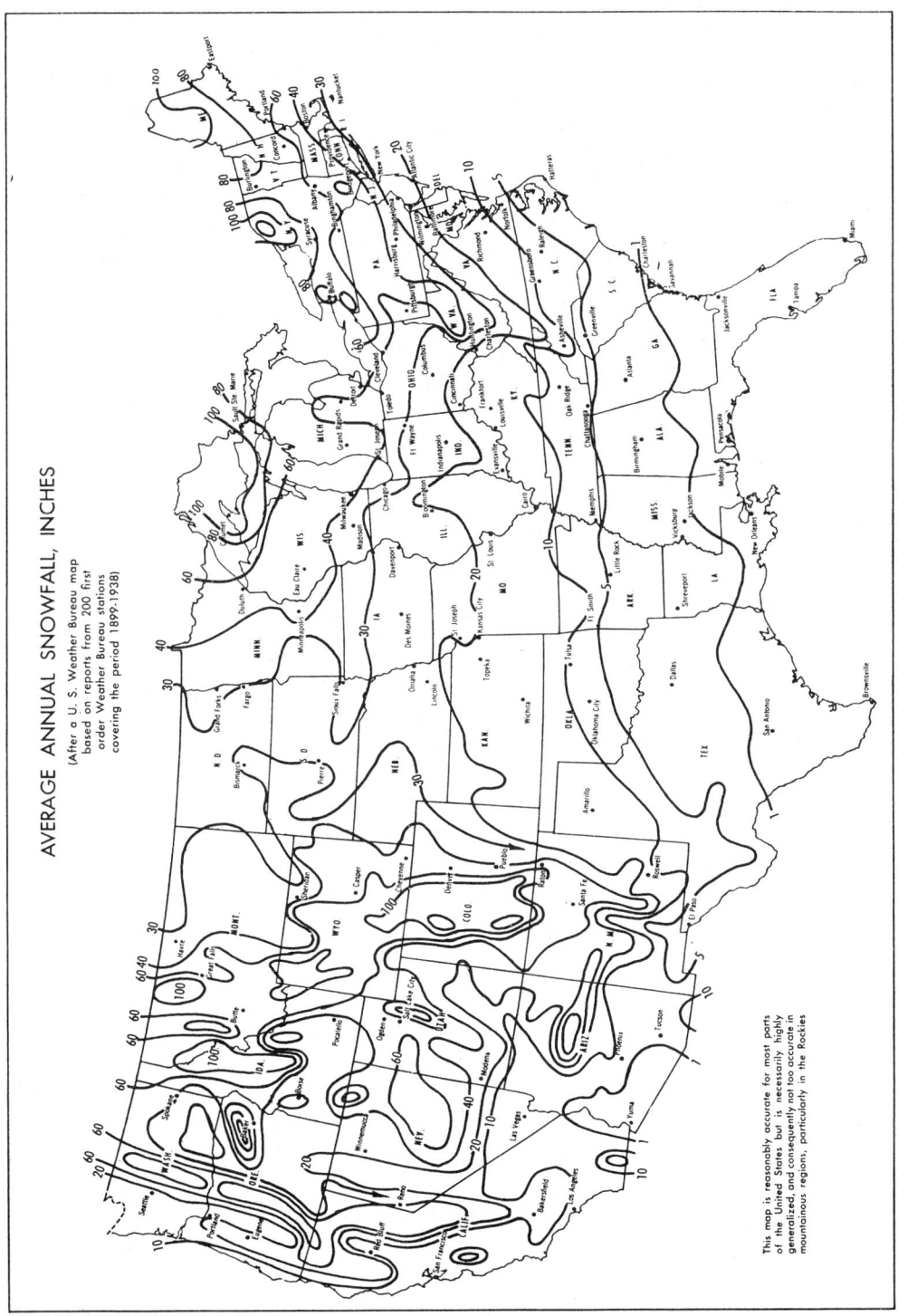

AVERAGE ANNUAL SNOWFALL, INCHES

(After a U. S. Weather Bureau map
based on reports from 200 first
order Weather Bureau stations
covering the period 1899-1938)

This map is reasonably accurate for most parts
of the United States but is necessarily highly
generalized, and consequently not too accurate in
mountainous regions, particularly in the Rockies

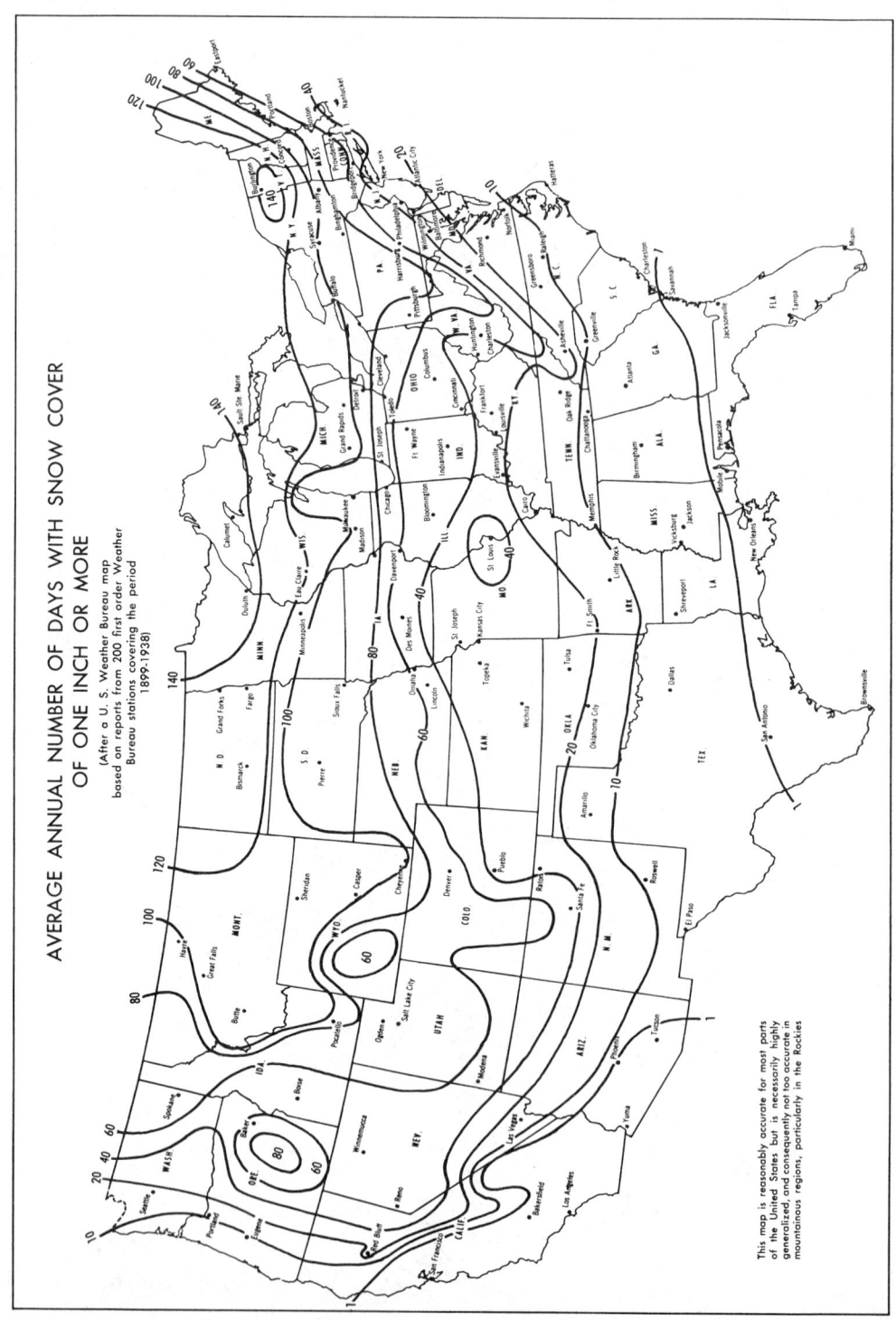

AVERAGE ANNUAL NUMBER OF DAYS WITH SNOW COVER
OF ONE INCH OR MORE

(After a U. S. Weather Bureau map
based on reports from 200 first order Weather
Bureau stations covering the period
1899-1938)

This map is reasonably accurate for most parts
of the United States but is necessarily highly
generalized, and consequently not too accurate in
mountainous regions, particularly in the Rockies

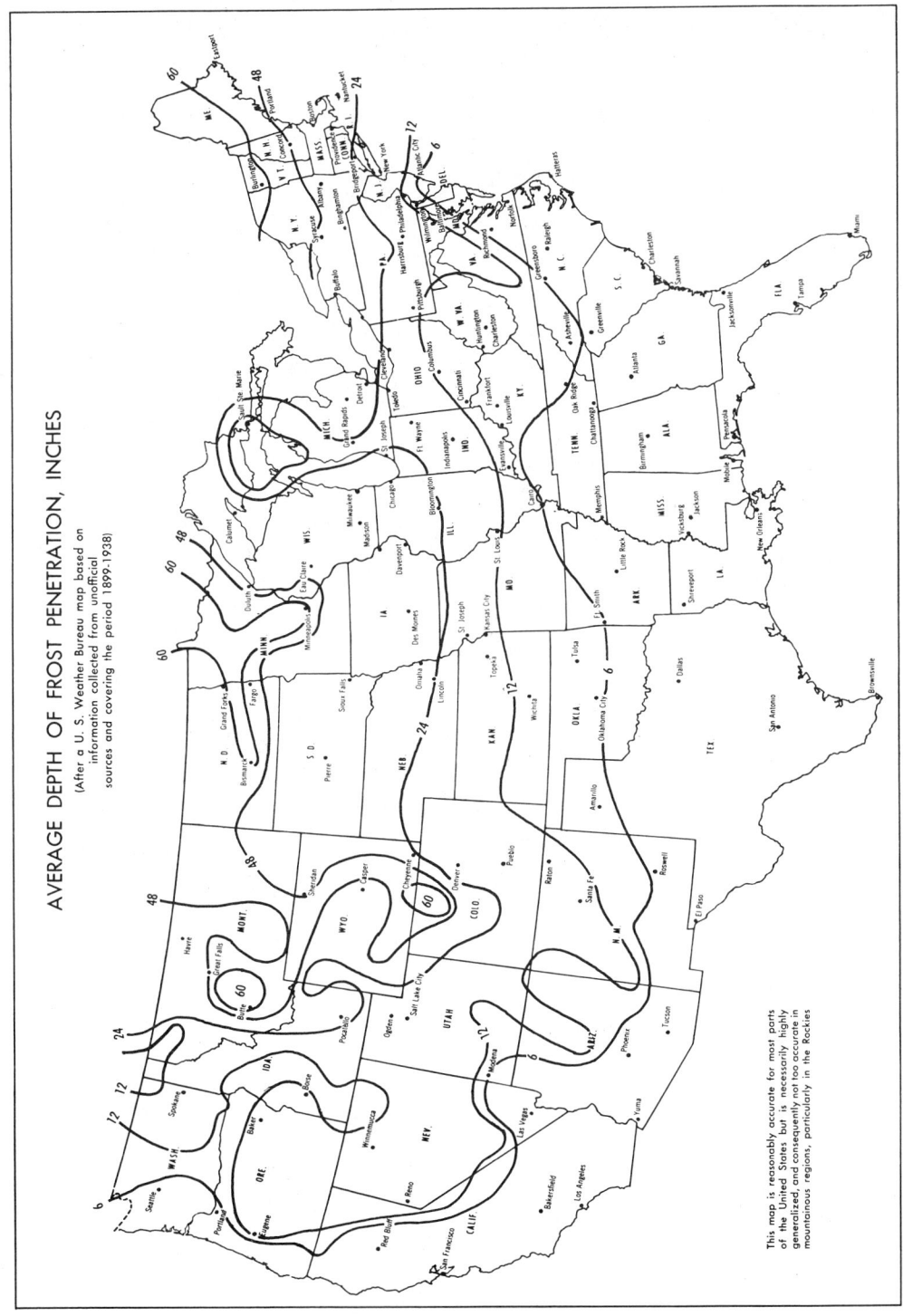

AVERAGE DEPTH OF FROST PENETRATION, INCHES

(After a U. S. Weather Bureau map based on information collected from unofficial sources and covering the period 1899-1938)

This map is reasonably accurate for most parts of the United States but is necessarily highly generalized, and consequently not too accurate in mountainous regions, particularly in the Rockies

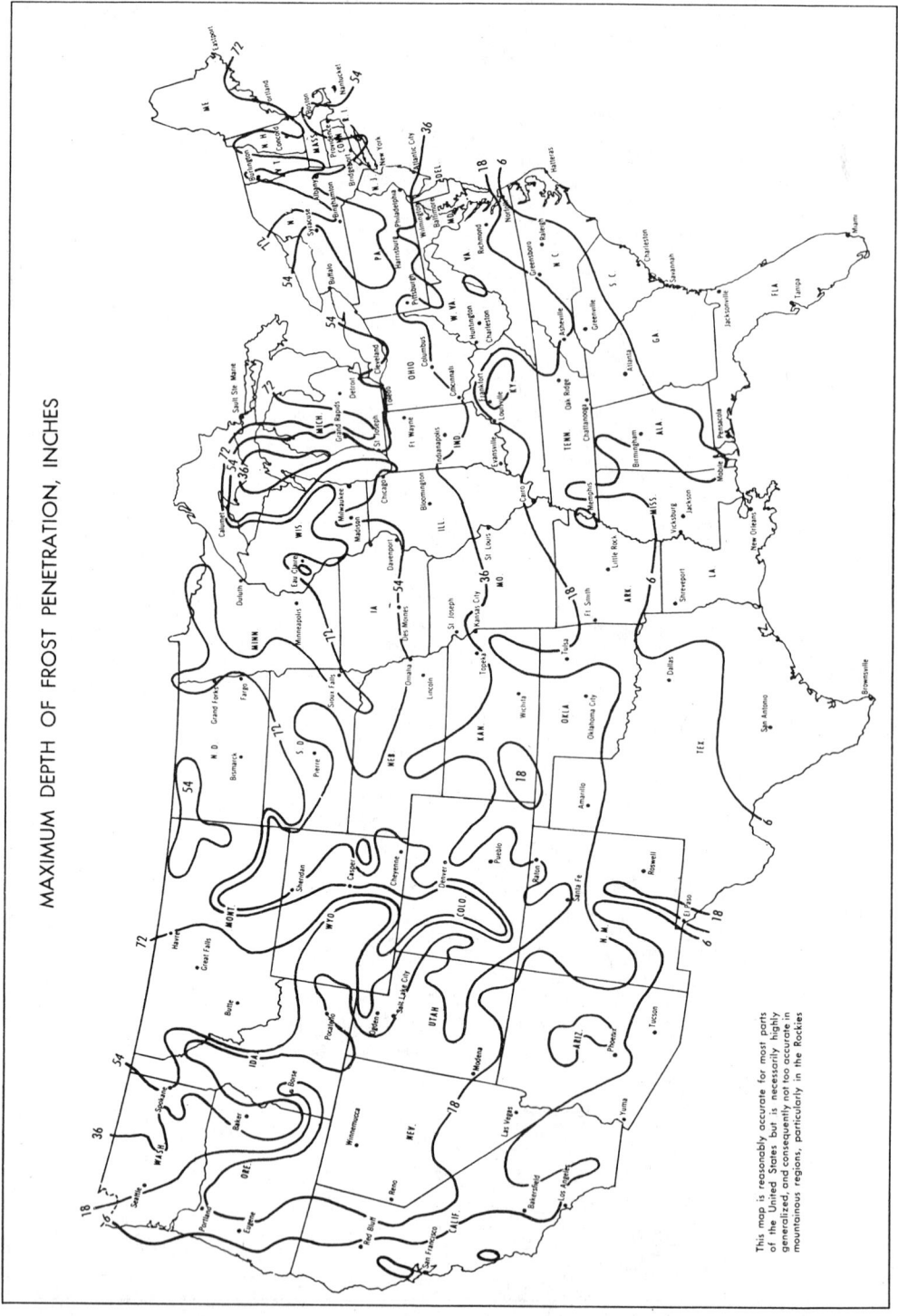

MAXIMUM DEPTH OF FROST PENETRATION, INCHES

This map is reasonably accurate for most parts of the United States but is necessarily highly generalized, and consequently not too accurate in mountainous regions, particularly in the Rockies

THE DEGREE-DAY

Experience has shown that, for buildings requiring an inside air temperature of approximately 70 deg, the amount of fuel or heat used per day is proportional to the number of degrees the average outside temperature falls below about 65 deg. The degree-day is based upon this principle. Thus the *number of degree-days (65 deg F base) per day is the difference between 65 deg and the daily mean temperature when the latter is less than 65 deg.*

The number of degree-days *for a given day* is thus equal to (65 deg − daily mean temperature for that day) × 1 (day), and the number of degree-days *for any longer period* is the sum of all such products for as many days as the period covers.

For example, the highest temperature recorded in Baltimore on December 12, 1931, was 70 deg and the minimum was 52 deg. The daily mean temperature was therefore (70 + 52) ÷ 2 or 61 deg. The number of degree-days for that day in Baltimore was thus (65 − 61) × 1 = 4. Carrying through this operation for each of the 31 days for December, 1931, it is found that the number of degree-days in Baltimore for that month is 601.

No attention is given to those days when the outside temperature averages above 65 deg.

The degree-day thus defined is now so widely used that when the unit is mentioned in the United States it is understood that the 65 deg base is referred to unless some other base or some other descriptive word is used with it.

As early as 1915, Eugene D. Milener, then an engineer with the gas utility company at Baltimore, found that the gas consumption of house-heating plants in that city varied with the number of degrees difference in outside temperature and 64 deg. Later studies indicated that when the figure was 65 deg the relationship was improved. As a result the 65 deg

figure was put into use and the name of degree-day was given to the unit related to this base temperature. Thus, from the inception of the unit, field study supported the conclusion that the proportionality between fuel consumption and temperature difference begins at a temperature of 65 deg in heating residential buildings.

65 Deg As the Base

Perhaps the earliest attempt to gather field data to confirm or reject the validity of the 65 deg value of the inside daily temperature took the form of noting in the field if there was any relation between the outside temperature and the time of starting up residential heating plants in the fall. It was found that when the mean daily outside temperature fell to about 65 deg, there was a considerable tendency to start the heating plant. Such observations tended to confirm the general idea that the base of proportionality was in the vicinity of 65 deg.

The general method of investigating the proportionality by direct field studies is illustrated in Fig. 1 where observed fuel consumption figures are plotted against daily mean outside temperature. In order to make such a study the fuel should be one that is easily measured, so that the exact quantity burned can be recorded without too much work, and the plant should be one in which the fuel measured is all used for the heating of the building and none for cooking or service water heating. In the example shown in Fig. 1, on a day when the mean temperature outside was 35 deg the plant required 14½ gallons of oil; when it was 40 deg, 13½ gallons; 0 deg, 32¼ gallons, etc. These points were plotted and a straight line drawn through them so that the points on either side of the line numbered about the same. In this case the line crosses the zero fuel consumption axis at 65 deg. In any particular plant, the straight line will cross at or a few degrees above or below 65 deg. If the variation is more than a few degrees some special conditions apply which must be taken into consideration. The point at which the axis is crossed fixes the value of the outside temperature at which fuel consumption starts, or the base of the proportionality between fuel consumption and temperature difference. Evidently, if a large number of separate plant operations could be studied the results would show whether or not the base line was generally at 65 deg or whether it was only accidental. Many such studies have now been made.

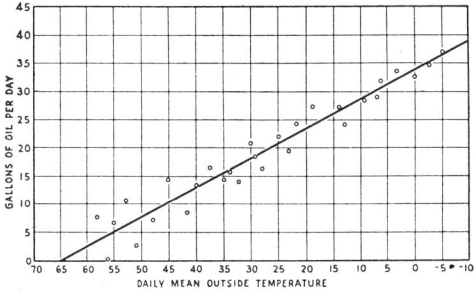

Fig. 1. Fuel consumption per day plotted against mean daily temperature. The curve best fitting the points shows a zero fuel consumption at 65°.

It might be supposed that, since the relationship of heat required and temperature was determined from an analysis of the performance of the heating plants in residences, a base temperature of other than 65 deg would apply to buildings with a different character of occupancy—such as office buildings and stores. When the inside temperature maintained is sharply different from 65 deg, as in certain industrial buildings, such a conclusion is correct. But buildings heated to 70 deg or about 70 deg show a relationship to the 65 deg base, even though their usage and character of occupancy may differ from that of the ordinary residence. A study made by the National District Heating Association of a number of buildings showed that the heat requirements of the buildings reached zero at the bases listed in Table 1. It will be noted that in most cases the base temperature indicated by these studies is at or near 65 deg.

While most types of building are operated with inside air temperatures approximating 70 deg during the heating season, there are important cases where the maintained temperature departs considerably from 70 deg. Prominent among such cases are industrial buildings of various kinds. In general, but little authentic information is in existence to show at what temperature the proportionality between fuel and temperature difference begins in such plants.

There are two types of degree-day figures for any locality. One is the number of degree-days which have accrued for that locality during a specific period, such as May, 1958, or the whole heating season of 1957-58. These can be termed *actual* degree-days. The other is the normal, or average, number of degree-days for that locality, on a monthly or yearly basis; the *normal* is the average for similar periods over many years.

The number of degree-days accumulated in a given locality for any period, such as in a month or a year, necessarily varies from month to month and year to year. The following analysis by Anthracite Institute illustrates this fluctuation clearly.

Weather Bureau records were analyzed back to the turn of the century, and the lowest and highest degree day totals for each month averaged for the cities of Boston, New York, Philadelphia, Scranton, Syracuse and Toronto. The years of warmest and coldest months did not coincide in many instances. For example, the coldest Januarys were in 1918 in New York, Philadelphia and Scranton, 1920 in Boston and Syracuse and 1912 in Toronto. On the other hand, 1917 marked the coldest December in each city, while the warmest December everywhere was in 1923.

Table 2 shows the average normals, the average degree days in the warmest and coldest months and deviations from normals.

The warmest months constituted a season 31.7% warmer than normal, the coldest 38.2% colder than normal. At the coldest, the heat demand would be more than double that in the warmest of record.

An opinion sometimes expressed is that the degree-day is inaccurate because the wind is not taken into account. This is not altogether true. The effect of wind on the heating of a building depends on (1) the velocity and (2) the temperature of the wind. Both are considered for maximum conditions in the design of the heating plant and reflected in its size. The degree-day in turn, by indicating the temperature of the wind, takes care of this variable through the heating season. Not accounted for is the fluctuation of wind velocity through the season. Use of the degree-day implies that the effect of the wind varies inversely as its temperature and that the velocity varies with degree-days. The former is true, the latter true at times, not true at others.

Table 1. — Base Temperature for Different Buildings as Determined by Field Tests

Type of Building	Base Deg.	Type of Building	Base Deg.
Offices	66.2	Apartments	68.8
Offices and banks	65.8	Residences	66.9
Banks	66.2	Clubs	65.5
Telephone exchanges	65.5	Theaters	67.6
Offices and stores	67.4	Warehouses and lofts	65.2
Stores	64.0	Manufacturing	65.4
Department stores	64.3	Average for 175 build-	
Hotels	66.5	ings of all types	66.0

Table 2. — Variations in Monthly Degree-Days in Six Cities, 1899-1953

(From an analysis by Anthracite Institute)

Month	Number of Degree-Days				
	Normal	Average Warmest	Average Coldest	Compared with Normal	
				Warmest	Coldest
Sept.	92	27	193	− 65	+ 101
Oct.	346	176	555	− 170	+ 209
Nov.	652	454	852	− 198	+ 200
Dec.	1,014	785	1,319	− 229	+ 305
Jan.	1,119	774	1,447	− 345	+ 328
Feb.	1,016	804	1,411	− 212	+ 395
Mar.	859	544	1,106	− 315	+ 247
Apr.	519	334	687	− 185	+ 168
May	212	102	413	− 110	+ 201
June	42	7	132	− 35	+ 90
Total	5,871	4,007	8,115	−1,864	+2,244

Uses of the Degree-Day

There are two principal uses of the degree-day: (1) as a means of eliminating the outside temperature variable in comparing fuel consumption data, and (2) for predicting fuel consumption. The first use is one with wide application for comparison of data for the same building. If a building used a given amount of fuel for a certain period during which there was a known number of degree-days, simple division gives the amount used per degree-day. If, in a subsequent period, this unit figure has sharply increased, then some reason must be looked for as accountable for the indicated drop in operating efficiency.

The Degree-Day as a Guide in Operation

The degree-day has been found increasingly useful as time goes on as a means of securing efficient operation of heating plants, for it provides a means of checking results with the weather variable eliminated from comparisons. Another wide use of the unit is closely akin to this one and consists of securing good use of fuel or heat on the part of the ultimate consumer.

One example is sufficient to indicate this application of the degree-day although there are a wide variety of applications and variations in use.

The method described is particularly applicable to commercial and institutional buildings where the primary use of fuel or steam is for the heating of buildings.

Based on past experience for a given building, a standard is set up of so many pounds of steam, pounds of coal, gallons of oil or therms of gas which were used for that building per degree-day. If a number of buildings are to be considered, this constant can be reduced further to amount of fuel or steam per degree-day per 1,000 cubic feet of building volume or square feet of floor space so that it would be possible to compare one building with another.

A record of degree-days is kept each day and from this the standard based on previous experience is indicated in amount of fuel or steam per degree-day. The figure for the week, for example, is then compared with the standard and if approximately the same unit figure is arrived at, the plant is being operated as efficiently as during the previous period. If the figure is lower than the standard, then it has been operated more efficiently than formerly and consideration should be given to taking this into account in setting a new standard, or par. On the other hand, if the unit figure is higher than the standard, then something has happened to affect the operating efficiency and steps should be taken to improve the operation.

Naturally, in such a system the standard should be one that represents really good operating practice with clean boilers, good fuels and careful efficient operation.

It often happens that when such a standard is set up, the unit figure in fuel used per degree-day gradually increases from week to week rather than showing a sharp increase in any one week. This is due to the

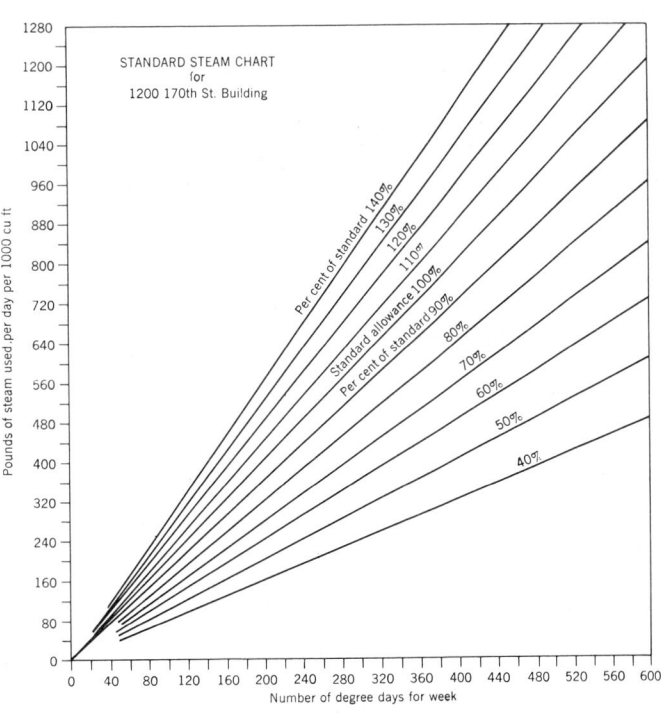

Fig. 2. Example of chart prepared for a specific building to check operating efficiency.

fact that as the boilers become dirty or the operator becomes more careless, the lessening efficiency is reflected in a gradual change in the unit figure.

An unusual but logical development in connection with the use of the degree-day for school buildings was reported from Salt Lake City a number of years ago. In this case, since schools are only open five days a week, a record was kept only of the degree-days on school days so that a comparison could be made on a more logical basis for this type of building.

Other uses of the degree-day include keeping a record of degree-days to show dissatisfied customers why their fuel bills are higher in one period than another.

For example, gas companies are frequently faced with this problem in explaining to customers why their gas bill goes up so sharply from November to December or from October to November. The consumer is aware, of course, that the weather is getting colder, but ordinarily the precise degree-day figures make more impression on him than his own awareness of low temperature. Such methods are used not only by gas companies, but by fuel oil dealers and district steam utilities as well. In one case, a utility company even prints the number of degree-days on its bills. In this case, the number of degree-days shown is for the period covered by the bill.

As an example of the use of the degree-day as a guide to operating efficiency the following is presented, not as an actual case, but as a composite of several actual uses. The figures used in examples are entirely fictitious and intended only to demonstrate that the method is workable. Actually, the forms used and details of working out such a method depend on the particular problem, number of buildings under one management, and the individual ideas of the person inaugurating the system.

Based on past records, a chart such as shown in Fig. 2 is prepared. In this example, a day is used as the time unit. As the ordinate, the quantity of steam or fuel used per day per 1000 cubic feet of building volume is used. If only one building is under consideration, the ordinate could omit the *per 1000 cubic feet*. Where a number of buildings are being compared, however, the cubic feet basis enables comparisons among different buildings of varying sizes.

After study of past records, a standard, or par, or 100% measure of performance is set up. From this, lines representing 90% of standard performance, 110%, 80%, 120%, and so on are plotted, as shown. This is drawn only once, and from then on the daily records are compared against the chart. Obviously,

if the standard has been properly set, readings over 100% represent substandard operation, readings under 100% good operation. Consistently substandard results should be checked into and operating procedures improved if possible. A wide variety of reasons can exist for below-standard or above-standard performance and every attempt should be made to determine what these are.

A second chart can be employed, as shown in Fig. 3, for recording the monthly performance of any building. Whereas Fig. 2 is drawn up once for each building, Fig. 3 is a blank chart, one being filled in for each building each month.

For each day of the month, the data for the bottom horizontal columns are filled in—only the top one, day of month, being printed in on the blank. The day of the week is filled in by initials, as shown; this is necessary since it will show up any variation from normal repeated on certain days of the week. For example, commercial buildings, such as offices, can be expected to show lower fuel or steam consumption on week ends when the building is only partly, if at all, occupied. The other columns are self-explanatory; the extreme bottom column is filled in by referring to Fig. 2.

The performance is then plotted in by days by drawing bars, as shown in the dashed lines for the first seven days. The purpose of this graphical portion of the chart is better to show up extreme or repetitive variations from normal.

Finally, the steam used for the whole month is obtained by totaling the daily figures, and this is also done for the degree-days. The monthly figures can then be inserted in the spaces at the top of the chart.

It can readily be seen that such a system or a similar one can easily be set up for almost any building where the necessary data can be obtained. Although steam or gas figures can be determined from meters every day, it is not too difficult to set up a system for weighing coal or measuring oil tank depletion.

Once the system is in operation, it can be kept in operation with ease and at very low cost. This cost may be far more than repaid when the system calls to the management's attention an increasing fuel or steam consumption due to faulty operation or equipment failure which otherwise might have gone on too long unnoticed.

The method shown here is a composite and can be varied widely. It may be that a weekly basis rather than a daily one would be more practicable. The system can even be a more simple one, such as the simple plotting on a graph of weekly energy

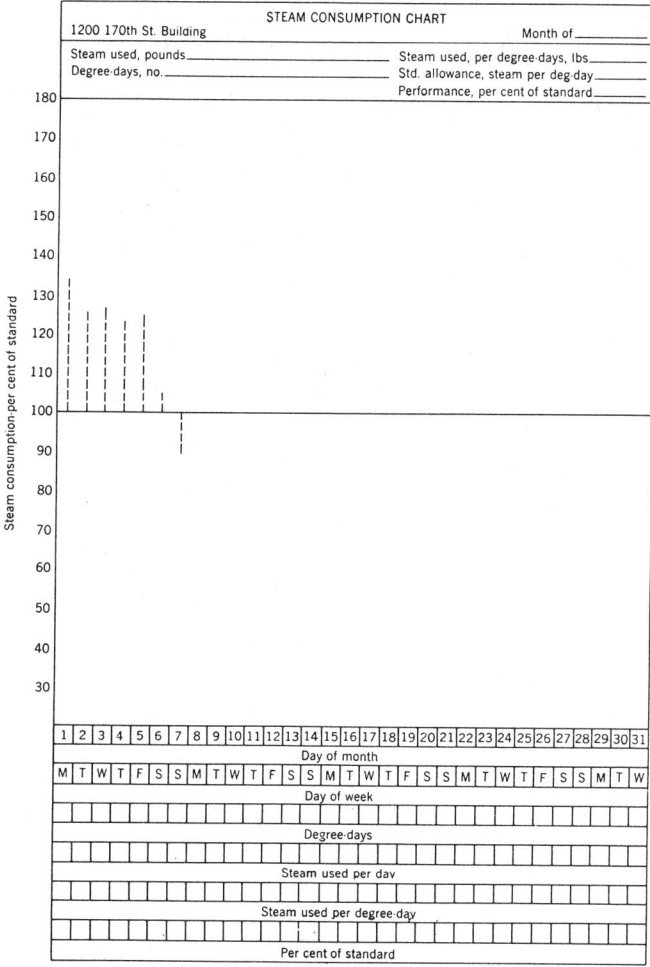

Fig. 3. Example of chart to record, on a monthly basis, operating results for a specific building.

consumption per degree-day. The details of the system depend on the size of the building and many other variables. The exact form and type of the system depends entirely on the enthusiasm and ingenuity of the person adopting the system.

Predicting Fuel Consumption

The use of the degree-day unit in predicting fuel consumption is of two different types: (1) predicting what the fuel consumption will be in an existing building for which performance data on fuel consumption during past periods are known, and (2) predicting fuel consumption for a new or proposed

building or one with a new heating plant for which no previous comparable fuel data are available.

One of the best examples of the use of the degree-day for predicting fuel consumption for an existing building is that used in the fuel oil delivery business where, by use of the degree-day, fuel deliveries are simplified and the fuel oil dealer knows by his records exactly when to deliver fuel oil just prior to the time when the oil in the tank becomes dangerously low.

As near as can be determined, the first use of the degree-day for this purpose was by Aetna Oil Service Co., Louisville, Ky., the system having been devised by W. R. Abbott of that company and reported by him in a talk before the American Society of Mechanical Engineers in 1928. The method used was somewhat as follows: a separate card was maintained for each customer and a constant K determined by dividing the gallons of oil used by that customer for a given period for which the number of degree-days was known. This K was then the gallons used per degree-day for that customer. Another constant K_1 was also kept, this being the gallons per degree-day per unit load, the unit load being assumed as 1,000 square feet of equivalent steam radiation. Also on the customer's card was entered the load on the boiler together with this equivalent load which was designated as C. K_1 was then determined by dividing K by the equivalent unit load C.

The constant K_1 was not necessary in connection with fuel deliveries, but was kept as an index of the overall efficiency of the plant and for comparing fuel consumption of that plant with other plants in the same city. It was of interest in evaluating the various types of building construction, insulation and weather-stripping.

Since a record was kept each day of the number of degree-days, it was possible at short intervals to go

through the cards and determine how much oil was used by each customer during that period and then subtract that from the previous known amount of fuel in the customer's tank. Consequently, it was quite easy to determine just when the customer's tank would become empty and anticipate this by a delivery of oil.

By adding the fuel consumption on all cards, it was possible to arrive at the number of gallons of oil needed by the company for its customers per degree-day so as to determine its own stocks. Obviously, too, truck schedules could be made up from information obtained from the cards so that the most efficient method of truck operation was possible.

Predicting Future Needs

An entirely different use of the degree-day unit is in predicting the future fuel (or steam or electrical energy) requirements in a building yet to be built, or in an existing building where, perhaps, a change in fuel or energy is contemplated. It is this particular use of the degree-day that originally led to the development of the degree-day years ago.

If the heat loss calculations for a building were accurate for the design conditions, and if the heat loss at any other outside temperature were proportional to that at design conditions, then the heat lost from the building for a whole heating season could be expressed by the equation

$$H = \frac{24hd(t_i - t_a)}{t_i - t_o} \qquad (1)$$

where H = seasonal heat loss in Btu,
h = hourly heat loss from the building for the design conditions, Btu,
t_i = inside design temperature, deg F,
t_o = outside design temperature, deg F,
24 = hours per day,
d = number of days in the heating season,
t_a = average outside temperature for the heating season.

For buildings where the inside design temperature is 70 deg, the formula becomes

$$H = \frac{24hd(70 - t_a)}{70 - t_o} \qquad (2)$$

This formula is rational and workable if figures for d (number of days in the heating season) and t_a (average heating season temperature) are available. While some reference books state that the heating season in the United States extends from October 1 to May 1 and that the average outside temperature for the period from October 1 to April 30 can be used for t_a, this is incorrect; the length of heating season varies widely throughout the United States and, unfortunately, such a simple assumption is not applicable over the whole country.

However, the degree-day makes it possible to determine the number of days in a normal, or average, heating season for any locality for which sufficient weather data are available, as well as the average outside temperature during the period comprising the heating season.

If the *daily* mean temperatures for any city are plotted for each day in a given year, the result appears something like Fig. 4. As the fall season approaches, the trend of the temperature is to get colder and colder until the middle of winter is reached, following which it gets warmer and warmer until the summer, when the cycle begins again. This is true in all parts of the United States, the only difference being that the swing is not so pronounced in some parts of the country as in others.

Although the trend in the fall is for the weather to get colder, the temperature does not get colder uniformly but fluctuates more or less violently, being relatively warm one day and cooler the next. If, however, we were similarly to plot for a city the *normal* daily mean temperature for all the years for which we have records (that is, if we were to average all of the April firsts, then all the April seconds, then all the April thirds, and so on) the result would be a smooth curve as shown in Fig. 5. This curve shows graphically the normal daily mean temperature for a typical city.

This normal daily temperature curve will, for most United States cities, cross the 65 deg line as shown, except in the extreme south, where the normal curve at all times is above the 65 deg line, and in the far north and west, where, in many localities, it is always below the 65 deg line.

The premise of the degree-day is that heating is required on days having degree-days, and heating is not required on days having no degree-days. Therefore, the heating season in a normal year begins on the date when the normal daily mean temperature crosses downward over the 65 deg line, and ends where it crosses upward over the same line. The length of the heating season is the number of days between the two crossings, and thus d (number of days in the heating season) in equations (1) and (2) can be determined for any city for which normal daily temperatures are available.

The degree-day is a product of time (in days) and

temperature, and when a graph such as Fig. 5 is plotted in days on the horizontal scale and daily mean temperature is on the vertical scale, the number of degree-days D in the heating season is indicated graphically by the shaded area in Fig. 6.

Referring to Fig. 6, if the shaded area represents the number of degree-days, then this area is equal to the base of the area, d, times the average altitude of the irregular area, t_z, or

$$D = dt_z \tag{3}$$

As shown on the graph,

$$t_z = 65 - t_a \tag{4}$$

so that

$$t_a = 65 - \frac{D}{d} \tag{5}$$

Consequently, the average outside temperature (t_a) can be found by equation (5).

Once d and t_a are available for any city the heat required can be calculated by equation (2). For example, what would be the heat required for a residence in New York where the design heat loss is 70,000 Btu per hour, $d = 241$ and $t_a = 44.0$ deg and $t_o = 0$ deg?

Substituting in equation (2)

$$H = \frac{24 \times 70{,}000 \times 241 \times (70 - 44)}{(70 - 0)}$$

$H = 150{,}384{,}000$ Btu, the heat loss for the whole heating season.

This can be converted into fuel units by dividing by the heat value per fuel unit and by the utilization efficiency. For example, assuming 535 Btu per cubic foot of gas at 80% efficiency, the gas consumption per season would be

$$\frac{150{,}384{,}000}{535 \times .80} = 351{,}364 \text{ cu ft}$$

However, equation (2) can be further simplified. Substituting equation (5) into equation (2) for t_a,

$$H = \frac{24hd}{(70 - t_o)}\left[70 - \left(65 - \frac{D}{d}\right)\right] \tag{6}$$

which reduces to

$$H = \frac{24(5d + D)h}{(70 - t_o)} \tag{7}$$

Since t_o is constant for a given city, as are d and D, the unit heat required per degree-day per Btu heat loss at design conditions is

$$\frac{H}{D} = \frac{24(5d + D)}{(70 - t_o)D}h \tag{8}$$

and since it is desirable to express the unit figure in terms of *thousands* of Btu heat loss

$$\frac{1000H}{D} = \frac{24(5d + D)1000h}{(70 - t_o)D} \tag{9}$$

The whole term $\dfrac{24(5d + D)1000}{(70 - t_o)D}$ can be calculated for any given city. Expressed in words, this term, which can be called K, is the unit heat requirement for that locality, in Btu per degree-day per 1000 Btu per hour heat loss at the design condition. The working formula is thus:

$$H = KD\,\frac{h}{1000} \tag{10}$$

where H = heat to be supplied, Btu per normal heating season,

$h/1000$ = heat loss *in thousands of Btu* per hour at design conditions,

D = normal number of degree-days per year,

K = heat to be supplied, Btu per degree-day per 1000 Btu per hour heat loss at design conditions, for the city in question. Note that K is a constant only for a given city or small locality.

Values of K and D for a large number of United States cities are given in the table titled "Data on Normal Heating Season in U. S. Cities."

Example. The New York residence previously cited, had a calculated heat loss of 70,000 Btu per hour. From the table, K for New York City is 424.7 and D is 5050, so that, using equation (10),

$$H = 424.7 \times 5050 \times 70$$
$$= 150{,}131{,}450 \text{ Btu per year}$$

Burning 535 Btu per cu ft gas at 80% efficiency,

$$\text{Cu ft gas required per season} = \frac{150{,}131{,}450}{535 \times .80}$$
$$= 350{,}774 \text{ cu ft}$$

In cases involving gas or steam, or in any cases where monthly figures are necessary, the table can be used to determine K; then for D use the number of degree-days in that locality for the given month and enter in equation (10). In this case, the answer will be in Btu to be supplied for that month.

Example. For the same residence, what would be the gas required for December?

The number of degree-days in a normal December

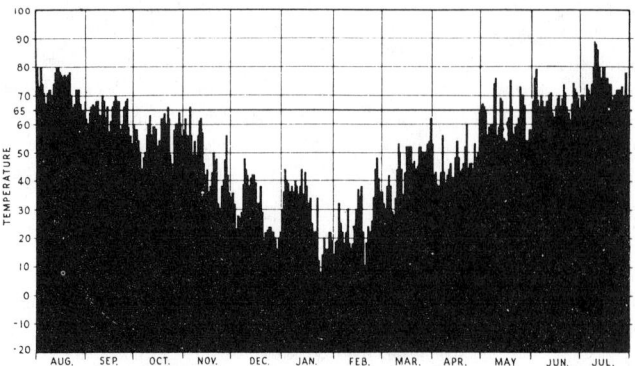

Fig. 4. Daily mean temperatures for a given year for a specific city.

quite useful when used in the locality in which they were determined. However, in the early days of the degree-day, it was found that fuel constants used in the northeast portion of the country were highly inaccurate in the south and southwest. The reason for this is that the term K is not a constant but varies widely throughout the country, as has been shown, due to the variation in d, D, and t_o.

in New York is 908. Using this in equation (10).

$$H_{Dec} = 424.7 \times 908 \times 70$$

$$H_{Dec} = 26{,}993{,}932 \text{ Btu}$$

and for 535 Btu gas at 80% efficiency

$$\text{Cu ft gas for Dec.} = \frac{26{,}993{,}932}{535 \times .80} = 63{,}069 \text{ cu ft}$$

Empirical Constants

The term K in equation (10) is Btu required per degree-day per 1000 Btu heat loss at design conditions. Constants of this general type, in different units, are widely used and, in fact, their use dates back to the earliest applications of the degree-day. Among such units which have been used are "cubic feet of gas per square foot of steam radiator per degree-day," "gallons of oil per square foot of steam radiator per degree-day," "pounds of steam per 1000 cubic foot of building volume" and a great many others.

These units were determined empirically and are

The exceedingly wide variation in value of K, which explains why one value of K cannot be used over a wide area, is shown graphically in the accompanying map titled "Unit Heat Requirements per Degree-Day." This map can be used to obtain values of K for cities not appearing in the table. Discretion must be observed in using the map, however, for mountainous localities.

The advantage of using K in terms of hourly heat loss at design conditions is that the Btu is the basic unit and the result is easily converted to other units (such as square foot of radiator).

Load Factor and Operating Hours

Refer to Fig. 6 which shows design temperature, t_o (which in this graph is at $+10$ deg), t_x, the average altitude of the shaded area, and $t_a = 65 - t_x$.

The Seasonal Load Factor (SLF) of a winter heating plant can be defined as the per cent of time during the heating season that the heating plant operates. Graphically, in Fig. 6, it is the ratio, expressed in per cent, of the shaded area to the area of the rectangle $ABCD$. Since the shaded area has already been shown to be equal to D, the number of degree-days during the heating season, and since $AB = 65 - t_o$ and $BC = d$, then

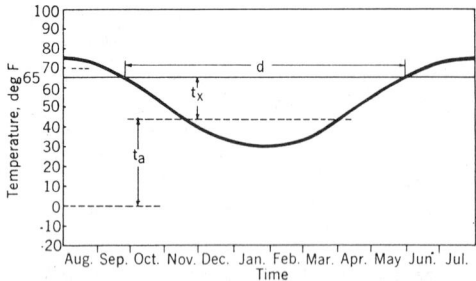

Fig. 5. Normal daily mean temperatures plot in a smooth curve as shown here.

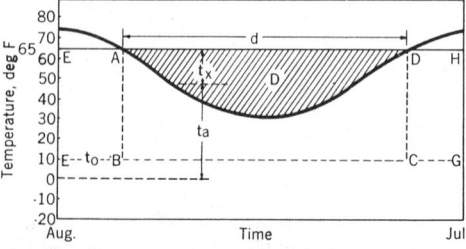

Fig. 6. Normal daily mean temperature curve showing graphically the number of degree-days.

$$SLF = \frac{D}{d(65 - t_o)} \times 100 \qquad (11)$$

But $D = dt_x$ so that

$$SLF = \frac{t_x}{65 - t_o} \times 100$$

Annual Load Factor (ALF) of the heating plant is the per cent of total time over the year that the heating plant operates. Graphically, it is the ratio of the shaded area in Fig. 6 to the area of the rectangle $EFGH$. Therefore, it can be shown that

$$ALF = \frac{D}{365(65 - t_o)} = \frac{dt_x}{365(65 - t_o)} \qquad (12)$$

A subsequent table gives values of the load factor for a wide range of cities.

The number of operating hours, or OH, that an intermittent or on-off heating plant operates during a normal year is thus

$$OH = ALF \times 24 \times 365 \qquad (13)$$

It is useful to know the number of hours of operation of a heating plant to determine the electrical consumption of burner motors, pumps, and similar equipment.

Limitations

The foregoing method of predicting fuel consumption and the supplementary data derived from this method are subject to limitations as follows:

(1) By definition, the method applies to spaces heated to 70 deg during daytime and evening hours, and will not apply to spaces heated to temperatures differing sharply from 70 deg, such as industrial buildings.

(2) The data apply to a normal year and are subject to variations from year to year as the number of degree-days varies.

(3) As H in equation (10) depends upon design heat loss, the results in terms of fuel can be no more accurate than the heat loss calculation. If the heat losses are figured liberally, on the conservative side, the fuel figure will likewise be too liberal and give too high values of fuel consumption.

(4) Since heat gains are not ordinarily calculated in winter heat loss, fuel estimates indicated by the degree-day method will be too high by the amount of heat gain. Heat gains in winter include those from (1) the chimney, (2) cooking and bathing, (3) body heat loss, (4) appliances, and (5), perhaps most important, solar heat. These added together can amount to as much as 25% of the heat loss and, if they are ignored, the fuel consumption will be estimated too liberally by a corresponding amount.

(5) The method is based on an outside design temperature (t_o) as listed in the table. If any other design temperature is used, refigure K by using equation (9). This is important because of increasing use of less severe design temperatures.

(6) The calculated fuel consumption, since it is based on space heating only, does *not* include the fuel for service hot water.

(7) Note that equation (10) does *not* include heating plant efficiency; this must be allowed for separately, as shown in the examples.

Degree-Days Abroad

Since considerably lower inside temperatures are maintained in European as compared to American buildings, the 65 deg base used for the degree-day is not applicable in countries abroad. In many of these countries a base of 60 deg F is used in calculating degree-days.

NORMAL MONTHLY AND ANNUAL DEGREE-DAYS BELOW 65F FOR CITIES IN THE UNITED STATES

(Reprinted, with permission, from 1976 ASHRAE Systems Handbook)

State	Station		Avg. Winter Temp^d	July	Aug.	Sept.	Oct.	Nov.	Dec.	Jan.	Feb.	Mar.	Apr.	May	June	Yearly Total
Ala.	Birmingham	A	54.2	0	0	6	93	363	555	592	462	363	108	9	0	2551
	Huntsville	A	51.3	0	0	12	127	426	663	694	557	434	138	19	0	3070
	Mobile	A	59.9	0	0	0	22	213	357	415	300	211	42	0	0	1560
	Montgomery	A	55.4	0	0	0	68	330	527	543	417	316	90	0	0	2291
Alaska	Anchorage	A	23.0	245	291	516	930	1284	1572	1631	1316	1293	879	592	315	10864
	Fairbanks	A	6.7	171	332	642	1203	1833	2254	2359	1901	1739	1068	555	222	14279
	Juneau	A	32.1	301	338	483	725	921	1135	1237	1070	1073	810	601	381	9075
	Nome	A	13.1	481	496	693	1094	1455	1820	1879	1666	1770	1314	930	573	14171
Ariz.	Flagstaff	A	35.6	46	68	201	558	867	1073	1169	991	911	651	437	180	7152
	Phoenix	A	58.5	0	0	0	22	234	415	474	328	217	75	0	0	1765
	Tucson	A	58.1	0	0	0	25	231	406	471	344	242	75	6	0	1800
	Winslow	A	43.0	0	0	6	245	711	1008	1054	770	601	291	96	0	4782
	Yuma	A	64.2	0	0	0	0	108	264	307	190	90	15	0	0	974
Ark.	Fort Smith	A	50.3	0	0	12	127	450	704	781	596	456	144	22	0	3292
	Little Rock	A	50.5	0	0	9	127	465	716	756	577	434	126	9	0	3219
	Texarkana	A	54.2	0	0	0	78	345	561	626	468	350	105	0	0	2533
Calif.	Bakersfield	A	55.4	0	0	0	37	282	502	546	364	267	105	19	0	2122
	Bishop	A	46.0	0	0	48	260	576	797	874	680	555	306	143	36	4275
	Blue Canyon	A	42.2	28	37	108	347	594	781	896	795	806	597	412	195	5596
	Burbank	A	58.6	0	0	6	43	177	301	366	277	239	138	81	18	1646
	Eureka	C	49.9	270	257	258	329	414	499	546	470	505	438	372	285	4643
	Fresno	A	53.3	0	0	0	84	354	577	605	426	335	162	62	6	2611
	Long Beach	A	57.8	0	0	9	47	171	316	397	311	264	171	93	24	1803
	Los Angeles	A	57.4	28	28	42	78	180	291	372	302	288	219	158	81	2061
	Los Angeles	C	60.3	0	0	6	31	132	229	310	230	202	123	68	18	1349
	Mt. Shasta	C	41.2	25	34	123	406	696	902	983	784	738	525	347	159	5722

State	City	Stn.[b]	Temp. (°F)	July	Aug.	Sept.	Oct.	Nov.	Dec.	Jan.	Feb.	Mar.	Apr.	May	June	Annual
	Oakland	A	53.5	53	50	45	127	309	481	527	400	353	255	180	90	2870
	Red Bluff	A	53.8	0	0	0	53	318	555	605	428	341	168	47	0	2515
	Sacramento	A	53.9	0	0	0	56	321	546	583	414	332	178	72	0	2502
	Sacramento	C	54.4	0	0	0	62	312	533	561	392	310	173	76	0	2419
	Sandberg	C	46.8	0	0	30	202	480	691	778	661	620	426	264	57	4209
	San Diego	A	59.5	9	0	21	43	135	236	298	235	214	135	90	42	1458
	San Francisco	C	53.4	81	78	60	143	306	462	508	395	363	279	214	126	3015
	San Francisco	C	55.1	192	174	102	118	231	388	443	336	319	279	239	180	3001
	Santa Maria	A	54.3	99	93	96	146	270	391	459	370	363	282	233	165	2967
Colo.	Alamosa	A	29.7	65	99	279	639	1065	1420	1476	1162	1020	696	440	168	8529
	Colorado Springs	A	37.3	9	25	132	456	825	1032	1128	938	893	582	319	84	6423
	Denver	A	37.6	6	9	117	428	819	1035	1132	938	887	558	288	66	6283
	Denver	C	40.8	0	0	90	366	714	905	1004	851	800	492	254	48	5524
	Grand Junction	A	39.3	0	0	30	313	786	1113	1209	907	729	387	146	21	5641
	Pueblo	A	40.4	0	0	54	326	750	986	1085	871	772	429	174	15	5462
Conn.	Bridgeport	A	39.9	0	0	66	307	615	986	1079	966	853	510	208	27	5617
	Hartford	A	37.3	0	12	117	394	714	1101	1190	1042	908	519	205	33	6235
	New Haven	A	39.0	0	12	87	347	648	1011	1097	991	871	543	245	45	5897
Del.	Wilmington	A	42.5	0	0	51	270	588	927	980	874	735	387	112	6	4930
D.C.	Washington	A	45.7	0	0	33	217	519	834	871	762	626	288	74	0	4224
Fla.	Apalachicola	C	61.2	0	0	16	153	319	347	260	180	33	0	0	0	1308
	Daytona Beach	A	64.5	0	0	0	75	211	248	190	140	15	0	0	0	879
	Fort Myers	A	68.6	0	0	0	24	109	146	101	62	0	0	0	0	442
	Jacksonville	A	61.9	0	0	12	144	310	332	246	174	21	0	0	0	1239
	Key West	A	73.1	0	0	0	0	0	28	40	31	9	0	0	0	108
	Lakeland	C	66.7	0	0	0	57	164	195	146	99	0	0	0	0	661
	Miami	A	71.1	0	0	0	0	0	65	74	56	19	0	0	0	214

[a] Data for United States cities from a publication of the United States Weather Bureau, *Monthly Normals of Temperature, Precipitation and Heating Degree Days*, 1962, are for the period 1931 to 1960 inclusive. These data also include information from the 1963 revisions to this publication, where available.

[b] Data for airport stations, A, and city stations, C, are both given where available.

[c] Data for Canadian cities were computed by the Climatology Division, Department of Transport from normal monthly mean temperatures, and the monthly values of heating degree days data were obtained using the National Research Council computer and a method devised by H. C. S. Thom of the United States Weather Bureau. The heating degree days are based on the period from 1931 to 1960.

[d] For period October to April, inclusive.

Normal Monthly and Annual Degree-Days for U. S. Cities (continued)

State	Station	Avg. Winter Temp[d]	July	Aug.	Sept.	Oct.	Nov.	Dec.	Jan.	Feb.	Mar.	Apr.	May	June	Yearly Total
Fla. (Cont'd)	Miami Beach C	72.5	0	0	0	0	0	40	56	36	9	0	0	0	141
	Orlando.............. A	65.7	0	0	0	0	72	198	220	165	105	6	0	0	766
	Pensacola.............. A	60.4	0	0	0	19	195	353	400	277	183	36	0	0	1463
	Tallahassee.............. A	60.1	0	0	0	28	198	360	375	286	202	36	0	0	1485
	Tampa.............. A	66.4	0	0	0	0	60	171	202	148	102	0	0	0	683
	West Palm Beach.............. A	68.4	0	0	0	0	6	65	87	64	31	0	0	0	253
Ga.	Athens .... A	51.8	0	0	12	115	405	632	642	529	431	141	22	0	2929
	Atlanta.............. A	51.7	0	0	18	124	417	648	636	518	428	147	25	0	2961
	Augusta.............. A	54.5	0	0	0	78	333	552	549	445	350	90	0	0	2397
	Columbus.............. A	54.8	0	0	0	87	333	543	552	434	338	96	0	0	2383
	Macon.............. A	56.2	0	0	0	71	297	502	505	403	295	63	0	0	2136
	Rome.............. A	49.9	0	0	24	161	474	701	710	577	468	177	34	0	3326
	Savannah.............. A	57.8	0	0	0	47	246	437	437	353	254	45	0	0	1819
	Thomasville.............. C	60.0	0	0	0	25	198	366	394	305	208	33	0	0	1529
Hawaii	Lihue.............. A	72.7	0	0	0	0	0	0	0	0	0	0	0	0	0
	Honolulu.............. A	74.2	0	0	0	0	0	0	0	0	0	0	0	0	0
	Hilo.............. A	71.9	0	0	0	0	0	0	0	0	0	0	0	0	0
Idaho	Boise.............. A	39.7	0	0	132	415	792	1017	1113	854	722	438	245	81	5809
	Lewiston.............. A	41.0	0	0	123	403	756	933	1063	815	694	426	239	90	5542
	Pocatello.............. A	34.8	0	0	172	493	900	1166	1324	1058	905	555	319	141	7033
Ill.	Cairo.............. C	47.9	0	0	36	164	513	791	856	680	539	195	47	0	3821
	Chicago (O'Hare).............. A	35.8	0	12	117	381	807	1166	1265	1086	939	534	260	72	6639
	Chicago (Midway).............. A	37.5	0	0	81	326	753	1113	1209	1044	890	480	211	48	6155
	Chicago.............. C	38.9	0	0	66	279	705	1051	1150	1000	868	489	226	48	5882
	Moline.............. A	36.4	0	9	99	335	774	1181	1314	1100	918	450	189	39	6408
	Peoria.............. A	38.1	0	6	87	326	759	1113	1218	1025	849	426	183	33	6025
	Rockford.............. A	34.8	6	9	114	400	837	1221	1333	1137	961	516	236	60	6830
	Springfield.............. A	40.6	0	0	72	291	696	1023	1135	935	769	354	136	18	5429

State	Station		Temp													Annual
Ind.	Evansville	A	45.0	0	0	66	220	606	896	955	767	620	237	68	0	4435
	Fort Wayne	A	37.3	0	0	105	378	783	1135	1178	1028	890	471	189	39	6205
	Indianapolis	A	39.6	0	0	90	316	723	1051	1113	949	809	432	177	39	5699
	South Bend	A	36.6	0	0	111	372	777	1125	1221	1070	933	525	239	60	6439
Iowa	Burlington	A	37.6	0	0	93	322	768	1135	1259	1042	859	426	177	33	6114
	Des Moines	A	35.5	0	6	96	363	828	1225	1370	1137	915	438	180	30	6588
	Dubuque	A	32.7	12	31	156	450	906	1287	1420	1204	1026	546	260	78	7376
	Sioux City	A	34.0	0	9	108	369	867	1240	1435	1198	989	483	214	39	6951
	Waterloo	A	32.6	12	19	138	428	909	1296	1460	1221	1023	531	229	54	7320
Kans.	Concordia	A	40.4	0	0	57	276	705	1023	1163	935	781	372	149	18	5479
	Dodge City	A	42.5	0	0	33	251	666	939	1051	840	719	354	124	9	4986
	Goodland	A	37.8	0	6	81	381	810	1073	1166	955	884	507	236	42	6141
	Topeka	A	41.7	0	0	57	270	672	980	1122	893	722	330	124	12	5182
	Wichita	A	44.2	0	0	33	229	618	905	1023	804	645	270	87	6	4620
Ky.	Covington	A	41.4	0	0	75	291	669	983	1035	893	756	390	149	24	5265
	Lexington	A	43.8	0	0	54	239	609	902	946	818	685	325	105	0	4683
	Louisville	A	44.0	0	0	54	248	609	890	930	818	682	315	105	9	4660
La.	Alexandria	A	57.5	0	0	0	56	273	431	471	361	260	69	0	0	1921
	Baton Rouge	A	59.8	0	0	0	31	216	369	409	294	208	33	0	0	1560
	Lake Charles	A	60.5	0	0	0	19	210	341	381	274	195	39	0	0	1459
	New Orleans	A	61.0	0	0	0	19	192	322	363	258	192	39	0	0	1385
	New Orleans	C	61.8	0	0	0	12	165	291	344	241	177	24	0	0	1254
	Shreveport	A	56.2	0	0	0	47	297	477	552	426	304	81	0	0	2184
Me.	Caribou	A	24.4	78	115	336	682	1044	1535	1690	1470	1308	858	468	183	9767
	Portland	A	33.0	12	53	195	508	807	1215	1339	1182	1042	675	372	111	7511
Md.	Baltimore	A	43.7	0	0	48	264	585	905	936	820	679	327	90	0	4654
	Baltimore	C	46.2	0	0	27	189	486	806	859	762	629	288	65	0	4111
	Frederick	A	42.0	0	0	66	307	624	955	995	876	741	384	127	12	5087
Mass.	Boston	A	40.0	0	9	60	316	603	983	1088	972	846	513	208	36	5634
	Nantucket	A	40.2	12	22	93	332	573	896	992	941	896	621	384	129	5891
	Pittsfield	A	32.6	25	59	219	524	831	1231	1339	1196	1063	660	326	105	7578
	Worcester	A	34.7	6	34	147	450	774	1172	1271	1123	998	612	304	78	6969

Normal Monthly and Annual Degree-Days for U. S. Cities (continued)

State	Station		Avg. Winter Temp^d	July	Aug.	Sept.	Oct.	Nov.	Dec.	Jan.	Feb.	Mar.	Apr.	May	June	Yearly Total
Mich.	Alpena	A	29.7	68	105	273	580	912	1268	1404	1299	1218	777	446	156	8506
	Detroit (City)	A	37.2	0	0	87	360	738	1088	1181	1058	936	522	220	42	6232
	Detroit (Wayne)	A	37.1	0	0	96	353	738	1088	1194	1061	933	534	239	57	6293
	Detroit (Willow Run)	A	37.2	0	0	90	357	750	1104	1190	1053	921	519	229	45	6258
	Escanaba	C	29.6	59	87	243	539	924	1293	1445	1296	1203	777	456	159	8481
	Flint	A	33.1	16	40	159	465	843	1212	1330	1198	1066	639	319	90	7377
	Grand Rapids	A	34.9	9	28	135	434	804	1147	1259	1134	1011	579	279	75	6894
	Lansing	A	34.8	6	22	138	431	813	1163	1262	1142	1011	579	273	69	6909
	Marquette	C	30.2	59	81	240	527	936	1268	1411	1268	1187	771	468	177	8393
	Muskegon	A	36.0	12	28	120	400	762	1088	1209	1100	995	594	310	78	6696
	Sault Ste. Marie	A	27.7	96	105	279	580	951	1367	1525	1380	1277	810	477	201	9048
Minn.	Duluth	A	23.4	71	109	330	632	1131	1581	1745	1518	1355	840	490	198	10000
	Minneapolis	A	28.3	22	31	189	505	1014	1454	1631	1380	1166	621	288	81	8382
	Rochester	A	28.8	25	34	186	474	1005	1438	1593	1366	1150	630	301	93	8295
Miss.	Jackson	A	55.7	0	0	0	65	315	502	546	414	310	87	0	0	2239
	Meridian	A	55.4	0	0	0	81	339	518	543	417	310	81	0	0	2289
	Vicksburg	C	56.9	0	0	0	53	279	462	512	384	282	69	0	0	2041
Mo.	Columbia	A	42.3	0	0	54	251	651	967	1076	874	716	324	121	12	5046
	Kansas City	A	43.9	0	0	39	220	612	905	1032	818	682	294	109	0	4711
	St. Joseph	A	40.3	0	6	60	285	708	1039	1172	949	769	348	133	15	5484
	St. Louis	A	43.1	0	0	60	251	627	936	1026	848	704	312	121	15	4900
	St. Louis	C	44.8	0	0	36	202	576	884	977	801	651	270	87	0	4484
	Springfield	A	44.5	0	0	45	223	600	877	973	781	660	291	105	6	4900
Mont.	Billings	A	34.5	6	15	186	487	897	1135	1296	1100	970	570	285	102	7049
	Glasgow	A	26.4	31	47	270	608	1104	1466	1711	1439	1187	648	335	150	8996
	Great Falls	A	32.8	28	53	258	543	921	1169	1349	1154	1063	642	384	186	7750
	Havre	A	28.1	28	53	306	595	1065	1367	1584	1364	1181	657	338	162	8700
	Havre	C	29.8	19	37	252	539	1014	1321	1528	1305	1116	612	304	135	8182

State	City		Temp	Jul	Aug	Sep	Oct	Nov	Dec	Jan	Feb	Mar	Apr	May	Jun	Annual
	Helena	A	31.1	31	59	294	601	1002	1265	1438	1170	1042	651	381	195	8129
	Kalispell	A	31.4	50	99	321	654	1020	1240	1401	1134	1029	639	397	207	8191
	Miles City	A	31.2	6	6	174	502	972	1296	1504	1252	1057	579	276	99	7723
	Missoula	A	31.5	34	74	303	651	1035	1287	1420	1120	970	621	391	219	8125
Neb.	Grand Island	A	36.0	0	6	108	381	834	1172	1314	1089	908	462	211	45	6530
	Lincoln	C	38.8	0	6	75	301	726	1066	1237	1016	834	402	171	30	5864
	Norfolk	A	34.0	9	0	111	397	873	1234	1414	1179	983	498	233	48	6979
	North Platte	A	35.5	0	6	123	440	885	1166	1271	1039	930	519	248	57	6684
	Omaha	A	35.6	0	12	105	357	828	1175	1355	1126	939	465	208	42	6612
	Scottsbluff	A	35.9	0	0	138	459	876	1128	1231	1008	921	552	285	75	6673
	Valentine	A	32.6	9	12	165	493	942	1237	1395	1176	1045	579	288	84	7425
Nev.	Elko	A	34.0	9	34	225	561	924	1197	1314	1036	911	621	409	192	7433
	Ely	A	33.1	28	43	234	592	939	1184	1308	1075	977	672	456	225	7733
	Las Vegas	A	53.5	0	0	0	78	387	617	688	487	335	111	6	0	2709
	Reno	A	39.3	43	87	204	490	801	1026	1073	823	729	510	357	189	6332
	Winnemucca	A	36.7	0	34	210	536	876	1091	1172	916	837	573	363	153	6761
N.H.	Concord	A	33.0	6	50	177	505	822	1240	1358	1184	1032	636	298	75	7383
	Mt. Washington Obsv.		15.2	493	536	720	1057	1341	1742	1820	1663	1652	1260	930	603	13817
N.J.	Atlantic City	A	43.2	0	0	39	251	549	880	936	848	741	420	133	15	4812
	Newark	A	42.8	0	0	30	248	573	921	983	876	729	381	118	0	4589
	Trenton	C	42.4	0	0	57	264	576	924	989	885	753	399	121	12	4980
N. M.	Albuquerque	A	45.0	0	0	12	229	642	868	930	703	595	288	81	0	4348
	Clayton	A	42.0	0	6	66	310	699	899	986	812	747	429	183	21	5158
	Raton	A	38.1	9	28	126	431	825	1048	1116	904	834	543	301	63	6228
	Roswell	A	47.5	0	0	18	202	573	806	840	641	481	201	31	0	3793
	Silver City	A	48.0	0	0	6	183	525	729	791	605	518	261	87	0	3705
N.Y.	Albany	A	34.6	0	19	138	440	777	1194	1311	1156	992	564	239	45	6875
	Albany	C	37.2	0	9	102	375	699	1104	1218	1072	908	498	186	30	6201
	Binghamton	A	33.9	22	65	201	471	810	1184	1277	1154	1045	645	313	99	7286
	Binghamton	C	36.6	0	28	141	406	732	1107	1190	1081	949	543	229	45	6451
	Buffalo	A	34.5	19	37	141	440	777	1156	1256	1145	1039	645	329	78	7062
	New York (Cent. Park)	C	42.8	0	0	30	233	540	902	986	885	760	408	118	9	4871
	New York (La Guardia)	A	43.1	0	0	27	223	528	887	973	879	750	414	124	6	4811

Normal Monthly and Annual Degree-Days for U. S. Cities (continued)

State	Station		Avg. Winter Temp[d]	July	Aug.	Sept.	Oct.	Nov.	Dec.	Jan.	Feb.	Mar.	Apr.	May	June	Yearly Total
	New York (Kennedy)	A	41.4	0	0	36	248	564	933	1029	935	815	480	167	12	5219
	Rochester	A	35.4	9	31	126	415	747	1125	1234	1123	1014	597	279	48	6748
	Schenectady	C	35.4	0	22	123	422	756	1159	1283	1131	970	543	211	30	6650
	Syracuse	A	35.2	6	28	132	415	744	1153	1271	1140	1004	570	248	45	6756
N. C.	Asheville	C	46.7	0	0	48	245	555	775	784	683	592	273	87	0	4042
	Cape Hatteras		53.3	0	0	0	78	273	521	580	518	440	177	25	0	2612
	Charlotte	A	50.4	0	0	6	124	438	691	691	582	481	156	22	0	3191
	Greensboro	A	47.5	0	0	33	192	513	778	784	672	552	234	47	0	3805
	Raleigh	A	49.4	0	0	21	164	450	716	725	616	487	180	34	0	3393
	Wilmington	A	54.6	0	0	0	74	291	521	546	462	357	96	0	0	2347
	Winston-Salem	A	48.4	0	0	21	171	483	747	753	652	524	207	37	0	3595
N. D.	Bismarck	A	26.6	34	28	222	577	1083	1463	1708	1442	1203	645	329	117	8851
	Devils Lake	C	22.4	40	53	273	642	1191	1634	1872	1579	1345	753	381	138	9901
	Fargo	A	24.8	28	37	219	574	1107	1569	1789	1520	1262	690	332	99	9226
	Williston	A	25.2	31	43	261	601	1122	1513	1758	1473	1262	681	357	141	9243
Ohio	Akron-Canton	A	38.1	0	9	96	381	726	1070	1138	1016	871	489	202	39	6037
	Cincinnati	C	45.1	0	0	39	208	558	862	915	790	642	294	96	6	4410
	Cleveland	A	37.2	9	25	105	384	738	1088	1159	1047	918	552	260	66	6351
	Columbus	A	39.7	0	6	84	347	714	1039	1088	949	809	426	171	27	5660
	Columbus	C	41.5	0	0	57	285	651	977	1032	902	760	396	136	15	5211
	Dayton	A	39.8	0	6	78	310	696	1045	1097	955	809	429	167	30	5622
	Mansfield	A	36.9	9	22	114	397	768	1110	1169	1042	924	543	245	60	6403
	Sandusky	C	39.1	0	6	66	313	684	1032	1107	991	868	495	198	36	5796
	Toledo	A	36.4	0	16	117	406	792	1138	1200	1056	924	543	242	60	6494
	Youngstown	A	36.8	6	19	120	412	771	1104	1169	1047	921	540	248	60	6417
Okla.	Oklahoma City	A	48.3	0	0	15	164	498	766	868	664	527	189	34	0	3725
	Tulsa	A	47.7	0	0	18	158	522	787	893	683	539	213	47	0	3860

State	Station		Temp													Annual
Ore.	Astoria	A	45.6	146	130	210	375	561	679	753	622	636	480	363	231	5186
	Burns	C	35.9	12	37	210	515	867	1113	1246	988	856	570	366	177	6957
	Eugene	A	45.6	34	34	129	366	585	719	803	627	589	426	279	135	4726
	Meacham	A	34.2	84	124	288	580	918	1091	1209	1005	983	726	527	339	7874
	Medford	A	43.2	0	0	78	372	678	871	918	697	642	432	242	78	5008
	Pendleton	A	42.6	0	0	111	350	711	884	1017	773	617	396	205	63	5127
	Portland	A	45.6	25	28	114	335	597	735	825	644	586	396	245	105	4635
	Portland	C	47.4	12	16	75	267	534	679	769	594	536	351	198	78	4109
	Roseburg	A	46.3	22	16	105	329	567	713	766	608	570	405	267	123	4491
	Salem	A	45.4	37	31	111	338	594	729	822	647	611	417	273	144	4754
Pa.	Allentown	A	38.9	0	0	90	353	693	1045	1116	1002	849	471	167	24	5810
	Erie	A	36.8	0	25	102	391	714	1063	1169	1081	973	585	288	60	6451
	Harrisburg	A	41.2	0	0	63	298	648	992	1045	907	766	396	124	12	5251
	Philadelphia	A	41.8	0	0	60	297	620	965	1016	889	747	392	118	40	5144
	Philadelphia	C	44.5	0	0	30	205	513	856	924	823	691	351	93	0	4486
	Pittsburgh	A	38.4	0	9	105	375	726	1063	1119	1002	874	480	195	39	5987
	Pittsburgh	C	42.2	0	0	60	291	615	930	983	885	763	390	124	12	5053
	Reading	C	42.4	0	0	54	257	597	939	1001	885	735	372	105	0	4945
	Scranton	A	37.2	0	19	132	434	762	1104	1156	1028	893	498	195	33	6254
	Williamsport	A	38.5	0	9	111	375	717	1073	1122	1002	856	468	177	24	5934
R. I.	Block Island	A	40.1	0	0	78	307	594	902	1020	955	877	612	344	99	5804
	Providence	A	38.8	0	16	96	372	660	1023	1110	988	868	534	236	51	5954
S. C.	Charleston	A	56.4	0	0	0	59	282	471	487	389	291	54	0	0	2033
	Charleston	C	57.9	0	0	0	34	210	425	443	367	273	42	0	0	1794
	Columbia	A	54.0	0	0	0	84	345	577	570	470	357	81	0	0	2484
	Florence	A	54.5	0	0	0	78	315	552	552	459	347	84	0	0	2387
	Greenville-Spartenburg	A	51.6	0	0	6	121	399	651	660	546	446	132	19	0	2980
S. D.	Huron	A	28.8	9	12	165	508	1014	1432	1628	1355	1125	600	288	87	8223
	Rapid City	A	33.4	22	12	165	481	897	1172	1333	1145	1051	615	326	126	7345
	Sioux Falls	A	30.6	19	25	168	462	972	1361	1544	1285	1082	573	270	78	7839
Tenn.	Bristol	A	46.2	0	0	51	236	573	828	828	700	598	261	68	0	4143
	Chattanooga	A	50.3	0	0	18	143	468	698	722	577	453	150	25	0	3254
	Knoxville	A	49.2	0	0	30	171	489	725	732	613	493	198	43	0	3494
	Memphis	A	50.5	0	0	18	130	447	698	729	585	456	147	22	0	3232

Normal Monthly and Annual Degree-Days for U. S. Cities (concluded)

State or Prov.	Station		Avg. Winter Temp[d]	July	Aug.	Sept.	Oct.	Nov.	Dec.	Jan.	Feb.	Mar.	Apr.	May	June	Yearly Total
	Memphis	C	51.6	0	0	12	102	396	648	710	568	434	129	16	0	3015
	Nashville	A	48.9	0	0	30	158	495	732	778	644	512	189	40	0	3578
	Oak Ridge	C	47.7	0	0	39	192	531	772	778	669	552	228	56	0	3817
Tex.	Abilene	A	53.9	0	0	0	99	366	586	642	470	347	114	0	0	2624
	Amarillo	A	47.0	0	0	18	205	570	797	877	664	546	252	56	0	3985
	Austin	A	59.1	0	0	0	31	225	388	468	325	223	51	0	0	1711
	Brownsville	A	67.7	0	0	0	0	66	149	205	106	74	0	0	0	600
	Corpus Christi	A	64.6	0	0	0	0	120	220	291	174	109	0	0	0	914
	Dallas	A	55.3	0	0	0	62	321	524	601	440	319	90	6	0	2363
	El Paso	A	52.9	0	0	0	84	414	648	685	445	319	105	0	0	2700
	Fort Worth	A	55.1	0	0	0	65	324	536	614	448	319	99	0	0	2405
	Galveston	A	62.2	0	0	0	6	147	276	360	263	189	33	0	0	1274
	Galveston	C	62.0	0	0	0	0	138	270	350	258	189	30	0	0	1235
	Houston	A	61.0	0	0	0	6	183	307	384	288	192	36	0	0	1396
	Houston	C	62.0	0	0	0	0	165	288	363	258	174	30	0	0	1278
	Laredo	A	66.0	0	0	0	0	105	217	267	134	74	0	0	0	797
	Lubbock	A	48.8	0	0	18	174	513	744	800	613	484	201	31	0	3578
	Midland	A	53.8	0	0	0	87	381	592	651	468	322	90	0	0	2591
	Port Arthur	A	60.5	0	0	0	22	207	329	384	274	192	39	0	0	1447
	San Angelo	A	56.0	0	0	0	68	318	536	567	412	288	66	0	0	2255
	San Antonio	A	60.1	0	0	0	31	204	363	428	286	195	39	0	0	1546
	Victoria	A	62.7	0	0	0	6	150	270	344	230	152	21	0	0	1173
	Waco	A	57.2	0	0	0	43	270	456	536	389	270	66	0	0	2030
	Wichita Falls	A	53.0	0	0	0	99	381	632	698	518	378	120	6	0	2832
Utah	Milford	A	36.5	0	0	99	443	867	1141	1252	988	822	519	279	87	6497
	Salt Lake City	A	38.4	0	0	81	419	849	1082	1172	910	763	459	233	84	6052
	Wendover	A	39.1	0	0	48	372	822	1091	1178	902	729	408	177	51	5778
Vt.	Burlington	A	29.4	28	65	207	539	891	1349	1513	1333	1187	714	353	90	8269

State	City		Temp														Annual
Va.	Cape Henry	C	50.0	0	0	0	112	360	645	694	633	536	246	53	0	3279
	Lynchburg	A	46.0	0	0	51	223	540	822	849	731	605	267	78	0	4166
	Norfolk	A	49.2	0	0	0	136	408	698	738	655	533	216	37	0	3421
	Richmond	A	47.3	0	0	36	214	495	784	815	703	546	219	53	0	3865
	Roanoke	A	46.1	0	0	51	229	549	825	834	722	614	261	65	0	4150
Wash.	Olympia	A	44.2	68	71	198	422	636	753	834	675	645	450	307	177	5236
	Seattle-Tacoma	A	44.2	56	62	162	391	633	750	828	678	657	474	295	159	5145
	Seattle	C	46.9	50	47	129	329	543	657	738	599	577	396	242	117	4424
	Spokane	A	36.5	9	25	168	493	879	1082	1231	980	834	531	288	135	6655
	Walla Walla	C	43.8	0	0	87	310	681	843	986	745	589	342	177	45	4805
	Yakima	A	39.1	0	12	144	450	828	1039	1163	868	713	435	220	69	5941
W. Va.	Charleston	A	44.8	0	0	63	254	591	865	880	770	648	300	96	9	4476
	Elkins	A	40.1	9	25	135	400	729	992	1008	896	791	444	198	48	5675
	Huntington	A	45.0	0	0	63	257	585	856	880	764	636	294	99	12	4446
	Parkersburg	C	43.5	0	0	60	264	606	905	942	826	691	339	115	6	4754
Wisc.	Green Bay	A	30.3	28	50	174	484	924	1333	1494	1313	1141	654	335	99	8029
	La Crosse	A	31.5	12	19	153	437	924	1339	1504	1277	1070	540	245	69	7589
	Madison	A	30.9	25	40	174	474	930	1330	1473	1274	1113	618	310	102	7863
	Milwaukee	A	32.6	43	47	174	471	876	1252	1376	1193	1054	642	372	135	7635
Wyo.	Casper	A	33.4	6	16	192	524	942	1169	1290	1084	1020	657	381	129	7410
	Cheyenne	A	34.2	28	37	219	543	909	1085	1212	1042	1026	702	428	150	7381
	Lander	A	31.4	6	19	204	555	1020	1299	1417	1145	1017	654	381	153	7870
	Sheridan	A	32.5	25	31	219	539	948	1200	1355	1154	1051	642	366	150	7680

NORMAL MONTHLY AND ANNUAL DEGREE-DAYS BELOW 65°F
FOR CANADIAN CITIES

(Data submitted by Donald W. Boyd, Meteorologist, National Research Council Canada)

City	Jan	Feb	Mar	Apr	May	Jun	Jul	Aug	Sep	Oct	Nov	Dec	Total
					ALBERTA								
Athabasca	1969	1661	1460	852	502	291	158	233	468	812	1308	1779	11493
Banff	1624	1364	1237	855	589	402	220	295	498	797	1185	1485	10551
Calgary	1636	1306	1259	805	494	282	126	189	415	711	1130	1446	9799
Edmonton	1843	1469	1327	775	413	209	96	162	411	728	1221	1624	10278
Fort McMurray	2220	1802	1525	954	536	291	133	242	504	880	1425	1950	12462
Grande Prairie	1984	1565	1439	861	462	253	142	211	441	793	1332	1756	11239
Jasper	1659	1319	1184	801	527	324	189	257	441	753	1182	1476	10112
Lethbridge	1497	1291	1159	696	403	213	056	112	318	611	1011	1277	8644
Medicine Hat	1696	1344	1203	665	348	148	36	66	300	603	1076	1441	8926
Red Deer	1773	1494	1321	786	446	267	118	189	423	750	1188	1547	10302
					BRITISH COLUMBIA								
Bull Harbour	818	709	722	588	474	351	276	260	333	484	636	766	6417
Crescent Valley	1311	1090	961	618	375	222	084	118	324	660	999	1184	7946
Estevan Point	763	675	704	573	456	330	260	239	294	450	591	701	6036
Fort Nelson	2322	1808	1518	923	480	210	116	212	518	964	1658	2182	12911
Kamloops	1293	1011	800	435	211	093	016	022	177	521	879	1119	6577
Kelowna	1290	1090	915	564	288	144	050	078	252	614	936	1141	7362
Kitimat	1221	975	868	648	440	249	146	155	300	608	873	1079	7562
Penticton	1180	910	817	518	279	103	26	40	212	536	824	1043	6488
Prince George	1653	1238	1126	765	482	273	184	255	463	761	1135	1442	9777
Prince Rupert	962	785	820	670	509	365	281	259	346	545	731	863	7136
Vancouver	888	707	698	506	329	168	73	85	221	460	659	850	5644
Victoria	799	641	649	499	359	247	174	179	235	424	602	733	5541
					MANITOBA								
Brandon	2006	1718	1463	810	403	183	040	084	345	722	1278	1776	10828
Churchill	2560	2284	2147	1576	1142	664	363	385	678	1073	1643	2247	16762
Flin Flon	2229	1842	1621	984	552	240	056	136	447	828	1473	2006	12414
The Pas	2269	1854	1639	998	585	239	77	142	453	801	1396	1994	12447
Winnipeg	2044	1730	1469	808	437	144	42	74	328	656	1224	1784	10740
					NEW BRUNSWICK								
Bathurst	1615	1452	1287	846	508	207	065	093	294	645	987	1463	9462
Chatham	1547	1366	1203	821	489	190	47	85	276	598	920	1337	8879
Fredericton	1516	1337	1146	760	423	162	42	71	267	572	891	1370	8557
Grand Falls	1686	1500	1311	831	450	180	096	112	315	651	984	1519	9635
Moncton	1462	1310	1177	820	485	208	52	87	273	568	863	1312	8617
Saint John	1430	1276	1147	806	514	254	108	122	295	568	842	1279	8641
					NEWFOUNDLAND								
Gander	1357	1245	1216	941	670	379	142	163	357	667	883	1207	9227
Goose	1935	1659	1491	1081	747	398	174	222	460	838	1181	1708	11894
Grand Bank	1166	1093	1107	885	688	429	202	140	288	570	765	1045	8378
St. John's (Torbay)	1225	1140	1152	926	727	452	190	176	345	621	797	1089	8840
					NORTHWEST TERRITORIES								
Cambridge Bay	2846	2684	2675	2157	1541	897	561	642	999	1652	2277	2697	21628
Coppermine	2616	2461	2465	1920	1336	807	505	555	852	1411	2067	2489	19484
Fort Reliance	2678	2333	2155	1521	967	552	313	329	639	1178	1740	2375	16780
Resolute	2830	2678	2768	2217	1597	960	766	856	1233	1835	2292	2641	22673
Yellowknife	2621	2238	2079	1424	800	335	149	247	617	1086	1748	2347	15691

City	Jan	Feb	Mar	Apr	May	Jun	Jul	Aug	Sep	Oct	Nov	Dec	Total
					NOVA SCOTIA								
Halifax	1228	1141	1061	770	519	255	76	61	202	466	717	1082	7578
Sydney	1264	1203	1151	871	596	299	81	81	246	523	756	1100	8171
Yarmouth	1163	1076	1005	739	505	273	114	113	241	468	690	1028	7415
					ONTARIO								
Chatham	1206	1096	967	573	257	051	012	019	096	375	744	1107	6503
Hamilton	1267	1131	996	598	320	73	10	20	142	403	734	1125	6819
Kapuskesing	2034	1732	1523	949	585	256	133	200	432	736	1211	1805	11596
Kenora	2012	1677	1409	852	479	173	55	93	367	685	1230	1788	10820
Kingston Ont Hydro	1451	1297	1113	654	329	075	019	040	147	481	831	1287	7724
Kirkland Lake	1965	1681	1522	963	533	192	115	174	384	760	1197	1783	11269
Kitchener	1342	1226	1101	663	322	066	016	059	177	505	855	1234	7566
London	1350	1210	1060	631	343	94	24	43	187	469	816	1215	7442
North Bay	1736	1492	1316	814	460	167	74	119	328	632	1027	1545	9710
Ottawa	1628	1409	1187	681	336	85	20	54	232	536	911	1442	8521
St. Catharines	1187	1096	986	597	285	039	003	028	108	394	726	1088	6537
Sudbury	1752	1509	1345	813	451	152	58	105	316	620	1020	1525	9666
Thunder Bay	1843	1587	1360	856	547	247	91	136	381	682	1120	1618	10468
Timmins	1965	1706	1486	931	561	240	124	194	419	729	1176	1767	11298
Toronto	1261	1119	988	588	304	65	8	17	141	405	730	1118	6744
Trenton	1420	1240	1066	636	332	82	15	33	184	473	800	1245	7526
Windsor	1254	1100	947	543	264	56	6	16	122	378	751	1123	6560
				PRINCE EDWARD ISLAND									
Charlottetown	1354	1282	1189	854	543	246	52	75	244	533	808	1225	8405
					QUEBEC								
Bagotville	1897	1620	1382	884	520	189	87	142	368	699	1067	1670	10525
Chicoutimi	1845	1585	1336	834	496	177	087	115	294	654	1050	1631	10104
Fort Chimo	2356	2088	1928	1491	998	597	363	437	687	1042	1443	2015	15445
Fort George	2275	2009	1888	1305	868	513	341	388	558	871	1230	1848	14094
Knob Lake	2306	1978	1814	1326	936	498	310	415	681	1073	1494	2049	14880
Megantic	1634	1475	1333	849	493	192	105	133	324	660	1011	1479	9688
Mont Joli	1640	1449	1308	894	567	243	121	146	354	688	1017	1497	9924
Montreal (Dorval)	1569	1376	1156	675	314	74	14	40	198	495	859	1387	8157
Nitchequon	2297	1952	1804	1296	911	468	260	341	609	989	1425	2046	14398
Quebec	1629	1469	1256	805	428	145	44	87	290	610	970	1492	9225
Roberval	1900	1641	1445	903	496	177	084	124	315	676	1089	1671	10521
Schefferville	2286	2008	1845	1359	964	527	320	420	682	1069	1455	2030	14965
Sept-iles	1795	1572	1392	998	700	361	182	231	466	797	1102	1587	11183
Sherbrooke	1584	1391	1179	729	384	127	43	89	267	562	901	1404	8660
Three Rivers	1696	1509	1293	780	378	114	040	084	255	623	990	1544	9306
					SASKATCHEWAN								
Moose Jaw	1829	1551	1349	744	378	174	059	059	309	670	1191	1581	9894
North Battleford	2024	1706	1482	813	406	207	065	130	384	756	1317	1792	11082
Prince Albert	2108	1763	1559	867	446	219	081	136	414	797	1368	1872	11630
Regina	1989	1661	1482	809	428	180	57	92	376	736	1272	1745	10827
Saskatoon	2066	1698	1505	816	428	179	58	107	385	744	1300	1801	11087
Swift Current	1800	1469	1366	787	449	210	68	108	370	684	1142	1562	10015
					YUKON TERRITORY								
Dawson	2561	2150	1838	1068	570	258	164	326	645	1197	1875	2415	15067
Mayo Landing	2427	1992	1665	1020	580	294	208	366	648	1135	1794	2325	14454
Teslin	2108	1701	1460	1041	663	372	276	378	597	995	1422	1885	12898
Whitehorse	2055	1599	1438	985	625	323	243	336	575	985	1478	1908	12550

NORMAL YEARLY TOTAL DEGREE-DAYS FOR 877 U. S. CITIES

The data which follow give the seasonal total number of degree days for the heating season for 877 towns, villages, and cities not covered in the previous table in which the data are given by months.

Figures in this table were compiled by using normal monthly mean temperatures for each station and are based on these monthly normal figures for periods as long as 50 years and at least 20 years. They are not official figures of the U. S. Weather Bureau, with certain exceptions, but were calculated from official publications of the Weather Bureau.

The method used in calculating these figures was to subtract the mean monthly temperature from 65

degrees and multiply by the number of days in that month.

Consequently, to arrive at the seasonal totals given, the monthly figures had to be calculated. The only reason these monthly figures are not presented here is due to lack of space.

The exceptions are those stations marked with a dagger for which the figures are those calculated by the U. S. Weather Bureau.

In using this table, first check to see whether the city is covered in the more complete tabulation which precedes this table and in which the figures are given for each month.

ALABAMA		ARKANSAS			
Ashville	2882	Arkadelphia	2500	Georgetown	3737
Bermuda	1816	Batesville	3429	Grass Valley	3754
Camp Hill	2251	Bentonville†	4036	Greenland Ranch	1088
Clanton	2408	Bergman	3957	Healdsburg	2728
Decatur	3047	Corning	3703	Hollister	2651
Eufaula	2147	Dardanelle	3014	Independence†	3834
Gadsden	2903	Dumas	2303	King City	2408
Hamilton	2864	Dutton	4167	Lick Observatory	4972
Healing Springs	2263	El Dorado	2295	Livermore	2473
Livingston	2373	Helena	2764	Lone Pine	3799
Ozark	1737	Mammoth Spring	3716	Macdoel	7732
Riverton	3497	Mena	2945	Madeline	8015
Scottsboro	3153	Mountain Home	3537	Marysville	2498
Tuscaloosa	2529	Okay	3430	McCloud	6053
Valley Head	3319	Osceola	3141	Mecca	1175
		Portland	2158	Merced	2473
ARIZONA		Stuttgart	2835	Mojave	2463
		Wynne	3066	Montague	5423
Alpine Ranger Sta.	7859			Mount Tamalpais	3867
Casa Grande	1283			Needles	1251
Chin Lee	5908	CALIFORNIA		Oceanside	1653
Clifton	2184			Oroville	2513
Congress	1666	Alturas	6673	Pasadena	2067
Douglas	2627	Bagdad	1141	Placerville	3959
Fort Apache	4922	Barstow	2299	Point Reyes†	4474
Granite Reef Dam	1418	Berkeley	3269	Pomona	2284
Kayenta	5317	Blythe	1740	Quincy	5724
Leupp	5311	Branscomb	4342	St. Helena	2869
Miami	2418	Burney	6163	San Bernardino	1879
Moccasin	5320	Cahuilla	4920	San Luis Obispo	2382
Nogales	2516	Calexico	1089	Santa Barbara	1995
Parker	1518	China Flat	3524	Santa Clara	3087
Payson	4496	Cloverdale	2703	Santa Cruz	2812
Pinto	5395	Crescent City	4711	Santa Monica	2121
Salome	2143	Cuyamaca	5245	Sterling	852
San Carlos	2257	Davis	2899	Stockton	2675
Seligman	5004	Delta	3088	Storey	2942
Sentinel	1181	Downieville	4873	Susanville	6140
Supai	2722	Emigrant Gap	5613	Tulare	2301
Truxton	3375	Eureka†	4758	Upper Lake	3367
Willcox	3320	Fort Bidwell	6680	Weaverville	4662
				West Branch	5040

Willows	2670	**FLORIDA**		**ILLINOIS**	
Yosemite	5345	Avon Park	198	Antioch	6897
COLORADO		Bradentown	354	Aurora	6661
		Cedar Keys	632	Bloomington	5693
Akron	6724	Fort Lauderdale	0	Danville	5391
Arriba	6392	Gainesville	732	Decatur	5485
Blue Valley Ranch	9759	Monticello	1129	Freeport	7052
Boulder	5535	Okeechobee	112	Griggsville	5275
Buena Vista	8852	St. Augustine	1116	Harrisburg	4076
Burlington	5875	Titusville	345	Havana	5365
Cheesman	6898			Henry	6089
Cheyenne Wells	5936	**GEORGIA**		Hillsboro	5009
Delta	6063	Americus	1895	La Harpe	5815
Dillon	11515	Bainbridge	1473	Mt. Carmei	4725
Dolores	6826	Brunswick	1268	Mt. Vernon	4691
Durango†	7143	Carrollton	2806	Palestine	5015
Fort Lewis	7646	Clayton	3611	Pontiac	5796
Fraser	12023	Concord	2597	Quincy	5302
Fruita	6205	Covington	2776	Rockford	6847
Garnett	8634	Diamond	3658	Urbana	5805
Grand Junction†	5613	Dublin	2106		
Grover	7305	Eastman	1816	**INDIANA**	
Gunnison	10218	Elberton	2728	Anderson	5441
Hamps	6939	Fitzgerald	1658	Angola	6349
Holly	5070	Fort Gaines	1802	Auburn	6522
Huerfano	6159	Gainesville	3239	Bedford	4682
Husted	6743	Gillsville	2969	Bloomington	4768
Julesburg	6435	Glennville	1623	Bluffton	5867
Lake City	9320	Griffin	2508	Columbus	5255
Las Animas	5595	Harrison	2183	Connersville	5452
Lay	8337	Marshallville	2019	Crawfordsville	5336
Leadville†	10678	Milledgeville	2335	Delphi	5909
Manassa	8285	Millen	1969	Greencastle	5389
Nast	10383	Ramhurst	3216	Greensburg	5274
Pagosa Springs	8512	St. George	1168	Hammond	6240
Rangely	7887	St. Marys	1190	Howe	6472
Redvale	7054	Statesboro	1769	Huntingburg	4372
Rifle	6642	Talbotton	2206	Huntington	5884
Salida	7167	Thomasville†	1513	Jeffersonville	4325
San Luis	8229	Valona	1484	Kokomo	5594
Steamboat Spr'gs	9630	Washington	2655	Lafayette	5816
Trinidad	5326	Waycross	1471	Laporte	5797
Two Buttes	1042	Waynesboro	2090	Logansport	5798
Wagon Wheel Gap	10833	West Point	2435	Madison	4472
Wray	5828	**IDAHO**		Marion	5715
CONNECTICUT		Ashton	8667	Plymouth	6149
		Avery	6947	Princeton	4639
Cream Hill	7136	Glenns Ferry	5684	Richmond	5757
N. Grosvenordale	6799	Grace	7551	Rome	4010
DELAWARE		Kooskia	6009	Royal Center†	6239
		Mackay	8609	Seymour	4817
Delaware City	4913	New Meadows	8525	Shelbyville	5106
Dover	4591	Oakley	6458	Valparaiso	6350
Milford	4303	Sandpoint	7326	Vincennes	4776
Millsboro	4527	Springfield	7477	Washington	4639
Seaford	4551				

Whiting	5993	Calhoun	2168	Kalamazoo	6655
Winona Lake	5896	Cameron	1281	Ludington†	7458
IOWA		Cheneyville	1607	Menominee	8043
		Collinston	1998	Mio	8335
Algona	7420	Dodson	1727	Pontiac	6924
Bonaparte	6041	Franklin	1198	Port Austin	7361
Clarinda	6285	Franklinton	1484	Port Huron	7195
Corydon	6087	Grand Cane	2099	Saginaw	7063
Cresco	8369	Houma	1078	St. Joseph	6203
Denison	6802	Melville	1471	Seney	9293
Guthrie Center	6573	Rayne	1268	**MINNESOTA**	
Independence	7177	Robeline	1997		
Keokuk†	5663	Tallulah	2057	Beardsley	8490
Marshalltown	6853	Tenmile	1454	Cass Lake	9445
Mason City	7566	**MAINE**		Fairmont	7849
Rockwell City	7008			Hallock	10285
Sibley	7912	Bar Harbor	7643	Luverne	7915
Washington	6322	Greenville†	9439	Montevideo	8169
Webster City	6781	North Brighton	7600	Moorhead†	9327
KANSAS		Orono	7934	Red Wing	7883
		Presque Isle	9440	Sandy Lake Dam	9620
Burr Oak	5657	**MARYLAND**		Thief River Falls	10044
Colby	5639			Tower	10440
Emporia	4694	Annapolis	4483	Wadena	9523
Fort Scott	4379	Cambridge	4245	**MISSISSIPPI**	
Hanover	5243	Clear Spring	5522		
Hays	5277	Cumberland	5196	Austin	2753
Healy	5275	Emmitsburg	5230	Batesville	2905
Hutchinson	4677	Fallston	5261	Columbia·	1739
Independence	4024	Grantsville	6493	Corinth†	3087
Iola†	4616	Princess Anne	4205	Crystal Springs	2013
Medicine Lodge	4269	Solomons	4202	Greenwood	2391
Phillipsburg	5301	Sudlersville	4676	Holly Springs	3298
Salina	4969	Takoma	4948	Kosciusko	2482
Toronto	4582	**MASSACHUSETTS**		Leakesville	1611
Ulysses	5009			Macon	2507
KENTUCKY		Amherst	6894	Nitta Yuma	2649
		Fitchburg†	6743	Palo Alto	2523
Alpha	3540	Williamstown	7225	Pearlington	1329
Bardstown	4178	**MICHIGAN**		Pecan	1346
Catlettsburg	4279			Porterville	2104
Farmers	4501	Ann Arbor	6877	Rosedale	2643
Frankfort	4211	Battle Creek	6585	Shoccoe	1932
Franklin	3570	Bay City	7051	Shubuta	2140
Greensburg	4369	Cadillac	7970	Tupelo	2781
Junction City	4113	Calumet	9297	Woodville	1565
Marion	3935	Charlevoix	7689	Yazoo City	2152
Middlesboro	3853	Cheboygan	8234	**MISSOURI**	
Owensboro	4175	Coldwater	6423		
Paducah	3654	Flint	7179	Bolivar	4330
Richmond	4126	Harrison	7749	Caruthersville	3439
Scott	4785	Hart	7040	Chillicothe	4902
LOUISIANA		Houghton†	9030	Dean	3942
		Iron River	9416	Glasgow	5235
Alexandria	1745	Ironwood	9126	Grant City	5661
Amite	1492	Jackson	6483	Hannibal†	5393
Antioch	2164			Hollister	3750

Jackson	4228	**NEW JERSEY**		Cooperstown	7756
Jefferson City	4874	Asbury Park	5272	Cortland	7388
Joplin	3904	Belvidere	5792	Cutchogue	5621
New Haven	4544	Bergen Point	5366	De Ruyter	7687
Poplar Bluff	3908	Boonton	5977	Elmira	6412
Potosi	5340	Bridgeton	4661	Franklinville	7455
Unionville	5929	Burlington	5009	Glens Falls	7244
Warsaw	4353	Cape May†	4870	Gloversville	7468
Willow Springs	4304	Clayton	5045	Indian Lake	8777
MONTANA		Dover	6270	Ithaca†	6914
Biddle	7786	Elizabeth	5302	Jamestown	6740
Bowning	9400	Hightstown	5360	Liberty	7676
Columbia Falls	8080	Indian Mills	5172	Little Falls	7604
Dillon	7796	Jersey City	5193	Lyons	6448
Glendive	8235	Lakewood	5399	Middlebury	6803
Harlowton	8781	Lambertville	5316	Mohonk Lake	6635
Hebgen Dam	10945	Long Branch	5396	Moira	8149
Lewistown	8230	Moorestown	5620	North Lake	8853
Livingston	7205	New Brunswick	5404	Number Four	8716
Malta	8892	Newton	6135	Ogdensburg	7688
Ovando	9452	Paterson	5365	Plattsburg	7897
Poplar	9241	Phillipsburg	5694	Port Jervis	6388
Roundup	7512	Plainfield	5535	Rome	7418
Thompson Falls	7347	Ridgefield	6091	Saranac Lake	8675
Upper Yaak River	8857	Sandy Hook†	5369	Setauket	5549
Victor	7637	Somerville	5586	South Canisteo	7200
NEBRASKA		South Orange	5745	Ticonderoga	7323
Bridgeport	6830	Sussex	6034	Utica	6796
Culbertson	5917	Vineland	5043	Wappingers Falls	6238
Drexel†	6611	Woodbine	5036	Watertown	7298
Dumas	7207	**NEW MEXICO**		Westfield	6548
Fort Robinson	7095	Aragon	6158	West Point	5882
Halsey	6646	Campana	4284	**NORTH CAROLINA**	
Hebron	5843	Carlsbad	2589	Andrews	3603
Hillside	6832	Chama	7946	Banners Elk	5746
Kearney	6050	Clovis	4197	Beaufort	2309
Kennedy	6922	Corona	5114	Brevard	4180
Kimball	6724	Elephant Butte	3401	Brewers	3911
Madrid	6456	Elizabethtown	9449	Charlotte	3120
St. Paul	6082	Fort Stanton	5069	Eagletown	3346
Santee	6766	Fort Sumner	3912	Hickory	3401
Springview	6872	Fort Union	6231	Hot Springs	3483
Stanton	6796	Fort Wingate	5638	Kinston	2768
Tecumseh	5773	Fruitland	5438	Louisburg	3518
NEVADA		Gage	3252	Manteo†	3109
Belmont	7399	Gallinas Springs	4552	Marion	3604
Beowawe	6622	Magdalena	5218	Monroe	3153
McGill	6980	Santa Fe†	6123	Mount Airy	4128
Quinn River R.	6673	Senorito	9031	Nashville*	3219
Wells	7575	**NEW YORK**		Pinehurst	2818
NEW HAMPSHIRE		Appleton	6775	Pittsboro	3619
Bethlehem	8773	Auburn	6747	Roxboro	3842
Hanover	8005	Canton†	8305	Salisbury	3222
Keene	7458	Cape Vincent	7775	Southport	2199
				Washington	2673

NORTH DAKOTA

Ashley	9471
Crosby	10091
Donnybrook	10006
Forman	8893
Fort Yates	8522
Grafton	10267
Grand Forks†	9871
Hannah	11018
Jamestown	9406
Manfred	9749
Mayville	9252
Minot	9581
New England	8961
Oakdale	8723
Washburn	9062
Westhope	10384

OHIO

Bangorville	5942
Bellefontaine	6036
Bowling Green	6233
Cambridge	5484
Cardington	6003
Clarington	5044
Clarksville	5231
Dover	6004
Green	4643
Greenville	5754
Hillhouse	6520
Kenton	5870
Lancaster	5272
Lima	5935
Marietta	4757
Marion	5685
Millport	6307
Montpelier	6411
New Bremen	5769
North Lewisburg	5802
Oberlin	6173
Orangeville	6465
Peebles	5088
Plattsburg	5688
Pomeroy	4828
Portsmouth	4411
Shenandoah	6343
Somerset	5314
Upper Sandusky	5804
Warren	6117
Waverly	4948

OKLAHOMA

Broken Arrow†	3826
Chattanooga	2962
Enid	3882
Hugo	2648
Kenton	4457

Keokuk Falls	3618
Muskogee	3616
Newkirk	3830
Rankin	3851
Ravia	2740
Shawnee	3240
Vinita	3681
Woodward	4008

OREGON

Corvalis	4746
Dayville	5513
Diamond	7306
Falls City	5368
Klamath Falls	6465
Lakeview	7094
McKenzie Bridge	5703
Parkdale	6552
Silver Lake	7455
Umatilla	4686
Vale	6268
Waldo	5648
Wallowa	7361
Warmspring	5687

PENNSYLVANIA

Altoona	6121
Bethlehem	5230
Bradford	6929
Canonsburg	5678
Carlisle	5710
Coatesville	5365
Corry	7190
Easton	5727
Emporium	6465
Franklin	6383
Gettysburg	5581
Gordon	6377
Greensburg	5796
Grove City	6489
Hanover	4943
Hyndman	5750
Johnstown	5539
Kennett Square	5388
Lancaster	5482
Lebanon	5585
Lewisburg	6395
Mount Pocono	7543
New Castle	5958
Pottstown	5153
State College	6297
Towanda	6540
Uniontown	5176
Vandergrift	5398
Warren	6740
Wellsboro	6893

West Chester	5441
York	5449

RHODE ISLAND

Narragansett Pier†	6397

SOUTH CAROLINA

Aiken	2311
Allendale	2107
Anderson	2732
Cheraw	2816
Dillon	2500
Due West†	2890
Ferguson	2245
Georgetown	1991
Heath Spring	2965
Newberry	2650
Paris Island	1601
Walhalla	3175
York	2725

SOUTH DAKOTA

Academy	7044
Dowling	7340
Faulkton	8260
Flandreau	8027
Forestburg	7889
Jefferson	8419
Milbank	8465
Murdo	7142
Oelrichs	7271
Pierre	7283
Pollock	8778
Spearfish	7072
Yankton	7144

TENNESSEE

Carthage	3587
Rogersville	4023
Savannah	3199
Springville	3654
Tullahoma	3717

TEXAS

Barstow	2224
Beaumont	1072
Blanco	1989
Bowie	2400
Brazoria	1008
Brownfield	3150
Childress	3045
Clarendon	3371
Claytonville	2533
Dalhart	4881
Dennison	2533
Eagle Pass	871
Fort Clark	1374
Fort Davis	2650

Fort Lancaster	2030	Hopewell	3663	New Martinsville	4778
Fort McIntosh	730	Ivanhoe	5132	Point Pleasant	4516
Haskell	2552	Newport News	3483	Sutton	4590
Hereford	4186	Quantico	4894	Upper Tract	5032
Knickerbocker	1959	Runnymede	3599	Wheeling	5218
Lampasas	1974	Salem	3789	White Sulphur Spgs.	5122
Liberty	1101	Saxe	3904	Williamson	4242
Lieb	4393	Staunton	4448	**WISCONSIN**	
Marshall	1923	Warsaw	4308		
Midland	2594	Williamsburg	3675	Appleton	7705
o2 Ranch	2160	Winchester	4784	Beloit	6876
Panter	2169	Woodstock	4804	Crandon	8978
Plainview	3332	Wytheville†	5103	Darlington	7420
Rock Island	1297			Deerskin Dam	9784
San Augustine	1758	**WASHINGTON**		Downing	8643
Sierra Blanca	2128			Eau Claire	7970
Somerville	1450	Aberdeen	5586	Florence	9115
Sonora	2091	Anacortes	5632	Fond du Lac	7617
Spur	3196	Bellingham	5417	Iron River	9195
Taylor†	1909	Blaine	6122	Kewaunee	7957
UTAH		Brewster	6396	Manitowoc	7730
		Buckley	5574	Medford	8564
Black Rock	6438	Centralia	5201	Oshkosh	7508
Cedar City	5679	Clearwater	5701	Park Falls	9459
East Portal	11129	Colfax	6411	Racine	6763
Emery	7071	Cusick	7323	Reedsburg	7277
Escalante	6193	Davis Ranch	6295	Rest Lake	9425
Green River	5674	Ephrata	5551	Sheboygan	7382
Ibapah	7569	Everett	5941	Spooner	8760
Kelton	6696	Glacier	6959	Sturgeon Bay	8201
Levan	6885	Kiona	4974	Superior	9750
Logan	6748	Lake Kachess	7689	Tomahawk	8659
Manila	8454	Longmires Sprgs.	7646	Valley Junction	7917
Modena†	6598	Lyle	5375	Waupaca	7892
Ogden	5538	Northport	7170	Wausau†	8494
St. George	3756	Oroville	6126	Whitehall	7850
VERMONT		Port Townsend	5510	Wisconsin Rapids	8028
		Puyallup	5431	**WYOMING**	
Enosburg Falls	8388	Sedro Woolley	5404		
Northfield†	8804	Sixprong	5062	Cody	7771
Wells	8037	Stehekin	6679	Cokeville	10020
VIRGINIA		Union City	5313	Colony	7793
		Vancouver	4748	Daniel	11118
Alexandria	4572	Wenatchee	6555	Dome Lake	11614
Big Stone Gap	4626	Zindel	4308	Ervay	8626
Birdsnest	3753	**WEST VIRGINIA**		Gillette	7518
Blacksburg	4977			Green River	8255
Bristol	4509	Bayard	6488	Griggs	7521
Callaville	3758	Cairo	4642	Kinnear	8114
Charlottesville	4223	Elkhorn	4411	Newcastle	7426
Columbia	4210	Fairmont	5047	Pathfinder	7402
Culpeper	4606	Green Sulphur Spgs.	5087	Saratoga	8630
Danville	3511	Lost Creek	5179	Snake River	11111
Fortress Monroe	3552	Marlinton	6220	South Pass City	11101
Fredericksburg	4242	Morgantown	5095	Worland	8058
Hampton	3092	New Cumberland	5664	Yellowstone Park†	9605

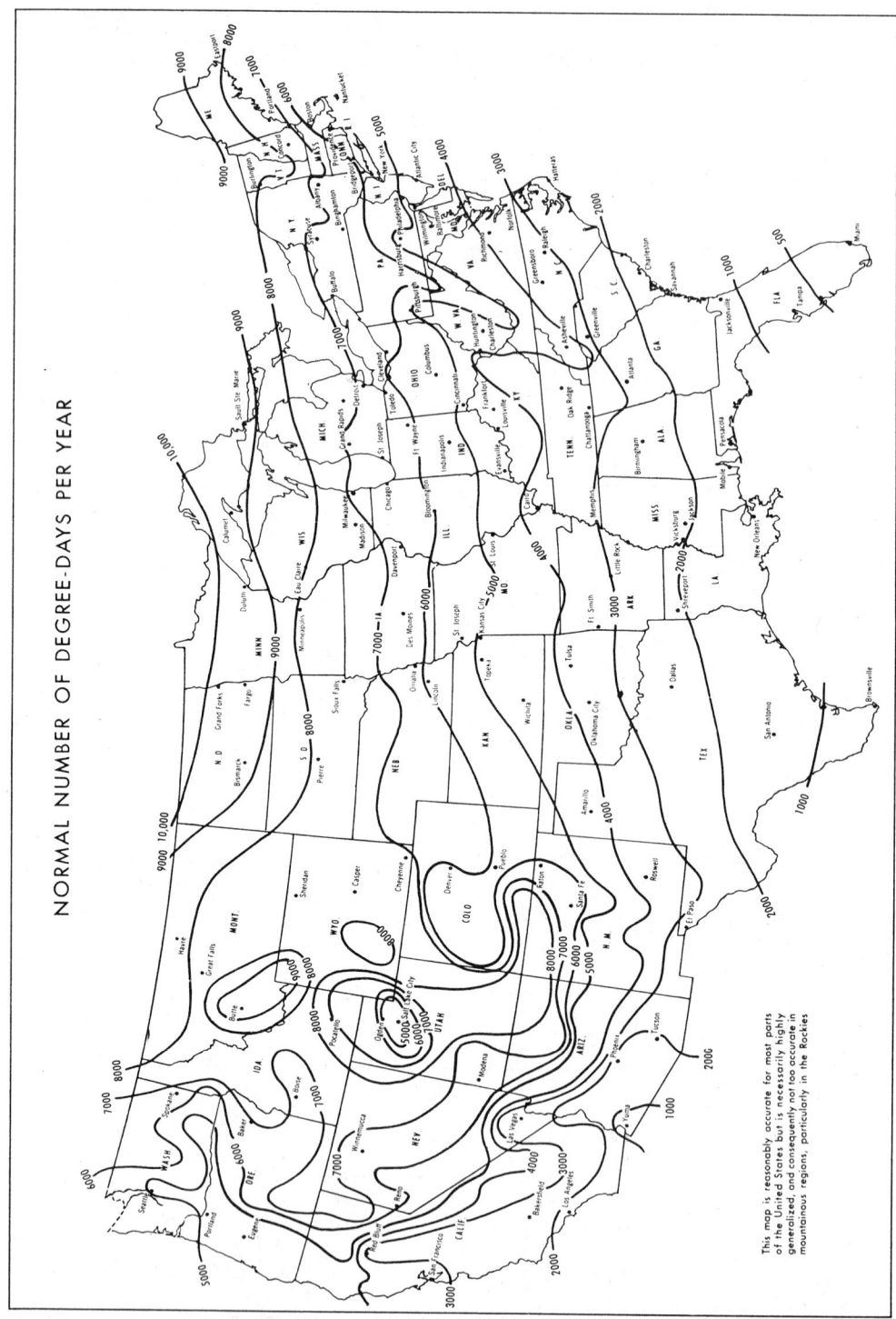

NORMAL NUMBER OF DEGREE-DAYS PER YEAR

This map is reasonably accurate for most parts of the United States but is necessarily highly generalized, and consequently not too accurate in mountainous regions, particularly in the Rockies

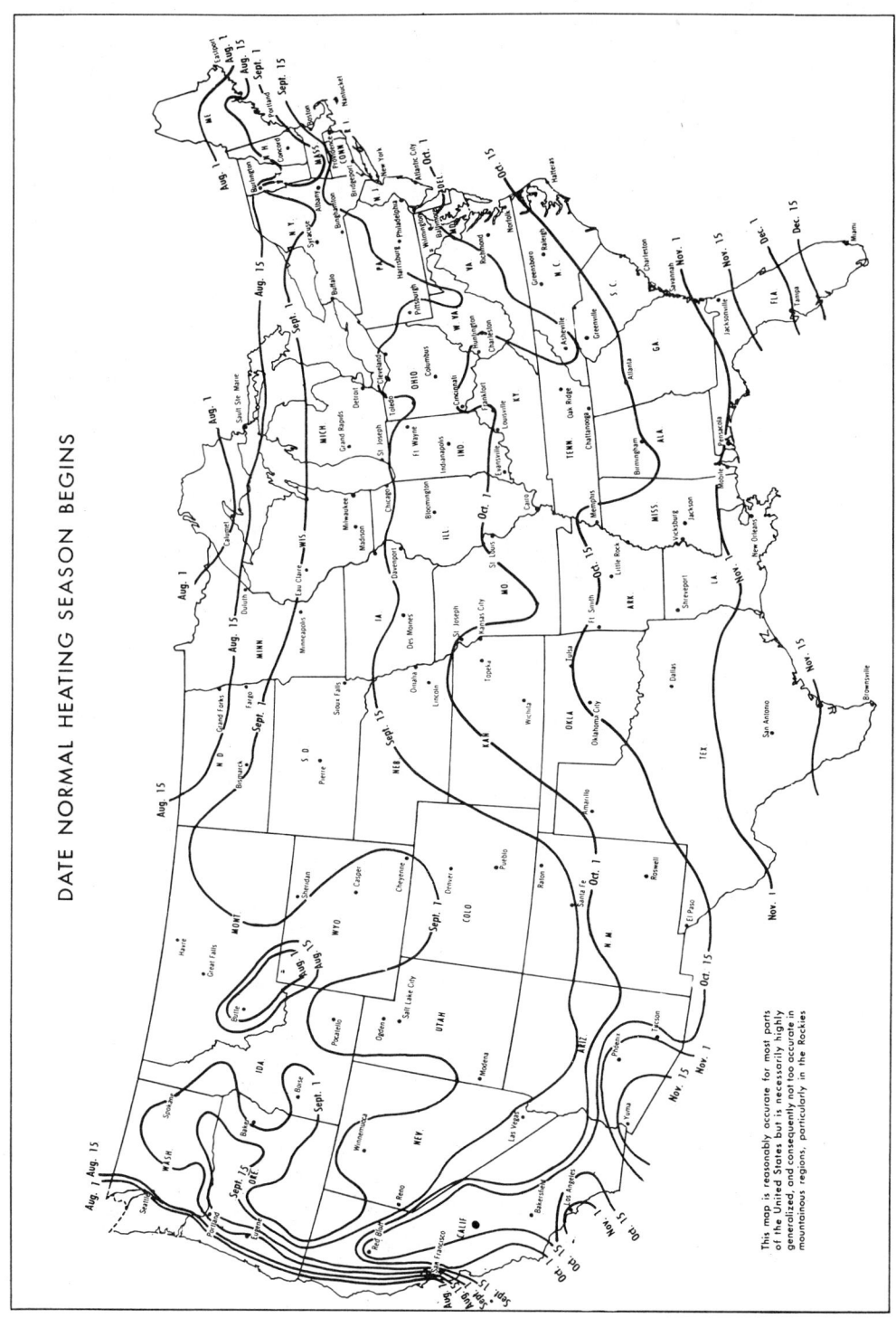

DATE NORMAL HEATING SEASON BEGINS

This map is reasonably accurate for most parts of the United States but is necessarily highly generalized, and consequently not too accurate in mountainous regions, particularly in the Rockies

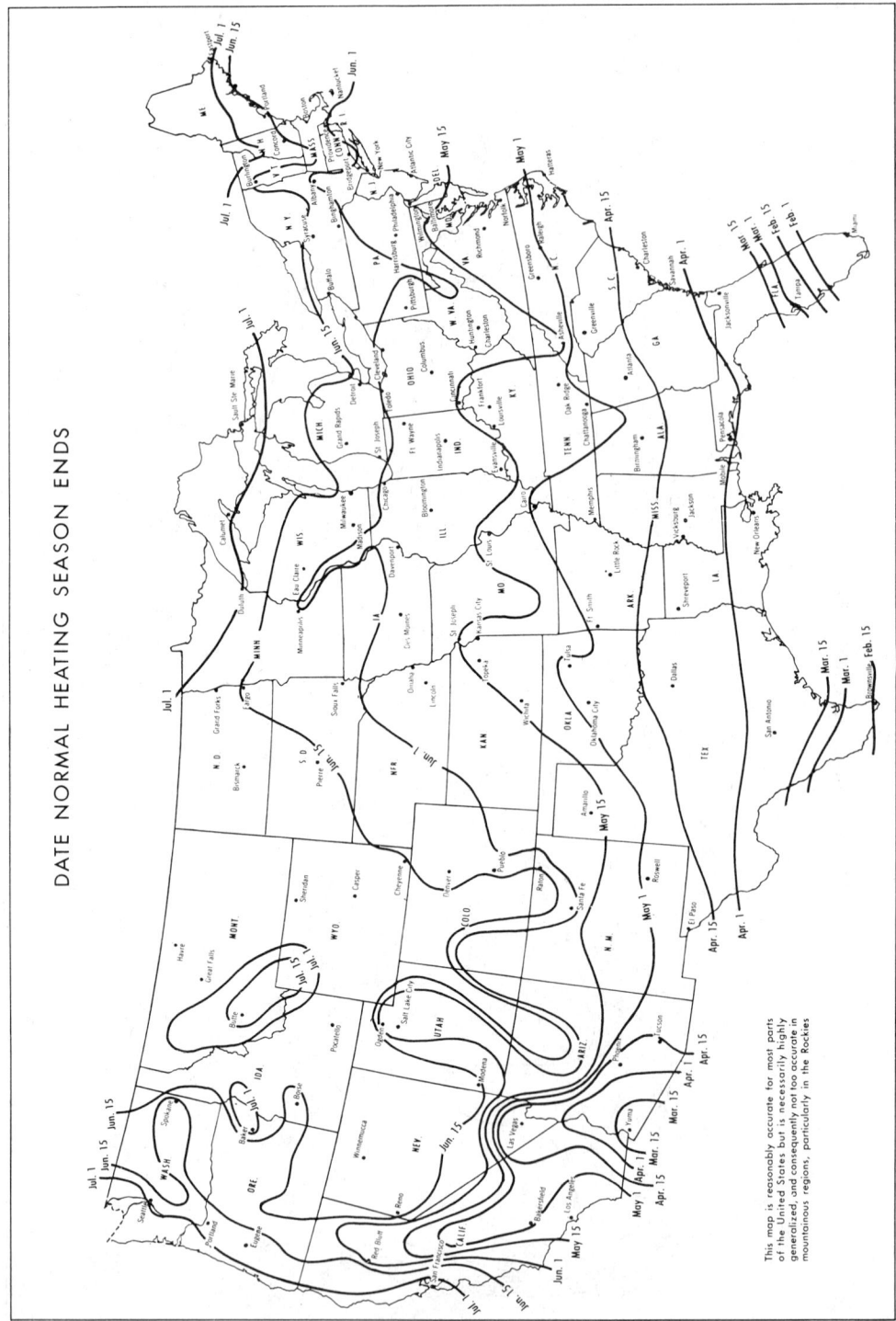

DATE NORMAL HEATING SEASON ENDS

This map is reasonably accurate for most parts of the United States but is necessarily highly generalized, and consequently not too accurate in mountainous regions, particularly in the Rockies

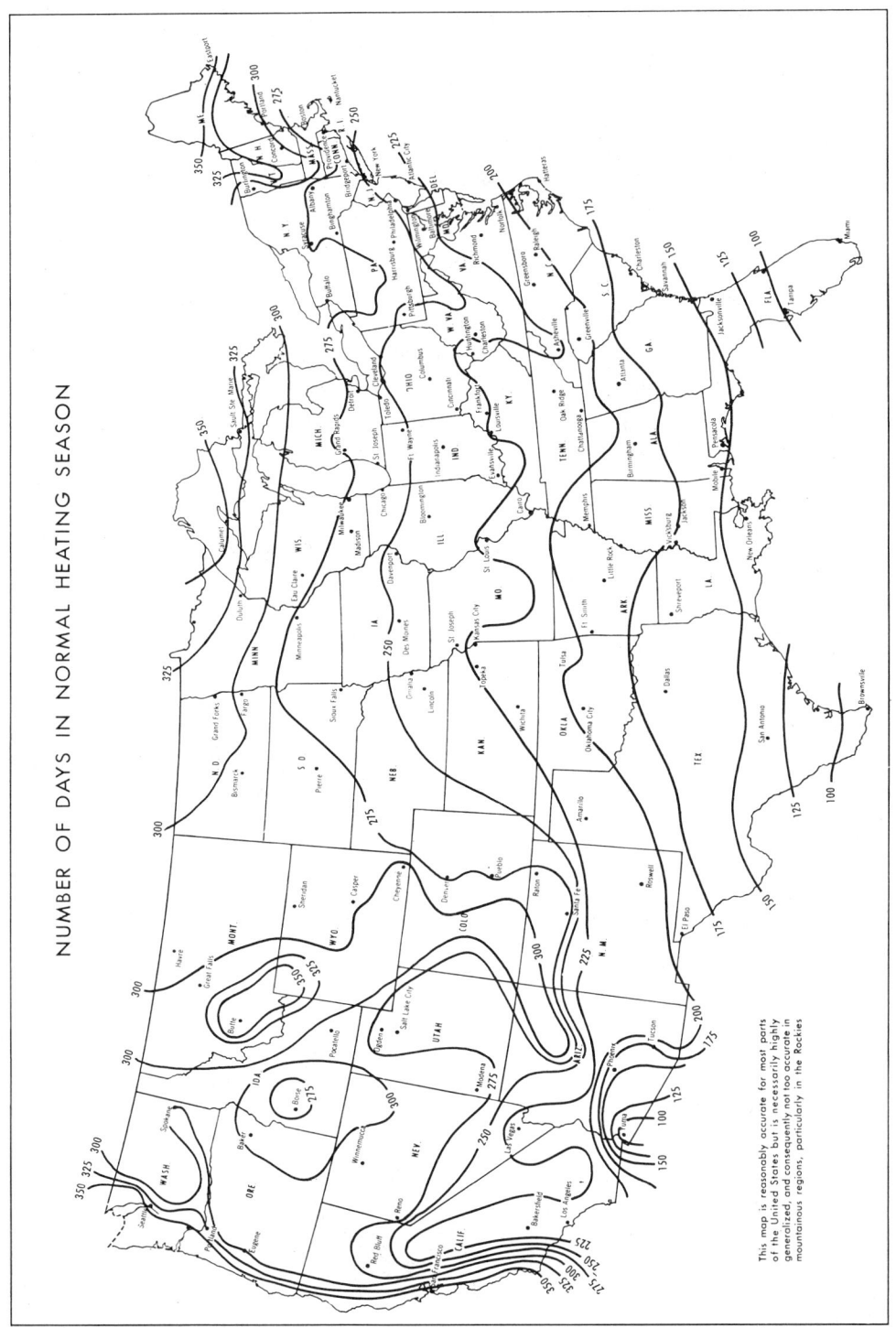

NUMBER OF DAYS IN NORMAL HEATING SEASON

This map is reasonably accurate for most parts of the United States but is necessarily highly generalized, and consequently not too accurate in mountainous regions, particularly in the Rockies

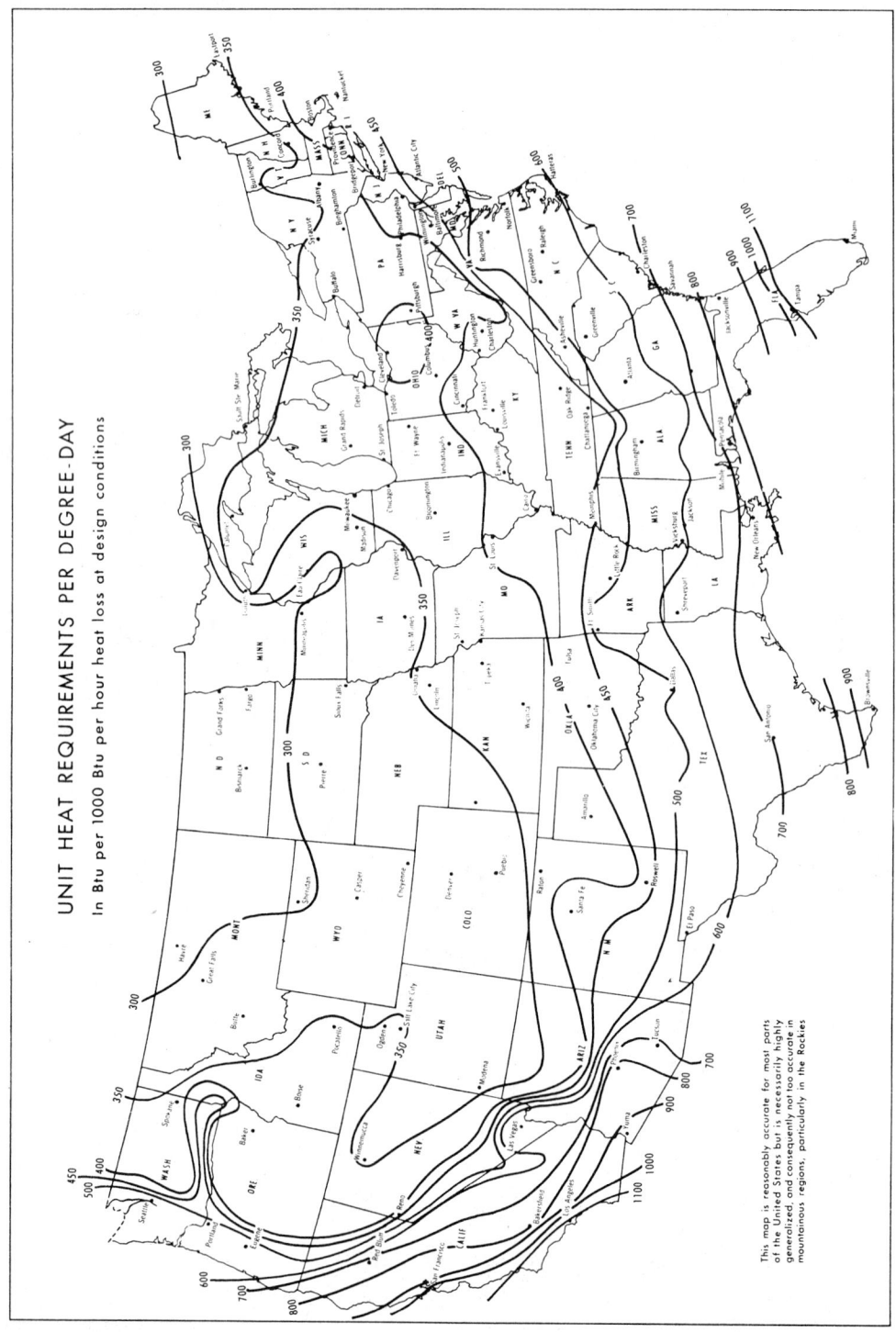

UNIT HEAT REQUIREMENTS PER DEGREE-DAY

In Btu per 1000 Btu per hour heat loss at design conditions

This map is reasonably accurate for most parts of the United States but is necessarily highly generalized, and consequently not too accurate in mountainous regions, particularly in the Rockies

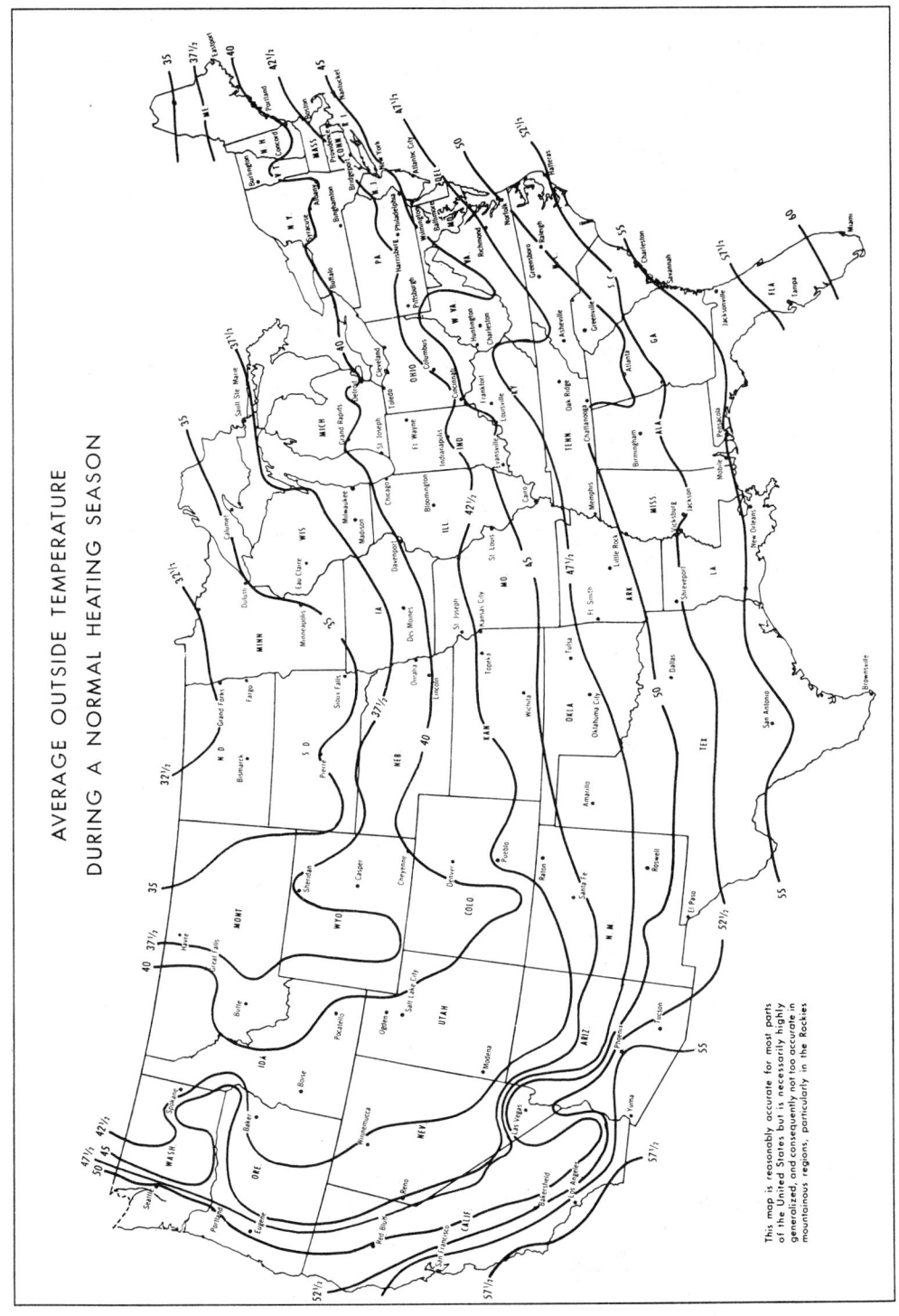

AVERAGE OUTSIDE TEMPERATURE
DURING A NORMAL HEATING SEASON

This map is reasonably accurate for most parts of the United States but is necessarily highly generalized, and consequently not too accurate in mountainous regions, particularly in the Rockies

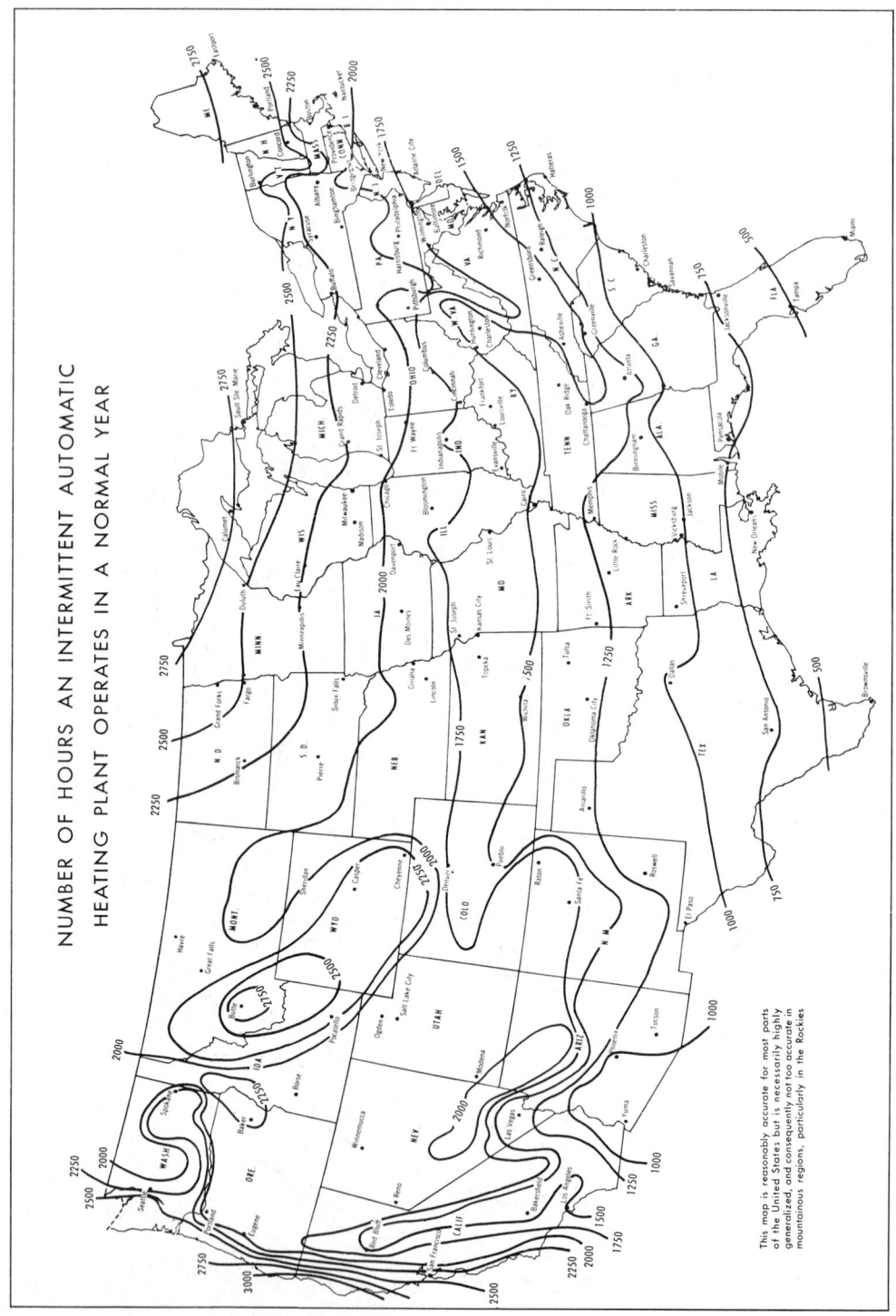

NUMBER OF HOURS AN INTERMITTENT AUTOMATIC
HEATING PLANT OPERATES IN A NORMAL YEAR

This map is reasonably accurate for most parts
of the United States but is necessarily highly
generalized, and consequently not too accurate in
mountainous regions, particularly in the Rockies

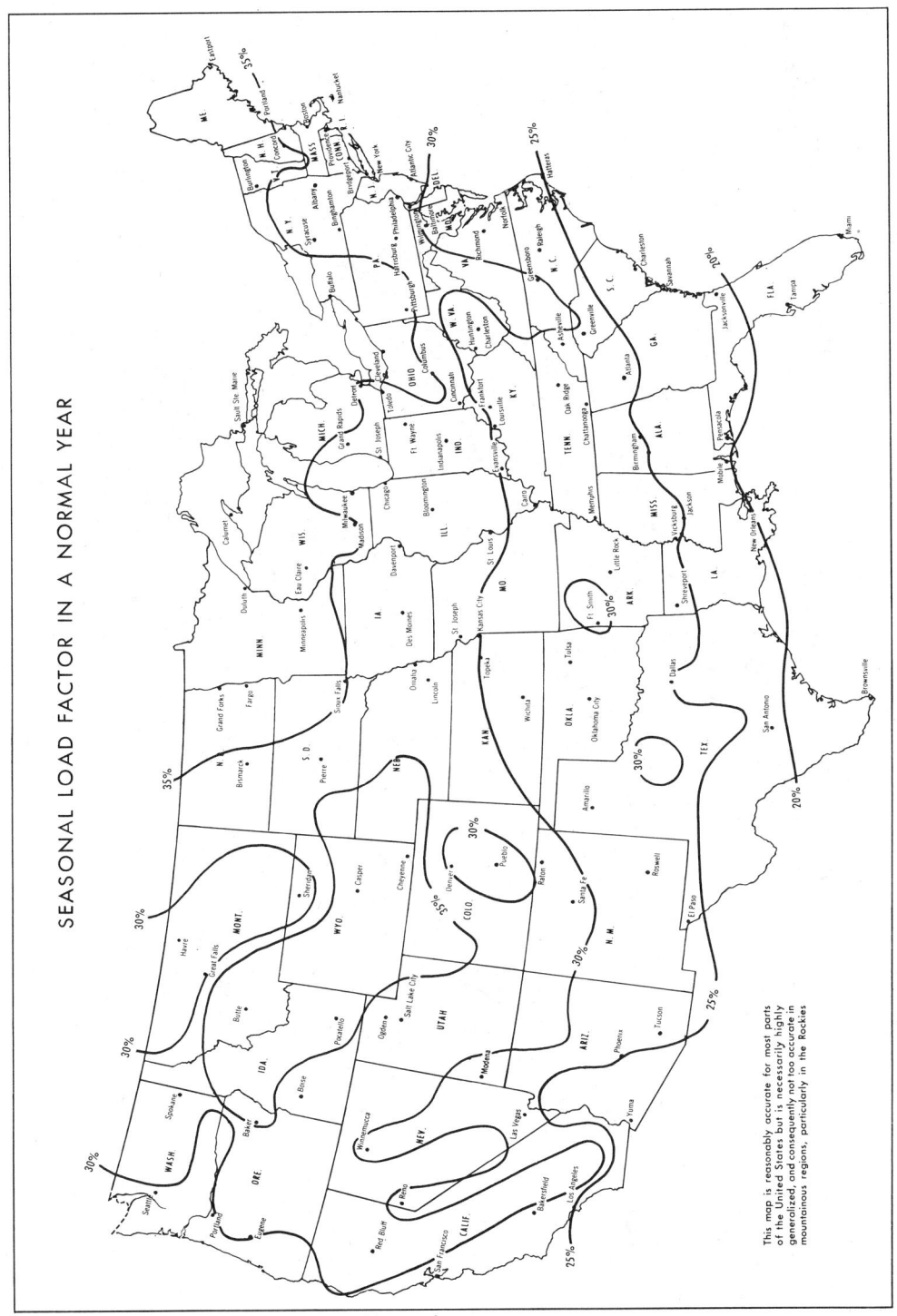

SEASONAL LOAD FACTOR IN A NORMAL YEAR

This map is reasonably accurate for most parts of the United States but is necessarily highly generalized, and consequently not too accurate in mountainous regions, particularly in the Rockies

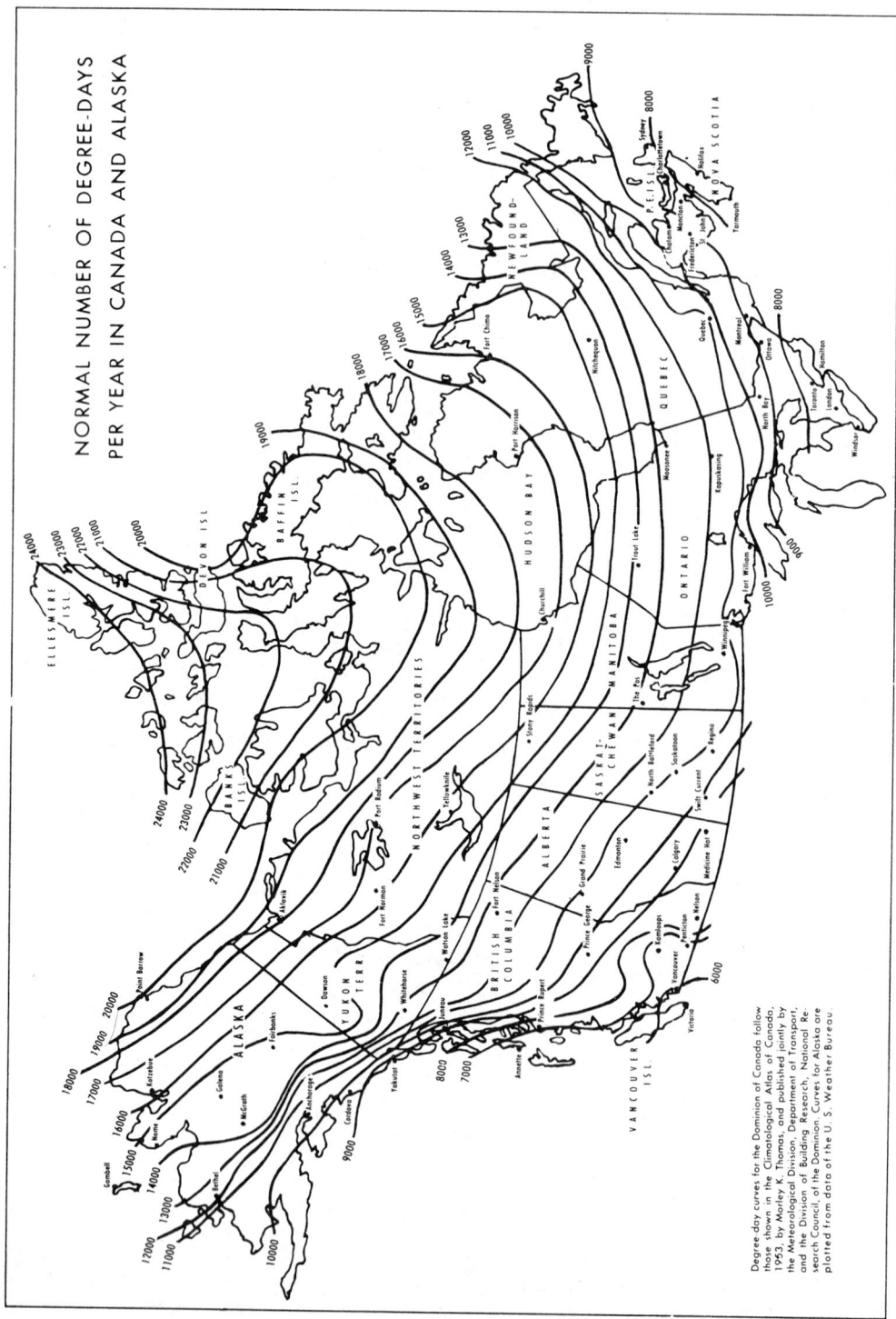

NORMAL NUMBER OF DEGREE-DAYS
PER YEAR IN CANADA AND ALASKA

Degree day curves for the Dominion of Canada follow those shown in the Climatological Atlas of Canada, 1953, by Morley K. Thomas, and published jointly by the Meteorological Division, Department of Transport, and the Division of Building Research, National Research Council of the Dominion. Curves for Alaska are plotted from data of the U. S. Weather Bureau.

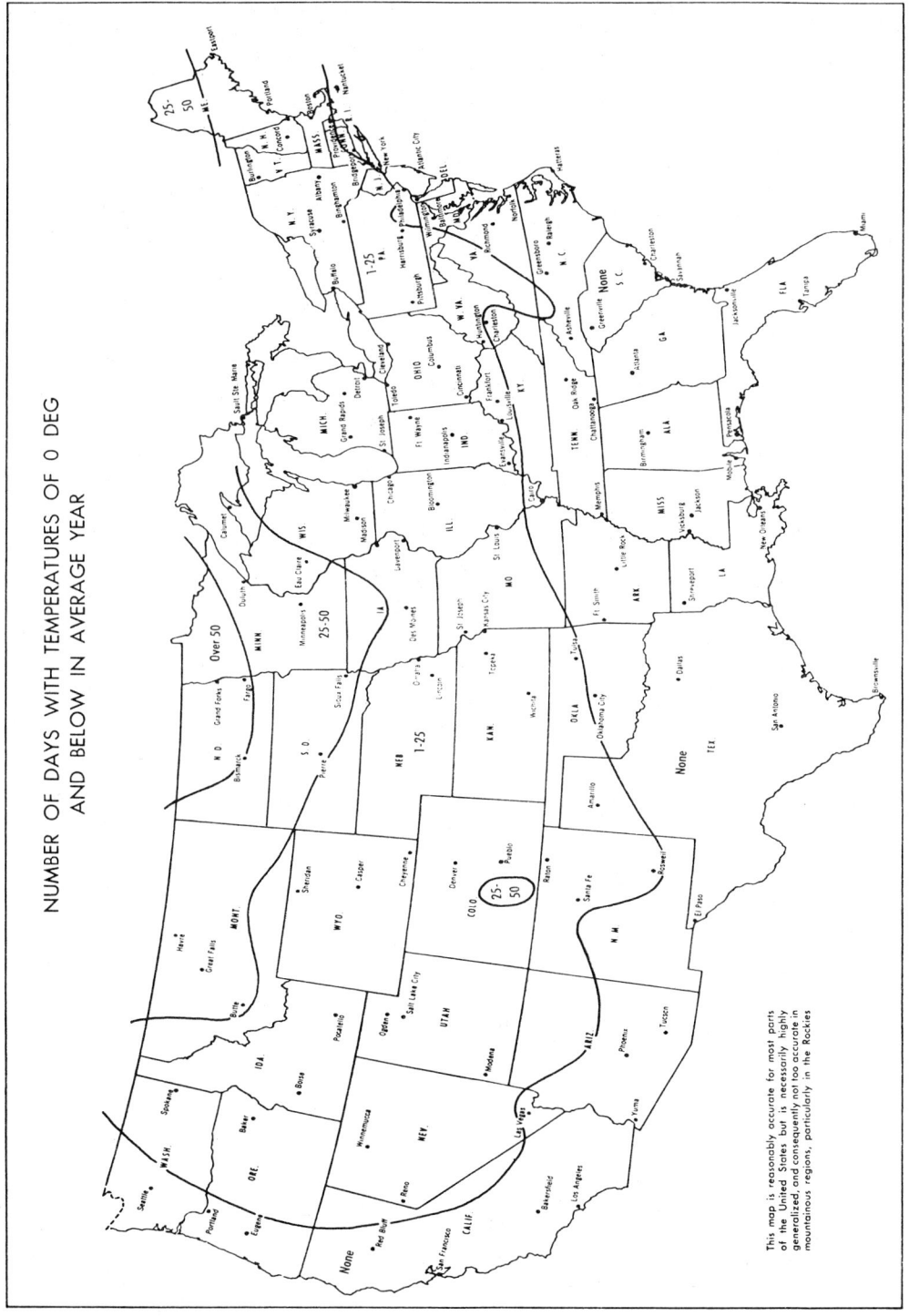

NUMBER OF DAYS WITH TEMPERATURES OF 0 DEG
AND BELOW IN AVERAGE YEAR

This map is reasonably accurate for most parts
of the United States but is necessarily highly
generalized and consequently not too accurate in
mountainous regions, particularly in the Rockies

DATA ON NORMAL HEATING SEASON IN U. S. CITIES

City	Average Date Season Begins	Average Date Season Ends	Days per Season	Average Temp., Deg. F.	K, Heat Required, Btu *	No. of Degree-Days
ALABAMA						
Anniston–A	Oct 15	Apr 24	192	50.3	494.6	2820
Birmingham–A	Oct 15	Apr 25	193	50.6	538.8	2780
Gadsden†	Oct 10	May 1	204	50.8	463.4	2903
Mobile–A	Oct 29	Apr 5	159	54.9	651.9	1612
Mobile–C	Nov 2	Apr 1	151	54.9	651.9	1529
Montgomery–A	Oct 21	Apr 14	176	52.9	564.8	2137
Montgomery–C	Oct 25	Apr 10	168	53.4	572.0	1954
Tuscaloosa†	Oct 14	Apr 22	191	51.8	508.5	2529
ARIZONA						
Flagstaff–A	Aug 12	Jul 5	328	42.1	365.4	7525
Nogales†	Oct 10	May 11	214	53.2	684.2	2516
Phoenix–A	Oct 30	Apr 3	155	54.0	808.1	1698
Phoenix–C	Nov 5	Apr 2	149	55.0	799.5	1492
Prescott–A	Sept 29	May 29	243	46.3	380.4	4533
Tucson–A	Oct 29	Apr 11	165	54.2	781.3	1776
Winslow–A	Sept 29	May 25	239	45.3	370.3	4702
Yuma–A	Nov 15	Mar 7	113	56.6	956.5	951
ARKANSAS						
Fort Smith–A	Oct 16	Apr 20	187	48.0	517.3	3188
Little Rock–A	Oct 17	Apr 21	187	49.1	484.9	2982
Texarkana–A	Oct 24	Apr 12	171	51.2	544.8	2362
CALIFORNIA						
Bakersfield–A	Oct 23	Apr 21	181	53.3	761.5	2115
Beaumont–C	Oct 12	May 29	230	52.7	843.0	2840
Bishop–A	Sept 29	May 23	239	47.3	615.8	4222
Burbank–A	Oct 20	May 18	211	56.4	—	1808
Eureka–C	—	—	365	52.3	837.5	4632
Fresno–A	Oct 16	Apr 25	192	51.8	735.5	2532
Independence†	Sept 25	May 23	241	49.1	—	3832
Los Angeles–A	Oct 11	Jun 13	246	56.8	1103.9	2015
Los Angeles–C	Nov 2	May 10	190	57.4	1134.8	1451
Montague†	Sept 6	Jun 18	286	46.0	—	5423
Mount Shasta–C	Aug 28	Jun 29	306	45.7	—	5913
Oakland–A	—	—	365	56.3	948.0	3163
Pasadena†	Oct 7	Jun 8	245	56.6	—	2067
Red Bluff–A	Oct 19	May 3	197	52.1	739.7	2546
Sacramento–A	Oct 13	May 16	216	51.9	829.8	2822
Sacramento–C	Oct 16	May 14	211	52.7	843.5	2600
San Diego–A	Oct 22	Jun 2	224	58.0	1173.9	1574
San Francisco–A	—	—	365	55.6	1053.1	3421
San Francisco–C	—	—	365	56.6	1095.7	3069
San Jose–C	Oct 7	Jun 7	244	55.1	803.0	2410
San Luis Obispo†	Sept 7	Aug 5	333	57.8	—	2382
Santa Catalina–A	Oct 11	Jun 27	260	56.3	841.5	2249
Santa Maria–A	—	—	365	57.0	1112.2	2934

A–Airport C–City station †Data not entirely comparable *per (degree-day) (Mbtu design heat loss)

City	Average Date Season Begins	Average Date Season Ends	Days per Season	Average Temp., Deg. F.	K, Heat Required, Btu	No. of Degree-Days
COLORADO						
Alamosa–A	Aug 2	Jul 10	343	39.8	—	8659
Boulder†	Sept 9	Jun 12	277	45.1	315.9	5535
Colorado Springs–A	Sept 8	Jun 10	276	41.4	308.5	6254
Denver–A	Sept 12	Jun 9	271	42.4	325.6	6132
Denver–C	Sept 15	Jun 7	266	43.7	329.1	5673
Grand Junction–A	Sept 25	May 23	241	41.0	341.1	5796
Pueblo–A	Sept 17	May 31	257	42.8	309.6	5709
CONNECTICUT						
Bridgeport–A	Sept 19	Jun 5	260	42.3	390.6	5896
Hartford–A	Sept 12	Jun 2	264	41.7	388.7	6139
New Haven–A	Sept 14	Jun 8	268	42.5	391.0	6026
DELAWARE						
Dover†	Sept 25	May 20	238	45.7	431.8	4591
Milford†	Sept 28	May 17	232	46.5	435.7	4303
Wilmington–A	Sept 27	May 19	235	44.1	424.9	4910
DISTRICT OF COLUMBIA						
Washington–A	Oct 2	May 14	225	45.7	432.0	4333
Washington–C	Oct 2	May 12	223	45.9	432.7	4258
Silver Hill Obs.	Sept 28	May 17	232	45.4	430.6	4539
FLORIDA						
Apalachicola–C	Nov 4	Apr 1	149	56.2	837.2	1307
Daytona Beach–A	Nov 19	Mar 20	122	57.9	—	868
Fort Myers–A	Jan 22	Jan 24	3	—	719.4	405
Jacksonville–A	Nov 5	Mar 25	141	56.2	835.7	1243
Jacksonville–C	Nov 10	Mar 21	132	56.6	849.5	1113
Lakeland–C	Nov 30	Feb 24	87	57.5	—	649
Melbourne–A	Dec 13	Feb 15	65	56.7	—	537
Miami–A	—	—	0	—	—	178
Miami–C	Dec 4	Mar 13	100	63.3	—	173
Miami Beach	—	—	0	—	—	123
Orlando–A	Nov 28	May 11	165	61.1	1555.9	650
Pensacola–C	Nov 3	Apr 2	151	55.5	732.4	1435
Tallahassee–A	Oct 27	Apr 1	157	55.3	809.0	1519
Tampa–A	Dec 1	Mar 3	93	57.8	1014.0	674
West Palm Beach–A	—	—	0	—	—	248
GEORGIA						
Albany–A	Oct 28	Apr 5	160	55.0	—	1763
Athens–A	Oct 15	Apr 23	191	50.3	536.3	2800
Atlanta–A	Oct 15	Apr 25	193	50.4	536.7	2826
Atlanta–C	Oct 14	Apr 25	194	50.5	538.0	2811
Augusta–A	Oct 23	Apr 12	172	52.6	560.8	2138
Columbus–A	Oct 20	Apr 17	180	51.7	550.3	2396
Macon–A	Oct 23	Apr 9	169	52.9	616.1	2049
Rome–A	Oct 12	Apr 28	199	49.2	526.8	3138
Savannah–A	Oct 28	Apr 9	164	54.6	710.4	1710
Valdosta–A	Oct 30	Apr 1	154	55.1	—	1525

Data on Normal Heating Season in U.S. Cities (continued)

City	Average Date Season Begins	Average Date Season Ends	Days per Season	Average Temp., Deg. F.	K, Heat Required, Btu	No. of Degree-Days
IDAHO						
Boise–A	Sept 9	Jun 12	277	43.7	370.5	5890
Lewiston–A	Sept 13	Jun 8	269	44.6	459.6	5483
Pocatello–A	Sept 4	Jun 20	290	40.9	386.4	6976
Salmon	Aug 24	Jun 26	307	39.2	—	7922
ILLINOIS						
Aurora†	Sept 10	Jun 2	266	40.0	359.9	6661
Bloomington†	Sept 22	May 21	242	41.5	363.8	5693
Cairo–C	Oct 11	Apr 29	201	46.3	434.4	3756
Chicago–A	Sept 21	May 29	251	39.9	359.8	6310
Danville†	Sept 23	May 17	237	42.3	390.3	5391
Decatur†	Sept 22	May 21	242	42.3	366.1	5485
Joliet–A	Sept 16	May 30	257	39.4	358.5	6578
Moline–A	Sept 20	May 25	248	39.3	358.5	6364
Peoria–A	Sept 22	May 27	248	40.5	361.3	6087
Rockford†	Sept 10	Jun 4	268	39.5	358.8	6847
Springfield–A	Sept 24	May 22	241	41.4	366.0	5693
Springfield–C	Oct 1	May 16	228	42.1	365.4	5225
INDIANA						
Evansville–A	Oct 4	Jun 15	255	47.9	443.0	4360
Fort Wayne–A	Sept 18	May 30	255	40.3	360.9	6287
Indianapolis–A	Sept 25	May 24	242	41.8	364.8	5611
Indianapolis–C	Oct 2	May 18	229	42.6	366.9	5134
South Bend–A	Sept 18	Jun 1	257	39.6	383.0	6524
Terre Haute–A	Sept 28	May 20	235	42.2	390.0	5366
IOWA						
Burlington–A	Sept 23	May 23	243	39.9	359.8	6101
Charles City–C	Sept 10	May 31	264	36.6	—	7504
Davenport–C	Sept 24	May 21	240	39.6	338.0	6091
Des Moines–A	Sept 22	May 26	247	38.9	336.6	6446
Des Moines–C	Sept 23	May 23	243	39.2	337.2	6274
Dubuque–A	Sept 10	Jun 4	268	37.9	315.8	7271
Marshalltown†	Sept 12	Jun 1	263	39.0	336.6	6853
Sioux City–A	Sept 16	May 26	253	37.3	314.7	7012
KANSAS						
Concordia–C	Oct 2	May 19	230	41.9	364.8	5323
Dodge City–A	Oct 2	May 20	231	43.1	368.4	5058
Goodland–A	Sept 16	Jun 3	261	40.6	—	6367
Salina†	Sept 29	May 18	232	43.6	370.0	4969
Topeka–A	Sept 30	May 18	231	42.5	366.6	5209
Topeka–C	Oct 4	May 13	222	42.8	367.8	4919
Wichita–A	Oct 7	May 14	220	44.2	372.3	4571
KENTUCKY						
Bowling Green–A	Oct 4	May 16	225	46.0	433.0	4279
Frankfort†	Sept 27	May 13	229	46.6	436.3	4211
Lexington–A	Oct 1	May 21	233	43.6	423.1	4979
Louisville–A	Oct 3	May 12	222	45.0	428.6	4439
Louisville–C	Oct 6	May 10	217	45.3	430.0	4279

City	Average Date Season Begins	Average Date Season Ends	Days per Season	Average Temp., Deg. F.	K, Heat Required, Btu	No. of Degree-Days
LOUISIANA						
Baton Rouge–A	Oct 30	Apr 4	157	54.8	716.2	1595
Burrwood	Nov 21	Mar 28	128	56.9	—	1033
Lake Charles–A	Nov 1	Mar 31	151	54.8	—	1543
New Orleans–A	Nov 5	Mar 29	145	55.9	744.0	1317
New Orleans–C	Nov 10	Mar 24	135	56.3	755.6	1175
Shreveport–A	Oct 26	Apr 6	163	52.0	710.0	2117
MAINE						
Bar Harbor†	Aug 10	Jun 15	310	40.4	360.9	7643
Caribou–A	Aug 3	Jul 12	344	35.4	—	10173
Eastport–C	—	—	365	42.4	366.3	8246
Orono†	Aug 24	Jun 27	308	39.3	341.4	7934
Portland–A	Aug 27	Jun 25	303	39.7	383.0	7681
MARYLAND						
Annapolis†	Sept 24	May 17	236	46.0	466.3	4483
Baltimore–A	Sept 28	May 19	234	45.3	430.0	4611
Baltimore–C	Sept 8	May 13	248	48.1	440.0	4203
Cambridge†	Oct 1	May 13	225	46.1	466.9	4245
Frederick–A	Sept 29	May 16	230	43.9	395.8	4854
MASSACHUSETTS						
Boston–A	Sept 17	Jun 5	262	42.9	420.3	5791
Fitchburg†	Sept 6	Jun 16	284	41.7	416.4	6632
Nantucket–A	Sept 6	Jun 29	297	44.5	426.1	6102
Pittsfield–A	Aug 24	Jun 20	301	39.4	358.8	7694
MICHIGAN						
Alpena–C	Aug 25	Jun 25	305	38.5	356.8	8073
Ann Arbor†	Sept 6	Jun 8	276	40.1	384.2	6877
Calumet†	—	—	365	39.5	337.8	9297
Detroit–A	Sept 17	Jun 2	259	40.3	360.6	6404
Escanaba–C	Aug 22	Jun 27	310	37.1	332.9	8657
Flint†	Sept 1	Jun 13	286	39.9	359.8	7179
Grand Haven†	Aug 31	Jun 19	293	41.4	—	6915
Grand Rapids–A	Sept 9	Jun 6	271	38.9	357.6	7075
Grand Rapids–C	Sept 16	Jun 2	260	40.1	360.3	6474
Houghton†	Aug 9	Jul 6	332	38.0	—	8964
Kalamazoo†	Sept 8	Jun 6	272	40.5	385.6	6655
Lansing–A	Sept 9	Jun 3	268	38.9	357.3	6982
Marquette–C	Aug 24	Jul 1	312	37.7	354.9	8529
Muskegon–A	Sept 6	Jun 13	281	39.8	359.4	7089
Port Huron†	Sept 4	Jun 17	287	39.9	383.8	7195
Saginaw†	Sept 6	Jun 7	275	39.3	358.4	7063
Sault Ste. Marie–A	Aug 8	Jul 15	342	37.3	314.6	9475
St. Joseph†	Sept 12	Jun 7	269	42.0	365.0	6203
Willow Run–A	Sept 17	May 31	257	39.8	—	6469

Data on Normal Heating Season in U.S. Cities (continued)

City	Average Date Season Begins	Average Date Season Ends	Days per Season	Average Temp., Deg. F.	K, Heat Required, Btu	No. of Degree-Days
MINNESOTA						
Duluth–A	Aug 21	Jul 1	315	33.5	292.6	9937
Duluth–C	Aug 21	Jul 5	319	35.3	295.2	9474
International Falls–A	Aug 12	Jul 3	326	32.5	—	10600
Minneapolis–A	Sept 11	May 31	263	35.1	311.2	7853
Moorhead†	Aug 23	Jun 16	298	34.0	—	9315
Rochester–A	Sept 5	Jun 5	274	35.5	311.7	8095
St. Cloud–A	Aug 30	Jun 11	286	33.9	293.4	8893
St. Paul–A	Sept 11	Jun 1	264	35.4	311.7	7804
MISSISSIPPI						
Jackson–A	Oct 21	Apr 13	175	52.4	609.6	2202
Meridian–A	Oct 17	Apr 14	180	52.0	554.3	2333
Vicksburg–C	Oct 26	Apr 8	165	52.9	565.2	2000
MISSOURI						
Columbia–A	Sept 29	May 17	231	42.9	367.8	5113
Kansas City–A	Oct 5	May 12	220	42.8	365.5	4888
St. Joseph–A	Oct 2	May 15	226	41.4	363.6	5336
St. Louis–A	Oct 4	May 13	222	43.8	423.8	4699
St. Louis–C	Oct 8	May 10	215	44.2	425.6	4469
Springfield–A	Oct 1	May 19	231	44.7	373.8	4693
MONTANA						
Billings–A	Sept 4	Jun 16	286	40.2	303.5	7106
Butte–A	—	—	365	38.3	317.9	9760
Glasgow–C	Aug 31	Jun 16	290	35.0	—	8690
Great Falls–A	Aug 26	Jun 28	307	40.4	320.0	7555
Havre–C	Aug 29	Jun 22	298	37.4	283.4	8213
Helena–A	Aug 24	Jun 30	311	38.5	316.7	8250
Helena–C	Aug 20	Jul 3	318	39.4	318.6	8126
Kalispell–A	Aug 17	Jul 5	323	40.1	320.0	8055
Miles City–A	Sept 7	Jun 13	280	37.1	269.2	7822
Missoula–A	Aug 24	Jul 1	312	39.8	—	7873
NEBRASKA						
Grand Island–A	Sept 23	May 26	246	39.3	337.2	6311
Lincoln–A	Sept 23	May 25	245	40.1	358.3	6104
Lincoln–C	Sept 27	May 21	237	40.3	360.6	5865
Norfolk–A	Sept 17	May 31	257	37.5	333.4	7065
North Platte–A	Sept 15	Jun 1	260	39.8	319.4	6546
Omaha–A	Sept 24	May 20	239	39.2	357.9	6160
Scotts Bluff–A	Sept 11	Jun 9	272	39.8	—	6841
Valentine–C	Sept 12	Jun 5	267	38.5	300.2	7075
NEVADA						
Elko–A	Aug 29	Jun 27	303	40.8	—	7335
Ely–A	Aug 26	Jun 25	304	40.5	—	7443
Las Vegas–C	Sept 18	Apr 11	206	53.2	683.6	2425
Reno–A	Sept 4	Jun 26	296	44.6	398.3	6036
Tonopah	Sept 17	Jun 12	269	43.4	454.3	5813
Winnemucca–A	Sept 4	Jun 17	287	42.8	345.9	6369

City	Average Date Season Begins	Average Date Season Ends	Days per Season	Average Temp., Deg. F.	K, Heat Required, Btu	No. of Degree-Days
NEW HAMPSHIRE						
Concord–A	Aug 29	Jun 15	291	38.8	336.2	7612
Hanover†	Aug 21	Jun 25	309	39.1	318.1	8005
Keene†	Aug 23	Jun 25	307	40.7	321.5	7458
NEW JERSEY						
Asbury Park†	Sept 20	Jun 5	259	44.7	459.9	5272
Atlantic City–C	Sept 29	May 30	244	45.6	464.0	4741
Belvidere†	Sept 14	May 30	259	42.6	367.1	5792
Bridgeton†	Sept 25	May 18	236	45.3	462.6	4661
Dover†	Sept 4	Jun 6	276	42.3	418.4	6270
Elizabeth†	Sept 20	May 24	247	43.5	422.8	5302
Jersey City†	Sept 22	May 25	246	43.9	424.1	5193
Long Branch†	Sept 18	Jun 10	266	44.7	460.2	5396
Newark–A	Sept 25	May 24	242	43.3	421.9	5252
New Brunswick†	Sept 17	May 30	256	43.9	424.1	5404
Paterson†	Sept 18	May 27	252	43.7	423.6	5365
Phillipsburg†	Sept 16	May 27	254	42.6	391.2	5694
Plainfield†	Sept 16	Jun 1	259	43.6	423.3	5535
Sandy Hook†	Sept 23	Jun 4	254	43.5	—	5470
Somerville†	Sept 18	May 29	254	43.1	392.7	5586
South Orange†	Sept 12	Jun 1	263	43.2	393.1	5745
Trenton–C	Sept 25	May 21	239	43.8	423.4	5068
NEW MEXICO						
Albuquerque–A	Oct 3	May 12	222	45.2	429.3	4389
Clayton–A	Sept 21	May 29	251	44.5	—	5138
Raton–A	Sept 4	Jun 15	285	42.5	391.0	6417
Roswell–A	Oct 8	May 3	208	48.5	390.9	3424
Santa Fe†	Sept 1	Jun 15	287	43.9	424.1	6063
NEW YORK						
Albany–A	Sept 9	Jun 3	268	39.0	357.6	6962
Albany–C	Sept 15	May 28	256	40.3	360.6	6319
Bear Mountain–C	Sept 10	Jun 6	270	40.9	—	6511
Binghamton–A	Aug 30	Jun 16	291	39.1	357.9	7537
Binghamton–C	Sept 10	Jun 4	268	40.5	361.3	6556
Buffalo–A	Sept 10	Jun 9	273	40.0	383.6	6838
Cortland†	Aug 19	Jun 25	311	41.3	363.1	7388
Elmira†	Sept 6	Jun 6	274	41.6	364.1	6412
Glens Falls†	Sept 1	Jun 9	282	39.3	358.4	7244
Ithaca†	Sept 4	Jun 9	279	40.9	362.3	6719
Jamestown†	Sept 3	Jun 17	288	41.6	364.1	6740
Little Falls†	Aug 27	Jun 22	300	39.7	319.2	7604
New York–LaGuard.–A	Sept 30	May 24	237	43.9	424.1	4989
New York–C	Sept 28	May 26	241	44.0	424.7	5050
New York–Central Park	Sept 30	May 22	235	43.9	423.7	4965
Ogdensburg†	Sept 2	Jun 17	220	30.1	301.3	7688
Oswego–C	Sept 9	Jun 15	280	40.1	360.3	6975
Port Jervis†	Sept 9	Jun 5	270	41.4	387.5	6388
Rochester–A	Sept 9	Jun 6	271	39.7	383.0	6863
Schenectady–C	Sept 10	Jun 1	265	38.4	356.1	7050
Syracuse–A	Sept 13	Jun 2	263	40.2	360.3	6520
Watertown†	Sept 2	Jun 13	285	39.4	337.5	7298

Data on Normal Heating Season in U.S. Cities (continued)

City	Average Date Season Begins	Average Date Season Ends	Days per Season	Average Temp., Deg.. F.	K, Heat Required, Btu	No. of Degree-Days
NORTH CAROLINA						
Asheville–C	Sept 29	May 19	233	47.5	440.9	4072
Charlotte–A	Oct 9	Apr 30	204	49.3	527.2	3205
Greensboro–A	Oct 4	May 9	218	47.5	514.3	3810
Hatteras–C	Oct 23	May 3	193	52.6	—	2392
Raleigh–A	Oct 8	May 6	211	49.0	525.2	3369
Raleigh–C	Oct 12	May 1	202	49.8	531.2	3075
Salisbury†	Oct 8	May 2	212	49.8	531.8	3222
Wilmington–A	Oct 19	Apr 24	188	52.6	612.7	2323
Winston-Salem–A	Oct 5	May 8	216	47.8	516.0	3721
NORTH DAKOTA						
Bismarck–A	Aug 31	Jun 16	290	33.9	278.4	9033
Devils Lake–C	Aug 27	Jun 22	300	31.9	276.0	9940
Fargo–A	Sept 1	Jun 14	287	32.7	291.7	9274
Grand Forks†	Aug 15	Jun 23	313	33.8	293.3	9764
Jamestown†	Aug 20	Jun 23	308	34.5	294.1	9406
Minot†	Aug 16	Jun 25	314	34.5	279.3	9581
Williston–C	Sept 1	Jun 19	292	33.9	265.3	9068
OHIO						
Akron-Canton–A	Sept 19	May 30	254	40.6	385.2	6203
Cincinnati–A	Sept 26	May 20	237	43.1	421.0	5195
Cincinnati–C	Oct 4	May 14	223	44.7	427.1	4532
Cincinnati–Abbe Obs.	Sept 30	May 18	231	43.9	424.1	4870
Cleveland–A	Sept 21	May 29	251	41.1	414.1	6006
Cleveland–C	Sept 23	May 29	249	42.0	417.3	5717
Columbus–A	Sept 23	May 25	245	42.1	365.4	5615
Columbus–C	Sept 27	May 22	238	42.8	367.5	5277
Dayton–A	Sept 25	May 25	243	42.0	417.3	5597
Lancaster†	Sept 17	May 25	251	44.0	396.0	5272
Lima†	Sept 13	Jun 1	262	42.4	390.6	5935
Marion†	Sept 19	May 27	251	42.4	366.3	5685
Sandusky–C	Sept 23	May 29	249	41.5	415.6	5859
Toledo–A	Sept 16	May 31	258	40.2	360.3	6394
Warren†	Sept 10	Jun 11	275	42.8	467.4	6117
Youngstown–A	Sept 18	Jun 1	257	41.0	386.4	6172
OKLAHOMA						
Muskogee†	Oct 8	May 6	211	47.9	443.0	3616
Oklahoma City–A	Oct 15	May 1	199	46.7	436.4	3644
Oklahoma City–C	Oct 16	Apr 27	194	46.9	437.1	3519
Tulsa–A	Oct 15	Apr 27	195	46.6	436.1	3584
OREGON						
Astoria–A	—	—	365	51.3	—	4995
Baker–C	Aug 25	Jul 4	314	42.4	390.6	7087
Burns–C	Aug 29	Jun 26	302	42.1	—	6918
Eugene–A	Sept 2	Jun 28	300	49.1	572.9	4779
Meacham–A	Aug 6	Jul 16	345	42.1	—	7888
Medford–A	Sept 17	Jun 13	270	48.2	478.3	4547

City	Average Date Season Begins	Average Date Season Ends	Days per Season	Average Temp., Deg. F.	K, Heat Required, Btu	No. of Degree-Days
OREGON (Concluded)						
Pendleton–A	Sept 18	Jun 4	260	45.0	352.7	5204
Portland–A	Sept 7	Jun 26	293	49.2	526.3	4632
Portland–C	Sept 16	Jun 18	276	50.0	532.2	4143
Roseburg–C	Sept 13	Jun 19	280	50.3	519.0	4122
Salem–A	Sept 5	Jun 23	292	49.3	575.6	4574
PENNSYLVANIA						
Allentown–A	Sept 18	May 26	251	41.6	415.9	5880
Altoona†	Sept 8	Jun 7	273	42.6	391.2	6121
Coatesville†	Sept 19	May 26	250	43.7	394.4	5365
Erie–C	Sept 20	Jun 4	258	41.3	387.1	6116
Franklin†	Sept 5	Jun 11	280	42.2	344.2	6383
Harrisburg–A	Sept 22	May 20	241	43.2	421.4	5258
Lancaster†	Sept 13	May 27	257	43.7	423.4	5482
Lebanon†	Sept 16	May 28	255	43.1	421.3	5585
New Castle†	Sept 12	May 30	261	42.2	418.1	5958
Philadelphia–A	Sept 28	May 19	234	44.2	425.1	4866
Philadelphia–C	Oct 2	May 17	228	45.2	429.3	4523
Pittsburgh-Gr. Pitt.–A	Sept 19	May 29	253	41.7	416.3	5905
Pittsburgh–C	Sept 28	May 19	234	43.4	422.0	5048
Reading–C	Sept 26	May 21	238	43.7	423.4	5060
Scranton–C	Sept 15	May 30	258	41.6	388.0	6047
Uniontown†	Sept 10	May 24	257	44.9	399.4	5176
Warren†	Sept 2	Jun 18	290	41.8	343.1	6740
Williamsport–A	Sept 18	May 28	253	41.7	388.4	5898
York†	Sept 16	May 26	253	43.5	394.2	5449
RHODE ISLAND						
Block Island–A	Sept 13	Jun 19	280	44.1	424.9	5843
Providence–A	Sept 11	Jun 8	271	42.4	418.6	6125
Providence–C	Sept 18	Jun 1	257	43.2	421.4	5607
SOUTH CAROLINA						
Charleston–A	Sept 27	Apr 15	201	55.2	658.5	1973
Charleston–C	Oct 28	Apr 10	165	54.3	639.6	1769
Columbia–A	Oct 18	Apr 17	182	51.6	549.3	2435
Columbia–C	Oct 19	Apr 17	181	52.4	558.4	2284
Florence–A	Oct 16	Apr 20	187	51.6	598.7	2507
Greenville–A	Oct 12	Apr 30	201	49.8	531.3	3060
Spartanburg–A	Oct 12	Apr 30	201	49.9	580.3	3044
SOUTH DAKOTA						
Huron–A	Sept 12	Jun 3	265	29.8	311.2	7902
Pierre†	Sept 4	Jun 17	287	39.2	—	7420
Rapid City–A	Sept 5	Jun 17	286	26.3	317.1	7535
Sioux Falls–A	Sept 10	Jun 4	268	29.3	312.0	7848
Yankton†	Sept 14	Jun 1	261	37.6	—	7144

Data on Normal Heating Season in U.S. Cities (continued)

City	Average Date Season Begins	Average Date Season Ends	Days per Season	Average Temp., Deg. F.	K, Heat Required, Btu	No. of Degree-Days
TENNESSEE						
Bristol–A	Sept 29	May 16	230	47.0	—	4148
Chattanooga–A	Oct 8	May 4	209	48.8	523.3	3384
Knoxville–A	Oct 8	May 4	209	47.8	442.7	3590
Memphis–A	Oct 15	Apr 23	191	48.6	447.3	3137
Memphis–C	Oct 19	Apr 22	186	48.8	449.0	3006
Nashville–A	Oct 11	May 2	204	47.8	442.4	3513
TEXAS						
Abilene–A	Oct 22	Apr 14	175	49.8	579.9	2657
Amarillo–A	Oct 2	May 17	228	45.9	378.6	4345
Austin–A	Nov 2	Apr 1	151	53.7	691.2	1713
Big Springs–A	Oct 25	Apr 14	172	50.6	538.4	2480
Corpus Christi–A	Nov 15	Mar 15	121	56.6	767.0	1011
Dallas–A	Oct 27	Apr 6	162	51.0	465.0	2272
Del Rio–A	Nov 5	Mar 22	138	54.8	650.1	1407
El Paso–A	Oct 19	Apr 19	183	50.6	538.4	2641
Fort Worth–A	Oct 27	Apr 10	166	50.8	540.4	2361
Galveston–A	Nov 13	Mar 28	136	55.9	744.4	1233
Galveston–C	Nov 13	Mar 29	137	56.2	751.2	1211
Houston–A	Nov 6	Mar 30	145	55.4	730.6	1388
Houston–C	Nov 8	Mar 25	138	55.8	739.2	1276
Lubbock–A	Oct 8	May 2	207	47.7	—	3587
Palestine–C	Oct 30	Apr 7	160	52.6	612.7	1980
Port Arthur–A	Nov 2	Apr 1	151	55.0	718.6	1517
Port Arthur–C	Nov 7	Mar 29	143	55.6	735.8	1340
San Antonio–A	Nov 5	Mar 29	145	54.1	700.4	1579
Victoria–A	Nov 14	Mar 18	125	56.0	—	1126
Waco–A	Oct 30	Apr 5	158	52.2	667.2	2025
Wichita Falls–A	Oct 19	Apr 20	184	48.6	—	3025
UTAH						
Blanding	Sept 11	Jun 10	273	42.5	—	6138
Logan†	Sept 4	Jun 20	290	41.7	343.1	6748
Milford–A	Sept 10	Jun 11	275	41.6	338.0	6445
Modena†	Sept 3	Jun 20	291	42.2	—	6635
Ogden†	Sept 13	May 28	258	49.4	369.9	5538
Salt Lake City–A	Sept 16	Jun 6	264	42.8	367.5	5866
Salt Lake City–C	Sept 20	May 29	252	43.3	369.0	5463
VERMONT						
Burlington–A	Sept 3	Jun 10	281	37.0	353.4	7865
Northfield†	Aug 8	Jul 1	328	38.6	—	8719
VIRGINIA						
Cape Henry–C	Oct 12	May 12	213	49.5	528.8	3307
Charlottesville†	Sept 17	May 12	238	47.3	512.8	4223

City	Average Date Season Begins	Average Date Season Ends	Days per Season	Average Temp., Deg. F.	K, Heat Required, Btu	No. of Degree-Days
VIRGINIA (Concluded)						
Danville†	Oct 8	May 6	211	48.4	520.3	3511
Lynchburg–A	Sept 29	May 12	226	46.6	469.6	4153
Norfolk–A	Oct 8	May 11	216	49.0	572.5	3454
Norfolk–C	Oct 13	May 7	207	49.9	580.7	3119
Richmond–A	Oct 4	May 12	221	47.1	558.1	3955
Richmond–C	Oct 6	May 9	216	47.8	562.9	3720
Roanoke–A	Oct 1	May 14	226	46.6	469.6	4152
Wytheville†	Sept 13	Jun 2	263	45.9	—	5022
WASHINGTON						
Aberdeen†	—	—	365	49.7	530.8	5586
Bellingham†	—	—	365	50.2	534.9	5417
Ellensburg–A	Sept 2	Jun 23	295	42.8	—	6542
Everett†	—	—	365	48.7	522.9	5941
Kelso–A	—	—	365	50.6	—	5239
North Head	—	—	365	50.7	648.0	5211
Olympia–A	—	—	365	49.9	580.7	5501
Seattle–C	Aug 28	Jul 2	309	50.6	588.1	4438
Seattle-Tacoma–A	Aug 10	Jul 18	343	49.6	578.1	5275
Spokane–A	Aug 31	Jun 27	301	42.2	365.8	6852
Tacoma–C	Aug 13	Jul 9	331	50.3	584.7	4866
Tatoosh Island–C	—	—	365	49.3	575.0	5724
Walla Walla–C	Sept 20	May 27	250	45.6	377.1	4848
Yakima–A	Sept 6	Jun 9	277	43.9	395.4	5845
WEST VIRGINIA						
Charleston–A	Oct 1	May 17	229	45.7	431.8	4417
Elkins–A	Sept 12	Jun 6	268	43.5	369.6	5773
Fairmont†	Sept 20	May 23	246	44.5	372.5	5047
Huntington–C	Oct 4	May 11	220	46.5	406.3	4073
Parkersburg–C	Sept 30	May 18	231	44.4	372.9	4750
Petersburg–C	Sept 23	May 22	242	44.5	—	4966
Wheeling†	Sept 19	May 21	276	46.1	404.6	5218
WISCONSIN						
Beloit†	Sept 10	Jun 5	269	39.4	358.6	6876
Eau Claire†	Sept 2	Jun 9	281	36.6	313.6	7970
Green Bay–A	Sept 2	Jun 12	284	35.9	312.2	8259
La Crosse–A	Sept 10	May 31	264	36.0	296.2	7650
Madison–A	Sept 10	Jun 4	268	37.3	333.2	7417
Madison–C	Sept 13	Jun 3	264	37.3	333.2	7300
Milwaukee–A	Sept 10	Jun 12	276	38.9	336.2	7205
Milwaukee–C	Sept 15	Jun 12	271	39.4	337.7	6944
Oshkosh†	Sept 5	Jun 11	280	38.1	335.1	7508
Sheboygan†	Sept 4	Jun 21	291	39.6	359.1	7382
WYOMING						
Casper–A	Aug 31	Jun 22	296	39.2	336.8	7638
Cheyenne–A	Aug 26	Jun 28	307	40.4	339.4	7562
Lander–A	Aug 29	Jun 22	298	37.1	314.4	8303
Rock Spring–A	Aug 27	Jun 29	307	37.4	—	8473
Sheridan–A	Sept 1	Jun 23	296	38.3	284.9	7903

HEATING PLANT OPERATING HOURS AND LOAD FACTOR

Location	No. of Hours Intermittent Heating Plant Operates per Normal Year	Seasonal Load Factor, Normal Year, Per-Cent	Location	No. of Hours Intermittent Heating Plant Operates per Normal Year	Seasonal Load Factor, Normal Year, Per-Cent
ALABAMA			**CONNECTICUT**		
Anniston–A	1129	24.5	Bridgeport–A	2022	32.4
Birmingham–A	1214	26.2	Hartford–A	2110	33.3
Mobile–A	771	20.2	New Haven–A	2065	32.1
Mobile–C	732	20.2	**DELAWARE**		
Montgomery–A	929	22.0			
Montgomery–C	851	21.1	Wilmington–A	1816	32.2
ARIZONA			**DISTRICT OF COLUMBIA**		
Flagstaff–A	2401	30.5			
Phoenix–A	1023	27.5	Silver Hill Obs.	1682	30.2
Phoenix–C	894	25.0	Washington–A	1604	29.7
Prescott–A	1452	24.9	Washington–C	1573	29.4
Tuscon–A	1069	27.0	**FLORIDA**		
Winslow–A	1509	26.3			
Yuma–A	651	24.0	Apalachicola–C	787	22.0
ARKANSAS			Jacksonville–A	744	22.0
			Jacksonville–C	665	21.0
Fort Smith–A	1387	30.9	Orlando–A	515	13.0
Little Rock–A	1189	26.5	Pensacola–C	863	23.8
Texarkana–A	1030	25.1	Tallahassee–A	916	24.3
CALIFORNIA			Tampa–A	460	20.6
Bakersfield–A	1273	29.3	**GEORGIA**		
Beaumont–C	1938	35.1			
Bishop–A	2254	39.3	Athens–A	1224	26.7
Eureka–C	3180	36.3	Atlanta–A	1863	26.5
Fresno–A	1521	33.0	Atlanta–C	1863	26.4
Los Angeles–A	1612	27.3	Augusta–A	929	22.5
Los Angeles–C	1154	25.3	Columbus–A	1045	24.2
Oakland–A	2181	24.9	Macon–A	982	24.2
Red Bluff–A	1527	32.3	Rome–A	1371	28.7
Sacramento–A	1939	37.4	Savannah–A	909	23.1
Sacramento–C	1777	35.1	**IDAHO**		
San Diego–A	1253	23.3			
San Francisco–A	2742	31.3	Boise–A	1888	28.4
San Francisco–C	2453	28.0	Lewiston–A	2195	34.0
San Jose–C	1452	24.8	Pocatello–A	2394	34.4
Santa Catalina–A	1360	21.8	**ILLINOIS**		
Santa Maria–A	2339	26.7			
COLORADO			Cairo–C	1389	28.8
			Chicago–A	2018	33.5
Colorado Springs–A	1735	26.2	Moline–A	2042	34.3
Denver–A	1730	26.6	Peoria–A	1946	32.7
Denver–C	1653	25.1	Springfield–A	1822	31.5
Grand Junction–A	1735	30.0	Springfield–C	1669	30.5
Pueblo–A	1523	24.7			

A–Airport station   C–City station
Seasonal Load Factor = No. Hrs. Plant Operates per Yr. ÷ No. Hrs. in Heating Season

Location	No. of Hours Intermittent Heating Plant Operates per Normal Year	Seasonal Load Factor, Normal Year, Per-Cent	Location	No. of Hours Intermittent Heating Plant Operates per Normal Year	Seasonal Load Factor, Normal Year, Per-Cent
INDIANA			**MICHIGAN**		
Evansville–A	1610	26.3	Alpena–C	2584	35.3
Fort Wayne–A	2013	32.9	Detroit–A	2045	32.9
Indianapolis–A	1795	30.9	Escanaba–C	2597	34.9
Indianapolis–C	1643	29.9	Grand Rapids–A	2263	34.8
South Bend–A	2239	36.3	Grand Rapids–C	2072	33.2
Terre Haute–A	1839	32.6	Lansing–A	2238	34.8
IOWA			Marquette–C	2726	36.4
			Muskegon–A	2266	33.6
Burlington–A	1954	33.5	Sault Ste. Marie–A	2676	32.6
Davenport–C	1832	31.8	**MINNESOTA**		
Des Moines–A	1933	32.6			
Des Moines–C	1884	32.3	Duluth–A	2646	35.0
Dubuque–A	2052	31.9	Duluth–C	2526	33.0
Sioux City–A	1979	32.6	Minneapolis–A	2222	35.2
KANSAS			Rochester–A	2282	34.7
			Saint Cloud–A	2375	34.6
Concordia–C	1700	30.8	Saint Paul–A	2205	34.8
Dodge City–A	1619	29.2	**MISSISSIPPI**		
Topeka–C	1577	29.6			
Wichita–A	1463	27.7	Jackson–A	1058	25.2
KENTUCKY			Meridian–A	1020	23.6
			Vicksburg–C	871	22.0
Bowling Green–A	1577	29.2	**MISSOURI**		
Lexington–A	1840	32.9			
Louisville–A	1641	30.8	Columbia–A	1635	29.5
Louisville–C	1578	30.3	Kansas City–A	1563	29.6
LOUISIANA			Saint Joseph–A	1709	31.5
			Saint Louis–A	1737	32.6
Baton Rouge–A	855	22.7	Saint Louis–C	1651	32.0
New Orleans–A	703	20.2	Springfield–A	1502	27.1
New Orleans–C	625	19.3	**MONTANA**		
Shreveport–A	1131	28.9			
MAINE			Billings–A	1894	27.6
			Butte–A	2751	31.4
Eastport–C	2637	30.1	Great Falls–A	2129	28.9
Portland–A	2625	36.1	Havre–C	2081	29.1
MARYLAND			Helena–A	2329	31.2
			Helena–C	2297	30.1
Baltimore–A	1702	30.3	Kalispell–A	2271	29.3
Baltimore–C	1548	26.0	Miles City–A	1875	27.9
Frederick–A	1662	30.1	**NEBRASKA**		
MASSACHUSETTS			Grand Island–A	1895	32.1
			Lincoln–A	1952	33.2
Boston–A	2138	34.0	Lincoln–C	1871	32.9
Nantucket–A	2245	31.5	Norfolk–A	2122	34.4
Pittsfield–A	2463	34.1	North Platte–A	1847	29.6

Heating Plant Operating Hours and Load Factor (continued)

Location	No. of Hours Intermittent Heating Plant Operates per Normal Year	Seasonal Load Factor, Normal Year, Per-Cent	Location	No. of Hours Intermittent Heating Plant Operates per Normal Year	Seasonal Load Factor, Normal Year, Per-Cent
NEBRASKA (Concluded)			**OHIO**		
Omaha–A	1973	34.4	Akron-Canton–A	2128	34.9
Valentine–C	1884	29.4	Cincinnati–A	1917	33.7
			Cincinnati–C	1670	31.2
NEVADA			Cincin.–Abbe Obs.	1802	32.5
			Cleveland–A	2217	36.8
Las Vegas–C	1295	26.2	Cleveland–C	2116	35.4
Reno–A	2067	29.1	Columbus–A	1793	30.5
Winnemucca–A	1915	27.8	Columbus–C	1691	29.6
			Dayton–A	2065	35.4
NEW HAMPSHIRE			Sandusky–C	2163	36.2
			Toledo–A	2050	33.1
Concord–A	2291	32.8	Youngstown–A	2116	34.3
NEW JERSEY			**OKLAHOMA**		
Atlantic City–C	1891	32.3			
Newark–A	1940	33.4	Oklahoma City–A	1347	28.2
Trenton–C	1870	32.6	Oklahoma City–C	1294	27.8
			Tulsa–A	1324	28.3
NEW MEXICO			**OREGON**		
Albuquerque–A	1625	30.5			
Roswell–A	1098	22.0	Baker–C	2434	32.3
			Eugene–A	2290	31.8
NEW YORK			Medford–A	1814	28.0
			Pendleton–A	1560	25.0
Albany–A	2232	34.7	Portland–A	2018	28.7
Albany–C	2021	32.9	Portland–C	1808	27.3
Binghamton–A	2409	34.5	Roseburg–C	1794	26.7
Binghamton–C	2103	32.7	Salem–A	2201	31.4
Buffalo–A	2339	35.7			
New York–LaG.–A	1849	32.5	**PENNSYLVANIA**		
New York–C	1868	32.3			
New York–Cent. Pk.	1833	32.5	Allentown–A	2169	36.0
Oswego–C	2231	33.2	Erie–C	2099	33.9
Rochester–A	2337	36.1	Harrisburg–A	1938	33.5
Schenectady–C	2258	35.5	Philadelphia–A	1797	32.0
Syracuse–A	2089	33.1	Philadelphia–C	1669	30.5
			Pitts. All. Airport	2055	35.1
NORTH CAROLINA			Pitts. Gr. Pitt. Air.	2174	35.8
			Pittsburgh–C	1865	33.2
Asheville–C	1504	26.9	Reading–C	1874	32.8
Charlotte–A	1395	28.5	Scranton–C	2068	33.4
Greensboro–A	1664	31.8	Williamsport–A	2022	33.3
Raleigh–C	1338	27.6			
Wilmington–A	1119	24.8	**RHODE ISLAND**		
Winston-Salem–A	1623	31.3			
NORTH DAKOTA			Block Island–A	2164	32.2
			Providence–A	2263	34.8
Bismarck–A	2276	32.7	Providence–C	2066	33.5
Fargo–A	2473	35.9			
Williston–C	2179	31.1			

Location	No. of Hours Intermittent Heating Plant Operates per Normal Year	Seasonal Load Factor, Normal Year, Per-Cent	Location	No. of Hours Intermittent Heating Plant Operates per Normal Year	Seasonal Load Factor, Normal Year, Per-Cent
SOUTH CAROLINA			**UTAH**		
Charleston–A	946	19.6	Milford–A	1934	29.3
Charleston–C	847	21.4	Salt Lake City–A	1875	29.6
Columbia–A	1066	24.4	Salt Lake City–C	1748	28.9
Columbia–C	995	22.9	**VERMONT**		
Florence–A	1203	26.8			
Greenville–A	1331	27.6	Burlington–A	2516	37.3
Spartanburg–A	1457	30.2	**VIRGINIA**		
SOUTH DAKOTA			Cape Henry–C	1442	28.2
			Lynchburg–A	1665	30.7
Huron–A	2232	35.1	Norfolk–A	1659	32.0
Rapid City–A	2121	30.9	Norfolk–C	1500	30.2
Sioux Falls–A	2219	34.5	Richmond–A	1899	35.8
			Richmond–C	1783	34.4
			Roanoke–A	1665	30.7
TENNESSEE			**WASHINGTON**		
Chattanooga–A	1480	29.5	North Head	2786	31.8
Knoxville–A	1329	26.5	Olympia–A	2646	30.2
Memphis–A	1155	25.2	Seattle–C	2136	28.8
Memphis–C	1112	24.9	Seattle-Tacoma–A	2535	30.8
Nashville–A	1297	26.5	Spokane–A	2355	32.6
			Tacoma–C	2336	29.4
			Tatoosh Isl.–C	2751	31.4
TEXAS			Walla Walla–C	1554	25.9
			Yakima–A	2001	30.1
Abilene–A	1277	30.4	**WEST VIRGINIA**		
Amarillo–A	1395	25.5			
Austin–A	910	25.1	Elkins–A	1846	28.7
Big Springs–A	1082	26.2	Huntington–C	1394	26.4
Brownsville–A	423	22.9	Parkersburg–C	1525	27.5
Corpus Christi–A	543	18.7	**WISCONSIN**		
Dallas–A	836	21.5			
Del Rio–A	676	20.4	Green Bay–A	2331	34.2
El Paso–A	1151	26.2	LaCrosse–A	2040	32.2
Fort Worth–A	1028	25.8	Madison–A	2225	34.6
Ft.Worth–A. Cart. Fld.	1022	25.8	Madison–C	2192	34.6
Galveston–A	659	20.2	Milwaukee–A	2159	32.6
Galveston–C	644	19.6	Milwaukee–C	2081	32.0
Houston–A	741	21.3	**WYOMING**		
Houston–C	676	20.4			
Palestine–C	952	24.8			
Port Arthur–A	805	22.2	Casper–A	2295	32.3
Port Arthur–C	717	20.9	Cheyenne–A	2269	30.8
San Antonio–A	842	24.2	Lander–A	2346	32.8
Waco–A	1076	28.4	Sheridan–A	1996	28.1

NORMAL INDUSTRIAL DEGREE-DAYS - 45 AND 55F BASES

The standard degree-day (65 deg. base) cannot be used with accuracy in estimates and records involving plants in which temperatures are maintained below 70 deg.

The table below gives the normal number of industrial degree-days on 45 and 55 deg. bases for many large U. S. cities. The industrial degree-day unit is highly inaccurate in the far South.

State and City	45F Base	55F Base	State and City	45F Base	55F Base	State and City	45F Base	55F Base
COLORADO			Houghton......	4029	6112	OKLAHOMA		
Denver.........	1548	3440	Lansing........	2537	4444	Oklahoma City.	600	1835
Grand Junction.	1757	3433	Sault Ste. Marie.	4049	6575	OREGON		
Pueblo.........	1499	3261	MINNESOTA			Baker.........	2321	4307
CONNECTICUT			Duluth.........	4419	6774	Portland......	373	1911
Meriden........	—	734	Minneapolis....	3309	5417	Roseburg	272	1868
New Haven.....	1769	3237	Moorhead......	4706	6572	PENNSYLVANIA		
DISTRICT OF			St. Paul........	2497	5497	Erie...........	2337	3837
COLUMBIA			MISSOURI			Harrisburg.....	1565	3236
Washington	1041	2487	Kansas City....	1463	2980	Philadelphia....	1122	2695
IDAHO			Saint Louis.....	1186	2745	Pittsburgh......	1377	3028
Boise..........	1045	2814	Springfield......	982	2423	Scranton.......	1938	3755
Lewiston.......	1034	2688	MONTANA			RHODE ISLAND		
Pocatello.......	2161	4140	Havre.........	3736	5874	Block Island....	871	3388
ILLINOIS			Helena.........	2843	5071	SOUTH DAKOTA		
Cairo..........	749	2119	Kalispell......	2874	5131	Yankton.......	2898	6045
Chicago........	1969	3743	NEBRASKA			TENNESSEE		
Springfield......	1677	3289	Lincoln........	3023	3850	Chattanooga....	242	1398
INDIANA			North Platte....	2291	4152	Knoxville......	431	1741
Evansville......	799	2335	Omaha........	2284	3982	Memphis.......	166	1284
Indianapolis....	1397	2829	Valentine.......	2833	4801	Nashville.......	419	1678
IOWA			NEVADA			UTAH		
Davenport......	2296	4142	Winnemucca....	1670	3468	Modena........	1978	3981
Des Moines.....	2440	4180	NEW HAMPSHIRE			Salt Lake City..	1475	3202
KANSAS			Concord........	2646	4640	VERMONT		
Dodge City.....	1385	2962	NEW JERSEY			Burlington.....	3014	4984
Topeka........	1518	1811	Atlantic City...	1123	2904	Northfield......	3652	7121
Wichita........	1152	2587	NEW YORK			VIRGINIA		
KENTUCKY			Albany........	2018	4302	Lynchburg.....	554	1928
Lexington......	—	2557	Binghamton....	2073	4296	Norfolk........	260	1496
Louisville.......	1073	2294	Buffalo........	2359	4316	Richmond......	549	1895
MAINE			Ithaca.........	2412	4023	WASHINGTON		
Eastport.......	2956	5236	New York......	1412	3089	North Head....	184	2062
Portland.......	2530	4572	Oswego........	2274	4363	Seattle.........	408	2185
MARYLAND			Rochester	2341	4231	Spokane.	1741	3672
Baltimore......	986	2491	NORTH DAKOTA			WEST VIRGINIA		
MASSACHUSETTS			Bismarck.......	3831	6468	Elkins.........	1506	3327
Boston.........	1787	3603	Williston.......	4616	6399	Parkersburg....	1147	2784
Nantucket......	1514	3419	OHIO			WISCONSIN		
MICHIGAN			Cincinnati......	1376	3003	Green Bay.....	3318	5331
Alpena.........	3131	5499	Cleveland......	1525	3795	La Crosse......	3034	3992
Detroit........	2240	4089	Columbus......	1600	3255	Madison.......	3067	4850
Escanaba......	3699	5918	Dayton........	1487	3147	Milwaukee.....	2657	4617
Grand Haven ..	2405	3435	Sandusky......	1949	3425	WYOMING		
Grand Rapids ..	2332	4177	Toledo........	1990	3757	Cheyenne.....	2500	4700
						Lander........	3208	5450

CITY WATER TEMPERATURES

State and City and Water Source (W=well; S=surface)	Month						Probable Monthly Max. Temp.
	April	May	Jun	Jul	Aug	Sept	
	Temperature, Deg F						
ALABAMA							
Anniston (W)	—	—	65.0	66.0	67.0	65.0	70
Birmingham	63.0	73.0	78.0	82.0	81.0	79.0	85
Gadsden (S)	—	—	76.0	78.0	82.0	80.0	85
Tuscaloosa (S)	—	—	64.0	69.0	70.0	71.0	74
ARKANSAS							
Fort Smith (S)	—	—	77.0	75.0	80.0	77.0	83
Little Rock (WS)	—	—	85.0	88.0	86.0	82.0	89
N. Little Rock (S)	—	—	80.0	82.0	85.0	80.0	88
Pine Bluff (W)	—	—	75.0	78.0	80.0	75.0	83
ARIZONA							
Tucson (W)	—	—	80.0	80.0	80.0	80.0	80
CALIFORNIA							
Alameda	—	—	59.0	62.0	64.0	64.0	67
Alhambra (W)	—	—	68.0	68.0	68.0	68.0	68
Berkeley	—	—	59.0	62.0	64.0	64.0	67
Fresno (W)	—	—	72.0	72.0	72.0	72.0	—
Glendale (WS)	—	—	68.0	68.2	68.2	67.3	71
Los Angeles	63.0	68.0	73.0	74.0	76.0	75.0	79
Oakland	55.0	57.0	59.0	62.0	64.0	64.0	67
Pasadena (WS)	—	—	68.0	73.0	74.0	74.0	77
Pomona (W)	—	—	67.0	—	—	—	—
Richmond	—	—	59.0	62.0	64.0	64.0	67
Riverside (W)	—	—	72.0	74.0	74.0	73.0	77
Sacramento (S)	—	—	70.7	79.7	80.6	77.0	83
San Bernardino (WS)	67.0	67.0	67.0	67.0	67.0	67.0	67
San Francisco	60.0	60.0	60.0	60.0	60.0	60.0	—
San Jose (WS)	—	—	68.0	73.0	73.0	73.0	76
Santa Ana (W)	—	—	69.0	69.0	69.0	69.0	69
Santa Barbara (S)	—	—	—	—	70.0	68.0	—
Stockton (W)	—	—	70.0	70.0	70.0	70.0	70
COLORADO							
Colorado Springs (S)	—	—	57.0	60.0	62.0	60.0	65
Denver	54.3	61.7	63.1	70.9	70.7	68.0	74
Pueblo (S)	—	—	69.0	74.0	74.0	70.0	77
CONNECTICUT							
Bridgeport	43.0	45.0	53.0	62.0	65.0	66.0	69
Hartford	44.0	54.0	59.0	66.0	70.0	69.0	73
New Haven	50.0	—	—	—	70.0	60.0	74
Waterbury	47.0	64.0	68.0	74.0	74.0	73.0	—
Stamford (S)	—	—	69.0	76.0	67.5	68.5	71
Stratford	—	—	61.0	63.0	64.0	64.0	67
DISTRICT OF COLUMBIA							
Washington	49.0	49.0	43.0	67.0	73.0	75.0	—

City Water Temperatures (continued)

State and City and Water Source (W=well; S=surface)	April	May	Jun	Jul	Aug	Sept	Probable Monthly Max. Temp.
				Temperature, Deg F			
DELAWARE							
Wilmington	52.0	68.0	73.0	78.0	79.0	73.0	82
FLORIDA							
Jacksonville	79.2	80.6	84.8	86.3	86.7	82.4	90
Miami			From 70 to 75 year round				—
Orlando (S)	—	—	84.0	86.0	87.0	86.0	90
Pensacola (W)	—	—	70.0	70.0	70.0	70.0	—
Tampa	80.0	85.0	87.0	85.0	85.0	83.0	—
West Palm Beach (S)	—	—	84.2	85.6	86.2	84.9	89
GEORGIA							
Atlanta	59.0	71.5	78.1	83.5	79.5	77.8	87
Augusta (S)	—	—	82.0	84.0	85.0	79.0	88
Columbus (S)	—	—	75.0	80.0	80.0	79.5	83
Rome (S)	—	—	77.0	78.0	77.0	72.0	81
Savannah (W)			72.0 year round				—
ILLINOIS							
Aurora (W)	—	—	60.0	60.0	60.0	60.0	—
Belleville	—	—	80.0	90.0	86.0	83.0	93
Bloomington (S)	—	—	69.8	78.8	78.8	68.0	82
Champaign (W)	—	—	58.0	58.5	56.0	56.0	—
Chicago	39.2	47.1	55.4	68.0	69.4	62.5	—
Danville (S)	—	—	79.5	84.0	83.0	75.0	87
Decatur (S)	—	—	75.9	78.6	83.3	73.8	86
E. St. Louis (S)	—	—	80.0	90.0	86.0	83.0	93
Elgin (W)	—	—	57.0	56.0	56.0	54.0	—
Evanston (S)	—	—	57.0	67.0	72.0	66.0	75
Freeport (W)	—	—	56.0	56.0	56.0	56.0	—
Granite City	—	—	80.0	90.0	86.0	83.0	93
Maywood (W)	—	—	60.0	60.0	60.0	60.0	—
Moline (S)	—	—	74.2	85.9	80.8	72.6	89
Oak Park (S)	—	—	68.2	70.0	75.0	73.0	78
Peoria	54.0	54.0	56.0	56.0	56.0	54.0	—
Quincy (S)	—	—	73.4	82.2	79.9	71.6	85
Rockford (W)	—	—	55.0	55.0	55.0	55.0	58
Springfield (S)	—	—	76.6	79.9	80.6	76.3	84
Waukegan (S)	—	—	54.5	65.6	69.8	64.7	—
INDIANA							
Elkhart (W)	—	—	57.0	58.0	60.0	58.0	—
Evansville	56.0	66.0	85.0	84.0	83.0	74.0	87
Fort Wayne	49.0	67.0	73.0	79.0	78.0	73.0	82
Gary	41.0	50.0	60.0	70.0	70.0	65.0	—
Hammond (S)	—	—	70.0	78.0	82.0	82.0	85
Indianapolis	53.0	68.0	73.0	80.0	82.0	77.0	85
Lafayette (W)	—	—	53.0	53.0	53.0	53.0	—
Marion (W)	—	—	54.0	54.0	55.0	55.0	—
Muncie (WS)	—	—	70.5	75.5	74.4	68.4	79
South Bend			Averages 60 year round				—
IOWA							
Burlington (S)	—	—	—	81.0	87.0	86.0	90

City Water Temperatures (continued)

State and City and Water Source (W = well; S = surface)	Month						Probable Monthly Max.
	April	May	Jun	Jul	Aug	Sept	
	Temperature, Deg F						
Cedar Rapids (S)	—	—	71.4	82.5	79.2	68.0	—
Council Bluffs (S)	—	—	76.0	80.5	81.5	72.7	85
Des Moines	44.1	49.1	58.2	65.7	72.9	71.1	77
Dubuque (W)	—	—	59.0	60.0	60.0	59.0	—
Fort Dodge (W)	—	—	52.0	52.0	52.0	52.0	—
Mason City	—	—	—	—	59.0	—	—
Ottumwa (S)	—	—	72.5	77.5	77.5	72.5	81
Sioux City (W)			Averages 54.0				—
Waterloo (W)			Averages 54.0				—
KANSAS							
Hutchinson (W)			Averages 60.0				—
Kansas City	63.0	78.0	84.0	93.0	91.0	85.0	—
Salina (W)	—	—	58.0	60.0	62.0	62.0	—
Topeka (WS)	—	—	76.0	84.0	80.0	74.0	87
KENTUCKY							
Ashland (S)	—	—	77.0	80.6	82.4	73.4	85
Covington (WS)	—	—	75.0	82.0	81.0	77.0	85
Louisville	49.0	69.0	77.0	82.0	82.0	77.0	85
Paducah (S)	—	—	87.0	89.0	87.0	80.0	92
LOUISIANA							
Alexandria (W)	—	—	85.0	86.0	86.0	86.0	89
New Orleans	66.0	77.0	86.0	89.0	90.0	90.0	93
Shreveport (S)	—	—	88.0	91.0	89.0	84.0	—
MAINE							
Portland (S)	—	—	56.3	64.0	66.0	64.0	69
MARYLAND							
Baltimore	47.0	53.0	61.0	66.0	70.0	64.0	73
Cumberland (S)	—	—	59.0	62.0	65.0	69.0	72
Hagerstown (S)	—	—	82.0	83.0	82.0	79.0	—
MASSACHUSETTS							
Brockton (S)	—	—	53.0	58.0	62.0	62.0	65
Brookline (W)	—	—	57.0	62.0	67.0	66.0	70
Cambridge	43.0	55.0	64.0	72.0	74.0	68.8	—
Chicopee (S)	—	—	61.5	66.4	66.7	62.1	70
Holyoke (S)	—	—	69.0	74.0	72.0	68.0	77
Leominster (S)	—	—	65.0	71.0	74.0	64.0	77
Lowell	50.0	50.0	50.0	50.0	50.0	50.0	—
New Bedford	42.0	48.0	60.0	68.0	71.0	69.0	74
Quincy (S)	—	—	65.0	74.0	73.0	69.0	77
Springfield	41.0	49.0	52.0	54.0	55.0	54.0	—
Taunton (S)	—	—	65.0	70.0	70.0	70.0	73
Westfield (S)	—	—	56.0	61.0	63.0	58.0	—
Weymouth (S)	—	—	68.0	73.4	75.2	68.0	78
Worcester	46.0	53.5	65.0	70.0	73.0	68.5	76
MICHIGAN							
Ann Arbor (W)	—	—	58.0	62.0	62.5	61.5	66
Battle Creek (WS)	—	—	52.0	51.0	52.0	52.0	—
Bay City (S)	—	—	70.0	78.0	75.0	67.0	81

City Water Temperatures (continued)

State and City and Water Source (W=well; S=surface)	Month						Probable Monthly Max. Temp.
	April	May	Jun	Jul	Aug	Sept	
	Temperature, Deg F						
Dearborn	—	—	64.0	75.0	74.0	68.0	78
Detroit	41.0	56.0	64.0	75.0	74.0	68.0	78
Grand Rapids	54.0	66.0	71.0	73.0	74.0	69.0	—
Highland Park	—	—	64.0	73.0	74.0	71.0	77
Jackson (W)	—	—	52.0	52.0	52.0	52.0	—
Kalamazoo (W)	—	—	52.0	52.0	52.0	52.0	55
Lansing (W)	—	—	57.5	58.0	59.0	59.0	62
Muskegon (S)	—	—	49.0	46.0	68.0	66.0	71
Pontiac (W)			Averages 55.0				—
Saginaw (S)	—	—	70.0	78.0	75.0	70.0	81
MINNESOTA							
Duluth	33.4	52.7	57.6	70.6	66.6	58.7	—
Minneapolis	40.5	61.2	69.3	80.2	73.0	67.6	83
Rochester (W)	—	—	53.0	55.0	58.0	56.0	—
St. Cloud (S)	—	—	68.0	77.0	70.0	64.0	80
Winona (W)			Averages 52.0				—
MISSISSIPPI							
Jackson (S)	—	—	80.0	82.0	74.0	74.0	85
Meridian (WS)	—	—	78.0	86.0	86.0	82.0	89
MISSOURI							
Hannibal (S)	—	—	78.0	80.0	80.0	75.0	83
Jefferson City (S)	—	—	65.0	70.0	80.0	76.0	83
Springfield (WS)	—	—	68.0	74.0	76.0	75.0	79
St. Louis	53.0	69.0	77.0	85.0	83.0	75.0	88
NEBRASKA							
Lincoln (W)	—	—	58.0	59.0	59.0	59.0	—
Omaha	50.8	55.8	68.0	80.9	79.4	69.1	84
NEW HAMPSHIRE							
Berlin (S)	—	—	58.0	66.0	63.0	56.0	69
Nashua (W)	—	—	52.0	58.0	67.0	65.0	70
NEW JERSEY							
Atlantic City (WS)	—	—	71.6	72.5	68.0	64.4	71
Elizabeth	43.3	48.8	51.7	54.7	61.3	66.7	70
Newark	45.0	56.0	64.0	68.5	71.5	70.5	75
Paterson	46.0	49.2	54.6	55.9	59.0	64.0	—
Woodbridge (S)			Averages 51.8				
NEW YORK							
Albany	40.0	52.0	60.0	56.0	66.0	65.0	69
Amsterdam (S)	—	—	44.0	50.0	57.0	57.0	60
Binghamton (S)			Ranges from 34.0 to 74.0				—
Buffalo	32.0	37.0	62.0	71.0	73.0	66.0	76
Ithaca (S)	—	—	61.1	71.6	70.8	67.5	75
Jamestown (W)			Averages 48.0				—
Lackawanna (S)	—	—	—	75.0	75.0	—	—
Mount Vernon (S)	—	—	60.0	64.0	67.0	68.0	71

City Water Temperatures (continued)

State and City and Water Source (W=well; S=surface)	April	May	Jun	Jul	Aug	Sept	Probable Monthly Max. Temp.
			Temperature, Deg F				
Newburgh (S)			Averages 54.0		70.0	69.0	—
New York	45.0	55.0	63.0	70.0	68.0	66.2	73
Niagara Falls (S)	—	—	59.9	69.8	68.0	66.2	73
N. Tonawanda (S)	—	—	66.0	67.0	78.0	73.0	81
Poughkeepsie (WS)	—	—	68.9	75.2	76.6	71.3	78
Rochester	41.9	52.8	62.3	68.0	68.9	64.6	72
Schenectady (W)	—	—	51.0	55.0	57.0	59.0	60
Syracuse	44.2	50.5	59.8	66.4	70.4	67.7	73
Troy (S)	—	—	67.5	70.0	72.5	69.5	76
Utica	43.2	53.6	63.0	70.0	70.2	67.6	73
Watertown (S)	—	—	69.0	72.0	71.0	67.0	75
White Plains (WS)	—	—	—	—	—	52.0	—
Yonkers	60.0	69.0	75.0	77.0	79.0	78.0	
NORTH CAROLINA							
Asheville (S)	—	—	68.0	69.0	74.0	76.0	79
Charlotte (S)	—	—	76.0	81.0	80.0	77.0	84
Durham (S)	—	—	72.0	74.7	76.2	74.4	79
High Point (S)	—	—	75.0	78.0	—	73.0	—
Raleigh (S)	—	—	77.0	83.4	81.8	79.0	—
Wilmington (S)	—	—	86.0	86.0	84.2	80.6	89
NEW MEXICO							
Albuquerque (W)			Averages 72.0				—
OHIO							
Akron	43.7	61.0	69.3	74.3	74.7	69.4	78
Alliance (WS)	—	—	64.4	68.0	68.0	64.4	71
Cincinnati	49.0	66.0	76.0	82.0	81.0	77.0	85
Cleveland	39.0	50.0	58.0	68.0	73.5	71.0	77
Cleveland Hts. (S)	—	—	58.0	68.0	73.5	71.0	77
Columbus	46.0	64.0	72.0	76.0	76.0	74.0	79
Elyria (W)	—	—	66.0	70.0	72.0	66.0	75
Findlay (WS)	—	—	63.0	61.7	60.8	59.0	64
Hamilton (W)	—	—	63.0	65.0	65.0	64.0	—
Lima (S)	—	—	71.6	74.7	77.4	73.4	80
Mansfield (W)			Averages 50.0				—
Newark (S)	—	—	67.0	70.0	70.0	64.0	73
Norwood (W)			Averages 56.0				—
Portsmouth (S)	—	—	77.0	80.6	78.8	75.2	82
Sandusky (S)	—	—	64.4	75.2	75.2	66.2	78
Steubenville (S)	—	—	60.0	63.0	68.0	67.0	71
Toledo	48.0	66.0	72.0	87.0	85.0	72.0	—
Warren (S)	—	—	73.2	78.6	76.5	71.2	82
Youngstown	43.0	50.5	58.5	62.5	66.0	68.5	72
OKLAHOMA							
Muskogee (S)	—	—	80.4	87.7	90.7	83.8	94
Oklahoma City	55.4	68.0	73.4	77.2	77.0	72.4	80
Tulsa	62.2	70.0	77.2	81.8	81.8	79.1	85
OREGON							
Portland	43.7	50.0	56.0	62.0	55.0	51.6	65

City Water Temperatures (continued)

State and City and Water Source (W=well; S=surface)	April	May	Jun	Jul	Aug	Sept	Probable Monthly Max. Temp.
PENNSYLVANIA							
Allentown (WS)	—	—	62.0	70.0	69.0	61.0	72
Bethlehem (WS)	—	—	70.6	74.0	75.0	62.7	78
Bradford (WS)	—	—	57.0	60.0	60.0	56.0	63
Butler (S)	—	—	55.0	65.0	70.0	70.0	73
Carbondale (S)	—	—	60.0	65.0	70.0	65.0	73
Duquesne (W)	Ranges from 56.0 to 68.0						—
Easton (S)	—	—	54.0	65.0	65.0	60.0	68
Erie	36.9	53.7	63.8	72.3	72.9	69.8	76
Hazleton (S)	Averages 65.0						—
Homestead	Ranges from 60.0 to 75.0						—
Johnstown (S)	—	—	53.5	57.7	59.5	58.7	63
Lancaster (S)	—	—	70.0	76.0	77.0	72.0	80
Norristown (S)	—	—	—	—	84.0	—	—
Philadelphia	40.0	68.0	71.0	79.0	77.0	72.0	82
Pittsburgh	46.4	66.2	75.2	80.6	80.6	75.2	84
Reading	46.4	59.9	70.7	78.8	76.1	71.6	82
Scranton	—	58.2	64.1	70.3	70.9	67.9	74
Sharon (S)	—	—	71.0	74.0	73.0	67.0	77
Uniontown (WS)	—	—	65.0	68.0	68.0	65.0	71
Wilkinsburg (S)	—	—	75.0	79.0	77.0	75.0	82
Williamsport (WS)	—	—	62.0	66.0	65.0	63.0	68
RHODE ISLAND							
Central Falls	—	—	70.0	74.0	74.0	67.0	77
Cranston	—	—	62.0	64.0	65.0	63.0	68
E. Providence (S)	—	—	50.0	55.0	60.0	60.0	63
Newport (S)	—	—	66.0	70.2	70.2	59.1	73
Pawtucket (S)	—	—	70.0	74.0	74.0	67.0	77
Providence	48.0	56.0	62.0	64.0	65.0	63.0	68
SOUTH CAROLINA							
Charleston (S)	—	—	81.0	83.0	85.0	80.0	—
Columbia	Averages 75.0						—
Greensville (S)	—	—	70.0	76.0	76.0	71.0	79
Spartanburg (S)	—	—	72.5	78.8	78.8	75.3	82
SOUTH DAKOTA							
Sioux Falls (W)	Averages 55.0						—
TENNESSEE							
Chattanooga	64.0	67.0	73.0	79.0	79.5	76.5	83
Jackson (W)	—	—	68.0	68.0	70.0	70.0	73
Knoxville	59.2	75.6	81.5	84.0	84.3	79.9	87
Nashville	61.0	76.0	84.0	88.0	88.0	84.0	91
TEXAS							
Abilene (WS)	—	—	70.0	75.0	80.0	80.0	83
Amarillo (W)	—	—	63.0	65.0	65.0	65.0	68
Austin (S)	—	—	84.0	84.0	85.0	79.0	88
Beaumont (S)	—	—	88.0	88.0	87.0	87.0	91
Brownsville (S)	—	—	84.0	86.0	87.0	89.0	92
Dallas	65.0	66.0	77.0	82.0	82.5	74.0	86

City Water Temperatures (continued)

State and City and Water Source (W = well; S = surface)	April	May	Jun	Jul	Aug	Sept	Probable Monthly Max. Temp.
				Temperature, Deg F			
El Paso	80.0	82.0	84.0	85.0	85.0	84.0	88
Fort Worth	62.0	72.0	81.0	83.0	83.0	81.0	86
Houston	84.0	84.0	84.0	84.0	84.0	84.0	—
Laredo (S)	—	—	84.2	83.3	84.6	78.5	88
Lubbock (S)	—	—	67.0	67.0	67.0	66.0	80
San Angelo (S)	—	—	75.0	78.0	80.0	77.0	83
San Antonio	76.0	76.0	76.0	76.0	76.0	76.0	—
Texarkana (WS)	—	—	77.0	83.0	86.0	86.0	89
Waco (S)	—	—	79.5	83.1	84.6	79.8	88
UTAH							
Ogden (WS).............			Averages 60.0				—
Salt Lake City	50.0	50.0	58.0	58.0	57.0	50.0	61
VERMONT							
Burlington (S)	—	—	58.0	63.0	66.0	68.0	71
VIRGINIA							
Lynchburg (S)	—	—	62.0	67.0	73.0	73.0	76
Newport News (WS)	—	—	77.0	82.0	82.0	76.0	85
Norfolk	62.0	70.0	77.5	83.0	83.0	79.5	—
Petersburg (S)	—	—	75.4	75.4	78.0	71.6	81
Richmond	56.8	69.6	74.7	80.4	79.5	73.4	83
WASHINGTON							
Everett (S)	—	—	54.0	60.0	62.0	50.0	65
Spokane			Averages 4, year round				—
Tacoma	49.0	51.0	53.0	59.0	61.0	60.0	—
Yakima (S)	—	—	56.0	61.5	65.0	62.0	68
WEST VIRGINIA							
Charleston (S)	—	—	79.0	83.0	82.0	77.0	85
WEST VIRGINIA (Cont'd)							
Clarksburg (S)	—	—	75.0	76.0	77.0	72.0	80
Wheeling (S)	—	—	75.8	80.9	75.5	76.4	—
WISCONSIN							
Appleton	—	—	67.0	76.0	72.0	67.0	79
Beloit (W)			Averages 57.0				—
Eau Claire (W)	—	—	50.0	51.0	53.0	62.0	65
Fond du Lac (W)	—	—	54.0	56.0	56.0	54.0	59
Kenosha (S)	—	—	56.0	59.0	67.0	60.0	70
La Crosse (W)			Averages 52.0				—
Madison (W)	—	—	53.0	52.0	53.0	52.0	56
Milwaukee	38.4	42.8	49.7	57.0	60.6	56.6	—
Manitowoc (W)	—	—	48.0	53.0	57.0	58.0	61
Superior (W)	—	—	56.2	61.5	64.4	62.9	67

The foregoing tabulation was compiled from data collected in a single year for many of the cities listed. While long term averages might show slightly different figures, it should be born in mind that well water temperatures usually reflect annual mean dry bulb temperatures of the air, while surface water temperatures vary more closely with seasonal wet bulb averages.

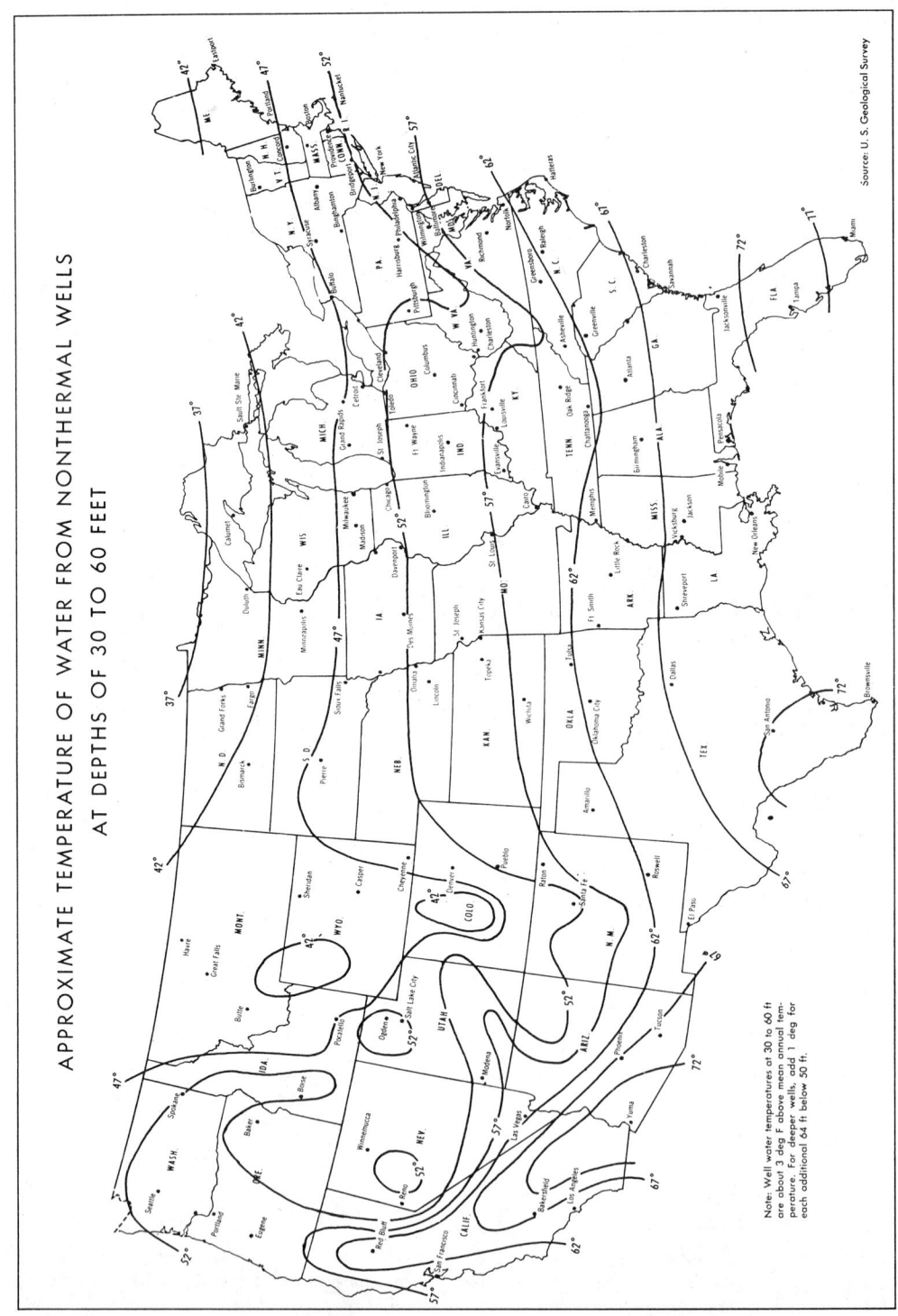

APPROXIMATE TEMPERATURE OF WATER FROM NONTHERMAL WELLS AT DEPTHS OF 30 TO 60 FEET

Source: U. S. Geological Survey

Note: Well water temperatures at 30 to 60 ft are about 3 deg F above mean annual temperature. For deeper wells, add 1 deg for each additional 64 ft below 50 ft.

Section 2

LOAD CALCULATIONS

Contents

Authorities

Section Editors: *Eugene Stamper* and *Richard L. Koral*

C.J. Baroczy, Nomograph for Free-Convection Heat Transfer with Air

William P. Chapman, Heat Requirements of Snow Melting Systems

William B. Foxhall, How to Use the Psychrometric Chart

Samuel J. Friedman, Heat Loss from Water Surface

R.W. Haines, High Altitude Psychrometric Charts

*Herbert Herkimer:*Apparatus Dew Point

E.W. Jerger, Condensation on Glass Surfaces

S. Konzo and *Wyman K. Ender:* Heat Gain through Uninsulated Ducts

Gordon W. Neal, Quick Method for Finding Air Sensible Heat Gain or Loss

*Stig Sylvan (*with *James R. Kayse)* High Temperature Psychrometric Chart

Nathan N. Wolpert, Design Conditions-Industrial Plants and Commercial Buildings

INTRODUCTION TO EVALUATING HEATING AND COOLING LOADS

The following information on evaluating heating and cooling loads is provided through the courtesy and permission of National Electrical Contractors Association, 7315 Wisconsin Ave., N.W,. Washington DC 20014.*

The primary purpose of heating and air conditioning is to provide thermal comfort for building occupants. Inside air temperature is used as the basis for calculation, whether there is heat gain or heat loss between indoors and outdoors. But even though air temperature is in the comfortable mid-70's, occupants still will be uncomfortable if there is excessive radiant loss or gain to the exterior wall or if there are cold downdrafts caused by cold surfaces. Discomfort can occur when there are large areas of single lights of glass and when the exterior wall has a high heat transmission factor (when it is poorly insulated). Thus the assumption of a "comfortable" air temperature and maintenance of the temperature by adding or removing heat in accordance with calculated values does not automatically assure comfort. Assuming that there is proper heat input or heat removal to maintain air temperature within recommended limits, comfort will depend upon the design of the structure and the heating and air-conditioning system, and upon integration of the two.

Heat loss/heat gain calculations are made for two basic reasons: (1) to size the heating and air-conditioning equipment and system, and (2) to provide data necessary for estimating operating costs. Depending upon the degree of accuracy desired and the building type, determination of heating or cooling demand and energy costs can range from the extremely simple to the highly complex.

Personal habits and preferences have a much greater effect upon air conditioning costs than heating costs. It has been shown that even with a number of similar residences equipped with identical air-conditioning systems, summer cooling costs can vary by a factor of two or more. This can be attributed to personal preferences in air temperature and periods of shutdown during days of absence and vacation. Thermostat settings for central residential air-conditioning systems have been shown in surveys to vary by as

much as 6F to 10F. The same surveys have indicated that as many as 30% of home-owners operate their systems manually, turning them on only when their houses have become excessively warm. Beyond this there is the question of whether windows and doors are kept closed, and at what outdoor temperature they are opened or closed. Further, the capacity required for the air-conditioning unit—and, of course, operating cost—will vary depending upon how much temperature swing is allowed. For example, if a 6F temperature swing (say from 74F to 80F) is permissible, then the capacity of the air-conditioning unit may need to be only 75 per cent of the heat gain calculation. Also, a 4½F swing might require a capacity from 85 to 90% of the design heat gain. In these cases, the low temperature would be reached only during the early morning hours, and the high only during "design" weather.

Next in the hierarchy of determining heating/cooling loads and operating cost is the apartment building, because of the multiplicity of similar spaces and space occupancies, and usually the simplicity of the heating-cooling system. Most power suppliers have power-consumption data that enables them to suggest fairly accurate correction factors to be applied to the degree-day formula. Similar experience information is generally available for small office buildings having a fairly uniform type of occupancy and also for small elementary schools.

Basically, steady-state heat loss/heat gain calculations themselves are simple and straightforward. With smaller types of buildings, the only aspects of heat loss calculations that may be important are transmission loss through the building structure, infiltration of outside air due to crackage, and outside air required for ventilation. Heat gain calculations, on the other hand, must take solar radiation into consideration, because this factor may have a considerable effect.

More comprehensive calculations will be called for in larger buildings having a wide variety of occupancy load and usage. Further, more detailed calculations generally are indicated for buildings using the larger air-source heat pumps because the operating efficiency and percentage of operation time at part-load capacity (instead of "full on-full off" operation) will vary in relation to the outdoor temperature. In such cases, a degree-hour approach generally will need to be used. With the

---

* Sources and credits for illustrations in this section are listed at the end of this section.

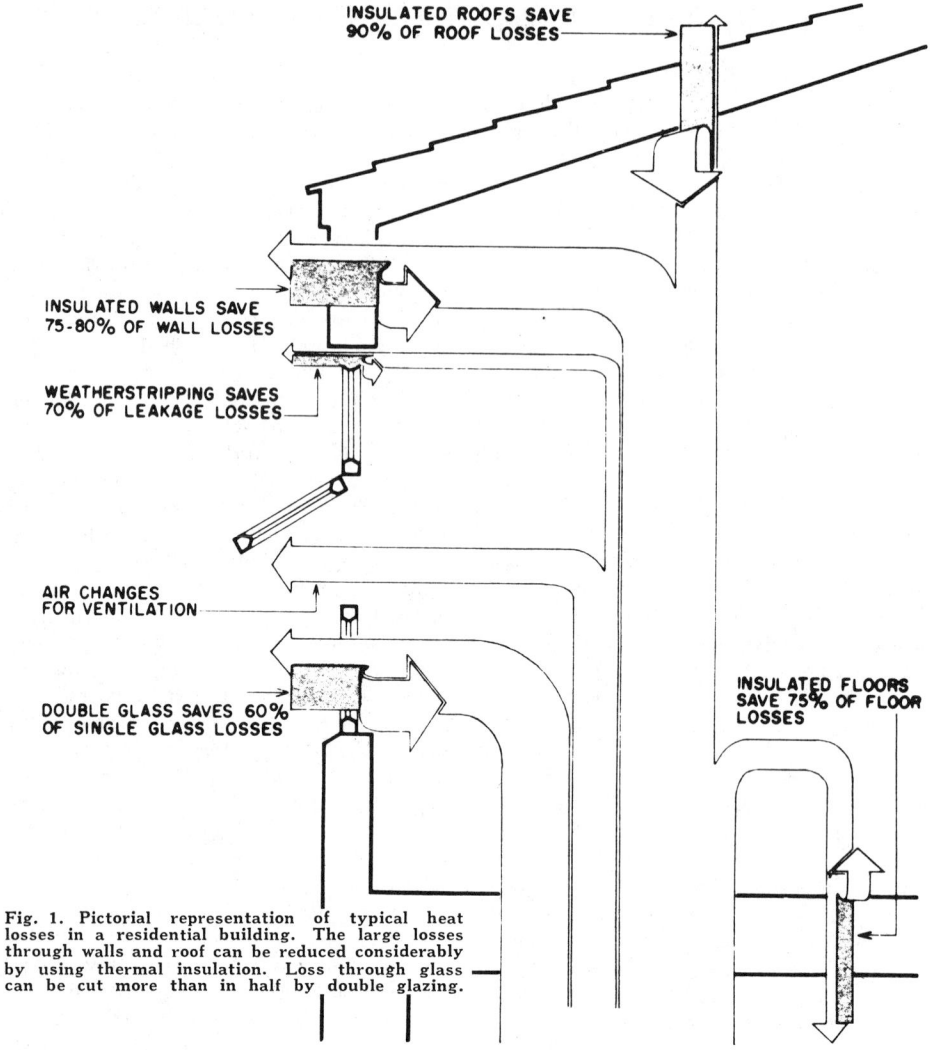

INSULATED ROOFS SAVE
90% OF ROOF LOSSES

INSULATED WALLS SAVE
75-80% OF WALL LOSSES

WEATHERSTRIPPING SAVES
70% OF LEAKAGE LOSSES

AIR CHANGES
FOR VENTILATION

DOUBLE GLASS SAVES 60%
OF SINGLE GLASS LOSSES

INSULATED FLOORS
SAVE 75% OF FLOOR
LOSSES

Fig. 1. Pictorial representation of typical heat losses in a residential building. The large losses through walls and roof can be reduced considerably by using thermal insulation. Loss through glass can be cut more than in half by double glazing.

degree-hour method using comprehensive weather data from the Weather Bureau, annual energy consumption estimates for heating can be made by taking into account hourly heat losses, amount of internal heat gain and degree-hours for occupied and non-occupied conditions.

The greatest refinement in determining heating-cooling loads and energy consumption is a computer program to integrate the large number of factors affecting these unknowns. Correlation between computer-calculated results and actual load and energy consumption figures obtained by

measurement in an actual building will be limited by accuracy of the weather data, building design parameters, assumed internal design conditions, and the response of the heating-cooling system to varying loads. Weather data will include hourly figures on temperature, humidity, wind, and solar radiation. Building design parameters will include building orientation, heat transmission factors and areas of wall, floor and roof materials, ventilation and infiltration, and shading. Internal design conditions will include the number of people, electric loads for lighting and equipment, and summer

and winter design temperatures and humidities. The accuracy of the computer approach also is limited to some extent by the fact that heat gain and heat loss do not take place under steady-state conditions. Nevertheless, theoretical methods have been developed that take into account non-periodic outdoor conditions and variable heat transfer coefficients (due to differences in wind, radiation, etc.). In this connection, two factors that can have a considerable effect on maximum heat gains and losses are thermal lag and storage. Lightweight materials have very little lag, massive materials a great deal. Solar load through clear glass is practically instantaneous. Effect of interior load from lights and people is not felt immediately because of the storage capacity of floors, interior partitions, furniture, and the like.

In the past, storage factors have not been taken into account in determination of heating load. They have, however, been built into data used for determining maximum cooling loads. The reason for this, no doubt, is that the maximum heating load of a building will occur when there is no sun. The maximum cooling load, however, is largely dependent upon solar radiation. Yet the effect of solar radiation on the system is not instantaneous but delayed because of storage. Generally this means that the system need not meet the load that would be indicated by steady-state, instantaneous, heat flow calculations, but probably quite a bit less than this. When all the zones are totaled up, the effect of storage plus load diversity will indicate a much smaller plant than the total of all calculated maximum instantaneous heat gains.

Requirements for Comfort

In order to make heat loss or heat gain calculations, an inside design air temperature that is expected to result in a comfortable thermal environment has to be assumed, and outdoor air temperatures have to be determined. The values depend upon whether the calculation is for determination of maximum heat gain or loss, or whether it is for heat gain or loss at various part load conditions. The latter, for example, would include load evaluation at all hours in a computer prediction of operating costs.

For most buildings, indoor air temperature for comfort has been taken to be close to 75F. (Recent practice in response to energy conservation criteria has set 68F for winter and 78F for summer.) With humidities and room air move- ment within the normal range, 75F has been demonstrated, by test and experience, to be preferred by people who are lightly clothed and engaged in sedentary activity such as office work. There will be variations from this temperature if people are more heavily clothed or are involved in more arduous activity.

To understand why maintenance of air temperature at a certain value does not by itself ensure comfort, we need to examine briefly what thermal comfort involves. This understanding is meaningful, not only in terms of the characteristics of the heating and air-conditioning system to be designed, but of the thermal-insulating characteristics of the building enclosure.

The purpose of a thermal environmental system is to allow the human body to lose the heat it produces so as to maintain a constant internal body temperature. This amount will vary, depending upon how much muscular activity a person is engaged in. (See Figs. 2 and 3.) If the thermal environment is such that a person's body would tend to lose more heat than it produces, various mechanisms automatically are brought into play to try to restore the balance.

The body loses heat by radiation, convection, and evaporation. When room temperature is below 70F, the skin loses most of its heat by exchange with lower-temperature room surfaces (radiation loss) and with lower-temperature air (convection). Only a small amount of heat is lost by evaporation of moisture from the skin and by respiration. Heat loss by radiation and convection decreases as the room air temperature rises (see Fig. 4), because there is less differential between the skin's temperature (98.6F) and that of the room air and room enclosure. At 80F, bodily heat loss is evenly divided between radiation and convection on the one hand, and evaporation on the other. When room temperature exceeds skin temperature, all heat has to be lost by evaporation.

Discomfort can be caused by certain local effects even though room air temperature is at a satisfactory level. This will occur, for example, if an inside surface of exposed wall, floor, or roof is too cold or too hot, as could happen if it were poorly insulated or had large areas of ordinary glass. The cold effects might be manifested in excessive radiation to the cold wall, or as drafts. The drafts could be counteracted by a current of warm air directed at the wall from the floor. The cold radiant loss could be lessened by double

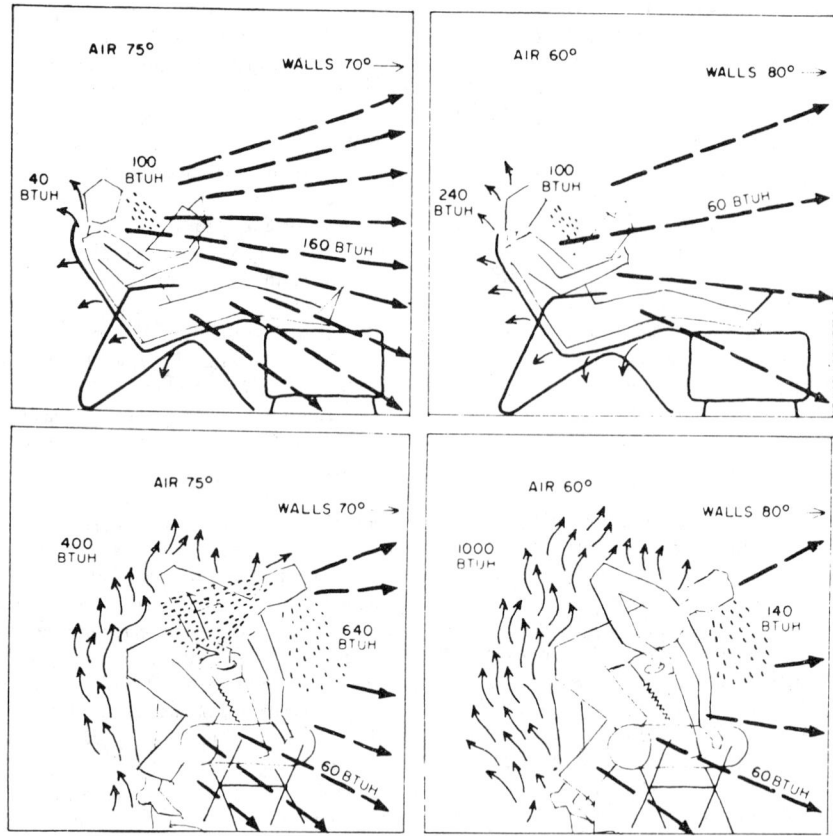

Fig. 2. Body heat is lost in light and active situations when air temperatures and surface temperatures have different relationships. The man reading is comfortable in either situation. The man sawing wood does not have to lose nearly as much heat by sweating when the air temperature is lower.

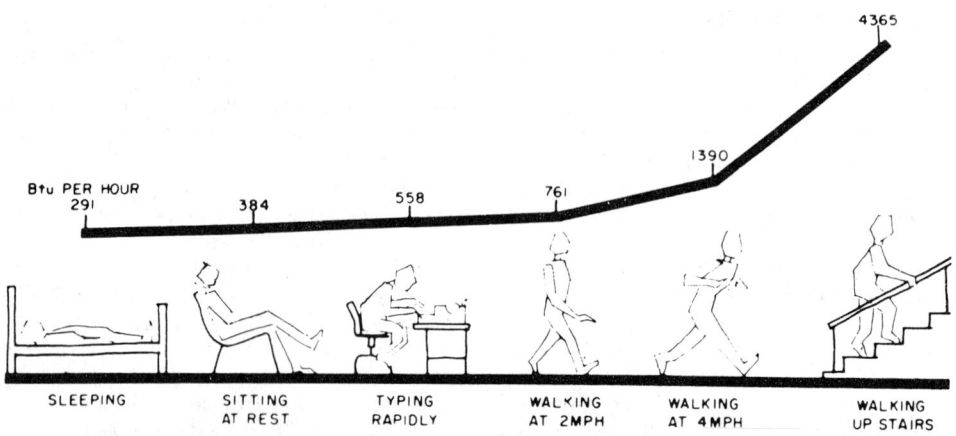

Fig. 3. The effect of different types of physical activity on heat production by the body.

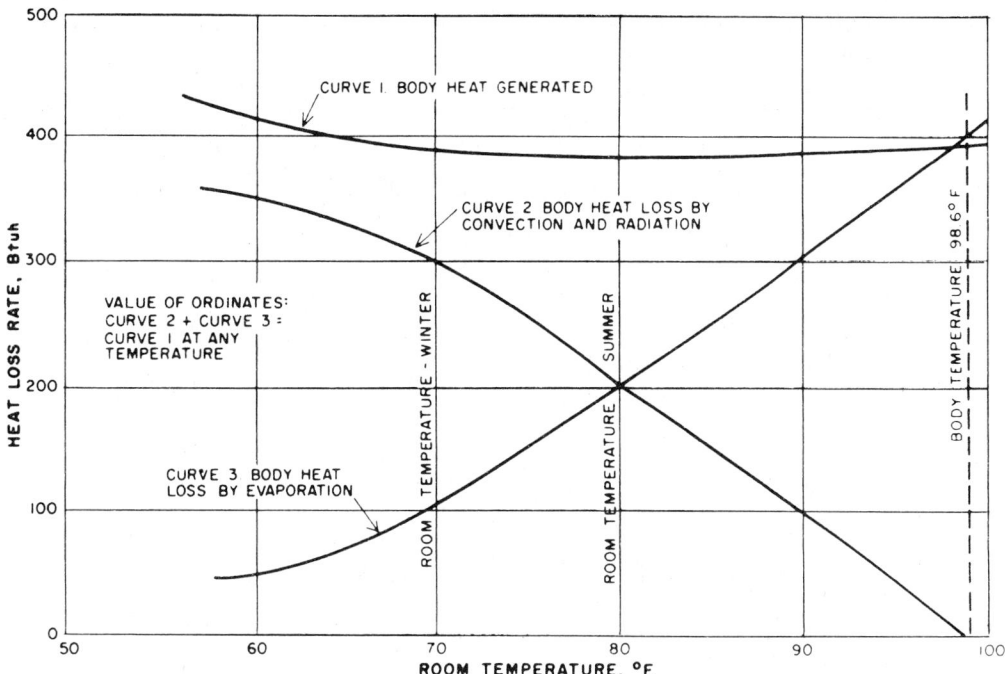

Fig. 4. At low environmental temperatures, most body heat loss is by convection and radiation, very little by moisture evaporation. At 70F, heat loss is about one-fourth by evaporation and three-fourths by radiation and convection. When air temperature is the same as skin temperature, evaporation of moisture is only source of heat loss.

glazing and by a sufficient flow of warm air. In severe situations, draperies or blinds might be necessary. The hot effects would be manifested in direct radiation of the sun and re-radiation by the wall and glass of absorbed solar energy. These effects are reduced by heat-absorbing and reflective glasses and by external and internal shading. (See Table 1.)

Another possible source of discomfort may arise from air motion, but this will depend upon the air temperature. Research studies have shown that air velocities below 40 ft per minute are not perceptible at ordinary room temperatures. In a hot environment of 86F, a velocity of 120 fpm is just perceptible. The ASHRAE comfort chart, Fig. 5, is based on air movement of from 15 to 25 fpm, and the air temperature comfort range is from 73F to 77F. Fig. 6 shows somewhat more of an effect of humidity on comfort.

In the past it was fairly common to assume winter indoor design temperature to be 70F, and summer design temperature to be 80F. The 1967

ASHRAE *Handbook of Fundamentals* recommends that 75F dry bulb temperature be used as the indoor design temperature for both heating and cooling seasons, based on research showing that, for inactive individuals, lightly clothed in still-air conditions, the line of optimum comfort ranges from 77.6F at 30% relative humidity to 76.5 at 85% rh. It also was determined that when the mean radiant temperature of room surfaces differs from the air temperature, the dry-bulb temperature should be increased 1.4F deg for each 1.0 deg MRT below air temperature and vice versa. Generally the MRT of a room will be higher than the air temperature in summer and lower in the winter. For this reason 75F was selected as the recommended summer indoor design temperature. The winter indoor design temperature was also set at 75F, even though the MRT is lower in winter, because it was assumed that people probably would wear heavier clothes in winter than the clothing worn by the subjects who participated in the comfort experiments. The ASHRAE *Handbook* further states that when the

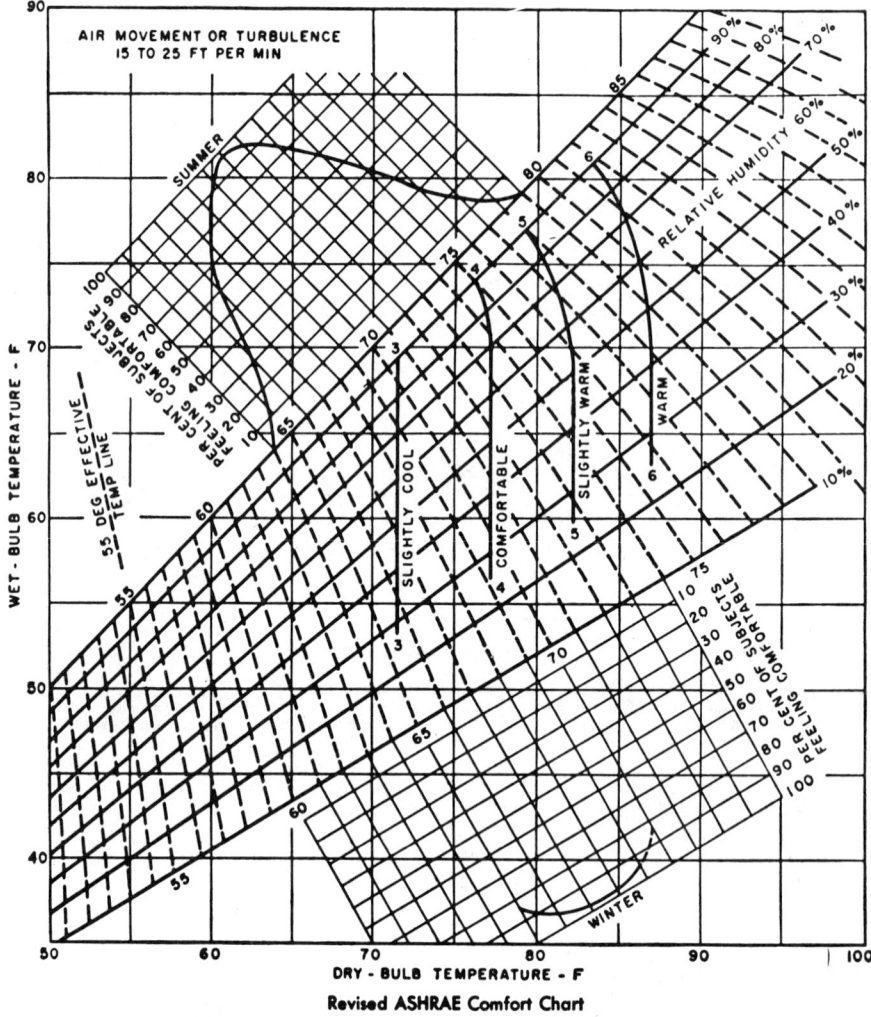

Revised ASHRAE Comfort Chart

Fig. 5. The ASHRAE Comfort Chart indicates the effect of different air temperatures and humidities on comfort. Studies showed little effect of humidity except at quite high humidities.

recommended design condition or 75F is applicable, the dry-bulb temperature may vary from 73F to 77F within the occupied zone. The recommendation for indoor relative humidity is 50% for summer cooling, with the qualification that 40% is probably more realistic for arid climates. No recommendation is given in the *Handbook* for winter indoor relative humidity, but it suggests that it should never exceed 60% to prevent material deterioration, and should not fall below 20% to prevent dryness of human nostrils and of

interior furnishings. The *Carrier Handbook of Air Conditioning Design* suggests a summer relative humidity of 45 to 50% for most applications and 50 to 55% for spaces with high latent load (restaurants, auditoriums, etc); for winter relative humidity the corresponding figures are 30 to 35% and 35 to 40%, assuming that artificial humidification is supplied. Newer ASHRAE research indicates that relative humidity can vary between 30 and 70% with no effect on thermal comfort.

Table 1. Effect of Positive or Negative Radiation from Glass

Relative		Solar Rad. Btuh F[1]	Low Temp. Rad. Btuh F[1]	Surf. Temp., deg F Glass (Large)	Out-Door Air deg F	Sensation of Warmth Vote	
Sun Inten- sity	Glass Temp.					Obs.	Calc.
High	Low	105	28	89	29	5.2	5.3
High	Med	118	33	98[1]	—	5.5	5.6
High	High	122	38	123[1]	35	5.7	5.8
Low	Low	19	−16	37	15	2.3	2.9
Low	Med	28	− 2	70	—	4.0	3.6
Low	High	7	18	119[1]	42	4.5	3.9
[1]Heated glass							

NOTE: Table 1 indicates the effect of positive or negative radiation from glass on the sensation of comfort. Low-temperature radiation is that due to surface temperature of the glass itself. The sensation of warmth was graded: 1, cold; 2, cool; 3, slightly cool; 4, neither warm nor cool; 5, slightly warm; 6, warm; 7, hot.

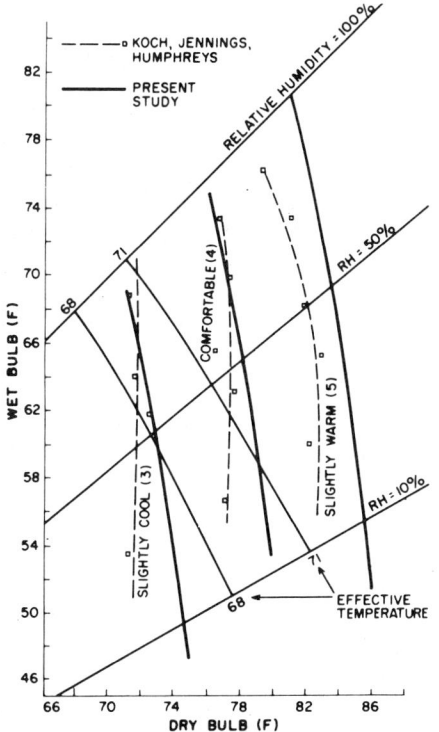

Fig. 6. Data obtained at Kansas State University for ASHRAE. The figure shows somewhat more influence of humidity on comfort.

Heat Flow Through a Structure

Sizing of the heating and cooling plant, distribution system, and terminal units is much easier than the determination of energy usage. The reason is that the main item of information needed to size the heating or cooling system is maximum heat loss or gain. Calculating energy usage is more complex because of variations in load conditions seasonally and daily. Further, the heating load is simpler to determine than the cooling load because the cooling load involves solar and internal heat gains, whose effects are partly instantaneous and partly delayed because of absorption.

The mechanisms of heat loss normally considered include (1) heat transmission from indoors to outdoors through the building enclosure, and (2) heat used by the system in warming infiltration and ventilation air. Heat gain can result from transmission, infiltration and ventilation air solar heat, lights and electrical equipment and people. (See Figs. 9 and 10.)

Heat flows through the elements of the building enclosure—walls, windows, roof, floor—whenever a temperature differential exists. The rate of heat flow depends partly upon the magnitude of the temperature differential between indoors and outdoors and the insulating value of the building section. Wind velocity and solar heat also affect how much heat will be transmitted.

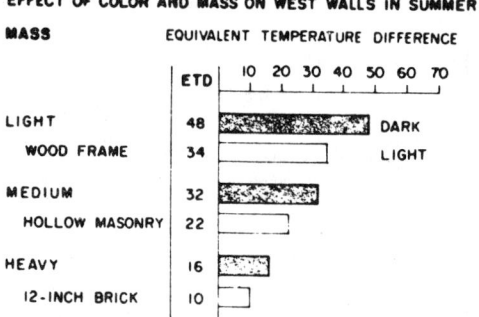

Fig. 7. Color of the building exterior as well as its mass affect the impact of solar load on a residential building, as seen on this chart.

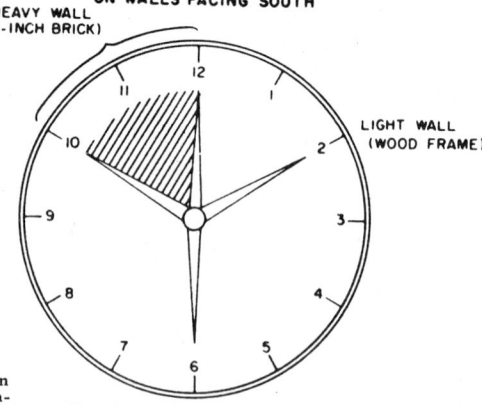

Fig. 8. Clock shows the length of time lags in heat flow that are encountered in residential construction.

Involved, too, is the relative mass and color of the material. (See Fig. 7.) It takes longer for a change in outdoor temperature to be felt through a heavy masonry wall than it does a lightweight wood frame wall—a difference of as much as four to eight hours. (See Figs. 8 and 9.)

Heat flow resulting from a difference between indoor and outdoor air temperatures can be reduced through use of thermal insulating materials, reflective surfaces, and air spaces. Still air offers an effective and practical barrier to heat flow. Relatively large air spaces which occur in some composite building sections have moderate insulating value. But more effective is the trapped air in the multitudes of interstices in various thermal insulations—tiny closed cells in foamed plastics, spaces between the particles of materials such as perlite and vermiculite, and films that cling to the masses

Table 2. Recommended Inside Design Conditions [1]—Summer and Winter

| | SUMMER | | | | | WINTER | | | | |
| | Deluxe | | Commercial Practice | | | With Humidification | | | Without Humidification | |
Type of Application	Dry-Bulb (F)	Rel. Hum. (%)	Dry-Bulb (F)	Rel. Hum. (%)	Temp. Swing[2] (F)	Dry-Bulb (F)	Rel. Hum. (%)	Temp. Swing[3] (F)	Dry-Bulb (F)	Temp. Swing[3] (F)
General Comfort Apt., House, Hotel, Office Hospital, School, etc.	74-76	50-45	77-79	50-45	2 to 4	74-76	35-30	−3 to −4	75-77	−4
Retail Shops (Short term occupancy) Bank, Barber or Beauty Shop, Dept. Store, Supermarket, etc.	76-78	50-45	78-80	50-45	2 to 4	72-74	35-30[4]	−3 to −4	73.75	−4
Low Sensible Heat Factor Applications (High Latent Load) Auditorium, Church, Bar, Restaurant, Kitchen, etc.	76-78	55-50	78-80	60-50	1 to 2	72-74	40-35	−2 to −3	74-76	−4
Factory Comfort Assembly Areas, Machining Rooms, etc.	77-80	55-45	80-85	60-50	3 to 6	68-72	35-30	−4 to −6	70-74	−6

[1] The room design dry-bulb temperature should be reduced when hot radiant panels are adjacent to the occupant and increased when cold panels are adjacent, to compensate for the increase or decrease in radiant heat exchange from the body. A hot or cold panel may be unshaded glass or glass block windows (hot in summer, cold in winter) and thin partitions with hot or cold spaces adjacent. An unheated slab floor on the ground or walls below the ground level are cold panels during the winter and frequently during the summer also. Hot tanks, furnaces or machines are hot panels.

[2] Temperature swing is above the thermostat setting at peak summer load conditions.

[3] Temperature swing is below the thermostat setting at peak winter load conditions (no lights, people or solar heat gain).

[4] Winter humidification in retail clothing shops is recommended to maintain the quality texture of goods.

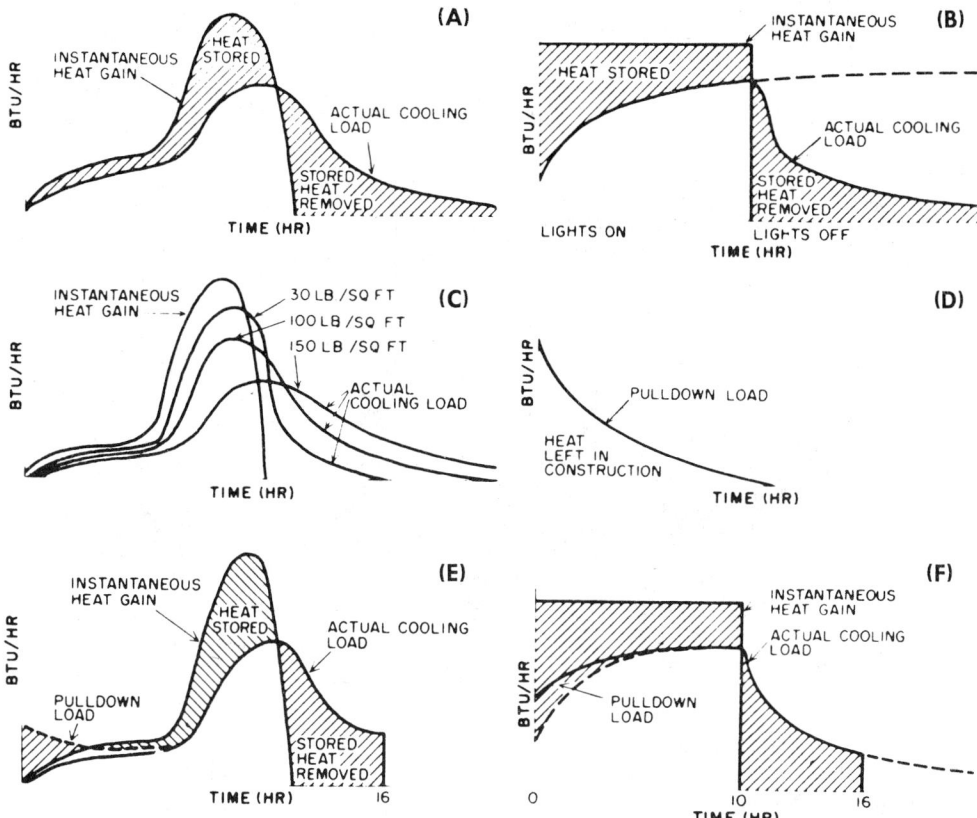

Fig. 9. Only part of the heat load from the sun and from lights is "felt" immediately by the system. The rest is stored in building materials and furnishings and released gradually. The effect of this phenomenon is to reduce the size of the peak load on the air-conditioning system and to delay it in time. (A) illustrates this for solar heat gain on a west exposure. (B) illustrates the effect of storage on cooling load due to lighting. As the construction of a building is made heavier, its stor-age capacity is increased; the effect is to both decrease the peak load and delay it further. See (C). When the air-conditioning system is shut off at the end of the operating day, some heat still may be stored by the building. This heat must be removed and is known as pulldown load. See (D). Both solar heat and lighting heat are stored; the extent of pulldown load for each is shown in (E) and (F).

of fine fibers of mineral wool insulations. For example, an ordinary air space (¾ in. with non-reflective surfaces) has a resistance to heat flow, R, greater than a ¾-in. sheet of plywood. A reflective air space theoretically has a resistance three times as large as an ordinary air space; an inch of mineral wool insulation has a resistance almost four times as large as an ordinary air space. Heat flow resulting from absorbed solar radiation can be reduced by use of light color for opaque materials and by absorptive or reflective glasses for windows.

Practically all building enclosures are comprised of composite sections, each of which has its own particular value of resistance to heat flow. The rate of heat flow through a homogeneous material under steady state conditions is called thermal conductivity, k, expressed in Btu per hr per sq ft of surface per degree F per in. of thickness. When the thickness of a particular material is known, the unit of inches can be factored out by dividing k by the thickness. Then the term is called conductance, C. It is sometimes used in the literature to indicate relative heat transmissions through building boards, masonry units, and the like. Conductance can be calculated for various thicknesses of homogeneous materials; for heterogeneous materials it is determined by test. In

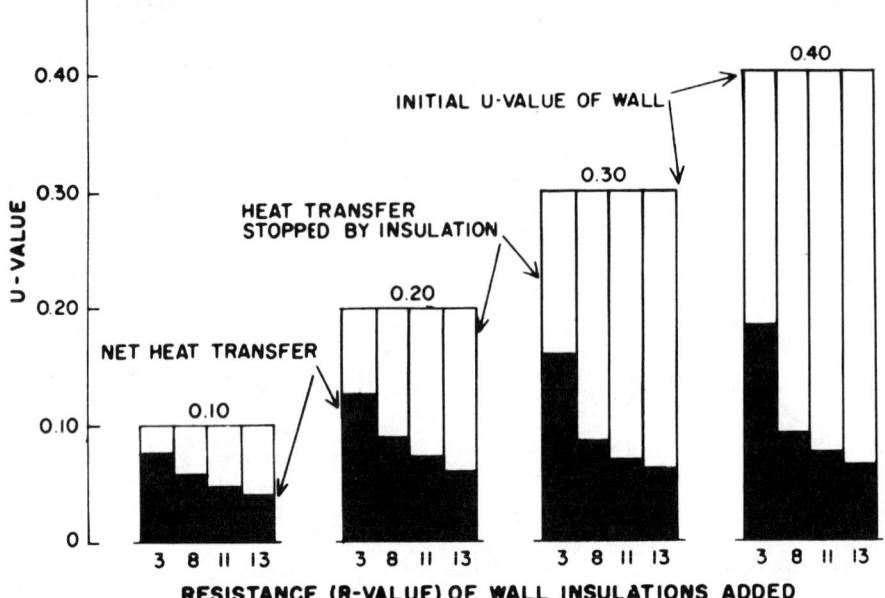

Fig. 10. Increasing the amount of insulation has more relative effect when the initial U value of the wall is high. Nevertheless, economics of insulation should, of course, be evaluated on a dollars and cents basis, not merely by comparison of U values.

practice the thermal transmittance of a building section is less than the sum of the conductances of its component materials because of inside and outside air films and air spaces, which add their respective resistances. When these resistances are added in, the applicable term becomes the overall coefficient of heat transmission, U. It is the reciprocal of the total resistance, R_t, of the section which can be found by totaling the resistances of all the elements as follows:

$$U = 1/R_t$$

Heat transfer terms and relations are covered in more detail later in this section under "Calculating Building Heat Losses."

Heat loss calculations for residences to determine system size include rates of heat loss through walls, roofs, windows and doors, basement walls and slabs, slabs on grades, floors over unheated spaces and infiltration.

For large buildings with high internal heat loads, these loads should be taken into account in the design since they offset heat loss, at least during occupied periods; sometimes during unoccupied periods, too, depending upon the design. (See Table 3.) For example, office building lights

Table 3. Breakdown of Heat Gain into Radiant and Convective Portions

Heat Gain Source	Radiant Heat	Convective Heat
Solar, without inside blinds	100%	—
Solar, with inside blinds	58%	42%
Fluorescent lights	50%	50%
Incandescent lights	80%	20%
People[1]	40%	20%
Transmission[2]	60%	40%
Infiltration and ventilation	—	100%
Machinery or appliances[3]	20-80%	80-20%

[1] The remaining 40% is dissipated as latent load.

[2] Transmission load is considered to be 100% convective load. This load is normally a relatively small part of the total load and for simplicity is considered to be instantaneous load on the equipment.

[3] The load from machinery or appliances varies, depending upon the temperature of the surface. The higher the surface temperature, the greater the radiant heat load.

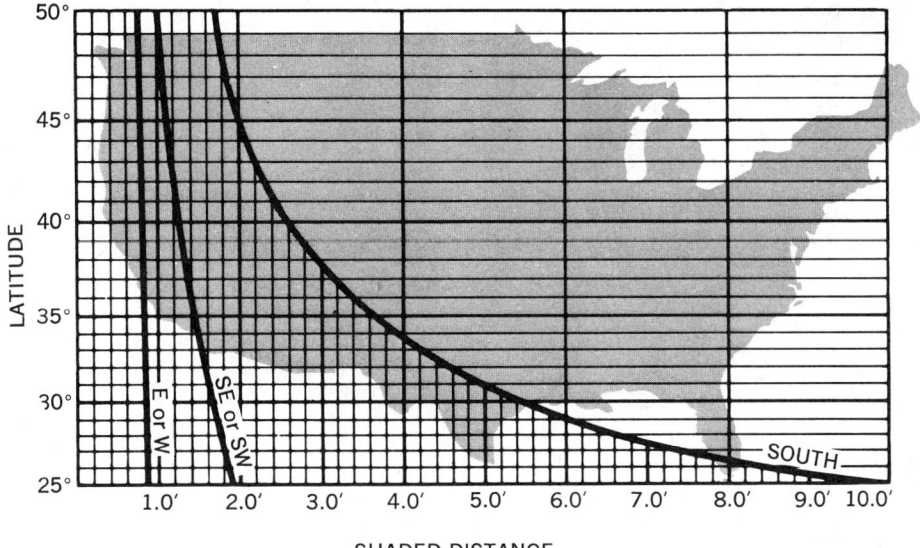

LATITUDE

SHADED DISTANCE

Fig. 11. This graph was developed to provide a quick method, yet with sufficient accuracy, for determining the shading of residential windows when calculating heat gains for the purpose of sizing air-conditioning equipment. The shaded distance per foot of overhang is the distance from the soffit to the shade line on the glass.

may be kept on for some time after working hours. Larger buildings will have ventilation-air heating loads during occupied periods. Solar load is not taken into account as a credit for residential load because it does not affect maximum load condition, which normally occurs during the early morning hours.

With heat gain calculations, on the other hand, solar load can have a big impact. In a modern residence, for example, solar load through windows alone might be one-third or more of the total. See Fig. 11. It can be considerably larger for non-residential buildings having a lot of glass. Solar load effect upon opaque surfaces is taken into account by a factor known as equivalent temperature difference, which combines the effects of mass, orientation, color of surfaces and degree of shading. Further, latent heat (moisture from people, cooking, and bathing) is taken into account in the cooling load. With houses it is roughly estimated to be 30% of total sensible heat gain.

Determination of design heat loss or heat gain by summing up the individual maximum losses or gains will give conservative results, mainly because of such influences as thermal storage and the fact that individual loads are not necessarily

coincident. This is not a particularly serious matter in regard to the sizing of most heating equipment because heating equipment by and large is less expensive per unit of capacity than cooling equipment. An exception is the heat pump which uses the refrigerant cycle for both heating and cooling and almost always is sized to meet cooling requirements. Energy cost estimates for heating based strictly on the degree-day method will also give highly conservative results. For this reason correction factors are frequently applied, particularly for all-electric residential buildings for which experience information is available. Such judgmental factors can be applied to other building types if they are of fairly standard design and sufficient operating cost data have been compiled to give a statistically accurate correction factor.

In the case of small unitary cooling equipment, there is an additional reason that cooling load calculations have been developed to a much more sophisticated state than heating load calculations. If the equipment has too much excess capacity, there can be long periods, when the refrigeration cycle is off, because the temperature criterion has been satisfied, and during which humidity can become excessive.

Available Data and Calculation Methods

Heat loss and gain calculations made for the purpose of sizing the terminal heating and cooling units, distribution lines, and the heating or cooling plant can be done using simple arithmetic. The times it takes to do them will depend upon the type and size of building, the building's internal arrangement and zoning, whether external solar shading is employed, and the relative accuracy and degree of comfort expected.

Computer solution of the calculations that relate solely to sizing considerations would seem to be indicated when the building is large and there are a large number and variety of space occupancies, internal gains, and alternate types of fenestration, wall construction, and building orientation. While it is not uncommon to size heating equipment conservatively, it is customary to size the cooling plant with greater precision, taking into account diversity of demand and storage factors. For instance, in a residential system, on-off cycling of an oversized compressor can result in poor temperature and humidity control.

There is no one standard calculation method suitable for all purposes. Generally speaking, high degrees of exactitude are not necessary for many types of buildings. By this is meant that refinements embracing every imaginable source of heat loss and heat gain will not be necessary for simpler buildings such as houses, apartment buildings, small conventional office buildings, and the like. Simplifications can be made as long as the major avenues of heat exchange are accounted for. The reason, basically, is that there is sufficient experience with these buildings to permit assuming simple straight-line factors for such components as latent heat gain (per cent of sensible load) infiltration (air changes per hour), etc. Further, the degree-day formula works with surprising accuracy in conjunction with experience factors for the estimation of heating energy usage.

There used to be many industry-promulgated calculation methods for cooling loads for residences, resulting in discrepancies of as much as 100% in estimated cooling load. Joint industry sponsored research at the University of Illinois Experiment Station on five research residences has produced a new calculation method that gives a much more accurate estimation of cooling load caused by solar energy transmission through both opaque walls and through glass with various types of shading—the main reason for the large discrepancies. (See Table 4.)

A procedure for determining heat loss, plus a method for estimating annual electrical energy usage, is presented in the National Electrical Manufacturers Association's *Manual for Electric Comfort Heating*. The heat transmission loss calculation is not done conventionally room by room, but areas of similar U factors are totaled. Then, total Btu/hr figures are calculated for similar U-factor building sections. Infiltration is assumed to be either $\frac{1}{2}$ or $\frac{3}{4}$ air change per hour. Installed heating capacity in each individual space is determined by weighting the relative linear feet of walls and square feet of windows and doors (combined infiltration effect and heat transmission effect) that each space has out of the total, then multiplying this factor by the total heat loss for the structure.

Energy consumption in the NEMA *Manual* method is determined through use of the degree-day formula, modified by an experience (C) factor. This experience factor, originally used in TVA areas to make subtractive adjustments between estimates (overages) given by the traditional degree-day formula and metered experience records, has been adopted for application to residential buildings, and, in some cases, to a few other uncomplicated building types. The C factor makes adjustments for: (1) living habits of the occupants, (2) orientation and shading effects, (3) variation in year-to-year weather, (4) workmanship in installing insulation and weatherstripping, and (5) type of equipment. The NEMA *Manual* suggests a C factor of 17 when the hourly air-change rate is $\frac{3}{4}$ and 18.5 when it is $\frac{1}{2}$. Annual kwhr are determined by the following formula:

$$\frac{\text{HL} \times \text{DD} \times \text{C}}{\text{DTD}}$$

where HL is design heat loss, DD is total degree days, C is the experience factor (in effect, a modified number of hours in a day—fewer than 24), and DTD is the difference between inside and outside design air temperature in degrees F.

Architects and engineers, asked to prepare time-period energy cost estimates in all-electric buildings, can rely upon the local electric utility for the proper C factor. This formula should be expected to work satisfactorily for buildings having fairly simple electric-heating systems and for which

Table 4. Comparison of Calculated and Measured Sensible Cooling Loads [1]

House	Method	Total Load	Glass Load
No. 1	A	12,747	5,311
	B	11,930	5,105
	Joint Industry	14,885	8,901
	Measured max.	15,080	
No. 2	A	14,751	5,306
	B	10,472	3,136
	Joint Industry	12,387	4,208
	Measured max.	11,890	
No. 3	A	21,516	8,500
	B	23,132	6,389
	C	25,059	8,801
	Joint Industry	23,270	11,657
	Measured max.	23,240	
No. 4	A	17,320	7,810
	B	14,325	7,150
	Joint Industry	16,830	8,950
	Measured max.	11,077	
No. 5	A	32,460	22,420
	B	19,975	10,825
	C	41,845	25,140
	Joint Industry	17,840	9,710
	Measured max.	19,850	

[1] Indoor air temperature of 80F used with Methods A and C. 75F in all other cases.

NOTE: Table shows the wide divergence in estimated cooling loads possible with calculation procedures recommended in the past by various industry associations. The principal reason for this was the different methods used for determining solar load. Because of this problem, and at the urging of FHA, a special Industry Heat Gain Joint Study Group was formed to develop a unified procedure. The results given by this procedure are listed in the table. The houses were all research residences at the University of Illinois Engineering Experiment Station.

there is sufficient experience information. It cannot be expected to work accurately for buildings having new types of heating systems and equipment for which metered data have not been accumulated, and for buildings in which there are complicated forms of heat-recovery systems and devices. Then a much more detailed analysis must be performed, involving estimates of heating system and heat-recovery device efficiencies, load coincidences, diversity factors, and the like. Computer analysis may be indicated because of the time-consuming aspect of the calculations. But even with a detailed computer analysis, energy usage estimates may vary from actual fact because of a lack of information regarding how the system responds to varying heating loads, from hour to hour and from one part of the building to the other. A fair amount of sophisticated information has been built into the heat transmission data in the ASHRAE *Handbook of Fundamentals*. But this accounts only for the response of the building enclosure to varying external loads. It tells nothing about how the building's thermal system responds. The reason for this situation is that, first of all, hardly any buildings are submetered sufficiently to indicate the details of power usage

of the heating and air-conditioning components. Further, manufacturers by and large do not furnish adequate minimum-to-maximum performance data under all possible operating conditions for many types of mechanical equipment, nor are such data that are available necessarily accurate at other than the rating point. When manufacturers furnish these data, computer programs will be developed, and those that exist will be refined, to reflect the characteristics of actual mechanical systems.

Evaluating Alternate Approaches to Design

Building professionals have become more knowledgeable and experienced in making fairly accurate judgmental decisions early in project development, regarding the feasibility of alternate systems for environmental control. Each has his own approach in arriving at guidelines, depending upon certain design preferences and past experience. Actual on-site building inspection and a critical examination of details of design, operating performance and cost data are the most worthwhile and valid means for aiding the judgment of those responsible for system selection and design.

The engineer's preliminary design activity should include specific discussion with the local utilities of watts per square foot for lighting in various areas, and of total installed load for equipment as well as estimated total miscellaneous load of appliances and business machines. On the basis of this information, the utilities should be asked for a specific estimate of monthly kilowatt demands and monthly energy usages resulting from these loads. The utilities have considerable experience data on this type of load breakdown. The designer's role here is to indicate to the utility any special considerations pertaining to design and operation that might require the utility to modify its data if the reference base is in any way different from the average installation.

The designer should study the utility's rate schedule in detail, making sure that he understands the various bases upon which rates are established for different situations. Having all this information in mind, along with a knowledge of various systems and equipment compatible with the overall concepts of the building being designed, the engineer can make some basic decisions regarding one or several design approaches that seem to be indicated for lighting, heating, and air-

conditioning systems. The goal is the least degree of complexity possible while achieving a certain standard of performance.

For buildings other than those of residential and small commercial occupancies, the reason for considering more than merely maximum heating and cooling requirements is that there is a wide range of possible "balancing temperatures." These will vary with type of building and occupancy, with type of heating and cooling system, and with the degree of heat recovery, redistribution, and storage. As a matter of fact, the 65F base used in determining degree days is, in effect the "balance temperature" for the typical non-electric residence, i.e., it is the assumed outdoor temperature above which no heating will be required. This base is too high where there are high insulation standards and high internal loads from appliances and lighting.

The lower the balance temperatures for various building operating periods, the less the energy used for heating and the lower the energy demand. When the designer has made a preliminary selection of system and equipment, he then can determine the outside temperatures and occupancy periods during which equipment will operate at various part-load capacities and efficiencies.

In the past there has not been too much concern over especially accurate energy usage determinations. Fuel costs were evaluated on a comparative basis, generally using degree-days for estimation purposes. Even now, one compares relative costs of various energy sources to help in system selection. It is essential that all important relative differences be taken into account. This means that the seasonal weather data being assumed can be for a hypothetical year and that various simplifying assumptions concerning temperature, wind, and solar effects are permissible, as long as these do not affect the relative difference in energy usage between systems.

But with the advent of the all-electric building, more definitive data on operating costs, because of the higher cost of the "raw" energy, are needed. This is more nearly possible today because of the availability of more detailed weather data resulting in growing use of temperature data on an hour-by-hour basis. With such data, one can break down required plant energy input on a temperature-level-bracket basis and know when these periods occur. In those cases where the electric utility has a monthly demand charge applicable to heating, for instance, one will

need to determine the lowest temperature bracket for each month in order to determine the demand charge.

In the making of energy-usage estimates, except for simple in-space resistance heating, the varying efficiencies of equipment under different load conditions must be taken into account. The most important reference base is, of course, outside temperature. This varying efficiency applies particularly to those systems involving air-source heat pumps and exhaust-air heat-recovery equipment.

Guidelines for Studying a Particular System

In evaluating the mechanical and electrical systems of a particular building, the following steps are useful:

1. Securing simplified floor plans, and adding the information necessary to indicate occupancies and usage of all spaces. This would include: occupancy densities; lighting, business machine, and other electrical equipment wattages; important design features with respect to type of lighting fixtures, air distribution supply and return, devices (including all terminal heating and cooling provisions at the perimeter with manufacturers' types and size ranges); zoning; equipment room areas with designations of principal pieces of equipment.
2. Listing all important construction design features: fenestration; insulation; type of structural frame, walls, partitions; sound and vibration isolation control features.
3. Listing design data on all heating and cooling design loads, heat recovery estimates, etc.
4. Listing sizes and types and manufacturers of all major equipment, and areas or functions served by them, including horsepower and energy consumption figures for each.
5. Drawing simplified schematic flow diagrams—air, water, electric, etc.
6. Drawing simplified automatic temperature control diagrams and noting all provisions for central control or monitoring.
7. Obtaining a description of operation of the various systems if they are available in some published form, and making detailed notations of variations from this description in the as-built installation. Give

details of part-load operation, system cycle changeover, and other features.
8. Obtaining detailed utility rate schedule breakdown and interpretation, as applicable to the installation.
9. Drawing schematic riser diagram indicating electric power demand and kwh metering.
10. Describing building space occupancies and usages by time of day, day of week, and season.
11. Obtaining breakdown of initial cost data on systems.
12. Looking for indications of particularly well thought-out, novel and/or economical solutions to the problem of physical integration or coordination of equipment or distribution systems into the structure and the overall architectural design.

Evaluation of a particular installation will be more worthwhile if site inspection of the building and systems under study are only part of the overall review. There will always be information gaps that have to be extrapolated, and, on the more complex installation, a complete picture can never be obtained. Before inspection of a building having relatively new equipment or areas of design, one should read up on the job to get the most out of the inspection trip. Additional inspections of parts of installations will often prove necessary so that one can draw his own conclusions. Also, it is preferable to visit installations at those times of the year when the installations are put to their most critical test. Such inspections should serve to validate any design assumptions about which there is some uncertainty.

When all of the basic data, as listed earlier, are down on paper, along with comments on system operating performance, then it is time to review the particular design approach and come to a reasoned judgment as to its appropriateness. One should consider making, in this after-inspection review, qualitative and quantitative commentary for future reference.

Computer Calculations

The computer is being applied to design of thermal environmental systems because of (1) the potential savings in manhours in doing repetitive, time-consuming calculations, (2) the promise of more exact solutions, and (3) the possibility of performing complex mathematical computations for these more exact solutions that would have

been impractical heretofore because of the inordinate amount of time required.

For thermal systems, the computer is employed to perform load calculations, energy-usage estimations, system simulation, prediction of system performance, minimum cost selections (see Fig. 12) and control of system operation. The six applications listed will find their greatest initial acceptance in connection with large jobs in which a large amount of money is at stake.

The question of when computer use is indicated is not easy to answer. For the simpler type of calculation such as steady-state heat loss, engineers find that while they can save time in getting the results they probably have not saved any money in doing so. On the other hand, they might find the computer appropriate if a large number of different constructions and materials for an enclosure are being considered or if the building is very large and has many different zones and types of construction, simply because of the large number of manhours involved in computing results by hand.

One type of computation for which the computer is especially well suited is that of nonperiodic heat flow caused either by varying outdoor and indoor temperatures or by varying solar load. With theoretical computational procedures relating heat loss and gain to energy consumption, there is no doubt that—accepting the premises taken for weather and system operation—precise results can be obtained. How precise depends on how accurate the weather data are, how well the thermal-conditioning system responds to changes in heating or cooling load, also how much inherent waste there is in the system caused by reheat or by mixing of hot and cold fluids to meet supply temperatures called for by room thermostats.

SOURCES AND CREDITS: Figs. 1, 7, 8 and 10, "Thermal Design of Buildings," Tyler S. Rogers, John Wiley & Sons, Inc.; Figs. 2 and 3, *Architectural Forum*, Nov. 1948; Fig. 4, "Mechanical and Electrical Equipment for Buildings," 4th ed., McGuinness, Stein, et al, John Wiley & Sons, Inc.; Fig. 5, "ASHRAE Handbook of Fundamentals"; Fig. 6, *ASHRAE Journal*, April 1966; Figs. 9, 11 and Tables 2 and 3, "Carrier Handbook of Air Conditioning System Design," McGraw-Hill Book Co.; Fig. 12, *Electrical Construction and Design*; Table 1, "A Subjective Evaluation of Effects of Solar Radiation and Reradiation from Windows on the Thermal Comfort of Women," Schutrum, Stewart and Nevins, ASHRAE, June 1968.

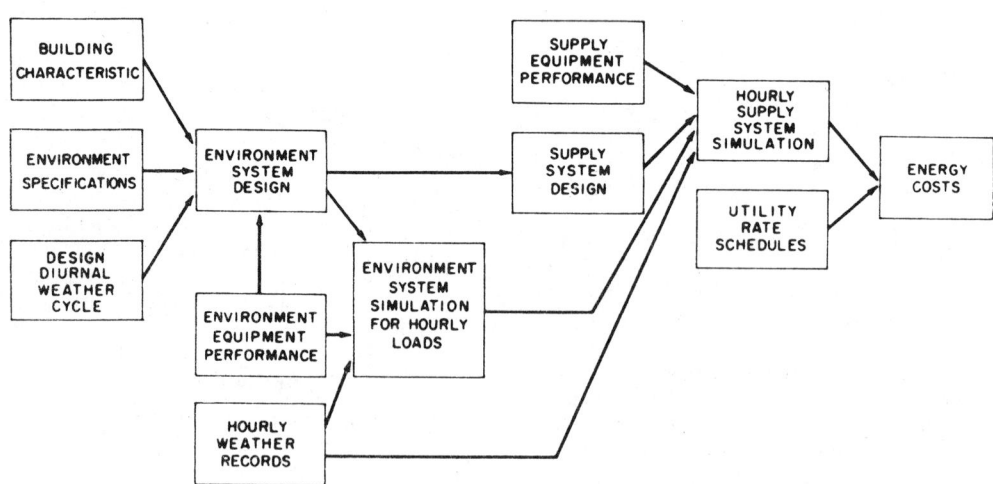

Fig. 12. Logic diagram for computer calculation of energy costs. Blocks indicate various inputs and subroutines involved.

HEAT AND MOISTURE IN AIR

Composition of Air

Atmospheric air is a mixture of gases and water vapor. Its main constituents are oxygen and nitrogen, but air also contains a portion of water vapor, some carbon dioxide, traces of other gases like argon, neon, etc.; also a certain amount of dust particles.

Dry air is air containing no water vapor. This term is used comparatively, since in nature there is always some water vapor included in atmospheric air.

Table 1 — Composition of Dry Air

Gas	Per Cent by Weight	Per Cent by Volume
Nitrogen	75.47	78.03
Oxygen	23.19	20.99
Argon	1.29	.94
Carbon Dioxide	.05	.03
Hydrogen	.00	.01
Xenon, krypton, and other gases	minute portions	

Table 1 represents one unit volume of dry air without any moisture whatsoever. The percentage of water vapor in reality varies considerably. For instance, when at 14.696 psia total pressure and in a saturated condition, the percentage of water vapor (by weight) is 0.1% at 5F, 1.0% at 57F, and 3.0% at 89F. Water vapor in atmospheric air is principally the result of evaporation from the surfaces of bodies of water and from the ground and other objects containing moisture. In addition, moisture is added to the air in buildings by cooking, baths, and human beings themselves.

Heat

Heat, according to the present theory, is a vibration of the molecules of a body. The standard unit commonly used in measuring heat in the United States and in England is the *British thermal unit,* commonly known as the Btu, which is the quantity of heat required to raise the temperature of 1 pound of water 1 degree Fahrenheit. Hence, 10 Btu will heat 1 pound of water 10F, or 10 pounds of water 1F.

Sensible heat is heat that is manifest to body senses, because changes in this form of heat are indicated by the sense of feeling. It may also be defined as heat which changes the temperature of a substance.

Latent heat. The heat that is required at a given pressure to produce changes in the physical state of a substance—to change it from solid to liquid, from liquid to gas, or the reverse of either, without changing its temperature—is called latent heat. Thus, the change of physical state may be either melting, freezing, vaporization (evaporation) or condensation. During melting and vaporization, heat must be supplied to the substance; during freezing and condensation, heat must be taken from it.

Latent heat of fusion is the amount of heat that is required for the fusion (melting) of a solid, such as ice, or the freezing of a liquid, such as water. Melting or freezing at constant temperature is generally characteristic of crystalline substances. Amorphous substances, such as tar and glass, pass imperceptibly from one state to another. It is customary to use one pound of a substance as the basis for comparing latent heats of fusion. A pound of ice at 32F requires 144 Btu to convert it into water at 32F. Hence, the latent heat of ice is said to be 144 Btu.

Latent heat of vaporization. If a quantity of water is boiled in the open air at sea level and its temperature is noted, it will be found that the temperature remains at 212F until the water entirely disappears. The water has absorbed a large quantity of heat while being converted into vapor without change of temperature. The amount of heat thus expended in converting a pound of water at the boiling point into steam at the same temperature is called the latent heat of vaporization of steam. The latent heat of steam at 212F is 970.4 Btu; that is, it requires 970.4 Btu to convert one pound of water at 212F to steam at the same temperature, under atmospheric pressure. The latent heat of vaporization of any liquid is the amount of heat required at a given pressure to change one pound of it from the liquid to the vapor state or vice versa without increase of temperature.

Specific heat. Specific heat is the number of Btu that must be added to, or subtracted from each pound of a substance to change its temperature one degree Fahrenheit. Different substances have dif-

ferent specific heats. Thus the specific heat of water at 15 degrees Centigrade is 1.000; of iron 0.113; of coal 0.314. The specific heat of air is the quantity of heat required to raise the temperature of air 1F at constant pressure, c_p or constant volume, c_r. Specific heat of air at constant pressure is 0.24 Btu at any condition likely to be encountered in air conditioning.

Heat of liquid is (in the case of water) the heat required to raise water at 32F to the temperature at which it evaporates. When referred to water vapor mixed with air, the temperature at which it evaporated is the same as that at which it would condense, which temperature is called the dew point.

Enthalpy is the term used for the total heat of the mixture of air and water vapor above a point of reference. Heat of air is measured above oF (usually above oC in the metric system). Heat of the water vapor is measured above the freezing point of water (at one atmosphere) or 32F (oC in the metric system).

The enthalpy of a mixture of air and water vapor is the sum of (1) the energy required to heat the air from oF to the mixture temperature, (2) to heat the water from 32F to the dew point or temperature of evaporation, (3) to evaporate the water (latent heat) at the dew point, and (4) to heat the water vapor from the dew point to the mixture temperature.

A term called the *sigma function*, denoting the sum of (1) and (3) only, is sometimes used.

Heat content of water vapor at saturation is very closely approximated by $1061 + 0.444t$, when t is in degrees F.

Enthalpy of a mixture of air and saturated steam is given by

$$h = 0.24 + W(1061 + 0.444t)$$

where h is the enthalpy in Btu per lb of dry air
0.24 is the specific heat of dry air in Btu per lb
W is the weight of moisture in lb per lb of dry air

Absolute temperature is the temperature of a substance measured above absolute zero. *Absolute zero* is the lowest temperature theoretically attainable. It corresponds to −273C and −459F. In the metric system, addition of 273 to the Centigrade temperature converts the temperature to the absolute scale, or degrees Kelvin (°K); addition of 459.6 (or 460) converts Fahrenheit temperature to the British absolute system, degrees Rankine (°R).

Pressure

Atmospheric pressure is the pressure exerted by the earth's atmosphere and is about 14.7 pounds per square inch at sea level. That means that a column of air, which has a cross-sectional area of one square inch and which extends from sea level to the upper limit of the earth's atmosphere, weighs 14.7 lbs, or more exactly, 14.696 lb. This pressure is often called *one atmosphere.*

Barometric pressure. The barometer is an instrument which operates upon the principle of equilibrium between the weight of a column of mercury and the weight of a column of atmospheric air of the same cross section. Pressure which is read on a barometer is called barometric pressure. This condition of equilibrium is attained when the mercury column falls to a height of 29.92 inches at sea level. Barometric pressure is affected by altitude above sea level, temperature and humidity.

Absolute pressure. Gage pressure indicates only the pressure above atmospheric pressure. Absolute pressure is the true total pressure. Consequently, absolute pressure (usually, pounds per square inch absolute, psia) equals atmospheric pressure plus gage pressure (usually, pounds per square inch gage, psig). Psig + 14.7 = psia at sea level.

Density is the weight of a unit volume of substance — that is, a body's weight divided by its volume. The weight and volume may be measured in pounds and the volume in cubic feet, the resultant density is expressed in pounds per cubic foot.

Pressure, temperature, volume relationships. Air, in the range considered for air conditioning, like most other gases not too near their critical temperature, acts very much like a perfect gas. Therefore, the gas volume change, corresponding to a pressure or temperature change, can be calculated in accordance with Boyle's and Charles' laws.

Boyle's law. At constant temperature, the volume of a given weight of gas varies inversely with its *absolute* pressure. Thus, if a particular weight of gas occupies one cu ft at one psia, it will occupy ½ cu ft at 2 psia and 2 cu ft at ½ psia.

Charles' law. If the pressure of a gas is kept constant, its volume varies directly with its absolute temperature. If its volume is kept constant, its absolute pressure varies directly with its absolute temperature.

These laws can be stated mathematically by the expression:

$$PV = RT$$

where

P = absolute pressure, lb per sq ft
V = specific volume, cu ft per lb
R = gas constant (53.3 for air)
T = absolute temperature (deg Rankine, or $F + 459.6$)

and

$$V = 386/m$$

where

V = volume, cu ft per lb at 29.92 inches Hg and 70F
m = molecular weight of gas

Table 2 gives values of R and m for various gases. The preceding expressions apply only when the particular materials are in gaseous form at the pressure and temperature being considered.

In connection with specific heat, where c_p or c_v is known and the other is desired, the following formulas are useful for reasonable approximations:

$$c_p = \frac{1.99}{m} + c_v$$

$$c_v = c_p - \frac{1.99}{m}$$

Dalton's law. In a mixture of gases and water (such as atmospheric air) the total pressure of the mixture is equal to the sum of the pressure which each would exert separately if it were alone in the space occupied by the mixture. (Dalton's law is rigidly accurate for ideal gases only, but the deviation from this law for actual gases is so slight, that it may be considered reasonably accurate for air and water vapor mixtures also at temperatures above −200F and pressures not over about 1,000 psi.) This principle, which is sometimes called the law of partial pressures, may be made clear with the following simple experiment: Place 1 cu ft of oxygen and 1 cu ft of nitrogen, at atmospheric pressure, together in a vessel of 1 cu ft capacity and the pressure of the mixture will (since atmospheric pressure equals 14.7 psia) become 14.7 × 2 = 29.4 psia.

Evaporation and Condensation

Vapor pressure is the pressure exerted by any vapor, whether by itself or mixed with other gases. When there is a mixture of gases, it refers to that part of the total pressure of the atmosphere which is exerted by the vapor of a particular substance. In air conditioning, the term usually refers to the water vapor pressure. It is always given in terms

Table 2 — Values of Gas Constant R, Molecular Weight m and Specific Heats of Various Gases

Gas	R	m	Specific Heat of Vapor, Btu per lb per F	
			Constant pressure c_p	Constant Volume c_v
Acetylene	59.34	26.02	0.36	0.29
Air	53.34	28.97	0.24	0.17
Ammonia	90.50	17.03	0.52	0.40
Carbon Dioxide	35.12	44.00	0.21	0.16
Carbon Monoxide	55.14	28.00	0.24	0.17
Refrigerant 12	12.77	120.92	0.15	0.13
Refrigerant 21	15.00	102.92	0.18	0.16
Ethane	51.38	30.05	0.41	0.35
Hydrogen	765.86	2.02	3.42	2.44
Isobutane	26.58	58.08	0.55	0.50
Methane	96.31	16.03	0.52	0.39
Methyl Chloride	30.59	50.48	0.24	0.20
Nitrogen	54.99	28.02	0.25	0.18
Oxygen	48.25	32.00	0.22	0.16
Propane	35.04	44.06	0.58	0.51
Steam	85.78	18.01	0.46	0.36
Sulphur Dioxide	24.10	64.07	0.15	0.12
Refrigerant 11	11.24	137.37	0.14	0.12
Water vapor	85.78	18.01	0.46	0.36

of absolute pressure and usually in pounds per square inch or inches of mercury (in. Hg).

Evaporation is the change in state of a substance from a liquid to a gas. It takes place when the liquid temperature corresponds to a saturation pressure greater than the vapor pressure of the same substance in vapor form which is in the space in contact with it. (Evaporation will occur even though there are other gases in the space, and the total pressure in the space exceeds the vapor pressure of the liquid.)

When the substance is in solid form and the vapor pressure of the same substance in the surrounding space is less than that corresponding to the temperature of the solid, some of the solid will change to gas. This change is called *sublimation*.

When the total pressure of the gases in contact with a substance in liquid form is less than the vapor pressure corresponding to the temperature of the substance, the evaporation takes place within the liquid, as well as at the surface. This type of evaporation is referred to as *boiling*.

Rate of evaporation is in direct proportion to the difference between the vapor pressure of the liquid and the lower vapor pressure of the same substance in the space adjoining the liquid surface. As liquid evaporates, two things happen:

1. Latent heat required for evaporation is drawn from the liquid and from any gases in the space in contact with the liquid, thus removing sensible heat from the liquid and gases and reducing their temperature. Thus, the sensation of cooling when part of one's skin is wet and surrounded by a mixture of air and of water vapor whose water vapor pressure is lower than that corresponding to skin temperature. It is still more apparent if alcohol or ether is dropped on the skin because of the higher vapor pressure of those liquids at the same temperature and the practically zero vapor pressure of alcohol or ether in the usual atmosphere.

2. The vapor pressure in the space adjoining the liquid increases.

It is evident that both of these changes tend to reduce the difference causing the evaporation. If evaporation is to continue, heat must be added to the liquid to replace what is removed and maintain the temperature. Also the vapor must be continually removed to prevent an increase in the vapor pressure in the space adjoining the liquid surface.

In many cases, both functions are taken care of by circulation of an air-vapor mixture over the liquid surface. The temperature of the mixture is higher than that corresponding to its vapor pressure. In circulating over the liquid surface, the mixture adds sensible heat to the evaporating liquid and displaces the accumulating vapors.

Product drying. If evaporation takes place at a temperature corresponding to a vapor pressure higher than the total pressure of the space adjoining the liquid surface, there is no necessity for mechanically displacing, by vapor at a lower pressure, the vapor which results from the evaporation, since the generated gas moves to lower pressure areas because of its own higher pressure. (That is, a vapor boiled off moves out under its own steam!)

An illustration of this is the drying of a product at sea level pressure (14.7 psia) and 240F. The pressure of water vapor at 240F saturated is 25.0 psia, but the pressure of the vapor cannot exceed the 14.7 psia of the ambient air. It is only necessary to add the sensible heat to replace the latent heat which is removed as liquid changes to vapor.

This heat is often added by forced circulation of a gas over the product being dried. The gas is recirculated after passing through a heating coil which raises its temperature from 240F to about 245F. Initially, the gas is air with a little water vapor. As the vapor pressure increases, the total pressure of the mixture increases above 14.7 psia and some of the air and water vapor leave. Since no new air is added, the only addition is water vapor.

With continual discharge of air and water vapor and addition of water vapor only, in time nearly pure water vapor is being circulated at 240-245F and 14.7 psia. This water vapor delivers the sensible heat required as it cools from 245F to 240F.

Because the specific heat of water vapor is nearly double that of air, and its density is about 60% of air at the same temperature and pressure, the vapor is about 20% more effective than air per unit of volume, for sensible heat transfer to the product.

When the liquid is in a material such as wood, cotton, etc., its vapor pressure is somewhat less than that corresponding to saturation for the free liquid at the same temperature. For this reason, all water would not be evaporated from such a material even if its temperature were maintained indefinitely at a value somewhat higher than that corresponding to the water vapor pressure in the space in contact with it.

Condensation. Condensation is the change in state from a gas or vapor to a liquid. If a vapor is in

contact with a surface (solid or liquid) at a temperature below that corresponding to saturation at the actual vapor pressure, some vapor will change to a liquid on the surface. This is illustrated by a cold glass of water on a summer day. Although the condensation on the glass is most apparent, condensation also takes place on the surface of the water. If the surface is at a temperature below freezing, the vapor will condense as ice.

If two quantities of vapor at different temperatures are mixed and if the temperature of the mixture is lower than that corresponding to saturation at the pressure of the resulting vapor, then condensation will occur in the form of minute droplets suspended in the space. This is called *fog* near the earth and *clouds* at high altitudes. It is incorrectly called steam as it leaves a kettle spout or a pan on a stove.

In fog form, water does not usually freeze to form frost just because it is at less than 32F. In fact, unless it is disturbed in some manner, fog will probably not form ice crystals (frost or snow) until it is about −40F. If a solid body passes through fog at less than 32F, ice forms on its surfaces. Examples are the icing up of automobile windshields and airplane wings.

Rate of Condensation on a surface is directly proportional to the difference between the pressure of the vapor and the vapor pressure corresponding to the temperature of the surface. As condensate forms on a surface, its latent heat is transferred to the surface and the vapor pressure decreases. Condensation continues at the same rate only if heat is removed, in some manner, from the surface and additional vapor brought to it to maintain the vapor pressure.

Air-Water Vapor Mixtures

Saturated vapor. A vapor is saturated when it is in equilibrium with the liquid of the same substance at the same temperature. (The equilibrium condition may be visualized as one in which neither evaporation nor condensation is taking place, although the true condition is probably one in which the two processes are occurring at equal rates.)

Saturated air. The term is often used to indicate that the space occupied by the air is also occupied by saturated water vapor. Actually, only the water vapor is saturated. (Contrary to the popular conception, air is in no sense a sponge. It cannot absorb water vapor.)

Dry air is air with no moisture present. When changing the condition of a mixture of air and vapor, moisture may be added or removed. The weight of the air is the only thing which remains constant. For this reason, everything is referred to the weight of dry air in the mixture.

Humidity is the moisture or water vapor occupying the same space as the air.

Absolute humidity is the weight of water vapor per unit volume, usually expressed in pounds or grains per cubic foot.

Specific humidity is the weight of water vapor per unit weight of dry air, usually expressed as pounds or grains of vapor per pound of dry air.

Specific humidity in pounds per pound $= 0.622\ P_w/P_a$

where $P_w =$ partial pressure of water vapor,
and
$P_a =$ partial pressure of dry air

Since $(P_w + P_a)$ is usually 29.92 inches of mercury, when P_w and P_a are in inches of mercury, the specific humidity in lb per lb

$$= 0.622\ P_w/(29.92\text{-}P_w)$$

Of course, the value in grains per pound is 7000 times the lb per lb value.

Relative humidity is the ratio of actual vapor pressure to saturated vapor pressure at the same temperature. It is also the ratio of the absolute humidity to the absolute humidity of saturated vapor at the same temperature. Note that no reference is made to air, which may or may not be present in the space. The term relative humidity has meaning even above 212F at 14.696 psia total pressure (the boiling point of water).

Percentage humidity is the ratio of the specific humidity to the saturated specific humidity at the same temperature. At temperature of 90F and less and 14.696 psia total pressure, the differences between the values of relative humidity and percentage humidity are less than 3%. However, the differences increase at higher temperatures, becoming infinite at 212F and 14.696 psia. At the temperature corresponding to saturated water vapor at any total pressure, percentage humidity is meaningless. "Humidity ratio" and "degree of saturation" have also been used with the same meaning as percentage humidity. This ratio is not as useful as relative humidity and it is being used less frequently.

Temperature and relative humidity. If the temperature of an air-vapor mixture is increased with no change in its moisture content, its relative humidity is reduced. For example, if outside air at 30F saturated (100% rh), which has 1.93 grains per cu ft, is heated to 70F, where saturation corresponds to 8.06 grains per cu ft, the relative humidity would be 1.93/8.06 or 24%.

If the temperature of an air-vapor mixture is reduced, the relative humidity will rise. If any surface, which is cooling the mixture, has a temperature lower than the dew point of the mixture, condensation will take place on the surface and sometimes as fog near the cool surface.

Measurement of Air Conditions

Dry-bulb temperature is the true temperature of the gas, or mixture of gases. It is measured with an accurate thermometer with any necessary corrections made for radiation.

Wet-bulb temperature is the temperature indicated by a wet-bulb thermometer constructed and used according to specifications. Basically, this is a thermometer with a thin cloth, wet with distilled water, covering the bulb and means for moving the air, or other gas, past the wet bulb at the prescribed rate.

If the water vapor in the sample, the wet-bulb of which is to be measured, is not saturated, evaporation takes place with a resultant lowering of temperature until equilibrium is reached. Thermodynamic wet-bulb temperature is the temperature of adiabatic saturation.

As the water vapor in an air-vapor mixture is saturated, the wet-bulb temperature remains constant. However, at any temperature above 32F, the enthalpy increases (very slightly at ordinary atmospheric conditions) because enthalpy includes the heat of the liquid (from the base 32F) whereas the water is added at the wet-bulb temperature.

Dew point temperature is the temperature at which condensation begins. If the temperature of a surface in contact with the air, is gradually lowered, the surface temperature when the first trace of a film of moisture appears on the surface is the dew point.

Obtaining relative humidity. If dry-bulb and wet-bulb temperatures of an air-vapor mixture have been measured, relative humidity can be obtained from a psychrometric table or chart or a table of wet-bulb depressions (differences between dry- and wet-bulb for a particular relative humidity at the actual dry-bulb temperature).

If the dew point and dry-bulb are known, the corresponding vapor pressure for each can be obtained from a psychrometric table. The vapor pressure at the dew point divided by the vapor pressure at the dry-bulb temperature is the relative humidity. Relative humidity can be read directly from a psychrometric chart when dry-bulb and dew point are known.

Saturation and 100% relative humidity define the same condition. At saturation, dry-bulb, wet-bulb and dew-point temperatures are the same. When fog is present, dry-bulb and wet-bulb are the same.

Comfort Conditions

Effective Temperature is an arbitrary index which combines into a single value the effect of temperature, humidity and air movement on the sensation of warmth or cold felt by the human body. The numercial value is that of the temperature of still, saturated air which would induce an identical sensation. Thus, an atmospheric condition is said to have an Effective Temperature of 70F when it induces the same sensation of warmth or cold as sensed in still, saturated air at 70F.

Experience has shown that the effect of humidity is not as great as indicated by the Effective Temperature scale when the air temperature and the activity of the person are such that little heat is lost by evaporation of perspiration. If an individual is performing sedentary activity and the temperature of the air is about 75F to 77F, a range of relative humidity from 30% to 65% has no effect on comfort. A higher rate of activity, at the same temperature will result in a feeling of warmth but a relative humidity of 40% will be more comfortable than 60% under such conditions. When people in the same room are engaged in different activities, such as sitting and dancing, a 75F temperature with 40% or less relative humidity would come closest to maintaining comfortable conditions for everyone.

Around 71F Effective Temperature (ET) has been a satisfactory summer condition for sedentary activity for the average person for some time. For winter, 66 ET was acceptable to the average person some years ago but 68 ET is now about average, probably because most people, at least in the United States, no longer wear much heavier clothing in winter than in summer.

PROPERTIES OF DRY AIR
(At 14.696 psia or 29.92 Inches of Mercury)

Temp., F	Volume, Cu. Ft. per Lb.	Density, Lb. per Cu. Ft.	Specific Heat, Btu per Lb. per F	Cu. Ft. Warmed 1F By 1 Btu	Btu to Raise 1 Cu. Ft. 1F
−50	10.328	.09682	0.240	43.034	.02324
−40	10.580	.09452	0.240	44.084	.02268
−30	10.832	.09232	0.240	45.134	.02216
−20	11.084	.09022	0.240	46.184	.02165
−10	11.336	.08821	0.240	47.234	.02117
0	11.589	.08629	0.240	48.289	.02071
1	11.614	.08610	0.240	48.392	.02066
2	11.639	.08592	0.240	48.496	.02062
3	11.664	.08573	0.240	48.600	.02058
4	11.689	.08555	0.240	48.705	.02053
5	11.715	.08536	0.240	48.813	.02049
6	11.740	.08518	0.240	48.917	.02044
7	11.765	.08500	0.240	49.021	.02040
8	11.790	.08482	0.240	49.125	.02036
9	11.816	.08463	0.240	49.234	.02031
10	11.841	.08445	0.240	49.338	.02027
11	11.866	.08427	0.240	49.442	.02023
12	11.891	.08410	0.240	49.546	.02018
13	11.917	.08391	0.240	49.655	.02014
14	11.942	.08374	0.240	49.759	.02010
15	11.967	.08356	0.240	49.863	.02005
16	11.992	.08339	0.240	49.967	.02001
17	12.017	.08322	0.240	50.071	.01997
18	12.043	.08304	0.240	50.180	.01993
19	12.068	.08286	0.240	50.284	.01989
20	12.093	.08269	0.240	50.388	.01985
21	12.118	.08252	0.240	50.492	.01981
22	12.143	.08235	0.240	50.596	.01976
23	12.169	.08218	0.240	50.705	.01972
24	12.194	.08201	0.240	50.809	.01968
25	12.219	.08184	0.240	50.913	.01964
26	12.244	.08167	0.240	51.017	.01960
27	12.269	.08151	0.240	51.121	.01956
28	12.295	.08133	0.240	51.230	.01952
29	12.320	.08177	0.240	51.334	.01948
30	12.345	.08100	0.240	51.438	.01944
31	12.370	.08084	0.240	51.542	.01940
32	12.396	.08067	0.240	51.650	.01936
33	12.421	.08051	0.240	51.755	.01932
34	12.446	.08035	0.240	51.859	.01928
35	12.471	.08019	0.240	51.963	.01924
36	12.496	.08003	0.240	52.067	.01921
37	12.522	.07986	0.240	52.175	.01917

PROPERTIES OF DRY AIR
(At 14.696 psia or 29.92 Inches of Mercury)

Temp., F	Volume, Cu. Ft. per Lb.	Density, Lb. per Cu. Ft.	Specific Heat, Btu per Lb. per F	Cu. Ft. Warmed 1F By 1 Btu	Btu to Raise 1 Cu. Ft. 1F
38	12.547	.07970	0.240	52.280	.01913
39	12.572	.07954	0.240	52.384	.01909
40	12.597	.07938	0.240	52.487	.01905
41	12.622	.07923	0.240	52.592	.01901
42	12.648	.07906	0.240	52.700	.01898
43	12.673	.07891	0.240	52.805	.01894
44	12.698	.07875	0.240	52.909	.01890
45	12.723	.07860	0.240	53.013	.01886
46	12.748	.07844	0.240	53.117	.01882
47	12.774	.07828	0.240	53.225	.01879
48	12.799	.07813	0.240	53.330	.01875
49	12.824	.07798	0.240	53.434	.01871
50	12.849	.07783	0.240	53.538	.01868
51	12.874	.07768	0.240	53.642	.01864
52	12.900	.07752	0.240	53.750	.01860
53	12.925	.07737	0.240	53.855	.01857
54	12.950	.07722	0.240	53.959	.01853
55	12.975	.07707	0.240	54.063	.01850
56	13.001	.07692	0.240	54.171	.01846
57	13.026	.07677	0.240	54.275	.01842
58	13.051	.07662	0.240	54.380	.01839
59	13.076	.07648	0.240	54.484	.01835
60	13.101	.07633	0.240	54.588	.01832
61	13.127	.07618	0.240	54.696	.01828
62	13.152	.07603	0.240	54.800	.01825
63	13.177	.07589	0.240	54.905	.01821
64	13.202	.07575	0.240	55.009	.01818
65	13.227	.07560	0.240	55.113	.01814
66	13.253	.07545	0.240	55.221	.01811
67	13.278	.07531	0.240	55.325	.01808
68	13.303	.07517	0.240	55.430	.01804
69	13.328	.07503	0.240	55.534	.01801
70	13.353	.07489	0.240	55.638	.01797
71	13.379	.07474	0.240	55.746	.01794
72	13.404	.07460	0.240	55.850	.01791
73	13.429	.07447	0.240	55.955	.01787
74	13.454	.07433	0.240	56.059	.01784
75	13.480	.07418	0.240	56.167	.01780
76	13.505	.07405	0.240	56.271	.01777
77	13.530	.07391	0.240	56.375	.01774
78	13.555	.07377	0.240	56.480	.01771
79	13.580	.07364	0.240	56.584	.01767
80	13.606	.07350	0.240	56.692	.01764

PROPERTIES OF DRY AIR

(At 14.606 psia or 29.92 Inches of Mercury)

Temp., F	Volume, Cu. Ft. per Lb.	Density, Lb. per Cu. Ft.	Specific Heat, Btu per Lb. per F	Cu. Ft. Warmed 1F By 1 Btu	Btu to Raise 1 Cu. Ft. 1F
81	13.631	.07336	0.240	56.796	.01761
82	13.656	.07323	0.240	56.900	.01757
83	13.681	.07309	0.240	57.005	.01754
84	13.706	.07296	0.240	57.109	.01751
85	13.732	.07282	0.240	57.217	.01748
86	13.757	.07269	0.240	57.321	.01745
87	13.782	.07256	0.240	57.425	.01741
88	13.807	.07243	0.240	57.530	.01738
89	13.832	.07230	0.240	57.634	.01735
90	13.858	.07216	0.240	57.742	.01732
91	13.883	.07203	0.240	57.846	.01729
92	13.908	.07190	0.240	57.950	.01726
93	13.933	.07177	0.240	58.055	.01723
94	13.959	.07164	0.240	58.163	.01719
95	13.984	.07151	0.240	58.267	.01716
96	14.009	.07138	0.240	58.371	.01713
97	14.034	.07126	0.240	58.475	.01710
98	14.059	.07113	0.240	58.580	.01707
99	14.085	.07100	0.240	58.688	.01704
100	14.110	.07087	0.240	58.792	.01701
101	14.135	.07075	0.240	58.896	.01698
102	14.160	.07062	0.240	59.000	.01695
103	14.185	.07050	0.240	59.105	.01692
104	14.211	.07037	0.240	59.213	.01689
105	14.236	.07024	0.240	59.317	.01686
110	14.362	.06963	0.240	59.842	.01671
115	14.488	.06902	0.240	60.367	.01656
120	14.614	.06843	0.240	60.892	.01642
125	14.740	.06784	0.240	61.417	.01628
130	14.866	.06727	0.241	61.685	.01621
135	14.992	.06670	0.241	62.208	.01608
140	15.118	.06615	0.241	62.731	.01594
145	15.244	.06560	0.241	63.253	.01581
150	15.370	.06506	0.241	63.776	.01568
155	15.496	.06453	0.241	64.299	.01555
160	15.622	.06401	0.241	64.822	.01543
165	15.748	.06350	0.241	65.345	.01530
170	15.874	.06300	0.241	65.868	.01518
175	16.001	.06250	0.241	66.395	.01506
180	16.127	.06201	0.241	66.917	.01494
185	16.253	.06153	0.241	67.440	.01483

PROPERTIES OF DRY AIR
(At 14.696 psia or 29.92 Inches of Mercury)

Temp., F	Volume, Cu. Ft. per Lb.	Density, Lb. per Cu. Ft.	Specific Heat, Btu per Lb. per F	Cu. Ft. Warmed 1F By 1 Btu	Btu to Raise 1 Cu. Ft. 1F
190	16.379	.06105	0.241	67.963	.01471
195	16.505	.06057	0.241	68.485	.01460
200	16.631	.06013	0.241	69.008	.01449
205	16.757	.05967	0.241	69.531	.01438
210	16.883	.05924	0.241	70.054	.01428
215	17.009	.05879	0.241	70.577	.01417
220	17.135	.05834	0.242	70.806	.01422
225	17.262	.05793	0.242	71.331	.01420
230	17.388	.05750	0.242	71.851	.01392
235	17.514	.05711	0.242	72.372	.01382
240	17.640	.05669	0.242	72.893	.01372
245	17.766	.05627	0.242	73.413	.01362
250	17.892	.05590	0.242	73.934	.01353
255	18.018	.05549	0.242	74.455	.01343
260	18.144	.05513	0.242	74.975	.01334
265	18.270	.05473	0.242	75.496	.01325
270	18.396	.05435	0.242	76.017	.01315
275	18.522	.05400	0.242	76.537	.01307
280	18.648	.05362	0.242	77.058	.01298
285	18.774	.05328	0.243	77.259	.01294
290	18.900	.05291	0.243	77.778	.01286
295	19.026	.05255	0.243	78.296	.01278
300	19.153	.05222	0.243	78.819	.01269
310	19.405	.05152	0.243	79.856	.01252
320	19.657	.05086	0.243	80.893	.01236
330	19.909	.05023	0.243	81.930	.01221
340	20.161	.04960	0.244	82.627	.01210
350	20.413	.04900	0.244	83.660	.01195
360	20.665	.04838	0.244	84.693	.01181
370	20.918	.04870	0.244	85.730	.01665
380	21.170	.04724	0.245	86.408	.01157
390	21.422	.04669	0.245	87.437	.01144
400	21.674	.04615	0.245	88.465	.01130
410	21.926	.04560	0.246	89.130	.01122
420	22.178	.04509	0.246	90.154	.01109
430	22.430	.04458	0.246	91.179	.01097
440	22.683	.04409	0.246	92.207	.01084
450	22.935	.04359	0.247	92.854	.01077
460	23.187	.04312	0.247	93.874	.01065
470	23.439	.04266	0.248	94.512	.01058
480	23.691	.04221	0.249	95.145	.01051
490	23.943	.04177	0.249	96.157	.01040
500	24.195	.04134	0.249	97.169	.01029

MISCELLANEOUS PROPERTIES OF WATER

Water consists of two parts of hydrogen and one of oxygen by volume, or two parts of hydrogen and sixteen of oxygen by weight. Its freezing point at 29.92 inches of mercury is 32F and its boiling point 212F. The boiling point varies with the pressure as shown in the table below, where the pressure is in inches of mercury. The boiling point at an altitude of 1 mile above sea level is 202F, and 192F at 2 miles altitude.

Water expands when heated in the range above 39.2F but contracts when heated in the range from 32F to 39.2F.

The specific heat of water can be taken as 1.00 up to 212F, 1.01 from 212F to 260F, 1.02 from 260F to 290F, 1.03 from 290F to 320F, and 1.04 from 320F to 350F.

In the solid form, ice, the specific heat is 0.46 to 0.50 per lb. One pound of ice at 32F has a volume of 30.07 cubic inches, and 1 cubic foot weighs 57.5 lb. Snow weighs from 5 to 12 lb. per cubic foot.

Boiling Point of Water at Various Pressures

Boiling Point, F	Tenths of Degrees									
	.0	.1	.2	.3	.4	.5	.6	.7	.8	.9
	Barometric Pressure, Inches of Mercury									
185	17.075	17.112	17.150	17.187	17.224	17.262	17.300	17.337	17.375	17.413
186	17.450	17.488	17.526	17.564	17.602	17.641	17.679	17.717	17.756	17.794
187	17.832	17.871	17.910	17.948	17.987	18.026	18.065	18.104	18.143	18.182
188	18.221	18.261	18.300	18.340	18.379	18.419	18.458	18.498	18.538	18.578
189	18.618	18.658	18.698	18.738	18.778	18.818	18.859	18.899	18.940	18.980
190	19.021	19.062	19.102	19.143	19.184	19.225	19.266	19.308	19.349	19.390
191	19.431	19.473	19.514	19.556	19.598	19.639	19.681	19.723	19.765	19.807
192	19.849	19.892	19.934	19.976	20.019	20.061	20.104	20.146	20.189	20.232
193	20.275	20.318	20.361	20.404	20.447	20.490	20.533	20.577	20.620	20.664
194	20.707	20.751	20.795	20.839	20.883	20.927	20.971	21.015	21.059	21.103
195	21.148	21.192	21.237	21.282	21.326	21.371	21.416	21.461	21.506	21.551
196	21.597	21.642	21.687	21.733	21.778	21.824	21.870	21.915	21.961	22.007
197	22.053	22.099	22.145	22.192	22.238	22.284	22.313	22.377	22.424	22.471
198	22.517	22.564	22.611	22.658	22.706	22.752	22.800	22.847	22.895	22.942
199	22.990	23.038	23.085	23.133	23.181	23.229	23.277	23.325	23.374	23.422
200	23.470	23.519	23.568	23.616	23.665	23.714	23.763	23.812	23.861	23.910
201	23.959	24.009	24.058	24.108	24.157	24.207	24.257	24.307	24.357	24.407
202	24.457	24.507	24.557	24.608	24.658	24.709	24.759	24.810	24.861	24.912
203	24.963	25.014	25.065	25.116	25.168	25.219	25.271	25.322	25.374	25.426
204	25.478	25.530	25.582	25.634	25.686	25.738	25.791	25.843	25.896	25.948
205	26.001	26.054	26.107	26.160	26.213	26.266	26.319	26.373	26.426	26.480
206	26.534	26.587	26.641	26.695	26.749	26.803	26.857	26.912	26.966	27.021
207	27.075	27.130	27.184	27.239	27.294	27.349	27.404	27.460	27.515	27.570
208	27.626	27.681	27.737	27.793	27.848	27.904	27.960	28.016	28.073	28.129
209	28.185	28.242	28.298	28.355	28.412	28.469	28.526	28.583	28.640	28.697
210	28.754	28.812	28.869	28.927	28.985	29.042	29.100	29.158	29.216	29.275
211	29.333	29.391	29.450	29.508	29.567	29.626	29.685	29.744	29.803	29.862
212	29.921	29.981	30.040	30.100	30.159	30.219	30.279	20.339	30.399	30.459
213	30.519	30.580	30.640	39.701	39.761	30.822	30.883	30.944	31.005	31.066
214	31.127	31.199	31.250	31.311	31.373	31.435	31.497	31.559	31.621	31.683

THERMAL PROPERTIES OF WATER

(Data on the specific volume are reprinted by permission from " Thermodynamic Properties of Steam" by Keenan and Keyes, published by John Wiley & Sons, Inc.)

Temp., F	Pressure Lb. per Sq. In. Abs.	Specific Volume, Cu. Ft. per Lb.	Density, Lb. per Cu. Ft.	Vapor Pressure, In. Mercury	Temp., F	Pressure Lb. per Sq. In. Abs.	Specific Volume, Cu. Ft. per Lb.	Density, Lb. per Cu. Ft.	Vapor Pressure, In. Mercury
32	0.08854	0.01602	62.418	0.180	76	0.4443	0.01607	62.252	0.896
33	0.09223	0.01602	62.420	0.187	77	0.4593	0.01607	62.244	0.926
34	0.09603	0.01602	62.422	0.195	78	0.4747	0.01607	62.235	0.957
35	0.09995	0.01602	62.423	0.203	79	0.4906	0.01608	62.225	0.989
36	0.10401	0.01602	62.425	0.211	80	0.5069	0.01608	62.216	1.022
37	0.10821	0.01602	62.425	0.219	81	0.5237	0.01608	62.207	1.056
38	0.11256	0.01602	62.426	0.228	82	0.5410	0.01608	62.197	1.091
39	0.11705	0.01602	62.426	0.237	83	0.5588	0.01609	62.187	1.127
40	0.12170	0.01602	62.426	0.247	84	0.5771	0.01609	62.177	1.163
41	0.12652	0.01602	62.426	0.256	85	0.5959	0.01609	62.167	1.201
42	0.13150	0.01602	62.425	0.266	86	0.6152	0.01609	62.156	1.241
43	0.13665	0.01602	62.424	0.277	87	0.6351	0.01610	62.146	1.281
44	0.14199	0.01602	62.423	0.287	88	0.6556	0.01610	62.135	1.322
45	0.14752	0.01602	62.421	0.298	89	0.6766	0.01610	62.124	1.364
46	0.15323	0.01602	62.419	0.310	90	0.6982	0.01610	62.113	1.408
47	0.15914	0.01603	62.417	0.322	91	0.7204	0.01611	62.102	1.453
48	0.16525	0.01603	62.415	0.334	92	0.7432	0.01611	62.090	1.499
49	0.17157	0.01603	62.412	0.347	93	0.7666	0.01611	62.079	1.546
50	0.17811	0.01603	62.409	0.360	94	0.7906	0.01612	62.067	1.595
51	0.18486	0.01603	62.406	0.373	95	0.8153	0.01612	62.055	1.645
52	0.19182	0.01603	62.403	0.387	96	0.8407	0.01612	62.043	1.696
53	0.19900	0.01603	62.399	0.402	97	0.8668	0.01612	62.031	1.749
54	0.20642	0.01603	62.395	0.417	98	0.8935	0.01613	62.019	1.803
55	0.2141	0.01603	62.391	0.432	99	0.9210	0.01613	62.006	1.859
56	0.2220	0.01603	62.386	0.448	100	0.9492	0.01613	61.994	1.916
57	0.2302	0.01603	62.382	0.465	101	0.9781	0.01614	61.981	1.975
58	0.2386	0.01604	62.377	0.482	102	1.0078	0.01614	61.968	2.035
59	0.2473	0.01604	62.372	0.499	103	1.0382	0.01614	61.955	2.097
60	0.2563	0.01604	62.366	0.517	104	1.0695	0.01615	61.942	2.160
61	0.2655	0.01604	62.361	0.536	105	1.1016	0.01615	61.929	2.225
62	0.2751	0.01604	62.355	0.555	106	1.1345	0.01615	61.915	2.292
63	0.2850	0.01604	62.349	0.575	107	1.1683	0.01616	61.902	2.360
64	0.2951	0.01605	62.343	0.595	108	1.2029	0.01616	61.888	2.431
65	0.3056	0.01605	62.336	0.616	109	1.2384	0.01616	61.874	2.503
66	0.3164	0.01605	62.330	0.638	110	1.2748	0.01617	61.860	2.576
67	0.3276	0.01605	62.323	0.661	111	1.3121	0.01617	61.846	2.652
68	0.3390	0.01605	62.316	0.684	112	1.3504	0.01617	61.831	2.730
69	0.3509	0.01605	62.309	0.707	113	1.3896	0.01618	61.817	2.810
70	0.3631	0.01606	62.301	0.732	114	1.4298	0.01618	61.802	2.891
71	0.3756	0.01606	62.294	0.757	115	1.4709	0.01618	61.788	2.975
72	0.3886	0.01606	62.286	0.783	116	1.5130	0.01619	61.773	3.061
73	0.4019	0.01606	62.278	0.810	117	1.5563	0.01619	61.758	3.148
74	0.4156	0.01606	62.269	0.838	118	1.6006	0.01620	61.743	3.239
75	0.4298	0.01607	62.261	0.866	119	1.6549	0.01620	61.728	3.331

THERMAL PROPERTIES OF WATER

Temp., F	Pressure Lb. per Sq. In. Abs.	Specific Volume, Cu. Ft. per Lb.	Density, Lb. per Cu. Ft.	Vapor Pressure, In. Mercury	Temp., F	Pressure Lb. per Sq. In. Abs.	Specific Volume, Cu. Ft. per Lb.	Density, Lb. per Cu. Ft.	Vapor Pressure, In. Mercury
120	1.6924	0.01620	61.713	3.425	166	5.461	0.01643	60.879	11.107
121	1.7400	0.01621	61.697	3.522	167	5.590	0.01643	60.859	11.369
122	1.7888	0.01621	61.681	3.621	168	5.721	0.01644	60.838	11.636
123	1.8387	0.01622	61.666	3.723	169	5.855	0.01644	60.817	11.909
124	1.8897	0.01622	61.650	3.827	170	5.992	0.01645	60.796	12.187
125	1.9420	0.01622	61.634	3.933	171	6.131	0.01645	60.775	12.470
126	1.9955	0.01623	61.618	4.042	172	6.273	0.01646	60.754	12.759
127	2.0503	0.01623	61.602	4.154	173	6.417	0.01647	60.732	13.054
128	2.1064	0.01624	61.585	4.268	174	6.565	0.01647	60.711	13.354
129	2.1638	0.01624	61.569	4.385	175	6.715	0.01648	60.689	13.660
130	2.2225	0.01625	61.552	4.504	176	6.868	0.01648	60.668	13.972
131	2.2826	0.01625	61.535	4.627	177	7.024	0.01649	60.646	14.289
132	2.3440	0.01626	61.519	4.752	178	7.183	0.01650	60.625	14.613
133	2.4069	0.01626	61.502	4.880	179	7.345	0.01650	60.602	14.943
134	2.4712	0.01626	61.485	5.011	180	7.510	0.01650	60.580	15.279
135	2.5370	0.01627	61.467	5.145	181	7.678	0.01650	60.558	15.621
136	2.6042	0.01627	61.450	5.282	182	7.850	0.01652	60.536	15.970
137	2.6729	0.01628	61.433	5.422	183	8.024	0.01653	60.514	16.325
138	2.7432	0.01628	61.415	5.565	184	8.202	0.01653	60.492	16.687
139	2.8151	0.01629	61.398	5.712	185	8.383	0.01654	60.469	17.055
140	2.8886	0.01629	61.380	5.862	186	8.567	0.01654	60.447	17.430
141	2.9637	0.01630	61.362	6.015	187	8.755	0.01655	60.424	17.812
142	3.0404	0.01630	61.344	6.171	188	8.946	0.01656	60.401	18.202
143	3.1188	0.01631	61.326	6.331	189	9.141	0.01656	60.378	18.598
144	3.1990	0.01631	61.308	6.395	190	9.339	0.01657	60.355	19.001
145	3.281	0.01632	61.289	6.662	191	9.541	0.01658	60.332	19.412
146	3.365	0.01632	61.271	6.832	192	9.746	0.01658	60.309	19.830
147	3.450	0.01633	61.252	7.006	193	9.955	0.01659	60.286	20.255
148	3.537	0.01633	61.233	7.184	194	10.168	0.01659	60.263	20.688
149	3.627	0.01634	61.215	7.366	195	10.385	0.01660	60.239	21.129
150	3.718	0.01634	61.196	7.552	196	10.605	0.01661	60.216	21.578
151	3.811	0.01635	61.177	7.742	197	10.830	0.01661	60.192	22.034
152	3.906	0.01635	61.158	7.936	198	11.058	0.01662	60.169	22.499
153	4.003	0.01636	61.139	8.133	199	11.290	0.01663	60.145	22.972
154	4.102	0.01636	61.119	8.335	200	11.526	0.01663	60.121	23.45
155	4.203	0.01637	61.100	8.541	202	12.011	0.01665	60.073	24.44
156	4.306	0.01637	61.080	8.752	204	12.512	0.01666	60.025	25.46
157	4.411	0.01638	61.061	8.966	206	13.031	0.01667	59.976	26.52
158	4.519	0.01638	61.041	9.186	208	13.568	0.01669	59.927	27.62
159	4.629	0.01639	61.021	9.409	210	14.123	0.01670	59.877	28.75
160	4.741	0.01639	61.001	9.637	212	14.696	0.01672	59.828	29.92
161	4.855	0.01640	60.981	9.870	214	15.289	0.01673	59.773	31.13
162	4.971	0.01640	60.961	10.108	216	15.901	0.01674	59.737	32.38
163	5.090	0.01641	60.941	10.350	218	16.533	0.01676	59.666	33.67
164	5.212	0.01641	60.920	10.597	220	17.186	0.01677	59.630	35.01
165	5.335	0.01642	60.900	10.850	222	17.861	0.01679	59.559	36.38

THERMAL PROPERTIES OF WATER

Temp., F	Pressure Lb. per Sq. In. Abs.	Specific Volume, Cu. Ft. per Lb.	Density, Lb. per Cu. Ft.	Vapor Pressure, In. Mercury	Temp., F	Pressure Lb. per Sq. In. Abs.	Specific Volume, Cu. Ft. per Lb.	Density, Lb. per Cu. Ft.	Vapor Pressure, In. Mercury
224	18.557	0.01680	59.524	37.79	324	94.84	0.01770	56.497	193.8
226	19.275	0.01682	59.453	39.26	328	100.26	0.01774	56.370	204.9
228	20.016	0.01683	59.418	40.77	332	105.92	0.01778	56.243	216.4
230	20.780	0.01684	59.382	42.34	336	111.84	0.01783	56.085	228.5
232	21.567	0.01686	59.312	43.94	340	118.01	0.01787	55.960	241.1
234	22.379	0.01688	59.242	45.59	344	124.45	0 01792	55.804	254.2
236	23.217	0.01689	59.207	47.31	348	131.17	0.01797	55.648	268.0
238	24.080	0.01691	59.137	49.08	352	138.16	0.01801	55.525	282.3
240	24.969	0.01692	59.102	50.89	356	145.45	0.01806	55.270	297.1
242	25.884	0.01694	59.032	52.76	360	153.04	0.01811	55.218	312.6
244	26.827	0.01696	58.962	54.69	364	160.93	0.01816	55.066	328.7
246	27.798	0.01697	58.928	56.67	368	169.15	0.01821	54.915	345.4
248	28.797	0.01699	58.858	58.71	372	177.68	0.01826	54.765	362.8
250	29.825	0.01700	58.824	60.82	376	186.55	0.01831	54.615	380.9
252	30.884	0.01702	58.754	62.98	380	195.77	0.01836	54.466	399.7
254	31.973	0.01704	58.685	65.20	384	205.33	0.01842	54.289	419.1
256	33.093	0.01705	58.651	67.48	388	215.26	0.01847	54.142	439.2
258	34.245	0.01707	58.582	69.85	392	225.56	0.01853	53.967	460.1
260	35.429	0.01709	58.514	72.26	396	236.24	0.01858	53.821	481.9
262	36.646	0.01710	58.480	74.75	400	247.31	0.01864	53.648	504.4
264	37.897	0.01712	58.411	77.31	405	261.71	0.01871	53.447	533.6
266	39.182	0.01714	58.343	79.93	410	276.75	0.01878	53.248	564.1
268	40.502	0.01715	58.309	82.63	415	292.45	0.01886	53.022	595.7
270	41.858	0.01717	58.241	85.41	420	308.83	0.01894	52.798	628.8
272	43.252	0.01719	58.173	88.25	425	325.92	0.01902	52.576	663.3
274	44.682	0.01721	58.106	91.18	430	343.72	0.01910	52.356	699.2
276	46.150	0.01722	58.072	94.18	435	362.27	0.01918	52.138	736.5
278	47.657	0.01724	58.005	97.25	440	381.59	0.01926	51.921	775.3
280	49.203	0.01726	57.937	100.41	445	401.68	0.01935	51.680	815.5
282	50.790	0.01728	57.870	103.66	450	422.6	0.0194	51.546	—
284	52.418	0.01730	57.803	106.99	455	444.3	0.0195	51.282	—
286	54.088	0.01732	57.737	110.41	460	466.9	0.0196	51.020	—
288	55.800	0.01733	57.703	113.91	465	490.3	0.0197	50.761	—
290	57.556	0.01735	57.637	117.50	470	514.7	0.0198	50.505	—
292	59.356	0.01737	57.571	121.18	475	539.9	0.0199	50.251	—
294	61.201	0.01739	57.504	124.94	480	566.1	0.0200	50.000	—
296	63.091	0.01741	57.438	128.81	485	593.3	0.0201	49.751	—
298	65.028	0.01743	57.372	132.78	490	621.4	0.0202	49.505	—
300	67.013	0.01745	57.307	136.8	495	650.6	0.0203	49.261	—
304	71.127	0.01749	57.176	145.2	500	680.8	0.0204	49.020	—
308	75.442	0.01753	57.045	154.1	550	1045.2	0.0218	45.872	—
312	79.96	0.01757	56.915	162.3	600	1542.9	0.0236	42.373	—
316	84.70	0.01761	56.786	173.0	650	2208.2	0.0268	37.313	—
320	89.66	0.01765	56.657	183.1	700	3093.7	0.0369	27.100	—

HOW TO USE THE PSYCHROMETRIC TABLE

The psychrometric table which follows this explanation and examples, and covering the range from —40F to 200F, can be used to solve almost any type of problem in psychrometry. Use of the table is illustrated by the problems which follow.

The table was calculated using vapor pressure data compiled by the U. S. Weather Bureau, the pressures below 32F being those over ice. Figures for heat content of saturated vapor, on which enthalpy data are based, are from the Keenan and Keyes steam tables.

The table gives important properties of a mixture of dry air and saturated steam. Such a mixture is commonly called "saturated air." It could equally well be called "air at 100% relative humidity." As a matter of fact it is the only air and steam mixture which contains saturated steam, for such mixtures at any relative humidity except 100% contain only superheated steam and no saturated steam. Also, such a mixture (saturated) of dry air and saturated steam contains the greatest weight of water vapor which can be mixed with a given weight of dry air at any particular temperature and total pressure.

As already mentioned, in the process of conditioning air, water vapor is likely to be added or removed. Thus, the only thing which is carried through the whole process, in all cases, is the air. This makes the *dry air* the most convenient base to refer everything to and, for this reason, *dry air* is the point of reference in the table and chart in this section and practically all presently used psychrometric tables and charts.

Columns (1) and (11) of the Psychrometric Table are dry-bulb and wet-bulb temperatures. These are the same because the table is based on saturated conditions.

Column (2) is the partial pressure of the saturated water vapor in inches of mercury.

Column (3) is the weight of the water vapor in grains per cubic foot. This value is nearly independent of the total pressure of the air-vapor mixture. (One pound = 7,000 grains).

Column (4) is the weight of the water vapor (in grains) which is in the space occupied by one pound of dry air.

Column (5) is the heat to raise one pound of dry air from 0F to the particular temperature. For example, the value for 100F = 0.24 (100-0) = 24.00.

Column (6) is the volume occupied by one pound of dry air when it is in the same space (mixed with) saturated vapor at the particular temperature. For example, the table of Properties of Dry Air shows the volume of one pound of dry air at 29.92 inches of mercury pressure and 70F as 13.353 cu ft per lb. If the same space contains water vapor as well as air, and the total pressure is 29.92 inches of mercury, the pressure on the air will be 29.92 inches less the vapor pressure. At lower pressure, the pound of dry air occupies a larger volume. Referring to column (6) for 70F saturated air, the volume of one pound of dry air is shown as 13.67 cu ft per lb.

Column (7) shows the weight (in grains) of dry air in one cubic foot of the mixture of air and saturated vapor at the particular temperature. The values in column (7) are 7000 times the reciprocal of those in column (6).

Column (8) is the weight (in grains) of dry air plus vapor in one cubic foot. The values in column (8) are the total of those in columns (3) and (7).

Column (9) is the heat of the liquid water above 32F plus the latent heat of evaporation for the weight of vapor in column (4).

Column (10) is the enthalpy of one pound of air plus that of the saturated vapor occupying the same space. The values of Column (10) are the sum of those in (5) and (9).

This psychrometric table makes it possible to solve any problems for air and water vapor mixture in the range of temperatures and the pressures for which it is made.

From tables such as this, psychrometric charts are made. These charts make the solution of most psychrometric problems very much simpler than it would be with the table alone. The chart permits easy plotting of any process. Direct solutions replace relatively long solutions or trial and error. The chart gives a visual display of the process.

This helps in the choice of the best process, provides additional means of checking the calculations and more confidence in the choice of process. It also gives greater assurance of satisfactory performance of the system installed on that basis.

The calculation of a few air-water vapor problems, by means of the psychrometric table will help toward a complete understanding of the subject. However, it is not recommended that one solve air conditioning problems in this way. In this Handbook and elsewhere, there are considerable data on the use of the psychrometric chart. The time spent on learning to use the chart will be generously rewarded by easier solutions and usually better and more accurate ones.

In the solutions to the following problems, total heat at saturation, which corresponds to a particular wet bulb temperature, is assumed to match the same wet-bulb temperature at lower than 100% relative humidity. This is not quite true but, up to about 90F wet bulb and 120F dry-bulb, the deviation is less than about 2% of the heat of the vapor. The correction would add much work to the solution. Some charts make the more accurate solution just as easy as the approximate.

All problems are based on a total pressure of 29.92 in. Hg or 14.696 psia.

Relative Humidity

Example 1. What is the relative humidity of air having a dry-bulb temperature of 80F and a wet-bulb temperature of 72F?

At 72F wet-bulb, the enthalpy in Col. 10 is 35.75 Btu per lb. The heat content of dry air at 80F (Col. 5) is 19.20 Btu per lb. The difference between the two is 16.55 Btu per lb which is the heat of the vapor. Referring to Col. 9, it will be noted that 16.03 Btu per lb corresponds to 0.6903 in. Hg vapor pressure (Col. 2) and 16.61 Btu per lb corresponds to 0.7144 in. Hg. By interpolation, 16.55 Btu per lb corresponds to 0.7119 in. Hg. At 80F saturated, the vapor pressure is 1.0321 in. Hg. Since the relative humidity is the ratio of the actual vapor pressure to that of saturated vapor at the same temperature, the relative humidity is 0.7119/1.0321 = 69.0%.

Example 2. What is the relative humidity for a dry-bulb temperature of 70F and a dew point of 59F?

The vapor pressure corresponding to 59F saturated is 0.5035 in. Hg. That at 70F is 0.7392 in. Hg. Therefore, its relative humidity is 0.5035/0.7392 = 68.1%.

Example 3. If the air in Example 2 is heated to 78F, what will be its new relative humidity?

Vapor pressure at 59F saturated was 0.5035 in. Hg. Vapor pressure at 78F saturated is 0.9666 in. Hg. The relative humidity is 0.5035/0.9666 = 52.1%.

Example 4. What is the relative humidity corresponding to 62F wet bulb and 54F dew point?

Enthalpy at 62F wet-bulb is 27.80 Btu per lb. The heat of vapor at 54F dew point is 9.62 Btu per lb. The difference, 27.80 − 9.62 = 18.18 Btu per lb, is the enthalpy of the dry air. The dry bulb temperature is the enthalpy of the dry air divided by the dry air specific heat (Btu per lb per deg F), 18.18/0.24 = 75.8F. (This could also be found by interpolation of Col. 5 and Col. 1.) The vapor pressure at 54F saturated is 0.4203 in. Hg; that at 75.8F saturated is 0.8986 in. Hg. The relative humidityy = 0.4203/0.8986 = 46.8%.

Example 5. What is the relative humidity corresponding to 80F db and 70 grains per lb?

The vapor pressure corresponding to 70 gr per lb (interpolating Col. 2 and Col. 4) is 0.4735 in. Hg. That for 80F saturated is 1.0321 in. Hg. The relative humidity is 0.4735/1.0321 = 45.9%.

Dry Bulb Temperature

Example 6. What is the dry-bulb temperature of air having a relative humidity of 76% and a dew point of 66F?

The vapor pressure corresponding to 66F saturated is 0.6442 in. Hg. With 0.6442 in. Hg. representing 76% relative humidity, 0.6442/0.76 = 0.8476 in. Hg. is the vapor pressure at saturation, which is practically the vapor pressure for 74F.

Example 7. What is the temperature of a mixture of 200 lb of dry air at 30F and 300 lb at 80F?

The temperature of the mixture is

$$\frac{(200)(30) + (300)(80)}{200 + 300} = 60F$$

(Actually, at dew points of less than about 90F, the temperature of the mixture would be within about one degree F of that as calculated above, even with moisture in one or both parts of the mixture, unless the total moisture of the mixture, in grains, divided by the total weight of air results in a value, in gr per lb, in excess of that corresponding to saturation at the calculated dry-bulb temperature of the mixture).

Dew Point Temperature

Example 8. What is the dew point of air having a dry-bulb temperature of 80F and a wet-bulb temperature of 72F?

The enthalpy, at 72F saturated, is 35.75 Btu per lb. The enthalpy of dry air at 80F is 19.20 Btu per lb. The heat of the vapor is 35.75 — 19.20 = 16.55 Btu per lb. By interpolation of Col. 9 and Col. 1, 16.55 Btu per lb corresponds to 68.9F.

Example 9. What is the dew point of air having a dry-bulb temperature of 80F with a relative humidity of 50%?

The vapor pressure at 80F saturated is 1.0321 in. Hg. At 50% rh, the vapor pressure is 0.5 × 1.0321 = 0.516 in. Hg. This corresponds to 59.7F dew point.

Wet-Bulb Temperature

Example 10. What is the wet-bulb temperature corresponding to 80F dry-bulb and 32.5% relative humidity?

The pressure of 80F saturated vapor is 1.0321 in. Hg. At 32.5% rh, it is 0.325 × 1.0321 = 0.3354 in. Hg, which corresponds to 47.9F dew point and 7.63 Btu per lb heat of vapor. The enthalpy of dry air at 80F is 19.20 Btu per lb. The resulting enthalpy is 19.20 + 7.63 = 26.83 Btu per lb, which corresponds to 60.6F wet-bulb.

Example 11. What is the wet-bulb corresponding to 72F dry-bulb and 56F dew point?

The enthalpy of the dry air at 72F is 17.28 Btu per lb. The enthalpy of the vapor at 56F saturated is 10.36 Btu per lb. Enthalpy is 17.28 + 10.36 = 27.64 Btu per lb, which corresponds to 61.8F wet-bulb temperature.

Example 12. What is the wet-bulb temperature corresponding to 66F dew point and 76% relative humidity?

The vapor pressure at 66F dew point is 0.6442, which corresponds to 76% rh. At saturation, the vapor pressure would be 0.6442/0.76 = 0.8476, corresponding to 74.2F. The enthalpy of vapor at 66F dew point is 14.93 Btu per lb. That of dry air at 74.2F is 17.81 Btu per lb. Enthalpy is 17.81 + 14.93 = 32.74 Btu per lb, which corresponds to 68.5F wet-bulb.

Example 13. Find the heat removal required to change air from 68F wet-bulb to 60F wet-bulb.

Enthalpy of air at 68F wet-bulb is 32.35 Btu per lb; that at 60F wet-bulb is 26.41 Btu per lb. The difference is 32.35—26.41 = 5.94 Btu per pound of dry air to be removed.

Density

Example 14. What is the weight of dry air per cubic foot at 70F saturated?

From Col. 6, saturated air at 70F has a volume of 13.67 cu ft per pound of dry air. The density is therefore 1/13.67 = 0.0732 lb dry air per cu ft.

It is also (Col. 7) 511.9 gr per cu ft ÷ 7000 = 0.0731 lb per cu ft.

Example 15. What is the weight of dry air per cu ft in air at 70F dry-bulb and 60F wet-bulb?

Enthalpy at 60F wet-bulb is 26.41 Btu per lb. Enthalpy of dry air at 70F is 16.80 Btu per lb. Enthalpy of the vapor is 26.41 — 16.80 = 9.61 cu ft per lb. This corresponds to a dew point of 54F and a vapor pressure of 0.4203 in. Hg. Pressure of the air is 29.92 — 0.42 = 29.50 in. Hg. From the Properties of Dry Air table, at 70F and 29.92 in. Hg, density of dry air is 0.07489 lb per cu ft. From Boyle's law, the density at 70F and 29.50 in. Hg is (0.07489) (29.50)/29.92 = 0.0738 lb dry air per cu ft.

Example 16. What is the weight of dry air in 2,000 cu ft of air at 70F and 40% rh?

At 70F saturated, the vapor pressure is 0.7392 in. Hg. At 40% rh, the vapor pressure is 0.4 × 0.7392 = 0.2957 in. Hg (0.30). The pressure of the air is 29.92 — 0.30 = 29.62 in. Hg. From the Properties of Dry Air table, the volume of dry air at 70F and 29.92 in. Hg is 13.353 cu ft per lb. At 70F and 29.62 in. Hg (from Boyle's law), the volume is

$$\frac{(13.353)(29.92)}{29.62} = 13.487 \text{ cu ft per lb.}$$

The weight of dry air in 2,000 cu ft is 2000/13.487 = 148.3 lb.

Heat

Example 17. How many Btu per hr of sensible heat is required to change 1,000 cu ft of air from 0F to 70F? (The 1,000 cu ft is measured at the entering condition.)

From the table, Properties of Dry Air, the density at 0F is 0.0863 lb per cu ft. The weight of air is 1000 × 0.0863 = 86.3 lb. Sensible heat required is 86.3 × 0.24 × (70-0) = 1,450 Btu.

Example 18. How many Btu of sensible heat must be removed to cool 1,000 lb of air from 100F to 80F?

Sensible heat to be removed is (1000) (0.24) (100-80) = 4800 Btu.

Example 19. What is the enthalpy in Btu per pound of dry air in a mixture of air and water vapor at 70F and 40% rh?

The enthalpy of the dry air at 70F is 16.80 Btu per lb. The vajor pressure at 70F saturated is 0.7392 in. Hg. At 40% rh the vapor pressure is 0.40 × 0.7392 = 0.2957 in. Hg. By interpolation of Col. 2 and Col. 9, the heat of vapor which corresponds to 0.2957 in. Hg is 6.71 Btu per lb. The enthalpy is 16.80 + 6.71 = 23.51 Btu per lb.

Example 20. How many Btus must be removed to change 1,000 lb of air from 100F and 55% rh to 76F and 50% rh?

At 100F saturated, the vapor pressure is 1.9325 in. Hg and the enthalpy of the dry air is 24.00 Btu per lb. At 55% rh, the vapor pressure is 0.55 × 1.9325 = 1.0629 in. Hg. By interpolation with Col. 2 and Col. 9, the enthalpy of the vapor is 25.14 Btu per lb. The enthalpy at 100F and 55% rh is 24.00 + 25.14 = 49.14 Btu per lb.

The vapor pressure at 76F saturated is 0.9046 in. Hg and the heat content of the dry air is 18.24 Btu per lb. At 50% rh, the vapor pressure is 0.50 × 0.9046 = 0.4523 in. Hg, which corresponds closely to 56F dew point where the vapor pressure is 0.4520 in. Hg and the heat of the vapor is 10.36 Btu per lb. The enthalpy at 76F and 50% rh is 18.24 + 10.36 = 28.60 Btu per lb.

The difference in the enthalpies is 49.14 — 28.60 = 20.54 Btu per lb and for 1,000 lb, this would be 20,540 Btu. However, this answer would not give an accurate idea of the process or the cooling capacity required. Except by means of a refrigerated absorbent, a mixture of air and water vapor cannot be changed from 100F and 55% rh to 76F and 50% rh. If a mixture has its enthalpy reduced by passing it over a surface at a lower temperature, its relative humidity must increase because a surface at a lower temperature than that of any air-water vapor mixture can only bring the vapor nearer to saturation. In this case, the initial relative humidity is 55% (and during the cooling it will increase) but 50% rh is required in the final condition.

The initial mixture must have its dew point reduced to 56F. To reduce the dew point to 56F, the mixture must be cooled to less than the final 76F required. Any temperature less than 76F makes it necessary first to remove more sensible heat than desired and then, second, to return that heat to the mixture to reach the final 76F.

Therefore, the 56F dew point should be attained with the mixture temperature no farther below 76F than necessary. The determination of the highest dry bulb temperature possible when 100F and 55% rh has the dew point reduced to 56F, is very difficult with only a psychrometric table. It is much easier if one refers to accurate manufacturers' data for cooling coils.

If the process is plotted on a psychrometric chart, the highest dry bulb possible is immediately visible. Actually, a mixture at 100F and 55% rh cannot have its dew point lowered to 56F, by a cooler surface, without reaching saturation. Therefore the problem requires changing the initial condition to 56F saturated where the enthalpy (Col. 10) is given as 23.80 Btu per lb.

The heat to be removed is therefore

(1,000) (49.14 — 23.80) = 25,340 Btu.

The heat to be added (reheat) is

(1,000) (28.60 — 23.80) = 4,800 Btu

(this could be calculated as (1,000) (0.24) (76 — 56) = 4,800 Btu).

The *net* heat removal is 25,340 —4,800 = 20,540 Btu, as calculated previously, but it is necessary to have the refrigeration capacity to remove 25,340 Btu and then replace 4,800 Btu to attain the particular change of condition specified.

Example 21. How much total heat and latent heat must be added to air at 40F and 60% rh to change its condition to 72F and 45% rh?

The vapor pressure at 40F saturated is 0.2478 in. Hg. At 40F and 60% rh, the vapor pressure is 0.6 × 0.2478 = 0.1487 in. Hg. This vapor pressure corresponds to 27.8F dew point and 3.34 Btu per lb heat of *moisture*.

The vapor pressure at 72F saturated is 0.7912 in. Hg. At 72F and 45% rh, the vapor pressure is 0.45 × 0.7912 = 0.3560 in. Hg. This vapor pressure corresponds to 49.5F dew point and 8.12 Btu per lb heat of *moisture*. The increase in heat of moisture is 8.12 — 3.34 = 4.78 Btu per lb. (It is not strictly correct to call this heat of moisture "latent heat".)

The dry air heat content must be increased by 0.24 (72-40) = 7.68 Btu per lb.

The total heat increase is 7.68 + 4.78 = 12.46 Btu per lb.

Example 22. What will be the enthalpy per pound of dry air of the resulting mixture when 200 lb of dry air at 30F and 80% rh is mixed with 400 lb of dry air at 80F and 50% rh?

The vapor pressure at 30F saturated is 0.164 in. Hg. At 30F and 80% rh, it is 0.80 × 0.164 = 0.131 in. Hg. This corresponds to 25.2F dew point and 2.98 Btu per lb heat of moisture. The enthalpy of the dry air is 7.20 Btu per lb (Col. 5) and the enthalpy is 7.20 + 2.93 = 10.13 Btu per lb.

The vapor pressure at 80F saturated is 1.0321 in. Hg. At 80F and 50% rh it is 0.50 × 1.0321 = 0.5160 in. Hg. This corresponds to a 59.7F dew point and 11.87 Btu per lb heat of moisture. The enthalpy of the dry air is 19.20 Btu per lb and the resulting enthalpy is 19.20 + 11.87 = 31.07 Btu per lb. Total enthalpy of the mixture is:

$$[(200 \times 10.13) + (400 \times 31.07)]/600 = 24.09$$
Btu per lb.

Example 23. What is the absolute humidity, or grains of moisture per cubic foot, of a mixture of air and water vapor at 70F dry-bulb and 56% rh?

At 70F saturated, absolute humidity is 8.065 gr per cu ft (Col. 3); at 70F and 56% rh, it is 0.56 × 8.065 = 4.516 gr per cu ft.

Example 24. What is the dry-bulb temperature, specific humidity and enthalpy of a mixture of 2 lb of air at 30F and 80% rh, and 3 lb of air at 70F and 50% rh?

Vapor pressure at 30F saturated is 0.164 in. Hg. At 30F and 80% rh, it is 0.80 × 0.164 = 0.131 in. Hg. This corresponds to 25.2F dew point, 19.15 gr moisture per lb and 2.93 Btu per lb heat of moisture. Heat of the dry air is 7.20 Btu per lb and total heat is 7.20 + 2.93 = 10.13 Btu per lb.

Vapor pressure at 70F saturated is 0.7392 in. Hg. At 70F and 50% rh, it is 0.50 × 0.7392 = 0.3696 in. Hg. This corresponds to 51.5F dew point, 54.46 gr moisture per lb and 8.43 Btu per lb heat of moisture. Heat of the dry air is 16.80 Btu per lb and total enthalpy is 16.80 + 8.43 = 25.23 Btu per lb.

The dry-bulb temperature is [(2 × 30) + (3 × 70)]/5 = 54F (unless fog formed). The specific humidity is [(2 × 19.15) + (3 × 54.46)]/5 = 40.34 gr per lb. (At 54F saturated the specific humidity is 62.03 gr per lb. Since the 40.34 gr per lb is not greater than the value at saturation, fog will not form and 54F is the dry-bulb temperature).

Enthalpy of the mixture is [(2 × 10.13) + (3 × 25.23)]/5 = 19.19 Btu per lb.

Example 25. What is the dry-bulb temperature, specific humidity and enthalpy of a mixture of equal weights of dry air at 25F with 30% rh and 75F with 90% rh?

The vapor pressure at 25F saturated is 0.130 in. Hg. At 25F and 30% rh, it is 0.30 × 0.130 = 0.039 in. Hg. This corresponds to 0.4F dew point, 5.68 gr per lb and 0.862 Btu per lb heat of moisture. The heat content of the dry air is 6.00 Btu per lb and total heat is 6.00 + 0.862 = 6.86 Btu per lb.

The vapor pressure at 75F saturated is 0.8750 in. Hg. At 75F and 90% rh it is 0.90 × 0.8750 = 0.7875 in Hg. This corresponds to 71.9F dew point, 117.69 gr per lb and 18.38 Btu per lb heat of moisture. The heat content of the dry air is 18.00 Btu per lb and the enthalpy is 18.00 + 18.38 = 36.38 Btu per lb.

The dry bulb temperature is (25 + 75)/2 = 50F (unless fog forms) and the specific humidity is (5.68 + 117.69)/2 = 61.68 gr per lb.

(At 50F saturated, specific humidity is 53.41 gr per lb. Since the 61.68 gr per lb specific humidity determined above is greater than the 53.41 gr per lb at 50F saturated, fog will form. Some of the vapor will condense, giving up its latent heat, and the dry bulb temperature will be above 50F).

Enthalpy of the mixture is (6.86 + 36.38)/2 = 21.62 Btu per lb and 21.62 Btu per lb corresponds to 52.4F wet-bulb.

The dry-bulb temperature will also be 52.4F. Vapor at 52.4F saturated is 58.43 gr per lb and that will be the weight of vapor in the mixture. 61.68 − 58.43 = 3.25 gr per lb will have condensed and be in the form of fog. If heat were added to the mixture (including the fog), when the enthalpy increased to about 22.49 Btu per lb, the dry-bulb and wet-bulb temperatures would be 53.8F and the fog would be evaporated because at 53.8F saturated, the specific humidity is 61.68 gr per lb.

Example 26. Assume that a quantity of air-vapor mixture which is brought into a building from the outside at a temperature of 32F and 50% rh is heated so as to leave the grilles at 172F and with sufficient moisture added so that its relative humidity is 45% when the mixture has cooled to 70F (room temperature). The air flow, measured as it leaves the grilles, is 9,000 cu ft per hr. How much moisture must be added?

At 70F saturated, the vapor pressure is 0.7392 in. Hg; at 70F and 45% rh it is 0.45 × 0.7392 =

0.3326 in. Hg. This corresponds to 47.7F dew point and 48.94 gr per lb.

At 32F saturated, the vapor pressure is 0.1803 in. Hg; at 32F and 50% rh, it is 0.50 × 0.1803 = 0.0902 in. Hg. This corresponds to 17.3F dew point and 13.16 gr per lb.

The air leaving the grilles has a vapor pressure of 0.3326 in. Hg. The pressure of the air is 29.92 — 0.33 = 29.59 in. Hg. From the table, Properties of Dry Air, air at 29.92 in. Hg. and 172F has a density of 0.06280 lb per cu ft. At 29.59 in. of Hg, the density is

0.06280 × 29.59/29.92 = 0.06211 lb per cu ft.

The weight of dry air circulated is

9000 × 0.06211 = 559 lb per hr.

The moisture to be added is

(559) (48.94 — 13.16)/7000 = 2.86 lb per hr. or 2.86 × 24 = 68.64 lb per 24 hr or 68.64/8.33 = 8.24 gal per day.

Example 27. How much moisture must be removed from 15,000 cu ft per hr (measured at standard conditions of 0.075 lb dry air per cu ft) entering a building at 91F db and 74F wb, when it is changed to the 75F db and 50% rh to be maintained in a conditioned space?

At 74F wb, enthalpy is 37.57 Btu per lb. The heat of dry air at 91F is 21.84 Btu per lb. The heat of the vapor is 37.57 — 21.84 = 15.73 Btu per lb. By interpolation from Col. 9 and Col. 4, the specific humidity is 100.91 gr per lb.

At 75F saturated, the vapor pressure is 0.8750 in. Hg. At 75F and 50% rh it is 0.50 × 0.8750 = 0.4375 in. Hg. By interpolation, from Col. 2 and Col. 4, the specific humidity is 64.61 gr per lb.

The moisture to be removed per pound of dry air is 100.91 — 64.61 = 36.30 grains.

The moisture to be removed from the 15,000 cu ft per hr is

[(15,000) (0.075) (36.30)/7000] = 5.8 lb per hr

Example 28. The sensible heat loss from a shop (at 70F inside and 10F outside) is 14,000 Btu per hr. Infiltration is insignificant, so that there is no moisture loss due to inleakage of low moisture content outside air. Ignore the small moisture output of occupants and any sensible heat from people, lights, etc. The inside condition is to be 70F and 40% rh when the outside condition is 10F and 50% rh. The air circulated is to be 25% from outside and 75% recirculated (by weight). The air leaves the grilles (in the shop) at 130F. The air flow is to be checked and adjusted at a time when the conditions are as given. What cubic feet per minute (cfm) should be measured, at the outside air intake, the recirculated air duct, in the mixed air duct before any heat or moisture is added? How much moisture must be added? What is the heat of vapor required? What is the enthalpy required?

Total air flow must deliver 140,000 Btu per hr to the shop while it cools from 130F to 70F. Therefore, the flow should be $\dfrac{140,000}{0.24\,(130\text{-}70)} = 9,720$ lb per hr (dry air).

Outside air is 0.25 × 9,720 = 2,430 lb per hr. Recirculated air is 0.75 × 9,720 = 7,290 lb per hr.

At 10F saturated, vapor pressure is 0.0631 in. Hg; at 10F and 50% rh, it is 0.50 × 0.0631 = 0.0316 in. Hg. This corresponds to —3.5F dew point, 4.60 gr per lb and 0.697 Btu per lb heat of vapor. The air pressure is 29.92 — 0.03 = 29.89 in. Hg.

From the table, Properties of Dry Air, at 10F and 29.92 in. Hg, volume is 11.841 cu ft per lb. At 29.89 in. Hg, it is (11.841 × 29.92)/29.89 = 11.853 cu ft per lb.

At 70F saturated, vapor pressure is 0.7392 in. Hg; at 70F and 40% rh, it is 0.40 × 0.7392 = 0.2957 in. Hg. This corresponds to 44.6F dew point, 45.13 gr per lb and 6.71 Btu per lb heat of vapor. The air pressure is 29.92 — 0.2957 = 29.62 in. Hg. From the table, Properties of Dry Air, at 70F and 29.92 in. Hg, the volume is 13.353 cu ft per lb. At 29.62 in. Hg, it is (13.353 × 29.92)/29.62 = 13.488 cu ft per lb.

When 25% of the 10F and 50% rh mixture is combined with 75% of the 70F and 40% rh mixture, the dry-bulb is (0.25 × 10) + (0.75 × 70) = 55F and the specific humidity is (0.25 × 4.60) + (0.75 × 45.13) = 35.00 gr per lb. This corresponds to 39.8F dew point and 0.2449 in. Hg vapor pressure. From the table, Properties of Dry Air, at 55F and 29.22 in. Hg, the volume is 12.975 cu ft per lb. The air pressure is 29.92 — 0.2449 = 29.68 in. Hg. At 55F and 29.68 in. Hg, the volume is (12.975 × 29.92)/29.68 = 13.080 cu ft per lb.

The outside air flow at 10F and 50% rh is

(2430 × 11.857)/60 = 480 cfm

The recirculated air flow at 70F and 40% rh is

(7290 × 13.488)/60 = 1638 cfm

The amount of moisture to be added is

$$\frac{2,430\,(45.13 - 4.60)}{7000} = 14.0 \text{ lb pr hr}$$

The heat of vapor required is

$$2430 \ (6.71 - 0.697) = 14,600 \text{ Btu per hr}$$

The enthalpy required is

$$140,000 + 14,600 + (2430)(0.24)(70\text{-}10) =$$
$$189,600 \text{ Btu per hr}$$

Example 29. The internal sensible and latent heat gains of a building are 40,000 Btu per hr and 7,500 Btu per hr respectively. What will be the amount of air to be handled and the amount of heat and moisture to be removed when the outside condition is 95F db and 70F wb, the inside condition is 75F and 50% rh and 75% (by weight) of the air is recirculated?

It is necessary first to determine the rate of air circulation required to remove both sensible and latent heat loads from the space. The larger the difference between the 75F space temperature and the temperature of the conditioned air, the lower the rate of circulation. However, the larger the temperature difference, the more difficult is the distribution in the space. Also if the temperature of the conditioned air is too far below the space dew point, more latent heat than desired may be removed, resulting in a lower space relative humidity than specified. The lower relative humidity may be unacceptable, which would make it necessary to use a higher temperature. If the lower relative humidity is acceptable, the result would be to increase the load per unit of outside air.

(For a problem of this type the psychrometric chart is strongly recommended because the possibilities are all graphically visible and the solution much easier).

A good place to start is with full details of the space condition. At 75F saturated, vapor pressure is 0.8750 in. Hg. At 75F and 50% rh; it is $0.50 \times 0.8750 = 0.4375$ in. Hg. This corresponds to 55.1F dew point, 64.61 gr per lb and 10.02 Btu per lb heat of moisture. The heat of dry air at 75F is 18.00 Btu per lb and enthalpy is $18.00 + 10.02 = 28.02$ Btu per lb.

Since the dew point is 55.1F and the conditioned air will have a lower dew point (it must absorb 7500 Btu per hr of latent heat), a 55F conditioned air temperature seems practical. Air entering a conditioned space at 15-20F less than space temperature is a frequent basis for design. On this basis, total circulation rate is

$$\frac{40.000}{(0.24)(75\text{-}55)} = 8330 \text{ lb dry air per hr.}$$

Since the same 8330 lb per hr must absorb 7500 Btu per hr, the heat of the vapor must be 7500/

8330 = 0.90 Btu per lb less than the space condition. The heat of the vapor for the conditioned air is $10.02 - 0.90 = 9.12$ Btu per lb, which corresponds to a 52.6F dew point and 58.28 gr per lb.

Outside air is at 95F db and 70F wb. Heat content of 95F dry air is 22.80 Btu per lb. Total enthalpy is 34.01 Btu per lb and heat of the moisture is $34.01 - 22.80 = 11.21$ Btu per lb. This corresponds to a dew point of 58.1F and 72.17 gr per lb.

Outside air $= 0.25 \times 8330 = 2082$ lb per hr. Required heat removal is

$$40,000 + 7,500 + 2082 \ (34.01 - 28.02) =$$
$$59,971 \text{ Btu per hr. Moisture removal is}$$
$$[2,082 \ (72.17 - 64.61) +$$
$$8,330 \ (64.61 - 58.28)]/7,000 = 9.78 \text{ lb per hr.}$$

Example 30. Air at 95F dry-bulb and 75F wet-bulb is to be cooled by an existing system having the capacity to remove 10 Btu per lb of dry air handled. If the air is treated by this system, what will be its final wet-bulb temperature? What, if any, is the minimum condensation in grains per pound?

At 95F, the enthalpy of dry air is 22.80 Btu per lb. Enthalpy at 75F wet-bulb is 38.51 Btu per lb. The heat of the vapor is $38.51 - 23.80 = 14.71$ Btu per lb, which corresponds to a dew point of 65.6F and 94.41 gr per lb (by interpolation of Columns 9, 1 and 4).

If 10 Btu per lb is removed, enthalpy will be $38.51 - 10.00 = 28.51$ Btu per lb, which corresponds to 63F wet-bulb. This will be the wet-bulb of the air leaving the system.

At 63F wet-bulb and saturation, the moisture content is 86.09 gr per lb and the heat of the vapor 13.40 Btu per lb. Since the original moisture ratio is 94.41 gr per lb and the highest vapor ratio at 63F is 86.09 gr per lb, the minimum condensation is $94.41 - 86.09 = 8.32$ gr per lb. How much more moisture will be removed and how much above 63F the leaving dry-bulb temperature will be, depends on the coil characteristics and the cooling system temperature.

The leaving air condition is very unlikely to be 63F saturated, with 95F db and 75F wb the entering condition, because that would require an unusual cooling system design. Although such a design would not be difficult, it would be more costly than the more common design. Details of this type are very difficult to handle without the psychrometric chart and will not be discussed further here.

PSYCHROMETRIC TABLE

(29.92 Inches of Mercury)

Dry Bulb Temp., F	Saturated Water Vapor Pressure Inches of Mercury	Saturated Water Vapor Weight Grains per Cu. Ft. of Vapor	Saturated Water Vapor Weight Grains per Pound of Dry Air	Dry Air, Btu per Lb. Above oF	Air Saturated with Water Vapor Volume Cu. Ft. of 1 lb. Dry Air and Vapor to Saturate It	Air Saturated with Water Vapor Weight, Grains Dry Air in 1 Cu. Ft. of Mixture	Air Saturated with Water Vapor Weight, Grains Air and Vapor in 1 Cu. Ft. of Mixture	Air Saturated with Water Vapor Heat Content per lb. Dry Air and Vapor to Saturate It Heat of Vapor Above 32F, Btu	Air Saturated with Water Vapor Heat Content per lb. Dry Air and Vapor to Saturate It Total En-thalpy, Btu	Wet Bulb Temp., F
(1)	(2)	(3)	(4)	(5)	(6)	(7)	(8)	(9)	(10)	(11)
−40	.0039	.054	.569	−9.60	10.58	661.4	661.5	.085	−9.515	−40
−39	.0041	.056	.597	−9.36	10.61	659.9	660.0	.089	−9.271	−39
−38	.0044	.060	.640	−9.12	10.63	658.3	658.4	.096	−9.025	−38
−37	.0046	.063	.669	−8.88	10.66	656.8	656.9	.100	−8.780	−37
−36	.0048	.065	.699	−8.64	10.68	655.2	655.3	.104	−8.536	−36
−35	.0051	.069	.742	−8.40	10.71	653.7	653.8	.111	−8.289	−35
−34	.0054	.073	.786	−8.16	10.74	652.1	652.2	.117	−8.043	−34
−33	.0057	.077	.830	−7.92	10.76	650.6	650.7	.124	−7.796	−33
−32	.0061	.082	.888	−7.68	10.79	650.0	650.1	.133	−7.547	−32
−31	.0065	.088	.946	−7.44	10.81	647.5	647.6	.142	−7.298	−31
−30	.0069	.093	1.004	−7.20	10.84	646.1	646.2	.150	−7.050	−30
−29	.0074	.099	1.077	−6.96	10.86	644.6	644.7	.161	−6.799	−29
−28	.0078	.104	1.135	−6.72	10.89	643.0	643.1	.170	−6.550	−28
−27	.0083	.111	1.209	−6.48	10.92	641.3	641.4	.181	−6.299	−27
−26	.0089	.118	1.295	−6.24	10.94	640.0	640.1	.194	−6.046	−26
−25	.0094	.125	1.368	−6.00	10.96	638.6	638.7	.205	−5.795	−25
−24	.0100	.132	1.456	−5.76	10.99	637.1	637.2	.218	−5.542	−24
−23	.0106	.140	1.543	−5.52	11.01	635.7	635.9	.232	−5.288	−23
−22	.0112	.148	1.630	−5.28	11.04	634.2	634.3	.245	−5.035	−22
−21	.0119	.157	1.732	−5.04	11.06	632.8	633.0	.260	−4.780	−21
−20	.0126	.165	1.834	−4.80	11.09	631.3	631.5	.276	−4.524	−20
−19	.0133	.174	1.936	−4.56	11.11	629.9	630.1	.291	−4.269	−19
−18	.0141	.184	2.052	−4.32	11.14	628.4	628.6	.309	−4.011	−18
−17	.0150	.196	2.184	−4.08	11.17	626.8	627.0	.329	−3.751	−17
−16	.0159	.207	2.315	−3.84	11.19	625.4	625.6	.349	−3.491	−16
−15	.0168	.218	2.446	−3.60	11.22	624.1	624.3	.368	−3.232	−15
−14	.0178	.231	2.592	−3.36	11.24	622.6	622.8	.391	−2.969	−14
−13	.0188	.243	2.737	−3.12	11.27	621.3	621.5	.413	−2.707	−13
−12	.0199	.257	2.898	−2.88	11.30	619.7	620.0	.437	−2.443	−12
−11	.0210	.270	3.058	−2.64	11.32	618.4	618.6	.461	−2.179	−11
−10	.0222	.285	3.233	−2.40	11.34	617.2	617.5	.488	−1.920	−10
−9	.0234	.300	3.407	−2.16	11.37	615.6	615.9	.514	−1.646	−9
−8	.0247	.316	3.597	−1.92	11.40	614.2	614.5	.544	−1.376	−8
−7	.0260	.331	3.787	−1.68	11.42	612.8	613.1	.572	−1.108	−7
−6	.0275	.350	4.005	−1.44	11.45	611.4	611.8	.606	−0.834	−6

PSYCHROMETRIC TABLE

(29.92 Inches of Mercury)

Dry Bulb Temp., F	Saturated Water Vapor			Dry Air, Btu per Lb. Above oF	Air Saturated with Water Vapor					Wet Bulb Temp., F
	Pressure	Weight			Volume	Weight, Grains		Heat Content per lb. Dry Air and Vapor to Saturate It		
	Inches of Mercury	Grains per Cu. Ft. of Vapor	Grains per Pound of Dry Air		Cu. Ft. of 1 lb. Dry Air and Vapor to Saturate It	Dry Air in 1 Cu. Ft. of Mixture	Air and Vapor in 1 Cu. Ft. of Mixture	Heat of Vapor Above 32F, Btu	Total Enthalpy, Btu	
(1)	(2)	(3)	(4)	(5)	(6)	(7)	(8)	(9)	(10)	(11)
−5	.0291	.370	4.239	−1.20	11.47	610.1	610.4	.641	−0.559	−5
−4	.0307	.389	4.472	−0.96	11.50	608.6	609.0	.677	−0.283	−4
−3	.0325	.412	4.734	−0.72	11.53	607.3	607.7	.717	−0.003	−3
−2	.0344	.434	5.001	−0.48	11.55	605.9	606.3	.759	+0.279	−2
−1	.0363	.457	5.289	−0.24	11.58	604.6	605.1	.802	0.562	−1
0	.0383	.481	5.58	0.00	11.60	603.2	603.7	.846	0.846	0
1	.0403	.505	5.87	0.24	11.63	601.8	602.3	.891	1.131	1
2	.0423	.529	6.16	0.48	11.66	600.5	601.1	.935	1.415	2
3	.0444	.554	6.47	0.72	11.68	599.2	599.7	.982	1.702	3
4	.0467	.581	6.81	0.96	11.71	597.8	598.4	1.03	1.994	4
5	.0491	.610	7.16	1.20	11.73	596.6	597.2	1.09	2.287	5
6	.0515	.638	7.51	1.44	11.76	595.2	595.9	1.14	2.581	6
7	.0542	.670	7.90	1.68	11.79	593.9	594.5	1.20	2.882	7
8	.0570	.703	8.31	1.92	11.81	592.5	593.2	1.26	3.184	8
9	.0600	.739	8.75	2.16	11.83	591.7	592.4	1.33	3.491	9
10	.0631	.775	9.19	2.40	11.86	590.4	591.1	1.40	3.799	10
11	.0665	.815	9.70	2.64	11.89	588.6	589.4	1.48	4.118	11
12	.0699	.855	10.19	2.88	11.92	587.3	588.2	1.55	4.433	12
13	.0735	.897	10.72	3.12	11.95	586.0	586.9	1.64	4.755	13
14	.0772	.941	11.26	3.36	11.97	584.7	585.7	1.72	5.078	14
15	.0810	.985	11.82	3.60	12.00	583.4	584.3	1.80	5.403	15
16	.0850	1.031	12.40	3.84	12.03	582.1	583.1	1.89	5.733	16
17	.0891	1.079	13.00	4.08	12.05	580.8	581.8	1.99	6.066	17
18	.0933	1.127	13.62	4.32	12.08	579.5	580.6	2.08	6.401	18
19	.0979	1.180	14.29	4.56	12.11	578.2	579.4	2.19	6.745	19
20	.103	1.239	15.04	4.80	12.14	576.8	578.0	2.30	7.099	20
21	.108	1.297	15.68	5.04	12.16	575.6	576.9	2.41	7.452	21
22	.113	1.354	16.51	5.28	12.19	574.2	575.6	2.53	7.806	22
23	.118	1.411	17.24	5.52	12.22	572.9	574.4	2.64	8.159	23
24	.124	1.480	18.12	5.76	12.25	571.7	573.1	2.78	8.535	24
25	.130	1.548	19.00	6.00	12.27	570.4	571.9	2.91	8.911	25
26	.136	1.616	19.88	6.24	12.30	569.1	570.7	3.05	9.289	26
27	.143	1.696	20.90	6.48	12.32	568.0	569.7	3.21	9.685	27
28	.150	1.775	21.94	6.72	12.36	566.5	568.3	3.37	10.09	28
29	.157	1.854	22.97	6.96	12.39	565.2	567.1	3.53	10.49	29

PSYCHROMETRIC TABLE

(29.92 Inches of Mercury)

Dry Bulb Temp., F	Saturated Water Vapor			Dry Air, Btu per Lb. Above oF	Air Saturated with Water Vapor					Wet Bulb Temp., F
	Pressure	Weight			Volume	Weight, Grains		Heat Content per lb. Dry Air and Vapor to Saturate It		
	Inches of Mercury	Grains per Cu. Ft. of Vapor	Grains per Pound of Dry Air		Cu. Ft. of 1 lb. Dry Air and Vapor to Saturate It	Dry Air in 1 Cu. Ft. of Mixture	Air and Vapor in 1 Cu. Ft. of Mixture	Heat of Vapor Above 32 F, Btu	Total Enthalpy, Btu	
(1)	(2)	(3)	(4)	(5)	(6)	(7)	(8)	(9)	(10)	(11)
30	.164	1.933	23.99	7.20	12.41	563.9	565.9	3.68	10.88	30
31	.172	2.023	25.17	7.44	12.44	562.6	564.6	3.87	11.31	31
32	.1803	2.117	26.39	7.68	12.47	561.5	563.6	4.06	11.74	32
33	.1878	2.201	27.50	7.92	12.49	560.4	562 6	4.23	12.15	33
34	.1955	2.289	28.63	8.16	12.52	559.1	561.3	4.40	12.56	34
35	.2035	2.375	29.81	8.40	12.55	557.7	560.1	4.59	12.99	35
36	.2118	2.467	31.04	8.64	12.58	556.5	559.0	4.78	13.42	36
37	.2203	2.562	32.29	8.88	12.60	555.4	557.9	4.97	13.85	37
38	.2292	2.660	33.61	9.12	12.64	554.0	556.6	5.18	14.30	38
39	.2383	2.760	34.95	9.36	12.66	552.8	555.6	5.39	14.75	39
40	.2478	2.864	36.36	9.60	12.69	551.4	554.3	5.61	15.21	40
41	.2576	2.971	37.81	9.84	12.73	550.1	553.1	5.83	15.67	41
42	.2677	3.082	39.30	10.08	12.75	548.9	552.1	6.07	16.15	42
43	.2782	3.196	40.86	10.32	12.78	547.6	550.8	6.31	16.63	43
44	.2891	3.314	42.48	10.56	12.82	546.2	549.5	6.56	17.12	44
45	.3004	3.437	44.15	10.80	12.84	545.0	548.4	6.82	17.62	45
46	.3120	3.564	45.87	11.04	12.87	543.8	547.3	7.09	18.13	46
47	.3240	3.694	47.66	11.28	12.90	542.5	546.2	7.37	18.65	47
48	.3364	3.828	49.51	11.52	12.93	541.3	545.1	7.66	19.18	48
49	.3493	3.967	51.43	11.76	12.97	539.9	543.9	7.96	19.72	49
50	.3626	4.110	53.41	12.00	13.00	538.7	542.8	8.27	20.27	50
51	.3764	4.257	55.47	12.24	13.03	537.3	541.6	8.59	20.83	51
52	.3906	4.409	57.59	12.48	13.06	535.9	540.4	8.92	21.40	52
53	.4052	4.565	59.77	12.72	13.09	534.7	539.2	9.26	21.98	53
54	.4203	4.727	62.03	12.96	13.12	533.4	538.1	9.62	22.58	54
55	.4359	4.893	64.37	13.20	13.16	532.1	537.0	9.98	23.18	55
56	.4520	5.064	66.78	13.44	13.19	530.8	535.8	10.36	23.80	56
57	.4686	5.240	69.27	13.68	13.22	529.5	534.8	10.75	24.43	57
58	.4858	5.422	71.85	13.92	13.25	528.2	533.6	11.16	25.08	58
59	.5035	5.609	74.52	14.16	13.29	526.8	532.4	11.58	25.74	59
60	.5218	5.801	77.27	14.40	13.32	525.5	531.3	12.01	26.41	60
61	.5407	5.999	80.12	14.64	13.36	524.1	530.1	12.46	27.10	61
62	.5601	6.203	83.05	14.88	13.39	522.9	529.1	12.92	27.80	62
63	.5802	6.414	86.09	15.12	13.42	521.5	528.0	13.40	28.52	63
64	.6009	6.631	89.23	15.36	13.46	520.2	526.8	13.89	29.25	64

PSYCHROMETRIC TABLE

(29.92 Inches of Mercury)

Dry Bulb Temp., F	Saturated Water Vapor			Dry Air, Btu per Lb. Above oF	Air Saturated with Water Vapor					Wet Bulb Temp., F
	Pressure	Weight			Volume	Weight, Grains		Heat Content per lb. Dry Air and Vapor to Saturate It		
	Inches of Mercury	Grains per Cu. Ft. of Vapor	Grains per Pound of Dry Air		Cu. Ft. of 1 lb. Dry Air and Vapor to Saturate It	Dry Air in 1 Cu. Ft. of Mixture	Air and Vapor in 1 Cu. Ft. of Mixture	Heat of Vapor Above 32F, Btu	Total Enthalpy, Btu	
(1)	(2)	(3)	(4)	(5)	(6)	(7)	(8)	(9)	(10)	(11)
65	.6222	6.853	92.46	15.60	13.49	518.9	525.7	14.40	30.00	65
66	.6442	7.082	95.80	15.84	13.53	517.5	524.6	14.93	30.77	66
67	.6669	7.318	99.25	16.08	13.56	516.1	523.4	15.47	31.55	67
68	.6903	7.560	102.82	16.32	13.60	514.7	522.3	16.03	32.35	68
69	.7144	7.810	106.50	16.56	13.64	513.3	520.9	16.61	33.17	69
70	.7392	8.065	110.28	16.80	13.67	511.9	520.0	17.21	34.01	70
71	.7648	8.329	114.21	17.04	13.71	510.5	518.9	17.83	34.87	71
72	.7912	8.601	118.25	17.28	13.75	509.1	517.7	18.47	35.75	72
73	.8183	8.879	122.42	17.52	13.79	507.7	516.6	19.12	36.64	73
74	.8462	9.165	126.71	17.76	13.83	506.3	515.5	19.81	37.57	74
75	.8750	9.460	131.16	18.00	13.87	504.9	514.3	20.51	38.51	75
76	.9046	9.762	135.73	18.24	13.91	503.4	513.2	21.23	39.47	76
77	.9352	10.07	140.47	18.48	13.94	502.0	512.1	21.98	40.46	77
78	.9666	10.39	145.35	18.72	13.99	500.5	510.9	22.75	41.47	78
79	.9989	10.72	150.37	18.96	14.03	499.0	509.8	23.55	42.51	79
80	1.0321	11.06	155.52	19.20	14.07	497.7	508.7	24.37	43.57	80
81	1.0664	11.40	160.88	19.44	14.11	496.1	507.5	25.21	44.65	81
82	1.1016	11.76	166.41	19.68	14.15	494.7	506.4	26.09	45.77	82
83	1.1378	12.12	172.11	19.92	14.20	493.1	505.2	26.99	46.91	83
84	1.1750	12.50	177.95	20.16	14.24	491.6	504.1	27.92	48.08	84
85	1.2133	12.88	184.00	20.40	14.29	490.0	502.9	28.88	49.28	85
86	1.2527	13.28	190.23	20.64	14.33	488.5	501.8	29.87	50.51	86
87	1.2931	13.68	196.63	20.88	14.37	487.0	500.7	30.89	51.77	87
88	1.3347	14.09	203.24	21.12	14.42	485.4	499.5	31.94	53.06	88
89	1.3775	14.52	210.08	21.36	14.47	483.4	498.3	33.03	54.39	89
90	1.4215	14.96	217.13	21.60	14.52	482.2	497.2	34.15	55.75	90
91	1.4667	15.40	224.40	21.84	14.57	480.5	495.9	35.31	57.15	91
92	1.5131	15.86	231.89	22.08	14.62	478.8	494.7	36.50	58.58	92
93	1.5608	16.34	239.60	22.32	14.67	477.3	493.6	37.73	60.05	93
94	1.6097	16.82	247.52	22.56	14.72	475.6	492.5	38.99	61.55	94
95	1.6600	17.31	255.75	22.80	14.77	473.9	491.2	40.30	63.10	95
96	1.7117	17.82	264.16	23.04	14.82	472.2	490.1	41.64	64.68	96
97	1.7647	18.34	272.87	23.28	14.88	470.5	488.8	43.03	66.31	97
98	1.8192	18.87	281.83	23.52	14.93	468.8	487.6	44.47	67.99	98
99	1.8751	19.42	291.06	23.76	14.99	467.0	486.4	45.94	69.70	99

PSYCHROMETRIC TABLE

(29.92 Inches of Mercury)

Dry Bulb Temp., F	Saturated Water Vapor			Dry Air, Btu per Lb. Above oF	Air Saturated with Water Vapor					Wet Bulb Temp., F
	Pressure	Weight			Volume	Weight, Grains		Heat Content per lb. Dry Air and Vapor to Saturate It		
	Inches of Mercury	Grains per Cu. Ft. of Vapor	Grains per Pound of Dry Air		Cu. Ft. of 1 lb. Dry Air and Vapor to Saturate It	Dry Air in 1 Cu. Ft. of Mixture	Air and Vapor in 1 Cu. Ft. of Mixture	Heat of Vapor Above 32F, Btu	Total Enthalpy, Btu	
(1)	(2)	(3)	(4)	(5)	(6)	(7)	(8)	(9)	(10)	(11)
100	1.9325	19.98	300.60	24.00	15.05	465.2	485.2	47.46	71.46	100
101	1.9915	20.55	310.44	24.24	15.10	463.4	484.0	49.03	73.27	101
102	2.0519	21.14	320.54	24.48	15.16	461.7	482.8	50.65	75.13	102
103	2.1138	21.75	330.95	24.72	15.22	459.9	481.6	52.31	77.03	103
104	2.1775	22.36	341.70	24.96	15.28	458.0	480.4	54.03	78.99	104
105	2.2429	22.99	352.80	25.20	15.34	456.1	479.1	55.81	81.01	105
106	2.3099	23.63	364.21	25.44	15.41	454.2	477.9	57.64	83.08	106
107	2.3786	24.30	375.97	25.68	15.47	452.4	476.7	59.52	85.20	107
108	2.4491	24.97	388.12	25.92	15.54	450.4	475.4	61.47	87.39	108
109	2.5214	25.67	400.61	26.16	15.61	448.5	474.2	63.47	89.63	109
110	2.5955	26.38	413.50	26.40	15.68	446.5	472.9	65.54	91.94	110
111	2.6715	27.10	426.82	26.64	15.75	444.4	471.5	67.68	94.32	111
112	2.7494	27.84	440.69	26.88	15.82	442.4	470.3	69.90	96.78	112
113	2.8293	28.61	454.69	27.12	15.89	440.4	469.0	72.15	99.27	113
114	2.9111	29.39	469.23	27.36	15.97	438.4	467.8	74.48	101.84	114
115	2.9948	30.19	484.21	27.60	16.04	436.4	466.6	76.89	104.49	115
116	3.0806	31.00	499.71	27.84	16.12	434.3	465.3	79.38	107.22	116
117	3.1687	31.83	515.69	28.08	16.20	432.1	463.9	81.95	110.03	117
118	3.2589	32.68	532.19	28.32	16.29	430.0	462.5	84.60	112.92	118
119	3.3512	33.54	549.13	28.56	16.37	427.6	461.1	87.33	115.89	119
120	3.4458	34.44	566.67	28.80	16.46	425.4	459.8	90.16	118.96	120
121	3.5427	35.35	584.70	29.04	16.55	423.2	458.5	93.06	122.10	121
122	3.6420	36.28	603.38	29.28	16.63	420.9	457.2	96.06	125.34	122
123	3.7436	37.23	622.62	29.52	16.73	418.6	455.8	99.17	128.69	123
124	3.8475	38.20	642.48	29.76	16.82	416.2	450.4	102.37	132.13	124
125	3.9539	39.19	662.90	30.00	16.92	413.9	453.0	105.67	135.67	125
126	4.0629	40.21	684.10	30.24	17.01	411.4	451.6	109.08	139.32	126
127	4.1745	41.24	705.91	30.48	17.12	409.0	450.2	112.60	143.08	127
128	4.2887	42.30	728.47	30.72	17.22	406.5	448.8	116.24	146.96	128
129	4.4055	43.38	751.72	30.96	17.33	404.0	447.4	120.00	150.96	129
130	4.5251	44.49	775.80	31.20	17.44	401.4	445.9	123.90	155.10	130
131	4.6474	45.62	800.57	31.44	17.56	398.9	444.5	127.90	159.34	131
132	4.7725	46.77	826.26	31.68	17.67	396.2	443.0	132.04	163.72	132
133	4.9005	47.95	852.73	31.92	17.79	393.6	441.6	136.34	168.26	133
134	5.0314	49.15	880.12	32.16	17.91	390.9	440.1	140.75	172.91	134

PSYCHROMETRIC TABLE

(29.92 Inches of Mercury)

Dry Bulb Temp., F	Saturated Water Vapor			Dry Air, Btu per Lb. Above oF	Air Saturated with Water Vapor					Wet Bulb Temp., F
	Pressure	Weight			Volume	Weight, Grains		Heat Content per lb. Dry Air and Vapor to Saturate It		
	Inches of Mercury	Grains per Cu. Ft. of Vapor	Grains per Pound of Dry Air		Cu. Ft. of 1 lb. Dry Air and Vapor to Saturate It	Dry Air in 1 Cu. Ft. of Mixture	Air and Vapor in 1 Cu. Ft. of Mixture	Heat of Vapor Above 32F, Btu	Total Enthalpy, Btu	
(1)	(2)	(3)	(4)	(5)	(6)	(7)	(8)	(9)	(10)	(11)
135	5.1653	50.38	908.42	32.40	18.04	388.2	438.6	145.34	177.74	135
136	5.3022	51.63	937.68	32.64	18.16	385.4	437.0	150.07	182.71	136
137	5.4421	52.91	967.89	32.88	18.30	382.7	435.6	154.97	187.85	137
138	5.5852	54.21	999.24	33.12	18.43	379.8	434.0	160.05	193.17	138
139	5.7316	55.55	1031.68	33.36	18.57	376.9	432.4	165.15	198.51	139
140	5.8812	56.91	1065.1	33.60	18.71	374.1	431.0	170.72	204.32	140
141	6.0341	58.29	1099.3	33.84	18.87	371.2	429.5	176.27	210.11	141
142	6.1903	59.71	1135.7	34.08	19.02	368.0	427.7	182.17	216.25	142
143	6.3500	61.16	1172.9	34.32	19.18	365.0	426.2	188.20	222.52	143
144	6.5132	62.63	1211.5	34.56	19.34	362.0	424.6	194.45	229.01	144
145	6.680	64.13	1251.4	34.80	19.51	358.7	422.9	200.96	235.76	145
146	6.850	65.67	1292.7	35.04	19.68	355.7	421.4	207.66	242.70	146
147	7.024	67.23	1335.6	35.28	19.86	352.4	419.6	214.63	249.91	147
148	7.202	68.82	1380.1	35.52	20.05	349.1	417.9	221.86	257.28	148
149	7.384	70.45	1426.4	35.76	20.25	345.7	416.2	229.39	265.15	149
150	7.569	72.11	1474.3	36.00	20.44	342.5	414.6	237.17	273.17	150
151	7.759	73.80	1524.9	36.24	20.66	338.8	412.6	245.41	281.65	151
152	7.952	75.53	1575.9	36.48	20.87	335.4	410.9	253.69	290.17	152
153	8.150	77.29	1629.8	36.72	21.10	331.9	409.2	262.47	299.19	153
154	8.351	79.08	1685.6	36.96	21.32	328.3	407.4	271.55	308.51	154
155	8.557	80.91	1743.8	37.20	21.56	324.8	405.7	281.03	318.23	155
156	8.767	82.76	1804.6	37.44	21.80	321.1	403.9	290.93	328.37	156
157	8.981	84.65	1867.3	37.68	22.06	317.3	402.0	301.14	338.82	157
158	9.200	86.59	1933.0	37.92	22.32	313.6	400.2	311.88	349.80	158
159	9.424	88.56	2001.8	38.16	22.60	309.7	398.3	323.08	361.24	159
160	9.652	90.57	2073.2	38.40	22.89	305.8	396.4	334.73	373.13	160
161	9.885	92.62	2148.0	38.64	23.19	301.8	394.4	346.92	385.56	161
162	10.122	94.70	2225.8	38.88	23.50	297.9	392.6	359.62	398.50	162
163	10.364	96.82	2307.2	39.12	23.83	293.7	390.6	372.92	412.04	163
164	10.611	98.97	2392.4	39.36	24.17	289.6	388.6	386.81	426.17	164
165	10.863	101.17	2481.6	39.60	24.53	285.4	386.5	401.04	440.64	165
166	11.120	103.41	2575.0	39.84	24.90	281.1	384.5	416.64	456.48	166
167	11.382	105.69	2673.0	40.08	25.29	276.8	382.5	432.65	472.73	167
168	11.649	108.03	2775.7	40.32	25.69	272.5	380.5	449.41	489.73	168
169	11.921	110.39	2883.4	40.56	26.12	267.9	378.4	467.02	507.58	169

PSYCHROMETRIC TABLE

(29.92 Inches of Mercury)

Dry Bulb Temp., F	Saturated Water Vapor			Dry Air, Btu per Lb. Above oF	Air Saturated with Water Vapor						Wet Bulb Temp., F
	Pressure	Weight			Volume	Weight, Grains		Heat Content per lb. Dry Air and Vapor to Saturate It			
	Inches of Mercury	Grains per Cu. Ft. of Vapor	Grains per Pound of Dry Air		Cu. Ft. of 1 lb. Dry Air and Vapor to Saturate It	Dry Air in 1 Cu. Ft. of Mixture	Air and Vapor in 1 Cu. Ft. of Mixture	Heat of Vapor Above 32F, Btu	Total Enthalpy, Btu	
(1)	(2)	(3)	(4)	(5)	(6)	(7)	(8)	(9)	(10)	(11)
170	12.199	112.79	2996.9	40.80	26.57	263.5	376.2	485.59	526.39	170
171	12.483	115.25	3116.6	41.04	27.04	258.9	374.1	505.18	546.22	171
172	12.772	117.74	3242.5	41.28	27.54	254.2	371.9	525.74	567.02	172
173	13.066	120.28	3375.0	41.52	28.06	249.5	369.7	547.43	588.95	173
174	13.366	122.94	3515.1	41.76	28.61	244.7	367.6	570.36	612.12	174
175	13.671	125.49	3662.8	42.00	29.19	239.8	365.3	594.50	636.50	175
176	13.983	128.18	3819.7	42.24	29.79	234.9	363.2	620.20	662.44	176
177	14.301	130.89	3986.1	42.48	30.46	229.8	360.7	647.46	689.94	177
178	14.625	133.66	4162.7	42.72	31.14	224.8	358.5	676.36	719.08	178
179	14.955	136.48	4350.5	42.96	31.88	219.6	356.1	707.09	750.05	179
180	15.291	139.36	4550.4	43.20	32.65	214.4	353.7	739.81	783.01	180
181	15.633	142.28	4763.7	43.44	33.48	209.1	351.3	774.77	818.21	181
182	15.982	145.26	4991.8	43.68	34.36	203.7	349.0	812.16	855.84	182
183	16.337	148.27	5236.1	43.92	35.32	198.2	346.5	852.23	896.15	183
184	16.699	151.35	5498.5	44.16	36.32	192.7	344.1	895.22	939.38	184
185	17.068	154.49	5781.2	44.40	37.42	187.1	341.6	941.59	985.99	185
186	17.443	157.66	6085.9	44.64	38.60	181.3	339.0	991.58	1036.22	186
187	17.825	160.88	6415.6	44.88	39.88	175.5	336.4	1045.68	1090.56	187
188	18.214	164.17	6773.5	45.12	41.26	169.7	333.8	1104.37	1149.49	188
189	18.611	167.50	7164.1	45.36	42.77	163.7	331.2	1168.46	1213.82	189
190	19.014	170.90	7589.5	45.60	44.41	157.6	328.5	1237.93	1283.53	190
191	19.425	174.35	8057.1	45.84	46.21	151.5	325.8	1314.92	1360.76	191
192	19.843	177.85	8571.9	46.08	48.20	145.2	323.1	1399.93	1446.01	192
193	20.269	181.44	9142.5	46.32	50.39	138.9	320.4	1493.07	1539.39	193
194	20.703	185.14	9777.7	46.56	52.85	132.5	317.5	1597.61	1644.17	194
195	21.144	188.73	10487.9	46.80	55.57	125.9	314.7	1714.04	1764.84	195
196	21.593	192.47	11287.8	47.04	58.64	119.4	311.8	1845.92	1892.96	196
197	22.050	196.30	12195.9	47.28	62.13	112.7	309.0	1994.41	2041.69	197
198	22.515	200.17	13234.8	47.52	66.12	105.9	306.0	2165.38	2212.90	198
199	22.987	204.08	14431.8	47.76	70.72	98.9	303.1	2361.62	2409.38	199
200	23.467	208.09	15828.9	48.00	76.07	92.0	300.1	2591.88	2638.88	200

HOW TO USE THE PSYCHROMETRIC CHART

A psychrometric chart is a graph of the properties of moist air at various conditions of temperature and humidity. Many such charts are in use; that in Fig. 1 was plotted from U. S. Weather Bureau data and other sources. It is typical of most such charts and is intended to serve as an illustration of the utility of the graphical method. For actual engineering practice, the same scale range over a larger area would be preferred.

Properties of Air. The variable properties of moist air pictured in Fig. 1 at a single pressure (standard atmospheric) are as follows:

Dry bulb temperature, degrees F, the horizontal scale, is the temperature read on an ordinary room thermometer.

Weight of water vapor in one pound of dry air, grains, the major vertical scale in grains per pound of dry air. A grain is 1/7000 of a pound. Quantities of air and moisture in weight units are used to simplify the presentation. The weight of vapor is the weight of water which has been evaporated into the air. The vapor is actually steam which occupies its own fraction of space and exerts its own pressure in proportion. Thus, in ordinary air, it is very low pressure superheated steam and has been evaporated into the air at a temperature corresponding to the boiling point of water at that very low pressure.

Pressure of water vapor, inches of mercury, shown in the scale at the left, is the absolute pressure of the steam, and *Dew point temperatures,* shown in the scale on the upper curved line, are the corresponding boiling points of water at those low pressures. If a single vapor coordinate in Fig. 1, say 96 grains, is followed from the right scale toward the left and across the decreasing dry bulb temperature coordinates, it is found that the steam will cool until it reaches the boiling point corresponding to its own (very low) pressure at 66 deg. The steam will then begin to condense back into water and fog will appear to the left of 66 deg. at the 96 grain level. This establishes the point where the 66 deg. dry bulb and 96 grains moisture coordinates meet at a condition called saturation because the amount of water in air at 66 deg. cannot be increased beyond 96 grains without condensation. Also, if there are 96 grains of moisture per pound of dry air at any temperature, no matter how high, that air will have to be cooled to 66 deg. before condensation takes place. Therefore the dew point temperature coordinates are also horizontal in Fig. 1. Each dry bulb temperature and weight of vapor combination has a corresponding saturation point which establishes the 100% relative

humidity or saturation curve.

Percent relative humidity curves show the number of grains of moisture actually in the air as a fraction of the total amount possible at the various dry bulb and vapor weight combinations.

Total heat, Btu per pound of dry air, is shown in Fig. 1 by a diagonal system of coordinates with the scale on the diagonal line separate from the body of the chart and above the saturation curve. A certain amount of heat (.24 Btu) is required to raise the temperature of a pound of air one degree. This is called sensible heat because its effect can be sensed by a dry bulb thermometer. Another definite amount of heat (1 Btu) is required to raise the temperature of one pound of water one degree. This is also sensible heat in that it affects a thermometer reading, but is called heat of the liquid. The third component of total heat in a given sample of moist air is the heat required to convert the water at the dew point to steam at the same temperature. Since no temperature change is involved, this component is called the latent heat of vaporization. Total heat on the chart is the sum of the heat required to warm a pound of air from 0 deg. to dry bulb, plus the heat to warm the water in a pound of air from 32 deg. to dew point, plus the latent heat required to vaporize that much water at dew point. Total heat is also called enthalpy.

Wet bulb temperatures are shown on diagonal coordinates coinciding with the total heat coordinates but with their scale shown on the saturation curve. Wet bulb is the temperature assumed by an evaporating film of water in air at constant dry bulb and dew point.

Any two known properties of air determine a point on the chart called a state point from which all other properties can be read. For example, to find the wet bulb of air at 78 deg. D.B. (dry bulb) and 50% R.H. (relative humidity), enter the chart at 78 deg on the bottom scale, as in Fig. 2, proceed up to the 50% curve and read diagonally upward to 64.8 deg. on the wet bulb scale as shown in Fig. 2. Construction of the chart and coordinates for a single state point are shown in Fig. 3.

Total Heat. The coordinates for total heat and wet bulb temperature are drawn to coincide in slope. Actually, at constant total heat the corresponding wet bulb may show slight variation over the range of humidity from dry air to saturation. For practical purposes, however, the difference is so slight as to be negligible in most air conditioning problems.

An increase in moisture content from zero to

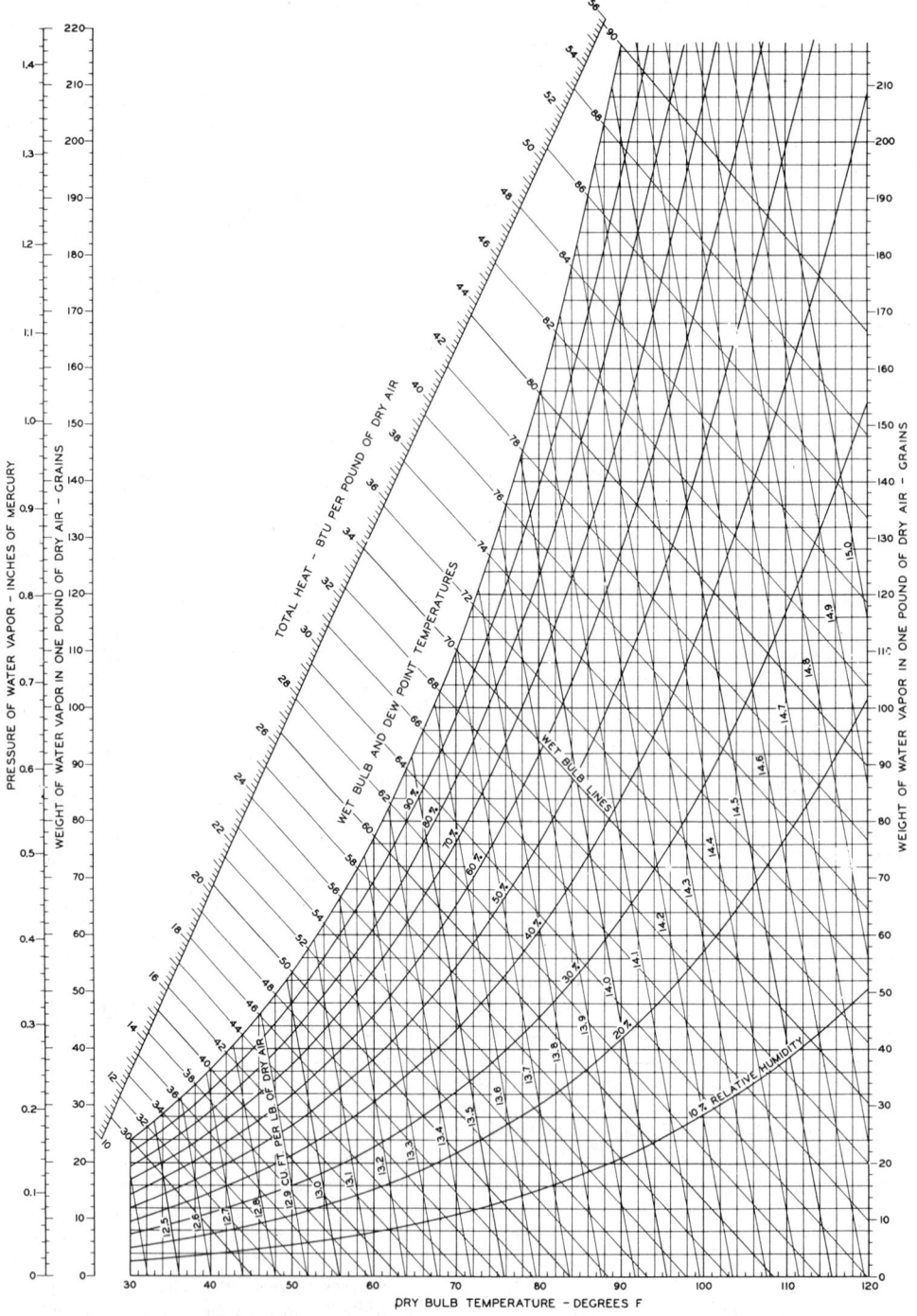

Fig. 1. Psychrometric properties of air at 29.92 inches of mercury absolute pressure.

saturation at constant dry bulb is accompanied by an increase in wet bulb temperature and in total heat at a rate depending on the enthalpy of the moisture added. Vapor enthalpy consists of the heat of the liquid from the datum temperature, 32 deg., to the dew point plus the latent heat of vaporization at the dew point. The slope of the total heat coordinates is, therefore, a function of the heat content of vapor in the mixture.

Since the datum temperature for calculating enthalpy of dry air is o deg., and the specific heat

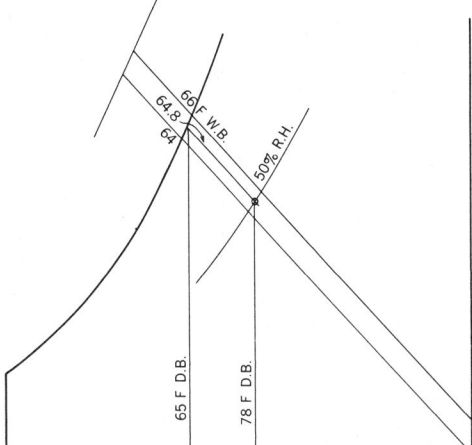

Fig. 2. How to find wet bulb from dry bulb and relative humidity. Interpolation between wet bulb coordinates is aided by the fact that wet bulb and dry bulb are identical at the saturation curve.

of air is .24 (for all practical purposes over the range of this chart) the product of dry bulb scale times .24 will be the sensible heat content per pound of dry air in any mixture, and dry bulb lines are lines of constant sensible heat of dry air.

Heat of Vapor. At any point other than along the saturation (100% relative humidity) line, the vapor in an air-vapor mixture is superheated steam. For absolute accuracy, therefore, total heat would include a small value allowed for sensible heat of the vapor in cooling from dry bulb to dew point. The specific heat of water vapor at constant pressure between 32 deg. and 212 deg. is about .45 Btu per (lb) (deg.). Thus, the allowance for sensible heat of superheated vapor would be .45/7000 = .00006 Btu per (gr) (deg.) above dew point. For most problems in air conditioning practice, this is a negligible amount.

Another component of total heat is the heat required to warm the liquid from the datum tem-

perature, 32 deg., to the dew point. For all practical purposes, this can be figured at 1 Btu per (lb) (deg.) or 1/7000 = .000143 Btu per (gr) (deg.).

Perhaps the most important component of heat in the vapor is the latent heat of evaporation. For total heat computations used in plotting the chart, enthalpy values of saturated vapor are from the Keenan and Keyes steam tables. The latent heat in any air-vapor sample can be determined by subtracting sensible heat of air and liquid from total heat. That is

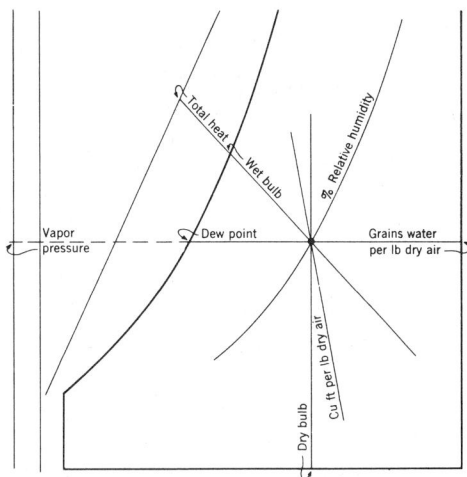

Fig. 3. Skeleton chart showing systems of coordinates for a single state point on the psychrometric chart.

$$L = T - .24t - .000143h(d_p - 32)$$

where L = latent heat, Btu per lb dry air
T = total heat, Btu per lb dry air
t = dry bulb temperature, deg. F.
h = absolute humidity, grains per lb dry air
d_p = dew point temperature, deg. F.

For example, the temperature and moisture conditions for air at 80 deg. D.B. and 50% R.H. are shown in Fig. 4. In addition to the values which can be read directly from the chart, the following properties can be derived as described:

	Btu per lb dry air
Sensible heat of dry air = 80 × .24 =	19.20
Enthalpy of vapor = 31.2 − 19.2 =	12.00
Heat of liquid = 76 × (59.7 − 32) × .000143 =	.30
Superheat of vapor = 76 × .00006 × (80 − 59.7) =	.10

Latent heat of evaporation =
$$12.00 - .30 = \qquad 11.70$$

Note that in the computation of total heat, the superheat of the vapor is not included and so is not subtracted from enthalpy of vapor in deriving latent heat. The amount of superheat in this case is less

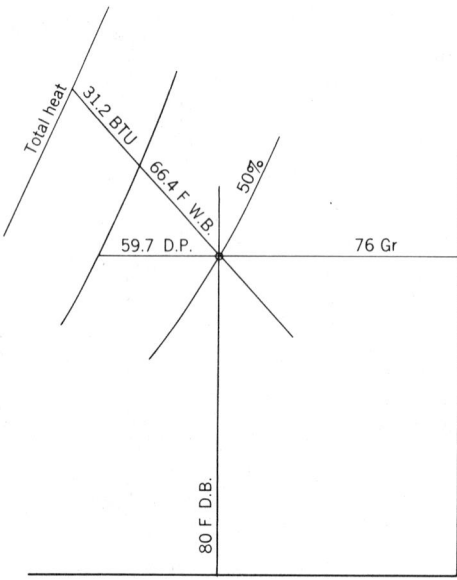

Fig. 4. Direct readings of heat and moisture properties at a single state point.

than 0.9% of vapor enthalpy and less than 0.4% of total heat.

Wet Bulb Temperature. Wet bulb temperature is the lowest temperature to which a film of water can be brought by free evaporation into air at constant dry bulb and dew point.

When water is in the presence of unsaturated air, there is always a boundary condition in which two opposing influences exist. (1) There is an exchange of heat by conduction to the cooler of the two media; and (2) there is a transfer of vapor to the medium of lower vapor pressure. Unsaturated air, by definition, has a vapor pressure lower than water at any temperature higher than the dew point temperature of the air.

A film of water, approaching the dry bulb of the air by conducted heat exchange, is evaporated at the dew point saturation pressure at a rate depending on the degree of saturation of the air; that is to say, on the magnitude of the vapor pressure difference. Removal of the latent heat of evaporation tends to

cool the water film, and equilibrium is reached at a point between the dry bulb and dew point of the air. This point is the wet bulb temperature. It is a directly measurable property of air, and is extremely useful as a limit in evaporative cooling problems.

As the dry bulb approaches the dew point, vapor pressures of both water film and air approach that of steam at the dew point saturation pressure.

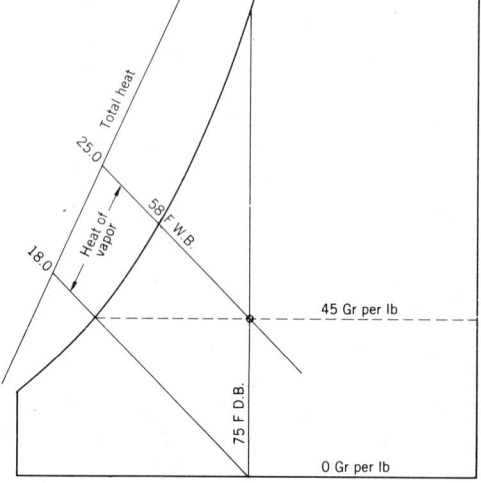

Fig. 5. Direct reading of vapor enthalpy (heat of vapor) as the difference between total heats of sample air and dry air at the same dry bulb.

Hence, at 100% saturation, dry bulb, dew point, and wet bulb temperatures coincide.

At constant dry bulb, any increase in moisture content of air results in an increase in total heat. Since the sensible heat of the air remains constant, such an increase in total heat must be entirely an increase in vapor enthalpy. The influence of conduction at a water interface is stable at constant dry bulb, because conduction is a function of temperature. Therefore, the influence of evaporation is affected exclusively and directly by an increase in moisture content; the wet bulb is thus a direct function of total heat. Wet-bulb lines and total heat lines have very nearly the same slope in a range up to about 90F wb. There is a slight deviation because, at constant wet-bulb, moisture is added merely by supplying the latent heat of evaporation at the wet-bulb temperature, while the total heat or enthalpy lines are based on latent heat plus heat of the liquid between the wet-bulb temperature and 32F. (At 32F wet-bulb, the lines do coincide.)

In the example shown in Fig. 5, the state point is at 75 deg. D.B. and 58 deg. W.B. Total heat at zero moisture is 18.0 Btu per lb and at the state point is 25.0. Enthalpy of vapor at the state point is therefore 25.0 − 18.0 = 7.0 Btu per lb dry air. This is in reasonable agreement with the calculated value based on 45 grains of moisture evaporated at 45.2 deg. (dew point).

Volume Lines. The volume of an air-vapor mixture per pound of dry air is indicated in Fig. 1 by a set of diagonal coordinates with the scale shown between the 20% and 30% relative humidity curves.

ordinates at the original moisture level. When a change in moisture content accompanies a temperature change, the intervening process between two state points becomes more complex, and the line joining the two points is an indication of the means by which the change can be brought about.

Cooling and Humidifying. For example, in Fig. 7, the change from point A (90 deg. D.B., 20% R.H.) to B (70 deg. D.B., 50% R.H.) represents a cooling and humidifying process accompanied by a

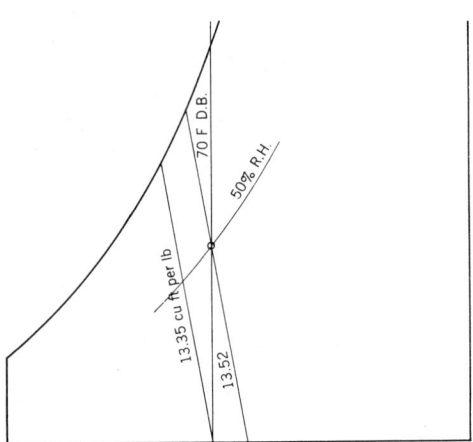

Fig. 6. Direct reading of vapor volume as the difference between volumes of actual state and of dry air at the same dry bulb temperature.

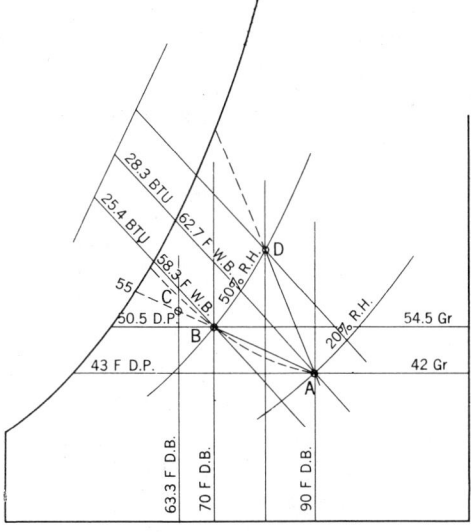

Fig. 7. Action of air washer and mixture of air streams in cooling and humidifying with water above the dew point of air.

In addition to the volume of mixture for any state point, the volume of dry air and of steam can be derived.

For example, as shown in Fig. 6, by reading the volume of dry air at the base of the dry bulb coordinate and subtracting from total volume at the state point, volume of steam is obtained. According to Fig. 6, the volume of mixture at 70 deg. D.B. and 50% R.H. is 13.52 cu ft per lb dry air and the volume of dry air at 70 deg. is 13.35. The volume of vapor at this state point, therefore, is .17 cu ft per lb dry air.

Use of the Chart. A principal use of the psychrometric chart is to visualize over-all changes in properties of air between one state point and another. Changes in dry bulb temperature without an accompanying change in moisture content (sensible heating or cooling) simply follow along the horizontal co-

net loss in total heat (28.3 − 25.4 = 2.9 Btu loss per lb dry air) while the line A–D represents sensible cooling and humidifying with a net gain in total heat. A net loss in total heat must indicate the presence of water at a temperature lower than the original wet bulb at A. Otherwise either thermal equilibrium or a net gain in total heat would obtain. Since there is a gain in humidity from A to B, the temperature (hence vapor pressure) of the water must be higher than at the dew point of the air. The change from A to B, therefore, is accomplished by passing the original air through a washer with water at a constant temperature between the dew point and wet bulb temperatures of the air.

To fix the theoretical, constant temperature of the washer water, simply extend the A–B line to the saturation curve and read the temperature at the point of intersection (55 deg. in this case). Although

impossible in practice (because the water will always change temperature toward the wet bulb of the air and vice versa) this theoretical procedure is useful to illustrate the point that, should the water temperature remain constant at 55 deg. and the washer prove to be 100% efficient (that is, capable of bringing the air to saturation at its own water temperature), the point B would move along to the saturation curve, and the ultimate condition would be 55 deg. saturated air. Thus the distance A-B, in relation to the total distance from A to saturation, is the efficiency of the washer.

In actual practice, the washer water will warm (or cool) toward the wet bulb of the air supplied to it. The condition line, then, would follow some curve similar to the dashed curve in Fig. 7.

Another means of changing air from condition A to B would be by mixing the original air with a specified amount of cooler air.

For example, if one part of air at condition A were mixed with three parts of cooler air, the dry bulb of the cooler air would be determined by the formula

$$t_c = \frac{70 - (.25 \times 90)}{.75} = 63.3$$

where t_c is the dry bulb temperature of the cooler air.

Note that the intersection of this dry bulb line with the extension of the A-B line fixes the state point of the cooler air at point C in Fig. 7. Also, the length of the line B-C is equal to one-third of the distance A-B, representing the one part of warmer air to three parts of cooler air.

The same reasoning can be applied to the mixing of any two samples of air in any proportion. The ultimate dry bulb is always the sum of the dry bulb temperatures of the contributing samples multiplied respectively by the fraction each contributes to the total quantity of air.

$$t_u = xt_1 + (1 - x)t_2$$

where t_u = dry bulb of mixture, deg. F.

x = fraction of mixture at dry bulb t_1

t_2 = dry bulb of second sample, deg. F.

The state point of the mixture is always the intersection of the ultimate dry bulb coordinate with a line joining the two state points of the components.

Quantities of air in most air conditioning problems are expressed in cubic feet per minute. For strict accuracy these quantities should be converted to weights of dry air, but for practical examples depending only on proportions mixed, the quantities expressed in cubic feet are sufficiently accurate.

Line A-D in Fig. 7 is the condition line for air at A in the presence of water at a temperature between the original wet bulb and dry bulb temperatures of the sample. It is of interest in showing a decrease in dry bulb accompanied by an increase in total heat.

If water is continuously recirculated in any air washer without being heated or cooled by any means other than action of the air, the water will assume the wet bulb temperature of the air and changes in state point of the air will follow directly along the wet bulb line of the original sample. The distance moved as a fraction of the total distance from the original point to saturation, is, again, the efficiency of the washer.

Actually, the effect is that of a mixture of two portions of the stream of air. One portion contacts the water and becomes saturated. The other portion by-passes the water and is unaffected. The degree of contact is affected by the number of banks of spray nozzles and the fineness (pressure) of the spray. Efficiencies of recirculating washers of average commercial construction are approximately:

Single bank	65%
Two banks	85%
Three banks	95%

Heating and Humidifying. Many commercial central systems combine the principles outlined in an arrangement of coils and air washer to heat and humidify air.

For example, if outside air at 34 deg. and 60% R.H. (point A in Fig. 8) is to be delivered to a room at 70 deg. and 50% R.H. (point D in Fig. 8), it can be introduced through a heating coil which will heat it to 85 deg. without adding moisture (line A-B). The heated air then goes through an 80% efficient recirculating washer in which the water is in thermal equilibrium at the wet bulb temperature of the air (55 deg.). Air cools and absorbs moisture at constant total heat and emerges at point C, 80% of the distance from B to the saturation curve. Air is now at 61 deg. D.B. and is reheated to 70 deg. in a second coil to point D.

It is apparent in Fig. 8 that an increase in the exit temperature at the first coil will raise the equilibrium temperature of the circulating water and the moisture level of the air emerging from the washer.

An alternate method is to decrease the exit temperature (and load) at the first coil to a point such as B^1 in Fig. 8. The washer water is then heated to some temperature above 61 deg. and heats and humidifies the air along a condition line such as B^1-C. The reheating coil functions as before.

Dehumidifying. Moisture can be removed from air by two processes; (1) by cooling below its dewpoint so that condensation occurs, and (2) by absorption through means of hygroscopic chemicals. The first can be accomplished by passing the air through a cooling coil the surface temperature of which is below the dew point of the air or through an air washer with chilled water at a temperature below the dew point of the air. Either of these methods removes both moisture and heat from the air. Chemical absorption is accomplished by passing air through a bed of absorptive material, like silica gel, or through a spray of absorptive solution, like lithium bromide. Chemical absorption takes place, theoretically, at constant total heat so that it is accompanied by an increase in dry bulb temperature of the air as the moisture gives up its latent heat. The two processes are shown in Fig. 9, where the line A–B represents dehumidification by cooling and condensing and the line A–C represents drying by hygroscopic chemicals. It should be noted that the line A–B in Fig. 9 does not show the progressive changes through the cooling medium, but only the terminal conditions.

by any air stream can be expressed

$$H_s = .24 \left(\frac{60}{13.34} Q \right) \Delta T = 1.1 Q \Delta T$$

and

$$Q = \frac{H_s}{1.1 \Delta T}$$

where H_s = sensible heat given up (or absorbed) in Btu per hr

Q = cubic feet of air delivered per minute (cfm)

ΔT = change in dry bulb temperature of the air.

The quantity of cooled air to be introduced into a room to maintain it at a given dry bulb against a given sensible heat load is therefore dependent only on the respective dry bulb temperatures of room and

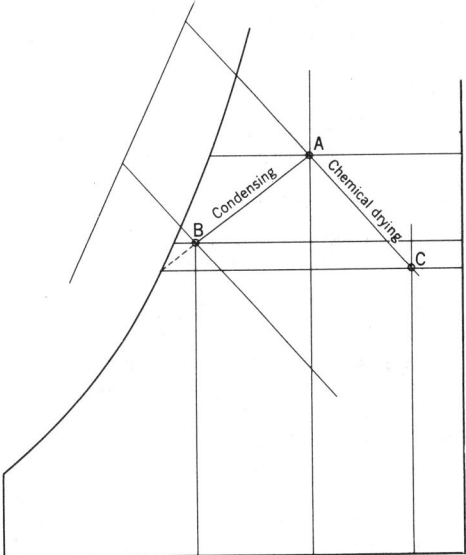

Fig. 9. Dehumidifying air by (1) cooling in contact with surface below dew point, line A-B, and (2) absorption of moisture by chemicals at constant total heat, line A-C.

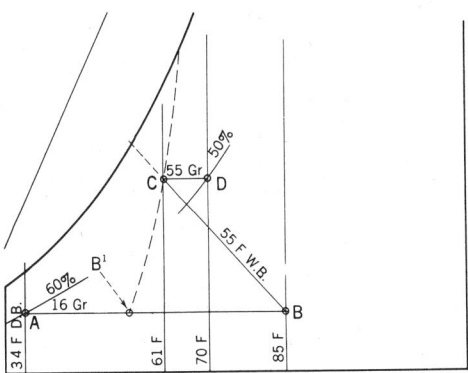

Fig. 8. Heating with coils, humidifying with washer. Washer water may be either unheated (solid line) or heated (broken line).

Psychrometric Calculations. Owing to the necessity of plotting the psychrometric chart on a weight basis and the convention of specifying air supply quantities in cubic feet (cfm), practical design practice is to avoid complicated conversions from weight to volume at various state points by assuming all air supply to be at standard specific volume of 13.34 cu ft per lb and standard specific heat of .24 Btu per (lb) (deg.).

In this way, the sensible heat delivered or removed

supply air. The condition lines of sensible heat change are horizontal and independent of the moisture level.

For example, to maintain a room at 80 deg. D.B. against a sensible heat gain of 20,000 Btu per hr by introducing air at 58 deg. D.B. would require

$$Q = \frac{20,000}{1.1 (80 - 58)} = 826 \text{ cfm}$$

If, now, the design condition in the room is fixed

at 50% R.H. (point A, Fig. 10), it is apparent in Fig. 10 that even saturated air at 58 deg. cannot be introduced to handle the sensible heat load without dehumidifying the space to about 47% along the line B–C — *unless there is an additional latent load of humidity gain in the room.*

If the humidity gain in the space is exactly enough to maintain 50% R.H. and thus enough to offset

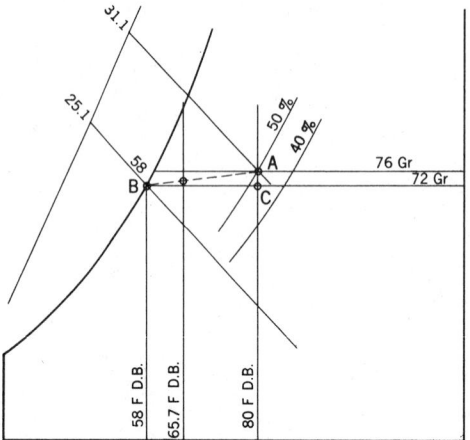

Fig. 10. Geometry of sensible heat ratio, expressed in total heat units as B-C/A-B.

the moisture deficiency in the proposed 826 cfm of saturated air introduced at 58 deg., the amount of moisture gain is seen to be represented on Fig. 10 by the distance A–C, or four grains of moisture per pound of supply. air. This moisture gain has the effect of establishing the slope of the line A–B.

It is apparent in the geometry of Fig. 10 that, since the line B–C represents sensible heat gain and the line A–B represents total heat gain, the ratio of sensible load to total load (cosine of the angle ABC) also establishes the slope of A–B, regardless of its actual position on the chart. Any space with the same sensible heat ratio would have its own state point and that of its compensating supply air on a condition line of identical slope.

In this example, the sensible heat gain can be read on the chart as 30.5 − 25.1 = 5.4 Btu per lb of supply air, and the total heat gain is 31.1 − 25.1 = 6.0 Btu per lb of supply air. The sensible heat ratio is therefore 5.4/6.0 = 0.9, and would be established in practice by the design sensible load of 20,000 Btu per hr plus a latent load of 2,200 Btu per hr in the ratio 20,000/22,200.

In this case, the introduction of saturated air presupposes a 100% efficient cooling apparatus at

the dew point 58 deg. The slope of the condition line characteristic of the sensible heat ratio intersects the saturation curve at the theoretically perfect *apparatus dew point.* This is the temperature on the saturation curve at which the smallest possible quantity of supply air (that is, saturated air on the required condition line) will handle the total load on the space. The position of the intersection (apparatus dew point) on the saturation curve is, of course, fixed by room air conditions and the slope of the condition line characteristic of the sensible heat ratio of the load.

Since all of the conditions along the line A–B from the room air state point to saturation will handle the characteristic load on the space, in accordance with the theory of air mixtures previously discussed, apparatus of any practical efficiency can be used to supply air having any state point along A–B with compensating adjustment of the amount of supply air to allow for the decrease in temperature difference. The by-pass factor of the apparatus (one minus the efficiency) thus poses a simple problem in air mixture such as has been discussed. That is, the efficiency point will lie on the condition line a proportionate distance to the right of the saturation curve and a compensating increase in supply air quantity will be made based on the smaller value of ΔT thus obtaining in the denominator of the formula $Q = H_s/1.1\Delta T$.

For example, assume that the apparatus in the previous example is 67% efficient. The state point of supply air will then move $1 - .67 = .33$ of the distance from saturation toward the room air state point. It will then be at a point intercepted by the 65.7 deg. dry bulb line as shown in Fig. 10, and the quantity of air required to handle 20,000 Btu per hr sensible load will increase from 826 cfm to

$$Q = \frac{20,000}{1.1\ (80 - 65.7)} = 1{,}270 \text{ cfm}$$

The geometry of Fig. 10 points up the fact that ΔT (that is, the range through which conditioned air is cooled) can be expressed as

$$\Delta T = \left[\left(\begin{array}{c}\text{Space}\\ \text{D.B.}\end{array}\right) - \left(\begin{array}{c}\text{Apparatus}\\ \text{Dew Point}\end{array}\right)\right] \times \left(\begin{array}{c}\text{Apparatus}\\ \text{Efficiency}\end{array}\right)$$

This quantity can be substituted in the basic formula for supply air quantity to give

$$Q = \frac{H_s}{1.1\ (t_1 - A_{dp})\ e}$$

where t_1 = space dry bulb, deg. F.
          A_{dp} = apparatus dew point
          e = efficiency

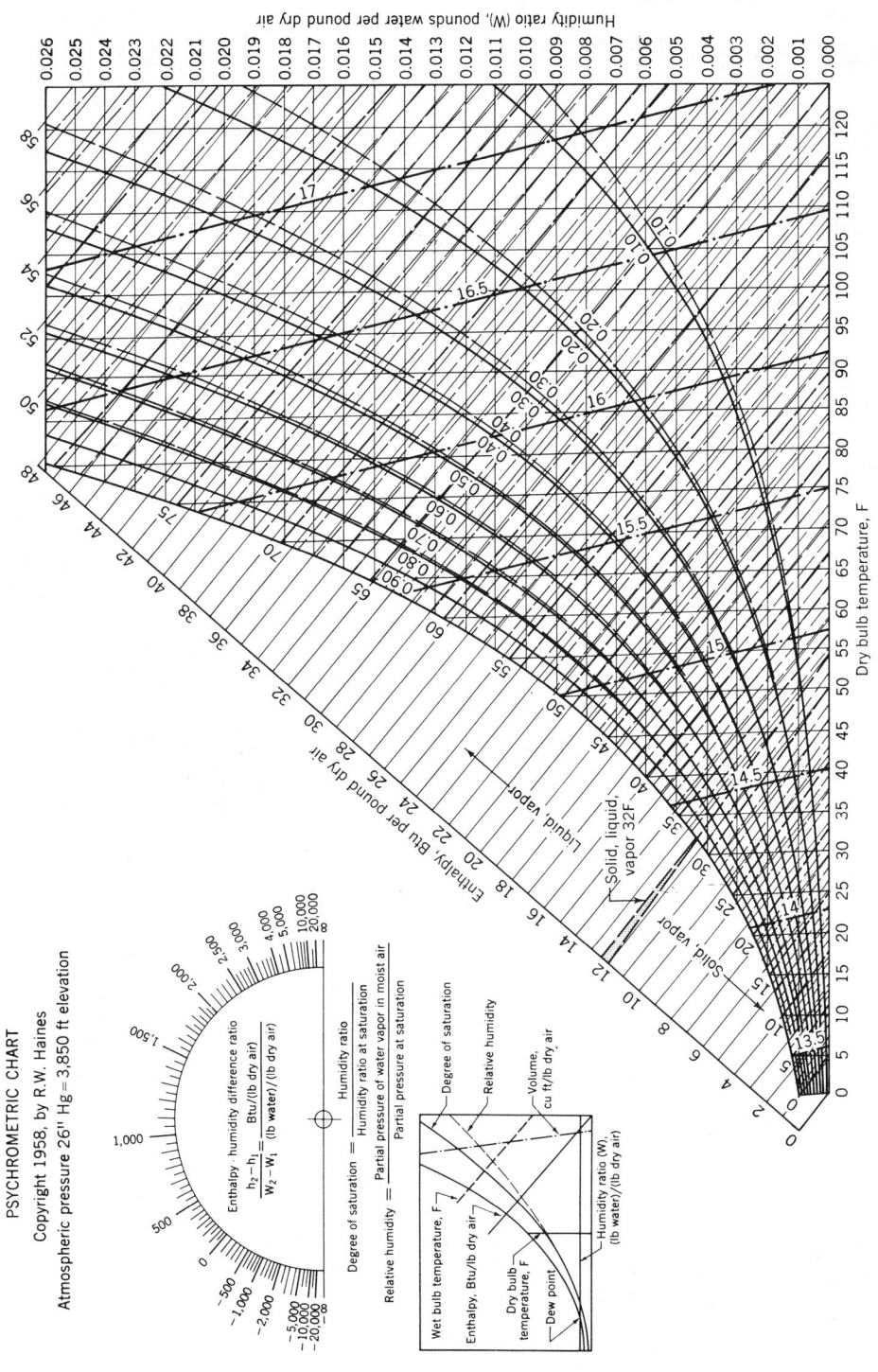

PSYCHROMETRIC CHART
Copyright 1958, by R.W. Haines
Atmospheric pressure 26" Hg = 3,850 ft elevation

Humidity ratio (W), pounds water per pound dry air

Dry bulb temperature, F

Enthalpy, Btu per pound dry air

Enthalpy - humidity difference ratio
$$\frac{h_2 - h_1}{W_2 - W_1} = \frac{\text{Btu/(lb dry air)}}{\text{(lb water)/(lb dry air)}}$$

Degree of saturation = $\dfrac{\text{Humidity ratio}}{\text{Humidity ratio at saturation}}$

Relative humidity = $\dfrac{\text{Partial pressure of water vapor in moist air}}{\text{Partial pressure at saturation}}$

Humidity ratio at saturation

Degree of saturation

Relative humidity

Volume,
cu ft/lb dry air

Wet bulb temperature, F

Enthalpy, Btu/lb dry air

Dry bulb
temperature, F

Dew point

Humidity ratio (W),
(lb water)/(lb dry air)

Liquid vapor

Solid, liquid,
vapor 32F

Solid vapor

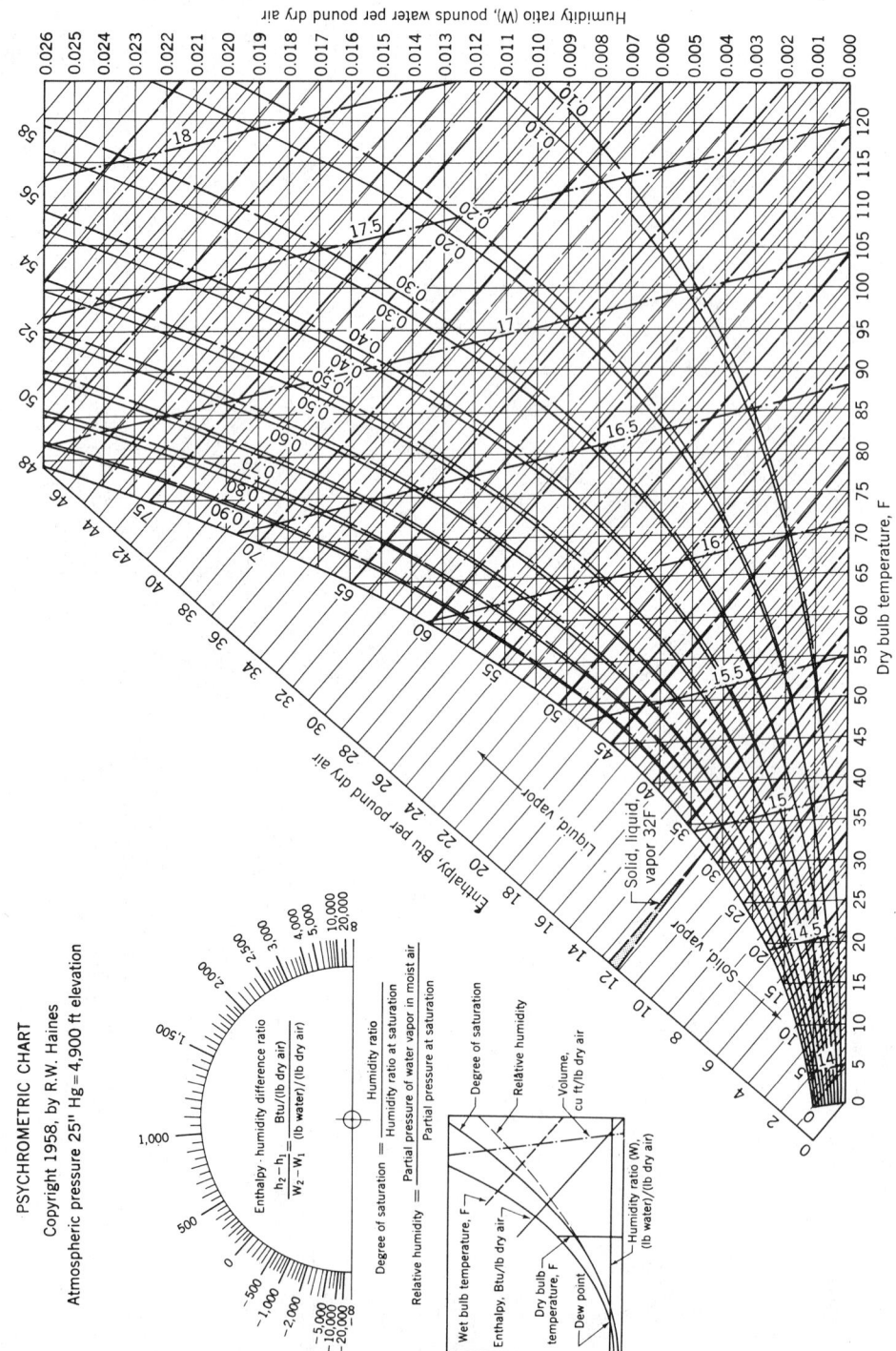

PSYCHROMETRIC CHART
Copyright 1958, by R.W. Haines
Atmospheric pressure 25" Hg = 4,900 ft elevation

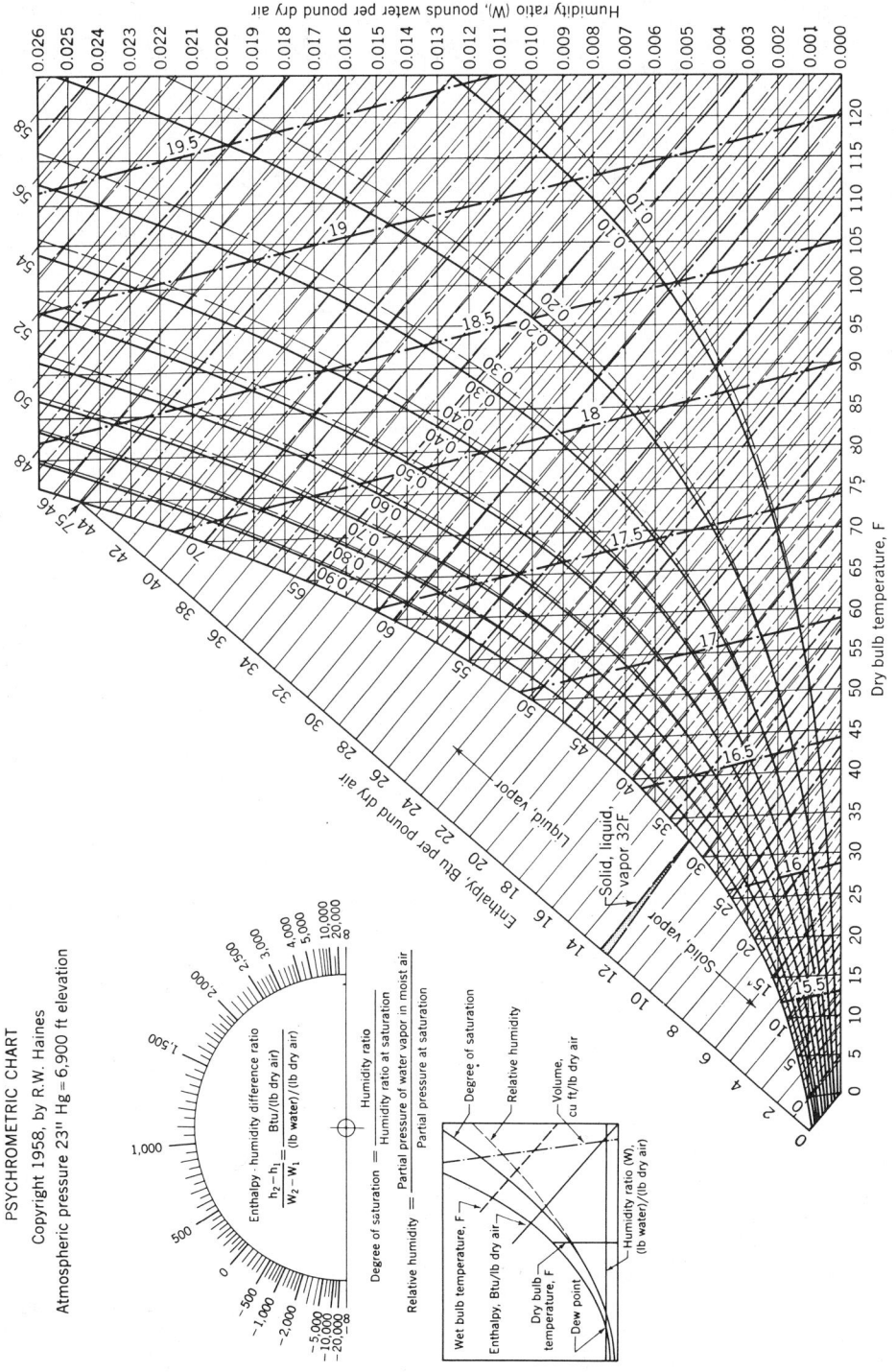

PSYCHROMETRIC CHART
Copyright 1958, by R.W. Haines
Atmospheric pressure 23" Hg = 6,900 ft elevation

APPARATUS DEW POINT

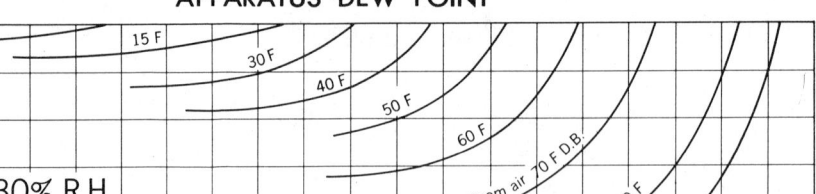

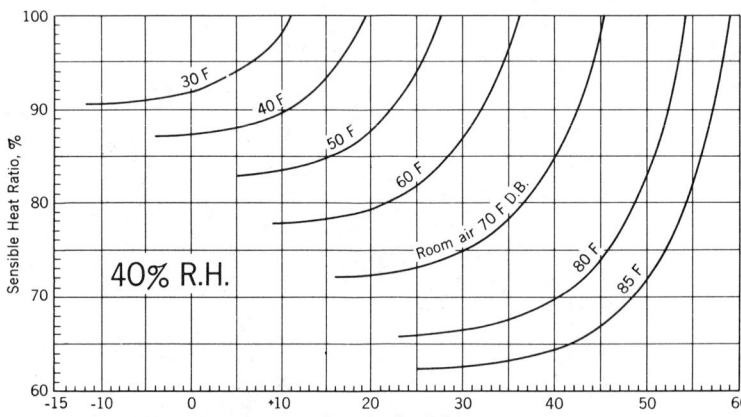

Fig. 1 (Above). Chart for finding apparatus dew point to maintain room air at 30% relative humidity at various dry bulb temperatures (marked on curves) with loads at various sensible heat ratios.

Fig. 2 (Left). Chart for finding apparatus dew point to maintain room air at 40% relative humidity at various dry bulb temperatures (marked on curves) with loads at various sensible heat ratios.

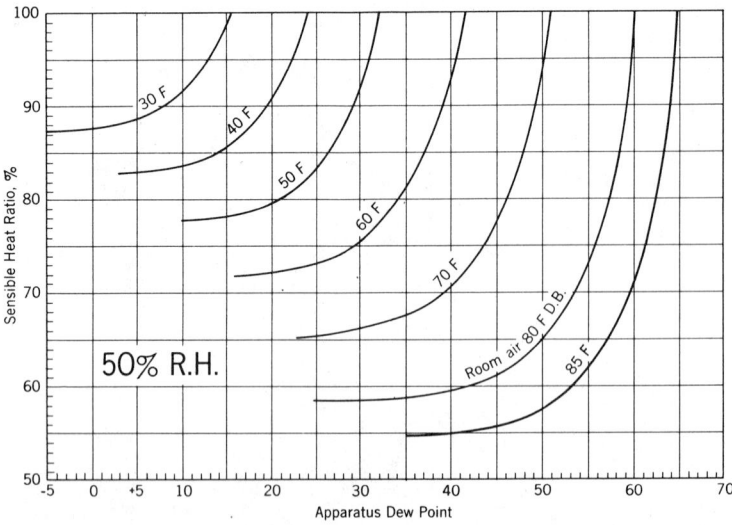

Fig. 3 (Left). Chart for finding apparatus dew point to maintain air at 50% relative humidity at various dry bulb temperatures (marked on curves) with loads at various sensible heat ratios.

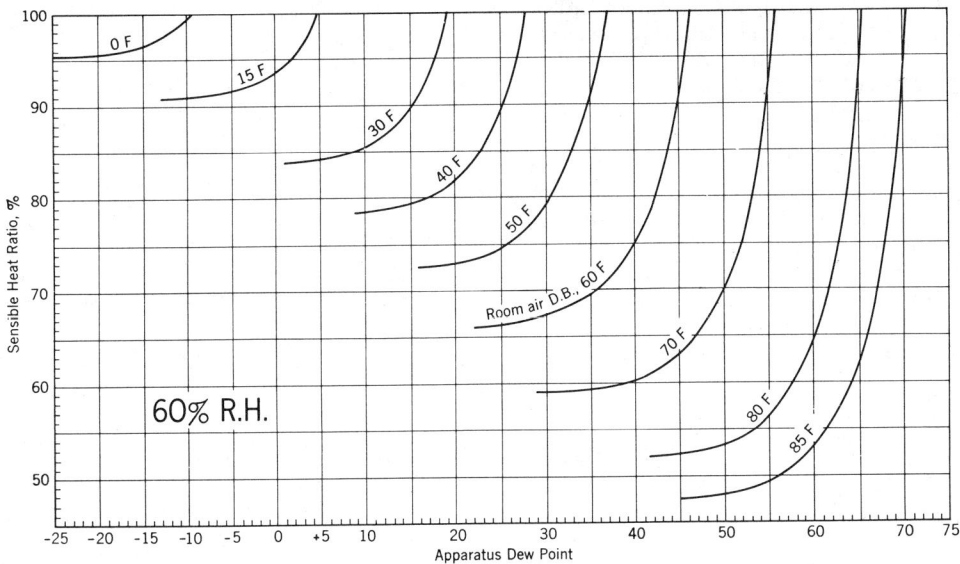

Fig. 4 (Above). Chart for finding apparatus dew point to maintain room at 60% relative humidity at various dry bulb temperatures (marked on curves) with loads at various sensible heat ratios.

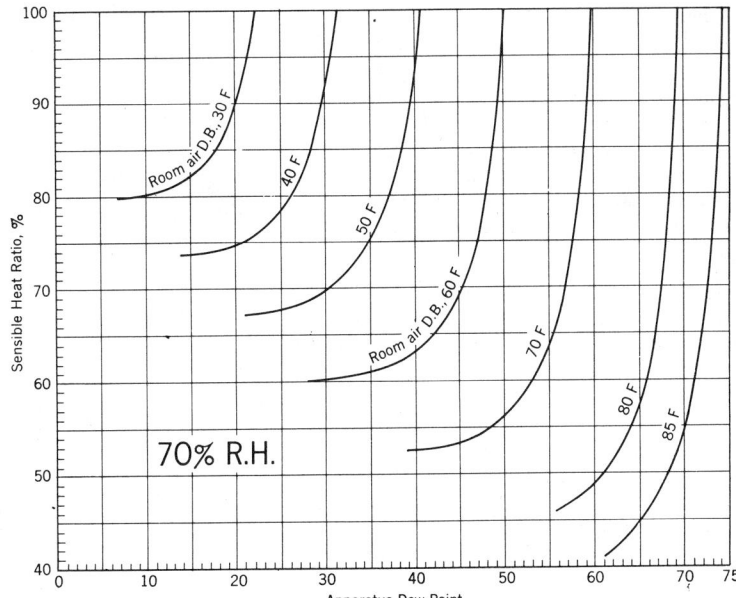

Fig. 5 (Right). Chart for finding apparatus dew point to maintain room air at 70% relative humidity at various dry bulb temperatures (marked on curves) with loads at various sensible heat ratios.

Apparatus dew point is the temperature of saturated supply air which will maintain specified conditions of temperature and humidity in a space against given latent and sensible load conditions. There are several methods of determining the apparatus dew point for a given space condition and load. Most methods are based on the existence of a characteristic slope of condition line for each sensible-to-total heat ratio. Protractor diagrams or special rules may be provided by means of which a line of proper slope can be drawn through the space condition to be maintained. The geometric method previously demonstrated can also be used.

The limitation of such methods lies not only in

the possibility of error inherent in geometric construction to close tolerances but also in the fact that, owing to differences in the latent heat of vaporization at different temperatures, the lines for a given sensible heat ratio may not be exactly parallel at all points of the chart.

Herbert Herkimer has taken this into account in calculating the apparatus dew points from thermodynamic data for literally thousands of combinations of space conditions and sensible heat ratios. The results of his calculations have been plotted on the charts shown in Figs. 1, 2, 3, 4, and 5. The usefulness of these charts in combination with the psychrometric chart is shown in the examples which follow.

Examples

(1) Given the space condition of 80 deg. D.B., 50% R.H., to be maintained against a sensible heat load of 20,000 Btu per hr and a latent load of 7,000 Btu per hr. Supply air is to be delivered at 58 deg. D.B.

Find (a) wet bulb and dew point of supply air, (b) quantity of supply air Q, and (c) efficiency of required coil or washer.

Solution: (a) The sensible heat ratio, $S_R = 20,000/$ 27,000 = .74. Reference to the apparatus dew point chart for 50% R.H. (Fig. 3) shows a value of 55.5 as the apparatus dew point for the problem conditions. The condition line is accordingly drawn as shown in Fig. 6. The intersection point of the condition line and 58 deg. D.B. coordinate establishes the condition of supply air to be 56.7 deg. W.B., 56 deg. D.P.

(b) The quantity of supply air (Q) is equal to the

(sensible heat load) + [1.1 × (dry bulb to be maintained − dry bulb of supply air)], so that

$$Q = \frac{20,000}{1.1\ (80-58)} = 826 \text{ cfm}$$

(c) $\Delta T = (t_1 - A_{dp})e$, so that

$$(80-58) = (80-55.5)e$$

$$e = \frac{22}{24.5} = 90\%$$

Note also that by direct measurement on Fig. 6

$$\frac{\text{length } AD}{\text{length } AC} = .9$$

(2) Given same conditions as in previous example except that sensible heat ratio is 60%.

Find (a) wet bulb and dew point of supply air, (b) quantity of supply air, and (c) efficiency of coil.

Solution: (a) According to Fig. 3, the apparatus dew point at S_R of 60% for 80 deg. D.B. and 50% R.H. is 42 deg. Draw line $A-E$, Fig. 6. Then point F on $A-E$ at 58 deg. D.B. shows W.B. = 53.7 deg.; D.P. = 50.6 deg.

(b) Q (not changed from previous example because D.B. is same) = 826 cfm.

(c) $$e = \frac{(80-58)}{(80-42)} = 58\%$$

(3) Given $H_s = 16,500$ Btu per hr, $H_L = 4,700$ Btu per hr, 1000 cfm is to maintain space at 85 deg.

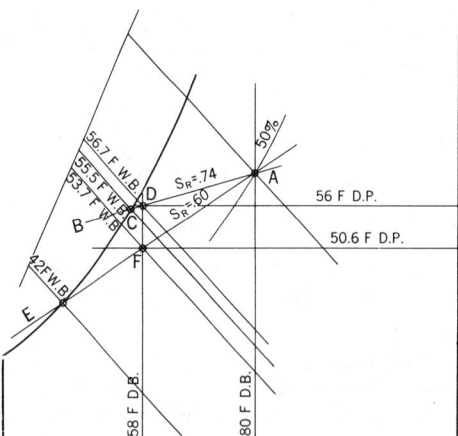

Fig. 6. Influence of sensible heat ratio on apparatus dew point and supply dew point at constant supply and room dry bulb (examples 1 and 2.)

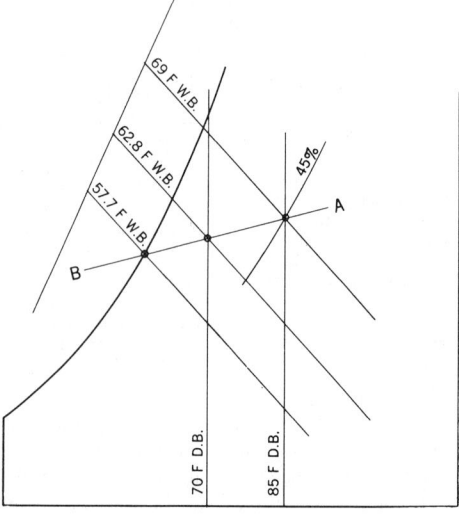

Fig. 7. Finding state point of supply air from specified volume (example 3).

D.B., 69 deg. W.B.

Find required wet bulb and dry bulb.

Solution:

$$\Delta T = \frac{H_s}{1.1Q} = \frac{16,500}{1.1 \times 1000} = 15$$

Required D.B. = 85 − 15 = 70 deg.

From a psychrometric chart find R.H. at 85 D.B., 69 W.B. = 45%. Note that $S_R = 16,500/21,200$ = 77.8%.

Read on apparatus dew point charts: A_{dp} at 40%, 85 deg. = 53.5. A_{dp} at 50%, 85 deg. = 62.0. By interpolation, A_{dp} at 45%, 85 deg. = 57.7.

Draw condition line $A-B$ as in Fig. 7. Intersection of condition line with 70 deg. D.B. coordinate shows supply at 62.8 deg. W.B.

(4) Given room condition 80 deg. D.B., 50% R.H., H_s = 75,000 Btu per hr, and H_L = 25,000 Btu per hr. Air leaves conditioning equipment at 90% R.H. Find condition and volume of air supply.

Solution: Plot point A, Fig. 8, at room condition on psychrometric chart. $S_R = 75,000 \div (75,000 + 25,000) = 75\%$.

From 50% R.H. apparatus dew point chart, Fig. 3, at 80 deg., 75%, sensible heat ratio, A_{dp} = 55.8. Draw condition line $A-B$ on Fig. 8. Locate point C on condition line at 90% R.H. Required condition of supply air is read at C, D.B. = 59.1 deg. W.B. = 57.5 deg.

Then

$$Q = \frac{H_s}{1.1T} = \frac{75,000}{1.1 (80 - 59)} = 3250 \text{ cfm.}$$

(5) Given room condition 80 deg. D.B., 50% R.H.,

H_s = 40,000 Btu per hr, and H_L = 10,000 Btu per hr. Total volume to be circulated, 3,000 cfm.

Find (a) condition of chilled air leaving coils if at 90% R.H., (b) volume of air to be cooled and dehumidified, (c) volume of recirculated air to be by-passed, (d) condition of mixture of chilled and by-passed air delivered to room.

Solution: (a) Plot point A, Fig. 9, at room condition on psychrometric chart. $S_R = 40,000 \div (40,000 + 10,000) = 80\%$. From 50% R.H. apparatus dew point chart, Fig. 3, at 80% sensible heat ratio, 80 deg., A_{dp} = 57.5.

Draw condition line $A-B$ and find point C on this line at 90% R.H. This fixes the condition of air leaving coils at 60.5 deg. D.B., 58.5 deg. W.B.

(b) $Q = \dfrac{40,000}{1.1 (80 - 60.5)} = 1,865$ cfm of chilled air

(c) Volume of return air by-passed is 3,000 − 1,865 = 1,135 cfm.

(d) Since a mixture of two air streams has a state point on a line joining the two conditions, the condition of the mixture of chilled and return air in this problem will be fixed by intersection of its dry bulb coordinate with the line $A-C$.

$$\Delta T = \frac{H_s}{1.1Q} = \frac{40,000}{1.1 \times 3,000} = 12.1$$

Hence, temperature of mixture is 80 − 12 = 68 deg. D.B., and its condition at point D, Fig. 9, is 61.8 deg. W.B.

Note that the same principles apply when it is desired to fix the temperature of cooled air supplied to a room.

(6) Given a room condition of 80 deg. DB.,

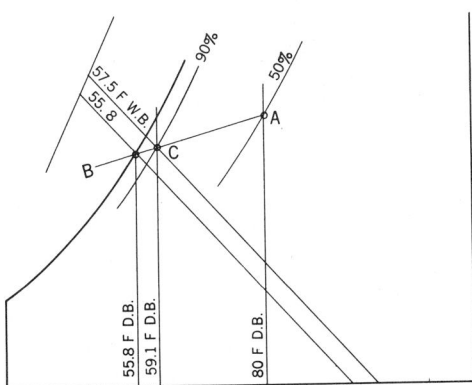

Fig. 8. Skeleton chart showing how to find supply condition from chilled air condition (example 4).

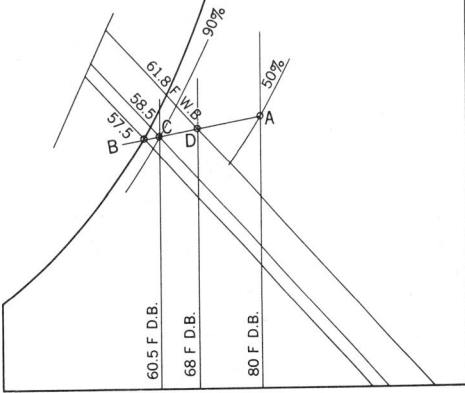

Fig. 9. Skeleton chart showing how to find amount of by-passed air (example 5).

67 deg. W.B. H_s = 50,000 Btu per hr, H_L = 17,500 Btu per hr. Outdoor air condition, 95 deg. D.B., 78 deg. W.B. Specified circulation, 3,000 cfm, of which 25% is outside air. Chilled air leaves coils at 90% R.H.

Find (a) W.B. and D.B. of air supplied to room, (b) W.B. and D.B. of return and outside air mixture ahead of cooling coils, (c) required conditions of chilled air leaving coils, (d) quantity of air to be chilled, (e) quantity of air to be by-passed.

Solution: (a) Plot room condition point A, Fig. 10, on psychrometric chart. S_R = 50,000 ÷ (50,000 + 17,500) = 74%. R.H. in room = 52.5% (read from psychrometric chart at A). A_{dp} for 50% R.H., 80 deg D.B., 74% S_R = 55.5 (read from apparatus dew point chart for nearest R.H., Fig. 3).

Draw slope line A'–B' for nearest condition, from 50% R.H., 80 deg. D.B. through 55.5 deg. on saturation curve. Then draw parallel line through point A to find A_{dp} (for 52.5%, 80 deg. D.B., 74% S_R) = 56.5.

Since 3,000 cfm will be supplied to room,

$$\Delta T = \frac{50,000}{1.1\,(3,000)} = 15 \text{ deg.}$$

This locates the dry bulb of supply air at (80 − 15) = 65 deg. and the state point of supply air at C where the wet bulb can be read as 60.7 deg. W.B.

(b) Locate outside air condition on chart at D, Fig. 10. Mixture of outside air and room air ahead of coils will then locate on line A–D.

Dry bulb of the mixture will be, according to previous discussion, 80 + .25 (95 − 80) = 83.75.

This locates the state point of return air ahead of coils at E, where D.B. = 83.75; W.B. = 70.

(c) The state point of chilled air leaving the cooling coils, since it is a component of a mixture of supply air and by-passed return air will lie on a line including points E and C. The intersection of this line with the 90% R.H. curve locates the condition of chilled air at F where D.B. = 59.4, W.B. = 57.6.

(d) The mixture of chilled (F) air and by-passed (E) air makes up the 3,000 cfm of supply air in proportions determined by the dry bulb intervals between them. The proportion of chilled air is therefore

$$\frac{83.75 - 65}{83.75 - 59.4} = \frac{18.75}{24.35} = 77\%$$

.77 × 3000 = 2310 cfm chilled air

(e) 3000 − 2310 = 690 cfm by-passed air.

It may happen that a room with an exceptionally large latent heat gain, such as a dance hall, will have a very low sensible heat ratio so that the slope of the supply condition line may be so steep that it does not intersect the saturation curve. In that event, the apparatus dew point can not be established directly from the chart for the room condition, because each point on those charts represents a point on the saturation curve.

In such cases, use is made of the fact that sensible heat ratio lines are approximately parallel for a given ratio when drawn for any state point on the chart. That is, if the apparatus dew point for the problem conditions can not be found on the saturation curve (and hence does not appear in the charts), any other room conditions with the same sensible heat ratio can be substituted to establish the slope of the supply air condition line. A line parallel to the substitute line can then be drawn through the desired room conditions.

If the proper line thus established is too steep to include the state point of chilled air leaving the coils, some means of reheating the chilled air must be devised, as in the following example.

(7) Given H_s = 141,000, H_L = 159,000 Room

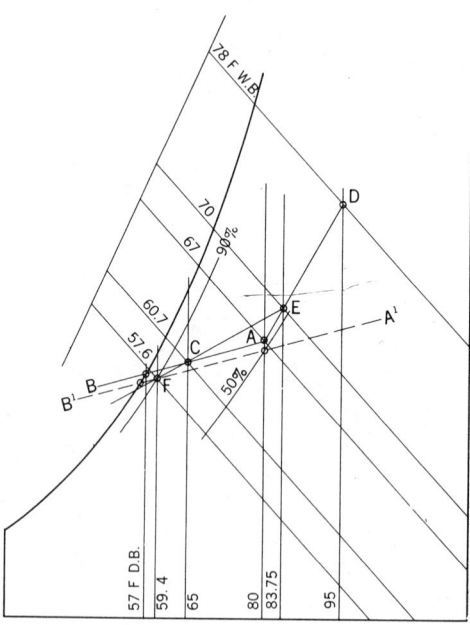

Fig. 10. Skeleton chart showing the mixing of outside and return air (example 6).

at 80 deg. D.B., 50% R.H. and air leaving coil is 90% R.H.

Find: (a) state of air leaving coil requiring the least possible amount of additional sensible heat, (b) state of supply to room, (c) volume of supply, and (d) sensible heat required for reheat.

Solution:

(a) $S_R = 47\%$.

This is not on the 50% chart, Fig. 3, but can be found on the 70% chart, Fig. 5. Therefore, find A_{dp} for 70% R.H., 80 deg. D.B., 47% S_R to be 58 deg. Draw line $A'-B'$, as in Fig. 11, representing slope of $S_R = 47\%$, and draw line $A-B$ parallel to $A'-B'$ and through the room condition.

This is the condition line of air supply to the room.

The point on the 90% rh curve requiring least addition of sensible heat to satisfy conditions on $A-B$ (with air entering coil at A) is on a tangent from A to the 100% curve. Draw tangent $A-C$ and locate chilled air condition at D, as 46F db with 44.7F wb. (It should be noted that, although this solution results in the least reheat, it may result in too low coil surface temperatures. This is so in this case because $A-C$ is tangent at about 32F, the intersection with the saturation curve is the (approximate average), temperature of the coil surface, and so the coil might freeze up.)

(b) Sensible heat will be added at constant humidity along the horizontal line from D to E on $A-B$. Supply air will then be at 58.2F db, 50.2F wb.

(c) $Q = \dfrac{141,000}{1.1\ (80 - 58.2)} = 5,880$ cfm

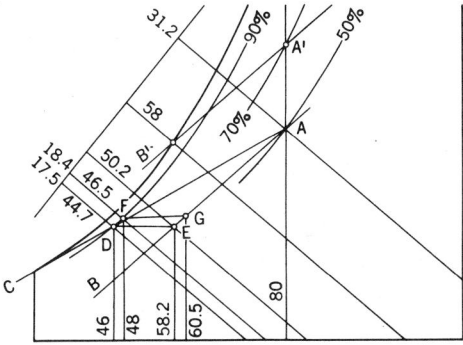

Fig. 11. Skeleton chart showing steep condition lines and reheat (example 7).

(d) Sensible heat for reheat will be $H_s = 1.1 \times 5,880\ (58.2 - 46) = 79,000$ Btu per hr.

Notice in Fig. 11 that any point on tangent $A-C$ results in this minimum reheat. If the air leaves the coil at 90% rh, D is the one point for the least reheat.

Since D as the leaving condition will result in freezing the coil, F on the 90% line is a better leaving coil condition and G on line $A-B$ is the reheated condition. The circulation rate is

$$Q = \frac{141,000}{1.1\ (80 - 60.5)} = 6,600 \text{ cfm}$$

The heat required for reheat is $H_s = 1.1 \times 6,600\ (60.5 - 48) = 91,000$ Btu per hr.

Refrigeration Capacity Required

The refrigeration plant is required to remove an amount of heat equal to the difference in total heat between the chilled air and the air supplied to the coil or washer.

Since the total heat coordinates in the psychrometric chart are in terms of Btu per pound of dry air and the air streams are calculated in cfm, it is convenient to convert again in terms of standard air at 13.34 cf ft per lb.

Thus the weight of air treated is

$$W = \frac{60}{13.34} Q = 4.5Q$$

and the total heat removed is

$$H_T = 4.5Qh_T$$

where h_T is the difference in total heat in Btu per pound of air before and after the coil.

The capacity of a plant may be expressed in tons of refrigeration using the factor, 1 ton = 12,000 Btu per hr. Thus, in tons, the capacity required is $H_T/12000$.

In the example illustrated by Fig. 11, the 6,600 cfm enters the coil at A and leaves at F, the refrigeration required is

$$H_T = (4.5)\ (6600)\ (31.2 - 18.4) \eqsim 380,000$$
Btu per hr.

This checks reasonably closely with the total of $H_s + H_L +$ reheat
$$= 141,000 + 159,000 + 91,000 = 391,000$$
Btu per hr.

Where the heat for reheating is extracted from return air by a means of a sensible heat exchanger (never by mixing the air streams) the total heat involved would be calculated between the points A and E in Fig. 11.

MULTI-PRESSURE PSYCHROMETRIC CHART

To allow rapid determination of psychrometric data for air conditioning problems involving barometric pressures that exist from 8000 feet above sea level to 8000 feet below sea level, G. E. McElroy of the Bureau of Mines has developed an all-pressure psychrometric chart representing a practical compromise between simplicity and absolute accuracy.

The effects of variations in atmospheric pressure upon psychrometric properties are quite pronounced. For problems involving considerable deviations from standard sea level barometric pressure, the standard psychrometric charts are not generally suitable.

Because of the practical limitations of existing data, it was desired to develop a single chart that would offer psychrometric data over a wide range of pressures. The major requirement was that heat and moisture contents should be determinable directly from the observed wet bulb temperatures, dry bulb temperatures, and barometric pressures. A second objective was the direct solution of problems involving the interchange of heat between air and water. A third objective was the determination of changes of wet bulb in downcast air shafts due to adiabatic compression.

The two secondary objectives required the use of heat content versus temperature as the primary coordinates while other useful variables such as dew-point temperature, relative humidity, and specific volume, were superimposed to as great an extent as was possible.

Formulas and tables of Goodman's Air Conditioning Analysis were used as basic data in the construction of the chart. Fig. 1 shows a small scale reproduction of the original chart with alternate lines omitted for clarity.

The total heats of the major coordinate block are figured from datum temperatures, and the general custom of using 0 deg. for the air and 32 deg. for the vapor has been followed. Absolute values are thus 10 Btu lower than values in Goodman's tables in which a value of 10 Btu is arbitrarily assigned to air at 0 deg.

The temperature scale is labeled both dry bulb temperature and water temperature. While water temperature carries throughout the height of the chart, dry bulb lines stop at saturation curve; that is, they apply to the clear area only and coincide with wet bulb lines in the fogged area.

The term total heat signifies the total of the sensible heat of air and liquid plus the latent heat of vaporization. The proportion each contributes

to the total heat of any air-vapor mixture can be determined readily from the chart.

Relative humidity curves, as plotted, are total-heat curves for unsaturated mixtures at a pressure of 30 inches of mercury, but are also total-heat curves for saturated mixtures at one particular pressure. The same statement in reverse applies to saturation curves. The relative humidity scale is therefore carried to 136.3 (30/22 × 100) on the 22 inch low-pressure saturation curve, and the scale values of the relative-humidity curves from 85 to 100 are set to conform to the previously drawn saturation curves, thus introducing a variable relative humidity scale in this range in order to avoid the confusion of closely spaced curves. Certain relative humidity curves (75 and 80% on Fig. 1) are also scaled as saturation curves for pressures greater than 35 inches of mercury.

There is a natural internal heat gain from compression as the pressure increases and a corresponding loss in adiabatic expansion as the pressure decreases. One Btu is added to each pound of mixture from this cause for every 778 feet of decrease in elevation or subtracted for the same increase in elevation. For dry air only, the change of dry bulb temperature is .00535 degree per foot or 1 degree for each 187 feet change of elevation. For air-vapor mixtures, without change of moisture content, the figure in feet per degree is slightly higher depending on the vapor content. If the vapor content is w pound water per pound of air, the change of dry bulb temperature is $\dfrac{1 + w}{(0.24 + 0.45w)778}$ degree per foot of change in elevation. Changes in total heat due to adiabatic compression or expansion are shown in a separate diagram in the left portion of the chart.

As several features of the chart have been drawn to a base pressure of 30 inches of mercury, it can be used as a simple sea-level-pressure chart for uniform-elevation conditions at or near sea-level elevation in the same way that single-pressure charts are now used and with the same degree of approximation. The saturation curve for 30 inches of mercury should, of course, be used (differences in values between 30 inches and 29.92 inches are not significant) at sea level. Wet bulb slope lines then serve directly as wet bulb temperature lines, and relative-humidity and specific-volume scales should not be corrected for pressure ratios. For many problems, small errors in absolute pressures are not important as the resulting errors in specific volumes

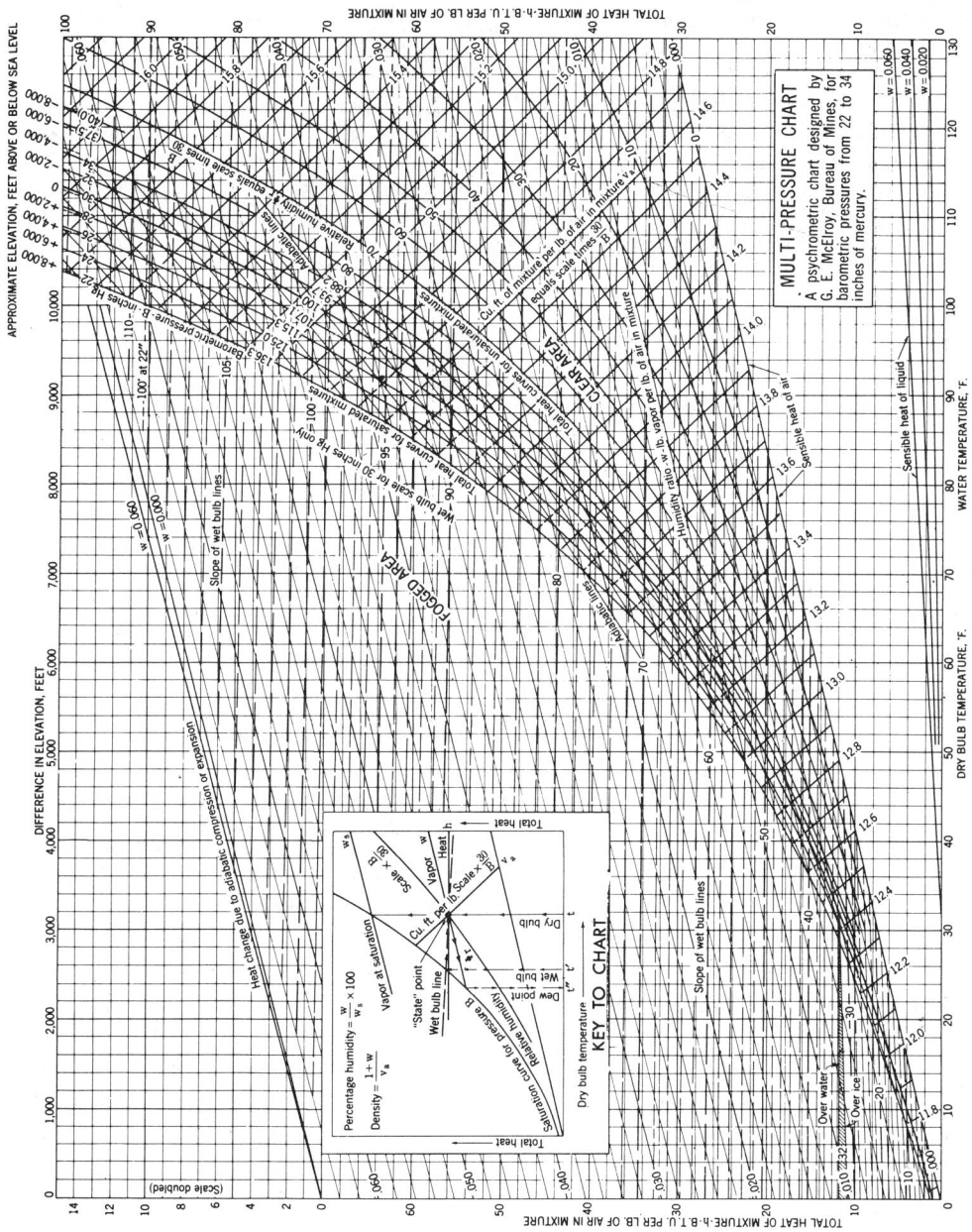

Fig. 1. Multi-pressure psychrometric chart designed by G. E. McElroy and covering barometric pressures from 22 to 34 in. of mercury.

almost completely compensate for corresponding errors in unit heat and moisture contents, and relative humidities and dew-point temperatures are largely dependent on wet-bulb depressions.

The major uses of the chart can be explained by six diagrams: A psychrometric diagram (Fig. 2), an

air-mixture diagram (Fig. 3), an adiabatic compression diagram (Fig. 4), an adiabatic expansion diagram (Fig. 5), and two water-air diagrams (Fig. 6 for parallel-flow conditions and Fig. 7 for counter-flow conditions).

Barometric Pressures. Points on saturation curves for intermediate pressures are obtained by vertical interpolation between adjacent curves. For any particular problem, the curve for the given pressure is used and the others are disregarded. For pressures greater than 34 inches of mercury, the relative humidity curve having a scale value of $30/B \times 100$ can be used as the saturation curve. Thus, for 40 inches use the $30/40 \times 100$, or 75% relative humidity curve. Relative humidity curves

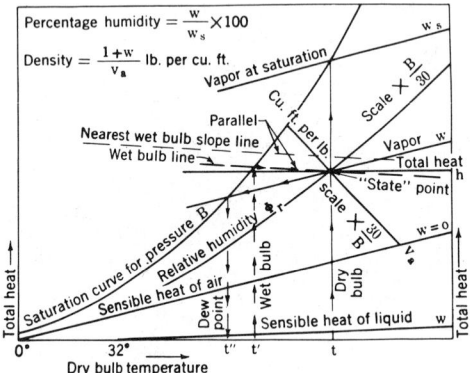

Fig. 2. Skeleton diagram of the McElroy psychrometric chart.

for intermediate scales are obtained by vertical interpolation between adjacent curves.

Wet Bulb Lines. For pressures other than 30 inches, only the slope of the wet bulb line is drawn in on the chart, and one point is required to fix the position of the line for a particular temperature and pressure. The required point is that point on the saturation curve for the given pressure where it is intersected by the dry bulb temperature line corresponding to the given wet bulb temperature (see Fig. 2). Lines of constant wet bulb temperature are straight lines parallel to the nearest wet bulb slope line (actual wet bulb line for 30 inches pressure) and extend throughout both fogged and clear areas. Adjacent wet bulb slope lines are essentially parallel across the width of the chart.

Dry Bulb Lines. Dry bulb temperature lines extend vertically from the lower edge to the satura-

tion curve but coincide with the wet bulb lines in the fogged area above the saturation curve for the given pressure.

State Point. The intersection of the wet bulb line with the dry bulb line determines the state point for the air-vapor mixture and serves to determine its unit-heat content, unit-vapor content, relative humidity, and (apparent) specific volume directly and, indirectly, its dew-point temperature, percentage humidity, and density.

Heat Content. The horizontal ordinate through the state point determines the total heat content on the scale on right or left edge. The relative proportion contributed by air and vapor can be scaled along the dry bulb line. The distance from the lower edge to the $w = 0$ line represents the heat of the air, and the distance from this point to the state point represents the heat of the vapor. The sensible heat of liquid evaporated or condensed is a relatively small amount closely approximating $w(t - 32)$ Btu per pound of air in the mixture, as represented to scale just above the lower edge of the chart. (See also Fig. 2).

Humidity Ratio. The humidity ratio or vapor content, w, is determined from the state point by interpolating vertically between adjacent humidity ratio lines, which are scaled within the chart along the right and left edges.

Relative Humidity. A scale value (actual for 30 inches pressure) for relative humidity is determined from the state point by interpolating vertically between adjacent relative humidity curves. The scale value is then multiplied by the given pressure divided by 30 to obtain the true relative humidity, Φ_r. The scale is carried to 136.3 on the 22-inch saturation curve, for which the relative humidity is $136.3 \times 22/30$ or 100%. Unequal spacings and scale differences are used for scale values greater than 85 to avoid the confusion otherwise caused by closely spaced curves.

Cubic Feet of Mixture Per Pound of Air. A scale value (actual value for 30 inches pressure) for cubic feet per pound is determined from the state point by interpolating horizontally between the two adjacent specific volume curves. The scale value is then multiplied by 30 and divided by the given pressure to obtain the true value of cubic feet of mixture per pound of air in the mixture. Like the dry bulb temperature lines, specific-volume curves extend only to the saturation curve for the given pressure and follow the slope of the nearest wet bulb slope line extending into the fogged area.

Dew-Point Temperature. The dew-point temperature is fixed by the point at which the humidity ratio line through the state point intersects the saturation curve for the given pressure. This intersection determines the dew-point temperature as that of the dry bulb temperature line passing through it. The very slight convergence of adjacent humidity lines needs to be taken into account only for differences between dry-bulb and dew-point temperatures greater than about 20 deg., as over short distances adjacent lines are essentially parallel.

Percentage Humidity. Percentage humidity, Φ_p, is obtained by dividing the humidity ratio, w, for the determined state point by the humidity ratio, w_s, for the point of intersection of the dry bulb line with the saturation curve for the given pressure and multiplying the result by 100; that is $\Phi_p = w/w_s \times 100$.

Mixture Density. The density of an air-vapor mixture is equal to one plus humidity ratio divided by cubic feet per pound of air (scale value corrected for pressure), or $(1 + w)/v_a$ pound per cubic foot.

Example 1

Given a barometric pressure 27 inches of mercury, wet bulb temperature 81 deg., dry bulb temperature 86 deg.

Find total heat content, vapor content, dew-point temperature, relative humidity, cubic feet of mixture per pound of air percentage humidity, and density of mixture.

Solution: Follow 81 deg. dry bulb line to intersection with saturation curve for 27 inches. Through this point, project a line parallel to nearest wet bulb slope line to intersection with the 86 deg. dry bulb line. This intersection is the state point for the air-vapor mixture. As it is on the 47.5 horizontal ordinate the total heat content is 47.5 Btu per pound of air. It is also on the 0.0244 humidity-ratio line, so that the vapor content is 0.0244 pound water per pound of air. A humidity-ratio line passing through the state point intersects the 27-inch saturation curve on the 79.6 dry bulb line, so that the dew-point temperature is 79.6 deg. The scale value of the relative humidity curve passing through the state point is 90.0, so that the actual relative humidity is $90.0 \times 27/30$ or 81.0%. The state point is located on the 14.26 cu ft per pound curve, so that $14.26 \times 30/27$ or 15.84 cu ft of mixture contains 1 pound of air. The intersection of the 86 deg. dry bulb line with the 27-inch saturation curve shows the humidity ratio for a saturated

mixture to be 0.0303. Dividing the actual humidity ratio of 0.244 by this figure and by multiplying by 100, gives the percentage humidity as $0.0244/0.0303 \times 100$ or 80.5%. The density of the mixture is $(1 + 0.0244) \div 15.84$, or 0.0647 lb per cu ft.

Air-Mixtures Diagram, Fig. 3

For a Mollier-type chart with heat content and vapor content per pound of air as the major coordinates, it can be proved that the state point of a mixture of two airstreams lies on a straight line connecting the state points of the individual airstreams and at distances from each inversely proportional to the part contributed by its airstream to the total weight of air in the mixture. Because of the slight convergence of the humidity-ratio lines

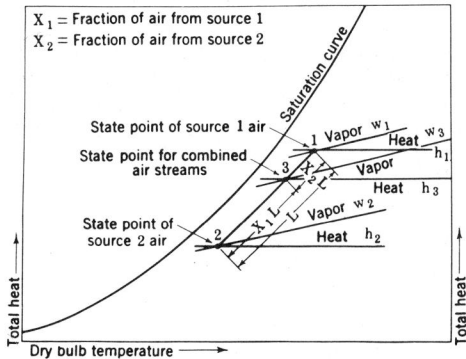

Fi 3. Skeleton chart showing use of chart for air mixtures.

of this chart the state point of the mixture (point 3 in Fig. 3) does not fall exactly on the straight line connecting the state points (points 1 and 2) of the individual components nor divide it in exact proportionality. The errors involved in following the same procedure are negligibly small, however, for most practical problems, and, in case of doubt, results by the chart are easily checked.

Example 2

Given two streams of moist air, one containing 750 pounds per minute of air with a heat content of 45.0 Btu per pound of air, and a vapor content of 0.0201 pound water per pound of air and another containing 250 pounds per minute of air with a heat content of 33.0 Btu per pound of air and a vapor content of 0.0136 pound water per pound of air.

Find the heat and vapor contents of a single

airstream formed by a junction of the two given airstreams.

Solution: Connect the two state points (points 1 and 2 of Fig. 3) of the given airstreams by a straight line. Spot a point (point 3 on the diagram) on this line one-fourth (250/1000) of its total length from the upper state point or three-fourths (750/1000) of its length from the lower state point. The point so spotted is the state point of the combined airstreams and gives the heat content as 42.0 Btu per pound of air and the humidity ratio, or vapor content, as 0.185 pounds of water per pound of air. Calculated values are 42.00 Btu and 0.018475 pounds of water.

Adiabatic Compression and Expansion Diagrams

The heat change, in Btu per pound of air, for adiabatic compression or expansion due to difference of elevation in a column of air-vapor mixture can be determined from the supplementary chart in the left of the main chart, Fig. 1. In assessing heat gains and losses due to sources other than compression or expansion, this figure is added to the heat

and pressure resulting from adiabatic compression due only to change in elevation, the heat of compression is determined from the supplementary chart, and the state point for the initial condition of temperatures and pressure (marked 1 in Fig. 4) is located ·as in the psychrometric diagram (Fig. 2). Adding the heat of compression to that for point 1 gives the heat content for the final condition. As the humidity ratio remains constant, the intersection of the heat ordinate for the final condition with the initial humidity-ratio line determines the state point (marked 2 on the diagram) for the final condition. The dry-bulb line through point 2 determines the final dry bulb temperature. A line parallel to the nearest wet-bulb slope line is then extended through state point 2 to intersection with an adiabatic line (obtained by interpolating between two adjacent lines of the chart) that passes through the intersection of the initial wet-bulb line and the saturation curve for the initial pressure. The point so obtained determines the final wet bulb temperature as that of the dry bulb line passing through it and the final pressure as that corresponding to the saturation curve passing through it, as obtained by interpolating

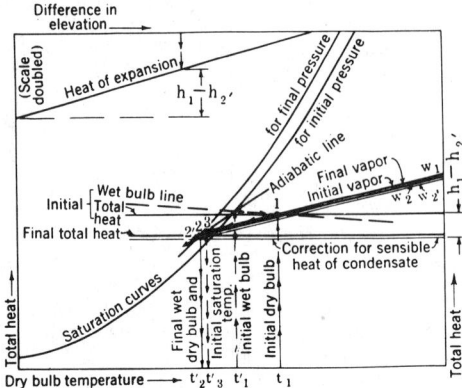

Fig. 4. Adiabatic compression diagram.

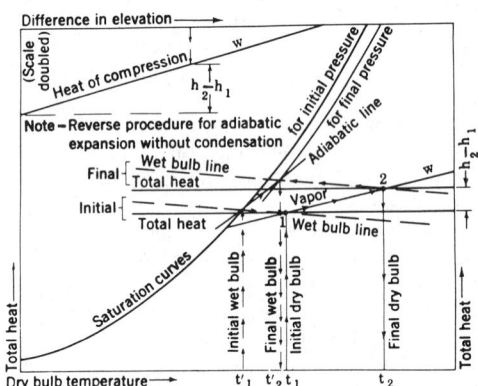

Fig. 5. Adiabatic expansion with condensation diagram.

content at the upper elevation for compression or subtracted from the heat content at the lower elevation for expansion.

The adiabatic lines drawn on the chart are constant-moisture wet-bulb temperatures that make it possible to determine adiabatic changes in pressure and wet bulb temperature that are caused by changes in elevation.

Adiabatic Compression

To determine the final conditions of temperature

between the two adjacent saturation curves.

Example 3

Given surface air conditions as in example 1 (that is, 81 deg. wet bulb and 86 deg. dry bulb air at 27 inches pressure) at the top of a vertical downcast air shaft 2,200 ft deep.

Find the theoretical temperatures and pressure of same air at the foot of the shaft due to adiabatic compression.

Solution: From example 1, the surface heat

content is 47.5 Btu per pound and the humidity ratio is 0.0244 pound of water per pound of air. From the supplementary chart (upper left of Fig. 1), the heat of compression for 2,200 feet and a humidity ratio of 0.0244 is 2.9 Btu per pound. The state point for the final condition is at the intersection of the 50.4 total-heat ordinate and the 0.0244 humidity ratio line. The dry bulb line through this point, 97.5 deg., is the final dry bulb temperature. A wet bulb line through this final state point intersects the adiabatic line, drawn through the intersection of the 81 deg. dry bulb line with the 27-inch saturation curve, on the 85.2 deg. dry bulb line and on the 29.0-inch saturation curve, so the final wet bulb temperature is 85.2 deg. and the final pressure is 29.0 inches.

Adiabatic Expansion

Where adiabatic expansion takes place without condensation, the procedure for determining the final conditions of temperature and pressure is the exact reverse of that used for compression. Point 2 in the diagram then represents the initial conditions, and point 1 the final conditions. In practice, expansion usually results in condensation, and this is particularly true of mine-upcast flow. For this case, as shown in the diagram (Fig. 5), a sufficiently accurate solution is obtained by following the initial humidity ratio from state point 1 to intersection with the adiabatic line at point 3 and then following down on this adiabatic line to its intersection with the ordinate of final heat content, as calculated by subtracting the heat change due to expansion from the original heat content. This intersection (point 2' in the diagram) is approximately the state point for the final condition, from which the approximate temperature, pressure, and humidity ratio of the saturated mixture may be obtained. If the difference between initial humidity ratio, w_1, and the final humidity ratio, w_2', so obtained is of significant amount, the final heat content should be increased by the sensible heat, $(t - 32)(w_1 - w_2')$, of the condensate that drops out of the mixture. Following back up the adiabatic line to this new final heat content determines a new final state condition at point 2 in the diagram, which in turn determines a new final temperature, pressure, and humidity ratio. Although the latter could be used successively for more accurate solutions, the sensible heat of the condensate is usually so small in practical problems that but one correction for condensate gives sufficient accuracy.

The intersection of the initial humidity ratio line with the adiabatic line at point 3 theoretically shows the temperature at which saturation occurs and condensation starts and, by the proportional decrease of heat content between the initial state point 1 and point 3, the approximate elevation at which it starts. However, in a vertical shaft, part of the condensate dropping through the unsaturated mixture below is reevaporated, and condensation actually starts at a higher temperature and lower elevation than indicated.

Example 4

Given 81 deg. wet bulb and 86 deg. dry bulb air at 27 inches pressure at the foot of a 2,200-foot vertical upcast air shaft.

Find the theoretical temperature and pressure of saturated air at the top of the shaft due to adiabatic expansion only.

Solution: From example 1, the initial heat content is 47.5 Btu per pound and the initial humidity ratio 0.0244 pound. The heat change for expansion is the same as for compression, which, from example 3, is 2.9 Btu per pound. Therefore the final approximate heat content is 44.6 Btu per pound. Extending the adiabatic line passing through 81 deg. on the 27-inch saturation curve downward to the left as in Fig 6, it is found that it intersects the humidity ratio line for 0.244 pound on the 78.2 deg. temperature line at the 45.6 Btu ordinate and intersects the 44.6 Btu heat ordinate on the 76.7 deg. temperature line and on the saturation curve for 25.1 inches. Consequently, saturation starts at 78.2 deg. and about $(47.5 - 45.6) \div 2.9 \times 2,200$, or 1,440 feet from the shaft bottom.

Subject to a possible small correction for condensed moisture, the temperature of saturated air at the surface is 76.7 deg. and the pressure is 25.1 inches. In this case, the correction of final heat content for the condensed moisture is negligible, as it amounts to only $(77 - 32)(0.0244 - 0.0239)$, or 0.02 Btu per pound.

The basic coordinates of this chart—total heat versus temperature—and the coordinate scales of 1 Btu = 1 deg. were selected primarily to facilitate the solution of air conditioning problems involving exchange of heat between air and water where the weight of moisture evaporated into the air, or condensed from the air, is negligible compared to the weight of water circulated.

Slope Reference Lines

Slope reference lines, expressing a continuous heat balance between the heat change in the air-vapor mixture and the heat change in the water, may be constructed as the hypothenuses of right triangles, of

which the bases are horizontal lines X coordinate units long and the altitudes are vertical lines XN units long in the same coordinate units, where N is pounds of water circulated per pound of air in the air-vapor mixture circulated. For parallel flow, such reference lines slope downward from left to right, whereas for counterflow they slope downward from right to left.

Parallel Flow

For parallel flow, draw a line such as 1–3 in the diagram, Fig. 6, parallel to the reference slope line from the intersection of the initial heat content of the mixture per pound of air and the initial temperature of the water, to the saturation curve for the given pressure. Any point on this line, such as 2, gives the heat content of the mixture and the corresponding water temperature at that stage in the process. The limiting condition, or 100% efficiency, is marked by point 3 where the slope line intersects the saturation curve, and defines the quality of saturated air temperature and water temperature possible of attainment. In the case where point 2 represents the final condition, the ratio of the distance 1–2 to the distance 1–3 measures the efficiency of the process.

Example 5

Given the shaft-bottom air conditions of example 3, that is, 85.2 deg. wet bulb and 97.5 deg. dry bulb air at 29.0 inches pressure, and parallel flow conditions.

Find (a) the lowest temperature (saturated) to which this air can be reduced by spraying with 0.25 pound of 60 deg. water per pound of air.

(b) What would be the saturated air temperature and the temperature of the water if the process were 85 percent efficient?

Note: A flow of 75,000 cfm at the surface air conditions of example 1 would change to 71,400 cfm at the shaft-bottom air conditions of example 3. At both positions, the airflow would contain 4,735 pounds of air per minute and require a circulation of $(4735 \times 0.25) \div 8.345$, or 142 gpm at the given rate of 0.25 pound water per pound of air.

Solution: (a) Draw a reference slope line in any near-by position, say through the 60 Btu-ttf coordinate intersection and through a point 20 units to the right and 20×0.25, or 5 units, down; that is, through the 55 Btu–75 deg. coordinate intersection. The initial heat content of the air, from example 3, is 50.4 Btu per pound. Through the intersection of the 60 deg. temperature line and the 50.4 Btu heat ordinate (point 1 to the left of the saturation curve

in the diagram) draw a line parallel to the slope reference line previously drawn and extend it to intersection with the 29.0-inch saturation curve (point 3 in the diagram). The point of intersection is on the 80.8 deg. temperature line, so that the lowest temperature to which the air can be reduced and the highest temperature to which the water can be raised is 80.8 deg.

(b) A point 85% of the distance from point 1 to point 3, such as point 2 in the diagram, represents the combination of total heat of the air and temperature of the water for a process that is 85% efficient. This point is on the 77.7 deg. temperature line, which represents the final temperature of the water. The total heat ordinate of point 2 is 46.0, which

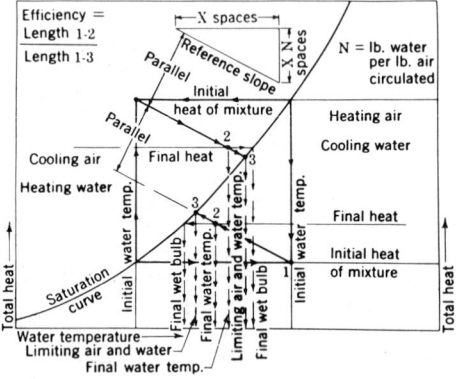

Fig. 6.   Water-air diagram for parallel flow processes.

intersects the 29.0-inch saturation curve on the 81.4 deg. temperature line, so that the final temperature of the saturated mixture is 81.4 deg.

Counterflow

For counterflow, draw in the initial total heat ordinate of the air from the line of initial water temperature to the saturation curve, such as line 1–3 in the diagram (Fig. 7) and draw in also the initial water temperature line from the initial heat of the air to the saturation curve, such as line 1–4 in the diagram. Then each line drawn parallel to the reference slope line across the right angle so formed represents a separate stage in the process. Limiting conditions are represented by that one particular line of proper slope that intersects one point on the saturation curve between terminal points 3 and 4 and lies wholly to the right of the saturation curve for the case of heating air or wholly to the left of it for the case of cooling air. Three conditions are possible for each case. In heating air, the slope

line for the limiting condition can pass through either terminal point 3 or 4 or be tangent to the saturation curve at some intermediate point. In cooling air, the slope line for the limiting condition can pass through either terminal point 3 or 4 or, for one particular slope, through both of these points.

In any case, the intersection of a slope line, such as 2–5, with the vertical line of initial water temperature (1–4) at point gives the heat content of the air for this stage of the process and the intersection of this new heat ordinate with the saturation curve gives the temperature of the (saturated) air. Also,

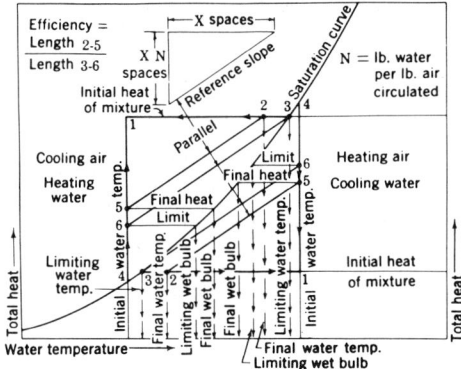

Fig. 7. Water-air diagram for counterflow processes.

the intersection of any slope line, such as 2–5, with the horizontal line 1–3 of initial heat content of the air at point 2 gives the corresponding temperature of the water. The length of a slope line representing a particular stage of the process, such as line 2–5, compared to the length of a line representing the limiting condition, such as line 3–6, determines the efficiency of the process.

Example 6

Given the shaft-bottom air conditions of example 3, that is, 85.2 deg. wet bulb and 97.5 deg dry bulb air at 29.0 inches pressure and counterflow conditions.

Find (a) the lowest temperature (saturated) to which this air can be reduced by spraying with 0.25 pound of 60 deg. water per pound of air, and to what temperature would the water be raised.

(b) What would the corresponding figures be if the process were 85% efficient?

Solution: (a) Draw a reference slope line in any near-by position, say through the 60 Btu–75 deg.

coordinate intersection and through a point 20 units to the left and 20 × 0.25, or 5 units, down, that is through the 55 Btu–55 deg. coordinate intersection. The initial heat content of the air, from example 3, is 50.4 Btu per pound. Draw in this ordinate from the 60 deg. temperature line to the saturation curve for 29.0 inches and draw in the 60 deg. temperature line from the 50.4 Btu ordinate to the saturation curve. Through the intersection of the 50.4 Btu ordinate with the 29.0-inch saturation curve (point 3 in the diagram) draw a line parallel to the reference slope line and extend it to intersection with the 60 deg. temperature line at point 6. The heat ordinate through this point of intersection (44.1 Btu per pound) intersects the 29.0-inch saturation curve on the 79.7 deg. temperature line, which is the lowest temperature to which the saturated air can be cooled. The intersection of the slope line 3–6 with the initial heat ordinate of 50.4 Btu occurs at point 3 on the saturation curve at the 85.1 deg. temperature line, which is the temperature to which the water would be raised.

(b) For 85% efficiency, draw in a parallel slope line, 2–5 in the diagram, just 85% as long as line 3–6. In the same way as in the preceding problem, the intersection of this slope line with the initial water temperature line at point 5 establishes the final heat content of the air as 45.1 Btu per pound, and the intersection of this heat ordinate with the saturation curve gives the temperature of saturated air as 80.6 deg. Intersection of the slope line 2–5 with the initial heat-content line at point 2 establishes the final water temperature as 81.4 deg.

Barometric Pressures at Various Altitudes at 70 Deg.

Altitude, Ft.	Barometer, In. of Mercury	Relative Density
15,000	16.88	.56
10,000	20.57	.69
8,000	22.12	.74
7,000	23.09	.77
6,000	23.98	.80
5,000	24.89	.83
4,000	25.84	.86
3,500	26.33	.88
3,000	26.81	.90
2,500	27.31	.91
2,000	27.82	.93
1,500	28.33	.95
1,000	28.85	.96
500	29.38	.98
Sea level	29.92	1.00

HIGH TEMPERATURE PSYCHROMETRIC CHART

In high temperature air handling, the effect of water vapor in an air stream has a substantial bearing on proper selection of exhauster size, speed and motor horsepower. This is especially true in cases involving dryers, kilns, wet type dust collectors, or cooling chambers. With the lower density of water vapor as compared to dry air at the same temperature, a mixture will weigh considerably less per cubic foot than will dry air. True density of the air stream must be known before exhauster size, speed and motor horsepower can be properly determined.

To assist in solving problems of high temperature moisture laden air, the accompanying chart, Fig. 1, has been prepared. The *humid volumes* shown in Fig. 1 *are expressed in cubic feet per pound of dry air and are actually the apparent specific volumes.* True specific volume, since it is expressed in cubic feet per pound of mixture, is used to compute the density of the air-vapor mixture and the density factor.

Apparent specific volume =

$$13.33 \left(\frac{460 + t}{530} \right) + 13.33 \left(\frac{460 + t}{530} \right) \left(\frac{p}{P - p} \right)$$

where 13.33 = specific volume of dry air at standard conditions, cubic feet per pound,

t = dry bulb temperature, deg F,

p = partial pressure of water vapor, inches of mercury,

P = standard atmospheric pressure, inches of mercury.

$$\text{True specific volume} = \frac{\text{apparent specific volume}}{1 + w}$$

where w = specific humidity, pounds of water vapor per pound of dry air

$$\text{Density of air stream} = \frac{1}{\text{true specific volume}}$$

$$\text{Density factor} = \frac{\text{density of air stream}}{0.075}$$

where 0.075 = density of standard air (70 deg and 29.92 inches of mercury)

Using the Chart

After obtaining the needed information from the chart, the correct way of applying the factors found should be known. In an air-vapor mixture, the density will be different from that of standard air and consequently a density factor should be applied to volume and pressure to be developed by the exhauster before proper selection can be made.

If the volume required is expressed in terms of standard air and operating conditions are different, the volume should be divided by the density factor to obtain the actual volume required of the exhauster.

$$CFM_{\text{(required)}} = \frac{SCFM}{\text{density factor}}$$

where $SCFM$ = air volume at standard conditions

Any changes in volume due to cooling or heating can be determined by multiplying the existing volume by the ratio of new humid volume over existing humid volume.

$$CFM_1 \times \frac{V_2}{V_1} = CFM_2$$

where V_1 = existing humid volume, cubic feet per pound of dry air,

V_2 = new humid volume, cubic feet per pound of dry air,

CFM_1 = existing air volume,

CFM_2 = new air volume.

When static or total pressure is expressed in terms of operating conditions, it is necessary to know what the pressure would be if standard air were being exhausted. This conversion is necessary before exhauster speed and horsepower required can be determined, because standard exhauster performance tables are usually based on standard air. A pressure expressed in terms of operating conditions should be divided by the density factor to obtain an equivalent pressure for standard conditions.

$$\begin{array}{c} \text{Pressure} \\ \text{(standard conditions)} \end{array} = \frac{\text{pressure (required)}}{\text{density factor}}$$

After obtaining the actual volume to be exhausted at operating conditions and the pressure to be developed in terms of standard conditions, the exhauster speed and horsepower required can be found. Since volume is a direct function of exhauster speed, the rpm read from the fan performance table is used because it was selected for the volume actually required. The brake horsepower, as read from the table, is for standard air and consequently is higher than necessary for less dense air. Actual horsepower required is obtained by multiplying the BHP obtained from the performance tables by the density factor.

$$BHP \text{ (required)} =$$
$$BHP \text{ (from tables)} \times \text{density factor}$$

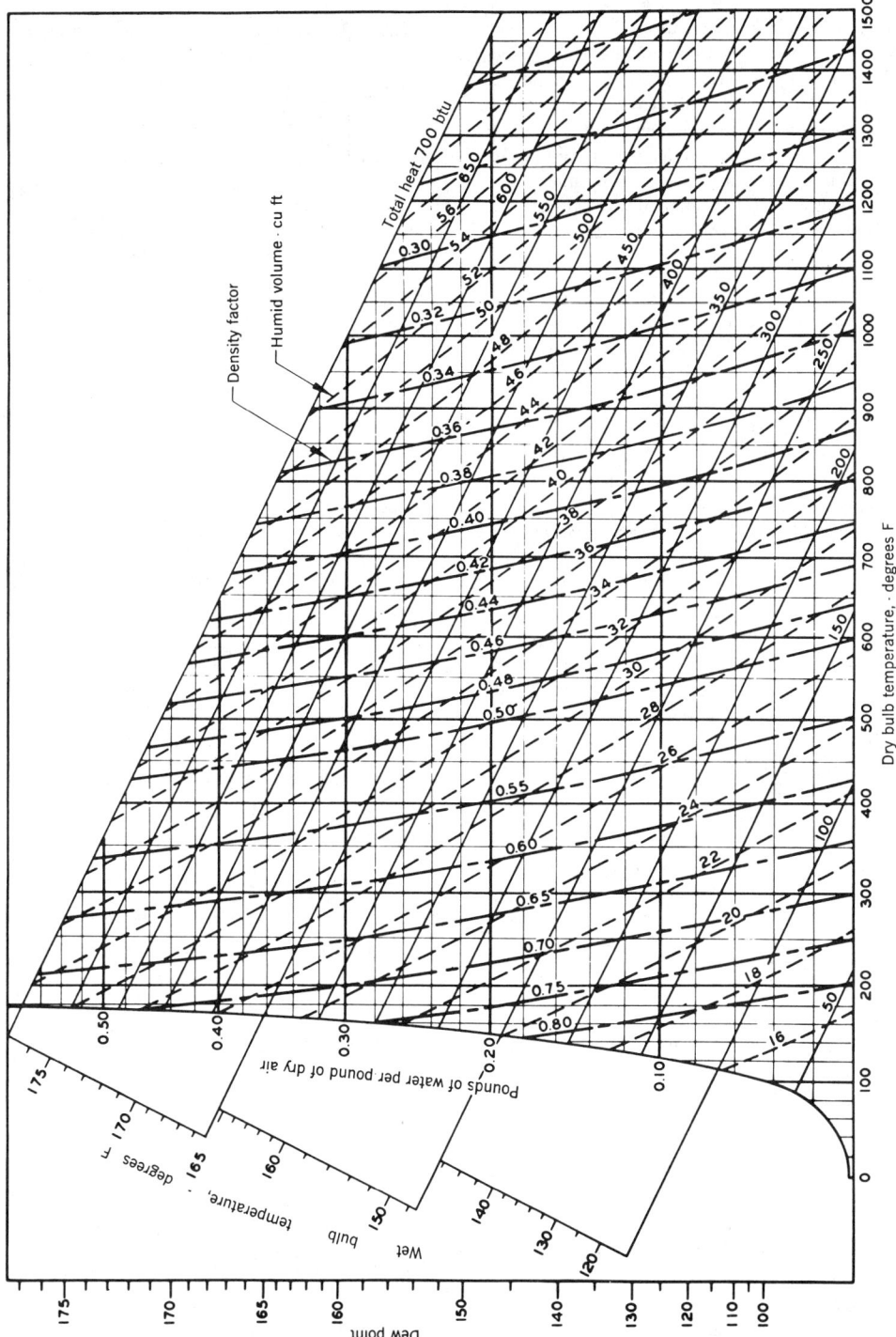

Fig. 1. High temperature psychrometric chart.

Where evaporative coolers are included in the air stream, it is necessary to know the humidifying efficiency before the outlet conditions can be determined; or, if a known outlet condition is desired, it is necessary to compute the humidifying efficiency and humidifying efficiency is defined as the ratio of the actual moisture gain through the machine to the moisture gain required to adiabatically saturate the air. By equation, this definition may be written as

$$N_h = \frac{W_2 - W_1}{W_s - W_1} \times 100$$

where N_h = humidifying efficiency, percent,

W_1 = weight of water vapor of inlet air,

W_2 = weight of water vapor of outlet air,

W_s = weight of water vapor of inlet air after adiabatically saturation. (All weights in any consistent units.)

Substituting the known values in the above equation, either humidifying efficiency (N_h) or outlet specific humidity (W_2), whichever is required, can be determined.

Example

Problem: The exit volume from a dryer is 35,000 cfm of moisture laden air at 600 deg F. The dryer has a feed rate of 60 tons per hour of material containing 15% moisture. The material leaving the dryer contains 5% moisture. The required static pressure on the dryer is 2.0 inches of water at operating conditions. The exhaust system has a dry centrifugal dust collector with 3.0 inches of water pressure drop at standard conditions and the calculated duct loss is 2.0 inches of water at standard condition. With the exhauster discharging directly to the atmosphere, determine exhauster speed and motor horsepower required.

Solution: To determine the density factor, it is necessary to know the specific humidity before using the psychrometric chart.

(1) Amount of moisture in pounds per minute released by dryer is:

$$\frac{60 \text{ tons/hr} \times 2000 \text{ lb/ton} \times (15\% - 5\%)}{60 \text{ min/hr}} = 200$$

(2) Table 1 shows that water vapor at 600 deg F has a specific volume of 42.9 cu ft per pound or it can be determined by:

$$\frac{386}{\text{molecular wt of water}} \times \left(\frac{460 + t}{460 + 70}\right) = \frac{386}{18} \times$$

$$\left(\frac{460 + 600}{460 + 70}\right) = 42.8 \text{ cu ft per lb}$$

where 386 is the volume in cubic feet of one

Table 1. Densities and Specific Volumes Dry Air and Water Vapor

225 to 900 Deg F; 29.96 Inches Mercury

Temperature, Deg F	Dry Air		Water Vapor	
	Specific Volume, Cu Ft per Lb	Density, Lb per Cu Ft	Specific Volume, Cu Ft per Lb	Density, Lb per Cu Ft
225	17.2	0.0581	27.7	0.0361
250	17.8	0.0562	28.7	0.0349
275	18.5	0.0541	29.7	0.0337
300	19.1	0.0524	30.8	0.0325
325	19.7	0.0508	31.8	0.0314
350	20.3	0.0493	32.7	0.0306
375	21.0	0.0476	33.8	0.0296
400	21.6	0.0463	34.8	0.0287
450	22.8	0.0439	36.8	0.0272
500	24.1	0.0415	38.8	0.0258
600	26.6	0.0376	42.9	0.0233
700	29.1	0.0344	46.9	0.0213
800	31.6	0.0316	51.0	0.0196
900	34.1	0.0293	55.0	0.0182

pound-mole of any gas at 70 deg F and 14.7 psia.

(3) Volume of water vapor = 42.9 cu ft per lb
× 200 lb per min = 8580 cfm

(4) Volume of dry air = Volume of mixture
— volume of vapor = 35,000 — 8580
= 26,420 cfm

(5) From Table 1, the specific volume of dry air at 600 deg is 26.6 cu ft per lb or it can be determined arithmetically by:

$$\frac{386}{\text{mol wgt}} \times \left(\frac{460 + t}{460 + 70}\right) = \frac{386}{29} \times \left(\frac{460 + 600}{460 + 70}\right)$$
$$= 26.6 \text{ cu ft per lb}$$

(6) Pounds of dry air in the system

$$= \frac{26,420 \text{ cfm}}{26.6 \text{ cu ft/lb}} = 990 \text{ lb per min}$$

(7) Specific humidity $= \frac{200 \text{ lb of water vapor}}{990 \text{ lb of dry air}}$
= .202 lb of vapor/lb dry air

(8) Knowing the specific humidity and dry bulb temperature, reference to the psychrometric chart indicates:

humid volume = 35.5 cu ft per lb of dry air
density factor = .455

(9) Pressure loss through the system at standard conditions is:

$$\text{Dryer loss} = \frac{2.0}{.455} = 4.4$$

Dust Collector loss = 3.0
Duct loss   = 2.0
    Total   = 9.4 inches of water

(10) Assuming the velocity pressures on the inlet and outlet of the exhauster are equal, the summation of the above values gives the total pressure of the system. As most exhauster performance tables are expressed in static pressure, it is necessary to subtract the fan inlet velocity pressure. This will be assumed as 1.0 inch of water. Therefore, the exhauster performance will be selected for 35,000 cfm at 8.4 inches of water.

From manufacturers' tables,

$$RPM = 720$$
$$BHP = 77.5$$

(11) The actual horsepower required equals

$$77.5 \times .455 = 35.3 \ hp$$

Extending the example problem by including an evaporative cooler and fabric dust collector in the system will demonstrate even further the importance of knowing the correct density factor for proper exhauster selection. With the exhauster located on the air leaving side of the fabric collector and discharging to the atmosphere, the cooling chamber having a pressure drop of 1.0 inch water at standard conditions, the fabric collector a 3.0-inch pressure drop at standard conditions, and the air leaving the cooling chamber at 200 deg, determine the correct exhauster speed, motor horsepower, humidifying efficiency and amount of water evaporated.

(1) From the inlet condition (600 deg F D.B.; $W_1 = 0.202$ lb water per lb dry air) on the chart, a line following a constant wet bulb temperature for adiabatic saturation should be drawn to the saturation curve. The intersection will be at $W_s = .33$ lb water vapor per lb dry air. Following the vertical line for a dry bulb temperature of 200 deg, it can be seen that the intersection of this vertical line with the constant wet bulb line is at the outlet condition from the cooling chamber with the air stream containing $W_2 = .32$ lb of vapor per lb of dry air. Then humidifying N_h is:

$$N_h = \frac{W_2 - W_1}{W_s - W_1} = \frac{.32 - .20}{.33 - .20} = 92.3\%$$

(2) Reference to the point on the chart indicated by intersection of 200 deg D.B. with the line of constant wet bulb from the inlet condition shows:

humid volume = 25.2 cu ft per lb of dry air
density factor =   .70

(3) Amount of water evaporated in the cooling chamber will be: $(W_2 - W_1) \times$ pounds of dry air $= (.32 - .20) \times 990 = 119$ lb per min

(4) The volume of air in the system after passing through the evaporative cooler can be found by using the equation for volume change:

$$CFM_1 \times \frac{V_2}{V_1} = 35,000 \times \frac{25.2}{35.5} = 24,800 \ cfm$$

(5) The pressure loss in the system corrected to standard condition (density factor at 200 deg = .70) is:

Dryer loss = $\frac{2.0}{.70}$ = 2.9
Dust Collector loss   = 3.0
Duct loss   = 2.0
Cooling chamber loss = 1.0
Fabric collector loss = 3.0
    Total   = 11.9 inches of water

(6) Subtracting 1 VP at the fan inlet, the exhauster will be selected for 24,800 cfm at 10.9 inches.

From manufacturers' tables:

$$RPM = 868$$
$$BHP = 62.5$$

(7) Actual horsepower required =
$$62.5 \times .70 = 43.8 \ hp$$

In this example, if only temperature reduction is considered without computing the effect on density by the addition of water vapor, an exhauster would be selected for 21,800 cfm at 10.5 inches of water which requires 843 rpm and a motor selected for 42.6 hp. This represents an error of 12% in volume, 4% in static pressure, 3% in exhauster speed, and 3% in horsepower required.

Conclusions

The example illustrates the importance of considering the effect of water vapor in an air stream and also the ease in which the psychrometric chart can be used for taking this water vapor into consideration. There are many other advantages in its use, namely, in the above example, the outlet temperature from the cooling chamber was made a constant and the humidifying efficiency was determined. If wet type dust collectors are used in an exhaust system employing high temperature air, it is very easy to predict the outlet temperature and volume from the unit if the humidifying efficiency is known. Knowing only two variables, dry bulb temperature and specific humidity or dew point temperature, enables anyone with the use of the psychrometric chart to accurately and easily find the proper density factor, humid volume, and wet bulb temperature.

DEW POINT TEMPERATURES FROM DRY AND WET BULB TEMPERATURES

Dry Bulb Temp., °F.	Wet Bulb Temperature, °F.									
	30	31	32	33	34	35	36	37	38	39
	Dew Point Temperature, °F.									
30	30.0									
31	28.5	31.0								
32	27.0	29.6	32.0							
33	25.3	28.2	30.7	33.0						
34	23.5	26.6	29.2	31.6	34.0					
35	21.6	24.9	27.8	30.2	32.6	35.0				
36	19.4	23.0	26.2	28.8	31.2	33.6	36.0			
37	17.1	21.0	24.4	27.3	29.8	32.2	34.7	37.0		
38	14.5	18.9	22.5	25.6	28.4	30.8	33.3	35.7	38.0	
39	11.5	16.4	20.5	23.8	26.8	29.5	31.9	34.4	36.8	39.0
40	8.0	13.7	18.3	21.9	25.1	28.0	30.5	33.0	35.5	37.8
41	4.0	10.6	15.8	19.8	23.3	26.4	29.1	31.6	34.1	36.6
42		7.0	12.9	17.5	21.3	24.6	27.6	30.2	32.7	35.3
43		2.7	9.7	14.9	19.2	22.8	26.0	28.8	31.3	33.9
44			6.0	12.0	16.7	20.7	24.2	27.3	30.0	32.5
45			1.4	8.6	14.1	18.5	22.3	25.6	28.5	31.1
46				4.6	11.1	16.1	20.3	23.8	27.0	29.8
47					7.6	13.3	18.0	21.9	25.3	28.3
48					3.4	10.2	15.5	19.8	23.5	26.7
49						6.5	12.6	17.5	21.5	25.0
50						2.1	9.4	14.9	19.4	23.2
51							5.5	12.0	17.0	21.2
52							0.9	8.6	14.4	19.0
53								4.6	11.4	16.6
54									7.9	14.0
55									3.8	10.9
56										7.4
57										3.1
58										
59										
60										
61										
62										
63										
64										
65										
66										
67										
68										
69										
70										
71										
72										
73										
74										
75										
76										
77										
78										
79										
80										

Dew Point Temperatures from Dry and Wet Bulb Temperatures (continued)

Dry Bulb Temp., °F.	Wet Bulb Temperature, °F. — Dew Point Temperature, °F.									
	40	41	42	43	44	45	46	47	48	49
40	40.0									
41	38.4	41.0								
42	37.7	39.9	42.0							
43	36.4	38.7	40.9	43.0						
44	35.1	37.6	39.8	42.0	44.0					
45	33.8	36.3	38.7	40.9	43.0	45.0				
46	32.3	35.0	37.5	39.8	42.0	44.0	46.0			
47	31.0	33.6	36.2	38.6	40.9	43.0	45.1	47.0		
48	29.6	32.2	34.9	37.4	39.8	42.0	44.1	46.1	48.0	
49	28.1	30.8	33.6	36.2	38.6	41.0	43.1	45.2	47.1	49.0
50	26.5	29.5	32.1	34.9	37.4	39.8	42.1	44.2	46.2	48.2
51	24.8	28.0	30.8	33.5	36.2	38.7	41.0	43.2	45.3	47.3
52	22.9	26.3	29.4	32.1	34.9	37.5	39.9	42.2	44.4	46.4
53	20.9	24.6	27.9	30.7	33.5	36.2	38.8	41.2	43.4	45.5
54	18.7	22.8	26.3	29.3	32.1	34.9	37.6	40.1	42.4	44.5
55	16.3	20.7	24.5	27.8	30.7	33.6	36.3	38.9	41.3	43.6
56	13.6	18.5	22.6	26.2	29.3	32.1	35.0	37.7	40.2	42.6
57	10.4	16.1	20.6	24.5	27.8	30.8	33.7	36.5	39.1	41.5
58	6.7	13.3	18.4	22.6	26.2	29.4	32.2	35.2	37.9	40.4
59	2.4	10.1	15.9	20.6	24.5	27.9	30.9	33.8	36.7	39.3
60		6.4	13.1	18.3	22.6	26.3	29.5	32.4	35.4	38.1
61		2.0	9.9	15.8	20.5	24.5	28.0	31.0	34.0	36.9
62			6.2	13.0	18.3	22.6	26.4	29.6	32.6	35.6
63			1.7	9.8	15.8	20.6	24.7	28.2	31.2	34.3
64				6.1	13.0	18.4	22.8	26.6	29.9	32.9
65				1.5	9.8	15.9	20.8	24.9	28.4	31.5
66					6.1	13.1	18.6	23.0	26.8	30.1
67					1.5	9.9	16.1	21.0	25.1	28.7
68						6.2	13.4	18.8	23.3	27.1
69						1.7	10.2	16.4	21.3	25.5
70							6.6	13.7	19.2	23.7
71							2.1	10.6	16.8	21.7
72								7.0	14.1	19.6
73								2.6	11.0	17.2
74									7.5	14.6
75									3.3	11.6
76										8.2
77										4.2
78										
79										
80										
81										
82										
83										
84										
85										
86										
87										
88										
89										
90										

Dew Point Temperatures from Dry and Wet Bulb Temperatures (continued)

Dry Bulb Temp., °F.	Wet Bulb Temperature, °F.									
	50	51	52	53	54	55	56	57	58	59
	Dew Point Temperature, °F.									
50	50.0									
51	49.2	51.0								
52	48.3	50.2	52.0							
53	47.5	49.4	51.2	53.0						
54	46.6	48.6	50.5	52.3	54.0					
55	45.7	47.7	49.7	51.5	53.3	55.0				
56	44.8	46.8	48.8	50.7	52.6	54.3	56.0			
57	43.8	45.9	48.9	50.0	51.8	53.6	55.3	57.0		
58	42.8	45.0	47.1	49.1	51.1	52.9	54.7	56.4	58.0	
59	41.7	44.1	46.2	48.3	50.3	52.1	54.0	55.7	57.4	59.0
60	40.7	43.1	45.3	47.4	49.5	51.4	53.2	55.0	56.7	58.4
61	39.5	42.0	44.4	46.5	48.6	50.6	52.5	54.3	56.1	57.8
62	38.4	41.0	43.4	45.6	47.8	49.8	51.8	53.6	55.4	57.1
63	37.2	39.9	42.3	44.7	46.9	49.0	51.0	52.9	54.7	56.5
64	35.9	38.7	41.3	43.7	46.0	48.2	50.2	52.2	54.0	55.8
65	34.6	37.5	40.2	42.7	45.1	47.3	49.4	51.4	53.3	55.2
66	33.2	36.3	39.1	41.7	44.1	46.4	48.6	50.6	52.6	54.5
67	31.8	35.0	37.9	40.6	43.1	45.5	47.7	49.9	51.9	53.8
68	30.3	33.6	36.7	39.5	42.1	44.5	46.8	49.0	51.1	53.1
69	29.0	32.1	35.4	38.3	41.0	43.6	45.9	48.2	50.3	52.3
70	27.5	30.8	34.0	37.1	39.9	42.7	45.0	47.3	49.5	51.6
71	25.8	29.4	32.6	35.8	38.8	41.5	44.0	46.4	48.7	50.8
72	24.1	27.9	31.2	34.5	37.6	40.4	43.0	45.5	47.8	50.0
73	22.1	26.3	29.8	33.1	36.3	39.3	42.0	44.6	47.0	49.2
74	20.1	24.5	28.4	31.7	35.0	38.1	41.0	43.6	46.1	48.4
75	17.8	22.7	26.8	30.3	33.7	36.9	39.9	42.6	45.1	47.5
76	15.2	20.6	25.1	28.9	32.2	35.6	38.7	41.5	44.2	46.6
77	12.3	18.4	23.3	27.3	30.9	34.3	37.5	40.5	43.2	45.7
78	9.0	15.9	21.3	25.7	29.5	32.9	36.3	39.3	42.2	44.8
79	5.1	13.1	19.1	24.0	28.0	31.5	35.0	38.2	41.1	43.8
80	0.8	10.0	16.7	22.0	26.4	30.1	33.6	36.9	40.0	42.8
81		6.3	14.1	19.9	24.7	28.6	32.1	35.7	38.9	41.8
82			11.0	17.6	22.8	27.1	30.8	34.3	37.7	40.7
83			7.5	15.0	20.8	25.4	29.4	32.9	36.4	39.6
84			3.3	12.1	18.6	23.6	27.9	31.5	35.1	38.4
85				8.7	16.1	21.7	26.3	30.2	33.8	37.2
86				4.8	13.3	19.6	24.5	28.7	32.3	36.0
87					10.2	17.2	22.7	27.2	31.0	34.6
88					6.5	14.6	20.6	25.5	29.6	33.3
89					2.5	11.6	18.4	23.7	28.1	31.8
90						8.2	15.9	21.8	26.5	30.5
91						4.2	13.1	19.7	24.8	29.1
92							10.0	17.3	22.9	27.5
93							6.3	14.7	20.9	25.9
94							1.7	11.8	18.7	24.2
95								8.4	16.3	22.2
96								4.3	13.6	20.2
97									10.4	17.9
98									6.8	15.3
99									2.4	12.5
100										9.2
101										5.3
102										0.6
103										

Dew Point Temperatures from Dry and Wet Bulb Temperatures (continued)

Dry Bulb Temp., °F.	Wet Bulb Temperature, °F.									
	60	61	62	63	64	65	66	67	68	69
	Dew Point Temperature, °F.									
60	60.0									
61	59.4	61.0								
62	58.8	60.4	62.0							
63	58.2	59.9	61.4	63.0						
64	57.6	59.3	60.9	62.5	64.0					
65	56.9	58.7	60.3	61.9	63.5	65.0				
66	56.3	58.1	59.7	61.4	63.0	64.5	66.0			
67	55.6	57.4	59.1	60.8	62.4	64.0	65.5	67.0		
68	55.0	56.8	58.5	60.2	61.9	63.5	65.0	66.5	68.0	
69	54.3	56.1	57.9	59.7	61.3	62.9	64.5	66.0	67.5	69.0
70	53.6	55.5	57.3	59.1	60.8	62.4	64.0	65.6	67.1	68.6
71	52.8	54.8	56.7	58.4	60.2	61.9	63.5	65.1	66.6	68.1
72	52.1	54.1	56.0	57.8	59.6	61.3	63.0	64.6	66.1	67.7
73	51.4	53.4	55.3	57.2	59.0	60.8	62.4	64.1	65.6	67.2
74	50.6	52.7	54.7	56.6	58.4	60.2	61.9	63.5	65.2	66.7
75	49.8	51.9	54.0	55.9	57.8	59.6	61.3	63.0	64.7	66.2
76	49.0	51.2	53.2	55.2	57.2	59.0	60.8	62.5	64.2	65.8
77	48.1	50.4	52.5	54.6	56.5	58.4	60.2	61.9	63.6	65.3
78	47.2	49.6	51.8	53.9	55.9	57.8	59.6	61.4	63.1	64.8
79	46.4	48.7	51.0	53.2	55.2	57.1	59.0	60.8	62.6	64.3
80	45.4	47.9	50.2	52.4	54.5	56.5	58.4	60.3	62.0	63.8
81	44.5	47.0	49.4	51.7	53.8	55.8	57.8	59.7	61.5	63.2
82	43.5	46.1	48.6	50.9	53.1	55.2	57.2	59.1	60.9	62.7
83	42.5	45.2	47.7	50.1	52.4	54.5	56.5	58.5	60.4	62.2
84	41.5	44.3	46.9	49.3	51.6	53.8	55.9	57.9	59.8	61.6
85	40.4	43.3	46.0	48.5	50.8	53.1	55.2	57.2	59.2	61.1
86	39.2	42.2	45.0	47.6	50.1	52.3	54.5	56.6	58.6	60.5
87	38.1	41.2	44.1	46.7	49.2	51.6	53.8	55.9	58.0	59.9
88	36.8	40.1	43.1	45.8	48.4	50.8	53.1	55.3	57.3	59.3
89	35.6	39.0	42.0	44.9	47.5	50.0	52.4	54.6	56.7	58.7
90	34.2	37.8	41.0	43.9	46.7	49.2	51.6	53.9	56.1	58.1
91	32.8	36.5	39.9	42.9	45.8	48.4	50.9	53.2	55.4	57.5
92	31.4	35.2	38.7	41.9	44.8	47.5	50.1	52.4	54.7	56.9
93	30.1	33.9	37.5	40.8	43.9	46.6	49.3	51.7	54.0	56.2
94	28.6	32.5	36.3	39.7	42.9	45.7	48.4	50.9	53.3	55.6
95	27.0	31.1	35.0	38.6	41.8	44.8	47.6	50.2	52.6	54.9
96	25.4	29.7	33.6	37.4	40.8	43.8	46.7	49.3	51.8	54.2
97	23.6	28.2	32.2	36.1	39.6	42.8	45.8	48.5	51.1	53.5
98	21.6	26.6	30.8	34.8	38.5	41.8	44.9	47.7	50.3	52.8
99	19.5	25.0	29.5	33.4	37.3	40.7	43.9	46.8	49.5	52.0
100	17.1	23.1	27.9	32.0	36.0	39.6	42.9	45.9	48.7	51.3
101	14.5	21.1	26.3	30.6	34.7	38.5	41.9	45.0	47.8	50.5
102	11.5	18.9	24.6	29.2	33.3	37.3	40.8	44.0	46.9	49.7
103	8.1	16.5	22.7	27.7	31.9	36.0	39.7	43.0	46.0	48.9
104	4.0	13.8	20.7	26.1	30.5	34.7	38.5	42.0	45.1	48.0
105		10.7	18.5	24.4	29.1	33.3	37.3	40.9	44.2	47.2
106		7.2	16.0	22.5	27.6	31.9	36.1	39.8	43.2	46.3
107		2.9	13.3	20.4	26.0	30.5	34.8	38.6	42.2	45.4
108			10.3	18.2	24.3	29.1	33.4	37.4	41.1	44.4
109			6.4	15.7	22.3	27.6	31.9	36.2	40.0	43.4
110			1.9	12.9	20.3	26.0	30.6	34.9	38.9	42.4

Dew Point Temperatures from Dry and Wet Bulb Temperatures (continued)

Dry Bulb Temp., °F.	Wet Bulb Temperature, °F.									
	70	71	72	73	74	75	76	77	78	79
	Dew Point Temperature, °F.									
70	70.0									
71	69.6	71.0								
72	69.1	70.6	72.0							
73	68.7	70.2	71.6	73.0						
74	68.2	69.7	71.2	72.6	74.0					
75	67.8	69.3	70.8	72.2	73.6	75.0				
76	67.3	68.9	70.4	71.8	73.2	74.6	76.0			
77	66.9	68.4	69.9	71.4	72.8	74.3	75.6	77.0		
78	66.4	68.0	69.5	71.0	72.4	73.9	75.3	76.7	78.0	
79	65.9	67.5	69.1	70.6	72.0	73.5	74.9	76.3	77.7	79.0
80	65.4	67.0	68.6	70.2	71.6	73.1	74.5	75.9	77.3	78.7
81	64.9	66.6	68.2	69.7	71.2	72.7	74.2	75.6	77.0	78.3
82	64.4	66.1	67.7	69.3	70.8	72.3	73.8	75.2	76.6	78.0
83	63.9	65.6	67.3	68.8	70.4	71.9	73.4	74.9	76.3	77.7
84	63.4	65.1	66.8	68.4	70.0	71.5	73.0	74.5	75.9	77.3
85	62.9	64.6	66.3	68.0	69.5	71.1	72.6	74.1	75.6	77.0
86	62.3	64.1	65.8	67.5	69.1	70.7	72.2	73.7	75.2	76.6
87	61.8	63.6	65.3	67.0	68.7	70.3	71.8	73.3	74.8	76.3
88	61.2	63.1	64.8	66.6	68.2	69.8	71.4	73.0	74.5	75.9
89	60.7	62.5	64.3	66.1	67.8	69.4	71.0	72.6	74.1	75.6
90	60.1	62.0	63.8	65.6	67.3	69.0	70.6	72.2	73.7	75.2
91	59.5	61.4	63.3	65.1	66.8	68.5	70.2	71.8	73.3	74.8
92	58.9	60.9	62.8	64.6	66.4	68.1	69.7	71.4	72.9	74.5
93	58.2	60.3	62.2	64.1	65.9	67.6	69.3	70.9	72.5	74.1
94	57.7	59.7	61.7	63.6	65.4	67.2	68.9	70.5	72.1	73.7
95	57.1	59.1	61.1	63.1	64.9	66.7	68.4	70.1	71.7	73.3
96	56.4	58.5	60.6	62.5	64.4	66.2	68.0	69.7	71.3	72.9
97	55.8	57.9	60.0	62.0	63.9	65.7	67.5	69.2	70.9	72.5
98	55.1	57.3	59.4	61.4	63.4	65.3	67.1	68.8	70.5	72.1
99	54.4	56.7	58.8	60.9	62.9	64.8	66.6	68.4	70.1	71.7
100	53.6	56.0	58.2	60.3	62.3	64.2	66.1	67.9	69.6	71.3
101	53.0	55.4	57.6	59.7	61.8	63.7	65.6	67.4	69.2	70.9
102	52.3	54.7	57.0	59.1	61.2	63.2	65.1	67.0	68.8	70.5
103	51.5	54.0	56.3	58.5	60.7	62.7	64.6	66.5	68.3	70.1
104	50.7	53.3	55.7	57.9	60.1	62.2	64.1	66.0	67.9	69.7
105	49.9	52.5	55.0	57.3	59.5	61.6	63.6	65.6	67.4	69.2
106	49.1	51.8	54.3	56.7	58.9	61.1	63.1	65.1	67.0	68.8
107	48.3	51.0	53.6	56.0	58.3	60.5	62.6	64.6	66.5	68.3
108	47.4	50.3	52.9	55.4	57.7	59.9	62.0	64.1	66.0	67.9
109	46.5	49.4	52.1	54.7	57.1	59.3	61.5	63.5	65.5	67.4
110	45.6	48.6	51.4	54.0	56.4	58.7	60.9	63.0	65.0	67.0

Dew Point Temperatures from Dry and Wet Bulb Temperatures (continued)

Dry Bulb Temp., °F.	Wet Bulb Temperature, °F.									
	80	81	82	83	84	85	86	87	88	89
	Dew Point Temperature, °F.									
80	80.0									
81	79.7	81.0								
82	79.4	80.7	82.0							
83	79.0	80.4	81.7	83.0						
84	78.7	80.1	81.4	82.7	84.0					
85	78.4	79.7	81.1	82.4	83.7	85.0				
86	78.0	79.4	80.8	82.1	83.4	84.7	86.0			
87	77.7	79.1	80.5	81.8	83.1	84.4	85.7	87.0		
88	77.4	78.8	80.2	81.5	82.8	84.2	85.5	86.7	88.0	
89	77.0	78.4	79.8	81.2	82.6	83.9	85.2	86.5	87.7	89.0
90	76.7	78.1	79.5	80.9	82.3	83.6	84.9	86.2	87.5	88.8
91	76.3	77.8	79.2	80.6	82.0	83.3	84.6	86.0	87.2	88.5
92	76.0	77.4	78.9	80.3	81.7	83.0	84.3	85.7	87.0	88.3
93	75.6	77.1	78.5	80.0	81.4	82.7	84.1	85.4	86.7	88.0
94	75.2	76.7	78.2	79.6	81.0	82.4	83.8	85.1	86.5	87.8
95	74.9	76.4	77.9	79.3	80.7	82.1	83.5	84.9	86.2	87.5
96	74.5	76.0	77.5	79.0	80.4	81.8	83.2	84.6	85.9	87.3
97	74.1	75.7	77.2	78.7	80.1	81.5	82.9	84.3	85.7	87.0
98	73.8	75.3	76.8	78.3	79.8	81.2	82.6	84.0	85.4	86.7
99	73.4	75.0	76.5	78.0	79.5	80.9	82.3	83.7	85.1	86.5
100	73.0	74.6	76.1	77.7	79.2	80.6	82.0	83.5	84.9	86.2
101	72.6	74.2	75.8	77.3	78.8	80.3	81.7	83.2	84.6	86.0
102	72.2	73.8	75.4	77.0	78.5	80.0	81.4	82.9	84.3	85.7
103	71.8	73.4	75.1	76.6	78.2	79.7	81.1	82.6	84.0	85.4
104	71.4	73.1	74.7	76.3	77.8	79.3	80.8	82.3	83.7	85.1
105	71.0	72.7	74.3	75.9	77.5	79.0	80.5	82.0	83.4	84.9
106	70.6	72.3	73.9	75.6	77.1	78.7	80.2	81.7	83.2	84.6
107	70.1	71.9	73.6	75.2	76.8	78.4	79.9	81.4	82.9	84.3
108	69.7	71.5	73.2	74.8	76.5	78.0	79.6	81.1	82.6	84.0
109	69.3	71.1	72.8	74.5	76.1	77.7	79.2	80.8	82.3	83.7
110	68.8	70.6	72.4	74.1	75.7	77.3	78.9	80.5	82.0	83.5

Dew Point Temperatures from Dry and Wet Bulb Temperatures (continued)

Dry Bulb Temp., °F.	Wet Bulb Temperature, °F.									
	90	91	92	93	94	95	96	97	98	99
	Dew Point Temperature, °F.									
90	90.0									
91	89.8	91.0								
92	89.5	90.8	92.0							
93	89.3	90.5	91.8	93.0						
94	89.0	90.3	91.5	92.8	94.0					
95	88.8	90.1	91.3	92.6	93.8	95.0				
96	88.5	89.8	91.1	92.3	93.6	94.8	96.0			
97	88.3	89.6	90.8	92.1	93.4	94.6	95.8	97.0		
98	88.1	89.3	90.6	91.9	93.1	94.4	95.6	96.8	98.0	
99	87.8	89.1	90.4	91.7	92.9	94.2	95.4	96.6	97.8	99.0
100	87.5	88.9	90.1	91.4	92.7	93.9	95.2	96.4	97.6	98.8
101	87.3	88.6	89.9	91.2	92.5	93.7	95.0	96.2	97.4	98.6
102	87.0	88.4	89.7	91.0	92.3	93.5	94.8	96.0	97.2	98.4
103	86.8	88.1	89.4	90.7	92.0	93.3	94.5	95.8	97.0	98.3
104	86.5	87.9	89.2	90.5	91.8	93.1	94.3	95.6	96.8	98.1
105	86.2	87.6	88.9	90.3	91.6	92.9	94.1	95.4	96.7	97.9
106	86.0	87.4	88.7	90.0	91.4	92.6	93.9	95.2	96.5	97.7
107	85.7	87.1	88.4	89.8	91.1	92.4	93.7	95.0	96.3	97.5
108	85.5	86.8	88.2	89.6	90.9	92.2	93.5	94.8	96.1	97.3
109	85.2	86.6	88.0	89.3	90.7	92.0	93.3	94.6	95.9	97.1
110	84.9	86.3	87.7	89.1	90.4	91.8	93.1	94.3	95.6	96.9

Dry Bulb Temp., °F.	Wet Bulb Temperature, °F.									
	100	101	102	103	104	105	106	107	108	109
	Dew Point Temperature, °F.									
100	100.0									
101	99.8	101.0								
102	99.6	100.8	102.0							
103	99.5	100.7	101.8	103.0						
104	99.3	100.5	101.7	102.8	104.0					
105	99.1	100.3	101.5	102.7	103.8	105.0				
106	98.9	100.1	101.3	102.5	103.7	104.8	106.0			
107	98.7	99.9	101.1	102.3	103.5	104.7	105.8	107.0		
108	98.5	99.8	101.0	102.2	103.3	104.5	105.7	106.9	108.0	
109	98.3	99.6	100.8	102.0	103.2	104.4	105.5	106.7	107.9	109.0
110	98.2	99.4	100.6	101.8	103.0	104.2	105.4	106.6	107.7	108.9

DEW POINT TEMPERATURES FROM DRY BULB AND HUMIDITY

Dry Bulb Temp., °F.	Relative Humidity, Per-Cent — Dew Point Temperature, °F.									
	2	4	6	8	10	12	14	16	18	20
30	−42.	−31.	−24.	−19.	−15.	−12.	−9.	−7.	−5.	−3.
35	−39.	−28.	−21.	−16.	−12.	−8.	−5.	−3.	−1.	1.3
40	−36.	−24.	−17.	−12.	−8.	−5.	−2.	.7	3.0	5.1
45	−33.	−21.	−14.	−9.	−5.	−1.	1.9	4.5	6.8	8.9
50	−30.	−18.	−10.	−5.	−1.	2.6	5.6	8.2	10.6	12.7
55	−26.	−14.	−7.	−2.	2.6	6.2	9.2	11.9	14.3	16.5
60	−23.	−11.	−4.	1.8	6.1	9.7	12.8	15.6	18.0	20.2
61	−23.	−11.	−3.	2.5	6.8	10.4	13.6	16.3	18.8	21.0
62	−22.	−10.	−2.	3.1	7.5	11.2	14.3	17.1	19.5	21.7
63	−22.	−9.	−2.	3.8	8.2	11.9	15.0	17.8	20.2	22.5
64	−21.	−9.	−1.	4.5	8.9	12.6	15.7	18.5	21.0	23.2
65	−20.	−8.	0.	5.2	9.6	13.3	16.5	19.2	21.7	23.9
66	−20.	−7.	.3	5.9	10.3	14.0	17.2	20.0	22.4	24.7
67	−19.	−7.	.9	6.6	11.0	14.7	17.9	20.7	23.2	25.4
68	−19.	−6.	1.6	7.3	11.7	15.4	18.6	21.4	23.9	26.2
69	−18.	−5.	2.3	7.9	12.4	16.1	19.3	22.1	24.7	26.9
70	−17.	−5.	3.0	8.6	13.1	16.8	20.1	22.9	25.4	27.7
71	−17.	−4.	3.6	9.3	13.8	17.6	20.8	23.6	26.1	28.4
72	−16.	−3.	4.3	10.0	14.5	18.2	21.5	24.3	26.9	29.2
73	−16.	−3.	4.9	10.6	15.2	19.0	22.2	25.1	27.6	29.9
74	−15.	−2.	5.6	11.3	15.9	19.6	22.9	25.8	28.3	30.6
75	−14.	−2.	6.3	12.0	16.6	20.4	23.6	26.5	29.1	31.4
76	−14.	−1.	6.9	12.7	17.3	21.1	24.3	27.2	29.8	32.1
77	−13.	0.	7.6	13.3	17.9	21.8	25.1	27.9	30.5	32.9
78	−13.	.3	8.2	14.0	18.6	22.5	25.8	28.7	31.2	33.8
79	−12.	.9	8.9	14.7	19.3	23.2	26.5	29.4	32.0	34.6
80	−11.	1.5	9.5	15.4	20.0	23.9	27.2	30.1	32.8	35.4
81	−11.	2.2	10.2	16.0	20.7	24.5	27.9	30.8	33.5	36.2
82	−10.	2.8	10.9	16.7	21.4	25.2	28.6	31.5	34.4	37.0
83	−10.	3.5	11.5	17.4	22.0	26.0	29.3	32.3	35.2	37.9
84	−9.	4.1	12.1	18.0	22.7	26.7	30.0	33.1	36.0	38.7
85	−8.	4.7	12.8	18.7	23.4	27.4	30.7	33.9	36.8	39.5
86	−8.	5.3	13.5	19.4	24.1	28.0	31.4	34.7	37.6	40.3
87	−7.	6.0	14.1	20.1	24.8	28.7	32.1	35.5	38.4	41.1
88	−7.	6.6	14.7	20.7	25.5	29.4	32.9	36.3	39.3	42.0
89	−6.	7.2	15.4	21.4	26.1	30.1	33.7	37.1	40.1	42.8
90	−5.	7.8	16.0	22.1	26.8	30.8	34.5	37.8	40.9	43.6
92	−4.	9.1	17.3	23.4	28.2	32.2	36.1	39.4	42.5	45.2
94	−3.	10.3	18.6	24.7	29.5	33.7	37.6	41.0	44.1	46.9
96	−2.	11.6	19.9	26.0	30.9	35.3	39.2	42.6	45.7	48.5
98	−1.	12.8	21.2	27.3	32.3	36.8	40.8	44.2	47.3	50.1
100	.3	14.0	22.5	28.7	33.8	38.4	42.3	45.8	48.9	51.8
102	1.4	15.2	23.7	30.0	35.3	39.9	43.9	47.4	50.5	53.4
104	2.6	16.5	25.0	31.3	36.8	41.4	45.4	49.0	52.1	55.0
106	3.8	17.7	26.3	32.6	38.2	42.9	47.0	50.6	53.7	56.6
108	4.9	18.9	27.6	34.1	39.7	44.5	48.5	52.1	55.3	58.3
110	6.0	20.1	28.8	35.6	41.2	46.0	50.1	53.7	57.0	59.9

Dew Point Temperatures from Dry Bulb and Relative Humidity (continued)

Dry Bulb Temp., °F.	Relative Humidity, Per-Cent									
	22	24	26	28	30	32	34	36	38	40
	Dew Point Temperature, °F.									
30	1.	.7	2.2	3.7	5.0	6.3	7.5	8.6	9.7	10.7
35	3.1	4.8	6.4	7.9	9.2	10.5	11.7	12.9	14.0	15.0
40	7.0	8.7	10.3	11.8	13.2	14.5	15.7	16.9	18.1	19.1
45	10.8	12.6	14.2	15.7	17.2	18.5	19.8	21.0	22.1	23.2
50	14.6	16.3	18.1	19.6	21.1	22.4	23.7	25.0	26.1	27.2
55	18.4	20.3	21.9	23.5	25.0	26.4	27.7	29.0	30.1	31.3
60	22.2	24.1	25.8	27.4	28.9	30.3	31.7	33.0	34.4	35.7
61	23.0	24.8	26.6	28.2	29.7	31.1	32.5	33.9	35.3	36.6
62	23.7	25.6	27.3	29.0	30.5	31.9	33.4	34.8	36.2	37.5
63	24.5	26.4	28.1	29.7	31.2	32.7	34.2	35.7	37.0	38.3
64	25.2	27.1	28.9	30.5	32.0	33.6	35.1	36.6	37.9	39.2
65	26.0	27.9	29.6	31.3	32.9	34.5	36.0	37.4	38.8	40.1
66	26.7	28.6	30.4	32.0	33.7	35.4	36.9	38.3	39.7	41.0
67	27.5	29.4	31.2	32.9	34.6	36.2	37.8	39.2	40.6	41.9
68	28.3	30.2	31.9	33.7	35.5	37.1	38.6	40.1	41.5	42.8
69	29.0	30.9	32.8	34.6	36.3	38.0	39.5	41.0	42.4	43.7
70	29.8	31.7	33.6	35.5	37.2	38.8	40.4	41.9	43.3	44.6
71	30.5	32.5	34.5	36.3	38.1	39.7	41.3	42.8	44.2	45.5
72	31.2	33.3	35.3	37.2	38.9	40.6	42.1	43.6	45.1	46.4
73	32.0	34.1	36.2	38.0	39.8	41.5	43.0	44.5	45.9	47.3
74	32.8	35.0	37.0	38.9	40.7	42.3	43.9	45.4	46.8	48.2
75	33.7	35.8	37.9	39.7	41.5	43.2	44.8	46.3	47.7	49.1
76	34.5	36.7	38.7	40.6	42.4	44.1	45.7	47.2	48.6	50.0
77	35.3	37.5	39.6	41.5	43.2	44.9	46.5	48.0	49.5	50.9
78	36.1	38.3	40.4	42.3	44.1	45.8	47.4	48.9	50.4	51.8
79	37.0	39.2	41.2	43.2	45.0	46.7	48.3	49.8	51.3	52.7
80	37.8	40.0	42.1	44.0	45.8	47.5	49.2	50.7	52.2	53.5
81	38.6	40.9	42.9	44.9	46.7	48.4	50.0	51.6	53.0	54.4
82	39.5	41.7	43.8	45.7	47.6	49.3	50.9	52.4	53.9	55.3
83	40.3	42.5	44.6	46.6	48.4	50.1	51.8	53.3	54.8	56.2
84	41.1	43.4	45.5	47.4	49.3	51.0	52.7	54.2	55.7	57.1
85	42.0	44.2	46.3	48.3	50.1	51.9	53.5	55.1	56.6	58.0
86	42.8	45.1	47.2	49.1	51.0	52.7	54.4	56.0	57.5	58.9
87	43.6	45.9	48.0	50.0	51.9	53.6	55.3	56.9	58.3	59.8
88	44.4	46.7	48.9	50.8	52.7	54.5	56.1	57.7	59.2	60.7
89	45.3	47.6	49.7	51.7	53.6	55.3	57.0	58.6	60.1	61.6
90	46.1	48.4	50.5	52.6	54.4	56.2	57.9	59.5	61.0	62.5
92	47.8	50.1	52.2	54.3	56.2	57.9	59.6	61.2	62.8	64.2
94	49.4	51.7	53.9	56.0	57.9	59.7	61.4	63.0	64.5	66.0
96	51.1	53.4	55.6	57.7	59.6	61.4	63.1	64.7	66.3	67.8
98	52.7	55.1	57.3	59.4	61.3	63.1	64.9	66.5	68.1	69.6
100	54.4	56.7	59.0	61.1	63.0	64.9	66.6	68.2	69.8	71.3
102	56.0	58.4	60.7	62.7	64.7	66.6	68.3	70.0	71.6	73.1
104	57.6	60.1	62.3	64.4	66.4	68.3	70.1	71.7	73.3	74.9
106	59.3	61.7	64.0	66.1	68.1	70.0	71.8	73.5	75.1	76.6
108	60.9	63.4	65.7	67.8	69.8	71.7	73.5	75.2	76.9	78.4
110	62.6	65.1	67.4	69.5	71.5	73.5	77.0	77.0	78.6	80.2

Dew Point Temperatures from Dry Bulb and Relative Humidity (continued)

Dry Bulb Temp., °F.	Relative Humidity, Per-Cent									
	42	44	46	48	50	52	54	56	58	60
	Dew Point Temperature, °F.									
30	11.7	12.7	13.6	14.4	15.3	16.1	16.9	17.6	18.3	19.1
35	16.0	17.0	17.9	18.8	19.7	20.5	21.3	22.1	22.8	23.5
40	20.1	21.1	22.1	23.0	23.8	24.7	25.5	26.3	27.0	27.8
45	24.2	25.2	26.1	27.1	28.0	28.8	29.6	30.4	31.2	32.0
50	28.3	29.3	30.3	31.2	32.1	33.1	34.1	35.0	35.8	36.7
55	32.4	33.6	34.7	35.7	36.7	37.7	38.6	39.6	40.5	41.4
60	36.9	38.1	39.2	40.3	41.3	42.3	43.3	44.3	45.2	46.1
61	37.8	39.0	40.1	41.2	42.2	43.3	44.3	45.2	46.2	47.1
62	38.7	39.9	41.0	42.1	43.2	44.2	45.2	46.2	47.1	48.0
63	39.5	40.7	41.9	43.1	44.1	45.1	46.1	47.1	48.0	48.9
64	40.5	41.7	42.9	44.0	45.1	46.1	47.1	48.0	49.0	49.9
65	41.4	42.6	43.8	44.9	45.9	47.0	48.0	49.0	49.9	50.8
66	42.3	43.5	44.6	45.7	46.8	47.9	48.9	49.9	50.8	51.8
67	43.2	44.4	45.6	46.7	47.8	48.8	49.8	50.8	51.8	52.7
68	44.1	45.3	46.5	47.6	48.7	49.7	50.8	51.8	52.7	53.7
69	45.0	46.2	47.4	48.5	49.6	50.7	51.7	52.7	53.7	54.6
70	45.9	47.2	48.3	49.4	50.5	51.6	52.6	53.6	54.6	55.5
71	46.8	48.0	49.2	50.3	51.4	52.5	53.5	54.5	55.5	56.4
72	47.7	48.9	50.1	51.3	52.4	53.5	54.5	55.4	56.4	57.4
73	48.6	49.8	51.0	52.2	53.3	54.4	55.4	56.4	57.4	58.3
74	49.5	50.8	52.0	53.1	54.2	55.3	56.3	57.3	58.3	59.2
75	50.4	51.6	52.9	54.0	55.1	56.2	57.3	58.3	59.2	60.2
76	51.3	52.5	53.8	54.9	56.0	57.1	58.2	59.2	60.2	61.1
77	52.2	53.5	54.7	55.8	57.0	58.1	59.1	60.1	61.1	62.1
78	53.1	54.4	55.6	56.7	57.9	59.0	60.1	61.1	62.0	63.0
79	54.0	55.3	56.5	57.7	58.8	59.9	61.0	62.0	63.0	63.9
80	54.9	56.2	57.4	58.6	59.7	60.8	61.9	62.9	63.9	64.9
81	55.8	57.1	58.3	59.5	60.6	61.7	62.8	63.8	64.8	65.8
82	56.7	58.0	59.2	60.4	61.5	62.7	63.7	64.8	65.8	66.8
83	57.5	58.8	60.1	61.3	62.5	63.6	64.6	65.7	66.7	67.7
84	58.5	59.8	61.0	62.2	63.4	64.5	65.6	66.6	67.6	68.6
85	59.3	60.7	61.9	63.1	64.3	65.4	66.5	67.5	68.6	69.6
86	60.3	61.6	62.8	64.0	65.2	66.3	67.4	68.5	69.5	70.5
87	61.2	62.5	63.7	65.0	66.1	67.2	68.3	69.4	70.4	71.4
88	62.0	63.4	64.6	65.9	67.0	68.2	69.3	70.3	71.4	72.4
89	62.9	64.3	65.6	66.8	67.9	69.1	70.2	71.3	72.3	73.3
90	63.9	65.2	66.5	67.7	68.9	70.0	71.1	72.2	73.2	74.2
92	65.6	67.0	68.3	69.5	70.7	71.9	73.0	74.1	75.1	76.1
94	67.4	68.8	70.1	71.3	72.5	73.6	74.7	75.9	77.0	78.0
96	69.2	70.6	71.9	73.1	74.3	75.5	76.7	77.8	78.8	79.9
98	71.0	72.4	73.7	75.0	76.2	77.3	78.5	79.6	80.7	81.7
100	72.8	74.2	75.5	76.8	78.0	79.2	80.3	81.5	82.6	83.6
102	74.6	75.9	77.3	78.6	79.8	81.0	82.2	83.3	84.4	85.5
104	76.3	77.8	79.1	80.4	81.6	82.9	84.0	85.2	86.3	87.3
106	78.1	79.5	80.9	82.2	83.5	84.7	85.9	87.0	88.1	89.2
108	79.9	81.3	82.7	84.0	85.3	86.5	87.7	88.9	90.0	91.1
110	81.7	83.1	84.5	85.8	87.1	88.4	89.6	90.7	91.8	92.9

Dew Point Temperatures from Dry Bulb and Relative Humidity (continued)

Dry Bulb Temp., °F.	Relative Humidity, Per-Cent									
	61	62	63	64	65	66	67	68	69	70
	Dew Point Temperature, °F.									
30	19.4	19.7	20.1	20.4	20.7	21.1	21.4	21.7	22.0	22.3
35	23.9	24.2	24.6	24.9	25.2	25.6	25.9	26.2	26.5	26.8
40	28.1	28.5	28.8	29.2	29.5	29.8	30.2	30.5	30.8	31.1
45	32.4	32.8	33.2	33.6	34.0	34.3	34.7	35.1	35.4	35.8
50	37.1	37.5	37.9	38.3	38.7	39.1	39.5	39.9	40.2	40.6
55	41.8	42.2	42.7	43.1	43.5	43.9	44.3	44.7	45.1	45.4
60	46.5	47.0	47.4	47.8	48.2	48.7	49.1	49.4	49.8	50.2
61	47.5	47.9	48.4	48.8	49.2	49.6	50.0	50.4	50.8	51.2
62	48.4	48.9	49.3	49.7	50.1	50.5	50.9	51.3	51.7	52.1
63	49.4	49.8	50.3	50.7	51.1	51.5	51.9	52.3	52.7	53.1
64	50.3	50.8	51.2	51.6	52.0	52.4	52.9	53.3	53.7	54.0
65	51.3	51.7	52.1	52.5	53.0	53.4	53.8	54.2	54.6	55.0
66	52.2	52.6	53.1	53.5	53.9	54.4	54.8	55.2	55.6	56.0
67	53.1	53.6	54.0	54.4	54.9	55.3	55.7	56.1	56.5	56.9
68	54.1	54.5	54.9	55.4	55.8	56.2	56.7	57.1	57.5	57.9
69	55.0	55.4	55.9	56.4	56.8	57.2	57.6	58.0	58.4	58.8
70	55.9	56.4	56.8	57.3	57.7	58.1	58.6	59.0	59.4	59.8
71	56.9	57.4	57.8	58.2	58.7	59.1	59.5	59.9	60.3	60.7
72	57.8	58.3	58.7	59.2	59.6	60.1	60.5	60.9	61.3	61.7
73	58.8	59.2	59.7	60.1	60.5	61.0	61.4	61.8	62.2	62.6
74	59.7	60.2	60.6	61.1	61.5	61.9	62.4	62.8	63.2	63.6
75	60.6	61.1	61.6	62.0	62.4	62.9	63.3	63.7	64.1	64.6
76	61.6	62.0	62.5	63.0	63.4	63.8	64.3	64.7	65.1	65.5
77	62.5	63.0	63.5	63.9	64.4	64.8	65.2	65.7	66.1	66.5
78	63.5	63.9	64.4	64.8	65.3	65.7	66.1	66.6	67.0	67.4
79	64.4	64.9	65.3	65.8	66.2	66.7	67.1	67.5	68.0	68.4
80	65.3	65.8	66.3	66.7	67.2	67.6	68.0	68.5	68.9	69.3
81	66.3	66.8	67.2	67.7	68.1	68.5	69.0	69.4	69.9	70.4
82	67.2	67.7	68.2	68.6	69.1	69.5	70.0	70.4	70.8	71.2
83	68.2	68.6	69.1	69.6	70.0	70.5	70.9	71.4	71.8	72.2
84	69.1	69.6	70.0	70.5	71.0	71.4	71.9	72.3	72.7	73.2
85	70.0	70.5	71.0	71.5	71.9	72.4	72.8	73.3	73.7	74.1
86	71.0	71.5	71.9	72.4	72.9	73.3	73.7	74.2	74.6	75.1
87	71.9	72.4	72.9	73.4	73.8	74.3	74.7	75.2	75.6	76.0
88	72.9	73.4	73.8	74.3	74.8	75.2	75.6	76.1	76.5	77.0
89	73.8	74.3	74.8	75.2	75.7	76.1	76.6	77.1	77.5	77.9
90	74.7	75.2	75.7	76.2	76.6	77.1	77.6	78.0	78.4	78.9
92	76.6	77.1	77.6	78.0	78.5	79.0	79.5	79.9	80.4	80.8
94	78.5	79.0	79.5	80.0	80.4	80.9	81.4	81.8	82.3	82.7
96	80.4	80.9	81.4	81.8	82.3	82.8	83.3	83.7	84.2	84.6
98	82.2	82.7	83.2	83.7	84.2	84.7	85.2	85.6	86.1	86.5
100	84.1	84.6	85.1	85.6	86.1	86.6	87.1	87.5	88.0	88.4
102	86.0	86.5	87.0	87.5	88.0	88.5	88.9	89.4	89.9	90.3
104	87.8	88.4	88.9	89.4	89.9	90.3	90.8	91.3	91.8	92.2
106	89.7	90.2	90.8	91.3	91.8	92.2	92.7	93.2	93.7	94.1
108	91.6	92.1	92.6	93.1	93.6	94.1	94.6	95.1	95.6	96.1
110	93.5	94.0	94.5	95.0	95.5	96.0	96.5	97.0	97.5	98.0

Dew Point Temperatures from Dry Bulb and Relative Humidity (continued)

Dry Bulb Temp., °F.	Relative Humidity, Per-Cent									
	71	72	73	74	75	76	77	78	79	80
	Dew Point Temperature, °F.									
30	22.6	22.9	23.2	23.5	23.8	24.0	24.3	24.6	24.9	25.2
35	27.2	27.5	27.8	28.0	28.3	28.6	28.9	29.2	29.5	29.7
40	31.4	31.8	32.1	32.4	32.7	33.1	38.4	33.7	34.0	34.3
45	36.2	36.5	36.9	37.2	37.6	37.9	38.2	38.6	38.9	39.2
50	40.9	41.3	41.7	42.0	42.4	42.8	43.1	43.4	43.7	44.1
55	45.8	46.2	46.5	46.9	47.2	47.6	48.0	48.3	48.6	49.0
60	50.6	51.0	51.3	51.7	52.1	52.4	52.8	53.1	53.5	53.9
61	51.5	51.9	52.3	52.7	53.1	53.4	53.7	54.1	54.4	54.8
62	52.5	52.9	53.3	53.7	54.0	54.4	54.7	55.1	55.4	55.8
63	53.5	53.9	54.2	54.6	54.9	55.3	55.7	56.1	56.4	56.7
64	54.4	54.8	55.2	55.6	55.9	56.3	56.7	57.0	57.4	57.7
65	55.4	55.8	56.1	56.5	56.9	57.3	57.6	58.0	58.3	58.7
66	56.4	56.7	57.1	57.5	57.9	58.2	58.6	59.0	59.3	59.7
67	57.3	57.7	58.1	58.5	58.9	59.2	59.6	59.9	60.3	60.6
68	58.3	58.6	59.0	59.4	59.8	60.2	60.5	60.9	61.3	61.6
69	59.2	59.6	60.0	60.4	60.7	61.1	61.5	61.9	62.2	62.6
70	60.2	60.6	61.0	61.4	61.7	62.1	62.5	62.8	63.2	63.6
71	61.1	61.5	61.9	62.3	62.7	63.1	63.4	63.8	64.2	64.5
72	62.1	62.5	62.9	63.3	63.7	64.0	64.4	64.8	65.2	65.5
73	63.1	63.5	63.8	64.2	64.6	65.0	65.4	65.7	66.1	66.5
74	64.0	64.4	64.8	65.2	65.6	66.0	66.3	66.7	67.1	67.5
75	65.0	65.4	65.7	66.1	66.5	66.9	67.3	67.7	68.0	68.4
76	65.9	66.3	66.7	67.1	67.5	67.9	68.3	68.6	69.0	69.4
77	66.9	67.3	67.7	68.1	68.5	68.9	69.2	69.6	70.0	70.4
78	67.8	68.2	68.6	69.0	69.4	69.8	70.2	70.6	70.9	71.3
79	68.8	69.2	69.6	70.0	70.4	70.8	71.2	71.6	71.9	72.3
80	69.8	70.2	70.6	71.0	71.4	71.7	72.2	72.5	72.9	73.3
81	70.7	71.1	71.5	71.9	72.3	72.7	73.1	73.5	73.9	74.2
82	71.7	72.1	72.5	72.9	73.3	73.7	74.1	74.5	74.8	75.2
83	72.6	73.1	73.5	73.9	74.3	74.7	75.0	75.4	75.8	76.2
84	73.6	74.0	74.4	74.8	75.2	75.6	76.0	76.4	76.8	77.2
85	74.6	75.0	75.4	75.8	76.2	76.6	77.0	77.4	77.8	78.1
86	75.5	75.9	76.3	76.7	77.1	77.5	77.9	78.3	78.7	79.1
87	76.4	76.9	77.3	77.7	78.1	78.5	78.9	79.3	79.7	80.1
88	77.4	77.8	78.3	78.7	79.1	79.5	79.9	80.3	80.7	81.0
89	78.4	78.8	79.2	79.6	80.0	80.4	80.8	81.2	81.6	82.0
90	79.3	79.8	80.2	80.6	81.0	81.4	81.8	82.2	82.6	83.0
92	81.2	81.7	82.1	82.5	82.9	83.3	83.7	84.1	84.5	84.9
94	83.2	83.6	84.0	84.4	84.8	85.3	85.7	86.1	86.5	86.9
96	85.1	85.5	85.9	86.4	86.8	87.2	87.6	88.0	88.4	88.8
98	87.0	87.4	87.8	88.3	88.7	89.1	89.5	90.0	90.4	90.8
100	88.9	89.3	89.8	90.2	90.6	91.1	91.5	91.9	92.3	92.7
102	90.8	91.2	91.7	92.1	92.5	93.0	93.4	93.8	94.2	94.6
104	92.7	93.1	93.6	94.0	94.5	94.9	95.3	95.7	96.2	96.6
106	94.6	95.1	95.5	95.9	96.4	96.8	97.3	97.7	98.1	98.5
108	96.5	97.0	97.4	97.9	98.3	98.8	99.2	99.6	100.0	100.5
110	98.4	98.9	99.4	99.8	100.2	100.7	101.1	101.6	102.0	102.4

Dew Point Temperatures from Dry Bulb and Relative Humidity (continued)

Dry Bulb Temp., °F.	Relative Humidity, Per-Cent									
	82	84	86	88	90	92	94	96	98	100
	Dew Point Temperature, °F.									
30	25.7	26.2	26.7	27.2	27.7	28.2	28.6	29.1	29.6	30.0
35	30.3	30.8	31.3	31.8	32.4	32.9	33.5	34.0	34.5	35.0
40	35.0	35.6	36.2	36.7	37.3	37.9	38.4	39.0	39.5	40.0
45	39.8	40.5	41.1	41.7	42.2	42.8	43.4	43.9	44.5	45.0
50	44.7	45.4	46.0	46.6	47.2	47.8	48.4	48.9	49.5	50.0
55	49.6	50.3	50.9	51.5	52.1	52.7	53.3	53.9	54.5	55.0
60	54.5	55.1	55.8	56.4	57.1	57.7	58.3	58.9	59.5	60.0
61	55.5	56.1	56.8	57.4	58.1	58.7	59.2	59.8	60.4	61.0
62	56.5	57.1	57.8	58.4	59.0	59.7	60.3	60.8	61.4	62.0
63	57.4	58.1	58.7	59.4	60.0	60.6	61.2	61.8	62.4	63.0
64	58.4	59.1	59.7	60.4	61.0	61.6	62.2	62.8	63.4	64.0
65	59.4	60.1	60.7	61.4	62.0	62.6	63.2	63.8	64.4	65.0
66	60.4	61.1	61.7	62.3	63.0	63.6	64.2	64.8	65.4	66.0
67	61.3	62.0	62.7	63.3	64.0	64.6	65.2	65.8	66.4	67.0
68	62.3	63.0	63.7	64.3	65.0	65.6	66.2	66.8	67.4	68.0
69	63.3	64.0	64.6	65.3	65.9	66.6	67.2	67.8	68.4	69.0
70	64.3	65.0	65.6	66.3	66.9	67.6	68.2	68.8	69.4	70.0
71	65.2	65.9	66.6	67.3	67.9	68.5	69.2	69.8	70.4	71.0
72	66.2	66.9	67.6	68.3	68.9	69.6	70.2	70.8	71.4	72.0
73	67.2	67.9	68.5	69.2	69.9	70.6	71.2	71.8	72.4	73.0
74	68.2	68.9	69.6	70.2	70.9	71.5	72.2	72.8	73.4	74.0
75	69.1	69.8	70.5	71.2	71.9	72.5	73.1	73.8	74.4	75.0
76	70.1	70.8	71.5	72.2	72.9	73.5	74.1	74.8	75.4	76.0
77	71.1	71.8	72.5	73.2	73.9	74.5	75.1	75.8	76.4	77.0
78	72.1	72.8	73.5	74.2	74.8	75.5	76.1	76.8	77.4	78.0
79	73.0	73.7	74.4	75.1	75.8	76.5	77.1	77.8	78.4	79.0
80	74.0	74.7	75.4	76.1	76.8	77.5	78.1	78.8	79.4	80.0
81	75.0	75.7	76.4	77.1	77.8	78.4	79.1	79.8	80.4	81.0
82	75.9	76.7	77.4	78.1	78.8	79.4	80.1	80.8	81.4	82.0
83	76.9	77.7	78.4	79.1	79.8	80.4	81.1	81.7	82.4	83.0
84	77.9	78.6	79.3	80.0	80.7	81.4	82.1	82.7	83.4	84.0
85	78.9	79.6	80.3	81.0	81.7	82.4	83.1	83.7	84.4	85.0
86	79.9	80.6	81.3	82.0	82.7	83.4	84.1	84.7	85.4	86.0
87	80.8	81.6	82.3	83.0	83.7	84.4	85.1	85.7	86.4	87.0
88	81.8	82.6	83.3	84.0	84.7	85.4	86.0	86.7	87.4	88.0
89	82.8	83.5	84.2	85.0	85.7	86.4	87.1	87.7	88.4	89.0
90	83.7	84.5	85.2	86.0	86.7	87.4	88.0	88.7	89.4	90.0
92	85.7	86.5	87.2	87.9	88.6	89.3	90.0	90.7	91.4	92.0
94	87.7	88.4	89.2	89.9	90.6	91.3	92.0	92.7	93.3	94.0
96	89.6	90.4	91.1	91.9	92.6	93.3	94.0	94.7	95.3	96.0
98	91.5	92.3	93.1	93.8	94.6	95.3	96.0	96.7	97.3	98.0
100	93.5	94.3	95.0	95.8	96.5	97.3	98.0	98.6	99.3	100.0
102	95.4	96.2	97.0	97.7	98.5	99.2	99.9	100.6	101.3	102.0
104	97.4	98.2	99.0	99.7	100.5	101.2	101.9	102.6	103.3	104.0
106	99.3	100.1	100.9	101.7	102.4	103.2	103.9	104.6	105.3	106.0
108	101.3	102.1	102.9	103.7	104.4	105.2	105.9	106.6	107.3	108.0
110	103.2	104.0	104.8	105.6	106.4	107.2	107.9	108.6	109.3	110.0

TABLES FOR DETERMINING RELATIVE HUMIDITY

Dry Bulb Temp., °F	Wet-Bulb Temperature, °F									
	30	31	32	33	34	35	36	37	38	39
	Relative Humidity, Per-Cent									
30	100.0									
31	89.1	100.0								
32	79.2	89.6	100.0							
33	70.4	80.2	90.3	100.0						
34	62.1	71.7	81.3	90.6	100.0					
35	54.4	63.6	72.8	81.7	90.8	100.0				
36	47.1	56.0	64.9	73.5	82.2	91.0	100.0			
37	40.5	49.0	57.5	65.7	74.1	82.6	91.2	100.0		
38	34.2	42.4	50.6	58.5	66.6	74.7	83.0	91.5	100.0	
39	29.1	36.2	44.1	51.7	59.5	67.3	75.3	83.4	91.6	100.0
40	22.9	30.5	38.1	45.4	52.9	60.4	68.1	75.9	83.8	91.9
41	17.9	25.2	32.5	39.5	46.7	53.9	61.3	68.8	76.4	84.2
42	13.2	20.2	27.2	32.8	40.9	47.8	55.0	62.2	69.5	76.9
43		15.6	22.3	28.8	35.4	42.2	49.0	56.0	63.0	70.1
44		11.2	17.7	24.0	30.4	36.8	43.4	50.1	56.9	63.8
45			13.5	19.5	25.7	31.9	38.2	44.7	51.1	57.8
46			9.5	15.3	21.3	27.2	33.3	39.5	45.8	52.2
47				11.4	17.1	22.9	28.8	34.7	40.8	46.9
48					13.3	18.9	24.5	30.3	36.1	42.0
49					9.7	15.1	20.5	26.0	31.6	37.3
50						11.5	16.8	22.1	27.5	33.0
51						8.2	13.3	18.4	23.6	29.0
52							10.1	15.0	20.0	25.1
53							7.0	11.8	16.6	21.6
54								8.8	13.5	18.2
55								6.0	10.5	15.1
56									7.7	12.2
57										9.4
58										6.9

Dry Bulb Temp., °F	Wet-Bulb Temperature, °F									
	40	41	42	43	44	45	46	47	48	49
	Relative Humidity, Per-Cent									
40	100.0									
41	92.0	100.0								
42	84.5	92.2	100.0							
43	77.4	84.8	92.4	100.0						
44	70.8	77.9	85.2	92.5	100.0					
45	64.5	71.4	78.4	85.5	92.7	100.0				
46	58.7	65.3	72.0	78.8	85.8	92.8	100.0			
47	53.2	59.5	66.0	72.6	79.2	86.0	93.0	100.0		
48	48.0	54.1	60.3	66.7	73.1	79.6	86.3	93.1	100.0	
49	43.1	49.0	55.0	61.1	67.3	73.6	80.1	86.6	93.2	100.0
50	38.6	44.3	50.1	55.9	61.9	68.0	74.2	80.5	86.9	93.4
51	34.3	39.8	45.3	51.0	56.8	62.6	68.6	74.7	80.8	87.1
52	30.3	35.6	40.9	46.4	51.9	57.6	63.3	69.2	75.1	81.2
53	26.5	31.6	36.8	42.1	47.4	52.8	58.4	64.0	69.8	75.5
54	23.0	27.9	32.9	38.0	43.1	48.4	53.7	59.1	64.7	70.3
55	19.7	24.4	29.3	34.1	39.1	44.2	49.3	54.6	59.9	65.3
56	16.6	21.2	25.8	30.5	35.3	40.2	45.1	50.2	55.3	60.6

Relative Humidity from Dry and Wet Bulb Temperatures (continued)

Dry Bulb Temp., °F	Wet-Bulb Temperature, °F									
	40	41	42	43	44	45	46	47	48	49
	Relative Humidity, Per-Cent									
57	13.7	18.1	22.6	27.1	31.8	36.4	41.2	46.1	51.1	56.1
58	11.0	15.2	19.6	24.0	28.4	32.9	37.6	42.3	47.0	51.9
59	8.5	12.6	16.7	21.0	25.3	29.6	34.1	38.6	43.2	47.9
60	6.1	10.1	14.1	18.2	22.3	26.5	30.8	35.2	39.6	44.2
61		7.7	11.6	15.5	19.5	23.6	27.7	32.0	36.3	40.6
62		5.5	9.3	13.1	16.9	20.8	24.9	28.9	33.1	37.3
63			7.1	10.8	14.5	18.3	22.1	26.1	30.0	34.1
64			5.0	8.6	12.2	15.8	19.6	23.4	27.2	31.2
65				6.6	10.0	13.5	17.2	20.8	24.6	28.4
66				4.6	8.0	11.4	14.9	18.4	22.0	25.7
67					6.1	9.4	12.8	16.2	19.7	23.2
68					4.3	7.5	10.8	14.1	17.4	20.9
69						5.7	8.9	12.1	15.3	18.6
70						4.1	7.1	10.2	13.4	16.6
71							5.5	8.5	11.5	14.6
72							3.9	6.8	9.7	12.7
73								5.3	8.1	11.0
74								3.8	6.6	9.3
75									5.1	7.8
76									3.7	6.4
77										5.0
78										3.7

Dry Bulb Temp., °F	Wet-Bulb Temperature, °F									
	50	51	52	53	54	55	56	57	58	59
	Relative Humidity, Per-Cent									
50	100.0									
51	93.5	100.0								
52	87.3	93.6	100.0							
53	81.5	87.6	93.7	100.0						
54	76.0	81.8	87.8	93.8	100.0					
55	70.8	76.4	82.2	88.0	93.9	100.0				
56	65.9	71.3	76.8	82.5	88.2	94.0	100.0			
57	61.2	66.5	71.8	77.2	82.8	88.4	94.1	100.0		
58	56.8	61.9	67.0	72.3	77.6	83.1	88.6	94.3	100.0	
59	52.7	57.6	62.5	67.6	72.7	78.0	83.3	88.8	94.3	100.0
60	48.8	53.5	58.3	63.2	68.1	73.2	78.4	83.6	89.0	94.4
61	45.1	49.6	54.2	58.9	63.7	68.6	73.6	78.7	83.9	89.1
62	41.6	46.0	50.4	55.0	59.6	64.3	69.1	74.0	79.0	84.1
63	38.3	42.5	46.8	51.2	55.7	60.2	64.9	69.6	74.4	79.3
64	35.2	39.2	43.4	47.6	51.9	56.3	60.8	65.4	70.1	74.8
65	32.2	36.2	40.2	44.3	48.4	52.7	57.0	61.4	65.9	70.5
66	29.4	33.2	37.1	41.1	45.1	49.2	53.4	57.7	62.0	66.4
67	26.8	30.5	34.2	38.1	41.9	45.9	49.9	54.1	58.3	62.5
68	24.3	27.9	31.5	35.2	39.0	42.8	46.7	50.7	54.7	58.9
69	22.0	25.4	28.9	32.5	36.1	39.8	43.6	47.4	51.4	55.4
70	19.8	23.1	26.5	30.0	33.5	37.0	40.7	44.4	48.2	52.0
71	17.7	20.9	24.2	27.5	30.9	34.4	37.9	41.5	45.2	48.9
72	15.8	18.9	22.0	25.3	28.5	31.9	35.3	38.7	42.3	45.9
73	13.9	16.9	20.0	23.1	26.3	29.5	32.8	36.1	39.6	43.1
74	12.2	15.1	18.0	21.1	24.1	27.2	30.4	33.7	37.0	40.4
75	10.5	13.4	16.2	19.1	22.1	25.1	28.2	31.3	34.5	37.8

Relative Humidity from Dry and Wet Bulb Temperatures (continued)

Dry Bulb Temp., °F	Wet-Bulb Temperature, °F									
	50	51	52	53	54	55	56	57	58	59
	Relative Humidity, Per-Cent									
76	9.0	11.7	14.5	17.3	20.2	23.1	26.1	29.1	32.2	35.4
77	7.6	10.2	12.9	15.6	18.3	21.2	24.1	27.0	30.0	33.0
78	6.2	8.7	11.3	14.0	16.6	19.4	22.2	25.0	27.9	30.9
79	4.9	7.4	9.9	12.4	15.0	17.7	20.4	23.1	25.9	28.8
80	3.7	6.1	8.5	11.0	13.5	16.0	18.7	21.3	24.0	26.8
81	2.6	4.9	7.2	9.6	12.0	14.5	17.0	19.6	22.2	24.9
82		3.7	6.0	8.3	10.7	13.1	15.5	18.0	20.6	23.1
83			4.9	7.1	9.4	11.7	14.1	16.5	18.9	21.5
84			3.8	6.0	8.2	10.4	12.7	15.1	17.4	19.9
85			2.8	4.9	7.0	9.2	11.4	13.7	16.0	18.4
86				3.9	5.9	8.0	10.2	12.4	14.6	16.9
87				2.9	4.9	7.0	9.0	11.2	13.3	15.5
88					3.9	5.9	7.9	10.0	12.1	14.2
89					3.0	5.0	6.9	8.9	10.9	13.0
90					2.2	4.0	5.9	7.9	9.8	11.9
91						3.2	5.0	6.9	8.8	10.7
92						2.4	4.2	6.0	7.8	9.7
93							3.3	5.1	6.9	8.7
94							2.6	4.3	6.0	7.8
95							1.8	3.5	5.2	6.9
96								2.8	4.4	6.1
97								2.1	3.6	5.3
98									2.9	4.5
99									2.3	3.8
100									1.7	3.1
101										2.5
102										1.9
103										1.3

Dry Bulb Temp., °F	Wet-Bulb Temperature, °F									
	60	61	62	63	64	65	66	67	68	69
	Relative Humidity, Per-Cent									
60	100.0									
61	94.5	100.0								
62	89.3	94.6	100.0							
63	84.3	89.5	94.7	100.0						
64	79.6	84.6	89.6	94.8	100.0					
65	75.2	80.0	84.8	89.8	94.8	100.0				
66	70.9	75.5	80.2	85.0	89.9	94.9	100.0			
67	66.9	71.4	75.9	80.5	85.3	90.1	95.0	100.0		
68	63.1	67.4	71.8	76.2	80.8	85.5	90.2	95.0	100.0	
69	59.4	63.6	67.8	72.1	76.6	81.1	85.6	90.3	95.1	100.0
70	56.0	60.0	64.1	68.3	72.5	76.9	81.3	85.9	90.5	95.2
71	52.7	56.6	60.5	64.6	68.7	72.9	77.2	81.6	86.0	90.6
72	49.6	53.3	57.1	61.1	65.0	69.1	73.3	77.5	81.8	86.2
73	46.6	50.2	53.9	57.7	61.6	65.5	69.5	73.6	77.8	82.0
74	43.8	47.3	50.9	54.5	58.3	62.1	65.9	69.9	73.9	78.1
75	41.1	44.5	48.0	51.5	55.1	58.8	62.5	66.3	70.3	74.3
76	38.6	41.9	45.2	48.6	52.1	55.7	59.3	63.0	66.8	70.6
77	36.2	39.3	42.6	45.9	49.3	52.7	56.2	59.8	63.4	67.2
78	33.9	36.9	40.1	43.3	46.5	49.9	53.3	56.7	60.3	63.9
79	31.7	34.7	37.7	40.8	43.9	47.2	50.5	53.8	57.2	60.7

Relative Humidity from Dry and Wet Bulb Temperatures (continued)

Dry Bulb Temp °F	Wet-Bulb Temperature, °F									
	60	61	62	63	64	65	66	67	68	69
	Relative Humidity, Per-Cent									
80	29.6	32.5	35.4	38.4	41.5	44.6	47.8	51.0	54.3	57.7
81	27.7	30.4	33.3	36.2	39.1	42.2	45.2	48.4	51.6	54.9
82	25.8	28.5	31.2	34.0	36.9	39.8	42.9	45.9	49.0	52.1
83	24.0	26.6	29.3	32.0	34.8	37.6	40.5	43.4	46.5	49.5
84	22.3	24.9	27.4	30.1	32.8	35.5	38.3	41.2	44.1	47.0
85	20.7	23.2	25.7	28.2	30.9	33.5	36.2	39.0	41.8	44.7
86	19.2	21.6	24.0	26.5	29.0	31.6	34.2	36.9	39.6	42.4
87	17.8	20.1	22.4	24.8	27.3	29.8	32.3	34.9	37.5	40.2
88	16.4	18.6	20.9	23.2	25.6	28.0	30.5	33.0	35.6	38.2
89	15.1	17.3	19.5	21.7	24.0	26.4	28.7	31.2	33.7	36.2
90	13.9	16.0	18.1	20.3	22.5	24.8	27.1	29.4	31.9	34.3
91	12.7	14.8	16.8	18.9	21.1	23.3	25.5	27.8	30.1	32.5
92	11.6	13.6	15.6	17.7	19.7	21.9	24.0	26.2	28.5	30.8
93	10.6	12.5	14.4	16.4	18.4	20.5	22.6	24.7	26.9	29.2
94	9.6	11.4	13.3	15.2	17.2	19.2	21.2	23.3	25.5	27.6
95	8.6	10.4	12.3	14.1	16.0	18.0	20.0	22.0	24.0	26.1
96	7.8	9.5	11.3	13.1	14.9	16.8	18.7	20.7	22.7	24.7
97	6.9	8.6	10.3	12.1	13.9	15.7	17.6	19.5	21.4	23.4
98	6.1	7.7	9.4	11.1	12.9	14.6	16.4	18.3	20.2	22.1
99	5.4	6.9	8.6	10.2	11.9	13.6	15.4	17.2	19.0	20.8
100	4.6	6.2	7.7	9.3	11.0	12.7	14.4	16.1	17.9	19.7
101	4.0	5.4	7.0	8.5	10.1	11.7	13.4	15.1	16.8	18.6
102	3.3	4.8	6.2	7.7	9.3	10.9	12.5	14.1	15.8	17.5
103	2.7	4.1	5.5	7.0	8.5	10.0	11.6	13.2	14.8	16.5
104	2.1	3.5	4.9	6.3	7.8	9.2	10.8	12.3	13.9	15.5
105	1.6	2.9	4.3	5.6	7.1	8.5	10.0	11.5	13.0	14.6
106		2.4	3.7	5.0	6.4	7.8	9.2	10.7	12.1	13.7
107		1.8	3.1	4.4	5.7	7.1	8.5	9.9	11.3	12.8
108		1.3	2.6	3.8	5.1	6.5	7.8	9.2	10.6	12.0
109			2.1	3.3	4.6	5.8	7.1	8.5	9.8	11.2
110			1.6	2.8	4.0	5.3	6.5	7.8	9.1	10.5

Dry Bulb Temp., °F	Wet-Bulb Temperature, °F									
	70	71	72	73	74	75	76	77	78	79
	Relative Humidity, Per-Cent									
70	100.0									
71	95.2	100.0								
72	90.7	95.3	100.0							
73	86.4	90.8	95.4	100.0						
74	82.3	86.6	91.0	95.4	100.0					
75	78.3	82.5	86.7	91.1	95.5	100.0				
76	74.6	78.6	82.7	86.9	91.2	95.5	100.0			
77	71.0	74.9	78.8	82.9	87.0	91.3	95.6	100.0		
78	67.6	71.3	75.2	79.1	83.1	87.2	91.4	95.7	100.0	
79	64.3	67.9	71.7	75.5	79.3	83.3	87.4	91.5	95.7	100.0
80	61.2	64.7	68.3	72.0	75.7	79.6	83.5	87.5	91.6	95.7
81	58.2	61.6	65.1	68.7	72.3	76.0	79.8	83.7	87.6	91.7
82	55.4	58.7	61.2	65.5	69.0	72.6	76.3	80.0	83.8	87.8
83	52.7	55.9	59.1	62.5	65.9	69.3	72.9	76.5	80.2	84.0
84	50.1	53.2	56.3	59.6	62.9	66.2	69.7	73.2	76.8	80.5
85	47.6	50.6	53.7	56.8	60.0	63.3	66.6	70.0	73.5	77.1
86	45.3	48.2	51.2	54.2	57.3	60.5	63.7	67.0	70.3	73.8

Relative Humidity from Dry and Wet Bulb Temperatures (continued)

Dry Bulb Temp., °F	Wet-Bulb Temperature, °F									
	70	71	72	73	74	75	76	77	78	79
	Relative Humidity, Per-Cent									
87	43.0	45.8	48.7	51.6	54.6	57.7	60.8	64.0	67.3	70.6
88	40.8	43.6	46.4	49.2	52.1	55.1	58.1	61.2	64.4	67.6
89	38.8	41.4	44.1	46.9	49.7	52.6	55.5	58.5	61.6	64.7
90	36.8	39.4	42.0	44.7	47.4	50.2	53.0	55.9	58.9	61.9
91	35.0	37.4	40.0	42.6	45.2	47.9	50.7	53.5	56.4	59.3
92	33.2	35.6	38.1	40.6	43.1	45.8	48.4	51.2	53.9	56.8
93	31.5	33.8	36.2	38.6	41.1	43.7	46.2	48.5	51.6	54.4
94	29.8	32.1	34.4	36.8	39.2	41.7	44.2	46.7	49.4	52.0
95	28.3	30.5	32.7	35.0	37.4	39.8	42.2	44.7	47.2	49.8
96	26.8	29.0	31.1	33.4	35.6	37.9	40.3	42.7	45.2	47.7
97	25.4	27.5	29.6	31.7	33.9	36.2	38.5	40.8	43.2	45.7
98	24.1	26.1	28.1	30.2	32.3	34.5	36.7	39.0	41.3	43.7
99	22.7	24.7	26.7	28.7	30.8	32.9	35.1	37.3	39.5	41.8
100	21.5	23.4	25.3	27.3	29.3	31.4	33.5	35.6	37.8	40.0
101	20.3	22.2	24.1	26.0	27.9	29.9	31.9	34.0	36.1	38.3
102	19.2	21.0	22.8	24.7	26.6	28.5	30.5	32.5	34.5	36.6
103	18.1	19.9	21.6	23.4	25.3	27.1	29.1	31.0	33.0	35.1
104	17.1	18.8	20.5	22.3	24.0	25.9	27.7	29.6	31.6	33.6
105	16.1	17.8	19.4	21.1	22.9	24.6	26.4	28.3	30.2	32.1
106	15.2	16.8	18.4	20.1	21.7	23.5	25.2	27.0	28.8	30.7
107	14.3	15.8	17.4	19.0	20.6	22.3	24.0	25.8	27.5	29.4
108	13.5	14.9	16.5	18.0	19.6	21.2	22.9	24.6	26.3	28.1
109	12.6	14.1	15.6	17.1	18.6	20.2	21.8	23.5	25.1	26.8
110	11.9	13.3	14.7	16.2	17.7	19.2	20.8	22.4	24.0	25.7

Dry Bulb Temp., °F	Wet-Bulb Temperature, °F									
	80	81	82	83	84	85	86	87	88	89
	Relative Humidity, Per-Cent									
80	100.0									
81	95.8	100.0								
82	91.8	95.8	100.0							
83	87.9	91.8	95.9	100.0						
84	84.2	88.0	91.9	95.9	100.0					
85	80.7	84.4	88.2	92.1	96.0	100.0				
86	77.3	80.9	84.6	88.3	92.1	96.0	100.0			
87	74.6	77.5	81.1	84.7	88.4	92.2	96.0	100.0		
88	70.9	74.3	77.7	81.2	84.8	88.5	92.2	96.1	100.0	
89	67.9	71.2	74.5	77.9	81.4	84.9	88.5	92.3	96.1	100.0
90	65.0	68.2	71.4	74.8	78.1	81.5	85.0	88.7	92.4	96.2
91	62.3	65.4	68.5	71.7	75.0	78.3	81.7	85.2	88.8	92.5
92	59.7	62.7	65.7	68.8	72.0	75.2	78.5	81.9	85.4	89.0
93	57.2	60.1	63.0	66.0	69.1	72.2	75.4	78.7	82.1	85.5
94	54.8	57.6	60.4	63.4	66.3	69.4	72.5	75.7	78.9	82.3
95	52.5	55.2	58.0	60.8	63.7	66.6	69.6	72.7	75.9	79.2
96	50.3	52.9	55.6	58.4	61.2	64.0	66.9	70.0	73.0	76.2
97	48.2	50.7	53.3	56.0	58.7	61.5	64.3	67.2	70.2	73.3
98	46.1	48.6	51.1	53.7	56.4	59.0	61.8	64.6	67.5	70.5
99	44.2	46.6	49.0	51.5	54.1	56.7	59.4	62.1	64.9	67.8
100	42.3	44.7	47.0	49.5	52.0	54.5	57.1	59.7	62.5	65.3
101	40.5	42.8	45.1	47.4	49.9	52.3	54.8	57.4	60.1	62.8
102	38.8	41.0	43.2	45.5	47.9	50.2	52.7	55.2	57.8	60.4

Relative Humidity from Dry and Wet Bulb Temperatures (continued)

Dry Bulb Temp., °F	Wet-Bulb Temperature, °F									
	80	81	82	83	84	85	86	87	88	89
	Relative Humidity, Per-Cent									
103	37.2	39.3	41.5	43.7	46.0	48.3	50.6	53.1	55.6	58.1
104	35.6	37.7	39.8	41.9	44.1	46.4	48.7	51.1	53.5	56.0
105	34.1	36.1	38.1	40.2	42.4	44.6	46.8	49.1	51.5	53.9
106	32.6	34.6	36.6	38.6	40.7	42.8	45.0	47.2	49.5	51.8
107	31.2	33.1	35.1	37.0	39.1	41.1	43.2	45.4	47.6	49.9
108	29.9	31.7	33.6	35.5	37.5	39.5	41.5	43.6	45.8	48.0
109	28.6	30.4	32.2	34.1	36.0	37.9	39.9	42.0	44.1	46.2
110	27.4	29.1	30.9	32.7	34.6	36.4	38.4	40.4	42.4	44.5

Dry Bulb Temp., °F	Wet-Bulb Temperature, °F									
	90	91	92	93	94	95	96	97	98	99
	Relative Humidity, Per-Cent									
90	100.0									
91	96.2	100.0								
92	92.6	96.3	100.0							
93	89.1	92.6	96.2	100.0						
94	85.7	89.1	92.7	96.3	100.0					
95	82.5	85.8	89.2	92.8	96.4	100.0				
96	79.4	82.6	85.9	89.4	92.8	96.4	100.0			
97	76.4	79.5	82.8	86.1	89.5	92.9	96.4	100.0		
98	73.5	76.6	79.7	82.9	86.2	89.5	92.9	96.4	100.0	
99	70.7	73.7	76.7	79.9	83.0	86.3	89.6	93.0	96.5	100.0
100	68.1	71.0	73.9	77.0	80.0	83.2	86.4	89.7	93.1	96.5
101	65.5	68.3	71.2	74.1	77.1	80.2	83.3	86.5	89.8	93.1
102	63.1	65.8	68.6	71.4	74.3	77.3	80.3	83.4	86.6	89.8
103	60.7	63.4	66.0	68.8	71.6	74.5	77.4	80.4	83.6	86.7
104	58.5	61.0	63.6	66.3	69.1	71.9	74.7	77.6	80.7	83.7
105	56.3	58.8	61.3	63.9	66.6	69.3	72.1	74.9	77.8	80.8
106	54.2	56.6	59.1	61.6	64.2	66.9	69.5	72.3	75.1	78.0
107	52.2	54.5	56.9	59.4	61.9	64.5	67.1	69.7	72.5	75.3
108	50.2	52.5	54.8	57.2	59.7	62.2	64.7	67.3	70.0	72.7
109	48.4	50.6	52.9	55.2	57.5	60.0	62.4	64.9	67.6	70.2
110	46.6	48.7	50.9	53.2	55.5	57.9	60.2	62.7	65.2	67.8

Dry Bulb Temp., °F	Wet-Bulb Temperature, °F									
	100	101	102	103	104	105	106	107	108	109
	Relative Humidity, Per-Cent									
100	100.0									
101	96.5	100.0								
102	93.1	96.5	100.0							
103	89.9	93.2	96.6	100.0						
104	86.8	90.0	93.3	96.6	100.0					
105	83.8	86.9	90.1	93.4	96.6	100.0				
106	80.9	84.0	87.1	90.2	93.4	96.7	100.0			
107	78.2	81.1	84.1	87.1	90.2	93.4	96.7	100.0		
108	75.5	78.3	81.2	84.2	87.2	90.3	93.4	96.7	100.0	
109	72.9	75.7	78.5	81.3	84.3	87.3	90.4	93.5	96.7	100.0
110	70.4	73.1	75.8	78.6	81.5	84.4	87.4	90.4	93.6	96.8

VAPOR PRESSURES FROM DRY AND WET BULB TEMPERATURES

Dry Bulb Temp., °F.	Wet Bulb Temperature, °F.									
	30	31	32	33	34	35	36	37	38	39
	Vapor Pressure, Inches of Mercury									
30	.1646									
31	.1538	.1726								
32	.1430	.1618	.1806							
33	.1323	.1510	.1698	.1880						
34	.1216	.1403	.1590	.1772	.1957					
35	.1107	.1295	.1483	.1664	.1849	.2036				
36	.0999	.1187	.1375	.1557	.1741	.1928	.2119			
37	.0892	.1079	.1267	.1449	.1633	.1820	.2011	.2204		
38	.0784	.0972	.1159	.1341	.1526	.1712	.1903	.2096	.2292	
39	.0676	.0864	.1052	.1233	.1418	.1605	.1795	.1988	.2184	.2384
40	.0568	.0755	.0944	.1125	.1310	.1497	.1687	.1880	.2076	.2276
41	.0461	.0649	.0836	.1017	.1202	.1389	.1580	.1772	.1968	.2168
42	.0353	.0540	.0728	.0909	.1094	.1281	.1472	.1665	.1860	.2060
43		.0433	.0620	.0802	.0986	.1173	.1364	.1557	.1752	.1952
44		.0325	.0512	.0694	.0878	.1065	.1256	.1449	.1644	.1844
45			.0405	.0586	.0771	.0957	.1148	.1341	.1536	.1736
46			.0297	.0478	.0663	.0849	.1040	.1233	.1429	.1628
47				.0370	.0555	.0741	.0932	.1125	.1321	.1520
48				.0263	.0447	.0634	.0824	.1017	.1213	.1412
49					.0339	.0526	.0716	.0909	.1105	.1304
50					.0231	.0418	.0608	.0801	.0997	.1196
51						.0310	.0500	.0693	.0889	.1089
52							.0393	.0585	.0781	.0981
53							.0285	.0477	.0673	.0873
54								.0369	.0565	.0765
55								.0261	.0457	.0657
56									.0349	.0549
57										.0441
58										.0333
59										
60										
61										
62										
63										
64										
65										
66										
67										
68										
69										
70										
71										
72										
73										
74										
75										
76										
77										
78										
79										
80										

Vapor Pressures from Dry and Wet Bulb Temperatures (continued)

Dry Bulb Temp., °F.	Wet Bulb Temperature, °F.									
	40	41	42	43	44	45	46	47	48	49
	Vapor Pressure, Inches of Mercury									
40	.2478									
41	.2370	.2576								
42	.2262	.2468	.2678							
43	.2154	.2360	.2570	.2783						
44	.2046	.2252	.2462	.2675	.2891					
45	.1938	.2145	.2354	.2567	.2783					
46	.1830	.2036	.2246	.2459	.2675	.2895				
47	.1722	.1928	.2138	.2351	.2567	.2787	.3012			
48	.1614	.1820	.2030	.2243	.2459	.2679	.2904	.3132		
49	.1506	.1712	.1922	.2135	.2351	.2571	.2796	.3024	.3256	
50	.1398	.1604	.1814	.2027	.2243	.2463	.2688	.2916	.3148	.3384
51	.1290	.1496	.1706	.1919	.2135	.2355	.2580	.2808	.3040	.3276
52	.1182	.1388	.1598	.1811	.2027	.2247	.2472	.2700	.2932	.3168
53	.1074	.1280	.1490	.1703	.1919	.2139	.2364	.2592	.2824	.3060
54	.0966	.1172	.1382	.1595	.1811	.2031	.2256	.2484	.2716	.2952
55	.0858	.1064	.1274	.1487	.1703	.1923	.2148	.2376	.2608	.2844
56	.0750	.0956	.1166	.1379	.1595	.1815	.2039	.2268	.2499	.2735
57	.0642	.0848	.1058	.1271	.1487	.1706	.1931	.2159	.2391	.2627
58	.0534	.0740	.0950	.1163	.1379	.1598	.1823	.2051	.2283	.2519
59	.0426	.0632	.0842	.1055	.1271	.1490	.1715	.1943	.2175	.2412
60	.0318	.0524	.0734	.0947	.1163	.1382	.1607	.1835	.2067	.2303
61		.0416	.0626	.0839	.1054	.1274	.1499	.1727	.1959	.2195
62		.0308	.0518	.0731	.0946	.1166	.1391	.1619	.1851	.2088
63			.0410	.0623	.0838	.1058	.1283	.1511	.1742	.1979
64			.0302	.0515	.0730	.0950	.1175	.1403	.1635	.1871
65				.0407	.0622	.0842	.1067	.1295	.1527	.1763
66				.0299	.0514	.0734	.0959	.1187	.1419	.1655
67					.0406	.0626	.0851	.1079	.1311	.1547
68					.0298	.0518	.0743	.0971	.1203	.1439
69						.0410	.0635	.0863	.1095	.1331
70						.0302	.0527	.0755	.0987	.1223
71							.0419	.0647	.0879	.1115
72							.0311	.0539	.0770	.1006
73								.0430	.0662	.0898
74								.0323	.0554	.0790
75									.0446	.0682
76									.0338	.0574
77										.0466
78										.0358
79										
80										
81										
82										
83										
84										
85										
86										
87										
88										
89										
90										

Vapor Pressures from Dry and Wet Bulb Temperatures (continued)

Dry Bulb Temp., °F.	Wet Bulb Temperature, °F.									
	50	51	52	53	54	55	56	57	58	59
	Vapor Pressure, Inches of Mercury									
50	.3624									
51	.3516	.3761								
52	.3408	.3653	.3903							
53	.3300	.3545	.3795	.4049						
54	.3192	.3437	.3687	.3941	.4200					
55	.3084	.3329	.3579	.3833	.4092	.4356				
56	.2976	.3221	.3471	.3725	.3984	.4248	.4517			
57	.2867	.3113	.3363	.3617	.3876	.4140	.4409	.4684		
58	.2759	.3004	.3255	.3509	.3768	.4032	.4301	.4576	.4855	
59	.2651	.2897	.3146	.3401	.3660	.3924	.4193	.4468	.4747	.5032
60	.2543	.2788	.3038	.3293	.3552	.3816	.4085	.4360	.4639	.4924
61	.2435	.2681	.2930	.3184	.3444	.3708	.3977	.4252	.4531	.4816
62	.2327	.2572	.2822	.3076	.3335	.3600	.3869	.4144	.4423	.4708
63	.2219	.2464	.2714	.2968	.3227	.3492	.3761	.4036	.4315	.4600
64	.2111	.2356	.2606	.2860	.3119	.3383	.3653	.3928	.4207	.4492
65	.2003	.2248	.2498	.2752	.3011	.3275	.3545	.3820	.4099	.4384
66	.1895	.2140	.2390	.2644	.2903	.3167	.3436	.3712	.3991	.4276
67	.1787	.2032	.2282	.2536	.2795	.3059	.3328	.3604	.3883	.4168
68	.1679	.1924	.2174	.2428	.2687	.2951	.3220	.3496	.3775	.4060
69	.1571	.1816	.2066	.2320	.2579	.2843	.3112	.3387	.3667	.3952
70	.1463	.1708	.1958	.2212	.2471	.2735	.3004	.3279	.3559	.3844
71	.1355	.1600	.1850	.2104	.2363	.2627	.2896	.3171	.3451	.3736
72	.1246	.1491	.1742	.1996	.2255	.2519	.2788	.3063	.3343	.3628
73	.1138	.1383	.1633	.1888	.2147	.2411	.2680	.2955	.3235	.3520
74	.1030	.1275	.1525	.1780	.2039	.2303	.2572	.2847	.3126	.3412
75	.0922	.1167	.1417	.1671	.1931	.2195	.2464	.2739	.3018	.3304
76	.0814	.1059	.1309	.1563	.1823	.2087	.2356	.2631	.2910	.3196
77	.0706	.0951	.1201	.1455	.1714	.1979	.2248	.2523	.2802	.3088
78	.0598	.0843	.1093	.1347	.1606	.1871	.2140	.2415	.2694	.2980
79	.0490	.0735	.0985	.1239	.1498	.1763	.2032	.2307	.2586	.2872
80	.0382	.0627	.0877	.1131	.1390	.1654	.1924	.2199	.2478	.2764
81	.0274	.0519	.0769	.1023	.1282	.1546	.1816	.2091	.2370	.2656
82		.0411	.0661	.0915	.1174	.1438	.1708	.1983	.2262	.2548
83			.0553	.0807	.1066	.1330	.1600	.1875	.2154	.2440
84			.0445	.0699	.0958	.1222	.1491	.1767	.2046	.2332
85			.0337	.0591	.0850	.1114	.1383	.1659	.1938	.2224
86				.0483	.0742	.1006	.1275	.1551	.1830	.2115
87				.0375	.0634	.0898	.1167	.1443	.1722	.2007
88					.0526	.0790	.1059	.1335	.1614	.1899
89					.0418	.0682	.0951	.1227	.1506	.1791
90					.0310	.0574	.0843	.1119	.1398	.1683
91						.0466	.0735	.1011	.1290	.1575
92						.0358	.0627	.0903	.1182	.1467
93							.0519	.0794	.1074	.1359
94							.0411	.0686	.0966	.1251
95							.0303*	.0578	.0858	.1143
96								.0470	.0750	.1035
97								.0362	.0642	.0927
98									.0534	.0819
99									.0426	.0711
100									.0318	.0603
101										.0495
102										.0387
103										.0279

Vapor Pressures from Dry and Wet Bulb Temperatures (continued)

Dry Bulb Temp., °F.	Wet Bulb Temperature, °F.									
	60	61	62	63	64	65	66	67	68	69
	Vapor Pressure, Inches of Mercury									
60	.5214									
61	.5106	.5403								
62	.4998	.5295	.5597							
63	.4890	.5187	.5489	.5798						
64	.4782	.5079	.5381	.5690	.6005					
65	.4674	.4971	.5273	.5582	.5897	.6218				
66	.4566	.4863	.5165	.5474	.5789	.6110	.6438			
67	.4458	.4755	.5057	.5366	.5681	.6002	.6330	.6664		
68	.4350	.4647	.4949	.5258	.5573	.5894	.6222	.6556	.6898	
69	.4242	.4539	.4841	.5150	.5465	.5786	.6114	.6448	.6790	.7139
70	.4134	.4431	.4733	.5042	.5357	.5679	.6007	.6341	.6682	.7031
71	.4026	.4323	.4625	.4934	.5250	.5571	.5899	.6233	.6575	.6923
72	.3918	.4215	.4517	.4827	.5142	.5463	.5791	.6125	.6467	.6816
73	.3810	.4107	.4409	.4719	.5034	.5355	.5683	.6017	.6359	.6708
74	.3702	.3999	.4301	.4611	.4926	.5247	.5575	.5909	.6251	.6600
75	.3594	.3891	.4193	.4503	.4818	.5139	.5467	.5801	.6143	.6492
76	.3486	.3783	.4085	.4395	.4710	.5031	.5359	.5694	.6036	.6385
77	.3378	.3675	.3978	.4287	.4602	.4923	.5252	.5586	.5928	.6277
78	.3270	.3567	.3870	.4179	.4494	.4815	.5144	.5478	.5820	.6169
79	.3162	.3459	.3762	.4071	.4386	.4708	.5036	.5370	.5712	.6061
80	.3054	.3351	.3654	.3963	.4278	.4600	.4928	.5262	.5604	.5954
81	.2946	.3243	.3546	.3855	.4170	.4492	.4820	.5154	.5497	.5846
82	.2838	.3135	.3438	.3747	.4062	.4384	.4712	.5047	.5389	.5738
83	.2730	.3027	.3330	.3639	.3955	.4276	.4604	.4939	.5281	.5630
84	.2622	.2919	.3222	.3531	.3847	.4168	.4496	.4831	.5173	.5523
85	.2514	.2811	.3114	.3423	.3739	.4060	.4389	.4723	.5065	.5415
86	.2406	.2703	.3006	.3315	.3631	.3952	.4281	.4615	.4958	.5307
87	.2298	.2595	.2898	.3207	.3523	.3844	.4173	.4508	.4850	.5199
88	.2190	.2487	.2790	.3099	.3415	.3736	.4065	.4400	.4742	.5092
89	.2082	.2379	.2682	.2991	.3307	.3629	.3957	.4292	.4634	.4984
90	.1974	.2271	.2574	.2884	.3199	.3521	.3849	.4184	.4526	.4876
91	.1866	.2163	.2466	.2776	.3091	.3413	.3741	.4076	.4419	.4768
92	.1758	.2055	.2358	.2668	.2983	.3305	.3634	.3968	.4311	.4661
93	.1650	.1947	.2250	.2560	.2875	.3197	.3526	.3860	.4203	.4553
94	.1542	.1839	.2142	.2452	.2767	.3089	.3418	.3753	.4095	.4445
95	.1434	.1731	.2034	.2344	.2660	.2981	.3310	.3645	.3987	.4337
96	.1326	.1624	.1926	.2236	.2552	.2873	.3202	.3537	.3880	.4229
97	.1218	.1516	.1818	.2128	.2444	.2765	.3094	.3429	.3772	.4122
98	.1110	.1408	.1710	.2020	.2336	.2658	.2986	.3321	.3664	.4014
99	.1002	.1300	.1602	.1912	.2228	.2550	.2879	.3213	.3556	.3906
100	.0894	.1192	.1494	.1804	.2120	.2442	.2771	.3106	.3449	.3798
101	.0786	.1084	.1386	.1696	.2012	.2334	.2663	.2998	.3341	.3691
102	.0678	.0976	.1278	.1588	.1904	.2226	.2555	.2890	.3233	.3583
103	.0570	.0868	.1170	.1480	.1796	.2118	.2447	.2782	.3125	.3475
104	.0462	.0760	.1062	.1372	.1688	.2010	.2339	.2674	.3017	.3367
105	.0354	.0652	.0954	.1264	.1580	.1902	.2231	.2566	.2910	.3260
106		.0544	.0846	.1156	.1472	.1794	.2123	.2459	.2802	.3152
107		.0436	.0739	.1048	.1364	.1687	.2016	.2351	.2694	.3044
108		.0328	.0631	.0941	.1257	.1579	.1908	.2243	.2586	.2936
109			.0523	.0833	.1149	.1471	.1800	.2135	.2478	.2829
110			.0415	.0725	.1041	.1363	.1692	.2027	.2371	.2721

Vapor Pressures from Dry and Wet Bulb Temperatures (continued)

Dry Bulb Temp., °F.	Wet Bulb Temperature, °F.									
	70	71	72	73	74	75	76	77	78	79
	Vapor Pressure, Inches of Mercury									
70	.7386									
71	.7278	.7642								
72	.7171	.7534	.7906							
73	.7063	.7427	.7798	.8177						
74	.6955	.7319	.7691	.8069	.8456					
75	.6847	.7211	.7583	.7962	.8348	.8744				
76	.6740	.7104	.7475	.7854	.8241	.8637	.9040			
77	.6632	.6996	.7368	.7747	.8133	.8529	.8933	.9345		
78	.6524	.6888	.7260	.7639	.8026	.8422	.8825	.9238	.9658	
79	.6417	.6781	.7153	.7532	.7918	.8314	.8718	.9130	.9551	.9981
80	.6309	.6673	.7045	.7424	.7811	.8207	.8610	.9023	.9443	.9874
81	.6201	.6565	.6937	.7316	.7703	.8099	.8503	.8916	.9336	.9767
82	.6093	.6458	.6830	.7209	.7596	.7991	.8395	.8808	.9229	.9659
83	.5986	.6350	.6722	.7101	.7488	.7884	.8288	.8701	.9122	.9552
84	.5878	.6242	.6614	.6994	.7381	.7777	.8181	.8593	.9014	.9445
85	.5770	.6135	.6507	.6886	.7273	.7669	.8073	.8486	.8907	.9338
86	.5662	.6027	.6399	.6778	.7166	.7562	.7966	.8379	.8800	.9230
87	.5555	.5919	.6292	.6671	.7058	.7454	.7858	.8271	.8692	.9123
88	.5447	.5811	.6184	.6563	.6951	.7347	.7751	.8164	.8585	.9016
89	.5339	.5704	.6076	.6456	.6843	.7239	.7644	.8057	.8478	.8909
90	.5232	.5596	.5969	.6348	.6735	.7132	.7536	.7949	.8370	.8802
91	.5124	.5488	.5861	.6241	.6628	.7024	.7429	.7842	.8263	.8694
92	.5016	.5381	.5753	.6133	.6520	.6917	.7321	.7735	.8156	.8587
93	.4908	.5273	.5646	.6025	.6413	.6809	.7214	.7627	.8049	.8480
94	.4801	.5165	.5538	.5918	.6305	.6702	.7107	.7520	.7941	.8373
95	.4693	.5058	.5430	.5810	.6198	.6594	.6999	.7413	.7834	.8266
96	.4585	.4950	.5323	.5703	.6090	.6487	.6892	.7305	.7727	.8158
97	.4478	.4842	.5215	.5595	.5983	.6380	.6784	.7198	.7619	.8051
98	.4370	.4735	.5108	.5487	.5875	.6272	.6677	.7090	.7512	.7944
99	.4262	.4627	.5000	.5380	.5768	.6165	.6569	.6983	.7405	.7836
100	.4154	.4519	.4892	.5272	.5660	.6057	.6462	.6876	.7298	.7729
101	.4047	.4412	.4785	.5165	.5553	.5950	.6355	.6768	.7190	.7622
102	.3939	.4304	.4677	.5057	.5446	.5842	.6247	.6661	.7083	.7515
103	.3831	.4196	.4569	.4949	.5338	.5735	.6140	.6554	.6976	.7408
104	.3724	.4089	.4462	.4842	.5230	.5627	.6032	.6446	.6868	.7300
105	.3616	.3981	.4354	.4734	.5122	.5520	.5925	.6339	.6761	.7193
106	.3508	.3874	.4247	.4627	.5015	.5412	.5817	.6232	.6654	.7086
107	.3400	.3766	.4139	.4519	.4907	.5305	.5710	.6124	.6546	.6979
108	.3293	.3659	.4031	.4412	.4800	.5197	.5603	.6017	.6439	.6871
109	.3185	.3550	.3924	.4304	.4692	.5090	.5495	.5910	.6332	.6764
110	.3077	.3443	.3816	.4196	.4585	.4982	.5388	.5802	.6225	.6657

Vapor Pressures from Dry and Wet Bulb Temperatures (continued)

Dry Bulb Temp., °F.	Wet Bulb Temperature, °F.									
	80	81	82	83	84	85	86	87	88	89
	Vapor Pressure, Inches of Mercury									
80	1.031									
81	1.021	1.066								
82	1.010	1.055	1.101							
83	.9993	1.044	1.090	1.137						
84	.9885	1.034	1.079	1.126	1.174					
85	.9778	1.023	1.069	1.116	1.163	1.212				
86	.9671	1.012	1.058	1.105	1.153	1.201	1.251			
87	.9564	1.001	1.047	1.094	1.142	1.191	1.240	1.292		
88	.9457	.9906	1.037	1.084	1.131	1.180	1.230	1.281	1.334	
89	.9350	.9799	1.026	1.073	1.121	1.169	1.219	1.271	1.323	1.377
90	.9242	.9692	1.015	1.062	1.110	1.159	1.208	1.260	1.313	1.366
91	.9135	.9585	1.005	1.052	1.099	1.148	1.198	1.249	1.302	1.356
92	.9028	.9478	.9938	1.041	1.089	1.137	1.187	1.239	1.291	1.345
93	.8921	.9371	.9830	1.030	1.078	1.127	1.176	1.228	1.281	1.334
94	.8814	.9264	.9724	1.019	1.067	1.116	1.166	1.217	1.270	1.324
95	.8707	.9157	.9617	1.009	1.057	1.105	1.155	1.207	1.260	1.313
96	.8600	.9050	.9510	.9980	1.046	1.095	1.144	1.196	1.249	1.303
97	.8492	.8943	.9403	.9873	1.035	1.084	1.134	1.186	1.238	1.292
98	.8385	.8836	.9296	.9766	1.024	1.073	1.123	1.175	1.228	1.281
99	.8278	.8729	.9189	.9659	1.014	1.063	1.112	1.164	1.217	1.271
100	.8171	.8622	.9082	.9552	1.003	1.052	1.102	1.154	1.206	1.260
101	.8064	.8514	.8975	.9446	.9924	1.041	1.091	1.143	1.196	1.249
102	.7957	.8407	.8868	.9339	.9817	1.031	1.080	1.132	1.185	1.239
103	.7849	.8300	.8761	.9232	.9710	1.020	1.070	1.122	1.174	1.228
104	.7742	.8193	.8654	.9125	.9603	1.009	1.059	1.111	1.164	1.217
105	.7635	.8086	.8547	.9018	.9497	.9985	1.048	1.100	1.153	1.207
106	.7528	.7979	.8440	.8911	.9390	.9878	1.038	1.090	1.142	1.196
107	.7421	.7872	.8333	.8804	.9283	.9772	1.027	1.079	1.132	1.186
108	.7314	.7765	.8226	.8697	.9176	.9665	1.016	1.068	1.121	1.175
109	.7207	.7658	.8119	.8590	.9069	.9558	1.006	1.058	1.111	1.164
110	.7099	.7551	.8012	.8483	.8963	.9451	.9950	1.047	1.100	1.154

Vapor Pressures from Dry and Wet Bulb Temperatures (continued)

Dry Bulb Temp., °F.	Wet Bulb Temperature, °F.									
	90	91	92	93	94	95	96	97	98	99
	Vapor Pressure, Inches of Mercury									
90	1.421									
91	1.410	1.466								
92	1.400	1.455	1.512							
93	1.389	1.445	1.501	1.560						
94	1.379	1.434	1.491	1.549	1.609					
95	1.368	1.424	1.480	1.539	1.598	1.659				
96	1.357	1.413	1.470	1.528	1.588	1.648	1.710			
97	1.347	1.402	1.459	1.518	1.577	1.638	1.700	1.763		
98	1.336	1.392	1.448	1.507	1.567	1.627	1.689	1.753	1.818	
99	1.325	1.381	1.438	1.497	1.556	1.617	1.678	1.742	1.808	1.874
100	1.315	1.371	1.427	1.486	1.546	1.606	1.668	1.731	1.797	1.864
101	1.304	1.360	1.417	1.475	1.535	1.596	1.657	1.721	1.787	1.853
102	1.294	1.349	1.406	1.465	1.524	1.585	1.647	1.710	1.776	1.843
103	1.283	1.339	1.395	1.454	1.514	1.575	1.636	1.700	1.765	1.832
104	1.272	1.328	1.385	1.444	1.503	1.564	1.626	1.689	1.755	1.822
105	1.262	1.318	1.374	1.433	1.493	1.553	1.615	1.679	1.744	1.811
106	1.251	1.307	1.364	1.422	1.482	1.543	1.605	1.668	1.734	1.801
107	1.240	1.296	1.353	1.412	1.472	1.532	1.594	1.658	1.723	1.790
108	1.230	1.286	1.342	1.401	1.461	1.522	1.584	1.647	1.713	1.780
109	1.219	1.275	1.332	1.391	1.450	1.511	1.573	1.637	1.702	1.769
110	1.209	1.264	1.321	1.380	1.440	1.501	1.562	1.626	1.692	1.759

Dry Bulb Temp., °F.	Wet Bulb Temperature, °F.									
	100	101	102	103	104	105	106	107	108	109
	Vapor Pressure, Inches of Mercury									
100	1.931									
101	1.921	1.990								
102	1.910	1.979	2.051							
103	1.899	1.969	2.041	2.113						
104	1.889	1.959	2.030	2.103	2.176					
105	1.879	1.948	2.020	2.092	2.166	2.241				
106	1.868	1.938	2.009	2.082	2.155	2.231	2.308			
107	1.858	1.927	1.999	2.071	2.145	2.220	2.298	2.377		
108	1.847	1.917	1.988	2.061	2.134	2.210	2.287	2.367	2.448	
109	1.837	1.906	1.978	2.050	2.124	2.199	2.277	2.356	2.438	2.520
110	1.826	1.896	1.967	2.040	2.114	2.189	2.267	2.346	2.427	2.510

HEAT REMOVED PER POUND OF DRY AIR

Since the wet-bulb temperature is a measure of the total heat of a mixture of dry air and vapor, the amount of heat to be removed in cooling a mixture is the difference between the total heats at the initial and final wet bulb temperatures, as given in these tables. These tables are on the basis of a pound of dry air plus the moisture.

As an example of the use of the tables, how much heat must be removed from air at 90F dry bulb, 70F wet bulb, to cool it to 70F dry bulb and 45F wet bulb?

The initial wet bulb is 70F, final wet bulb 45F. Therefore, under 70F initial wet bulb and opposite 45F final wet bulb find the answer to be 16.39 Btu to be removed from each pound of dry air (plus moisture) to be cooled.

Final Wet Bulb Temp., F	Initial Wet Bulb Temperature, F															
	60	61	62	63	64	65	66	67	68	69	70	71	72	73	74	75
	Heat Removed per Lb. Dry Air, Btu															
40	11.20	11.89	12.59	13.31	14.04	14.79	15.56	16.34	17.14	17.96	18.80	19.66	20.54	21.43	22.36	23.30
41	10.74	11.43	12.13	12.85	13.58	14.33	15.10	15.88	16.68	17.50	18.34	19.20	20.08	20.97	21.90	22.84
42	10.26	10.95	11.65	12.37	13.10	13.85	14.62	15.40	16.20	17.02	17.86	18.72	19.60	20.49	21.42	22.36
43	9.78	10.47	11.17	11.89	12.62	13.37	14.14	14.92	15.72	16.54	17.38	18.24	19.12	20.01	20.94	21.88
44	9.29	9.98	10.68	11.40	12.13	12.88	13.65	14.43	15.23	16.05	16.89	17.75	18.63	19.52	20.45	21.39
45	8.79	9.48	10.18	10.90	11.63	12.38	13.15	13.93	14.73	15.55	16.39	17.25	18.13	19.02	19.95	20.89
46	8.28	8.97	9.67	10.39	11.12	11.87	12.64	13.42	14.22	15.04	15.88	16.74	17.62	18.51	19.44	20.38
47	7.76	8.45	9.15	9.87	10.60	11.35	12.12	12.90	13.70	14.52	15.36	16.22	17.10	17.99	18.92	19.86
48	7.23	7.92	8.62	9.34	10.07	10.82	11.59	12.37	13.17	13.99	14.83	15.69	16.57	17.46	18.39	19.33
49	6.69	7.38	8.08	8.80	9.53	10.28	11.05	11.83	12.63	13.45	14.29	15.15	16.03	16.92	17.85	18.79
50	6.14	6.83	7.53	8.25	8.98	9.73	10.50	11.28	12.08	12.90	13.74	14.60	15.48	16.37	17.30	18.24
51	5.58	6.27	6.97	7.69	8.42	9.17	9.94	10.72	11.52	12.34	13.18	14.04	14.92	15.81	16.74	17.68
52	5.01	5.70	6.40	7.12	7.85	8.60	9.37	10.15	10.95	11.77	12.61	13.47	14.35	15.24	16.17	17.11
53	4.43	5.12	5.82	6.54	7.27	8.02	8.79	9.57	10.37	11.19	12.03	12.89	13.77	14.66	15.59	16.53
54	3.83	4.52	5.22	5.94	6.67	7.42	8.19	8.97	9.77	10.59	11.43	12.29	13.17	14.06	14.99	15.93
55	3.23	3.92	4.62	5.34	6.07	6.82	7.59	8.37	9.17	9.99	10.83	11.69	12.57	13.46	14.39	15.33
56	2.61	3.30	4.00	4.72	5.45	6.20	6.97	7.75	8.55	9.37	10.21	11.07	11.95	12.84	13.77	14.71
57	1.98	2.67	3.37	4.09	4.82	5.57	6.34	7.12	7.92	8.74	9.58	10.44	11.32	12.21	13.14	14.08
58	1.33	2.02	2.72	3.44	4.17	4.92	5.69	6.47	7.27	8.09	8.93	9.79	10.67	11.56	12.49	13.43
59	.67	1.36	2.06	2.78	3.51	4.26	5.03	5.81	6.61	7.43	8.27	9.13	10.01	10.90	11.83	12.77
60		.69	1.39	2.11	2.84	3.59	4.36	5.14	5.94	6.76	7.60	8.46	9.34	10.23	11.16	12.10
61			.70	1.42	2.15	2.90	3.67	4.45	5.25	6.07	6.91	7.77	8.65	9.54	10.47	11.41
62				.72	1.45	2.20	2.97	3.75	4.55	5.37	6.21	7.07	7.95	8.84	9.97	10.71
63					.73	1.48	2.25	3.03	3.83	4.65	5.49	6.35	7.23	8.12	9.05	9.99
64						.75	1.52	2.30	3.10	3.92	4.76	5.62	6.50	7.39	8.32	9.26
65							.77	1.55	2.35	3.17	4.01	4.87	5.75	6.64	7.57	8.51
66								.78	1.58	2.40	3.24	4.10	4.98	5.87	6.80	7.74
67									.80	1.62	2.46	3.32	4.20	5.09	6.02	6.96
68										.82	1.66	2.52	3.40	4.29	5.22	6.16
69											.84	1.70	2.58	3.47	4.40	5.34
70												.86	1.74	2.63	3.56	4.50
71													.88	1.77	2.70	3.64
72														.89	1.82	2.76
73															.93	1.87
74																.94

HEAT REMOVED PER POUND OF DRY AIR

Final Wet Bulb Temp., F.	Initial Wet Bulb Temperature, F.														
	76	77	78	79	80	81	82	83	84	85	86	87	88	89	90
	Heat Removed per Lb. Dry Air, Btu														
40	24.26	25.25	26.26	27.30	28.36	29.44	30.56	31.70	32.87	34.07	35.30	36.56	37.85	39.18	40.54
41	23.80	24.79	25.80	26.84	27.90	28.98	30.10	31.24	32.41	33.61	34.84	36.10	37.39	38.72	40.08
42	23.32	24.31	25.32	26.36	27.42	28.50	29.62	30.76	31.93	33.13	34.36	35.62	36.91	38.24	39.60
43	22.84	23.83	24.84	25.88	26.94	28.02	29.14	30.28	31.45	32.65	33.88	35.14	36.43	37.76	39.12
44	22.35	23.34	24.35	25.39	26.45	27.53	28.65	29.79	30.96	32.16	33.39	34.65	35.94	37.27	38.63
45	21.85	22.84	23.85	24.89	25.95	27.03	28.15	29.29	30.46	31.66	32.89	34.15	35.44	36.77	38.13
46	21.34	22.33	23.34	24.38	25.44	26.52	27.64	28.78	29.95	31.15	32.38	33.64	34.93	36.26	37.62
47	20.82	21.81	22.82	23.86	24.92	26.00	27.12	28.26	29.43	30.63	31.86	33.12	34.41	35.74	37.10
48	20.29	21.28	22.29	23.33	24.39	25.47	26.59	27.73	28.90	30.10	31.33	32.59	33.88	35.21	36.57
49	19.75	20.74	21.75	22.79	23.85	24.93	26.05	27.19	28.36	29.56	30.79	32.05	33.34	34.67	36.03
50	19.20	20.19	21.20	22.24	23.30	24.38	25.50	26.64	27.81	29.01	30.24	31.50	32.79	34.12	35.48
51	18.64	19.63	20.64	21.68	22.74	23.82	24.94	26.08	27.25	28.45	29.68	30.94	32.23	33.56	34.92
52	18.07	19.06	20.07	21.11	22.17	23.25	24.37	25.51	26.68	27.88	29.11	30.37	31.66	32.99	34.35
53	17.49	18.48	19.49	20.53	21.59	22.67	23.79	24.93	26.10	27.30	28.53	29.79	31.08	32.41	33.77
54	16.89	17.88	18.89	19.93	20.99	22.07	23.19	24.33	25.50	26.70	27.93	29.19	30.48	31.81	33.17
55	16.29	17.28	18.29	19.33	20.39	21.47	22.59	23.73	24.90	26.10	27.33	28.59	29.88	31.21	32.57
56	15.67	16.66	17.67	18.71	19.77	20.85	21.97	23.11	24.28	25.48	26.71	27.97	29.26	30.59	31.95
57	15.04	16.03	17.04	18.08	19.14	20.22	21.34	22.48	23.65	24.85	26.08	27.34	28.63	29.96	31.32
58	14.39	15.38	16.39	17.43	18.49	19.57	20.69	21.83	23.00	24.20	25.43	26.69	27.98	29.31	30.67
59	13.73	14.72	15.73	16.77	17.83	18.91	20.03	21.17	22.34	23.54	24.77	26.03	27.32	28.65	30.01
60	13.06	14.05	15.06	16.10	17.16	18.24	19.36	20.50	21.67	22.87	24.10	25.36	26.65	27.98	29.34
61	12.37	13.36	14.37	15.41	16.47	17.55	18.67	19.81	20.98	22.18	23.41	24.67	25.96	27.29	28.65
62	11.67	12.66	13.67	14.71	15.77	16.85	17.97	19.11	20.28	21.48	22.71	23.97	25.26	26.59	27.95
63	10.95	11.94	12.95	13.99	15.05	16.13	17.25	18.39	19.56	20.76	21.99	23.25	24.54	25.87	27.23
64	10.22	11.21	12.22	13.26	14.32	15.40	16.52	17.66	18.83	20.03	21.26	22.52	23.81	25.14	26.50
65	9.47	10.46	11.47	12.51	13.57	14.65	15.77	16.91	18.08	19.28	20.51	21.77	23.06	24.39	25.75
66	8.70	9.69	10.70	11.74	12.80	13.88	15.00	16.14	17.31	18.51	19.74	21.00	22.29	23.62	24.98
67	7.92	8.91	9.92	10.96	12.02	13.10	14.22	15.36	16.53	17.73	18.96	20.22	21.51	22.84	24.20
68	7.12	8.11	9.12	10.16	11.22	12.30	13.42	14.56	15.73	16.93	18.16	19.42	20.71	22.04	23.40
69	6.30	7.29	8.30	9.34	10.40	11.48	12.60	13.74	14.91	16.11	17.34	18.60	19.89	21.22	22.58
70	5.46	6.45	7.46	8.50	9.56	10.64	11.76	12.90	14.07	15.27	16.50	17.76	19.05	20.38	21.74
71	4.60	5.59	6.60	7.64	8.70	9.78	10.90	12.04	13.21	14.41	15.64	16.90	18.19	19.52	20.88
72	3.72	4.71	5.72	6.76	7.82	8.90	10.02	11.16	12.33	13.53	14.76	16.02	17.31	18.64	20.00
73	2.83	3.82	4.83	5.87	6.93	8.01	9.13	10.27	11.44	12.64	13.87	15.13	16.42	17.75	19.11
74	1.90	2.89	3.90	4.94	6.00	7.08	8.20	9.34	10.51	11.71	12.94	14.20	15.49	16.82	18.18
75	.96	1.95	2.96	4.00	5.06	6.14	7.26	8.40	9.57	10.77	12.00	13.26	14.55	15.88	17.24
76		.99	2.00	3.04	4.10	5.18	6.30	7.44	8.61	9.81	11.04	12.30	13.59	14.92	16.28
77			1.01	2.05	3.11	4.19	5.31	6.45	7.62	8.82	10.05	11.31	12.60	13.93	15.29
78				1.04	2.10	3.18	4.30	5.44	6.61	7.81	9.04	10.30	11.59	12.92	14.28
79					1.06	2.14	3.26	4.40	5.57	6.77	8.00	9.26	10.55	11.88	13.24
80						1.08	2.20	3.34	4.51	5.71	6.94	8.20	9.49	10.82	12.18

HEAT REMOVED PER CUBIC FOOT OF AIR

The accompanying tables show the amount of heat removed from one cubic foot of air and vapor in cooling from one wet bulb temperature to a lower wet bulb temperature. The figures are based on the specific volume of the initial air-vapor mixture.

As an example of the use of the tables, how much heat must be removed from 60,000 cubic feet of air at 70F wet bulb to cool it to 45F wet bulb?

Under 70F and opposite 45F find 1.198 Btu to be removed from one cubic foot. The answer is thus 60,000 × 1.198 or 71,880 Btu to be removed from 1 cu ft of the mixture.

Final Wet Bulb Temp., F	Initial Wet Bulb Temperature, F															
	60	61	62	63	64	65	66	67	68	69	70	71	72	73	74	75
	Heat Removed per Cubic Foot of Air Being Cooled															
40	.841	.891	.940	.992	1.043	1.096	1.150	1.204	1.260	1.316	1.374	1.433	1.493	1.554	1.617	1.680
41	.807	.856	.906	.957	1.009	1.062	1.116	1.170	1.226	1.283	1.341	1.400	1.460	1.520	1.583	1.647
42	.771	.820	.870	.922	.973	1.026	1.080	1.135	1.191	1.248	1.306	1.365	1.425	1.486	1.549	1.612
43	.734	.784	.834	.886	.938	.991	1.045	1.100	1.155	1.212	1.270	1.330	1.390	1.451	1.514	1.578
44	.698	.748	.798	.849	.901	.954	1.009	1.063	1.120	1.176	1.235	1.294	1.354	1.415	1.479	1.542
45	.660	.710	.760	.812	.864	.918	.972	1.027	1.083	1.140	1.198	1.258	1.318	1.379	1.442	1.506
46	.622	.672	.722	.774	.826	.880	.934	.989	1.045	1.102	1.161	1.220	1.281	1.342	1.406	1.469
47	.583	.633	.684	.735	.788	.841	.896	.951	1.007	1.064	1.123	1.182	1.243	1.304	1.368	1.432
48	.543	.593	.644	.696	.748	.802	.857	.912	.968	1.025	1.084	1.144	1.205	1.266	1.330	1.394
49	.502	.553	.604	.656	.708	.762	.817	.872	.928	.986	1.045	1.104	1.165	1.227	1.291	1.355
50	.461	.512	.562	.615	.667	.721	.776	.831	.888	.946	1.004	1.064	1.125	1.187	1.251	1.315
51	.419	.470	.521	.573	.626	.679	.735	.790	.847	.905	.963	1.024	1.085	1.146	1.210	1.275
52	.376	.427	.478	.530	.583	.637	.692	.748	.805	.863	.922	.982	1.043	1.105	1.169	1.234
53	.333	.383	.435	.487	.540	.594	.650	.705	.762	.820	.879	.940	1.001	1.063	1.127	1.192
54	.288	.339	.390	.443	.496	.550	.605	.661	.718	.776	.836	.896	.957	1.019	1.084	1.149
55	.243	.294	.345	.398	.451	.505	.561	.617	.674	.732	.792	.852	.914	.976	1.040	1.105
56	.196	.247	.299	.352	.405	.459	.515	.571	.628	.687	.746	.807	.869	.931	.996	1.061
57	.149	.200	.252	.305	.358	.413	.469	.525	.582	.641	.700	.761	.823	.885	.950	1.015
58	.100	.151	.203	.256	.310	.365	.420	.477	.534	.593	.653	.714	.776	.838	.903	.968
59	.050	.102	.154	.207	.261	.316	.372	.428	.486	.545	.605	.666	.728	.790	.855	.921
60		.052	.104	.157	.211	.266	.322	.379	.437	.496	.556	.617	.679	.742	.807	.872
61			.052	.106	.160	.215	.271	.328	.386	.445	.505	.566	.629	.692	.757	.823
62				.054	.108	.163	.219	.276	.334	.394	.454	.515	.578	.641	.706	.772
63					.054	.110	.166	.223	.282	.341	.401	.463	.526	.589	.654	.720
64						.056	.112	.170	.228	.287	.348	.410	.473	.536	.602	.668
65							.057	.114	.173	.232	.293	.355	.418	.481	.547	.614
66								.057	.116	.176	.237	.299	.362	.426	.492	.558
67									.059	.119	.180	.242	.305	.369	.435	.502
68										.060	.121	.184	.247	.311	.377	.444
69											.061	.124	.188	.252	.318	.385
70												.063	.126	.191	.257	.324
71													.064	.128	.195	.262
72														.065	.132	.199
73															.067	.135
74																.068

HEAT REMOVED PER CUBIC FOOT OF AIR

Final Wet Bulb Temp., F.	Initial Wet Bulb Temperature, F.														
	76	77	78	79	80	81	82	83	84	85	86	87	88	89	90
	Heat Removed per Cubic Foot of Air Being Cooled														
40	1.744	1.810	1.878	1.946	2.016	2.087	2.161	2.232	2.307	2.385	2.464	2.545	2.623	2.707	2.793
41	1.711	1.777	1.845	1.914	1.984	2.055	2.128	2.199	2.275	2.353	2.432	2.513	2.591	2.676	2.762
42	1.677	1.743	1.810	1.879	1.950	2.021	2.094	2.166	2.241	2.319	2.398	2.479	2.558	2.642	2.728
43	1.642	1.709	1.776	1.845	1.915	1.987	2.060	2.132	2.208	2.286	2.365	2.446	2.525	2.609	2.695
44	1.607	1.673	1.741	1.810	1.881	1.952	2.026	2.097	2.173	2.251	2.331	2.412	2.491	2.575	2.662
45	1.571	1.638	1.705	1.775	1.845	1.916	1.990	2.062	2.138	2.216	2.296	2.377	2.456	2.541	2.627
46	1.534	1.601	1.669	1.738	1.809	1.880	1.954	2.026	2.102	2.181	2.260	2.341	2.421	2.506	2.592
47	1.497	1.564	1.632	1.701	1.772	1.843	1.917	1.990	2.066	2.144	2.224	2.305	2.385	2.470	2.556
48	1.459	1.526	1.594	1.663	1.734	1.806	1.880	1.952	2.029	2.107	2.187	2.268	2.348	2.433	2.520
49	1.420	1.487	1.555	1.625	1.696	1.768	1.842	1.914	1.991	2.069	2.149	2.231	2.310	2.396	2.482
50	1.380	1.448	1.516	1.607	1.657	1.729	1.803	1.875	1.952	2.031	2.111	2.192	2.272	2.358	2.445
51	1.340	1.407	1.476	1.546	1.617	1.689	1.763	1.836	1.913	1.992	2.072	2.153	2.234	2.319	2.406
52	1.299	1.367	1.435	1.505	1.576	1.648	1.723	1.796	1.873	1.952	2.032	2.114	2.194	2.280	2.367
53	1.258	1.325	1.394	1.464	1.535	1.607	1.682	1.755	1.832	1.911	1.991	2.073	2.154	2.236	2.327
54	1.214	1.282	1.351	1.421	1.492	1.565	1.640	1.713	1.790	1.869	1.950	1.962	2.112	2.198	2.285
55	1.171	1.239	1.308	1.378	1.450	1.522	1.597	1.671	1.748	1.827	1.908	1.990	2.071	2.157	2.244
56	1.127	1.195	1.263	1.334	1.406	1.478	1.553	1.627	1.704	1.784	1.864	1.947	2.028	2.114	2.201
57	1.081	1.149	1.218	1.289	1.361	1.434	1.509	1.583	1.660	1.740	1.820	1.903	1.984	2.070	2.158
58	1.035	1.103	1.172	1.243	1.315	1.388	1.463	1.537	1.615	1.694	1.775	1.858	1.939	2.025	2.113
59	.987	1.055	1.125	1.196	1.268	1.341	1.416	1.490	1.568	1.648	1.729	1.812	1.893	1.980	2.068
60	.939	1.007	1.076	1.149	1.220	1.293	1.369	1.443	1.521	1.601	1.682	1.765	1.847	1.933	2.022
61	.889	.958	1.027	1.099	1.171	1.244	1.320	1.395	1.473	1.553	1.634	1.717	1.799	1.886	1.974
62	.839	.908	.977	1.049	1.121	1.195	1.270	1.345	1.424	1.504	1.585	1.668	1.751	1.837	1.926
63	.787	.856	.926	.997	1.070	1.144	1.220	1.295	1.373	1.453	1.535	1.618	1.701	1.788	1.876
64	.735	.804	.874	.945	1.018	1.092	1.168	1.243	1.322	1.402	1.484	1.567	1.650	1.737	1.826
65	.681	.750	.820	.892	.965	1.039	1.115	1.190	1.269	1.350	1.432	1.515	1.598	1.685	1.774
66	.626	.695	.765	.837	.910	.984	1.061	1.136	1.215	1.296	1.378	1.462	1.545	1.632	1.721
67	.569	.639	.709	.781	.855	.929	1.005	1.081	1.160	1.241	1.323	1.407	1.491	1.578	1.667
68	.512	.581	.652	.724	.798	.872	.949	1.025	1.104	1.185	1.268	1.352	1.435	1.523	1.612
69	.453	.523	.593	.666	.739	.814	.891	.967	1.047	1.128	1.210	1.295	1.378	1.467	1.556
70	.393	.462	.533	.606	.680	.754	.831	.908	.988	1.069	1.152	1.236	1.320	1.408	1.498
71	.331	.401	.472	.545	.619	.693	.771	.848	.927	1.009	1.092	1.176	1.261	1.349	1.439
72	.267	.338	.409	.482	.556	.631	.708	.786	.866	.947	1.030	1.115	1.200	1.288	1.378
73	.203	.274	.345	.419	.493	.568	.645	.723	.803	.885	.968	1.053	1.138	1.227	1.317
74	.137	.207	.279	.352	.427	.502	.580	.658	.738	.820	.903	.988	1.073	1.162	1.253
75	.069	.140	.212	.285	.360	.435	.513	.591	.672	.754	.838	.923	1.008	1.097	1.188
76		.071	.143	.217	.292	.367	.445	.524	.604	.687	.771	.856	.942	1.031	1.122
77			.072	.146	.221	.297	.375	.454	.535	.617	.701	.787	.873	.963	1.053
78				.074	.149	.225	.304	.383	.464	.547	.631	.717	.803	.893	.984
79					.075	.152	.230	.310	.391	.474	.558	.644	.731	.821	.912
80						.077	.156	.235	.317	.400	.484	.571	.658	.748	.839

QUICK METHOD FOR FINDING AIR SENSIBLE HEAT GAIN OR LOSS

Often, when one has to determine the amount of sensible heat gain or loss for a quantity of air when it is changed from one temperature to another, one knows only the rate of air flow, the initial temperature, and the final temperature. A solution involving only these three factors would save considerable computation time, since most other methods require either a transfer of flow from a volume to a weight basis or an adjustment based on the absolute temperature of the air.

Sensible Heat Only

Tables 1 and 2 provide factors for calculating sensible heat change over the range of temperatures ordinarily encountered in heating, ventilating, and air conditioning work. *The factors do not apply to changes involving cooling below the dew point.* The errors that would result are of particular significance when the initial temperature is in the range covered by Table 2.

Factors apply to the formula:

$$H = (Q)(\Delta t)(F)$$

where H = heat gain or loss, Btuh,

Q = air flow, cfm,

Δt = air temperature change, deg F, and

F = factor from Tables 1 or 2, corresponding to the temperature at which Q is taken.

Values in Table 1 are based upon an average of dry air and saturated air. As air temperature rises, moisture content plays an increasingly important part in the amount of sensible heat required to change the temperature. For the range of temperatures of Table 1 the maximum error in computed sensible heat, which could arise from moisture content variations, is less than 1%.

Relative Humidity

For the range of temperatures in Table 2, variations in moisture content may be considered significant, depending upon the precision required. Thus, sensible heat change factors for air in increments of 20% relative humidity are given. Linear interpolation will give values at smaller increments.

Example

Assume that it is desired to size a coil for heating 1000 cfm of 50 F air to 100 F. From Table 1, $F = 1.12$; therefore,

$$H = (1000)(100 - 50)(1.12) = 56{,}000 \text{ Btuh}$$

On the other hand, if the 1000 cfm were the quantity of air desired at 100 F, then $F = 1.03$; and

$$H = (1000)(100 - 50)(1.03) = 51{,}500 \text{ Btuh}$$

Table 1—Factors for Air to 110 Deg F

Air Temperature, F	Factor, Btuh per (cfm) (deg F)	Air Temperature, F	Factor, Btuh per (cfm) (deg F)
—20	1.30	50	1.12
—10	1.27	60	1.10
0	1.24	70	1.08
10	1.22	80	1.06
20	1.19	90	1.05
30	1.17	100	1.03
40	1.15	110	1.01

Table 2—Factors for Air from 110 to 210 Deg F

Air Temperature, F	Percent Relative Humidity					
	0	20	40	60	80	100
	Factor, Btuh per (cfm) (deg F)					
110	1.005	1.008	1.011	1.014	1.017	1.020
120	0.988	0.992	0.996	1.000	1.004	1.009
130	0.971	0.977	0.983	0.988	0.994	0.999
140	0.955	0.962	0.970	0.978	0.986	0.994
150	0.941	0.951	0.961	0.971	0.981	0.991
160	0.926	0.939	0.952	0.965	0.979	0.992
170	0.911	0.929	0.946	0.964	0.982	0.999
180	0.898	0.920	0.943	0.966	0.988	1.011
190	0.884	0.913	0.943	0.972	1.002	1.031
200	0.871	0.909	0.948	0.986	1.024	1.062
210	0.859	0.908	0.956	1.005	1.053	1.102

NOMOGRAPH FOR FREE-CONVECTION HEAT TRANSFER WITH AIR

Many simplified expressions and charts are available for free-convection heat-transfer calculations, however, they are generally limited in either accuracy or scope. Simplified expressions which assume a mean temperature for application with any temperature difference can predict heat transfer coefficients which are in substantial error. Charts which are appropriate for different fluids are usually limited to a single heat-transfer region and to a particular surface shape and position. A group of generalized curves are presented in nomograph form, which permit rapid, accurate determination of the free-convection heat transfer coefficient for any surface in air, under heating or cooling conditions.

The curves (a) encompass both laminar and turbulent regions; (b) are accurate, in that ambient air temperature, air properties, and surface temperature are considered in both heating and cooling over a wide temperature range; and (c) cover all surface shapes and positions for which data are available.

Example 1. Determine the free-convection heat transfer coefficient for a 6.62-inch diameter horizontal pipe at 250 F (T_s) in air at 70 F (T_a).

1. From the *shape constant table* accompanying the nomograph, see that, for horizontal cylinders, characteristic surface dimension, L, is diameter, D.

$$L = D = 6.62/12 = 0.55 \text{ ft}$$

2. Mean temperature,
$$T_M = (T_a + T_s)/2 = (250 + 70)/2 = 160 \text{ F}$$

3. Since T_s is greater than T_a, the solid curves in the center (A) portion of the nomograph are used. Starting from the base scale with $T_M = 160$ F, proceed vertically to $T_a = 70$ F, interpolating between the 60 F and 75 F solid T_s curves; then read out on extreme left hand scale, $(a\Delta T)^{1/3} = 530$.

4. Multiply $(a\Delta T)^{1/3}$, from Step 3, by L, from Step 1: $530 \times 0.55 = 292$. Since this value falls within the 10-1000 range (laminar region), the B curves at the upper right apply.

5. Move horizontally and right from the location determined in Step 3 (530 in the A curves) until $T_M = 160$ F is reached, then down to the D curves until $L = 0.55$ ft is reached, and, finally, horizontally to the right and read out $h/C = 2.02$.

6. From the shape constant table, $C = 0.45$ for a horizontal pipe, therefore,

$$h = (h/C) \, C = 2.02 \times 0.45$$
$$= 0.91 \text{ Btu per (hr)(sq ft)(deg F)}$$

Nomenclature

C = shape and position constant
h = free-convection heat-transfer coefficient, Btu per (hr) (sq ft) (deg F)
L = characteristic surface dimension, ft
P_x = ambient air pressure, psia
T_a = ambient air temperature, F
T_M = mean temperature, F = $(T_a + T_s)/2$
T_s = surface temperature, F
ΔT = temperature difference, F = $T_s - T_a$ or $T_a - T_s$, as appropriate

Example 2. Determine the free-convection heat-transfer coefficient for the pipe in Example 1 if the *pipe* temperature is 70 F and the *air* temperature is 250 F.

$$L = 0.55 \text{ ft}; \ T_a = 250 \text{ F}; \ T_s = 70 \text{ F}$$
$$T_M = (T_a + T_s)/2$$
$$= (250 + 70)/2 = 160 \text{ F}$$

Since T_s is less than T_a, the dashed A curves are used. With $T_M = 160$ F, a value of 530 on the extreme left hand scale is obtained from the curves. This is identical to the value obtained in Example 1. This follows, as T_M and ΔT are identical in both cases. The balance of the steps, and the resultant h are also the same.

Example 3. Determine the free-convection heat-transfer coefficient for Example 1 if the pipe diameter is increased to 24 inches ($L = 2$ ft).

1. For $T_M = 160$ F and $T_a = 70$ F, the solid A curves ($T_s > T_a$) give a value of 530 for $(a\Delta T)^{1/3}$.

$$(a\Delta T)^{1/3} L = 530 \times 2 = 1060$$

Since this value is larger than 1000, indicating turbulent flow, the C curves at the left apply.

2. Move horizontally to the C curve until (interpolating) $T_M = 160$ F is reached. Then move vertically downward and read out $h/C = 8.3$.

3. From the shape constant table, $C = 0.11$.

$$h = (h/C) \, C = 8.3 \times 0.11$$
$$= 0.91 \text{ Btu per (hr) (sq ft) (deg F)}$$

Total Must Include Radiation

Since, with free-convection heat-transfer, radiation is significant even at low-temperature levels, it must not be neglected. Total heat transfer is the sum of free convection and radiation heat transfer.

Nomograph for Free-Convection Heat Transfer with Air

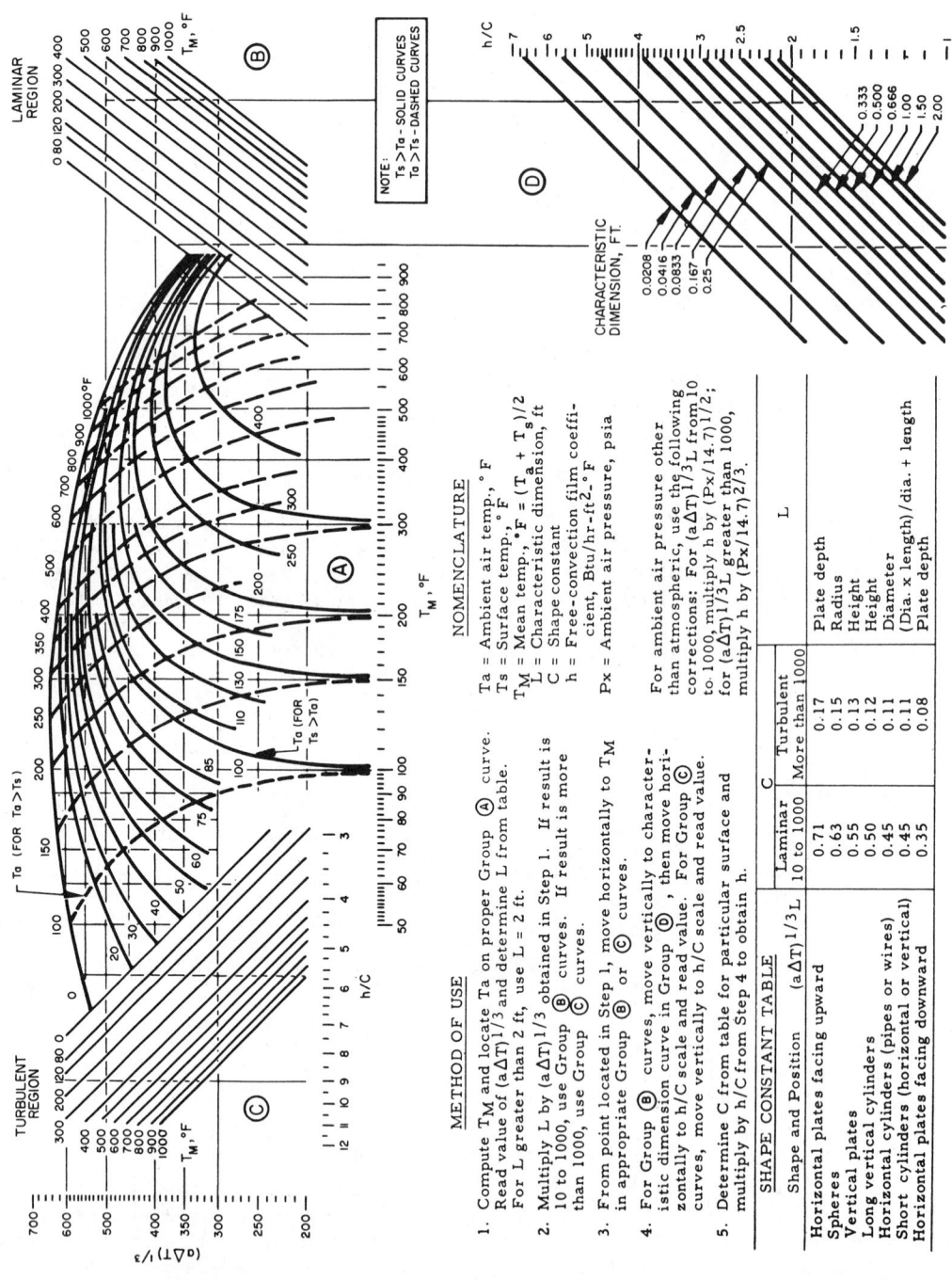

EFFECT OF AIR CONDITIONING ON OFFICE WORKERS

No better index of the effect of air conditioning on office workers has been provided than the results of a five-months study made public in 1959 by the Office of Buildings Management, General Services Administration of the U. S. Government.

The building in which the test was conducted is a wing-type structure, with the head house in the center. The wings were of open construction, without definite corridors. The west wing was selected as the test area, and the center wing as the control area. Both were principally occupied by file cabinets where the searches are performed.

All space was heated by direct radiators under the windows. Ventilation depended on opening the windows, with the usual disadvantages.

Identical painting (walls, a flat light green, ceilings, flat white) and lighting (semi-indirect fluorescent) were provided so that, with the exception of air conditioning in the test area, all employees would be working under the same conditions. The air conditioning system installed for the test area was operated to hold a uniform temperature of 75 F (± 2 deg), and no more than 50% rh.

Increase in Productivity

Records for the test period show that work production of the group in the air conditioned space varied from 6.5 to 13.5% higher than that of the employees in the control area. Weather Bureau data for the period furnished no definite clues to this varied production curve. Neither could the occupying agency pinpoint the reason for this variation.

Despite the "unknowns" in this study, the 9.5% average production increase in the air conditioned test area over the rate of production in the non-air conditioned control area is very significant and exceeded the agency's expectations. GSA knew, from previous studies, that a seven-minute saving in time per employe per day will more than pay for the expense of an air conditioning system. In other words, if work production increases only 1.5% per year, the system will pay for itself through savings in salaries.

During the test period the best work production in the non-air conditioned area occurred in June and September; in the air conditioned area production was highest in June, July and September. The low point in the air conditioned space was in August, and probably was related to the outside weather, (which was hot and humid), to which the employees were subject during off-duty hours.

There was no significant difference in the numbers of errors made in the air conditioned and non-air conditioned areas.

Decrease in Absenteeism

Disregarding absences due to sickness positively not associated in any way with air conditioning or the lack of it (fractured bones, operations, abscesses, skin rashes, bronchitis, etc.), leaving only minor ailments, such as headaches and upset stomache, to be considered, GSA suggests that no definite conclusions be drawn from this test, but pointed out that in the non-air conditioned area the absenteeism was 2.5% higher than in the air conditioned space. This is the more striking considering that the groups mixed with each other and strangers at lunch time, rest periods, in stores, on streets, and in public conveyances for the major portion of the day, and therefore were exposed to all kinds of diseases prevailing in the community.

Intangible Effects

All employees, with the exception of one girl, preferred to work in the air conditioned space and hoped they would not be transferred out of it.

The most important reasons for this preference, from their point of view, are as follows (no attempt has been made to list them in the order of their importance):

(1) Considerably less consumption of chilled drinking water.

(2) The girls were pleased with the noticeable saving in money on make-up and hairdos, attributed to the absence of electric fans and relief from the high humidity which prevails in Washington. It not only saved them money, it boosted their morale to know they still looked their best at the end of a busy day.

(3) Elimination of the nuisances associated with wall fans.

(4) Less soiling of clothing.

(5) The majority felt that the heat after working hours did not affect them as much as formerly.

It is safe to assume that in any other type of office work, where the rate of production is not standardized, or where no minimum requirement must be met consistently, the increase in productivity would be more than enough to pay for air conditioning through savings in employee salaries.

HEAT GAINS FROM APPLIANCES

In order to size cooling equipment correctly, it is necessary to determine the amount of heat added to an air-conditioned space by gas and electric appliances during the period of peak load from all sources. There are several methods for estimating peak appliance load, all of which require knowledge of the appliance nameplate rating (in Btu per hour) and factors affecting the rate at which nameplate energy enters the conditioned space at the design hour.

To assist in preparing estimates, the tables which follow have been prepared. In some cases nameplate ratings are given, while in others some of the factors for applying the data are presented.

The column marked "Use Factor" is a guide to (a) average nameplate ratings and (b) the actual heat load imposed on a space by an appliance in normally active service. Nameplate ratings are total loads divided by use factors. Note that the use factor does not take into account the actual time of peak service, which may vary from one appliance to another with relation to the peak meal-serving hour. For instance, the food preparation appliances may deliver their loads considerably in advance of food warmers and holders.

Before attempting to use the tables the estimator should be sure that he understands the nature of this information. The first point to be noted is that, while the ratings given represent typical appliances, they do not hold for all appliances since these devices have not been fully standardized. If accurate information is necessary concerning some particular appliance it must be found individually. However, many appliances are nearly enough standardized to make the table useful in practically all cases excepting a very few where extreme accuracy is desired.

The second point to be noted is that in the case of gas-fired appliances the data are based on Btu per hour contained in the gas fed to the appliances and not the Btu per hour added to the room air. The quantities also assume steady operation of the appliances with gas being fed at a constant rate for a full hour. Since gas appliances operate intermittently or with throttled gas feed oftener than they operate steadily, it is necessary to make allowances for this.

The third point is that the way in which the appliance is installed must also be considered. In some cases the flue gases are released directly into the room air, and in other cases they are vented to the outside. When appliances are both vented and hooded, tabulated values for gas appliances can be reduced by 50 to 60%. Hooding alone reduces the load by 10 to 40%.

The figures in the tables when modified by these factors permit finding the maximum amount of heat which a given appliance may add to the room air if operated steadily. What is necessary, however, is knowledge of the amount of heat which all of these appliances operating together do actually add at the time of the peak load.

There are two general methods of finding the required information. One is to go directly into the building or room and observe which appliances are operating there at various hours. Knowing the hour of the day when the system operates at peak it would be possible to determine how much energy is being consumed and from this calculate how much of an allowance to make in the estimate for cooling equipment capacity.

The second method makes use of data collected from meter readings taken at short intervals in the

Heat Gains from Steam Heated Restaurant Appliances
(Unhooded)[1]

Appliance	Lowest Working Pressure, psig	Pounds Steam per Hour	Heat Gain in Space, Btu per Hr	
			Latent	Sensible
Steam jacketed kettle	15	2.0 per gal contents	420 per gal contents	515
Steam table	5	1.6 per sq ft table surface	345 per sq ft surface	425
Bain Marie (vegetable & sauce warmer)	5	3.2 per sq ft table surface	690 per sq ft surface	845
Plate warmer	5	1.5 per cu ft of vol	325 per cu ft vol	395
Urn	15	3.0 per gal contents	710 per gal contents	710
Vegetable steamer	15	2.0 per gal contents	375 per gal contents	560
Egg boiler	15	5.0 per compartment	900 per compartment	1400
Dishwasher	5 to 15	.33 per person served	630 per person served	950
Traywasher	5	.166 per person served	300 per person served	500
Cupwasher	5	.083 per person served	150 per person served	250

[1] Hooding cuts sensible heat gain 50%, latent gain 40%

building to be cooled. These readings are then tabulated to show the ratio of the actual fuel consumptions for any particular time to the maximum possible consumption if all appliances were in operation. For instance, if the meter reading showed that during the peak hour only 40% as much energy had been used as if all the appliances had been operating steadily, then the ratio of the maximum meter reading to connected load is 0.40, and the diversity of the load is 0.4.

Heat Gains from Unhooded[1] Appliances

Appliance	Use[2] Factor	Heat Gain, Btu per Hour					
		Electric			Gas		
		Sensible	Latent	Total	Sensible	Latent	Total
Restaurant Appliances							
Coffee maker—							
½ gal brewer	.60	850	350	1200	1400	500	1900
½ gal warmer	1.00	200	100	300	400	150	550
2 gal urn	.25	2000	1420	3420	2200	2200	4400
3 gal urn	.25	2400	1600	4000	2800	2800	5600
5 gal urn	.30	3300	2200	5500	4000	4000	8000
10 gal urn	.30	5000	3500	8500	5000	5000	10000
Egg boiler, 2-cup	.50	1000	800	1800	1000	1000	2000
Water heater, 5 gal	1.00	4500	3700	8200	8000	7000	15000
10 gal	1.00	7500	6200	13700	8500	7500	16000
Food warmer, water pan, per burner	.50	400	500	900	700	800	1500
Fryer, deep, per lb fat	.47	140	220	360	310	110	420
Griddle, per sq ft	.35	1800	950	2750	3250	1750	5000
Grill, small	.60	5500	2500	8000	7000	3000	10000
large	.60	10000	5000	15000	14400	3600	18000
Hot plate, simmer	.33	1200	800	2000	1200	800	2000
full on	1.00	3000	3000	6000	4500	4500	9000
Ovens, per cu ft	1.00	3750	1250	5000	5600	2400	8000
Ranges, per burner	1.00	3500	3500	7000	4250	4250	8500
Toaster, continuous, 2 slices wide.	.64	5100	1300	6400	7700	3300	11000
Miscellaneous Appliances							
Domestic—							
Range, per burner	1.00	2400	2400	4800	4000	4000	8000
Toaster	.22	800	200	1000	1500	500	2000
Waffle iron	1.00	1800	1200	3000	2700	1300	4000
Coffee maker	.50	600	400	1000	800	700	1500
Flat iron	.50	700	800	1500	—	—	—
Beauty shop hair dryer—							
blower type	.50	2300	400	2700	—	—	—
helmet type	.57	1870	330	2200	3000	800	3800
Permanent wave machine	.20	850	150	1000	—	—	—

[1] Where appliances are hooded use 50% of tabulated values.
[2] Use factors have been applied to average nameplate ratings to bring tabulated values into line with usual expected loads in active service. Total load divided by use factor equals nameplate rating.

Heat Gain from Electric Motors

Nameplate Rating of Motor in Horsepower	Average Motor Efficiency in Continuous Operation	Btu per Hr to Room Air per Rated Hp. of Motor		
		Motor Outside of Room, Driven Device Inside Room	Motor in Room; Driven Device Outside of Room	Motor and Driven Device Both Inside Room
⅛ to ½	0.60	2,546	1,700	4,246
½ to 3	0.69	2,546	1,100	3,646
3 to 20	0.85	2,546	400	2,946

General Rule for Motors: if H_m = Btu per hr of motor input,

$$H_m = \frac{2,546 \times Hp. \text{ (connected load)}}{\text{Motor Efficiency}}$$

Note: Where possible obtain actual value of motor efficiency. Where not possible: for motors use average efficiencies as listed above; for motor generators use average efficiency sets up to 3 hp. as 0.55; for larger sets use average as listed above; for motor generators use average effciency sets up to 3 hp. as 0.55; for larger sets use average efficiency as 0.80.

Electric Motor-Driven Appliances

(Motor and Driven Appliance Both in Same Room)

Fans (Blade Diameters, Inches)		Btu per Hr	Appliances	Btu per Hr
Ceiling	32	340	Clock	7
	52	410	Hair dryer	1,900
	56	600	Drink mixer	240
Desk or Wall	8	120	Sewing machine (domestic)	220
	10	140	Vacuum cleaner (domestic)	250
	12	200	Hair clipper	78
	16	300	Vibrator (beauty)	11
Figures are thermal equivalent of nameplate rating corrected for motor efficiency.			Figures are thermal equivalent of nameplate ratings.	

Electric Refrigerators

With well insulated cabinet in 80F air and 40F inside cabinet

Electric Motor and Air Cooled Condenser in Cabinet in Room Air		Cold Cabinet in Room Compressor and Condenser Remote	
Cabinet Volume (Cu Ft)	Btu per Hr (Thermal Equivalent of Motor Input)	Cabinet Volume (Cu Ft)	Btu per Hr per Cu Ft of Cabinet Volume
2-4	530	2-4	100 to 75
5	710	5-6	70 to 65
6-10	850	7-10	60 to 55
12-18	1,060	12, 14, 16	55 to 50
		20, 25, 30	50, 45, 40

Miscellaneous Electric Appliances

Type of Appliance	Btu per Hr	Type of Appliance	Btu per Hr
Domestic electric irons		Radio (6 to 8 tubes)	340
small	—	Infra red lamp (heat)	850
medium	2,300	Sun lamp	1,350
large	3,400	Neon lights (15 mm), per ft	12
Tumbler water heaters	1,200	Neon lights (11 mm), per ft	18
Television set	680	Permanent waver (beauty)	4,240
Figures are thermal equivalent of nameplate rating of typical example of appliance listed.			

INFILTRATION LOADS IN SUMMER

All air that enters a cooled structure from the outside adds to the air conditioning load, and this part of the load must be calculated as carefully as possible.

Operating in favor of the air conditioning load is the fact that at design conditions in summer the wind velocity is usually low. Consequently, the quantity of air that enters through cracks around windows and doors is perhaps less than that entering in winter. The same data as used for heating can be used as a basis for calculating this part of the infiltration load.

At the other extreme is the open door, frequently encountered in air cooled stores where it is considered good business to make the entrance of the customer as easy as possible. If there is an appreciable breeze, the load thus caused can be practically impossible to estimate.

However, when the air is calm and a door is open on one outside wall, an infiltration of 800 cubic feet of air per minute can be assumed for the conventional three foot door.

The heat loss tables following, are useful for estimating the load placed on air cooling systems due to the infiltration of air through doors. Practically every time a door leading to an air conditioned space is opened, some outside air will be admitted, the amount depending upon the number of factors, such as wind velocity, size, location and type of doors, and the length of time the doors remain open.

In calculating the heat load due to infiltration through doors it is first necessary to select the type of application; that is, the type of store the installation is to be made in, as a bank, drug store or restaurant. Table 3 shows the number of entrance passages to be expected each hour for each occupant for various applications.

The amount of air entering each time the doors are opened depends on the type of door and whether there are doors in one wall or more than one wall. Tables 1 and 2 have been set up to give the amount of air per passage for single swing doors, swing doors with vestibules and revolving doors. Table 1 is for applications where the doors are in one wall only while Table 2 covers applications with doors in more than one wall.

The use of these tables may be explained by the following example: Suppose that we have a drug store which has an average occupancy of ten persons. Table 3 shows that there will be eight entrance passages per hour for each occupant in the drug store. Since there are ten occupants, the total number of passages per hour would be eighty. If this door is a single swinging door and there are entrances in one wall only, we would use Table 1 and in the line of 100 passages per hour and in the column of infiltration per passage for single swing doors, we would find 110 cu. ft. per passage. In other words, each person entering or leaving would allow 110 cu. ft. of air into the store. Since there are eighty passages per hour, the total infiltration will be 8,800 cu. ft. per hour, and the total heat gain per hour per degree temperature difference would be $1.98 \times 80 = 158.4$ Btu per hr.

In many stores there are exhaust fans to rid the space of objectionable odors, etc., and in such cases it is obvious that the air exhausted must be replaced by warm outside air, so that the capacity of such fans should be added to the infiltration through doors.

It should be kept in mind that if air enters the conditioned space from outside it is first cooled to a point near the temperature of the space; then, on entering the cooling coils, it is cooled to the temperature of air leaving the coils. The coils are calculated on the basis of a given quantity of air at a given temperature returning, so that the load from any ventilating or infiltration air is thus that due to cooling it to room conditions — namely, to the wet bulb temperature of the conditioned space.

Table 1. Infiltration through Entrances with Door in One Wall Only

No. of Passages per Hour, Up to	Single Swing Doors		Swing Doors-Vestibule		Revolving Doors†	
	Infiltration per Passage, Cu. Ft.	Heat Gain per Passage per Deg. Differ.	Infiltration per Passage, Cu. Ft.	Heat Gain per Passage per Deg. Differ.	Infiltration per Passage, Cu. Ft.	Heat Gain per Passage per Deg. Differ.
300	110	1.98	83	1.49	30	0.54
500	110	1.98	83	1.49	29	0.52
700	110	1.98	82	1.47	27	0.49
900	109	1.96	82	1.47	25	0.45
1100	109	1.96	82	1.47	23	0.41
1200	108	1.95	82	1.47	21	0.38
1300	108	1.95	82	1.47	19	0.34
1400	108	1.95	81	1.46	18	0.32
1500	108	1.95	81	1.46	17	0.31
1600	108	1.95	81	1.46	16	0.29
1700	107	1.93	81	1.46	15	0.27
1800	105	1.89	80	1.44	14	0.25
1900	104	1.87	80	1.44	13	0.23
2000	100	1.80	79	1.42	12	0.22
2100	96	1.73	79	1.42	11	0.20

Table 2. Infiltration through Entrances with Doors in More than One Wall

No. of Passages per Hour, Up to	Single Swing Doors		Swing Doors-Vestibule		Revolving Doors†	
	Infiltration per Passage, Cu. Ft.	Heat Gain per Passage per Deg. Differ.	Infiltration per Passage, Cu. Ft.	Heat Gain per Passage per Deg. Differ.	Infiltration per Passage, Cu. Ft.	Heat Gain per Passage per Deg. Differ.
300	168	3.02	125	2.25	48	0.86
500	168	3.02	125	2.25	46	0.83
700	168	3.02	125	2.25	43	0.78
900	168	3.02	125	2.25	39	0.70
1100	168	3.02	125	2.25	33	0.59
1200	168	3.02	125	2.25	30	0.54
1300	168	3.02	125	2.25	28	0.50
1400	168	3.02	125	2.25	25	0.45
1500	168	3.02	125	2.25	23	0.41
1600	167	3.01	125	2.25	21	0.38
1700	163	2.93	124	2.23	19	0.34
1800	159	2.86	122	2.20	18	0.32
1900	156	2.81	120	2.16	17	0.30
2000	152	2.74	118	2.12	16	0.29
2100	147	2.64	115	2.07	15	0.27

† Data are for 7 x 7 ft. door. For 6 x 7 ft. door, deduct 25%.

Table 3. Entrance Passages per Occupant per Hour

Banks	8	Dress shops	3	Office buildings	2
Barber shops	4	Drug stores	8	Professional offices	4
Brokers offices	8	Furriers	3	Public buildings	3
Candy and soda stores	6	Hospital rooms	4	Restaurants	3
Cigars and tobacco stores	25	Lunchrooms	6	Shoe stores	4
Department stores	8	Men's shops	4	Variety stores	12

HEAT GAINS FROM HUMAN BEINGS

Especially in spaces where a number of people gather, the heat given off by human beings is a factor which must be taken into account in the air conditioning estimate.

The human body, being a living machine, is obliged to expend a certain measure of energy simply to keep alive. The measure of this resting energy consumption is the measure of one's basal metabolism, which is simply a base line from which to determine the extra energy required in the performance of particular tasks calling for physical or mental effort. Although the basal metabolism of every individual is different, the heat production for the average man when lying quietly in bed before breakfast, is about one kilogram calorie per minute. This is equivalent to about 238 Btu per hr. Just sitting up increases the energy consumption, and consequently, the heat output, by 5%. Standing results in an increase of 10%, a brisk walk will increase the base figure by 200%, and work to the limit of human endurance will increase it 1000% or more.

It must be borne in mind that a heavy man gives off more heat in support of basal metabolism than a thin man of the same height. A tall man expends more than a short man of the same weight. A man of 25 years of age expends more than a man of 70.

The sex, too, is important in these calculations, since a man's basal metabolism runs about 10% higher than that of a woman, even though the weight, height, and age are the same. Consequently, in calculating the air conditioning load, consideration should be given to these factors, if the human load consists of predominantly one or the other of the above mentioned types rather than of average types.

The accompanying table gives the heat emission of human beings engaged in various occupations, as reported under different conditions.

In the table the classifications following the first cover a broad range. Sitting down, for example, may mean almost complete relaxation, or it may mean furiously typewriting; similarly brisk activity covers the whole range from a fairly rapid walk to boxing or swimming. Consequently, the range of figures is quite large.

Values in the first column labeled "post-absorptive stage," are from Carnegie Institution tests made on people 12 hours after the last meal; in other words, the heat emission is probably at its lowest.

The second column is based on reports of various investigations, including those of the Research Laboratory of the American Society of Heating and Air Conditioning Engineers, and covering the extreme high values in each category.

Tests by Carnegie Institution showed that mental effort, contrary to popular belief, does not consume energy rapidly and, consequently, the heat output is not increased greatly over the basal rate. The increase was on an average of about 4%.

Representative values of heat emission by male adults in an ambient temperature of 80 deg F may be used as a guide to selecting values from the table. A man seated at rest in a theater, for instance, may emit a total of 390 Btu per hr, of which about 40% is latent heat. Seated at moderately active office work he may give off 475 Btu per hr with 53% latent; walking slowly about as in shopping, 550 Btu, 55% latent; ball room dancing, 900 Btu, 70% latent; exercising actively, 1500 Btu, 72% latent, etc. A cooler ambient temperature will increase the proportion of sensible heat, about 10% for two degrees, 15% for 10 degrees, etc. Adult women produce about 10% less heat per pound than men, but average lighter in weight so that a reasonable allowance would be 15% less per person; children about 25% less.

Heat Emission from Human Beings

Activity	Average Emission, Btu per Hour per Person						
	Post Absorptive Stage		Possible Maximum		Average		
	Sensible	Latent	Sensible	Latent	Sensible	Latent	Total
Lying in bed	140	98	165	125	150	110	260
Sitting down	145	105	280	370	195	205	400
Standing	150	112	285	665	240	360	600
Brisk Activity	215	500	650	1350	390	910	1300
Limit of Endurance	650	1968	1350	3450	975	2525	3500

HEAT EMISSION OF ANIMALS

In connection with air conditioning animal houses in research laboratories, veterinary establishments, and zoos, it is frequently desirable to estimate the heat load of various animals. Data on this are available but widely scattered.

It was known for many years that metabolism was a function of the surface area of the body and several different methods were proposed for determining this area; from the area, the metabolism was calculated.

However, the surface area is related to body weight, so that, knowing the weight, surface area could be determined. This appeared to be needlessly roundabout to Max Kleiber, of the University of California, who, in 1932, studied the direct relationship between body weight and metabolism. He demonstrated that, with reasonable accuracy, the basal metabolism of warm-blooded animals, from ring doves and mice to steers, including man, can be expressed by the equation

$$H = 6.6W^{0.75}$$

where H is the heat in Btu per hour per animal, and W is the weight of the animal in pounds. The formula applies to man as well as other animals.

The accompanying graph is a plot of this formula, so that, knowing the weight of the animal in pounds, the value of H can be read directly.

Since the basal metabolism is that for a state of utter relaxation, the figures read from the chart must be increased in proportion to the degree of activity of the animal. A multiplying factor of 2 is suggested for animals moderately active.

There are other variables affecting metabolism in addition to degree of activity. For example, Benedict showed that young men have a higher metabolism than old men, the decrease per year being 0.4% of the metabolism at 30 years of age. Further, although somewhat open to question, it has been stated that, due to radiation, dark colored animals have a higher metabolism than those of light color. For estimating purposes, however, these points have little bearing on the air conditioning engineer's problem.

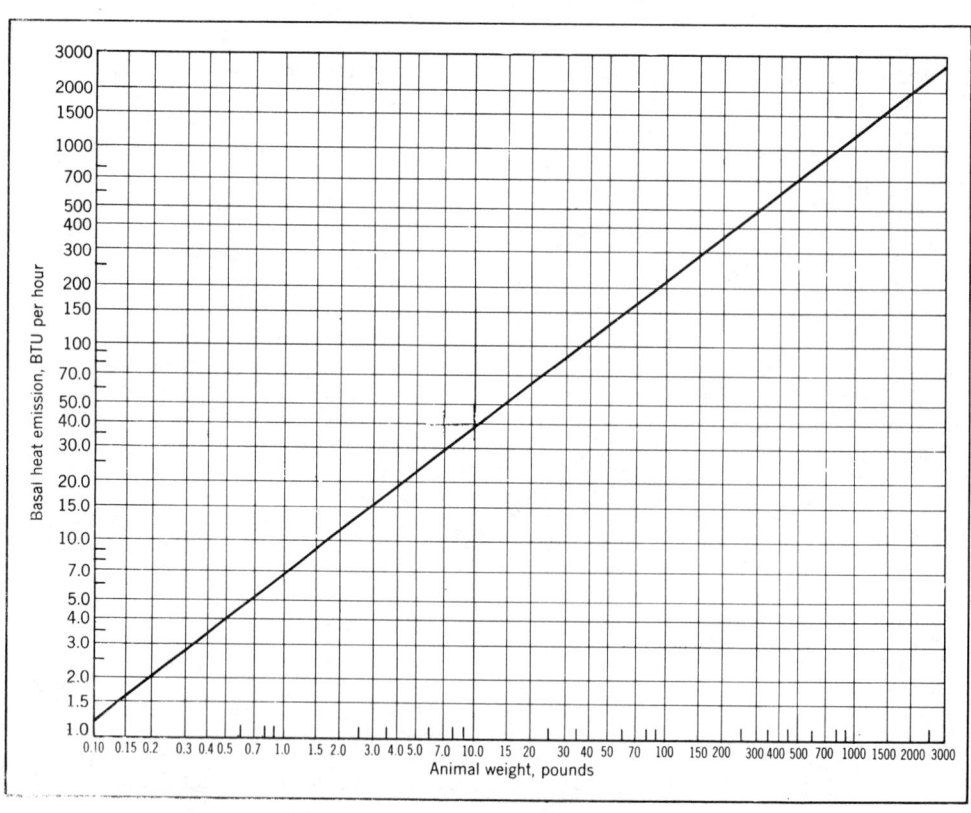

HEAT GAIN THROUGH UNINSULATED DUCTS

The quantity of air to be supplied through a register to a room is

$$Q = CH \qquad (1)$$

where Q is the volume of air leaving the register in cfm, H is the sensible heat gain of the room in Btu per hour, and C is a coefficient depending on the difference between the air entering the return grille and that entering through the inlet grille, and given in Table 1.

Table 1—Value of C for Use in Formula 1.

Temp. Difference, F (Return grille temp. minus inlet register temp.)	Coefficient C
7	0.132
8	0.115
9	0.102
10	0.0917
11	0.0832
12	0.0761
13	0.0701
14	0.0650
15	0.0605
16	0.0566
17	0.0532
18	0.0501
19	0.0474
20	0.0450
21	0.0427
22	0.0407
23	0.0389
24	0.0372
25	0.0356
26	0.0342
27	0.0329
28	0.0316
29	0.0305
30	0.0294
31	0.0284
32	0.0274
33	0.0266
34	0.0257
35	0.0249
36	0.0242
37	0.0235
38	0.0228
39	0.0222
40	0.0216

Example—The calculated sensible heat gain of a room is 4,500 Btu per hour, return register air 75 deg F, air supply at 65 deg. What conditioned air quantity is required?

Solution — The temperature difference is $(75-65) = 10$ deg. From Table 1 for 10 deg difference, $C = .0917$. Using Formula 1

$$Q = 4,500 \times .0917 = 413 \text{ cfm}$$

Four charts are presented as Figs. 1 to 4 covering air velocities of 600, 800, 1,000 and 1,200 fpm. All can be used for ambient air temperatures of 60 to 90 deg. The charts are based on a temperature difference of 25 deg between the temperature of the air entering the duct and the ambient air temperature. If the temperature is less than 25 deg, correction must be made as explained later. The charts apply to uninsulated galvanized steel ducts with an emissivity of 0.25.

Duct size is on the basis of equivalent diameter. For round ducts, duct diameter is numerically equal to equivalent diameter; for square ducts, the length of one side is numerically equal to equivalent diameter; for rectangular ducts, equivalent diameter $E = 2\,ab/(a+b)$ where E is the equivalent diameter, a the length of one side, b the other side, all in inches. Thus the equivalent diameter of an 8×10 inch duct is $(2 \times 8 \times 10) \div (8 + 10) = 8.9$ inches.

Use of Charts

If the temperature difference between air entering the duct and the ambient air is 25 deg: Find the equivalent diameter, enter base scale at length of duct, move upward to curve for equivalent diameter, thence left to temperature difference. Subtract from 25 to find the temperature rise. This, added to the entering air temperature gives the inlet register temperature.

Example—Air flows at 600 fpm through a 10×14 inch duct 200 ft long to a register. Air leaves the conditoner at 50F, air about the duct is 75F. What is the temperature rise in the duct?

Solution—For a 10×14 duct, $E = (2 \times 10 \times 14) \div (10 + 14) = 11.7$. The temperature difference is 25 and, since Fig. 1 applies to 600 fpm velocities, enter Fig. 1 at 200 ft, move vertically and interpolate between the 10 and 12 equivalent diameter curves, thence left and read 12.8F temperature difference. The temperature difference was 25, and $25 - 12.8 = 12.2$ temperature rise. The final air temperature is thus $50 + 12.2 = 62.2$ deg F.

If the temperature difference between air entering the duct and ambient air is less than 25: In this case, enter the chart at the required temperature difference, left scale, move right to the proper equivalent diameter curve, thence down and find the distance along the duct at which that temperature difference exists. Add to this, actual length of

duct, enter base scale at that point, up to diameter curve, and find at the left the temperature difference. Subtract from the required temperature difference to find temperature rise. Add to entering air temperature.

Example—Air enters a 60-ft length of 10-inch round duct at 55 deg and flows at 600 fpm with 75-deg ambient air. What is the inlet register temperature?

Solution—The equivalent diameter is 10 inches.

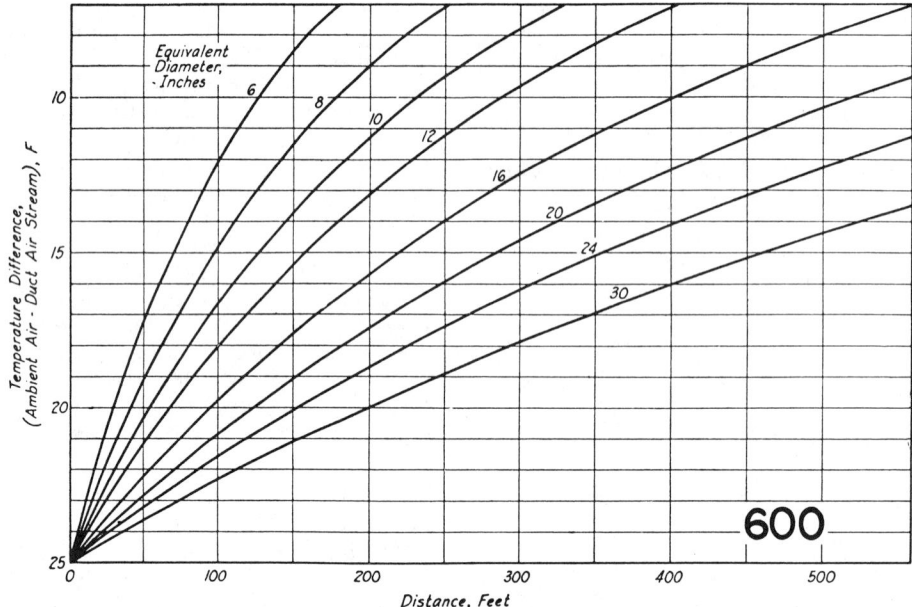

Fig. 1. Chart for 600 fpm air velocity.

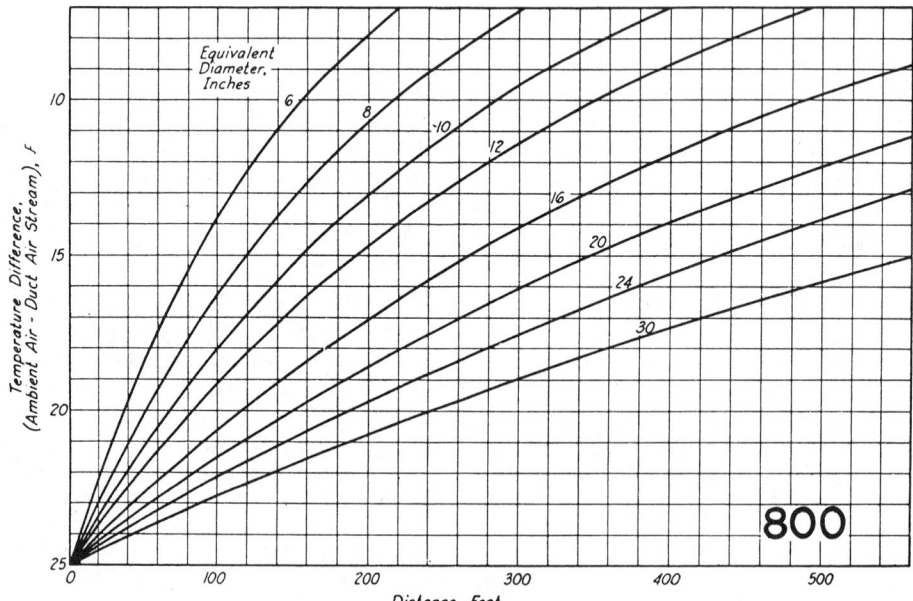

Fig. 2. Chart for 800 fpm air velocity.

The temperature difference is $75 - 55 = 20$. Enter Fig. 1 at 20 on the left scale, move horizontally to the right to the intersection with the 10 equivalent diameter curve, and find that a 20-deg temperature difference exists at 55 ft. Add length of duct, 60, to 55, and enter chart at 115 ft upward to the 10 inch diameter curve, thence left and find 15.6 deg. Then $20 - 15.6 = 4.4$F temperature rise, so that the air temperature at the end of the duct is $55 + 4.4 = 59.4$F.

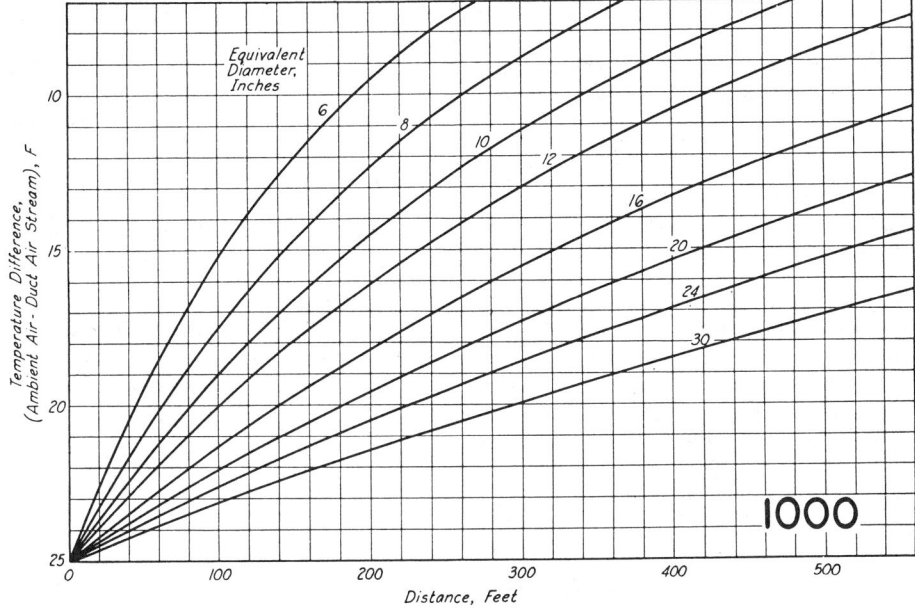

Fig. 3. Chart for 1000 fpm air velocity.

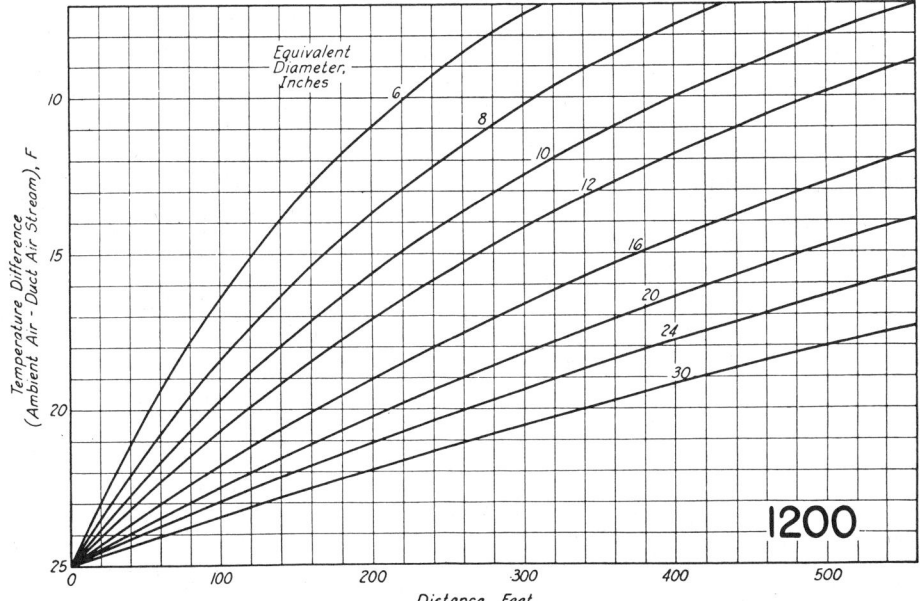

Fig. 4 Chart for 1200 fpm air velocity.

AIR CONDITIONING ESTIMATING DATA

It is frequently desirable to have at hand estimating data for use especially in working up preliminary figures. The following unit data on air conditioning were compiled by Gayle B. Priester, Air Conditioning Engineer, Consolidated Gas, Electric Light and Power Company of Baltimore.

From the large number of air conditioning installations that have already been made, sufficient operating data are available for arriving at general figures (Table 1) for classes of commercial installations. The information is not intended to replace the careful heat gain calculations necessary in a particular air conditioning design. These data, however, do enable an engineer to obtain a broad general picture with reference to the areas and number of persons generally served, the range of size of unit commonly installed for such service, and the required power consumption. These data can also be used as check figures to substantiate particular design calculations.

The information was obtained from analyses of actual load estimates and from other sources. In most cases, outside design conditions were 95F dry bulb and 75 or 78F wet bulb; inside design conditions were 76 to 80F dry bulb and 50% relative humidity.

When the watts per square foot are higher than average, or the square feet per person are lower than average, the total load usually is higher than average.

Since people, lights, and outside air usually account for the greatest part of the cooling load, except for living quarters or industrial applications, the data in Table 2 are given to show the important effect that a change in these items has on the total cooling load.

For the average application, approximately 1 hp per ton of refrigeration is required for the refrigerant compression equipment. When the predominant load is due to people, as in theaters and taverns, a lower refrigerant temperature is required to remove the high latent heat load caused by moisture from the occupants. Therefore, more than 1 hp per ton is required. Similarly, when a large part of the cooling load is sensible heat due to transmission, solar radiation, or lights, a higher refrigerant temperature can be used and less than 1 hp per ton is needed. The supply air fan power requirement is about 0.1 hp per ton of refrigeration.

Table 3 gives electric power consumption data

Table 1. Design Data for Commercial Air Conditioning

Application	Btu per Hr per Sq Ft			Persons per Ton			Cfm per Sq Ft			Sq Ft per Person			Watts per Sq Ft		
	Low	Avg.	High	Low	Avg.	High	Low	Avg.	High	Low	Avg.	High	Low	Avg.	High
Apartments and hotel guest rooms	13	20	30	1.4	1.7	2.2	0.5	0.7	0.9	100	175	325	0.2	0.6	0.9
Art museums, libraries	30	51	75	2.5	4.3	8.3	0.9	1.6	2.1	40	60	80	0.5	1.0	2.0
Bank areas	35	53	75	2.5	4.7	7.5	1.1	1.8	2.5	26	53	80	0.9	2.9	4.4
Barber shops	45	73	112	2.9	4.9	7.4	1.3	2.6	4.4	20	40	60	0.6	1.4	4.6
Beauty shops	49	75	114	2.5	4.3	7.1	1.6	2.3	3.0	17	42	75	2.7*	4.2*	9.3*
Clothing stores, children's	39	41	42	1.4	3.1	5.9	1.1	1.8	3.2	48	96	130	1.1	1.6	2.5
Clothing stores, men's	33	45	85	1.2	3.0	6.2	0.9	1.4	1.8	60	118	205	1.0	2.2	4.4
Clothing stores, women's	30	43	65	2.5	5.7	11.0	0.8	2.4	6.9	22	61	107	0.8	3.3	7.4
Clothing stores, general	29	44	68	3.2	5.2	7.0	0.9	1.4	2.1	27	65	111	1.5	2.2	3.5
Department store basements	20	30	39	6.2	8.0	15.0	0.5	0.8	1.2	20	30	95	0.8	2.4	3.9
Department stores, main floors	25	42	62	2.0	6.0	7.0	0.9	1.3	2.0	16	35	90	0.7	2.5	5.2
Doctors' and dentists' offices	33	51	68	1.3	4.0	7.0	1.2	1.7	2.4	29	75	160	1.4	1.7	3.4
Drug stores	35	70	109	1.3	4.5	6.9	1.1	1.9	3.4	17	39	92	0.2	1.6	3.9
Food stores	44	82	142	3.0	5.3	7.9	1.3	2.5	4.8	12	36	72	0.9	2.6	5.0
Offices, private and general	22	43	72	1.2	3.5	6.3	0.7	1.4	2.2	32	105	278	0.6	2.0	4.8
Restaurants	62	115	260	3.4	7.0	11.1	0.8	2.1	3.8	9	18	32	0.2	1.4	6.8
Specialty stores	22	52	179	1.1	3.1	5.5	0.8	1.9	5.9	20	90	192	0.9	3.9	12.9
Taverns and night clubs	25	80	165	6.6	8.6	10.7	0.8	1.4	2.8	8	18	75	0.2	1.1	2.2
Theaters	74	92	115	10.4	16.0	19.0	15.0†	20.0†	30.0†	6	8	12	0.1	0.3	0.8

* For lights and equipment. † Cfm dehumidified air per person.

Table 2. Cooling Load Calculations

Application or Item	Btu per hr = (1) × (2)	
	Quantity (1)	Factor (2)
Theaters	No. persons	350
Quiet offices or living quarters	No. persons	400
Active offices or retail stores	No. persons	450
Drug stores or restaurants	No. persons	500
Lights	Watts on in room	3.4
Outside air (at 95F db and 78F wb)	cfm	43

based on actual installations analyzed for 95F dry bulb and 78F wet bulb conditions at about 40 deg latitude and near sea level. For different conditions and length of cooling season, these figures would have to be adjusted accordingly. Spaces are left to insert comparable figures for a local territory. Equivalent full load operating hours are included.

In addition to electric power costs, there usually are water costs. If a water-cooled condenser is used, then from 1 to 2 gpm of water per ton of refrigeration are required. The colder the water, the less refrigeration tonnage is needed. The total water consumption cost for the season would be obtained by multiplying the total cooling load in tons by the operating hours from Table 3 times an average figure

of 1.5 gpm times 60 minutes per hour times the cost per gallon (7.5 gal = 1 cu ft). In other words,

$$\text{Average water cost} = \text{tons cooling load} \times \text{hours} \times 1.5 \times 60 \times \text{cost of water per gal}$$

or

$$= \text{tons} \times \text{hours} \times 12 \times \text{cost of water per cu ft}$$

Where restrictions on the use of water exist or where water is expensive, a water-cooling tower or an evaporative type refrigerant condenser should be used. These require about 10% more connected electric load and power consumption, but reduce the water consumption by at least 90%.

The installation cost of most air conditioning systems ranges from $300 to $1,000 per ton, the average cost varying for different applications and different localities. The cost per sq ft of floor area varies from about $1 to $4.

When air conditioning is not practical, some means of ventilation is necessary, particularly in the summer, to prevent the heat and odors in a space from accumulating and becoming objectionable, or to remove smoke and vapors.

Roughly, 1 hp per 9,000 to 18,000 cfm is required by a direct-driven propeller fan with free delivery. The actual value depends upon the fan design and operating speed. One hp per 3,000 to 6,000 cfm is required under reasonable conditions for a blower operating against 1/2 inch static pressure. Noise must be considered in many cases, and a large enough fan must be selected to deliver air quietly.

Table 3. Power Consumption Data for 2 Consecutive Years

Installation			Hp Range	Annual Kwhr per Hp			Equiv. Full Load Operating Hours	
No. Used	Class			Range	2-Yr Avg.	Local	Avg.	Local
9	Restaurants		12–36	850–1,870	1,425		1,620	
4	Cafeterias		11–21	770–1,365	1,124		1,270	
15	Drug stores		3–13	516–2,101	1,021		1,160	
5	Jewelry stores		5–9	565–1,180	840		950	
8	Night clubs		15–47	410–1,190	827		940	
9	Dress shops		5–45	354–1,085	770		870	
11	Residences		3–11	111–1,229	661		750	
11	Theaters		35–60	327–1,300	618		700	
4	Large offices		45–60	490–1,018	547		620	
7	Ten-cent stores		40–140	344–978	527		600	
6	Shoe stores		all 3.5	270–782	526		600	
7	Beauty shops		5–8	287–745	510		580	
4	Small offices		all 5.5	264–580	441		500	
6	Recreation spaces		25–65	276–651	421		480	
9	Funeral parlors		10–18	108–572	187		210	

NEMA ROOM AIR CONDITIONER COOLING LOAD FORM

(from Standard CN1, National Electric Manufacturers Association)

A. This cooling load estimate form is suitable for estimating the cooling load for comfort air-conditioning installations which do not require specific conditions of inside temperature and humidity.

B. The form is based on an outside design temperature of 95 F dry bulb and 75 F wet bulb. It can be used for areas in the continental United States having other outside design temperatures by applying a correction factor for the particular locality as determined from the map.

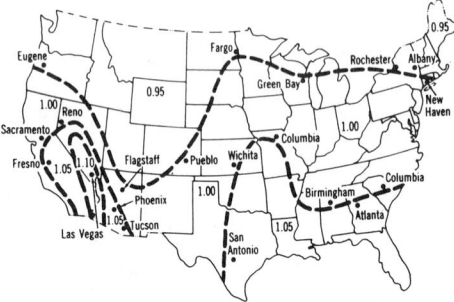

C. The form includes "day" factors for calculating cooling loads in rooms where daytime comfort is desired (such as living rooms, offices, etc.), as well as "night" factors for calculating cooling loads in rooms where only nighttime comfort is desired (such as bedrooms). "Night" factors should be used only for those applications where comfort air-conditioning is desired during the period from sunset to sunrise.

D. The numbers of the following paragraphs refer to the correspondingly numbered item on the form:

1. Multiply the square feet of window area for each exposure by the applicable factor. The window area is the area of the wall opening in which the window is installed. For windows shaded by inside shades or venetian blinds, use the factor for "Inside Shades." For windows shaded by outside awnings or by both outside awnings and inside shades (or venetian blinds), use the factor for "Outside Awnings." "Single Glass" includes all types of single-thickness windows, and "Double Glass" includes sealed airspace types, storm windows, and, glass block. Only one number should be entered in the right-hand column for item 1, and this number should represent *only the exposure with the largest load.*

2. Multiply the total square feet of *all* windows in the room by the applicable factor.

3a. Multiply the total length (linear feet) of all walls exposed to the outside by the applicable factor. Doors should be considered as being part of the wall. Outside walls facing due north should be calculated separately from outside walls facing other directions.

Walls which are permanently shaded by adjacent structures should be considered as being "North Exposure." Do not consider trees and shrubbery as providing permanent shading. An uninsulated frame wall or a masonry wall 8 inches or less in thickness is considered "Light Construction." An insulated frame wall or a masonry wall over 8 inches in thickness is considered "Heavy Construction."

3b. Multiply the total length (linear feet) of all inside walls between the space to be conditioned and any unconditioned spaces by the given factor. Do not include inside walls which separate other air-conditioned rooms.

4. Multiply the total square feet of roof or ceiling by the factor given for the type of construction most nearly describing the particular application. (Use one line only.)

5. Multiply the total square feet of floor area by the factor given. Disregard this item if the floor is directly on the ground or over a basement.

6. Multiply the number of people who normally occupy the space to be air-conditioned by the factor given. Use a minimum of 2 people.

7. Determine the total number of watts for lights and electrical equipment, except the air conditioner itself, that will be *in use* when the room air-conditioning is operating. Multiply the total wattage by the factor given.

8. Multiply the total width (linear feet) of any doors or arches which are continually open to an unconditioned space by the applicable factor.

NOTE—Where the width of the door or arches is more than 5 feet, the actual load may exceed the calculated value. In such cases, both adjoining rooms should be considered as a single large room, and the room air-conditioner unit or units should be selected according to a calculation made on this new basis.

9. Total the loads estimated for the foregoing 8 items.

10. Multiply the sub-total obtained in Item 9 by the proper correction factor, selected from the map, for the particular locality. The result is the total estimated design cooling load in Btu per hour.

E. For best result a room air-conditioner unit or units having a cooling capacity rating (determined in accordance with the NEMA Standards Publication for Room Air Conditioners, CN 1-1960) as close as possible to the estimated load should be selected. In general, a greatly oversized unit which would operate intermittently will be much less satisfactory than one which is slightly undersized and which would operate more nearly continuously.

F. Intermittent loads such as kitchen and laundry equipment are not included in this form.

NEMA ROOM AIR CONDITIONER COOLING LOAD FORM

(from Standard CN1, National Electric Manufacturers Association)

Heat Gain From	Quantity	Factors Night	Factors Day				Btu/Hr (Quantity x Factor)
			No. Shades*	Inside Shades*	Outside Awnings*	(Area x Factor)	
1. Windows: Heat gain from sun.							
Northeast	___ sq ft	0	60	25	20	___	___
East	___ sq ft	0	80	40	25	___	___
Southeast	___ sq ft	0	75	30	20	___	___
South	___ sq ft	0	75	35	20	___	___
Southwest	___ sq ft	0	110	45	30	___	___
West	___ sq ft	0	150	65	45	___	___
Northwest	___ sq ft	0	120	50	35	___	___
North	___ sq ft	0	0	0	0	___	___

(Use only the largest load)

2. Windows: Heat gain by conduction. (Total of all windows.)					
Single glass	___ sq ft	14	...... 14		___
Double glass or glass block	___ sq ft	7	...... 7		___

			Light Construction	Heavy Construction	
3. Walls: (Based on linear feet of wall.) a. Outside walls					
North exposure	___ ft	30	30	20	___
Other than north exposure	___ ft	30	60	30	___
b. Inside walls (between conditioned and unconditioned spaces only)	___ ft	30	30		___

4. Roof or Ceiling: (Use one only.)				
a. Roof, uninsulated	___ sq ft	5	19	___
b. Roof, 1 inch or more insulation	___ sq ft	3	8	___
c. Ceiling, occupied space above	___ sq ft	3	3	___
d. Ceiling, insulated with attic space above	___ sq ft	4	5	___
e. Ceiling, uninsulated, with attic space above	___ sq ft	7	12	___
5. Floor: (Disregard if floor is directly on ground or over basement.)	___ sq ft	3	3	___
6. Number of People:	___	600	600	___
7. Lights and Electrical Equipment in Use:	___ watts	3	3	___
8. Doors and Arches Continuously Open to Unconditioned Space: (Linear feet of width.)	___ ft	200	300	___
9. Sub-Total	x x x x x	x x x x x	x x x x x	___

10. Total Cooling Load: (Btu per hour to be used for selection of room air-conditioner(s).)____ (Item 9) X _____(Factor from Map) = _____

* These factors are for single glass only. For glass block, multiply the above factors by 0.5; for double-glass or storm windows, multiply the above factors by 0.8.

PRELIMINARY COOLING ESTIMATES

Approximate Heat Loads in Summer in Various Types of Small Buildings
(For rough estimates only)

PERSONAL SERVICE SHOPS

Floor Space: Allow one chair per 90 sq ft of floor space for barber shops; one chair per 100 sq ft for beauty shops.

Hair Dryers: Allow 500 Btu per hr for each hair dryer.

Hot Water: Allow 200 Btu per hr per chair sensible heat, and 400 Btu per hr per chair latent heat.

Heat Load per Operator's Chair, Btu per Hr

	10° Cooling	15° Cooling
Sensible heat	3000	3800
Latent heat	1000	1200
Total heat	4000	5000

SMALL PRIVATE OFFICES

Heat Load per Cu Ft of Cooled Space, Btu per Hr

S = Sensible Heat / T = Total Heat	10° Cooling		15° Cooling	
	S	T	S	T
Surrounding Space Uncooled				
Top Floor				
Two exposures	5¾	7	7	8¾
One exposure ....!	4½	5¾	5¾	7¼
One exposure, best conditions..	2½	3½	3½	4¾
Intermediate Floor				
Two exposures	3	4	4	5¼
One exposure	2½	3¼	3¼	4¼
One exposure, best conditions..	1¾	2½	2½	3½
Adjacent Floor Space Cooled				
Top Floor				
Two exposed walls	4¼	5½	5½	7
One exposed wall	3½	4½	4½	5½
One exposure, best conditions..	2¾	3½	3½	4½
Intermediate Floor				
Two exposed walls	2½	3¼	3¼	4½
One exposed wall	1¾	2½	2½	3½
One exposure, best conditions..	1½	2	2	2¾

RESTAURANTS

Floor Space: Allow 14 sq ft per customer chair.

No. of Occupants: Add employees to number of chairs.

Outside Air: Allow not less than 7 to 10 cfm per chair.

Hot Food: Allow 30 Btu sensible heat, and 30 Btu per hr latent heat, per meal served.

Coffee Urn: 1 burner, 10 gal., allow total heat of 12,000 Btu per hr., of which 5,000 sensible and 7,000 latent. If fully ventilated and exhausted allow 2,500 Btu per hr sensible, no latent.

Steam Table: 1 burner, allow 1,500 Btu per hr of which 750 sensible and 750 latent. If fully ventilated and exhausted, allow 200 Btu sensible, no latent.

Waffle Irons: Allow 75% of input rating. If fully ventilated allow 50% of input rating.

Heat Load per Chair, Btu per Hr

Sensible heat	900
Latent heat	300
Total heat	1200

ROOMS IN RESIDENCES

Rooms Exposed On Two Or Three Sides
Heat Load per Cu Ft Of Cooled Space, Btu per Hr

S = Sensible heat / T = Total heat	10° Cooling		15° Cooling	
	S	T	S	T
2nd floor room..	3	4	4	5½
1st floor room ..	2½	3½	3½	5
Same, with surrounding rooms also cooled ...	2	2¾	3	4

SMALL THEATERS

Floor Space: Allow 7½ sq ft per seat.

Outside Air: Allow not less than 7½ to 10 cfm of outside air per seat.

Heat Load per Seat. Btu per Hr

	10° Cooling	15° Cooling
Sensible heat	500	600
Total heat	800	900

AIR CONDITIONING DESIGN CONDITIONS—INDUSTRIAL PLANTS

Abrasives

In the manufacture of abrasive grinding wheels, an attempt is made to hold the room temperature to 78 deg F and 55% relative humidity.

Baking

Steam is delivered to bread ovens at 5 to 15 lb pressure. The air conditioning equipment installed not only filters but washes the air and this holds down the dust particles carrying mold spores. In some plants, cakes are passed under ultra-violet rays to sterilize the baking product from mold spores before wrapping. Table 1 presents temperatures and humidities used in various stages of baking and the storage of supplies.

Bakery Products

Heat generated in the dough mixers used in bakeries will cause burning unless the mixer is cooled by chilled air or brine at 33 to 38 deg.

In fermentation rooms, air conditions should be held at 78 to 80 deg and 75% relative humidity. Temperatures below 80 deg will reduce yeast cell development. It is important that a relative humidity of 75% be maintained to prevent absorption or release of moisture by the dough. High relative humidity will cause slimy dough and stickiness when handled. Air of low relative humidity will

Table 1. Air Conditions for Bakeries

Room or Process	Temperature, Deg F	Relative Humidity %
Storage		
Flour	70-75	65
Yeast	40-45	60
Milk and sugar	45-55	60
Shortening	50-55	60
Canned goods	40	50
Fresh fruits	35	60
Eggs	35	50
Waxed paper	50-60	—
Ingredient water	34-38	—
Bread dough in mixers	78-80	—
Fermentation room	78-82	75
Make-up room	82-84	65
Proof box	95-98	80-85
Bread cooling	80-85	75
Bread wrapping	65	65
Cake mixing	70-75	—
Cake mixing—		
sponge type	95-110	—
Bread ovens	375-450	—
Cake ovens	300-430	—

cause crustness, prevent the release of gas, and will form blow holes and lumps. Air must be so distributed to the fermentation room that drafts, cold spots and overheated areas are eliminated.

For proofing, the usual satisfactory condition is 90 deg and 85% relative humidity. Loss of product weight can be prevented by cooling for 80 to 120 minutes at a temperature of 70 to 75 deg and 65% relative humidity.

While best storage conditions for flour is 78 deg and 75% relative humidity, a humidity of 13% is best for the maturing or aging period. Malt, yeast and shortening is best stored at 40 deg, while the waxed paper for wrapping should be held in stock at 50-60 deg F.

Beer and Ale

One brewer blows air through coils of cold brine and the cooled air is then delivered to the front of aging and fermenting tanks to maintain temperatures as low as 32 deg. An air cooling system is used in preference to a system of circulating cooled brine to eliminate the inconvenience that results from frost accumulation on pipe and blower type evaporators. In areas where the aging and fermenting tanks are located, water is used for washing floors and tanks. As a result, with a direct expansion system of air conditioning, considerable frost would accumulate. The air cooling system succeeds in bringing down relative humidity although no attempt is made to control this moisture to a definite range.

Butter Processing

Desirable air conditions for butter processing ranges from a minimum of 65 deg and 35% relative humidity in winter to 75F and 50% relative humidity in summer.

These conditions are in the range of normal comfort for people working at machines and also within the range necessary to maintain the quality, body and texture of the product. Although it is possible to exceed 50% relative humidity, excessive condensation will result in some of the packaging machines, and a contaminating influence may be promoted.

Brewing

While the variables of physical location, equipment, raw materials and processes all affect the conditions in any one brewery, average figures for steps in brewing are given in Table 2, furnished by one authority.

Table 2. Air Conditions for Brewing

Room or Process	Temperature. Deg F	Relative Humidity. %
Fermentating cellar		
Lager	40-45	
Ale	55	
Storage cellar		75 min.*
Beer or ale	32-35	75 min.*
Racking cellar		
Keg filling and storage	32-35	75 min.†
Hop storage	30-32	
		75 min.
Liquid yeast storage	32-34	55-62
Grain storage—		70 max.
When storing		
for considerable period	80 or less	60 or less

* If wooden tankage is used, relative humidity must be 75% min.; otherwise humidity is maintained to prevent condensation on walls and ceilings.
† If wooden tankage is used.

Wort is cooled after removal from kettle in the coolship and open cooler room. Here the air may be heated to between 90 to 110 deg to reduce the relative humidity to promote evaporative cooling. Air to the fermenting cellar must be filtered if fermentation is carried on in open tanks. Filtered air is also supplied to the room for storing liquid yeast and to the coolship and open cooler room. It is not required for the other processes.

The following representative data on air conditions in various departments of a brewery were supplied by another authority. For fermenting cellars, temperature should be 38 to 50 deg and the relative humidity 60 to 85%; bacteriological control is desirable. For storage and finishing cellars, temperature should be 30-34 deg and relative humidity 60 to 85%; bacteriological control is desirable for rooms with open tanks. While of less importance in rooms with closed tanks, it is still favored as a means of controlling mold growths on ceilings and walls. Low humidity and dust control are important where malt and other grains are ground and where materials are handled.

Candy

Data on conditions maintained in confectionery plants are given in Tables 3 and 4—each based on data from a different source.

Dust control is employed for handling starch. Ventilation systems are installed to remove heat and vapor from many of the candy furnaces. The use of bacteriological control is increasing and some candy companies employ outside firms that specialize in such sanitation.

While Table 4 and information in the following paragraphs may vary with the preceding data, a study of the combined figures may help to arrive at suitable air conditioning design conditions.

This second candy authority states that jellies and gums are cast into starch molds and are dried in air conditioned hot rooms maintained at 120 to 140 deg F. While humidity is important (13-19% relative humidity for temperatures given) close control is not necessarily needed, since a large amount of personal experience governs the drying procedure.

Cooling tunnels are employed in the cooling of cast chocolate bars, coated centers, and hard candy. Basically, this tunnel is a long insulated box placed around a conveyor in such a manner that the product travels through it. Cold air is supplied at the end of the tunnel at which the product leaves so as to utilize the counterflow principle for the cold air. In general, air is supplied at velocities up to 1,500 fpm and at temperatures of 35 to 45 deg. The high air velocity creates a turbulent condition in the tunnel to promote better cooling. In chocolate enrobing, the aim is to secure an even chocolate with some luster. Since the best luster is

Table 3. Air Conditions for Candy Manufacture

Prescribed by Source A		
Room or Process	Temperature, Deg F	Relative Humidity, %
Chewing gum		
Manufacture	77	33
Rolling	68	63
Stripping	72	53
Breaking	74	47
Wrapping	74	58
Hard Candy		
After drops are formed	60	40-60
Cooling, mixing,		
packing	75-80	40-45
Chocolate		
Candy centers for		
coating	75-80	
Hand dipping	63	45
Packing	68	45
Finished chocolate		
storage	55	50
Nougat, fudge, caramel,		
toffee		
Panned specialty		
rooms	70-75	45
General candy storage	65	45
Nut Storage	50-55	45

Table 4. Air Conditions for Candy Manufacture

Prescribed for Source B		
Room or Process	Temperature, Deg F	Relative Humidity %
Hand dipping	60-65	50-55
Enrober room	75	60
Jellies and gums—drying	120-140	13-19
Cooling tunnel	35-45	
Packing		
Chocolate	65	55
Hard candy	70-75	35-45
Storage		
Finished candy	70	40-50
Nuts	30-32	75-80

obtained with rather slow cooling, a slow production rate must be followed or excessively long cooling tunnels used.

Although most finished goods will keep at conditions shown in Table 4, special attention must be given to nuts so that excessive dehydration of the nuts is avoided, mold growth is retarded, and insect eggs are kept dormant.

Starch or sugar dust is normally picked up by the return air of the cooling cycle and accumulates with condensed moisture on the cooling coils. This reduces the cooling capacity of the coil and provides a maintenance problem. Therefore, the use of filters is recommended for both outside and return air.

Ceramics

Comfort for the worker is the chief cause for temperature control in the ceramic industry, although exact temperature conditions are important where decalcomania are transferred onto whiteware.

In the refractories industry, temperature can range from a comfortable working temperature of about 65 deg to whatever the maximum summer temperature might be — 95 or 98 deg. Relative humidity does not have to be controlled for it has no effect upon the product. However, dust control is important and the dust count must be held down to 4 million particles per cubic foot due to the danger of silicosis. Since the material might contain over 9% free silica, the dust content must be controlled.

In the manufacture of whiteware, air temperature is not important in most departments except as it affects the comfort of the men. One of several spots in the whiteware plant where temperature must be controlled is the decorating shop.

Here, where paper is used to transfer patterns from a machine or decalcomania onto ware, reasonable temperature and humidity controls must be exercised. Otherwise the paper will contract or expand to interfere with color registration on the ware. Another critical spot where temperature and humidity control is mandatory is the decalcomania production room. Because temperature and moisture effect the size of the paper backing, it also governs registration of colors.

Conditions in the enamel industry are about the same as in the pottery field. Temperature is controlled only for the comfort of the men. Although relative humidity is an important factor in the firing, up to this time no one has done anything to control it. Dust control is necessary because of the inherent danger due to silicosis and because of the possible damage that dirt can do to porcelain enameled steel used in refrigerators and appliances. Much of the air in enamel plants is filtered before it enters the rooms where spraying and particular handling is done.

In glass plants conditions are about the same as in the whiteware plants except that some precautions are taken to control the heat in the machine room where glass bottles and ware are made. Generally this is done by blowers to establish proper air movement in the factory. The dust count must be held down due to the possibility of silicosis.

Chocolate

In producing fine ground chocolate, a supply of cooling water is important to counteract the heat produced by friction during grinding. For a 50 hp motor, about 20 gpm cooling water at 35 deg F is required in an 80 deg F room. Where the room temperature is above 80 deg F, add about 5 gpm to this amount. Water should not be colder than 55 deg F nor warmer than 75 deg F to keep the rolls at the proper temperature. Should the rolls be too cold, they will shrink in size so that some of the chocolate will pass through the rolls without refining. If the rolls run too warm, the metal will expand and the rolls will become too tight, resulting in crystallization of the sugar content, followed by scorching of the chocolate.

Chocolate Candy

For cooling chocolates, there should be little temperature differential between the temperature of the candy and the air in the cooling tunnel. If the temperature is held below 40 deg, the chocolate may become discolored due to moisture condensing from the warmer air. Air flow should be calculated to

prevent more than a 10 deg rise between the entering air and the outlet air. Best results are produced where the direction of air flow is opposite the candy travel; when the warmer air strikes the warmest candy as it enters the cooler, the finish of the chocolate is improved.

Cigars

Air conditions to be maintained in the manufacture of cigars are listed in Table 5.

Table 5. Air Conditions for Cigar Manufacture

Room or Process	Temperature, Deg F	Relative Humidity, %
Packing and shipping	76	65
Cigar making (machine)	76	73
Tobacco stripping (machine)	75	75
Wrapper tobacco conditioning and storage	75	75
Filler tobacco storage and preparation	78	70
Filler tobacco casing and conditioning	75	75

Cosmetics

Manufacturing throughout the cosmetic industry is done at room temperatures of 65 to 70 deg with no humidity control. Often the windows are open to outside air. Although there is some dust control where powders are made, the practice is for the cosmetic companies to buy powders from the few companies that specialize in making them. In most of the cosmetic plants there is no regulation of dust. Bacteriological control is obtained by germicides used in the manufacture and included in the products.

Crackers and Biscuits

One manufacturer of crackers and biscuits has found it helpful to keep the packaging department temperature at 70 to 72F, with the relative humidity from 50 to 53%. Not only do these air conditions maintain the products in a fresh and palatable condition, but they produce comfort for the workers.

Distilling

In the aging of whiskey in warehouses, the whiskey is stored in 50-gal charred oak barrels, in rooms with temperatures between 65 to 72 deg and 50 to 60% relative humidity. Fermentation is carried on in enclosed tanks; heat generated during this process is carried off and the temperature controlled by heat exchangers.

Dust control is essential in the milling department where grain is received, ground and processed. Two methods are generally employed—the use of a fan to discharge the dust to filters; use of cyclone dust collectors.

Bacterial control is essential although the industry is converting to enclosed fermenters.

Electrical Manufacturing

A large number of products and processes are covered by the generic term electrical manufacturing. It includes not only coil winding, machine operations, assembling and testing, but a certain amount of laboratory work where specified conditions must be maintained.

In a room where refrigerator units are assembled, filtered air is supplied under positive pressure to keep out dirt from adjoining areas. Cooling is for comfort purposes only and no attempt is made to hold the relative humidity to any fixed amount. For performance testing of refrigerator units, definite air conditions are maintained. Although the air is filtered, no effort is made to hold the dust count to a special figure. In testing of refrigerator units, temperature and humidity control is employed for the return air.

The room where refrigerator valves are manufactured is held to as low a relative humidity as possible to prevent corrosion of valve steel.

The amount of rejects of coil and high tension transformer windings for electronic work has been definitely reduced as the result of keeping moisture content and dust out of the windings.

Maintenance of constant temperature is a primary consideration to insure uniform calibration of thermal circuit breakers.

Humidity is held down to prevent rusting during the internal assembly of electronic tubes and in the lapping operation listed in Table 6.

Table 6. Electrical Manufacturing

Room or Process	Temperature. Deg F	Relative Humidity, %
Engineering, drafting, and sales	73	40-50
Instruments Manufacturing	70 ± 1½	Comfort
Laboratory	70 ± 1½	Comfort
Electronic and X-ray Coil winding and Coil winding and high tension transformer winding	72	15
Assembly of tubes	68	40

(Concluded on next page)

Table 6. Electrical Manufacturing (Concluded)

Room or Process	Tempera-ture, Deg F	Relative Humidity, %
Refrigeration		
Manufacture of		
refrigeration valves	75	40
Assembly of refrigerators	75	—
Testing	65 and 82	47 ± 2
Testing refrigeration		
units in traveling,		
electrified chain		
conveyors	65 and 82	47 ± 2
Spray painting	75	—
Thermostat assembly and		
calibration	76 ± 1½	Comfort
Multiplying and wrapping		
cotton and glass yarns		
on copper conductors	75	65-70
Small mechanisms—		
assembly to close		
tolerances	72	40-45
Assembly of lightning-		
arresters	68	20-40
Thermal circuit breaker—		
assembly and test	76 ± ½	30-60
Lapping and thrust		
runners for vertical		
water wheel generation	70	Under 50
Meters—assembling and		
testing	74-76	60-63
Switchgear		
Winding of capacitor		
sections, storage of		
paper, assembly of		
fuses and cutouts	73	50
Rectifiers—processing		
selenium and copper		
oxide plates	74 ± 2	30-40

Fur Storage

Members of the Certified Cold Fur Storage Association have reached the conclusion that the best conditions for storing furs in a vault is to maintain the temperature at 40 deg or lower, and the relative humidity from 50 to 65%. In fact, the American Institute of Refrigeration will not certify a vault as being a cold storage vault unless it can maintain a temperature of 40 deg in the vault.

Many members of this association store furs just below freezing, chiefly because they store in a wholesale way for many retail stores. Since the price for storage is based on the quantity of furs stored, it is not economical for the storer to inspect the fur. The below-freezing temperature is depended on to prevent any moth eggs or larvae, with which the fur may be infested, from hatching or becoming active.

A shock system for eradication of any insect infestations is also used. This treatment consists of dropping the temperature to 18 or 20 deg, then raising it to 60 or 70 deg before the temperature is lowered to 40 deg, which is the holding temperature. While in the north only one shock treatment is given during a storage season, southern fur storers may employ this system once or twice a season.

Glass and Glass Yarn

In the manufacture of laminated or safety glass, a sheet of plastic is used as a filler or binder between two sheets of glass. It is important that the two pieces of glass be dust-free to produce a clear product, and that the plastic binder be properly dried to remove intercellular moisture.

The plastic (vinyl resin) is slightly hygroscopic under humid conditions so that it may take up enough moisture to prevent it serving as a satisfactory binder for the glass sheets. At ordinary room temperature it is limp and tacky but at temperatures below 65 deg it becomes fairly stiff and it loses its stickiness.

Air conditions to be maintained in this industry are listed in Table 7.

All air supplied to the dehumidifier serving the cutting room is double filtered to protect the alumina beds for contamination.

Glass coolers on each assembly conveyor cool the glass from 110 deg, the temperature as it leaves the washers, to 60 deg F as it enters the laminating room. It is important to keep the air in the cooler at low humidity to pick up any surface moisture that may remain on the glass from the washer, and to prevent condensation on the cooling coils.

Fusing in the manufacture of optical lenses is carried out under air conditions of 75 deg and 45% relative humidity. Electrostatic filters are used to hold down the dust count.

In the manufacture of fiber glass strands which form the basis of fiber glass yarn, small glass marbles are melted in an electric furnace. Diameter of filaments products are controlled by the temperature of the molten glass and the speed at which the filaments are drawn.

The very fine filaments are passed through a spray of a special lubricant and then through a metal V-shaped notch that draws the fibers into a single strand. The lubricant oils the filaments and

Table 7. Air Conditions for Laminated Glass Manufacture

Room or Process	Temperature, Deg F	Relative Humidity, %
Cutting room	65	15
Vinyl inspection and laminating room	55	25
Vinyl drying ovens	142	5

Table 8. Conditions for Drying Macaroni and Spaghetti

Stage of Drying	Long Goods in Batch Dryers		Short Goods, Continuous Drying		Noodles— Continuous Drying	
	Dry Bulb, Deg F	Wet Bulb, Deg F	Dry Bulb, Deg F	Wet Bulb, Deg F	Dry Bulb, Deg F	Wet Bulb, Deg F
Preliminary	90	80	100	85	90	82
Mid-period	90	84	95	87	100	88
End	85	73	90	78	100	88

serves as a temporary binder to make the filaments adhere to each other in subsequent fabricating processes. The single strand that is formed is drawn to a high speed winder. These strands are so fine that at least two strands are twisted together to form a thread suitable for weaving. The strands are twisted to form yarn which is then wound on cones for shipping.

Where the continuous filaments are produced, it is necessary to maintain a temperature of about 75 deg and a relative humidity of 65% to prevent the lubricant from drying too rapidly. It is also necessary to pass large amounts of air through this room to carry away the heat given off by the electric furnaces. Temperature on the electric furnace floor is about 85 deg.

Hygroscopic Material

Substances as hygroscopic as dried molasses, or dried orange and lemon powders, must be handled and packaged in atmospheres not exceeding 20% relative humidity. Soluble coffee should be packed and handled in atmospheres containing not over 30% relative humidity.

Macaroni and Spaghetti

Control of air conditions in the macaroni industry is a very important phase of production. Drying of both long macaroni and spaghetti is carried on normally in enclosed rooms and requires from 30 to 96 hours per batch, depending on the degree of control exercised. For the shorter drying rates, a very close control is required and the control temperatures are held constant throughout the drying process. For example, see Table 8, which gives both dry bulb and wet bulb temperatures for three products.

Macaroni, spaghetti and noodles are hygroscopic materials and will pick up or lose moisture, depending on the moisture content of the surrounding air. For quality control, a moisture content of about 11% is desirable, corresponding to an equilibrium condition of 50% relative humidity. Therefore, in

the packaging room, and in the storage rooms, a relative humidity of 50% should be maintained. Unfortunately, not too many macaroni plants make any attempt to control humidity in the packaging rooms.

Another reason why humidity should be maintained in packaging rooms is that the cellophane and cardboard cartons are easier to handle and stand up better.

In the manufacturing area where the semolina flour and water are mixed, and the macaroni is extruded, it is important to keep this room warm and relatively humid—somewhere between 45 and 55% relative humidity.

Laundry and Dry Cleaning

In laundry work, the problem is to make conditions as comfortable for the workers as possible. The most important phase of the laundry industry is dry cleaning which is conducted under conditions of 65 to 70 deg and 65% relative humidity. At present, there is a tendency to filter the air supplied to the drying rooms instead of the air supplied to drying tumblers. Bacterial control is not considered necessary. In rooms used for drying, after rugs have been washed, air is maintained at 175 to 180 deg and 35-40% relative humidity.

Leather

For drying heavy leather, such as vegetable-tanned sole leather, the tanned hides are hung in an atmosphere having a temperature of about 70 deg, with 75% relative humidity. Over a period of ten days or so, these conditions are changed to 90 deg, with a relative humidity of 35% or less. While these conditions are not closely controlled and they may vary greatly in different tanneries, a difference of 10 units in these figures would not be surprising. Control largely depends on the judgment of a skilled operator.

Most light leathers, such as chrome-tanned shoe upper leathers, are dried under more closely controlled conditions in a drying tunnel. A typical set-up calls for 120F and 45% relative humidity.

Leather is sometimes moistened to prepare it for rolling or stretching in manufacturing operations. This is usually done by dipping or spraying with water until the required weight of water is absorbed. Then the leather is allowed to stand in an atmosphere at room temperature or slightly higher and with a relative humidity of 95% for a period long enough to permit the water to become properly distributed throughout the leather. In some shoe-making operations, part of the shoe may be moistened by spraying with steam.

Leather is usually stored in warehouses without temperature or humidity control. However, it is necessary to keep the relative humidity below 85% to avoid mildew. For long time storage, 50 to 60F and 40 to 60% relative humidity would be considered desirable.

No special control of dust is followed through in the industry.

Until 1948, laboratories for physical testing of leather were maintained at 70 deg and 65% relative humidity. These conditions are now being changed to the new standard which is 73.4 deg and 50% relative humidity,

Linoleum

Precise control of temperatures and humidity is required in the first basic step in the manufacture of linoleum—the mechanical oxidizing of linseed oil which results in the product that forms the linoleum binder. This is a major operation which is conducted in the presence of air that has been conditioned at a temperature of 90 to 100F and 20 to 28% relative humidity. If temperatures during this oxidizing operation were not controlled, the batch would continue to get hotter and hotter until it became unusable. Eventually, it would burst into flame.

For this reason, the rate of flow and temperature of the cooling water in the jacket surrounding the tank is measured at regular intervals. The volume of air used in the progress is correspondingly measured.

After the oxidized linseed oil has been cured for several weeks, it is mixed with the cork flour, minerals and color pigments used in making the various types of linoleum.

Stoving processes for the maturing of linoleum involve heated air and are conducted at temperatures between 160 to 250 deg. Relative humidity of the air involved is naturally low at these high temperatures. In terms of absolute humidity, moisture content of 60 grains or less per pound of air is desirable.

While bacteriological control is not a factor in manufacturing operations, some operations are stabilized against the possibility of mold growth.

Machinery Manufacture

Controlled air conditions for various rooms in the manufacture of automobiles are listed in Table 9. These conditions are maintained by a major automobile manufacturer.

In offices, drafting rooms and gear matching rooms where employes spend the entire day, the dry bulb temperature is maintained higher during the summer months to avoid the possible health hazard resulting when one moves from conditioned zones to high temperatures that may prevail outdoors.

Another manufacturer employer air conditioning for—

(1) Precision machining of parts.
(2) Accurate gaging and inspection.
(3) Manufacture and adjustment of precision gages.
(4) Processing of certain kinds of material which can be satisfactorily worked only if their moisture content and temperature are correct.

Conditions maintained for operations (1) and (2) are generally 75 deg and 45 to 50% relative humidity. For (3), conditions are generally 68F and 50% relative humidity, although at times 75

Table 9. Air Conditions for Automobile Manufacture

Room or Process	Temperature, Deg F	Relative Humidity, %
Engineering offices and drafting rooms		
Summer	80	40
Winter	75	35
Spectrographic analysis room		
Summer	80	50
Winter	75	45
Gear matching and special assembly rooms		
Summer	80	40
Winter	75	35
Gasket storage		
All seasons	100	50
Cement and glue storage		
All seasons	65	40

deg and 50% relative humidity may be permitted. For (4), a great deal depends upon the operation. In one case, a year-round temperature of 65 deg and 10 to 15% relative humidity is necessary.

For general manufacturing and assembly areas, no attempt is made to control conditions during hot weather. An ample supply of fresh air and air motion are relied upon to provide personal comfort.

Machine Assembly

Controlled air conditions are important in the assembly rooms and in some machining operations. Machining tolerances and finishes of parts for precision gear drives are held to such specifications that assembly under constant temperature conditions is essential. Dust counts are held down fairly well in the room reserved for assembly and run-in of compressors. See Table 10.

Table 10. Air Conditions Maintained in Machine Assembly

Room or Process	Temperature, Deg F	Relative Humidity, %
Assembly of precision gear drives	72	42-50
Assembly and run-in of air conditioning compressors	70-76	45-55

Matches

Generally speaking, air conditioning in match manufacture is used only in the drying of the match heads. In older installations, the entire room was air conditioned but in some of the newer ones, only the individual machine is air conditioned. Drying is speeded by keeping the relative humidity low, about 40%, and the temperature 70 to 75 deg.

Meat and Meat Products

Aside from the information given in Table 12, which is a tabulation for an entire industry, one meat packer has found the conditions listed in Table 11 satisfactory for their processing and packaging departments.

During hog chilling, while the final cooler temperature is 28 deg and the minimum relative humidity is 85%, the relative humidity is much higher during the first part of the cooling cycle. Throughout the cooler, there are from 1 to 1.5 air changes per minute.

In areas other than those listed in Table 11, satisfactory humidities are maintained to avoid excessive dryness. However, reheat is necessary to

Table 11. Air Conditions for Meat Processing

Process Room or Area	Temperature, Deg F	Relative Humidity, %
Drying room for dry sausage	53-54	72-73
Bacon slicing and packaging	55	60*
Beef holding cooler	33-34	85
Fresh pork cooler	34-38	85
Sausage wrapping	55	60*
Hog chilling—final temperature	28	88

* Steps are being taken to reduce this to 40% R. H.

provide air of low humidity and comfort for employes in the wrapping rooms and in the dry sausage drying rooms.

Air conditioning is necessary for a wide variety of processes and products in meat packing plants. Each type of meat requires special air conditions. Table 12 lists the suggested air conditions for various processes and rooms in meat packing plants.

Table 12. Air Conditions in Meat Packing Plants*

Room or Process	Temperature, Deg F	Relative Humidity, %
Beef		
Chill	30-32	85-90
Curing	36-38	85-90
Car shipping	33-35	84-87
Aging	33-35	85-87
Boning	40-45	80-85
Cutting	33-36	80-85
Bacon		
Chilling	10-28	80-85
Molding	32-34	80-85
Slicing	50-55	50-55
Calf cooler	34-36	85-90
Cooked ham cooler	35-40	80-85
Canned ham cooler	32-38	75-80
Canning meats chill	20-38	75-80
Dried beef cooler	36-38	60-65
Dry salt curing cellar	36-38	85-90
Hogs		
Chill	28-32	85-90
Cutting	50-60	60-70
Cuts grading	45-50	60-75
Lard		
Shipping of shortening	48-50	65-70
Packaging	50-55	60-65
Leak-lard cooler	28-30	80-85

(Concluded on next page)

Table 12. Air Conditions in Meat Packing Plants* (Concluded)

Room or Process	Temperature. Deg F	Relative Humidity. %
Loading dock—shipping	40-45	65-70
Offal cooler	33-35	85-90
Oleo shipping	48-50	65-70
Pigs feet, curing	40-45	85-90
Pork cuts, fresh	26-30	80-85
Sausage, dry		
Meat curing	38-40	80-85
Stuffing	55-60	65-70
Grinding	40-45	80-85
Sausage, fresh		
Manufacture	55-60	65-70
Packaging cooler	40-45	80-85
Sausage		
Grinding	40-45	80-85
Stuffing	55-60	65-70
Pre-chill	38-40	80-85
Packaging	40-45	80-85
Sheep-lamb chill	34-36	85-90
Smoked meats wrapping and shipping	60-65	60-70
Veal chilling	34-36	85-90

* From Meat Packers Guide.

Table 13. Air Conditions for Munitions

Room or Process	Temperature, Deg F	Relative Humidity, %
Metal percussion elements		
Drying parts	180-200	—
Drying paint	110	—
Black powder—drying	120-130	—
Powder type fuse		
Conditioning powder	70	40
Powder loading	70	40
Pelleting room for tracer loading	80	35-40

Munitions

Air conditioning for munitions plants is used with good results in the manufacture of time ring fuses. It is important in permitting close gage tolerances and in holding down the moisture content of the explosive charge. Air conditioning is also important in holding down moisture condensation on ceilings, walls and windows during the winter months.

Because of the danger of explosive dusts carrying a fire from one room to another, usually duct work is not allowed to pass from one explosive room to another, regardless of whether or not outlets are taken off in each room.

Filters in return air ducts can become exceedingly hazardous when powder accumulates on them. The amount of explosive pick-up varies and there is the chance of human error in deciding how often to clean the filter and in judging the amount of powder present.

In rooms where powders are pressed, air distribution must be uniform and very accurately controlled so as to disturb the powders as little as possible. Since the type of items manufactured in air conditioned rooms may change frequently, it is important that the air distribution system be flexible enough to take care of such changing conditions.

Table 13 lists some of the air conditions that must be maintained for various rooms.

The metal parts of percussion elements are rinsed in clear water, after washing, and dried by air at 180 to 200 deg. It is important that the paint in the flash holes of an artillery primer dries from the inside out without shrinking. In the first section of the drier, air is maintained at 110 deg; when the solvents have been removed from the paint, 90 deg air is blown over the paint surface.

In drying black powder for artillery primers, moisture content is maintained below ¾ of 1% to procure uniform burning of the powder. Black powder is very dusty and is very sensitive to friction. Therefore, it is necessary to supply 100% fresh outside air.

Two types of fuses are used for projectiles: (1) mechanical, which, similar to a watch. requires close mechanical tolerances, and requires air conditioned air in its manufacture to hold down perspiration of the operators; and (2) powder train type, the powder for which must be accurately controlled. Relative humidity is held to within 1% so that the powder moisture content can be controlled to within 0.1% moisture content by weight. More moisture gives a slower timing.

Pharmaceuticals

For the purpose of this tabulation, grouped under the general term of pharmaceuticals is the manufacture of tablets, powders, capsules, ampules, toxines, and the maintenance of small animals for research work.

Conditions maintained in the various air conditioned departments are summarized in Table 14.

Rooms in which filled gelatin capsules are stored should be at 40 to 50% relative humidity at 75F, according to an authority other than quoted in Table 14, and from whom the following is abstracted.

In the manufacture of uncoated pills, the temperature and relative humidity must be kept accurately controlled for if the humidity is too high, the pills and tablets will not dry at the proper rate, and the powder that may be used as a surfacing will not adhere. If the humidity is too low, a hard outer layer may form before the inside of the pill has had a chance to dry thoroughly. Ideal conditions for a tablet compressing room are from 35% relative humidity at 80 deg to 50% relative humidity at 75 deg.

Conditions in rooms where tablets are coated should be close to 35% relative humidity at 80 deg.

Conditioning of filling and packing rooms should be on a definite basis—not a differential. In most cases 35% relative humidity at 78 to 80 deg is ideal. An excellent general condition for many purposes is 50% at 75 to 80 deg, although many products can be handled most efficiently at 5 to 10% relative humidity at 80 deg.

Drug manufacturers use large numbers of animals in testing and research work and also for production. While the animals used include guinea pigs, mice, rats, rabbits, monkeys, dogs, cats, chickens, hogs, horses and cows, the ones most frequently used in testing and research work are the smaller animals—mice, guinea pigs and rabbits. Ideal conditions for these animals are 50% relative humidity at 77 to 80 deg. While 60% relative humidity is satisfactory, the temperature must be uniform.

In the research room, it is only necessary to maintain comfort conditions. However, in the microanalysis room, 80 deg and 50% relative humidity are maintained throughout the year.

Table 14. Air Conditions for Manufacture of Pharmaceuticals

Room or Process	Temperature, Deg F	Relative Humidity, %
Tablet compressing department	77	40
Powder storage and packing areas	75	35
Milling room	80	35
Ampule manufacturing department	80	35
Biological manufacturing department	80	35
Liver extract	80	10.5
Capsule storage	75	35-40
Serums	74-48	50
Animal room	80	40
Small animal room	75-78	47-48

Photographic Film and Paper Manufacture

Air control is most important in the manufacture of photographic film and paper. Steps are taken to provide dust-free conditions and to prevent the development of microscopic growths. Air conditions maintained in the various rooms are given in Table 15.

All air is washed with water sprays to which an algae inhibitor is added. Air is filtered through glass wool, air mat paper, or electrostatic filters. All make-up air is filtered. Make-up air that is used for film drying is passed through activated carbon filters to absorb any harmful fumes that may be present in the air.

Photographic Film Storage

Temperature and humidity conditions in motion picture film storage vaults should be such as to insure maximum life for photographic emulsions which, because of their nature, undergo gradual deterioration as they age. Vaults should be cool and dry. Temperatures of 65 to 70F with a relative humidity of 40 to 55% meet the storage requirements of most films. While lower temperatures and humidity would be conducive to longer film life, the conditions mentioned are considered satisfactory except for special products for which the manufacturer may recommend other air conditions. These figures are for both raw stock storage and for the storage of developed film. Unless properly stored, developed film may become brittle.

By maintaining general laboratory conditions the same as those under which the raw stock is stored, there is no time lag in bringing the film to equilibrium with laboratory conditions before use. However, other factors may influence design conditions. Operator comfort suggests a slightly higher temperature, say 70-75 deg. It may be necessary to maintain relative humidity at 65 to 70% to minimize the static which is sometimes encountered in handling film in a too-dry atmosphere. Because of the static electricity problem, a relative humidity below 40% should be avoided.

Table 15. Air Conditions in Photographic Film and Paper Manufacture

Room or Process	Temperature, Deg F	Relative Humidity, %
Drying	20-125	40-80
Cutting and packing	65-75	40-70
Film base and paper storage	70-75	40-65
Coated paper and film storage	70-75	40-65

Table 16. Maximum Recommended Temperatures for Film Storage

Type of Film	Storage Periods Up to 1 Month	Storage Periods Up to 6 Months
	Maximum Temperature. Deg F	
Black and white negative and sound	65	50
Black and white positive	75	65
Color	65	50

Another authority recommended that motion picture laboratories and film storage vaults be kept at 70 deg and 55-65% relative humidity. He points out that while vault temperature could be lower, it should not go low enough to cause condensation on the film when it is brought out to an area having normal room conditions.

A third authority suggests the figures given in Table 16.

Plywood

Building up the laminations for plywood may be done by cold pressing or hot pressing. For cold pressing, the air should be maintained at 90 deg and 15 to 25% relative humidity. For hot pressing with resin film, it should be 90 deg and 60% relative humidity.

Poultry Incubation

According to one well versed in poultry husbandry, most of the large forced draft incubators for incubation of chicken, duck and turkey eggs operate at a temperature of 99.5 to 100 deg and a relative humidity of 54 to 56%. At hatching time, the temperature is usually dropped one-half a degree or a little better, and the humidity is allowed to rise.

Fumigation is used about two to three times during the last two days of incubation—that is, on the twentieth and twenty-first days. A mixture of formaldehyde and potassium permanganate is used to control pullorum disease and omphalitis or navel ill.

Table 17 was drawn from data supplied by another poultry specialist.

No mention was made of air conditions in laying quarters because such conditions vary with the seasons. In summer months, it is desirable to have a maximum of ventilation, regardless of temperature; in winter, to avoid condensation on the interior walls, an attempt is made to hold the temperature to

Table 17. Air Conditions for Poultry Incubation

Room or Process	Temperature, Deg F	Relative Humidity, %
Incubator room	70-75	30-60
Brooder house		
1st week turkeys and chickens	70-75	60
After 1st week	50-60	60
Battery rooms		
Starting chicks or turkeys	70	70-75
Growing chicks or turkeys	50-60	50-60
Egg storage rooms	45-60	70 & up

10 deg over or below the outside temperature. This procedure applies to houses which do not have forced ventilation systems.

Precision Manufacturing

It is important to maintain standard air conditions in gage rooms so that parts manufactured in several plants can be assembled to form one machine or unit. Most of the industrial nations have adopted 68 deg F or 20 deg C as the standard temperature for gage rooms. This figure was selected because it is a reasonable room temperature that is fairly easy to maintain; it is expressed as a whole number on whichever of the two temperature scales that is used; and it was adopted in some plants turning out metric gages.

This standard condition for gages has made possible a very close tolerance. For example, during World War I, 0.002 inch was considered a close tolerance; during World War II, a tolerance of 0.0002 inch was not unusual and was required on many government specifications.

Relative humidity is maintained at 40 to 50%, and dust control is necessary for a quality product.

Air conditions are controlled in all rooms where watches are assembled. Temperature is held within the range of 72 to 75 deg, with a maximum relative humidity of 50%. Filters are used in the air circulating system to keep down dust. In the manufacturing divisions, the only rooms air conditioned are the mainspring department and the dial department. Temperature is held within a range of 72 to 75 deg mainly for comfort. However, in the manufacture of mainsprings, relative humidity is held to below 50% to control corrosion. By keeping down the relative humidity to 50% in the room where watch dials are lacquered and printed, many manufacturing

troubles are eliminated. Previous to air conditioning, there were difficulties due to high humidity, particularly during the summer.

Printing

Difficulties experienced in printing establishments operating under normal conditions are primarily caused by use of hygroscopic materials, although there are equipment problems, too.

The exact air conditions that are employed are not as important as the consistency with which such conditions are maintained. Whatever conditions are selected must be maintained both during winter and summer. While a lower temperature and relative humidity would be more economical in winter, such values would be more difficult to maintain in summer. Where winter temperature is so low that excessive sweating may occur at 40% relative humidity, double glass windows or glass block windows have proven effective. One engineer selected 77 deg and 50% relative humidity as the best year-round condition. Another printing authority suggests 80 deg and 50% relative humidity as the condition to be maintained when it is 90 deg outside.

Not only is it important to have correct consistent conditions so that accurate registration may result in color printing, but controlled conditions help wood-mounted engravings to be at their correct printing height at all times.

Air humidification in winter is best achieved by spray type air washers; silica gel for dehumidification.

For multi-color printing, air should be at 82 deg and 45% relative humidity. For photo-gravure printing, it should be 72 deg and 50% relative humidity.

Textiles—Knitted Wear

Recommended relative humidities for knitting mills are given in Table 18.

For knitting nylon, a relative humidity of 48% is recommended. If relaxation of the yarn is a problem, then 70% relative humidity should be maintained.

The relative humidity percentages of Table 18 are based on a temperature of 70 deg.

According to U. S. Government tests, when the relative humidity of the air is over 45%, static cannot exist. This allows material to run freely on high speed machinery without the danger of jumping due to electrification. When the yarns run moist they are pliable and tend to run without cracking off of the small fibers.

Table 18. Recommended Relative Humidities for Knitting Mills

Area	Textile		
	Wool	Cotton	Rayon
	Relative Humidity, %		
Stock, storage	65	60	55
Warping	65	60	55
Winding	60	60	55
Knitting	65	68	60

Rust will not set in when the atmosphere is kept under 70% relative humidity. Relative humidity can be kept well below this figure and still high enough for better working of the yarns. The change-over from haphazard atmospheric conditions to controlled air conditions is claimed to lessen the amount of breakages and seconds in a mill at least 25%, and the productivity of the mill increased from 3 to 8%.

Textiles—Cotton, Wool, Silk Processing

Table 19 lists conditions to be maintained in the processing of cotton, linen, woolens and silk.

Table 19. Air Conditions for Textile Manufacture

Prescribed by Source A		
Room or Process	Temperature, Deg F	Relative Humidity %
Cotton		
Opening	75	55
Picking	75	55
Carding	75	55
Drawing	80	60
Roving	80	60
Ring Spinning		
Conventional	80-85	60-70[1]
Long draft	80-85	55[1]
Spooling and warping	80	65
Weaving	80	70-85[2]
Cloth room	75	65-70
Linen		
Carding	75-80	60
Spinning	75-80	60
Weaving	80	80
Woolen Processing		
Pickers	80-85	60
Carding	80-85	65-70
Spinning	80-85	50-60[3]
Dressing	75-80	60
Weaving	80-85	60-75[4]

(Concluded on next page)

Table 19. Air Conditions for Textile Manufacture (Continued)

Room or Process	Tempera-ture, Deg F	Relative Humidity, %
Worsteds		
Carding	80-85	65
Combing	80-85	65-70
Gilling	80-85	65-70
Top storage	80-85[5]	75-80
Drawing	80-85	65
Cap spinning	80-85	50-55
Spooling and winding	75-80	65-70
Weaving	80	55-70[6]
Finishing	75-80	60
Silk		
Preparatory	80	60-65
Weaving	80	60-70

[1] Relative humidity maintained in ring spinning depends upon the staple, twist, and whether the spinning is long draft with leather aprons for conveying the stock. The aprons readily absorb moisture so that the cotton sticks to the aprons when 55% is the limit of the relative humidity. If conventional, 3- or 4-roll spinning is run, relative humidities can be as high as 70%, but that depends upon the draft, twist and staple.
[2] Relative humidity carried in cotton weaving depends on the cloth construction.
[3] Relative humidities to be maintained for the mules is generally 55%; conditions over frame spinning, 55-60%. Both types of spinning depend upon the class of stock spun, also the regain in the roving.
[4] Woolen weaving generally calls for 60% relative humidity. However, on heavy goods such as 32 oz overcoatings, highest production is experienced with 75% relative humidity.
[5] This depends upon whether cellar long period conditioning at low temperature or quick conditioning at high temperature is used.
[6] This depends upon staple quality and construction.

Animal fibers best draw and twist by surrendering regain which accounts for the lower relative humidities maintained in spinning than in drawing.

One objective of processing textiles is to get the proper weight of material per given length. All machinery is mechanical and delivers a definite amount of stock per given length of time. Weight is influenced by the moisture content of the air surrounding the processed stock—namely, increase or decrease of regain.

Therefore, regardless of whether the textile fibers are synthetic, animal or vegetable, controlled humidification is of great importance in the processing of such textiles.

All textile machinery generates static electricity. This is materially allayed by use of 50-55% relative humidity.

High relative humidities in the various processes also serves considerably to lay dust. This is attested to by the fact that in blue denim weave sheds, where weaving is done under low relative humidities, after one spends an hour's time in such a shed, there is a definite blue ring around the nostrils and lips due to dust. Such a condition is not experienced in these sheds when high relative humidities prevail.

Table 20. Air Conditions for Textile Manufacture

Prescribed by Source B		
Room or Process	Tempera-ture, Deg F	Relative Humidity, %
Cotton		
Opening and picking	75	50-60[1]
Carding	75	50-55[2]
Combing	75	60-65
Drawing	75	50-65
Roving	75	50-60
Spinning	75-80	50-65
Spooling	—	70
Weaving	75-78[3]	70-80
Warp tying	—	60-70
Cloth room	—	65-70
Laboratory	70[4]	65[4]
Woolens		
Bradford		
Carding	75	65-70
Combing	—	60-65
Drawing	75	50-60
Spinning	75	50-55
Weaving	75	50-55
French		
Carding	75	65
Combing	75	65
Drawing	75	70-80
Spinning	80-85	75-80
Weaving	75	50-55

[1] If air is recirculated and cleaned, then relative humidity should be 60%.
[2] It is essential in late fall, winter and early spring, particularly on windy days, to keep the relative humidity above 40%. Under dry conditions, the card is hard to start.
[3] Relative humidity can go as high as 90% when many automatic machines are to be tended and warps are heavily sized. In winter, temperature should be 75 to 78 deg; in summer, 83 deg at 90% relative humidity; 85 deg at 80% relative humidity; 87 deg at 70% relative humidity.
[4] Both temperature and relative humidity are held to ± 2.

Data in Table 19 were obtained from another textile authority. In some items, they vary with the information given in Table 20, but a study of the two tables helps one to arrive at suitable average conditions.

Textiles—Manufacture of Rayon

Although it is possible to produce rayon from many substances, wood pulp and cotton are used almost exclusively because they have the purest cellulose in the most readily available state. Raw material in the form of sheets is dipped in a caustic soda solution. At the end of the steeping operation, the sheets are broken up into a fine matted crumb. These operations are carried on in a room maintained at 70 deg and 55% relative humidity.

The crumb is dumped into tanks for aging. Although this room is maintained at 73 deg, it is not necessary to control the humidity.

In the churn room, the sodium cellulose is converted to cellulose xanthate. Temperatures are maintained at 75 to 80 deg, and since the cellulose is not exposed to the air, humidity control is not important. During summer, room temperatures are held below 85 deg.

After ripening, the viscose solution is pumped to spinning machines. The many filaments that are produced by fine jets are twisted into thread and collected in a revolving circular box. In the spinning operation, it is desirable to limit the temperature to 90 deg during the summer and to keep the relative humidity at least 70% or higher. Variations in temperature and humidity may cause crystallization of the solution which in turn results in breakage of the thread. Winter conditions are 75 deg and 70% relative humidity.

The centrifugal force of the revolving box in spinning produces an annular cake. This cake is removed from the box and is held in storage until needed for later processing. This storage room is kept at 85 deg and 100% relative humidity.

From the storage room, the cake goes through the following processes: first washing, desulfurizing, acidification, bleaching and final washing. Cakes are loaded on skid-type racks and moved into the drying room. After cakes are dried, they are left in the drier under controlled conditions of 100 deg and 65% relative humidity to bring the rayon back to the proper regain.

The final operation takes place in the coning room where winding machines place the rayon yarn on cones. As rayon yarn is sold by weight, definite humidity control is needed. Temperature is kept at a maximum of 80 deg to prevent perspiration from workers' hands from soiling the yarn, and relative humidity is kept at 55% so that the yarn will have the legal moisture.

Textiles—Processing Synthetic Fibers

A number of twisting and winding operations, called throwing, up-spinning, twisting, winding, coning, warping, copping and quilling, are required to prepare rayon filament yarn for weaving or knitting. Air conditioning serves to condition the yarn so that it is workable at high speed with a minimum of breakages and irregularities. See Tables 21 and 22.

Temperature, too, is important. It should be high enough to prevent hardness in the rayon and combined with adequate humidity to minimize static. During cool weather, temperature should be high enough to lessen the variations between cool and hot weather operation. Temperature is important in the weave room. Many of the troubles formerly attributed to variations in humidity are now known to be caused also by variations in temperature. A minimum temperature of 80 deg helps to prevent hardness in the yarn. Uniform pliability and tension are also favored by a fixed, uniform temperature. A moderate amount of humidity is looked upon as insurance against static.

The A.S.T.M. has standardized testing and laboratory conditions at 70 deg and 65% relative humidity based on four hours' exposure.

Table 21. Air Conditions for Rayon and Synthetic Fiber Processing

Prescribed by Source A		
Room or Process	Temperature. Deg F	Relative Humidity. %
Viscose		
Preparatory	80	60
Weaving	80	60
Celanese		
Preparatory	80	70
Weaving	80	70-75
Nylon		
Preparatory	80	50-60
Weaving	80	50-60

Table 22. Air Conditions for Rayon Processing

Prescribed by Source B		
Room or Process	Temperature, Deg F	Relative Humidity %
Throwing	80[1]	55-65
Picking	—	50-60
Carding, drawing, roving	80-90	50-60
Spinning	80-90	50-60
Weaving		
Regenerated rayons	80	50-65
Acetate	80	55-60[2]
Spun rayons	80	80[3]
Knitting		
Viscose or cuprammonium type	80-85	65
Acetate	80-85	70

[1] There is a growing tendency to increase the humidity and some manufacturers have tried 65% relative humidity with satisfactory results. In the South, a minimum temperature of 85 deg is preferred.
[2] The minimum range of relative humidity is 55 to 60% and the maximum, 70 to 75%.
[3] Maximum condition.

AIR CONDITIONING DESIGN CONDITIONS—COMMERCIAL BUILDINGS

Banks

Packaged air conditioning systems are commonly employed for offices in country locations, while the central system is installed in the larger city banks. It used to be that an effort was made to maintain a 15 to 20 deg differential between inside and outside temperatures, and to hold the relative humidity at 40 to 50%. The modern trend is to maintain 73-77F, 50-60 rh.

Dentists' Offices

Air conditioning in dentists' offices not only promotes the comfort of the patients, but it increases the efficiency of the dental operators. Unpleasant temperature and humidity conditions greatly multiply the discomfort suffered from the simplest form of dental operation.

Several materials that are commonly used, such as zinc oxide cements, silicates, and various impression materials, are more or less affected by temperature and humidity conditions in their manipulation. Air conditioning stabilizes the technic of handling these materials.

In lieu of more specific information, design conditions may be considered as 78 deg and 50% relative humidity.

Department Stores

Research and practical experience has long established that the conditions most satisfactory to both customer and employee is 78 to 80 deg and a relative humidity of 50%. Temperatures above 80 deg have not been satisfactory, while higher humidities tend to make odors noticeable and to make necessary a greater amount of outdoor air for odor dilution. Too low a temperature is not desirable, for it is no longer necessary to impress customers with the fact that the store is air conditioned.

Comfort should not be stressed to the customer; rather, the customer should be unaware that the air conditioning system exists. The paramount thing is to have a system that functions—one that is definitely not uncomfortable in its results.

In department stores, it is impossible to have the temperature indoors change to the same degree as the outdoor temperature. The heat storage of the building simply does not follow the predictions made. It is far more desirable to hold a temperature at a point that will prevent customers from perspiring. So-called temperature shock is merely an awareness, as one enters, that the place is cooled.

Hospitals

Air conditioned space in hospitals is generally limited to the operating room, recovery room, and labor and delivery rooms of the maternity wing. Air conditioning has been used in rooms for the care of premature infants, treatment of fevers, and a variety of ailments that are common during periods of intense summer heat.

The operating room is the one that is most generally air conditioned. For this and also for the labor room, temperature is from 78 to 80 deg and relative humidity from 55 to 60%. The high relative humidity is necessary to eliminate possible explosions due to anesthetic gases. Because of this high relative humidity, steps must be taken to prevent condensation and frosting on windows during cold weather.

Nothing more than ordinary filtering is practiced because ordinances in some cities prohibit air recirculation. However, activated carbon has been helpful in removing odors and traces of anesthesia gases.

Libraries

Record material in libraries such as paper, leather, cloth, thread and adhesives are susceptible to decay brought about by unfavorable conditions of temperature and humidity. Extreme dryness makes these materials so brittle that books and manuscripts show signs of repeated handling. Paper, cotton and leather, under certain conditions, are subject to mildew which propagates by spores nearly always present in air. These grow slowly below 40 deg and very rapidly at the optimum temperature which varies with the species of growth. Since an abundance of water is required for their growth, if the relative humidity is kept below 80%, paper, cotton and leather will not contain enough moisture for such growth, but glue and starch will still be susceptible. Therefore, if these materials are protected in the bookbinding process by adding a bactericide such as betanaphthol, mildew will not form.

It is fortunate that air conditions for personal comfort are similar to air conditions favorable for the storage of records and books in libraries. Suitable values are: summer temperature, 80 deg;

winter temperature, 70 deg; relative humidity, 40 to 50%. However, in extremely cold weather, a lower relative humidity is necessary to prevent moisture condensation on windows. Under the conditions stated, the strength and flexibility of record materials are not adversely affected, and growth of molds and fungi is inhibited.

Motion Picture Theaters

The accepted standard for air conditioning motion picture theaters is to maintain summer conditions of 78 to 80 deg and 50% relative humidity, based on an outside temperature of 95 deg. However, in the South it is held at 80 deg, while in Cuba, for example, it is 85 deg.

More than anything else, excessive relative humidity will lessen the sensation of comfort of occupants seated in the theater. While it is expected that the next advance in theater air conditioning will be the development of an independent means for dehumidification, at present reduction in humidity is obtained by lowering the temperature of the air supply to a point required for theater cooling, and then to raise the temperature of air leaving the cooling coil by using the heat in the return air from the theater, or by depending on steam reheating coils.

Tests have shown that the desired differential between inside and outside temperatures should be from 12 to 15 deg. Some theaters have inside-outside controls.

Clean air is considered desirable to protect theater furnishings and decorations and also from the health standpoint. While electrostatic filters are considered best, they are also the costliest type of air filter. Types generally used are the throwaway type or those that can be washed and reused again. Most operators consider filters an operating nuisance. Where the air conditioning system employs washers, the washers are relied upon to remove dirt from the supply air.

Photographic Studios

Both personal comfort and industrial air conditioning must be considered in the design of commercial photographic studios. The studio proper, dressing rooms, and offices are maintained at 74 deg and 30 to 40% relative humidity, as listed in Table 1.

In the printing and developing rooms, not only must air conditions be controlled, but it is desirable to maintain developing solutions and rinse water at proper temperatures, for the time of development

Table 1. Design Conditions for Commercial Photographic Studios

Room or Process	Temperature, Deg F	Relative Humidity, %
Studio, dressing rooms and offices	74	30-40
Developing of film	70-75	60
Film drying	75-80	50
Printing	70	70
Cutting	72	65
Film and paper storage	60	45

and the quality of the final product is directly related to a controlled solution and wash-water temperature. Heat is liberated during the printing, enlarging and drying processes and this must be removed. Controlled conditions are also necessary for the storage of photo-sensitive materials such as film and paper.

Restaurants

Three areas must be considered in air conditioning restaurants—the dining room proper, the space reserved for the bar, and the kitchen.

In the dining areas, conditions are held at 76 deg and 50% relative humidity. Cafeterias, which have the problem of the open steam tables, maintain temperatures one or two degrees lower than those followed in conventional dining areas. On warm days, an attempt is made to maintain a differential of 12 to 14 deg between inside and outside conditions. It is important that a positive pressure be created as compared with the kitchen in order to prevent kitchen air and odors from entering the dining areas.

Because of the higher metabolic rate of persons drinking at the bar, as compared with those dining in the restaurant, it is necessary to supply a greater amount of cool air to create the necessary coolness.

One large restaurant chain keeps its kitchens at 76 to 80 deg with a relative humidity of 55%. The summer condition is 80 deg. It is rather expensive to air condition kitchens because of the large heat gain produced in cooking; therefore, air-cooled kitchens are not too common. From the cost standpoint, it is advantageous when establishing design conditions to figure on temperatures in the kitchen being 5 deg higher than those in the dining areas. However, the dew point in both places is the same. This slightly raises the wet bulb temperature in the kitchen over that in the restaurant.

Retail Food Stores

Almost all retail stores are fully climate controlled on a year-round basis. This is the result not only of high internal heat loads such as high lighting levels, both interior and at display windows, but from high people and ventilation loadings as well. A special factor contributing to year-round control is that retail food-store customers tend to expect climate control. Also, special heat-emitting sources such as boilers, must not be overlooked.

As a rule, packaged unitary systems—roof-mounted where possible—are the rule. However, owners of large shopping centers will often supply their tenants with both chilled and hot water, frequently on a metered basis. *York Budget and Data Book* recommends a budget of 400-500 sq ft per ton of refrigeration. The *ASHRAE Handbook* suggests a lighting load of one watt per square foot, and two air changes per hour, for a small grocery store.

Open refrigerated display cases are another special problem that must be considered. They contribute 24-hour heating load, year-round. If ignored, cold air spilling into aisleways from such refrigerated cases will require the use of new energy—even in summer. One energy-conserving solution collects the heat from remotely located refrigeration compressors, using it to supply the heating requirement.

To eliminate the stratified cold air that lays near the floor around refrigerated cases, it is recommended that return air be taken at or near the floor and as close to the cases as possible. An alternative is placing circulating fans behind the cases. These draw air from near the floor and discharge it into upper space levels.

Television Studios

It is considered good practice in the air conditioning of television studios to have a maximum inside-outside temperature differential of 20 deg. The chief problem is to provide conditions that would not endanger the health of the performers because of wide variations in ambient temperatures.

In general, winter air conditions should be 72 deg and 45 to 50% relative humidity, with maximum summer conditions of 80 deg and 50 to 60% relative humidity, depending on outside conditions.

Women's Specialty Stores

In a women's specialty store, the selling area is maintained at 76 deg and 50% relative humidity during the summer; for the winter months, temperature is 74 deg with little attention paid to humidity. The fitting room requires a slightly higher temperature by 2 to 3 deg. Sufficient outside air must be circulated to prevent odors. Exhaust air from the fitting room and should be discarded.

Other Applications

INSIDE COMFORT DESIGN CONDITIONS —SUMMER***

The inside design conditions listed in Table 2 are recommended for types of applications listed. These conditions are based on experience gathered from many applications, substantiated by ASHRAE tests.

The optimum or deluxe conditions are chosen where costs are not of prime importance and for comfort applications in localities having summer outdoor design dry-bulb temperatures of 90F or less. Since all of the loads (sun, lights, people, outdoor air, etc.) do not peak simultaneously for any prolonged periods, it may be uneconomical to design for the optimum conditions.

The commercial inside design conditions are recommended for general comfort air conditioning applications. Since a majority of people are comfortable at 75F or 76F db and around 45% to 50% rh, the thermostat is set to these temperatures, and these conditions are maintained under partial loads. As the peak loading occurs (outdoor peak dry-bulb and wet-bulb temperatures, 100% sun, all people and lights, etc.), the temperature in the space rises to the design point, usually 78F db.

INSIDE COMFORT DESIGN CONDITIONS —WINTER***

For winter season operation, the inside design conditions listed in Table 2 are recommended for general heating applications. With heating, the temperature swing (variation) is below the comfort condition at the time of peak heating load (no people, lights, or solar gain, and with the minimum outdoor temperature). Heat stored in the building structure during partial load (day) operation reduces the required equipment capacity for peak load operation in the same manner as it does with cooling.

***From *Handbook of Air Conditioning System Design,* by Carrier Air Conditioning Co., © 1965 by McGraw-Hill Inc. Used with permission.

Table 2. Recommended Inside Design Conditions*—Summer and Winter

TYPE OF APPLICATION	SUMMER					WINTER				
	Deluxe		Commercial Practice			With Humidification			Without Humidification	
	Dry-Bulb (F)	Rel. Hum. (%)	Dry-Bulb (F)	Rel. Hum. (%)	Temp. Swing† (F)	Dry-Bulb (F)	Rel. Hum. (%)	Temp. Swing‡ (F)	Dry-Bulb (F)	Temp. Swing‡ (F)
GENERAL COMFORT Apt., House, Hotel, Office Hospital, School, etc.	74-76	50-45	77-79	50-45	2 to 4	74-76	35-30	−3 to −4	75-77	−4
RETAIL SHOPS (Short term occupancy) Bank, Barber or Beauty Shop, Dept. Store, Supermarket, etc.	76-78	50-45	78-80	50-45	2 to 4	72-74	35-30**	−3 to −4	73-75	−4
LOW SENSIBLE HEAT FACTOR APPLICATIONS (High Latent Load) Auditorium, Church, Bar, Restaurant, Kitchen, etc.	76-78	55-50	78-80	60-50	1 to 2	72-74	40-35	−2 to −3	74-76	−4
FACTORY COMFORT Assembly Areas, Machining Rooms, etc.	77-80	55-45	80-85	60-50	3 to 6	68-72	35-30	−4 to −6	70-74	−6

*The room design dry-bulb temperature should be reduced when hot radiant panels are adjacent to the occupant and increased when cold panels are adjacent, to compensate for the increase or decrease in radiant heat exchange from the body. A hot or cold panel may be unshaded glass or glass block windows (hot in summer, cold in winter) and thin partitions with hot or cold spaces adjacent. An unheated slab floor on the ground or walls below the ground level are cold panels during the winter and frequently during the summer also. Hot tanks, furnaces or machines are hot panels.

†Temperature swing is above the thermostat setting at peak summer load conditions.

‡Temperature swing is below the thermostat setting at peak winter load conditions (no lights, people or solar heat gain).

**Winter humidification in retail clothing shops is recommended to maintain the quality texture of goods.

HEATING AND HEAT TRANSMISSION

THERMOMETRIC SCALES AND CONVERSION

Two thermometer scales are in general use: the Fahrenheit (F) and the Celsius (°C or degree C). The Fahrenheit scale has been common to most English speaking countries. Nevertheless, the Celsius scale (formerly Centigrade) is making increasing inroads, impelled by the international changeover to the SI Metric System.

In the Fahrenheit thermometer, the freezing point of water is marked at 32 degrees on the scale and the boiling point, at atmospheric pressure, at 212 degrees. The distance between these two points is divided into 180 degrees. On the Celsius scale, the freezing point of water is at 0 degrees and the boiling point at 100 degrees. The following formulas may be used for converting temperatures given on any one of the scales to the other scales:

$$\text{Degrees Fahrenheit} = \frac{9 \times \text{degrees C}}{5} + 32$$
$$= \frac{9 \times \text{degrees R}}{4} + 32.$$

$$\text{Degrees Celsius} = \frac{5 \times (\text{degrees F} - 32)}{9}$$
$$= \frac{5 \times \text{degrees R}}{4}.$$

Tables follow for converting degrees Celsius into degrees Fahrenheit. The tables can, of course, be conveniently used in the reverse order. As an example of the use of the tables, it will be seen that 1040 degrees Celsius equals 1904 degrees Fahrenheit, that — 130 degrees Celsius equals — 202 degrees Fahrenheit, and that 79 degrees C. equals 174.2 degrees Fahrenheit.

Absolute Zero and Absolute Temperature Scales. A point has been determined on the thermometer scale, by theoretical considerations, which is called the absolute zero and beyond which a further decrease in temperature is inconceivable. This point is located at — 273.2 degrees Celsius or — 459.7 degrees F. A temperature reckoned from this point, instead of from the zero on the ordinary thermometers, is called absolute temperature. To find the absolute temperature when the temperature in degrees F is known, add 459.7 to the number of degrees F. For example, find the absolute temperature of the freezing point of water (32 degrees F).

$$459.7 + 32 = 491.7 \text{ absolute temperature Fahrenheit.}$$

The temperature on the Celsius scale but measured from absolute zero is also expressed as degrees kelvin or K. Degrees kelvin is thus °C + 273.2. Similarly, the Fahrenheit scale measured from absolute zero is called the Rankine scale, and degrees Rankine is thus F + 459.7.

Example. What is the Fahrenheit equivalent of 543.2K.

Solution. $K = 273.2 + °C$, so that $°C = 543.2 - 273.2 = 270$

$$F = \frac{9}{5}°C + 32 \text{ or } \frac{9}{5}(270) + 32$$
$$= 518 \text{ deg.}$$

The final step can also be solved by use of the conversion tables.

Example. What is the Celsius equivalent of 764.7R.

Solution. $R = 459.7 + F$, so that $F = 764.7 - 459.7 = 305.0$

$$°C = \frac{5}{9}(F - 32) \text{ or } \frac{5}{9}(305 - 32)$$
$$= 151.7$$

The final step can also be solved by use of the conversion tables.

FAHRENHEIT-CELSIUS CONVERSION

The numbers in the center column, in bold face type, refer to the temperature in either Fahrenheit or Celsius degrees. If it is desired to convert from Fahrenheit to Celsius degrees, consider the center column as a table of Fahrenheit temperatures and read the corresponding Celsius temperature in the column at the left. If it is desired to convert from Celsius to Fahrenheit degrees, consider the center column as a table of Celsius values, and read the corresponding Fahrenheit temperature on the right.

Interpolation factors are given for use with that portion of the table in which the center column advances in increments of 10. To illustrate, suppose it is desired to find the Fahrenheit equivalent of 314 deg C. The equivalent of 310 deg C, found in the body of the main table, is seen to be 590 deg F. The Fahrenheit equivalent of a 4-deg C difference is seen to be 7.2, as read in the table of interpolating

Interpolation Factors

Deg C		Deg F	Deg C		Deg F
0.56	1	1.8	3.33	6	10.8
1.11	2	3.6	3.89	7	12.6
1.67	3	5.4	4.44	8	14.4
2.22	4	7.2	5.00	9	16.2
2.78	5	9.0	5.56	10	18.0

factors. The answer is the sum or 597.2 deg F.

Above 1000 in the center column, the table increases in increments of 50. To convert 1462 deg C to Fahrenheit, for instance, add to the Fahrenheit equivalent of 1400 deg C ten times the interpolation factor for 6, and the interpolation factor for 2, or 2552 + 108 + 3.6, which equals 2663.6.

For conversions not covered in the table, the following formulas are used:

$$F = 1.8 \ C + 32$$
$$C = (F-32) \div 1.8$$

Deg C		Deg F	Deg C		Deg F	Deg C		Deg F	Deg C		Deg F
−273	−459.4	...	−101	−150	−238	− 8.3	17	62.6	9.4	49	120.2
−268	−450	...	− 96	−140	−220	− 7.8	18	64.4	10.0	50	122.0
−262	−440	...	− 90	−130	−202	− 7.2	19	66.2	10.6	51	123.8
−257	−430	...	− 84	−120	−184	− 6.7	20	68.0	11.1	52	125.6
−251	−420	...	− 79	−110	−166	− 6.1	21	69.8	11.7	53	127.4
−246	−410	...	− 73	−100	−148	− 5.6	22	71.6	12.2	54	129.2
−240	−400	...	− 68	− 90	−130	− 5.0	23	73.4	12.8	55	131.0
−234	−390	...	− 62	− 80	−112	− 4.4	24	75.2	13.3	56	132.8
−229	−380	...	− 57	− 70	− 94	− 3.9	25	77.0	13.9	57	134.6
−223	−370	...	− 51	− 60	− 76	− 3.3	26	78.8	14.4	58	136.4
−218	−360	...	−46	− 50	−58	− 2.8	27	80.6	15.0	59	138.2
−212	−350	...	−40	−40	−40	− 2.2	28	82.4	15.6	60	140.0
−207	−340	...	−34	−30	−22	− 1.7	29	84.2	16.1	61	141.8
−201	−330	...	−29	−20	− 4	− 1.1	30	86.0	16.7	62	143.6
−196	−320	...	−23	−10	14	− 0.6	31	87.8	17.2	63	145.4
−190	−310	...	−17.8	0	32—	0—	32	89.6	17.8	64	147.2
−184	−300	...	−17.2	1	33.8	0.6	33	91.4	18.3	65	149.0
−179	−290	...	−16.7	2	35.6	1.1	34	93.2	18.9	66	150.8
−173	−280	...	−16.1	3	37.4	1.7	35	95.0	19.4	67	152.6
−169	−273	−459.4	−15.6	4	39.2	2.2	36	96.8	20.0	68	154.4
−168	−270	−454	−15.0	5	41.0	2.7	37	98.6	20.6	69	156.2
−162	−260	−436	−14.4	6	42.8	3.3	38	100.4	21.1	70	158.0
−157	−250	−418	−13.9	7	44.6	3.9	39	102.2	21.7	71	159.8
−151	−240	−400	−13.3	8	46.4	4.4	40	104.0	22.2	72	161.6
−146	−230	−382	−12.8	9	48.2	5.0	41	105.8	22.8	73	163.4
−140	−220	−364	−12.2	10	50.0	5.6	42	107.6	23.3	74	165.2
−134	−210	−346	−11.7	11	51.8	6.1	43	109.4	23.9	75	167.0
−129	−200	−328	−11.1	12	53.6	6.7	44	111.2	24.4	76	168.8
−123	−190	−310	−10.6	13	55.4	7.2	45	113.0	25.0	77	170.6
−118	−180	−292	−10.0	14	57.2	7.8	46	114.8	25.6	78	172.4
−112	−170	−274	− 9.4	15	59.0	8.3	47	116.6	26.1	79	174.2
−107	−160	−256	− 8.9	16	60.8	8.9	48	118.4	26.7	80	176.0

FAHRENHEIT-CELSIUS CONVERSION TABLE (CONT'D)

Deg C		Deg F	Deg C		Deg F	Deg C		Deg F	Deg C		Deg F
27.2	81	177.8	58.3	137	278.6	89.4	193	379.4	304.4	580	1076
27.8	82	179.6	58.9	138	280.4	90.0	194	381.2	310.0	590	1094
28.3	83	181.4	59.4	139	282.2	90.6	195	383.0	315.6	600	1112
28.9	84	183.2	60.0	140	284.0	91.1	196	384.8	321.1	610	1130
29.4	85	185.0	60.6	141	285.8	91.7	197	386.6	326.7	620	1148
30.0	86	186.8	61.1	142	287.6	92.2	198	388.4	332.2	630	1166
30.6	87	188.6	61.7	143	289.4	92.8	199	390.2	337.8	640	1184
31.1	88	190.4	62.2	144	291.2	93.3	200	392.0	343.3	650	1202
31.7	89	192.2	62.8	145	293.0	93.9	201	393.8	348.9	660	1220
32.2	90	194.0	63.3	146	294.8	94.4	202	395.6	354.4	670	1238
32.8	91	195.8	63.9	147	296.6	95.0	203	397.4	360.0	680	1256
33.3	92	197.6	64.4	148	298.4	95.6	204	399.2	365.6	690	1274
33.9	93	199.4	65.0	149	300.2	96.1	205	401.0	371.1	700	1292
34.4	94	201.2	65.6	150	302.0	96.7	206	402.8	376.7	710	1310
35.0	95	203.0	66.1	151	303.8	97.2	207	404.6	382.2	720	1328
35.6	96	204.8	66.7	152	305.6	97.8	208	406.4	387.8	730	1346
36.1	97	206.6	67.2	153	307.4	98.3	209	408.2	393.3	740	1364
36.7	98	208.4	67.8	154	309.2	98.9	210	410.0	398.9	750	1382
37.2	99	210.2	68.3	155	311.0	99.4	211	411.8	404.4	760	1400
37.8	100	212.0	68.9	156	312.8	100.0	212	413.6	410.0	770	1418
38.3	101	213.8	69.4	157	314.6	104.4	220	428.0	415.6	780	1436
38.9	102	215.6	70.0	158	316.4	110.0	230	446.0	421.1	790	1454
39.4	103	217.4	70.6	159	318.2	115.6	240	464.0	426.7	800	1472
40.0	104	219.2	71.1	160	320.0	121.1	250	482.0	432.2	810	1490
40.6	105	221.0	71.7	161	321.8	126.7	260	500.0	437.8	820	1508
41.1	106	222.8	72.2	162	323.6	132.2	270	518.0	443.3	830	1526
41.7	107	224.6	72.8	163	325.4	137.8	280	536.0	448.9	840	1544
42.2	108	226.4	73.3	164	327.2	143.3	290	554.0	454.4	850	1562
42.8	109	228.2	73.9	165	329.0	148.9	300	572.0	460.0	860	1580
43.3	110	230.0	74.4	166	330.8	154.4	310	590.0	465.6	870	1598
43.9	111	231.8	75.0	167	332.6	160.0	320	608.0	471.1	880	1616
44.4	112	233.6	75.6	168	334.4	165.6	330	626.0	476.7	890	1634
45.0	113	235.4	76.1	169	336.2	171.1	340	644.0	482.2	900	1652
45.6	114	237.2	76.7	170	338.0	176.7	350	662.0	487.8	910	1670
46.1	115	239.0	77.2	171	339.8	182.2	360	680.0	493.3	920	1688
46.7	116	240.8	77.8	172	341.6	187.8	370	698.0	498.9	930	1706
47.2	117	242.6	78.3	173	343.4	193.3	380	716.0	504.4	940	1724
47.8	118	244.4	78.9	174	345.2	198.9	390	734.0	510.0	950	1742
48.3	119	246.2	79.4	175	347.0	204.4	400	752.0	515.6	960	1760
48.9	120	248.0	80.0	176	348.8	210	410	770.0	521.1	970	1778
49.4	121	249.8	80.6	177	350.6	215.6	420	788	526.7	980	1796
50.0	122	251.6	81.1	178	352.4	221.1	430	806	532.2	990	1814
50.6	123	253.4	81.7	179	354.2	226.7	440	824	537.8	1000	1832
51.1	124	255.2	82.2	180	356.0	232.2	450	842	565.6	1050	1922
51.7	125	257.0	82.8	181	357.8	237.8	460	860	593.3	1100	2012
52.2	126	258.8	83.3	182	359.6	243.3	470	878	621.1	1150	2102
52.8	127	260.6	83.9	183	361.4	248.9	480	896	648.9	1200	2192
53.3	128	262.4	84.4	184	363.2	254.4	490	914	676.7	1250	2282
53.9	129	264.2	85.0	185	365.0	260.0	500	932	704.4	1300	2372
54.4	130	266.0	85.6	186	366.8	265.6	510	950	732.2	1350	2462
55.0	131	267.8	86.1	187	368.6	271.1	520	968	760.0	1400	2552
55.6	132	269.6	86.7	188	370.4	276.7	530	986	787.8	1450	2642
56.1	133	271.4	87.2	189	372.2	282.2	540	1004	815.6	1500	2732
56.7	134	273.2	87.8	190	374.0	287.8	550	1022	1093.9	2000	3632
57.2	135	275.0	88.3	191	375.8	293.3	560	1040	1648.9	3000	5432
57.8	136	276.8	88.9	192	377.6	298.9	570	1058	2760.0	5000	9032

HEAT TRANSFER BY RADIATION

The Stefan-Boltzmann law states that the radiation from a black body to another body is proportional to the difference between the fourth power of the absolute temperatures of the two bodies. When the radiating body is not a perfect black one,

$$Q_r = \frac{0.174 A e(T_h{}^4 - T_c{}^4)}{10^8} \cdots \quad (1)$$

where Q_r = radiation in Btu per hour
A = area in square feet
e = emissivity of radiating body
T_h = absolute temperature of radiating body
T_c = absolute temperature of receiving body

When $A = 1$, and when the receiving body is at absolute zero, the equation reduces to

$$q = \frac{0.174 e T_h{}^4}{10^8}$$

Table 1 is a tabular solution of this equation. It is useful in determining the unit radiation by giving the radiation to absolute zero of the hot body, then that to absolute zero by the receiving body and subtracting to find the difference which is the unit radiation when the emissivity is 1.0. Table 2 gives values for e.

Example. What is the radiation from a rough plaster ceiling at 100F to surrounding surfaces at a temperature of 50F?

Solution. From Table 1, q for the 100F ceiling to absolute zero surroundings is 169.7, that from the 50F surroundings is 116.6, or $(169.7 - 116.6) = 53.1$ Btu per sq. ft. per hr. at an e of 1. From Table 2, e for rough plaster is 0.91, so that the radiation is $53.1 \times .91 = 48.3$ Btu per sq. ft. per hour.

The accompanying drawing is a graphical solution of equation (1) for special conditions, such as with radiant heating, and when $e = 1.0$. Note that neither the table nor the curve includes the heat emitted by convection.

Table 1. — Radiation from Surfaces to Surroundings at Absolute Zero
(When emissivity = 1.0)

Surface Temperature, F	Radiation, Btu per Sq. Ft. per Hr.	Surface Temperature, F	Radiation, Btu per Sq. Ft. per Hr.	Surface Temperature, F	Radiation, Btu per Sq. Ft. per Hr.	Surface Temperature, F	Radiation, Btu per Sq. Ft. per Hr.
30	99.3	57	123.0	84	150.8	111	183.3
31	100.0	58	124.0	85	151.9	112	184.7
32	101.0	59	125.1	86	153.1	113	186.0
33	101.7	60	125.9	87	154.1	114	186.9
34	102.3	61	126.9	88	155.4	115	188.5
35	103.3	62	127.7	89	156.5	116	189.9
36	104.2	63	128.8	90	157.7	117	191.2
37	105.0	64	130.0	91	159.0	118	192.5
38	105.9	65	130.8	92	160.0	119	193.9
39	106.7	66	131.8	93	161.2	120	194.6
40	107.6	67	132.8	94	162.4	121	195.9
41	108.3	68	133.9	95	163.5	122	197.3
42	109.3	69	134.8	96	164.6	123	198.8
43	110.1	70	135.9	97	165.8	124	200.0
44	111.1	71	136.8	98	167.0	125	201.4
45	112.2	72	137.9	99	168.5	126	202.8
46	113.1	73	139.0	100	169.7	127	204.2
47	113.9	74	140.0	101	170.9	128	205.5
48	114.7	75	141.1	102	172.2	129	206.9
49	115.5	76	142.2	103	173.3	130	208.4
50	116.6	77	143.3	104	174.4	135	215.3
51	117.6	78	144.4	105	175.5	140	222.8
52	118.4	79	145.4	106	176.6	145	230.4
53	119.4	80	146.5	107	176.9	150	238.1
54	120.2	81	147.6	108	179.3	175	279.6
55	121.1	82	148.6	109	180.6	200	326.3
56	122.1	83	149.6	110	182.0		

Table 2. — Emissivity of Various Surfaces

Asbestos board	0.96	Oil paints, all colors	0.92–0.96
Asbestos paper	0.93–0.945	Aluminum paint	0.27–0.67
Brick, red	0.93	Paper	0.924–0.944
Enamel, white, on iron	0.897	Plaster, rough	0.91
Glass, smooth	0.92–0.94	Absolute black surface	1.00
Gypsum	0.903	Dark paint	0.92
Marble, light gray, polished	0.903	Rough concrete	0.94
Oak, planed	0.895	Lime wash	0.91
Black lacquer on iron	0.875	Polished aluminum	0.03

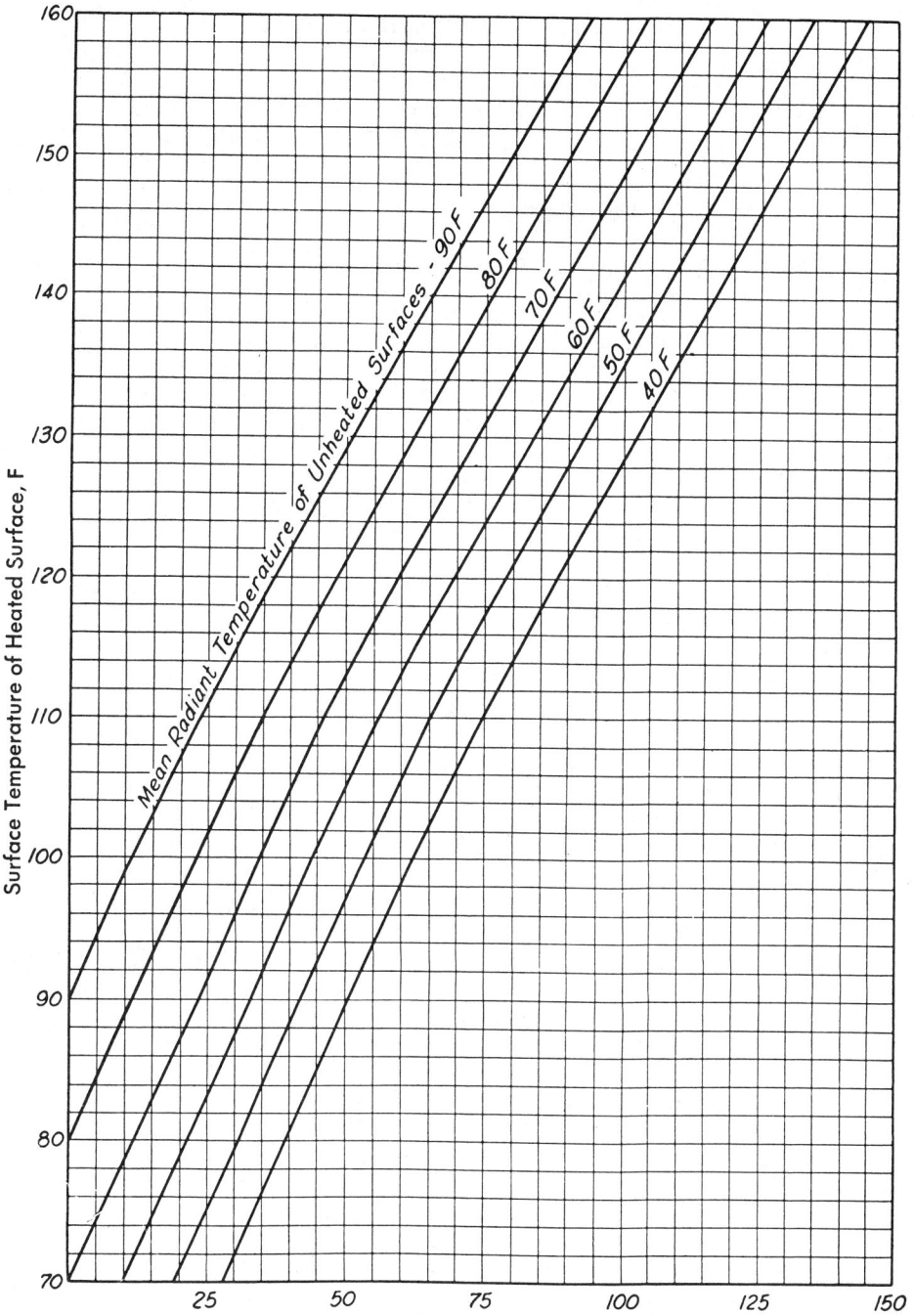

Heat emission by radiation from flat surfaces with emissivity = 1.0.

HEAT LOSS FROM WATER SURFACE

Heat loss from the exposed liquid surface of an open tank is by convection and radiation from the surface and by evaporation of the liquid. The last, especially, is a rather involved phenomenon to measure, and the formula for the sum of these three different phases of heat loss is extremely complex.

The accompanying table is a solution of the formula for the common case of water where the relative humidity of the surrounding air is 70%, and for various water temperatures, air temperatures, and air velocities. Note that air velocities are in feet per *second*.

Example: A 4 ft wide × 12 ft long dye vat maintained at 200 deg F is located in a room where the average air temperature is 80 deg F, and the air velocity is estimated at 0.5 ft per second. What is the heat lost from the water surface if the relative humidity of the air is 70%?

Solution: Referring to the accompanying table in the section devoted to 80 deg F air find, opposite 200 deg F and under 0.5 velocity, a heat loss of 4850 Btu per (sq ft) (hour). Since the tank has a water surface area of 4 × 12, or 48 sq ft, the heat loss from the water surface is 48 × 4850 = 232,800 Btu per hour.

Heat Loss from Water Surface at Various Air and Water Temperatures and Air Velocities

(When Relative Humidity Is 70%)

Water Temp., Deg F	Air Velocity, Feet per Second							
	0	0.5	1	2	5	10	20	50
	Heat Loss, Btu per (Hour) (Square Foot)							
Air Temperature 100 Deg F								
100	35	70	82	105	130	200	330	700
110	105	170	200	250	315	480	800	1600
120	200	295	350	420	550	810	1380	2750
130	325	460	520	650	820	1250	2050	4200
140	500	670	800	950	1230	1800	3000	6200
150	730	930	1100	1300	1700	2600	4300	8700
160	1000	1290	1500	1750	2400	3600	5900	12,100
170	1380	1750	2050	2450	3300	4950	8200	17,000
180	1830	2450	2800	3300	4600	7000	11,500	24,000
190	2400	3500	4050	4800	6600	10,000	16,500	34,000
200	3100	5000	5800	7100	9400	14,300	24,000	48,000
Air Temperature 80 Deg F								
80	17	34	40	49	66	106	185	330
100	58	96	113	135	185	280	460	920
110	210	285	320	380	540	780	1250	2550
120	315	420	470	560	760	1150	1850	3800
130	460	580	660	780	1090	1620	2600	5200
140	640	780	900	1060	1480	2200	3600	7200
150	860	1020	1200	1420	1970	2950	4800	9700
160	1150	1330	1590	1800	2550	3800	6300	13,000
170	1500	1730	2050	2500	3400	5200	8500	17,300
180	2000	2400	2750	3400	4700	7100	12,000	24,000
190	2550	3400	3900	4800	6600	10,000	17,000	34,000
200	3300	4850	5600	6900	9500	14,000	24,000	50,000

Water Temp., Deg F	Air Velocity, Feet per Second							
	0	0.5	1	2	5	10	20	50
	Heat Loss, Btu per (Hour) (Square Foot)							
Air Temperature 60 Deg F								
60	7	18	21	26	34	56	95	180
70	33	52	60	78	105	170	270	580
80	78	110	125	150	210	315	510	1050
90	130	180	210	240	330	490	780	1600
100	205	270	320	350	480	710	1150	2300
110	290	370	420	480	670	1000	1600	3300
120	400	500	570	660	930	1350	2200	4400
130	550	780	750	890	1240	1800	3000	6000
140	710	870	970	1150	1600	2400	4000	8000
150	950	1130	1260	1510	2100	3200	5300	10,500
160	1230	1450	1600	1850	2700	4050	6900	13,700
170	1600	1900	2100	2600	3700	5200	9100	18,500
180	2050	2600	2900	3550	4900	7200	12,500	25,000
190	2600	3550	4000	4950	6900	10,300	17,500	35,000
200	3300	4800	5500	7000	9700	14,600	25,000	50,000
Air Temperature 40 Deg F								
50	25	36	42	47	60	100	160	320
60	58	75	86	94	130	190	310	620
70	90	115	130	150	210	300	490	970
80	135	170	195	225	310	450	720	1460
90	190	230	265	310	440	640	1030	2100
100	260	315	360	420	600	860	1400	2800
110	350	420	470	550	790	1130	1880	3700
120	460	550	610	720	1020	1500	2500	5000
130	600	700	770	930	1300	1950	3150	6400
140	800	880	970	1200	1760	2500	4200	8300
150	1020	1130	1430	1530	2150	3200	5300	10,600
160	1300	1450	1570	1900	2800	4200	7000	14,000
170	1700	1900	2100	2600	3700	5500	9400	18,800
180	2200	2550	2750	3500	4900	7400	12,500	25,000
190	2750	3400	3750	4700	6800	10,300	17,000	34,000
200	3400	4750	5400	6700	9700	15,000	25,000	50,000
Air Temperature 0 Deg F								
40	77	93	102	113	155	235	360	700
50	105	121	134	153	215	315	490	940
60	135	155	170	195	270	410	630	1210
70	175	200	225	255	360	520	840	1610
80	225	255	280	320	450	680	1080	2100
90	285	320	350	410	580	860	1400	2700
100	360	400	440	510	720	1100	1760	3450
110	450	500	540	630	920	1350	2200	4400
120	560	610	680	800	1150	1700	2800	5600
130	700	750	840	1000	1440	2150	3550	7000
140	890	930	1030	1240	1800	2700	4450	9000
150	1100	1160	1300	1560	2250	3450	5700	11,300
160	1400	1460	1600	2000	2800	4400	7300	14,600
170	1800	1900	2100	2600	4700	5800	9700	19,400
180	2200	2500	2750	3400	5000	7700	13,000	26,000
190	2800	3400	3800	4700	7000	10,000	17,500	35,000
200	3500	4700	5300	6600	9800	15,000	25,000	50,000

HEAT TRANSFER TO AIR BY CONVECTION

The heat transferred to air by convection is given by the formula

$$H_c = h_c A (t_1 - t_2)$$

where H_c = heat transferred to the air by convection, Btu per hour,

h_c = convection coefficient, Btu per hour per square foot of surface per degree temperature difference,

t_1 = temperature of the hot surface, deg F,

t_2 = ambient air temperature, deg F,

A = area of hot surface, square feet.

The table inserted in the accompanying graph gives values of h_c for common cases where the formulas are well established. Values of H_c per square foot for six of the cases listed can be read directly from the graph.

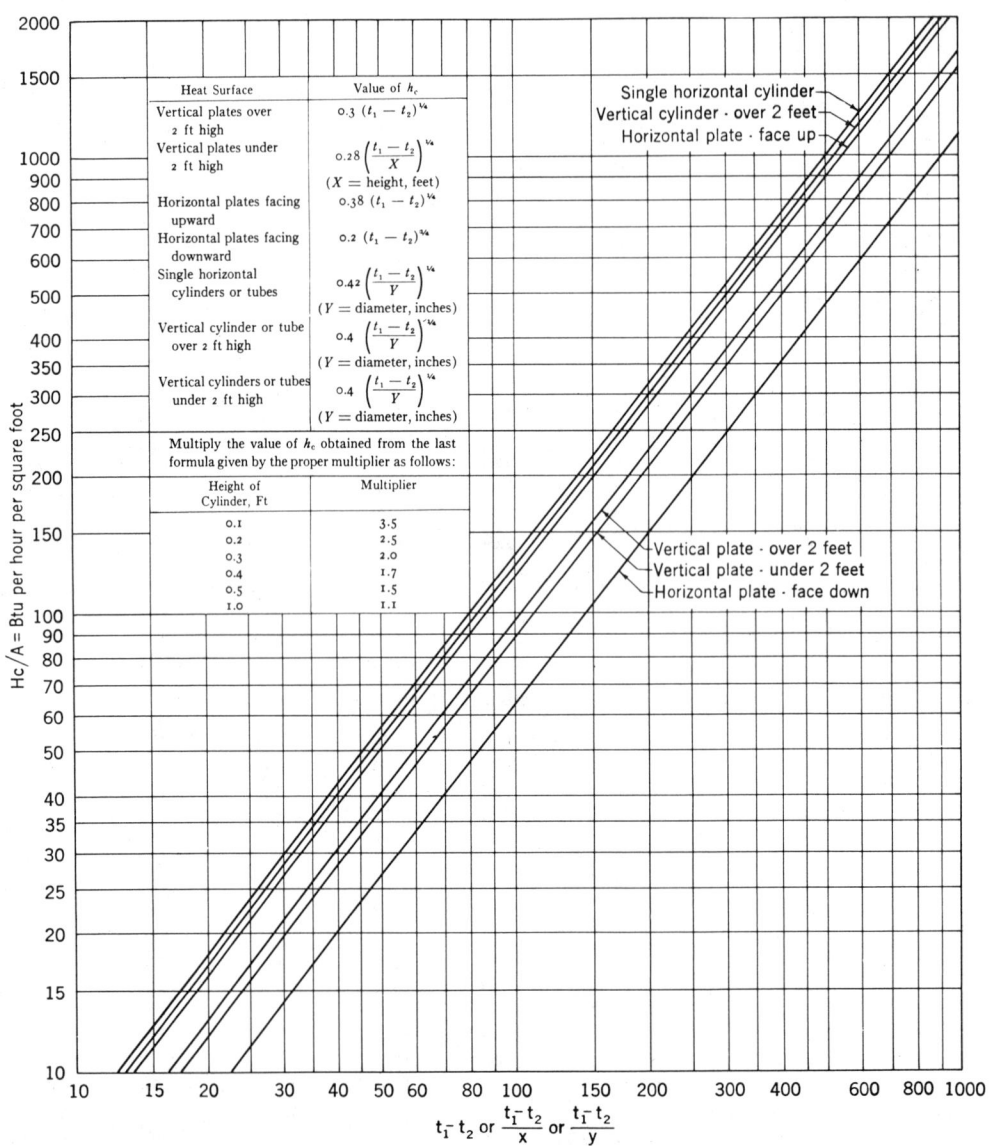

Heat Surface	Value of h_c
Vertical plates over 2 ft high	$0.3 (t_1 - t_2)^{1/4}$
Vertical plates under 2 ft high	$0.28 \left(\dfrac{t_1 - t_2}{X}\right)^{1/4}$ (X = height, feet)
Horizontal plates facing upward	$0.38 (t_1 - t_2)^{1/4}$
Horizontal plates facing downward	$0.2 (t_1 - t_2)^{1/4}$
Single horizontal cylinders or tubes	$0.42 \left(\dfrac{t_1 - t_2}{Y}\right)^{1/4}$ (Y = diameter, inches)
Vertical cylinder or tube over 2 ft high	$0.4 \left(\dfrac{t_1 - t_2}{Y}\right)^{1/4}$ (Y = diameter, inches)
Vertical cylinders or tubes under 2 ft high	$0.4 \left(\dfrac{t_1 - t_2}{Y}\right)^{1/4}$ (Y = diameter, inches)

Multiply the value of h_c obtained from the last formula given by the proper multiplier as follows:

Height of Cylinder, Ft	Multiplier
0.1	3.5
0.2	2.5
0.3	2.0
0.4	1.7
0.5	1.5
1.0	1.1

Single horizontal cylinder
Vertical cylinder · over 2 feet
Horizontal plate · face up

Vertical plate · over 2 feet
Vertical plate · under 2 feet
Horizontal plate · face down

H_c/A = Btu per hour per square foot

$t_1 - t_2$ or $\dfrac{t_1 - t_2}{x}$ or $\dfrac{t_1 - t_2}{y}$

CALCULATING BUILDING HEAT LOSSES

The quantity of heat which must be supplied to a building during the coldest periods in the winter is called the design heat loss. Ordinarily it is made up of two basic quantities—the heat which is lost through the walls, roof, window surface, etc., called transmission losses, and the heat which is needed to raise the temperature of the cold air which is leaking in around the window and door cracks, etc., termed infiltration loss. These pages are concerned with the former—the heat transmitted through the walls, roof doors, windows, and so on. Infiltration losses are discussed elsewhere.

As temperatures shift inside and outside a building the rate of heat flow, and even the direction of flow, changes. With massive structures especially, the heat storage of the structure is such that the heat may not even be flowing in the direction it is believed to be. For purposes of calculating winter heat losses, however, a "steady state" of heat flow from inside to outside is assumed.

This heat is transmitted from the warm interior to the cold exterior by combinations of conduction, convection, and radiation. Of these, the simplest is conduction.

Conductivity. This is the rate of heat flow per unit area per unit time per unit of temperature difference between the two *surfaces* (inside and outside) and *not* between the air temperatures inside and outside. It applies only to heat transferred by conduction and is expressed per unit of thickness. It is designated as the k value, and is expressed in btu per (hour) (square foot) (deg. F temperature difference per inch of thickness). Note that conductance is less the greater the thickness.

Conductance. Heat flow by conduction is also given in terms of *conductance* (C), a related unit

(cont. on page 2-174)

Summary of Heat Transfer Terms

U Overall heat transmission coefficient, or quantity of heat in Btu transmitted per hour through one square foot of a building section (wall, roof, window, floor, etc.) for each degree F of temperature difference between the air on the warm side and the air on the cold side of the building section.

$1/U$ Overall resistance to the flow of heat through a building section between the air on the warm side and air on the cold side of the section, and expressed in (hours) (square feet) (deg F temperature difference) per Btu.

k Thermal conductivity, the quantity of heat in Btu transmitted by conduction per hour through one square foot, one inch thick, of a homogenous material for each degree F temperature difference between the *surfaces* (not the air on each side) of the material.

C Thermal conductance. Same as conductivity except that it is per stated thickness and not per inch thickness.

$1/k$ Resistivity, or amount of resistance offered by one inch of a material to heat flow between the two surfaces. It is expressed in (hours) (square foot) (deg F temperature difference per inch) per Btu.

x/k Resistance of a given thickness x of a homogenous material. Same as k except that it applies to a stated thickness x other than 1 inch.

h Surface or film conductance. The quantity of heat in Btu transmitted one hour from (or to) one square foot of surface to (or from) the air adjacent to the surface. The symbol h_o is used for the outside surface and h_i for the inside surface. For most winter calculations $h_i = 1.46$ and $h_o = 6.00$.

ϵ Surface emittance. The ratio of radiant energy emitted by or absorbed by a surface to that emitted or absorbed by a perfect black surface under the same temperature conditions.

E Effective emittance. The ratio of radiant energy transmitted across an air space from a warm to a cold surface to the radiant energy that would be transmitted across the same air space and under the same temperature conditions if the surfaces were both perfect black bodies. E is further defined in footnote (b), Table 2.

R Resistance to flow of heat. Reciprocal of conductance.

Table 1 Surface Conductances and Resistances for Air

All conductance values expressed in Btu/(hr·ft²·F).

A surface cannot take credit for both an air space resistance value and a surface resistance value. No credit for an air space value can be taken for any surface facing an air space of less than 0.5 in.

SECTION A. Surface Conductances and Resistances[a,b]

Position of Surface	Direction of Heat Flow	Non-reflective ε = 0.90		Reflective ε = 0.20		Reflective ε = 0.05	
		h_i	R	h_i	R	h_i	R
STILL AIR							
Horizontal	Upward	1.63	0.61	0.91	1.10	0.76	1.32
Sloping—45 deg	Upward	1.60	0.62	0.88	1.14	0.73	1.37
Vertical	Horizontal	1.46	0.68	0.74	1.35	0.59	1.70
Sloping—45 deg	Downward	1.32	0.76	0.60	1.67	0.45	2.22
Horizontal	Downward	1.08	0.92	0.37	2.70	0.22	4.55
		h_0	R	h_0	R	h_0	R
MOVING AIR (Any Position)							
15-mph Wind (for winter)	Any	6.00	0.17				
7.5-mph Wind (for summer)	Any	4.00	0.25				

SECTION B. Reflectivity and Emittance Values of Various Surfaces and Effective Emittances of Air Spaces

Surface	Reflectivity in Percent	Average Emittance ε	Effective Emittance E of Air Space	
			One surface emittance ε; the other 0.90	Both surfaces emittances ε
Aluminum foil, bright	92 to 97	0.05	0.05	0.03
Aluminum sheet	80 to 95	0.12	0.12	0.06
Aluminum coated paper, polished	75 to 84	0.20	0.20	0.11
Steel, galvanized, bright..	70 to 80	0.25	0.24	0.15
Aluminum paint	30 to 70	0.50	0.47	0.35
Building materials: wood, paper, masonry, nonmetallic paints		0.90	0.82	0.82
Regular glass	5 to 15	0.84	0.77	0.72

[a] For ventilated attics or spaces above ceilings under summer conditions (heat flow down) see Table 6.
[b] Conductances are for surfaces of the stated emittance facing virtual blackbody surroundings at the same temperature as the ambient air. Values are based on a surface-air temperature difference of 10 deg F and for surface temperature of 70 F.

Reprinted, with permission, from ASHRAE Handbook of Fundamentals. 1977.

Table 2 Thermal Resistances of Plane Air Spaces [c,d]*

All resistance values expressed in (hour)(square foot)(degree Fahrenheit temperature difference) per Btu
Values apply only to air spaces of uniform thickness bounded by plane, smooth, parallel surfaces with no leakage of air to or from the space.
Thermal resistance values for multiple air spaces must be based on careful estimates of mean temperature differences for each air space.

Position of Air Space	Direction of Heat Flow	Mean Temp,[b] (F)	Temp Diff,[b] (deg F)	0.5-in. Air Space Value of E [a,b]					0.75-in. Air Space[c] Value of E [a,b]				
				0.03	0.05	0.2	0.5	0.82	0.03	0.05	0.2	0.5	0.82
Horiz.	Up	90	10	2.13	2.03	1.51	0.99	0.73	2.34	2.22	1.61	1.04	0.75
		50	30	1.62	1.57	1.29	0.96	0.75	1.71	1.66	1.35	0.99	0.77
		50	10	2.13	2.05	1.60	1.11	0.84	2.30	2.21	1.70	1.16	0.87
		0	20	1.73	1.70	1.45	1.12	0.91	1.83	1.79	1.52	1.16	0.93
		0	10	2.10	2.04	1.70	1.27	1.00	2.23	2.16	1.78	1.31	1.02
		-50	20	1.69	1.66	1.49	1.23	1.04	1.77	1.74	1.55	1.27	1.07
		-50	10	2.04	2.00	1.75	1.40	1.16	2.16	2.11	1.84	1.46	1.20
45° Slope	Up	90	10	2.44	2.31	1.65	1.06	0.76	2.96	2.78	1.88	1.15	0.81
		50	30	2.06	1.98	1.56	1.10	0.83	1.99	1.92	1.52	1.08	0.82
		50	10	2.55	2.44	1.83	1.22	0.90	2.90	2.75	2.00	1.29	0.94
		0	20	2.20	2.14	1.76	1.30	1.02	2.13	2.07	1.72	1.28	1.00
		0	10	2.63	2.54	2.03	1.44	1.10	2.72	2.62	2.08	1.47	1.12
		-50	20	2.08	2.04	1.78	1.42	1.17	2.05	2.01	1.76	1.41	1.16
		-50	10	2.62	2.56	2.17	1.66	1.33	2.53	2.47	2.10	1.62	1.30
Vertical	Horiz.	90	10	2.47	2.34	1.67	1.06	0.77	3.50	3.24	2.08	1.22	0.84
		50	30	2.57	2.46	1.84	1.23	0.90	2.91	2.77	2.01	1.30	0.94
		50	10	2.66	2.54	1.88	1.24	0.91	3.70	3.46	2.35	1.43	1.01
		0	20	2.82	2.72	2.14	1.50	1.13	3.14	3.02	2.32	1.58	1.18
		0	10	2.93	2.82	2.20	1.53	1.15	3.77	3.59	2.64	1.73	1.26
		-50	20	2.90	2.82	2.35	1.76	1.39	2.90	2.83	2.36	1.77	1.39
		-50	10	3.20	3.10	2.54	1.87	1.46	3.72	3.60	2.87	2.04	1.56
45° Slope	Down	90	10	2.48	2.34	1.67	1.06	0.77	3.53	3.27	2.10	1.22	0.84
		50	30	2.64	2.52	1.87	1.24	0.91	3.43	3.23	2.24	1.39	0.99
		50	10	2.67	2.55	1.89	1.25	0.92	3.81	3.57	2.40	1.45	1.02
		0	20	2.91	2.80	2.19	1.52	1.15	3.75	3.57	2.63	1.72	1.26
		0	10	2.94	2.83	2.21	1.53	1.15	4.12	3.91	2.81	1.80	1.30
		-50	20	3.16	3.07	2.52	1.86	1.45	3.78	3.65	2.90	2.05	1.57
		-50	10	3.26	3.16	2.58	1.89	1.47	4.35	4.18	3.22	2.21	1.66
Horiz.	Down	90	10	2.48	2.34	1.67	1.06	0.77	3.55	3.29	2.10	1.22	0.85
		50	30	2.66	2.54	1.88	1.24	0.91	3.77	3.52	2.38	1.44	1.02
		50	10	2.67	2.55	1.89	1.25	0.92	3.84	3.59	2.41	1.45	1.02
		0	20	2.94	2.83	2.20	1.53	1.15	4.18	3.96	2.83	1.81	1.30
		0	10	2.96	2.85	2.22	1.53	1.16	4.25	4.02	2.87	1.82	1.31
		-50	20	3.25	3.15	2.58	1.89	1.47	4.60	4.41	3.36	2.28	1.69
		-50	10	3.28	3.18	2.60	1.90	1.47	4.71	4.51	3.42	2.30	1.71

Table 2   Thermal Resistances of Plane Air Spaces [c,d]*

Position of Air Space	Direction of Heat Flow	Mean Temp.[b] (F)	Temp Diff.[b] (deg F)	1.5-in. Air Space[c] Value of E[a,b] 0.03	0.05	0.2	0.5	0.82	3.5-in. Air Space[c] Value of E[a,b] 0.03	0.05	0.2	0.5	0.82
Horiz	Up	90	10	2.55	2.41	1.71	1.08	0.77	2.84	2.66	1.83	1.13	0.80
		50	30	1.87	1.81	1.45	1.04	0.80	2.09	2.01	1.58	1.10	0.84
		50	10	2.50	2.40	1.81	1.21	0.89	2.80	2.66	1.95	1.28	0.93
		0	20	2.01	1.95	1.63	1.23	0.97	2.25	2.18	1.79	1.32	1.03
		0	10	2.43	2.35	1.90	1.38	1.06	2.71	2.62	2.07	1.47	1.12
		-50	20	1.94	1.91	1.68	1.36	1.13	2.19	2.14	1.86	1.47	1.20
		-50	10	2.37	2.31	1.99	1.55	1.26	2.65	2.58	2.18	1.67	1.33
45° Slope	Up	90	10	2.92	2.73	1.86	1.14	0.80	3.18	2.96	1.97	1.18	0.82
		50	30	2.14	2.06	1.61	1.12	0.84	2.26	2.17	1.67	1.15	0.86
		50	10	2.88	2.74	1.99	1.29	0.94	3.12	2.95	2.10	1.34	0.96
		0	20	2.30	2.23	1.82	1.34	1.04	2.42	2.35	1.90	1.38	1.06
		0	10	2.79	2.69	2.12	1.49	1.13	2.98	2.87	2.23	1.54	1.16
		-50	20	2.22	2.17	1.88	1.49	1.21	2.34	2.29	1.97	1.54	1.25
		-50	10	2.71	2.64	2.23	1.69	1.35	2.87	2.79	2.33	1.75	1.39
Vertical	Horiz.	90	10	3.99	3.66	2.25	1.27	0.87	3.69	3.40	2.15	1.24	0.85
		50	30	2.58	2.46	1.84	1.23	0.90	2.67	2.55	1.89	1.25	0.91
		50	10	3.79	3.55	2.39	1.45	1.02	3.63	3.40	2.32	1.42	1.01
		0	20	2.76	2.66	2.10	1.48	1.12	2.88	2.78	2.17	1.51	1.14
		0	10	3.51	3.35	2.51	1.67	1.23	3.49	3.33	2.50	1.67	1.23
		-50	20	2.64	2.58	2.18	1.66	1.33	2.82	2.75	2.30	1.73	1.37
		-50	10	3.31	3.21	2.62	1.91	1.48	3.40	3.30	2.67	1.94	1.50
45° Slope	Down	90	10	5.07	4.55	2.56	1.36	0.91	4.81	4.33	2.49	1.34	0.90
		50	30	3.58	3.36	2.31	1.42	1.00	3.51	3.30	2.28	1.40	1.00
		50	10	5.10	4.66	2.85	1.60	1.09	4.74	4.36	2.73	1.57	1.08
		0	20	3.85	3.66	2.68	1.74	1.27	3.81	3.63	2.66	1.74	1.27
		0	10	4.92	4.62	3.16	1.94	1.37	4.59	4.32	3.02	1.88	1.34
		-50	20	3.62	3.50	2.80	2.01	1.54	3.77	3.64	2.90	2.05	1.57
		-50	10	4.67	4.47	3.40	2.29	1.70	4.50	4.32	3.31	2.25	1.68
Horiz.	Down	90	10	6.09	5.35	2.79	1.43	0.94	10.07	8.19	3.41	1.57	1.00
		50	30	6.27	5.63	3.18	1.70	1.14	9.60	8.17	3.86	1.88	1.22
		50	10	6.61	5.90	3.27	1.73	1.15	11.15	9.27	4.09	1.93	1.24
		0	20	7.03	6.43	3.91	2.19	1.49	10.90	9.52	4.87	2.47	1.62
		0	10	7.31	6.66	4.00	2.22	1.51	11.97	10.32	5.08	2.52	1.64
		-50	20	7.73	7.20	4.77	2.85	1.99	11.64	10.49	6.02	3.25	2.18
		-50	10	8.09	7.52	4.91	2.89	2.01	12.98	11.56	6.36	3.34	2.22

[a] Interpolation is permissible for other values of mean temperature, temperature differences, and effective emittance E. Interpolation and moderate extrapolation for air spaces greater than 3.5 in. are also permissible.

[b] Effective emittance of the space E is given by $1/E = 1/e_1 + 1/e_2 - 1$, where e_1 and e_2 are the emittances of the surfaces of the air space (See section B of Table

c Credit for an air space resistance value cannot be taken more than once and only for the boundary conditions established.

d Resistances of horizontal spaces with heat flow downward are substantially independent of temperature difference.

e Thermal resistance values were determined from the relation $R = 1/C$, where $C = h_c + Eh_r$, h_c is the conduction-convection coefficient, Eh_r is the radiation coefficient $\cong 0.00686\,E\,[(460 + t_m)/100]^3$, and t_m is the mean temperature of the air space. For interpretation from Table 2 to air space thicknesses less than 0.5 in. (as in insulating window glass), assume $h_c = 0.795\,(1 + 0.0016)$ and compute R-values from the above relations for an air space thickness of 0.2 in.

*Based on National Bureau of Standards data presented in Housing Research Paper No. 32, Housing and Home Finance Agency 1954, U. S. Government Printing Office, Washington 20402.

Table 3A Thermal Properties of Typical Building and Insulating Materials—(Design Values)[a]

(For Industrial Insulation Design Values, see Table 3B). These constants are expressed in Btu per (hour) (square foot) (degree Fahrenheit temperature difference). Conductivities (k) are per inch thickness, and conductances (C) are for thickness or construction stated, not per inch thickness. **All values are for a mean temperature of 75 F, except as noted by an asterisk (*) which have been reported at 45 F.** The SI units for Resistance (last two columns) were calculated by taking the values from the two Resistance columns under Customary Unit, and multiplying by the factor $1/k$ (r/in.) and $1/C$ (R) for the appropriate conversion factor.

Description	Density (lb/ft3)	Conductivity (k)	Conductance (C)	Customary Unit Resistance[b] (R) Per inch thickness (1/k)	For thickness listed (1/C)	Specific Heat, Btu/(lb)(deg F)	SI Unit Resistance[b] (R) (m·K)/W	(m2·K)/W
BUILDING BOARD[c]								
Boards, Panels, Subflooring, Sheathing Woodboard Panel Products								
Asbestos-cement board............ 0.125 in.	120	4.0	—	0.25	—	0.24	1.73	
Asbestos-cement board............ 0.25 in.	120		33.00		0.03			0.005
Asbestos-cement board............	120		16.50		0.06			0.01
Gypsum or plaster board .... 0.375 in.	50		3.10		0.32			0.06
Gypsum or plaster board ..... 0.5 in.	50		2.22		0.45	0.26		0.08
Gypsum or plaster board .... 0.625 in.	50		1.78		0.56			0.10
Plywood (Douglas Fir)...........	34	0.80	—	1.25	—	0.29	8.66	
Plywood (Douglas Fir)........... 0.25 in.	34		3.20		0.31			0.05
Plywood (Douglas Fir).......... 0.375 in.	34		2.13		0.47			0.08
Plywood (Douglas Fir)...........0.5 in.	34		1.60		0.62			0.11
Plywood (Douglas Fir).......... 0.625 in.	34		1.29		0.77			0.19

Reprinted, with permission, from ASHRAE Handbook of Fundamentals, 1977.

Table 3A Thermal Properties of Typical Building and Insulating Materials—(Design Values)[a]

Description	Density (lb/ft³)	Customary Unit					SI Unit	
		Conductivity (k)	Conductance (C)	Resistance[b] (R) Per inch thickness (1/k)	Resistance[b] (R) For thickness listed (1/C)	Specific Heat, Btu/(lb)(deg F)	Resistance[b] (R) (m·K)/W	Resistance[b] (R) (m²·K)/W
Plywood or wood panels. . . . . 0.75 in.	34	—	1.07	—	0.93	0.29	—	0.16
Vegetable Fiber Board								
Sheathing, regular density. . . . . . 0.5 in.	18	—	0.76	—	1.32	0.31		0.23
. . . . 0.78125 in.	18	—	0.49	—	2.06			0.36
Sheathing intermediate density. . . 0.5 in.	22	—	0.82	—	1.22	0.31		0.21
Nail-base sheathing. . . . . 0.5 in.	25	—	0.88	—	1.14	0.31		0.20
Shingle backer. . . . . 0.375 in.	18	—	1.06	—	0.94	0.31		0.17
Shingle backer. . . . . 0.3125 in.	18	—	1.28	—	0.78			0.14
Sound deadening board. . . . . 0.5 in.	15	—	0.74	—	1.35	0.30		0.24
Tile and lay-in panels, plain or acoustic	18	0.40	—	2.50	—	0.14	17.33	
. . . . 0.5 in.	18	—	0.80	—	1.25			0.22
. . . . 0.75 in.	18	—	0.53	—	1.89			0.33
Laminated paperboard.	30	0.50	—	2.00	—	0.33	13.86	
Homogeneous board from repulped paper	30	0.50	—	2.00	—	0.28	13.86	
Hardboard								
Medium density	50	0.73	—	1.37	—	0.31	9.49	
High density, service temp. service underlay	55	0.82	—	1.22	—	0.32	8.46	
High density, std. tempered	63	1.00	—	1.00	—	0.32	6.93	
Particleboard								
Low density	37	0.54	—	1.85	—	0.31	12.82	
Medium density	50	0.94	—	1.06	—	0.31	7.35	
High density	62.5	1.18	—	0.85	—	0.31	5.89	
Underlayment. . . . . 0.625 in.	40	—	1.22	—	0.82	0.29		0.14
Wood subfloor. . . . . 0.75 in.		—	1.06	—	0.94	0.33		0.17
BUILDING MEMBRANE								
Vapor—permeable felt.	—	—	16.70	—	0.06			0.01
Vapor—seal, 2 layers of mopped 15-lb felt	—	—	8.35	—	0.12			0.02
Vapor—seal, plastic film	—	—	—	—	Negl.			

FINISH FLOORING MATERIALS									
Carpet and fibrous pad	—	—	0.48	—	*2.08*	0.34		*0.37*
Carpet and rubber pad	—	—	0.81	—	*1.23*	0.33		*0.22*
Cork tile . . . 0.125 in.	—	—	3.60	—	*0.28*	0.48		*0.05*
Terrazzo . . . 1 in.	—	—	12.50	—	*0.08*	0.19		*0.01*
Tile—asphalt, linoleum, vinyl, rubber	—	—	20.00	—	*0.05*	0.30		*0.01*
vinyl asbestos						0.24		
ceramic						0.19		
Wood, hardwood finish . . . 0.75 in.			1.47		*0.68*			0.12
INSULATING MATERIALS								
Blanket and Batt								
Mineral Fiber, fibrous form processed from rock, slag, or glass								
approx.[e] 2–2.75 in.	0.3–2.0	—	0.143	—	*7*[d]	0.17–0.23		*1.23*
approx.[e] 3–3.5 in.	0.3–2.0	—	0.091	—	*11*[d]			*1.94*
approx.[e] 3.50–6.5	0.3–2.0	—	0.053	—	*19*[d]			*3.35*
approx.[e] 6–7 in.	0.3–2.0		0.045		*22*[d]			*3.87*
approx.[d] 8.5 in.	0.3–2.0		0.033		*30*[d]			*5.28*
Board and Slabs								
Cellular glass	8.5	0.38	—	*2.63*	—	0.24	*18.23*	
Glass fiber, organic bonded	4–9	0.25	—	*4.00*	—	0.23	*27.72*	
Expanded rubber (rigid)	4.5	0.22	—	*4.55*	—	0.40	*31.53*	
Expanded polystyrene extruded Cut cell surface	1.8	0.25	—	*4.00*	—	0.29	*27.72*	
Expanded polystyrene extruded Smooth skin surface	2.2	0.20	—	*5.00*	—	0.29	*34.65*	
Expanded polystyrene extruded Smooth skin surface	3.5	0.19	—	*5.26*	—		*36.45*	
Expanded polystyrene, molded beads	1.0	0.28	—	*3.57*	—	0.29	*24.74*	
Expanded polyurethane[f] (R-11 exp.)	1.5	0.16	—	*6.25*	—	0.38	*43.82*	
(Thickness 1 in. or greater)	2.5							
Mineral fiber with resin binder	15	0.29	—	*3.45*	—	0.17	23.91	
Mineral fiberboard, wet felted								
Core or roof insulation	16–17	0.34	—	*2.94*	—		20.38	
Acoustical tile	18	0.35	—	*2.86*	—	0.19	19.82	
Acoustical tile	21	0.37	—	*2.70*	—		18.71	
Mineral fiberboard, wet molded								
Acoustical tile[g]	23	0.42	—	*2.38*	—	0.14	16.49	
Wood or cane fiberboard								
Acoustical tile[g] . . . 0.5 in.	—	—	0.80	—	*1.25*	0.31		0.22
Acoustical tile[g] . . . 0.75 in.	—	—	0.53	—	*1.89*			0.33
Interior finish (plank, tile)	15	0.35	—	*2.86*	—	0.32	19.82	
Wood shredded (cemented in preformed slabs)	22	0.60	—	*1.67*	—	0.31	11.57	

Table 3A Thermal Properties of Typical Building and Insulating Materials—(Design Values)[a]

Description	Density (lb/ft³)	Customary Unit					SI Unit	
		Conductivity (k)	Conductance (C)	Resistance[b] (R) Per inch thickness (1/k)	Resistance[b] (R) For thickness listed (1/C)	Specific Heat, Btu/(lb)(deg F)	Resistance[b] (R) (m·K)/W	Resistance[b] (R) (m²·K)/W
LOOSE FILL								
Cellulosic insulation (milled paper or wood pulp)	2.3–3.2	0.27–0.32	—	3.13–3.70	—	0.33	21.69–25.64	
Sawdust or shavings	8.0–15.0	0.45	—	2.22	—	0.33	15.39	
Wood fiber, softwoods	2.0–3.5	0.30	—	3.33	—	0.33	23.08	
Perlite, expanded[d]	5.0–8.0	0.37	—	2.70	—	0.26	18.71	
Mineral fiber (rock, slag or glass)								
approx.[e] 3.75–5 in.	0.6–2.0	—	—		11	0.17		1.94
approx.[e] 6.5–8.75 in.	0.6–2.0	—	—		19			3.35
approx.[e] 7.5–10 in.	0.6–2.0	—	—		22			3.87
approx.[e] 10.25–13.75 in.	0.6–2.0	—	—		30			5.28
Vermiculite, exfoliated	7.0–8.2	0.47	—	2.13	—	3.20	14.76	
	4.0–6.0	0.44	—	2.27	—,		15.73	
ROOF INSULATION[h]								
Preformed, for use above deck								
Different roof insulations are available in different thicknesses to provide the design C values listed.[h] Consult individual manufacturers for actual *thickness of their material.*			0.72 to 0.12		1.39 to 8.33		— —	0.24 to 1.47
MASONRY MATERIALS								
CONCRETES								
Cement mortar	116	5.0	—	0.20	—		1.39	
Gypsum-fiber concrete 87.5% gypsum, 12.5% wood chips	51	1.66	—	0.60	—		4.16	
Lightweight aggregates including expanded shale, clay or slate; expanded slags; cinders; pumice; vermiculite; also cellular concretes	120	5.2	—	0.19	—	0.21	1.32	
	100	3.6	—	0.28	—		1.94	
	80	2.5	—	0.40	—		2.77	
	60	1.7	—	0.59	—		4.09	
	40	1.15	—	0.86	—		5.96	
	30	0.90	—	1.11	—		7.69	
	20	0.70	—	1.43	—		9.91	
Perlite, expanded	40	0.93		1.08		0.32	7.48	
	30	0.71		1.41			9.77	
	20	0.50		2.00			13.86	

Material								
Sand and gravel or stone aggregate (oven dried)	140	9.0	—	0.11	—	0.22	*0.76*	
Sand and gravel or stone aggregate (not dried)	140	12.0	—	0.08	—		*0.55*	
Stucco	116	5.0	—	0.20	—		*1.39*	
MASONRY UNITS								
Brick, common[i]	120	5.0	—	0.20	—	0.19	*1.39*	
Brick, face[i]	130	9.0	—	0.11	—		*0.76*	
Clay tile, hollow:								
1 cell deep .... 3 in.			1.25		0.80	0.21		*0.14*
1 cell deep .... 4 in.			0.90		1.11			*0.20*
2 cells deep .... 6 in.			0.66		1.52			*0.27*
2 cells deep .... 8 in.			0.54		1.85			*0.33*
2 cells deep .... 10 in.			0.45		2.22			*0.39*
3 cells deep .... 12 in.			0.40		2.50			*0.44*
Concrete blocks, three oval core:								
Sand and gravel aggregate ... 4 in.			1.40		0.71	0.22		*0.13*
8 in.			0.90		1.11			*0.20*
12 in.			0.78		1.28			*0.23*
Cinder aggregate ... 3 in.			1.16		0.86	0.21		*0.15*
4 in.			0.90		1.11			*0.20*
8 in.			0.58		1.72			*0.30*
12 in.			0.53		1.89			*0.33*
Lightweight aggregate ... 3 in.			0.79		1.27	0.21		*0.22*
(expanded shale, clay, slate ... 4 in.			0.67		1.50			*0.26*
or slag; pumice) ... 8 in.			0.50		2.00			*0.35*
12 in.			0.44		2.27			*0.40*
Concrete blocks, rectangular core.*[j]								
Sand and gravel aggregate								
2 core, 8 in. 36 lb.k*			0.96		1.04	0.22		*0.18*
Same with filled cores[j]*			0.52		1.93	0.22		*0.34*
Lightweight aggregate (expanded shale, clay, slate or slag; pumice):								
3 core, 6 in. 19 lb.k*			0.61		1.65	0.21		*0.29*
Same with filled cores[j]*			0.33		2.99			*0.53*
2 core, 8 in. 24 lb.k*			0.46		2.18			*0.38*
Same with filled cores[j]*			0.20		5.03			*0.89*
3 core, 12 in. 38 lb.k*			0.40		2.48			*0.44*
Same with filled cores[j]*			0.17		5.82			*1.02*
Stone, lime or sand.		12.50	—	0.08	—	0.19	*0.55*	
Gypsum partition tile:								
3 × 12 × 30 in. solid			0.79		1.26	0.19		*0.22*
3 × 12 × 30 in. 4-cell			0.74		1.35			*0.24*
4 × 12 × 30 in. 3-cell			0.60		1.67			*0.29*

Table 3A Thermal Properties of Typical Building and Insulating Materials—(Design Values)[a]

Description	Customary Unit						SI Unit	
	Density (lb/ft³)	Conductivity (k)	Conductance (C)	Resistance[b] (R) Per inch thickness (1/k)	Resistance[b] (R) For thickness listed (1/C)	Specific Heat, Btu/(lb)(deg F)	Resistance[b] (R) (m·K)/W	Resistance[b] (R) (m²·K)/W
METALS								
(See Chapter 37, Table 3)								
PLASTERING MATERIALS								
Cement plaster, sand aggregate	116	5.0	—	0.20	—	0.20	1.39	
..........................0.375 in.			13.3	—	0.08	0.20		0.01
..........................0.75 in.	—		6.66	—	0.15	0.20		0.03
Gypsum plaster:								
Lightweight aggregate..................0.5 in.	45	—	3.12	—	0.32			0.06
Lightweight aggregate................0.625 in.	45	—	2.67	—	0.39			0.07
Lightweight agg. on metal lath........0.75 in.		—	2.13	—	0.47			0.08
Perlite aggregate	45	1.5	—	0.67	—	0.32	4.64	
Sand aggregate	105	5.6	—	0.18	—	0.20	1.25	
Sand aggregate........................0.5 in.	105		11.10	—	0.09			0.02
Sand aggregate......................0.625 in.	105		9.10	—	0.11			0.02
Sand aggregate on metal lath..........0.75 in.	—		7.70	—	0.13			0.02
Vermiculite aggregate	45	1.7	—	0.59	—		4.09	
ROOFING								
Asbestos-cement shingles	120	—	4.76	—	0.21	0.24		0.04
Asphalt roll roofing	70	—	6.50	—	0.15	0.36		0.03
Asphalt shingles	70	—	2.27	—	0.44	0.30		0.08
Built-up roofing......................0.375 in.	70	—	3.00	—	0.33	0.35		0.06
Slate.................................0.5 in.	—	—	20.00	—	0.05	0.30		0.01
Wood shingles, plain and plastic film faced	—	—	1.06	—	0.94	0.31		0.17
SIDING MATERIALS (On Flat Surface)								
Shingles								
Asbestos-cement	120	—	4.75	—	0.21	0.31		0.04
Wood, 16 in., 7.5 exposure	—	—	1.15	—	0.87	0.28		0.15
Wood, double, 16-in., 12-in. exposure	—	—	0.84	—	1.19	0.31		0.21
Wood, plus insul. backer board, 0.3125 in.	—	—	0.71	—	1.40			0.25
Siding								
Asbestos-cement, 0.25 in., lapped	—	—	4.76	—	0.21	0.24		0.04
Asphalt roll siding	—	—	6.50	—	0.15	0.35		0.03
Asphalt insulating siding (0.5 in. bed.)	—	—	0.69	—	1.46	0.35		0.26
Wood, drop, 1 × 8 in.	—	—	1.27	—	0.79	0.28		0.14
Wood, bevel, 0.5 × 8 in., lapped	—	—	1.23	—	0.81	0.28		0.14

Material								
Wood, bevel, 0.75 × 10 in., lapped	—	—	0.95	—	1.05	0.28		0.18
Wood, plywood, 0.375 in., lapped	—	—	1.59	—	0.59	0.29		0.10
Wood, medium density siding, 0.4375 in.	40	1.49	—	0.67		0.28	4.65	
Aluminum or Steel[m], over sheathing								
Hollow-backed	—	—	1.61	—	0.61	0.29		0.11
Insulating-board backed nominal								
0.375 in.	—	—	0.55	—	1.82	0.32		0.32
Insulating-board backed nominal								
0.375 in., foil backed	—	—	0.34	—	2.96	0.20		0.52
			10.00		0.10			0.02
Architectural glass	—	—	—	—	—			
WOODS								
Maple, oak, and similar hardwoods	45	1.10	—	0.91	—	0.30	6.31	0.17
Fir, pine, and similar softwoods	32	0.80	—	1.25	0.94	0.33	8.66	0.33
Fir, pine, and similar softwoods . . . . 0.75 in.	32	—	1.06	—	1.89	0.33		0.60
. . . . 1.5 in.			0.53	—	3.17			0.75
. . . . 2.5 in.			0.32	—	4.35			
. . . . 3.5 in.			0.23	—				

Notes for Table 3A

[a] Representative values for dry materials were selected by ASHRAE TC4.4, Insulation and Moisture Barriers. They are intended as design (not specification) values for materials in normal use. For properties of a particular product, use the value supplied by the manufacturer or by unbiased tests

[b] Resistance values are the reciprocals of C before rounding off C to two decimal places.

[c] Also see Insulating Materials, Board.

[d] Does not include paper backing and facing, if any. Where insulation forms a boundary (reflective or otherwise) of an air space, see Tables 1 and 2 for the insulating value of air space for the appropriate effective emittance and temperature conditions of the space.

[e] Conductivity varies with fiber diameter. Insulation is produced by different densities; therefore, there is a wide variation in thickness for the same R-value among manufacturers. No effort should be made to relate any specific R-value to any specific thickness. Commercial thicknesses generally available range from 2 to 8.5.

[f] Values are for aged board stock.

[g] Insulating values of acoustical tile vary, depending on density of the board and on type, size, and depth of perforations.

[h] The U. S. Department of Commerce, *Simplified Practice Recommendation for Thermal Conductance Factors for Preformed Above-Deck Roof Insulation*, No. R 257-55, recognizes the specification of roof insulation on the basis of the C-values shown. Roof insulation is made in thicknesses to meet these values.

[i] Face brick and common brick do not always have these specific densities. When density is different from that shown, there will be a change in thermal conductivity.

[j] Data on rectangular core concrete blocks differ from the above data on oval core blocks, due to core configuration, different mean temperatures, and possibly differences in unit weights. Weight data on the oval core blocks tested are not available.

[k] Weights of units is approximately 7.625 in. high and 15.75 in. long. These weights are given as a means of describing the blocks tested, but conductance values are all for 1 ft² of area.

[l] Vermiculite, perlite, or mineral wool insulation. Where insulation is used, vapor barriers or other precautions must be considered to keep insulation dry.

[m] Values for metal siding applied over flat surfaces vary widely, depending on amount of ventilation of air space beneath the siding; whether air space is reflective or nonreflective; and on thickness, type, and application of insulating backing-board used. Values given are averages for use as design guides, and were obtained from several guarded hotbox tests (ASTM C236) or calibrated hotbox tests (BSS 77) on hollow-backed types and types made using backing-boards of wood fiber, foamed plastic, and glass fiber. Departures of ±50% or more from the values given may occur.

Reprinted, with permission, from ASHRAE Handbook of Fundamentals, 1977.

Table 3B Thermal Conductivity (k) of Industrial Insulation (Design Values)[a] (For Mean Temperatures Indicated)

Expressed in Btu per (hour)(square foot)(degree Fahrenheit temperature difference per in.)

Form — Material Composition	Accepted Max Temp for Use, F*	Typical Density (lb/ft³)	Typical Conductivity k at Mean Temp F													
			-100	-75	-50	-25	0	25	50	75	100	200	300	500	700	900
BLANKETS & FELTS																
MINERAL FIBER (Rock, slag or glass)																
Blanket, metal reinforced	1200	6-12									0.26	0.32	0.39	0.54		
	1000	2.5-6									0.24	0.31	0.40	0.61		
Mineral fiber, glass																
Blanket, flexible, fine-fiber organic bonded	350	0.65				0.25	0.26	0.28	0.30	0.33	0.36	0.53				
		0.75				0.24	0.25	0.27	0.29	0.32	0.34	0.48				
		1.0				0.23	0.24	0.25	0.27	0.29	0.32	0.43				
		1.5				0.21	0.22	0.23	0.25	0.27	0.28	0.37				
		2.0				0.20	0.21	0.23	0.23	0.25	0.26	0.33				
		3.0				0.19	0.20	0.21	0.22	0.23	0.24	0.31				
Blanket, flexible, textile-fiber organic bonded	350	0.65				0.27	0.28	0.29	0.30	0.31	0.32	0.50	0.68			
		0.75				0.26	0.27	0.28	0.29	0.31	0.32	0.48	0.66			
		1.0				0.24	0.25	0.26	0.27	0.29	0.31	0.45	0.60			
		1.5				0.22	0.23	0.24	0.25	0.27	0.29	0.39	0.51			
		3.0				0.20	0.21	0.22	0.23	0.24	0.25	0.32	0.41			
Felt, semirigid organic bonded																
Laminated & felted	400	3-8	0.16	0.17	0.18	0.19	0.20	0.24	0.25	0.26	0.27	0.35	0.44			
Without binder	850	3						0.21	0.22	0.23	0.24	0.35	0.55			
	1200	7.5											0.35	0.45	0.60	
VEGETABLE & ANIMAL FIBER																
Hair Felt or Hair Felt plus Jute	180	10						0.26	0.28	0.29	0.30					
BLOCKS, BOARDS & PIPE INSULATION																
ASBESTOS																
Laminated asbestos paper	700	30									0.40	0.45	0.50	0.60		
Corrugated & laminated asbestos Paper																
4-ply	300	11-13								0.54	0.57	0.68				
6-ply	300	15-17								0.49	0.51	0.59				
8-ply	300	18-20								0.47	0.49	0.57				
MOLDED AMOSITE & BINDER	1500	15-18									0.32	0.37	0.42	0.52	0.62	0.72
85% MAGNESIA	600	11-12									0.35	0.38	0.42			
CALCIUM SILICATE	1200	11-13									0.38	0.41	0.44		0.62	0.72
	1800	12-15													0.74	0.95
CELLULAR GLASS	800	9			0.32	0.33	0.35	0.36	0.38	0.40	0.42	0.48	0.55			
DIATOMACEOUS SILICA	1600	21-22												0.64	0.68	0.72
	1900	23-25												0.70	0.75	0.80

The following table (rotated on the page) gives representative thermal conductivity (k) values for insulating materials. For each material the "Max Temp*" (°F) and "Density" (lb/ft³) columns are followed by k-values at increasing mean temperatures.

Material	Max Temp*	Density	k values (increasing mean temperature)
MINERAL FIBER			
Glass,			
Organic bonded, block and boards	400	3–10	0.16, 0.17, 0.18, 0.19, 0.20, 0.22, 0.24, 0.25, 0.26, 0.33, 0.40, 0.52, 0.62, 0.74
Nonpunking binder	1000	3–10	0.26, 0.31, 0.38
Pipe insulation, slag or glass	350	3–4	0.24, 0.29
Inorganic bonded-block	500	3–10	0.26, 0.33, 0.40
	1000	10–15	0.33, 0.38, 0.45, 0.55
	1800	15–24	0.32, 0.37, 0.42, 0.52
Pipe insulation slag or glass	1000	10–15	0.33, 0.38, 0.45, 0.55
MINERAL FIBER			
Resin binder		15	0.23, 0.24, 0.25, 0.26, 0.28, 0.29
RIGID POLYSTYRENE			
Extruded, Refrigerant 12 exp	170	3.5	0.16, 0.16, 0.15, 0.16, 0.16, 0.17, 0.18, 0.19, 0.20
Extruded, Refrigerant 12 exp	170	2.2	0.16, 0.16, 0.17, 0.16, 0.17, 0.18, 0.19, 0.20
Extruded	170	1.8	0.17, 0.18, 0.19, 0.20, 0.21, 0.23, 0.24, 0.25, 0.27
Molded beads	170	1	0.18, 0.20, 0.21, 0.23, 0.24, 0.25, 0.26, 0.28
POLYURETHANE**			
Refrigerant 11 exp	210	1.5–2.5	0.16, 0.16, 0.17, 0.17, 0.18, 0.17, 0.17, 0.16, 0.16, 0.17, 0.17
RUBBER, Rigid Foamed	150	4.5	0.20, 0.20, 0.21, 0.22, 0.23, 0.23
VEGETABLE & ANIMAL FIBER			
Wool felt (pipe insulation)	180	20	0.28, 0.30, 0.31, 0.33
INSULATING CEMENTS			
MINERAL FIBER			
(Rock, slag, or glass)			
With colloidal clay binder	1800	24–30	0.49, 0.55, 0.61, 0.73, 0.85
With hydraulic setting binder	1200	30–40	0.75, 0.80, 0.85, 0.95
LOOSE FILL			
Cellulose insulation (milled pulverized paper or wood pulp)		2.5–3	0.19, 0.21, 0.23, 0.25, 0.26, 0.27, 0.29
Mineral fiber, slag, rock or glass		2–5	0.26, 0.28, 0.31
Perlite (expanded)		5–8	0.29, 0.30, 0.32, 0.34, 0.35, 0.37, 0.39
Silica aerogel		7.6	0.13, 0.14, 0.15, 0.15, 0.16, 0.17, 0.18
		7–8.2	0.39, 0.40, 0.42, 0.44, 0.45, 0.47, 0.49
Vermiculite (expanded)		4–6	0.34, 0.35, 0.38, 0.40, 0.42, 0.44, 0.46

aRepresentative values for dry materials as selected by ASHRAE TC 4.4, Insulation and Moisture Barriers. They are intended as design (not specification) values for materials of building construction for normal use. For thermal resistance of a particular product, use the value supplied by the manufacturer or by unbiased tests.

*These temperatures are generally accepted as maximum. When operating temperature approaches these limits follow the manufacturer's recommendations.

**Values are for aged board stock.

Note: Some polyurethane foams are formed by means which produce a stable product (with respect to k), but most are blown with refrigerant and will change with time.

Reprinted, with permission, from ASHRAE Handbook of Fundamentals, 1977.

Table 4A Coefficients of Transmission (U) of Frame Walls

These coefficients are expressed in Btu per (hour) (square foot) (degree Fahrenheit) difference in temperature between the air on the two sides, and are based on an outside wind velocity of 15 mph

Replace Air Space with 3.5-in. R-11 Blanket Insulation (New Item 4)

Construction	Resistance (R)			
	1		2	
	Between Framing	At Framing	Between Framing	At Framing
1. Outside surface (15 mph wind)	0.17	0.17	0.17	0.17
2. Siding, wood, 0.5 in.× 8 in. lapped (average)	0.81	0.81	0.81	0.81
3. Sheathing, 0.5-in. asphalt impregnated	1.32	1.32	1.32	1.32
4. Nonreflective air space, 3.5 in. (50 F mean; 10 deg F temperature difference)	1.01	—	11.00	—
5. Nominal 2-in. × 4-in. wood stud	—	4.38	—	4.38
6. Gypsum wallboard, 0.5 in.	0.45	0.45	0.45	0.45
7. Inside surface (still air)	0.68	0.68	0.68	0.68
Total Thermal Resistance (R)	R_i=4.44	R_s=7.81	R_i=14.43	R_s=7.81

Construction No. 1: $U_i = 1/4.44 = 0.225$; $U_s = 1/7.81 = 0.128$. With 20% framing (typical of 2-in. × 4-in. studs @ 16-in. o.c.), $U_{av} = 0.8$ (0.225) + 0.2 (0.128) = 0.206 (See Eq 9)

Construction No. 2: $U_i = 1/14.43 = 0.069$; $U_s = 0.128$. With framing unchanged, $U_{av} = 0.8(0.069) + 0.2(0.128) = 0.081$

Table 4B Coefficients of Transmission (U) of Solid Masonry Walls

Coefficients are expressed in Btu per (hour) (square foot) (degree Fahrenheit) difference in temperature between the air on the two sides), and are based on an outside wind velocity of 15 mph

Replace Furring Strips and Air Space with 1-in. Extruded Polystyrene (New Item 4)

Construction	Resistance (R)		
	1		2
	Between Furring	At Furring	
1. Outside surface (15 mph wind)	0.17	0.17	0.17
2. Common brick, 8 in.	1.60	1.60	1.60
3. Nominal 1-in. ×3-in. vertical furring	—	0.94	—
4. Nonreflective air space, 0.75 in. (50 F mean; 10 deg F temperature difference)	1.01	—	5.00
5. Gypsum wallboard, 0.5 in.	0.45	0.45	0.45
6. Inside surface (still air)	0.68	0.68	0.68
Total Thermal Resistance (R)	$R_i = 3.91$	$R_s = 3.84$	$R_i = 7.90 = R_s$

Construction No. 1: $U_i = 1/3.91 = 0.256$; $U_s = 1/3.84 = 0.260$. With 20% framing (typical of 1-in. × 3-in. vertical furring on masonry @ 16-in. o.c.) $U_{av} = 0.8 (0.256) + 0.2 (0.260) = 0.257$

Construction No. 2: $U_i = U_s = U_{av} = 1/7.90 = 0.127$

Table 4C Coefficients of Transmission (U) of Frame Partitions or Interior Walls

Coefficients are expressed in Btu per (hour) (square foot) (degree Fahrenheit difference in temperature between the air on the two sides), and are based on still air (no wind) conditions on both sides

Replace Air Space with 3.5-in. R-11 Blanket Insulation (New Item 3)

Construction	Resistance (R)			
	1		2	
	Between Framing	At Framing	Between Framing	At Framing
1. Inside surface (still air)	0.68	0.68	0.68	0.68
2. Gypsum wallboard, 0.5 in.	0.45	0.45	0.45	0.45
3. Nonreflective air space, 3.5 in. (50 F mean; 10 deg F temperature difference)	1.01	—	11.00	—
4. Nominal 2-in. × 4-in. wood stud	—	4.38	—	4.38
5. Gypsum wallboard 0.5 in.	0.45	0.45	0.45	0.45
6. Inside surface (still air)	0.68	0.68	0.68	0.68
Total Thermal Resistance (R)	$R_i = 3.27$	$R_s = 6.64$	$R_i = 13.26$	$R_s = 6.64$

Construction No. 1: $U_i = 1/3.27 = 0.306$; $U_s = 1/6.64 = 0.151$. With 10% framing (typical of 2-in. × 4-in. studs @ 24-in. o.c.), $U_{av} = 0.9(0.306) + 0.1(0.151) = 0.290$

Construction No. 2: $U_i = 1/13.26 = 0.075$; $U_s = 1/6.64 = 0.151$. With framing unchanged, $U_{av} = 0.9(0.075) + 0.10(0.151) = 0.083$

Table 4D Coefficients of Transmission (U) of Masonry Walls

Coefficients are expressed in Btu per (hour) (square foot) (degree Fahrenheit difference in temperature between the air on the two sides), and are based on an outside wind velocity of 15 mph

Replace Cinder Aggregate Block with 6-in. Light-weight Aggregate Block with Cores Filled (New Item 4)

Construction	Resistance (R)			
	1		2	
	Between Furring	At Furring	Between Furring	At Furring
1. Outside surface (15 mph wind)	0.17	0.17	0.17	0.17
2. Face brick, 4 in.	0.44	0.44	0.44	0.44
3. Cement mortar, 0.5 in.	0.10	0.10	0.10	0.10
4. Concrete block, cinder aggregate, 8 in.	1.72	1.72	2.99	2.99
5. Reflective air space, 0.75 in. (50 F mean; 30 deg F temperature difference)	2.77	—	2.77	—
6. Nominal 1-in. × 3-in. vertical furring	—	0.94	—	0.94
7. Gypsum wallboard, 0.5 in., foil backed	0.45	0.45	0.45	0.45
8. Inside surface (still air)	0.68	0.68	0.68	0.68
Total Thermal Resistance (R)	$R_i = 6.33$	$R_s = 4.50$	$R_i = 7.60$	$R_s = 5.77$

Construction No. 1: $U_i = 1/6.33 = 0.158$; $U_s = 1/4.50 = 0.222$. With 20% framing (typical of 1-in. × 3-in. vertical furring on masonry @ 16-in. o.c.), $U_{av} = 0.8(0.158) + 0.2(0.222) = 0.171$

Construction No. 2: $U_i = 1/7.60 = 0.132$; $U_s = 1/5.77 = 0.173$. With framing unchanged, $U_{av} = 0.80(0.132) + 0.2(0.173) = 1.40$

Table 4E   Coefficients of Transmission (U) of Masonry Cavity Walls

Coefficients are expressed in Btu per (hour) (square foot) (degree Fahrenheit difference in temperature between the air on the two sides), and are based on an outside wind velocity of 15 mph

Replace Furring Strips and Gypsum Wallboard with 0.625-in. Plaster (Sand Aggregate) Applied Directly to Concrete Block-Fill 2.5-in. Air Space with Vermiculite Insulation (New Items 3 and 7.

Construction	Resistance (R)			
	1		2	
	Between Furring	At Furring	Between Furring	At Furring
1. Outside surface (15 mph wind)	0.17	0.17	0.17	0.17
2. Common brick, 8 in.	0.80	0.80	0.80	0.80
3. Nonreflective air space, 2.5 in. (30 F mean; 10 deg F temperature difference)	1.10*	1.10*	5.32**	5.32**
4. Concrete block, stone aggregate, 4 in.	0.71	0.71	0.71	0.71
5. Nonreflective air space 0.75 in. (50 F mean; 10 deg F temperature difference)	1.01	—	—	—
6. Nominal 1-in. × 3-in. vertical furring	—	0.94	—	—
7. Gypsum wallboard, 0.5 in.	0.45	0.45	0.11	0.11
8. Inside surface (still air)	0.68	0.68	0.68	0.68
Total Thermal Resistance (R)	$R_i = 4.92$	$R_s = 4.85$	$R_i = R_s = 7.79$	

Construction No. 1: $U_i = 1/4.92 = 0.203$; $U_s = 1/4.85 = 0.206$. With 20% framing (typical of 1-in. × 3-in. vertical furring on masonry @16-in. o.c.), $U_{av} = 0.8(0.203) + 0.2(0.206) = 0.204$

Construction No. 2: $U_i = U_s = U_{av} = 1.79 = 0.128$

* Interpolated value from Table 2.
** Calculated value from Table 3.

Table 4F   Coefficients of Transmission (U) of Masonry Partitions

Coefficients are expressed in Btu per (hour) (square foot) (degree Fahrenheit difference in temperature between the air on the two sides), and are based on still air (no wind) conditions on both sides

Replace Concrete Block with 4-in. Gypsum Tile (New Item 3)

Construction	1	2
1. Inside surface (still air)	0.68	0.68
2. Plaster, lightweight aggregate, 0.625 in.	0.39	0.39
3. Concrete block, cinder aggregate, 4 in.	1.11	1.67
4. Plaster, lightweight aggregate, 0.625 in.	0.39	0.39
5. Inside surface (still air)	0.68	0.68
Total Thermal Resistance(R)	3.25	3.81

Construction No. 1: $U = 1/3.25 = 0.308$
Construction No. 2: $U = 1/3.81 = 0.262$

Table 4G Coefficients of Transmission (U) of Frame Construction Ceilings and Floors

Coefficients are expressed in Btu per (hour) (square foot) (degree Fahrenheit difference between the air on the two sides), and are based on still air (no wind) on both sides

Assume Unheated Attic Space above Heated Room with Heat Flow Up—Remove Tile, Felt, Plywood, Subfloor and Air Space—Replace with R-19 Blanket Insulation (New Item 4)

	1		2	
	Heated Room Below		Unheated Space	
	Resistance (R)			
Construction (Heat Flow Up)	Between Floor Joists	At Floor Joist	Between Floor Joists	At Floor Joists
1. Bottom surface (still air)	0.61	0.61	0.61	0.61
2. Metal lath and lightweight aggregate, plaster, 0.75 in.	0.47	0.47	0.47	0.47
3. Nominal 2-in. × 8-in. floor joist	—	9.06	—	9.06
4. Nonreflective airspace, 7.25-in.	0.93*	—	19.00	—
5. Wood subfloor, 0.75 in.	0.94	0.94	—	—
6. Plywood, 0.625 in.	0.78	0.78	—	—
7. Felt building membrane	0.06	0.06	—	—
8. Resilient tile	0.05	0.05	—	—
9. Top surface (still air)	0.61	0.61	0.61	0.61
Total Thermal Resistance (R)	$R_i = 4.45$	$R_s = 12.58$	$R_i = 20.69$	$R_s = 10.75$

Construction No. 1 $U_i = 1/4.45 = 0.225$; $U_s = 1/12.58 = 0.079$. With 10% framing (typical of 2-in. joists @ 16-in. o.c.), $U_{av} = 0.9\,(0.225) + 0.1\,(0.079) = 0.210$

Construction No. 2 $U_i = 1/20.69 = 0.048$; $U_s = 1/10.75 = 0.093$. With framing unchanged, $U_{av} = 0.9\,(0.048) + 0.1\,(0.093) = 0.053$

Table 4H   Coefficients of Transmission (U) of Flat Masonry Roofs with Built-up Roofing, with and without Suspended Ceilings[a]. (Winter Conditions, Upward Flow)

These Coefficients are expressed in Btu per (hour) (square foot) (degree Fahrenheit difference in temperature between the air on the two sides), and are based upon an outside wind velocity of 15 mph

Add Rigid Roof Deck Insulation, $C = 0.24$ ($R = 1/C$) (New Item 7) Construction (Heat Flow Up)	1	2
1. Inside surface (still air)	0.61	0.61
1. Metal lath and lightweight aggregate plaster, 0.75 in.	0.47	0.47
3. Nonreflective air space, greater than 3.5 in. (50 F mean; 10 deg F temperature difference)	0.93*	0.93*
4. Metal ceiling suspension system with metal hanger rods	0**	0**
5. Corrugated metal deck	0	0
6. Concrete slab, lightweight aggregate, 2 in.	2.22	2.22
7. Rigid roof deck insulation (none)	—	4.17
8. Built-up roofing, 0.375 in.	0.33	0.33
9. Outside surface (15 mph wind)	0.17	0.17
Total Thermal Resistance (R)	4.73	8.90

Construction No. 1: $U_{av} = 1/4.73 = 0.211$
Construction No. 2: $U_{av} = 1/8.90 = 0.112$

[a] To adjust U values for the effect of added insulation between framing members, see Table 5 or 6.
* Use largest air space (3.5 in.) value shown in Table 2.
** Area of hanger rods is negligible in relation to ceiling area.

Table 4I Coefficients of Transmission (U) of Wood Construction Flat Roofs and Ceilings (Winter Conditions, Upward Flow)

Coefficients are expressed in Btu per (hour) (square foot) (degree Fahrenheit difference in temperature between the air on the the two sides), and are based upon an outside wind velocity of 15 mph

Replace Roof Deck Insulation and 7.25-in. Air Space with 6-in. R-19 Blanket Insulation and 1.25-in. Air Space (New Items 5 and 7)

Construction (Heat Flow Up)	Resistance (R)			
	1		2	
	Between Joists	At Joists	Between Joists	At Joists
1. Inside surface (still air)	0.61	0.61	0.61	0.61
2. Acoustical tile, fiberboard, glued, 0.5 in.	1.25	1.25	1.25	1.25
3. Gypsum wallboard, 0.5 in.	0.45	0.45	0.45	0.45
4. Nominal 2-in. × 8-in. ceiling joists	—	9.06	—	9.06
5. Nonreflective air space, 7.25 in. (50 F mean; 10 deg F temperature difference)	0.93*	—	1.05**	—
6. Plywood deck, 0.625 in.	0.78	0.78	0.78	0.78
7. Rigid roof deck insulation, c = 0.72, (R = 1/C)	1.39	1.39	19.00	—
8. Built-up roof	0.33	0.33	0.33	0.33
9. Outside surface (15 mph wind)	0.17	0.17	0.17	0.17
Total Thermal Resistance (R)	R_i=5.91	R_s=14.04	R_i=23.64	R_s=12.65

Construction No. 1 U_i = 1/5.91 = 0.169; U_s = 1/14.04 = 0.071. With 10% framing (typical of 2-in. joists @ 16-in. o.c.), U_{av} = 0.9 (0.169) + 0.1 (0.071) = 0.159

Construction No. 2 U_i = 1/23.64 = 0.042; U_s = 1/12.65 = 0.079. With framing unchanged, U_{av} = 0.9 (0.042) + 0.1 (0.079) = 0.046

*Use largest air space (3.5 in.) value shown in Table 2.
**Interpolated value (0 F mean; 10 deg F temperature difference).

Table 4J Coefficients of Transmission (U) of Metal Construction Flat Roofs and Ceilings (Winter Conditions, Upward Flow)

Coefficients are expressed in Btu per (hour) (square foot) (degree Fahrenheit difference in temperature between the air on the two sides), and are based on upon outside wind velocity of 15 mph

Replace **Rigid Roof Deck Insulation** ($C = 0.24$) **and Sand Aggregate Plaster with Rigid Roof Deck Insulation**, $C = 0.36$ **and Lightweight Aggregate Plaster (New Items 2 and 6)**

Construction (Heat Flow Up)	1	2
1. Inside surface (still air)	0.61	0.61
2. Metal lath and sand aggregate plaster, 0.75 in	0.13	0.47
3. Structural beam	0.00*	0.00*
4. Nonreflective air space (50 F mean; 10 deg F temperature difference)	0.93**	0.93**
5. Metal deck	0.00*	0.00*
6. Rigid roof deck insulation, $C = 0.24(R = 1/c)$	4.17	2.78
7. Built-up roofing, 0.375 in.	0.33	0.33
8. Outside surface (15 mph wind)	0.17	0.17
Total Thermal Resistance (R)	6.34	5.29

Construction No. 1: $U = 1/6.34 = 0.158$
Construction No. 2: $U = 1/5.29 = 0.189$

*Total R. Full scale testing of a suitable portion of the construction is preferable.
**Use largest air space (3.5 in.) value shown in Table 2.

Table 4K Coefficients of Transmission (U) of Pitched Roofs[a].

Coefficients are expressed in Btu per (hour) (square foot) (degree Fahrenheit difference in temperature between the air on the two sides), and are based on an outside wind velocity of 15 mph for heat flow upward and 7.5 mph for heat flow downward

Find U_{av} for same Construction 2 with Heat Flow Down (Summer Conditions)

Construction 1 (Heat Flow Up) (Reflective Air Space)	1		2	
	Between Rafters	At Rafters	Between Rafters	At Rafters
1. Inside surface (still air)	0.62	0.62	0.76	0.76
2. Gypsum wallboard 0.5 in., foil backed	0.45	0.45	0.45	0.45
3. Nominal 2-in. × 4-in. ceiling rafter	—	4.38	—	4.38
4. 45 deg slope reflective air space, 3.5 in. (50 F mean, 30 deg F temperature difference)	2.17	—	4.33	—
5. Plywood sheathing, 0.625 in.	0.78	0.78	0.78	0.78
6. Felt building membrane	0.06	0.06	0.06	0.06
7. Asphalt shingle roofing	0.44	0.44	0.44	0.44
8. Outside surface (15 mph wind)	0.17	0.17	0.25**	0.25**
Total Thermal Resistance (R)	$R_i=4.69$	$R_s=6.90$	$R_i=7.07$	$R_s=7.12$

Construction No. 1: $U_i=1/4.69= 0.213$; $U_s = 1/6.90 = 0.145$. With 10% framing (typical of 2-in. rafters @16-in. o.c.), $U_{av}= 0.9 (0.213) + 0.1 (0.145) = 0.206$

Construction No. 2: $U_i=1/7.07 = 0.141$; $U_s = 1/7.12 = 0.140$. With framing unchanged, $U_{av} = 0.9 (0.141) + 0.1 (0.140) = 0.141$

Find U_{av} for same Construction 2 with Heat Flow Down (Summer Conditions)

Construction 1 (Heat Flow Up) (Non-Reflective Air Space)	3		4	
	Between Rafters	At Rafters	Between Rafters	At Rafters
1. Inside surface (still air)	0.62	0.62	0.76	0.76
2. Gypsum wallboard, 0.5 in.	0.45	0.45	0.45	0.45
3. Nominal 2-in. × 4-in. ceiling rafter	—	4.38	—	4.38
4. 45 deg slope, nonreflective air space, 3.5 in. (50 F mean; 10 deg F temperature difference)	0.96	—	0.90*	—
5. Plywood sheathing, 0.625 in.	0.78	0.78	0.78	0.78
6. Felt building membrane	0.06	0.06	0.06	0.06
7. Asphalt shingle roofing	0.44	0.44	0.44	0.44
8. Outside surface (15-mph wind)	0.17	0.17	0.25**	0.25**
Total Thermal Resistance (R)	$R_i=3.48$	$R_s=6.90$	$R_i=3.64$	$R_s=7.12$

Construction No. 3: $U_i = 1/3.48 = 0.287$; $U_s = 1/6.90 = 0.145$. With 10% framing typical of 2-in. rafters @ 16-in. o.c.), $U_{av} = 0.9 (0.287)+ 0.1 (0.145) = 0.273$

Construction No. 4: $U_i = 1/3.64 = 0.275$; $U_s = 1/7.12 = 0.140$. With framing unchanged, $U_{av} = 0.9 (0.275) + 0.1 (0.140) = 0.262$

a Pitch of roof—45 deg.
* Air space value at 90 F meann, 10 F dif. temperature difference.
** 7.5-mph wind.

Reprinted, with permission, from ASHRAE Handbook of Fundamentals, 1977.

Table 5A Determination of U-Value Resulting from Addition of Insulation to the Total Area[e] of any Given Building Section

Given Building Section Property[a,b]		Added R[c,d,e]						
		$R = 4$	$R = 6$	$R = 8$	$R = 12$	$R = 16$	$R = 20$	$R = 24$
U	R	U	U	U	U	U	U	U
1.00	1.00	0.20	0.14	0.11	0.08	0.06	0.05	0.04
0.90	1.11	0.20	0.14	0.11	0.08	0.06	0.05	0.04
0.80	1.25	0.19	0.14	0.11	0.08	0.06	0.05	0.04
0.70	1.43	0.18	0.13	0.11	0.07	0.06	0.05	0.04
0.60	1.67	0.18	0.13	0.10	0.07	0.06	0.05	0.04
0.50	2.00	0.17	0.13	0.10	0.07	0.06	0.05	0.04
0.40	2.50	0.15	0.12	0.10	0.07	0.05	0.04	0.04
0.30	3.33	0.14	0.11	0.09	0.07	0.05	0.04	0.04
0.20	5.00	0.11	0.09	0.08	0.06	0.05	0.04	0.03
0.10	10.00	0.07	0.06	0.06	0.05	0.04	0.03	0.03
0.08	12.50	0.06	0.05	0.05	0.04	0.04	0.03	0.03

[a] For U- or R-values not shown in the table, interpolate as necessary.

[b] Enter column 1 with U or R of the design building section.

[c] Under appropriate column heading for added R, find U-value of resulting design section.

[d] If the insulation occupies previously considered air space, an adjustment must be made in the given building section R-value.

[e] If insulation is applied between framing members, determine average U-value.

Table 5B Determination of U-Value Resulting from Addition of Insulation to Uninsulated Roof Deck

U-Value of Roof without Roof-Deck Insulation[a]	Conductance C of Roof-Deck Insulation					
	0.12	0.15	0.19	0.24	0.36	0.72
	U	U	U	U	U	U
0.10	0.05	0.06	0.07	0.07	0.08	0.09
0.15	0.07	0.08	0.08	0.09	0.11	0.12
0.20	0.08	0.09	0.10	0.11	0.13	0.16
0.25	0.08	0.09	0.11	0.12	0.15	0.19
0.30	0.09	0.10	0.12	0.13	0.16	0.21
0.35	0.09	0.11	0.12	0.14	0.18	0.24
0.40	0.09	0.11	0.13	0.15	0.19	0.26
0.50	0.10	0.12	0.14	0.16	0.21	0.30
0.60	0.10	0.12	0.14	0.17	0.23	0.33
0.70	0.10	0.12	0.15	0.18	0.24	0.35

[a] Interpolation or mild extrapolation may be used.

Reprinted, with permission, from ASHRAE Handbook of Fundamentals, 1977.

Table 6   Effective Resistance of Ventilated Attics[a]—(Summer Condition)

PART A. NONREFLECTIVE SURFACES

Ventilation Air Temp., F	Sol-Air[c] Temp., F	No Ventilation		Natural Ventilation		Ventilation Rate, cfm/ft²		Power Ventilation[d]			
		0		0.1		0.5		1.0		1.5	
		10	20	10	20	10	20	10	20	10	20
		1/U Ceiling Resistance, R[b]									
80	120	1.9	1.9	2.8	3.4	6.3	9.3	9.6	16	11	20
	140	1.9	1.9	2.8	3.5	6.5	10	9.8	17	12	21
	160	1.9	1.9	2.8	3.6	6.7	11	10	18	13	22
90	120	1.9	1.9	2.5	2.8	4.6	6.7	6.1	10	6.9	13
	140	1.9	1.9	2.6	3.1	5.2	7.9	7.6	12	8.6	15
	160	1.9	1.9	2.7	3.4	5.8	9.0	8.5	14	10	17
100	120	1.9	1.9	2.2	2.3	3.3	4.4	4.0	6.0	4.1	6.9
	140	1.9	1.9	2.4	2.7	4.2	6.1	5.8	8.7	6.5	10
	160	1.9	1.9	2.6	3.2	5.0	7.6	7.2	11	8.3	13

PART B. REFLECTIVE SURFACES[e]

Ventilation Air Temp., F	Sol-Air[c] Temp., F	No Ventilation		Natural Ventilation		Ventilation Rate, cfm/ft²		Power Ventilation[d]			
		0		0.1		0.5		1.0		1.5	
		10	20	10	20	10	20	10	20	10	20
80	120	6.5	6.5	8.1	8.8	13	17	17	25	19	30
	140	6.5	6.5	8.2	9.0	14	18	18	26	20	31
	160	6.5	6.5	8.3	9.2	15	18	19	27	21	32
90	120	6.5	6.5	7.5	8.0	10	13	12	17	13	19
	140	6.5	6.5	7.7	8.3	12	15	14	20	16	22
	160	6.5	6.5	7.9	8.6	13	16	16	22	18	25
100	120	6.5	6.5	7.0	7.4	8.0	10	8.5	12	8.8	12
	140	6.5	6.5	7.3	7.8	10	12	11	15	12	16
	160	6.5	6.5	7.6	8.2	11	14	13	18	15	20

[a] The term *effective resistance* is used when there is attic ventilation. A value for no ventilation is also included. The effective resistance of the attic may be added to the resistance (1/U) of the ceiling (Table 4G) to obtain the effective resistance of the combination based on sol-air and room temperature. These values apply to wood frame construction with a roof deck and roofing having a conductance of 1.0 Btu/(hr · ft² · F).

[b] *Resistance* is one (hr · ft² · F)/Btu. Determine ceiling resistance from Tables 4G and 5A, and adjust for framing by Eq 9. Do not add the effect of a reflective surface facing the attic to the ceiling resistance from Table 4G, as it is accounted for in Table 6, Part B.

[c] Roof surface temperature rather than sol-air temperature (see Chapter 28) may be used if 0.25 is subtracted from the attic resistance shown.

[d] Based on air discharging outward from attic.

[e] Surfaces with effective emittance E of 0.05 between ceiling joists facing the attic space.

Table 7 Estimated Heat Lost from Building by Infiltration

The tabulated factors, when multiplied by room or building volume (ft³), will result in estimated heat loss (Btu/hr) due to infiltration and does not include the heat needed to warm ventilating air

Room or Building Type	No. of Walls with Windows	Temp. Difference, deg F				Room or Building Type	No. of Walls with Windows	Temp. Difference, deg F			
		25	50	75	100			25	50	75	100
A	None	0.23	0.45	0.68	0.90	B	Any	1.35	2.70	4.05	5.40
	1	0.34	0.68	1.02	1.36	C	Any	0.90-1.35	1.80-2.70	2.70-4.05	3.60-5.40
	2	0.68	1.35	2.02	2.70	D	Any	0.45-0.68	0.90-1.35	1.35-2.02	1.80-2.70
	3 or 4	0.90	1.80	2.70	3.60	E	Any	0.68-1.35	1.35-2.70	2.03-4.05	2.70-5.40

A = Offices, apartments, hotels, multistory buildings in general.
B = Entrance halls or vestibules.
C = Industrial buildings.
D = Houses, all types, all rooms except vestibules.
E = Public or institutional buildings.

Reprinted, with permission, from ASHRAE Handbook of Fundamentals, 1977.

Table 8 Coefficients of Transmission (U) of Windows, Skylights, and Light Transmitting Partitions

These values are for heat transfer from air to air, Btu/(hr · ft² · F). To calculate total heat gain, include solar transmission.

PART A—VERTICAL PANELS (EXTERIOR WINDOWS, SLIDING PATIO DOORS, AND PARTITIONS)— FLAT GLASS, GLASS BLOCK, AND PLASTIC SHEET

Description	Exterior[a]		Interior
	Winter	Summer	
Flat Glass[b]			
single glass	1.10	1.04	0.73
insulating glass—double[c]			
0.1875-in. air space[d]	0.62	0.65	0.51
0.25-in. air space[d]	0.58	0.61	0.49
0.5-in. air space[e]	0.49	0.56	0.46
0.5-in. air space, low emittance coating[f]			
e = 0.20	0.32	0.38	0.32
e = 0.40	0.38	0.45	0.38
e = 0.60	0.43	0.51	0.42

PART B—HORIZONTAL PANELS (SKYLIGHTS)— FLAT GLASS, GLASS BLOCK, AND PLASTIC DOMES

Description	Exterior[a]		Interior[f]
	Winter[i]	Summer[j]	
Flat Glass[e]			
single glass	1.23	0.83	0.96
insulating glass—double[c]			
0.1875-in. air spaced	0.70	0.57	0.62
0.25-in. air spaced	0.65	0.54	0.59
0.5-in. air space[e]	0.59	0.49	0.56
0.5-in. air space, low emittance coating[f]			
e = 0.20	0.48	0.36	0.39
e = 0.40	0.52	0.42	0.45
e = 0.60	0.56	0.46	0.50

Glass Block[h]

Description			
11 × 11 × 3 in. thick with cavity divider	0.53	0.35	0.44
12 × 12 × 4 in. thick with cavity divider	0.51	0.34	0.42
Plastic Domes[k]			
single-walled	1.15	0.80	—
double-walled	0.70	0.46	—

PART C—ADJUSTMENT FACTORS FOR VARIOUS WINDOW AND SLIDING PATIO DOOR TYPES (MULTIPLY U VALUES IN PARTS A AND B BY THESE FACTORS)

Description	Single Glass	Double or Triple Glass	Storm Windows
Windows			
All Glass[l]	1.00	1.00	1.00
Wood Sash—80% Glass	0.90	0.95	0.90
Wood Sash—60% Glass	0.80	0.85	0.80
Metal Sash—80% Glass	1.00	1.20[m]	1.20[m]
Sliding Patio Doors			
Wood Frame	0.95	1.00	—
Metal Frame	1.00	1.10[m]	—

Description		
insulating glass—triple[c]		
0.25-in. air spaces[d]	0.39	0.38
0.5-in. air spaces[g]	0.31	0.30
storm windows		
1-in. to 4-in. air space[d]	0.50	0.44
Plastic Sheet		
single glazed		
0.125-in. thick	1.06	0.98
0.25-in. thick	0.96	0.89
0.5-in. thick	0.81	0.76
insulating unit—double[c]		
0.25-in. air space[d]	0.55	0.56
0.5-in. air space[c]	0.43	0.45
Glass Block[h]		
6 × 6 × 4 in. thick	0.60	0.57
8 × 8 × 4 in. thick	0.56	0.54
—with cavity divider	0.48	0.46
12 × 12 × 4 in. thick	0.52	0.50
—with cavity divider	0.44	0.42
12 × 12 × 2 in. thick	0.60	0.57

[a] See Part C for adjustment for various window and sliding patio door types.
[b] Emittance of uncooled glass surface = 0.84.
[c] Double and triple refer to the number of lights of glass.
[d] 0.125-in. glass.
[e] 0.25-in. glass.
[f] Coating on either glass surface facing air space; all other glass surfaces uncoated.
[g] Window design: 0.25-in. glass—0.125-in. glass—0.25-in. glass.
[h] Dimensions are nominal.
[i] For heat flow up.
[j] For heat flow down.
[k] Based on area of opening, not total surface area.
[l] Refers to windows with negligible opaque area.
[m] Values will be less than these when metal sash and frame incorporate thermal breaks. In some thermal break designs, U-values will be equal to or less than those for the glass. Window manufacturers should be consulted for specific data.

Reprinted, with permission, from ASHRAE Handbook of Fundamentals, 1977.

differing from conductivity only in that the latter is per inch of thickness while conductance is for some stated thickness. For example, the conductivity k in Table 3A for asbestos-cement board is given as 4.0. The conductance for a-⅛-inch asbestos-cement board is 33.0. Conductance, then, is expressed in Btu per (hour) (square foot) (deg. F temperature difference per stated thickness).

Surface Conductance. Conductivity is concerned with the difference in temperature between two surfaces. In a building which is heated there is a difference in temperature between the air outside and the surface temperature of the outside of the heated building; similarly, between the room air temperature and the inside wall surface temperature. The rate of heat flow between the room air and the wall surface temperature and adjacent air is termed the surface or film conductance, usually expressed as f_o (for outside wall) or f_i (for inside wall). Both are in btu per (hour) (square foot) (deg. F temperature difference). The temperature difference is that between the air and the surface involved.

Air Space Conductance. The heat flow across an air space is complex, involving conduction, convection, and radiation. This cannot be calculated, but must be measured by actual test, and this has been done by the National Bureau of Standards for a wide variety of conditions. It is usually symbolized by a, and is the heat flow in Btu per (hour) (square foot) (deg F temperature difference between the surfaces bounding the space).

Thermal Resistance. This is the reciprocal of conductivity or conductance, usually called R, and is in (hours) (square feet) (deg. F temperature per inch or stated thickness) per Btu. Its convenience will be seen later.

Emissivity. The emissivity, frequently symbolized by the Greek letter epsilon (ϵ), is the ratio of the radiant energy emitted by a surface to that emitted by an ideal black surface at the same temperature. Bright polished surfaces have a low emissivity, dull dark ones a high emissivity.

Radiation is across an air space, and the amount of heat transmitted depends on the emissivities of both sides of the air space. The product of the two emissivities gives the *effective emissivity* (E) of the air space.

The commonly used emissivities are: for most building materials, 0.90, for bright aluminum foil. 0.05, for aluminum coated paper, 0.20. Combina-

tions, or effective emissivities of both sides are thus: for cases where one side has an ϵ of 0.90, the other of 0.90, the effective emissivity E is 0.90 x 0.90 or approximately 0.82; one side, 0.90, the other side 0.05, 0.90 x 0.05 = 0.045, or 0.05; one side 0.90, the other 0.20, E = 0.20 x 0.90 = 0.18 (but 0.20 is usually used); both sides 0.05, E is usually taken as 0.05.

U Values. Knowing the values of k or C for all common building materials, and of f and h for surfaces and air spaces, all of which are given in Tables 1-8, the U values, or heat transmission coefficients can be calculated. The U values are the overall units of heat transfer from inside air to outside air through an almost infinite number of combinations of building materials and air spaces. U-values are in Btu per (hour) (square foot) (deg. F difference in temperature, inside to outside air.)

Due to the large number of combinations for which U values are required, combinations are not ordinarily tested to obtain U. Instead, the k and other values being known rather precisely for the individual components, U can be calculated with considerable accuracy. The basic formula for deriving U is, consequently, the reciprocal of the sum of all the individual resistances of the component materials, air spaces and the inside and outside surface resistances, thus:

$$U = \cfrac{1}{\cfrac{1}{h_1} + \cfrac{x}{k} + \cfrac{x_n}{k_n} + \cfrac{1}{C} + \cfrac{1}{C_n} + \cfrac{1}{h} + \cfrac{1}{h_n} + \cfrac{1}{h_o}}$$

All of these symbols have been defined except x which is the thickness in inches of some material whose conductivity is k. There will be, in an actual case, as many values of $\dfrac{x}{k}$ as there are materials in the construction to be solved; the same is true of $\dfrac{1}{C}$ and $\dfrac{1}{h}$. There will always be two values of $1/h$; in the case of partitions, both values are that of h_1.

In defining the symbols, it was shown that the reciprocals of k, C, f, and h can be given as R. Therefore, the foregoing equation can be expressed as

$$U = \cfrac{1}{R_1 + R_2 + R_3 + R_4 + R_5 + R_6} = \cfrac{1}{R_t}$$

where R_t = the sum of the resistances of the individual components. Note, however, that for any

component whose k and thickness are known, R_x $= x\left(\dfrac{1}{k}\right) = xR$. For example, if one of the components is a perlite aggregate gypsum plaster $\frac{1}{2}$ inch thick, find in Table 3A "Plastering Materials," k to be 1.5, $R = 0.67$ for 1 inch. For $\frac{1}{2}$ inch thickness, $\dfrac{x}{k} = 0.5 \times 0.67 = 0.34$. Note that, in most cases, $\left(\dfrac{1}{k}\right)$ is also given.

Values for U for winter heat loss are usually calculated on the basis of a 15 mph wind velocity *outside* and still air inside. For these conditions, $f_o = 6.00$ and $f_1 = 1.46$. For summer conditions, a value of $f_o = 4.00$ is used, a $7\frac{1}{2}$ mph wind velocity being assumed in summer.

Example. What would be the U value of a wall made of 2 x 4-inch studing with pine wood $\frac{3}{4}$ inch thick on the outside, $\frac{3}{4}$-inch plywood on the inside?

Solution. From Table 3A, under Woods, find that pine has a k of 0.80, so that $\dfrac{x}{k} = \dfrac{0.75}{0.80}$.

Therefore, the R for the pine is 0.94. Under Building Board, $\frac{3}{4}$-inch plywood has a C of 1.07, an R of 0.93. In Table 2, for vertical wall, 50F mean temperature and 30F temperature difference (one-half total temperature difference, inside to outside air), for 3.5-in. air space where $E = 0.82$ (from Table 1), find that $R = .91$, and for the outside surface of a wall, winter, to be 6.00, or $R = .17$ (with 15 mph wind). Similarly, (h_i) for the inside wall (still air) is 1.46, or $R = 0.68$. Then

Total $R_t = R_1$ (pine) $+ R_2$ (air space) $+ R_3$ (plywood) $+ R$ (outside) $+ R$ (inside)

$R_t = 0.94 + 0.91 + 0.93 + 0.17 + 0.68$

$R_t = 3.63$

But $U = \dfrac{1}{R_t}$ so that $U = 1/3.63 = 0.28$

Example. What would be the heat loss coefficient U for a frame consisting of 0.5-inch x 8-inch lapped wood siding, 0.5-inch vegetable fiber board sheathing, air space between studding, with .625-inch gypsum lath? Find the values for outside and inside surface resistances, the x/k values for the various materials, the air space resistance, and total them as follows:

Solution.

Component	R
Outside surface (15 mph wind)	0.17
$\frac{1}{2}$-inch wood siding, 8-inch lapped	0.85
Fiber board sheathing	1.32
Air space	0.91
.625-inch gypsum board	0.56
Inside surface (still air)	0.68
Total Resistance	4.49

$$U = 1/R_t = 1/4.49 = 0.22$$

Tables 1-8 give the needed values of k, C, R, etc., needed for virtually all calculations. Tables 4A-8, inclusive, give U values for all the commonly used walls, ceilings, floors, and roofs, where air-to-air heat transmission is involved. Heat loss from room air to ground, as with slab floors laid on earth, are treated elsewhere.

The possible combinations of all these constructions with insulation are so numerous that tables are given if insulation is to be added. Tables 5A and 5B are for this purpose.

Example. What would be the U value of the previous example if $3\frac{3}{4}$ inches of rock wool were to be used between the studs?

Solution. The U value, as calculated without insulation was 0.22. In Table 3A, find insulation $R = 11$. Added $R = 11 - .9 = 10.1$. Interpolating in Table 5A for $U = .23$ and between $R = 8$ and $R = 12$, the new $U = 0.072$.

Example. What would be the U value for the same construction ($U = 0.23$) if one side of the air space were lined with bright aluminum foil?

Solution. From Table 1, Sec. B, one side would have an ϵ of 0.90, the other 0.05, and the combined E would be 0.05. Therefore, from Table 2, the new $R = 1.25$, an increase of 0.34. Interpolating in Table 5A, the new $U = 0.156$.

If the air space in the same construction had been broken up into three air spaces using aluminum foil, the U would have been 0.088.

Doors and Windows. Worked-out tables for all common sizes of doors and windows appear on other pages.

Wind Velocity. The effect of air movement on the surface conductance is to increase it. If, for any reason, it is required to know the U value at other wind velocities of any construction for which the

U value for 15 mph wind is known, use the formula

$$U_x = \cfrac{1}{\cfrac{1}{h_x} + \left(\cfrac{1}{U_{15}} - .17\right)}$$

where U_x is the U value at some other wind velocity x, $h_x =$ the outside surface conductance at velocity x, $U_{15} =$ the U value at 15 mph. Values of h_x for other wind velocities are as shown at the right.

For example, a certain construction has a U value of 0.24 at 15 mph outside. What would U be for a 30-mile wind.

The formula would be

$$U_{30} = \cfrac{1}{\cfrac{1}{h_x} + \left(\cfrac{1}{U_{15}} - .17\right)}$$

so that

$$U_{30} = \cfrac{1}{\cfrac{1}{10.00} + \left(\cfrac{1}{0.24} - .17\right)}$$

$$= \cfrac{1}{.1 + 4.16 - .17} = \cfrac{1}{4.09} = 0.244$$

Wind Velocity, mph	Value of h_x
0	1.46
5	3.20
7½	4.00
10	4.60
15	6.00
20	7.30
25	8.60
30	10.00

Summer Coefficients. As has been mentioned, the outside surface resistance for summer calculations is on a 7½-mph wind velocity basis, rather than 15 mph as for winter. In the case of walls this makes little difference so that winter coefficients for walls are used likewise for summer.

In the case of ceilings and roofs, however, the heat flow upward (in winter) is greater than the heat flow downward (in summer). The difference is primarily in the effective emittance of the air spaces, its mean temperature and temperature difference. Use Table 2 to correct for summer conditions.

Calculating Heat Losses

The total winter heat loss for a building is the transmitted heat, through walls, roofs, etc., plus the heat required to raise the temperature of air leaking into the building from the outside (design) temperature to the temperature maintained inside (inside design temperature), less the amount of heat from lights, motors, people, etc., on the inside. This heat "gain" is usually neglected in the case of residences. The inside design temperature is usually (but not always) taken at 70 deg F, while the outside design temperature (discussed elsewhere) varies from locality to locality.

It can be seen, then, that the *transmitted* loss depends on (1) the difference between the design inside and design outside temperatures, (2) area of transmitting surface, (3) the U values of the building section involved.

Special Cases. There are five special cases involved in most heat loss calculations:

(1) Spaces adjacent to the room whose heat loss is being calculated and which are either to be maintained at the same or some other temperature. If the adjacent room is to be at the same temperature, there is no heat flow through the partition. If it is to be maintained at a lower temperature, heat will flow to that room, and the temperature difference used in calculating should be that between the two rooms.

(2) Unheated crawl spaces under, or attics over the room being calculated.

In the case of crawl spaces under floors, the temperature of the air in the crawl space tends to approach the outside temperature (a) the more open the space is to the outside, (b) the better the floor insulation, and (c) if there is no pipe or duct in the space giving off heat. Judgment must be used in selecting a temperature of a crawl space, this varying from the extreme, the case where the temperature will be only slightly above the outside design temperature, to the other extreme where the crawl space temperature is the mean between the inside and outside design temperatures.

The same reasoning applies to attic spaces. If the attic is well ventilated, with no heat, and with an

insulated ceiling, a design temperature slightly above the outside design temperature can be selected. If it is tightly sealed, with no ceiling insulation, the attic temperature can be taken as the mean between the inside and outside design temperatures. Table 6 provides effective resistances for attics, in summer only.

(3) The third common case is that of basements. Usually such basements containing the heating plant which keeps the basement reasonably warm. In such cases, no heat loss is ordinarily assumed to the basement through the floor above. If the basement has no heat, the basement can be assumed to be at a temperature which is the mean between the inside and outside design temperatures.

(4) The fourth case is a garage adjacent to a heated space. Since the garage doors are frequently open, calculate the garage temperature as being at the outside design temperature.

(5) The final case is that of a slab floor laid directly on the ground. This is treated elsewhere in this section.

Unheated, Unventilated Spaces. For unheated and *unventilated* areas adjacent to heated rooms,

Heat lost by heated space to unheated space = Heat lost by unheated space to outside, or

$$A_1 U_1 (t_h - t_u) = A_a U_a (t_u - t_o)$$

where A_1 = area of surface between heated and unheated space

U_1 = U for that area,

A_a = area of surface between unheated space and outside

U_a = U for that area,

t_h = temperature of heated space

t_u = temperature of unheated space

t_o = outside temperature.

There are almost always more than one area and one U value involved, but it can be shown that

$$t_u = \frac{t_h (A_1 U_1 + A_2 U_2 ..) + t_o (A_a U_a + A_b U_b ..)}{A_1 U_1 + A_2 U_2 .. + A_a U_a + A_b U_b ..}$$

where the inferior numerals denote surfaces between the heated and unheated spaces and inferior letters surfaces between the unheated space and the outside.

Example: A corner room, 15 x 12 ft in plan, with a 9-ft ceiling and with 3 windows each 30 x 54 inches is to be calculated. The outside design temperature is 0 deg. Rooms overhead and below are

heated, as are adjacent rooms. Outside walls are face brick veneer with ½-inch insulation board sheathing, with ½-inch plaster on wood lath inside. What are the transmission losses (infiltration not to be calculated)?

Solution. Outside gross wall area is $(12 \times 9) + (15 \times 9) = 243$ sq ft. From this must be deducted the window area. Window area $= (2\frac{1}{2} \times 4\frac{1}{2}) \times 3 = 33.75$ sq ft, approximately 34. The net wall area is then $243 - 34 = 209$ sq ft. From Table 2, the U value for windows is 1.13. The total transmitted losses are thus

$$209 \times 0.25 \times 70 = 3658 \text{ Btu per hour}$$
$$34 \times 1.13 \times 70 = \underline{2689}$$
$$6347 \text{ Btu per hour}$$

Example. What would be the transmission losses from this room if an adjacent space along the long wall (partition = 135 sq ft) is unheated and unventilated with outside surfaces totaling 200 sq ft with a U value of 0.29 and if the partition between has a U value of 0.52 and the outside design temperature is 10 deg.

Solution. The temperature of the unheated space would be

$$t_u = \frac{(70 \times 135 \times .52) + 10 (200 \times .29)}{(135 \times .52) + (200 \times .29)}$$

$$= \frac{4914 + 580}{70 + 58} = \frac{5494}{128} = \text{appr. } 43 \text{ deg.}$$

The heat lost through the 135 sq ft of partition is thus $(70 - 43) \times .52 \times 135 = 1895$ Btu per hr. (Check: The heat lost through the partition equals the loss through the outside walls: This loss is $(43 - 10) (200 \times .29) = 1914$ Btu per hr (close enough to 1895).

The outside design temperature is 10 instead of 0, so that the wall and window losses are 60/70 of what they were in the previous example, or

$$(60/70 \times 6347 = 5440 \text{ Btu per hr})$$

which, plus the 1895 Btu lost through the partition equals a transmitted loss of 7335 Btu per hr.

Infiltration can be calculated in either of two ways. The crack method is described under Heat Losses through Wood Windows and tables for wood, steel and aluminum windows, showing both infiltration loss and transmission losses are given on pages devoted to these windows. The second method, the air change method, is explained under Estimating Heat Loss by Infiltration.

ESTIMATING HEAT LOST BY INFILTRATION

There are two methods of estimating the amount of heat lost by heating to room temperature the cold air entering the building around doors and windows.

The first, and more accurate, is to determine the lineal feet of crack in windows and doors and multiply such footage by an appropriate figure representing the cubic feet per hour leakage per foot of crack for the type of window or door used. This product, in turn, is multiplied by the specific heat of air, temperature range through which the air is raised, and the density of air. Worked-out tables for a wide variety of windows appear elsewhere in this book.

With the crack method, it is usually assumed that air can enter on only two sides of a rectangular building, since heated air must leave from the opposite sides. Consequently, the combination of those two adjacent sides which give the greatest heat load is commonly used as the basis of estimating infiltration. In no case, however, should less than half of the total crack be used.

Air Change Method

The second method of computing infiltration is the air change method. This assumes that enough air infiltrates to replace the whole volume of air in the room from $\frac{1}{2}$ to some higher number of times per hour. Since the air volume is in cubic feet, the volume is multiplied by the number of air changes assumed per hour, and this in turn multiplied by the temperature range, density, and specific heat, as with the crack method.

Expressed as a formula, the infiltration loss by the air change method is

$$H = 0.24\ d\ (t_i - t_o)\ n\ v \qquad (1)$$

where $H =$ infiltration loss, Btu per hour,

$n =$ number of air changes per hour,

$v =$ volume of room, cubic feet,

$d =$ density of air, usually 0.075 lb per cu ft,

$t_i =$ inside design air temperature, deg F,

$t_o =$ outside design air temperature, deg F.

Table 1 gives values of n for use in this equation. Table 7, in the preceding section, provides a rapid method for arriving at infiltration heat loss.

Infiltration is, in the methods described, the loss due to heating cold air which is unavoidably introduced. Ventilation air, which is introduced intentionally, is a separate matter.

Air can and does also infiltrate through walls themselves. Ordinarily, however, this is neglected in heat loss calculations.

An additional cause of infiltration is due to so-called chimney effect which is proportional to the height of the building and which is caused by an influx of cold air at all possible points at the lower floors to replace the heated air leaving the upper floors. This is a separate and important problem in multi-story buildings but does not constitute a problem in residences.

Still another cause of infiltration is the mechanical withdrawal of air through exhaust systems, such as in industrial buildings. This causes a reduction of air pressure in the building and increases the rate of cold air flow into the building at doors and windows.

Table 1—Air Changes (n) for Use in Equation 1

Room or Building Type	No. of Sides of Room Having Windows	No. of Air Changes per Hour (n)	$.24d\ (t_i - t_o)n$ when $(t_i - t_o) = 70F$
Offices, apartments, hotels, multi-story buildings in general.......	none	$\frac{1}{2}$	0.63
	1	$\frac{3}{4}$	0.95
	2	$1\frac{1}{2}$	1.89
	3	2	2.52
	4	2	2.52
Entrance halls or vestibules........	any	3	3.78
Industrial buildings.............	any	2 to 3	2.52 to 3.78
Houses, all types, all rooms except vestibules..................	any	1 to $1\frac{1}{2}$	1.26 to 1.89
Public or institutional buildings....	any	$1\frac{1}{2}$ to 3	1.89 to 3.78

ESTIMATING THE AMOUNT OF VENTILATION AIR

In order to arrive at figures showing the heat quantity involved in the ventilation air deliberately passed through a building, use is usually made of the formula:

$$V = \frac{C_v(t_i - t_0)}{55}$$

where V = heat for ventilation air, Btu per hour.

C_v = volume of ventilation air flowing, cubic feet per hour.

t_i = temperature of indoor air, F

t_0 = temperature of outdoor air, F

$1/55$ = 0.018 = heat to raise 1 cu. ft. of air at 70F through 1F.

Fixing suitable values for C_v is always troublesome. It is important that they be fixed as definitely as possible for the value enters into the sizing of equipment and greatly affects its cost. There is a variety of procedures used in fixing the quantity. The table on the next page gives values of C_v for three different methods of attack.

The amount of ventilation air can be computed on the basis of (1) number of air changes per hour; (2) volume supplied per occupant, or (3) volume required per square foot of floor space.

For the first method $C_v = n \times V_R$, where n is the number of air changes per hour and V_R the volume of the room in cubic feet. Values in the first two columns of the table opposite are values of n for substitution in the foregoing formula to find C_v.

For the second method, $C_v = A \times N$ where A is the cubic feet of ventilation air per hour per occupant and N the number of occupants. Values of A are given in the table.

For the third method, $C_v = B \times S$ where B is the air volume required per square foot of floor surface and S the area of floor in square feet. Values for B are given in the table.

The three methods of arriving at values of C_v are indicated at the top of each column in the table. The three figures are alternates. Select one method to use depending on your judgment of which is most suitable. Then fix the upper and lower limiting values of C_v. Select one in between these limits and use it as C_v in the formula for V previously given.

Another point about the figures in the table is the great range covered by many of them. For instance, in the case of paint booths the figures show that from zero to 600 air changes per hour are required. With such a range there must be some additional information before deciding where a particular job

falls in between such wide limits. Therefore use the figures with the greatest care and only after careful thought to select one which fits the case. An intelligent selection will necessarily require experience. In other words the table, while it can be made helpful, in no way sets up a specific rule of practice — it merely serves as a guide based on the findings of others. It is thought to be the most complete compilation of ventilation air quantities of this kind put together in a single table.

The table itself was prepared by carefully reviewing and tabulating values as reported in various texts, handbooks, catalogs and codes. Note that it applies only to air deliberately introduced for ventilation and does not purport to cover infiltration air, which will be subsequently discussed.

Recirculation. In estimating the heat lost by ventilation air by the foregoing methods, the designer should keep in mind that the introduction of outdoor air for ventilation is an expensive process, in either winter or summer, and the amount of such outside ventilation air introduced should be kept at a minimum. In general, the purpose of ventilation is to carry away heat, moisture, odors or dust or reduce the concentration of any of these. If, then, the objective can be attained without introducing *outside* air more economical operation will result.

In other words, the possibilities of removing the dust, odors or moisture from the air, then recirculating it, should be studied and the cost weighed against the cost of heating the outside air. Filters, cyclones, odor adsorbers and other methods of air recovery should be studied.

Consider a kitchen of 5500 cu. ft. volume in a New York City restaurant and where odors are being generated so that at least thirty air changes per hour are estimated as necessary to alleviate the condition. If outside air is introduced for 16 hours per day, and since New York has 250 days per heating season with an average outside temperature of 43.6F, the heat required is

$$\frac{(5500 \times 30)\,(70 - 43.6)}{55} \times 16 \times 250 =$$

$$316.8 \text{ million Btu}$$

At 1.00 per million Btu for oil, the cost of heating the ventilation air alone would be $316.80 per year. The summer costs, if air conditioning is involved, would add sharply to this, so that recirculation through odor adsorbers should be considered.

VOLUME OF VENTILATION AIR FOR VARIOUS BUILDINGS

Type of Building	No. of Changes per Hour, (n)		Cu Ft per Occupant per Hour, (A)		Cu Ft per Sq Ft Floor Surface per Hour, (B)	
	Min.	Max.	Min.	Max.	Min.	Max.
Commercial						
Garages	6	12	—	—	—	—
Offices	1½	12	—	—	—	—
Waiting rooms (public)	4	6	—	—	—	—
Restaurants (dining)	4	20	—	—	—	—
Restaurants (kitchen)	4	60	—	—	240	240
Stores (retail)	6	12	—	—	—	—
Farm Buildings						
Cow	—	—	3000	6000	—	—
Horse	—	—	3600	4200	—	—
Hog	—	—	1200	1500	—	—
Sheep	—	—	600	900	—	—
Hospitals						
Dining rooms	6	12	—	—	90	90
Kitchens	20	60	—	—	240	240
Operating rooms	—	—	3000	3000	—	—
Toilets	7.5	30	—	—	120	120
Wards (ordinary)	—	—	2100	4500	60	60
Wards (contagious)	—	—	2400	6000	—	—
Hotels						
Barber shops	7.5	7.5	—	—	—	—
Cafes	7.5	7.5	—	—	—	—
Dining rooms	4	20	—	—	90	90
Guest rooms	3	5	—	—	—	—
Kitchens	4	60	—	—	240	240
Lobbies	3	4	—	—	—	—
Lounges	6	6	—	—	—	—
Toilets	10	12	—	—	—	—
Prisons						
Single cells	9	12	—	—	—	—
Sleeping cells	6	6	—	—	—	—
General quarters	6	6	1800	1800	—	—
Residences						
Bathrooms and toilets	1	5	—	—	—	—
Halls	1	3	—	—	—	—
Kitchens	1	40	—	—	—	—
Living rooms	1	2	—	—	—	—
Sleeping rooms	0	1	—	—	—	—
Various Public Spaces						
Auditoriums, churches, dance halls	4	30	600	4000	90	120
Billiard and bowling	6	20	—	—	—	—
Classrooms (colleges)	—	—	1500	2400	—	—
Classrooms (schools)	—	—	1800	2400	120	120
Corridors	4	4	—	—	30	30
Dining rooms	5	20	—	—	90	90
Gymnasiums	12	12	—	—	90	90
Kitchens	15	60	—	—	120	120
Laboratories	6	20	—	—	—	—
Locker rooms	2	10	—	—	120	120
Projection booths	30	30	—	—	90	90
Reading rooms	3	5	—	—	—	—
Toilets	10	30	—	—	120	120
Engine and Boiler Room	3	12	—	—	—	—

HEAT LOSSES FROM FLOOR SLABS

Increasing use of concrete floors laid directly on the ground has spurred research into the thermal characteristics of such floors. Further investigations were prompted by the installation in such slabs of warm air perimeter ducts and hot water coils, transforming the slab floor into a radiant panel. The material for this summary is drawn from work done at the University of Illinois, at various stages in cooperation with the National Bureau of Standards, the National Warm Air Heating and Air-Conditioning Association, and the Institute of Boiler and Radiator Manufacturers.

Test results show that heat loss from the center of the slab is practically constant, while loss from the peripheral portion varies with outside temperature. Therefore, for estimating purposes it is convenient to consider first a ribbon of concrete 3 ft wide, the thickness of the floor slab, and the length of the slab's perimeter. Losses through this portion depend only upon its length (perimeter) and the temperature difference across it. Losses through the interior portion of the slab are approximately 2 Btu per (hr) (sq ft). Therefore,

$$Q = F_1 P (t_i - t_o) + 2A_i \qquad (1)$$

where

Q = heat loss through floor, Btu per hr,
P = length of edge of floor adjacent to exposed wall of building
t_i = design inside temperature, deg F,
t_o = design outside temperature, deg F,
F_1 = heat loss factor based on heat loss through peripheral portion of slab within 3 ft of edge, Btu per hour per lineal foot of exposed edge per deg F,
A_i = floor area within a 3 foot border along exposed edge, sq ft.

A simpler method of estimating heat losses through concrete floors for use where the ratio of the slab area (in square feet) to the length of the perimeter (in feet) does not exceed 12 is as follows:

$$Q = F_2 P(t_i - t_o) \qquad (2)$$

where Q, P, t_i, and t_o are as previously defined, and F_2 is a heat loss factor expressed in the same units as F_1.

Values of F_1 and F_2 are given in Table 3, for the various types of floor construction illustrated in Fig. 1.

For cases where the slab construction differs from

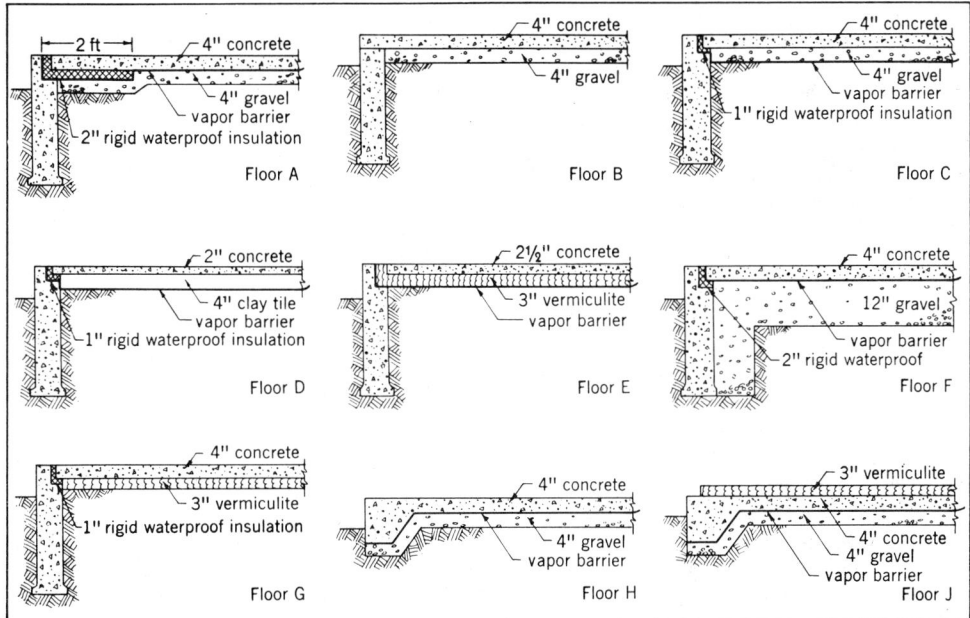

Fig. 1. Various floor constructions, heat loss data through which are given in Table 3.

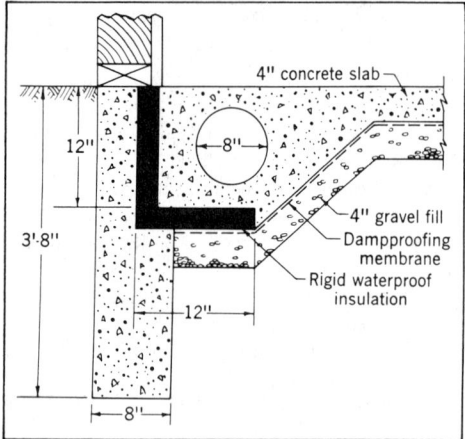

Fig. 2(a). Typical 2-inch L-type floor slab edge construction with warm air perimeter heating duct.

those shown in Fig. 1, and also for checking purposes, Table 2 is included, from work done by the National Bureau of Standards. This gives values of F_2 for both heated and unheated slabs for use where the conductance of the insulation is known.

Since the greatest portion of the heat loss is near the edge of the slab, it is necessary to insulate the edge of the slab and to extend this insulation either down along the concrete foundation or else under the floor for a distance of at least 2 ft. Two-inch rigid waterproof insulation is recommended.

Heat loss through unheated slabs may represent about 10% of the total heat loss for the average small house. Insulation is of importance primarily to help raise floor temperatures which otherwise will be uncomfortably low in winter. A gravel fill is helpful in drainage control but is not an effective insulator.

Heated Slabs

Warmer floors obtained with a warm air perimeter heating system or with embedded hot water coils increase heat loss from the subfloor. Losses of 14 to 24% are to be expected, or about double for that of unheated slabs. Table 1 gives data for estimating heat loss for several outside design temperatures, assuming 70 deg F inside design, and for two types of construction illustrated in Fig. 2a and 2b. Figures obtained from Table 1 are in Btu per hr per linear foot of exposed edge. Therefore, when multiplied by feet of exposed edge the result is Btu per hr. *Do not multiply the figures in Table 1 by temperature difference.*

It is found that fuel savings resulting from insulating the entire underside of floor panels are too small to warrant the additional cost. This is true whether a warm air or hot water system is used.

Table 1. Subfloor Heat Loss, Warm Air Perimeter Heating

Outdoor Design Temperature, Deg F	Type of Edge Construction	
	2-inch L Type. Fig. 2(a)	1-inch Vertical. Fig. 2(b)
	Heat Loss, Btu per (Hr) (Linear Ft Exposed Edge)	
25	40	55
10	55	75
0	65	85
—10	75	95
—20	85	105

Note: These values are for a room width of 12 ft. For rooms over 12 ft wide, add 2 Btu for each additional foot of width to the above values.

From "Warm-Air Perimeter Heating, Heat Losses from Floor Slab—Part III," J. R. Jamieson, R. W. Roose, and S. Konzo, University of Illinois, Vol. 58, Trans. of ASHAE, 1952.

Effect of Floor Coverings

The effect of floor coverings on performance of hot water floor panels was investigated. Any insulation on top of the slab necessitates a raise in temperature of the hot water circulating in order that radiation into the space be undiminished. There is a corresponding increase of heat flow to the ground. Asphalt or rubber tile have little effect on loss to the ground. A light weight carpet with rubber pad in-

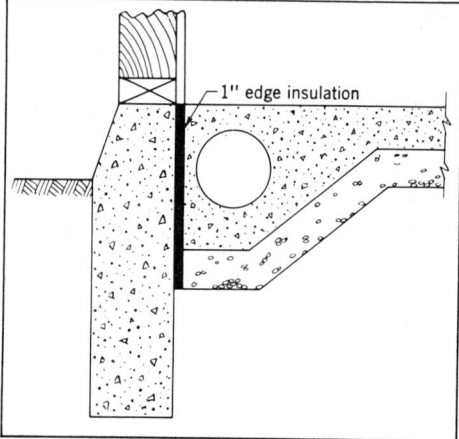

Fig. 2(b). Typical 1-inch vertical floor slab construction with warm air perimeter heating duct.

Table 2. Heat Loss Factors for Use in Equation (2)

Conductance of Insulation	Total Horizontal and/or Vertical Width on Insulation		
	1 Foot	1.5 Feet	2.00 Feet
	Values of F_2*		
Unheated Slabs			
0.15	0.29	0.26	0.25
0.20	0.38	0.35	0.33
0.25	0.49	0.44	0.42
0.30	0.58	0.52	0.50
0.35	0.67	0.61	0.59
0.40	0.77	0.70	0.67
Heated Slabs			
0.15	0.32	0.28	0.27
0.20	0.43	0.39	0.37
0.25	0.57	0.58	0.48
0.30	0.70	0.62	0.59
0.35	0.83	0.75	0.71
0.40	1.00	0.88	0.83

* In the case of unheated slabs, temperature difference is design indoor minus design outdoor temperature. In the case of heated slabs, temperature difference is the temperature of the heating medium on the slab minus the outside design temperature.

creases the heat loss about 50%; a heavy carpet with 40-oz pad increases the losses 150%. However, much of the heat lost is stored in the earth beneath the slab. During periods of milder weather, a reverse flow will take place and much of the "lost" heat will be recovered. It was found that carpeting had little effect on the seasonal fuel bill.

Essentials of Floor Slab Construction

In order that insulation be effective, good drainage must be provided. The floor slab must be above the level of the surrounding land and the ground must slope away from the building. Pockets, where roof water or drainage might collect, are to be avoided. A 4-inch fill of coarse washed gravel or crushed rock, spread over the entire area where the floor is to be laid, is recommended by the University of Illinois Small Homes Council. Structural clay tile may also be used.

A vapor barrier must be provided to prevent the rise of moisture from the ground into the slab. If, instead of rigid waterproof insulation, insulating concrete is used on the edge of the slab, it should be at least 4-inches thick to give the same resistance as 2 inches of the other.

Table 3. Values of F_1 and F_2 for Use in Equations (1) and (2)

Floor Symbol (Fig. 1)	Indoor-Outdoor Temperature Difference						
	50	55	60	65	70	75	80
Values of F_1							
A	0.35	0.34	0.33	0.33	0.32	0.32	0.32
B	0.69	0.66	0.63	0.60	0.58	0.57	0.55
C	0.54	0.52	0.49	0.48	0.46	0.45	0.44
D	0.71	0.68	0.64	0.62	0.59	0.57	0.56
E	0.55	0.53	0.50	0.48	0.47	0.46	0.44
F	0.89	0.83	0.78	0.75	0.72	0.68	0.65
G	0.51	0.48	0.47	0.46	0.45	0.44	0.43
H	0.62	0.69	0.57	0.55	0.54	0.53	0.51
J	0.33	0.32	0.31	0.30	0.30	0.29	0.28
Values of F_2							
A	0.71	0.67	0.63	0.60	0.58	0.56	0.54
B	1.05	0.98	0.93	0.88	0.84	0.80	0.77
C	0.90	0.89	0.79	0.75	0.72	0.68	0.66
D	1.06	1.00	0.94	0.90	0.85	0.82	0.78
E	0.91	0.85	0.80	1.03	0.98	0.93	0.89
F	1.25	1.17	1.10	0.74	0.70	0.68	0.65
G	0.88	0.82	0.77	0.83	0.79	0.75	0.72
H	0.98	0.93	0.87	0.58	0.56	0.53	0.52
J	0.69	0.65	0.62				

HEAT LOSS THROUGH WOOD WINDOWS

Heat is lost via windows in two distinctly different ways. The first is by conduction through the glass, the second is by leakage of cold air around the cracks of the movable sash or sections of windows. There is also some leakage around the frames of the windows but usually this is not substantial and may be ignored in most heat loss calculations.

Loss through Glass

Ordinarily the U value for vertical glass windows is taken at 1.13 Btu per (sq ft) (hr) (deg temperature difference) based on a 15 mph wind velocity. In the rare cases where the glass is horizontal, the U value is 1.40 when the heat flow is upward. Where two vertical sheets are separated by a ¼ inch air space, $U = 0.61$ and if the air space is ½ inch, $U = 0.55$. Values of U for hollow glass block range from 0.55 to 0.60.

Strictly speaking, the area of the window opening occupied by the wood sash should be calculated separately from the area occupied by glass. Actually, it is almost universal practice to calculate the loss through the entire window sash opening as if it were all occupied by glass, and this has been done in the tables which follow.

Loss by Infiltration

In determining the heat lost by infiltration around windows, it is customary to size the heating equipment for each room based on infiltration through the worst possible combination of infiltration from two adjacent walls. On the other hand, the capacity for the whole building need not necessarily be the sum of capacities needed for individual spaces, on the theory that if there is infiltration on one or two sides of a building, there must be exfiltration on other sides. With mechanical air systems, however, where the interior of the building may be under slight pressure, there can be exfiltration even on the windward side.

The heat lost per hour by infiltration is

$$q \times .075 \times .24 \times (t_i - t_o) \times L \quad (1)$$

where q is the cubic feet per hour of leakage per foot of crack, .075 is the density of air in pounds per cubic foot, .24 is the specific heat of air, Btu per pound per degree, t_i is the inside air temperature, t_o the outside air temperature, and L the lineal feet of crack.

Tabular Data

The tables which follow are tabular solutions of equation 1 for standard size double-hung wood windows and wood casement windows for a temperature difference of 70 deg. The standard sizes are those given in Commercial Standard 163-52 of the Department of Commerce and covering standard stock Ponderosa pine windows and sash.

The following points should be noted by those using the tables:

(1) The tables give the heat loss for a 0° to 70° temperature difference. It is a simple matter to correct for other temperature differences in totals by applying correction factors given in Table 1.

Table 1. Correction Factors for Temperature Differences Other than 70 Deg

Design Temperature Difference, Deg. F	Multiply Table Value By
40	.57
50	.71
60	.86
70	1.00
80	1.13
90	1.29
100	1.43

(2) The window opening has been calculated as being all glass, rather than calculating the wood sash separately. The error here is, of course, on the conservative side.

(3) A value of 39 for average non-weather stripped wood windows and 24 for weatherstripped wood windows has been used for q in equation 1.

(4) The value of 1.13 for U was used in calculating the heat loss through the glass.

(5) The total values to the right of the tables are the total of the transmission loss and the infiltration loss.

(6) The tables for both casement and double hung wood windows apply to all combinations of number of panes for the given size opening. The lineal feet of crack for casements is the perimeter of the opening; for double hung windows it is the perimeter of the opening plus the width of the opening to provide for the crack at the meeting rail.

HEAT LOSSES (0° TO 70°) THROUGH WOOD CASEMENT WINDOWS

Opening, Ft. and In.		Area, Sq. Ft.	Heat Loss through Glass, Btu per Hr.	Crack, Ft.	Infiltration Loss, Btu per Hr.		Total Window Loss, Btu per Hr.	
Width	Height				Weather-stripped	Not Weather-stripped	Weather-stripped	Not Weather-stripped
0′11½″	2′6″	2.40	190	6.92	208	339	398	529
	2′10″	2.72	215	7.58	227	371	442	586
	3′2″	3.04	240	8.26	248	405	488	645
	3′6″	3.36	266	8.92	268	437	534	703
	3′10″	3.68	291	9.58	287	469	578	760
	4′2″	4.00	316	10.26	308	503	624	819
	4′6″	4.32	342	10.92	328	535	670	877
	4′10″	4.64	367	11.58	347	567	714	934
	5′2″	4.96	392	12.26	368	601	760	993
1′3½″	2′6″	3.23	255	7.58	227	371	482	626
	2′10″	3.65	289	8.26	247	405	536	694
	3′2″	4.09	324	8.92	268	437	592	761
	3′6″	4.52	358	9.58	287	469	645	827
	3′10″	4.94	391	10.26	308	503	699	894
	4′2″	5.38	426	10.92	328	535	754	961
	4′6″	5.81	460	11.58	347	567	807	1027
	4′10″	6.23	493	12.26	368	601	861	1094
	5′2″	6.67	528	12.92	388	633	916	1161
1′7½″	2′6″	4.05	320	8.26	248	405	568	725
	2′10″	4.58	362	8.92	268	437	630	799
	3′2″	5.14	407	9.58	287	469	694	876
	3′6″	5.67	448	10.26	308	503	756	951
	3′10″	6.20	490	10.92	328	535	818	1025
	4′2″	6.76	535	11.58	347	567	882	1102
	4′6″	7.29	577	12.26	368	601	945	1178
	4′10″	7.82	619	12.92	388	633	1007	1252
	5′2″	8.38	663	13.58	407	665	1070	1328
1′11½″	2′6″	4.90	388	8.92	268	437	656	825
	2′10″	5.55	439	9.58	287	469	726	908
	3′2″	6.21	491	10.26	308	503	799	994
	3′6″	6.86	543	10.92	328	535	871	1078
	3′10″	7.51	594	11.58	347	567	941	1161
	4′2″	8.17	646	12.26	368	601	1014	1247
	4′6″	8.82	698	12.92	388	633	1086	1331
	4′10″	9.47	749	13.58	407	665	1156	1414
	5′2″	10.13	801	14.26	428	699	1229	1500
2′3½″	2′6″	5.73	453	9.58	287	469	740	922
	2′10″	6.48	513	10.26	308	503	821	1016
	3′2″	7.26	574	10.92	328	535	902	1109
	3′6″	8.02	634	11.58	347	567	981	1201
	3′10″	8.77	694	12.26	368	601	1062	1295
	4′2″	9.55	755	12.92	388	633	1143	1388
	4′6″	10.31	816	13.58	407	665	1223	1481
	4′10″	11.06	875	14.26	428	699	1303	1574
	5′2″	11.84	937	14.92	448	731	1385	1668

HEAT LOSSES (0° to 70°) THROUGH DOUBLE-HUNG WOOD WINDOWS

Opening, Ft. and In.		Area, Sq. Ft.	Heat Loss through Glass, Btu per Hr.	Crack, Ft.	Infiltration Loss, Btu per Hr.		Total Window Loss, Btu per Hr.	
Width	Height				Weather-stripped	Not Weather-stripped	Weather-stripped	Not Weather-stripped
1' 8"	3' 2"	5.29	418	11.34	340	556	758	974
	3' 6"	5.85	462	12.00	360	588	822	1050
	3' 10"	6.40	506	12.66	380	620	886	1126
	4' 2"	6.96	550	13.34	400	654	950	1204
	4' 6"	7.52	594	14.00	420	686	1014	1280
	4' 10"	8.07	638	14.66	440	718	1078	1356
	5' 2"	8.63	682	15.34	460	752	1142	1434
	5' 6"	9.19	726	16.00	480	784	1206	1510
	5' 10"	9.74	770	16.66	500	816	1270	1586
	6' 2"	10.30	814	17.34	520	850	1334	1664
	6' 6"	10.86	859	18.00	540	882	1399	1741
2' 0"	2' 6"	5.00	396	11.00	330	539	726	935
	2' 10"	5.66	448	11.66	350	571	798	1019
	3' 2"	6.34	501	12.34	370	605	871	1106
	3' 6"	7.00	554	13.00	390	637	944	1191
	3' 10"	7.66	606	13.66	410	669	1016	1275
	4' 2"	8.34	660	14.34	430	703	1090	1363
	4' 6"	9.00	712	15.00	450	735	1162	1447
	4' 10"	9.66	764	15.66	470	767	1234	1531
	5' 2"	10.34	818	16.34	490	801	1308	1619
	5' 6"	11.00	870	17.00	510	833	1380	1703
	5' 10"	11.66	922	17.66	530	865	1452	1787
	6' 2"	12.34	976	18.34	550	899	1526	1875
	6' 6"	13.00	1028	19.00	570	931	1598	1959
2' 4"	2' 6"	5.83	461	12.00	360	588	821	1049
	2' 10"	6.59	521	12.66	380	620	901	1141
	3' 2"	7.39	585	13.34	400	654	985	1239
	3' 6"	8.16	645	14.00	420	686	1065	1331
	3' 10"	8.92	706	14.66	440	718	1146	1424
	4' 2"	9.72	769	15.34	460	752	1229	1521
	4' 6"	10.49	830	16.00	480	784	1310	1614
	4' 10"	11.25	890	16.66	500	816	1390	1706
	5' 2"	12.05	953	17.34	520	850	1473	1803
	5' 6"	12.82	1014	18.00	540	882	1554	1896
	5' 10"	13.58	1074	18.66	560	914	1634	1988
	6' 2"	14.38	1137	19.34	580	948	1717	2085
	6' 6"	15.15	1198	20.00	600	980	1798	2178
2' 8"	2' 10"	7.56	598	13.66	410	669	1008	1267
	3' 2"	8.46	669	14.34	430	703	1099	1372
	3' 6"	9.35	740	15.00	450	735	1190	1475
	3' 10"	10.23	809	15.66	470	767	1279	1579
	4' 2"	11.13	880	16.34	490	801	1370	1681
	4' 6"	12.02	951	17.00	510	833	1461	1784
	4' 10"	12.90	1020	17.66	530	865	1550	1885
	5' 2"	13.80	1092	18.34	550	899	1642	1991
	5' 6"	14.69	1162	19.00	570	931	1732	2094
	5' 10"	15.57	1232	19.66	590	963	1822	2195
	6' 2"	16.47	1303	20.34	610	997	1913	2300
	6' 6"	17.36	1373	21.00	630	1029	2003	2402

HEAT LOSSES (0° to 70°) THROUGH DOUBLE-HUNG WOOD WINDOWS

Opening, Ft. and In.		Area, Sq. Ft.	Heat Loss through Glass, Btu per Hr.	Crack, Ft.	Infiltration Loss, Btu per Hr.		Total Window Loss, Btu per Hr.	
Width	Height				Weather-stripped	Not Weather-stripped	Weather-stripped	Not Weather-stripped
3' 0''	2' 10''	8.49	672	14.66	440	718	1112	1390
	3' 2''	9.51	752	15.34	460	752	1212	1504
	3' 6''	10.50	831	16.00	480	784	1311	1615
	3' 10''	11.49	909	16.66	500	816	1409	1725
	4' 2''	12.51	990	17.34	520	850	1510	1840
	4' 6''	13.50	1068	18.00	540	882	1608	1950
	4' 10''	14.49	1146	18.66	560	914	1706	2060
	5' 2''	15.51	1227	19.34	580	948	1807	2175
	5' 6''	16.50	1305	20.00	600	980	1905	2285
	5' 10''	17.49	1383	20.66	620	1012	2003	2395
	6' 2''	18.51	1464	21.34	640	1046	2104	2510
	6' 6''	19.50	1542	22.00	660	1078	2202	2620
3' 4''	2' 10''	9.42	745	15.66	470	767	1215	1512
	3' 2''	10.56	835	16.34	490	801	1325	1636
	3' 6''	11.66	922	17.00	510	833	1422	1755
	3' 10''	12.75	1009	17.66	530	865	1539	1874
	4' 2''	13.89	1099	18.34	550	899	1649	1998
	4' 6''	14.99	1186	19.00	570	931	1756	2117
	4' 10''	16.08	1272	19.66	590	963	1862	2235
	5' 2''	17.22	1362	20.34	610	997	1972	2359
	5' 6''	18.32	1449	21.00	630	1029	2079	2478
	5' 10''	19.41	1535	21.66	650	1061	2185	2596
	6' 2''	20.55	1626	22.34	670	1095	2296	2721
	6' 6''	21.65	1713	23.00	690	1127	2403	2840
3' 8''	3' 6''	12.85	1016	18.00	540	882	1556	1898
	3' 10''	14.06	1112	18.66	560	914	1672	2026
	4' 2''	15.30	1210	19.34	580	948	1790	2158
	4' 6''	16.52	1307	20.00	600	980	1907	2287
	4' 10''	17.73	1402	20.66	620	1012	2022	2414
	5' 2''	18.97	1501	21.34	640	1046	2141	2547
	5' 6''	20.19	1597	22.00	660	1078	2257	2675
	5' 10''	21.40	1693	22.66	680	1110	2373	2803
	6' 2''	22.64	1791	23.34	700	1144	2491	2935
	6' 6''	23.86	1887	24.00	720	1176	2607	3063
4' 0''	3' 6''	14.00	1107	19.00	570	931	1677	2038
	3' 10''	15.32	1212	19.66	590	963	1802	2175
	4' 2''	16.68	1319	20.34	610	997	1929	2316
	4' 6''	18.00	1424	21.00	630	1029	2054	2453
	4' 10''	19.32	1528	21.66	650	1061	2178	2589
	5' 2''	20.68	1636	22.34	670	1095	2306	2731
	5' 6''	22.00	1740	23.00	690	1127	2430	2867
	5' 10''	23.32	1845	23.66	710	1159	2555	3004
	6' 2''	24.68	1952	24.34	730	1193	2682	3147
	6' 6''	26.00	2057	25.00	750	1225	2807	3282
4' 4''	4' 6''	19.49	1542	22.00	660	1078	2202	2620
	5' 2''	22.39	1771	23.34	700	1144	2471	2915
	5' 10''	25.24	1996	24.66	740	1208	2736	3204
	6' 6''	28.15	2227	26.00	780	1274	3007	3501

HEAT LOSS THROUGH STEEL WINDOWS

The accompanying tabulations cover heat loss by transmission and infiltration through various types of steel windows on the 0 to 70 deg. temperature difference basis. The same principles apply as to wood windows as explained under "Heat Loss through Wood Windows."

For steel windows, the value of q (cubic feet per hour leakage per foot of crack) varies with the particular type of window. Values of q (based on a 15 mph wind velocity) used were:

Double-hung steel windows, weatherstripped..... 32
Double-hung steel windows, not weatherstripped 72
Steel casement windows, 1/64 in. crack 33
Steel casement windows, 1/32 in. crack............ 52
Industrial pivoted steel windows, 1/16 in. crack..176
Industrial pivoted steel windows, 1/32 in. crack . 88
Commercial projected steel windows,
  1/16 in. crack 176
Commercial projected steel windows,
  1/32 in. crack 88
Architectural projected steel windows,
  1/64 in. crack 62
Architectural projected steel windows,
  1/32 in. crack 88
Combination side-hinged and projected
  steel windows, 1/64 in. crack 18
Combination side-hinged and projected
  steel windows, 1/32 in. crack 38

HEAT LOSSES (0° TO 70°), ARCHITECTURAL PROJECTED STEEL WINDOWS

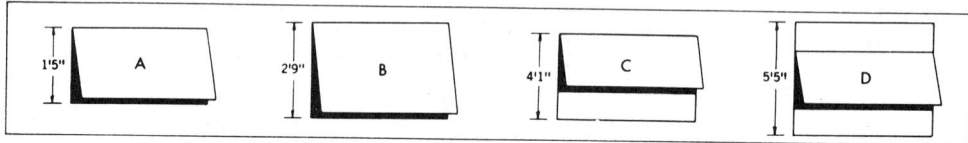

Symbol (See Drawing Above)	Opening, Ft. and In.		Area, Sq. Ft.	Heat Loss through Glass, Btu per Hr.	Crack, Ft.	Infiltration Loss, Btu per Hr.		Total Window Loss, Btu per Hr.	
	Width	Height				1/64 In. Crack	1/32 In. Crack	1/64 In. Crack	1/32 In. Crack
A	2'0⅞"	1'5"	2.94	233	7.02	548	779	781	1012
	2'8⅞"		3.89	308	8.32	649	924	957	1232
	3'4⅞"		4.84	383	9.66	753	1072	1136	1455
	3'8⅞"		5.31	420	10.32	805	1146	1225	1566
	4'0⅞"		5.78	457	10.98	856	1219	1313	1676
	4'8⅞"		6.73	532	12.32	961	1368	1493	1900
B	2'0⅞"	2'9"	5.69	450	9.66	753	1072	1203	1522
	2'8⅞"		7.54	596	10.98	856	1219	1452	1815
	3'4⅞"		9.38	742	12.32	961	1368	1703	2110
	3'8⅞"		10.29	814	12.98	1012	1441	1826	2255
	4'0⅞"		11.19	885	13.64	1064	1514	1949	2399
	4'8⅞"		13.04	1031	14.98	1168	1663	2199	2694
C	2'0⅞"	4'1"	8.45	668	9.66	753	1072	1421	1740
	2'8⅞"		11.18	884	10.98	856	1219	1740	2103
	3'4⅞"		13.91	1100	12.32	961	1368	2061	2468
	3'8⅞"		15.26	1207	12.98	1012	1441	2219	2648
	4'0⅞"		16.61	1314	13.64	1064	1514	2378	2828
	4'8⅞"		19.34	1530	14.98	1168	1663	2698	3193
D	2'0⅞"	5'5"	11.22	888	9.66	753	1072	1641	1960
	2'8⅞"		14.85	1175	10.98	856	1219	2031	2394
	3'4⅞"		18.48	1462	12.32	961	1368	2423	2830
	3'8⅞"		20.27	1603	12.98	1012	1441	2615	3044
	4'0⅞"		22.06	1745	13.64	1064	1514	2809	3259
	4'8⅞"		25.69	2032	14.98	1168	1663	3200	3695

The drawings shown in certain tables are not to scale and for many cases not in proportion; their purpose is to show general arrangement. The direction of projection (in or out) is, in the drawings, all one way, since this is of no significance to the heating engineer.

The values of q as given are not necessarily the values which might apply to a given installation, since the quality of workmanship both in manufacture and installation can affect this value considerably. It should also be noted that the infiltration loss applies only to the crack opening around movable sections of the window and does not include any infiltration around the overall window frame; in other words, the table assumes perfect caulking around the frame.

For design temperature differences other than 0 to 70 deg., multiply the Btu values in the tables by the correction factors as given under "Heat Loss through Wood Windows."

In all cases, the window opening has been calculated as all glass, with a U value of 1.13.

HEAT LOSSES, ARCHITECTURAL PROJECTED STEEL WINDOWS (Ctd.)

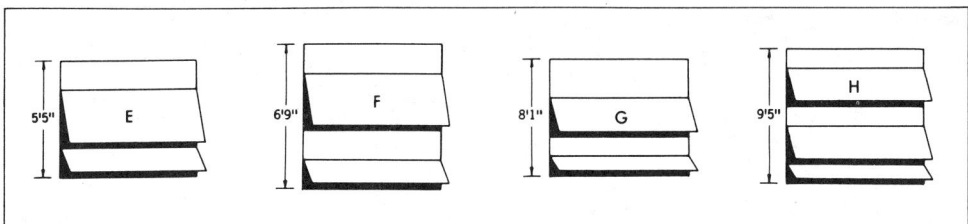

Symbol (See Drawing Above)	Opening, Ft. and In.		Area, Sq. Ft.	Heat Loss through Glass, Btu per Hr.	Crack, Ft.	Infiltration Loss, Btu per Hr.		Total Window Loss, Btu per Hr.	
	Width	Height				$\frac{1}{64}$ In. Crack	$\frac{1}{32}$ In. Crack	$\frac{1}{64}$ In. Crack	$\frac{1}{32}$ In. Crack
E	2'0⅞"	5'5"	11.22	888	14.34	1119	1592	2007	2480
	2'8⅞"		14.85	1175	16.35	1275	1815	2450	2990
	3'4⅞"		18.48	1462	18.36	1432	2038	2894	3500
	3'8⅞"		20.27	1603	19.35	1509	2148	3112	3751
	4'0⅞"		22.06	1745	20.34	1587	2258	3322	4004
	4'8⅞"		25.69	2032	22.35	1743	2481	3775	4513
F	2'0⅞"	6'9"	13.97	1105	16.35	1275	1815	2380	2920
	2'8⅞"		18.50	1463	19.35	1509	2148	2972	3611
	3'4⅞"		23.02	1821	21.74	1696	2413	3517	4234
	3'8⅞"		25.25	1997	23.06	1799	2560	3796	4557
	4'0⅞"		27.47	2173	24.38	1902	2706	4075	4879
	4'8⅞"		32.00	2531	27.06	2111	3004	4642	5535
G	2'0⅞"	8'1"	16.73	1323	16.35	1275	1815	2598	3138
	2'8⅞"		22.14	1751	19.35	1509	2148	3260	3899
	3'4⅞"		27.55	2179	21.74	1696	2413	3875	4592
	3'8⅞"		30.22	2390	23.06	1799	2560	3189	4950
	4'0⅞"		32.89	2602	24.38	1902	2706	4504	5308
	4'8⅞"		38.30	3030	27.06	2111	3004	5141	6034
H	2'0⅞"	9'5"	19.50	1542	23.66	1845	2626	3387	4168
	2'8⅞"		25.81	2042	27.06	2111	3004	4153	5046
	3'4⅞"		32.12	2541	30.36	2368	3370	4909	5911
	3'8⅞"		35.23	2787	32.01	2497	3553	5284	6340
	4'0⅞"		38.34	3033	33.66	2625	3736	5658	6769
	4'8⅞"		44.65	3532	37.01	2887	4108	6419	7640

HEAT LOSSES (0° TO 70°) THROUGH STEEL CASEMENT WINDOWS

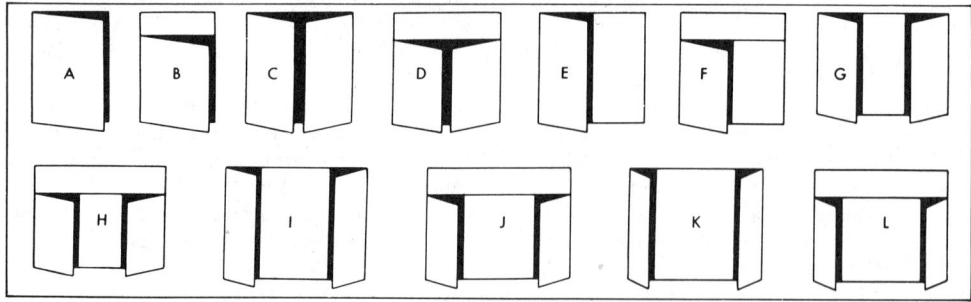

Symbol (See Key Above)	Opening Ft. and In.		Area, Sq. Ft.	Heat Loss through Glass, Btu per Hr.	Crack, Ft.	Infiltration Loss, Btu per Hr.		Total Window Loss, Btu per Hr.	
	Width	Height				1/64 In. Crack	1/32 In. Crack	1/64 In. Crack	1/32 In. Crack
A	1′7⅛″	2′2″	3.47	274	7.54	314	494	588	768
		3′2⅜″	5.12	405	9.60	399	629	804	1034
		4′2⅝″	6.75	534	11.64	484	762	1018	1296
B	1′7⅛″	4′2⅝″	6.75	534	9.53	396	624	930	1158
		5′3″	8.40	664	11.60	483	760	1147	1424
C	3′1″	2′2″	6.68	528	14.84	617	972	1145	1500
		3′2⅜″	9.86	780	18.96	789	1242	1569	2022
		4′2⅝″	13.00	1028	23.04	958	1509	1986	2537
D	3′1″	4′2⅝″	13.00	1028	18.82	783	1233	1811	2239
		5′3″	16.17	1279	22.96	955	1504	2234	2783
E	3′1″	2′2″	6.68	528	7.42	309	486	837	1014
		3′2⅜″	9.86	780	9.48	394	621	1174	1401
		4′2⅝″	13.00	1028	11.52	479	755	1507	1783
F	3′1″	4′2⅝″	13.00	1028	9.41	391	616	1419	1644
		5′3″	16.17	1279	11.48	478	752	1757	2031
G	4′5⅛″	2′2″	9.61	760	14.60	607	956	1367	1716
		3′2⅜″	14.18	1122	18.72	779	1226	1901	2348
		4′2⅝″	18.69	1478	22.80	948	1493	2426	2971
H	4′5⅛″	4′2⅝″	18.69	1478	18.58	773	1217	2251	2695
		5′3″	23.26	1840	22.72	945	1488	2785	3328
I	5′9⅜″	2′2″	12.54	992	14.46	602	947	1594	1939
		3′2⅜″	18.50	1463	18.58	773	1217	2236	2680
		4′2⅝″	24.39	1929	22.66	943	1484	2872	3413
J	5′9⅜″	4′2⅝″	24.39	1929	18.44	767	1208	2696	3137
		5′3″	30.35	2401	22.58	939	1479	3340	3880
K	7′7½″	2′2″	16.56	1310	14.78	615	968	1925	2278
		3′2⅜″	24.42	1932	18.90	786	1238	2718	3170
		4′2⅝″	32.20	2547	22.98	956	1505	3503	4052
L	7′7½″	4′2⅝″	32.20	2547	18.76	780	1229	3327	3776
		5′3″	40.06	3169	22.90	953	1500	4122	4669

HEAT LOSSES (0° TO 70°) THROUGH STEEL COMBINATION WINDOWS

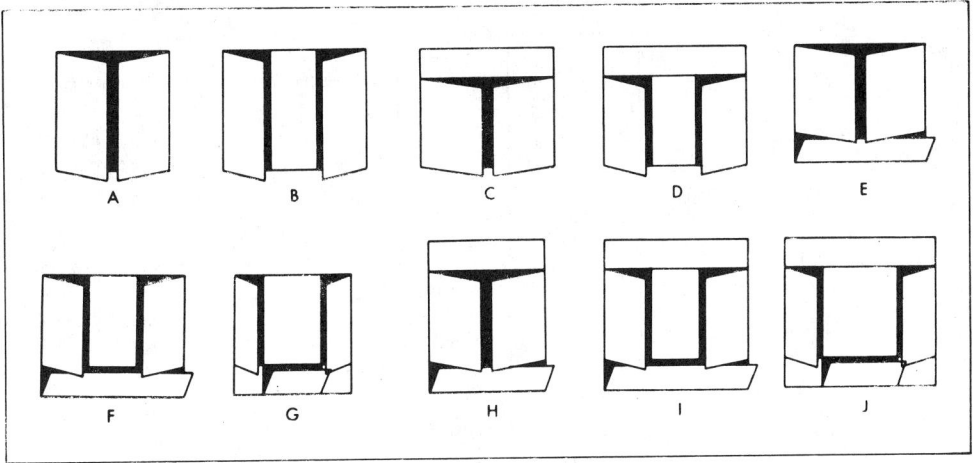

Symbol (See Drawing Above)	Opening, Ft. and In.		Area, Sq. Ft.	Heat Loss through Glass, Btu per Hr.	Crack, Ft.	Infiltration Loss, Btu per Hr.		Total Window Loss, Btu per Hr.	
	Width	Height				1/64 In. Crack	1/32 In. Crack	1/64 In. Crack	1/32 In. Crack
A	3'4⅞"	2'9"	9.38	742	15.07	347	723	1089	1465
	4'0⅞"		11.19	885	16.39	377	787	1262	1672
	3'4⅞"	4'1"	13.91	1100	19.06	438	915	1538	2015
	4'0⅞"		16.61	1314	20.38	469	978	1783	2292
B	5'0⅞"	2'9"	13.94	1103	17.76	408	852	1511	1955
	6'8⅞"		18.54	1467	20.00	460	960	1927	2427
	5'0⅞"	4'1"	20.69	1637	21.39	492	1027	2129	2664
	6'8⅞"		27.50	2175	23.06	530	1107	2705	3282
C	3'4⅞"	5'5"	18.48	1462	19.02	437	913	1899	2375
	4'0⅞"		22.06	1745	20.34	468	976	2213	2721
D	5'0⅞"	5'5"	27.48	2174	26.40	607	1267	2781	3441
	6'8⅞"		36.53	2890	29.74	684	1428	3574	4318
E	3'4⅞"	5'5"	18.48	1462	25.14	578	1207	2040	2669
	4'0⅞"		22.06	1745	27.12	624	1302	2369	3047
F	5'0⅞"	5'5"	27.48	2174	34.18	786	1641	2960	3815
G	6'8⅞"	5'5"	36.53	2890	24.32	559	1167	3449	4057
H	3'4⅞"	6'9"	23.02	1821	25.08	577	1204	2398	3025
	4'0⅞"		27.47	2173	27.06	622	1299	2795	3472
	3'4⅞"	8'1"	27.55	2179	27.74	638	1332	2817	3511
	4'0⅞"		32.89	2602	29.72	684	1427	3286	4029
I	5'0⅞"	6'9"	34.22	2707	30.40	699	1459	3406	4166
		8'1"	40.97	3241	36.14	831	1735	4072	4976
J	6'8⅞"	6'9"	45.50	3599	30.36	698	1457	4297	5056
		8'1"	54.46	4308	36.10	830	1733	5138	6041

HEAT LOSSES (0° TO 70°) THROUGH DOUBLE-HUNG STEEL WINDOWS

Size of Window in Ft. and In.		Area, Sq. Ft.	Heat Loss through Glass, Btu per Hr.	Crack, Ft.	Infiltration Loss, Btu per Hr.		Total Loss, Btu per Hr.	
Width	Height				Weather-stripped	Not Weather-stripped	Weather-stripped	Not Weather-stripped
2' 0"	3' 1½"	6.34	501	12.34	492	1107	993	1608
	3' 9½"	7.66	606	13.66	545	1226	1151	1832
	4' 5½"	9.00	712	15.00	599	1348	1311	2060
	5' 1½"	10.34	818	16.34	652	1467	1470	2285
	5' 9½"	11.66	922	17.66	705	1586	1627	2508
	6' 5½"	13.00	1028	19.00	758	1706	1786	2734
	7' 1½"	14.34	1131	20.34	813	1829	1944	2960
2' 4"	3' 1½"	7.39	585	13.34	532	1196	1117	1782
	3' 9½"	8.92	706	14.66	585	1316	1291	2022
	4' 5½"	10.49	830	16.00	638	1436	1468	2266
	5' 1½"	12.05	953	17.34	692	1557	1648	2510
	5' 9½"	13.58	1074	18.66	745	1676	1819	2750
	6' 5½"	15.15	1198	20.00	798	1796	1996	2994
	7' 1½"	16.70	1321	21.34	853	1919	2174	3240
2' 8"	3' 1½"	8.46	669	14.34	572	1287	1241	1956
	3' 9½"	10.23	809	15.66	625	1406	1434	2215
	4' 5½"	12.02	951	17.00	678	1526	1629	2477
	5' 1½"	13.80	1092	18.34	732	1647	1824	2739
	5' 9½"	15.57	1232	19.66	785	1766	2017	2998
	6' 5½"	17.36	1373	21.00	840	1890	2213	3263
	7' 1½"	19.10	1511	22.34	893	2009	2404	3520
3' 0"	3' 1½"	9.51	752	15.34	612	1377	1363	2129
	3' 9½"	11.49	909	16.66	665	1496	1574	2405
	4' 5½"	13.50	1068	18.00	718	1616	1786	2684
	5' 1½"	15.51	1227	19.34	771	1735	1998	2962
	5' 9½"	17.49	1383	20.66	825	1856	2208	3239
	6' 5½"	19.50	1542	22.00	878	1976	2420	3518
	7' 1½"	21.50	1701	23.34	933	2099	2634	3800
3' 4"	3' 1½"	10.56	835	16.34	652	1467	1487	2302
	3' 9½"	12.75	1009	17.66	705	1586	1714	2595
	4' 5½"	14.99	1186	19.00	758	1706	1944	2892
	5' 1½"	17.22	1362	20.34	811	1825	2173	3187
	5' 9½"	19.41	1535	21.66	865	1946	2400	3481
	6' 5½"	21.65	1713	23.00	918	2066	2631	3779
	7' 1½"	23.80	1883	24.34	973	2189	2856	4072
3' 8"	3' 1½"	11.60	918	17.34	694	1562	1612	2480
	3' 9½"	14.06	1112	18.66	745	1676	1857	2788
	4' 5½"	16.52	1307	20.00	798	1796	2105	3103
	5' 1½"	18.97	1501	21.34	851	1915	2352	3416
	5' 9½"	21.40	1693	22.66	904	2034	2597	3729
	6' 5½"	23.86	1887	24.00	958	2156	2845	4043
	7' 1½"	26.30	2080	25.34	1013	2279	3093	4359
4' 0"	3' 1½"	12.60	997	18.34	733	1649	1730	2646
	3' 9½"	15.32	1212	19.66	785	1766	1997	2978
	4' 5½"	18.00	1424	21.00	838	1886	2262	3310
	5' 1½"	20.68	1636	22.34	891	2005	2527	3641
	5' 9½"	23.32	1845	23.66	944	2124	2789	3969
	6' 5½"	26.00	2057	25.00	998	2246	3055	4303
	7' 1½"	28.60	2262	26.34	1053	2369	3315	4631

HEAT LOSSES (0° TO 70°) THROUGH PIVOTED STEEL WINDOWS

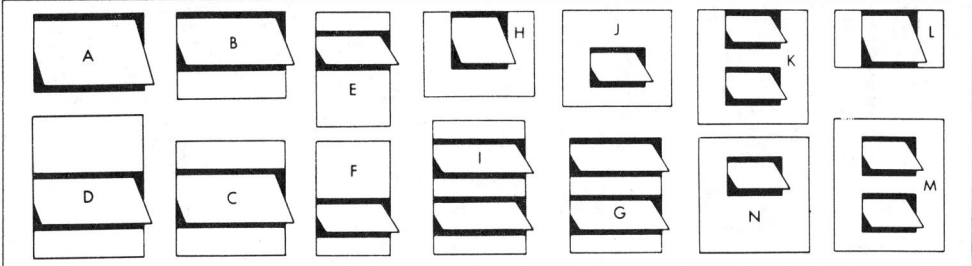

Symbol (See Key Above)	Opening Ft. and In.		Area, Sq. Ft.	Heat Loss through Glass, Btu per Hr.	Crack, Ft.	Infiltration Loss, Btu per Hr.		Total Window Loss, Btu per Hr.	
	Width	Height				1/32 In. Crack	1/16 In. Crack	1/32 In. Crack	1/16 In. Crack
A	1'8⅞"	2'9"	4.79	379	9.00	1998	999	2377	1378
	3'8⅞"		10.29	814	13.00	2886	1443	3700	2257
	5'0⅞"		13.94	1103	15.50	3441	1721	4544	2824
B	1'8⅞"	4'1"	7.14	565	9.00	1998	999	2563	1564
	3'8⅞"		15.39	1210	13.00	2886	1443	4096	2653
	5'0⅞"		20.69	1637	15.50	3441	1721	5078	3358
C	1'8⅞"	5'5"	9.43	746	9.00	1998	999	2744	1745
	3'8⅞"		20.27	1603	13.00	2886	1443	4489	3046
	5'0⅞"		27.48	2174	15.50	3441	1721	5615	3895
D	1'8⅞"	6'9"	11.75	929	9.00	1998	999	2927	1928
	3'8⅞"		25.25	1997	13.00	2886	1443	4883	3440
	5'0⅞"		34.22	2707	15.50	3441	1721	6148	4428
E	3'8⅞"	6'9"	25.25	1997	13.00	2886	1443	4883	3440
	5'0⅞"		34.22	2707	15.50	3441	1721	6148	4428
F	3'8⅞"	8'1"	30.22	2390	13.00	2886	1443	5276	3833
	5'0⅞"		40.97	3241	15.50	3441	1721	6682	4962
	3'8⅞"	9'5"	35.23	2787	13.00	2886	1443	5673	4230
	5'0⅞"		47.76	3778	15.50	3441	1721	7219	5499
G	3'8⅞"	8'1"	30.22	2390	26.00	5772	2886	8162	5276
	5'0⅞"		40.92	3241	31.00	6882	3441	10123	6682
H	6'8⅞"	4'1"	27.50	2175	12.25	2720	1360	4895	3535
I	3'8⅞"	9'5"	35.23	2787	26.00	5772	2886	8559	5673
	5'0⅞"		47.76	3778	31.00	6882	3441	10660	7219
J	6'8⅞"	5'5"	36.53	2890	12.25	2720	1360	5610	4250
		6'9"	45.50	3599	12.25	2720	1360	6319	4959
		8'1"	54.46	4308	12.25	2720	1360	7028	5668
		9'5"	63.49	5022	12.25	2720	1360	7744	6382
K	6'8⅞"	8'1"	54.46	4308	24.50	5440	2720	9748	7028
L	6'8⅞"	2'9"	18.54	1467	12.25	2720	1360	4187	2827
M	6'8⅞"	9'5"	63.49	5022	24.50	5440	2720	10462	7744
N	6'8⅞"	6'9"	45.50	3599	12.25	2720	1360	6319	4959

HEAT LOSSES (0° TO 70°), COMMERCIAL PROJECTED STEEL WINDOWS

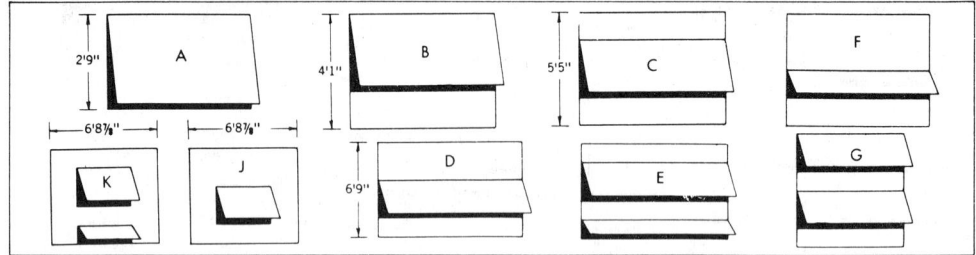

Symbol (See Drawing Above)	Opening, Ft. and In.		Area, Sq. Ft.	Heat Loss through Glass, Btu per Hr.	Crack, Ft.	Infiltration Loss, Btu per Hr.		Total Window Loss, Btu per Hr.	
	Width	Height				1/16 In. Crack	1/32 In. Crack	1/16 In. Crack	1/32 In. Crack
A	1'8⅞"	2'9"	4.79	379	8.95	1986	993	2365	1372
	3'8⅞"		10.29	814	12.93	2870	1435	3684	2249
	5'0⅞"		13.94	1103	15.60	3463	1732	4566	2835
B	1'8⅞"	4'1"	7.10	562	8.95	1986	993	2548	1555
	3'8⅞"		15.26	1207	12.93	2870	1435	4077	2642
	5'0⅞"		20.69	1637	15.60	3463	1732	5100	3369
C	1'8⅞"	5'5"	9.43	746	8.95	1986	993	2732	1739
	3'8⅞"		20.27	1603	12.93	2870	1435	4473	3038
	5'0⅞"		27.48	2174	15.60	3463	1732	5637	3906
D	1'8⅞"	6'9"	11.75	929	8.95	1986	993	2915	1922
	3'8⅞"		25.25	1997	12.93	2870	1435	4867	3432
	5'0⅞"		34.22	2707	15.60	3463	1732	6170	4439
E	3'8⅞"	6'9"	25.25	1997	23.04	5114	2557	7111	4554
	5'0⅞"		34.22	2707	28.35	6294	3147	9001	5854
	3'8⅞"	8'1"	30.22	2390	23.04	5114	2557	7740	4947
	5'0⅞"		40.97	3241	28.35	6294	3147	9535	6388
	3'8⅞"	9'5"	35.23	2787	23.04	5114	2557	7901	5344
	5'0⅞"		47.76	3778	28.35	6294	3147	10072	6925
F	3'8⅞"	8'1"	30.22	2390	12.93	2870	1435	5260	3825
	5'0⅞"		40.97	3241	15.60	3463	1732	6740	4973
	3'8⅞"	9'5"	35.23	2787	12.93	2870	1435	5657	4222
	5'0⅞"		47.76	3778	15.60	3463	1732	7241	5510
G	3'8⅞"	8'1"	30.22	2390	25.73	5712	2856	8102	5246
	5'0⅞"		40.97	3241	31.05	6893	3447	10134	6688
	3'8⅞"	9'5"	35.23	2787	25.73	5712	2856	8499	5643
	5'0⅞"		47.76	3778	31.05	6893	3447	10671	7225
J	6'8⅞"	5'5"	36.53	2890	12.14	2695	1348	5585	4238
		6'9"	45.50	3599	12.14	2695	1348	6294	4947
		8'1"	54.46	4308	12.14	2695	1348	7003	5656
		9'5"	63.49	5022	12.14	2695	1348	7717	6370
K	6'8⅞"	6'9"	45.50	3599	21.56	4786	2393	8385	5992
		8'1"	54.46	4308	21.56	4786	2393	9094	6501
		9'5"	63.49	5022	21.56	4786	2393	9808	7415

HEAT LOSSES (0° TO 70°) THROUGH ALUMINUM AWNING WINDOWS

(For other types of aluminum windows, refer to corresponding prototype in steel windows)

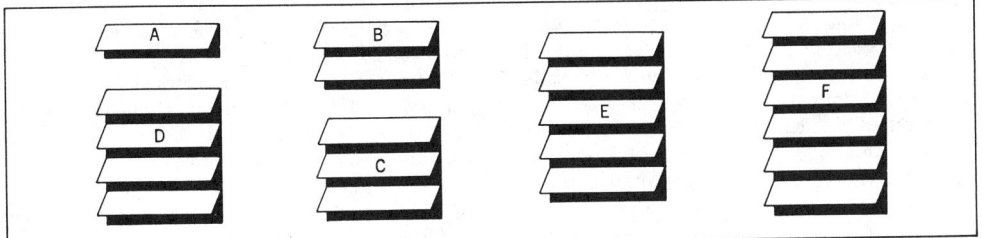

Symbol (See Drawing Above)	Opening, Ft. and In.		Area, Sq. Ft.	Heat Loss through Glass Btu per Hr.	Crack, Ft.	Infiltration Loss, Btu per Hr.		Total Window Loss, Btu per Hr.	
	Width	Height				1/64 In. Crack	1/32 In. Crack	1/64 In. Crack	1/32 In. Crack
A	2'0⅞"	1'5"	2.92	231	6.96	543	773	774	1004
	2'8⅞"		3.86	305	8.29	647	920	952	1225
	3'4⅞"		4.81	380	9.62	750	1068	1130	1448
	4'0⅞"		5.74	454	10.95	854	1215	1308	1669
	4'8⅞"		6.68	528	12.28	958	1363	1486	1991
B	2'0⅞"	2'9"	5.69	450	11.71	913	1300	1363	1750
	2'8⅞"		7.54	596	13.71	1069	1522	1665	2118
	3'4⅞"		9.38	742	15.71	1225	1744	1967	2486
	4'0⅞"		11.19	885	17.71	1381	1966	2266	2851
	4'8⅞"		13.04	1031	19.71	1537	2188	2568	3219
C	2'0⅞"	4'1"	8.45	668	16.44	1282	1825	1950	2493
	2'8⅞"		11.18	884	19.11	1491	2121	2375	3005
	3'4⅞"		13.91	1100	21.78	1699	2418	2799	3518
	4'0⅞"		16.61	1314	24.45	1907	2714	3221	4028
	4'8⅞"		19.34	1530	27.12	2115	3010	3645	4540
D	2'0⅞"	5'5"	11.20	886	21.17	1651	2350	2537	3216
	2'8⅞"		14.82	1172	24.51	1912	2721	3084	3893
	3'4⅞"		18.45	1459	27.85	2172	3091	3631	4550
	4'0⅞"		22.02	1742	31.19	2433	3462	4175	5204
	4'8⅞"		25.64	2028	34.53	2693	3833	4721	5863
E	2'0⅞"	6'9"	13.97	1105	25.92	2022	2877	3127	3982
	2'8⅞"		18.50	1463	29.92	2334	3321	3800	4784
	3'4⅞"		23.02	1821	33.92	2646	3765	4467	5567
	4'0⅞"		27.47	2173	37.92	2958	4209	5131	6382
	4'8⅞"		32.00	2531	41.92	3270	4653	5801	7184
F	2'0⅞"	8'1"	16.73	1323	30.65	2391	3402	3714	4725
	2'8⅞"		22.14	1751	35.32	2755	3921	4506	5672
	3'4⅞"		27.55	2179	39.99	3119	4439	5298	6618
	4'0⅞"		32.89	2602	44.66	3483	4957	6085	7559
	4'4⅞"		38.30	3030	49.33	3848	5476	6878	8506

HEAT LOSS THROUGH DOORS

In the accompanying table, the heat loss through wooden doors has been tabulated. The left-hand portion of the table is devoted to transmission loss, and for this part of the table a *U*-value of 0.48 Btu per (hour) (degree) (sq ft) has been used, this being the value for a 1¾ inch (nominal) wood door, as suggested by the ASHAE *Guide* and is still current. Also following this authority, a coefficient of 0.31 was used as an overall *U* value for the door and storm door calculation. In both cases, the loss was calculated for a 70 deg. temperature differential; for other temperature differences, use window heat loss correction factor.

The effect of weatherstripping doors is believed to be a reduction of 50% infiltration loss a storm door is likewise believed to have the same effect. Consequently, in the table, the no-storm-door with no-weatherstripping combination (last column) has been calculated on the same basis as a loose fitting double-hung wood window—that is, at 111 cu ft per hour infiltration. The values for the non-weatherstripped but with storm door combination in the next-to-last column is taken as 50% of that in the last column. Similarly, the weatherstripped but no-storm-door values are 50% of those in the last column. Finally, the addition of the storm door to the weatherstripping (fourth from last column) is calculated to effect a further 50% reduction.

In an elaborate investigation of air flow through swinging doors in tall buildings by the ASHAE (Winter Infiltration through Swinging Door Entrances in Multi-Story Buildings, by T. C. Min, ASHAE Trans., 1958) the author concluded that such air flow depended not only on the door size but type of entrance, arrangement of doors, direction of swinging, depth of vestibule, size of cracks, traffic rate, height of building (chimney effect) and other variables. For rough estimating the author gives 900 cubic feet of infiltration per person per passage for single bank door entrances, and 550 cu ft for vestibule type entrances, both *for a 30 story building*.

For revolving doors, Air Conditioning & Refrigeration Institute, Inc., gives a value of 60 cubic feet per passage infiltration (72 inch door) as an average and with 75 and 40 cu ft for infrequent and heavy usage, respectively.

The American Society of Refrigerating Engineers gives 110 cubic feet per passage as average for single swing doors where the doors are in one wall only, and a range from 11 to 30 cubic feet per passage for revolving doors.

Heat Loss Through Wood Doors at 70 Deg. Differential

Size of Opening, Inches		Area, Sq. Ft.	Transmission Loss, Btu per Hour		Lin. Ft. of Crack	Infiltration Loss, Btu per Hour			
						Weatherstripped		Not Weatherstripped	
Width	Height		No Storm Door	With Storm Door		With Storm Door	No Storm Door	With Storm Door	No Storm Door
Single Doors									
24	78	13.0	385	274	17.0	595	1190	1190	2380
24	80	13.3	394	281	17.3	605	1211	1211	2422
28	78	15.2	450	321	17.7	619	1239	1239	2478
28	80	15.6	462	329	18.0	630	1260	1260	2520
30	78	16.2	480	342	18.0	630	1260	1260	2520
30	80	16.7	494	352	18.3	640	1281	1281	2562
32	80	17.8	527	376	18.8	658	1316	1316	2632
34	82	19.3	571	407	19.3	675	1351	1351	2702
36	80	20.0	592	422	19.3	675	1351	1351	2702
36	84	21.0	622	443	20.0	700	1400	1400	2800
Double (French) Doors									
48	78	26.0	770	549	27.5	962	1925	1925	3850
48	80	26.7	790	563	28.0	980	1960	1960	3920
48	84	28.0	829	591	29.0	1015	2030	2030	4060
60	78	32.5	962	686	29.5	1032	2065	2065	4130
60	80	33.0	986	703	30.0	1050	2100	2100	4200
60	84	35.0	1036	739	31.0	1085	2170	2170	4340

INSIDE WALL SURFACE TEMPERATURES

The resistance to heat flow through a wall including the inside surface conductance is $\frac{1}{U}$, where U is the coefficient of heat transmission. The resistance to heat flow from the inside air temperature to the inside wall surface consists only of the inner surface conductance $\frac{1}{f_i}$, usually taken as $\frac{1}{1.65}$. It can be shown that

$$\frac{t_i - t_s}{t_i - t_o} = \frac{1/1.65}{1/U}$$

so that

$$t_s = t_i - \frac{U}{1.65}(t_i - t_o)$$

where t_s = temperature of the inside surface of the wall,

t_i = inside air temperature,

t_o = outside air temperature.

The accompanying chart is a solution of this equation when $t_i = 70$ deg.

For example, what is the inner wall temperature of a wall having a U of 0.30 when the outside air temperature is 10 deg and the inside air temperature is 70 deg.

From 0.30 on the horizontal scale move upward to the diagonal line for 10 deg, thence to the left and read 59 deg.

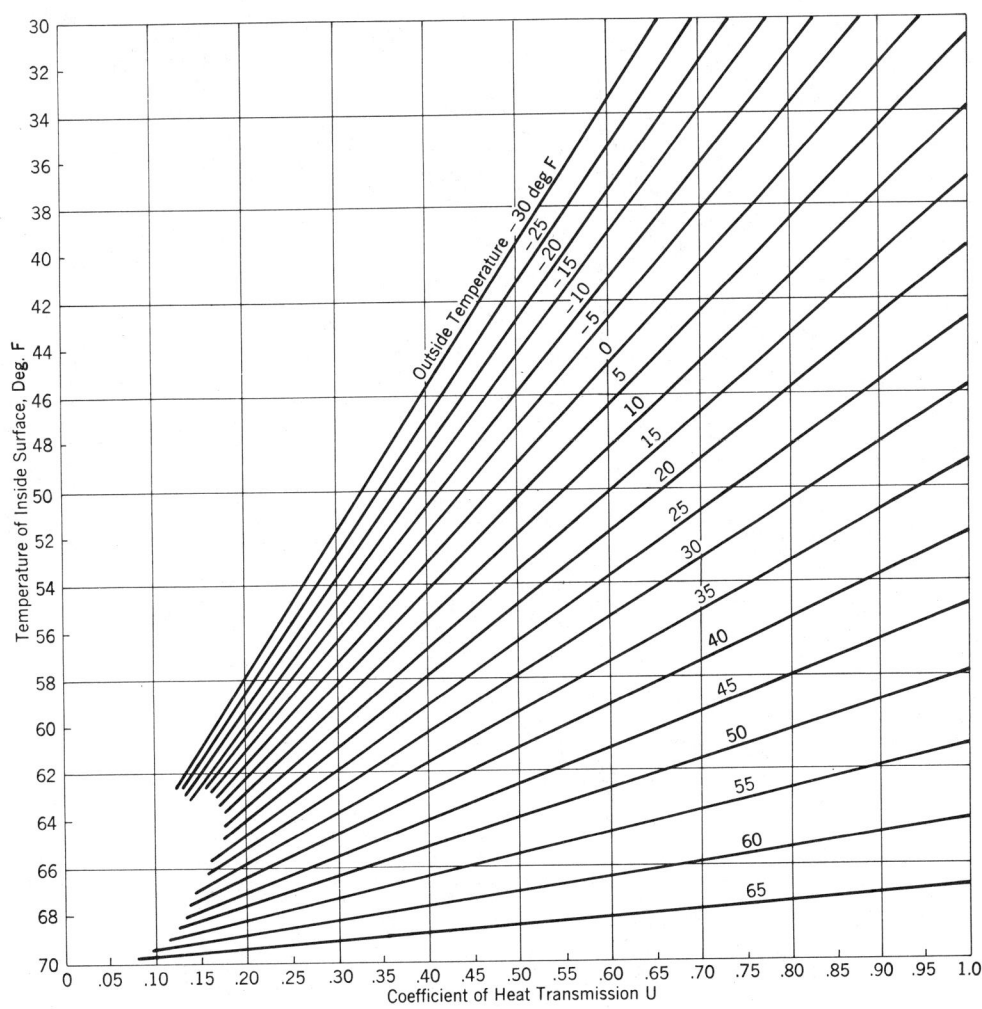

HEAT WASTED BY HIGH INSIDE TEMPERATURES

The accompanying charts show the percentage of heat wasted by carrying excessive inside temperatures. They can be used for either daily or seasonal calculations. Since they are based on both 55F and 65F daily average temperatures, they are usable for most installations.

Example: An apartment normally heated to 70F during the day and evening hours but with a 24-hour average inside temperature of 65F, is being heated to an average inside temperature (24 hr) of 75F. The temperature outside averages 50F on the day in question. What per cent of the heat is being wasted? *Solution:* Refer to Fig. 1 and locate 50F on the lower scale, move vertically to the curved 75F line, thence horizontally to read 40%. That is, on this day, 40% of the heat is wasted.

Note that the answer is in per cent of the heat

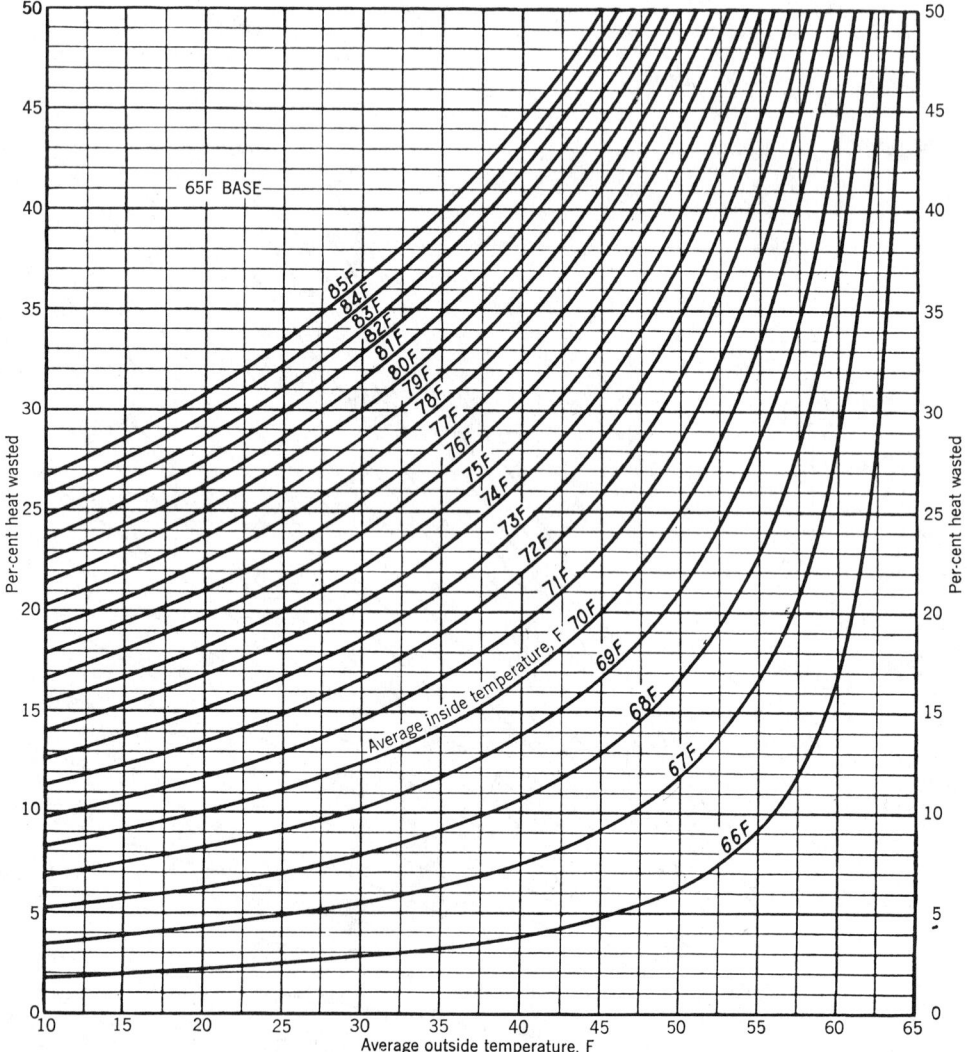

Fig. 1. Per-cent of heat wasted by high inside temperatures in buildings where 24-hour average inside temperature is normally 65F.

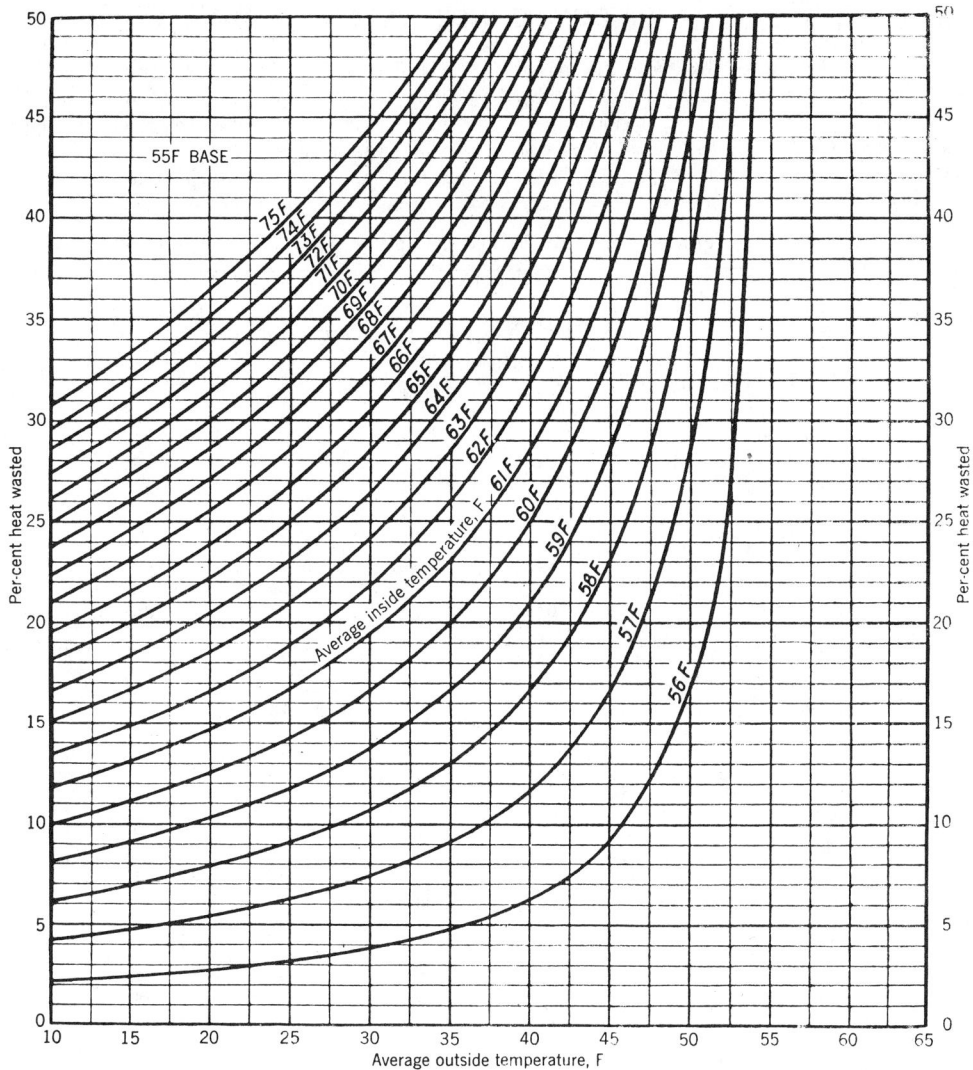

Fig. 2. Per-cent of heat wasted by high inside temperatures in buildings where 24-hour average inside temperature is normally 55F.

load under the excessive condition. In the foregoing example 40% of the heat is wasted; this does not mean the 40% more heat is being used than necessary.

The excess heat $= X \div (100 - X)$, where X is the percent of heat being wasted (from the charts). In this example, then $40 \div (100 - 40) = 66.7\%$ more heat is being used than would normally be used.

Example: An industrial plant, normally operated at 60 deg F is held for a day at 65 deg while the outside temperature averages 40 deg. What is the percent of heat wasted? *Solution:* Refer to Fig. 2 and find 40 deg on the lower scale. Move vertically to the 65 F curve and note that it intersects the horizontal line for 40% heat wasted at that point. Thus for a five-degree excess in inside temperature 40% of all heat supplied was wasted.

WINDOW CONDENSATION

The accompanying tables give information on the combinations of temperature and relative humidity which will result in condensation forming on windows. The data are given for both single and double windows, and for each of these for both a no-wind condition and with wind.

The data were arrived at by equating the dew point corresponding to the inside condition and the inside surface temperature, based on the type window, wind and temperature conditions.

For example, assume an inside temperature of 65 deg and a relative humidity of 40%. The dew

Table 1. Outside Temperatures at Which Moisture Will Deposit on Windows

Inside Relative Humidity, %	Inside Temperature, Deg F				
	60	65	70	75	80
	Outside Temp. at Which Moisture Will Deposit				
	Single Windows, Still Air				
10	−48	−45	−44	−43	−40
20	−20	−17	−14	−13	−10
30	−2	1	4	7	12
40	12	15	20	23	26
50	22	27	30	35	40
60	32	37	40	45	50
70	40	45	50	55	58
80	48	53	58	61	66
90	54	59	64	69	74
100	60	65	70	75	80
	Single Windows, Wind				
10	−21	−17	−15	−14	−10
20	0	4	7	9	13
30	14	17	20	24	29
40	24	28	33	36	40
50	31	37	40	45	50
60	39	44	48	53	58
70	45	50	55	60	64
80	51	56	61	65	70
90	55	60	65	70	75
100	60	65	70	75	80
	Double Windows, Still Air				
10	−156	−155	−158	−161	−160
20	−100	−99	−98	−101	−100
30	−64	−63	−62	−61	−56
40	−36	−35	−30	−29	−28
50	−16	−11	−10	−5	0
60	4	9	10	15	20
70	20	25	30	35	36
80	36	41	46	47	52
90	48	53	58	63	68
100	60	65	70	75	80
	Double Windows, Wind				
10	−120	−118	−120	−122	−120
20	−73	−72	−70	−72	−70
30	−43	−41	−40	−38	−33
40	−20	−18	−13	−11	−10
50	−3	2	3	8	13
60	13	18	20	25	30
70	27	31	37	42	43
80	40	45	50	52	57
90	50	55	60	65	70
100	60	65	70	75	80

Values shown in italics indicate that moisture will be in the form of frost.

Table 2. Inside Relative Humidities at Which Moisture Will Deposit on Windows

Deg F Outside Temperature,	Inside Temperature, Deg F				
	60	65	70	75	80
	Inside Rel. Hum. at Which Moisture Will Deposit				
Single Windows, Still Air					
−50	9	9	9	8	8
−40	12	12	11	10	10
−30	16	14	14	13	13
−20	20	18	18	17	16
−10	25	23	22	21	20
0	32	29	27	25	24
10	39	36	33	31	29
20	47	43	40	37	35
30	57	52	50	45	42
40	70	63	60	53	50
Single Windows, Wind					
−50	4	3	3	3	2
−40	5	4	4	4	3
−30	7	6	6	6	5
−20	10	9	9	8	7
−10	14	13	12	11	10
0	20	18	16	15	13
10	28	24	22	20	18
20	36	32	30	26	24
30	48	41	38	34	30
40	62	54	49	43	40
Double Windows, Still Air					
−50	34	34	34	33	32
−40	38	38	36	36	35
−30	42	42	41	40	38
−20	47	46	46	44	42
−10	52	50	49	48	46
0	57	55	55	52	50
10	62	60	59	57	54
20	69	66	63	62	59
30	76	73	71	68	65
40	84	80	79	74	71
Double Windows, Wind					
−50	28	26	26	26	25
−40	32	30	30	29	28
−30	36	34	34	32	31
−20	41	39	38	35	35
−10	46	45	42	42	39
0	52	49	47	46	44
10	58	56	53	52	49
20	65	63	59	58	54
30	72	70	66	64	60
40	80	78	73	70	67

Values shown in italics indicate that moisture will be in the form of frost.

point temperature for this condition is 40 deg. The inside surface temperature (with wind) is found by the equation

$$t_s = t_i - \frac{U}{f_i} (t_i - t_o)$$

where t_s = inside glass surface temperature,

t_i = room air temperature,

U = heat transmission coefficient,

f_i = inside surface conductance,

t_o = outside air temperature.

Single glass, with wind, has a U value of 1.13, and f_i = 1.65. Therefore, making the surface temperature equal the dewpoint,

$$40 = 65 - \frac{1.13}{1.65} (65 - t_o)$$

$$t_o = 28.4$$

CONDENSATION ON GLASS SURFACES

The accompanying chart, designed by E. W. Jerger, associate professor, University of Notre Dame, shows, for a variety of glass walls and windows, the combination of inside relative humidity and outside temperature which will produce condensation on the inside of the glass. For example, with an indoor relative humidity of 40% and a single glass window, condensation will occur at and below a temperature of 33 deg. outside.

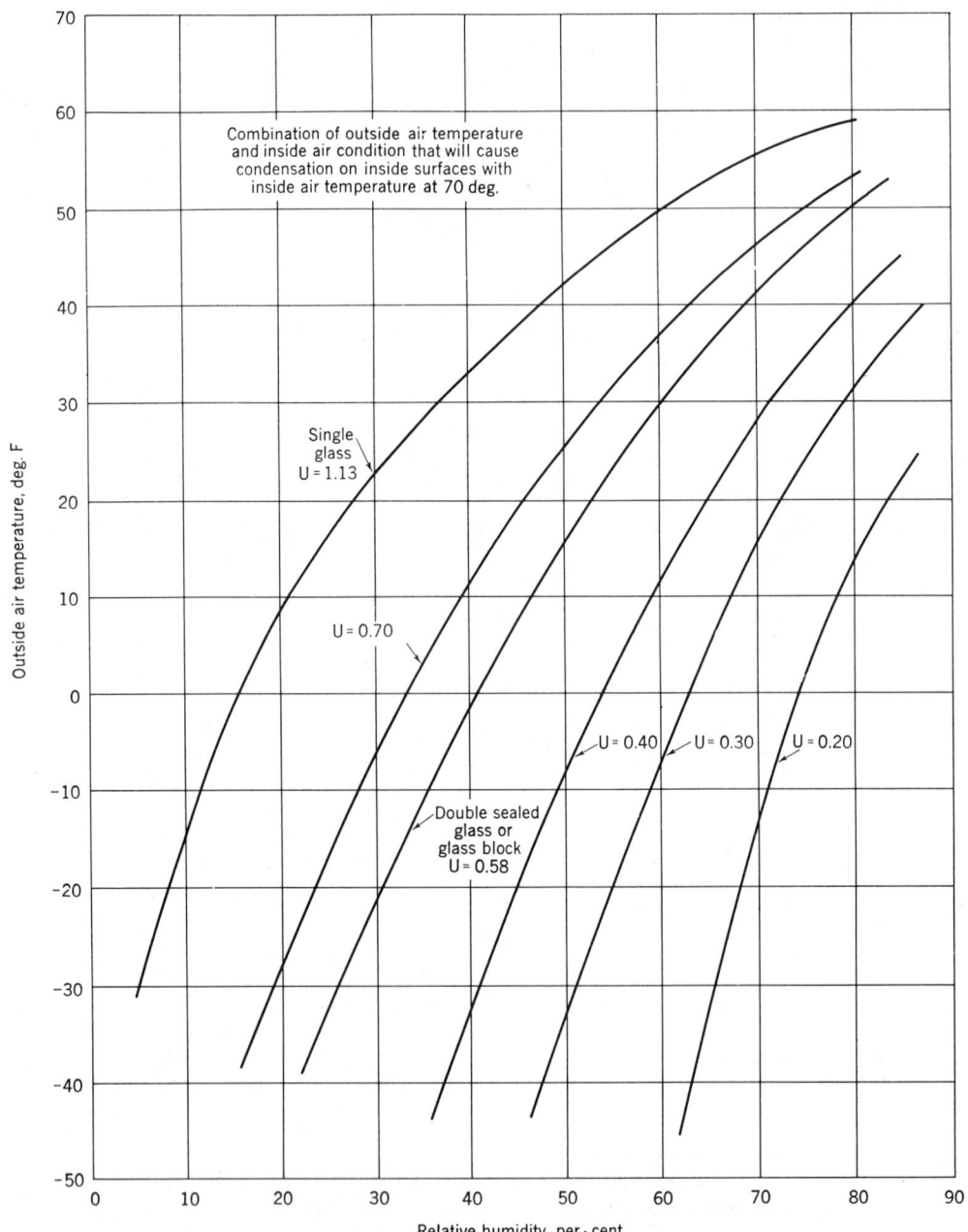

Combination of outside air temperature and inside air condition that will cause condensation on inside surfaces with inside air temperature at 70 deg.

Single glass U = 1.13

U = 0.70

U = 0.40 U = 0.30 U = 0.20

Double sealed glass or glass block U = 0.58

Outside air temperature, deg. F

Relative humidity, per-cent

SPECIFIC HEATS OF VARIOUS SOLIDS, LIQUIDS AND GASES

Specific Heats of Various Solids and Metals

Material	Temp., Deg F	Spec. Heat	Material	Temp., Deg F	Spec. Heat	Material	Temp., Deg F	Spec. Heat
Alloys			Earth	22-210	.440	Pine	32-212	.670
Red brass	32	.090	Fir	32-212	.650	Porcelain	—	.260
Yellow brass	32	.088	Glass, common	—	.200	Potassium	35	.177
80 Cu + 20 Sn	57-208	.086	Granite	50-125	.192	Pyrex	—	.200
Monel metal	68-2400	.127	Ice	—112	.350	Pyroxlin plastics	—	.360
Aluminum	32	.209	Ice	—40	.434	Quartz	54-212	.188
Antimony	59	.049	Ice	—4	.465	Rubber	—	.480
Asbestos	65-200	.195	Ice	32	.490	Silk	—	.330
Bakelite, asbestos fill.	—	.380	Iron	100	.109	Silver	32	.056
Bakelite, wood filler	—	.330	Iron, cast	60-212	.119	Steel	—	.120
Brick	32-212	.220	Iron, wrought	60-212	.115	Stoneware	—	.190
Carbon	52	.160	Lead	60-500	.032	Sandstone	32-212	.250
Cast iron	60-212	.119	Limestone	59-212	.216	Talc	62-208	.209
Cement, dry	—	.370	Marble	32-212	.210	Tar	—	.350
Chromium	32	.104	Nickel	65-215	.109	Tin	32	.054
Clay	—	.220	Oak	32-212	.570	Tile, hollow	—	.150
Coal	68-1900	.315	Plaster, neat gypsum	—	.250	Vanadium	32-212	.115
Concrete	32-212	.156	Plaster, sand	—	.190	Vulcanite	68-212	.331
Copper	60-212	.095	Plaster, perlite	—	.200	Zinc	32-212	.095
Cork	22-210	.485	Plaster, vermiculite	—	.200			

Specific Heats of Gases and Vapors

Gas or Vapor	Specific Heat, Btu per Lb per F		Gas or Vapor	Specific Heat, Btu per Lb per F	
	At Constant Pressure	At Constant Volume		At Constant Pressure	At Constant Volume
Air, dry, 70F	.24	.172	Hydrogen sulphide	.25	.189
Air, saturated, 70F	.85	—	Methane	.53	.405
Alcohol, C_2H_5OH	.45	.398	Nitrogen	.24	.170
Alcohol, CH_3OH	.46	.366	Nitric oxide	.23	.166
Ammonia	.54	.422	Nitrous oxide	.21	.166
Benzene, C_6H_6	.33	.236	Oxygen	.22	.157
Bromine	.06	.047	Steam, 32F	.47	.370
Carbon dioxide	.20	.156	Steam, 212F	.42	.316
Carbon monoxide	.24	.172	Steam, 356F	.51	.389
Chlorine	.11	.082	Sulphur dioxide	.15	.119
Chloroform	.15	.131	Water vapor, 32F	.47	.370
Hydrogen	3.41	2.43	Water vapor, 212F	.42	.316
			Water vapor, 356F	.51	.389

Specific Heats of Liquids

Liquid	Specific Heat, Btu per Lb per F	Liquid	Specific Heat, Btu per Lb per F
Acetone	.51	Fuel oil, sp. gr. = .91	.44
Ammonia, 32F	1.10	Fuel oil, sp. gr. = .86	.45
Ammonia, 104F	1.16	Fuel oil, sp. gr. = .81	.50
Ammonia, 176F	1.29	Gasoline	.53
Ammonia, 212F	1.48	Kerosene	.50
Ammonia, 238F	1.61	Sea water, sp. gr. = 1.024	.94
Calcium chloride, sp. gr. = 1.20	.73	Water	See "Properties of Water"
Fuel oil, sp. gr. = .96	.40		

Specific Heats of Foodstuffs

Product	Specific Heat, Btu per Lb per F		Product	Specific Heat, Btu per Lb per F	
	Above Freezing	Below Freezing		Above Freezing	Below Freezing
Apples	.87	.42	Dandelion greens	.88	.43
Apricots, fresh	.88	.43	Dates	.20	.007
Artichokes	.87	.42	Eels	.77	.39
Asparagus	.94	.45	Eggs	.76	.40
Avocados	.72	.37	Eggplant	.94	.45
Bananas	.80	.40	Endive	.95	.45
Barracuda	.80	.40	Figs, fresh	.82	.41
Bass	82	.41	Figs, dried	.39	.26
Beef, carcass	.68	.48	Figs, candied	.37	.26
Beef, flank	.56	.32	Flounders	.86	.42
Beef, loin	.66	.35	Flour	.38	.28
Beef, rib	.67	.36	Frogs legs	.88	.44
Beef, round	.74	.38	Garlic	.79	.40
Beef, rump	.62	.34	Gizzards	.78	.39
Beef, corned	.63	.34	Goose	.61	.34
Beer	.89	—	Gooseberry	.86	.42
Beets	.90	.43	Granadilla	.84	.41
Blackberries	.87	.42	Grapefruit	.91	.44
Blueberries	.87	.42	Grapes	.86	.42
Brains	.84	.41	Grape juice	.82	.41
Broccoli	.92	.44	Guavas	.86	.42
Brussels sprouts	.88	.43	Guinea hen	.75	.38
Butter	.65	.34	Haddock	.85	.42
Butterfish	.77	.39	Halibut	.80	.40
Cabbage	.94	.45	Herring, smoked	.71	.37
Candy	.93	—	Horseradish, fresh	.79	.40
Carp	.82	.41	Horseradish, prepared	.88	.43
Carrots	.91	.44	Ice cream	.74	.45
Cauliflower	.93	.44	Kale	.89	.43
Celery	.94	.45	Kidneys	.81	.40
Chard	.93	.44	Kidney beans, dried	.28	.23
Cheese	.65	—	Kohlrabi	.92	.44
Cherries, sour	.88	.43	Kumquats	.85	.41
Chicken, squab	.80	.40	Lamb, carcass	.73	.38
Chicken, broilers	.77	.39	Lamb, leg	.71	.37
Chicken, fryers	.74	.38	Lamb, rib cut	.61	.34
Chicken, hens	.65	.35	Lamb, shoulder	.67	.35
Chicken, capons	.88	.44	Lard	.54	.31
Clams, meat only	.84	.41	Leeks	.91	.44
Coconut, meat and milk	.68	.36	Lemons	.91	.44
Coconut, milk only	.95	.45	Lemon juice	.92	.44
Codfish	.86	.42	Lettuce	.96	.45
Cod Roe	.76	.39	Lima beans	.73	.38
Cowpeas, fresh	.73	.39	Limes	.89	.43
Cowpeas, dry	.28	.22	Lime juice	.93	.44
Crabs	.84	.41	Lobsters	.82	.41
Crab apples	.85	.41	Loganberries	.86	.42
Cranberries	.90	.43	Loganberry juice	.91	.44
Cream	.90	.38	Milk, cow	.94	.47
Cucumber	.97	.45	Mushrooms, fresh	.93	.44
Currants	.97	.45	Mushrooms, dried	.30	.23

The four accompanying tables give the specific heat of a wide variety of substances as collected from numerous sources. Where no temperature is given, the specific heat is that at approximately room temperature. Liquid ammonia is at saturation.

By definition specific heat is a ratio and consequently not expressed in units. However, the engineer commonly uses the term "specific heat" to

Specific Heats of Foodstuffs, concluded

Product	Specific Heat, Btu per Lb per F		Product	Specific Heat, Btu per Lb per F	
	Above Freezing	Below Freezing		Above Freezing	Below Freezing
Muskmelons	.94	.45	Rabbit	.76	.39
Nectarines	.86	.42	Radishes	.95	.45
Nuts	.28	.24	Raisins	.39	.26
Olives, green	.80	.40	Raspberries, Black	.85	.41
Onions	.90	.43	Raspberries, red	.89	.43
Onions, Welsh	.91	.44	Raspberry juice, black	.91	.44
Oranges, fresh	.90	.43	Reindeer	.73	.37
Orange juice	.89	.43	Rhubarb	.96	.45
Oysters	.84	.41	Rutabagas	.91	.44
Peaches, Georgia	.87	.42	Salmon	.71	.37
Peaches, N. Carolina	.89	.43	Sapote	.73	.37
Peaches, Maryland	.90	.43	Sauerkraut	.93	.44
Peaches, New Jersey	.91	.44	Sausage, beef and pork	.56	.32
Peach juice, fresh	.89	.43	Sausage, bockwurst	.71	.37
Pears, Bartlett	.89	.43	Sausage, bologna	.71	.37
Pears, Beurre Bosc	.85	.41	Sausage, frankfurt	.69	.36
Pears, dried	.39	.26	Sausage, salami	.45	.28
Peas, young	.85	.41	Sardines	.77	.39
Peas, medium	.81	.40	Shrimp	.83	.41
Peas, old	.88	.43	Spanish mackerel	.73	.39
Peas, split	.28	.23	Strawberries	.95	.45
Peppers, ripe	.91	.44	String beans	.91	.44
Perch	.82	.41	Sturgeon, raw	.83	.41
Persimmons	.72	.37	Sugar apple, fresh	.79	.39
Pheasant	.75	.38	Sweet potatoes	.75	.38
Pickerel	.84	.41	Swordfish	.80	.40
Pickles, sweet	.82	.41	Terrapin	.80	.40
Pickles, sour and dill	.96	.45	Tomatoes	.95	.45
Pickles, sweet mixed	.78	.29	Tomato juice	.95	.45
Pickles, sour hixed	.95	.45	Tongue, beef	.74	.38
Pig's feet, pickled	.50	.31	Tongue, calf	.79	.40
Pike	.84	.41	Tongue, lamb	.76	.38
Pineapple, fresh	.88	.43	Tongue, pork	.74	.39
Pineapple, sliced	.82	.41	Tripe, beef	.83	.41
Pineapple juice	.90	.43	Tripe, pickled	.89	.43
Plums	.89	.43	Trout	.82	.41
Pomegranate	.85	.41	Tuna	.76	.39
Pompano	.77	.39	Turkey	.67	.35
Porgy	.81	.40	Turnips	.93	.44
Pork, bacon	.36	.25	Turtle	.84	.41
Pork, ham	.62	.34	Veal, flank	.65	.35
Pork, loin	.66	.35	Veal, loin	.75	.38
Pork, shoulder	.59	.33	Veal, rib	.73	.37
Pork, spareribs	.62	.34	Veal, shank	.77	.39
Pork, smoked ham	.65	.35	Veal, quarter	.74	.38
Pork, salted	.31	.24	Venison	.78	.39
Potatoes	.82	.41	Watercress	.95	.45
Prickly pears	.91	.43	Watermelons	.94	.45
Prunes	.81	.40	Whitefish	.76	.39
Pumpkin	.92	.44	Wines	0.9	—
Quinces	.88	.43	Yams	.78	.39

indicate the quantity of heat necessary to raise a given mass one degree in temperature; this works out satisfactorily numerically since in both the English and metric systems the heat quantity to raise one unit mass of water one degree is one unit.

Many of the values in the table on foodstuffs have been calculated from the formula $S = .008M + 0.2$ where M is the percentage of moisture in the food.

LOGARITHMIC MEAN TEMPERATURE DIFFERENCE

The basic heat transfer formula is

$$Q = A \, U \, D_m$$

where Q = heat transmitted, Btu per hr
   A = heat transfer surface, sq ft
   U = transmission coefficient, Btu per (hr) (sq ft) (deg F), and
   D_m = logarithmic mean temperature difference.

Where temperature on each side of the heat transfer surface is uniform, as in the case of heat loss through a building wall, D_m is merely the arithmetic difference in temperatures.

When moving fluids are involved, however, as in a shell and tube condenser, steam condenser, or liquid-to-liquid heat exchanger, there is a steadily rising temperature on one side, and a steadily falling temperature on the other. Then,

$$D_m = \frac{t_2 - t_1}{\log_e (t_2 \div t_1)}$$

where t_2 = the larger of the two temperature differences, and
   t_1 = the smaller of the two temperature differences at one or the other ends of the exchanger, as illustrated.

The tables that follow contain values of D_m for various values of t_2 and t_1. It makes no difference whether the vertical or horizontal columns are chosen for t_1.

When $t_2 - t_1$ is small, the arithmetic mean is a good approximation of the logarithmic mean. Thus, if the large temperature difference is 10, and the small temperature difference is 9, the sum of 9 and 10 divided by 2 is 9.50. The tabular value is 9.49.

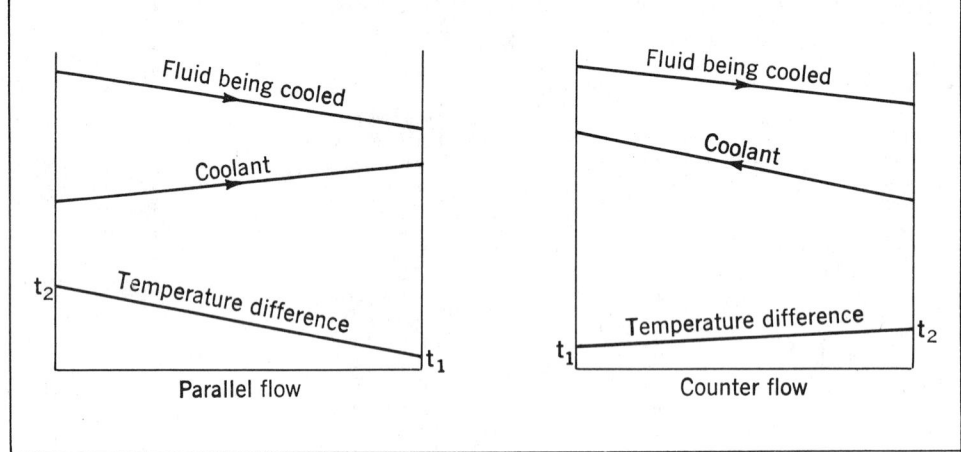

Temperature relations in a heat exchanger. When heat is being extracted from a gas undergoing a change of state, the "fluid being cooled" line is horizontal.

Example:

A refrigerant vapor is liquefying in a water-cooled condenser at a temperature of 110 deg. Water enters at 75 deg, leaves at 100 deg.

A change of state takes place on the refrigerant side of the condenser, but not a change of temperature. Thus, hot gas enters at 110 deg, and hot liquid leaves at 110 deg. Water enters at 75 deg, a difference of 35 deg from the entering gas. Water leaves at 100 deg, a difference of 10 deg from the leaving hot liquid. Proceeding to the right from *35* in the first vertical column until we intersect the vertical column headed *10*, we then find that the

logarithmic mean temperature difference is 19.96.

Example:

Water enters a heat exchanger at 190 deg and leaves at 178 deg. A salt solution enters in the opposite direction (counterflow) at 153 deg and leaves at 175 deg. Again we are interested in the temperature differences at the two ends of the exchanger; in this case, two inlet-outlet combinations, because of counterflow. On one end there is a temperature difference of (190 − 175), or 15 deg; on the other, a temperature difference of (178 − 153), or 25 deg. The logarithmic mean temperature difference is seen to be 19.57.

LOGARITHMIC MEAN TEMPERATURE DIFFERENCES

Temp. Diff. at One End	Temperature Difference at Other End									
	1	2	3	4	5	6	7	8	9	10
	Logarithmic Mean Temperature Difference									
1	1.00	1.44	1.82	2.16	2.49	2.79	3.08	3.37	3.64	3.91
2	1.44	2.00	2.47	2.89	3.27	3.64	3.99	4.33	4.65	4.97
3	1.82	2.47	3.00	3.48	3.92	4.33	4.72	5.10	5.46	5.81
4	2.16	2.89	3.48	4.00	4.48	4.93	5.36	5.77	6.17	6.55
5	2.49	3.27	3.92	4.48	5.00	5.49	5.94	6.38	6.81	7.21
6	2.79	3.64	4.33	4.93	5.49	6.00	6.49	6.96	7.40	7.83
7	3.08	3.99	4.72	5.36	5.94	6.49	7.00	7.49	7.96	8.41
8	3.37	4.33	5.10	5.77	6.38	6.96	7.49	8.00	8.49	8.97
9	3.64	4.65	5.46	6.17	6.81	7.40	7.96	8.49	9.00	9.49
10	3.91	4.97	5.81	6.55	7.21	7.83	8.41	8.97	9.49	10.00
11	4.17	5.28	6.16	6.92	7.61	8.25	8.86	9.42	9.98	10.49
12	4.43	5.58	6.49	7.28	8.00	8.66	9.28	9.86	10.44	10.97
13	4.68	5.88	6.82	7.64	8.37	9.05	9.69	10.30	10.89	11.44
14	4.93	6.17	7.14	7.98	8.74	9.44	10.10	10.72	11.33	11.89
15	5.17	6.45	7.46	8.32	9.10	9.82	10.50	11.14	11.74	12.33
16	5.41	6.73	7.77	8.66	9.46	10.19	10.89	11.54	12.16	12.77
17	5.65	7.01	8.07	8.99	9.81	10.56	11.27	11.94	12.58	13.19
18	5.88	7.28	8.37	9.31	10.15	10.92	11.65	12.33	12.98	13.61
19	6.11	7.55	8.67	9.63	10.49	11.28	12.02	12.72	13.38	14.02
20	6.34	7.82	8.96	9.94	10.82	11.63	12.38	13.10	13.78	14.43
21	6.57	8.08	9.25	10.25	11.15	11.97	12.74	13.47	14.17	14.83
22	6.79	8.34	9.54	10.56	11.47	12.31	13.10	13.84	14.55	15.22
23	7.02	8.60	9.82	10.86	11.80	12.65	13.45	14.20	14.92	15.61
24	7.24	8.85	10.10	11.16	12.11	12.98	13.80	14.56	15.29	15.99
25	7.46	9.11	10.38	11.46	12.43	13.31	14.14	14.92	15.66	16.37
26	7.67	9.36	10.65	11.75	12.74	13.64	14.48	15.27	16.02	16.75
27	7.88	9.61	10.92	12.05	13.05	13.96	14.82	15.62	16.38	17.12
28	8.10	9.85	11.19	12.33	13.35	14.28	15.15	15.97	16.74	17.48
29	8.32	10.10	11.46	12.62	13.65	14.60	15.48	16.31	17.09	17.85
30	8.53	10.34	11.73	12.90	13.95	14.91	15.80	16.65	17.44	18.21
31	8.74	10.58	11.99	13.19	14.25	15.22	16.13	16.98	17.79	18.56
32	8.95	10.82	12.25	13.47	14.55	15.53	16.45	17.31	18.13	18.91
33	9.15	11.06	12.51	13.74	14.84	15.84	16.77	17.64	18.47	19.26
34	9.36	11.30	12.77	14.02	15.13	16.14	17.08	17.97	18.81	19.61
35	9.56	11.53	13.03	14.29	15.41	16.44	17.40	18.29	19.14	19.96
36	9.77	11.76	13.28	14.56	15.70	16.74	17.71	18.62	19.48	20.30
37	9.97	12.00	13.53	14.83	15.99	17.04	18.02	18.94	19.81	20.64
38	10.17	12.23	13.79	15.10	16.27	17.34	18.32	19.25	20.14	20.97
39	10.37	12.56	14.04	15.37	16.55	17.63	18.63	19.70	20.46	21.31
40	10.57	12.69	14.28	15.64	16.83	17.92	18.93	19.88	20.78	21.96

LOGARITHMIC MEAN TEMPERATURE DIFFERENCES (Cont'd)

Temp. Diff. at One End	Temperature Difference at Other End									
	11	12	13	14	15	16	17	18	19	20
	Logarithmic Mean Temperature Difference									
1	4.17	4.43	4.68	4.93	5.17	5.41	5.65	5.88	6.11	6.34
2	5.28	5.58	5.88	6.17	6.45	6.73	7.01	7.28	7.55	7.82
3	6.16	6.49	6.82	7.14	7.46	7.77	8.07	8.37	8.67	8.96
4	6.92	7.28	7.64	7.98	8.32	8.66	8.99	9.31	9.63	9.94
5	7.61	8.00	8.37	8.74	9.10	9.46	9.81	10.15	10.49	10.82
6	8.25	8.66	9.05	9.44	9.82	10.19	10.56	10.92	11.28	11.63
7	8.86	9.28	9.69	10.10	10.50	10.89	11.27	11.65	12.02	12.38
8	9.42	9.86	10.30	10.72	11.14	11.54	11.94	12.33	12.72	13.10
9	9.98	10.44	10.89	11.33	11.74	12.16	12.58	12.98	13.38	13.78
10	10.49	10.97	11.44	11.89	12.33	12.77	13.19	13.61	14.02	14.43
11	11.00	11.48	11.91	12.43	12.89	13.33	13.79	14.22	14.64	15.06
12	11.48	12.00	12.43	12.95	13.44	13.92	14.35	14.80	15.24	15.65
13	11.96	12.43	13.00	13.48	13.96	14.44	14.90	15.35	15.80	16.26
14	12.43	12.95	13.48	14.00	14.58	14.96	15.47	15.90	16.38	16.81
15	12.89	13.44	13.96	14.58	15.00	15.42	16.02	16.46	16.90	17.40
16	13.33	13.92	14.44	14.96	15.42	16.00	16.63	16.98	17.41	17.93
17	13.79	14.35	14.90	15.47	16.02	16.63	17.00	17.44	17.93	18.50
18	14.22	14.80	15.35	15.90	16.46	16.98	17.44	18.00	18.36	19.00
19	14.64	15.24	15.80	16.38	16.90	17.41	17.93	18.36	19.00	19.36
20	15.06	15.65	16.26	16.81	17.40	17.93	18.50	19.00	19.36	20.00
21	15.47	16.08	16.69	17.26	17.83	18.36	18.95	19.43	20.03	20.50
22	15.87	16.50	17.11	17.71	18.27	18.84	19.40	19.95	20.45	20.99
23	16.27	16.90	17.53	18.13	18.73	19.27	19.85	20.38	20.90	21.46
24	16.66	17.31	17.94	18.56	19.15	19.73	20.29	20.88	21.42	21.94
25	17.05	17.72	18.35	18.97	19.57	20.15	20.73	21.30	21.85	22.41
26	17.44	18.10	18.76	19.39	20.01	20.60	21.20	21.77	22.34	22.87
27	17.82	18.50	19.15	19.80	20.42	21.01	21.62	22.20	22.77	23.33
28	18.20	18.89	19.56	20.20	20.82	21.44	22.05	22.62	23.20	23.78
29	18.57	19.26	19.94	20.60	21.24	21.85	22.47	23.07	23.66	24.22
30	18.94	19.64	20.33	20.99	21.64	22.27	22.88	23.48	24.08	24.66
31	19.31	20.02	20.71	21.39	22.04	22.67	23.29	23.92	24.50	25.10
32	19.67	20.39	21.09	21.77	22.44	23.08	23.72	24.35	24.94	25.53
33	20.03	20.76	21.47	22.16	22.83	23.48	24.13	24.76	25.36	25.96
34	20.38	21.13	21.85	22.54	23.21	23.88	24.53	25.16	25.79	26.38
35	20.73	21.48	22.22	22.92	23.61	24.27	24.92	25.57	26.17	26.80
36	21.08	21.85	22.58	23.30	23.99	24.66	25.32	25.97	26.60	27.22
37	21.43	22.20	22.95	23.66	24.36	25.04	25.72	26.36	27.02	27.63
38	21.78	22.55	23.31	24.04	24.75	25.43	26.11	26.77	27.41	28.04
39	22.12	22.91	23.67	24.40	25.12	25.81	26.50	27.16	27.81	28.45
40	22.47	23.26	24.02	24.77	25.49	26.19	26.88	27.56	28.21	28.85

LOGARITHMIC MEAN TEMPERATURE DIFFERENCES (Cont'd)

Temp. Diff. at One End	Temperature Difference at Other End									
	21	22	23	24	25	26	27	28	29	30
	Logarithmic Mean Temperature Difference									
21	21.00	21.50	21.98	22.47	22.94	23.41	23.87	24.33	24.79	25.23
22	21.50	22.00	22.50	22.99	23.47	23.94	24.41	24.88	25.34	25.77
23	21.98	22.50	23.00	23.50	23.99	24.47	24.95	25.42	25.91	26.35
24	22.47	22.99	23.50	24.00	24.50	25.00	25.47	25.95	26.42	26.89
25	22.94	23.47	23.99	24.50	25.00	25.50	25.99	26.47	26.95	27.43
26	23.41	23.94	24.47	25.00	25.50	26.00	26.50	26.99	27.47	27.95
27	23.87	24.41	24.95	25.47	25.99	26.50	27.00	27.50	27.99	28.47
28	24.33	24.88	25.42	25.95	26.47	26.99	27.50	28.00	28.50	28.99
29	24.79	25.34	25.91	26.42	26.95	27.47	27.99	28.50	29.00	29.50
30	25.23	25.77	26.35	26.89	27.43	27.95	28.47	28.99	29.50	30.00
31	25.68	26.24	26.80	27.35	27.89	28.43	28.95	29.47	29.99	30.50
32	26.12	26.69	27.25	27.81	28.36	28.90	29.43	29.96	30.48	30.99
33	26.55	27.13	27.70	28.26	28.82	29.36	29.90	30.43	30.96	31.48
34	26.98	27.57	28.14	28.71	29.27	29.82	30.37	30.90	31.43	31.96
35	27.41	28.00	28.58	29.16	29.72	30.28	30.83	31.37	31.91	32.44
36	27.83	28.43	29.02	29.60	30.17	30.73	31.28	31.83	32.37	32.91
37	28.25	28.85	29.45	30.03	30.61	31.18	31.74	32.29	32.84	33.38
38	28.66	29.27	29.88	30.47	31.05	31.62	32.19	32.75	33.30	33.84
39	29.08	29.69	30.30	30.90	31.48	32.06	32.63	33.20	33.75	34.30
40	29.49	30.11	30.72	31.32	31.91	32.50	33.08	33.64	34.21	34.76

Temp. Diff. at One End	Temperature Difference at Other End									
	31	32	33	34	35	36	37	38	39	40
	Logarithmic Mean Temperature Difference									
31	31.00	31.50	31.99	32.48	32.96	33.44	33.91	34.38	34.85	35.31
32	31.50	32.00	32.50	32.99	33.48	33.96	34.44	34.91	35.38	35.85
33	31.99	32.50	33.00	33.50	33.00	34.48	34.96	35.44	35.92	36.39
34	32.48	32.99	33.50	34.00	34.50	34.99	35.48	35.96	36.44	36.92
35	32.96	33.48	33.99	34.50	35.00	35.50	35.99	36.48	36.96	37.44
36	33.44	33.96	34.48	34.99	35.50	36.00	36.50	36.99	37.48	37.96
37	33.91	34.44	34.96	35.48	35.99	36.50	37.00	37.50	37.99	38.48
38	34.38	34.91	35.44	35.96	36.48	36.99	37.50	38.00	38.50	38.99
39	34.85	35.38	35.92	36.44	36.96	37.48	37.99	38.50	39.00	39.50
40	35.31	35.85	36.39	36.92	37.44	37.96	38.48	38.99	39.50	40.00

Note: Redundant tabular values have been eliminated on this page. For instance, values from 1 to 20, under the column headed "Temp. Diff. at One End", are omitted because they are identical to the values from 21 to 30 in the third quarters of the first two tables. Up to this point, all values are covered up to and including a 40-deg temperature difference at one end and a 40-deg temperature difference at the other end. Linear interpolation will yield results within the accuracy of the tables.

HEAT REQUIREMENTS OF SNOW MELTING SYSTEMS

Total heat output of a snow melting slab consists of the heat required to (1) raise the temperature of the snow to the melting point, (2) melt the snow (heat of fusion), (3) evaporate a portion of the melted snow and (4) make up the heat transferred by convection and radiation from that portion of the slab free of snow.

Total heat requirements of a snow melting slab can be expressed in the basic equation:

$$q_t = q_s + q_m + A_r (q_e + q_h),$$

where q_t = total heat,
q_s = sensible heat,
q_m = heat to melt the snow,
q_e = heat to evaporate the melted snow,
q_h = heat transferred by radiaion and convecion,
(all q values are Btuh per sq ft)
A_r = free area ratio.

In the foregoing equation, note that the last two items making up the equation, q_e and q_h, have no value while the slab is entirely covered by snow, since the melted snow beneath cannot evaporate nor can the slab lose heat by radiation and convection. When the slab is partially freed of snow, q_e and q_h have a value depending on the ratio of free area (melted area) to the total slab area. This ratio is termed free and a ratio, expressed by

$$A_r = A_f / A_t$$

where A_r = free area ratio, A_f = area of slab free of snow, in sq ft, A_t = total area of slab.

It can be shown that there are just three practical values for A_r, namely, zero, 0.5 and 1.0. When the free area ratio, A_r, is equal to zero, the slab is covered completely with snow. Although this condition can be defined as a failure of the system it may be tolerable in some instances for short periods of time. When the free area ratio, A_r, is equal to unity, the slab is completely free of snow. This situation requires the maximum output for a given weather condition.

Total heat output of a snow melting system depends on the rate at which it melts the snow, and this rate in turn depends on the relative importance of rapid snow removal among different types of applications. Snow melting systems can be classified as follows:

Class I (minimum): Residential walks or driveways and interplant areaways;

Class II (moderate): Commercial (store and office) sidewalks and driveways and steps of hospitals; and

Class III (maximum): Toll plazas of highways and bridges, and aprons and loading areas of airports.

These classifications depend upon the allowable rate of snow melting.

Table 1

Table 1 contains the operating information of a snow melting system. For freezing temperatures (32 F and below) but no snowfall the system may be "idling." That is, the designer may want to supply some heat to the slab so that there will be immediate melting when snow starts to fall. The column on mean temperature gives the average temperature during *freezing* temperatures. This temperature is used in calculating the "idling" load. The other term need to calculate idling load is the wind velocity during the period of freezing temperatures.

The column "Hours of Snowfall" indicates the number of hours that snow is falling at rates equal to or greater than 0.01 inch of water equivalent per hour. Roughly speaking there are snowfalls of "trace" quantities about twice as often as there are for measurable quantities of 0.01 inch or more. It is assumed that these light falls can be handled by the idling load.

The remaining columns of Table 1 represent the frequency distribution of required heat output. This distribution is based on the solution to the basic equation for two values of the free area ratio, A_r. This distribution represents the heart of the analysis and is also the basis for Tables 2 and 3.

Table 2

Table 2 contains the *design* heat requirement based on classification of snow melting systems. Under Class I systems, the figures in parentheses are *idling rates* and, since they exceed the Class I design rates, should be taken as design ouput for this classification. Design rates may be altered by the designer if he feels that a particular job should have a different design criterion from those given in the footnote. Any change in design selection should be based on the frequency distribution given in Table 1.

Table 1. Data for Determining Operating Characteristics of Melting Systems[1]

City	Period of No Snowfall — Air temperature, F[2] Over 32 (% of winter hours with no snow at above temperatures)	Below or equal to 32	Mean temperature during freezing period[6]	Wind velocity, freezing period,[6] mph	Hours of snowfall[2] Per cent	Hr per year	Free area ratio, A_r	Period of Snowfall[3] — Required Output, Btu per (hr) (sq ft)[4] 0-49	50-99	100-149	150-199	200-249	250-299	300-349	350-399	400-up	Maximum output, Btu per (hr) (sq ft)
Albuquerque, N. M.	74.7	24.7	26.2	8.5	0.6	22	1	62.0	25.4	7.6	4.2	0.0	0.8	—	—	—	259
							0	94.1	5.9	—	—	—	—	—	—	—	82
Amarillo, Tex.	73.1	26.0	24.6	13.3	0.9	33	1	33.7	35.4	15.4	10.7	3.0	1.8	—	—	—	260
							0	88.1	10.1	1.8	—	—	—	—	—	—	143
Boston, Mass.	64.6	31.4	24.7	14.2	4.0	145	1	51.5	30.0	12.3	4.3	1.2	0.6	0.1	—	—	320
							0	83.2	14.0	2.0	0.3	0.3	0.1	—	0.2	—	370*
Buffalo-Niagara Falls, N. Y.	46.5	46.9	23.9	10.8	6.6	240	1	50.7	32.6	11.2	3.7	1.4	0.2	0.2	—	—	309
							0	95.9	3.4	0.2	0.5	—	—	—	—	—	192
Burlington, Vt.	39.0	54.5	19.6	10.8	6.5	236	1	53.7	29.9	13.2	2.5	0.6	0.1	—	—	—	280
							0	91.8	7.6	0.6	—	—	—	—	—	—	142
Caribou-Limestone, Me.	21.4	70.6	16.5	10.0	8.0	290	1	35.0	39.7	16.0	5.7	2.0	1.0	0.5	0.1	—	378
							0	92.0	7.5	0.5	—	—	—	—	—	—	138
Cheyenne, Wyo.	46.4	49.8	21.5	15.3	3.8	138	1	16.5	26.2	19.4	13.1	8.6	4.7	4.2	4.7	2.6	499
							0	94.3	5.4	0.3	—	—	—	—	—	—	129
Chicago, Ill.	45.4	50.9	21.4	11.5	3.7	134	1	45.8	37.4	11.4	3.1	1.4	0.6	0.2	0.1	—	368
							0	91.5	8.1	0.3	0.1	—	—	—	—	—	165
Colorado Springs, Colo.	54.3	43.6	22.1	11.5	2.1	76	1	26.8	36.3	19.0	7.5	4.4	5.5	0.5	—	—	311
							0	98.4	1.6	—	—	—	—	—	—	—	63
Columbus, Ohio	59.0	38.1	24.5	10.0	2.9	105	1	65.8	22.4	8.0	1.7	1.7	0.4	—	—	—	261
							0	97.7	2.3	—	—	—	—	—	—	—	72
Detroit, Mich.	47.0	49.3	24.1	10.6	3.7	134	1	60.4	27.7	9.3	1.5	0.8	0.3	—	—	—	278
							0	95.9	3.5	0.6	—	—	—	—	—	—	140
Duluth, Minn.	12.6	80.5	14.5	12.0	6.9	250	1	23.7	32.9	20.6	13.7	4.3	2.5	1.7	0.6	—	382
							0	94.8	4.7	0.0	0.3	0.2	—	—	—	—	206
Falmouth, Mass.	68.5	29.5	25.5	12.8	2.0	73	1	50.0	33.9	14.2	1.6	0.3	—	—	—	—	204
							0	91.5	7.4	1.1	—	—	—	—	—	—	144
Great Falls, Mont.	49.0	46.2	16.5	14.4	4.8	174	1	26.2	27.6	16.7	16.4	7.5	4.6	0.3	0.5	0.2	451
							0	94.6	4.8	0.6	—	—	—	—	—	—	138
Hartford, Conn.	56.4	38.9	24.4	8.2	4.7	171	1	48.4	34.6	11.2	4.3	0.8	0.7	—	0.1	—	396
							0	80.4	16.7	2.2	0.5	—	0.1	—	0.1	—	383
Lincoln, Neb.	45.0	52.5	20.8	10.1	2.5	91	1	32.7	26.2	20.0	13.9	5.7	1.5	—	—	—	293
							0	97.2	2.6	0.0	0.0	0.2	—	—	—	—	202
Memphis, Tenn.	87.2	12.5	27.0	11.5	0.3	11	1	48.4	28.3	6.7	13.3	3.3	—	—	—	—	227
							0	85.0	8.3	6.7	—	—	—	—	—	—	144
Minneapolis-St. Paul, Minn.	23.6	70.8	16.9	11.1	5.6	203	1	28.4	31.4	21.7	14.1	3.5	0.6	0.3	—	—	313
							0	96.5	3.1	0.3	0.1	—	—	—	—	—	155
Mt. Home, Idaho	56.3	42.6	24.9	9.5	1.1	40	1	74.2	21.9	3.9	—	—	—	—	—	—	143
							0	98.1	1.9	—	—	—	—	—	—	—	90
New York, N. Y.	55.7	42.2	24.2	11.8	2.1	76	1	53.1	31.8	9.4	2.2	1.5	1.7	—	0.3	—	385
							0	87.6	9.6	1.5	0.7	0.3	0.3	—	—	—	298
Ogden, Utah	50.0	45.6	24.3	9.4	4.4	160	1	64.6	29.2	5.8	0.3	0.1	—	—	—	—	216
							0	88.8	9.4	1.4	0.3	0.1	—	—	—	—	216*
Oklahoma City, Okla.	79.0	19.8	24.6	15.8	1.2	44	1	27.8	18.7	17.0	12.6	14.3	5.9	2.7	1.0	—	394
							0	95.7	4.3	—	—	—	—	—	—	—	81
Philadelphia, Pa.	75.8	22.6	26.7	9.7	1.6	58	1	62.3	23.6	10.4	2.3	0.9	0.5	—	—	—	296
							0	84.3	14.0	1.1	0.2	0.4	—	—	—	—	229
Pittsburgh, Pa.	55.2	39.8	24.3	11.6	5.0	182	1	53.6	30.8	8.4	4.6	1.9	0.7	—	—	—	282
							0	93.3	5.9	0.7	0.1	—	—	—	—	—	157
Portland, Ore.	92.9	6.1	28.9	8.4	1.0	36	1	78.0	16.9	5.1	—	—	—	—	—	—	125
							0	91.5	8.5	—	—	—	—	—	—	—	97
Rapid City, S. D.	45.2	51.6	19.3	12.9	3.2	116	1	29.7	29.0	16.0	8.4	6.3	3.6	1.9	2.0	3.1	581
							0	97.6	2.2	0.2	—	—	—	—	—	—	102
Reno, Nev.	56.0	41.6	24.3	5.6	2.4	87	1	82.6	15.4	1.8	0.2	—	—	—	—	—	152
							0	90.2	8.0	1.6	0.2	—	—	—	—	—	154*
St. Louis, Mo.	68.7	30.4	25.0	11.5	0.9	33	1	42.9	31.4	16.7	7.1	1.9	—	—	—	—	225
							0	85.2	11.6	2.6	0.6	—	—	—	—	—	152
Salina, Kan.	60.0	38.5	23.3	10.9	1.5	54	1	44.9	31.9	12.7	7.6	2.2	0.7	—	—	—	286
							0	93.5	6.2	0.3	—	—	—	—	—	—	120
Sault Ste. Marie, Mich.	21.3	69.2	18.6	9.4	9.5	345	1	45.7	32.8	14.3	5.7	1.4	0.1	—	—	—	262
							0	97.9	2.0	0.1	—	—	—	—	—	—	144
Seattle-Tacoma, Wash.	88.0	10.8	28.5	5.9	1.2	44	1	86.3	12.3	1.4	—	—	—	—	—	—	137
							0	90.0	8.1	0.9	—	—	—	—	—	—	128
Spokane, Wash.	48.5	46.1	25.7	10.7	5.4	196	1	62.6	28.7	7.4	1.1	0.2	—	—	—	—	205
							0	92.0	7.8	0.2	—	—	—	—	—	—	127
Washington, D. C.	77.9	21.2	26.8	9.6	0.9	33	1	59.0	29.8	10.6	0.6	—	—	—	—	—	154
							0	85.7	11.8	2.5	—	—	—	—	—	—	121

[1] The period covered by this table is from Nov. 1 to March 31, incl. with February taken as a 28¼ day month. Total hours in period = 3630.
[2] The percentage in Columns 2 and 3 plus the percent under hours of snowfall total 100%. Note that "Hours of Snowfall" does not include idling time, and is not actual operating time.
[3] Snowfalls of trace amounts are not included; hence, "Hours of Snowfall" includes only those hours of 0.01 inches water equivalent per hour snowfall.
[4] Output does not include allowance for back or edge losses since these depend on slab construction.
[5] Percentage total 100% of the number of hours of snowfall.
[6] "Freezing Period" is that during No Snowfall when the air temperature is 32 F or below.
* When heat output for $A_r = 0$ equals or exceeds the heat output for $A_r = 1$, the heat transfer q_h is **from** the air **to** the slab. This occurs when snowfall is at temperatures above 32 F.

Table 2. Design Data for Three Classes of Snow Melting Systems

(Where idling rate is greater than Class I design rate, idling rate is given in parentheses and should be used as Class I design output)

City	Design output, Btu per (hr)(sq ft)			City	Design output. Btu per (hr)(sq ft)		
	Class I system[1]	system[2] Class II	Class III system[3]		Class I system[1]	Class II system[2]	Class III system[3]
Albuquerque, N. M.	71	82	167	Minneapolis-			
Amarillo, Tex.	98	143	241	St. Paul, Minn.	63(95)	155	254
Boston, Mass.	107	231	255	Mt. Home, Idaho	50	90	140
Buffalo-Niagara				New York, N. Y.	121	298	342
Falls, N. Y.	80	192	307	Ogden, Utah	98	216	217
Burlington, Vt.	90	142	244	Oklahoma City,			
Caribou-				Okla.	66	81	350
Limestone, Me.	89(93)	138	307	Philadelphia, Pa.	97	229	263
Cheyenne, Wyo.	83	129	425	Pittsburgh, Pa.	89	157	275
Chicago, Ill.	89	165	350	Portland, Ore.	86	97	111
Colorado Springs,				Rapid City, S. D.	58(86)	102	447
Colo.	49(63)	63	293	Reno, Nev.	98	154	155
Columbus, Ohio	52	72	253	St. Louis, Mo.	122	152	198
Detroit, Mich.	69	140	255	Salina, Kan.	85	120	228
Duluth, Minn.	83(114)	206	374	Sault Ste.			
Falmouth, Mass.	93	144	165	Marie, Mich.	52(78)	144	213
Great Falls, Mont.	84(112)	138	372	Seattle-Tacoma,			
Hartford, Conn.	115	254	260	Wash.	92	128	133
Lincoln, Neb.	64(67)	202	246	Spokane, Wash.	87	127	189
Memphis, Tenn.	134	144	212	Washington, D. C.	117	121	144

[1] For Class I (residential) Systems, the design output is set at that required heat output (see Table 1) when $A_r = 0$ at the 98th percentile of the frequency distribution; that is, where 98% of the hours have this output or less.

[2] For Class II (commercial) systems, the design output is the maximum output when $A_r = 0$ in Table 1 (last column).

[3] For Class III (industrial) systems, the design ouput is determined by the following four requirements: (1) Output

is never exceeded for two consecutive hours; (2) Output for $A_r = 1$; Table 1, is at least 1 Btu per hour per sq ft greater than maximum output for $A_r = 0$; (3) A_r is greater than or equal to 0.5 maximum requirement shown in Table 2 for $A_r = 1$. That is $q_d = q_s + q_m + 0.5 (q_e + q_h)$ for the conditions where $q_t = q_s + q_m + q_h + q_e$ are a maximum; (4) The free area ratio, A_r, is unity for at least 98% of the hours listed in Table 1.

Table 3

Table 3 contains the data required to estimate the operating cost of a snow melting system. The footnotes of Table 3 explain the various items.

Fluid Temperature

$$q = (t_w - t_f) / R$$

where q = heat emitted by slab,

t_w = fluid temperature in the pipe,

t_f = temperature at the slab surface, and

R = thermal resistance of slab, pipe to surface.

For a snow melting system with the equivalent of 3 inches of concrete above the panels, take $R = 0.50$. Slab surface temperature is usually taken at 70 F. Hence

$$t_w = 0.5 q + 70$$

Example 1. Assume that engineers A and B have been assigned to select the required output for a snow melting system for automobile traffic in Buffalo, N. Y. It is decided that 25% will be adequate for back and edge losses, Engineer A submits a figure of 250 Btu per (hr) (sq ft) and B submits a figure of 375.

A argues that automobiles will not be out in heavy snowstorms, hence a Class II system is adequate; furthermore an output of 200 is adequate for 98.6% of the snowfall in Buffalo (see Table 1). An additional 50 Btu per (hr) (sq ft) provides 25% back loss.

B contends that public conveyances will be out in the worst weather and, in addition, so will such vehicles as fire engines, ambulances and other vehicles on emergency calls. Hence, B's estimate of 300 plus 75 for back losses. In addition, B points

Table 3. Yearly Operating Data

City	Idling Time[1] hr per year	Idling Rate[2] Btu per (hr) (sq ft)	Idling Output[3] Btu per (year) (sq ft)	Melting Time[4] hr per year	Class[5]	Melting Output[6] Btu per (sq ft) (year)
Albuquerque, N. M.	897	32.5	29100	22	I / II / III	908 / 969 / 1150
Amarillo, Tex.	944	51.1	48200	33	I / II / III	2150 / 2520 / 2770
Boston, Mass.	1140	52.1	59400	145	I / II / III	8000 / 9080 / 9100
Buffalo-Niagara Falls, N. Y.	1702	50.2	85500	240	I / II / III	11800 / 14600 / 14900
Burlington, Vt.	1978	76.9	152000	236	I / II / III	11800 / 13500 / 13800
Caribou-Limeston, Me.	2563	93.0	238000	290	I / II / III	17800* / 20600 / 22500
Cheyenne, Wyo.	1808	77.7	140000	138	I / II / III	9730 / 13200 / 20200
Chicago, Ill.	1848	67.8	125000	134	I / II / III	7200 / 8390 / 8700
Colorado Springs, Colo.	1583	63.4	100000	76	I / II / III	3960* / 3960 / 7390
Columbus, Ohio	1383	45.0	62200	105	I / II / III	3590 / 4180 / 5350
Detroit, Mich.	1790	49.0	87600	134	I / II / III	5540 / 6850 / 7070
Duluth, Minn.	2922	113.8	332000	250	I / II / III	33200* / 38100 / 39500
Falmouth, Mass.	1071	44.2	47400	73	I / II / III	3830 / 4250 / 4290
Great Falls, Mont.	1677	111.6	187000	174	I / II / III	13700 / 15400 / 19100
Hartford, Conn.	1412	41.9	59200	171	I / II / III	9830 / 10800 / 10810
Lincoln, Neb.	1906	67.2	128000	91	I / II / III	4750* / 8350* / 8520
Memphis, Tenn.	454	32.0	14500	11	I / II / III	702 / 721 / 792
Minneapolis-St. Paul, Minn.	2570	95.1	244000	203	I / II / III	14200* / 17600 / 18400
Mt. Home, Idaho	1546	41.9	64800	40	I / II / III	1260 / 1530 / 1590
New York, N. Y.	1532	50.7	77700	76	I / II / III	4180 / 4690 / 4710
Ogden, Utah	1655	44.6	73800	160	I / II / III	7050 / 7370 / 7370
Oklahoma City, Okla.	719	56.2	40400	44	I / II / III	2380 / 2800 / 5400
Philadelphia, Pa.	820	31.4	25700	58	I / II / III	2710 / 3100 / 3110
Pittsburgh, Pa.	1445	49.5	71500	182	I / II / III	9050 / 10700 / 11100
Portland, Ore.	221	17.4	3840	36	I / II / III	1300 / 1330 / 1360
Rapid City, S. D.	1873	86.4	162000	116	I / II / III	7450* / 8250 / 13400
Reno, Nev.	1510	36.9	55700	87	I / II / III	2970 / 3030 / 3030
St. Louis, Mo.	1104	44.8	4950	33	I / II / III	2190 / 2290 / 2380
Salina, Kan.	1398	53.9	75400	54	I / II / III	2920 / 3370 / 3810
Sault Ste. Marie, Mich.	2512	77.7	195000	345	I / II / III	17600* / 22200 / 23200
Seattle-Tacoma, Wash.	392	17.2	6750	44	I / II / III	1410 / 1430 / 1430
Spokane, Wash.	1673	39.1	65500	196	I / II / III	8650 / 9350 / 9560
Washington, D. C.	770	30.6	23600	33	I / II / III	1650 / 1660 / 1690

[1] From Table 1, Column 3 $\times$ 3630 (hr per year).

[2] Rate when idling. Btu per (hr) (sq ft) = $(0.27V + 3.3)(32 - t)$, where V = wind velocity from Column 5, Table 1, and t = air temperature from Column 4, Table 1.

[3] Product of the two preceding columns.

[4] Hours of snowfall, from Column 7, Table 1.

[5] See footnote for Table 2.

[6] Based on the condition that surface temperature is maintained at 33 F until required output exceeds designed output, at which time design output is used regardless of required output. Distribution of required output based on Table 1.

* Based on idling rate rather than design rate.

out that operating costs for a Class III system are 14,900 Btu per (year) (sq ft) and 14,600 for a Class II. Surely such a slight increase in operating costs are warranted for the added protection of emergency travel.

In fact, 250 is adequate for Buffalo. A careful examination of Table 1 indicates that an output of 200 Btu per (hr) (sq ft) always will provide a free area ratio greater than zero. In other words, snow will not accumulate beyond that amount required to give $A_r = 0$; furthermore, frequencies of 0.002 for rates above 240 indicate that a Class III rate of 307 is based on rule 1, footnote 3 of Table 2, and probably for one unusual storm every four years. True, two consecutive hours would be bad, but since the snow cannot be accumulating, A_r must still be above zero but less than unity. In other words, the rate of 307 is due to high winds and low temperatures rather than heavy snowfall. Emergency vehicles would have to be driven carefully, but they would not be stalled. It seems adequate, then, to select the design rate of 200 Btu per (hr) (sq ft) and allow for back and edge losses.

Example 2. A plant engineer for a company located just outside Duluth, Minnesota, has decided that a snow melting system would increase plant shipments during the winter. There are two choices. Run a pipe line from a remote station or put in a boiler just for the snow melting system. The water from the remote station is being wasted at the present time. If he uses it as a heat source for a snow melting system, he can reclaim the energy that is being wasted now. His choice is governed by the economics of each system. If he runs the pipe line he will have an almost unlimited source of good, non-corrosive hot water that will develop a temperature of 160 F in the anti-freeze fluid of his snow melting system.

The problem is to compare the cost of the pipe line (including maintenance) to the cost of the boiler and the annual fuel consumption. In one case there is high initial cost but low operating costs; in the other case there is just the opposite. First, he must decide on the performance of a system operating with 160 F fluid, and second he must decide on the annual operating cost of such a system.

A system with a fluid temperature (t_w) of 160 F and a surface temperature (t_f) of 33 F should have an output of

$$q = \frac{t_w - t_f}{R} = \frac{160 - 33}{0.50}$$
$$= 254 \text{ Btu per (hr) (sq ft)}.$$

where $R =$ thermal resistance of slab, pipe to surface, (deg) (sq ft) (hr) per Btu.

Assume an output of 250 Btu per (hr) (sq ft). From Table 1, for Duluth, it is seen that $2.5 + 1.7 + 0.6 = 4.8\%$ of the snowfalls will require an output in excess of 250 for $A_r = 1.0$. On the other hand the maximum output for $A_r = 0$ is shown as 206 Btu per (hr) (sq ft). In addition, the maximum output for $A_r = 1$ is only 382; therefore, an output for 250 should develop a free area ratio of at least 0.25, but this is adequate for plant vehicular traffic, especially since $A_r = 1$ can be expected for 95% of the time.

It seems likely that a fluid temperature of 160 F will be satisfactory. The remaining problem is to compute the fuel consumption of a system such as this. The engineer has decided to control the surface temperature at 33 F and maintain $A_r = 1$ up to the capacity of the system. He can assume that the frequency distribution in each cell of Table 1 is linear, so that the system output for $A_r = 1$ up to $q = 250$ and $t_f = 33$ could be calculated as follows:

$$
\begin{aligned}
0.237 \times 25 &= 5.93 \\
0.329 \times 75 &= 24.68 \\
0.206 \times 125 &= 25.75 \\
0.137 \times 175 &= 24.00 \\
0.043 \times 225 &= \underline{9.68} \\
& 90.04 \text{ Btu per (hr) (sq ft)}
\end{aligned}
$$

For the remaining 4.8% of the cases where the required output is in excess of 250 Btu per (hr) (sq ft) and $A_r < 1$, the system output could be calculated as:

$$0.048 \times 250 = 12.00 \text{ Btu per (hr) (sq ft)}$$

Since Duluth has an average snowfall of 250 hours per year, the annual system output would be:

$$
\begin{aligned}
(\, 90.04 + 12.00 \,) &\times 250 \\
&= 25,510 \text{ Btu per (yr) (sq ft)}
\end{aligned}
$$

To this figure must be added the estimated back and edge losses.

With these data and the fuel costs for a boiler the engineer is in a position to compare the economics of the systems. One advantage of the remote system is that it can "idle" whenever there is no snow. The only cost is the pumping cost. With such an operation, the slab will be warm before the snow starts falling. In this way, there will never be an accumulation of snow.

BUILDING ENERGY CONSERVATION CODES AND STANDARDS

When energy was relatively inexpensive, the primary criteria for standard engineering design of building HVAC systems concerned itself with the economy of installation and the flexibility of temperature and humidity control. However, the increased cost of energy has provided the incentive for a change in design philosophy. This change now reflects life-cycle costing, which takes into account operating as well as installation costs.

Lately, energy consumption has become the concern of government at both the national and state level. As of this writing, many jurisdictions already have or are in the process of enacting regulations that supersede the builders' private concerns in the building design process by mandating energy conservation criteria.

Basically, there are two approaches to most current codes or standards. In type they can be described as "prescriptive" or "performance." A prescriptive code details the method to be employed to achieve the desired goal. A mandated thickness of insulation in walls is an example of a prescription. A requirement that a building of a certain category consume no more than say, 55,000 Btu per sq ft/yr is a performance type requirement.

Currently, the most important document in building energy conservation is Standard 90, "Energy Conservation in New Building Design" of the American Society of Heating, Refrigerating and Air-Conditioning Engineers (ASHRAE), which is a development from original drafts prepared by National Bureau of Standards of the U. S. Department of Commerce. Since ASHRAE Standard 90 will be updated at frequent intervals, the reader is advised to obtain the most recent edition from the Society at 345 East 47th Street, New York, N.Y. 10017. At present, Standard 90 is before the American National Standard Institute (ANSI) for possible adoption as an ANSI Standard. Also, as of this writing, ASHRAE is preparing a standard for the retrofitting of existing buildings.

ASHRAE Standard 90 is a mixture of prescriptive and performance approaches and can be subdivided into three parts. This was done by Arthur D. Little, Inc. (ADL), for the purpose of analysis of the standard's impact on energy consumption (See: "Energy Conservation in New Building Design," Conservation Paper No. 43A,

issued by the Federal Energy Administration, for sale by the U. S. Government Printing Office.) :

- *Standard Prescriptive/Performance Approach* (*Sections 4-9*): Well-defined performance criteria, based upon an element-by-element design analysis, are applied in this approach to the selection of building materials and systems.
- *Systems Analysis Approach* (*Section 10*): An alternative to the above, this approach allows for compliance if the building's energy consumption is to be equal to or lower than that achievable through the standard prescriptive/performance approach of Sections 4 through 9.
- *Energy Augmentation Approach* (*Section 11*): This alternative allows for the use of solar- or wind-powered systems, or other nondepletive energy sources, to supplement the energy usage of the building by including a "credit" for the energy supplied by such systems.

ADL analyzed the effect that the Standard would have on energy consumption on five building types:

Single-Family Detached Residence
Low-Rise Apartment Building
Office Building
Retail Store
School Building

1. Impact on Building Energy Consumption

Under a strict interpretation of the standard, ASHRAE 90 is very effective in reducing annual energy consumption in all building types and locations. The unweighted average reduction in annual energy consumption relative to 1973 construction and operational practices across the four locations investigated were as follows:

Single-Family Residence	—11.3%
Low-Rise Apartment Building	—42.7%
Office Building	—59.7%
Retail Store	—40.1%
School Building	—48.1%

The lowest unit-energy consumption that was obtainable in the prototypical buildings after the prescriptive/performance approach (Sections 4 through 9) had been made was on the order of

67,000 to 72,000 Btu per square foot/yr. This consumption is considerably greater than the General Services Administration announced "goal" of 55,000 Btu per square foot per year. ADL stated that it does not believe the implementation of ASHRAE 90 alone would suffice to meet the GSA goal for any building type similar to those investigated.

If measured in terms of energy reduction potential, the most "effective" sections in the document will vary by type of building. With few exceptions, all of the sections have some influence on the reduction in annual energy consumption. Changes in winter design conditions (Section 5) and supplied domestic hot water temperatures (Section 7) appear to be the most effective parameters for the single-family residence. Those chapters dealing with HVA/C equipment, systems, and control (Sections 5 and 6) appear to be most effective in non-residential construction.

2. Impact on Physical Characteristics

The application of ASHRAE 90 brought about the following physical differences in the conventional versus ASHRAE 90 modified buildings:

Exterior Glass—Glass area (percent fenestration) was reduced in approximately two-thirds of the buildings. Reductions were as much as 30%, but most were less than 20%. One region—the North Central—required reductions in glass area for all buildings.

Exterior Wall—Decreases in glass area were balanced by increases in net wall area; virtually all increases were less than 8%.

Insulation—Additional insulation requirements for residential construction varied from 80 to 300 pounds per unit. Increased requirements for insulation in commercial construction were even greater than those needed in residential construction.

Lighting—Reductions in lamps and lamp fixtures varied by building type, and averaged 24% and 22%, respectively, for nonresidential construction.

HVA/C System Capacities—Reductions in heating system capacities were significant, averaging 42%, while reductions in cooling systems were generally less, averaging 31%. The greatest reductions were found in the school building. Auxiliary HVA/C equipment, including pumps, towers, fans, supply fans, etc., also

showed a significant reduction, averaging 44% in rated kilowatt capacity.

3. Impact on Building Economics

Based on 1975 energy costs compiled by ADL, annual savings in operating costs ranged between $0.05 and $1.05 per square foot, but were generally within the range between $0.20 and $0.70 per square foot. Savings in single-family residences ($0.05 to $0.14 per square foot) were lower and less broad than those for commercial construction ($0.12 to $1.05 per square foot). Percent savings in annual energy costs ranged from 9-15% in the single-family house to 30-45% for commercial buildings.

These savings may be large enough to induce building owners to follow the standard on a voluntary basis, providing they had such decision information available to them and providing financial institutions recognize that the loan quality is improved.

The initial construction costs of those buildings modified under the standard prescriptive/performance approach were shown to be *less* than those of conventional buildings. Unit savings range from $0.04 to $0.94 per square foot, with the greatest savings experienced in office buildings.

ASHRAE 90 generally increases the cost of the exterior wall, floors, roof, and domestic hot water system. Glazing costs may be higher or lower depending upon building type. Unit costs for lighting, and particularly HVA/C equipment and distribution systems, were significantly lower and tended to offset the increase in other costs.

Average changes in unit costs are as follows:

	Dollars Per Square Foot
Single-Family Residence	—0.02
Multi-Family Residence	—0.41
Office Building	—0.63
Retail Store	—0.18
School Building	—0.44

For the prototypical buildings investigated, the cost of additional design effort was found to be between $0.09 and $0.36 per square foot of floor area. With the exception of the single-family residence, the straight payback of design services due to energy cost savings was found to be less than one year, and less than six months in most cases. Average additional design costs were as follows:

	Dollars Per Square Foot		
	Annual Energy Savings	Additional Design Services	Straight Payback
Single-Family Detached Residence	0.07	0.24	2.9 years
Low-Rise Apartment	0.31	0.09	3.4 months
Office Building	0.40	0.16	2.5 months
Retail Store	0.68	0.09	7.6 months
School	0.70	0.15	4.6 months

Thus, savings in initial cost can be offset by increased design fees; consequently it appears that the ASHRAE 90 modified buildings should cost no more to build and will have significantly less annual energy costs. Furthermore, even if total initial cost did increase, the savings in operating cost (over those of conventionally-designed buildings) would more than recover such costs in a few months.

ADL's approach to assessing the impact of Standard 90 in modifying conventional buildings was based upon professional design judgment as to what the architect/engineer would be most likely to do, and what the client would be most likely to permit aesthetically. In applying the standard, recommended numerical values listed in the standard were assumed to be targets, and as such, only the barest minimum of modification was undertaken to meet the targets. The philosophy was to meet—but not purposely exceed—the standard. It was felt that the client and design community would not choose to adopt major modifications in either building appearance or system performance. Typically, such decisions are not controlled by life-cycle economics.

Using ASHRAE 90 on a case-by-case basis, changes were made in indoor and outdoor design conditions, exterior wall and roof heat transmittance, lighting levels, window area and type, etc. For each modification, actual materials and/or different HVA/C equipment were selected which later formed the basis for determining the impact on the selected industry subsectors.

Tables 1 through 5 summarize the major physical changes made in each building type for each of four regions, Northeast, North Central, South, and West. For reference purposes, the appropriate section of ASHRAE 90 is shown for each design parameter changed. Other critical assumptions may be summarized as follows:

Type of Fuel: The same fuel was considered for both the conventional and ASHRAE 90 modified buildings, i.e., alternative fuel sources were not evaluated. The fuels selected for the various regions were predominant in these regions.

HVA/C Systems: Systems selected for the various building types were as follows:
Single Family—Hot air furnace with split system cooling, direct expansion (DX) coil. No economizer cycle, no humidification, no night setback.

Low-Rise Apartment—Fan coil units with air-cooled reciprocating chiller; hot water boiler; two-pipe system with chilled/hot water; 100% recirculated air, and ventilation provided by infiltration. No economizer cycle, no humidification, no night setback.
Office Building—Constant volume, low-pressure air system with terminal reheat; perimeter radiation and with economizer cycle; centrifugal chiller with constant condenser water temperature; cooling tower; hot-water boiler. Winter humidification for 30% rh. No night setback.

Retail Store—Rooftop constant volume, low-pressure air system, with economizer cycle; hot-water boiler. DX and hot-water heating coil (both in the unit), air-cooled condenser, no perimeter radiation. No humidification, no night setback.

School—Unit ventilators with four-pipe system using chilled water and hot water; centrifugal chiller; hot-water boiler; cooling tower; and economizer cycle. No humidification, no night setback.

The same systems are used for the ASHRAE 90 buildings, except simulation is modified for controls to include reset by maximum demand and to allow for "deadband" requirements.

Hot-Water Demand: Residential, domestic, hot-water demand was assumed to be 20 gallons per person per day. Other maximum loads were assumed to be:

	Maximum Peak Load (Gallons Per Hour)
Low-Rise Apartment Building	230
Office Building	160
Retail Store	40
School Building	720

Temperature Rise/Drop: To establish flow rates, the same temperature rise/drop was used in both conventional and ASHRAE 90 modified buildings:

	°F.
Chilled Water	10
Hot Water	20
Condenser Water	10
Air, Retail Store	17
Air, All Others	20

Safety Factor: As per conventional practice, a safety factor of 10% is used for conventional building load calculations and zero % for ASHRAE 90 buildings. The boilers were 20% oversized in both cases.

Table 1. Single Family Residence: Summary of Changes in Design Parameters, Conventional vs. ASHRAE 90-75, Modified Prototypical Structure

Design Variable	Applicable Section of ASHRAE 90-75	Northeast		North Central		South		West	
		Conv.	90-75	Conv.	90-75	Conv.	90-75	Conv.	90-75
Design Conditions:									
Summer Outdoor, °F DB/°F WB	4.2.5, 5.3.2.1, 5.3.2.2	91/77	87/76	97/79	94/78	95/78	92/77	96/66	94/65
Indoor, °F DB/ %RH max		75/50	78/60	75/50	78/60	75/50	78/60	75/50	78/60
Winter Outdoor, °F DB		12	21	-12	-1	14	23	6	17
Indoor, °F DB		75	72	75	72	75	72	75	72
Exterior Envelope:									
Glass Area (percent of gross wall area) North	4.3.2.1	15	15	15	14.4	15	15	15	14.8
East		15	15	15	14.4	15	15	15	14.8
South		15	15	15	14.4	15	15	15	14.8
West		15	15	15	14.4	15	15	15	14.8
Glass U (Btu/hr ft² °F)		1.13	1.13	1.13	1.13	1.13	1.13	1.13	1.13
Wall U (Btu/hr ft² °F)	4.3.2.2	0.087	0.063	0.087	0.063	0.087	0.087	0.087	0.087
Overall Wall "Uo" (Btu/hr ft² °F)		0.24	0.22	0.24	0.21	0.24	0.24	0.24	0.24
Roof U (Btu/hr ft² °F)	4.3.2.3, 4.3.2.4	0.074	0.050	0.074	0.050	0.048	0.045	0.048	0.045
Floor Perimeter U (Btu/hr ft² °F)									
Lighting/Power (Watts/sq ft)	9.3, 9.4	1.0	0.8	1.0	0.8	1.0	0.8	1.0	0.8
Ventilation (cfm/sq ft)	4.5.3, 5.3.2.4								
Infiltration (Air Change/hr)		1.0	0.93	1.0	0.93	1.0	0.93	1.0	0.93
Domestic Hot Water Temperature Rise (°F)	7.3	100	70	100	70	100	70	100	70

SOURCE: Kling-Lindquist, Inc., based on strict interpretation of ASHRAE 90-75.

Table 2. Low-Rise Apartment Building: Summary of Changes in Design Parameters, Conventional vs. ASHRAE 90-75, Modified Prototypical Structure

Design Variable	Applicable Section of ASHRAE 90-75	Northeast		North Central		South		West	
		Conv.	90-75	Conv.	90-75	Conv.	90-75	Conv.	90-75
Design Conditions:									
Summer Outdoor, °F DB/°F WB	4.2.5, 5.3.2.1, 5.3.2.2	91/77	87/76	97/79	94/78	95/78	92/77	96/66	94/65
Indoor, °F DB/ %RH max		75/50	78/60	75/50	78/60	75/50	78/60	75/50	78/60
Winter Outdoor, °F DB		12	21	-12	-1	14	23	6	17
Indoor, °F DB		75	72	75	72	75	72	75	72
Exterior Envelope:									
Glass Area (percent of gross wall area) North	4.3.2.1	30	26.7	30	23.6	30	30	30	28.2
East		0	0	0	0	0	0	0	0
South		30	26.7	30	23.6	30	30	30	28.2
West		0	0	0	0	0	0	0	0
Glass U (Btu/hr ft² °F)		1.13	1.13	1.13	1.13	1.13	1.13	1.13	1.13
Wall U (Btu/hr ft² °F)		0.093	0.068	0.093	0.068	0.370	0.072	0.370	0.072
Overall Wall "Uo" (Btu/hr ft² °F)	4.3.2.2	0.340	0.290	0.340	0.264	0.550	0.334	0.550	0.305
Roof, U (Btu/hr ft² °F)		0.070	0.045	0.070	0.045	0.070	0.045	0.070	0.045
Floor Perimeter U (Btu/hr ft² °F)	4.3.2.3, 4.3.2.4	0.20	0.16	0.20	0.14	0.41	0.21	0.41	0.17
Lighting/Power (watts/sq ft)	9.3, 9.4	1	.8	1	.8	1	.8	1	.8
Ventilation (cfm/sq ft)	5.3.2.3	.05	.025	.05	.025	.05	.025	.05	.025
Infiltration (Air Change/hr)	4.5.3, 5.3.2.4	.5	.3	.5	.3	.5	.3	.5	.3
Domestic Hot Water Temperature Rise (°F)	7.3	100	70	100	70	100	70	100	70

SOURCE: Kling-Lindquist, Inc., based on strict interpretation of ASHRAE 90-75

Table 3. Office Building: Summary of Changes in Design Parameters, Conventional vs. ASHRAE 90-75, Modified Prototypical Structure

Design Variable	Applicable Section of ASHRAE 90-75	Northeast		North Central		South		West	
		Conv.	90-75	Conv.	90-75	Conv.	90-75	Conv.	90-75
Design Conditions:									
Summer Outdoor, °F DB/°F WB	4.2.5, 5.3.2.1, 5.3.2.2	91/77	87/76	97/79	94/78	95/78	92/77	96/66	94/65
Indoor, °F DB/ %RH max		75/50	78/60	75/50	78/60	75/50	78/60	75/50	78/60
Winter Outdoor, °F DB		12	21	-12	-1	14	23	6	17
Indoor, °F DB		75/30	72/30	75/30	72/30	75/30	72/30	75/30	72/30
Exterior Envelope:									
Glass Area (percent of gross wall area) North	4.4.2.1, 4.4.3.1	30	25	30	29	50	34.7	50	35.9
East		30	25	30	29	50	34.7	50	35.9
South		30	25	30	29	50	34.7	50	35.9
West		30	25	30	29	50	34.7	50	35.9
Glass U (Btu/hr ft^2 °F)		1.13	0.65	1.13	0.65	1.13	0.65	1.13	0.65
Wall U (Btu/hr ft^2 °F)		0.34	0.168	0.34	0.108	0.20	0.113	0.20	0.113
Overall Wall "Uo" (Btu/hr ft^2 °F)	4.4.2.2, 4.4.3.2	0.580	0.290	0.580	0.265	0.665	0.300	0.665	0.300
Roof U (Btu/hr ft^2 °F)		0.14	0.079	0.14	0.079	0.18	0.089	0.18	0.089
Lighting/Power (Watts/sq ft)	9.3, 9.4	4.5	3.5	4.5	3.5	5.5	3.5	5.5	3.5
Ventilation (cfm/sq ft)	5.3.2.3	.25	.148	.25	.148	.25	.148	.25	.148
Infiltration (Air Change/hr)	4.5.3, 5.3.2.4	.5	.3	.5	.3	.5	.3	.5	.3
Domestic Hot Water Temperature Rise (°F)	7.3	100	70	100	70	100	70	100	70

SOURCE: Kling-Lindquist, Inc., based on strict interpretation of ASHRAE 90-75.

Table 4. Retail Store: Summary of Changes in Design Parameters, Conventional vs. ASHRAE 90-75, Modified Prototypical Structure

Design Variable	Applicable Section of ASHRAE 90-75	Northeast		North Central		South		West	
		Conv.	90-75	Conv.	90-75	Conv.	90-75	Conv.	90-75
Design Conditions:									
Summer Outdoor, °F DB/°F WB	4.2.5, 5.3.2.1, 5.3.2.2	91/77	87/76	97/79	94/78	95/78	92/77	96/66	94/65
Indoor, °F DB/ %RH max		75/50	78/60	75/50	78/60	75/50	78/60	75/50	78/60
Winter Outdoor, °F DB		12	21	-12	-1	14	23	6	17
Indoor, °F DB		75	72	75	72	75	72	75	72
Exterior Envelope:									
Glass Area (percent of gross wall area) North	4.4.2.1, 4.4.3.1	0	0	0	0	0	0	0	0
East		0	0	0	0	0	0	0	0
South		60	60	60	52	60	60	60	60
West		0	0	0	0	0	0	0	0
Glass U (Btu/hr ft² °F)		1.13	1.13	1.13	1.13	1.13	1.13	1.13	1.13
Wall U (Btu/hr ft² °F)		0.29	0.135	0.29	0.135	0.29	0.135	0.29	0.135
Overall Wall "Uo" (Btu/hr ft² °F)		0.416	0.284	0.416	0.265	0.416	0.284	0.416	0.284
Roof U (Btu/hr ft² °F)	4.4.2.2, 4.4.3.2	0.14	0.079	0.14	0.065	0.14	0.089	0.14	0.089
Floor Perimeter U (Btu/hr ft² °F)	4.4.2.4	0.41	0.16	0.41	0.14	0.41	0.21	0.41	0.17
Lighting/Power (Watts/sq ft)	9.3, 9.4	6.0	4.5	6.0	4.5	6.0	4.5	6.0	4.5
Ventilation (cfm/sq ft)	5.3.2.3	.3	.216	.3	.216	.3	.216	.3	.216
Infiltration (Air Change/hr)	4.5.3, 5.3.2.4	.5	.3	.5	.3	.5	.3	.5	.3
Domestic Hot Water Temperature Rise (°F)	7.3	100	70	100	70	100	70	100	70

SOURCE: Kling-Lindquist, Inc., based on strict interpretation of ASHRAE 90-75.

Table 5. School Building: Summary of Changes in Design Parameters, Conventional vs. ASHRAE 90-75, Modified Prototypical Structure

Design Variable	Applicable Section of ASHRAE 90-75	Northeast		North Central		South		West	
		Conv.	90-75	Conv.	90-75	Conv.	90-75	Conv.	90-75
Design Conditions:									
Summer Outdoor, °F DB/°F WB	4.2.5, 5.3.2.1, 5.3.2.2	91/77	87/76	97/79	94/78	95/78	92/77	96/66	94/65
Indoor, °F DB/ %RH max		75/50	78/60	75/50	78/60	75/50	78/60	75/50	78/60
Winter Outdoor, °F DB		12	21	-12	-1	14	23	6	17
Indoor, °F DB		75	72	75	72	75	72	75	72
Exterior Envelope:									
Glass Area (percent of gross wall area) North	4.4.2.1, 4.4.3.1	20	18.5	20	16	20	20	20	20
East		20	18.5	20	16	20	20	20	20
South		20	18.5	20	16	20	20	20	20
West		20	18.5	20	16	20	20	20	20
Glass U (Btu/hr ft² °F)		1.13	1.13	1.13	1.13	1.13	1.13	1.13	1.13
Wall U (Btu/hr ft² °F)		0.10	0.10	0.10	0.10	0.30	0.10	0.30	0.10
Overall Wall "Uo" (Btu/hr ft² °F)	4.4.2.2, 4.4.3.2	0.306	0.265	0.306	0.265	0.466	0.306	0.466	0.306
Roof U (Btu/hr ft² °F)		.14	.078	.14	.065	.23	.09	.23	.079
Lighting/Power (Watts/sq ft)	9.3, 9.4	4.0	3.5	4.0	3.5	4.0	3.5	4.0	3.5
Ventilation (cfm/sq ft)	5.3.2.3	.50	.25	.50	.25	.50	.25	.50	.25
Infiltration (Air Change/hr)	4.5.3, 5.3.2.4	.50	.30	.50	.30	.50	.30	.50	.30
Domestic Hot Water Temperature Rise (°F)	7.3	100	70	100	70	100	70	100	70

SOURCE: Kling-Lindquist, Inc., based on strict interpretation of ASHRAE 90-75.

Section 3

AIR CONDITIONING SYSTEMS
AND COMPONENTS

Contents

Authorities

Section Editors: *Eugene Stamper* and *Richard L. Koral*

Louis Austerweil, Optimized Data for Heat Pump Systems

John D. Constance, Check List for Air Conditioning Surveys

James G. Deflon, Mechanical Draft and Atmospheric Cooling Towers

Harrison D. Goodman, Winterizing Chilled Water Systems

F.E. Ince, Air Conditioning Equipment Maintenance

Herbert Kunstadt, Selecting Air Handling Units

David Rickelton, Variable Volume Air Conditioning Systems

N.S. Shataloff, High Velocity Dual Duct Systems

Jim Shih, Automatic Control Applications

K.E. Sontag, Servicing the Cooling Plant for Summer Use

E. Sternberg, The Roof as a Location for Air Conditioning Equipment

Stephen W. Trelease, Electrohydronic Heat Recovery

UNITARY AIR CONDITIONING SYSTEMS

A unitary air conditioner is an apparatus consisting of one or more factory-made assemblies, which normally include an evaporator, blower, a compressor and a condenser, designed to be used together and often performs a heating as well as a cooling function. In other words, a unitary air conditioner is a packaged, complete refrigeration cycle.

Classification

Air-Conditioning and Refrigeration Institute has found the following basic order of classification the most convenient way of identifying the many types of unitary air conditioners:

Single-package air conditioner—
SP-A: air cooled
SP-E: evaporatively-cooled
SP-W: water-cooled

Air conditioner with remote condenser—
RC-A: air-cooled condenser
RC-E: evaporative condenser
RC-W: water-cooled condenser

Refrigeration chassis—
RCH-A: air-cooled
RCH-E: evaporatively-cooled
RCH-W: water-cooled

Refrigeration chassis with remote condenser—
RCH-RCA: air-cooled condenser
RCH-RCE: evaporative condenser
RCH-RCW: water-cooled condenser

Year-round single-package air conditioner—
SPY-A: air-cooled
SPY-E: evaporatively-cooled
SPY-W: water-cooled

Year-round single-package air conditioner with remote condenser—
RCY-A: air-cooled condenser
RCY-E: evaporative condenser
RCY-W: water-cooled condenser

Year-round air conditioner, remote condensing unit—
RCUY-A-CB: air-cooled condenser
RCUY-E-CB: evaporative condenser
RCUY-W-CB: water-cooled condenser

Split system with condensing unit and coil alone—
RCU-A-C: air-cooled condenser
RCU-E-C: evaporative condenser
RCU-W-C: water-cooled condenser

Split system with condensing unit and coil with blower—
RCU-A-CB: air-cooled condenser
RCU-E-CB: evaporative condenser
RCU-W-CB: water-cooled condenser

Split system with remote water chiller and coil alone—
RWC-A-C: air-cooled

Split system with remote water chiller and coil with blower—
RWC-A-CB: air-cooled

G: Preceding any type designation, indicates gas-fired, heat-operated unit.

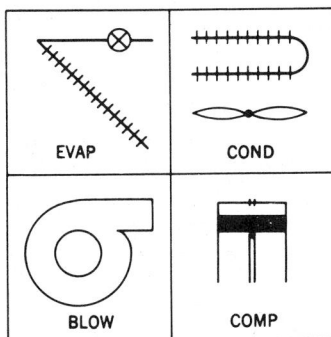

Fig. 1. So-called "self-contained" unitary air conditioner has all components in one package.

Single-Package Units (SP)— The single-package, or so-called "self-contained" air conditioner is shown symbolically in Fig. 1. All four basic components of the refrigeration cycle have been placed in a single "package" or housing.

Vertical single-package equipment is usually designed for inside-the-building application. Units often are equipped with line voltage controls including integral thermostat, all factory pre-wired. Return-air filter and grille are standard, with supply-air duct flanges usually at the top of the unit. Most have water-cooled condensers which can be used with either city water or cooling tower water. Figure 2 is typical. Where it is necessary to apply the unit without ductwork, a plenum and grille may be fitted to the unit for free air delivery.

Vertical units are available, in many lines, with either decorative cabinets, for installation in the space (the traditional "store cooler" or with more

utilitarian cabinets (sometimes designated "commercial") for remote location, especially where space and noise factors assume extra importance. Commercial units are available in capacities to 50 tons and higher.

Horizontal single-package units (Fig. 3) are designed for both indoor and outdoor installation, with the majority of units incorporating air-cooled or evaporatively-cooled condensers. The most common design has supply and return air connections located in the same face, with horizontal air flow. Since horizontal units are designed for remote location, they are usually equipped with low-voltage thermostat control. Filters are not usually supplied with the unit.

Horizontal units are manufactured in capacities from 2 to 7½ tons with many modifications for use in residential and light commercial applications. Their outstanding virtue is flexibility, being capable of installation indoors, with ducted air for condensing; outdoors, with ducted conditioned air supply and return; and through the wall. Indoors, they can be floor, wall, or ceiling mounted, and placed in the attic, basement, or in the conditioned space.

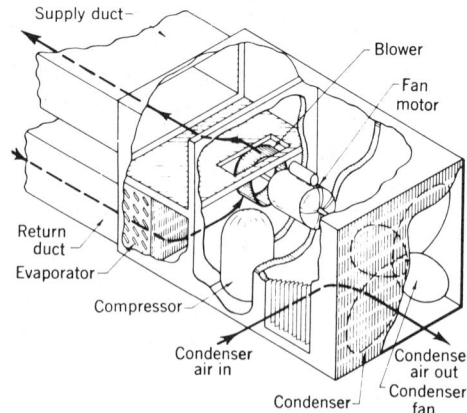

Fig. 3. Simplified view of horizontal self-contained air conditioner.

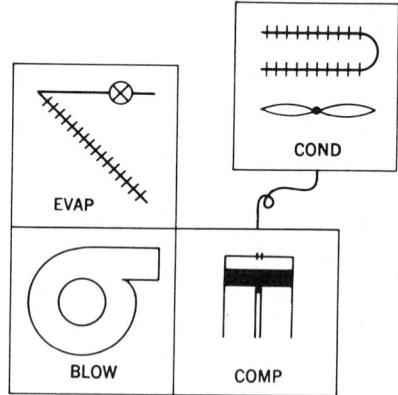

Fig. 4. Unit with remote condenser, the converted "store cooler." Refrigerant piping is applied in the field.

Units with Remote Condensers (RC)— Unitary air conditioners with a remote condenser, piped and charged in the field, are basically the store cooler with remote air-cooled condensing substituted for the integral water-cooled condenser. See Fig. 4.

Refrigeration Chassis (RCH, RCH-RC)— Units in this category are designed to be used with a warm air heating system, whose blower supplies the air for the unit's cooling coil. See Fig. 5.

In one version, the compressor-condenser is connected to the evaporator only by several feet of precharged refrigerant line for not-too-remote location. See Fig. 6 and 7. These units are somewhat of a cross between single-package and split systems.

Refrigeration chassis are available, also, with (truly) remote air-cooled condensers.

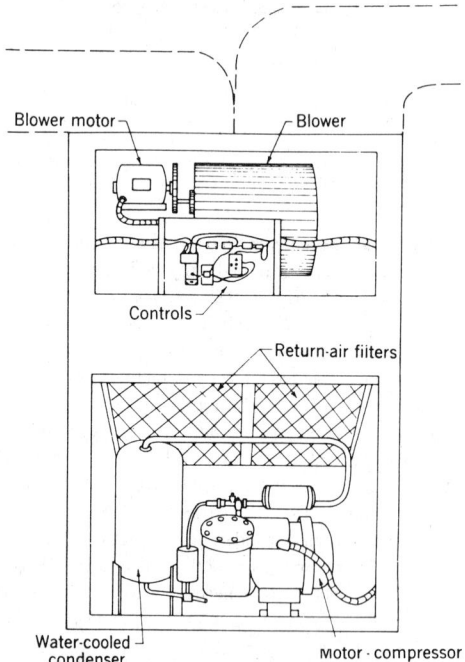

Fig. 2. Simplified rear view of typical vertical self-contained unit.

Single-Package Year-Round Units (SPY)—Operated by a factory wired, interconnected control system through a single heating-cooling thermostat, these units contain a fuel-fired heat exchanger and a factory sealed and charged refrigeration cycle. See Fig. 8. There are water cooled models designed for installation indoors, but by far the majority are designed for roof mounting with air condensing. See Fig. 9. Units are available with natural-gas engine compressor drive.

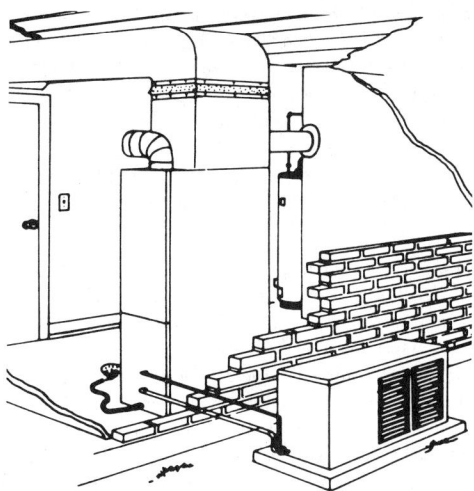

Fig. 7. Same refrigeration chassis as in Fig. 6, added to garden apartment warm air heating system.

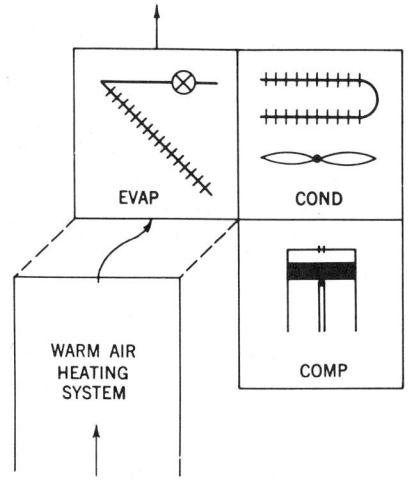

Fig. 5. Refrigeration chassis, with warm air heating system, makes year-round installation.

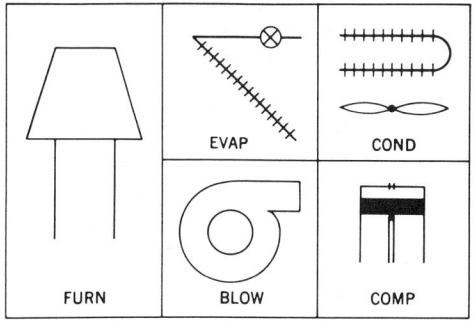

Fig. 8. Above, single package year-round unit is complete refrigeration system plus interconnected heating apparatus.

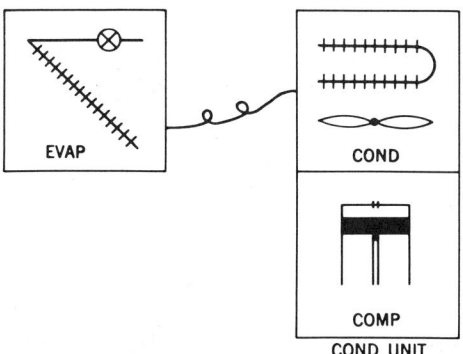

Fig. 6. One type of refrigeration chassis consists of an evaporator-condensing unit combination with several feet of interconnecting tubing that permits not-too-remote location.

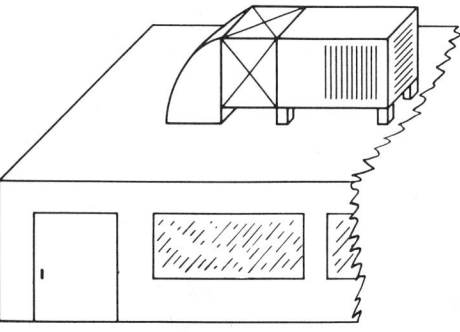

Fig. 9. Year-round self-contained units are most popular for roof mounting, installed in multiples to achieve flexible zone control.

Year-Round, Remote Units (RCY, RCUY)—These units are the SP (single-package) and RC (single-package with remote condenser) units with heating units attached. See Fig. 10.

The RCUY unit, year-round air conditioner with remote condensing unit (See Fig. 11) makes the real jump to the "split system." In the case of both these units, interconnecting refrigerant piping is usually field supplied.

Remote Condensing Units (RCU-C, RCU-CB)—A split-system air conditioner is a unitary conditioner whose components are assembled in more than one factory-made enclosure, and in which the cooling coil is remote from the condensing unit. Because of their similarity to single-package air conditioners, units whose condensers only are remote and refrigeration chassis, with or without remote condenser, have already been discussed in conjunction with single-package air conditioners, although they are

not truly "single-package."

Split-system components include electric and engine-driven condensing units, remote, gas-fired water chillers, cooling coils (evaporators without blowers), and cooling coils with blowers. There is more split system equipment available than any other type.

Condensing units contain compressor, condenser and electrical controls. Air-cooled and evaporatively cooled condenser models also contain a fan and fan motor. Units are normally designed for outdoor installation, but some models are designed for indoor or through-the-wall application. See Fig. 12 and 13.

Condensing units in the unitary category are available in capacities from about 2 to 15 tons. (Of course, units in much larger capacities are manufactured and are used to supply multiple cooling coils. However, the engineering of such "split" systems is more complicated and requires a degree of sophistication on the part of the installer that puts the whole system in the "engineered systems" category. The essence of the unitary approach, let it be re-

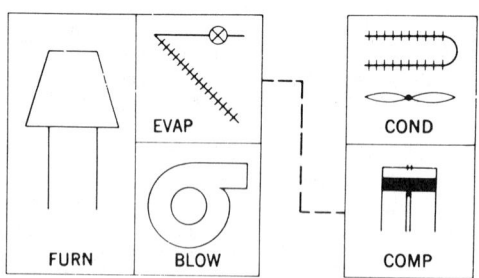

Fig. 11. Year-round air conditioner with remote condensing unit. This is truly a split system.

Fig. 10. Year-round air conditioner with remote air-cooled condenser.

Fig. 12. Above, left and right, two applications of remote condensing units.

membered, is pre-matched components that can be installed with a minimum of technical skills, quickly and economically. Where total requirements are large, the unitary approach calls for multiple installations of complete unitary systems.)

Gas-engine driven condensing units are available in 3 and 4½ tons, air cooled, and 3½ to 5 tons water cooled.

Evaporator coils (see Fig. 14) come in a multitude of sizes and designs. Coils for use with warm air heating systems are available for upflow, downflow, and horizontal air patterns, that is, in bonnet units, under-the-furnace units, and duct units.

Coils with blowers (see Fig. 15) for applications requiring separate cooling supply air systems are available for horizontal, vertical or downflow air discharge. Some manufacturers offer models in building-block design for increased flexibility. See Fig. 16. Both horizontal and vertical coil-with-blower units are available with decorative grilles for free-air delivery. See Fig. 17.

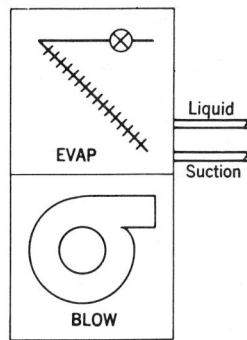

Fig. 15. Coil-with-blower units are used in cooling-only split systems.

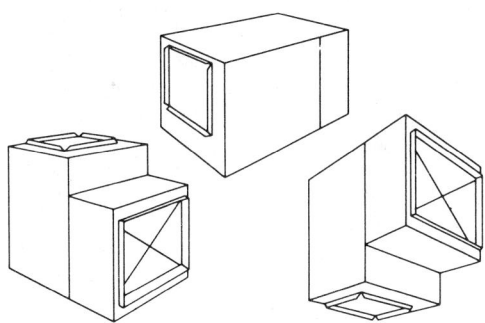

Fig. 16. Coil-with-blower units for upflow, downflow, or horizontal ductwork are piped to condensing units, completing refrigeration cycle.

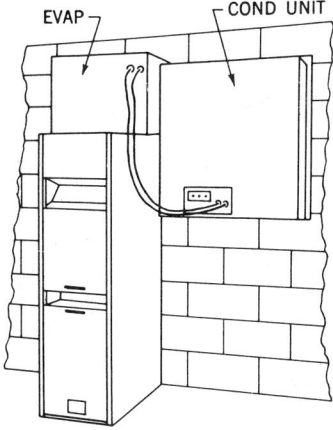

Fig. 13. Pre-charged split system with flush-wall condensing unit.

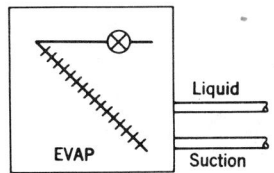

Fig. 14. Evaporator coils are used in conjunction with warm air furnaces: above them, below them, or in the air supply duct.

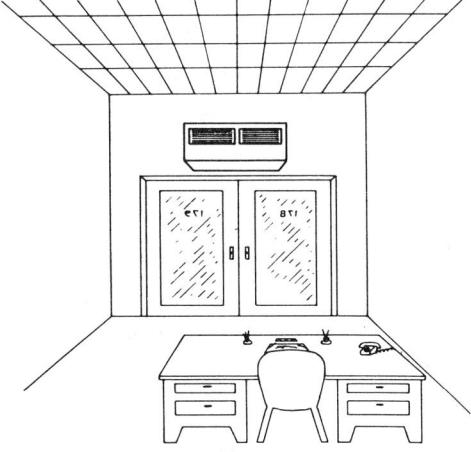

Fig. 17. Horizontal coil-with-blower unit with decorative front grille is wall mounted to condition small commercial space.

Absorption Split Systems (RWC)— Gas-fired water chillers are available for use with cooling coils in connection with winter air conditioning systems and with coil-with-blower units for summer air conditioning only. The absorption chiller, complete with controls and circulating water pump, is located outside the building and is air cooled. With polyethylene plastic water connections between chiller and cooling coil, the units are said to be unaffected by freezing, hence never require draining. Fig. 18 shows the chiller, Fig. 19, its application.

Multizone Units—A relatively recent addition to available unitary equipment is the package multizone air conditioner. It consists of an otherwise conventional self-contained combination in which the fan is located so as to blow through the coil or a bypass around the coil. The face of the coil is divided into compartments extending into the bypass area with a damper located in the bypass whereby the proportion of air that is bypassed may be controlled.

The multi-zone unitary air conditioner makes possible, from a single unit, the simultaneous individual control of the conditions in many zones. However, while they provide all the advantages of unitary equipment, they do require an appreciable amount of engineering in the determination of the zoning and design of the duct system supplied by the unit.

Unitary air conditioners will be found almost anywhere in a building. Because they are available in such variety, with so many air outlet patterns, and because they are relatively so compact, it is fair to say that their chief virtue is flexibility of location. If an available niche within the building is unobtainable, then the unitary equipment may be located in an outside wall or with the bulk of it outside the building.

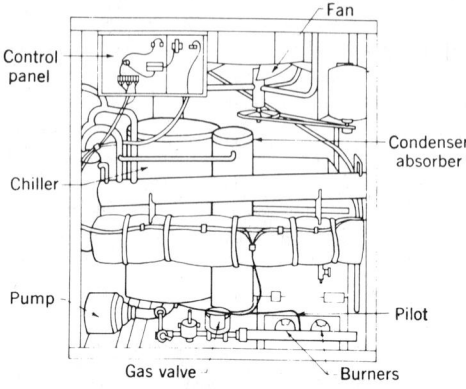

Fig. 18. Components of gas-fired water chiller, including water pump.

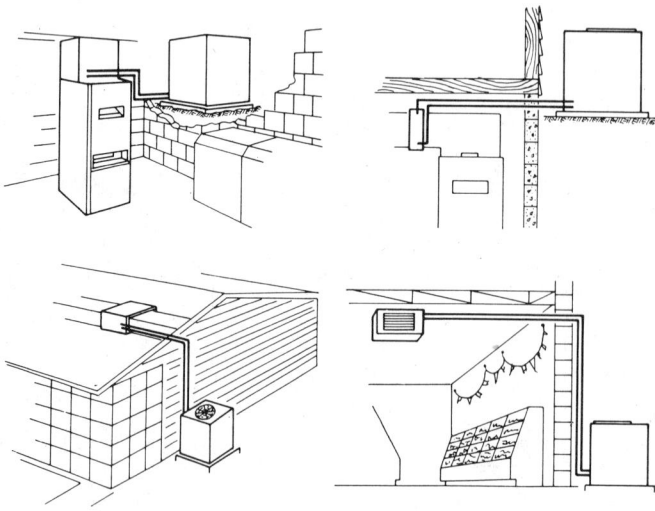

Fig. 19. Application of gas-fired water chiller with cooling coil or coil-with blower unit in residential and commercial buildings.

Single-Package Installations

Within the Space—For small areas with no obstructions or partitions, installation within the conditioned space of units arranged for free delivery will often prove to be the most economical. For free delivery applications, an accessory discharge plenum and grille can be added to self-contained units.

A system of ductwork must be designed for applications involving partitioned or obstructed areas, installation outside the conditioned space, or any other condition that cannot be handled properly with discharge plenums. Fig. 20 illustrates a typical duct system for a vertical self-contained unit.

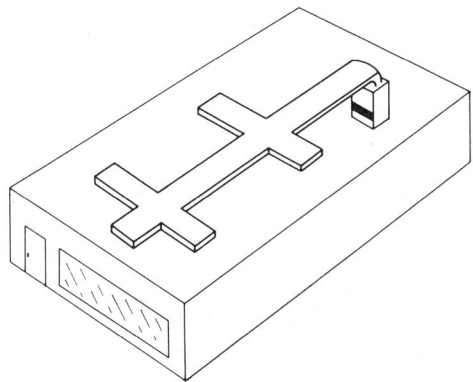

Fig. 20. Commercial application of vertical self-contained unit.

Attic or Crawl Space— Location of self-contained units in an attic or crawl space has been a popular method of installing air-conditioning in small homes and in some smaller commercial buildings. However, because of service problems and some restrictive codes, this type of installation appears to be declining in popularity, although the amount of ductwork required is minimal and the unit is completely out of sight, or visible only as a screened and louvered opening.

Because the attic and crawl spaces are relatively inaccessible, some method for remote reset is required when the pressure switch kicks out in the event that head pressure exceeds its maximum safe limit. (Underwriters' Laboratories requires a manual reset to prevent rapid cycling of the compressor that would occur should the condenser fan fail.)

Because of the constricted space in most attics and crawl spaces, these installations are frequently difficult to service. And they do need service from

time to time, particularly since drains should be checked periodically, for when they become plugged the first indication will be water dripping through the ceiling, and the cost of water damage in such a situation can run high. Because of this, FHA and some city codes require a secondary drain. That is, in addition to the primary drain pan at the bottom of the unit, a second pan, about 2 inches deep and slightly wider and deeper than the unit, is set underneath the unit, to collect any overflow from the primary drain or any condensation on the outside of the unit that drips down. Both drains must be independently piped (¾-inch is recommended) down to the sewer. While the secondary drain is often part of the unit, the double piping is an expense.

Also to be considered is that condenser air may be restricted, or hot attic air may be the only cooling source for the condenser, resulting in high head pressures, high operating costs, and increased need for service. More-or-less satisfactory solutions include: mounting of unit on outside wall of attic; mounting unit through attic wall; installing booster fan or blower in condenser air duct-work to increase air flow through the condenser, and ducting in of air at one end of attic, discharging at the other end.

Finally, noise and vibration from the compressor and fan will be transmitted to the conditioned space unless extreme care is used in mounting the unit and running the ductwork. Spring mounts are effective if proper springs are used. Rubber shock mounts have been applied successfully, and a 2-inch high-density fiberglass batt supporting the entire unit is worth considering.

Through-the-Wall—Small commercial and some residential buildings without forced air heating are the most likely applications of through-the-wall installations. See Fig. 21. Through-the-wall units may be used on any or all floors of a multi-story building. With their weight borne by a solid, load-bearing wall, no footings or other special supports are required.

Units may be placed high enough in the wall so that there can be no damage from passersby. Location of moving parts outside the space tends to reduce noise level inside, although noise and vibration may be transmitted to all parts of the structure unless properly installed and mounted.

An important factor in favor of through-the-wall installation is that units are available for service from the outside, which can be an advantage if units are not too high off the ground, while some

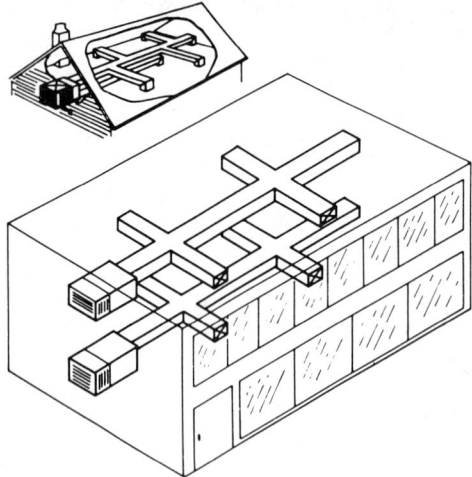

Fig. 21. Through-the-wall self-contained units can be installed at any level.

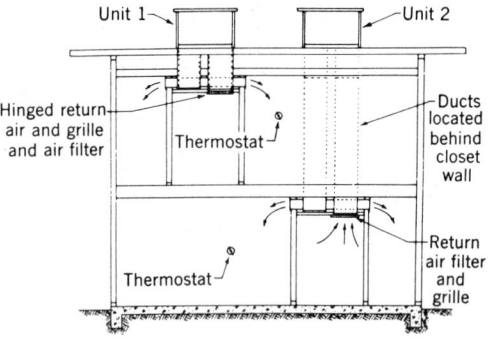

Fig. 22. Unitary packages on roof ducted to first and second floors, a technique successfully applied to both residential and small commercial office buildings.

are available that can be serviced from the inside, where outside service is not feasible.

One must consider whether a replacement unit will be available to fit the wall sleeve in future years.

Roof-Top—Frequently, roof top units are the most economical way to air condition single-story commercial and factory buildings. See Fig. 22-24.

In commercial installations, it has been found that a number of units may be used (as in shopping centers), with a minimum of duct-work in addition to excellent zone control, plus the fact that in most cases the equipment is completely hidden from the view of persons on the ground. Another advantage is to be found in the fact that all "noise-producing" components are outside the conditioned space (although here, again, vibration may cause noise unless unit is properly installed and mounted, and objection to noise by neighboring tenants must be considered, especially if the neighboring building is higher). There is usually no need to run a condensate drain, as condensate can be dumped on the roof or a short line run to a gutter.

In installing this type of unit, it is well to remember that the roof of an existing building may require reinforcement (new buildings designed for this equipment may have their roofs made of adequate strength during construction), and that a relatively large hole in the roof must be made for ductwork, which, if not properly sealed, will cause trouble.

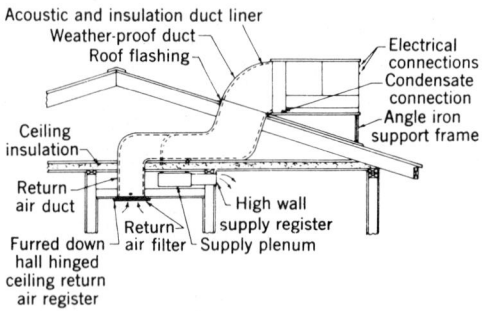

Fig. 23. Self-contained units mounted on gabled roof may not be visible from front if front is at left.

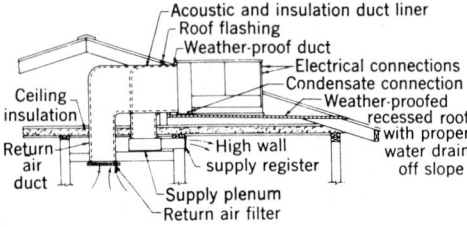

Fig. 24. Dormer recess installation is less obtrusive.

Here, again, the possibility of inconvenience of servicing should be considered, as well as the fact that condenser air rejection may be adversely affected by the wind.

Slab or Ground Level—Mounting of packaged units at ground level outside the conditioned space has become a standard method of installation in many areas, particularly where the heat pump has gained acceptance. See Fig. 25, 26, and 27.

The reasons for this are fairly obvious: (1) The only connection between the conditioned space and the unit itself are the supply and return ducts, thus virtually eliminating transmission of vibrations and reducing noise to a minimum; (2) the unit is very accessible for servicing, thus tending to reduce service cost and start-up time; (3) ground-level installation requires no hoist or crane to install, and (4) if the condensate drain plugs, no damage is done; in fact, in some areas a condensate drain line is not even required. (Sometimes, a small rock-filled dry well is constructed next to the unit to take care of condensate.)

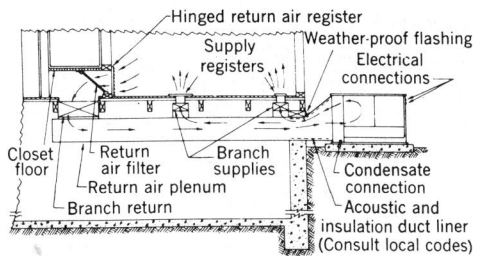

Fig. 27. Unit on slab where building has basement.

Furnace Mounting—One finds self-contained units mounted directly on warm air furnaces, usually where first cost is the prime consideration. Many manufacturers have designed units to match specific heating equipment (see Fig. 28); others have modified existing packaged equipment by removing the evaporator blower and adding sheet metal parts (this is the "refrigeration chassis"—see prior section, "Classification") so that the furnace blower may be used for both heating and cooling distribution.

Installations of this type are limited almost entirely to new construction, as this equipment usually requires an outside utility room and fairly large holes in the outside wall. A few are installed in a central utility room or basement, with ducted condenser air.

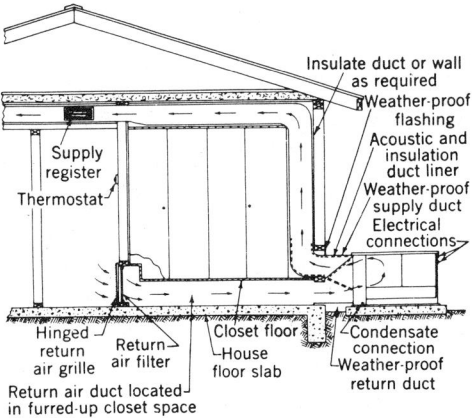

Fig. 25. Self-contained unit mounted on slab.

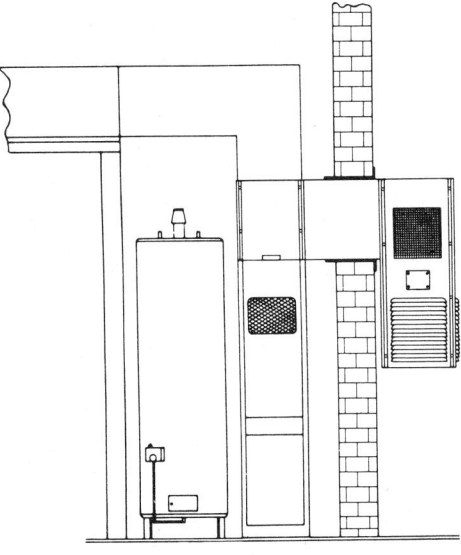

Fig. 28. There are units especially designed to attach to warm air furnaces. This is but one of many arrangements.

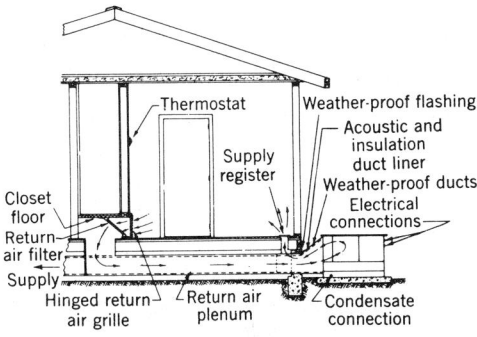

Fig. 26. Ducts run through crawl space.

Installation of Split Systems

Slab or Ground Level—Slab or ground-level installations of split systems are the most common for residential and small commercial and apartment buildings with only one or two floors. Fig. 29 is typical.

While such installations place the major noise-producing components outside the conditioned space, which is good from the sound-level standpoint, the condenser becomes subject to possible damage from passersby, trucks, cars, etc., and here again, as with single-package units, the condenser blower tends to pick up dirt and requires more frequent cleaning. Also, condenser air flow may be obstructed by surrounding shrubbery, or trash placed near the air intake or discharge openings.

However, these installations are convenient to install, service or replace, and they add no load to any part of the building—roof or walls. An added advantage is that they are generally protected by the building itself from strong winds which might affect air flow through the condenser.

To prevent flashing of hot liquid refrigerant, the evaporator should not be installed more than 20 feet higher than the condensing unit, which restricts the application to one-and-two-story buildings, unless extensive ductwork is used.

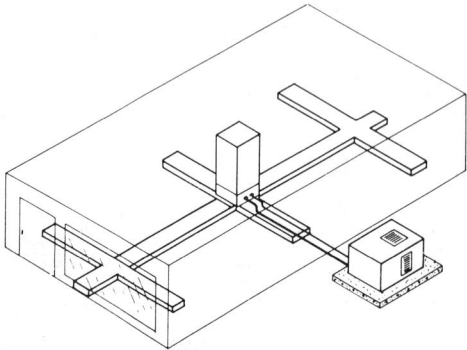

Fig. 29. Split system shown applied to commercial structure. Furnace (downflow here) makes the system serve year-round.

"Through the Roof"—Roof-top split-system installations are becoming increasingly popular in commercial buildings (see Fig. 30), as well as apartments and industrial (factory) buildings for a number of reasons, although they usually require a crane for installation and may call for reinforcement of the roof in some cases.

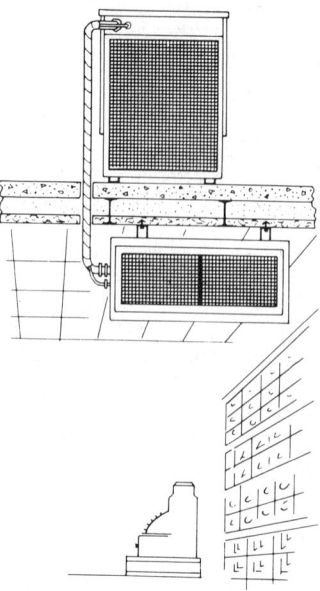

Fig. 30. Split system with condensing unit on roof is often used in stores and other one-story, flat-roofed buildings.

However, they require a minimum of connecting tubing (particularly if the evaporator is located directly below the condensing unit), and frequently a minimum of ductwork. Also they have the advantage that the condensing unit is out of the way of street or yard traffic.

Through-the-Wall Condensing Units—Split systems with condensing units through the wall are particularly well suited for individual cooling and heating in highrise apartments. When installed flush with outside walls, do not mar the appearance of the building.

Zoning Unitary Installations

Single-Zone—There are two basic ways to apply the unitary approach to air conditioning a large single zone. A decision has to be made whether to employ a large single unit or smaller multiple units.

First to be settled is whether the area under consideration is truly a single zone. A single zone has uniform type and periods of occupancy, uniform exposure, and uniform lighting load. Seasonal variations in one department could create zones within an area generally considered homogeneous. Con-

versely, areas that are actually physically separated may be similarly enough exposed and insulated to consider them as a single zone.

If the building is new, will exposed ductwork be aesthetically acceptable, and if not, can it be hidden? If old, is it easier to run multiple condenser piping or the duct work required with a single unit?

Many states require that units over a certain size or containing a maximum refrigerant charge must have an operating engineer. Obviously, this requirement would be obviated by utilizing multiple smaller units.

Loss of refrigerant can be costly. Use of multiple circuits or multiple units can minimize this loss, and, further, will reduce the inconvenience and cost resulting from such loss.

Multiple units offer greater standby capacity than a single large unit with one compressor—but a single large unit with multiple compressors is even better in this respect than multiple units. The entire zone will still be covered in the event of failure of one compressor, although at reduced capacity.

Once a decision for multiple units has been made, the engineer must consider thermostat locations carefully. Do not place the thermostat for one unit where it might be affected by the air delivery of another unit. It should only be within the lineal confines of the distribution pattern of the air delivered by the unit it controls.

Multiple Zones— Here again we find the option of single unit versus multiple units within the unitary approach. The single unit would be considered a central unit in this case, but it is still "unitary" in every sense of the definition (Fig. 31, 32 and 33.)

Many of the factors listed previously also apply in the case of multiple zones, and these should be considered—new or old construction and the structural and aesthetic factors they imply—desirability of an operating engineer if unit size makes one necessary—operating economy and standby capacity.

Also a factor is the quality of the installing labor. A central system is decidedly more complex than the multiple unit system, and whether the available labor is adequate to the task of installation is a question that has to be resolved in each instance.

If the decision is made to use a central unitary system, a further choice has to be made between two possible methods:

1. A large unit with single or multiple compressors, connected to a duct system for proper air distribution. Zoning in this instance will be accom-

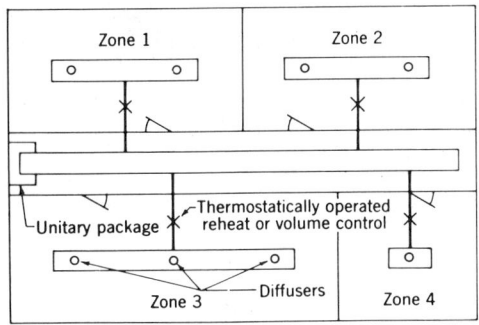

Fig. 31. Here a single unit serves multiple zones, and zoning is accomplished by means of reheat or throttled air. The latter will probably require static pressure regulation in main trunk duct.

plished either by applying conventional reheat or (less desirable) some means of throttling air in the individual zones for temperature control (Fig. 31).

2. A direct-expansion multizone unit incorporating steam or hot water heating coils. Here hot and cold air may be blended according to each zone's needs by face and bypass dampers within the unit and delivered in individual ducts to the zones (Fig. 32). Or, hot and cold air can be distributed through a double duct system and blended at the zone in accordance with temperature requirements (Fig. 33)

Reheat or Volume Control— The reheat device, whether steam, hot water or electric, should be installed in the duct take-off to a zone and should be thermostatically controlled from the zone.

If air volume control is to be used instead, this can be done at the diffuser outlet if there is only one diffuser serving a zone. If, however, the zone is covered by more than one outlet, air delivery should be throttled by a thermostatically operated damper-

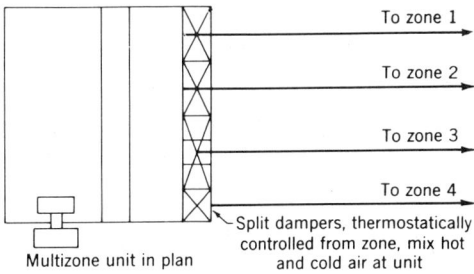

Fig. 32. A multizone unitary package can serve multiple zones through individual ducts, shown here, or through a double duct system shown in Fig. 33.

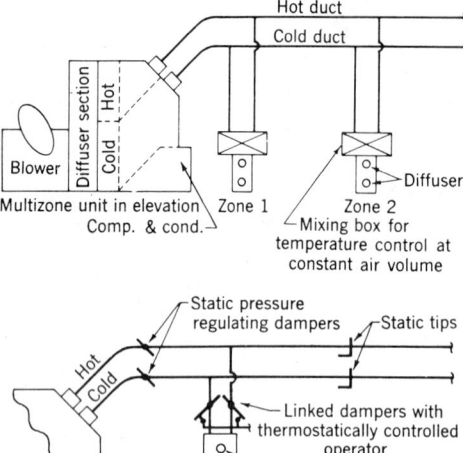

Fig. 33. Two methods of zoning with a multizone unitary package serving a double duct system are shown schematically. Above, temperature and air volume are regulated in a mixing box for each zone. Below, mechanically linked dampers mix warm and cold air to meet temperature requirements of zone.

ing system installed in the duct take-off to the zone, as seen in Fig. 31.

In certain applications, such as sporadically-occupied conference rooms, it will be most economical to use a manual volume control for comfort regulation. And in other applications, such as a split office within one zone, manual volume control provides some measure of independent control at minimum expense.

Multizone or Double Duct— In a multizone air conditioning unit, the heating and cooling coils are placed in parallel with respect to air flow. With the individual duct method of zone control, Fig. 32, face and bypass dampers divide the flow of air over the two coil surfaces in the right proportion to control the temperature and humidity of the air delivered to the zone. For year-round humidity control, a humidifier should be included for winter operation.

If an application should be critical so far as noise is concerned (and system resistance is low due to short ducts and large diffusers) fluctuations in air delivery could cause noise problems, and static pressure regulation would be in order.

This contrasts with the double duct method, where some form of static pressure regulation is mandatory. Two variations of this system are shown in Fig. 33.

Controls

Multiple unitary air-conditioner installations are often advantageously provided with a central control panel. This panel should include controls for starting and stopping unit fans (preferably by time clock with manual override); fresh air damper adjustments; manual summer-winter switches, and fan speed control switches, if used. Optional features that may be included on the panel are; remote adjustment of thermostat set points; remote space temperature reading; pilot light indication of fan, refrigeration or burner operation, and alarms for fan failure, burner failure or reset lockout, refrigeration overload or failure, condenser fan failure or dirty filters.

The key to control centralization is simplicity. Only functions that are actually needed working tools for the particular situation should be on the panel. For example, if the unitary systems are roof-mounted and thus not easily accessible, more indication is needed than for units mounted in an equipment room. Again, if the client employs regular maintenance personnel, less indication and alarms are required than for clients depending on outside help. In such a case, a centralized control panel to help a non-technical building manager diagnose the trouble and decide whom to call will pay for itself in a short time.

Locating the Panel—Location of the centralized panel is important. In a store or other commercial establishment the manager's office is generally the best place. In a factory, the maintenance office may be best. The area should be well lighted and accessible, yet closed to unauthorized persons.

Ventilation Air— Because of varying ventilation codes and individual customer requirements, outside and return air plenums are generally fabricated in the field when the manufacturer's standard provisions for ventilation will not meet the need. The most simple configuration of a ventilation system (Fig. 34) has a spring-return motor on the outside air damper. Whenever the fan is started, the outside air damper opens to admit a fixed percentage of outside air. When the fan stops, the outside air damper closes.

A more complex system (Fig. 35) replaces the two-position damper motor with a modulating motor The percentage of outside air is varied by a positioning switch on the control panel. While the operator can open the outside air damper to 100 per cent,

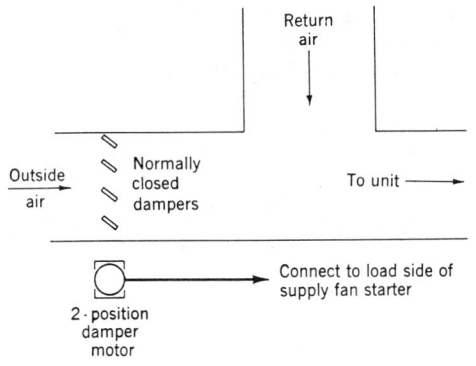

Fig. 34. Most simple ventilation system.

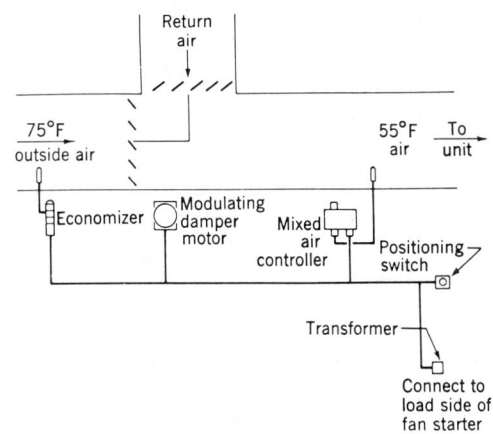

Fig. 36. Outside air controlled from mixed-air temperature.

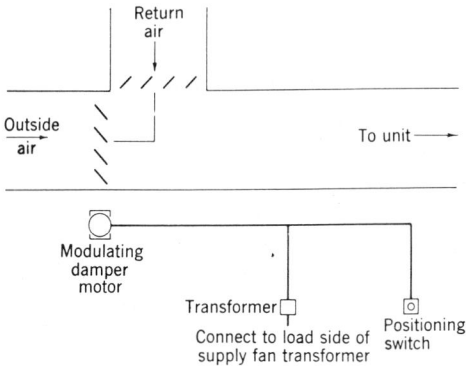

Fig. 35. More complex system has modulating damper motor.

the modulating motor automatically closes the outside air damper whenever the fan is stopped.

An automatic system of ventilation control (Fig. 36) uses a mixed air controller to operate the damper motor, modulating fresh and return-air dampers to maintain a constant temperature input to the unitary air conditioner.

Economical cooling operation is assured by an outside thermostat which overrides the mixed air thermostat in the cooling season. The dampers are controlled by the centrally located minimum position switch. In the heating season, this switch provides minimum ventilation required by code, or extra ventilation if required for any other reason.

Locating the Thermostat— Nothing works better than a room thermostat properly mounted in the air-conditioned space. However, in large spaces with no columns, or where store shelves cover all the wall space, the return-air thermostat is the best. A return-air thermostat should offer high sensitivity, fast response, low mass and excellent repeatability.

One possibility is electronic controls. A typical transistorized amplifier is no larger than a conventional relay, but has a performance that outshines the best bulb-type controller.

An added bonus with electronic control is remote temperature reset. In the case of rooftop units, the problem of resetting return-air thermostats can be thorny. A central panel with remote temperature reset can solve this problem handily.

Humidity Control— From a control point of view, a slightly undersize unit is preferable to an oversize unit. The undersize unit will keep the cold evaporator running, necessary for good latent heat removal. An oversize unit, cycling even in the hottest weather, will re-evaporate moisture into the air and increase the already-high humidity. No direct humidity control should be attempted unless summer reheat is available. (See "Reheat or Volume Control" under "Zoning," above.)

Selection Procedure

Recommended selection procedures vary somewhat with different makes and models of self-contained units, but all procedures are basically similar. The following discussion and selection of a vertical, self-contained unit will illustrate the method.

Before actual selection begins, a job survey and cooling load estimate must have been made, and these seven basic items of necessary data compiled: (1) outdoor design temperatures, (2) indoor design temperatures, (3) ventilation requirement, (4) total cooling load, (5) condensing medium and temperature, (6) air delivery requirements, and (7) external static pressure. Once these data are available, the self-contained unit is selected as follows:

1. First a *tentative* unit selection, based on the total cooling load, is made. This selection is made from the manufacturer's catalogued nominal unit capacities (Fig. 37).

2. Next the rated cfm delivery for the tentative unit is determined from a table of unit specifications, and compared to the air delivery requirement. If the rated cfm delivery is considerably larger or smaller than the required cfm but still within 20% either way, the selection should proceed on the basis of the *required cfm. If the two cfm values differ by more than 20%, a special built-up system probably should be considered.*

3. Temperatures of the air entering the unit are then determined. Entering dry-bulb and wet-bulb temperature can be calculated directly, or they can be obtained from a prepared table. In either case, outdoor and indoor design temperatures and ventilation air requirement must be known.

4. For units with water-cooled condensers, condensing temperature must be determined. A tentative condensing temperature is established as approximatly 30F higher than the entering water temperature.

5. Next, total unit capacity and sensible heat capacity at the given conditions are determined for the selected unit. This information is obtained from the manufacturer's capacity table for the appropriate type of unit (Fig. 37).

(a) For water-cooled condenser models, the evaporator entering air dry-bulb and wet-bulb temperatures and the tentative condensing temperature are necessary.

(b) For air-cooled condenser models, the outdoor design dry-bulb temperature and the dry-bulb and wet-bulb temperatures of the air entering the evaporator are needed. The smaller of the two recommended condensers should be used at this stage in the selection.

(c) For evaporative condenser models, the average maximum wet-bulb temperature of the air entering the condenser, together with the dry-bulb and wet-bulb temperatures of the air entering the evaporator, are required.

COOLING CAPACITIES—WATER-COOLED MODELS

Unit Model	Entering Air DB, Deg F	Refrigeration, Tons	Condensing Temperature, Deg F											
			95				105				115			
			Entering Air WB, Deg F											
			64	67	70	73	64	67	70	73	64	67	70	73
VW3	75	Total	3.22	3.40	3.60	3.78	2.98	3.18	3.38	3.56	2.73	2.93	3.14	3.32
		% Sens	71	58	46	35	73	60	47	35	75	61	48	35
	80	Total	3.22	3.40	3.60	3.78	2.98	3.18	3.38	3.56	2.73	2.93	3.14	3.32
		% Sens	81	71	61	49	83	73	63	50	85	75	65	51
	85	Total	3.22	3.40	3.60	3.78	2.98	3.18	3.38	3.56	2.73	2.93	3.14	3.32
		% Sens	90	82	75	63	93	84	77	65	95	86	79	67
VW7	75	Total	6.93	7.43	7.98	8.60	6.53	7.00	7.54	8.13	6.10	6.59	7.13	7.72
		% Sens	71	58	46	35	73	60	47	35	75	61	48	35
	80	Total	6.93	7.43	7.98	8.60	6.53	7.00	7.54	8.13	6.10	6.59	7.13	7.72
		% Sens	81	71	61	49	83	73	63	50	85	75	65	51
	85	Total	6.93	7.43	7.98	8.60	6.53	7.00	7.54	8.13	6.10	6.59	7.13	7.72
		% Sens	90	82	75	63	93	84	77	65	95	86	79	67
	90	Total	7.20	7.43	7.98	8.60	6.80	7.00	7.54	8.13	6.35	6.59	7.13	7.72
		% Sens	100	93	84	75	100	96	87	77	100	98	89	79

Fig. 37. Use this type table first for tentative selection.

CAPACITY CORRECTION FACTORS

To Be Applied to Cooling and Heating Capacities

Cfm°Compared to Rated Quantity	−40%	−30%	−20%	−10%	Std.	+10%	+20%
Cooling Capacity Multiplier	.88	.93	.96	.98	1.00	1.02	1.03
Sensible Capacity Multiplier	.80	.85	.90	.95	1.00	1.05	1.10
Heating Capacity Multiplier	.78	.83	.89	.94	1.00	1.06	1.12

Fig. 38. For use where cfm per ton is other than 400.

CONDENSER WATER REQUIREMENTS, GPM

Unit Model	Load, Tons	Condensing Temperature, Deg F											
		95				105				115			
		Entering Water Temperature, Deg F											
		65	70	75	80	70	75	80	85	80	85	90	95
VW3	2.5	3.5	4.6	6.8	10.6	2.9	3.5	4.6	6.8	2.9	3.9	4.6	6.8
	3.0	4.2	5.5	8.1	12.7	3.5	4.2	5.5	8.1	3.5	4.2	5.5	8.1
	3.5	4.9	6.5	9.5	14.9	4.1	4.9	6.5	9.5	4.1	4.9	6.5	9.5
VW5	4.0	5.6	7.4	10.8	17.0	4.6	5.6	7.4	10.8	4.6	5.6	7.4	10.8
	5.0	7.0	9.2	13.5	21.2	5.8	7.0	9.2	13.5	5.8	7.0	9.2	13.5
	6.0	8.4	11.1	16.2	25.4	6.9	8.4	11.1	16.2	6.9	8.4	11.1	16.2
VW7	6.5	9.1	12.0	17.6	27.3	7.5	9.1	12.0	17.6	7.5	9.1	12.0	17.6
	7.5	10.5	13.9	20.3	31.8	8.7	10.5	13.9	20.3	8.7	10.5	13.9	20.3
	8.5	11.9	15.7	23.0	36.0	9.8	11.9	15.7	23.0	9.8	11.9	15.7	23.0

Fig. 39. Use after condensing temperature is finally established and total cooling load, in tons, is fixed.

CONDENSER WATER PRESSURE DROP

Unit Model		Flow Rate, Gpm, and Pressure Drop (PD), Ft of Water								
VW3	GPM	3	4	5	6	7	8	9	10	11
	P. D.	1.7	2.9	4.5	6.3	8.4	10.8	13.0	16.5	19.3
VW5	GPM	5	6	7	8	10	12	14	16	18
	P. D.	1.8	2.6	3.3	4.1	6.1	8.6	11.3	14.3	17.9
VW7	GPM	6	8	10	12	15	18	21	24	27
	P. D.	2.2	3.9	5.5	7.8	11.5	16.0	21.2	27.6	33.5
VW10	GPM	10	12	14	16	20	24	28	32	36
	P. D.	1.8	2.6	3.3	4.1	6.1	8.6	11.3	14.3	17.9

Fig. 40. Pressure drop of water through condenser can be read directly from manufacturers' tables like this one.

SPECIFICATIONS AND GENERAL INFORMATION

Unit Model	Compressor				Evaporator Fan				
	Type	Qty	Hp	Rpm	No.	DWDI Size	Cfm	Motor Hp	Rp
VW3 VA3	Herm	1	3	3450	1	10″	1200	½	34
VW5 VA5	Herm	1	5	3450	1	12″	2000	¾	34
VW7 VA7	Herm	1	7½	3450	1	15″	3000	1½	34
VW10	Herm	2	5	3450	2	12″	4000	2	34

Fig. 41. Use in Step 2 of hypothetical example to find cfm of model air conditioner tentatively selected.

In each case, total unit capacity and per cent sensible heat are read directly from the tables. It will be necessary to interpolate to get the exact capacities if any of the given conditions fall between the values listed on the capacity table. The capacities given in these tables are for rated (standard) air delivery only. If quantities other than standard are being used, correction factors will have to be applied to the tabulated capacities. These correction factors are given in a table such as the one shown in Fig. 38.

6. The final total unit capacity should not be less than the total cooling load unless a compromise in design temperatures can be accepted. If additional capacity is required, a new selection will have to be made.

(a) For units with *water-cooled condensers,* the tentative condensing temperature should be lowered 5 or 10 F and new capacities determined before a larger unit is considered. If the load still cannot be met, the next size larger unit will have to be used.

(b) If the capacity is insufficient with the initial *air-cooled condenser* selection, the larger of the two recommended condensers should be tried. If capacity is still short, the next size larger air conditioning unit will have to be selected.

(c) Since the condensing temperature and condenser size are more or less fixed with *evaporative condenser* models, the next size larger unit will have to be selected unless minor adjustments in conditioned air fan speed will provide for the increase in capacity.

7. After actual unit capacity has finally been established, sensible heat capacity is compared to sensible heat load. If sensible capacity is not equal to or greater than the load, a new system will have to be selected.

8. For water-cooled condenser units, water flow and pressure drop through the condenser must be be determined. Using the condensing temperature finally established and the total cooling load in tons, the gpm requirement is obtained from a condenser water requirement table (Fig. 39). When gpm through the condenser is known, the pressure drop can be read directly from a condenser water pressure drop table (Fig. 40).

9. The next items to be determined are fan speed and fan motor horsepower necessary to deliver the required air quantity against the calculated external static pressure. These values are read directly from the fan performance table. If the fan motor horse-power exceeds the recommended limits, a larger fan motor should be substituted.

Because vertical self-contained units are available in fairly large increments of capacity, many situations arise in which the actual load falls between two available capacities. For example, suppose the cooling load estimate sheet indicates a 12½ ton load. The units which most nearly match this load have either 10 or 15 tons of capacity. However, a 7½ ton unit plus a 5-ton unit can be installed to meet the load exactly, but the cost will be higher for the two-unit installation. With an eye on possible future expansion of the area, the single 15-ton unit may be preferable. Or, if first cost is the major criterion, it may be best to tolerate a few warm days and accept the undersized 10-ton unit.

Whatever the case may be, it is always wise to check actual sensible and latent heat capacities of the unit against the estimated load. From a comfort standpoint, it is often most important that the unit be capable of handling the entire sensible load.

If the ratio of sensible heat gains to latent heat gains (sensible heat percentage) is unusually high or low, a more detailed check should be made. This detailed check can be accomplished by use of the manufacturer's published capacity table.

To illustrate the application of the nine-point selection procedure outlined above, a hypothetical example is given below. Estimation of the cooling load is omitted from the discussion, it being assumed that ARI's convenient form or equivalent has been used.

Example: Select a vertical self-contained unit with water-cooled condenser for use in a general office in Dallas, Tex. Condensing water will be supplied from a cooling tower, and a free discharge plenum will be added in place of ductwork. No air delivery requirements are specified.

Data from Cooling Load Estimate:

Outdoor design temperatures:	100F db; 78F wb
Indoor design temperatures:	78F db; 65F wb
Ventilation requirement:	300 cfm
Total cooling load:	6.97 tons

Condensing medium and temperature: cooling tower water @ 85F

Air delivery requirements:
400 cfm per ton = 2788 cfm

External static pressure	none

FAN PERFORMANCE—WATER & AIR-COOLED UNITS

Unit Model	Standard Fan Speed Range, Rpm	Cfm	External Pressure, Inches of Water						
			3/16		1/4		3/8		1/2
			Rpm	Bhp	Rpm	Bhp	Rpm	Bhp	Rpm
VW3	728—	1000	790	.24	740	.20	840	.28	920
		1100	845	.29	800	.26	890	.33	970
		1200	890	.35	850	.32	930	39	1020
	1113	1300	950	.44	910	.41	980	.46	1060
VW7	628—	2400	655	.61	640	.60	670	.63	700
		2700	700	.80	715	.91	740	.96	770
		3000	800	1.25	785	1.21	810	1.28	833
VA7	904	3300	870	1.62	860	1.57	880	1.67	905
		3600	945	2.10	935	2.05	955	2.14	975
VW10	753—	3200	735	.76	705	.70	770	82	810
		3600	805	1.01	785	.94	825	1.08	880

Fig. 42. Use in Step 9 of hypothetical example to find fan speed and fan motor horsepower for unit selected.

Selection:

1. From **Fig.** 37 tentatively select a Model VW7 unit to meet the 6.97 ton load.

2. From **Fig.** 41 determine a rated cfm delivery of 3000 cfm for a Model VW7 unit. This is close enough to the required cfm.

3. Calculate the dry-bulb and wet-bulb temperatures of the air entering the self-contained unit.

(a) $\dfrac{\text{vent requirement}}{\text{std. cfm delivery}} = \dfrac{300 \text{ cfm}}{3000 \text{ cfm}}$ = 1/10 or 10% outside air.

(b) ent. db = (0.10 × 100F) + (0.90 × 78F) = 80.2F db.

(c) wet-bulb temperature at 80.2F db with given mixture is 66.4 (from psychrometric chart).

4. Select a tentative condensing temperature: 85F + 30 = 115F condensing temperature.

5. From Fig. 37 determine total unit capacity and the sensible heat percentage. In this case, sufficient accuracy can be obtained by using 80F entering db and 67F entering wb for a VW7 unit at 115F condensing temperature. Using the above figures, total unit capacity is 6.59 tons and the sensible heat percentage is 75%.

6. This final total unit capacity is less than the calculated total cooling load by 0.38 tons. By re- ducing the condensing temperature to 105F and re-selecting, a capacity of 7.00 tons can be obtained.

7. Since there is no specified latent heat load, sensible heat percentage need not be considered for this selection.

8. Determine water flow requirement and pressure drop through the condenser.

(a) By interpolating between 6.5 and 7.5 in the "Load in Tons" column for a Model VW7 unit in the "Condenser Water Requirements" table (Fig. 39), determine a gpm water flow rate at 105F condensing—85F entering water temperature of 18.9 gpm.

(b) From the "Condenser Water Pressure Drop" table (Fig. 40), determine a water pressure drop through the condenser of 17.6 feet of water for 18.9 gpm. Next, size the cooling tower pump from the pump manufacturer's literature.

9. Finally, from the "Fan Performance—Water and Air-Cooled Models" table (Fig. 42), determine a fan speed of 800 rpm and a fan horsepower of 1.25 hp for a Model VW7 unit at 3000 cfm with 3/16-inch external static pressure corresponding to use of a free discharge plenum. Since this is the rated cfm and there is no unusually high external static pressure, it will not be necessary to check the required fan horsepower against an "Electrical Characteristics" table.

EVAPORATIVE AIR CONDITIONING

Refrigeration provides the best type of cooling, serving well where people are closely spaced in well constructed or relatively insulated structures. However, its first and operating costs bar it from the hottest commercial, industrial and residential buildings.

Evaporative cooling is an economical substitute in many regions. It is 60% to 80% cheaper to buy and operate. Thus, it is practical for lower income groups and where summers are short. Moreover, it cheaply cools hot, thinly-constructed mills, factories, workshops, foundries, powerhouses, farm buildings, canneries, etc., where refrigerated cooling is prohibitively expensive.

Evaporative air conditioning includes air cooling by evaporation of water. When water evaporates into the air being cooled, it creates *direct* evaporative cooling, the oldest and most common form. When evaporation occurs separately, and the air is cooled without humidity gain, the process is *indirect* evaporative cooling.

Permissible Air Motion

The closer entering air temperatures approximate room conditions, the higher the induced air velocities can be. When temperature differences are negligible, as with fan circulation, or outdoors, velocities of 880 fpm (a 10-mph breeze) are acceptable. Thus, since evaporative cooling usually provides air only 3F to 6F below room temperature, permissible room velocities are higher than is the case with refrigerated air conditioning.

Evaporatively cooled air inherently creates few draft problems. When it enters, although 3F to 6F cooler, its humidity exceeds that of the room air and it evaporatively cools skin proportionally less for equal velocity. Hence, its stray currents create fewer chill sensations, and higher velocities become permissible. Of course, full outlet velocity should not strike people directly, not rustle or blow papers in offices, etc. Thus, velocities under 200 fpm are recommended for sedentary workers, with higher ones used where necessary for manual and physical workers.

The designer of an evaporative cooling system can take advantage of the cooling effect of higher air velocities. However, it is difficult to predict how much the sensible temperature and relative humidity can be raised above those values considered comfortable at 15-25 fpm for a given increase in air motion. Table 1 attempts to define this relationship in terms of Effective Temperature. One degree ET is approximately equal to a unit of the Temperature-Humidity Index, defined by the equation

$$THI = 0.4(t_d + t_w) + 15$$

where THI = temperature-humidity index

t_d = dry-bulb temperature, F

t_w = wet-bulb temperature, F

Table 1—Effect of Air Velocity on Effective Temperature at 50% RH.[1,2]

Air Velocity, Feet per Minute	Room Effective Temperature	ET Reduction Due to Velocity	Room Effective Temperature	ET Reduction Due to Velocity	Room Effective Temperature	ET Reduction Due to Velocity	Room Effective Temperature	ET Reduction Due to Velocity
15-25	65	—	70	—	75	—	80	—
50	64.2	0.8	69.4	0.6	74.7	0.3	79.7	0.3
100	63.3	1.7	68.7	1.3	74.0	1.0	79.0	1.0
150	62.2	2.8	68.0	2.0	73.3	1.6	78.5	1.5
200	61.1	3.9	67.0	3.0	72.8	2.2	78.2	1.8
250	60.4	4.6	66.5	3.5	72.2	2.8	77.7	2.3
300	59.7	5.3	66.0	4.0	71.9	3.1	77.2	2.8

[1] For other humidities, the above ET reductions should be multiplied by the following approximate factors:

Relative Humidity	Multiplier
40%	0.95
60	1.05
70	1.13
80	1.25

[2] Table was computed from Fig. 9, Effective Temperature Chart for Normally Clothed Persons Doing Sedentary or Light Work, ASHRAE Guide and Data Book, 1961, p. 109.

At 70 or 71 ET (or THI) most people are comfortable; at 75, at least half will be uncomfortably warm; and at 80, almost all will certainly be very uncomfortable.

Table 1 indicates, for instance, that if a maximum THI of 72 is required, and if air velocity can be 300 fpm, one can design as if for THI = 75, choosing any suitable combination of wet- and dry-bulb temperatures that will satisfy the equation. The designer is cautioned, however, that the Effective Temperature and Temperature-Humidity Indexes are not always true indices of comfort. In the case of persons at rest or seated and performing light work, there is practcially no difference in the sensation of comfort, at moderate temperatures, between 30% and 70% relative humidity. Above 80F, however, increasing relative humidities produce discomfort at an accelerating pace.

Permissible Outdoor Conditions

Evaporative cooling performs well only where summer provides adequately dry air. The minimum required weather differs with the type. Unstaged indirect systems usually require the lowest humidity, while textile mill evaporative cooling and animal cooling and "spot" cooling for factories tolerate the most. Between these extremes, lie thousands of conventional direct systems.

Comfort cooling in all cases requires a minimum wet-bulb depression (difference between dry-bulb and wet-bulb temperature) of 22F and a maximum wet-bulb temperature which depends upon the Effective Temperature considered comfortable by the people in a given area of the country. A comfort cooling system is here defined as one that will produce comfort 80-90% of all hot hours.

Relief cooling occurs whenever evaporative coolers maintain air and surroundings cooler than skin temperature (87-94F) and circulate it rapidly regardless of drafts or humidity. It requires a minimum permissible outdoor wet-bulb depression of 13F.

Table 2, Geographic Performance of Direct Evaporative Cooling, adopted from Watt, "Evaporative Air Conditioning," Table IX (The Industrial Press, 1963) tells whether a particular city is suitable for comfort or relief cooling by evaporative techniques and is based on available weather data for these locations.

Table 2—Performance of Direct Evaporative Cooling

City	Evaluation*	City	Evaluation*	City	Evaluation*
ALABAMA		Livermore	Comfort	Santa Ana	Comfort
Anniston	Possible	Long Beach	NR	Santa Barbara	Possible
Birmingham	NR	Los Angeles	Relief	Santa Maria	Relief
Mobile	NR	Madera	Comfort	Santa Monica	NR
Montgomery	NR	Marysville	Comfort	Thermal	Comfort
		Maywood	Relief	Victorville	Comfort
ARIZONA	Comfort	Merced	Comfort	Williams	Comfort
		Muroc	Comfort		
ARKANSAS		Needles	Comfort	COLORADO	Comfort
Fort Smith	Possible	Oakland	Possible		
Little Rock	Relief	Oxnard	Possible	CONNECTICUT	NR
Texarkana	Possible	Palmdale	Comfort		
		Pasadena	Comfort	DELAWARE	NR
CALIFORNIA		Pleasanton	Comfort		
Bakersfield	Comfort	Pomona	Comfort	FLORIDA	NR
Blythe	Comfort	Red Bluff	Comfort		
Burbank	Comfort	Redding	Comfort	GEORGIA	
Culver City	Relief	Riverside	Comfort	Albany	Possible
Daggett	Comfort	Sacramento	Comfort	Atlanta	Possible
El Centro	Comfort	San Bernardino	Comfort	Augusta	Relief
El Toro	Comfort	Sandberg	Comfort	Macon	Possible
Eureka	Comfort	San Diego	NR	Marietta	Possible
Fairfield	NR	San Francisco	NR	Savannah	NR
Fresno	Comfort	San Jose	Comfort	Valdosta	NR
Laguna	NR	San Rafael	Possible	Warner-Robins	Possible

*NR = Not Recommended. See end of table for explanation.

Table 2 (Continued)—Performance of Direct Evaporative Cooling

City	Evaluation*	City	Evaluation*	City	Evaluation*
IDAHO	Comfort	MASSACHUSETTS		NEVADA	Comfort
		Bedford	NR		
		Boston	NR	NEW HAMPSHIRE	
ILLINOIS		Chicopee Falls	Possible	Concord	Possible
Belleville	Possible	East Lynn	NR	Manchester	NR
Cairo	Relief	Falmouth	NR	Portsmouth	NR
Chicago	Possible	Nantucket	NR		
Moline	Relief	Springfield	Possible	NEW JERSEY	
O'Hare	Possible	West Lynn	NR	Atlantic City	NR
Peoria	Relief			Camden	NR
Rantoul	NR			Fort Dix	NR
Springfield	Relief	MICHIGAN		Navesink	NR
		Alpena	NR	Newark	NR
		Detroit	Possible	Trenton	Possible
INDIANA		Grand Rapids	Relief		
Columbus	Possible	Lansing	Possible	NEW MEXICO	Comfort
Evansville	Relief	Marquette	NR		
Fort Wayne	Relief	Sault Ste Marie	NR	NEW YORK	
Helmer	NR			Albany	Possible
Indianapolis	Possible			Binghamton	NR
South Bend	Possible	MINNESOTA		Buffalo	NR
Terre Haute	Relief	Duluth	NR	Canton	NR
		Minneapolis	Possible	Elmira	Possible
		Rochester	Possible	Hempstead	NR
IOWA		St. Cloud	Possible	Newburgh	NR
Burlington	Possible	St. Paul	Possible	New York City	NR
Davenport	Relief			Niagara Falls	NR
Des Moines	Relief			Rochester	NR
Dubuque	Possible	MISSISSIPPI		Syracuse	Possible
Keokuck	Relief	Biloxi	NR	Watertown	NR
Sioux City	Relief	Greenville	Possible	White Plains	NR
		Jackson	Possible		
		Gulfport	NR	NORTH CAROLINA	
KANSAS		Meridian	Possible	Asheville	NR
Dodge City	Comfort	Vicksburg	NR	Charlotte	Relief
Goodland	Comfort			Greensboro	NR
Topeka	Relief			Raleigh	NR
Wichita	Comfort	MISSOURI		Wilmington	NR
		Columbia	Relief	Winston-Salem	NR
		Kansas City	Relief		
KENTUCKY		Kirksville	Relief		
Lexington	Relief	Springfield	Possible	NORTH DAKOTA	
Louisville	Possible	St. Joseph	Possible	Bismarck	Comfort
		St. Louis	Relief	Devil's Lake	Relief
LOUISIANA				Dickinson	Relief
Alexandria	NR	MONTANA	Comfort	Fargo	Relief
Baton Rouge	NR			Grand Forks	Relief
Lake Charles	NR			Pembina	Relief
New Orleans	NR	NEBRASKA		Williston	Comfort
Shreveport	Possible	Grand Island	Relief		
		Lincoln	Relief		
		North Platte	Relief	OHIO	
MAINE	NR	Omaha	Relief	Akron	NR
		Scottsbluff	Comfort	Canton	NR
MARYLAND	NR	Valentine	Relief		

Table 2 (Continued)—Performance of Direct Evaporative Cooling

City	Evaluation*	City	Evaluation*	City	Evaluation*
OHIO (Cont'd)		SOUTH DAKOTA		VERMONT	
Cincinnati	Possible	Gettysburg	Relief	Burlington	Possible
Cleveland	Possible	Huron	Relief		
Columbus	Possible	Rapid City	Comfort	VIRGINIA	
Dayton	Possible	Sioux Falls	Relief	Cape Henry	NR
Sandusky	Possible			Fort Eustis	NR
Toledo	Possible			Hampton	NR
Youngstown	NR	TENNESSEE		Lynchburg	Relief
		Bristol	Possible	Norfolk	NR
		Chattanooga	Relief	Petersburg	NR
OKLAHOMA		Knoxville	Possible	Quantico	NR
Ardmore	Relief	Memphis	Possible	Richmond	NR
Enid	Relief	Nashville	Relief	Roanoke	Possible
Muskogee	Possible				
Oklahoma City	Relief				
Tulsa	Relief			WASH., D.C.	NR
Waynoka	Comfort	TEXAS			
		Abilene	Comfort	WASHINGTON	
		Amarillo	Comfort	Ellensburg	Comfort
OREGON		Austin	Relief	Everett	NR
Arlington	Comfort	Big Spring	Comfort	Moses Lake	Comfort
Baker	Relief	Brownsville	NR	North Bend	NR
Eugene	Relief	Bryan	Possible	Seattle	NR
Klamath Falls	Relief	Corpus Christi	NR	Spokane	Comfort
Medford	Comfort	Dallas	Relief	Tacoma	NR
Pendleton	Comfort	Del Rio	Relief	Vancouver	Possible
Portland	Possible	El Paso	Comfort	Yakima	Comfort
Roseburg	Relief	Fort Worth	Relief		
Salem	Possible	Galveston	NR	WEST VIRGINIA	
		Harlingen	NR	Charleston	Possible
		Houston	NR	Elkins	NR
PENNSYLVANIA		Killeen	Relief	Parkersburg	Possible
Allentown	Relief	Laredo	Relief		
Curwensville	NR	Lubbock	Comfort		
Erie	NR	Midland	Comfort	WISCONSIN	
Harrisburg	Possible	Mineral Wells	Relief	Green Bay	NR
Philadelphia	Possible	Monahans	Comfort	La Crosse	NR
Pittsburgh	Possible	New Braunfels	Relief	Madison	Possible
Reading	Relief	Palestine	Possible	Milwaukee	NR
Scranton	Relief	Port Arthur	NR		
Sunbury	NR	Randolph Field	Relief		
Williamsport	Relief	San Angelo	Comfort	WYOMING	Comfort
		San Antonio	Relief		
		San Marcos	Relief		
RHODE ISLAND	NR	Sherman	Relief		
		Sweetwater	Comfort		
		Victoria	NR		
SOUTH CAROLINA		Waco	Relief		
Charleston	NR	Wichita Falls	Relief		
Columbia	NR	Wink	Comfort		
Greenville	Relief				
Myrtle Beach	NR				
Sumter	Possible	UTAH	Comfort		

*COMFORT: Climate is suitable for maintaining comfort conditions with direct evaporative cooling.

RELIEF: Climate is suitable for maintaining indoor conditions in the relief cooling range (no more than 2 deg ET above comfort range).

POSSIBLE: Climatic data for area is only slightly unfavorable or, if favorable, is judged doubtful because of proximity to coast.

NOT RECOMMENDED: Area unsuited to *direct* evaporative cooling techniques, but note, indirect evaporative cooling might be feasible.

VARIABLE VOLUME A.C. SYSTEMS

Central air distribution systems for air conditioning are either constant or variable volume. In *Constant Volume Systems,* the volumetric rate of flow of air to a conditioned space remains the same, while air temperature is varied to maintain space conditions. In *Variable Volume Systems* (sometimes called Variable Air Volume—VAV), air temperature may be held constant over a range of room requirements while the amount of air is changed to satisfy conditioning needs.

The variable volume system is truly an energy conservation system. Within certain limitations, only the actual amount of air required to do the heating or cooling is circulated through the system. Savings in system initial cost and operating cost are evident. When air volume reduction occurs as a function of system control, good system design dictates equipment selection based on system diversity. Present day load calculation methods using computers permit fairly accurate determination of system diversity and therefore equipment selection. When equipment is selected on this basis the initial size reduction can be as much as 20 to 30% compared with constant volume systems. Initial equipment size reduction is reflected in initial dollars saved and in lower operation costs. In an all-air system, initial cost savings are almost directly proportional to system size reduction.

High lighting levels designed into today's buildings dictate requirements for cooling even in very cold weather. Solar and lighting loads may require full cooling, at a particular exposure, at any time of year. The Variable Volume concept handles this type of control problem very well. Where Dual Duct or Reheat Terminals are used for perimeter areas, supply air volume is reduced as the solar load diminishes. At the preset minimum volume, some form of heat is added. Minimum volume is usually determined by heating requirements, although ventilation codes or air movement standards may take precedence.

For properly designed interior spaces, lighting loads automatically set the minimum volume. Except for space warm-up, a source of heat for interior spaces is not necessary. As lights are usually turned on while the building is occupied, interior space load fluctuations are caused simply by people migration. Volume reduction will generally suffice to offset people load changes.

In an attempt to calculate energy savings by applying variable volume, consider the effect of reducing flow on the operating characteristics of a system as applied to a specific building. A load profile for the southern exposure of this building is shown in Fig. 1. Maximum cooling load occurs at 2 P.M. and is 33 Btu/hr per sq ft of space. Maximum heating load is 30 Btu/hr per sq ft with building unoccupied and lights off.

Assume that designs are being made of a Dual Duct System and a Reheat System for perimeter areas, and that a comparison is to be made

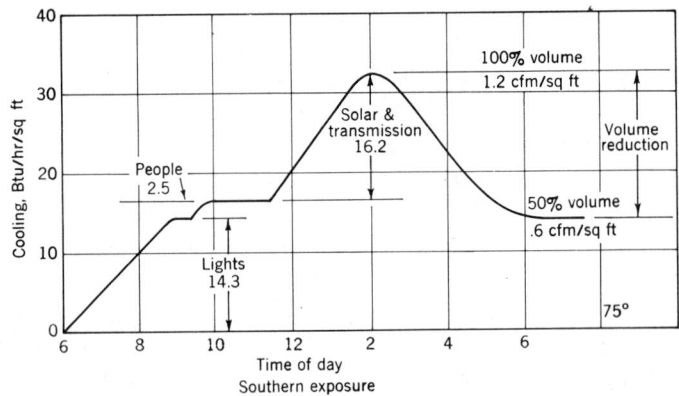

Fig. 1. Load profile.

between a constant volume and a variable volume design for both systems. Following are the design requirements for this building:

		SUMMER	WINTER
Room Temp. 75°F	Outside Temp.	95°F DB 78°F WB	0°F DB
Supply Air Temp.	Cold Air	50°F	60°F
Supply Air Volume	cfm/sq ft	1.2	.6
Supply Air Temp.	Warm air	80	95

Lighting Load 6 watts/sq ft (30% of this load is picked up through light troffers and doesn't enter the space).

People—1 per 100 sq ft

From an energy standpoint a temperature reset schedule for central apparatus is important. It is general practice to neglect occupied operating conditions and the reset schedule is determined with maximum temperature selected for un-occupied conditions. This means that for Dual Duct system design, the temperature of the warm air pumped to the space is higher than necessary, causing the mixing unit to provide an increased amount of cold air to offset this higher tempera-ture. Energy is wasted when refrigeration is in operation. Duct transmission losses are higher particularly if warm ducts are uninsulated and heat losses due to mixing unit damper leakage are increased.

A better procedure is to select the maximum warm air temperature with occupied conditions as the basis for selection. For example—The max-imum unoccupied load is 30 Btu/hr/sq ft. Occu-pied load due to lights and people is 16.8 Btu/hr/sq ft. Subtracting 16.8 from 30 leaves 13.2 Btu/hr/sq ft as the maximum heat loss when the building is occupied. See Fig. 2 for a com-parison of reset schedules. The reset schedule for an occupied building begins at 0°F outside and 95°F warm temperature and reduces to 80°F at 25°F outside. At 25°F outside, the transmission loss is −20 Btu/hr/sq ft, or −3.2 Btu/hr/sq ft when the building is occupied. For the Variable Volume Dual Duct system the supply air tempera-ture required is:

$$\Delta t = \frac{3.2}{.60 \times 1.1} = 4.85°,$$

$$t_s = 4.85° + 75° = 79.85°$$

To be practical we should allow some duct heat loss, mixing damper leakage, plus possible un-predictable building leakage conditions. A slightly higher reset schedule would probably be provided. (See Fig. 2.). Warm up conditions require higher

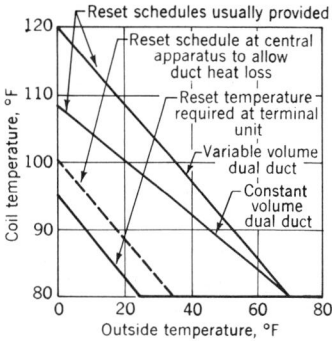

Fig. 2. Reset schedules compared.

temperatures. The heating coil should be selected for unoccupied load plus any leakage of outside air through dampers, plus infiltration, plus cold mixing unit leakage. Maximum coil temperature could be made available by manual or automatic control for the warm up period only. For inter-mediate weather operation (50°F to 70°F) an unnecessarily high warm air temperature will require additional energy. For example, if the reset schedule usually provided for variable vol-ume systems was applied, the warm air tem-perature, as indicated in Fig. 2, would be 92°F at 50°F outside. After refrigeration is on, the in-creased energy required, when compared to 80°F warm air temperature, is 1.8/Btu/hr/sq ft, or an increase of approximately 16% in cooling and heating energy.

Initial Costs

A comparison of energy in Btu required by the two systems (constant and variable volume) to satisfy the load profile of Fig. 1 is shown in Fig. 3. It is assumed that exterior and interior areas

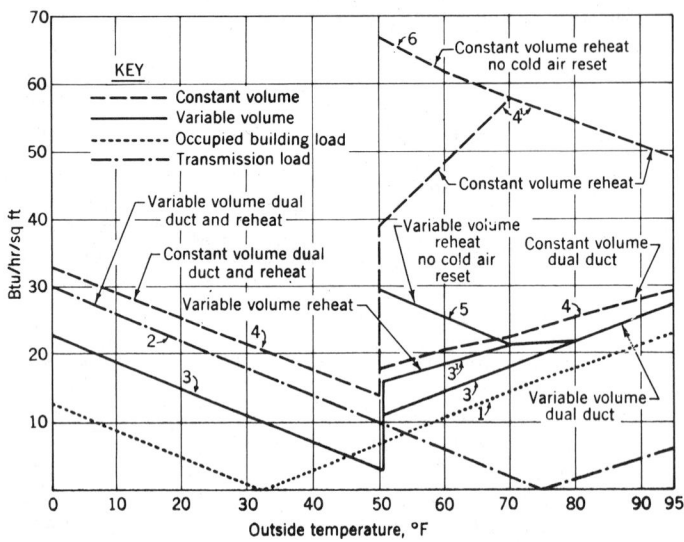

Fig. 3. Energy requirements in Btu. Conditions—no sunload. Southern exposure.

are handled by the same apparatus. This is generally done, particularly in small buildings, in order to keep the initial cost as low as possible. To keep energy costs down the cold air temperature is reset from 50°F to 60°F as the outside temperature falls from 70°F to 50°F. Figure 3 shows energy in Btu/hr/sq ft required to maintain 75°F in the space while the building is occupied and as the outside temperature changes from 0°F to 95°F. Where heating and cooling occur simultaneously, they are added in Fig. 3. Line 1 is the occupied building load plus transmission loss or gain. Line 2 is unoccupied transmission loss or gain. Line 3 gives Btu requirements for variable volume systems, dual duct and reheat. Line 4 indicates Btu requirements for constant volume systems, dual duct and reheat. Included in Lines 3 and 4 is the energy required to heat or cool ventilation air. The increase in energy from Line 3 to Line 4 below 50°F may be accounted for in the increased outside air volume required to maintain 60°F cold supply temperature for the constant volume system. If ventilation codes required a minimum outside air volume greater than the volume required to produce 60°F cold air, then the variable volume system energy requirements would increase by that amount. The reheat system deviates from dual duct both for constant and variable air volume after refrigeration is applied and outside dampers

are closed to the minimum position. The comparison demonstrates the high energy requirements of the constant volume systems, particularly the reheat system. (Line 4'.) The reason for increased energy requirements of the constant volume reheat system is the heat applied to maintain dry bulb conditions at reduced load. Notice the reduction in energy requirements below and above 70°F on Line 4'. Above 70°F less heat is needed to offset cold supply air temperature of 50°F. Energy for the constant volume reheat system drops off below 70°F due to resetting of cold air temperature from 50 to 60°F. If the cold air temperature is not reset, energy requirements would increase to 66.8 Btu/hr/sq ft at 50°F with the refrigeration system running. This is .7 times the energy required by transmission losses. The constant volume dual duct system captures space heat for control purposes and is therefore considerably more economical above 50°F. Variable volume systems generally follow space load conditions. For the dual duct variable volume system additional energy required over and above load conditions is simply ventilation load requirements. At some point below 80°F the load is reduced sufficiently so that the volume of air supplied to the space reaches minimum. At this point heat is added to the variable volume reheat unit while the mixing unit for the dual duct system begins to mix warm air from the space to

offset load changes. The additional heat required by the reheat unit below 80°F is indicated by Line 3'. Lines 5 and 6 indicate additional heat required by the reheat systems if cold air temperature is not reset to 60°F. Total energy requirements for the various systems are shown in Table 1.

The calculation procedure for obtaining energy usage for a condition of 50°F with refrigeration on, and no sun load on this exposure, is shown in Table 2. At 50°F transmission load is −10 Btu/hr. This, subtracted from occupied building load of +16.8, is +6.8 Btu/hr/sq ft. For the variable volume systems a supply air temperature of 64.7°F is required, and for the constant volume systems, it is 69.88°F.

The 7½°F heating rise required for dual duct systems is the result of return air mixing with minimum outside air (25% minimum and assuming perfect mixing conditions). Neglecting supply fan temperature rise, 72½°F is the temperature of the air entering the heating coil. The variable volume dual duct system supplies one-half the volume of air required by the constant volume

Table 1—Summary of Energy Requirements at Outside Temperatures Indicated

OUTSIDE Temp. °F	CONSTANT VOLUME DUAL DUCT			VARIABLE VOLUME DUAL DUCT			CONSTANT VOLUME REHEAT			VARIABLE VOLUME REHEAT		
	1	2	3	1	2	3	1	2	3	1	2	3
0			33.0			23.0			33.0			23.0
10			29.0			19.0			29.0			19.0
20			25.2			15.2			25.2			15.2
30			21.2			11.2			21.2			11.2
40			17.2			7.2			17.2			7.2
* 50	14.0		14.0	3.1		3.1	14.0		14.0	3.1		3.1
**50	9.4 + 12.4		21.8	5.0 + 10.1		15.1	14.0 + 26.4		40.4	3.1 + 13.2		16.3
60	6.6 + 17.6		24.2	2.6 + 14.1		16.7	15.6 + 33.0		48.6	2.4 + 16.5		18.9
70	2.0 + 21.4		23.4	1.0 + 18.1		19.1	18.2 + 39.6		57.8	1.7 + 19.8		21.5
80			25.1			22.0	14.5 + 39.6		54.1			22.0
90			27.5			25.7	11.2 + 39.6		50.8			25.7
95			29.6			27.5	10.0 + 39.6		49.6			27.5

NOTE: 1. Heating 2. Cooling 3. Totals
  * Refrigeration Off ** Refrigeration On

Table 2—Heating-Cooling Calculations 50°F to Refrigeration on— No Sun for Southern Exposure Only

			BTU/HR/SQ. FT.		
			HEAT	COOL	TOTAL
1. C. V. DUAL DUCT	$\Delta t = \dfrac{6.8}{1.2 \times 1.1} = 5.12,$ $\quad 75 - 5.12 = 69.88$ Heating (.5 × 1.2) × 7.5 × 1.1 = Cooling (.50 × 1.2) × 20 × 1.1 =		5.0	13.2	18.2
2. V. V. DUAL DUCT	$\Delta t = \dfrac{6.8}{.60 \times 1.1} = 10.3,$ $\quad 75 - 10.3 = 64.7$ Heating (.235 × .60) × 7.5 × 1.1 = Cooling (.765 × .60) × 20 × 1.1 =		1.2	10.1	11.3
3. C. V. REHEAT	$\Delta t = \dfrac{6.8}{1.2 \times 1.1} = 5.12,$ $\quad 75 - 5.12 = 69.88$ Heating 1.2 × (69.88 − 60) × 1.1 = Cooling 1.2 × 20 × 1.1 =		13.0	26.4	39.4
4. V. V. REHEAT	$\Delta t = \dfrac{6.8}{.60 \times 1.1} = 10.3,$ $\quad 75 - 10.3 = 64.7$ Heating .60 × (64.7 − 60) × 1.1 = Cooling .60 × 20 × 1.1 =		3.1	13.2	16.3

system at this condition. Approximately 75% less air is heated and 25% less air is cooled. As the reheat system is unable to take advantage of building heat (unless heat recovery cycles are used), this system requires 2½ times the heating energy of the constant volume dual duct system and twice the cooling energy. Even the variable volume reheat system does not compare favorably with the variable volume dual duct system. The variable volume reheat system uses one-half the air volume and one-half the temperature rise for heating compared with constant volume reheat, reducing heating energy approximately 4 times. One-half the air volume is cooled and cooling energy usage is reduced 100%.

Although the analysis applies to only one zone, energy usage should be similar around the perimeter of the building when sun load is discounted. Winter energy needs are reduced from Fig. 3 loading as sun load increases for all systems. After refrigeration is turned on energy requirements increase for all but the constant volume reheat system. As the solar load increases less reheat is required, and therefore less total energy is needed. Fig. 4 shows energy used with full sun load on the exposure. The variable volume dual duct and reheat systems, Line 7; follow the same energy line at full sun load as no volume reduction takes place under this condition. Lines 8 and 9 show constant volume systems energy

requirements at full sun load. Reheat system Line 9 and dual duct system Line 8. Conditions of Fig. 3 are superimposed on Fig. 4 in light lines to permit comparison.

Cooling Considerations

Energy requirements in the building interior are purely cooling considerations. Cooling energy usage stops when the refrigeration system is off and outside air is provided. Depending on system design, this occurs between 50°F and 60°F outside temperature. When the building is occupied heat available within the building is sufficient to raise the outside air temperature to the design cold air temperature of 60°F. A mixture of 25% outside air and 75% return will produce 60°F. No additional heating for outside air is needed and free cooling is provided for the interior.

Some care must be exercised in equipment layout to prevent stratification of the two air streams at the central apparatus. Where stratification is anticipated and the design is for cold climates, some additional energy in the form of outside air preheat may be required. 25% outside air is generally enough for ventilation purposes, provided of course that this volume is sufficient to meet ventilation codes. If necessary a means of insuring minimum volumes at reduced flow may be applied at the central apparatus. Energy cost to provide interior cooling in winter is limited

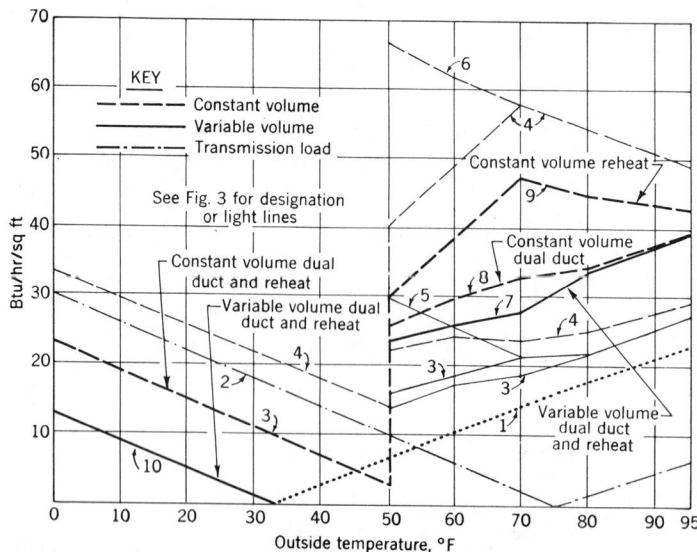

Fig. 4. Energy requirements in Btu. Condition—full sunload. Southern exposure.

to the cost of pumping the air to the space.

Load fluctuations in interior spaces are principally caused by people moving from place to place. Interior load changes seldom exceed 20% of the total interior space load. Variable Volume units can handle this load change with no difficulty and will produce some energy savings as the volume is reduced. Variable Volume terminal units for the interior may be Dual Duct, Reheat, or Shut-off type. Since the development of variable volume regulating devices, dual duct or reheat types are seldom installed for interior spaces. While the system is operating on outside air, no cooling energy is required for the shut-off type. Volume reduction takes care of load fluctuations. Cooling energy is required after refrigeration is on and then only that energy required to exactly offset the load.

When the building is unoccupied outside dampers should be closed and system fans cycled to maintain a predetermined minimum temperature. For winter, energy requirements will be lower than Lines 3 or 4, depending upon the amount of outside air leaking through closed dampers. The system is not generally cycled during summer operation.

Consider first the system using radiation at the perimeter. The usual procedure for most economical operation is to zone the radiation by exposure and control from outside dry bulb temperature. Two types of terminals are available for the air system. One type shuts off air flow completely and the other permits a minimum flow setting. Where the shut-off terminal is installed, radiation is sized to offset transmission loss while the lighting load is considered sufficient to keep the "shut-off" terminal open.

Some variable volume system designs provide either a separate source of heating at the perimeter in the form of radiation or a separate air heating and cooling system at the perimeter. The perimeter system is usually designed to handle transmission loads only. Solar loads are handled by the interior system terminals sized to offset solar loads when they occur, thereby providing a minimum volume at the perimeter. Figure 5 indicates energy requirements for this design approach, and once again is superimposed over previous figures. As this is principally an interior zone system it is economical to continue the outside air cycle as long as possible, and eliminate resetting of cold air temperature. 55°F is selected as the changeover point. Heating requirements follow Line 2 (Transmission Loss Line) up to 55°F. Heat required above the variable volume system (either dual duct or reheat) is the amount necessary to keep the interior terminals "open." This difference could be minimized by balancing. This, however, is seldom done, as it is normally desired to heat the building perimeter with the perimeter

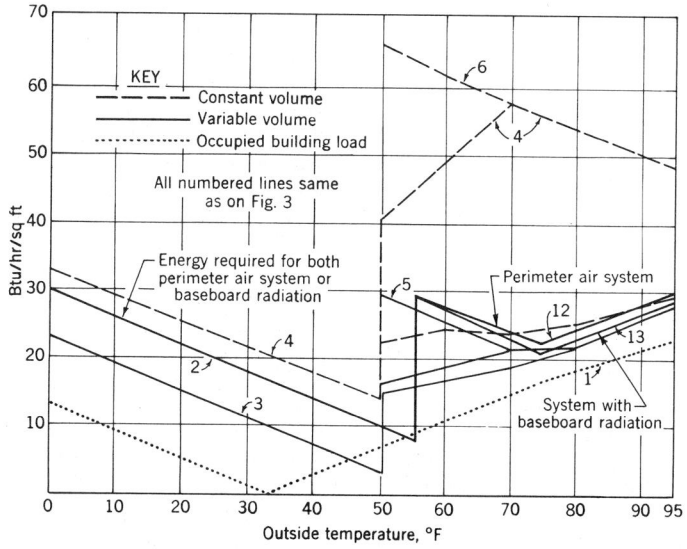

Fig. 5. Energy requirements in Btu. Condition—no sunload. Southern exposure.

system during unoccupied hours, thereby permitting shutdown of the interior system. Line 13, Fig. 5, shows energy required for the system using perimeter radiation and an interior variable volume system. The increase in energy below 75°F occurs only at the perimeter terminals and is the overlap caused by radiation in operation below 75°F. When the building is unoccupied energy requirements are at a minimum as fan systems are off and radiation handles the heating requirements.

The design with a separate perimeter air system will perform in a similar manner except that minimum flow may be provided in the perimeter system permitting interior terminals to cut off if the loads so dictate. With cooling capacity available at the perimeter, interior terminal size may be reduced. Since perimeter design is to offset transmission loads only, no zoning of the perimeter supply temperature is usually provided. The perimeter system is generally a constant volume system and temperature is programmed using the constant volume reset schedule shown in Fig. 2. Figure 5, Line 2, indicates winter energy requirement and Line 12 shows the energy required above 55°F for this approach.

Overlapping

There is some overlapping of energy requirements unless the perimeter system is zoned by exposure. When the perimeter system is controlled by outside temperature and not zoned, the system is unable to take advantage of sun load shifting and tends to overheat areas exposed to the sun. Actually, no overheating takes place as an additional volume of cold air enters the space from the interior terminal. From a control standpoint the system works quite well. The increase in energy, Line 12 over 13, is the increased volume of air handled by the perimeter air system and the energy required to cool the additional heat in the return air supplied to the perimeter air system.

This discussion is somewhat theoretical. The analysis is for only one load profile at one set of varying temperatures, however energy requirements will be similar around the building at no sun load. To go beyond this by manual calculation is an endless task. Only proven calculation procedures and the computer will simplify the task.

As previously stated, many factors affect an analysis of this type. Comfort and economy of operation are not always compatible. In fact, some overlapping of heating may be desirable to insure comfort at perimeter areas. Capturing building heat and distributing it to areas where required will pay dividends in improved comfort and reduced costs. Large glass areas and/or poor building construction tend to defeat attempts to capture space heat and will add to operating costs.

To reach the conclusion that Btu savings are proportional to energy savings would be incorrect. Type of energy used, cost of energy, equipment performance under varying load conditions, etc., must be taken into account to obtain cost savings.

Varying volume permits operating cost savings as described herein. The most uneconomical point of operation occurs when refrigeration is first required. The Dual Duct Variable Volume System in combination with Single Duct Variable Volume for interior control is the most economical to operate of the systems discussed. The Reheat System for comfort applications is extremely uneconomical to operate *unless* the variable volume concept is applied. With the exception of special applications such as hospitals or laboratories, constant volume is unnecessary for average comfort applications.

A single terminal providing heating and cooling as needed, has the capability to provide the energy required, and only that energy to compensate for changing conditions at that particular zone or room. A split system (perimeter heating and/or cooling, with cooling system only for areas just inside of the perimeter) must overlap energy requirements for proper performance. There are several reasons for the overlap—1. It is not possible to exactly offset transmission losses and, to be safe, a perimeter system is usually oversized. 2. It is desired to insure that inside terminals are open. 3. Exposure zoning is expensive as solar compensators must be used, therefore systems are seldom controlled by exposure. Perimeter systems, however, provide greater comfort and are mandatory in severe climates. The ultimate solution for comfort and operating economy is the single terminal installed at the perimeter, either mixing unit or reheat unit, controlled from the space being served, with "cooling only" variable volume terminals for interior spaces.

The variable volume concept is a *very simple* energy conservation approach. It can be used with conventional heating systems or heat re-

covery refrigeration systems. The engineer must weigh cost of heating outside air from 55°F, 60°F to approximately 95°F, 100°F (including warmup costs) against energy required for the refrigeration machine plus the heating for ventilation air. It is certainly a question in one's mind if it is economical to run a refrigeration machine to cool a building interior and areas exposed to the sun when the same work can be done with outside air. An exception to this would be warmer climates, with high wet bulbs where refrigeration demand time is quite high, and/or special considerations where outside air might not be desirable.

HEAT RECOVERY

Integrated Lighting-Heating-Cooling Systems

The following material on integrated systems and heat recovery has been excerpted, with permission, from the *Electric Heating and Cooling Handbook,* Edison Electric Institute.

The evolution of environment control has made the ceiling an especially functional element of the structure. As shown in Fig. 1, it has become the principal distribution point for the elements of a controlled environment—lighting, cleaned air supply, air motion, comfortable temperature and humidity, and noise control. Thus it provides the opportunity for integration of the lighting, heating, and cooling systems for an optimum in design.

A properly designed system will prevent much of the heat from the lighting fixtures from entering the occupied space and thereby reduce the cooling load and air handling requirements for the space. If this control over the lighting heat results in cooling the lamps and fixtures, the lamps will operate at higher efficiency, the ballast life will be lengthened, and infrared radiation into the space will be reduced. In addition, lighting heat can be used to supply some portion of the space heating requirement. The magnitude of the portion which the lighting system can supply will depend on the lighting density and the climatic design parameters.

Many systems have been designed for environmental control using integrated lighting-heating-cooling systems. Basically, they are divided into air systems and air-water systems.

Air Systems

A typical fluorescent lamp emits 19 percent of its energy input as light and 31 percent as radiant heat. The balance, or 50 percent, is convected or conducted heat. The energy represented in the visible light itself also is converted to heat. Figure 2 shows the energy distributed from one type of fixture. In the air system, most of the heat generated is carried away from the occupied space by

Fig. 1. The modern office environment includes pleasingly controlled lighting, temperature, humidity, and air movement.

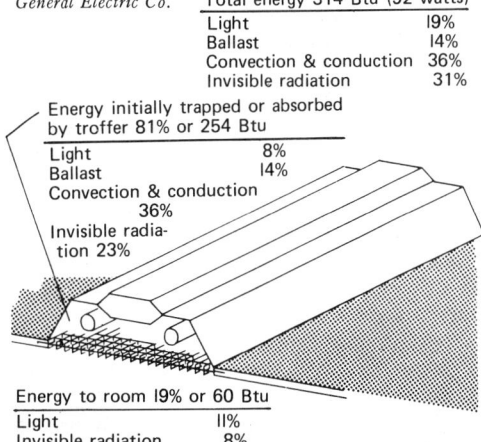

General Electric Co.

Total energy 314 Btu (92 watts)	
Light	19%
Ballast	14%
Convection & conduction	36%
Invisible radiation	31%

Energy initially trapped or absorbed by troffer 81% or 254 Btu	
Light	8%
Ballast	14%
Convection & conduction	36%
Invisible radiation	23%

Energy to room 19% or 60 Btu	
Light	11%
Invisible radiation	8%

Fig. 2. The theoretical dissipation of heat with two 40-watt fluorescent lamps and ballast (assumed luminaire efficiency—60 percent).

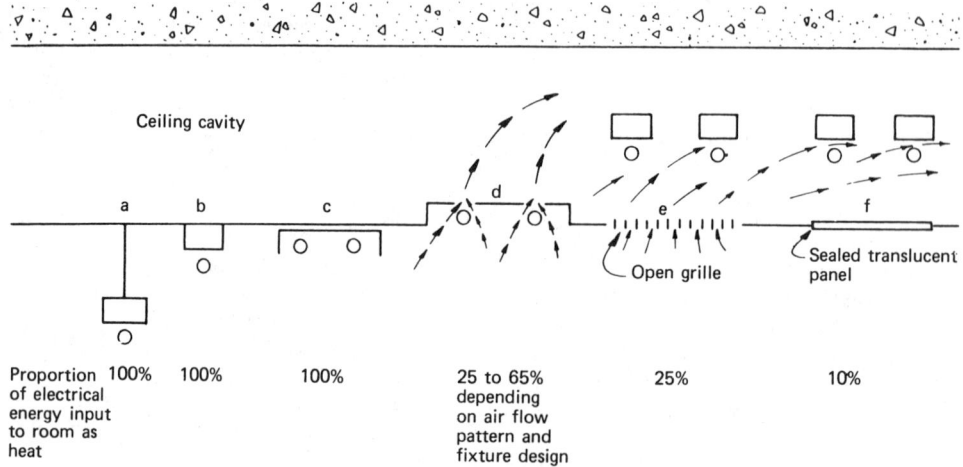

Fig. 3. **Proportions of energy input to lighting fixtures which appear in the room as heat.**

air induced to flow from the room, through the fixture, and into the ventilating system for heat recovery, if warranted, or for exhaust, if desired. Figure 3 shows a number of typical fixtures and the air-flow patterns which they employ to prevent most of the heat from entering the room.

Typically, the 75°F room air drawn from the room through these fixtures will be heated to about 85 to 95°F, depending on air-flow rate. Under many circumstances, the heat in this air can be used elsewhere in the building. If the heat thus absorbed in the air can be confined to a minimum volume of air giving a relatively high temperature, heat recovery may be relatively economical and simple. On the other hand, if the heat is diffused in a large volume of low-temperature air, heat recovery may not be justified.

Two basic air systems are currently used, with many variations. One general method supplies conditioned air to the office module and removes room air through the troffer and, thence, to the air handling and conditioning unit for recirculation or exhaust. Another method, the bleed-off system, wastes the return air if no heat is needed, or extracts the heat with a heat pump or other means if heat is needed for perimeter offices. Figures 4, 5, and 6 show basic ceiling variations with these systems.

Figure 4 shows an arrangement in which air admission and return air are separate from the luminaires.

The dual ducts—one hot, one cold—supply hot

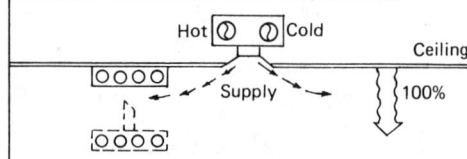

a. Flush mounted or suspended fixture (100% energy enters room)

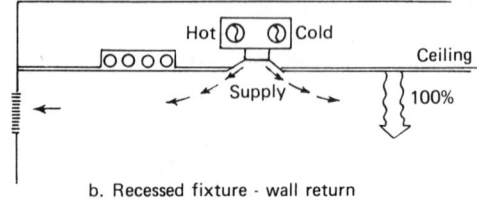

b. Recessed fixture - wall return (100% energy enters room)

c. Recessed fixture - plenum return (35% energy enters room)

IES Lighting Handbook, 4th ed.

Fig. 4. **Proportionate heat distribution from ceiling arrangement using air diffusers separate from lighting fixtures.**

and cold air to a mixing box above the ceiling. A room thermostat controls the proportions of hot and cold air supplied to provide the desired room temperature.

If the luminaire is suspended from the ceiling, 100 percent of the energy from the lights enters the room as heat.

If the luminaire is recessed in the ceiling, and the ceiling space serves as a return plenum, then only about 35 percent of the heat from the light enters the room. The exact amount of heat entering or removed will depend on air flow rate and fixture design.

If the return is through the wall, then about 45 percent of the heat will enter the room from a recessed luminaire.

In Fig. 5 mixed air at a controlled temperature is supplied from the hot and cold ducts to the recessed luminaire which serves as the point of air distribution. Other luminaires in the room can serve as return outlets or return can be through a wall outlet. Most of the heat from a supply luminaire will enter the room. Less will enter from the ventilated luminaire used as a return.

In Fig. 6, the light source is in the cavity above either a louvered or translucent luminous ceiling; all of the heat of the lights enters the room. But if the air enters from a diffuser and returns by way of the cavity, then only about 25 percent of the light heat reaches the room.

Manufacturers of luminaires will provide data on the air-handling capacity of each of their products. The proportion of input energy withdrawn by the return air before it has an oppor-

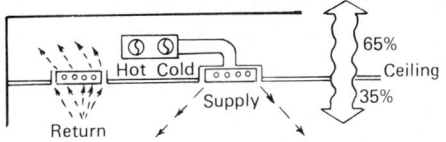

a. Separate fixtures for supply and return (35 % energy enters room)

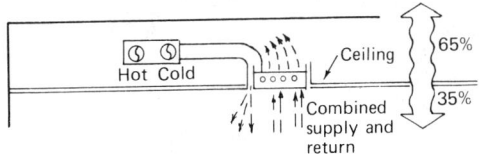

b. Combined fixture for both supply and return (35% energy enters room)

Fig. 5. Proportionate heat distribution from ceiling arrangement using luminaires for conditioned air supply distributions and return air removal in same ceiling system.

tunity to enter the room will also be provided by the manufacturers. Figure 7 shows the heat distribution for a typical fixture at various return-air flow rates.

Calculation of Air Flow Required in a Typical Office Building

Selection of operating conditions for a combined lighting-heating-cooling system is illustrated in the following example:

In a multistory office building, each floor is to

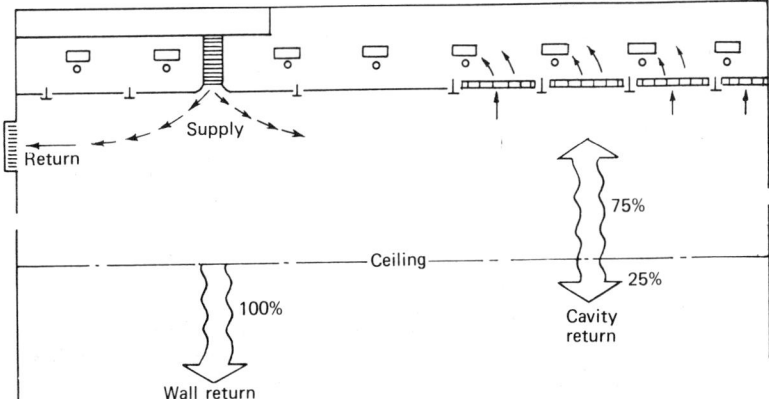

IES Lighting Handbook, 4th ed.

Fig. 6. Proportionate heat distribution through luminous or louvered ceiling.

be provided with an illumination level of 150 foot-candles. This will require a lighting input of 63.3 kw to each floor.

To remove as much of this heat from the fixtures as possible, so that it does not add to the heat load in the room when not needed, return air from the room must be drawn into the fixture. Thus, after the illumination level, 150 footcandles, and the fixture type and arrangement required to achieve that level satisfactorily, are decided, the supply air rates required can be determined.

The supply air will be through separate diffusers in the ceiling supplied by insulated ducts. Return air will be through the lighting troffers to the return air plenum above the suspended ceiling.

The following calculation method was provided by W. S. Fisher, General Electric:

Assume outdoor design temperatures as:
   Summer: 95°F
   Winter: −5°F
Indoor design temperature: 75°F
Area per floor: 11,500 sq ft
Perimeter of building: 500 ft
Electrical lighting load per floor: 63.3 kw
Other heat gains (human, solar, motors): 190,000 Btu per hr
Return air rate per troffer: To be calculated
Ratio of return air through troffers to by-pass air: 4:1
   (By-pass air is that air which flows upward into the return air plenum without passing through the troffers.)
Return air volume for four separate floating panel luminaires per floor: 2,530 cfm
   (This to be added to volume of air assumed for troffer return.)
Ventilation exhaust rate through toilet rooms: 1,500 cfm
   (Not included as troffer exhaust but included in total air supply.)

Figure 8 shows a typical plenum arrangement.

The required air volume to remove as much of the heat from the lights as practicable to prevent its passing from the plenum into the occupied space is calculated as follows:

Assume a return air rate through each light troffer as 40 cfm
Thus:

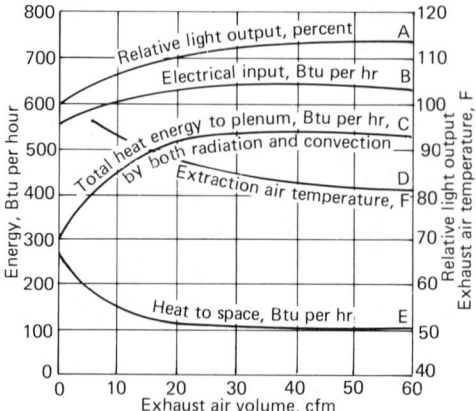

Day-Brite Lighting Div. of Emerson Electric Co.

Fig. 7. Manufacturer's data on heat distribution for a typical four-lamp troffer:

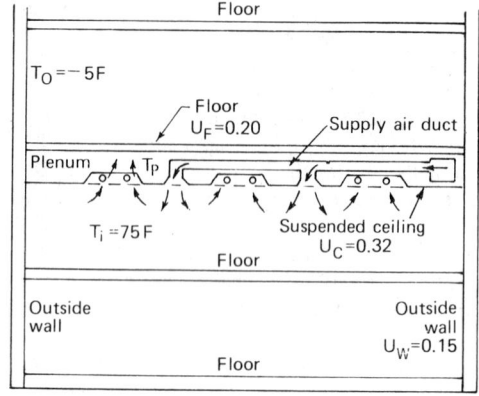

A_F = Area of floor
A_W = Area of wall
F_P = Fraction of lighting heat into plenum (a function of troffer flow rate, from manufacturer's data)
D_F = Duct heat loss factor (0.03)
H_{RS} = Sensible cooling load
Q_R = Quantity of return air (cfm)
T_I = Indoor temperature
T_P = Plenum temperature
T_O = Outdoor temperature
W_L = Energy input to lighting, watts
U_F, U_C, U_W = Heat transfer coefficients, Btu per sq ft-hr-F

Fig. 8. Typical plenum arrangement.

Troffer return: 40×295 troffers $= 11,800$ cfm

Bypass (4:1)	2,950
Floating panel return	2,530
Return air volume (Q_R)	17,280
Vent exhaust from toilet rooms	1,500
Supply air volume, Q_S (estimated)	18,780 cfm

The resulting heat balance equation is:

Heat gain by plenum = Heat losses from plenum

$$H_1 + H_2 = H_3 + H_4 + H_5 + H_6 + H_7$$

The objective of this calculation is to determine the rate of flow of return air required to enable the heat removed from the plenum to just equal the heat input to the plenum:

Heat Gains by Plenum:

Gain from lighting	$H_1 = 3.415 W_L F_P$
Gain from wall	$H_2 = U_W A_W (T_O - T_P)$

Heat Losses from Plenum:

Loss to floor above	$H_3 = U_F A_F (T_P - 75)$
Loss through suspended ceiling	$H_4 = U_C A_C (T_P - 75)$
Loss through wall	$H_5 = U_W A_W (T_P - T_O)$
Loss to return air	$H_6 = 1.08 Q_R (T_P - 75)$
Loss to air supply ducts	$H_7 = H_{RS} D_F$

$$\underset{H_1}{3.415 W_L F_P} + \underset{H_2}{U_W A_W (T_O - T_P)} = \underset{H_3}{U_F A_F (T_P - 75)} + \underset{H_4}{U_C A_C (T_P - 75)} + \underset{H_5}{0} + \underset{H_6}{1.08 Q_R (T_P - 75)} + \underset{H_7}{0}$$

$$3.415(63,300)(0.67) + 0.28(1500)(95 - T_P) = 0.20(11,500)(T_P - 75) + 0.32(9140)(T_P - 75) + 1.08(17,280)(T_P - 75)$$

$$145,000 + 40,000 - 420 T_P = (T_P - 75)(2300 + 2920 + 18,660)$$
$$185,000 - 420 T_P = (T_P - 75)(23,880)$$
$$185,000 - 420 T_P = 23,880 T_P - 1,790,000$$
$$24,300 T_P = 1,975,000$$
$$T_P = 81.2°F$$

The corresponding amount of heat to the return air is:

$$H_6 = 1.08 Q_R (T_P - 75) = 1.08(17,280)(6.2)$$

$$H_6 = 116,000 \text{ Btu per hr}$$

But the total heat from the light is:

$$3.415 W_L = 3.415(63,300) = 216,000 \text{ Btu per hr}$$

Hence, the heat into the occupied space from the lights is the total heat liberated less the amount removed by the return air, as follows:

$$216,000 - 116,000 = 100,000 \text{ Btu per hr}$$

Other heat gains (human, solar, motors) $= 190,000 \text{ Btu per hr}$

Sensible heat gain to occupied space $= 290,000 \text{ Btu per hr}$

The supply air quantity to remove this is:

$$Q_s = \frac{\text{Sensible Gain}}{1.08(T_O - 75)} = \frac{290,000}{1.08(95 - 75)} = 13,400 \text{ cfm}$$

However, this value for Q_S is considerably less than the 18,780 cfm estimated at the start of this calculation. Hence, a new and lower value of troffer return rate must be selected and the calculation repeated.

A second calculation was then made assuming only 25 cfm per troffer instead of the 40 cfm as first assumed. This turned out to give too high a supply air quantity (Q_S) compared to that assumed, hence a third and then a fourth calculation were made. This fourth calculation gave good agreement based on a troffer air rate of 28 cfm per troffer. The corresponding supply air rate, Q_S, of 14,300 cfm is the amount required by this particular floor of the building. The sum of the air supplies needed for each of the floors, determined similarly, establishes the size of the air-handling equipment needed for summer design conditions for the building. For large, complex building requirements, the above calculation is best carried through with the aid of a computer.

Under winter design conditions, it will often be found that the offices at the exterior walls will require some additional heat even at the minimum air-ventilation rate.

Controls

Control systems used can be composed of a wide variety of available components and can vary considerably in complexity, degree of control, and in flexibility to meet the desires of individual occupants. Through control of air-mixing dampers the individual room thermostats can control the proportionate air flow to and from each room and, thus, control the heat extraction rate from each room. The disposition of recovered heat, whether to re-use, to store, or to waste, can be controlled from appropriate sensors and controls at the mechanical equipment room.

Evaluation of the degree and the frequency of need for heat by the exterior offices under various weather conditions and of the cost of various methods is most readily made by means of a computer study conducted by or under the direction of the consulting engineer.

Integrated Lighting-Heating-Cooling Systems Using Water and Air for Control of Heat Distribution

As with the use of air alone, there are also many possible variations of systems using air and water. The advantages of using water are:

1. Heat transfer to water tubes or channels near or attached to the metallic luminaire surfaces is very effective.

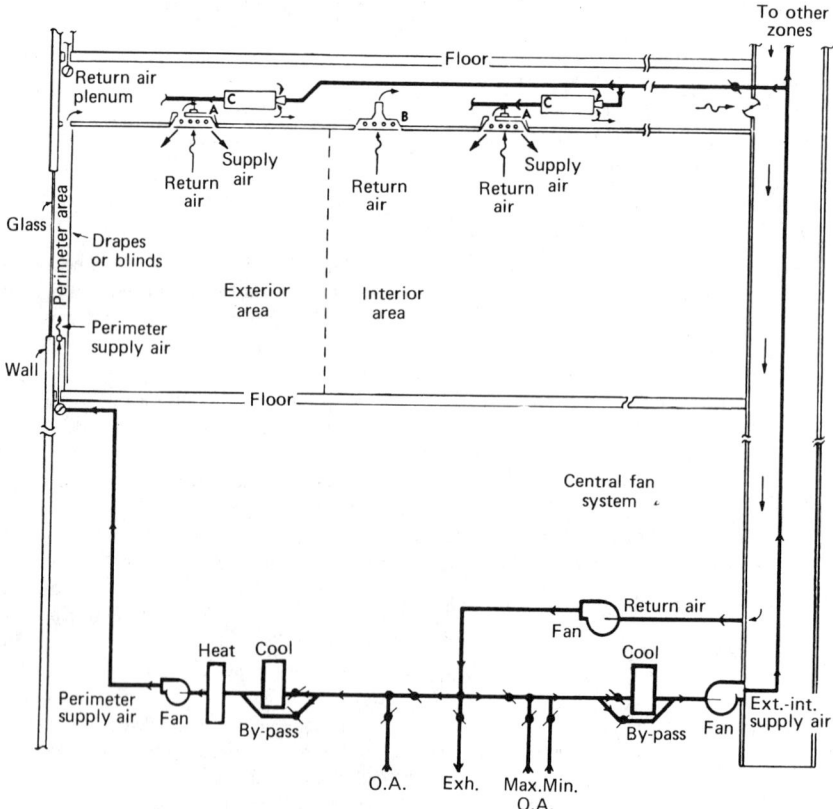

Courtesy of Barber-Colmán Co.

**Fig. 9. Diagram of typical air system using induction-type mixing units shown in Fig. 10.
A. Combination heat extractor and supply air light fixtures. B. Heat extractor light fixtures. C. Jetronic mixing units.**

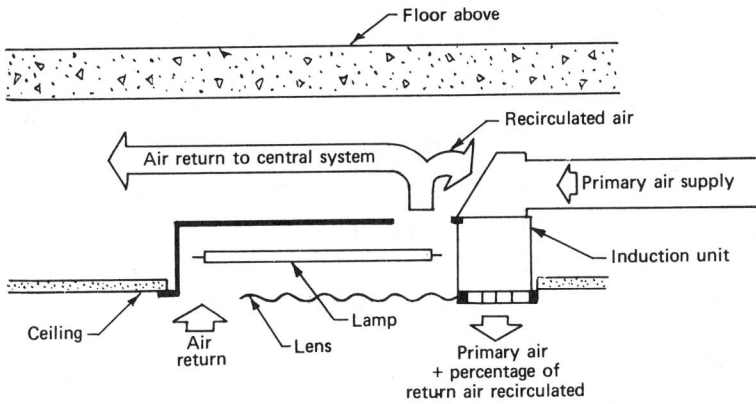

Heating, Piping and Air Conditioning

Fig. 10. In variation of air-cooled luminaire application, warm return air from fixture is induced into primary cold air supply in response to temperature changes in interior spaces. Balance is returned to fan station coils.

2. The heat absorbed in the water is readily available to:
   a. Be utilized at some other location in the building.
   b. Serve as a heat source for a heat pump.
   c. Simply be dissipated in a cooling tower.
3. Space requirements for heat-removal systems are less.
4. The plenum space is cooler, hence less heat is dissipated to the floor above.

With the water system there is still the need for air motion in the office space, hence some minimal ventilation air *must* be supplied.

Examples of Integrated Systems Using Air Alone

One system, using air alone, which has been successfully applied to many office buildings, is shown diagrammatically in Fig. 9. Essential to its operation is the induction-type mixing box shown in Fig. 10. The single duct system supplies cold air, either filtered outdoor air or refrigerated air, to the air-flow induction unit. This flow-inducing unit, which is also the luminaire, induces the flow of warm air into the cool air by the jet action of the accelerated jet of cool air passing through the unit. The resulting mixture of tempered air is then distributed through the ceiling to the room.

Figure 11 illustrates the action, viewed from the room, of the air leaving and entering the room

under the action of the air-induction unit. Where full cooling is needed in the space, all of the warm room air is returned to the main ventilating fan, carrying with it most of the heat from the lights. When needed, this warm air can be circulated to the perimeter offices to supply the heat lost through the walls and windows.

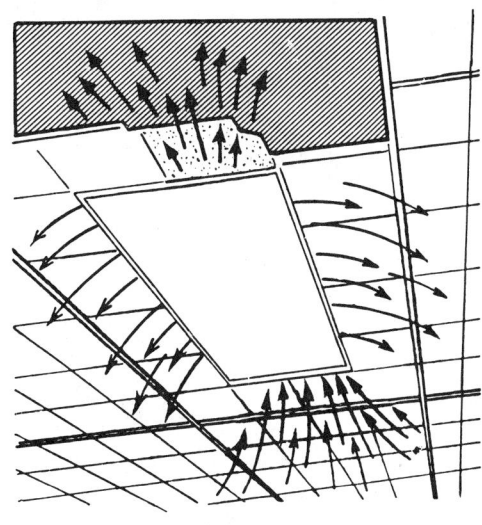

Air Conditioning, Heating and Ventilating

Fig. 11. Combination supply and return air and lighting luminaire.

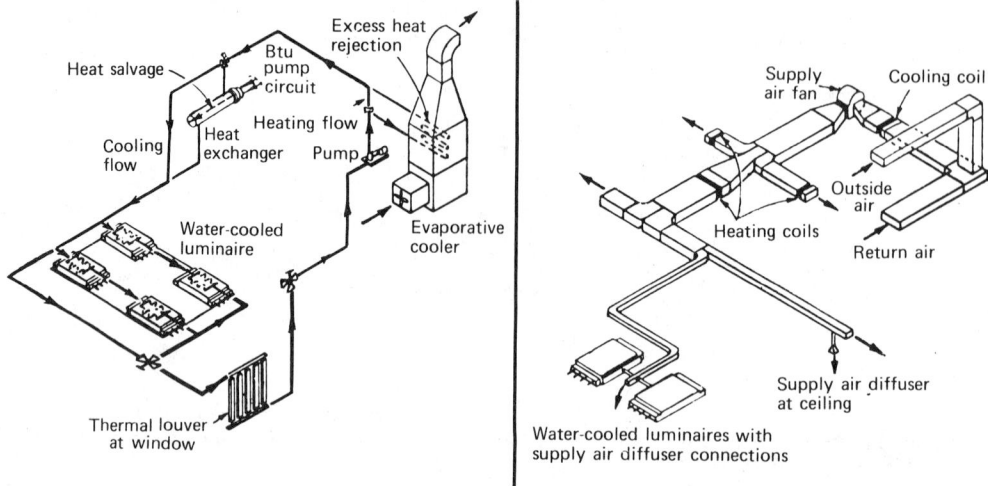

Electric Construction and Maintenance

Fig. 12. Simplified diagram of integrated system operation. Recirculated water circuit through fixtures and thermal louvers is at left. Duct system providing heated or cooled air to areas is shown at right.

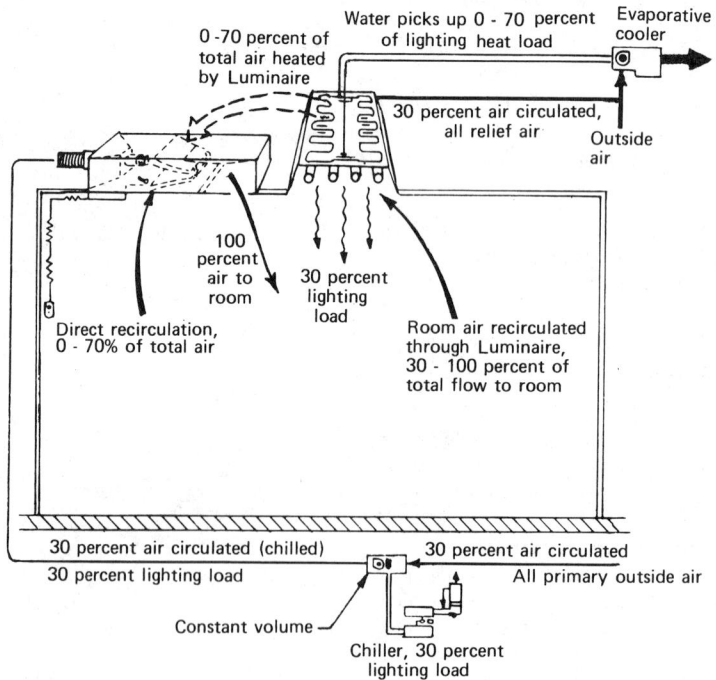

Gershon Meckler

Fig. 13. Patterns of heat flow, air flow, and water flow for system using induction mixing box to modulate air recirculation.

Air-Water Systems

The use of non-refrigerated water to cool luminaires designed for that method enables reduction of conditioned air flow to that quantity needed for removal of the heat released by occupants, office machines, solar load, and the reduced heat from the lamps and fixtures.

Figure 12 shows simplified diagrams of a typical air-water system. On cold days the warmed water can be raised in temperature by an electric resistance heater, or in a heat pump from which the hot water can be circulated to the perimeter for heating. On warm days, the heat in the water from the luminaires and thermal louvers can be dissipated from the evaporative cooler.

The conditioned ventilating air can be introduced to the offices through separate diffusers or through the water-cooled luminaires. This air can also be used to induce mixing with recirculated room air, as shown in Fig. 13.

Figure 14 shows typical air and water connections for a general office area.

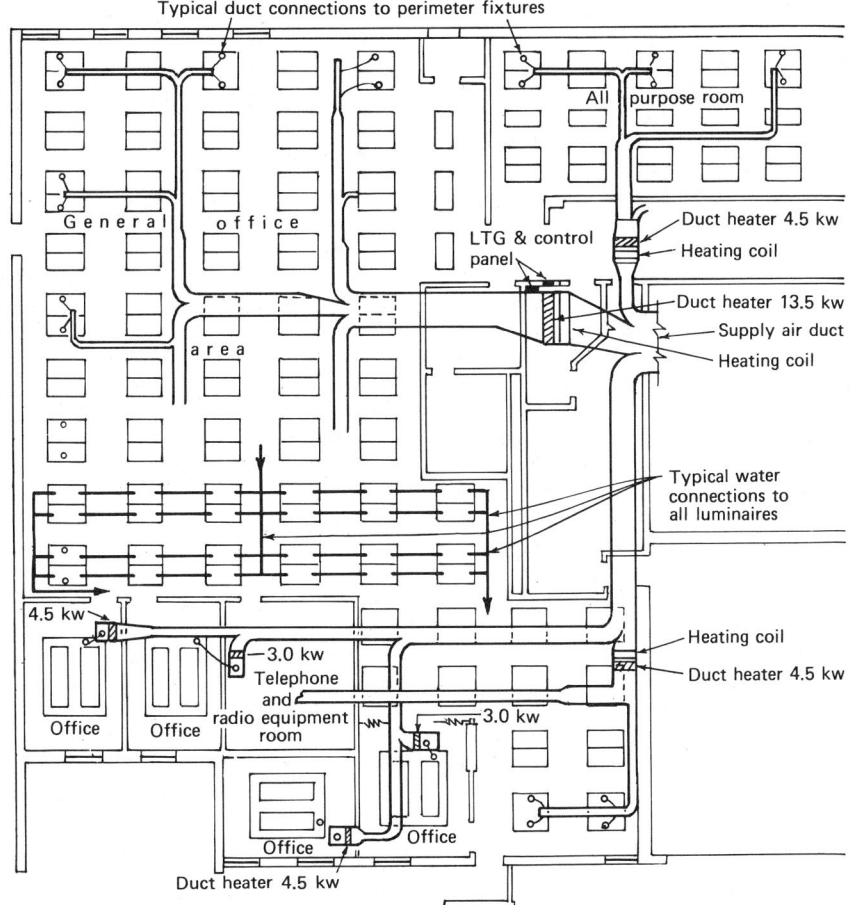

Electric Construction and Maintenance

Fig. 14. Plan of office area integrated heat-by-light installation with portions of supply air duct system showing typical connections to fluorescent fixtures. Note duct coils using recirculated water from fixture; also electric duct heaters for night heating of building. Special fixtures in private offices are dual water-cooled.

SOURCES OF INTERNAL HEAT

Some of the heat energy released by lights, appliances, and production machinery, or otherwise made available in buildings, can often be recovered or redistributed for reuse at points in the building which need additional heat. Thus, the heat released from lighting fixtures, as described previously, is frequently used to provide the energy for heating the perimeter areas of the building. Also, the heat in the exhaust air from ventilating systems can be recovered by various methods for return to the building at points where heat is needed.

The principal sources of internal heat in modern buildings are lights, office and production machines, and occupants. In addition, windows exposed to sunlight act as heat traps because the solar radiation passes inward through the glass much more readily than that heat can be re-radiated outward. If this solar heat, and/or internally generated heat, results in an excess of heat in the affected rooms, cooling must be provided to maintain comfort conditions and the heat which is thus removed may then be stored or utilized elsewhere in the building.

The tables in Section 2 include typical heat release rates from various sources of internal heat gain.

The internal heat gains within buildings are usually absorbed by circulating air, circulating water or, in some cases of small buildings, by direct-expansion refrigeration.

After the internal heat has been absorbed in circulating air, water, or refrigerant, it must be utilized, stored, or discharged. A wide variety of systems is possible for handling these alternatives. In some cases, the heat may be recovered by a simple regenerative air heater or stored in enclosed reservoirs of water. However, the heat collected from the internal heat sources by means of circulating air, water, or refrigerant cannot be directly utilized for comfort heating because its temperature usually is too low. For example, the air leaving luminaires is usually heated to the range of 85 to 95°F. In some winter climates this temperature range is too low for use in comfort heating directly. However, the temperature of the air or water can be increased by means of a refrigeration system acting on the heat pump principle.

Thus, wherever a building requires refrigeration for summer cooling, and hence must have refrigeration equipment installed, by appropriate design the central refrigeration system can be used during the winter to raise the temperature of the water or air carrying the recovered heat. Furthermore, where a building requires simultaneous heating and cooling because of intense solar heating on the south side of the building on clear winter days, the central refrigeration system can simultaneously serve both the heating and cooling needs of the building without requiring any additional energy beyond that required to drive the refrigeration equipment.

Example of Heat Recovery Using Air or Water to Absorb the Internal Heat Gains

Figure 15 is a schematic diagram of the system installed in an office building in which the heat from the internal heat gains is first absorbed in air, then is transferred to water by means of a duct coil for return to the refrigeration condenser to be heated to a useful temperature level. The heated water is then circulated to heating coils serving the perimeter areas of the building.

If some of the internal heat in the structure is absorbed by water, such as in water-cooled lighting fixtures or in water-cooled machinery, the warmed water can be pumped directly to the condenser for further heating. From the condenser the heated water can be distributed to points in the perimeter zone needing heat.

Under intermediate or summer weather conditions where the building requires heating only part of the 24-hour day, the internal heat collected can be circulated in water to a cooling tower or pond to be dissipated outdoors or to a water storage tank to be stored for later use for nighttime or week-end heating of the building.

Figure 16 is a diagram of the heat-recovery system for a two-story, 95,000-square-foot office-laboratory building. The heat from the central core is collected by return air passing through the ceiling luminaires. Water coils in the return duct system extract this heat and the warmed water then conveys the recovered heat to either of two refrigeration condensers where it is heated to as high as 130°F. The heated water is then pumped to the peripheral zones for heating, or to the

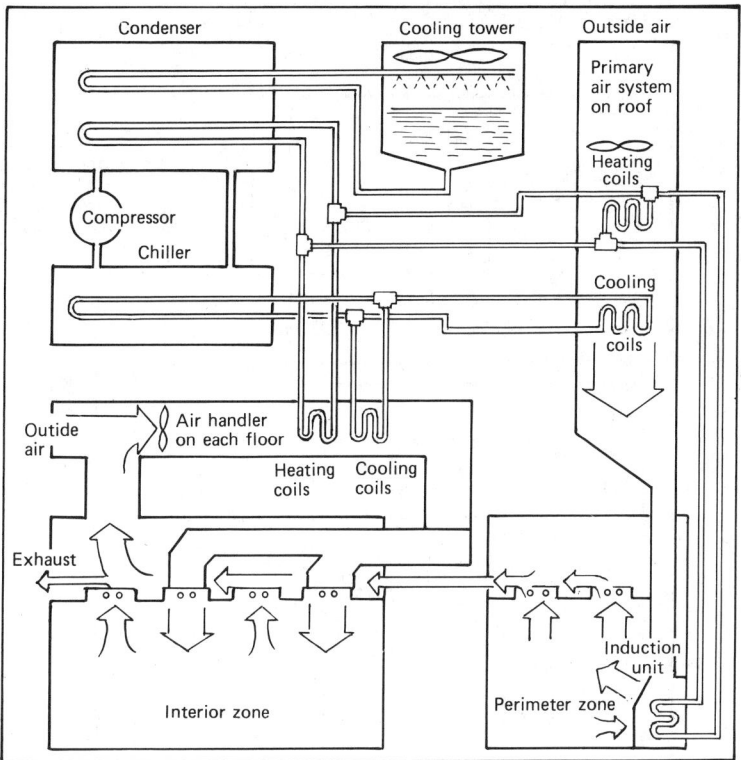

Actual Specifying Engineer

Fig. 15. Schematic diagram of heat recovery system using split condenser as installed in an office building.

150,000-gallon storage tank to be held available for later use.

The operating temperature range for the storage tank is considered to be from a high of 125°F down to 45°F. This 80-degree range represents a total heat storage capacity of 100 million Btu. The designer arranged to utilize the first 25 million Btu of this stored heat directly, by circulating the stored water to the peripheral zone until the storage-tank water temperature is reduced to 105°F. Below that temperature the water is first passed through the refrigeration condenser to be heated to at least 105°F before being used for comfort heating.

This use of the refrigeration system to lift the temperature level of the water from the storage tank to a point suitable for comfort heating is sometimes referred to as operating the refrigeration system with a "false load." In effect, in the system described, this "false loading" technique not only makes available the full 100 million Btu held in the storage tank, but adds to it the heat of compression, making additional heat available for off-peak night-time and week-end heating when most of the unoccupied building's lighting system may be off.

Exhaust Air Heat Recovery Systems

If the exhaust air from a building is discharged from only a few concentrated points, the heat ordinarily lost in such a discharge can largely be recovered by the use of an air-to-air heater to heat the incoming ventilating air. Or in the summer, the same exchanger can be used to partially cool the incoming ventilating air. The heat exchanger can be a static or rotating air-to-air type, or it can consist of two liquid-air heat

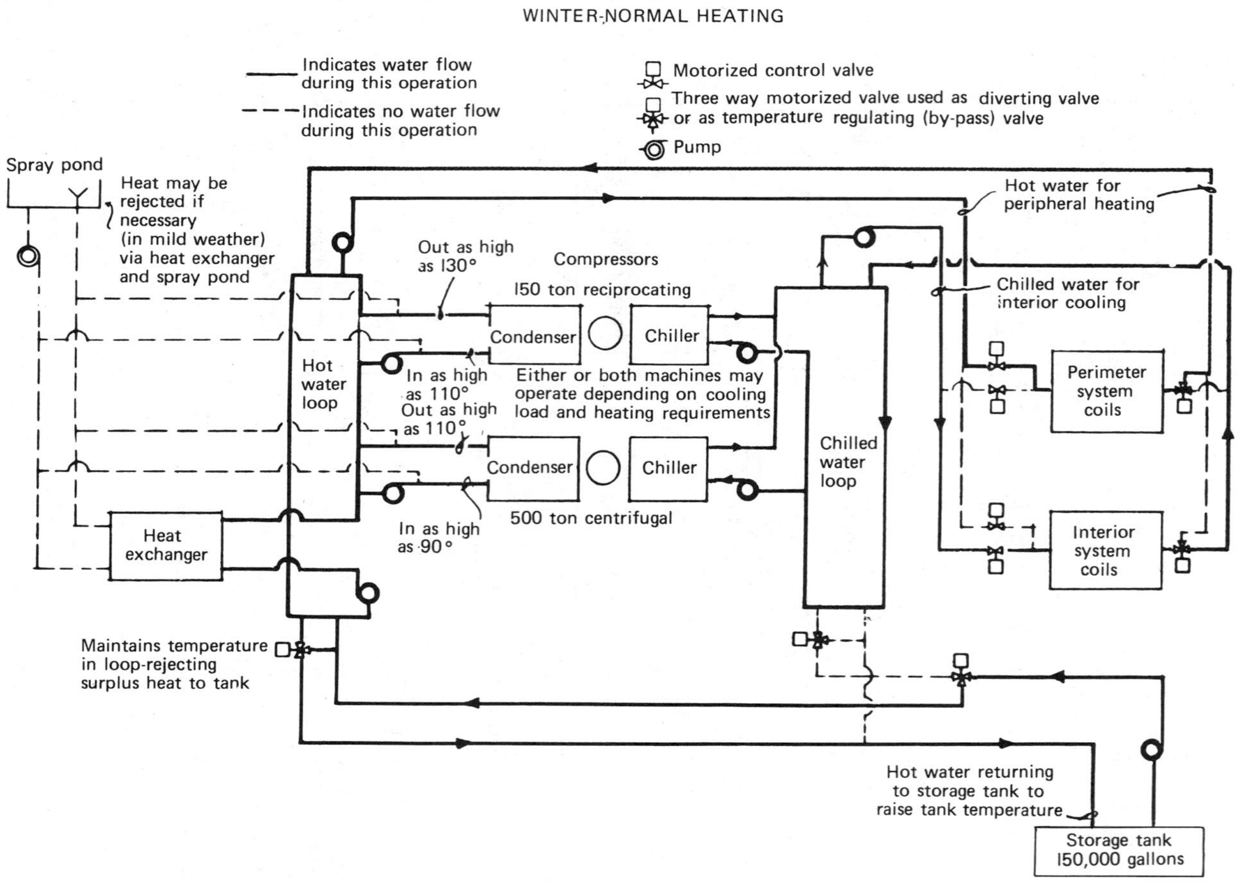

WINTER-NORMAL HEATING

☐ Motorized control valve

⋈ Three way motorized valve used as diverting valve or as temperature regulating (by-pass) valve

◯ Pump

—— Indicates water flow during this operation

---- Indicates no water flow during this operation

Spray pond

Heat may be rejected if necessary (in mild weather) via heat exchanger and spray pond

Hot water for peripheral heating

Chilled water for interior cooling

Perimeter system coils

Interior system coils

Chilled water loop

Hot water returning to storage tank to raise tank temperature

Storage tank 150,000 gallons

Compressors

150 ton reciprocating

Chiller

Condenser

Either or both machines may operate depending on cooling load and heating requirements

Chiller

Condenser

500 ton centrifugal

Out as high as 130°

In as high as 110°
Out as high as 110°

In as high as 90°

Hot water loop

Heat exchanger

Maintains temperature in loop-rejecting surplus heat to tank

Actual Specifying Engineer

Fig. 16. Piping system for heat recovery and storage for a 95,000-sq-ft, two-story office-laboratory building.

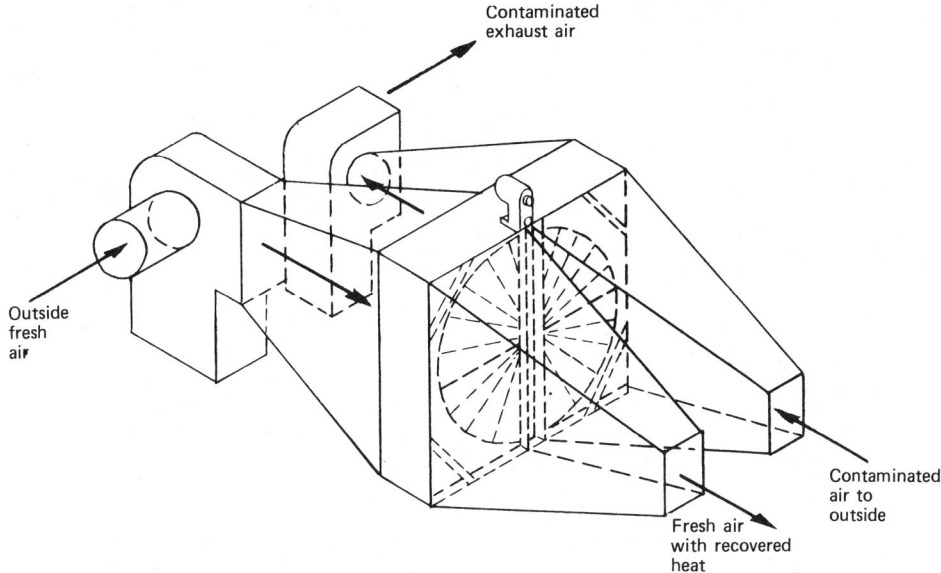

Contaminated
exhaust air

Outside
fresh
air

Contaminated
air to
outside

Fresh air
with recovered
heat

Heat Recovery Inc.

Fig. 17. Diagram of regenerative heat recovery system.

exchangers, one in each air stream with liquid circulating between them.

Figure 17 indicates the principle of the rotary air-to-air heat exchanger. The slowly rotating wheel is packed with corrosion-resistant metallic fabric or fibers through which the air is forced by ventilating fans. As the warm exhaust air leaves the building in the winter through one-half of the packing in the wheel, it gives up most of its heat to the metal fibers. Then, as these fibers are carried by the rotation of the wheel into the duct through which the incoming outdoor air is flowing, the fibers release their heat to raise the temperature of that incoming air. Heat-reclaiming efficiencies up to 80 percent can be attained by these devices. By means of closely fitting flexible seals separating the two air passages, the air leakage between them can be held to less than 2 percent. Experience has shown that such equipment can be safely applied to heat recovery from hospital exhaust air without risk of significant air contamination.

Figure 18 shows schematically the application of a rotary regenerative exchanger. In the winter, outdoor air is taken in at 10°F and heated by the rotating wheel to 62°F. In the summer, 95°F intake air is cooled by the same wheel to 79°F.

Further cooling and the necessary dehumidification then occurs in the conventional duct coil.

The following tabulation shows the estimated design load saving that can be achieved through the application of the recovery system shown in Fig. 18.

	Temp. Rise	Heat Required, Btu per Hr per 1000 cfm
Ventilation Air Heating, Winter		
Without regenerator	65°F	70,200
With regenerator	13°F	14,000
Saving in design heating load:		56,200 Btu per hr per 1000 cfm
Ventilation Air Cooling, Summer		
Without regenerator	20°F	21,600
With regenerator	4°F	4,320
Saving in design cooling load:		17,280 (equivalent of 1.44 tons saving per 1000 cfm of outdoor ventilating air)

The above calculation is based on the saving in

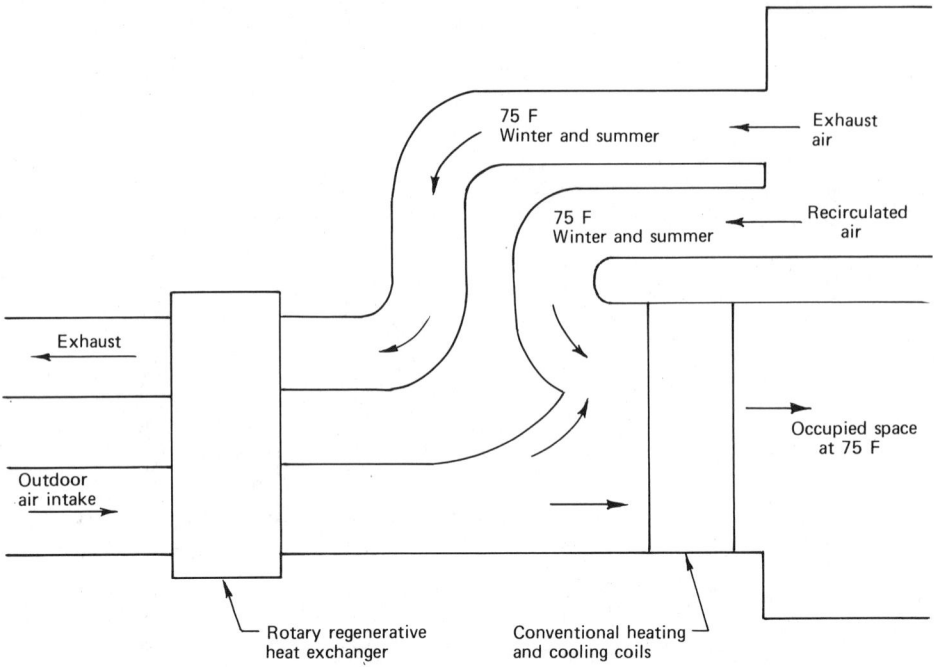

Fig. 18. Diagram of heat recovery system using rotary heat exchanger.

sensible heat only. The use of a regenerator having a chemical moisture adsorbent coating on its surfaces will increase the potential saving by transferring moisture between the two air streams, resulting in retaining moisture within the system in the winter and reducing the incoming air moisture in the summer.

An important requirement in the economic application of rotary regenerators to heat recovery for occupied buildings is that the intake and exhaust ports must be near to each other so that extensive special duct-work is not needed to bring the two air streams to the wheel. However, a closed system which avoids this limitation is known as the "runaround" system. In this arrangement the heat is extracted from the one air stream by means of an air-to-liquid heat exchanger. Then, by means of a pump, the heat exchange liquid is passed to a similar coil located in the other airstream. By the use of heavy thermal insulation on the connecting piping and pump, the loss of recovered heat is minimal. To permit operation during freezing conditions, the runaround system is filled with antifreeze solution.

Although the runaround system permits a more flexible air-duct arrangement, the introduction of two air-to-liquid heat transfer units in the heat-recovery path may require a large heat-transfer surface to provide sufficient heating or cooling of the ventilating air. Thus, in applying the runaround principle for heat recovery, the data of the coil manufacturer should be considered regarding air-to-liquid heat transfer and pressure-loss characteristics to assure that ample coil and air-blower capacities are selected.

Heat from Service Refrigeration

Supermarkets and similar enterprises using large amounts of refrigeration supplied from central condensing units are uniquely fitted for space heating by means of the heat discharged from the condensers. This heat can be absorbed in circulating water for distribution to unit heaters, convectors, duct coils, or baseboard radiation for comfort heating.

Figure 19 shows a diagram for a heat-recovery installation in a supermarket. In some cases the recovery is through a single central condensing

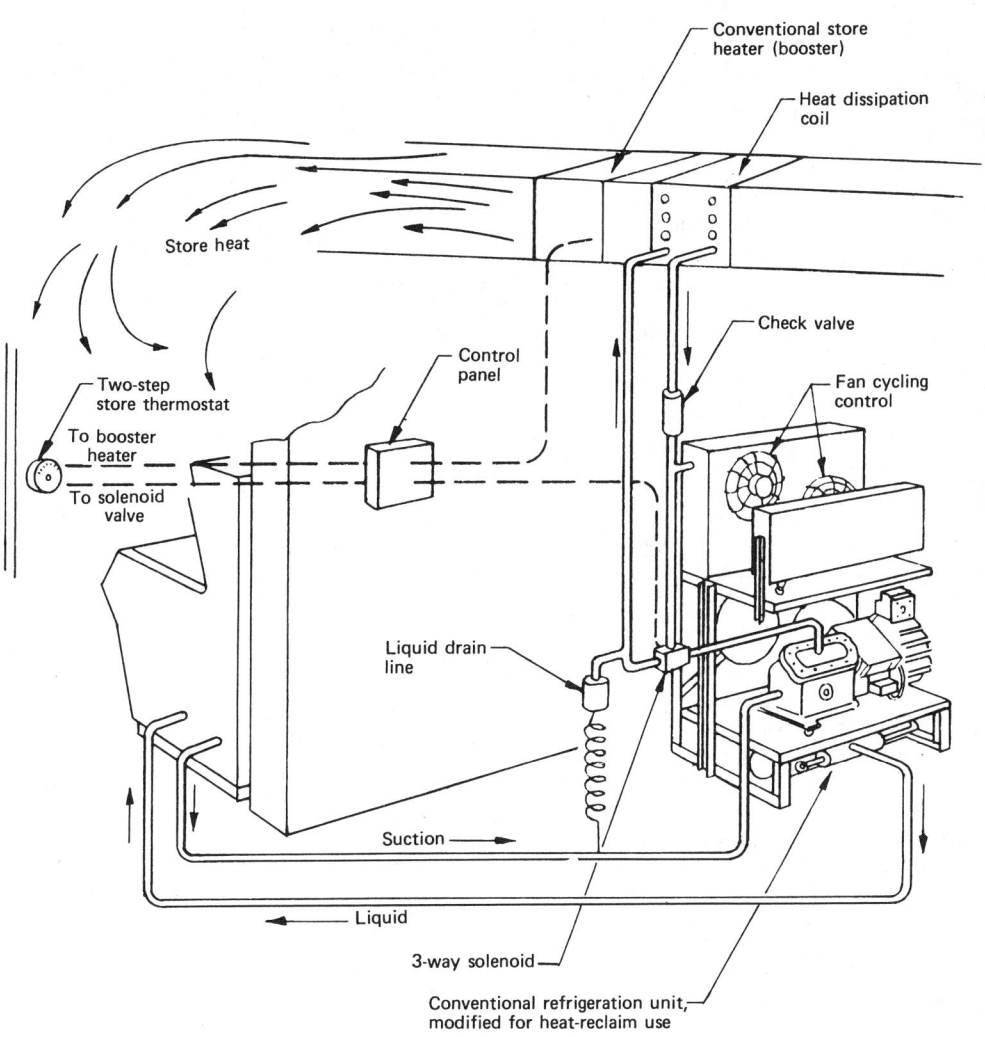

Hussman Manufacturing Co.

Fig. 19. Diagram of refrigeration heat recovery system for supermarket.

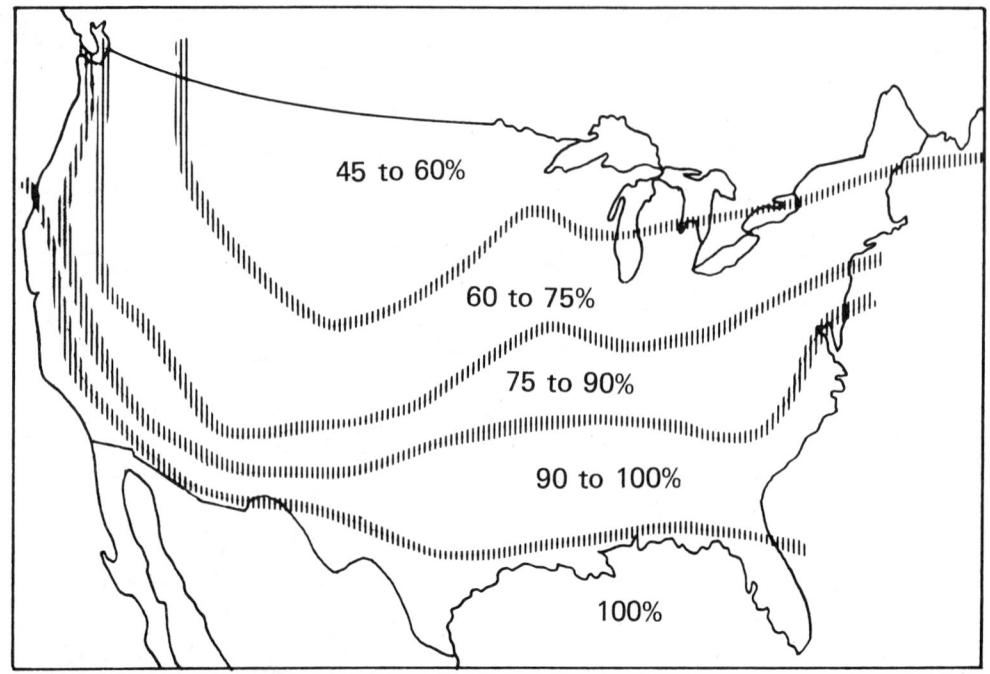

Hussman Manufacturing Co.

Fig. 20. Approximate savings possible from heat recovery system as percent of total heating requirements in various zones.

unit. In other cases where the refrigeration load is handled by a series of condensing units, the condenser of each unit is connected to the heat-recovery water loop.

In some cases the recovered heat is sufficient to *supply all the comfort heating needs;* in others, supplementary heat must be provided. Figure 20 shows the approximate percentage of the heating load which can be supplied by recovered heat for supermarkets in various zones.

HEAT PUMPS *

A heat pump consists of the same basic components as a refrigerating machine:

1. An element which extracts heat from or rejects heat to a medium outside a conditioned space.
2. An indoor element which adds heat to or extracts heat from a conditioned space.
3. A compressor for pumping the refrigerant containing sensible and/or latent heat in the desired direction so that the heat may be rejected for cooling or used for heating.

4. Controls and other items as required for satisfactory operation.

Reverse-Cycle Principle

For many years engineers have known that the functions of the evaporator and condenser could be interchanged by reversing the direction of flow of the refrigerant from the compressor. When customers' needs and economics justified the addition of components in a refrigeration system so that it performed as a heating or a cooling device, the public began to hear about reverse-cycle units and heat pumps. Both terms have been used to

---

* Excerpted, with permission, from the *Electrical Heating and Cooling Handbook,* Edison Electric Institute.

identify the same system. Today, most people refer to a heating-and-cooling refrigeration system as a heat pump.

Many heat pump systems are assembled with a refrigerant changeover valve or valves, depending upon the capacity and operation of the installation. The changeover valve (or valves) is installed in the refrigerant circuit at or near the compressor. When the structure calls for cooling, the valve directs the flow of the refrigerant so that the indoor coil functions as the evaporator, thereby absorbing heat from the interior. When heating is required, the valve changes the direction of flow of the refrigerant so the indoor coil acts as a condenser, thereby releasing heat to the interior. The internal positioning of the changeover valve (or valves) is automatically controlled by the indoor thermostat or temperature-control system. Figure 1 is a schematic sketch of refrigerant flow in a heat pump system for heating and cooling. In addition to a refrigerant-changeover heat-pump system, there are air-changeover and water-changeover systems. In these units the refrigerant flow is not changed and heating or cooling is obtained by controlling the flow of air or water. In air-changeover systems, ductwork is installed to direct air across the condenser for heating or across the evaporator for cooling before passing to the air-conditioned

space. This system requires rather complicated ductwork and dampers.

In the water-changeover system the direction of flow of the water leaving the condenser and chiller (evaporator section) is controlled by valves to provide heating or cooling in the conditioned area as required.

Water-changeover and air-changeover systems can be installed to provide heating and cooling for different areas at the same time. This can also be accomplished with extra heat exchangers in the refrigerant-changeover system.

Coefficient of Performance and Performance Factor

When operating as a cooling unit, electric heat pumps remove more heat energy from the structure than the electrical energy required to run the equipment. Likewise, when operating as a heating unit, electric heat pumps usually provide more heat to the structure than they use or consume in performing the function.

The ratio of the useful heat delivered (in Btuh or kw) to the equivalent heat (in Btuh or kw) used to operate the entire system at a fixed operating condition is known as the Coefficient of Performance (C.O.P.). The C.O.P. may also be specified with respect to the compressor only or

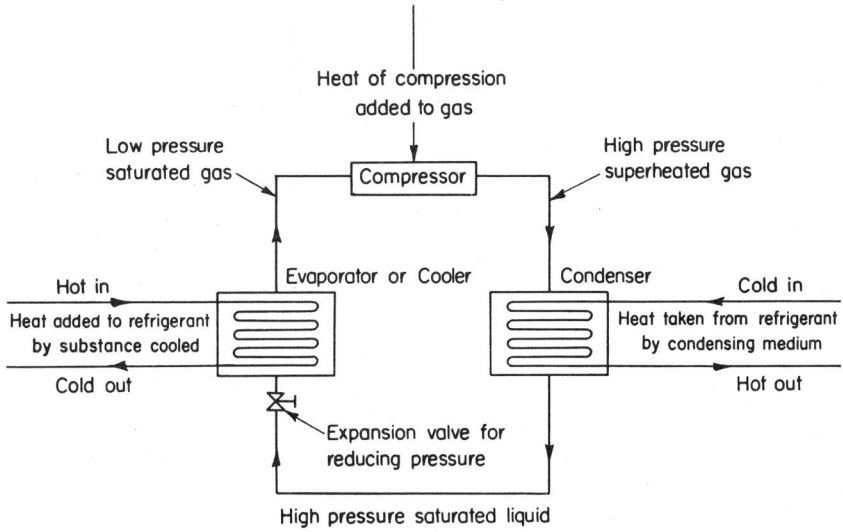

Fig. 1. Schematic sketch of refrigerant flow in a heat pump system for heating and cooling.

any given portion of the system. The following examples are typical for air-to-air pumps.

Cooling Cycle

A 5-ton air-to-air heat pump, delivering 60,000 Btuh, with a total input of 8 kw, has a C.O.P. of 2.2. (A.R.I. rating with outdoor air at 95°F dry bulb and 78°F wet bulb and indoor return air at 80°F.)

$$\text{Cooling C.O.P.} = \frac{60{,}000}{8 \times 3413} = 2.20$$

Heating Cycle

Operating in an outdoor temperature of 45°F, the same heat pump has a heating capacity of 55,800 Btuh, with a total of 6.28 kw. At this rating the indoor-return-air temperature is 70°F at the unit. The heating C.O.P. is 2.6 at these operating conditions.

$$\text{Heating C.O.P.} = \frac{55{,}800}{6.28 \times 3413} = 2.6$$

When C.O.P. figures are used to compare various types of heat pump systems and/or makes of equipment, care must be exercised to insure proper evaluations. The C.O.P.'s should be based on the same operating conditions, i.e., outdoor-indoor temperatures, complete systems or equal portions thereof, etc.

The C.O.P. of a heat pump varies with the temperature difference between the heat source and heat sink; the larger this difference, the lower the C.O.P. will be.

Figure 2 shows how the C.O.P. of an air-to-air heat pump varies with outdoor temperatures during the heating season.

Because the C.O.P. varies with operating conditions, it cannot be used to compare systems operating over a period of time, such as a day, week, month, or season.

The ratio of the total heating or cooling energy (heat) delivered during a given period to the total equivalent energy used by the system in the same period is indicative of the overall operating conditions. This ratio is known as the Performance Factor (P.F.). It is expressed as a function of time and operating conditions, such as heating season performance factor or cooling season performance factor. Its usage is associated with the heating characteristics of heat pump systems more

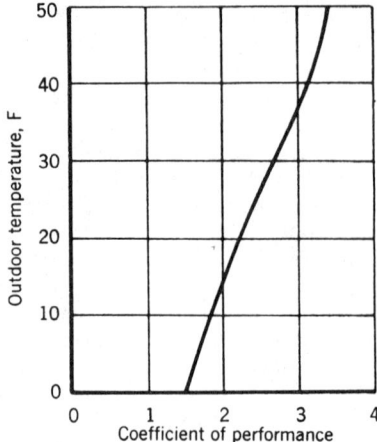

Fig. 2. Coefficient of performance of air-to-air heat pumps for heating. (This varies somewhat with size and make of unit.)

often than with cooling operations of mechanical refrigeration systems. The following example shows how a typical heating season performance factor was determined.

Heating Season Performance Factor

During the 1962-63 heating season, an air-to-air pump system consumed 14,000 kwhr while heating a building requiring 115,000,000 Btu. The 14,000 kwhr includes power used to drive all components and supplementary resistance heaters as needed. The heat pump system operated with a heating season performance factor of 2.4.

$$\text{P.F.} = \frac{115{,}000{,}000}{(14{,}000)(3413)} = 2.4$$

Performance factor values vary with seasonal climatic conditions, heat pump systems, application, and sizing of the equipment, as well as with the specific characteristics of the structures and their internal loads.

Because of a higher temperature heat source, residential heat pump installations in the southern regions operate with higher seasonal performance factors than like installations in the northern regions.

Because of more efficient motors and compressors, commercial and industrial installations may operate with higher seasonal performance factors than residential installations in the same area.

When performance factors are used to compare

heat pump systems, care must be exercised to insure that all components or the same proportion of components are included.

Types of Heat Pumps

Heat pump systems are designated by the heat source or sink and the type of distribution system in the building, for example, air-to-air, water-to-air, air-to-water, water-to-water, etc., systems. The medium designated first is the external heat source and/or sink. The medium identified second is that which comes in contact with the indoor refrigerant coil. Some larger heat-pump systems are installed with more than one heat source and with both water- and air-distribution systems.

An air-to-air system uses outdoor air as the heat source and/or sink. Heated or cooled air is supplied through a duct system or air-distribution system within the building.

An air-to-water system uses air as the heat source and/or sink. Heating and cooling is provided within the structure by a water-distribution system and appropriate convectors to transfer the heat between the water and the air.

From the standpoint of temperature, ground water is an excellent heat source and heat sink, but it is not readily available in all areas and is in limited supply in many regions during some periods. Air is used predominantly as the heat source and/or sink in residential and small commercial installations. It is becoming popular for larger installations.

Heat pump systems may also use the earth as a heat source and/or sink. This has not proven to be too practical, and few, if any, ground-source units are being considered today.

Other than air and water, a few sources have been utilized for the heat source for a heat pump. These are waste heat from various industrial processes, ventilating-air exhausts, solar energy, and heat extracted from refrigerated spaces. These sources are generally used in addition to rather than as replacements of the basic heat source.

Air-to-Air Heat Pumps

Atmospheric air is the most universally available heat source and heat sink; hence, it is the one most commonly utilized. Air-to-air heat pumps use outdoor air as a heat sink or heat source with the heat being delivered to or re-moved from the indoor air through the use of a direct refrigerant.

The air-to-air-type heat pump is generally the smaller-tonnage unit in integral packaged form for residential and commercial application up to 30 tons. However, built-up systems of any size may be constructed. The models that are presently being marketed of the air-to-air type are either self-contained in one unit or split into two sections. Both types offer definite advantages, depending mainly on the type of application.

An advantage with using air as the heat source is that as temperatures moderate, efficiency improves, thus allowing the heat pump to operate in its more efficient range for most of heating use.

The same economic limitations on size of low-velocity duct systems that apply to cooling systems usually limit air-to-air heat pumps to approximately 50 tons per low-velocity duct system.

Generally, a heat pump is selected for the design cooling load. If the design heating load exceeds the heating capacity of the selected heat pump, the necessary supplementary heat is generally supplied by auxiliary electric resistance heaters, which may be required to operate during the infrequent periods of maximum requirements.

The annual kilowatt-hour consumption of the supplementary heaters depends upon the application, the capacity of the heat pump in relation to the heat loss of the structure, the climatic conditions, and the control system. In residental applications, supplementary heaters seldom account for more than 10 to 15 percent of the total heating season kilowatt-hours used by the system. In southern areas, use of electricity by supplemental heaters may be negligible. In commercial and industrial installations, lighting loads, body loads, etc., reduce the need for and/or use of the supplementary heaters; therefore, they may account for a very small percentage of the total heating season kilowatt-hours.

Various methods have been conceived to get the system through the periods of peak heat demand; however, none of these have yet approached the resistance heating method for practicability and minimum investment.

Another item that requires consideration on this type of heat pump is the outdoor-air coil defrosting system. Continuous operation of an air-source heat pump with outdoor temperature below 45°F may cause frost accumulation on the outdoor-air coils. This frost must be periodically removed;

otherwise, the coil would become completely blocked with frost, permitting no air-flow through it, and the heat pump system would then be ineffective.

Depending upon a number of variables, the frequency of defrost is most severe at outside temperatures between approximately 28 and 40°F and when high relative humidities are experienced. At lower outside-air temperatures, the ability of the air to hold moisture is limited and fewer defrosts are required. At higher temperatures, the refrigerant temperature in the evaporator will be above the point where frost will form on the coil, and water condensing on the coil surface is easily drained from the unit.

Most air-to-air heat pumps are equipped with automatic defrosting controls. These controls may be timers, devices which measure the difference in temperature between the outdoor air and refrigerant, or devices which react to differences in air pressure-drop across the outdoor coil. Since the defrosting operation accounts for relatively few kilowatt-hours, how it is accomplished is unimportant as long as the device is reliable in performing the operation.

In most air-source heat-pump systems, the outdoor coil is defrosted by reversing the refrigerant cycle so that the system uses the building as a source of heat for defrosting. Supplementary electric heaters located in the air stream may be energized to provide a source of heat to prevent chilling of inside air below comfort levels.

Under special circumstances and in some older systems, heat for defrosting can be provided by resistance elements in the outdoor coil or by water sprays. These systems are most widely used in commercial refrigeration applications for defrosting evaporators in systems without reversing valves.

With electric resistance heating, probes are inserted in the outdoor coil in spaces provided between the refrigerant tubes. To defrost the coil the compressor is shut down and with electricity, the coil is heated directly to melt the frost. This provides a slight improvement over the first method since the compressor is idle, and no compressor and motor losses are encountered.

General considerations in design and use of air-to-air units may be summarized as follows:

1. Since air-source units do not use water, problems of piping, corrosion, condenser water disposal, etc., do not exist.

2. Air-to-air units may operate as in-space units to supply filtered warm and cool air, as needed, at a nominal cost.

3. Air-to-air systems are available as packaged integral or remote (split) systems. They are also available as custom-built systems.

4. Air-to-air units are generally installed with an interior distribution system (ductwork) supplying filtered warm and cool air as needed.

5. The cooling capacity drops off a slight amount and the input to the compressor motor increases slightly as the outdoor temperature increases above rated conditions.

6. The heating capacity, as well as the input to the compressor motor, decreases with lowering outdoor temperature. Because of this characteristic, air-to-air systems are generally supplemented with resistance heaters that supply the additional heating requirements under extreme weather conditions. They are controlled via indoor and outdoor thermostats, or other means, to limit their usage to actual need.

7. Frost will form on the outdoor coil during certain outdoor temperature and humidity conditions, necessitating the use of defrosting means and controls.

Water-to-Water Heat Pumps

The heat pump that uses water from wells, lakes, or other heat sources, with the heat being delivered to the indoor area through the use of a secondary refrigerant—in this case, water—is generally referred to as a water-to-water type.

Water-to-water heat pumps in this country have been used since the early thirties at least, and, for the most part, have operated quite successfully. The growth of this type of system has been relatively slow and has been surpassed by the air-to-water type, which had its infancy, except on an experimental basis only, in the years following World War II.

The reason for the slower growth of the water-source heat pumps compared to the air-source units is simple: there are few populated areas with generous supplies of clean water. Well water provides an excellent source of heat. Water from this source is occasionally returned to the ground in order not to deplete underground water supplies. Disposal of the water also becomes a

problem, with many communities charging high sewer rates.

Well water has an advantage over air as a heat source in that it is at a more nearly constant temperature and is a more efficient heat-transfer medium than air. This is particularly advantageous when the peak heating load occurs, since its temperature is considerably higher than the temperature of outdoor air at this peak condition. This temperature difference also permits the water-source heat pump to have a higher C.O.P. than a comparable air-source type.

Another problem is water quality. Soluble minerals such as calcium salts, magnesium salts, and iron will deposit themselves on heat exchanger surfaces and retard heat transfer. On the other hand, water without mineral content (soft water) can be corrosive. A neutral water supply with a pH-value of 7.0 is seldom found.

The water-to-water heat pump may operate with either refrigerant or water-reversing valves. With the water-reversing system the water flow through the heat exchangers is reversed by automatically operated valves. Therefore, whenever the system operation is reversed from heating to cooling, or vice versa, the indoor conditioned water circuit becomes contaminated with the water supplied, whether it is well, lake, or even sea water in more southerly climates. This condition aggravates the water-treating problem of the conditioned-water circuit which is exposed to each air-handling-unit coil throughout the building. For this reason, units with refrigerant-reversing valves are preferred except where near-perfect water conditions prevail.

Treatment of the supply water from the heat sink is impractical because of the large quantities involved. Therefore, use of unsuitable water will result in excessive maintenance of heat-exchange surfaces and components.

Where design requires a water-changeover system in the face of undesirable water, the problem may be eliminated or minimized by one or more of the following means:

1. Use of admiralty metal or cupro-nickel heat exchangers in all system components.
2. Use of a large heat exchanger to separate well water from condenser-chiller water.
3. Provide means to positively drain the condenser when changing from cooling to heating and the chiller when changing from heating to cooling. In this way, contamination

of the indoor circuit is eliminated and the loss of interior circulating-system water is limited to the volume of the condenser and piping common to both circuits. This allows for proper treatment of the conditioned water.

In some areas the sand content in the well water, if not completely separated from the water supplied to the system, can greatly shorten the life of the pumps, heat exchangers, and valves. Increased maintenance of the equipment of the system is also required.

The water-to-water heat pump has several advantages. The yearly electrical operating costs are frequently substantially lower than for any of the fossil fuels commonly used in industrial and commercial establishments. The electricity required will usually be less than that for the air-source heat pump for year-round operation, due to the more efficient refrigeration cycle. A further advantage is that the first cost of the system generally will be lower than a conventional heating system provided with a second system for summer cooling.

In brief, the water-to-water heat pump can be justified in first cost and operating cost if an abundant supply of approximately 50°F, or warmer, water is available, free from foreign matter and not conducive to excessive corrosion or scale build-up on heat-transfer surfaces. Water-to-water systems are available as packaged or custom-built systems. They offer considerable flexibility in commercial and industrial applications.

There is a renewed interest in water-to-water heat pumps used in conjunction with (closed circuit) flat plate solar collectors as a heat source, and a water-storage tank as the heat sink. This application is discussed later in more detail.

Water-to-Air Heat Pumps

A heat pump which uses well or lake water or any other available water source as a heat sink or source, with the heat being delivered to the indoor area through the use of a direct refrigerant, is commonly referred to as a water-to-air type. The market for this type of unit is similar to the air-to-air type in the 2- to 50-ton range.

The water-to-air type heat pump has some of the same basic problems as a water-to-water type in regard to availability, quality, and disposal means for the water.

In water-to-air heat pumps the refrigerant cycle is reversed to change between summer and winter cycles. This eliminates all conditioned-water problems which can occur in water-to-water units.

General considerations in use and design of water-to-air heat pumps may be summarized as follows:

1. They operate generally with a higher seasonal performance factor than air-source units. This advantage is dependent upon the availability of an ample amount of water above 50°F during the entire heating season and upon the amount of power required to supply the water.

2. Integral water-to-air units have greater application flexibility than integral air-to-air units in that they are not tied down to the need for an outdoor air supply.

3. They may be operated as free-standing, in-space units.

4. Water-to-air units are available as packaged units and as custom-built units.

5. Water-to-air units may be operated with a duct system supplying warmed or cooled filtered air as needed.

6. An ample volume of water must be available to assure sufficient flow in the evaporator section to avoid freezing on the heating cycle.

7. Water availability and disposal may become a problem at some future date.

8. Corrosion problems may create serious maintenance costs unless equipment selected has ability to use available water.

Air-to-Water Heat Pumps

A heat pump which uses the outdoor air as a heat source or sink, and which delivers the heat to the indoor area to be conditioned through the use of a secondary medium—in this case, water—is referred to as an air-to-water type.

This type is marketed in sizes of 3 tons and up. No upper limit is anticipated with jobs being designed to compete with conventional heating-cooling systems of all sizes.

If the design heat load exceeds the heating capacity of the selected heat pump, the necessary supplementary heat is generally supplied by auxiliary electric resistance heaters which may be required to operate during the infrequent periods of maximum requirements. The resistance heaters may be installed in the individual air-handling units. This may be preferable, since it permits higher heat pump efficiencies and provides heat to colder spaces before system changeover to heating.

With additional first costs the efficiency of a heat pump can be high at low temperatures by providing compound compression when the pumping differential pressure across the compressors becomes high (Patented). Compound compression is a means of pumping gas through several stages of compression to limit the pressure differential and compression ratio across each stage. Two-stage compression reduces the average pumping differential across each compressor by approximately 50 percent and decreases the compression ratio by approximately 67 percent.

By using compound compression, two things are accomplished toward improving system efficiency. First, the theoretical refrigeration cycle is greatly improved, since only about 75 percent of the total refrigerant flow need be pumped by the low-stage compressor, with the remaining 25 percent pumped by only the high stage. This is made possibly by the use of an inter-cooler which expands some of the refrigerant to intermediate pressure while subcooling the liquid refrigerant flowing to the evaporator. Secondly, the volumetric efficiency of the reciprocating-type compressor varies inversely with the compression ratio; when the compression ratio is reduced 67 percent, a sizable gain in compressor volumetric efficiency results.

Typical figures indicate that two-stage compression can produce about 65 percent more heat at −20°F evaporator temperature than the same displacement compressors operating single-stage. Furthermore, the electrical requirement for the single-stage compressors would be approximately 65 percent greater than for the two-stage for each unit of heat output.

Systems can be arranged to operate single-stage with gas flow in parallel for summer cooling, and can be converted to compound compression with series gas flow to improve the operating efficiency at lower outside-air temperatures. This is accomplished by providing a single, automatically operated gas valve plus two check valves.

Thus, if the ratio of heating load to cooling load is favorable, the same compressors used for summer cooling may provide sufficient capacity, when compounded, for outdoor-air temperatures of even −10°F without the use of supplementary electric heaters.

Air-to-water systems have the following advantages and limitations:

1. They present the same advantages and limitations as air-to-air units insofar as the heat source and sink are concerned.

2. Warm and cold water are piped throughout the building, thereby eliminating the need for large supply- and return-duct systems.

3. In large central-system installations, rooms and areas may be thermostatically controlled more easily than with large supply and return air-distribution systems.

Ground-Source Heat Pumps

The ground-coil method for a heat source is mentioned here for general information. The problems arising from this type of heat pump as well as recent developments in other types of heat pumps have practically eliminated the desirability of any further developments of this type.

The availability of heat in the earth for a heat pump system on peak-demand days will be better than atmospheric air but poorer than well water at the same location. If ground coils are used, either a direct-expansion or a secondary-refrigerant system may be used. The problems that arise if leaks develop in the underground coil are very serious and expensive.

Uncertainties arise in designing the underground coil. These stem from the variation of ground temperatures in a relatively small area. Even if test borings that give accurate temperatures are made, the fact that these temperatures will vary somewhat with changes in ground moisture content makes design difficult and uncertain.

Difficulties with moisture content in the soil, chemical content and make-up of the earth (clay, gravel, sand, etc.), plus expansion and contraction of the buried piping, discourage heat pump installers from using the earth as a heat source and/or sink.

Special Heat Sources

A few other sources besides air and water have been utilized as the heat source for a heat pump. These sources are generally used as additions rather than replacements of the basic heat source. These additional sources supply heat from operations or processes that exist in areas close enough to the desired conditioned area to be used to economic advantage.

On most large commercial and industrial buildings, a positive means of exhausting air from the building is mandatory to obtain good air distribution. Exhausting this air at one or more central points may be advantageous from the standpoint that heat may be extracted from it. One economical method of removing this heat and returning it to the portions of the building requiring heat is by the use of a direct-expansion coil located in the exhaust-air stream.

Another method of extracting this heat for beneficial use is by means of the "run-around" system. Coils are located in both the ventilation supply and exhaust air ducts, with an antifreeze solution such as ethylene glycol circulating between the two coils. Heat is removed from the exhaust air and absorbed by incoming ventilation air to provide a reduction in heating requirements of the mechanical equipment. Care should be exercised in the design of the run-around system to prevent excessive frost formation on the coil in the exhaust-air stream. A bypass valve, thermostatically controlled, can be located in the antifreeze-solution piping circuit to maintain solution temperatures not lower than approximately 28°F to the exhaust-air coil as a means of preventing the formation of frost.

A third method of extracting this heat, for useful heating as well as for reducing the compressor motor electrical consumption, is to locate a refrigerant-liquid subcooling coil in the ventilation-air plenum chamber, providing preheat to the incoming fresh air while subcooling the liquid refrigerant as it flows to the evaporator. On single-stage systems, this arrangement also provides an increase in heating capacity for a given displacement compressor. In effect, useful subcooling of the refrigerant liquid increases the refrigerant effect of each pound circulated to the evaporator, thereby reducing the quantity of refrigerant that the compressor must handle to absorb a given amount of heat. The horsepower is also reduced for each unit of useful work in the same proportion that the flow rate may be decreased. This coil must be bypassed for summer operation.

Other means of subcooling the liquid refrigerant, such as disposing of the heat to the outdoor air or preheating the air circulated to the outdoor evaporator, would actually result in a net loss over a conventional system without sub-

cooling and should not be used. Rejecting this heat to the outside air would be wasteful. Pre-heating air at the outdoor evaporator coil increases the air temperature a small amount in proportion to the quantity of heat given up by the system.

The use of subcooling coils is limited in application to buildings where a large portion of the ventilation-supply air is introduced to the building at a central point near the compressor room. It becomes impractical to run long refrigerant lines to scattered coils located throughout the building because of the first cost (including insulation), the leak hazard with expensive refrigerants, the increased maintenance, and the design problems brought about by increased pressure-drop in the refrigerant circuit. Also, it should be kept in mind that a system provided with liquid subcooling means will only be beneficial when the normal supply of fresh air is furnished to the building. When the fresh-air supply is shut down during the unoccupied periods, the single-stage system must operate without benefit of the extra capacity. On the two-stage systems, this disadvantage can be offset by locating an intercooler in parallel with the subcooling coil, and no reduction in system capacity will be experienced.

Numerous other building functions and system applications offer means of economically capturing heat from within by a heat pump system. Telephone exchanges, calculating and tabulating machine rooms, and other business machines generate considerable quantities of heat which must be removed to maintain expensive equipment at the necessary temperature level as well as to provide a comfortable temperature for the occupants. Another source is in industrial plants where heat is removed from process work, friction-generating equipment, transformer vaults, and exhaust fumes. Still another source is the recovery of heat from refrigeration-condensing units or packaged air coolers.

An example of condensing units being available is in supermarkets where many units are used for the various refrigerated cabinets in the market. Heat from this source is generally discarded to the atmosphere but may be economically captured by a heat pump system and made available where heating is necessary.

Operating and Installation Factors

Residential heat pump installations and commercial-industrial heat pump installations must be examined on the basis of application considerations and operating cost factors applicable to the area and type of installation. Like other heating and cooling systems, they do have some common considerations such as:

1. Rule-of-thumb applications sometimes result in uneconomical and/or unsatisfactory installations.
2. The sizing of the complete system is dependent upon reliable heat-gain and heat-loss calculations. Design conditions recommended in this handbook or in the current *ASHRAE Handbook* should be followed.
3. Operating cost estimates should be reviewed with the local electric utility. This is especially important with respect to commercial and industrial jobs.
4. Care should be exercised not to oversell the heat pump. The heat pump is not necessarily a panacea for all air-conditioning problems.
5. Proper installation of insulation, double glass, ductwork, and other components is most important.

When a heat pump system fails to meet the owner's ideas and/or requirements, he generally blames the heat pump. The owner seldom appreciates the fact that other considerations, such as lack of insulation or poor ductwork, may have created a problem.

Heat pump manufacturers usually provide the following helpful information on the units:

1. Instructions recommending the manner in which their equipment should be installed.
2. Heating and cooling capacities and kilowatt input at various heat-source and heat-sink temperatures.
3. Service requirements with respect to installation and location of the equipment.

Outdoor Temperature Effects

The operation of air-to-air heat pumps is influenced by outdoor temperatures, and their effects are best explained by means of a graph. Figure 3 presents typical heating characteristics of an air-to-air unit.

With decreasing outdoor temperature the heating capacity decreases, as shown by Line A, while the heat loss (heating load) increases, as shown by Line B. The balance point is the temperature at which the heat loss is equal to the heating

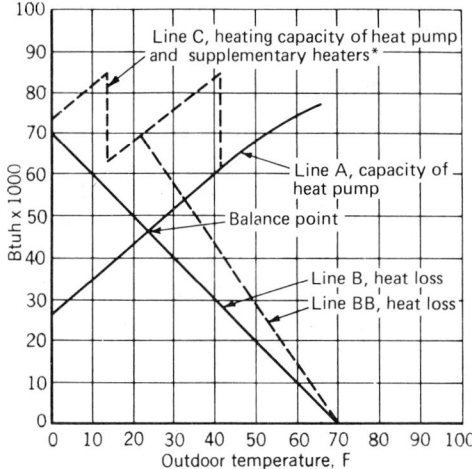

* Resistance heaters may be installed in 2-kw, 3-kw, 5-kw or 6.9-kw banks. They may be in larger banks in commercial and industrial installations. In this example, two 6.9-kw banks are shown.

Fig. 3. Typical air-to-air heat pump operation with various outdoor temperatures.

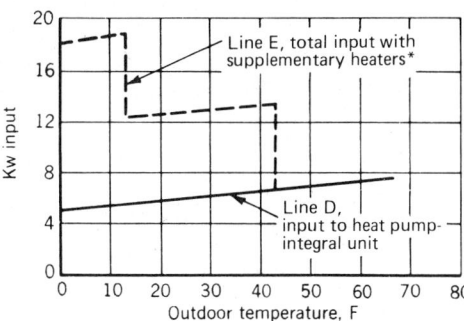

* Supplementary heater remarks on Fig. 3 apply to this diagram.

Fig. 4. Electrical input for various outdoor temperatures. (Input to heat pump includes input to compressor motor, indoor and outdoor fan motors, and control circuit.)

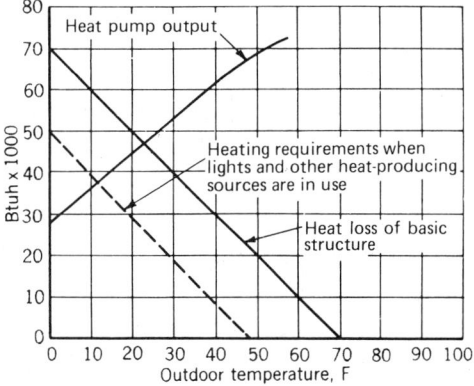

Fig. 5. Effect of interior heat sources on the heating load and balance point.

capacity of the pump. This is the point where Lines A and B cross and, for the example shown, this point is 23°F.

The balance point varies with the application and is a function of capacity of the unit in relationship to the heat loss of the structure. For example, using a heat pump with the capacity as shown in Fig. 3 in a home with a heat loss of 70,000 Btuh at 20°F (Line BB), the balance point would be 32°F.

Supplementary resistance heaters are generally used to augment the capacity of the unit when temperatures fall below the balance point. These are available in fixed amounts (such as 5-kw banks) and are installed in the supply plenum. Line C in Fig. 3 represents a typical application of resistance heaters.

Line D in Fig. 4 represents the instantaneous input in kilowatts versus outdoor temperatures. This value varies, since the load on the compressor motor changes with outdoor temperatures. Line D includes the total input to the compressor motor, the outdoor-fan motor, the indoor-fan motor, and the controls. Line E shows how the input to the heaters adds to that of the unit.

Figure 5 shows how lighting loads, body loads, etc., shift the balance point in commercial and industrial buildings. With the internal heat sources the balance point would be 12°F, com-

pared with 23°F for the basic structure with no internal heat sources.

Thermostats

Heat pumps operate automatically to provide controlled air conditioning under varying conditions of weather, internal heat sources, and demands of the occupants. Thermostats and other control components are so set up that the electric power consumption of the heat pump and electric resistance heaters is held to a minimum. Figure 6 shows a typical control circuit for a residential heat pump with two banks of electric resistance heaters.

Contacts 1 and 2 control the heating functions of the system. Contact 2 is in series with outdoor

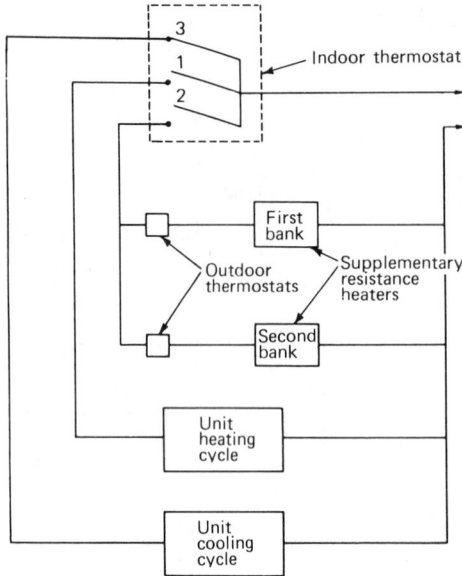

Fig. 6. Control circuit for residential heat pump.

thermostats. These function to block unneeded use of the supplementary heaters. When Contact 1 closes, the heat pump operates as a heating device.

Contact 2 is mechanically locked to Contact 1. It will close when the indoor temperature drops about 1½°F below the setting of Contact 1. When Contact 2 closes, it energizes the supplementary-heater control circuit up to the outdoor thermostats. If the outdoor thermostat are closed, the supplementary heaters are energized.

The setting of the outdoor thermostats depends upon the heat loss of the building, the design temperature conditions, and the capacity of the heat pump. Most manufacturers include outdoor thermostat settings for applications of their units in their installation manuals.

Contact 3 on the indoor thermostat controls the cooling operations of the heat pump. In most thermostats the heating and cooling settings are mechanically blocked from coming within less than 4 to 5°F of each other. For example, when Contact 1 (heating) is set for 70°F, then Contact 3 (cooling) cannot be set below 74°F.

In commercial and industrial installations, various thermostat-control arrangements may be used. They may be interconnected with the lighting to block the use of the supplementary heaters when

certain amounts of lighting are in use. In multiple-unit installations where units are located throughout the building, each unit may be controlled by its own thermostat.

As with any heating and cooling system, the location of the thermostat or thermostats is most important in assuring good comfort control. There are many do's and don'ts in locating the thermostat. With respect to the don'ts, thermostats should not be mounted:

1. where heated or cooled air will blow on them;
2. behind an inside door which will usually be in the open position;
3. in front of a hot or cold pipe within the wall;
4. on an exterior wall.

In general, thermostats should be located in the space most often used by the occupants. This, frequently, is the space having the greatest heat gain or heat loss per unit volume.

Heat Anticipators

Heating anticipators used for heat pumps should be designed for the lower thermal storage of heat pump systems. Proper anticipation will prevent excessive cycling of the compressor above the balance point. The moderate anticipation of the two-stage heat pump thermostat should not cause cycling of the compressor below the balance point.

Equipment Arrangement

In residential installations, the heat pump should be located so all supply-duct and return-duct runs are as short as possible. This will minimize duct losses and help to assure good rated air flows. To avoid transmission of noise and vibration by the duct system, nonmetallic, flexible connections should be installed between the unit and ductwork. In some installations, acoustical treatment may be required in the return and/or supply duct to reduce the noise level.

The situation may arise where an integral air-to-air unit may require outdoor-air-circuit ductwork. This section handles more air than the indoor section. Large ducts are required in order to provide proper air flow with the available fan power. This ductwork must be insulated and vapor sealed in the conditioned space to avoid

condensation in the winter and to minimize increased heating and cooling requirements in the conditioned space.

In general, it is always best to locate integral-air-source units adjacent to exterior walls. A foam rubber gasket may be used around the edges of the outdoor-air circuit to isolate the unit from the structure and to seal the outdoor-air circuits.

Some commercial and industrial installations operate with supply- and return-air-duct systems. Others operate with heating and cooling piped to fan coil units. Still other installations may comprise multiple units operating in free space with less piping and ductwork.

In examining any application, the designer should consider integral units versus remote (split) units with respect to simplification of indoor supply- and return-air ductwork. Remote units eliminate the need for outdoor ductwork at times needed for integral applications. They do, however, require refrigerant piping which, if improperly installed, may be a noise problem. Manufacturers' instructions should be carefully followed regarding the use of mufflers in the refrigerant circuit and the method of supporting the refrigerant lines between the outdoor and indoor sections.

Manufacturers' application and equipment data sheets should be consulted in locating the equipment within a home or structure. All types of heating and cooling equipment require preventive maintenance as well as annual service. The heat pump is no exception. Consideration should be given to the fact that fans, fan motors, water pumps—if used, and other equipment must be lubricated, and that filters must be cleaned or changed periodically. Integral, remote, and custom-built heat pumps are designed with definite space requirements for servicing. These cannot be ignored in locating the equipment.

Good indoor air distribution contributes as much or more to the success of a heat pump system as any other factor. Therefore, care should be used to assure ample air distribution throughout the structure, and to assure operation with a reasonable noise level. Conformance to good duct-sizing practice is imperative to accomplish this.

The heating coefficient of performance and the heating season performance factor decrease when the indoor air flow is much below rated conditions. Therefore, it is important to maintain approximate rated indoor air flow during the heating season to avoid increases in the estimated operating cost. Registers and ducts should be sized for maximum air flow to the area.

Heating efficiency and comfort are improved if the winter air return is at or near floor level.

ELECTROHYDRONIC HEAT RECOVERY

Some air conditioning systems used regularly in low-rise apartments or hotels become impractical or uneconomical in high-rise buildings, while other systems on the contrary, become increasingly practical with high-rise structures. The Electrohydronic System utilizing water-to-air heat pumps, with a common heat sink, is used in both high- and low-rise buildings. Generally, the installed cost, in dollars per square foot, is less with high-rise structures.

In virtually all apartments and hotels, the need for each occupant to control the temperature in his own area is basic. The most critical period of maintaining uniform comfort control may not occur at design summer or winter conditions, but rather at outdoor temperatures of 40 to 70°F. Under such conditions, the traditional central system chiller and boiler can be running simultaneously.

Some building areas, particularly of hotels, are subjected to varying heat gains and losses to a much greater extent than other interior areas. Therefore, they require a system with greater flexibility and quicker response than does a typical interior zone.

The electrohydronic system has only four major components: a device for rejecting excess heat, a water heater, circulating pumps and the air conditioners. Although these components are well-known, there are important differences in the way they function in this system.

The system, (see Fig. 1) uses unitary water-to-air, reverse cycle, air conditioners. These contain the same components as traditional air conditioners with one important addition: A reversing valve diverts hot refrigerant gas to the water coil when the air conditioner is cooling, or to the air coil for heating. Note, however, that the direction of refrigerant flow through the compressor is *not* reversed.

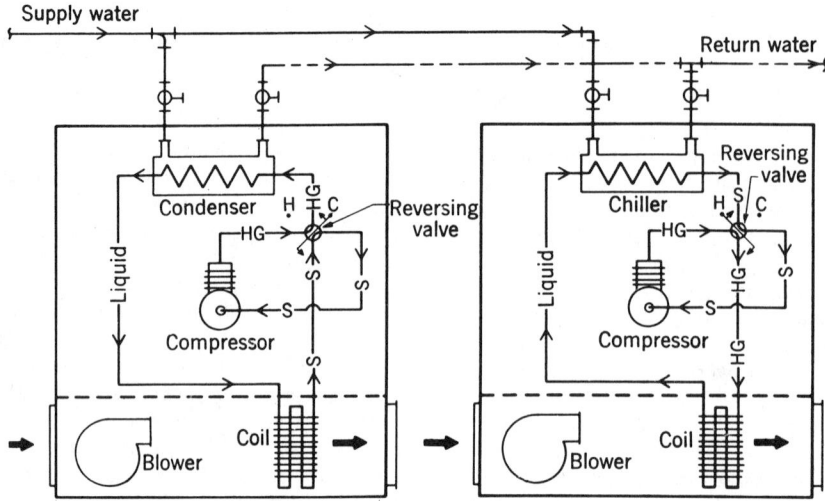

Fig. 1. Water is recirculated in loop as a heat reservoir, providing a closed loop water-to-air heat pump. Unit No. 1, left (shown cooling), Unit No. 2, right (shown heating).

The air conditioner is connected to a closed circuit, non-refrigerated water loop. The water is recirculated in this loop as a heat reservoir, providing a closed loop water-to-air heat pump.

Cooling Cycle

On demand of the thermostat for cooling, the reversing valve guides the hot gas from the compressor to the water coil which serves during the cooling cycle as a condenser, removing heat from it and condensing it into a liquid. The heat picked up is circulated in the closed water loop. The liquid refrigerant then flows through an expansion device to the air coil which acts as an evaporator. The boiler refrigerant absorbs heat from the room air passing across the evaporator coil. Refrigerant vapor then flows through the reversing valve and into the compressor, thus completing the cooling cycle.

When the thermostat calls for heat, the reversing valve reverses the flow of refrigerant so that hot gas from the compressor flows through the tubes of the air coil which now serves as a condenser. Air to be heated passes over the tubes, absorbing heat from the refrigerant gas and causing it to condense. This warm air heats the room. Liquid refrigerant then flows through the expansion device to the water coil, which now functions as a water cooler or chiller. As the refrigerant vaporizes, it absorbs heat from the closed water loop, which is a heat reservoir. The refrigerant gas is conducted back through the reversing valve to the compressor, completing the heating cycle.

Maintenance of Loop Temperature Limits

Note that cooling or heating of room air is done by transfer with the *refrigerant, not the water* in the closed loop. This water, ordinarily between 60°F and 90°F, accepts heat on the cooling cycle and provides heat on the heating cycle.

When enough units are on cooling to raise the water temperature above 90°F, the excess heat must be rejected from the water loop. This is done most frequently by a closed circuit evaporative cooler, a small factory designed and assembled package with a coil through which the warm system water flows. This system water is cooled by the evaporation of spray water which is recirculated over the outside of the coil. (See Fig. 2.)

A small water heater is included to maintain the loop water temperature at 60°F when almost all the units are on heating. This heater can be about two-thirds the capacity required by conventional systems.

On moderate days, units serving the side of the building exposed to the sun may be cooling, while those on the shady side may simultaneously be

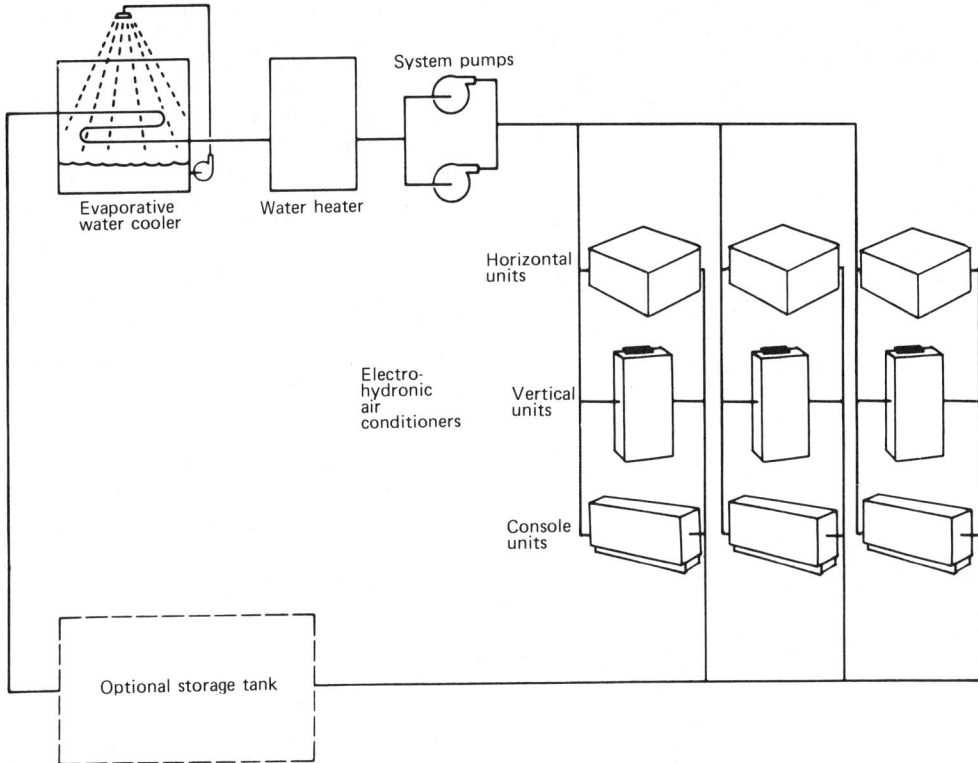

Fig. 2. System water is cooled by evaporation of spray water which is recirculated over the outside of the coil.

heating. This situation, which occurs during much of the year, greatly reduces the hours of operation and costs for the water heater and the heat rejector. Even on cold days, interior zones with high heat gain from people, lights, and equipment, requiring cooling, often provide sufficient heat to the water loop for the units warming the perimeter zones so that it is not necessary to operate the supplementary heater.

In addition to effective recovery and utilization of heat where simultaneous heating and cooling is required, the system takes advantage of the "fly-wheel effect" in many buildings where net cooling is required during the day and net heating at night. It is possible to calculate building heat requirements throughout any given day as outdoor temperatures and building usage vary. It is also possible to determine the internal heat gains within the building during that same period of time.

If these heat gains and losses are superimposed, it is possible to see how this system can transfer heat within the building and also store excess heat from day-time hours for utilization at night when it would otherwise be necessary to add supplementary heat. During a 24-hour period, the system water temperature may never drop low enough to require supplementary heat, nor reach a high enough temperature to reject excess heat.

Controls are so arranged that, if loop water temperature falls to 50°F, the heater automatically operates. If it rises to 90°F, the cooler operates. If the temperature is between 60 and 90 degrees, neither device is required. At no time do the cooler and heater operate simultaneously.

Among the factors which may influence design and installations are the following:

1. If a unit requires service, it can be quickly removed and replaced without affecting other apartments, or the rest of the system.

2. Less ductwork may mean less overall building height for the same number of stories which, of course, reduces building costs.

3. Non-insulated pipes reduce costs of material and labor. The water in the closed loop is not hot enough for heat loss or cold enough to cause condensation on the pipes. Plastic pipe is being used with increasing frequency.

4. The central water core can be installed and then individual conditioners furnished later as apartments are sold or leased. This is also particularly applicable to modernization projects because the conditioners can be added one floor at a time.

5. The conditioners may be individually metered. This is very desirable for rental and leasing situations or condominiums.

6. No complex control system is required. The services of an operating engineer normally are not needed. A simple panel to indicate extreme operating temperatures or lack of flow in the closed water loop is all that is necessary.

System Design

Building heat gains and heat losses should be calculated in accordance with normal procedures. Water piping for the system should provide a balanced two-pipe reverse return arrangement wherever possible. Since system water flows continuously to provide either cooling or heating, two pumps should be installed—one on standby.

Supply and return water connections to each air conditioner should include a shut-off valve and a means of disconnecting piping at the air conditioner to facilitate removal should servicing be required. These connections may be made with a good quality hose. The hose will simplify installation and servicing as well as minimize any possible vibration transmission through the piping.

Polyvinyl chloride (PVC) hose may be used as a condensate drain line. It requires no insulation and can be installed very economically.

It is essential that the piping system be thoroughly flushed of foreign material. The recommended means of doing this is to loop over the roughed-in piping from supply to return at each unit. It is especially convenient to do this when hose connections are being used. The pump can then be operated to flush the system prior to

installation of any air conditioners.

Excess heat may be rejected from the water loop through a heat exchanger into a cooling tower, a well, river water, a lake or any other readily available source of coolant suitable to prevent the system water from exceeding the maximum desired temperature.

In order to keep the water system clean and to maintain peak performance, the water loop must be a closed system. The heat exchanger selected for rejection of excess heat may be a water-to-water exchanger or when a cooling tower is to be used, a closed circuit evaporative cooler.

The building's peak heat rejection load will be less than the sum of rated heat rejection for all the air conditioning units plus any additional heat from other sources. That portion (diversity factor) of maximum load which can actually be required at any given time must be determined for each job but it will generally be less than unity. This is based on the premise that not all rooms in an entire building will simultaneously need maximum cooling (room fully occupied, lights and other heat sources all operating, sun shining brightly through the window, etc.). Although it will vary with different applications, the use of a diversity factor of 0.8 for comfort air conditioning will generally be reasonably accurate.

Should a closed circuit evaporative cooler or cooling tower be installed where it may be exposed to freezing temperatures, provision must be made to avoid damage from freezing. If the entire tower cannot be located inside where temperatures can be maintained above 40°F, perhaps the tower sump and makeup water line could. Otherwise, they may be equipped with an immersion heater in the sump and electric heating tape around the makeup water line. The heaters should be controlled to prevent the water temperature from dropping below about 40°F.

It is recommended that full system water flow be maintained through the cooling coil of an evaporative cooler during the entire year. To minimize heat loss in cold climates, any exposed piping to the cooler should be insulated and a positive closure damper installed at the evaporative cooler air discharge. Buildings having an exhaust system which can conveniently be directed into the cooler will benefit during all seasons. The exhaust air will improve warm weather performance of the cooler and will reduce heat loss during cold weather.

Supplementary Heat

A source of supplementary heat must be provided for any extended cold weather periods when there is inadequate heat for the system to recover. This heat source must add whatever heat is removed from the water loop by the air conditioners once the water temperature has dropped to its minimum temperature of about 60°F. Since the air conditioner's heat of compression adds about 30% of the total unit's heat output, the capacity of the supplementary heat source need be only 70% of the heat loss from the area served by the system. Add to this any additional heat that may be required for other areas or purposes.

Fuel-fired boilers, electric boilers or immersion elements, steam converters or other heat sources are suitable. They are sized and controlled to add whatever heat is required to prevent system water from dropping below its minimum allowable temperature.

OPTIMIZED DATA FOR HEAT PUMP SYSTEMS

While there are substantial operating economies with heat pump systems, it is often a misconception that the first costs of such systems are determined by the capacity of the chiller only. However, the following shows that first costs can be optimized with the proper choice of other operating parameters.

Certain factors, including those of supply- and return-water temperatures and total water flow, which have some influence on chiller selection—at least on its heating condenser—have a major influence on the heating system itself.

The heating capacity of any heating unit—air heating coil, fan-coil unit, induction unit or radiator—is directly proportional to supply water temperature and to the rate of flow of water passing through it. In the case of the heating system, the solution is usually routine: water is supplied at 190°F or 200°F and flow is fixed by selection of a 20 deg. F temperature drop across the unit. In the case of heat pump systems, however, available supply temperatures are much lower and temperature drop is also limited. The reason is that maximum condenser temperature and, thus, the temperature of water leaving it, cannot be higher than the temperature corresponding to the surge point of the refrigerant.

Optimum conditions occur at that temperature and flow which results in the lowest first costs for the heat distributing system (not including chiller or auxiliary boiler).

The following approach is based on the assumption that a required heating capacity for a building will be provided by the necessary number of identical heating units with identical capacities and operating under identical conditions (this will result in fractional numbers as shown in Table 1, Column 3).

The capacity of a unit with heating coil and

Table 1. Fan-Coil Unit Data

GPM	Unit Capacity, MBh*	No. Units to Serve 1000 MBh Load
1	35.4	28.2
2	52.9	18.9
4	70.4	14.2
6	79.2	12.6
7	82.2	12.16
8	84.5	11.8
9	86.4	11.57
10	88.0	11.26

*Based on data for Trane Force-Flow #12 unit.

fan at a given airflow rate and supplied with 190°F water varies with the rate of flow of water through it, as indicated in Table 1. [At water temperatures, T, other than 190°F, the number of units required for a 1000 MBh load as given in Table 1, Column 3, can be corrected by multiplying by the reciprocal of $1 - 0.008$ $(190 - T)$].

The amount of water flowing through the system and the chiller's heating bundle is the product of the number of heating units and the appropriate water flow through each unit.

The assumed cost of each heating unit is approximately $350. This includes enclosure, two 5-ft lengths of copper piping with a control valve, a shut-off valve, and a balancing valve.

Piping cost can also be estimated easily (See Table 5). Water flow is taken at 50% of maximum flow on each system (every floor with a separate horizontal distribution system or every riser with a vertical distribution shall be treated

as a separate system). Average pipe size is selected to produce a friction of more than 3.5 ft per 100 ft of piping. Table 5 shows the cost of piping for a total length of piping of 1000 ft as a function of actual maximum water flow.

With the help of these cost figures and with the assumption that this function is continuous, a curve can be established. Then, for any water flow, costs can be accurately interpolated. This curve is shown on Fig. 1. For different lengths of piping (L), the costs shall be multiplied by the ratio $L/1000$.

Development of Equations

Let Q_H be the total heating load in thousands of Btu per hr (MBh); Q_c, minimum cooling capacity required during the heating season; W, total water flow through the system in gallons per minute; T_1, supply-water temperature (the temperature at which the water is discharged from the auxiliary boiler); and T_2, return-water temperature entering the heating bundle of the condenser.

Total heating capacity equals total flow (taken at 50% of maximum flow) multiplied by total temperature rise and the specific heat of water (taken as 1).

$$Q_H = 0.5W(T_1 - T_2) \qquad (1)$$

Heat transferred to the water in the condenser equals approximately $1.3Q_c$. This amount of heat shall not raise the temperature of the water from T_2 above the temperature corresponding to the refrigerant surge point, which we assume to be 105°F. Then,

$$1.3Q_c = 0.5W(105 - T_2) \qquad (2)$$

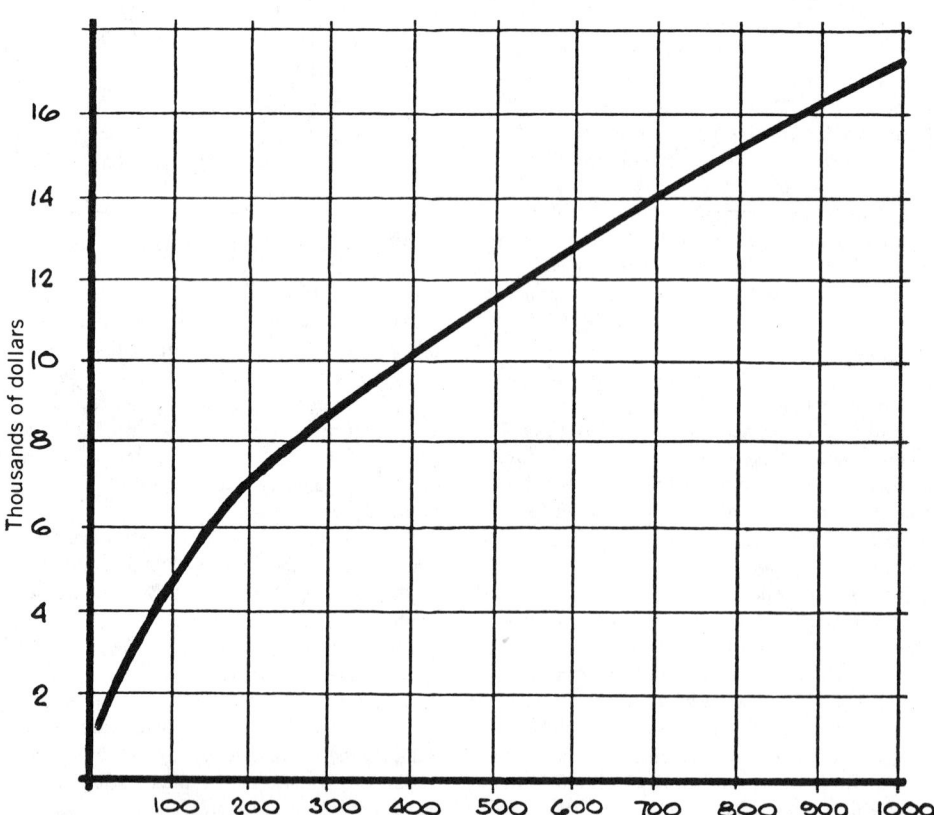

Fig. 1. Cost of piping per 1000 ft as a function of gpm.

Table 2. Possible Water Flow/per Unit (GPM) at Different Q_C/Q_H Values and Supply Temperatures.

$\dfrac{Q_c}{Q_H}$	Supply Water Temperature, F						
	108	110	112½	115	117½	120	125
.20	15.89	9.26	5.92	4.18	3.26	2.60	1.77
.25	14.33	8.27	5.23	3.63	2.72	2.20	1.44
.30	12.76	7.28	4.54	3.08	2.34	1.79	1.11
.35	11.19	6.30	3.85	2.54	1.75	1.39	.78
.40	9.62	5.31	3.15	1.99	1.42	.99	.46
.45	8.05	4.33	2.46	1.44	.875	.59	.13
.50	6.48	3.34	1.77	.89	.505	.19	—
.60	3.24	1.37	.38	—	—	—	—
.70	.20	—	—	—	—	—	—

Table 3. Number of Units for Possible Conditions and for 1000 MBh Heating Load.

$\dfrac{Q_c}{Q_H}$	Supply Water Temperature, F						
	108	110	112½	115	117½	120	125
.20	—	32.3	33.27	35.32	37.94	39.62	43.76
.25	—	32.85	34.82	37.78	40.97	41.74	50.16
.30	—	33.86	36.17	40.93	43.10	47.23	56.57
.35	—	34.97	38.22	44.11	49.64	55.73	71.90
.40	33.57	36.85	42.57	47.07	57.82	64.65	91.10
.45	34.27	39.06	46.85	60.29	75.23	90.25	110.90
.50	36.06	44.12	55.34	78.20	99.27	116.65	—
.60	46.48	69.34	93.10	—	—	—	—
.70	154.00	—	—	—	—	—	—

It has been found that total water flow is a linear function of unit water flow, (W_0) at supply water temperature of 190°F, as follows:

$$W = (Q_H/1000)(9.5W_0 + 18.7) \qquad (3)$$

At supply water temperatures (T) different from 190°F, total water flow is:

$$W = \left[\frac{Q_H}{1000}\right]\left[\frac{9.9W_0 + 18.7}{1 - 0.008(190 - T_1)}\right] \qquad (4)$$

Combining Equations (1), (2) and (4),

$$\frac{(1 - 1.3Q_c/Q_H) \times 1000}{0.5(9.5W_0 + 18.7)} = \frac{T_1 - 105}{1 - 0.008(190 - T_1)} \qquad (5)$$

which will provide, for any selected ratio (Q_c/Q_H) and any selected supply temperature (T_1), a single possible value of (W_0); providing, also, the only possible total water flow (W) from Equation (3).

Table 4. Cost of Heating Units for Possible Conditions and for 1000 MBh Heating Load.

	Supply Water Temperature, F						
	108	110	112½	115	117½	120	125
.20	—	11,305	11,644½	12,362	13,279	13,867	15,316
.25	—	11,497½	12,187	13,223	14,339½	14,609	17,556
.30	—	11,851	12,654½	14,325½	15,085	16,530½	19,799½
.35	—	12,239½	13,377	15,438½	17,374	19,505½	25,165
.40	11,749½	12,997½	14,849½	16,474½	20,237	22,629½	31,885
.45	11,994½	13,671	16,397½	21,101½	20,330½	31,587½	38,815
.50	12,621	15,442	19,369	27,370	34,744½	40,827½	—
.60	16,268	24,269	32,583	—	—	—	—
.70	53,900	—	—	—	—	—	—

Table 5. Cost of Piping for a Total of L = 1000 Ft.

Total Water Flow, Gpm	45	140	220	400	900
Cost of Piping, $	3,000	6,200	7,600	10,000	16,100

Development of Tables

With the help of Equation (5), Table 2 has been established showing the possible unitary water flow values for different (Q_c/Q_H) ratios and (T_1) temperatures. From Table 2 and Table 1, the number of units required for a heating load of 1000 MBh and their cost can be figured. The results are shown in Tables 3 and 4. For heating loads (Q_H) different from 1000 MBh, the values should be multiplied by $(Q_H/1000)$. See Table 5 for typical piping cost estimates.

Table 2 and Equation (3) will provide total water flow data for any combination of (Q_H/Q_c) and (T_1) values and, with the help of Fig. 1, cost of piping can be established.

Costs Plotted

The sum of the costs for heating units and piping provides the total cost of the heating system as a function of supply water temperature for any given (Q_H/Q_c) ratio and given Q_H heating load. These total cost data are shown graphically in the accompanying curves. Figs. 2 to 9.

Because each curve generally has a minimum, the temperature which results in the minimum cost can be established for any given (Q_H) heating load, (Q_c/Q_H) ratio and given length of piping. For this optimum temperature, Table 2 will furnish the unitary water flow and, with Equation (3), the total water flow can be calculated. The temperature drop is given by Equation (1).

Every individual system should be calculated separately, i.e., on multistory systems served by a vertical main riser, every floor should be calculated separately and, for a system with risers and horizontal mains, every riser is a separate system. The cost curves of all the individual systems are added and the resulting minimum will provide the optimum water temperature.

As previously shown, with the proper selection of heating water temperatures, an optimum first cost can be achieved for hot water piping and heating units. However, to these savings can be added those achieved by maintaining the pressure of the chiller plant below the limit where certain codes require licensed operating personnel.

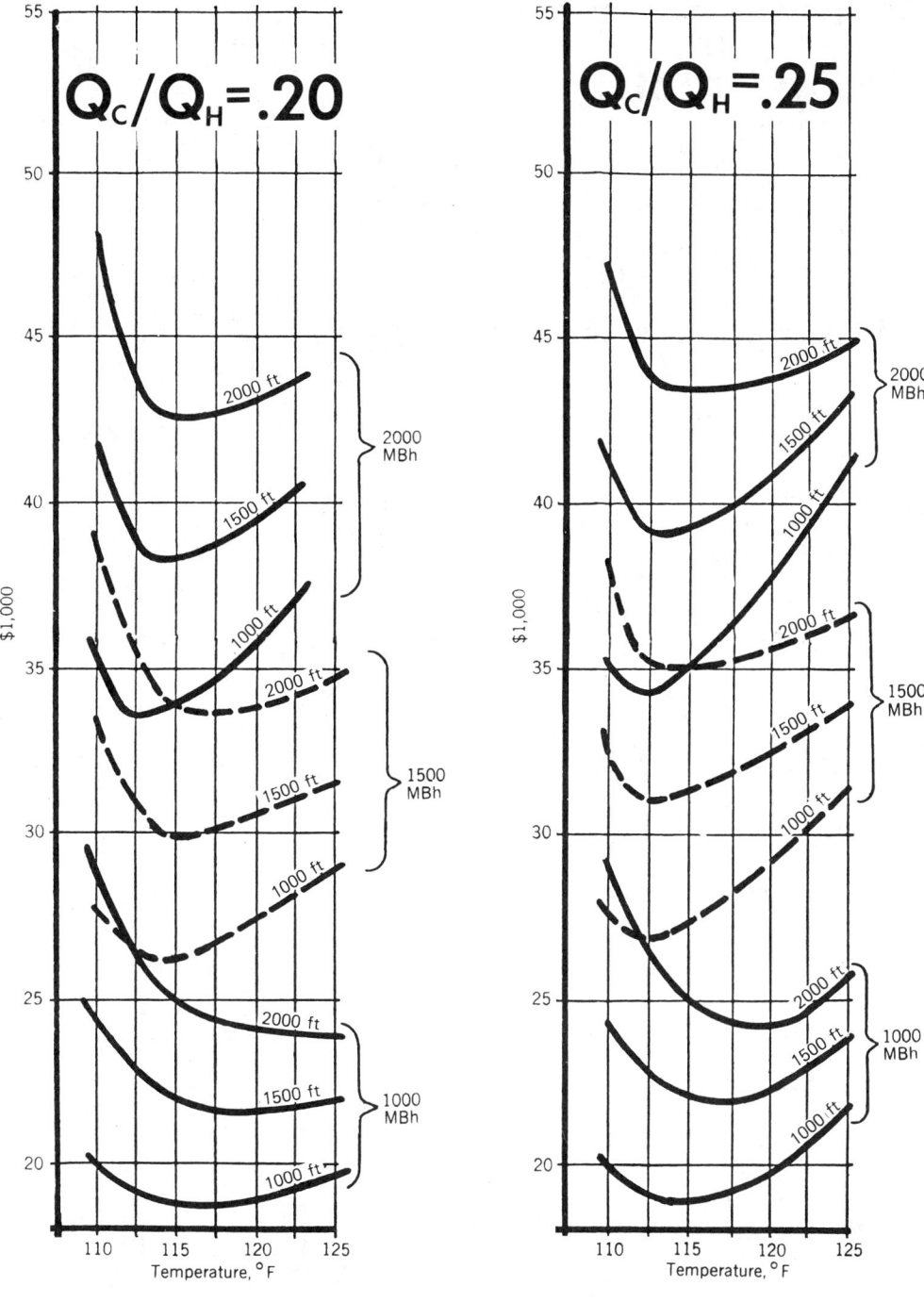

Fig. 2

Fig. 3

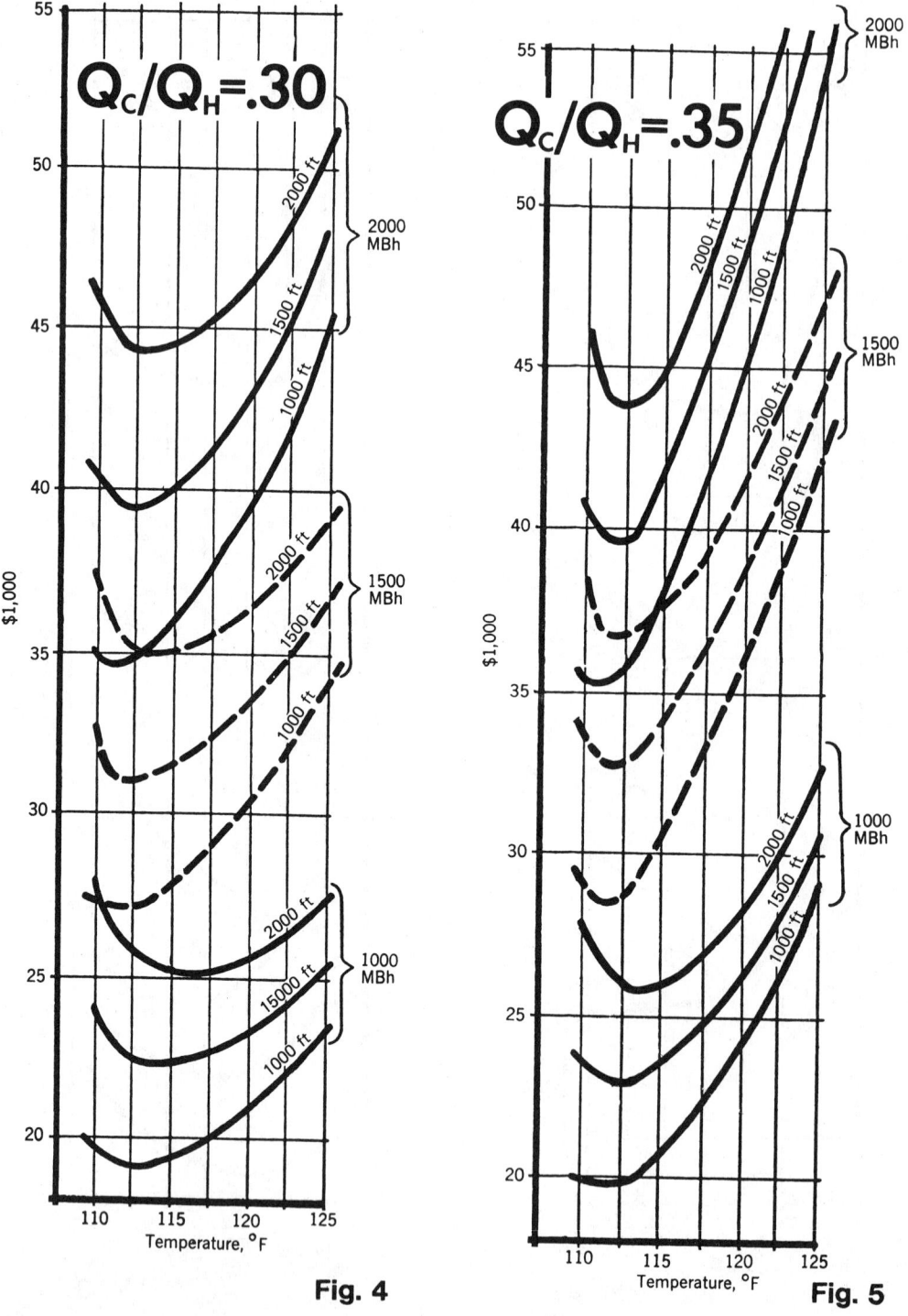

Fig. 4

Fig. 5

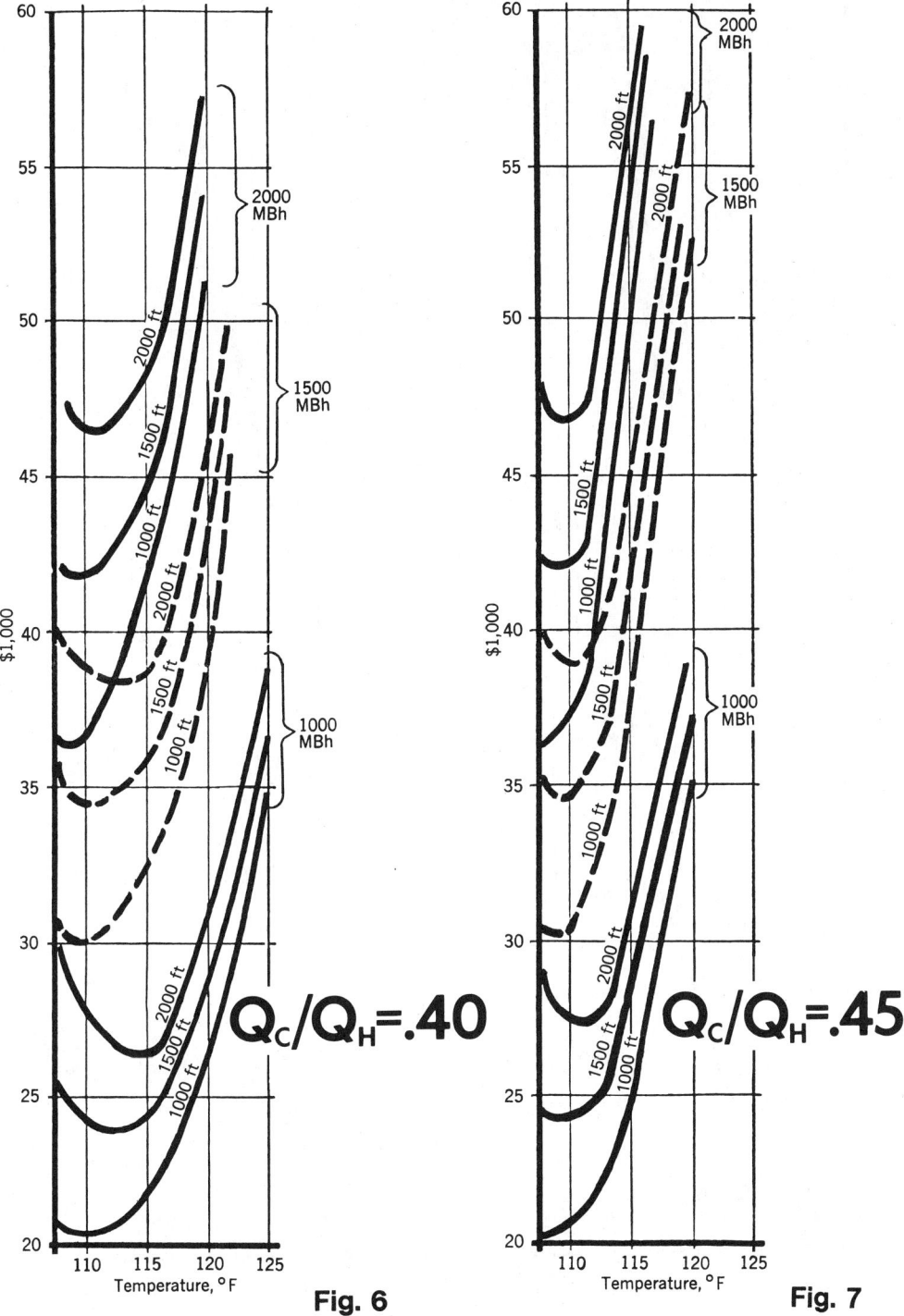

Fig. 6

Fig. 7

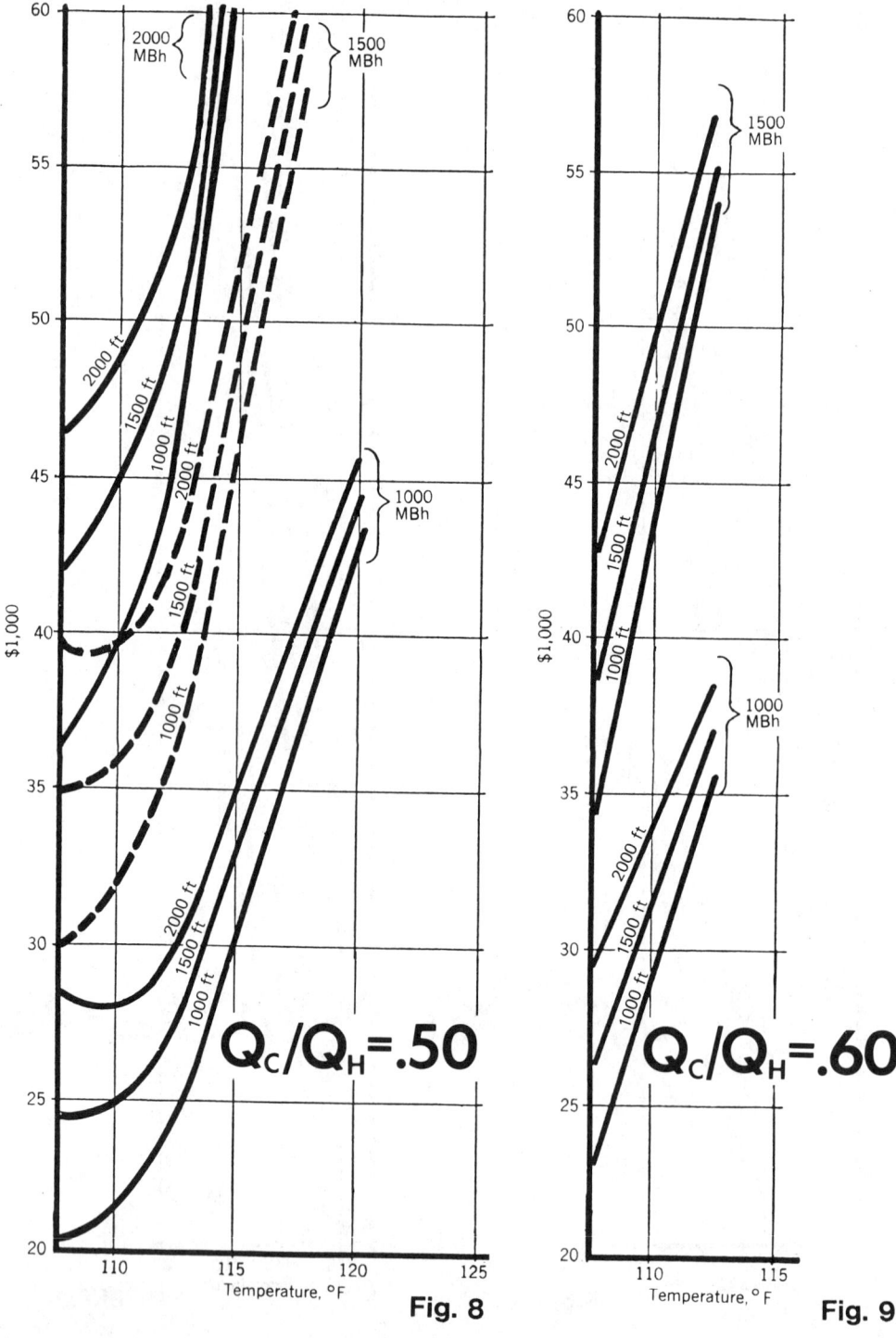

Fig. 8

Fig. 9

SELECTING AIR HANDLING UNITS

The data that follow can be used in a simple fashion for the quick and accurate selection of air-handling units, without resorting to manufacturer's literature. All data are based on a survey of equipment that is commercially available, and will be helpful for schematic, preliminary, and final layouts:

CHART FOR DOUBLE WIDTH FAN SELECTION

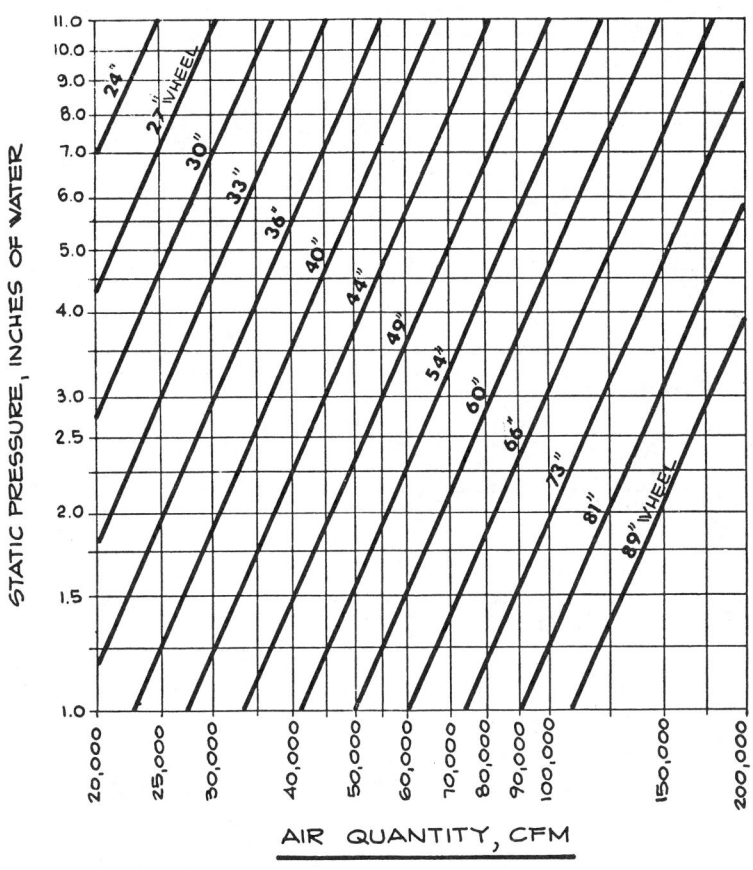

NOTE: This chart is applicable to
all double width centrifugal
fans built in accordance
with AMCA Standard 2401-66

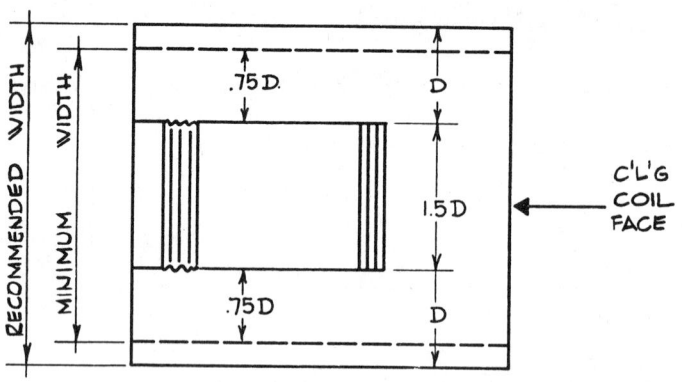

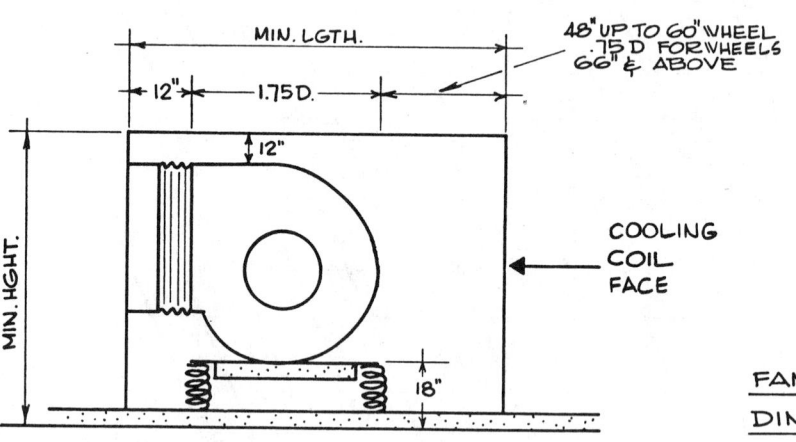

FAN PLENUM

DIMENSIONS

Wheel Diam-eter, D, Inches	Casing Width				Minimum Height, Ft-In.	Minimum Length, Ft-In.
	Minimum (3D)		Rec. Min (3.5 D)			
	Inches	Ft-In.	Inches	Ft-In.		
27	81	6-9	95	8-0	7-0	9-0
30	90	7-6	105	8-9	7-6	9-6
33	99	8-3	116	9-8	8-0	9-10
36	108	9-0	126	10-6	8-6	10-3
40	120	10-0	140	11-9	9-3	10-10
44	132	11-0	154	12-10	9-10	11-6
49	147	12-3	171	14-2	10-8	12-2
54	162	13-6	189	15-9	11-6	13-0
60	180	15-0	210	17-6	12-6	13-9
66	198	16-0	231	19-3	13-6	14-9
73	219	18-4	256	21-3	14-6	16-3
81	243	20-3	284	23-3	16-0	17-10
89	267	22-3	312	26-0	17-6	19-8

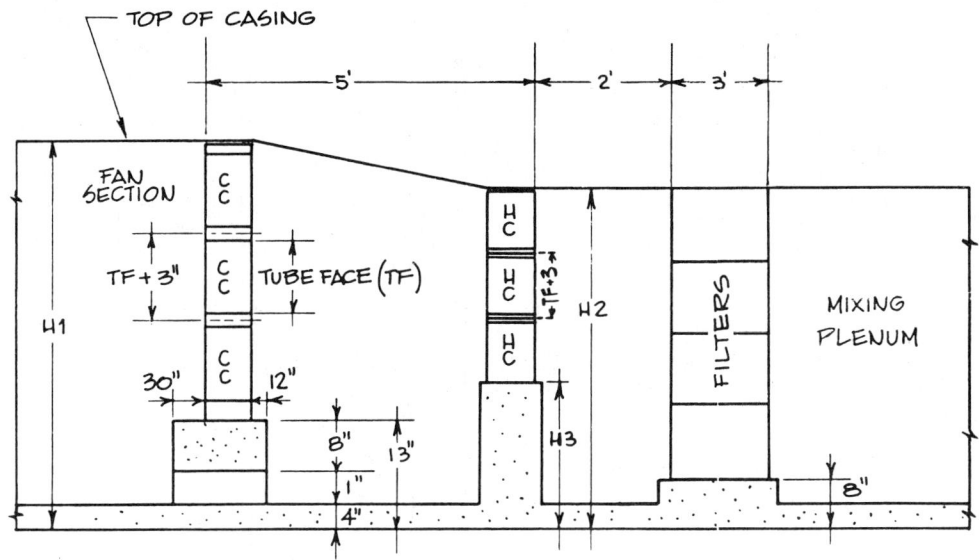

AIR HANDLING UNIT
CASING HEIGHT DATA
(BASED ON COOLING COIL SELECTIONS)

Casing Height H_1, Feet and Inches [$H_1 = N(TF + 3) + 13$]						
9-4	10-10	12-4	14-1	15-11	16-1	17-4
Number of Cooling Coils (N) and TF Dimension, Inches						
(3) 30	(3) 36	(3) 42	(4) 36	(4) 39	(4) 42	(5) 36
Number of Heating Coils (N) and TF Dimension, Inches						
(2) 30	(2) 36	(2) 42	(3) 33	(3) 36	(3) 39	(3) 42
Height of Heating Coil Section, Inches [$N(TF + 3)$]						
66	78	90	108	117	126	135
Height to Bottom of Heating Coil, H_3, Inches						
38	52	40	44	35	50	41
Casing Height, $H_2 = N(TF + 3) + H_3$, Inches						
104	130	130	152	152	176	176
Number of (24-Inch) Filters High						
4	5	5	6	6	7	7

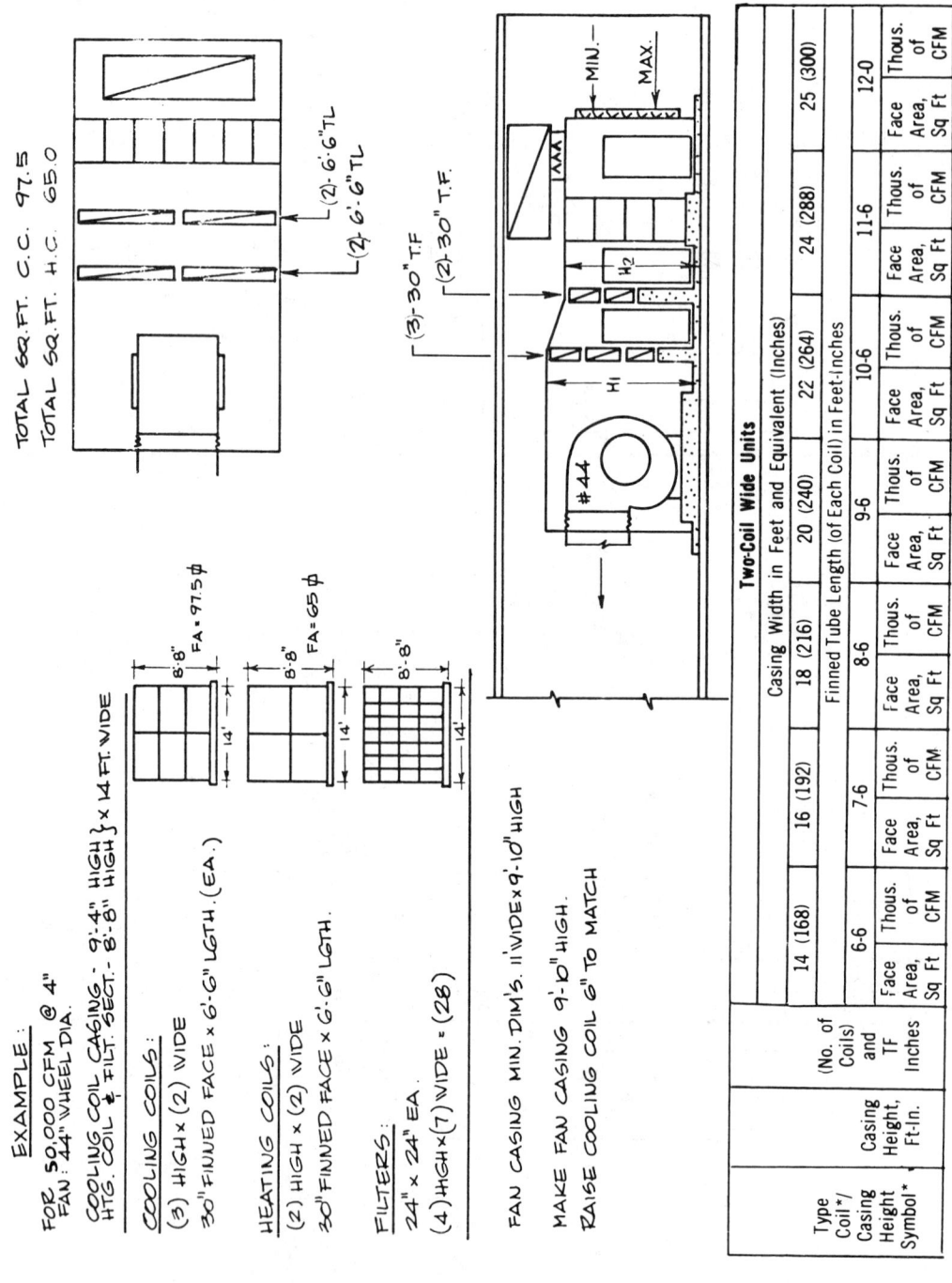

Air Handling Unit Casing Data

CC/H₁	9-4	(3) 30	97.5	48.7	112.5	56.2	127.5	64.0	142.5	71.2	157.5	79.0	172.5	86.0	—	—
HC/H₂	8-8	(2) 30	65	—	75	—	85	—	95	—	105	—	115	—	—	—
CC/H₁	10-10	(3) 36	117	58.5	135	67.5	153	76.5	171	85.5	189	94.5	207	103.5	—	—
HC/H₂	10-10	(2) 36	78	—	90	—	102	—	114	—	126	—	138	—	—	—
CC/H₁	12-4	(3) 42	136.5	68.2	151.5	80.0	178.5	89.0	199.5	100.0	220	110.0	242	121.0	—	—
HC/H₂	10-10	(2) 42	91	—	105	—	119	—	133	—	147	—	161	—	—	—
CC/H₁	14-1	(4) 36	156	78.0	180	90.0	204	102.0	228	114.0	252	126.0	288	144.0	—	—
HC/H₂	12-8	(3) 33	107	—	124	—	140	—	157	—	173	—	190	—	—	—
CC/H₁	15-1	(4) 38	169	84.5	195	97.5	221	110.0	247	123.5	273	136.5	299	149.5	—	—
HC/H₂	12-8	(3) 36	117	—	135	—	153	—	171	—	189	—	207	—	—	—
CC/H₁	16-1	(4) 42	182	91.0	210	105.0	238	119.0	266	133.0	294	147.0	322	161.0	336	168.0
HC/H₂	14-8	(3) 39	127	—	146	—	166	—	135	—	205	—	224	—	117	—
CC/H₁	17-4	(5) 36	—	—	—	—	—	—	—	—	315	157.0	345	172.0	360	180.0
HC/H₂	14-8	(3) 42	—	—	—	—	—	—	—	—	220	—	242	—	126	—
CC/H₁	19-10	(5) 42	—	—	—	—	—	—	—	—	368	184.0	402	201.0	420	210
HC/H₂	16-8	(4) 36	—	—	—	—	—	—	—	—	252	—	276	—	288	—

One-Coil Wide Units

			Casing Width, Feet and Equivalent (inches)							
			9 (108)		10 (120)		11 (132)		12 (144)	
			Tube Length (Coil Width) in Feet-Inches							
			8-6		9-6		10-6		11-6	
CC/H₁	8-7	(2) 42	59.5	29.8	86.5	33.2	73.5	36.8	80.5	40.2
HC/H₂	8-7	(2) 30	42.5	—	47.5	—	52.5	—	57.5	—
CC/H₁	9-4	(3) 30	63.8	32.0	71.3	35.6	78.8	39.6	86.3	42.1
HC/H₂	8-8	(2) 30	42.5	—	47.5	—	52.5	—	57.5	—
CC/H₁	10-10	(3) 36	75.5	38.3	85.5	42.8	94.5	47.3	103.5	51.8
HC/H₂	10-10	(2) 36	50.5	—	57	—	63	—	69	—
CC/H₁	12-4	(3) 42	89	44.6	99.6	49.9	110	55.0	121	60.4
HC/H₂	10-10	(2) 42	59.5	—	66.5	—	73.5	—	80.5	—
CC/H₁	14-1	(4) 36	104	52.0	114	57.0	126	63.0	138	69.0
HC/H₂	12-8	(3) 33	70	—	78.5	—	87	—	101	—

Note: data based on

500 fpm coil face velocity

2200 fpm max. fan outlet velocity

*CC — Cooling Coil

HC — Heating Coil

H₁, H₂ — See Fig. 3

Adjust H₁ (casing ht. at cooling coil) to match fan plenum height.

AIR HANDLING UNIT SCHEMATIC
PLAN LAYOUT

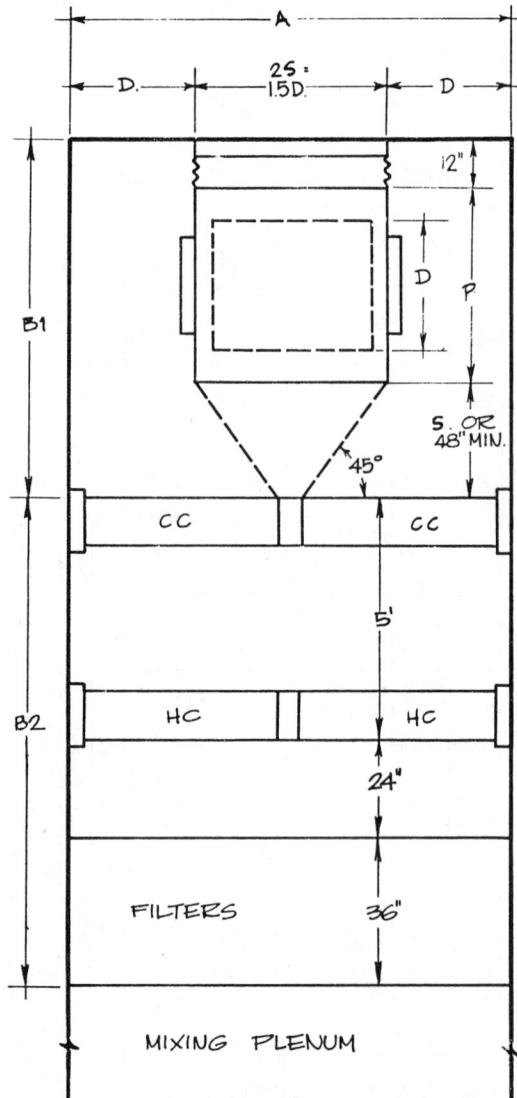

FAN HEIGHT = F
MIN. HEIGHT = F + 3'

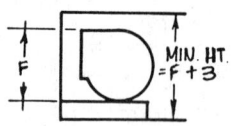

B1 = 12" + P + S (OR 48")
A = 2S + 2D
S = .75D
A = 1.5D + 2D = 3.5 D.
B2 = 10'-0"

Casing Width, W, Feet					
24	22	20	18	16	14
Net Coil Tube Length, Feet (2 Coils Wide)					
23	21	19	17	15	13
Tube Length of Each Coil, Feet-Inches					
11-6	10-6	9-6	8-6	7-6	6-6
Number of 24-Inch Filters Wide					
12	11	10	9	8	7

WELL WATER FOR AIR CONDITIONING

When groundwater is cold enough, a direct supply to cooling coils will eliminate the need for refrigeration in air-conditioning systems. Well water is usually a little too warm to accomplish the dehumidification required. In that case, the water may be used advantageously for precooling outside air prior to refrigerated cooling and dehumidification, thus reducing the amount of refrigeration needed. In addition, the leaving water from the precooling coil is still cold enough to be an excellent source of condensing water, eliminating the need for—and thermodynamically superior to—a cooling tower or air-cooled con-

denser for the refrigeration system.

Both of these techniques, plus use of a heat recovery wheel, were employed in a 100 percent outside-air laboratory air-conditioning system at Pratt Institute, Brooklyn, New York. Figure 1, a schematic of the system, shows the supply and recharge wells, well-water precooling coil, direct-expansion refrigerant coil, the wheel, fans, and compressor. An analysis by its designer, Richard Shaw, P.E., shows that the wheel/well-water system dropped the required refrigeration tonnage from 388 to 60 for that facility, with a comparable saving in operating energy.

HEAT PUMP/SOLAR ENERGY APPLICATION

As part of an energy management research and development program, the Georgia Power Company commissioned its demonstration project, "Atlanta Answer House," to illustrate the combined uses of electric heat pump, heat storage, and solar energy systems.

Atlanta Answer House contains 1984 sq ft of living space. The design is country ranch containing three bedrooms, two baths, great room and fireplace, kitchen, two-car garage, and expandable attic. The structure is 2" x 6" frame construction with wood siding exterior, wood floor over crawl space and partial basement.

Ceiling construction uses raised trusses, with 6" batt and 8" blown insulation, foilbacked sheetrock, giving an R-value of 39. The wall construction is 2" x 6" studs, 24" o.c., 1" exterior foam sheeting, 6" friction fit batts, foilbacked sheetrock, giving an R-value of 24. Floors are constructed using wood and carpet, 6" batt insulation, giving an R-value of 22. Windows are double pane with storm windows, representing only 9 percent of the floor area. Special equipment includes solar heating and water heating with heat pump thermal storage cooling, high efficiency appliances, and fresh air intake type fireplace.

Heat loss for the Atlanta Answer House was 6.1 kw *vs.* 15.9 kw for a minimum property standard house of this type. Heat gain was 12,298 Btuh *vs.* 25,493 Btuh for a minimum property standard house. In order to meet these requirements, a special solar heating and off-peak

storage cooling system was designed.

Solar Heating and Off-Peak Storage Cooling System

The system designed for the Atlanta house integrates: (a) Direct solar heat, (b) Solar-augmented electric-heat pump, (c) Off-peak electric cooling, and (d) Dehumidification.

The characteristics of the system are: (a) Lower electric heat consumption, (b) Load management, with (1) Low winter/summer peak demands, and (2) Off-peak demands, (c) Multipurpose equipment utilization, (d) Pre-engineered package, (e) Simple installation.

The controls involved are a room thermostat, room humidistat, manual heat/cool switch, manual winter/summer switch. The system has been built for ease of installation by integrating the compressor, evaporator, condenser, and all pumps, valves and controls in one package, 2' x 2' x 3'.

The solar collector is a low-cost collector with low installation cost and efficiency equivalent to conventional non-selective coated collectors. The concept is simply to flow a non-toxic, non-flammable, transparent fluid with good wetting characteristics and a very low vapor pressure over a black sheet metal pan under a single sheet of fiberglass glazing. The collector is built into the roof rafter structure with the fiberglass glazing serving as the weatherproof covering. There are approximately 700 sq ft of collectors on the roof.

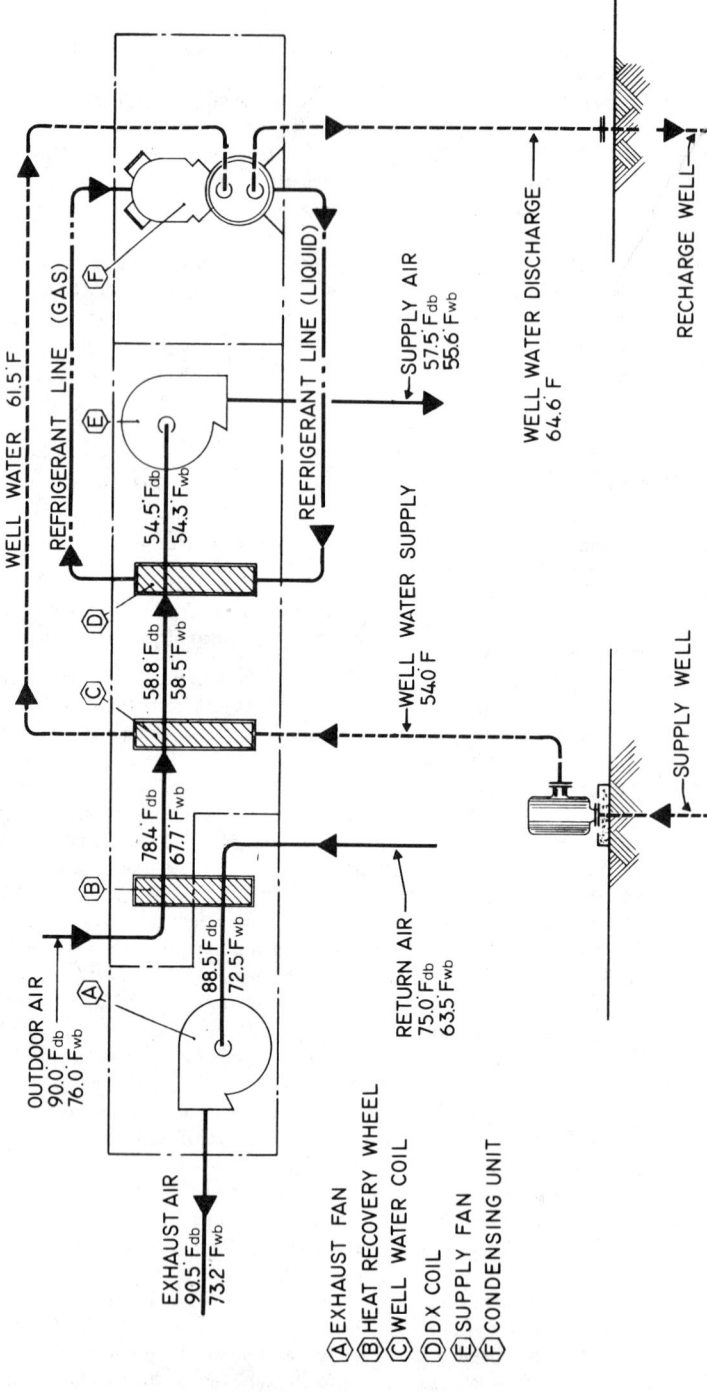

Fig. 1. Schematic of well water precooling and refrigerant condensing system. Tests proved that a reliable source at 54F would be available.

System Description and Operation

A schematic of the components of the system is shown in Fig. 1. They include: (1) Central energy management package, (2) Closed circuit outside water-to-air coil, (3) Domestic solar hot water storage tank, (4) Air handler with hot and chilled water coils and emergency resistance duct heater, (5) Solar collector heat exchanger, (6) Solar collector, and (7) Main water storage tank with immersion heater.

The system operates in nine modes. The proper mode is determined by the input information from: (1) House thermostat, (2) House humidistat, (3) Collector temperature sensor, (4) Main storage tank temperature sensor, (5) Domestic hot water tank temperature sensor, and (6) Position of winter/summer switch (changed once a year). The modes are shown in concept by Figs. 2 to 10.

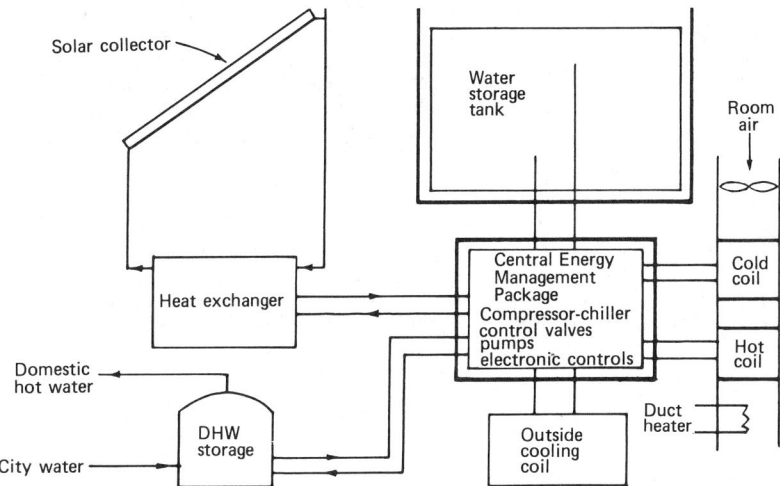

Fig. 1. Component layout.

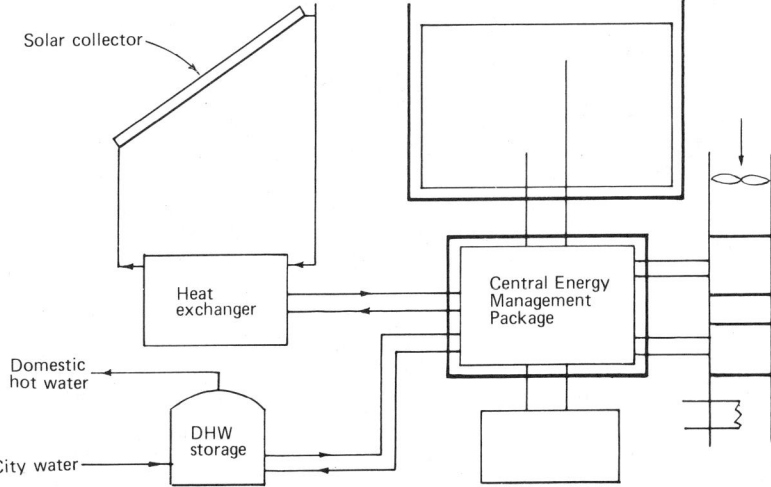

Fig. 2. (Mode 1) Solar heating of domestic hot water. Hot water from the collector heat exchanger is circulated through a coil in the bottom of the domestic hot water storage tank. This mode takes priority over the following mode.

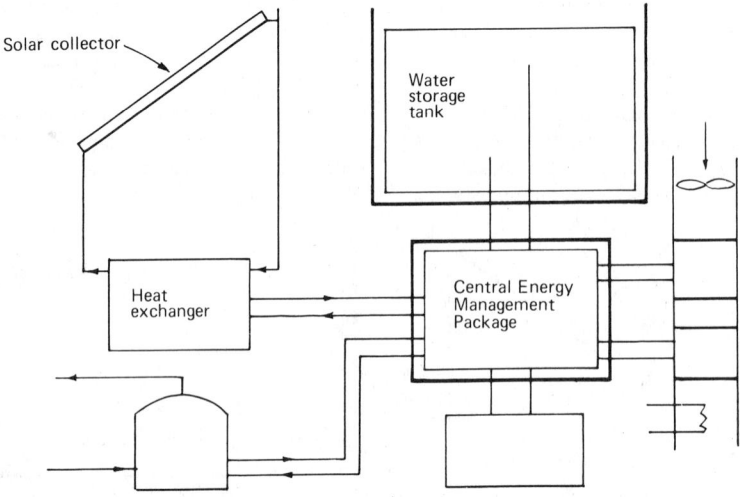

Fig. 3. (Mode 2) Solar heating of main storage tank. This occurs only during the winter and is accomplished by the storage tank water being circulated through the collector heat exchanger.

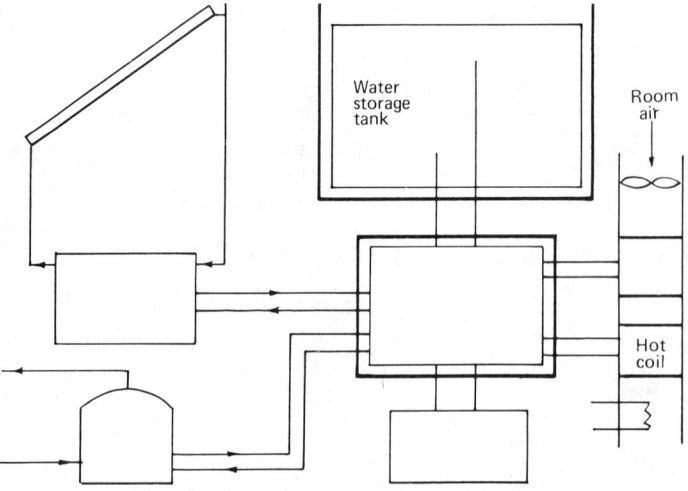

Fig. 4. (Mode 3) Direct solar heating. When the house calls for heating and the tank is above 100°F, hot water from the main storage tank is circulated through the hot water coil in the air handler.

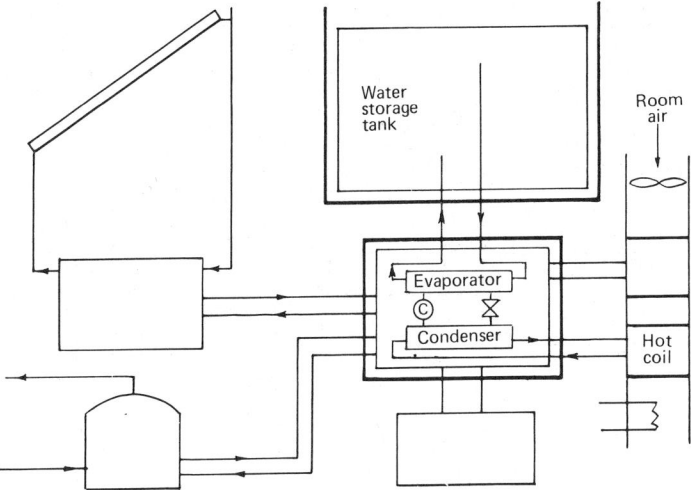

Fig. 5. (Mode 4) Solar augmented heat pump heating. If the main storage tank temperature is less than 100°F, or if the house thermostat goes to its 2nd stage, indicating inadequate heat is available from the direct solar heating mode, the compressor comes on, which acts as water-to-water heat pump, cooling the main storage tank and producing 105°F hot water from the condenser. This hot water is circulated through the hot water coil in the air handler.

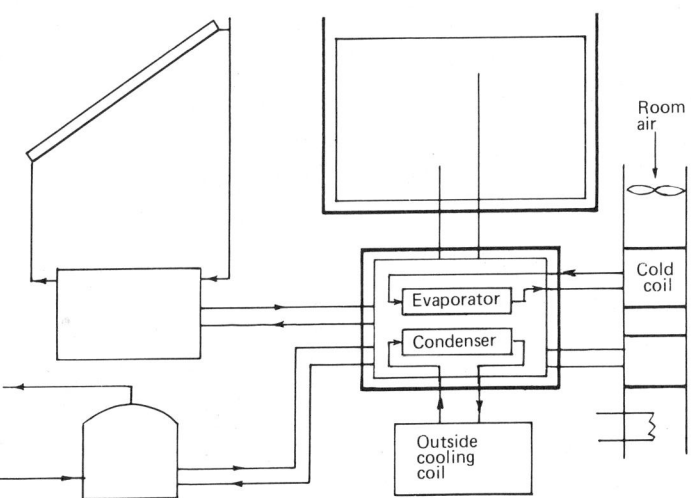

Fig. 6. (Mode 5) Utility off-peak space cooling. During the late fall, winter, and early spring, as well as daily off-peak summer nights, the house is cooled when necessary by the compressor acting as a conventional air conditioner, circulating chilled water through the cold coil in the air handler while the hot condenser water is cooled by the outside water air cooling coil. (This coil is automatically drained to the lower main storage tank when the circulating pump or valve is turned off, to prevent winter freezing.)

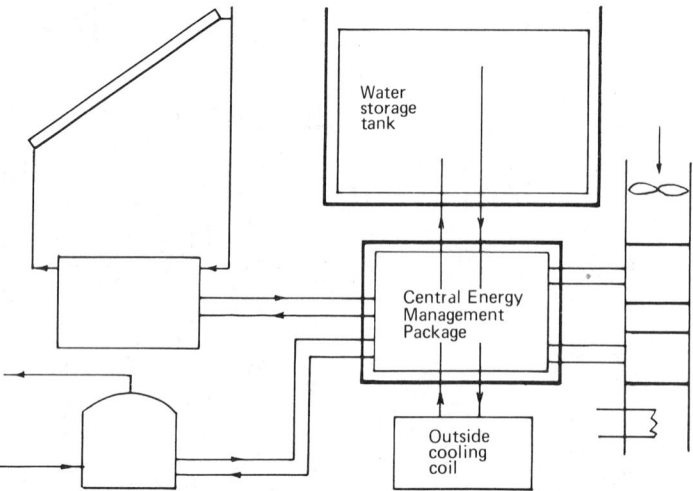

Fig. 7. (Mode 6) Winter to Summer Tank Transition—When the manual winter/summer switch is turned to the summer position, the tank water will generally be hot and is automatically cooled to outside ambient temperature by circulating water to the outside water-to-air cooling coil.

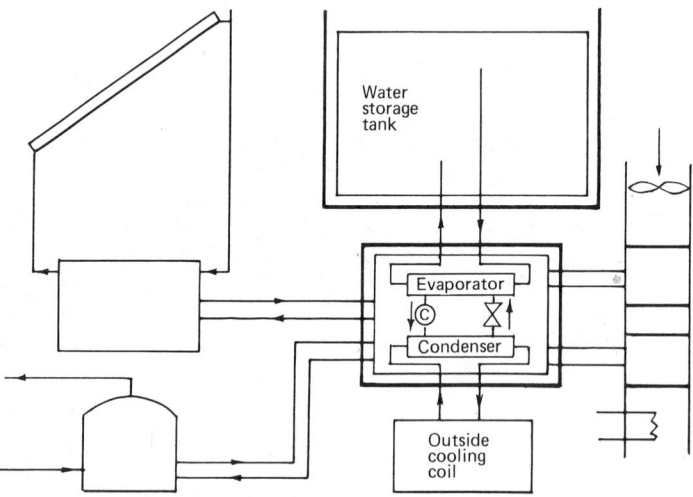

Fig. 8. (Mode 7) Cold storage. During low electric utility demand periods in the summer (nights), a time clock in the central energy management package allows the compressor to run, to cool the tank. When the tank temperature reaches 40°F, the compressor shuts off.

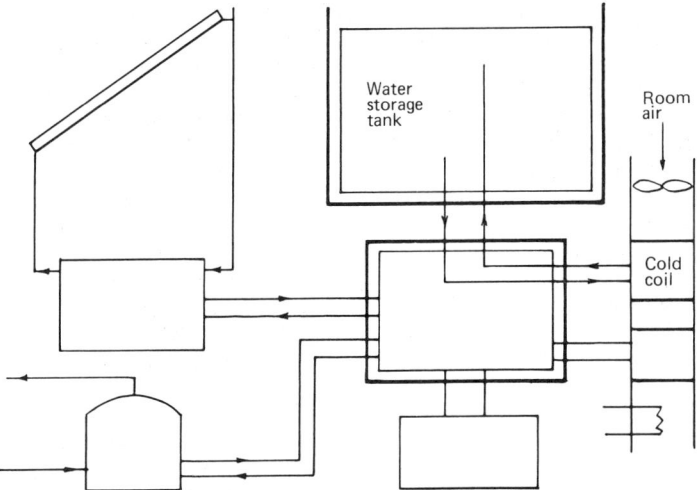

Fig. 9. (Mode 8) Utility on-peak space cooling. During high electrical utility demand periods (hot summer afternoons), the compressor is not allowed to run. When the house thermostat calls for cooling, chilled water from the main tank is circulated through the chilled water coil in the air handler.

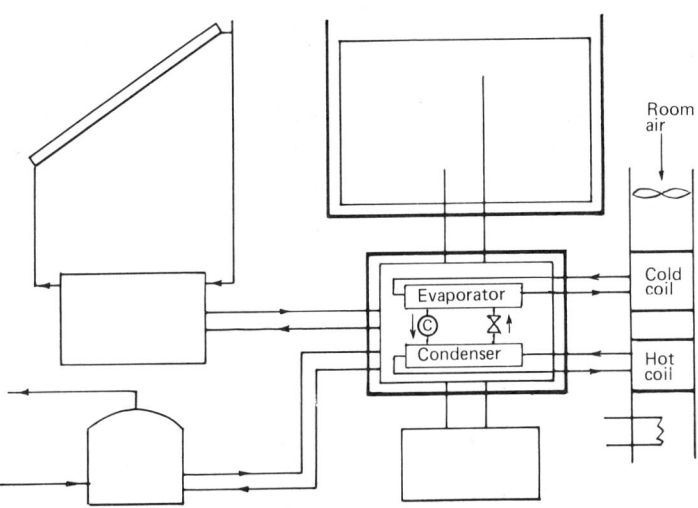

Fig. 10. (Mode 9) Dehumidification. When the house humidistat calls for dehumidification, the compressor comes on, producing chilled water from the evaporator, which is circulated through the chilled water coil to cool and dehumidify the house air. Hot water from the condenser is circulated through the hot water coil to reheat the air before it is returned to the house.

HIGH VELOCITY DUAL DUCT SYSTEMS

Dual duct systems as referred to in the discussion which follows are air conditioning systems for multi-zone applications. They employ two parallel ducts, one carrying warm air, the other cold air, thus providing a constant air source for heating *and* cooling at all times.

In each conditioned space or zone, a mixing valve responsive to a room thermostat mixes the warm and cold air in proper proportions to satisfy the instantaneous heat load of the space. In high and medium pressure systems the volumetric delivery of air is controlled at terminal points in each air mixing unit. In systems with low pressures, the rise in air delivery during partial load operation may sometimes be restricted by controlling air pressures at various points in the duct system.

Advantages and Disadvantages

The principal advantages of dual duct air conditioning systems are: (1) all conditioning equipment is centrally located, simplifying maintenance and operation; (2) the system can cool and heat simultaneously without the necessity of central zoning or seasonal changeover; (3) on all-air dual duct systems there is no water and steam piping, electrical equipment or wiring in conditioned spaces; (4) the system can utilize outdoor air for cooling purposes in intermediate seasons; (5) the system can be used in combination with direct radiation or other conditioning methods without interference or conflict among the several types of systems. This quality of dual duct systems is especially valuable in existing buildings because no change or removal of existing heating system is required; (6) the system has great flexibility to meet specific design or cost objectives on any particular project.

Limitations of dual duct systems in large buildings include: (1) to make the system mechanically stable, a large number of control points must be inserted in the air stream of the distributing system for control of volumetric delivery; (2) the space limitations and the necessity of running two ducts instead of one requires higher velocities and pressures in the duct system than with conventional systems; (3) the arrangement of two parallel ducts with crossovers to terminal mixing units requires special attention and study on the part of designer and installer; (4) accessibility to terminal mixing devices demands close cooperation between architectural, mechanical, and structural designers.

The necessity of using higher pressures and velocities in dual duct conduits for conditioning large buildings introduces additional problems which are actually inherent to all high pressure systems. These problems are: (1) Dynamic losses in high velocity fittings and take-offs are not yet determined with sufficient degree of accuracy; (2) temperature gains and losses in high velocity ducts and flexible tubing also can be determined only approximately and more basic research is needed on this subject; (3) high velocities and pressures introduce acoustical problems. No standard method of presenting acoustical ratings for component parts of high velocity air handling system has yet been established. This requires the designer to rely partly on empirical information for use in achieving desired acoustical results in conditioned spaces.

Dual Duct Cycles

Basic dual duct cycles applicable to multi-zone installations are shown by diagram in Figs. 1, 2, 3, 4 and 5. Modification to the arrangements shown may be dictated by design conditions. These cycles actually represent no new concept in air conditioning process. Thermodynamically, all dual duct

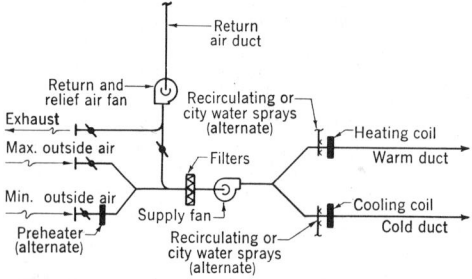

Fig. 1. Single fan dual duct cycle with blow-through dehumidifier.

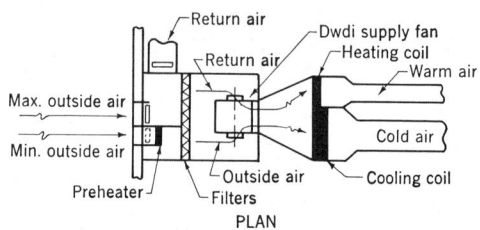

Fig. 2. Single fan dual duct cycle with stratified air.

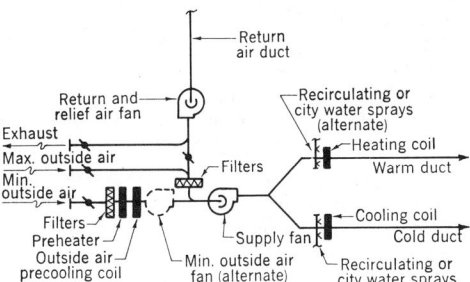

Fig. 3. Single fan dual duct cycle with additional minimum outside air dehumidifier.

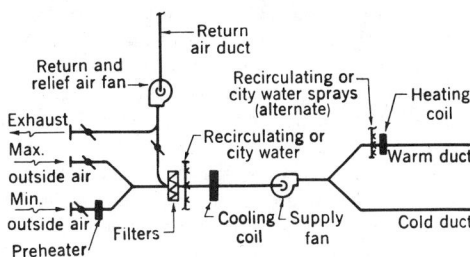

Fig. 5. Single fan dual duct cycle with push-through dehumidifier.

cycles are equivalent to conventional single duct systems, with face and bypass dampers at the cooling coils. Functionally, the cold and warm duct dampers in dual duct mixing units are the same as the face and bypass dampers in a conventional single duct system, except that the dual duct system possesses flexibility automatically to zone each area independently.

Since most buildings are not centrally zoned for dual duct systems, consideration must be given to maintenance of conditions in lightly loaded zones or no-load zones. The ability of the system to maintain conditions in no-load zones might be considered a criterion of adequacy of design for a particular installation. Though it is conceivable that on systems with selective temperature controls even more severe loading conditions may occur, experience has shown that when dual duct systems for comfort application are designed to satisfy the conditions of no-load zones, such systems in actual operation will maintain the desired room conditions at all times.

The no-load zones are represented on psychrometric charts of Figs. 6 and 7 by point 2. Thermo-

stats controlling no-load zones will demand a mixture of air from cold and warm ducts in such proportion as to make the temperature of supply air and room air the same. Practically speaking, provisions for maintenance of conditions in no-load zones will be met if cold air temperatures are kept at a sufficiently low level. The temperature range of 50 to 55 deg F in the cold duct is standardized in dual duct systems.

If the volumetric delivery of supply air is controlled and the ceiling outlets are properly selected,

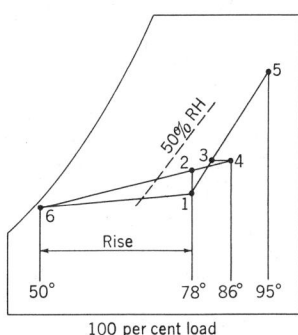

Fig. 6. Psychrometric chart for dual duct system, Fig. 1, full load.

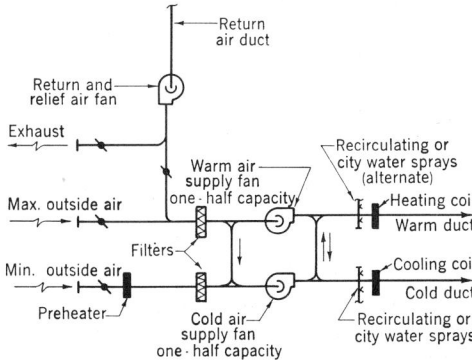

Fig. 4. Two-fan dual duct cycle with blow-through dehumidifier.

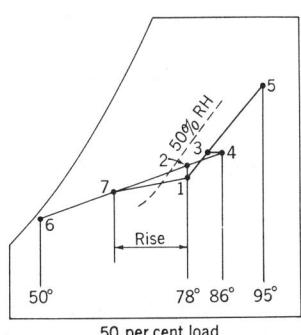

Fig. 7. Psychrometric chart for dual duct system, Fig. 1, 50% load.

temperature differentials (room air temperature minus supply air temperature) as high as 25-30 F can be used. High temperature differentials also can be used if the air from underwindow type units is discharged through high-velocity, high-entrainment type outlets or slots.

Outdoor air dampers at the central apparatus are usually split into two sections, one for minimum, the other for maximum outdoor air volume. The maximum outdoor air damper is interlocked with the return and relief dampers and used whenever outdoor air is favorable for cooling purposes. In climates where outdoor air is not suitable for cooling, only minimum outdoor air need be provided for ventilating requirements. Return fans shown may be eliminated on small installations if provisions are made to relieve the excess outdoor air from the conditioned spaces. The main function of cold chamber sprays is to effect evaporative cooling during intermediate seasons for operating economy. The cold sprays are arranged to operate prior to the opening of the maximum outdoor dampers, thus decreasing the load on the heating coil in the warm chamber. Warm chamber sprays are only installed when control of humidity is required for winter operation.

In recent years, usage of water sprays on dual duct comfort installations has greatly diminished, especially in large cities, because of operating and maintenance problems associated with sprays.

An important point in the design of dual duct systems is the close relationship between the first cost of the system and the total amount of air handled. The first cost and annual owning and operating cost will always be reduced by using lower temperature in the cold air stream—a wide differential between room temperature and cold air supply temperature.

Cycle of Figs. 1 and 2

Fig. 1 shows the simplest, least costly, and most compact apparatus arrangement for dual duct air conditioning.

Thermostats located in each zone control the dry bulb temperature by mixing warm and cold air to give desired results. Close control of relative humidity is not essential and is never attempted for systems installed to produce comfort conditions. It has been established that humidity variations between 35 and 55% are adequate for summer operation with room dry bulb temperatures between 74 and 78 F.

Cold air stream temperature is normally set at 50 F, with the air nearly saturated. The warm air stream, for summer operation, is maintained at not less than 5 F above the average return air temperature. In winter the warm air temperature varies with that of the outdoor air.

Thermodynamically, this cycle is equivalent to a single duct system having face and bypass dampers at the cooling coil, arranged to bypass a mixture of outdoor and recirculated air in response to a zone thermostat, as internal heat load fluctuates. With falling internal heat load and prevailing high outdoor dewpoint temperatures, the cycle will produce a rise in room relative humidities unless reheat is added to the warm air.

Psychrometric representation of this cycle for specific design conditions is shown in Figs. 6 and 7.

In these two figures, the numbers on the psychrometric charts represent: 1. Average room air; 2. No-load zone conditions; 3. Average mixture of room and 25% outdoor air; 3-4. Heat of compression of supply fan; 4. Warm air; 5. Outdoor air; 6. Cold air; 7. Mixture of cold and warm air entering room at 50% load.

Fig. 6 shows the state of air when internal heat gain is maximum. Fig. 7 shows the conditions when internal heat gain falls to 50% of maximum. Note that both points 1 and 2 representing conditions of average room air and no-load zone are higher when the system is operating at partial load. If maximum outdoor design conditions, as shown by point 5 in Figs. 6 and 7 remain constant through the entire range of operation from 0 to 100% of internal heat load, the resulting average room conditions can be shown by curve B in Fig. 8. Corresponding con-

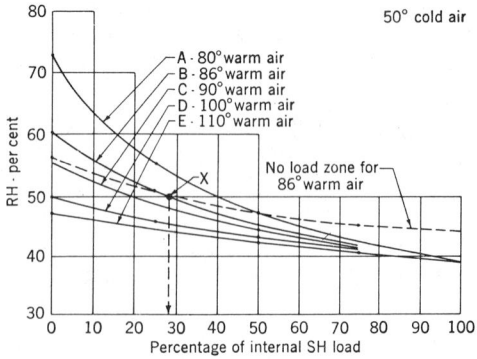

Fig. 8. Performance over range of internal sensible heat load; System of Fig. 1 and Fig. 2 for 50-degree cold air.

ditions of no-load zone could be represented by the dotted line. Additional curves *A, C, D,* and *E* are plotted for arbitrarily selected warm air temperatures and show average resulting room conditions with different temperatures of the warm air.

To limit the rise of humidity in conditioned spaces to, say, 50% rh, the systems shown would require reheat in the warm air when the internal sensible heat load reduces to 28% of maximum, as shown by point *X.* If this system were designed with 55 F air in the cold duct, the conditions of average room air and no-load zone over the whole range of internal sensible heat loads would be as shown in Fig. 9. In such a system reheat would be required when the internal sensible load falls to approximately 64% of maximum. Furthermore, it would be impossible to prevent rise of humidities in no-load zones above 50% rh under any condition. Since, in actual operation, internal partial loads coincide with lower outdoor dewpoint temperatures, the build-up of relative humidities in conditioned spaces can never be as critical as theoretical calculations indicate.

The graphs of Figs. 8 and 9 are based on 25% outside air at 95 F dry bulb, 75 F wet bulb, 98 grains moisture per pound of dry air, and room air temperature of 78 F dry bulb.

This cycle can be used for most multi-zone comfort installations which are operated during regular hours and located in moderate climates. Because there is direct bypass of outdoor air into the warm air duct, the application of this cycle is usually limited to systems which do not require more than 40% of full fan capacity for ventilation of minimum outside air.

The dual duct cycle in Fig. 1 will behave as

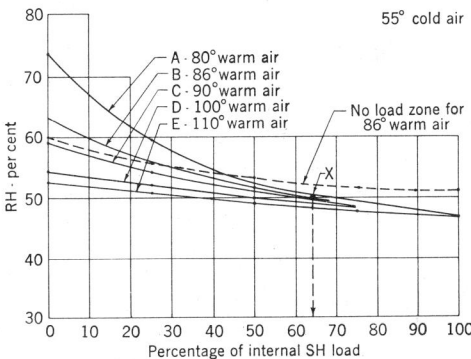

Fig. 9. Performance for same system as Fig. 8 but for 55 deg cold air.

described only if there is a perfect mixture of air on the leaving side of the supply fan. If the air stratifies, as it may happen when double-width, double-inlet fans are used, the performance of the cycle alters. When the apparatus is arranged as shown in Fig. 2, the cycle performance becomes similar in performance to the arrangements shown in Fig. 4. In Fig. 2 the fan discharges the moist outdoor air to the cooling coil and return air to the warm duct coil. This arrangement of the cycle is applicable even to structures located in humid climates because only a small part of moist outdoor air will be bypassed in the warm duct. If the position of return air and minimum outdoor air were reversed, the cycle would be objectionable for all applications. If the apparatus is arranged as shown in Fig. 2 the control instruments in the cold chamber should be of averaging types and care should be exercised not to connect separate mains to opposite sides of the cold or warm apparatus chamber.

It is recommended that a "stratified arrangement" in the apparatus room be used only on smaller installations in warmer climates. Stratified air in larger plants on installations located in colder climates is inadvisable because of danger of freeze-ups of cooling coils and because of control and operating difficulties.

Cycle of Fig. 3

The arrangement in Fig. 3 overcomes the objection of bypassing excessively humid outdoor air to the warm duct on partial loads. The minimum outdoor air is normally precooled to a point which will not necessitate the application of reheat in the warm duct on high summer loads. Precooling with well water or with water leaving main cooling coils is ideally suitable for application to this cycle. Otherwise, the performance of this cycle is similar to that of the cycle illustrated in Fig. 1, but with reduced rate of humidity built up on partial loads.

Again, care must be exercised so as not to stratify the air in cold and warm apparatus chambers, especially on large installations in colder climates.

This cycle could be applied to all systems which might be called to operate at low internal loads while high humidity conditions prevail outdoors. The systems serving hotels, hospitals or apartment buildings fall into this classification.

Cycle of Fig. 4

The two-fan system shown in Fig. 4 is the most economical to operate and will permit very close

control of room humidities when less than half of total air is handled by the warm duct. During this time the system will be equivalent to a single duct system with face and bypass dampers at the cooling coils where only return air is bypassed around the cooling coil. On a further drop in the internal load, which will increase the flow of air in the warm duct over approximately half of the total system capacity, the performance of this cycle will resemble more and more that of the cycle of Fig. 1. The application of reheat to the warm duct coil still will be necessary to depress the relative humidities in the conditioned spaces during partial load operation when dewpoint temperature of outdoor air is high.

The general behavior of the system is shown in Fig. 10. Note that the rate of build-up of humidity is relatively low, while internal SH (sensible heat) load of the system decreases from 100 to 50%. Actually, in this range, no-load curve and average room condition curve, curve *A*, coincide. On further decrease in load, the humidity of both average zone and no-load zone will rise rapidly until it reaches a point where addition of reheat will be necessary. The point of application of reheat is shown at point *X* which occurs when internal SH load is 30%. Note also that the curves of Fig. 10 are plotted on the assumption that maximum outdoor air design conditions will prevail through entire range of operation of the system. As pointed out, in actual practice the system will operate under more favorable conditions and the demand for the reheat will not be as critical as indicated in Fig. 10. This diagram is based on 25% outside air at 95 F dry bulb and 75 F wet bulb with 98 grains moisture per pound of dry air and average room temperature is 80 F dry bulb.

Application of this cycle should be limited to smaller installations located in warmer climates. As stated previously, a stratified arrangement of apparatus on larger installations in colder climates may cause cooling coil freeze-ups and will be difficult to control and to operate.

Cycle of Fig. 5

Fig. 5 shows an arrangement that will permit very close control of relative humidities in the conditioned spaces through the entire range of operation. However, the operating cost of the cycle, because of continuous demand for reheat, is relatively high. Further, additional refrigeration must be provided in most installations to treat the air which will be bypassed to the warm air duct under conditions of maximum loading. For reasons of economy, use of this cycle is not justified for the majority of comfort installations and is restricted only to special applications.

System Design

The design of a dual duct system does not involve any new concepts. All basic principles, procedures and standards are fully developed and appear in the technical literature. The accepted codes and practices of all recognized groups are equally applicable to the design of a dual duct system. The objective of this discussion is to point out and to discuss only the specific problems and practices associated with handling air in two parallel streams.

Calculating Air Quantities

Air quantities required for a dual duct air handling system are derived from equations 1 to 7.

$$Q_c^1 = H_{ss}^1/1.08 \ (t_r - t_c) \tag{1}$$
$$Q_w^1 = H_{sw}^1/1.08 \ (t_w - t_r) \tag{2}$$
$$Q_v^1 = Q_o^1/P_o \tag{3}$$

(The ventilating requirements may also be expressed in terms of air changes per hour, cfm per square foot of floor area, or cfm per person.)

$$Q = Q_z^1 + Q_z^2 + \ldots + Q_z^{n-1} + Q_z^n \tag{4}$$

(Supercripts refer to an arbitrary zone number.)

$$Q = Q_c + Q_w \tag{5}$$

For summer design conditions:

$$H_{ss} + 1.08 \ Q_w \ (t_w - t_r) = 1.08 \ Q_c \ (t_r - t_c) \tag{6}$$

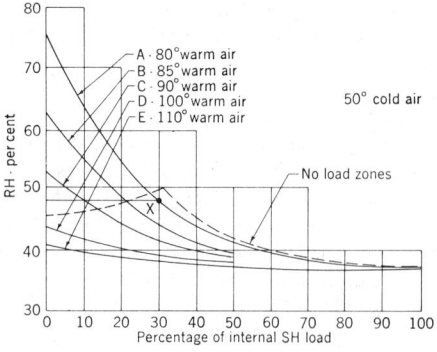

Fig. 10. Performance over whole range of internal sensible heat load for dual duct system of Fig. 4 for 50 deg cold air.

For winter design conditions:

$$H_{sw} + 1.08\,Q_c\,(t_r - t_c) = 1.08\,Q_w\,(t_w - t_r) \quad (7)$$

where:

t_r = room air dry bulb temperature, F

t_c = cold air dry bulb temperature, F

t_w = warm air dry bulb temperature, F

H_{ss} = total internal sensible heat load on a system during summer peak: Btu per hour.

H_{sw} = total internal sensible heat load on system during winter peak, Btu per hour.

H_{ss}^1, H_{ss}^2, etc. = internal sensible heat load of a particular zone during summer peak, Btu per hour.

H_{sw}^1, H_{sw}^2, etc. = internal sensible heat load of a particular zone during winter peak, Btu per hour.

Q = total air handled by the supply fan, cfm.

Q_c = total cold air required by the system during summer or winter peak, cfm.

Q_w = total warm air required by the system during summer or winter peak, cfm.

Q_c^1, Q_c^2, etc. = cold air required in a particular zone for summer peak, when zone receives cold air only, cfm.

Q_w^1, Q_w^2, etc. = warm air required in a particular zone for winter peak, assuming that the zone receives warm air only, cfm.

Q_o^1, Q_o^2, etc. = outdoor ventilating air required in a particular zone, cfm.

Q_v^1, Q_v^2, etc. = total air required by a particular zone to satisfy ventilating conditions, cfm.

Q_z^1, Q_z^2

Q_z^n = the value of Q_c, Q_w or Q_v, whichever is greatest, for each particular zone as shown by the superscript, cfm.

P_o = ratio of outside to total air in the system.

Since all zones of a dual duct system are supplied by a substantial constant volume of air at all times, the air requirements for each zone must be checked for demands of summer and winter peaks and for ventilating requirements. *The highest value is used to establish the zone cfm.*

Equations (1), (2) and (3) are used to establish these demands.

In structures where heating is done by perimeter radiation, only equations (1) and (3) are used for this purpose.

The total air volume to be supplied by the system must be equal to the sum of the peaks of all zones, as stated by equation (4). Peaks for each zone or room are established by equations (1), (2) or (3).

Equations (5) and (6) determine the total cold and warm air quantities for summer peak operation. The second term of equation (6) represents heat bypassed in the warm duct under maximum load conditions. The solution of equations (5) and (7) will establish the cold and warm air demand for winter peak operation. The second term of equation (7) will represent the net cooling effect due to bypass of cold air in the cold duct during the winter peak. The second terms of both equations (6) and (7) actually do not constitute a "system penalty" but represent part of the ventilating load and heat of compression of the supply fan.

Example

The following illustrates how to determine system air quantities by using equations (5), (6), and (7).

Q is given as 60,000 cfm, having been calculated from equations 1, 2, 3 and 4. Also given are:

t_r (summer) = 78 F
t_c (summer) = 50 F   See Fig. 6
t_w (summer) = 86 F
H_{ss} (summer) = 1,400,000 Btu per hr
t_r (winter) = 75 F
t_c (winter) = 55 F
t_w (winter) = 125 F
H_{sw} (winter) = 2,100,000 Btu per hr

Solution

Determine cold and warm quantities for peak summer and winter operation:

For Summer Operation: (Eq 6)

$$1,400,000 + 1.08\,Q_w\,(86\text{-}78) = 1.08\,Q_c\,(78\text{-}50)$$
$$60,000 = Q_c + Q_w \quad (5)$$

Solving equations (5) and (6):

Q_c (summer) = 49,500 cfm
Q_w (summer) = 10,500 cfm

For Winter Operation: (Eq 7)

$$2{,}100{,}000 + 1.08\,Q_c\,(75\text{-}55) = 1.08\,Q_w\,(125\text{-}75)$$
$$60{,}000 = Q_c + Q_w \qquad\qquad (5)$$

Solving equations (7) and (5):

Q_w (winter) $= 45{,}000$ cfm
Q_c (winter) $= 15{,}000$ cfm

Air Quantities

After the air quantities needed for each zone are established by equations (1), (2) and (3), the exact determination of air flow in the cold and warm ducts becomes a rather time-consuming process, especially on larger installations. In cold ducts, for example, it will be necessary to calculate flow at two or three periods of the day to establish the highest rate of flow in different branches. The calculated rate of flow then has to be corrected for the warm air bypass if the design warm air temperatures are higher than the average room design temperatures. During the heating season, actual demand for maximum warm air in each zone has to be calculated and then corrected to compensate for cold air bypass in the cold duct. At the same time, the flow through the warm ducts should be analyzed, considering partial load operation during the cooling cycle.

For a system which feeds both exterior and interior zones, a simplified procedure may be used as follows, to establish air quantities for sizing cold and warm ducts. For most comfort installations this procedure, if followed, will introduce a slight safety factor in the design of the duct system as compared to exact calculations. In general, the steps are:

1. Size the cooling coil for the quantity of cold air calculated from equations (5) and (6).
2. Establish the ratio of total summer cold air to the total system air. In the example given, it will be $49{,}500/60{,}000 = 0.83$.
3. Using Table 1, select a corresponding factor from column B and use this as a multiplier for all zone air quantities as established by equations (1), (2) or (3). The factor for the example would be 0.90. The air quantities established in this manner can be considered to be simultaneous peak requirements for design of cold air ducts.
4. After the cold duct sizes are determined, select the corresponding area ratio of warm duct to cold duct from columns C or D of Table 1 and use this ratio as a multiplier in determining warm duct sizes. The ratio in the example will either be 0.75 (Col C) or 0.85 (Col D), depending upon the type of the system.
5. Select the dual duct mixing units and size low pressure ducts leaving the units, using the total zone cfm as established by equations (1), (2) or (3).

As an alternate to the procedure described, the cold air quantities and cold duct sizes could be determined by any desired method and warm ducts then sized by using factors from columns C or D of Table 1. The alternate procedure may be more desirable when the central systems are zoned or when the duct mains supply air to either interior or exterior zones.

The practice of undersizing warm ducts has been the cause, in many cases, of serious acoustical problems. A system may be "designed" to handle a relatively small quantity of warm air. However, there are many sets of operating conditions which may cause much higher air flow in the warm duct than calculated, thus generating rumble and other noise in the duct system.

Duct Sizing Technique

After the general scheme of the ductwork is established, and the air to be supplied to each section of the system is calculated, the actual sizing of the ductwork does not differ greatly from that used for other systems.

In dual duct systems where volume regulation is accomplished at terminal points, the duct sizing technique becomes even less critical than for other systems, and extreme precision in duct sizing becomes superfluous. The ducts may be sized by equal friction, static regain or velocity methods, or a combination of all three, if desired. Volume regulators in mixing units will absorb all pressure unbalance caused by the initial design, partial load operation or future unavoidable changes in load distribution and will produce a mechanically stable system under all operating conditions, irrespective of the method or procedure used for duct sizing.

On dual duct multi-zone installations there will be an infinite set of operating conditions which will create rates of air flow in the cold and warm ducts entirely different from those used in the original design. For this reason, the possibility of achieving uniformity in duct pressures, for the purpose of volumetric control by some program of duct sizing, is very limited.

Large vs. Small Ducts

Every effort should be made in designing the ductwork to reduce the total power needed to produce the required rate of flow in the supply system. This will assure a quieter system, reduce duct leakages and will approach in the majority of cases the point of maximum owning cost economy.

The basic objective in designing a distribution system is to obtain minimum owning cost by balancing the increment cost of power against the increment cost of ductwork, including all associated costs such as insulation, motors, starters, wiring, etc. For most dual duct systems, the analysis of cost factors indicates economy of larger ducts if the cost of the building structure is disregarded. However, the relative economy of a larger duct as compared to power consumption is not of such magnitude as to justify extensive cost calculation in ordinary design work. Economy will be achieved if efforts are made to keep the total resistance of the system to a practical minimum by utilizing most effectively the space available.

Design Velocity

The chief justification for high duct velocity is space limitation created by past and present architectural and structural practices. By necessity, in a majority of multizone installations, relatively high velocities and high friction rates must be used in some parts of the distribution system. On the other hand, where space is not at a premium, the use of high friction rates is not economically justifiable. In most installations there will always be a point or points of greatest space restriction where the duct velocity must be high. As soon as these points are passed and space becomes less critical, the friction rates should be gradually reduced toward the low end of the duct system.

For example: The longest run may be started with a friction rate of 0.75 inch per 100 ft and then gradually reduced toward the end to say 0.15 inch per 100 ft, giving an average friction rate of 0.45 inch per 100 ft. Such a procedure may involve additional work in estimating the total duct friction but will in general satisfy the requirements of economy. The shorter duct branches, if desired, may be equalized in the usual manner by using the equal friction method of sizing. However, in so doing, care should be exercised to avoid excessively high velocities in smaller ducts. Excessively high velocities in duct fittings, tap-offs, and flexible connections may cause noisy operation.

For installations where no space restrictions exist, the ductwork may be sized by the equal friction method, using low friction rates of, say, 0.20 to 0.30 inch per 100 ft. The shorter branches again can be equalized as previously described. The straight, long risers may be sized effectively by the static regain method, in this manner achieving equal static pressures at the takeoffs for each floor under specified set of operating conditions.

A substantial saving in duct friction is usually achieved by a careful study of the duct mains adjacent to the apparatus room. In order to conserve energy and to prevent regeneration of noise, the connections to the cold and warm plenums should be made as streamlined as possible to reduce the entrance loss to the high velocity duct. In actual layout this is usually accomplished by using a funnel shaped fitting between the high velocity ductwork and the plenum. In many cases the space conditions at the apparatus room are not critical and a large duct can be used in this area. The duct velocity can then be stepped up gradually when the point of space restriction is reached.

The inefficiency of static and velocity pressure conversions must be carefully evaluated because in high velocity systems it may represent an appreciable part of the total resistance to be imposed on the supply fan.

In the main ductwork, it is safe to assume that only 75% of the velocity pressure will be regained. For example, if the highest velocity in the duct is 4000 fpm and the terminal velocity is 2000 fpm, the total loss will be $[(4000/4005)^2 - (2000/4005)^2] \times (1 - 0.75) = 0.19$ inches.

Velocity pressures in the round connection to the last mixing unit can not be credited with regain. This pressure must be considered as the part of the total energy required to create the needed rate of flow through the unit. Rating tables for mixing units are normally based on the static pressures at the entrance of the unit, which means that velocity pressures should be added to this value to establish total pressure loss through the unit.

Maximum Velocity

Maximum velocities recommended for cold and warm ducts are given in Table 2.

For duct elbows having velocities less than 2500 fpm, ordinary turning vanes may be used. For higher velocities the vanes should be streamlined, having no sharp edges. It is also recommended that vanes of streamlined construction be used on all

Table 1—Factors for Duct Design

COL. A	COL. B	COL. C	COL. D
Ratio of system summer cold air to system total air $= Q_c/Q$	Ratio of zone cold air to total zone air for purpose of sizing cold duct	System with supplementary heating at perimeter of building	100% air system
		Ratio of warm duct area to cold duct area	
1.0 to 0.9	1.0	0.70	0.80
.89 to .89	0.95	0.70	0.80
.84 to .80	0.90	0.75	0.85
.79 to .75	0.85	0.75	0.85
.74 or smaller	0.80	0.80	0.90

Table 2—Recommended Maximum Velocities in Ducts

Duct Carries, cfm	Max Velocity, fpm
60,000 to 40,000	6,000
40,000 to 25,000	5,000
25,000 to 15,000	4,500
15,000 to 10,000	4,000
10,000 to 6,000	3,500
6,000 to 3,000	3,000
3,000 to 1,000	2,500

sharp offsets and transformers handling high velocity air. In square elbows the turning vanes should be extended at least 5 inches on the downstream side of the vanes. Extreme care should be exercised in designing vanes for fittings placed in series and located close to each other. With high velocity air, two unvaned or poorly vaned fittings in series might create air turbulence and rumble which may nullify all sound attenuating provisions made in the air handling system.

For better acoustical results in connection with design of high velocity ducts:*

Avoid the use of dampers for adjustment of distribution. They are practically useless in a dual duct system with its inherently unstable pressure characteristics. In systems with relatively high velocities, they are a source of noise, leakage and an unnecessary expense.

Avoid splitters of any kind. They are a source of noise and can cause distributional troubles.

Avoid unvaned rectangular elbows. (All vanes should be of streamline construction.)

Avoid undersizing small ducts and take-offs from mains to mixing units.

---

* See also, Acoustical Problems in High Velocity Air Distribution Systems, elsewhere in this Handbook.

Avoid offsets if possible. When not possible, the transformations and bends should be kept easy; they usually require turning vanes.

Avoid undersizing of warm duct mains under *all* design conditions. (See Table 1.)

Never reduce or restrict warm connections to mixing units.

Do not use tape for sealing high pressure ductwork. Avoid large 90° tap-offs to risers and mains (say, over 64 sq in.). If this is not possible, use straightening vanes at the entrance to avoid air turbulence and regeneration of rumble and other noise.

Do not have less than 5 equivalent diameters between two fittings in series. If this is not psosible, use straightening vanes.
When space conditions permit, use round ductwork.

Do not use tap-off fittings which have sharp edges at the entrance. This applies to both rectangular and round ductwork.

The maximum total pressure which may be used in designing a dual duct system will depend chiefly on the characteristics of the mixing units used. Constant volume units are commercially available which will absorb up to 8 inches of static pressure.

Sizing High Pressure Ducts—Example

A system supplying 60,000 cfm of cold air is diagrammatically shown in Fig. 11. The space conditions are critical at the top of the riser and at the

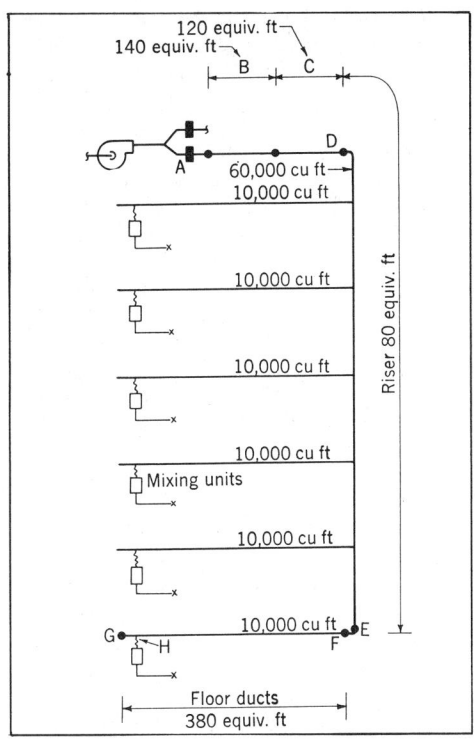

Fig. 11. Dual duct system showing cold duct only.

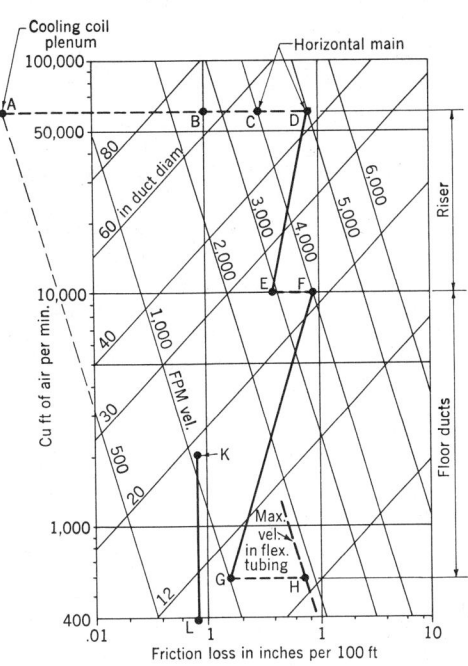

Fig. 12. Air friction chart; design conditions of dual duct system of Fig. 11.

take-off to each floor, as designated by points D and F. The maximum permissible duct sizes are determined at these points and corresponding velocities are plotted on the air friction chart in Fig. 12. The increase in air velocity from point A, in the plenum chamber, to point D is accomplished in two steps. Part of the horizontal main is sized for 2600 fpm and part for 4000 fpm, as indicated by points B and C. These velocities are again determined from consideration of space conditions. The vertical riser sizes are selected on line DE. This line closely resembles that static regain line, assuming that 75% of the velocity pressure is regained in the riser resulting in 25% loss. The horizontal ducts on each floor are sized on line FG. The point G is arbitrarily picked for a relatively low friction rate. The velocity in the flexible connections to the mixing units are stepped up to point H, as required by the inlet size of the selected unit. The low pressure ducts leaving the mixing units are sized on equal friction line KL.

The total resistance of the system is calculated as follows:

Item	Inches of Water
Apparatus	1.25
Horizontal main B —	
140 equiv. ft @ 0.1 inch per 100 ft	.014
Horizonel Main C —	
120 equiv. ft @ 0.23 inch per 100 ft	0.28
Riser DE —	
80 equiv. ft @ 0.56 inch per 100 ft (aver.)	0.45
Vel. head @ 5000 fpm = 1.56 inch	
Vel. head @ 2800 fpm = 0.49 inch	
_____	
1.07 @ 25% loss	0.27
Floor Ducts FH —	
380 equiv. ft @ 0.40 inch per 100 ft (aver.)	1.50
Vel. head @ 4000 fpm = 1.00 inch	
Vel. head @ 1000 fpm = 0.25 inch	

.75 @ 25%   0.19

Loss through flexible conn. and tap-off *H*   0.25

Last dual duct unit   1.00

Low pressure ductwork and outlet   0.15

Total inches of water 5.46

After the cold duct sizes are determined, it is more practical to size the warm duct by using factors from Table 1. It is obvious from the example considered that no set procedure can be followed in sizing the duct system. The judgment of the design engineer and space conditions of each project will always be the chief factors in selecting duct sizes.

After the duct sizes are determined by a method such as described, careful examination of the entire duct system is recommended and further changes in the duct sizes can be made by: (1) Eliminating unnecessary transformations, thus keeping the number of duct sizes to a minimum; (2) standardizing the duct sizes in order to have as many duplicate sizes as possible; (3) increasing, by inspection, the end run sizes of longer branches or those branches where load may be expected to increase in the future; (4) eliminating unnecessary fittings, crossovers, and so forth.

This procedure may increase the weight of the metal slightly but it will reduce the actual cost of the installation and assure a more flexible, more quiet and tighter distributing system.

Return Air Ducts

Return air ducts are sized at low velocities as in any conventional system. Attempts to pressurize the return duct have not as yet proved to be entirely successful; clogging of high pressure throttling valves by lint and dust still presents a problem. If pressure reducing valves are used in the return air ductwork, easy access should be provided for cleaning. One scheme to simplify the arrangement of the dual duct system is to use hung-ceilings or corridors for return plenums and collect all return air at central points on each floor. In some localities there are code restrictions on this method.

Low Pressure Ductwork

Low pressure ducts leaving the mixing units are sized as any other conventional ductwork. Normally this ductwork is sized for a friction rate of 0.08 inch to 0.10 inch per 100 ft based on ASHRAE duct friction charts, which will give an approximate friction rate of 0.10 inch to 0.12 inch per 100 ft

if ductwork is lined. Since, in the majority of cases, the dual duct systems are designed for low cold air supply temperatures, the low pressure ductwork must be either lined or insulated externally. The internal lining is preferable in all cases for acoustical reasons.

Experience has indicated that the least expensive and the safest way to assure a quiet installation is to have some length of lined ductwork on the leaving side of the mixing unit. The lined ductwork, especially when it contains one or two elbows, is a very effective sound attenuator.

The lined ductwork will also provide a necessary safety factor in acoustical design if noise regeneration should occur in the air distributing system because of poorly constructed ducts, fittings and tap-offs.

Basic Arrangement

Development of the basic scheme of air distribution suitable for a particular structure is the most important function in the design of a dual duct system. The first cost, owning and operating cost, and future flexibility will always reflect the skill with which the basic planning of the system is executed. In every installation, several schemes of air distribution will be feasible and a careful evaluation of all schemes must be made before selecting the one most suitable for the installation.

The cost of the duct system forms a relatively large part of the cost of the total installation. If the system is simple, the weight of metal in the duct system may be, say, 1 lb per square foot of conditioned area. If the system is complex, with long runs of low capacity, this weight may rise to 3 or 4 lb. Assuming the erected cost of metal with its associated components to be $1 per lb of metal, the penalty for a complicated arrangement becomes obvious.

Special effort should always be made to achieve simplicity in the duct system, by utilizing structural and architectural features of the building, judicious location of risers, mains, and apparatus room.

As pointed out previously, overpressurizing of the system should be avoided in all installations. High pressures are always associated with higher owning and operating costs and, in addition, magnify noise and vibration problems. Bear in mind that the actual increment cost of a larger duct is not an average cost of the ductwork but consists mainly of the added cost of the metal and can be assumed to be only 15-20 cents per pound.

Zoning. The cost of the installation is greatly affected by the general zoning arrangement of the system. If the structure is subdivided into many small zones, when larger zones would meet the objectives of the design, the increase in the installed cost of the system represents a very substantial amount of money.

However, on an average comfort installation, little will be gained by using excessively large units (zones). As it can be seen in Fig. 30, the point of diminishing return will be reached when the size of the unit will be increased to, say, 1200-1400 cfm. Under some design conditions, the layout based on large units may be even costlier than the layout where the spaces are served by units of medium capacity.

In existing buildings, it will be more economical to leave the existing perimeter radiation system as it is and to condition all spaces with overhead units.

In some installations only part of the structure must be conditioned, but provisions for treating the remainder must be incorporated in the design. Existing buildings with large numbers of tenants are often treated in this manner. A dual duct system with automatic volume regulation at terminal points is ideally suited for installations of this type. If the central apparatus and duct mains are installed, any part of the building may be provided with mixing units and the system put into operation prior to its entire completion. When additional units are installed, no rebalancing of the original units will be necessary, as all changes in duct pressures will automatically be compensated for at the terminal points.

Typical Floor Layouts, Existing Office Buildings. Floor layouts shown in Fig. 13-15 are applicable to existing office buildings or to new office buildings where only limited amount of glass is used in the facade of the building. The modular arrangement of buildings shown is such that each modulus on the perimeter wall can be subdivided by partitions. This is usually the case when the moduli are 7 ft or larger.

Fig. 13 shows a typical floor of an office building

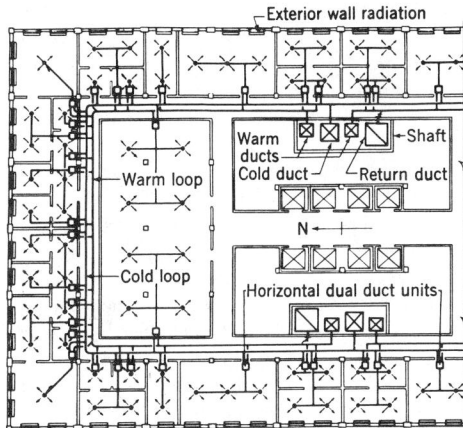

Fig. 14. Typical office building dual duct layout with supplementary radiation at exterior walls, zoned to provide selective temperature control for each enclosure.

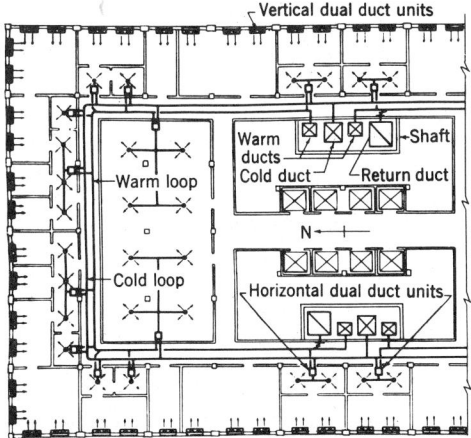

Fig. 15. Typical office building dual duct layout, 100 per cent air system, zoned to provide selective temperature control for each enclosure with vertical dual duct units at exterior walls.

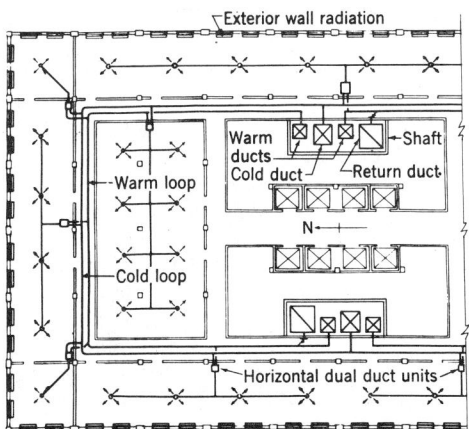

Fig. 13. Typical office building dual duct layout with supplementary radiation at exterior wall with minimum number of zones.

arranged for a minimum recommended number of zones. The basic minimum number of zones in a building of rectangular shape should be as follows: (a) Four exposure zones; (b) four corner rooms with windows in two walls; (c) interior zones as required by type of loading or occupancy.

In modern buildings, the need for zoning of interior areas with heavy population and business machinery loads is actually greater than in exterior areas of the same exposure. The basic need for zoning exterior areas exists only when one side of the building is partially shaded by the adjacent structures or selective temperature control is needed for each room. The arrangement shown in Fig. 13 is supplemented by perimeter radiation for winter operation, which may be controlled independently or in conjunction with the air conditioning system. The system arranged in this manner obviously will be the lowest in first cost and could be applied to both new and existing buildings for conditioning open floors or non-executive areas. Each large zone may consist of a separate mixing unit or of two or more units discharging air into a common duct, as shown on the west side of the building. If the ductwork is arranged in the manner shown, any change in future zoning arrangement can be easily accomplished by installing additional mixing units where required.

Figure 14 shows the same basic layout as indicated on Fig. 13, with the exception that all spaces are equipped with additional mixing units for selective temperature control. If the horizontal duct mains are properly arranged, future changes in mixing units and zone arrangement are easily made without disturbing the main system. When mixing units are equipped with volume regulators, no rebalancing of units will be required if the pattern of air distribution is changed. In some cases, the mixing units are arranged on a modular basis providing one mixing unit for each exterior modulus. A design executed in this manner has the merit of greater flexibility for future partition changes and should be considered for installations where this is expected.

In buildings located in warmer climates, exterior wall radiation can be omitted and the entire cooling and heating load can be handled by the air supplied from overhead. The air can be distributed through ceiling and sidewall outlets or through a combination of air diffusers and light troffers. In Europe, especially in Switzerland and Italy, there are very successful dual duct installations with air distribution through perforated ceilings.

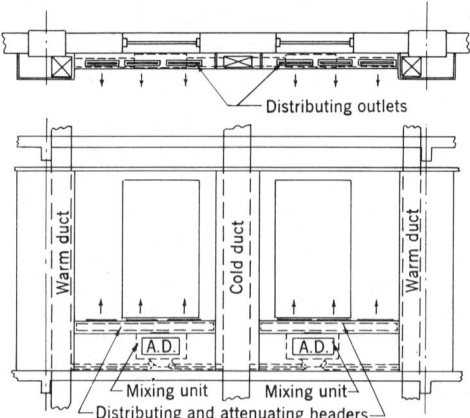

Fig. 16. Arrangement for vertical underwindow dual duct units connected to risers—one unit per riser.

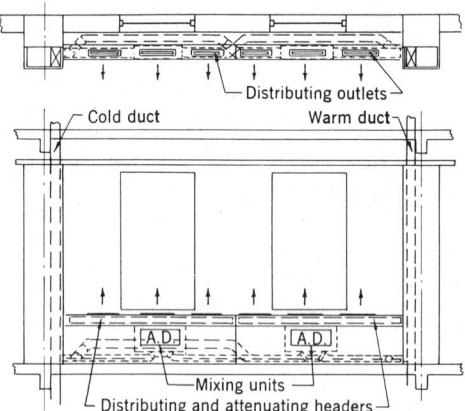

Fig. 17. Arrangement for vertical underwindow dual duct units—two units per riser.

Figure 15 illustrates a typical layout with exterior zones supplied by underwindow air mixing units. The units may be connected to vertical risers, as shown in Fig. 16 and 17, or to horizontal mains located at the ceiling or on the floor below, as shown in Fig. 18. The vertical risers for exterior units are usually collected on one or more floors to form a separate distributing header or loop. The exterior zones might be supplied by a separate air handling system or be a part of the interior system.

When space is not available under the windows for the installation of self-contained units or when the exterior wall, for reasons of economy, is subdivided into large moduli, the exterior can be supplied by horizontal units located at the ceiling or on

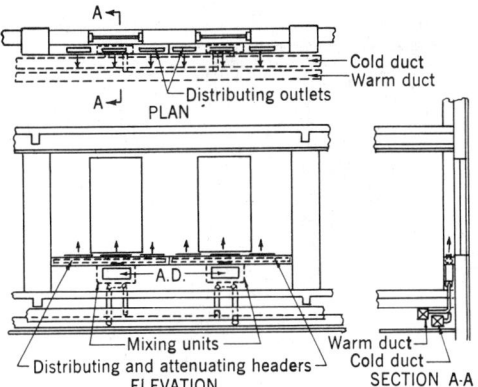

Fig. 18. Arrangement for vertical underwindow dual duct unit connected to horizontal mains.

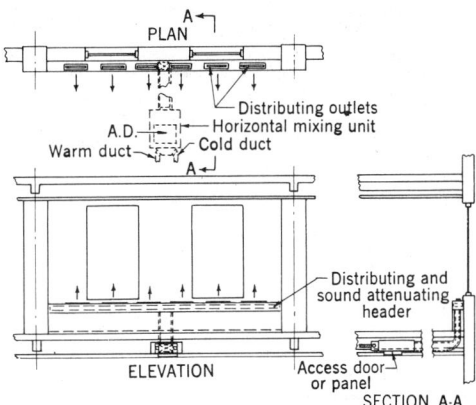

Fig. 19. Perimeter header fed by horizontal unit.

the floor below, as shown in Fig. 19. The air from the unit is discharged into a distributing and sound attenuating header located at the perimeter of the building. The chief objective of the distributing headers, as shown in the layout, is to prevent down drafts during cold weather.

The high pressure horizontal supply ducts feeding individual floors may be arranged in a conventional manner or form a system of loops, as shown in Fig. 13-15. The loop system has an obvious advantage when the floors are supplied by more than one central air handling apparatus. If the air flows freely through the ends of the loops from one side of the building to the other to satisfy the sun load, appreciable economy is effected in designing the cold circuits of the central apparatus. With the loop system, the combined coil capacity of all apparatus is sized to take care of the peak of the entire building. If two separate apparatus feed the building, one the east and another the west side, each apparatus must be designed to provide for the peak load of its side of the building.

With more than one apparatus supplying a building, combined by looping the supply ducts, each central apparatus can be the same size, thus simplifying the layout work and providing a saving in the first cost of the installation by a duplication of equipment. The looped system also permits partial conditioning of the building by operating only half of the central fans for night, week-end operation, or during a breakdown.

Experience has shown that the best location for high pressure ductwork and mixing units is in corridors or in areas adjacent to the service core of the building. If space for the installation of units is not available in the corridor, the unit can be moved to the adjacent spaces.

Ceiling Plenum. A dual duct layout will be greatly simplified if the return air system is arranged as in Fig. 13-15. If the space above the hung ceiling is used for a return air plenum, all return air may be drawn into the ceiling plenum through conventional grilles and registers or combination light fixtures and return grilles and collected at central points near the risers. In some designs, it is sufficient to use as return plenums only the space above corridor ceilings where, normally, high pressure ducts and units are located. Under some design and layout conditions, it is possible to use both the corridors and the hung ceiling spaces for collecting and conveying of return air. However, local codes in some cities prohibit or restrict the use of hung ceilings and corridors for return plenum chambers.

The most critical points in a dual duct system will generally be the areas near the main shaft where the ducts enter a floor and branch off into horizontal mains. A careful study of space conditions must be made in these areas and the ducts sized at the lowest velocity possible. To avoid crossovers of horizontal mains, the risers can be rearranged by dividing the warm riser and placing a cold riser between two warm risers, as shown in Fig. 13-15.

Modular Type Office Buildings

When new, modular type buildings with large percentages of glass in their facades are conditioned by perimeter mixing units, these can be arranged according to several alternatives as shown on Schemes A, B, C, D, or E in Fig. 20.

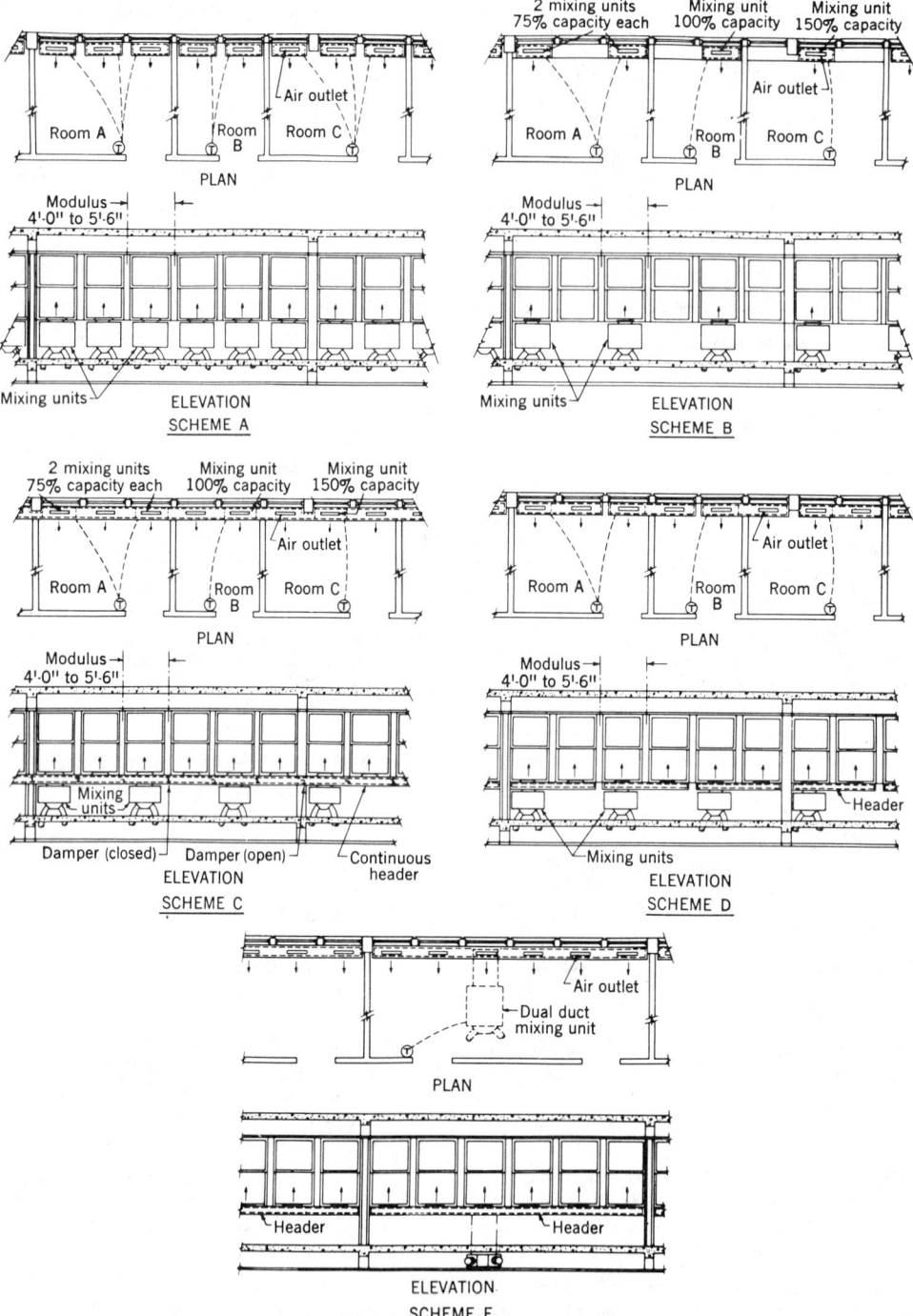

Fig. 20. Modular arrangements of dual duct mixing units. (Reprinted by permission from ASHRAE Guide and Data Book, 1964.)

It is obvious that infinite flexibility for future partition changes could be obtained if one mixing unit is placed in each building modulus, as shown by Scheme *A*. By discharging a fixed quantity of air upward to blanket the window in each modulus, proper air distribution will be obtained in all conditioned spaces and the possibility of downdrafts during cold weather operation will be eliminated. Since each unit must be sized to take care of the maximum load of the particular modulus it serves, rebalancing of units will not be required with future partition changes. The installed cost of the system based on this scheme will be the highest, and for this reason it should be used only when the building moduli are large enough to be individually partitioned.

If the building moduli are small (4-5½ ft) and at no time will less than two moduli be partitioned off, Schemes *B*, *C*, or *D* can be selected. Although almost identical design objectives will be accomplished by Schemes *A, B, C,* and *D,* the installed cost of the last three schemes will be lower by approximately 70 cents to one dollar per square foot of exterior area on an average office building.

In warmer climates, or in moderate climates if windows are protected by double glazing, units at the perimeter can be arranged according to Scheme *B*. All mixing units for this scheme should be sized initially to take care of three moduli. Thus, the units may have to be rebalanced as required by future partition changes. Each unit may be called upon to supply 75, 100, or 150% of air needed for two moduli, as indicated by location of partitions for rooms *A, B,* and *C,* respectively.

Because of cold downdrafts at unprotected windows, this scheme is not applicable to buildings located in colder climates. If it is desired to eliminate the possibility of cold downdrafts at the perimeter glass in buildings located in colder climates and at the same time to provide infinite flexibility for future partition changes, Scheme *C* can be selected. With this scheme, the air from the mixing units is discharged into a common header with an air outlet in each building modulus. Complete separation of two adjacent spaces can be obtained by dampers (or friction type baffles) installed in headers as indicated on the elevation drawing. As with Scheme *B*, the mixing units for Scheme *C* should be sized initially to provide air for three moduli, if that is required, by partition arrangement. Each perimeter unit may be called upon to supply 75, 100, or 150% of the air required for two moduli. Scheme *C* is especially suitable for

areas where two adjacent spaces may have different load intensities or where the reversal of load in the adjacent spaces is possible. This condition is usually met when the building is either self-shaded or is shaded by adjacent buildings.

If the character of the internal load for two adjacent spaces on each exposure of the building is subject to only moderate variations, the arrangement of units as indicated in Scheme *D* will suffice. The headers for this arrangement will extend over two moduli only and future partition changes will not require rebalancing of air mixing units.

For general areas which will not be partitioned, or where selective temperature control in each cubicle is not required by the design criteria, one mixing unit per bay should be provided as shown by Scheme *E*. Little will be saved by serving more than one bay with one mixing unit.

In all schemes illustrated in Fig. 20, it is assumed that exterior spaces are partitioned from the interior areas. In such cases, it is possible to install room thermostats on the back walls, adjacent to doors or return grilles as shown. When the exterior and interior areas are not separated by partitions, room thermostats and thermostatic bulbs should be mounted inside the perimeter type mixing units or in the air distributing headers. One of the methods of mounting and actuating the thermostatic bulb inside the mixing unit using an air sampler is shown in Fig. 21 and 22.

When room air samplers are used, it is recommended that high-velocity high-entrainment type grilles or slots for discharge of air from perimeter mixing units be installed. This will assure that representative air from the occupied zone is drawn

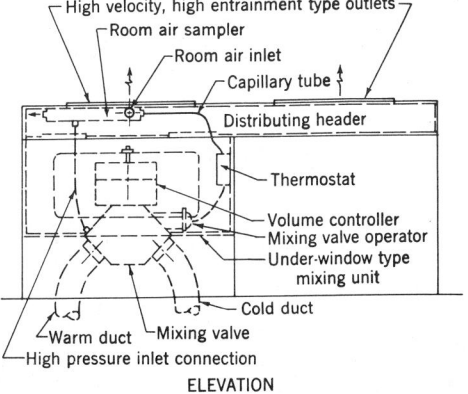

ELEVATION

Fig. 21. Method of mounting room air sampler in mixing unit.

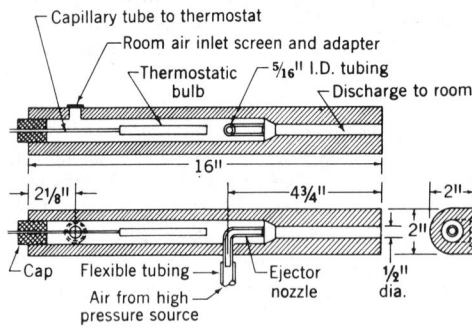

Fig. 22. Room air sampler tube.

toward the air inlet of the room air sampler. Locating room thermostats on exterior walls or columns should be avoided.

When using Schemes C and D of Fig. 20, further savings in installed cost of the system can be achieved by providing only one mixing unit per three or four moduli in all corner rooms.

Schemes A, B, C, D, and E in Fig. 20 can be applied also for overhead distribution of air. For example, in Figs. 23 and 24, an overhead air distribution arrangement is shown for Schemes C and D, respectively. The arrangement of outlets shown is schematic and any type of air distributing devices can be adapted with the illustrated schemes.

Many patterns of duct distributing systems are

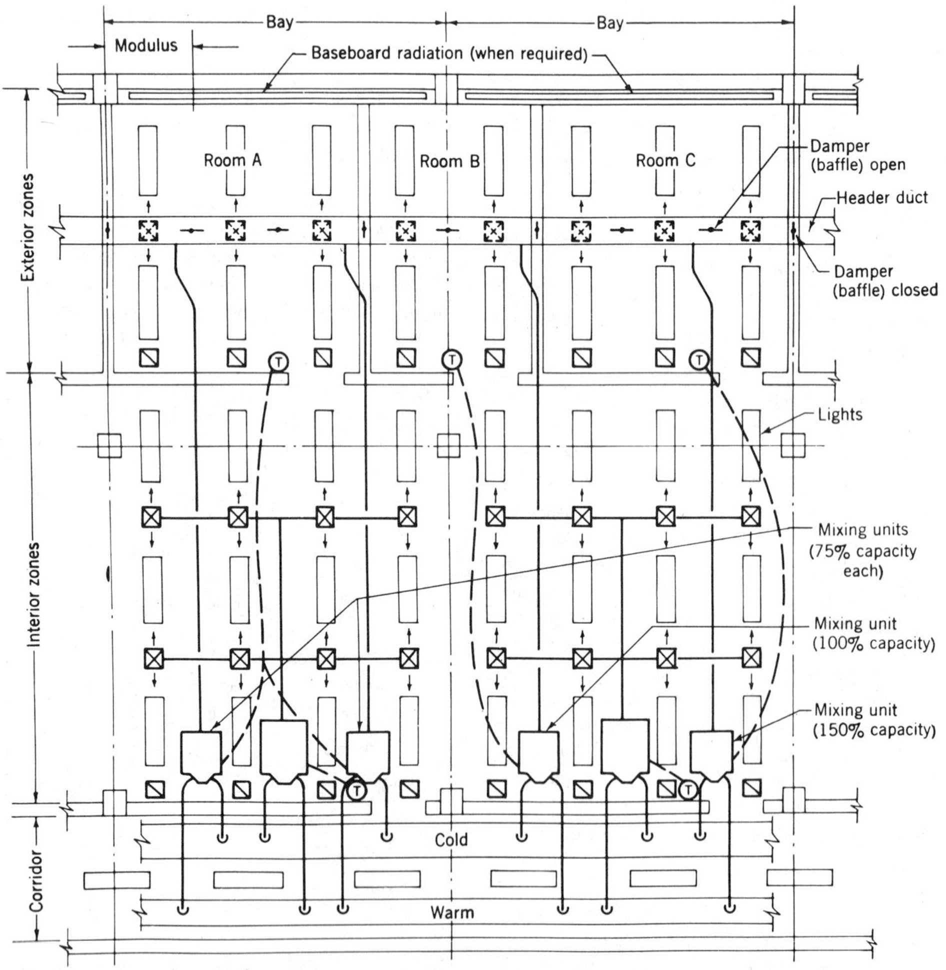

Fig. 23. Modular arrangement for overhead dual duct system, Scheme C.

possible with dual duct systems to meet a variety of design objectives for a particular installation. Cost and space requirements for distributing systems can be minimized by properly organizing the element of distribution. The emphasis on organization has resulted in the modern trend toward integrated designs, where structural elements of the building are used for distribution of air, thereby reducing the cost of sheet metal and building space normally occupied by the ductwork.

Guides to Layout. In laying out distribution systems one does not simply follow the technique used in low pressure layouts by running high pressure ductwork over an entire conditioned area. The ex-

tent of high pressure mains on each floor should be limited to well defined areas and air distributed to individual rooms by low-pressure, internally lined ductwork. The high pressure ductwork is considered a high velocity plenum which remains intact for the life of the structure. In order to have maximum future flexibility and to avoid the danger of obsolescence, all future changes due to shifting loads and partition changes should be confined to low pressure ductwork and mixing units.

Layouts consisting of large numbers of small dual duct mixing units should be avoided unless dictated absolutely by the design conditions. A layout made in this manner will be the highest in first cost.

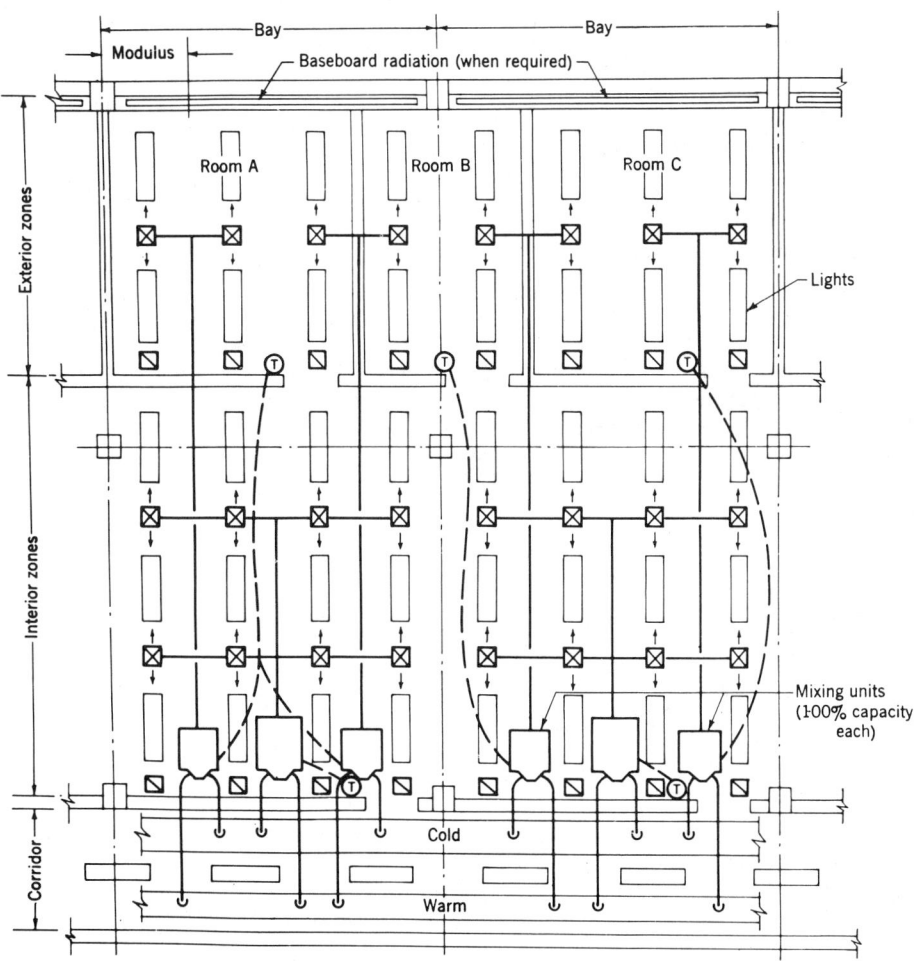

Fig. 24. Modular arrangement for overhead dual duct system, Scheme D.

In making alternate air distribution schemes for a particular project, do not rule out a scheme consisting of perimeter radiation and overhead air system. On many projects such a scheme will be the least expensive in first cost.

Overhead mixing units of small capacity, unless absolutely necessary for zoning requirements, should be avoided. In order to decrease the cost of installation, use of a larger unit is better with air distributed to individual outlets by low pressure lined ductwork. However, little will be gained by increasing sizes of the units over, say, 1200-1500 cfm.

Mixing units located as close to the high pressure mains as possible will facilitate future changes and servicing.

The high pressure duct system should be designed (and shaped) from consideration of more than present loads only. Internal loads may increase or change in the future and the supply fans will have to be speeded up or a new booster air handling system connected to the distributing system. With high pressure dual duct systems, the provisions for future adequacy of the installation can be made at a very small additional cost.

Constant Volume Mixing Units

At the present time there are several manufacturers of dual duct air mixing units who offer equipment with automatic control of delivered volume. Arrangement of the mechanical components vary greatly among the manufacturers but the basic method of automatic control of volume and temperature falls into only two categories. These two methods are shown in diagrams of Fig. 25 and 26.

For this discussion they will be referred to as Control Method *A* and *B*, respectively.

Method A is a direct control of both temperature and volume. A thermostat controls the mixture of cold and warm air and a self-contained, spring-balanced, volume regulator controls the volume at the discharge of a 3-way air mixing valve. With this arrangement, as long as there is sufficient air pressure at both warm and cold inlets of the mixing valve to overcome the resistance interposed by the mixing unit, the volume will remain constant and the thermal quality of the air delivered will be responsive to the requirement of the thermostat.

If the pressure at either inlet of the valve should fall below the point necessary to overcome unit resistance, the thermostat will throttle or close off the other inlet and the volume delivered by the unit may fall below the specified quantity but the temperature control of the reduced quantity will not be affected. This is especially important under special design and operating conditions during the heating season which will be discussed later.

When using Method *A* control, care should be exercised to avoid creating conditions where air from one duct can be by-passed to the other through the mixing units. If such conditions do exist or may be anticipated, air check valves should be installed in the inlet of the units. Introduction of air check valves may be required:

(1) When air mixing dampers are hand operated.

(2) When room thermostats controlling the units are not properly located or adjusted.

(3) When a smooth operation of air mixing dampers is prevented by such mechanical defects as

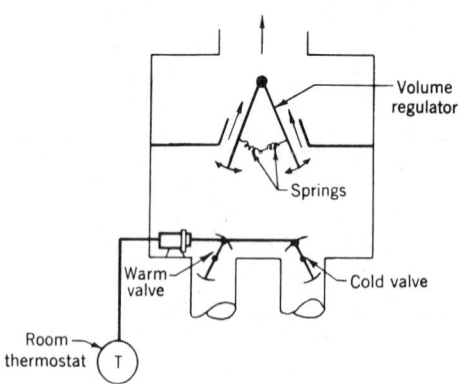

Fig. 25. Type A dual duct constant volume unit.

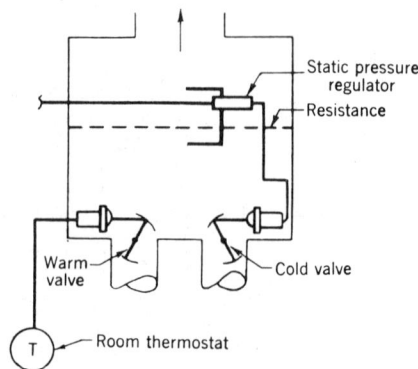

Fig. 26. Type B dual duct constant volume unit.

(a) folded or leaky diaphragm operator of pneumatic motor.

(b) mixing valve linkages out of adjustment.

(c) any other mechanical defect which may cause binding of the valve in an intermediate position.

(4) When one room thermostat controls mixing units of different sizes and when movement of mixing valves in such units is not properly synchronized. It is imperative that volume regulators in Method A-type units have a positive means to eliminate possible fluttering of the units under conditions of extreme turbulence.

Method B is indirect in controlling both temperature and volume. Method B units may be actuated by compressed air or by the air pressure of the system. The results are the same. A space thermostat controls only the warm air inlet and a static pressure regulator controls the flow through the cold air inlet. The pressure drop across a resistance plate is kept constant to give fixed outlet volume. As the thermostat throttles the warm air, the static pressure regulator opens the cold air inlet, maintaining constant volume regardless of the resulting temperature of the mixture. If the warm air pressure is insufficient to overcome the resistance of the mixing unit, the thermostat will open the warm air inlet wide when calling for heat, but low pressure will prevent full flow of warm air and the static pressure regulator will therefore open the cold air inlet until the volume control is satisfied. Thus, insufficient pressure in the warm air duct can produce cooling of the conditioned space when the thermostat is calling for heat.

Several normal design and operating conditions can produce low pressure in some parts of the system. Some of the normal reasons for insufficient pressure in warm air ducts are:

(1) In starting up after shutdown in cold weather, if the warm ducts are substantially smaller than the cold ducts.

(2) When supply fans are run at low speed for night heating. Normally night and week-end operation do not require maintenance of exact conditions and low speed operation of fans saves power.

(3) When one of several fans feeding the same duct system is non-operative, as during night heating, servicing or break-downs.

(4) When the warm duct temperature falls substantially below the design temperature due to faulty action of central thermostats, failure of steam pressure, etc.

(5) When warm ducts are sized without due consideration of partial heating loads or intermediate season operation while low temperatures are maintained in the warm ducts.

(6) When an excessive temperature drop occurs in uninsulated warm ducts.

(7) When a partial or complete obstruction occurs in the warm duct.

(8) When the air distributing system is used in conjunction with perimeter radiation there will be an infinite set of operating conditions which may cause a drop of pressures in the warm ducts. This may be caused by improper timing in applying radiation or by inadequate temperatures in the warm ducts or by a combination of both.

(9) The type B method may overcool even during the summer cycle under normal operation. For example, when the temperature of outdoor air, warm duct air, and conditioned spaces are about the same; the internal load in the building is negligible and there is no sun. This condition may be met in summer during partial occupancy of the building. Under this condition of operation, all system air will tend to flow in the warm duct, producing low or negligible pressures at the end of the warm duct. Consequently the spaces at the far end of the duct system will be receiving cold air although room thermostats do not call for it.

It should be kept in mind, also, that Method B control may introduce acoustical problems. When the room thermostat calls for all warm air, the volume control function of the unit will be lost completely. During such times, the increased air flow through the unit may generate objectionable noise, or low frequency rumble.

Air Handling Apparatus

In a majority of cases the dual duct central apparatus is based on one of the four basic cycles previously described. Minor modifications of the cycles and equipment might be dictated by design conditions. There are no general or special rules which should be followed in the selection of the equipment or shaping of the apparatus casings. The judgment of the design engineer and manufacturers' recommendations will determine the type, size and quality of the component parts forming the central apparatus. All mechanical equipment is of standard manufacture as used at present by the industry. However, there are several points which should be given special consideration in designing a push-through high pressure apparatus of dual duct type. These will now be discussed in detail.

Apparatus Floor Area. Apparatus floor area con-

Table 3. Floor Area Required by Air Handling Equipment Room (Fig. 27)

Net Area of Building Served by Equip. Room, Sq Ft	Total Area Required by Equip. Room, Sq Ft	Clear Head Room, Ft	Preferable Shape of Equipment Room		Percent of Net Area Served by Equipment Room
			"A", Ft	"B", Ft	
50,000	1100	8	30	37	2.2
	1000	10	30	33	2.0
	950	12	30	32	1.9
	850	14	30	27	1.7
100,000	2100	8	32	66	2.1
	1900	10	32	60	1.9
	1700	12	32	53	1.7
	1500	14	32	47	1.5
150,000	2700	10	34	80	1.8
	2400	12	34	70	1.6
	2100	14	34	62	1.4

Notes: 1. Net usable area assumed to be 80% of gross building area.
2. Areas required for refrigeration plant or miscellaneous exhaust systems not included.
3. Total air handled assumed to be 1.2 cfm per square foot of net area.

stitutes a large part of the total owning and operating cost of any type of air conditioning system. An improperly planned apparatus room may involve waste in floor area which could represent an element of cost greater than the cost of all mechanical equipment in it.

A well planned dual duct apparatus room can be fitted into a space no more than $1\frac{1}{2}$ to $2\frac{1}{2}\%$ of the net area it conditions. Such economy in floor area is due to the following factors:

(1) Zoning and multiplicity of apparatus systems is not required for dual duct systems and a larger central apparatus may be used which will always take less space than several small systems of same total capacity.

(2) By supplying the cold air at lower temperatures, the total air supplied is greatly reduced. With 50 deg air in the cold duct, the size of the central apparatus is only 60 to 70% of one using 60 deg air.

(3) Use of vertically split cold and warm chambers (shown in Fig. 27) allows the piping to the cooling coils to be placed in the passage for warm air, saving $2\frac{1}{2}$ to $3\frac{1}{2}$ ft of space on each side of the bank of cooling coils.

(4) By using the full height of rooms for banks of coils and filters, floor area is saved.

All air in the dual duct system is conditioned by a central air handling plant. It is arranged so that return air can be recirculated or, in suitable climates, 100% outdoor air can be used for cooling purposes. Additional cooling effect could be achieved by evaporative cooling in the cold chamber.

Figure 27 shows a typical layout of a room containing a plant of large capacity.

The amount of floor space required varies approximately between 1.4% and 2.2% of the net area served by the plant.

Table 3 shows the floor area required for the air handling plant related to net areas served by the plant. Dimensions A and B are those of Fig. 7.

It should be noted that Table 3 is compiled on the basis of total air flow, assumed to be 1.2 cfm per sq ft of net area and that the net usable area is assumed to be 80% of the gross area.

While a single plant is very desirable, it may be that in large buildings two plants may prove more practical and economical. These may be arranged on the same floor or one could be in the basement, say, and one on the roof. The latter arrangement could effect economies in shaft sizes for ducts.

Back-Lash and Carry-Over. Due to unequal distribution of velocity over the face of the cooling coils in the push-through type apparatus, the carry-over of moisture in the cold chamber and back-lash from the cooling coil represent a serious problem in the design of apparatus rooms. Experience has shown that the most effective way to eliminate carry-over and back-lash is to install the following: distributing plates over the inlet of both heating and cooling coils. The distributing plate is usually a perforated metal sheet with $\frac{1}{2}$ inch round holes, having 40-50% free area. This plate should be

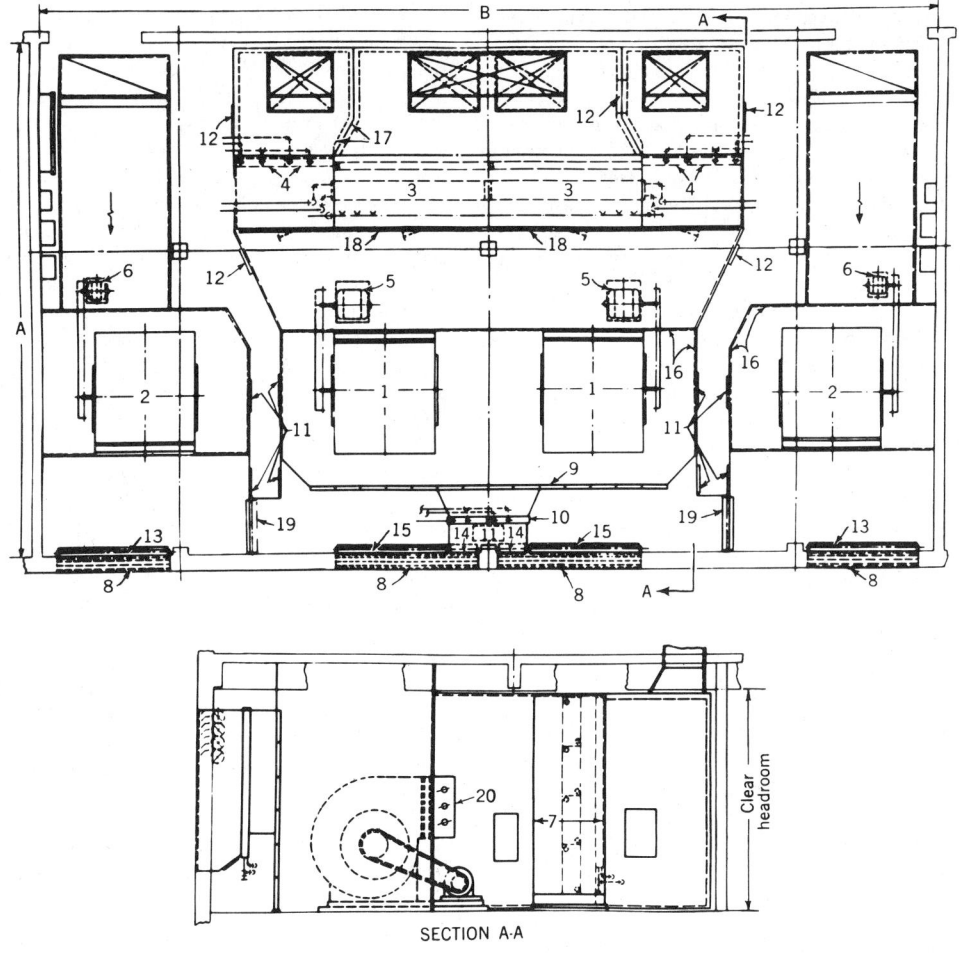

SECTION A·A

Fig. 27. Typical large capacity dual duct apparatus room.

1. Supply fans	6. Return fan motors	11. Low pressure access doors	16. One-inch acoustical lining
2. Return and relief fan	7. High pressure dehumidifier	12. High pressure access doors	17. Four-inch acoustical lining
3. Cooling coils	8. Outdoor louvers and screens	13. Relief dampers	18. Perforated distributing plate
4. Heating coils	9. Filters	14. Minimum outside air dampers	19. Return air dampers
5. Supply fan motors	10. Preheaters	15. Maximum outside air dampers	20. Shut-off dampers

amply reinforced, made removable and provided with access panels to coil chambers and sprays.

In addition, corrosion resistant eliminator plates should be provided on the leaving side of the cooling coils. Three-bend eliminators spaced $1\frac{1}{8}$ inch apart have been found to give good results for dual duct installations. If sprays are used in the warm chamber, eliminator plates should also be installed on the leaving side of the warm duct heating coil.

Apparatus Casing. Utmost care should be exercised in the construction, bracing and sealing of the high pressure apparatus casing. For practical reasons, it is almost impossible to pressure test the casing chamber before the system is put in actual operation. Consequently, any error in construction, assembly and sealing is extremely difficult to correct after the system is completed.

Construction standards developed for low pressure apparatus casings are inadequate for large high-pressure and medium-pressure systems and their application has created serious acoustical problems on many such installations. It is recommended that

structural framing for apparatus casings subjected to higher pressures be carefully designed for the imposed loads. In addition, casings should be amply braced and stiffened to prevent vibration and generation of noise. On larger apparatus casings handling over, say, 40,000 cfm, one of the following should be considered: double-wall factory insulated panels, cellular casings, or masonry casings. Masonry casings, internally insulated and made airtight, will provide the optimum construction to achieve desired acoustical results.

The vertical split of cooling and heating chambers will be more desirable for the following reasons: the heating chamber will be more accessible for inspection and repair; provisions for humidifying devices in the warm air chamber are more easily made; floor area required for the apparatus will be saved if the pipe space for the cooling coils is utilized for the warm air chamber; in many installations the connections to cold and warm mains are simpler; and less air turbulence and casing vibration will be caused.

Insulation and Sound Lining. It is important that all sound lining installed inside casings be mechanically attached to the casing walls in addition to cementing. The metal clips and washers used for this purpose should not be spaced on more than 1 ft centers. All joints between insulation sheets should be filled with cement or protected by 1 inch x $\frac{1}{8}$ inch flat bars to prevent air erosion. If lining is installed on the bottom of enclosures, it should be protected by heavy gage screen or perforated metal.

Fans. Careful consideration should be given to the proper selection of fan and motor vibration absorbing bases. In addition to selecting the base for maximum efficiency, it should be adjustable—a feature which is usually incorporated in spring type bases.

In larger air handling plants, it might be advisable to have two or more supply fans in parallel discharging air to a common plenum. See Fig. 27. When such an arrangement is used, the fans are equipped with normally closed inlet vanes or discharge dampers operated by electro-pneumatic switches.

Noise in High Velocity Systems. Any air conditioning or air handling system is a potential source of noise which may enter the conditioned area through supply, return or exhaust ducts, by transmission through the building framework, or by communication from zone to zone through interconnecting ducts. Generally, low-velocity conventional systems present relatively few acoustical design problems. Considerable work has been done in acoustical ratings of fans, terminal devices and acoustical treatment of distribution systems by equipment manufacturers. The design engineer, with a knowledge of the noise spectrum generated by the air moving device and the attenuating factors of the low-velocity system can, with reasonable assurance, predict an end environment for the spaces served by the system.

It is not sufficiently understood that high velocity systems cannot be treated in the same manner. Therefore, it is not uncommon to find that a design engineer has provided what might appear to be a conservative acoustical design but that system noise is none the less transmitted to occupied spaces. The excessive system noise is usually generated by the turbulence of the air stream and is not noise originating at the fan and transmitted through the distribution system. Although there is very little published information regarding this aspect of noise control, it is, by far, the more serious and costly to rectify.

Various recommendations for acoustical considerations in design and layout of high-velocity systems can be found elsewhere in this Handbook under the title "Acoustical Problems in High-Velocity Air Distribution."

Installed Costs

Cost data presented here are offered for the use of the design engineer in a form that will allow him to establish budget information quickly, without making detailed estimates and partial plans which must be discarded and replaced before final figures are produced. These figures also show the areas in which savings may be found to reduce costs to meet predetermined budget allowance.

All cost data are based on the net, or usable, area of the building. The non-usable areas, stairwells, elevators, toilets, main corridors, etc. have little or no effect on the air conditioning cost unless these spaces contribute to the complexity of the distribution system.

Data given were compiled from detailed cost analyses of various high pressure dual duct systems with total refrigeration capacity of over 40,000 tons, ranging in size from 50-2500 tons each. Special effort has been made to arrange the information so the design engineer has a tool to guide his planning.

Variables Affecting Costs. The main variables which affect the installed cost of dual duct systems are:

1. Initial TD (temperature difference, equal to room temperature minus cold air supply temperature) used in design of the system.
2. Size of individual zones (capacity of mixing units).
3. Size of air handling apparatus rooms.
4. Type of dual duct cycle selected.
5. Basic arrangement of air handling system.
6. Degree of future flexibility required.
7. Number, location and arrangement of riser shafts.
8. Type of construction of rectangular duct-work used.
9. Structural and architectural features of the building.
10. Effect of interior zones on cost of exterior zones.
11. Ratio of floor area of exterior zones to interior zones.
12. Type of terminal units used.

Records of recent installations indicate an extremely wide cost range for high velocity dual duct systems. For example, on new office buildings in 1964 this range is approximately $2.75 to $5.50 per square foot of floor area (net). This variance in costs could be traced directly to effect of the above mentioned cost variables.

In preliminary design of dual duct systems, the above variables may be grouped and combined conveniently into four basic items:

1. TD (Room—cold air supply temperature)
2. Size of central apparatus (capacity of supply fan)
3. Average size of control zones (capacity of mixing units)
4. Complexity of distribution duct system

Cost analyses as are shown in Figures 28 to 31 apply to an average multi-story office building of standard construction and normal shape. Wage rates prevailing in northeastern states in 1960 were used in calculating total erection costs. The central apparatus and distributing system were assumed to be in accordance with good engineering practice. Buildings with architectural and structural designs deviating from normal will introduce some degree of error in costs shown on curves, but these curves will still be useful to indicate cost tendencies and are valuable in developing basic schemes for any project.

All data pertaining to the cost of dual duct sys-tems are combined in the four curves of Figs. 28-31. Costs are given per cfm of total supply air. This may not be as graphic as the customarily used units, cost per square foot of floor area, or cost per ton of refrigeration, but it is more indicative of the true cost of the system.

Figure 28 shows the relationship between TD and cfm per sq ft of usable floor area. A TD decrease from 30 to 20 deg represents a 50% increase in size of the air distributing system. After considering the variables represented in Figs. 29-31, the saving due to higher TD is obvious. Any extra cost of outlets, insulation or other minor factors associated with higher TD is only a fraction of the savings obtained by reduction in capacity of the air handling system. This statement applies to both first cost and annual owning and operating costs. High TDs can be safely used if the air is distributed by suitably designed outlets with relatively high entrainment ratio. The effect of TD on cost is not shown directly in Fig. 28, but is reflected by Figs. 29-31.

The cost of refrigeration is shown on the right vertical scale of Fig. 28. The refrigeration is calculated for average conditions:

1. One ton per 330 sq ft of net conditioned area; this is average for office buildings having 35% of glass in the outside walls, with outside ventilating air of 0.25 cfm per square foot and outside design conditions of 95 F dry bulb, 78 F wet bulb.

2. The cost of refrigeration and piping has been assumed to be $250 per ton.

The cost of refrigeration includes the refrigeration cycle, cooling tower, water cooler, condensate water and cold water circulating pumps, interconnecting piping, motors and starters. It is assumed that cooling tower, condenser, water cooler and dehumidifier are reasonably close to each other. If refrigeration apparatus is remote from either, or both, cooling tower or chilled water coils, an additional allowance must be made for extra piping.

Figure 29 shows the cost of central apparatus, in terms of its size, for the conditions listed. Special features, such as more expensive filters, should be added to this cost. Costs of sheet metal casings and internal lining included in the curve represent a carefully planned apparatus room, using not over 2% of the total conditioned area, and having approximately 0.2 lb of sheet metal casings per cfm of total air.

Figure 30 indicates the cost of zoning expressed in terms of cfm required by the average zone, or the average capacity of the mixing unit. The relatively

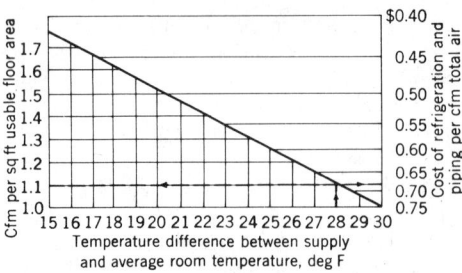

Fig. 28. Ratios of total air to floor area and cost of refrigeration.

Cost of refrigeration includes compressor, compressor motor, condenser, water chiller, cooling tower, cold water and condenser water pumps, automatic controls, all interconnecting piping, insulation, and starters for all motors.

Air volume and refrigeration is based on design conditions of 95 F db, 78 F wb; an average electric load of 4 watts per sq ft; glass area of 35% of gross outside wall; one occupant per 100 sq ft; 0.25 cfm per sq ft outside air for ventilation.

For electric load other than 4 watts per sq ft, correct cfm value by 0.1 cfm per 1-watt change.

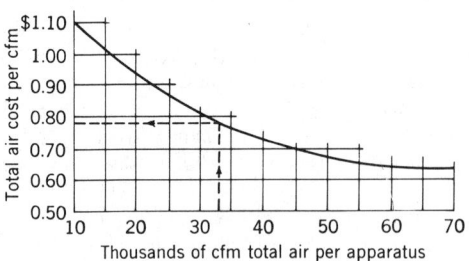

Fig. 29. Cost of central apparatus.

Cost of central apparatus includes Class II supply fan, Class I return air fan, fan motors, motor starters, variable pitch drives, vibration absorbing bases, high pressure dehumidifier, 8-row cooling coil, 2-row heating coil, 1-row preheat coil, spray pumps and piping, apparatus casing, simple filters, automatic control, specialties, doors, etc.

Capacity of central apparatus is based on 100% total air fan capacity, 80% total air dehumidifier capacity, and 70% total air reheater capacity.

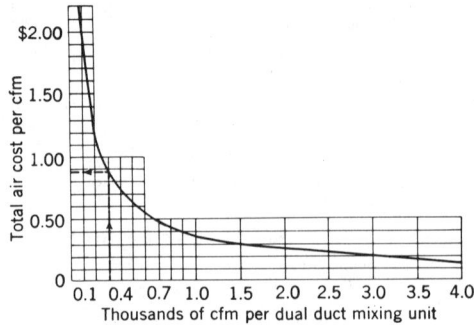

Fig. 30. Cost of zoning.

Cost of zoning includes air mixing unit, thermostat installed, duct connection to high-pressure mains, 2 tap-offs on high pressure mains, insulation of cold flexible duct.

Note that, if one thermostat is installed to control more than one air mixing unit, $80. should be deducted for each thermostat thus omitted.

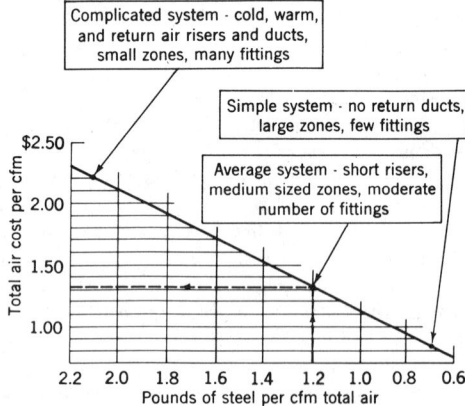

Fig. 31. Cost of general distribution duct system.

Cost of system includes sheet metal mains and risers, insulation of cold air ducts and risers, metal and lining of low-pressure ductwork, air outlets and grilles, miscellaneous vanes, caulking compounds, fire dampers, etc.

high costs of smaller zones are clearly demonstrated, and such zones should be used only when required by design conditions.

Figure 31 shows the cost of the distribution system. It is difficult to find a definite unit which reflects the combined cost of the distribution system, especially in the preliminary stages of design. Analysis shows that two criteria may be used to establish the cost of the distributing ducts.

1. The weight of steel per cfm of total air handled.
2. The complexity of the layout, which may be

classified as complicated, average, or simple.

By applying either of these criteria, a suitable point on the horizontal scale of Fig. 31 can be selected and the cost predicted with reasonable accuracy. Some familiarity with the art of dual duct system design is necessary for proper use of this graph, if an attempt is made to apply the curve without making a preliminary duct layout.

The weight of sheet metal per cfm is an accurate indication of the cost, providing proper allowance is made for the way the duct is designed and constructed and its average size. The cost of duct fit-

tings is three to four times that of straight duct of the same weight and size. A given run of ductwork containing a large number of fittings and transformations may cost 50-75% more than a similar run of well designed duct with standardized sizes and a limited number of fittings.

Use of Cost Curves. *Example:* A multi-story office building of average shape and with average design conditions, having 90,000 sq ft of net usable area and 120,000 sq ft gross area, is to be conditioned by a high pressure dual duct system. It is decided that the system will have three central fan units and 330 zones. The budget cost of the installation is required.

Step 1. Assume a temperature difference (TD) of 28 F. From Fig. 28, read 1.1 cfm per sq ft required on vertical scale at left. Total air to be distributed = 90,000 sq ft × 1.1 cfm per sq ft = 99,000 cfm. From Fig. 28, vertical scale at right, cost of refrigeration and piping is $0.68 per cfm; thus, total cost for refrigeration and piping = 99,000 cfm × 0.68 dollars per cfm = $67,320. Assuming the average for office buildings with 35% glass, etc., of one ton of refrigeration for each 330 sq ft of net conditioned area, total refrigeration = 90,000/330 = 272 tons.

Step 2. Each central fan system will distribute 99,000/3 = 33,000 cfm. From Fig. 29, cost of central apparatus is $0.78 per cfm; thus, total cost of apparatus = 99,000 cfm × 0.78 = $77,220.00.

Step 3. Each average zone requires 99,000/330 = 300 cfm. From Fig. 30, cost of zoning is $0.89 per cfm, or 99,000 × 0.89 = $87,110.

Step 4. From consideration of a tentative scheme, it is decided that the distribution system will be of average complexity. From Fig. 31, cost of average distribution system is $1.32 per cfm or 99,000 × 1.32 = $130,680.00.

Step 5. Add costs from Steps 1-4, total cost of system (to contractor) is $362,680, which, when divided by 99,000 cfm is $3.67 per cfm, when

divided by 272 tons is $1,336 per ton, when divided by 90,000 sq ft gives $4.04 per sq ft of usuable area, and by 120,000 sq ft gives $3.03 per sq ft of gross area.

These costs do *not* include (1) power wiring, (2) building construction work, painting, etc.; (3) miscellaneous exhaust and special systems, or (4) boilers and steam piping outside of apparatus rooms.

Automatic Control

Basic function of automatic controls for dual duct cycle of Figs. 1 and 2 is shown on Fig. 32. The system is assumed to provide for humidity control in winter time and a preheat coil in the minimum outdoor air duct. Minor modifications in control arrangement would be required for cycles of Figs. 3, 4 and 5.

When the supply fan is started, relay E-1, actuated by the fan motor starter, supplies compressed air to all apparatus and room control instruments. Minimum outdoor damper D-1 opens when the supply fan runs. Damper D-1 could be arranged to be closed manually or automatically when ventilation is not required, as for night operation, warm-up, or cool-down periods. Summer-winter switches S-1 and S-2 can be manual or automatic.

Summer Operation. With summer-winter switch S-1 in summer position, cold duct thermostat T-1 controls chilled water valve V-3 to maintain desired temperature in the cold deck. The practice of eliminating chilled water control valves or controlling only water temperature leaving the coil is not recommended because of danger of having excessively low temperatures in the cold duct on partial loads.

Sub-master warm duct thermostat T-2 is reset through cumulator C-1 by summer humidistat H-1 and controls steam or hot water valves V-4 and V-5. Minimum setting of thermostat T-2 should be approximately 5 deg above average room temperature and maximum setting equal to the temperature required for maximum heating load. In order to avoid hunting of warm duct control at low loads, it is recommended to provide two valves, V-4 and V-5, in parallel approximately of one third and two thirds capacity respectively. Summer humidistat H-1 adds heat to the warm duct only when low internal loads occur during humid weather (see Fig. 8, 9 and 10). During high summer load operation, control of temperature in warm duct is not

essential and source of heat might be shut off or the controls made inoperative.

Winter Operation. When refrigeration is not required, the summer-winter switch S-1 is placed in winter position allowing thermostat T-1 to modulate dampers D-2, D-3 and D-4 to maintain cold duct temperature by drawing cool air from outdoors. For economy, to minimize the amount of reheat in the warm duct, city water sprays or recirculating pump sprays could be put into operation ahead of dampers D-2, D-3 and D-4 through relay C-2. Evaporative cooling of 8 to 12 deg could be achieved. Contamination of air in industrial cities makes use of recirculating sprays inadvisable for prolonged use. Hand-off-automatic switch, if recirculating pump is used, should be provided to stop sprays during very cold weather. In order to reduce the amount of outdoor air heated during cold weather, thermostat T-1 is usually reset upward in winter, for operating economy. Experience indicates proper winter setting of thermostat T-1 can be determined only by actual operation of the particular system.

Warm duct temperature is controlled by submaster thermostat T-2 as reset by master thermostat T-3. Humidity control can be accomplished by winter humidistat H-2 controlling city water spray valve V-2 or a recirculating spray pump. When preheater in minimum outdoor air is used, it is controlled by thermostat T-1 to maintain minimum temperature of the air drawn into the system.

On new buildings, where temporary heat is to be provided by a dual duct system, room thermostats T-R should be direct acting with warm air connections in room mixing units normally open. If cold connections in the mixing units normally open, room thermostats should be reverse acting.

Construction Details

On many multi-zone installations due to space limitations, rectangular ducts must be used to make installation feasible. At present the high pressure contruction for rectangular ducts is not yet fully standardized and several types of duct construction have been developed by the industry to conform to local practices. The construction shown in Fig. 33 and Table 4 has been used successfully for fifteen years and is now extensively used in the United States and abroad. (See, also, details of tap-off fittings for high pressure ductwork under "Acoustical Problems in High Velocity Air Distribution" elsewhere in this Handbook.)

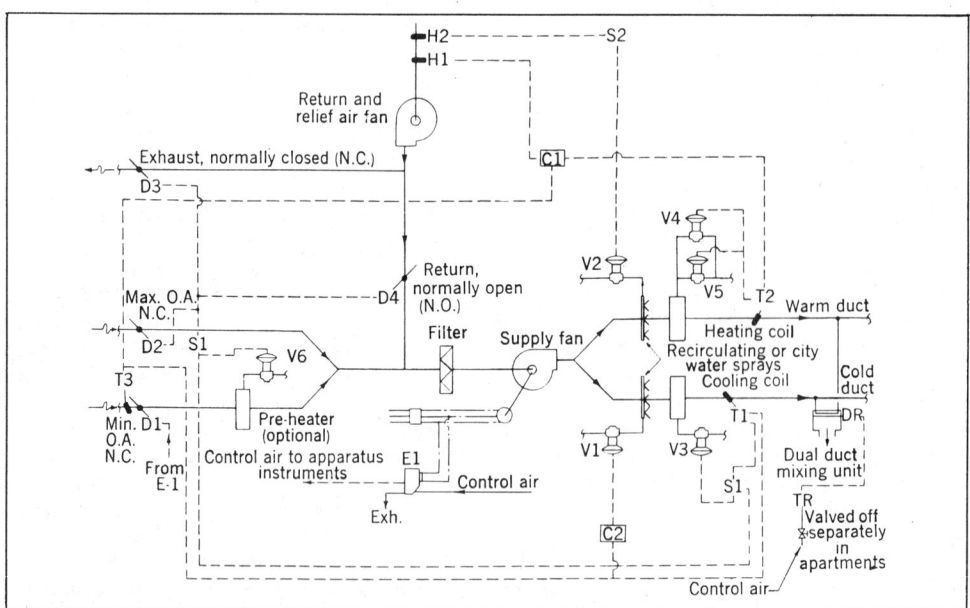

Fig. 32. Typical automatic control diagram for dual duct system.

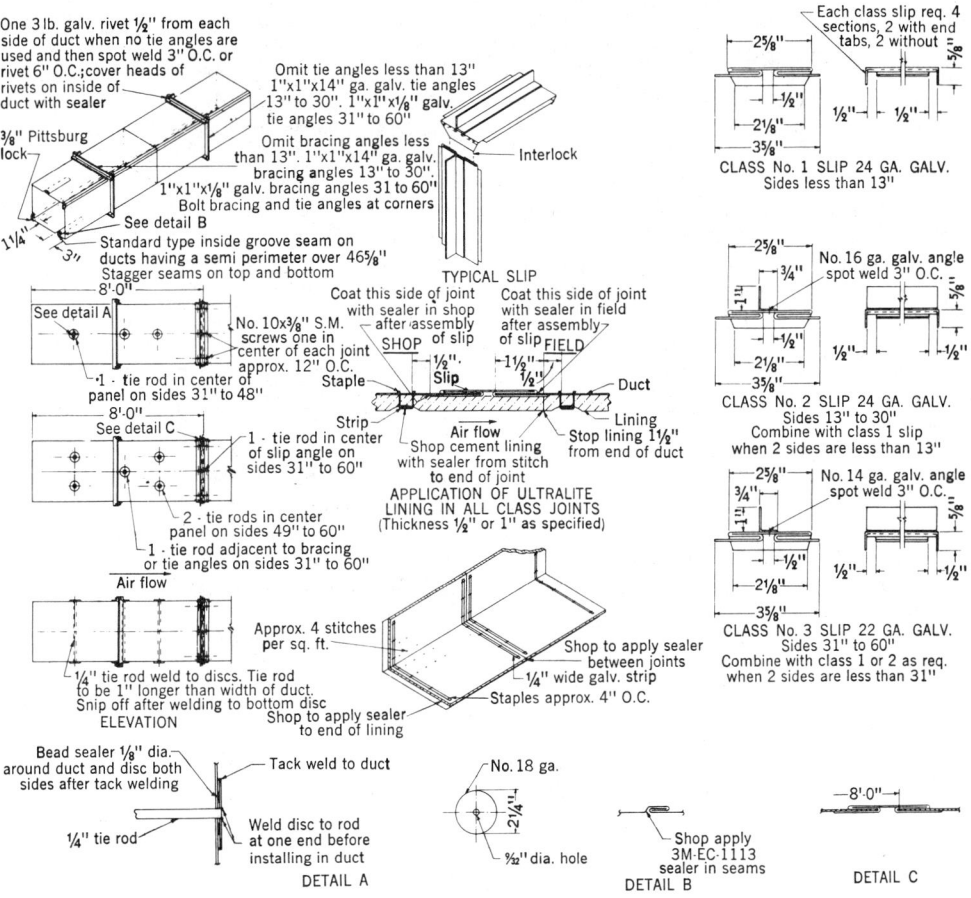

Fig. 33. Construction of high pressure rectangular ducts.

Table 4. Data for Rectangular High Pressure Duct

Duct Size, Inches	Gage	Joint Length, Feet	Slip Class	Tie Rod Reinforcing	Bracing Angle (Top and Bottom)	Tie Angles (Sides)
1-12	24	8	1	None	None	None
13-25	24	8	2	None	1 x 1 inch x 14 ga. angle irons 2 per joint	1 x 1 inch x 14 ga. angle irons 2 per joint
26-30	22	8	2	None	Same as above	Same as above
31-45	22	8	3	4—¼ inch rods	1 x 1 inch x ⅛ inch angle irons 2 per joint	1 x 1 inch x ⅛ inch angle irons 2 per joint
46-48	20	8	3	Same as above	Same as above	Same as above
49-60	20	8	3	6—¼ inch rods	Same as above	Same as above

Several types of sealing compounds have been developed for making up the joints in high pressure ductwork. They all have a synthetic rubber base with a bonding strength that does not decrease with age and retains good elasticity.

Sealing Duct Joints. Shop procedure for sealing high pressure ducts is as follows:

(1) Before fittings and joints are assembled, sealer is applied to rivets, grooved seams, and tap-off collars on the internal side of the metal. Pittsburgh lock pocket must be flooded with sealer, and the duct assembled; (2) sealer is brushed around reinforcing rod washers, corners, rivets, notches and tap-off collars after ducts are assembled. A double S-slip is installed on the air-leaving-side of the duct and fastened in place, using metal screws on 6-inch centers. Sealer is brushed into connecting lap and corner joints of an S-slip; (3) inside of connecting lap of S-slip and duct surface is coated with sealer. Where possible, sealing should be done on inside of the ductwork.

Field procedure for sealing duct joints is as follows:

(1) Sealer is spread on the inside of the double S-slip and the joints of the duct assembled. Immediately after joints are assembled, holes are drilled through the S-slip and metal screws inserted on 6-inch centers. Sealer is applied over the screw heads and the joint; (2) after 24 hours, a second coat of sealer is spread over the joints and allowed to dry for 24 hours before testing; (3) where joints are not accessible for proper sealing, hand holes should be cut in the duct and the joints sealed from the inside. Special care should be taken to seal all duct corners; (4) when testing ducts for leaks, leaks should be marked and resealed without pressure in the duct and allowed to dry for 24 hours.

A very similar rectangular duct construction was recently developed by Sheet Metal and Air Conditioning Contractors National Association Inc. (SMACNA) and is illustrated in SMACNA "Duct Manual & Sheet Metal Construction for Ventilating and Air Conditioning Systems, Section II—High Velocity Systems."

AUTOMATIC CONTROL APPLICATIONS

The following briefly describes applications of commonly used control systems in building air conditioning: rooftop multizone units, multizone units, dampers, unit ventilators, hot water systems, mixing boxes, and rotary air-to-air heat exchangers.

Basic categories of automatic controls include controls for primary equipment such as boilers, chillers, and packaged units, central fan systems, including damper control; valves for steam or hot and chilled water coils and terminal units.

Special applications of control systems are too numerous to be completely discussed. Special cases that are not included can be found in manufacturers' technical bulletins and control manuals. Control manufacturers offer to engineers, assistance in control methodology.

Growing interest in rooftop and other packaged units has focused attention on methods to control these units. Their control cycles are more complicated than those of a simple reheat coil, for example. Energy conservation devices, such as air-to-air heat exchanger controls, are also of increasing interest to HVAC engineers.

Rooftop Multizone Units

The multizone is a constant volume, variable temperature, central system (see Fig. 1). Outdoor air and return air are mixed and drawn through the filter section by the blower. Air leaving the blower is divided in parallel paths (the hot and cold decks). After leaving the heating and cool-

ing sections, the two air flows are mixed in proportions necessary to maintain individual zone temperatures.

Most manufacturers provide eight or more zone dampers on each unit. When fewer zones are required, two or more sets of zone dampers can be linked together to give the required number and size. Rooftop units are equipped with belt-driven supply fans for easy adjustment of air flow rates. Linked zone dampers and adjustable air flow rates combine to make rooftop multizones very flexible.

Mixed air section: Damper arrangements in multizones vary considerably. Generally, rooftop units are furnished with an outdoor and return air damper linked together. The exhaust damper is controlled in any of three ways: (1) by an additional damper motor, (2) by linkage to the outdoor and return air dampers, and (3) by being opened by exhaust fan static pressure (spring or gravity closed).

Rooftop units often employ the same thermostat for both mixed air and cold deck control. Placed downstream of the DX coil, the thermostat controls the mixed air damper motor and cooling pneumatic-electric switches.

With an economizer cycle, usual with rooftop units, the exhaust fan runs only when the outdoor air damper is open past the minimum setting. The exhaust damper should begin opening slightly before the exhaust fan starts and should not completely close until the fan has stopped.

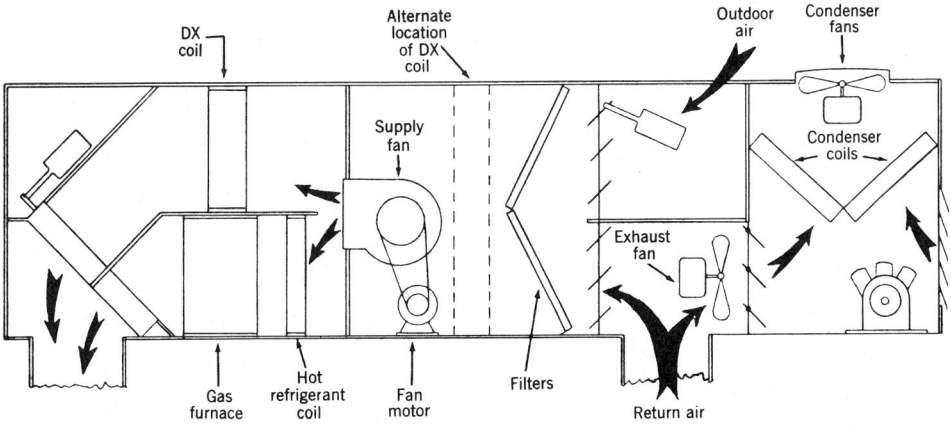

Fig. 1. Multizone rooftop control.

The exhaust fan, exhaust damper and outside and return air dampers are so sequenced that the building will maintain a positive static pressure. This minimizes the possibility of dust infiltration and non-conditioned air passing into the building.

Figure 2 shows typical cold deck—mixed air control. When outdoor air temperature is below 55°F, T-3 (thermostat No. 3) locks out cooling and T-1 modulates the damper to maintain 55°F cold deck (mixed air) temperature. The exhaust fan starts when the dampers move above minimum position. Between 55 and 65°F outdoor temperature, the exhaust fan will be on and the outdoor and exhaust dampers will be fully opened.

Cold deck control: The cooling coil is found in either of two locations (see Fig. 2): in the cold deck, or immediately upstream of the supply fan. The latter is preferred for applications involving high latent loads, as it provides dehumidification of all air passing through the unit.

When the coil is located in the cold deck, air passing through the hot deck will not be dehumidified. In this application, space humidity can be lowered by manually or automatically increasing the hot deck temperature, increasing the volume of dehumidified cool air and decreasing the volume of warm, humid air required to main-

tain zone dry-bulb temperature. Chilled water coils are controlled with either a two- or three-way valve.

The compressor will run to keep cold deck temperature at 55°F when one or more zones call for full cooling (via highest pressure R-3 [relay No. 3] and positive relay R-2). Above 65°F outdoor temperature, outdoor and exhaust dampers go to their minimums and the exhaust fan stops. The compressor continues to maintain 55°F cold deck temperature. When the supply fan stops, EP-1 (electro-pneumatic switch No. 1) is de-energized, which stops the compressor and exhaust fan. All dampers then go to normal position.

An alternate control method places T-1 in the mixed air stream (upstream of the cooling coil) and controls PE-2 from a controller mounted in the cold deck.

The following cold deck control is recommended: The compressor is locked out until one or more zones call for full cooling. At that time, the compressor may start if the cold deck temperature is above the thermostat set point. The cold deck thermostat cycles the compressor *on* and *off.* Compressor capacity control is achieved through one or two steps of unloading, and a hot-gas bypass system. (See Fig. 3.)

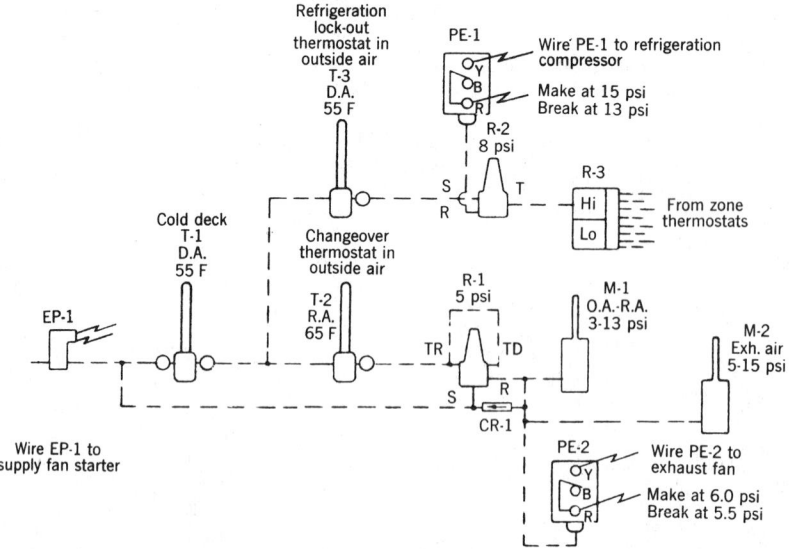

Fig. 2. Mixed-air control.

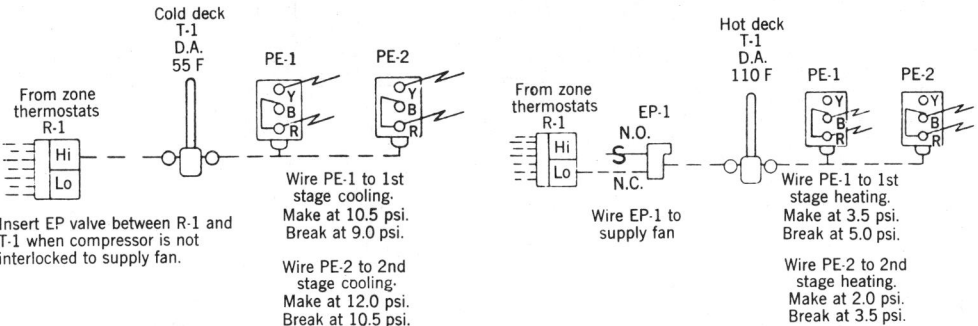

Fig. 3. Cold deck control.   Fig. 4. Hot deck control.

The refrigeration control circuit is always factory wired with the exception of the cold-deck thermostat (or pneumatic-electric switches). Each unit should have a low-ambient cooling lockout thermostat, and a short cycle time-delay relay and compressor interlock with supply fan. The control contractor should make certain these devices are in the refrigeration control circuit.

Hot Deck: The type of heating most widely used for rooftop units, the gas-fired furnace, comes with on-off, two-stage or modulating control. A pneumatic hot deck thermostat and PE switch is used for on-off or two-stage control. Upon a call for heat, the vent motor starts. When this motor is running, a centrifugal switch or air-flow switch closes, allowing the pre-purge timer to start. After the pre-purge period, the pilot solenoid valve opens and the electric igniter lights the pilot. Once the flame is proven, the low-stage solenoid valve opens, and on a further call for heat, the second-stage gas valve opens. Figure 4 shows typical hot-deck control of a gas-fired furnace. It is good practice to prevent the furnace from starting until one or more zones call for full heat.

Gas burner modulating control is similar to the two-stage control described above. Once the first stage is on (approximately 40 to 60 percent full flame), the modulating valve can increase gas input to 100 percent.

On larger units, where one gas furnace will not span the entire width of the hot deck, two heat exchangers can be placed side-by-side and operated simultaneously.

Several rooftop manufacturers offer a hot-refrigerant coil located in the hot deck. Hot liquid refrigerant leaves the condenser and flows to the hot refrigerant coil, then to the thermal expansion valve, evaporator, and back to the compressor. The hot-refrigerant coil warms hot-deck air whenever the compressor is running, providing free heating during mild and warm weather. If more heat is needed than delivered by the coil, the hot-deck thermostat energizes the hot gas solenoid. This allows the gas to flow directly from the compressor to the hot-refrigerant coil.

Most rooftop manufacturers carry a line of roof-mounted hot water or steam boilers which can serve one or more multizone units.

Multizone Unit Control

As its name implies, the multizone unit is an air conditioning unit which supplies air to a number of zones at varying temperatures. A source of hot and cold air is available at the unit and, by mixing the two air streams, the unit can easily satisfy the needs of each zone.

Figure 5 illustrates the controls for a *typical* multizone unit. The controls may vary, depending upon the individual job requirements.

Control of mixed air: Thermostat T-1 controls motors M-1 (motor No. 1) and M-2 to maintain 55°F mixed air temperature when the outside air is below 55°F. As outside air rises above 55°F, the outside air damper fully opens, and return-air damper closes. When the outside air rises above 75°F, T-2, through the minimum pressure relay R-1, position the outside air damper to its minimum open position. On shut-down, EP-1, which is wired to the fan starter, allows the outside air damper to close and the return air damper to open.

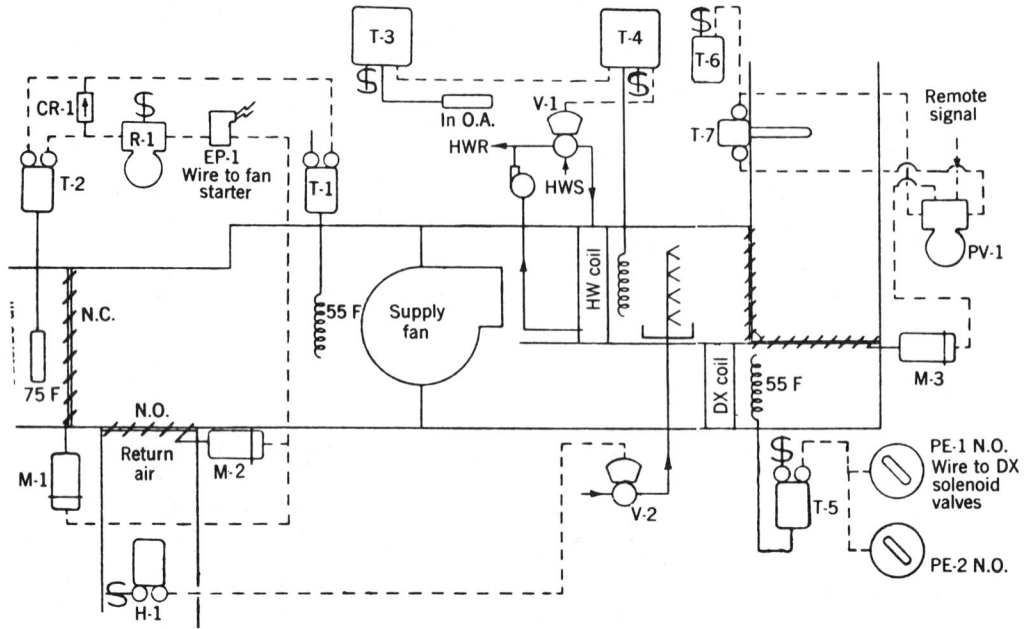

Fig. 5. Control of Typical Multizone Unit.

Control of hot air plenum: Control for hot water heating is also shown in Fig. 5. The plenum is reset in accordance with outside temperature by master thermostat *T-3*. The reset range for *T-4* will vary, depending upon the job location. Ranges of 110°F to 70°F are typical. These ranges are for outside temperatures of minus 10°F to plus 65°F in northern areas, and 30°F to 70°F in southern areas. By varying the hot plenum temperature, there is an attempt to balance the supply air with load. This allows the zone dampers to operate in their mid-positions.

While not shown in Fig. 5, a pressure-electric switch, operated by the master thermostat, is often used to stop the circulating pump when the outside temperature rises to a specified value.

A circulating pump is connected to the leaving side of the coil to insure constant circulation through the coil. This pump is necessary because, if one tried to reduce coil output by reducing flow through the coil, one would find that the water would take a greater temperature drop and coil output would not be reduced sufficiently.

In areas where it is necessary to protect against coil freeze-up, the averaging bulb of a thermostat is located right after the heating coil. The thermostat is set for approximately 40°F and

When humidification is required in all areas, the humidifier is located in the hot air plenum and the humidity controller (*H-1*) in the return air. *H-1* then controls valve *V-2* or a step switch for electric immersion heaters. If humidification is needed only in the zone, humidifier is located in the particular zone duct and controlled from a zone humidistat.

Control of cold air plenum: Figure 5 shows capacity control with liquid line solenoid valves. Thermostat *T-5* controls a number of normally open pressure-electric switches which, when closed, energize the solenoid valves. The pressure-electric switches operate in sequence.

Figure 6 illustrates control with cylinder unloaders. A reverse acting thermostat with averaging bulb controls the cylinder unloaders. On a rise in temperature above the set point of about 50°F to 55°F, the cylinders cut in gradually to give greater capacity.

Figure 7 shows control with a suction (back) connected to the fan starter. Then, whenever the temperature falls below this value, the supply fan stops, and it requires inspection of the unit by maintenance personnel before the freeze detection thermostat is manually reset.

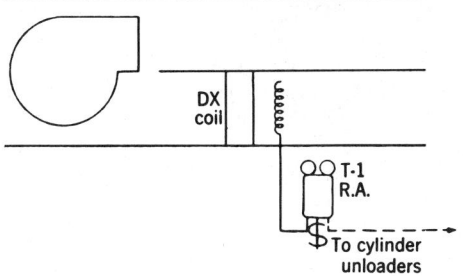

Fig. 6. Control of cold plenum with cylinder un-
loader.

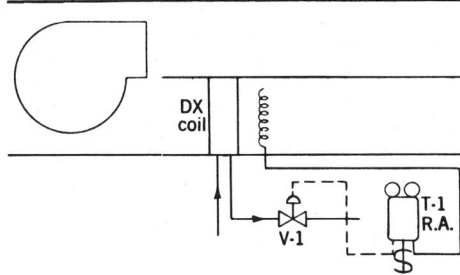

Fig. 7. Same, with suction pressure regulator
valve.

pressure regulator valve. As the temperature rises, the valve opens gradually to reduce the coil suction pressure and, therefore, the temperature.

The compressor must have a wide range of capacity reduction in order to handle the varying load conditions. Otherwise, there may be cycling of the zone mixing dampers. Cycling of the dampers is usually caused by varying temperatures across the coil. This occurs when one of the zone dampers changes position. Unless the suction pressure and, hence, coil temperature, is carefully controlled, the variation in air temperature may result in damper cycling. Cylinder unloaders offer very suitable capacity reduction.

To obtain the best humidity control, the coil should be operated as cold as possible during mild weather. During change in seasons, a considerable amount of air bypasses the cooling coil through the hot air plenum. This allows air to enter the zone without being dehumidified. By operating the coil at a low temperature, there may be adequate dehumidification. The bypass air is then used for reheat. If possible the coil should be operated at a colder temperature during mild weather than during peak load. This can be done using master-submaster instruments.

Refrigeration can be started by the zone requiring cooling, as shown in Fig. 8. The zone

Fig. 8. Refrigeration starting.

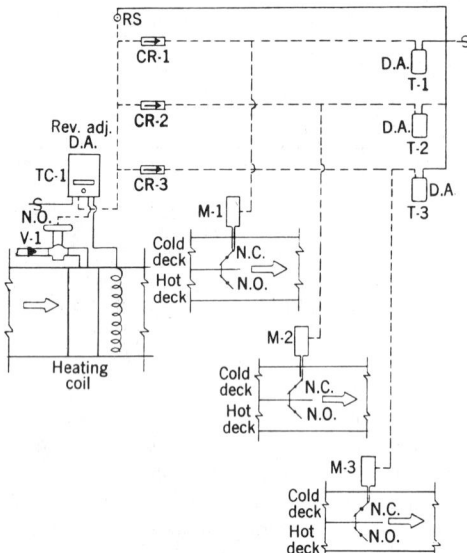

Fig. 9. Zone mixing damper.

thermostats are connected through check valves to a pressure-electric switch which is wired to the refrigeration starter. On a call for cooling by any one zone, refrigeration is started.

Control of zone mixing dampers: Room thermostat $T-6$ (Fig. 8) positions zone mixing dampers to maintain desired zone temperature. With the increase in temperature, the thermostat increases branch pressure to close the hot deck damper and open the cold deck damper. With a decrease in temperature, the thermostat decreases branch pressure to open the hot deck and close the cold deck.

During the heating season in northern climates, it is usually desirable to provide low limit control for certain zones. $PV-1$ (Pilot Valve No. 1) is used to place controller $T-7$ in the control line to the damper motor during heating season and bypass it during cooling season.

Damper Control

Zone mixing damper control: With hot deck reset from coldest zone: As shown in Fig. 9, room thermostats $T-1$, $T-2$ and $T-3$ control multizone damper motors $M-1$, $M-2$ and $M-3$, respectively. The lowest control pressure from any room thermostat resets $TC-1$. $TC-1$ controls the multizone hot deck discharge temperature through $V-1$.

As temperature decreases, the zone thermostat will decrease the control pressure to its damper motor. The hot deck zone damper will open and the cold deck will close. The zone thermostat with the lowest control pressure will reset $TC-1$. As this reset pressure drops, the $TC-1$ set point will rise. Thus, at direct-acting controller $TC-1$ control pressure decreases, which opens $V-1$.

When the temperature in the coldest zone increases, control pressure from the thermostat in that zone increases and resets $TC-1$ to a lower set point. This increases $TC-1$ control pressure and closes heating coil valve $V-1$. The increase in thermostat control pressure gives modulation with the hot deck zone damper closed, and the cold deck damper open.

Mixed-air control: With minimum fresh-air damper: As shown in Fig. 10, the minimum outdoor air-intake damper motor $M-1$ will open fully at any time the supply fan is operating. Mixed air control $T-1$ will modulate outdoor air and return air dampers $M-2$ and $M-3$ as required to maintain the desired mixed-air temperature. Electropneumatic valve $EP-1$ will prevent damper operation unless the supply fan is running.

The system is capable of providing a two-position minimum damper control to open the minimum outdoor air intake damper fully at any time the fan system is operating. The mixed-air control modulates the maximum outdoor air damper and return air damper in unison, to maintain a constant 55°F mixed-air temperature. When the fan system is not operating, the outdoor air dampers are closed and the return air damper is open.

Damper operation during day cycle only: This is a positive control system for the outdoor return and exhaust air dampers to prevent damper operation during night cycle, or whenever the fan system is not running. The outdoor and exhaust air dampers shall be normally closed and the return air damper shall be normally open. Referring to Fig. 11:

Day Cycle—During day operation, relay $R-1$ will connect ports B and D and outdoor, return air, and exhaust damper motors $M-1$, $M-2$, and $M-3$, respectively, will be controlled by the air signal from the controlling device. Electropneumatic valve $EP-1$ will prevent damper operation unless the fan system is running.

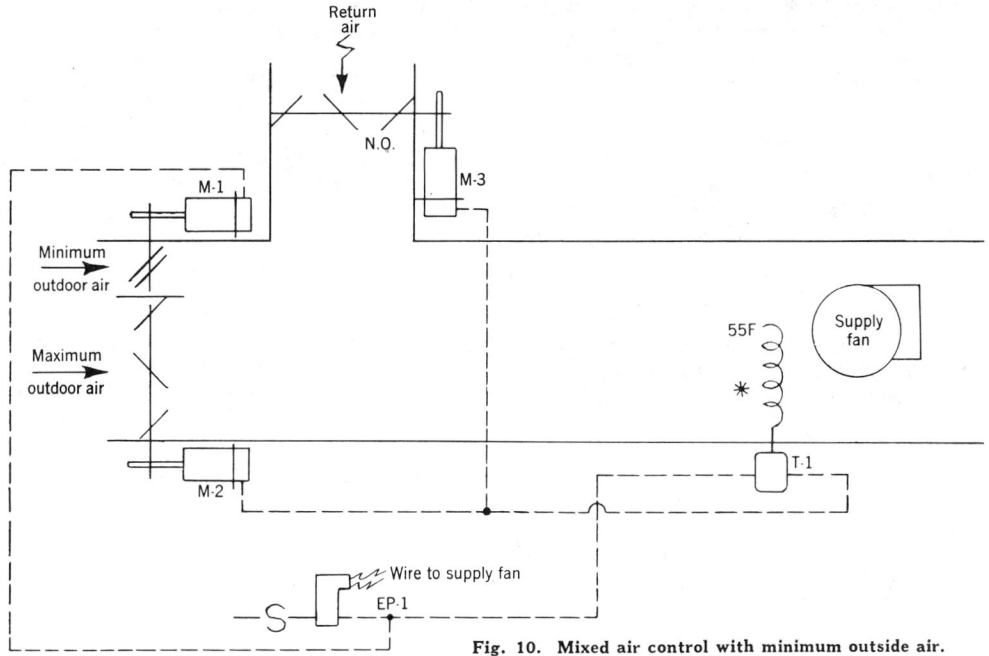

Fig. 10. Mixed air control with minimum outside air.

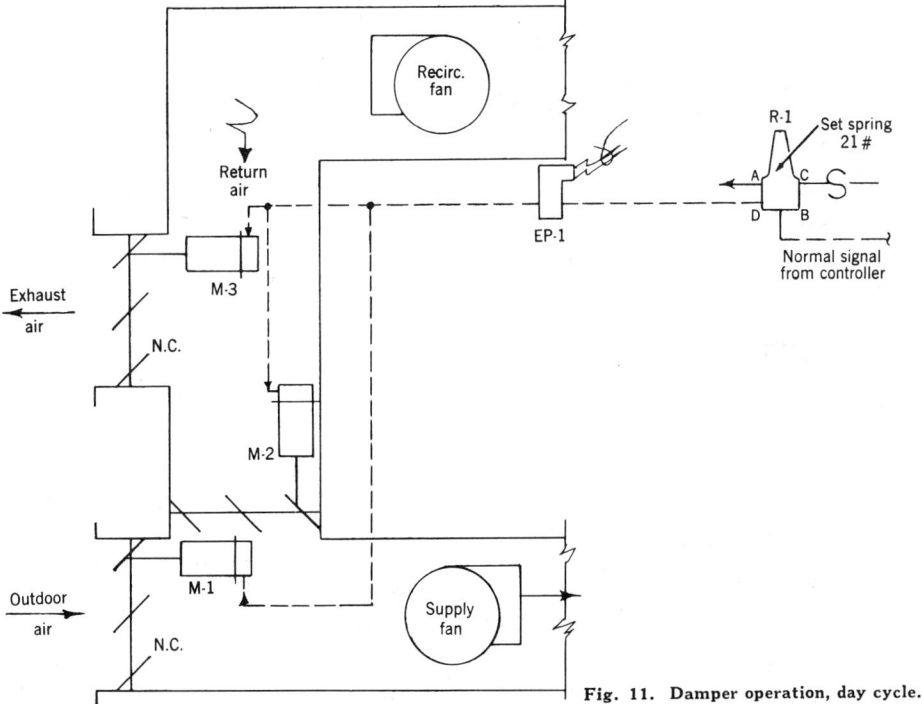

Fig. 11. Damper operation, day cycle.

Night Cycle—During night operation relay R-1 will connect ports A and D and the outdoor return and exhaust air dampers will assume their normal position. This allows for night operation of the fan system without opening the outdoor air and exhaust dampers.

Economizer control cycle: With changeover and minimum damper in summer only: This is a mixed air control to proportion automatically the outdoor and return air dampers as required to maintain the desired mixed air temperature. It provides a changeover controller to reduce outdoor air intake to minimum requirements whenever the outdoor air temperatures rise above a designated temperature. Referring to Fig. 12:

Winter Cycle—Mixed air controller *T-1* will modulate outdoor and return air damper motors *M-1* and *M-2* to maintain the desired mixed air

temperature. Valve EP-1 will prevent damper operation when fan is not running.

Summer Cycle—Whenever the outdoor air temperature exceeds the setting of changeover control *T-2*, the outdoor air intake will be limited to the minimum position. Relay *R-1* provides field adjustment (panel mounting if desired) of the minimum damper position.

Unit Ventilator Control

Large Units: Generally, the space thermostat is required to modulate in sequence the heating valve and the outdoor and return air dampers to maintain the desired space conditions. Under normal operation, the outdoor and return air damper motors will maintain a minimum fresh-air intake and the space thermostat will control temperature by modulating the heating valve or introducing

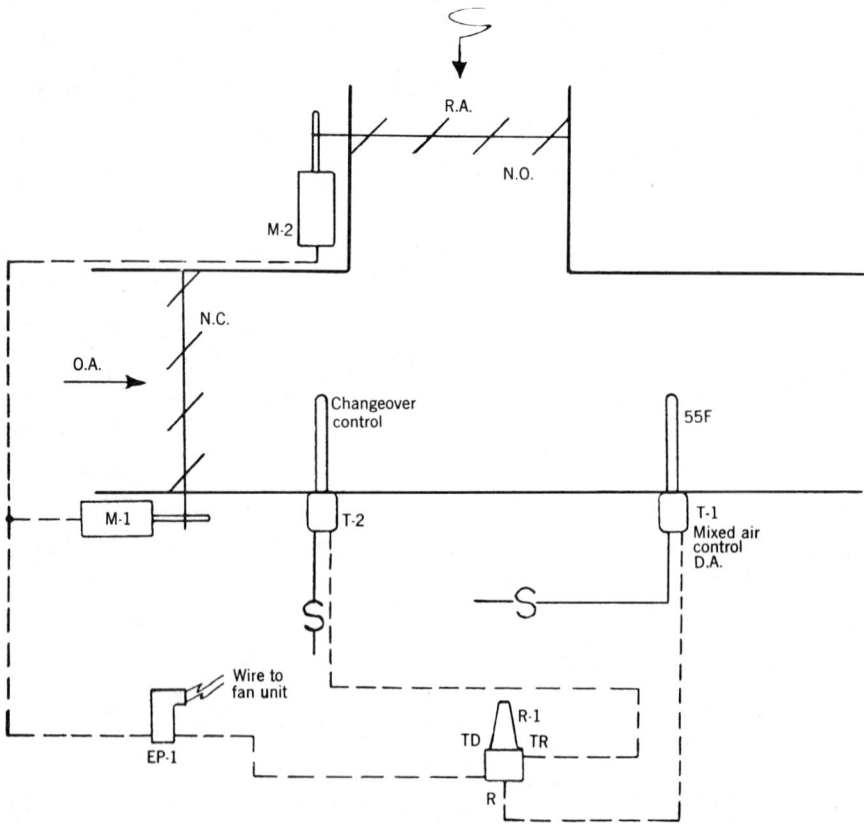

Fig. 12. Changeover control.

additional outdoor air. However, if the space temperature drops below the thermostat setting, the thermostat should be able to reduce the minimum fresh-air intake, as required, in order to maintain space conditions. Discharge low limit control is frequently used to allow reduction of fresh air intake and opening of the heating valves required to maintain a minimum discharge-air condition. When the system is not operating, the outdoor air damper should remain fully closed. The heating valve will remain under control of the space thermostat or discharge low limit.

Referring to Fig. 13, at any time that space temperature is below the setting of space thermostat, *T-1*, valve *V-1* will remain in the fully open position and the unit will operate on 100% recirculated air for rapid space warm-up. As the space temperature approaches the setting of *T-1*, outdoor and return air damper motors *M-1* and *M-2* will modulate to a minimum damper position. A further rise in space temperature will actuate heating valve *V-1* into a closed position. A continued rise in space temperature will modulate

outdoor and return air dampers *M-1* and *M-2* for increased outdoor ventilation, as required to maintain space conditions. Low limit control *T-2*, located in the fan discharge will vary the outdoor air damper to the minimum position and open heating valve *V-1* as required to maintain a minimum discharge air temperature. *EP-1* will prevent damper operating when the fan is not running.

Zone day-night operation: Referring to Fig. 14, when room temperature is below setting of thermostat *T-1*, the valve serving the heating coil is open and the outside air damper is closed. As the temperature approaches the setting of the thermostat, the outside air damper will move to a predetermined minimum open position. On a further rise in room temperature, the valve closes. On a still further rise, the outside air damper gradually moves to the fully open position. The airstream thermostat *T-2* overrides the room thermostat and repositions the damper and valve in order to maintain a minimum unit discharge

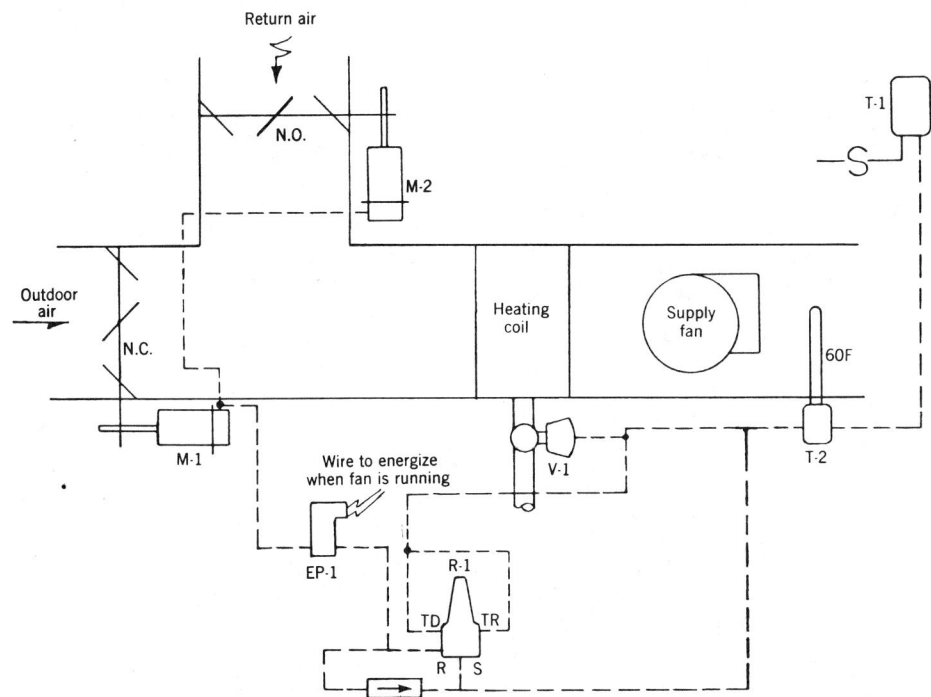

Fig. 13. Unit ventilator control—large unit.

temperature of 60°F. An electropneumatic valve closes the outside air damper on fan shutdown.

The controls function, as described above, during the "Occupied" cycle of operation. During the "Unoccupied" cycle of operation, a zone thermostat actuates a pressure-electric switch, PE-1 to cycle the unit fan. When the fan is de-energized, the electropneumatic valve EP-1 opens the valve (V-1) and closes the outside air damper M-2 to maintain a reduced "Unoccupied" temperature of 60°F.

Hot water coil: The simpler unit ventilator control cycle is a *Day Only* room thermostat con-

trolled valve and damper. Referring to Fig. 15, with fan motor on, electropneumatic valve EP-1 is energized to open the control line to damper motor M-1 and control valve V-1. Room thermostat T-1 gradually positions M-1 and V-1 to maintain the desired temperature according to the control schedule shown as part of Fig. 15. If the discharge temperature drops below 60°F, discharge thermostat T-2 vents pressure from motor M-1 and valve operator V-1 to maintain the minimum discharge temperature.

With fan motor off, valve EP-1 is de-energized, motor M-1 closes and valve V-2 opens, permitting full water flow for freeze protection.

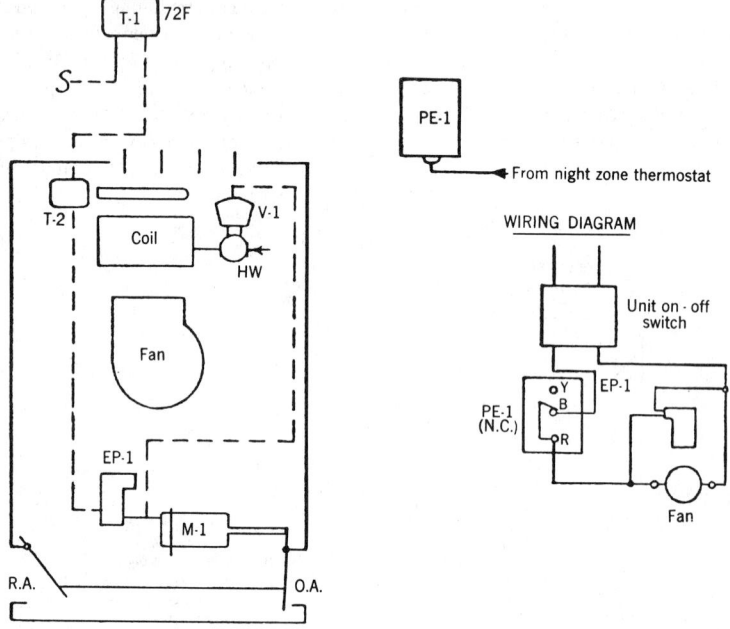

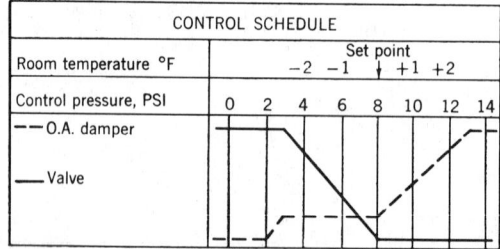

Fig. 14. Zone day-night operation.

Hot Water System Control

Hot water temperature control with outside air reset: For the entire system or zone, an outdoor thermostat resets supply water temperature according to a predetermined reset schedule. Control is maintained when the heat output of the various heat exchange units closely approximates the load, as reflected in the reset schedule.

For the heating unit itself, a three-way water mixing valve provides a variable water temperature by mixing supply water with system return water. Figure 16 shows the general arrangement.

Preheat control: Figure 17 shows the general arrangement of a preheat control. This layout requires proper sizing of the coil so that freezing can be prevented. A freeze protection thermostat is required. The first coil operates at 40°F and the second coil operates when outdoor temperature drops to 15°F. The damper motor and freeze detection thermostat are all wired to the fan motor for freeze protection.

Face and bypass control: Heat transfer to the air stream, proportional to the amount of air passing through the coil face, is controlled by an automatic face and bypass damper. A control valve stops water flow after the face damper is fully closed. The outside air thermostat will open the coil valve at temperatures below 40°F, while the discharge thermostat will position the face

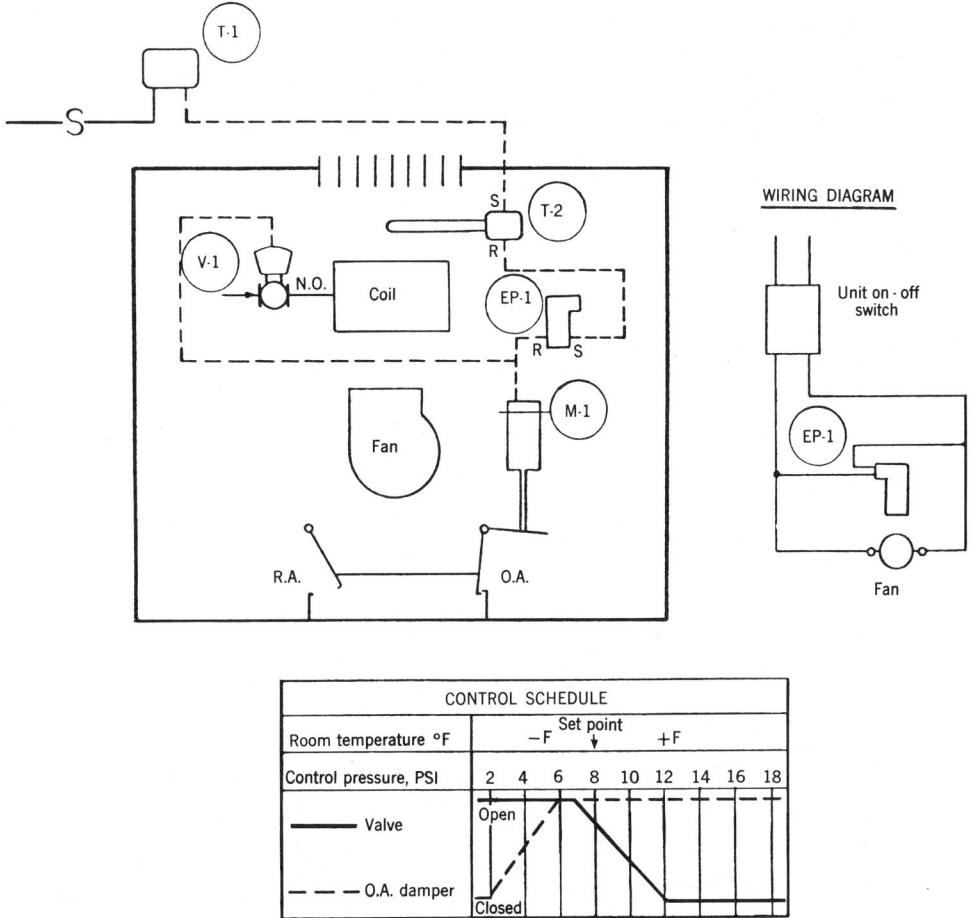

CONTROL SCHEDULE									
Room temperature °F		−F		Set point ↓		+F			
Control pressure, PSI	2	4	6	8	10	12	14	16	18
——— Valve	Open								
− − − O.A. damper	Closed								

Fig. 15. Unit ventilator control—hot water.

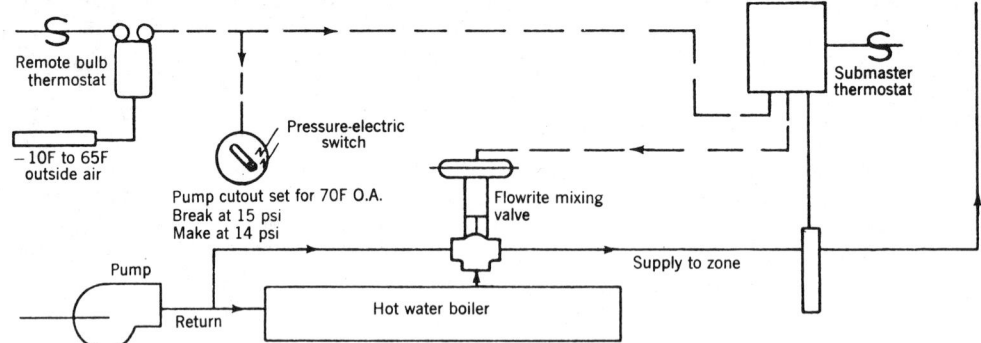

Fig. 16. Scheduled hot water control.

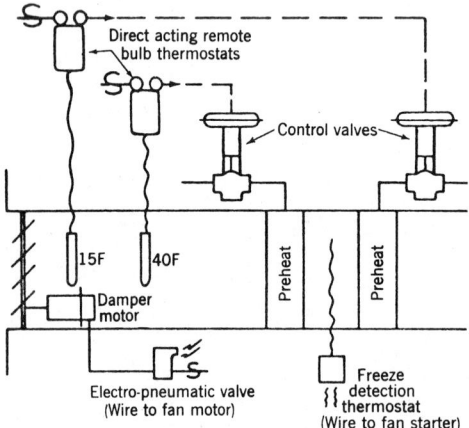

Fig. 17. Hot water preheat coil control.

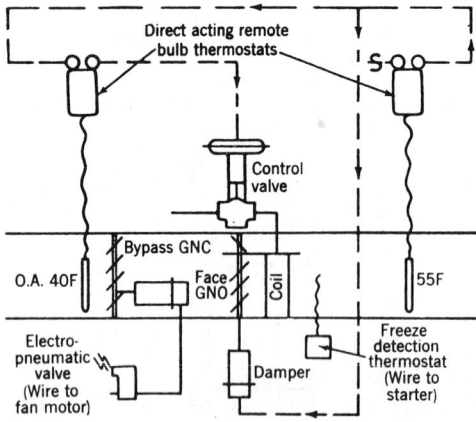

Fig. 18. Face and bypass control.

and bypass damper to maintain 55°F for necessary ventilation cooling. Further reheating is added in the air handling unit or zone ducts to satisfy room conditions. A freeze detection thermostat with coiled bulb is recommended on the leaving side of heating coil to insure fan and outside air damper shut-down in case of system failure. (See Fig. 18.)

Pressure control: Figure 19(a) illustrates a sloping pump characteristic curve. Without pressure control of the system, pressure will rise when control valves throttle water flow. This causes a shift of system curve from *A* to *B* and increases the pressure drop across the valves, adversely affecting control.

It is desirable to maintain a constant system pressure. Most of the pressure variation is caused by water flow change. A bypass around the pump or piping system is indicated.

1. A pressure bypass can be used around the pump, providing the pump chosen has a steep head curve. See Fig. 19(b). To prevent motor overload, caution should be taken in limiting the flow when the valve is fully open.

2. A pressure reducing valve is recommended when the pump curve is flat. See Fig. 19(c).

3. A system bypass valve can be employed at the end of a system or zone. Care should be taken in sizing supply and return mains to insure adequate pressure control. This can apply to either flat or steep head pump curves. See Fig. 19(d).

Hot water reheat: Figure 20 shows a commonly used reheat system with room thermostat *T-1* control. Preconditioned air from the fan discharge enters the reheat coil at approximately

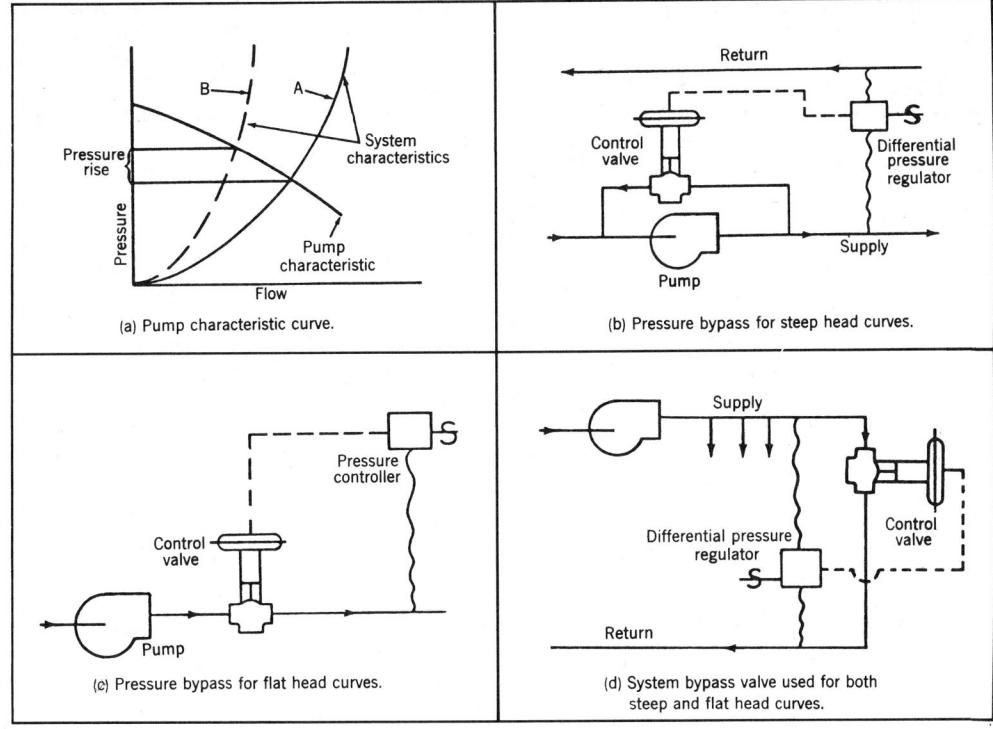

(a) Pump characteristic curve.

(b) Pressure bypass for steep head curves.

(c) Pressure bypass for flat head curves.

(d) System bypass valve used for both steep and flat head curves.

Fig. 19. Hot water pressure control.

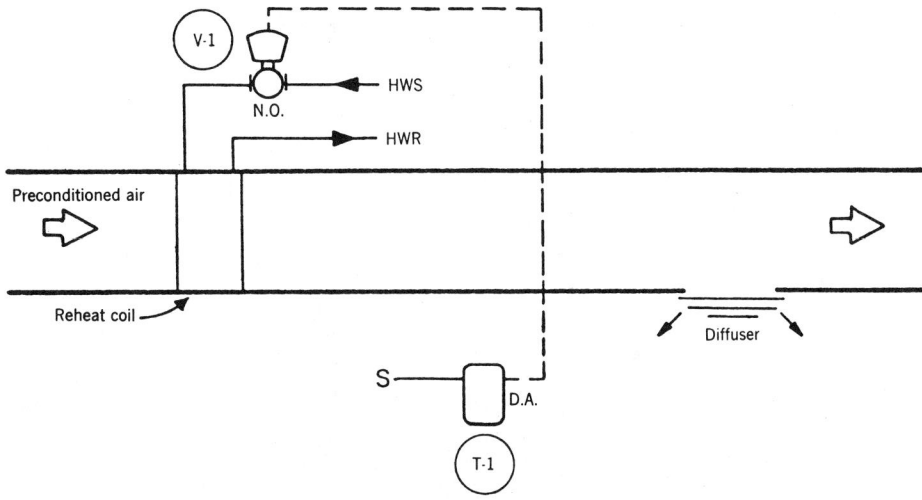

Fig. 20. Hot water reheat control.

55°F. *T-1* modulates valve *V-1* to maintain desired room temperature.

Hot water converter control: Outside air temperature resets the hot water supply temperature. The signal from the outside air transmitter, *TT-1,* resets submaster controller, *TC-1,* in accordance with a preset schedule. *TC-1* positions the normally opened steam valve, *V-1,* to maintain scheduled hot water temperature. Pressure-electric switch *PE-1* will shut off the pump when outside air temperature reaches 65°F. The schedule will depend on the load of the system. Figure 21 shows the control diagram.

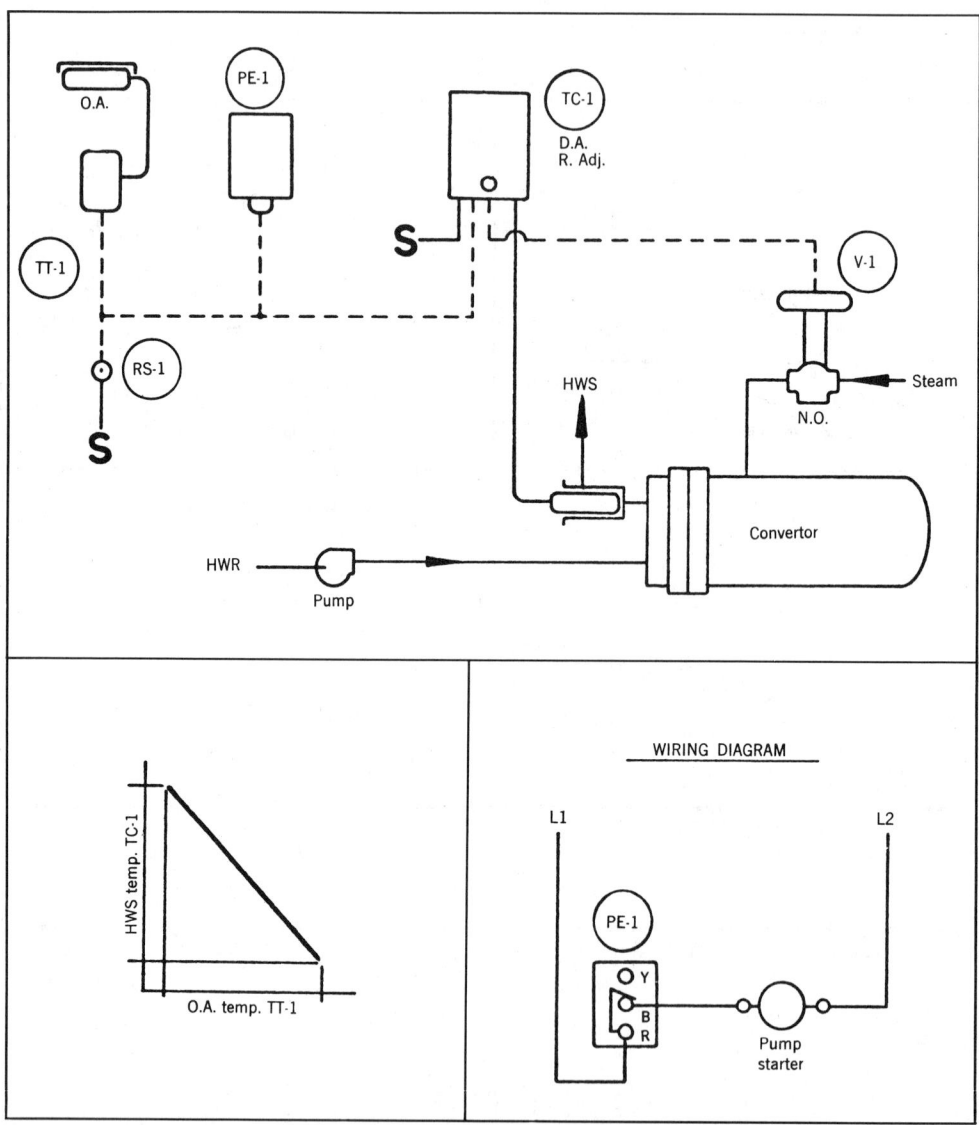

Fig. 21. Converter control with reset.

Mixing Box Control

Mixing boxes of various designs are generally used for local (room) comfort control. The unit usually has two air dampers for mixing hot and cold air. Damper motors may be mounted inside or outside of the unit. The dampers may be linked together or separately controlled. Following are a few typical control systems for mixing boxes:

Dual duct variable volume: Referring to Fig. 22, room thermostat *T-1* gradually positions the hot and cold duct valves through motor *M-1* and regulator through motor *M-2* to maintain the desired temperature according to the control schedule shown.

Dual duct constant volume—high velocity: Referring to Fig. 23, the constant air volume (CAV) regulator controls the pressure drop across the diffusers at a constant value. Static and velocity pressure will increase or decrease simultaneously as the flow of the unit increases or decreases in the system whereas resistance downstream from the measuring tip is constant. This allows the use of static or total pressure measurement. The room

thermostat controls the mixing ratio of hot and cold air for its load. The CAV regulator keeps the total flow steady.

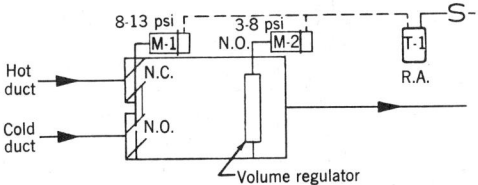

CONTROL SCHEDULE						
Room temperature	+3	+2	+1	SP	−1	−2
Control pressure	2	4	6	8	10	12
Hot duct valve flow	100% Open / 0% Closed					
Cold duct valve flow	100% Open / 0% Closed					
Volume regulator flow	100% Open / 0% Min.					

Fig. 22. Dual duct variable volume mixing box.

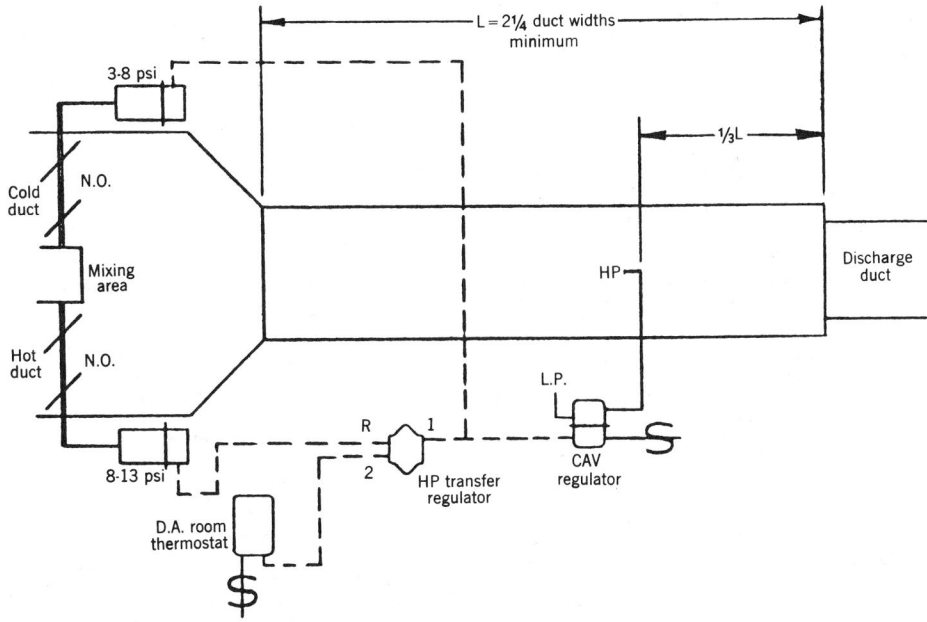

Fig. 23. Dual duct constant volume mixing box.

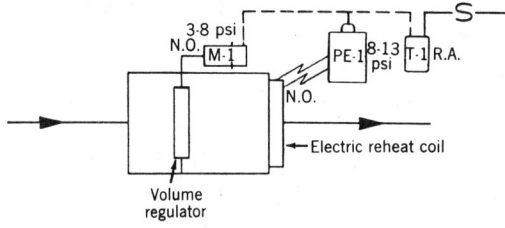

CONTROL SCHEDULE						
Room temperature		+3	+2	+1	SP	−1	−2
Control pressure		2	4	6	8	10	12
P.E. switch	on	Open					
	off						
Volume regulator flow	100%						
	0%	Min.					

Fig. 24. Single duct variable volume mixing box with electric reheat.

Single duct variable volume: Referring to Fig. 24, thermostat *T-1* gradually positions regulator through motor *M-1* and pressure-electric switch or step controller *PE-1* to maintain the desired temperature according to the control schedule shown.

Rotary Air-to-Air Heat Exchanger Control

Since the concept of energy conservation is of interest to this industry, air-to-air heat exchanger control is described:

Face and bypass control: Figure 25 shows the control diagram for a rotary air-to-air heat exchanger. The unit runs continuously at maximum speed. Reverse acting thermostat *T-1* is set for maximum sensitivity. When outside air temperature is below 75°F, thermostat *T-1* allows full control pressure from thermostat *T-2* to damper motor *M-2*. *T-2* controls the face and bypass dampers to maintain 55°F incoming air tempera-

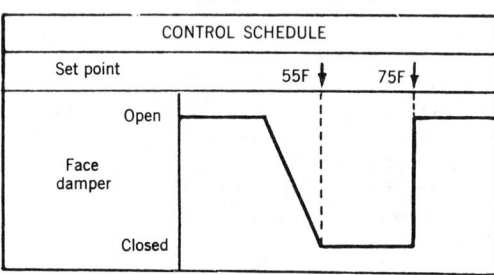

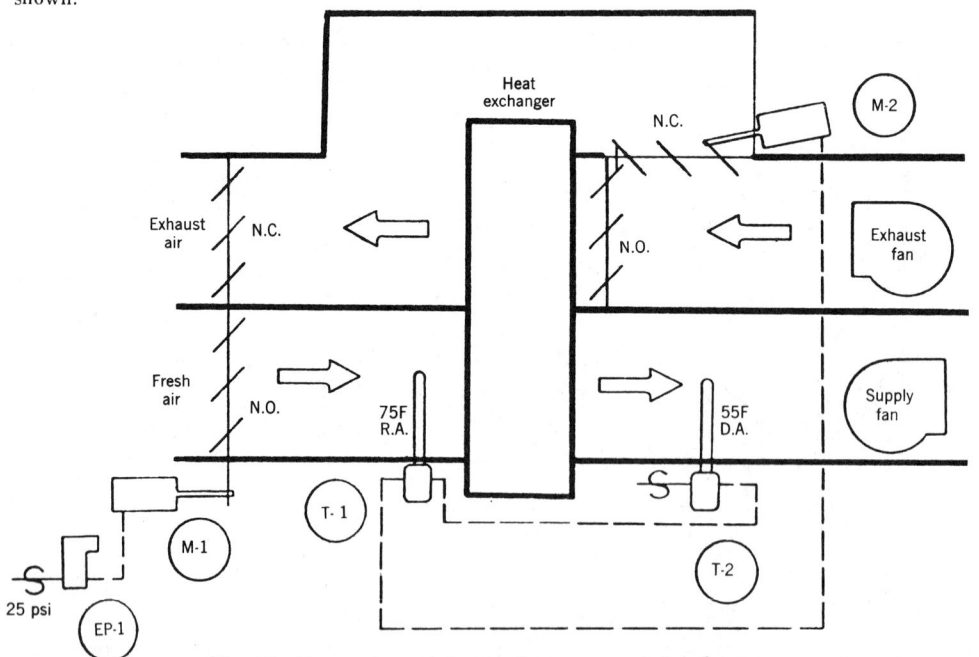

Fig. 25. Rotary air-to-air heat exchanger control, face-bypass.

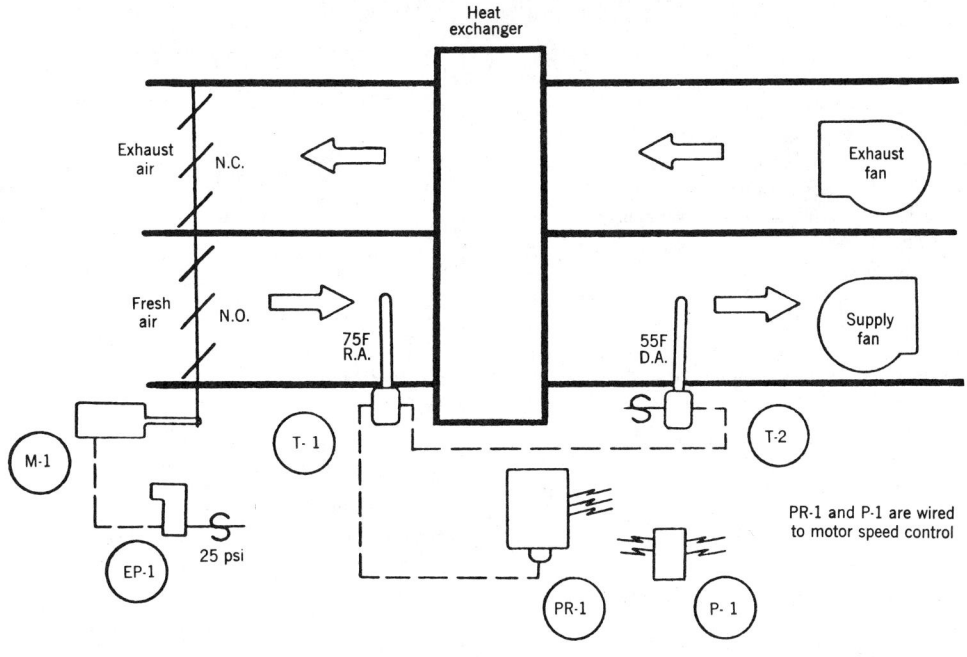

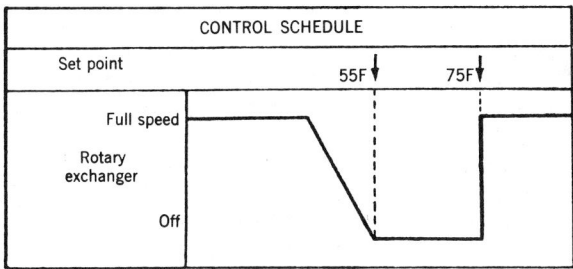

Fig. 26. Rotary air-to-air heat exchanger control, variable speed.

ture. When the outside air temperature rises above 75°F, *T-1* stops control pressure from *T-2* and bleeds air from motor *M-2* to allow full air flow through the rotary heat exchanger. Electropneumatic valve *EP-1* is wired to the supply fan to close the outside air and exhaust dampers on fan shutdown.

Variable speed control: Referring to Fig. 26, the reverse acting thermostat *T-1* is set for maximum sensitivity. When outside air temperature is below 75°F, *T-1* allows full control pressure from *T-2* to pressure controller *PR-1*. *T-2* controls the speed of the rotary heat exchanger to maintain 55°F incoming air temperature. When outside air temperature rises above 75°F, *T-1* stops control pressure from *T-2* and bleeds air from pressure regulator *PR-1* to maintain maximum speed of the rotary heat exchanger. Valve *EP-1* is wired to the supply fan to close outside air and exhaust dampers on fan shutdown.

WINTERIZING CHILLED WATER SYSTEMS

During winter operation of year-round air conditioning systems, there is the danger of frozen chilled water coils. A common practice to prevent this hazard is to drain these coils and to circulate anti-freeze in them temporarily, or to leave them filled with the anti-freeze for the winter.

In very large buildings with more than one central station system, the cost of labor and anti-freeze can be significant. Also, use of anti-freeze entails special precautions to prevent leakage through valve packings. Finally, most of the common aqueous solutions with a low freezing point are more viscous than chilled water at low temperatures. As a result, streamline flow might develop in the cooling coils and almost certainly would develop in the secondary water coils of any perimeter units on the system. Streamline flow sharply reduces coil capacity, perhaps better than 50%.

Water Circulation To Prevent Freeze-Up

Another common method of winterizing chilled water systems lies simply in circulating the water in the system when outside temperature drops below 32 F. However, most chilled water coils, 4 to 6 rows deep, are connected for counterflow operation to obtain good cooling efficiency. In Fig. 1, which shows a typical cooling coil counterflow arrangement, chilled water enters the last row downstream at (1) and leaves the first or upstream row at (7). Cold air contacts first the coldest water (at 7), which is an invitation to freeze-up, not a precaution against it.

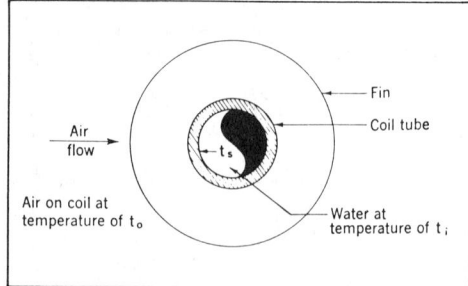

Fig. 2. Section through a typical chilled water coil tube.

What is actually required to prevent freeze-up is that the warmest water first meet the coldest air, that is, parallel flow. In Fig. 1, this would be accomplished by water entering at (7) and leaving at (1).

To prevent freeze-up, how warm should the water in the coil tubes be? By referring to Fig. 2, the following equation gives the heat transferred from the water to the air:

$$h_o' (t_s - t_o) = h_i (t_i - t_s). \qquad (1)$$

where

h_o'—Outside surface coefficient referred to inside surface, Btu per (hr)(sq ft)(deg F).

t_s—Temperature at the inside wall of the coil tube, deg F.

t_o—Temperature of the air upstream of the coil tube, deg F.

t_i—Temperature of the water inside of the coil, deg F.

h_i and h_o—Inside and outside film coefficients, Btu per (hr)(sq ft)(deg F).

Inside tube surface temperature, t_s, must be 32 deg or higher to avert ice formation. The air temperature, t_o, should generally be taken as the outside design temperature for the winter. The inside film coefficient, h_i, for water at a specified velocity and estimated temperature is found in most heat transfer books.

The outside film coefficient, h_o, for the air can be found from the equation:

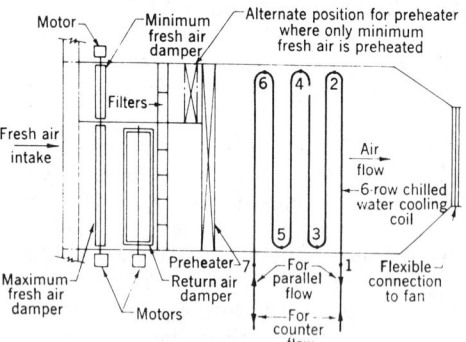

Fig. 1. Schematic of a large central station air conditioning system.

$$h_o' = \left[\frac{(\omega \times A_F) + A_o}{A_i} \right] h_o. \qquad (2)$$

where

ω—Finned surface efficiency of coil, dimensionless.

A_F—Surface of coil fins (both sides) per linear foot of coil length, sq ft per foot.

A_o—Outside surface of coil tube exposed to air per linear foot of coil length, sq ft per foot.

A_i—Inside surface of coil tube per linear foot of coil length, sq ft per foot.

The engineer can obtain ω and h_o from the coil manufacturer or approximate them from references 1 and 3. The water temperature, t_i, can now be found from the above equations. If it is greatly different from that used to find h_i, then trial and error solutions should be made until the estimated and calculated values of t_i are in close agreement.

For example, assume that the preheater fails and outside air at 0 F and 500 fpm face velocity should enter over part of the upstream row of the chilled water coil because of air stratification. From the previous equations, a water temperature of approximately 55 F with a velocity of 1 fps would be required, under these circumstances, to prevent freeze-up in a commercial coil with a high finned surface efficiency. Based on a water temperature of 55 F, evaluation of parallel flow vs. counterflow can now be completed with some interesting conclusions.

Counterflow and Parallel Flow

In counterflow, a 6-row coil with air and water conditions as cited would require an inlet water temperature of approximately 180 F to prevent freezing in the upstream row on the air side. Aside from destroying the air conditioning potential of all systems concerned, such a high temperature would make an oven of the building.

Parallel flow winterizing, on the other hand, is feasible with moderate inlet water temperatures, such as 55 F, which will not destroy or even reduce the air conditioning potential in any system in the building.

Steps in Design

There are essentially three steps to design a parallel flow winterizing system.

1. Calculate inlet water temperature required to prevent freeze-up as the water proceeds from the first row upstream to that row at which the air temperature is warm enough to prevent freezing the water.

2. Design a control system and a warning system which will guarantee that the water pumps actually operate during sub-freezing weather with water at the temperature calculated in step (1).

3. Design and layout the piping and equipment necessary for changing to parallel flow in cold weather.

As for the first step, the engineer should be able to ascertain this water temperature with the heat transfer equations given and in references 1 and 3. The coil manufacturer, however, is in a better position to make this check and should do so whenever possible. Step 2 is self-explanatory. The sales engineer for the temperature controls is in the best position to help design a suitable control system for parallel flow winterizing.

Piping Arrangement

The piping arrangement to achieve parallel flow through the coils in cold weather may be accomplished at one of two locations of the water system: (1) At each coil, or (2) at the central chilled water pumping station. Under scheme (2) it is necessary to have a manual switch to open the normally closed chilled water control valves. Under neither scheme is it advisable to circulate the water through the refrigerating units during winter shutdown because it will unnecessarily pressurize the evaporator. This might cause leakage of the primary refrigerant. Fig. 3 shows a typical piping arrangement to effect parallel flow at each coil.

Parallel flow at the central pumping station requires reversing the connections at the pumps with bypass valves. The exact method of piping a system for parallel flow will depend on initial cost and the ease for personnel to make this changeover.

Operating Procedure

Operating procedure for parallel flow when outside air temperature is expected to drop below 32 F usually involves the following steps:

1. Set all valving for parallel flow.

2. Start pump or pumps when outside air temperature drops to 35 F. Water velocity in coils should be kept at approximately 2 fps for good turbulence and low pumping cost. Since it is of vital importance to keep the pumps in continuous operation in sub-freezing weather, some guarantee must be made to assure this, such as automatic controls actuated by an outside air thermostat. When shut-

down of the plant occurs, as over week-ends, and operating personnel are usually reduced to a minimum, continuous operation of the pumps is most

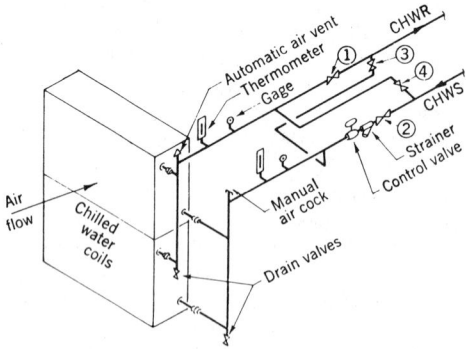

Fig. 3. Isometric of the piping for chilled water coils for both counterflow and parallel flow.

advisable, even though the outside temperature may not fall below 32 F for most of this time.

3. As the outside air temperature falls to subfreezing, it may become necessary to heat the water in order to attain that inlet water temperature calculated to prevent freeze-up. This would require installation of a small converter, if none were already available for this purpose. Fortunately, this is probably not required for most installations, since the vast mass of water will usually come to an equilibrium temperature well above that required to prevent ice formation.

REFERENCES

1. Donald Q. Kern, *Process Heat Transfer*, McGraw-Hill Book Co., Inc.
2. Ingersoll and Zobel, *Heat Conduction*, McGraw-Hill Book Co., Inc.
3. Tamami Kusuda, *Coil Performance Solutions Without Trial and Error*, AIR CONDITIONING, HEATING AND VENTILATING, January 1960.
4. Engineering Department Bulletin No. 2-59, 1959, John J. Nesbitt, Inc.
5. Engineering Department Bulletin No. 5-59, 1959, John J. Nesbitt, Inc.

MECHANICAL DRAFT COOLING TOWERS

There are two types of cooling towers in general use — the atmospheric and the mechanical draft. The spray pond and natural draft chimney tower have been largely replaced by those two types of cooling towers. Objection to the spray pond is the limited performance available and the nuisance created by the high water loss during certain seasons of the year. Objection to the natural draft tower is the high initial cost and the serious reduction in performance experienced during periods of hot weather.

There are two types of mechanical draft towers — the forced draft and the induced draft. The forced draft tower has its fan mounted at its base and the air is forced in at the bottom and discharged through the top at low velocity. In the induced draft tower, the fan is mounted on the roof of the structure and air is pulled upward and discharged at a high velocity.

Except for fan location, the structural and operational features of the two types of mechanical draft towers are essentially the same. A cross-section of the induced-draft tower with the various parts labeled is shown in the accompanying drawing. Entrained moisture is removed from the exhaust air by the drift eliminator just above the spray chamber and below the fan. Water is pumped to the main header at the top of the tower and from there distributed to the various nozzles. This water is sprayed up in a manner similar to that used in a spray pond and is intimately mixed with the exhaust air before dropping to decks below. In performance, the upspray distributing system represents the equivalent of adding 8 or 9 ft. to the height of the cooling tower over that of the gravity-type system. Slat-type grids interrupt the water as it flows counter-currently to the air. In flowing counter-currently, the coldest water contacts the driest air and the warmest water contacts the most humid air. Maximum performance is thus obtained.

Performance of a given type cooling tower is governed by the ratio of of weights of air to water and the time of contact between water and air. In commercial practice, variation in the ratio of air to water is first obtained by keeping the air velocity constant at about 350 fpm per sq ft of active tower cross-section and varying the water concentration (gpm per sq ft).

Time of contact between water and air is governed largely by varying the tower height. Should contact time be insufficient, no amount of increase in the ratio of air to water will produce the desired cooling.

Approach is the difference between the cold water temperature and the wet bulb temperature. Cooling range is the difference between the hot water temperature and the cold water temperature.

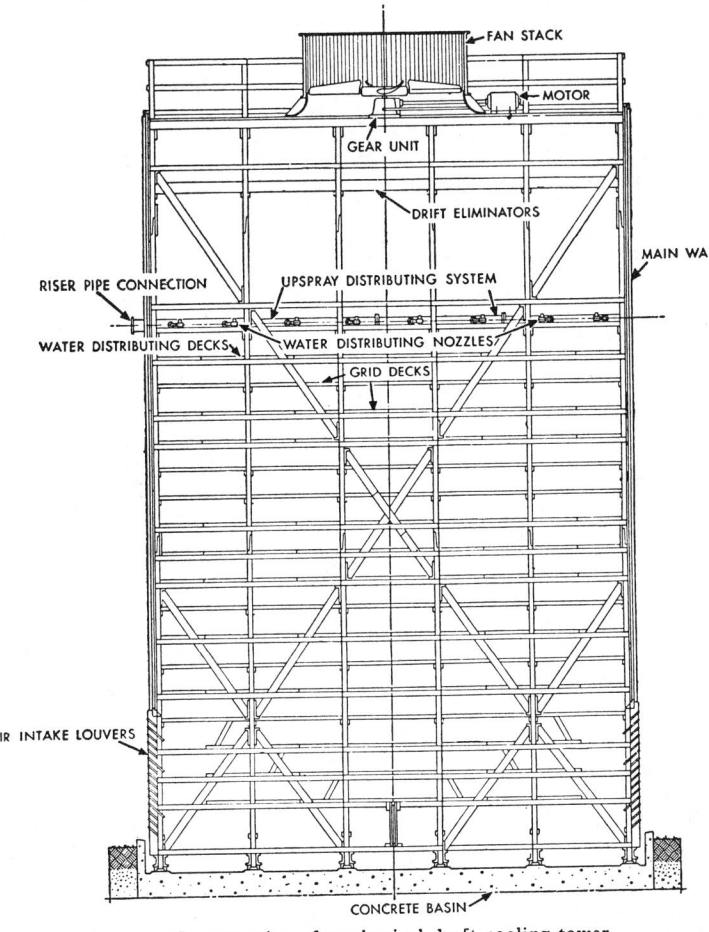

Cross-section of mechanical draft cooling tower.

ESTIMATING DATA FOR MECHANICAL DRAFT COOLING TOWERS

Fig. 1 shows the relationship of the hot water, cold water and wet bulb temperatures to water concentration. From this, the minimum area required for a given performance of a well-designed counter-flow induced draft cooling tower can be obtained. The horsepower per square foot of tower area required for a given performance is given in Fig. 2. These curves do not apply to parallel or cross-flow cooling; also they do not apply where the approach to the cold water temperature is less than 5F. They should be considered approximate and for preliminary estimates only. Many factors not shown in the graphs must be included in the computation and hence the manufacturer should be consulted for final design recommendations.

3. Quantity of water to be cooled;
4. Wet bulb temperature;
5. Air velocity through the cell; and
6. Tower height.

To illustrate use of the charts, assume the following cooling conditions:

Hot water temperature	102F
Cold water temperature	78F
Wet bulb temperature (T_{wb})	70F
Water quantity, gpm	2000

Place a straight edge on Fig. 1 to connect points representing the design water and wet bulb temperature, and find that a water concentration of 2 gal. per

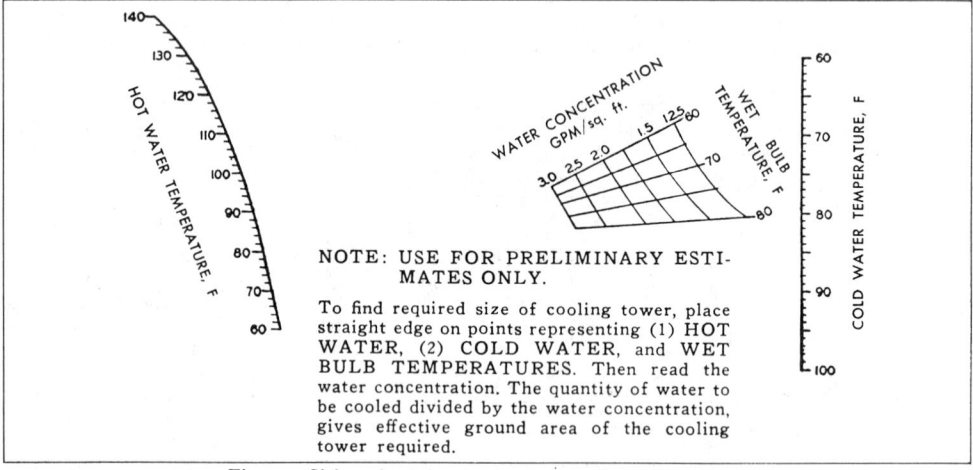

NOTE: USE FOR PRELIMINARY ESTI-MATES ONLY.

To find required size of cooling tower, place straight edge on points representing (1) HOT WATER, (2) COLD WATER, and WET BULB TEMPERATURES. Then read the water concentration. The quantity of water to be cooled divided by the water concentration, gives effective ground area of the cooling tower required.

Fig. 1. Sizing chart for mechanical draft cooling tower.

The cooling performance of any tower containing a given depth of filling varies with the water concentration. It has been found that the maximum contact and performance are obtained with a tower having a water concentration of 2 to 3 gpm per sq. ft. of ground area. Thus, the problem of calculating the size of a cooling tower becomes one of determining the proper concentration of water required to obtain the desired results. A higher tower will be required if the water concentration falls below 1.6 gal. per sq. ft. Should the water concentration exceed 3 gal. per sq. ft., a cooling tower of less height may be used. Once the necessary water concentration is obtained, the tower area can be calculated by dividing the gallons per minute circulated by the water concentration in gallons per square foot. Required tower size then is a function of the following:

1. Cooling range (hot water temperature minus cold water temperature);

2. Approach to wet bulb temperature (cold water temperature minus wet bulb temperature);

sq. ft. is required. Dividing the quantity of water circulated by the water concentration, find that the theoretical area of the tower is 1,000 sq. ft.

To obtain the theoretical fan horsepower, use Fig. 2. Connect points representing the 100% of standard tower performance with the turning point and find that it will require 0.041 hp per sq. ft. of actual effective tower area. Multiplying this by the tower area of 1,000 sq. ft., find that 41.0 fan horse-power are required to perform the necessary cooling.

Suppose that the commercial tower size is such that the actual tower area is 910 sq. ft. The cooling equivalent to 1,000 sq. ft. of standard tower area can still be obtained by increasing the air velocity through the tower. Within reasonable limits, the actual area shortage can be compensated for by an increase in air velocity through the tower which, in turn, requires a higher fan horsepower. The problem then becomes one of increasing the performance of the smaller tower by 10%. From Fig. 2, by connecting the points representing 110% of standard tower performance and the turning point,

the fan horsepower is found to be 0.057 hp per sq. ft. of actual tower area, or $0.057 \times 910 = 51.9$ hp.

On the other hand, suppose the commercial tower size is such that the actual tower is 1,110 sq. ft., the cooling equivalent to 1,000 sq. ft. of standard tower area can be accomplished with less air and less fan horsepower. By the use of Fig. 2 the theoretical fan horsepower for a tower doing only 90% of standard performance is found to be 0.031 per sq. ft. of actual tower area or 34.5 hp.

This illustrates how sensitive the fan horsepower is to small changes in tower area. The importance of designing a tower which is slightly oversize in ground area becomes immediately apparent.

concentration does not change since the volume of water and the tower area remain constant. With the water concentration at 2.0 and the wet bulb temperature at 60F, by adjusting the angle of the straight edge in Fig. 2 until a 30F differential is obtained between the hot water and cold water temperatures, find the hot water temperature to be 103F, and the cold water temperature to be 73F.

Suppose that the above designed tower had 1,500 gpm flowing through it, and the total heat load remained constant, what would the cold water temperature be when the wet bulb temperature is 65F?

Design heat load was $1000 \times 8.33 \times 33 = 250,000$ Btu per minute. The new cooling range (heat load

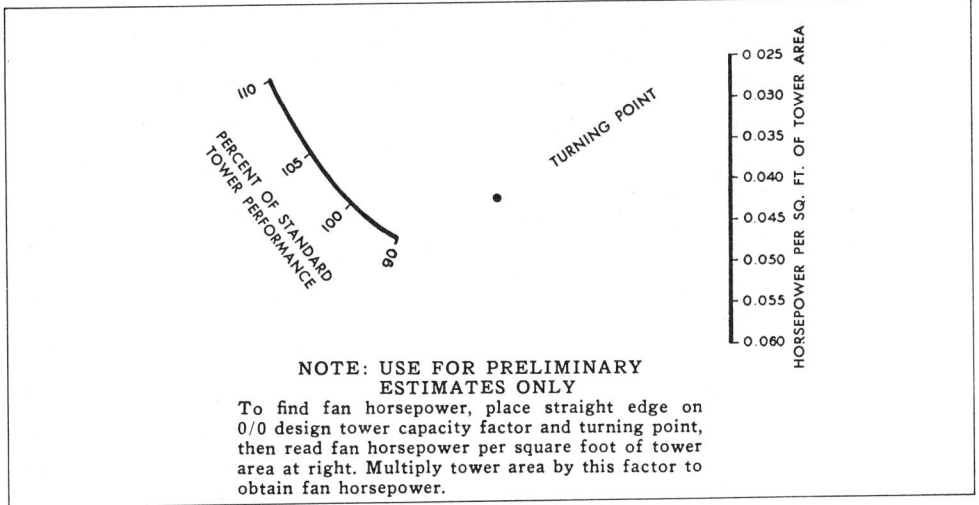

NOTE: USE FOR PRELIMINARY
ESTIMATES ONLY
To find fan horsepower, place straight edge on 0/0 design tower capacity factor and turning point, then read fan horsepower per square foot of tower area at right. Multiply tower area by this factor to obtain fan horsepower.

Fig. 2. Fan horsepower curve for mechanical draft cooling towers.

Assume the same cooling range and approach as used in the first example, except that the wet bulb temperature is lower. The design conditions would then be: gpm, 2,000; range, 24F; approach, 8F; T_1, 92F; T_2, 68F; wet bulb temperature, (T_{wb}) 60F.

From Fig. 1, find the water concentration required to perform the cooling is 1.75, giving a theoretical tower area of 1,145 sq. ft. as compared with 1,000 sq. ft. for a 70F wet bulb temperature. This shows that the lower the wet bulb temperature for the same cooling range and approach, the larger the tower area required, and, therefore, the more difficult the cooling job.

Estimating the performance of an existing tower, at other than design conditions, is often required. For example, assume a tower designed for the following conditions: gpm, 1,000; range, 30F; approach, 10F; T_1, 110F; T_2, 80F; wet bulb temperature, 70F. What will the cold water temperature T_2 be when the wet bulb temperature T_{wb} drops to 60F, providing that the heat load and water quantity remain constant?

From Fig. 1, find that the water concentration is 2.0 gal. per sq. ft. at design conditions. This water

remaining constant) when recirculating 1,500 gpm over the tower would be

$$\frac{250,000}{1500 \times 8.33} = 20.0F$$

Theoretically, the design area of the tower from Fig. 1 was 500 sq. ft. (1000 gpm/2.0 gal. per sq. ft. = 500 sq. ft.). Water concentration when circulating 1,500 gpm is

$$\frac{1500}{500} = 3 \text{ gal. per sq. ft.}$$

Now, referring to Fig. 1, with a water concentration of 3.0 gal. per sq. ft. and 65F wet bulb temperature, adjust the straight edge until a difference of 20F exists between the hot water and cold water temperatures. This shows the hot water temperature to be 100F and the cold water temperature to be 80F. This indicates that the possibility of a lower cold water temperature obtained by the lower existing wet bulb temperature was lost due to the adverse effect of the increased water quantity.

ATMOSPHERIC (NATURAL DRAFT) COOLING TOWERS

In atmospheric towers, water is pumped to the top of the tower, where it is discharged through a distributing system. Cross-section of a typical atmospheric cooling tower is shown in the drawing. As the water begins its downward flow, it is broken up and redistributed by the decks that comprise the filling of the tower. This continually creates newly exposed cooling surface for air to encounter passing horizontally through the tower. The redistribution insures even concentration of water throughout the tower during its entire fall.

Although the initial cost for an atmospheric cooling tower designed for a 3-mile wind is about the same as that for a mechanical draft tower, there are certain important limitations governing its performance. The tower must be located broadside to the prevailing wind in an exposed area. Any surrounding structures, hills or other barriers would tend to block the wind.

The cooling capacity of any tower, with a given wet bulb temperature and wind velocity, varies with the water concentration. Therefore, the problem of calculating tower size becomes one of obtaining the correct water concentration for a tower of chosen height that will operate with a certain wind velocity and wet bulb temperature. Once this water concentration factor is obtained, the area of a given height tower can easily be calculated by dividing the gallons per minute circulated by the concentration factor.

The concentration required to produce desired cooling depends primarily on the following conditions:

(1) Temperature range $(T_1 - T_2)$
(2) Approach to wet bulb temperature $(T_2 - T_{wb})$
(3) Tower height
(4) Wind velocity
(5) Wet bulb temperature (T_{wb})

Variation in items (3), (4) or (5) affect the water concentration, as follows:

Wind Velocity. The higher the wind velocity, the greater the amount of air that goes through the tower. This results in greater cooling. Therefore, when the wind velocity is higher, the concentration can be greater and still obtain equal cooling.

Bear in mind that there is a period occurring during the day and night in which the prevailing wind shifts. When this occurs, there is a short period when there is little or no air movement. At such a time, the tower water temperatures will rise from 2 to 5F, depending on the duration of the calm and the design wind velocity, and should be evaluated in determining a design wind velocity. Many atmospheric towers are designed to operate successfully at a zero wind condition.

Tower Height. In general, with atmospheric as well as mechanical draft towers, the greater the cooling range and the closer the approach to the wet bulb temperature, the higher will be the tower required to give sufficient time of contact between water and air to accomplish the desired cooling. In atmospheric towers, the performance is limited by both maximum and minimum water concentrations. Should the water concentration fall below 1 gpm per sq. ft. of tower area, it will be necessary to employ the next higher tower size. Should the water concentration exceed 3 gpm per sq. ft. of tower area, it will then be necessary to choose the next size lower tower.

Wet Bulb Temperature. Theoretically, a cooling tower cannot cool water to a temperature lower than the prevailing wet bulb temperature. With this limit, one is more interested in the economic approach of the cold water temperature to the wet bulb temperature. Air has a greater capacity for absorbing heat at the higher wet bulb temperatures. At the lower wet bulb, air in passing through the tower must have a greater temperature rise to accomplish the same cooling. Therefore, to obtain the same approach at the lower wet bulb temperatures, it is necessary to reduce the concentration.

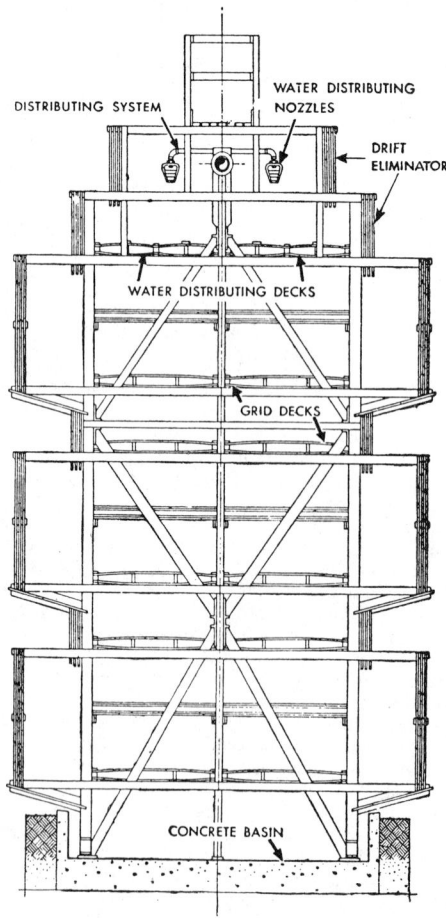

Section of atmospheric cooling tower.

ESTIMATING DATA FOR ATMOSPHERIC COOLING TOWERS

To calculate the size of an atmospheric type cooling tower with effective width of 12 ft., the following general formula may be used:

$$L = \frac{Q \times W}{C \times 12 \times C_w \times C_h} \qquad (1)$$

when L = length of tower in feet,
$\quad Q$ = quantity of water in gallons per minute,
$\quad W$ = wind correction factor,
$\quad C$ = concentration of water per sq. ft. of cooling tower area,
$\quad C_w$ = wet bulb correction factor,
$\quad C_h$ = tower height correction factor,
$\quad T_1$ = inlet temperature,
$\quad T_2$ = outlet temperature,
$\quad (T_1 - T_2)$ = temperature range,
$\quad T_{wb}$ = wet bulb temperature, and
$\quad (T_2 - T_{wb})$ = approach to wet bulb temperature.

Fig. 1 gives capacity curves for determining the water concentration (capacity) C for various temperature drops $T_1 - T_2$ (left scale) for various approaches to wet bulb temperature $T_2 - T_{wb}$ (curves). The accompanying table shows the values of the correction factors C_w, W, and C_h for cases where the wet bulb is other than 70F, the wind other than 3 mph and the height other than 35 ft.

Example. Determine the length of a tower 49 ft. high to cool 1500 gpm from 90F to 75F with a 70F wet bulb and a 5 mph wind.

From these conditions the approach $(T_2 - T_w)$ = 5F and the range $(T_1 - T_2)$ = 15F. From Fig. 1, the intersection of the 5° curve and the horizontal 15F line gives, on the bottom scale, a value of C = 1.19. Since the height is 49 ft., find in the table C_h = 1.60 and for 70F W.B., C_w = 1.0. Referring to the table, a wind of 5 mph has a correction factor of .83. Substituting in formula 1

$$L = \frac{1500 \times .83}{1.19 \times 12 \times 1.60 \times 1} = 54.5 \text{ ft.}$$

Once a tower is installed the problem often arises of determining what cold water temperature one can expect under conditions differing from those for which the tower was designed. For example, assume the tower is 35 ft. high, 12 ft. wide, and 107 ft. long. What cold water temperature T_2 can be expected when the wet bulb temperature is 60 F, wind velocity 4 mph, cooling range 15F, and water circulation 2000 gpm?

The wet bulb temperature correction factor C_w from the table for a wet bulb of 60F is 0.71. The wind velocity correction factor (W) for a 4 mph wind is 0.90. By substituting those values in the general formula and solving for water concentration, find:

$$C = \frac{2000 \times .90}{107 \times 12 \times .71 \times 1}$$
$$= 1.97 \text{ gal per sq. ft.}$$

Referring to Fig. 1 find that when the cooling range is 15F and the water concentration is 1.9, the approach to the wet bulb temperature is 9F. This means that the cold water temperature will be 60F + 9F = 69F.

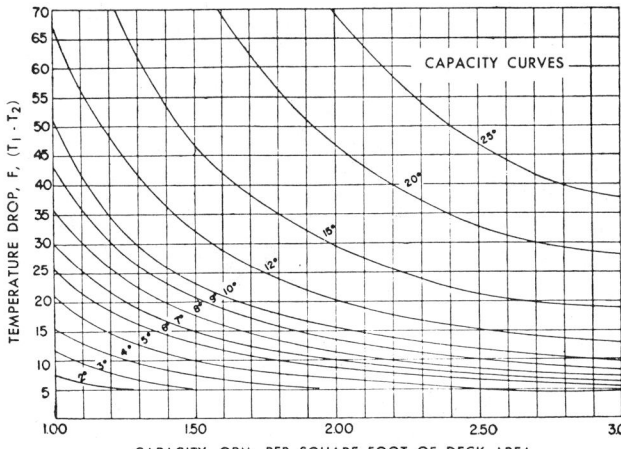

CAPACITY, GPM, PER SQUARE FOOT OF DECK AREA
BASED ON 70 F W.B.; 35-FT HIGH, 12-DECK TOWER; AND 3-MPH WIND
Fig. 1. Capacity curves for atmospheric cooling towers.

Correction Factors for Formula 1

Wet bulb temperature, F........	50	55	60	65	70	75	80	85	90	—
Wet bulb correction factor C_w.....	.56	.61	.71	.86	1.00	1.16	1.30	1.45	1.60	—
Tower height, feet.............	19	24	29	34	35	39	44	49	—	—
Height correction factor C_h......	.55	.68	.82	.98	1.00	1.15	1.36	1.60	—	—
Wind velocity, mph.............	1	2	3	4	5	6	7	8	9	10
Wind correction factor W........	1.28	1.13	1.00	.90	.83	.77	.72	.68	.64	.61

QUANTITY OF COOLING WATER REQUIRED

As shown in the table of compressor heat, elsewhere in this section, for every 200 Btu per minute (1 ton of refrigeration) absorbed by the mechanical refrigeration cycle there are approximately 40 Btu absorbed from work done by the compressor, or a total of 240 Btu per minute per ton to be rejected by the condenser where Freon-12 is the refrigerant. The quantity of condenser (cooling) water required, then, depends on the temperature rise of the cooling water.

The accompanying table gives the cooling water requirements for various temperature rises, in gallons per minute. The table is arranged with columns showing the quantity of water (gpm) necessary to conduct away from the refrigerant the heat quantities represented by the two units commonly used in air cooling practice, namely: the Mbh and the ton of refrigeration capacity. The table indicates these quantities at the increase in temperature which the water experiences in passing through the apparatus over the range usually covered in practice.

The following example illustrates the use of the table.

Example. How much water must be circulated through a water cooled refrigeration condenser of 11 ton capacity rating if the temperature rise of the water is 10F?

Solution. Enter the table at 10 in the first column and move horizontally to column 2 and find the figure 2.88. This is the volume (gpm) per ton and consequently multiplying it by 11 gives 31.68

gpm as the quantity which must be circulated. It should be noted that if the temperature rise were 20F instead of 10F, the quantity of water required would be one-half of 31.68 or 15.84 gpm.

Clear distinction should be made between the figures given in the table and the water consumption by refrigeration condensers, or of the total quantity of water required over a period of time per unit of heat removed. The figures in the table are for the quantity of water which must be circulated and are not necessarily the quantity of water consumed. In other words, there is a very definite difference between the water quantity which must be circulated and the water quantity which may be consumed. This is made obvious by keeping in mind that if the equipment is so arranged that the water which circulates through a refrigerant condenser can be re-cooled and used over again, time after time, the actual water consumption of the condenser can be zero, even though the amount of water circulated is large. Estimates of the water consumption required can be arrived at only by knowing the actual performance of the installed equipment. Generally, for air conditioning applications in many sections of the country, maximum water demands for refrigeration condensers are estimated on the basis of 2 gpm being required per ton of refrigeration effect, if there is any re-circulation of water.

With well designed evaporative cooling equipment, it is possible to reduce the water consumption of refrigerant condensers to from 5 to 10% of the amounts necessary without their use.

Water Circulation Necessary with Various Rises in Water Temperature in Cooling Surfaces

Rise in Water Temperature through Cooling Surface, F	Required per 1000 Btu per Hour of Cooling Load	Required per Ton of Refrigeration (12,000 Btu per Hour)	Rise in Water Temperature through Cooling Surface, F	Required per 1000 Btu per Hour of Cooling Load	Required per Ton of Refrigeration (12,000 Btu per Hour)
	Gallons per Minute			Gallons per Minute	
0.5	4.800	57.60	13.5	0.178	2.14
1.0	2.400	28.80	14.0	0.172	2.06
1.5	1.600	19.20	14.5	0.166	2.00
2.0	1.200	14.40	15.0	0.160	1.92
2.5	0.960	11.52	15.5	0.155	1.86
3.0	0.800	9.60	16.0	0.150	1.80
3.5	0.686	8.23	16.5	0.146	1.75
4.0	0.600	7.20	17.0	0.142	1.70
4.5	0.535	6.42	17.5	0.137	1.64
5.0	0.480	5.76	18.0	0.133	1.60
5.5	0.437	5.24	18.5	0.130	1.56
6.0	0.400	4.80	19.0	0.127	1.52
6.5	0.370	4.14	19.5	0.123	1.48
7.0	0.343	4.12	20.0	0.120	1.44
7.5	0.320	3.84	20.5	0.117	1.40
8.0	0.300	3.60	21.0	0.115	1.38
8.5	0.283	3.40	21.5	0.112	1.34
9.0	0.276	3.31	22.0	0.109	1.31
9.5	0.253	3.04	22.5	0.109	1.31
10.0	0.240	2.88	22.5	0.107	1.28
10.5	0.229	2.75	23.0	0.104	1.25
11.0	0.218	2.62	23.5	0.102	1.22
11.5	0.209	2.51	24.0	0.100	1.20
12.0	0.200	2.40	24.5	0.098	1.18
12.5	0.193	2.32	25.0	0.096	1.15
13.0	0.185	2.22	25.5	0.094	1.13

THE ROOF AS A LOCATION FOR AIR CONDITIONING EQUIPMENT

The accompanying discussion of the advantages and disadvantages of roofs and basements as locations for air conditioning equipment was prepared by Edwin Sternberg, of Fred L. Moesel Associates, New York, N. Y.

At one time, the invariable practice of designers was to locate the refrigeration system for the air conditioning installation in a basement or sub-basement machine room, irrespective of refrigeration capacity required. With basement installations, foundations for the main pieces of equipment such as refrigeration compressor and driving motor or turbine, chilled water and condenser water circulating pumps could rest on bedrock underlying the building foundations. This was a logical location then, due, in part, to the available equipment which was bulky, heavy in weight and noisy. In the case of slow speed reciprocating compressors, vibration and pounding had to be counteracted. In addition, operation of the equipment was generally manually controlled. Therefore, a basement machine room was convenient for the full attention of the operating engineer.

The development of automatic, compact refrigeration equipment and the evaporative condenser for water conservation, have changed this situation with the result that today the equipment can be located anywhere in the building, providing adequate means of isolation are installed. Basement machine rooms are not a must in the overall design of a system; economic and space considerations should be determining factors in the final selection of the machinery location.

Both roof and basement locations for the necessary equipment have distinct advantages as well as disadvantages. The capacity in tons of refrigeration for the proposed plant is an important consideration in arriving at a satisfactory decision.

Advantages of Roof

Use of Evaporative Condensers. When the refrigeration compressors are located within a machine room on the top floor of a building or a penthouse designed for this purpose, evaporative condensers can be located within a very short distance of the compressors. Since such condensers can be located on the compressor room roof above or alongside of the penthouse, usually there is no problem of air supply and discharge for the equipment. This arrangement eliminates the need for water-cooled condensers, cooling tower and circulating pump and also reduces the cost of refrigerant piping. The compressor brake horsepower per ton of refrigeration can be kept to a minimum by the installation of adequate size evaporative condensers, thereby reducing the actual cost of operation.

The physical size of the evaporative condenser is usually smaller than that of an equivalent capacity cooling tower. Maintenance of the evaporative condenser depends on its construction; a well designed condenser should have upkeep similar to a cooling tower. Water treatment for both types is desirable, especially in contaminated air surroundings, to protect condensing coils and condenser tubes. Prime surface condensing coils with corrosion-protected metal such as hot-dip galvanizing or non-ferrous tubes and corrosion-resistant fans are very desirable.

Evaporative condensers can be enclosed in a louvered shell similar to cooling towers to improve their appearance.

Multiple Units. It is economically feasible to divide the total required refrigeration capacity between several compressors. The evaporative condenser installation can likewise be made with separate coils or a separate unit for each refrigeration compressor. This arrangement provides flexibility in the operation of the plant under varying load requirements and can readily be accomplished with the roof installation due to the close spacing of all equipment.

Automatic Control. The development of automatically operated refrigeration equipment lends itself to the roof installation. Refrigeration compressors and evaporative condensers can be readily controlled automatically by a means such as the chilled water temperature. This arrangement provides maximum economy of operation in matching the required load and does not require continual supervision by the operating engineer. Periodic supervision by the operating engineer to check machine operation, oiling and pressures, can readily be combined with other duties. A bell or light in the operating engineer's office connected to the high pressure cut-out and to the low temperature chilled water thermostatic cut-out, will give added protection to the system in case of faulty operation.

Size of System. The total capacity of the refrigeration system is a paramount factor in the determination of location. Small and medium tonnage installations that lend themselves to the use of reciprocating compressors, either singly or in multiple, have greater economic advantages when located in roof machinery rooms than do large tonnage installations which generally require centrifugal or axial-flow compressors. The latter are built with integral water-cooled condensers and therefore require a cooling tower. The high weight concentration of the cooling tower over a small area calls for heavier structural framing and supports. Usually this is not the case with the smaller capacity, lighter weight reciprocating compressors. It would be impossible to set up an arbitrary tonnage figure as a boundary between reciprocating and centrifugal compressors, for each installation should be analyzed from a cost standpoint and decided on its own merits.

Ventilation. Adequate ventilation of machinery rooms is often a problem to the designer. It is obvious that the roof installation can be readily ventilated at a minimum cost due to its close proximity to outside air. Ductwork required for the

basement location can be eliminated and it is not difficult to locate the fresh air intake and exhaust louvers. In some roof installations, the evaporative condenser can be set at one end of the machinery room with the outside air intake louvers placed at the opposite end. The condenser fans will then augment the machinery room ventilation by drawing air across the compressor area.

Location of Air Conditioning Equipment. The location of the air conditioning equipment can plan an important part in arriving at the best location for the refrigeration equipment. When the air conditioning equipment is located in the upper floor or penthouse fan rooms of a building, it is obvious that a most economical installation can be made should the refrigeration equipment be located nearby. With the development of primary air conditioning systems used in conjunction with individual room conditioners, a pent-house or top floor fan room has decided advantages, and lends itself to such an arrangement. There are other less obvious advantages such as the lower static water pressure on the chilled water circulating pump casing which exists when the pump is located at the top of the closed circulating system. This arrangement minimizes stuffing-box seal maintenance problems.

Machinery Space. The location of equipment presents an economic problem, especially in rental type buildings. For example, in large cities where the building cost per cubic foot is high, the basement area, which is usually reserved for refrigeration equipment, can yield a good return when rented for restaurant or similar use. This basement use for a restaurant is feasible if good air conditioning and ventilation are provided and arrangements made for a ground floor entrance. In some cases, the elimination of costly rock excavation required for a subbasement equipment room can materially change the construction cost. With the increasing use of district steam, the basic need for high ceiling boiler rooms is eliminated and this tends to further reduce excavation cost. The exclusion of a high ceiling refrigeration equipment room can again help reduce these costs. It must be remembered that the refrigeration capacity of the installation, as discussed previously, plays an important part in the analysis.

Disadvantages of Roof

Supporting Structure. The roof installation obviously requires both a supporting structure for the dead weight of the equipment and adequate bracing for unbalanced loads. The additional steel requirements in the form of columns, girders and beams is a function of several variables so no arbitrary standards can be established. For example, the installation of multiple smaller capacity refrigeration compressors and evaporative condensers does not present the same structural problem that would be encountered with one or two large capacity units. In the former case, by judicious location of the equipment with reference to the main load sup-

porting columns and beams, the cost of additional framing can be minimized; in the latter case, the supporting structure must be designed to meet the expected loadings which may result in deep section girders and beams.

This situation can increase the floor to floor heights to maintain proper headroom under the structural supports. The building height directly affects the cost of additional steel for columns. A skyscraper type of building, with multiple setbacks and with high refrigeration requirement, obviously is not as well suited to a roof installation as a building of lower height and with uniform floor areas. The designer should analyze each type of structure on its own merits. In many cases, the floor loading selected by the structural engineer for fan rooms is suited to the loading imposed by small tonnage refrigeration equipment without additional steel framing.

Vibration. Reciprocating compressors located on a building roof present problems of vibration isolation to prevent transmission and telegraphing of this motion to the occupied portions of the building. Small multicylinder compressors do not create the same situations as large compressors of slower speed. Several types of vibration-absorbing bases and foundation are available, each engineered to the particular type of installation. Specialists in vibration-control should be consulted.

It is not costly to isolate a small compressor and the same type of isolation would generally be used irrespective of the equipment location. Large units require spring-supported floating structural subbases which in turn must be properly supported by the building. In some cases, weight must be incorporated in the base to absorb unbalanced forces.

It is desirable to isolate interconnecting piping from the building structure by the use of isolated pipe hangers and flexible seamless tube type connections at the equipment.

Services to Equipment. Electric power must be carried from the main service location, usually located in the basement, to the equipment on the roof. The cost of this service will depend on the distance involved and the wire size. When the fan room is located nearby, both power requirements can be combined; this is less costly than where a separate service is run to each. Separate metering and switches can be located in a service center.

Since the water supply must be brought to the evaporative condenser or cooling tower, irrespective of where the design places the refrigeration equipment, no new situation is established. Water can be taken from the main water supply for the water-cooled stuffing-box seals, make-up water for the chilled water system, and for similar needs.

Installation of Equipment. The cost of rigging refrigeration equipment to a roof location will usually be greater than for a basement location. Rigging costs should be estimated for each location and this figure used in the cost analysis.

SERVICING THE COOLING PLANT FOR SUMMER USE

Following are practical suggestions for putting the summer air conditioning system into operation:

Because all air conditioning systems are practically "tailor-made," it is impossible to give explicit instructions which will be applicable to every installation. Therefore, while the points listed are general points, they are important whether they require the adjustment of 20 valves or one valve to make the system operate properly.

Water System

(1) Check coils in the cooling system. Plugs may have been removed and system drained for winter. Replace plugs and close valves to the coils. Be certain that all valves to coils are closed before cracking the main water supply valve to allow gradual filling of system. If coils are located in several equipment rooms on different floors, start at the bottom of the system venting coils in each equipment room from first floor to top of system.

(2) If a deep well pump is part of the system, it should be checked for proper lubrication and electrical connections. After all valves are open, trip pump switch for momentary operation of pump. Do this two or three times. If operation is normal, throw switch for regular operation.

(3) After the system has operated for two days, recheck coils for air binding.

Air Handling System

The procedure for preparing air handling equipment for operation is the same whether the system uses water or mechanical refrigeration as the source of cooling. Here again there are a number of important steps in setting the system in operation.

(1) Check the motors for proper lubrication and electrical connections.

(2) Check the fan, bearings and belts to see that they are in operating condition.

(3) Do not forget the filters. While checking the filters for cleanliness also check the surface of the coil to see that it is clean. One building owner had his serviceman start up the system after a check, but he got no cooling—in fact, he got no air distribution at all. The filters were clogged with dust and dirt to such an extent that air could not be drawn into the system.

If it is possible to manually operate the air handling equipment on the spot, check it for noise and general operating characteristics.

Compressor Oil

Before proceeding further, check the oil pressure in the crankcase of the compressor. The normal operation of shutting down a refrigeration plant for the season is to leave 2 lb pressure in the system. Now, before opening any valves, observe this pressure to determine if there was a loss of pressure during the shutdown period. A lower pressure indicates a leak in the system. Remember that both Freon and oil will escape through the defective part. How to search out these leaks is described later.

Condenser

If there is an evaporative condenser in the system, it should be next on the check list. In a good many installations, the evaporative condenser is mounted on the roof. The service suggestions here are simple to accomplish, provided the unit was serviced properly when the system was shut down at the end of the season the previous fall.

(1) Here again motors, fans and belts are first on the list for checking and servicing. V-belts should be free from wear and have a live feeling when struck with the hand. They should not be "fiddle-string" tight.

(2) For outdoor installations, the next step is to clean out the sump. If there is water in the sump, start looking for trouble. This, very possibly, indicates the unit was not properly drained and lines in the system may have frozen and ruptured. Before going ahead with the checkup, inspect the piping that is exposed to the elements or where freeze-up may occur.

(3) Close the drain on the sump.

(4) Open the water supply valve to the unit and at the same time check the operation of the float valve in the sump.

(5) Check the operation and for leaks in pump. First, turn the pump over by hand to check freedom of operation. Then with the pump operating electrically, check for leaks. There should be a slight leak at the packing for the pump to be operating at its best. This leak should amount to a rather infrequent drip and not a steady flow from the packing.

(6) Examine the coil for corrosion or fouling. If the coil is fouled with scale, it should be cleaned with an inhibited acid. After using the inhibited acid, the coils should be well flushed with clear water to remove the cleaner.

(7) If the liquid receiver is mounted near the evaporative condenser, now is the time to open both valves in the liquid receiver.

On some installations, there is a shut off valve in the hot gas line on the evaporative condenser. This valve should be opened as the last step in getting this unit ready for operation.

If the system uses a shell and tube condenser instead of an evaporative condenser, there are two inspections to make. (1) Open the valve for the water supply to the condenser. (2) Use a screw driver to momentarily open the water regulating valve on the condenser: This will flush the valve mechanism and put it in readiness for operation.

Refrigeration Unit

All parts of the system are important, but the refrigeration unit is the most vital. Every precaution and check must be made to certify the correct operation of this unit.

(1) First of all, make the usual inspections for proper motor lubrication, electrical connections and condition of belts, if belt-driven. If direct-driven, inspect the coupling for signs of wear.

(2) Open all valves on the unit. These include the liquid valve, suction valve, and hot gas discharge valve. It's a good idea to double check this operation. Things can happen very fast if one or more valves remain shut during operation of the compressor.

(3) Check electrical connections to the compressor. The main fuse disconnect switch has probably been locked in the open position to guarantee non-operation of the unit during the seasonal shut-down. This should be unlocked and closed at this time.

(4) Next, throw compressor switch in and out two or three times. Allow compressor to run a few seconds each time the switch is thrown in to get oil on the seal. This short periodic operation also helps to separate the Freon from the oil.

(5) Put the system into operation. This includes starting up the air handling equipment, evaporative condenser (if included in the system) and compressor. If there is a gage on the crankcase, observe the oil pressure. It should be above the pressure in the suction line. Observe the oil in the crankcase. If low, one of two things has happened. Oil may have mixed with the Freon and be held in another part of the system, or oil has escaped through a Freon leak. To determine which of these malfunctions has occurred, run the compressor intermittently for several seconds to try and separate the oil and Freon. If, after several tries, the oil level still remains low, there may be leaks in the system. Check Freon charge through the sight glass on the liquid line. If it bubbles while the compressor is in operation, the charge is short. This is another signal that troubles are present.

(6) Shut down the system and close valves. Leaks may be discovered in two different manners. A halide torch may be used for this purpose. Escaping Freon will turn the color of the torch flame for it is extremely sensitive to the presence of Freon. The other method is to examine the piping system for oil droplets hanging to the underside of the piping. While the Freon escapes into the air, the oil usually clings to the piping in the vicinity of the leak.

Check Oil

There is another possibility, in case the leak cannot be discovered. A previous leak may have been repaired, Freon added, but the oil forgotten. To correct this requires only the addition of compressor oil.

When the system is in shape again, add the necessary refrigerant and oil. Always remember to add both Freon and oil. When there is a leak in the system, both the lubricating oil and the refrigeration charge will escape.

Compressor Now Ready

A few final inspections remain before placing the unit in operation for the summer. You are now ready to start the air handling unit and the compressor. Run the compressor for approximately 30 minutes. At the end of this time observe the oil in the crankcase and check the condition of the Freon in the sight glass. If everything appears correct, continue to run the equipment while checking the entering and leaving air in the air handling unit. If these temperatures are in accordance with design conditions, check the operation of the controls and upon completion of this step, the air conditioning system should be in tip-top shape for summer-long operation.

It is important that this "spring tune-up" be carried out as fully as possible. The investment in the air conditioning system is well worth the few hours required to set it in top condition for its job during the hot days ahead. While these suggestions are necessarily general, they cover the majority of installations and their form should lend them to easy modification to fit your particular installation.

AIR CONDITIONING EQUIPMENT MAINTENANCE

Maintenance of air conditioning equipment includes the following:

Air Handling Equipment. Treating water for washers, replacing or cleaning air filters, checking fan belts, lubrication of pumps and fans, coils, electrical repairs and checking of controls.

Air Distribution Equipment. Cleaning air ducts and grilles, repairing insulation on ducts, painting ducts, and checking dampers.

Water-Using Equipment. Treatment of water, lubrication of pumps and fans, checking of piping and valves, painting, and cleaning.

Cooling Equipment. Recharging refrigerant, lubrication, checking of V-belts, maintenance of compressors, checking of motors and controls, checking of valves and piping.

Air conditioning equipment is designed to perform a particular function in the most efficient manner possible and to give good mechanical performance. But before the expected results can be achieved, it is necessary to (1) so install the unit that there are no conditions that are detrimental to good operation; (2) operate and maintain the unit so that a continuance of satisfactory operation is assured.

The discussion which follows was prepared by F. Edward Ince of Marlo Coil Company. It is concerned with cooling towers, evaporative coolers and evaporative condensers, and also with air conditioning units.

One point that many engineers slight is that not only must equipment have enough space for installation, but there must be sufficient clearance around the unit to enable proper servicing and operation. There should be sufficient space to permit oiling of bearings, removal of coils, fans and shaft, eliminators and wetted deck surface, servicing the spray water pump and strainer and for the air conditioning unit; enough room for servicing fan and pump motors, and the belts. Space should be allowed to permit the use of brushes and other equipment to clean the coils in the unit.

It is important that the units are lined up horizontally and vertically so that there will be proper slope for satisfactory drainage.

Air discharge from water-saving devices should be above the roof line, and well above and as far as possible from the air intake. Since dirt and fumes carried into the unit aggravate the problem of keeping clean the heat transfer surfaces, air entering the apparatus should be kept as free as possible from dirt, fumes and steam. Simple upward direction of the air discharge is not sufficient, and the back-wash effect of air currents flowing over buildings should be taken into consideration. Both the air inlet and outlet should be kept free from obstructions that might impede the free flow of air. Runs of discharge duct inside the building should be insulated to reduce condensation on the inside and the ducts should be made water tight.

Before starting equipment, remove the belts and be sure that fans and pump rotate freely. Fill the oil cups on the fan and motor bearings before the electric wiring is completed. Then check the motor and replace the belts. Belts should be adjusted to prevent slipping when starting. However, too much tension will cause rapid wear on both belts and bearings and an unnecessary load on the motor. It is important to check the motor for correct rotation. Directional arrows serve as a guide.

Fill the pump with water and then check and adjust the float valve to maintain proper water level—not more than one inch below the overflow pipe. Place the waste line from the spray header inside the overflow cap. Next open the waste line valve at the spray header about two turns, and with the aid of a gallon bucket, check the flow rate. Set the waste at about 2 gph per ton of refrigeration or per 15,000 Btu per hr heat rejection, and then adjust as required under operation.

Since the spray water system is designed to maintain a running film of water on all heat transfer surfaces, it is necessary to maintain the spray pumping unit in good operation. As one check, see that pump and motor bearings are properly lubricated, that the stuffing box is not binding the shaft, and that the fan and pump belts are maintained in proper tension and alignment. It is advisable to waste enough water so that the mineral concentration does not exceed that point at which scale will form. One must remember that in operation about one pound of water is evaporated per 1,000 Btu and that the mineral content of this water is left after evaporation takes place.

Some mineral salts are more scale-forming than others, and it may be advisable to treat the water with compounds which form soluble and non-scaling type of salts after combining with whatever minerals are present. Where wasting of water does not solve the problem, water treatment by some competent company is recommended.

From the cleaning standpoint, it is important that the sump tank should be drained and cleaned often enough to prevent large accumulations of sediment that are carried into the unit through the air stream. Pump suction strainer should be cleaned and the

spray nozzles checked to see that each is producing a full spray cone and that all parts of coil or wetted deck surface are flooded. Air intake screens should be kept free of leaves and paper, and a check made that boxes or crates do not obstruct free flow of air into the unit.

It is important to note (1) the in-going and outgoing temperatures of the fluids to be cooled, (2) temperature of the spray water, (3) the in-going and outgoing air temperature, and occasionally (4) the volume of the air. Readings that deviate from normal, indicate the need for closer inspection of the apparatus and the equipment served by it.

All systems using a direct expansion refrigerant should be thoroughly tested for leaks before and after the system is charged with the refrigerant. Particular attention should be given to the thermal valve packing. Avoid any condition that creates gas in the liquid line but if such a condition is unavoidable, provide adequate subcooling of the liquid with a heat exchanger to prevent gasification.

Water coils, sump pan, and various water lines, should be adequately drained and blown out for winter operation to prevent freezing of these components. During winter operation, if the unit is shut down for any period, the steam coils should also be drained to prevent freezing. Water and steam coils are circulated and sloped in the unit at the factory for proper drainage. Precautionary methods should be taken at all times during winter operation to prevent freezing in the steam coils. Adequate filters should be placed on water, steam and humidifier lines to prevent dirt accumulation in the coils.

The pump should be located as close as possible to the unit sump tank and one should always avoid a high suction lift. In no case should the total suction lift, including friction in the pipe, be more than 15 ft. If the pump is to handle hot water, it must be so located that water will flow freely into the suction.

On units with face and bypass dampers, make certain that the dampers are properly linked for full opening and closing.

Periodic maintenance and inspections should be enforced to insure long and dependable service. About every 60 days inspect the damper bearings and linkage, and check fan and motor bearings. A suitable oil should be added. Fans and scrolls should be free and clean of all foreign matter.

Once a year, it is good policy to shut down the unit and thoroughly clean all surfaces, remove all scale and rust deposits and protect surfaces with a good corrosion resistant paint where necessary. One should remember that the misuse of strong acids or alkali solutions is dangerous and can result in serious damage to this equipment. Interior parts should be brushed and freed of loose particles. As a safeguard, paint bare metal before using any cleaning solutions. After using any chemical, wash, brush clean all surfaces, and repaint as necessary for further protection. However, do not paint coils on evaporative coolers.

Pumps require attention. Clean, flush and inspect the pump, checking water passages and impeller clearance; clean and regrease pump bearings; and every six months, the pump should be greased. All worn or damaged parts should be replaced. In addition, inspect the motor and motor mounting brackets and at the same time clean and oil the motor. However, keep oil away from fan belts and pulleys. Inspect pulleys for worn or damaged grooves and make such replacements as are necessary.

Maintenance resolves itself into a program of cleaning and inspection. Every 60 days, the fan and motor bearings should be checked and the proper type of oil added. Also periodically check the tightness and alignment of belts.

Clean the strainer and the pipe in the spray piping as often as necessary. There is no set rule for the periodic cleaning of spray piping and nozzles, but when any nozzle does not have a full cone spray, it should be cleaned. The constant waste valve and waste line should be checked periodically to insure free flow of waste water to the drain. Examine the float valve for proper function, and replace the valve seat if it is worn.

At regular intervals, inspect the water humidifier and steam jet humidifier nozzles, and clean if necessary. On vapor pan humidifiers check for the proper functioning of the float valve. Because of water evaporation, these pans will accumulate salt scale deposits so that it is important to flush the pans at least once a month during the winter season.

All coils should be carefully washed with a hose once a month to remove accumulated lint or foreign matter carried to the unit by the air stream. Fans and scrolls should also be washed or brushed down to remove the accumulated dirt. Incidentally, clean or replace air filters when dirty.

Before a unit is to be shut down for a long period of time, at the end of either the cooling or heating season, it should be thoroughly drained, well flushed out and cleaned, belt tension relieved, and the unit given a coat of corrosion-resistant paint, especially in spots where rust might appear.

The sump pump on spray type dehumidifier units should be drained and thoroughly flushed at least once every two weeks. If scale deposits appear on the cooling coils, the water should be treated.

AIR CONDITIONING MAINTENANCE SCHEDULE

The schedule which follows is designed to serve as a guide for maintenance inspection for air conditioning equipment. In general, inspection should be divided into daily, monthly, and yearly periods.

Daily inspections are those that have to do with checking water levels, water conditions, heating up of parts, hand-oiling and whatever may be termed a daily once-over.

Monthly inspections are usually those specified by the manufacturer as routine checks on oil levels in reservoirs, checks for leaks of the refrigerant, oil, and water, cleaning of electrical apparatus, filters, spray nozzles, collection pans, and checking of control devices, belt tensions, scale traps, and purging.

Yearly inspections require the opening up of condensers for examination, dismantling of machines, and any inspections which require the apparatus to be shut down for prolonged periods. The grouping in the check list has been arranged in what is considered a logical order for efficient work.

It might appear that the schedules outlined are an attempt to reduce inspection to a rigid routine. No such intention was in mind. On the contrary, the grouping is definitely arranged to permit flexibility for physical reasons unique to any given plant. Most failures of equipment are, without question, preventable. A pulley that causes frequent replacements of belts is being neglected by somebody and the relief valve that fails to open the one time it is required to has been a forgotten piece of equipment for some time. Even in an extreme case it may take a dollar's worth of maintenance to prevent a dollar repair; the maintenance dollar, however, is the wiser one because the failure that necessitated the repair may be accompanied by intangibles.

The following list is a schedule, as explained. The list entitled Air Conditioning Maintenance Procedure, relates to *what* is to be done as compared to the accompanying list covering *when* things are to be done. Both are suggested practice.

UNIT AIR CONDITIONERS

Unit Air Conditioners—self-contained air conditioning systems that are built in compact units.

Daily—Since these units are as near "fool proof" as it is possible to make them, inspection becomes co-incidental with "operation."

Monthly—The system should be checked for leaks with a standard halide gas leak detector. Belts should be inspected each month for signs of abnormal wear or pulley misalignment. The builder's instructions on purging should be followed. Oil levels are checked each month. Knocks should be investigated.

Yearly—A check should be made on the water being used to determine its condition. The service of the small unit does not warrant other than reasonable precautions to see that clean water is at all times supplied to the system. At the annual inspection the water in the system should be drained and the various parts examined for corrosion and scale. Safety devices shoud be removed and checked and their condition noted. An over-all "condition" record should be made for the year. In some cases it is recommended that a general check-up or inspection be made of unit systems after a stipulated number of service hours; quite often the number of service hours is much longer than a year. The recommendations based on service hours should always be followed because the maintenance requirements are a function of hours of actual operation which determines the life of individual parts of the unit.

CENTRAL SYSTEMS

Rotating Apparatus—those parts of the air conditioning system, such as pumps, compressors, blowers, motors, and the controlling devices which convert or assist in the conversion of electrical energy into refrigeration.

Daily—Motors should be given a daily "once over" with particular attention being given to automatic starters and contactors. Any arcing should be noted. Such equipment properly cared for seldom gives trouble.

Monthly—Note the condition of brushes, protective devices and oil levels. Belt tensions should be checked and any apparent misalignment or abnormal wear should be fully investigated. Blower bearings should be checked for proper adjustment. Pump packings should be inspected monthly for leakage. The casing covers of motors used with induced draft cooling towers should be removed once a month for inspection. Where gears are used on this type fan they should be inspected once a month.

Refrigerant Circuit Controls—the part of the system through which the refrigerant circulates including the compressor, connected piping and valves, condenser receiver, and the evaporator. The controls consist of such devices as the thermostatic expansion valves, switches and regulators with automatic cutouts and solenoid valves for water and refrigerant.

Daily—System should be inspected for overheating of the moving parts, "knocking," excessive pressures in the system of piping and vessels. The proper cutting in and cutting out points of machines should be noted. Any evidence of sticking of the expansion valves should be noted since such sticking is usually caused by moisture in the system. Any signs of leakage should be traced to their source.

Monthly—On monthly inspections the refrigerant circuit should be checked for leaks of the refrigerant. The standard halide refrigerant gas leak detector is recommended for this purpose. Also look for oil leaks around packings and water leaks throughout the system. Scale traps should be checked. Purge valve should be opened or the system should be purged through the means provided. Check thermostatic gas-filled bulbs for tightness against pipes and moisture between the bulb and the pipe; moisture may freeze under certain conditions and break the contact necessary for good operation of such devices. Remove the covers from all pressure switches and such control mechanisms at least once a month as loose screws, springs, and contacts will cause trouble if not checked in time.

Yearly—Safety devices such as relief valves should be removed and tested unless it is required more frequently by local ordinances. Valves suspected of leaking should be examined for wear. Packings should be inspected. It is assumed that at the time of the yearly inspection the system is out of service so that parts normally inaccessible can be dismantled to check for wear. A summation of the year's record should be made at this time including the notes made of the yearly inspection.

Condensing Water Circuit—the part of the system through which the condenser cooling water circulates. It consists of condensers, water piping, floats and valves. If the water is recirculated it may include spray or cooling ponds, cooling towers, settling tanks, and collection pans; or it may include an evaporative type condenser.

Daily—The condition of the water should be checked for corrosive tendencies. Temperatures should be taken daily since they often are indications of the cleanliness of the system. Quantity of makeup should be recorded daily when means are at hand to do so. Float valves should be inspected daily since improper functioning of floats may result in waste of water. Leaks, even though not serious, should be noted.

Monthly—Sumps, tanks, and collection pans should be checked for slime formation and algae growth, and for general cleanliness. Water samples should be taken for chemical analysis. Spray nozzles should be inspected for plugging. Accessible parts of evaporative type condensers should be given a thorough inspection each month for accumulations in the air and water passages.

Yearly—Water should be drained from all parts of the system so that all internal parts can be properly inspected. The circulating system should be checked for clogging and partial stoppages. Any valves suspected of leaking during operation should be dismantled for inspection at this time. All internal surfaces should be carefully scrutinized for corrosive action and scale formation. There should be a general summing up of the yearly record at the time of annual inspections. All items carried over from the other inspections will not become a part of the yearly report.

Cooling Water System—the circuit through which the refrigerated water passes. Includes storage tanks, evaporators, piping, spray units or washers, and collection pans and heating coils.

Daily—The storage tank should be checked for water level and condition. Floats should be inspected daily. On new systems the spray nozzles in the units should be inspected daily for plugging as small particles of dirt find their way to the nozzles for the first month or so of operation.

Monthly—The storage tank should be drained monthly for inspection; the large amount of dirt washed out of the air may cause plugging if a careful check is not made. After daily inspection of the spray nozzles has been dropped it should be made monthly. A water sample should be taken for chemical analysis. Drains should be inspected.

Yearly—Water should be drained from the cooling water circuit for the annual inspection. This inspection should include a check of evidences of scale formation and corrosion.

Filters and Ducts—the part of the system which cleans and carries the air of the system. Because of their structural nature they require less attention than other parts of the system.

Daily—Static pressure drop across filters should be checked.

Monthly—Filters should be inspected for dust accumulations. Sections should be removed at random from sectional filters each for checking. Insulated duct work should be inspected monthly for traces of moisture getting between insulation and duct walls.

Yearly—The general condition of filters should be determined. Ducts are inspected at this time for defects and leakages. Dirt accumulations are noted, and any abnormal conditions investigated.

AIR CONDITIONING MAINTENANCE PROCEDURE

CENTRAL SYSTEMS

Refrigerant Circuit and Controls—the part of the systems through which the refrigerant circulates including the compressor, connected piping and valves, condensers, receiver, and evaporator. The controls consist of such devices as thermostatic expansion valves, switches and regulators with automatic cutouts and solenoid valves for water and refrigerant.

Leaking Glands—When repairing compressors, valves and fittings, leaking glands should be tightened only to the point where the leak stops. If it is necessary to replace the packing, the system may have to be pumped down.

Pumping Down—If the leaks are on the low side, the liquid line valve is closed and the compressors operated until a vacuum of 20 to 25 in. is reached. This forces the refrigerant into the receiver. In some cases where the system is to be open only for a short period it is possible to prevent leakage of air into the system, or leakage of refrigerant out by lowering the pressure to atmospheric pressure before the system is opened. After the repair is completed carbon dioxide may be used for testing for tightness. Where air is allowed to get into the system the system will have to be purged after operation is resumed. In making a repair to the high side of the system the refrigerant is pumped into the low side. In most cases it is not possible to pump all the refrigerant into the low side and therefore it is necessary to draw some of the refrigerant into drums temporarily. Only drums intended for such storage should be used.

Refrigerant Storage in Drums—The drum is laid on a slant and connected to the refrigerant valve in the receiver with a flexible connection. The refrigerant flows into the drum by gravity. If the drum is kept in an ice bath, refrigerant will be speeded. The drum should be filled only to its stamped capacity. The amount of refrigerant in drum can be checked by placing the drum on a scale and noting the weight increase. After refrigerant is replaced and operation resumed the system should be purged.

Purging System—The best way to purge the refrigerant circuit is to first settle out the system to permit all foreign gases and air to rise to the high, or purging point of the system. With pressure on the system the purge valve is opened until refrigerant shows; it is then closed.

Replacing Refrigerant—When additional refrigerant must be added to the system this is done by connecting a refrigerant drum to the suction side of the system in such a way that no liquid refrigerant can enter into the compressor. When conditions permit a warm bath around the drum will speed up the charging. The amount of refrigerant necessary can be determined by watching the head pressure.

Scale Trap Cleaning—System should be pumped down, if necessary, scale trap cleaned and the system should be put back into normal operation. After a short period of operation, air should be purged.

Expansion Valves—To check the operation of expansion valves, the valve should be removed and connected to a cylinder of the refrigerant with the bulb hanging free. Refrigerant will be passed through the outlet of the valve when refrigerant drum is open. Placing the expansion valve bulb in the refrigerant stream should cause the valve to close.

Moisture—To remove moisture from the refrigerant a dehydrator of activated alumina, silica gel, or similar substance, should be installed. Dehydrators should be removed after about two weeks because they serve no purpose after the moisture has been removed.

Safety Devices—Fusible plug in liquid receiver should be renewed at least every two years. Relief valves can be tested by removing them from the system and checking them with compressed air or water. If water is used it must be perfectly clean and free from solid matter which might injure the valve. It should be noted whether safety devices of all types comply with local ordinances covering such devices. The operator should always be sure that protective devices will pass inspection.

Condensing Water Circuit—the part of the system through which the condenser cooling water circulates. It consists of condensers, water piping, floats and valves. If the water is recirculated it may include spray or cooling ponds, cooling towers, settling tanks, and collection pans.

Strainers—Devices to catch foreign matter; they are often a protective device. Strainers should be examined for signs of wearing or warping in such a way that they might become an obstruction.

Valves—Periodical flushing with clean water under pressure is necessary to keep the valves clean even though strainers are placed ahead of the valves. When valves become pitted they should be ground in with a fine abrasive. After repairs valves should be adjusted to close when unit is shut down.

Condenser—Condenser should be cleaned mechanically once a year and thoroughly washed with clean water. Tools which are likely to damage the tubes should not be used, nor should the tubes be bulged. A cleaning "brush" should be used for mechanical cleaning. For prolonged lay-up periods condenser should be left completely dry on the water side.

Tanks and Pans—Any tank and pan used in a condensing water system should be kept as clean as conditions will permit. During lay-up periods tanks should be first cleaned then laid up dry. Where tanks show signs of deterioration or corrosion, a coat of corrosion resisting paint should be applied.

Evaporative Condensers—If ordinary flushing of the tubes does not completely clean them, other methods may have to be used. A caustic solution forced through a spray gun in which air pressure is used to spray, or in which steam is mixed with the caustic, will effectively remove stubborn scale. There are a number of commercial solutions, with caustic bases, which need no preparation.

Cooling Water System—the circuit through which the refrigerated water passes. Includes storage tanks, evaporators, piping, spray units or washers, and collection pans and heating coils.

Cooling Water—Water should be changed frequently otherwise it may develop odors that will be carried into the air conditioning spaces. In using deodorants, care should be taken to prevent an overdose. The pH concentration of the water should be watched to maintain it at the proper point. Addition of chemicals to the water may be necessary to establish the proper pH.

Spray Nozzles—Nozzles must be kept clean. To do this, nozzles, or a complete bank of nozzles including the header, should be removed and blown out backwards with compressed air.

Filters and Ducts—the part of the system which cleans and carries the air of the system. Because of their structural nature they require less attention than other parts of the system.

Filters—Oil dipping filters should be cleaned with hot caustic solution; they are then washed with hot water or blown with clean, live steam. After cleaning they must be dried well. The oil dipping should give a thin film over all the surfaces after a period of draining to remove the excess oil. After cleaning filters should be thoroughly washed with clean water and dried and are then ready for dipping in an oil. Filters must be well drained before they are replaced in a filter bank.

Ducts—Ducts can be blown out but vacuum cleaning is preferred because of better control of cleaning. Unusually large accumulations of dust and dirt in a particular part of an otherwise clean duct may indicate improper contour at that point. Dirt-laden ducts mean poor filtering and washing.

Dampers—Dampers should operate smoothly, which means that piston drives must be free from air leaks that might cause sluggish operation. Dirty parts of such drives should be washed in a solution such as carbon tetrachloride to clean them. Orifices and parts should not be scraped since they may be enlarged by such treatment.

Rotating Apparatus—those parts of the air conditioning system, such as pumps, compressors, blowers, motors, and the controlling devices which convert or assist in the conversion of electrical energy into refrigeration.

Compressors—Only lubricants recommended by manufacturer should be used. Oil taken from separators should not be replaced in the system until it has been properly filtered. Packing glands should be tightened only enough to prevent leakage and only soft packing should be used. Where conditions permit, bearings should be opened once a year and the oil grooves cleaned out. Oil piping also should be cleaned. Cross head and connecting rods should be carefully checked for cracks and for signs of unusual wear. The whitewash test is highly useful for locating cracks. The part to be tested is first soaked in oil and then cleaned dry. Whitewash is applied and left to dry. After complete drying, the part is tapped with a hammer and oil will ooze from any cracks. Repairs should be made by welding.

Belts—During long lay-up periods belts should be removed and wrapped in a protective covering. They should be stored away from the heat and in a dry place. No dressing should be applied unless specifically called for by manufacturer's instruction. Unusual wear of belts means misalignment.

Motors—Cooling air passages through motors should be kept clean. Broken brushes are an indication of high spots on the commutator. Motors should be kept covered when not in use. Starters on motors should have all contacts free from arcing during the starting cycle.

Water Pumps—Small water drips from pump packings should be disregarded. As a matter of fact many manufacturers specify this condition for good operation. Pump packing which is too tight may injure the shaft. When pump is to be idle for an extended period, packing on pumps should be eased off and all exposed parts should be covered with grease.

Gears—Gear boxes should be well lubricated with the proper grade of oil. They should be washed out and oil renewed each year unless otherwise specified. Attention should be paid to the drives to see that they are kept in good condition.

UNIT AIR CONDITIONERS

Unit Air Conditioners—self-contained air conditioning systems that are built in compact units.

Piping—Flared joints in the piping or tubing should be protected from moisture formation by a coating of grease such as vaseline. Moisture which freezes will cause cracked nuts or flared ends of tubing and result in leaks.

Motors and Fans—In replacing motor bearing oil, unless otherwise recommended, SAE No. 10 or No. 20 should be used on small motors. Cleaning of small motors is of considerable importance. Belt tension should be maintained at the proper point.

Cooling Coil—Cooling coil should be kept clean at all times since the dirt will not only reduce the efficiency but also may cause odors. Coils may be cleaned by washing with a caustic soda solution and rinsing with clean water.

Condensers—Air cooled condensers should be cleaned frequently by compressed air. Tower type condensers may be cleaned by water under pressure.

Air Filters—Air filters should be replaced when they become dirty and a static pressure gauge should be used to determine when filters should be renewed.

CHECK LIST FOR AIR CONDITIONING SURVEYS

In making air conditioning surveys it is important that care be taken so that no items are left out. The accompanying check list has been in use for a number of years, and is believed to have all the important items included. This list furnishes a systematic method of checking all the important information needed in making an air conditioning survey.

Since the preparation of air conditioning drawings is closely related to the survey, it is suggested that reference also be made to the check lists on drawings appearing elsewhere in this book.

Particular attention is called to the last heading in the check list under which is listed items of work which may or may not be performed by or at the expense of the air conditioning contractor, and which, consequently, should be clearly understood; in some cases, such items may total enough to throw

what might otherwise be a profitable contract into a loss proposition.

The use of such a check as given does not, however, entirely compensate for a surveyor's lack of observation on a site. Instances which might not be duplicated in a thousand jobs do arise and no check list can possibly foresee them. The cost of returning to a location once the survey is completed may be prohibitive, or at least costly, and it is highly desirable that all data be collected on one visit.

In this connection, one point in relation to existing motors: Instead of attempting to copy only what seems to be pertinent data from the motor nameplate, the surveyor should make a pencil rubbing on a sheet of paper of the whole nameplate; the advantage is that all the information is then available if needed, without the trouble of copying it all, with the possibility of error.

Hot Water Heating Supply		
Location of boiler Size of boiler Excess boiler capacity	Location of storage tank Capacity of storage tank Possible location of hot water pump	Distance, tank to heating coil Distance, heating coil to tank Temperature of hot water
Electric Power Facilities		
Electric power voltage Electric power phase Electric power cycles	Location of power panel Horsepower connected at present Horsepower to be added	Permissible starting current in-rush Electric power cost Name of utility company
Refrigeration Facilities		
Name of compressor manufacturer Model number of compressor Operating condensing pressure	Operating suction pressure Refrigerant used Is 15 lb. air available continuously	Quantity flow of chilled water or brine available if system is indirect Excess refrigeration available
Drain and Sewer Facilities		
Size of present sewer Location of sewer Condition of present sewer Is sewer vented	Size of present slop sink* Location of slop sink Conditon of slop sink Head for draining condensate pan	Distance, condenser to slop sink Distance, condensate drip line to sink Is condensate pump necessary Must condensate drip line be insulated

*A slop sink, properly vented and provided with a supply of water to maintain water seal in trap is required. Sinks used for food washing or provided with a plug receptacle and strainer are not acceptable.

General Information

Date of survey	Who must approve job	Must job conform to local code
Name of prospect	New building or alteration	Must job conform to Under-
Location of job	Length of lease	writers code
Owner	Does owner agree to installation	Must job conform to any state
		code
		Are other trades union or non-
		union

Space Conditions

To what use is space put	Floor area	What equipment is hooded
Is smoking allowed	Ceiling height	Are hoods effective
Any unusual odors generated	Cubical contents	Are other hoods desirable
Loose or tight building con-	Area of doors and windows to	Any unusual dust condition
struction	be open	

Cooling and Heating Load Quantities

Number of occupants	Horsepower of motors	Moisture load
Degree of activity of occupants	Heat generated by gas-burners	Exposed window and wall area
Total light wattage	Any other heat generating	Compass rose
Light wattage on sunny day	equipment	Winter conditions to be main-
Are awnings and blinds on ex-	Product load, if any	tained
posed windows	Time phases on component	Summer conditions to be main-
Are blinds of venetian type	loads	tained
Is load such as to require zoning	Period spaces are occupied	Outside winter design conditions
	Areas with concentrated loads	Outside summer design condi-
		tions

Water Facilities

Size of city main connection	Is well water available	Length of water line to con-
Location of possible tap-in	Well water flow, quantity	denser
City water temperature	Well water temperature	Length of water line to spray
City water pressure	Well water pressure	nozzles
Size of present water meter	Well water location	Must water lines to condenser
Is constant quantity flow	Corrosion properties of well	be insulated
assured	water	Cost of water
		Hardness of water
		Head available using house tank

Steam Supply Facilities

Location of steam supply	Excess capacity of boiler	Is return trap or condensate
Kind of steam supply available	Boiler operating pressure	pump needed
Steam pressure available	Boiler water level, operating	Distance, steam source to heat-
Is there constant supply year-	Head available to drain heating	ing coil
round	coil	Length condensate return run to
Boiler location	How is boiler fired	boiler
Boiler capacity	Location of controlling thermo-	Location steam main nearest
Type of existing heating system	stat	heat coil
	Type of heating system control	

Air Distribution

Exhaust duct sizes	Air quality, exhaust system	Supply fan foundation data
Exhaust fan data	Exhaust grille sizes	Supply fan motor foundation
Exhaust fan motor data	Supply duct sizes	data
Exhaust fan foundation data	Supply fan data	Air quantity, supply system
Exhaust fan motor foundation	Supply fan motor data	Supply system grille sizes
		Grille locations

Locating Equipment

Does available space for ducts indicate a central duct system or room units	Is elevator available	Is water-proof equipment floor necessary
	Check loading and size of elevator	Visualize equipment set in place and check location for accessibility
Passage space for delivery of equipment	Is rigging or special handling needed	
Check for possible changes to partitions between survey and installation	Is spring loaded compressor foundation needed due to upper floor mounting	Check the heating of the proposed equipment room in winter to prevent the freeze-up of condenser if water cooled
Noise considerations	Check outside window area with fire code	

Duct System

Noise considerations	Locate exhaust to atmosphere and place on side of building opposite to that on which air intake is located	What duct work is to be insulated
Locate fresh air intake so as to prevent contamination of air supply		Locate fire walls and partitions
		Possible location of control equipment

Miscellaneous Considerations

Bonded roof to be cut	Is job location such that treatment of air washer water is needed	Length of service of nearby washers
Economizer to run year round		Present condition of nearby washers
Working hours — straight time in what spaces, overtime in what spaces	Are other washers operating nearby	Do nearby washers use treated water
	If so, of what material are they	Bar corrosion experience with nearby plants

Under the Contract Who Will Perform Following

Cutting and patching	Make water connections	Lay foundations
Install water meter	Make refrigerating connections	Carpenter work
Electric wiring	Make steam connections	Masonry work
Clean existing ducts	Insulate steam piping	Excavating
Finish painting of equipment	Insulate ducts	Plastering
Remove or relocate present wiring	Insulate cold water mains	Structural supports for equipment
Remove existing electric fixtures	Insulate hot water mains	
Remove existing piping	Install access openings	Waterproof floors
Install and connect drain mains	Install grilles and plaques	Build equipment room partitions
		Cut holes for pipes, ducts, grilles

REFRIGERATION FOR AIR CONDITIONING

Contents

Authorities

Section Editors: *Eugene Stamper* and *Richard L. Koral*

E. R. Ambrose, The Heat Pump
W. N. Fitzcharles, Chillers in Parallel and Series
Sidney N. Ludwig, Capacity of Direct Expansion Evaporator

J. Partington, *Jr.*, Making Industrial Refrigeraton Surveys
William A. Pennington, Refrigerants
M. A. Ramsey, Refrigerant Piping

AIR CONDITIONING AND REFRIGERATION ESTIMATING GUIDE

These data were compiled to aid in the determination of the economic soundness of air conditioning or refrigeration installations with a minimum of effort. There is a risk of over-simplification always present in their use; the size and cost of any specific installation will almost always be more or less than that predicted on the basis of generalized information. Properly used and tempered with judgment, however, this information involves but a small risk. *However, when designing as accurately as possible for energy efficiency, a detailed analysis is necessary in those cases.*

Table 3—Approximate Turbine Steam Rates

Inlet pressure	125 psi	125 psi	5 psi
Exhaust pressure	5 psi	26 in.*	26 in.*
Steam rate, lb/hp/hr	40	15	30

*In. Hg, vacuum condensing

Table 4—Approximate Condenser Water Rates

Refrigerant	Condenser Cooling Rate, gpm/ton	
Ammonia	Cooling towers	4-5
	City water	3
R-12, R-22	Cooling towers	2-3
(25F evap.)	City water	2
(105F cond)	55 F well water	1
LiBr absorb	Cooling towers	3-3.6
All	30 gal-deg/min-ton (25/105F ev/cn)	

Table 1—Air Conditioning and Refrigeration Load Factors *

Load	Factor/ton of Refrigeration
People	30 people
Lights	3.5 kw
Motors	4.7 hp
Walls, floors, and ceilings	2000 sq ft (Δt = 20F)
Glass--single pane	N=375 NE,NW=91 E,W=71
	S=150 SE,SW=94sq ft
Circulated air	400 cfm
Refrigeration service factor	20% of heat gain
Refrigeration product factor	50% of heat gain

* Reprinted with permission: *York Budget and Data Book,* 1973, York Div. of Borg Warner.

Table 5—Approximate Condensing Surface

Type of Condenser		Surface (sq ft/ton)
Evaporative	(ammonia)	12-20
Horizontal	(ammonia)	5-10
Vertical	(ammonia)	10
Evaporative	(R-12,R-22)	8-11
Horizontal	(R-12,R-22)	6-10

Table 2—Compressor Horsepower-Evaporator Temperature Ratio (100 F cond.)

Evaporator temperature, F	40	25	-5	-35	-65	-95
Bhp/ton of refrigeration	1.0	1.3	2.0	3.0	4.5	6.0

Table 6—Approximate Chilled Water and Brine Flow Rates

Coolant	Application	Rate, gpm/ton
Chilled Water Brine	Air conditioning	1½- 3
	General purpose refrigeration	4-7
	Flaked ice mach. skating rink	10-14
Chilled water	24 gal-deg/min-ton	
Brine	28 " " " "	

Table 7—Approximate Conversion from Refrigerant 12 to Refrigerant 22 for a Given Compressor Displacement

tons(R-22 system) = tons(R-12 system) × 1.61
bhp(R-22 system) = tons(R-12 system) × 1.48

Table 8—Approximate Surface for Horizontal Coolers for Refrigerants

Refrigerant	Surface, square feet per ton
Ammonia	8-10
R-12 or R-22	7-10

Table 9 *—R-12 Discharge and Liquid Line Capacities [1]

	Discharge Lines	Liquid Lines	
Line [2] size in.	Temp.,175F press.drop 1psi/100ft	Condenser to receiver Velocity, 100 fpm	Receiver to exp. valve Press. drop, 2psi/100ft
	Capacity, tons		
OD			
1/2	0.33	–	–
5/8	0.62	3.18	4.23
3/4	1.06	4.77	7.25
7/8	1.62	6.61	11.2
1-1/8	3.30	11.2	23.1
1-3/8	5.72	17.1	40.0
1-5/8	9.10	24.3	64.0
2-1/8	18.8	42.3	133
ips			
2-1/2	26.6	65.5	197
3	46.2	101	350
4	94.7	174	712
5	172	–	–
6	280	–	–
8	573	–	–
10	1030	–	–
12	1625	–	–

[1] Based on fluid flow at 105°F saturated condensing temperature and 40°F saturated evaporated temperature.
[2] "ips" data based on Schedule 40 steel piping except that liquid lines 1½" and smaller are Schedule 80. "OD" data based on Type L copper tubing.
* Tables 9 to 12 reprinted with permission: *York Budget and Data Book*, 1973, York Div. of Borg Warner.

Table 10 *—Brake Horsepower Requirements

Water pump	bhp= gpm × ft head × spec grav/2800
Fan,61% eff.	bhp= cfm × total press. (in.H$_2$0)/4000

Table 11 *—Floor Area of Cooling Tower

	Forced Draft
R-12,-22 system	1.28 sq ft per ton
Ammonia system	1.92 sq ft per ton

Table 12 *—R-12 Suction Line Capacities [1]

Line [2] Size, in.	Saturated Suction Temperature, F			
	-20	0	20	40
	Capacity, tons			
OD				
5/8	–	–	0.32	0.40
3/4	0.22	0.29	0.54	0.68
7/8	0.34	0.45	0.84	1.05
1-1/8	0.69	0.91	1.69	2.14
1-3/8	1.22	1.61	2.94	3.74
1-5/8	1.92	2.54	4.70	6.02
2-1/8	4.01	5.28	9.75	12.3
ips				
2-1/2	6.02	7.86	14.1	17.6
3	10.6	13.8	24.9	30.7
4	21.8	28.6	51.8	64.4
5	39.9	51.7	93.0	117
6	64.5	83.5	151	187
8	130	170	306	381
10	239	310	553	689
12	382	490	882	1093
	Pressure Drop, psi/100 ft			
	.50	.50	1.00	1.00

[1] Based on fluid flow at 105°F saturated condensing temperature.
[2] "ips" data based on Schedule 40 piping. "OD" data based on Type L copper tubing.

Pressure Drop Correction Factors

Tons capacity at these pressure drops, psi		0.50			1.00	
Times these factors, equals	1.45	2.10	2.60	1.45	1.80
Tons capacity at these pressure drops, psi	1	2	3	2	3

MAKING INDUSTRIAL REFRIGERATION SURVEYS

The quickest way to learn what we want to find out is to ask questions. Aside from asking questions, the first step in making a survey is to listen to what the prospect has to say, and then to evaluate the information. In many cases, he will have a fairly definite idea of what he desires performance-wise, although his ideas may be purely theoretical or based on some empirical observations.

He may say, for example, "I have no trouble in the winter when I use city water for cooling. In the summer I have to reduce my output 50% and sometimes shut down the operation entirely." While this establishes the problem, it affords insufficient data on which to base an intelligent solution. What then do we need to know?

A breakdown of the energy input must be obtained to calculate the heat gain which the system must overcome. In addition to the conduction and sun effect components of the load, which are calculated in the normal manner, care must be exercised to determine accurately the number of people, lights, and motor horsepower, as well as the energy given off by retorts, solder pots, furnaces, bunsen burners and the like. Moisture release from gas appliances, quench tanks and plating operations must also be carefully enumerated, especially where close humidity control is required.

In addition to the more usual heat gain components, what further information must be gathered? The check list of Table 1 covers most of the items, why such information is necessary, and where such data may be obtained.

Knowing these answers, or at least to those which are pertinent to the particular problem, you are now equipped to attack the problem intelligently so as to submit a sound engineered proposal for its solution. In this connection, it is frequently advisable to prepare a reasonable number of alternate plans should the problem lend itself to such treatment. Budgets are not made of rubber and a compromise solution may have to be accepted.

During the survey, whenever any large items of energy input are encountered, immediate steps should be taken to learn whether such loads can be diminished. Factful inquiries may determine that motors could be located outside of the air conditioned space by running motor shafts through the wall. Hoods can frequently be installed to remove heat at its source before it can spread to the entire space. Ovens can sometimes be relocated so that only their doors are within the area to be conditioned. Vents can be installed to remove noxious fumes at the source and thus reduce excessive quantities of ventilation air. Frequently, furnaces can be enclosed separately. Since money spent on structural changes is often less than expenditures for added cooling capacity otherwise required, finding ingenious ways to reduce the total heat gain pays off in more installations and better satisfied customers.

When the prospect states he wants certain conditions maintained, find out the operating range limits—what tolerance he can accommodate. Must conditions be held right on the button or will a definite operating range be satisfactory? Learn whether or not the limits are one-sided.

It is not always possible to accept the prospect's stated design requirements at face value. Extreme diplomacy is required, especially if the prospect's product has to be tested in accordance with a set of specifications provided by the prospect's customer. It is doubly difficult if this customer is Uncle Sam, although, here again, the question routine proves invaluable and affords an opening for suggestions.

Table 1. Check List of Questions to Ask

Item	Why Necessary	Where Obtained
Brief story of process	Orient design	Mfg mgr or plant engr
How vital is job?	To find if spares are required	Mfg mgr or plant engr
Is production rate likely to increase?	Influences equipt choice	Mfg manager
Is load constant?	Influences equipt size	Production mgr
Range and rate of load variation	Influences system design	Production mgr
Operating range limits	Influences control selection	Quality engr
Energy input	Determines heat gain	Plant engr
Cost of electricity	Influences system design	Plant engr
Electrical characteristics	Influences equipt selection	Plant engr
Is cooling water available?	Influences system design	Plant engr
Cost of water	Influences system design	Plant engr
Max summer water temp	Influences equipt selection	Plant engr
Min winter water temp	Influences equipt selection	Plant engr
Max safe water temp	Influences system design	Prod mgr
Water quality	Influences equipt selection	Water dept
Ventilation air—quality and source	Influences equipt design	Plant engr
Electric power source and characteristics	Influences equipt selection	Plant engr and power co.
Architectural plans	Influences equipt location	Plant engr
Can work be done in normal working hours?	Influences price	Prod mgr or plant engr

Table 2. Industrial Air Conditioning and Refrigeration Form

Customer

....Buyer...............................Address.............................
Type of Application.................Engineer............................Survey by...........Date...........
Extreme Outdoor Temperatures: Summer.........Fdb.........Fwb   Winter.........Fdb.........Fwb
Extreme Ambient Temperatures: Summer.........Fdb.........Fwb   Winter.........Fdb.........Fwb

Gas (or Air) Cooling

Maker's Designation.................Specific Heat............ @ Constant Volume .... @ Constant Pressure
Pressure.............. Altitude.............. Quantity.............. cfm Cooled from..........F to..........F
Moisture Removal from.....Grains to.....Grains   Tolerance ±.....F, ...± Grains or ±.....Relative Humidity
Ventilation Air Quantity ....Cfm   Impurities ....

Liquid Cooling

Liquid .... Vendor's Designation ....Quantity.............. Gpm Pressure..............
Specific Heat.......Btu/Lb F   Conductivity............Viscosity @ .... F; @ .... F; @ .... F
Density........Cu Ft per lb   Cooled from....F to ....F.   Tolerance ±   F. Corrosive Properties........
Preferred   Materials...

Solid Cooling

Material................. Vendor's Designation ....Specific Heat...... Btu/Lb F   Conductivity....
Density........ Cut Ft per Lb   Unit Volume........Unit Surface...Sq Ft Quantity, Lb/Hr Cooled from..F to ..F
Method of Handling..
Method of Cooling: Contact.....Gas Convection.....Radiation.....Liquid Convection.....Immersion......Spray......

Thermal Properties

Boiling Point......F   Freezing Point......F   Latent Heat of Vaporization........   Latent Heat of Fusion........

Load Characteristics

Constant........Variable........Range........to........%   Expanding Production........ Key Process........

Controls

Pneumatic........Electrical........Fully Automatic........ Semi-Automatic........ Manual........ Sequence........

Utilities or Services

Extreme Water Temp: Winter.....F   Summer.....F   Cost...../M Gal   Available Quantity.....Gpm   Quality.....
Electrical Characteristics............. Volts............ Phase............ Cycles............ Cost............¢/Kwh
Steam Pressure.........Lb/Sq in Gage   Quantity.......... Availability..........Hr/Day   Superheat..........F
Drain Capacity.........Storm.........Sanitary Compressed Air Available..........Pressure..........Lb/Sq In Gage

General

Plans Available.............Max Floor Loading.............Lb/Sq Ft   Foundations Suitable....................
Work Schedule............Work by Owner.....................Largest Door to Equipment......... × ....

Comments

Date of Next Appointment..

When the process involves fluids other than water, or gases other than air, find out by questioning the specific gravity, specific heat, conductivity and, in the case of liquids, the viscosity. It is not necessary to quibble over the units in which the answers are given (conversion factors are obtainable) as long as the answers are labeled. If the prospect does not have these data, try to determine the source of supply of the fluid.

In such low temperature applications where pulldown time is important, if the objects to be cooled are composed of heterogeneous materials, it is vital to know the proportions by weight of each of the components. Should some of the components be proprietary secrets, at least obtain the specific heats of elements X, Y and Z, and the quantity of each.

The presence of abrasives, inflammable or explosive vapors or dusts and corrosive fumes should be carefully noted, since their effect on the life of equipment or the safety of the people may be extremely harmful. The action of salt air on aluminum ducts and coil fins or the reaction of copper pipe to ammonia fumes point up the necessity for caution. Here again, the people who live with such hazards every day are in the best position to advise about suitable precautionary measures.

After the many steps have been taken in making an industrial application survey, you should possess the following information:

(1) A clear understanding of the problem.

(2) Most of the necessary data required for its solution (obtained by asking questions).

(3) An evaluation of the answers which has formed a basis for the tentative solution.

(4) A general idea of the equipment required and a rough cost estimate.

(5) A determination of the prospect's qualifications as a source of business.

(6) A definite idea as to what further steps should be taken.

If any of this information is lacking or incomplete, further questions are indicated.

CAPACITY OF DIRECT EXPANSION EVAPORATOR

Presented below is a direct method for determining the cooling and dehumidifying capacities of a direct expansion evaporator. Physical characteristics of the evaporator, entering air conditions, air flow rate, and refrigerant evaporating temperature are the only pre-determined data necessary to complete the calculation.

The method, outlined for plate fin type coils with Refrigerants 12 or 22 evaporating within the tubes, is easily adapted to other fin configurations and refrigerants. The method may also be applied to water coils by considering successive portions of the coil rather than the coil as a whole.

Each complete calculation is based on a specific refrigerant evaporating temperature and specific entering air temperatures. Only a small amount of recalculation is necessary to determine coil capacities at other temperature conditions, because the coil characteristic, K, is essentially a function of air flow rate and physical characteristics of the coil only. Actually, the coil characteristic is also a function of air properties such as specific heat, thermal conductivity, specific volume, and absolute viscosity, which do vary with air temperature. Only a small error is introduced, however, if these properties are assumed constant over the normal air conditioning temperature range. The coil characteristic, therefore, remains constant as temperature conditions vary. It is then possible to obtain a complete rating table for a specific evaporator in a relatively short period of time.

In addition, by plotting refrigerant evaporating temperature versus the corresponding total wet coil capacity, it is possible to obtain a "coil line" for this same evaporator, which, when superimposed upon a set of standard compressor capacity lines (usually plotted for parameters of compressor discharge pressure), will yield the balanced refrigerant evaporating temperature and total coil capacity for that specific system.

Using procedures outlined herein, capacities of many direct expansion evaporators at varying temperature and air flow rate conditions have been obtained which, when compared to results with laboratory test runs for the same evaporators at corresponding conditions of service, indicate that calculated values are accurate to within 5% of total coil capacity and to within 8% of sensible coil capacity for sensible heat factors less than 0.90.

Accuracy of calculated sensible coil capacity decrease as the sensible heat factor increases from 0.90 to 1.00, probably due to discontinuities in the condensate film near the "dry coil" region. When operating in this region, therefore, it is recommended that the step-by-step procedure be applied to successive increments of coil (e.g., row by row) rather than to the coil as a whole. Here the coil characteristic per row is assumed constant, and the air temperatures entering each row are set equal to the corresponding temperatures leaving the preceding row. This procedure, which tends to separate the "wet" and "dry" portions of the coil, thus decreasing the error in calculated sensible coil capacity near the "dry coil" range, is especially valuable for deep coils having overall CF values above 0.85.

ILLUSTRATIVE EXAMPLE

Given:

L = finned coil length, 50 inches
H = finned coil height, 22.5 inches
N = number of tubes high, 18
n = number of rows deep, 3
S_1 = tube center-to-center distance normal to air flow, 1.25 inches
S_2 = row center-to-center distance, parallel to air flow, 1.00 inch
δ = coil fin thickness, 0.01 inch
m = number of fins per inch, 10
OD = tube outside diameter, 0.508 inches
D_o = $OD + 2\delta = 0.508 + 2(0.01)$
    = 0.528 inch

D_i = tube inside diameter, 0.474 inch
q_{cfm} = air volume flow rate, standard air, 3000 cfm
V_F = air face velocity, 144 q_{cfm}/LH
    = $(144)(3000)/(50)(22.5)$
    = 384 fpm
t_{wb}^e = entering air wet-bulb, 67 F
t_{db}^e = entering air dry-bulb, 80 F
t_{dp}^e = entering air dew-point, 60 F
h_a^e = entering air enthalpy, 31.62 Btu per lb
t_r = refrigerant evaporating temperature, 40 F
    Staggered tube arrangement

Step-by-Step Solution

1. Find total coil surface per foot of tube, S
$$S = (m/6)(S_1 S_2 - \tfrac{1}{4}\pi D_o{}^2) + (\pi D_o/12)(1 - m\delta)$$
$$S = (10/6)[(1.25)(1.00) - \tfrac{1}{4}\pi(0.528)^2] + (\pi/12)(0.528)[1 - 10(0.01)] = 1.84 \text{ sq ft per ft}$$

2. Find total external coil surface area, A_o
$$A_o = S(L/12)(N)(n)$$
$$A_o = 1.84(50/12)(18)(3) = 414 \text{ sq ft}$$

3. Determine f_a, air film heat transfer coefficient from Table 1
$$f_a = 10.9 \text{ Btu per (hr)(sq ft)(F)}$$

4. Determine y, a dimensionless parameter for determining CF
$$y = (f_a A_o/1.08\,q_{cfm})$$
$$y = (10.9)(414)/(1.08)(3,000) = 1.39$$

5. Determine CF, wet coil contact factor (dimensionless) from Fig. 2
$$CF = 0.750$$

6. Determine r_m, fin metal heat transfer resistance, from Table 2 (but see Fig. 3)
$$r_m = 0.0120 \text{ (hr)(sq ft)(F) per Btu}$$

7. Determine overall heat transfer resistance from inner fluid to outer coil surface, R
$$R = (S/79\,D_i) + r_m$$
$$R = [(1.84)/(79)(0.474)] + 0.0120 = 0.0611 \text{ (hr)(sq ft)(F) per Btu}$$

8. Determine K, coil characteristic (dimensionless)
$$K = 4.5(q_{cfm})(CF)(R)(1/A_o)$$
$$K = (4.5)(3,000)(0.750)(0.0611)(1/414) = 1.50$$

9. Determine the quantity $K h_a{}^e + t_r$
$$K h_a{}^e + t_r = (1.50)(31.62) + 40 = 87.4$$

10. Determine t_s from Fig. 4
$$t_s = 53.7 \text{ F} \qquad (h_s = 22.44 \text{ Btu per lb})$$
$$(\therefore t_s < t_{dp}{}^e)$$

Wet Coil Only $(t_s < t_{dp}{}^e)$

11. Determine total wet coil capacity, Q_T
$$Q_T = 4.5(q_{cfm})(CF)(h_a{}^e - h_s)$$
$$Q_T = (4.5)(3,000)(0.750)(31.62 - 22.44) = 93,000 \text{ Btu per hr}$$

12. Determine sensible wet coil capacity, Q_s
$$Q_s = 1.08(q_{cfm})(CF)(t_{db}{}^e - t_s)$$
$$Q_s = (1.08)(3,000)(0.750)(80 - 53.7) = 64,000 \text{ Btu per hr}$$

13. Determine latent heat factor, F_{LH} (dimensionless)
$$F_{LH} = 1 - Q_s/Q_T$$
$$F_{LH} = 1 - (64,000/93,000) = 0.31$$

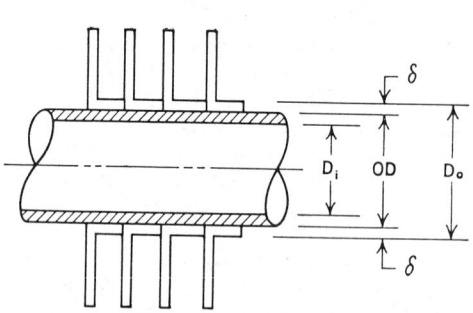
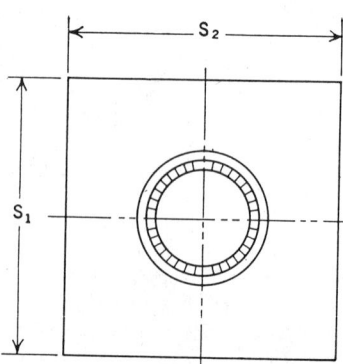

Fig. 1. Diagram of finned tubing.

Dry Coil Only $(t_s > t_{dp}{}^e)$

11. Determine overall heat transfer coefficient from inner fluid to outside air stream, U

$$U = f_a/(Rf_a + 1) \text{ Btu per (hr)(sq ft)(F)}$$

12. Determine y', a dimensionless parameter for determining CF'

$$y' = (UA_o/1.08\, q_{cfm})$$

13. Determine CF' from Fig. 2.

14. Determine total dry coil capacity, Q_{dry}

$$Q_{dry} = 1.08(q_{cfm})(CF')(t_{db}{}^e - t_r) \text{ Btu per hr}$$

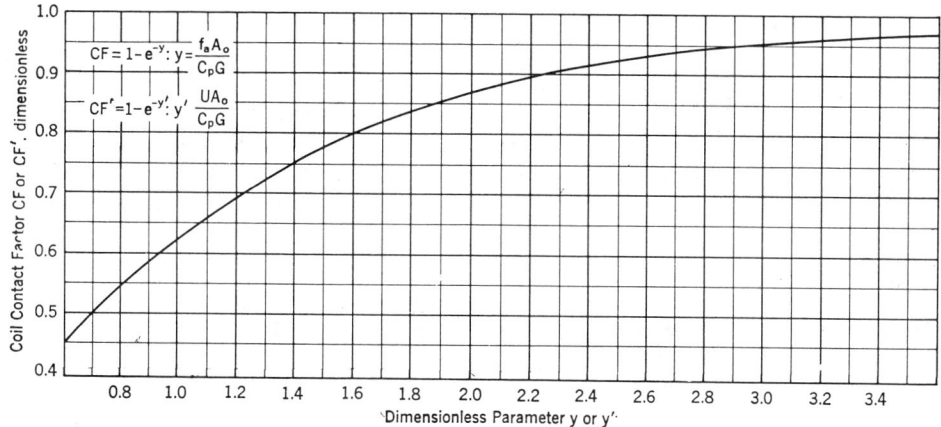

Fig. 2. Contact factor.

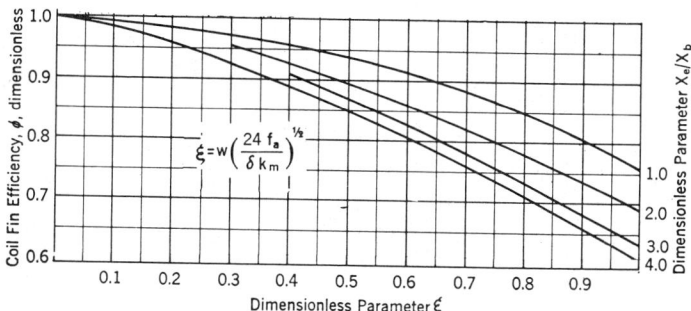

Fig. 3. Fin efficiency.

W. H. Carrier and S. W. Anderson give the following formula for fin metal resistance, r_m

$$r_m = (1/f_a)\,[(1 - \phi)\,/\,(\phi + S_P/S_F)]$$

where f_a = air film heat transfer coefficient, Btu per (hr) sq ft)(deg F), ϕ = coil fin efficiency, dimensionless, S_P = Primary coil surface per foot of tube, sq ft per ft, and S_F = Secondary coil surface per foot of tube, sq ft per ft.

This equation, derived for use with dry extended surfaces, may be applied directly to most wet coil configurations with only small error.

K. A. Gardner, in "The Efficiency of Extended Surface," *Trans. ASME*, November, 1945, presented a series of curves for determining fin efficiency, ϕ, for various fin configurations. The curves above are based on Gardner's data for circumferential fins of rectangular cross section. A reasonably close approximation of the efficiency of a square or rectangular plate fin is obtained by assuming a flat circular fin of equal area. Above, ϕ is plotted against ξ for parameters of X_e/X_b where

$$\xi = w(24 f_a/\delta k_m)^{1/2}$$
$$X_e/X_b = 1 + (24 \, w/D_o)$$

where δ = coil fin thickness, inches, k_m = fin metal thermal conductivity, Btu per (hr)(sq ft)(deg F) per ft, D_o = OD + 2 δ and w, equivalent circular fin width, in feet

$$= 1/12\,[S_1\, S_2\pi)^{1/2} - D_o/2]$$

For convenience, values of r_m for the more common coil configurations are listed in Table 2.

Table 1. Air Film Coefficient*

Air Face Velocity, V_F, Ft per Min	In-Line		Staggered		
	Tube Outside Diameter, Inches				
	½	⅝	¾	½	⅝
300	8.0	7.2	10.8	9.5	8.6
350	8.8	8.0	11.8	10.4	9.4
400	9.6	8.7	12.7	11.2	10.1
450	10.3	9.4	13.6	12.0	10.8
500	11.0	10.0	14.5	12.7	11.5
550	11.7	10.6	15.3	13.4	12.2
600	12.4	11.2	16.1	14.0	12.8
650	13.0	11.8	16.9	14.6	13.3
700	13.6	12.4	17.7	15.2	13.8
750	14.2	12.9	18.4	15.8	14.3
800	14.8	13.4	19.0	16.4	14.8

*Based on the work of E. D. Grimison, "Correlation and Utilization of New Data on Flow Resistance and Heat Transfer for Cross Flow of Gases over Tube Banks," *Trans. ASME*, Vol. 59, 1937

Table 2. Fin Metal Resistance

Tube Outside Diameter, Inches	Tube Spacing $S_1 \times S_2$*	Heat Transfer Resistance, r_m (hr)(sq ft)(F)/Btu
⅜	1.00 × .866	0.0100
	1.00 × 1.00	0.0135
½	1.00 × 1.00	0.0085
	1.25 × 1.00	0.0120
	1.50 × 1.50	0.0310
	1.50 × 1.75	0.0400
⅝	1-19/32 × 1⅜	0.0225
	1.50 × 1.50	0.0235
	1.50 × 1.75	0.0315

*S_1 = Tube center distance, inches, normal to air flow.
S_2 = Row center distance, inches, parallel to air flow.

Note: The minor effect of change in face velocity (between 300 and 800 fpm) on r_m has been neglected in the tabulation. In addition, the table is based on 0.008 inch aluminum fin thickness. A small error is introduced if these data are used for other fin thicknesses. However, since r_m is generally less than 30% of the overall coil resistance, R, the effect of these errors on the final evaporator capacity is extremely small.

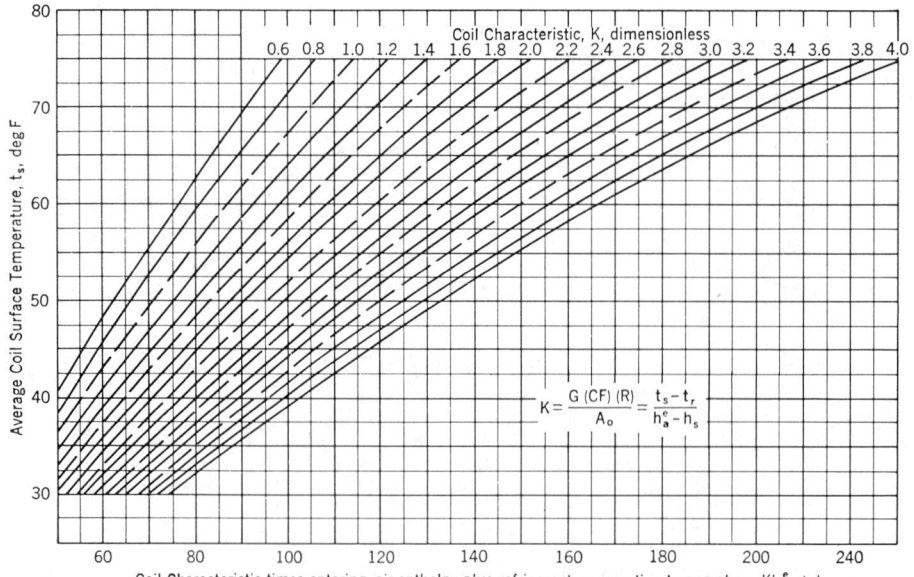

$$K = \frac{G\,(CF)\,(R)}{A_o} = \frac{t_s - t_r}{h_a^e - h_s}$$

Coil Characteristic times entering air enthalpy plus refrigerant evaporating temperature, $Kh_a^e + t_r$

Fig. 4. Wet coil surface temperature.

(The value K, derived by T. Kusuda, is called the *coil characteristic* since it contains all the physical characteristics of a given fan-coil. Fig. 4 is a series of direct-reading curves for the solution of Mr. Kusuda's equation, given above, obtained by plotting $Kh_a^e + t_r$ vs. t_s for parameters of K.)

CHILLERS IN PARALLEL AND SERIES

These charts show unloading and power requirements for water chillers used in parallel, series (two methods shown) and combined parallel series hookup of three water chillers.

Tonnage of each system was selected so that each combination would provide approximately the same capacity. The series combinations, by this selection, have a slightly smaller capacity with the same size

chilling units. This is a result of a 4% reduction in capacity imposed by the one-pass chiller required.

Control of (series-connected) System 3 requires a rather complex arrangement. One solution is to provide a reset control where return-water-sensing stat T_1, acting as a master controller, resets a submaster controller, T_2, for a lower or higher leaving water temperature.

System 1—Parallel Connected Chillers
(T_1, T_2 and T_3 are indicators. T_4 is a controller.)

Percent System Load	System Tonnage	Percent Load	Percent Capacity	Percent Load	Percent Capacity	Water Temperature, F				Power Consumption, Kw		
		Chiller No. 1		Chiller No. 2		T_1	T_2	T_3	T_4	No. 1	No. 2	Total
100	488.4	50	100	50	100	52	42	42	42	199	199	398
90	438	45	90	45	90	51	42	42	42	175	175	350
80	390	40	80	40	80	50	42	42	42	151	151	302
70	341	35	70	35	70	49	42	42	42	129.5	129.5	259
60	293	30	60	30	60	48	42	42	42	109.5	109.5	219
50	244	25	50	25	50	47	42	42	42	91.5	91.5	183
40	195	20	40	20	40	46	42	42	42	74.8	74.8	149.6
30	146.4	30	60	0	0	45	39	45	42	109.5	0	109.5
20	97.5	20	40	0	0	44	40	44	42	75.5	0	75.5
10	48.8	20	20	0	0	43	41	43	42	47.8	0	47.8
5	24.4	5	10	0	0	42.5	41.5	42.5	42	39.8	0	39.8

System 2—Series Connected Chillers
(T_1 is an indicator. T_2 and T_3 are controllers.)

Percent System Load	System Tonnage	Percent Load	Percent Capacity	Percent Load	Percent Capacity	Water Temperature, F			Power Consumption, Kw		
		Chiller No. 1		Chiller No. 2		T_1	T_2	T_3	No. 1	No. 2	Total
100	485.7	52	100	48	100	52	46.8	42	199	199	398
90	437	46.5	91.2	53.5	100	51	46.8	42	179	199	378
80	387	40	77	60	100	50	46.8	42	139	199	338
70	349	31.2	60	68.8	100	49	46.8	42	109	199	308
60	297	20	38.5	80	100	48	46.8	42	71.5	199	270.5
50	242	4*	7.7*	96	100	47	46.8	42	39.8	199	238.8
40	194	0	0	40	83.3	46	46	42	0	159	159
30	145	0	0	30	62.5	45	45	42	0	115	115
20	97.2	0	0	20	41.7	44	44	42	0	74.8	74.8
10	48.6	0	0	10	20.8	43	43	42	0	47.8	47.8
5	24.3	0	0	5	10.4	42.5	42.5	42	0	39.8	39.8

* Theoretical unloading. NOTE: Chiller No. 1 is unloaded as the load decreases, and chiller No. 2 is unloaded as the load falls below 50% of total connected capacity.

System 3—Series Connected Chillers

(T_1 is a master controller. T_3 is a controller. T_2 is a sub-master controller.)

Percent System Load	System Tonnage	Percent Load	Percent Capacity	Per cent Load	Percent Capacity	Water Temperature, F			Power Consumption, Kw		
		Chiller No. 1		Chiller No. 2		T_1	T_2	T_3	No. 1	No. 2	Total
100	485.7	52	100	48	100	52	46.8	42	199	199	398
90	437	58.7	100	41.3	76.2	51	45.8	42	199	143	342
80	387	63.4	100	36.6	57.0	50	44.9	42	199	103	302
70	349	69.5	100	30.5	43.7	49	44.0	42	199	80	279
60	291	82.5	100	17.5	20.1	48	43.1	42	199	50	249
50	253	97.2	100	2.6*	2.9*	47	42.2	42	199	36	235
40	194	40	82.7	0	0	46	42	42	157	0	157
30	145	30	61.8	0	0	45	42	42	112	0	112
20	97.2	20	41.5	0	0	44	42	42	74.8	0	74.8
10	48.6	10	20.7	0	0	43	42	42	47.8	0	47.8
5	24.3	5	10.4	0	0	42.5	42	42	39.8	0	39.8

* Theoretical unloading. NOTE: The table above shows the unloading of series chillers allowing the first chiller No. 1 to handle as much of the load as possible while the second chiller No. 2 is unloaded.

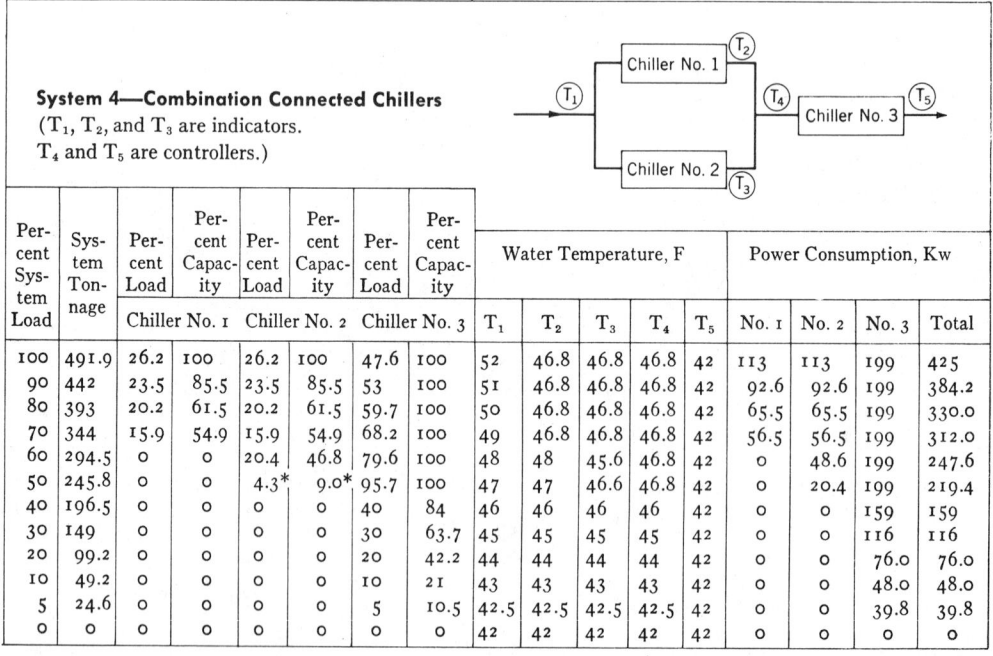

System 4—Combination Connected Chillers

(T_1, T_2, and T_3 are indicators. T_4 and T_5 are controllers.)

Percent System Load	System Tonnage	Percent Load	Percent Capacity	Per cent Load	Percent Capacity	Per cent Load	Percent Capacity	Water Temperature, F					Power Consumption, Kw			
		Chiller No. 1		Chiller No. 2		Chiller No. 3		T_1	T_2	T_3	T_4	T_5	No. 1	No. 2	No. 3	Total
100	491.9	26.2	100	26.2	100	47.6	100	52	46.8	46.8	46.8	42	113	113	199	425
90	442	23.5	85.5	23.5	85.5	53	100	51	46.8	46.8	46.8	42	92.6	92.6	199	384.2
80	393	20.2	61.5	20.2	61.5	59.7	100	50	46.8	46.8	46.8	42	65.5	65.5	199	330.0
70	344	15.9	54.9	15.9	54.9	68.2	100	49	46.8	46.8	46.8	42	56.5	56.5	199	312.0
60	294.5	0	0	20.4	46.8	79.6	100	48	48	45.6	46.8	42	0	48.6	199	247.6
50	245.8	0	0	4.3*	9.0*	95.7	100	47	47	46.6	46.8	42	0	20.4	199	219.4
40	196.5	0	0	0	0	40	84	46	46	46	46	42	0	0	159	159
30	149	0	0	0	0	30	63.7	45	45	45	45	42	0	0	116	116
20	99.2	0	0	0	0	20	42.2	44	44	44	44	42	0	0	76.0	76.0
10	49.2	0	0	0	0	10	21	43	43	43	43	42	0	0	48.0	48.0
5	24.6	0	0	0	0	5	10.5	42.5	42.5	42.5	42.5	42	0	0	39.8	39.8
0	0	0	0	0	0	0	0	42	42	42	42	42	0	0	0	0

* Theoretical unloading.

COLD WATER FRICTION CHART – ROUGH PIPE

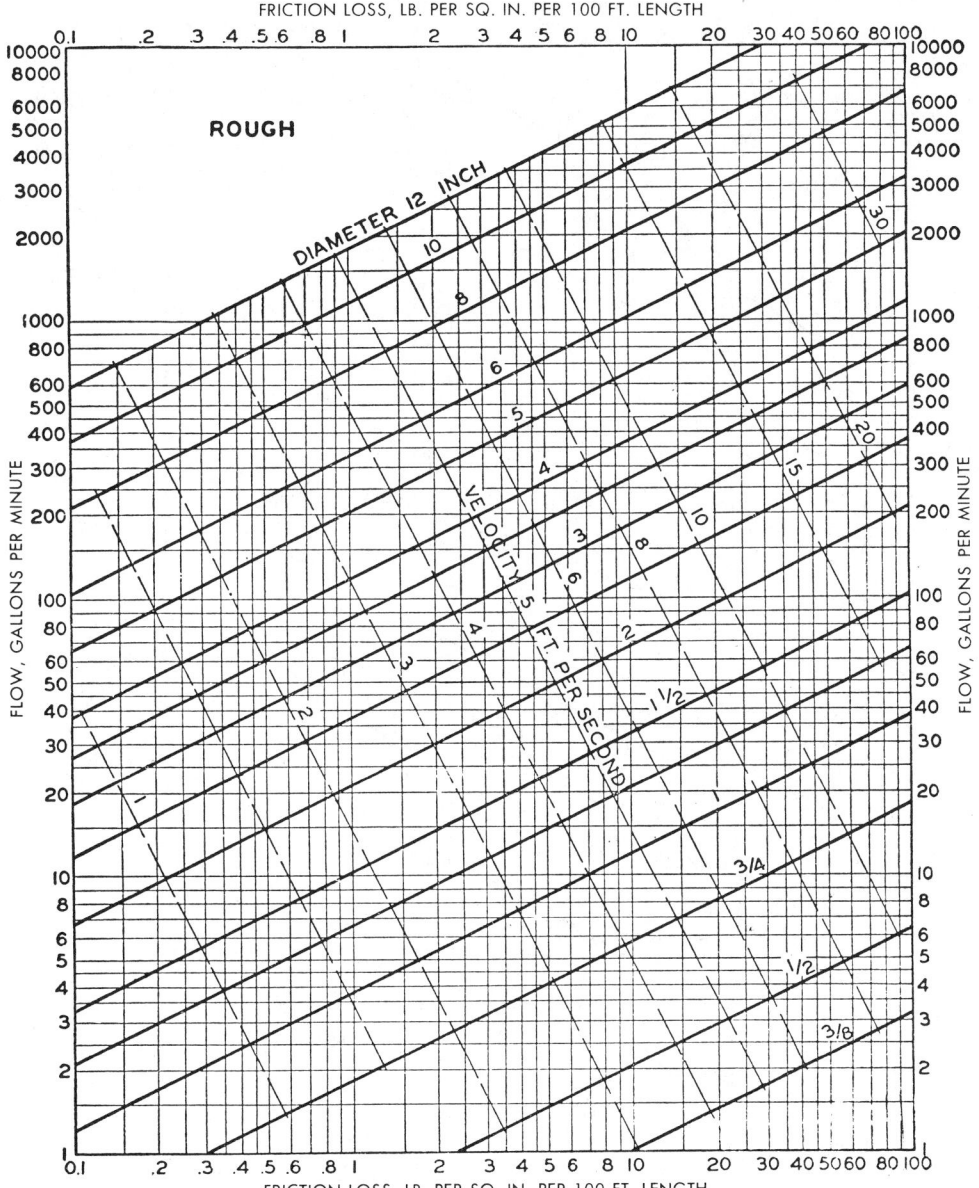

FRICTION LOSS, LB. PER SQ. IN. PER 100 FT. LENGTH

FLOW, GALLONS PER MINUTE

The above chart applies to iron or steel pipe which has been corroded or caked handling water at 50°F It is based on the formula

$$h = 0.010v^2/d$$

where h is friction loss in feet of water per foot of pipe, v is velocity in feet per second, and d is actual inside diameter of pipe in inches.

COLD WATER FRICTION CHART – FAIRLY SMOOTH PIPE

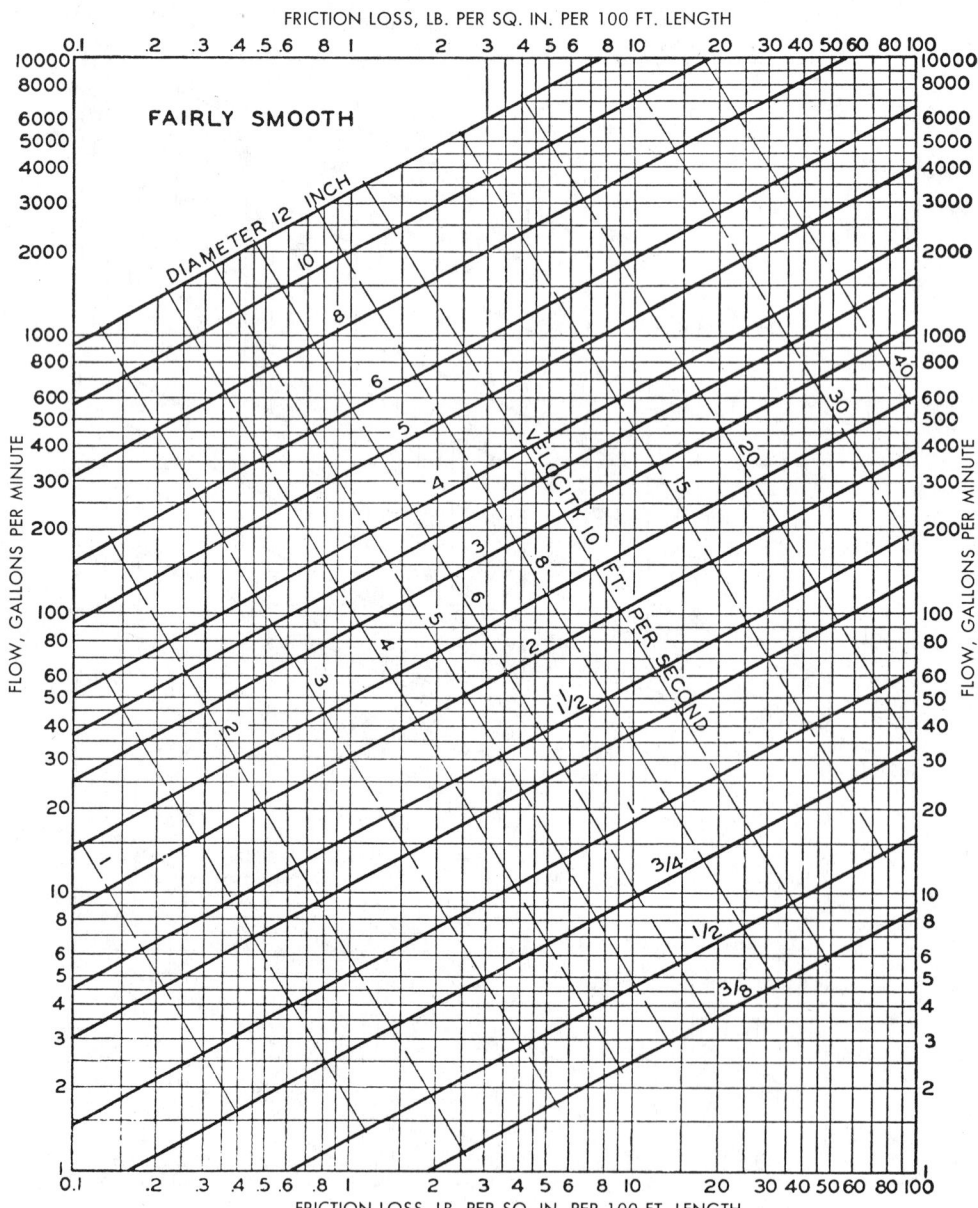

FRICTION LOSS, LB. PER SQ. IN. PER 100 FT. LENGTH

The above chart applies to fairly smooth new steel or iron pipe handling water at 50°F. It is based on the formula

$$h = 0.0073 v^{1.83}/d^{1.17}$$

where h is friction loss in feet of water per foot of pipe (1 lb per sq inch per 100 ft = 1.92 milinches water per foot) v is velocity in feet per second, d is actual inside diameter of pipe in inches.

COLD WATER FRICTION CHART – SMOOTH PIPE OR TUBING

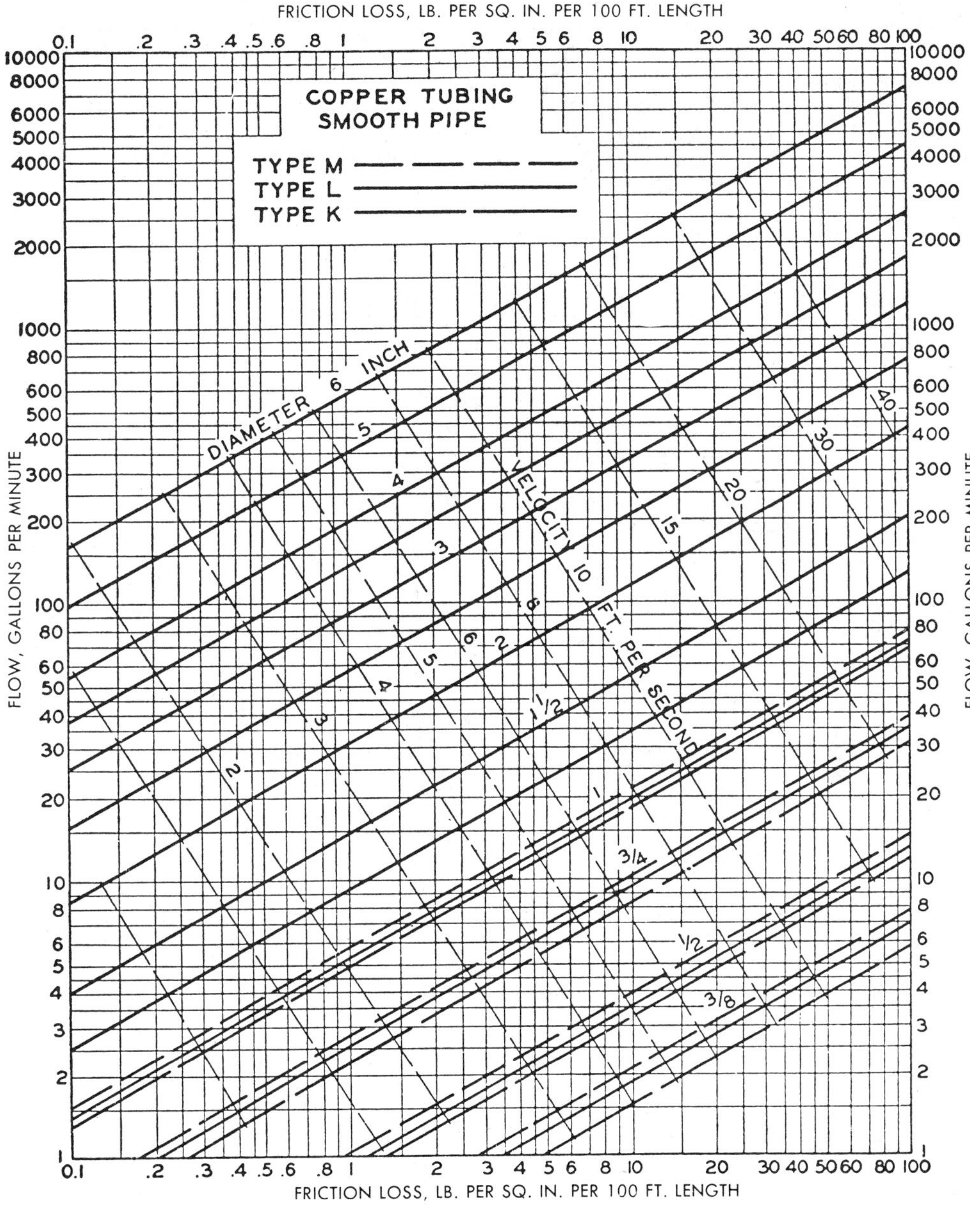

The above chart is for smooth copper pipe with streamlined fittings, designed on the basis of actual diameters, but with nominal sizes indicated, and on the basis of water temperature of 50°F. Can also be used for brass pipe (IPS) with recessed fittings, by taking into account difference in diameters.

Chart is based on the formula

$$h = 0.00682 v^{1.75}/d^{1.25}$$

where h = friction loss in feet of water per foot of pipe

v = velocity, feet per second, and

d = inside diameter of pipe, inches.

ASHRAE REFRIGERANT NUMBERING SYSTEM †

The accompanying table presents the refrigerant numbering system of The American Society of Heating, Refrigerating and Airconditioning Engineers as published in ASHRAE Standard 34-57. For purposes of this standard a refrigerant is the medium of heat transfer in a refrigerating system which picks up heat by evaporating at a low temperature and pressure, and gives up heat on condensing at a higher temperature and pressure. The principle of this standard is to assign an identifying number to each refrigerant. This identifying number can be used either alone or in combination with the word "refrigerant" as "refrigerant 10 , " etc.

ASHRAE Number	Chemical Name	Formula	Mol. Wt.	Boiling Point. Deg F
		Halocarbon Compounds		
10	Carbontetrachloride	CCl_4	153.8	170.2
11	Trichloromonofluoromethane	CCl_3F	137.4	74.8
12	Dichlorodifluoromethane	CCl_2F_2	120.9	— 21.6
13	Monochlorotrifluoromethane	$CClF_3$	104.5	—114.6
13B1	Monobromotrifluoromethane	$CBrF_3$	148.9	— 72.0
14	Carbontetrafluoride	CF_4	88.0	—198.4
20	Chloroform	$CHCl_3$	119.4	142
21	Dichloromonofluoromethane	$CHCl_2F$	102.9	48.1
22	Monochlorodifluoromethane	$CHClF_2$	86.5	— 41.4
23	Trifluoromethane	CHF_3	70.0	—119.9
30	Methylene Chloride	CH_2Cl_2	84.9	105.2
31	Monochloromonofluoromethane	CH_2ClF	68.5	48.0
32	Methylene Fluoride	CH_2F_2	52.0	— 61.4
40	Methyl Chloride	CH_3Cl	50.5	— 10.8
41	Methyl Fluoride	CH_3F	34.0	—109
50	(Methane)[2]	CH_4	16.0	—259
110	Hexachloroethane	CCl_3CCl_3	236.8	365
111	Pentachloromonofluoroethane	CCl_3CCl_2F	220.3	279
112	Tetrachlorodifluoroethane	CCl_2FCCl_2F	203.8	199.0
112a	Tetrachlorodifluoroethane	CCl_3CClF_2	203.8	195.8
113	Trichlorotrifluoroethane	CCl_2FCClF_2	187.4	117.6
113a	Trichlorotrifluoroethane	CCl_3CF_3	187.4	114.2
114	Dichlorotetrafluoroethane	$CClF_2CClF_2$	170.9	38.4
114a	Dichlorotetrafluoroethane	CCl_2FCF_3	170.0	38.5
114B2	Dibromotetrafluoroethane	$CBrF_2CBrF_2$	259.9	117.5
115	Monochloropentafluoroethane	$CClF_2CF_3$	154.5	— 37.7
116	Hexafluoroethane	CF_3CF_3	138.0	—108.8
120	Pentachloroethane	$CHCl_2CCl_3$	202.3	324
123	Dichlorotrifluoroethane	$CHCl_2CF_3$	153	83.7
124	Monochlorotetrafluoroethane	$CHClFCF_3$	136.5	10.4
124a	Monochlorotetrafluoroethane	CHF_2CClF_2	136.5	14
125	Pentafluoroethane	CHF_2CF_3	120	— 55
133a	Monochlorotrifluoroethane	CH_2ClCF_3	118.5	43.0
140a	Trichloroethane	CH_3CCl_3	133.4	165
142b	Monochlorodifluoroethane	CH_3CClF_2	100.5	12.2
143a	Trifluoroethane	CH_3CF_3	84	— 53.5
150a	Dichloroethane	CH_3CHCl_2	98.9	140
152a	Difluoroethane	CH_3CHF_2	66	— 12.4
160	Ethyl chloride	CH_3CH_2Cl	64.5	54.0
170	(Ethane)[2]	CH_3CH_3	30	—127.5
218	Octafluoropropane	$CF_3CF_2CF_3$	188	— 36.4
290	(Propane)[2]	$CH_3CH_2CH_3$	44	— 44.2

(Footnotes on next page at conclusion of table)

ASHRAE Refrigerant Numbering System (Concluded)

ASHRAE Number	Chemical Name	Formula	Mol. Wt.	Boiling Point. Deg F
	Miscellaneous Organic Compounds			
	Cyclic Organic Compounds			
C316	Dichlorohexafluorocyclobutane	$C_4Cl_2F_6$	233	140
C317	Monochloroheptafluorocyclobutane	C_4ClF_7	216.5	77
C318	Octafluorocyclobutane	C_4F_8	200	21.1
	Azeotropes			
500	Refrigerants 12/152a 73.8/26.2 wt %	CCl_2F_2/CH_3CHF_2	99.29	— 28.0
501	Refrigerants 22/12 75/25 wt %	$CHClF_2/CCl_2F_2$	93.1	— 42
502	Refrigerants 22/115 48.8/51.2 wt %	$CHClF_2/CClF_2CF_3$	112	— 50.1
	Hydrocarbons			
50	Methane	CH_4	16.0	—259
170	Ethane	CH_3CH_3	30	—127.5
290	Propane	$CH_3CH_2CH_3$	44	— 44.2
600	Butane	$CH_3CH_2CH_2CH_3$	58.1	31.3
601	Isobutane	$CH(CH_3)_3$	58.1	14
1150	(Ethylene)[3]	$CH_2=CH_2$	28.0	—155.0
1270	(Propylene)[3]	$CH_3CH=CH_2$	42.1	— 53.7
	Oxygen Compounds			
610	Ethyl ether	$C_2H_5OC_2H_5$	74.1	94.3
611	Methyl formate	$HCOOCH_3$	60.0	89.2
	Sulfur Compounds			
620				
	Nitrogen Compounds			
630	Methyl amine	CH_3NH_2	31.1	20.3
631	Ethyl amine	$C_2H_5NH_2$	45.1	61.8
	Inorganic Compounds			
717	Ammonia	NH_3	17	— 28.0
718	Water	H_2O	18	212
729	Air		29	—318
744	Carbon dioxide	CO_2	44	—109 (subl.
744A	Nitrous oxide	N_2O	44	—127
764	Sulfur dioxide	SO_2	64	14.0
	Unsaturated Organic Compounds			
1112a	Dichlorodifluoroethylene	$CCl=CF_2$	133	67
1113	Monochlorotrifluoroethylene	$CClF=CF_2$	116.5	— 18.2
1114	Tetrafluoroethylene	$CF_2=CF_2$	100	—105
1120	Trichloroethylene	$CHCl=CCl_2$	131.4	187
1130	Dichloroethylene	$CHCl=CHCl$	96.9	118
1132a	Vinylidene fluoride	$CH_2=CF_2$	64	—119
1140	Vinyl chloride	$CH_2=CHCl$	62.5	7.0
1141	Vinyl fluoride	$CH_2=CHF$	46	— 98
1150	Ethylene	$CH_2=CH_2$	28.0	—155.0
1270	Propylene	$CH_3CH=CH_2$	42.1	— 53.7

† Reprinted, by permission, from ASHRAE Standard 34-57.

[2] The compounds methane, ethane, and propane appear in the Halocarbon section in their proper numerical positions, but in parentheses since they are not halocarbons.

[3] The compounds ethylene and propylene appear in the Hydrocarbon section as parenthetical items in order to indicate that these compounds are hydrocarbons. Ethylene and propylene are properly identified in the Unsaturated Organic Compounds section.

REFRIGERANTS

A refrigerant is the medium of heat transfer in a refrigerating system which picks up heat by evaporating at a low temperature, and gives up heat on condensing at a higher temperature and pressure.

Types of Refrigerants

Until the early nineteen thirties the principal refrigerants were ammonia, carbon dioxide, sulfur dioxide, methyl chloride, and the hydrocarbons. Thermodynamically they are all good refrigerants for commercial use in reciprocating machines, except carbon dioxide; its critical temperature is too low although it has been employed rather extensively. As far as chemical properties are concerned they all fall far short. Ammonia, methyl chloride, and the hydrocarbons will burn and under certain conditions, where mixed with air, will explode. They are all toxic. Methyl chloride will react directly with aluminum to produce aluminum trimethyl, a gas which will catch fire spontaneously in the presence of air.

With the development of centrifugal refrigeration by W. H. Carrier about 1920 there arose a need for a high molecular weight compound which would have a normal boiling near room temperature. Dilene ($C_2H_2Cl_2$) was obtained from a German firm in 1921. Methylene chloride (CH_2Cl_2) came into use in 1926 giving approximately twice the refrigeration capacity for a given impeller at the same speed, twice the power being required.

In 1930, Midgley presented his paper, Organic Fluorides as Refrigerants, before the American Chemical Society. Carrier was interested in the subject and immediately began testing "Freon-11" (CCl_3F) with a normal boiling point of 74.7 deg F. It was a much improved refrigerant over methylene chloride and was soon to attain considerable popularity. Today it is used extensively in centrifugal machines with refrigeration capacities from 100 tons upward. More recently the compound CCl_2FCClF_2 with a boiling point of 117.6 deg is coming into use, making lower capacity machines practicable.

The first commercial azeotropic refrigerant, Carrene 7, containing 26.2% CH_3CHF_2 and 73.8% CCl_2F_2, was put on a restricted market in 1950 to be used in special situations to get about 20% more refrigeration capacity than CCl_2F_2 with a given compressor running at the same speed. As far as is known, it is the only commercial azeotropic refrigerant today.

The halocarbons are so much safer to use than ammonia, sulfur dioxide, and the other refrigerants used before 1930 that they have almost completely replaced the less desirable refrigerants.

There are so many names bearing close resemblance to each other, such as trichlorofluoromethane and chlorotrifluoromethane, that it becomes imperative to have a simple number designation system. The first refrigerant mentioned immediately above has a normal boiling point of 74.7 deg; the other, −73.6 deg, a difference of about 158 deg—quite a difference.

Designating Systems

The E. I. du Pont de Nemours Company has given its number system to the industry. This "Freon" system of numbers originally applied to saturated halocarbons and was later extended to include unsaturated halocarbons.

For the saturated halocarbons, there are two or three digits in the number. If there are two digits, the number should be regarded as having three digits, the first being zero. One may obtain the formula by adding "one" to the first digit to get the number of carbon atoms, subtracting "one" from the second digit to get the number of hydrogen atoms, and neither adding nor subtracting anything from the third digit to get the number of fluorine atoms. Any valencies not satisfied by hydrogen or fluorine will be filled with chlorine atoms.

While the same numbers are now used, the word "Freon" being a trade name, cannot be used to apply to a material if it is not made by du Pont. The generic term adopted by The American Society of Heating, Refrigerating and Airconditioning Engineers is "Refrigerant." Under the present ASHRAE Standard on refrigerant numbering, CCl_2F_2 is "Refrigerant 12" (or simply "12," if it is understood that a refrigerant is intended).

Refrigerant 12 is in reality, then, "Refrigerant 012." The number can be decoded "102" to designate directly one carbon atom, zero hydrogen atoms, and two fluorine atoms.

The number system for halocarbons was not offered until it was extended to cover many other refrigerants. The extended system covers saturated halocarbons with numbers up to, but not including, 500; the five hundred series is for azeotropes; the six hundred series certain hydrocarbons with oxygen, sulfur, and nitrogen compounds; the seven hundred series for inorganic compounds; the thou-

Table 1. Comparative Refrigerant Characteristics†
(Based on a Standard Ton)

Refrigerant Number	Formula	Boiling Point, Deg F	Compression Ratio	Coefficient of Performance	Horsepower per Ton	Discharge Temp., Deg F
13B1	CBrF$_3$	− 73.6	3.36	4.25	1.109	124
290	C$_3$H$_8$	− 44.2	3.70	4.58	1.029	97
22	CHClF$_2$	− 41.4	4.06	4.66	1.012	131
717	NH$_3$	− 28.0	4.94	4.76	0.990	210
500	—	− 28.0	4.12	4.61	1.023	105
12	CCl$_2$F$_2$	− 21.7	4.07	4.70	1.003	100
40	CH$_3$Cl	− 10.8	4.48	4.90	0.962	172
764	SO$_2$	14.0	5.63	4.87	0.968	191
600	n-C$_4$H$_{10}$	31.3	5.07	4.95	0.952	88
114	CClF$_2$CClF$_2$	38.4	5.42	4.64	1.016	86
21	CHCl$_2$F	48.0	5.96	5.05	0.933	142
11	CCl$_3$F	74.7	6.24	5.09	0.926	112
113	CCl$_2$FCClF$_2$	117.6	8.02	4.92	0.958	86

† Based on a table by Reed, *Air Conditioning and Refrigeration News*, 1951.

sand series for unsaturated hydrocarbons, the last three digits having the same significance as with the saturated halocarbons.

For the seven hundred series, the last two digits give the molecular weight, water becoming Refrigerant 718.

Refrigerant 500, the only known commercial azeotropic refrigerant, contains 26.2% of 152 and 73.8% of 12. It does not have two azeotropic competitors as might be inferred from the other designations. Commercially it is in a class by itself.

Desirable Properties in Refrigerants

Not only is it important to have good thermodynamic properties in a refrigerant, but it is also imperative to have certain desirable physical and chemical properties.

In the first place the liquid must have a fairly low freezing point so that solid will not be formed in the evaporator. Water, which is a good refrigerant in many ways, cannot be employed to attain temperatures lower than 32 deg. The critical temperature should not be too low or otherwise the efficiency of operation will be extremely low. Carbon dioxide, with a critical temperature of 87.8 deg, gives a poor performance for any ordinary refrigeration. Ammonia, on the other hand, with a critical temperature of 271.4 deg, gives very satisfactory performance.

For a good many years nearly all refrigerants used were toxic, flammable, or explosive where mixed with the proper amount of air; some had all three disadvantages. Methyl chloride not only has these disadvantages, but reacts with aluminum to form aluminum trimethyl, a spontaneously combustible gas. Therefore it cannot be used in machines where it will come in contact with aluminum.

The halocarbon refrigerants have gone a long way toward replacing carbon dioxide, methyl chloride, sulfur dioxide, the hydrocarbons, methylene chloride, dielene, and, even ammonia, because many of them are nontoxic, nonflammable, and will not explode where mixed with air. Refrigerants 11, 12, 22, 113, 114, and 500 are the most important commercial refrigerants today, having these desirable properties.

The matter of flammability is not so important where the refrigerant is used in a place such as an oil refinery. Everything the plant makes is flammable, so that engineers would not be adding much to the hazards by using propane as a refrigerant. The importance of nonflammability for small units has been far overplayed. After all, gasoline is quite flammable, but plenty of it is used around every home.

Chemical stability is important in any refrigerant. If it decomposes, one naturally has to be concerned about the products. There are so many stable refrigerants on the market that one does not need to take a chance with a questionable one.

Ammonia is quite stable except where used with a metal such as copper. Particularly in the presence of water, its corrosion rate on copper is very high. This has been learned well through the years. However, the characteristics of ammonia are so well known that it can be used to a great advantage in some large commercial applications which operate

under the direction of an engineer—to a great advantage because thermodynamically it is an excellent fluid to use in compression refrigeration.

In general, the molecular weight should be low for compression refrigeration and high for centrifugal refrigeration. One often sees a reference to the effect that the compression ratio should be low. On the contrary, it should be high. It is the molar latent heat of vaporization that produces the refrigeration. The greater the compression ratio, the greater will be the molar latent heat and the greater will be the percentage of the latent heat used to cool the work.

The molar specific heat of the liquid should be low. Otherwise, too much cooling is required for the hot liquid entering the evaporator. The specific heat of the vapor should be high, but only if this does not automatically call for high specific heat of the liquid. Of the two, that of the liquid is more important. Since they go hand in hand, the specific heats should be low.

There are other properties such as water solubility, viscosity, and heat transfer that have their importance even though the range on each can be rather broad.

Ideal Refrigeration Cycle

Through a consideration of the second law of thermodynamics one may arrive at the ideal refrigeration cycle, so called because it is reversible and gets the maximum amount of cooling for work expended.

From Carnot's cycle, which is described in any text on thermodynamics, it may be found that a heat engine, taking in heat Q_2 at a temperature T_2 and rejecting Q_1 at T_1, will show the following for

the relation of the work W done because of the introduction of the heat Q_2:

$$\frac{W}{Q_2} = \frac{T_2 - T_1}{T_2} , \qquad (1)$$

the temperatures being absolute.

If the cycle be reversed, as in refrigeration, interest focuses on the amount of heat Q_1 "picked up" at T_1. Since W in Equation (1) is $Q_2 - Q_1$, by substitution and proper rearrangement, it may be shown that

$$\frac{Q_1}{W} = \frac{T_1}{T_2 - T_1} \qquad (2)$$

The quotient Q_1W is, of course, the ratio of the heat pumped out of the evaporator to the work required for the pumping. This quotient is called the coefficient of performance, *COP*, that of the Carnot cycle being a maximum for a given temperature lift.

In an actual refrigeration cycle, reversibility does not exist and therefore there will be losses causing the *COP* to be less than that for the ideal cycle.

Standard Ton Conditions

In order to compare refrigerants as to performance, either the operating conditions should be the same as nearly as possible or the same machine should be used and let the conditions change as they will. In most cases the former is used.

In other words, it is assumed that a different machine is designed for each refrigerant such that the same evaporator temperature will be maintained; likewise, the same condensing temperature will be maintained. Once these conditions are set, the superheat temperature and a good many other properties are fixed depending upon the nature of the two refrigerants.

Before discussing further standard conditions, it becomes necessary to define the term "ton of refrigeration." It is the removal in the evaporator of 200 Btu per min, the concept being derived from the fact that this is the rate that heat must be removed from 32 deg liquid water to produce one ton of ice at 32 deg in 24 hours.

Where the evaporator temperature is held at 5 deg and the condenser at 86 deg, the condition is called the "standard ton." Ice could well be made where such conditions exist. It is significant that, for most refrigerants, one horsepower will produce about one ton of refrigeration under these conditions.

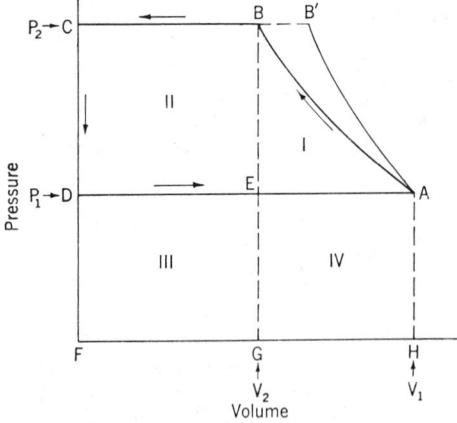

Fig. 1 Simple refrigeration thermodynamic cycle.

Table 2. Reliability of the Refrigerant Performance Equation†

Refrigerant Number	Normal Boiling Point, Deg F	Coefficient of Performance		Deviation, %
		Observed	Calculated	
290	− 44.2	4.58	4.551	+0.63
22	− 41.4	4.66	4.671	−0.24
115	− 36.4	—	4.367	—
218	− 34.3	—	4.214	—
500	− 28.0	4.61	4.643	−0.72
12	− 21.7	4.70	4.700	0.00
40	− 10.8	4.90	4.902	−0.04
600	31.3	4.95	4.838	+2.26
114	38.4	4.64	4.652	−0.26
21	48.0	5.05	5.037	+0.26
11	74.7	5.09	5.090	0.00
C_4F_{10}	84.8	—	4.527	—
113	117.6	4.92	4.953	−0.91

† Largely data by Reed, Data Book, 1953-54, The American Society of Refrigerating Engineers.

Simple Refrigeration Cycle

A realistic refrigeration cycle is somewhat complicated and it is rather difficult to fill in all the details for a complete understanding of a cycle for a given refrigerant.

In trying to get a general understanding of the phenomenon, it pays to make a simple model and see how well it fits the engineering facts. In other words, one should predict from the simple model and make the proof with a refrigeration calorimeter.

To set up a simple model let us first assume that the cycle is started at A in Fig. 1 with an isentropic compression to B. This assumption is tantamount to saying that no heat is exchanged between the compressor and the fluid. The gas at B will have a temperature at least as high as the condensing liquid, generally higher depending on the nature of the gas. For standard ton conditions Refrigerant 22

will be at 131 deg at B; Refrigerant 114, at 86 deg, which is identical with the condensing temperature.

If there is superheat, as for 22, it will be rejected to the heat sink until the condensation temperature is reached, at which point the latent heat of vaporization will likewise be rejected, resulting in condensation of the fluid. The condensation, the second step in the cycle, will give a great reduction in volume. As a matter of fact, the volume of the liquid is being neglected in order to build up the desired simple model. The condensed liquid is represented by C where the pressure is p_2 and the temperature t_2. The third step involves the transfer of the liquid through an expansion valve or capillary tube where it is cooled to t_1 and p_1 at D. This cooling is effected at the expense of some of the latent heat of vaporization. The fourth step, going

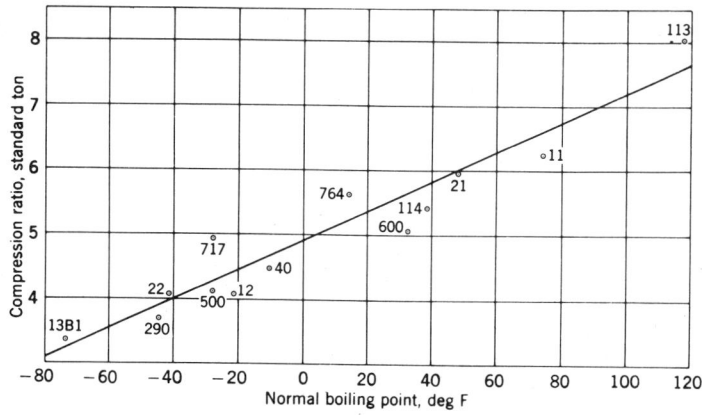

Fig. 2. Effect of boiling point on the compression ratio of some refrigerants.

Table 3. Standard Ton Refrigeration Capacity of Some Refrigerants

Refrigerant Number	Boiling Point, Deg F	Pressure at 5 deg F	Tons
113	117.6	0.980	5.77
11	74.7	2.931	16.00
21	48.0	5.243	28.47
114	38.4	6.772	29.69
12	− 21.7	26.51	100.00
500	− 28.0	31.08	116.3
218	− 34.3	35.85	—
115	− 36.4	—	—
22	− 41.4	43.02	161.7

from D to A, involves the boiling of the refrigerant to get again vapor at A thus completing the cycle.

The work done in the cycle is represented by the sum of Areas I and II which can be evaluated in two fairly simple ways.

In one method use can be made of the fact that the work of compression, $- \int p\,dv$, is the sum of I and IV, but IV is $- p_1 (v_2 - v_1)$ and this must be subtracted in order to get Area I. The area of II is $- v_2 (p_1 - p_2)$. The cyclic work then is

$$W = - \int_{v_1}^{v_2} p\,dv + p_1 (v_2 - v_1) - v_2 (p_1 - p_2) \tag{3}$$

Each of the three terms can be evaluated, but the integration may require a disproportionate amount of time, depending on the degree of complication of the equation of state.

In the second method inasmuch as the volume is assumed to go to zero (neglecting the volume of the liquid), the area of I and II may be given as $+ \int_{p_1}^{p_2} v\,dp$. Hence, the cyclic work is

$$W = + \int_{p_1}^{p_2} v\,dp \tag{4}$$

which is just as difficult to evaluate as the previous integral.

Since the compression process approaching a

Table 4. Low Temperature Refrigerants

Refrigerant Number	Formula	Boiling Point, Deg F	Critical Temp., Deg F	Status
13B1	$CBrF_3$	− 73.6	153.5	S
116	CF_3CF_3	−108.8	—	—
13	$CClF_3$	−114.6	83.9	C
23	CHF_3	−119.9	—	D
14	CF_4	−198.4	− 49.9	S

condition of constant entropy and, indeed, has been assumed to be isentropic it behooves us to examine the combined statements of the first and second law of thermodynamics involving the differential dS for entropy. These equations are

and

$$dE = TdS - pdv \tag{5}$$

$$dH = TdS + vdp. \tag{6}$$

For constant entropy,

$$dE = - pvd \text{ and} \tag{7}$$

$$dH = vdp. \tag{8}$$

Enthalpy values are of greater significance generally in thermodynamic calculations involving refrigerants and consequently are tabulated among the useful thermodynamic data. This fact leads to a preference for Equation (8). It is immediately obvious on combining Equations (4) and (8) that

$$W = H_2 - H_1. \tag{9}$$

The change in enthalpy for the compression process is identical with the work for the entire cycle.

For a given refrigerant then, one may use a Mollier chart where pressure is plotted against enthalpy and go out along a constant entropy line to find H_2, having started at a point on the saturation curve giving H_1. The discharge temperature can also be read from the chart. Because the charts are difficult to read unless they are made very large, it is often desirable to make use of the tabular data and interpolate for a more precise value of the change in enthalpy and, hence, the work of the cycle.

It is abundantly clear now why the simple model is so important—it is so easy to get the work of the cycle from tabulated, thermodynamic data. Sight must not be lost of the fact that one does not only want something simple; he also wants something accurate. Calorimeter tests involving refrigerants in action have demonstrated that the simple model fits the real situation well enough to justify its use in comparing the merits of various refrigerants.

In evaluating the COP (Q_1/W) for a refrigerant one uses

$$\frac{Q_1}{W} = \frac{- (h_1)_{t_2} + (h_g)_{t_1}}{(h_g)_{t_3} - (h_g)_{t_1}}, \tag{10}$$

where h, t, l, and g denote enthalpy, temperature, liquid, and gas, respectively. t_1, t_2, and t_3 are the

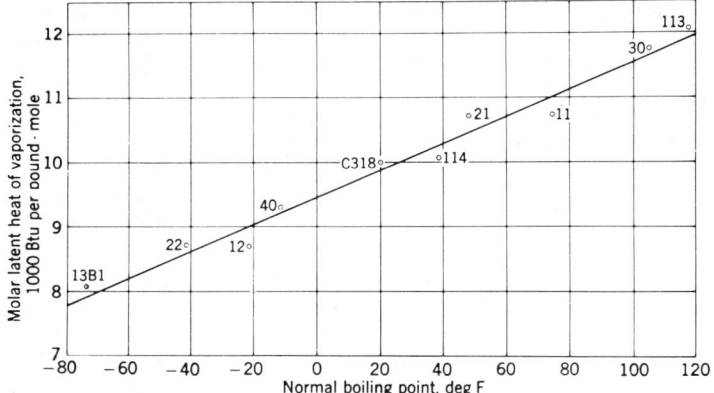

Fig. 3. Effect of boiling point on the molar latent heat of some refrigerants.

evaporator, condenser, and discharge temperatures, respectively.

Q_1 is often referred to as the net refrigeration effect. In the case of Refrigerant 12 for the standard ton conditions, it is 50.036 Btu per lb and yet the latent heat of vaporization is 68.204 Btu per lb. This means that 26.64% of the latent heat cooling power must be used to cool the hot liquid refrigerant coming from the condenser. In seeking more desirable refrigerants, it certainly looks as though one should attempt to find one where the greatest percentage of the latent heat is used for useful cooling.

The concept of "latent heat efficiency" has been defined in the literature as the fraction or percentage of the latent heat used to get useful cooling. On the percentage basis

$$L_v = \frac{100 Q_1}{h_g - h_1} \qquad (11)$$

Obviously, if L_v is zero there can be no useful cooling whatsoever. In other words, all the latent heat cooling power is used to cool the refrigerant. This condition can be visualized where the evaporator is surrounded by a perfect insulator. If such a system is started, using Refrigerant 12, and the condenser maintained at 210 deg, the evaporator will be at —130 deg and dry. All the heat rejected at 210 deg is the heat of compression, none of it being net heat pumped from a level of —130 deg to a level of 210 deg. It would be impossible then to pump heat from —130.1 deg to heat a chemical solution to 210 deg using Refrigerant 12. Hypothetically it could be done from —129.9 deg but, of course, would be exceedingly inefficient.

Returning to the standard ton conditions where L_v for 12 is 73.36, one may surmise that some other or new refrigerant whose latent heat efficiency is, say 80, will be more desirable. In general this is quite true, but one must be careful in coming to

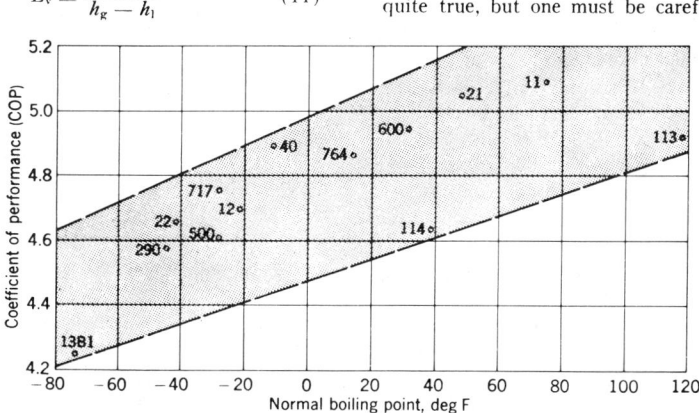

Fig. 4. Effect of boiling point on the COP for some refrigerants.

such a conclusion. This can be borne out through a speculation involving Refrigerant 12.

Carnot's cycle for the standard ton conditions gives

$$\frac{Q_1}{W} = \frac{T_1}{T_2 - T_1} = \frac{464.72}{81} = 5.737.$$

Calculations for Refrigerant 12 using Equation (10) give

$$Q_1/W = 4.70,$$

which means that if Q_1 is 200, W is 42.55, the work each minute in doing refrigeration at a capacity of one ton.

It has already been shown the L_v is 73.36, meaning that the 200 Btu could be boosted to 272.6 Btu if all the latent heat could be used. However, this could not be done unless the specific heat of the liquid is zero. While that of Refrigerant 12 is not zero, some other refrigerant might be found with a lower value nearer this limiting value. The limiting condition, even for the imagination, is zero so the situation can be analyzed where the specific heat of Refrigerant 12 is regarded as zero. Further assume that this has no other effect on the cycle as given in Fig. 1. In other words, the same work, 42.55 Btu, will now pump 272.6 Btu to the 86 deg level. Q_1/W will now be 6.407 which is considerably better than the ideal Carnot cycle. This is a gross violation of the second law of thermodynamics.

Since there is no question about the pressures or the beginning point A in Fig. 1, B must be shifted for the new situation. There must be more work to bring the COP down below 5.737 and, therefore, the sum of the areas of I and II must be increased as indicated by moving B to B'. The refrigerant would get hotter in the compression if the specific heat of the liquid were less. It naturally follows that the specific heat of the vapor would be less if that of the liquid were less.

The speculation indulged in the foregoing demonstrates that there is an advantage in having a low specific heat in the liquid. The lower the specific heat of the vapor the greater will be the discharge temperature for an otherwise constant set of conditions and, consequently, the more work will have to be done in the compression. The advantage of the low specific heat in the liquid outweighs the disadvantage of the low specific heat of vapor so that there is a net gain by having low specific heat in the liquid even though it is accompanied with a low specific heat in the vapor.

Again in considering the compression ratio and claims that a low ratio is desired, examine the situation quantitatively. The limiting ratio downward is unity, even for the imagination. On examining Fig. 1, it is found that W is zero and, hence, Q_1/W is infinite, trampling the second law into oblivion, unless Q_1 is also zero. As long as Q_1/W is constant, it does not matter about the compression ratio. For the same evaporation pressure and temperature it is true that more work will be done per stroke if the compression ratio is high, but so will more latent heat be involved (Clausius-Clapeyron). Perhaps the best way to regard the compression ratio is in terms of latent heat efficiency. The higher the compression ratio, between the same temperature conditions, the higher the molar latent heat will be, meaning that if the specific heat remains the same, the latent heat efficiency is greater, and therefore the more efficient the refrigerant will be.

Comparative Characteristics

The data given in Table 1 are presented to depict the relations existing between various characteristics of refrigerants.

The values for horsepower per ton were calculated from the COP. Obviously,

$$H_p = K(W/Q_1), \tag{12}$$

where H_p is the horsepower per ton, W/Q_1 is the reciprocal of the COP, and K is a constant. The values in the table values for 13B1 and 21 are not consistent with Equation (12), which is a precise relationship.

The normal boiling point affects the compression ratio as shown in Fig. 2; the relation is linear. There is no question that the compression ratio increases with the normal boiling point. Likewise, the molar latent heat of vaporization at the normal boiling point is, on the average, a straight line function of the normal point, as shown in Fig. 3. The data fitted by the law of least squares give

$$y = 9446 + 20.89x, \tag{13}$$

where x is the boiling point in deg F and y is the molar latent heat of vaporization in Btu per pound mole. The COP is also affected by the normal boiling, the relation again being linear as shown in Fig. 4, where a least squares fitting gives

$$y = 4.749 + 0.003223x, \tag{14}$$

x being the boiling point and y being the COP.

On the introduction of the concept of latent heat efficiency into the literature, it was shown that, for standard ton conditions, the product of the horsepower per ton (H_p) and latent heat efficiency (L_v) is a constant:

$$H_p L_r = 74.703. \tag{15}$$

Table 5. Characteristics of Some Cascade Systems†

Combination	Refrigerant Number	Stage	Temperature, Deg F		Pressure, psi	
			Evaporator	Condenser	Evaporator	Condenser
	12	High	— 30	100	—	—
1	13	Low	—125	— 20	10.5	126.5
2	13B1	Low	—114	— 20	4.1	48.9
	22	High	— 50	100	—	—
3	13	Low	—134	— 40	7.3	87.4
4	13B1	Low	—127	— 40	2.7	32.4

† From *Refrigerating Engineering*, 64, No. 2, 1956.

The relation is shown graphically in Fig. 5. The data fall fairly close to the straight line except those for CH_3C1 and 114.

An equation similar to Equation (15) for the condition where the evaporator is at 0 deg and the condenser is at 100 deg was used to estimate the horsepower per ton for refrigerant 115 from limited data. It was found to be about 35% higher than that for 22. The data in Fig. 4 are referred to for an estimate for the standard ton.

Obviously the scatter is such as to suggest a classification as to types of molecules. This classification has been made and the results appear graphically in Fig. 6. The equations of the straight lines appear in the figure and were obtained in each case from two points.

In a given class, it is obvious that increasing the number of carbon atoms lowers the COP and increasing the number of hydrogen atoms increases the COP. In other words, more carbon atoms per molecule increase the horsepower; more hydrogen atoms decrease it.

If the 113-114 line is extrapolated to a boiling point of —36 deg F for Refrigerant 115, it will be found that the COP is 4.377 giving by Equation (12), using Refrigerant 12 to evaluate the constant.

a value of 1.077 as the horsepower per ton for 115. This value is only 6.4% higher than that for Refrigerant 22. The agreement with the relative value of 35% given in a previous publication for the cycle 0-100 deg is not at all good. The former estimate of the horsepower requirement for 115 is not realistic. Refrigerant 115 appears to be a much better refrigerant than was previously indicated.

With a little mathematical manipulation into the empirical, it has been possible to develop, for the standard ton, the following relation from the data for Refrigerant 11, 12, 21, 22, 113, and 114:

$$C_\pi = (1.011)^H (1.00000 - 0.19599 \log C)$$
$$(4.788 — 0.00405 B_p), \qquad (16)$$

where C_π is the COP, H is the number of hydrogen atoms per molecule, C is the number of carbon atoms per molecule, and B_p is the normal boiling point in deg F. This question is called the "refrigerant performance equation".

Data are given in Table 2 to demonstrate what has been wrested in the way of scientific classification out of the somewhat chaotic picture shown in Fig. 4. Aside from Refrigerant 600 (C_4H_{10}) the calculated and observed values agree as well as could ever be expected.

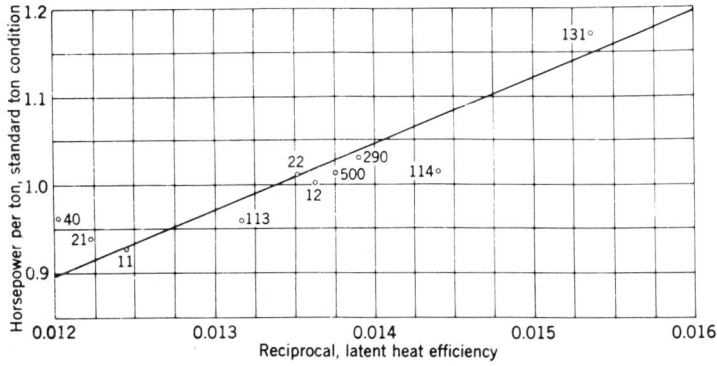

Fig. 5. Effect of latent heat efficiency on horsepower per ton for some refrigerants.

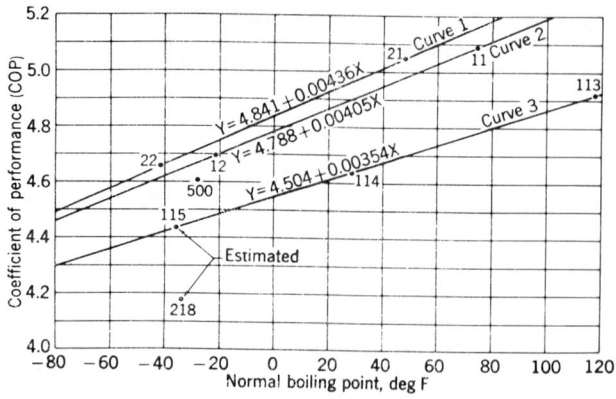

Fig. 6. Effect of boiling point on COP for various classes of refrigerants.

Calculated values for three other refrigerants which may be employed in the future have also been put in the table; reference will be made to them later.

Refrigeration Capacity

Different refrigerants, where used in identical machines with appropriate motors, will yield different refrigeration capacity in each case. Advantage has been taken of this fact to use the same compressor in air conditioning for several applications. A 30-ton unit designed for Refrigerant 12 will have a capacity of 35 tons with Refrigerant 500 and 45 tons with Refrigerant 22. Using the different refrigerants in this way results in some genuine savings for the manufacturer of air conditioning and refrigeration equipment. A new refrigerant is needed to fill the gap between Refrigerant 500 and 22; one that can give 40 tons capacity in the above unit designed to give 30 tons with 12.

Table 3 shows how the standard ton refrigeration capacity of some refrigerants varies with the normal boiling point and also with the pressure. A machine which will give 100 tons with Refrigerant 12 has been considered in the calculations with the speed of the compressor constant for all refrigerants. Sufficient data are not at hand to analyze Refrigerant 115 and 218, but their boiling points indicate that one of them may serve as the intermediate between 500 and 22. It is feared, however, that they may be so inefficient as to have capacity values much lower than the pressures would indicate. Incidentally, the values given in Table 3 indicate that the capacity is close to a linear function of the evaporator pressure. See Fig. 7 in which the line has merely been placed by inspection.

One can visualize the importance, in air conditioning, of using different refrigerants with the same compressor when he considers that a line of compressors 3, 5, 10, 15, 25, 35, and 50 tons in capacity, where Refrigerant 12 is used, will give 3.6, 6, 12, 18, 30, 42.5, and 60 tons where Refrigerant 500 is used. Thus the number of sizes are doubled without any redesign and new tooling.

Low Temperature Refrigerants

In order to cool to a temperature of, say—125 deg F, it is important to have a refrigerant with adequate pressure so that the compressor will not need to be unduly large. For such a service, some of the refrigerants listed in Table 4 may be used: In the column headed "status," D denotes in the development stage, S semicommercial, and C commercial.

The critical temperature is so low in many of these refrigerants that they cannot be condensed in the vicinity of 100 deg where heat is finally rejected from so many commercial systems. It becomes expedient to use a cascade system involving two refrigerants. Refrigerant 13 can be used to pump heat from an evaporator at —125 deg up to —20 deg where its heat of condensation is transferred to boiling Refrigerant 12 which condenses at 100 deg transferring the heat to cooling water. Missimer describes systems using 12 or 22 as the high stage refrigerant and 13 or 13B1 as the low stage refrigerant. Table 5 contains data from his paper.

High Molecular Weight Refrigerants

While high molecular weight should be avoided in refrigerants used in compression refrigeration, just the opposite holds in centrifugal refrigeration. As a matter of fact, the greater the molecular weight the fewer stages are required in the impeller. It is

now visualized that a single stage impeller may be used if suitable refrigerants can be found with molecular weights around 300. Three such refrigerants with boiling points of 75, 100, and 125 deg could probably find immediate application. Table 6 contains some high molecular weight compounds in which interest has been shown.

Badylkes has proposed CF_2ClBr, CBr_2F_2 and $(CBrF_2)_2$ for use in single stage compressors, stating that an evaporator temperature of 15 deg and a condensing temperature as high as 160 deg can be used.

Azeotropic Refrigerants

An azeotropic solution is one (containing at least two components) that boils at some given pressure without change in composition. The term "azeotrope" refers to the solution with the exact composition which is azeotropic. An azeotropic system is a chemical system in which an azeotropic solution exists for some temperature.

To illustrate, a solution of Refrigerants 152a and 12 containing 26.2% 152a is an azeotropic solution at 32 deg F. The solution of this composition is also called the azeotrope at 32 deg. The chemical system, Refrigerants 152a and 12, is an azeotropic system whether one knows the azeotropic composition or not. One may use a solution in an azeotropic system without that solution being the azeotrope. A maximum pressure azeotropic system can be fractionated in a distilling column to give the azeotrope at the top and one of the pure constituents at the bottom; which one depends upon which side of the azeotropic composition the feed solution is on. Homoazeotrope refers to an azeotrope in a single liquid phase.

Refrigerant 500 is an azeotropic mixture (32 deg) containing 26.2% Refrigerant 152 and 73.8% 12. It was developed to be used commercially in a machine, designed for Refrigerant 12 and 60 cycle current, to give the same capacity where 50 cycle current is employed. Another encouragement toward its development was the extension and filling in of sizes in a given line of compressors. While it is the only azeotropic mixture known to the writer to be used commercially, there are many other azeotropes not yet made commercial. Some of these are given in Table 7. The boiling points are given in the order of appearance in the combination column, No. 1 always being the higher boiling constituent. The last column contains the weight percentage of the higher boiling constituent.

The pressure-composition for the system, Refrigerants 152a and 12, at 32 deg is shown in Fig. 8 and a portion of the Refrigerants 12 and 22 system is given in Fig. 9. Both of these systems are azeotropic. Note the great increase in pressure of the azeotrope in the first system over the higher pressure constituent. In the 12-22 system, the azeotrope has a pressure only slightly above that of Refrigerant 22. The maximum (azeotrope) in the 12-22 system comes at about 2.1% 12.

The pressure for the 12-22 system at 32 deg may be given by

$$p = 72.36x^a + 44.77 (1 - x)^b \qquad (17)$$

where p is pressure in psia and x is the mole fraction of Refrigerant 22, $a = 0.8593 - 0.0825x$ and $b = 0.9305 - 0.1308 (1 - x)$.

There is no inherent advantage or disadvantage of an azeotrope or any other mixture insofar as thermodynamic properties are concerned. There is a feeling among some mechanical engineers that a mixture, other than an azeotrope, requires more horsepower per ton just because it is a mixture. This is not true, as is shown in Fig. 10, where curves are given for various mixtures showing the results of calorimeter tests. The mixtures 21-22 and 114-22 were of such composition as to give the same pres-

Table 6. High Molecular Weight Refrigerants

Refrigerant Number	Formula	Molecular Weight	Boiling Point, Deg F
227	C_3HF_7	170	3.3
—	C_4F_{10}	238	28.9
12B2	CBr_2F_2	210	73.6
—	C_5F_{12}	288	84.9
319	C_4ClF_9	254.5	85.7
216	$C_3Cl_2F_6$	221	95

Table 7. Azeotropic Refrigerants

Combination	Boiling Point, Deg F			Percent Weight (1)
	Of (1)	Of (2)	Of Azeo.	
152-12	—11.4	—21.7	—28.0	26.2
12-22	—21.7	—41.4	—41.45	2
40-12	—10.8	—21.7	—25.6	22
115-22	—36.4	—41.4	—50	50
22-290	—41.4	—44.2	—49	68
21-114	48.0	38.4	34.5	25
218-22	—34.4	—41.4	—45	54
227-12	3.3	—21.7	—22	13.5
$(CH_3)_2O$-12	—10.5	—21.7	—22	10
152-115	—11.4	—36.4	—42	16

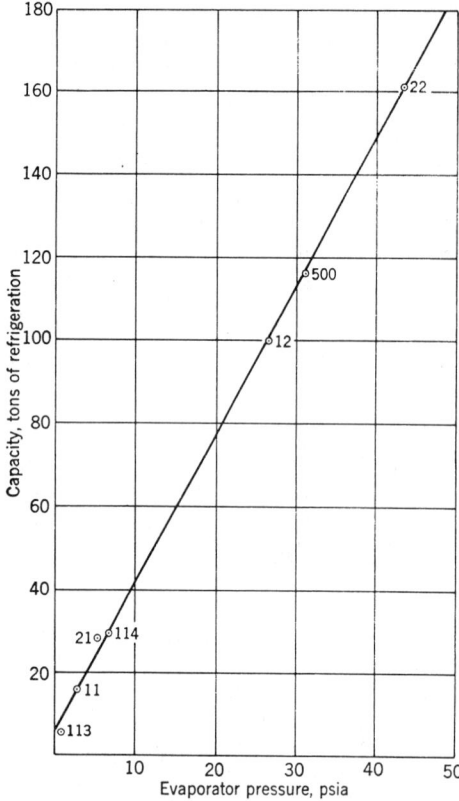

Fig. 7. Refrigeration capacity as related to evaporator pressure for some refrigerants.

sure at 32 deg as 12 has. Three points were found for each composition. The abscissa gives the Btu per min transferred from an electric heater to the boiling refrigerant. The ordinate is the rate of heat transfer per watt hour for the motor.

Obviously 21-22 is more efficient than 12 and 114-22 is less efficient. This is to be expected almost quantitatively from the following facts:

Refrigerant	Latent Heat Efficiency, Standard Ton
12	75.51
22	74.03
114	69.54
21	81.77

Since 22 and 12 have almost identical values for L_v and 114 has a lower value, a mixture of 22-114 would be expected to require more horsepower per ton, not because it is a mixture, but because it is 22-114. Refrigerant 21 has the best value for latent heat efficiency. Thus, it is not surprising that more heat can be transferred with the same work using

21-22; the horsepower per ton will be less for the mixture than where 12 is used.

Advantages of Nonazeotropic Mixtures

Already mixtures of 12 and 22 far from the azeotropic composition are being used as refrigerants primarily because the addition of 12 increases the oil solubility and makes it possible to get much better oil return to the compressor. One can use up to about 25% (by weight) of Refrigerant 12 without getting a great deal of change in capacity.

From time to time other mixtures may be employed for specific reasons. Food freezers have been operated with a mixture of 114-22 in a very satisfactory manner. The refrigerant at the beginning of the evaporator may be −40 deg and at the end near 0 deg while the motor is running. Once the motor is cut off by the thermostat, the liquid refrigerant will come from the high side into the evaporator, the last portion being as high as 50 deg. The heat in this "hot" refrigerant has not the slightest chance of doing any damage on the refrigerated food around it. The temperature of this part of the refrigerant is very quickly far below zero because of the enormous amount of sensible heat required to raise the temperature of the food. The average temperature at various spots in the cabinet were quite close to what they were where Refrigerant 12 was used. While such a mixture can be employed in a food freezer there seems to be no advantage whatsoever in using it.

Refrigerants for Absorption Refrigeration

Were there sufficient water at, say, 45 deg F, just under the surface of the earth, it could be used directly to give all the air conditioning that would be desired. There is no such supply. However, there is a considerable amount at such temperatures so that it can be employed to maintain temperatures as low as 110 deg or lower.

Here is where absorption refrigeration comes in. Water is used to keep a solution such as 61% lithium bromide in water at a temperature not exceeding 110 deg. This salt solution, in turn, keeps the refrigerant, boiling water, at a temperature near 40 deg which then can be pumped around a circuit to collect heat, thereby producing air conditioning. The lithium bromide solution and liquid water have the same vapor above them, namely, water vapor. A 61% LiBr solution has the same vapor pressure at 110 deg that liquid water has at 38 deg. Consequently, if chambers containing the two are in communication and held at the above mentioned temperatures, respectively, there will be no flow of

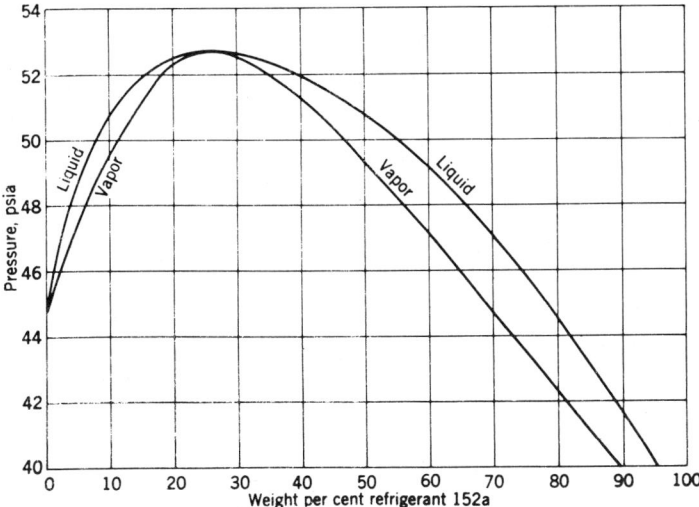

Fig. 8. Equilibrium pressure-composition diagram, Refrigerants 152a and 12 at 32 deg F.

vapor. If the water were at 40 deg its vapor pressure will be higher and vapor will flow into the salt solution making it more dilute. This is exactly what happens in the lithium bromide absorption air conditioning machine. The condensation of the vapor into the salt solution takes the place of the compressor. In a working machine there must be a driving force so that the 61% LiBr solution cannot cool water down to 38 deg except at zero load. At full load the water may be 10 deg higher.

The absorbing solution pecomes more dilute and will become ineffective unless it is concentrated. To accomplish this it is pumped into a generator where steam is used to boil off water concentrating the solution which is returned to the absorber. The water boiled off is condensed through the use of cooling water and thus the refrigerant is recovered; it is returned to the evaporator. The simple absorption cycle is illustrated in Fig. 11.

The absorption cycle referred to is in practical use in air conditioning in machines from some few up to perhaps 1000 tons capacity. This cycle can truly be appreciated when it is realized that it would require a reciprocating compressor with about 6400 cylinders 4 inches in diameter with a 6-inch stroke to give a 1000 tons if the crankshaft were rotating at 1760 revolutions per minute. Of course the cylinders might be much larger. Reciprocating compression is impractical in any event where water is the refrigerant. The absorption cycle makes it possible to use water, a very good refrigerant thermodynamically. The water vapor velocity in parts of the absorption machine is probably as much as 300 miles an hour.

Water of course, is also used as the refrigerant in a steam jet steam where water vapor from the evaporator is entrained in a fast moving jet of steam.

Ammonia is also used in absorption refrigeration. The gas-fired home refrigerator uses an ammonia-water solution as the absorbent and ammonia as the refrigerant. Ammonia has also been tested in a large pilot model where a concentrated solution of lithium nitrate in anhydrous ammonia is the absorbent and ammonia is the refrigerant. Technically it works, but may not prove to be economically feasible.

Among chemical systems which have been proposed for absorption systems are those listed in Table 8; the two systems just discussed are also included. *DETG* is used to designate dimethyl ether of tetramethylene glycol.

Table 8. Some Absorbent-Refrigerant Combinations

No.	Absorbent	Refrigerant	Refrigerant Boiling Point, Deg F
1	LiBr	H_2O	212
2	$(CH_3)_2SO$	H_2O	212
3	H_2O	NH_3	— 28
4	$LiNO_3$	NH_3	— 28
5	DETG	21	48
6	DETG	22	— 41.4

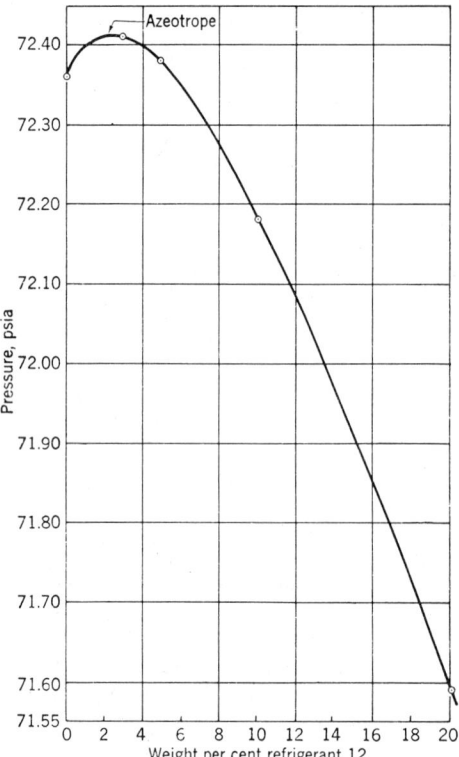

Fig. 9. Pressure-composition diagram, Refrigerants 12 and 22 at 32 deg F.

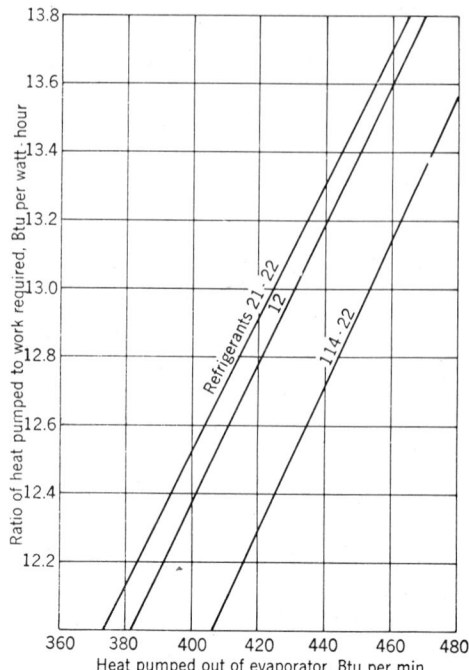

Fig. 10. Capacity tests on nonazeotropic mixtures.

The absorbing solution is actually a concentrated solution of the absorbent and the refrigerant.

Oil Solubility

All hydrocarbons and the halocarbons containing no hydrogen and much chlorine are completely miscible in the oils employed normally in refrigeration systems. This is a good characteristic.

When the motor stops in a refrigeration system the liquid refrigerant evaporates, over a time, and condenses in the oil in the basin in the housing containing the compressor. On the start-up, the oil-refrigerant mixture will foam and some of this goes over into the condenser and thence into the evaporator. If the oil and refrigerant are completely miscible at the low temperature, the oil will return and a separation will be made with the warm-up, leaving the oil where it belongs.

If there is considerable fluorine or fluorine with hydrogen and little or no chlorine in the molecule, the oil ordinarily used is not miscible to a large extent with the refrigerant. Refrigerant 13 ($CClF_3$)

and 14 (CF_4) are almost completely inmiscible; so is 218 (C_3F_8) and 152a (CH_3CHF_2). Intermediate compositions, such as Refrigerant 22 ($CHClF_2$) are completely miscible at room temperature, but have limited solubility at low temperature.

Figure 12 is the behavior of Refrigerant 22 with Suniso 3G oil. Below zero there are two phases in the oil-refrigerant mixture for certain compositions. At −10 deg F, 8% oil is all that is needed to form two liquids. This is why, in a low temperature system, oil separators are needed.

There has been a marked trend toward the use of Refrigerant 22, instead of 12, in the last several years because of its having a higher pressure at the

Table 9. Solubility (ppm) of Water in Refrigerants 12 and 22

Tem-perature. Deg F	Refrigerant 12		Refrigerant 22	
	Liquid	Vapor	Liquid	Vapor
− 40	1.7	29.1	120	24.4
0	8.3	110.4	308	92.7
40	32	323.2	690	268.4
80	98	—	1350	—
100	165	—	1800	—

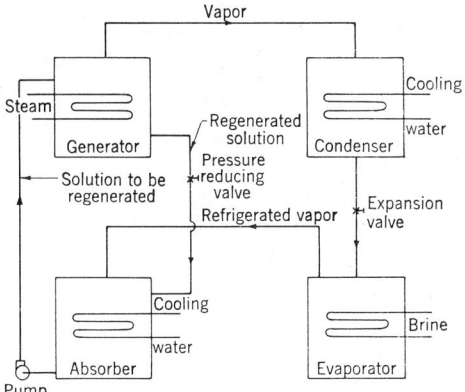

Fig. 11. Simple absorption refrigeration cycle.

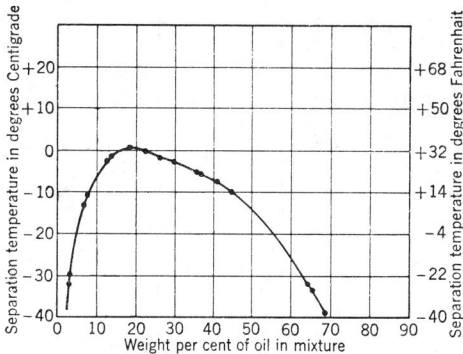

Fig. 12. Equilibrium diagram, Refrigerant 22 and
Suniso 3G oil (after Walker *et al*).

same temperatures. Many difficulties have arisen in trying to make the change and this is the primary reason for using the mixture of 12 and 22 containing 25% 12. The mixture should be used if a refrigerant capacity must be had in this range, but calling it an azeotrope, when it is not, does not enhance its properties.

Steinle has reported that secondary butyl polysilicate esters can be used as oils in Refrigerant 22 systems. It has a viscosity of 55 centistokes at 68 deg and is miscible in all proportions with 22 down to —108 deg (dry ice temperature).

Water Solubility

Moisture inside the circulating system has caused a great deal of difficulty in refrigeration systems. Great pains are taken to dehydrate the systems, particularly the hermetic units where the motor is on the inside of the refrigerant system. For years this dehydration was effected with vacuum at 275 deg F. In later years, some use an air sweep instead of vacuum. The latter method should be seriously questioned for low temperature systems because of the tendency to oxidize the small amount of oil put in for the short time initial run in test of the motor and compressor.

Table 9 shows the solubility of water in Refrigerants 12 and 22 in parts per million. Most other halocarbons used have solubilities not too different. Refrigerant 12 has more water in the vapor; Refrigerant 22, in the liquid. A pressure-composition diagram for 12 and water would be of the type shown in Fig. 13. A homoazeotrope exists in the system unless the solubility runs out before it is reached.

Refrigerant 12 is autodrying, meaning that the liquid becomes drier during distillation; 22 is auto-

wetting. Where moisture is determined by absorption in phosphorus pentoxide, it is not necessary to sweep out the sample cylinder, if it is made of proper materials, after the distillation.

For moisture analyses of Refrigerant 22, it is imperative to have a dry gas sweep or otherwise some of the moisture will be left in the cylinder and low values will be obtained.

One great advantage in being able to analyze a refrigerant for moisture accurately is to learn indirectly about a great many things in a refrigeration system. A refrigerant sample taken from a

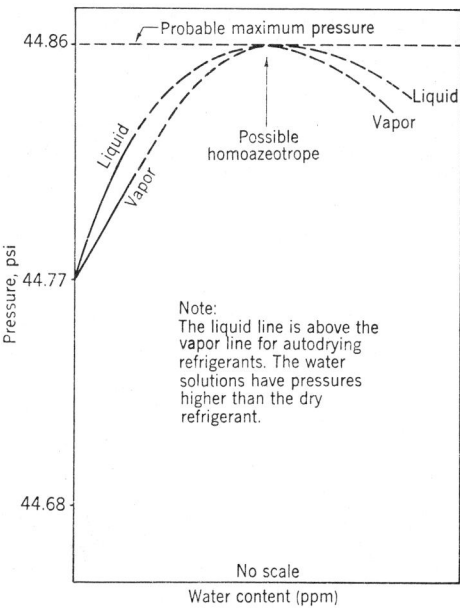

Fig. 13. Type pressure-composition diagram for
Refrigerant 12 and water, at 32 deg F.

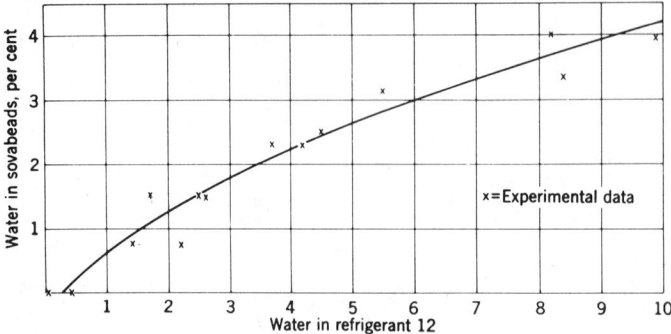

Fig. 14. Moisture equilibrium relation with drier at 100 deg F.

system containing a desiccant and which has been run for a few hours can be used to learn how much water is in the desiccant. Fig. 14 shows how much water is in the desiccant Sovabeads, a type of silica gel, where the temperature is 100 deg for various moisture contents in Refrigerant 12.

If one gets 3 ppm in a new unit just run in and then after a year of operation gets 9 ppm, he can know if there is 25 grams of the gel in the drier that roughly 0.6 grams of water have gone into the drier within the year. Did it come from improper dehydration or from a breakdown of insulation in the motor? The writer has run several units for a yearly period and found no difference in the moisture content of the refrigerant. Such results prove conclusively that the dehydration was perfect and there was no motor breakdown.

All low temperature systems ought to have a drier in them and, in general, it should be placed where the refrigerant has the lowest, not the highest, solubility for water. It will always be the low temperature side. Whether it should be at the beginning of the evaporator or in the vapor space beyond the evaporator depends on several factors, the solubility

of water in the different phases being among them. The data in Table 9 show conclusively that a drier at the beginning of the evaporator for a Refrigerant 12 system will have the greatest capacity. One must take the fouling with oil into consideration before putting the drier in the vapor space.

Data in Table 10 show the capacity of silica gel type driers at different locations and under different operating conditions in a Refrigerant 12 food freezer. The absorbing capacity of silica gel at the beginning of the evaporation is nine times as much as it is in the high side liquid at 130 deg.

Manufacturers guarantee less than 10 ppm in many refrigerants. This number is taken far too seriously. It is not the few parts per million entering the system that causes trouble, it is the milliliters that are not pumped out during evacuation because the valve to the system is never opened or it is the large amount that may leak in where the unit is being tested under water.

In a food freezer there is about 450 grams of refrigerants. If it contained 50 ppm, there would be only 0.02 grams water introduced in the refrigerant, a mere trifle. A 25-gram drier where the unit is operated as Condition No. 1 will handle 9 grams which is 450 times as much.

Air Solubility

Air or any other noncondensible present in a refrigerant can best be determined through the use of a reflux apparatus where the gas can be driven into a comparatively small tube above the condensing refrigerant.

Air solubility has been determined by a rather complicated method using extremely low temperature.

One can determine quite accurately by simply measuring the total pressure at a given temperature. Since the solubility is low, Henry's Law works as perfectly as it ever would for anything.

Table 10. Drier Capacity in a Refrigerant 12 Food Freezer

Con-dition No.	Temperature, Deg F			Relative Percent Liquid	Water in Gel. Percent
	Drier	Evaporator	Condenser		
1	— 10	—10	95	80	36
2	— 10	—10	95	60	26
3	0	—10	95	0	14
4	0	0	130	75	35.6
5	20	20	100	75	34
6	70	0	130	100	8
7	95	—10	95	100	8
8	130	0	130	100	4

From ASRE Circular 8.

PROPERTIES OF REFRIGERANT-12

Dichlorodifluoromethane (CCl_2F_2), a halogen derivative of methane, is a colorless gas, heavier than air (specific gravity = 4.26). At ordinary temperatures it is a liquid when under a pressure of 75 lb per sq in. gage.

The extended table of thermodynamic properties of saturated Refrigerant-12 which follows was copyrighted by the makers in 1955 and 1956. The original tables were issued in 1931 and added to in 1942 and subsequently. The new tables are more consistent, based on additional experimental measurements, and calculated from equations which apply throughout the whole range of temperature, density, and pressure. The complete tables cover the range from −152 F to 233.6 F; the data which follow begin at −40 deg.

Characteristics of Refrigerant-12, based on the new data, including the characteristics at an evaporator temperature of 5 F and a condenser temperature of 86 F, the conventional temperatures of the standard ton, are given in the table below. Saturation properties of Refrigerant-12 follow on the next page.

Characteristics of Refrigerant-12

Compression ratio, 86 deg/5 deg	4.08
Net refrigerating effect, Btu per pound	50.035
Refrigerant circulated per ton, pound per minute	3.9972
Refrigerant liquid circulated per ton, cubic inches per minute	85.621
Compressor displacement per ton, cubic feet per minute	5.8279
Refrigeration per cubic foot of compressor displacement, Btu	34.318
Heat of compression, Btu per pound	10.636
Temperature of compressor discharge, deg F	100.84
Coefficient of performance	4.704
Horsepower per ton	1.0023
Critical pressure, pound per square inch absolute	596.9
Critical temperature, deg F	233.6
Critical volume, cubic feet per pound	0.0287
Molecular weight	120.93
Melting point, deg F	−252
Boiling point, deg F	−21.62
Flammability	Nonflammable and Noncombustible
Toxicity	Very low—Group 6

	At 86 deg F	At 5 deg F
Pressure, pound per square inch absolute	108.04	26.483
Latent heat of vaporization, Btu per pound	58.052	68.204
Net refrigerating effect, Btu per pound	50.035	51.07
Specific volume, liquid, cubic feet per pound	0.0124	0.01109
Specific volume, vapor, cubic feet per pound	0.3766	1.4580
Heat of liquid, Btu per pound	27.769	9.6005
Heat in vapor, Btu per pound	85.821	77.805
Entropy, liquid, Btu per pound per degree	0.05730	0.02165
Saturated density of vapor, pound per cubic foot	2.6556	0.68588
Specific heat, liquid	0.235	0.216
Specific heat, vapor (C_p), 1 atmosphere	0.1482	0.1398
Specific heat, vapor (C_v), 1 atmosphere	0.1302	0.1202
Specific gravity, liquid (water = 1)	1.292	1.442

SATURATION PROPERTIES OF REFRIGERANT-12

Tem-pera-ture, Deg. F.	Pressure, Lb. per Sq. In.		Volume, Cu. Ft. per Lb.		Density, Lb. per Cu. Ft.		Enthalpy, Btu per Lb.			Entropy, Btu per Lb. per Deg. Abs.	
Deg. F.	Abs.	Gage	Liq.	Vapor	Liq.	Vapor	Liq.	Latent	Vapor	Liq.	Vapor
−40	9.3076	10.9709*	0.010564	3.8750	94.661	0.25806	0	72.913	72.913	0	0.17373
−39	9.5530	10.4712*	0.010575	3.7823	94.565	0.26439	0.2107	72.812	73.023	0.000500	0.17357
−38	9.8035	9.9611*	0.010586	3.6922	94.469	0.27084	0.4215	72.712	73.134	0.001000	0.17343
−37	10.059	9.441*	0.010596	3.6047	94.372	0.27741	0.6324	72.611	73.243	0.001498	0.17328
−36	10.320	8.909*	0.010607	3.5198	94.275	0.28411	0.8434	72.511	73.354	0.001995	0.17313
−35	10.586	8.367*	0.010618	3.4373	94.178	0.29093	1.0546	72.409	73.464	0.002492	0.17299
−34	10.858	7.814*	0.010629	3.3571	94.081	0.29788	1.2659	72.309	73.575	0.002988	0.17285
−33	11.135	7.250*	0.010640	3.2792	93.983	0.30495	1.4772	72.208	73.685	0.003482	0.17271
−32	11.417	6.675*	0.010651	3.2035	93.886	0.31216	1.6887	72.106	73.795	0.003976	0.17257
−31	11.706	6.088*	0.010662	3.1300	93.788	0.31949	1.9003	72.004	73.904	0.004469	0.17243
−30	11.999	5.490*	0.010674	3.0585	93.690	0.32696	2.1120	71.903	74.015	0.004961	0.17229
−29	12.299	4.880*	0.010685	2.9890	93.592	0.33457	2.3239	71.801	74.125	0.005452	0.17216
−28	12.604	4.259*	0.010696	2.9214	93.493	0.34231	2.5358	71.698	74.234	0.005942	0.17203
−27	12.916	3.625*	0.010707	2.8556	93.395	0.35018	2.7479	71.596	74.344	0.006431	0.17189
−26	13.233	2.979*	0.010719	2.7917	93.296	0.35820	2.9601	71.494	74.454	0.006919	0.17177
−25	13.556	2.320*	0.010730	2.7295	93.197	0.36636	3.1724	71.391	74.563	0.007407	0.17164
−24	13.886	1.649*	0.010741	2.6691	93.098	0.37466	3.3848	71.288	74.673	0.007894	0.17151
−23	14.222	0.966*	0.010753	2.6102	92.999	0.38311	3.5973	71.185	74.782	0.008379	0.17139
−22	14.564	0.270*	0.010764	2.5529	92.899	0.39171	3.8100	71.081	74.891	0.008864	0.17126
−21	14.912	0.216	0.010776	2.4972	92.799	0.40045	4.0228	70.978	75.001	0.009348	0.17114
−20	15.267	0.571	0.010788	2.4429	92.699	0.40934	4.2357	70.874	75.110	0.009831	0.17102
−19	15.628	0.932	0.010799	2.3901	92.599	0.41839	4.4487	70.770	75.219	0.010314	0.17090
−18	15.996	1.300	0.010811	2.3387	92.499	0.42758	4.6618	70.666	75.328	0.010795	0.17078
−17	16.371	1.675	0.010823	2.2886	92.399	0.43694	4.8751	70.561	75.436	0.011276	0.17066
−16	16.753	2.057	0.010834	2.2399	92.298	0.44645	5.0885	70.456	75.545	0.011755	0.17055
−15	17.141	2.445	0.010846	2.1924	92.197	0.45612	5.3020	70.352	75.654	0.012234	0.17043
−14	17.536	2.840	0.010858	2.1461	92.096	0.46595	5.5157	70.246	75.762	0.012712	0.17032
−13	17.939	3.243	0.010870	2.1011	91.995	0.47595	5.7295	70.141	75.871	0.013190	0.17021
−12	18.348	3.652	0.010882	2.0572	91.893	0.48611	5.9434	70.036	75.979	0.013666	0.17010
−11	18.765	4.069	0.010894	2.0144	91.791	0.49643	6.1574	69.930	76.087	0.014142	0.16999
−10	19.189	4.493	0.010906	1.9727	91.689	0.50693	6.3716	69.824	76.196	0.014617	0.16989
− 9	19.621	4.925	0.010919	1.9320	91.587	0.51759	6.5859	69.718	76.304	0.015091	0.16978
− 8	20.059	5.363	0.010931	1.8924	91.485	0.52843	6.8003	69.611	76.411	0.015564	0.16967
− 7	20.506	5.810	0.010943	1.8538	91.382	0.53944	7.0149	69.505	76.520	0.016037	0.16957
− 6	20.960	6.264	0.010955	1.8161	91.280	0.55063	7.2296	69.397	76.627	0.016508	0.16947
− 5	21.422	6.726	0.010968	1.7794	91.177	0.56199	7.4444	69.291	76.735	0.016979	0.16937
− 4	21.891	7.195	0.010980	1.7436	91.074	0.57354	7.6594	69.183	76.842	0.017449	0.16927
− 3	22.369	7.673	0.010993	1.7086	90.970	0.58526	7.8745	69.075	76.950	0.017919	0.16917
− 2	22.854	8.158	0.011005	1.6745	90.867	0.59718	8.0898	68.967	77.057	0.018388	0.16907
− 1	23.348	8.652	0.011018	1.6413	90.763	0.60927	8.3052	68.859	77.164	0.018855	0.16897
0	23.849	9.153	0.011030	1.6089	90.659	0.62156	8.5207	68.750	77.271	0.019323	0.16888
1	24.359	9.663	0.011043	1.5772	90.554	0.63404	8.7364	68.642	77.378	0.019789	0.16878
2	24.878	10.182	0.011056	1.5463	90.450	0.64670	8.9522	68.533	77.485	0.020255	0.16869
3	25.404	10.708	0.011069	1.5161	90.345	0.65957	9.1682	68.424	77.592	0.020719	0.16860
4	25.939	11.243	0.011082	1.4867	90.240	0.67263	9.3843	68.314	77.698	0.021184	0.16851
5	26.483	11.787	0.011094	1.4580	90.135	0.68588	9.6005	68.204	77.805	0.021647	0.16842
6	27.036	12.340	0.011107	1.4299	90.030	0.69934	9.8169	68.094	77.911	0.022110	0.16833
7	27.597	12.901	0.011121	1.4025	89.924	0.71300	10.033	67.984	78.017	0.022572	0.16824
8	28.167	13.471	0.011134	1.3758	89.818	0.72687	10.250	67.873	78.123	0.023033	0.16815
9	28.747	14.051	0.011147	1.3496	89.712	0.74094	10.467	67.762	78.229	0.023494	0.16807
10	29.335	14.639	0.011160	1.3241	89.606	0.75523	10.684	67.651	78.335	0.023954	0.16798
11	29.932	15.236	0.011173	1.2992	89.499	0.76972	10.901	67.539	78.440	0.024413	0.16790
12	30.539	15.843	0.011187	1.2748	89.392	0.78443	11.118	67.428	78.546	0.024871	0.16782
13	31.155	16.459	0.011200	1.2510	89.285	0.79935	11.336	67.315	78.651	0.025329	0.16774
14	31.780	17.084	0.011214	1.2278	89.178	0.81449	11.554	67.203	78.757	0.025786	0.16765
15	32.415	17.719	0.011227	1.2050	89.070	0.82986	11.771	67.090	78.861	0.026243	0.16758

*Inches of mercury below one atmosphere

Saturation Properties of Refrigerant-12 (continued)

Temperature, Deg. F.	Pressure, Lb. per Sq. In.		Volume, Cu. Ft. per Lb.		Density, Lb. per Cu. Ft.		Enthalpy, Btu per Lb.			Entropy, Btu per Lb. per Deg. Abs.	
	Abs.	Gage	Liq.	Vapor	Liq.	Vapor	Liq.	Latent	Vapor	Liq.	Vapor
15	32.415	17.719	0.011227	1.2050	89.070	0.82986	11.771	67.090	78.861	0.026243	0.16758
16	33.060	18.364	0.011241	1.1828	88.962	0.84544	11.989	66.977	78.966	0.026699	0.16750
17	33.714	19.018	0.011254	1.1611	88.854	0.86125	12.207	66.864	79.071	0.027154	0.16742
18	34.378	19.682	0.011268	1.1399	88.746	0.87729	12.426	66.750	79.176	0.027608	0.16734
19	35.052	20.356	0.011282	1.1191	88.637	0.89356	12.644	66.636	79.280	0.028062	0.16727
20	35.736	21.040	0.011296	1.0988	88.529	0.91006	12.863	66.522	79.385	0.028515	0.16719
21	36.430	21.734	0.011310	1.0790	88.419	0.92679	13.081	66.407	79.488	0.028968	0.16712
22	37.135	22.439	0.011324	1.0596	88.310	0.94377	13.300	66.293	79.593	0.029420	0.16704
23	37.849	23.153	0.011338	1.0406	88.201	0.96098	13.520	66.177	79.697	0.029871	0.16697
24	38.574	23.878	0.011352	1.0220	88.091	0.97843	13.739	66.061	79.800	0.030322	0.16690
25	39.310	24.614	0.011366	1.0039	87.981	0.99613	13.958	65.946	79.904	0.030772	0.16683
26	40.056	25.360	0.011380	0.98612	87.870	1.0141	14.178	65.829	80.007	0.031221	0.16676
27	40.813	26.117	0.011395	0.96874	87.760	1.0323	14.398	65.713	80.111	0.031670	0.16669
28	41.580	26.884	0.011409	0.95173	87.649	1.0507	14.618	65.596	80.214	0.032118	0.16662
29	42.359	27.663	0.011424	0.93509	87.537	1.0694	14.838	65.478	80.316	0.032566	0.16655
30	43.148	28.452	0.011438	0.91880	87.426	1.0884	15.058	65.361	80.419	0.033013	0.16648
31	43.948	29.252	0.011453	0.90286	87.314	1.1076	15.279	65.243	80.522	0.033460	0.16642
32	44.760	30.064	0.011468	0.88725	87.202	1.1271	15.500	65.124	80.624	0.033905	0.16635
33	45.583	30.887	0.011482	0.87197	87.090	1.1468	15.720	65.006	80.726	0.034351	0.16629
34	46.417	31.721	0.011497	0.85702	86.977	1.1668	15.942	64.886	80.828	0.034796	0.16622
35	47.263	32.567	0.011512	0.84237	86.865	1.1871	16.163	64.767	80.930	0.035240	0.16616
36	48.120	33.424	0.011527	0.82803	86.751	1.2077	16.384	64.647	81.031	0.035683	0.16610
37	48.989	34.293	0.011542	0.81399	86.638	1.2285	16.606	64.527	81.133	0.036126	0.16604
38	49.870	35.174	0.011557	0.80023	86.524	1.2496	16.828	64.406	81.234	0.036569	0.16598
39	50.763	36.067	0.011573	0.78676	86.410	1.2710	17.050	64.285	81.335	0.037011	0.16592
40	51.667	36.971	0.011588	0.77357	86.296	1.2927	17.273	64.163	81.436	0.037453	0.16586
41	52.584	37.888	0.011603	0.76064	86.181	1.3147	17.495	64.042	81.537	0.037893	0.16580
42	53.513	38.817	0.011619	0.74798	86.066	1 3369	17.718	63.919	81.637	0.038334	0.16574
43	54.454	39.758	0.011635	0.73557	85.951	1.3595	17.941	63.796	81.737	0.038774	0.16568
44	55.407	40.711	0.011650	0.72341	85.836	1.3823	18.164	63.673	81.837	0.039213	0.16562
45	56.373	41.677	0.011666	0.71149	85.720	1.4055	18.387	63.550	81.937	0.039652	0.16557
46	57.352	42.656	0.011682	0.69982	85.604	1.4289	18.611	63.426	82.037	0.040091	0.16551
47	58.343	43.647	0.011698	0.68837	85.487	1.4527	18.835	63.301	82.136	0.040529	0.16546
48	59.347	44.651	0.011714	0.67715	85.371	1.4768	19.059	63.177	82.236	0.040966	0.16540
49	60.364	45.668	0.011730	0.66616	85.254	1.5012	19.283	63.051	82.334	0.041403	0.16535
50	61.394	46.698	0.011746	0.65537	85.136	1.5258	19.507	62.926	82.433	0.041839	0.16530
51	62.437	47.741	0.011762	0.64480	85.018	1.5509	19.732	62.800	82.532	0.042276	0.16524
52	63.494	48.798	0.011779	0.63444	84.900	1.5762	19.957	62.673	82.630	0.042711	0.16519
53	64.563	49.867	0.011795	0.62428	84.782	1.6019	20.182	62.546	82.728	0.043146	0.16514
54	65.646	50.950	0.011811	0.61431	84.663	1.6278	20.408	62.418	82.826	0.043581	0.16509
55	66.743	52.047	0.011828	0.60453	84.544	1.6542	20.634	62.290	82.924	0.044015	0.16504
56	67.853	53.157	0.011845	0.59495	84.425	1.6808	20.859	62.162	83.021	0.044449	0.16499
57	68.977	54.281	0.011862	0.58554	84.305	1.7078	21.086	62.033	83.119	0.044883	0.16494
58	70.115	55.419	0.011879	0.57632	84.185	1.7352	21.312	61.903	83.215	0.045316	0.16489
59	71.267	56.571	0.011896	0.56727	84.065	1.7628	21.539	61.773	83.312	0.045748	0.16484
60	72.433	57.737	0.011913	0.55839	83.944	1.7909	21.766	61.643	83.409	0.046180	0.16479
61	73.613	58.917	0.011930	0.54967	83.823	1.8193	21.993	61.512	83.505	0.046612	0.16474
62	74.807	60.111	0.011947	0.54112	83.701	1.8480	22.221	61.380	83.601	0.047044	0.16470
63	76.016	61.320	0.011965	0.53273	83.580	1.8771	22.448	61.248	83.696	0.047475	0.16465
64	77.239	62.543	0.011982	0.52450	83.457	1.9066	22.676	61.116	83.792	0.047905	0.16460
65	78.477	63.781	0.012000	0.51642	83.335	1.9364	22.905	60.982	83.887	0.048336	0.16456
66	79.729	65.033	0.012017	0.50848	83.212	1.9666	23.133	60.849	83.982	0.048765	0.16451
67	80.996	66.300	0.012035	0.50070	83.089	1.9972	23.362	60.715	84.077	0.049195	0.16447
68	82.279	67.583	0.012053	0.49305	82.965	2.0282	23.591	60.580	84.171	0.049624	0.16442
69	83.576	68.880	0.012071	0.48555	82.841	2.0595	23.821	60.445	84.266	0.050053	0.16438
70	84.888	70.192	0.012089	0.47818	82.717	2.0913	24.050	60.309	84.359	0.050482	0.16434

Saturation Properties of Refrigerant-12 (continued)

Tem-perature, Deg. F.	Pressure, Lb. per Sq. In.		Volume, Cu. Ft. per Lb.		Density, Lb. per Cu. Ft.		Enthalpy, Btu per Lb.			Entropy, Btu per Lb. per Deg. Abs.	
	Abs.	Gage	Liq.	Vapor	Liq.	Vapor	Liq.	Latent	Vapor	Liq.	Vapor
70	84.888	70.192	0.012089	0.47818	82.717	2.0913	24.050	60.309	84.359	0.050482	0.16434
71	86.216	71.520	0.012108	0.47094	82.592	2.1234	24.281	60.172	84.453	0.050910	0.16429
72	87.559	72.863	0.012126	0.46383	82.467	2.1559	24.511	60.035	84.546	0.051338	0.16425
73	88.918	74.222	0.012145	0.45686	82.341	2.1889	24.741	59.898	84.639	0.051766	0.16421
74	90.292	75.596	0.012163	0.45000	82.215	2.2222	24.973	59.759	84.732	0.052193	0.16417
75	91.682	76.986	0.012182	0.44327	82.089	2.2560	25.204	59.621	84.825	0.052620	0.16412
76	93.087	78.391	0.012201	0.43666	81.962	2.2901	25.435	59.481	84.916	0.053047	0.16408
77	94.509	79.813	0.012220	0.43016	81.835	2.3247	25.667	59.341	85.008	0.053473	0.16404
78	95.946	81.250	0.012239	0.42378	81.707	2.3597	25.899	59.201	85.100	0.053900	0.16400
79	97.400	82.704	0.012258	0.41751	81.579	2.3951	26.132	59.059	85.191	0.054326	0.16396
80	98.870	84.174	0.012277	0.41135	81.450	2.4310	26.365	58.917	85.282	0.054751	0.16392
81	100.36	85.66	0.012297	0.40530	81.322	2.4673	26.598	58.775	85.373	0.055177	0.16388
82	101.86	87.16	0.012316	0.39935	81.192	2.5041	26.832	58.631	85.463	0.055602	0.16384
83	103.38	88.68	0.012336	0.39351	81.063	2.5413	27.065	58.488	85.553	0.056027	0.16380
84	104.92	90.22	0.012356	0.38776	80.932	2.5789	27.300	58.343	85.643	0.056452	0.16376
85	106.47	91.77	0.012376	0.38212	80.802	2.6170	27.534	58.198	85.732	0.056877	0.16372
86	108.04	93.34	0.012396	0.37657	80.671	2.6556	27.769	58.052	85.821	0.057301	0.16368
87	109.63	94.93	0.012416	0.37111	80.539	2.6946	28.005	57.905	85.910	0.057725	0.16364
88	111.23	96.53	0.012437	0.36575	80.407	2.7341	28.241	57.757	85.998	0.058149	0.16360
89	112.85	98.15	0.012457	0.36047	80.275	2.7741	28.477	57.609	86.086	0.058573	0.16357
90	114.49	99.79	0.012478	0.35529	80.142	2.8146	28.713	57.461	86.174	0.058997	0.16353
91	116.15	101.45	0.012499	0.35019	80.008	2.8556	28.950	57.311	86.261	0.059420	0.16349
92	117.82	103.12	0.012520	0.34518	79.874	2.8970	29.187	57.161	86.348	0.059844	0.16345
93	119.51	104.81	0.012541	0.34025	79.740	2.9390	29.425	57.009	86.434	0.060267	0.16341
94	121.22	106.52	0.012562	0.33540	79.605	2.9815	29.663	56.858	86.521	0.060690	0.16338
95	122.95	108.25	0.012583	0.33063	79.470	3.0245	29.901	56.705	86.606	0.061113	0.16334
96	124.70	110.00	0.012605	0.32594	79.334	3.0680	30.140	56.551	86.691	0.061536	0.16330
97	126.46	111.76	0.012627	0.32133	79.198	3.1120	30.380	56.397	86.777	0.061959	0.16326
98	128.24	113.54	0.012649	0.31679	79.061	3.1566	30.619	56.242	86.861	0.062381	0.16323
99	130.04	115.34	0.012671	0.31233	78.923	3.2017	30.859	56.086	86.945	0.062804	0.16319
100	131.86	117.16	0.012693	0.30794	78.785	3.2474	31.100	55.929	87.029	0.063227	0.16315
101	133.70	119.00	0.012715	0.30362	78.647	3.2936	31.341	55.772	87.113	0.063649	0.16312
102	135.56	120.86	0.012738	0.29937	78.508	3.3404	31.583	55.613	87.196	0.064072	0.16308
103	137.44	122.74	0.012760	0.29518	78.368	3.3877	31.824	55.454	87.278	0.064494	0.16304
104	139.33	124.63	0.012783	0.29106	78.228	3.4357	32.067	55.293	87.360	0.064916	0.16301
105	141.25	126.55	0.012806	0.28701	78.088	3.4842	32.310	55.132	87.442	0.065339	0.16297
106	143.18	128.48	0.012829	0.28303	77.946	3.5333	32.553	54.970	87.523	0.065761	0.16293
107	145.13	130.43	0.012853	0.27910	77.804	3.5829	32.797	54.807	87.604	0.066184	0.16290
108	147.11	132.41	0.012876	0.27524	77.662	3.6332	33.041	54.643	87.684	0.066606	0.16286
109	149.10	134.40	0.012900	0.27143	77.519	3.6841	33.286	54.478	87.764	0.067028	0.16282
110	151.11	136.41	0.012924	0.26769	77.376	3.7357	33.531	54.313	87.844	0.067451	0.16279
111	153.14	138.44	0.012948	0.26400	77.231	3.7878	33.777	54.146	87.923	0.067873	0.16275
112	155.19	140.49	0.012972	0.26037	77.087	3.8406	34.023	53.978	88.001	0.068296	0.16271
113	157.27	142.57	0.012997	0.25680	76.941	3.8941	34.270	53.809	88.079	0.068719	0.16268
114	159.36	144.66	0.013022	0.25328	76.795	3.9482	34.517	53.639	88.156	0.069141	0.16264
115	161.47	146.77	0.013047	0.24982	76.649	4.0029	34.765	53.468	88.233	0.069564	0.16260
116	163.61	148.91	0.013072	0.24641	76.501	4.0584	35.014	53.296	88.310	0.069987	0.16256
117	165.76	151.06	0.013097	0.24304	76.353	4.1145	35.263	53.123	88.386	0.070410	0.16253
118	167.94	153.24	0.013123	0.23974	76.205	4.1713	35.512	52.949	88.461	0.070833	0.16249
119	170.13	155.43	0.013148	0.23647	76.056	4.2288	35.762	52.774	88.536	0.071257	0.16245
120	172.35	157.65	0.013174	0.23326	75.906	4.2870	36.013	52.597	88.610	0.071680	0.16241
121	174.59	159.89	0.013200	0.23010	75.755	4.3459	36.264	52.420	88.684	0.072104	0.16237
122	176.85	162.15	0.013227	0.22698	75.604	4.4056	36.516	52.241	88.757	0.072528	0.16234
123	179.13	164.43	0.013254	0.22391	75.452	4.4660	36.768	52.062	88.830	0.072952	0.16230
124	181.43	166.73	0.013280	0.22089	75.299	4.5272	37.021	51.881	88.902	0.073376	0.16226
125	183.76	169.06	0.013308	0.21791	75.145	4.5891	37.275	51.698	88.973	0.073800	0.16222

Saturation Properties of Refrigerant-12 (continued)

Temperature, Deg. F.	Pressure, Lb. per Sq. In.		Volume, Cu. Ft. per Lb.		Density, Lb. per Cu. Ft.		Enthalpy, Btu per Lb.			Entropy, Btu per Lb. per Deg. Abs.	
	Abs.	Gage	Liq.	Vapor	Liq.	Vapor	Liq.	Latent	Vapor	Liq.	Vapor
125	183.76	169.06	0.013308	0.21791	75.145	4.5891	37.275	51.698	88.973	0.073800	0.16222
126	186.10	171.40	0.013335	0.21497	74.991	4.6518	37.529	51.515	89.044	0.074225	0.16218
127	188.47	173.77	0.013363	0.21207	74.836	4.7153	37.785	51.330	89.115	0.074650	0.16214
128	190.86	176.16	0.013390	0.20922	74.680	4.7796	38.040	51.144	89.184	0.075075	0.16210
129	193.27	178,57	0.013419	0.20641	74.524	4.8448	38.296	50.957	89.253	0.075501	0.16206
130	195.71	181.01	0.013447	0.20364	74.367	4.9107	38.553	50.768	89.321	0.075927	0.16202
131	198.16	183.46	0.013476	0.20091	74.209	4.9775	38.811	50.578	89.389	0.076353	0.16198
132	200.64	185.94	0.013504	0.19821	74.050	5.0451	39.069	50.387	89.456	0.076779	0.16194
133	203.15	188.45	0.013534	0.19556	73.890	5.1136	39.328	50.194	89.522	0.077206	0.16189
134	205.67	190.97	0.013563	0.19294	73.729	5.1829	39.588	50.000	89.588	0.077633	0.16185
135	208.22	193.52	0.013593	0.19036	73.568	5.2532	39.848	49.805	89.653	0.078061	0.16181
136	210.79	196.09	0.013623	0.18782	73.406	5.3244	40.110	49.608	89.718	0.078489	0.16177
137	213.39	198.69	0.013653	0.18531	73.243	5.3965	40.372	49.409	89.781	0.078917	0.16172
138	216.01	201.31	0.013684	0.18283	73.079	5.4695	40.634	49.210	89.844	0.079346	0.16168
139	218.65	203.95	0.013715	0.18039	72.914	5.5435	40.898	49.008	89.906	0.079775	0.16163
140	221.32	206.62	0.013746	0.17799	72.748	5.6184	41.162	48.805	89.967	0.080205	0.16159
141	224.00	209.30	0.013778	0.17561	72.581	5.6944	41.427	48.601	90.028	0.080635	0.16154
142	226.72	212.02	0.013810	0.17327	72.413	5.7713	41.693	48.394	90.087	0.081065	0.16150
143	229.46	214.76	0.013842	0.17096	72.244	5.8493	41.959	48.187	90.146	0.081497	0.16145
144	232.22	217.52	0.013874	0.16868	72.075	5.9283	42.227	47.977	90.204	0.081928	0.16140
145	235.00	220.30	0.013907	0.16644	71.904	6.0083	42.495	47.766	90.261	0.082361	0.16135
146	237.82	223.12	0.013941	0.16422	71.732	6.0895	42.765	47.553	90.318	0.082794	0.16130
147	240.65	225.95	0.013974	0.16203	71.559	6.1717	43.035	47.338	90.373	0.083227	0.16125
148	243.51	228.81	0.014008	0.15987	71.386	6.2551	43.306	47.122	90.428	0.083661	0.16120
149	246.40	231.70	0.014043	0.15774	71.211	6.3395	43.578	46.904	90.482	0.084096	0.16115
150	249.31	234.61	0.014078	0.15564	71.035	6.4252	43.850	46.684	90.534	0.084531	0.16110
151	252.24	237.54	0.014113	0.15356	70.857	6.5120	44.124	46.462	90.586	0.084967	0.16105
152	255.20	240.50	0.014148	0.15151	70.679	6.6001	44.399	46.238	90.637	0.085404	0.16099
153	258.19	243.49	0.014184	0.14949	70.500	6.6893	44.675	46.012	90.687	0.085842	0.16094
154	261.20	246.50	0.014221	0.14750	70.319	6.7799	44.951	45.784	90.735	0.086280	0.16088
155	264.24	249.54	0.014258	0.14552	70.137	6.8717	45.229	45.554	90.783	0.086719	0.16083
156	267.30	252.60	0.014295	0.14358	69.954	6.9648	45.508	45.322	90.830	0.087159	0.16077
157	270.39	255.69	0.014333	0.14166	69.770	7.0592	45.787	45.088	90.875	0.087600	0.16071
158	273.51	258.81	0.014371	0.13976	69.584	7.1551	46.068	44.852	90.920	0.088041	0.16065
159	276.65	261.95	0.014410	0.13789	69.397	7.2523	46.350	44.614	90.964	0.088484	0.16059
160	279.82	265.12	0.014449	0.13604	69.209	7.3509	46.633	44.373	91.006	0.088927	0.16053
161	283.02	268.32	0.014489	0.13421	69.019	7.4510	46.917	44.130	91.047	0.089371	0.16047
162	286.24	271.54	0.014529	0.13241	68.828	7.5525	47.202	43.885	91.087	0.089817	0.16040
163	289.49	274.79	0.014570	0.13062	68.635	7.6556	47.489	43.637	91.126	0.090263	0.16034
164	292.77	278.07	0.014611	0.12886	68.441	7.7602	47.777	43.386	91.163	0.090710	0.16027
165	296.07	281.37	0.014653	0.12712	68.245	7.8665	48.065	43.134	91.199	0.091159	0.16021
166	299.40	284.70	0.014695	0.12540	68.048	7.9743	48.355	42.879	91.234	0.091608	0.16014
167	302.76	288.06	0.014738	0.12370	67.850	8.0838	48.647	42.620	91.267	0.092059	0.16007
168	306.15	291.45	0.014782	0.12202	67.649	8.1950	48.939	42.360	91.299	0.092511	0.16000
169	309.56	294.86	0.014826	0.12037	67.447	8.3080	49.233	42.097	91.330	0.092964	0.15992
170	313.00	298.30	0.014871	0.11873	67.244	8.4228	49.529	41.830	91.359	0.093418	0.15985
171	316.47	301.77	0.014917	0.11710	67.038	8.5394	49.825	41.562	91.387	0.093874	0.15977
172	319.97	305.27	0.014963	0.11550	66.831	8.6579	50.123	41.290	91.413	0.094330	0.15969
173	323.50	308.80	0.015010	0.11392	66.622	8.7783	50.423	41.015	91.438	0.094789	0.15961
174	327.06	312.36	0.015058	0.11235	66.411	8.9007	50.724	40.736	91.460	0.095246	0.15953
175	330.64	315.94	0.015106	0.11080	66.198	9.0252	51.026	40.455	91.481	0.095709	0.15945
176	334.25	319.55	0.015155	0.10927	65.983	9.1518	51.330	40.171	91.501	0.096172	0.15936
177	337.90	323.20	0.015205	0.10775	65.766	9.2805	51.636	39.883	91.519	0.096636	0.15928
178	341.57	326.87	0.015256	0.10625	65.547	9.4114	51.943	39.592	91.535	0.097102	0.15919
179	345.27	330.57	0.015308	0.10477	65.326	9.5446	52.252	39.297	91.549	0.097569	0.15910
180	349.00	334.30	0.015360	0.10330	65.102	9.6802	52.562	38.999	91.561	0.098039	0.15900

Saturation Properties of Refrigerant-12 (continued)

Temperature, Deg. F.	Pressure, Lb. per Sq. In.		Volume, Cu. Ft. per Lb.		Density, Lb. per Cu. Ft.		Enthalpy, Btu per Lb.			Entropy, Btu per Lb. per Deg. Abs.	
	Abs.	Gage	Liq.	Vapor	Liq.	Vapor	Liq.	Latent	Vapor	Liq.	Vapor
180	349.00	334.30	0.015360	0.10330	65.102	9.6802	52.562	38.999	91.561	0.098039	0.15900
181	352.76	338.06	0.015414	0.10185	64.877	9.8182	52.874	38.697	91.571	0.098509	0.15891
182	356.55	341.85	0.015468	0.10041	64.649	9.9587	53.188	38.391	91.579	0.098982	0.15881
183	360.38	345.68	0.015524	0.098992	64.418	10.102	53.504	38.081	91.585	0.099457	0.15871
184	364.23	349.53	0.015580	0.097584	64.185	10.248	53.822	37.767	91.589	0.099933	0.15861
185	368.11	353.41	0.015637	0.096190	63.949	10.396	54.141	37.449	91.590	0.10041	0.15850
186	372.02	357.32	0.015696	0.094810	63.711	10.547	54.463	37.127	91.590	0.10089	0.15839
187	375.96	361.26	0.015756	0.093443	63.470	10.702	54.786	36.800	91.586	0.10138	0.15828
188	379.94	365.24	0.015816	0.092089	63.225	10.859	55.111	36.469	91.580	0.10186	0.15817
189	383.94	369.24	0.015878	0.090747	62.978	11.020	55.439	36.133	91.572	0.10235	0.15805
190	387.98	373.28	0.015942	0.089418	62.728	11.183	55.769	35.792	91.561	0.10284	0.15793
191	392.05	377.35	0.016006	0.088101	62.475	11.351	56.101	35.447	91.548	0.10333	0.15780
192	396.14	381.44	0.016073	0.086796	62.218	11.521	56.435	35.096	91.531	0.10382	0.15768
193	400.27	385.57	0.016140	0.085502	61.958	11.696	56.772	34.739	91.511	0.10432	0.15755
194	404.44	389.74	0.016209	0.084218	61.694	11.874	57.111	34.377	91.488	0.10482	0.15741
195	408.63	393.93	0.016280	0.082946	61.426	12.056	57.453	34.009	91.462	0.10532	0.15727
196	412.86	398.16	0.016352	0.081683	61.155	12.242	57.797	33.636	91.433	0.10583	0.15713
197	417.12	402.42	0.016426	0.080431	60.879	12.433	58.144	33.256	91.400	0.10634	0.15698
198	421.41	406.71	0.016502	0.079188	60.599	12.628	58.494	32.869	91.363	0.10685	0.15683
199	425.73	411.03	0.016580	0.077953	60.315	12.828	58.847	32.476	91.323	0.10737	0.15667
200	430.09	415.39	0.016659	0.076728	60.026	13.033	59.203	32.075	91.278	0.10789	0.15651
201	434.48	419.78	0.016741	0.075511	59.732	13.243	59.562	31.668	91.230	0.10841	0.15634
202	438.91	424.21	0.016826	0.074301	59.433	13.459	59.924	31.252	91.176	0.10894	0.15617
203	443.36	428.66	0.016912	0.073099	59.128	13.680	60.290	30.828	91.118	0.10947	0.15599
204	447.85	433.15	0.017002	0.071903	58.818	13.908	60.659	30.396	91.055	0.11001	0.15580
205	452.38	437.68	0.017094	0.070714	58.502	14.141	61.032	29.955	90.987	0.11055	0.15561
206	456.94	442.24	0.017188	0.069531	58.179	14.382	61.409	29.505	90.914	0.11109	0.15541
207	461.53	446.83	0.017286	0.068353	57.849	14.630	61.790	29.045	90.835	0.11164	0.15521
208	466.16	451.46	0.017387	0.067179	57.513	14.886	62.175	28.574	90.749	0.11220	0.15499
209	470.82	456.12	0.017492	0.066009	57.168	15.149	62.565	28.092	90.657	0.11276	0.15477
210	475.52	460.82	0.017601	0.064843	56.816	15.422	62.959	27.599	90.558	0.11332	0.15453
211	480.25	465.55	0.017713	0.063679	56.455	15.704	63.359	27.093	90.452	0.11390	0.15429
212	485.01	470.31	0.017830	0.062517	56.084	15.996	63.764	26.573	90.337	0.11448	0.15404
213	489.82	475.12	0.017952	0.061355	55.703	16.299	64.174	26.040	90.214	0.11506	0.15377
214	494.65	479.95	0.018079	0.060193	55.312	16.613	64.591	25.490	90.081	0.11566	0.15349
215	499.53	484.83	0.018212	0.059030	54.908	16.941	65.014	24.925	89.939	0.11626	0.15320
216	504.44	489.74	0.018351	0.057864	54.492	17.282	65.444	24.341	89.785	0.11687	0.15290
217	509.38	494.68	0.018497	0.056694	54.062	17.639	65.881	23.738	89.619	0.11749	0.15257
218	514.36	499.66	0.018651	0.055518	53.616	18.012	66.327	23.113	89.440	0.11813	0.15223
219	519.38	504.68	0.018814	0.054334	53.153	18.405	66.782	22.465	89.247	0.11877	0.15187
220	524.43	509.73	0.018986	0.053140	52.670	18.818	67.246	21.790	89.036	0.11943	0.15149
221	529.52	514.82	0.019169	0.051934	52.167	19.255	67.722	21.086	88.808	0.12010	0.15108
222	534.65	519.95	0.019365	0.050711	51.638	19.720	68.209	20.350	88.559	0.12079	0.15064
223	539.82	525.12	0.019576	0.049468	51.082	20.215	68.711	19.575	88.286	0.12150	0.15017
224	545.02	530.32	0.019804	0.048200	50.494	20.747	69.228	18.757	87.985	0.12223	0.14966
225	550.26	535.56	0.020053	0.046900	49.868	21.322	69.763	17.888	87.651	0.12298	0.14911
226	555.54	540.84	0.020327	0.045559	49.196	21.949	70.320	16.958	87.278	0.12377	0.14850
227	560.85	546.15	0.020632	0.044166	48.468	22.642	70.904	15.953	86.857	0.12459	0.14782
228	566.20	551.50	0.020978	0.042702	47.669	23.418	71.519	14.854	86.373	0.12545	0.14705
229	571.60	556.90	0.021378	0.041140	46.778	24.307	72.177	13.629	85.806	0.12638	0.14617
230	577.03	562.33	0.021854	0.039435	45.758	25.358	72.893	12.229	85.122	0.12739	0.14512
231	582.50	567.80	0.022450	0.037492	44.544	26.672	73.696	10.553	84.249	0.12852	0.14380
232	588.01	573.31	0.023262	0.035041	42.988	28.538	74.651	8.335	82.986	0.12987	0.14191
233.6 (Critical)	596.9	582.2	0.02870	0.02870	34.84	34.84	78.86	0	78.86	0.1359	0.1359

CHARACTERISTICS OF REFRIGERANT-12

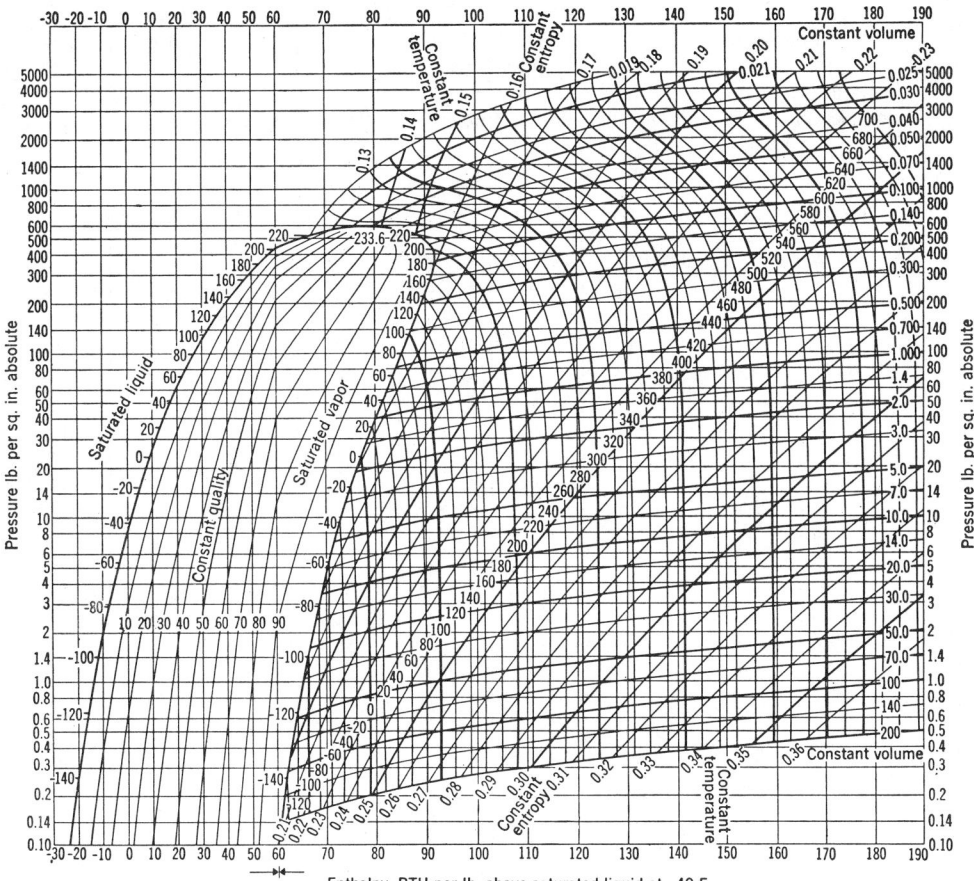

Scale change Enthalpy, BTU per lb. above saturated liquid at −40 F

The Mollier chart on Refrigerant-12 covers an exceptionally wide range. It is based on the properties of this refrigerant compiled and published by the manufacturer, Kinetic Chemicals Division, E. I. duPont de Nemours, copyrighted by that company, and reproduced here by permission.

In the saturated vapor region in the left portion of the chart, the horizontal figures interpolated between the pressure coordinates and skirting the entire saturated vapor area, are temperatures, in deg. F. In this same area, the horizontal figures on the lines slanting upward to the right designate per-cent quality lines.

In the superheat region, occupying all of the chart to the right and by far the greater part of the whole chart, the lines sloping slightly upward to the right are lines of constant specific volume and ranging in value (at the extreme right) from 200 at the bottom to 0.019 at the top. In the same region, the lines sloping to the left from vertical are temperatures.

Lines in the superheat region sloping upward to the right at about a 45° angle are lines of constant entropy with the scales given near the lower axis (from 0.21 to 0.36) or near the top of the chart (0.13 to 0.23). Horizontal lines are, of course, pressure and vertical lines are enthalpy.

SUPERHEAT DATA—REFRIGERANT-12

Temp. of Saturated Vapor, Deg.	Degree of Superheat								
	10.			15			20		
	Vol- ume, Cu. Ft. per Lb.	Enth- alpy, Btu per Lb.	Ent- ropy, Btu per Deg. R.	Vol- ume, Cu. Ft. per Lb.	Enth- alpy, Btu per Lb.	Ent- ropy, Btu per Deg. R.	Vol- ume, Cu. Ft. per Lb.	Enth- alpy, Btu per Lb.	Ent- ropy, Btu per Deg. R.
−10	2.0260	77.61	.1730	2.0522	78.23	.1745	2.0781	78.85	.1761
−5	1.8392	78.16	.1725	1.8630	78.83	.1740	1.8866	79.50	.1759
0	1.6525	78.72	.1720	1.6739	79.44	.1735	1.6952	80.16	.1750
5	1.5063	79.26	.1715	1.5229	79.99	.1731	1.5454	80.72	.1746
10	1.3602	79.81	.1711	1.3779	80.55	.1726	1.3957	81.29	.1741
12	1.3140	80.03	.1709	1.3311	80.77	.1725	1.3483	81.51	.1740
14	1.2678	80.24	.1708	1.2843	80.99	.1723	1.3009	81.73	.1738
16	1.2215	80.46	.1706	1.2376	81.21	.1721	1.2536	81.95	.1737
18	1.1753	80.68	.1705	1.1908	81.43	.1720	1.2062	82.18	.1735
20	1.1291	80.89	.1703	1.1440	81.65	.1718	1.1588	82.40	.1733
22	1.0922	81.11	.1702	1.1067	81.86	.1717	1.1210	82.62	.1732
24	1.0554	81.32	.1700	1.0693	82.08	.1715	1.0833	82.84	.1731
26	1.0185	81.53	.1699	1.0320	82.30	.1714	1.0455	83.06	.1729
28	.9816	81.75	.1697	.9947	82.51	.1713	1.0078	83.28	.1728
30	.9447	81.96	.1696	.9573	82.73	.1711	.9699	83.50	.1727
32	.9149	82.17	.1695	.9272	82.94	.1710	.9394	83.72	.1725
34	.8851	82.38	.1694	.8970	83.16	.1709	.9089	83.93	.1724
36	.8553	82.59	.1692	.8668	83.37	.1708	.8783	84.15	.1723
38	.8255	82.80	.1691	.8366	83.59	.1706	.8478	84.37	.1722
40	.7956	83.01	.1690	.8064	83.80	.1705	.8172	84.59	.1720
42	.7714	83.22	.1689	.7819	84.00	.1704	.7924	84.80	.1719
44	.7472	83.43	.1688	.7574	84.22	.1703	.7676	85.01	.1718
46	.7230	83.63	.1687	.7329	84.43	.1702	.7428	85.23	.1717
48	.6988	83.84	.1685	.7084	84.64	.1701	.7180	85.44	.1716
50	.6745	84.05	.1684	.6839	84.85	.1700	.6932	85.66	.1715
52	.6547	84.25	.1683	.6638	85.06	.1699	.6728	85.87	.1714
54	.6348	84.45	.1682	.6436	85.27	.1698	.6525	86.08	.1713
56	.6149	84.66	.1681	.6235	85.47	.1697	.6321	86.29	.1712
58	.5950	84.86	.1681	.6034	85.68	.1696	.6118	86.50	.1711
60	.5751	85.06	.1680	.5833	85.89	.1695	.5914	86.71	.1710
62	.5587	85.26	.1679	.5666	86.09	.1694	.5746	86.92	.1709
64	.5423	85.46	.1678	.5500	86.29	.1693	.5577	87.12	.1709
66	.5259	85.66	.1677	.5334	86.50	.1692	.5409	87.33	.1708
68	.5094	85.86	.1676	.5168	86.70	.1691	.5241	87.54	.1707
70	.4930	86.06	.1675	.5001	86.90	.1691	.5072	87.75	.1706
80	.4245	87.02	.1671	.4309	87.89	.1687	.4372	88.75	.1702
90	.3670	87.96	.1668	.3728	88.85	.1683	.3785	89.73	.1699
100	.3187	88.87	.1664	.3238	89.78	.1680	.3289	90.69	.1696
120	.2423	90.57	.1658	.2466	91.54	.1674	.2509	92.50	.1690
140	.1860	92.09	.1651	.1897	93.11	.1668	.1934	94.14	.1684
160	.1433	93.35	.1643	.1467	94.46	.1660	.1500	95.57	.1678
180	.1105	94.22	.1636	.1136	95.46	.1653	.1168	96.70	.1669

Data interpolated from tables copyrighted by Kinetic Chemicals Div., E. I. duPont de Nemours & Co., Inc.

Superheat Data—Refrigerant-12 (continued)

Temp. of Saturated Vapor, Deg.	Degree of Superheat								
	25			30			35		
	Volume, Cu. Ft. per Lb.	Enthalpy, Btu per Lb.	Entropy, Btu per Deg. R.	Volume, Cu. Ft. per Lb.	Enthalpy, Btu per Lb.	Entropy, Btu per Deg. R.	Volume, Cu. Ft. per Lb.	Enthalpy, Btu per Lb.	Entropy, Btu per Deg. R.
−10	2.1041	79.65	.1776	2.1300	80.45	.1791	2.1555	81.17	.1805
−5	1.9170	80.27	.1770	1.9472	81.03	.1785	1.9637	81.76	.1800
0	1.7299	80.89	.1765	1.7645	81.62	.1780	1.7719	82.35	.1795
5	1.5715	81.46	.1761	1.5976	82.19	.1776	1.6097	82.93	.1791
10	1.4132	82.03	.1756	1.4307	82.77	.1771	1.4475	83.51	.1786
12	1.3653	82.25	.1755	1.3822	83.00	.1770	1.3985	83.74	.1785
14	1.3173	82.48	.1753	1.1337	83.23	.1768	1.3496	83.97	.1783
16	1.2694	82.70	.1752	1.2852	83.45	.1767	1.3006	84.21	.1781
18	1.2214	82.93	.1750	1.2367	83.62	.1765	1.2517	84.44	.1780
20	1.1735	83.16	.1748	1.1882	83.91	.1763	1.2027	84.67	.1778
22	1.1353	83.38	.1747	1.1495	84.14	.1762	1.1636	84.90	.1777
24	1.0970	83.60	.1746	1.1108	84.36	.1761	1.1244	85.13	.1775
26	1.0588	83.82	.1744	1.0722	84.59	.1759	1.0853	85.36	.1774
28	1.0206	84.05	.1743	1.0335	84.81	.1758	1.0461	85.58	.1773
30	.9824	84.27	.1742	.9948	85.04	.1756	1.0070	85.81	.1771
32	.9515	84.49	.1740	.9635	85.26	.1755	.9753	86.04	.1770
34	.9205	84.71	.1739	.9322	85.49	.1754	.9437	86.27	.1769
36	.8893	84.93	.1738	.9010	85.71	.1753	.9121	86.49	.1768
38	.8587	85.15	.1737	.8697	85.94	.1752·	.8805	86.72	.1766
40	.8278	85.37	.1735	.8384	86.16	.1750	.8488	86.95	.1765
42	.8027	85.59	.1734	.8130	86.38	.1749	.8231	87.17	.1764
44	.7776	85.81	.1733	.7876	86.60	.1748	.7975	87.39	.1763
46	.7525	86.02	.1732	.7622	86.82	.1747	.7718	87.62	.1762
48	.7274	86.24	.1731	.7368	87.04	.1746	.7461	87.84	.1761
50	.7023	86.46	.1730	.7115	87.26	.1745	.7204	88.06	.1759
52	.6817	86.67	.1729	.6906	87.48	.1744	.6994	88.28	.1759
54	.6611	86.89	.1728	.6698	87.70	.1743	.6783	88.50	.1758
56	.6405	87.10	.1727	.6490	87.91	.1742	.6572	88.72	.1757
58	.6199	87.31	.1726	.6281	88.13	.1741	.6362	88.95	.1756
60	.5993	87.53	.1725	.6073	88.35	.1740	.6151	89.17	.1755
62	.5823	87.74	.1725	.5901	88.56	.1740	.5977	89.38	.1754
64	.5653	87.95	.1724	.5729	88.77	.1739	.5803	89.60	.1754
66	.5483	88.16	.1723	.5556	89.99	.1738	.5629	89.82	.1753
68	.5312	88.37	.1722	.5384	90.20	.1737	.5454	90.03	.1752
70	.5142	88.58	.1721	.5212	89.41	.1736	.5280	90.25	.1751
80	.4434	89.61	.1717	.4496	90.46	.1733	.4556	91.31	.1747
90	.3840	90.61	.1714	.3895	91.48	.1729	.3949	92.35	.1744
100	.3339	91.58	.1711	.3389	92.48	.1726	.3437	93.37	.1741
120	.2550	93.44	.1706	.2591	94.38	.1721	.2630	95.32	.1737
140	.1969	95.14	.1700	.2003	96.14	.1716	.2037	97.12	.1732
160	.1531	96.64	.1695	.1562	97.72	.1711	.1591	98.76	.1727
180	.1196	97.87	.1687	.1225	99.04	.1704	.1251	100.16	.1721

Superheat Data—Refrigerant-12 (continued)

Temp. of Saturated Vapor, Deg.	Degree of Superheat								
	40			45			50		
	Volume, Cu. Ft. per Lb.	Enthalpy, Btu per Lb.	Entropy, Btu per Deg. R.	Volume, Cu. Ft. per Lb.	Enthalpy, Btu per Lb.	Entropy, Btu per Deg. R.	Volume, Cu. Ft. per Lb.	Enthalpy, Btu per Lb.	Entropy, Btu per Deg. R.
−10	2.1810	81.89	.1820	2.2064	82.62	.1835	2.2318	83.34	.1849
−5	1.9801	82.48	.1815	2.0032	83.21	.1830	2.0262	83.95	.1844
0	1.7793	83.08	.1810	1.8000	83.81	.1825	1.8207	84.55	.1839
5	1.6222	83.67	.1805	1.6412	84.41	.1820	1.6601	85.15	.1834
10	1.4652	84.26	.1801	1.4824	85.00	.1815	1.4995	85.75	.1830
12	1.4156	84.49	.1799	1.4322	85.24	.1814	1.4487	85.99	.1828
14	1.3660	84.72	.1798	1.3820	85.48	.1812	1.3979	86.23	.1826
16	1.3163	84.96	.1796	1.3318	85.71	.1811	1.3472	86.47	.1825
18	1.2667	85.19	.1794	1.2816	85.95	1809	1.2964	86.71	.1823
20	1.2171	85.43	.1793	1.2314	86.19	.1807	1.2456	86.95	.1822
22	1.1775	85.66	.1791	1.1914	86.42	.1806	1.2051	87.18	.1820
24	1.1379	85.89	.1790	1.1513	86.65	.1805	1.1647	87.42	.1819
26	1.0983	86.12	.1789	1.1113	86.89	.1803	1.1242	87.66	.1818
28	1.0587	86.35	.1787	1.0712	87.12	.1802	1.0837	87.89	.1816
30	1.0191	86.58	.1786	1.0312	87.36	.1800	1.0433	88.13	.1815
32	.9871	86.81	.1785	.9989	87.59	.1799	1.0106	88.37	.1814
34	.9552	87.04	.1784	.9665	87.82	.1798	.9779	88.60	.1812
36	.9232	87.27	.1782	.9342	88.05	.1797	.9452	88.84	.1811
38	.8912	87.50	.1781	.9019	88.29	.1796	.9125	89.07	.1810
40	.8592	87.73	.1780	.8695	88.52	.1794	.8798	89.31	.1809
42	.8333	87.96	.1779	.8433	88.75	.1793	.8533	89.54	.1808
44	.8073	88.18	.1778	.8170	88.98	.1792	.8267	89.77	.1807
46	.7813	88.41	.1777	.7908	89.21	.1791	.8002	90.00	.1806
48	.7554	88.64	.1776	.7645	89.44	.1790	.7736	90.24	.1804
50	.7294	88.86	.1775	.7382	89.67	.1789	.7471	90.47	.1803
52	.7081	89.09	.1774	.7167	89.89	.1788	.7253	90.70	.1802
54	.6868	89.31	.1773	.6952	90.12	.1787	.7035	90.93	.1802
56	.6655	89.54	.1772	.6736	90.35	.1786	.6817	91.16	.1801
58	.6442	89.76	.1771	.6521	90.57	.1785	.6600	91.39	.1800
60	.6229	89.98	.1770	.6305	90.80	.1784	.6382	91.62	.1799
62	.6053	90.20	.1769	.6127	91.02	.1784	.6202	91.84	.1798
64	.5877	90.42	.1768	.5949	91.24	.1783	.6022	92.07	.1797
66	.5701	90.64	.1767	.5771	91.47	.1782	.5842	92.29	.1796
68	.5525	90.86	.1767	.5593	91.69	.1781	.5662	92.52	.1796
70	.5349	91.08	.1766	.5415	91.91	.1780	.5482	92.75	.1795
80	.4616	92.16	.1761	.4662	93.01	.1776	.4734	93.86	.1791
90	.4003	93.22	.1759	.4055	94.09	.1774	.4107	94.95	.1788
100	.3485	94.25	.1756	.3532	95.14	.1771	.3579	96.02	.1786
120	.2670	96.25	.1752	.2708	97.17	.1767	.2747	98.09	.1781
140	.2071	98.11	.1747	.2103	99.07	.1762	.2135	100.04	.1777
160	.1620	99.80	.1743	.1648	100.82	.1758	.1676	101.84	.1774
180	.1277	101.28	.1738	.1302	102.36	.1753	.1327	103.45	.1769

SATURATION PROPERTIES OF REFRIGERANT-22

Temperature, Deg. F.	Pressure, Lb. per Sq. In.		Volume, Cu. Ft. per Lb.		Density, Lb. per Cu. Ft.		Enthalpy, Btu per Lb.			Entropy, Btu per Lb. per Deg. Abs.	
	Abs.	Gage	Liq.	Vapor	Liq.	Vapor	Liq.	Latent	Vapor	Liq.	Vapor
0	38.79	24.09	0.01192	1.373	83.90	0.7282	10.63	94.39	105.02	0.0240	0.2293
2	40.43	25.73	0.01195	1.320	83.68	0.7574	11.17	94.07	105.24	0.0251	0.2289
4	42.14	27.44	0.01198	1.270	83.45	0.7877	11.70	93.75	105.45	0.0262	0.2285
5	43.02	28.33	0.01200	1.246	83.34	0.8034	11.97	93.59	105.56	0.0268	0.2283
6	43.91	29.21	0.01201	1.221	83.23	0.8191	12.23	93.43	105.66	0.0274	0.2280
8	45.74	31.04	0.01205	1.175	83.01	0.8514	12.76	93.11	105.87	0.0285	0.2276
10	47.63	32.93	0.01208	1.130	82.78	0.8847	13.29	92.79	106.08	0.0296	0.2272
12	49.58	34.88	0.01211	1.088	82.55	0.9191	13.82	92.47	106.29	0.0307	0.2268
14	51.59	36.89	0.01215	1.048	82.32	0.9545	14.36	92.14	106.50	0.0319	0.2264
16	53.66	38.96	0.01218	1.009	82.09	0.9911	14.90	91.81	106.71	0.0330	0.2260
18	55.79	41.09	0.01222	0.9721	81.86	1.029	15.44	91.48	106.92	0.0341	0.2257
20	57.98	43.28	0.01225	0.9369	81.63	1.067	15.98	91.15	107.13	0.0352	0.2253
22	60.23	45.53	0.01229	0.9032	81.39	1.107	16.52	90.81	107.33	0.0364	0.2249
24	62.55	47.85	0.01232	0.8707	81.16	1.149	17.06	90.47	107.53	0.0375	0.2246
26	64.94	50.24	0.01236	0.8398	80.92	1.191	17.61	90.12	107.73	0.0379	0.2242
28	67.40	52.70	0.01239	0.8100	80.69	1.235	18.17	89.76	107.93	0.0398	0.2239
30	69.93	55.23	0.01243	0.7816	80.45	1.280	18.74	89.39	108.13	0.0409	0.2235
32	72.53	57.83	0.01247	0.7543	80.21	1.326	19.32	89.01	108.33	0.0421	0.2232
34	75.21	60.51	0.01250	0.7283	79.97	1.373	19.90	88.62	108.52	0.0433	0.2228
36	77.97	63.27	0.01254	0.7032	79.73	1.422	20.49	88.22	108.71	0.0445	0.2225
38	80.81	66.11	0.01258	0.6791	79.49	1.473	21.09	87.81	108.90	0.0457	0.2222
40	83.72	69.02	0.01262	0.6559	79.25	1.525	21.70	87.39	109.09	0.0469	0.2218
42	86.69	71.99	0.01266	0.6339	79.00	1.578	22.29	86.98	109.27	0.0481	0.2215
44	89.74	75.04	0.01270	0.6126	78.76	1.632	22.90	86.55	109.45	0.0493	0.2211
46	92.88	78.18	0.01274	0.5922	78.51	1.689	23.50	86.13	109.63	0.0505	0.2208
48	96.10	81.40	0.01278	0.5726	78.26	1.747	24.11	85.69	109.80	0.0516	0.2205
50	99.40	84.70	0.01282	0.5537	78.02	1.806	24.73	85.25	109.98	0.0528	0.2201
52	102.8	88.10	0.01286	0.5355	77.77	1.868	25.34	84.80	110.14	0.0540	0.2198
54	106.2	91.5	0.01290	0.5184	77.51	1.929	25.95	84.35	110.30	0.0552	0.2194
56	109.8	95.1	0.01294	0.5014	77.26	1.995	26.58	83.89	110.47	0.0564	0.2191
58	113.5	98.8	0.01299	0.4849	77.01	2.062	27.22	83.41	110.63	0.0576	0.2188
60	117.2	102.5	0.01303	0.4695	76.75	2.130	27.83	82.95	110.78	0.0588	0.2185
62	121.0	106.3	0.01307	0.4546	76.50	2.200	28.46	82.47	110.93	0.0600	0.2181
64	124.9	110.2	0.01312	0.4403	76.24	2.271	29.09	81.99	111.08	0.0612	0.2178
66	128.9	114.2	0.01316	0.4264	75.98	2.346	29.72	81.50	111.22	0.0624	0.2175
68	133.0	118.3	0.01320	0.4129	75.72	2.422	30.35	81.00	111.35	0.0636	0.2172
70	137.2	122.5	0.01325	0.4000	75.46	2.500	30.99	80.50	111.49	0.0648	0.2168
72	141.5	126.8	0.01330	0.3875	75.20	2.581	31.65	79.98	111.63	0.0661	0.2165
74	145.9	131.2	0.01334	0.3754	74.94	2.664	32.29	79.46	111.75	0.0673	0.2162
76	150.4	135.7	0.01339	0.3638	74.68	2.749	32.94	78.94	111.88	0.0684	0.2158
78	155.0	140.3	0.01344	0.3526	74.41	2.836	33.61	78.40	112.01	0.0696	0.2155
80	159.7	145.0	0.01349	0.3417	74.15	2.926	34.27	77.86	112.13	0.0708	0.2151
82	164.5	149.8	0.01353	0.3313	73.89	3.019	34.92	77.32	112.24	0.0720	0.2148
84	169.4	154.7	0.01358	0.3212	73.63	3.113	35.60	76.76	112.36	0.0732	0.2144
86	174.5	159.8	0.01363	0.3113	73.36	3.213	36.28	76.19	112.47	0.0744	0.2140
88	179.6	164.9	0.01368	0.3019	73.09	3.313	36.94	75.63	112.57	0.0756	0.2137
90	184.8	170.1	0.01374	0.2928	72.81	3.415	37.61	75.06	112.67	0.0768	0.2133
92	190.1	175.4	0.01379	0.2841	72.53	3.520	38.28	74.48	112.76	0.0780	0.2130
94	195.6	180.9	0.01384	0.2755	72.24	3.630	38.97	73.88	112.85	0.0792	0.2126
96	201.2	186.5	0.01390	0.2672	71.95	3.742	39.65	73.28	112.93	0.0803	0.2122
98	206.8	192.1	0.01396	0.2594	71.65	3.855	40.32	72.69	113.00	0.0815	0.2119
100	212.6	197.9	0.01402	0.2517	71.35	3.973	40.98	72.08	113.06	0.0827	0.2115
102	218.5	203.8	0.01408	0.2443	71.05	4.094	41.65	71.47	113.12	0.0839	0.2111
104	224.6	209.9	0.01414	0.2370	70.74	4.220	42.32	70.84	113.16	0.0851	0.2107
106	230.7	216.0	0.01420	0.2301	70.42	4.347	42.98	70.22	113.20	0.0862	0.2104
108	237.0	222.3	0.01426	0.2233	70.11	4.479	43.66	69.58	113.24	0.0874	0.2100
110	243.4	228.7	0.01433	0.2167	69.78	4.614	44.35	68.94	113.29	0.0886	0.2096
112	249.9	235.2	0.01440	0.2104	69.45	4.752	45.04	68.30	113.34	0.0898	0.2093
114	256.6	241.9	0.01447	0.2043	69.12	4.896	45.74	67.64	113.38	0.0909	0.2089
116	263.4	248.7	0.01454	0.1983	68.78	5.043	46.44	66.98	113.42	0.0921	0.2085
118	270.3	255.6	0.01461	0.1926	68.44	5.192	47.14	66.32	113.46	0.0933	0.2081
120	277.3	262.6	0.01469	0.1871	68.10	5.345	47.85	65.67	113.52	0.0945	0.2078

Data copyrighted by Freon Products Div., E. I. duPont de Nemours & Co., Inc.

PRESSURE ENTHALPY CHART—REFRIGERANT-22

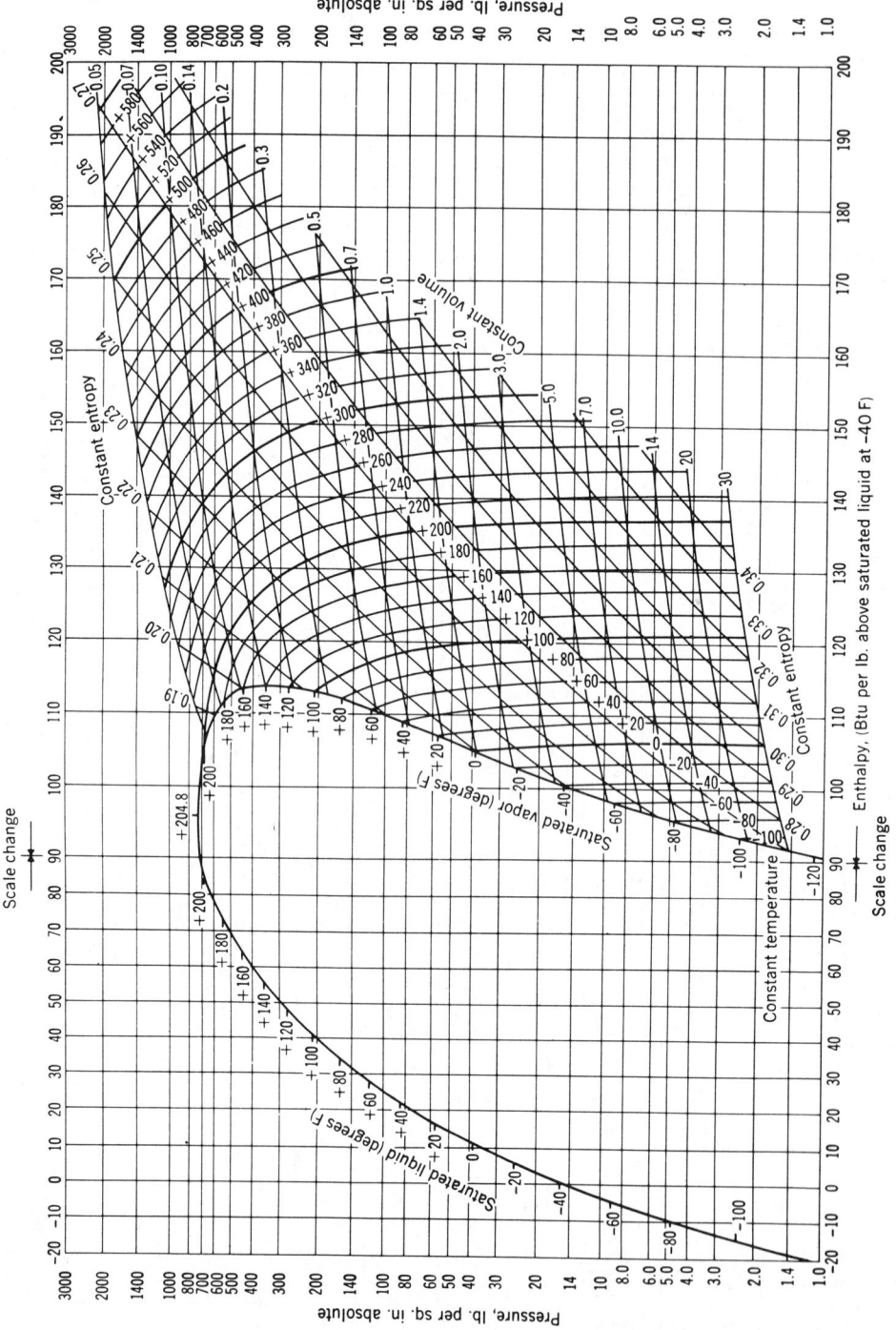

REFRIGERATION HEAT QUANTITIES

In the simple refrigeration cycle the essential equipment consists of two heat exchangers—the condenser and the evaporator—compressor, liquid receiver, expansion valve, and connecting piping. The warm refrigerant gas is compressed by the compressor, the work done by the compressor raising the temperature of the refrigerant vapor. This hot gas is cooled in the condenser, the rejected heat being carried away by air or water. As its name implies, the condenser condenses the refrigerant gas and the liquid is collected in a receiver from which it flows to the expansion valve. This valve allows only as much liquid to enter the evaporator as that heat exchanger can evaporate. In the expansion valve there is a change of pressure but theoretically no change in total heat. In the evaporator, heat is exchanged from the surrounding air (or water) to the refrigerant which is changed from a liquid to a saturated vapor at the same pressure. The vapor then passes to the compressor, completing the cycle.

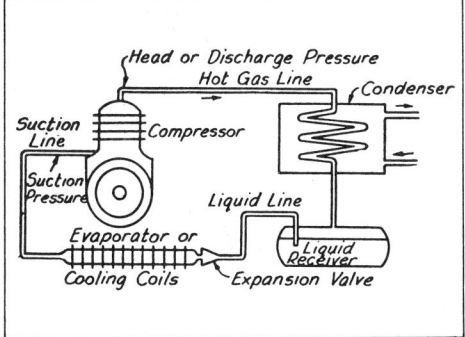

Heat Added to System. Heat is added to the refrigerant in two major ways: (1) the compression of the gas in the compressor requires work which appears as heat in the gas; (2) the heat flowing into the system in the evaporator. This total heat must be removed by the condenser. The foregoing neglects heat gained or lost in piping, etc.

The quantity of heat added to each pound of refrigerant in the evaporator is termed refrigerating effect. The refrigerating effect is equal to the heat in the vapor leaving the evaporator less the heat in the liquid entering the evaporator, both on a Btu per pound basis. The actual value of the refrigerating effect depends on the (a) properties of the refrigerant used, (b) temperature of refrigerant entering the expansion valve and (c) the temperature at which the refrigerant is evaporating in the coils. Knowing (a), (b), and (c) the refrigerating

effect can be calculated from tables of refrigerant properties found elsewhere in this section of the Handbook.

$$E = h_{lg} - h_l$$

where E = refrigerating effect, Btu per lb refrigerant,

h_{lg} = enthalpy of refrigerant vapor leaving evaporator, Btu per lb,

h_l = enthalpy of liquid refrigerant entering expansion valve, Btu per lb.

The quantity of refrigerant that must be circulated, W_t is

$$W_t = 200/E$$

where W_t = circulating refrigerant, lb per ton per min,

200 = Btu-per-min equivalent of 1 ton.

The charts, Flow Rate per Ton of Refrigeration for R-12, and for R-22, are solutions of this equation.

Compressor capacity, C_t is

$$C_t = d/W_t$$

where C_t = compressor capacity, tons per cu ft piston displacement per min,

d = density of refrigerant vapor at evaporator temperature, lb per cu ft.

Example: How much Refrigerant-12 must be circulated to supply 12 tons of refrigeration, if the evaporator temperature is 30F and the temperature of the refrigerant ahead of the expansion valve is 80F?

Solution: Enthalpy of the 80-F liquid, from the table, Saturation Properties of Refrigerant-12, is 26.365 Btu per lb. Enthalpy of saturated vapor at 30F, from the same table, is 80.419. The difference, 54.05 Btu per lb, is E, the refrigerating effect.

The amount of refrigeration to be circulated, W_t, is 200/54.05 = 3.7 lb per min per ton, which checks with the chart of flow rates for R-12. For 12 tons, the flow rate is 44.4 lb per min.

Density of R-12 vapor at the evaporator temperature, 30F, from the table, Saturation Properties of Refrigerant-12, is 1.0884 lb per cu ft. Compressor capacity, C_t, is 1.0884/3.7 = 0.272 tons per cu ft piston displacement per min, which gives 12/0.272 = 44 cu ft per min for a 12-ton system.

The compressor adds heat to the refrigerant in an amount theoretically equivalent to the work done,

$$h_c = (h_{g1} - h_{g2})$$

where h_c = heat added (or work done), Btu per min per lb,

h_{g1} = enthalpy of refrigerant gas leaving compressor, Btu per lb,

h_{g2} = enthalpy of refrigerant gas entering compressor, Btu per lb.

Solution of this equation is more difficult. The value of h_{g2} can be assumed to be the enthalpy of the saturated gas leaving the evaporator, which has been determined above. However, the gas may have been superheated by picking up heat in the suction line from the ambient air or in a heat exchanger designed to sub-cool the hot liquid on the way to the expansion valve. In these cases, the degree of superheat has to be calculated and the tables of refrigerant superheat data consulted.

The value of h_{g2}, the enthalpy of gas leaving the compressor, is obtained from the pressure-enthalpy charts (Mollier charts) for refrigerants. Since compression of the gas is practically isentropic (at constant entropy), one enters the chart at the enthalpy of the entering gas, proceeds upward to the appropriate condition (the simplest would be saturation), then along the line of constant entropy until the condensing temperature is reached, then vertically down to read out the leaving gas enthalpy.

In addition to heat added to the gas by the work done on it (the heat of compression), there is heat dissipated due to friction of the cylinders.

Note that $h_c/42.4$ is compressor horsepower.

Neglecting heat lost or gained by piping, total heat added by compressor and evaporator is rejected at the condenser. Since heat added by the evaporator per ton is, by definition, 200 Btu per min, then heat to be dissipated at the condenser is h_c + 200. The value of h_c over the usual air conditioning range is close to 40 Btu per min per ton, so that the condenser ordinarily must dissipate 200 + 40 = 240 Btu per min per ton. The quantity of condenser water required, then, is:

$$G = \frac{240}{(t_1 - t_2)8.33}$$

where G = condenser water required, gpm

$(t_1 - t_2)$ = temperature rise of condenser water, deg F, and

8.33 = pounds of water per gallon.

For determining the charge for a new system, it is desirable to know the weight of refrigerant in a system. The receiver contains liquid refrigerant, as does the liquid line, while refrigerant gas occupies the suction line and hot gas line.

Tables appear elsewhere in this section of the Handbook for estimating volume of R-12 in piping. R-22 volume would be slightly less.

In calculating the amount of refrigerant in the receiver, some engineers use a rough figure of one third as that part of the diameter of the receiver occupied by the liquid refrigerant, while others arbitrarily set a figure in pounds and size the receiver 20% above the volume for such a charge.

Refrigerant in the condenser can be calculated as nearly filled with vapor at the condensing temperature but with, say, 10% liquid on the bottom. The volume of the vapor and the liquid at the condensing temperature can be determined from the tables of refrigerant properties. Evaporator coil length being known, the volume is estimated as being half liquid and half vapor at evaporator temperature.

In the simplified calculations given it has been assumed that the condenser removes only the latent heat from the gas and reduces it to a liquid at the same temperature. The cooling of the liquid to a point below the condensing temperature is called *sub-cooling*. Sub-cooling is an advantage if the cooling is by some source external to the refrigerant cycle, since it reduces the quantity of flash gas at the expansion valve.

In the foregoing discussion it has been assumed that the vapor leaving the evaporator is saturated. This is by no means always the case. The refrigerant, after vaporizing, may continue to absorb heat and become *superheated;* if this absorption takes place in the evaporator the refrigerant effect per pound of refrigerant is increased. On the other hand, since superheated gas occupies more volume per pound than saturated gas, the compressor capacity is reduced; the net result must be computed for each case to determine whether the superheating is a net advantage or disadvantage.

Due to clearance between the piston and the cylinder head, to the expansion of the cold gas in the hot cylinder, and to the friction of the compressor valves, the actual volume of gas handled by the compressor must be greater than the theoretical volume. The term *volumetric efficiency* is used to express the relationship between the two, volumetric efficiency being the actual weight of refrigerant in a cylinder on one stroke divided by the theoretically computed weight, and the result expressed in per cent.

REFRIGERANT PIPING

It is impossible to give several, or even a hundred rules which would cover all of the problems arising during the design and installation of refrigeration or air conditioning piping. However, there are several basic rules which should always be followed in laying out piping to provide a neat, serviceable installation. These basic rules are:

1. Design refrigerant lines to be as short and direct as possible.

2. Locate refrigerant piping so that access to system components is not hindered, and so that refrigerant flow sight glasses can be easily observed.

3. Avoid locations where copper tube will be exposed to possible injury. If this is not feasible, enclose tube in a protective conduit. Some codes require this.

4. Use as few joints and fittings as possible. Fittings should be of the wrought copper, long radius type. When possible, use wye fittings in preference to 90° connections.

5. Use a continuous length of soft copper tube bent around obstructions in inaccessible locations where it would be difficult to install tube with fittings or where obstructions would require an unusually large number of fittings. Soft tube so used must be properly supported and should never be bent more than 22½° without the use of bending tools.

6. Pitch all horizontal lines a minimum, of ¼ inch and preferably ½ inch in 10 ft in the direction of refrigerant flow, and avoid unnecessary traps.

7. Do not expose piping to external sources of heat, if it can be avoided.

8. If space will not allow a single large line, use two or more smaller lines in parallel.

Once equipment locations and approximate line locations have been determined, the next problem is to size and detail the suction, discharge (hot gas) and liquid lines. An improperly sized or connected line can result in system failure.

The sizing and location of refrigerant piping is complicated by the fact that there is usually oil, in liquid form, with the refrigerant vapor in discharge and suction lines. At any usual liquid temperature in air conditioning applications, oil is sufficiently miscible with almost all liquid refrigerants that possible separation in the liquid line need not be considered.

Refrigerant piping is a means ·for conducting refrigerant liquid and gas and any liquid oil with the gas. It cannot perform this function and at the same time keep refrigerant from moving from one part of a system to another, such as from an evaporator to a compressor. In the past, and occasionally even in present literature, there have been indications that the use of traps or loops would be of some value in keeping refrigerant liquid out of the compressor, particularly during a period when the compressor is not operating. Of course, any loops or traps for this purpose are, at best, a waste of money. In some cases they are harmful, as when they collect liquid, only to release it, all at once and in large quantity, to the compressor, or in giving a false notion that the compressor has been protected by the loop or trap.

During compressor operation, liquid reaches the compressor because of:

1. Sudden increases in *load* or of compressor *capacity*. In either case, the liquid is caused to boil more rapidly, increasing the volume it occupies in the evaporator, which can cause it to boil over.

When the *load* increases, the evaporating (suction) pressure is likely to rise somewhat, which causes a superheat controlled expansion valve to move toward the *closed* position, thus reducing the chance that the more rapidly boiling liquid will carry over into the suction line.

When compressor *capacity* increases without a simultaneous load increase of the same or greater magnitude, evaporating pressure will fall. A sudden fall will move a superheat controlled expansion valve toward the *open* position. This increases the chance that liquid will reach the suction line.

When the compressor starts up, there is a capacity increase. If there is much liquid left in the evaporator from the previous operation, it may boil over. This is the reason for "pumpdown," which guarantees an empty evaporator when the compressor starts up.

2. A poorly designed flooded evaporator may continually slop over or deliver a mist of liquid with the suction gas.

3. Liquid feed valves (expansion valves) may be poorly selected, poorly adjusted or, like any mechanism, may occasionally fail.

If any of the things just described can result in liquid damaging the compressor, it is wise to install

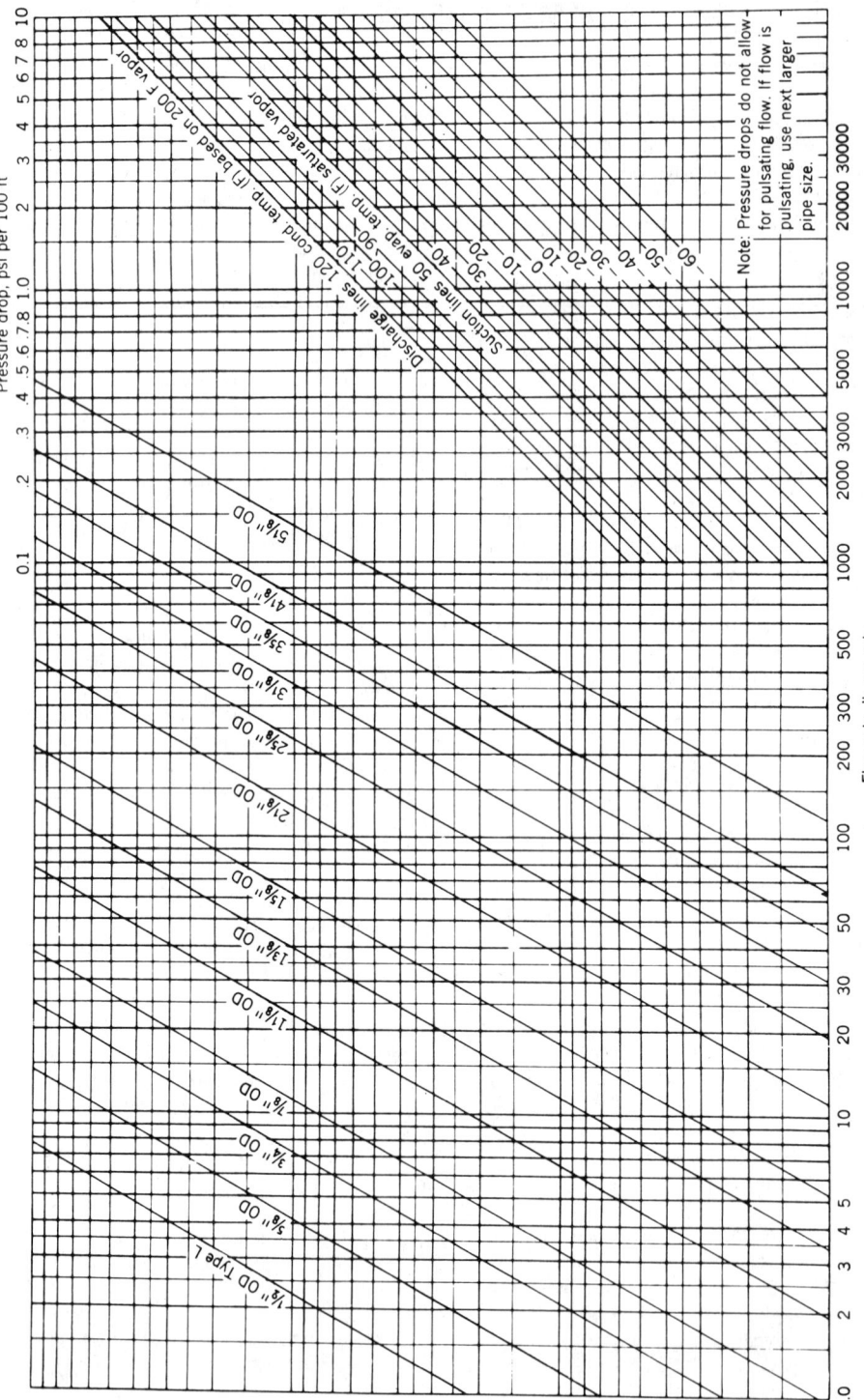

Fig. 1. Pressure drop in copper tubing for Refrigerant-22 vapor. Reprinted, by permission, from "Refrigerant Piping Data" published by Air-Conditioning and Refrigeration Institute.

an adequate and properly designed heat exchanger, suction trap or knockout drum.

When a compressor is idle, liquid reaches it due to temperature differences. With refrigerants like R-12 or R-22, a few degrees difference results in pressure differences corresponding to many feet of liquid head. Thus, it is unlikely that a trap of significant size can be installed. Much of the time, liquid boils in the evaporator, condenser or receiver and condenses elsewhere, passing through the piping as a gas making a liquid trap useless. If the liquid is at a higher temperature at one point in the idle system than at any other point in that system not already filled with liquid, it will boil at the higher temperature point, resulting in a higher pressure at that point and flow of gas to the lower temperature point where it will condense.

Refrigerant gas often flows through the compressor, from suction to discharge (the compressor valves, of course, impose no appreciable resistance to flow in that direction).

Control of temperatures or valving that *effectively* isolates sections of the circuit are the only means to keep the liquid where it is desired. Definite means to assure that the compressor is at a higher temperature than any other part of the system, before starting, is the surest way to assure that there is no liquid in it at that time.

Suction Lines

The following discussion of refrigerant piping will start at the suction line and proceed through the system in the direction of gas flow until the full refrigeration circuit has been traced.

The suction line conveys refrigerant gas from the evaporator to the compressor with the lowest economical pressure loss and, unless there is other provision to take care of the lubricating oil, it returns to the compressor the oil which has left the evaporator along with the refrigerant gas.

Normally, oil should be taken back at the same rate it leaves the evaporator. It should not be allowed to accumulate and return intermittently in large quantities ("slugs") that will damage the compressor. If this cannot be done, a knockout drum and separate oil return system is required.

The greater the pressure loss in the suction line, the lower is the pressure of the gas at the compressor intake; hence, the greater is the volume of gas the compressor must handle. The combination of greater volume and larger pressure differential significantly reduces compressor capacity and increases power

consumption per ton of refrigeration. Thus, the higher the suction loss, the greater are the initial and operating costs of the refrigeration system. Of course, increasing the size of the suction line to reduce the pressure loss will itself increase initial cost, so that the proper size is dictated by an economic balance. Usually, a pressure loss roughly corresponding to a 2F temperature difference is reasonable. At 40F evaporating temperature, a 2F difference corresponds to about 2 psi in the case of R-12, and 3 psi with R-22. At lower evaporating pressures, these pressure differences are smaller. They can be found from the appropriate refrigerant table.

Complete tables and charts are available for friction losses in pipe, fittings and valves. Figure 1 is a chart for R-22 vapor (or gas) in copper tubing, taken from "Refrigerant Piping Data," published by Air-Conditioning and Refrigeration Institute. Complete data for R-12 and R-22, as well as ammonia, are to be found in that booklet.

When, as is usually the case, oil must be carried with the gas, this factor also may affect the pipe size. If the pipe is horizontal, or if it pitches down in the direction of flow, the oil will be carried along with the gas regardless of its velocity. However, if flow is upward, line size must be checked.

Although a fixed minimum velocity has often been considered the principal criterion, this ignores other important factors. Table 1 contains friction values which should be maintained or exceeded when the flow rate is at its lowest value in any line pitching up in the direction of flow.

Tables 2 and 3 are derived from Table 1 for refrigerants R-12 and R-22, and copper and steel pipe. These tables give values in tons of refrigeration for different sizes and conditions since this is usually the most convenient form. The following illustrates the sizing of suction lines containing vertical rises.

Table 1. Minimum Friction for Gas-Oil Risers

Saturation Temperature, F	Pipe Size, Inches	
	< 2, 2	> 2
	Minimum Friction for Oil Return with Gas Flow is Upward, Psi per 100 Ft	
50, > 50	0.25	0.15
0 — 50	0.35	0.20
—50 — 0	0.45	0.25

Table 2. Minimum Recommended Load in Vertically Upward Refrigerant-12 Gas Lines*

Gas Condition[1]	Copper Tube Outside Diameter, Inches												
	5/8	3/4	7/8	1 1/8	1 3/8	1 5/8	2 1/8	2 5/8	3 1/8	3 5/8	4 1/8	5 1/8	6 1/8
	Minimum Recommended Load for Satisfactory Oil Lift, Tons[2]												
175F & 132 psia	0.30	0.48	0.75	1.5	2.7	4.3	8.8	12	18	27	40	70	110
50F sat.	0.21	0.35	0.55	1.1	2.0	3.2	6.5	8.8	14	21	29	52	85
40F sat.	0.20	0.32	0.53	1.05	1.9	3.0	6.2	8.5	13	20	28	50	80
30F sat.	0.19	0.30	0.50	1.0	1.8	2.8	5.8	7.8	12	19	26	48	75
0F sat.	0.15	0.25	0.4	0.8	1.4	2.4	4.8	6.2	10	15	21	38	62
—50F sat.	0.10	0.16	0.25	0.5	0.9	1.5	3.0	3.5	6.2	9	13	23	38

Gas Condition[1]	Steel Pipe Nominal Diameter, Inches												
	1/2	3/4	1	1 1/4	1 1/2	2	2 1/2	3	4	5	6	8	10	12
	Minimum Recommended Load for Satisfactory Oil Lift, Tons[2]												
175F & 132 psia	0.35	0.75	1.85	3.0	4.0	8.0	10	18	35	65	105	212	400	620
50F sat.	0.25	0.52	1.0	2.1	3.0	6.0	7.5	13	28	49	78	163	287	475
40F sat.	0.24	0.50	0.98	2.0	2.9	5.8	7.0	12	26	46	75	153	275	425
30F sat.	0.22	0.48	0.92	1.9	2.8	5.5	6.5	12	24	42	70	145	262	400
0F sat.	0.19	0.40	0.78	1.6	2.4	4.5	5.5	10	20	38	58	120	215	350
—50F sat.	0.12	0.26	0.50	1.0	1.5	3.0	3.5	6.2	12	22	35	75	140	225

[1] Tables, as noted are based on saturated condition for temperatures that are normal for suction, and superheated at pressure corresponding to 100F for a typical discharge condition. An increase in superheat (other conditions equal) increases tendency of gas to carry oil with it.

[2] Tons are based on: Refrigerant 12—Four lb per min per ton. For other conditions, divide by actual lb per min per ton and multiply by 4.

* Table reprinted by permission of Worthington Air Conditioning Co.

Example—Sizing Suction Lines With Vertical Rises.

Assume an R-22 evaporator operating at 40F evaporating temperature and that liquid reaches the expansion valve at 100F. Design capacity is 75 tons, but the evaporator must operate at as low as 25-ton capacity.

The suction line is about 90 ft long. The equivalent length of its fittings is estimated to be 60 ft, giving a 150 ft total.

Pressure loss is not to exceed 3 psi, so that the suction line size will be selected for not over (3) (100)/150 = 2 psi per 100 ft. The suction line contains a 10-ft vertical rise.

Refer to the chart, elsewhere in this section of the Handbook, Flow Rate per Ton for Refrigerant-12. For the conditions in this problem, the flow rate is 2.94 lb per min per ton. Thus, total flow at design conditions is (2.94)(75) = 220 lb per min.

From Fig. 1, the pressure drop in 3 1/8-inch copper tubing is 2 psi per 100 ft when flow is about 230 lb per min; thus 3 1/8-inch tubing is satisfactory for the full capacity.

Next, the suction line must be checked as to its oil return function at 25 tons. Reference to Table 3 shows that, at 40F evaporating, a 3 1/8-inch line will be satisfactory at as little as 20 tons.

What if the evaporator is to operate at as low as 15 tons? From Table 3, it is seen that a 3 1/8-inch line is not acceptable, but note that a 2 5/8-inch line is suitable down to 12 tons. The 10-ft vertical riser section, and its two elbows, would then have to be reduced to 2 5/8 inches dia. The equivalent length of this portion of the suction line would then be nearly 25 ft.

Table 3. Minimum Recommended Load in Vertically Upward Refrigerant-22 Lines*

Gas Condition[1]	Copper Tube Outside Diameter, Inches													
	½	⅝	¾	⅞	1⅛	1⅜	1⅝	2⅛	2⅝	3⅛	3⅝	4⅛	5⅛	6⅛
200F & 213 psia	Minimum Recommended Load for Satisfactory Oil Lift, Tons[2]													
	0.22	0.40	0.70	1.2	2.2	4.0	6.0	13	17	27	41	56	103	166
50F sat.	0.17	0.30	0.55	0.85	1.7	3.0	4.8	10	13	21	31	44	80	123
40F sat.	0.16	0.29	0.53	0.80	1.6	2.9	4.5	9.3	12	20	29	42	75	118
30F sat.	0.15	0.27	0.50	0.75	1.5	2.7	4.2	8.7	12	18	27	40	70	107
0F sat.	0.12	0.22	0.40	0.62	1.2	2.2	3.3	7.3	9.0	15	22	32	55	90
−50F sat.	0.08	0.14	0.26	0.40	0.8	1.4	2.2	4.7	5.7	9.3	14	20	37	57
Gas Condition[1]	Steel Pipe Nominal Diameter, Inches													
	½	¾	1	1¼	1½	2	2½	3	4	5	6	8	10	12
200F & 213 psia	Minimum Recommended Load for Satisfactory Oil Lift, Tons[2]													
	0.50	1.0	1.9	4.3	6.0	12	14	25	50	92	150	310	530	870
50F sat.	0.37	0.80	1.50	3.2	4.8	8.7	10.7	19	34	68	113	240	433	670
40F sat.	0.35	0.73	1.43	3.0	4.4	8.3	10.3	18	33	66	110	230	400	630
30F sat.	0.32	0.68	1.33	2.7	4.0	8.0	9.7	17	32	64	106	220	367	600
0F sat.	0.27	0.57	1.10	2.3	3.3	6.7	8.0	14	28	52	83	177	307	500
−50F sat.	0.18	0.37	0.73	1.5	2.2	4.3	5.0	8.7	17	32	50	107	197	310

[1] Tables, as noted are based on saturated condition for temperatures that are normal for suction, and superheated at pressure corresponding to 100F for a typical discharge condition. An increase in superheat (other conditions equal) increases tendency of gas to carry oil with it.

[2] Tons are based on: Refrigerant 22—Three lb per min per ton. For other conditions, divide by actual lb per min per ton and multiply by 3.

* Table reprinted by permission of Worthington Air Conditioning Co.

From Fig. 1, the pressure drop of 2⅝-inch tubing with 220 lb per min flow is about 4.3 psi per 100 ft. Thus, the loss in the 25-ft (equivalent) section is now about 1.1 psi, instead of the (25/100)(2) = 0.5 psi when it had been assumed to be 3⅛-inch dia, which is 0.6 psi more than desired.

Three alternative solutions to this problem present themselves:

1. *Accept the 0.6 psi additional loss.*

2. *Use larger size tubing for the horizontal* (125 ft equivalent length) portion of the suction line, so that, instead of a net drop of 3.0-0.5 = 2.5 psi in this portion of the line, the pressure drop will be 3.0 − 1.1 = 1.9 psi, or (1.9)(100)/125 = 1.5 psi per 100 ft.

Referring to Fig. 1, a 3⅝-inch line carries 290 lb per min per ton at this pressure loss.

3. *A double suction riser may be used.* This is shown in Fig. 2. When flow rate drops to the point that oil no longer returns up both risers, oil begins to accumulate in the trap, cutting off flow in the trapped riser. This leaves the flow adequate for oil return in the untrapped riser at the lowest capacity

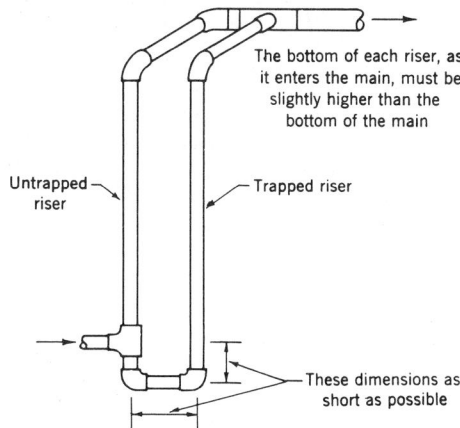

The bottom of each riser, as it enters the main, must be slightly higher than the bottom of the main

Untrapped riser

Trapped riser

These dimensions as short as possible

Fig. 2. Double suction (or discharge) riser.

of operation. At higher capacity, flow takes place in both risers at an acceptable pressure loss.

In this example, a 2⅝-inch line was found to be satisfactory for oil return at a 15-ton minimum

capacity, so that if a double suction riser were used, the untrapped riser would be sized 2⅝-inch dia.

From Fig. 1, the flow rate through a 2⅝-inch line at 2 psi per 100 ft loss is about 145 lb per min. If the loss at 220 lb per min is to be limited to 2 psi per 100 ft, the trapped line must have that pressure loss with a flow rate of 220 − 145 = 75 lb per min. Fig. 1 shows that a 2⅛-inch line satisfies this condition.

Double suction risers can be slug producers. After oil has sealed the trap, it will finally fill to just above the bottom of the tee inlet. In the trapped riser, the level will be as much above that point as will balance the pressure loss through the untrapped riser with 2.5 feet difference corresponding to 1 psi loss.

If the load increases gradually, so that increased pressure loss does not lower the level at the entrance to the trap more rapidly than new oil leaves the evaporator, the only limits to the height of oil in the trapped riser is the pressure loss at greatest flow rate through the untrapped riser or the height of the trapped riser.

If there is a sudden load increase or a sudden capacity increase (as when a compressor switches from 50% to 75% capacity) the increased loss through the untrapped line will probably be more than the height of the trap. This will break the seal and most of the accumulated oil will be carried into the suction main immediately. This has damaged compressors.

Keeping trap dimensions and the trapped line sizes as small as possible reduces the size of the oil accumulation. If there is a possibility that the compressor could be damaged by the return of this mass of oil, a properly designed knockout drum or suction trap of adequate size should be installed in the suction line. This would collect the oil and feed it to the compressor gradually.

Evaporators

Refrigerant evaporaters can be divided into two main categories: direct expansion and flooded.

Direct expansion evaporators are so arranged that refrigerant passes through the inside of tubes. There may be a single tube circuit or there may be any number of parallel-tube, equal-load circuits. The number of parallel circuits may be the same throughout the length of the evaporator or it may vary. For example, refrigerant may enter one end of each of twelve tubes and continue in the twelve parallel circuits until they all join the common suction line or it may enter 8 tubes and, on leaving those, branch to 12 tubes which branch to 16 tubes which finally terminate in the common suction line.

This type of evaporator may be cross feed, as in a coil having horizontal air flow. It may be up or down feed in the case where the fluid to be cooled passes up or down through the evaporator. Upfeed is usually the most desirable. (The direction of flow, upfeed or downfeed, is not a criteria which determines whether an evaporator is direct expansion or flooded.)

The refrigerant enters one end of each of the evaporator circuits and all of the liquid becomes a gas before reaching the end of its circuit. It is this characteristic of the evaporator which makes it the direct expansion type.

A direct expansion evaporator is usually fed by a superheat control (often called thermostatic) expansion valve. In special cases, a constant pressure valve, a capillary tube or even a high side float may be used. Except for many small evaporators using capillary tubes, the superheat control valve is used almost exclusively.

Most direct expansion evaporators are too large to have only a single circuit and therefore are divided into a number of parallel circuits. It is most important that all circuits fed by one expansion valve have the same capacity because the refrigerant distributor will feed approximately the same quantity to each. If the circuits fed by a single valve and its distributor do not have equal loading, the refrigerant feed will be controlled by the least loaded circuit and thus all other circuits fed by the same valve will have an effective capacity equal to that of the one with the least load. This will result in the heavier loaded circuits being starved in proportion to their greater loading. Unequal loading (such as results from uneven air distribution) is often mistakenly diagnosed as poor refrigerant distribution.

If there are not too many different loadings of circuits, a separate expansion valve may feed each circuit or each group of equally loaded circuits. But if there are too many unequally loaded circuits to make this practical, or if the relative circuit loading is not accurately predictable or if the loadings of too many circuits vary with reference to each other during operation, the use of a direct expansion evaporator will not be practical and a flooded evaporator or circulation of a chilled liquid must be used.

The flooded type of evaporator may be in several forms.

It may be of the shell and tube type with liquid refrigerant in the shell space and surrounding the tubes. The tubes may occupy the lower ½ to ¾ of the shell, with the liquid level well below the top of the shell (but covering the tubes) so that the space in the shell and above the tubes serves as a separation space. The tubes may occupy the whole shell and the outlet from the top join a surge drum where the separation of gas and liquid takes place. A line is usually taken from the bottom of the surge drum to return liquid to the bottom of the shell.

It may be a tubular type in which the liquid boils in the tubes. To be a flooded evaporator however, the refrigerant must enter one end of any tube but a liquid and gas mixture must leave the tube and enter a surge drum in which the gas separates from the liquid. The gas goes to the suction line and the liquid drains, with make-up liquid, to the tube for further evaporation.

Thus, in a flooded chiller, a particular part of the refrigerant will usually pass over the same part of the evaporator surface more than once before it is all changed to vapor. This may be by recirculation through tubes from the surge drum or by internal circulation over the tubes in the flooded shell.

A pump may be used to assist the refrigerant circulation. In the case of some large shell and tube units, a pump may take the liquid from a drain well at the bottom of the shell and pump it to sprays which wet the tube surface. This reduces the quantity of refrigerant necessary. Whenever refrigerant pumps are used, it is extremely important that the full net positive suction head is available. This requires a liquid level, above the pump, somewhat greater than the NPSH required for the pump and a low friction loss suction line so designed as to assure no cavitation.

The feed of refrigerant to a flooded evaporator is controlled by a valve in response to the liquid refrigerant level in the evaporator or surge drum. If there is only one evaporator in the system, the feed is sometimes controlled by the level at the condenser outlet (high-pressure float valve) thus merely keeping all of the liquid possible out of the condenser where it does no good and in the evaporator where it is useful.

Connections to direct expansion evaporators.
To assure steady oil return and that the expansion valve bulb for one circuit is not affected by flow from another circuit, suction connections from the evaporator should be properly made.

Figure 3 shows connections on a single evaporator controlled by a single expansion valve. Usually, only one connection from the suction header is used. The coil manufacturer's instructions should determine whether one or both are used. In either case, it is desirable, possibly necessary, to take the connection from the bottom, as shown.

Where two or more coils are stacked, the connections should be made as shown in A of Fig. 4. This helps assure that oil return will be steady and that liquid from an upper coil will not drain down and affect the control bulb of the lower coil as might be the case with B of Fig. 4.

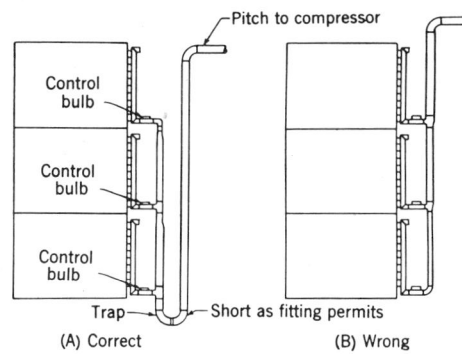

Fig. 4 Suction connections to multiple DE coils.

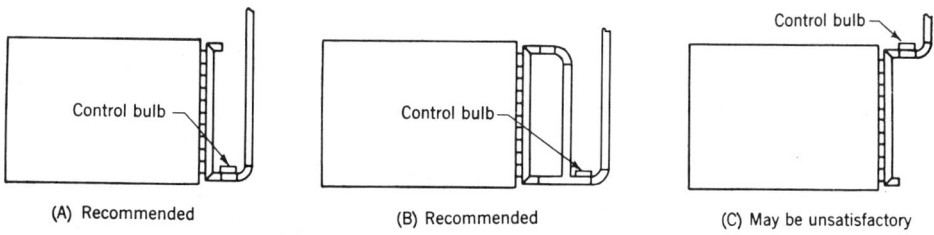

Fig. 3. Suction connections to single direct-expansion coils.

If the coils are to operate at lower capacity, the vertical upflow suction line must be sized in accordance with the instructions already given, including the decision as to whether or not to use a double suction riser as shown in *B* of Fig. 5.

If lower capacity is attained by stopping the liquid feed to one or more of the coils, the arrangement shown in *A* of Fig. 5 may be used to reduce the possibility of a requirement for the double suction riser as shown in *B*.

If, as is usual, the main is larger than the risers, the risers may enter the side or end of the main instead of the top. This saves an elbow in each riser. The reason for either arrangement is to assure that a little liquid (oil or refrigerant) on the bottom of the main will not drain into a riser having a low or zero flow rate.

In Figs. 3, 4 and 5, all suction lines are shown rising from the evaporators. This is probably the most frequent arrangement because the suction line is likely to be carried overhead. In some cases the most convenient route is down from the evaporator to a suction main or the compressor. This simplifies the connection since the suction line merely continues down. If the suction connections from two or more evaporators join, the connections should be made so that the flow from one evaporator will not affect the control bulb of another. This is most important when all evaporators do not operate together. Separate connections, as shown in *A* of Fig. 5, may be desirable, with the suction of each half of the coil bank dropping individually to a main below the bottom coil outlet.

Some expansion valve manufacturers would rec-

ommend a small trap ahead of the riser, as shown in Fig. 6, instead of the direct rise shown in *A* and *B* of Fig. 3. This trap reduces the chance of liquid accumulation at the control bulb.

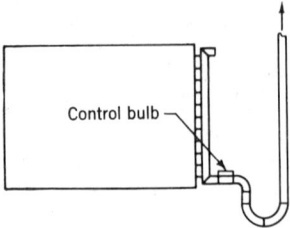

Fig. 6. Trap in suction line.

Many installations have been made, as shown in *A* and *B* of Fig. 3, without the trap. When the suction riser has been sized as previously described and the short horizontal from the coil is no larger than the riser, most, if not all of these installations have had perfectly satisfactory expansion valve operation. It is probably preferable, if the trap is omitted, to have the length of the horizontal from the coil at least two pipe diameters more than the bulb length. The bulb can then be located that distance away from the riser and hence, from the small amount of liquid that often remains at the bottom of the ell.

If an individual installer believes there is any possibility that he may later require help from the manufacturer in connection with the expansion valve operation, he would be wise to include the trap, unless the particular manufacturer approves its omission.

Oil Return

At the evaporating temperatures usually used in refrigeration systems for air conditioning, there is not likely to be any problem with oil return from a direct expansion evaporator if the suction line is designed in accordance with the preceding instructions. An oil separator in the discharge line is probably unnecessary.

If a flooded evaporator is used, the rate of oil leaving the evaporator will be less than the rate oil enters with the refrigerant liquid. The oil proportion in the evaporator will increase to an unacceptable value unless means are taken to recover it. This is done by taking a sample of liquid from the evaporator, boiling off the refrigerant and returning the remaining oil to the compressor.

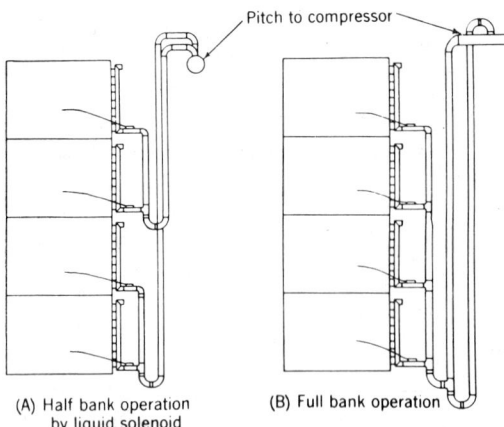

(A) Half bank operation by liquid solenoid

(B) Full bank operation

Fig. 5. Suction connections to multiple DE coils with capacity reduction.

Heat from the relatively warm liquid refrigerant before the expansion valve is usually used for the evaporation because this results in no loss of refrigeration capacity.

An oil separator in the discharge line is usually desirable with a flooded evaporator because any oil removed at that point reduces equally that to be recovered at the evaporator, thus lowering the cost of the equipment for this recovery. Fig. 7 shows an oil recovery arrangement for a flooded evaporator.

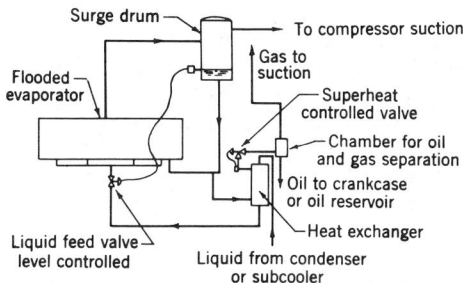

Fig. 7. Flooded evaporator with oil recovery. Arrangement is for refrigerants of greater density than the oil.

If there is much chance of unacceptable amounts of liquid in the suction line, or extra assurance is desired against liquid which might get into the suction line because of failure of expansion valves or other controls, a knockout drum may be installed in the suction line near the compressor. Fig. 8 shows a tangential entrance knockout drum, which is very effective. These are frequently installed without insulation and with a pan underneath to catch and drain any water which condenses on its outside surface.

If the quantities of liquid to the knockout drum are small and infrequent, heat from the surroundings of the vessel or even heat leaking to the drain line may be adequate to evaporate any liquid refrigerant. If not, these vessels may have a coil in their lower parts for the warm liquid. With this type knockout drum, the flow of gas is mainly in the upper part of the vessel so that a coil in the bottom adds very little superheat but is very effective for liquid evaporation.

Liquid draining from the knockout drum should pass through a filter which catches material down to micron size. A valve should be used to maintain the greatest possible flow rate such that the temperature of the liquid, before it enters the compressor crankcase, is 10F or more above that corresponding to saturation at crankcase pressure. Use

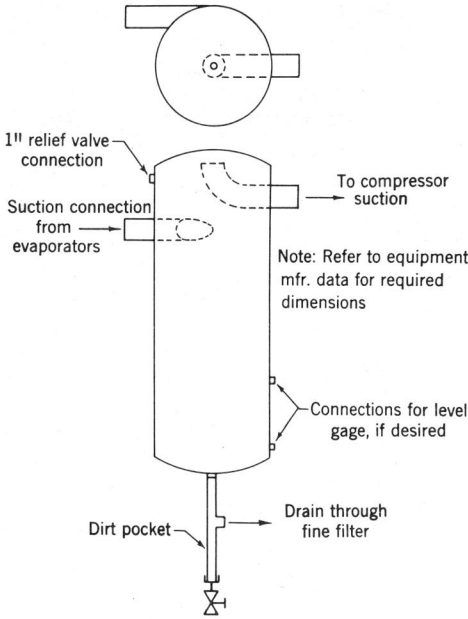

Fig. 8. One design of a knockout drum.

of a large pipe, or even finned tubing, for the drain line will help transfer heat to its contents.

Suction lines to parallel compressors. There is no problem in the suction connection to a single compressor. However, when there are two or more in parallel, the connections should be as symmetrical as possible to help keep the pressures in each compressor as nearly equal as possible and to get the oil to return equally to each. Figs. 9 and 10 show some arrangements for parallel connections.

Crankcase equalizer lines, one above the oil level and one below it, are shown connecting the compressors in these figures.

An equalizing line never completely equalizes; it does bring pressures closer together. If there is a difference in pressure between two points connected by a so-called equalizer, there will be a

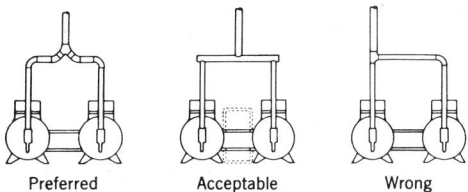

Fig. 9. Suction connections to two compressors in parallel.

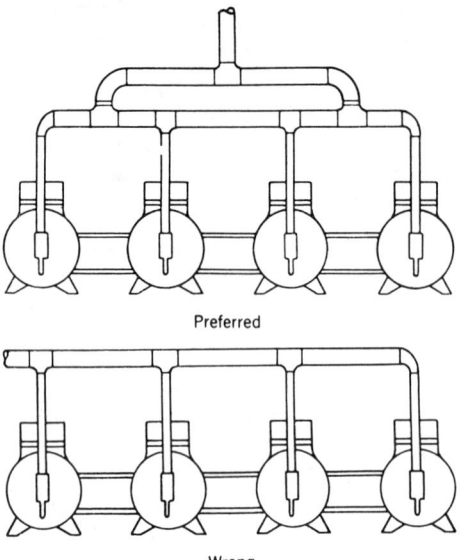

Preferred

Wrong

Fig. 10. Connections from single suction main to parallel compressors.

flow through the equalizing line with a corresponding pressure loss. The larger the equalizing line, the smaller the pressure difference. Equalizing lines often have valves for isolation purposes. The importance of small pressure difference for a small liquid level difference is indicated by the fact that 30 inches of oil level or about 24 inches of R-22 liquid correspond to one pound per square inch.

There are other methods of maintaining oil levels in the crankcases of several parallel compressors. Some maintain the proper level in each crankcase even though the compressors themselves may be on different levels, even one above the other. The compressor manufacturer should be consulted about such systems.

Discharge Lines

The function of the discharge line is to convey gas from compressor to condenser with lowest economical pressure loss and to carry any oil which is mixed with the gas along with the gas.

Pressure loss in the discharge line adds an equal amount to the compressor discharge pressure, increasing compressor power requirements. It has some affect on compressor capacity because an increase in compression ratio affects volumetric efficiency. The decision on pipe size is an economic one. A pressure loss corresponding to about 2F

temperature difference is reasonable. At 100F, this represents about 4 psi for R-12 and 6 psi for R-22.

The charts cited previously in the discussion of suction lines includes, also, data for discharge lines. Fig. 1, in this section, covers copper tubing with R-22.

Any upflow discharge lines carrying oil with the gas must also be sized so that, at the lowest flow rate, the oil will be carried by the gas. Table 1 gives values for vertical upward flow lines for any refrigerant; if R-12 or R-22, Tables 2 or 3 may be more useful. In those few cases where flow rate, at full capacity, results in excessive pressure loss through a vertical line sized for oil lift at minimum capacity, double discharge risers may be used. These are shown in Fig. 2 and explained under "Suction Lines."

It is best for the discharge line to pitch down to the condenser, eliminating the chance that oil or refrigerant liquid will drain back into the compressor discharge space. However, the relative locations of compressor and condenser frequently do not permit this.

When there is a considerable vertical rise from the compressor, any oil which was being carried up the discharge line will tend to drain back when the compressor stops. This might then drain into the compressor discharge space. Depending on the compressor construction, this may or may not do any harm. In an attempt to keep this oil from reaching the compressor, it has been frequent practice to provide a trap in a vertical up flow discharge line as shown in Fig. 11. A rule for determining the depth of this trap with any accuracy cannot be given but a value which seems reasonable (probably generous) is one inch for each foot of rise.

It is very important to realize that a trap of this type has no value for keeping liquid refrigerant out of the compressor discharge space. If there is liquid refrigerant in a condenser or receiver which are at

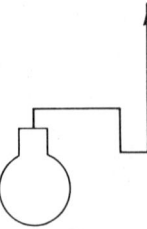

Fig. 11. Trap in discharge line keeps oil from draining into compressor.

a higher temperature than the compressor, the refrigerant will evaporate at the higher temperature point, the gas will pass down the discharge line, bubble through any oil and condense in the compressor discharge space.

If the discharge line temperature is lower than that of the compressor and the temperature of both is less than that of a condenser or receiver containing liquid, condensation will take place in the discharge line and the liquid will mix with the oil, making the trap of no value at all under such conditions.

If the compressor head temperature is higher than that of the discharge line, condenser and receiver, the smallest trap which keeps gas from moving away from the compressor, while liquid flows through the same pipe in the opposite direction, will keep liquid or oil from the compressor discharge space.

(If the discharge line is cooler than the compressor head and full of liquid, the liquid column increases the pressure in the compressor discharge space. This is equivalent to raising the condenser and receiver temperature to a value corresponding to saturation at a pressure as much higher as the equivalent of the column. The effect of this is illustrated by the fact that a 30 foot column of R-22 will raise an 80F condenser temperature to the equivalent of 86F).

A check valve in the discharge line, preferably near the condenser, will help keep liquid refrigerant out of the discharge line and compressor discharge space. Of course, its effectiveness depends on how tightly it closes.

Oil separators. There are many factors to be considered in making a decision to use or not to use an oil separator. However, these factors are not discussed here.

When a separator is used, it should be as near as convenience will allow to the compressor. The most desirable location is one where the discharge line dips down to it, but there is no objection to a 6-8 ft rise to it if the vertical line is sized to carry the oil.

The drain from the oil separator should go to the compressor crankcase unless there is a separate oil return system. In the drain line between the oil separator and the crankcase should be a float valve, a temperature or superheat control valve (responsive to conditions in the bottom of the separator) and a filter which will remove material down to micron size. Hand valves for isolation are useful.

A ¼ inch OD or ⅜ inch OD drain is adequate for any but the largest systems. Fig. 12 shows this oil drain line.

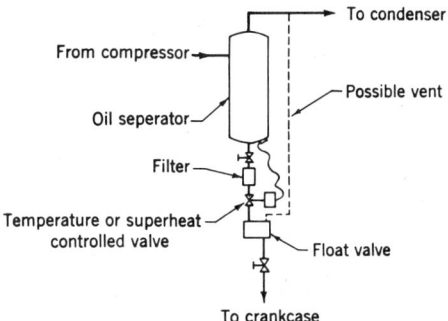

Fig. 12. Oil separator, showing how it is drained.

Many separators are made with the float valves inside them and, in that case the one shown in Fig. 12 would not be used. The inside float valve is inaccessible for cleaning or elimination of anything which causes it to stick. This makes the location shown preferable. The vent shown is usually found to be unnecessary.

When the compressor is not operating, the temperature of the separator may be *lower* than that corresponding to saturation at the pressure within it. Under these conditions, the separator becomes a condenser and the liquid resulting may drain into the crankcase, with probably serious results, unless a valve closes off the drain when the compressor is idle. If the temperature at the bottom of the vessel, where any liquid would collect, is several degrees *above* that corresponding to saturation at the internal pressure, there will be no free refrigerant liquid and what is left will be oil suitable for delivery to the crankcase.

The preferred valve for this service is responsive to superheat (like the standard valve for feeding liquid to the evaporator). However, the high temperatures during operation usually make the standard valves unusable. Since the minimum operating gas temperature is considerably above that corresponding to saturation at the highest condensing temperature, the result can usually be attained by the use of a temperature controlled valve. This can be a regular expansion valve controlled by *temperature only* or a solenoid valve and thermostat. The thermostat setting should be 10F to 20F above the highest expected condensing temperature.

Insulating a separator does not stop liquid accumulation. It merely retards the rate and is not

recommended. Insulation is sometimes applied to a separator, at low level, to protect against burning an individual who comes in contact with it.

Mufflers. These are frequently required to reduce discharge gas pulsations. (When an oil separator is used, it also serves to some extent as a muffler and often will make unnecessary an otherwise required muffler.) Flow through a muffler should be downward so that oil will not accumulate in it. Horizontal flow may be acceptable but manufacturers' instructions show be consulted.

Parallel Compressors. The discharge lines from two or more compressors should be reasonably symmetrical where they join the common discharge. Two discharge lines should never connect to opposite branches of a tee. The bottom of the discharge main should be lower than the bottom of the branches so that oil or refrigerant liquid cannot drain from the main to the branches.

A check valve (designed for compressor discharge service) should be installed in each branch just before it joins the main. If an oil separator happens to be used in each branch, the check valves should be between the separators and the main.

Since check valves cannot always be counted on to be perfectly tight, a good practice is to install a small ($\frac{1}{4}$ or $\frac{3}{8}$ inch) bypass with solenoid valve between each compressor discharge and suction. The solenoid closes when the particular compressor starts and opens when it stops. (In some cases, with the individual separators in each discharge branch and pumpdown control, all solenoids may be closed when no compressor is operating.) With this arrangement, everything from the check valve back, in an idle compressor circuit, will be at suction pressure. At any normal machine room or evaporating temperature, this assures that there will be no liquid in any part of the circuit. Such pressure conditions also make starting easy.

Condenser connections. If a single condenser is used, whether with single or multiple compressors, there is nothing special about its piping. The discharge and liquid lines connect to the points indicated by the maker of the particular condenser. If a so-called single condenser has multiple circuits which are not all previously joined so as to leave a single discharge and liquid connection, the instructions for multiple condensers which follow apply, unless the condenser manufacturer has specific and different instructions.

Multiple Condensers

Water piping connections. Two or more condensers may get their cooling water from a single water source but may not be a part of the same refrigerant circuit. The source may be a cooling tower, city water or water to be heated by the heat rejected from the condenser. When the condensers are not a part of a common refrigerant circuit, the water may pass through the condensers in parallel or series. Series connections will result in a lower condensing temperature for part of the refrigeration and thus result in more tons per horsepower for part of the system and probably for the system as a whole.

Series connection will complicate the control of water flow if it is to be kept at the lowest possible rate as normally required if the supply is city water. It results in increased total pressure drop unless the condenser can be purchased for, or changed so as to have an appropriate number of circuits.

Care must be taken in selecting water flow through condensers to avoid excessive water velocities in the tubes. More than about 10 ft per sec tube velocity is likely to result in tube erosion problems.

Series water connections are permissible when the condensers are part of the same refrigerant circuit if the refrigerant passes through the condensers in series. In such a case, the water and refrigerant flows should be countercurrent.

When condensers are part of the same refrigerant circuit and connected in parallel (connected to a common discharge and common liquid line) the water circuits of those condensers should also be parallel so that about the same temperature water enters each. The water flow rate through each should be approximately equal if the condensers are of equal capacity or approximately proportional to the respective capacity if the condenser capacities are unequal.

Refrigerant connections. When the refrigerant circuits of two or more condensers are connected in parallel, the piping sizes and arrangement must be selected carefully to assure proper operation.

Pressure drop through a shell and tube condenser is so small as to be practically negligible but there is an appreciable pressure loss through the entering and leaving connections, which usually have elbows and valves. Some other types of condensers, such as air-cooled (dry or evaporative), have a considerable pressure drop through the condenser itself, as well as the connecting lines.

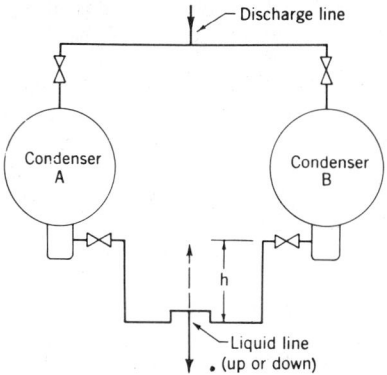

Fig. 13. Parallel hook-up of condensers of equal capacity.

Figure 13 shows what looks like and may be an ideal arrangement. The condensers are intended to be alike and the piping as near as possible to completely symmetrical. The pressure drop from the common discharge to the common liquid line, through either parallel circuit with equal flow, will always be the same. Suppose the water flow rate through condenser A is appreciably lower than that through B, so that less heat is removed in A. This results in less gas flow in the discharge and liquid branches for A and consequently less pressure drop in those branches than in the ones for B. To compensate for this difference the level of liquid in the B circuit must be higher above the junction than that in the A circuit. If the h dimension is not sufficient for a liquid head equal to or greater than the difference between the pressure losses in the two branches, the necessary additional head will be attained by a rise in the level above the bottom of the condenser. Any rise into the actual condenser has two possible disadvantages:

1. A rise in the condenser shell requires a considerable quantity of liquid for a small head so that a considerable amount of refrigerant is removed from use.

2. If the level rises much in the condenser it will cover some of the heat transfer surface and make it ineffective as condensing surface.

To avoid these problems, the distance h must be more than enough to result in a liquid head equal to the greatest difference expected between the pressure drops through the respective condenser circuits. The number of feet of head to equal one psi is $144/\rho$ where ρ is the refrigerant liquid density in pounds per cubic foot at the appropriate temperature, which may be taken as the expected condensing temperature. It will be noted that the head to balance one psi is around 2 ft for R-12 and R-22.

The lines from each condenser should be sized in the normal manner for liquid lines. It is better if the junction of the liquid lines is at a slightly higher head level than the bottom of any condenser drop leg as shown in the Fig. 13. This arrangement can be attained by the use of fittings or by pitching the lines slightly up toward the junction, as shown in Fig. 16, until the bottom of the pipe at that point is above the top of the pipe at the point where it leaves the vertical.

Liquid line connections are shown differently in Figs. 13, 15 and 16. These are merely alternate methods of joining the liquid lines. The dimension h is the important factor. Fig. 15 illustrates a method of obtaining the required head without having the lines to the junction low where they might be an obstruction. The limitation on the rise after the drop to attain the required h is the same as the limitation on the rise of any liquid line, due to flashing if there is not enough sub-cooling. Also, the longer these lines are, before the junction, the more friction to be used in calculating h.

Compressors and condensers in parallel can be

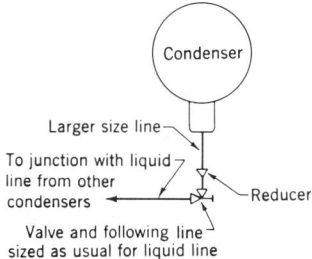

Fig. 14. Arrangement to reduce flashing at valve when liquid from condenser is saturated due to design and existing conditions.

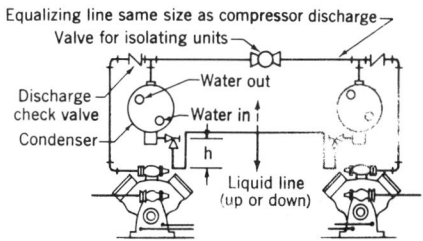

Fig. 15. Multiple condensing units operating in parallel.

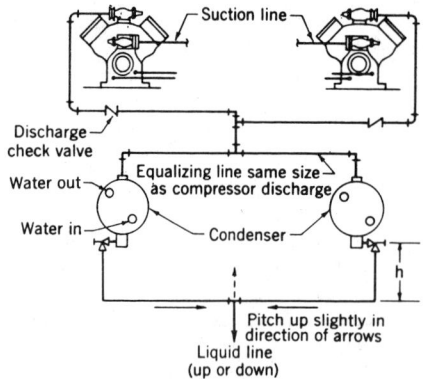

Fig. 16. Multiple compressors and condensers operating in parallel.

connected in the same manner as shown in Fig. 15 instead of with a junction of the discharge lines connecting to a common condenser equalizing line as shown in Fig. 16. The connection shown in Fig. 16 reduces the unbalance if a single compressor is running, probably reducing the size of h.

The dimension h, as mentioned previously, must be such as to allow the formation of a column large enough to balance the greatest difference in pressure drops which could otherwise exist between the parallel circuits. In calculating this difference, consideration must be given to the method of operation.

If a single discharge line from two or more compressors joins a gas header to two or more condensers as in Figs. 13 or 16, operation of one or more compressors while others are idle will have no effect on the relative flow to the various condensers.

If the connections for compressors and condenser are made as shown in Fig. 15, operation of one compressor while the other is idle will result in gas entering one condenser directly and the other through a so-called "equalizing" line. It is important to remember that connecting two units by an "equalizing" line does not make the two pressures equal. It merely makes them more nearly equal. If, without the connecting line, the pressure difference would be very large, the flow through the connecting line will be relatively great, resulting in an appreciable pressure drop in that line. Only by making this connecting line a larger pipe size, shorter, or containing less fittings or valves, can the pressure loss through it be decreased. Although this pressure drop may be reduced to a small amount, it cannot reach zero. In considering the size of this small pressure drop and the resulting difference in pressure between the points to which this "equaliz-

ing" line connects, it must be realized that $\frac{1}{4}$ psi corresponds to about a 6 inch column of R-12 or R-22.

If one compressor is to operate while the other is idle, making the connections with the tees as shown in Fig. 15 will give better equalization than entering the condensers directly with ells and connecting the equalizing line to the branch of a tee in each line. However the connections are made, h must be large enough to compensate for any difference.

With water-cooled condensers in the same refrigerant circuit, water will usually go through all condensers even though only a part of the compressor capacity is effective, because keeping all of the condenser surface in use either reduces the condensing pressure or, if the pressure is held constant by an automatic water valve, reduces the water consumption. In this case, the refrigerant pressure losses through all condensers will probably be made about equal by proper adjustment of the water flow and the necessary h will be small when connections are made as in Figs. 13 and 16.

If, for any reason, water may not flow through all the condensers in the same parallel circuit, h for the drain leg of any condenser must be equivalent to the pressure loss (at full condenser capacity) through that condenser and its circuit. This is true because the flow through a condenser not being cooled will be close to zero and the pressure drop will be nearly zero.

If two or more condensers discharge to a common liquid line which feeds a single evaporator circuit, a high side float on each condenser drain may be the best control. Such a control eliminates the need of the drain leg.

The illustrations and discussion, up to this point, have been based on shell and tube condensers. In general, they apply equally to evaporative, air-cooled or other condensers having appreciable refrigerant pressure loss. The only difference is that the pressure loss through the condenser itself must be added to the loss through its parallel circuit when determining the value of h.

The pressure loss through most air-cooled or evaporative condensers is likely to be 2-7 psi or higher. This pressure loss, corresponding to the rate of heat transfer for which it is to be used, should be obtained from the manufacturer of the particular condenser.

If the principles outlined above are properly applied, any type of condenser may be connected in parallel with any other type, if physical location

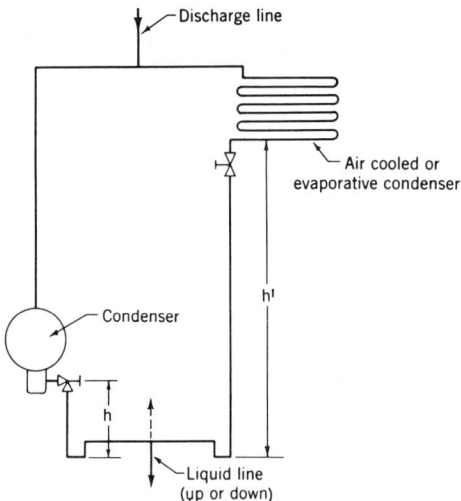

Fig. 17. Parallel connections for two types of condensers.

and spaces permit drain legs long enough to provide a liquid column of sufficient height for each condenser.

Figure 17 shows parallel connections for two types of condensers. The dimension h must result in a liquid column corresponding to the pressure drop between the discharge junction, through the shell and tube condenser, to the liquid junction, when that condenser is taking the *largest* load it is expected to take, less the pressure drop, between the same junctions, through the other condenser when it has the *least* load.

The dimension h' must correspond to the pressure loss through the air-cooled or evaporative condenser circuit when it is transferring the *most* heat, less the pressure loss through the other condenser circuit while it is transferring the *least* heat.

It is usually not wise to assume both condensers operating at design heat transfer rates at all times. In the case illustrated by Fig. 17, the water for the shell and tube condenser might be at 85F. The other condenser may be evaporative and its ambient on a cool night might be 50F wet-bulb. The sizes might be such that all the heat could be transferred by the outdoor condenser at 80F condensing temperature. Under such conditions, the shell and tube condenser would do nothing and the pressure drop through its circuit would be zero. Under other conditions, the shell and tube condenser might transfer all the heat while the other transfers nothing.

With such possibilities, the safe way is to have the h and h' correspond to the pressure loss through the respective condenser when it is transferring the most heat at any conditions anticipated. No deduction is made for the pressure drop through the other condenser circuit because of the possibility that it might be transferring no heat at that time. This may result in greater h or h' dimensions than necessary but the intelligent design on any other basis requires accurate information on all the operating conditions which might occur.

Occasionally there are cases when it is practical and possibly, even preferable, to connect the condensers into the refrigerant circuit in series. However, it must be realized that there is a pressure loss in each condenser and its connections and the condensing temperature will therefore be lower in each successive condenser in the direction of refrigerant flow. The water or other condenser cooling medium must therefore be at a lower temperature in each succeeding condenser or there will be a smaller temperature difference for heat transfer. If the pressure drop through each condenser and its connections is small, the temperature difference at which condensation takes place in each condenser may be so small as to be nearly negligible. If the temperature of the cooling medium in any condenser might at any time be above the temperature corresponding to the pressure of the gas and liquid entering from a preceding condenser, re-evaporation or flashing would occur and the result will be most unsatisfactory.

There are other possible arrangements and connections for condensers and all cannot be described here but an attempt has been made to explain the principles. If these are properly applied, with full knowledge of all operating conditions, the performance will be satisfactory.

Liquid Lines

Within reasonable limits, pressure loss in liquid lines has no effect on system capacity or power requirements. Pressure loss in liquid lines may be the result of a column of liquid when flow is upward. (When flow is downward, the liquid column is, of course, a pressure gain.)

The factors to be considered are:

1. Pressure loss must not result in liquid pressure dropping below that corresponding to saturation at its existing temperature, at any point before the expansion valve. If the pressure is lower, evaporation or "flashing" will occur at that point, causing

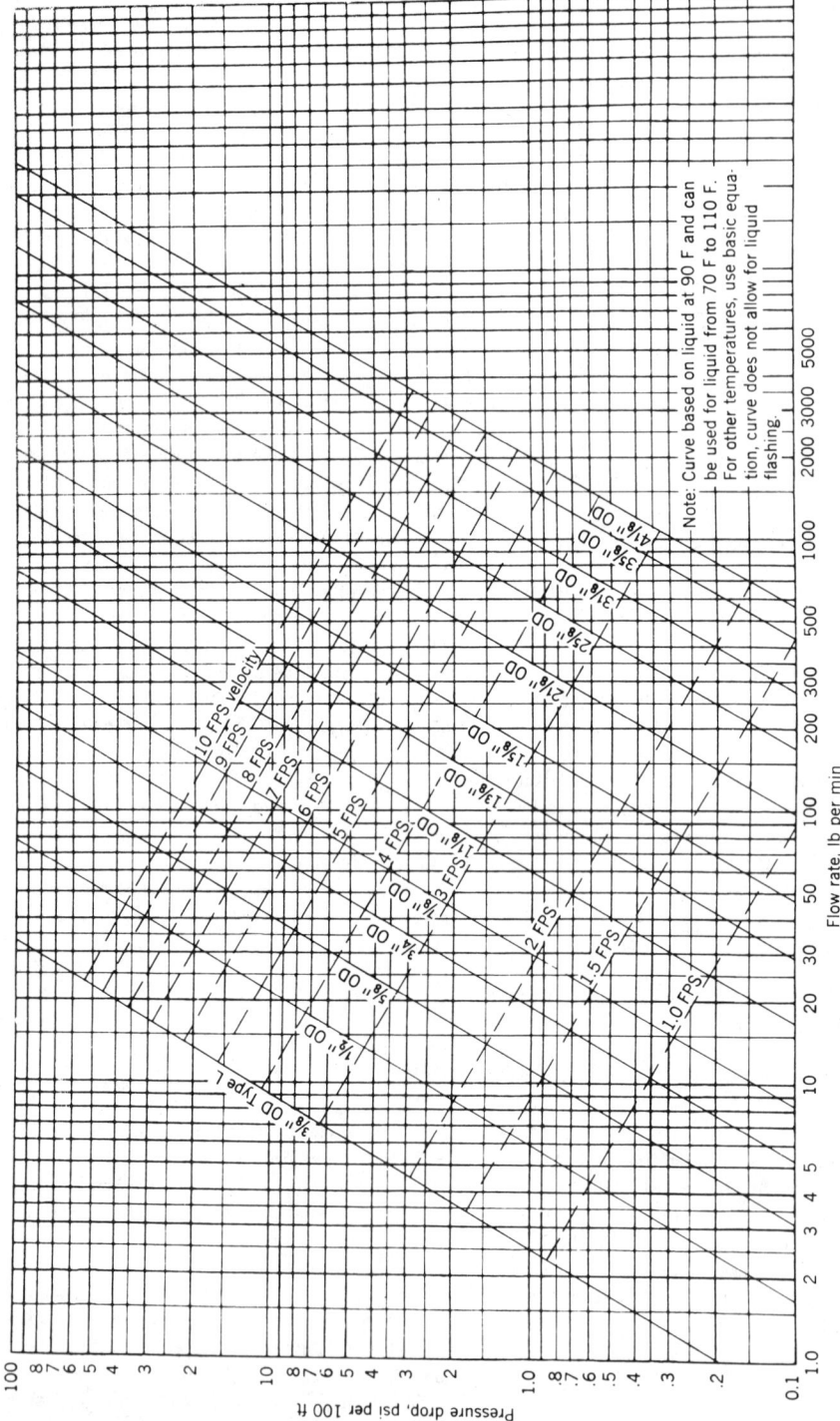

Fig. 18. Velocity and pressure drop in copper tubing for R-22 liquid. Reprinted, by permission, from "Refrigerant Piping Data" published by Air-Conditioning and Refrigeration Institute.

higher flow resistance, so that liquid may not be able to get to the evaporator at the desired rate.

2. If quick-closing solenoid valves are used, the velocity in the liquid lines should not exceed about 5 ft per sec.

3. If the line passes through places where the temperature is higher than that of the liquid, insulation may be desirable; sometimes it is necessary.

4. Pressure and temperature of the liquid at the entrance to the expansion valve must be such as to result in the desired rate of flow through the existing expansion valve at the desired evaporator pressure.

Fig. 18 gives pressure loss and velocities for R-22 in copper tubing.

Refrigerant liquid leaves all condensers with some subcooling. The amount of subcooling depends on the design of the condenser and the difference between the temperatures of the condensing refrigerant and the coolant.

For example, in air-cooled or evaporative condensers of usual construction, subcooling is about 25% of the difference between the condensing temperature and the entering-air temperature (dry-bulb for air-cooled, wet-bulb for evaporative cooled). If a given air-cooled condenser is operated with a loading such that the condensing temperature is 110F with 95F air, subcooling will be about 0.25 (110-95) or about 4F. If the loading is such that the condensing temperature is 125F with 95F air, the subcooling will be about 0.25 (125-95) or 8F. (These values presume no condenser flooding.)

In the standard shell and tube condenser, subcooling probably is not much less than 25% of the difference between the condensing temperature and entering water temperature. If the condenser is of a type with a lot of space and without tubes in its lower part, some of the subcooling potential may be lost due to gas heating of the liquid. In some shell and tube condensers the tubes are carried to the bottom and a baffle keeps the gas from contacting the liquid as it passes over those tubes so that there is more subcooling.

Subcooling can be increased further by passing the liquid through additional cooled surface, even a refrigerated cooler if necessary. Except in special cases, refrigerated subcoolers are not used.

Since the liquid leaving most average condensers, during full capacity operation has 4F to 7F subcooling (corresponding, at 100F condensing, to 11-20 psi in the case of R-22 and to 7-13 psi in the case of R-12) there is usually no problem if the liquid line is taken directly from the condenser outlet without going through a receiver. Unless there is a considerable rise in elevation to the evaporator, the available pressure drop is enough to overcome the losses through the solenoid valve, the filter (or filter-drier) and the liquid line.

With air cooled or evaporative condensers, it is possible to increase subcooling considerably, with no loss in performance, by adding enough liquid charge to equal about 6% of the condenser volume. This will result in about double the subcooling with about 6% greater difference between the condensing and entering dry or wet-bulb temperatures. Thus, these two types of condensers offer considerable flexibility.

Receivers

Receivers have been a frequent source of operating problems, although as a source of trouble they are often not suspected.

A receiver is basically a bulge in the liquid line in which an attempt is made to keep gas in its upper part while liquid flows freely through its lower part. Since this gas can condense, or some of the liquid evaporate and raise its pressure, depending on the surrounding and liquid temperatures, the consequences are not always as desired.

A receiver should be used in a system only if required for a particular purpose. Use of a receiver when one is not required is likely to be detrimental to system operation. If a receiver is not properly connected or is not correctly located (the correct location is sometimes below the condenser, but there are many cases where operation is poorer or impossible when so located), it may seriously hinder system performance.

The function of a receiver is to provide a reservoir for removal of refrigerant from some other part of the system. The following are probably the only valid reasons for using a receiver:

1. It may be desirable to remove refrigerant from the entire system so that it can be isolated in a place not likely to leak when the isolation is for prolonged periods. (If the receiver for this purpose is one through which the refrigerant flows, as will be explained later, it will usually be desirable to arrange to bypass it during normal operation.)

2. It may not be possible to pump the whole charge of refrigerant into the condenser either for automatic pumpdown or if it is desirable to remove refrigerant from a part of the system to do work on the evaporator, clean a strainer, replace

a drier, etc. Before installing a receiver for this purpose, the pumpdown capacity of the condenser should be checked. This is available for most condensers.

3. It may be desirable to operate the system with the refrigerant in only part of the evaporator while maintaining full condenser capacity; for example, in a low temperature system where, to avoid excessive heat rejection during pulldown, it is desirable to have only part of the evaporator capacity while retaining the full condenser capacity.

4. The head pressure control may be one which floods the condenser to reduce its effective surface. There must be a receiver to contain this refrigerant when full condenser capacity is required.

Most receivers are connected so that the liquid runs through them. If the relative levels of condenser and receiver permit, the receiver can be connected as a surge vessel and when so connected, it does not have the disadvantages of the "flow through" receiver.

The use of a flow through receiver may result in problems for two reasons.

1. Liquid leaves such a receiver practically saturated, even though it had been considerably subcooled when it left the condenser.

2. If the air surrounding such a receiver has a temperature much above the condensing temperature, the pressure in the receiver becomes higher than that in the condenser.

The first problem reduces the possibility of liquid leaving the receiver and in a number of cases has reduced the rate of flow to the point where system operation was at a small fraction of what was desired.

The second makes it impossible for liquid to flow from the condenser to the receiver. If the condenser is a type that has a large cross-section for gas flow (such as a shell and tube condenser) and thus a very small pressure difference in the gas space, the difficulty can be easily overcome. The receiver is mounted below the condenser. The liquid line between the vessels is made larger than required for the liquid flow and arranged so as to have no traps. Any globe valve is mounted so that its stem is horizontal to assure that it does not create a trap. With this arrangement, gas can readily flow from the receiver to the condenser while still permitting liquid to drain, in the same pipe, from the condenser to the receiver.

Since, with water cooled condensers, the temperature of the air in the machine room around the receiver has been often above the condensing tem-

perature, this has been a common arrangement. The liquid line between condenser and receiver was not sized for pressure loss but for a velocity of 100-150 ft per min.

There is a better method of connecting the condenser and receiver which is described below but, where the arrangement is as above, the liquid line must be sized as indicated. However, this is not a general rule for sizing lines from any condenser to a receiver. It applies only to the condenser having an insignificant internal pressure loss, with the receiver below it and the possibility that the temperature of the air surrounding the receiver may be higher than the condensing temperature. In all other cases, the size should be based on pressure loss and the velocity as it affects the pressure energy equal to the velocity energy.

Sometimes shell and tube condensers have been connected to receivers as shown in Fig. 19. This is a better method. The liquid line need not be oversized, but a vent must be added. The vent line size must be a little larger than required for the generally recommended arrangement of Fig. 20 because the exposed surface of the subcooled liquid continually passing through serves to condense the gas and increase flow through the vent.

As shown in Fig. 19, liquid must pass through the receiver where it losses some (probably nearly all) of its subcooling as it condenses gas, thus making the flow in the liquid line, after the receiver, more critical. Also this requires a minimum level of liquid in the receiver to match the flow rate at full capacity. This liquid is not a reserve because it is the least that can be in the receiver during operation. Only liquid above this level is a reserve. That below is a useless investment, since proper connections eliminate any need for it.

Figure 20 is the preferred method for connecting the receiver. It requires no liquid in the receiver

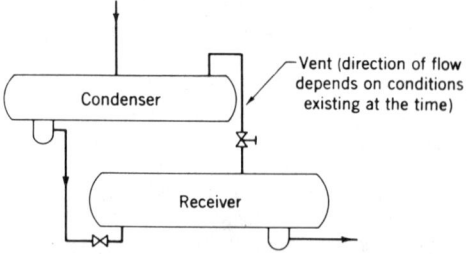

Fig. 19. A method of connecting condenser and receiver, but method shown in Fig. 20 is better.

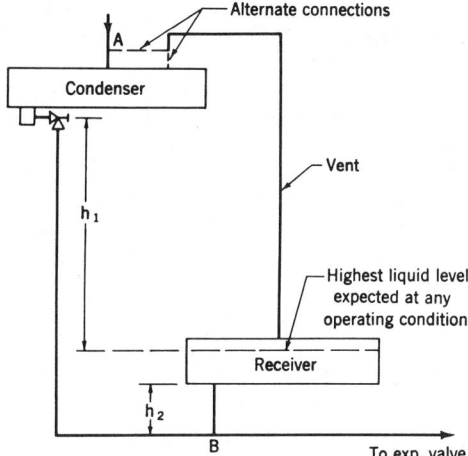

Fig. 20. Recommended method of connecting condenser and receiver. The relative elevations are critical.

Table 4. Rate of Flow in Receiver Vent of Fig. 20

(Flow is from receiver when temperature of air is higher than the condensing temperature, and vice versa.)

Difference Between Condensing Temperature and Surrounding Air Temperature, F	R-12	R-22
	Flow, lb per min per 100 sq ft of Receiver Surface	
10	1.0	0.8
20	2.0	1.7
30	3.0	2.5
40	4.0	3.3

for operation (any in it is all usable reserve) and does not lose any subcooling because the liquid does not pass through it.

The value of h must be such that a liquid column of that height is at least equal to the pressure loss from A, through the condenser and liquid line, to B, at the greatest capacity, plus the loss in the vent line when the temperature of the air surrounding the receiver is the highest value above the condensing temperature ever expected during operation. Since the pressure loss through a shell and tube condenser is small, the required levels are available practically without exception. For an air-cooled or evaporative condenser, the pressure loss is usually between 2 and 7 psi. This value can generally be obtained from the manufacturer. It will enable the determination of h. A decision can then be made as to whether or not the levels are available.

The vent must also be adequate when the temperature of the air is lower than the condensing temperature so that gas condenses in the receiver, lowering the pressure such that there is a flow in the vent line toward the receiver. The value of h_2 (Fig. 20) must represent a liquid head at least equal to the vent line pressure loss at this condition, minus the pressure loss from A to B through the condenser and liquid line. Since at very low loads this latter loss is small, it seems wise to ignore it.

Table 4 gives reasonable rates of flow through the vent line assuming reasonably still air around an uninsulated receiver. If the receiver is insulated with the equivalent of about one inch of cork, the values in Table 4 would be about 10% of those given.

As an example of vent selection, assume a receiver with 80 sq ft of uninsulated surface, containing R-22, with surrounding air temperature varying from 30F below to 20F above the condensing temperature. The equivalent length of the vent (including valve and fittings) is about 50 ft. The height of h_2 (Fig. 20) is not to exceed 6 inches.

As possible liquid temperature from the condenser, about 2.2 ft may correspond to 1 psi, so that the permissible pressure loss is $6/(2.2)(12)$ = 0.23 psi for the 50 ft length or 0.45 psi per 100 ft.

The flow rate to the receiver is used for selecting h_2 and it is based on a 30F difference. The flow rate is $(80/100)(2.5) = 2.0$ lb per min. The pressure of the gas in the vent line may correspond closely to a low condensing temperature. Therefore, approximately 80F will be used when selecting the vent size from Fig. 1.

From Fig. 1, for 0.45 psi per 100 ft and 2.0 lb per hr, a ¾ inch OD tube is the smallest suitable.

When the temperature of the air is 20F above the condensing temperature, the flow in the vent line from the receiver is about $(80/100)(1.7)$ = 1.36 lb per min. The loss through the ¾ inch OD line is about 0.14 psi per 100 ft or 0.07 psi in 50 ft. The corresponding liquid column is $(0.07)(2.2)$ = 0.154 ft.

The value for h is the greatest pressure loss from A to B (Fig. 20) through the condenser and liquid line, plus 0.154 ft. It may be desirable to check the loss from A to B at two or three expected operating conditions. It is likely to be greatest when the condensing temperature is lowest.

The line from the condenser to B is sized, like any liquid line, for the pressure loss permissible for that particular length. The line from receiver to B may be the same size or a little smaller.

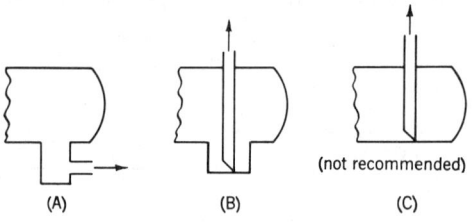

(not recommended)

(A)   (B)   (C)

Fig. 21. Types of receiver outlets.

Flow Through Receivers

If a receiver is required and the equipment cannot be so located so that their relative positions are as in Fig. 20, the flow-through receiver must be used. Satisfactory operation requires:

1. That the air temperature around the receiver be not much above (in some cases, less than) the condensing temperature. For greatest flexibility in locating the receiver, it should be colder than the subcooled liquid.

2. The rise in elevation of, and the friction loss in, the liquid line, from the receiver to the expansion valve, should be very small. This includes the entrance loss as it leaves the receiver.

The first requirement is likely to be met automatically in the case of an air-cooled condenser because the receiver is likely to be surrounded by the same or cooler air than air fed to the ccondenser. At worst, it is usually in the air leaving the condenser.

With this first requirement met, there is no necessity that the receiver be below the condenser. The liquid will flow from a condenser up to a receiver just as well as it will flow to the expansion valve, up to the point where it is no longer subcooled or the air temperature around the receiver is higher than the liquid temperature.

If the evaporator is above the receiver, operation will be improved if the rise is between the condenser and receiver, rather than between the receiver and the expansion valve.

In the case of the evaporative condenser, the first requirement is frequently not met because, in many localities and particularly at low capacities, air temperature around the receiver is likely to be higher than the condensing temperature, which is controlled by the wet-bulb temperature.

Every effort should be made to avoid using a flow-through receiver in the circuit during operation. If one must be used, its surface should be kept wet when the system is in operation.

Vents, from the receiver to the outlet of an evaporative condenser, are almost never of any value.

Since the liquid leaving a receiver is very slightly subcooled, it is not only important to keep friction loss and liquid elevation small, but to make it as easy as possible for liquid to leave the receiver. The velocity in the liquid line leaving it should be kept low. The head loss required for a particular velocity is found from

$$h = v^2/2g$$

where h = liquid height, ft

g = 32.2 ft per sec per sec, and

v is the liquid velocity in ft per sec.

Since the liquid probably forms a vena contracta upon entering the pipe, it is safest to take the velocity based on about 0.6 of the cross-section of the pipe.

Fig. 21 shows three types of receiver outlets: B is as good as A if the liquid line rises from the receiver; C requires an extra refrigerant charge and is therefore not recommended.

Accessories

Subcoolers are used where advantageous for economy or when necessitated by the existing liquid line pressure loss. When a receiver is used, they are always installed between receiver and expansion valve.

Expansion valves. The superheat control type only will be discussed here and only so far as it affects piping and bulb location.

The pressure under the diaphragm should be the pressure in the suction line at the point where the bulb is located. If there is an extremely small pressure drop through the evaporator, so that the pressure at the expansion valve is practically equal at all times (even at maximum load) to the pressure in the suction line at the point where the bulb is located, a valve in which the space under the diaphragm is directly open to the discharge space of the valve may be used. This is known as an *internally equalized valve*.

Normally however, there is a considerable pressure drop through distributors and also through the evaporator itself, so that there is a considerable (and probably variable) difference between the pressure at the valve discharge and that in the suction line. This makes it necessary to use a valve in which the space under the diaphragm is not open

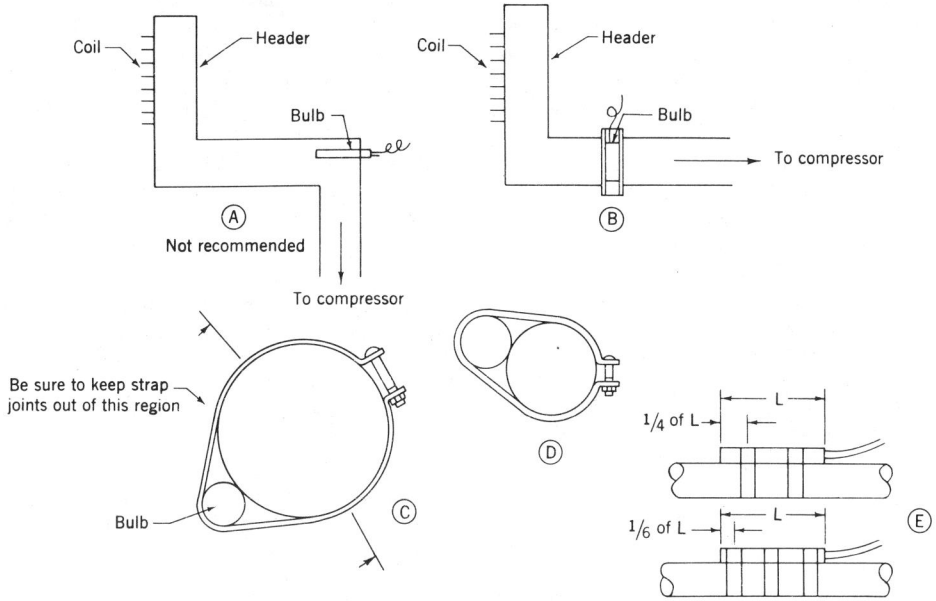

Fig. 22. Bulb locations and applications.

to the valve discharge. A separate small line then connects the space under the diaphragm with the suction line at the point where the bulb is to be located. (Usually, the equalizing line is connected to the suction line at a point immediately following the bulb. This small line is called an *external equalizer*.

If there is any doubt as to whether or not it is necessary, the external equalizer should be used. It will never be harmful.

If the valve is arranged for external equalizing, one must be used.

To control properly, the bulb should be at exactly the same temperature as the gas in the suction line at the point leaving the evaporator (or the superheat source if it is separate from the evaporator). If liquid reaches that point, the temperature of the bulb should immediately fall to saturation temperature so that the valve will close. Since the suction gas temperature is likely to keep changing and heat transfer from and to the bulb is required to change its temperature, it is important to have the best possible thermal path between the bulb and the suction gas and the poorest possible thermal contact between the bulb and any other surroundings.

A bulb mounted in a well in the suction line can give very satisfactory results but it is most important that the bulb fit the well very closely or have other means of good thermal contact with the well. A well filled with any liquid for contact purposes has the disadvantage of adding more mass and making the bulb lag more in following the suction gas temperature. The well should be so mounted in the suction line that refrigerant liquid cannot pass along one side of the pipe without touching the well. In the well shown in Fig. 22(a), liquid can pass along the bottom of the pipe without touching the well. A type of well shown at (b) is open at both ends and welded in the pipe so that it passes completely through. It is of such an inside diameter that the bulb fits it quite closely. This type meets the requirements very well.

A properly applied, externally mounted bulb will also perform well and may even be better than some internal well mountings. The real problem of external mounting is that the bulb and pipe are usually cylindrical and two cylinders can at best have only line contact, which results in an extremely small contact. To obtain good heat transfer, the bulb is attached with straps which should be of generous cross section, of a metal conducting heat

well and soft enough to wrap well and have good contact with the bulb and the pipe. Both the bulb and the pipe should be clean and have a good smooth surface so that the straps can make good contact. The joints of the straps should not be in the space where heat transfer is important. See Fig. 22(c). The bulb should never be attached with tape or wire.

Consideration should be given as to which side of the pipe liquid would contact if it passes the end of the evaporator. It will be assumed to be on the bottom of the pipe, since that will usually be the point of contact, but should not always be assumed so because, in some cases, inertia may carry it to the top or sides. There are times when it may be necessary to locate the bulb on a vertical line. On the basis of any liquid being on the bottom, Fig. 22(c) and (d) show good bulb locations.

The general rule is to locate the bulb as near as possible to the point where liquid will be expected to contact the pipe while still having strap contact at that point. This results in a different bulb location for different pipe sizes, as is clear from Fig. 22(c) and (d).

The straps should not be at the ends of the bulb. If two straps are used, they should be about ¼ of the bulb length from the ends of the bulb. If three are used, one should be in the center and the other two about 1/6 of the length from the ends. This keeps the heat transfer path as short as possible. See Fig. 22(e). Extra straps contribute to good operation.

Better distribution, particularly at part capacity, is likely if the expansion valve discharges vertically downward. Vertically upward is the second choice, and horizontal discharge the least desirable.

Strainers or filters. An appropriate filter should precede anything which could be damaged or have its operation interfered with by particles of foreign matter. The filter should be selected to remove particles down to the size which can interfere with operation. Very fine filters should be used at least for the operating parts of any automatic valve.

Sight glasses and moisture indicators. At least one of these should be installed in the liquid line of every system, to indicate if any gas is entrained in the liquid, and to indicate the moisture content of the refrigerant, so that proper steps will be taken to dry it if necessary.

A sight glass is also desirable in such places as drain lines from oil separators and knockout drums.

Suction-liquid heat exchangers can be of several types. The simplest is a liquid line soldered to a suction line. In another, called a double pipe, liquid passes through the annular space between two concentric pipes and the gas through the inner pipe.

Exchangers are commercially available in shell and coil construction with large gas side cross-sections, and in other forms where the cross-section of the gas side is about the same as the suction line.

Suction-liquid heat exchangers are installed for several reasons:

1. To evaporate any liquid which is carried with the suction gas. Most shell and coil exchangers are not very good for this purpose because the liquid can pass through without touching much of the coil surface. Some other types, including the double pipe, are better for this purpose.

2. To increase the degree of superheat in the suction gas. This may be for reasons of economy in the case of refrigerants like R-12, or because more superheat is needed for other reasons. When it is being done for reasons of economy, it should be ascertained whether the additional suction pressure drop, due to operation of the exchanegr, entails a loss greater than what is gained by the superheating.

3. To provide additional liquid subcooling which may be desirable to increase the available liquid line pressure loss.

The expansion valve bulb should, almost without exception, be mounted on the suction line between the evaporator and the heat exchanger.

On very rare occasions, it is desirable to cool a fluid with refrigerant liquid whose evaporating temperature is very little below the initial fluid temperature. When the difference is too small to provide the superheat for the expansion valve, one way to obtain this superheat is from the relatively warm liquid about to enter the expansion valve.

For this purpose, a heat exchanger is installed about a foot or so from the evaporator on each section fed by a single expansion valve. The bulb and external equalizer are on the suction line next to the exchanger on the compressor side. For this to provide satisfactory operation, several rules must be followed:

1. The exchanger should be of the type whose cross-sectional area for gas is about the same as the suction line. No liquid or oil can be trapped in this type of exchanger.

2. Liquid and gas flows should be actually countercurrent.

3. If flow is not vertically upward, the exchanger must be pitched upward in the direction of flow.

4. If the evaporator has more than one section, there must be a separate exchanger on the suction from each section. The liquid line must divide into the same number of branches and each branch must take the liquid through the exchanger which corresponds to the section it feeds. There must be no connection between the suction lines of the various circuits until they join after the point where the bulbs are located. They should join, as previously described for separate evaporators, so that nothing from one branch can affect the other.

Double pipe, and some commercial types of exchangers, meet the requirements given. The shell and coil exchanger does not and should not be used for this purpose.

Construction

Refrigerant lines may be either of copper tube or steel pipe.

Copper Tube. Commercial hard drawn copper tube is suitable for refrigerant lines. Of the various wall thicknesses available, only type *K* or *L* type should be used. The heavier wall type *K* is particularly recommended because of its greater strength. Soft copper tube of type *K* or *L* wall thickness is also suitable and should be used where ease of bending or forming is required. To conform to standard practice, copper tube orders should specify type and O.D. dimenson, although I.D. or nominal dimension and wall thickness may be given. To minimize the amount of cleaning required, it is also advisable to specify deoxidized tube with sealed ends.

Fittings for Copper Tube. The required elbows, tees, couplings, reducers, plugs and miscellaneous fittings may be either forged brass or wrought copper solder type fittings. Flare type compression fittings are adaptable to connections for smaller size soft tube such as is used on gage and control lines. These fittings are ordered by O.D. size, specifying male or female connection. Cast fittings are generally not as satisfactory for refrigerant applications because they may be porous and may crack if overheated.

Steel Pipe. Steel pipe is gaining in acceptance as a piping material for air conditioning and refrigeration installations and is quite satisfactory if applied and installed correctly. Particularly, it must be thoroughly cleaned and precautions taken to minimize the formation of scale during welding. Seamless or welded steel pipe conforming to specification A-53 of the American Society for Testing and Materials is recommended. Pipe should be Grade-A with wall thickness complying with Schedule 40 formulated by the American National Standards Institute. Nominal size, grade and schedule number should be specified when ordering steel pipe. Fittings for steel pipe, forged, seamless steel, butt welding type fittings conforming to the ANSI B31.5-1974 Standard A-106 specifications are recommended.

Various types and sizes of fittings may be fabricated on the job by making mitered cuts in copper tubing or steel pipe and brazing or welding the cut ends together. Also, various types of joints can be fabricated in copper tubing by swedging and brazing. However, it is advisable to use standard wrought or forged fittings.

Expansion

In order to obtain adequate flexibility to absorb changes in length due to linear expansion and contraction of piping with temperature change, long, straight runs should have U-bends or other conventional expansion loops and flexible hangers. Table 5 gives basic data necessary for the fabrication of such loops with copper tube. For steel pipe the same general loop dimensions as shown for a double offset bend may be employed, but the loop fabricated of straight pipe and ell fittings. This takes full advantage of the transverse flexibility of the straight pipe lengths to provide adequately for expansion or contraction. Expansion loops must be supported in a manner that will allow the expansion loop to take the strain, and should be in the horizontal plane to prevent the formation of a trap. It is generally advisable to pre-stress the piping when it is anchored so that it will not be stressed when under normal operating conditions.

Support

On horizontal pipe runs, supports should be substantial enough to prevent high bending stresses in the pipe. Table 6 gives hanger spacing dimensions. Rod-type hangers, perforated metal strips or wall-brackets which are all commercially available are satisfactory and simple means of support. On long horizontal runs, flexible hangers should be used to permit the free expansion and contraction of the piping. Spring hangers, roll hangers or pipe roll

Table 5. Bending Radius of Expansion Loops

Nominal Pipe Size, Inches	Expansion Absorbed by Loop, Inches								
	½	I	1½	2	2½	3	4	5	6
	Radius-R, Inches								
¾	10	15	19	22	25	27	30	34	38
I	11	16	20	24	27	29	33	38	42
1¼	11	17	21	26	29	32	36	42	47
1½	12	18	23	28	31	35	39	46	51
2	14	20	25	31	34	38	44	51	57
2½	16	22	27	32	37	42	47	56	62
3	18	24	30	34	39	45	53	60	67
4	20	28	34	39	44	48	58	66	75
5	22	31	39	44	49	54	62	70	78
6	24	34	42	48	54	59	68	76	83

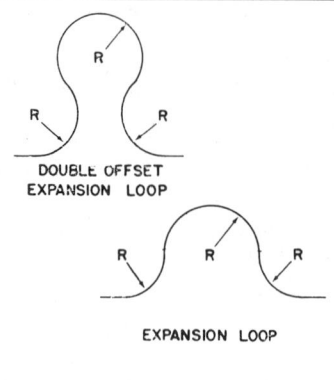

DOUBLE OFFSET EXPANSION LOOP

EXPANSION LOOP

Bends of radius less than 37 inches can be made from 20 ft or less of pipe. All bends made from type K pipe. Bends requiring more than 20 ft pipe are made in sections and assembled with couplings or flanges.

stands are typical supports designed for this purpose.

The weight of vertical piping may be either supported with riser clamps bearing on structural members of the building, or by a platform at the bottom of the riser. A horizontal pipe run connected to a vertical riser must be well supported so that its additional weight is not added to the vertical riser.

When the compressor is supported by vibration absorbing mounts, refrigerant lines should not be anchored rigidly at the unit, but at some distance away so a length of pipe can absorb the motion without being overstressed. Similarly, an accessory such as an oil separator should not be mounted securely to the floor and close coupled to a flexibly mounted compressor, but should be mounted on the compressor itself. Piping flexibility may be obtained by using spring type hangers, by including loops in the lines, or by installing a length of flexible tubing.

Vibration

Flexible hose is often used as a means of reducing vibration transmission along piping. While this method is effective with small size pipe, to be effective with larger diameters, long lengths of hose

are required which may make it uneconomical. Also, the larger sizes must be designed to withstand the larger forces of internal pressure and are less flexible. In using flexible tubing to compensate for compressor motions, install a length of flexible hose at right angles to the direction of compressor motion; horizontally if the motion is vertical, vertical if the motion is horizontal; and both horizontal and vertical if the motion is both vertical and horizontal. If flexible hose is used, it should be of the best grade available, since this is a weak point in a refrigerant system and a potential trouble source.

Vibration is frequently caused by something as simple as a poorly balanced belt sheave. Large sheaves, particularly those with several belts, should by dynamically balanced.

In some cases, severe vibration is caused by the fact that the natural period of vibration of the piping is the same, or an even multiple of the RPM of the motor, compressor or other piece of equipment.

This can often be corrected by changing its mass, unsupported length, etc. If one has no special competence in this field it is usually wise to consult someone who has; there are a number of specialists in this field.

Table 6. Maximum Spacing Between Pipe Supports for Copper or Steel Pipe

Nom. Pipe Size, Ins.	I	1½	2	2½	3	3½	4	5	6	8	10	12	14	16	18	20	24
Maximum Span, Ft	7	9	10	11	12	13	14	16	17	19	22	23	25	27	28	30	32

FLOW RATE PER TON OF REFRIGERATION FOR R-12

Reprinted from "Refrigerant Piping Data" published by Air-Conditioning and Refrigeration Institute.

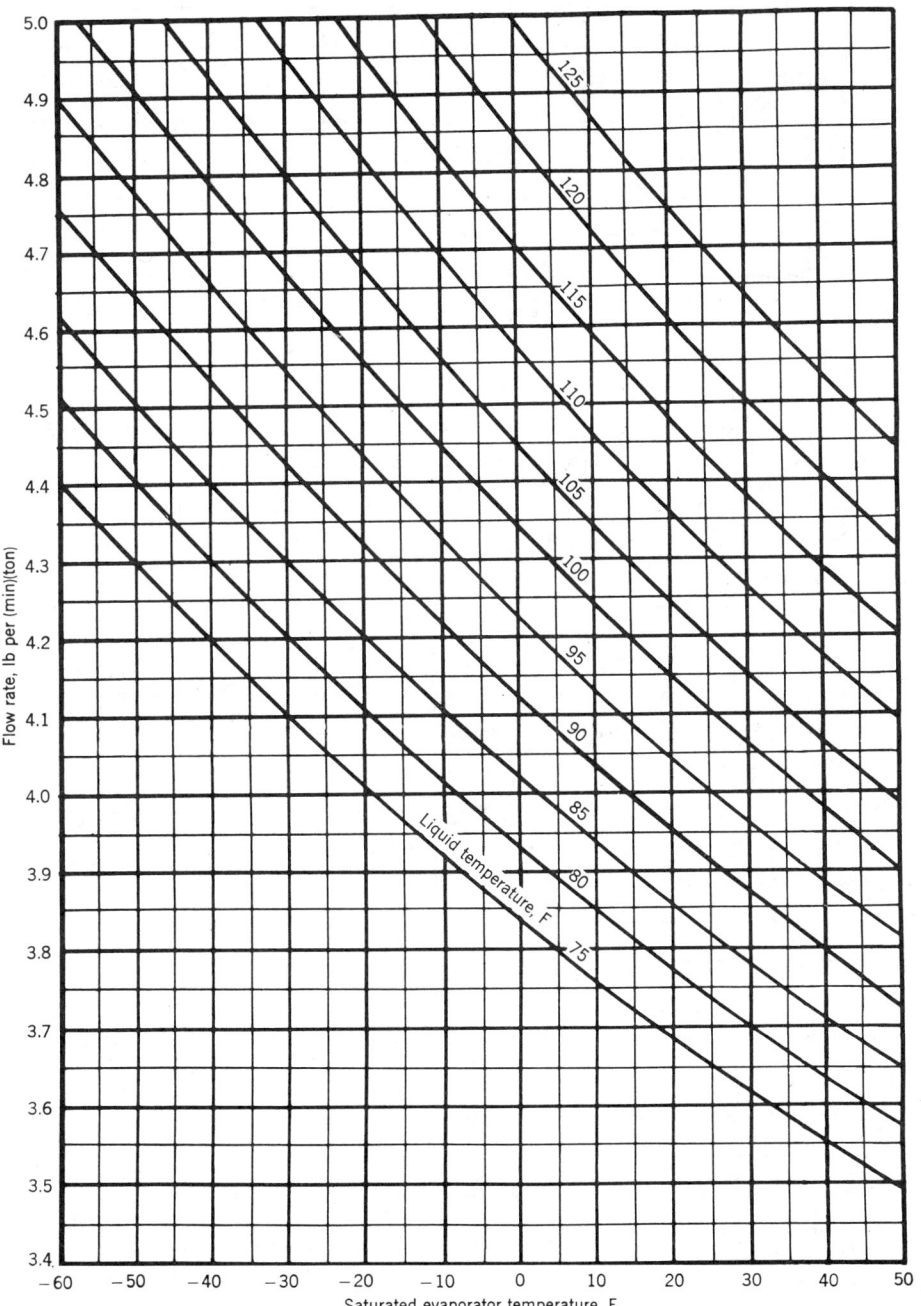

FLOW RATE PER TON OF REFRIGERATION FOR R-22

Reprinted from "Refrigerant Piping Data" published by Air-Conditioning and Refrigeration Institute.

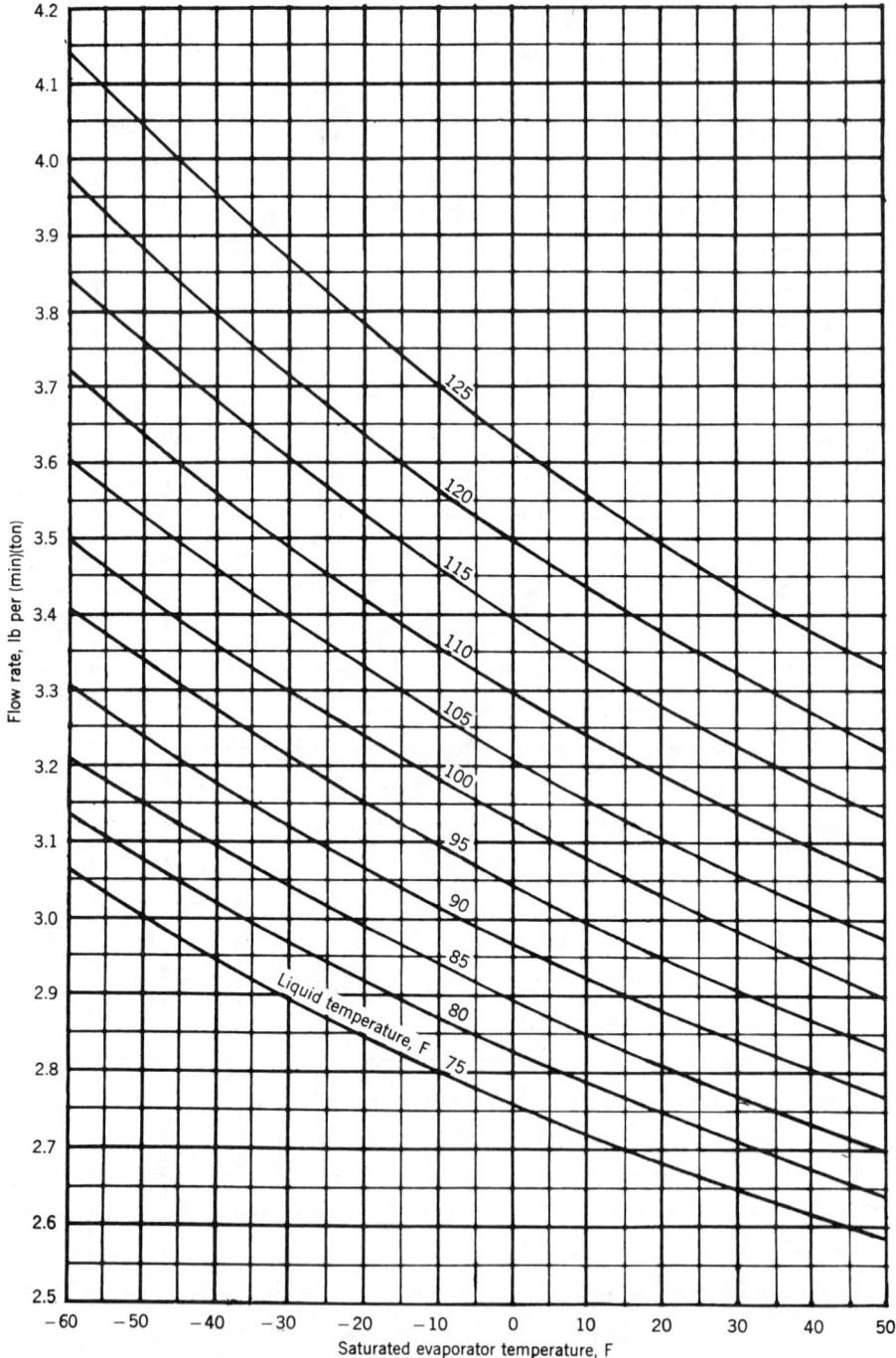

WEIGHT OF REFRIGERANT-12 IN TYPE K LINES IN OPERATION

SUCTION LINE

Nominal Dia., In.	Outside Dia., In.	SUCTION TEMPERATURE, DEG. F								
		50	40	30	20	10	0	−10	−20	−30
		WEIGHT OF REFRIGERANT-12 PER FOOT OF TYPE L TUBE, LB.								
3/8	1/2	.0013	.0011	.0009	.0008	.0007	.0005	.0004	.0004	.0003
1/2	5/8	.0022	.0019	.0016	.0013	.0011	.0009	.0008	.0006	.0005
5/8	3/4	.0034	.0029	.0025	.0021	.0017	.0014	.0012	.0009	.0008
3/4	7/8	.0045	.0038	.0032	.0027	.0022	.0019	.0015	.0012	.0010
1	1 1/8	.0080	.0068	.0058	.0048	.0040	.0033	.0027	.0022	.0017
1 1/4	1 3/8	.0125	.0106	.0090	.0075	.0063	.0052	.0042	.0034	.0027
1 1/2	1 5/8	.0178	.0151	.0127	.0107	.0089	.0073	.0060	.0048	.0039
2	2 1/8	.0311	.0264	.0223	.0187	.0155	.0128	.0104	.0085	.0068
2 1/2	2 5/8	.0480	.0408	.0344	.0288	.0239	.0197	.0161	.0131	.0105
3	3 1/8	.0684	.0582	.0491	.0411	.0341	.0281	.0230	.0186	.0149
3 1/2	3 5/8	.0928	.0789	.0666	.0557	.0463	.0382	.0312	.0253	.0202
4	4 1/8	.1205	.1025	.0864	.0724	.0600	.0496	.0405	.0328	.0263
5	5 1/8	.1869	.1590	.1341	.1123	.0932	.0769	.0628	.0509	.0408
6	6 1/8	.2668	.2269	.1913	.1603	.1330	.1098	.0897	.0726	.0582

HOT GAS LINE

Nominal Dia., In.	Outside Dia., In.	CONDENSING TEMPERATURE, DEG. F								
		80	85	90	95	100	105	110	115	120
		WEIGHT OF REFRIGERANT-12 PER FOOT OF TYPE K TUBE, LB.								
3/8	1/2	.0021	.0022	.0024	.0026	.0028	.0030	.0032	.0034	.0037
1/2	5/8	.0036	.0038	.0041	.0044	.0047	.0051	.0055	.0059	.0063
5/8	3/4	.0055	.0059	.0063	.0068	.0073	.0078	.0084	.0090	.0097
3/4	7/8	.0071	.0077	.0082	.0089	.0095	.0102	.0109	.0118	.0126
1	1 1/8	.0127	.0137	.0147	.0158	.0169	.0182	.0195	.0210	.0225
1 1/4	1 3/8	.0199	.0214	.0230	.0247	.0265	.0285	.0305	.0329	.0352
1 1/2	1 5/8	.0281	.0303	.0325	.0350	.0375	.0403	.0432	.0465	.0498
2	2 1/8	.0492	.0531	.0569	.0613	.0656	.0706	.0755	.0814	.0872
2 1/2	2 5/8	.0760	.0820	.0879	.0946	.1013	.1090	.1167	.1257	.1347
3	3 1/8	.1084	.1169	.1254	.1349	.1444	.1554	.1663	.1792	.1920
3 1/2	3 5/8	.1470	.1585	.1700	.1830	.1959	.2108	.2256	.2430	.2604
4	4 1/8	.1909	.2058	.2207	.2375	.2543	.2736	.2928	.3155	.3380
5	5 1/8	.2962	.3193	.3425	.3685	.3946	.4246	.4544	.4895	.5245
6	6 1/8	.4228	.4558	.4889	.5261	.5633	.6060	.6486	.6987	.7487

LIQUID LINE

Nominal Dia., In.	Outside Dia., In.	LIQUID TEMPERATURE, DEG. F								
		40	50	60	70	80	90	100	110	120
		WEIGHT OF REFRIGERANT-12 PER FOOT OF TYPE K TUBE, LB.								
3/8	1/2	.0758	.0747	.0737	.0727	.0716	.0705	.0693	.0682	.0669
1/2	5/8	.1300	.1283	.1265	.1247	.1229	.1210	.1190	.1170	.1148
5/8	3/4	.1998	.1971	.1944	.1916	.1888	.1859	.1828	.1797	.1764
3/4	7/8	.2609	.2574	.2539	.2503	.2466	.2427	.2388	.2347	.2303
1	1 1/8	.4649	.4587	.4524	.4460	.4395	.4326	.4255	.4183	.4105
1 1/4	1 3/8	.7275	.7177	.7079	.6980	.6877	.6769	.6659	.6545	.6424
1 1/2	1 5/8	1.0298	1.0159	1.0020	.9879	.9734	.9581	.9424	.9264	.9092
2	2 1/8	1.8012	1.7769	1.7527	1.7280	1.7027	1.6759	1.6485	1.6205	1.5903
2 1/2	2 5/8	2.7828	2.7453	2.7078	2.6696	2.6305	2.5892	2.5468	2.5035	2.4570
3	3 1/8	3.9666	3.9132	3.8597	3.8054	3.7496	3.6907	3.6303	3.5686	3.5022
3 1/2	3 5/8	5.3804	5.3079	5.2354	5.1617	5.0861	5.0061	4.9242	4.8405	4.7505
4	4 1/8	6.9844	6.8903	6.7962	6.7005	6.6024	6.4985	6.3923	6.2836	6.1667

Nominal diameters are 1/8 inch less than actual outside diameters.

WEIGHT OF REFRIGERANT-12 IN TYPE L LINES IN OPERATION

SUCTION LINE

Nominal Dia., In.	Outside Dia., In.	Suction Temperature, Deg. F								
		50	40	30	20	10	0	−10	−20	−30
		Weight of Refrigerant-12 per Foot of Type L Tube, Lb.								
3/8	1/2	.0015	.0013	.0011	.0009	.0008	.0006	.0005	.0004	.0003
1/2	5/8	.0024	.0020	.0017	.0014	.0012	.0010	.0008	.0006	.0005
5/8	3/4	.0036	.0031	.0026	.0022	.0018	.0015	.0012	.0010	.0008
3/4	7/8	.0050	.0042	.0036	.0030	.0025	.0021	.0017	.0014	.0011
1	1 1/8	.0085	.0072	.0061	.0051	.0042	.0040	.0029	.0023	.0019
1 1/4	1 3/8	.0130	.0110	.0093	.0078	.0065	.0053	.0044	.0035	.0028
1 1/2	1 5/8	.0183	.0156	.0131	.0110	.0091	.0075	.0067	.0050	.0040
2	2 1/8	.0319	.0271	.0229	.0192	.0159	.0131	.0107	.0087	.0070
2 1/2	2 5/8	.0492	.0418	.0353	.0295	.0245	.0202	.0165	.0134	.0107
3	3 1/8	.0702	.0597	.0503	.0422	.0350	.0289	.0236	.0191	.0153
3 1/2	3 5/8	.0950	.0808	.0681	.0571	.0474	.0391	.0319	.0259	.0207
4	4 1/8	.1234	.1050	.0885	.0742	.0615	.0508	.0415	.0336	.0269
5	5 1/8	.1924	.1636	.1380	.1156	.0959	.0792	.0647	.0524	.0420
6	6 1/8	.2765	.2352	.1983	.1661	.1378	.1138	.0930	.0753	.0603

HOT GAS LINE

Nominal Dia., In.	Outside Dia., In.	Condensing Temperature, Deg. F								
		80	85	90	95	100	105	110	115	120
		Weight of Refrigerant-12 per Foot of Type L Tube, Lb.								
3/8	1/2	.0024	.0026	.0027	.0030	.0032	.0034	.0036	.0039	.0042
1/2	5/8	.0038	.0041	.0044	.0047	.0050	.0054	.0058	.0062	.0067
5/8	3/4	.0057	.0061	.0066	.0071	.0076	.0081	.0087	.0094	.0101
3/4	7/8	.0079	.0085	.0091	.0098	.0105	.0113	.0121	.0130	.0140
1	1 1/8	.0135	.0145	.0156	.0167	.0180	.0193	.0207	.0222	.0239
1 1/4	1 3/8	.0205	.0221	.0237	.0255	.0273	.0293	.0315	.0338	.0363
1 1/2	1 5/8	.0291	.0313	.0336	.0361	.0387	.0415	.0446	.0479	.0515
2	2 1/8	.0505	.0544	.0584	.0628	.0673	.0722	.0775	.0833	.0895
2 1/2	2 5/8	.0779	.0839	.0901	.0968	.1038	.1114	.1196	.1284	.1380
3	3 1/8	.1112	.1197	.1286	.1382	.1482	.1590	.1707	.1834	.1970
3 1/2	3 5/8	.1505	.1620	.1741	.1869	.2006	.2152	.2310	.2481	.2666
4	4 1/8	.1956	.2105	.2262	.2429	.2606	.2796	.3001	.3224	.3464
5	5 1/8	.3049	.3281	.3526	.3786	.4062	.4357	.4677	.5025	.5399
6	6 1/8	.4382	.4715	.5067	.5441	.5838	.6263	.6723	.7222	.7760

LIQUID LINE

Nominal Dia., In.	Outside Dia., In.	Liquid Temperature, Deg. F								
		40	50	60	70	80	90	100	110	120
		Weight of Refrigerant-12 per Foot of Type K Tube, Lb.								
3/8	1/2	.0867	.0858	.0846	.0834	.0822	.0809	.0795	.0782	.0768
1/2	5/8	.1386	.1368	.1349	.1330	.1310	.1290	.1269	.1247	.1224
5/8	3/4	.2083	.2056	.2027	.1999	.1970	.1939	.1907	.1875	.1840
3/4	7/8	.2893	.2854	.2815	.2775	.2735	.2692	.2648	.2603	.2554
1	1 1/8	.4934	.4867	.4801	.4733	.4664	.4590	.4515	.4438	.4356
1 1/4	1 3/8	.7508	.7407	.7306	.7203	.7097	.6986	.6871	.6755	.6629
1 1/2	1 5/8	1.0633	1.0490	1.0347	1.0201	1.0052	.9894	.9732	.9566	.9388
2	2 1/8	1.8494	1.8245	1.7996	1.7742	1.7483	1.7208	1.6926	1.6638	1.6329
2 1/2	2 5/8	2.8594	2.8209	2.7823	2.7431	2.7030	2.6605	2.6169	2.5724	2.5246
3	3 1/8	4.0708	4.0160	3.9611	3.9053	3.8481	3.7876	3.7257	3.6623	3.5942
3 1/2	3 5/8	5.5087	5.4345	5.3602	5.2847	5.2073	5.1254	5.0416	4.9559	4.8638
4	4 1/8	7.1575	7.0611	6.9646	6.8665	6.7660	6.6595	6.5506	6.4392	6.3195

Section 5

AIR HANDLING
AND VENTILATION

Contents

Authorities

Section Editor: *Harrison D. Goodman,* Jos. R. Loring & Associates, Inc.

Morton A. Bell, Air Filters and Dust Collectors
F. Caplan, Nomographs for Fan HP and Capacity
Elliot Godes, High Velocity System Design

F. W. Hutchinson, Determining Required Air Volume
Knowlton J. Kaplan, Air Flow in Ducts

FAN TERMINOLOGY AND DEFINITIONS *

A fan is a gas flow producing machine which operates on the same basic principles as a centrifugal pump or compressor. Each of these devices, including the fan, converts rotational mechanical energy, applied to their shafts, to total pressure increase of the moving gas. This conversion is accomplished by changing the momentum of the fluid.

The American Society of Mechanical Engineers' power test codes limit the fan definition to machines which increase the density of the gas by no more than 7% as it travels from inlet to outlet. This is a rise of about 30 inches of water pressure based on standard air. For pressure higher than 30 in. WG, the air-moving device is a compressor, or "pressure blower."

Fans for heating, ventilating and air conditioning, even on high velocity, high-pressure systems, rarely encounter more than 10-12 inches of water pressure.

Terminology, Abbreviations and Definitions

Definitions of terms common in fan technology follow.

Standard Air—Air at a temperature of 70° F dry bulb and a barometric pressure of 29.92 inches of mercury with a density of .075 lb/cu ft.

Water Gauge (WG) (Fig. 1)—The measure of pressure above atmospheric expressed as the height of a column of water in inches (atmospheric at sea level equals 407.1 inches of water or 33.97 feet of water).

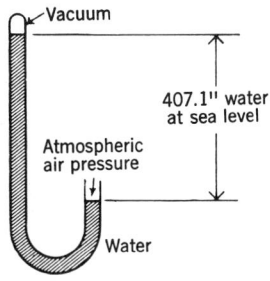

Fig. 1. Atmospheric pressure.

---

* R. O. Hunton, P.E., "Review of Typical Fan Performance," *Fan Seminar*. (La Crosse, Wisc.: The Trane Co.)

CFM and SCFM = the cubic feet per minute (cfm) of air produced by a fan in a given system is independent of the air density.

  CFM—Cubic feet per minute of air handled by a fan at any air density.
  SCFM—Cubic feet per minute of standard air (.075 lb/ft) handled by a fan.

Fan Total Pressure (TP)—Fan total pressure is the difference between the total pressure at the fan outlet and the total pressure at the fan inlet. The fan total pressure is the measure of the total mechanical energy added to the air or gas by the fan. How this is measured is illustrated in Fig. 2.

Fan Static Pressure (SP)—Fan static pressure (Fig. 3) is the fan total pressure less the fan velocity pressure. It can be calculated by subtracting the total pressure at the fan inlet from

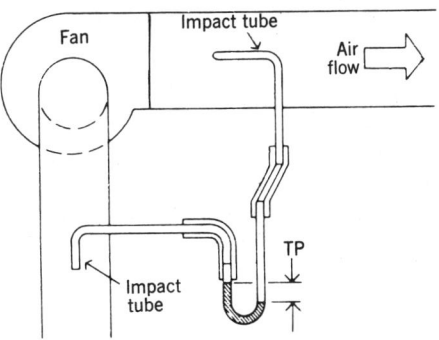

Fig. 2. Fan total pressure.

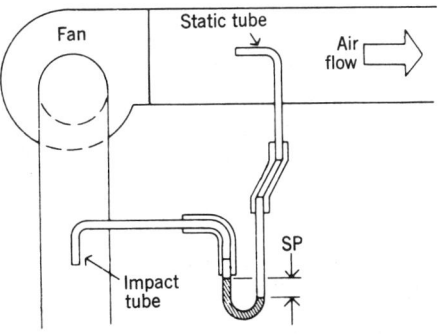

Fig. 3. Fan static pressure.

the static pressure at the fan outlet. This is a source of some confusion within the industry but, by definition, fan SP equals fan TP (outlet) −TP (inlet) −VP (outlet). Also, we have TP (outlet) −SP (outlet) =VP (outlet) and, substituting, results in Fan SP=SP (outlet) −TP (inlet). Thus, in taking field measurements, care must be exercised when a duct inlet is employed to measure the total pressure at the inlet rather than just static pressure.

Fan Velocity Pressure (VP)—Fan velocity pressure (Fig. 4) is the pressure corresponding to the fan outlet velocity. It is the kinetic energy per unit volume of flowing air.

Air Horsepower (AHP)—Assuming 100% efficiency, it is the horsepower required to move a given volume of air against a given pressure.

$$\text{Static AHP} = \frac{\text{CFM} \times \text{SP}}{6356} =$$

$$\frac{\text{ft}^3/\text{min} \times \text{in. water} \times \dfrac{14.7\ \text{lb/in.}^2}{407.1\ \text{in. water}} \times \dfrac{144\ \text{in.}^2}{\text{ft}^2}}{33,000\ \text{ft}-\text{lb/min/AHP}}$$

$$\text{Total AHP} = \frac{\text{CFM} \times \text{TP}}{6356}$$

Brake Horsepower (BHP)—The actual horsepower a fan requires. It is greater than air horsepower, because no fan is actually 100% efficient. It may include power absorbed by V-belt drives, accessories, and any other power requirements, in addition to power input to the fan.

$$\text{BHP} = \frac{\text{CFM}}{6356} \times \frac{\text{TP}}{\text{Fan Total Efficiency}}$$

Static Efficiency (S.E.)—The static air horsepower divided by the power input to the fan.

$$\text{S.E.} = \frac{\text{Power Output}}{\text{Power Input}} = \frac{\text{CFM} \times \text{SP}}{6356 \times \text{BHP}}$$

Mechanical Efficiency (M.E.)—Also called total efficiency (T.E.). Ratio of power output over power input.

$$\text{M.E.} = \frac{\text{CFM} \times \text{TP}}{6356 \times \text{BHP}}$$

Application Range (Fig. 5)—The range of operating volumes and pressures, determined by the manufacturer, at which a fan will operate satisfactorily.

Blocked Tight Static Pressure (BTSP) (Fig. 6)—Operating condition when the fan outlet is completely closed, resulting in no air flow.

Free Delivery (Fig. 7)—Also called wide open CFM (WOCFM). At this operating condition, static pressure across the fan is zero.

Wide Open Brake Horsepower (WOBHP)—The horsepower consumed when the fan is operating at free delivery. Frequently, fan character-

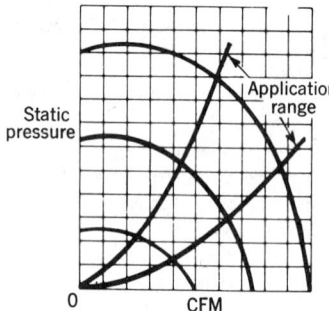

Fig. 5. Application range.

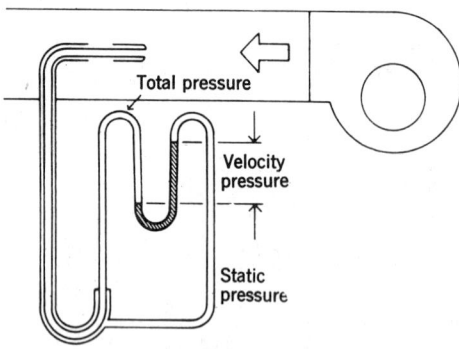

Velocity pressure=total pressure=static pressure

Fig. 4. Fan velocity pressure.

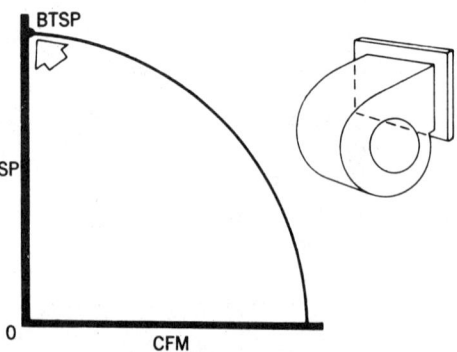

Fig. 6. Blocked tight static pressure.

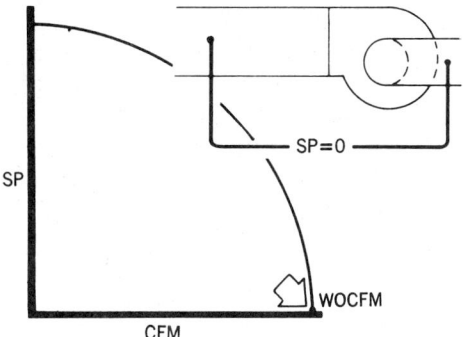

Fig. 7. Free delivery.

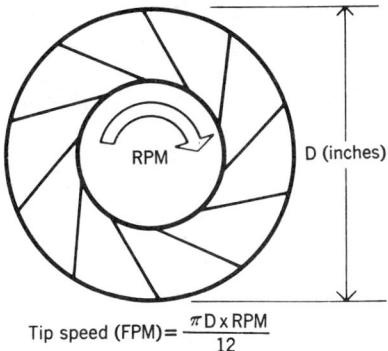

$$\text{Tip speed (FPM)} = \frac{\pi D \times RPM}{12}$$

Fig. 8. Tip speed.

istics are referred to in terms of the percent of wide open CFM (percent WOCFM) which for a given fan then fixes the corresponding percent blocked tight static pressure (percent BTSP) and percent WOBHP.

Tip Speed (TS) (Fig. 8)—Also called periph-eral velocity, equals the circumference of the fan wheel times the RPM of the fan and is ex-pressed in ft/min.

$$TS = \frac{\pi D \times RPM}{12} \quad \begin{array}{l} \text{(when D is expressed} \\ \text{in inches)} \end{array}$$

FAN LAWS

The fan laws relate the performance variables for any homologous series of fans, and apply to both centrifugal and axial flow types. The vari-ables involved are fan size [related to wheel diameter (D), rotating speed in revolutions per minute (RPM), gas density (d), capacity (CFM), static pressure (SP), brake horsepower (BHP), sound-power level and efficiency]. All ten fan laws are summarized in the Buffalo Forge *Fan Engineering* handbook; but for practical purposes, only the laws relating to RPM, fan size, and density effects are illustrated below. The fan laws will be accurate for geometrically proportioned fans; however, because tolerances are usually not proportioned, slightly better per-formance is generally obtained when projecting from a given fan size to a larger one.

Effect of RPM Change (Fan Law 1)

First considered are the fan laws applying to a change only in RPM (constant system) with a given fan and a given system handling air at a given density. Fan law 1 is illustrated in Fig. 1.

CFM varies directly as the RPM.

$$\frac{CFM_1}{CFM_2} = \left(\frac{RPM_1}{RPM_2}\right)$$

Static pressure varies as the square of the RPM

$$\frac{SP_1}{SP_2} = \left(\frac{RPM_1}{RPM_2}\right)^2$$

Horsepower varies as the cube of the RPM

$$\frac{BHP_1}{BHP_2} = \left(\frac{RPM_1}{RPM_2}\right)^3$$

Efficiency will not change.

Point of operation (percent of wide open CFM) will not change for a given system. Changing the speed will not change the percent WOCFM, static efficiency, etc.

Note that the SP change is greater than the CFM change.

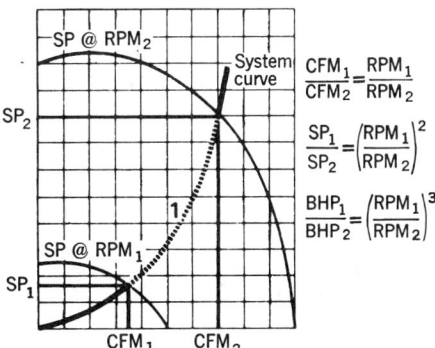

Fig. 1. Fan law No. 1—RPM change.

Effect of Fan Size Change

Fan laws 2 and 3 account for changes in performance due to proportioned changes in fan size. They are based on either constant tip speed or constant RPM.

Fan law 2 is illustrated in Fig. 2; with constant tip speed, air density, fan proportions, and fixed operating point, fan size is the variable.

CFM and BHP vary as the square of the wheel diameter.

$$\frac{CFM_1}{CFM_2} = \frac{BHP_1}{BHP_2} = \left(\frac{D_1}{D_2}\right)^2$$

Static pressure remains constant.

$$SP_1 = SP_2$$

RPM varies inversely as the wheel diameter.

$$\frac{RPM_1}{RPM_2} = \frac{D_2}{D_1}$$

Fan law 2 is used mostly by fan designers and rarely has application in the field.

Fan law 3 is illustrated in Fig. 3; with constant RPM, air density, fan proportions, and fixed operating point.

CFM varies as the cube of the fan diameter.

$$\frac{CFM_1}{CFM_2} = \left(\frac{D_1}{D_2}\right)^3$$

Static pressure varies as the square of wheel diameter.

$$\frac{SP_1}{SP_2} = \left(\frac{D_1}{D_2}\right)^2$$

Tip speed varies directly as wheel diameter.

$$\frac{TS_1}{TS_2} = \frac{D_1}{D_2}$$

Brake horsepower varies as the fifth power of wheel diameter.

$$\frac{BHP_1}{BHP_2} = \left(\frac{D_1}{D_2}\right)^5$$

Fan law 3, with constant RPM and density, is used by fan manufacturers to generate performance data for geometrically proportioned "families" of fans.

Density Effects

Considered next is the effect of change in air density on fan performance. Three fan laws apply in this situation.

Fan law 4 is illustrated in Fig. 4; with constant volume (CFM), system, fan size, and RPM.

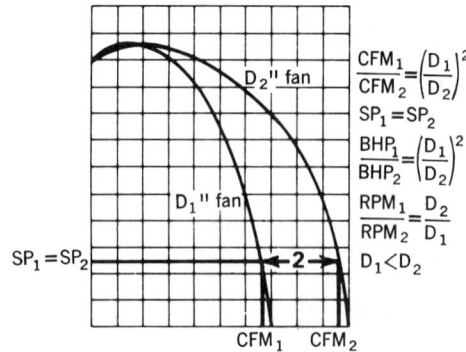

Fig. 2. Fan law No. 2—Change in wheel diameter (constant tip speed).

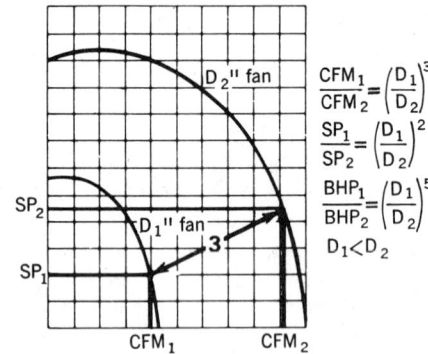

Fig. 3. Fan law No. 3—Change in wheel diameter (constant RPM).

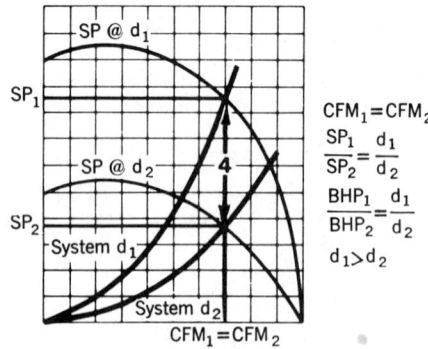

Fig. 4. Fan law No. 4—Effect of density change (constant volume).

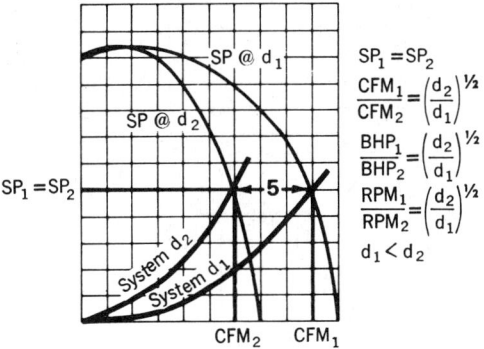

$$SP_1 = SP_2$$
$$\frac{CFM_1}{CFM_2} = \left(\frac{d_2}{d_1}\right)^{\frac{1}{2}}$$
$$\frac{BHP_1}{BHP_2} = \left(\frac{d_2}{d_1}\right)^{\frac{1}{2}}$$
$$\frac{RPM_1}{RPM_2} = \left(\frac{d_2}{d_1}\right)^{\frac{1}{2}}$$
$$d_1 < d_2$$

Fig. 5. Fan law No. 5—Density change (constant static pressure).

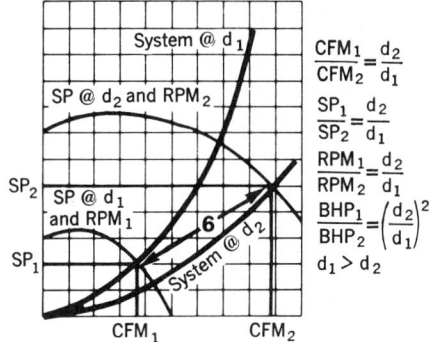

$$\frac{CFM_1}{CFM_2} = \frac{d_2}{d_1}$$
$$\frac{SP_1}{SP_2} = \frac{d_2}{d_1}$$
$$\frac{RPM_1}{RPM_2} = \frac{d_2}{d_1}$$
$$\frac{BHP_1}{BHP_2} = \left(\frac{d_2}{d_1}\right)^2$$
$$d_1 > d_2$$

Fig. 6. Fan law No. 6—Density change (constant mass flow).

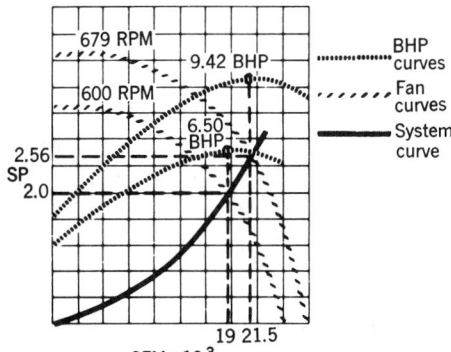

Fig. 7. Example No. 1—Fan law No. 1 (RPM change).

The fan volume, in CFM, will not change with density. A fan is a constant volume machine and will produce the same CFM no matter what the air density may be.

$$CFM_1 = CFM_2$$

The static pressure and BHP developed by the fan will vary in direct proportion to the density. In other words, the heavier the air, the more pressure will be produced and the more horsepower will be required.

$$\frac{SP_1}{SP_2} = \frac{BHP_1}{BHP_2} = \frac{d_1}{d_2}$$

Fan law 5 is illustrated in Fig. 5; with constant pressure, system, and fan size. Variable RPM.

CFM, RPM, and brake horsepower vary inversely as the square root of density.

$$\frac{CFM_1}{CFM_2} = \frac{RPM_1}{RPM_2} = \frac{BHP_1}{BHP_2} = \left(\frac{d_2}{d_1}\right)^{\frac{1}{2}}$$

Static pressure remains constant.

$$SP_1 = SP_2$$

Fan law 6 is illustrated in Fig. 6; with constant mass flow rate, constant system and fixed fan size. Variable RPM.

CFM, RPM, and static pressure (SP), vary inversely with density.

$$\frac{CFM_1}{CFM_2} = \frac{RPM_1}{RPM_2} = \frac{SP_1}{SP_2} = \frac{d_2}{d_1}$$

Brake horsepower varies inversely with square of density.

$$\frac{BHP_1}{BHP_2} = \left(\frac{d_2}{d_1}\right)^2$$

Fan laws 4 and 6 are the basis for selecting fans for other than standard air density using the catalog fan tables which are based on standard air.

The following examples illustrate the application of fan laws.

Example No. 1 (Fig. 7)

An air-conditioning supply fan is operating at 600 RPM against 2" SP and requiring 6.50 BHP. It is delivering 19,000 CFM at standard conditions. In order to handle an air-conditioning load heavier than originally planned, more air is desired. In order to increase the CFM to 21,500, what are the new RPM, SP, and BHP? The solution to this problem is found by applying Fan law 1, as illustrated in Fig. 7.

$$CFM_2 = CFM_1 \; (RPM_2/RPM_1)$$

Rearranging, $RPM_2 = (CFM_2/CFM_1) \; RPM_1$

$$RPM_2 = \frac{21,500}{19,000} \; (600) = 679$$

$$SP_2 = SP_1 \left(\frac{RPM_2}{RPM_1}\right)^2$$

$$SP_2 = (2) \left(\frac{679}{600}\right)^2 = 2.56 \; inches$$

$$BHP_2 = (BHP_1) \left(\frac{RPM_2}{RPM_1}\right)^3$$

$$BHP_2 = (6.50) \left(\frac{679}{600}\right)^3 = 9.42$$

Example No. 2 (Fig. 8)

A fan is operating at 2714 RPM on 70°F air against 3" SP. It is delivering 3,560 CFM and requires 2.84 BHP. A 5 hp motor is powering the fan. The system is short capacity but the owner doesn't want to spend any money to change the motor. What is the maximum capacity from this system with the existing 5 hp motor? What is the allowable speed increase? What will CFM and SP be under the new conditions? The solution to this problem is found by applying Fan law 1.

$$RPM_2 = RPM_1 \left(\frac{BHP_2}{BHP_1}\right)^{1/3}$$

$$RPM_2 = (2714) \left(\frac{5.00}{2.84}\right)^{1/3} = 3280$$

$$CFM_2 = CFM_1 \left(\frac{RPM_2}{RPM_1}\right)$$

$$CFM_2 = 3560 \left(\frac{3280}{2714}\right) = 4300$$

$$SP_2 = SP_1 \left(\frac{RPM_2}{RPM_1}\right)^2$$

$$SP_2 = 3 \left(\frac{3280}{2714}\right)^2 = 4.4 \; inches$$

Example No. 3 (Fig. 9)

A fan manufacturer wishes to project data obtained for a 30-inch-diam fan to a 60-inch-diam fan. At one operating point the 30 inch fan delivers 7,755 CFM of 70°F air against 3 inches SP. This requires 694 RPM (tip speed = 5450 FPM) and 1.77 BHP. What will the projected CFM, SP, BHP, and tip speed (TS) be for a 60 inch fan at the same RPM? The solution to this problem is found by applying Fan law 3.

$$CFM_2 = CFM_1 \left(\frac{D_2}{D_1}\right)^3$$

$$CFM_2 = 7,755 \left(\frac{60}{30}\right)^3 = 62,000$$

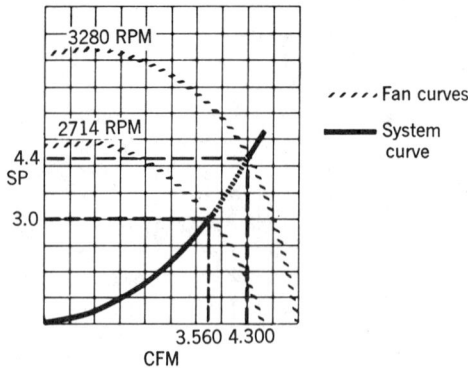

Fig. 8. Example No. 2—Fan law No. 1 (RPM change).

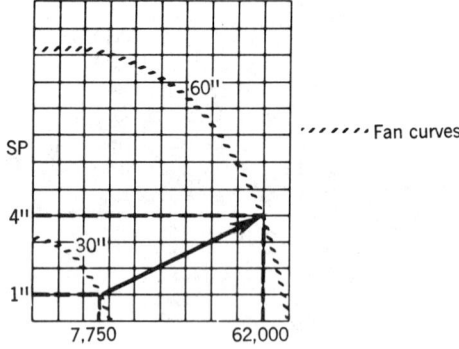

Fig. 9. Example No. 3—Fan law No. 3 (diameter change).

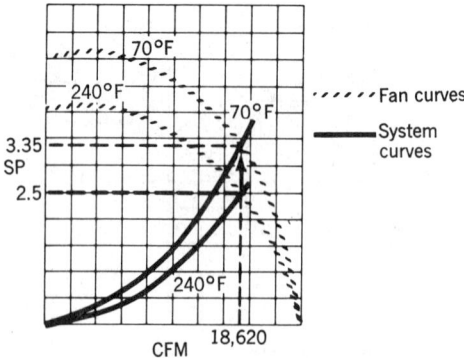

Fig. 10. Example No. 4—Fan law No. 4 (density change).

$$SP_2 = SP_1 \left(\frac{D_2}{D_1}\right)^2$$

$$SP_2 = 3 \left(\frac{60}{30}\right)^2 = 12 \text{ inches}$$

$$BHP_2 = 1.77 \left(\frac{60}{30}\right)^5 = 56.6$$

$$TS_2 = TS_1 \left(\frac{D_2}{D_1}\right)$$

$$TS_2 = 5450 \left(\frac{60}{30}\right) = 10,900$$

This, plus Fan law 1, are the fan laws used to project catalog data for many diameters and speeds from a test on a single fan at one speed.

Example No. 4 (Fig. 10)

A fan drawing air from an oven is delivering 18,620 CFM of 240°F air against 2½ inches SP. It is operating at 796 RPM and requires 9.90 BHP. Assume the oven loses its heat and the air is at 70°F. What happens to the SP and BHP required? The solution to this problem is found by applying Fan law 4.

Density of 70°F air = .075 lb/ft³
Density of 240°F air = .056 lb/ft³

$$CFM_2 = CFM_1 = 18,620$$

$$SP_2 = SP_1 \left(\frac{d_2}{d_1}\right)$$

$$SP_2 = (2.5'') \left(\frac{.075}{.056}\right) = 3.35''$$

$$BHP_2 = BHP_1 \frac{d_2}{d_1}$$

$$BHP_2 = (9.90) \left(\frac{.075}{.056}\right) = 13.3$$

This example illustrates why the fan motor should always be selected on the BHP at the maximum density, which would be at the lowest air temperature expected.

Example No. 5 (Fig. 11)

An engineer specifies that he wants 15,200 CFM at 2 inches SP, 120°F and 1000 feet altitude. Determine the RPM and BHP.

There are two ways to solve this problem. Fan law 4 is used in the first solution method.

In order to enter in the manufacturer's catalog fan tables which are based on standard air, we must determine the SP that would be required with standard air.

From a chart of air density ratios we would find

$$\frac{\text{d-Actual}}{\text{d-Standard}} = .88$$

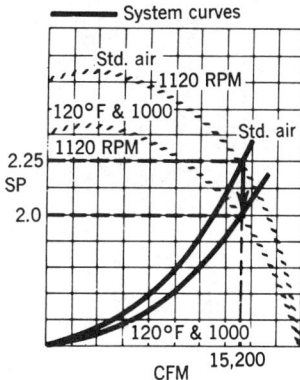

Fig. 11. Example No. 5—Fan law No. 4 (density change).

$$SP \text{ Std.} = SP \text{ Act.} \left(\frac{d \text{ Std.}}{d \text{ Act.}}\right)$$

$$SP \text{ Std.} = 2''/.88 = 2.27 \text{ inches, say } 2.25$$

From the catalog fan table we find to deliver 15,200 CFM against 2.25'' will require 1120 RPM. The BHP required is 8.07. The RPM is correct at 1120, but since the fan is handling less dense air, then:

$$BHP \text{ Act.} = BHP \text{ Std.} \frac{d \text{ Act}}{d \text{ Std}} = 8.07 \,(.88) = 7.1$$

Note also from this example that the static pressure resistance of the system varies directly with air density.

The other method of solution utilizes Fan law 6, as illustrated in Fig. 12.

In this case, assume that operating condition is standard to determine an RPM and BHP in the catalog. Then the catalog BHP and SP will be corrected according to Fan law 6.

$$CFM \text{ Std} = CFM \text{ Act} \frac{d \text{ Act}}{d \text{ Std}}$$
$$= 15,200 \,(.88) = 13,400$$

$$SP \text{ Std} = SP \text{ Act} \frac{d \text{ Act}}{d \text{ Std}}$$
$$= 2.0 \,(.88) = 1.76'', \text{ say } 1.75$$

The fan will deliver 13,400 CFM against 1.75'' when operating at 988 RPM. Required BHP = 5.55. Correcting the RPM for density according to Fan law 6, we obtain:

$$RPM \text{ Act} = RPM \text{ Std} \left(\frac{d \text{ Std}}{d \text{ Act}}\right) = \frac{988}{.88} = 1120$$

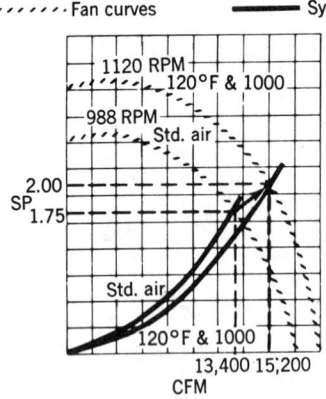

········ Fan curves ——— System curves

Fig. 12. Example No. 5—Fan law No. 6 (density change).

$$BHP\ Act = BHP\ Std \left(\frac{d\ Std}{d\ Act}\right)^2 = \frac{5.55}{(.88)^2} = 7.1\ BHP$$

As would be expected, the answer comes out the same with either solution.

Example No. 6—Effect of Changing the Resistance of a Fan System

Assume that a fan is handling 41,280 CFM at $1\frac{1}{2}''$ SP, running at 418 RPM and requiring 14.99 BHP. If the speed remains constant at 418 RPM,

but an additional resistance of $\frac{1}{2}''$ (based on existing velocities) is placed in the system, the static pressure would be 2 inches if the capacity, 41,280 CFM, remains the same. From the fan manufacturer's rating table, it is seen that the speed would have to be increased to 454 RPM and would require 18.7 BHP. This new fan rating must be reduced to the predetermined speed of 418 RPM along the new duct resistance curve by use of Fan law 1.

$$CFM_1 = CFM_2 \left(\frac{RPM_1}{RPM_2}\right)$$
$$= 41,280 \left(\frac{418}{454}\right) = 38,000$$

$$SP_1 = SP_2 \left(\frac{RPM_1}{RPM_2}\right)^2$$
$$= 2'' \left(\frac{418}{454}\right)^2 = 1.70''$$

$$BHP_1 = BHP_2 \left(\frac{RPM_1}{RPM_2}\right)^3$$
$$= 18.7 \left(\frac{418}{454}\right)^3 = 14.6$$

This example, taken from the 6th ed. of *Fan Engineering*, by the Buffalo Forge Co., is useful in those cases where added resistance, such as absolute filters, is inserted in the fan system and thereby raises its static pressure beyond the fan manufacturer's cataloged ratings.

AIR DENSITY CONSIDERATIONS

(1) Standard Density = .075 lb/cu ft at 29.92" Hg (Atmos. Press.) and 70° F *Dry Air*.

(2) Density varies inversely as absolute temperature: i.e., at 120°F

$$Density = .075 \times \frac{460+70}{460+120} = .075 \times \frac{530}{580}$$
$$= .0685\ lb/cu\ ft\ Dry\ Air$$

(3) Density is reduced by elevation: i.e.,

Sea Level— 29.92" Hg.
1000'   — 28.85" Hg.
2000'   — 27.80" Hg.

See chart for other elevations; i.e., at 120°F and 2000' elevation

$$Density = .075 \times \frac{530}{580} \times \frac{27.80}{29.92}$$
$$.0637\ lb/cu\ ft\ Dry\ Air$$

(4) Density is reduced when an inlet duct is

used, and this reduction may be significant. For a 20" H_2O inlet suction (atmospheric pressure is 407" H_2O) the density correction is $\frac{407-20}{407} = \frac{387}{407} = .951$; i.e., @ 120°F, 2000' elevation and 20" suction,

$$Density = .075 \times \frac{530}{580} \times \frac{27.80}{29.92} \times \frac{387}{407} = .0605\ lb/cu\ ft$$

(5) Density is reduced when water vapor is added to air, as in wet scrubber exhaust systems. For saturated air, determine density and proceed per Steps 3 and 4, above, as required, i.e., at 120°F saturated, (Read 0.0655 from Fig. 21B), 2000' elevation and 20" suction.

$$Density = .0655^* \times \frac{27.80}{29.92} \times \frac{387}{407} = .0578\ lb/cu\ ft$$

* For densities of non-saturated air, refer to psychrometric charts. With wet bulb and dry bulb temperatures, determine air volume (cu ft/lb). Density = 1 ÷ air volume.

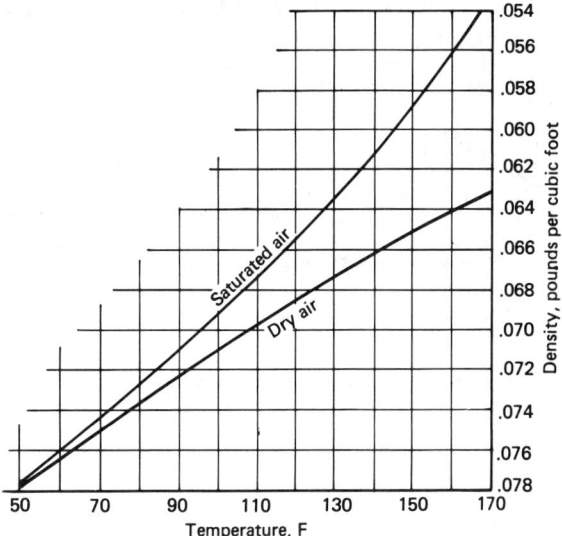

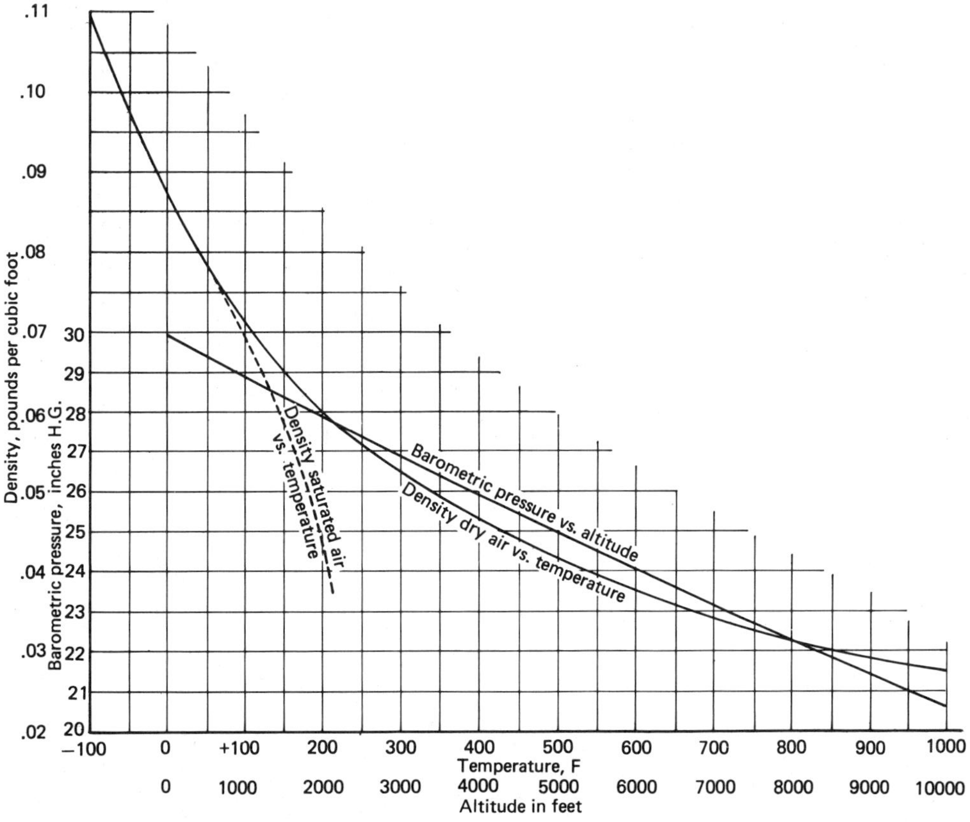

Density versus temperature and pressure versus elevation.

RELATION OF AIR CHARACTERISTICS AND TEMPERATURE

The table below shows the effects of temperature on air volume, weight, and pressure, as well as power required to move the air, all based on air at 70 deg. F, for which the ratio is 1.00 in all cases.

Column (2) gives the relative volume of air of the same weight while Column (3) shows the relative velocities required to produce the same pressure. Column (4) contains the relative pressures required to pass the same weight of air through the same orifice and Column (5) the relative fan speeds required to move the same weight of air through the same orifice. Column (6) shows the relative power required to move the same weight of air through the same orifice. Column (7) is the relative power needed to move the same volume of air against the same resistance, Column (8) the relative power to move the same weight of air against the same resistance, and Column (9) is the power to move the same volume of air at the same velocity.

The chart is presented here by permission of American Blower Corporation from its publication "Air Conditioning and Engineering."

Temperature, Deg. F. (1)	Volume of Same Weight (2)	Velocity to Produce Same Pressure (3)	Pressure to Pass Same Weight (4)	Fan Speed, Same Weight (5)	Power, Same Weight, Orifice (6)	Power, Same Volume, Resistance (7)	Power, Same Weight, Resistance (8)	Power, Same Volume, Velocity (9)
0	.86	.932	.86	.86	.74	1.00	.86	1.15
10	.88	.942	.88	.88	.77	1.00	.88	1.13
20	.90	.952	.90	.90	.81	1.00	.90	1.10
30	.92	.961	.92	.92	.85	1.00	.92	1.08
40	.94	.971	.94	.94	.88	1.00	.94	1.06
50	.96	.981	.96	.96	.92	1.00	.96	1.04
60	.98	.990	.98	.98	.96	1.00	.98	1.02
70	1.00	1.000	1.00	1.00	1.00	1.00	1.00	1.00
80	1.02	1.009	1.02	1.02	1.04	1.00	1.02	.98
90	1.04	1.018	1.04	1.04	1.08	1.00	1.04	.96
100	1.06	1.028	1.06	1.06	1.12	1.00	1.06	.94
125	1.11	1.051	1.11	1.11	1.23	1.00	1.11	.90
150	1.15	1.073	1.15	1.15	1.34	1.00	1.15	.87
175	1.20	1.094	1.20	1.20	1.45	1.00	1.20	.83
200	1.25	1.114	1.25	1.25	1.56	1.00	1.25	.80
225	1.29	1.135	1.29	1.29	1.67	1.00	1.29	.77
250	1.34	1.156	1.34	1.34	1.79	1.00	1.34	.75
275	1.39	1.177	1.39	1.39	1.92	1.00	1.39	.72
300	1.43	1.197	1.43	1.43	2.06	1.00	1.43	.70
325	1.48	1.216	1.48	1.48	2.20	1.00	1.48	.68
350	1.53	1.235	1.53	1.53	2.34	1.00	1.53	.66
375	1.58	1.254	1.58	1.58	2.48	1.00	1.58	.64
400	1.62	1.273	1.62	1.62	2.63	1.00	1.62	.62
425	1.67	1.291	1.67	1.67	2.79	1.00	1.67	.60
450	1.72	1.309	1.72	1.72	2.95	1.00	1.72	.58
475	1.77	1.327	1.77	1.77	3.11	1.00	1.77	.56
500	1.81	1.345	1.81	1.81	3.28	1.00	1.81	.55
525	1.86	1.362	1.86	1.86	3.46	1.00	1.86	.54
550	1.91	1.379	1.91	1.91	3.64	1.00	1.91	.52
575	1.96	1.396	1.96	1.96	3.82	1.00	1.96	.51
600	2.00	1.414	2.00	2.00	4.00	1.00	2.00	.50
625	2.05	1.430	2.05	2.05	4.20	1.00	2.05	.49
650	2.10	1.446	2.10	2.10	4.40	1.00	2.10	.48
675	2.15	1.462	2.15	2.15	4.60	1.00	2.15	.47
800	2.19	1.478	2.19	2.19	4.79	1.00	2.19	.46
800	2.38	1.542	2.38	2.38	5.65	1.00	2.38	.42
1000	2.76	1.660	2.76	2.76	7.59	1.00	2.76	.36
1200	3.14	1.770	3.14	3.14	9.85	1.00	3.14	.32
1500	3.70	1.920	3.70	3.70	13.70	1.00	3.70	.27
2000	4.65	2.160	4.65	4.65	21.60	1.00	4.65	.21

FAN PERFORMANCE CURVES

Since each type and size of fan has different characteristics, fan performance curves must be developed by the fan manufacturers. Generally, these curves are determined by laboratory tests, conducted according to an appropriate industry test standard. In the United States and Canada, the test procedures for fan testing have been standardized by the Air Moving and Conditioning Association (AMCA).

It is important to note that the test setup required by AMCA standards is nearly ideal. For this reason, the performance curves for static pressure and brake horsepower versus CFM, are those obtained under ideal conditions, which rarely exist in practice. To make allowances for deviations from the laboratory test setup, see "Fan Inlet Connections" and "Fan Discharge Connections," later in this section.

The "Fan Laws" are used to determine the brake horsepower and performance characteristics at other speeds and fan sizes; normally, as mentioned before, only one fan size and speed must be tested to determine the capacity for a given "family" of fans.

SYSTEM RESISTANCE CURVES

System resistance is the sum total of all pressure losses through filters, coils, dampers, and ductwork. The system resistance curve (illustrated in Fig. 1) is simply a plot of the pressure that is required to move air through the system.

For fixed systems, that is, with no changes in damper settings, etc., system resistance varies as the square of the airflow (CFM). The resistance curve for any system is represented by a single curve. For example, consider a system handling 1000 CFM with a total resistance of 1 inch SP.

If the CFM is doubled, the SP resistance will increase to 4 inches, as shown by the squared value of the ratio given in Fig. 1. This curve changes, however, as filters load with dirt, coils start condensing moisture, or when outlet dampers are changed in position. For a more detailed discussion on fan systems, the reader is referred to *Fan Engineering,* 6th edition, by the Buffalo Forge Co.

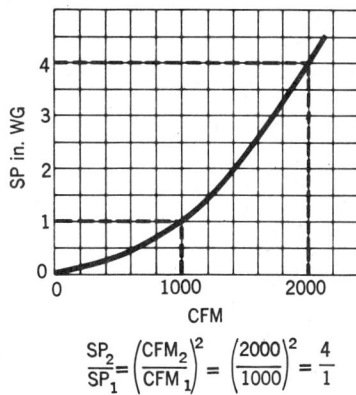

$$\frac{SP_2}{SP_1} = \left(\frac{CFM_2}{CFM_1}\right)^2 = \left(\frac{2000}{1000}\right)^2 = \frac{4}{1}$$

Fig. 1. System resistance curve.

Operating Point

The operating point (Fig. 2) at which the fan and system will perform is determined by the intersection of the system resistance curve and fan performance curve. Note that every fan operates only along its performance curve. If the system resistance designed is not the same as the resistance in the system installed, the operating point will change and the static pressure and volume delivered will not be as calculated.

Note in Fig. 3 that the actual system has more pressure drop than predicted in the design. Thus, CFM is reduced and static pressure is increased.

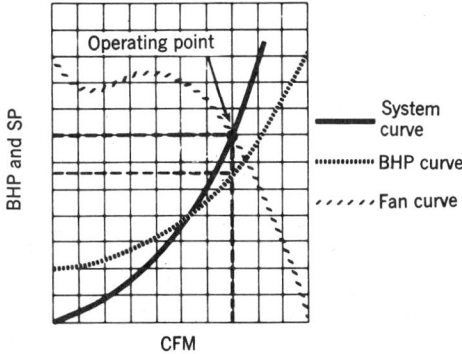

Fig. 2. Operating point.

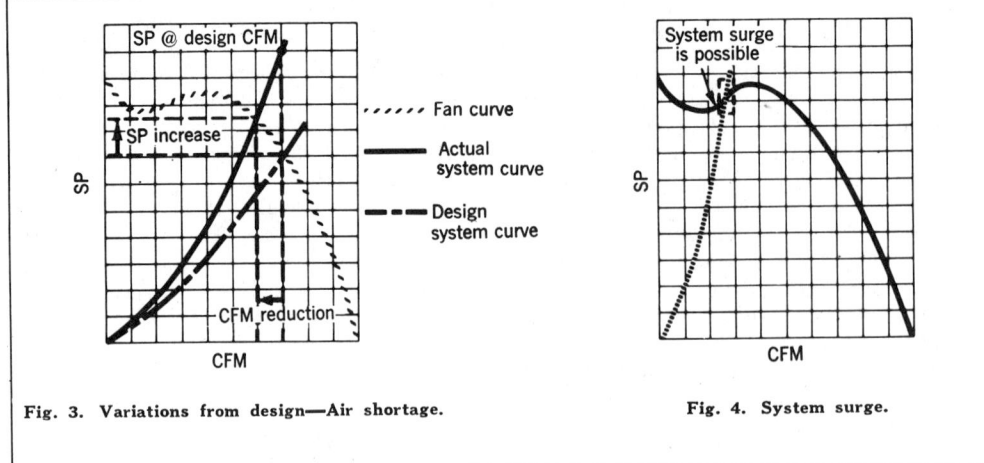

Fig. 3. Variations from design—Air shortage.

Fig. 4. System surge.

The shape of the HP curve typically would result in a reduction in BHP. Typically, the RPM would then be increased and more BHP would be needed to achieve the desired CFM. In many cases where there is a difference between actual and calculated fan output, it is due to a change in system resistance rather than any shortcomings of the fan or motor. Frequently the mistake is made of taking the static pressure reading across the fan and concluding that if it is at or above design requirements the CFM is also at or above design requirements. Figure 3 shows why the assumption is completely invalid.

SYSTEM SURGE, FAN SURGE, AND PARALLELING

The three main reasons for unstable airflow in a fan system are:

(1) System Surge
(2) Fan Surge, and
(3) Paralleling.

System surge (Fig. 4), occurs when the system resistance and fan performance curves do not intersect at a distinct point but rather over a range of volumes and pressures. This situation does not occur with backward inclined (BI), airfoil (AF), and radial fans. However, it can occur with a forward curve centrifugal fan when operating, as shown in Fig. 4, to the left of the peak of its performance curve.

In this situation, because the fan curve and system curve are almost parallel, the operating point can be over a range of CFM and static pressures. This will result in unstable operation known as system surge, pulsation, or pumping.

System surge should not be confused with "paralleling," which can only occur when two fans are installed in parallel.

Fan surge (Fig. 5), on the other hand, is different than system surge. They may or may not occur at the same time.

For any fan, *the point of minimum pressure occurs at the center of rotation of the fan wheel* and the *maximum pressure occurs just at the discharge side of the wheel.* If the wheel were not turning and this pressure differential existed, flow would be from the high pressure point to the low pressure point. This is opposite from the direction air normally flows through the fan. The only thing that keeps the air moving in the proper direction *is the whirling of the blades.* Stall occurs unless there is sufficient air entering the fan wheel to completely fill the space between the blades.

This shows up in Fig. 6 as fluctuation in CFM

Fig. 5. Fan surge explanation.

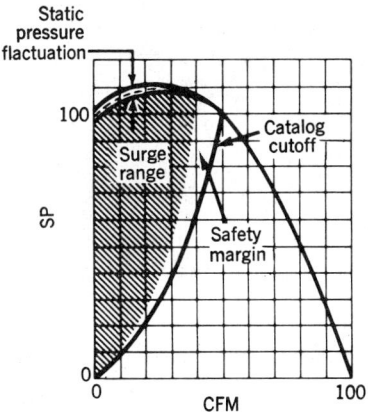

Fig. 6. Fan surge.

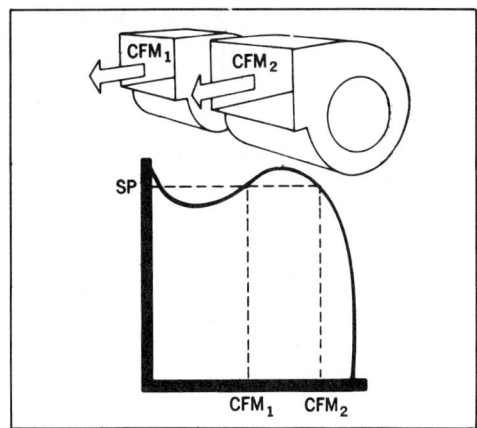

Fig. 7. Unbalanced parallel operation.

and pressure. This surge can be both felt and heard and occurs in nearly all fan types, to varying degrees, as block-tight static pressure is approached. The radial blade is a notable exception. While the magnitude of surge varies for different types of fans, (being greatest for airfoil and least for forward curve), the pressure fluctuation close to block-tight may be on the order of 10%. For example, a fan in surge developing about 6 inches of total static pressure might have pressure fluctuation of 6/10 of an inch. This explains why a large fan in surge is intolerable. Equipment room walls have been cracked from the vibration of ducts serving a fan in surge.

Selections should not be made to the left of the "surge point" on the fan curve. This point, which defines a system curve when all operating speeds of the fan are considered, varies for different fan installations. For instance, stable operation can be obtained much further to the left when the fan is installed in an ideal laboratory type situation. These conditions, of course, are seldom encountered in field applications. Consequently, *most* manufacturers do not catalog operating ranges all the way to the surge line.

However, since the catalog cut-off point is basically one of engineering judgment, conservative catalog performance data will provide operating ranges which will allow stable operation with any reasonable field ductwork design.

The third cause for unstable operation is paralleling, (Fig. 7), which can occur only in a multiple fan installation connected with either a common inlet or common discharge, or both.

In addition, the fan curve must have a characteristic shape where two different CFM's are possible at the same SP. This allows a paralleling situation to exist. It can be theoretically illustrated by drawing a "sum of the differences curve" or "difference curve" for short. See Fig. 8.

If the curve indicates that the paralleling situation can exist, then corrective measures must be taken to achieve satisfactory performance.

The sum of the differences curve is determined by first plotting the combined two-fan curve and the system curve along with the curve of one fan shown alone on the same sheet. Then, the difference curve (abbreviation for sum of the differences curve) is plotted by taking the CFM difference at constant static pressure between the system curve and the single fan curve.

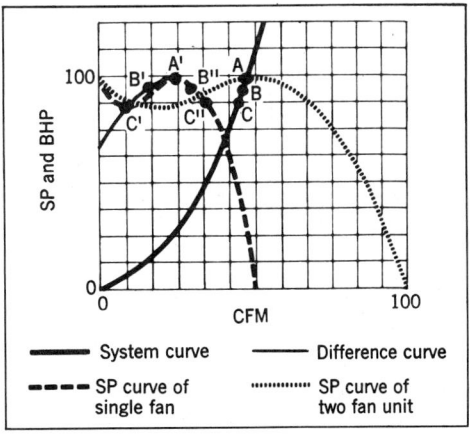

| — System curve | — Difference curve |
| - - - SP curve of single fan | .... SP curve of two fan unit |

Fig. 8. Paralleling.

For example, from Fig. 8, the CFM difference between C and C″ is plotted as point C′ at the same static pressure. Then another point to develop the difference curve is shown by the CFM difference B and B″ and is plotted as B′. Finally, A′ is exactly half the CFM of A and is the intended operating point for each of the two fans.

Every point where the difference curve intersects the single fan curve indicates a possible point of operation. Intersection at B′ or C′ in Fig. 8 indicates a paralleling situation. Each of these points then defines a CFM and static pressure falling on the system curve which both fans combined can satisfy; one fan operating on one side of its curve and the other fan operating on the other side.

In this example, the difference curve intersects the single fan curve at C′ and B′ in addition to the intended operating point at A′. Thus the operating point can be not only at the intended point A but also at points B and C. The result is that the fan will not operate at any one of these points, but will bounce between them, causing a fluctuation in static pressure, CFM, and noise level.

This requires the installation of scroll volume (outlet volume) dampers (Fig. 9). A scroll volume damper serves to change the shape of the fan scroll and thus, for each position of the damper, there is a corresponding different performance curve.

The effect of outlet volume dampers is shown in Fig. 10. The fan curve resulting from various positions of the outlet volume dampers is also shown. The purpose, of course, is to change the fan curve sufficiently such that the sum of the

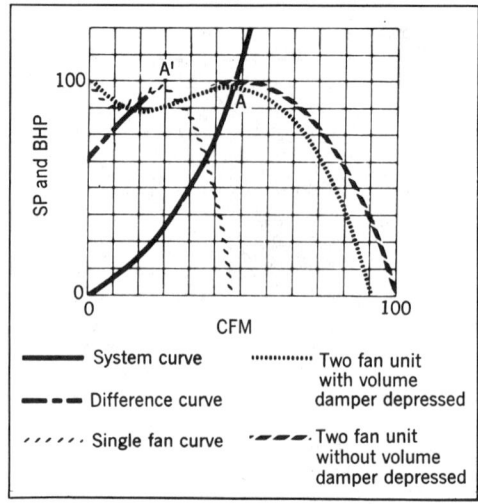

Fig. 10. Elimination of paralleling.

—— System curve	·········· Two fan unit with volume damper depressed
– – – Difference curve	
′′′′′′ Single fan curve	·–▬–▬· Two fan unit without volume damper depressed

differences curve will intersect the single fan curve at A′ and provide stable operation.

The performance may be reduced slightly and a corresponding increase in RPM should be made to achieve the specified conditions. However, this is rarely done since the difference is typically negligible.

To correct, the scroll volume damper is merely pushed down on both fans until the static pressure and noise level pulsation disappear. Generally, they are then left in this position permanently. The curve generated by the damper at this point is so shaped that the sum of the differences curve intersects at only one point.

FAN TYPES

There are two general types of fans—*centrifugal* and *axial*. Flow within the centrifugal fan is substantially radial through the wheel. In an axial fan, flow is parallel to the shaft.

Centrifugal fans are divided into four general classifications: Forward Curve (FC), Backward Inclined (BI), Radial Blade, and Tubular Centrifugal.

The RPM for a given type centrifugal fan wheel is determined by the tip speed necessary to produce the required absolute particle velocity (Fig. 11). This absolute particle velocity vector relative to ground (S) has two components, one radial (r) and the other tangential (t) to the wheel.

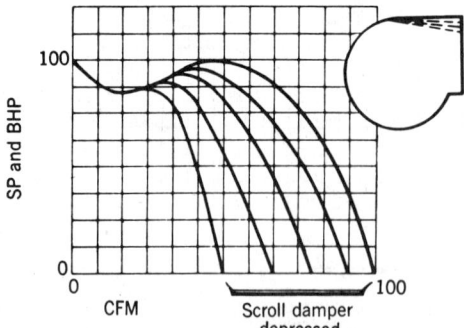

Fig. 9. Effect of scroll dampers.

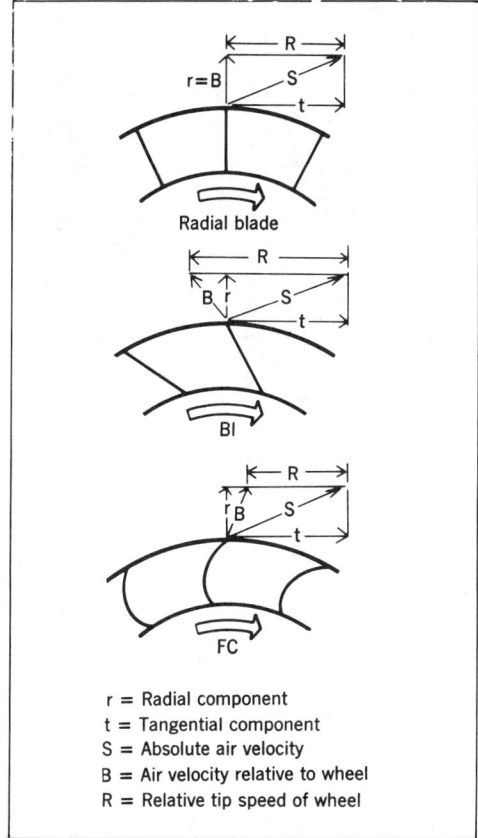

Fig. 11. Centrifugal fan wheels.

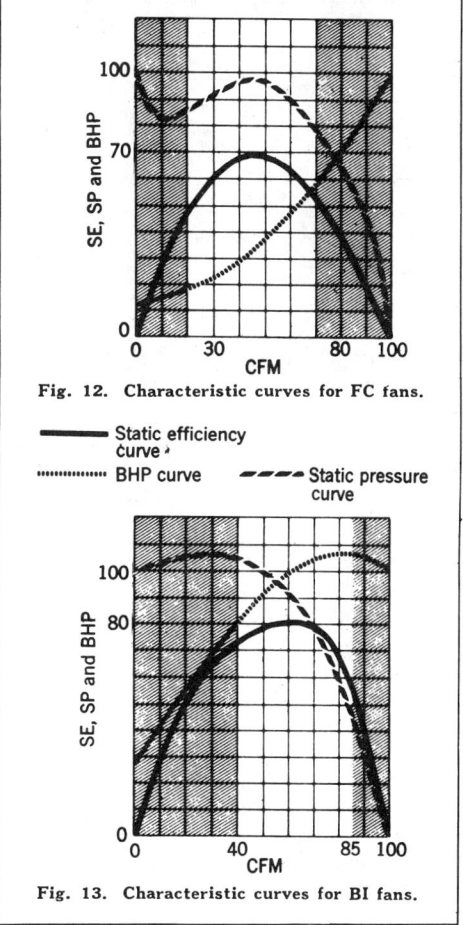

Fig. 12. Characteristic curves for FC fans.

—— Static efficiency curve

·············· BHP curve   ▬ ▬ ▬ Static pressure curve

Fig. 13. Characteristic curves for BI fans.

The velocity of the air relative to the blade is indicated by the blade vector (B) which is nearly tangential to the blade though some slip occurs. The length of the tip speed vector (R) as shown in the diagram, indicates the relative wheel RPM to produce a given capacity. By examining the relative length of the R vector, it can be seen that the FC fan requires the lowest tip speed for a given capacity, while the BI requires the highest tip speed.

Forward Curve Centrifugal Fan

The forward curve centrifugal fan travels at a relatively slow speed and generally is used for producing high volumes at low static pressure.

Typical operating range of this type of fan is from 30% to 80% wide-open volume. (Fig. 12.) The maximum static efficiency of 60-68% generally occurs slightly to the right of peak static pressure. The horsepower curve has an increasing slope and is referred to as an "overloading type."

The FC fan will surge but the magnitude is typically less than for other types.

Advantages of the FC fan are low cost, slow speed which minimizes shaft and bearing size, and wide operating range. The disadvantages are the shape of its performance curve which allows the possibility of paralleling, and an overloading horsepower curve. Thus, overloading of the motor may occur if system static pressure decreases. Also, it is not suitable for material handling because of its blade configuration. It is inherently weak structurally. Therefore, forward curved fans are generally not capable of the high speeds necessary for developing higher static pressures.

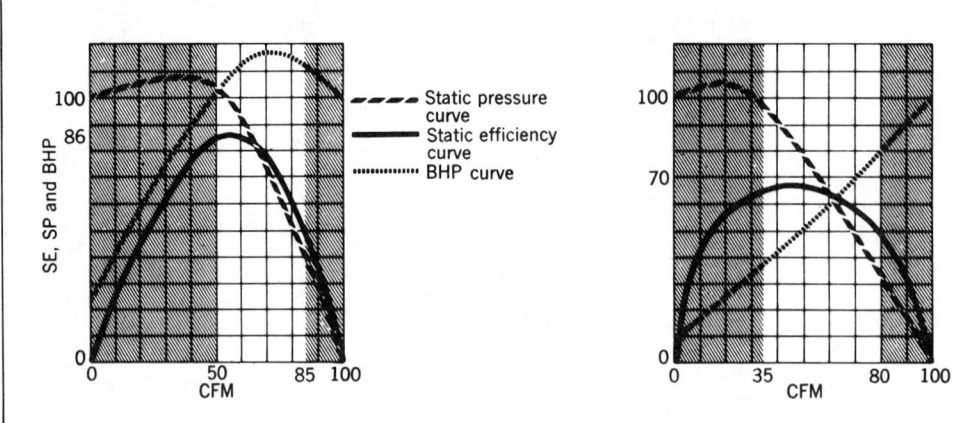

Fig. 14. Characteristic curves for airfoil fans. Fig. 15. Characteristic curves for radial blade fans.

Backward Inclined Fan

Backward inclined fans travel about twice the speed of the forward curved fan as previously indicated by the velocity vector diagram. The normal selection range of the backward inclined fan is approximately 40-85% of wide open CFM. (See Fig. 13.) Maximum static efficiency of about 80% generally occurs close to the edge of its normal operating range. Generally, the larger the fan the more efficient for a given selection.

The magnitude of surge for a BI fan is greater than for an FC fan.

Advantages of the BI fan are higher efficiency and nonoverloading horsepower curve. The horsepower curve generally reaches a maximum in the middle of the normal operating range, thus overloading is normally not a problem. Inherently stronger design makes it suitable for higher static pressure operation.

The BI fan's disadvantages include first, the higher speed which requires larger shaft and bearing sizes and places more importance on proper balance, and secondly, unstable operation occurs as block-tight static pressure is approached. This fan is also unsuitable for material handling.

A refinement of the flat-blade, backward inclined fan makes use of airfoil shaped blades. This improves the static efficiency to about 86% and reduces noise level slightly. The magnitude of surge also increases with the airfoil blades. Characteristic curves for airfoil fans are shown in Fig. 14.

Radial Blade Fans

Radial blade fans (Fig. 15) are generally narrower than other types of centrifugal fans. Consequently, they require a larger diameter wheel for a given capacity. This increases the cost and is the main reason why they are not used for air conditioning duty.

The radial blade fan is well suited for handling low volumes at relatively high static pressures and for material handling. Absence of surge and a nearly straight horsepower curve with linear relationship with CFM are its other advantages. This proportional relationship allows capacity control to be actuated from motor power input.

Disadvantages of this type of fan are higher cost and lower efficiency.

Tubular Centrifugal Fans

Tubular centrifugal fans, illustrated in Fig. 16, generally consist of a single width airfoil wheel arranged in a cylinder to discharge air radially against the inside of the cylinder. Air is then deflected parallel with the fan shaft to provide straight-through flow. Vanes are used to recover static pressure and straighten air flow.

Characteristic curves are shown in Fig. 17. The selection range is generally about the same as the scroll type BI or airfoil-blade wheel, 50-85% of wide open volume. However, because there is no housing of the turbulent air flow path through the fan, static efficiency is

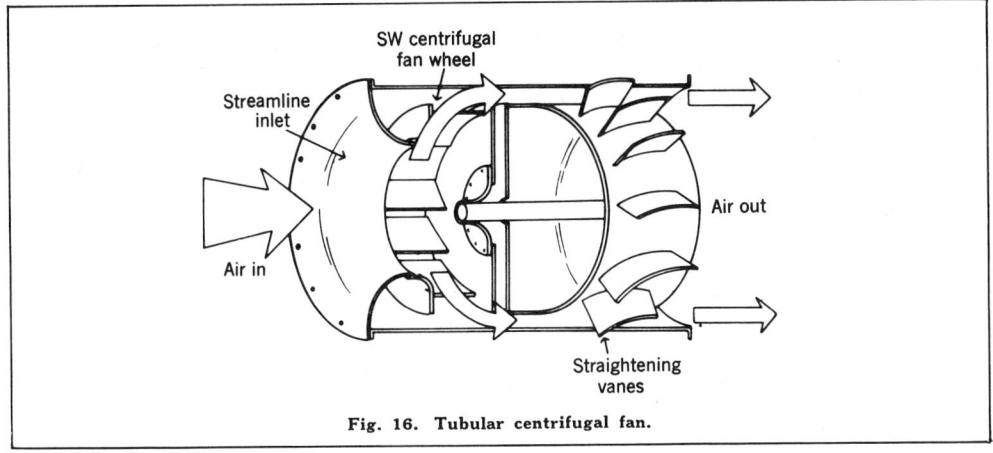

Fig. 16. Tubular centrifugal fan.

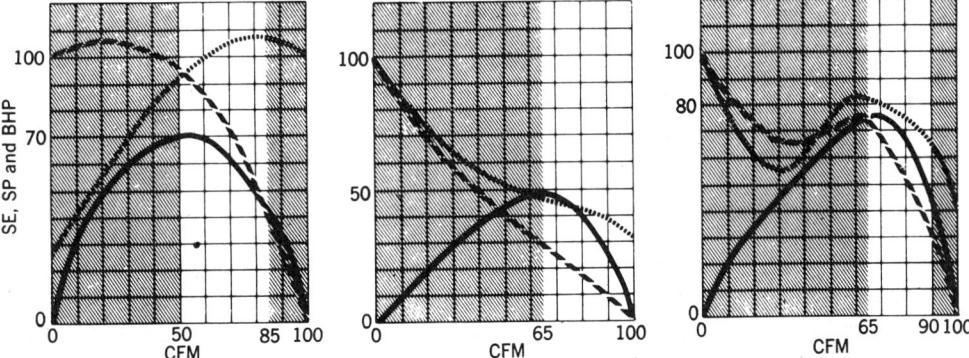

Fig. 17. Characteristic curves for tubular centrifugal fans.

Fig. 18. Characteristic curves for propeller fans.

Fig. 19. Characteristic curves for vaneaxial fans (high performance).

reduced to a maximum of about 72% and the noise level is increased.

Frequently, the straight-through flow results in significant space savings. This is the main advantage of tubular centrifugal fans.

Axial Fans

Axial fans are divided into three groups—propeller, tubeaxial, and vaneaxial.

The propeller fan (Fig. 18) is well suited for high volumes of air at little or no static pressure differential. Tubeaxial and vaneaxial fans, (Fig. 19), are simply propeller fans mounted in a cylinder and are similar except for vane type straighteners on the vaneaxial. These vanes remove much of the swirl from the air and improve efficiency. Thus, a vaneaxial fan is more efficient

than a tubeaxial and can reach higher pressures. Note that with axial fans the BHP is maximum at block-tight static pressure. With centrifugal fans the BHP is minimum at block-tight static pressure.

Advantages of tube and vaneaxial flow fans are the reduced size and weight and straight-through air flow which frequently eliminates elbows in the ductwork. The maximum static efficiency of an industrial vaneaxial fan is approximately 85%. The operating range for axial fans is from about 65% to 90%.

The disadvantages of axial fans are high noise level and lower efficiency than centrifugal fans.

In recent years, more sophisticated design of vaneaxial fans has made it possible to use these fans at pressures comparable to those developed by the airfoil backward inclined fans, with

equal overall efficiency. These fans have variable pitch blades which can be activated by an external operator. For large-size fans requiring motor horsepowers above 100 hp, it is comparatively simple to change the fan characteristics by using either a manual or pneumatic controller. The disadvantage of these fans is their high noise level; sound traps are generally required both upstream and downstream. Despite this added acoustical requirement, the initial cost of these fans compares favorably with the airfoil BI fans.

CLASS LIMITS FOR FANS

The Air Moving and Conditioning Association (AMCA) has adopted a standard which defines operating limits for the various classes of centrifugal fans used in general ventilation applications.

The standard uses limits based on "mean brakehorsepower per square foot of outlet area," expressed in terms of outlet velocity and static pressure. (See Figs. 1 to 3.)

When a fan is designated as meeting the requirements of a particular class, as defined by the standard, it must be physically capable of operating safely at any point inside the "minimum performance" limits for that class.

To assist fan users and consulting engineers, members of AMCA's Centrifugal Fan Division have agreed on a standard system of shading multirating tables in catalogs which should simplify selection of the appropriate fan class for each application. Where the new class limits are used in a catalog, the manufacturer will state that these are "in accordance with revised AMCA Standard 2408-69."

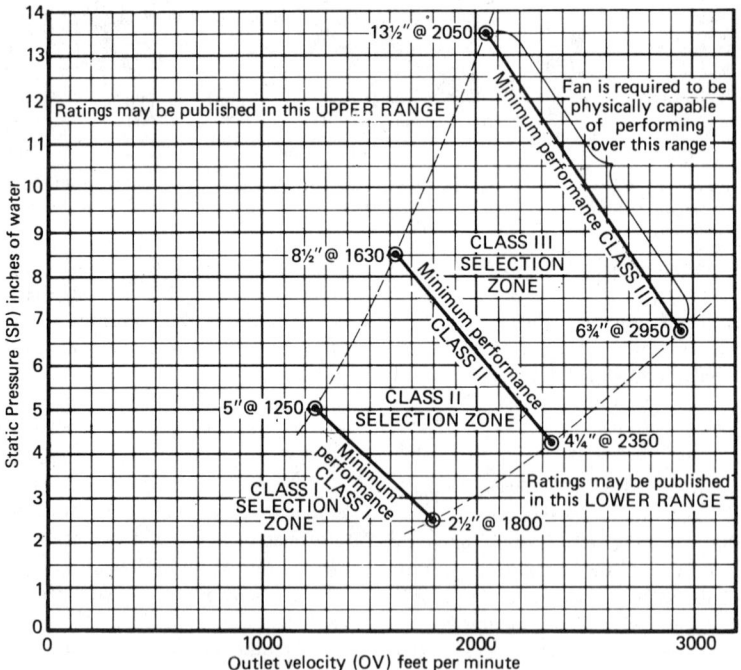

Fig. 1. Operating limits for centrifugal fans—tubular. (*Source: Air Moving and Conditioning Assn., Inc.*)

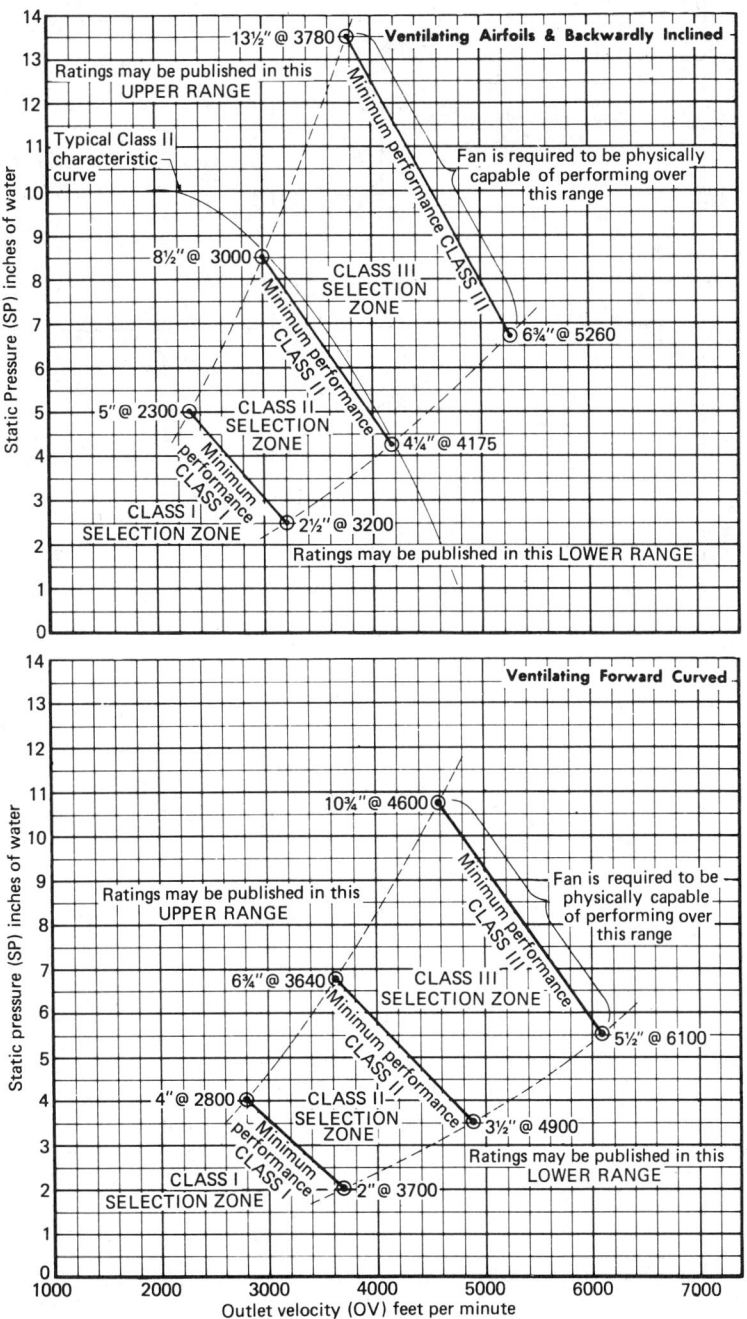

Fig. 2. Operating limits for single width centrifugal fans. (*Source: Air Moving and Conditioning Assn., Inc.*)

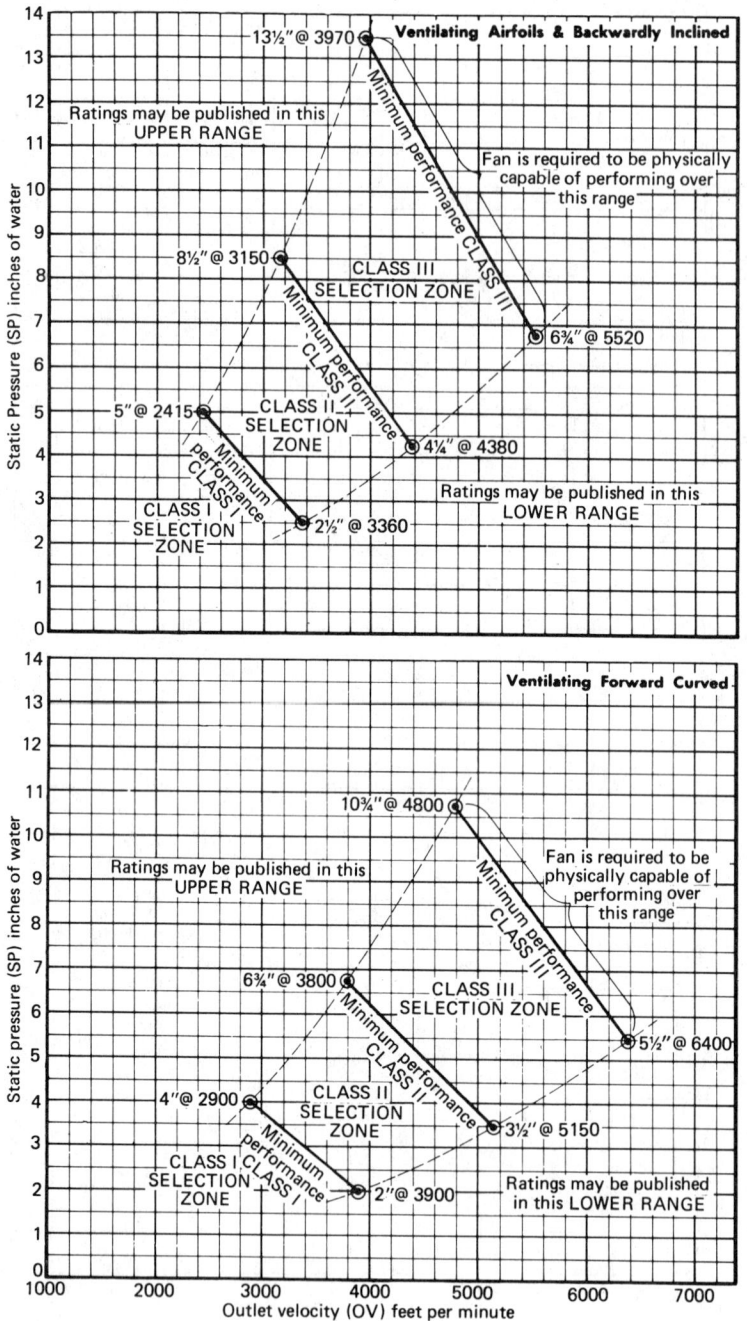

Fig. 3. Operating limits for double width centrifugal fans. (*Source: Air Moving and Conditioning Assn., Inc.*)

CENTRIFUGAL FAN, DRIVE AND INLET BOX ARRANGEMENTS

Reprinted by permission of Air Moving and Conditioning Association, Inc.

Arrangements of Drive

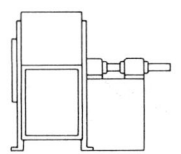 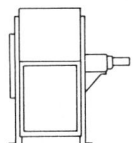 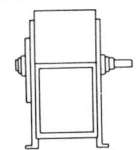 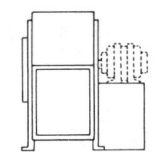 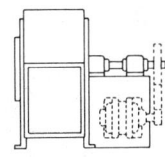

No. 1, SW, SI.
For belt drive or direct connection. Wheel overhung. Two bearings on base.

No. 2, SW, SI.
For belt drive or direct connection. Wheel overhung. Bearings in bracket supported by fan housing.

No. 3, SW, SI.
For belt drive or direct connection. One bearing on each side and supported by fan housing. Not recommended in sizes 27″ diam. and smaller.

No. 4, SW, SI.
For direct drive. Wheel overhung on prime mover shaft. No bearings on fan. Base mounted or an integrally direct connected prime mover.

No. 9, SW, SI.
For belt drive. Arrangement No. 1 designed for mounting prime mover on side of base.

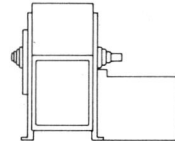 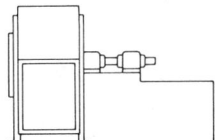 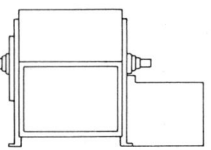 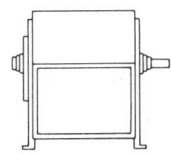

No. 7, SW, SI.
For belt drive on direct connection. Arrangement No. 3 plus base for prime mover. Not recommended in sizes 27″ diameter and smaller.

No. 8, SW, SI.
For belt drive or direct connection. Arrangement No. 1 plus base for prime mover.

No. 7, DW, DI.
For belt drive or direct connection. Arrangement No. 3 plus base for prime mover.

No. 3, DW, DI.
For belt drive or direct connection. One bearing on each side and supported by fan housing.

SW indicates single width, DW double width
SI indicates single inlet, DI double inlet

Designation of Direction of Rotation and Discharge

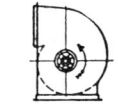 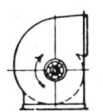 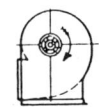 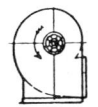

Counter-Clockwise Top Horizontal   Clockwise Top Horizontal   Clockwise Bottom Horizontal   Counter-Clockwise Bottom Horizontal

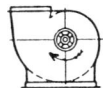 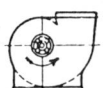 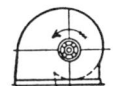 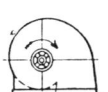

Clockwise Up Blast   Counter-Clockwise Up Blast   Counter-Clockwise Down Blast   Clockwise Down Blast

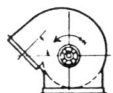 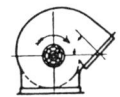 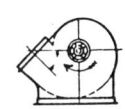 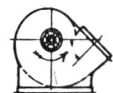

Counter-Clockwise Top Angular Down   Clockwise Top Angular Down   Clockwise Bottom Angular Up   Counter-Clockwise Bottom Angular Up

 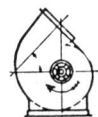

Counter-Clockwise Top Angular Up   Clockwise Top Angular Up

Direction of Rotation is determined from drive side for either single or double width, or single or double inlet fans. (The driving side of a single inlet fan is considered to be the side opposite the inlet regardless of actual location of the drive.) For fan inverted for ceiling suspension, Direction of Rotation and Discharge is determined when fan is resting on floor.

Designation of Position of Inlet Boxes

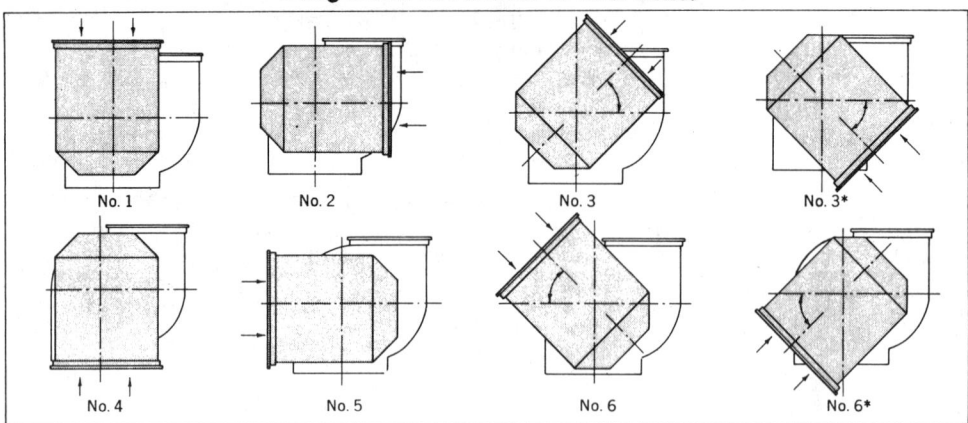

No. 1   No. 2   No. 3   No. 3*

No. 4   No. 5   No. 6   No. 6*

Definitions

Reference line is a horizontal line through center of fan shaft.

Air entry to inlet box is determined from drive side of fan.

On single inlet fans the drive side is always considered as the side opposite the fan inlet.

When drives are on both ends of fan shaft, the drive side is that side having the higher horsepower driving unit and is the same side from which the fan rotation is designated.

Air Entry Position Designation

1. Top Intake
2. Horizontal Right Intake
3. (Number of degrees) Above or below horizontal center line on right.
4. Bottom Intake
5. Horizontal Left Intake
6. (Number of degrees) Above or below horizontal center line on the left.

* It will be found in some cases that this arrangement interferes seriously with the framing of the floor structure by the amount of floor space required.

Motor Position, Belt or Chain Drive

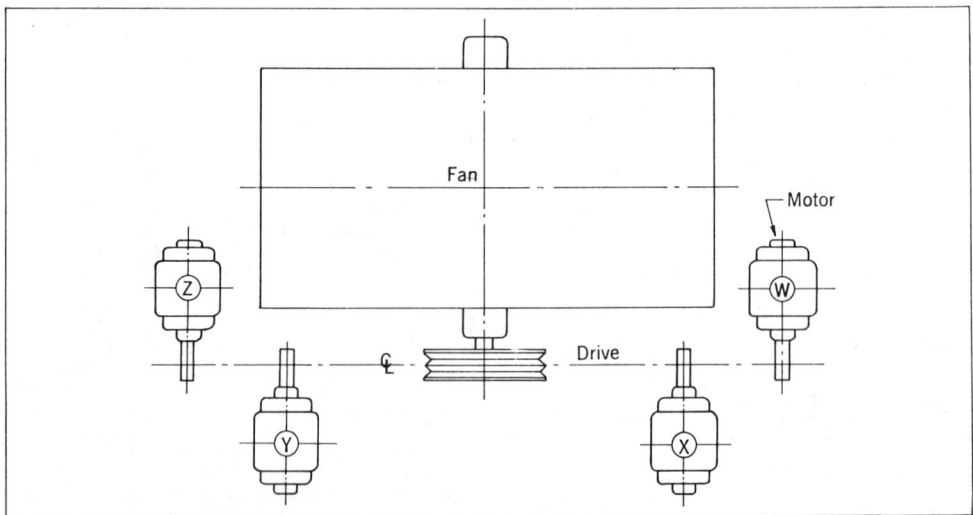

Location of motor is determined by facing the drive side of fan and designating the motor position by letters W, X, Y, or Z as the case may be.

FAN SELECTION

In any fan system, two basic parameters are required for fan selection: rate of flow or capacity, generally expressed in cubic feet per minute (CFM), and the potential required to move the air through the system, expressed either as static or total pressure (SP or TP).

The CFM is determined by the system designer at a specified temperature and barometric pressure at the system inlet. If the system requirement is given in pounds per minute, CFM can be found by the following equation:

$$CFM = \frac{\text{lb per min flowing}}{\text{lb per cu ft at inlet (density)}}$$

The fan performance is a function of air density at the fan inlet. This density not only determines the volumetric capacity for a given weight of flow, but the developed pressure of the fan as well. Factors which affect air density are: Barometric pressure, temperature, and relative humidity. Whenever these conditions are not specified, the fan supplier usually assumes air at standard conditions (70°F and 29.92" Hg. barometer).

Definitions of static, velocity, and total fan pressures are given earlier in this section, under "Terminology, Abbreviations, and Definitions." Fan manufacturers in the United States have customarily cataloged fan static pressure as the single basis for fan selection. This practice has led to confusion and very often poor fan selection. For this reason, a growing number of fan manufacturers now are showing their fan performance in terms of total pressure, since total pressure is actually what the fan must deliver.*

While a fan of almost any size, and either centrifugal or axial, could be selected for a given air flow and system resistance, the realistic possibilities are limited by practical engineering and economic considerations:

1. Space for the fan and its driving mechanism
2. Application, such as material handling, temperature of air, parallel operation, pressure range, and other factors listed under "Fan Types," above
3. Fan first cost versus its operating cost (fan

horsepower energy and maintenance)
4. Type and loudness of noise produced by fan
5. Effect of system dampering on fan performance
6. Fan drive mechanism and its reliability, particularly vee-belts versus direct drive for large fans
7. Expected life of fan versus first cost. This is closely tied to the fan construction and class.

While it is possible to select a fan by the "Equivalent Air Method of Rating" by using the fan laws,* the same results can be more rapidly obtained by resorting to the tables or selection charts published by the fan manufacturer. Multirating tables are the most common type of such data, usually based on standard air. After space requirements, fan application, expected life, and other such considerations have been established, the optimum fan selection is either at, or just to the right of, the peak efficiency point on the performance curve. This results in a slightly undersize fan. However, selection in this range provides a more stable operation than would an oversize fan. In fact, oversize fans should be selected only where future increase in capacity is expected, and great care must be exercised not to select one in the unstable range of the curve.

Peak efficiency can be determined from either the fan performance curves, or from multirating tables, by noting which fan meets the design requirements with minimum BHP. There is only one fan size in any type which will meet these requirements. If the design requirements do not exactly match the cataloged values of CFM or SP, linear interpolation of these values will give accurate results. The tabulated value of RPM is the required operating speed. However, the listed value of BHP must be multiplied by the ratio of actual density to standard density, to obtain the required operating horsepower.

Multirating charts are also very useful for fan selection; their main advantage is the graphic depiction of the performance for a family of similar fans. To obtain a better understanding of how these curves are constructed and used, several excellent references are available.

A second method of fan selection is the "Specific

---

* John W. Markert, "Use Total Pressure in Air System Design, Fan Selection," *Heating, Piping and Air Conditioning*, October, 1969.
Erich Brandt, "Fan Pressure Distributions in a Ducted Air System," *Heating, Piping and Air Conditioning*, May, 1969.

* R. Jorgensen, *Fan Engineering*, 6th ed., (Buffalo, N.Y.: Buffalo Forge Co.).

DESCRIPTION	% LOSS IN CFM IF NOT CORRECTED	% INCREASE NEEDED IN FAN SP TO COMPENSATE
3 piece elbow R/D = .5 1.0 2.0 6.0	12 6 5 5	30 13 11 11
4 piece elbow R/D = 1.0 2.0 8.0	6 4 4	13 9 9
5 or more piece R/D = 1.0 elbow 2.0 8.0	5 4 4	11 9 9
Mitered elbow	16	42
Square Ducts with Vanes No Vanes A B C D	17 8 6 5 4	45 18 13 11 9
Round to Square to Round	8	18

4 Pc Elbow

D   R

Rectangular Elbows without Vanes*

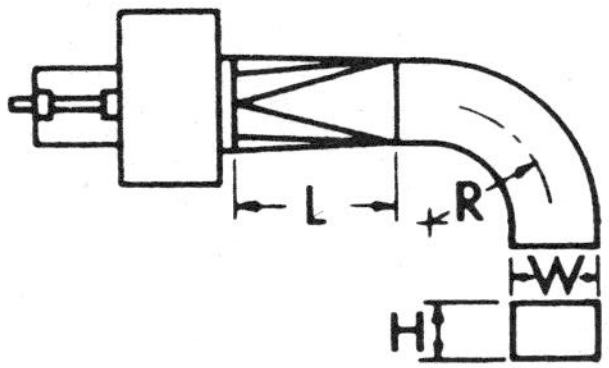

$\frac{H}{W}$	$\frac{R}{W}$		
$\frac{H}{W} = .25$, &	$\frac{R}{W} = .5$	7	15
	1.0	4	9
	2.0	4	9
$\frac{H}{W} = 1.00$, &	$\frac{R}{W} = .5$	12	30
	1.0	5	11
	2.0	4	9
$\frac{H}{W} = 4.00$, &	$\frac{R}{W} = .6$	15	39
	1.0	8	18
	2.0	4	9

*In all cases use of 3 long, equally spaced vanes will reduce loss and needed sp increase to 1/3 the values for unvaned elbows.

The maximum included angle of any element of the transition should never exceed 30". If it does, additional losses will occur. If angle is less than 30° and L is not longer than the fan inlet diameter, the effect of the transition may be ignored. If it is longer, it will be beneficial because elbow will be farther from the fan.

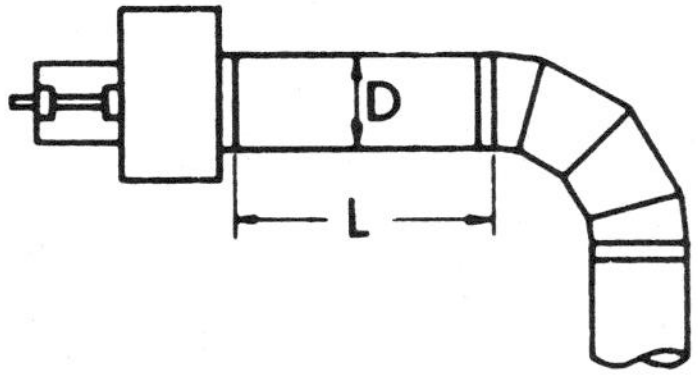

Each 2½ diameters of straight duct between fan and elbow or inlet box will reduce the adverse effect approximately 20%. For example, if an elbow that would cause a loss of 10% in CFM or an increase of 23% in fan SP, if on the fan inlet, is separated from the fan by straight duct the effect of the duct may be tabulated thus:

No duct	Loss = 10%	SP needed = 23%
L/D = 2½	Loss = 8%	SP needed = 19%
5	Loss = 6%	SP needed = 13%
7½	Loss = 4%	SP needed = 9%
10	Loss = 2%	SP needed = 4%

Fig. 1. Probable effects of various inlet connections. The losses given do not include friction losses. (Source: Industrial Ventilation Manual, 10th ed., pp. 10-13.)

Speed Method of Rating." This method is commonly used to select larger fans with direct drives. Since alternating-current electric motors are not available with standard rotating speeds, such as 3600, 1800, 1200 rpm, etc., selection of a motor speed that will produce the most efficient fan selection is a matter of trial assumptions of the standard motor speeds available. From these, the corresponding specific speeds may be calculated and then used with the base performance curves to select fan CFM and efficiency for a given SP and air density. This method is generally not recommended for fans with variable-speed drives, such as variable sheave, vee-belt drives commonly used for most HVAC systems.

Regardless of which method is used to select a fan, there is generally a possible selection of two or more suitable fans. Economics is usually the determining factor in the final selection. The initial cost of each fan which includes all required accessories, acoustical attenuators and vibration isolators, must be determined. To these component costs must be added the cost of installation. The first cost can be translated into an annual owning cost, to which is added the annual energy cost for running the fan and the annual maintenance cost. The fan which has the lowest annual owning and operating cost would then be the logical selection.

Fan noise and vibration are important considerations and are influenced by the size and type of fan, its rotating speed, and its efficiency. Generally, axial fans require acoustical treatment on both the suction and discharge sides of the fan, while centrifugal fans usually will need minimal treatment and then only on the discharge. For medium- and high-pressure fan systems, the advice of an acoustical consultant is recommended. Some manufacturers publish certified sound-rating data for their fans, and these should be consulted when available. In addition, the section on "Noise and Radiation Control" provides helpful information for estimating the sound-power levels of a fan.

FAN INLET CONNECTIONS

Cataloged fan performance is based on lab tests performed under ideal conditions that almost never occur at the fan inlet. This deviation from the ideal produces fan losses which reduce, often seriously, cataloged performance data. There are three basic causes, or various combinations of the three, for fan inlet losses:

1. Nonuniform flow into suction of fan
2. Swirl or vorticity
3. Flow blockage or inlet restrictions.

Due to the infinite variety of inlet conditions at each fan installation, it is difficult to assign specific loss values to these basic causes of fan inlet losses. However, some general guidelines will be useful in reducing them. While bad inlet conditions adversely affect the performance of axial fans, centrifugal fans are extremely susceptible to these conditions. For this reason, most of the ensuing discussion on inlet conditions pertains to centrifugal fans only.

Nonuniform flow into the suction of a fan is typically caused by an elbow installed too close to the fan inlet. The probable effects of various inlet connections are shown in Fig. 1.

Inlet swirl, or vorticity, is a frequent cause of reduction in fan performance. If the spin is imparted in the direction of wheel rotation, a situation corresponding to the use of inlet vanes arises: the fan volume, pressure, and horsepower are lower than expected. If the air spin is counter to wheel rotation, the volume and static pressure will be greater than expected and the brake horsepower will also be greater. In either case, spin always reduces efficiency. These conditions are readily overcome by installing vanes and a splitter at the fan inlet, as is graphically shown in Fig. 2.

Fan inlet blockage or restrictions may be encountered because of field installation conditions. In these cases, a loss in static pressure will result. This will require an increase in fan speed, with a corresponding increase in brake horsepower, to correct this situation.

Under some conditions, a fan may have a relatively short straight inlet duct starting in a plenum, through a wall, or in a flanged pipe. In some cases, the duct ends abruptly. (See Fig. 3.) Where the duct terminates in a plenum, through a wall, or flanged pipe, there is a pressure loss of $\frac{1}{2}$ the inlet duct velocity head. Where the duct ends abruptly, the pressure loss is $\frac{9}{10}$ of the inlet duct velocity head. In all these cases, a bell-mouth inlet would reduce the inlet loss to $\frac{1}{20}$ of the inlet duct velocity head.

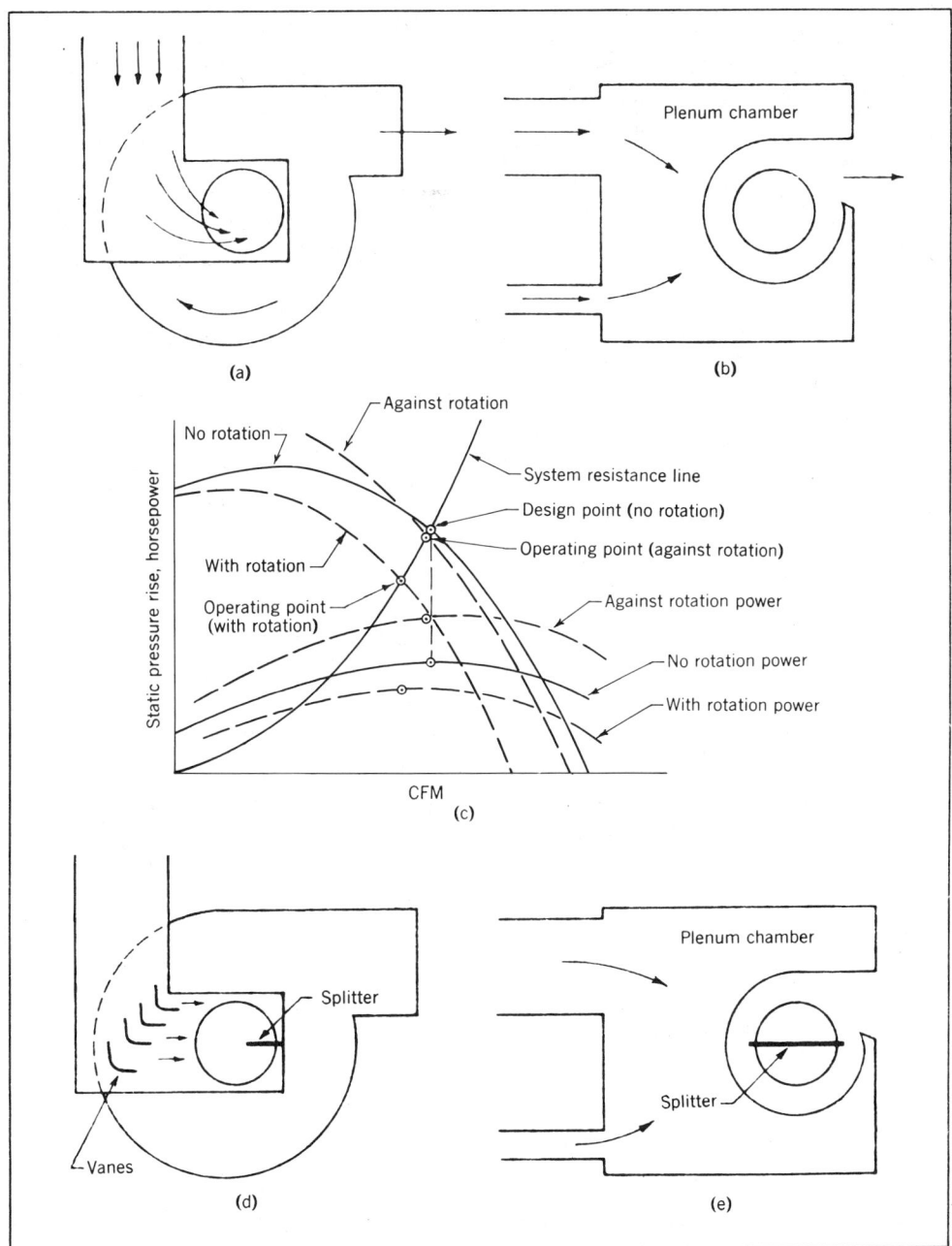

Fig. 2. Fan inlet swirl. (a) Inertia of air tends to crowd it to bottom of inlet, setting up swirl. (b) With two unequally sized inlets to plenum chamber, imbalance is set up, causing swirl at fan inlet. (c) Effect of inlet swirl on fan performance. (*Source: D. G. Traver, "System Effects on Centrifugal Fan Performance," ASHRAE paper presented at Symposium on "Fan Application—Testing and Selection," ASHRAE Semiannual Meeting, January, 1970.*) (d) Vanes and splitter prevent swirl in inlet box. (e) A splitter overcomes the imbalance that is due to unequal inlets. (*Source: a, b, d, and e; Air Conditioning, Heating and Ventilating, December, 1961, p. 79.*)

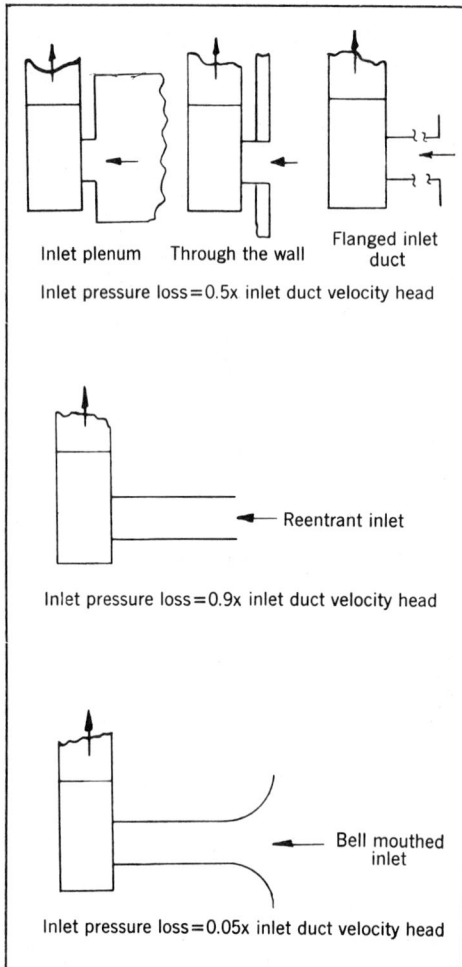

Inlet plenum   Through the wall   Flanged inlet duct

Inlet pressure loss=0.5x inlet duct velocity head

← Reentrant inlet

Inlet pressure loss=0.9x inlet duct velocity head

← Bell mouthed inlet

Inlet pressure loss=0.05x inlet duct velocity head

Fig. 3. Straight inlet duct losses. (*Source: D. G. Traver, "System Effects on Centrifugal Fan Performance," ASHRAE paper presented at Symposium on "Fan Application—Testing and Selection," ASHRAE Semiannual Meeting, January, 1970.*)

In some applications, fans are installed in plenum chambers, with open inlets. Occasionally, the wall of the plenum may be close enough to the fan inlet as to restrict its flow. Walls or similar obstructions should be kept at a minimum distance (A) of ½ a fan wheel diameter. (See Fig. 4.) A spacing of ⅓ wheel diameter will reduce pressure and flow about 10%.

Fan installations employing variable inlet guide vanes frequently result in an additional resistance to flow which decreases cataloged performance. There is an increasing fan industry trend to

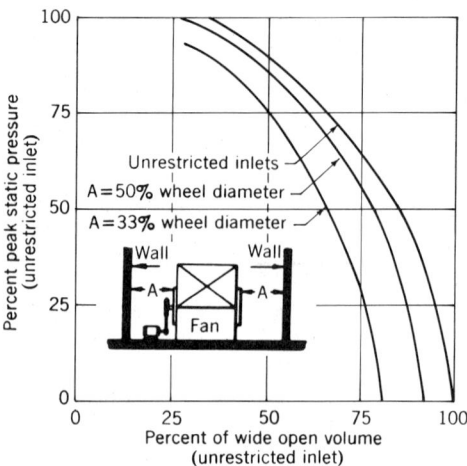

Fig. 4. Effect of space on fan performance. (*Courtesy of the Trane Co.*)

mount variable inlet guide vanes in the inlet bell of the fan, as contrasted with the practice of mounting an accessory set of vanes just upstream of the fan inlet in a larger diameter, lower velocity flow field. The vanes in the inlet bell often have their actuating mechanism at the center and this partially obstructs the flow as do the blades themselves. This blockage is percentage-wise more on smaller fans so the performance loss is proportionally greater. (See Fig. 5.) For example, a 30-in.-diameter fan must run at 4% higher rpm to meet catalog rating with a corresponding increase in hp of about 12%.

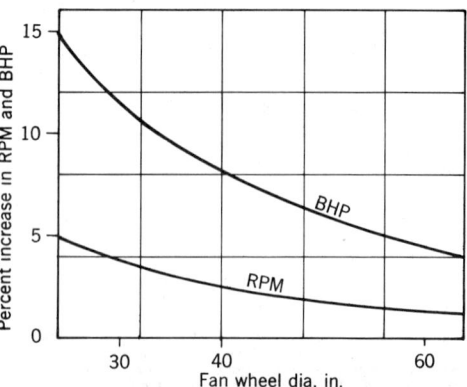

Fig. 5. Inlet guide vane restriction. (*Courtesy of the Trane Co.*)

Single inlet (SI) fans are often tested for rating purposes in an arrangement without any inlet bearing; hence, the performance of SI fans with an inlet bearing will be slightly less than the cataloged value. The performance reduction will be proportionately greater for smaller fans than for larger fans due to the relatively greater blockage area. The reduction will be greater for higher-pressure fans than for low-pressure fans due to the larger bearing and bearing support.

Double inlet (DI) fans are often rated using an extension drive shaft. This eliminates the blockage effect of the drive pulley and belts. Cataloged performance is slightly reduced by normal belt drives. This reduction is greater on higher-pressure fans due to the wider belts and pulley.

The comment about the effects of bearing support for SI fans applies here as well. Consult a fan supplier for specific correction factors for SI, DI, and drive effects on performance. These effects are usually less than 4% on speed or flow and 12% on hp.

In addition to belt and drive blockage, additional horsepower is required when heavy-duty bearings, heavy-duty grease, and belt drives are used. Belt losses are a function of belt tension, number of belts, and the type of belts. Typical belt drive losses are from 2 to 6 percent and can be significantly greater with small fans at slow speeds. When selecting a motor at or near its nameplate rating, this should be taken into account.

FAN DISCHARGE CONNECTIONS

While fan discharge conditions cannot alter the performance characteristics of a fan in the same way that inlet swirl does, fan outlet conditions can be responsible for system losses that are often sizable. Basically, these losses are a result of one or both of the following.

1. Reduction in Static Pressure Recovery

Air leaving a centrifugal type fan impeller is discharged with a radial velocity component (see Fan Types, page 5-15), which results in air discharge vortices. In addition, fan discharge velocity is not uniform across this discharge area, but is peaked from the concentration of air at the outside radius of the scroll. The resulting air flow from a fan outlet is, therefore, one of a non-uniform, spiral nature that does not fill the fan discharge area.*

When fans are performance tested, they typically have many equivalent diameter lengths of constant area duct attached to the fan discharge, including a flow straightener. As a result, there is ample distance for the flow to redistribute itself and the spiral flow will disappear partly of its own accord and partly due to the straightener. As a result, at the measuring station the flow will be very uniform, exhibiting a typical fully turbulent

flow velocity profile. Some of the dynamic energy is converted into static pressure, and the fan manufacturer then tabulates the fan's performance data, derived from these ideal discharge conditions. Unfortunately, these outlet conditions are almost never obtained in practice. Either the system design should attempt to use straight ductwork for 3 to 5 equivalent duct diameters downstream of the fan discharge and thereby realize static regain, or if this is not possible, provide added brake horsepower to accommodate the dynamic losses.

When straight discharge ducts are used, it is not recommended that any sudden transition to larger areas be used. It is recommended that transition to a larger area duct be accomplished with a taper having an included angle of no more than 15 degrees to minimize losses. This is common, good, duct-design practice. (See Fig. 1.)

When fans discharge into a plenum (commonly called *bulkheading*), as they do in many fan systems, a loss occurs due to the sudden enlargement in flow area. Theoretically, if the fan had a uniform velocity over its entire discharge area, the discharge pressure loss would be one velocity head, based on the fan discharge velocity. This is the velocity which is tabulated in fan catalogs. However, the actual velocity is not at all equal to the tabulated value; actual average discharge velocities are from 120 to 180 percent of the catalog value. This results in actual discharge pressure losses of 150 to 300 percent of what one

* D. G. Traver, "System Effects on Centrifugal Fan Performance," ASHRAE paper presented at Symposium on "Fan Application—Testing and Selection," ASHRAE Semiannual Meeting, January, 1970.

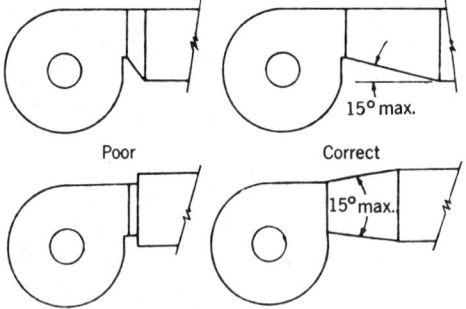

15° max.

Poor   Correct

15°max.

Fig. 1. Duct transition pieces for centrifugal fan discharge. *Top:* Right and wrong transition piece with expansion all on one side of duct. *Bottom:* Right and wrong transition piece with expansion in two directions. (*Source: Air Conditioning, Heating and Ventilating, February, 1962, p. 70*).

would compute from the catalog outlet velocity. The addition of a short discharge duct of only one or two equivalent diameters in length will significantly reduce this sudden enlargement loss. Even this short distance permits a significant redistribution of the velocity with its corresponding static regain. The discharge loss will then be considerably reduced.

It is recommended that the bulkheading loss be obtained from the fan manufacturer, because the magnitude of this loss varies with the type of fan. The Trane Co. has suggested the following approximate increases in fan RPM and BHP to the cataloged values when there is no discharge ductwork: *

Fan Type	Percent Increase, in RPM	Percent Minimum, BHP
Forward Curved	6	20
Backward Inclined	4	13
Airfoil	3	9

Fan discharge into an after-filter plenum must, in addition to the bulkheading losses, take into account the damage to the filters, especially the bag type, by hitting them with high velocity air from a short distance. For this reason either 2-inch-thick, cleanable metal prefilters or a perforated metal "splash" plate, set in front of the after-filter bank will protect the after-filters

* Trane Engineering Bulletin, "Effect of Accessories and Installation on· Cataloged Fan Performance," Section EB-FAN16, March 24, 1969.

from damage. Unfortunately, this also adds a slight pressure loss for the fan to overcome. The pressure loss for 2-inch-thick, cleanable filters would generally not exceed 0.10 inch WG, while a perforated metal splash plate's resistance depends on the perforation size and free area. This loss can be obtained by referring to air performance data of perforated grilles or small lattice grilles in the catalog of any air-grille manufacturer.

2. Duct Turns

The following has been excerpted from ASHRAE Bulletin SF-70-8, "Fan Application—Testing and Selection," 1972.

Duct turns, immediately off the fan discharge, create a higher-than-expected static pressure drop due to turbulence and to the velocity profile existing at the discharge.

When an elbow must be used at the fan discharge, it is not recommended that a short-radius elbow be attached directly to the fan discharge. Preferably, a medium-radius elbow should be used (minimum mean radius—1.5 × equivalent duct diameter) or a one-diameter length of straight duct followed by a cascade square elbow will give a lower loss but only if the catalog outlet velocity is below 2000 fpm to minimize noise generation problems.

Assuming that a medium-radius elbow of rectangular cross-section is fitted to the fan discharge, it can turn the air in any one of four directions. If the fan discharge velocity were uniform, one could readily calculate the loss in the elbow and whatever direction the elbow faced would be immaterial. With a spiraling non-uniform discharge velocity, one cannot apply any of the normal elbow and duct friction factors appearing in Tables 9, 10, and 21, which are in the *ASHRAE Handbook,* or other references. These only apply when the flow is uniform across the duct without any spiral. If the flow were uniform, such an elbow would have a pressure loss of 0.25 × fan outlet velocity.

The flow in a fan discharge elbow differs in each of the four positions, for both single inlet and double inlet fans. (See Fig. 2 for an illustration of the four positions.) For position A, the high flow velocity is on the same side of the scroll and the elbow. This will result in the lowest loss of the four positions. It should be used whenever possible. Assume a loss equal to

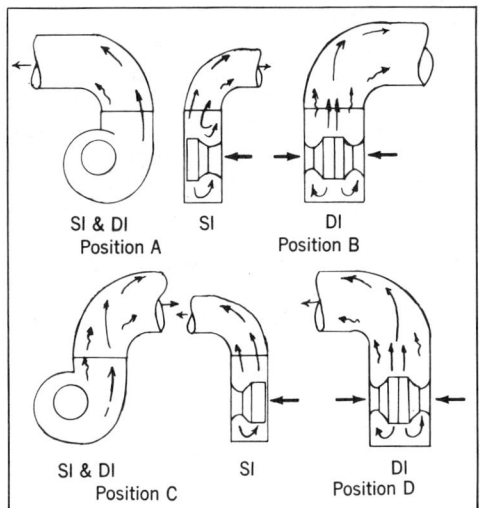

Fig. 2. Discharge elbow flow patterns (*Source: D. G. Traver, "System Effects on Centrifugal Fan Performance," "ASHRAE paper presented at Symposium on "Fan Application—Testing and Selection," ASHRAE Semiannual Meeting, January, 1970.*)

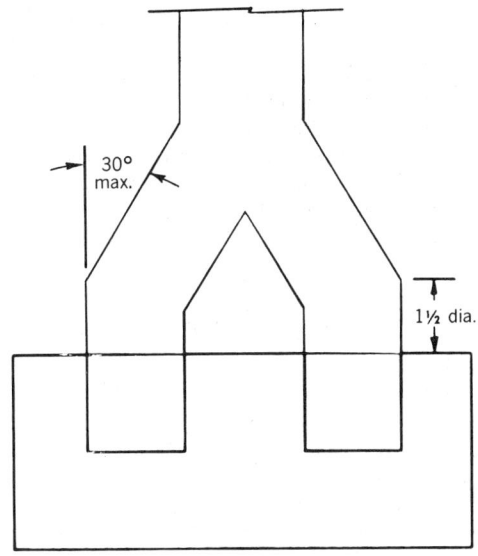

Fig. 3. Proper construction of a "pants" connection on a two-fan unit. (*Courtesy of the Trane Co.*)

0.5 × cataloged outlet velocity head for both SI and DI fans. For position B, this is the second lowest loss position for a SI fan, the high velocity out of the scroll continues to the outside of the elbow. Assume a loss equal to 0.60 × cataloged outlet velocity head for SI fans and 0.75 × cataloged outlet velocity head for DI fans. The DI loss is higher than the SI loss because the peak velocity out of the fan discharge is at the center and it must be diverted to the outside of the elbow. Energy must be expended to shift the flow and, consequently, an additional loss is introduced.

For position C, assume a loss equal to 1.0 × cataloged outlet velocity for both type fans. In SI and DI fans, the peak discharge velocity is on the opposite side of the elbow than normal. The resulting flow redistribution results in high losses. This position is the most unfavorable of the four.

For position D, assume 0.90 × cataloged outlet velocity head for SI fans and 0.75 × cataloged outlet velocity for DI fans. The SI fan has its peak discharge velocity from the fan discharge at the opposite side from that which is normal for an elbow. As a result the velocity redistribution in this situation results in a higher loss than for DI fans where the peak discharge velocity is centered as in position B. These loss factors are approximate only, but they do establish an approximate level of loss for design purposes.

Fans are frequently installed in factory-fabricated rectangular enclosures called "cabinets." Cabinet fans often have two fans discharging into a common duct by a *pants* connection. Figure 3 shows the requirements of the pants connection, to achieve catalog performance. There should be 1½ equivalent fan diameters of straight duct before the transition, with the convergence angle a maximum of 30° on each side. If these design parameters cannot be met, the fan discharge is treated as though it were a free discharge into a plenum, and bulkheading losses are then used.*

---

* Trane Engineering Bulletin, "Effect of Accessories and Installation on Cataloged Fan Performance," Section EB-FAN16, March 24, 1969.

FAN PERFORMANCE MODULATION

Some fan systems have changing air requirements during operation, such as variable air volume systems, while others have changing pressure requirements; both airflow and pressure are often altered during operation. To accommodate these changes, some form of fan performance modulation is required. The types of modulation typically used in ducted applications are:

1. Scroll Volume Control

This is discussed under "System Surge, Fan Surge, and Paralleling," on page 5-14.

Scroll volume dampers are sometimes used on small, single-fan utility sets as a means of quickly adjusting the air delivery. However, it is not regarded as a good means of capacity control. Efficiency is reduced and the very nature of outlet volume control makes it difficult to operate automatically off a static pressure sensing device. Thus, while the scroll volume damper serves a useful purpose in controlling the paralleling of fans, it is not recommended for capacity modulation.

2. Inlet Dampers

The primary purpose of inlet dampers, or face dampers, as they are better known in central station units, is for the prevention of backdraft and air circulation when the unit is shut down.

Inlet dampers merely add resistance to the system and cause a corresponding change in static pressure at the fan to vary the CFM.

There are two basic drawbacks to inlet dampers. First, they allow little capacity modulation without forcing the fan to operate in an unstable part of its performance range. Secondly, since they are frequently mounted in front of an outside air opening or in front of a coil bank, they are much larger in size than the fan inlet. Thus, the static pressure differential across the damper is spread over a large area.

Because of this second drawback, care must be taken to make sure that the fan is not capable of producing a static pressure sufficient to warp or cave-in the dampers. Static pressure differential across most face dampers used on central station air-handling units should not exceed 4 inches, total. If the fan is capable of developing more than this static pressure at the operating RPM,

care must be taken to insure that face dampers cannot be closed while the fan is operating. If the dampers are used for trimming the system, a manual stop can be put in the damper linkage to prevent them from closing completely. For an on-off shutoff to prevent air circulation, damper motors should be installed to close the dampers only after the fan motor has shut off; conversely, the fan motor should not start until these dampers are at least partly opened. This can be done by an end switch on the dampers, which precludes fan motor operation when the dampers are completely closed and will permit fan operation only when the dampers are sufficiently opened to avert high suction static pressure.

3. Discharge Dampers

Discharge dampers are a method of varying the CFM over a rather narrow performance range.

Since discharge dampers are typically mounted on the fan discharge, the area of the dampers is relatively small. Thus, there is generally no need to worry about excessive static pressure damaging the dampers. They will operate satisfactorily at block-tight conditions unless fan static pressure exceeds the structural capability of the dampers. Normally, the damper strength should withstand at least 4 inches of static pressure.

Figure 1 shows fan performance with discharge dampers. These dampers increase system static pressure to modulate the CFM. Discharge dampers do not change the unstable area of the fan. Thus, they should not be used for CFM modulation with AF centrifugal fans below about 50% of wide open CFM, as this figure indicates.

Neither discharge dampers nor inlet dampers have much effect on the system noise level in the wide open position on low- and medium-pressure applications. However, they do increase the noise level as they near a closed position. The magnitude of the increase is a function of the air velocity and static pressure differential.

4. Inlet Vanes

Inlet vanes are sometimes given the misnomer of *vortex dampers*. Actually, these vanes are not dampers; their sole purpose is to impart a swirl to the air in the direction of rotation as it enters the fan. The resulting vorticity results in

a reduction in CFM, static pressure, and brake horsepower. Moreover, for every position of the inlet vanes, separate curves for static pressure and brake horsepower versus CFM are generated.

As these vanes are modulated, the brake horsepower curve generated is lower than the brake horsepower curve with the vanes wide open. Therefore, inlet vanes do provide some operating cost savings. The magnitude of these savings is generally about 20 to 30 percent, if the vanes are operated a majority of the time in the range of 60 to 80 percent of design CFM. Since inlet vanes cost two to three times as much as parallel-blade discharge dampers, it doesn't pay to use them unless capacity reduction is at least 50 percent for long periods of time, since its horsepower savings over parallel and opposed blade dampers average about 25 percent under these conditions.

Quite apart from economics, inlet vanes are useful for capacity reduction on large centrifugal fans requiring more than 100 BHP, which are equipped with direct drive. This results from the difficulty of using vee-belt, variable-speed drives on such large fans.

There are three drawbacks in using inlet vanes for capacity modulation: First, the fan can be forced to operate in an unstable range with inlet vanes. This is most likely to occur when the vanes are used to modulate a constant static pressure system. The resultant noise and vibration has been known to shake an entire floor.

Secondly, capacity reduction also occurs when the inlet blades are in the wide-open position, as is shown in Fig. 5, page 5-28. Construction of the vanes with the hub and turning mechanism located in the center creates a pressure drop, the magnitude of which is a function of the fan size. For very small fans, the hub is a relatively large percentage of the total inlet area. Thus, the capacity reduction is substantial. On the other hand, with very large fans, the hub area is a very small percentage of the total and the reduction is negligible. With belt drive applications, this does not present any particular problem since the fan speed can be readily increased to com-

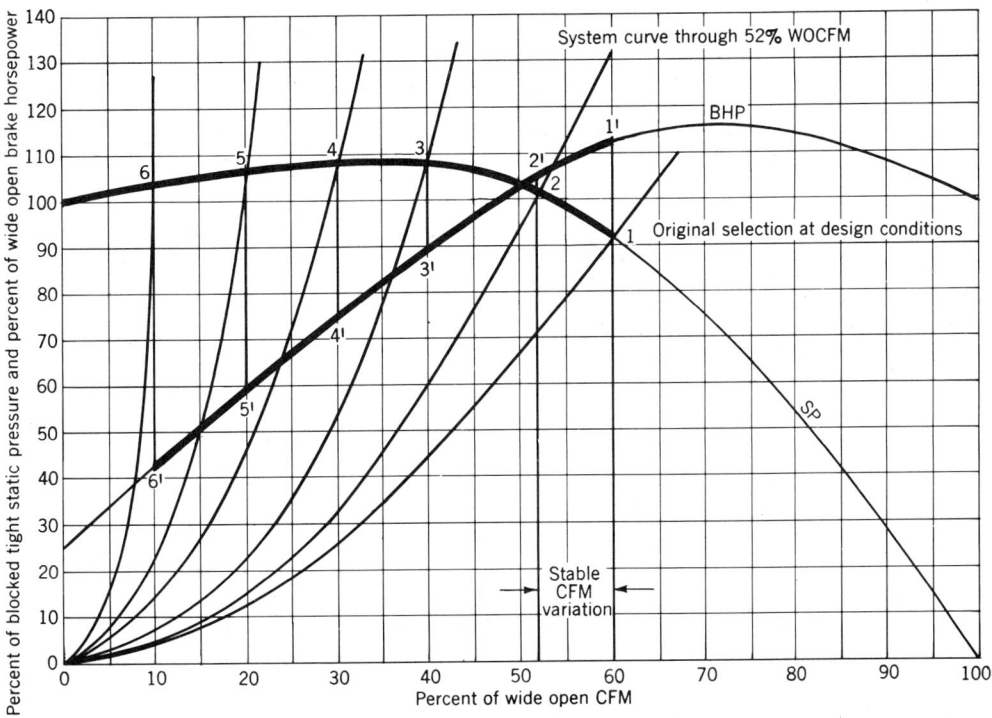

Fig. 1. Discharge damper performance for air foil fan. (*Courtesy of the Trane Co.*)

pensate. However, brake horsepower also goes up.

For example, the RPM must be increased approximately 3%, with a 36-inch-diameter wheel, to achieve full load capacity with inlet vanes in wide-open position. This increases the brake horsepower approximately 9.3% which could be a problem if the brake horsepower is very close to the nameplate motor horsepower.

In direct drive units, however, the use of inlet vanes becomes more of a problem. Fairly accurate means must be available for estimating the capacity reduction for various size fans.

Thirdly, inlet vanes will increase the fan's noise level, even at wide-open position. Because test data is limited, a good rule to follow is to add 5 db to the fan noise level when using inlet vanes.

Before using inlet vanes, the fan manufacturer should be consulted for information regarding the unstable range of operation, the capacity reduction due to inlet area restriction, and the resultant noise levels.

5. Speed Modulation

Speed variation in fans can be accomplished in a number of ways, including: Multispeed motors; fluid drives; mechanical speed reducers; and solid-state devices.

Speed modulation is not generally used in air conditioning applications and will not be discussed in detail. Typically, the cost is greater and requires more elaborate control.

Solid state devices have some merit on fractional horsepower motors and smaller integral horsepower motors. However, the contol must be closely matched with the motor to operate properly.

All of these devices affect fan performance in accordance with the following fan laws:

$$\left(\frac{CFM_2}{CFM_1}\right) = \left(\frac{RPM_2}{RPM_1}\right)$$

$$\left(\frac{SP_2}{SP_1}\right) = \left(\frac{RPM_2}{RPM_1}\right)^2$$

$$\left(\frac{BHP_2}{BHP_1}\right) = \left(\frac{RPM_2}{RPM_1}\right)^3$$

Care must be exercised in using this type of modulation in systems requiring constant static pressure either at the fan or at remote distribution boxes, as the static pressure at the fan reduces proportionally to the square of the RPM reduction.

6. Fan Blade Pitch Variation

Vaneaxial fans are available with adjustable pitch blades to permit varying the fan's performance. This may be used to increase or decrease system capacity on direct drive fans, depending upon the original selection. On belt-driven fans it may allow some increase in efficiency if the static pressure was grossly over-estimated when the original selection was made.

This form of capacity modulation will generally reduce brake horsepower more than any of the previous methods for a given CFM and static pressure. It also obviates the vee-belt drive problem for larger fans requiring more than 100 BHP, since control modulation can be accomplished fairly easily.

One method of fan-blade pitch variation allows for a change in pitch while the fan is in operation. This makes the fan very adaptable for such applications as automatic static pressure control for variable air volume systems.

Since the vaneaxial fan generally must have acoustic treatment anyway, noise generation due to change in fan pitch is easily handled. For this reason, the greatest drawback in this type of fan modulation is the added cost of the device. The more sophisticated the modulation and its controls, the greater will be the cost premium. These fans may still be cheaper, however, in both initial and operating costs than the centrifugal fans with either dampers or inlet vanes.

USEFUL FAN FORMULAS

$$Q = A \times V, \qquad (1)$$

where

$Q =$ volume per unit time, usually cubic feet per minute

$A =$ cross-sectional flow area, usually square feet

$V =$ average air velocity, usually feet per minute

$$CFM = \frac{lb/hr \ (Air \ or \ Gas)}{60 \times Density} \qquad (2)$$

$$Density = .075 \frac{(460° + 70°)}{(460° + Elev. \ Temp)}$$

$$= .075 \frac{(°R \ Base)}{(°R \ Actual)} \qquad (3)$$

[Air at 29.92 in. Hg and 70°F at Sea Level.] (4)

Atmospheric Pressure

$\qquad$ = 14.7 psig at Sea Level and 70°F

$\qquad$ = 407 in. H₂O at Sea Level and 70°F

$\qquad$ = 29.92 in Hg at Sea Level and 70°F

$\qquad$ = 33.94 ft H₂O at Sea Level and 70°F

$$TP = SP + VP \quad \text{(Total Pressure = Static Pressure + Velocity Pressure)} \qquad (5)$$

$$°R = 460° + °F = \text{Absolute Temperature or °Rankine} \qquad 6)$$

$$V = 1096.2 \sqrt{\frac{VP}{\text{Density}}} \qquad (7)$$

$$V = 4005 \sqrt{VP} \quad \text{(at .075 lb/cu ft Density)} \quad (8)$$

$$AHP = \frac{Q \times TP}{6356} = \frac{CFM \times TP}{6356} \qquad (9)$$

$$BHP = \frac{Q \times TP}{6356 \times TE} = \frac{CFM \times TP}{6356 \times \text{Total Effic.}} \qquad (10)$$

$$BHP = \frac{Q \times SP}{6356 \times SE} = \frac{CFM \times SP}{6356 \times \text{Static Effic.}} \qquad (10a)$$

$$OV = \frac{CFM}{\text{Outlet Area}} \qquad (11)$$

$$°F = \left(°C \times \frac{9}{5} \right) + 32 \qquad (12)$$

$$T = \frac{W \times D}{55} \qquad (13)$$

where:

T (Starting Torque, in pound feet)

W (Weight of wheel plus shaft, in pounds)

D (Diameter of fan bearings, in inches)

$\qquad$ [The above is for sleeve bearings]

Flywheel effect (Polar Moment of Inertia): WR^2 (fan wheel) = Whl Wgt × (Rad. Gyration)² +

$$\text{Shaft Wgt.} \times \left(\frac{\text{Shaft Hub Dia}}{3} \right)^2 \qquad (14)$$

[Use for direct connected fans and motors]

$$WR^2 = WR^2 \text{ Fan wheel} \times \left(\frac{\text{Fan RPM}}{\text{Motor RPM}} \right)^2$$

[Use for V-belt-driven fans and motors]

$$SHP = \frac{\text{Amps} \times \text{Volts} \times \sqrt{3} \times \text{Effic} \times \text{Power Factor}}{746} \qquad (15)$$

[SHP = Shaft Horse Power]

$$\text{Operating Torque} = \frac{SHP \times 33000}{2\pi r \times RPM}$$

$$= 5250 \times \frac{SHP}{RPM} \qquad (16)$$

[Operating Torque in foot-pounds]

$$\text{Air Temp. Rise through a fan} = \frac{.00278 \times SP}{SE \times \text{Density}}$$

$$= \frac{42.5 \times BHP}{\text{lb/Min} \times .241} \qquad (17)$$

where:

$\qquad$ SP = Static Pressure (inches H₂O)

$\qquad$ SE = Static Efficiency

$\qquad$ .241 = Specific Heat of Air

$\qquad$ 42.5 = Constant

$\qquad$ .00278 = Constant

NOMOGRAPHS FOR FAN HORSEPOWER AND ACTUAL CAPACITY

The nomographs on this and the following page are for use in solving problems in air handling.

In either nomograph, should a quantity be involved that exceeds the range of any of the scales, it may be divided by any number (10, 100, and 1000 are most convenient) and the reduced number used. For each time that a quantity has been reduced by division, the answer must be multplied by the same number.

The fan horsepower nomograph on this page is a solution of the equation:

$$BPH = (0.01573)(P)(CFM)/E$$

Example: What brake horsepower is required to drive a 95% efficient blower in order to deliver 150,000 cfm at a pressure of 3 inches of water?

Since 3 inches of water is outside the range of the P scale, divide by 10 and use 0.3. Then, align $E = 95\%$ with $P = 0.3$ (dashed line 1), marking its intersection with the pivot line. Since 150,000 cfm is outside the range of the CFM scale, divide it, also, by ten and use 15,000. Through the intersection of line 1 and the pivot line, draw a line through CFM = 15,000, and read BHP = 0.745 (line 2). Because two quantities were divided by ten, the answer must be multiplied twice by ten. Therefore, the answer is 74.5 brake horsepower.

The second nomograph is based on the following equation:

$$SCFM = \frac{(ACFM)(68 + 460)(P_g + 14.7)}{(t + 460)(14.7)}$$

Example: If a dryer can handle 41,000 standard cubic feet of air per minute (SCFM) at 66 psig and

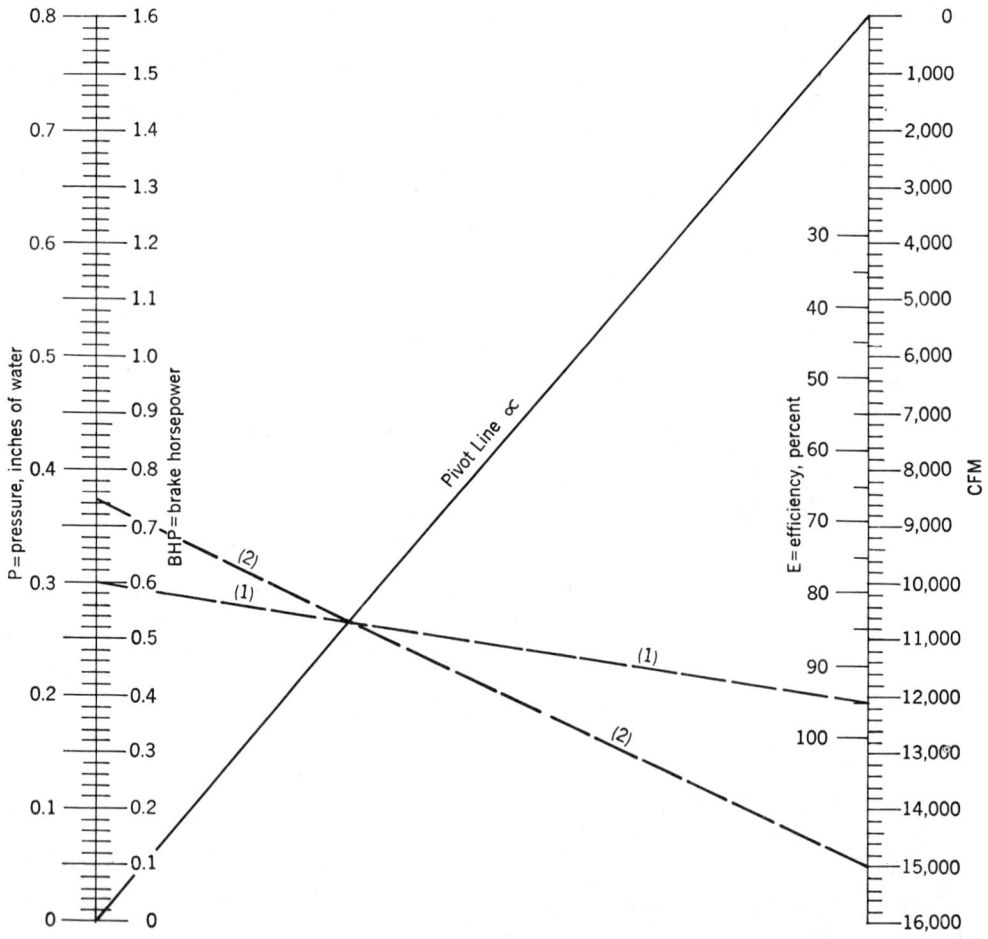

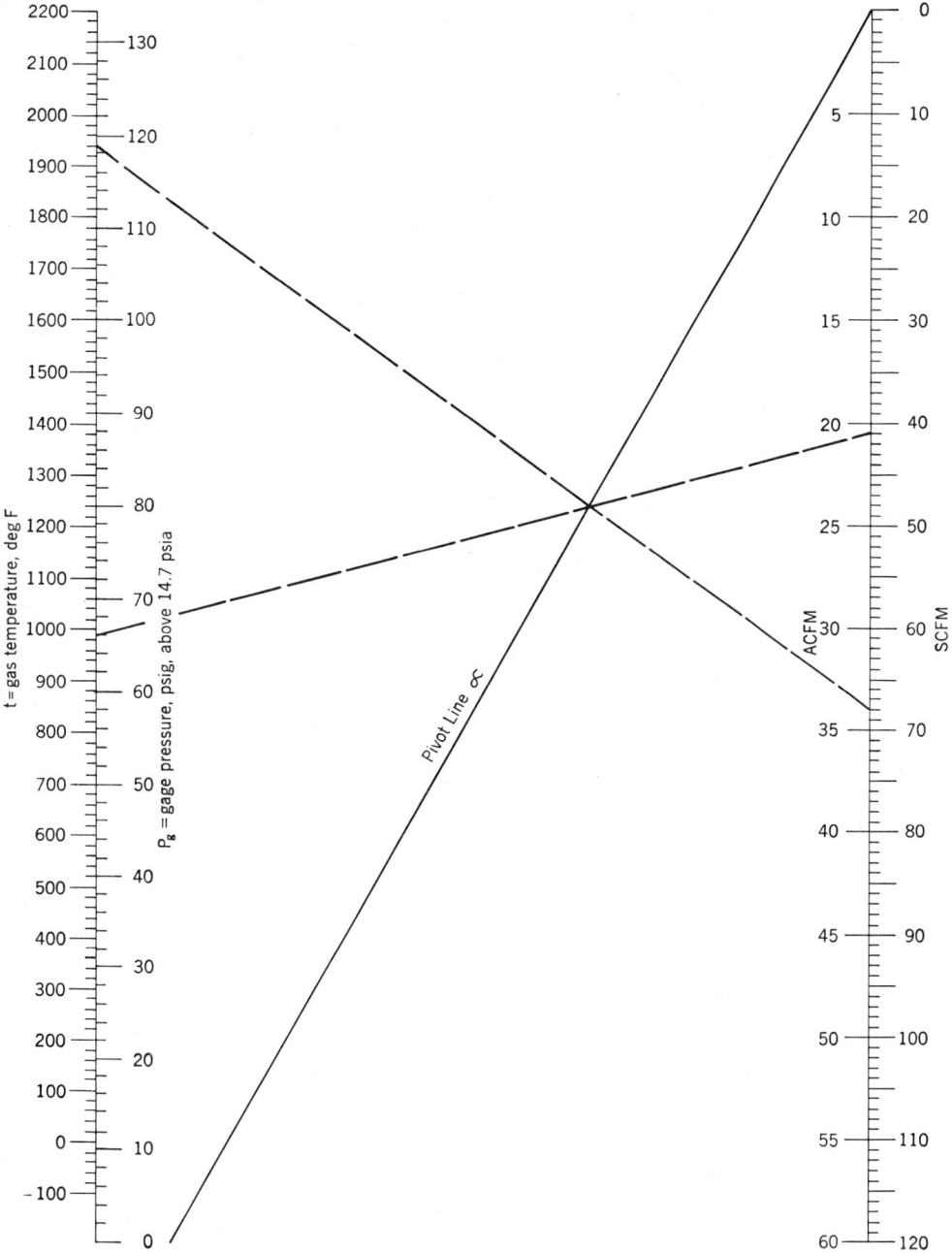

t = gas temperature, deg F

P_g = gage pressure, psig, above 14.7 psia

Pivot Line ∝

ACFM

SCFM

1940 F, what is the actual cfm (ACFM) it is handling at those conditions?

In order to bring the given data within the scales of the nomograph, SCFM is divided by 1000. Align

SCFM = 41 with P_g = 66, (dashed line). Through the intersection of this line and the pivot line, draw another through t = 1940. Read ACFM = 1000 × 34 = 34,000.

FAN SELECTION QUESTIONNAIRE

To assist the designer in a systematic procedure for selection of the proper fan and drive, a fan selection questionnaire, as given below, lists a logical progression of steps in the choice of fan and drive for a specific application.

1. **Fan Selection:**
   Manufacturer: _____
   Air Temperature and Density: _____
   Fan CFM at Design Temperature and Density: _____
   Fan CFM Corrected to Standard Conditions: _____
   Fan SP Corrected to Standard Conditions: _____ in. WG
   Type of Fan: _____ (Centrifugal, Roof-Top, Axial Flow, Tubular Centrifugal)
   Fan Speed (Corrected for Inlet and Discharge Factors): _____ RPM
   Wheel Diameter: _____
   Tip Speed: _____ FPM
   Outlet Velocity: _____ FPM
   Fan Width and Inlet: _____
   Class: _____ (I, II, III)
   Arrangement: _____
   Discharge: _____
   Rotation: _____
   Fan Curve: Is operating point on the negative slope? _____
   Fan Efficiency: _____

2. **Fan Motor:**
   BHP (Corrected for Inlet and Discharge Factors): _____
   Motor HP: _____
   Voltage: = _____ (Open Drip-proof, Spash-proof, TE, TE—Explosion Proof)
   NEMA Frame No.: _____
   Single Speed or 2-Speed: _____
   2-Speed RPM's: _____
   Fan's Polar Moment of Inertia, WR^2—for acceleration characteristic of motor drive:
   _____
   Special Motor (Description): _____

3. **Fan Drive:**
   Vee-Belt:
   Direct:
   Special:

4. **Fan Construction:**
   Type of Bearings: _____
   Housing Material: _____
   Wheel Material: _____
   Birdscreen—Mesh and Material: _____
   Drain in Fan Housing: _____
   Adjustable Inlet Vanes: _____
   Gravity Discharge Dampers: _____
   Cleanout Doors: _____

5. If fan is operating in parallel, show curves for single and parallel operation. Also, select motor for non-overloading capability.

6. Is fan flywheel effect (WR^2) large enough to require special starting?
   (See section on "Motors and Starters" later in this Handbook, for proper motor and starter selection.)

7. Type of Fan Performance Modulation Control: _____

FAN COMPARISON EXAMPLE

The following example is intended to show the factors that are taken into account when a fan selection is made. In this case, the engineer is looking for a fan which will deliver 75,000 cfm of standard air against a static pressure of 8.0 inches WG. This fan will have a free inlet and bulkheaded discharge into a sound-attenuated plenum. Capacity control is required.

The following fans are considered as feasible selections:

Fan No.	Type	Arrangement No.	Type of Capacity Control
1	Airfoil centrifugal—DWDI (Double Width, Double Inlet	4	Inlet vanes
2	Airfoil centrifugal—DWDI	3	" "
3	Inline tubular centrifugal	9	" "
4	Vaneaxial	4	Controllable pitch

Note: Fans 1 and 4 have direct drive and fans 2 and 3 are driven by vee-belts. All fans are Class III fans. Fan 4 has a bell-mounted inlet and a sound trap on the suction side.

Flywheel Effect

Fan motors and drives for most small fans can be selected on the basis of full-load horsepower and speed alone. However, in larger fans, it is advisable to examine the starting characteristics of the fan-drive combination. Since this involves breakaway torques, speed-torque relations, flywheel effect, acceleration time, temperature effects, etc., it is beyond the scope of this handbook to cover these items. However, the fan manufacturer can readily supply all necessary information required for fan load characteristics. The engineer can also find several excellent references on this subject.

Since there is a dramatic difference in flywheel effect between the centrifugal fans and the vaneaxial fan in this example, the polar moment of inertia for each fan rotor is:

Fan No.	WR^2
1	1910
2	1910
3	8100
4	210

The above flywheel effects are for the fan rotors only, not the drives.

As can be seen from this example, the retardant effect of the polar moment of inertia is high for the tubular centrifugal fan and very low for the vaneaxial; or conversely, the vaneaxial fan can come to rated speed much faster than can the centrifugal fans.

Fan Sound Data

	Fan 1	Fan 2	Fan 3	Fan 4
Octave Band, cps	Certified Sound-Power Level, decibels			
1-63	117	113	120	101
2-125	118	116	113	98
3-250	109	111	111	107
4-500	106	106	107	120
5-1000	103	103	103	116
6-2000	97	98	98	114
7-4000	94	95	93	109
8-8000	92	92	88	105

Note: These are AMCA certified values without any attenuation. Fans 1, 2, and 3 have comparable sound-power spectrums, with higher values at the low frequencies which the human ear can better tolerate. Fan 4 has somewhat lower values in the low frequencies, but significantly higher values in the higher frequencies. This is what gives the vaneaxial fan its shrill sound, akin to a jet engine. For this reason, it is usually necessary to acoustically treat both the inlet and discharge of the vaneaxial fan with sound trapping. Such sound treatment will make the vaneaxial fan as quiet as, or quieter than, Fans 1, 2, and 3.

The comparative cost of each of these four types of fan is a difficult area to evaluate because of the wide range of fan quality. For larger fans (such as those in the example), the tubular centrifugal would probably be about twice the cost of the specified vaneaxial. Fans 1 and 2 would be higher (somewhere around 35 to 50 percent) in installed cost than the vaneaxial, but considerably lower than the tubular centrifugal.

Fan Performance Data

Performance Requirement	Fan 1	Fan 2	Fan 3	Fan 4
	Performance Data			
System resistance, inches WG	8.0	8.0	8.0	8.0
Velocity pressure, including "bulkhead," inches WG	0.375	0.375	0.125	1.10
Added sound attenuation	...	...	...	0.22
Total pressure, inches WG	8.375	8.375	8.125	9.32
CFM of standard air	75,000	75,000	75,000	75,000
RPM for rated conditions	1140	1041	800	1770
BHP for rated conditions and RPM	123	123	126	132
Vee-belt drive loss on fan RPM, percent	...	4	4	...
RPM required because of Vee-belts	1140	1081	832	1770
BHP required because of Vee-belts	123	135	138	132
Inlet vane loss on fan RPM, percent	...	3	3	...
RPM required because of inlet vanes	1140	1113	857	1770
Inlet vane loss on fan BHP, percent	15	9	9	...
BHP required because of inlet vanes	141	147	150	132
Static efficiency, percent	67.0	64.0	63.0	73.0
Overall total efficiency, percent	70.5	67.5	64.0	83.5

Fan Weights

Fan No.	Weight less Motor, lbs	Installed Weight with Motor, Inertia Block and Isolators, lbs
1	3800	13,000
2	3800	13,000
3	7200	17,000
4	970	3,000

For smaller fan sizes, the relative installed cost differential would tend to lessen.

Conclusions

1. Based on economics alone, the vaneaxial fan (Fan number 4) would have an advantage both in operating cost (because of a better overall efficiency) and in first cost.
2. Based on sound data, the vaneaxial fan is the least attractive selection and would certainly require sound attenuation both upstream and downstream. However, even with the added sound trap, Fan number 4 might still cost less initially.
3. Other considerations such as direction of fan

Installed Cubage

Fan No.	Volume, Cubic Feet
1	880
2	1080
3	1152
4	245

discharge, may dictate the use of Fans 1 and 2 over Fans 3 and 4.

4. While these fans were selected for a free inlet and a "bulkheaded" discharge, this is not the most general requirement for HVAC. Usually, there are filters, coils, etc., upstream of the fan, and very often a duct connection on the discharge. This set of parameters may give Fans 1, 2, or 3 a decided advantage over Fan 4 because of sound-trapping requirements.
5. For variable volume control, Fan number 4 is the best in terms of efficiency, and probably also best in terms of ease and range of control. Inlet vane control simply cannot match controlled pitch.

The main theme of this example is that there are many considerations in properly selecting a fan, particularly a large fan. The engineer must evaluate them all for any particular application.

AIR FLOW IN DUCTS

Air flowing in ducts or pipes is characterized by three pressures; static pressure, velocity pressure, and total pressure.

Static pressure, P_s, is the pressure exerted normal to the direction of flow. Atmospheric pressure is usually taken as the zero datum, so static pressures less than atmospheric are negative. Static pressure is analogous to potential energy.

Velocity pressure, P_v, is the pressure in the direction of flow equivalent to the kinetic energy of motion of the gas. Since velocity pressure is always measured in the direction of flow, numerical values are always positive.

Total pressure, P_t, is the algebraic sum of P_s and P_v.

The basic equation for the velocity of a gas is

$$V = 1096.5 \sqrt{\frac{P_v}{W}} \qquad (1)$$

where V is the velocity in feet per minute, P_v is the velocity pressure in inches water gage, and W is the gas density in pounds per cubic foot. For standard air, at 29.92 inches of mercury and 70 deg F, $W = 0.07495$, and formula (1) becomes

$$V = 4005 \sqrt{P_v}. \qquad (2)$$

Table 1 is a solution of equation (2).

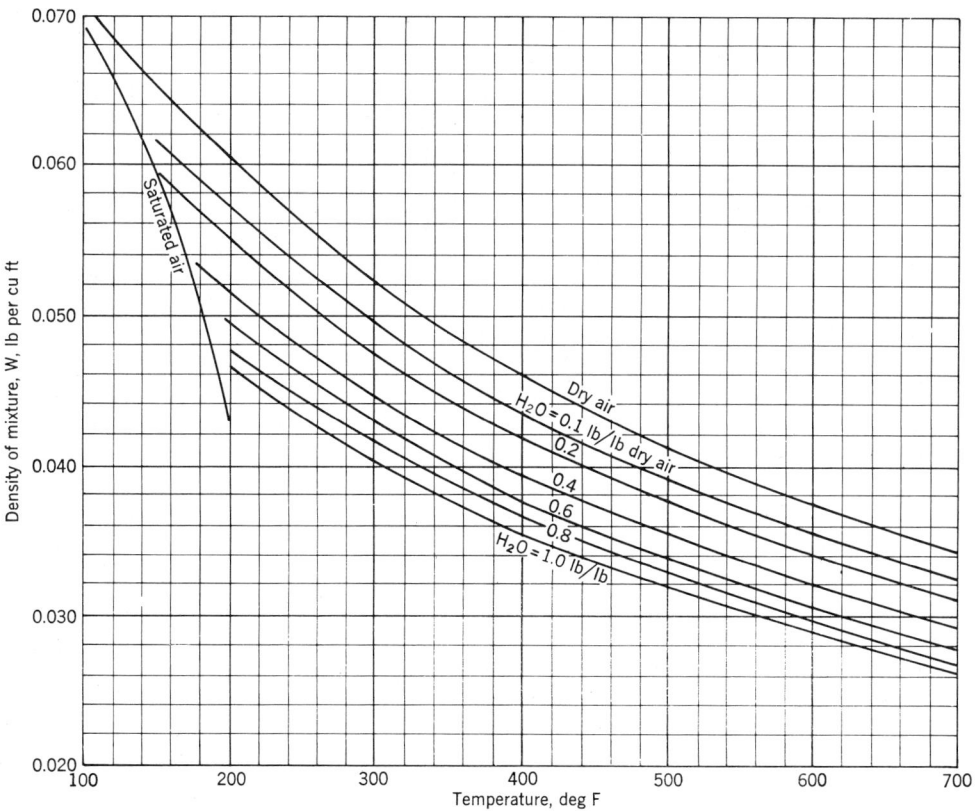

Fig. 1. Density of air-water-vapor mixtures, pounds per cubic foot of mixture, 100 to 700 deg. F.

Table 1. Velocity Equivalents of Velocity Pressures

(For conversion of velocity pressure gage readings in inches of water to velocity in feet per minute when air is at 70 deg and 29.92 inches barometric pressure.)

Pressure, P_v, Inches of Water	Velocity, V, Feet per Min.	Pressure, P_v, Inches of Water	Velocity, V, Feet per Min.	Pressure, P_v, Inches of Water	Velocity, V, Feet per Min.	Pressure, P_v, Inches of Water	Velocity, V, Feet per Min.
.01	400.5	.50	2,831.9	.99	3,985.0	1.48	4,872.3
.02	566.3	.51	2,859.0	1.00	4,005.0	1.49	4,888.7
.03	693.7	.52	2,888.4	1.01	4,021.0	1.50	4,905.1
.04	801.0	.53	2,915.6	1.02	4,044.8	1.51	4,921.4
.05	895.5	.54	2,942.9	1.03	4,064.6	1.52	4,937.7
.06	980.8	.55	2,970.1	1.04	4,084.3	1.53	4,953.9
.07	1,059.7	.56	2,996.9	1.05	4,103.9	1.54	4,970.1
.08	1,132.6	.57	3,023.8	1.06	4,123.4	1.55	4,986.2
.09	1,201.5	.58	3,050.2	1.07	4,142.8	1.56	5,002.2
.10	1,266.4	.59	3,076.2	1.08	4,162.1	1.57	5,018.3
.11	1,328.5	.60	3,102.3	1.09	4,181.3	1.58	5,034.2
.12	1,387.3	.61	3,127.9	1.10	4,200.5	1.59	5,050.1
.13	1,444.2	.62	3,153.5	1.11	4,219.5	1.60	5,066.0
.14	1,498.7	.63	3,178.8	1.12	4,238.5	1.61	5,081.8
.15	1,551.1	.64	3,204.0	1.13	4,257.4	1.62	5,097.5
.16	1,602.0	.65	3,228.8	1.14	4,276.2	1.63	5,113.2
.17	1,651.3	.66	3,253.7	1.15	4,294.9	1.64	5,128.9
.18	1,699.3	.67	3,254.1	1.16	4,313.5	1.65	5,144.5
.19	1,745.8	.68	3,302.5	1.17	4,332.1	1.66	5,160.0
.20	1,791.0	.69	3,327.0	1.18	4,350.6	1.67	5,175.5
.21	1,835.5	.70	3,351.0	1.19	4,368.9	1.68	5,191.1
.22	1,870.3	.71	3,374.6	1.20	4,387.3	1.69	5,206.5
.23	1,920.8	.72	3,398.2	1.21	4,405.5	1.70	5,221.9
.24	1,962.0	.73	3,421.9	1.22	4,423.7	1.71	5,237.2
.25	2,002.5	.74	3,445.1	1.23	4,441.7	1.72	5,252.5
.26	2,042.1	.75	3,468.3	1.24	4,459.8	1.73	5,267.7
.27	2,081.0	.76	3,491.6	1.25	4,477.7	1.74	5,283.0
.28	2,119.4	.77	3,514.4	1.26	4,495.6	1.75	5,298.1
.29	2,156.7	.78	3,537.2	1.27	4,513.4	1.76	5,313.2
.30	2,193.5	.79	3,559.6	1.28	4,531.1	1.77	5,328.3
.31	2,230.0	.80	3,582.1	1.29	4,548.8	1.78	5,343.4
.32	2,265.6	.81	3,604.5	1.30	4,566.4	1.79	5,358.3
.33	2,300.9	.82	3,626.5	1.31	4,583.9	1.80	5,373.3
.34	2,335.3	.83	3,648.6	1.32	4,601.4	1.81	5,388.2
.35	2,369.4	.84	3,670.6	1.33	4,618.8	1.82	5,403.0
.36	2,403.0	.85	3,692.6	1.34	4,636.1	1.83	5,417.8
.37	2,436.2	.86	3,714.2	1.35	4,653.4	1.84	5,432.7
.38	2,468.7	.87	3,735.5	1.36	4,670.6	1.85	5,447.4
.39	2,573.2	.88	3,757.1	1.37	4,687.7	1.86	5,462.1
.40	2,533.2	.89	3,778.3	1.38	4,704.8	1.87	5,476.8
.41	2,564.4	.90	3,799.5	1.39	4,721.8	1.88	5,491.4
.42	2,595.6	.91	3,820.4	1.40	4,738.8	1.89	5,506.0
.43	2,626.1	.92	3,841.6	1.41	4,755.7	1.90	5,520.5
.44	2,656.5	.93	3,862.4	1.42	4,772.5	1.91	5,535.0
.45	2,686.6	.94	3,882.8	1.43	4,789.3	1.92	5,549.5
.46	2,716.2	.95	3,903.7	1.44	4,806.0	1.93	5,563.9
.47	2,745.8	.96	3,924.1	1.45	4,822.7	1.94	5,578.3
.48	2,774.7	.97	3,944.5	1.46	4,839.2	1.95	5,592.7
.49	2,803.5	.98	3,964.5	1.47	4,855.8	1.96	5,607.0

Pressure, P_v, Inches of Water	Velocity, V, Feet per Min.	Pressure, P_v, Inches of Water	Velocity, V, Feet per Min.	Pressure, P_v, Inches of Water	Velocity, V, Feet per Min.	Pressure, P_v, Inches of Water	Velocity, V, Feet per Min.
1.97	5,621.3	2.32	6,100.2	2.67	6,544.2	3.02	6,959.9
1.98	5,635.5	2.33	6,113.4	2.68	6,556.5	3.03	6,971.5
1.99	5,649.7	2.34	6,126.5	2.69	6,568.7	3.04	6,983.0
2.00	5,663.9	2.35	6,139.5	2.70	6,580.9	3.05	6,994.4
2.01	5,678.0	2.36	6,152.6	2.71	6,593.1	3.06	7,005.9
2.02	5,692.2	2.37	6,165.6	2.72	6,605.2	3.07	7,017.3
2.03	5,706.2	2.38	6,178.6	2.73	6,617.3	3.08	7,028.7
2.04	5,720.3	2.39	6,191.6	2.74	6,629.4	3.09	7,040.1
2.05	5,734.3	2.40	6,204.5	2.75	6,641.5	3.10	7,051.5
2.06	5,748.3	2.41	6,217.4	2.76	6,653.6	3.11	7,062.9
2.07	5,762.2	2.42	6,230.3	2.77	6,665.6	3.12	7,074.2
2.08	5,776.1	2.43	6,243.2	2.78	6,677.7	3.13	7,085.6
2.09	5,789.9	2.44	6,256.0	2.79	6,689.7	3.14	7,096.9
2.10	5,803.8	2.45	6,268.9	2.80	6,701.6	3.15	7,108.2
2.11	5,817.6	2.46	6,281.6	2.81	6,713.6	3.16	7,119.4
2.12	5,831.4	2.47	6,294.3	2.82	6,725.6	3.17	7,130.7
2.13	5,845.1	2.48	6,307.0	2.83	6,737.5	3.18	7,142.0
2.14	5,858.8	2.49	6,319.8	2.84	6,749.3	3.19	7,153.2
2.15	5,872.5	2.50	6,332.5	2.85	6,761.2	3.20	7,164.3
2.16	5,886.1	2.51	6,345.1	2.86	6,773.1	3.21	7,175.6
2.17	5,899.7	2.52	6,357.7	2.87	6,784.9	3.22	7,186.7
2.18	5,913.3	2.53	6,370.4	2.88	6,796.7	3.23	7,197.9
2.19	5,926.8	2.54	6,382.9	2.89	6,808.5	3.24	7,209.0
2.20	5,940.4	2.55	6,395.5	2.90	6,820.3	3.25	7,220.1
2.21	5,953.9	2.56	6,408.0	2.91	6,832.0	3.26	7,231.2
2.22	5,967.3	2.57	6,420.5	2.92	6,843.7	3.27	7,242.3
2.23	5,980.7	2.58	6,433.0	2.93	6,855.4	3.28	7,253.4
2.24	5,994.1	2.59	6,445.4	2.94	6,867.1	3.29	7,264.4
2.25	6,007.5	2.60	6,457.9	2.95	6,878.8	3.30	7,275.4
2.26	6,020.8	2.61	6,470.3	2.96	6,890.5	3.31	7,286.5
2.27	6,034.1	2.62	6,482.7	2.97	6,902.1	3.32	7,297.5
2.28	6,047.4	2.63	6,495.0	2.98	6,913.7	3.33	7,308.4
2.29	6,060.6	2.64	6,507.4	2.99	6,925.3	3.34	7,319.4
2.30	6,073.9	2.65	6,519.7	3.00	6,936.9	3.35	7,330.4
2.31	6,087.0	2.66	6,532.0	3.01	6,948.4	3.36	7,341.3

Air density varies with temperature, pressure, and water vapor content. Velocity equivalents of velocity pressures for non-standard air may be found from equation (1). However, in order to speed computation and eliminate extraction of the square root, a correction factor method has been developed. If CF, the correction factor, is defined as the square root of the ratio, 0.07495 divided by W, equation (1) becomes

$$V = (4005 \sqrt{P_v}) \, (CF). \qquad (3)$$

The quantity in the first parenthesis of equation (3) is given in Table 1. Values of CF are given in Table 2 for various densities. Air densities may be obtained from psychrometric charts and tables for most conditions. For moisture-laden air beyond this range, densities may be obtained from Fig. 1.

Air Temperature, Deg. F	Constant in Eq. (2)
35	3872
40	3891
45	3910
50	3932
55	3952
60	3970
65	3987
70	4005
75	4025
80	4045
85	4064
90	4081
95	4101
100	4117
150	4298

Table 2. Correction Factors to be Applied to Table 1 for Non-Standard Air

Gas Density, lb per cu ft	Correction Factor
0.01	2.74
0.02	1.94
0.03	1.58
0.04	1.37
0.05	1.22
0.06	1.12
0.07	1.03
0.075	1.000
0.08	0.968
0.09	0.913
0.10	0.866
0.11	0.825
0.12	0.790
0.13	0.759
0.14	0.732
0.15	0.707

Dry air density, W_d, is given by the equation

$$W_d = 0.07495 \left(\frac{530}{\text{deg F} + 460} \right) \left(\frac{P_a}{29.92} \right) \quad (4)$$

where P_a is the absolute pressure in inches of mercury.

The procedures to be followed for non-standard conditions are as follows:

(1) *To find P_v for a known velocity,* divide known velocity by CF from Table 2 to determine equivalent velocity. Determine P_v from equivalent velocity using Table 1.

Example: What is the velocity pressure of air at 560 deg F, 0.2 lb moisture per lb dry air, if its velocity is 3,000 fpm?

Solution: From Fig. 1, $W = 0.0355$. From Table 2, CF = 1.46 (by interpolation between $W = 0.03$ and $W = 0.04$). Dividing the known velocity, 3,000 fpm, by the correction factor, the equivalent velocity of 2040 fpm is found. From Table 1, the velocity pressure is seen to be 0.026 inches of water.

(2) *To determine V for a known P_v,* determine equivalent velocity for P_v from Table 1. Multiply equivalent velocity by CF to determine actual velocity.

Example: What is the velocity of saturated air at 180 deg, if the velocity pressure reading is 1.42?

Solution: From Table 1, if $P_v = 1.42$, $V = 4772.5$ fpm for standard air. From Fig. 1, $W = 0.050$, and from Table 1, CF = 1.22. Dividing V by CF, the velocity is found to be 3912 fpm.

(3) *To determine average velocity* from a number of P_v readings as in a Pitot traverse (discussed below), determine the equivalent velocities using Table 1, then average the equivalent velocities. Multiply the average equivalent velocity by the correction factor, CF, from Table 2.

Pitot Traverse

Air volumes flowing in a duct are frequently determined by traversing the duct with a standard Pitot tube. The Pitot tube reads P_v directly in inches of water. A Pitot traverse should be made only at a point where relatively uniform velocities exist, such as seven or more duct diameters downstream from the nearest elbow, branch entry, damper, or like flow disturbance. For rectangular duct, the cross-section of the duct should be hypothetically divided into 12 or more equal areas and a Pitot reading taken at the center of each area. For round ducts, Table 3 give the distance inward form the side wall for each traverse point in order that readings be taken at the center of equal annular areas. If readings are taken at such equal-area points, the average velocity may be determined

Table 3. Pitot Traverse Points for Round Duct
(10-Point Traverse)

Point	Percent of Diameter	Duct Diameter, Inches						
		12	16	20	24	30	36	48
		Distance Inward from Side Wall, Inches						
1	1.95	1/4	5/16	3/8	7/16	9/16	11/16	15/16
2	8.15	1	1- 5/16	1- 5/8	1-15/16	2-7/16	2-15/16	3-15/16
3	14.65	1- 3/4	2- 5/16	2-15/16	3- 1/2	4-3/8	5- 1/4	7- 1/16
4	22.60	2-11/16	3- 5/8	4- 1/2	5- 7/16	6-3/4	8- 1/8	10- 7/8
5	34.20	4- 1/8	5- 1/2	6-13/16	8- 3/16	10-1/4	12- 5/16	16- 7/16
6	65.80	7- 7/8	10- 1/2	13- 3/16	15-13/16	19-3/4	23-11/16	31- 9/16
7	77.40	9- 5/16	12- 3/8	15- 1/2	18- 9/16	23-1/4	27- 7/8	37- 1/8
8	85.35	10- 1/4	13-11/16	17- 1/16	20- 1/2	25-5/8	30- 3/4	41
9	91.85	11	14-11/16	18- 3/8	22- 1/16	27-9/16	33- 1/16	44- 1/16
10	98.05	11- 3/4	15-11/16	19- 5/8	23- 1/2	29-7/16	35- 5/16	47- 1/16

from the simple arithmetic average of velocities at each point. *Do not average the velocity pressures.* For small ducts, of 6 inches in diameter or less, average velocities may be estimated by taking 0.81 times the centerline velocity.

Friction Losses

Duct friction varies directly with duct length, inversely with duct diameter, as the square of the velocity of the gas flowing, and with a friction factor. Because the friction factor varies in a

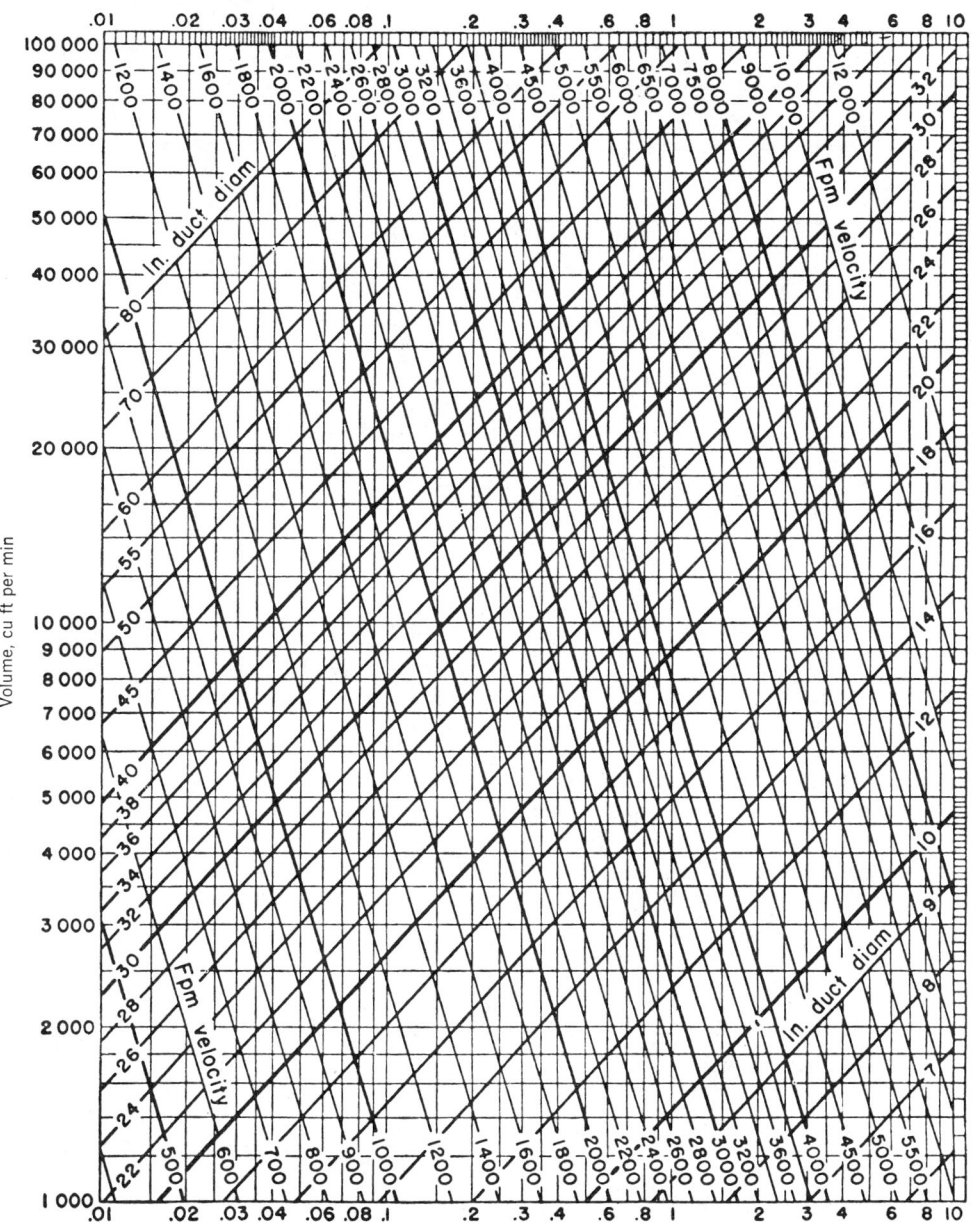

Fig. 2a. Friction of air in straight galvanized ducts,
for volumes of 1000 to 100,000 cfm.
(Same basis and source as Fig. 2b).

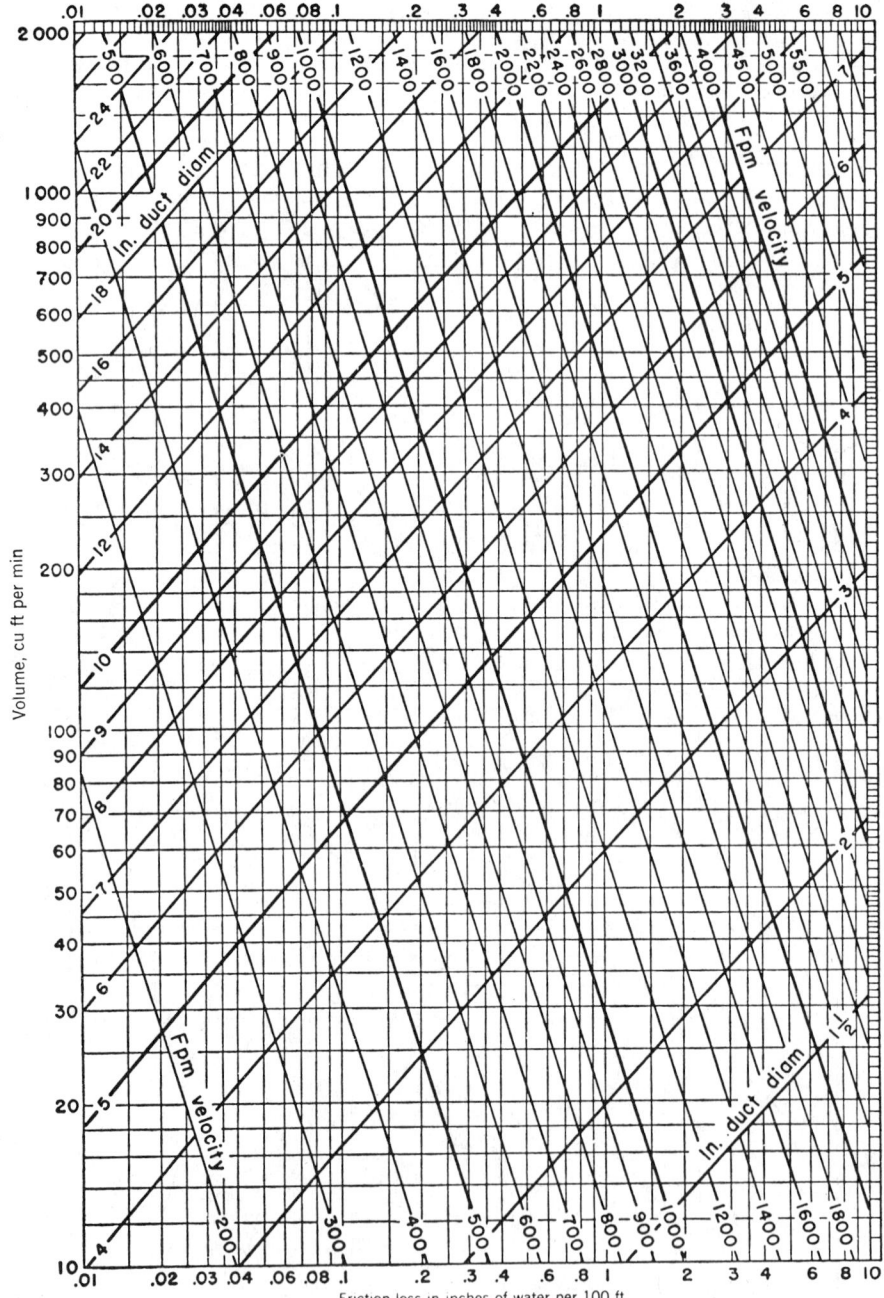

Fig. 2b. Friction of air in straight galvanized ducts,
for volumes of 10 to 2000 cfm.

(Based on standard air of 0.075 lb. per cu ft density flowing through clean round galvanized ducts having approximately 40 joints per 100 ft. Reproduced by permission from *Heating Ventilating and Air Conditioning Guide*, 1958, Chapter 31).

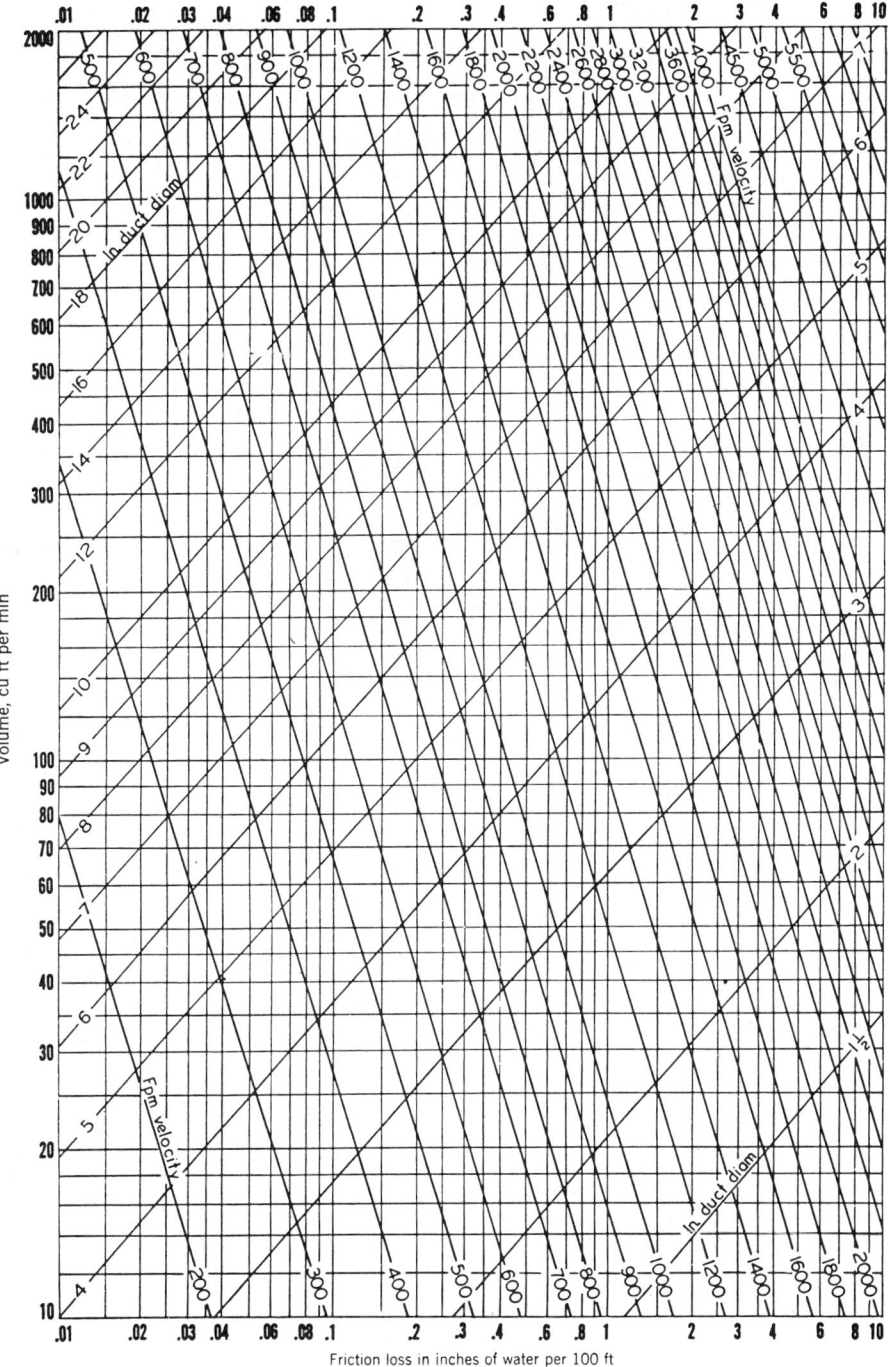

Fig. 3a. Friction of air in straight aluminum ducts
for volumes of 10 to 2000 cfm.
(Same basis and source as Fig 3b).

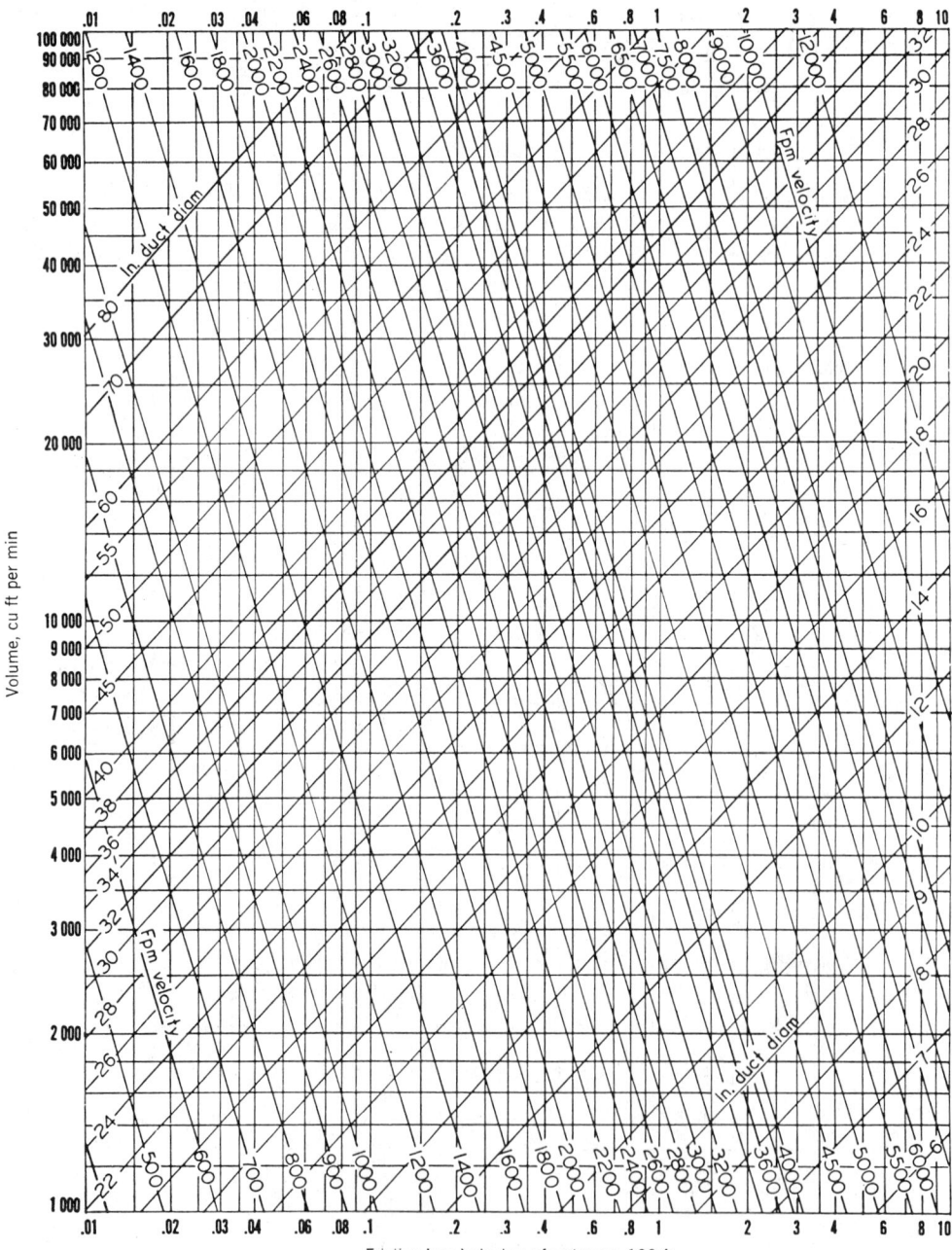

Friction loss in inches of water per 100 ft

Fig. 3b. Friction of air in straight aluminum ducts
for volumes of 1000 to 100,000 cfm.

(Based on standard air of 0.075 lb per cu ft density flowing through clean round aluminum ducts having approximately 40 joints per 100 ft. Reproduced by permission from *The Design of Aluminum Duct Systems*, prepared for Kaiser Aluminum & Chemical Sales, Inc., by F. W. Hutchinson).

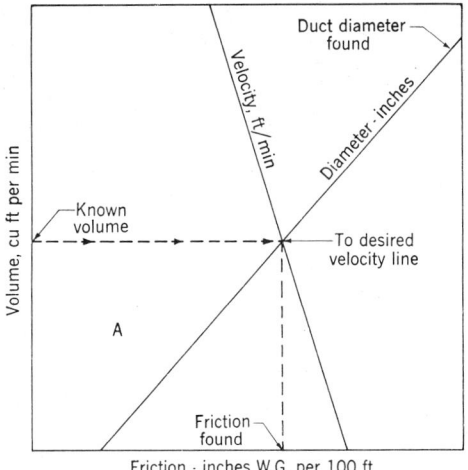

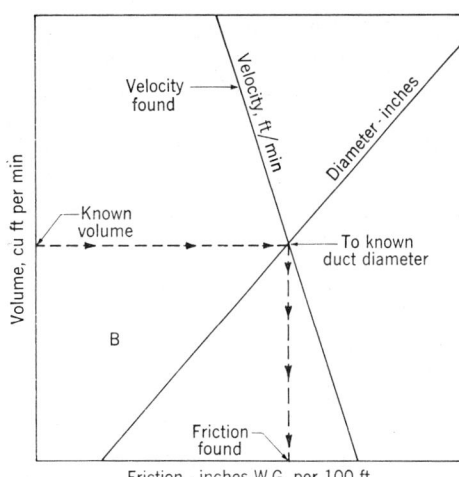

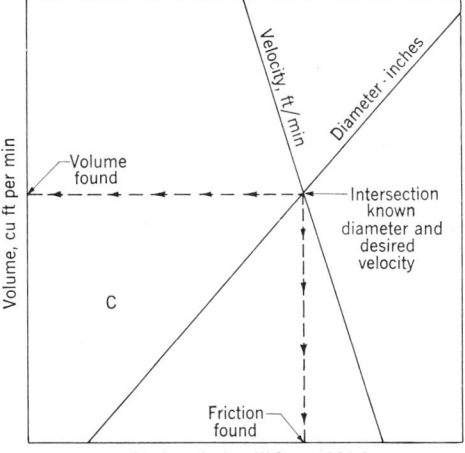

Fig. 4. Uses of friction chart:
(A) Finding diameter and friction from known volume and velocity;
(B) Finding velocity and friction from known volume and diameter;
(C) Finding volume and friction from known velocity and diameter.

rather complex way with roughness of the duct, with Reynolds number, and with viscosity of the gas flowing, and because these factors are not constants but are in turn dependent upon other factors, the formulas that describe gas flow are time-consuming in their application. Therefore, the application engineer depends upon friction charts, such as are presented here. Figure 2 gives friction losses in standard round galvanized duct for standard air. Fig. 3 gives friction losses in round aluminum ducts. Knowing any two of the three factors (1) duct diameter, (2) air velocity, and (3) air flow rate, the third factor and the friction loss may be determined. Examples of friction chart manipulations are shown in Fig. 4.

Table 4. Degrees of Roughness in Various Ducts

Pipe	Degree of Roughness
Drawn tubing	Very smooth
New steel or wrought iron	Medium smooth
Asphalted cast iron	
Galvanized iron	Average
Wood stave	
Average concrete	Medium rough
Riveted steel	Very rough

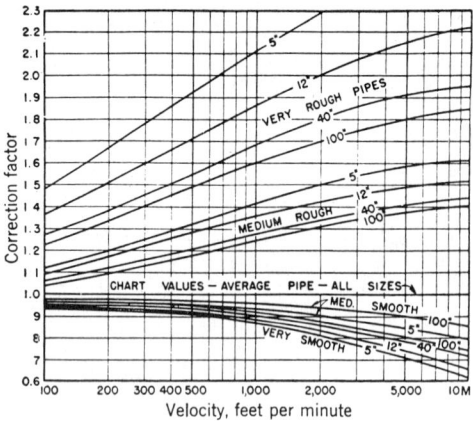

Fig. 5. Correction factors for pipe roughness, to be applied to values read from Fig. 2.

Correction for Roughness

For duct materials other than galvanized iron or aluminum, roughness correction factors given in Fig. 5 should be applied. Table 4 may be used as a guide to roughness of various pipes.

Rectangular Duct

Figures 2 and 3, the friction charts for round duct, may be used for rectangular duct. Table 5 presents the equivalent size of round duct for equal friction at equal volume flow (*not* equal velocity) for various rectangular duct sizes. It is a solution of the formula,

$$d = 1.265 \sqrt[5]{\frac{(ab)^3}{a+b}}$$

where d = equivalent round duct diameter, inches,

a = one side of rectangular duct, inches,

b = other side of rectangular duct, inches.

Example: (Refer to Table 5.) A 10-inch (left margin) by 6-inch (top line) rectangular duct has the same friction losses at the same volume flow as an 8.4-inch (body of table) round duct.

Correction for Density

For the normal range of temperature, pressure, and moisture content encountered in air handling,

Table 5. Equivalent Circular and Rectangular Ducts with Equal Friction

One Side Rect'lar Duct, Inches	22	24	26	28	30	32	34	36	38	40	42	44	46	48
	\multicolumn{14}{Diameter of Equivalent Circular Duct, Inches}
22	24.2													
24	25.2	26.4												
26	26.3	27.5	28.6											
28	27.3	28.5	29.7	30.8										
30	28.2	29.5	30.7	31.9	33.0									
32	29.1	30.5	31.7	32.9	34.1	35.2								
34	30.0	31.3	32.7	33.9	35.1	36.3	37.4							
36	30.8	32.2	33.7	34.9	36.1	37.3	38.5	39.6						
38	31.5	33.1	34.6	35.9	37.1	38.4	39.5	40.7	41.8					
40	32.4	33.9	35.3	36.7	38.0	39.3	40.5	41.7	42.9	44.0				
42	33.0	34.5	36.0	37.6	39.0	40.3	41.5	42.7	44.0	45.1	46.2			
44	33.7	35.3	36.9	38.5	39.9	41.2	42.5	43.7	44.9	46.1	47.2	48.4		
46	34.6	36.2	37.8	39.3	40.8	42.2	43.5	44.8	46.0	47.2	48.4	49.5	50.6	
48	35.2	37.0	38.5	40.0	41.5	43.0	44.4	45.6	46.9	48.1	49.3	50.5	51.6	52.8
50	35.9	37.6	39.2	40.8	42.3	43.8	45.2	46.5	47.9	49.1	50.4	51.6	52.9	54.0
52	36.5	38.3	40.0	41.6	43.1	44.7	46.1	47.5	48.9	50.1	51.3	52.5	53.8	55.0
54	37.2	38.9	40.7	42.4	44.0	45.5	47.0	48.4	49.9	51.1	52.3	53.5	54.8	56.0
56	37.8	39.6	41.3	43.0	44.6	46.2	47.7	49.1	50.6	52.0	53.3	54.6	55.9	57.0
58	38.4	40.3	42.1	43.8	45.4	47.0	48.5	50.0	51.5	52.9	54.2	55.5	56.8	58.0
60	39.1	40.9	42.7	44.5	46.1	47.8	49.3	50.9	52.3	53.8	55.0	56.4	57.7	58.9
62	39.6	41.6	43.4	45.1	46.8	48.4	50.0	51.7	53.0	54.5	55.9	57.2	58.5	59.7
64	40.2	42.2	44.0	45.8	47.5	49.2	50.9	52.4	53.9	55.4	56.8	58.1	59.4	60.6
66	40.8	42.8	44.7	46.5	48.2	50.0	51.6	53.1	54.7	56.2	57.6	59.1	60.4	61.6

Table 5 (Continued). Equivalent Circular and Rectangular Ducts with Equal Friction

One Side Rect'lar Duct, Inches	Other Side of Rectangular Duct, Inches													
	3	4	5	6	7	8	9	10	11	12	14	16	18	20
	Diameter of Equivalent Circular Duct, Inches													
3½	—	—	—	—	—	5.7	6.0	6.3	6.6	6.9	7.4	7.8		
4	3.8	4.4	4.9	5.4	5.8	6.1	6.5	6.8	7.1	7.4	7.9	8.4		
4½	4.0	4.7	5.2	5.8	6.1	6.5	6.9	7.2	7.6	7.9	8.5	9.0		
5	4.3	4.9	5.5	6.3	6.5	6.9	7.3	7.7	8.0	8.3	8.9	9.5		
5½	4.4	5.2	5.9	6.4	6.8	7.3	7.7	8.1	8.5	8.8	9.5	10.1		
6	4.6	5.4	6.3	6.6	7.0	7.6	8.0	8.4	8.8	9.2	9.9	10.5		
7	5.0	5.8	6.5	7.0	7.7	8.2	8.7	9.2	9.6	10.0	10.8	11.4		
8	5.3	6.1	6.9	7.6	8.2	8.8	9.3	9.8	10.2	10.7	11.5	12.3		
9	5.5	6.5	7.3	8.0	8.7	9.3	9.9	10.4	10.9	11.4	12.3	13.1		
10	5.8	6.8	7.7	8.4	9.2	9.8	10.4	11.0	11.5	12.0	12.9	13.8		
11	6.1	7.1	8.0	8.8	9.6	10.2	10.9	11.5	12.1	12.6	13.6	14.5		
12	6.3	7.4	8.3	9.2	10.0	10.7	11.4	12.0	12.6	13.2	14.3	15.2		
13	6.4	7.6	8.7	9.6	10.4	11.1	11.8	12.5	13.1	13.7	14.8	15.8		
14	6.7	7.9	8.9	9.9	10.8	11.5	12.3	12.9	13.6	14.3	15.4	16.5		
15	7.0	8.2	9.2	10.2	11.1	11.9	12.7	13.4	14.1	14.7	16.0	17.1		
16	7.2	8.4	9.5	10.5	11.4	12.3	13.1	13.8	14.5	15.2	16.5	17.6		
17	7.3	8.6	9.8	10.8	11.8	12.6	13.5	14.2	15.0	15.7	17.0	18.2		
18	7.5	8.9	10.0	11.1	12.1	13.0	13.8	14.6	15.4	16.1	17.4	18.7	19.8	
19	7.7	9.1	10.3	11.4	12.4	13.3	14.2	15.0	15.8	16.5	17.9	19.2	20.4	
20	7.9	9.3	10.5	11.6	12.7	13.6	14.5	15.4	16.2	17.0	18.4	19.7	20.9	22.0
22	8.2	9.7	11.0	12.1	13.2	14.2	15.2	16.1	16.9	17.8	19.2	20.6	21.9	23.1
24	8.5	10.0	11.4	12.6	13.8	14.8	15.8	16.8	17.6	18.5	20.0	21.5	22.8	24.0
26	8.8	10.4	11.8	13.1	14.3	15.4	16.4	17.3	18.3	19.2	20.8	22.3	23.8	25.1
28	9.1	10.8	12.2	13.5	14.8	15.9	17.0	18.0	19.0	19.8	21.5	23.1	24.6	26.0
30	9.2	11.0	12.6	13.9	15.2	16.4	17.5	18.5	19.5	20.5	22.2	23.9	25.4	26.8
32	9.6	11.3	12.9	14.3	15.6	16.9	18.0	19.1	20.1	21.1	22.9	24.6	26.2	27.7
34	9.9	11.6	13.2	14.7	16.1	17.3	18.5	19.6	20.7	21.6	23.5	25.3	26.9	28.5
36	10.0	11.9	13.6	15.1	16.4	17.7	19.0	20.1	21.2	22.2	24.2	26.0	27.7	29.3
38	—	12.2	13.9	15.4	16.8	18.2	19.4	20.6	21.7	22.8	24.8	26.7	28.4	30.0
40	—	12.5	14.3	15.7	17.2	18.6	19.8	21.1	22.2	23.3	25.4	27.3	29.1	30.8
42	—	12.7	14.5	16.1	17.6	19.0	20.3	21.6	22.7	23.8	25.9	27.9	29.8	31.4
44	—	13.0	14.8	16.4	18.0	19.4	20.7	22.0	23.1	24.3	26.5	28.5	30.3	32.1
46	—	13.3	15.1	16.7	18.4	19.8	21.1	22.4	23.6	24.8	27.0	29.1	31.0	32.8
48	—	13.5	15.4	17.0	18.7	20.1	21.5	22.8	24.1	25.2	27.5	29.6	31.6	33.4
50	—	13.7	15.7	17.3	19.0	20.4	21.9	23.2	24.5	25.7	28.0	30.3	32.2	34.1
52	—	13.9	15.9	17.6	19.2	20.8	22.2	23.6	24.9	26.2	28.5	30.7	32.9	34.7
54	—	14.1	16.1	17.9	19.6	21.1	22.6	24.0	25.3	26.6	29.0	31.2	33.4	35.3
56	—	14.3	16.3	18.2	19.9	21.5	22.9	24.4	25.7	27.0	29.5	31.7	33.9	35.9
58	—	14.6	16.6	18.4	20.2	21.8	23.3	24.7	26.1	27.4	30.0	32.2	34.4	36.4
60	—	14.7	16.8	18.7	20.4	22.1	23.6	25.1	26.5	27.8	30.5	32.7	34.9	37.1
62	—	15.0	17.0	19.0	20.7	22.4	24.0	25.5	26.9	28.2	30.9	33.2	35.4	37.7
64	—	15.1	17.3	19.2	21.0	22.7	24.3	25.9	27.3	28.6	31.3	33.7	35.9	38.2
66	—	15.3	17.5	19.5	21.2	23.0	24.6	26.2	27.7	29.0	31.7	34.2	36.4	38.7

variations in air viscosity are small enough to be neglected. Friction chart values may be corrected for variation in air density from standard by the following methods:

(A) If actual air volumes or velocities at operating conditions are used in reading the friction chart, then to get the actual friction loss per 100 ft, multiply the friction loss per 100 ft from the chart by either air density divided by 0.075, or by an equivalent ratio, 13.35 divided by actual specific volume.

Example: Dry air at 300 deg F is conveyed through 12-inch duct at 2000 cfm (actual). From the friction chart, Fig. 2, friction loss per 100 ft is 0.75. From Fig. 1, density is 0.052. Then the actual friction loss per 100 ft is 0.75 (0.052/0.075) = 0.52.

(B) If standard air volumes are used in reading the friction chart, then the actual friction loss per

100 ft will be the loss from the chart multiplied by the reciprocal of the factor in (A), 0.075

Table 6. Losses in 90° Round Elbows

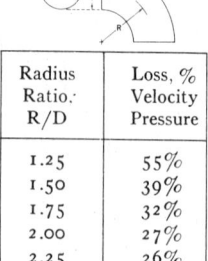

Radius Ratio. R/D	Loss, % Velocity Pressure
1.25	55%
1.50	39%
1.75	32%
2.00	27%
2.25	26%
2.50	22%
2.75	26%

Table 7. Percent of 90° Elbow Loss for Various Bends

Bend. degrees	Loss, % 90° bend
30°	33%
45°	52%
60°	67%
75°	85%
90°	100%
105°	115%
120°	127%
135°	140%
150°	150%
165°	158%
180°	165%

Table 8. Loss in Rectangular 90-Deg. Elbows, Percent Velocity Pressure *

Aspect Ratio. H/W	No. Splitters	0.0 (mitre elbow)	0.50	0.75	1.0	1.5	2.0	3.0	4.0
					Friction Loss, Percent of Velocity Pressure				
0.25	0	150	136	55	46	28	24	24	28
	1	135	77	22	26	24	26	—	—
	2	120	49	16	22	26	—	—	—
	3	109	31	12	22	—	—	—	—
0.5	0	132	120	48	28	18	15	15	18
	1	119	67	19	16	15	16	—	—
	2	104	43	14	13	16	—	—	—
	3	96	27	10	13	—	—	—	—
1.0	0	115	105	42	21	13	11	11	13
	1	104	59	17	12	11	12	—	—
	2	92	38	12	10	12	—	—	—
	3	84	24	9	10	—	—	—	—
2.0	0	104	95	38	21	13	11	11	13
	1	94	53	15	12	11	12	—	—
	2	83	34	11	10	12	—	—	—
	3	76	22	11	10	—	—	—	—
3.0	0	83	76	30	20	13	11	11	13
	1	75	42	12	12	11	12	—	—
	2	67	27	9	10	12	—	—	—
	3	61	18	9	10	—	—	—	—
4.0	0	61	55	22	19	11	10	10	11
	1	55	31	12	11	10	11	—	—
	2	49	20	9	9	11	—	—	—
	3	45	13	9	9		—	—	—

Radis Ratio, R/W

* For radius elbow with splitters illustrated later in Fig. 6, use radius ratio R/W, of 1.0 with appropriate number of splitters to determine pressure loss. When throat radius of this elbow is squared off, use "O" splitters and R/W = 1.0 for pressure-loss calculation.

Table 9. Total Pressure Losses in Duct Elements

Element	Illustration	Conditions	Loss, Fraction of Velocity Pressure	Equivalent Length of Round Duct, Diameters
Weather Cap		$H/D = 1.00$	.10	6
		.75	.18	11
		.65	.30	18
		.55	.56	34
		.50	.73	44
		.45	1.00	60
Expansion		$a = 5°$	.17	10
		$10°$	.28	17
		$20°$	.45	27
		$30°$	.59	36
		$40°$	.73	44
		Loss referred to (hv_1-hv_2)		
Contraction		$a = 30°$	.02	1
		$45°$	.04	2
		$60°$	.07	4
		Loss referred to V_2		
Transition Piece			.15	9
Abrupt Entrance			.50	30
Abrupt Exit			1.0	60
Bell Mouth Entrance			.03	2
Bell Mouth Exit			1.0	60
Re-entrant Entrance			.85	51
Sharp Edge Round Orifice Exit		$A_2/A_1 = 0.00$	2.8	170
		.25	2.4	140
		.50	1.9	115
		Loss referred to A_2 .75	1.5	90
		1.00	1.0	60
Abrupt Contraction		$V_1/V_2 = 0.00$	.50	30
		.25	.45	27
		Loss referred to V_2 .50	.32	19
		.75	.18	11

Element	Illustration	Conditions	Loss, Fraction of Velocity Pressure	Equivalent Length of Round Duct, Diameters
Abrupt Expansion		$V_1/V_2 = 0.00$	1.0	60
		.20	.64	39
		Loss .40	.36	22
		referred .60	.16	9
		to V_1 .80	.04	2
Double Elbows	R=1.5D	Loss for both elbows		
		$L = 0$	.43	26
		$L = D$	.31	19
		Elbows vaned	.15	9
	R=1.5D	$L = 0$	.62	38
		$L = D$	.68	40
		Elbows vaned	.19	12
Branch Entry		Angle a, Degrees		
		10	.06	4
		15	.09	5
		20	.12	7
	2D min.	25	.15	9
		30	.18	11
		35	.21	13
		40	.25	15
		45	.28	17
		50	.32	19
		60	.44	26
		90	1.00	60
		Loss referred to branch velocity		
Pipe Running Through Duct		$E/D = .10$	.20	12
		.25	.55	33
		.50	2.00	120
Bar Running Through Duct		$E/D = .10$	.7	40
		.25	1.4	80
		.50	4.0	240
Streamlined Covering over Obstruction		$E/D = .10$	.07	4
		.25	.23	14
		.50	.90	54

divided by gas density, or its equivalent, specific volume divided by 13.35.

Example: An air flow of 5000 cfm (standard air) has been specified, and a duct size of 18 inches diameter has been chosen. Before the air reaches the duct it passes through a spray chamber and emerges at 100 deg, saturated. From Fig. 2, for 5,000 cfm and an 18-inch duct, friction loss per 100 ft is 0.56. From Fig. 1, density is 0.069. Therefore, the actual friction loss in the duct is 0.56 (0.075/0.069) = 0.61.

Losses in Elbows and Fittings

When an air stream undergoes a change of either direction or velocity, a dynamic loss occurs. Unlike friction losses in straight duct, fitting losses are due to internal turbulence rather than skin friction. This turbulence destroys the forward motion of some elements of the air stream and therefore requires energy to reestablish its forward motion. Hence, roughness of material has but slight effect over a wide range of moderately smooth materials.

Fitting losses can be expressed as equivalent length of straight duct; or as a fraction of velocity pressure; or directly in inches water gage.

Losses as a percent of velocity pressure and as equivalent length of round duct in diameters, are shown in Tables 7, 8, and 9. The losses for a variety of fittings are shown in equivalent feet of duct in the section on Warm-Air Heating. Losses are shown in the following section on Duct System Design as "inches wg (water gage)."

The equivalent-feet-of-duct method of determining turbulent losses has great appeal to the designer of small air systems, which predominate in the residential warm-air heating industry. Here, the University of Illinois and ASHRAE have done much research on small duct fittings, particularly those used in warm-air, low-velocity heating systems. Therefore, the equivalent-feet method will give very satisfactory results. For larger commercial and institutional air conditioning systems, the designer is advised to use the fraction of velocity pressure or the tabulated loss readings for various fittings in preference to the equivalent-length method. In high-velocity systems (over 2000 fpm air velocity), use of the equivalent-feet method will often result in errors on the low side, because large, high-velocity ducts usually have fairly low friction losses against which the equivalent feet for a particular fitting is applied. A recent study on high-velocity duct design showed that even the loss coefficients for elbows (such as those shown in Table 8), vary with velocity, but fortunately there is, at most, an 8% difference over the air velocity range of 2000 to 6000 fpm.*

For duct fittings used aboard ships, the U.S. Dept. of the Navy has published an extensive list of pressure losses for these fittings, "Pressure Losses of Ventilation Fittings, Navships Design Data Sheet DDS 3801-2." This publication contains data on pressure losses for very complex elbow arrangements and may be obtained from National Technical Information Service, Operations Division, 5285 Port Royal Road, Springfield, Va. 22151.

Good Duct Turns

Air will not turn of its own volition; it must be turned. Even with a standard elbow (throat radius equals duct width), the air cannot make the turn, and high velocity occurs at the outside with an eddy zone at the throat. This nonuniform condition, as shown in Fig. 6a does not permit outlets or branches to be located near the elbow on the throat side. Ductwork can be designed to accommodate this nonuniform condition. Elbows *b* through *g* will cause difficulties. Elbow *f* is fair because the air can expand along the vanes, but this arrangement should be used with care and a good section of straight duct ahead of the vanes. For elbow *g*, the air will jam on the vanes at the top and cause excessive energy loss. Elbow *h* should be used where the depth of the duct changes also. Note that the transition accommodates the air flow as ·compared to the poor condition in *d*. Elbows *i, j,* and *k* are needed when an outlet or branch is located less than six duct widths from an outlet. Even with vanes or splitters, some distance must be allowed for improved uniformity and better performance. For elbow *k,* keep in mind that the splitter should extend beyond the tangency point in the direction of flow to straighten the air stream. As a minimum, L should equal W and preferably should be 1.5 to 2W. If space conditions make it necessary for the throat radius to be eliminated, i.e., $R_1 = O$, then W_1 should be a minimum of 3 inches, $W_2 = 6$ inches, $W_3 = 12$ inches, etc. The venturi elbow, shown in Fig. 7, will also provide a smooth expansion.

* Rick Eschman and Wayne E. Long, "A Critical Assessment of High-Velocity Duct Design Information," Transaction No. 2137, ASHRAE Transactions, 1970, Part I, Vol. 76, p. 157.

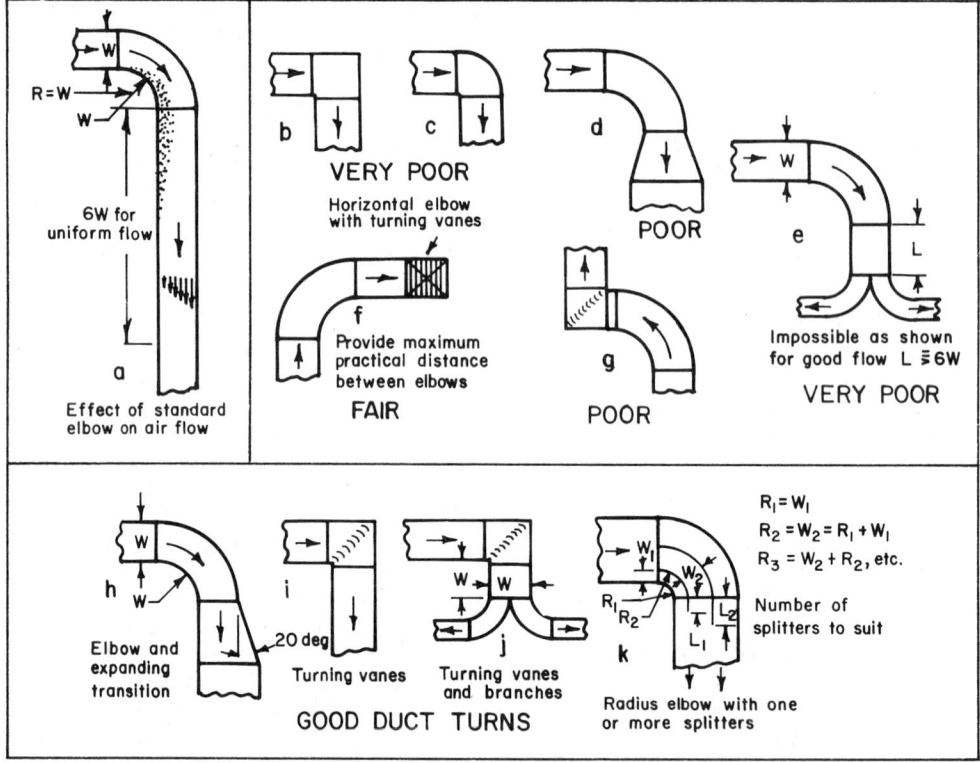

Fig. 6. Duct turns. *Source: J. H. Clarke, "The Art of Duct Hunting," Heating, Piping & Air Conditioning,* April, 1971, page 104.

Good Aerodynamics for Duct Systems

Since air does not turn by itself, as mentioned previously, it must be guided by fittings and elbows whose configuration should make for a good aerodynamic path. From Table 10, the designer should use those fittings, where possible, which produce the lowest loss fraction of velocity pressure. Abrupt transitions and compound elbows should be avoided unless space conditions make them absolutely necessary.

It is beyond the scope of this handbook to discuss theoretical aspects of aerodynamics. However, some practical guidelines can be given to the design of turns and elbows in rectangular ducts, as shown in Fig. 6.* The two worst cases of duct turns in Fig. 6 are shown in (b) and (c); these can readily be converted to good duct turns by turning vanes (i) and splitters (k), respectively.

---

* John H. Clarke, "The Art of Duct Hunting," *Heating, Piping and Air Conditioning,* April, 1971, p. 104.

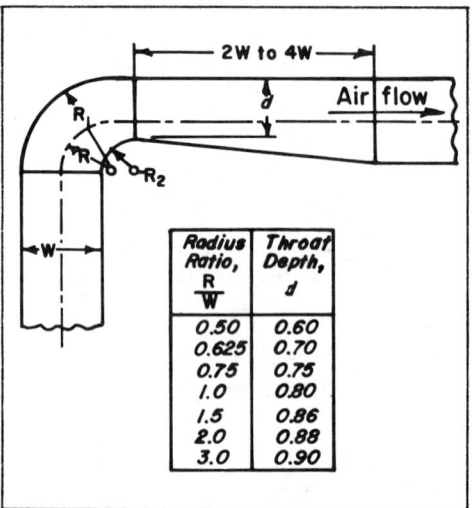

Radius Ratio, $\frac{R}{W}$	Throat Depth, d
0.50	0.60
0.625	0.70
0.75	0.75
1.0	0.80
1.5	0.86
2.0	0.88
3.0	0.90

Fig. 7. Venturi elbow. *Source: J. L. Alden and J. M. Kane, Design of Industrial Exhaust Systems, 4th Ed., New York: Industrial Press Inc., 1970.*

Until recently, it was thought that better aerodynamics resulted from use of double-thickness turning vanes in square or rectangular elbows, as against the single-thickness type. This theory has been completely contradicted in a recent study, which indicates that single-thickness vanes result in lower losses than double-thickness vanes. Moreover, the dynamic losses in single-thickness vaned elbows can be reduced even more by extending the trailing edge of each vane. Turning-vane loss coefficients are a function of air velocity, vane radius, and vane spacing; the lowest loss coefficient for the several vane types and geometry tested was a constant 0.12, based on the duct velocity range of 500 to 5,000 feet per minute. This particular vane was single-thickness type, 2-inch radius, 1.5-inch space between vanes on the diagonal, trailing edge extended 0.75 inch. This study further showed that the best double-

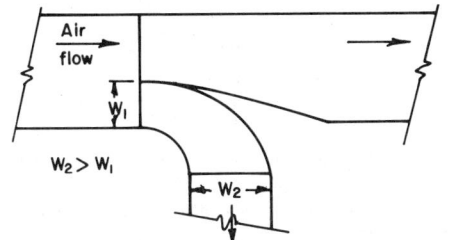

Fig. 8. "Warped" elbow. (Note: Depth is greater at inlet than at outlet of this elbow; this gives the "warped" appearance to the elbow.)

thickness vane had a loss coefficient that varied from 0.34 at 500 fpm to 0.16 at 5,000 fpm.

One type of elbow, shown in Fig. 8, should be avoided; this is the "warped" elbow which changes in both depth and width at the outlet. It has poor aerodynamics and is expensive to manufacture because of the labor involved.

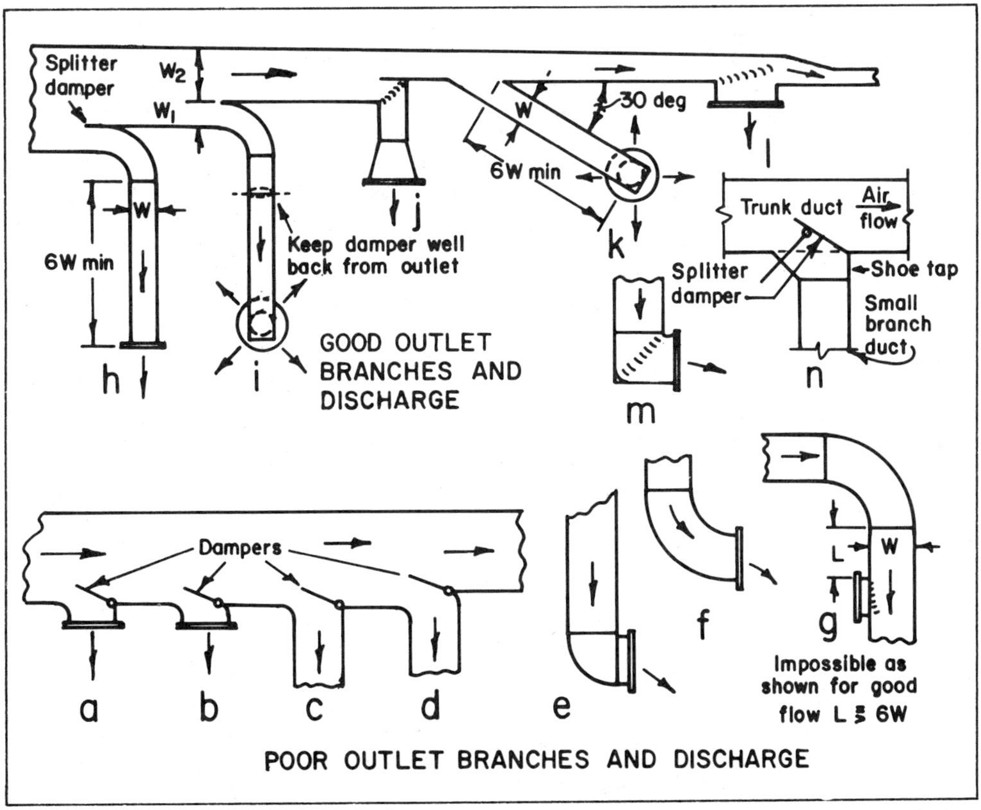

Fig. 9. Outlet branches and discharge. *Source: J. H. Clarke, "The Art of Duct Hunting," Heating, Piping & Air Conditioning, April, 1971.*

For branch duct connections to the trunk duct, some means must be provided to assure good approach at the outlets. Otherwise, uneven flow will occur from the outlet, which can cause draft or prevent the desired air motion. The throw may be greatly reduced, perhaps to as little as one-third of the catalog value. Poor flow will always result from the outlet connections and branches shown by *a* through *g* in Fig. 9. Noise, higher friction, and balancing difficulty will also occur if these configurations are used. For arrangement *g,* inflow is more probable than outflow at the grille.

Full tap branch connections should always have shoe tap connections and splitter dampers.

Cost is always a factor, but savings are of no value if outlets do not perform satisfactorily. Outlets and branches *a, b, c,* and *d* in Fig. 9 should not be used. Impossible grille performance will result with *a* and *b* because the air will discharge through the far end. Recirculation will exist at the upstream end. Branches *c* and *d* will produce extremely nonuniform flow, which may cause difficulties with an outlet or branch downstream. All of these dampers will be noisy and because of turbulence downstream will make it more difficult to balance downstream branches and outlets. Outlets *e* and *f* will be greatly affected by the nonuniformity that occurs with these elbows. The throws may be as little as 35 percent of catalog values. The outlet in *g* cannot be expected to perform unless it is intended to aspirate air into the duct. This is unlikely. If an outlet must be near an elbow, provide room for flow uniformity to be established, or provide turning vanes as shown in *h* through *m.* In all branches, keep dampers far enough back from outlets so outlets can perform as intended, with good approach. Branch *k* is used frequently, particularly where a small branch is taken off a large duct. In industrial areas, it may be straight, as shown, or if interferences prevent this, the duct may be turned to provide a branch similar to *h* or *i.* Splitter dampers should never be less than 4 inches wide, and should generally have a width equal to 1.5 W_1 (where W_1 is small neck size). Note that neck sizes W_1 and W_2, in Fig. 9, are in direct proportion to air flow through each neck, except that minimum neck size, W, in this case, should never be less than three inches.

Where air registers in low-velocity systems are connected into the side of the trunk duct, an extractor should be used to obtain uniform air

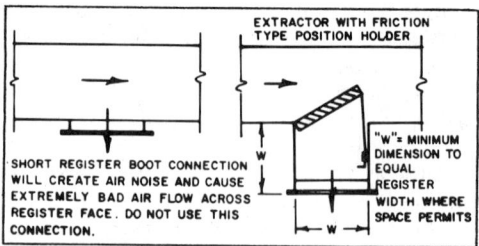

Fig. 10. Top register connection to trunk duct.

flow, as shown in Fig. 10. Also, to preclude noise generation, the trunk duct velocities should not exceed those recommended by the register manufacturer.

Air Balancing and Air Turning Hardware

For air balancing, proper hardware should be provided at branch ducts to divert or throttle air flow. Most air diffusers and registers have their own terminal dampers, usually the opposed blade type which are key operated, and these alone would seem to be sufficient. Unfortunately, this is not true, because any throttling required to "kill" excess static pressure, above 1/4" WG, virtually closes these terminal dampers and will thereby cause either a whistle or a whirring noise. For this reason, local branch dampers are required upstream of the outlet or outlets. These branch dampers are manually operated and for small ducts are of the single-blade type; while for larger duct sizes they are the opposed-blade type. For high-velocity duct systems, factory fabricated dampers with good aerodynamic design (often called "air pressure reducing valves") are required to reduce the high pressures encountered in high-velocity design, to low pressures suitable for conventional low-velocity duct construction. If air pressure reducing valves are not used, dampers for high-velocity systems should certainly be the "airfoil" type, preferably factory made where close quality control can guarantee against defective damper configuration, burrs, poor leading edges, etc.

Where supply trunk ducts split into branches, splitter dampers are recommended since sizing of the necks does not automatically assure that the air goes through the branches in the designed quantities. In effect, the splitter damper offers a means of adjusting the neck size. It is true that manual volume dampers will do the same thing,

i.e., by adding resistance to the branch which supplies more air than the design flow. However, manual dampers will add resistance to the system, unlike splitter dampers.

Air flow into exhaust or return ducts follows the same rules of fluid flow as it does for supply ducts. The inlets in Fig. 11 a, b, and c will cause severe turbulence in the duct and interfere with the flow from upstream inlets the same as if a damper were located about halfway, or more, into the air stream. Full flow will not occur in the length of the grilles, so much of the length is wasted. The inlets in d through h will provide good flow. In industrial areas, appearance is seldom of great importance, so inlets such as d and e can and should be used. These will provide the best flow. Balancing can be improved by proper sizing of the branches. For better performance, the grilles in i and j should have straight vanes as opposed to the 45-degree setting of the grille blades in a and b. However; if curved or deflected vanes are used to hide the interior of the duct, they should deflect in the direction of air flow.*

Exhaust or return ducts and inlets suffer in the same manner and for much the same reasons as do supply ducts. Air is not pulled into ducts; it moves because of a pressure differential, as does supply air. It also has mass and momentum. Air entering an exhaust or return duct perpendicularly

at a branch or inlet creates considerable impact and turbulence because it cannot make the turn, as shown by a, b, and c in Fig. 11. This acts as a block, or choke, for flow from inlets upstream. For reduced turbulence and friction, the air should enter the main stream as close to parallel as is practical and at nearly the same velocity. Straight inlets such as d and e, in Fig. 11, should be the first choice. Arrangements f, g, h, i, and j also provide good flow. It should be noted that the branch channel in g is extended into the duct by the length L. The rule here is that at an entrance or elbow turn, the splitter or branch should be extended to avoid serious impingement on the main stream or adjacent streams, and the air should be fully turned to provide good flow. L should at least equal W. Better flow results when L equals $1.5W$. Where it is very important that the air stream be turned for particularly good flow conditions, L should equal $2W$.*

Full taps for small exhaust or return-duct branch connections into trunk ducts should not be used unless there is an internal boot, or at least a splitter or a splitter damper. Tests have shown that even a full tap with a "shoe tap" connection to the trunk duct, will result in turbulent losses four times those with boots or

---

* John H. Clarke, "The Art of Duct Hunting," *Heating, Piping and Air Conditioning*, April, 1971, p. 104.

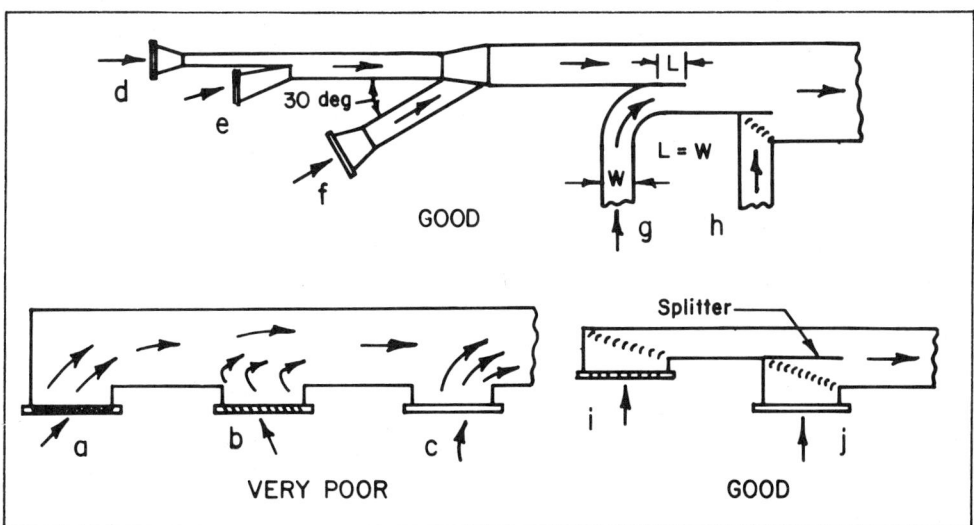

Fig. 11. Air flow into ducts. *Source: J. H. Clarke, "The Art of Duct Hunting," Heating, Piping & Air Conditioning, April, 1971.*

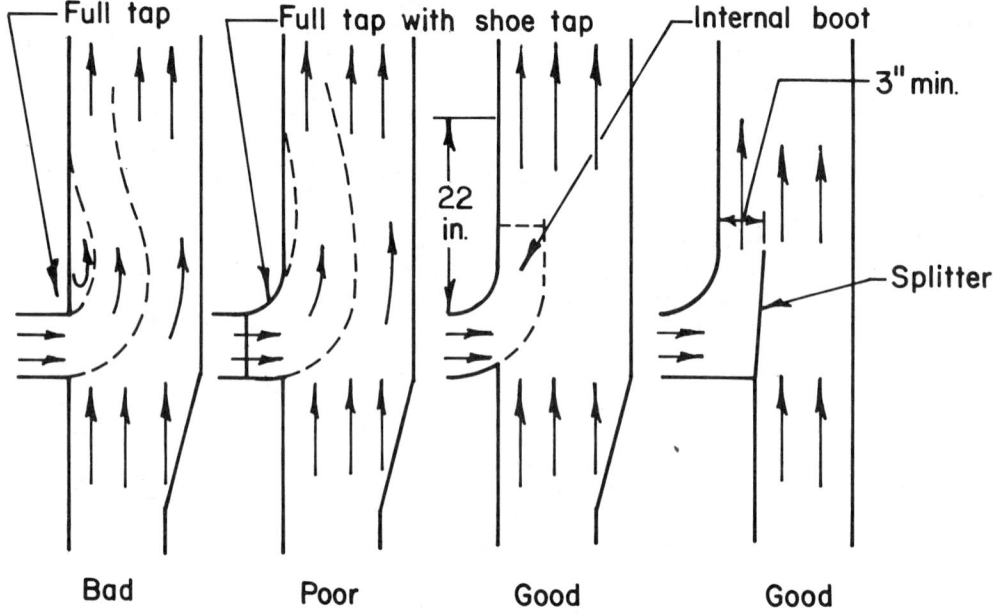

Fig. 12. Four types of arrangements of return or exhaust air-duct branch fittings.

splitters.† This is shown in Fig. 12.

A splitter damper may be used in lieu of a fixed splitter; however, its only purpose in a return or exhaust duct is better aerodynamics, not air balancing. For this reason, the added cost of the splitter damper may not justify it over a fixed splitter.

Return Air Plenums

It has long been known that return air plenums, often called "mixing boxes," where return air and outside air are brought together, do not really mix the air. Rather, stratification generally results when the return air and outside air have widely different temperatures. Short of using a propeller fan inside the "mixing box" or a bulky factory-fabricated air blender to achieve a homogeneous mixture, there is really no effective way to mix the two air streams of widely different temperatures. However, some duct arrangements at the mixing box are better than others; these are shown in Figs. 13 and 14.

Uniform flow and thorough mixing of outside

† Carlyle M. Ashley, "Return Duct System Design," *Heating, Piping and Air Conditioning*, February, 1969, p. 100.

and return air are essential for good equipment operation. In Fig. 13(a), air directed by inlet louvers to the top of the casing results in backflow through bottom filters, condensate blow-off from cooling coils, and uneven heating or possibly freezing of heaters. Similarly in b, the air cannot make the turn and flows to the top. In c, the flow depends on the damper setting. With a high percentage of return air, return air flows to the opposite side, causing recirculation on the near side. Only a thin stream of outside air is allowed to enter. With mostly outside air, return air flows to the opposite side, allowing a thin stream of outside air in any damper position. Stratification is extreme for d, and this arrangement should not be used; particularly if outside air enters under the return air. Better arrangements for uniform flow and mixing are shown by e, f, and g. In e, two-position dampers are arranged to act as louver extensions to turn the air. In f, the inlet hood splitters are extended to turn the air. The extension should be at least as long as the channel width. Better flow will occur if the extension is $1.5W$, and if it is important that the air be fully turned, the extension should be $2W$. This will not provide instant

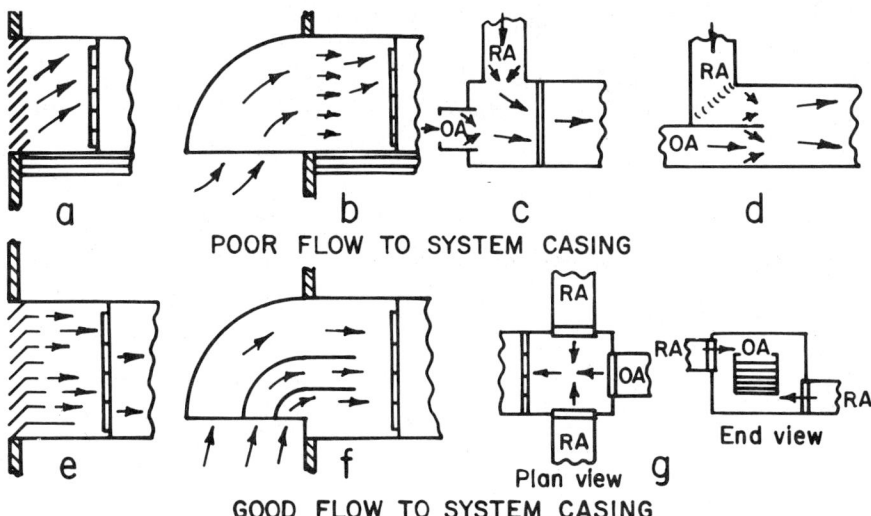

Fig. 13. Flow to system casing. *Source: J. H. Clarke, "The Art of Duct Hunting," Heating, Piping & Air Conditioning, April, 1971.*

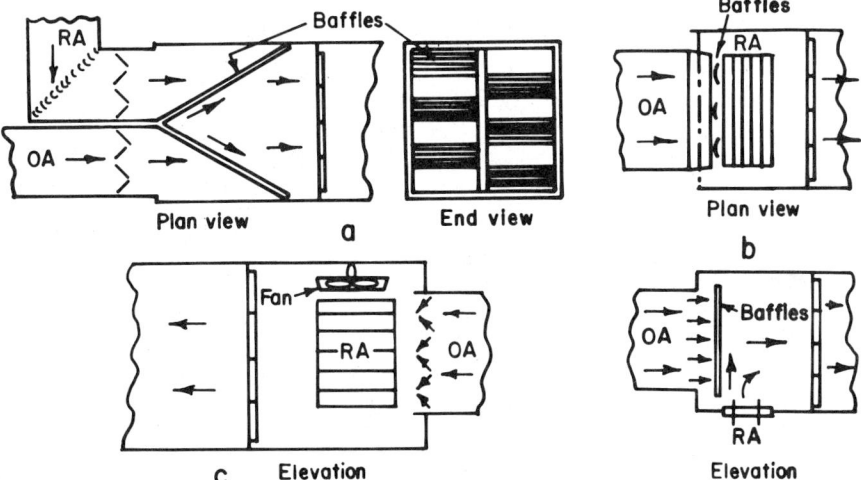

Fig. 14. Mixing outside and return air. *Source: J. H. Clarke, "The Art of Duct Hunting," Heating, Piping & Air Conditioning, April, 1971.*

uniform flow, but it will help greatly. Good mixing and flow are encouraged in g by returning the air in an offset pattern, which causes rotation and mixing of the streams.

Outside and return air have been successfully mixed by various added means. In Fig. 14a, outside and return air should not be brought into the casing in parallel; but if this cannot be avoided, alternate baffles, as shown, will create alternate streams of outside and return air, following the baffles to the opposite side, to encourage mixing. In b, similar baffles will cause mixing for air streams entering at right angles. In this case, the dampers do not modulate: but where they do, they should be set for opposed operation. Another excellent way of mixing air stream is by means of a propeller fan in the mixing compartment, as in c. This was used effectively in one large building to eliminate freezing caused by stratification.

AIR DISTRIBUTION

At the ends of the air supply duct system, air supply outlets must be selected to provide proper air motion in the occupied spaces. Table 1 can be used as a general guide in this respect.

In recent years, both the Air Diffusion Council of the United States; and the American Society of Heating, Refrigerating and Air-Conditioning Engineers (ASHRAE)—through its experimental programs at Kansas State University—have done much to upgrade the criteria for proper air motion through the use of air diffusers. This has led to rating air distribution systems by use of a

Table 1. Acceptable Air Motion in the Occupied Zone

(Courtesy of the Carrier Corp.)

Room Air Velocity, FPM	Recommended Application
0-16	Should not be used because of complaints about "stuffiness" (stagnant air).
25	Ideal for all commercial air-conditioning applications.
25-50	Acceptable range for commercial air conditioning where occupants are seated.
50-75	Retail and department store air conditioning where people are moving about.
75-300	Factory air conditioning (spot cooling).

Table 2. Characteristics of Air Supply Outlets

Outlet	Characteristics
1. All outlet Air Steams	Act as jet pumps. Entrain many times their primary volumes. They slow down and approach room temperature slowly
2. Circular (or similar) Diffusers	Short throw, high entrainment for a flat pattern
3. Round or Square (low aspect ratio) outlets	Long throw, moderate entrainment
4. Long slots	Moderate throw, high entrainment
5. Perforated sheet	Short throw, low entrainment

Table 3. Applications for Air Supply Outlets

Outlet	Applications
Conventional diffusers, grilles	For fully air conditioned spaces or nominal ventilation not requiring heat relief or velocity control.
Directional grilles	For general area relief ventilation systems or spot cooling; must provide direction and velocity control.
Directional diffusers	For general area relief ventilation but better suited for aisle ventilation and spot cooling; must provide direction and velocity control.
Nozzles	For spot cooling (large sizes are needed) and high local air velocities; must provide direction and velocity control.
Two-way damper sections	For heat recovery and conservation, to permit directing adjustable air supply to ceiling in winter and work zone in summer; these also prevent volume reduction of supply air for makeup systems.
Slots	For crowded areas or areas requiring good mixing and air motion for close environmental control; can release high volume in small space.
Perforated panels	For high air change rates and where very low air motion is required because of hood or process requirements.

valid, single number called the Air Diffusion Performance Index (ADPI). This index is based on comfort criteria at which 80% of test occupants felt comfortable, and a maximum allowable draft velocity of 70 fpm.

With the increased concern for energy conservation and the use of the variable-air volume system to maximize energy efficiency, the ADPI methodology will assist engineers in designing a draft-free environment. Further information can be obtained from the Air Diffusion Council, 435 North Michigan Avenue, Chicago, Illinois 60611.

Outlets vary widely in their throw and performance characteristics. Manufacturers provide a wealth of data on air throws, noise levels, etc., for various types of air supply diffusers, registers.

and grilles. The ASHRAE Guides contain detailed recommendations for particular applications; however, a very general guide to air outlet characteristics is indicated in Table 2.*

Having determined the type of performance required for a specific space, the outlet types can now be selected with the help of Table 3. Care should be taken in laboratories against using conventional diffusers and grilles ordinarily employed in air conditioned spaces. They may undermine the safety of nearby fume hoods. Such outlets are not good selections for laboratories with hoods, particularly if high air-change rates are involved.

---

* J. H. Clarke, "The Art of Duct Hunting," *Heating, Piping and Air Conditioning*, April, 1971, p. 104.

FIRE DAMPERS AND FIRE PROTECTION

Every fire prevention code—Federal, state or local—uses fire-stopping devices at fire walls and partitions, and fire-rated ceilings or floors; these safety devices are called *fire dampers*. The current edition of NFPA 90A requires that the HVAC designer show and specify all fire dampers and fire doors required by this standard. Moreover, these dampers and doors must have an Underwriter's Laboratories rating for a give time span such as one hour, under this standard. This has given rise to the use of factory-made fire dampers and doors with the U.L. label. It has also stimulated the greater use of the "shutter" or "guillotine" type of fire damper, over the "butterfly"

type. The latter is less compact in size and sustains more pressure loss from air turbulence than the shutter type.

The NFPA standards are replete with information on fire prevention or fire stopping for special exhaust systems, such as kitchen ventilation, hospital operating rooms, labs, etc. The designer should always consult the most current edition of these standards, available from National Fire Protection Assn., 470 Atlantic Ave., Boston, Mass. 02210. They are revised annually to meet the increasingly stringent requirements of current fire prevention.

DUCT SYSTEM DESIGN

(1) *Constant Pressure Drop*—With this method, pressure loss in each foot of duct length is constant. The choice of this constant (in inches of water column per 100 feet of duct) is based largely on noise and economical limitations. Tables 1 and 2 contain normal and maximum velocities which will serve as a guide in planning the design. As a first step, the velocity to be assigned to the main duct is chosen from the tables. Then, with the main duct volume known and its velocity assigned, Figs. 2a, 2b, 3a, 3b on pages 5-45 to 5-48 are used to determine the pressure loss per 100 ft. This loss is then uniformly assigned to the rest of the duct system except for losses in transition pieces, air inlets and outlets, filters, coils, etc. Because, for a given pressure loss, larger ducts carry proportionately larger air volumes at higher velocities,

it follows that velocities automatically diminish as the ducts become smaller.

This system is most applicable where the lengths of branch runs are symmetrical and of nearly uniform length. Any differences in pressure at the take-offs are made up by dampers at each branch.

In systems with unequal branches, mains and branches of equal length may be sized by the constant-pressure-drop method. Each unequal branch may then be sized in accordance with the pressure at its take-off. This is a refinement not always followed but one which permits branches near the fan to carry air at a higher pressure drop than the uniform drop used throughout the rest of the system. Care should be taken, however, to make certain that branch velocities do not exceed the maximum velocities shown in Table 1.

Table 1. Recommended and Maximum Duct Velocities for Low Velocity Systems

Designation	Recommended Velocities, FPM		
	Residences	Schools, Theaters, Public Buildings	Industrial Buildings
Outdoor Air Intakes [a]	500	500	500
Filters [a]	250	300	350
Heating Coils [a]	450	500	600
Cooling Coils [a]	450	500	600
Air Washers [a]	500	500	500
Fan Outlets	1000–1600	1300–2000	1600–2400
Main Ducts	700–900	1000–1300	1200–1800
Branch Ducts	600	600–900	800–1000
Branch Risers	500	600–700	800
	Maximum Velocities, FPM		
Outdoor Air Intakes [a]	800	900	1200
Filters [a]	300	350	350
Heating Coils [a]	500	600	700
Cooling Coils [a]	450	500	600
Air Washers [a]	500	500	500
Fan Outlets	1700	1500–2200	1700–2800
Main Ducts	800–1200	1100–1600	1300–2200
Branch Ducts	700–1000	800–1300	1000–1800
Branch Risers	650–800	800–1200	1000–1600

[a] These velocities are for total face area, not the net free area; other velocities in table are for net free area.

An example of the constant-pressure-drop method is given below for the system shown in Fig. I.

Example: Given, a public building in which a total air volume of 3,700 cfm is moved. The main duct velocity chosen from Table I is 1,300 fpm. From Fig. 2a (page 5-45) the volume of 3,700 cfm and the velocity of 1,300 fpm intersect at 0.09 inches WG per 100 ft of duct. This loss is used in sizing all ducts from *A*, via *C*, *D*, *E* and *I* to *L*, not including transitions at filters, fan, and coil. Other portions of duct, such as *G* to *M* and *E* to *P*, are often sized on the same uniform pressure drop as the main run; in this case, 0.09 inches per 100 ft. It is possible, however, to design a more economical system, and one more readily balanced, if the pressure drop from *G* to *M* is made the same as from *G* to *L* and the pressure drop from *E* to *P*

Table 2. Recommended Maximum Duct Velocities for High Velocity Systems

CFM Carried by the Duct	Maximum Velocities, FPM
60,000 to 40,000	6,000
40,000 to 25,000	5,000
25,000 to 15,000	4,500
15,000 to 10,000	4,000
10,000 to 6,000	3,500
6,000 to 3,000	3,000
3,000 to 1,000	2,500

the same as from *E* to *L*. This must be done with caution, as high velocities may result in branch circuits, causing noisy operation.

Assuming the same register and take-off losses, the duct pressure loss from G to L for the 90 ft length is $(90/100)(.09)=0.081$. If this drop is applied to the 60 ft length from G to M, the design pressure loss per 100 ft for G to M would be $(.081)(100)/60=0.123$ per 100 ft. To size the branch E to P, the 120 ft from E to L would have a pressure loss of $(120/100)(.09)=0.108$. If this is applied to the 40 ft length from E to P, the design pressure loss per 100 ft for E to P would be $(.108)(100)/40=0.27$ per 100 ft. If we now apply these pressure losses per 100 ft against the respective air volumes, we obtain for branches G to M and E to P the values shown in the lower part of Table 3.

For small branches, such as G-M and E-P,

the designer will usually find that it generally does not pay in sheet metal costs to size them by the higher pressure drops, 0.123 and 0.27, as shown in the example. This derives from the fact that most duct work is priced the same per foot up to certain sizes, such as a 36-inch total duct circumference (e.g., 12 in. x 6 in.-cross section); therefore, these branches could more readily be sized at 0.09 in. WG/100 ft with very little additional cost. If this is done, a volume damper would be inserted at the entrance to each branch, to handle the pressure difference.

Total pressure loss in the system and, hence, the static pressure developed by the fan is obtained as follows: (table numbers refer to section on high velocity system design following):

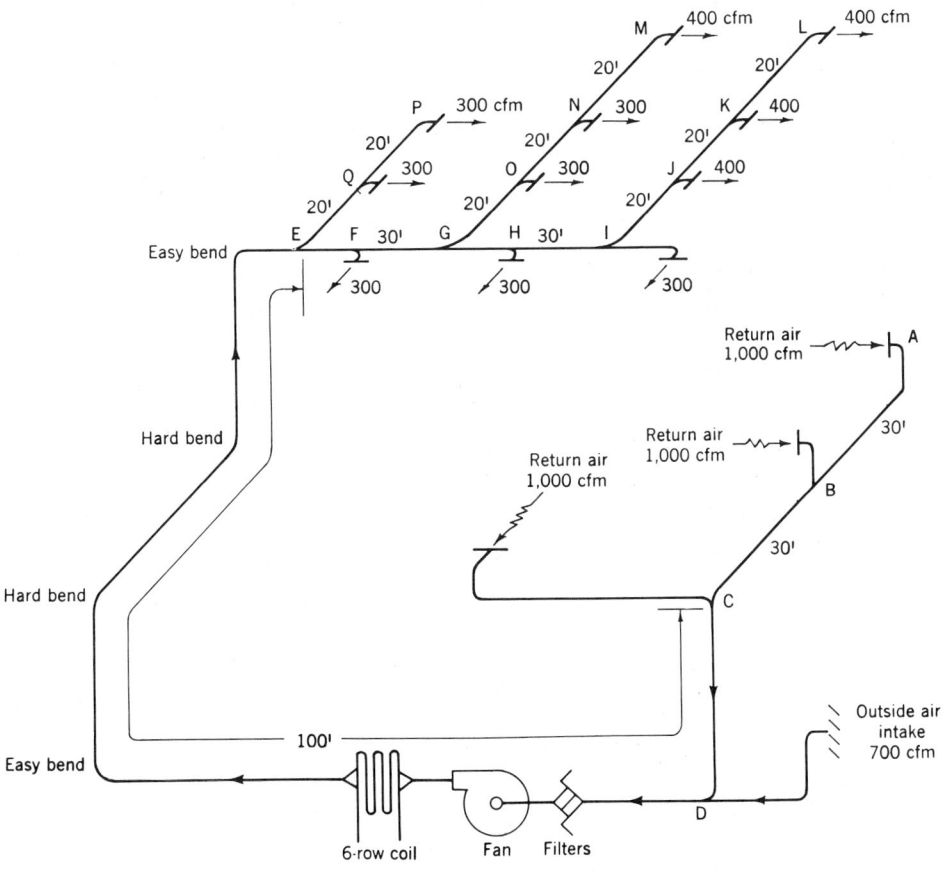

Fig. 1. Schematic layout of system employing constant pressure drop method.

Total straight-line duct work *A* to *L* 280'
Pressure loss in ductwork, inches WG
  (280/100) (.09)   0.25
2 hard bends at .02 (Table 5:
  H/W=0.41; R/W=1.0)   .04
2 easy bends at .01 (Table 5:
  H/W=2.4; R/W=1.5)   .02
Branch take-off at (Table 4.
  Based on I-J)   .03
Loss thru 6-row coil (approx.)
  8 fins per inch at 500 fpm   .34*
Filters—assumed dirty   .25*
Loss in supply register   .05*
Loss in return-air grille   .15*
Transitions—fan and filters   .10

    Static pressure loss in system,
      inches WG   1.23

* Refer to manufacturers' data.

(2) *Static Regain*—The intent of this method is to maintain the static pressure at practically a constant value throughout the system. The advantage of doing this is that static pressure determines the rate of discharge through outlets; hence, if the static pressure remains constant, the size of outlet for given volume of discharge would also be constant. For installations such as hotels or hospitals, there would be an obvious advantage in having the same size outlet in a series of like rooms.

The static regain method utilizes a velocity reduction at the end of each section of duct; the magnitude of the reduction being sufficient to provide a loss of velocity pressure equal to the loss of total pressure that occurred in the preceding section of duct. Thus, for a given length of duct the method of application would be as follows:

(a) Determine the friction loss per 100 ft from friction charts.

(b) Calculate the static drop in the given length of duct by multiplying the friction loss (from "a") by the duct length divided by 100.

(c) For known velocity in the duct section read the velocity pressure, VP_u from a table of velocity equivalents (Table 1 in the preceding section on Air Flow in Ducts).

(d) Calculate the velocity pressure in the down-

Table 3. Calculations for Duct Design Example

Duct Section	Air Volume, Cfm	Pressure Drop, Inches/100 Ft	Round Duct Diam., Inches	Round Duct Vel., Fpm	Equiv. Rect. Duct, Inches	Rect. Duct Vel., Fpm
A-B	1000	.09	14.2	950	14 X 12	860
B-C	2000	.09	18.5	1100	24 X 12	1000
C-D	3000	.09	21.7	1200	34 X 12	1050
D-E	3700	.09	23.4	1300	34 X 14	1120
E-F	3100	.09	22.	1200	34 X 12	1090
F-G	2800	.09	21.	1200	32 X 12	1050
G-H	1800	.09	17.9	1080	22 X 12	980
H-I	1500	.09	16.6	1030	19 X 12	945
I-J	1200	.09	15.3	970	16 X 12	905
J-K	800	.09	13.1	890	12 X 12	820
K-L	400	.09	10.0	750	12 x 8	600
For branch *G* to *M* designed for a higher pressure loss of 0.123 inch per 100 ft, we obtain						
G-O	1000	0.123	13.4	1050	12 X 12	1000
O-N	700	0.123	11.7	960	10 X 12	840
N-B	400	0.123	9.4	850	10 x 8	720
For branch *E* to *P* designed for a higher pressure loss of 0.27 inch per 100 ft, we obtain						
E-Q	600	0.27	9.5	1250	10 x 8	1080
Q-P	300	0.27	7.2	1100	6 x 8	895

stream section from the equation,

$$VP_d = (VP_u) - 2SP$$

where SP is the friction loss as determined in step "b";

VP_u is the velocity pressure from step "c". The coefficient reflects the assumption that the regain occurs at 50% efficiency. VP_d is the necessary velocity in the downstream section of duct to provide complete regain of the static pressure lost in the upstream section.

(e) Knowing VP_d go to the velocity equivalent table and read the necessary velocity (thus fixing diameter) in the downstream section of the duct.

The static regain method is likely to be more time-consuming than other duct design precedures, but in many cases it will be found to justify the added effort through more effective air distribution.

(3) *Simplified Static Regain*—It will generally be found that the constant-pressure-drop method produces a constant friction drop of roughly 0.15 in. wg per 100 ft of duct for large trunk ducts, carrying 10,000 to 15,000 cfm. By dividing the longest equivalent duct run into approximately

three parts, each third can be sized at a descending equal friction rate, in the direction of air flow, say 0.15, 0.12, and 0.09 in WG per 100 ft of duct run. These rates can be determined in the same manner as the example above.

By designing with descending equal friction rates, the duct system realizes static regain, which makes the system easier to balance and less wasteful of fan energy, while making the design procedure considerably more simple than pure static regain.

Branch sizing can be done as in the above example (Table 3); however, it is usually satisfactory for short branches to simply use the same friction rate used for the trunk duct.

(4) *Total Pressure Method (High-Velocity Method)*—The following coverage on High-Velocity System Design includes all the elements for design of large systems at any velocity, as well as the special considerations for design at high velocity. It will serve as an example of a design method which can be modified to suit any situation, whether more or less rigorous. The 1972 *ASHRAE Handbook of Fundamentals* alludes to this method as the "Total Pressure Method."

HIGH VELOCITY SYSTEM DESIGN

High velocity duct design is based on the same basic laws of fluid flow as the design of conventional duct systems. There are a few important differences, however:

Smaller ducts are used and, therefore, higher friction and dynamic losses are incurred.

More extensive use is made of standardized fittings.

At high velocities, static pressure regain due to velocity changes becomes significant.

Therefore, the following points should be borne in mind:

Selection of High Velocity Units—High velocity units and diffusers are selected from manufacturers' catalog ratings according to volume flow rate, permissible sound level, and desired location.

Main Duct—The critical duct run is selected on the basis of the system layout, not necessarily on the basis of length alone. Duct runs with greater air flow rate, with more fittings (elbows, transitions, branch take-offs, etc.), and/or with a high velocity terminal having greater minimum resis-

tance are often more appropriate as a basis for duct design.

Duct Velocity and Friction—An initial main duct velocity of approximately 4000 to 4500 fpm and a constant friction loss of about 1.0 inch water gage per 100 ft of run are recommended. Note that, since round ducts of standard diameter are generally used, it is not always possible to maintain an exactly constant friction loss factor. The recommended duct velocities for high-velocity systems were listed earlier in Table 2, page 5-64.

Dynamic Losses—Static pressure drop caused by branch take-offs, elbows, and duct fittings depends upon both fitting design details and air velocity. Tables 1 to 5 permit direct reading of pressure drop due to commonly used fittings.

Static Pressure Loss and Regain—Velocity drop in a system can produce regain of static pressure, and, conversely, a velocity rise will result in a loss of static pressure. These changes occur in connection with conversion of velocity into static pressure and vice versa. They are important because their

Table 1. Static Regain or Loss in Round Transition Pieces

Values for velocity increase (in italics) are static losses.
Values for velocity reductions are static regain.
All values in inches water gage.
Note: Table 1 values do not apply where airstream changes direction as at elbow or in branch at take-off.

Up-stream Velocity V_U, fpm	Downstream Velocity V_D, fpm												
	600	800	1000	1200	1400	1600	1800	2000	2100	2200	2300	2400	2500
	Static Loss (italics) or Regain, Inches wg												
600	.001	.02	.043										
800	.015	.002	.026	.055	.088								
1000	.035	.018	.003	.032	.066	.106							
1200	.06	.043	.022	.005	.038	.078	.122	.172	.20				
1400	.09	.072	.051	.026	.006	.046	.09	.14	.167	.195	.22	.26	.29
1600	.123	.116	.084	.058	.03	.008	.05	.102	.129	.16	.185	.22	.25
1800	.161	.144	.123	.096	.066	.033	.01	.06	.087	.115	.143	.175	.21
2000	.20	.186	.165	.139	.106	.073	.036	.013	.04	.068	.096	.128	.16
2100	.23	.21	.188	.162	.132	.096	.058	.016	.014	.042	.070	.102	.134
2200	.25	.23	.21	.186	.156	.12	.081	.04	.016	.015	.043	.075	.107
2300	.27	.25	.23	.21	.18	.144	.104	.062	.043	.016	.016	.048	.08
2400	.30	.28	.26	.24	.21	.172	.131	.089	.065	.045	.017	.018	.05
2500	.33	.31	.29	.26	.23	.198	.158	.114	.093	.068	.047	.017	.02
2600	.36	.34	.32	.29	.26	.23	.188	.143	.119	.097	.071	.049	.017
2700	.39	.37	.35	.32	.29	.26	.22	.172	.147	.124	.10	.074	.051
2800	.42	.40	.38	.35	.32	.29	.25	.21	.178	.152	.129	.103	.077
2900	.45	.43	.41	.39	.36	.32	.28	.24	.21	.185	.158	.134	.107
3000	.48	.47	.45	.42	.39	.35	.31	.27	.24	.22	.192	.164	.139
3100	.52	.50	.48	.45	.42	.39	.35	.30	.28	.25	.23	.198	.17
3200	.55	.54	.52	.49	.46	.42	.38	.34	.31	.29	.26	.23	.21
3300	.58	.57	.55	.53	.50	.46	.42	.37	.35	.32	.30	.27	.24
3400	.63	.61	.59	.56	.53	.50	.46	.41	.39	.36	.34	.31	.28
3500	.67	.65	.63	.60	.57	.54	.50	.45	.43	.40	.38	.35	.32
3600	.71	.69	.67	.64	.61	.58	.54	.49	.47	.44	.42	.39	.36
3700	.75	.73	.71	.68	.65	.62	.58	.53	.51	.48	.46	.43	.40
3800	.79	.77	.75	.72	.69	.66	.62	.58	.55	.52	.50	.47	.44
3900	.83	.81	.79	.76	.73	.70	.66	.62	.59	.57	.54	.51	.48
4000	.88	.86	.84	.81	.78	.74	.70	.66	.64	.61	.58	.55	.52
4100	.92	.91	.88	.86	.83	.79	.75	.71	.68	.66	.63	.60	.57
4200	.97	.95	.93	.90	.87	.84	.80	.76	.73	.70	.68	.65	.62
4300	1.02	1.0	.98	.95	.92	.88	.84	.81	.78	.75	.73	.70	.67
4400	1.07	1.05	1.03	1.00	.97	.93	.89	.85	.83	.80	.78	.75	.72
4600	1.17	1.15	1.13	1.10	1.07	1.04	1.00	.95	.93	.90	.88	.85	.82
4800	1.27	1.25	1.23	1.20	1.17	1.14	1.10	1.05	1.03	1.00	.98	.95	.92
5000	1.38	1.37	1.34	1.32	1.28	1.25	1.21	1.17	1.14	1.11	1.09	1.06	1.03
5200	1.50	1.48	1.46	1.43	1.40	1.36	1.33	1.28	1.26	1.23	1.20	1.18	1.15
5400	1.61	1.60	1.58	1.55	1.52	1.48	1.44	1.40	1.37	1.35	1.32	1.29	1.27
5600	1.73	1.72	1.70	1.67	1.64	1.61	1.57	1.53	1.50	1.47	1.45	1.42	1.39
5800	1.86	1.85	1.83	1.80	1.77	1.74	1.70	1.66	1.63	1.60	1.57	1.55	1.52
6000	1.99	1.98	1.96	1.93	1.91	1.87	1.83	1.78	1.76	1.73	1.71	1.68	1.65

Table 1 (continued). Static Regain or Loss in Round Transition Pieces

Upstream Velocity V_U, fpm	Downstream Velocity V_D, fpm												
	2600	2700	2800	2900	3000	3100	3200	3300	3400	3500	3600	3700	3800
	Static Loss (italics) or Regain, Inches wg												
1600	.28	.32											
1800	.24	.27	.30	.35									
2000	.194	.23	.26	.30	.34	.38	.42						
2100	.168	.20	.24	.28	.31	.35	.40	.44					
2200	.141	.17	.21	.25	.29	.33	.37	.41	.46				
2300	.114	.148	.183	.22	.26	.30	.34	.38	.43	.47			
2400	.084	.118	.153	.191	.23	.27	.31	.35	.40	.44	.49		
2500	.054	.088	.123	.161	.20	.24	.28	.32	.37	.41	.46	.51	
2600	.021	.055	.09	.128	.168	.21	.25	.29	.34	.38	.43	.47	.52
2700	.018	.023	.058	.096	.135	.175	.22	.26	.30	.35	.39	.44	.49
2800	.052	.018	.025	.062	.102	.141	.183	.23	.27	.31	.36	.41	.46
2900	.080	.054	.019	.026	.065	.105	.147	.189	.23	.28	.32	.37	.42
3000	.11	.083	.056	.019	.028	.068	.11	.152	.196	.24	.29	.34	.38
3100	.144	.114	.087	.057	.02	.03	.072	.114	.158	.20	.25	.30	.35
3200	.176	.149	.118	.09	.059	.02	.032	.074	.118	.163	.21	.26	.31
3300	.21	.183	.153	.122	.093	.06	.021	.034	.078	.123	.169	.22	.27
3400	.25	.22	.189	.158	.126	.096	.061	.021	.036	.081	.127	.176	.22
3500	.29	.26	.22	.194	.162	.129	.099	.062	.022	.038	.084	.133	.182
3600	.33	.30	.26	.23	.199	.166	.132	.102	.063	.022	.041	.087	.137
3700	.37	.34	.31	.27	.24	.20	.171	.136	.105	.064	.023	.043	.091
3800	.41	.38	.35	.31	.28	.25	.21	.175	.139	.108	.065	.023	.045
3900	.45	.42	.39	.36	.32	.29	.25	.22	.18	.142	.111	.066	.024
4000	.49	.46	.43	.40	.36	.33	.29	.26	.22	.184	.146	.114	.067
4100	.54	.51	.48	.45	.41	.37	.34	.30	.27	.23	.189	.149	.117
4200	.59	.56	.53	.49	.46	.42	.38	.34	.31	.27	.24	.193	.153
4300	.64	.61	.57	.54	.50	.47	.43	.39	.35	.32	.28	.24	.198
4400	.69	.66	.62	.59	.55	.52	.48	.44	.40	.36	.33	.28	.25
4600	.79	.76	.72	.69	.65	.62	.58	.54	.50	.46	.42	.38	.34
4800	.89	.86	.83	.79	.75	.72	.68	.65	.60	.56	.52	.48	.44
5000	1.00	.97	.94	.90	.87	.83	.79	.76	.72	.68	.63	.59	.55
5200	1.12	1.09	1.05	1.02	.98	.95	.91	.87	.83	.79	.75	.71	.66
5400	1.24	1.20	1.17	1.14	1.10	1.07	1.03	.99	.95	.91	.87	.82	.78
5600	1.36	1.33	1.30	1.26	1.23	1.19	1.15	1.11	1.08	1.03	.99	.95	.91
5800	1.49	1.46	1.43	1.39	1.36	1.31	1.28	1.24	1.20	1.16	1.12	1.08	1.03
6000	1.62	1.59	1.56	1.52	1.49	1.45	1.41	1.37	1.33	1.29	1.25	1.21	1.17

calculation may result in ((1) reduction of required fan horsepower; (2) calculation of correct pressure available at a branch take-off; (3) determination of correct friction losses in the branch and hence selection of the proper take-off fitting.

Table 1 permits direct reading of static regain, or transition loss, by relating downstream velocity to upstream velocity. The values in the table apply equally to reducer transitions with included angle between 45° and 60°. *Caution: Table 1 values do not apply to velocity changes that occur with a change in direction of the airstream, as at an elbow or a branch take-off.*

Primary Equipment Losses—To determine the overall static pressure drop for fan selection, it is essential to consider also the losses due to primary equipment, such as coils, filters, fresh-air intakes, transitions between fan and plenum, heat-exchange

Table 1 (concluded). Static Regain or Loss in Round Transition Pieces

Upstream Velocity V_U, fpm	Downstream Velocity V_D, fpm													
	3900	4000	4100	4200	4300	4400	4600	4800	5000	5200	5400	5600	5800	6000
	Static Loss (italics) or Regain, Inches wg													
2700	.54													
2800	.51	.56	.61	.67										
2900	.47	.52	.58	.63	.69									
3000	.44	.48	.54	.59	.65	.71								
3100	.40	.45	.50	.56	.61	.67	.79							
3200	.36	.41	.46	.52	.57	.63	.75	.87						
3300	.32	.37	.42	.48	.53	.59	.71	.83	.96					
3400	.28	.33	.38	.43	.49	.55	.67	.79	.91	1.05				
3500	.23	.28	.34	.39	.45	.50	.62	.74	.87	1.00				
3600	.188	.24	.29	.35	.40	.46	.58	.70	.83	.96	1.10			
3700	.143	.194	.25	.30	.36	.41	.53	.65	.78	.92	1.05			
3800	.094	.145	.20	.26	.31	.37	.49	.61	.74	.87	1.00			
3900	.048	.098	.153	.21	.26	.32	.44	.56	.69	.82	.96	1.10		
4000	.024	.05	.102	.16	.22	.27	.39	.51	.64	.78	.91	1.06		
4100	.068	.025	.053	.106	.162	.22	.34	.46	.59	.72	.86	1.00	1.15	
4200	.12	.068	.025	.055	.111	.167	.29	.41	.54	.66	.81	.95	1.10	
4300	.159	.123	.069	.026	.058	.114	.23	.35	.48	.62	.76	.90	1.05	
4400	.20	.164	.126	.07	.026	.061	.179	.30	.43	.56	.70	.84	.99	1.15
4600	.30	.25	.21	.174	.129	.071	.066	.188	.32	.45	.59	.73	.88	1.04
4800	.40	.36	.31	.27	.22	.18	.072	.072	.20	.33	.47	.62	.76	.92
5000	.50	.46	.41	.37	.32	.28	.186	.074	.078	.21	.35	.49	.64	.80
5200	.62	.57	.52	.48	.43	.39	.29	.192	.076	.084	.22	.37	.51	.67
5400	.73	.69	.64	.59	.54	.49	.40	.30	.20	.078	.091	.23	.38	.54
5600	.86	.81	.76	.72	.66	.61	.52	.42	.31	.21	.079	.098	.25	.40
5800	.99	.94	.89	.85	.79	.74	.64	.54	.43	.32	.22	.08	.106	.26
6000	1.12	1.07	1.02	.98	.92	.87	.77	.66	.55	.44	.34	.22	.081	1.15

apparatus, etc. Return duct losses must also be added unless a separate return air fan is provided. Table 6 gives losses or regain in commonly used primary equipment.

Branch Trunk Ducts—Generally, branch ducts are sized on the basis of an equal friction loss of 1.0 inch per 100 ft of run. Take-offs and other fittings are selected according to the static pressure available at take-off less branch friction and terminal pressure resistance. Thus, excess available pressure may be partly absorbed by choosing fittings with higher resistance. The first branch take-off, and frequently the last branch take-off, are considered fittings of the branch trunk duct.

Single Branch Lines—Generally, required air vol-ume flow rate and inlet size of the terminal unit also determine branch duct velocity and diameter. Again, fitting details are selected according to available static pressure at the take-off.

Dual Duct Systems—Dual duct systems are designed in the same manner as single duct systems, with the following two exceptions:

First, the high velocity units are selected on the basis of sound level ratings corresponding to an inlet static pressure of 2.5 inches of water. However, the minimum inlet pressures of the selected units are used for the purpose of duct sizing and fan horsepower determination.

Secondly, to improve stability throughout the system, branch ducts and fittings are designed for low static pressure resistance and excess available

static pressure is allowed to be absorbed at the inlet valves of the high velocity unit.

Design air quantities for cold ducts are usually 100% of the required cfm. Hot ducts are generally based on 50% to 75% of the cold duct cfm, depending on the expected load variations, building heat losses, and the required comfort conditions.

Economies in duct sizing can be realized by taking advantage of the probability that 100% of design air quantity will rarely be required even through cold air ducts. One method is to divide the system into four sections of equal design capacity. Then the cold air ducts of the section farthest from the fan are sized for 100% of design capacity; the preceding quarters are sized for 90%, 80% and 70%, decreasing towards the fan. The hot air ducts are then designed for 75% of the cold duct capacities.

Combined Duct Systems—Sometimes high velocity ducting ends in a multiple (octopus) or end-discharge unit from which rectangular or round ducts carry air to the supply outlets at conventional velocities. Static pressure losses in these downstream portions of the system are determined in the same manner, with certain modifications noted later in this article.

Rectangular Ducts—Use of rectangular ducting in high velocity systems is generally avoided because of its greater weight per volume of air carried, greater cost, lower rigidity, and leakage hazards. However, when space and structural reasons require the use of high-velocity rectangular ducts, their dimensions are determined either by the desired air velocity through the duct or by the equivalent diameter tables for round ducts (Table 5 in the preceding section on Air Flow in Ducts). The average air velocity in a rectangular duct can be found by dividing the CFM by the duct's cross-sectional area in square feet. Do not use the velocity for the equivalent round duct size found in Figs. 2a, 2b, 3a and 3b also in the preceding section, as it will obviously always be higher than the average air velocity through the equivalent rectangular duct for the same air flow.

When calculating *dynamic air losses* in rectangular duct fittings, always use the average air velocity, not that of the equivalent round duct, to determine the velocity pressure. On the other hand, with rectangular ducts, the equivalent round-duct sizes should be used to obtain friction losses from Figs. 2a, 2b, 3a, or 3b in the preceding section.

Pressure Losses in Branch Take-offs, Elbows, and Fittings in High-Velocity Rectangular Ductwork

As stated in "Good Aerodynamics for Duct Systems," above, some of the preferred elbow types for rectangular ductwork in high-velocity systems are venturi elbows, easy-bend radius elbows and radius elbows with elongated splitters. These are shown in Figs. 7, 6h, and 6k, in the preceding section.

Where ample space for installation is available, venturi elbows or radius elbows with very easy bends (aspect ratios of 3.0 and above, and radius ratios of 1.5 and above) are preferred. Venturi elbows do not require as much turning space as easy-bend radius elbows, and have an excellent loss factor; roughly 0.2 loss fraction of velocity pressure.* The losses for easy-bend elbows are given in Table 5.

Turning space frequently is unavailable for either venturi or easy-bend radius elbows, in which case square elbows with double-thickness turning vanes might be considered. However, these are *NOT* recommended as they tend to increase resistance and become noise generators in high-velocity systems. Instead, the designer should use radius elbows with splitters elongated to 2W as in Fig. 6k in the preceding section on Air Flow in Ducts, to be constructed according to the SMACNA High Velocity Duct Manual to preclude rattling and fluttering. While a minimum throat radius of 3 inches is desirable for a radius elbow with these splitters, the throat radius (R_1 in Fig. 6k) can be "squared off" in extreme cases where tight installation conditions prevail; hence these elbows can be installed in the same space which accommodates a square elbow with turning vanes. Dynamic loss fractions, given as a percent of velocity pressure, are tabulated in Table 8 of the preceding section on Air Flow in Ducts.

For 90° branch take-offs and elbows in rectangular high-velocity ducts, the designer is advised to use very easy bends where space conditions permit. The losses for these branch take-offs are found in Tables 4 and 5. The loss fractions for various other high-velocity duct fittings, such as transition pieces, are tabulated in Tables 6 and 9 of the preceding Air Flow Section.

Care should be taken to allow an absolute minimum of five equivalent diameters of straight duct run between successive 90° turns. If this is

---

* John L. Alden, *Design of Industrial Exhaust Systems,* 2nd ed., The Industrial Press, pp. 114-118.

Table 2A. Static Pressure Losses in 90° Branch Take-offs in Round Ducts

Branch Duct Velocity, V_B, fpm	90° Cylindrical	90° "Short Cone" Type Take-off					90° "Long Cone" Type Take-off						90° "Long" or "Short" Take-off		
		.8	.9	1.0	1.25	1.5	.8	.9	1.0	1.25	1.5	1.75	2.0	2.5	3.0
		Static Pressure Loss, Inches wg													
600	.034	.016	.017	.019	.02	.021	.007	.009	.011	.015	.018	.021	.023	.024	.025
800	.06	.028	.03	.033	.036	.038	.013	.016	.02	.027	.032	.037	.04	.042	.044
1000	.094	.043	.047	.052	.057	.059	.02	.025	.031	.042	.051	.058	.063	.066	.069
1200	.134	.062	.068	.075	.082	.086	.029	.036	.045	.06	.073	.084	.09	.095	.099
1400	.183	.084	.092	.101	.111	.116	.039	.049	.061	.082	.099	.113	.122	.128	.134
1600	.24	.11	.12	.133	.146	.152	.051	.065	.08	.107	.13	.149	.16	.168	.176
1800	.30	.139	.152	.168	.184	.192	.065	.082	.101	.135	.164	.188	.20	.21	.22
2000	.37	.172	.182	.21	.23	.24	.080	.101	.128	.167	.20	.23	.25	.26	.27
2100	.41	.19	.21	.23	.25	.26	.088	.111	.138	.184	.22	.26	.28	.29	.30
2200	.45	.21	.23	.25	.28	.29	.097	.122	.151	.20	.25	.28	.30	.32	.33
2300	.49	.23	.25	.27	.30	.31	.105	.133	.165	.22	.27	.31	.33	.35	.36
2400	.54	.25	.27	.30	.33	.34	.115	.145	.18	.24	.29	.33	.36	.38	.40
2500	.58	.27	.29	.32	.35	.37	.124	.158	.195	.26	.32	.36	.39	.41	.43
2600	.63	.29	.32	.35	.38	.40	.135	.171	.21	.28	.34	.39	.42	.44	.46
2700	.68	.31	.34	.38	.41	.43	.145	.184	.23	.30	.37	.42	.45	.47	.50
2800	.73	.34	.37	.41	.44	.46	.156	.198	.24	.33	.40	.45	.49	.51	.54
2900	.79	.36	.39	.44	.48	.50	.168	.21	.26	.35	.43	.49	.52	.55	.58
3000	.84	.39	.42	.47	.51	.53	.18	.23	.28	.38	.46	.52	.56	.59	.62
3100	.90	.41	.45	.50	.55	.57	.19	.24	.30	.40	.49	.56	.60	.63	.66
3200	.96	.44	.48	.53	.58	.61	.20	.26	.32	.43	.52	.59	.64	.67	.70
3300	1.02	.47	.51	.56	.62	.65	.22	.27	.34	.45	.55	.64	.68	.71	.74
3400	1.08	.50	.54	.60	.66	.69	.23	.29	.36	.48	.58	.67	.72	.76	.79
3500	1.15	.53	.57	.63	.70	.73	.24	.31	.38	.51	.62	.71	.76	.80	.84
3600	1.21	.56	.61	.67	.74	.77	.26	.33	.40	.54	.65	.75	.81	.85	.89
3700	1.28	.59	.64	.71	.76	.81	.27	.35	.42	.57	.69	.79	.85	.90	.94
3800	1.35	.62	.68	.75	.82	.86	.29	.37	.45	.60	.73	.84	.90	.95	.99
3900	1.42	.65	.71	.79	.86	.90	.30	.39	.48	.63	.77	.88	.95	1.00	1.04
4000	1.49	.68	.74	.82	.90	.95	.32	.41	.50	.66	.80	.92	.99	1.05	1.09
4100	1.57	.72	.78	.87	.95	1.00	.34	.43	.52	.70	.85	.97	1.05	1.10	1.15
4200	1.65	.76	.83	.91	1.00	1.05	.35	.45	.55	.73	.89	1.02	1.10	1.16	1.21
4300	1.73	.80	.87	.96	1.05	1.10	.37	.47	.58	.77	.93	1.07	1.15	1.21	1.27
4400	1.81	.84	.91	1.00	1.10	1.15	.39	.49	.60	.80	.98	1.12	1.21	1.27	1.33
4600	1.98	.91	.99	1.10	1.20	1.25	.42	.54	.66	.88	1.07	1.23	1.32	1.39	1.45
4800	2.15	.99	1.08	1.19	1.31	1.36	.46	.58	.72	.96	1.16	1.34	1.44	1.51	1.58
5000	2.34		1.17	1.29	1.42	1.48		.63	.78	1.04	1.26	1.45	1.56	1.64	1.71
5200	2.53		1.26	1.40	1.53	1.60		.68	.84	1.13	1.37	1.57	1.69	1.77	1.85
5400	2.72		1.36	1.50	1.65	1.72		.73	.91	1.21	1.47	1.69	1.81	1.90	1.99
5600	2.93			1.62	1.78	1.86			.98	1.31	1.58	1.82	1.95	2.05	2.15
5800	3.14			1.74	1.91	1.99			1.05	1.40	1.70	1.95	2.10	2.20	2.31
6000	3.37			1.86	2.04	2.13			1.12	1.50	1.82	2.09	2.24	2.36	2.47

Table 2B. Static Pressure Losses in Round Branch Take-offs at Different Angles

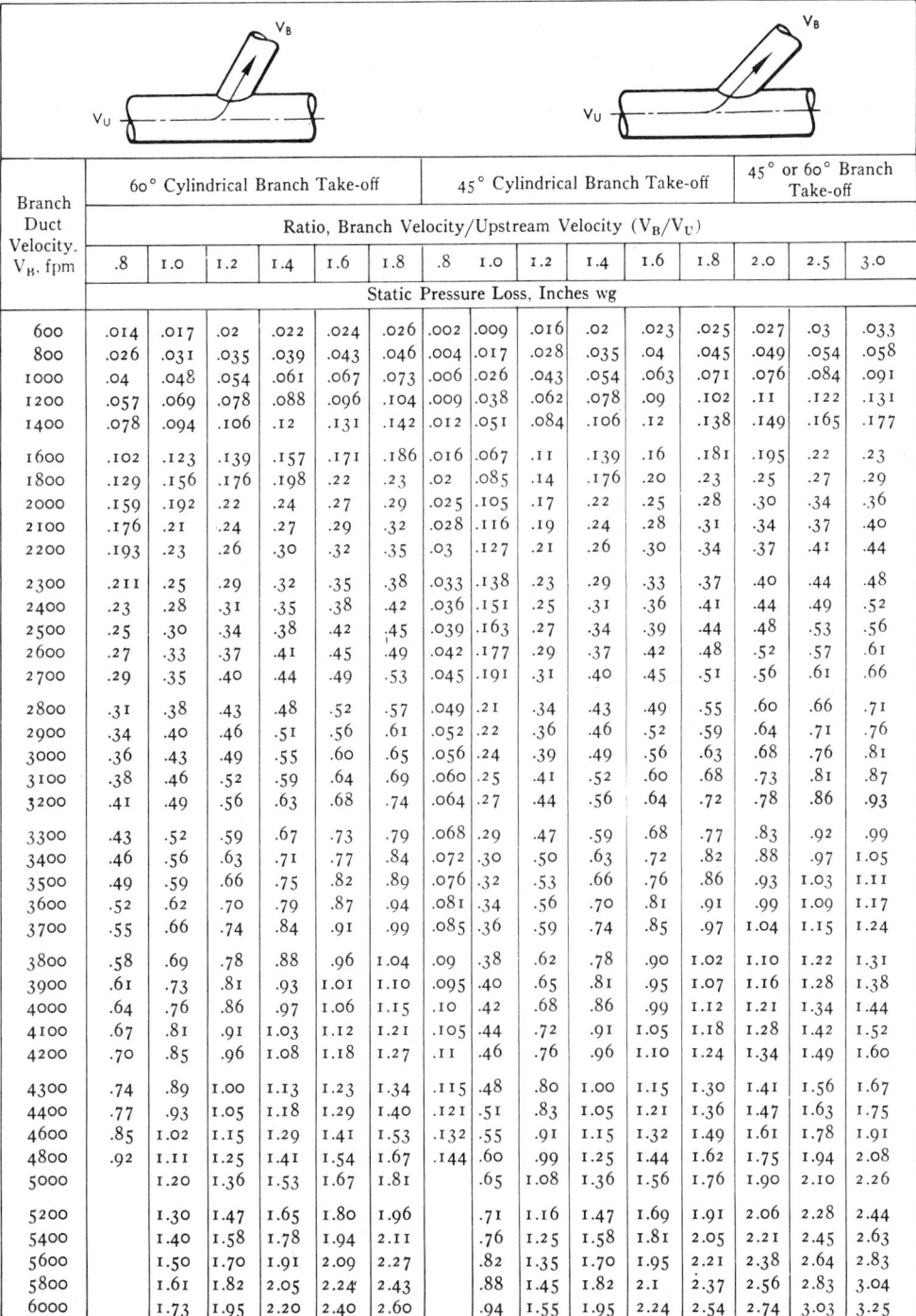

Branch Duct Velocity. V_B. fpm	60° Cylindrical Branch Take-off						45° Cylindrical Branch Take-off						45° or 60° Branch Take-off		
	Ratio, Branch Velocity/Upstream Velocity (V_B/V_U)														
	.8	1.0	1.2	1.4	1.6	1.8	.8	1.0	1.2	1.4	1.6	1.8	2.0	2.5	3.0
	Static Pressure Loss, Inches wg														
600	.014	.017	.02	.022	.024	.026	.002	.009	.016	.02	.023	.025	.027	.03	.033
800	.026	.031	.035	.039	.043	.046	.004	.017	.028	.035	.04	.045	.049	.054	.058
1000	.04	.048	.054	.061	.067	.073	.006	.026	.043	.054	.063	.071	.076	.084	.091
1200	.057	.069	.078	.088	.096	.104	.009	.038	.062	.078	.09	.102	.11	.122	.131
1400	.078	.094	.106	.12	.131	.142	.012	.051	.084	.106	.12	.138	.149	.165	.177
1600	.102	.123	.139	.157	.171	.186	.016	.067	.11	.139	.16	.181	.195	.22	.23
1800	.129	.156	.176	.198	.22	.23	.02	.085	.14	.176	.20	.23	.25	.27	.29
2000	.159	.192	.22	.24	.27	.29	.025	.105	.17	.22	.25	.28	.30	.34	.36
2100	.176	.21	.24	.27	.29	.32	.028	.116	.19	.24	.28	.31	.34	.37	.40
2200	.193	.23	.26	.30	.32	.35	.03	.127	.21	.26	.30	.34	.37	.41	.44
2300	.211	.25	.29	.32	.35	.38	.033	.138	.23	.29	.33	.37	.40	.44	.48
2400	.23	.28	.31	.35	.38	.42	.036	.151	.25	.31	.36	.41	.44	.49	.52
2500	.25	.30	.34	.38	.42	.45	.039	.163	.27	.34	.39	.44	.48	.53	.56
2600	.27	.33	.37	.41	.45	.49	.042	.177	.29	.37	.42	.48	.52	.57	.61
2700	.29	.35	.40	.44	.49	.53	.045	.191	.31	.40	.45	.51	.56	.61	.66
2800	.31	.38	.43	.48	.52	.57	.049	.21	.34	.43	.49	.55	.60	.66	.71
2900	.34	.40	.46	.51	.56	.61	.052	.22	.36	.46	.52	.59	.64	.71	.76
3000	.36	.43	.49	.55	.60	.65	.056	.24	.39	.49	.56	.63	.68	.76	.81
3100	.38	.46	.52	.59	.64	.69	.060	.25	.41	.52	.60	.68	.73	.81	.87
3200	.41	.49	.56	.63	.68	.74	.064	.27	.44	.56	.64	.72	.78	.86	.93
3300	.43	.52	.59	.67	.73	.79	.068	.29	.47	.59	.68	.77	.83	.92	.99
3400	.46	.56	.63	.71	.77	.84	.072	.30	.50	.63	.72	.82	.88	.97	1.05
3500	.49	.59	.66	.75	.82	.89	.076	.32	.53	.66	.76	.86	.93	1.03	1.11
3600	.52	.62	.70	.79	.87	.94	.081	.34	.56	.70	.81	.91	.99	1.09	1.17
3700	.55	.66	.74	.84	.91	.99	.085	.36	.59	.74	.85	.97	1.04	1.15	1.24
3800	.58	.69	.78	.88	.96	1.04	.09	.38	.62	.78	.90	1.02	1.10	1.22	1.31
3900	.61	.73	.81	.93	1.01	1.10	.095	.40	.65	.81	.95	1.07	1.16	1.28	1.38
4000	.64	.76	.86	.97	1.06	1.15	.10	.42	.68	.86	.99	1.12	1.21	1.34	1.44
4100	.67	.81	.91	1.03	1.12	1.21	.105	.44	.72	.91	1.05	1.18	1.28	1.42	1.52
4200	.70	.85	.96	1.08	1.18	1.27	.11	.46	.76	.96	1.10	1.24	1.34	1.49	1.60
4300	.74	.89	1.00	1.13	1.23	1.34	.115	.48	.80	1.00	1.15	1.30	1.41	1.56	1.67
4400	.77	.93	1.05	1.18	1.29	1.40	.121	.51	.83	1.05	1.21	1.36	1.47	1.63	1.75
4600	.85	1.02	1.15	1.29	1.41	1.53	.132	.55	.91	1.15	1.32	1.49	1.61	1.78	1.91
4800	.92	1.11	1.25	1.41	1.54	1.67	.144	.60	.99	1.25	1.44	1.62	1.75	1.94	2.08
5000		1.20	1.36	1.53	1.67	1.81		.65	1.08	1.36	1.56	1.76	1.90	2.10	2.26
5200		1.30	1.47	1.65	1.80	1.96		.71	1.16	1.47	1.69	1.91	2.06	2.28	2.44
5400		1.40	1.58	1.78	1.94	2.11		.76	1.25	1.58	1.81	2.05	2.21	2.45	2.63
5600		1.50	1.70	1.91	2.09	2.27		.82	1.35	1.70	1.95	2.21	2.38	2.64	2.83
5800		1.61	1.82	2.05	2.24	2.43		.88	1.45	1.82	2.1	2.37	2.56	2.83	3.04
6000		1.73	1.95	2.20	2.40	2.60		.94	1.55	1.95	2.24	2.54	2.74	3.03	3.25

Table 3. Static Pressure Losses in 90° Elbows in Round Ducts

 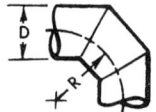 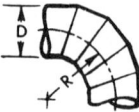 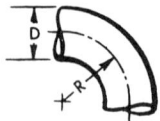

Duct Velocity fpm	Miter Elbows	3 pc. Elbows				5 pc. & Smooth Elbows				
		Ratio, Bending Radius/Diameter (R/D)								
		.5	.75	1.0	1.5-3.0	.5	.75	1.0	1.5	2.0
		Static Pressure Loss, Inches wg								
600	.029	.022	.012	.009	.007	.02	.01	.007	.005	.004
800	.052	.038	.021	.016	.013	.036	.018	.013	.01	.008
1000	.081	.06	.033	.025	.021	.056	.028	.021	.015	.012
1200	.117	.086	.048	.036	.03	.081	.041	.03	.022	.017
1400	.159	.117	.065	.049	.04	.11	.055	.04	.029	.023
1600	.21	.154	.085	.064	.053	.144	.072	.053	.038	.03
1800	.26	.194	.107	.081	.067	.182	.091	.067	.048	.038
2000	.32	.24	.132	.10	.082	.22	.112	.082	.06	.047
2100	.36	.26	.146	.11	.091	.25	.124	.091	.066	.052
2200	.39	.29	.16	.121	.10	.27	.136	.10	.072	.057
2300	.43	.32	.174	.132	.109	.30	.148	.109	.079	.063
2400	.47	.34	.19	.144	.118	.32	.162	.118	.086	.068
2500	.51	.37	.21	.156	.128	.35	.175	.128	.093	.074
2600	.55	.41	.22	.169	.139	.38	.19	.139	.101	.08
2700	.59	.44	.24	.182	.15	.41	.20	.15	.109	.086
2800	.63	.47	.26	.195	.161	.44	.22	.161	.117	.093
2900	.68	.50	.28	.21	.173	.47	.24	.173	.126	.10
3000	.73	.54	.30	.22	.185	.51	.25	.185	.135	.107
3100	.78	.58	.32	.24	.198	.54	.27	.198	.144	.114
3200	.83	.61	.34	.26	.21	.58	.29	.21	.153	.121
3300	.88	.64	.36	.27	.22	.61	.31	.22	.163	.129
3400	.94	.69	.38	.29	.24	.65	.32	.24	.173	.137
3500	.99	.73	.40	.31	.25	.69	.34	.25	.183	.145
3600	1.05	.78	.43	.32	.27	.73	.36	.27	.194	.154
3700	1.11	.82	.45	.34	.28	.77	.38	.28	.20	.162
3800	1.17	.86	.48	.36	.30	.81	.41	.30	.22	.171
3900	1.23	.91	.50	.38	.31	.85	.43	.31	.23	.18
4000	1.30	.96	.53	.40	.33	.90	.45	.33	.24	.189
4100	1.36	1.01	.56	.42	.35	.94	.47	.35	.25	.20
4200	1.43	1.06	.58	.44	.36	.99	.50	.36	.26	.21
4300	1.50	1.11	.61	.46	.38	1.04	.52	.38	.28	.22
4400	1.57	1.16	.64	.48	.40	1.09	.54	.40	.29	.23
4600	1.72	1.27	.70	.53	.44	1.19	.59	.44	.32	.25
4800	1.87	1.38	.76	.57	.47	1.29	.65	.47	.35	.27
5000	2.03	1.50	.83	.62	.51	1.40	.70	.51	.38	.30
5200	2.19	1.62	.89	.67	.56	1.52	.76	.56	.41	.32
5400	2.36	1.74	.96	.73	.60	1.63	.82	.60	.44	.34
5600	2.54	1.88	1.04	.78	.65	1.76	.88	.65	.47	.37
5800	2.72	2.01	1.11	.84	.69	1.89	.94	.69	.50	.40
6000	2.92	2.15	1.19	.90	.74	2.02	1.01	.74	.54	.43

Example: Determine static pressure drop of 12-in. dia, 90-deg, 5-piece elbow where R = 18 in. It handles 1570 cfm.
Solution: Velocity (cfm ÷ duct area) = 2000 fpm. R/ = 18/12 = 1.5. Static pressure drop = 0.06 inches wg.
Note: For elbows other than 90 deg, multiply static pressure loss for 90-deg elbow by the angle divided by 90.

Table 4. Static Pressure Losses in 90° Branch Take-offs in Rectangular Ducts

Branch Velocity, V_B, fpm	Aspect Ratio: H/W									
	.25—.50		.75—3.0			4—6				
	Velocity Ratio: V_B/V_U									
	1.2	1.3	1.1	1.2	1.3	.9	1.0	1.1	1.2	1.3
	Area Ratio: A_B/A_U									
	.7—1.2		.4—.6			.15—.3				
	Static Pressure Loss, Inches wg									
600	.011	.013	.007	.011	.018	.008	.011	.015	.018	.02
800	.02	.024	.013	.02	.032	.015	.02	.027	.032	.036
1000	.032	.038	.020	.032	.051	.023	.031	.042	.05	.057
1200	.046	.055	.028	.046	.073	.033	.045	.061	.073	.082
1400	.062	.074	.038	.062	.099	.045	.061	.082	.098	.111
1600	.081	.097	.05	.081	.129	.059	.08	.108	.129	.145
1800	.102	.123	.063	.102	.163	.074	.101	.136	.163	.183
2000	.126	.151	.078	.126	.20	.091	.125	.168	.20	.23
2100	.139	.167	.086	.139	.22	.101	.138	.185	.22	.25
2200	.153	.184	.095	.153	.24	.111	.151	.20	.24	.27
2300	.166	.20	.104	.166	.27	.121	.165	.22	.27	.30
2400	.182	.22	.113	.182	.29	.131	.18	.24	.29	.33
2500	.197	.24	.122	.197	.31	.142	.195	.26	.31	.35
2600	.21	.26	.133	.21	.34	.154	.21	.28	.34	.38
2700	.23	.28	.143	.23	.37	.166	.23	.31	.37	.41
2800	.25	.30	.153	.25	.39	.179	.24	.33	.39	.44
2900	.27	.32	.165	.27	.42	.192	.26	.35	.42	.48
3000	.28	.34	.176	.28	.45	.21	.28	.38	.45	.51
3100	.30	.36	.188	.30	.48	.22	.30	.40	.48	.54
3200	.32	.39	.20	.32	.52	.23	.32	.43	.52	.58
3300	.34	.41	.21	.34	.55	.25	.34	.46	.55	.62
3400	.36	.44	.23	.36	.58	.26	.36	.49	.58	.65
3500	.39	.46	.24	.39	.62	.28	.38	.51	.62	.69
3600	.41	.49	.25	.41	.65	.30	.40	.54	.65	.73
3700	.43	.52	.27	.43	.69	.31	.43	.58	.69	.78
3800	.46	.55	.28	.46	.73	.33	.45	.61	.73	.82
3900	.48	.58	.30	.48	.77	.35	.47	.64	.77	.86
4000	.51	.61	.31	.51	.81	.36	.50	.67	.81	.91
4100	.53	.64	.33	.53	.85	.38	.52	.71	.85	.95
4200	.56	.67	.35	.56	.89	.40	.55	.74	.89	1.00
4300	.58	.70	.36	.58	.93	.42	.58	.78	.93	1.05
4400	.61	.73	.38	.61	.97	.44	.60	.81	.97	1.10
4600	.67	.80	.41	.67	1.07	.48	.66	.89	1.06	1.20
4800	.73	.87	.45	.73	1.16	.53	.72	.97	1.16	1.30
5000	.79	.95	.49	.79	1.26	.57	.78	1.05	1.26	1.41
5200	.85	1.03	.53	.85	1.36	.62	.84	1.14	1.36	1.53
5400	.92	1.11	.57	.92	1.47	.67	.91	1.22	1.46	1.65
5600	.99	1.19	.61	.99	1.58	.72	.98	1.32	1.57	1.77
5800	1.06	1.27	.66	1.06	1.69	.77	1.05	1.41	1.69	1.90
6000	1.14	1.36	.70	1.14	1.81	.82	1.12	1.51	1.81	2.04

Example: Upstream duct, 24 by 12 inches, handling 5000 cfm; branch duct, 12 by 12, handling 3000 cfm; $A_B/A_U = 0.5$; $H/W = 1.0$; $V_B/V_U = 3000/2500 = 1.2$. Static pressure drop for 3000-fpm branch velocity = 0.28 inches wg.

Note: Radius ratio (R/W) should equal or exceed unity. For branch take-offs with lower velocity ratios (V_B/V_U) uses values for elbows of comparable aspect ratio (H/W) and radius ratio of 1.0. See Table 5.

Table 5. Static Pressure Losses of 90° Elbows in Rectangular Ducts

Hard Bend

Easy Bend

Duct Ve-locity fpm	Aspect Ratio: H/W															
	.25				.50				.75-3.0				4-6			
	Radius Ratio: R/W															
	.75	1.0	1.5	2-3	.75	1.0	1.5	2-3	.75	1.0	1.5	2-3	.75	1.0	1.5	2-3
	Static Pressure Loss, Inches wg															
600	.012	.010	.006	.005	.01	.006	.004	.003	.008	.005	.003	.002	.007	.004	.003	.002
800	.022	.018	.011	.009	.018	.011	.007	.006	.014	.008	.005	.004	.013	.007	.005	.004
1000	.034	.028	.017	.015	.028	.017	.011	.009	.023	.013	.008	.007	.020	.011	.007	.006
1200	.049	.04	.025	.021	.04	.025	.015	.013	.032	.019	.012	.01	.029	.017	.01	.009
1400	.066	.054	.033	.028	.054	.034	.021	.018	.044	.026	.016	.013	.039	.023	.014	.012
1600	.086	.07	.044	.037	.07	.044	.027	.023	.058	.034	.021	.018	.051	.030	.018	.016
1800	.109	.089	.055	.047	.089	.056	.035	.029	.072	.042	.026	.022	.064	.037	.023	.02
2000	.13	.11	.068	.058	.11	.069	.043	.036	.09	.052	.032	.027	.079	.046	.028	.024
2100	.149	.121	.075	.064	.121	.076	.047	.04	.099	.058	.036	.030	.087	.051	.031	.026
2200	.163	.133	.083	.07	.133	.084	.052	.044	.109	.063	.039	.033	.096	.056	.034	.029
2300	.178	.145	.09	.076	.145	.091	.056	.048	.118	.069	.043	.036	.104	.061	.038	.032
2400	.194	.158	.098	.083	.158	.10	.061	.052	.129	.075	.047	.039	.114	.066	.041	.035
2500	.21	.17	.107	.09	.17	.108	.067	.056	.14	.082	.051	.043	.123	.072	.044	.038
2600	.23	.186	.116	.098	.186	.114	.072	.061	.152	.089	.055	.046	.134	.078	.048	.041
2700	.25	.20	.124	.105	.20	.126	.078	.066	.163	.095	.059	.050	.144	.084	.052	.044
2800	.26	.21	.134	.113	.21	.136	.083	.071	.176	.102	.063	.054	.155	.09	.056	.047
2900	.28	.23	.144	.122	.23	.146	.090	.076	.189	.11	.068	.058	.166	.097	.060	.051
3000	.30	.25	.154	.13	.25	.156	.096	.081	.20	.118	.073	.062	.178	.104	.064	.054
3100	.32	.26	.164	.139	.26	.167	.102	.087	.22	.126	.078	.066	.19	.111	.068	.058
3200	.35	.28	.175	.148	.28	.178	.109	.093	.23	.134	.083	.07	.20	.118	.073	.062
3300	.37	.30	.186	.158	.30	.189	.116	.098	.24	.143	.088	.075	.22	.126	.077	.066
3400	.39	.32	.198	.167	.32	.20	.123	.105	.26	.151	.094	.079	.23	.133	.082	.07
3500	.41	.34	.21	.177	.34	.21	.131	.111	.28	.16	.099	.084	.24	.141	.087	.074
3600	.44	.36	.22	.187	.36	.22	.138	.117	.29	.17	.105	.089	.26	.149	.092	.078
3700	.46	.38	.23	.198	.38	.24	.146	.124	.31	.179	.111	.094	.27	.158	.097	.083
3800	.49	.40	.25	.21	.40	.25	.154	.131	.32	.189	.117	.099	.29	.167	.103	.087
3900	.51	.42	.26	.22	.42	.26	.162	.138	.34	.199	.123	.104	.30	.176	.108	.092
4000	.54	.44	.27	.23	.44	.28	.17	.144	.36	.21	.129	.109	.32	.184	.113	.097
4100	.57	.46	.29	.24	.46	.29	.179	.152	.38	.22	.136	.115	.33	.194	.12	.102
4200	.59	.48	.30	.26	.48	.31	.188	.16	.40	.23	.143	.121	.35	.20	.125	.107
4300	.62	.51	.32	.27	.51	.32	.197	.167	.42	.24	.15	.127	.37	.21	.131	.112
4400	.65	.53	.33	.28	.53	.34	.21	.175	.43	.25	.157	.133	.38	.22	.138	.117
4600	.71	.58	.36	.31	.58	.37	.23	.191	.48	.28	.172	.145	.42	.24	.15	.128
4800	.78	.63	.39	.33	.63	.40	.25	.21	.52	.30	.187	.158	.46	.27	.164	.139
5000	.84	.69	.43	.36	.69	.43	.27	.23	.56	.33	.20	.171	.49	.29	.178	.151
5200	.91	.74	.46	.39	.74	.47	.29	.24	.61	.35	.22	.185	.53	.31	.192	.164
5400	.98	.80	.50	.42	.80	.51	.31	.26	.65	.38	.24	.20	.58	.34	.21	.176
5600	1.06	.86	.54	.45	.86	.54	.33	.28	.70	.41	.25	.21	.62	.36	.22	.19
5800	1.13	.92	.57	.49	.92	.58	.36	.30	.75	.44	.27	.23	.66	.39	.24	.20
6000	1.21	.99	.61	.52	.99	.62	.38	.33	.81	.47	.29	.25	.71	.42	.26	.22

Table 5. (concluded). Static Pressure Losses of 90° Elbows in Rectangular Ducts

Example: An elbow in a 24 by 8 inch duct with 12-inch center radius handles 2800 cfm. Velocity is 2400 (previously calculated). H/W = 24/8 = 3.0; R/W = 12/8 = 1.5; and static pressure drop = 0.047 inches wg.
Note: Rectangular elbows with radius ratio below 0.5 not recommended for high velocity work. Aspect ratios below 0.25 and above 6.0 should be avoided.
Elbows other than 90 deg cause a drop in static pressure equal to tabular value times the angle divided by 90.
Static pressure losses of rectangular elbows with low radius ratio can be reduced by use of turning vanes. Correction factors are given in table below. Values not shown indicate no static pressure drop reduction due to vanes.

R/W	Number of Turning Vanes		
	1	2	3
	Correction Factor		
.75	.43	.32	.25
1.0	.56	.46	—
1.5	.85	—	—

Table 6. Static Pressure Loss and Regain, Fan-Plenum-Duct Transitions, Inches WG

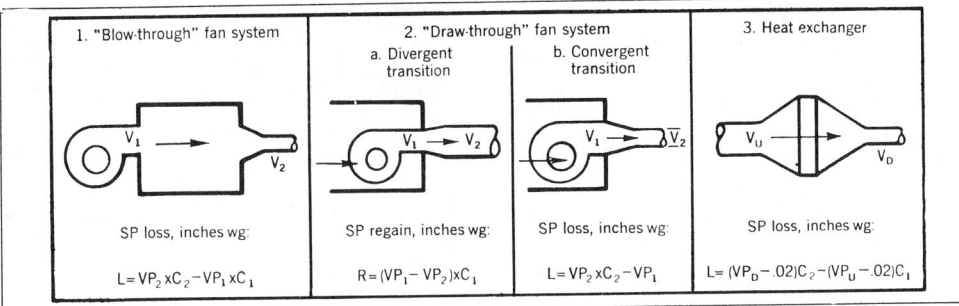

1. "Blow-through" fan system	2. "Draw-through" fan system		3. Heat exchanger
	a. Divergent transition	b. Convergent transition	
SP loss, inches wg:	SP regain, inches wg:	SP loss, inches wg:	SP loss, inches wg:
$L = VP_2 \times C_2 - VP_1 \times C_1$	$R = (VP_1 - VP_2) \times C_1$	$L = VP_2 \times C_2 - VP_1$	$L = (VP_D - .02)C_2 - (VP_U - .02)C_1$

Velocity V, fpm—Velocity Pressure VP, inches wg

"C" Factor Table

V	VP	V	VP	V	VP	V	VP	Transition	C_1	C_2
600	.023	2300	.33	3300	.68	4300	1.15	Abrupt enlargement	.18	—
800	.04	2400	.36	3400	.72	4400	1.21	Abrupt contraction	—	1.47
1000	.062	2500	.39	3500	.77	4600	1.32			
1200	.09	2600	.42	3600	.81	4800	1.44	Gradual enlargement 30°	.44	—
1400	.122	2700	.46	3700	.85	5000	1.56	(Divergent transition), 45°	.33	—
1600	.16	2800	.49	3800	.90	5200	1.69	Included Angle: 60°	.26	—
1800	.20	2900	.53	3900	.95	5400	1.82	Gradual contraction		
2000	.25	3000	.56	4000	1.00	5600	1.96	(Convergent transition)		
2100	.28	3100	.60	4100	1.05	5800	2.10	30°—60° Incl. Angle	—	1.07
2200	.30	3200	.64	4200	1.10	6000	2.25	"Bellmouth"	—	1.05

Examples

I   "Blow-Through" Fan
Fan outlet velocity, V_1 = 3000 fpm; main duct velocity, V_2 = 4400 fpm. Both transitions (fan-plenum and plenum duct) of 60-deg included angle. VP_1 = .56, VP_2 = 1.21, C_1 = .26, C_2 = 1.07; static pressure loss = 1.21 × 1.07—.56 × .26 = 1.15 inches wg.

II-a   "Draw-Through" Fan
V_1 = 5000 fpm, V_2 = 4200 fpm, 45-deg divergent transition (fan-duct). VP_1 = 1.56, VP_2 = 1.10, C_1 = .33; regain (1.56—1.1) × .33 = .152 inch wg.

II-b   "Draw-Through" Fan
V_1 = 3000 fpm, V_2 = 4200 fpm, 45 deg convergent transition (fan-duct). VP_1 = .56 VP_2 = 1.10, C_2 = 1.07; static pressure loss: 1.1 × 1.07—.56 = .617 inch wg.

III   Reheater Coil Section
V_U = 3800 fpm, V_D = 4600 fpm. Both transitions at 45-deg included angles. VP_U = .9, VP_D = 1.32, C_1 = .33; C_2 = 1.07; static pressure loss: (1.32—.02) × 1.07 — (.9—.02) × .33 = 1.1 inch wg.

Note: Static pressure losses due to conventional return duct systems are calculated in the customary manner. Static pressure losses due to equipment, such as filters, coils, etc are determined according to manufacturer's literature. To determine fan motor hp all pressure losses must be considered.

not possible, use either 45° elbows or radius elbows with elongated splitters. A good arrangement for close distances between turns is shown in Fig. 1.*

For duct main shaft take-offs, conical fittings (for circular duct), and prism shaped shoe fittings (for rectangular duct), should be used; 90° duct main or shaft take-offs should not be tolerated for larger sizes (say, over 8 inches diameter) and high velocities (say, over 2,000 fpm). All efforts, where large duct panels are involved, at shaft take-off and main duct branch take-offs, should be made to reduce the velocity even further. See Fig. 2.*

Where shaft take-off conditions do not permit adequate space conditions to provide conical or prism type taps, it is recommended that 10-gage duct be installed to permit the fire damper to be located further downstream of the tap within the 10-gage duct. The shaft wall-thickness may then be utilized to permit the installation of conical or prism type tap and thereby provide better aerodynamic take-off and approach conditions to the fire damper. See Fig. 3.

When possible, the fire damper should be constructed with blades out of the take-off air stream (shutter type). Under no circumstances should the blades be permitted to extend into the shaft air stream.

Pressure Losses for Divided Flow Fittings in Round, High-Velocity Ductwork

While Tables 2A and 2B, above, give tabulated losses for several types of round, divided flow fittings, the *ASHRAE Handbook of Fundamentals* (1972 edition) gives the loss data for several additional shapes of these fittings. These are round fittings which are commercially available, manufactured by large sheet steel factories. Local sheet metal shops can make these fittings, but the labor involved in their fabrication will usually make their pricing prohibitive. It is advisable to check with the local sheet metal union whose contract may expressly prohibit the use of factory-made fittings.

A modified type of round, high-velocity duct is that with an oval cross-section, generally called "flat oval." This type of duct is very satisfactory for high-velocity systems where limited height is available for installation. However, this duct-work should be factory-fabricated, not shop-fabricated. Only factory-fabricated fittings are recommended for "flat oval" ducts.

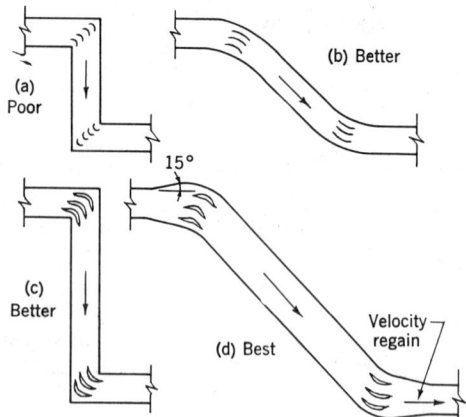

Fig. 1. Air stream turns between successive fittings.

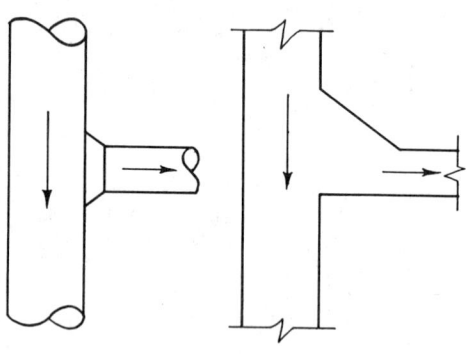

Round duct   Rectangular duct

Fig. 2. Tap-off from riser.

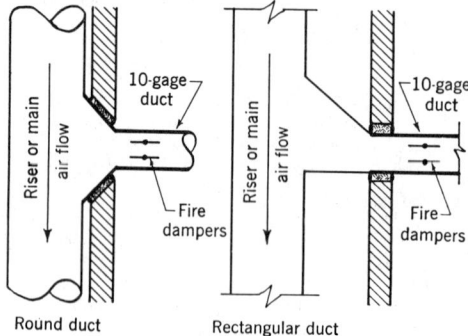

Round duct   Rectangular duct

Fig. 3. Where shaft take-off space is inadequate, use extra-heavy duct.*

---

* V. V. Cerami and N. S. Shataloff, "Quiet High-Velocity Air Distribution," *Air Conditioning, Heating and Ventilating*, August, 1964.

Losses in Tap-Off Fittings in High-Velocity Rectangular-Duct Systems

The pressure losses in the tap-off fittings shown in the charts in Fig. 4 are experimentally determined for the following set of conditions:

1. The fittings are constructed and installed as shown later, in Fig. 3, *"Duct Construction."*
2. The ratio of main duct area to branch duct area was greater than four.
3. The mains were of rectangular construction.

4. The mains on the upstream side of the tap-off fittings had approximately 8 diameters of straight pipe.

Only static pressure losses through the tap-off fittings are shown in the charts. When the branch ducts from the tap-off connections feed high-pressure dual-duct or single-duct units, the velocity pressure in the branch duct cannot be credited with regain. This velocity pressure must be considered as a part of the total energy required to create the needed rate-of-flow through

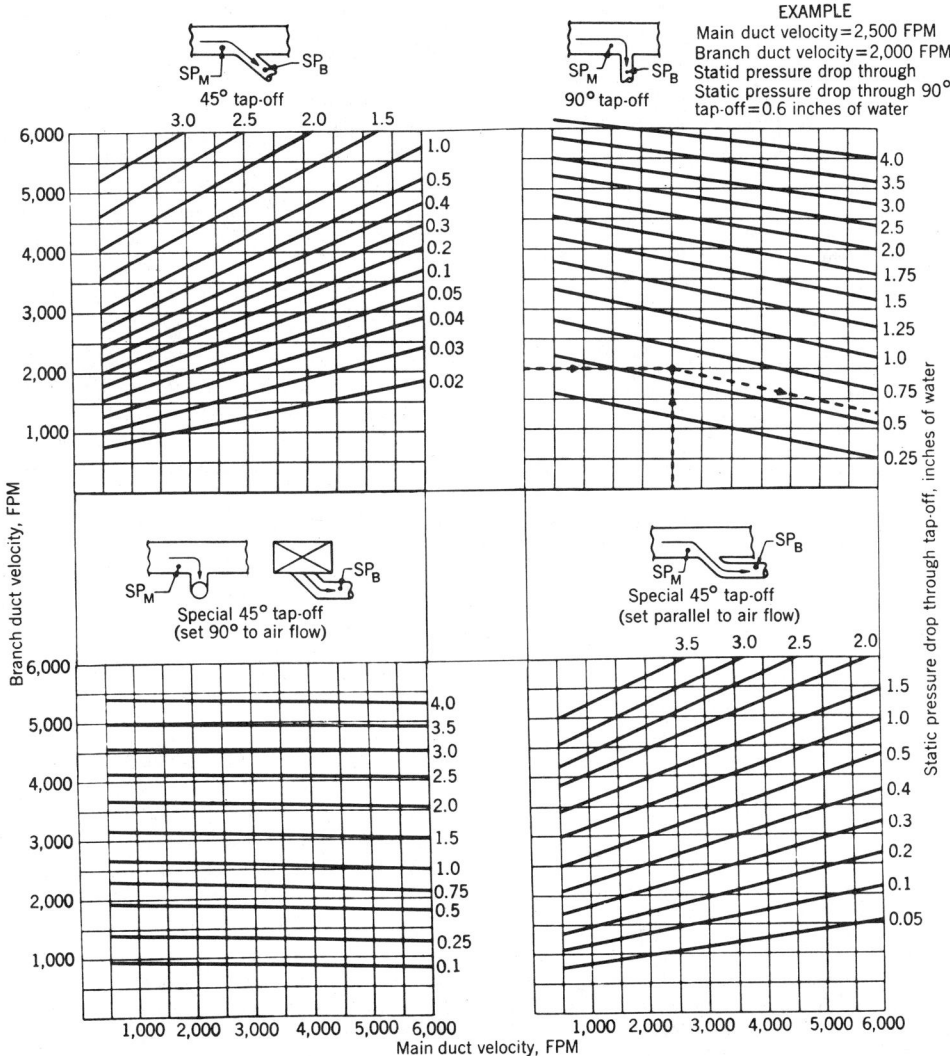

Fig. 4. Static pressure losses through four types of tap-off connections in high-velocity duct systems

the unit. Rating tables for the high-pressure units are normally based on the static pressures at the entrance of the units, which means that velocity pressure should be added to these values to establish the total pressure loss through the unit.

If it is desired to establish the true loss through the tap-off fittings, the velocity pressure in the main duct should be added to, and the velocity pressure in the branch duct should be deducted from, the values given in the illustrations. In these figures, SP_M is the static pressure in the main duct and SP_B, static pressure in the branch duct.

One should keep in mind that the losses indicated in the chart apply only to well-constructed fittings as specified. Fittings with raw edges at the main duct will have losses much greater than those indicated in the chart and will have different acoustical characteristics at higher air velocities. For proper construction of these fittings, see "Construction of Tap-off Fittings for High-Velocity Ductwork," on page 5-79.

Flexible Round Ducting for Connections between High-Velocity Ductwork and Terminal Units

Flexible duct is available in aluminum, plastic, and steel and can stand pressures within the usual ranges of high-velocity duct systems. Manufacturers of flexible duct should be consulted as to the suitability of the material and pressure selected for a particular application; flexible metal ducts, for example, may not be sufficiently airtight to meet high-velocity duct specifications. The prevailing practice has been to use flexible ducts to connect high-velocity ductwork to the terminal units, such as dual-duct mixing boxes, in diameters up to 8 inches and in lengths up to 12 feet.

Operating temperature limitations for metallic ducts are:

Aluminum	350°F
Steel	650°F
Stainless Steel	1400°F

Single-ply, aluminum flexible duct is good for a maximum working pressure 3 in. wg, whereas double-ply aluminum can be used for negative or positive pressure as high as 15 to 20 in. wg.

Steel flexible duct is usually galvanized and is stronger than aluminum for the same ply. Single-ply steel is made for negative or positive working pressures as high as 15 in. wg.

It is recommended that flexible duct be U.L. approved for fire safety; this may be required by some codes and should be checked.

In some areas, the sheet metal trade unions have agreements which limit the length of flexible ducting that can be used in terminal connections. For metallic, flexible, round-duct friction losses, consult manufacturer's literature, which will generally be very reliable.

STEP-BY-STEP DESIGN EXAMPLE

A 14-step method for design of high-velocity systems is demonstrated by its application to a hypothetical 3-story office building. Schematic layout of the system is shown in Fig. 5. The work sheet is Fig. 6. Air requirements and duct layout for each of the three floors are assumed to be identical.

Main Duct

Step 1: Select critical high velocity unit and duct run. Consider capacity and pressure loss characteristics of units and fittings, as well as duct length. Include branch take-off before critical unit.

Referring to the schematic layout, Fig. 5, note that Unit A is farthest from riser, but has low minimum static pressure, only one elbow fitting, and may possibly benefit from static regain at upstream transition fittings. Unit B has shorter duct run but higher minimum SP, both an elbow and a take-off fitting, and less potential benefit from static regain. Unit B is, therefore, selected as the critical unit. For similar reasons, the ducting serving the second floor is selected as part of the main duct run.

Step 2: Starting at the critical high velocity unit and working upstream toward the fan, identify and enter in the work sheet, Fig. 6, Columns 1, 2, and 5, the main duct fittings, their cfm and section lengths. (Refer to sections 1 to 5 in the layout, Fig. 5.)

Step 3: Starting with the duct section handling the most cfm, determine duct velocities, diameters, and SP per 100 ft (Columns 3, 4, and 6) by means of a standard air duct friction chart (Figs. 2a, 2b, 3a and 3c in the preceding section on Air Flow in Ducts). For example:

Main duct, Section 5 .... 7650 cfm;
Initial velocity (Column 3) .... 4400 fpm;
Duct diameter (Column 4) .... 18 inches;
SP per 100 ft (Column 6) .... 1.3 in. wg.

Step 4: Determine preliminary SP losses (Col 8).

(a) Refer to manufacturers' literature for high velocity units' minimum SP resistance.

(b) Multiply duct length by friction SP loss per 100 ft and divide by 100. For example, Section 5:

Length (Column 5) .... 16 ft;
SP per 100 ft (Column 6) .... 1.3 in. wg;
SP loss, (Column 8) .... 0.208 in. wg.

(c) Refer to Table 3 for elbow losses. For example, Duct Section 1:
Velocity, (Column 3) .... 2800 fpm;
90°, 5-piece elbow, R/D = 1.5,
   SP loss (Column 8) .... 0.117 in. wg.

(d) Refer to Table 2-A or 2-B for branch take-off losses. For example:

Duct Section 1, branch velocity,
  V_B .... 2800 fpm;
Duct Section 2, upstream velocity,
  V_U .... 2500 fpm;
Velocity ratio, V_B/V_U (Column 7) 1.12;
Short cone take-off loss (Column
  8) .... 0.425 in. wg.

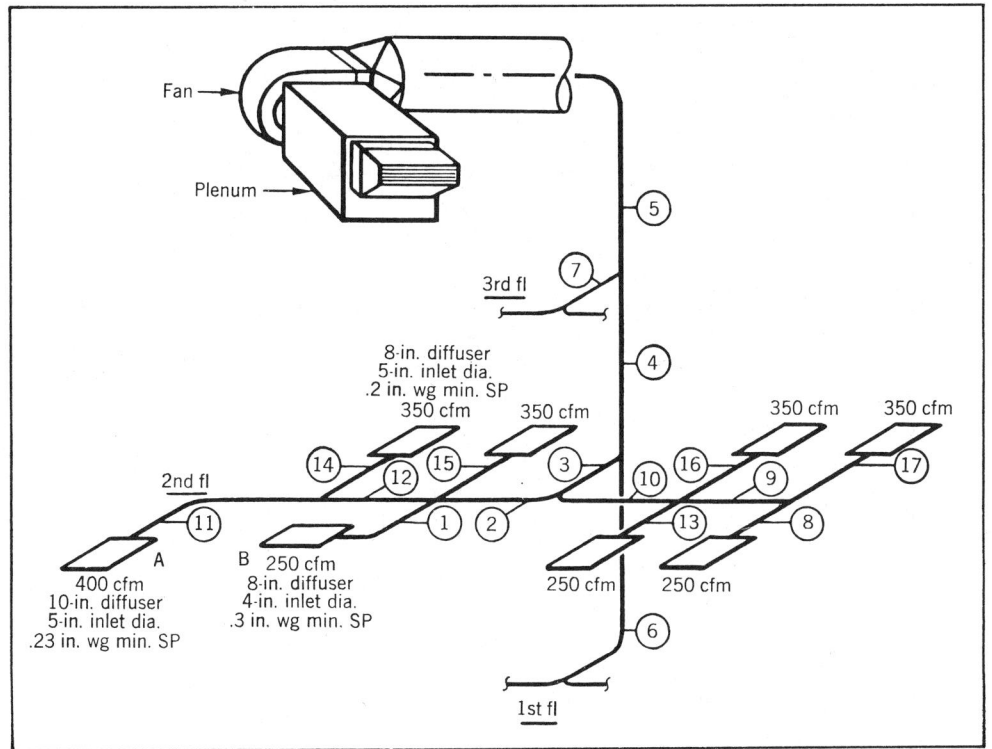

Section	Cfm	Length, Ft	Section	Cfm	Length, Ft	Section	Cfm	Length, Ft
1	250	9	6	2550	18	11	400	12
2	1350	10	7	2550	6	12	750	10
3	2550	6	8	250	6	13	250	6
4	5100	12	9	600	10	14-15	350	6
5	7650	16	10	1200	10	16-17	350	6

Fig. 5. Layout and specifications chart for duct system in a 3-story office building.

DUCT DESIGN WORK SHEET

1	2	3	4	5	6	7	8	9	10	11	12	13	14	15
DUCT SECTION OR FITTING DETAIL	CFM	FPM	DIA. IN.	SECT. LGTH. FT.	SP 100 FT. "WG	Vs/Vu	STATIC PRESSURE, "WG					BRANCH-OFF		REMARKS
							PRELIM.	VEL. CONVERS'N LOSS	REGAIN	SECT'L.	CUMUL.	AT	AVAIL. SP	
8" Diffuser,	250						.3							Mfr's Ratgs
1	250	2800	4	9	3.5		.315	-*		} 1.157	1.157			
90°L,R=1.5D, 5 Pc.		2800	4				.117							Table III
90°T Short Cone	250	2800				1.12	.425							Table II-A
2	1350	2500	10	10	.9		.09	-*						
45°L,R=1.5D, 3Pc."		2500	10				.047			} .176	1.333			
45° T	1350	2500				.80	.039							Table II-B
3	2550	3200	12	6	1.2		.072	-*		} .312	1.645			
90°T, Long Cone		2550	3200				.86	.24						
4	5100	3700	16	12	1.1		.132	.28		} .148	1.497			
5	7650	4400	18	16	1.3		.208	-	-	} .498	1.995			
90°L,R=1.5D, 5Pc."		4400	18				.29							
Draw-Thru Fan System, 30° Incl. Angle Transition,									.154					Table
V₁=5000 fpm; V₂=4400 fpm										} .496	2.491			Mfr's Ratgs
Primary Equipment & Air Intake							.65							& Sep.Calc.
Static Pressure Recheck of Other Branch Trunk Ducts:														
1st Floor Equal to Main Duct Sections 1,2										1.333		3	1.645	(Longest duct run)
6	2550	3200	12	18	1.2		.216	.171		} .198	1.531			OK
90°L,R=1.5D, 5Pc."		3200	12				.153							
3rd Floor Equal to Main Duct Sections 1,2										1.333		4	1.497	(Lowest available take-off)
7	2550	3200	12	6	1.2		.072	-*		} .072	1.405			
90°T, Long Cone		2550	3200				.725	-						OK
Select fan on basis of 2.491, say 2.5" w.g. overall S.P. resistance.														
8" Diffuser,	250						.3			}		2	1.333	
8	250	2800	4	6	3.5		.21	-*		.95	.95			
(90°)T (Short Cone)	250	2800				1.27	(.44)							
9	600	2200	7	10	1.2		.12			} .124	.004	.946		
10	1200	2700	9	10	1.2		.12	-*						
(45°)L,R=1.5D, 3Pc."		2700	9				(.055)			} .26	1.206			OK
(45°) T	1200	2700				.845	(.085)							
Preliminary S.P.=							.75	-	.124	.626;				
Available for fittings: 1.333 - .626 = .707, Selected: Section 8 - 90°T, Short Cone = .44)														
" 10 - 45°L,R=1.5D, 3 Pc.=.055) .58"														
45° T =.085) w.g.														
Excess available S.P. absorbed at terminal inlet valve: .707 - .58=1.333-1.206= .127" w.g.														
10" Diffuser.	400						.23			}		1	1.157	
11	400	3000	5	12	3.0		.36	.29		1.015	1.015			
90°L(R=1.5D)	400	3000					(.135)							
12	750	2200	8	10	.9		.09	.068		.022	1.037			OK
13	250	Equal to Section 8			1.04	.51+.41	-*			.92		9	.946	OK
8" Diffuser,	350						.2			}				
14	350	2600	5	6	2.25		.135	-*		.705		11	1.015	**)
90°T, Short Cone	350	2600				1.18	.37							
15	350	Equal to Section 14			1.04	.335+.36	-*			.695		1	1.157	**)
16	350					.965	.335+.34	-*		.675		9	.946	
17	350					1.18	.335+.37	-*		.705		8	.95	

Notes: -* Table I values do not apply, when change in velocity is combined with change in direction of flow (elbow or branch take-off)

** 90° cyl. branch take-off fittings could be used for sections 14 and 15, but short cone type was selected throughout for reasons of uniformity and safety.

Fig. 6. Work sheet for design example.

(e) Refer to manufacturers' literature for SP ratings of primary equipment, such as coils and filters, and determine losses due to fresh-air intakes, return ducts, etc., by separate calculation.

Step 5: Determine SP losses (Column 9) and regain (Column 10) due to velocity conversion.

(a) Refer to Table 1 for the duct system proper. For example, Duct Section 4:

Upstream velocity in Section 5
(Column 3) V_U .... 4400 fpm;
Downstream velocity in Section 4
(Column 3) V_D .... 3700 fpm;
Regain (Column 10) .... 0.28 in. wg.

(b) Refer to Table 6 for fan-plenum-duct transitions. For example, fan-plenum Arrangement II-a:

Static regain, $R = (VP_1 - VP_2) \times C_1$ in. wg.;
Fan discharge velocity $V_1 = 5000$ fpm;
From the table, $VP_1 = 1.56$ in. wg.;
Initial duct velocity, $V_2 = 4400$ fpm;
From the table $VP_2 = 1.21$ in. wg.;
From C-Factor table
C_1 for 30° divergent transition = 0.44;
Enter static regain (Column 10) = 0.154 in. wg.

Step 6: Add up pressure losses (Columns 8 and 9) and deduct regain (Column 10) for each section to obtain sectional SP losses (Column 11).

Step 7: Determine cumulative SP losses (Column 12) by adding up sectional losses step by step, starting with Section 1. The highest value obtained represents the tentative SP drop for the entire system, subject to recheck of other branch trunk ducts, as outlined in Step 8.

Branch Trunk Ducts

Step 8: Determine the static pressure available for designing the branch duct. It is equal to the pressure in the main duct at the junction and, therefore, also to the sum of pressure losses downstream of the take-off.

From Columns 1 and 12 of the main duct calculations, identify and enter in the work sheet the main duct section following the branch take-off (Column 13) and available SP at the take-off (Column 14). For example:

(Refer to layout, Fig. 5.) Main duct Section 2 follows junction of branch trunk duct, Sections 8 to 10.

(Refer to work sheet, Fig. 6.) Available SP at take-off = cumulative SP for duct Section 2 = 1.333 inches wg.

Step 9: Branch trunk ducts, resembling a portion of the main duct in layout and capacity, often require rechecking of only one fitting and/or duct section against available SP. Check particularly for:

(a) Low available SP ..... Check main duct, Column 12;

(b) High V_B/V_U ratio .... Check Columns 3 and 7;

(c) Longest duct run .... Check Column 5.

For example, referring to layout and work sheet for Section 7:

Duct layout and cfm for 3rd floor are equal to previously calculated main duct Sections 1 to 3.
Available SP (Column 14) = 1,497 in. wg. Note that this is a lower value than that for the 2nd floor.
Branch SP losses (Column 12) = 1.405 in. wg. due to low V_B/V_U ratio, take-off losses are disregarded. See Table 2-A.
Available SP exceeds branch duct losses. Therefore, no changes are required. (Should branch duct losses exceed available SP, increase fan SP rating by a corresponding amount or select fittings or high velocity terminal with lower resistance.

Step 10: Design branch trunk ducts with different cfm and dimensions, as follows: Determine location of take-off and available SP, as before, (Columns 13 and 14). Enter on work sheet the branch duct sections, fittings, and high velocity unit, their cfm and section lengths, as before (Columns 1, 2, and 5) but omit fitting details. An example is provided on the work sheet for Duct Sections 8 to 12.

Step 11: Determine duct velocities, diameters, SP per 100 ft, V_B/V_U ratios, friction losses, conversion losses and regain, and high velocity unit minimum SP resistances, as before, (Columns 3, 4, 6, 7, 8, 9, and 10) but omit fitting losses. Add up values, entered in Columns 8 and 9 so far, and deduct those in Column 10. Subtract the result from the available SP, (Column 14) to obtain the pressure available for the fittings.

Step 12: Select fitting details from Tables 2-A, 2-B and/or 3. Enter them and their losses on the work

sheet (Columns 1 and 8). Add up sectional and cumulative SP losses, as before (Columns 11 and 12). As an example, refer to work sheet for duct sections 8 to 10.

Single Branch Lines

Step 13: Enter take-off location and available SP from previously established data (Columns 13 and 14); also terminal, duct section, and duct length, according to layout (Columns 1, 2, and 5); duct velocity, diameter, SP per 100 ft, and V_B/V_U, according to the high velocity unit selected and an air duct friction chart (Columns 3, 4, 6, and 7).

Step 14: Determine duct friction and minimum high velocity SP loss (Column 8), also conversion loss or regain wherever applicable (Columns 9 and 10). Finally, check fitting details and SP losses (Columns 1 and 8) and sectional pressure losses (Column 11) against available SP (Column 14).

Alternate Procedure for High-Velocity Duct Design

The preceding example suggests that trial duct sizes in the longest run can be obtained by finding the equivalent friction loss of the main trunk duct (Section 5), and using this static pressure

RESUME OF PROCEDURE

Step	Description	Ref. or Table	Work Sheet Column
	MAIN DUCT		
1	Select "critical" high velocity unit and duct run.	Layout	—
2	Enter duct sections and fittings, cfm and duct section lengths, starting at high velocity unit.	Layout	1, 2, 5
3	Enter duct fpm, dia. and SP/100 ft, starting at fan.	Friction Chart	3, 4, 6
4	Enter preliminary SP losses of:		
	a) High Velocity Unit.	Catalog	8
	b) Duct sections.	Col. 5, 6	8
	c) Fittings	Tables 2-5	8
	(for branch take-offs, determine V_B/V_U ratio).		7
	d) Fan-plenum-duct transitions.	Table 6	9, 10
	e) Primary equipment.	Mf'rs. Ratings	8
	f) Air intakes and/or return ducts.	Separ. Calcul.	8
5	Enter velocity conversion losses and regain.	Table 1	9, 10
6	Add up SP losses, deduct regain for each section.	Col. 8, 9, 10	11
7	Add up sectional SP losses, step-by-step.	Col. 11	12
	BRANCH TRUNK DUCT		
8	Enter main duct section following take-off and available cumulative SP.	Main Duct Col. 1, 12	13, 14
9	With branch ducts resembling portion of main duct, check take-off fitting losses against available SP.	Col. 14	7, 8
10	With branch trunk ducts in general, enter duct sections and fittings, cfm and lengths; omit fitting details.	Layout	1, 2, 5
11	Enter fpm, dia., SP/100 ft V_B/V_U ratio, friction losses, conversion losses and regain and high velocity unit resistance. Add values in Col. 8 and 9, deduct values in Col. 10, and deduct result from available SP.	Friction Chart Table 1 Catalog Col. 14	3, 4, 6, 7, 8 9, 10 8
12	Difference is SP available for fittings; select fittings and enter their SP losses.	Tables 2-5	1, 8
	Add up sectional and cumulative losses.	Col. 8, 9, 10	11, 12
	SINGLE BRANCH LINE		
13	Proceed as under Step 8 and 9 (or 10).		1, 2, 5, 13, 14
	Enter duct fpm, dia., SP/100 ft, V_B/V_U ratio.		3, 4, 6, 7
14	Enter SP losses and regain as before.	Col. 14	8, 9, 10
	Subtract their sum from available SP and select fitting details accordingly.	Tables 2-5	1, 8, 11, 12

loss per 100 ft, go on to size each succeeding duct section. This method would result in excessively high fan pressures for larger and more sophisticated systems than that shown in the example. In preference to the example's method, the designer is advised to take a standard duct friction chart (Figs. 2a, 2b, 3a or 3b of the preceding section on Air Flow in Ducts) and continue as follows:

1. Select a velocity in the trunk duct leaving the fan. Table 2 will be useful for showing the limiting velocities in trunk ducts.

2. After a trunk duct velocity is selected for the maximum CFM, it will almost invariably result in a duct friction loss below 1.0 inches wg per 100 ft. If so, base all trial trunk-duct sizes on the trunk duct velocity selected in Step 1, until the duct velocity line crosses the constant friction line of 1.0 in. wg per 100 ft. It will not be possible to base all duct sizes on exactly the velocity selected in Step 1 since round ducts come in standard diameters and rectangular ducts are usually made to the nearest inch; however, the designer should attempt to keep velocity changes in this section to the minimum or lowest range possible, because the dynamic losses for each velocity change become significant at higher velocities. Trial sizes for rectangular ducts in this section must be based on the average velocity in the duct cross section, not the velocity of an equivalent round duct.

3. At the point where the constant velocity line from Step 1 crosses the constant friction line of 1.0 in. wg per 100 ft., the CFM for this duct size and all trial sizes for downstream sections thereafter, will be sized on an equal duct friction of 1.0 in. wg per 100 ft. When sizing on a constant friction line, all ducts with rectangular cross sections are sized from the equivalent round-duct size.

The constant pressure drop was arbitrarily set at 1.0 in. wg per 100 feet, which should give reasonably conservative trial duct sizes. The designer can use 1.5 in. wg per 100 ft. for smaller trial sizes, or select trial sizes based on a band between 1.5 in. and 1.0 in. wg. For very conservative trial duct sizes, a constant pressure of 0.75 in. wg per 100 ft. is recommended.

After the trial sizes have been selected, the designer can proceed to calculate the duct friction losses, static losses, and static regain, all as shown in the preceding example. Where there are large losses due to friction or static losses (from velocity increase), the designer can change the trial size to a larger size, and reduce these losses.

In the end, the designer must not come out with static pressures above 6 in. wg if he expects to use high velocity, medium-pressure duct construction. Moreover, he must stay below 10 in. wg in the main trunk duct since this is the limiting pressure for high-velocity, high-pressure duct construction. If the pressure limits for high-velocity duct construction are exceeded, the ducts may very well "balloon" or blow apart at the seams.

The smallest duct sizes recommended for high-velocity systems are 4 inches dia for round ducts and 6 in. square for rectangular ducts.

DUCT DESIGN BY COMPUTER

Recently, great strides have been made in the computerization of duct design. This includes programs which design duct systems on the equal friction method, on static regain, or on total pressure methods. It further includes high- and low-velocity duct design, and both supply and return (or exhaust) systems.

Most programs give the following items on the printout:

1. Dimensions of each duct section,
2. Velocity in each duct section,
3. Static pressure required at fan discharge,
4. Pressure drop through each duct section,
5. Total pounds of sheet metal required for the whole duct system,
6. Air flow rate at each air device,
7. Air supply temperature at each air device for cases where duct is bare,
8. Static pressure at each branch take-off.

Some programs are set up to optimize duct design so that the available static pressure is dissipated completely in each run. Others permit

the designer to make fitting substitutions so that fan static pressures can be reduced. In general, the computer can do in seconds what a designer may never be able to do in a short time—i.e., determine which run in a system has the greatest pressure drop. While the computer can do marvels in duct design, the major problem is to set the system up so that its input data can be readily fed into the computer's program.

For example, one program has been set up in conversational English. It progressively asks the designer all pertinent questions required by the program, such as, "What are the air flows for each diffuser?" "How long are the duct lengths for each section?" "What type of fitting has been chosen?" "What is the friction rate to be used for constant friction design?" "What are the velocity limits, the type of duct construction desired, etc.?" For this program, it is necessary to set up the duct system on an architectural sepia, in single line. From this, nodes can be numbered, duct section lengths estimated, and input sheets made out. This input data may require many hours for a large, complicated duct system. But, the results are rewarding: All answers are in conversational English, and they give a clear picture of the whole system, as designed, by the assumed parameters. The designer has only to transcribe the duct sizes on the single-line sepia, prints from which the final duct work drawing can now be made by a draftsman or draftsmen.

There is, however, one word of caution about the use of computer results. They are based on ideal construction of the duct system, which rarely ever happens. For this reason, the static pressure for the system, as calculated by the computer, may be on the light side. To be safe for fan selection, a healthy safety factor should be applied to the calculated static pressure for the duct system, and the fan motor should be able to accommodate a fan static pressure of 30 to 50 percent more than the calculated static pressure.

There are many companies, technical institutions, and trade organizations which offer ductwork programs for computers, as well as any number of arrangements for renting or buying these programs. The reader is referred to the "Use of Computers for Environmental Engineering Related to Buildings," Building Science Series 39, published by the National Bureau of Standards, for sale by the Superintendent of Document, U.S. Government Printing Office, Catalog No. C13.29/2:39. Also, the APEC organization in the United States has developed computer programs for HVAC design, which can be rented by APEC members for a nominal fee. For particulars, APEC (Automated Procedures for Engineering Consultants) can be reached through its executive office at Grant-Deneau Tower, Suite M-15, Fourth and Ludlow Streets, Dayton, Ohio 45402.

FIBROUS GLASS DUCT CONSTRUCTION

Fibrous glass duct construction is used for low-velocity, low-pressure systems, and enjoys an increasing popularity, particularly in smaller duct systems for homes, small offices, soundproof areas such as studios, etc. The advantages of this duct material are:

1. It combines the conduit capability of carrying air with thermal insulation and sound absorption. On the other hand, sheet metal ductwork requires two separate materials to accomplish the same result.
2. Because it is taped at the construction joints, air leakage is almost nil and rated air delivery is dependable.
3. Fibrous glass duct systems are approximately 75 percent lighter in weight than insulated

sheet-metal systems of comparable size. This greatly aids both duct delivery to the job site and field assembly. For example, duct sections 16 to 20 feet long can be easily assembled, handled, and installed as one unit, whereas sheet metal ducts are limited to 8 ft lengths.

4. Duct condensation is not a problem; the vapor barrier jackets on the preformed sections are excellent for the warm-side barriers on air conditioning systems. With properly taped and sealed joints, no condensation will occur.

5. At one time, the interior surfaces of fibrous glass ductwork were rough and required

greater fan horsepower to overcome the added skin friction. Advances in fibrous glass duct-board technology have resulted in smooth interior surfaces, essentially equal to those of sheet metal.

6. By the use of color-coded hand tools, fibrous glass ductwork can be rapidly grooved and notched in the field or on the shop bench. The mechanic need not possess unusual manual skills to perform these simple operations.

The current SMACNA * standard for fibrous glass duct construction limits this material to positive or negative static pressures in the duct of 2 in. wg and an air velocity of 2000 fpm. There are other limitations on the use of fibrous glass for ductwork:

1. Maximum air temperature inside duct of 250°F and maximum ambient air temperature outside the duct of 150°F.

2. Fibrous glass cannot be used for kitchen or exhaust ducts, or to convey solids or corrosive gases.

3. Ducts of fibrous glass should not be buried below grade or in concrete.

4. These ducts are limited to indoor use, unless weather protection and proper reinforcement are provided.

5. Fibrous glass cannot be used for casings and housings.

6. High-temperature heating coils should not be installed immediately adjacent to the fibrous glass material.

7. For other limitations on fibrous glass ductwork, such as flexural rigidity, maximum allowable deflection, board fatigue, moisture, and safety standards, the reader should consult the SMACNA standard which may be had by writing them at 8224 Old Courthouse Rd., Tyson's Corner, Vienna, Va. 22180.

8. If a fibrous duct system is to be used for air conditioning only, positive-closing type air diffusers and return-air grilles should be used. They must be closed during off-

season periods, to minimize the entrance of heat and moisture.

·While the past market for fibrous glass ductwork has been primarily in the smaller systems, the technology for this material has advanced so rapidly that it is now used on larger installations. In the current SMACNA standard, maximum duct dimensions of 96 inches are indicated. One fibrous glass duct manufacturer shows duct dimensions up to 10 ft, using heavy duct board. Sample specifications are available from the SMACNA standard, as well as from the duct board manufacturers.

Duct board is one-inch thick, but comes in three classifications, based on flexural rigidity as determined by ASTM Test D-1037-64: 475EI (light), 800EI (medium), and 1400EI (heavy). These are so marked by the manufacturers. Small duct sizes may be made without reinforcement; however, in larger sizes, steel tee or channel reinforcement is essential for duct strength.

The current SMACNA standard shows all reinforcement requirements condensed on a single page, with color accent to emphasize ranges of reinforcement spacing relative to duct size and to each of the three types of duct board. Reinforcement spacings are listed in four categories: Not required and maximum longitudinal spacings of 48 in., 24 in., and 16 in. These spacings are further tabulated on the basis of 3 static pressure ranges: 0"-½", over ½"-1", and over 1"-2".

To simplify reinforcement construction, the SMACNA standard shows only two alternative shapes (tee or channel) of reinforcement members; they come in only two gages of metal (22 and 18) and four sizes.

Finally, the current SMACNA standard has added some new features over the previous issue:

1. New plates are added for fire damper connections, transverse joints, and ducted electric heater installations. In addition, plates for hangers, access doors, and reinforcements are revised.

2. Thermal conductivity, air-flow friction loss, and acoustical absorption data are now included.

3. Shiplap joint assembly methods are illustrated, as well as V-Groove methods.

* Sheet Metal & Air Conditioning Contractors' National Assn.

DETERMINING REQUIRED AIR VOLUME

The first step in the design of a heating or a cooling system is to determine the heat and moisture loads due to transmission and infiltration losses or gains, and to internal heat and moisture sources. Once the load has been determined the next step is to calculate the volume of air which must be delivered in order to carry the load. Fig. 1 provides a graphical means of evaluating air volume without calculation. Although the sensible heat scale (Fig. 1) goes only to 10,000 Btu per hr and the volume scale to 1000 cfm the figure can be used for larger values by merely multiplying *both* scale readings by 10 or 100 as may be necessary.

Example: The load on a room which is to be heated is 8000 Btu per hr. Room temperature is to be held at 72 deg F and air is to be supplied at 92 deg F.

The temperature difference available for supplying the sensible heat losses amount to 92 − 72 = 20 deg. Enter Fig. 1 at 8000 load and rise to intersection with the radial line for a 20 deg F temperature difference; from this point move horizontally left to read the required air volume as 370 cfm.

Example: The sensible heat load on a house is 78,000 Btu per hr and warm air is to be supplied with a temperature 70 deg F greater than room temperature. Enter the graph at 7800 rise to the 70-deg line and move left to read 100 cfm. But the load is ten times greater than 7800 so the required air volume will be 10 x 100 or 1000 cfm.

Handling Moisture Loads

Figure 2 provides a graphical means of determining air volume for systems which provide either

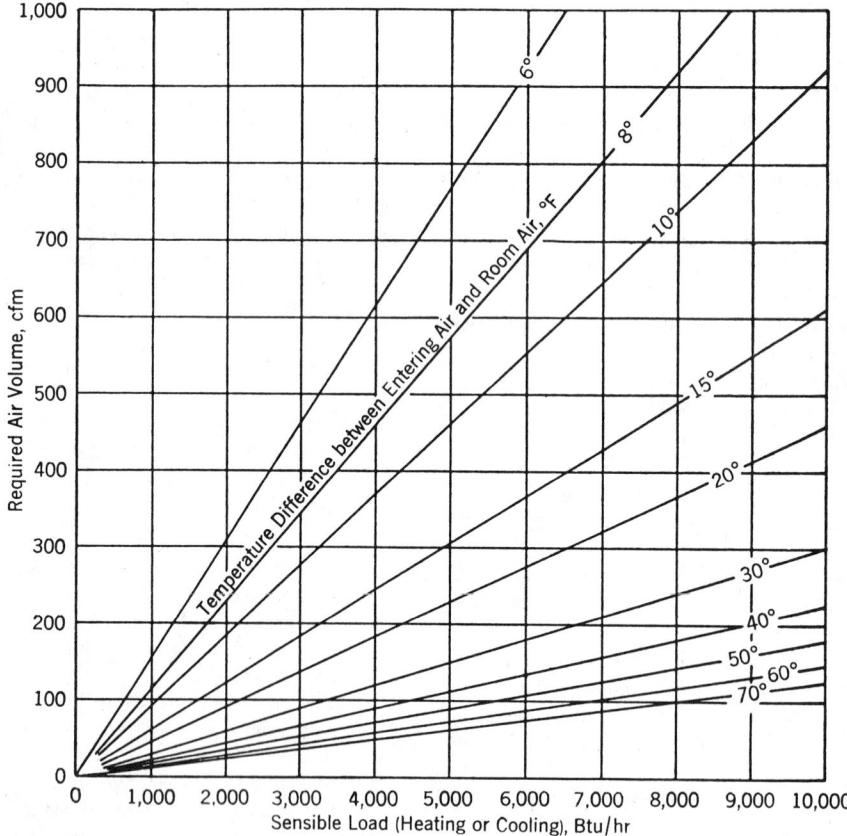

Fig. 1. Air volume required to handle sensible heating or cooling loads. Reprinted by permission from "The Design of Aluminum Duct Systems" prepared for Kaiser Aluminum & Chemical Sales, Inc., by F. W. Hutchinson.

humidification or dehumidification. The latent heat load is first determined by the usual methods, the humidity difference between entering air and room air selected and the required air volume then read from the figure. Note that with Fig. 2 as with Fig. 1 the graphs can be used beyond their scale range by merely multiplying both scales by 10.

In many heating systems the total load will include both sensible and latent fractions. In such a case the designer can select either the temperature or the humidity of the supply air and then use Figures 1 and 2 as follows:

(1) If the humidity of the supply air is known:
  (a) From Fig. 2, for known latent load and known humidity difference, read the required air volume.
  (b) Enter Fig. 1 at known sensible load, then re-enter at the air volume from step (a) and at the intersection of these two lines read the necessary temperature difference.

(2) If the temperature of the supply air is fixed:
  (a) From Fig. 1, for known sensible load and known temperature difference, read the required air volume.
  (b) Enter Fig. 2 at known air volume from step (a), then re-enter for known latent heat load; the intersection of these two lines will give the required humidity difference.
  (c) Add (or subtract) the humidity difference to (or from) room humidity to determine the required humidity of the supply air.

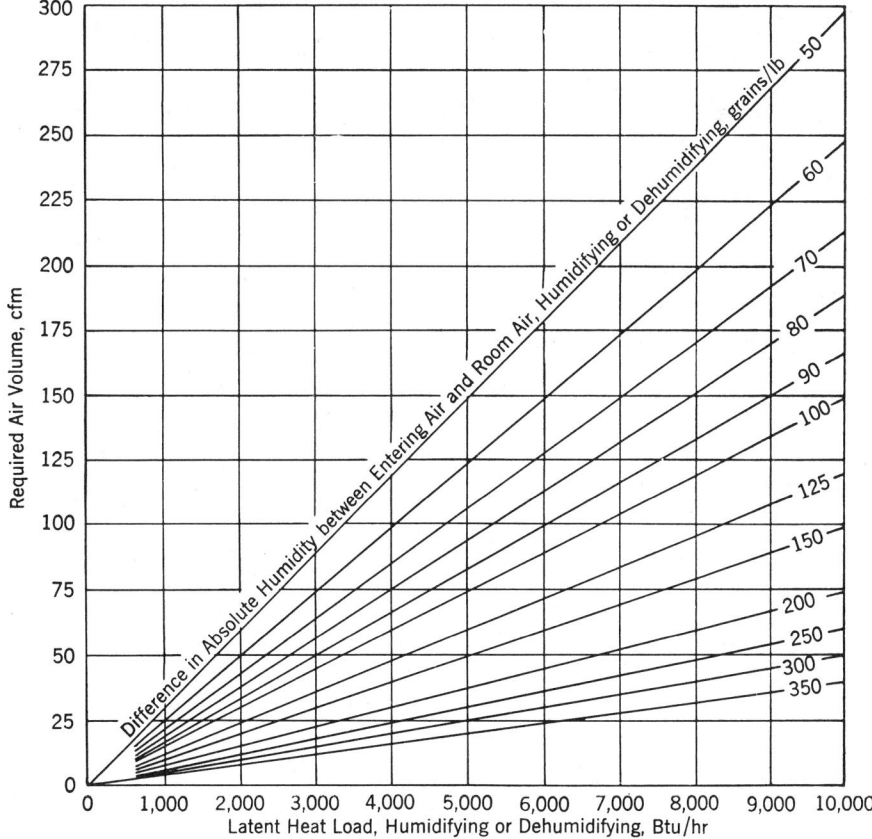

Fig. 2. Air volume required to handle humidifying or dehumidifying loads. Reprinted by permission from "The Design of Aluminum Duct Systems" prepared for Kaiser Aluminum & Chemical Sales, Inc., by F. W. Hutchinson.

WEIGHTS OF GALVANIZED STEEL DUCTS—1 × 1 TO 15 × 15 IN.

Weight of Galvanized Steel Duct per Lineal Foot

Gage																Gage
16	1.13	1.57	2.03	2.48	2.93	3.38	3.83	4.28	4.73	5.18	5.63	6.08	6.53	7.20	7.65	16
18	.92	1.28	1.65	2.02	2.38	2.75	3.12	3.48	3.85	4.22	4.58	4.95	5.32	5.87	6.23	18
20	.71	.99	1.28	1.56	1.84	2.13	2.41	2.69	2.98	3.26	3.54	3.83	4.11	4.53	4.82	20
22	.59	.82	1.05	1.29	1.52	1.75	1.99	2.22	2.45	2.68	2.92	3.15	3.38	3.73	3.96	22
24	.50	.70	.90	1.10	1.30	1.50	1.70	1.90	2.10	2.30	2.50	2.70	2.90	3.20	3.40	24
26	.38	.52	.68	.83	.97	1.13	1.28	1.42	1.58	1.73	1.87	2.03	2.18	2.40	2.54	26

Square Feet of Duct Surface per Lineal Foot of Duct

Width of Duct, Inches															Width of Duct, Inches
.33	.50	.67	.83	1.00	1.17	1.33	1.50	1.67	1.83	2.00	2.17	2.33	2.50	2.67	
															1
															2
															3
															4
															5
															6
															7
															8
															9
															10
															11
															12
															13
															14
															15

Height of Duct, Inches (triangular index, read along the diagonal):

1, 2, 3, 4, 5, 6, 7, 8, 9, 10, 11, 12, 13, 14, 15

Square Feet of Duct Surface per Lineal Foot of Duct

2.83	3.00	3.17	3.33	3.50	3.67	3.83	4.00	4.17	4.33	4.50	4.67	4.83	5.00

Weight of Galvanized Steel Duct per Lineal Foot

Gage															Gage
26	2.70	2.85	3.00	3.15	3.30	3.45	3.60	3.75	3.90	4.05	4.20	4.35	4.50	4.73	26
24	3.60	3.80	4.00	4.20	4.40	4.60	4.80	5.00	5.20	5.40	5.60	5.80	6.00	6.30	24
22	4.20	4.43	4.66	4.90	5.13	5.36	5.60	5.83	6.06	6.30	6.52	6.75	7.00	7.35	22
20	5.10	5.38	5.67	5.95	6.23	6.52	6.80	7.08	7.37	7.65	7.93	8.22	8.50	8.93	20
18	6.60	6.97	7.33	7.70	8.07	8.43	8.80	9.17	9.53	9.90	10.27	10.63	11.00	11.55	18
16	8.10	8.55	9.00	9.45	9.90	10.35	10.80	11.25	11.70	12.15	12.60	13.05	13.50	14.18	16

WEIGHTS OF GALVANIZED STEEL DUCTS—1 × 16 TO 15 × 30 IN.

Weight of Galvanized Steel Duct per Lineal Foot

Gage	2.83	3.00	3.17	3.33	3.50	3.67	3.83	4.00	4.17	4.33	4.50	4.67	4.83	5.00	5.17
16	8.10	8.55	9.00	9.45	9.90	10.35	10.80	11.25	11.70	12.15	12.60	13.05	13.50	14.18	14.63
18	6.60	6.97	7.33	7.70	8.07	8.43	8.80	9.17	9.53	9.90	10.27	10.63	11.00	11.55	11.92
20	5.10	5.38	5.67	5.95	6.23	6.52	6.80	7.08	7.37	7.65	7.93	8.22	8.50	8.93	9.21
22	4.20	4.43	4.66	4.90	5.13	5.37	5.60	5.83	6.06	6.30	6.53	6.76	7.00	7.35	7.59
24	3.60	3.80	4.00	4.20	4.40	4.60	4.80	5.00	5.20	5.40	5.60	5.80	6.00	6.30	6.50
26	2.70	2.85	3.00	3.15	3.30	3.45	3.60	3.75	3.90	4.05	4.20	4.35	4.50	4.73	4.88

Square Feet of Duct Surface per Lineal Foot of Duct — Height of Duct, Inches

Width of Duct, Inches	2.83	3.00	3.17	3.33	3.50	3.67	3.83	4.00	4.17	4.33	4.50	4.67	4.83	5.00	5.17
1	16	17	18	19	20	21	22	23	24	25	26	27	28	29	30
2		16	17	18	19	20	21	22	23	24	25	26	27	28	29
3			16	17	18	19	20	21	22	23	24	25	26	27	28
4				16	17	18	19	20	21	22	23	24	25	26	27
5					16	17	18	19	20	21	22	23	24	25	26
6						16	17	18	19	20	21	22	23	24	25
7							16	17	18	19	20	21	22	23	24
8								16	17	18	19	20	21	22	23
9									16	17	18	19	20	21	22
10										16	17	18	19	20	21
11											16	17	18	19	20
12												16	17	18	19
13													16	17	18
14														16	17
15															16

Square Feet of Duct Surface per Lineal Foot of Duct — Height of Duct, Inches (continued)

Width of Duct, Inches	5.33	5.50	5.67	5.83	6.00	6.17	6.33	6.50	6.67	6.83	7.00	7.17	7.33	7.50
2	30													
3	29	30												
4	28	29	30											
5	27	28	29	30										
6	26	27	28	29	30									
7	25	26	27	28	29	30								
8	24	25	26	27	28	29	30							
9	23	24	25	26	27	28	29	30						
10	22	23	24	25	26	27	28	29	30					
11	21	22	23	24	25	26	27	28	29	30				
12	20	21	22	23	24	25	26	27	28	29	30			
13	19	20	21	22	23	24	25	26	27	28	29	30		
14	18	19	20	21	22	23	24	25	26	27	28	29	30	
15	17	18	19	20	21	22	23	24	25	26	27	28	29	30

Weight of Galvanized Steel Duct per Lineal Foot

Gage	5.33	5.50	5.67	5.83	6.00	6.17	6.33	6.50	6.67	6.83	7.00	7.17	7.33	7.50
26	5.02	5.18	5.33	5.47	5.63	5.78	5.92	6.08	6.23	6.37	6.53	6.68	6.82	7.05
24	6.70	6.90	7.10	7.30	7.50	7.70	7.90	8.10	8.30	8.50	8.70	8.90	9.10	9.40
22	7.82	8.05	8.28	8.51	8.75	9.00	9.22	9.45	9.66	9.92	10.15	10.40	10.62	10.97
20	9.49	9.78	10.06	10.34	10.63	10.91	11.19	11.48	11.76	12.04	12.33	12.61	12.89	13.32
18	12.28	12.65	13.02	13.38	13.75	14.12	14.48	14.85	15.22	15.58	15.95	16.32	16.68	17.23
16	15.08	15.53	15.98	16.43	16.88	17.33	17.78	18.23	18.68	19.13	19.58	20.03	20.48	21.15

WEIGHTS OF GALVANIZED STEEL DUCTS — 1 × 31 TO 15 × 45 IN.

Weight of Galvanized Steel Duct per Lineal Foot / Square Feet of Duct Surface per Lineal Foot of Duct

Width of Duct, Inches	Sq Ft of Duct Surface per Lin Ft	Gage 16	Gage 18	Gage 20	Gage 22	Gage 24	Gage 26
1	5.33	15.08	12.28	9.49	7.82	6.70	5.02
2	5.50	15.53	12.65	9.78	8.05	6.90	5.18
3	5.67	15.98	13.02	10.06	8.29	7.10	5.33
4	5.83	16.43	13.38	10.34	8.52	7.30	5.47
5	6.00	16.88	13.75	10.63	8.75	7.50	5.63
6	6.17	17.33	14.12	10.91	8.99	7.70	5.78
7	6.33	17.78	14.48	11.19	9.22	7.90	5.92
8	6.50	18.23	14.85	11.48	9.45	8.10	6.08
9	6.67	18.68	15.22	11.76	9.67	8.30	6.23
10	6.83	19.13	15.58	12.04	9.92	8.50	6.37
11	7.00	19.58	15.95	12.33	10.15	8.70	6.53
12	7.17	20.03	16.32	12.61	10.40	8.90	6.68
13	7.33	20.48	16.68	12.89	10.62	9.10	6.82
14	7.50	21.15	17.23	13.32	10.97	9.40	7.05
15	7.67	21.60	17.60	13.60	11.20	9.60	7.20

Height of Duct, Inches

(Rows: Width of Duct, Inches. Columns: Square Feet of Duct Surface per Lineal Foot.)

Width	7.67	7.50	7.33	7.17	7.00	6.83	6.67	6.50	6.33	6.17	6.00	5.83	5.67	5.50	5.33
1	45	44	43	42	41	40	39	38	37	36	35	34	33	32	31
2	44	43	42	41	40	39	38	37	36	35	34	33	32	31	45
3	43	42	41	40	39	38	37	36	35	34	33	32	31	45	44
4	42	41	40	39	38	37	36	35	34	33	32	31	45	44	43
5	41	40	39	38	37	36	35	34	33	32	31	45	44	43	42
6	40	39	38	37	36	35	34	33	32	31	45	44	43	42	41
7	39	38	37	36	35	34	33	32	31	45	44	43	42	41	40
8	38	37	36	35	34	33	32	31	45	44	43	42	41	40	39
9	37	36	35	34	33	32	31	45	44	43	42	41	40	39	38
10	36	35	34	33	32	31	45	44	43	42	41	40	39	38	37
11	35	34	33	32	31	45	44	43	42	41	40	39	38	37	36
12	34	33	32	31	45	44	43	42	41	40	39	38	37	36	35
13	33	32	31	45	44	43	42	41	40	39	38	37	36	35	34
14	32	31	45	44	43	42	41	40	39	38	37	36	35	34	33
15	31	45	44	43	42	41	40	39	38	37	36	35	34	33	32

Weight of Galvanized Steel Duct per Lineal Foot / Square Feet of Duct Surface per Lineal Foot of Duct

Sq Ft of Duct Surface per Lin Ft	Gage 26	Gage 24	Gage 22	Gage 20	Gage 18	Gage 16
7.83	7.35	9.80	11.43	13.88	17.97	22.05
8.00	7.50	10.00	11.67	14.17	18.33	22.50
8.17	7.65	10.20	11.90	14.45	18.70	22.95
8.33	7.80	10.40	12.13	14.73	19.07	23.40
8.50	7.95	10.60	12.37	15.02	19.43	23.85
8.67	8.10	10.80	12.60	15.30	19.80	24.30
8.83	8.25	11.00	12.84	15.58	20.17	24.75
9.00	8.40	11.20	13.08	15.87	20.53	25.20
9.17	8.55	11.40	13.30	16.15	20.90	25.65
9.33	8.70	11.60	13.53	16.43	21.27	26.10
9.50	8.85	11.80	13.77	16.72	21.63	26.55
9.67	9.00	12.00	14.00	17.00	22.00	27.00
9.83	9.15	12.20	14.23	17.28	22.37	27.45
10.00	9.38	12.50	14.60	17.71	22.92	28.13

WEIGHTS OF GALVANIZED STEEL DUCTS — 16 × 16 TO 30 × 30 IN.

Weight of Galvanized Steel Duct per Lineal Foot

Column headings are the Square Feet of Duct Surface per Lineal Foot of Duct.

Gage	5.33	5.50	5.67	5.83	6.00	6.17	6.33	6.50	6.67	6.83	7.00	7.17	7.33	7.50	7.67
16	15.08	15.53	15.98	16.43	16.88	17.33	17.78	18.23	18.68	19.13	19.58	20.03	20.48	21.15	21.60
18	12.28	12.65	13.02	13.38	13.75	14.12	14.48	14.85	15.22	15.58	15.95	16.32	16.68	17.23	17.60
20	9.49	9.78	10.06	10.34	10.63	10.91	11.19	11.48	11.76	12.04	12.33	12.61	12.89	13.32	13.60
22	7.82	8.05	8.29	8.52	8.75	8.99	9.22	9.45	9.67	9.92	10.15	10.40	10.62	10.97	11.20
24	6.70	6.90	7.10	7.30	7.50	7.70	7.90	8.10	8.30	8.50	8.70	8.90	9.10	9.40	9.60
26	5.02	5.18	5.33	5.47	5.63	5.78	5.92	6.08	6.23	6.37	6.53	6.68	6.82	7.05	7.20

Square Feet of Duct Surface per Lineal Foot of Duct: 5.33, 5.50, 5.67, 5.83, 6.00, 6.17, 6.33, 6.50, 6.67, 6.83, 7.00, 7.17, 7.33, 7.50, 7.67

Height of Duct, Inches (for the table above)

Rows = Width of Duct, Inches; columns = Square Feet per Lineal Foot.

Width \ Sq Ft	5.33	5.50	5.67	5.83	6.00	6.17	6.33	6.50	6.67	6.83	7.00	7.17	7.33	7.50	7.67
16	16	17	18	19	20	21	22	23	24	25	26	27	28	29	30
17		16	17	18	19	20	21	22	23	24	25	26	27	28	29
18			16	17	18	19	20	21	22	23	24	25	26	27	28
19				16	17	18	19	20	21	22	23	24	25	26	27
20					16	17	18	19	20	21	22	23	24	25	26
21						16	17	18	19	20	21	22	23	24	25
22							16	17	18	19	20	21	22	23	24
23								16	17	18	19	20	21	22	23
24									16	17	18	19	20	21	22
25										16	17	18	19	20	21
26											16	17	18	19	20
27												16	17	18	19
28													16	17	18
29														16	17
30															16

Height of Duct, Inches (for the table below)

Rows = Width of Duct, Inches; columns = Square Feet per Lineal Foot.

Width \ Sq Ft	7.83	8.00	8.17	8.33	8.50	8.67	8.83	9.00	9.17	9.33	9.50	9.67	9.83	10.00
17	30													
18	29	30												
19	28	29	30											
20	27	28	29	30										
21	26	27	28	29	30									
22	25	26	27	28	29	30								
23	24	25	26	27	28	29	30							
24	23	24	25	26	27	28	29	30						
25	22	23	24	25	26	27	28	29	30					
26	21	22	23	24	25	26	27	28	29	30				
27	20	21	22	23	24	25	26	27	28	29	30			
28	19	20	21	22	23	24	25	26	27	28	29	30		
29	18	19	20	21	22	23	24	25	26	27	28	29	30	
30	17	18	19	20	21	22	23	24	25	26	27	28	29	30

Square Feet of Duct Surface per Lineal Foot of Duct: 7.83, 8.00, 8.17, 8.33, 8.50, 8.67, 8.83, 9.00, 9.17, 9.33, 9.50, 9.67, 9.83, 10.00

Weight of Galvanized Steel Duct per Lineal Foot

Gage	7.83	8.00	8.17	8.33	8.50	8.67	8.83	9.00	9.17	9.33	9.50	9.67	9.83	10.00
26	7.35	7.50	7.65	7.80	7.95	8.10	8.25	8.40	8.55	8.70	8.85	9.00	9.15	9.38
24	9.80	10.00	10.20	10.40	10.60	10.80	11.00	11.20	11.40	11.60	11.80	12.00	12.20	12.50
22	11.43	11.67	11.90	12.13	12.37	12.60	12.84	13.08	13.30	13.53	13.77	14.00	14.23	14.60
20	13.88	14.17	14.45	14.73	15.02	15.30	15.58	15.87	16.15	16.43	16.72	17.00	17.28	17.71
18	17.97	18.33	18.70	19.07	19.43	19.80	20.17	20.53	20.90	21.27	21.63	22.00	22.37	22.92
16	22.05	22.50	22.95	23.40	23.85	24.30	24.75	25.20	25.65	26.10	26.55	27.00	27.45	28.13

WEIGHTS OF GALVANIZED STEEL DUCTS—16 × 31 TO 30 × 45 IN.

Weight of Galvanized Steel Duct per Lineal Foot

Gage	7.83	8.00	8.17	8.33	8.50	8.67	8.83	9.00	9.17	9.33	9.50	9.67	9.83	10.00	10.17
16	22.05	22.50	22.95	23.40	23.85	24.30	24.75	25.20	25.65	26.10	26.55	27.00	27.45	28.13	28.58
18	17.97	18.33	18.70	19.07	19.43	19.80	20.17	20.53	20.90	21.27	21.63	22.00	22.37	22.92	23.28
20	13.88	14.17	14.45	14.73	15.02	15.30	15.58	15.87	16.15	16.43	16.72	17.00	17.28	17.71	17.99
22	11.43	11.67	11.90	12.13	12.37	12.60	12.84	13.08	13.30	13.53	13.77	14.00	14.23	14.60	14.81
24	9.80	10.00	10.20	10.40	10.60	10.80	11.00	11.20	11.40	11.60	11.80	12.00	12.20	12.50	12.70
26	7.35	7.50	7.65	7.80	7.95	8.10	8.25	8.40	8.55	8.70	8.85	9.00	9.15	9.38	9.52

Square Feet of Duct Surface per Lineal Foot of Duct: 7.83, 8.00, 8.17, 8.33, 8.50, 8.67, 8.83, 9.00, 9.17, 9.33, 9.50, 9.67, 9.83, 10.00, 10.17

Height of Duct, Inches (Width of Duct, Inches at left)

Width	7.83	8.00	8.17	8.33	8.50	8.67	8.83	9.00	9.17	9.33	9.50	9.67	9.83	10.00	10.17
16	31	32	33	34	35	36	37	38	39	40	41	42	43	44	45
17		31	32	33	34	35	36	37	38	39	40	41	42	43	44
18			31	32	33	34	35	36	37	38	39	40	41	42	43
19				31	32	33	34	35	36	37	38	39	40	41	42
20					31	32	33	34	35	36	37	38	39	40	41
21						31	32	33	34	35	36	37	38	39	40
22							31	32	33	34	35	36	37	38	39
23								31	32	33	34	35	36	37	38
24									31	32	33	34	35	36	37
25										31	32	33	34	35	36
26											31	32	33	34	35
27												31	32	33	34
28													31	32	33
29														31	32
30															31

Height of Duct, Inches (Width of Duct, Inches at left) — continued

Width	10.33	10.50	10.67	10.83	11.00	11.17	11.33	11.50	11.67	11.83	12.00	12.17	12.33	12.50
17	45													
18	44	45												
19	43	44	45											
20	42	43	44	45										
21	41	42	43	44	45									
22	40	41	42	43	44	45								
23	39	40	41	42	43	44	45							
24	38	39	40	41	42	43	44	45						
25	37	38	39	40	41	42	43	44	45					
26	36	37	38	39	40	41	42	43	44	45				
27	35	36	37	38	39	40	41	42	43	44	45			
28	34	35	36	37	38	39	40	41	42	43	44	45		
29	33	34	35	36	37	38	39	40	41	42	43	44	45	
30	32	33	34	35	36	37	38	39	40	41	42	43	44	45

Square Feet of Duct Surface per Lineal Foot of Duct: 10.33, 10.50, 10.67, 10.83, 11.00, 11.17, 11.33, 11.50, 11.67, 11.83, 12.00, 12.17, 12.33, 12.50

Weight of Galvanized Steel Duct per Lineal Foot

Gage	10.33	10.50	10.67	10.83	11.00	11.17	11.33	11.50	11.67	11.83	12.00	12.17	12.33	12.50
26	9.68	9.84	9.98	10.11	10.29	10.41	10.58	10.72	10.88	11.02	11.18	11.31	11.48	11.70
24	12.90	13.10	13.30	13.50	13.70	13.90	14.10	14.30	14.50	14.70	14.90	15.10	15.30	15.60
22	15.05	15.28	15.51	15.75	16.00	16.20	16.45	16.68	16.92	17.15	17.38	17.61	17.86	18.20
20	18.28	18.56	18.84	19.13	19.41	19.69	19.98	20.26	20.54	20.83	21.11	21.39	21.68	22.10
18	23.65	24.02	24.38	24.75	25.12	25.48	25.85	26.22	26.58	26.95	27.32	27.68	28.05	28.60
16	29.03	29.48	29.93	30.38	30.83	31.28	31.73	32.18	32.63	33.08	33.53	33.98	34.43	35.10

WEIGHTS OF GALVANIZED STEEL DUCTS—31 × 31 TO 45 × 45 IN.

Weight of Galvanized Steel Duct per Lineal Foot

Square Feet of Duct Surface per Lineal Foot of Duct (column headers)

Gage	10.33	10.50	10.67	10.83	11.00	11.17	11.33	11.50	11.67	11.83	12.00	12.17	12.33	12.50	12.67
16	29.03	29.48	29.93	30.38	30.83	31.28	31.73	32.18	32.63	33.08	33.53	33.98	34.43	35.10	35.55
18	23.65	24.02	24.38	24.75	25.12	25.48	25.85	26.22	26.58	26.95	27.32	27.68	28.05	28.60	28.97
20	18.28	18.56	18.84	19.13	19.41	19.69	19.98	20.26	20.54	20.83	21.11	21.39	21.68	22.10	22.38
22	15.05	15.28	15.52	15.75	15.99	16.21	16.45	16.68	16.92	17.15	17.39	17.60	17.85	18.20	18.42
24	12.90	13.10	13.30	13.50	13.70	13.90	14.10	14.30	14.50	14.70	14.90	15.10	15.30	15.60	15.80
26	9.68	9.84	9.98	10.11	10.29	10.41	10.58	10.72	10.88	11.02	11.18	11.31	11.48	11.70	11.85

Square Feet of Duct Surface per Lineal Foot of Duct

Height of Duct, Inches (body entries); column headers = Square Feet

Width of Duct, Inches	10.33	10.50	10.67	10.83	11.00	11.17	11.33	11.50	11.67	11.83	12.00	12.17	12.33	12.50	12.67
31	31	32	33	34	35	36	37	38	39	40	41	42	43	44	45
32		31	32	33	34	35	36	37	38	39	40	41	42	43	44
33			31	32	33	34	35	36	37	38	39	40	41	42	43
34				31	32	33	34	35	36	37	38	39	40	41	42
35					31	32	33	34	35	36	37	38	39	40	41
36						31	32	33	34	35	36	37	38	39	40
37							31	32	33	34	35	36	37	38	39
38								31	32	33	34	35	36	37	38
39									31	32	33	34	35	36	37
40										31	32	33	34	35	36
41											31	32	33	34	35
42												31	32	33	34
43													31	32	33
44														31	32
45															31

Square Feet of Duct Surface per Lineal Foot of Duct

Height of Duct, Inches (body entries); column headers = Square Feet

Width of Duct, Inches	12.83	13.00	13.17	13.33	13.50	13.67	13.83	14.00	14.17	14.33	14.50	14.67	14.83	15.00
31														
32	45													
33	44	45												
34	43	44	45											
35	42	43	44	45										
36	41	42	43	44	45									
37	40	41	42	43	44	45								
38	39	40	41	42	43	44	45							
39	38	39	40	41	42	43	44	45						
40	37	38	39	40	41	42	43	44	45					
41	36	37	38	39	40	41	42	43	44	45				
42	35	36	37	38	39	40	41	42	43	44	45			
43	34	35	36	37	38	39	40	41	42	43	44	45		
44	33	34	35	36	37	38	39	40	41	42	43	44	45	
45	32	33	34	35	36	37	38	39	40	41	42	43	44	45

Weight of Galvanized Steel Duct per Lineal Foot

Square Feet of Duct Surface per Lineal Foot of Duct (column headers)

Gage	12.83	13.00	13.17	13.33	13.50	13.67	13.83	14.00	14.17	14.33	14.50	14.67	14.83	15.00
26	12.00	12.15	12.30	12.45	12.60	12.75	12.90	13.05	13.20	13.35	13.50	13.65	13.78	13.97
24	16.00	16.20	16.40	16.60	16.80	17.00	17.22	17.40	17.60	17.80	18.00	18.20	18.40	18.70
22	18.66	18.90	19.12	19.35	19.60	19.85	20.05	20.28	20.53	20.76	21.00	21.24	21.46	21.80
20	22.67	22.95	23.23	23.52	23.80	24.08	24.37	24.65	24.93	25.22	25.50	25.78	26.07	26.49
18	29.33	29.70	30.07	30.43	30.80	31.17	31.53	31.90	32.27	32.63	33.00	33.37	33.73	34.28
16	36.00	36.45	36.90	37.35	37.80	38.25	38.70	39.15	39.60	40.05	40.50	40.95	41.40	42.08

WEIGHTS OF GALVANIZED STEEL DUCTS—31 × 46 TO 45 × 60 IN.

Gage	Weight of Galvanized Steel Duct per Lineal Foot																Gage
16		36.00	36.45	36.90	37.35	37.80	38.25	38.70	39.15	39.60	40.05	40.50	40.95	41.40	42.08	42.53	16
18		29.33	29.70	30.07	30.43	30.80	31.17	31.53	31.90	32.27	32.63	33.00	33.37	33.73	34.28	34.65	18
20		22.67	22.95	23.23	23.52	23.80	24.08	24.37	24.65	24.93	25.22	25.50	25.78	26.07	26.49	26.78	20
22		18.66	18.90	19.12	19.35	19.60	19.85	20.08	20.28	20.53	20.76	21.00	21.24	21.46	21.80	22.05	22
24		16.00	16.20	16.40	16.60	16.80	17.00	17.20	17.40	17.60	17.80	18.00	18.20	18.40	18.70	18.90	24
26		12.00	12.15	12.30	12.45	12.60	12.75	12.90	13.05	13.20	13.35	13.50	13.65	13.78	13.97	14.18	26
Width of Duct, Inches	Square Feet of Duct Surface per Lineal Foot of Duct																Width of Duct, Inches
		12.83	13.00	13.17	13.33	13.50	13.67	13.83	14.00	14.17	14.33	14.50	14.67	14.83	15.00	15.17	
	Height of Duct, Inches																
31		46	47	48	49	50	51	52	53	54	55	56	57	58	59	60	31
32	60		46	47	48	49	50	51	52	53	54	55	56	57	58	59	32
33	59	60		46	47	48	49	50	51	52	53	54	55	56	57	58	33
34	58	59	60		46	47	48	49	50	51	52	53	54	55	56	57	34
35	57	58	59	60		46	47	48	49	50	51	52	53	54	55	46	35
36	56	57	58	59	60		46	47	48	49	50	51	52	53	54	55	36
37	55	56	57	58	59	60		46	47	48	49	50	51	52	53	54	37
38	54	55	56	57	58	59	60		46	47	48	49	50	51	52	53	38
39	53	54	55	56	57	58	59	60		46	47	48	49	50	51	52	39
40	52	53	54	55	56	57	58	59	60		46	47	48	49	50	51	40
41	51	52	53	54	55	56	57	58	59	60		46	47	48	49	50	41
42	50	51	52	53	54	55	56	57	58	59	60		46	47	48	49	42
43	49	50	51	52	53	54	55	56	57	58	59	60		46	47	48	43
44	48	49	50	51	52	53	54	55	56	57	58	59	60		46	47	44
45	47	48	49	50	51	52	53	54	55	56	57	58	59	60		46	45
Gage	Square Feet of Duct Surface per Lineal Foot of Duct																Gage
	15.33	15.50	15.67	15.83	16.00	16.17	16.33	16.50	16.67	16.83	17.00	17.17	17.33	17.50			
	Weight of Galvanized Steel Duct per Lineal Foot																
26	14.33	14.47	14.63	14.78	14.92	15.08	15.24	15.37	15.53	15.68	15.82	15.98	16.13	16.35			26
24	19.10	19.30	19.50	19.70	19.90	20.10	20.30	20.50	20.70	20.90	21.10	21.30	21.50	21.80			24
22	22.30	22.51	22.75	23.00	23.22	23.45	23.70	23.92	24.15	24.38	24.60	24.85	25.10	25.44			22
20	27.06	27.34	27.63	27.91	28.19	28.48	28.76	29.04	29.33	29.61	29.89	30.18	30.46	30.88			20
18	35.02	35.38	35.75	36.12	36.48	36.85	37.22	37.58	37.95	38.32	38.68	39.05	39.42	39.97			18
16	42.98	43.43	43.88	44.33	44.78	45.23	45.68	46.13	46.58	47.03	47.48	47.93	48.38	49.05			16

WEIGHTS OF GALVANIZED STEEL DUCTS—46 × 46 TO 60 × 60 IN.

Weight of Galvanized Steel Duct per Lineal Foot

Gage	15.33	15.50	15.67	15.83	16.00	16.17	16.33	16.50	16.67	16.83	17.00	17.17	17.33	17.50	17.67
16	42.98	43.43	43.88	44.33	44.78	45.23	45.68	46.13	46.58	47.03	47.48	47.93	48.38	49.05	49.50
18	35.02	35.38	35.75	36.12	36.48	36.85	37.22	37.58	37.95	38.32	38.68	39.05	39.42	39.97	40.33
20	27.06	27.34	27.63	27.91	28.19	28.48	28.76	29.04	29.33	29.61	29.89	30.18	30.46	30.88	31.17
22	22.30	22.51	22.75	23.00	23.22	23.45	23.70	23.92	24.15	24.38	24.60	24.85	25.10	25.44	25.65
24	19.10	19.30	19.50	19.70	19.90	20.10	20.30	20.50	20.70	20.90	21.10	21.30	21.50	21.80	22.00
26	14.33	14.47	14.63	14.78	14.92	15.08	15.24	15.37	15.53	15.68	15.82	15.98	16.13	16.35	16.50

Square Feet of Duct Surface per Lineal Foot of Duct — Height of Duct, Inches

Upper section (Square Feet headers 15.33 to 17.67):

Width of Duct, Inches	15.33	15.50	15.67	15.83	16.00	16.17	16.33	16.50	16.67	16.83	17.00	17.17	17.33	17.50	17.67
46	46	47	48	49	50	51	52	53	54	55	56	57	58	59	60
47		46	47	48	49	50	51	52	53	54	55	56	57	58	59
48			46	47	48	49	50	51	52	53	54	55	56	57	58
49				46	47	48	49	50	51	52	53	54	55	56	57
50					46	47	48	49	50	51	52	53	54	55	56
51						46	47	48	49	50	51	52	53	54	55
52							46	47	48	49	50	51	52	53	54
53								46	47	48	49	50	51	52	53
54									46	47	48	49	50	51	52
55										46	47	48	49	50	51
56											46	47	48	49	50
57												46	47	48	49
58													46	47	48
59														46	47
60															46

Lower section (Square Feet headers 17.83 to 20.00):

Width of Duct, Inches	17.83	18.00	18.17	18.33	18.50	18.67	18.83	19.00	19.17	19.33	19.50	19.67	19.83	20.00
47	60													
48	59	60												
49	58	59	60											
50	57	58	59	60										
51	56	57	58	59	60									
52	55	56	57	58	59	60								
53	54	55	56	57	58	59	60							
54	53	54	55	56	57	58	59	60						
55	52	53	54	55	56	57	58	59	60					
56	51	52	53	54	55	56	57	58	59	60				
57	50	51	52	53	54	55	56	57	58	59	60			
58	49	50	51	52	53	54	55	56	57	58	59	60		
59	48	49	50	51	52	53	54	55	56	57	58	59	60	
60	47	48	49	50	51	52	53	54	55	56	57	58	59	60

Weight of Galvanized Steel Duct per Lineal Foot

Gage	17.83	18.00	18.17	18.33	18.50	18.67	18.83	19.00	19.17	19.33	19.50	19.67	19.83	20.00
26	16.65	16.80	16.95	17.10	17.25	17.40	17.55	17.70	17.85	18.00	18.15	18.30	18.45	18.68
24	22.20	22.40	22.60	22.80	23.00	23.20	23.40	23.60	23.80	24.00	24.20	24.40	24.60	24.90
22	25.90	26.15	26.36	26.60	26.82	27.05	27.30	27.52	27.77	28.00	28.28	28.45	28.70	29.05
20	31.45	31.73	32.02	32.30	32.58	32.87	33.15	33.43	33.72	34.00	34.28	34.57	34.85	35.28
18	40.70	41.07	41.43	41.80	42.17	42.53	42.90	43.27	43.63	44.00	44.37	44.75	45.10	45.65
16	49.95	50.40	50.85	51.30	51.75	52.20	52.65	53.10	53.55	54.00	54.45	54.90	55.35	56.03

ESTIMATING WEIGHT OF METAL IN CIRCULAR DUCTS

The accompanying table is for use in determining the weight of galvanized steel sheet metal used in fabricating ducts of circular cross section. Also included are data on the surface area of circular ducts for use in estimating insulation areas.

Recommended gages for circular ducts for heating and general ventilating work are:

Diameter, Inches	Gage
6 to 19	26
20 to 29	24
30 to 39	22
40 to 49	20
over 49	18

For industrial exhaust systems the following table of gages and diameters is good practice:

Diameter, Inches	Gage
Under 9	24
9 to 14	22
15 to 20	20
21 to 30	18
over 30	16

The weights given in the table include allowance for seams on the basis that the ducts are cut from sheets 30 in. wide, and that one lap of 1 inch is required for each sheet. Thus, ducts up to 9 inches diameter contain one lap and so on up to ducts of 72 inches diameter which contain 8 laps. No allowance is made for waste, bracing or hanging.

Areas and Weights of Circular Galvanized Steel Ducts

Diameter, In.	Area per Ft., Sq.Ft.	Gage 26	Gage 24	Gage 22
		Weight per Ft., Lb.		
4	1.05	1.02	1.36	1.59
5	1.31	1.25	1.67	1.95
6	1.57	1.49	1.98	2.32
7	1.83	1.72	2.30	2.69
8	2.09	1.96	2.61	3.06
9	2.36	2.20	2.93	3.42
10	2.62	2.51	3.34	3.91
11	2.88	2.74	3.66	4.28
12	3.14	2.98	3.97	4.64
13	3.40	3.21	4.28	5.01
14	3.67	3.45	4.60	5.38
15	3.93	3.68	4.91	5.75
16	4.19	3.92	5.23	6.12
17	4.45	4.16	5.54	6.48
18	4.72	4.39	5.85	6.85
19	4.98	4.63	6.17	7.22
20	5.24	4.94	6.58	7.70
21	5.50	5.18	6.90	8.07
22	5.76	5.41	7.21	8.44
23	6.02	5.64	7.53	8.80
24	6.28	5.88	7.84	9.17
25	6.54	6.12	8.15	9.54
26	6.80	6.35	8.47	9.91

Diameter, In.	Area per Ft., Sq.Ft.	Gage 24	Gage 22	Gage 20	Gage 18
		Weight per Ft., Lb.			
27	7.07	8.78	10.27	12.47	16.07
28	7.33	9.10	10.64	12.92	16.65
29	7.59	9.41	11.01	13.36	17.22
30	7.85	9.83	11.50	13.95	17.98
31	8.11	10.14	11.86	14.40	18.55
32	8.38	10.45	12.23	14.84	19.13
33	8.65	10.77	12.60	15.29	19.70
34	8.91	11.08	12.96	15.74	20.28
35	9.17	11.40	13.33	16.18	20.85
36	9.43	11.71	13.70	16.63	21.43
37	9.69	12.02	14.07	17.07	22.00
38	9.95	12.34	14.44	17.52	22.58
39	10.21	12.65	14.80	17.97	23.15
40	10.47	13.07	15.29	18.55	23.91
41	10.73	13.39	15.66	19.00	24.49
42	10.99	13.70	16.02	19.45	25.06
43	11.26	14.01	16.39	19.89	25.64
44	11.52	14.32	16.76	20.34	26.21
45	11.78	14.64	17.13	20.78	26.79
46	12.04	14.95	17.49	21.23	27.36
47	12.30	15.27	17.86	21.68	27.94
48	12.56	15.58	18.23	22.12	28.51
49	12.83	15.89	18.60	22.57	29.09

Diameter, In.	Area per Ft., Sq.Ft.	Gage 20	Gage 18	Gage 16
		Weight per Ft., Lb.		
50	13.09	23.16	29.84	36.69
51	13.35	23.60	30.42	37.40
52	13.61	24.05	30.99	38.11
53	13.87	24.50	31.57	38.81
54	14.13	24.94	32.14	39.52
55	14.39	25.39	32.72	40.23
56	14.65	25.83	33.29	40.93
57	14.92	26.28	33.87	41.64
58	15.19	26.73	34.44	42.35
59	15.45	27.17	35.02	43.05
60	15.71	27.76	35.78	43.99
61	15.97	28.21	36.35	44.69
62	16.23	28.65	36.93	45.40
63	16.50	29.10	37.50	46.11
64	16.76	29.54	38.07	46.81
65	17.02	29.99	38.65	47.52
66	17.29	30.44	39.23	48.23
67	17.55	30.88	39.80	48.94
68	17.81	31.33	40.38	49.64
69	18.07	31.78	40.95	50.35
70	18.33	32.36	41.71	51.28
71	18.60	32.81	42.28	51.99
72	18.86	33.26	42.86	52.70

ESTIMATING WEIGHT OF METAL IN DUCTS

American Wire or Brown & Sharpe Gage No.	Thick- ness, Inch	Approximate Weight, Pounds per Square Foot				
		Copper*	Yellow Brass	Tobin Bronze	5 Per Cent Phosphor- Bronze	Everdur 1010
0000	0.4600	21.33	20.27	20.14	21.20	20.40
000	0.4096	18.99	18.05	17.93	18.88	18.17
00	0.3648	16.92	16.07	16.41	16.81	16.18
0	0.3249	15.06	14.32	14.23	14.98	14.41
1	0.2893	13.41	12.75	12.67	13.33	12.83
2	0.2576	11.94	11.35	11.28	11.87	11.43
3	0.2294	10.64	10.11	10.04	10.57	10.17
4	0.2043	9.473	9.002	8.943	9.414	9.061
5	0.1819	8.434	8.015	7.963	8.382	8.068
6	0.1620	7.512	7.138	7.092	7.465	7.185
7	0.1443	6.691	6.358	6.317	6.649	6.400
8	0.1285	5.958	5.662	5.625	5.921	5.699
9	0.1144	5.304	5.041	5.008	5.272	5.074
10	0.1019	4.725	4.490	4.461	4.696	4.519
11	0.0907	4.206	3.997	3.971	4.180	4.023
12	0.0808	3.747	3.560	3.537	3.723	3.584
13	0.0720	3.338	3.173	3.152	3.318	3.193
14	0.0641	2.972	2.825	2.807	2.954	2.843
15	0.0571	2.648	2.516	2.500	2.631	2.532
16	0.0508	2.355	2.238	2.223	2.341	2.253
17	0.0453	2.100	1.996	1.983	2.087	2.009
18	0.0403	1.869	1.776	1.764	1.857	1.787
19	0.0359	1.665	1.582	1.572	1.654	1.592
20	0.0320	1.484	1.410	1.401	1.475	1.419
21	0.0285	1.321	1.256	1.248	1.314	1.264
22	0.0253	1.178	1.119	1.112	1.170	1.127
23	0.0226	1.048	0.9958	0.9893	1.041	1.002
24	0.0201	0.9320	0.8857	0.8799	0.9263	0.8915
25	0.0179	0.8300	0.7887	0.7836	0.8248	0.7939
26	0.0159	0.7373	0.7006	0.6960	0.7327	0.7052
27	0.0142	0.6584	0.6257	0.6216	0.6544	0.6298
28	0.0126	0.5842	0.5552	0.5516	0.5806	0.5588
29	0.0113	0.5240	0.4979	0.4947	0.5207	0.5012
30	0.0100	0.4637	0.4406	0.4377	0.4608	0.4435
31	0.0089	0.4127	0.3922	0.3897	0.4102	0.3947
32	0.0080	0.3709	0.3525	0.3502	0.3686	0.3548
33	0.0071	0.3292	0.3129	0.3109	0.3272	0.3149
34	0.0063	0.2921	0.2776	0.2758	0.2903	0.2794
35	0.0056	0.2597	0.2468	0.2452	0.2581	0.2484
36	0.0050	0.2318	0.2203	0.2189	0.2304	0.2218
37	0.0045	0.2087	0.1983	0.1970	0.2074	0.1996
38	0.0040	0.1855	0.1763	0.1752	0.1844	0.1774
39	0.0035	0.1623	0.1542	0.1532	0.1613	0.1552
40	0.0031	0.1437	0.1366	0.1357	0.1429	0.1375

* Copper sheets can also be obtained in fractional-inch thicknesses varying by sixteenths of an inch from 1/16 to 2 inches.

WEIGHTS OF NON-FERROUS METAL SHEETS

American Wire or Brown & Sharpe Gage No.	Thickness, Inch	Approximate Weight, Pounds per Square Foot				
		S.A.E. Aluminum Alloys Nos. 26 and 27	S.A.E. Aluminum Alloy No. 28	Aluminum Commercially Pure (99 to 99.4 Per Cent)	Nickel Silver 18%*	Nickel Silver 20%-30%
0000	0.4600	6.680	6.410	6.490	20.93	21.20
000	0.4096	5.950	5.710	5.780	18.64	18.88
00	0.3648	5.290	5.090	5.140	16.60	16.81
0	0.3249	4.720	4.530	4.580	14.78	14.97
1	0.2893	4.200	4.030	4.080	13.16	13.33
2	0.2576	3.738	3.591	3.632	11.72	11.87
3	0.2294	3.329	3.198	3.234	10.44	10.57
4	0.2043	2.964	2.848	2.880	9.296	9.414
5	0.1819	2.640	2.536	2.565	8.277	8.382
6	0.1620	2.351	2.258	2.284	7.372	7.466
7	0.1443	2.094	2.012	2.034	6.566	6.649
8	0.1285	1.865	1.792	1.812	5.847	5.921
9	0.1144	1.660	1.595	1.613	5.206	5.272
10	0.1019	1.479	1.420	1.437	4.637	4.696
11	0.0907	1.316	1.264	1.279	4.127	4.179
12	0.0808	1.172	1.126	1.139	3.677	3.724
13	0.0720	1.045	1.004	1.015	3.276	3.318
14	0.0641	0.930	0.894	0.904	2.917	2.954
15	0.0571	0.829	0.796	0.805	2.598	2.631
16	0.0508	0.737	0.708	0.716	2.312	2.341
17	0.0453	0.657	0.631	0.639	2.061	2.087
18	0.0403	0.585	0.562	0.568	1.834	1.857
19	0.0359	0.5210	0.5010	0.5060	1.634	1.655
20	0.0320	0.4640	0.4460	0.4510	1.456	1.474
21	0.0285	0.4140	0.3970	0.4020	1.297	1.313
22	0.0253	0.3671	0.3527	0.3567	1.156	1.171
23	0.0226	0.3280	0.3150	0.3186	1.028	1.041
24	0.0201	0.2917	0.2802	0.2834	0.9146	0.9262
25	0.0179	0.2597	0.2495	0.2524	0.8145	0.8248
26	0.0159	0.2307	0.2216	0.2242	0.7235	0.7327
27	0.0142	0.2060	0.1980	0.2002	0.6462	0.6544
28	0.0126	0.1828	0.1756	0.1776	0.5734	0.5807
29	0.0113	0.1640	0.1575	0.1593	0.5142	0.5207
30	0.0100	0.1451	0.1394	0.1410	0.4550	0.4608
31	0.0089	0.1296	0.1245	0.1259	0.4050	0.4101
32	0.0080	0.1154	0.1108	0.1121	0.3640	0.3686
33	0.0071	0.1027	0.0987	0.0998	0.3231	0.3272
34	0.0063	0.0914	0.0878	0.0888	0.2867	0.2903
35	0.0056	0.0814	0.0782	0.0791	0.2548	0.2580
36	0.0050	0.0726	0.0697	0.0705	0.2275	0.2304
37	0.0045	0.0646	0.0620	0.0627	0.2048	0.2074
38	0.0040	0.0576	0.0553	0.0560	0.1820	0.1843
39	0.0035	0.0512	0.0492	0.0498	0.1593	0.1613
40	0.0031	0.0456	0.0438	0.0443	0.1411	0.1429

* Multiply weights in this column by 0.9905 for 10 per cent nickel-silver and by 0.9937 for 15 per cent nickel-silver.

APPARATUS CASING CONSTRUCTION

Built-up air conditioning units must house the "mixing box," air filters, heating and cooling coils, and for medium- or high-pressure design, the fan or fans. Since most built-up units are for large air systems—which cannot be accommodated by factory-built units—the physical size of these built-up units make their casing construction a very substantial structural problem. An internal positive pressure of 8 in. wg exerts a pressure of 41.5 psf; this requires that the casing walls and roof be of reinforced metal construction.

Casing construction for low-, medium-, and high-pressure units is specified and illustrated in the SMACNA standards for low- and high-velocity ductwork. All construction details, such as access doors, insulation, fan, and duct openings are clearly illustrated in these standards. Casing floors are usually poured concrete.

Attempts to use unreinforced concrete blocks for casing walls have been successful only for low-pressure units, because concrete block construction cannot tolerate pressures in excess of 3.0 in. wg (8'-0" wall height) or 1.3 in. wg (12'-0" wall height). Concrete block walls, for this reason, do make an economical fresh air intake plenum since the pressures in such a plenum rarely exceed 1.0 in. wg. Typical design loads for plenum or casing wall construction are listed in Table 1, and proper mortar mixes in Table 2.

When the values in Table 1 are compared with the commonly used design pressures of 6", 8", or 10" in the fan plenum of today's buildings, it can readily be seen that an 8" unreinforced block wall is not a satisfactory type of construction for these plenums.

Casing construction for low pressures (max. of 2 in. wg) is adequate for either positive or negative pressures; however, medium- and high-pressure casings are rated for 6 in. wg and 10 in. wg maximum positive pressures, respectively, in the SMACNA high-velocity-duct standard. They are not rated for negative pressures, but experience has shown that they will not tolerate negative pressures nearly as high as the maximum positive pressures.

If the fan is started on either a medium- or

Table 1. Typical Design Lateral Loads for Unreinforced Concrete Block Walls

Wall Height	Vertical Load From Roof or Ceiling	Type N Mortar, In. WG	Type M or S Mortar, In. WG
8'-0"	0	2.6	3.7
8'-0"	25 lb/ft.	3.0	4.4
10'-0"	0	1.6	2.3
10'-0"	25 lb/ft.	2.0	2.9
12'-0"	0	1.0	1.5
12'-0"	25 lb/ft.	1.3	1.9

Note: The above pressures are all based on allowable design loads on the wall, as published by the National Concrete Masonry Association (1971).

Table 2. Mortar Mixes for Concrete Block Walls

Mortar Types	Portland Cement	Masonry Cement	Masonry Sand
Type M	1	1	3
Type S	½	1	3
Type N	...	1	3

Note: These types of mortars are in conformance with ASTM Standard #C270-68.

high-pressure system with both the outside air dampers and return air damper in the closed position, it is possible that the extreme negative pressure will collapse the casing. To preclude this possibility, it is recommended that system start-up first open the outside air damper to minimum position. An end switch on the open damper will then start the return fan. The supply fan should be interlocked with the return fan so that it starts after the return fan. During system construction and initial start-up, the automatic control system will probably be inoperative. Until the controls are functioning properly, the contractor is well advised to block open the outside-air, or return-air, damper.

CONDENSATE DRAINS FOR AIR-CONDITIONING UNITS

One of the most serious problems with air conditioning casings is the proper drainage of the cold condensate off the cooling coils. Firstly, the drain pan must be large enough to hold the condensate under maximum latent load. If the pan is long, say 10 ft, it is advisable to drain the pan at each end. Also, if the pan is too shallow, the water will tend to "slop" over the downstream edge of the pan because of the air flow across the water surface.

The second problem associated with drainage is the failure to design a proper trap for the drain tubing, which carries the cold condensate from the pan to the floor drain. This drain tubing should be Type "L" copper tubing, minimum diameter of 1⅜ inches O.D., with a trap designed as in Fig. 1, for a "draw-through" unit.

Determine the design negative static pressure within the fan plenum. Note that this pressure is *not* the same as the fan total pressure, which includes the pressure losses *downstream* as well as *upstream* of the supply fan. Assume the worst conditions, such as having the air filters fully loaded.

Differential 1 (Fig. 1a) must be equal to, or larger than, the plenum negative static pressure at design operating conditions.

To store enough water to prevent losing the trap seal, Differential 2 (Fig. 1b) must be equal to, or larger than, one-half the plenum maximum negative static pressure.

Differential 3 (Fig. 1c) is equal to the maximum negative static pressure in the plenum. This condition probably occurs at start-up, before the air distribution system is brought up to operating pressure. However, tests of actual fan installations indicate that Differential 3 is practically the same as Differential 1, the plenum negative static pressure at design operating conditions. Evidently, the initial surge of negative pressure at

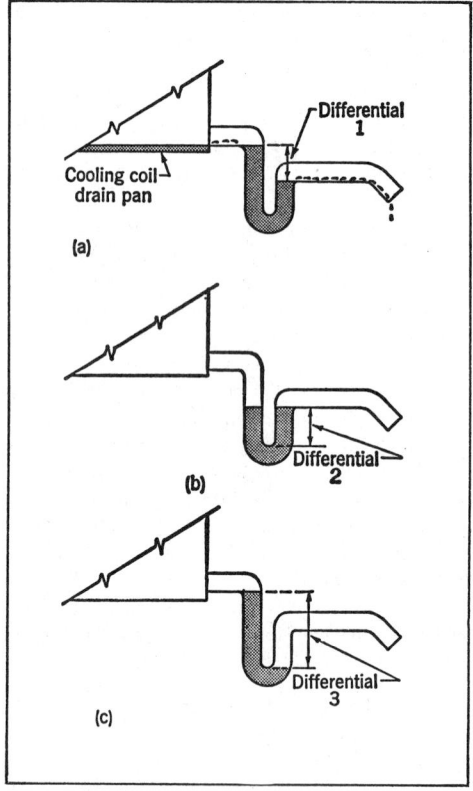

Fig. 1. Properly designed trap, shown here, does not lose its seal under various fan conditions. (a) Fan running and condensate draining: (b) Fan "off;" (c) Trap condition when fan starts. (*Source:* Warren Ng, "Proper Trap Solves the Problem," *Building Systems Design Magazine,* October, 1969.)

fan start-up is not noticeably greater than the normal negative pressure within the plenum, and can be ignored.

Therefore, having determined Differentials 1 and 2, a proper condensate trap for draw-through air conditioning units can be designed.

BALANCING AIR FLOW IN AIR CONDITIONING SYSTEMS

The duct system of an air conditioning installation must be properly balanced to insure satisfactory operation. Engineers of York Corporation compiled these easy step-by-step instructions for use in balancing small, medium or large systems.

Balancing Small Air Conditioning Systems

To balance small air conditioning systems, proceed as follows:

(1) See that all the filters installed as part of the system are clean.

(2) Start the fan. With forward curved blade fans and variable pitch pulleys see that the motor is not overloaded by excessive fan speed.

(3) Examine all supply and return grilles to see that they are each getting air, and that there are no restrictions.

(4) Block the face damper so it is open and the bypass damper closed.

(5) By means of velocity measurements on the return and outside air inlets, set these two air quantities so that they are in the same proportion as the job conditions require (for example, 75% return air and 25% outside air) regardless of what their actual values may be. The air velocities may be read by means of anemometer, Velometer, or pitot tube, as convenience dictates.

(6) Take anemometer readings at the grilles and convert these to cfm to get the total air quantity handled by the fan. If this is within 10% of the estimate, adjustment of the fan speed is unnecessary. If the cfm differs more than this from the required value, change the fan speed in the proportion of cfm required over cfm obtained. Note: The cfm may be obtained also by pitot tube readings in the fan discharge, if convenient.

(7) Select one outlet and take an accurate velocity reading (record for further reference).

(8) Open the by-pass damper and partially close the face damper.

(9) Again take a velocity reading at the selected outlet or grille. If the velocity read is below that observed under (7), the face damper should be opened until the velocity corresponds .within 10% to the value obtained under (7). If the observed velocity is higher than obtained under (7), move the face damper toward the closed position. Mark the damper position finally obtained. Note: In-

stead of observing grille velocities, the above procedure can be followed by using in the same manner, the static pressure in the fan discharge.

(10) Adjust the damper motor linkage so that with the damper motor at one end of its travel, the by-pass will be closed and the face damper wide open; and at the other end of its travel, the bypass will be wide open and the face damper will be in the position determined under (9).

(11) Set the splitters and dampers so that each grille delivers the required cfm by means of anemometer or Velometer readings (generally readings within 10% are close enough).

(12) Place the system under automatic control.

(13) Check the room temperatures for uniformity, and adjust the grilles further if necessary.

(14) Mark all damper positions finally obtained so that if a complaint arises later, it can be determined whether the positon of the damper (or dampers) has been moved.

Balancing Medium and Large Systems

General—While these systems may vary considerably in size or in the extent of the duct distribution, the same general procedure for balancing may be followed on all jobs. In any case, it should be done in a systematic manner by keeping a complete set of records of every operation, and the results obtained. Furthermore, the larger the job, the more difficult it is for an engineer to assimilate the details of the installation in a short time. For this reason, it is good practice, as well as economical, to have someone available who is familiar with the location of all dampers and splitters.

As mentioned before, the apparatus should be balanced first. This procedure would be as follows:

(1) Examine the entire duct system to make sure all grilles, dampers and splitters are open for both supply and return.

(2) Start the fan. (If it is a forward curved blade fan, see that the fan motor is not overloaded.) Note: If the ducts are dirty, it may be necessary to cover the outlets temporarily with cheesecloth to catch the dirt.

(3) Examine the filters to see whether they are dirty or not, as testing should be done with clean filters.

Setting Outside and Return Air Dampers—The following steps are taken in checking dampers:

Table 1—Record of Data in Making Check of Air Quantity.

Item	Est. Cfm	Test 1	Test 2	Test 3	Test 4
By-pass damper		Closed	Closed	Open	Open
Return damper		Open	Closed	Closed	Open
Outside dampers, minimum		Open	Closed	Open	Open
Outside dampers, maximum		Closed	Open	Open	Closed
Face damper		None	None	None	None
Maximum washer cfm	17,800	18,000	18,000		
Minimum washer cfm	9,500			9,600	9,500
Fan speed	428		430		
Fan discharge static pressure, inches ..		0.62	0.63	0.62	0.63
Velocity at eliminator		750	750	400	400
Minimum by-pass cfm					
Maximum by-pass cfm	8,300			8,400	8,400
Minimum outside air cfm	5,000				5,200
Maximum outside air cfm	17,000		18,000		

(4) Open the minimum outside air damper to its approximate setting, and close the maximum outside damper.

(5) Close the by-pass damper and open the return air dampers. By means of velocity measurements on return and minimum outside air dampers, set these two air quantities so that they are in the same proportion as the job conditions require (such as 75% return air and 25% outside air, for example) regardless of what their actual values may be. The air velocities may be read by means of anemometer, Velometer, or pitot tube, as convenience dictates.

(6) Take a static pressure reading on the fan discharge.

(7) Close the by-pass, open the maximum outside air damper, and close the return damper.

(8) Take a static pressure reading as in (6). If this static pressure is lower than in (6) the outside air intake is restricted. If the resistance cannot be removed, block the return air damper partially open so that the fan discharge static pressure is equal to that observed in (6). Install a permanent stop on the damper. If the fan discharge static pressure is higher than in (6), block the outside air damper partially closed so that the fan discharge static pressure is equal to that observed in (6), and install a permanent stop on the damper. Note: By means of this procedure, outside and return air dampers have been set so that the same total air quantity will be handled with the dampers in the positions corresponding to maximum and

minimum outside air. The next step is to set the by-passes.

Setting By-Pass Dampers—There are two cases, washers with and without face dampers. If face dampers are used, it is possible to set the face and by-pass dampers so as to keep the total cfm constant for any position of these dampers, and at the same time to give the designed maximum and minimum cfm through the washer. The procedure is as follows:

(9) With the by-pass closed, and face, return, and minimum outside air dampers opened, read and record the fan discharge static pressure.

(10) With the by-pass, minimum outside air and return air dampers open, move the washer face damper to such a position that the fan discharge static pressure reads the same as in (9).

(11) By means of anemometer, Velometer, or pitot tube readings, determine the cfm through the bypasses and the washer when the dampers are in the positon determined by (10). If the portion by-passed exceeds the design quantity, add resistance to the by-passes. If the proportion by-passed falls below that required, by-pass is restricted and must be relieved.

(12) By trial and error, repeat steps (9), (10) and (11), until the by-pass resistance and face damper position are found which give the design proportion of by-passed and washer air when the by-pass is wide open and at the same time give the same fan discharge static pressure with the dampers in positions described in (9) and (10).

(13) Adjust the damper motor linkage so that with the damper motor at one end of its travel, the by-pass will be closed and the face damper will be in its minimum open position as determined in (12); and with the damper motor at the other end of its travel, the by-pass is closed and the face damper open. Note: When the washer is not fitted with face damper, it is not possible to keep the total air quantity constant with the by-pass damper in open and closed position, the cfm variation depending on the portion of the total system resistance represented by the washer. In such cases, the procedure is as follows:

(14) With the by-pass closed and reutrn and minimum outside air damper open, observe the fan discharge static pressure.

(15) Open the by-pass and osberve the fan discharge static pressure. If this exceeds the static pressure observed in (14) by more than about 20%, resistance must be added to the by-pass until the static pressure increase at the fan discharge falls within 20% of the value in (14).

(16) By velocity readings over the face of the washer and in the by-pass, determine the proportion of by-passed to washer air with the by-pass open. If the proportion of by-pass air exceeds the design value, add resistance to the by-pass by trial until the design proportion is reached. If the proportion of by-passed to washed air falls below that required, the by-pass is restricted and must be relieved. The limit of relieving the bypass is fixed by (15) above. If after the limit of static pressure given in (15) has been reached and there is still insufficient by-passed air, a washer face damper must be used, in which case the adjustment is as described in (9) to (13) inclusive.

Setting Total Air Quantity—In establishing the setting for the total air quantity, proceed as follows:

(17) Take readings of the fan discharge velocity to determine accurately the cfm handled. If this cfm differs from the design value, change the fan speed in the proportion of cfm required over cfm observed.

(18) As a final check, read and record fan discharge static pressure and air quantities, and record as tabulated in Table 1. If the air quantities are not within 10% of the design volume, the previously described procedure should be checked through to find errors.

As in balancing the apparatus, the testing of the duct outlets should be done in a careful and systematic manner.

Balancing Duct Distribution

The simplest procedure is to take the blue-prints of the duct layout, and number each grille, splitter and damper. On the larger jobs, it is good practice to mark also the grilles, dampers, etc. In marking the prints, where th supply duct is divided into two or more branches, each branch should be numbered or lettered as well as the grilles, etc. For example, supply S-1 will have grille S-1-1, S-1-2, etc., and the return R-1-1, R-1-2, etc.

Tabulation of test results is very important and it will be found that careful recording will save considerable time.

(1) When the cfm from the fan is correct, for all damper positions, determine the amount of air to be handled by each branch.

(2) Measure each branch at a suitable point, and from the setimated cfm figure the velocity of air in the duct.

(3) Convert velocity into pressure in inches of water.

(4) Take velocity readings in the branch ducts and set the splitter so that the velocity pressure will be as figured under (3). Note: It is not necessary to take a complete transverse of the duct, but for this purpose three or four readings is all that is required in large ducts.

(5) When the cfm to all the branch supplies are approximately correct, select one branch and take reading of all grilles, record and compare these with the estimated velocities.

(6) From the tabulations of the grille test, note which ones have higher velocity than estimated, and set the dampers or valves to reduce the air quantity to these grilles. This will automatically increase the cfm of the grilles which have low velocity.

(7) Repeat the readings as under (5) and (6) until all are within 10% of estimate.

(8) Proceed in a similar manner on the remaining branches.

(9) Make up a tabulation of the return grilles similar to the tabulation for the supply grilles.

(10) By means of an anemometer, determine the cfm on the return grilles and set the dampers so each handles the correct amount.

(11) Take wet and dry bulb readings in various areas to determine the uniformity of temperature and humidity and adjust the supply grilles if necessary.

Balancing Systems Using Booster Fans

On installations using booster fans, the by-pass air is handled at each zone fan, and there is no by-pass at the main apparatus. The balancing of such a system is done substantially in the following manner:

Setting of Outside and Return Air Dampers—
The following three steps should now be followed:

(1) Close the by-pass and open the dehumidified air dampers on all the booster fans.

(2) Proceed as (1) to (6) under Balancing Medium or Large Air Conditioning Systems. Note: Under "Starting the Fan," the booster fans should also be started.

(3) Close the maximum outside air damper and open the return damper on the main supply apparatus.

Setting Total Air Quantities—This is a more extensive procedure and calls for the following steps:

(4) Close the dehumidified air and open the by-pass dampers on the booster fans.

(5) By means of pitot tube readings on the booster fan discharges, determine the actual cfm being handled by each fan. If this cfm differs from the design value, change the fan speed in the proportion of the cfm required over the cfm observed.

(6) When the cfm of each booster is correct, take the static pressure in the booster fan discharge.

(7) Open the dehumidified air and close the by-pass dampers on the booster fans.

(8) Set the splitters in the supply duct from the main fan so that each booster fan receives the correct proportion of the total cfm regardless of whether the total is correct or not. To do this, take the pitot tube readings and determine the total cfm from the main fan, and then take readings at each branch duct to the booster fans and set the splitters accordingly.

(9) Take a static pressure reading in the booster fan discharge. If this reading is 10% above or below the static pressure taken at (6), change the speed of the main supply fan by:

$$\text{New rpm} = \text{old rpm} \sqrt{\frac{\text{S.P. taken under (6)}}{\text{S.P. taken under (9)}}}$$

This procedure will be satisfactory when the cfm of the booster fan is equal to the cfm of dehumidified air from the main supply fan. When the booster fan cfm is greater than the cfm of dehumidified air supplied to it, such as when a constant amount of by-pass air is handled, use steps (1) to (8).

(10) With the by-pass dampers open and the dehumidified air dampers closed, take a static pressure reading in the by-pass air duct near the damper.

(11) Open the dehumidified air and close the by-pass dampers on all booster fans. If a hand by-pass damper is provided, open it slightly to an approximate position to handle the constant by-pass; if not, block the automatic dampers partially open.

(12) By means of an anemometer, Velometer or pitot tube, determine the cfm of the constant by-pass air and place a stop on the dampers when the correct position is found.

(13) Open the dehumidified air dampers and close the by-pass dampers to the stop.

(14) Take a static pressure reading in the dehumidified air supply duct, as near as possible to the damper. If this static pressure is more than 10% above or below the static pressure taken at (10), change the main supply fan speed accordingly. Note: While the static pressure taken under item (10) may not be the same on each fan, the difference between the static pressure taken at (10) and (14) should be the same for each fan.

(15) With the static pressure taken under (10) and (14) equal, the total cfm from the main supply fan should be correct. However, take pitot tube readings in the main dehumidified air supply duct to check. If it is not correct, the indication is that either the by-pass air dampers were not set correctly or the booster fans were not receiving the correct proportion of air as set under item (8).

(16) Take a static pressure reading in the discharge of the main supply fan and set the static regulator to maintain a constant static pressure (if a static pressure regulator is furnished).

(17) Open the by-pass damper and close the dehumidified air damper to the approximate minimum position.

(18) With a pitot tube, check the dehumidified air supply to each booster fan and set the dampers to the correct minimum position, and install stops.

There are two important reasons for checking the performance of air conditioning installations as soon as they are placed in operation. They are:

(1) To see that each part is operating correctly.

(2) To see that the various parts and their functions are coordinated.

Air Balancing by Balancing and Testing Engineers

In recent years, balancing of both air and water HVAC systems has become a profession.

Associated Air Balance Council, 2146 Sunset Boulevard, Los Angeles, Calif. 90026, certifies that its member firms will test and balance all systems in accordance with the HVAC plans and specifications. This certification guarantees that the mechanical contractor or owner will get a satisfactory job.

Also, the mechanical and sheet metal contractors have joined forces in the National Environmental Balancing Bureau (NEBB) to train and certify its members to test and balance HVAC systems. NEBB is located at 8224 Old Courthouse Rd., Tyson's Corner, Vienna, Va. 22180.

The Associated Air Balance Council has also prepared publications on air balancing; these are available at a small fee. The "Engineer's Project Check List," taken from the *Engineer's Project Design Check List and Design Guide Manual,* Volume 1, published by the Associated Air Balance Council, is quoted below (with some editing), to assure the "balanceability" of the system prior to putting plans and specifications out for bids:

Basic Design Items

1. Supply fan designed for 10% over required air to allow for leakage.
2. Return fan included if system contains "economy cycle" (100% outside air for winter, spring, and fall cooling).
3. Adjustable pressure relief system included (in lieu of #2) to limit building pressurization to .05 in. wg if system contains "economy cycle."
4. Fan capacities selected with correction factor for elevation above sea level.
5. Fan motor selected to allow for additional static due to field conditions.
6. Fan curve checked to provide safety factor against higher operating static pressures.
7. Ceiling diffusers selected for proper application according to the manufacturer's recommendations (i.e., high induction type to satisfy high air volume requirements).
8. Select air outlets which have damper mechanisms readily accessible.

Specification Items

1. Variable-pitch (adjustable) sheaves specified for fan motors—supply, return, and exhaust.
2. Provision made for complete change of filters just prior to balancing.
3. Provision for single-blade manual damper with locking quadrant hardware in each zone duct of multizone system near unit.
4. Provision for manual damper at each branch duct take-off from main duct (return or supply air).
5. Provision for manual damper at each duct drop to diffusers from main duct.
6. Provision for opposed blade dampers at each diffuser and register or grille.
7. Provision for manual damper downstream of mixing dampers at each zone of double duct system.
8. Provision for turning vanes in all square elbows (double-wall type or extended-edge type) in low velocity systems.
9. Provision for perforated static pressure plates at fan discharge of blow through air handling units.
10. Provision for access holes for tachometer readings in all belt guards.
11. No duct liner at duct connections to high-velocity air handlers or multizone units.
12. Provision for 1/2" mesh screen on intakes to prevent clogging.
13. Provision for access doors of adequate size within working distance of all volume dampers, fire dampers, pressure reducing valves, reheat coils, mixing boxes, blenders, constant volume regulators, etc.

14. Provision of extractors at all 90° boot type branch duct connections to main duct.
15. Provision for all duct seams, duct connections, casing and plenum connections to be sealed to minimize leakage factor to approx. 5%.
16. Provision for pressure testing of high pressure duct runs (i.e., spiral duct) for casing wall leakage.
17. Requirement for mixing-box manufacturers to set mechanical volume controllers to plus or minus 5% of design CFM by actual air-flow test methods rather than measured spring adjustments.
18. Requirement for duct-liner in discharge duct after leaving mixing box according to manufacturer's recommendations.
19. Provision for mixing baffles (perforated plate, etc.) at the outlet of all mixing boxes to prevent temperature stratification in supply duct.
20. Requirement for mixing-box dampers not leaking more than 3% of design CFM when unit is operating at the design static-pressure conditions.
21. Provision for round volume dampers to be installed in hot and cold duct take-offs to mixing boxes.
22. Provision for control manufacturer to be responsible for setting and testing all controls in conjunction with air-balance work.

Plan Items

1. Be absolutely certain that all dampers required in "specification item section" above, are distinctly shown in their proper locations on the plans. If a choice is necessary, the best place to cover dampers is on the plans rather than just in the specs.
2. For purposes of locating duct obstructions which reduce air flow to various outlets or zones, all damper locations, including fire dampers, should be shown.
3. Indicate damper locations at accessible points and, whenever possible, at a distance from a duct transition or fitting.
4. Provide manual, opposed-blade type dampers in outside and return air ducts at the entrance to the mixed-air plenum in addition to the automatic proportioning dampers.
5. Splitter dampers to be used as diverters only and not to control air volume; use only in low velocity systems.
6. Where proportional type take-offs are employed, a splitter blade is also recommended to allow adjustment of proportion in the event of higher or lower duct pressures. This does not eliminate the necessity for volume control dampers in the resulting branch ducts.
7. Prevent short-circuiting of discharge air from cooling towers, condensing units, relief exhausts, roof exhausters, etc., to any fan system outside air intake due to close proximity or wind.
8. Connections of outside air or return air ducts should not be made at only one side of a fan inlet plenum.
9. Avoid placement of return-air inlet, in or adjacent to, the return air plenum.
10. Avoid the use of masonry or composition wall vertical shafts for exhaust systems on multistory buildings without interconnecting ductwork. Inherent leakage prohibits the removal of exhaust air from designated locations within the occupied spaces.
11. Avoid locating diffusers, registers, or grilles directly into the bottom or sides of a main duct as no amount of adjustment will decrease the generated noise level.
12. Avoid locating troffer light diffusers on the same duct run or zone with standard type diffusers or registers due to difference in static pressure imposed by each. The higher statics of the troffers will require the standard outlets to be throttled severely and will cause objectionable noise levels at these outlets.
13. Avoid long duct runs containing large volume diffusers on main ducts with small diffusers or registers on branch ducts from the same main duct.

14. Avoid locating ceiling diffusers on the same duct run with registers due to differences in static pressure imposed by each.
15. Provision for extractors (adjustable type) at each boot type branch-duct take-off.
16. Length of duct drops to diffusers should be two times the duct diameter and should contain a volume control damper.
17. Avoid the passage of return air from one space or zone through that of another to reach a return air register.
18. Avoid the use of door louvers for passage of return air when supply air system operates at low pressure (i.e., ceiling plenum supply system).
19. Avoid the use of combination supply and return outlets due to inability to determine amount of short circuiting.
20. Avoid short discharge ducts between mixing boxes and supply registers as they will cause excessive discharge velocities and air noise at face of register.
21. Avoid short, abrupt connections from mixing-box outlet to proportional take-offs or branch ducts.
22. Provision of temperature and pressure gages, flow meters, etc.
23. Duct sizing increased at mixing damper and fire-damper locations so as to provide same free area as in adjoining ductwork.
24. For high-velocity systems, manual volume dampers should be avoided where possible because they are noise generators. If manual throttling is required in these systems, factory-made pressure reducing valves with airfoil-shaped vanes should be used. They are available with either manual or motor operators.

AIR FILTERS AND DUST COLLECTORS

Devices employed to separate solid particulate from a moving air system may be categorized as either (1) air filters, or (2) dust collectors.

Generally speaking, air filters are used in conjunction with warm air heating systems, air conditioning and ventilating systems, and at the intake of process air systems. Applications suitable for air filters are limited usually to contaminant concentrations found to occur naturally in the atmosphere. These concentrations normally will cover a range from as low as 0.03 grains (7,000 grains = 1 lb) per 1,000 cubic feet of air to a high of 0.20 grains per 1,000 cubic feet of air. While these numbers appear to have a low order of magnitude it is obvious that a system operating on a continuous basis and bringing in 50,000 cfm (cubic feet per minute) of air containing a concentration of 0.10 grains per cubic foot will introduce about 30 lbs of airborne dirt per month into the space receiving the air.

Dust collectors are most frequently applied to the exhaust from dust control systems and from processes which result in a contaminant of some form being dispersed into the ambient surrounding the process. Systems employed primarily to capture dust to provide a satisfactory in-plant environment for industrial workers are classified as "dust control systems," whereas those which are intended to prevent the escape of contaminants into the atmosphere are defined as "air-pollution control systems." Dust collectors are called upon to entrap contaminant concentrations which vary from a low of 0.5 grains per cubic foot of exhaust air or gas, to as high as 20 grains per cubic foot. A simple exercise in arithmetic will show that contaminant concentrations handled by dust collectors are approximately 100,000 times greater than those to which air filters may be successfully applied.

Air Filters *

Three basic types of air filters are in common use today: viscous impingement, dry, and electronic. The principles employed by these filters in removing airborne solids include viscous impingement, interception, impaction, diffusion, and electrostatic precipitation. Some filters utilize only one of these principles; others employ them in combination.

---

* Mr. Daniel Lapedes, ed., McGraw-Hill Encyclopedia of Science and Technology, 3rd. ed. (New York: McGraw-Hill Book Co., 1971).

For selected heavy duty applications, where high concentrations are present, inertial separators merit consideration as first-stage precleaners.

Viscous impingement filters.—Viscous impingement filters are made up of a relatively loosely arranged medium, usually consisting of spun glass fibers, metal screens, or layers of crimped expanded metal. The surfaces of the medium are coated with a tacky oil, generally referred to as an adhesive. The arrangement of the filter medium is such that the air stream is forced to change direction frequently as it passes through the filter. Solid particles, because of their momentum, are thrown against, and adhere to, the viscous coated surfaces. Larger airborne particles, having greater mass, are filtered in this manner, whereas small particles tend to follow the path of the air stream and escape entrapment. Operating characteristics of viscous impingement filters are shown in Table I as Category I.

Dry filters.—Dry air filters constitute the broadest category of air filters in terms of the variety of designs, sizes, and shapes in which they are manufactured. The commonest filter medium is glass fiber. Other materials used are cellulose, paper, cotton, and polyurethane and other synthetics. Glass fiber is used extensively because of its relatively low cost and the unique ability to control the diameter of the fiber in manufacture (in general, the finer the fiber diameter, the higher will be the air-cleaning efficiency of the medium).

Dry filters employ the principles of interception, in which particles too large to pass through the filter openings are literally strained from the air stream; impaction, in which particles strike and stick to the surfaces of the glass fibers because of natural adhesive forces (even though the fibers are not coated with a filter adhesive), and diffusion, in which very small particles behave as true gases and move toward a filter fiber to replace those particles in the immediate vicinity of the fiber which have impacted on its surface. Through the process of diffusion a filter is able to separate from the air stream particles much smaller than the openings in the medium itself. Operating characteristics of dry type filters are shown as Categories IIa and IIb in Table I.

Electronic air-cleaners.—Used primarily on applications requiring high air-cleaning efficiency, these devices operate on the principle of passing the air stream through an ionization field where a 12-14 kV potential imposes a positive charge on all airborne particles. The ionized particles are then passed between aluminum plates, alternately grounded and connected to a 6-8 kV source, and are precipitated onto the grounded plates.

The original design of the electronic air-cleaner utilizes a water-soluble adhesive coating on the plates, which holds the dirt deposits until the plates require cleaning. The filter is then de-energized, and the dirt and adhesive film are washed off the plates. Fresh adhesive is applied before the power is turned on again. Newer versions of electronic air-cleaners are designed so that the plates serve as agglomerators; the agglomerates of smaller particles are allowed to slough off the plates and to be trapped by viscous impingement or dry-type filters downstream of the electronic unit. Operating characteristics of electronic air cleaners are shown as Category III in Table I.

Testing and rating.—Filter testing and rating has been recently standardized for the first time on an industry-wide basis with the implementation of ASHRAE Test Standard 52-76. Applicable both to nominal efficiency filters, where Arrestance efficiency is determined, and to high efficiency filters, where a dust spot efficiency on atmospheric air is established, the ASHRAE procedure requires the reporting of initial (clean) efficiency, dust holding capacity with a standardized test dust, and simultaneous determination of average efficiency and dust holding capacity. To insure the accuracy of published data there must be independent laboratory certification of the manufacturer's performance claims. For high efficiency particulate air (HEPA) filters, as used in special-purpose applications involving protection from radioactive contamination, clean rooms, etc., the dioctylphthalate (DOP) test is employed. This test uses a specially generated smoke of uniform 0.3 micron particles.

Filter selection.—Unlike other components of mechanical ventilation and air conditioning systems the selection of the proper air filter for a specific application involves judgment more than the use of clearly defined parameters. The coordinates of air filtration do not lend themselves to selection charts and tables as do other system components such as fans and heat ex-

changers. However, there are important guidelines to proper selection. These are: Efficiency requirements; initial and operating cost factors; maintenance characteristics; effect on system of air friction loss; and equipment space requirements.* A generalized analysis of how these

guidelines apply to the more commonly used types of air filters is shown in Table 2. Since Operating Cost can be a major variable it is good practice to prepare a detailed cost study for any sizable installation.

* Lapedes, Ibid.

Table 1. Characteristics of Air Filters *

Filter Categories	Principles of Operation	Type of Filter Surface	Media Velocity versus Face Velocity fpm	Air Friction Loss [1]	Efficiency Test Methods (See *Testing and rating* in text.)	Efficiency %	AFI Dust Holding Capacity, Grams Per Sq Ft
I. Viscous Coated	Straining & Impingement	Coarse Fiber, Loosely Spaced, or Metal Screens	300-600 vs. 300-600	Variable, Low, 0.10-0.50	Arrestance (ASHRAE)[5]	Low (70-85)	High 75-150 [4]
IIa. Dry	Straining & Interception	Medium to Fine Fiber, Moderately to Closely Spaced	20-50 vs. 250-625	Variable, Moderate 0.20-1.0	Arrestance or Dust Spot (ASHRAE)[5]	Moderate (85-95) to High (85-95)	Moderate, 5-15
IIb. HEPA	Interception & Diffusion	Ultra Fine Fiber, Very Closely Spaced	5-6 vs. 250-300	Variable, High, 1.0-2.0	DOP	Ultra High (99.97)	Low, 1-2
III. Electro-static	Precipitation &/or Agglomeration	Plates	400-600 vs. 400-600	Constant & Low,[2] or Variable & Moderate,[3] 0.45, or 0.5-1.0	Dust Spot (ASHRAE)[5]	High (85-97)	

[1] Values shown are in inches of water column and range from initial (clean) to final (dirty).
[2] Automatic renewable media storage section
[3] Cartridge storage section
[4] Dynamic test value shown for upper limit.
[5] Per ASHRAE Test Standard 52-76

(Source: American Air Filter Co., Inc.)

Table 2. Guide to Air Filter Selection *

Filter Types—

A—Throwaway (I)
B—Cleanable (I)
C—Replaceable pad (I)
D—Automatically renewable media (I)
E—Cartridge, 30% NBS (I or II)
F—Cartridge, 40% NBS (I or II)
G—Cartridge, 50-55% NBS (IIa)

H—Cartridge, 80-85% NBS (IIa)
J—Cartridge, 90-97% NBS (IIa)
K—Electronic agglomerator with automatically renewable media storage (III)
L—Electronic agglomerator with replaceable cartridge storage section, 93-97% NBS (III)
M—Electronic air cleaner, washable plates (III)

System Selection Factor	Filter Type Ratings		
	Nominal efficiency	Medium efficiency	High efficiency
Air friction loss—			
Constant	D	—	K, M
Variable	A, B, C	E, F, G	H, J, L
Initial cost—			
Low	A, B, C	E, F, G	—
Moderate	D	—	H, J
High	—	—	K, L, M
Operating cost—			
Low	D	F	K, M
Moderate	B, C	E, G	L
High	A	—	H, J
Maintenance skills required—			
Minimal	A, B, C	—	—
Average	D	E, F, G	H
Trained	—	—	J, K, L, M

* Filters of various types are rated with respect to selection factors of air friction loss, initial cost, operating cost, and maintenance skills required for nominal-, medium-, and high-efficiency applications. The filter category from Table 1 is shown in parentheses after each type commonly used. (*Courtesy of Heating, Piping and Air Conditioning,* Sept. 1970.)

Dust Collectors

Dust collectors may be divided into four major categories:

1. Dry Mechanical Collectors
2. Wet Collectors and Scrubbers
3. Fabric Collectors
4. Electrostatic Precipitators.

Industrial dusts emanate from three major sources:

1. Abrading
2. Combustion
3. Materials Handling

A typical industrial process, such as metal grinding, will produce metallic dusts in the range from 10 to 100 microns.

Differences in size and concentration of particulate matter dictate a considerable difference in the performance requirements of dust collectors, as compared with air filters. The family of curves shown in Fig. 1 may be used as an unrefined method of arriving at the recommended type of dust collector for a given application.

The usual method of rating dust collectors, even those with a high efficiency characteristic, is on a weight basis. By comparison, only low efficiency air filters are rated on a weight method

while high efficiency filters are always rated by the dust spot or DOP methods.

Major differences between rating air filters and dust collectors on the weight method is that an artificial test dust is used for rating air filters, whereas on dust collectors the rating is determined by actual performance against the specific dust which is to be collected.

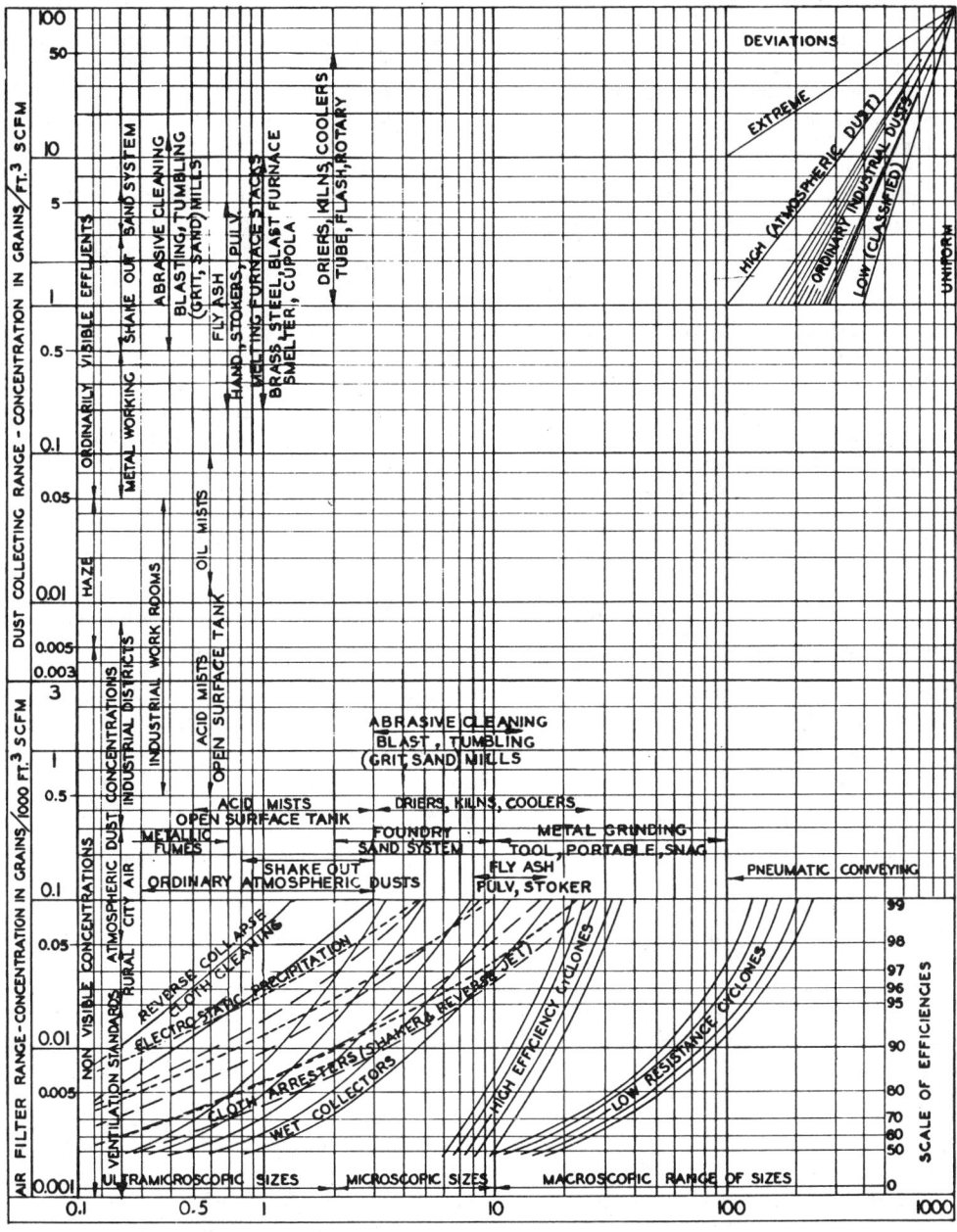

Fig. 1. Range of particle sizes, concentration, and collector performance. (*Courtesy American Air Filter Co., Inc., 1962.*)

Dry Centrifugal Collectors.—Dry centrifugal collectors are the least costly and for many years were the most frequently used type of dust-collection equipment. They are effective on almost all types of medium- to coarse-size granular dusts and are commonly used as either primary collectors or as precleaners to more efficient final collectors. They have the advantage of being low in cost, inexpensive to operate and maintain, and of being relatively small in size.

Because of today's more stringent emission standards, the use of the dry centrifugal collector is limited almost exclusively to product recovery applications that are upstream of more efficient secondary collectors.

Dry centrifugal collectors are commercially available in a wide variety of designs, but all depend on centrifugal force to separate dust particles from the air stream.

The magnitude of the centrifugal force which can be exerted on a particle is a function of the particle mass and the angular acceleration. The latter is dependent on the air velocities in the collector. As velocities increase, energy losses due to friction and turbulence also increase, and the pressure drop across the collector rises. In general, dust collection efficiency becomes greater as collector pressure-drop increases.

Unfortunately, collection efficiency cannot be improved without limit by increasing the energy input and collector pressure drop. Although higher velocities result in a larger centrifugal force vector, which causes the particle path to diverge from that of the air stream, the aerodynamic forces acting to resist path divergence are also increased. This effect is particularly pronounced on smaller particles because they have a greater surface area per unit mass. Every dry centrifugal collector, therefore, has an upper performance limit (which can be expressed conveniently in terms of pressure drop) beyond which its collection efficiency for a given dust cannot, for practical purposes, be increased. Collector sizing and selection is largely a matter of choosing an optimum operating point—the point selected should yield the highest possible return, in terms of collection efficiency, for power input.

Wet Collectors.—Wet dust collectors provide a comparatively simple, moderate-cost solution to many dust control and air pollution problems. Space requirements are generally less than for other collector types. Because equipment size is small in relation to air cleaning capacity, most collectors can be shipped from the manufacturer completely assembled or in major sub-assemblies, simplifying installation and reducing erection costs.

Capable of cleaning hot, moist gases which are difficult or even impossible to handle with other collector types, wet collectors also are often able to eliminate or substantially reduce the hazards associated with the collection of explosive or highly flammable materials. In addition, since solids are collected in a wetted form, secondary dust problems during material disposal are avoided. They are commercially available in a wide variety of designs, shapes, and sizes. The collection principles employed are centrifugal force, impaction, and impingement, either separately or in combination.

Independent investigators studying wet collector performance have developed the Contact Power Theory, which states that for well-designed equipment, collection efficiency is a function of the energy consumed in the air-to-water contact process, and is independent of the collector design. On this basis, well designed collectors operating at or near the same pressure drop can be expected to exhibit comparable performance.

All wet collectors have a fractional efficiency characteristic; that is, their cleaning efficiency varies directly with the size of the particle being collected. In general, collectors operating at a very low pressure-loss will remove only medium- to coarse-size particles. High-efficiency collection of fine particles requires increased energy input, which will be reflected in higher collector-pressure loss.

High-efficiency wet collection of submicron particulate, fume, and smoke has been made possible largely by the development of the high-energy venturi type collector. Venturi designs are now used on a large number of applications formerly limited to fabric or electrostatic collectors. In accordance with the Contact Power Theory, venturi type collectors require substantial energy input to achieve high collection efficiency on submicron particles. A recent new concept in wet collectors which appears to present a solution to a long-standing problem employs a mobile bed contactor for scrubbing of SO_2 fumes from the products of combustion of power generating and large industrial boilers.

Collector water requirements represent a con-

tinuing operating cost which must be evaluated when selecting specific equipment. Further considerations are necessary when applying wet collectors to certain fume scrubbing processes since a solution of water and a chemical additive may be required to achieve the desired objective. When required water rates are high, substantial savings can usually be realized by using a recirculating water system. Such systems usually employ a settling tank or pond to separate the collected material by gravity. Since the water returned to the collector will invariably contain some solids, it is advantageous in these applications to choose a collector which does not require spray nozzles or other small water orifices.

Corrosive substances are often present in typical wet collector applications. Modern construction materials are capable of providing satisfactory protection against nearly all corrosive agents, but the chemical compounds present must be correctly anticipated and identified in order to make the proper material selection.

Fabric Collectors.—Fabric collectors represent one of the oldest and most successful methods of dust collection. They are commonly used when particle size is small and a high degree of cleaning efficiency is required. They also have the added capability of reclaiming maximum quantities of valuable process materials in a dry state. Newly enacted governmental regulations placing limitations on process emissions has created an even stronger demand for the use of fabric collection.

All fabric collectors employ the same method of separating particulate from the air stream. Dust-laden air flows through a cloth tube or envelope, where particles larger than the fabric interstices are deposited by simple sieving action. A mat or "cake" of dust is quickly formed on the air-entering surface of the fabric. The dust cake acts as a highly efficient filter, capable of removing submicron dusts and fumes, while the fabric serves principally as a supporting structure for the cake.

There are many different fabric collector designs in current use. Basically, the collectors differ in the method used to remove the deposited material from the fabric surface. Common methods include mechanical shaking, reverse-air collapse, and pulse-jet.

Selection of the method of fabric reconditioning (removal of dust accumulation from the fabric surface) is primarily a function of the nature of the dust, whether the process is of a continuous or intermittent nature, and the choice of fabric. Fiberglass cloth, for example, is often selected for higher temperature applications (up to 500 F). However it is generally limited to reverse-air equipment since the glass fibers would fail if subjected to the flexing imposed by a shaker mechanism. Shaker collectors are frequently used for intermittent-process applications, while the pulse-jet type is a more common choice for continuous processes. As with most capital equipment, the final selection generally resolves itself into a matter of economics where there are no other overriding factors.

Fabric options include almost all of the organic and synthetic fibers. The choice of fabric is usually a function of the temperature and fume content of the exhaust gases, and the chemical composition of the dust being captured.

Electrostatic Precipitators.—These are capable of attaining a dust collecting efficiency equal to that of a fabric collector. The frequency with which they are used is somewhat limited, however, due to both high initial installed cost, which prohibits their use on small and medium air volumes, and the need for more sophisticated engineering personnel to maintain them properly. The high voltage electrostatic precipitator should not be confused with the low voltage, small dust holding capacity designs used for filtration in air conditioning systems.

The principle of collection relies on the ability to impart a negative charge to particles in the gas stream causing them to move and adhere to the grounded, or positively charged, collector plates. Most precipitators are made for horizontal air flow with velocities of 100 to 600 fpm. The collecting plates are parallel elements, typically on 9-inch centers, and constructed in various ways including corrugated or perforated plates or rod curtains. The weighted wire or rigid electrodes are centered between the collector plates. Voltage difference between electrode and plate is 60,000 to 75,000 volts in most designs. Collector plates of cylindrical shapes surrounding the electrode rod are provided where water is used to wash off collected material and where the gas stream is under high pressure or vacuum.

Removal of the collected material is obtained by rapping or vibrating the elements, either continuously or at predetermined intervals. Vibration or unloading usually takes place without stopping

air flow through the precipitator, although some loss to the effluent air can be expected in most applications during the cleaning cycle.

Pressure drop is negligible. Collection efficiency is high and nearly uniform, regardless of particle size including submicron particles. Space is relatively large and cost is high where small gas volumes (below 50,000 cfm) are involved, due to the cost of high-voltage equipment. Efficiency is improved with increased humidity of the air stream as a change takes place in the dielectric properties of the dust. Heavy concentrations, on the other hand, cause a reduction in collection as the space charge on numerous particles blankets the corona effect from the electrodes.

Under these circumstances a precleaner, often of the dry centrifugal type, is introduced for the purpose of reducing the inlet concentration entering the precipitator.

Electrostatic precipitators have been extensively used in high-temperature gas cleaning from equipment such as blast furnaces, open hearth furnaces, and central station pulverized fuel boilers. The chemical industry has many applications including sulfuric acid plants, carbon black, cement kilns, and soda ash from paper mill black-liquor furnaces. As voltage setting is close to the spark-over potential, application is limited to materials that are not explosive or combustible in nature unless the carrier gas stream is an inert gas.

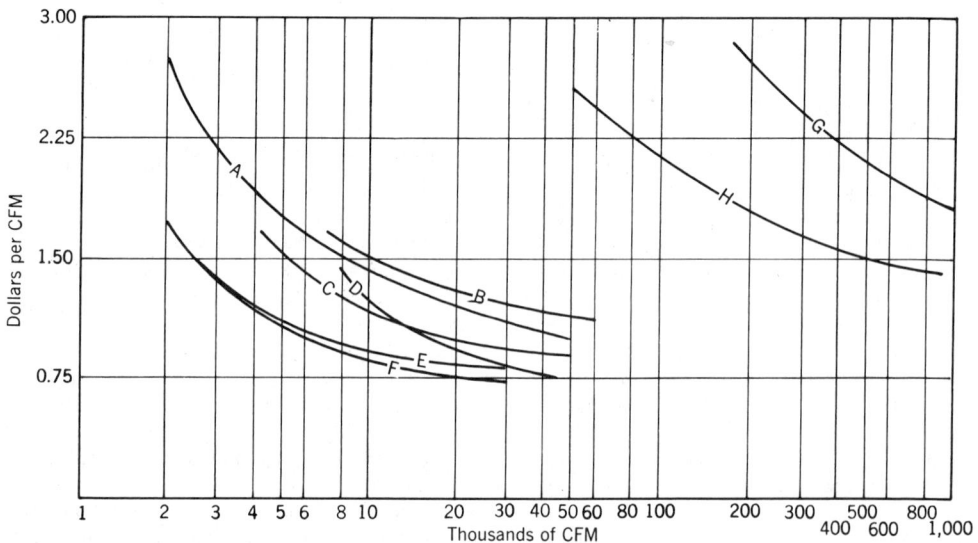

A— Pulse-jet fabric collectors (continuous duty)
B— Wet collectors and separator (venturi type)
C— Intermittent duty fabric collectors
D— High efficiency centrifugal collectors
E— Wet collectors (moderate pressure drop)
F— Low pressure drop cyclones
G— Electrostatic precipitators (99 + % collection efficiency)
H— Electrostatic precipitators (90% collection efficiency)

Note 1: Cost data for A-F based on collector and exhauster only. Cost data for electrostatic precipitators does not include fan. Costs do not include ducting, water requirement or power requirement.
Note 2: Cost of continuous duty sectional fabric collector approaches cost of pulse-jet continuous duty collector.
Note 3: Price of electrostatic precipitators will vary with the contact time and the electrical equipment required. Prices shown are for fly ash installations when high velocities of 200 to 400 fpm are usual. Precipitators for metallurgical fumes, etc., will be considerably higher in cost per cfm.
Note 4: Curves based on data available in January, 1975.

Fig. 2. Cost Estimates of Dust Collecting Equipment. (*Source:* American Air Filter Co.)

Dust Collector Selection

As with air filters, the proper dust collector selection involves both judgment as well as a knowledge of the operating conditions of the particular system to which the equipment will be applied. The primary parameters affecting selection consist basically of the following:

—efficiency requirements
—initial and operating cost factors
—maintenance characteristics
—effect on system of variable friction loss
—equipment space requirements
—availability of water disposal considerations
—humidity of exhaust air or gas stream operating temperatures.

These factors have all been considered in the compilation of the data in Table 3. Initial cost of the various types of collectors is plotted against air volume curves in Fig. 2. It is essential that operating cost be calculated, since this may well influence the final decision, which is based on fan and pump hp considerations as well as on maintenance costs obtainable from recognized equipment manufacturers.

After Filter Placement.—When after filters are used for labs, hospitals, etc., the fan discharge is directly into the filter bank in most cases. This has two adverse effects:

1. The structural strength of the filter media may not be able to resist the high air velocities, and will be torn apart from air impact.

2. Only those filters which are in the direct stream of the air blast will accomplish any filtering since most of the air goes through these.

To prevent filter destruction and unequal distribution over the face of the filter bank, a perforated stainless steel "splash plate" should be installed in front of those filters on which the air blast is mainly directed. This would appear to introduce an additional dynamic loss for the air. However, when a fan discharges into a plenum space, such as an after-filter bank, the velocity pressure of the air is reduced to an insignificant figure anyway, hence, the splash plate actually contributes very little aerodynamic loss that was not already in the system. For a discussion of this problem, see "Fan Discharge Connections."

BREECHING DESIGN AND CONSTRUCTION

Boiler breeching for HVAC installations must handle flue gases at elevated temperatures; for HVAC boilers, these temperatures vary in a range from 400 to 500°F. For the correct flue gas temperature the boiler manufacturer should be consulted. These elevated temperatures affect breeching design as explained in the following.

Expansion

The sheet steel breeching is erected at room ambient temperatures, so that it expands when hot flue gases are carried to the chimney flue. Expansion joints are required to absorb relative movement between fixed sections of breeching or between a section of breeching and the chimney flue.

As a "rule-of-thumb" for expansion joint location, use the following:

For very short breeching runs of 10 feet or less between fixed points, such as connections to boilers, expansion joints are required for combined axial and lateral movements over 1/4 inch.

For longer runs of breeching between fixed points, the maximum total relative axial and lateral movement which can be tolerated is 1/2 inch. If this movement exceeds 1/2 inch, an expansion joint is required.

To find the total relative axial and lateral movement, multiply the linear coefficient of expansion of steel (0.0000702 inch/inch/deg. F) by the expected temperature rise (flue gas temperature minus ambient temperature at installation, usually taken at 40°F) ; this gives the breeching movement per inch of run. Next, multiply the total runs of breeching between fixed points, based on centerline distances (including offsets) in inches, by the movement per inch of run. If this total expansion exceeds 1/4 inch for very short runs or 1/2 inch for longer runs, an expansion joint is required in this section.

Individual breechings must have individual expansion joints to take care of the temperature variance of each duct causing different expansions at different times, as shown in Fig. 1.

Expansion joint construction is based on either

Table 3. Comparison of Dust Collector Characteristics

Type	Higher Efficiency Range On Particles Greater Than Mean Size, In Microns	Pressure Loss, Inches of water (WG)	H₂O Gal., Per 1000 CFM	Space	Sensitivity To CFM Change — Pressure	Sensitivity To CFM Change — Efficiency	Humid Air Influence	Maximum Temperature, F. Standard Construction
Electrostatic	0.25	½	—	Large	Negligible	Yes	Improves Efficiency	500
Fabric:								
Conventional	0.25	3-6	—	Large	As CFM	Negligible	May Make Reconditioning Difficult	180, Note 1
Reverse Jet	0.25	3-8	—	Moderate	As CFM	Negligible		180
Glass, Reverse Flow	0.25	3-8	—	Large	As CFM	Negligible		550
Wet:								
Packed Tower	1-5	1.5-3.5	5-10	Large	As CFM	Yes	None	Unlimited
Wet Centrifugal	1-5	2.5-6	3-5	Moderate	As (cfm)²	Yes		
Wet Dynamic	1-2	Note 2	½ to 1	Small	Note 2	No		
Orifice Types	1-5	2½-6	10-40	Small	As CFM	Varies With Design		
Higher Efficiency:								
Fog Tower	0.5-5	2-4	5-10	Moderate	As (cfm)²	Slightly Yes	None	Note 3
Venturi	0.5-2	10-100	5-15	Moderate	As (cfm)²	Yes		Unlimited
Dry Centrifugal:								
Low Pressure Cyclone	20-40	0.75-1.5	—	Large	As (cfm)²	Yes	May Cause Condensation And Plugging	750
High Efficiency Centrifugal	10-30	3.6	—	Moderate	As (cfm)²	Yes		750
Dry Dynamic	10-20	Note 2	—	Small	Note 2	No		750

Note 1: 180°F based on cotton fabric. Synthetic fabrics may be used to 275°F.
Note 2: A function of the mechanical efficiency of these combined exhausters and dust collectors.
Note 3: Precooling of high temperature gases will be necessary to prevent rapid evaporation of fine droplets.
(*Source: Industrial Ventilation Manual,* 14th ed.)

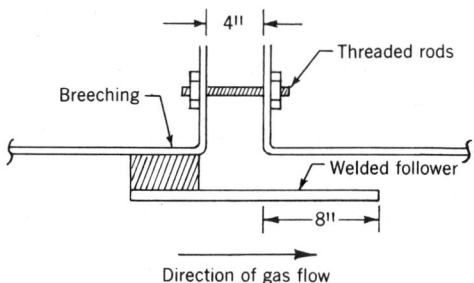

Fig. 1. Expansion joint.

a slip joint, illustrated in Fig. 2, or a welded bellows joint ("accordion" type). The slip joint is easier to fabricate and less expensive, and though not airtight, will be satisfactory for positive pressures encountered in most HVAC installations. The welded bellows joint is used for large boilers, such as power plant installations.

Aerodynamics

Good aerodynamics are essential for breeching to prevent eddy currents which cause vibrations and pressure losses. There are, however, some special considerations for breeching design:

1. If a boiler uses natural draft, which is not common practice, the maximum attainable flue gas velocity is 1500 fpm. The engineer should check local code requirements which may not accept velocities that low.

2. If a boiler uses mechanical draft, either forced or induced, a flue gas velocity of 2000 to 3000 fpm in the breeching is recommended to prevent soot or ash dropout; however, flue gas velocities as high as 4000 fpm are permissible, provided the draft is available. Again, the engineer should check local code requirements.

3. In calculating flue gas draft requirements,

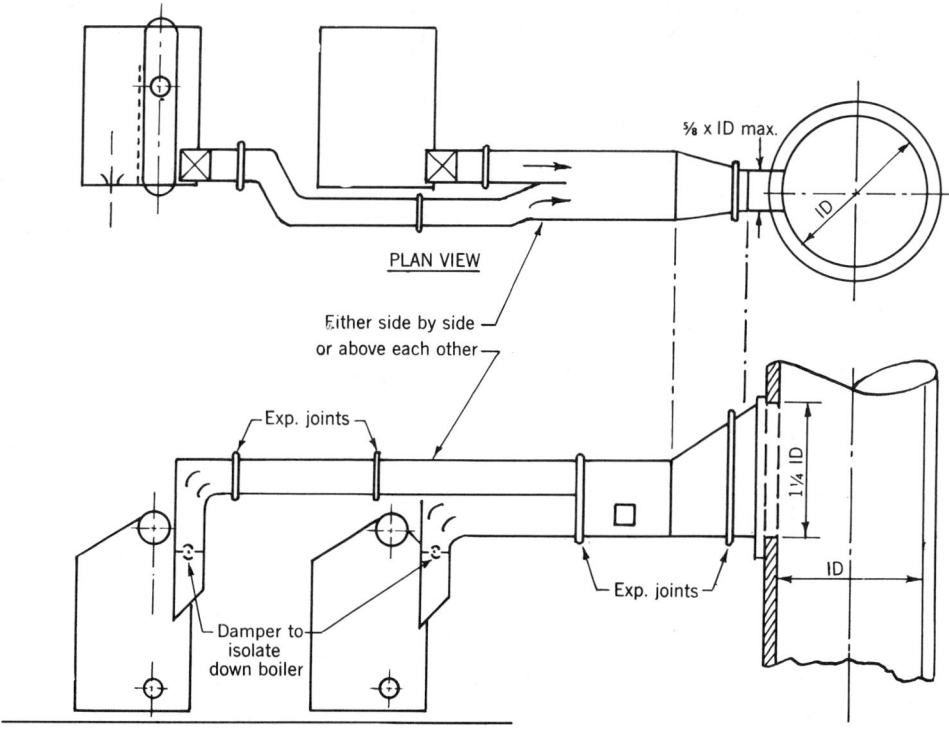

Fig. 2. Typical breeching flow arrangements and stack entry.

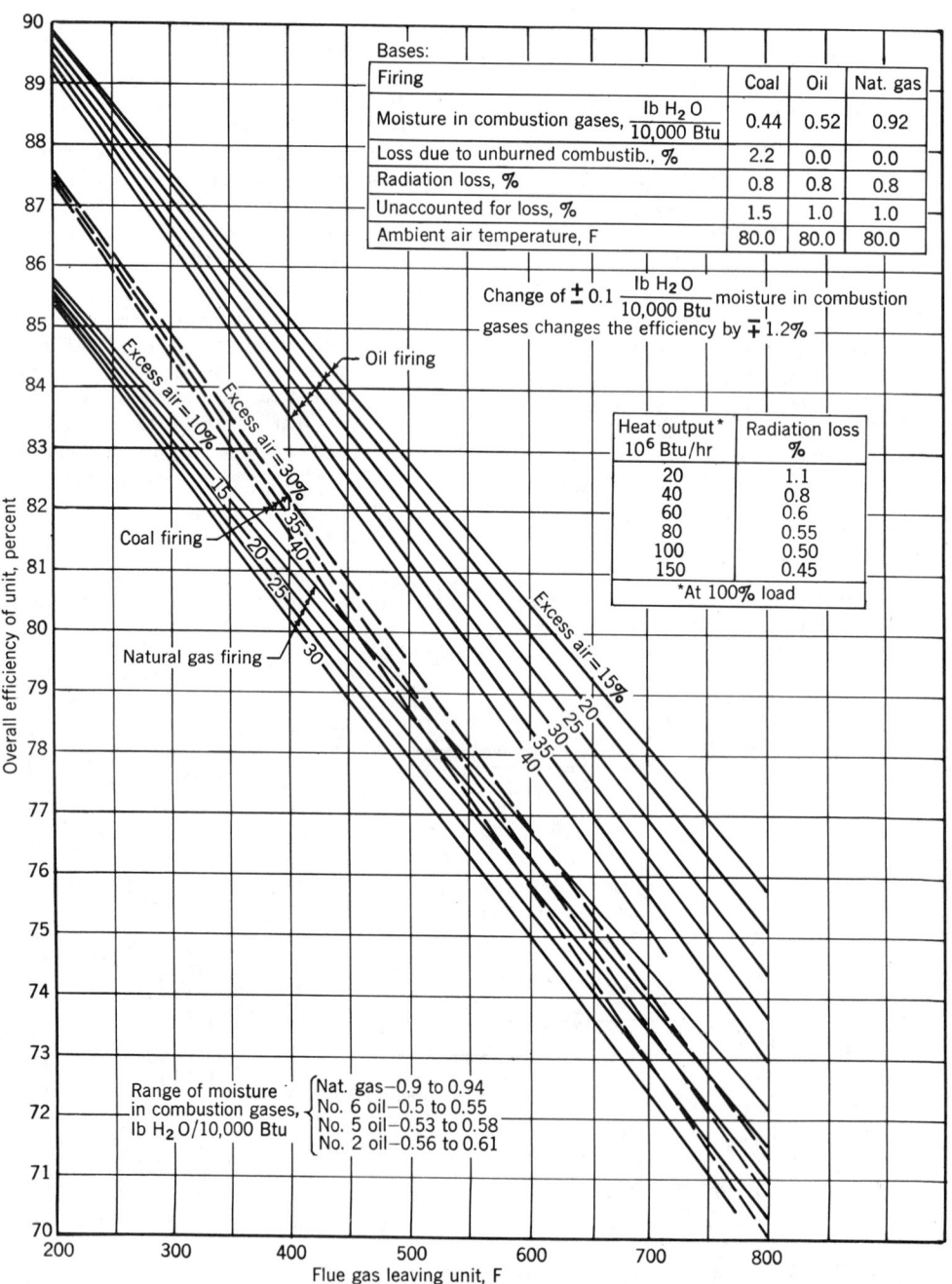

Fig. 3. Overall efficiency of steam generating unit. *Courtesy of Cleaver-Brooks Co., Milwaukee, Wisc.*

breeching and chimney losses can be calculated in the same manner as ductwork losses, except that these losses must be based on the actual flue gas temperature. This affects the gas density, hence the losses. Fig. 3 gives the overall efficiency of a steam generating unit vs. leaving flue gas temperature. In the absence of manufacturer's data, this will assist the designer in selecting the leaving flue gas temperature leaving the boiler.

4. The type of draft arrangement depends on the boiler and burner selected. Some boilers have induced or forced draft only, others have both; and a few use natural draft only. The draft requirements through the boiler are set by the boiler manufacturer; however, those in the breeching and chimney are calculated by the engineer. The forced and/or induced draft head available, should be adequate to guarantee maximum flue gas flow requirements. In recent years, air pollution control agencies have demanded that exiting velocity from the chimney be a minimum of 3000 fpm at 60% of the total load. This means a stack exiting velocity of 5000 fpm when all boilers are firing at their maximum rate. To get this high terminal velocity, a tapered cone is installed on the chimney outlet. This flue gas pressure to velocity conversion requires a large increase in the draft pressure, and must be added to the other calculated draft losses for a proper fan selection. The height of the reducing cone should be at least two diameters of the stack before the taper starts.

5. Transition pieces in breeching should not be abrupt. For *decreasing* cross-sectional area of the breeching (converging section), the total angle of transition should not exceed 60°. For *increasing* cross-sectional area of the breeching (diverging section), the total angle of transition should not exceed 30°.

6. Multiple gas flows from more than one boiler into a common breeching should flow parallel into the breeching without one flow disturbing or even cutting off the flow from the other boiler *as this can lead to explosions during start up.* This is shown in Figs. 1 and 5.

7. Where shoe tops are made into the bottom

of the breeching, due to tight ceiling heights, splitters must be used in the breeching—otherwise, the upward jet of the boiler's flue gas into the main stream will result in high dynamic losses.

8. Entering a round stack with two opposite breechings requires a partition plate that should be arranged tangentially or diagonally so as to prevent bounce-back eddy currents when a breeching discharges at a 90 degree dead impact on the partition plate. The available stack draft would then not be reduced, provided that the flow areas are correct. This is illustrated in Fig. 4.

9. If directional flow changes are at an angle of more than 30°, they should have vane guidance plates, approximately 12″ apart so that a man can get through the clearance between the vane blades. These turning vanes must be stiff and connected to the duct sides by continuous strength weld for the full edge length of the vane plate to eliminate the possibility of having them tear off the duct walls. If the vane blades are over 4 ft long, a center support for the vanes (made of flats or pipes) should be provided. The vanes should be made of 16 gage, minimum thickness, with smooth leading and trailing edges. The trailing edge should be extended at least 12 inches on a straight run, to obtain better regain. See Fig. 5.)

10. Round breeching cross-section is preferred to rectangular cross section for two reasons:

    1. It is a more efficient cross section for gas flow.
    2. It has rigidity from its shape, whereas the rectangular cross section usually requires external or internal stiffeners.

Unfortunately, tight headroom conditions may preclude round breeching cross-section; in this case, the rectangular cross-section should be as close to square as is possible.

Access

Access must be provided to all parts of the breeching for the purpose of conducting inspection maintenance and repair. The means of obtaining this access are doors attached to the breeching plate or, in some cases, entry from

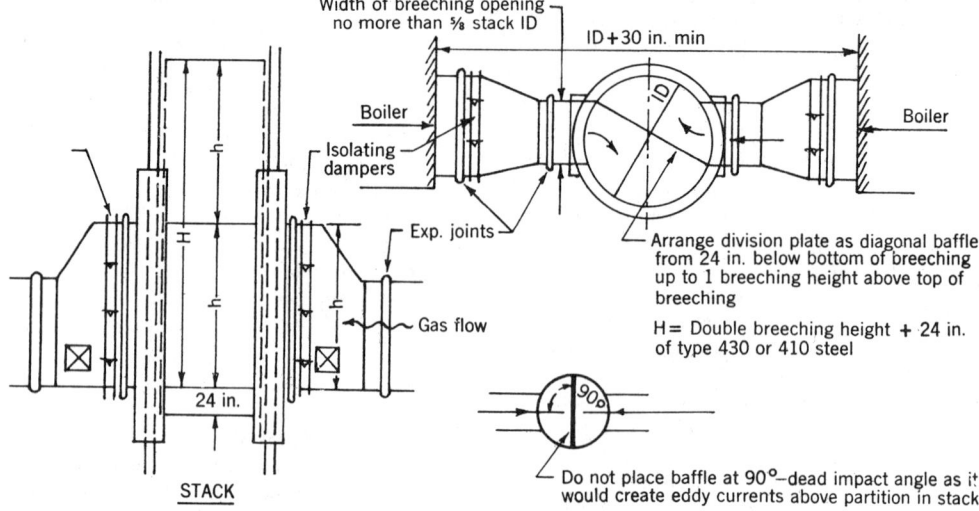

Width of breeching opening — no more than ⅝ stack ID

ID + 30 in. min

Boiler

Isolating dampers

Exp. joints

Boiler

Arrange division plate as diagonal baffle from 24 in. below bottom of breeching up to 1 breeching height above top of breeching

Gas flow

H = Double breeching height + 24 in. of type 430 or 410 steel

24 in.

STACK

Do not place baffle at 90°—dead impact angle as it would create eddy currents above partition in stack

Fig. 4. Opposite stack entry.

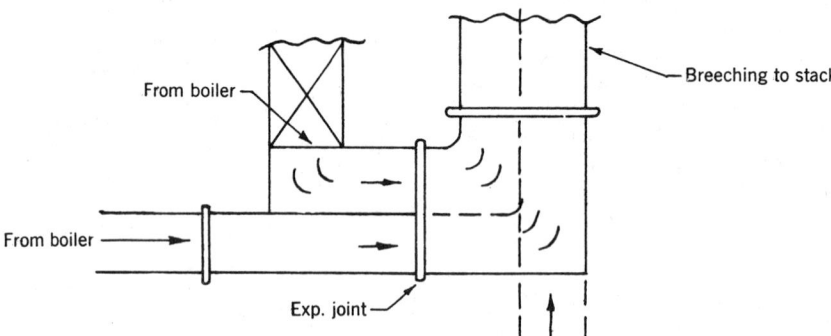

From boiler

Breeching to stack

From boiler

Exp. joint

Fig. 5. Breeching entries for joining gas flows parallel. *Courtesy of Cleaver Brooks Co., Milwaukee, Wisc.*

some other area that is provided with a door. Location and means of access should be carefully considered in order to ensure the following:

1. Convenience—Doors must be located where they may be easily entered from platforms or ladders.

2. Interference—Do not locate doors within 12 inches of corners, expansion joints, or dampers.

3. Code requirements—A minimum clear space of 15 inches must be maintained around breeching for access to its surfaces.

4. Access doors on both sides (certainly one side) of dampers are advisable.

In addition to access doors, 2-inch drain nipples with gate valves should be welded to the bottom of depressed breeching sections, to permit draining condensate during start-up, etc.

Openings in the breeching must be provided for draft gage connections, smoke density sensors, thermometers, and carbon dioxide sampling connections.

Round Breeching Construction

Round breeching is made of rolled black steel plate (0.15 to 0.25 carbon), all welded longitudinal seams. End joints are either welded or made with companion rolled angle flanges (2½ x ¼ inch) with asbestos gasket.

A schedule of steel gages for round breeching is tabulated below:

Breeching Diameter	Steel Plate Gage
12 inches, and below	18
13 to 24 inches	16
25 to 36 inches	14
37 to 60 inches	12
60 inches, and above	10

If the fuel has more than ½% sulfur content and flue gas temperatures are expected to drop below the dew point generally below 270°F) at low loads, the breeching steel should be corrosion resistant (COR-TEN or equal). This will be more likely to happen to uninsulated breeching; therefore, breeching inside the boiler plant should be insulated to keep flue gases above the dew point and assure comfort for the operating engineers. In this case, insulation retaining clips should be specified.

Rectangular Breeching Construction

Rectangular breeching is made of black steel plate (0.15 to 0.25 carbon) all welded construction. End joints for each section are 2 inch flanges with open corners, fully welded all around, as shown in Fig. 6.

For small rectangular breeching cross-sections, maximum side-width of 48 inches, the construction shown in Fig. 6 is satisfactory, provided the design pressures do not exceed 3 in. wg. While 10 gage steel plate is recommended for this construction, very small breeching sizes (up to 18″ max. width) can be made with 16 gage steel plate.

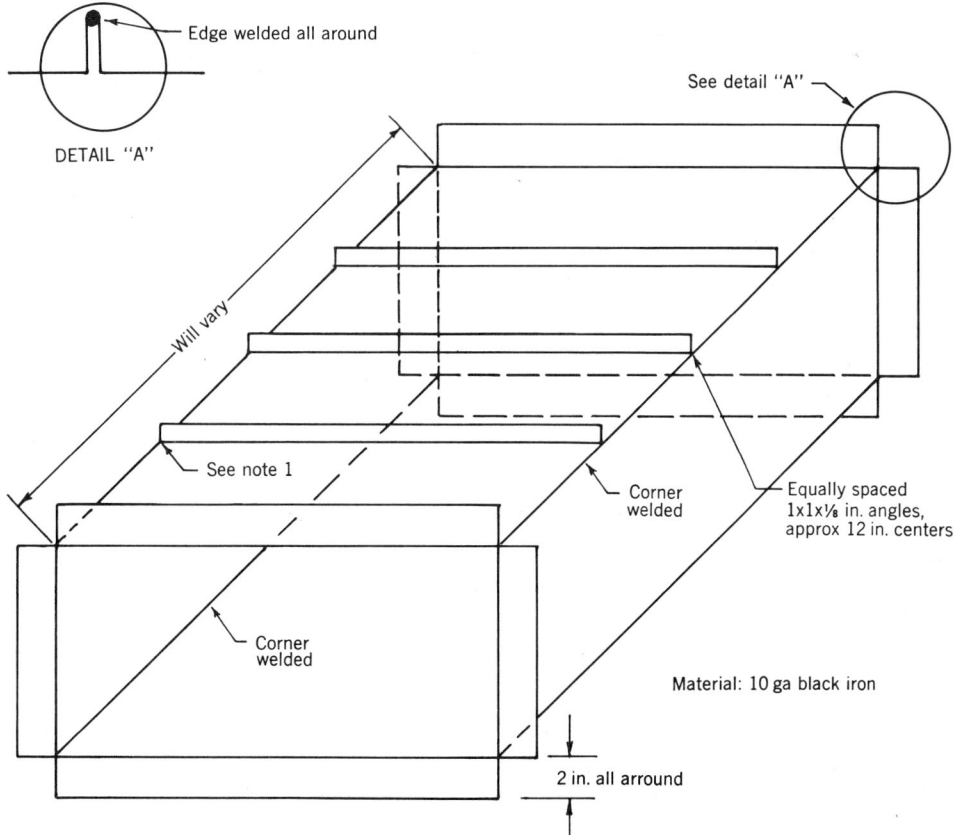

Fig. 6. Typical rectangular breeching for low pressure.

A schedule for steel gages for rectangular breeching construction is tabulated below for higher pressures and larger cross-sections than those shown in **Fig. 6**:

Largest cross-section Dimension	Design Pressure	
	up to 15 in. WG	16 in. to 27 in. WG
up to 6 ft.	10 gage	. . .
6 ft. to 10 ft.	3/16 inch	3/16 inch
over 10 ft.	1/4 inch	1/4 inch

A schedule for reinforcement of the flat areas of large breeching to prevent pulsations and fluttering, is given in Table 1.

Welded duct joints are preferred, but removable duct sections may have to be used for access to parts behind the duct. These removable sections should be companion flanged with a composition gasket material to withstand the high temperatures of the flue gases. Bolted flanges of joints should be stiff; 3" x 3" x 3/8" angles, or larger, may be required. If the long side of the duct is 3 ft, or less, 2½" x 2½" x 1/4" angles may do. The bolts should not be less than ½" and should be 5/8" where static pressures ± of more than 6"

Table 1. Reinforcements for Flat Areas of Breechings and Ducts *

Max. Design Pressure H₂O	Max. Sq. Ft. Reinf. Area	Maximum Stiffener Length				
		5'-0"	7'-0"	9'-0"	11'-0"	13'-0"
0" to 6"	16 Sq. Ft.	2" x 3/8"	2½" x 3/8"	3" x 3/8"	4" x 3/8"	4" x 15/8"
6.1" to 10"	14 Sq. Ft.	2½" x 3/8"	3" x 3/8"	4" x 3/8"	4" x 15/8"	4" x 2½"
10.1" to 12"	13 Sq. Ft.	3" x 3/8"	4" x 3/8"	4" x 15/8"	4" x 2½"	6" x 2" or 6" High Bent Shape
12.1" to 18"	11 Sq. Ft.	4" x 3/8"	4" x v5/8"	4" x 2½"	6" x 2" or 6" High Bent Shape	6" x 2½" or 6" High Bent Shape
18.1" to 27"	8 Sq. Ft.	4" x 15/8"	4" x 15/8"	4" x 2½"	6" x 2½" or 6" High Bent Shape	6" x 2½" or 6" High Bent Shape

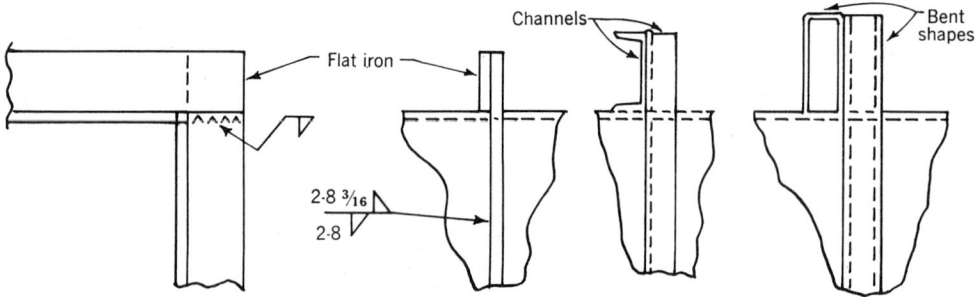

* Reinforcements for flat areas of breeching and ducts should be tack welded 2" every 9" using the following as guidelines.
NOTE: All breeching to be 3/16" minimum thickness.
    For 1/4" plates increase maximum allowable unreinforced area 25%.
    Stiffeners to run the shortest spans possible.
    Joints of ducts or breeching to equipment must be stiffened also to suit, as in table above.
(*Source: Cleaver Brooks Co., Milwaukee, Wis.*)

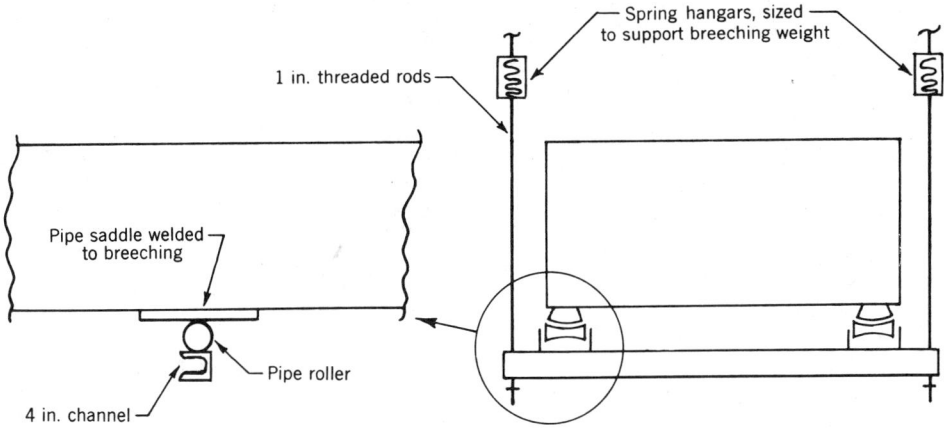

Note: Hangars positioned as shown on plans.

Fig. 7. Breeching hanger supports.

are encountered. Bolts should be spaced about 2½" inches, or a maximum of 3 inches apart.

Metal selection for rectangular breeching should also consider corrosion and wear; see "Round Breeching Construction" above. Shop painting generally is required and specifications may spell out: "Bare metal before painting; shot or sandblasting; primer and finish coat suitable for the temperatures involved."

Detail for expansion joint (for either round or rectangular cross-section) is shown in Fig. 2.

Breeching hanger supports are required to support each section of breeching. Typical hanger detail is shown in Fig. 7. Spacing between supports is dictated by the beam strength of the breeching. Usually, the supports are located next to the joints, which are normally strong enough to permit hanging spacing at 6- to 10-ft intervals.

Dampers at flue gas outlet of each boiler in a multiboiler installation are recommended to prevent injury to personnel from back puffs. These dampers must be tied in with the burner controls so that the burner cannot operate unless the dampers are fully open.

Breeching construction, flue gas velocities, flue gas dampers, and breeching access are often subject to legal or code requirements, which may supersede those hereinabove. Check all pertinent codes before designing the boiler breeching.

Section 6

FUELS AND COMBUSTION

Contents

Authorities

Section Editor: *Richard J. Trieste*, The Brooklyn Union Gas Company

F. W. Hutchinson, Pound-Mole Method for Combustion Calculations

C. C. McRae, Fuel Equivalent Chart

Everett F. Opperman, Fuel Oil Storage Tanks

C. George Segeler, Venting Gas-Fired Equipment

Leslie Silverman, Chimney Draft and Velocities

FUELS AND COMBUSTION

COMBUSTION THEORY

Definition. Combustion is the rapid oxidation of elements or combination of elements which liberate heat during their combination with oxygen. The ordinarily used fuels are those which contain in easily usable form a high proportion of elements or compounds which will so combine with oxygen.

Basic Fuels. The commonly used fuels can be classified as solid, liquid or gaseous, but from the standpoint of the theory of combustion this grouping is of little use. All fuels, however complex in formation, are composed, either in whole or in part, of basic fuels: (1) solid carbon, (2) carbon monoxide and hydrogen; (3) gaseous hydrocarbons. In some cases the fuel as supplied is a mixture of these basic fuels, while in other cases the original fuel breaks down into the basic fuels during combustion.

Elementary Chemistry. The combination of any element (or compound of elements) with oxygen is a chemical process which follows certain known laws. Since matter is never destroyed, all of the elements entering the process are found in the products of combustion following combustion, but they are, except for certain inactive elements, rearranged in combination with each other. The combination liberates a definite amount of heat, depending on the elements involved and the quantity of the elements, but independent of the time taken for the process.

Before the combination of the fuel and oxygen takes place the fuel must be brought to a certain ignition temperature, which varies with the fuel. After combustion has begun the heat liberated by the process raises the temperature of the remaining supply of fuel to the ignition temperature so that combustion continues.

For combustion processes with which the heating engineer deals, the oxygen is furnished by air, which contains 23.23% of oxygen (by weight). The remaining 76.77% of air is nitrogen and other inert gases which are unchanged during the combustion process, except for temperature change.

The reactions of oxygen with carbon, various hydrocarbons and other substances entering into combustion problems, are shown in Table 1.

Combustion of Carbon. When carbon is heated in the presence of air, the oxygen in the air combines with the carbon to form carbon dioxide, some carbon monoxide, and some complex compound of carbon and oxygen C_xO_y, whose exact formation is not known or understood but which in turn breaks down to form CO_2 and CO. The CO combines with more oxygen to form CO_2; if sufficient oxygen (or air containing oxygen) is supplied the resultant product of combustion consists entirely of CO_2. If insufficient air is supplied the products of combustion will include some CO.

From the standpoint of heating value of the fuel it does not matter whether the carbon burns to CO_2 immediately or to CO and finally to CO_2. In brief, if sufficient air is supplied during the process, each pound of carbon will liberate 14,542 Btu during its combination with oxygen to form CO_2. If insufficient oxygen is supplied, less heat will be liberated, for a pound of carbon burning to CO gives up but 4,451 Btu, and the heat released by carbon burning in insufficient air depends on the relative quantity of the carbon burning to CO_2 to that burning to CO.

Combustion of Hydrogen. When hydrogen is burned in the presence of oxygen the resultant is water (H_2O) and the heat liberated is 61,000 Btu per lb. provided (and this applies to all of the heats of reaction) the resulting product of combustion is cooled down at constant pressure to the initial temperature of the fuel.

Since the temperature of combustion is always relatively high, the water thus formed is immediately evaporated into steam, so that in actual practice the heat liberated by a pound of hydrogen is less than that given, due to the absorption of the water of sufficient heat to evaporate the water into steam and to raise the temperature of the steam to the temperature of the leaving products of combustion.

The one value is referred to as the *lower heat value*, while the other is the *higher* heat value. These are often also called the *net* and the *gross* heating values, respectively. Any fuel containing hydrogen, consequently, has a high and low heating value.

The net (or lower) heat value is defined as the quantity of heat produced by combustion at any stated temperature, usually 60F, if the flue gases are cooled back to the initial temperature of the fuel and air, but with the water vapor uncondensed. The gross (or higher) heat value is the heat developed if the flue gas water vapor is condensed and the flue products cooled back to the initial temperature.

Combustion of Hydrocarbons. Hydrocarbons are compounds containing only carbon and hydrogen. When these are burned, the theory is that the oxygen combines with the hydrocarbon molecule to form unstable compounds which in turn break down until eventually CO_2 and H_2O are formed. For example, methane (CH_4) is believed to burn as follows: (the equations are not balanced):

$$CH_4 + O_2 \rightarrow CH_3OH \text{ (methyl alcohol)}$$
$$\rightarrow CH_2(OH)_2 \text{ (formaldehyde and water)}$$
$$\rightarrow H_2O + CO_2$$

This process is called hydroxylation.

Some hydrocarbons may, at high temperatures, decompose rapidly into soot (carbon) and hydrogen — a "cracking" process. This may occur in the presence of insufficient oxygen, and if this action should occur before hydroxylation, the flame will be yellowish and sooty.

Combustion of Sulphur. Sulphur is almost always present in fuels, seldom, though, as a free element. Theoretically, sulphur should burn to sulphur dioxide (SO_2) and liberate 3,940 Btu per lb. In some cases, however, sulphur trioxide (SO_3) is formed, depending on conditions. All of the sulphur in the fuel may not burn in the case of coal and may reappear in the ash. High sulphur content is objectionable in a fuel: in oil it may corrode the tank or pipes; in coal it aids spontaneous combustion; in the products of combustion it may combine with other elements to form corrosive acids. Also it is doubtful if the full heating value of sulphur is ever realized in actual combustion.

Table 1. — Heating Value of Common Fuel Components, Dry, at 60F and 30 in. of Mercury

Substance	Heating Value			
	Btu per Lb.		Btu per Cu. Ft.	
	Gross	Net	Gross	Net
Carbon (to CO_2).....	14,150	14,150	—	—
Carbon (to CO).....	3,960	3,960	—	—
Carbon monoxide....	4,367	4,367	322	322
Hydrogen..........	61,000	51,500	323	273
Sulphur...........	3,940	—	—	—
Hydrogen sulphide...	7,479	—	672	—
Methane..........	23,912	21,533	1,015	914
Ethane...........	22,215	20,312	1,781	1,638
Hexane...........	20,526	18,976	4,667	4,315
Octane*..........	20,542	10,444	6,239	3,170
Pentane..........	20,908	19,322	3,981	3,679
Ethylene.........	21,884	20,525	1,631	1,530
Propylene........	21,042	19,683	2,336	2,185
Propane..........	21,564	19,834	2,572	2,371
Butylene.........	20,840	19,481	3,135	2,884
Acetylene........	21,572	20,840	1,503	1,453
Benzene..........	18,150	17,418	3,741	3,590
Toluene..........	18,129	17,301	4,408	4,206
Naphthalene	17,298	—	5,859	—
Ethyl alcohol*.....	12,804	—	1,548	—
Methyl alcohol*....	9,603	—	818	—
Xylene...........	18,410	—	5,155	—

Source — Miscellaneous. Hydrocarbons from "Combustion" except alcohols and octane which are from the National Bureau of Standards.
*Saturated with water, not dry

TEMPERATURES AND EFFICIENCY OF COMBUSTION

Ignition Temperatures. The ignition temperature of a substance is the lowest temperature at which, in the presence of oxygen, combustion will be independent of a supply of external heat. Representative approximate values of this temperature for various solids, liquids and gases are given in Table A.

Table A. — Approximate Ignition Temperatures of Solids, Liquids and Gases in Air

Substance	Ignition Temperature, F.
Anthracite	1112†
Bituminous coal	850†
Coke	1123†
Lignite	979†
Semi-bituminous coal	980†
Sulphur	470
Kerosene	563
Gas oil	637
Ethyl alcohol	900
Propyl alcohol	941
Pennsylvania crude oil	601
Carbon	800
Methane	1300
Hydrogen	1070
Ethane	1050
Ethylene	950
Acetylene	685
Hexane	570
Decane	865
Benzol	1364
Toluol	1490
Phenol	1319
Carbon monoxide	1200

†Glow point

Flame Temperature. It is practically impossible to measure the actual temperature of a flame accurately. For purposes of comparison, the "theoretical flame temperature" is frequently referred to. This is the temperature which the flame would attain if combustion was instantaneous and perfect and if no heat were lost to the surroundings; that is, if all the heat of combustion were used to heat the products of combustion. If this condition could be attained, then

$$H = pc(t_2 - t_1)$$

where H = Btu per lb liberated per lb of fuel,
  p = products of combustion in pounds,
  c = specific heat of the products of combustion,
  t_2 = temperature, F, of products of combustion, final, and
  t_1 = temperature, F, of products of combustion, initial

so that t_2 would also be the theoretical flame temperature and

$$t_2 = \frac{H}{pc} + t_1$$

Consequently, the theoretical flame temperature is equal to the term (H/pc) plus the initial temperature of the fuel and air, so that preheating the air will result in a higher flame temperature.

The theoretical flame temperature is never reached due to excess air which must be heated to flame temperature; to losses of heat by radiation

and conduction; and to dissociation of CO_2 and H_2O, and other causes. The dissociation of CO_2 and H_2O is a reaction which absorbs heat just as the combination of C and H_2 with O_2 liberates heat.

Efficiency of Combustion. The thermal efficiency of a combustion process is $(t_2 - t_s)/t_2$ where t_2 is the flame temperature, t_s is the stack temperature; $(t_2 - t_s)$ measures the heat used and t_2 the heat input. Thus the higher the flame temperature (t_2) the higher the thermal efficiency.

Table B. — Approximate Theoretical Flame Temperatures of Representative Fuels

Fuel	Flame Temperature, F.*
Acetylene	4160
Carbon monoxide	3850
Fuel oil	3850
Hydrogen	3920
Natural gas	3600
Manufactured gas	3630
Carbon	4946

*Figures are with no excess air, corrected for dissociation.

Table B gives some approximate theoretical flame temperatures of several fuels. The actual temperatures of a coal, oil or gas flame in practice are between 1500 and 2600F, depending on the composition of the fuel and the amount of excess air.

Dissociation. Heat is given off in the ordinary combustion reactions; that is, the action is exothermic. The chemical action can, however, be reversed, and in this case heat is necessary to complete the reaction (this action is termed endothermic). For example, when carbon burns to carbon dioxide, heat is given up; when the molecule of CO_2 is broken up into carbon and oxygen again, heat is absorbed.

When the temperature of combustion has reached the equilibrium temperature, the breaking up of molecules is taking place at such a rate that the heat absorbed is equal to that given up by the combination. Consequently, the equilibrium temperature is a limiting temperature beyond which a higher temperature cannot be obtained. The breaking up of the molecules at the equilibrium temperature is known as dissociation.

Complete Combustion. There are two terms in common use for ideal combustion: *perfect combustion*, where all combustible in the fuel is burned with no excess air, so that the products contain only CO_2, H_2O (plus SO_2 if there is sulphur in the fuel) and N_2; *complete combustion* where all combustible is burned, but with excess air, so that the flue products contain free O_2 in addition to CO_2, H_2O and N_2, but with no CO. The former is a theoretical unattainable ideal, the latter is easily maintained only with fairly high proportions of excess air. The most successful fuel burning equipment from an operating standpoint is that which attains complete combustion with a minimum of excess air, CO and smoke with the minimum of attention.

AIR REQUIRED FOR PERFECT COMBUSTION

Each of the combustible elements requires a certain quantity of oxygen in order to burn completely and complete the reaction, and this quantity varies with the elements according to the molecular weight of the element. Similarly, since the combustible in fuels is composed of combinations of combustible elements, the quantity of oxygen necessary for combustion is determined by the proportion of each element present.

Tables 1 and 2 list the reactions of common combustibles with oxygen and tabulate the quantities of oxygen — and air — needed for perfect combustion, as well as the quantities of flue products resulting. An example of how such data are arrived at follows:

Compute the combustion constants for ethane gas:

Table 1. Ethane (C_2H_6) has a molecular weight of 2 (molecular weight of C) + 3 (mol. wt. of H_2) = $2 \times 12.00 + 3 \times 2.016 = 30.05$ (Col. 3, Table 1). The weight of ethane per cubic foot (Col. 4) is determined by weighing, and Col. 5 is the reciprocal of Col. 4.

Col. 6, the weight of carbon per pound of gas, is found as follows: There is, in C_2H_6, 2 molecules of carbon for every 3 molecules of hydrogen and the molecular weight of C is 12, that of C_2H_6 is 30.05 (Col. 3). Then the the ratio of C_2H_6 which is carbon is

$$2 \times \frac{12.00}{30.05} = 0.799$$

so that there is 0.799 lb of carbon present in 1 lb. as given in Col. 6. By subtraction there is 0.201 lb. of hydrogen present in 1 lb. of C_2H_6.

Col. 7 is the product of Col. 4 and Col. 6, or 0.0801 lb. per cu. ft. of $C_2H_6 \times 0.799$ lb. of carbon per lb. of $C_2H_6 = 0.064$ lb. of C per cu. ft. of C_2H_6. From Col. 8 the combustion of C_2H_6 is:

$$2\ C_2H_6 + 7\ O_2 = 4\ CO_2 + 6\ H_2O$$

Table 2. Entering the molecular weights from Col. 3, Table 1, into the foregoing equation:

$$2(30.05) + 7(32) = 4(44.0) + 6(18.01)$$

$$60.1\ C_2H_6 + 224\ O_2 = 176\ CO_2 + 108.1\ H_2O$$

That is, 60.1 lb. of ethane will combine with 224 lb. of oxygen to form 176 lb. of CO_2 and 108.1 lb. of water, or 1 lb. of ethane will combine with $\frac{224}{60.1}$ lb. or 3.73 lb. of oxygen (Col. 2, Table 2) to form $\frac{176}{60.1}$ lb. or 2.93 lb. of CO_2 (Col. 4) and $\frac{108.1}{60.1}$ lb. or 1.80 lb. of water (Col. 5)

Since air consists of 23.23% oxygen and 76.77% nitrogen by weight, the air required in pounds for burning 1 lb. of gas (Col. 3) is Col. 2 ÷ .2323 or $\frac{3.73}{.2323} = 16.06$ lb. (Col. 3). The weight of nitrogen entering into the process but which is uncombined, is 16.06 − 3.73 = 12.33 lb. This appears in the flue products (Col. 6.)

Since 3.73 lb. of O_2 are needed to burn 1 lb. of C_2H_6, and since (opposite O_2 in Col. 5, Table 1) there are 11.85 cu. ft. of O_2 per lb., then it will require 11.85×3.73 cu. ft. of O_2 to burn 1 lb. of C_2H_6. But 1 lb. of C_2H_6 occupies 12.48 cu. ft. (Col. 5, Table 1), so that 1 cu ft of C_2H_6 will require $\frac{11.85 \times 3.73}{12.48} = 3.5$ cu ft. of O_2 for combustion (Col. 7, Table 2).

Air consists of 20.93% O_2 and 79.07% N_2 by volume, so that Col. 8 is Col. 7 ÷ .2093 or $\frac{3.5}{.2093} = 16.72$ cu. ft. air per cu. ft. of gas. Then Col. 11 is 16.72 − 3.5 = 13.22 cu. ft. of N_2.

Col. 9 and 10 are computed similarly to Col. 7.

Since each cubic foot of ethane gas requires 16.72 cu. ft. of air (Col. 8), the total mixture occupies (1.00 + 16.72) = 17.72 cu. ft. The proportion of gas and air (Col. 12 and 13) are thus $\frac{1}{17.72} = 5.6\%$ and $\frac{16.72}{17.72} = 94.4\%$, respectively.

Air Required — Common Fuels. From Table 2 the weight of air necessary for the perfect combustion of the combustible elements is given in Col. 3. Figures from this column may be combined to arrive at the weight of air necessary to burn a fuel composed of combinations of the elements.

Assuming that the oxygen in the fuel is free to combine with the hydrogen, then

$$P = 11.48C + 34.19\left(H_2 - \frac{O_2}{8}\right) + 4.30S$$

where the figures are from Col. 3 and C, H_2, O_2 and S are the percentages of carbon, hydrogen, oxygen and sulphur present in the ultimate analysis and expressed as a decimal, and P is the weight of air required per pound of fuel.

If hydrocarbons are present, the formula becomes

$$P = 11.48C + 34.19\left(H_2 - \frac{O_2}{8}\right) + 4.30S +$$

$$k_1(Z_1) + k_2(Z_2) \cdots k_n(Z_n)$$

where k is the value from Col. 3, Table 2, for the proportion of each hydrocarbon Z in Col. 1.

Example. What is the weight of air necessary to burn 1 lb. of a hydrocarbon consisting of 60% heptane and 40% pentylene by weight?

There is no C, H_2 or S present so that the formula becomes

$$P = k_1 Z_1 + k_2 Z_2$$

and, referring to Col. 3, Table 2, the formula is

$$P = .60 \times 15.15 + .40 \times 14.72$$

$$P = 9.09 + 5.89 = 14.98 \text{ lb. of air per lb.}$$

of fuel.

Example. What is the weight of air necessary to burn 1 lb. of coal whose analysis is 86.0% carbon,

Table 1. — Combustion Constants for the Common Gases, Vapors, and Solids

Substance	Formula	Molecular Weight	Lb. of Gas per Cu. Ft. at 30 in. Hg.	Cu. Ft. of Gas per Lb. at 30 in. Hg.	Lb. of Carbon in 1 Lb. of Gas	Lb. of Carbon in 1 Cu. Ft. of Gas at 30 in. Hg.	Combustion Reaction
(1)	(2)	(3)	(4)	(5)	(6)	(7)	(8)
Hydrogen.......	H_2	2.016	0.0053	188.7	—	—	$2 H_2 + O_2 = 2 H_2O$
Methane........	CH_4	16.03	0.0423	23.64	0.749	0.0317	$CH_4 + 2 O_2 = CO_2 + 2 H_2O$
Ethane.........	C_2H_6	30.05	0.0801	12.48	0.799	0.0640	$2 C_2H_6 + 7 O_2 = 4 CO_2 + 6 H_2O$
Propane........	C_3H_8	44.06	0.1162	8.61	0.817	0.0949	$C_3H_8 + 5 O_2 = 3 CO_2 + 4 H_2O$
Butane.........	C_4H_{10}	58.08	0.1535	6.51	0.826	0.1268	$2 C_4H_{10} + 13 O_2 = 8 CO_2 + 10 H_2O$
Pentane........	C_5H_{12}	72.10	0.1902	5.26	0.832	0.1582	$C_5H_{12} + 8 O_2 = 5 CO_2 + 6 H_2O$
Hexane.........	C_6H_{14}	86.11	0.2271	4.40	0.836	0.1899	$2 C_6H_{14} + 19 O_2 = 12 CO_2 + 14 H_2O$
Heptane........	C_7H_{16}	100.13	0.2642	3.79	0.839	0.2217	$C_7H_{16} + 11 O_2 = 7 CO_2 + 8 H_2O$
Octane.........	C_8H_{18}	114.14	0.3010	3.32	0.841	0.2531	$2 C_8H_{18} + 25 O_2 = 16 CO_2 + 18.H_2O$
Decane..... ..	$C_{10}H_{22}$	142.18	0.3750	2.67	0.844	0.3165	$2 C_{10}H_{22} + 31 O_2 = 20 CO_2 + 22 H_2O$
Duodecane......	$C_{12}H_{26}$	170.21	0.4489	2.23	0.846	0.3798	$2 C_{12}H_{26} + 37 O_2 = 24 CO_2 + 26 H_2O$
Ethylene........	C_2H_4	28.03	0.0739	13.53	0.856	0.0633	$C_2H_4 + 3 O_2 = 2 CO_2 + 2 H_2O$
Propylene.......	C_3H_6	42.05	0.1109	9.02	0.856	0.0949	$2 C_3H_6 + 9 O_2 = 6 CO_2 + 6 H_2O$
Butylene........	C_4H_8	56.06	0.1479	6.76	0.856	0.1266	$C_4H_8 + 6 O_2 = 4 CO_2 + 4 H_2O$
Pentylene.......	C_5H_{10}	70.08	0.1849	5.41	0.856	0.1583	$2 C_5H_{10} + 15 O_2 = 10 CO_2 + 2 H_2O$
Acetylene.......	C_2H_2	26.02	0.0692	14.45	0.922	0.0638	$2 C_2H_2 + 5 O_2 = 4 CO_2 + 2 H_2O$
Benzol..........	C_6H_6	78.05	0.2059	4.86	0.922	0.1898	$2 C_6H_6 + 15 O_2 = 12 CO_2 + 6 H_2O$
Toluol.........	C_7H_8	92.06	0.2429	4.12	0.912	0.2215	$C_7H_8 + 9 O_2 = 7 CO_2 + 4 H_2O$
Xylol..........	C_8H_{10}	106.08	0.2798	3.57	0.905	0.2532	$2 C_8H_{10} + 21 O_2 = 16 CO_2 + 10 H_2O$
Hydrogen sulphide	H_2S	34.09	0.0909	11.00	—	—	$2 H_2S + 3 O_2 = 2 H_2O + 2 SO_2$
Methyl alcohol...	CH_4O	32.03	0.0845	11.83	0.375	0.0317	$2 CH_4O + 3 O_2 = 2 CO_2 + 4 H_2O$
Ethyl alcohol....	C_2H_6O	46.05	0.1215	8.23	0.521	0.0633	$C_2H_6O + 3 O_2 = 2 CO_2 + 3 H_2O$
Carbon to CO_2...	C	12.00	—	—	—	—	$C + O_2 = CO_2$
Carbon to CO....	C	12.00	—	—	—	—	$2 C + O_2 = 2 CO$
Carbon monoxide	CO	28.00	0.0738	13.55	0.429	0.0317	$2 CO + O_2 = 2 CO_2$
Sulphur.........	S	32.07	—	—	—	—	$S + O_2 = SO_2$
Carbon dioxide..	CO_2	44.00	0.1167	8.57	0.273	0.3019	—
Nitrogen........	N_2	28.00	0.0739	13.53	—	—	—
Oxygen.........	O_2	32.00	0.0844	11.85	—	—	—
Air............	—	29.00[a]	0.0764	13.09	—	—	—
Sulphur dioxide..	SO_2	64.07	0.1729	5.78	—	—	—
Water vapor....	H_2O	18.01	0.0476	21.02	—	—	—

[a]Equivalent molecular weight.

4.8% hydrogen, 1.5% nitrogen, 5.5% oxygen, and 2.2% sulphur?

$$P = 11.48 \times .86 + 34.19\left(.048 - \frac{.055}{8}\right) + 4.30 \times .022$$

$$= 11.33 \text{ lb. of air per lb. of coal.}$$

In the foregoing equation the term $\left(H_2 - \frac{O_2}{8}\right)$ is due to the fact that when hydrogen is burned a portion of the heat is lost due to the formation of water.

Table 2. — Weights and Volumes of Gases Involved in Perfect Combustion

Substance	Lb. of Oxygen per Lb. of Gas	Lb. of Air per Lb. of Gas[b]	Lb. of Products of Combustion per Lb. of Gas			Cu. Ft. Oxygen per Cu. Ft. of Gas	Cu. Ft. Air per Cu. Ft of Gas	Cu. Ft. of Products of Combustion per Cu. Ft. of Gas			Per Cent by Volume, for Perfect Combustion	
			CO_2	H_2O	N_2			CO_2	H_2O	N_2	Gas	Air
(1)	(2)	(3)	(4)	(5)	(6)	(7)	(8)	(9)	(10)	(11)	(12)	(13)
Hydrogen........	7.94	34.19	—	8.94	26.25	0.5	2.39	—	1.0	1.89	29.5	70.5
Methane........	3.99	17.17	2.74	2.25	13.18	2.0	9.56	1.0	2.0	7.56	9.5	90.5
Ethane........	3.73	16.06	2.93	1.80	12.33	3.5	16.72	2.0	3.0	13.23	5.6	94.4
Propane........	3.63	15.63	3.00	1.64	12.00	5.0	23.89	3.0	4.0	18.90	4.0	96.0
Butane........	3.58	15.41	3.03	1.55	11.83	6.5	31.06	4.0	5.0	24.57	3.1	96.9
Pentane........	3.55	15.28	3.05	1.50	11.73	8.0	38.22	5.0	6.0	30.24	2.5	97.5
Hexane........	3.53	15.20	3.07	1.46	11.67	9.5	45.39	6.0	7.0	35.91	2.2	97.8
Heptane........	3.52	15.15	3.08	1.44	11.62	11.0	52.56	7.0	8.0	41.58	1.9	98.1
Octane..:......	3.50	15.07	3.08	1.42	11.57	12.5	59.72	8.0	9.0	47.25	1.65	98.35
Decane........	3.49	15.02	3.09	1.39	11.53	15.5	74.06	10.0	11.0	58.59	1.47	98.53
Duodecane......	3.48	14.98	3.10	1.38	11.50	18.5	88.39	12.0	13.0	69.93	1.33	98.67
Ethylene........	3.42	14.72	3.14	1.29	11.30	3.0	14.33	2.0	2.0	11.34	6.5	93.5
Propylene.......	3.42	14.72	3.14	1.29	11.30	4.5	21.50	3.0	3.0	17.01	4.4	95.6
Butylene........	3.42	14.72	3.14	1.29	11.30	6.0	28.67	4.0	4.0	22.68	3.4	96.6
Pentylene.......	3.42	14.72	3.14	1.29	11.30	7.5	35.83	5.0	5.0	28.35	2.7	97.3
Acetylene.......	3.07	13.22	3.38	0.69	10.15	2.5	11.94	2.0	1.0	9.45	7.7	92.3
Benzol..........	3.07	13.22	3.38	0.69	10.15	7.5	35.83	6.0	3.0	28.35	2.7	97.3
Toluol..........	3.13	13.47	3.35	0.78	10.34	9.0	43.00	7.0	4.0	34.02	2.3	97.7
Xylol..........	3.17	13.65	3.32	0.85	10.48	10.5	50.17	8.0	5.0	39.69	2.0	98.0
Hydrogensulphide	1.41	6.07	1.88[†]	0.53	4.66	1.5	7.17	1.0[*]	1.0	5.67	12.2	87.8
Methyl alcohol...	1.50	6.46	1.37	1.12	4.96	1.5	7.17	1.0	2.0	5.67	12.2	87.8
Ethyl alcohol....	2.08	8.95	1.91	1.17	6.87	3.0	14.33	2.0	3.0	11.34	6.5	93.5
Carbon to CO_2...	2.67	11.48	3.67	—	8.81	1.0	4.78	1.0	—	3.78	—	—
Carbon to CO...	.33	5.74	2.33[‡]	—	4.41	0.5	2.39	1.0[c]	—	1.89	—	—
Carbon monoxide	0.57	2.45	1.57	—	1.88	0.5	2.39	1.0	—	1.89	29.5	70.5
Sulphur.........	1.00	4.30	2.00[†]	—	3.30	1.0	4.78	1.0[*]	—	3.78	—	—
Carbon dioxide..	—	—	—	—	—	—	—	—	—	—	—	—
Nitrogen........	—	—	—	—	—	—	—	—	—	—	—	—
Oxygen.........	—	—	—	—	—	—	—	—	—	—	—	—
Air.............	—	—	—	—	—	—	—	—	—	—	—	—
Sulphur dioxide..	—	—	—	—	—	—	—	—	—	—	—	—
Water vapor.....	—	—	—	—	—	—	—	—	—	—	—	—

[b]Column (3) obtained by assuming air to equal 20.93% O_2 and 79.07% N_2 by volume or 23.23% and 76.77% by weight.
[†]Figure is in pounds at SO_2, not CO_2.
[‡]Figure is in pounds of CO, not CO_2.
[*]Figure is in volumes of SO_2, not cubic feet of CO_2.
[c]Figure is in volumes of CO, not cubic feet of CO_2.

APPROXIMATE COMBUSTION DATA FOR TYPICAL FUELS

It is frequently necessary or desirable to estimate air requirements or flue gas quantities approximately for rough calculations or where detailed fuel analyses are not available. For this purpose the accompanying table and formula will prove useful.

Approximations. Where detailed fuel analyses are not available or accuracy not essential, the following approximate formulas are useful:

Air required for perfect combustion in cubic feet per pound of solid or liquid fuel = heat content of fuel per pound × 0.0095.

Air required for perfect combustion in pounds per pound of solid or liquid fuel = heat content of fuel per pound × .00076.

Air required for perfect combustion in cubic feet per cubic foot of gas = (gross heat content of gas per cubic foot ÷ 100) − 0.5.

Flue gas volume in cubic feet per pound of solid or liquid fuel or per cubic foot of gas = heat content of fuel per pound or per cubic foot of gas × 0.01.

Combustion Data for Typical Fuels

(Units of Fuel are: Solid Fuels, Pounds; Liquid Fuels, Gallons; Gaseous Fuels, Cubic Feet)

Fuel	With Perfect Combustion				CO_2 Attained in Actual Practice
	Pounds of Air Required Per Unit of Fuel	Pounds of Flue Gases per Unit of Fuel	Cubic Feet of Flue Gases per Unit of Fuel	Ultimate CO_2 in Dry Flue Gas, %	
Coke..............	11.5	12.5	150.7	21.0	11–14
Anthracite...........	11.4	12.4	151.9	19.8	11–14
Semi-anthracite........	11.5	12.4	153.5	19.4	11–14
Semi-bituminous.......	11.3	12.3	153.0	18.7	10–13
Bituminous...........	10.8	11.7	145.4	18.7	10–13
Sub-bituminous........	9.7	10.6	131.3	18.9	9–13
Lignite..............	8.4	9.1	112.1	19.4	9–13
Wood..............	5.8	6.4	77.1	20.7	—
No. 2 Fuel oil..........	104	106	1394	15.2	8–12
No. 3 Fuel oil.........	107	110	1434	15.3	8–12
No. 5 Fuel oil.........	110	113	1471	16.0	9–13
No. 6 Fuel oil.........	114	117	1524	16.1	9–14
Natural gas (Penna.)....	.72	.77	10.5	11.8	7–11
Natural gas (Georgia)...	.81	.86	11.7	12.1	7–11
Natural gas (California)..	.80	.90	11.6	12.1	7–11
Natural gas (Kansas)....	.70	.74	10.2	11.9	7–11
Mixed natural-water gas.	.34	.38	5.2	15.3	9–13
Water gas............	.26	.21	2.7	18.0	10–15
Coke oven gas.........	.38	.42	5.8	11.2	7–11
Coal gas..............	.38	.39	5.4	10.7	6–10
Carbureted water gas ...	.35	.37	5.2	17.1	10–15
Blast furnace gas........	.05	.14	1.5	25.5	15–22
Butane (bottled)........	2.33	2.47	32.9	14.0	9–13
Propane (bottled).......	1.82	1.94	26.0	13.7	9–13

PRACTICAL AIR REQUIREMENTS

Air Supply to Boiler Rooms. Outside air must be admitted to the space where the burner is located for three possible reasons: 1. In every installation air must be supplied for combustion; 2. Additional air may be required for ventilation; and 3. In flue arrangements with barometric draft regulators or gas diverters, air drawn up the flue through these devices must be replaced.

The admission of air to the boiler room is usually handled by specifying the open area. Where the burner is located in a room or basement of loose construction, all or part of the air may filter in through building openings. Air requirements introduced in this manner are usually sufficient for small residential combustion equipment. When the Btu input of the equipment exceeds approximately 300,000 Btu per hour, outside air openings are necessary.

As previously mentioned, a supply of air is required to provide combustion air, to dilute flue products, and to control the ambient temperature of the boiler room. Boiler rooms should then be provided with permanent openings to the outdoors—equivalent in free (net, open) area to the boiler(s) breeching area but to be not less than 1 square inch per 10,000 Btu per hour input. Additional ventilation may be required to cool confined boiler rooms which otherwise would develop a high ambience. The openings to the outdoors may be louvered but otherwise must be unobstructed. If screens are utilized they should not be finer than $\frac{1}{4}$-inch mesh.

It might be desirable on some installations to equip the air openings to the outside with motor-operated louvers to close the openings while the burner is not running. Provision should be made in these cases to open the louvers prior to starting the burner cycle. These louvers should be interlocked with the combustion programmer and provided with an automatic end switch to prove they are open before the combustion cycle is allowed to continue.

Exhaust fans must never be installed in a boiler room as they will retard the flow of air required for combustion and could possibly cause a flue reversal.

Combustion Air. A burner is a device for the final conveyance of the fuel, or a mixture of fuel and air, to the combustion zone. This definition introduces the two factors involved in the combustion process, fuel and air.

Besides the theoretical quantity of air required for combustion, excess air must also be supplied. The amount of excess air required, in practice, depends upon several factors. These factors usually are the uniformity of air-distribution and mixing, the direction of fuel travel from the burner, and the height and temperature of the combustion chamber.

The heat lost in excess air represents waste heat, and proper burner design will help reduce this loss to a practical minimum. In practice, 15 to 30% excess air is found to provide reasonable assurance of satisfactory combustion without seriously lowering burner efficiency. The operation of burners at lower percentages of excess air is not practical in most heating installations. The small improvement in efficiency which can be realized may not offset the hazardous and troublesome conditions created.

The combustion equipment usually utilized in setting up burner equipment readily checks for carbon dioxide, carbon monoxide, and smoke to give an indication of burner operation. For this reason, burner operation is often specified as producing a given percentage of carbon dioxide without carbon monoxide, for gas, and carbon dioxide without smoke, for oil.

With an appropriate amount of excess air, the flue analysis will determine the effectiveness of the combustion process. For natural gas, the acceptable minimum carbon dioxide should be about 8.5% without any carbon monoxide; for oil, the acceptable minimum carbon dioxide should be about 10% without any smoke. These figures will vary slightly with different natural gases and oils and are given, therefore, as approximations.

Other factors besides burner design can affect minimum allowable amounts of excess air. Installations which do not properly control and direct secondary air will cause shortages of air in some sections of the burner flame. With the burning of gas, this produces carbon monoxide and with the burning of oil, smoke is produced. This may be overcome at a sacrifice of efficiency by increasing the total amount of secondary air to the point where no part of the burner flame is short of air. A similar condition can be found when a gas flame impinges on a cold surface or when particles of oil strike a cold surface. Poor oil atomization may also cause the same problem. Therefore, it is imperative that the burner manufacturer's installation instructions be faithfully followed to minimize excess air requirements.

WHAT THE FLUE GAS ANALYSIS MEANS

For each fuel there is an "ultimate" CO_2; that is, the percentage of the flue gas which would be CO_2 if combustion were perfect. If combustion were perfect, the flue gases would consist largely of N_2 and CO_2, some water vapor and SO_2. Too small an air supply would produce CO, H_2 and, depending on the fuel, some hydrocarbons in the flue gases. If there is excess of air supply, O_2 will be present in the flue gas. The following points, then, outline the meaning of the flue gas analysis:

1. Presence of N_2, CO_2, H_2O and SO_2 only indicates perfect combustion;
2. Presence of O_2 indicates excess air;
3. Per cent of CO_2 depends on the fuel used. Relative to the ultimate CO_2 of the fuel, a high CO_2 indicates a small amount of excess air but not necessarily good combustion;
4. Relative to the ultimate CO_2, low CO_2 indicates (a) large excess air supply or (b) improper mixing of air and fuel;
5. Presence of CO, H_2 or hydrocarbons indicates

incomplete combustion and (a) if there is no O_2, air supply is insufficient; (b) if there is free O_2 with CO, H_2 or hydrocarbons, fuel and air have not been mixed in the combustion zone, temperature of combustion is too low, or there has not been time for combustion to be completed.

The flue gas analysis gives results on a volume basis. If it is desired to convert these to a weight basis, use the following procedure:

1. Multiply the per cent CO_2, CO, O_2 and N_2 each by its molecular weight;
2. Total the products;
3. Divide the individual product from (1) by the total (2). The result is the percent on a weight basis.

The following example, purposely simplified and not typical, shows the method.

Example. A flue gas analysis shows 12% CO_2, 1% CO, 7% O_2 and, by difference, 80% N_2. What are the proportions on a weight basis?

The molecular weights are found on page 2-5:

Gas	Analysis, by Volume, %	Mol. Wt.	Product	Analysis, by Weight, %
CO_2	12	44	528	528/3020 = 17.5%
CO	1	28	28	28/3020 = 0.9%
O_2	7	32	224	224/3020 = 7.4%
N_2	80	28	2240	2240/3020 = 74.2%
Total	100%		3020	100.0%

The flue gas data resulting from an Orsat analysis are used in connection with suitable charts for determining the per-cent excess air and, together with information on flue gas temperatures, to determine the heat lost in the chimney. Charts for these purposes appear later in this section.

Heat Balance. Useful formulas for determining the losses in combustion follow. The formulas are all on the basis of a unit (pounds, gallons, or cubic feet) of fuel burned.

Heat lost in dry flue gas, $H_s = .24W\,(t_s - t_a)$ where W is the weight of flue gas per unit of fuel burned, t_s is the flue gas temperature, F, and t_a is the air temperature in the boiler room, F.

The weight of dry chimney gases per unit of fuel burned, W, can be computed as follows:

$$W = \frac{11\,CO_2 + 8\,O_2 + 7(N_2 + CO)}{3(CO_2 + CO)} \times (C + .545S)$$

where CO_2, O_2, N_2 and CO are the per cent of those gases in the flue gas by volume, expressed as a decimal, and C and S are the per cent of carbon and sulphur in the fuel by weight, expressed as a decimal. This loss usually ranges from 5 to 25%.

Heat lost through combustion of hydrogen is

$$H_H = 9H(1090.7 - t + .455\,t_s)$$

where H is the percent of hydrogen in the fuel by weight expressed as a decimal; t the temperature of the fuel, F, and t_s the flue gas temperature. This loss ranges usually from 2.5% to 8%.

Heat lost due to incomplete burning of carbon, is

$$H_c = \frac{10160 \times CO \times C}{CO_2 + CO}$$

where CO and CO_2 are the percent of those gases in the flue gas by volume, expressed as a decimal, and C is the percent of carbon in the fuel by weight, expressed as a decimal. Note that if there is no CO, H_c becomes zero. This loss is rarely over 2%.

Heat lost due to moisture in the fuel, is

$$H_M = W(1090.7 - t + .455\,t_s)$$

where W is the percent of free moisture per unit of fuel expressed as a decimal, t is the temperature of the fuel, F, and t_s the flue gas temperature, F. This loss ranges from zero to 0.3%.

Heat lost due to moisture in the air, is

$$H_a = .48W_v\,(t_s - t_a)$$

where W_v is the weight of water vapor in the air required per unit of fuel burned, t_s is the flue gas temperature, F, and t_a the boiler room temperature, F. The determination of the quantity of air burned per unit of fuel burned has been covered elsewhere in this book while the weight of moisture for any known quantity and humidity of air is covered in Section 1 of this book. This loss usually runs between 0.1 and 0.5%.

The sum of the foregoing losses plus the radiation and unaccounted losses when subtracted from 100% gives the heat usefully applied or boiler efficiency.

HOW TO ANALYZE FLUE GASES

For a hydrocarbon fuel of a given analysis, calculations will show the relative volumes of carbon dioxide and water vapor present in the flue gases when the fuel is perfectly burned. In actual practice there is always excess air supplied to the fuel, and in some cases of imperfect combustion there is carbon monoxide present in the products of combustion.

If the flue gases are measured dry, with the water vapor condensed out, the quality of combustion can then be measured by the relative volumes of the gases present in the products of the combustion. With incomplete combustion the gases will contain carbon monoxide and carbon and, depending on the fuel, hydrocarbons. Excess air will contribute free oxygen to the flue gases.

In order to properly examine the performance of a burner, the flue gases must be analyzed. These gases should be analyzed for CO_2, O_2, CO, and smoke. Without this information it is impossible to properly adjust a burner for its peak performance.

There are various instruments used for analyzing flue gases. Two of the most commonly used in the field are the Orsat and Fyrite.

The Orsat instrument, in its standard form, analyzes only for CO_2, O_2, and CO, in that order, with the three chemical reagents contained in a single, portable housing. The remaining gas is assumed to be N_2. The test for smoke, however, requires a separate instrument. The main disadvantages of the Orsat instrument are the high degree of skill required to use it and the amount of care needed in handling it.

The Fyrite checks only CO_2 and O_2, utilizing two separate analyzers. Like the Orsat, it uses the volumetric measurement involving chemical absorption of the gas sample. The CO and smoke then, would have to be measured with separate instruments. The chief advantages of using these instruments are the ease of taking readings, the short time required to take these readings, and the ruggedness of the equipment. The Fyrite analyzers, along with a carbon monoxide and smoke tester, are the instruments most commonly used in the field.

Points to Watch in Using Flue Analyzers. (1) The flue gas sampled should be a typical or average sample taken from the center of the stream and at a place where there is very little possibility of dilution because of air infiltration. It is desirable that more than one sample be taken so as to obtain average results. (2) When ground-glass cocks are used in the Orsat they should be well lubricated with petroleum jelly in thin films so that they are gas-tight. (3) Rubber tubes and connections should be checked to insure that they are not cracked or loose. (4) Absorption reagents should be fresh, since each solution is capable of taking in a definite quantity of a specific gas. If the solutions are weak, the absorption will be exceedingly slow or the results erratic. (5) Carbon filters must be used with all monoxide testers. (6) The manufacturer's instructions on the analyzers must be followed faithfully.

Results of the Flue Analysis. In analyzing the flue gases, the goal is the maximum CO_2 attainable with the particular fuel being burned, with the minimum O_2, no CO, and no smoke. In normal practice the ultimate CO_2 is nearly attained and the CO should never be over 0.005%. The smoke should never exceed 0 on a smoke scale. Smoke is an indication of free carbon which collects in boiler passages, thereby inhibiting heat transfer and reducing overall efficiency of the heating equipment.

DEW POINT OF FLUE GASES

Water vapor in the flue gases will begin to condense out at the dew point temperature of the mixture. This temperature is actually the boiling point (or condensing point) of the steam in the flue gases, which depends on the pressure of the steam. This pressure, in turn, depends on the amount of steam in the flue gases.

Each of the constituents of the flue gas exerts its own pressure independent of the other gases (Dalton's law) and the total of these partial pressures equals the pressure of the mixture. The partial pressure of each is proportional to its relative volume. If, for example, a flue gas is composed of 14.9% (by volume) of water vapor, the partial pressure of the water vapor when the flue gas is at atmospheric pressure will be .149 × 30.0 = 4.47 inches of mercury. The boiling point of steam at this pressure is 129F, so that when the steam is cooled to this point it begins to condense and this is the dew point for the mixture.

Table 1 gives this dew point temperature for various volumes of water vapor in the flue gases.

It is emphasized that the dew point is not the temperature at which all vapor condenses out. It is the temperature where such condensation begins. As the steam is condensed out, the volume of steam in the mixture is reduced, its partial pressure is thus lower, and the dew point consequently lower. Condensation begins at the dew point and continues as the temperature drops, with the steam continuing to condense out. At any given temperature below the dew point, the per cent of water vapor still uncondensed equals

$$\frac{\text{Water vapor, \% by volume}}{\text{opposite given temperature}} \times 100$$
$$\frac{}{\text{Water vapor, \% by volume,}}{\text{in flue products}} \times 100$$

Example. 17% of a flue gas, by volume is water. What is the dewpoint of the flue gas, and how much of the water will be uncondensed at 120F?

In Table 1, opposite 17% in the first column, find the dewpoint in the second column to be 135F. Opposite 120F in the second column, find in the first column by interpolation 11.3. The per cent uncondensed at 120F = $\frac{11.3}{17} \times 100 = 66.5\%$.

Table 1. — Approximate Dew Point Temperature of Flue Gases

Water Vapor in Flue Products, % by Volume	Dew Point Temp., F
1	45
2	64
3	76
4	85
5	92
6	98
7	103
8	108
9	111
10	116
11	119
12	122
13	125
14	128
15	130
16	132
17	135
18	137
19	139
20	131
21	143
22	145
23	146
24	148
25	150
26	151
27	153
28	154
29	156
30	157
31	158
32	160
33	161
34	162
35	163

Morgan, in a circular of the University of Illinois Engineering Experiment Station, gives the average temperature at which condensation takes place with common fuels as shown in Table 2.

Table 2. — Average Dew Point of Flue Gases for Typical Fuels
(From Univ. of Ill. Eng. Exp. Sta. Bull. Circ. 22, Condensation of Moisture in Flues, by William R. Morgan.)

Fuel	Per Cent Excess Air				
	0	25	50	75	100
	Average Dew Point, F				
Anthracite........	81	77	73	70	68
Semi-bituminous...	100	95	90	87	84
Bituminous.......	111	105	100	97	93
Oil..............	123	117	111	106	103
Natural Gas.......	139	132	127	122	118
Manufactured Gas.	148	141	137	132	128

FLUE GASES WITH PERFECT COMBUSTION

Tables 1 and 2 under Air Required for Perfect Combustion give the weight of flue gas constituents for perfect combustion.

Example. What would be the composition of the flue gas resulting from the perfect combustion of a bituminous coal consisting of 86.0% carbon, 4.8% hydrogen, 1.5% nitrogen, 5.5% oxygen, and 2.2% sulphur?

Solution. Refer to Col. 4, 5, and 6 of Table 2 previously referred to and tabulate:

Gas	CO_2	H_2O	N_2	SO_2
From C........	3.67×.86=3.16	—	8.81×.86= 7.58	—
From H_2......	—	$8.94 \times \left(.048 - \frac{.055}{8}\right) = .367$	$26.25 \times \left(.048 - \frac{.055}{8}\right) = 1.08$	—
From N........	—	—	Unchanged=	—
From S........	—	—	3.30×.022= .075	.073 2.00×.022=.044
From O_2.......	—	—	.073	—
Total weight of flue products..	3.16	0.37	8.75	.04

The total is thus 3.16 + 0.37 + 8.75 + .04 = 12.32 lb. Since the air required for this fuel, using the previously given formula and data from Column 3 of Table 1, was 11.33 lb. of air per lb. of fuel, the total weight of flue gases should be 11.33 + 1.0 = 12.33 lb. This checks closely with the 12.30 just obtained by the method outlined above.

However, flue gases are analyzed on a volume basis, so that the flue gas data are often needed in terms of volume. This can be computed from Col. 5 in Table 1 under Air Required for Perfect Combustion. The volume is at standard conditions:

Gas	Lb. of Gas, Col. 5	Lb. from Example	Cu. Ft. of Flue Gas	Per Cent by Volume, Moist	Per Cent by Volume, Dry
	(a)	(b)	(a) × (b) = (c)	(c) ÷ 153.47	(c) ÷ (153.47 − 7.78)
CO_2	8.57	3.16	27.08	17.65	18.59
H_2O	21.02	0.37	7.77	5.06	—
N_2	13.53	8.75	118.39	77.14	81.26
SO_2	5.78	0.04	0.23	0.15	0.15
Total	—	12.32	153.47	100.00%	100.00%

In actual practice, flue gases are measured dry. These figures differ slightly from those at actual flue gas temperatures. To obtain the volume at a given flue gas temperature:

Volume at actual temperature = Volume at

$$\text{standard condition} \times \frac{460 + \text{actual temperature}}{460}$$

The previous example worked out for a flue gas temperature of 700F. follows

Gas	Volume, Cu. Ft. per Lb. at 60F.	Volume, Cu. Ft. per Lb. at 700F.
CO_2	27.00	68.0
H_2O	7.78	19.6
N_2	118.25	298.0
SO_2	.23	.6

Note that the *ratio* of volumes has not changed.

Ultimate CO_2. The per cent of CO_2 in the flue gases which would be obtained with perfect com-bustion, or in other words, the maximum possible CO_2, is called the ultimate CO_2. The ultimate CO_2 can be determined by either of the following formulas.

$$\text{Ult.}CO_2,\% =$$

$$\frac{\begin{array}{c}\text{Cubic feet of } CO_2 \text{ per cubic}\\ \text{foot (or pound) of fuel}\end{array}}{\begin{array}{c}\text{Cubic feet of total dry flue gas}\\ \text{per cubic foot (or pound) of fuel}\end{array}} \times 100$$

when calculated from the analysis for perfect com-bustion at standard conditions.

Thus, for the example given, the ultimate CO_2 is

$$\frac{27.00}{153.26 - 7.78} = 18.5\%$$

The ultimate CO_2 can be computed from an actual flue gas analysis if no CO is present. CO_2 and O_2 from the same sample.

$$\text{Ult.}CO_2,\% = \frac{\% \ CO_2 \text{ in flue gas}}{100 - \text{Per cent } O_2 \text{ in flue gas}}$$

.21

HEAT LOST IN FLUE GAS

Heat is lost to the chimney in (1) the sensible heat in the chimney gases and (2) in the latent heat of the water vapor in the chimney gases. Both of these can be computed from the flue gas analysis if the flue gas temperature is known.

Sensible Heat. The sensible heat in the flue gases can be computed on the basis of a unit of fuel burned; that is, per pound of coal, gallon of oil or cubic foot of gas. The heat lost per unit of fuel burned = weight of flue gas per unit of fuel burned × (temperature of flue gases − temperature of air supply) × mean specific heat of flue gases.

By use of the data in Table A the mean specific heat of the flue gas can be calculated. For example, what would be the heat lost in a flue gas consisting of, by weight, 25.6% CO_2, 3.0% water vapor, 71.0% nitrogen and 0.4% sulphur, with 12.31 lb. of flue gas per pound of fuel burned, and a flue temperature of 700F.

The mean specific heat of the flue gas using Table A is (assuming a value of .227 for SO_2):

$$\frac{.256 \times .227 + .03 \times .227 + .71 \times .253 + .004 \times .227}{1.000}$$
$$= 0.247.$$

Therefore, the sensible heat lost above 60F per pound of fuel burned is

$$12.31 (700 - 60) .247 = 1946 \text{ Btu per lb.}$$

Table A. — Mean Specific Heat of Flue Gas Constituents Between 60F and Given Temperature

Gas	Temperature, F			
	300	500	700	900
	Mean Specific Heat, Btu per Lb. per F			
Hydrogen.....	3.48	3.50	3.52	3.54
Nitrogen......	0.252	0.252	0.253	0.254
Carbon dioxide.	0.218	0.221	0.227	0.235
Carbon monox-ide.........	0.251	0.252	0.255	0.257
Oxygen.......	0.219	0.220	0.221	0.223
Air..........	0.242	0.243	0.244	0.246
Water vapor...	0.218	0.221	0.227	0.235

Latent Heat. The loss in latent heat in flue gases is that due to evaporating the water formed by the combustion of hydrogen. At standard conditions (60F) the latent heat per pound of water vapor is

1057.8 Btu or 50.3 Btu per cubic foot. Therefore the latent heat loss is

Cubic feet of water vapor × 50.3 or

Pounds of water vapor × 1057.8

For the foregoing example, assuming 0.37 lb. of water vapor per pound of fuel, the latent heat loss would be

$$0.37 \times 1057.8 = 391.4 \text{ Btu per lb.}$$

The total chimney loss above 60F would then be

$$1946 + 391 = 2337 \text{ Btu per lb. of fuel burned}$$

Table B is the basis of another method. This table gives the sensible heat content per pound of the various flue gas constituents above 60F for various flue gas temperatures. The quantity of each constituent is multiplied by the heat content from Table B and the products added. For the example, with a flue gas temperature of 700F:

Gas	Lb. of Gas per Lb. of Fuel		Heat per Lb. at 700F above 60F		Sens. Heat Loss, Btu per Lb. Fuel
CO_2	3.19†	×	139.5	=	445
H_2O	0.37	×	290.0	=	107
N_2	8.75	×	159.8	=	1398
		Sensible heat loss		=	1950

†Includes the SO_2.

The foregoing checks closely with the 1946 Btu found by the previous method.

Another method, based on the volume of flue gas at standard conditions (60F), can be used by reference to Table C. The method is the same as the previous one but based on volume instead of weight. Using the same example, and referring to Table C:

Gas	Cu. Ft. of Gas per Lb. of Fuel		Heat per Cu. Ft. at 700F above 60F		Sens. Heat Loss Btu per Lb. Fuel
CO_2	27.23†	×	16.3	=	444
H_2O	7.78	×	13.8	=	107
N_2	118.39	×	11.8	=	1397
		Sensible heat loss		=	1948

†Includes the SO_2.

Table B. — Sensible Heat in Flue Gases, Btu per Lb. Above 60F

Temp., F	Carbon Dioxide	Water Vapor	Nitrogen and Carbon Monoxide	Oxygen	Hydrogen
200	28.3	63.0	33.8	29.6	472
300	49.8	107.1	59.6	52.1	830
400	71.2	153.4	84.0	73.5	1170
500	93.5	199.5	109.9	96.0	1528
600	116.9	244.0	135.4	118.5	1887
700	139.5	290.0	159.8	140.0	2227
800	163.9	338.0	185.5	162.3	2585
900	188.0	384.5	211.0	185.0	2944
1000	212.5	432.5	237.0	207.5	3302
1250	276.0	554.5	300.3	263.0	4189
1500	342.5	680.0	365.5	320.0	5095
2000	479.5	950.0	498.0	435.5	6944
2500	623.0	1250	635.0	555.5	8850
3000	770.0	1590	776.0	680.0	10831

Table C. — Sensible Heat in Flue Gases, Btu per Cu. Ft. Above 60F

Temp., F	Gas			
	Carbon Dioxide	Water Vapor	Nitrogen, Oxygen, and Carbon Monoxide	Hydrogen
200	3.3	3.0	2.5	2.5
300	5.8	5.1	4.4	4.4
400	8.3	7.3	6.2	6.2
500	10.9	9.5	8.1	8.1
600	13.6	11.6	10.0	10.0
700	16.3	13.8	11.8	11.8
800	19.1	16.1	13.7	13.7
900	21.9	18.3	15.6	15.6
1000	24.8	20.6	17.5	17.5
1250	32.2	26.4	22.2	22.2
1500	39.9	32.4	27.0	27.0
2000	55.9	45.2	36.8	36.8
2500	72.6	59.5	46.9	46.9
3000	89.9	75.6	57.4	57.4

Excess Air. The ratio of the quantity of air actually supplied to the combustion process over and above that theoretically necessary for perfect combustion, expressed as per cent, is termed excess air. In equation form

$$\% \text{ excess air} = \frac{(\text{Air actually supplied} - \text{Theoretical air required})}{\text{Theoretical air required}} \times 100$$

Rarely is it possible to measure the air volume supplied so that the foregoing formula is of little use. However, for cases where the fuel contains no nitrogen, the excess air can be calculated from data resulting from the flue gas analysis by the following formula:

$$\text{Per cent excess air} = \left[\frac{O_2 - \frac{1}{2}CO}{.266N_2 - (O_2 - \frac{1}{2}CO)} \right] \times 100$$

where O_2, N_2 and CO are the per cent of oxygen, nitrogen and carbon monoxide from the flue gas analysis, expressed as a decimal or a whole number.

Example. What would be the excess air if the flue gas analysis shows 6% O_2, 2% CO, 10% CO_2 and, by difference, 82% N_2?

Solution. Excess air =

$$\frac{6.0 - \frac{1}{2} \times 2.0}{.266 \times 82.0 - (6.0 - \frac{1}{2} \times 2.0)} \times 100 = \frac{5.0}{16.8} \times 100$$

$$= 29.8\%$$

PER CENT EXCESS AIR FROM CO₂ READING

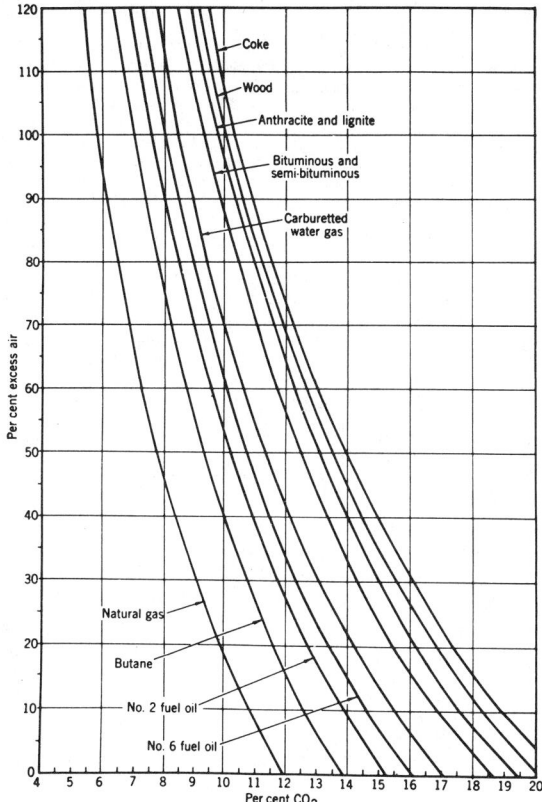

HEAT LOSS IN FLUE GAS

The following example illustrates the use of the flue gas charts that follow:

Assume a natural gas of the analysis shown on the upper half of page 6-22. Assume that a flue gas analysis shows 8% CO_2, no CO, and the flue gas temperature is 400 F. Find per cent excess air and sensible heat loss up the flue.

Enter chart on the right-hand scale at 8. Move horizontally to the left to the Excess Air curve, then vertically down. Read on the bottom scale approximately 52% excess air.

From the intersection of the Excess Air line and the 8% CO_2 line, move vertically down to the 400 F curve, thence left, reading 9+%, sensible heat loss in the flue gas.

HEAT LOSS IN FLUE GAS—WOOD AND LIGNITE

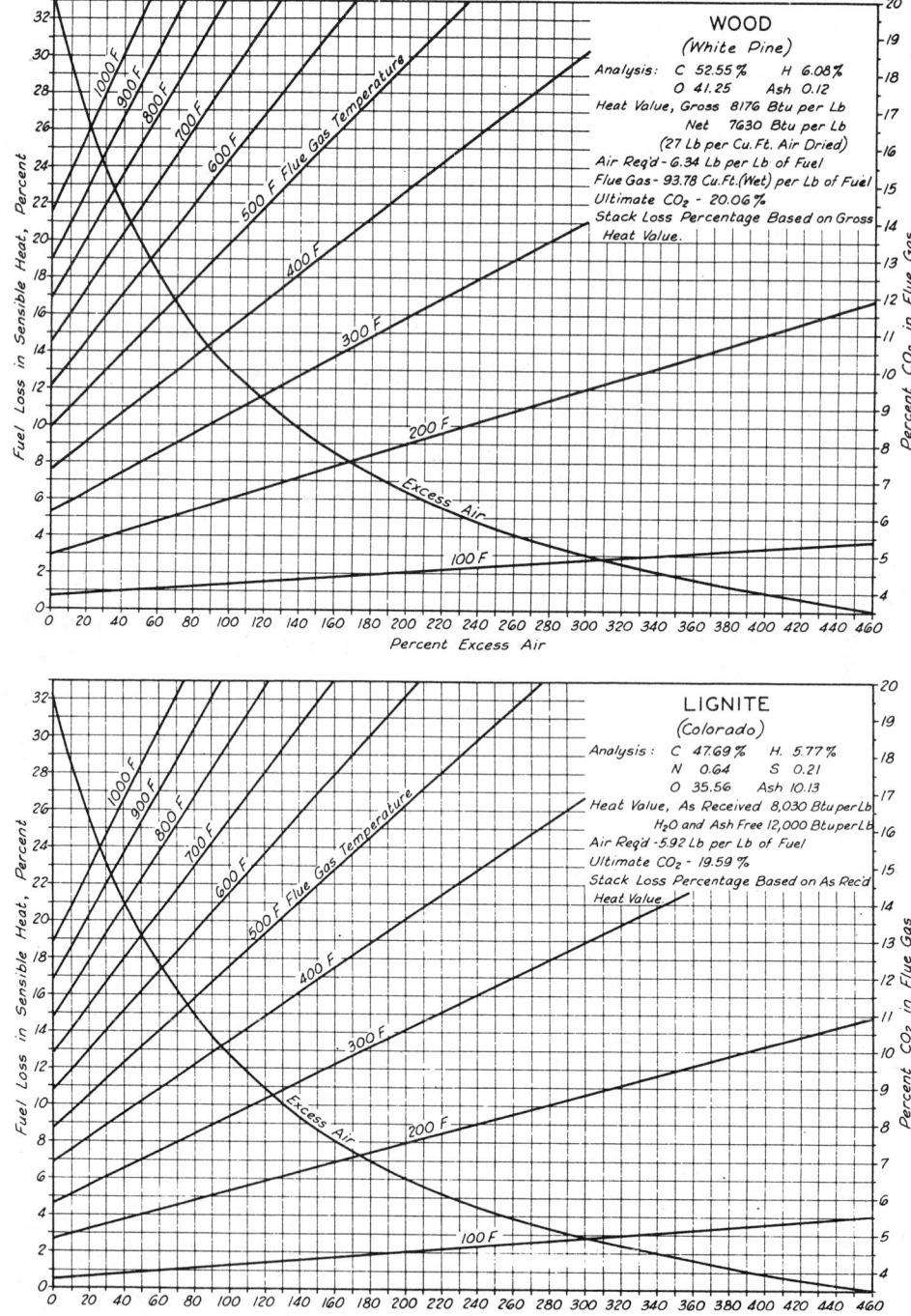

HEAT LOSS IN FLUE GAS—SEMI-BITUMINOUS AND COKE

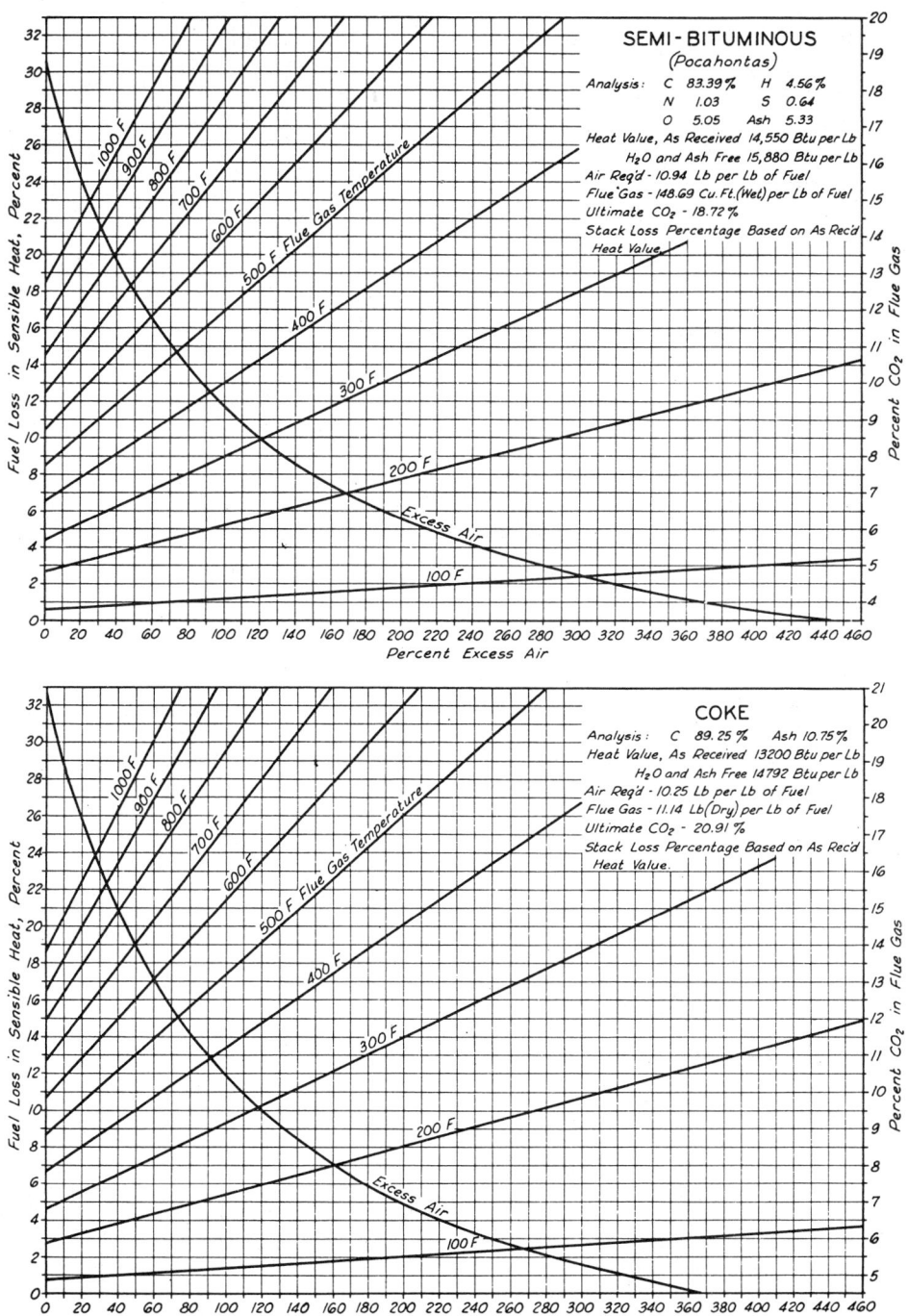

SEMI-BITUMINOUS
(Pocahontas)

Analysis: C 83.39%   H 4.56%
N 1.03   S 0.64
O 5.05   Ash 5.33
Heat Value, As Received 14,550 Btu per Lb
H₂O and Ash Free 15,880 Btu per Lb
Air Req'd - 10.94 Lb per Lb of Fuel
Flue Gas - 148.69 Cu.Ft.(Wet) per Lb of Fuel
Ultimate CO₂ - 18.72%
Stack Loss Percentage Based on As Rec'd Heat Value.

COKE

Analysis: C 89.25%   Ash 10.75%
Heat Value, As Received 13200 Btu per Lb
H₂O and Ash Free 14792 Btu per Lb
Air Req'd - 10.25 Lb per Lb of Fuel
Flue Gas - 11.14 Lb(Dry) per Lb of Fuel
Ultimate CO₂ - 20.91%
Stack Loss Percentage Based on As Rec'd Heat Value.

HEAT LOSS IN FLUE GAS—BITUMINOUS COALS

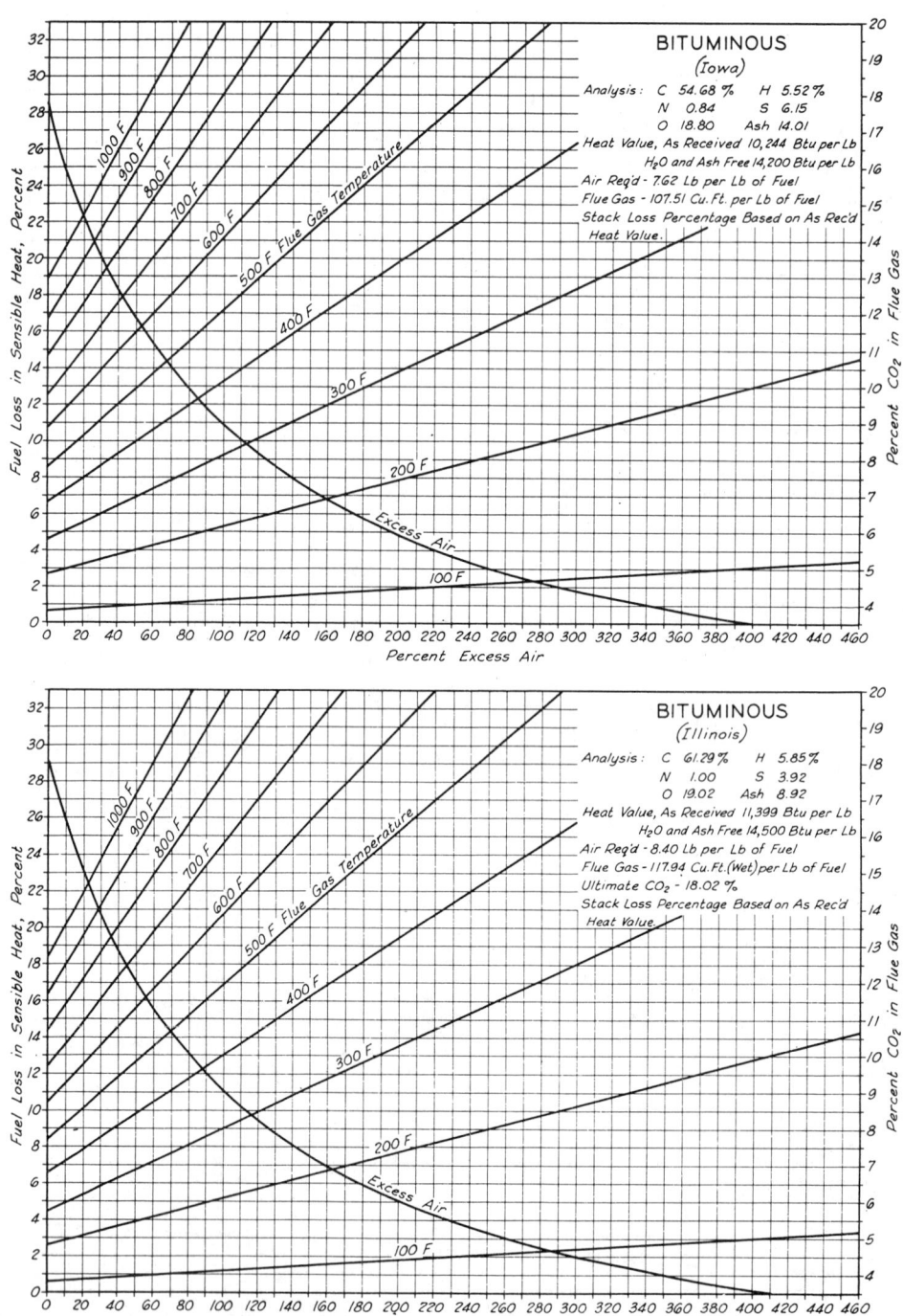

BITUMINOUS
(Iowa)
Analysis: C 54.68 %   H 5.52 %
N 0.84   S 6.15
O 18.80   Ash 14.01
Heat Value, As Received 10,244 Btu per Lb
H₂O and Ash Free 14,200 Btu per Lb
Air Req'd - 7.62 Lb per Lb of Fuel
Flue Gas - 107.51 Cu. Ft. per Lb of Fuel
Stack Loss Percentage Based on As Rec'd
Heat Value.

BITUMINOUS
(Illinois)
Analysis: C 61.29 %   H 5.85 %
N 1.00   S 3.92
O 19.02   Ash 8.92
Heat Value, As Received 11,399 Btu per Lb
H₂O and Ash Free 14,500 Btu per Lb
Air Req'd - 8.40 Lb per Lb of Fuel
Flue Gas - 117.94 Cu. Ft. (Wet) per Lb of Fuel
Ultimate CO₂ - 18.02 %
Stack Loss Percentage Based on As Rec'd
Heat Value.

HEAT LOSS IN FLUE GAS—BITUMINOUS AND ANTHRACITE

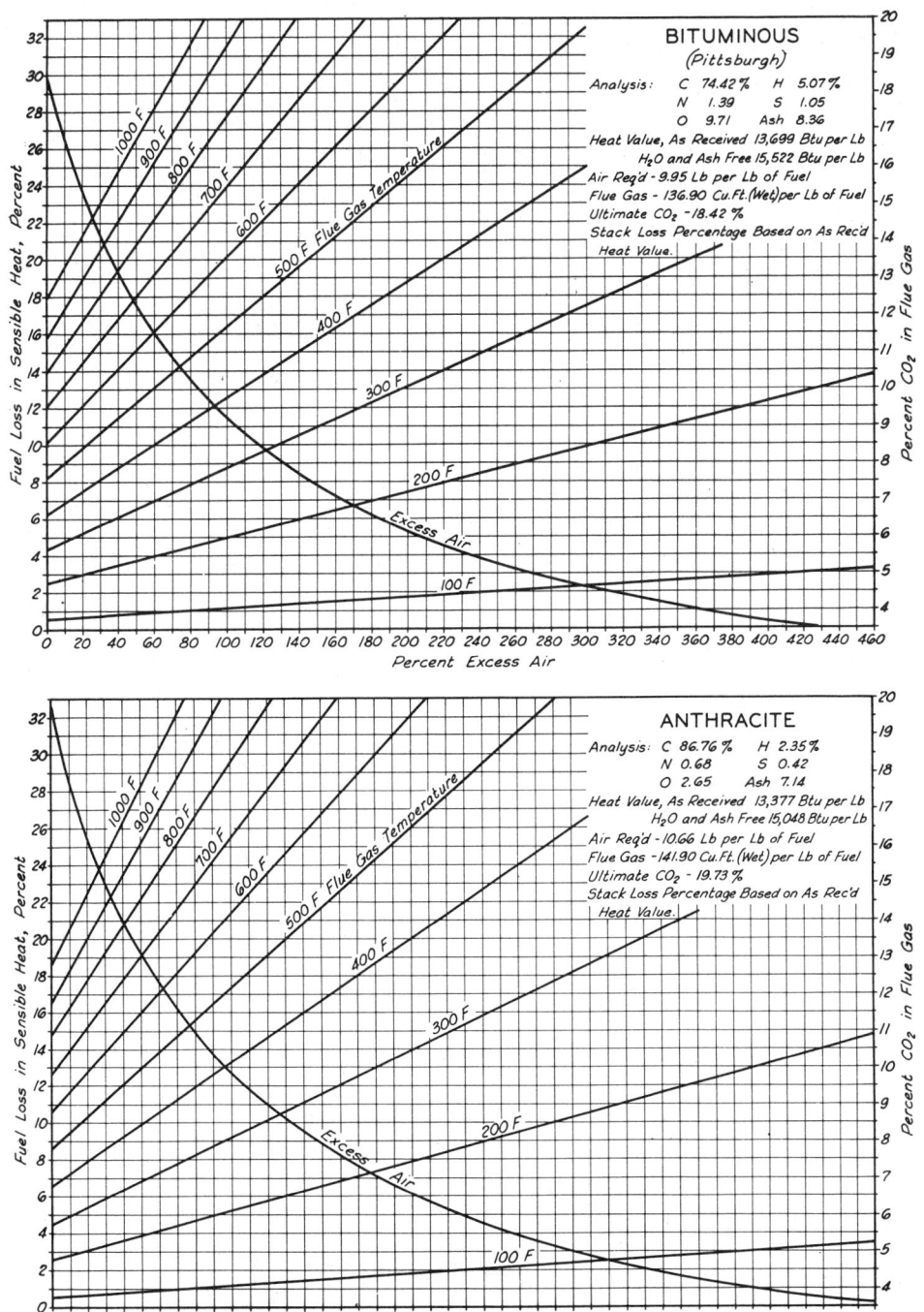

BITUMINOUS
(Pittsburgh)

Analysis: C 74.42% H 5.07%
N 1.39 S 1.05
O 9.71 Ash 8.36
Heat Value, As Received 13,699 Btu per Lb
H₂O and Ash Free 15,522 Btu per Lb
Air Req'd - 9.95 Lb per Lb of Fuel
Flue Gas - 136.90 Cu.Ft.(Wet)per Lb of Fuel
Ultimate CO₂ - 18.42 %
Stack Loss Percentage Based on As Rec'd
Heat Value.

ANTHRACITE

Analysis: C 86.76% H 2.35%
N 0.68 S 0.42
O 2.65 Ash 7.14
Heat Value, As Received 13,377 Btu per Lb
H₂O and Ash Free 15,048 Btu per Lb
Air Req'd - 10.66 Lb per Lb of Fuel
Flue Gas - 141.90 Cu.Ft.(Wet) per Lb of Fuel
Ultimate CO₂ - 19.73 %
Stack Loss Percentage Based on As Rec'd
Heat Value.

HEAT LOSS IN FLUE GAS—FUEL OILS

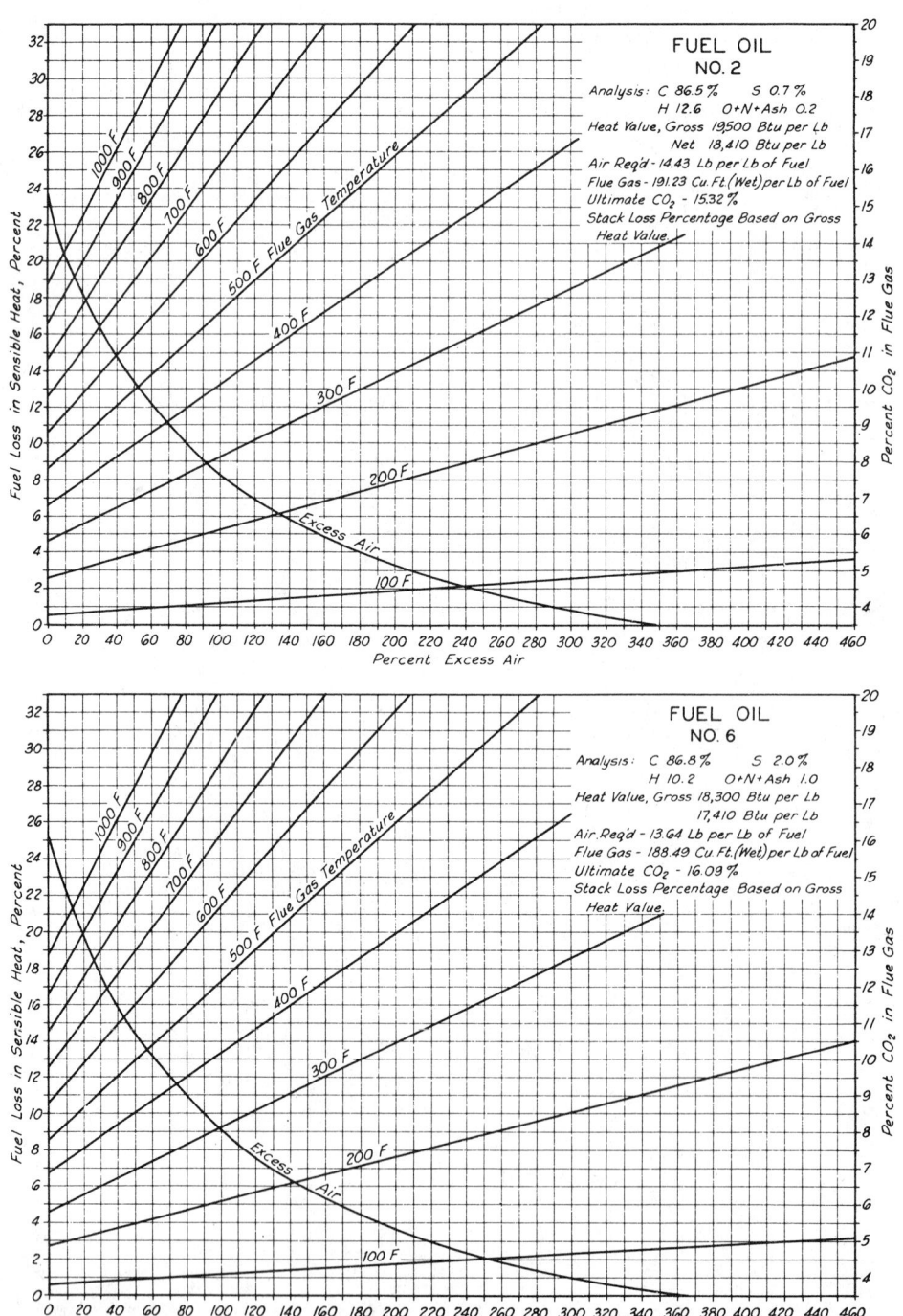

HEAT LOSS IN FLUE GAS—NATURAL GASES

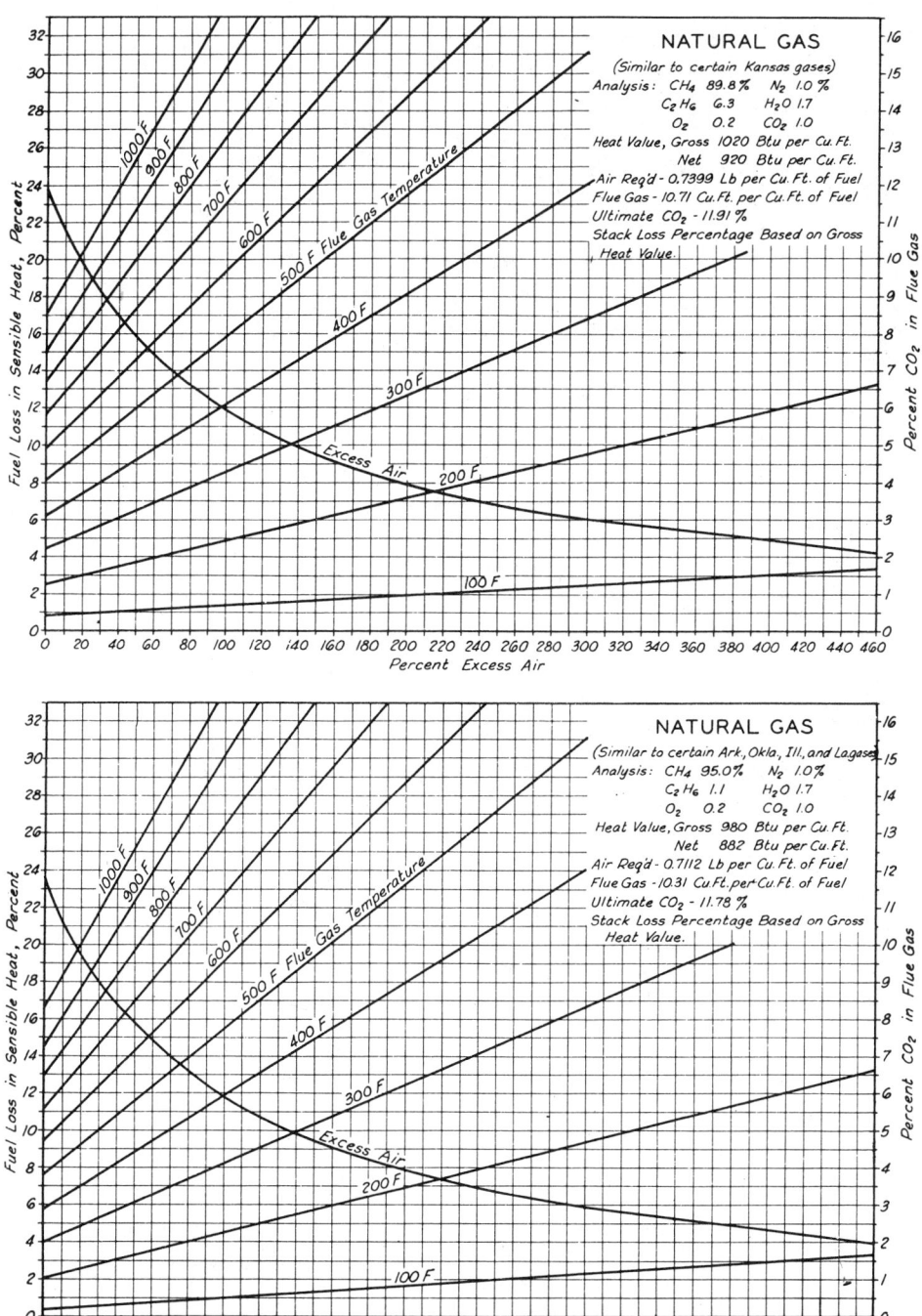

HEAT LOSS IN FLUE GAS—NATURAL GASES

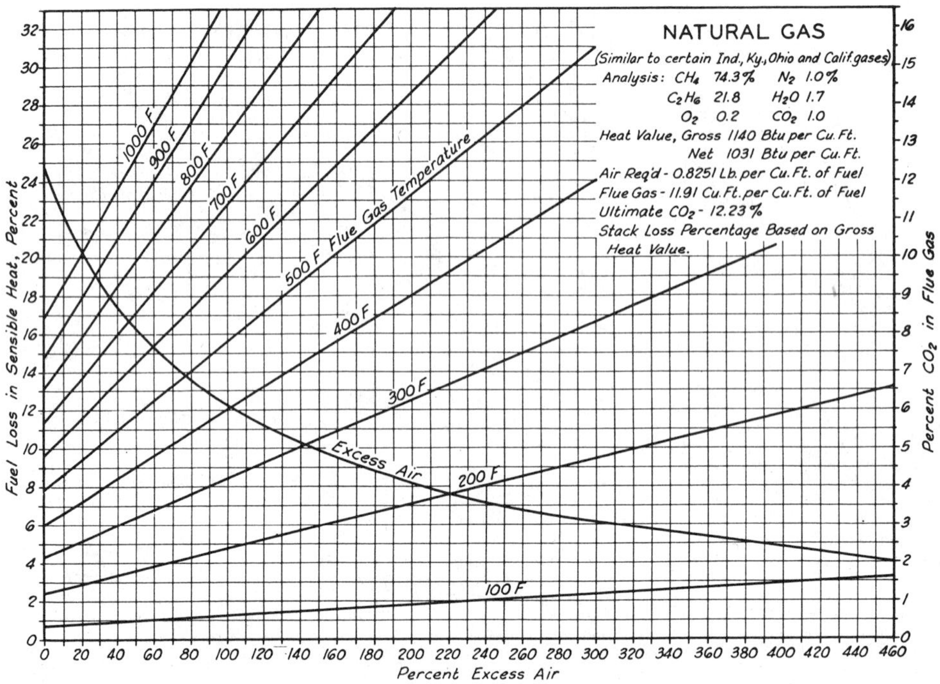

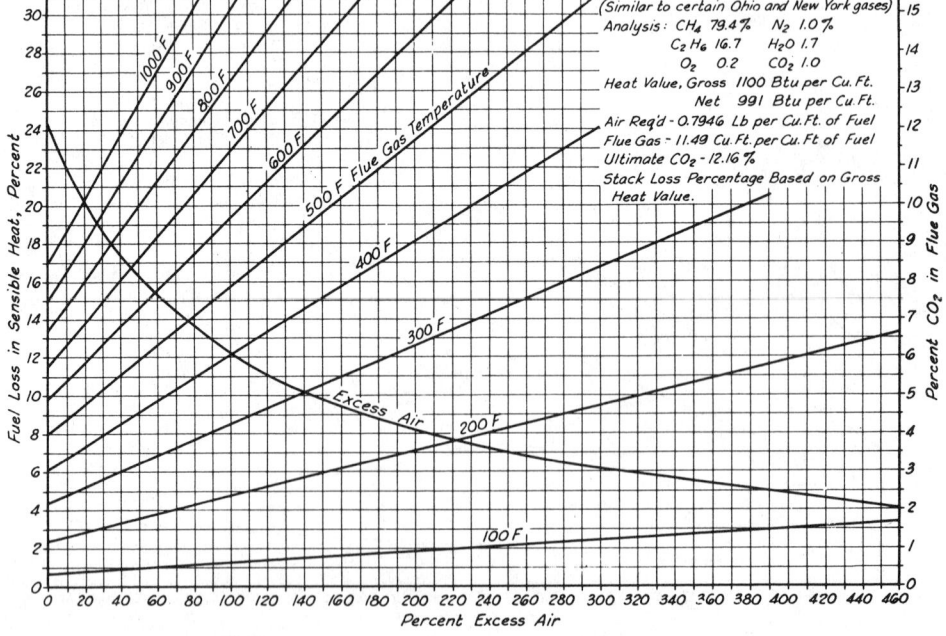

HEAT LOSS IN FLUE GAS—NATURAL GASES

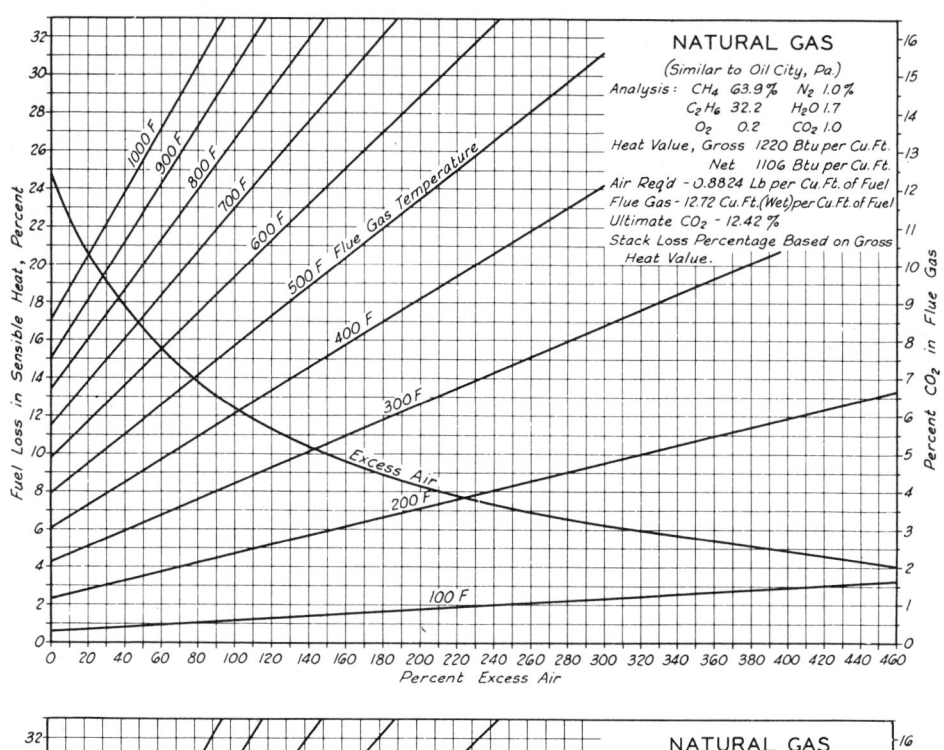

NATURAL GAS

(Similar to Oil City, Pa.)
Analysis: CH₄ 63.9%   N₂ 1.0%
C₂H₆ 32.2   H₂O 1.7
O₂ 0.2   CO₂ 1.0
Heat Value, Gross 1220 Btu per Cu.Ft.
Net 1106 Btu per Cu.Ft.
Air Req'd - 0.8824 Lb per Cu.Ft. of Fuel
Flue Gas - 12.72 Cu.Ft.(Wet)per Cu.Ft. of Fuel
Ultimate CO₂ - 12.42 %
Stack Loss Percentage Based on Gross
Heat Value.

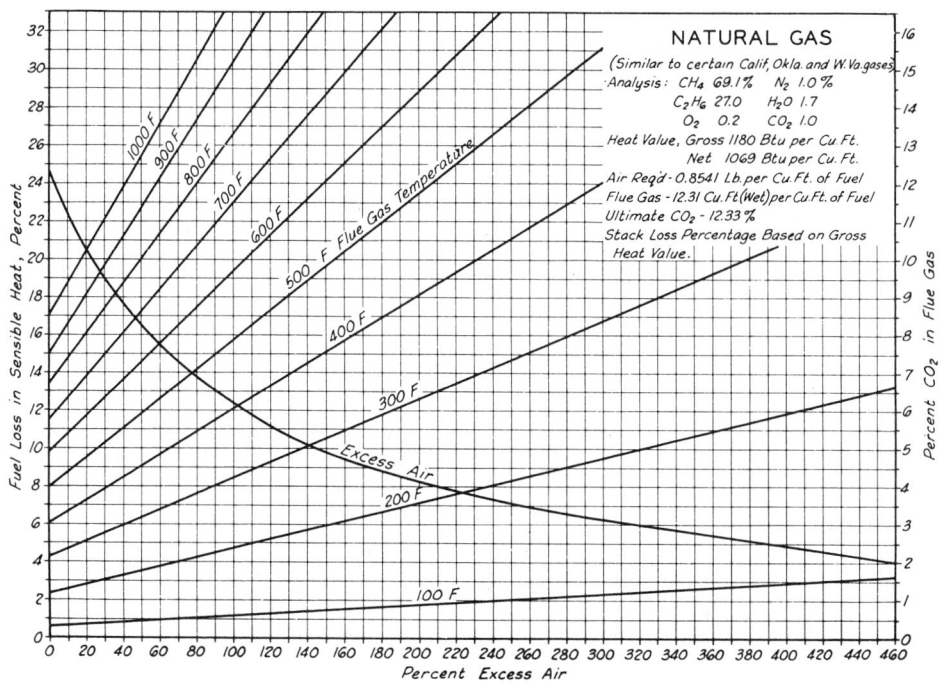

NATURAL GAS

(Similar to certain Calif., Okla. and W.Va.gases)
Analysis: CH₄ 69.1%   N₂ 1.0%
C₂H₆ 27.0   H₂O 1.7
O₂ 0.2   CO₂ 1.0
Heat Value, Gross 1180 Btu per Cu.Ft.
Net 1069 Btu per Cu.Ft.
Air Req'd - 0.8541 Lb.per Cu.Ft. of Fuel
Flue Gas - 12.31 Cu.Ft.(Wet)per Cu.Ft. of Fuel
Ultimate CO₂ - 12.33 %
Stack Loss Percentage Based on Gross
Heat Value.

HEAT LOSS IN FLUE GAS—NATURAL GAS AND COAL GAS

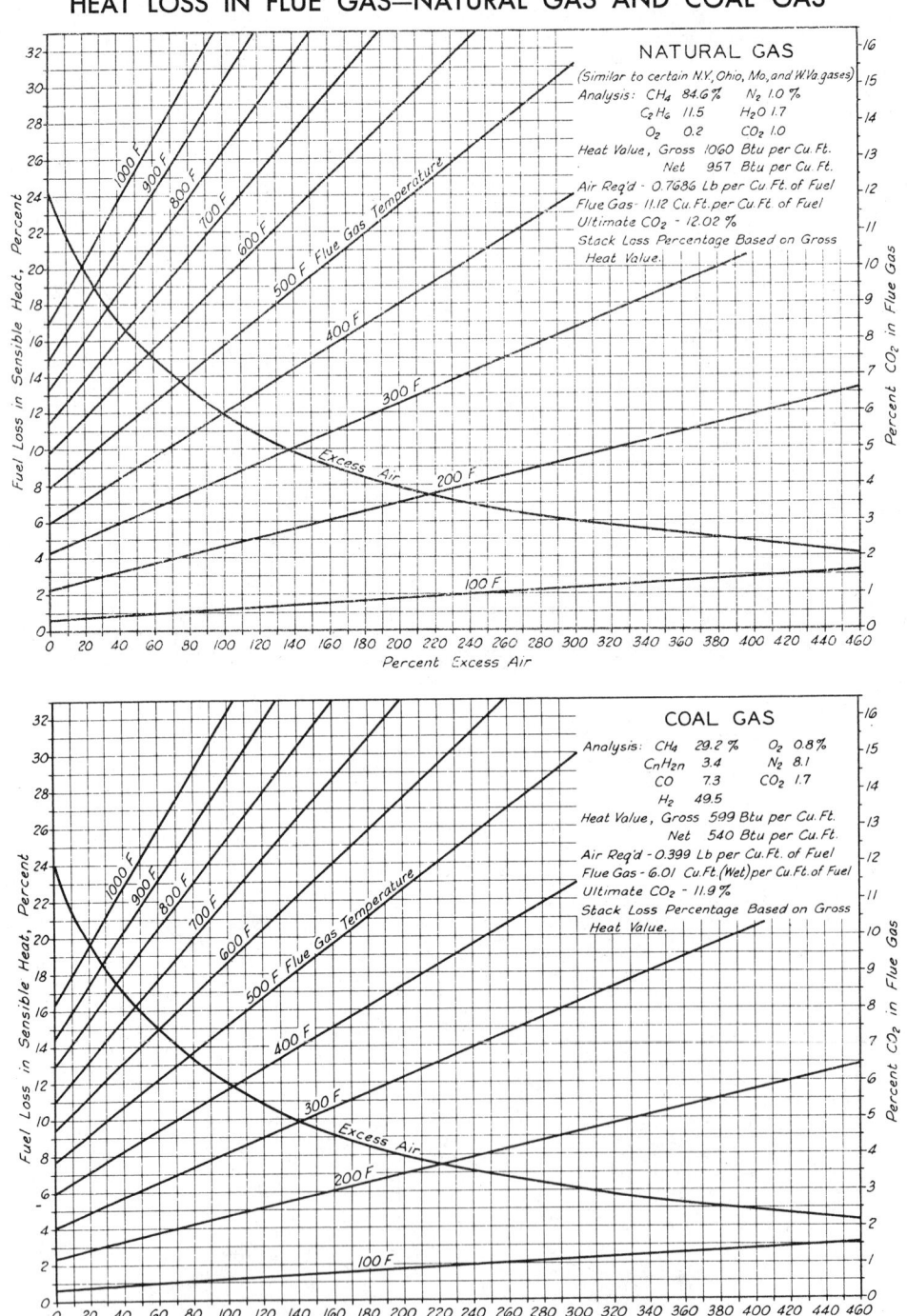

HEAT LOSS IN FLUE GAS—BUTANE AND PROPANE

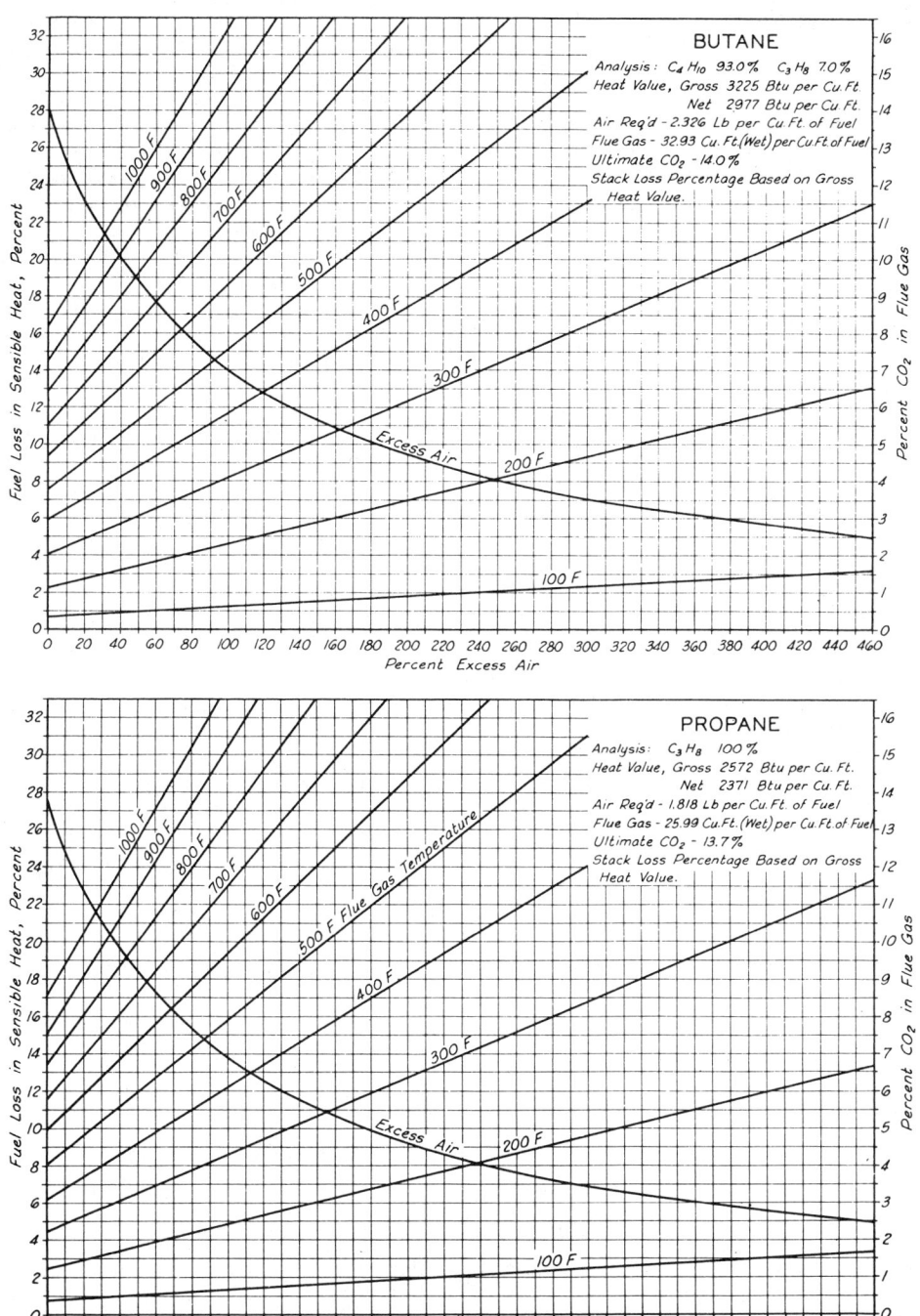

BUTANE

Analysis: C_4H_{10} 93.0% C_3H_8 7.0%
Heat Value, Gross 3225 Btu per Cu.Ft.
              Net 2977 Btu per Cu.Ft.
Air Req'd - 2.326 Lb per Cu.Ft. of Fuel
Flue Gas - 32.93 Cu.Ft.(Wet) per Cu.Ft. of Fuel
Ultimate CO_2 - 14.0%
Stack Loss Percentage Based on Gross
Heat Value.

PROPANE

Analysis: C_3H_8 100%
Heat Value, Gross 2572 Btu per Cu.Ft.
              Net 2371 Btu per Cu.Ft.
Air Req'd - 1.818 Lb per Cu.Ft. of Fuel
Flue Gas - 25.99 Cu.Ft.(Wet) per Cu.Ft. of Fuel
Ultimate CO_2 - 13.7%
Stack Loss Percentage Based on Gross
Heat Value.

POUND-MOLE METHOD FOR COMBUSTION CALCULATIONS

Once the user has become accustomed to its use, the pound-mole method of calculation is found to have great advantages. The detailed explanation of the pound-mole method which follows was prepared by F. W. Hutchinson, Professor of Mechanical Engineering, University of California.

The fundamental unit of all combustion calculations is the pound-mole which is defined as that mass of any material which has a weight in pounds equal to the molecular weight of the substance. The materials most often considered in combustion are carbon, C, with a molecular weight of 12, hydrogen, H_2, with a molecular weight of 2, oxygen, O_2, with 32, and nitrogen, N_2, with 28. For all but the most exact combustion calculations the molecular weight and other properties of the mixture of gases, other than oxygen, which make up air can be taken as identical with those of nitrogen; hence, for combustion analyses, air is treated as though it consisted of but two gases: 21% by volume of oxygen and 79% by volume of nitrogen (or 23.3% by weight oxygen and 76.7% by weight nitrogen).

$$\text{Pound-mole of air} = (0.21 \times 32) + (0.79 \times 28) = 28.84 \text{ lb}$$

The practical usefulness of the pound-mole arises from the fact that for material in gaseous form the volume of a pound-mole at any given temperature and pressure is the same for all materials, and the number of molecules in a pound-mole is likewise the same for all materials; at standard atmospheric pressure and 32F the volume of one pound-mole of any substance in gaseous form is 358.6 cubic feet. Thus when combustion equations are written in terms of pound-moles the number of molecules going into the reaction are indicated by the number of pound-moles involved (though this number need not be the same on both sides of the equation) whereas the actual weights of material entering and leaving are, of course, necessarily the same.

In writing combustion equations various units are often used and in some cases more than one type of unit may be used in a single equation. To avoid confusion, careful differentiation must be noted in the use of symbols representing various constituents of both the fuel and of the products of combustion. The following nomenclature is used:

(1) Italics, CO_2, represents per cent by weight.
(2) Bold-face, **CO_2**, represents per cent by volume.
(3) Subscripts f and a refer, respectively, to fuel and air: Thus $(CO_2)_f$ would indicate percentage by volume of carbon dioxide in the fuel whereas $(N_2)_a$ would indicate percentage by weight of nitrogen in the air.
(4) Primed italics, C', represents weight fraction of the constituent in the fuel ($= C_f/100$).
(5) Standard type, CO_2, represents the constituent without units. Thus total weight in pounds would be shown as 80 lb CO_2. Total volume (or weight) in moles would be shown as $80_m CO_2$.
(6) Combustion equations for the Orsat analysis are often based on 100 pounds of dry exit gases; in this case the italicized symbol, CO_2, indicates the weight in pounds of the constituent present in 100 pounds of dry exit gases.

Complete Combustion of Carbon

Utilizing the pound-mole concept, the basic equation for the complete combustion of carbon can be written as,

$$1 \text{ pound-mole carbon} + 1 \text{ pound-mole oxygen} = 1 \text{ pound-mole carbon dioxide}$$

or,

$$1_m C + 1_m O_2 = 1_m CO_2 \tag{1}$$

Then substituting the weights of a pound-mole of each substance, equation (4) can be re-written as

$$1(12) \text{ lb C} + 1(32) \text{ lb } O_2 = 1(12 + 32) \text{ lb } CO_2$$

$$12 \text{ lb C} + 32 \text{ lb } O_2 = 44 \text{ lb } CO_2$$

$$1 \text{ lb C} + 2.667 \text{ lb } O_2 = 3.667 \ CO_2 \tag{2}$$

which shows that 2.667 pounds of oxygen are required to bring about the complete combustion of one pound of free carbon.

In most thermodynamic cycles, however, the oxygen is supplied as part of atmospheric air. Since air consists of 21% oxygen by volume and since equal gas volumes—at the same temperature and pressure—represent

equal fractions of a pound-mole it follows that $0.79/0.21 = 3.76$ pound-moles of nitrogen will accompany each pound-mole of oxygen and 4.76 pound-moles of air will have to be supplied to provide one pound-mole of oxygen. Equation (1) then becomes,

$$1_mC + 4.76_m(\text{Air}) = 1_mCO_2 + 3.76_mN_2$$

or

$$1_mC + (1_mO_2 + 3.76_mN_2) = 1_mCO_2 + 3.76_mN_2 \tag{3}$$

Substituting in the weights of pound-moles,

$$1(12) \text{ lb C} + (32 + 3.76 \times 28) \text{ lb Air} = 1(12 + 32) \text{ lb } CO_2 + (3.76 \times 28) \text{ lb } N_2$$

$$12 \text{ lb C} + 137.3 \text{ lb Air} = 44 \text{ lb } CO_2 + 105.3 \text{ lb } N_2$$

$$1 \text{ lb C} + 11.44 \text{ lb Air} = 3.66 \text{ lb } CO_2 + 8.78 \text{ lb } N_2 \tag{4}$$

Hence 11.44 lb of air are needed to burn 1 lb of carbon.

Complete combustion cannot be achieved unless some excess air is provided. If $X\%$ excess air is supplied equation (3) becomes,

$$1_mC + 4.76_m(1 + X/100)_m(\text{Air}) = 1_mCO_2 + 3.76(1 + X/100)_mN_2 + (X/100)_mO_2 \tag{5}$$

giving,

$$1 \text{ lb C} + 11.44(1 + X/100 \text{ lb})(\text{Air}) = 3.66 \text{ lb } CO_2 + 8.78(1 + X/100) \text{ lb } N_2 + (X/100)2.667 \text{ lb } O_2 \tag{6}$$

Hence, for these generalized combustion conditions, the weight of air required per pound of free carbon is $11.44(1 + X/100)$ and the total weight of the products of combustion is equal to the sum of the weight of fuel and air or,

$$[11.44(1 + X/100) + 1] \text{ lb products per lb carbon} \tag{7}$$

The per cent by weight of the materials present in the products of combustion is then,

$$CO_2 = 100 \frac{3.66}{11.44(1 + X/100) + 1}\% \tag{8}$$

$$O_2 = 100 \frac{(X/100)2.667}{11.44(1 + X/100 + 1)}\% \tag{9}$$

$$N_2 = 100 \frac{8.78(1 + X/100)}{11.44(1 + X/100) + 1}\% \tag{10}$$

For the special case of zero excess air the above equations reduce to,

$$CO_2 = 100(3.66 \div 12.44)\% = 29.42\% \tag{11}$$

$$O_2 = 0\% \tag{12}$$

$$N_2 = 100(8.78 \div 12.44)\% = 70.58\% \tag{13}$$

In many practical cases analyses on a volume basis are of greater value than on a weight basis; from the above weight equations for combustion of free carbon:

(1) The volume of air required per pound of carbon burned (based on the specific volume, 13.35 cu ft per lb of standard air) is,

$$\text{Cu ft per lb of carbon} = (13.35)(11.44)(1 + X/100) \tag{14}$$

which, for zero excess air, becomes,

$$\text{Cu ft per lb of carbon} = 153 \text{ cu ft} \tag{15}$$

(2) The volume of the products of combustion is determined by the number of pound-moles which they represent. From equation (8) the total pound-moles in the products of combustion amount to,

$$1 + 3.76(1 + X/100) + X/100 = 4.76(1 + X/100)$$

pound-moles per pound-mole of C. Since one pound-mole of any material occupies the same volume as one pound-mole of any other material the percentages by volume of the materials making up the products of combustion are,

$$CO_2 = 100 \frac{1}{4.76(1 + X/100)} \% \qquad (16)$$

$$O_2 = 100 \frac{X/100}{4.76(1 + X/100)} \% \qquad (17)$$

$$N_2 = 100 \frac{3.76(1 + X/100)}{4.76(1 + X/100)} \% \qquad (18)$$

and for the special case of zero excess air,

$$CO_2 = 21\% \qquad O_2 = 0\% \qquad N_2 = 79\% \qquad (19)$$

The total volume of the products of combustion can be evaluated from the perfect gas law for partial pressure of P psfa, absolute temperature of T deg R = (deg F + 460), and the universal gas constant of $R' = 1544$; thus,

$$V = \frac{mR'T}{P} \qquad (20)$$

where m = number of moles of material entering into the reaction. The partial pressure is equal to the ratio of the pound-moles of the particular constituent divided by the pound-moles of all the products of combustion. Then selecting CO_2 as the particular constituent, the partial pressure [using mole values from equation (5)] is,

$$\frac{1}{1 + 3.76(1 + X/100) + X/100} (14.7)(144)$$

psfa. Hence the volume of the products of combustion of 1 mole of carbon at a temperature of 2000 deg R is

$$V = \frac{(1)(1544)(2000)}{\dfrac{1}{1 + 3.76(1 + X/100) + X/100} (14.7)(144)} \text{ cu ft} \qquad (21)$$

and for the special case of zero excess air,

$$V = \frac{3088000}{\dfrac{(14.7)(144)}{4.76}}$$

$$V = \frac{3088000}{445} = 6940 \text{ cu ft per mole carbon}$$

and the volume of the flue gas of 1 lb of carbon is $\frac{1}{12}$ of this or

$$V = \frac{6940}{12} = 578.3 \text{ cu ft per lb of carbon} \qquad (22)$$

Incomplete Combustion of Carbon

By analogy with the equations for complete combustion, the equations can be written for burning of carbon to carbon monoxide.

$$1_mC + 0.5_mO_2 = 1_mCO \qquad (23)$$

$$12 \text{ (lb C)} + 0.5(32) \text{ (lb O}_2) = 28 \text{ (lb CO)}$$

$$1 \text{ (lb C)} + 1.33 \text{ (lb O}_2) = 2.335 \text{ (lb CO)} \qquad (24)$$

which shows that 1.33 pounds of oxygen are needed to burn carbon to monoxide. For burning with excess air the equation becomes,

$$1 \text{ (lb C)} + 5.72(1 + X/100) \text{ (lb Air)}$$
$$= 2.335 \text{ (lb CO)} + 4.39(1 + X/100) \text{ (lb N}_2) + (X/100)(1.33) \text{ (lb O}_2) \quad (25)$$

Combustion of Hydrogen

By analogy with the equations for free carbon, the combustion equations for hydrogen can be written as follows:

$$2_m H_2 + 1_m O_2 = 2_m H_2O \tag{26}$$

$$2(2) \text{ lb } H_2 + 1(32) \text{ lb } O_2 = 2(2 + 16) \text{ lb } H_2O$$

$$4 \text{ lb } H_2 + 32 \text{ lb } O_2 = 36 \text{ lb } H_2O$$

$$1 \text{ lb } H_2 + 8 \text{ lb } O_2 = 9 \text{ lb } H_2O \tag{27}$$

which shows that 8 pounds of oxygen are required to accomplish the complete combustion of one pound of hydrogen.

With air,

$$2_m H_2 + 4.76_m (\text{Air}) = 2_m H_2O + 3.76_m N_2 \tag{28}$$

or,

$$2_m H_2 + (1_m O_2 + 3.76_m N_2) = 2_m H_2O + 3.76_m N_2 \tag{29}$$

$$2(2) \text{ lb } H_2 + (32 + 3.76 \times 28) \text{ lb Air} = 2(2 + 16) \text{ lb } H_2O + 3.76 \times 28 \text{ lb } N_2$$

$$4 \text{ lb } H_2 + 137.3 \text{ lb Air} = 36 \text{ lb } H_2O + 105.3 \text{ lb } N_2$$

$$1 \text{ lb } H_2 + 34.32 \text{ lb Air} = 9 \text{ lb } H_2O + 26.32 \text{ lb } N_2 \tag{30}$$

With X% excess air,

$$2_m H_2 + 4.76(1 + X/100)_m (\text{Air}) = 2_m H_2O + 3.76(1 + X/100)_m N_2 + (X/100)_m O_2 \tag{31}$$

$$1 \text{ lb } H_2 + 34.32(1 + X/100) \text{ lb Air} = 9 \text{ lb } H_2O + 26.32(1 + X/100) \text{ lb } N_2 + (X/100)8 \text{ lb } O_2 \tag{32}$$

When the fuel contains oxygen it is customary to assume that this oxygen combines with the hydrogen; the above reaction equations are then applicable to the combustion process with $(H_2 - O_2/8)$ substituted for H_2 as indicative of the weight of hydrogen that is free to combine with oxygen of the air.

The per cent by weight of the materials present in the products of combustion is,

$$H_2O = 100 \frac{9}{1 + 34.32(1 + X/100)} \% \text{ by weight} \tag{33}$$

$$O_2 = 100 \frac{(X/100)8}{1 + 34.32(1 + X/100)} \% \text{ by weight} \tag{34}$$

$$N_2 = 100 \frac{26.32(1 + X/100)}{1 + 34.32(1 + X/100)} \% \text{ by weight} \tag{35}$$

For the special case of zero excess air the above equations reduce to,

$$H_2O = 100 \frac{9}{35.32} = 25.5\% \tag{36}$$

$$O_2 = 0\% \tag{37}$$

$$N_2 = 100 \frac{26.32}{35.32} = 74.5\% \tag{38}$$

On a volume basis:

(1) The volume of standard air required per pound of hydrogen burned is,

$$\text{Cu ft per lb of hydrogen} = (13.35)(34.32)(1 + X/100) \tag{39}$$

which for zero excess air becomes,

$$\text{Cu ft per lb of hydrogen} = 458 \text{ cu ft} \tag{40}$$

(2) The percentage volumes of the products of combustion are,

$$H_2O = 100 \frac{2}{1 + 4.76(1 + X/100)} \% \tag{41}$$

$$O_2 = 100 \frac{X/100}{1 + 4.76(1 + X/100)} \% \tag{42}$$

$$N_2 = 100 \frac{3.76(1 + X/100)}{1 + 4.76(1 + X/100)} \% \tag{43}$$

and for the special case of zero excess air,

$$H_2O = 34.7\%; \quad O_2 = 0\%; \quad N_2 = 65.3\% \tag{44}$$

The total volume of the products of combustion of 2 moles of hydrogen based on the volume of the water vapor at its partial pressure and for a temperature of 2000R, is,

$$V = \frac{(2)(1544)(2000)}{\dfrac{2}{1 + 4.76(1 + X/100)}(14.7)(144)} \text{ cu ft per 2 moles of hydrogen or } \tfrac{1}{4} \text{ of this volume per pound of hydrogen} \tag{45}$$

and for the special case of zero excess air,

$$V = \frac{6176000}{\dfrac{2}{5.76}(14.7)(144)} = 8400 \text{ cu ft per 2 moles of } H_2 \text{ or 2100 cu ft per lb of } H_2 \tag{46}$$

Combustion Equations for Hydrocarbon

Combining the equations already developed for combustion of free carbon or of hydrogen, the equations for any hydrocarbon, having the formula C_xH_y, can be written

$$1_m C_xH_y + (x + y/4)_m O_2 = (y/2)_m H_2O + x_m CO_2$$

$$(12x + y) \text{ lb } C_xH_y + (x + y/4)(32) \text{ lb } O_2 = (y/2)(2 + 16) \text{ lb } H_2O + (x)(12 + 32) \text{ lb } CO_2 \tag{47}$$

$$1 \text{ lb } C_xH_y + \frac{32(x + y/4)}{(12x + y)} \text{ lb } O_2 = \frac{9y}{(12x + y)} \text{ lb } H_2O + \frac{44x}{(12x + y)} \text{ lb } CO_2 \tag{48}$$

For combustion with air,

$$1_m C_xH_y + 4.76(x + y/4)_m \text{Air} = (y/2)_m H_2O + x_m CO_2 + 3.76(x + y/4)_m N_2 \tag{49}$$

$$1 \text{ lb } C_xH_y + \left(11.44 \frac{12x}{12x + y} + 34.32 \frac{y}{12x + y}\right) \text{ lb Air} = \frac{9y}{(12x + y)} \text{ lb } H_2O + \frac{(3.66)(12)(x)}{(12x + y)} \text{ lb } CO_2$$

$$+ \left[8.78 \frac{12x}{(12x + y)} + 26.32 \frac{y}{(12x + y)}\right] \text{ lb } N_2 \tag{50}$$

For combustion with excess air the air and nitrogen terms in the above equations are multiplied by $(1 + X/100)$ and the oxygen term is multiplied by $(X/100)$,

$$1_m C_xH_y + 4.76(1 + X/100)(x + y/4)_m (\text{Air}) = (y/2)_m H_2O + x_m CO_2$$

$$+ 3.76(1 + X/100)(x + y/4)_m N_2 + (X/100)(x + y/4)_m O_2 \tag{51}$$

and,

$$1 \text{ lb } C_xH_y + (1 + X/100)\left[11.44 \frac{12x}{(12x + y)} + 34.32 \frac{y}{(12x + y)}\right] \text{ lb Air}$$

$$= \frac{9y}{(12x + y)} \text{ lb } H_2O + 3.66 \frac{12x}{(12x + y)} \text{ lb } CO_2$$

$$+ (1 + X/100)\left[8.78 \frac{12x}{(12x + y)} + 26.32 \frac{y}{(12x + y)}\right] \text{ lb } N_2 + (X/100)32 \frac{(x + y/4)}{(12x + y)} \text{ lb } O_2 \tag{52}$$

which is of the form,

$$1 \text{ lb } C_xH_y + A \text{ lb Air} = B \text{ lb } H_2O + C \text{ lb } CO_2 + D \text{ lb } N_2 + E \text{ lb } O_2 \qquad (53)$$

where the coefficients A,B,C,D, and E have fixed values for a given fuel when burned with a given percentage by weight of excess air. The per cent by weight of the materials present in the products of combustion is,

$$H_2O = 100 \frac{B}{B + C + D + E} \%$$

$$= 100 \frac{B}{1 + A} \% \text{ by weight} \qquad (54)$$

$$CO_2 = 100 \frac{C}{B + C + D + E} \%$$

$$= 100 \frac{C}{1 + A} \% \text{ by weight} \qquad (55)$$

$$N_2 = 100 \frac{D}{B + C + D + E} \%$$

$$= 100 \frac{D}{1 + A} \% \text{ by weight} \qquad (56)$$

$$O_2 = 100 \frac{E}{B + C + D + E} \%$$

$$= 100 \frac{E}{1 + A} \% \text{ by weight} \qquad (57)$$

The volume relationships for this fuel are obtained from the basic combustion equations in exactly the same way as was used in the sections on carbon and on hydrogen, thus,

$$H_2O = 100 \frac{y/2}{y/2 + x + 3.76(1 + X/100)(x + y/4) + (X/100)(x + y/4)} \% \text{ by volume} \qquad (58)$$

$$CO_2 = 100 \frac{x}{y/2 + x + 3.76(1 + X/100)(x + y/4) + (X/100)(x + y/4)} \% \text{ by volume} \qquad (59)$$

$$N_2 = 100 \frac{3.76(1 + X/100)(x + y/4)}{y/2 + x + 3.76(1 + X/100)(x + y/4) + (X/100)(x + y/4)} \% \text{ by volume} \qquad (60)$$

$$O_2 = 100 \frac{(X/100)(x + y/4)}{y/2 + x + 3.76(1 + X/100)(x + y/4) + (X/100)(x + y/4)} \% \text{ by volume} \qquad (61)$$

and for the special case of zero excess air,

$$H_2O = 100 \frac{y/2}{y/2 + x + 3.76(x + y/4)} \% \text{ by volume} \qquad (62)$$

$$CO_2 = 100 \frac{x}{y/2 + x + 3.76(x + y/4)} \% \text{ by volume} \qquad (63)$$

$$N_2 = 100 \frac{3.76(x + y/4)}{y/2 + x + 3.76(x + y/4)} \% \text{ by volume} \qquad (64)$$

$$O_2 = 0\% \text{ by volume} \qquad (65)$$

The total volume of the products of combustion based on the volume of carbon dioxide at its partial pressure and for 2000R is,

$$V = \cfrac{(x)(1544)(2000)}{\cfrac{x}{y/2 + x + 3.76(1 + X/100)(x + y/4) + (X/100)(x + y/4)}(14.7)(144)} \quad \text{cu ft per mole of } C_xH_y \quad (66)$$

or $1 \div (12x + y)$ of this volume per pound of C_xH_y

For the special case of zero excess air the above equation becomes,

$$V = \cfrac{3088000}{\cfrac{(14.7)(144)}{y/2 + x + 3.76(x + y/4)}} \quad \text{cu ft per mole of } C_xH_y \quad (67)$$

or $1 \div (12x + y)$ of this volume per pound of C_xH_y

Combustion Analyses from Orsat Readings

In most actual combustion processes the determination of air-fuel ratio cannot be made by direct measurement of entering air since various leakages through auxiliary openings and through the furnace setting will be responsible for a substantial increase in total air over that entering at the burner. Thus for practical purposes the air-fuel ratio must usually be determined by calculation from data available on analysis of the products of combustion. The usual Orsat analysis gives carbon dioxide, carbon monoxide and oxygen and from these data —together with a knowledge of the N_2 and O_2 content of the fuel—it is possible to carry through a satisfactorily complete analysis of combustion conditions and to determine—if this is not already known—the carbon and hydrogen content of the fuel. All analyses of this type utilize the concept of a mass balance, and by establishing such balances on nitrogen, carbon, hydrogen, and oxygen complete combustion information can be obtained.

When the ultimate analysis of the fuel is known, or when the percentage by either volume or weight of the carbon or the hydrogen in the fuel is known, the analysis in terms of Orsat data requires fewer steps than for the generalized case in which the only information available concerning the composition of the fuel is the percentage of N_2 and of O_2 that it contains. Since the latter case is the most complex and since all other cases can be treated as special and simplified forms of the same analysis, the general procedure will be established in detail.

Consider that some unknown fuel is burned under combustion conditions such that the Orsat reading, in volumetric percentages on a dry gas basis, is,

$$\text{Carbon dioxide} \quad = CO_2\% \text{ by volume}$$

$$\text{Carbon monoxide} = CO \% \text{ by volume}$$

$$\text{Oxygen} \quad = O_2\% \text{ by volume}$$

If the fuel contained sulphur the resultant sulphur dioxide would appear as part of the Orsat reading for carbon dioxide. Since a pound-mole of either carbon (12 pounds) or of sulphur (32 pounds) will react with the pound-mole of oxygen it follows that the volume of sulphur dioxide in the exit gases will be exactly the same as the volume of carbon dioxide that would have resulted if an equal fraction of a mole of carbon were burned in place of the sulphur. When the fuel contains only a very small amount of sulphur the analysis can be made with reasonable accuracy by assuming that all of the indicated carbon dioxide is due to combustion of carbon alone; when a larger, and known, percentage of sulphur is present in the fuel an approximate analysis can be made by adding 12/32 of the known weight of sulphur to the weight of carbon and thereafter continuing the analysis as though no sulphur were present. If greater accuracy is needed and if the weight of sulphur in the fuel is known, the equivalent amount of sulphur dioxide can be deducted from the Orsat reading for carbon dioxide.

For the burning, with adequate air, of any commercial fuel in a furnace there will not be any significant amount of hydrogen or of other gaseous combustible in the products of combustion. For burning of hydrocarbons in an internal combustion engine, however, combustibles will almost always be present in the exit gases. Extensive research reported by the National Advisory Committee of Aeronautics (Reports 476 and 616) shows that for either gasoline or diesel fuel the amount of hydrogen present in the exit gases bears a fixed

relationship to the amount of carbon monoxide and can be taken as 51% of the volumetric fraction of carbon monoxide reported in Orsat reading. The same source gives 0.22% as the amount of methane, by volume on a dry gas basis, that is usually found in the exit gases from an internal combustion engine. If there is evidence that other hydrocarbons are present in the products a modified type of Orsat analysis can be used to determine the fractions of saturated and unsaturated hydrocarbons in an exit-gas sample; aside, however, from research studies, it is rarely necessary to carry out the more complicated modified Orsat analysis.

Then adjusting the original Orsat readings and assuming—with good accuracy—that the undetermined remaining volume of the products of combustion consists of nitrogen, we have

$$\text{Carbon dioxide} = CO_2$$
$$\text{Carbon monoxide} = CO$$
$$\text{Oxygen} = O_2$$
$$\text{Hydrogen} = 0.51 CO$$
$$\text{Methane} = 0.22$$
$$\text{Nitrogen} = N_2$$

Since most of the desired results are on a weight rather than volume basis a first step is to convert the volumetric data on exit gas to a gravimetric basis; thus,

$$\text{Carbon dioxide} = \frac{CO_2(12 + 32)}{W}$$
$$= \frac{CO_2(44)}{W} = CO_2\% \text{ by weight} \tag{68}$$

$$\text{Carbon monoxide} = \frac{CO(12 + 16)}{W}$$
$$= \frac{CO(28)}{W} = CO_2\% \text{ by weight} \tag{69}$$

$$\text{Oxygen} = \frac{O_2(32)}{W} = O_2\% \text{ by weight} \tag{70}$$

$$\text{Hydrogen} = \frac{CO(2 \times 0.51)}{W}$$
$$= \frac{CO(1.02)}{W} = H_2\% \text{ by weight} \tag{71}$$

$$\text{Methane} = \frac{0.22(12 + 4)}{W}$$
$$= \frac{3.5}{W} = CH_4\% \text{ by weight} \tag{72}$$

$$\text{Nitrogen} = \frac{N_2(28)}{W} = N_2\% \text{ by weight} \tag{73}$$

$$W = \frac{44CO_2 + 28CO + 32O_2 + 1.02CO + 3.5 + 28N_2}{100} \tag{74}$$

where W is the total molecular weight of exit gas.

A reaction equation can then be written in the form,

$$(\text{lb C} + \text{lb H}_2 + \text{lb O}_2 + \text{lb N}_2)_f + (\text{lb O}_2 + \text{lb N}_2)_a = 100 + \text{lb H}_2O \tag{75}$$

where each term is in units of pounds per 100 pounds of dry products of combustion and the subscripts f and a refer to fuel and air, respectively; terms in the form "lb x" represent unknowns and it will be noted that in its present form equation contains seven such. In order to completely solve the reaction equation two of the unknowns must be given; usually the fuel is known to be free of, or to possess known percentages by weight of oxygen and nitrogen. A complete generalized solution can be obtained by considering that the fractions by weight of nitrogen and oxygen in the fuel, N' and O', respectively, are known.

Let Z equal the weight of fuel burned per 100 pounds of dry products of combustion; then

$$Z = (lb\ C + lb\ H_2 + lb\ O_2 + lb\ N_2)_f$$

$$= (lb\ C + lb\ H_2 + O'Z + N'Z)_f$$

and solving for Z

$$= \frac{(lb\ C + lb\ H_2)_f}{1 - O' - N'} \tag{76}$$

Now establish a mass-balance on the nitrogen entering and leaving the combustion space,

$$(lb\ N_2)_a + N'Z = N_2$$

or

$$(lb\ N_2)_a = N_2 - N'Z \tag{77}$$

hence

$$(lb\ N_2 + lb\ O_2)_a = (1/0.769)(N_2 - N'Z) \tag{78}$$

Four of the original unknowns in the reaction equation can now be eliminated; substituting from equations (76) and (78) into (75) and simplifying,

$$\frac{(lb\ C + lb\ H_2)_f}{1 - O' - N'}\left(1 - \frac{N'}{0.769}\right) + \frac{N_2}{0.769} = 100 + (lb\ H_2O) \tag{79}$$

Equation (79) contains three unknowns (lb C)$_f$, (lb H$_2$)$_f$, and (lb H$_2$O); two of these can be eliminated by establishing mass balance on the hydrogen and the carbon:

A mass-balance on the hydrogen gives,

$$(lb\ H_2)_f = (4/16)CH_4 + H_2 + (2/18)\ (lb\ H_2O)$$

$$= 0.25CH_4 + H_2 + 0.111\ (lb\ H_2O) \tag{80}$$

where (lb H$_2$O) is the only unknown on the right side of the equation.

A mass-balance on the carbon gives,

$$(lb\ C)_f = (12/44)CO_2 + (12/28)CO + (12/16)CH_4$$

$$= 0.2727CO_2 + 0.429CO + 0.75CH_4 \tag{81}$$

which does not contain any unknowns on the right side.

Substituting from equations (80) and (81) into equation (79),

$$\frac{[0.272CO_2 + 0.429CO + 0.25CH_4 + H_2 + 0.111\ (lb\ H_2O)](1 - N'/0.769)}{(1 - O' - N')} + N_2/0.769 = 100 + (lb\ H_2O)$$

which simplifies to

$$(lb\ H_2O)$$

$$= \left[\frac{(0.2727CO_2 + 0.429CO + CH_4 + H_2)(1 - N'/0.769)}{(1 - O' - N')} + \frac{N_2}{0.769} - 100\right]\left[\frac{1 - O' - N'}{0.889 - O' - 0.857N'}\right] \tag{82}$$

in which there are no unknowns on the right side; thus equation (82) provides a solution for the weight of water vapor in the exit gases per 100 pounds of *dry* exit gases as determined by Orsat analysis. Knowing (lb H$_2$O) the weight of carbon and of hydrogen in the fuel can be calculated from equations (80) and (81), the weight of fuel from equation (76), and the weight of air from equation (78); the actual fuel-air ratio can then be determined by simple division.

The air required for perfect combustion (no excess air) is given by,

$$\text{(lb Air)}/\text{(lb Fuel)} = 11.44C' + 34.32H'_2 - (1.0/0.231)O'_2 \tag{83}$$

where C', H'_2, and O'_2 are the fractions by weight in the fuel on a moisture and ash free basis; these fractions are readily calculable from data available from the above analysis. Then by division of the air-fuel ratio for perfect combustion into the air-fuel ratio for actual combustion the fraction of excess air used can be readily determined.

In practical application of the above method of analysis it must be noted that the calculated weight of carbon in the fuel, (lb C), is equal to that weight of carbon which undergoes combustion to form either CO_2 or CO. If any carbon leaves the system in smoke or with the ash it will *not* show up in the analysis and the derived value of the per cent by weight of carbon in the moisture and ash-free fuel will then be less than the actual value. If information is available as to either the weight of carbon lost as smoke or in ash, or the fraction of total carbon from the fuel which is so lost, the data resulting from the analysis can be revised to give the actual per cent by weight of each constituent of the moisture-free and ash-free fuel. If information is available concerning the per cent moisture and the per cent ash in the fuel the data from the analysis can be further developed to permit evaluation of the ultimate analysis of the fuel on an as-received basis. The steps in these various extensions of the analysis are as follows:

(1) To determine the *per cent* by weight of constituents in the moisture-free and ash-free fuel (assuming no loss of carbon in smoke or ash).

$$\% \text{ Carbon} = \text{(lb C/Z)}100 = 100C' \tag{84}$$

$$\% \text{ Hydrogen} = 100 \text{ (lb } H_2/Z) = 100H'_2 \tag{85}$$

$$\% \text{ Oxygen} = 100(O'_2) \tag{86}$$

$$\% \text{ Nitrogen} = 100.N'_2 \tag{87}$$

(2) To determine the per cent by weight *of constituents* in the moisture-free and ash-free fuel when the combined loss of carbon in the smoke or with the ash is known to be C_{loss} pounds per hundred pounds of fuel fired.

$$\% \text{ Carbon} = 100(100C' + C_{loss})/(100 + C_{loss}) \tag{88}$$

$$\% \text{ Hydrogen} = 100H'_2/(100 + C_{loss}) \tag{89}$$

$$\% \text{ Oxygen} = 100(O'_2)/(100 + C_{loss}) \tag{90}$$

$$\% \text{ Nitrogen} = 100.N'_2/(100 + C_{loss}) \tag{91}$$

(3) To determine the per cent by weight of constituents in the fuel on an as-received basis; the per cents by weight of moisture and of ash are known and are respectively equal to 100M and 100A.

$$\% \text{ Carbon} = 100(100C' + C_{loss})/(100 + C_{loss} + 100M + 100A) \tag{92}$$

$$\% \text{ Hydrogen} = 100H'_2/(100 + C_{loss} + 100M + 100A) \tag{93}$$

$$\% \text{ Oxygen} = 100(O'_2)/(100 + C_{loss} + 100M + 100A) \tag{94}$$

$$\% \text{ Nitrogen} = 100.N'_2/(100 + C_{loss} + 100M + 100A) \tag{95}$$

$$\% \text{ Moisture} = 100M$$

$$\% \text{ Ash} = 100A$$

Combination Analysis for Simplified Cases

The generalized solution established in the preceding section covers the most involved applications of Orsat data to fuel-and-combustion analysis. In by far the greater number of practical combustion problems, a number of simplifying conditions greatly facilitate the numerical work needed in obtaining a solution. Thus for steady state combustion processes (as in furnaces) no correction need be made to the Orsat reading for either methane or hydrogen. In many cases the fuel will be known to be free of either oxygen or nitrogen

or both and in many other cases the percentage by weight of carbon in the fuel may be known. Analyses for the various simplified cases can be made either by direct derivation or by reduction of the generalized solution. The latter procedure will be illustrated for some of the more common cases:

Case I. Fuel contains no oxygen. Equation (82) reduces to,

$$(\text{lb } H_2O) = \left[\frac{(0.2727CO_2 + 0.429CO + CH_4 + H_2)\left(1 - \dfrac{N'}{0.769}\right)}{1 - N'} + \frac{N_2}{0.769} - 100 \right]\left[\frac{1 - N'}{0.889 - 0.857N'} \right] \quad (96)$$

and equation (76) reduces to,

$$Z = (\text{lb } C + \text{lb } H_2)_f / (1 - N') \quad (97)$$

Equations (80) and (81) remain unchanged.

Case II. Fuel contains no nitrogen. Equation (82) reduces to,

$$(\text{lb } H_2O) = \left[\frac{0.2727\ CO_2 + 0.429\ CO + CH_4 + H_2}{1 - O'} + \frac{N_2}{0.769} - 100 \right]\left[\frac{1 - O'}{0.889 - O'} \right] \quad (98)$$

and equation (76) reduces to,

$$Z = \frac{(\text{lb } C + \text{lb } H_2)_f}{1 - O'} \quad (99)$$

Equations (80) and (81) remain unchanged.

Case III. Fuel contains neither oxygen nor nitrogen. Equation (82) becomes,

$$(\text{lb } H_2O) = \frac{(0.2727CO_2 + 0.429CO + CH_4 + H_2) + (N_2/0.769) - 100}{0.889} \quad (100)$$

and equation (76) reduces to,

$$Z = (\text{lb } C + \text{lb } H_2) \quad (101)$$

Equations (80) and (81) remain unchanged.

Case IV. Carbon content of the fuel is known. When the fraction by weight, C', of carbon in the fuel is known the procedure is to calculate the weight of carbon from equation (81) and then determine the weight of fuel as,

$$Z = (1/C')\,(\text{lb } C) \quad (102)$$

For the known O' and N' the weights of oxygen and nitrogen can then be calculated and the weight of hydrogen in the fuel (per 100 pounds of dry products of combustion) obtained by subtraction. Equation (78) then gives the weight of air supplied and the weight of water vapor formed during combustion is known as 9 times the weight of hydrogen in the fuel.

Case V. Carbon content of fuel known, no oxygen or nitrogen in the fuel, and no hydrogen or methane in the products of combustion. The weight of carbon in the fuel is obtained from a simplified form of equation (81) as,

$$\text{lb } C = 0.2727CO_2 + 0.429CO \quad (103)$$

and the weight of hydrogen in the fuel is then,

$$\text{lb } H_2 = \frac{1 - C'}{C'}\,(\text{lb } C) \quad (104)$$

The weight of air is $(N_2/0.769)$ and the weight of water vapor is 9 $(\text{lb } H_2)$.

Examples of Combustion Calculations

The following numerical examples illustrate the method of applying the procedures developed in the preceding sections.

Example 1. An anthracite coal having 1.3% nitrogen and 3.9% oxygen (on a gravimetric ash and moisture free basis) is burned without smoke and without loss of free carbon in the ash. The products of combustion, by Orsat analysis, contain 14.6% carbon dioxide, 0.2% carbon monoxide, 5.5% oxygen, and—by difference—

79.7% nitrogen; the fuel is assumed to contain no appreciable quantity of sulphur. Determine the percentages of carbon and hydrogen in this coal, determine the air-fuel ratio, and evaluate the supply of excess air to the furnace.

Solution: The volumetric Orsat percentages will first be changed to a gravimetric basis; from equation (77),

$$W = \frac{44(14.6) + 28(0.2) + 32(5.5) + 28(79.7)}{100}$$

$$= \frac{642 + 5.6 + 176 + 2230}{100} = 30.54$$

hence,

$$CO_2 = 44(14.6/30.54) = 21.0\% \text{ by weight}$$

$$CO = 28(0.2/30.54) = 0.2\% \text{ by weight}$$

$$O_2 = 32(5.5/30.54) = 5.8\% \text{ by weight}$$

$$N_2 = 28(79.7/30.54) = 73.0\% \text{ by weight}$$

Substituting in equation (82)

$$(\text{lb } H_2O) = \left[\frac{[(0.2727)(21.0) + 0.429(0.2)]\left(1 - \dfrac{0.013}{0.769}\right)}{1 - 0.039 - 0.013} + \frac{73.0}{0.769} - 100 \right]\left[\frac{1 - 0.039 - 0.013}{0.889 - 0.039 - 0.857(0.013)} \right]$$

$$= 1.07 \text{ lb.}$$

From equation (80),

$$(\text{lb } H_2)_t = 0.111(1.07) = 0.119 \text{ lb}$$

From equation (81),

$$(\text{lb } C)_t = 0.2727(21.0) + 0.429(0.2) = 5.81 \text{ lb}$$

From equation (76),

$$\text{Weight of fuel} = Z = \frac{5.81 + 0.119}{1 - 0.012 - 0.029} = 6.25 \text{ lb}$$

From equation (78),

$$\text{Weight air} = (1/0.769)[73.0 - 0.013(6.25)]$$

$$= 93.65$$

$$(\text{Air-fuel ratio})_{\text{actual}} = 93.65/6.25 = 15.0$$

On an ash and moisture free basis the percentages by weight of constituents in the fuel are:

$$
\begin{aligned}
C &= (5.81/6.25)100 = 92.9\% \text{ by weight in fuel} \\
H_2 &= (0.119/6.25)100 = 1.9\% \text{ by weight in fuel} \\
N_2 &= 1.3\% \text{ by weight in fuel} \\
O_2 &= \underline{3.9\% \text{ by weight in fuel}} \\
& 100.0\%
\end{aligned}
$$

The required air-fuel ratio for perfect combustion is then, from equation (87),

$$(\text{Air-fuel ratio})_{\text{perfect}} = 11.44(0.929) + 34.32(0.019) - [(1.0/0.231)](0.039)$$

$$= 11.1 \text{ lb}$$

The actual air supply is therefore $15.0/11.1 = 1.35$ times the requirements for perfect combustion; excess air is thus equal to 35%.

Example 2. A gaseous fuel is known to contain 70.4% nitrogen by weight and 24.4% oxygen. When burned in a furnace the Orsat analysis shows 83.9% nitrogen, 9.1% oxygen, and 7.0% carbon dioxide. Determine the carbon and hydrogen content of the fuel (on an ash and moisture free basis), the air-fuel ratio and the percentage of excess air supplied under the actual conditions of combustion.

Solution: Changing the volumetric Orsat percentages to a gravimetric basis, from equation (74),

$$W = \frac{44(7.0) + 32(9.1) + 28(83.9)}{100}$$

$$= \frac{308 + 292 + 2350}{100} = 29.5$$

hence,

$$
\begin{aligned}
CO_2 &= 44(7.0/29.5) &= 10.4\% \text{ by weight} \\
O_2 &= 32(9.1/29.5) &= 10.0\% \text{ by weight} \\
N_2 &= 28(83.9/29.5) &= \underline{79.6\%} \text{ by weight} \\
& & 100.0\%
\end{aligned}
$$

Substituting in equation (82),

$$(\text{lb } H_2O) = \left[\frac{[(0.2727)(10.4)]\left(1 - \dfrac{0.704}{0.769}\right)}{1 - 0.244 - 0.704} + \frac{79.6}{0.769} - 100 \right]\left[\frac{1 - 0.244 - 0.704}{0.889 - 0.244 - 0.857(0.704)} \right]$$

$$= 10.1 \text{ lb}$$

From equation (80),

$$(\text{lb } H_2)_f = 0.111(10.1) = 1.122 \text{ lb}$$

From equation (81),

$$(\text{lb } C)_f = 0.2727(10.4) = 2.837 \text{ lb}$$

From equation (76),

$$\text{Weight of fuel} = Z = \frac{2.837 + 1.122}{1 - 0.244 - 0.704} = 76.1 \text{ lb}$$

From equation (78),

$$\text{Weight air} = \left(\frac{1.0}{0.769}\right)[79.6 - 0.704(76.1)]$$

$$= 33.8 \text{ lb}$$

$$\text{Air-fuel ratio} = 33.8/76.1 = 0.444$$

On an ash and moisture free basis the percentages by weight of the constituents of the fuel are:

$$
\begin{aligned}
C &= (2.837/76.1) = & 3.73\% \text{ by weight in fuel} \\
H_2 &= (1.122/76.1) = & 1.47\% \text{ by weight in fuel} \\
O_2 &= & 24.4\% \text{ by weight in fuel} \\
N_2 &= & \underline{70.4\%} \text{ by weight in fuel} \\
& & 100.0\%
\end{aligned}
$$

The air-fuel ratio required for perfect combustion is then, from equation (83),

$$(\text{Air-fuel ratio})_{\text{perfect}} = 11.44(0.0373) + 34.32(0.0147) - (1/0.231)(0.244)$$
$$= 0.931 - 1.057 = -0.126 \text{ lb}$$

which indicates that the oxygen content of the fuel is more than sufficient to permit complete combustion hence no air is required.

The above results can be checked by noting that all of the oxygen in the air must appear in the exit gases and an additional 0.126 pound of oxygen will appear in the exit gases for each pound of fuel burned. The total weight of free oxygen in the exit gases is then,

$$\text{Free exit gas oxygen} = 0.231(33.8) + 0.126(0.231)(76.1) = 10.0 \text{ lb}$$

But the weight of dry exit gases is 100 lb hence the oxygen content is 10% by weight which is in agreement with the percentage determined from the Orsat data.

Although the knowledge of constituents in the fuel which is realized from the above analysis is of interest it does not permit evaluation of the heating value since no information can be gained as to the chemical arrangements which exist among the elements. Thus the carbon can be partially present in a free state or in

the form of carbon monoxide, carbon dioxide, or methane. Additional data concerning the fuel would be therefore necessary before its heating value could be determined.

Example 3. An anthracite coal has the following composition on an ash and moisture free basis:

$$C = 92.9\%$$
$$H_2 = 1.9\%$$
$$N_2 = 1.3\%$$
$$O_2 = 3.9\%$$

If this coal is burned without smoke and without loss of carbon to the ash determine the volumetric percentage of constituents in the actual products of combustion. Determine the percentages of nitrogen, carbon dioxide, and oxygen that would be shown in an Orsat test of the exit gases. Assume that 35% excess air is supplied.

Solution: The reaction equation for the carbon (based on 100 pounds of fuel) is obtained by substitution into equation (9),

$$92.9[1 \text{ (lb C)} + 11.44(1.35) \text{ (lb air)}] = 92.9[3.66 \text{ (lb CO}_2) + 8.78(1.35) \text{ (lb N}_2) + 2.667(0.35) \text{ (lb O}_2)]$$
$$= 92.9 \text{ (lb C)} + 1435 \text{ (lb air)}$$
$$= 340 \text{ (lb CO}_2) + 1101 \text{ (lb N}_2) + 86.8 \text{ (lb O}_2)$$

The reaction equation for the hydrogen is obtained by substitution in equation (32), and noting that the net weight of hydrogen which receives oxygen from the air is $H_2 - O_2/8$,

$$\left(1.9 - \frac{3.9}{8}\right)[1 \text{ (lb H}_2) + 34.32(1.35) \text{ (lb air)}]$$
$$= \left(1.9 - \frac{3.9}{8}\right)[9 \text{ (lb H}_2O) + 26.32(1.35) \text{ (lb N}_2) + 8(0.35) \text{ (lb O}_2)]$$

$$1.41 \text{ (lb H}_2) + 65.4 \text{ (lb air)} = 12.7 \text{ (lb H}_2O) + 50.2 \text{ (lb N}_2) + 3.95 \text{ (lb O}_2)$$

Adding the reaction equations for carbon and hydrogen and adding also nitrogen and oxygen that is in the coal and the hydrogen that combines directly with oxygen,

$$92.9 \text{ (lb C)} + 1.9 \text{ (lb H}_2) + 1.3 \text{ (lb N}_2) + 3.9 \text{ (lb O}_2) + 1500 \text{ (lb air)}$$
$$= 340 \text{ (lb CO}_2) + [12.7 + 3.9(9/8)] \text{ (lb H}_2O) + 1152.5 \text{ (lb N}_2) + 90.8 \text{ (lb O}_2)$$

The relative volumes in the products of combustion will correspond to the relative number of pound-moles:

$$
\begin{array}{llr}
\text{Carbon dioxide} = & 340/44 = & 7.73 \text{ lb moles} \\
\text{Water vapor} = & 17.1/18 = & 0.95 \\
\text{Nitrogen} = & 1152.5/28 = & 41.15 \\
\text{Oxygen} = & 90.8/32 = & 2.84 \\
\hline
& & 52.67
\end{array}
$$

Hence the volumetric percentage of constituents in the exit gases is,

$$CO_2 = 100(7.73/52.7) = 14.7\% \text{ by volume}$$
$$H_2O = 100(0.95/52.7) = 1.8\% \text{ by volume}$$
$$N_2 = 100(41.15/52.7) = 78.1\% \text{ by volume}$$
$$O_2 = 100(2.84/52.7) = 5.42\% \text{ by volume}$$

The volumetric percentages indicated by an Orsat analysis would be,

$$CO_2 = 100(14.7/98.2) = 15.0\% \text{ by volume}$$
$$N_2 = 100(78.1/98.2) = 79.6\% \text{ by volume}$$
$$O_2 = 100(5.4/98.2) = 5.4\% \text{ by volume}$$

Example 4. For the coal of the above example consider that 1.35% of the total carbon in the fuel burns to carbon monoxide and determine the volumetric percentage of constituents in the actual products of combustion and in the products as indicated by an Orsat analysis.

Solution: The weight of carbon incompletely burned, per 100 pounds of fuel, is equal to,

$$0.0135(92.9) = 1.25 \text{ lb}$$

and the weight which undergoes complete combustion is $92.9 - 1.25 = 91.65$ lb. The two reaction equations for carbon are, from equations (6) and (25):

$$91.65[1 \text{ (lb C)} + 11.44(1.35) \text{ (lb air)}] = 91.65 [3.66 \text{ (lb CO}_2\text{)} + 8.78(1.35) \text{ (lb N}_2\text{)} + 2.667(0.35) \text{ (lb O}_2\text{)}]$$

$$1.25[1 \text{ (lb C)} + 5.72(1.35) \text{ (lb air)}] = 1.25[2.33 \text{ (lb CO)} + 4.39(1.35) \text{ (lb N}_2\text{)} + 0.35(1.33) \text{ (lb O}_2\text{)}]$$

The reaction equation for the hydrogen is exactly as in example 3. Simplifying the carbon equations and adding all three reaction equations, together with the nitrogen and oxygen in the fuel, and the hydrogen that combines directly with the oxygen,

$$92.9 \text{ (lb C)} + 1.9 \text{ (lb H}_2\text{)} + 1.3 \text{ (lb N}_2\text{)} + 3.9 \text{ (lb O}_2\text{)} + 1490 \text{ (lb air)}$$
$$= 335.5 \text{ (lb CO}_2\text{)} + 2.9 \text{ (lb CO)} + 17.1 \text{ (lb H}_2\text{O)} + 1145 \text{ (lb N}_2\text{)} + 90.3 \text{ (lb O}_2\text{)}$$

The relative volumes in the products of combustion will correspond to the relative number of pound-moles:

$$
\begin{array}{llll}
\text{Carbon dioxide} & = & 335.5/44 = & 7.62 \\
\text{Carbon monoxide} & = & 2.91/28 = & 0.10 \\
\text{Water vapor} & = & 17.1/18 = & 0.95 \\
\text{Nitrogen} & = & 1145/28 = & 40.88 \\
\text{Oxygen} & = & 90.3/32 = & 2.82 \\
& & & \overline{52.37}
\end{array}
$$

Hence the volumetric percentages of constituents in the exit gases are:

$$
\begin{array}{llll}
\mathbf{CO_2} & = 100(7.62/52.37) = & 14.6 & \% \\
\mathbf{CO} & = 100(0.10/52.37) = & 0.2 & \% \\
\mathbf{H_2O} & = 100(0.95/52.37) = & 1.8 & \% \\
\mathbf{N_2} & = 100(40.9/52.37) = & 78.2 & \% \\
\mathbf{O_2} & = 100(2.8 \ /52.37) = & \underline{5.4} & \% \\
& & 100.0 + \% &
\end{array}
$$

The volumetric percentages indicated by an Orsat analysis would be,

$$
\begin{array}{ll}
\mathbf{CO_2} = 100(14.65/98.1) = & 14.9\% \\
\mathbf{CO} = 100(0.002/98.1) = & 0\% \\
\mathbf{N_2} = 100(78.2 \ /98.1) = & 79.6\% \\
\mathbf{O_2} = 100(5.4 \ /98.1) = & \underline{5.5\%} \\
& 100.0\%
\end{array}
$$

Example 5. Hexane (C_6H_{14}) is burned under conditions of perfect combustion. Determine the air-fuel ratio.

Solution: By substitution in equation (50) for the combustion of a hydrocarbon,

$$1 \text{ lb } C_6H_{14} + \left[11.44 \frac{(12)(6)}{(12)(6) + 14} + 34.32 \frac{14}{(12)(6) + 14} \right] \text{ lb air}$$

$$= 9 \frac{(14)}{(12)(6) + 14} \text{ (lb H}_2\text{O)} + 3.66 \frac{(12)(6)}{(12)(6) + 14} \text{ lb CO}_2 + \left[8.78 \frac{(12)(6)}{(12)(6) + 14} + 26.32 \frac{14}{(12)(6) + 14} \right] \text{ lb N}_2$$

$$1 \text{ (lb } C_6H_{14}\text{)} + 15.17 \text{ (lb air)} = 1.465 \text{ (lb H}_2\text{O)} + 3.06 \text{ (lb CO}_2\text{)} + 11.53 \text{ (lb N}_2\text{)}$$

The air-fuel ratio for this hydrocarbon under conditions of perfect combustion is thus slightly greater than 15 to 1.

CHIMNEY DRAFT AND VELOCITIES

The amount of natural convective draft theoretically available in a chimney is given by Leslie Silverman, Harvard School of Public Health, as

$$D_r = 2.96 \, BH \left[\frac{W_a}{T_a} - \frac{W_s}{T_s} \right] \qquad (1)$$

where D_r = draft in inches of water,
  B = atmospheric pressure in inches of mercury (30 usually taken as normal barometer),
  H = height of stack in feet,
  W_a = density of the atmosphere at oF and sea level pressure, pounds per cubic foot,
  W_s = density of stack gas under the same conditions,
  T_a = absolute air temperature, F absolute = 460 + t_a (t_a = air temperature, F), and
  T_s = absolute gas temperature, F absolute = 460 + t_s (t_s = gas temperature, F).

When the velocity V is in feet per minute, Silverman shows that

$$V = 1890 \sqrt{BH \left(\frac{1}{T_a} - \frac{W_s}{W_a T_s} \right)} \qquad (2)$$

and when t_a = oF and B = 30,

$$V = 484 \sqrt{H \left(1 - \frac{W_s T_a}{W_a T_s} \right)} \qquad (3)$$

Since gas densities are directly proportional to molecular weight M at the same pressure and temperature and $\frac{H_1}{12}$ (inches) = H we can place equation (3) in a form which will permit calculations to be made for unconfined convection currents where chimney or convective columns are of lower magnitude:

$$V = 140 \sqrt{H_i \left(1 - \frac{M_s T_a}{M_a T_s} \right)} \qquad (4)$$

where M_s is the molecular weight of the stack gas and M_a the molecular weight of air.

Equation (4) may be used to predict the maximum possible velocity under theoretical conditions for a given instance. For example, what is the maximum velocity that a column of steam will attain in a frictionless chimney at a point 12 inches above a boiling pan or vat in ambient air at 70F, assuming that the steam temperature is 212F at this height. (Because of the high latent heat of steam this is a reasonable assumption.) Substituting in (4)

$$V = 140 \sqrt{12 \left[1 - \frac{18(460 + 70)}{29(460 + 212)} \right]}$$

$$V = 344 \text{ fpm}$$

Equations (1) and (4) assume that no loss has taken place through friction or flow resistance. In the actual case, however, the resistance to flow will reduce the velocity considerably. In the case of confined convection currents this can be evaluated since the resistance loss is subtracted from the theoretical draft. In mathematical form the actual draft in a confined condition is defined by

$$D_r = \text{theoretical draft} - \text{resistance loss} \qquad (5)$$

In terms of actual chimneys or stacks

$$D_r = 2.96 \, BH \left[\frac{W_a}{T_a} - \frac{W_s}{T_s} \right] - \frac{0.184 \, f W_s BH V^2}{T_s D}$$

where in addition to the terms defined in equation (1) we have

  f = friction factor and
  D = minimum internal diameter of chimney

of stack in feet = $0.288 \sqrt{\dfrac{WT_s}{BW_s V}}$

where W = weight of gas flowing in pounds per second. Hence, if the friction factor f, gas velocity, and density are known, the loss in theoretical draft can be computed from Equation (6).

For the average chimney f is equal to 0.016. It should be recognized that f depends upon the nature of the chimney surface as well as the viscosity and temperature of the discharging gas. The value of 0.016 corresponds to chimney gases at 500F. If the stack is to discharge gases only slightly above room temperature or at an appreciably lower density than the ambient air, then the second term of the equation should be replaced by a value of friction loss corresponding to the type of stack construction and the gas density involved. If heavy-gauge galvanized iron pipe is employed, such as that used in industrial exhaust systems, then the ordinary friction factor charts can be used with a correction for gas density. Such a value can also be obtained from equation (7):

$$\text{Resistance loss} = \frac{L}{CD} \left(\frac{V}{4000} \right)^2 \qquad (7)$$

where resistance loss is in inches of water for air at 70F; for other gases multiply by ratio of

$$\frac{\text{density}}{0.075} = \frac{W_a}{0.075}$$

and L = length of pipe in feet,
  D = pipe diameter in feet,
  V = velocity in feet per minute,
  C = 60 for perfectly smooth pipe,
  C = 50 for heating and ventilating ducts,
  C = 45 for smooth tile ducts, brick, and
  C = 40 for rough tile ducts, concrete.

For other than round pipe the equivalent diameter is computed from the hydraulic radius.

Direct measurement of the velocity in the stack or chimney can be accomplished by means of the Pitot tube. The static pressure tube when used alone will read the draft pressure. The chief limitation of the Pitot tube is that traverses of equal area must be made for determining average velocity if the discharged quantity is to be calculated. Another limitation is that on velocity. The velocity is determined from the Pitot tube by means of the formula:

$$V = 1098 \sqrt{\frac{h_i}{W_a}}$$

where V = velocity in feet per minute,

h_i = manometer reading in inches of water,

W_a = air density, pounds per cubic foot.

If the air temperature is 70F, the formula becomes

$$V = 4000 \sqrt{h_i}$$

The lowest readings which can be made accurately on an inclined draft gage is approximately 0.025 in. of water. That is, on a 1 to 10 inclined gage a reading of 0.25 in. of water is obtained. This corresponds to a velocity of 600 fpm at 70F. For velocities less than this amount other devices must be used or a micromanometer such as the Wahlen gage, which will read as low as 0.001 in. of water may be employed.

The measurement of convection currents in uncon-fined instances requires that both qualitative and quantitative indicators be used. The most common direction indicator is smoke, either chemical or that produced by burning tobacco, etc. Small fine threads, candle flames and fine powders are also used for direction indicators. One handy type of smoke generator is a device which employs tin or titanium tetrachloride which hydrolyzes on contact with the moisture in air forming a dense white smoke. By means of this smoke tube, the direction of convection currents may be ascertained. The degree of turbulence of the air may also be estimated by the rate at which the smoke dissipates and diffuses as it moves with the air current. Quantitative measurements of low air flows can be made by timing the travel of a puff of smoke over a measured distance.

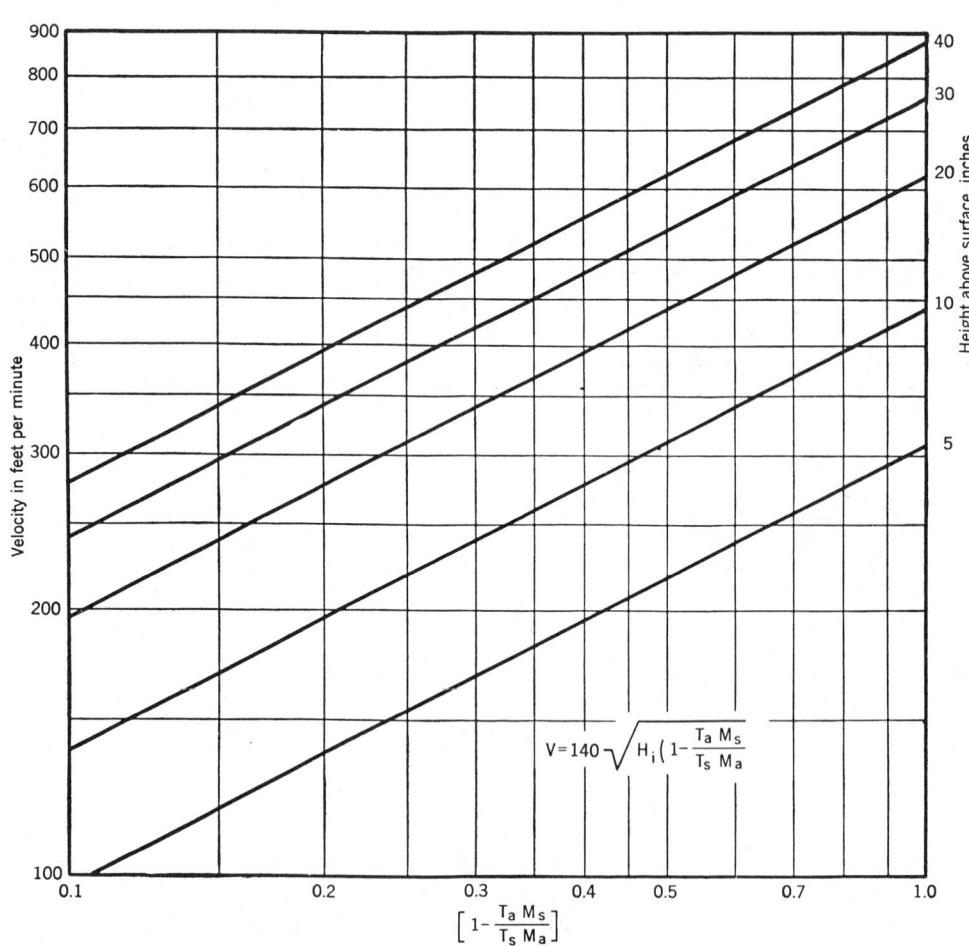

$$V = 140 \sqrt{H_i \left(1 - \frac{T_a\,M_s}{T_s\,M_a}\right)}$$

$$\left[1 - \frac{T_a\,M_s}{T_s\,M_a}\right]$$

Calculation chart for determining theoretical convection velocity.

FORCED DRAFT AND DRAFT CONTROL

Draft in a heating system refers to the pressure difference which causes a current of air or gases to flow through a combustion chamber, flue, or chimney. Natural draft is the draft obtained without any mechanical means. It is the heat of the combustion processes that creates the differential pressure causing the draft. Mechanical draft is created by fans which either force or pull the air or combustion products through the system. When the combustion products are forced through the system, the system is designated as a *Forced Draft System.* When the combustion products are pulled through the system, the system is designated as an *Induced Draft System.*

The overall problem of obtaining proper combustion of fuel includes, in addition to the design and installation of the actual firing equipment, a consideration of the entire path of the gas travel, from the air openings into the boiler room, through the burner, the combustion area, the heating surfaces, the breeching, and the vent or chimney.

To obtain this flow of air and gas, some force is required to provide the movement and to overcome the friction through the burner, boiler, and flue passages. Different firing systems employ one or more of the following means of providing this force:

1. Natural draft chimney
2. Forced draft fan
3. Induced draft fan.

Natural draft is used on rotary oil burners in both the air-register models and those with air openings in the combustion chamber floor. While these units have a primary air fan, its function is merely to atomize the oil or premix some air with the fuel gas, leaving the bulk of the required combustion air to be supplied by natural draft. Natural draft is also used in combination with forced draft on pressure and air atomizing oil burners and combination gas-oil burners. In these cases the fan provides the force to move the combustion air through the burner and the chimney pull provides the forces to move the combustion products through the boiler, breeching, and flue.

Forced draft burners are fired with gas, oil, or gas-oil in combination and the types of oil burners are usually of the pressure or air atomizing type. In forced draft equipment, the burner fan has sufficient power to overcome the resistance of the boiler passages. Under certain conditions there is even sufficient power to overcome the resistance of the boiler breeching and a short stub stack. This type of burner equipment is usually found on packaged sealed boilers. The forced draft type burner can be utilized on a natural draft boiler that is not sealed. In this case the natural draft chimney is sufficiently negative in pressure to pull the products of combustion through the boiler and breeching.

Induced draft fans are not usually furnished as an integral part of a burner. However, they can be used as auxiliary equipment with any burner where the chimney does not provide adequate draft. Some package boilers fired with rotary burners will utilize an induced draft fan to overcome the resistance of the boiler.

Larger heating plants will require draft at the boiler outlet in order to maintain satisfactory combustion. Obviously, equipment of this type cannot be operated with a draft diverter. However, some sort of draft control is necessary, and one of the simplest devices used for draft control is the barometric draft regulator. These are usually sized to have a free opening equal to the breeching to which they are attached. On gas or combination gas-oil fired installations they should be of the double swing type to relieve downdrafts or pressure due to a blocked chimney. A thermal spill switch should also be installed to shut down the burner equipment in the event of spillage.

Where chimneys are especially high—usually over 100 feet—or the firing rates must be varied widely in conjunction with large inputs, sequence draft control may be required. In a sequence draft control system, the draft is controlled by positioning a damper in the breeching, which automatically maintains a specified draft throughout the firing range of the combustion equipment.

In general, some sort of draft control is required whenever there is a negative pressure in the boiler breeching. Even on so-called forced draft equipment this rule holds true. It must be made clear, however, that a forced draft burner has 100% of the combustion air supplied by the burner fan; this burner does not normally have the capability of overcoming the resistance of the boiler flue passages, breeching, and chimney.

If draft control is not provided there will usually be instability in the main burner or pilot burner on both gas and oil. The result of this will be nuisance shutdowns.

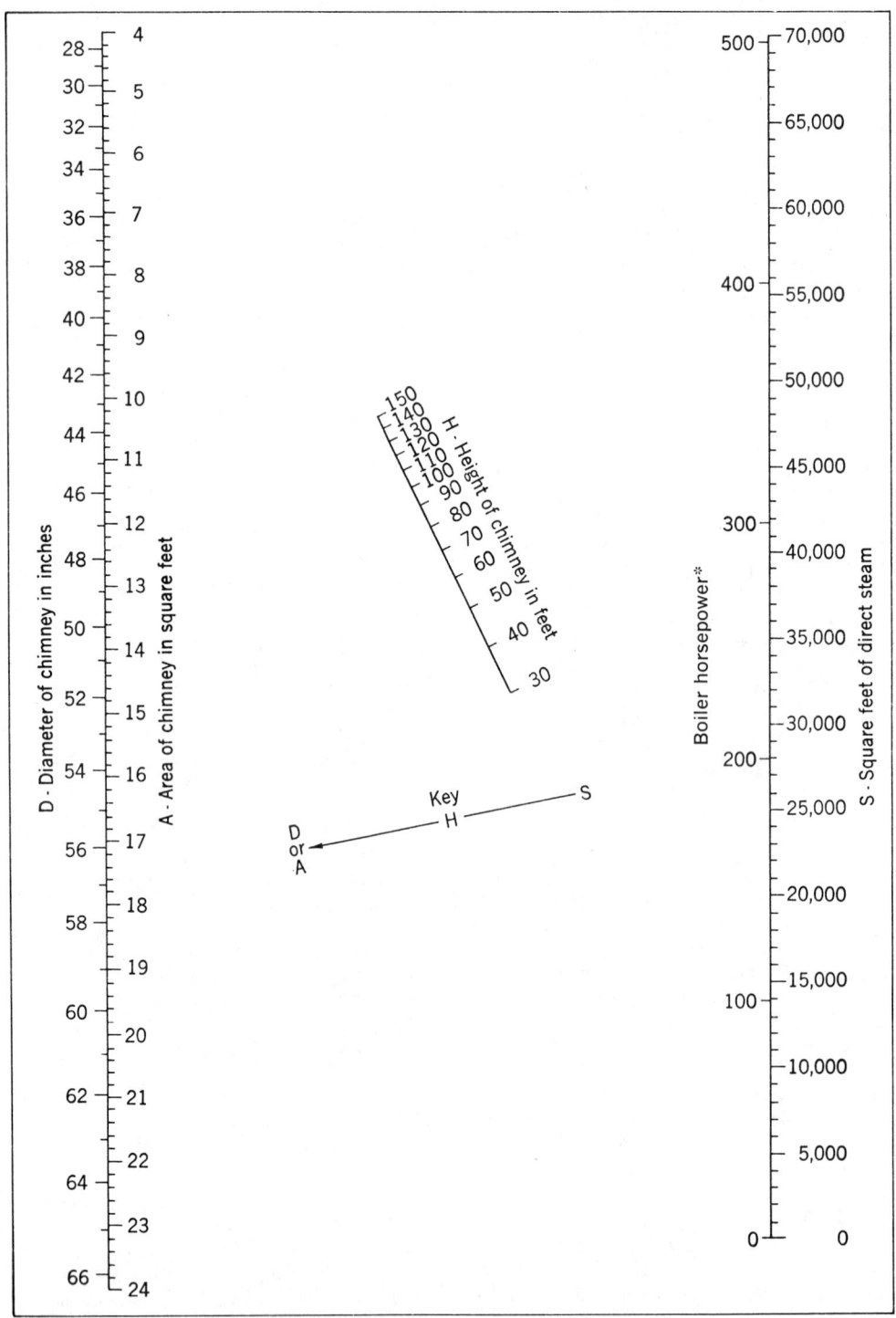

*Nominal boiler horsepower equals 33,500 Btu per hour

SIZING OF LARGE CHIMNEYS

The charts shown on these pages, designed by J. N. Arnold, Purdue University, are for the determination of chimney sizes when the nominal boiler horsepower or square feet of radiation is known.

For example, assume a 1000-boiler horsepower load and a tentative chimney height of 150 ft. What must be the area of the chimney? Refer to the chart on this page. Through the 1000 point on the horsepower scale and 150 on the diagonal scale, draw a line to intersect the Q-axis. About this point on the Q-axis adjust the straight edge until identical readings are obtained on the two A scales, in this case, 27.5 sq. ft. area. The diameter can be read directly opposite the area on the left-hand scale; in this case a 71-in. diameter chimney would be required.

To use the chart on the opposite page, assume a chimney 150 ft. high for a 55,000 sq. ft. load.

Through the 55,000 point on the S-scale and 150 on the diagonal scale, draw a straight line until it intersects the A and D scales. Find the diameter, to be 46 in. with an area of 11.5 sq. ft.

The chart on this page is a solution of the formula

$$HP = 3.33(A - 0.6 \sqrt{A}) \sqrt{H}$$

where HP = nominal boiler horsepower,*
A = chimney area, square feet,
H = chimney height, feet.

The chart on the next page is a graphical solution of the formula

$$S = 464(A - 0.6 \sqrt{A}) \sqrt{H}$$

where S = square feet of direct steam radiator surface,
A = chimney area in square feet,
H = chimney height, feet.

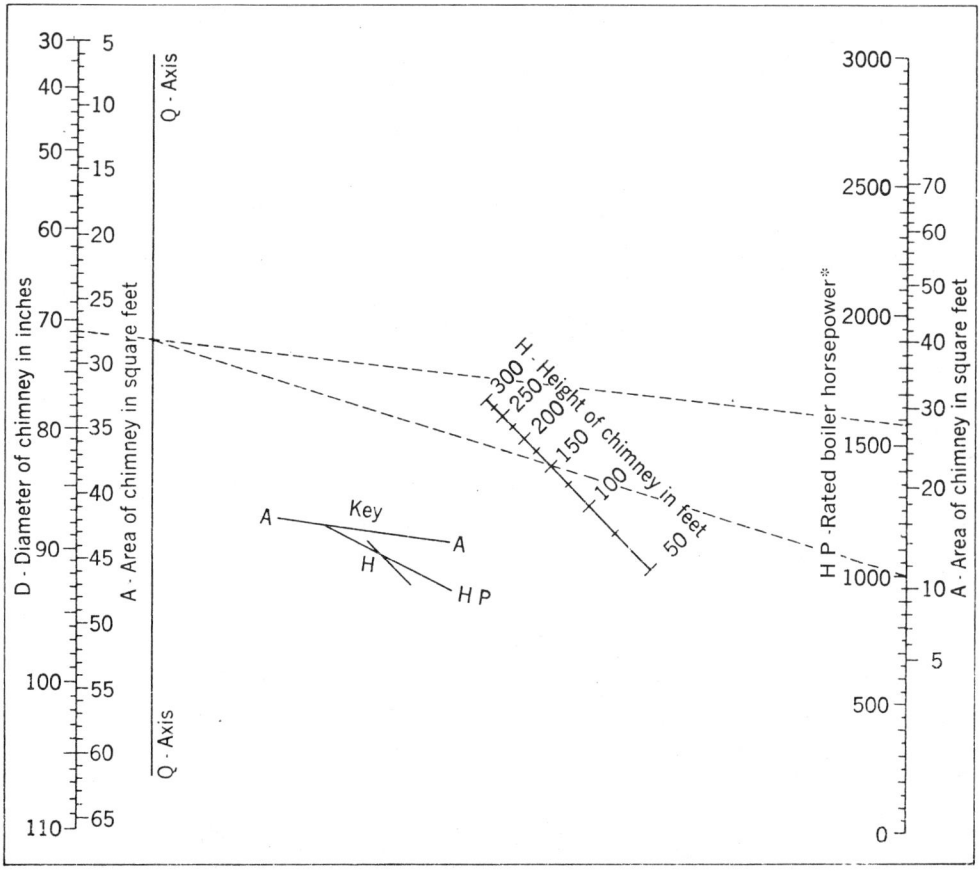

*Nominal boiler horsepower equals 33,500 Btu per hour

CHIMNEY DESIGN AND CONSTRUCTION

In connection with chimneys, the question of *area* arises, since this is related to the volume of flue gases.

The cross-sectional area is given by Severns as:

$$A = \frac{Q}{V \times K}$$

where A = chimney area, sq ft,

Q = flue gas volume,

K = velocity coefficient of roughness (from .30 to .50)

and V = theoretical gas velocity, ft per sec.

The draft required from a chimney varies with the fuel. Table 1 gives approximate data for hand-fired coal.

The maximum attainable (not theoretical) draft per foot of chimney height is given by Sefing as tabulated in Table 2. Note that the draft increases with the flue gas temperature.

Tests by the National Bureau of Standards on very short chimneys of 6-inch bare smoke pipe show results about 80% of those in Table 2.

Draft required for oil burning domestic heating equipment is about as follows: mechanical draft oil burners, 0.03 to 0.06 inches of water; space heaters, 0.04 to 0.08 inches of water.

Fig. 1 shows the draft for oil-fired steel heating boilers at various flue gas temperatures as given by Steel Boiler Institute in connection with its test code for "Table 3" boilers and boiler-burner units.

Allowable gas input to gas appliances for various sizes of circular flues is given in Table 3. This table is based on a 150 deg F flue temperature, 60 deg F outside air temperature, 100% excess air, and 100% dilution at draft hood, terra cotta flues, and short horizontal breeching. Reduce capacity in table by 3.5% for each 1000 feet above sea level.

The National Board of Fire Underwriters' code contains numerous provisions relating to chimneys, including the following:

Table 1. Draft Required for Burning Coal

Fuel	Burning Rate, pounds Coal per Square Foot Grate per Hour						
	5	10	15	20	25	30	35
	Draft, Inches of Water						
Anthracite No. 3, buckwheat	0.15	0.40	0.75	1.24	—	—	—
Anthracite No. 1, buckwheat	0.10	0.24	0.44	0.68	1.00	—	—
Anthracite, pea	0.06	0.16	0.30	0.45	0.65	0.90	1.20
Semi-bituminous	0.05	0.10	0.18	0.26	0.35	0.45	0.58
Bituminous	0.04	0.09	0.15	0.22	0.28	0.38	0.45
Bituminous, run of mine	0.04	0.05	0.08	0.10	0.14	0.16	0.20
Bituminous, on chain grate	0.05	0.12	0.15	0.23	0.31	0.44	0.57

Table 2. Maximum Attainable Draft per Foot of Chimney

Average Flue Gas Temperature, Deg F	Outside Air Temperature, Deg F					
	0	20	40	60	80	100
	Draft per Foot of Chimney Height, Inches of Water					
150	.00360	.00283	.00225	.00167	.00113	.00062
200	.00460	.00385	.00325	.00267	.00213	.00162
250	.00543	.00468	.00408	.00350	.00286	.00245
300	.00615	.00540	.00480	.00422	.00368	.00335
350	.00679	.00604	.00545	.00466	.00432	.00399
400	.00735	.00660	.00600	.00542	.00488	.00455
450	.00787	.00712	.00652	.00595	.00540	.00507
500	.00831	.00756	.00696	.00638	.00584	.00551

1. *Flue Connections Required.* Every heating apparatus or heat producing appliance requiring a flue connection shall be connected with a flue conforming to the provisions of this article. This shall not include electric appliances; gas appliances, except as specifically required in this article; nor oil fired appliances especially designed for use without flue connection.

2. *Use of Nonconforming Flues.* Flues not conforming to the requirements of this article for chimneys, metal smokestacks or vents for gas appliances, shall not be used unless listed by *Underwriters' Laboratories, Inc.*, installed in full compliance with the listing and the manufacturer's instructions, and approved for such use by the building official.

3. *Smoke Pipe Connections.*

a. No flue shall have smoke pipe connections in more than one story of a building, unless provision is made for effectively closing smoke pipe openings with devices made of noncombustible materials whenever their use is discontinued temporarily, and completely closing them with masonry when discontinued permanently.

b. Two or more smoke pipes shall not be joined for a single connection, unless the smoke pipes and flue are of sufficient size to serve all the appliances thus connected.

c. The smoke pipe of a heating appliance shall not be connected into the flue of an incinerator which has the rubbish chute identical with the smoke flue.

4. *Construction of Chimneys.*

a. Chimneys hereafter erected within or attached to a structure shall be constructed in compliance with the provisions of this section.

b. Chimneys shall extend at least 3 ft above the highest point where they pass through the roof of the building, and at least 2 ft higher than the ridge within 10 ft of such chimney.

c. Chimneys shall be wholly supported on masonry or self-supporting fireproof construction.

d. No chimney shall be corbeled from a wall more

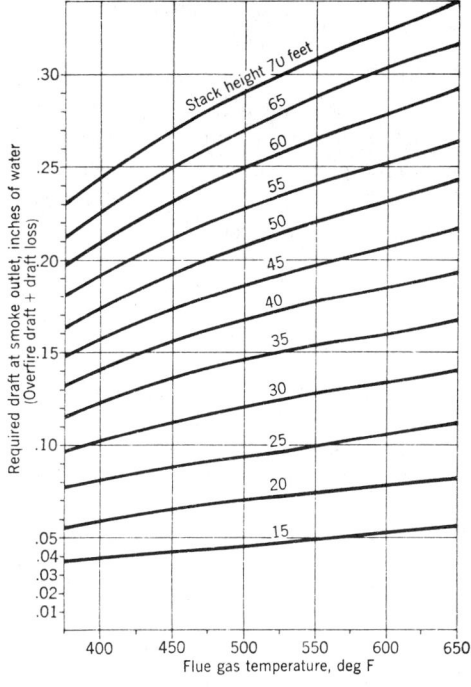

Fig. 1. Chimney draft versus flue gas temperature for various height chimneys.

than 6 in.; nor shall a chimney be corbeled from a wall which is less than 12 in. in thickness, unless it projects equally on each side of the wall; provided that in the second story of 2-story dwellings corbeling of chimneys on the exterior of the enclosing walls may equal the wall thickness. In every case the corbeling shall not exceed 1 in. projection for each course of brick projected.

Table 3. Capacities of Chimneys for Gas Appliances with Draft Hoods

Inner Diameter of Flue, Inches	Height of Chimney, Feet					
	10	15-20	30	40	50	60
	Gas Input, Thousands of Btu per Hour					
3	24	27	30	32	—	—
4	45	50	57	61	67	—
5	74	82	93	100	110	120
6	110	120	135	145	160	180
7	150	165	200	215	230	250
8	210	230	270	300	325	350
9	260	300	350	380	410	450
10	335	370	440	500	520	600
11	420	470	590	620	650	730
12	500	560	690	740	800	870

FUEL EQUIVALENT CHART

The purpose of the accompanying fuel equivalent chart, designed by C. C. McRae, is to determine the quantity of one fuel burned at a given efficiency equivalent to another fuel of different type, of different efficiency, and in different quantity units. That is, the chart is a solution of the equation:

$$X = \frac{M \times H_m \times E_m}{H_x \times E_x}$$

where X = required equivalent;
M = quantity of known fuel;
H_m = heating value of known fuel;
E_m = operating efficiency of known fuel;
H_x = heating value of required equivalent; and
E_x = operating efficiency of required equivalent.

The efficiencies are to be expressed as a decimal. The resulting answers, just as on a slide rule, do not indicate the position of the decimal points.

Table 1 is to assist in determining the location of the decimal point. To use Table 1, note that, first column of figures, 1 ton of coal at 100% efficiency is equivalent to 164 gal. of 152,000 Btu oil, 172 gal. of 145,000 Btu oil, 21,700 cu. ft. of 1150 Btu natural gas, 25,000 cu. ft. of 1,000 Btu per cu. ft. gas, 29,412 cu. ft. of 850 Btu per cu. ft. gas, and so on. To apply the table and use the chart, refer to the following examples:

Example: For a given process 9.5 gallons of diesel oil (145,000 Btu per gal.) are burned at an estimated 60% efficiency. How much 1150 Btu per cu. ft. natural gas will be required for this process if burned at 70% efficiency?

Solution: Locate 9.5 on horizontal scale, move vertically to the diesel oil curve, horizontally to the 60% efficiency curve, vertically to the 70% curve and horizontally to the 1150 Btu natural gas curve, thence downward and locate 10.3 M cu. ft. or 10,300 cu. ft. uncorrected for decimal place. From Table 1 note that 1 gal. of 145,000 Btu per gal. oil is equivalent to 126 cu. ft. of natural gas at 100% efficiency so that 10 gal. is equivalent to 1260 cu. ft.; therefore the answer is 1030.0 cu. ft.

Example: A 60% efficiency coal system burning 52 tons of 12,500 Btu per lb. coal is to be replaced by a 70% efficiency system burning 136,000 Btu per gal. oil. How much oil will be required?

Solution: Ignoring decimal points, move upward from 5.2 on the horizontal scale to the coal curve, horizontally to the 60% curve, vertically to the 70% curve, horizontally to the 136,000 oil curve thence down and read 815 gal., uncorrected for decimal point. From Table 1, each ton is equivalent to 184 gal. so 50 tons would be equivalent to approximately 9000 gal. at 100% efficiency, so that the answer is 8150 gal.

Table 1. — Approximate Equivalents, Based on 100% Efficiency, For Determining Position of Decimal Point

Tons of 12,500 Btu/lb. coal	1							
Gal. of 152,000 Btu/gal. oil	164	1						
Gal. of 145,000 Btu/gal. oil	172	1.05	1					
Gal. of 136,000 Btu/gal. oil	184	1.12	1.07	1				
Kilowatt-hours of elect.	7,327	44.5	42	40	1			
Cu. ft. of 1,150 Btu/cu. ft. gas	21,700	132	126	118	2.97	1		
Cu. ft. of 1,000 Btu/cu. ft. gas	25,000	152	145	136	3.41	1.15	1	
Cu. ft. of 850 Btu/cu. ft. gas	29,412	179	170	160	4.02	1.35	1.18	1

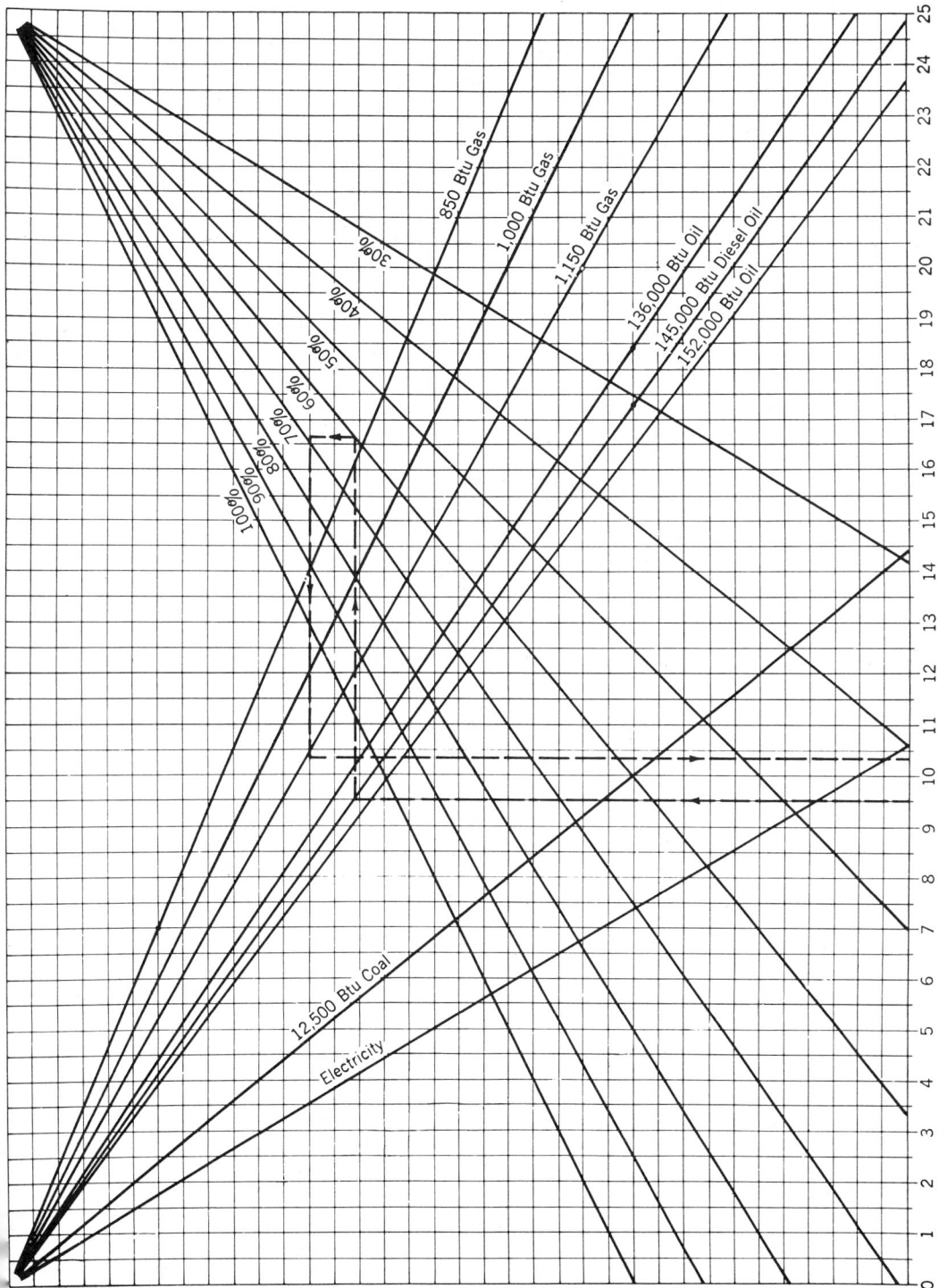

Fuel Consumption: Kw.-hr. of Electricity; Tons of Coal; Gal. of Oil; Thousands of Cu. Ft. of Gas

CHARTS FOR COMPARING FUEL COSTS

The accompanying charts are designed for finding, on a common left-hand scale, the cost per million Btu of useful heat from coal, oil, natural gas, electricity, or purchased steam. In the case of the last two, an efficiency of 100% is assumed, so the cost of useful heat is found by moving horizontally to the left opposite the unit cost. Unit heating values used in calculating the charts are: 141,000 Btu per gallon of oil; 13,000 Btu per lb. for coal, 2,000 lb. per ton; 1,000 Btu per cubic foot for natural gas; 1,000 Btu per lb. of steam; and 3,413 Btu per kw-hr for electricity.

For energy rates higher than those indicated on the curves even multiples can be used.

For gas, steam, and electricity, the unit cost is an average of the total cost, including monthly standby or services and demand charges.

Examples of the use of the charts are as follows:

Example: What would be the cost per million Btu of fuel oil burned at 80% efficiency and costing 36 cents per gallon?

Solution: Using the 18 cents per gallon curve at 80% efficiency costs $1.60 per million Btu. Since 18 cents per gallon is ½ the price of 36 cents per gallon, a multiplier of two is required. Therefore, 36 cents per gallon oil costs $1.60×2 or $3.20 per million Btu.

Example: What would be the cost per million Btu of electricity costing 2 cents per kw-hr?

Solution: Using the 0.2 cents per kw-hr on the right index is the same as 2 cents when it is multiplied by ten. The left scale opposite 0.2 cents is $.59 or $.59×10=$5.90 per million Btu.

Example: What would be the cost per million Btu of fuel oil burned at 65% overall efficiency and costing 11 cents per gallon?

Solution: Enter the oil chart at the bottom scale at 65 and move upward to the intersection of the efficiency line with the curve for 11 cents per gallon, thence left, and find the cost to be $1.20 per million Btu.

Example: What would the natural gas rate have to be, assuming it is to be burned at 75% efficiency, to produce heat at the same cost as oil burned at 70% efficiency and costing 10 cents per gallon?

Solution: On the oil chart, move upward from the 70 on the lower scale to the curve for 10 cents, thence horizontally to the right. Move upward on the natural gas scale at the extreme right and upward from the 75% efficiency line to the intersection with the previous horizontal line and find that a .75 cents per thousand cubic foot rate would be necessary.

Example: Electricity is available at 1.5 cents per kw-hr in a certain community. How does this compare with gas to be burned at 80% efficiency and costing $1.20 per 1,000 cu. ft.?

Solution: Using the extreme right-hand scale of 0.15 move to the left scale finding the cost per million Btu to be $.44×10=$4.40. For gas sold on a 1,000 cubic foot basis move upward from the 80 intersection with the 120 line and find the cost with gas would be $1.49 per million Btu.

Example: What would be the cost of oil, gas, and coal at 70% efficiency in their units equal to electricity at 1.5 cents a kw-hr?

Solution: Using the right-hand scale at 1.5, move to left-hand scale. Cost of electricity per million Btu is $4.40. On the oil curve, move up from 70 to the 11 cent curve then over to the left scale $1.12×4=$4.48. On the gas curve, move up from 70 to the 80 cent curve then over to the left scale $1.14×4=$4.56. On the coal curve, move up from 70 to the $20 curve then over to the left scale $1.10×4=$4.40. Therefore, electricity at 1.5 cents per kw-hr is comparable to oil at 40-44 cents per gallon, gas at $3.00-3.20 per 1,000 cubic foot, and coal at $80 to $88 per ton.

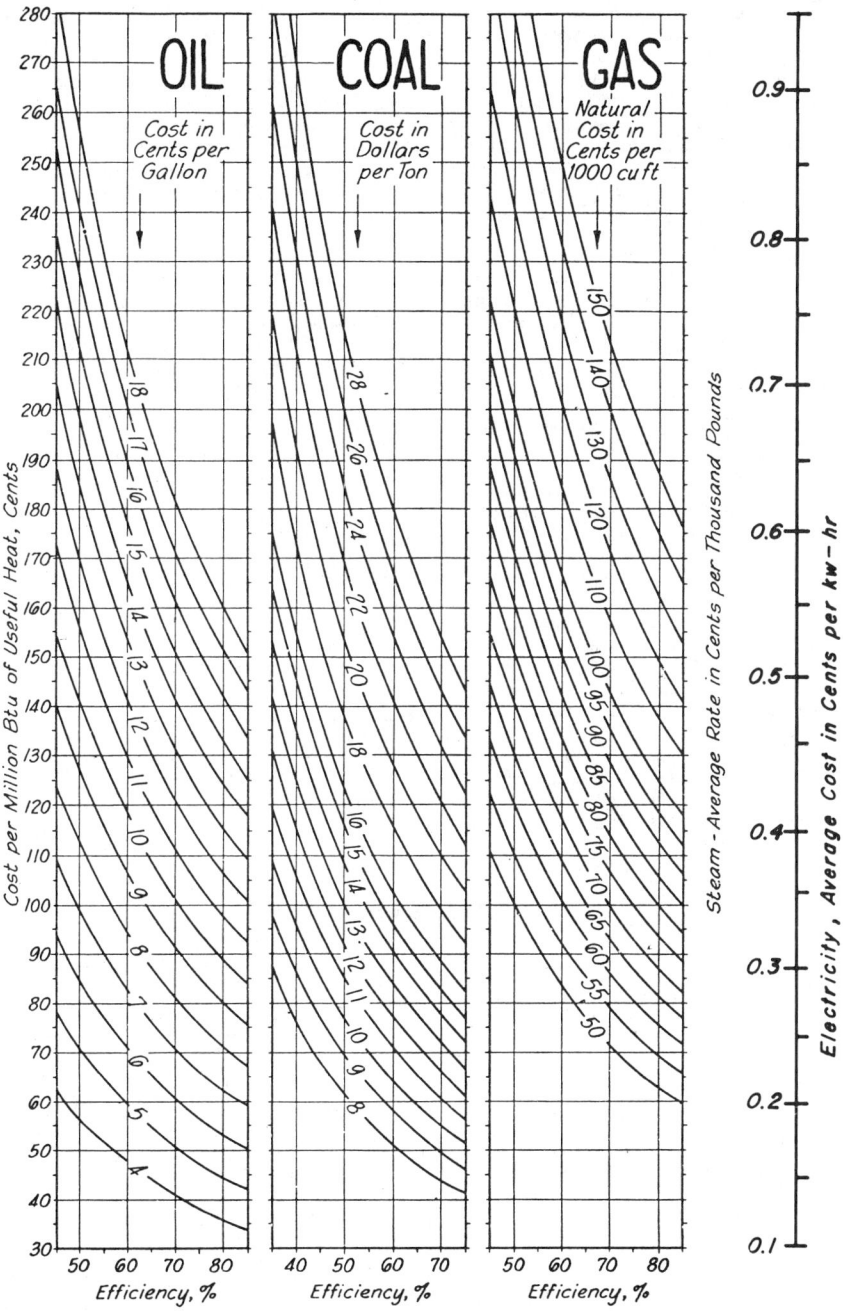

CLASSIFICATION AND PROPERTIES OF COALS

The accepted theory of the origin of coal is that it is formed from vegetable matter and transformed into coal in two stages. The first stage, which took place at a time when the rate of plant growth was more rapid than it is today, consisted of an accumulation of vegetable matter which was subjected to a bacterial decay. This stage was brought to an end by toxic substances which killed the bacteria, leaving a remainder which we today call peat. Peat which was subjected to the second stage was devolatilized by a tremendous pressure of the earth's crust. This pressure converted the peat, in successive stages, to lignites, sub-bituminous, bituminous, semi-bituminous and anthracite coals. The last is a high grade fuel devolatilized by pressure to a greater degree than the lower grade lignite at the other end of the scale. It is estimated that over 10,000 years were required for the formation of anthracite. In addition to the differences among fuels caused by the difference in pressures to which they were subjected, the nature of the vegetable matter and the extent of the decay result in differences in the properties of various coals.

Properties of Coals. The foregoing discussion of classification shows that there is no definite line of demarcation between the various ranks of coals. Regardless of the particular method of classification, however, the following general remarks summarize roughly the characteristics of the ranks of coals:

Lignite is a brown, non-caking woody fuel with high moisture content and low heating value. It disintegrates rapidly and is liable to spontaneous combustion, but is clean to handle. It is ordinarily used only in localities where it is mined.

Sub-bituminous coals are also called black lignites. Not so woody as lignites, these coals have a higher heating value. Burns with a medium length flame, free-burning and non-caking. Higher in volatile than lignites, but form little smoke.

Bituminous coals are, unlike lignitic fuels, little affected by weathering. This group covers the widest range of fuels found in the U. S. Commonly called "soft" coals, these fuels are generally brittle with a vitreous, greasy or resinous luster. The composition is principally fixed carbon and bitumen. This is a mixture of hydrocarbons which, when heated, breaks down into gases, oils, and tars. These coals burn with a yellow flame and smoke while burning. There are two principal types — caking coals and non-caking or free-burning. The former are high in hydrocarbons and used in gas manufacture; the latter do not fuse, but hold their shape, burn freely and are valuable for steam generating.

Semi-bituminous coals are mis-named; they should be called super-bituminous coals. They are softer than anthracite, but brighter in appearance than bituminous coals. These coals are free-burning, relatively free from smoke, and have a high heating value. It is sometimes called smokeless coal, but since it is soft and fusible it is not so clean to handle as anthracite.

Semi-anthracite is not so hard as anthracite but has less volatile matter than semi-bituminous. The volatile it does have, however, makes it more free-burning than anthracite. There is little of this fuel available.

Anthracite, or "hard" coal, is a clean, hard and dense fuel which is practically all fixed carbon. Consequently, it is hard to ignite but burns freely when well started. It burns with a short flame and without smoke. It is mined principally in Pennsylvania and its use is principally in domestic heating plants not too far from its source.

Classification. The classification of coals is not a definite matter. The various methods which have been proposed are very numerous, and are based on such properties of the fuel as burning characteristics, heating value, composition, flame, ratios of components, and so on. An old and arbitrary classification is that of Frazer, as follows:

Name	Ratio, $\dfrac{\text{Fixed Carbon}}{\text{Volatile Matter}}$
Anthracite	100–12
Semi-anthracite	12–8
Semi-bituminous	8–5
Bituminous	5–0

The American Society of Testing and Materials has adopted the classification shown in Table 1.

Table 1. — Classifications of Coals

(According to the American Society for Testing and Materials)

General Class	Group	Fixed Carbon, %, Dry	Volatile Matter, %, Dry	Btu. per Lb., Moist	Physical Properties
Anthracite	Meta-anthracite	98 or more	2 or less	—	Non-agglomerating (does not adhere in a mass). If agglomerating, classify as low volatile bituminous.
	Anthracite	92–98	2–8	—	
	Semi-anthracite	86–92	8–14	—	
Bituminous	Low volatile bituminous	78–86	14–22	—	
	Medium volatile bituminous	69–78	22–31	—	
	High volatile A bituminous	Under 69	Over 31	Over 14,000	
	High volatile B bituminous	—	—	13,000–14,000	
	High volatile C bituminous	—	—	11,000–13,000	Either non-agglomerating or non-weathering
Sub-bituminous	Sub-bituminous A	—	—	11,000–13,000	Weathering and non-agglomerating
	Sub-bituminous B	—	—	9,500–11,000	
	Sub-bituminous C	—	—	8,300–9,500	
Lignitic	Lignite	—	—	Under 8,300	Consolidated
	Brown coal	—	—	Under 8,300	Unconsolidated

ULTIMATE AND PROXIMATE ANALYSES OF COAL

The composition of coal furnishes a basis of comparison among coals, and the composition can be determined by either of two methods — an *ultimate* analysis or a *proximate* analysis. The former is obtained by a chemical process, while the latter is primarily a physical process. Both require a suitable sample, the obtaining of which should be carefully done and for which a definite procedure has been adopted by the A.S.M.E., A.S.T.M., American Chemical Society and Bureau of Mines.

Ultimate Analysis. This consists of a relative quantitive determination of the ultimate constituents — carbon, hydrogen, oxygen, nitrogen, ash, and sulphur. This is done by the ordinary methods of organic quantitative analysis, customarily by a standard procedure. The ultimate analysis is usually given in per cent of these elements present after the moisture has been determined and reported separately. The coal as originally sampled is termed "as received" or "as fired"; when the moisture is driven off it is "dry." Occasionally the per cent of both moisture and ash are reported on the coal as received, with the ultimate analysis on a "moisture and ash free" basis; in this case the remainder is "combustible," although neither oxygen nor nitrogen are combustible.

Proximate Analysis. The proximate analysis is easier and quicker to make and is one frequently encountered. This procedure has also been standardized. A proximate analysis has as its purpose the determination of the percentages of moisture, ash, fixed carbon and volatile matter. Such an analysis enables the engineer to determine in advance to a certain extent the burning characteristics of the coal.

The "ash" as determined from an analysis after the fuel is as completely burned as possible differs from the ash in the pit of a boiler, since the latter contains some combustibles. A better name for the latter is "refuse." The ash consists of slate, iron pyrites, clay, sand and stone. A large percentage of ash is undesirable; it is paid for on a weight basis, reduces the heat value of a fuel, increases handling costs, produces clinker and increases the draft necessary. Moisture is undesirable since it increases handling costs and absorbs heat for evaporation.

The carbon which, combined with hydrogen and other gaseous compounds, is driven off by heat is termed the "volatile." The uncombined carbon which remains is "fixed carbon."

Coals from a given bed have practically the same analysis when expressed in a moisture, ash and sulphur free basis.

Conversion of Analysis. Analyses are frequently given in terms other than desired. The following rules are for converting analyses (ultimate or proximate) to a "free" basis from an "as received" basis.

(1) to convert an "as fired" analyses to a combustible (or moisture free) basis, divide each of the components other than moisture by (100 − per cent moisture).

(2) to convert an "as fired" analysis to a combustible (or moisture and ash free) basis, divide each of the components other than moisture by [100 − (per cent moisture + per cent ash)].

(3) to convert an "as fired" analysis to an ash, moisture and sulphur free basis, divide each of the components other than ash, moisture or sulphur by [100 − (per cent moisture + per cent ash + per cent sulphur)].

Example: The proximate analysis of a coal as received is: Fixed carbon, 52.0%; Volatile matter, 30.0%; Ash, 10.2%; Moisture, 7.8%. What is the analysis on a moisture free basis?

Solution: Divide each of the first three percentages by (100% − 7.8%) or .922. Carbon, 52.0% ÷ .922 = 56.4%; Volatile Matter, 30.0% ÷ .922 = 32.5%; Ash, 10.2% ÷ .922 = 11.1%.

Analysis of Typical Coals. On a straight analysis basis, Tables 1, 2 and 3 show typical coals of each class with both proximate and ultimate analyses.

Table 1. — Typical Analyses of Various Ranks of Coals

Fuel	Proximate (As received)				Ultimate (As received)						Heating Value, Btu per Lb.
	Moisture, %	Volatile Matter, %	Fixed Carbon, %	Ash, %	Ash, %	Sulphur, %	Hydrogen, %	Carbon, %	Nitrogen, %	Oxygen, %	
Peat............	56.7	26.1	11.2	6.0	6.0	0.6	8.3	21.0	1.1	63.0	3,586
Lignite.........	34.5	35.3	22.9	7.3	7.3	1.1	6.6	42.4	0.6	42.0	7,090
Sub-bituminous ..	24.3	27.6	44.8	3.3	3.3	0.4	6.1	55.3	1.1	33.8	9,376
Bituminous......	3.2	27.1	62.5	7.2	7.2	1.0	5.2	78.0	1.3	7.3	13,919
Semi-bituminious.	2.0	14.5	75.3	8.2	8.2	2.3	4.1	80.0	1.3	4.1	14,081
Semi-anthracite ..	3.4	8.5	76.7	11.4	11.4	0.6	3.6	78.4	1.0	5.0	13,156
Anthracite.......	2.8	1.2	88.2	7.8	7.8	0.9	1.9	84.4	0.6	4.4	13,298

Detailed Analyses. The three tables (Tables 1, 2 and 3) are helpful in a general way, but the broad data given cannot be applied to a specific sample of coal. For cases where a highly accurate analysis of the shipment is not necessary and where the origin of the coal is definitely known, reference can be made to some of the many Bureau of Mines publications.

Table 2. — Proximate Analyses of Typical Coals of Various Ranks on an Ash-Free Basis

Fuel	Fixed Carbon, %	Volatile Matter, %	Moisture, %	Heating Value, Btu per Lb.
Lignite....................	37.8	18.8	43.4	7,400
Sub-bituminous............	42.4	34.2	23.4	9,720
Low-rank bituminous........	47.0	41.4	11.6	12,880
Medium-rank bituminous.....	54.2	40.8	5.0	13,880
High-rank bituminous........	64.6	32.2	3.2	15,160
Low-rank semi-bituminous....	75.0	22.0	3.0	15,480
High-rank semi-bituminous...	83.4	11.6	5.0	15,360
Semi-anthracite............	83.8	10.2	6.0	14,880
Anthracite................	95.6	1.2	3.2	14,440

These publications give specimen analyses exemplifying the analysis of coal mined in each coal-producing county, or, where it was feasible, coal from each bed in each county; also, ranges of analyses, within which the composition of most of the coal represented will fall, are given where there were enough analyses to permit satisfactory determination of the ranges. In addition, the degree of metamorphism from lignite toward anthracite of much of the coal is indicated by the classification by rank. This classification reveals to more experienced users of coal some of the general properties of coal, that is, whether the flame will be long or short, whether the coal will cake in the fire, and how it is likely to behave on exposure and in storage.

Table 3. — Analyses of Typical American Coals

Coal	Proximate, %				Ultimate, %					High Heat Value, Btu per Lb.	
	Mois-ture	Vola-tile Matter	Fixed Car-bon	Ash	Sul-phur	Hydro-gen	Car-bon	Nitro-gen	Oxy-gen	Air Re-ceived	Moisture and Ash Free
Pennsylvania anthracite	3.43	6.79	78.25	11.53	0.46	2.52	78.85	0.77	5.87	12,782	15,030
Pocahontas	2.8	14.5	77.4	5.33	0.64	4.56	83.39	1.03	5.05	14,550	15,800
New River	2.94	20.11	73.1	3.85	1.12	5.04	82.60	1.46	5.93	14,582	15,625
Pittsburgh	3.39	31.79	56.46	8.36	1.05	5.07	74.42	1.39	9.71	13,699	15,522
Alabama	4.83	31.40	52.40	11.37	0.74	5.25	71.61	1.23	9.79	12,780	15,250
Illinois	12.39	36.89	41.80	8.92	3.92	5.85	61.29	1.00	19.02	11,399	14,500
Iowa	13.88	36.94	35.17	14.01	6.15	5.52	54.68	0.84	18.8	10,244	14,200
Kansas	4.99	32.68	49.36	12.97	4.28	4.98	67.34	1.08	9.35	12,242	14,900
Arkansas	2.91	12.65	66.93	17.51	3.12	3.60	70.88	1.17	3.72	12,312	15,440
Olkahoma	4.61	37.0	47.25	11.14	3.63	4.92	67.36	1.48	11.46	12,319	14,600
Colorado anthracite	3.0	3.0	86.5	7.5	0.69	2.67	83.20	1.48	4.48	13,500	15,100
Colorado bituminous	4.06	34.48	52.84	8.62	0.65	5.34	71.18	1.24	12.97	12,888	14,750
Colorado lignite	23.07	31.20	35.60	10.13	0.21	5.77	47.69	0.64	35.56	8,030	12,000
Wyoming sub-bituminous	22.38	31.85	39.42	6.35	1.16	6.32	52.25	1.19	32.73	9,247	12,900

HEATING VALUE OF COALS

The heating value of coal is expressed in Btu per pound with the basis usually specified as to whether the pound was ash and/or moisture free. The heating value is determined by burning a weighed sample in oxygen in a calorimeter. Alternative methods are to calculate heating value from the ultimate analysis, or less accurately from calculations based on the proximate analysis.

Calculations of Heating Value. If the ultimate analysis is known the heating value can be calculated, as shown in the following examples.

Example. What is the heating value per pound of a Pocahontas coal with an ultimate analysis as follows: Carbon, 83.14%; Hydrogen, 4.58%; Oxygen, 4.65%; Nitrogen, 1.02%; Suphur, 0.75%; Ash, 5.86%.

Solution by Dulong's Formula. Use Dulong's formula for approximate values. This formula, substantially correct only for solid fuels except those high in hydrogen (cannel coal), and not recommended for lignite, wood or other solids high in oxygen, follows:

$$H = 14{,}544C + 62{,}028\left(H_2 - \frac{O}{8}\right) + 4{,}050S$$

where H = heating value, Btu per lb. and C, H_2, O, and S are the per cent of these elements expressed as a decimal. In this case

$$H = 14{,}544 \times .831 + 62{,}028\left(.046 - \frac{.047}{8}\right) + 4{,}050 \times .0075$$

$H = 14{,}610$ Btu per lb.

The heating value of this fuel as measured by a calorimeter was 14,672 Btu per lb.

Heating Value from Proximate Analysis. No accurate formula for determining the heating value of coal from the proximate analysis is known. The following formula by Goutal is believed, however, to be sufficiently accurate for most heating engineers' calculations:

$$H = 148C + xV$$

where H = heating value, Btu per lb.
C = per cent fixed carbon, dry and ash free
V = per cent volatile matter, dry and ash free
x = a variable depending on the per cent of volatile matter as follows:

Per cent Volatile Matter	Value of x
5	260
10	234
15	210
20	196
25	185
30	176
35	169
40	144

Example. A West Virginia coal yields a proximate analysis, as received, of 16.22% volatile, 76.35% fixed carbon, 3.32% moisture and 4.11% ash. What is the heating value?

Solution: The moisture and ash total 7.43%, the combustible (100.0% − 7.43%) = 92.57%. Dividing the volatile and carbon by 0.9257, the per cent of volatile and fixed carbon on an ash and moisture free basis are, respectively, 17.5% and 82.5%. Interpolating for x opposite 15 and 20% volatile, $x = 203$, so that

$$H = 148 \times 82.5 + 203 \times 17.5 = 15{,}770 \text{ Btu per lb.}$$

ash and moisture free. On an as-received basis, the heating value would be 15,770 × .9257 = 14,600 Btu per lb.

Evans (of the Engineering Experiment Station of Cornell University) gives the following formulas for determining heating value of coals from the proximate analysis. In these formulas V is the fraction of volatile in the coal on an ash and moisture free basis; H is the high heating value in Btu per lb:

For coals from Illinois and Michigan: $H = 16062 - 3830V$

For coals from other states $\begin{cases} V \text{ less than } 16\% & H = 14550 + 7810V \\ V \text{ 16 to 36\%} & H = 16160 - 2250V \\ V \text{ over 36\%} & H = 18750 - 9440V \end{cases}$

Effect of Weathering on Heating Value. Numerous tests have been run to determine the effect of storage on the heating value of coal. The results show: (1) there is no loss in the heating value of anthracite when stored, (2) bituminous coals lose heating value from 1 to 5% on weathering; Eastern soft coals lose less than the midwest coals and (3) sub-bituminous and lignitic coals lose more heating value than bituminous coals.

Bituminous coal is sometimes stored under water to prevent spontaneous combustion; this is not recommended as a means of preventing loss of heating value.

Spontaneous Combustion. The self-igniting of coal in piles is caused by its slow oxidation under circumstances which do not permit the dissipation of the heat generated, until the oxidation finally reaches a burning temperature. Oxidation depends on the surface exposed so that fine coal may spontaneously ignite more quickly than larger sizes. Anthracite is not liable to spontaneous combustion. Precautions for preventing spontaneous combustion of soft coal include:

1. Do not pile over 12 feet, nor so that any point in the interior will be over 10 feet from an air-cooled surface.

2. If possible, store only screened nut coal.

3. Keep out the dust as much as possible; reduce the handling to a minimum.

4. Pile so that the lumps and the fine are distributed as evenly as possible, not allowing lumps to roll to the bottom, forming air passages.

5. Re-handle and screen after two months.

6. Do not store near external sources of heat, even though the heat transmitted be moderate.

7. Allow six weeks seasoning after mining and before storing.

8. Avoid alternate wetting and drying.

9. Avoid air transmission to the pile through interstices around foreign objects, brickwork, etc., or through porous bottoms, such as coarse cinders.

10. Do not ventilate with pipes, which may do more harm than good.

SIZES OF COAL AND COKE

Commercial Sizes of Coals. Anthracite is sized by screening. The official Anthracite Institute sizes are as given in Table 1.

Table 1. — Sizes of Anthracite

Trade Name	Will Pass through Mesh of, In.	Will Not Pass through Mesh of, In.	Min. Size Grate, In.	Min. Depth Fuel Bed, In.	Heat Value Btu per Lb., Dry	Per Cent Ash
Barley......	$\frac{3}{16}$	$\frac{3}{32}$	—	—	12,858	13.9
Rice........	$\frac{5}{16}$	$\frac{3}{16}$	—	—	12,895	13.7
Buckwheat..	$\frac{9}{16}$	$\frac{5}{16}$	—	—	12,926	13.5
Pea........	$\frac{13}{16}$	$\frac{9}{16}$	Any	—	13,168	12.2
Nut........	$1\frac{5}{8}$	$\frac{13}{16}$	Any	—	13,395	10.7
Stove.......	$2\frac{7}{16}$	$1\frac{5}{8}$	16	12	13,395	10.7
Egg........	$3\frac{1}{4}$	$2\frac{7}{16}$	24	16	13,164	9.5
Broken.....	$4\frac{3}{8}$	$3\frac{1}{4}$	—	—	—	—

Coke sizes are not so well standardized, although the data in Table 2 are widely used.

Table 2. — Sizes of Coke

Trade Name	Size, Inches	Trade Name	Size, Inches
Foundry..........	Over $2\frac{1}{2}$	Nut..............	$\frac{3}{4}$–$1\frac{3}{8}$
Egg.............	$1\frac{7}{8}$–$2\frac{1}{2}$	Pea..............	$\frac{1}{2}$– $\frac{3}{4}$
Stove...........	$1\frac{3}{8}$–$1\frac{7}{8}$	Buckwheat........	Under $\frac{1}{2}$

There is no official size for soft coal, but the A S M E standard is as given in Table 3.

Table 3. — Sizes of Soft Coal

Trade Name	Dia., In., of Opening over or through which Coal Will Pass		Trade Name	Dia., In., of Opening over or through which Coal Will Pass	
	Through	Over		Through	Over
EASTERN COALS			WESTERN COALS		
Run of mine.......	As Mined		Run of mine.......	As Mined	
Lump............	—	$1\frac{1}{4}$	6 in. lump........	—	6
Nut.............	$1\frac{1}{4}$	$\frac{3}{4}$	3 in. lump........	—	3
Slack............	$\frac{3}{4}$	—	$1\frac{1}{4}$ in. lump......	—	$1\frac{1}{4}$
			3 in. nut.........	3	$1\frac{1}{4}$
			$1\frac{1}{4}$ in. nut......	$1\frac{1}{4}$	$\frac{3}{4}$
			$\frac{3}{4}$ in. nut........	$\frac{3}{4}$	$\frac{5}{8}$
			Screenings........	$1\frac{1}{4}$	—

STORAGE SPACE FOR SOLID FUELS

The accompanying Table 1 gives the space required for storage of anthracite and bituminous coals. The space occupied by coal varies with the specific gravity, size, moisture content and degree of settling. Other things being equal, the coal with the higher moisture content will weigh more per cubic foot than a coal of lower moisture on a dry basis.

Table 2 gives the storage space and heat content of various woods and other solid fuels, including coke.

Table 1. — Storage Space Required for Various Coals

Coal	Weight, Lb. per Cu. Ft.	Volume, Cu. Ft. per Short Ton
Anthracite, furnace	52–55	36.2–38.4
Anthracite, egg	53–58	34.5–37.7
Anthracite, chestnut	52.5–56.5	35.3–38.1
Anthracite, pea	53.5–54.5	36.7–37.3
Anthracite, buckwheat	49–52	38.4–40.8
Bituminous, lump	40–45	44.5–50
Bituminous, domestic	40–44	45.5–50
Bituminous, slack	50–55	36.2–40

Table 2. — Storage Space Required for and Heat Contents of Woods and Miscellaneous Solid Fuels

Name	Space Required, Cu. Ft. per Lb.	Heat Value, Btu per Lb., Dry
WOOD		
Ash	.0337	5400
Beech	.0328	5350
Birch	.0320	5200
Chestnut	.0475	5800
Elm	.0412	6000
Hickory	.0278	5400
Maple, red	.0400	5950
Oak, white	.0298	5550
Pine, white	.0581	5850
Pine, yellow	.0555	9200
Walnut	.0320	4650
MISCELLANEOUS SOLID FUELS		
Coke	.03	12,500 to 15,000
Charcoal	.07 to .10	11,000 to 14,000
Peat	.016 to .14	7,000 to 10,000
Hog fuel	.0256 to .0512	Same as wood
Sawdust	—	from which made
Wet tan bark	—	2,500 to 3,000
Straw	.05	5,450 to 6,750
Bagasse	—	8,000 to 8,700
Lignite	.0185 to .0208	11,000 to 12,500
Sawdust briquettes	.0125 to .0133	8,000 to 8,200
Shelled corn	—	7,800 to 8,500
Cottonseed hulls	.06 to .07	7,000 to 7,200
City garbage	.023 to .025	7,000 to 9,000

Angles of Sliding and Repose. In connection with the storage of coal, the angle at which coal will begin sliding is sometimes of use. Table 3 shows the angle with the horizontal at which coal will begin sliding on bright steel chutes. The angle of repose in piles can be assumed at about 2 degrees above the minimum sliding angle.

Storage Space for Season. The storage space required for coal for a season's operation of a heating plant, not including service hot water, for installations in most parts of the United States can be determined by the formula

$$T = \frac{H \times D}{(70 - t_0) \times E \times F_c \times 67}$$

where T = seasonal coal consumption, short tons,

H = estimated hourly heat loss under design (coldest) conditions, in Btu,

D = normal seasonal total of degree days for locality in question,

t_0 = outside design temperature for locality in question,

E = overall efficiency of heating plant, expressed as a decimal, and

F_c = heating value of coal used, per pound.

To the quantity of fuel obtained from the above should be added the fuel required for service hot water.

The formula is not applicable to the deep South, extreme northern part of northern states, or the Pacific Coast states.

Table 3. — Angle of Sliding for Coals

Coal	Angle with Horizontal			
	Sliding Starts		Sliding Continues	
BITUMINOUS				
	Deg.	Min.	Deg.	Min.
Illinois 6 in. lump...........................	21	50	20	40
Illinois 3 × 6 in. egg........................	21	00	19	10
Illinois 3 × 2 in. lump.......................	20	30	19	00
Illinois 1¼ × 2 in. nut.......................	20	40	19	20
Illinois ¾ × 1¼ in. nut.......................	21	00	19	40
Illinois slack................................	25	40	22	30
Oklahoma screenings..........................	24	00	22	30
Oklahoma chestnut............................	21	00	19	00
Kentucky egg.................................	21	50	20	20
ANTHRACITE				
	Deg.	Min.	Deg.	Min.
Pennsylvania egg.............................	15	40	14	00
Pennsylvania chestnut........................	16	40	15	10

COKE

(Sliding angles below are for coke on coke, not on chutes.
High values are for dry coke, lower values for wet coke)

Size	Sliding Angle	Angle of Repose
	Degrees	Degrees
Furnace......................................	43 to 45	45 to 47
Stove..	39 to 41	43 to 44
Breeze.......................................	37 to 40	41 to 44

CLASSIFICATION AND PROPERTIES
OF GAS FUELS

The design information and data required by the engineer when specifying combustion equipment that will use gas fuel can be obtained readily from the local gas utility company. For the most part, many of the manufactured fuel gases such as coal gas, blue water gas, blast-furnace and coke-oven gases, are no longer in service because of their low heating values and poor economy when related to other fuels, among other factors. Gas fuels, in present use, tend to burn cleanly and maintain a consistent efficiency in combustion over long periods of time. Because of their good burning characteristics, gas fuels no longer cause soot deposits which can build up on equipment surfaces and cause a reduction in heat transfer.

It is desirable, however, to know what different types of gas fuels are currently employed or are enjoying increasing usage. These different types will be described in the following sections.

Natural Gas

Natural gas obtained its name because it is a naturally formed gaseous mixture of hydrocarbons and non-hydrocarbons occurring in porous formations within the crust of the earth. It is often found along with natural deposits of crude petroleum. Natural gas is principally methane with small quantities of other hydrocarbons and may sometimes contain nitrogen or carbon dioxide. Before this gas is delivered to the pipelines it is usually stripped of the heavier hydrocarbons. Gathering systems from the producing gas fields collect the gas from wells in many fields for delivery to the pipelines. The pipelines will then carry the gas to the distant markets where it is utilized in the same state and condition in which it is delivered. The gas at the customer's meter will usually have a heating value of about 1,000 to 1,050 Btu per cubic foot and a specific gravity between 0.58 to 0.61 compared to air. For design considerations it is usually practical to use a heating value of 1,000 Btu per cubic foot and a specific gravity of 0.60.

Synthetic Natural Gas (SNG)

SNG is a manufactured gas whose chemical and physical characteristics match very closely those of natural gas. SNG is principally methane, free of other hydrocarbons unless they are deliberately added, and containing small quantities of hydrogen, nitrogen, and carbon dioxide. This manufactured product is usually made from naphtha and is completely and easily interchangeable with natural gas. Since the SNG generated has a heating value of about 985 Btu per cubic foot it may be spiked with heavier hydrocarbons, such as propane, to raise the Btu level. This synthetic gas is utilized by the natural gas distribution companies to augment their normal pipeline gas. For design considerations in combustion equipment it is handled as if it were the normal natural gas.

Liquefied Natural Gas (LNG)

LNG is natural gas that has been converted to a liquid state for storage purposes only. In this state natural gas has its volume reduced by 600%; that is, one cubic foot of LNG will produce approximately 600 cubic feet of natural gas when it is returned to the gaseous state. Since LNG is natural gas, the design considerations in combustion equipment are handled as if it never was a liquid. Like SNG, LNG is utilized by the natural gas distribution companies to augment their normal pipeline gas.

Liquefied Petroleum (LP) Gases

The so called L.P. gases are propane, butane, or pentane and are more commonly referred to as bottled gas. Propane has a gross heating value of 2,572 Btu per cubic foot and butane 3,225 Btu per cubic foot. Usually, regardless of the name, L.P. gases are sold as mixtures of the three gases mentioned. These gases are all obtained from natural gas. They are stripped out of the gas before the natural gas is delivered to the pipelines.

CLASSIFICATION AND PROPERTIES OF LIQUID FUELS

Petroleum, from which fuel oils are derived, is a mixture of hydrocarbons, the mixture varying depending on the field where found. Petroleums from the Appalachian field are of the paraffin series, those from the Gulf and West Coast are of the asphalt series, and so on. The types of hydrocarbons present are shown in Table 1.

There are many data available on the analysis and characteristics of petroleums, but since fuel oils are the result of refining operations they vary widely.

Table 1. — Hydrocarbons in Crude Oils

Hydrocarbon Series	Formula	Lowest in Series
Paraffin or methane..................................	C_nH_{2n+2}	CH_4
Unsaturated olefines of saturated naphthenes..............	C_nH_{2n}	C_2H_4
Unsaturated acetylene.............................	C_nH_{2n-2}	C_2H_4
Asphalt..	C_nH_{2n-4}	—
Aromatics or benzene.............................	C_nH_{2n-6}	C_6H_6
Naphthaline.......................................	C_nH_{2n-12}	$C_{10}H_8$

Generally, fuel oils consist of about 85% carbon and 12% hydrogen, with small amounts of oxygen, nitrogen and sulphur. The specific gravity is an index to the heating value of fuel oil but although used roughly in practice it is not in itself a sufficient basis for classification. The Department of Commerce commercial standard for fuel oils is the basis of classification widely used; however, there is considerable variation in actual practice among oils designated by these numbers.

Table 2 gives typical heating values and approximate ultimate analyses of these various grades.

Table 2. — Approximate Gravity, Heat Value and Analyses of Fuel Oils

Commercial Standard No.	Deg. API, Approx.	Heat Value, Btu per Gal.	Approximate Analysis Per Cent by Weight
1	38–40	136,000	
2	34–36	138,500	Typical analysis would be
3	28–32	141,000	85% carbon, 14% hydrogen
4	24–26	145,000	and 1% water and sediment
5	18–22	148,500	
6	14–16	152,000	

Miscellaneous Properties of Fuel Oils. The following formulas are useful in determining various properties of fuels derived from petroleum. In all these formulas

d = specific gravity at 60F;
t = temperature, F;
p = pressure, lb. per sq. in.

Heat of Combustion

Btu per lb. = $22,320 - 3780d^2$

Thermal Conductivity

Btu per in. per sq. ft. per hr. per F =

$$\frac{0.813}{d}[1 - 0.0003(t - 32)]$$

Specific Heat of Liquid

Btu per lb. per F = $\frac{1}{\sqrt{d}}(0.388 + 0.00045t)$

Btu per gal. per F = $\frac{1}{\sqrt{d}}(3.235 + 0.000375t)$

Latent Heat of Vaporization ·

Btu per lb. = $\frac{1}{d}(110.9 - 0.09t)$

Btu per gal. = $925 - 0.75t$

Heat Content of Liquid

Btu per gal. = $\sqrt{d}(3.235t + 0.001875t^2 - 105.5)$

Heat Content of Vapor

Btu per gal. = Heat of liquid + $925 - 0.75t$

Specific Volume of Vapor

Cu. ft. per lb. = $\frac{0.242(t + 460)}{p} \times \frac{1.03 - d}{d}$

Kerosenes. This fuel is composed of a number of hydrocarbons averaging about $C_{12}H_{26}$. Its specific gravity ranges from 0.78 to 0.82 (50° to 41° Bé.) and its higher heating value from 19,800 to 20,100 Btu per lb., according to the following formula:

$$H = 18,440 + 40(\text{Deg. Bé} - 10)$$

where H is the high heat value in Btu per pound.

Gasoline. The chief constituents of this fuel are chiefly paraffin hydrocarbons. The higher heating value per pound is $18320 + 40$ (deg. Bé − 10) and averages about 20,200 Btu per lb., with only 1.5% difference over a range of specific gravities from 0.687 to 0.745 (74 to 58 deg. API). The heating value per gallon is 116,500 Btu for a specific gravity of 0.687 and 124,000 Btu at 0.745 sp. gr.

WEIGHT AND VOLUME EQUIVALENTS FOR FUEL OIL AT 60F

Deg. API	Deg. Baumé	Relative Weight Water = 1.00	Weight, Lb. per Cu. Ft.	Weight, Lb. per Gal.
10	10.00	1.000	62.37	8.34
11	10.99	0.993	61.93	8.28
12	11.98	0.986	61.50	8.22
13	12.97	0.979	61.07	8.16
14	13.96	0.973	60.65	8.11
15	14.95	0.966	60.24	8.05
16	15.94	0.959	59.83	8.00
17	16.93	0.953	59.43	7.94
18	17.92	0.947	59.03	7.89
19	18.91	0.940	58.64	7.84
20	19.89	0.934	58.25	7.79
21	20.88	0.928	57.87	7.74
22	21.87	0.922	57.49	7.69
23	22.86	0.916	57.12	7.64
24	23.85	0.910	56.75	7.59
25	24.84	0.904	56.39	7.54
26	25.83	0.898	56.03	7.49
27	26.83	0.893	55.68	7.44
28	27.81	0.877	55.33	7.40
29	28.80	0.882	54.98	7.35
30	29.79	0.876	54.64	7.31
31	30.78	0.871	54.31	7.26
32	31.77	0.865	53.98	7.22
33	32.76	0.860	53.65	7.17
34	33.75	0.855	53.32	7.13
35	34.74	0.850	53.00	7.09
36	35.73	0.845	52.69	7.04
37	36.71	0.840	52.37	7.00
38	37.70	0.835	52.06	6.96
39	38.69	0.830	51.76	6.92
40	39.68	0.825	51.46	6.88
41	40.67	0.820	51.16	6.84
42	41.66	0.816	50.86	6.80
43	42.65	0.811	50.57	6.76
44	43.64	0.806	50.28	6.72
45	44.63	0.802	49.50	6.68
46	45.62	0.797	49.72	6.65
47	46.61	0.793	49.44	6.61
48	47.60	0.788	49.16	6.57
49	48.59	0.784	48.89	6.54
50	49.58	0.780	48.62	6.50

STORAGE SPACE FOR LIQUID FUELS

The storage space for fuel oil and other liquid fuels are given in Table 1.

Table 1. — Storage Space and Heat Content of Liquid Fuels

Fuel	Cubic Feet per Gal.	Heat Content, Btu per Gal.	Fuel	Cubic Feet per Gal.	Heat Content, Btu per Gal.
FUEL OIL			FUEL OIL		
No. 1.............	0.13368	136,000–139,000	No. 4.............	0.13368	144,000–145,500
No. 2.............	0.13368	139,500–140,500	No. 5.............	0.13368	146,000–148,500
No. 3.............	0.13368	141,000–143,500	No. 6.............	0.13368	149,000–152,000

Fuel	Cubic Feet per Pound	Heat Content, Btu per Pound	Fuel	Cubic Feet per Pound	Heat Content, Btu per Pound
MISCELLANEOUS LIQUID FUELS			MISCELLANEOUS LIQUID FUELS		
Coal tar............	.0128–.0169	16,000–17,300	Methyl (wood) alcohol	.0203	9,600
Benzene[1]...........	.0181–.0185	18,170	Gasoline............	.0212–.0224	20,600
Ethyl (grain) alcohol.	.0202	12,810	Kerosene.......'...	.0197	20,000–20,500

[1]Not to be confused with benzine (gasoline).

Seasonal Storage Space. The storage capacity for a whole season can be calculated for fuel oil by means of the following formula to which two restrictions apply: (1) it is not applicable to the deep South, the extreme northern portions of the most northerly states nor to the Pacific Coast states; (2) it does *not* include allowance for fuel needed to supply service hot water:

$$G = \frac{30 \times H \times D}{(70 - t_0) \times E \times F_0}$$

where G = seasonal oil consumption in gallons, heating only,
H = estimated hourly heat loss under design (coldest) conditions,
D = normal seasonal total of degree days for locality in question,
t_0 = outside design temperature for locality in question,
E = overall efficiency of heating plant, expressed as a decimal, and
F_0 = heating value of oil per gallon.

To the quantity of fuel obtained from the above should be added the fuel required for service hot water.

Tank Capacity. The capacity in U. S. gallons of cylindrical oil tanks with plane ends is $0.0034d^2L$, where d is the diameter in inches and L the length in inches. For bumped end cylindrical tanks, bumped outward at both ends, and with a bumped radius equal to the diameter, the capacity in barrels is $d^2(0.14l + 0.019d)$ where d is the diameter (and radius of bumps) in feet and l the length excluding convex ends.

Table 2 gives the sizes of oil tanks which are fairly well standardized.

Table 2. — Fuel Oil Tanks Commonly Used

Diameter, Inches	Length, Inches	Capacity, Gallons	Diameter, Inches	Length, Inches	Capacity, Gallons
CYLINDRICAL TANKS			CYLINDRICAL TANKS		
42	93¾	560	64	363	5,000
36	127	550	72	284	5,000
42	167½	1,000	84	212	5,000
45½	146¼	1,030	102	142	5,000
64	108	1,500	84	420	10,000
64	144	2,000	96	329	10,000
64	180	2,500	102	284	10,000
64	215	3,000	120	206	10,000
64	286	4,000			
OBLOID TANKS					
26 × 42	66⅞	275			

FUEL OIL STORAGE TANKS

Fuel oil storage tanks are generally of steel conconstruction, although occasionally concrete is used. Steel is the most practical material for tank construction and tanks so constructed are by far the least expensive Since steel has the most flexible characteristics, it is not limited by tank size or the grade or weight of the fuel to be stored. It is thoroughly adapted to small domestic installations as well as to the largest industrial requirements.

There are definite rules and regulations covering the construction of fuel oil storage tanks. While local rules and regulations take precedence, where these are nonexistent, it is customary to install a tank constructed according to the National Board of Fire Underwriters requirements.

Concrete Tanks

There are special rules and regulations for the construction of concrete tanks. The National Board of Fire Underwriters regulations states that:

(1) Concrete may be used as a material for storage tank construction when the fuel oil to be stored is heavier than 35 deg A.P.I.

(2) Selection of sites shall be governed by rules for steel tanks.

(3) All concrete tanks must be provided with a gas-tight concrete roof.

(4) Rules for piping shall in general be the same as for steel tanks.

(5) Plans for concrete tanks must be approved by the Inspection Department of the National Board of Fire Underwriters before the construction is started, and the entire erection must be under the charge of a fully competent engineer.

According to the Portland Cement Association, the maximum working stress of concrete in storage tanks should not exceed 8,000 to 10,000 lb per sq in. The concrete shall always be reinforced with steel, as concrete does not possess the strength in tension that it has in compression, and steel must, therefore, provide the tensile strength. Steel in tension in the walls and floor shall be designed for a safe working stress of 10,000 lb per sq in. whether in circumferential tension or in tension due to bending moments.

Tanks shall be designed to provide for all internal and external loads to include: (1) full hydrostatic pressure of contents upon floor and walls, (2) external pressure upon floor and walls, (3) earth pressure on walls, and (4) live and dead loads on roof

Column loads shall be distributed over a proper area by providing adequate footings and floor reinforcements. The minimum thickness of walls shall be 8 inches at the top and 10 inches at the bottom.

Steel Tanks

Steel tanks are constructed in three standard shapes now in general use, cylindrical, obround, and rectangular. Cylindrical tanks are always used for underground service. Large inside tanks are either rectangular or cylindrical. Where conditions require that tanks be fabricated on the premises they are generally rectangular in shape. Obround tanks are mostly used for domestic installations and are supported off the floor. They generally do not exceed 275 gal capacity.

Table 1 is a schedule of commercial sizes of cylindrical tanks for underground service built to National Board Standards.

Table 1. Dimensions of Fuel Oil Storage Tanks

(Actual dimensions will vary slightly depending on the manufacturer)

Capacity Gallons	Diameter. Inches	Length, Feet and Inches	Shell	Head	Weight, Pounds
			Thickness, Inch		
550	48	6-0	3/16	3/16	800
1,000	48	10-10	3/16	3/16	1,300
1,100	48	11-10	3/16	3/16	1,400
1,500 {	48	15-8	3/16	3/16	1,650
	65	9-0	3/16	3/16	1,500
2,000	65	11-10	3/16	3/16	2,050
2,500	65	14-10	3/16	3/16	2,275
3,000	65	17-8	3/16	3/16	2,940
4,000	65	23-8	3/16	3/16	3,600
5,000 {	72	23-8	1/4	1/4	5,800
	84	17-8	1/4	1/4	5,400
7,500 {	84	26-6	1/4	1/4	7,150
	96	19-8	1/4	1/4	6,400
8,500	108	18-0	1/4	5/16	7,250
10,000 {	96	26-6	1/4	5/16	8,540
	120	17-0	1/4	5/16	8,100
12,000 {	96	31-6	1/4	5/16	10,500
	120	20-8	1/4	5/16	9,500
15,000 {	108	31-6	5/16	5/16	13,300
	120	25-6	5/16	5/16	12,150
20,000	120	34-6	5/16	5/16	15,500
25,000	120	42-6	3/8	3/8	22,300
30,000	120	51-3	3/8	3/8	28,000

The tanks may be fabricated in the field, or in the basement or vault when outside space is limited. Tanks so fabricated are best constructed by electric weld, although acetylene weld is used when electric welding equipment is not available. If it is possible, the seams should be welded inside and outside; however, where conditions prevent this procedure, the welding m·y be limited to the inside seams. Rectangular tanks should always be crossbraced internally.

NBFU Requirements

The National Board of Fire Underwriters has set up the following minimum requirements for the construction of underground tanks and tanks inside buildings. Tanks of 181 to 275 gal capacity, installed in buildings with enclosure, shall be 14 gage minimum. Steel or wrought iron, thinner than No. 7 gage (3/16 or 0.188 in.), used in construction of underground and enclosed tanks, shall be galvanized. Otherwise, tanks shall be constructed according to Table 2.

If adequate internal bracing is provided, then tanks of 12,001 to 30,000 gal capacity may be built of 1/4 in. plate. For tanks larger than 1,100 gal capacity, a tolerance of 10% in capacity is permitted.

Sizing the Tank

The size of the fuel oil storage tank, or tanks, does not follow any single rule but is the outcome of: (1) the rate at which the fuel will be burned, (2) the space limitations available for storage capacity, (3) the availability or nearness of the sources of supply, (4) the method of delivery (by truck or tank car), (5) the possibility of transportation difficulties, and (6) the possibility of unit cost reduction on large purchases.

In sizing the fuel oil storage tank, it is first necessary to determine the rate at which the fuel will be consumed. Where the load is for heating purposes only, it is necessary to first obtain the annual fuel consumption. The greatest percentage of fuel consumption will occur in January and will amount to approximately 20% of the annual consumption. One-fourth of the January figure will give a week's supply which is the minimum tank size. Before selecting the tank, however, approximately 10% should be added to the minimum tank size to allow for the suction stub clearance.

The annual fuel consumption may be approximated by (1) the calculated heat loss method, or

Table 2. Thickness of Metal for Fuel Oil Tanks

Tank Capacity, Gallons	Thickness, Minimum	Weight of Plate, Lb per Sq Ft
7 to 285	16 gage	2.50
286 to 560	14 gage	3.125
561 to 1,100	12 gage	4.375
1,101 to 4,000	7 gage	7.50
4,001 to 12,000	1/4 in. nominal	10.00
12,001 to 20,000	5/16 in. nominal	12.50
20,001 to 30,000	3/8 in. nominal	15.00

(2) the degree day method. In the case of a conversion installation, reliable records of past heating operation or fuel consumption will usually produce the most trustworthy estimates for future consumption.

Example

As an example, assume that an approximated annual fuel consumption shows 80,000 gal of fuel oil are required. During the month of January, approximately 16,000 gal will be consumed. The weekly consumption will, therefore, be approximately 4,000 gal. Adding 10% for stub clearance, will result in the selection of a 5,000 gal tank which, according to Table 1, is nearest to the minimum requirements.

If the steam requirements are for processing, laundry or other similar load not of a building heating nature, the preceding methods will not be useful. The following method may be used.

Assume a six-day operating week and a ten-hour operating day at which time the plant will operate at peak consumption. As an example, assume that the maximum hourly consumption is 50 gal. The daily fuel consumption will be 500 gal, and 3,000 gal for the six-day week. Adding 10% for stub clearance will result in a minimum tank capacity of 3,300 gal. The tank selected would be 4,000 gal.

While these methods of sizing the storage tank will result in the minimum storage tank capacity, additional capacity must be allowed to compensate for the availability or distance to the source of supply, and the method of delivery. If the fuel oil is delivered by rail, the fuel oil storage tank should be sized to take a full tank car plus one week's fuel consumption. Tank cars range from 8,000 to 12,000 gal. In the preceding two examples, the

minimum storage tank size for tank delivery would be 17,000 gal. This could be met by the selection of two 8,500 gal tanks or one 20,000 gal tank.

Inside Tanks

The NBFU states that inside tanks larger than 60 gal capacity shall be located at the lowest level in a building whether it be the cellar or basement or at the first floor level. Unenclosed inside storage tanks, Fig. 1, shall not be located closer than 7 ft horizontally to any fire or flame. Individual oil supply tanks inside a building shall not exceed 275 gal capacity, or 550 gal aggregate capacity, unless enclosed. Enclosures shall consist of at least a 6-inch thickness of reinforced concrete, or of brick at least 8-inches thick. Between the tank and the enclosure, the space shall be filled completely with sand. If the floor above the tank is fire-resistive and has a load carrying capacity rated at least 150 lb per sq ft, the walls of the enclosure and the sand should be carried one foot above the tank; otherwise the tank must be covered with a 5-inch concrete slab or equivalent. As an alternate to this enclosure, the tank may be completely encased in 6 inches of concrete applied directly to the tank without air space.

In non-fire-resistive buildings, inside storage up to 5,000 gal is permitted. In fire-resistive buildings, up to 15,000 gal of oil may be stored inside. If special provisions are made in a fire-resistive building, up to 50,000 gal may be stored, but not more than 25,000 gal in a single tank.

Underground Tanks

NBFU recommends that oil supply tanks should be located outside of buildings and underground with the top of the tank below the level of all piping to which the tank is connected, to prevent discharge of oil through a broken pipe or connection by siphoning. An underground tank should be buried so that the top of the tank is at least two feet below grade. Otherwise the top of the tank shall be covered with not less than one foot of earth on top of which shall be placed a slab of reinforced concrete not less than 4 inches thick. The slab should be set on a firm, well tamped earth foundation and shall extend at least one foot beyond the tank in all directions. Where tanks are buried underneath buildings, such a concrete slab shall be provided in every instance. Underground tanks should be painted with heavy asphaltum or rust-resisting paint and should be set into excavation as level as possible.

Storage tanks should never be installed in or covered with cinders. Cinders contain sulfur and when wet with water will produce very rapid steel corrosion.

In many localities, considerable quantities of ground water may be encountered. Where such con-

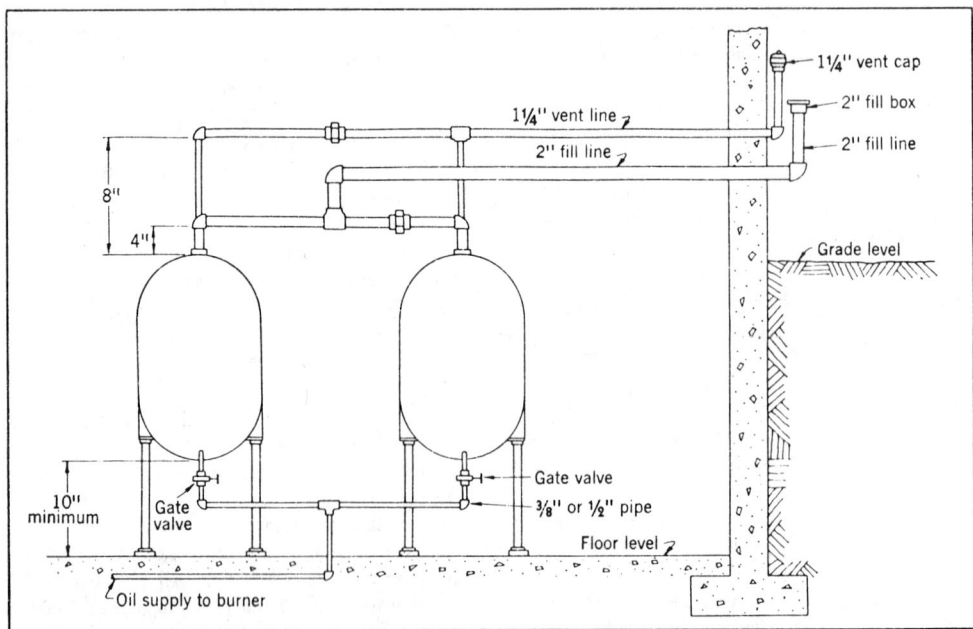

Fig. 1. A typical hook-up for two 275 gal basement tanks showing valves and pipe lines.

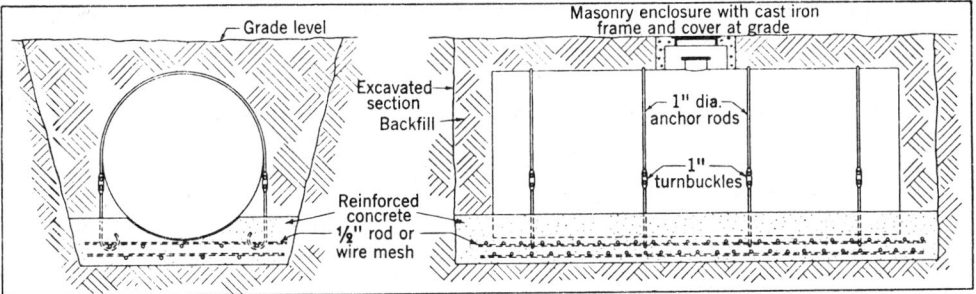

Fig. 2. Method of securing large underground storage tank when surface water condition exists.

ditions are found, it is necessary to provide an anchor, Fig. 2, to secure the tank in place. The general procedure is to set the tank on a poured concrete slab and attach the tank to this slab with steel straps or rods. The concrete slab must be reinforced with ½ inch diameter steel rods placed on centers not greater than 12-inch. The rods should be placed near the bottom of the slab where the slab is in tension. Reinforcing rods are placed crosswise in the concrete. When the concrete is poured, the anchor rods are set in place and are anchored into the slab by looping the end around horizontal rods embedded lengthwise in the concrete on either side of the tank. One-inch diameter steel anchor rods, equipped with turnbuckles for adjustment, should be placed on approximately 5 ft centers along the tank length.

Anchor Design

The weight of the concrete necessary to secure the tank should be between the maximum weight necessary to sink the tank in clear water and the weight to offset the maximum amount of ground water may be encountered in the special locality.

The weight of concrete in air varies from 110 to 155 lb per cut ft, according to the material used. Commonly used averages are 150 lb per cu ft for stone concrete. However, the actual effective weight of concrete, when immersed in water, is equal to the difference in weight between one cubic foot of concrete at 150 lb and one cubic foot of water at 62.5 lb. The effective weight of concrete is therefore 87.5 lb per cu ft. The tank is buoyed up by the weight of the displaced water. This is equal to the capacity of the tank in gallons multiplied by 8.3 lb. From this total displaced weight there may be deducted the weight of the tank, the weight of the concrete manhole enclosure, if present, the effective weight of the earth fill on top of the tank, and in

the judgment of the designer any equipment that may lend its weight for this purpose. When these items have been deducted, the resulting net weight of displaced water divided by 87.5 will give the cubic feet of concrete necessary.

When the concrete is placed over the tank, the full weight of 150 lb per cu ft may be used in calculating the volume of concrete necessary. The top of the concrete would be in tension and the reinforcing rods should, therefore, be located near the upper surface of the slab. The concrete slab should extend a minimum of one foot on all sides of the tank. The slab should, therfore, be at least two feet greater in length and two feet greater in width than the tank. Also, concrete slab may be used to prevent settling.

Manholes

Practically all underground storage tanks of 3,000 gal and greater capacity are equipped with access manholes for inspection, cleaning, and repair. Standard manholes are 16 inches in diameter and are made up of a manhole flange welded to the tank, and a steel cover plate which is bolted to the tank flange. An oil-tight gasket should separate the two surfaces. When the storage tank is to be heated by a tank coil, or suction bell, a second manhole is placed near the access manhole. The piping to the coil, including the oil suction and return stubs, are welded to the steel cover plate, and the entire unit is installed as one.

To provide access to a buried tank, it is customary to install a masonry or poured concrete enclosure around the manholes and to extend this enclosure to the grade level for receiving a heavy cast iron frame and cover. The inside dimensions of the masonry enclosure should be approximately 3 ft wide and 6 ft long. The cast iron manhole at the grade level should be placed directly over the

tank manhole containing the heating coil so that this entire unit may be installed and removed easily.

Above-Ground Tanks

In general, the use of outside above-ground tanks for oil burner work should be discouraged. This type of oil storage is subject to the sudden changes of climate and the extreme cold of winter. Residual oils, when stored above-ground, will be subject to numerous temperature changes making the flow and control of such oils a very difficult problem.

Where inside storage tank space is not available, and outside conditions are such that an above-ground tank is the last resort, then the grade of fuel oil to be burned should be limited to the distillate oils which are not too greatly affected by climactic changes. However, before such an installation is made, it is best to consult with the local authorities having jurisdiction and with the Fire Underwriters' representative.

Tank Heating Arrangements

Fuel tanks that are to store Bunker C fuel oil, and some of the more viscous grades of No. 5 oil, must be equipped with some means by which the oil can be heated to make it possible to pump these very viscous fluids.

Hot Well

Several heating arrangements are used for fuel oil tank heating. Probably the least expensive method, as far as the installation cost is concerned, is the hot well, Fig. 3. This method does not provide any heating in the tank itself but employs a confining chamber into which the fuel oil suction and return stubs terminate not greater than 12 inches apart. The oil is heated at the boiler by an indirect heater, using either steam or hot water as the medium, and the oil, in excess of the burning rate, is bled back to the storage tank where it is mixed in the hot well to give a desired pumping temperature.

The hot well is generally constructed on the job, using sheet iron or a lightweight steel plate. While it may be of most any shape, it must have holes for entrance of the fuel oil in the sides or ends, holes on the top for suction and return stubs, and one or two small vent holes to allow any entrained air to escape. The holes where the suction and return stubs enter should be cut one pipe size larger than the actual size of the stubs themselves.

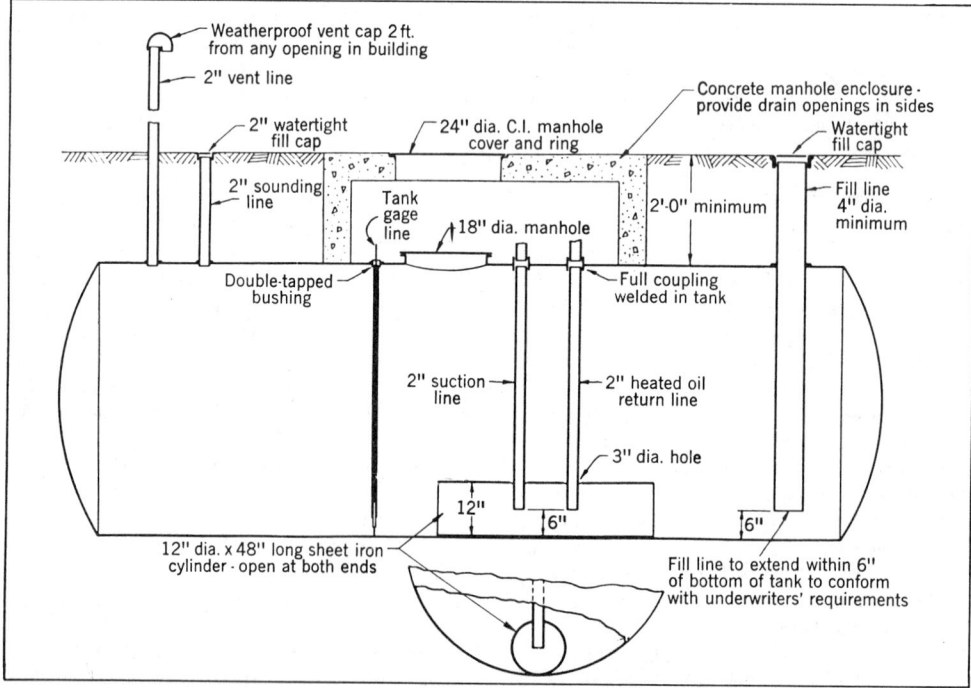

Fig. 3. Fuel oil storage tank with a hot well.

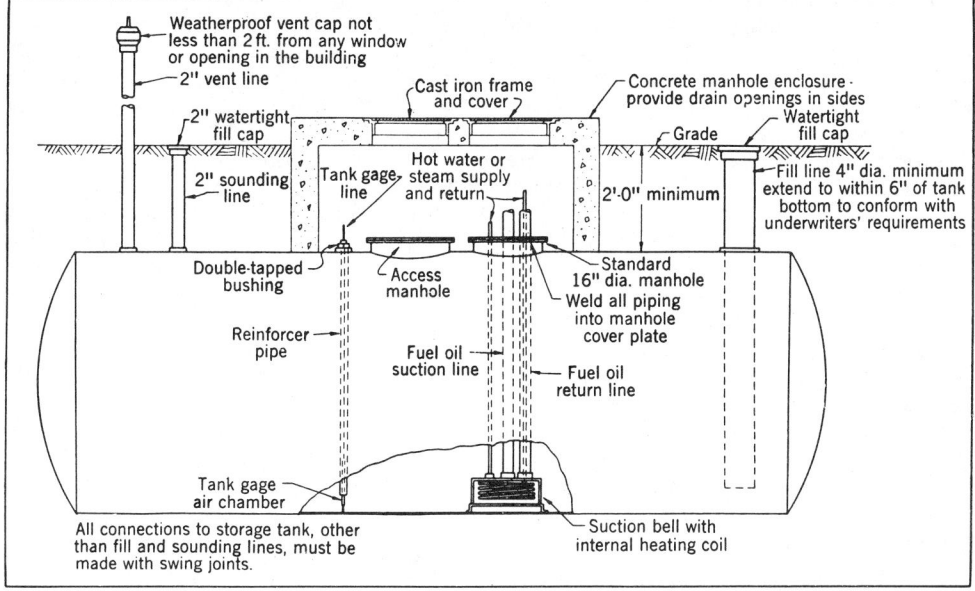

Fig. 4. Detail of storage tank with suction bell.

In lieu of the hot well, it is possible to purchase a suction bell. This type of unit may be obtained from several sources of manufacture and is constructed in two types. Of the two types, one consists of a cast iron shell and a cast iron heater with extended fins. The other, Fig. 4, consists of a sheet iron shell enclosing a pipe coil. This unit may be furnished without a pipe coil and may be employed as a hot well.

Heating Medium

In general, the suction bell is furnished with the heating element. Steam is generally used as the heating medium when it is of sufficient pressure to force the condensate back to the boiler, or where a vacuum return line pump is used. Where neither high pressure steam nor a vacuum return pump is available, forced circulation hot water may be employed. The boiler water may be pumped through the suction bell or an indirect heater may be attached to the boiler which in turn will heat the water pumped through the coil.

Use of the indirect heater lessens the possibility of oil entering the boiler should a leak occur. In any event, the capacity of the suction bell heating coil is not generally sufficient in itself and it is necessary to bleed back through the main fuel oil preheater or to employ a manifold return line heater as used with the hot well.

A second and much preferred method of storage tank heating is the use of the tank coil. This method, too, consists of two general types: (1) the horizontal grid coil, and (2) the vertical helical coil. The horizontal grid coil consists of a series of heating pipes running the length of the storage tank and covering the greater part of the tank bottom. This type of coil is generally made up of 1¼- or 1½-inch steel pipe, supported approximately 12 to 18 inches above the tank bottom, and extending the full width of the tank at this point. The horizontal grid coil is best suited where it is desired that the entire tank is to be heated.

Coil Fabrication

The vertical helical coil is generally constructed of 1- or 1¼-inch seamless steel tubing. Coils are generally fabricated with an outside diameter of 15 inches so that they may be inserted or removed through a standard 16-inch diameter manhole. Coils should be spaced on approximately 2-inch centers for 1-inch tube construction and 2½-inch centers when 1¼-inch tubing is used. The overall length of the coil itself should be a minimum of one-third the diameter of the storage tank. The coil should be set approximately 6 inches above the bottom of the storage tank and the fuel oil suction and return stubs carried down through the center of the coil.

The vertical helical coil, which is not intended for heating the entire tank, is used to heat the oil around the suction and return stubs for pumping only. The vertical helical coil has the advantage over the horizontal grid coil in that it may be installed and removed or replaced very easily through the tank manhole provided for this purpose. Steam or forced circulation hot water is used as the heating medium. There are several pipe fabricators that will make up a coil of this kind to suit any size or condition. There is at least one standard size coil available that can be used in most jobs.

With the various described types of heating devices in storage tanks, using steam or hot water for heating purposes, protection should be given where the pipe is buried underground. In such cases, the suction and return and the steam or hot water supply and return are all wrapped together and insulated with a waterproof covering.

Electric Heaters

Electricity is another means that has been used with success to heat the oil between the oil pump and the storage tank. In this system the suction and return oil piping actually form a low voltage circuit to heat the oil. The oil temperature is thermostatically controlled. The entire system consists of a step-down transformer, adjustable thermostat for temperature control, special insulated pipe flanges, and a special storage tank unit incorporating the suction and return stubs. Where the system is insufficient to give the desired oil temperature rise, conventional electric preheaters are inserted in the suction or return lines to give the necessary capacity.

In some installations, a complete system of steam or hot water preheaters are installed along with this system to take over after the plant is started up, and thus reduce the electrical operating cost. On some very rare occasions, insertion type electric heating elements have been inserted directly in the storage tank itself. The operating cost of heating oil by electricity has reduced its use. In general, electricity is used for cold start-up only and is supplemented by steam or hot water heating once the plant is in operation.

Tank Gages

All fuel oil storage tank installations should be equipped with some means by which the quantity of oil contained in the tank may be determined. Tank gaging equipment, or liquid-level indicating devices are made to indicate the vertical depth of the fuel oil in tanks of all descriptions, including cylindrical, rectangular, obround, horizontal, vertical, steel, and concrete. External tubular type glass gages as used on water tanks are not allowed for fuel oil storage tanks.

The simplest type of tank gage is the calibrated stick which is inserted through the fill line or a sounding line when provided. Calibrated sticks may be fabricated of wood or metal. This method, however, is limited to outside storage tanks. The NBFU states that test wells shall not be installed inside buildings, and where permitted for outside service shall be closed tight, when not in use, by a metal cover designed to prevent tampering. Also, the gaging of inside tanks by means of measuring sticks is a pronounced hazard and should not be permitted. Tank sticks must be calibrated and marked for the particular tank in which it is to be used in order to obtain a reasonable percentage of accuracy.

Automatic Gages

There are several types of automatic gages for measuring the depth of fuel oil in storage tanks that are used with oil burner installations. Small inside domestic tank installations make use of the float type gage. Various methods of operation employ the float gages. One gage consists of a series of small floats linked together, and of sufficient length to reach close to the bottom of the tank. The top end is attached to a spring loaded indicator with indicating dial which is screwed into one of the tank tappings. When the tank is full, all of the floats are buoyed up and exert no force on the indicator. As the tank is emptied, each float becomes suspended in air which in turn exerts its weight against the spring to move the indicating dial. There are other arrangements of the float gage consisting of lever arms, gearing, and other parts.

The disadvantage of the float gage is that the float becomes coated with sludge and tar and loses some of its buoyancy. At best the float gage is not accurate or dependable but merely offers a means of judging when the tank supply is low enough to require an oil delivery. The gage indicator is marked to show when the tank is full, $\frac{3}{4}$, $\frac{1}{2}$, $\frac{1}{4}$, and empty. Float gages should be limited to inside domestic installations.

Distant reading gages are employed to determine the tank contents regardless of the tank shape or location. The gage indicator may be placed at any desired point regardless of the storage tank location, up to $\frac{1}{4}$ mile distance.

The hydrostatic gage is the most common type of remote-reading liquid-level indicating device that is used with oil-burner installations. In general, the hydrostatic gage consists of an air chamber which rests on the bottom of the storage tank, a small copper tube that is attached to the air chamber and is extended to a desired point within the building, and an indicating instrument attached to the end of the tubing. The hydrostatic head of the liquid in the tank exerts pressure on the air trapped in the air chamber which is in turn transmitted by the copper tubing to the indicating device. Usually there is a small pump located at the indicating instrument to replenish the air in the system that is sometimes expelled by small leaks in the tubing and connections. Hydrostatic gages are sufficiently accurate for use with distances up to $\frac{1}{4}$ mile from the tank to indicator. When selecting a hydrostatic gage it is necessary to know the depth or diameter of the storage tank, the specific gravity of the oil, and the length of the communicating tubing.

Hydrostatic gage indicators are of several different types: Those with red indicator fluid in a glass tube, those using mercury liquid in a glass tube, and those using a dial indicator. In the first two of these types of indicators, the liquid in the gage glass is supported at the proper level by the direct action of the air pressure corresponding to the hydrostatic head of oil in the tank. The third type is by action of the air pressure on a bellows that is converted into pointed indication through knife-edge bearings.

Hydraulic Gages

Another type of remote-gage is the float hydraulic gage, a completely self-contained unit that can be tested for proper operation before it is installed. This system consists of a tank unit made up of a large copper float which is supported in the tank by a lever arm mechanism. This, in turn, is attached to and operates a system of bellows. Another system of bellows is contained in the indicating device placed within the building. The two units are connected together by tubing filled with a liquid. As the depth of liquid in the tank changes, the float moves and thus operates the bellows at the tank unit. This motion is transmitted by the liquid in the tubing and thus positions the bellows at the indicating device which in turn operates the indicating pointer. When selecting this type of gage, the inside diameter of the tank and the length of the transmitting tubing must be known.

Tank gages should be installed as close to the center of the tank as possible to minimize the effect of uneven installation or uneven settling of the storage tank. A reinforcer pipe, one inch or larger, with a double tapped bushing, should be used with each hydrostatic gage tank assembly to insure that the brass air chamber is resting on the tank bottom. The reinforcer pipe should extend down to approximately 4 inches from the tank bottom. The transmitting tubing or conduit should be run in a protective pipe which should extend from the storage tank, underground, and through the building foundation. Protective piping diameters will range from 1 to 2 inches, depending upon the type of gage used.

Hydrostatic and mechanical gages offer a reasonably accurate means of determining a tank's contents. When properly installed they will give an accuracy within 1%. However, there are conditions beyond the accuracy of the gage that will result in very erroneous readings.

Fuel oil storage tanks assume a slight change when full from that when they are empty. The sides and ends of rectangular tanks buckle outward, thus increasing the tanks actual capacity. While this is also true for the ends of horizontal cylindrical tanks, the change in capacity is less affected. Errors up to 3 or 4% can result from this condition. Underground tanks are generally not set level to start with, and, in addition, one end of the tank will probably settle more than the other after the installation is made. In addition, a tolerance of 10% in capacity for tanks over 1,100 gallons is allowed at the time of fabrication. It is, therefore, evident that absolute accuracy in tank measurement is not to be expected.

Another type of indicating device that has been used successfully is installed completely independent of the storage tank. The instrument is actuated by the plus or minus pressure of a column of oil on a large metallic bellows, the movement of which is transmitted to an indicator needle. Oil level in the tank is indicated in feet on a large dial. Installation cost of this type of gage is practically nothing. A tee is inserted in the fuel suction line and the gage is screwed into it. No other connections are necessary. Before selecting this type of gage, one should know the specific gravity of the fuel oil, the depth of the storage tank, and vertical distance that the gage will be located above or below the storage tank.

The foregoing discussion was prepared by Everett F. Opperman.

GAGING CHARTS FOR CYLINDRICAL TANKS

The accompanying charts were designed by the late Sybren R. Tymstra, University of Washington, for determining the contents of partially filled horizontal, spherical and vertical tanks. The charts are reproduced by permission of the Engineering Experiment Station, University of Washington.

The charts cover tanks with the types of heads shown in Fig. 1. Fig. 2, from the known information, yields Factors A and B. By using Factor A in Fig. 3, the contents of the heads are found; similarly, using Factor B in Fig. 4 gives the contents of the cylinder itself. Adding the two results gives the total contents.

Horizontal Tanks

The charts cover horizontal tanks from 5 ft to 10 ft in diameter and up to 20 ft long, and with any depth of contents from empty to full.

Smaller horizontal tanks are gaged by assuming that their dimensions and their depth of contents are doubled. The actual volume of contents will be one-eighth that of the chart reading.

Larger horizontal tanks are gaged by assuming that their dimensions and their depth of contents are halved. The actual volume of contents will be eight times that of the chart reading.

Horizontal tanks whose diameters fall within the range of the chart but whose lengths are greater are gaged as follows: First, find the contents of the heads. Second, assuming the cylindrical

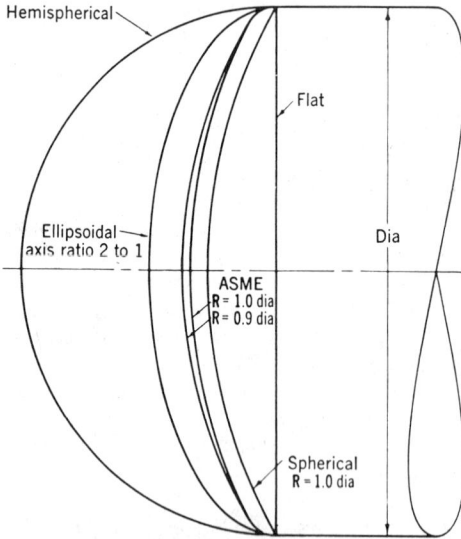

Fig. 1. Types of tank heads covered by the accompanying charts.

part to be only one-half, one-third, or one-quarter its actual length, find its contents. Double, triple, or quadruple this reading and add it to the contents of the heads.

Horizontal tanks, installed with a slight slope, can be gaged fairly accurately at mid-length unless they are nearly empty, or nearly full.

Example. A horizontal cylindrical tank with ellipsoidal heads of 2 to 1 axis ratio is 9 ft in diameter, 16 ft long, and has a depth of contents of 62½ inches. What is the volume of contents?

Contents of Heads—From the 62½-inch depth-of-contents point at left of Fig. 2, follow dashed line horizontally to the 9-ft diameter line, then vertically to the ellipsoidal-head curve, then horizontally. Read Factor $A = 831$.

Enter Factor A at left in Fig. 3 and go horizontally to the 9-ft diameter line, then down. Read contents of both heads = 880 gal.

Contents of Cylinder—From the 62½-inch depth-of-contents point at the left of Fig. 2, follow dashed line horizontally to the 9-ft diameter line, then down to read Factor $B = 422$.

Enter Factor B at top of Fig. 4, and go downward to cylinder line, then horizontally to 9-ft diameter line, upward to 16-ft length-of-cylinder line, and finally horizontally to the right to read contents of cylinder = 4560 gal.

Contents of Tank—The volume of the contents of the tank is the sum of contents of cylinder and heads, or 880 + 4560 = 5440 gal.

A similar tank, 4½ ft in diameter and 8 ft long, having a depth of contents of 31¼ inches, would have a volume of contents of 5440 ÷ 8 = 680 gal.

A similar tank 18 ft in diameter and 32 ft long, having a depth of contents of 125 inches, would have a volume of contents of 5440 × 8 = 43520 gal.

A similar tank 9 ft in diameter and 64 ft long, having a depth of contents of 62½ inches, would have a volume of contents of 880 + 4(4560) = 19,120 gal.

The contents of nearly all ASME heads can be gaged by means of the two curves representing this type of head, either directly or by interpolation.

Spherical Tanks

Spherical tanks are equivalent to two hemispherical heads and therefore can be gaged either directly or by size proportioning.

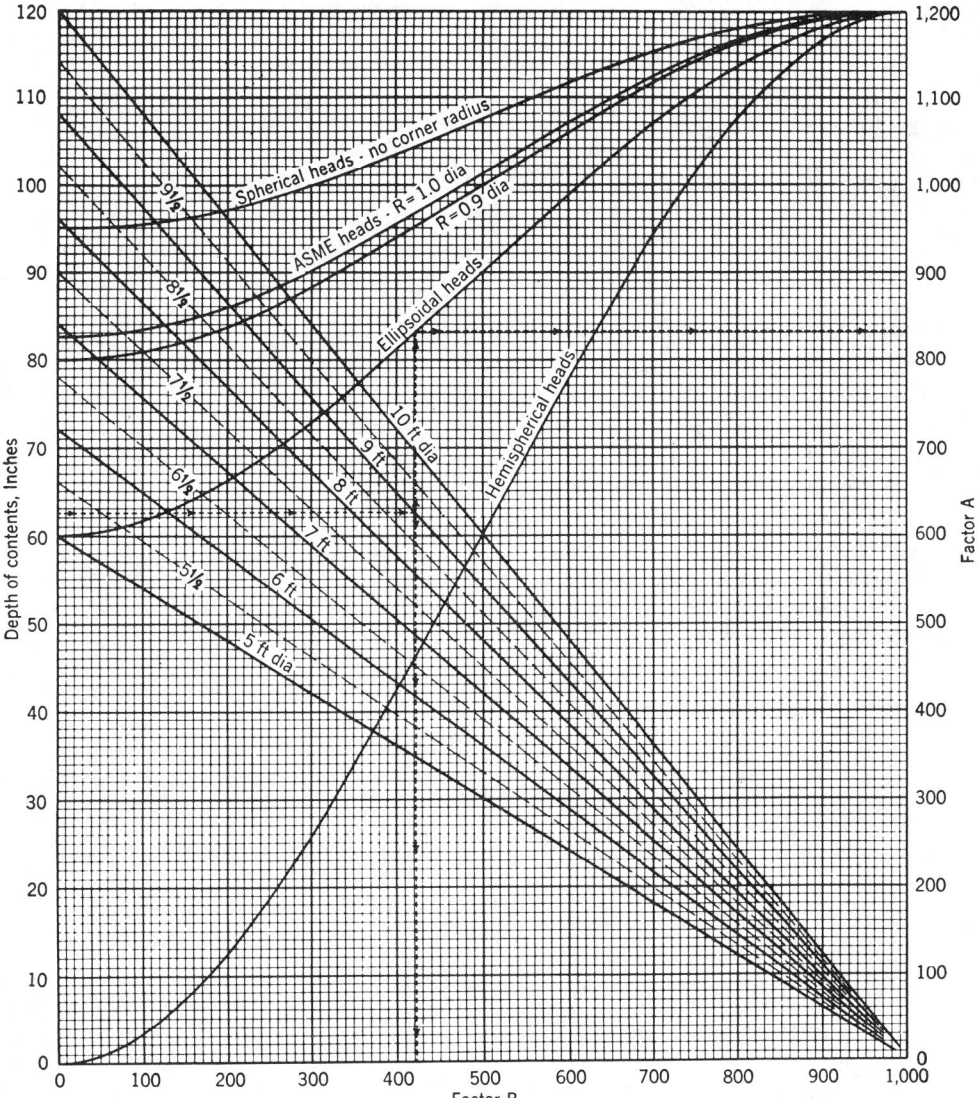

Fig. 2. Knowing the tank diameter, length, type of head, and depth of contents, this chart gives Factors A and B.

Example. A spherical tank 36 ft in diameter, having a depth of contents of 250 inches, would be gaged as follows: Consider a tank one-quarter (½ × ½) the size, i.e., with 9-ft diameter and depth of contents of 62½ inches. Entering Fig. 2, as before, and going horizontally to the 9-ft diameter line, then down to the hemispherical heads curve and horizontally to the right, read Factor *A* = 460. Entering Factor *A* at the left of Fig. 3, and proceeding horizontally to the 9-ft diameter line, then down, read 1760 for contents of both heads in gallons. Multiply the answer by the number eight, twice (because tank dimensions were halved (twice) to get 112,600 gal.).

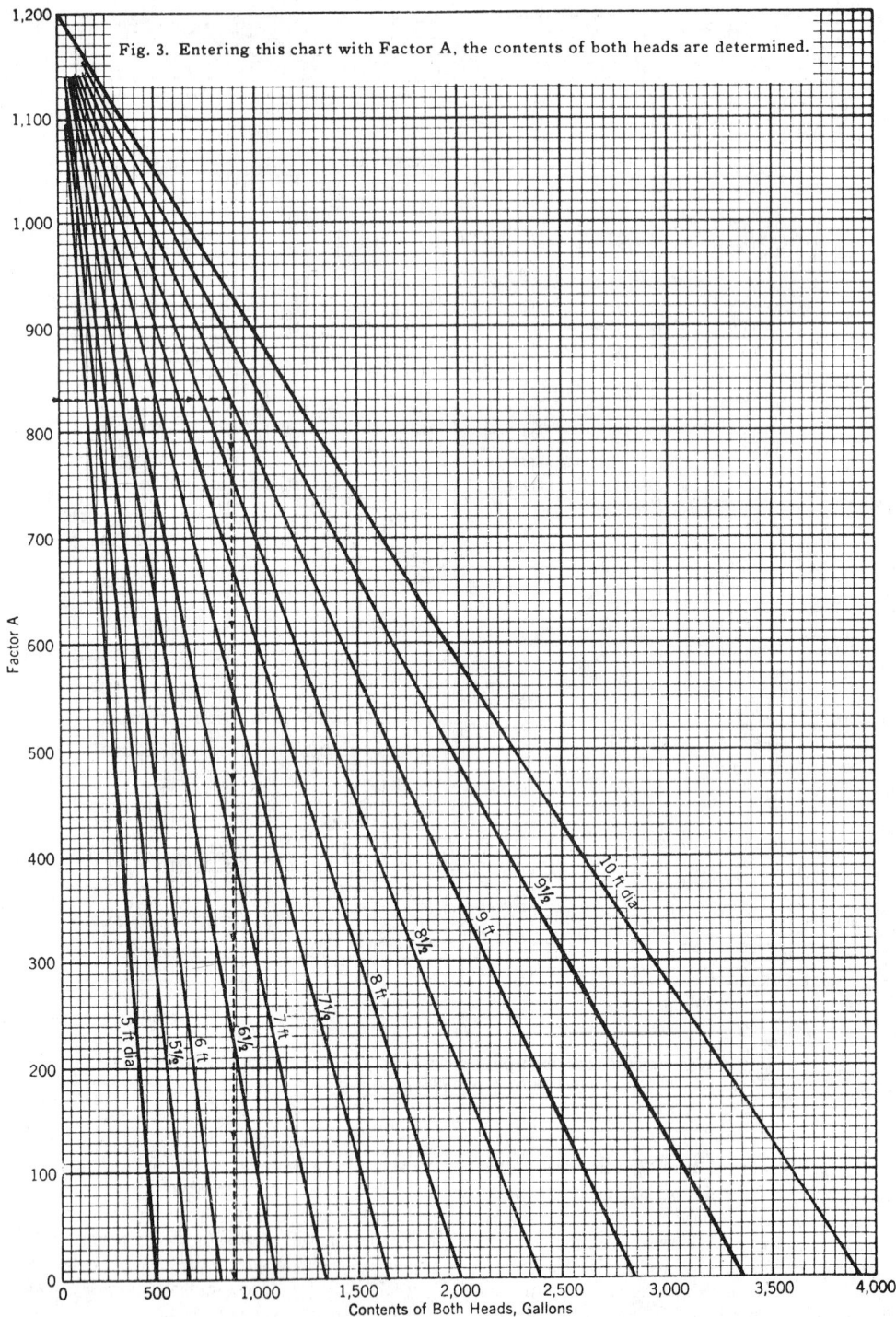

Fig. 3. Entering this chart with Factor A, the contents of both heads are determined.

Factor A (vertical axis)

Contents of Both Heads, Gallons (horizontal axis)

5 ft dia · 5½ · 6 ft · 6½ · 7 ft · 7½ · 8 ft · 8½ · 9 ft · 9½ · 10 ft dia

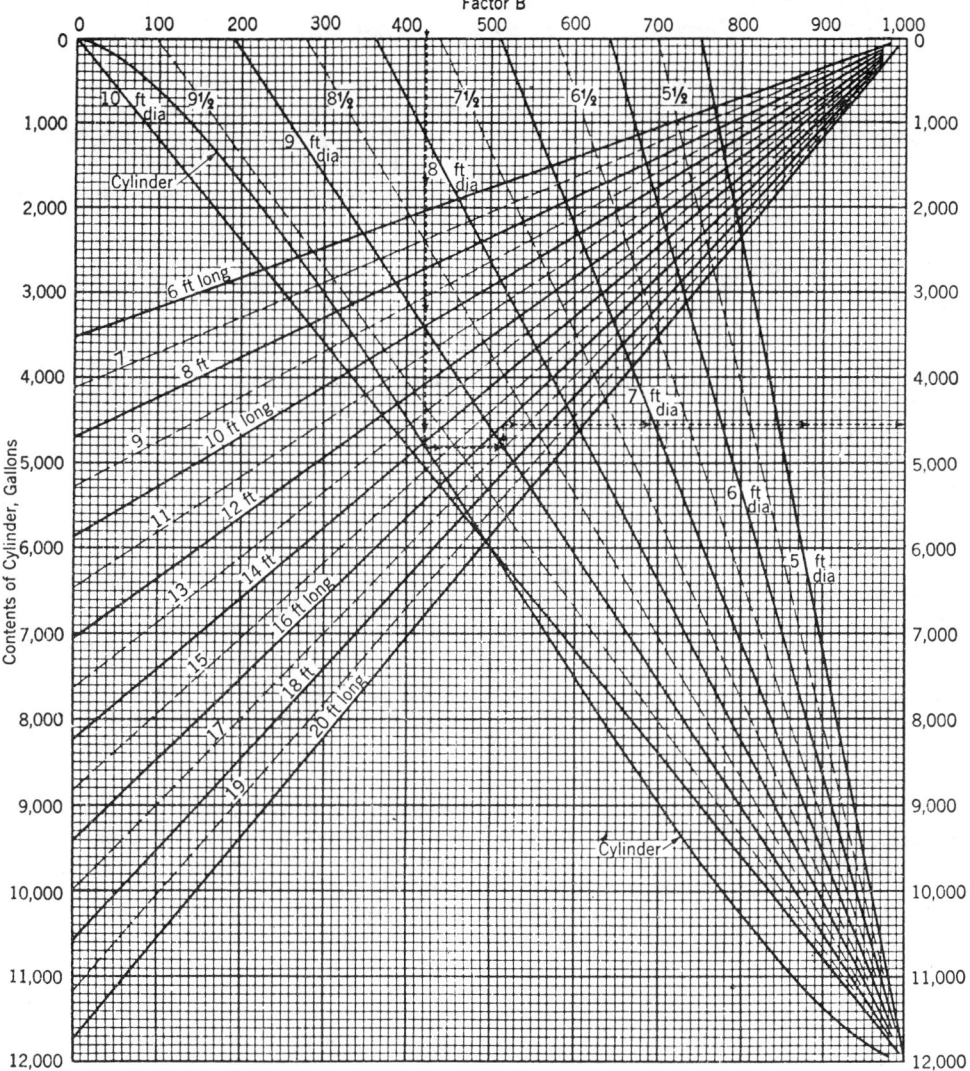

Fig. 4. Entering this chart with Factor B, the contents of the cylinder are determined.

Vertical Tanks

Vertical tanks with flat bottoms are equivalent to full cylindrical tanks of proper length and are therefore easily gaged.

Hemispherical and spherical bottoms of vertical tanks are gaged by considering them as partly filled spheres. (Radius of hemispherical bottom equals radius of tank; radius of spherical bottom equals diameter of tank.)

ASME bottoms of vertical tanks are gaged by considering them as spherical ones (with radius of dish equal to the tank diameter or 0.9 thereof) within 10% of their total depth. The upper 10% of the bottom is then considered an extension of the cylinder.

Ellipsoidal bottoms (with axis ratio of 2 to 1) of vertical tanks are gaged by doubling their depth of contents and then taking half the corresponding volume of contents of a hemispherical bottom.

VENTING GAS-FIRED EQUIPMENT

Venting is the removal of combustion products of flue gases to the outdoors through a system of piping, ducts, flues, vents, chimneys, or stacks especially designed for that purpose. Gas equipment designed for use with *draft hoods* are intended to operate without additional chimney draft. Therefore, the functions of a draft hood are: (1) to break chimney draft by introducing dilution air; and (2) to provide for safe combustion in the event of downdraft or blocked flue. Other gas equipment requires chimney draft for its proper functioning. In such cases, *barometric dampers* are used to control draft.

Among the methods for venting gas equipment are: (1) site-constructed chimneys of masonry, reinforced concrete, or metal; (2) prefabricated chimneys, UL-listed for use where flue gases do not exceed 1000 F continuously nor 1400 F except for infrequent brief periods; (3) UL listed Type *B* gas vents, consisting of noncombustible, corrosion-resistant piping, for flue gases up to 550 F; (4) venting through louver or roof venting, as, for example, of radiant heaters in industrial plants mounted 15-30 ft above the area heated; and (5) power venting.

For details on dimensions and installation of draft hoods, refer to National Fuel Gas Code, ANSI Z223.1.

Barometric dampers are frequently used on units having input rates over 400,000 Btuh in a single combustion chamber. Draft control is provided through a balanced butterfly gate located outside the flue gas flow on a tee fitting. It may control updraft only, or control updraft and relieve downdraft. Protection against flue stoppage or downdraft may be obtained through a thermal switch mounted where hot downdraft gases cause it to actuate a burner shutoff.

Natural-Draft Appliances

Table 3 provides vent design capacities for single appliances with draft hoods when connected to various vertical flues.

Recommended minimum internal masonry chimney diameter is one inch larger than the vent connector diameter.

A vent pipe diameter equal to, or slightly larger than, the draft hood outlet diameter should satisfactorily vent single gas appliances (excluding masonry chimneys). Insulated, factory-built chimneys

may be sized according to double metal pipe specifications, assuming use of a double-wall vent connector. If single-wall connector is used, or if double-wall vent pipe is used with a single-wall connector, the same specifications may be used if permissible lateral length is reduced 50%.

A semi-empirical equation for calculating vent capacity has been derived from more exact equations. For typical approved gas appliances, and for vents and stacks of 6 inches and larger, it can be simplified to the following form which may be used when tabulated sizing data does not cover a particular case:

$$I = 4650(H/R)^{0.5}[A - (0.031)(U)^{1.5}(H + 4L)] \qquad (1)$$

where

I = sea level input, Btuh,
H = total height, ft,
R = flow resistance, velocity heads, Table 1,
A = area of passageway, sq in.,
U = heat transfer coefficient, Btu per (hr) (sq ft) (deg F), Table 2,
L = lateral or connector length, ft. (Note: The importance of the lateral length is shown by $4L$.)

Example: Find the venting capacity of a steel stack 40 inches in diameter and 60 ft high, with a 10-ft breeching serving a large, efficient gas designed boiler.

From Table 1,

Acceleration head	1.0
Draft hood	1.0
Tee in breeching	1.4
70 ft piping, $0.4(H+L)/D$	0.7
$R =$	4.1

From Table 2, $U = 1.3$
Substituting in Equation (1)

$$I = (4650)(60/4.1)^{0.5}$$
$$[1256 - (0.031)(1.3)^{1.5}(60 + 40)]$$
$$= (4650)(3.82)(1256 - 4.6)$$
$$= 22,200,000 \text{ Btuh input.}$$

Note that heat losses reduce capacity by less than 0.5%, even with an uninsulated stack. Had barometric control been used, capacity shown above would be slightly conservative, depending on draft setting.

Table 1. Vent and Chimney Resistances

Item or Fitting	Design Resistance, Velocity Heads
Acceleration head	1.0
Draft hood	1.0
Elbow, 90°	0.75
Elbow, 45°	0.3
Round tee, 90° in individual vent, or chimney breeching	1.4
Vent caps, tops and cowls*	0.5
Round-to-oval tee fitting	2.5
Piping losses, metal and asbestos-cement. $(L + H) =$ total vertical and lateral piping, ft. D, in inches, up to 24	$0.5 \ (L + H)/D$
D greater than 24 inches	$0.4 \ (L + H)/D$
Piping losses, clay tile masonry and block, L, H, D as above	$0.8 \ (L + H)/D$
Increaser, $A_1 =$ inlet area, $A_2 =$ outlet area, velocity head at A_2	$\left(\dfrac{A_2 - A_1}{A_1}\right)^2$

* Use of tops with resistance over 1.0 velocity head can seriously reduce vent capacity below tabulated values. Caps can be designed with diffuser sections to produce negative resistance or recovery of the acceleration head.

Input for zero draft, with barometric draft control, may be found for the given example by reducing total flow resistance by one velocity head. Then I would be 25,600,000 Btuh.

Multiple Vents. A multiple vent serves two or more units on a single floor through a common vent system. Operation of only one, usually the smallest, unit produces the most critical phase of performance, for dilution air enters the draft hoods of the inoperative unit, lowering flue gas temperature.

Greater vent connector rise is the most effective method for increasing vent capacity. See Table 4. (Also refer to Figs. 1 and 2 for a fuller understanding of terms used in the venting tables.)

Short vent connector laterals are desirable and the common vent height should be greater than the vent connector lateral. Flow resistance of the laterals may also be compensated for by increasing vent connector diameter, and/or somewhat by the total height of the combined vent. Tables 4, 5 and 6 contain recommended sizes for vent connectors and common vertical vents. Regardless of results

Table 2. Chimney and Vent Heat Transfer Coefficients

Material	Coefficient*	Remarks
Industrial steel stacks	1.3	Wet wind outdoors
Clay or iron sewer pipe	1.3	Used as single-wall material
Asbestos cement gas vent	1.2	Tested per UL Std. 441
Black or painted steel stove pipe	1.2	Comparable to weathered galvanized steel
Single-wall galv. steel	1.0	Varies with surface and exposure
Single-wall pure unpainted aluminum	1.0	1100 or other bright surface aluminum alloy
Brick chimney, tile lined	1.0	Residental construction as used for gas appliances
Double-wall gas vent, 1/4-inch air space	0.6	Galv. steel outer, pure aluminum inner pipe, tested per UL Std. 441
Double-wall gas vent, 1/2-inch air space	0.5	Same
Insulated prefab chimney	0.3	Solid insulation tested per UL Std. 103

* Design coefficient, U Btu per (hr)(sq ft)(deg F) referred to inner surface area for use in Equation (1).

obtained from these tables, the common vent must be at least as large as the largest connector.

Since elbows are a principal source of flow resistance in vent connectors, an equal number of turns in each such connector is important in the interest of balanced connectors, particularly with the larger unit of a multiple venting system. In judging balance, a draft hood with side outlet is considered equivalent to one with a top outlet and a 90° elbow.

Flow resistance of the fittings, and lower common-vent flue-gas temperatures, offset any advantages of multiple-appliance vent systems with a common vent height (not total height) under 5 ft. For heights less than 5 ft, single-appliance vents perform better.

Table 3. Capacity of Vents for Single Appliances

Total Height, H in Feet	Total Lateral Length, L in Feet	Double-Wall Metal Vent Pipe			Single-Wall or Asbestos-Cement			Masonry Chimneys	
		Vent Diameter, D, Inches[1]							
		10	12	24	10	12	24	10	12
		Maximum Allowable (Appliance Rated) Input, 1000 Btuh							
6	0	540	790	3520	500	750	—	—	—
	2	450	660	2670	415	620	—	400	580
	5	442	650	2655	390	600	—	375	560
8	0	610	920	4010	542	815	—	—	—
	2	510	770	3050	451	680	—	445	650
	5	504	760	3040	430	648	—	422	638
	10	480	732	3020	406	625	—	400	598
10	0	695	1030	4450	606	912	—	—	—
	2	580	859	3390	505	760	—	490	722
	5	572	846	3365	480	724	—	465	710
	10	545	816	3340	455	700	—	441	665
	15	534	792	3295	432	666	—	421	634
15	0	800	1220	5300	684	1040	4910	—	—
	2	665	1020	4080	570	865	4100	560	840
	5	655	1010	4056	540	825	—	531	825
	10	625	970	4024	514	795	3960	504	774
	15	602	940	3980	488	760	3850	481	738
	20	585	920	3935	466	726	—	459	706
20	0	900	1370	6000	770	1190	5750	—	—
	2	750	1140	4700	641	990	4800	625	950
	5	740	1125	4650	610	945	—	594	930
	10	705	1090	4600	571	910	4630	562	875
	15	678	1050	4575	550	870	—	536	835
	20	660	1030	4550	525	832	4430	510	800
30	0	1030	1570	7060	878	1370	6850	—	—
	2	860	1310	5600	730	1140	5700	715	1110
	5	848	1290	5550	694	1080	—	680	1090
	10	808	1250	5470	656	1050	5500	644	1020
	15	778	1210	5390	625	1000	—	615	975
	20	756	1180	5310	596	960	5260	585	932
	30	705	1110	5250	540	890	4950	544	865
50	0	1140	1750	8540	980	1550	7800	—	—
	2	955	1460	6950	820	1290	6500	810	1240
	5	942	1440	6870	780	1230	—	770	1220
	10	895	1400	6850	730	1190	6260	728	1140
	15	865	1350	6750	705	1130	—	695	1090
	20	841	1310	6700	670*	1080	6000	660	1040
	30	783	1240	6550	605*	1010	5850	610	970

[1] Capacities at other diameters may be estimated by assuming that they are proportional to the vent cross-sections, or D^2. For example, when $D=16$ (and $H=20$, $L=0$) capacity for double-wall vent is approximately $(16/24)^2 \times 6000 \simeq 2660$.

Data in table for 10 and 12-inch diameter vents are extracted from AGA Laboratories Research Report No. 1319.

Data for 24-inch double-wall metal vent pipe are interpolated values from gas vent tables of William Wallace Co. These particular data may also be used for masonry chimneys with 24-inch size vent connectors. For smaller sizes, reduce by $(22 - D)\%$. Masonry chimneys should be one size larger than vent connector.

Data for 24-inch single-wall or asbestos-cement vents are from unpublished AGA Laboratories investigation (1962). Work is being continued to confirm the results.

* Possibility of continuous condensation. Consult local utility and /or local codes.

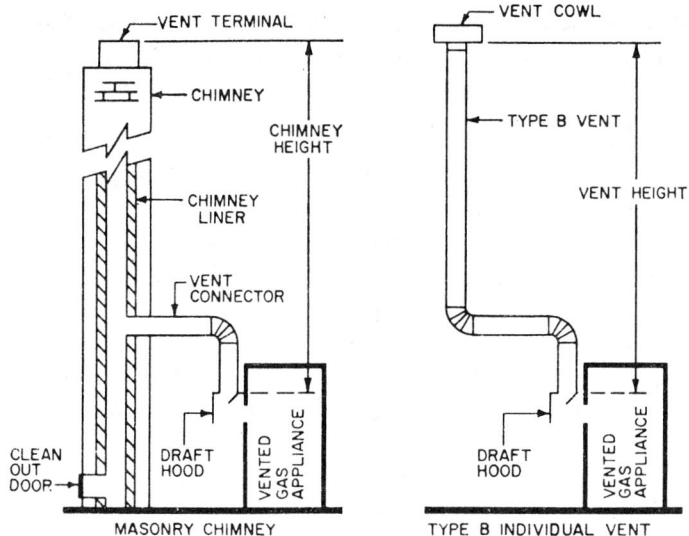

Fig. 1. Basic components of an individual natural draft vent system.

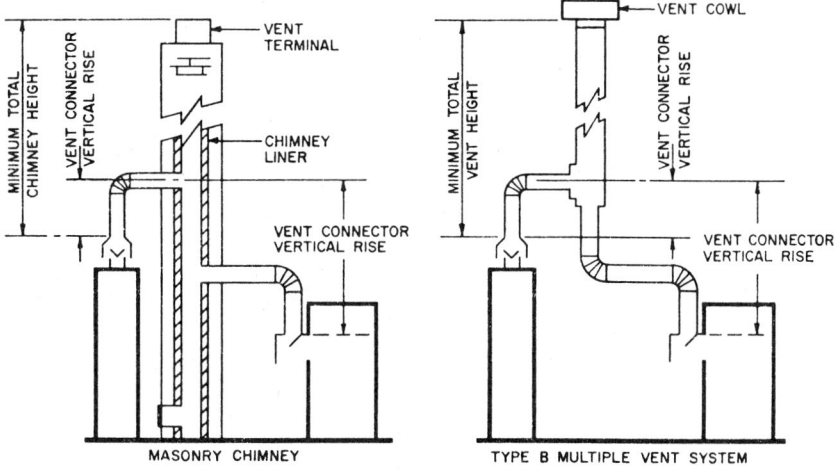

Fig. 2. Basic components of multiple vent system.

Vent Manifolds. Two or more units on the same level may be vented into a common horizontal duct or manifold which is joined to a common vertical vent. Constant-diameter and tapered manifolds perform about the same and may be sized using Table 4. Tapered manifolds of such pipe may be sized for total input to each portion also using capacities in Table 4. If manifolds must slope, care should be taken to allow sufficient rise in closed motor and the squirrel-cage the vent connector to the unit farthest from the common vent; otherwise, severe draft-hood spillage at this appliance may occur.

Table 4. Vent Connector Capacity, Double-Wall Metal Pipe

For single-wall metal or asbestos-cement pipe, reduce maximum heat inputs by 10%.

Draft Hood Exit Opening Diameter, Inches	Vent Connector Configuration		Common Vent Diameter, Inches									
			Up to 12				14 to 18			20 to 24		
	Vertical Rise, Feet	Diameter, Inches	Minimum Total Vent Height, Feet									
			6-8	15	30	50	15	30	50	15	30	50
			Maximum Allowable Heat Input to Vent Connector, Thousands of Btu Per Hour									
3	E	3	NR	NR	NR	NR	NR	NR	NR	NR	NR	NR
	1		22	24	26	27	20	20	20	20	20	20
	2		29	31	32	33	25	25	30	25	25	25
	3		34	35	37	38	30	30	35	30	30	30
	E	4	28	32	37	40	30	30	30	30	30	30
	1		33	38	43	45	35	35	35	35	35	35
	2		45	47	50	52	40	45	45	40	45	45
	3		50	54	57	59	50	55	55	50	50	55
4	E	4	34	40	46	48	35	35	35	35	35	35
	1		42	47	52	54	40	45	45	40	40	45
	2		54	57	61	63	50	55	55	50	55	55
	3		61	65	69	71	60	65	65	60	60	65
	E	5	53	62	72	77	55	60	60	50	55	60
	1		62	70	79	83	60	65	65	55	60	65
	2		78	83	90	93	75	80	80	70	75	80
	3		89	93	98	103	90	90	90	90	90	90
5	E	5	58	69	80	86	60	65	70	55	60	65
	1		70	80	88	92	65	70	75	65	70	75
	2		88	94	101	105	85	90	90	80	85	90
	3		100	105	110	115	100	100	100	100	100	100
	1	6	97	111	125	131	95	100	110	90	95	100
	2		121	131	142	148	120	120	130	115	120	125
	3		141	147	157	163	135	140	145	130	135	140
6	1	6	106	121	136	143	105	110	120	100	105	110
	2		132	142	155	161	130	130	140	125	130	135
	3		153	160	171	177	145	150	155	140	145	150
	1	7	142	165	189	200	135	150	170	130	140	150
	2		172	194	213	227	165	180	195	160	170	180
	3		197	214	234	248	195	200	215	185	190	200
7	1	7	150	175	199	212	145	160	180	140	150	160
	2		182	205	225	240	175	190	205	170	180	190
	3		209	226	248	262	205	210	230	195	200	210
8	1	8	210	247	282	310	205	220	245	190	205	220
	2		242	275	310	327	245	260	280	230	245	260
	3		283	310	345	365	270	285	305	260	270	285
10	1	10	300	350	400	500	300	350	400	250	300	350
	2		350	400	450	550	350	400	450	350	350	400
	3		400	450	500	600	400	450	500	400	400	450

See footnotes at end of table, next page.

Table 4. Vent Connector Capacity, Double-Wall Metal Pipe (Cont'd)

For single-wall metal or asbestos-cement pipe, reduce maximum heat inputs by 10%.

Draft Hood Exit Opening Diameter, Inches	Vertical Rise, Feet	Diameter, Inches	Up to 12				14 to 18			20 to 24		
			6-8	15	30	50	15	30	50	15	30	50
			Maximum Allowable Heat Input to Vent Connector, Thousands of Btu Per Hour									
12	1	12	400	500	650	750	450	550	650	400	450	550
	2		500	600	700	800	550	600	700	500	550	600
	3		600	650	750	850	600	700	750	550	600	650
14	1	14					650	800	950	550	700	800
	2						750	900	1050	700	800	900
	3						850	1000	1100	800	850	950
16	1	16					850	1050	1200	800	950	1100
	2						1000	1150	1300	950	1050	1200
	3						1100	1250	1400	1050	1150	1300
18	1	18					1100	1450	1650	1050	1350	1550
	2						1300	1550	1750	1250	1450	1650
	3						1400	1650	1900	1350	1550	1750
20	1	20								1300	1700	2000
	2									1600	1800	2200
	3									1700	2000	2300
22	1	22								1600	2100	2500
	2									1900	2300	2800
	3									2100	2500	2900
	1									2000	2600	3100
24	2	24								2300	2800	3300
	3									2600	3000	3500

Common vent diameter must be at least as large as the largest vent connector.

NR—not recommended

E Indicates vertical rise curves for draft hoods with horizontal outlets and no additional rise, or vertical outlets with a 90° elbow and no additional rise.

Power Venting

Many special problems are solved by power venting. Such problems usually involve gas appliances with draft hoods, designed to operate under natural draft, or those designed to operate at some minimum draft provided through barometric draft controls. Conditions responsible for venting difficulties include presence of adjacent high buildings which produce erratic draft performance, absence of chimneys sufficiently high to produce adequate draft, and use of exhaust fans creating negative pressure around the vented appliance.

Fans found acceptable for power venting include the mushroom type wall or roof fan with scroll type housing with its impeller mounted on the motor shaft. The mushroom type, requiring inlet connections only, is installed outdoors at the point of discharge. The scroll housing type usually is installed indoors with its outlet connected to the discharge ductwork. A weather-type scroll type fan may be installed on the roof, if desired.

The fan selected should have the capacity to handle both combustion products and dilution air. The following capacity formula has been found

Table 5. Common Vent Capacity, Double-Wall Pipe

Common Vent Diameter, Inches	Minimum Total Vent Height, Measured From Draft Hood Exit Opening, Feet							
	6	8	10	15	20	30	50	80
	Maximum Allowable Heat Input, 1,000 Btuh (combined appliance rated input)							
4	52	60	65	76	86	97	——	——
5	84	95	104	122	140	158	185	——
6	120	136	151	178	204	237	275	——
7	165	190	207	246	284	328	395	——
8	220	250	272	325	380	442	550	——
10	345	395	440	520	605	705	890	——
12	NR	570	640	760	890	1060	1320	——
14	NR	750	850	1000	1200	1400	1750	2100
16	NR	1000	1100	1350	1500	1800	2250	2700
18	NR	NR	1350	1650	1850	2250	2800	3400
20	NR	NR	1600	2000	2250	2650	3400	4050
22	NR	NR	NR	2450	2700	3250	4050	5000
24	NR	NR	NR	2850	3200	3900	4950	6000

NR—not recommended

sufficiently accurate:

$$Q = 0.02 \, ACDT \qquad (2)$$

where

Q = fan capacity against 0.375 in. w.g. external static pressure, standard cfm,

A = required combustion air, 10 cu ft per 1000 Btu gas,

C = gas input, cu ft per hr,

D = dilution factor = 2 with draft hood, otherwise 1.25 to 1.5, and

T = factor for correcting vent gas volume to temperature condition at fan inlet, normally 1.25.

Substituting, for draft hood installations,

$$Q = 0.5C$$

and for other installations,

$$Q = 0.3C \text{ to } 0.4C$$

For installations with vent runs longer than 10 ft and for duct manifolds venting several heaters through one fan, double the capacity obtained from the formulae above.

Ducts for power venting should be designed similarly to those for handling heating air. Runs from more than one appliance should be designed for equal pressure loss in each section between appliances, according to the quantity of diluted products handled as given by Equation (2). Total loss should not exceed 0.1 in. w.c.

Draft on each appliance should be regulated to prevent excess draft. For draft-hood equipped appliances, a single-acting barometric control may be installed at least one foot downstream from the draft hood. Draft is satisfactory when match flame, held around various points of relief opening, is drawn inward. For appliances requiring a draft, a double-acting barometric control is sufficient. Adjustment for proper combustion may be made with a flue-gas analyzer.

Fig. 3 is a diagram of a power vent system and its wiring. Diaphragm operated switch, sensitive to 0.05-in. w.c. draft, shuts off gas if exhaust system fails.

Vent Installation

Standards for venting gas appliances are presented in National Fuel Gas Code, ANSI Z223.1. Included are major installation features as proper sizing, required clearances to combustible materials, and means of interconnection when serving gas multiple appliances, or a gas appliance and one using other fuel.

Vents for Large Appliances. Draft requirements for burner and boiler assemblies vary widely, in some cases reaching 0.40 in. In order to establish

Table 6. Common Vent Capacity for Single-Wall or Asbestos-Cement Pipe

Common Vent Diameter, Inches	Minimum Total Vent Height, Measured From Draft Hood Exit Opening, Feet							
	6	8	10	15	20	30	50	80
	Maximum Allowable Heat Input, 1,000 Btuh (combined appliance rated input)							
4	48	55	59	71	80	NR	NR	——
5	78	89	95	115	129	147	NR	——
6	111	128	136	168	186	215	NR	——
7	155	175	190	228	260	300	306	——
8	205	234	250	305	340	400	490	——
10	320	365	395	480	550	650	810	——
12	NR	505	560	690	790	940	1190	——
14	NR	700	800	950	1100	1300	1600	1950
16	NR	950	1050	1250	1400	1700	2150	2600
18	NR	NR	1300	1600	1800	2200	2700	3300
20	NR	NR	1550	1950	2200	2600	3350	4000
22	NR	NR	NR	2450	2700	3250	4050	5000
24	NR	NR	NR	2850	3200	3900	4950	6000

NR—not recommended

Table 7. Potential Static Draft in Venting System

Height of Stack Above Level of Draft Measurement, Feet	Average Temperature Rise Above Ambient (60 deg F) of Gases in Vertical Part of Chimney, deg F							
	150	200	300	400	500	600	750	1000
	Theoretical or Potential Static Draft, Inches of Water Column							
5	0.017	0.020	0.027	0.032	0.036	0.039	0.043	0.049
10	0.033	0.041	0.054	0.064	0.072	0.079	0.087	0.097
15	0.049	0.061	0.081	0.096	0.108	0.118	0.130	0.145
20	0.066	0.082	0.11	0.13	0.14	0.16	0.17	0.19
25	0.083	0.10	0.13	0.16	0.18	0.20	0.22	0.24
30	0.10	0.12	0.16	0.19	0.22	0.24	0.26	0.29
40	0.13	0.16	0.22	0.26	0.29	0.32	0.35	0.39
60	0.20	0.25	0.32	0.38	0.43	0.47	0.52	0.58
80	0.26	0.33	0.44	0.51	0.58	0.63	0.70	0.78
100	0.33	0.41	0.54	0.64	0.72	0.79	0.87	0.97
150	0.50	0.61	0.81	0.96	1.08	1.18	1.30	1.45
200	0.66	0.82	1.07	1.28	1.44	1.58	1.74	1.94
300	0.99	1.23	1.61	1.92	2.16	2.37	2.61	2.91

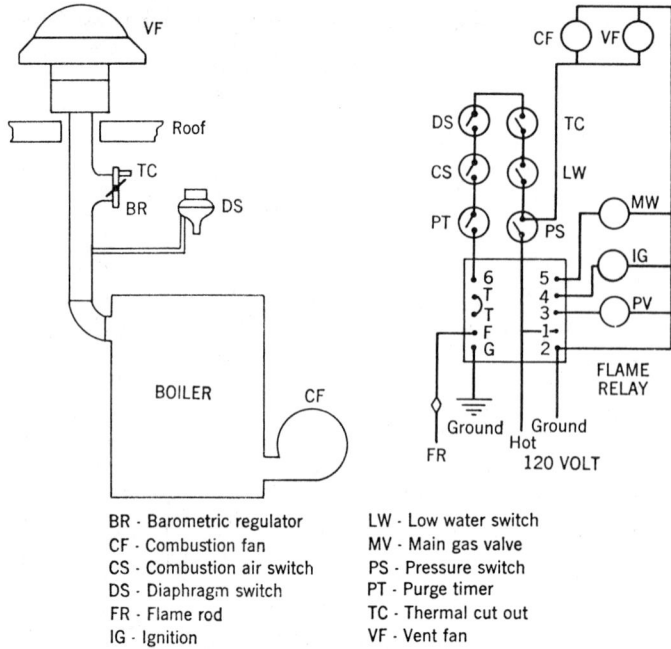

BR - Barometric regulator
CF - Combustion fan
CS - Combustion air switch
DS - Diaphragm switch
FR - Flame rod
IG - Ignition

LW - Low water switch
MV - Main gas valve
PS - Pressure switch
PT - Purge timer
TC - Thermal cut out
VF - Vent fan

Fig. 3. Power vent system and wiring diagram for gas boiler with power burner.

the criteria for proper venting, the boiler burner draft requirements should be known. Then these values are deducted from the theoretical draft available, as shown later.

Vent design for gas equipment needing draft may be separated into a two-step procedure:

1. From required draft, based on expected flue gas temperature and such equipment characteristics as efficiency, over-fire static pressure, firing rate, steam pressure or temperature, etc., choose from Table 7 a stack height to yield a theoretical draft sufficiently large to take care of required draft and the vent system friction losses. The difference between the theoretical and required draft is the available draft.

Such other considerations as building height, fume and gas dispersal, location of adjacent structures, etc., may require greater height than that theoretically called for. A look at Tables 4, 5, and 6 will show maximum vent capacities available under most favorable conditions for appliances requiring no draft. Thus, other appliances need larger or higher vent systems.

2. From vent operating conditions, choose an input loading at which the sum of the static draft pressure losses plus the available draft is less than the theoretical draft.

Tables 7 and 9 are used as follows to calculate height and size of vents for equipment requiring draft:

Start with Table 7 to obtain theoretical draft solely as a function of height and gas temperature. Next, find draft losses, using Table 9. The vent is

Table 8. Area to Diameter Conversion

(from area in square inches to round vent diameter in feet: $A = 113D^2$)

Area, A, Sq. in.	Dia. D, Feet	Area, A, Sq. in.	Dia. D, Feet
50	0.67	4,070	6
113	1	5,540	7
254	1.5	7,230	8
453	2	9,250	9
707	2.5	11,300	10
1,020	3	25,400	15
1,800	4	45,300	20
2,820	5		

Table 9. Static Draft Pressure Losses in Venting System with 8% CO_2 in Flue Gases

(applicable only if there is no draft hood; a barometric draft control may be used)

Average Temperature Rise Above Ambient of Gases in Vertical Part of Stack, Deg F Above 60	Input Loading: Thousands of Btuh per Sq In. of Vent or Chimney Area, I/A	L/D; System Length to Diameter Ratio, Feet of Connector and Stack per Foot of Diameter— Vertical Only							
		45	50	60	70	80	100	120	140
		L/D; System Length to Diameter Ratio, Feet of Connector and Stack per Foot of Diameter— Connector Used							
		5	10	20	30	40	60	80	100
		Static Pressure Losses, Inches of Water Column							
150*	6	0.005	0.006	0.007	0.008	0.009	0.011	0.013	0.015
	8	0.010	0.011	0.012	0.014	0.016	0.020	0.024	0.028
	10	0.015	0.017	0.019	0.022	0.025	0.031	0.037	0.043
	15	0.034	0.037	0.044	0.050	0.057	0.070	0.083	0.096
	20	0.061	0.066	0.078	0.090	0.101	0.125	0.148	0.171
	25	0.094	0.104	0.122	0.140	0.159	0.195	0.232	0.270
	30	0.136	0.149	0.175	0.201	0.227	0.280	0.332	0.384
300	6	0.007	0.007	0.009	0.010	0.011	0.014	0.016	0.019
	8	0.012	0.013	0.015	0.018	0.020	0.025	0.029	0.034
	10	0.019	0.020	0.024	0.027	0.031	0.038	0.045	0.052
	15	0.042	0.046	0.054	0.062	0.070	0.086	0.102	0.118
	20	0.074	0.081	0.096	0.110	0.124	0.153	0.182	0.209
	25	0.116	0.127	0.150	0.171	0.195	0.239	0.284	0.328
	30	0.166	0.182	0.208	0.246	0.278	0.342	0.406	0.470
500*	6	0.008	0.009	0.011	0.012	0.014	0.017	0.020	0.023
	8	0.015	0.016	0.019	0.022	0.025	0.031	0.036	0.042
	10	0.023	0.025	0.030	0.034	0.039	0.047	0.056	0.065
	15	0.052	0.057	0.067	0.077	0.087	0.107	0.127	0.146
	20	0.092	0.101	0.119	0.137	0.155	0.190	0.226	0.260
	25	0.145	0.158	0.186	0.213	0.242	0.297	0.354	0.408
	30	0.215	0.235	0.277	0.317	0.360	0.443	0.526	0.610
750*	6	0.010	0.011	0.013	0.015	0.017	0.021	0.025	0.029
	8	0.018	0.020	0.024	0.027	0.030	0.038	0.045	0.052
	10	0.029	0.031	0.037	0.042	0.048	0.059	0.070	0.081
	15	0.064	0.070	0.083	0.095	0.118	0.133	0.158	0.182
	20	0.114	0.125	0.148	0.170	0.192	0.236	0.281	0.324
	25	0.180	0.196	0.231	0.265	0.300	0.370	0.440	0.508
	30	0.256	0.281	0.322	0.381	0.430	0.530	0.628	0.730
1000*	6	0.012	0.014	0.016	0.018	0.021	0.025	0.030	0.035
	8	0.022	0.024	0.028	0.033	0.037	0.045	0.054	0.062
	10	0.034	0.038	0.044	0.051	0.057	0.071	0.084	0.097
	15	0.077	0.084	0.099	0.114	0.129	0.159	0.189	0.212
	20	0.127	0.150	0.177	0.204	0.230	0.283	0.336	0.390
	25	0.215	0.235	0.277	0.317	0.360	0.443	0.526	0.610
	30	0.307	0.337	0.386	0.456	0.515	0.634	0.753	0.870

* Table was developed from 300 F data by use of Table 10 factors.

so designed that draft produced will be greater than that needed by a margin sufficient to insure proper operation under all flow conditions.

Design application of Tables 7 and 9 requires cut-and-try calculation procedures. With the help of some auxiliary tables, effects of operating variables can be estimated and vent height and size computed with minimum effort.

Terms and table headings are explained below:

Friction: This is represented by the ratio L/D where L is the total piping length of connector plus stack and D the diameter of the connector and stack assumed to be the same, both in feet. The greater the system length for a given diameter, the greater the energy losses. Values given for systems having a connector include effects of energy losses in a tee. Where connector and stack sizes differ, the smaller area is used to find D, and L/D.

Input and cross-sectional area: This is represented by the ratio I/A where I is gas input rate, Btuh, and A the area of connector, or stack, sq in. This ratio is frequently used as a basis for gas chimney design and serves to relate heat input to flow velocity. Table 8 converts area to diameter. Where connector and stack area differ, the smaller should be used.

Average gas temperature in vertical stack. Table 9 shows temperature rises of 150 to 1000 F in the left marginal blocks. Energy losses increase with temperature due to greater flow velocity. Precise work also requires compensating average temperature for cooling according to Table 10. For vent design, ambient temperature is chosen as 60 F. Boiler room temperature does not influence stack operation. If design is adequate for 60 F operation, any drop in outdoor temperature improves stack action because of greater difference between stack and ambient temperature. Choice of a lower operating temperature than actually expected yields a conservative design, with extra overload capacity. Boilers required to operate at full capacity in summer may require somewhat greater vent capacity to compensate for high ambient temperatures.

Gas composition or CO_2 concentration (dry basis): Flue gas flow velocity for a given heat input increases as the CO_2 decreases, thereby increasing the energy losses at higher amounts of excess air.

Table 10. Effect of Chimney Gas Temperature upon Relative Draft Losses

(Based on 300 F temperature rise above 60 F ambient)

Av. Temp Rise of Chimney Gases. F	Relative Draft Loss	Av. Temp Rise of Chimney Gases, F	Relative Draft Loss
50	.70	700	1.49
100	.76	800	1.61
200	.88	900	1.73
300	1.00	1000	1.85
400	1.12	1200	2.10
500	1.24	1500	2.46
600	1.37	2000	3.08

Cooling of gases in connector, breeching and vertical section may be estimated as:
1. Steel connector and stack—4% per 10 diameters.
2. Masonry chimney—1% per 10 diameters.

Table 11. Effect of CO_2 Concentration Upon Relative Draft Losses

CO_2 in Flue Gases. %	Relative Draft Loss	CO_2 in Flue Gases, %	Relative Draft Loss
1	53.0	7	1.26
2	13.6	8	1.00
3	6.27	9	0.803
4	3.58	10	0.667
5	2.35	11	0.563
6	1.63	12	0.491

Table 12. Altitude Correction Factors

(for theoretical draft and for draft losses)

Altitude, Feet	Multiple Draft and Losses by	Altitude. Feet	Multiple Draft and Losses by
Sea level	1.00	6000	0.81
2000	0.93	8000	0.75
4000	0.86	10000	0.70

Table 9, for 8%, represents a common operating condition required to produce draft; 8% CO_2 is a suitable basis for boilers. For operation at other CO_2 values, Table 11 can be used to correct the loss in Table 9 for 8% CO_2. For operation at other than sea level, consult Table 12.

SPACE HEATING

Contents

Authorities

Section Editor: *Arthur H. Fox*, Tippets-Abbett- McCarthy-Stratton

Robert E. Bolaffio, Sizing of Vertical Flash Tanks, Sizing of Steam Traps at Air Heating Coils

Alford G. Canor, Check List for Heating System Servicing

John F. Collins, Jr., Check List for Conservation of Fuel

John D. Constance, Steam-Supplied Unit Heater Performance

Richard S. Dawson, Sizing Hot Water Expansion Tanks, Boiler Output Known

Stuart D. Distelhorst, Industrial and Commercial Steam Requirements

William B. Foxhall, Flash Steam Calculations; Miscellaneous Heating and Cooling Media

George J. Jankauskas, Sizing Hot Water Expansion Tanks, System Water Capacity Known

Owen S. Lieberg, High Temperature Water Systems

C. George Segeler, Gas-Fired Radiant (Infrared) Heaters; Unit Heaters and other Gas-Fired Heaters

STEAM HEATING SYSTEM DESIGN

Part 1 of this section of the Handbook pertains mainly to the design of large commercial, industrial, and institutional steam heating systems.

Part 2 is chiefly concerned with design of one-pipe

systems in the residential size range.

Part 3 applies to design of two-pipe steam heating systems of various types for residential and larger installations.

PART 1—LARGE SYSTEMS

Vacuum pumps are essential in large steam heating systems to assure quick and uniform distribution of steam to all parts of the system and to obtain prompt return of condensate to the boiler. Basic system components are shown in Fig. 1, 2, and 3.

Steam supply to the heating units may be upfeed or downfeed, as desired. When downfeed supply of steam is desired, the mains are located above the heating units and supplied by main risers. Steam is distributed by downfeed risers from the mains. When upfeed supply of steam is employed, the mains are located below the heating surfaces and steam is supplied through upfeed risers. A combination of downfeed and upfeed supply may be employed depending on the building construction.

Steam mains must pitch down in the direction of steam flow so as to provide concurrent flow of steam and of condensate occurring in the mains. Diagrams in a later section on Supply and Return Piping Connections illustrate the basic connections between mains and risers.

The heating units condense the steam, changing it to liquid or condensate. In this change of state,

latent heat is given up by the steam at the rate required by the heating units and is distributed by the heating units to the space being heated.

Traps permit condensate and air to pass into the return piping of the system, but prevent the passage of steam. It is essential that no connection be made between the supply side and the return side of the system at any point except through a suitable trap.

The return piping conducts the condensate and air back to the pump and must pitch down toward the pump. This piping must provide unrestricted gravity flow of condensate to the pump receiver.

As the condensate and air reach the pump they are separated, the gases discharged to atmosphere, and the condensate returned to the boiler or boiler feed system.

These same basic components may be used in the design of a vacuum or a subatmospheric system. The vacuum heating pump maintains a vacuum in the return piping under all operating conditions, assisting the supply steam in overcoming the resistance of the system piping and assuring rapid and uniform steam distribution. The subatmospheric system per-

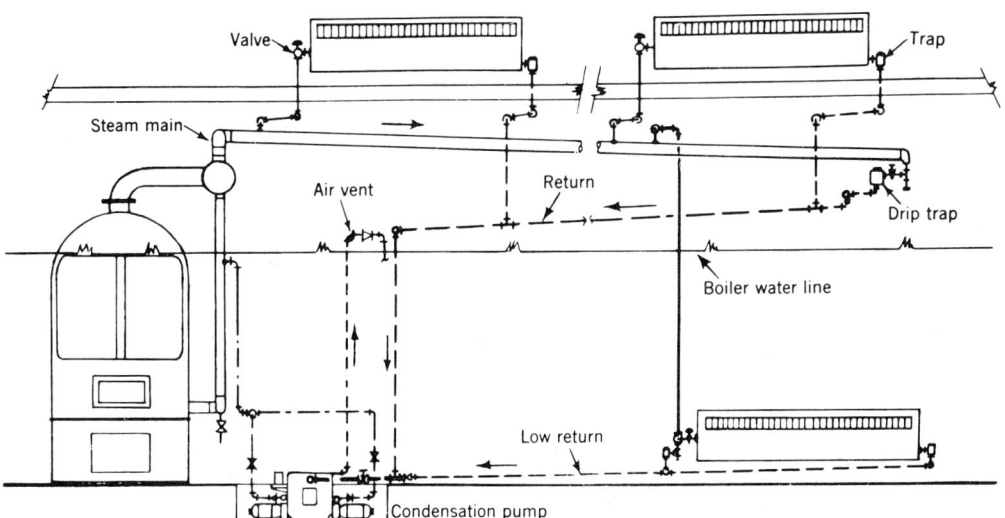

Fig. 1. Diagram illustrates the basic components of a steam heating system.

mits control of building temperature by variation of steam temperature and output of heating units.

Figure 2 illustrates the six basic components of a system which includes a vacuum heating pump.

When it is necessary to install returns below the level of the pump inlet, a mechanical lift should be installed to handle the low returns, as illustrated in Fig. 3.

Equivalent Direct Radiation (EDR). Equivalent direct radiation is a unit measure of heat output from a heating surface mounting to 240 Btu per hr. This is approximately equal to the latent heat given up when condensing ¼ lb steam per hr. For pounds steam per hour, multiply sq ft EDR by 0.25. To obtain MBH (1000 Btu per hr), multiply sq ft EDR by 0.24.

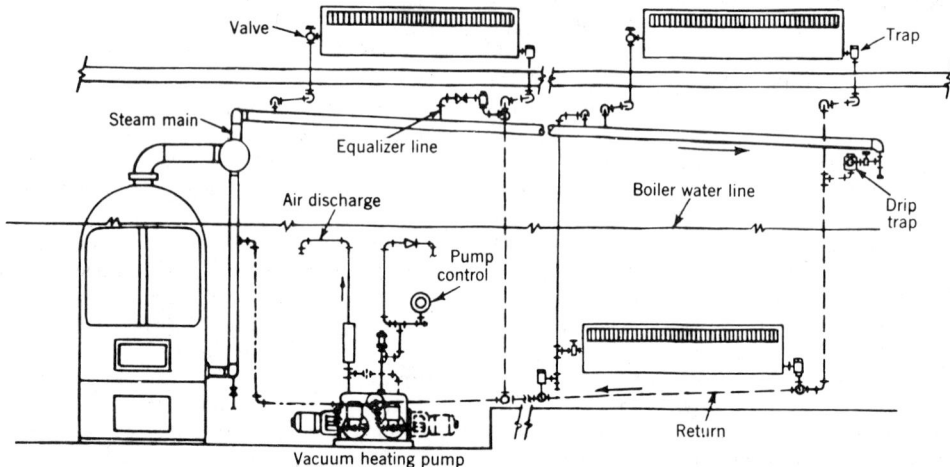

Fig. 2. Use of a vacuum heating pump maintains a vacuum in the return piping under all operating conditions.

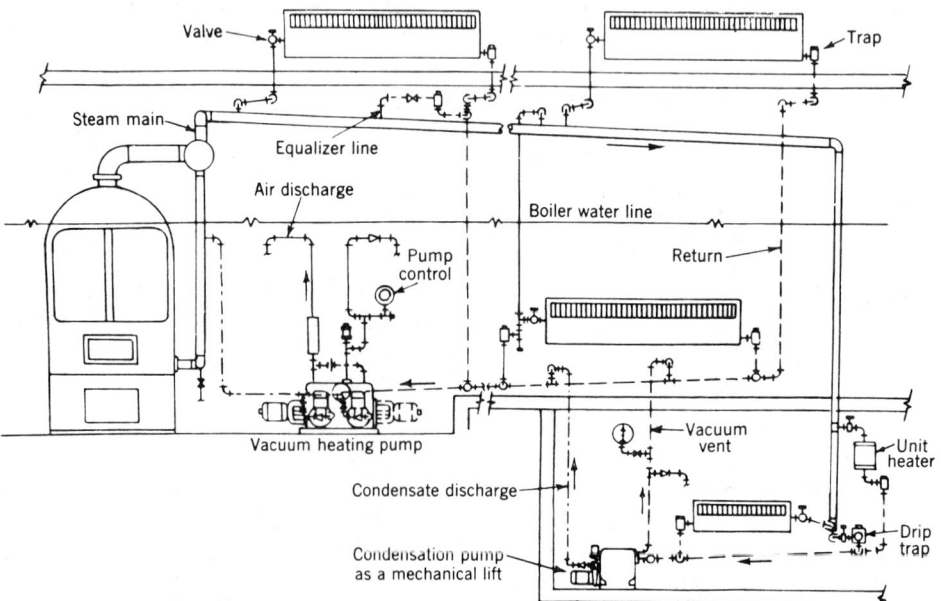

Fig. 3. A mechanical lift should be installed to handle returns below the level of the pump inlet.

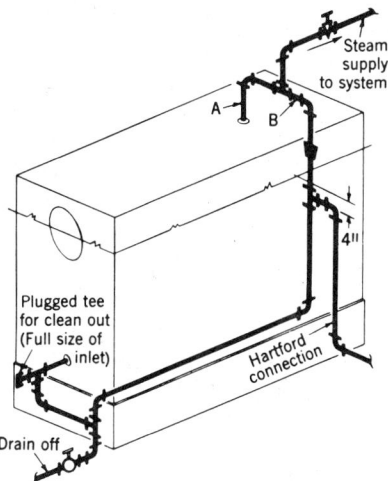

Fig. 4. Single boiler with Hartford connection.

Piping Connections to Boilers

Fundamentals applying to connections of supply and return piping to boilers are illustrated in Fig. 4, 5, 6, 7, and 8. Piping at the boiler including the Hartford Return Connection is shown for a single boiler in Fig. 4.

Steam is supplied by the boiler through the boiler nozzle. The size of the uptake A, Fig. 4, should never be less than the nozzle opening. The steam supply header B must be at least the same size as the uptake A. The connection between the end of the steam supply header B and the return inlet of the boiler is the equalizer and steam header drip. This connection is reduced in size depending upon the capacity of the boiler but should not be smaller than the size shown in Table 1 (later in this section).

The supply header piping is arranged to direct the flow of steam from the boiler toward the equalizer. This assures a pressure balance in the piping between the supply and return connections to the boiler.

Note the plugged tee at the connection to the boiler return inlet for cleaning and the drain valve at the bottom of the equalizer. The horizontal piping connection to the boiler inlet should be the same size as the inlet.

The condensate return to the boiler is connected to the equalizer at a point 4 inches below the recommended boiler water level. This constitutes the "Hartford connection." It is made with a short nipple to eliminate undesirable vapor accumulation,

a cause of water hammer. The vertical rise in the condensate return to the Hartford connection should be the same size as the equalizer line. This assures low velocity flow of condensate into the equalizer.

When the length of run from the condensate pump or boiler feeder to the boiler exceeds 25 ft, a check valve should be installed close to the base of the vertical rise to the Hartford connection. A check valve is always installed close to the pump discharge. When the pump discharge is overhead, a spring loaded check valve should be installed at the base of the vertical rise to the Hartford connection.

The pump discharge pipe should be sized to limit the velocity of flow to the boiler. The following table shows correct pipe sizes for velocities not to exceed 4 ft per sec.

Pipe Size, Inches	Flow, Gpm	Pipe Size, Inches	Flow, Gpm
1	10	2½	60
1¼	18	3	90
1½	25	3½	120
2	42	4	160

When a condensate boiler feed unit is vented to atmosphere and employed with a vacuum heating system, a check valve should be installed in the vent. Also, an equalizer line with thermostatic trap must be installed between the boiler supply header and the receiver of the boiler feed unit. A check valve is also required in the overflow line from the receiver. This will allow equalization between the receiver of the boiler feed unit and the boiler whenever an induced vacuum or vacuum caused by vacuum pump operation occurs in the boiler. Omission of these features can cause boiler flooding on shutdown.

Single Boiler with Direct Return Connection.

When the condensate return is connected directly to the boiler without the use of a Hartford connection, it is connected directly to the return inlet of the boiler (see Fig. 5). The connection is made with a check valve, gate valve, and a valved drain-off at the boiler inlet.

The direct return of condensate to the return inlet of the boiler has the advantage of delivering condensate to the coolest portion of the boiler. This advantage is important when a building is heated intermittently, for example, in a school with night and weekend shutdown, when the initial condensate returned after a shutdown is relatively cool.

Two Boilers with Common Return Header and Hartford Connection. In Fig. 6 the drop supply provides swing connections from each boiler to the common header. It also assures the separation of entrained condensate which flows to the supply header drip without entering the heating system piping with the steam. The common drop supply header may be provided with blank flanges for the addition of future boilers. Supply mains are taken from the top of the header.

The common supply header should be sized for the total connected load of the system based on the total developed length of the longest run.

The drip from the header is taken from the bottom and connected to the common return header. The horizontal run in the drip line must be at least 24 in. above the water line of the boilers to prevent surging of condensate in the line. The size of this line should not be less than the size of the equalizer line on each boiler.

The common return header should be sized for the total connected load of the entire system based on the total length of the longest return main.

The condensate return to the boilers is connected to the drip line at a point 4 in. below the water level of the boilers to constitute a Hartford connection.

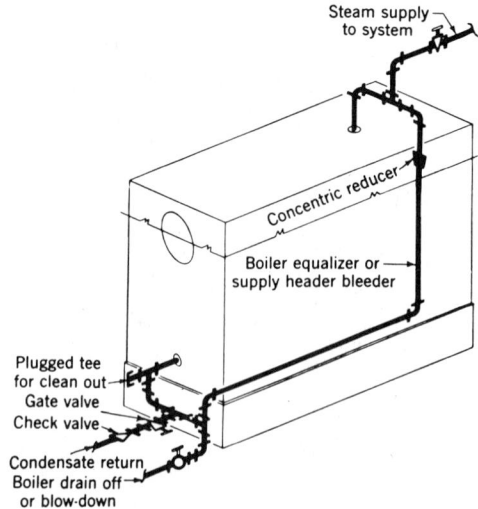

Fig. 5. Single boiler with direct return connection.

The boiler piping arrangement described eliminates the necessity of drip traps or other mechanical devices for dripping the steam headers and provides equalization of pressure in all the boiler piping.

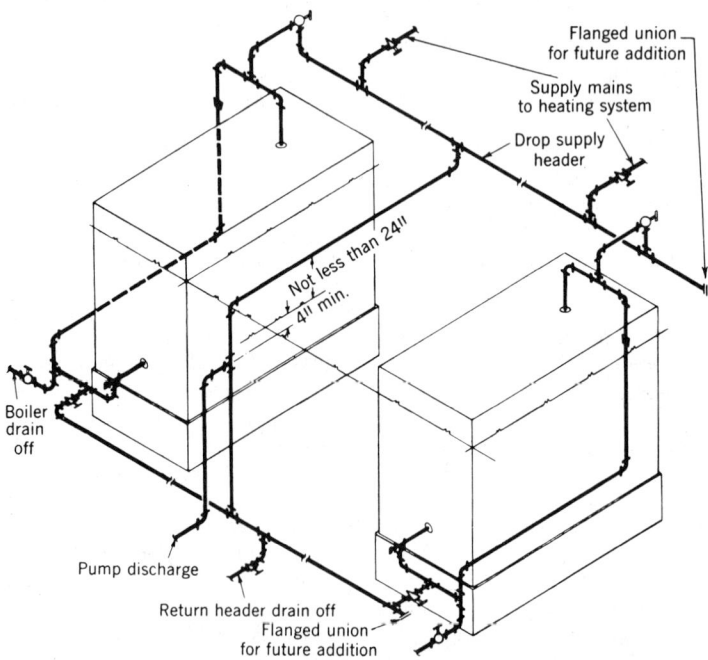

Fig. 6. Two boilers with common return header and Hartford connection.

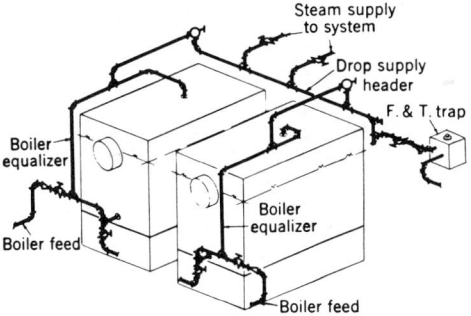

Fig. 7. Two boilers with separate direct return connections from below.

Two Boilers with Separate Direct Return Connections from Below. The drop supply header must be dripped through a float and thermostatic trap to the return main of the heating system since the boilers are not equalized. This arrangement is shown in detail in Fig. 7.

The supply outlet of each boiler is equalized with its return inlet. Since the boilers have no common return header, a separate boiler feed is made direct to the return inlet of each boiler. The connection is made with a check valve, gate valve, and a valved drain-off at the inlet.

Two Boilers with Separate Direct Return Connections from Overhead. When returns to the boilers are from overhead, as shown in Fig. 8, a spring loaded check valve should be installed at each boiler connection to prevent drainage in the overhead lines. This eliminates the possibility of water hammer on following pump operations. Note the discharge line enters the return header at each boiler so that direction of flow is toward the boiler return inlet.

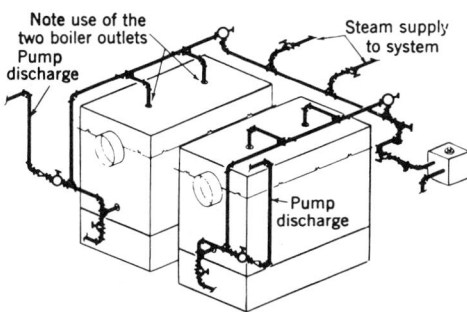

Fig. 8. Two boilers with separate direct return connections from overhead.

Connections to Steam-Using Equipment

Proper methods of connecting piping to the steam-using equipment, controls, and traps are shown in Figs. 9 to 15.

Convector Piping Detail. Piping connections for convectors are shown in Fig. 9 for both downfeed and upfeed risers.

When an automatic control valve is employed with a heating unit installed below the steam main, it is necessary to drip the downfeed riser to prevent accumulation of condensate in the riser when the valve is closed.

A control valve should be installed above the heating element of a convector to prevent accumulation of condensate in the supply end of the convector.

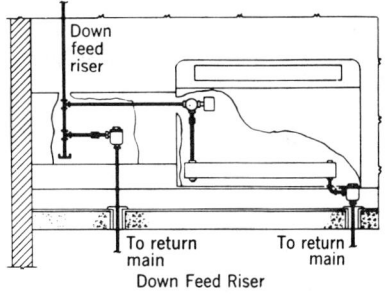
Down Feed Riser

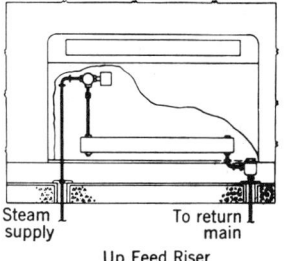
Up Feed Riser

Fig. 9. Convector piping connections.

Fin-Tube Piping Detail. Piping details for a fin-tube heater are shown in Fig. 10.

Note that the downfeed riser is provided with a drip to prevent the accumulation of condensate in the riser.

The installation of the fin-tube heaters requires consideration of the relation between length of element and steam distribution. Note a two tier fin-

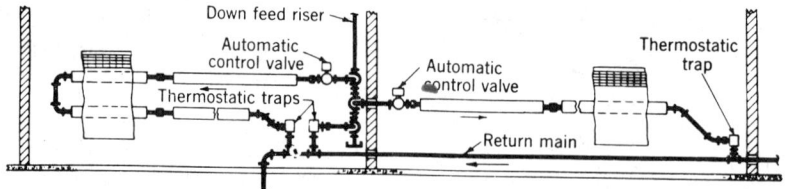

Fig. 10. Fin-tube heater piping connections.

tube heating element with a return bend has a length of run equal to twice the length of the unit. An approximation of the pressure drop through a long run of fin-tube can be made by referring to Table 3, friction losses for steam flow through branches, for the load involved.

Expansion joints are necessary to prevent expansion noises and element distortion.

Unit Heater Piping Details. Installation of a unit heater with propeller fan and large air capacity should be made with careful consideration to support of the unit and to prevention of expansion strains. The unit should be supported separately from the piping and allowance should be made for expansion of supply piping by use of swing joints. See piping details in Fig. 11.

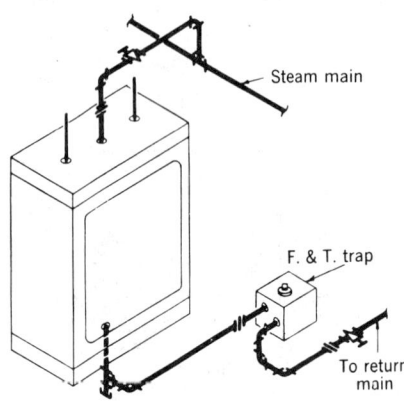

Horizontal Unit Heater

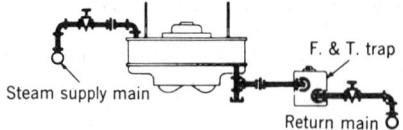

Vertical Delivery Unit Heater

Fig. 11. Unit heater piping details.

Industrial Type Unit Heater Piping Details. This heater, which has centrifugal fans, is capable of delivering air under sufficient pressure for distribution through duct work or with discharge flow directed to selected areas. The unit has capacity for heating large quantities of air relative to the space it occupies. The unit requires the use of a heavy-duty float and thermostatic trap. Automatic control of the steam supply is rarely used but thermostatic control of the fan operation is often employed. Piping details are shown in Fig. 12.

When thermostatic control of the steam supply is employed the supply and return should be arranged the same as shown for an auditorium type unit with equalizer and auxiliary air vent bypass; see Fig. 14.

Unit Ventilator Piping Details. The steam supply to the heating coils is normally automatic. The capacity range of the coils usually permits the use of a thermostatic trap rather than a float and thermostatic trap. Unit ventilators are usually operated with continuous fan operation during room occupancy. Piping details for unit ventilators are shown in Fig. 13.

An equalizing line with check valve is employed by some unit manufacturers, as illustrated, on units that have the heating element near bottom of cabinet. This assists coil drainage.

Auditorium Type Unit Ventilator Piping Details. An auditorium type unit ventilator having large air capacity is used when ventilation as well as heating is required for large spaces. Ventilation is accomplished by adding either a proportion of outdoor air with recirculated air from the space being heated or with 100% outdoor air. Equipped with centrifugal type fans, it is suitable for delivering air under sufficient pressure for distribution through duct work. Piping details are shown in Fig. 14.

A heavy-duty float and thermostatic trap is used. The auxiliary air vent bypass F may not be required when pressure steam is always available to the coils on automatic operation. The bypass should be em-

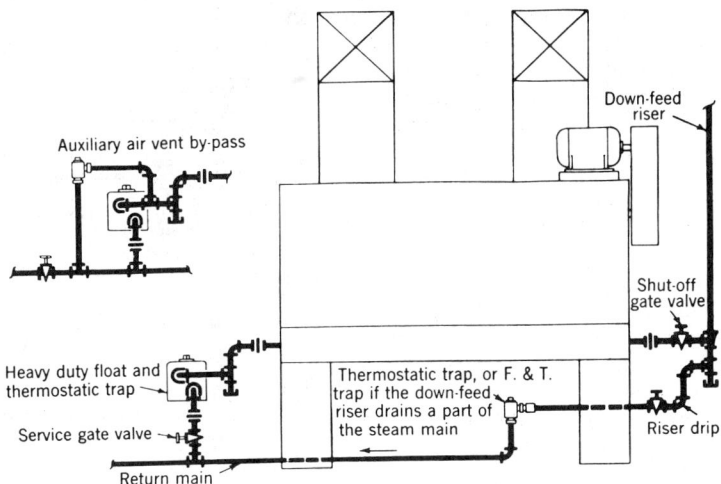

Fig. 12. Piping connections for industrial type unit heaters.

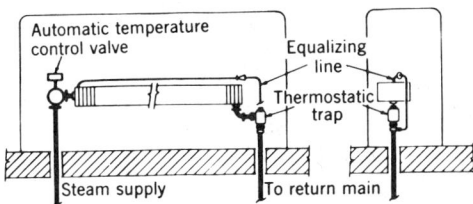

Fig. 13. Unit ventilator piping details.

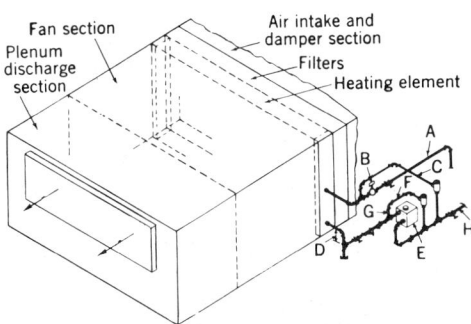

Fig. 14. Auditorium ventilator piping connections.

ployed with all subatmospheric systems to provide adequate venting of coils under this condition. This will assure removal of air inleakage from the supply mains and compensate for reduction of venting and condensate capacity in the float and thermostatic trap at low subatmospheric pressures. The bypass

connection should be made as close to the float and thermostatic trap inlet as possible and must be left uncovered to assure a cooling leg for proper thermostatic trap operation.

The equalizer line C operates as an induced vacuum breaker for the coils. When the automatic temperature control valve throttles, a high induced vacuum can occur in the coils and retard the drainage of condensate, particularly on conventional low vacuum systems due to frequent supply valve closing.

Supply valves for individual units installed on a subatmospheric system should be sized for not more than 1 psi pressure drop at design conditions.

Hot Water Generator Piping Details. Piping details for a hot water generator are shown in Fig. 15.

When the steam supply to a hot water generator comes from overhead, it is essential to drip the supply ahead of the control valve. This prevents

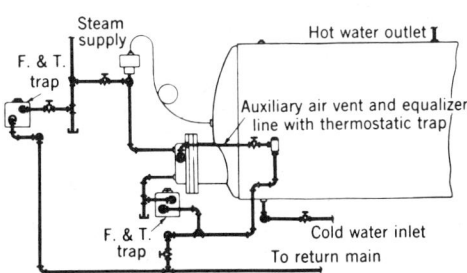

Fig. 15. Piping connections to hot water generator.

accumulation of condensate in the supply line when the automatic supply valve is closed. Accumulation of condensate ahead of this valve can cause water hammer. When the run of steam supply piping to the generator is short, a thermostatic trap may be used in place of the float and thermostatic trap shown.

The auxiliary air vent with thermostatic trap from the steam coil of the heater to the vacuum return line provides venting of the coils on heating up and relief of the induced vacuum caused by closing of the automatic control valve. This back venting allows complete drainage of the steam coils. The float and thermostatic trap is a heavy-duty trap.

When the returns from a hot water generator are close to a vacuum return pump provision should be made for cooling this condensate. This may be done with a section of fin-tube heating element.

Supply and Return Piping Connections

In laying out piping for a steam system particular attention must be given to connections and locations of traps for venting of air and removal of condensate. Means must be provided for permitting expansion of piping without incurring a change of alignment that may interfere with flow of condensate. Condensate should flow promptly and smoothly to the return piping. Proper pitch is needed for all mains and branches. Where changes in elevation or direction of mains are required, drips should be used at low points. The methods of making connections to assure proper system operation are shown in Figs. 16 to 36 which follow.

Boiler Supply Header Drip. Provision should be made for cooling the hot condensate from a boiler supply header when the drip connection is close to a vacuum return pump. This is accomplished as shown in Fig. 16 with a section of fin-tube radiation beyond the float and thermostatic trap. The 1/2 in. uncovered air line with thermostatic trap stabilizes the flow of condensate through the cooling coil by separating air and noncondensable gases from the liquid.

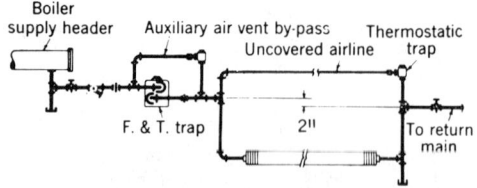

Fig. 16. Boiler supply header drip.

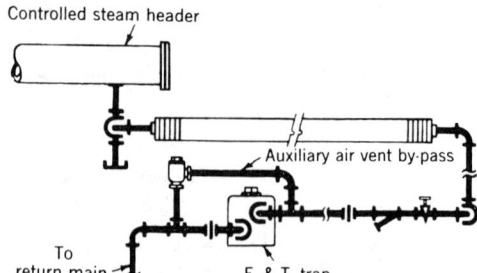

Fig. 17. Controlled steam header drip.

The auxiliary air vent bypass around the float and thermostatic trap is recommended to assure rapid venting of the supply header. It also prevents vapor binding of the float and thermostatic trap on this relatively severe service.

A priming or foaming boiler can carry boiler water into the header by entrainment with the steam. This can exceed the trap capacity. The excessive hot drips caused by this entrained boiler water can adversely affect the return side of a vacuum heating system.

Controlled Steam Header Drip (Vacuum, Subatmospheric System). The controlled steam header of a subatmospheric system contains superheated steam most of the time. This is due to rapid reduction in steam pressure caused by the throttling of the control valves admitting steam from the boiler supply header. The thermostatic element of the float and thermostatic trap for this controlled steam header must therefore be protected from this superheat which would cause excessive closing pressure of the thermostatic element. This is accomplished, as shown in Fig. 17, by means of a section of fin-tube radiation inserted ahead of the float and thermostatic trap.

Steam Main Rise and Drip. Whenever it is necessary to re-establish the elevation of a steam main to provide proper pitch in the main, a float and thermostatic trap is employed as shown in Fig. 18. The trap provides venting and drainage of the steam main ahead of the rise in the main.

On extended mains having several reductions in pipe size a rise and drip should be provided to relieve the end sections of the main from carrying all the condensate occurring on warm-up in the entire length of main.

On short connections between a trap and the return piping a swing joint is recommended to relieve pipe strain due to expansion of the main.

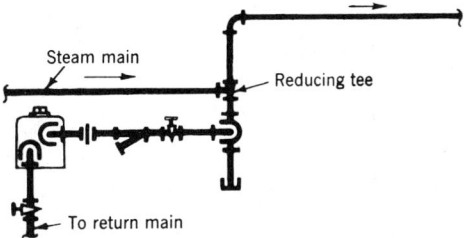

Fig. 18. Steam main rise and drip connections.

End of Main Drip. A float and thermostatic trap is used at the end of a steam main to provide venting and drainage of the main. This arrangement, shown in Fig. 19, with static head before the trap and a drop leg on the discharge assists drainage of the trap at all times.

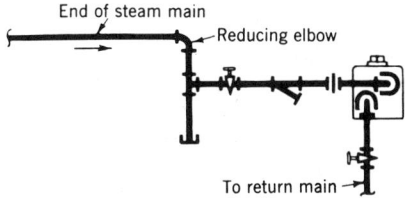

Fig. 19. Detail of end of main drip.

End of Main Drip with Minimum Elevation Between Steam Main and Return Main. When the relative elevation between a steam main and return main must be at a minimum, the arrangement in Fig. 20 with a float and thermostatic trap may be used for venting and draining the end of the steam main. With this arrangement, however, condensate flow may be periodically retarded because of insufficient differential across the trap. The arrangement shown in Fig. 19 above should be used whenever possible.

How to Split a Single Steam Main into Two Mains with a Drip. The arrangement shown in

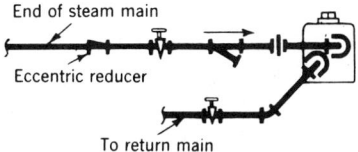

Fig. 20. Detail of end of main drip with minimum elevation between steam main and return main.

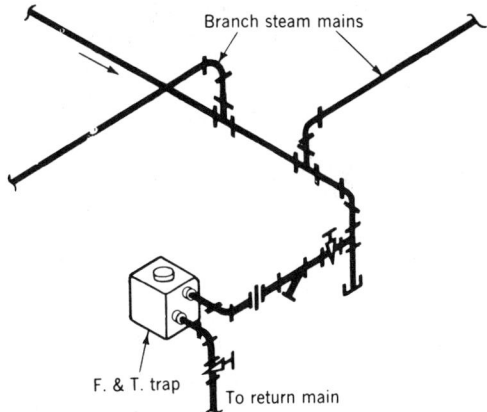

Fig. 21. Splitting a single steam main into two mains with a drip.

Fig. 21 is recommended whenever it is necessary to divide a main into two branches. It may be used at any point and the float and thermostatic trap serves to vent and drain the common section of the supply mains. An additional advantage of the arrangement is that it provides a means for establishing a new elevation for the branches.

Two Methods of Splitting a Single Steam Main into Two Mains without a Drip. The arrangements A and B in Fig. 22 are convenient whenever

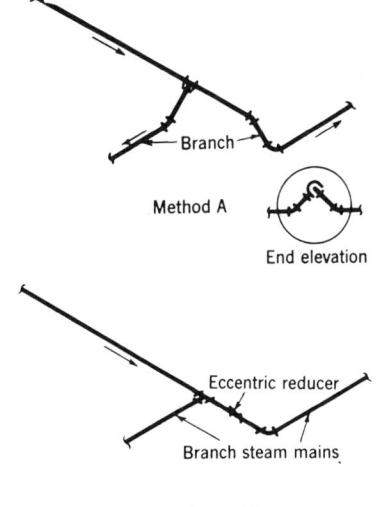

Fig. 22. Two methods of splitting a single steam main without a drip.

it is necessary to divide a main into two branches. Method A should be used when the common section of the main, beyond the last drip, is relatively short or when the main must be split close to the boiler header. This eliminates the necessity of a float and thermostatic trap close to the header where the service would be most severe. With Method B, most of the condensate drains into one branch.

Main Supply Riser Drip. A float and thermostatic trap is always used at the base of a main supply riser as shown in Fig. 23. This trap provides drainage of the steam main ahead of the riser as well as drainage of the riser itself.

Connections for Branch Mains. When the branch main is concealed in the ceiling, a 90° connection is preferred due to the ceiling opening. The 90° connection also serves to provide better conditions for expansion of the main. See Fig. 24.

End of Branch Main Drip with Thermostatic Trap. A float and thermostatic trap should be used at the end of a branch main as shown in Fig. 25 the same as shown in Fig 19 for the end of a main. However, a suitable thermostatic trap may be used when the load on the branch does not exceed 1500 sq ft EDR and the difference in elevation between the supply and return piping mains is limited. The use of a straightway, or side outlet thermostatic trap, permits the higher return line connection.

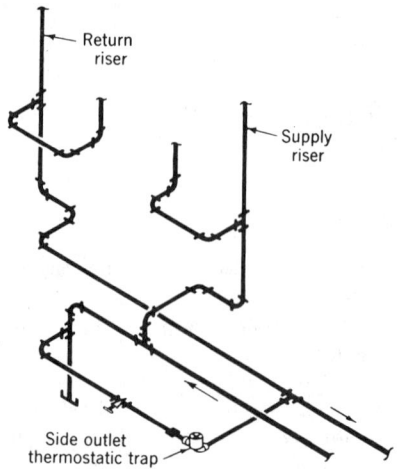

Fig. 25. End of branch main drip with thermostatic trap.

Supply Riser Drips. The arrangements in Fig. 26 may be employed whenever both the riser and spring piece are to be drained by the same drip trap. They

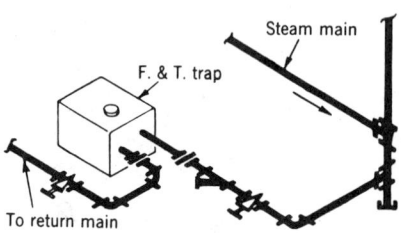

Fig. 23. Detail of main supply riser drip.

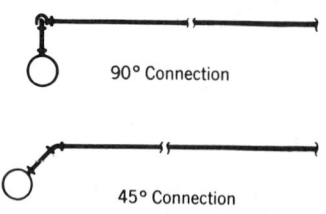

Fig. 24. Connections for branch mains.

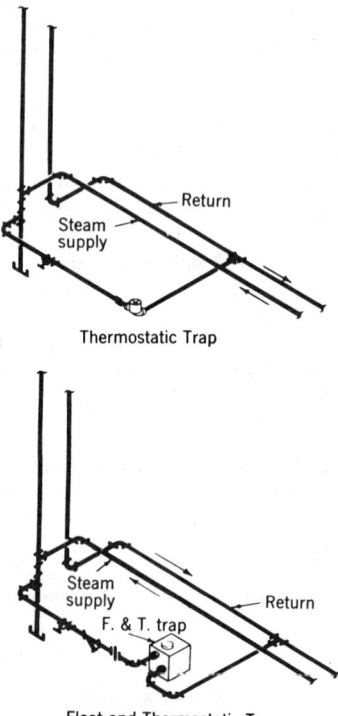

Fig. 26. Supply riser drip connections.

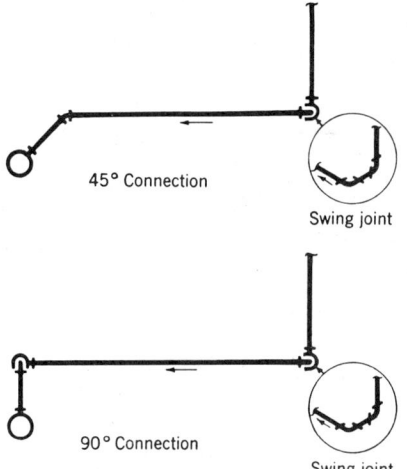

Fig. 27. Connections to first floor heating units or risers.

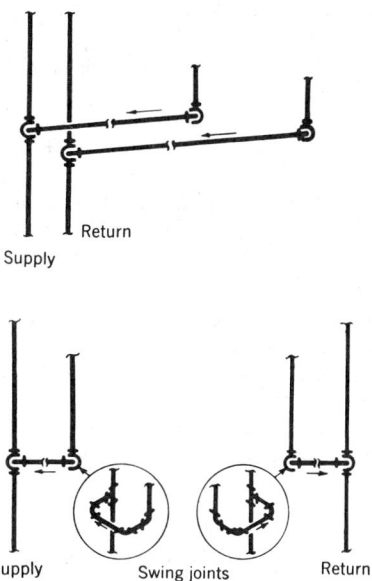

Fig. 29. Runout connections from risers to heating units.

should be used whenever a spring piece exceeds 16 ft of run. A float and thermostatic trap should be used for heavy loads. A thermostatic trap may be used when the load on the riser does not exceed 1500 sq ft EDR.

Connections to First Floor Heating Units or Risers. The 45° connection shown in Fig. 27 is preferred because, with counterflow of condensate and steam, it lessens entrainment of condensate with the steam. Note that the spring piece should be one pipe size larger than the riser due to counter flow.

Connections from Mains to Downfeed Risers. The 45° connection shown in Fig. 28 is preferred because it provides better separation of condensate and steam. The concurrent flow does not require an increase in size of spring piece over riser.

Runout Connections from Risers to Heating Units. The runout connections shown in Fig. 29 must pitch at least ½ in. per foot of length after allowing for the expansion of the riser. This assures positive drainage of the runout under all conditions of operation.

All runouts from risers must include swing joints.

Offset Connections in Risers. Offset connections in risers as shown in Fig. 30 must pitch at least ½ in. per foot of length after allowing for the expansion of the riser sections.

All offset connections in risers must include swing joints.

Note that the supply offset connection must be one size larger than the largest section of riser.

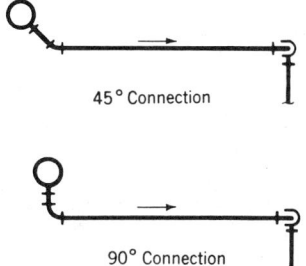

Fig. 28. Connections from mains to down feed risers.

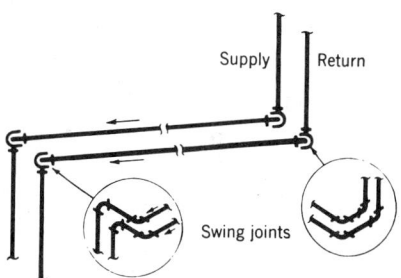

Fig. 30. Offset connections in risers.

Expansion Joints and Loops. The expansion joint and expansion loop shown in Fig. 31 are two general methods of allowing for expansion in low pressure steam mains. The expansion joint allows a greater amount of expansion than the loop with less strain.

Provision for expansion must be made in any length of steam pipe. Allowance for expansion in risers is important to prevent dislocation of the heating units and to maintain proper grade in the run-outs. Provision must also be made for expansion of the main where the riser is joined. See Fig. 29.

Swing Type Expansion Loop Combined with Rise and Drip of Main. Whenever it is necessary to re-establish the elevation of a steam main, it is possible to combine the rise and drip with a swing type expansion loop (see Fig. 32). The swing type expansion loop is less rigid than the expansion loop shown in Fig. 31 and allows for greater expansion.

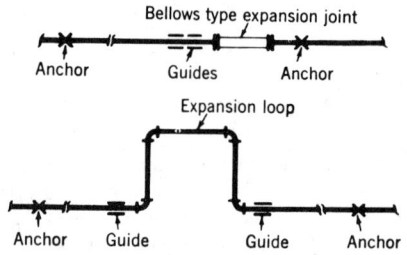

Fig. 31. Expansion joint detail.

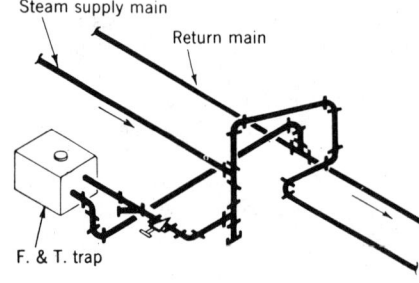

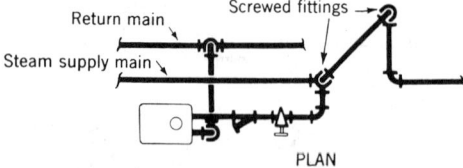

Fig. 32. Swing type expansion loop combined with main rise and drip.

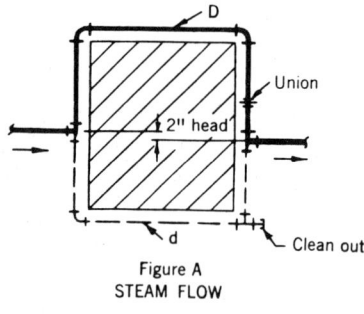

Figure A
STEAM FLOW

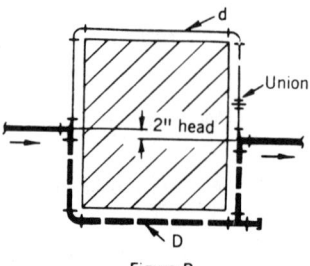

Figure B
CONDENSATE AND AIR FLOW

Fig. 33. Detail of piping around obstacle.

Piping Around a Door or Obstacle. The method of piping around a vertical obstacle is shown in Fig. 33.

Diagram A shows piping for a steam main. The steam will flow through the upper line D. Any condensate flowing with the steam will flow by gravity through the lower line d.

Diagram B shows piping for a return main. The condensate will flow by gravity through the lower line D. Any air and vapor will flow through the upper line d.

In Diagram A, line D remains the same size as the steam main. In Diagram B, line D remains the same size as the return main. Lines d in Diagrams A and B may be reduced.

Pipe Anchor Details. The best method of anchoring steam pipes will depend on the building construction. Two methods are shown in Fig. 34.

Piping Recessed Below Floor. Figure 35 shows a method of supporting pipe in a horizontal recess. In the recess shown the cover plate provides a semi-permanent flooring above the recess, permitting access to the piping during the life of the building.

When the elevation of the exterior grade permits, transverse recesses can be arranged with cover plates for access from the outside of the building.

Reducing Main or Branch through Eccentric Reducer. When the load on a steam main or branch main is reduced, and it is possible to reduce the pipe size, an eccentric reducer should be employed. This permits free flow of condensate from the larger to the smaller pipe size as shown in Fig. 36. The reducer should be installed a distance from the last outlet in the larger pipe equal to not less than 3 pipe diameters.

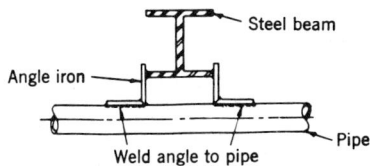

Note: Two angle irons may be used on each side of beam, depending on size of pipe, temperature differential and distance between anchors

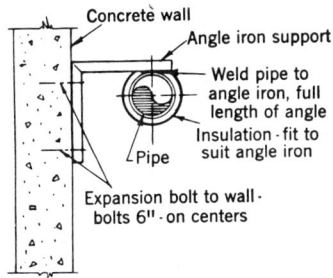

Fig. 34. Anchoring steam supply piping.

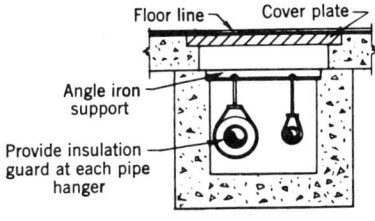

Fig. 35. Detail of pipe recess section.

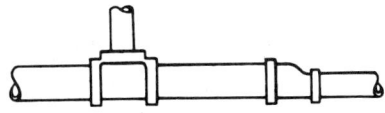

Fig. 36. Eccentric reducer.

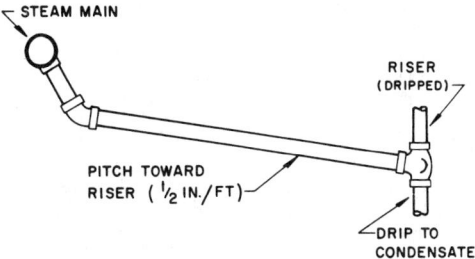

Fig. 37. Connection to Dripped Riser.

Steam Riser

The piping and trap material which follows has been prepared, with permission, from "Piping Design," *Carrier System Design Manual,* Carrier Corp.

Figures 37 and 38 illustrate steam supply risers connected to mains with runouts. The runout in Fig. 37 is connected to the bottom portion of the main and is pitched toward the riser to permit condensate to drain from the main. *This layout is used only when the riser is dripped.* If a dry return is used, the riser is dripped thru a steam trap. If a wet return is used, the trap is omitted.

Fig. 38 shows a piping diagram when the steam riser is not dripped. In this instance the runout is connected to the upper portion of the steam main and is pitched to carry condensate from the riser to the main.

Prevention of Water Hammer

If the steam main is pitched incorrectly when the riser is not dripped, water hammer may occur as illustrated in Fig. 39. Diagram "*a*" shows the runout partially filled with condensate but with enough space for steam to pass. As the amount of condensate increases and the space decreases, a wave motion is started as illustrated in diagram "*b*". As the wave or slug of condensate is driven against the turn in the pipe (diagram "*c*"), a

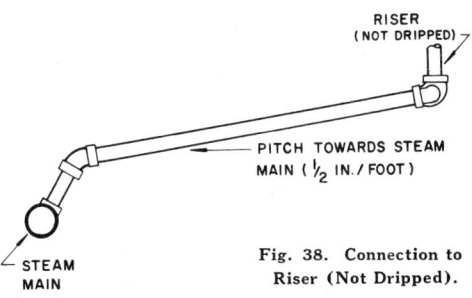

Fig. 38. Connection to Riser (Not Dripped).

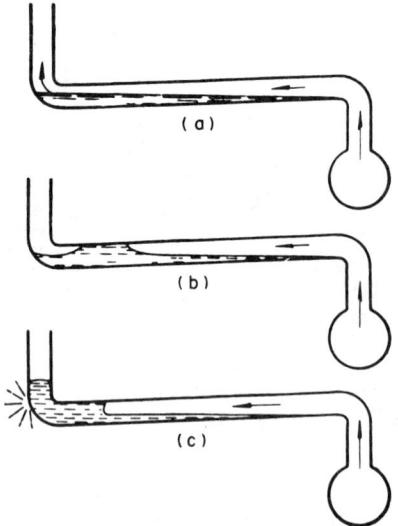

Fig. 39. Water Hammer.

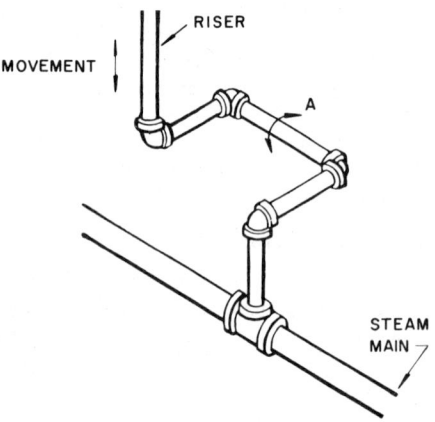

Fig. 40. Riser Connected to Allow For Expansion.

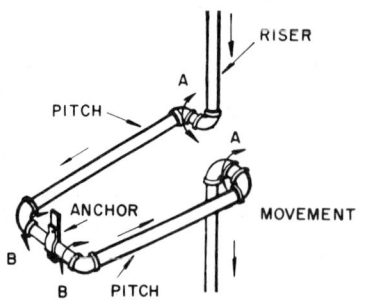

Fig. 41. Riser Anchor.

hammer noise is caused. This pounding may be of sufficient force to split pipe fittings and damage coils in the system.

The following precautions must be taken to prevent water hammer:

1. Pitch pipes properly.
2. Avoid undrained pockets.
3. Choose a pipe size that prevents high steam velocity when condensate flows opposite to the steam.

Expansion and Contraction

Where a riser is two or more floors in height, it should be connected to the steam main as shown in Fig. 40. Point (A) is subject to a twisting movement as the riser moves up and down.

Figure 41 shows a method of anchoring the steam riser to allow for expansion and contraction. Movement occurs at (A) and (B) when the riser moves up and down.

Obstructions

Steam supply mains may be looped over obstructions if a small pipe is run below the obstruction to take care of condensate as illustrated in Fig. 42. The reverse procedure is followed for condensate return mains as illustrated in Fig. 43. The larger pipe is carried under the obstruction.

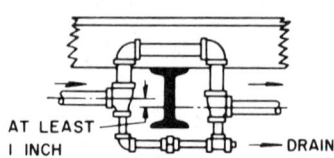

Fig. 42. Supply Main Loops.

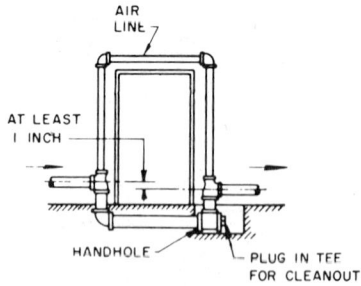

Fig. 43. Return Main Loop.

Dripping Riser

A steam supply main may be dropped abruptly to a lower level without dripping if the pitch is downward. When the steam main is raised to a higher level, it must be dripped as illustrated in Fig. 44. This diagram shows the steam main dripped into a wet return.

Figure 45 is one method of dripping a riser thru a steam trap to a dry return. The runout to the return main is pitched toward the return main.

Vacuum Lift

As described under vacuum systems, a lift is sometimes employed to lift the condensate up to the inlet of the vacuum pump. Figs. 46 and 47 show a one-step and two-step lift, respectively. The one-step lift is used for a maximum lift of 5 ft. For 5 to 8 ft a two-step lift is required.

Steam Coils

Figures 48 thru 57 show methods of piping steam coils in a high or low pressure or vacuum steam piping system. The following general rules are applicable to piping layout for steam coils used in all systems:

1. Use full size coil outlets and return piping to the steam trap.
2. Use thermostatic traps for venting only.
3. Use a 15° check valve only where indicated on the layout.
4. Size the steam control valve for the steam load, not for the supply connection.
5. Provide coils with air vents as required, to eliminate non-condensable gases.
6. Do not drip the steam supply mains into coil sections.
7. Do not pipe tempering coils and reheat coils to a common steam trap.

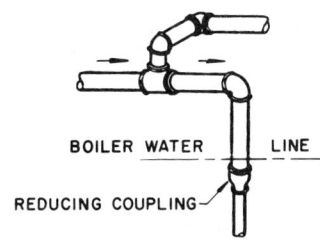

Fig. 44. Dripping Steam Main.

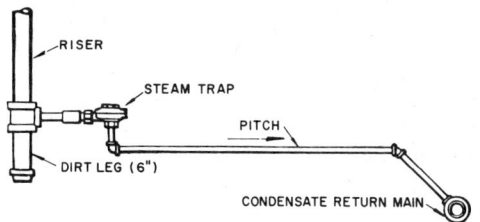

Fig. 45. Riser Dripped to Dry Return.

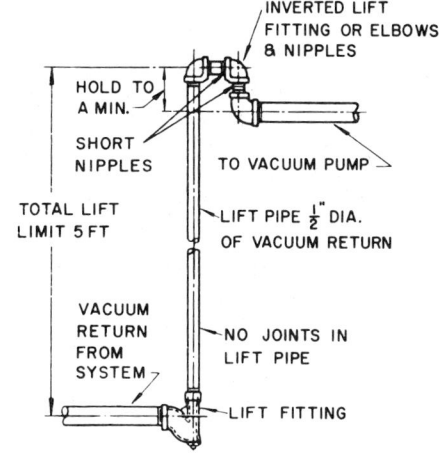

Fig. 46. One-Step Condensate Lift.

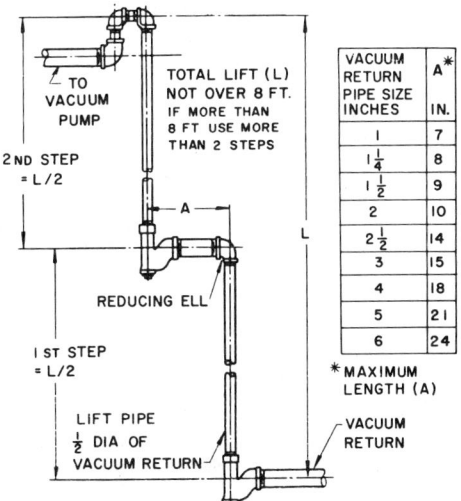

VACUUM RETURN PIPE SIZE INCHES	A* IN.
1	7
$1\frac{1}{4}$	8
$1\frac{1}{2}$	9
2	10
$2\frac{1}{2}$	14
3	15
4	18
5	21
6	24

* MAXIMUM LENGTH (A)

Fig. 47. Two-Step Condensate Lift.

8. Multiple coils may be piped to a common steam trap if they have the same capacity and the same pressure drop and if the supply is regulated by a control valve.

Piping Single Coils

Figure 48 illustrates a typical steam piping diagram for coils used in either a high or medium pressure system. If the return line is designed for low pressure or vacuum conditions and for a pressure differential of 5 psi or greater from steam to condensate return, a flash trap may be used.

Low pressure steam piping for a single coil is illustrated in Fig. 49. This diagram shows an open air relief located after the steam trap close to the unit. This arrangement permits non-condensable gases to vent to the atmosphere.

Figure 50 shows the piping layout for a steam coil in a vacuum system. A 15° check valve is used to equalize the vacuum across the steam trap.

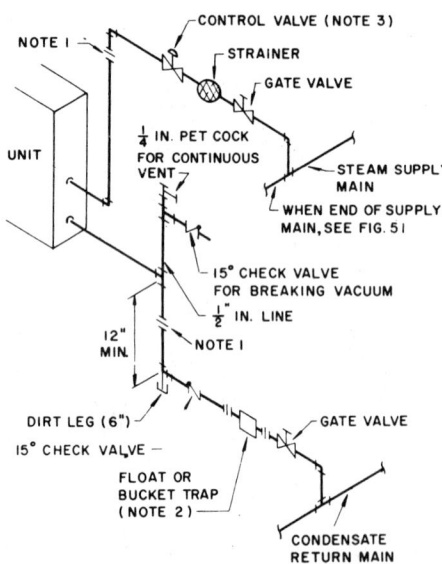

NOTES:
1. Flange or union is located to facilitate coil removal.
2. Flash trap may be used if pressure differential between steam and condensate return exceeds 5 psi.
3. When a bypass with control is required, see Fig. 52.
4. Dirt leg may be replaced with a strainer. If so, tee on drop can be replaced by a reducing ell.
5. The petcock is not necessary with a bucket trap or any trap which has provision for passing air. The great majority of high or medium pressure returns end in hot wells or deaerators which vent the air.

Fig. 48. High or Medium Pressure Coil Piping.

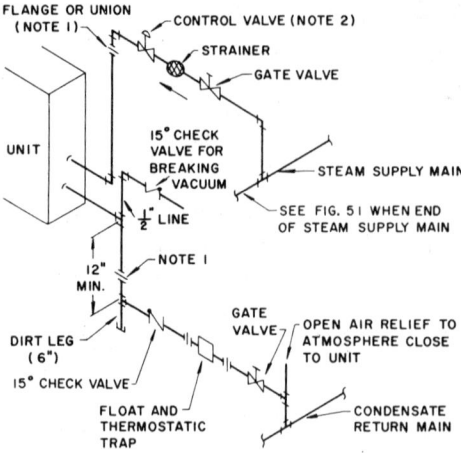

NOTES:
1. Flange or union is located to facilitate coil removal.
2. When a bypass with control is required, see Fig. 52.
3. Check valve is necessary when more than one unit is connected to the return line.
4. Dirt pocket is the same size as unit outlet. If dirt pocket is replaced by a strainer, replace tee with a reducing ell from unit outlet to trap size.

Fig. 49. Single Coil Low Pressure Piping Gravity Return.

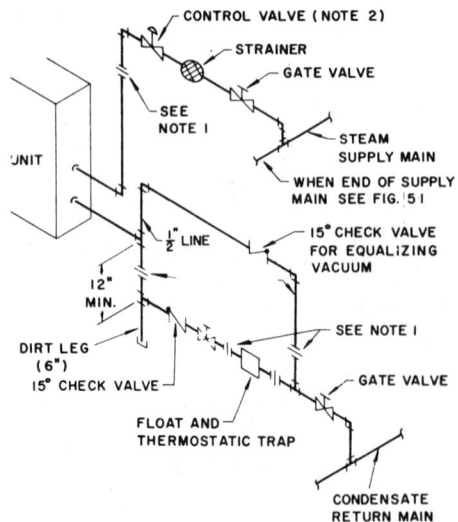

NOTES:
1. Flange or union is located to facilitate coil removal.
2. When a bypass with control is specified, see Fig. 52.
3. Check valve is necessary when more than one unit is connected to the return line.

Fig. 50. Vacuum System Steam Coil Piping.

Dripping Steam Supply Main

A typical method of dripping the steam supply main to the condensate return is shown in Fig. 51.

Steam Bypass Control

Frequently a bypass with a manual control valve is required on steam coils. The piping layout for a control bypass with a plug type globe valve as the manual control is shown in Fig. 52.

Lifting Condensate to Return Main

A typical layout for lifting condensate to an overhead return is described in Fig. 53. The amount of lift possible is determined by the pressure differential between the supply and return sides of the system. The amount of lift is not to exceed one foot for each pound of pressure differential. The maximum lift should not exceed 8 ft.

Piping Multiple Coils

Figures 54 thru 57 show piping layouts for high pressure, low pressure and vacuum systems with multiple coils. If a control valve is not used, each coil must have a separate steam trap as illustrated in Fig. 57. This particular layout may be used for a low pressure or vacuum system.

If the coils have different pressure drops or capacities, separate traps are required with or without a control valve in the system.

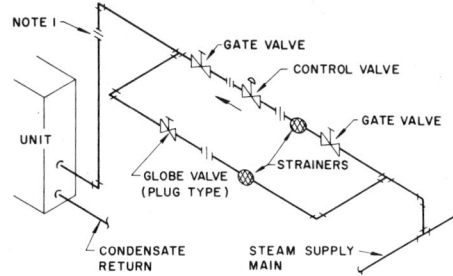

NOTES:
1. Flange or union is located to facilitate coil removal.
2. A bypass is necessary around valves and strainer when continuous operation is necessary.
3. Bypass to be the same size as valve port but never less than ½ inch.

Fig. 52. Bypass With Manual Control.

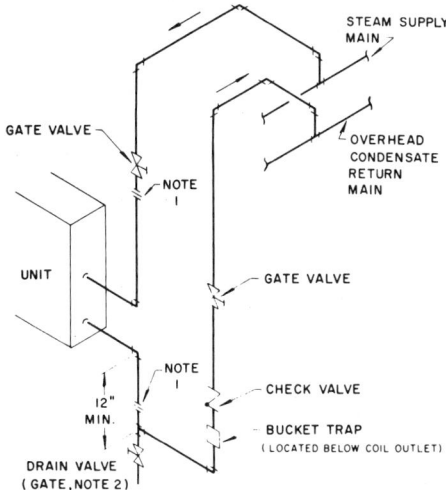

NOTES:
1. Flange or union is located to facilitate coil removal.
2. To prevent water hammer, drain coil before admitting steam.
3. Do not exceed one foot of lift between trap discharge and return main for each pound of pressure differential.
4. *Do not use this arrangement for units handling outside air.*

Fig. 53. Condensate Lift to Overhead Return.

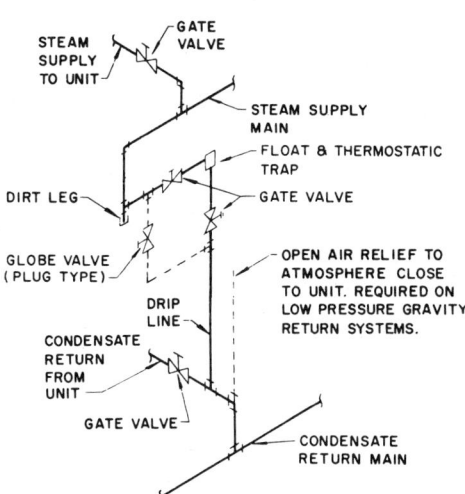

NOTES:
1. A bypass is necessary around trap and valves when continuous operation is necessary.
2. Bypass to be the same size as trap orifice but never less than ½ inch.

Fig. 51. Dripping Steam Supply To Condensate Return.

Boiler Piping

Figure 58 illustrates a suggested layout for a steam plant. This diagram shows parallel boilers and a single boiler using a "Hartford Return Loop."

FREEZE-UP PROTECTION

When steam coils are used for tempering or preheating outdoor air, controls are required to prevent freezing of the coil.

In high, medium, low pressure and vacuum systems, an immersion thermostat is recommended to protect the coil. The protection device controls the fan motor and the outdoor air damper. The immersion thermostat is actuated when the steam supply fails or when the condensate temperature drops below a predetermined level, usually 120 F to 150 F. The thermostat location is shown in Fig. 59.

The 15° check valve shown in the various piping diagrams provides a means of equalizing the pressure within the coil when the steam supply shuts off. This check valve is used in addition to the immersion thermostat. The petcock for continuous venting removes non-condensable gases from the coil. Non-condensable gases can restrict the flow of condensate, causing coil freeze-up.

On a low pressure and vacuum steam heating system, the immersion thermostat may be replaced by a condensate drain with a thermal element (Fig. 60). The thermal element opens and drains the coil when the condensate temperature drops

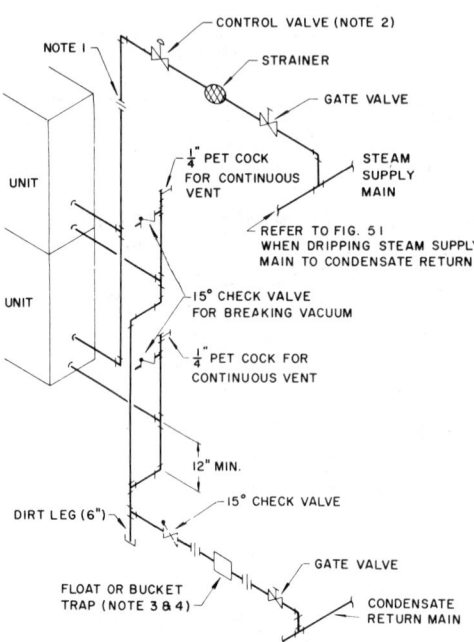

NOTES:
1. Flange or union is located to facilitate coil removal.
2. When bypass control is required, see Fig. 52.
3. Flash trap can be used if pressure differential between supply and condensate return exceeds 5 psi.
4. *Coils with different pressure drops require individual traps. This is often caused by varying air velocities across the coil bank.*
5. Dirt pocket may be replaced by a strainer. If so, tee on drop can be replaced by a reducing ell.
6. The petcock is not necessary with a bucket trap or any trap which has provision for passing air. The great majority of high pressure return mains terminate in hot wells or deaerators which vent the air.

Fig. 54. Multiple Coil High Pressure Piping.

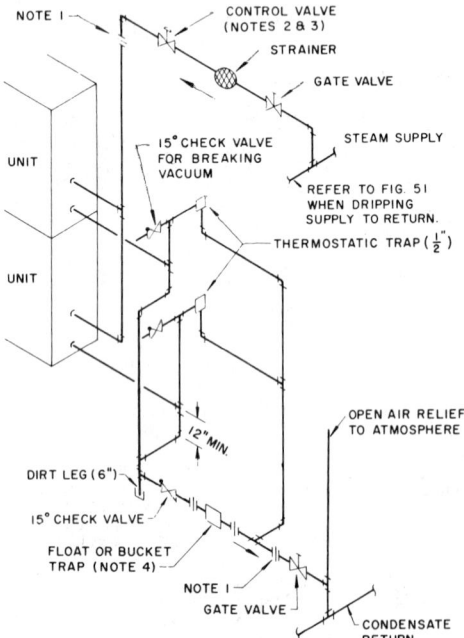

NOTES:
1. Flange or union is located to facilitate coil removal.
2. See Fig. 57 when control valve is omitted on multiple coils in parallel air flow.
3. When bypass control is required, see Fig. 52.
4. *Coils with different pressure drops require individual traps. This is often caused by varying air velocities across the coil bank.*

Fig. 55. Multiple Coil Low Pressure Piping Gravity Return.

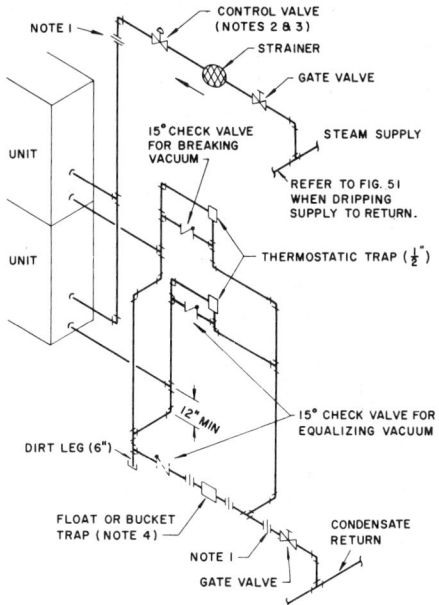

NOTES:
1. Flange or union is located to facilitate coil removal.
2. See Fig. 57 when control valve is omitted on multiple coils in parallel air flow.
3. When bypass control is required, see Fig. 52.
4. *Coils with different pressure drops require individual traps. This is often caused by varying air velocities across the coil bank.*

Fig. 56. Multiple Coil Low Pressure Vacuum System Piping.

below 165 F. Condensate drains are limited to a 5 lb pressure.

The following are general recommendations in laying out systems handling outdoor air below 35 F:

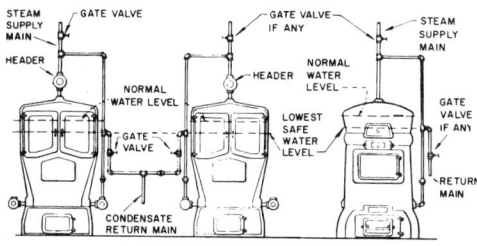

Fig. 58. "Hartford" Return Loop.

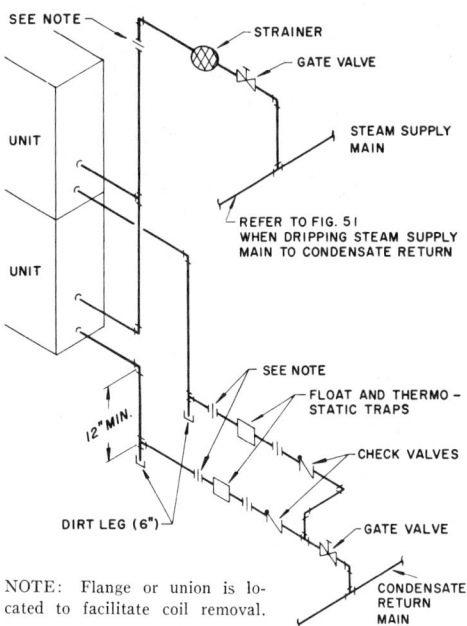

NOTE: Flange or union is located to facilitate coil removal.

Fig. 57. Low Pressure or Vacuum System Steam Traps.

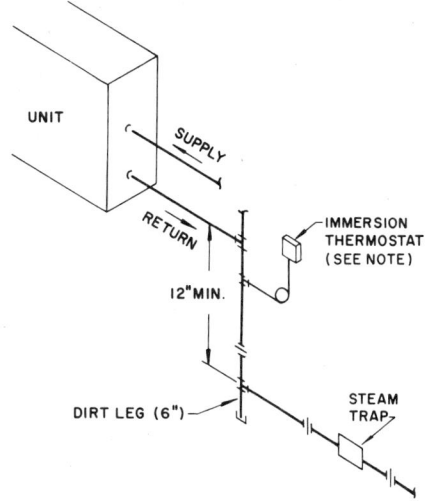

NOTE: Immersion thermostat is for control of outdoor air dampers and fan motor. Thermostat closes damper and shuts off fan when condensate temperature drops below a predetermined level.

Fig. 59. Freeze-up Protection for High, Medium, Low Pressure, and Vacuum Systems.

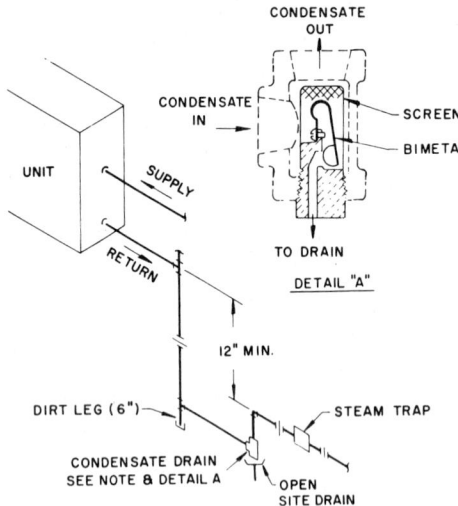

NOTE: Condensate drain drains coil when condensate temperature drops below a predetermined level.

Fig. 60. Freeze-up Protection for Low Pressure and Vacuum Systems.

1. Do not use overhead returns from the heating unit.
2. Use a strainer in the supply line to avoid collection of scale or other foreign matter in distributing orifices of nonfreeze preheater.

SIZING SUPPLY PIPING

Steam piping differs from other systems because it usually carries three fluids—steam, water and air. For this reason, steam piping design and layout require special consideration.

GENERAL SYSTEM DESIGN

Steam systems are classified according to piping arrangement, pressure conditions, and method of returning condensate to the boiler. These classifications are discussed in the following paragraphs.

PIPING ARRANGEMENT

A one- or two-pipe arrangement is standard for steam piping. The one-pipe system uses a single pipe to supply steam and to return condensate. Ordinarily there is one connection at the heating unit for both supply and return. Some units have two connections which are used as supply and return connections to the common pipe.

A two-pipe steam system is more commonly used in air conditioning, heating, and ventilating applications. This system has one pipe to carry the steam supply and another to return condensate. In a two-pipe system, the heating units have separate connections for supply and return.

The piping arrangements are further classified with respect to condensate return connections to the boiler and direction of flow in the risers:

1. Condensate return to boiler
   a. Dry-return—condensate enters boiler above water line.
   b. Wet-return—condensate enters boiler below water line.
2. Steam flow in riser
   a. Up-feed—steam flows up riser.
   b. Down-feed—steam flows down riser.

PRESSURE CONDITIONS

Steam piping systems are normally divided into five classifications—high pressure, medium pressure, low pressure, vapor and vacuum systems. Pressure ranges for the five systems are:

1. High pressure—100 psig and above
2. Medium pressure—15 to 100 psig
3. Low pressure—0 to 15 psig
4. Vapor—vacuum to 15 psig
5. Vacuum—vacuum to 15 psig

Vapor and vacuum systems are identical except that a vapor system does not have a vacuum pump, but a vacuum system does.

CONDENSATE RETURN

The type of condensate return piping from the heating units to the boiler further identifies the steam piping system. Two arrangements, gravity and mechanical return, are in common use.

When all the units are located above the boiler or condensate receiver water line, the system is described as a gravity return since the condensate returns to the boiler by gravity.

If traps or pumps are used to aid the return of condensate to the boiler, the system is classified as a mechanical return system. The vacuum return pump, condensate return pump and boiler return trap are devices used for mechanically returning condensate to the boiler.

CODES AND REGULATIONS

All applicable codes and regulations should be checked to determine acceptable piping practice for the particular application. These codes usually dictate piping design, limit the steam pressure, or qualify the selection of equipment.

WATER CONDITIONING

The formation of scale and sludge deposits on the boiler heating surfaces creates a problem in generating steam. Scale formation is intensified since scale-forming salts increase with an increase in temperature.

Water conditioning in a steam generating system should always be under the supervision of a specialist.

Table 1. Recommended Hanger Spacings for Steel Pipe

NOM. PIPE SIZE (in.)	DISTANCE BETWEEN SUPPORTS (FT)		
	Average Gradient		
	$1''$ in 10'	$\frac{1}{2}''$ in 10'	$\frac{1}{4}''$ in 10'
$\frac{3}{4}$	9	—	—
1	13	6	—
$1\frac{1}{4}$	16	10	5
$1\frac{1}{2}$	19	14	8
2	21	17	13
$2\frac{1}{2}$	24	19	15
3	27	22	18
$3\frac{1}{2}$	29	24	19
4	32	26	20
5	37	29	23
6	40	33	25
8	—	38	30
10	—	43	33
12	—	48	37
14	—	50	40
16	—	53	42
18	—	57	44
20	—	50	47
24	—	64	50

NOTE: Data is based on standard wall pipe filled with water and average number of fittings. Source: Crane Co. See pp. 8-146, 147.

PIPING SUPPORTS

All steam piping is pitched to facilitate the flow of condensate. Table 1 contains the recommended support spacing for piping pitched for different gradients. The data is based on Schedule 40 pipe filled with water, and an average amount of valves and fittings.

PIPING DESIGN

A steam system operating for air conditioning comfort conditions must distribute steam at all operating loads. These loads can be in excess of design load, such as early morning warmup, and at extreme partial load, when only a minimum of heat is necessary. The pipe size to transmit the steam for a design load depends on the following:

1. The initial operating pressure and the allowable pressure drop thru the system.
2. The total equivalent length of pipe in the longest run.
3. Whether the condensate flows in the same direction as the steam or in the opposite direction.

The major steam piping systems used in air conditioning applications are classified by a combination of piping arrangement and pressure conditions as follows:

1. Two-pipe high pressure
2. Two-pipe medium pressure
3. Two-pipe low pressure
4. Two-pipe vapor
5. Two-pipe vacuum
6. One-pipe low pressure

ONE-PIPE SYSTEM

A one-pipe gravity system was used primarily on residences and small commercial establishments. Fig. 61 shows a one-pipe, upfeed gravity system. The steam supply main rises from the boiler to a high point and is pitched downward from this point around the extremities of the basement. It is normally run full size to the last take-off and is then reduced in size after it drops down below the boiler water line. This arrangement is called a wet return. If the return main is above the boiler water line, it is called a dry return. Automatic air vents are required at all

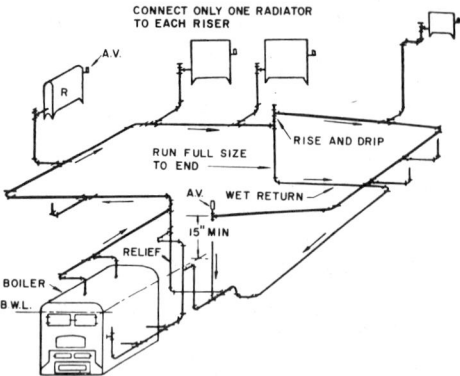

Fig. 61. One-Pipe, Upfeed Gravity System.

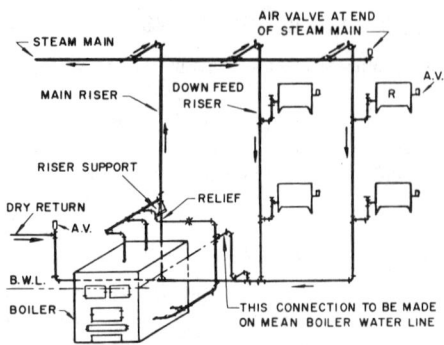

Fig. 63. One-Pipe, Downfeed Gravity System.

high points in the system to remove non-condensable gases. In systems that require long mains, it is necessary to check the pressure drop and make sure the last heating unit is sufficiently above the water line to prevent water backing up from the boiler and flooding the main. During operation, steam and condensate flow in the same direction in the mains, and in opposite direction in branches and risers. This system requires larger pipe and valves than any other system.

The one-pipe gravity system can also be designed as shown in Fig. 62, with each riser dripped separately. This is frequently done on more extensive systems.

Another type of one-pipe gravity system is the down-feed arrangement shown in Fig. 63. Steam flows in the main riser from the boiler to the

building attic and is then distributed throughout the building.

TWO-PIPE SYSTEM

A two-pipe gravity system is shown in Fig. 64. This system is used with indirect radiation. The addition of a thermostatic valve at each heating unit adapts it to a vapor or a mechanical vacuum system. A gravity system has each radiator separately sealed by drip loops on a dry return or by dropping directly into a wet return main. All drips, reliefs and return risers from the steam to the return side of the system must be sealed by traps or water loops to insure satisfactory operation.

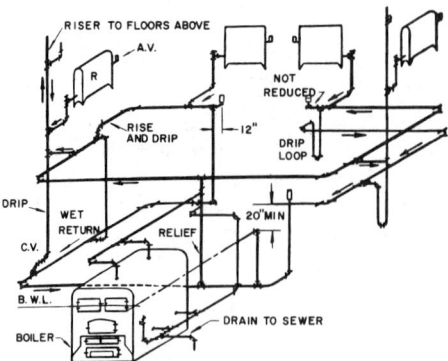

Fig. 62. One-Pipe Gravity System With Dripped
Risers.

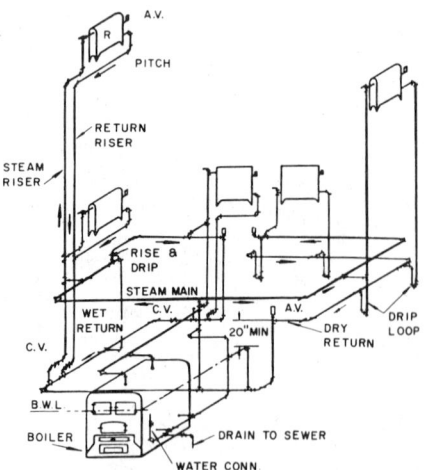

Fig. 64. Two-Pipe Gravity System.

If the air vent on the heating unit is omitted, and the air is vented thru the return line and a vented condensate receiver, a vapor system as illustrated in Fig. 65 results.

The addition of a vacuum pump to a vapor system classifies the system as a mechanical vacuum system. This arrangement is shown in Fig. 66.

PIPE SIZING

Tables have been developed which are used to select the proper pipe to carry the required steam rate at various pressures.

Figure 67 is a universal chart for steam pressure of 0 to 200 psig and for a steam rate of from 5 to 100,000 pounds per hour. However, the velocity as read from the chart is based on a steam pressure of 0 psig and must be corrected for the desired pressure from Fig. 68. The complete chart is based on the Moody friction factor

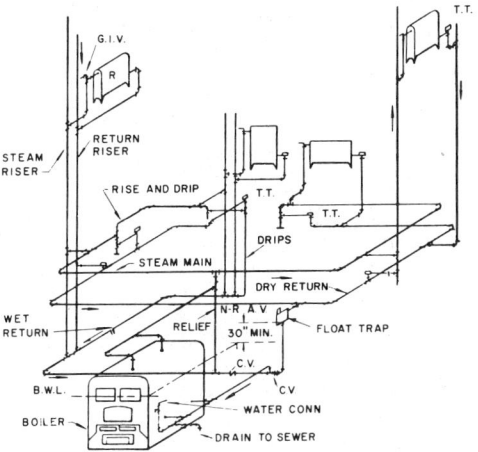

Fig. 65. Vapor System.

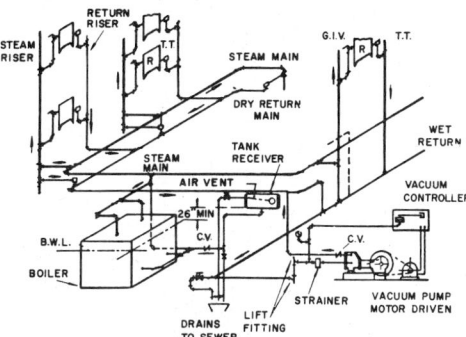

Fig. 66. Mechanical Vacuum System.

Table 2. High Pressure System Pipe Capacities (150 psig)

Pounds Per Hour

PIPE SIZE (in.)	PRESSURE DROP PER 100 FT							
	⅛ psi (2 oz)	¼ psi (4 oz)	½ psi (8 oz)	¾ psi (12 oz)	1 psi (16 oz)	2 psi (32 oz)	5 psi	10 psi
	SUPPLY MAINS AND RISERS				130 - 180 psig — Max Error 8%			
¾	29	41	58	82	116	184	300	420
1	58	82	117	165	233	369	550	790
1¼	130	185	262	370	523	827	1,230	1,720
1½	203	287	407	575	813	1,230	1,730	2,600
2	412	583	825	1,167	1,650	2,000	3,410	4,820
2½	683	959	1,359	1,920	2,430	3,300	5,200	7,600
3	1,237	1,750	2,476	3,500	4,210	6,000	9,400	13,500
3½	1,855	2,626	3,715	5,250	6,020	8,500	13,100	20,000
4	2,625	3,718	5,260	7,430	8,400	12,300	19,200	28,000
5	4,858	6,875	9,725	13,750	15,000	21,200	33,100	47,500
6	7,960	11,275	15,950	22,550	25,200	36,500	56,500	80,000
8	16,590	23,475	33,200	46,950	50,000	70,200	120,000	170,000
10	30,820	43,430	61,700	77,250	90,000	130,000	210,000	300,000
12	48,600	68,750	97,250	123,000	155,000	200,000	320,000	470,000
	RETURN MAINS AND RISERS				1 - 20 psig - Max Return Pressure			
¾	156	232	360	465	560	890		
1	313	462	690	910	1,120	1,780		
1¼	650	960	1,500	1,950	2,330	3,700		
1½	1,070	1,580	2,460	3,160	3,800	6,100		
2	2,160	3,300	4,950	6,400	7,700	12,300		
2½	3,600	5,350	8,200	10,700	12,800	20,400		
3	6,500	9,600	15,000	19,500	23,300	37,200		
3½	9,600	14,400	22,300	28,700	34,500	55,000		
4	13,700	20,500	31,600	40,500	49,200	78,500		
5	25,600	38,100	58,500	76,000	91,500	146,000		
6	42,000	62,500	96,000	125,000	150,000	238,000		

From Heating Ventilating Air Conditioning Guide, 1959. Used by permission.

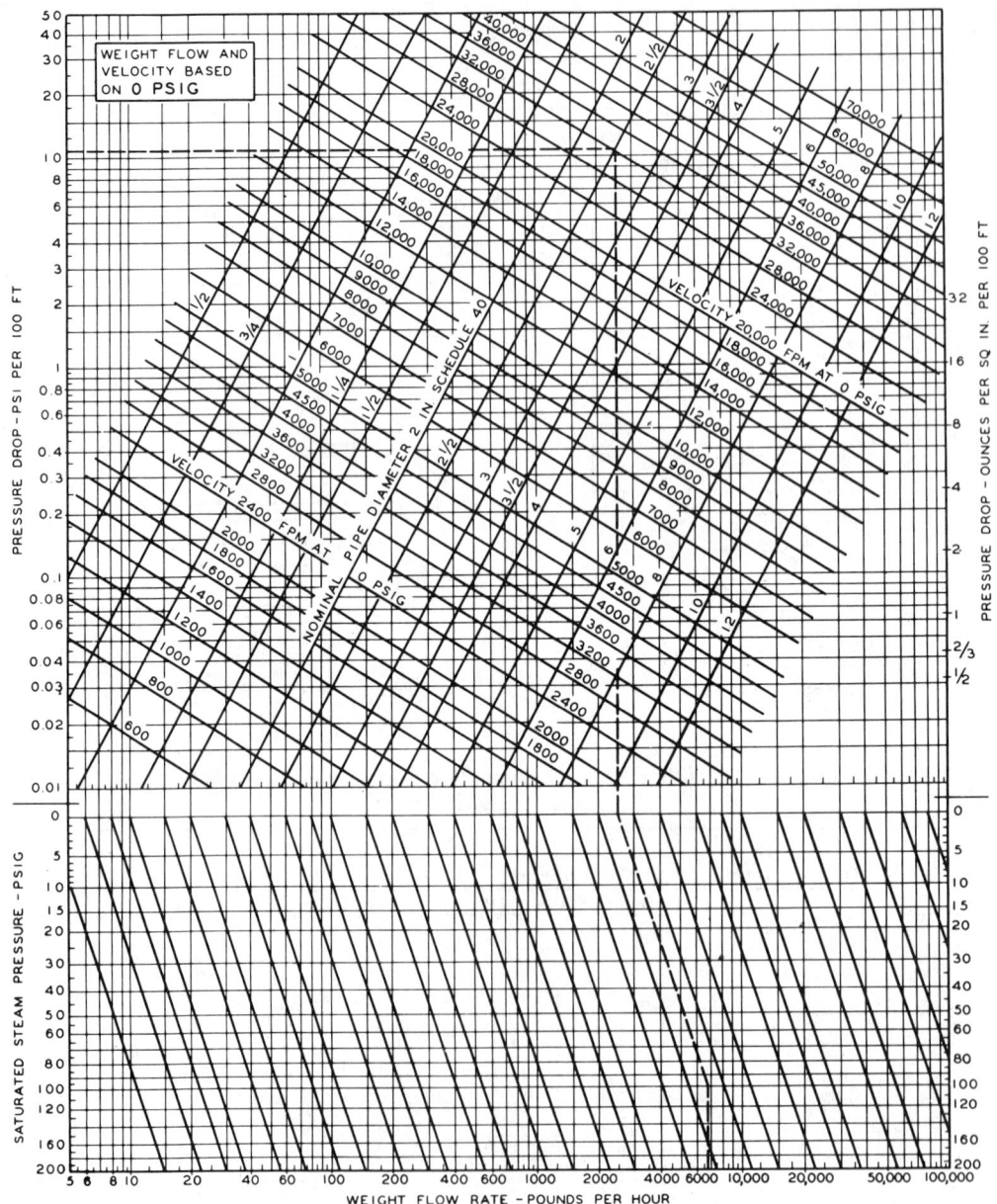

Based on Moody Friction Factor where flow of condensate does not inhibit the flow of steam.

Fig. 67. Pipe Sizing.*

* Use Fig. 68 to determine steam velocity at initial saturated pressures other than O psig. *ASHRAE Handbook of Fundamentals* (1972), p. 502.

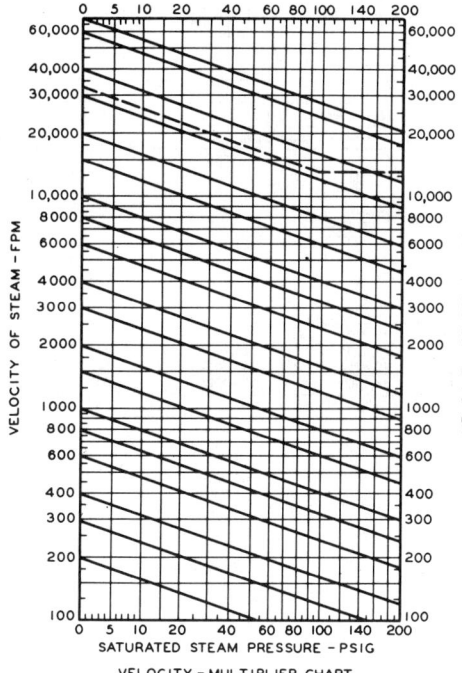

Fig. 68. Velocity Conversion.*

* See Example 3, page 7-28, for use of figure. *ASHRAE Handbook of Fundamentals* (1972), p. 503.

and is valid where condensate and steam flow in the same direction. See also, p. 8-29.

Tables 2 thru 7 are used for quick selection at specific steam pressures. Figure 67 has been used to tabulate the capacities shown in Tables 2 thru 4. The capacities in Tables 5 thru 7 are the results of tests conducted in the ASHRAE laboratories. Suggested limitations for the use of these tables are shown as notes on each table. In addition, Table 7 shows the total pressure drop for two-pipe low pressure steam systems. For equivalent resistance of valves and fittings, refer to Chapter 8, Page 8-5.

RECOMMENDATIONS

The following recommendations are for use when sizing pipe for the various systems:

Two-Pipe High Pressure System

This system is used mostly in plants and occasionally in commercial installations.

1. Size supply main and riser for a maximum drop of 25-30 psi.
2. Size supply main and risers for a maximum friction rate of 2-10 psi per 100 ft of equivalent pipe.
3. Size return main and riser for a maximum pressure drop of 20 psi.
4. Size return main and riser for a maximum friction rate of 2 psi per 100 ft of equivalent pipe.
5. Pitch supply mains ¼ in. per 10 ft away from boiler.
6. Pitch return mains ¼ in. per 10 ft toward the boiler.
7. Size pipe from Table 2.

Two-Pipe Medium Pressure System

This system is used mostly in plants and occasionally in commercial installations.

1. Size supply main and riser for a maximum pressure drop of 5-10 psi.
2. Size supply mains and risers for a maximum friction rate of 2 psi per 100 ft of equivalent pipe.
3. Size return main and riser for a maximum pressure drop of 5 psi.
4. Size return main and riser for a maximum friction rate of 1 psi per 100 ft of equivalent pipe.
5. Pitch supply mains ¼ in. per 10 ft away from the boiler.
6. Pitch return mains ¼ in. per 10 ft toward the boiler.
7. Size pipe from Table 3.

Two-Pipe Low Pressure System

This system is used for commercial, air conditioning, heating and ventilating installations.

1. Size supply main and risers for a maximum pressure drop determined from Table 7, depending on the initial system pressure.
2. Size supply main and riser for a maximum friction rate of 2 psi per 100 ft of equivalent pipe.
3. Size return main and riser for a maximum pressure drop determined from Table 7, depending on the initial system pressure.
4. Size return main and riser for a maximum friction rate of ½ psi per 100 ft of equivalent pipe.

Table 3. Medium Pressure System Pipe Capacities (30 psig)

Pounds Per Hour

PIPE SIZE (in.)	PRESSURE DROP PER 100 FT					
	1/8 psi (2 oz)	1/4 psi (4 oz)	1/2 psi (8 oz)	3/4 psi (12 oz)	1 psi (16 oz)	2 psi (32 oz)
	SUPPLY MAINS AND RISERS		25 - 35 psig – Max Error 8%			
3/4	15	22	31	38	45	63
1	31	46	63	77	89	125
1 1/4	69	100	141	172	199	281
1 1/2	107	154	219	267	309	437
2	217	313	444	543	627	886
2 1/2	358	516	730	924	1,033	1,460
3	651	940	1,330	1,628	1,880	2,660
3 1/2	979	1,414	2,000	2,447	2,825	4,000
4	1,386	2,000	2,830	3,464	4,000	5,660
5	2,560	3,642	5,225	6,402	7,390	10,460
6	4,210	6,030	8,590	10,240	12,140	17,180
8	8,750	12,640	17,860	21,865	25,250	35,100
10	16,250	23,450	33,200	40,625	46,900	66,350
12	25,640	36,930	52,320	64,050	74,000	104,500
	RETURN MAINS AND RISERS		0 - 4 psig – Max Return Pressure			
3/4	115	170	245	308	365	
1	230	340	490	615	730	
1 1/4	485	710	1,025	1,285	1,530	
1 1/2	790	1,155	1,670	2,100	2,500	
2	1,575	2,355	3,400	4,300	5,050	
2 1/2	2,650	3,900	5,600	7,100	8,400	
3	4,850	7,100	10,250	12,850	15,300	
3 1/2	7,200	10,550	15,250	19,150	22,750	
4	10,200	15,000	21,600	27,000	32,250	
5	19,000	27,750	40,250	55,500	60,000	
6	31,000	45,500	65,500	83,000	98,000	

Table 4. Low Pressure System Pipe Capacities

Pounds Per Hour

CONDENSATE FLOWING WITH THE STEAM FLOW

NOM. PIPE SIZE (in.)	PRESSURE DROP PER 100 FT													
	1/16 psi (1 oz)		1/8 psi (2 oz)		1/4 psi (4 oz)		1/2 psi (8 oz)		3/4 psi (12 oz)		1 psi		2 psi	
	SATURATED PRESSURE (PSIG)*													
	3.5	12	3.5	12	3.5	12	3.5	12	3.5	12	3.5	12	3.5	12
3/4	9	11	14	16	20	24	29	35	36	43	42	50	60	73
1	17	21	26	31	37	46	54	66	68	82	81	95	114	137
1 1/4	36	45	53	66	78	96	111	138	140	170	162	200	232	280
1 1/2	56	70	84	100	120	147	174	210	218	260	246	304	360	430
2	108	134	162	194	234	285	336	410	420	510	480	590	710	850
2 1/2	174	215	258	310	378	460	540	660	680	820	780	950	1,150	1,370
3	318	380	465	550	660	810	960	1,160	1,190	1,430	1,380	1,670	1,950	2,400
3 1/2	462	550	670	800	990	1,218	1,410	1,700	1,740	2,100	2,000	2,420	2,950	3,450
4	726	800	950	1,160	1,410	1,690	1,980	2,400	2,450	3,000	2,880	3,460	4,200	4,900
5	1,200	1,430	1,680	2,100	2,440	3,000	3,570	4,250	4,380	5,250	5,100	6,100	7,500	8,600
6	1,920	2,300	2,820	3,350	3,960	4,850	5,700	7,000	7,200	8,600	8,400	10,000	11,900	14,200
8	3,900	4,800	5,570	7,000	8,100	10,000	11,400	14,300	14,500	17,700	16,500	20,500	24,000	29,500
10	7,200	8,800	10,200	12,600	15,000	18,200	21,000	26,000	26,200	32,000	30,000	37,000	42,700	52,000
12	11,400	13,700	16,500	19,500	23,400	28,400	33,000	40,000	41,000	49,500	48,000	57,500	67,800	81,000

* The weight-flow rates at 3.5 psig can be used to cover sat. press. from 1 to 6 psig, and the rates at 12 psig can be used to cover sat. press. from 8 to 16 psig with an error not exceeding 8 percent.

Tables 2 thru 4 from Heating Ventilating Air Conditioning Guide 1959. Used by permission.

5. Pitch mains 1/4 in. per 10 ft away from the boiler.
6. Pitch return mains 1/4 in. per 10 ft toward the boiler.
7. Use Tables 4 thru 6 to size pipe.

Two-Pipe Vapor System

This system is used in commercial and residential installations.

1. Size supply main and riser for a maximum

Table 5. Return Main and Riser Capacities for Low Pressure Steam Systems—Pounds Per Hour

PRESSURE DROP PER 100 FT. — column groups: 1/32 psi (1/2 oz), 1/24 psi (2/3 oz), 1/16 psi (1 oz), 1/8 psi (2 oz), 1/4 psi (4 oz), 1/2 psi (8 oz)

RETURN MAINS

PIPE SIZE (in.)	1/32 Wet	1/32 Dry	1/32 Vac	1/24 Wet	1/24 Dry	1/24 Vac	1/16 Wet	1/16 Dry	1/16 Vac	1/8 Wet	1/8 Dry	1/8 Vac	1/4 Wet	1/4 Dry	1/4 Vac	1/2 Wet*	1/2 Dry	1/2 Vac
¾	125	62		145	71	42	175	80	100	250	103	142	350	115	200		120	283
1	213	130		248	149	113	300	168	175	425	217	249	600	241	350		255	494
1¼	338	206		393	236	244	475	265	300	675	340	426	950	378	600		385	848
1½						388			475			674			950			1,340
2	700	470		810	535	815	1,000	575	1,000	1,400	740	1,420	2,000	825	2,000		830	2,830
2½	1,180	760		1,580	868	1,360	1,680	950	1,680	2,350	1,230	2,380	3,350	1,360	3,350		1,410	4,730
3	1,880	1,460		2,130	1,560	2,180	2,680	1,750	2,680	3,750	2,250	3,800	5,350	2,500	5,350		2,585	7,560
3½	2,750	1,970		3,300	2,200	3,250	4,000	2,500	4,000	5,500	3,230	5,680	8,000	3,580	8,000		3,780	11,300
4	3,880	2,930		4,580	3,350	4,500	5,500	3,750	5,500	7,750	4,830	7,810	11,000	5,380	11,000		5,550	15,500
5	6,090	4,600		7,880	5,250	7,880	9,680	5,870	9,680	13,700	7,560	13,700	19,400	8,420	19,400		8,880	27,300
6	8,820	6,670		12,600	7,620	12,600	15,500	8,540	15,500	22,000	10,990	22,000	31,000	12,200	31,000		12,800	43,800
8	15,200	11,480		21,700	13,120	21,700	26,700	14,700	26,700	37,900	18,900	37,900	53,400	22,800	53,400		25,500	75,500
10	24,000	18,100		34,300	20,800	34,300	42,200	23,200	42,200	59,900	29,800	59,900	84,400	33,300	84,400		35,400	114,400

RETURN RISERS

PIPE SIZE (in.)	1/32 Wet	1/32 Dry	1/32 Vac	1/24 Wet	1/24 Dry	1/24 Vac	1/16 Wet	1/16 Dry	1/16 Vac	1/8 Wet	1/8 Dry	1/8 Vac	1/4 Wet	1/4 Dry	1/4 Vac	1/2 Wet*	1/2 Dry	1/2 Vac
¾		48			48	113		48	175		48	249		48	350		48	494
1		113			113	244		113	300		113	426		113	600		113	848
1¼		248			248	388		248	475		248	674		248	950		248	1,340
1½		375			375	815		375	1,000		375	1,420		375	2,000		375	2,830
2		750			750	1,360		750	1,680		750	2,380		750	3,350		750	4,730
2½		1,230			1,230	2,180		1,230	2,680		1,230	3,800		1,230	5,350		1,230	7,560
3		2,250			2,250	3,250		2,250	4,000		2,250	5,680		2,250	8,000		2,250	11,300
3½		3,230			3,230	4,480		3,230	5,500		3,230	7,810		3,230	11,000		3,230	15,500
4		4,830			4,830	7,880		4,830	9,680		4,830	13,700		4,830	19,400		4,830	27,300
5		7,560			7,560	12,600		7,560	15,500		7,560	22,000		7,560	31,000		7,560	43,800
6		10,990			10,990	21,700		10,990	26,700		10,990	37,900		10,990	53,400		10,990	75,500
8		18,900			18,900	34,300		18,900	42,200		18,900	59,900		18,900	84,400		18,900	114,400
10		29,800			29,800			29,800			29,800			29,800			29,800	

* Vac values may be used for wet return mains.　　From Heating Ventilating Air Conditioning Guide, 1959. Used by permission.

pressure drop of 1/16-1/8 psi.

2. Size supply main and riser for a maximum friction rate of 1/16-1/8 psi per 100 ft of equivalent pipe.

3. Size return main and supply for a maximum pressure drop of 1/16-1/8 psi.

4. Size return main and supply for a maximum friction rate of 1/16-1/8 psi per 100 ft of equivalent pipe.

5. Pitch supply 1/4 in. per 10 ft away from the boiler.

6. Pitch return mains 1/4 in. per 10 ft toward the boiler.

7. Size pipe from Tables 4 thru 6.

Two-Pipe Vacuum System

This system is used in commercial installations.

1. Size supply main and riser for a maximum pressure drop of 1/8-1 psi.

2. Size supply main and riser for a maximum friction rate of 1/8-1/2 psi per 100 ft of equivalent pipe.

3. Size return main and riser for a maximum pressure drop of 1/8-1 psi.

Table 6. Low Pressure System Pipe Capacities

Pounds Per Hour

CONDENSATE FLOWING AGAINST STEAM FLOW

PIPE SIZE (in.)	TWO-PIPE SYSTEM Vertical	TWO-PIPE SYSTEM Horizontal	ONE-PIPE SYSTEM Up-feed Supply Risers	ONE-PIPE SYSTEM Vertical Connectors	ONE-PIPE SYSTEM Riser Run-outs
A	B*	C†	D‡	E	F
¾	8	—	6		7
1	14	9	11	7	7
1¼	31	19	20	16	16
1½	48	27	38	23	16
2	97	49	72	42	23
2½	159	99	116	—	42
3	282	175	200	—	65
3½	387	288	286	—	119
4	511	425	380	—	186
5	1,050	788	—	—	278
6	1,800	1,400	—	—	545
8	3,750	3,000	—	—	—
10	7,000	5,700	—	—	—
12	11,500	9,500	—	—	—
16	22,000	19,000	—	—	—

* Do not use Column B for pressure drops less than 1/16 psi/100 ft of equivalent length. Use Fig. 67, page 7-23.

† Pitch of horizontal runouts to riser should be not less than 1/2 in./ft. Where this pitch cannot be obtained, runouts over 8 ft in length should be one pipe size larger than called for in this table.

‡ Do not use Column D for pressure drops less than 1/24 psi/100 ft of equivalent run except on sizes 3 in. and larger. Use Fig. 67. From Heating Ventilating Air Conditioning Guide, 1959. Used by permission.

4. Size return main and riser for a maximum friction rate of ⅛-½ psi per 100 ft of equivalent pipe.
5. Pitch supply mains ¼ in. per 10 ft away from the boiler.
6. Pitch return mains ¼ in. per 10 ft toward the boiler.
7. Size pipe from Tables 4 thru 6.

One-Pipe Low Pressure System

This system is used on small commercial and residential systems.

1. Size supply main and riser for a maximum pressure drop of ¼ psi.
2. Size supply main and risers for a maximum friction rate of ¹⁄₁₆ psi per 100 ft of equivalent pipe.
3. Size return main and risers for a maximum pressure drop of ¼ psi.
4. Size return main and risers for a maximum friction rate of ¹⁄₁₆ psi per 100 ft of equivalent pipe.
5. Pitch supply main ¼ in. per 10 ft away from the boiler.
6. Pitch return main ¼ in. per 10 ft toward the boiler.
7. Size supply main and dripped runouts from Table 4.
8. Size undripped runouts from Table 6, Column F.
9. Size upfeed risers from Table 6, Column D.
10. Size downfeed supply risers from Table 4.
11. Pitch supply mains ¼ in. per 10 ft away from boiler.
12. •Pitch return mains ¼ in. per 10 ft toward the boiler.

Use of Table 7

Example 1—Determine Pressure Drop for Sizing Supply and Return Piping

Given:
  Two-pipe low pressure steam system
  Initial steam pressure—15 psig
  Approximate supply piping equivalent length—500 ft
  Approximate return piping equivalent length—500 ft

Table 7. Total Pressure Drop for Two-Pipe Low Pressure Steam Piping Systems

INITIAL STEAM PRESSURE (psig)	TOTAL PRESSURE DROP IN SUPPLY PIPING (psi)	TOTAL PRESSURE DROP IN RETURN PIPING (psi)
2	½	½
5	1¼	1¼
10	2½	2½
15	3¾	3¾
20	5	5

Find:
1. Pressure drop to size supply piping
2. Pressure drop to size return piping

Solution:
1. Refer to Table 7 for an initial steam pressure of 15 psig. The total pressure drop should not exceed 3.75 psi in the supply pipe. Therefore, the supply piping is sized for a total pressure drop of 3.75 or ¾ psi per 100 ft of equivalent pipe.
2. Although ¾ psi is indicated in Step 1, Item 4 under the two-pipe low pressure system recommends a maximum of ½ psi for return piping. Therefore, use ½ psi per 100 ft of equivalent pipe.

$$\text{Return main pressure drop} = \frac{1}{2} \times \frac{500}{100} = 2.5 \text{ psi}$$

Friction Rate

Example 2 illustrates the method used to determine the friction rate for sizing pipe when the total system pressure drop recommendation (supply pressure drop plus return pressure drop) is known and the approximate equivalent length is known.

Example 2—Determine Friction Rate

Given:
  Four systems
  Equivalent length of each system—400 ft
  Total pressure drop of system—½, ¾, 1, and 2 psi

Find:
  Friction rate for each system.

Solution:

System Number	System Equiv. Length (ft)	Total System Press. Drop. (psi)	Friction Rate For Pipe Sizing (per 100 ft)
1	400	½	(400/100) (x) = ½ x = ⅛
2	400	¾	(400/100) (x) = ¾ x = 3/16
3	400	1	(400/100) (x) = 1 x = ¼
4	400	2	(400/100) (x) = 2 x = ½

Use of Charts in Figs. 67 and 68.

Example 3—Determine Steam Supply Main and Final Velocity

Given:

Friction rate—2 psi per 100 ft of equivalent pipe
Initial steam pressure—100 psig
Flow rate—6750 lb/hr

Find:

1. Size of largest pipe not exceeding design friction rate.
2. Steam velocity in pipe.

Solution:

1. Enter bottom of Fig. 67 at 6750 lb/hr and proceed vertically to the 100 psig line (dotted line in Fig. 67). Then move obliquely to the 0 psig line. From this point proceed vertically up the chart to the smallest pipe size not exceeding 2 psi per 100 ft of equivalent pipe and read 3½ in.

2. The velocity of steam at 0 psig, as read from Fig. 67, is 16,000 fpm. Enter the left side of Fig. 68 at 16,000 fpm. Proceed obliquely downward to the 100 psig line and horizontally across to the right side of the chart (dotted line in Fig. 68). The velocity at 100 psig is 6100 fpm.

Example 4 illustrates a design problem for sizing pipe on a low pressure, vacuum return system.

Example 4—Sizing Pipe for a Low Pressure, Vacuum Return System

Given:

Six units
Steam requirement per unit—72 lb/hr
Layout as illustrated in Figs. 69 thru 71
Threaded pipe and fittings
Low pressure system—2 psi

Find:

Size of pipe and total pressure drop
Note: Total pressure drop in the system should never exceed one-half the initial pressure. A reasonably small drop is required for quiet operation.

Solution:

Determine the design friction rate by totaling the pipe length and adding 50% of the length for fittings:

$$115 + 11 + 133 = 259$$
$$259 \times .50 = 130$$
$$\overline{389} \text{ ft equiv length}$$

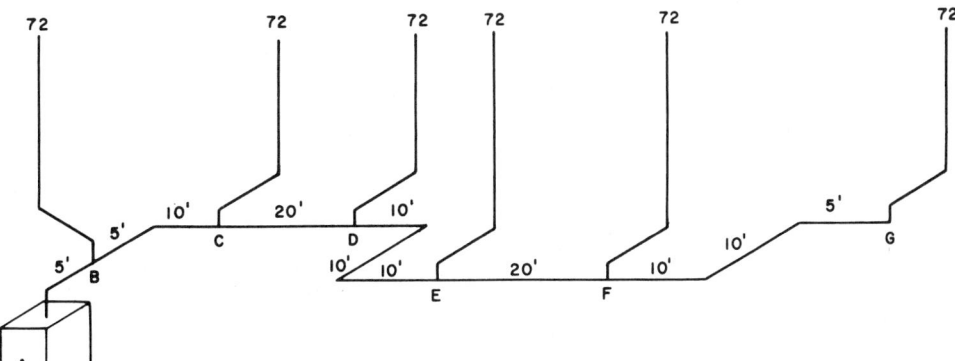

Fig. 69. Low Pressure Steam Supply Main.

Check pipe sizing recommendations for maximum friction rate from "Two-Pipe Vacuum System," Item 2, ⅛-½ psi. Check Table 7 to determine recommended maximum pressure drop for the supply and return mains (½ psi for each). Design friction rate $= 1/3.89 \times (1/2 + 1/2) =$ 1/4 psi per 100 ft. The supply main is sized by starting at the last unit "G" and adding each additional load from unit "G" to the boiler; use Table 4. The following tabulation results:

Section	Steam Load (lb/hr)	Pipe Size (in.)
F-G	72	1¼
E-F	144	2
D-E	216	2
C-D	288	2½
B-C	360	2½
A-B	432	3

Convert the supply main fittings to equivalent lengths of pipe and add to the actual pipe length, page 8-5.

Equivalent pipe lengths

1-1¼ in. side outlet tee	7.0
2-1¼ in. ells	4.6
1-2 in. reducing tee	4.7
1-2 in. run of tee	3.3
2-2 in. ells	6.6
1-2½ in. reducing tee	5.6
1-2½ in. run of tee	4.1
1-2½ in. ell	6.0
2-3 in. ells	15.0
1-3 in. reducing tee	7.0
Actual pipe length	115.0
Total equivalent length	178.9 ft

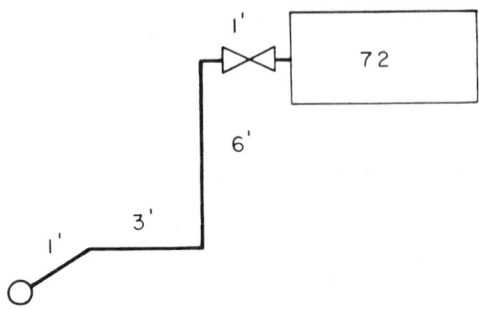

Fig. 70. Low Pressure Runout and Riser.

Pressure drop for the supply main is equal to the equivalent length times pressure drop per 100 ft:

$$178.9 \times .25/100 = .447 \text{ psi}$$

This is within the recommended maximum pressure drop (1 psi) for the supply.

The branch connection for Fig. 70 is sized in a similar manner at the same friction rate.

From Table 6 the horizontal runout pipe size for a load of 72 lb is 2½ in. and the vertical riser size is 2 in.

Convert all the fittings to equivalent pipe lengths, and add to the actual pipe length.

Equivalent pipe lengths

1-2½ in. 45° ell	3.2
1-2½ in. 90° ell	4.1
1-2 in. 90° ell	3.3
1-2 in. gate valve	2.3
Actual pipe length	11.0
Total equivalent length	23.9 ft

Pressure drop for branch runout and riser is

$$23.9 \times .25/100 = .060 \text{ psi}$$

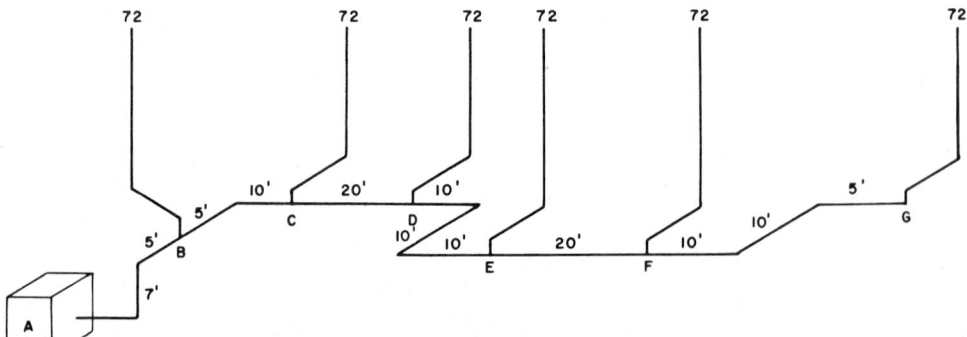

Fig. 71. Low Pressure Vacuum Return.

The vacuum return main is sized from Table 5 by starting at the last unit "G" and adding each additional load between unit "G" and the boiler. Each riser—72 lb per hr, ¾ in.

Section	Steam Load (lb/hr)	Pipe Size (in.)
F-G	72	¾
E-F	144	¾
D-E	216	1
C-D	288	1
B-C	360	1¼
A-B	432	1¼

Convert the return main fittings to equivalent pipe lengths and add to the actual pipe length, page 8-5.

Equivalent pipe lengths
1-¾ in. run of tee	1.4
5-¾ in. 90° ells	7.0
1-1 in. reducing tee	2.3
1-1 in. run of tee	1.7
2-1 in. 90° ells	3.4
1-1¼ in. reducing tee	3.1
3-1¼ in. 90° ells	6.9
1-1¼ in. run of tee	2.3
Actual pipe length	133.0
Total equivalent length	161.1 ft

Pressure drop for the return equals

$$161.1 \times .25/100 = .403 \text{ psi}$$

Total return pressure drop is satisfactory since it is within the recommended maximum pressure drop (⅛ — 1 psi) listed in the two-pipe vacuum return system. The total system pressure drop is equal to

$$.447 + .060 + .403 = 0.91 \text{ psi}$$

This total system pressure drop is within the maximum 2 psi recommended (1 psi for supply and 1 psi for return).

PIPING APPLICATION

The use and selection of steam traps, and condensate and vacuum return pumps are presented in this section.

Also, various steam piping diagrams are illustrated to familiarize the engineer with accepted piping practice.

STEAM TRAP SELECTION

The primary function of a steam trap is to hold steam in a heating apparatus or piping system and allow condensate and air to pass. The steam remains trapped until it gives up its latent heat and changes to condensate. The steam trap size depends on the following:

1. Amount of condensate to be handedl by the trap, lb/hr.
2. Pressure differential between inlet and discharge at the trap.
3. Safety factor used to select the trap.

Amount of Condensate

The amount of condensate depends on whether the trap is used for steam mains or risers, or for the heating apparatus.

The selection of the trap for the steam mains or risers is dependent on the pipe warm-up load and the radiation load from the pipe. Warm-up load is the condensate which is formed by heating the pipe surface when the steam is first turned on. For practical purposes the final temperature of the pipe is the steam temperature. Warm-up load is determined from the following equation:

$$C_1 = \frac{W \times (t_f - t_i) \times .114}{h_l \times T}$$

where:
- $C_1 =$ Warm-up condensate, lb/hr
- $W =$ Total weight of pipe, lb
- $t_f =$ Final pipe temperature, F (steam temp)
- $t_i =$ Initial pipe temperature, F (usually room temp)
- $.114 =$ Specific heat constant for wrought iron or steel pipe (.092 for copper tubing)
- $h_l =$ Latent heat of steam, Btu/lb (from steam tables)
- $T =$ Time for warm-up, hr

The radiation load is the condensate formed by unavoidable radiation loss from a bare pipe. This load is determined from the following equation and is based on still air surrounding the steam main or riser:

$$C_2 = \frac{L \times K \times (t_f - t_i)}{h_l}$$

where:
- $C_2 =$ Radiation condensate, lb/hr
- $L =$ Linear length of pipe, ft

K = Heat transmission coefficient, Btu/ (hr) (linear ft) (deg F diff between pipe and surrounding air)

t_f, t_i, h_i explained previously

The radiation load builds up as the warm-up load drops off under normal operating conditions. The peak occurs at the mid-point of the warm-up cycle. Therefore, one-half of the radiation load is added to the warm-up load to determine the amount of condensate that the trap handles.

Pressure Differential

The pressure differential across the trap is determined at design conditions. If a vacuum exists on the discharge side of the trap, the vacuum is added to the inlet side pressure to determine the differential.

Safety Factor

Good design practice dictates the use of safety factors in steam trap selection. Safety factors from 2 to 1 to as high as 8 to 1 may be required, and for the following reasons:

1. The steam pressure at the trap inlet or the back pressure at the trap discharge may vary. This changes the steam trap capacity.

2. If the trap is sized for normal operating load, condensate may back up into the steam lines or apparatus during start-up or warm-up operation.

3. If the steam trap is selected to discharge a full and continuous stream of water, the air could not be vented from the system.

The following guide is used to determine the safety factor:

Design	Safety Factor
Draining steam main	3 to 1
Draining steam riser	2 to 1
Between boiler and end of main	2 to 1
Before reducing valve	3 to 1
Before shut-off valve	
(closed part of time)	3 to 1
Draining coils	3 to 1
Draining apparatus	3 to 1

When the steam trap is to be used in a high pressure system, determine whether or not the system is to operate under low pressure conditions at certain intervals such as nighttime or weekends. If this condition is likely to occur, then an additional safety factor should be considered to account for the lower pressure drop available during nighttime operation.

Example 5 illustrates the three concepts mentioned previously in trap selection—condensate handled, pressure differential and safety factor.

Example 5—Steam Trap Selection for Dripping Supply Main to Return Line

Given:

Steam main—10 in. diam steel pipe, 50 ft long

Steam pressure—5 psig (227 F)

Room temperature—70 F db (steam main in space)

Warm-up time—15 minutes

Steam trap to drip main into vacuum return line (2 in. vacuum gage design)

Find:

1. Warm-up load.
2. Radiation load.
3. Total condensate load.
4. Specifications for steam trap at end of supply main.

Solution:

1. The warm-up load is determined from the following equation:

$$C_1 = \frac{W \times (t_f - t_i) \times .114}{h_i \times T}$$

where: W = 40.48 lb/ft $\times$ 50 ft

t_f = 227 F

t_i = 70 F

h_i = 960 Btu/lb (from *steam tables*)

T = .25 hr

$$C_1 = \frac{2024 \times (227 - 70) \times .114}{960 \times .25}$$

= 150 lb/hr of condensate

2. The radiation load is calculated by using the following equation:

$$C_2 = \frac{L \times K \times (t_f - t_i)}{h_i}$$

where: L = 50 ft

K = 6.41 Btu/(hr) (linear foot) (deg F diff between pipe and air)

$t_f = 227$ F

$t_i = 70$ F

$h_i = 960$ Btu/lb (from *steam tables*)

$$C_2 = \frac{50 \times 6.41 \times (227 - 70)}{960}$$

$= 52$ lb/hr of condensate

3. The total condensate load for steam trap selection is equal to the warm-up load plus one half the radiation load.

Total condensate load $= C_1 + (\frac{1}{2} \times C_2)$

$= 150 + (\frac{1}{2} \times 52)$

$= 176$ lb/hr

4. Steam trap selection is dependent on three factors: condensate handled, safety factor applied to total condensate load, and pressure differential across the steam trap.

The safety factor for a steam trap at the end of the main is 3 to 1 from the table on this page. Applying the 3 to 1 safety factor to the total condensate load, the steam trap would be specified to handle 3×176 or 528 lb/hr of condensate.

The pressure differential across the steam trap is determined by the pressure at the inlet and discharge of the steam trap.

Inlet to trap $= 5$ psig

Discharge of trap $= 2$ in. vacuum (gage)

When the discharge is under vacuum conditions, the discharge vacuum is added to the inlet pressure for the total pressure differential.

Pressure differential $= 6$ psi (approx)

Therefore the steam trap is selected for a differential pressure of 6 psi and 528 lb/hr of condensate.

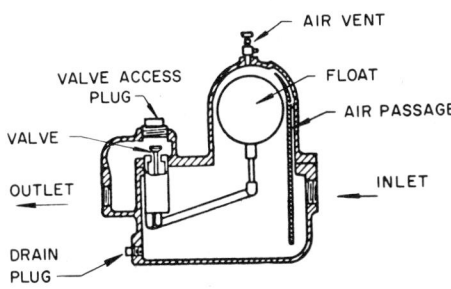

Fig. 72. Float Trap.

STEAM TRAP TYPES

The types of traps commonly used in steam systems are:

Float	Flash
Thermostatic	Impulse
Float & thermostatic	Lifting
Upright bucket	Boiler return or
Inverted bucket	alternating receiver

The description and use of these various traps are presented in the following pages.

Float Trap

The discharge from the float trap is generally continuous. This type (Fig. 72) is used for draining condensate from steam headers, steam heating coils, and other similar equipment. When a float trap is used for draining a low pressure steam system, it should be equipped with a thermostatic air vent.

Thermostatic Trap

The discharge from this type of trap is intermittent. Thermostatic traps are used to drain condensate from radiators, convectors, steam heating coils, unit heaters and other similar equipment.

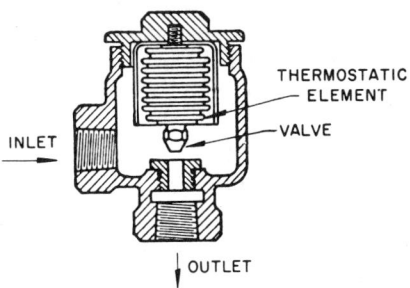

Fig. 73. Thermostatic Trap, Bellows Type.

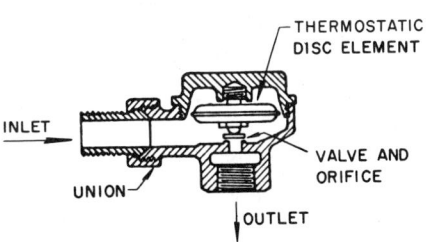

Fig. 74. Thermostatic Trap, Disc Type.

Strainers are normally installed on the inlet side of the steam trap to prevent dirt and pipe scale from entering the trap. On traps used for radiators or convectors, the strainer is usually omitted. Fig. 73 shows a typical thermostatic trap of the bellows type and Fig. 74 illustrates a disc type thermostatic trap.

When a thermostatic trap is used for a heating apparatus, at least 2 ft of pipe are provided ahead of the trap to cool the condensate. This permits condensate to cool in the pipe rather than in the coil, and thus maintains maximum coil efficiency.

Thermostatic traps are recommended for low pressure systems up to a maximum of 15 psi. When used in medium or high pressure systems, they must be selected for the specific design temperature. In addition, the system must be operated continuously at that design temperature. *This means no night setback.*

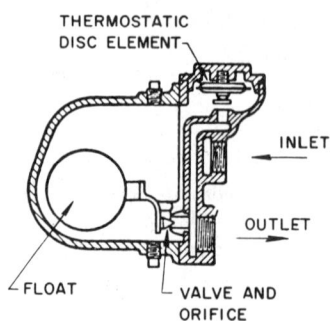

Fig. 75. Typical Float and Thermostatic Trap.

Float and Thermostatic Trap

This type of trap is used to drain condensate from blast heaters, steam heating coils, unit heaters and other apparatus. This combination trap (Fig. 75) is used where there is a large volume of condensate which would not permit proper operation of a thermostatic trap. Float and thermostatic traps are used in low pressure heating systems up to a maximum of 15 psi.

For medium and high pressure systems, the same limitations as outlined for thermostatic traps apply.

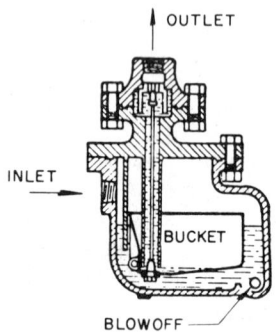

Fig. 76. Upright Bucket Trap.

Upright Bucket Trap

• The discharge of condensate from this trap (Fig. 76) is intermittent. A differential pressure of at least 1 psi between the inlet and the outlet of the trap is normally required to lift the condensate from the bucket to the discharge connection. Upright bucket traps are commonly used to drain condensate and air from the blast coils, steam mains, unit heaters and other equipment. This trap is well suited for systems that have pulsating pressures.

Inverted Bucket Trap

The discharge from the inverted bucket trap (Figs. 77 and 78) is intermittent and requires a differential pressure between the inlet and discharge of the trap to lift the condensate from the

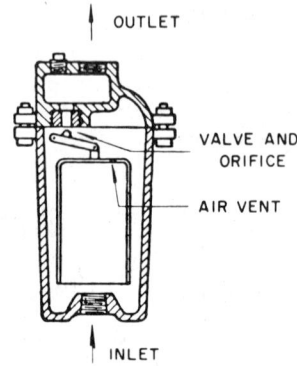

Fig. 77. Inverted Bucket Trap.

bottom of the trap to the discharge connection.

Bucket traps are used for draining condensate and air from blast coils, unit heaters and steam

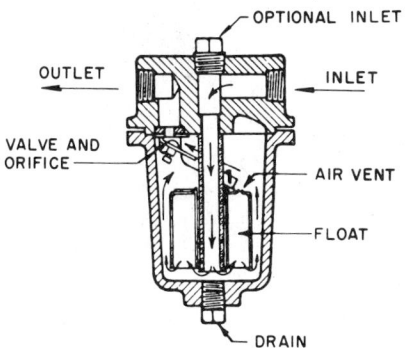

Fig. 78. Inverted Bucket Trap With Guide.

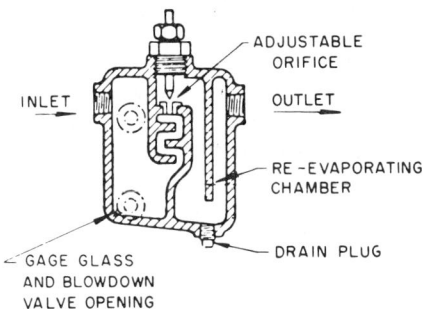

Fig. 79. Flash Trap.

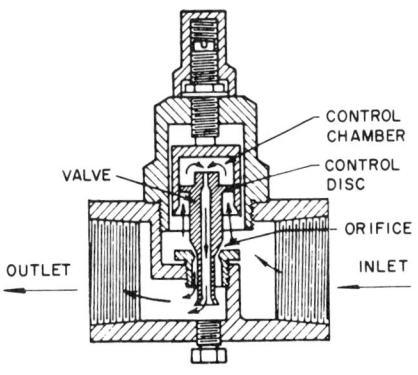

Fig. 80. Impulse Trap.

heating coils. Inverted bucket traps are well suited for draining condensate from steam lines or equipment where an abnormal amount of air is to be discharged and where dirt may drain into the trap.

Flash Trap

The discharge from a flash trap (Fig. 79) is intermittent. This type of trap is used only if a pressure differential of 5 psi or more exists between the steam supply and condensate return. Flash traps may be used with unit heaters, steam heating coils, steam lines and other similar equipment.

Impulse Trap

Under normal loads the discharge from this trap (Fig. 80) is intermittent. When the load is heavy, however, the discharge is continuous. This type of trap may be used on any equipment where the pressure at the trap outlet does not exceed 25% of the inlet pressure.

Lifting Trap

The lifting trap (Fig. 81) is an adaptation of the upright bucket trap. This trap can be used on all steam heating systems up to 150 psig. There is an auxiliary inlet for high pressure steam, as illustrated in the figure, to force the condensate to a point above the trap. This steam is normally at a higher pressure than the steam entering at the regular inlet.

Boiler Return Trap or Alternating Receiver

This type of trap is used to return condensate to a low pressure boiler. The boiler return trap (Fig. 82) does not hold steam as do other types, but is an adaptation of the lifting trap. It is used in conjunction with a boiler to prevent flooding

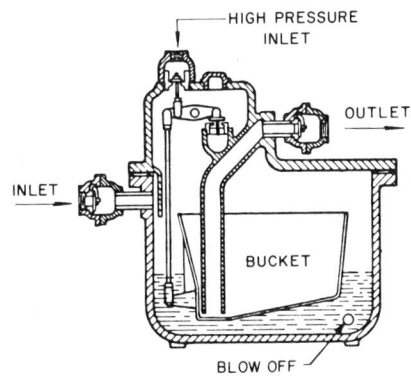

Fig. 81. Lifting Trap.

return mains when excess pressure prevents condensate from returning to the boiler by gravity. The boiler trap collects condensate and equalizes the boiler and trap pressure, enabling the condensate in the trap to flow back to the boiler by gravity.

CONDENSATE RETURN PUMP

Condensate return pumps are used for low pressure, gravity return heating systems. They are normally of the motor driven centrifugal type and have a receiver and automatic float control. Other types of condensate return pumps are the rotary, screw, turbine and reciprocating pump.

The condensate receiver is sized to prevent large fluctuations of the boiler water line. The storage capacity of the receiver is approximately 1.5 times the amount of condensate returned per minute, and the condensate pump has a capacity of 2.5 to 3 times normal flow. The relationship of pump and receiver to the condensate takes peak condensation load into account.

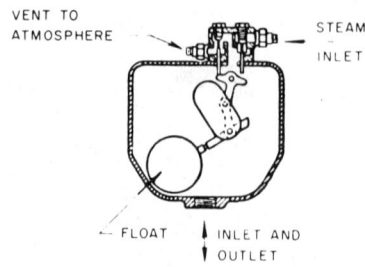

Fig. 82. Boiler Return Trap or Alternating Receiver.

VACUUM PUMP

Vacuum pumps are used on a system where the returns are under a vacuum. The assembly consists of a receiver, separating tank and automatic controls for discharging the condensate to the boiler.

Vacuum pumps are sized in the same manner as condensate pumps for a delivery of 2.5 to 3 times the design condensing rate.

INDUSTRIAL AND COMMERCIAL STEAM REQUIREMENTS

Following are data from many sources concerning steam requirements for a variety of specific industrial and commercial processes for use in gaging the adequacy of an existing steam plant in the case of a proposed expansion of steam-using facilities or for the purpose of sizing the boiler plant for a new facility.

Published Boiler Ratings

Boiler output is specified in a wide variety of systems, including square feet of equivalent direct radiation, steam or hot water, Btu per hr (or Btuh), pounds per hour of steam, and boiler horsepower. Table 1 permits conversion from one system to the other.

Table 1. — Conversion of Standard Boiler Ratings

Boiler Horse-power (BHP)	Lb per Hr Steam, from and at 212 F	MBh or 1000 Btu/hr	Sq Ft EDR (Steam)	Sq Ft EDR (Water)
1	34.5	33.5	139.55	223.30
0.02898	1	0.971	4.044	6.472
0.02985	1.0298	1	4.165	6.665
0.007165	0.2472	0.2400	1	1.600
0.004478	0.1545	0.1502	0.6249	1

Boiler horsepower is a theoretical figure, by definition equivalent to generating 34.5 lb per hr of steam at zero psig (212 F) from feedwater at 212 F.

In comfort heating, the square feet of equivalent direct radiation (EDR) is still in use. When the term was developed, in the days of less efficient gravity systems, a square foot of cast iron radiation was considered capable of radiating 240 Btu per hour when heated with steam or 150 Btu per hour if heated with lower temperature hot water. With modern improvements in heating radiators, making the definition generally obsolete, Btu per hr has become a more acceptable term.

The basic term in boiler output is Btu per hour, the fundamental unit of heat required to raise a pound of water one degree Fahrenheit. A boiler's capacity in Btu per hr is independent of makeup or return temperature or output pressure or temperature. It is the form in which heat losses in buildings are calculated. It is readily convertible into other more arbitrary rating systems.

Pounds per hour of steam is the most generally accepted method of rating industrial steam-using equipment. Steam ratings of boilers as published in lb per hr "from and at 212F" are inaccurate for practical application since feedwater is rarely if ever admitted as high as 212F and steam is never utilized at 212F (zero psig).

Pressure, Temperature Corrections

As a guide to more accurately estimating steam boiler capacity in lb per hr under specific conditions, the following table should be consulted. Note that with excessively low feedwater temperature, a boiler generating steam at reasonably high pressure can have an actual output in lb per hr of steam of only 80% or even less than that indicated in published tables which are "from and at 212F" figures.

Table 2. — Fraction of Boiler Rating Under Operating Conditions

Feedwater Temperature F	Boiler Operating Pressure, Psig						
	10	25	50	100	150	200	250
	Percent of "From and at 212 F" Rating						
40	84	83	82	82	81	81	81
60	86	85	84	83	83	82	82
80	87	86	85	85	84	84	84
100	89	88	87	86	86	85	85
120	90	89	89	88	88	87	87
140	92	91	90	89	89	89	89
160	94	93	92	91	91	91	91
180	95	95	94	93	93	92	92
200	97	97	96	95	94	94	94

Specific Applications

Following, in alphabetical order, are steam consumption data for various applications.

Asphalt Plants. Asphalt, for the preparation of road surfacing materials at bituminous central plants, must be kept hot to insure its ability to flow. A 125-psi boiler operating pressure is normally recommended. Steam at 125 psi will heat asphalt to 300F while steam at 150 psi will heat asphalt to 311F. In general, 50 MBh * is required for each 1000 gal of asphalt to be heated and 170 MBh is

required for each 100 ft of 3-inch steam jacketed pipe. A 10,000-gal tank car, therefore, requires 500 MBh for heating plus whatever heat may be dissipated in the steam jacketed piping.

Baths (Steam and Turkish). Turkish bath houses require 5-10 psi steam. The usual inside temperature is 110F. Allow 33.5 MBh per 100 sq ft of steam radiation. This estimate includes compensation for all losses.

Canning. Retorts, blanchers, steamers, preheaters, cookers and other vegetable canning machinery requiring steam are widely variable in size, design and purpose because of the specialized needs and whims of individual canners, equipment manufacturers and the vegetables themselves.

Because of batch operations, initial steam condensing rates of such equipment are usually high; thus a great deal of judgment is required for even reasonably accurate estimating.

For instance, scalders and blanchers, ranging from 6 to 20 ft in length and 4 to 6 ft wide, require 500-840 MBh of steam. Retorts for cooking filled cans generally fall in the same range. Some types of preheating and scalding equipment—such as required to loosen tomato skins for easy removal—use considerably less steam.

Cheese Vats. An equipment manufacturer has indicated the ratings in Table 3 for jacketed cheese vats. A boiler of 100-psi design pressure is recommended. The usual temperature rise is 60F to 105F.

Table 3. — Steam for Cheese Vats

Vat Capacity, Lb	Steam, MBh *
1,500	107
4,000	211
7,500	402
9,300	486
14,000	704

* An MBh equals 1000 Btu/hr.

Concrete Block. Concrete block, after about a two-hour atmospheric presetting period, is ordinarily cured in kilns with a start-up preheat cycle of high steam demand of perhaps three hours duration during which a maximum temperature of 140-155F is attained using 125-psi steam. This is followed by a three-hour or longer holding period of considerably less steam demand. Frequently, live steam up to 10 psi is injected initially into the kiln in combination with indirect air heating. Good curing practice requires a rate of temperature rise approximat-

ing 60F per hour up to the final 150F.

Such kilns are operated in batteries and are loaded and started one at a time so that the high start-up steam demand of more than one kiln is never imposed on the boiler at one time. Normally, some kilns will be unloading and some will be at holding temperature. However, it is well to establish the actual operating cycle at each specific concrete block operation.

Steam requirements can be safely estimated on the basis of a production rate of 1.5 standard blocks (8 x 8 x 16 inches) per MBh.

As an example, a plant with a block-making machine turning out 800 blocks per hour probably would have six 2500-block high-temperature kilns. On an 8-hr curing cycle, which is standard in the industry, the plant can thus be operated on a two or three-shift basis.

A maximum production load might be three kilns operating at one time with the other kilns being loaded and unloaded so that 7500 blocks (3 kilns x 2500-block capacity) might be produced in an 8-hr shift, requiring a 5025 MBh boiler.

Concrete Ready-Mix. Concrete ingredients contain varying amounts of moisture from outdoor storage and must be dried in both summer and winter weather before mixing to keep the amount of moisture in the finished concrete within proper values required by the mixing formulas. Steam is ordinarily used at high pressure, on the high side of the 100-150 psi range. For average conditions, aggregate is assumed to contain 5% moisture.

Table 4 gives steam required to dry one ton of aggregate with original moisture content up to 10% and at initial temperatures of 20F through 50F. At higher storage temperatures aggregate moisture content is proportionately higher, requiring increasing amounts of steam for adequate drying.

Table 4. — Steam Required to Dry Aggregate

Percent Moisture	Initial Temperature of Aggregate, F			
	20	30	40	50
	Lb per Hr Steam per Ton of Aggregate			
2	14.4	17.9	21.4	24.9
4	20.6	24.1	27.6	31.1
6	26.6	30.1	33.6	37.1
8	32.7	37.2	39.7	43.2
10	39.0	42.5	46.0	49.5

Steam is used to heat water for mixing with cold aggregate in winter climates to provide the required

60F wet concrete temperature for proper setting. For example, with aggregate at 35F and using 30 gal of water per cubic yard of concrete, water temperature must be 140F to produce 60F concrete. At higher aggregate temperatures and in summer climates, heated water is not required.

Table 5 presents water temperatures necessary to produce 60F concrete when the aggregate is at initial temperatures of 35F or 40F, using amounts of water varying between 15 gal and 40 gal per cu yd.

Table 5. — Mixing Water Temperature for 60 F Concrete

Aggregate Temperature, F	Water Required, Gal per Cu Yd					
	15	20	25	30	35	40
	Water Temperature, F					
35	—	180	155	140	130	120
40	190	155	135	126	115	110

Calculations of steam required for heating water are available elsewhere. A convenient rule of thumb is to assume 1 Btu required per degree of temperature rise per pound of water, using 8.3 lb of water per gallon.

Cookers, Coil. Submerged coils in open kettles are used widely in chemical and food processing operations. For instance, tomato puree is concentrated one-third for catsup in coil-heated open cookers. There are a number of coil arrangements, such as single and double spirals and various combinations of series and parallel connections. Variations in steam requirements for these different designs are inconsequential. Piping used for the coils is normally 3-inch dia.

Steam is usually utilized in the 100-150 psi range. A 1000-gal cooker requires about 6700 MBh during the 3 or 4 minutes when the watery puree is being brought to a boil and about 3400 MBh to maintain evaporation during the remainder of the cycle—about 40 min in tomato concentration. These cookers range from about 750 gal to 1500 gal capacity. Occasionally, coil cookers down to 250 gal capacities are found. Usually, steam jacketed kettles are used for smaller capacities.

In batteries, a quarter of the cookers may be starting up, a quarter may be boiling, another quarter may be washing up, while the final quarter may be reloading for another cycle. Therefore, on batch operations of this kind, peak load is always considerably less than connected load.

Cookers (Jacketed Kettles). Jacketed kettles are generally used where smaller volumes per batch are processed. For cooking jams, jellies, preserves, and various types of candies and confections, a 60-gal or 70-gal kettle is usually used. These also may be equipped with electrically driven mixers, evacuation devices for vacuum cooking and other accessories. Steam pressure rarely exceeds 70 psi.

A 150-gal to 300-gal jacketed kettle may be figured at 838-1675 MBh while the smaller 60-gal size will require 335-503 MBh. As with coil cookers, any battery of kettles will consist of about one quarter condensing at the maximum startup rate while half may be considered as cooking normally at the minimum load. The remainder would be in the midst of filling, dumping or cleaning-up.

Cookers, Rendering. Rendering cookers are jacketed, horizontal, cylindrical vessels, usually 5 or 6 ft in diameter and about 12 ft long, for recovering fat for soap and chemical manufacturing. Some method of rotation is employed, usually an internal agitator. Pressures are normally in the 60-70 psi range and the units condense at a rate of about 2000 MBh.

One manufacturer's cooker has a steam consumption of about 3/4 lb steam per pound of raw material. Thus, an 8000-lb capacity cooker requiring a 3-hr cycle would condense approximately 6000 lb of steam at a rate of 2000 lb per hr, corresponding roughly to 2350 MBh.

Distilleries. Distilling equipment is varied and somewhat complicated. In one typical arrangement, live steam is bubbled into the product through a perforated sparge pipe. In making complete calculations on a heat-input-equals-heat-output basis, the constants in Table 6 may be used:

Table 6. — Thermal Data for Water and Alcohol

Liquid	Specific Heat	Specific Gravity	Heat of Vaporization, Btu per Lb
Water	1.0	1.0	970
Alcohol	0.6	0.8	363

A good average is about 24 lb of steam required per wine gallon of 190-proof, in which distilling capacity is rated. Thus, a still producing 500

wine gallons per hour would require 12,000 lb steam per hr.

Drycleaning. Because more delicate fabrics are handled, drycleaning operations are usually performed at lower steam pressures and temperatures —70 psi—than laundry operations—100 psi. Principal operation in drycleaning is the solvent still. Cleaning solvents are used over and over and require the use of filters and stills to purify the solvent after each cleaning cycle. Depending on nominal still size, steam requirements range from about 50 to 500 MBh. See Table 7.

Table 7. — Dry Cleaning Still Steam Requirements

Size, Gal	Steam, MBh	Size, Gal	Steam, MBh
25	50	125-150	168
50	67	175-200	200
75	100	250	300
100	134	350-400	470

The small amounts of live steam used in hand-spotting operations and some finishing operations generally are inconsequential.

Steam requirements of tumblers and finishing equipment are about the same as for laundries and will be found elsewhere in this section under those specific types of equipment.

Dryers, Double-Drum. Double-drum dryers are frequently used to dry milk, brewer's yeast, various other foods, and chemicals. Steam is usually utilized at less than 100 psi. The dryers consist of two cylindrical, steam-filled rolls on parallel horizontal axes revolving slowly toward each other. In the center, between the two rolls, the fluid to be dried is sprayed thinly on the two rolls. After about three-quarters of a revolution the dried residue is removed by long doctor blades and the powdered product drops down below the rolls where it is picked up by an integral screw conveyor for continuous removal.

Table 8 presents steam requirements of several popular dryer sizes based on the size of *one* of the twin drying rolls.

Since operation is continuous rather than in batches, multiple units in a battery could all be conceivably in service simultaneously. The plant's specific cycle of operation should be explored to arrive at as accurate an estimate of steam load as possible.

Table 8. — Double-Drum Dryer Steam Requirements

Drum Size, Inches	Steam, MBh	Drum Size, Inches	Steam, MBh
24 x 60	1000	36 x 100	3350
28 x 72	1500	36 x 120	4200
36 x 84	2345	—	—

Dryers, Revolving Steel Tubular. Revolving steel tubular dryers consist of stationary steel piping inside a rotating steel shell. Sloped for continuous feeding and removal of dried product, they are used for drying grain, pigments, and various forms of granular solids. In this way, spent distillery and brewery mash is pressed dry, then dried for high-vitamin-content stock feed.

A 6-ft dia by 30-ft long unit, operated at 50 psi, condenses about 3350 MBh. During the start-up, the steam requirement is only slightly greater. In batteries, of course, all units may be operating simultaneously if the process is continuous rather than batch.

Dryers, Spray. The convenience to manufacturers of other foods of handling dried milk and the acceptance of reconstituted dry milk solids as a home beverage have led to tremendous increases in milk drying operations. Several methods are available for recovering the dry solids from milk, which is nearly 90% water. One of the most popular methods is spray drying, in which the milk is sprayed into a steam heated atmosphere, the water thus evaporated, and the white crystalline nutrients recovered by gravity (see, also, Dryers, Double-Drum).

Spray dryers for recovering dried milk solids can generally be estimated as requiring about a quarter of the pound-per-hour milk input capacity in boiler horsepower. Thus, a spray dryer rated to handle 1000 lb of raw milk per hour would require 8375 MBh of steam capacity.

Dyeing. The process of dyeing consists of dipping fabrics or yarns in a dye solution (called liquor) at boiling temperature, rinsing and drying. In the ordinary dye vat, steam jets in the tank bottom heat the liquor to boiling temperature. The turbulence from the jets agitates the liquor and mixes the solution thoroughly.

The following formula provides a practical method of computing the MBh required for a dye vat:

Btu/hr = (gal capacity) (8.33 lb/gal) (temp. rise to 212F)

Assume a vat of 400-gal capacity and water supply temperature of 60F:

(400 gal) (8.3 lb/gal) (212F — 60F)
= 505,000 Btu/hr or 505 MBh

This formula is based on raising the temperature of the liquor to the boiling point in one hour. If the vat is to be brought up to temperature in 30 min, multiply the result by 2, if 20 minutes multiply by 3, etc.

After the cloth has been dyed it must be dried, usually in a heated compartment or cabinet. The load can be determined from the size of the pipe as indicated under *Kilns*, below.

Feed Mills. In pelleting various ground grains with vitamins and other fortifying agents for stock feed, live steam at 10-30 psi (240-275 F) is admitted to the feed, both for heating and as a moistening agent as part of the process of compressing the material into solid lozenges from very small diameters for baby chicks to ¾-inch range cubes for cattle. Frequently, because of production problems, baby chick feed is processed into relatively large diameters and then recrumbled or reground into appropriate sizes for such small animals.

Steam consumption is normally estimated at about 50 lb of steam per 1000 lb of animal feed at the input. Some feeds, alfalfa for instance, may require as little as 30 lb of steam per 1000 lb of feed. Since there may be about 5-10% loss of fines in the processing, if output figures are used to measure the size of the production machinery, increase the output capacity by 10% for use in estimating steam requirements.

When a low-pressure boiler is used, steam pressure available at the steam controls on the pellet mill usually will be around 8-10 psi due to line losses, even with 13-15 psi boiler pressure. A high-pressure boiler, with steam reduced and regulated to 20-30 psi at the pellet mill, is more flexible and will give better control of the processing. Accurate temperature regulation is of primary importance for quality control. Effectiveness of the steam pressure regulation system should be carefully studied to avoid later complaints against the boiler.

In estimating the steam requirements, a convenient rule of thumb therefore can be 5% of the rated input capacity to the pellet mill. Thus, a pellet mill producing 10 tons (20,000 lb) per hour

should be increased by 10% to 22,000 lb per hr to arrive at feed input compensating for loss of fines in processing and multiplied by 5% to arrive at a steam consumption figure of 1100 lb of steam per hour. Since steam is condensed in the feed and not returned to the boiler, this is a 100% makeup application. Installation of a 1350 MBh boiler is better than attempting to squeeze by with a 1000 MBh boiler. Cold makeup would reduce the 1000 MBh boiler steaming capacity under actual operating conditions to a figure well below the required 1100 lb of steam per hour.

Garbage Cooking. To prevent the spread of disease, garbage used for hog feed is cooked at 212 F for at least 30 min, usually by bubbling live steam through the batch via a perforated or nozzled piping system. Ordinarily, steam is used at about 10 psi. 276 lb steam per hr has been established as the steam requirement to cook a ton of garbage. Garbage, for purposes of estimated calculations, is considered to weigh 4.6 lb per gal and 1000 lb per cu yd.

Assume a garbage load in a truck body 14 ft long, 7½ ft wide, and 2½ ft deep:

(14)(7½)(2½) cu ft/27 cu ft per yd = 10 cu yd

(10 cu yd)(1000 lb per cu yd) = 10,000 lb = (5 tons) (276) = 1380 lb/hr

(5 tons)(8 bhp per ton) = 40 bhp

Most states have laws requiring cooking of garbage used for hog feed, establishing minimum temperatures and time limits as is here given.

Greenhouses and Soil Sterilization. Fresh vegetables are frequently grown under glass, even in the coldest climates, year-round. Tomatoes, for instance, may be grown at the rate of three or four crops a year at temperatures of 55-58F, increased to 62-65F in milder weather. Greenhouse heating loads may be estimated at about 4200 MBh per acre (43,560 sq ft) although greenhouse sources recommend 5000 MBh.

Steam for soil sterilization to kill weeds between crops is usually admitted through buried porous pipes or tiles to maintain a section under treatment at 212-215F for 4 to 6 hours.

One method of controlling soil sterilization is to include a raw potato under the tarpaulins with which the ground being sterilized is covered to retard the escape of steam. When the potato is cooked, the soil is sterilized.

Sterilizing steam requirements are 4 or 5 times

greater than the space heating requirement and a boiler sized for normal space heating can be used for soil sterilization on the basis of one section at a time. This is also practicable since the remainder of the growing area is usually producing plants in various growth stages.

Kilns (Bare Pipe Radiation). For bare pipe radiation as utilized in brick, lumber, veneer and wallboard dryers and many other similar kiln-type applications, a factor of 0.3 lb per hr of steam for each square foot of pipe surface is a fairly accurate approximation.

For instance, assume an installation contains 1500 ft of 3/8-inch finned copper tubing and 1800 ft of 1½-inch steel pipe in each of five kilns. The radiating area may be calculated from the pipe OD (converted to feet) in each case:

$$\pi D(0.675 \text{ inch}/12) \times 1500 \text{ ft} = 264 \text{ sq ft}$$
$$\pi D(1.900 \text{ inch}/12) \times 1800 \text{ ft} = 895 \text{ sq ft}$$

Total radiating area per kiln = 264 + 895 = 1159 sq ft

Total radiating area = 1159 sq ft/kiln × 5 kiln = 5795 sq ft

5795 sq ft × 0.3 lb per hr per sq ft = 1738.5 lb per hr of steam

Laundries, Commercial. Commercial laundries require substantial quantities of steam for washing, drying and finishing operations. Steam is generally utilized at pressures slightly above 100 psi (338 F) to provide sufficiently high temperatures for finishing operations.

Heating of wash water to 180 F is probably the peak source of boiler load. Drying and finishing operations usually require less steam and are staggered with the actual washing although the individual laundry's operating procedure should be carefully checked in each instance. The number of wash wheels and their dimensions determine hot water load with reasonable accuracy. See Table 9.

A double-check is based on the poundage of dry, dirty clothes being laundered either by the laundry owner's estimate or rapid calculation based on two rules-of-thumb:

1. One cubic foot of wash wheel volume contains 4-6 lb of dry dirty clothes.

Example: 36-inch dia x 54-inch wash wheel holds

$$\frac{\pi}{4}\left(\frac{36}{12}\right)^2 \text{ft}^2 \times \left(\frac{54}{12}\right)\text{ft} \times \left(5\right)\frac{\text{lb}}{\text{ft}^3} = 160 \text{ lb}$$

2. About 3 gal of hot water are required per pound of dry dirty clothes washed.

Example: Dry clothes may be assumed to absorb between 1/4 and 1/3 gal of water per pound. Individual laundrymen have their own preferences as to specific washing cycles but typically one might be four suds with the wash wheel filled to a 3-in. level (about 21 gal each for a 36-inch x 54-inch unit) and five rinses filled to a 9-in. level (50 gal):

Absorption	0.33 gal per lb × 160 lb =	53 gal
Suds:	4 x 21 gal @ 3 inches =	84
Rinses:	5 x 50 gal @ 9 inches =	250
		387 gal

387 gal per 160 lb is 2.41 gal hot water per pound laundry; 3 gal per lb of dry dirty clothes is a safely conservative figure to use in estimating.

The sum of the individual capacities of the washers will give the maximum gallons of hot water required. In actual practice, a wash round can be completed in less than an hour. However, allowing for loading and unloading the wheels, starching, and the use of low-temperature water on certain rounds, the preceding estimate is reasonable.

Given four 42-inch wheels x 96-inch wheels requiring (from Table 9) 4 x 720 gph to be heated from 40-180 F.

$$4(720 \text{ gal/hr})(8.3 \text{ lb/gal})(1 \text{ Btu/lb/F})(180-40 \text{ F}) = 3,340,000 \text{ Btuh}$$

This is approximately equivalent to a 100 hp boiler requirement.

Some laundries are equipped with reclaimers to preheat incoming water with otherwise wasted heat in the laundry wash water going to the sewer and thus reduce the temperature differential.

Another factor that may reduce the amount of

Table 9. — Wash Wheel Hot Water Requirements

Wheel Dia, Inches	Hot Water, Gal	Wheel Dia, Inches	Hot Water, Gal
30 x 30	110	42 x 108	810
30 x 36	130	44 x 36	300
30 x 48	190	44 x 54	450
36 x 36	200	44 x 60	495
36 x 48	270	44 x 72	595
36 x 54	300	44 x 84	695
36 x 64	350	44 x 96	790
42 x 36	270	44 x 108	900
42 x 54	405	44 x 120	990
42 x 64	475	54 x 84	870
42 x 72	540	54 x 96	990
42 x 84	630	54 x 108	1110
42 x 96	720	54 x 120	1230

boiler capacity required is a hot water storage tank, smoothing peak hot water requirements.

Laundries, Self-Service. The increased popularity of automatic self-service laundries has resulted in new type of laundry equipment. This hot water load can generally be estimated at 38 gal of 140 F water per hour for each machine on the basis of a 9-lb clothes load, a 30-50 minute washing cycle, and a requirement of 25 gal of 140 F water per cycle.

Laundry and Drycleaning—Finishing Equipment. Flatwork ironers or mangles are the largest consumers of steam in the area of finishing equipment. They are used for pressing tablecloths, sheets, pillowcases, napkins and other flatwork. These machines have a large production output when properly operated with sufficient temperature. Wet multiple thickness hems of sheets and pillowcases are usually the first indication of inadequate pressing temperature (and steam pressure at the ironer inlet). Steam requirements per flatwork ironer range from 100 to 670 MBh/hr. See Table 10.

Table 10. — Flatwork Ironer Steam Requirements

Ironer Size	Length of Rolls, Inches	
	100	120
	MBh per hour Requirements	
2-roll	104	127
4-roll	211	255
6-roll	300	330
8-roll	430	476
12-roll	—	650

Large single- or dual-cylinder (24-48-inch dia.) mangles have largely been replaced by the more modern multiple cylinder steam chest flatwork ironers. Where the old type machines are still in use, a 335 or 500 MBh figure can be used.

Drying rooms or cabinets are used for drying curtains, blankets and bedspreads. These are usually heated by radiators or pipe coils. An estimate of 2 bhp is usually adequate. In unusual cases of special designs, radiation load can be computed as for comfort heating or as described under *Kilns*. However, these cabinets are not in frequent use and usually do not increase the maximum load.

Shirt finishing units, consisting of several pieces of specialized pressing equipment—collar, yoke, bosom, sleeves, cuffs—are designated as "2-girl" or "4-girl" units and require from 165 to 330 MBh per installation.

Individual apparel presses range from small pony, handkerchief or hosiery form presses, using from 1 to about 7 MBh. Such presses cannot be totally disregarded but they can be lumped together and estimated roughly at a total of 130 or 170 MBh since they do not increase the steam requirement greatly in comparison with the other larger laundry or drycleaning equipment.

Laundry and Drycleaning — Tumbler Dryers. Since most garments are pressed in a damp condition, the extractor plays the biggest part in the drying operation of a laundry. The extractor removes excess moisture from clothing by centrifugal action and does not require any steam. Further drying is provided by tumbler dryers for some of the delicate garments and for fabrics that must be dry before finishing.

The tumbler consists of a cylinder with perforated outer wall mounted inside a steam-heated cabinet. The cylinder is rotated while warm air circulates through the cabinet in the same general way that an electric or gas heated home laundry dryer operates. Size of the tumbler dryer is determined by the dimensions of the cylinder. Ratings for the more popular sizes are given in Table 11.

Table 11. — Tumbler Dryer Steam Requirements

Tumbler Size, Inches	Steam, MBh	Tumbler Size, Inches	Steam, MBh
36 x 24	124	42 x 40	218
36 x 30	130	42 x 60	268
36 x 42	150	42 x 90	400
36 x 48	201	42 x 120	536

Because of more delicate fabrics in drycleaning applications, fewer coils are sometimes used, thus reducing the steam requirements. However, Table 11 data are adequate for estimating purposes.

Ovens, Proof Boxes (Bakeries). Specialized requirements for steam in bakeries include ovens and proof boxes. Other steam uses include jacketed kettles (described elsewhere) for confections, hot water for washing, and space or comfort heating.

Proof boxes are used to speed the process of raising yeast dough before baking in ovens. The proof box is similar to a steam room, since steam is injected directly into the box to obtain a warm, humid atmosphere. The amount of steam released is regulated in accordance with the amount of dough to be raised.

Table 12. — Steam Requirement for Typical Commercial Ovens

Revolving Ovens			Traveling Ovens		
Number and Size of Trays, Inches	Capacity in 1-Lb Loaves	MBh Steam	Number and Size of Trays, Inches	Capacity in 1-Lb Loaves	MBh Steam
4-20 X 54	80	107	10-24 X 96	320	430
4-20 X 80	128	170	10-24 X 108	360	482
6-20 X 54	120	161	10-26 X 98	320	430
6-20 X 80	192	258	10-26 X 110	400	536
6-23 X 92	216	291	12-24 X 96	384	515
6-23 X 110	264	355	12-24 X 108	432	580
6-23 X 128	288	385	12-24 X 120	480	643
6-23 X 146	352	450	12-26 X 98	384	515
8-23 X 92	384	385	12-26 X 110	480	643
8-23 X 110	352	470	15-24 X 108	540	725
8-23 X 128	384	515	15-26 X 110	600	805
8-23 X 146	448	600	16-24 X 120	640	855
10-23 X 110	440	590	16-26 X 110	576	775
10-23 X 128	480	643	18-24 X 108	648	870
10-23 X 146	560	750	18-26 X 110	648	870
12-23 X 128	576	775	20-24 X 120	800	1070
12-23 X 146	672	900	22-24 X 120	880	1180

Steam requirements for proof boxes are normally inconsequential in comparison with other larger steam loads. Commercial types of proof boxes are rated by the number of racks they contain on the basis of 10 MBh per rack. The standard proof rack is about 30 inches wide by 6 ft long and holds 8-10 trays of bread. Proofing time varies from 45 to 60 min.

Baking ovens are of three types—the simple hearth oven, the traveling oven, and the revolving oven. In all types, steam is injected into the baking chamber to cure the bread or to glaze the surfaces of certain breads and rolls. The normal baking time for bread is 30 min, so that bakeries usually average two bakes per hour. Regardless of type of oven, a practicable rule of thumb is to assume about 67 MBh of steam required per 100 lb of bread baked per hour. Thus, an oven designed to hold 320 one-pound loaves could bake 640 lb of bread per hour, requiring about 430 MBh of steam capacity.

Hearth type oven capacities can be calculated from the number and dimensions of trays used.

Revolving ovens may contain from 4 to 12 trays which revolve on a wheel somewhat like a ferris wheel.

Traveling ovens are in continuous motion, being loaded with, and discharging, bread simultaneously, at opposite ends of a 100-ft long tunnel oven.

Table 12 gives typical commercial oven sizes, indicating the bread capacity and steam requirement of each.

Paper Corrugators. Paper corrugators are integrated machines that take rough paper filler, spray it with steam for softening, and corrugate it over toothed rolls. Following this, heavier liners are glued to one or both sides and the assembled sheet is passed rapidly over steam heated plates. Steam-filled preheating cylinders are frequently used ahead of the corrugating operation. The entire process may run at speeds of 200 to 400 ft per min. Steam is used at up to 175 psi. Corrugating machines condense from 2240-5175 lb/hr steam although a specific machine can be checked with the manufacturer.

A convenient rule of thumb is to use 40 lb of steam per 1000 sq ft of corrugated board produced. About 15% of the steam is directly condensed in softening the filler sheet although the remainder is easily recovered as condensate. With 60-inch board running at 300 fpm, this comes to 3600 lb per hr of steam.

Additional processing loads and plant and office heating should also be considered in sizing the boiler.

Paper Making. Paper is made from a slurry of ground wood pulp, waste paper or rags spread out and drained on a wire screen or a felt blanket and finally dried by passing continuously over the surface of steam-heated rotating drum dryers.

Steam is ordinarily utilized at relatively low pressures, under 50 psi, although newer machines are being constructed to utilize pressures up to 175 psi. Frequently, two or three successively increasing pressures are used in consecutive banks of drying rolls which may total anywhere from about 25 to 100 in number. The lower pressure (and temperature) steam is applied first to heat the new paper slowly. Higher pressure and temperature, if applied first, might case-harden the outer surface of the paper sheet. Moisture sealed inside later explodes and wrinkles the finished sheet.

In the absence of flow meter readings, a figure of 3½ or 4 lb steam per pound of paper produced may be used for paper machine dryers. Thus, a mill producing 30 tons of paper per 24-hour day would be producing 1¼ tons or 2500 lb of paper per hour requiring about 10,000 lb per hr of steam.

Occasionally, water for the pulp slurry will be preheated. Other steam process loads and plant and office heating loads should be calculated before selecting the boiler size.

Pasteurization. Conventional pasteurization in the dairy industry consists of raising milk to a predetermined temperature well short of boiling and holding it at that temperature for approximately 30 min to kill disease-carrying bacteria. The process is also employed in canning and in bottling fruit juices.

Some equipment is sized in terms of gallons of product per minute, other types may be sized in total gallons per 20-min or 30-min cycle. In the latter case, this can be converted readily into a gpm figure. A 400-gal pasteurizer for a 30-min cycle can be said to be a 13-1/3-gpm size.

A convenient estimate of steam need in boiler horsepower is to use 1½ times the nominal equipment size rating as specified in gallons per minute. Thus, a 400-gal pasteurizer for a 30-min cycle would require about 20 bhp of steam.

Platen Presses. Platen presses, used in the production of plastic parts and components, veneer plywood, and other manufactured products, consist of two or more steam-filled plates or platens which are brought together under high mechanical pressure for forming under combined heat and pressure as a part of each manufacturing cycle.

One convenient rule-of-thumb is to estimate about 30 lb steam per hr or 1 bhp for each 12-15 sq ft of exposed platen surface.

Process Heating. There are many variations to the application of heat by steam in the process industries. Frequently, determination of steam requirements must be accomplished by primary heat and steam formulas.

In the basic approach to the measurement of steam required for process heating, the following factors must be considered:

(*a*) Heat to raise temperature of substance.

(*b*) Heat to raise temperature of heating vessel.

(*c*) Heat to overcome radiation losses.

The greatest amount of steam is consumed in raising the temperature of the substance and the heating vessel to the operating temperature of the process. This is the maximum load which must be provided by the boiler. Maintenance of operating temperature, in batch operations, merely requires replacement of losses due to radiation, convection and conduction. They can be safely ignored in these rough calculations.

The heat required, in Btu, is equal to the weight of the substance in pounds times its specific heat plus the weight of the vessel in pounds times its specific heat, all multiplied by the temperature rise in degrees Fahrenheit:

$$\text{Btu required} = (W'S' + w''s'') \times (\deg \text{F temperature rise})$$

This is the total heat required per batch and must be related to time to get the steam needs. If the cycle is accomplished in one hour, the resulting total Btu requirement can be converted directly into pounds per hour of steam or boiler horsepower. If the cycle is only 30 min, or 15 min, the Btu figure must be doubled or quadrupled to get an equivalent Btu per hr rate for conversion to lb per hr or bhp steam need.

Specific heats of some typical vessel materials and substance materials are given in Table 13. Other specific heats can be found in standard reference sources.

Obviously, in practical plant situations, a great deal of guesswork is necessary in the selection of approximate specific heat figures as well as the weight of the vessel or heat exchanger. Weight of the substance being processed can usually be more accurately estimated from plant production figures. Erring high in estimates of weights and in selection of specific heat figures will give higher than necessary steam load figures and provide adequate safety factors for these rough estimates.

In continuous operations, only rate of heating of the substance really need be considered. Normal load then is the amount of heat added to the substance at its flow rate.

Table 13. — Specific Heats of Various Materials

Vessel Material		Substance Material	
Aluminum	0.22	Ammonia (gas)	0.50
Brass	0.09	Acetic acid	0.47
Copper	0.09	Air	0.24
Iron	0.11	Ethyl alcohol	0.58
Lead	0.03	Glycerin	0.57
Monel	0.13	Hydrogen peroxide	0.58
Magnesium alloy	0.25	Olive oil	0.47
Stainless steel	0.11	Paraffin	0.69
—	—	Sugar (solid)	0.28
—	—	Sulphuric acid	0.34
—	—	Turpentine	0.41

For example, assume a continuous flow of a vegetable oil through a copper tube heat exchanger in a pharmaceutical, food or cosmetic process at a rate of 30 to 50 gpm, or 400 lb per min—24,000 lb per hr. You might estimate the heat exchanger weighs 1000 lb. The temperature rise undoubtedly would be well documented by plant officials, say, from 60 F to 150 F. Thus, maximum start-up Btu per hr load becomes:

$$\frac{\left[(1000 \text{ lb})(0.09) + (24,000 \text{ lb})(0.50) \right] (90 \text{ F})}{(90 + 12,000)(90)}$$
$$1,088,100 \text{ Btu/hr}$$

This amounts to about 1100-1200 lb per hr of steam. Note that the weight and the specific heat of the vessel are insignificant in establishing the steam requirement of the process.

Restaurants. Because of lighting, cooking, and body heat, the comfort heating load requirements of a restaurant are on the low side. But, because restaurants are an integral part of more and more office buildings, industrial plants, apartments, hospitals, hotels and motels, and other installations the restaurant's process steam load must be considered.

Standard steam tables may be estimated to require about 30 lb steam per hr per ft of length. A bain-marie (hot water filled "steam" table) is considered at half of a steam-table heating requirement.

Jacketed holding kettles for stock, ranging from 10-60 gal capacity, require from 15-60 lb steam per hr.

Hot water loads are uniformly high because the sanitary use of hot water is spread out over the entire day. Restaurant dishwashers may use 2-8 bhp of steam. Restaurants serving only two, or even one, meal a day would have relatively higher peak hot water demands over a shorter time period—say, only 2 hr—and consequently, in the selection of equipment, require greater hot water storage or quicker recovery.

Snow Removal. Snow and ice is removed from sidewalks, driveways, airports, loading ramps and other paved areas by circulating 140 F water through a piping grid embedded in the pavement. A convenient rule of thumb is to assume a 100-Btu heat requirement to remove a square foot of 1-inch snow or 0.1-inch of ice. This works out to 335 sq ft of 1-inch snow or 0.1-inch ice per bhp. Thus, in figuring on removing a 2-inch snowfall from a filling station driveway measuring 100 ft on each side, an additional heating load of 2,000 MBh must be considered.

Sterilizers, Autoclaves. Sterilizers are used by the medical profession for surgical dressings and hospital supplies and industrially in various phases of the manufacture of pharmaceuticals, packed food products, textiles, plastics, hair and bristle products and allied processing operations.

The steam consumption figures in Table 14 are

Table 14. — Steam Requirements for Sterilizers

Size Sterilizer, Inches	Fabric Loads	Solution Loads
	Steam Consumption, Lb per Hr	
24 x 24 x 36	55	60
24 x 24 x 48	65	70
24 x 24 x 60	70	80
24 x 36 x 48	80	90
24 x 36 x 60	85	100
30 x 42 x 84	125	155
30 x 48 x 84	135	175
36 x 42 x 84	145	185
42 x 48 x 96	210	265
48 x 54 x 96	260	340
48 x 60 x 96	285	375
60 x 66 x 96	395	515
60 x 66 x 156	700	900
60 x 66 x 228	1000	1200

based on uninsulated sterilizers. If 2-inch magnesia insulation or its equivalent is applied, steam consumption figures can be reduced by 25%.

The operating cycle assumes starting with a cold fully loaded machine operating through a complete cycle in one hour (15 min slow heating of jacket to 250 F and 15 psi pressure, 30 min exposure of load to 240-250 F, 15 min for drying during which jacket is maintained at 15 psi).

In battery operations, sterilizers should be put into operation on a staggered schedule. To insure adequate steam pressure within the sterilizer, it is recommended that 25% to 50% be added to the steam load to compensate for line losses and that 40-psi steam pressure be available and reduced at the sterilizer inlet.

Tire Recapping. Tire recappers are increasingly important users for steam generating and steam handling equipment, both in this country and overseas. The best source of steam requirement information is the manufacturer of the tire recapping vulcanizers under consideration. In Table 15, based on tire sizes, are some reasonably accurate estimates of steam needs.

Table 15. — Steam Consumption for Tire Recapping

Tire Size	Steam, MBh
600-16, 650-15, 700-15	33.5
600-20, 700-20, 900-16	50
750-20, 1000-20	67
1200-24, 1400-20	84
(off-the-road equipment)	335

Vacuum Pans. Condensed or evaporated milk is produced by evaporating part of the moisture from whole or skim milk. The evaporating or condensing takes place in vacuum pans which are operated under approximately 25 inches of vacuum. The raw milk is usually preheated to about 210 F before entering the vacuum pan. Depending on the ratio of concentration, Table 16 is a guide to the steam requirements for vacuum pans of various diameters.

Table 16. — Steam Requirements for Vacuum Pans

Vacuum Pan Dia., Inches	Pan Output, Lb per Hr	Steam, lb/hr
16	21-165	100-200
26	56-450	250-590
36	131-1050	600-1360
42	248-1980	1170-2450
48	416-3330	1960-4300
60	574-4590	2700-5900
72	979-7830	4600-10,000

Washers, Bottle. Many types of bottle washers are in use by dairies, breweries, soft drink bottlers and other industries. The manufacturer of each particular machine normally provides the information regarding steam need which is essential for heating wash water to the required 180 F-200 F in a specified time from an assumed minimum water supply temperature.

There is little difference in the hot water consumption related to bottle size—half-pint, pint, quart, half-gallon, gallon—but the total amount of water required should be determined carefully and the steam load to produce it then added to the steam needs of other specialized plant process and comfort heating equipment to arrive at the total boiler load.

Bottle washers are identified in terms of bottles-per-minute capacity. If nameplates are obliterated the machine operator or plant management can estimate the operating rate. A consensus of manufacturers data indicates maximum steam need and water supply required as given in Table 17, related to the holding capacity and rate of bottle washing.

Table 17. — Bottle Washer Steam Consumption

Bottle Capacity	Bottles per Minute	Fresh Water Required, Gpm	Steam, MBh
5000	400	72	2000
4000	320	60	1675
3000	240	50	1350
2500	200	45	1200
2000	160	40	1000
1500	120	32	670
1000	80	21	500
500	50	15	335

FLASH STEAM CALCULATIONS

Where high temperature condensate or continuous boiler blowdown is available, there may be opportunities to make controlled use of flash steam. The energy in water at high pressure and near saturation temperature produces flash steam instantaneously when pressure is reduced below saturation. The amount of steam so produced depends on the enthalpy of water and steam at each temperature and pressure.

As an example, if saturated water at 70 psia and 303 deg F with enthalpy of 272.6 Btu per lb is discharged into a tank maintained at 20 psia, it must immediately assume the saturation conditions of the lower pressure, 228 deg and 196.2 Btu per lb. Some of the energy which had formerly held the temperature at 303 deg is converted into latent heat

of steam at the new temperature. The amount of energy so converted is 272.6 − 196.2 = 76.4 Btu per lb. Since the latent heat of steam at 20 psia is 960.1 Btu per lb, this excess heat is enough to produce 76.4/960.1 or 0.08 lb of steam per lb of high temperature condensate, or, in other words, 8% of the condensate will flash into steam at the lower pressure. A similar 50 lb drop in another range of the pressure scale, say from 100 to 50 psia, would

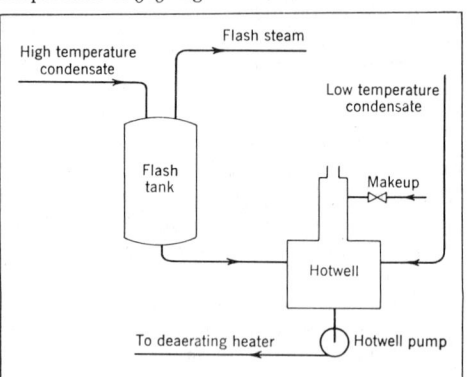

Fig. 1. Flow diagram for generation of flash steam from high temperature condensate returns.

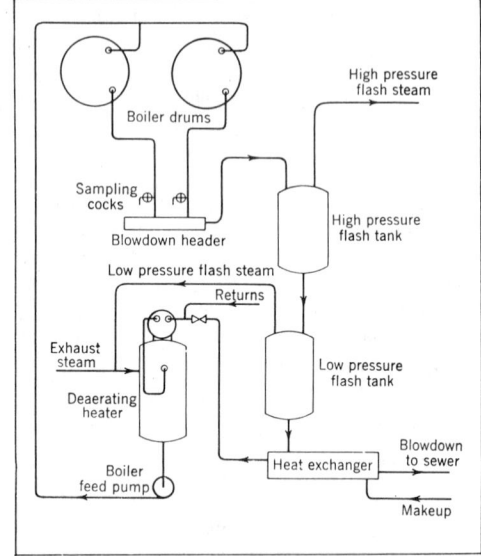

Fig. 2. Flow diagram for two-stage flash system based on heat in continuous boiler blowdown.

Flash Tank Dimensions for Various Steam Capacities[1]
(Tanks Provided with Tangential Inlet[2])

Tank No.	Dia. of Tank, Ft.	Height Over Heads, Ft.	Flash Area, Sq. Ft.	Volume, Cu. Ft.	Vent Dia., In.	Flashed Steam	
						Cfm.	Lb.
1	2.00	4.25	3.1	13	3	627	26
2	2.67	5.50	5.6	28	4	1,058	45
3	3.33	6.67	8.7	58	5	1,645	70
4	4.00	7.83	12.5	98	6	2,390	101
5	4.50	8.16	15.9	116	8	4,113	174
6	5.00	8.50	19.6	160	10	6,658	281
7	5.50	8.75	23.8	200	12	9,400	397
8	6.00	9.00	28.3	222	14	11,360	480
9	7.00	10.50	38.5	400	16	15,667	662
10	8.00	12.00	50.2	600	18	19,192	811
11	9.00	13.50	63.6	830	20	23,892	1010
12	10.00	15.00	78.5	1010	24	34,858	1474

[1] Quantities are for 2 lb gage pressure.
[2] For tanks with top inlet, multiply quantity of steam flash by 0.7.

QUANTITY OF HEAT IN FLASH STEAM

High Pressure System			Pressure in Low Pressure System, Lb. per Sq. In. Abs.										
Sat. Pressure Lb. per Sq. In. Abs.	Saturation Temperature, Deg. F.	Enthalpy of Sat. Liquid, Btu per Lb.	1	5	10	14.7	20	25	30	35	40	45	50
			Total Heat in Flash Steam, Btu per Lb. of H.P.* Condensate										
5	162.2	130.1	60.4	—	—	—	—	—	—	—	—	—	—
10	193.2	161.2	91.5	31.1	—	—	—	—	—	—	—	—	—
14.7	212.0	180.1	110.4	50.0	18.9	—	—	—	—	—	—	—	—
20	228.0	196.2	126.5	66.1	35.0	16.1	—	—	—	—	—	—	—
25	240.1	208.4	138.7	78.3	47.2	28.3	12.2	—	—	—	—	—	—
30	250.3	218.8	149.1	88.7	57.6	38.7	22.6	10.4	—	—	—	—	—
35	259.3	227.9	158.2	97.8	66.7	47.8	31.7	19.5	9.1	—	—	—	—
40	267.3	236.0	166.3	105.9	74.8	55.9	39.8	27.6	17.2	8.1	—	—	—
45	274.4	243.4	173.7	113.3	82.2	63.3	47.2	35.0	24.6	15.5	7.4	—	—
50	281.0	250.1	180.4	120.0	88.9	70.0	53.9	41.7	31.3	22.2	12.9	6.2	—
55	287.1	256.3	186.6	126.2	95.1	76.2	60.1	47.9	37.5	28.4	20.3	12.9	6.2
60	292.7	262.1	192.4	132.0	100.9	82.0	65.9	53.7	43.3	34.2	26.1	18.7	12.0
65	298.0	267.5	197.8	137.4	106.3	87.4	71.3	59.1	48.7	39.6	31.5	24.1	17.4
70	302.9	272.6	202.9	142.5	111.4	92.5	76.4	64.2	53.8	44.7	36.6	29.2	22.5
75	307.6	277.4	207.7	147.3	116.2	97.3	81.2	69.0	58.6	49.5	41.4	34.0	27.3
80	312.0	282.0	212.3	151.9	120.8	101.9	85.8	73.6	63.2	54.1	46.0	38.6	31.9
85	316.3	286.4	216.7	156.3	125.2	106.3	90.2	78.0	67.6	58.5	50.4	43.0	36.3
90	320.3	290.6	220.9	160.5	129.4	110.5	94.4	82.2	71.8	62.7	54.6	47.2	40.5
95	324.1	294.6	224.9	164.5	133.4	114.5	98.4	86.2	75.8	66.7	58.6	51.2	44.5
100	327.8	298.4	228.7	168.3	137.2	118.3	102.2	90.0	79.6	70.5	62.4	55.0	48.3
110	334.8	305.7	236.0	175.6	144.5	125.6	109.5	97.3	86.9	77.8	69.7	62.3	55.6
120	341.3	312.4	242.7	182.3	151.2	132.3	116.2	104.0	93.6	84.5	76.4	69.0	62.3
130	347.3	318.8	249.1	188.7	157.6	138.7	122.6	110.4	100.0	90.9	82.8	75.4	68.7
140	353.0	324.8	255.1	194.7	163.6	144.7	128.6	116.4	106.0	96.9	88.8	81.4	74.7
150	358.4	330.5	260.8	200.4	169.3	150.4	134.3	122.1	111.7	102.6	94.5	87.1	80.4
160	363.5	335.9	266.2	205.8	174.7	155.8	139.7	127.5	117.1	108.0	99.9	92.5	85.8
170	368.4	341.1	271.4	211.0	179.8	161.0	144.9	132.7	122.3	113.2	105.1	97.7	91.0
180	373.1	346.0	276.3	215.9	184.8	165.9	149.8	137.6	127.2	118.1	110.0	102.6	95.9
190	377.5	350.8	281.1	220.7	189.6	170.7	154.6	142.4	132.0	122.9	114.8	107.4	100.7
200	381.8	355.4	285.7	225.3	194.2	175.3	159.2	147.0	136.6	127.5	119.4	112.0	105.3
225	391.8	366.1	296.4	236.0	204.9	186.0	169.9	157.7	147.3	138.2	130.1	122.7	116.0
250	401.0	376.0	306.3	245.9	214.8	195.9	179.8	167.6	157.2	148.1	140.0	132.6	125.9
275	409.4	385.2	315.5	255.1	224.0	205.1	189.0	176.8	166.4	157.3	149.2	141.8	135.1
300	417.3	393.8	324.1	263.7	232.6	213.7	197.6	185.4	175.0	165.9	157.8	150.4	143.7
350	431.7	409.7	340.0	279.6	248.5	229.6	213.5	201.3	190.9	181.8	173.7	166.3	159.6
400	444.6	424.0	354.3	293.9	262.8	243.9	227.8	215.6	205.2	196.1	188.0	180.6	173.9
450	456.3	437.2	367.5	307.1	276.0	257.1	241.0	228.8	218.4	209.3	201.2	193.8	187.1
500	467.0	449.4	379.7	319.3	288.1	269.3	253.2	241.0	230.6	221.5	213.4	206.0	199.3
1000	544.6	542.4	472.7	412.3	381.2	362.3	346.2	334.0	323.6	314.5	306.4	299.0	292.3

* High pressure.

Quantity of Heat in Flash Steam (Concluded)

High Pressure System			Pressure in Low Pressure System, Lb. per Sq. In. Abs.										
Sat. Pressure, Lb. per Sq. In. Abs.	Saturation Temperature, Deg. F.	Enthalpy of Sat. Vapor, Btu per Lb.	55	60	65	70	75	80	85	90	95	100	110
			Total Heat in Flash Steam, Btu per Lb. of H.P.* Condensate										
60	292.7	915.5	5.8	—	—	—	—	—	—	—	—	—	—
65	298.0	911.6	11.2	5.4	—	—	—	—	—	—	—	—	—
70	302.9	907.9	16.3	10.5	5.1	—	—	—	—	—	—	—	—
75	307.6	904.5	21.1	15.3	9.9	4.8	—	—	—	—	—	—	—
80	312.0	901.1	25.7	19.9	14.5	9.4	4.8	—	—	—	—	—	—
85	316.3	897.8	30.1	24.3	18.9	13.8	9.0	4.4	—	—	—	—	—
90	320.3	894.7	34.3	28.5	23.1	18.0	13.2	8.6	4.2	—	—	—	—
95	324.1	891.7	38.3	32.5	27.1	22.0	17.2	12.6	8.2	4.0	—	—	—
100	327.8	888.8	42.1	36.3	30.9	25.8	21.0	16.4	12.0	7.8	3.8	—	—
110	334.8	883.2	49.4	43.6	38.2	33.1	28.3	23.7	19.3	15.1	11.1	7.3	—
120	341.3	877.9	56.1	50.3	44.9	39.8	35.0	30.4	26.0	21.8	17.8	14.0	6.7
130	347.3	872.9	62.5	56.7	51.3	46.2	41.4	36.8	32.4	28.2	24.2	20.4	13.1
140	353.0	868.2	68.5	62.7	57.3	52.2	47.4	42.8	38.4	34.2	30.2	26.4	19.1
150	358.4	863.6	74.2	68.4	63.0	57.9	53.1	48.5	44.1	39.9	35.9	32.1	24.8
160	363.5	859.2	79.6	73.8	68.4	63.3	58.5	53.9	49.5	45.3	41.3	37.5	30.2
170	368.4	854.9	84.8	79.0	73.6	68.5	63.7	59.1	54.7	50.5	46.5	42.7	35.4
180	373.1	850.8	89.7	83.9	78.5	73.4	68.6	64.0	59.6	55.4	51.4	47.6	40.3
190	377.5	846.8	94.5	88.7	83.3	78.2	73.4	68.8	64.4	60.2	56.2	52.4	44.1
200	381.8	843.0	99.1	93.3	87.9	82.8	78.0	73.4	69.0	64.8	60.8	57.0	49.7
225	391.8	833.8	109.8	104.0	98.6	93.5	88.7	84.1	79.7	75.5	71.5	67.7	60.4
250	401.0	825.1	119.7	113.9	108.5	103.4	98.6	94.0	89.6	85.4	81.4	77.6	70.3
275	409.4	816.9	128.9	123.1	117.7	112.6	107.8	103.2	98.8	94.6	90.6	86.8	79.5
300	417.3	809.0	137.5	131.7	126.3	121.2	116.4	111.8	107.4	103.2	99.2	95.4	88.1
350	431.7	794.2	153.4	147.6	142.2	137.1	132.3	127.7	123.3	119.1	115.1	111.3	104.0
400	444.6	780.5	167.7	161.9	156.5	151.4	146.6	142.0	137.6	133.4	129.4	125.6	118.3
450	456.3	767.4	180.9	175.1	169.7	164.6	159.8	155.2	150.8	146.6	142.6	138.8	131.5
500	467.0	755.0	193.1	187.3	181.9	176.8	172.0	167.4	163.0	158.8	154.8	151.0	143.7
1000	544.6	649.4	286.1	280.3	274.9	269.8	265.0	260.4	256.0	251.8	247.8	244.0	236.7

* High pressure.

produce only a 5.2% conversion to flash steam because of the non-linearity of the enthalpy function.

Several tables are presented pertaining to this matter. One, showing the amount of heat given up by condensate to flash steam at various pressure reductions, will save the time and tedium of individual calculations. The second table, converting this heat into percent of condensate flashed to steam, leaves nothing for calculation except interpolation when required. Both tables are based on absolute pressures.

Fortunately, experience has contributed to the design of flash tanks that work. The accompanying table gives flash tank dimensions for various steam capacities. The table is based on tanks provided with tangential inlets, flash pressure of 2 psig, and vent pipe diameters giving steam velocities leaving in the order of 11,000 fpm. Water storage space is

PER–CENT OF CONDENSATE FLASHED TO STEAM

High Pressure System			Pressure in Low Pressure System, Lb. per Sq. In. Abs.										
			1	5	10	14.7	20	25	30	35	40	45	50
			Latent Heat in Flash Steam, Btu per Lb.										
Sat. Pressure, Lb. per Sq. In. Abs.	Saturation Temperature, Deg.F	Enthalpy of Sat. Liquid Btu per Lb.	1036	1001	982.1	970.3	960.1	952.1	945.3	939.2	933.7	928.6	924.0
			Volume of Steam, Cu. Ft. per Lb.										
			333.6	73.5	38.4	26.8	20.1	16.3	13.7	11.9	10.5	9.4	8.5
			Percent of High Pressure Condensate Flashed at Low Pressure										
5	162.2	130.1	5.8	—	—	—	—	—	—	—	—	—	—
10	193.2	161.2	8.8	3.1	—	—	—	—	—	—	—	—	—
14.7	212.0	180.1	10.7	5.1	1.9	—	—	—	—	—	—	—	—
20	228.0	196.2	12.2	6.6	3.6	1.7	—	—	—	—	—	—	—
25	240.1	208.4	13.4	7.8	4.8	2.9	1.3	—	—	—	—	—	—
30	250.3	218.8	14.4	8.9	5.9	4.0	2.4	1.1	—	—	—	—	—
35	259.3	227.9	15.2	9.8	6.8	4.9	3.3	2.1	1.0	—	—	—	—
40	267.3	236.0	16.0	10.6	7.6	5.8	4.4	2.9	1.8	0.9	—	—	—
45	274.4	243.4	16.7	11.3	8.4	6.5	4.9	3.7	2.6	1.7	0.8	—	—
50	281.0	250.1	17.4	12.0	9.1	7.2	5.6	4.4	3.3	2.4	1.4	0.7	—
55	287.1	256.3	18.0	12.6	9.7	7.9	6.3	5.0	4.0	3.0	2.2	1.4	0.7
60	292.7	262.1	18.5	13.2	10.3	8.4	6.9	5.6	4.6	3.6	2.6	2.0	1.3
65	298.0	267.5	19.1	13.7	10.8	9.0	7.4	6.2	5.2	4.2	3.4	2.6	1.9
70	302.9	272.6	19.5	14.3	11.3	9.5	8.0	6.7	5.7	4.8	3.8	3.1	2.4
75	307.6	277.4	20.0	14.7	11.8	10.0	8.5	7.3	6.2	5.3	4.4	3.7	3.0
80	312.0	282.0	20.4	15.2	12.3	10.5	8.9	7.7	6.7	5.8	4.9	4.2	3.5
85	316.3	286.4	20.9	15.6	12.8	11.0	9.4	8.2	7.2	6.2	5.4	4.6	3.9
90	320.3	290.6	21.3	16.1	13.2	11.4	9.8	8.6	7.6	6.7	5.8	5.1	4.4
95	324.1	294.6	21.7	16.5	13.6	11.8	10.3	9.1	8.0	7.1	6.3	5.5	4.8
100	327.8	298.4	22.1	16.8	14.0	12.2	10.7	9.5	8.4	7.5	6.7	5.9	5.2
110	334.8	305.7	22.8	17.5	14.7	12.9	11.4	10.2	9.1	8.2	7.4	6.6	6.0
120	341.3	312.4	23.4	18.2	15.4	13.6	12.1	10.9	9.9	9.0	8.2	7.4	6.8
130	347.3	318.8	24.0	18.9	16.0	14.3	12.8	11.6	10.6	9.7	8.9	8.1	7.4
140	353.0	324.8	24.6	19.3	16.6	14.9	13.4	12.2	11.2	10.3	9.5	8.8	8.1
150	358.4	330.5	25.2	20.0	17.2	15.5	14.0	12.8	11.8	10.9	10.2	9.4	8.7
160	363.5	335.9	25.7	20.6	17.8	16.0	14.5	13.4	12.4	11.5	10.7	10.0	9.3
170	368.4	341.1	26.2	21.1	18.3	16.6	15.1	14.0	12.9	12.1	11.3	10.5	9.9
180	373.1	346.0	26.7	21.6	18.8	17.1	15.6	14.5	13.4	12.6	11.8	11.0	10.4
190	377.5	350.8	27.1	22.1	19.3	17.6	16.1	15.0	14.0	13.1	12.3	11.5	10.9
200	381.8	355.4	27.6	22.5	19.7	18.0	16.6	15.5	14.4	13.6	12.8	12.1	11.4
225	391.8	366.1	28.6	23.6	20.8	19.1	17.7	16.6	15.6	14.7	14.0	13.2	12.6
250	401.0	376.0	29.8	24.6	21.8	20.2	18.7	17.6	16.6	15.8	15.0	14.3	13.6
275	409.4	385.2	30.4	25.5	22.8	21.1	19.7	18.6	17.6	16.7	16.0	15.3	14.6
300	417.3	393.8	31.3	26.4	23.6	22.0	20.6	19.5	18.5	17.7	16.9	16.2	15.5
350	431.7	409.7	32.8	28.0	25.2	23.6	22.2	21.1	20.2	19.4	18.6	17.9	17.3
400	444.6	424.0	34.2	29.4	26.8	25.1	23.7	22.6	21.6	20.9	20.1	19.4	18.9
450	456.3	437.2	35.4	30.7	28.0	26.4	25.1	24.0	22.8	22.2	21.5	20.8	20.1
500	467.0	449.4	36.6	31.9	29.3	27.7	26.1	25.3	24.4	23.6	22.8	22.2	21.6
1000	544.6	542.4	45.6	41.2	38.8	37.2	36.0	35.1	34.2	33.5	32.8	32.2	31.7

Per-Cent of Condensate Flashed to Steam (Concluded)

High Pressure System			Pressure in Low Pressure System, Lb. per Sq. In. Abs.										
			55	60	65	70	75	80	85	90	95	100	110
			Latent Heat in Flash Steam, Btu per Lb.										
Sat. Pressure Lb. per Sq. In. Abs.	Saturation Temperature, Deg. F.	Enthalpy of Sat. Vapor, Btu per Lb.	919.6	915.5	911.6	907.9	904.5	901.1	897.8	894.7	891.7	888.8	883.2
			Volume of Steam, Cu. Ft. per Lb.										
			7.8	7.2	6.7	6.2	5.8	5.5	5.2	4.9	4.7	4.4	4.05
			Percent of High Pressure Condensate Flashed at Low Pressure										
60	292.7	1177.6	0.6	—	—	—	—	—	—	—	—	—	—
65	298.0	1179.1	1.2	0.6	—	—	—	—	—	—	—	—	—
70	302.9	1180.6	1.8	1.1	0.6	—	—	—	—	—	—	—	—
75	307.6	1181.9	2.3	1.7	1.1	0.5	—	—	—	—	—	—	—
80	312.0	1183.1	2.8	2.2	1.6	1.0	0.5	—	—	—	—	—	—
85	316.3	1184.2	3.3	2.7	2.1	1.5	1.0	0.5	—	—	—	—	—
90	320.3	1185.3	3.7	3.1	2.5	2.0	1.5	1.0	0.5	—	—	—	—
95	324.1	1186.2	4.2	3.5	3.0	2.4	1.9	1.4	0.9	0.4	—	—	—
100	327.8	1187.2	4.6	4.0	3.1	2.8	2.3	1.8	1.3	0.9	0.4	—	—
110	334.8	1188.9	5.7	4.7	4.2	3.6	3.1	2.6	2.1	1.7	1.2	0.8	—
120	341.3	1190.4	6.1	5.5	4.9	4.4	3.9	3.4	2.9	2.4	2.0	1.6	0.8
130	347.3	1191.7	6.8	6.2	5.6	5.1	4.6	4.1	3.6	3.2	2.7	2.3	1.5
140	353.0	1193.0	7.5	6.8	6.3	5.7	5.2	4.7	4.3	3.8	3.4	3.0	2.2
150	358.4	1194.1	8.1	7.5	6.9	6.4	5.9	5.4	4.9	4.5	4.0	3.6	2.8
160	363.5	1195.1	8.7	8.1	7.5	7.0	6.5	6.0	5.5	5.1	4.6	4.2	3.4
170	368.4	1196.0	9.2	8.6	8.1	7.5	7.0	6.6	6.1	5.6	5.2	4.8	4.0
180	373.1	1196.9	9.8	9.2	8.6	8.1	7.6	7.1	6.6	6.2	5.8	5.4	4.6
190	377.5	1197.6	10.3	9.7	9.1	8.6	8.1	7.6	7.2	6.7	6.3	5.9	5.1
200	381.8	1198.4	10.8	10.2	9.6	9.1	8.6	8.1	7.7	7.2	6.8	6.4	5.6
225	391.8	1199.9	11.9	11.4	10.8	10.3	9.8	9.3	8.9	8.4	8.0	7.6	6.8
250	401.0	1201.1	13.0	12.4	11.9	11.4	10.9	10.4	10.0	9.5	9.1	8.7	8.2
275	409.4	1202.1	14.0	13.4	12.9	12.4	11.9	11.5	11.0	10.6	10.2	9.8	9.0
300	417.3	1202.8	15.0	14.4	13.9	13.3	12.9	12.4	12.0	11.5	11.1	10.7	10.0
350	431.7	1203.9	16.7	16.1	15.6	15.1	14.6	14.2	13.8	13.3	12.9	12.5	11.8
400	444.6	1204.5	18.2	17.7	17.2	16.7	16.2	15.8	15.3	14.9	14.5	14.1	13.4
450	456.3	1204.6	19.6	19.1	18.6	18.1	17.7	17.2	16.8	16.4	16.0	15.6	14.9
500	467.0	1204.4	21.0	20.5	20.0	19.5	19.0	18.6	18.2	17.7	17.4	17.0	16.3
1000	544.6	1191.8	31.1	30.6	30.2	29.7	29.3	28.9	28.5	28.1	27.8	27.5	27.0

one-third of the tank volume.

Assume a condensate return system returning 2750 lb of condensate per minute at 45 psia and flashing down to the pressure of a heating system at 5 psig or, approximately, 20 psia. From the second table, it is found that 4.9% of condensate flashes at this pressure. The specific volume of steam at 5 psig is 20.4 cu ft per lb. Hence, 2750 × 20.4 × 4.9% = 2750 cfm of steam flashed. In order to use the flash tank table for flash steam at 2 psig, difference in specific volumes must be accounted for. The ratio is 23.9/20.4 = 1.17, and 2750/1.17 = 2350 cfm. Tank 4 provides the required dimensions to satisfy this capacity.

Figures 1 and 2 show typical flow diagrams for flash steam systems. Figure 1 is a possible arrangement for generation of flash steam from high temperature condensate returns. Figure 2 is a two-stage flash system based on the heat in continuous boiler blowdown.

SIZING OF VERTICAL FLASH TANKS

Following are data for sizing of vertical flash tanks for steam and the float traps that drain them.

To Size Flash Tank

To obtain the proper size of a vertical flash tank, the following steps should be taken:

1) Calculate maximum amount of flash steam from Fig. 1.

2) Find, from Fig. 2, the size area in which both condensation rate and flash steam quantity lie.

For Example:

A unit maintained at 80 psig condenses 3000 lb per hr of steam. Flash steam from this is to be used in another unit at 5 psig.

From Fig. 1 find that condensate at 80 psig produces just over 0.1 lb of flash per lb when dropped to 5 psig. So the total flash is 300 lb per hr.

From Fig. 2 we see that 3000 lb of condensate per hour meets the line of 300 lb flash per hour in Size 4 area. So a Size 4 flash vessel must be chosen for the duty.

See Table 1 for full dimensions for each size number of vertical flash tank.

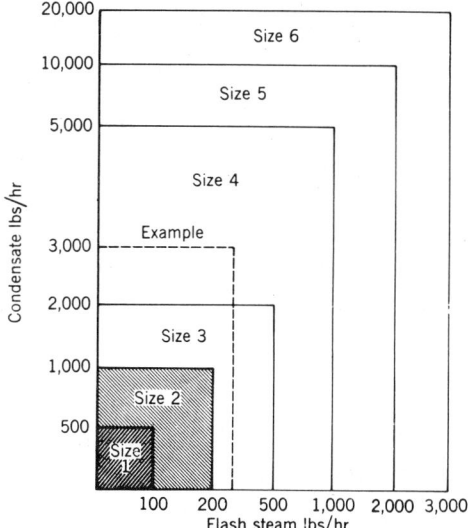

Fig. 2. Chart for determining size number of tank.

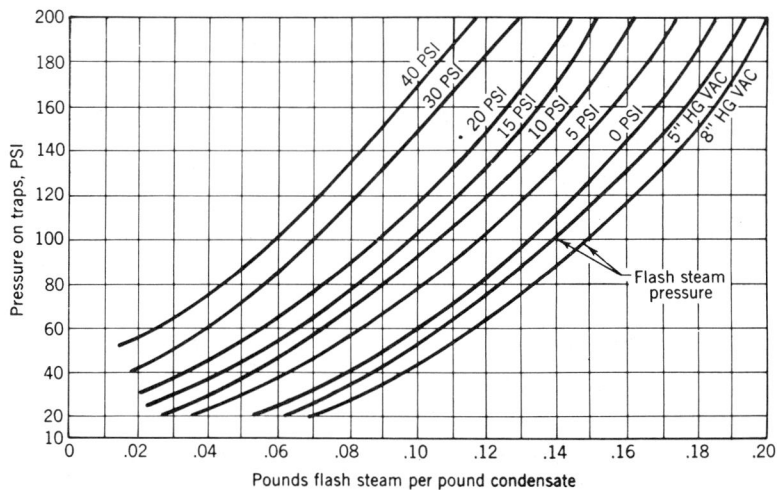

Fig. 1. Pounds flash steam per pound condensate at various pressure drops.

Table 1. Dimensions of Vertical Flash Tanks (See Fig. 3)

Size No.	A	B	C	D	E	F	G	I
	Approximate Dimensions (Inches)							
1	4	32	5	10	1	$1\frac{1}{4}$	$1\frac{1}{4}$	$\frac{1}{2}$
2	5	32	$5\frac{1}{2}$	10	$1\frac{1}{2}$	2	$1\frac{1}{2}$	$\frac{1}{2}$
3	8	34	$5\frac{1}{2}$	12	$1\frac{1}{2}$	$2\frac{1}{2}$	2	$\frac{1}{2}$
4	10	36	$5\frac{1}{2}$	12	$1\frac{1}{2}$	3	$2\frac{1}{2}$	$\frac{3}{4}$
5	14	36	$5\frac{1}{2}$	12	$1\frac{1}{2}$	3*	$2\frac{1}{2}$*	$1\frac{1}{2}$
6	16	42	6	14	2	6*	3*	2

* Flanged connection.

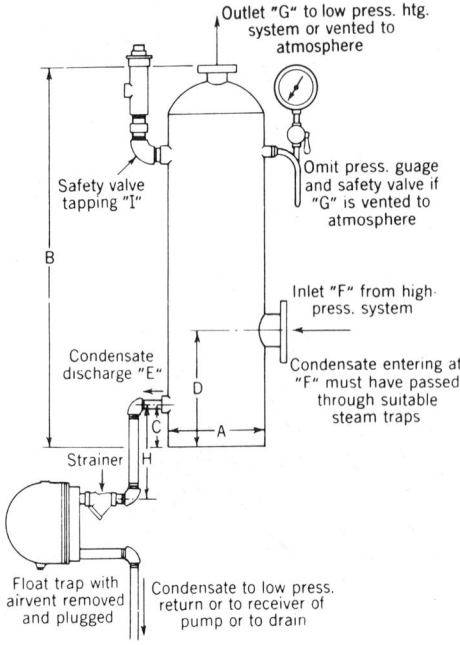

Fig. 3. Elevation—vertical flash tank.

To Size Float Trap

Select a trap having capacity to discharge given quantity of condensate (pounds per hour) at pressure corresponding to static head H, plus pressure differential (if any) between tank and return main (or atmosphere), as the case may be.

Airbinding

Air entrapped in a steam system retards or even stops the flow of steam. The condition is known as *airbinding*. Lack of a proper vent in the right place may be one cause; or the trap (usually a bucket trap) may have insufficient venting capacity. A water trap may be the result of locating the trap too far from the appliance being drained. It prevents the air from reaching the point where it can be vented. Illustration below shows a common cause, and a cure, for airbinding.

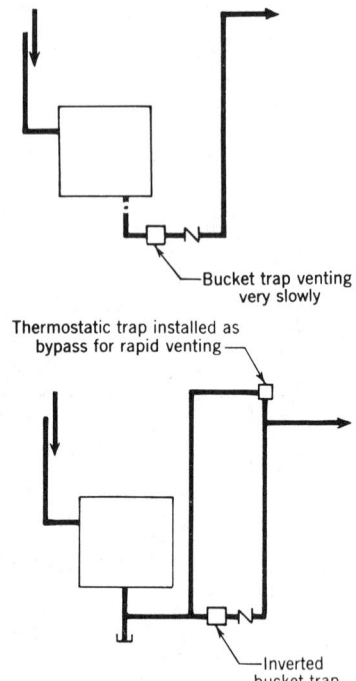

Cause and Cure of Airbinding

ESTIMATING FRICTION IN HOT WATER PIPING

Friction charts for cold water (see Piping section) were designed using the following formulas:

Smooth . . . . $h = 0.00682v^{1.75}/d^{1.25}$
Fairly Smooth . . . . $h = 0.0073v^{1.83}/d^{1.17}$
Fairly Rough . . . . $h = 0.008v^{1.92}/d^{1.08}$
Rough . . . . $h = 0.0101v^2/d$

where

h = friction loss in feet of water per foot of pipe;

v = velocity, feet per second; and

d = inside diameter of pipe, inches.

Following the same theory as outlined on *cold* water and using the basic formulas above, but substituting a value for the viscosity of water at 175F instead of that for 50F water, the following formulas result:

(1) Smooth: . . . . $h = 60v^{1.75}/d^{1.25}$
(2) Fairly Smooth: . . . . . . $h = 72v^{1.83}/d^{1.17}$
(3) Fairly Rough: . . . . . . $h = 84v^{1.92}/d^{1.08}$
(4) Rough: . . . . $h = 120v^2/d$

where

h = friction head in milinches per foot of pipe;

v = velocity in feet per second; and

d = actual inside diameter in inches.

Note that part of the difference between the coefficients in the two sets of formulas is due to a change in units of h.

The hot water friction charts which follow are graphical plots of formulas (1) and (3), the former being used for copper tubing. The fairly rough formula was used for iron and steel pipe so that, as the pipe ages, corrodes and scales, the friction values will be ample.

The hot water friction charts, then, show graphically the friction, as well as the heat carried, in steel pipe and Type L copper tubing when the fluid is water at 175F temperature and with a 20F temperature drop. The temperature of 175F was chosen as meeting the requirements of both forced and gravity systems with reasonable accuracy, so that the charts are suitable for all forms of hot water heating.

Use of Charts

Use of the charts is best illustrated by examples as follows:

Example: A 46 ft. branch, steel pipe, must have a loss of 3680 milinches to balance, and, based on a 20F drop, carry 150,000 Btu per hr. What size pipe and velocity are required?

Solution: The friction loss per foot of pipe must be 3680/46 = 80.0 milinches Locate the intersection of 150 on the side scale and 80 on the lower scale, and find that a 2 inch pipe and a velocity of 1.5 ft. per second meet the conditions.

The copper tubing chart is used in exactly the same manner.

Use of these charts is not, of course, limited to cases where the temperature drop is 20F, if correction is made. If the temperature drop is other than 20F, multiply the heat carried as found on the vertical scale by a correction factor of

Actual Temp. Drop ÷ 20

In other words, in the example above, if other conditions were the same but the heat to be carried were 150,000 Btu per hour, based on a 15F temperature drop, the correction factor would be 15 ÷ 20 = .75. Therefore 150,000 ÷ .75 = 200,000 Btu to be found on the vertical scale. Locate the intersection of 200 on the side scale and 80 on the lower scale, and find 2½ in. pipe and velocity of 1.5 ft. per second velocity, which would be somewhat large; however, 2 inch pipe would have been too small.

HOT WATER FRICTION CHART — TYPE L COPPER TUBING *

FRICTION LOSS IN MILINCHES (.001") PER FOOT OF PIPE

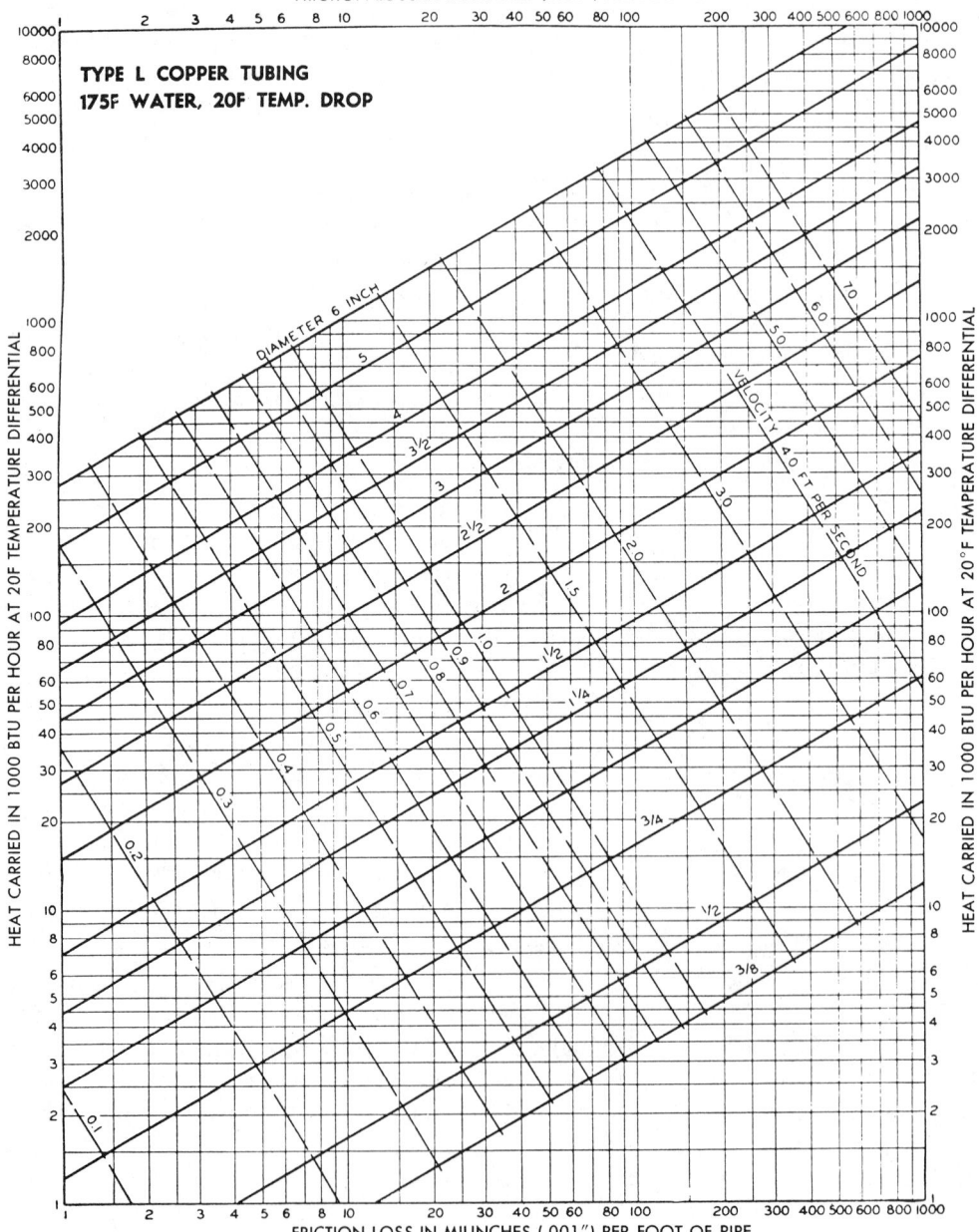

TYPE L COPPER TUBING
175F WATER, 20F TEMP. DROP

* For temperature drops other than 20°F, and for friction losses in feet per 100 ft, flow rate in gallons per minute or pounds per hour, please refer to tables on pages 8-8 through 8-28.

The above chart is based on the formula

$$h = 60v^{1.75}/d^{1.25}$$

where h is friction loss in milinches per foot of tube, v is velocity in feet per second, d is actual inside diameter of pipe in inches. (1 milinch of water = 0.000036 lb per sq inch.)

HOT WATER FRICTION CHART — IRON OR STEEL PIPE *

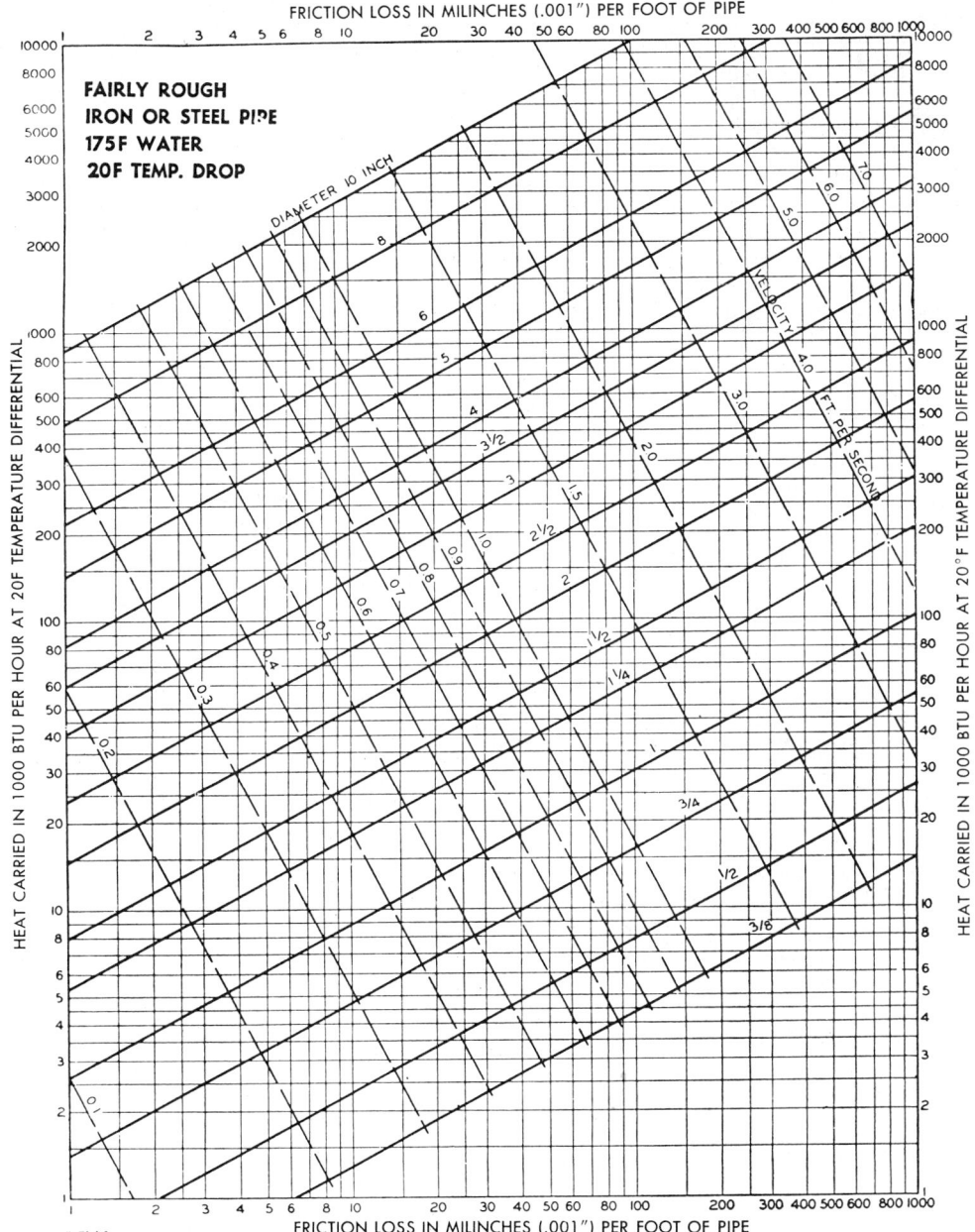

FRICTION LOSS IN MILINCHES (.001″) PER FOOT OF PIPE

FAIRLY ROUGH
IRON OR STEEL PIPE
175F WATER
20F TEMP. DROP

HEAT CARRIED IN 1000 BTU PER HOUR AT 20F TEMPERATURE DIFFERENTIAL

HEAT CARRIED IN 1000 BTU PER HOUR AT 20°F TEMPERATURE DIFFERENTIAL

FRICTION LOSS IN MILINCHES (.001″) PER FOOT OF PIPE

* Ibid.

The above chart is based on the formula

$$h = 84v^{1.92}/d^{1.08}$$

where h is friction loss in milinches per foot of pipe, v is velocity in feet per second, d is actual inside diameter of pipe in inches. (1 milinch of water = 0.000036 lb per sq inch.)

HOT WATER HEATING SYSTEMS

Practically all new hot water heating systems are designed for forced circulation because: circulation is positive, good circulation to all parts of the system is assured, system temperatures may be readily varied to suit the load, pipe sizes may be small and piping need not be pitched in direction of flow. Another feature of forced circulation systems (particularly valuable in residential and other small systems) is the practicability of obtaining service water from a coil in the boiler.

Although we are here concerned chiefly with forced circulation systems, the principles given can be applied to gravity circulated systems, as explained later.

Service Water Heating. If a flow control valve is provided at the boiler outlet, boiler water temperature may be maintained continuously so that a coil inserted in the water may be used to heat service water. The coil may be small, in which case it is used with a storage tank, or the coil may have enough surface so that water is heated to the required service temperature by one passage through the coil. This latter type is advantageous where a storage tank may overheat the basement and because the tank may corrode rapidly if the water is corrosive. Year-round operation of the system keeps the boiler warm continuously and keeps any soot deposits dry, thereby preventing corrosion of the boiler surfaces or chimney connection pipe.

In residence heating, the load imposed by the instantaneous water heater may be disregarded in determining boiler size because it is of such short duration. In some cases, however, the water-heating load of the instantaneous heater, 3-5 gpm heated 100 degrees, may be larger than the heating load and will therefore determine the boiler or burner capacity required.

Operating Water Temperature. Systems for heating with hot water are flexible in regard to operating temperature. For instance, if the system is designed for 250 F operating temperature it is easy to use a lower temperature in Spring and Fall by setting the boiler thermostat at a temperature that will satisfy the service water heater, say 160 F. This lowered setting at reduced heating loads results in more continuous pump operation and more uniform space temperature than with operation at 250 F. Service water temperature of 140 F is usually satisfactory for residences.

Air Removal from System. To facilitate movement of air to high points in the piping system for venting, it is common practice to give supply piping a slight upward pitch ($\frac{1}{4}$ inch in 25 ft) in the direction of flow of water. Separation of air from water is greatest where the water is hottest, which is in the top of the boiler.

In a closed heating system, air is admitted only to replace that absorbed from the expansion tank by the water passing in and out of the tank. Absorption of air from the tank occurs mostly during the off cycle when the water cools. This cannot be prevented because water when cool can absorb more air than when heated. Some of this absorbed air can be returned to the expansion tank if there is space at the top of the boiler (in which the water is heated) where the air released by the water may collect and be returned through a pipe to the tank. Special air-release fittings are sometimes used to separate air from the circulating water and return it to the tank. If air carried by the water collects in heating units or other points from which it must be vented there will be a loss of air from the expansion tank that must be replaced either from a source of compressed air or by air admitted to the tank after the tank has been drained.

Water Flow Velocity. The velocity of flow in any hot water heating system will be that at which the resistance of the system is equal to the head produced by the pump. In other words, when the pump is started, the velocity will build up until the friction of flow in the system uses up the head produced by the pump at whatever speed the pump is run by the pump motor. The pump head produced depends on the pump design characteristics as well as speed. Performance curves are available from the pump manufacturer; they should be considered when selecting the pump. If a system is found after installation to be noisy it may be necessary to reduce pump size or size of motor in order to reduce the velocity.

Quiet operation is expected from hot water heating systems. For residential and commercial buildings, it is common practice to limit velocities to 4 fps in small piping. In piping above 2 inch diameter, this velocity may be increased somewhat. In industrial buildings, friction loss in the piping is the limiting factor rather than the noise due to flow. Noises due to expansion can be prevented by designing the piping for necessary movement.

Prevention of Freezing. When freezing temperatures occur, there should be sufficient circulation of water to prevent lowering of water temperature at any point in the system to the freezing point. Dampers may be necessary to prevent incoming air from striking water coils before they have become heated. Complete drainage of all parts of the system should be provided if there is a shutdown long enough to cause freezing in cold weather.

Water Circulation Below Mains. Circulation of water to units placed below mains is practicable if special fittings are used to divert necessary flow to such units. Some special fittings are proprietary. Whether these fittings are required on both supply and return connections at the main will depend on the fitting design and the distance of the unit below the main. Capacities of diversion fittings and instructions for their use are available from manufacturers.

Limitation of Pressure. In tall buildings (six stories and higher) it may be advisable to divide the system into vertical zones, not exceeding six stories each, and have each zone supplied with water from a heat exchanger. By this arrangement the pressure in the boiler can be held within the common legal limit of 30 psig for low pressure water boilers. In some states the law demands that a full time licensed engineer operate equipment if pressures are over the legal low. At current prices, an engineer attendant is an expensive employee, and if there must also be a night attendant, the operating cost of the high-pressure system may seem excessive. Beyond the hydraulic limit of 30 psig on low-pressure classifications, boiler prices go up and there may be difficulty in obtaining convectors and baseboards for the high-pressure service.

System Adaptability. The variety of piping arrangements that can be used is an attractive feature of hot water heating systems.

Commercial and industrial structures frequently require several distinct climates in their various areas. If these various climates can be obtained with heat from a single heat source, the economy of the job generally is promoted. Thus, hot water from a central boiler can be sent to rooms heated by convectors; then beyond the partitions of these rooms to unit heaters or to a radiant panel floor or ceiling; and, still farther along the hot water main; to a fan coil for warming up air for local duct distribution. Each area can be heated with the apparatus that fits it best, with hot water as the basic heat source.

Fig. 1 illustrates this flexibility.

Use of Waste Steam Heat. It should be noted that if waste steam is available from some plant process, economic comparisons are likely to favor the waste steam over a hot water system for these installations. However, one point that must be considered is the effect of the source or prior use of the steam on the steam itself. To explain, low pressure steam taken from an engine exhaust header is almost certain to bear traces of oil; consequently, if adequate steam purification is not accomplished before the steam enters the heating system, a gradual fouling of the piping, traps and other components of the system is rather certain to follow. If waste steam is used in a heat exchanger to heat the circulating water in a hot water system the troubles due to oil will not occur in the heating system. Arrangements can be made with little difficulty to remove the oil reaching the exchanger.

Heat from District Steam System. When district steam is available as a heat source it may be used with a heat exchanger to transfer heat to the circulating water. Typical steam and water piping and controls between the steam service and the pump circulating hot water are shown in Fig. 2. If a condensate meter is used, the flow meter orifice and connections are omitted.

Summer Cooling. Summer cooling is readily incorporated in a hot water installation by providing a water chiller. It's just as easy to pump cold water in summer as it is to pump hot water in winter. This arrangement is the basis of some notably efficient all-year designs, with fan-coil cabinets strategically located.

However, chilled water is of no practical value when circulated through convectors, baseboards, or radiant panels. In the convectors and baseboard, heat exchange is insignificant unless the ambient air is forced to circulate over the cold surfaces. For the radiant panels, surface condensation is a hazard in most areas of the country, although in some parts of the southwest United States where relative humidities vary between 15 and 25 percent, cool panels probably may be suitable as a cooling means.

In general, some form of auxiliary cooling is necessary whenever fan coil apparatus is not a part of the area heating scheme.

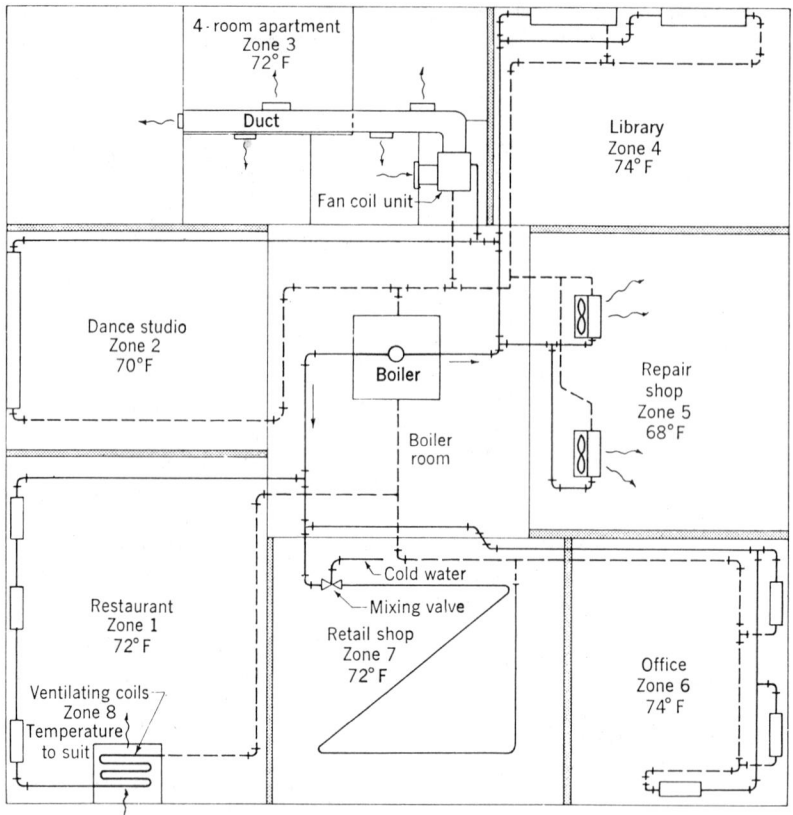

Fig. 1. Application of hot water system for production of different space requirements. Zone 1—Convectors, 1 pipe series loop; Zone 2—Baseboard, single pipe; Zone 3—Fan-coil duct system; Zone 4—Fan-coil room units; Zone 5—Unit heaters; Zone 6—Convectors, 2-pipe direct return; Zone 7—Panel ceiling; Zone 8—Fan-coil, for heating outdoor air for restaurant.

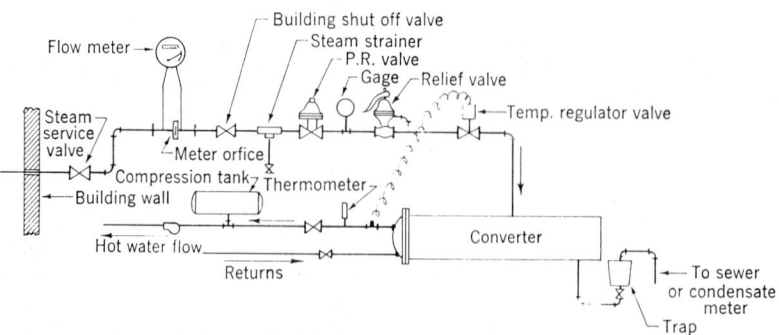

Fig. 2. Use of district steam as a heat source.

TYPES OF WATER HEATING SYSTEMS

The designations of the various types of hot water heating systems are based on the piping arrangement. Whether the final heat emitter is a convector, radiant panel, fan-coil unit or baseboard, the hot water system that serves it is described by one of the following designations:

(a) *One-pipe series* system, in which the water flows through each of the heat emitting units in turn before returning to the boiler.

Care must be exercised in sizing the successive heat emitting units to compensate for the progressively cooler water. The practice of sizing the whole loop on an average water temperature has given rise to overheating in some areas, underheating in others. For best results, the heating demands on each unit should be determined accurately, and the heating surface apportioned accordingly. Fig. 3 illustrates this system.

Because of its simplicity, a series layout can be very appropriate and useful as a branch or zone when incorporated in an installation of multiple requirements.

(b) *One-pipe diversion* system, in which the water is diverted from the main to the heat emitting unit, and then returns to the main for eventual reheating in the boiler. Fig. 4 illustrates this system.

As in the series system, it is necessary to increase the size of the heating units successively to compensate for lower water temperatures toward the end of the circuit.

(c) *Two-pipe direct return*, an arrangement of the piping that causes each heat emitter to discharge the cooled water back to the boiler by the shortest possible route. To use this system for balanced heating, it is necessary to control the water that flows through each heating unit circuit, otherwise short-circuiting takes place and some areas will be overheated, others underheated. The use of balancing cocks is recommended, a cock being placed on the return from each unit and set to permit the proper flow for that unit and no more. This system is illustrated in Fig 5.

(d) *Two-pipe reversed return*, a design that endeavors to make circuits to all heat emitters of equal length, so that the pipe friction to and from each unit is approximately the same. This system is illustrated in Fig. 6.

This system is balanced by replacing the cocks of a direct return arrangement with a return main that is calculated to make all of the heat emitter circuits of equal length, and consequently of equal friction. This scheme has the advantage of eliminating the individual balance cocks, and the disadvantage of increasing the amount of pipe which must be used on a job.

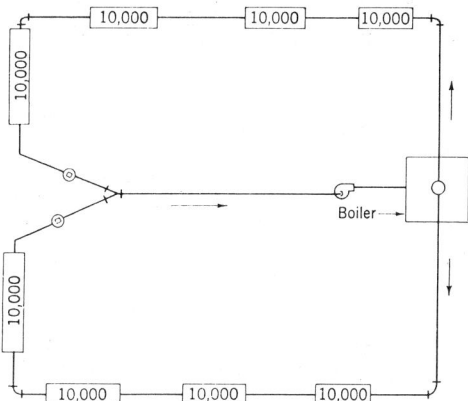

Fig. 3.   One-pipe series loop system.

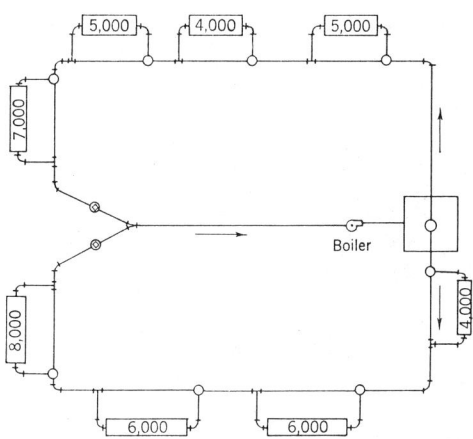

Fig. 4.   One-pipe diversion loop system.

nating the individual balance cocks, and the disadvantage of increasing the amount of pipe which must be used on a job.

For the large installations, as a general rule, either (c) or (d) is common; however in many cases where multiple zoning is involved, individual zones in the form of (a) or (b) may be found to best suit the requirements of a particular zone. In other words, there are four forms of hot water circulation available to be used as circumstances indicate. This is another illustration of the flexibility characteristic of hot water heating.

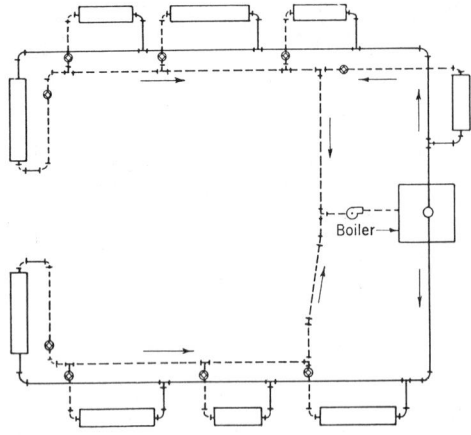

Fig. 5. Two-pipe direct return system.

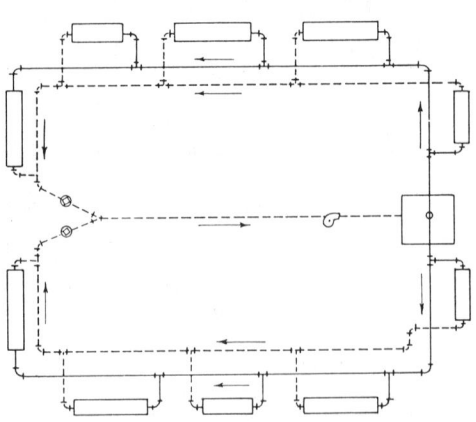

Fig. 6. Two-pipe reversed return system.

DESIGN RECOMMENDATIONS FOR HOT WATER SYSTEMS

Water Velocity. Water velocity should be limited in order to prevent noise in piping and to keep friction loss within acceptable limits. Generally accepted limits for velocity are 4 to 10 fps with 4 fps applying to pipe up to 2 inches diameter size and 10 fps to pipe of 6 inches and over. The 4 fps limitation should be observed for residences and commercial space such as offices. Friction loss, rather than noise, will be a limiting factor in industrial buildings because it affects the size of pump required and the cost of pumping. Common friction loss design values are 100 to 500 milinches per foot of pipe (0.83 to 4.1 ft of head per 100 ft of pipe).

Pump Location. Place the pump to *draw* water from the boiler for large installations, for those with high pumping heads, and for those cases where the radiation is even with or below the top of the boiler. When the radiation is below the boiler or at the same elevation, and the boiler is on the suction side of the pump, the pressure on the system piping is increased, a condition that helps a great deal in venting. Another benefit is less likelihood of pump cavitation, since the air separates largely· in the boiler rather than later in the pump.

For high head pumping, and for building more than three stories high, pumping out of the boiler becomes practically obligatory to avoid opening of the relief valves on the boiler, and to escape some occurrence of boiling at the high points. Why this is so is made clear in the discussion of effect of pump location on system performance.

Air Venting. Be careful to prevent air entry and pocketing, and make certain that air is vented. Venting is essential at the high points of the circuit, and is recommended additionally at the boiler outlet. An air separator, Fig. 7, is effective since the separation and removal of the air takes place in an area of high water temperature, and where separation probably is easiest.

Two air separators are shown in Fig. 7 to suggest a technique for handling two circuits with flow in opposite directions. One separator would usually suffice if placed in the piping before it is divided into two circuits.

To avoid air pockets, use eccentric reducers wherever the horizontal piping changes size. Fig. 8 illustrates this point.

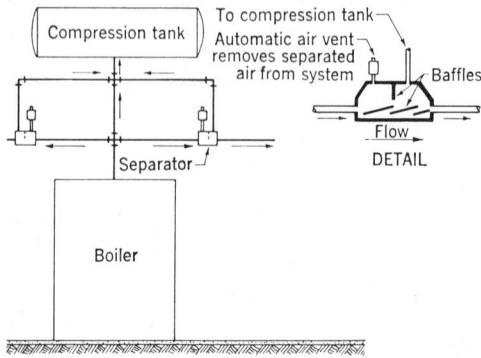

Fig. 7. Use of air separators for air removal.

Balancing Circuits. Use adjustable cocks to balance water flow through circuits of unequal loading whenever these circuits merge into a common main. Otherwise the circuit of lesser resistance will obtain too large a share of the circulating water.

Filling Pressure. In filling a low-pressure system with cold water, the pressure should not exceed 19 psig because the expansion of the water in being heated in the boiler from, for instance, 50 to 200 F would increase the pressure approximately 10 psi which, added to the fill pressure of 19 psig, would almost open the relief valve set at 30 psig. A slight increase in pressure due to pump operation would open the relief valve.

A commonly used cold fill pressure is 12 psig, but this may not be enough to fill a high system. When pressures exceed 30 psi, the system is no longer "low pressure" and, consequently, the boiler and equipment must be suitable for the more severe conditions.

Preventing Back-Flow. If two or more pumps discharge into a common header, a check valve or flow control valve must be used in each discharge to prevent backflow into non-operating circuits or to prevent unauthorized operation of any circuit. This is illustrated in Fig. 9.

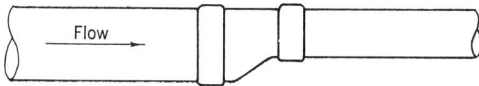

Fig. 8. Use of eccentric reducers to prevent air pockets at change of pipe size.

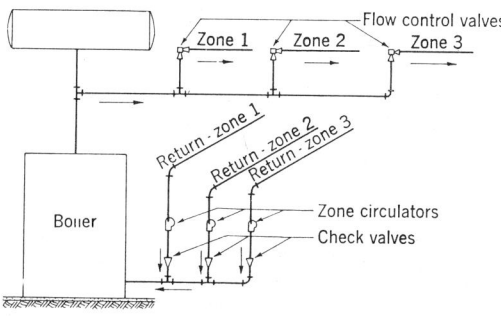

Fig. 9. Schematic diagram shows proper use of both check valves and flow control valves.

Connecting Returns to Boiler. Satisfactory ways for connecting a return to the boiler are illustrated in Fig. 10.

In general, when making boiler connections to both inlet and outlet, the paths of flow through the boiler should be equalized. Complaints of inadequate capacity sometimes are due to an incorrect arrangement of boiler connections, the trouble being that the water finds a short route to the outlet nozzle and thus leaves the boiler at less than the intended temperature. This situation is aggravated if the water temperature control is mounted in a hot spot which is out of the path of water circulation and can consequently shut down the firing equipment prematurely. When the boiler and system appear to be of adequate capacity but the building occupants are dissatisfied, the trouble may be due to the water flow pattern in the boiler.

Locating the Circulating Pump. Wherever the compression tank is connected into the system, this is the one point on the whole circuit whose pressure remains the same when the pump begins to operate. When the pump is started, pressure from the pump discharge to the tank rises above system static pressure; pressure from the pump suction to the tank falls below. The amount of rise and fall is a maximum at the pump and decreases in each direction, finally disappearing at the point of connection to the expansion tank. (The algebraic sum of the rise and fall at the pump is the pumphead, which equals total system friction.) Thus, location of pump and tank determines how system pressures will vary through the circuit.

The compression tank is most commonly located above the boiler, connected to the supply piping. Although not critical in small residential systems, it is recommended that it pump *out* of the boiler. With this arrangement, the entire circuit, from the pump discharge to the boiler, is under positive pressure, reducing the possibility, at any point in the circuit, but especially at the highest elevations, that pressure will fall below saturation pressure and cause boiling of the hot water. Lack of positive pressure in the system can also prevent proper venting of air. In addition, if pumping *into* the boiler is desirable for other reasons, other factors should be checked to guarantee that pressures in the boiler will not exceed the setting of the pressure relief valve.

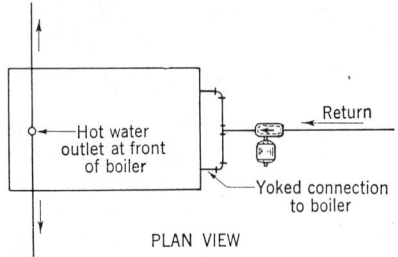

PLAN VIEW

This is good but outlet must be at front of boiler. Do not move back.

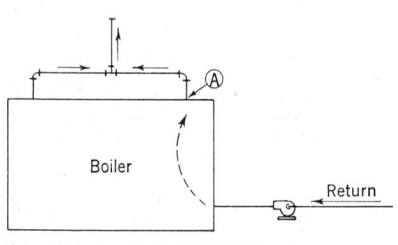

SIDE ELEVATION

Dotted lines show possible short circuit to rear boiler outlet. Install valve and throttle at point A to balance resistance with front outlet.

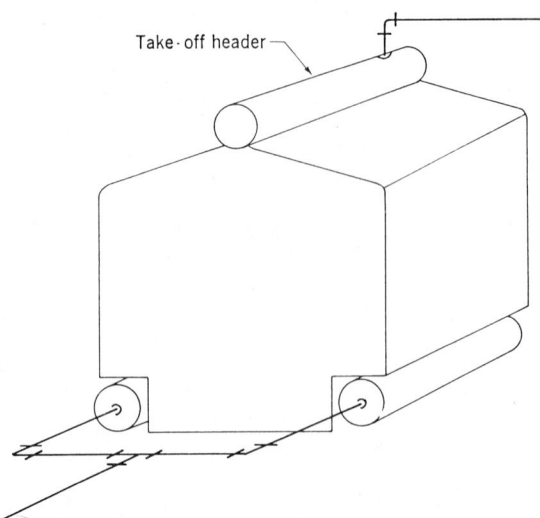

Return yoke is good. Hot water must be taken from front as shown.

Fig. 10. Satisfactory ways of connecting a return line to boiler.

Sizing the Expansion Tank. The simplest expansion tank is an open tank set sufficiently above the highest point in the system to produce a head of at least four feet to ensure adequate venting. Adding a few more feet of head may be desirable to prevent boiling. In addition, if the effect of pump location is to reduce system pressure between tank and pump suction, additional head may be required to offset this drop. Capacity can be calculated from the numerator of the formula below. It allows for the offsetting effect of pipe expansion on excess water volume when system temperature rises.

Compression tanks may be sized by the following ASME formula applicable for operating temperatures between 160F and 280F:

$$V_t = \frac{(0.00041t - 0.0466)V_s}{(P_a/P_f) - P_a/P_o}$$

where V_t = Minimum volume of tank, gal
V_s = System volume, gal

t = Maximum average operating temperature, F
P_a = Pressure in tank when water first enters, feet of water, absolute (usually atmospheric = 34 ft)
P_f = Initial fill or minimum pressure at tank, feet of water, absolute
P_o = Maximum operating pressure at tank, feet of water, absolute

System volume can be found from water capacities of boilers and heat consuming apparatus in manufacturers' catalogs and from tables of dimensional and capacity data for piping elsewhere in this volume.

The higher the compression tank is situated in the system, the smaller is P_f, minimum pressure at the tank, hence the smaller the tank. For this reason, compression tanks are often located in the attic.

Two graphical methods for sizing compression tanks are given elsewhere in this section of the *Handbook*.

Compressed Air to Reduce Tank Size. By injecting air under pressure into the compression tank, it is possible to economize on tank size. This is worthwhile if compressed air is available or if a hand pump can be kept conveniently in the area. Tank must be equipped with an air fill valve, a pressure gage, and a water column to allow an intelligent estimate of conditions within the tank at all times.

To determine tank size using compressed air, the value of P_a in the formula above should reflect the addition of air. Thus, if 5 psig air is used, $P_a = 34 + 5(2.31) = 45.55$ feet of water, absolute.

Piping Details. The following suggestions pertain to the layout of piping for a hot water heating system:

1. When a run of piping must cross a beam or other obstruction at a conflicting level, route the piping *under* the beam, not over it. Fig. 11 shows how this idea prevents the formation of an air pocket at this point.

If the pipe crosses *over* the beam, that point becomes a high point on the run and provides a collecting place for bubbles of air. Trouble will be almost certain to develop eventually unless the over-the-beam arrangement is vented at that point.

2. Do not connect the supply pipe to the *upper* part of a radiator when the supply originates at or under the floor, because air, accumulating in the top of the radiation, would gradually choke off circulation. Making both supply and return connections low doesn't stop air from accumulating at the top, but it does keep most of the radiation warm. Fig. 12 illustrates this type of installation.

3. When the radiation is *below* the main, be careful to connect the supply into the top of the radiation, the return to the bottom. This permits air to escape upwards to the main. This path of flow also allows the gradually-cooling water to sink to the bottom of the radiation, doing all the heating possible before returning to the main above. See Fig. 13.

4. When unit heaters are supplied with hot water from a main below the heaters, each must carry its own automatic vent. Otherwise the heater becomes a natural air pocket.

5. When changing the size of horizontal mains use eccentric reducers as shown in Fig. 8 in order to cause the top of the pipe to pitch smoothly to the desired high point. Concentric reducers create air pockets.

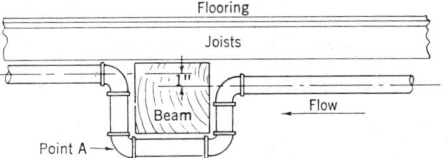

Fig. 11. Method of crossing a beam.

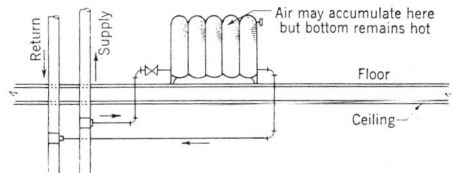

Fig. 12. Preferred connection when radiation is above supply and return.

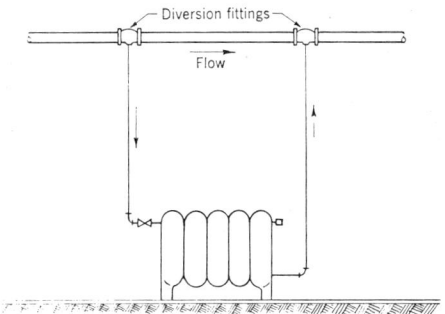

Fig. 13. Preferred connection when radiation is below supply and return.

6. In making up screwed pipe joints, use pipe compound on the male threads only. This keeps the compound from being pushed into the pipe where it can be picked up by the circulating water and deposited in undesirable locations.

7. Experience indicates that the upside-down placement of valves is best avoided. A valve in this position develops leaks and needs service earlier than when installed upright.

8. A temperature-pressure relief valve should be so connected that it is in intimate contact with the hottest water in the boiler. It should not be in a spot where incrustation can form over the relief orifice. A valve mounted several inches above the boiler does not register true temperatures of the boiler water and may not open when it should. Valves that must be so mounted can be obtained with bulb that extends into the boiler water, and therefore are desirable from a safety viewpoint.

DESIGN OF PIPING SYSTEMS FOR HOT WATER HEATING

To determine the pipe or tube sizes required in the various sections of a hot water heating system:

1. Calculate hourly heat loss of the structure for the indoor and outdoor design temperatures currently used in the locality.

2. Determine the sizes of heating units required to deliver the calculated heat loss at the selected system water temperature. Supply water temperatures in forced circulation systems are usually within the range of 180F to 250F. Temperature drop is usually assumed to be 20F deg. Output of heating units are obtained from manufacturers' catalogs.

3. Select the best locations of the heating units for proper heating of the space. Lay out the piping from the boiler to all heating units so that the length of all piping can be measured for both supply and return.

4. Measure total length of the piping circuit that appears to be the longest, from the boiler to the farthest heating unit and back to the boiler. Add 50% as an allowance for fitting-friction loss in terms of length. The total will be the length used to determine the friction loss per foot or per 100 ft that will be the basis for preliminary pipe sizes of all sections of the system.

5. Determine rate of flow in gallons per minute that must be maintained at the pump.

$$\text{Flow, gpm} = \frac{\text{Hourly heat loss of structure, Btuh}}{\text{temperature drop x } 8.33 \text{ x } 60}$$

(8.33 is the weight of one gallon of water; 60 is the number of minutes per hour.) If a temperature drop of 20F is chosen, the denominator becomes $20 \times 8.33 \times 60 = 9996$, which may be taken as 10,000.

For example, if the structure heat loss is 104,000 Btuh, and system temperature drop is selected as 20F, the required pumping rate is 104,000/10,000 = 10.4 gpm.

6. Select pump from a manufacturer's catalog that will deliver water at the required rate, noting

Table 1—Pressure Drop through 90° Screwed Elbow

Nominal Pipe Size, Inches	Water Velocity, Feet per Second										
	K^1	1	2	3	4	5	6	7	8	9	10
	Pressure Drop through 90° Elbow, Milinches[2]										
½	182	182	728	1,640	2,910	4,550	6,550	8,920	11,600	14,700	18,200
¾	176	176	704	1,580	2,820	4,400	6,340	8,620	11,300	14,300	17,600
1	170	170	680	1,530	2,720	4,250	6,120	8,330	10,900	13,800	17,000
1¼	162	162	648	1,460	2,590	4,050	5,830	7,940	10,400	13,100	16,200
1½	156	156	624	1,400	2,500	3,900	5,620	7,640	9,980	12,600	15,600
2	147	147	588	1,320	2,350	3,680	5,290	7,203	9,410	11,900	14,700
2½	140	140	560	1,260	2,240	3,500	5,040	6,860	8,960	11,300	14,000
3	132	132	528	1,190	2,110	3,300	4,750	6,470	8,450	10,700	13,200
3½	127	127	508	1,140	2,030	3,180	4,570	6,220	8,130	10,300	12,700
4	120	120	480	1,080	1,920	3,000	4,320	5,880	7,680	9,720	12,000
5	111	111	444	999	1,780	2,780	4,000	5,440	7,100	8,990	11,100
6	103	103	412	927	1,650	2,580	3,710	5,050	6,590	8,340	10,300
8	90	90	360	810	1,440	2,250	3,240	4,410	5,760	7,290	9,000
10	82	82	328	738	1,310	2,050	2,950	4,020	5,250	6,640	8,200
12	78	78	312	702	1,250	1,950	2.810	3,820	4,990	6,320	7,800

[1] From F. E. Giesecke, ASHRAE *Transactions*, Vol. 59, p. 52 (1953)
[2] Pressure drop $= Kv^2$

Table 2—Pressure Drop through Fittings

Fitting	Pressure Drop in Equivalent 90° Elbows
90° ell, screwed	1.0
90° ell, flanged	0.8
45° ell, screwed	0.5
45° ell, flanged	0.4
180° open return bend, screwed	1.0
180° close return bend, screwed	2.0
180° close return bend, flanged	1.5
tee (flow through side), screwed	2.0
tee (flow through side), flanged	1.5
tee (flow through run), screwed	0.6
tee (flow through run), flanged	0.5
90° welding elbows and smooth bends: (R = bend radius, d = pipe dia.)	
$R/d = 1$	0.5
$R/d = 1\frac{1}{2}$	0.4
$R/d = 2$	0.3
$R/d = 4$	0.3
$R/d = 6$	0.3
$R/d = 8$	0.4
gate valve, full open	0.3
globe valve, full open	11
angle valve, full open	6
angle radiator valve, full open	3
swing check valve, full open	3
radiator or convector	4*
boiler	4*

* For rough calculation. Check catalog specification.

permissible head in feet of water at pump outlet. Divide permissible head by equivalent length of pipe, as found in Item 4, and multiply the result by (a) 100, to obtain permissible loss in feet per 100 ft of pipe, or by (b) 12,000, to obtain permissible loss in milinches per foot of pipe.

If the pump selected will deliver 10.4 gpm at a head of 6.6 ft and total equivalent length of pipe is 230 ft, the design friction loss is $(6.6/230)(100)$ = 2.87 ft per 100 ft or $(6.6/230)(12,000)$ = 344 milinches per foot.

7. Make a table, entering each section of the circuit previously selected as the longest. Then determine pipe size. Two hot water friction charts, for Type L copper tubing and iron or steel pipe, will be found elsewhere in this section of the Handbook. One enters the chart, for each section of piping to be sized, with the "heat" carried in that section of piping, in Btuh, as determined from the piping layout in Item 3, selecting that commercial pipe size whose friction loss per foot is closest to the permissible drop per foot selected in Item 6. Note velocity. If excessive, choose a larger pipe size. The partially completed table should then show pipe section (with convenient letter designations), heat carried in that section, pipe size, friction loss per foot based on pipe size selected, and velocity.

8. Multiply friction loss per foot for each section by actual length of section in feet and enter friction loss of pipe for that section of table.

9. Determine the exact fittings required for each section of pipe tabulated in Item 7 and find the friction loss for each from Tables 1 and 2. Table 1 gives pressure drops through 90° screwed elbows for various nominal pipe sizes and velocities from one to 20 fps. Table 2 gives losses through other fittings in terms of the number of 90° elbows. Thus, from Table 2, pressure drop through a 45° screwed elbow is 0.5 elbows, and from Table 1, the pressure drop for a 1½-inch elbow at 3 fps is 1400 milinches. Therefore, pressure drop through a 1½-inch, 45° screwed elbow is $1400 \times 0.5 = 700$ milinches. These values are very approximate. For pipe fittings below 2 inches diameter, these values may vary above and below the values given by 40%, depending on construction and installation details. Enter fittings losses in their appropriate sections of the table.

10. Total straight pipe and fittings friction losses to obtain friction loss for each section of piping. Total friction losses for all sections to obtain total friction loss against which the pump will operate.

If total friction loss for the circuit exceeds the head which the pump can handle, an inspection of the pipe sizes tabulated may show that, by some changes in pipe size, the head can be reduced to an acceptable value. If total friction loss is too small, it may be desirable to reduce the pipe sizes in some sections.

Water velocity limitations as previously given in this chapter in the section Design Recommendations should be followed.

PIPING DESIGN FOR TWO-PIPE REVERSED RETURN SYSTEM

The procedure outlined for determining pipe sizes is illustrated in the following Example 1.

Example 1. Select pipe sizes for two-pipe forced circulation reversed return hot water heating system for a structure having a design heat loss of 104,000 Btuh. There are 10 radiators so arranged that the system may be laid out in two symmetrical equal circuits, as shown in Fig. 14, each having a total load of 52,000 Btuh.

Circuit to Radiator No. 5. The piping to Radiator No. 5 is selected as the longest circuit. Table 3 is constructed to facilitate its sizing. Section designations, corresponding to the letter symbols in Fig. 14, are listed in Column 1 and their actual lengths in Column 2. The amount of radiation carried by each section, in thousands of Btu, are set down in Column 3, having been taken from Fig. 14.

Total measured length of the circuit (Column 2) is 153 ft. It is common practice to increase actual length by 50% for preliminary sizing purposes. Therefore, it is assumed that equivalent length is 153 × 1.5 = 230 ft.

If a 20F drop is selected, flow through the pump will be, approximately, 104,000/10,000 = 10.4 gpm. From catalog data, it is found that a 10.4-gpm capacity will operate against a head of 6.6 ft of water. Since the circuit is 230 ft, permissible friction loss would average (6.6/230)(12,000) = 344 milinches per ft.

For each section of pipe, Table 3, the friction chart is entered at the Mbtu value in Column 3 and a pipe size selected as close to 344 milinch per ft pressure drop as possible. In each case actual friction drop for the pipe selected is recorded in Column 5 and the water velocity in feet per second to the nearest integer recorded in Column 6.

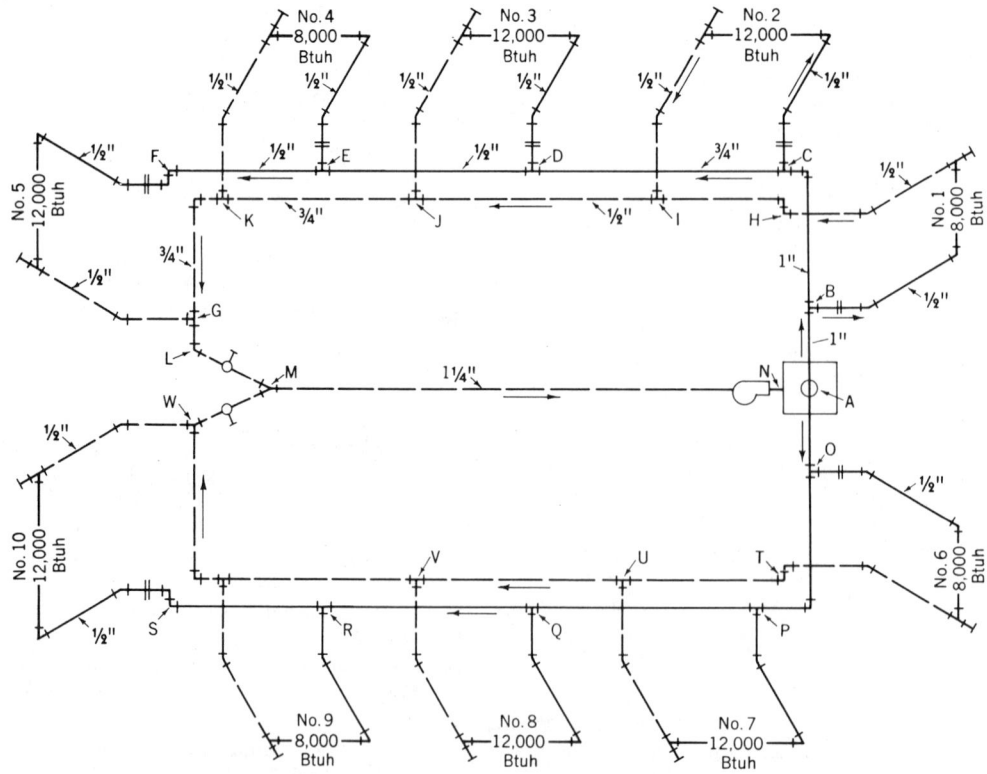

Fig. 14. Layout for two-pipe reversed return hot water heating system with forced circulation (for example 1).

Table 3—Sizing Circuit to Radiator No. 5, Example 1

I	2	3	4	5	6	7	8	9	10	11	12
Section Desig- nation	Actual Length, Ft	Mbtu Carried	Pipe Size	Friction, Milinches Per Ft	Vel- ocity, Fps	Loss in Pipe, Milinches (Col 2 × Col. 5)	Fittings	Equiv- alent Elbows	Loss in One Elbow, Mil- inches	Loss Thru Fittings, Milinches (Col. 9 × Col. 10)	Loss Thru Section, Milinches (Col. 7 Plus Col. 11)
A-B	15	52	1	300	2	4,500	90° ell tee	1.6	680	1,100	5,600
B-C	16	44	1	200	2	3,200	90° ell tee	1.6	680	1,100	4,300
C-D	13	32	¾	350	2	4,550	tee	0.6	704	420	4,970
D-E	14	20	½	600	2	8,400	tee	0.6	728	440	8,840
E-F	16	12	½	200	1	3,200	90° ell 6-90° ells 45° ell	1.0	182	180	3,380
F-G	20	12	½	200	1	4,000	rad. valve tee rad.	15.5	182	2,800	6,800
G-L	5	52	1	300	2	1,500	90° ell 1 cock	1.0	680	680	2,180
L-M	12	52	1	300	2	3,600	1-Y	1.0	680	680	4,280
M-N	38	104	1¼	275	2	10,450	2-90° ells 2-90° ells	2.0	648	1,300	11,750
N-A	4	104	1¼	275	2	1,100	2-tees flow valve boiler	21.0	648	1,200	23,600
Total	153										65,700

Losses in straight pipe sections are calculated by multiplying actual pipe length, Column 2, by friction per foot, Column 5, and the product recorded in Column 7.

The entire circuit, Fig. 14, is then carefully examined and the fittings in each section recorded in Column 8. The equivalent number of 90° elbows, from Table 2, are set down in Column 9 and the loss per elbow, from Table 1, recorded in Column 10. Loss through fittings is the product of Columns 9 and 10, recorded in Column 11.

Total loss in each section is the sum of straight piping loss, Column 7, and loss through fittings, Column 11. Their sum is recorded in Column 12. Column 12 is summed up, giving the approximate actual friction loss (in Table 3) of 65,700 milinches, which, when divided by 12,000, is 5.5 ft.

The total of 5.5 ft of head compares to 6.6 ft against which the pump selected must work in order to produce the proper flow. The discrepancy reflects the very rough approximations that were made at the beginning of the calculations and the very approximate nature of all the data, especially for loss through fittings, valves and equipment. An equivalent resistance of 11 elbows was assigned to the flow valve, assumed full open. Closing it partially will bring the resistance of the circuit up to the required amount if, indeed, the system, as installed, has the calculated resistance.

Had the calculated resistance of the circuit been significantly greater than 6.6 ft, it would have been necessary to increase some of the pipe sizes, thereby reducing the friction head to the allowable amount.

Table 4—Sizing Circuit to Radiator No. 1, Example 1

1	2	3	4	5	6	7	8	9	10	11	12
Sec-tion	Flow		Pipe size, inches	Vel., fps	Loss, mi/ft	Fittings			Meas. Pipe, ft	Total Eqiv. Pipe, ft	Total Loss, Milinches
	Btuh	gpm				No., Type	Ells	Eq. ft			
A-B	52,000	5.2	1	1.9	250	1-90° ell 1-tee	3	7.2	15	22.2	5,550
B-H	8,000	0.8	½	0.84	105	1-tee 1-45° ell 4-90° ell 1-rad vlv 1-rad	11.7	13.	20	33.0	3,460
H-I	8,000	0.8	½	0.84	105	1-90° ell 1-tee	3	3.3	13	16.3	1,710
I-J	20,000	2.0	½	2.05	550	1-tee	2	2.8	14	16.8	9,250
J-K	32,000	3.2	¾	1.88	330	1-tee	2	3.6	16	19.6	6,480
K-G	40,000	4.0	¾	2.40	500	1-90° ell	1	1.9	16	17.9	8,950
G-L	52,000	5.2	1	1.9	250	1-90° ell	1	2.4	5	7.4	1,850
L-M	52,000	5.2	1	1.9	250	1-cock 1-Y	1.7	4.1	12	16.1	4,020
M-N	104,000	10.4	1¼	2.3	240	2-90° ell	2	6.6	38	44.6	10,700
N-A	104,000	10.4	1¼	2.3	240	2-90° ell 2-tees flow vlv boiler	29	95.0	4	99.0	23,800
Totals								145.3	153	295.9	75,770 = 6.3 ft

Col. 1: From Fig. 14
Col. 2: From Fig. 14
Col. 3: Col. 2/10,000
Col. 4: From 1964 ASHRAE Guide and Data
Book, friction chart, p. 136
Col. 5: From friction chart
Col. 6: From friction chart

Col. 7: From Fig. 14
Col. 8: Ibid., Table 6, p. 140
Col. 9: Ibid., Table 5, p. 139
Col. 10: From Fig. 14
Col. 11: Col. 9 + Col. 10
Col. 12: Col. 6 × Col. 11

Circuit to Radiator No. 1. Table 4 contains the working data for sizing the circuit to Radiator No. 1 by a slightly different method. Resistance of fittings are recorded in equivalent feet of pipe from data in the ASHRAE Guide and Data Book, Applications volume, Chapter 8, Hot Water Heating Systems. Equivalent length is added to actual pipe length for each section and the sum multiplied by the friction loss per foot as found in the friction charts, giving friction loss for that section.

Note that the friction loss for the circuit is 6.32 ft. Since the friction loss is the circuit to Radiator No. 5 will be adjusted to approximately 6.6 ft, the difference may be eliminated by addition of a square head cock in Section H-I for achieving balance, or the angle valve at Radiator No. 1 can be partially closed. However, such a close balance is not necessary in a system with a reversed return.

Final Check of Pipe Sizes. The friction losses of all remaining radiator circuits should be obtained and tabulated as for Radiators No. 1 and No. 5 so that, if necessary, corrections can be made by changing some pipe sizes or by adding adjustable cocks, orifices, or valves to obtain a balanced flow through all circuits.

PIPING DESIGN FOR TWO-PIPE DIRECT RETURN SYSTEM

In a two-pipe *direct* return system, the procedure for selecting pipe sizes is the same as for a reversed return system. To the measured length of the longest circuit is added 50% to obtain the equivalent length. For a 20 deg temperature drop, the water flow through the pump is found by dividing the heat loss of the structure by 10,000. A pump is selected from a manufacturer's catalog and the available head is noted. The available head is divided by the equivalent length of the circuit to obtain a design unit friction loss for the system.

Preliminary pipe sizes are found in the friction chart for hot water and listed for all sections of the circuit as was done in Table 3 for the reversed return system in Example 1. The actual unit friction loss per foot, the velocity, and the total friction loss for the longest circuit are tabulated. If the total fric-tion loss does not exceed the rated pump head and if the velocities are within limits suggested in previous chapter section on Design Recommendations the selected pipe sizes are satisfactory.

If necessary, corrections are made in the piping of some parts to obtain a balance of the total circuit friction and the available pump head.

The procedure just outlined for the longest circuit is repeated for circuits to all heating units so that the friction losses in all circuits are equal. However, it may not be possible to achieve balance by changes in pipe sizes and number of fittings because there may be too much difference in circuit lengths, and therefore, it may be necessary to add adjustable cocks or orifices in circuits that have too low friction loss.

PIPING FOR ONE-PIPE DIVERSION SYSTEM

In a one-pipe system the same principles apply to design of piping as explained in Example 1 for a two-pipe system but, since special fittings are generally used to induce or proportion flow from the main through the heating units and back to the main, it is necessary to obtain from the manufacturer the capacities and resistances of these fittings at various rates of flow in the main for different fitting sizes.

Sizing Piping for Main. After the heat loss of the structure has been determined and the proper heating units have been selected and their location established, a layout should be made showing all piping and fittings.

To the measured length of the main add 50% to obtain the equivalent length of main. If there is more than one main circuit, this applies to one such circuit from boiler to the end of the main at the boiler return connection.

Assume a design temperature drop of 20 F.

Since the water taken from the main to any heating unit returns to the main, the flow through the main is the same throughout the circuit. The main may therefore be uniform in size except in parts, if any, where flow is combined with that from another circuit.

Next, select a pump from a manufacturer's catalog that will deliver the gallons per minute required. This flow is found by dividing the heat loss of the structure by 10,000.

Note the head produced by the pump at the rated flow. Divide the pump head by the equivalent main length found, to obtain the design unit friction loss for the circuit.

From friction chart, select a pipe size that will have approximately the unit design friction loss at the flow rate required.

Sizing Piping for Branches. Since the flow required through the piping from a main to a heating unit and back to the main is known (because it is determined from the heat required by the heating unit and the temperature drop) it is convenient to select, from the tables provided by the manufacturer of the flow-inducing fitting, the size of fitting required for a desired flow in gallons per minute for a given equivalent length of branch piping.

As an example, the capacity table for special fittings of one manufacturer shows:

For a 1¼-inch fitting in a 1¼-inch main, when the branch has an equivalent length of 30 ft, the connection to the fitting should be ¾-inch and the flow in the main should be 10 gpm if the flow through the branch is to be 1.9 gpm. This manufacturer states that the special fitting should be in the main at the return branch connection and that the tee at the connection of the supply branch should be a regular pipe tee. It is also stated that only one special fitting is required.

For these fittings the manufacturer recommends that the distance along the main between supply

and return connections should be at least 18 inches.

It should be noted that manufacturers show a reduction in flow inducing or diverting capacities of special fittings if the heating unit is below the level of the main. This reduction is due to the opposition of the gravity head in the branch. This gravity head is greatest when the pump starts because at that time the water in the branch is at room temperature. Opposition decreases as the water in the branch becomes warmer but, nevertheless, continues at a decreased rate as long as the pump is in operation.

The manufacturer of another special fitting that may be used singly or in pairs to divert the proper amount of water from the main to a heating unit circuit and back to the main shows the capacities of fittings having various sizes of outlet to branch for various milinch pressure drops for heating units one, two, and three floors above the main, using either one or a pair of the fittings. For circulation to heating units below the main, a pair of these special fittings are always required and consequently, capacities are shown only for pairs of fittings for this use.

Pipe Size Check. When finding the actual friction loss of mains and branches after the preliminary pipe sizes have been found, it is necessary to know the friction at the given flow in the main through

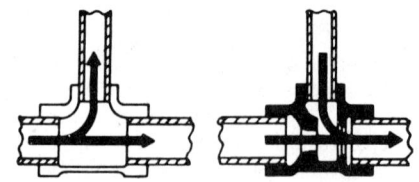

Fig. 15. Flow through Taco venturi fitting and standard pipe tee.

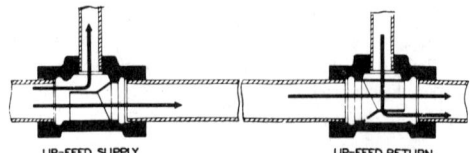

UP-FEED SUPPLY   UP-FEED RETURN

Fig. 16. Flow through Bell & Gossett Monoflo fitting, up-feed supply and up-feed return.

the special fitting. The manufacturer therefore shows the pressure drop at various rates of flow in single fittings, or for fittings in pairs for fittings of various sizes. The recommendation is given by one manufacturer that only 40% of the given pressure loss should be considered when determining the friction loss in the main due to the special fitting.

The operating principles of the special fittings mentioned will be evident from Figs. 15 and 16.

PIPING FOR ONE-PIPE SERIES SYSTEM

In the one-pipe series system, the same water flows progressively through the heating units in any circuit and consequently the water temperature entering any unit is lower than in the preceding unit. It is therefore common practice to limit the number of heating units in any one circuit so that the water temperature at the last units on the circuit will not be so low as to necessitate an increase of heating unit size. A common temperature drop used for these systems is 20 F. A drop of 5 F per heating unit in any circuit is sometimes used. An increase in velocity can be used to lower the temperature drop in the system but will be limited by noise considerations, pump head, or cost of pumping.

In order to size the piping, a layout is made and the total flow rate (in gallons per minute) in the system is obtained by dividing the design heat loss of the structure by 10,000, when a temperature drop of 20 F is used. A preliminary selection of pump

may now be made for the flow rate required. The available circulation head of the pump should be noted.

The equivalent length of piping should be the measured length plus 50% added length for fittings plus the length of the heating units.

The available pump head should be divided by the equivalent pipe length to obtain the design unit friction loss.

The pipe size may now be found from the friction chart for the required flow rate and the design unit friction loss.

The actual friction loss of the system is found by adding to the loss through piping the losses through heating units and fittings (see Tables 1 and 2) or by adding (1) the actual piping length, (2) the heating unit lengths, and (3) the equivalent length of boiler and pipe fittings and multiplying this total by the actual unit friction loss.

This total system friction loss should be compared with the available pump head. It may be necessary to change the pump size or pipe size if the two losses are not in balance.

In systems having baseboard units, preliminary pipe sizes are sometimes selected equal to the tappings or connection sizes provided by the manufacturer. The actual system friction loss is then determined as outlined above for comparison with the available pump head.

If the system has a large number of baseboard heating units, the system may be designed with more than one loop or circuit. In this case, the piping will be designed for the required flow in each circuit and it will be necessary to equalize the friction of the different circuits.

COMBINATION OF PIPING SYSTEMS

While the different piping systems for hot water heating have been discussed separately, it will be found practicable in many a system to use more than one type of piping in different parts of a system. The principles for obtaining proper water circulation, the balancing of friction drop in parallel circuits, the limiting of friction loss to the available pump head, and the limiting of velocities to accepted values, can be applied to assure proper functioning of all parts.

SIZING HOT WATER EXPANSION TANKS
(System Water Capacity Known)

The accompanying chart is for use where the quantity of water in the system is known. Use of the chart is explained by the following examples.

Example. A hot water system has a head of 53 feet and a system water volume (*W*) of 1,000 gallons. What size expansion tank is to be used, what pressure will be developed in it, and what would be the minimum size tank that could be used?

Solution. From 53 on the horizontal scale move vertically to the optimum reference line, thence right and read 0.30*W*. Since 0.30*W* = 300 gallons, this is the recommended size tank. The intersection of the curve and the 53 line also shows the working pressure to be 50 pounds gage.

The minimum tank size is found to be 0.163*W* or 163 gallons, in which case the working pressure would be 100 pounds gage.

If a predetermined working pressure is selected, read to the right. For example, if 60 pounds pressure is desired, move up the 53 line to 60, thence right and read 0.24*W*, or 240 gallons as the tank capacity.

The safety valve pressure should be 10 pounds above the working pressure.

If the system is connected to the water supply by means of a self-filling pressure reducing valve, the reducing valve must be adjusted to the system's static pressure. If it is not, a higher working pressure will be developed in the system.

Example. What will be the working pressure in the given example with a tank size 0.3*W*, if the self-filling pressure reducing valve is adjusted to 30 pounds gage?

Solution. The system's effective static head will be:

$$H = \frac{30}{0.434} = 70 \text{ feet.}$$

With 70 feet of head and 0.3*W* tank capacity in the system, 71 psi gage pressure will be developed instead of 50 psi gage. This is shown by the intersection of vertical 70 feet head with the horizontal 0.3*W*.

It is well to bear in mind that the pressures given are maximum pressures developed by the expansion of a system, filled cold and then heated.

The optimum tank size (volume) reference line is based on the volume of the tank capacity in gallons being equal to *kW*, where *W* is the water volume in the system in gallons, and *k* is 0.1 for a one-story building, and 0.13, 0.17 and 0.23 for 2, 3 and 4 story buildings, respectively.

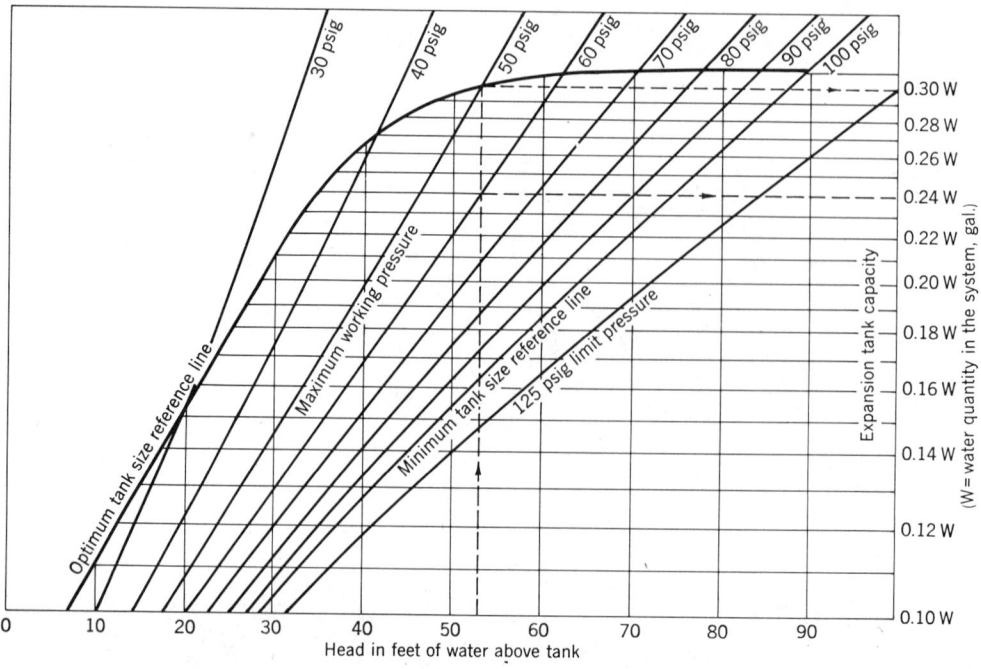

SIZING HOT WATER EXPANSION TANKS
(Boiler Output Known)

Use of this chart should be limited to conventional hot water heating systems. It should *not* be used for multi-story buildings, water temperatures above 220 deg F, or specially designed industrial installations. The selections it indicates are based on standard, commercially available compression tank sizes.

The following steps should be followed in use of the chart:

1. Determine net boiler output and locate on left-hand scale.

2. From this point, draw horizontal line to intersect the appropriate system curve. The choice is made from flash or conventional boiler source, hot water heating systems utilizing radiant panels or baseboard, convectors, or thin tube cast iron pipe.

3. From this intersection, extend vertical line downward through both maximum and minimum base lines.

4. Select compression tank size as indicated where vertical line intersects dotted, diagonal line. If vertical line does not intersect a dotted line, use next larger tank size to right.

5. Apply correction factor, if necessary, from the accompanying table to the tank size.

Correction Factors

(Standard conditions are 12 psig initial pressure, 30 psig final pressure, and 18.5 ft maximum system height above tank. For single-pipe systems, deduct 13-20% from tank size given for conventional boilers; 18-22% from tank size given for flash boilers)

Initial Pressure, Psig	Static Height, Ft	Multiply Tank Size as found by
12	18.5	1.00
16	27.5	1.43
20	36.5	2.18
24	46.0	3.82

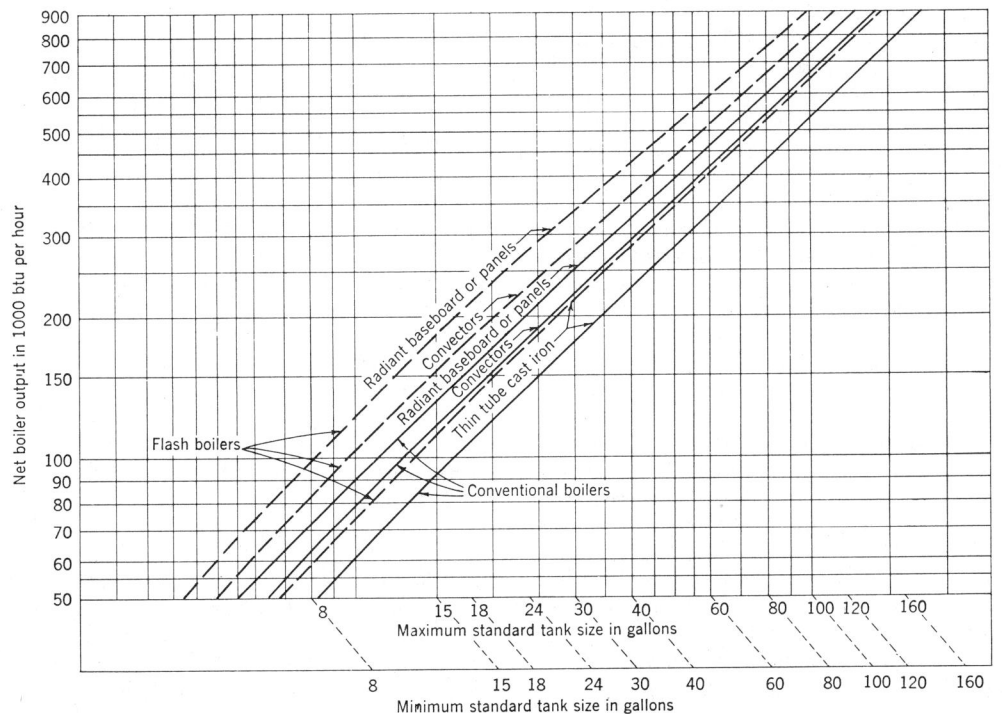

HIGH TEMPERATURE WATER SYSTEMS

High temperature water (HTW) systems have grown in popularity due to their adaptability for supplying heat at various temperatures for industrial and space heating purposes. The system supply temperature is governed by the heating unit or process requiring the highest temperature but lower temperatures, if required for other loads in the system, are readily provided by drawing mixed water from supply and return mains, by restricting flow to certain units, or by use of water-to-water heat exchangers. If a process in the system requires steam, this may be generated in a heat exchanger if the saturation temperature of the steam required is sufficiently below the supply water temperature to permit the transfer of heat to produce the steam.

Conditions Affecting Design

The designer of a HTW system is immediately concerned with problems due to the following conditions, some of which introduce complications not found in the design of low-temperature water heating systems:

1. Water temperature is high, often 300 F to 400 F. A common temperature is 350 F.

2. Pressures must be high enough to prevent formation of steam in any part of the system.

3. Expansion of pipe and metal parts is greater than with lower temperature systems.

4. Corrosion must be limited to a minimum by special precautions for excluding oxygen. Loss or removal of water should be limited in order to prevent introduction of oxygen when makeup water is added to replace loss of water.

5. Piping, pumps, boilers, heating units, controls, and other equipment must be constructed for operation at the high temperatures and pressures required in the system.

6. Pump characteristics and performance must be carefully considered to assure operation at economical rates for the capacity pumped and the friction loss of the piping.

7. Means for obtaining water at different temperatures for various uses in the equipment attached to the system will require heat exchangers or water mixing devices, and controls for such purpose.

8. A method of producing and maintaining the required pressure must be selected; the usual mediums being steam or nitrogen.

9. Steam formation may occur due to loss of pressure at elevated points in the piping or at the pump suction. Water hammer due to steam formation and its subsequent contact with cooled water must be avoided.

10. Condensation on boiler tubes may occur, causing them to corrode, if too-cool return water flows through the boiler.

11. Due to the large storage of heat in the system it is not advisable to operate the circulation system intermittently, or to try to make rapid changes in water temperature. A relatively constant temperature in the mains with separate control of heating units will, therefore, be necessary for maintaining or controlling space temperatures.

12. Pumps are always used to produce circulation in HTW systems and consequently the mains need not be pitched in the direction of flow. They may be run at any desired level to follow the contour of the land or to pass obstacles.

13. HTW systems are usually designed for a temperature drop of 100 to 150 F instead of the 20 F drop commonly used for low-temperature systems. While the design temperature drop may be as low as 80 F, it is more economical to use a 150 F drop. In some instances it can be shown that a drop of 180 F or even 200 F is practicable and economical.

Many of the conditions mentioned in the preceding thirteen items will be treated more in detail in later text.

Purpose of High Temperature Drop

High-temperature drops between supply and return mains are used in HTW systems in order to reduce the flow of water that must be circulated and thereby permit reduction of pipe sizes to a minimum practical limit. For instance, if a pipe line is to transmit 1,000,000 Btuh through a mile of pipe at a 50 F temperature drop and a velocity of 4 fps, a 2½-inch pipe will be required. For the same system, if the temperature drop is 120 F, the pipe size may be reduced to 1½ inches. Similarly, if the heat transmitted is to be 10,000,000 Btuh, the pipe size required is 6 inches for a 50 F drop or 4 inches for a 120 F drop.

Comparison with Steam

In heating units or processes the heat available from steam is the latent heat due to condensation only, while the heat available from water is the sensible heat given up by cooling of the water. Condensation of steam at 175 psia will yield 852.8 Btu

per lb of steam or 256 Btu per cu ft of steam. One cubic foot of water at the same temperature, if cooled through 100 deg (the minimum temperature drop) to 271 F will give up the difference between the fluid heats at 371 F and 271 F, or (18830—14002) = 4828 Btu. Therefore, water will give up, under equivalent design conditions almost 19 times the heat of an equivalent volume of condensing steam.

The high heat content of the water in the HTW distribution system increases the response of the system to demands for heat and facilitates the maintenance of uniform temperatures in the heated space.

Heat Storage

A HTW system contains a large quantity of stored heat in the circulating water. The stored heat may also be increased during periods of light load by increasing the temperature of the water in the return main. An example of large storage capacity is found in a system having a total length of 20 miles in the supply and return mains, which has a storage capacity equal to a 24-hour reserve of heat.

Limitation of Corrosion

In determining the method of system pressurization or provision of space for the expansion of water as temperature changes, it is of prime importance to limit corrosion within the system. It is therefore necessary to limit addition of oxygen which is the main cause of corrosion. Oxygen enters as a constituent of air (if air is used for pressurization) or as a gas dissolved with air in the makeup water and released as the water is heated.

It is advisable not to draw water from a system for outside purposes because this would necessitate additional makeup water which would introduce oxygen unless the water is given special treatment before being added.

PRESSURIZATION OF HTW SYSTEM

In all water heating systems there must be sufficient pressure on the surface of the water at all points of the system to prevent formation of steam. This pressure must be above the saturation pressure corresponding to the temperature of the water at any point. Some of the most critical points in the system are: (1) the pump suction (where cavitation may occur), (2) the top tubes in the boiler (where the tubes may burn out if not covered by water or filled with water), (3) the expansion tank or other high points in the system, and (4) heat exchangers or converters.

Extensive damage may be done in the system to pumps and other equipment if water hammer results from steam formation.

The medium for pressurizing is mainly steam, air, or an inert gas, usually nitrogen.

Steam Pressurization

The principle of steam pressurization is shown in Fig. 1 in which it will be noted that the expansion tank provides (1) space for expansion of system water, (2) means for pressurization by steam in the upper part of the tank, and (3) storage space for heated water from the boiler. The boilers shown in Fig. 1, schematically, are of the water-tube type with the expansion tank serving also as the steam drum. Note that the supply water for system circulation is taken from the expansion tank and that a bypass is provided to permit mixing some return water with the heated water in order to reduce water temperature at the pump suction and thereby reduce the possibility of cavitation at the pump.

Steam pressurized systems are very critical in operation with regard to fluctuations in pressure and therefore require precise control of pressure.

Pressure fluctuation in steam pressurized systems can be prevented by supplying additional steam to the expansion chamber from a separate steam generator. By this means any desirable pressure can be maintained in the system at all times regardless of

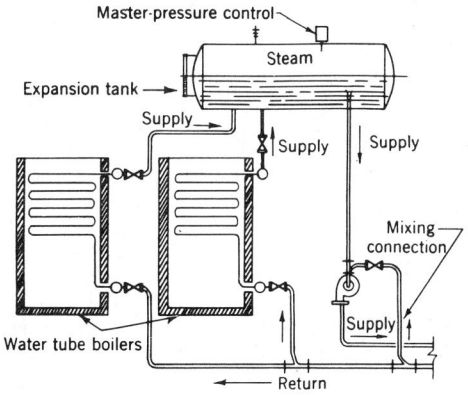

Fig. 1. HTW system pressurized by steam.

the load or water temperature. The heat required by the separate generator would be equivalent only to the heat loss from the expansion chamber surface.

When a boiler has a large steam space within its shell (such as a Scotch Marine or other fire-tube boiler) this steam space may be used for both pressurization space and for space for expansion of the water in the system. However, a separate expansion tank placed high above the boiler is generally preferred because it increases net positive suction head and prevents cavitation at the pump.

Expansion tank volume should be equal to the sum of (1) volume required for steam space, (2) volume required for expansion of the water in the system and (3) volume required or occupied by internal headers and piping and an allowance for sludge and reserve.

Steam space required will be about 20% of the sum of items (2) and (3). Item (2), the space required for expansion of the water in the system, will be the change in volume of the water content from minimum to maximum operating temperature, assuming that water will be drawn off as required after the initial filling of the system to limit the pressure in the expansion tank to the pressure corresponding to the lowest average operating temperature.

Item (3), volume allowance for internal headers, sludge, etc., may be equivalent to 40% of item (2) for small systems up to 80% of item (2) for large systems.

Gas Pressurization

Pressurization of a HTW system by means of gas overcomes the difficulties occurring in steam pressurized systems due to fluctuations of pressure which cause cavitation, water hammer, etc.

Air and nitrogen have been used but nitrogen is preferred because it is inert while air introduces oxygen which is very corrosive, especially where metal surfaces are in a moist condition.

An advantage of gas pressurization is that the pressure in the system can readily be maintained at the desired level and, consequently, the tendency for steam formation is eliminated. The expansion tank can be placed where convenient. Also, since it is considered as an unfired pressure vessel it need only meet the requirements of the ASME Unfired Pressure Vessel Code. Vertical expansion tanks are frequently used to save floor space.

Air Pressurization

Use of air for pressurization is shown schematically in Fig. 2. Air escapes through a relief valve to the atmosphere when pressure due to the expansion of the water exceeds the pressure setting of the valve. Air for replacement is supplied by an air compressor and pressure tank and enters the expansion tank under control of a regulating valve when the pressure falls. Since the interior of the tank is moist, corrosion within the tank may be extensive.

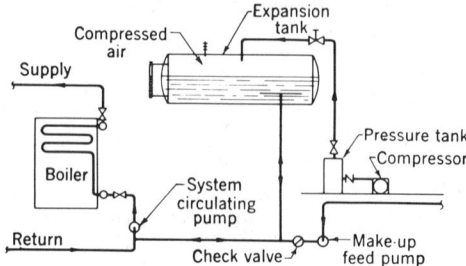

Fig. 2. Method of pressurizing HTW system by air.

Nitrogen Pressurization

There are three basic methods of pressurization with nitrogen using (a) an overhead tank, Fig. 3, (b) an overflow or spill tank at boiler floor level for expansion and contraction of the system water, with a hot well for make-up, Fig. 4, and (c) elimination of the hot well. In (a), nitrogen is recycled by a compressor, thereby eliminating gas losses except by absorption, which is small. Note that with this earlier method three drums are necessary: one expansion drum and two nitrogen drums. (In earlier

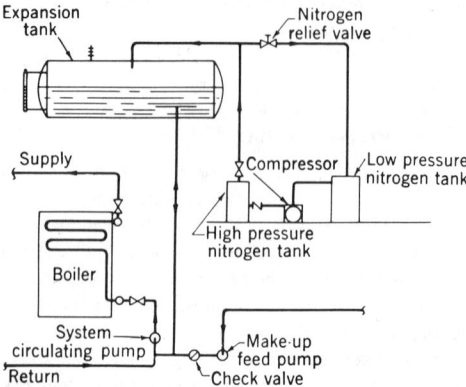

Fig. 3. Inert gas pressurized HTW system using nitrogen.

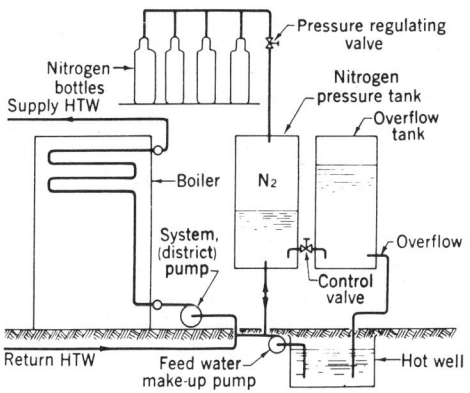

Fig. 4. Nitrogen pressurization HTW system without
overhead tank.

systems, nitrogen was released to the atmosphere
when excessive pressures were built up, as thse pressures became the system pressures and could cause
damage to equipment.)

The more modern method of pressurizing with an
inert gas is to combine the feed-water make-up with
the expansion and contraction of the circulating
water. (Some of these designs are proprietary, including those of completely packaged pressurization
units.) By eliminating the overhead pressure tank,
no elaborate and costly structural steel is needed
to support it and further, the height of the boiler
house can be considerably reduced, again reducing
capital costs.

Pressurization is accomplished in a tank or tanks
of minimum size at the boiler floor level and the
small losses of nitrogen that occur in these systems
are more than offset by the considerable capital and
other savings effected.

EXPANSION TANKS

Expansion Conditions

It is impossible to take up the expansion of
heated water even within a shell type boiler in
large HTW systems. An independent expansion tank
is necessary. Further, since this tank serves also as
a pressure vessel to maintain the operating pressure
on the system, it must be large enough for both
duties. By absorbing expansion and contraction, the
original water is retained and only a minimum quantity of feedwater makeup is required. This reduces
the amount of chemical water treatment necessary.
In a good system, daily makeup water should not
exceed 1% of the total water in the system.

Initial expansion of water to fill the system as it
is heated from fill temperature (assumed to be at
50 F) to the mean temperature of the heated water
at 300 F, can be drawn off the system through the
blowdown valve so that expansion during operation
begins at 300 F. The blowdown valve is closed when
the water level at 300 F is at the center of the expansion tank. Further expansion of water to the
maximum temperature is then contained in the expansion tank.

The following fittings are needed: safety valves
and relief valves constructed according to applicable
codes, high- and low-water-level alarms, and controls.

Determining Expansion Tank Size

The first step in calculating tank size is to establish design pressure. In steam pressurized systems
it is the saturation pressure corresponding to the
temperature of the water. In gas pressurized systems, gas pressure is usually maintained at about
40 psi above design saturation pressure.

The second step is to determine the total volume
of water in the system. It is calculated from the
length and size of all piping plus the water content
of all heat exchangers, drums, boilers, fittings, etc.
It is *not* a function of rated plant output.

The third step is to calculate the daily mean water
temperature change. This is a rather indefinite variable from the design standpoint. Its value is the
average of system flow and return temperatures at
daily maximum and minimum load conditions. The
daily load cycle, however, is a function of the type
of load, that is, whether it is strictly a heating load,
a process load, or a combination of both. From
studies of operating records at various heating installations, actual daily load swings of 25% of rated
output are about average and should be expected
during most of the heating season. In a process
application, daily load swings of 50% or more of rated
output are usual.

To illustrate, assume: system water capacity of 500,000 lb, supply temperature at 350 F, return temperature at 250 F and makeup boiler feed water at 50 F. Values for specific volume, in cubic feet per pound of water, at 50 F, 250 F, 300 F, and 350 F are, respectively 0.01603, 0.01700, 0.01745, and 0.01799 (from Table 1). Multiplying each by 500,000 lb gives the system volume at the four temperatures as 8015, 8500, 8725, and 8995 cu ft, respectively.

For expansion from 300 F to 350 F, the increase in volume is (8995—8725) or 270 cu ft. If a 24-inch (2-ft) rise of water in the expansion tank is considered, the area of tank required at the water line should be 270/2 = 135 sq ft. Referring to a cylindrical tank or drum placed horizontally as in Fig. 5 the surface area at the center should be satisfactory if the tank is 8 ft in diameter and 18 ft long.

If expansion had been computed from 50 F to 350 F, the expansion would have been (8995—8015) or 980 cu ft, which is about 3½ times the expansion found for a temperature rise from 300 F to 350 F.

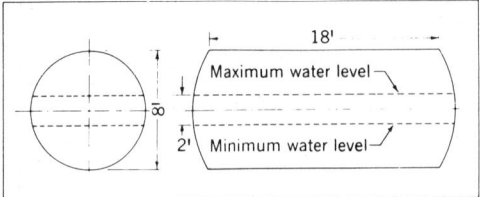

Fig. 5. Horizontal steam pressurizing tank.

Location of Steam Pressurizing Tank

The expansion tank for steam pressurization should be placed well above the boiler to allow for expansion loops in the pipe lines between the boiler header and tank. It does not have to be directly above the boiler. For piping design reasons, it is best to locate the tank above the circulating pumps.

Care must be taken when designing inlet and outlet connections from the expansion tank. The supply outlet leading from the tank to the system circulating pumps should be so placed that no vortex can be formed where the velocity is highest and where, consequently, steam can be drawn with the water. If a dip tube is used for hot water supply to the circulating line of the system, its entrance should

be well below the surface of the water (where steam bubbles collect) so that the bubbles cannot be drawn into the supply line to cause water hammer and troubles at the pump.

Where two expansion tanks are connected, they must be fitted with steam and water balance pipes to keep the two tanks in equilibrium. These steam and water pipes must be used solely to maintain equilibrium and should not be connected into any other lines.

Nitrogen Pressurizing Tanks

In nitrogen pressurized systems, gas pressure is usually maintained at about 10% above design saturation pressure. Vertical drums are preferred for nitrogen pressurization. As an example, assume that the drum is to be sized for the same conditions as used in the previous illustration for a steam pressurized system. The increase in volume of the water when heated from 300 F to 350 F is the same, 270 cu ft. Assume that the tank as illustrated in Fig. 6 is 8 ft in diameter, which provides a tank area of approximately 50 sq ft. The water will then change 5½ ft in level. If a distance of 5 ft is allowed from the maximum water level to the top and 4 ft from the minimum level to the bottom of the tank, the required tank height will be 12½ ft.

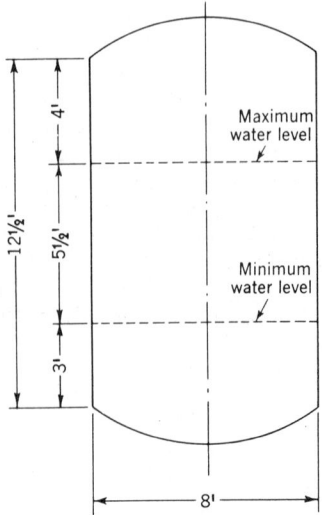

Fig. 6. Vertical nitrogen pressurizing tank.

Table 1. Selected Thermal Properties of Water and Steam

Extracted, by permission, from "Thermodynamic Properties of Steam" by Joseph H. Keenan and Frederick G. Keyes, copyright 1937 by the authors and published by John Wiley & Sons, Inc., New York.

Temp., Deg F	Saturation Pressure		Saturated Water				Heat of Vapori- zation, Btu/lb	Saturated Steam	
	Psia	Psig	Specific Volume, cu ft/lb	Den- sity, lb/cu ft	Specific Heat, Btu/ lb/deg	Enthalpy, Btu/lb		Specific Volume, cu ft/lb	Enthalpy, Btu/lb
32	0.089	—	0.01602	62.422	1.01	0.00	1075.8	3306	1075.8
50	0.178	—	0.01603	62.383	1.00	18.07	1065.6	1703.2	1083.7
80	0.507	—	0.01608	62.189	1.00	48.02	1048.6	633.1	1096.6
100	0.949	—	0.01613	61.996	1.00	67.97	1037.2	350.4	1105.2
110	1.275	—	0.01617	61.843	1.00	77.94	1031.6	265.4	1109.5
120	1.692	—	0.01620	61.728	1.00	87.92	1025.8	203.27	1113.7
130	2.223	—	0.01625	61.538	1.00	97.90	1020.0	157.34	1117.9
140	2.889	—	0.01629	61.387	1.00	107.89	1014.1	123.01	1122.0
150	3.718	—	0.01634	61.200	1.00	117.89	1008.2	97.07	1126.1
160	4.741	—	0.01639	61.013	1.00	127.89	1002.3	77.29	1130.2
170	5.992	—	0.01645	60.790	1.00	137.90	996.3	62.06	1134.2
180	7.510	—	0.01651	60.569	1.00	147.92	990.2	50.23	1138.1
190	9.339	—	0.01657	60.350	1.00	157.95	984.1	40.96	1142.0
200	11.526	—	0.01663	60.132	1.01	167.99	977.9	33.64	1145.9
210	14.123	—	0.01670	59.880	1.01	178.05	971.6	27.82	1149.7
212	14.696	0.000	0.01672	59.809	1.01	180.07	970.3	26.80	1150.4
220	17.186	2.490	0.01677	59.630	1.01	188.13	965.2	23.15	1153.4
230	20.780	6.084	0.01684	59.382	1.01	198.23	958.8	19.382	1157.0
240	24.969	10.273	0.01692	59.102	1.01	208.34	952.2	16.323	1160.5
250	29.825	15.129	0.01700	58.824	1.02	216.48	945.5	13.821	1164.0
260	35.429	20.733	0.01709	58.514	1.02	228.64	938.7	11.763	1167.3
270	41.858	27.162	0.01717	58.241	1.02	238.84	931.8	10.061	1170.6
280	49.203	34.507	0.01726	57.937	1.02	249.06	924.7	8.645	1173.8
290	57.556	42.860	0.01735	57.637	1.02	259.31	917.5	7.461	1176.8
300	67.013	52.317	0.01745	57.307	1.03	269.59	910.1	6.466	1179.7
310	77.68	62.98	0.01755	56.980	1.03	279.92	902.6	5.626	1182.5
320	89.66	74.96	0.01765	56.657	1.04	290.28	894.9	4.914	1185.2
330	103.06	88.36	0.01776	56.306	1.04	300.68	887.0	4.307	1187.7
340	118.01₁	103.31	0.01787	55.960	1.05	311.13	879.0	.788	1190.1
350	134.63	119.93	0.01799	55.586	1.05	321.63	870.7	3.342	1192.3
360	153.04	138.34	0.01811	55.218	1.05	332.18	862.2	2.957	1194.4
370	173.37	158.67	0.01823	54.855	1.05	342.79	853.5	2.625	1196.3
380	195.77	181.07	0.01836	54.466	1.06	353.45	844.6	2.335	1198.1
390	220.37	205.67	0.01850	54.054	1.07	364.17	835.4	2.0836	1199.6
400	247.31	232.61	0.01864	53.648	1.09	374.97	826.0	1.8633	1201.0
410	276.75	262.05	0.01878	53.248	1.09	385.83	816.3	1.6700	1202.1
420	308.83	294.13	0.01894	52.798	1.10	396.77	806.3	1.5000	1203.1
430	343.72	329.02	0.01910	52.356	1.11	407.79	796.0	1.3499	1203.8
440	381.59	366.89	0.01926	51.921	1.12	418.90	785.4	1.2171	1204.3

Table 1. Selected Thermal Properties of Water and Steam (Cont'd)

Temp., Deg F	Saturation Pressure		Saturated Water				Heat of Vapori- zation, Btu/lb	Saturated Steam	
	Psia	Psig	Specific Volume, cu ft/lb	Den- sity, lb/cu ft	Specific Heat, Btu/ lb/deg	Enthalpy, Btu/lb		Specific Volume, cu ft/lb	Enthalpy, Btu/lb
450	422.6	407.9	0.0194	51.55	1.12	430.1	774.5	1.0993	1204.6
460	466.9	452.2	0.0196	51.02	1.14	441.4	763.2	0.9944	1204.6
470	514.7	500.0	0.0198	50.51	1.15	452.8	751.5	0.9009	1204.3
480	566.1	551.4	0.0200	50.00	1.16	464.4	739.4	0.8172	1203.7
490	621.4	606.7	0.0202	49.51	1.17	476.0	726.8	0.7423	1202.8
500	680.8	666.1	0.0204	49.02	1.19	487.8	713.9	0.6749	1201.7
520	812.4	797.7	0.0209	47.85	1.22	511.9	686.4	0.5594	1198.2
540	962.5	947.8	0.0215	46.51	1.26	536.6	656.6	0.4649	1193.2
560	1133.1	1118.4	0.0221	45.25	1.31	562.2	624.2	0.3868	1186.4
580	1325.8	1311.1	0.0228	43.86	1.37	588.9	588.4	0.3217	1177.3
600	1542.9	1528.2	0.0236	42.37	1.45	617.0	548.5	0.2668	1165.5
620	1786.6	1771.9	0.0247	40.49	1.53	646.7	503.6	0.2201	1150.3
640	2059.7	2045.0	0.0260	38.46	1.66	678.6	452.0	0.1798	1130.5
660	2365.4	2350.7	0.0278	35.97	1.93	714.2	390.2	0.1442	1104.4
680	2708.1	2693.4	0.0305	32.79	2.47	757.3	309.9	0.1115	1067.2
705.4	3206.2	3191.5	0.0503	19.88	—	902.7	0	0.0503	902.7

APPLICATION OF HTW FOR PROCESS STEAM

Although HTW has many advantages over steam, it is not a substitute for steam in all cases. There are some processes for which steam is essential because of the type of equipment used, or by the nature of the operation. In such cases HTW can always be used as the prime heat source. Instead of piping steam over long distances, with attendant trap and grading problems, the HTW, with its simple circulating system, is brought to the equipment requiring steam. There, either below or adjacent to the unit, dry steam can be made available at a constant pressure and with adequate reserve.

It is not unusual for several pieces of equipment, all requiring steam, to be operated from the same HTW supply line; each steam-consuming unit working at a different pressure. If steam were used instead of hot water as the distributing medium from the central boiler house, pressure reducing valves would have to be fitted to every unit, and, apart from the somewhat complicated condensate return line with receivers and condensate pumps, there would not be the same adequate reserve of steam

(as with HTW) without a much larger boiler plant. Further, uniform control of pressure would, to a great extent, be dependent entirely on manual observation and control to make certain that large steam consuming units are not suddenly thrown on the line, as can often happen in a factory. Where each steam-consuming unit has its own heat exchanger, not only will the overall efficiency of the plant be higher, but waste can almost be eliminated and the cost of steam considerably reduced.

Heat exchangers that produce steam from HTW are often called converters. Steam temperatures range from 320 to 400 F, but more often from 337 to 353 F which correspond to 100 psig and 125 psig, respectively. Typical applications are: steam cooking equipment, sterilizers, calender rolls, radiators, finned coils, and many pieces of hospital equipment such as dressing sterilizers, kitchen and cafeteria equipment, and laundry calenders. Many of these applications require steam at different pressures but can still be connected to the same supply and return HTW lines which serve the space heating units.

CIRCULATING PUMPS

The circulating pump is the heart of the HTW system. Careful design and selection is of vital importance because all heat distribution depends upon it. It must be selected not only for the quantity of water to be pumped at the head or heads required, but also for the high temperatures at which it will be operating. If a pump is designed to deliver 700 gpm against a 200 ft head, it will deliver a greater gallonage against 160 ft head, and will absorb more power, too. It is, therefore, important for the design engineer and pump manufacturer to determine the point of maximum efficiency and allow for a non-overload characteristic so that if the pressure head drops, the horsepower required will be within the capacity of the driving motor.

In steam-pressurized systems, proper practice is to install circulating pumps in the supply line because this maintains the pressure above the flash point of the hottest water and prevents steam formation in the system.

Another safeguard against flashing or cavitation at the pump in steam-pressurized systems is a mixing connection as shown in Fig. 7, which bypasses some of the cooler water from the return into the supply line at the pump suction. This reduces the supply water temperature 5 or 10 F, enough to prevent flashing but without diminution of system pressure. For example, if the temperature of the water leaving the expansion tank is 350 F at 120 psig (near saturation, an unstable condition) a reduction of 5 F by admixture of cooler return water will result in 345 F, 120-psig water, which will not flash.

Further precautions against flashing include the location of the circulating pumps well below the level of the boiler outlet and the use of a large connecting suction pipe dropping vertically to the pumps. By thus increasing the static head on the pump suction, long suction lines are avoided and the suction piping is arranged so that any sudden acceleration in flow is avoided. The total circulating head equals the sum of the head due to pump pressure and the head due to gravity. It should be noted, however, that in design calculations, the head due to gravity is usually neglected when it is small in comparison with the pumping head.

In HTW systems *pressurized with air or inert gas*, the system circulating pump can be placed in the return line because the system pressure is always maintained well above the flashpoint of the water. The pumps may also be placed in the supply line if design conditions make this advisable.

Pumps for HTW Systems

The centrifugal pump is used to circulate water in HTW systems. A description of the characteristics of such pumps, their operation, performance, and methods of converting performance for changes in speed, head, horsepower, impeller size, and also a discussion of pump selection are given in the Piping and Plumbing section of this handbook. The discussion of pumps here will, therefore, be limited to features and factors relating to pump applications for HTW systems.

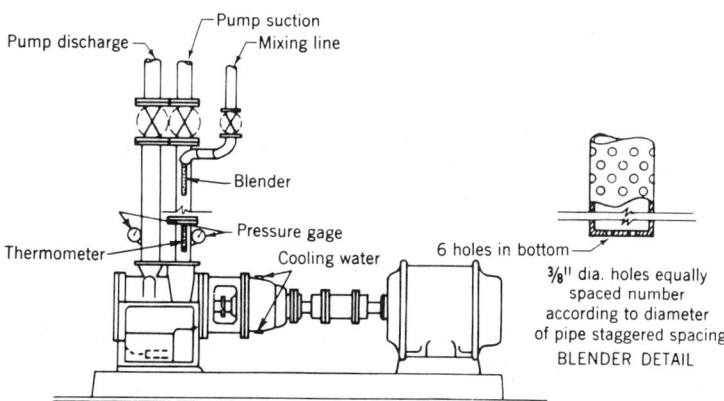

Fig. 7. Mixing connection for steam-pressurized HTW system.

Manufacturers' Information

A typical manufacturer's form will contain the following information:

Conditions of Service

Capacity (gpm)	Liquid
Total head (feet)	Pumping temperature (F)
Speed (rpm)	Specific gravity
Efficiency (%)	Viscosity
Brake horsepower	Vapor pressure
Suction pressure	NPSH above vapor pressure required

Materials of Construction:

Casing and nozzle	Impeller
Case and nozzle rings	Base plate
Shaft sleeve	Coupling
Shaft	

Stuffing Box Sealing
    Type

Pump Specifications

When writing a specification or calling for bids for the centrifugal pump, the pump manufacturer needs to have the following information:

1. The number of units required.

2. The temperature, vapor pressure, and specific gravity of the liquid pumped.

3. The required capacity of the pump, including the minimum and maximum amount of liquid the pump will ever be called upon to deliver.

4. The suction conditions. Is there a suction lift or a suction head? What are length and diameter of the suction pipe?

5. The discharge conditions. The static head. Is it constant or variable? The friction head. The maximum discharge pressure against which the pump must deliver liquid.

6. The total head which will vary as items 4 and 5, above vary.

7. Whether service is continuous or intermittent.

8. The type of power available to drive the pump and the characteristics of this power.

9. Space weight, or transportation limitations involved.

10. The location of the installation; its geographic location, elevation above sea level, whether indoors or outdoors, range of ambient temperatures.

11. Any special requirements or marked preferences with respect to the design, construction, or performance of the pump.

It is not necessary, in writing pump specifications, to state the required total head as long as the dis-charge pressure and the prevailing suction conditions are given. As a matter of fact, it is safer to specify the latter two only, rather than the suction pressures and total heads. The reason for this suggestion is that the specification writer generally works in terms of psi and will give more accurate conditions thereby with less effort.

Net Positive Suction Head (NPSH)

In any centrifugal pump, there is a definite friction and velocity head loss that takes place before the liquid enters the impeller vanes to a degree sufficient for it to receive energy from the pump. When water enters the first-stage impeller of a centrifugal pump, a certain amount of pressure energy is, therefore, converted into kinetic energy, the exact amount of conversion depending upon various features of the suction passage and impeller inlet design.

In order to prevent flashing of the water before it has a chance to receive additional pressure energy from the impeller, it is necessary to avoid lowering the pressure energy at the impeller vanes below the vapor pressure.

It is somewhat difficult to measure the static pressure at the impeller entrance proper and therefore pump designers specify the required pressure as measured at the pump suction flange and referred to the pump horizontal centerline. This pressure is determined for a given pump and for a given capacity on the basis of the calculated or predicted friction head losses. This is the NPSH.

In boiler pump terminology, the net positive suction head (NPSH) represents the net suction over the pump suction referred to the pump centerline *over and above* the vapor pressure of the water. The reason for this definition is that incipient cavitation must be expected to take place as soon as the static pressure at some one point within the pump has dropped to the value of the vapor pressure so that vaporization begins. This definition of suction head renders it automatically independent of any variations in temperature and vapor pressure of the water.

Effect of Cavitation within Pump

If the static pressure at the impeller vanes should fall below the vapor pressure corresponding to its temperature, a portion of the water contained in the impeller suction would immediately flash into steam. This flashing will form vapor-filled cavities within the body of the flowing water and cause cavitation,

the immediate effect of which is to prevent a further increase of the pump capacity because the pumping space normally allotted to the flow of water is occupied by steam.

There are other serious effects of cavitation. As the steam bubbles pass on to regions of the impeller under somewhat higher pressure, they condense in a sudden collapse with all the characteristics of an explosion. This violent collapse produces water hammer pressure of great intensity on small localized areas of the impeller.

Pump Construction for HTW Systems

Above a temperature of 150 F, the pump casing must always be steel. Bearings, housing, glands, and stuffing boxes must be water cooled. The suction inlet should be shaped to avoid sudden changes in flow and velocity and to avoid cavitation. The casing and head should be steel or cast iron with a high tensile strength, the impeller should be of chrome steel or high tensile ferrous alloy, the shaft of stainless steel, the sleeves and bushings of chrome steel, and bearings should be of the ball-bearing type, oil lubricated and water cooled. Couplings should be designed to permit dismantling of the pump without breaking the suction or discharge piping. Otherwise the entire pump must be dismantled.

Circulating Pump Seals

One of the most troublesome problems in the handling of HTW is the sealing of the rotating shaft. While packed stuffing boxes are not condemned, the use of mechanical seals for the higher pressures is strongly recommended. The mechanical seal, when installed by the pump manufacturer at the factory and not left to an inexperienced mechanic in the field, has many advantages compared to conventional stuffing boxes.

Mechanical seals are available from several manufacturers. It is of greatest importance that a mechanical seal properly installed should never be touched by inexperienced mechanics except for periodic inspection as *directed by the manufacturer*.

Since a great number of HTW systems are being installed in remote areas where the maintenance facilities and the ability of the personnel are very limited, the packed box pump could be more suited than the mechanical seal type. Most pump mechanics can handle packing gland problems but only the best and more experienced can handle properly

the installation and maintenance of a mechanical seal.

The mechanical seal is highly efficient and very effective for the duty required and has proved to be superior to the conventional asbestos or metallic stuffing box packing. Mechanical seals are especially suitable where pressures and shaft speeds are extremely high, or where corrosive conditions exist. The mechanical seal will reduce leakage to a minimum and will operate over long periods without the deterioration of the conventional packing. Shaft sleeve replacement, a regular repair with packed boxes, is almost eliminated.

The cooled gland is recommended for pumping temperatures of 400 F or over. Whereas cooling piping is standard for all HTW pump bearings, the cooling piping to the mechancal seal-gland cooling jacket must be of flexible tubing to prevent piping movement from affecting seal operation by deflecting the gland out of alignment.

In large installations, the outlet or hot line from the pump bearing can be used for many purposes, such as providing domestic or service water requirements.

Boiler Recirculating Pump

The boiler manufacturer should be consulted to ascertain the minimum flow requirements for the boiler. In some boilers the manufacturer may find 50% of full load flow to be safe while for others the manufacturer may insist that flow be not reduced below 85% of full load flow. Some manufacturers of forced circulation boilers recommend a minimum flow rate of 7,000 lb of water per million Btu maximum boiler output. This flow is equivalent in a 50,000,000-Btu boiler to 350,000 lb per hr. Under some conditions, especially during very light (summer) loads, the use of a separate boiler recirculating pump in addition to the system circulating pump might have advantages. But the additional cost of piping, complicated valve circuits, and fittings will increase the initial cost of the system to little advantage. In most cases it is practicable and efficient to install a separate system pump for summer loads and eliminate the boiler recirculation pump altogether.

However, where the friction head loss or boiler flow resistance is too high for one pump, i.e.,combining both boiler and distribution circuits, then a separate boiler recirculating pump must be used. If the boiler flow resistance exceeds 35 ft head loss one authority would recommend a two-pump system.

BOILERS FOR HTW SYSTEMS

Boilers for HTW systems, in addition to meeting construction requirements suitable for the operating pressure of the system, must be so constructed that proper circulation is assured in all parts of the boiler. To assure such circulation, some boilers are equipped with circulation pumps and distribution means to proportion the water to the heat absorbing capacity of the tube surface, which varies in heat absorption in different parts of the boiler. Boiler bypass connections must not create a condition that reduces circulation of water or supply to any part of the heating surface.

In some boilers (controlled circulation type) water is distributed to the various tubes through screens which practically meter the water to each tube circuit. This feature is of importance also when boilers are operated in parallel because it assists in obtaining equal flow distribution to each boiler.

Water velocity in a controlled circulation boiler is set for a minimum pressure loss which does not exceed 10 psi even for large boilers. A high pressure loss in boilers can easily require an increase of motor size by 10 hp. Since some motors may operate through the 24-hr day, it is important to use boilers that have a low pressure drop.

Boiler Emergency Protection

Many types of emergency valves have been designed to prevent loss of water in case of pipe breakage, but of these, only those controlled by flow have proved to be satisfactory. Such valves must be of a type that does not close too quickly, otherwise the shock will produce water hammer.

One valving arrangement for emergency operation that will prevent loss of water in the boilers is shown diagrammatically in Fig. 8. If a break should occur in the distribution line, the drop in flow will open the emergency control valve and bypass the water from the pump discharge header to the return water (boiler inlet) header. Thus the boilers will never be without water.

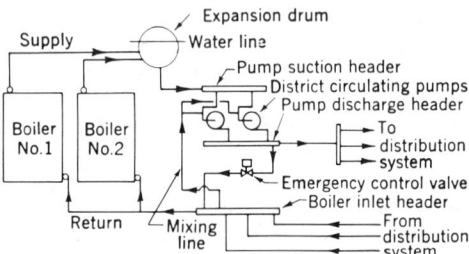

Fig. 8. Piping arrangement for emergency control valve.

PIPE, VALVES, AND FITTINGS FOR HTW SYSTEMS

HTW systems can be compared to steam systems of equivalent pressure in that valves designed for steam pressures of 150, 200, or 250 psig can be used in most HTW systems. Since the design or test pressure of any valve is always much greater than its operating pressure, a steam valve sold for a nominal 150 psig can be selected for HTW systems operating at 350 F, equivalent to 150-190 psig, depending on the method of pressurization. For HTW systems operating at 400 F, equivalent to 260-300 psig, and where process heating by steam or hot water is required, valves must be of the 300-psig class. All pipe, valves, and fittings for HTW systems should comply with the requirements of the *American National Standard for Pressure Piping*.

Generally speaking, HTW valves are restricted to the 150 and 300 psig classes, with preference for the former due to cost. This means that unless higher pressures are otherwise required, HTW systems for all forms of space heating should be limited to from 325 to 340 F, or pressures below 150 psig.

The importance of well made tight fitting valves cannot be overstressed because most leakage occurs at the valves. Leakage from HTW systems flashes into steam and is, therefore, not detected except by deposits of scale or salt formation on the surface at point of leakage.

Since it is usually not possible to control the difference in pressure between supply and return mains within narrow limits in the distribution system, the pressure available to force the water through the control valve may vary as much as 5 to 40 psi. Therefore, the valve has to be sized to pass the correct amount of hot water at both extremes. To do this the valve must have a large

number of positions, a high lift, an equal-percentage throttling plug, and an actuator powerful enough to hold the valve in position with the changes in water supply pressure.

Valve Installation

Valves in HTW service must always be located on the downstream or return side of converters or steam generators and similar equipment to be controlled because this allows the valve to operate at the lower temperature. If installed on the upstream side, a large pressure drop across the valve could occur when the valves starts to close thereby causing flashing of water into steam and consequent water hammering.

Welded Joints

To assure tightness of the HTW system all joints to valves and fittings for sizes above $1\frac{1}{4}$ in. should be welded except in the boiler house where flanges should be used to facilitate maintenance or changes in piping to equipment.

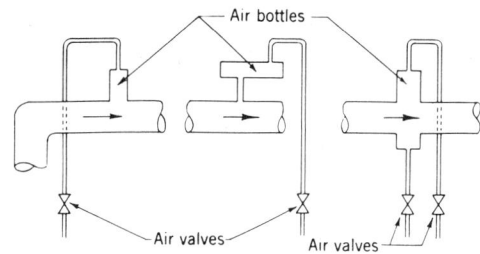

Fig. 9. Collecting chambers for venting of air.

Venting of Piping

Provision of air collecting chambers at all high points in the piping should be made as shown in Fig. 9. These chambers should have vent pipes run to valves conveniently located for manual operation. The accumulation of air is considerable when the system is first put in operation and for the period when the entrained air escapes from the heated circulating water. Thereafter, only periodic opening of the vent valves will be necessary.

EFFECT OF LOAD VARIATION ON OPERATION

Considerable differences occur between design temperatures and pressures as well as between winter and summer design loads. Sometimes in design, the extreme conditions for winter are overstressed without taking into account that extremes are usually of short duration and can be handled readily by either the storage capacity of the system or the ability of the boilers to operate above rating for such periods. In HTW systems, heat is available promptly from the main in which the water is always up to design temperature. For the brief extreme periods, space heating units can be operated as necessary to prevent large drops in space temperature during periods of non-occupancy and, consequently, reduce peak demands such as occur when the space is being brought up to occupancy temperature.

It is of most importance that the summer low load, usually a night load, be established carefully in order to plan boiler and pump operation that will maintain balanced system conditions. If only one boiler is put on the line, a bypass line will greatly reduce the load on the boiler when the system load is dropping to minimum conditions. Sum-

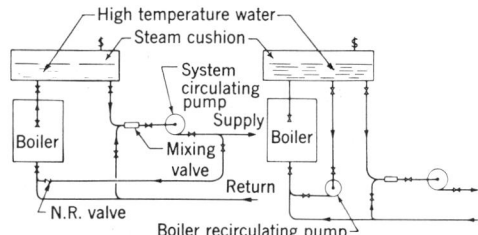

Fig. 10. HTW system with boiler recirculating.

mer low load conditions may be met by having a boiler circulating pump at each boiler to recirculate the boiler water or, having a bypass from the system circulating line to the boiler inlet (see Figs. 10 and 11).

In the case of one HTW system having three water tube boilers, each of 25,000,000 Btuh capacity, it was found necessary with an extremely low summer load, especially at night, to use a bypass from the discharge side of the circulating pump to the return connection of each boiler to obtain a balanced condition at this low load.

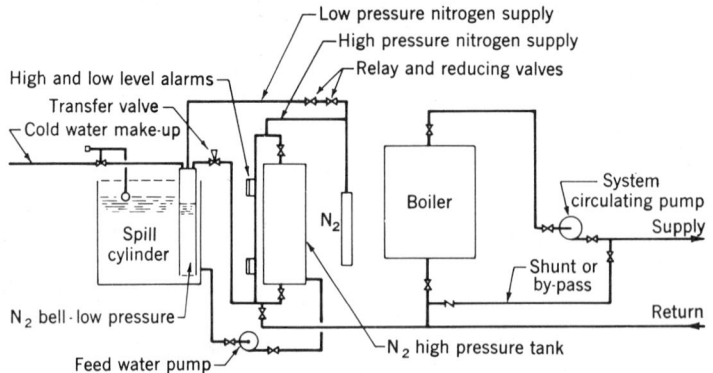

Fig. 11. HTW system with bypass for light-load conditions.

PIPE SIZING FOR HTW SYSTEMS

Pipe sizes for HTW systems are usually selected from charts or tables which show for some average temperature such as 200 or 300 F: (1) rate of water flow, (2) size of pipe, (3) water velocity, and (4) the unit friction loss in the piping. The units in which the items (1) to (4) are expressed vary with the preferences of the user. Since friction loss varies with the different water temperatures, the charts or tables are often accompanied by factors for correcting the values obtained from the charts or tables.

Approximate heat carrying capacities of pipe lines for HTW systems, based on 100 deg temperature drop, on 0.1 to 0.2 in. water friction loss per foot of pipe, and velocities of 1.5 to 4.5 fps are shown in Table 2. This table is indicative only and should not be substituted for the more precise tables and charts which follow.

Figure 13 is a convenient chart for sizing pipe for HTW systems. Elsewhere in this handbook will be found a complete table of pipe friction losses for 60F, 180F and 300F water in Schedule 40 pipe. Unit friction losses obtained from Fig. 13 and from this table for 300F water may be used, with the help of the correction factors in Fig. 12, for temperatures other than 300F.

For a discussion of calculation of friction loss in piping, including data for losses through valves and fittings, the reader is referred to the text on hot water heating systems elsewhere in this handbook.

Table 2. Heat Capacity of Pipes

	Pipe Dia., inches	Btu per hr Transmitted
	¾	100,000
	1	250,000
Data based on 100-F	1¼	400,000
temperature drop and	1½	700,000
friction loss of 0.1 to 0.2	2	1,500,000
inches wg per ft.	2½	2,500,000
	3	4,000,000
Velocities will vary from	3½	6,500,000
1.5 to 4.5 fps.	4	8,000,000
	5	12,000,000
	6	20,000,000

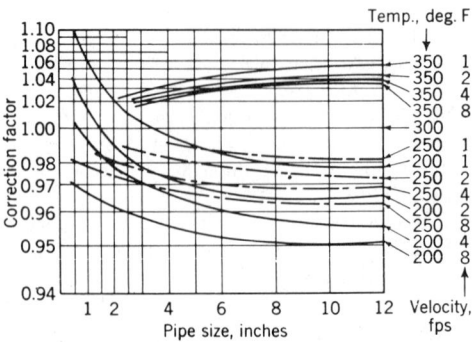

Fig. 12. Correction factors for friction loss from Fig. 13.

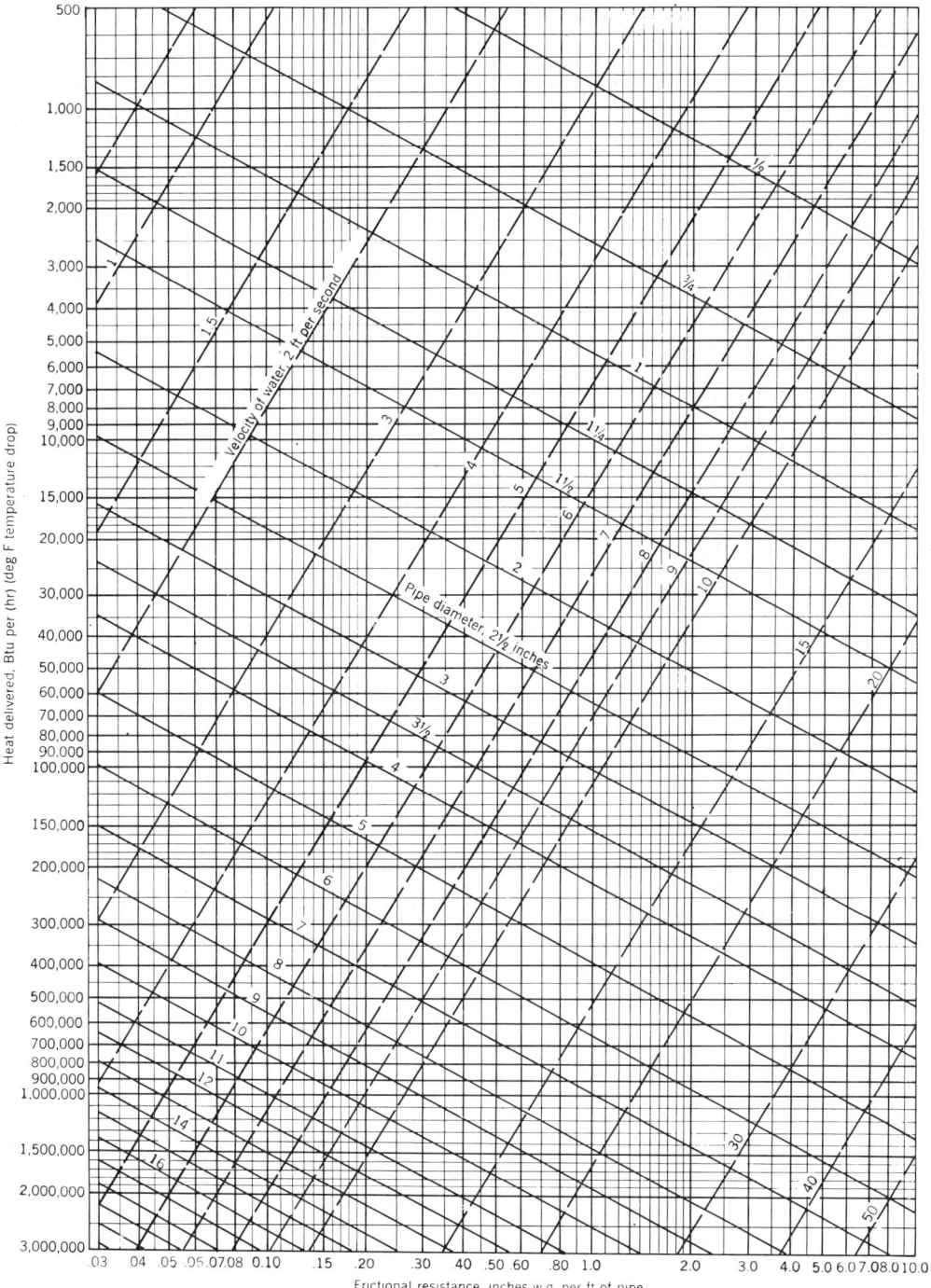

Fig. 13. Frictional resistance of 300 F water in Schedule 40 steel pipe.

RATINGS OF STEEL BOILERS

Steel heating boiler ratings given on these pages are those of the Steel Boiler Institute (SBI), from whose Rating Code (seventh edition, 1958) the accompanying tables and text are abstracted by permission.

GENERAL

The code applies to all designs of steel boilers designated as Table 1 Steel Boilers, Table 2 Steel Boilers, and Table 3 Steel Boilers.

Table 1 steel boilers are those containing the heating surfaces shown in Table 1 and the ratings of which have been determined in accordance with the Institute's code.

Table 2 steel boilers are those containing not more than 294 square feet of heating surface and having catalog SBI net ratings (steam) which are not more than 5000 square feet mechanically fired and 4120 square feet hand-fired.

Table 3 steel boilers are those with gross outputs and net ratings listed in Table 3. Table 3 boilers have no limits of minimum heating surface or furnace volume.

Ratings

One square foot of steam radiation shall be considered equal to the emission of 240 Btu per hour, and one square foot of water radiation shall be considered equal to the emission of 150 Btu per hour.

A boiler horsepower is the evaporation of 34.5 pounds of water per hour into dry steam from and at 212 deg F.

The SBI rating indicates the estimated design load that the boiler will carry as determined by the sum of items A, B and C given below under Connected Loads.

Table 1a. Ratings for Table 1 Steel Boilers, Mechanically-Fired

SBI Rating			SBI Net Rating			SBI Gross Output, 1000 Btu per Hr (7)	Minimum Furnace Volume, Cu Ft† (8)	Minimum Furnace Height, In.* (9)	Minimum Heating Surface, Sq Ft (10)
Sq Ft Steam (1)	Sq Ft Water (2)	1000 Btu per Hr (3)	Sq Ft Steam (4)	Sq Ft Water (5)	1000 Btu per Hr (6)				
2,190	3,500	526	1,800	2,880	432	648	15.7	26	129
2,680	4,280	643	2,200	3,520	528	792	19.2	28	158
3,160	5,050	758	2,600	4,160	624	936	22.6	29¼	186
3,650	5,840	876	3,000	4,800	720	1,080	26.1	29½	215
4,250	6,800	1,020	3,500	5,600	840	1,260	30.4	30	250
4,860	7,770	1,166	4,000	6,400	960	1,440	34.8	30½	286
5,470	8,750	1,313	4,500	7,200	1,080	1,620	39.1	31¼	322
6,080	9,720	1,459	5,000	8,000	1,200	1,800	43.5	31¾	358
7,290	11,660	1,750	6,000	9,600	1,440	2,160	52.1	32¾	429
8,500	13,600	2,040	7,000	11,200	1,680	2,520	60.8	34	500
10,330	16,250	2,479	8,500	13,600	2,040	3,060	73.8	35½	608
12,150	19,440	2,916	10,000	16,000	2,400	3,600	86.8	37½	715
15,180	24,280	3,643	12,500	20,000	3,000	4,500	108.5	40¼	893
18,220	29,150	4,373	15,000	24,000	3,600	5,400	130.2	43	1,072
21,250	34,000	5,100	17,500	28,000	4,200	6,300	151.8	46	1,250
24,290	38,860	5,830	20,000	32,000	4,800	7,200	173.5	48¾	1,429
30,360	48,750	7,286	25,000	40,000	6,000	9,000	216.9	54½	1,786
36,430	58,280	8,743	30,000	48,000	7,200	10,800	260.3	60½	2,143
42,500	68,000	10,200	35,000	56,000	8,400	12,600	303.6	66	2,500
48,570	77,710	11,657	40,000	64,000	9,600	14,400	346.9	71¾	2,857
54,640	87,420	13,114	45,000	72,000	10,800	16,200	390.3	77½	3,214
60,710	97,140	14,570	50,000	80,000	12,000	18.000	433.6	83½	3,571

† Oil, gas, or bituminous stoker-fired coal.
* Bituminous stoker-fired. Does not apply to Scotch type boilers.

The SBI net rating indicates the net load the boiler will carry as determined by the sum of items A and B below under Connected Loads except that if the heat loss from the piping system exceeds 20% of the installed radiation the excess shall be considered as additional net load.

The SBI gross output for Table 1 boiler indicates a heat output at the boiler nozzle of 5040 Btu per hour per square foot of catalog heating surface. For Table 2 boilers the SBI gross output indicates a heat output of 6120 Btu per hour per square foot of catalog heating surface. For Table 3 boilers the SBI gross output is the output tabulated in Table 3 for a boiler which is operating within all the limits stipulated in the code. The SBI gross output indicates the load the boiler will carry as determined by the sum of items A, B, C and D below, under Connected Loads.

Connected Loads

A. The estimated normal heat emission at design temperature of the connected radiation required

to heat the building as determined by accepted practice, expressed in square feet of radiation or in Btuh.

B. The estimated maximum heat required by water heater or other apparatus connected to the boiler, expressed in square feet of radiation or in Btu per hour.

C. The estimated heat emission at design temperature of piping connecting radiation and other apparatus to the Boiler, expressed in square feet of radiation or in Btu per hour.

D. The estimated increase in normal load in Btu per hour due to heating the boiler and contents to operating temperature and heating up cold radiation and piping.

Heating Surface

Heating surface is expressed in square feet and includes those surfaces in the boiler which are exposed to the products of combustion on one side and water on the other. The outer surface of tubes is used.

Table 1b. Ratings for Table 1 Steel Boilers, Hand-Fired

SBI Rating			SBI Net Rating			Minimum Grate Area Sq Ft†	Minimum Heating Surface. Sq. Ft.
Sq Ft Steam (11)	Sq Ft Water (12)	M-Btuh (13)	Sq Ft Steam (14)	Sq Ft Water (15)	M-Btuh (16)	(17)	(18)
1,800	2,880	432	1,500	2,400	360	7.9	129
2,200	3,520	528	1,830	2,930	439	8.9	158
2,600	4,160	624	2,170	3,470	521	9.7	186
3,000	4,800	720	2,500	4,000	600	10.5	215
3,500	5,600	840	2,920	4,670	700	11.4	250
4,000	6,400	960	3,330	5,330	800	12.2	286
4,500	7,200	1,080	3,750	6,000	900	13.4	322
5,000	8,000	1,200	4,170	6,670	1,000	14.5	358
6,000	9,600	1,440	5,000	8,000	1,200	16.4	429
7,000	11,200	1,680	5,830	9,330	1,400	18.1	500
8,500	13,600	2,040	7,080	11,330	1,700	20.5	608
10,000	16,000	2,400	8,330	13,330	2,000	22.5	715
12,500	20,000	3,000	10,420	16,700	2,500	25.6	893
15,000	24,000	3,600	12,500	20,000	3,000	28.4	1,072
17,500	28,000	4,200	14,580	23,330	3,500	30.9	1,250
20,000	32,000	4,800	16,670	26,670	4,000	33.2	1,429
25,000	40,000	6,000	20,830	33,330	5,000	37.4	1,786
30,000	48,000	7,200	25,000	40,000	6,000	41.2	2,143
35,000	56,000	8,400	29,170	46,670	7,000	44.7	2,500
40,000	64,000	9,600	33,330	53,300	8,000	48.0	2,857
45,000	72,000	10,800	37,500	60,000	9,000	51.0	3,214
50,000	80,000	12,000	41,670	66,700	10,000	54.0	3,571

† Does not apply to Scotch type boilers.

Furnace Volume

The furnace volume for furnaces in which solid fuel is burned is the cubical content of the space between the bottom of the fuel bed and the first plane of entry into or between the tubes or flues.

The furnace volume for furnaces in oil-fired Table 2 and Table 3 steel boilers is the furnace volume used in running SBI rating tests and does not include the volume occupied by refractory material, insulation, or fill used in the construction of the combustion chamber.

The furnace volume for furnaces in which pulverized fuel, liquid or gaseous fuel is burned is the cubical content of the space between the hearth and the first plane of entry into or between the tubes or flues.

Heating Value of Coal

The ratings for hand-fired steel boilers are based on coal having a heating value of 13,000 Btu per pound as fired. When selecting hand-fired steel boilers for coal of less heating value, the estimated design load should be multiplied by the proper factor selected from Table A.

Table A. Multiplying Factors for Coal Boilers

Heat Value Btu Per Lb.	Factor	Heat Value Btu Per Lb.	Factor
13000	1.00	10500	1.24
12500	1.04	10000	1.30
12000	1.08	9500	1.37
11500	1.13	9000	1.44
11000	1.18		

Grate Area

Grate area is expressed in square feet and is measured in the plane of the top surface of the grate. For double grate boilers, the grate area is the area of the upper grate plus one fourth of the area of the lower grate.

TABLE 1 RATINGS

The SBI rating of a boiler—expressed in square feet of steam radiation—in which solid fuel hand-fired is burned, is equal to fourteen times the heating surface of the boiler in square feet.

The SBI rating of a boiler in square feet of steam radiation in which solid fuel is mechanically fired or in which oil or gas is burned, is equal to seventeen times the heating surface of the boiler in square feet.

Grate Area

The grate area is not less than the values shown in Table 1 which have been determined by the following formulae:

For boilers with SBI ratings of 1800 sq ft to 4000 sq ft of steam radiation

$$G = \sqrt{\frac{R\text{-}200}{25.5}}$$

For boilers with SBI ratings of 4000 sq ft of steam radiation and larger

$$G = \sqrt{\frac{R\text{-}1500}{16.8}}$$

where G is grate area in square feet and R is SBI rating in square feet of steam radiation. The minimum grate requirement does not apply to Scotch type boilers.

Furnace Volume

The furnace volume for a boiler in which pulverized fuel, oil or gas, or stoker-fired bituminous coal is burned is not less than one cubic foot for every 140 sq ft of steam rating. No minimum furnace volume is specified for mechanically fired boilers burning anthracite.

Furnace Height

Furnace height is the vertical distance from the bottom of the water leg to the crown sheet measured midway between the sidewalls and midway between the front and back (or bridgewall if used) of the furnace.

The minimum furnace height requirement does not apply to Scotch Type Boilers.

TABLE 2 RATINGS

Oil-Fired Boilers

The SBI net rating expressed in square feet of steam radiation is seventeen times the minimum heating surface shown in Column 6, Table 2.

The net rating for water boilers, expressed in Btu per hour, is 270 times the SBI net rating in square feet of steam radiation.

In determining Table 2 ratings, the following factors were applied.

The burner was set to produce 10% CO_2 ($\pm$ 0.2%) in the flue gases, the O_2 in the flue gases was not less than 7%, the flue gas temperature did not exceed 600 deg F when the boiler operated at the gross output. Further the overall efficiency of the boiler and burner was not less than 70% when operating at the gross output. Also, the difference between the draft in the breeching and the draft in the firebox when the boiler is operating at the gross output was not greater than that determined by the following formula:

$$DL = \frac{NR}{200} + 4$$

where DL is the draft loss in hundredths of an inch of water, and NR is the net rating in square feet of steam radiation.

This limitation does not apply to integral boiler burner units regularly cataloged and marketed complete with boiler burner and refractory designed to operate with higher draft losses and the means are provided to develop sufficient draft to overcome the high draft losses.

Furnace Volume

The furnace volume is not less than one cu ft for every 110 sq ft of steam net rating.

This limitation shall not apply to integral boiler-burner units regularly cataloged and marketed complete with boiler, burner and refractory.

Stoker-Fired and Gas-Fired Boilers

The SBI net rating is not greater than the oil-fired gas-fired boiler ratings. If baffles or turbulators are used in determining the oil-fired ratings they may be omitted from boilers furnished for stoker-firing.

Table 2. Net Rating Data for Table 2 Steel Boilers, Oil Fired

| SBI Net Rating | | SBI Gross Output, 1000 Btu per Hr | Minimum Furnace Volume, Cu Ft | Minimum Heating Surface, Sq Ft |
| Steam | | Water, 1000 Btu per Hr | | |
Sq Ft (1)	1000 Btu per Hr (2)	(3)	(4)	(5)	(6)
275	66	74	99	2.5	16
320	77	87	115	2.9	19
400	96	108	144	3.6	24
550	132	149	198	5.0	32
700	168	189	252	6.4	41
900	216	243	324	8.2	53
1,100	264	297	396	10.0	65
1,300	312	351	468	11.8	77
1,500	360	405	540	13.6	88
1,800	432	486	648	16.4	106
2,200	528	594	792	20.0	129
2,600	624	702	936	23.6	153
3,000	720	810	1,080	27.3	177
3,500	840	945	1,260	31.8	206
4,000	960	1,080	1,440	36.3	236
4,500	1,080	1,215	1,620	40.9	265
5,000	1,200	1,350	1,800	45.4	294

Hand-Fired Boilers

The net rating in square feet of steam radiation is not greater than fourteen times the heating surface.

The grate area is not less than that determined by the formula:

$$G = \sqrt{\frac{NR + 600}{9}} - 8$$

where G is the grate area in square feet and NR is the SBI net rating in square feet of steam radiation.

The firebox volume for the purpose of determining the coal storage capacity of a hand-fired boiler is that volume between the grate and the crown sheet. The volume of the uptake, the upper combustion chamber and that beyond the front face of the bridgewall is not included.

The firebox volume is not less than that determined by the formula:

$$FV = \sqrt{\frac{NR + 90}{5}} - 5.5$$

where FV is the firebox volume in cubic feet and NR is the SBI net rating in square feet of steam radiation.

Boiler Nameplate

Minimum data to be shown on a boiler are:
(a) Manufacturer's name and address.
(b) Boiler number and type.
(c) SBI symbol.
(d) Heating surface.
(e) SBI net rating in Btu per hour for each type of firing recommended.
(f) ASME Code symbol.
(g) Whether or not baffles or turbulators or other means of directing or retarding flue gases are used.

TABLE 3 STEEL BOILERS

Table 3 boilers include boiler-burner units defined as integral units, regularly cataloged complete with boiler, burner and combustion chamber.

Oil-Fired Boilers

In rating Table 3 boilers, the burner is set to produce 10% CO_2 ($\pm$ 0.2%) in the flue gases. This limit does not apply to boiler-burner units but, if the CO_2 produced during the tests of such units exceeds 10.2%, the flue gas temperature and overall efficiency figures determined from the tests are adjusted as follows:—

For each 0.1% CO_2 over 10.2% 2.5 deg F is added to the recorded stack temperature and 0.175% deducted from the calculated test efficiency.

The O_2 percent by volume in the flue gases is not less than calculated according to the following formula:

$$7 (3 - 0.2\ CO_2) = \text{Minimum permissible}$$

Where CO_2 is the percent CO_2 recorded in the test.

With certain exceptions the flue gas temperature does not exceed 600 deg F when the boiler is operating at the gross output. When the CO_2 is over 10.2% the adjusted flue gas temperature shall meet this requirement.

Smoke Test

Smoke density is required to be not more than No. 2 on the Shell Bacharach scale when boiler is operating at 100% of the SBI gross output. One smoke test is made within 15 minutes after start, one after about one hour and one within five minutes of end of recorded test and all three tests must meet this requirement.

Flue Gas Temperature Adjustment

A boiler or a boiler burner unit is allowed an increase of one degree F over the 600 deg F flue gas temperature limit for each 0.1% increase in test efficiency achieved over 70%. For boiler burner units the allowance is above the calculated adjusted stack temperature limit. The maximum allowance in either case is 30 deg F.

Table 3. Net Rating Data for Table 3 Steel Boilers

SBI Gross Output, 1000 Btu per Hr	SBI Net Ratings		
	Water, 1000 Btu per Hr	Steam, 1000 Btu per Hr	Steam, Sq Ft
99	74	66	275
115	87	77	320
144	108	96	400
198	149	132	550
252	189	168	700
324	243	216	900
396	297	264	1,100
504	378	336	1,400
648	486	432	1,800
792	594	528	2,200
1,008	756	672	2,800
1,260	945	840	3,500
1,512	1,134	1,008	4,200
1,800	1,350	1,200	5,000

Table B. Minimum Stack Dimensions, Table 3 Boilers and Boiler Burner Units

Rectangular Stack		
Firing Rate, GPH, Not Over	Nominal Dimensions, Inches	Inside Dimensions of Liner, Inches
2.1	8 x 8	6¾ x 6¾
3.5	8 x 12	6½ x 10½
5.3	12 x 12	9¾ x 9¾
7.7	12 x 16	9½ x 13½
11.5	16 x 16	13¼ x 13¼
15.0	16 x 20	13 x 17
21.0	20 x 20	16¾ x 16¾
26.0	20 x 24	16½ x 20½

Round Stack	
Firing Rate, GPH, Not Over	Inside Diameter of Liner, Inches
1.3	6
1.8	7
2.5	8
4.7	10
5.6	11
7.0	12
12.0	15
19.0	18

The overall efficiency of the boiler and burner is not less than 70% when operating at the SBI gross output.

There are no furnace volume limitations for Table 3 Boilers.

Draft Loss

The difference between the draft in the breeching and the draft in the firebox when the boiler is operating at SBI gross output is not greater than determined by the following formula:—

$$DL = \frac{\text{SBI Gross Output (Thousand Btu)}}{75} + 4$$

where DL is the draft loss in hundredths inches of water. This limitation does not apply to boiler-burner units regularly cataloged and marketed complete with boiler, burner and combustion chamber, designed to operate with higher draft losses and where means are provided to develop sufficient draft to overcome the high draft loss.

Minimum stack dimensions for Table 3 boilers are given in Table B.

I-B-R * RATINGS FOR CAST IRON BOILERS

(As established by the Institute of Boiler & Radiator Manufacturers [now Hydronics Institute]; abstracted by permission)

Gross I-B-R Output, 1000 Btu	Hand-Fired						Automatic-Fired			
	Net I-B-R Rating		Piping and Pickup Factor	Time Available Fuel Will Last, Hour	Maximum Stack Height, Feet	Minimum Stack Area, Sq In	Net I-B-R Rating		Piping and Pickup Factor	Maximum Allowable Draft Loss
	Steam and Water, 1000 Btu	Steam, Sq Ft					Steam and Water, 1000 Btu	Steam, Sq Ft		
30	—	—	—	—	—	—	22.5	94	1.333	.043
40	—	—	—	—	—	—	30.0	125	1.333	.044
50	21.2	88	2.360	7.50	29.0	50.0	37.5	156	1.333	.046
100	42.4	177	2.360	7.09	31.0	50.0	75.0	313	1.333	.052
200	87.5	365	2.285	6.41	36.0	50.0	150.0	625	1.333	.065
300	136.8	570	2.193	5.95	39.5	50.0	225.1	938	1.333	.078
400	190.1	792	2.104	5.58	43.0	62.5	300.1	1250	1.333	.090
500	245.9	1025	2.033	5.29	46.0	84.5	375.1	1563	1.333	.102
600	304.0	1266	1.974	5.02	48.5	106.5	450.1	1875	1.333	.114
700	363.1	1513	1.928	4.80	51.0	128.0	525.1	2188	1.333	.126
800	423.3	1764	1.890	4.59	53.5	149.0	600.1	2501	1.333	.137
900	484.4	2018	1.858	4.42	55.5	169.5	675.2	2813	1.333	.149
1000	546.7	2278	1.829	4.31	57.5	190.5	750.2	3126	1.333	.161
1100	611.1	2546	1.800	4.22	59.0	211.5	825.2	3438	1.333	.172
1200	677.2	2822	1.772	4.14	61.0	232.5	900.2	3751	1.333	.183
1300	745.4	3106	1.744	4.07	63.0	254.0	978.9	4079	1.328	.195
1400	815.4	3397	1.717	4.00	64.5	275.0	1061.4	4423	1.319	.207
1500	886.5	3694	1.692	4.00	66.5	295.5	1145.0	4771	1.310	—
1600	960.4	4002	1.666	4.00	68.0	315.0	1227.9	5116	1.303	—
1700	1036.0	4316	1.641	4.00	69.5	333.0	1311.7	5466	1.296	—
1800	1113.9	4641	1.616	4.00	71.0	350.0	1394.3	5809	1.291	—
1900	1193.5	4973	1.592	4.00	72.5	366.0	1474.0	6142	1.289	—
2000	1274.7	5311	1.569	4.00	73.5	381.0	1552.8	6470	1.288	—
2500	1707.6	7115	1.464	4.00	80.0	450.0	1941.0	8087	1.288	—
3000	2142.9	8929	1.400	4.00	86.5	512.0	2329.2	9705	1.288	—
3500	2500.0	10417	1.400	4.00	92.5	572.0	2717.4	11322	1.288	—
4000	2857.1	11905	1.400	4.00	98.5	630.0	3105.6	12940	1.288	—
4500	3214.3	13393	1.400	4.00	104.0	688.0	3493.8	14557	1.288	—
5000	3571.4	14881	1.400	4.00	109.0	746.0	3882.0	16175	1.288	—
5250	3750.0	15625	1.400	4.00	111.5	775.0	4076.1	16984	1.288	—
5500	3928.6	16369	1.400	4.00	113.5	803.0	4270.2	17792	1.288	—
5750	4107.1	17113	1.400	4.00	115.5	831.0	4464.3	18601	1.288	—
6000	4285.7	17857	1.400	4.00	117.5	858.0	4658.4	19410	1.288	—
6250	4464.3	18601	1.400	4.00	119.0	884.0	4852.5	20219	1.288	—
6500	4642.9	19345	1.400	4.00	120.0	900.0	5046.6	21027	1.288	—
7000	5000.0	20833	1.400	4.00	120.0	900.0	5434.8	22645	1.288	—
7190	5135.7	21399	1.400	4.00	120.0	900.0	5582.3	23260	1.288	—
7300	5214.3	21726	1.400	4.00	120.0	900.00	—	—	—	—

The accompanying table is from the 6th edition of the I-B-R Testing and Rating Code for Low Pressure Cast Iron Boilers, effective June, 1958.

The Gross I-B-R Output is obtained by test. For hand-fired boilers, the minimum efficiency, time available fuel will last, chimney size, and draft are subject to limitations set by the code. For automatically-fired boilers, flue gas temperature and analysis, draft loss and heat output of the combustion chamber are similarly subject to code limitations. Net I-B-R Rating is obtained from Gross Rating by applying Piping and Pick-up Factors.

* Hydronics Institute.

HEAT EMISSION OF CAST IRON RADIATORS

The accompanying data make readily available the information necessary in the field to determine the net heat given off by direct cast iron radiators. The data given are not intended to be used in design but are for surveys of existing plants.

Each of the tables can be used alone if desired. To determine the combined effect of the different variables affecting the heat given off by radiators, refer to the formula given later.

Table 1 — Heat emissions in equivalent square feet (1 EDR = 240 Btu per hr.) of bare, unenclosed, direct, cast iron radiators standing in still air at 70F and with 215F heating medium temperature.

Table 2 — Gives the relative heat emissions of radiators under varying radiator and room temperatures. In general it may be assumed that the radiator surface temperature will be the same as the temperature of the heating medium. The table shows that, for example, a radiator with 210F steam in an 80F room will use 0.87 as much steam as one standing in 70F air using 215F steam. Similarly, a radiator supplied with hot water at 170F in 70F air has a heat output 0.62 as much as a steam radiator in still air at 70 with a 215F surface. The factors given in this table when multiplied by the ratings in Table 1 give the equivalent square feet (EDR) for the room and heating medium conditions given.

Table A — Painting of a radiator affects its heat output. This table gives the relative effect of commonly used finishes as compared with a bare cast iron radiator.

Table B — Gives the relative effect of some common enclosures on radiators as compared with bare, unenclosed, cast iron direct radiators.

The combined effect of all of these variables can be expressed as

$$R = H \times N \times F \times P \times E,$$

where

R = Net heat given off in EDR (240 Btu per hr.)
H = Rated EDR as given in Table 1.
N = Number of sections of the given radiator.
F = Factor from Table 2 depending on the room and heating medium temperature.
P = Multiplier from Table A, for finish.
E = Multiplier from Table B depending on the type of enclosure used.

Example — What is the net heat given off by a 12-section tubular radiator with 5 tubes, 22 in. high, 1 lb. per sq. in. steam pressure? The radiator is painted with aluminum bronze, has a shield above it, and is in a room maintained at 80F.

Solution — Find the net heat given off by substituting in the formula

$$R = H \times N \times F \times P \times E$$

The section on tubular radiators of Table 1 shows that H (the rated EDR per section) for a 5-tube, 22-in., bare cast iron radiator is 3.00.

N (number of sections) is given as 12.

F (correction factor for room and surface temperature) for an 80F room temperature and 1 lb. per sq. in. steam pressure is 0.92 in Table 2.

P (multiplier depending on the radiator finish) is 0.91 for an aluminum finish from Table A.

E (multiplier for the enclosure) is found in Table B to be 0.90 for a shield above a radiator.

Substituting these figures in the formula

$$R = 3.00 \times 12 \times 0.92 \times 0.91 \times 0.90$$
$$= 27.1 \ EDR$$

Radiators which use hot water for a heating medium frequently have a lower EDR per section because of the lower surface temperatures maintained. Hot water temperatures will range anywhere from 170F for a gravity installation to 220F or higher for a forced pressure system. Of course, when surface temperatures of 215F or over are encountered, the EDR will be equal to or higher than the rating under standard conditions.

Table A.—Relative Effect of Radiator Finishes on Heat Emission

Finish	Multiplier (P)
Bare cast iron, foundry finish....	1.00
Gold bronze	0.90
Aluminum bronze.............	0.91
White paint..................	1.02
Cream paint.................	1.04
Red lacquer.................	1.00
Dull green paint.............	0.96
Brown paint.................	1.05

Table B. — Relative Effect of Enclosures on Radiator Heat Emission

Description of Enclosure	Multiplier (E)
None, bare cast iron radiator....	1.00
None, shield above radiator....	0.90
Open front and ends at bottom. Large free area in outlet, front only, upper half.................	0.90
Solid top, grilled front and ends, with large free area..........	0.86
Solid top, grilled front and ends, small free area..............	0.85
Solid top and ends, full grilled front. Factor depends on free area of grille; drops off rapidly as free area decreases........	0.70–0.87
Solid top and ends, upper half of front grilled.............	0.83

Table 1. Capacity of Unenclosed Bare Cast Iron Radiators

(With 70 deg. air and 215 deg. surface temperatures)

Height, Floor to Top, Inches	Column Type—Number of Columns							Tubular Type—Number of Tubes				
	1	2	3	4	5	6	Solid	3	4	5	6	7
	Heat Emission (H) per Section, Sq. Ft. Equiv. Steam Radiator											
12	—	—	—	—	—	3.00	—	—	—	—	—	—
13	—	—	—	—	3.00	—	—	—	—	—	—	2.50
14	—	—	—	—	—	3.25	—	—	—	—	—	2.50
15	—	1.50	—	—	—	—	—	—	—	—	—	—
16	—	—	—	—	3.50	3.75	—	—	—	—	—	2.50
17	—	—	—	—	—	4.00	—	—	—	2.25	—	3.00
18	—	—	2.25	3.00	4.00	4.33	—	—	—	—	—	3.50
20	1.50	2.00	—	—	5.00	5.25	3.25	1.75	2.25	2.67	3.00	3.67
22	—	—	3.00	4.00	6.00	—	—	2.00	—	3.00	—	4.00
23	1.67	2.33	—	4.67	—	—	4.00	2.00	2.50	3.00	3.50	4.25
26	2.00	2.67	3.75	5.00	7.00	—	4.50	2.33	2.75	3.50	4.00	5.00
28	—	—	—	—	—	—	5.00	—	—	—	—	—
30	—	—	—	—	—	—	—	3.00	3.50	4.00	5.00	5.50
31	—	—	—	—	—	—	—	3.00	—	4.25	—	—
32	2.50	3.33	4.50	6.50	8.50	—	5.75	3.00	3.50	4.33	5.00	6.25
36	—	—	—	—	—	—	—	3.50	4.25	5.00	6.00	6.75
37	—	—	—	—	—	—	—	—	4.25	5.00	6.00	—
38	3.00	4.00	5.00	8.00	10.00	—	7.00	3.50	4.25	5.00	6.00	7.50
45	—	5.00	6.00	10.00	—	—	—	—	—	—	—	—

Ultra-Slender Tubular				Bathroom Radiators		
Height, Floor to Top, Inches	Number of Tubes			Short Side, Inches	Long Side, Inches	
	3	4	5		17	20½
	Heat Emission (H) per Section, EDR				Heat Emission (H) per Radiator, EDR	
19	1.1	1.4	1.8	8	—	3.50
22	1.3	1.6	2.1	9	3.50	—
25	1.5	1.8	2.4	12	4.25	—

Flat Front Wall Radiators				Wall Radiators						
Number of Tubes	Height, Inches			Dimension, One Side, Inches	Dimension, Other Side, Inches					
	17	20	23		16½	18½	21	22-24	26-29	38
	Heat Emission (H) per Section in EDR				Heat Emission (H) per Radiator in EDR					
1	1.75	2.00	2.25	2½	—	—	1.4	—	1.8	—
2	3.33	—	4.25	12	—	—	—	5.0	7.0	9.0
3	2.00	2.33	2.67	12½	5.0	—	6.0	7.0	9.0	—
4	2.67	3.00	3.33	13¼	5.0	7.0	—	7.0	9.0	—
5	3.25	—	4.00							

Table 2. Relative Heat Emission of Bare Radiators at Various Temperatures

Room Temperature, Deg. F.	Surface Temperature of Radiator, Deg. F.									
	150	160	170	180	190	200	210	215	250	275
	Multipliers (F) for Bare Cast Iron Radiators									
50	.62	.70	.78	.87	.96	1.04	1.15	1.20	1.54	1.82
60	.54	.62	.70	.78	.87	.96	1.04	1.10	1.45	1.69
65	.50	.57	.66	.74	.83	.92	1.00	1.04	1.41	1.64
70	.46	.54	.62	.70	.78	.87	.96	1.00	1.35	1.59
75	.42	.50	.57	.66	.74	.83	.92	.96	1.30	1.54
80	.39	.46	.54	.62	.70	.78	.87	.92	1.25	1.49

Typical Ratings for Baseboard Radiation

Hot Water Ratings (65F Entering Air) — Avg. Water Temp, F: 170	180	190	200	Steam Ratings (1 psi)	Heating Element Material and Tube Dia, Inches	Finned Surface: Material	Finned Surface: Size, Inches	Overall Height from Floor, Inches	Nominal Length of Sections, Feet
Btu per Lin Ft per Hr at 1 gpm Flow Rate				Btu per Hr per Lin Ft					
500	560	630	700	820	¾ Copper	Alum	2x2¾x0.010	9	3, 4, 5, 8
350	390	430	480	560	Cast Iron	C.I.	——	7	——
520	590	660	720	830	Cast Iron	C.I.	——	9⅞	1½-6
590	660	730	800	990	¾ Copper	Alum	2½x3½x0.017	9 9/16	4, 5, 6, 8
543	618	702	785	930	1 IPS	Steel	2¾x4	10	1-6
620	706	802	898	1060	1 Copper	Alum	2¾x4	10	1-8
330	365	430	510	—	⅝ Copper	Alum	1⅝x1¾	9½	8
410	455	500	550	—	⅝ Copper	Alum	1⅝x1¾	12	8
520	585	640	705	—	⅝ Copper	Alum	1⅝x1⅝	9½	8
650	720	795	865	—	⅝ Copper	Alum	1⅝x1⅝	12	8

Typical Capacities of Unit Ventilators*

Standard Air Rating, Cfm[1]	Anemometer Rating, Cfm[2]	Heat for Ventilation 1000 Btuh[3]	Heating Medium: Steam	Hot Water	Electric
			Total Heat, 1000 Btuh[4]		
500	750	9.45	40.6	32.0	31.6
750	1000	14.2	59.8	48.0	47.6
1000	1260	18.9	76.5	64.0	63.9
1250	1560	23.6	96.6	80.0	79.6
1500	1860	28.3	117.7	96.0	95.5

* Adapted from ASHRAE Guide and Data Book, 1963.
[1] Base on 218.5F steam, 0F entering air, leaving air converted to standard air at 70F.
[2] Rating specified in some school codes.
[3] Based on 25% outdoor air at 0F, 70F room temperature.
[4] Heat for ventilation plus heat available to offset room heat loss. The latter is called "surplus heat."

Heat Emission of Pipe Coils*

(Wall mounted, pipes horizontal, 215F steam, air at 70F)

Number of Rows	Pipe Size, Inches: 1	1¼	1½
	Heat Emission, Btuh per Foot of Coil (Not Pipe)		
1	132	162	185
2	252	312	348
4	440	545	616
6	567	702	793
8	651	796	907
10	732	907	1020
12	812	1005	1135

* Reprinted from ASHRAE Guide and Data Book, 1963.

Typical Propeller Unit Heat Capacities

(200F water, 60F entering air)

Heat Emission, 1000 Btuh	Final Air, F	Water Rate, Gpm	Water Friction, Ft of Head
10F Water Temperature Drop			
23	107	4.5	0.14
45	110	9.0	0.40
90	110	18.0	0.73
210	107	41	1.89
20F Water Temperature Drop			
20	102	2	0.08
45	104	4.5	0.14
90	98	9.0	0.21
225	98	22.5	0.75

Typical Propeller Unit Heater Capacities

(2 psi steam, 60F entering air)

Heat Emission, 1000 Btuh	Final Air, F	Air Rate, Cfm	Motor HP
Standard 2-Row Coil Units			
18	105	370	1/30
50	118	790	1/20
100	124	1450	1/8
200	128	2700	1/6
High Air Capacity Units			
26	101	590	1/30
55	101	1230	1/8
100	101	2260	1/6
130	100	2970	1/6

MISCELLANEOUS HEATING AND COOLING MEDIA

Following are data on heating and cooling media other than water.

Brine

The introduction of certain salts into water alters the boiling and freezing point of the solution in a specific manner according to the concentration of salt. It is interesting to note that the freezing point of water is gradually lowered as salt is added up to a concentration above which the freezing point rises and may eventually be higher than that of water. The actual boiling and freezing points depend on the nature of the salt as well as the concentration of salt in the solution. Calcium chloride is a salt commonly used in brine solutions for heat transfer purposes.

Table 1 shows the boiling and freezing points of calcium chloride solutions.

Table 1. Boiling and Freezing Points of Calcium Chloride Solutions

Concentration Per Cent, Weight	Specific Gravity at 60F	Freezing Point, F	Boiling Point, F
40	1.410	+56	248
30	1.295	−51	237
20	1.186	0	221
10	1.087	+22	214

The characteristics of calcium chloride brine are such as to make it attractive for cooling systems but less attractive for heating systems.

Glycerine

Glycerine can be readily handled for both heating and cooling applications. In high concentrations it has a high boiling point. Both the boiling point and freezing point are lowered when glycerine is

Table 2. Boiling and Freezing Points of Aqueous Solutions of Glycerine

Concentration, Per Cent, Weight	Specific Gravity at 60F	Freezing Point, F	Boiling Point, F
98.2	1.261	+56	554
95.0	1.253	+46	332
90.0	1.240	+29	281
80.0	1.213	−5	250
70.0	1.185	−38	237
66.7	1.178	−51	234
60.0	1.157	−30	228
50.0	1.129	−9	223
40.0	1.102	+4	219
30.0	1.075	+15	217

diluted with water, the freezing point reaching a minimum of −51F when the water content is about a third of the weight of solution. Table 2 shows the characteristics of aqueous solutions of glycerine.

Glycol

Ethylene glycol has some of the characteristics of glycerine. It is heavier than water, has no marked corrosive effect on iron and steel. On some other metals and alloys it has a marked corrosive effect

Table 3. Boiling and Freezing Points of Aqueous Solutions of Ethylene Glycol

Concentration, Per Cent, Volume	Specific Gravity at 60F	Freezing Point, F	Boiling Point, F
100	1.113	+10	386
90	1.105	−17	288
80	1.098	−45	260
70	1.088	−47	245
60	1.078	−54	235
50	1.061	−34	229
40	1.045	−13	225
30	1.032	+3.2	221

which can be modified by inhibitors. It should not be used with zinc or galvanized iron. Characteristics of ethylene glycol solutions are shown in Table 3.

Other Media

Aroclor is a trade name applied to a group of chlorinated biphenyls. Aroclor 1248 has a boiling point of 644F and a freezing point less than 20F. Its specific heat at 86F is .29 and increases to .44 at 248F. Viscosity at 86F is 112 centipoises.

Dowtherm A is a trade name of diphenyl and diphenyl oxide, produced as a heating medium at high temperature. Table 4 shows properties.

A heat transfer oil developed specifically for snow melting systems by Socony-Vacuum Oil Co. is S/V Sovaloid S. Characteristics of two S/V oils are shown in Table 5.

Certain organic silicates are liquids having high boiling and low melting points. By chemical combination and mixing of compounds, liquids of a common generic type but having different physical properties are produced within a liquid range of −60 to 800F. Tables 6 and 7 give some of the properties of aryl and cresyl silicates.

Physical properties of mercury are shown in Table 8. It is the only metal which liquefies at atmospheric temperatures. It can be used with iron and steel but dissolves brass and other metals.

Table 4. Physical Properties of Dowtherm A

Boiling point		496F
Freezing point		54F
Flash point		228F
Fire Point		246F
Auto-ignition temperature exceeds		932F
Viscosity, centipoises, at:	60F	5
	250F	0.8
	450F	0.4
Specific heat, at	54F	0.37
	250F	0.47
	450F	0.6
Density, lb per cu ft, at:	54F	66.9
	250F	61.1
	450F	54.8
Heat Content, Btu per (cu ft) (F), at:	54F	24.7
	250F	28.6
	450F	32.8

Table 5. Physical Properties of Heat Transfer Oils

Property	S/V Sovaloid S	S/V Heat Transfer Oil
Gravity, API at 60F/60F	10.0	11.0
Specific Gravity at 60F/60F	1.0	.9930
Flash point, F, open cup	260 Minimum	355
Pour point, F	—40	10
Viscosity, kinematic centistokes, at:		
450F		.86
300F		1.95
175F	1.7	8.25
100F	3.9	50.00
50F	10.0	450.00
0F	48.0	....
—10F	70.0	....
Coefficient of Expansion, per deg F	.00035	....
Distillation range		
Initial boiling point, F	500 Minimum	615
Final boiling point, F	640 Maximum	735 (90%)
Specific heat, Btu per (lb) (F), at:		
450F	..	.966
300F	..	.910
180F	.47	.864
100F	.43	.835
60F	.42	.820
0F	.39	..

Table 6. Properties of Tetra Aryl Silicates

Viscosity, centipoises, at:	0F	740.00
	50F	52.00
	300F	1.55
	600F	0.43
Specific heat, at:	50F	0.37
	300F	0.48
	600F	0.74
Density, lb per cu ft, at:	50F	72.20
	300F	65.0
	600F	56.2
Heat content, Btu per (cu ft) (F), at:	50F	26.7
	300F	31.1
	600F	41.5

Table 7. Properties of Tetra Cresyl Silicate

Boiling point, F	815
Specific gravity, at:	
77F	1.125
392F	.982
572F	.908
Specific heat, Btu per (lb) (F), at:	
77F	.40
392F	.530
572F	.620
Viscosity, centistokes, at:	
77F	41
392F	1.4
Coefficient of expansion, per deg F	.000775 to .000977
Density, lb per cu ft, at 60F	69.3
Flash point, F	325 minimum
Pour point, F	1.0

Table 8. Physical Properties of Mercury

Boiling point, F		675
Melting point, F		—38
Viscosity, centipoises, at:	—4F	1.85
	68F	1.56
	212F	1.22
	572F	0.93
Specific heat, at:	32F	0.0333
	212F	0.0326
	392F	0.0318
Density, lb per cut ft, at:	32F	845
	212F	830
	392F	816
Heat content, Btu per (cu ft) (F), at:	32F	28.2
	212F	27.0
	392F	25.9

WARM AIR HEATING

If a warm-air furnace is defined as a heat-transfer device in which heat is released on one side of the heat-exchanger surface and heat is absorbed by circulating air on the other side, a large number of devices can be included in this category, some of which are illustrated in schematic form in Fig. 1.

Early Types

(a) Stove. The earliest ancestor of the present-day warm-air furnace is the parlor stove, sometimes referred to as the pot-bellied stove. Aside from the cheery aspects of the flame that could be viewed through the isinglass openings, the stove was a utility that was tolerated but not aesthetically admired.

(b) Pipeless Furnace. Warmed air rose by gravity action and was delivered into the room above through a large grilled opening in the floor.

(c) Gravity Hot-Air Furnace (Pre-1918). These early systems were installed without benefit of engineering knowledge. Air flow was restricted, and the furnaces were aptly described as "hot-air" furnaces, since the air temperatures were in excess of 200 deg F.

(d) Gravity Warm-Air Furnace (Pre-1930). The early experimental work at the University of Illinois by Professors A. C. Willard and A. P. Kratz in the period following 1918 showed that, for example, markedly improved results could be obtained by the use of streamlined, amply sized return-air ducts, large warm-air stacks, and registers with ample free area. In these improved systems the actual operating register-air temperature was much less than the assumed design value of 175 deg F, even under design weather loads.

(e) Early Forced-Air Conversion Models (Pre-1940). Even in the greatly improved gravity warm-air furnace system the surfaces had to be extensive and the furnaces were large in size. The obvious step to increase the heat transfer rate was to use some mechanical means for circulating the air. Propeller fans located in the return-air boot or in the bonnet were found to be lacking in pressure and capacity. When centrifugal fans were placed in a compartment attached to the casing, it was found that furnace casings had to be made smaller to prevent by-passing the heating surface. Later, casings became rectangular and smaller; fans, filters, burners, and controls were integrally built into the casing, and a whole new line of products was introduced. In the trade, the centrifugal fans used for warm-air furnaces are referred to as "blowers."

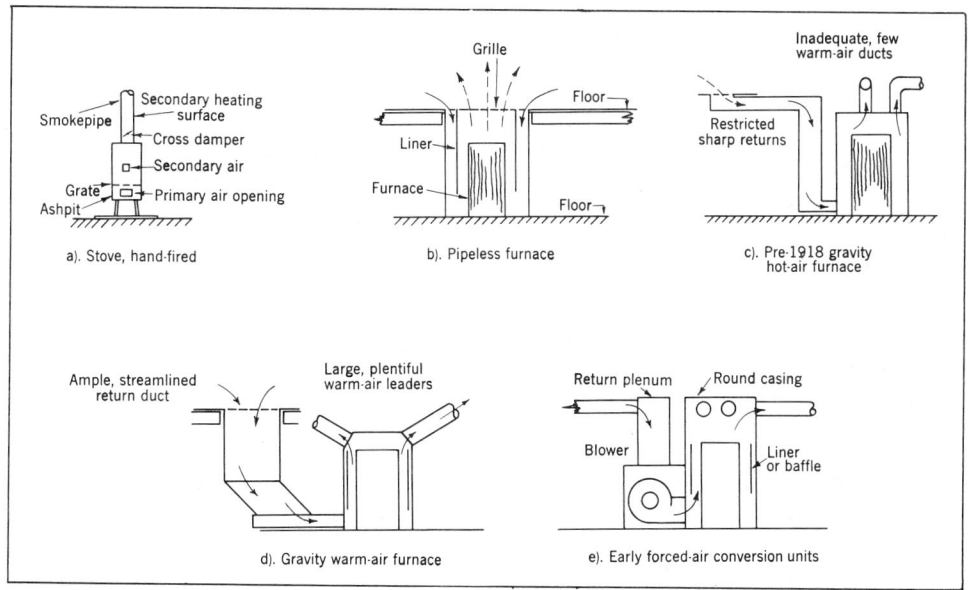

a). Stove, hand-fired

b). Pipeless furnace

c). Pre-1918 gravity hot-air furnace

d). Gravity warm-air furnace

e). Early forced-air conversion units

Fig. 1. Developmental stages of forced-air furnace.

Modern Types

(f) Current Forced-Air Furnace Types. The current types of furnaces, some of which are illustrated in Fig. 2, show a great diversity of shapes and arrangements. Among the principal types used for residences and small buildings the following arrangements are common:

(1) Low-boy arrangement with furnace and blower in separate compartments, and with air discharged upwards. Used for basement installation, but also usable with furnace closets located on the first story.
(2) High-boy arrangement with up-flow of air, in which the blower is located at the bottom of the casing and the heat exchanger is located at the top. Used for either basement installation or for first-story installation.
(3) High-boy arrangement with down-flow of air, in which the blower is located at the top of the casing and the heat exchanger is located below the blower. Used mainly for first-story installation in which warm air is discharged downward.
(4) Horizontal arrangement in which the blower and the heat exchanger are located side by side and the warm air is discharged horizontally. Used for basement or attic installations, or crawl-space installations.

The current types are integrally designed and coordinated units in which the blower and control equipment are selected and arranged for the specific unit. In fact, in the smaller size of high-boy units,

the furnace package consists of a factory-wired and factory-installed assembly of heat exchanger, blower, blower motor, filter, controls, and humidifier.

Special Types

(g) Other Current Types of Equipment. The dividing line between a warm-air furnace and a device which is not a warm-air furnace is difficult to define sharply. The following are included here as special types of warm-air furnaces (see Fig. 3):

(1) *Space Heater.* This reminder of the early parlor stove still persists. Most space heaters are gravity circulating systems, but some are provided with propeller fans to stir the room air. No ducts are attached. Oil-fired and gas-fired space heaters are available, some of them with thermostatic control equipment.
(2) *Floor Furnace.* This device is reminiscent of the early pipeless furnace, but on a much smaller scale and with refinements of automatic firing and controls. The unit is usually attached to the ceiling joists of the basement and is ductless. The circulating air temperature is extremely high with gravity circulating units.
(3) *Wall Furnace.* This device is a space-saving unit that is installed in the wall of a room and is usually a gravity circulation, pipeless unit. Some are tall and some are low enough to be installed just above the baseboard.
(4) *Direct-Fired Unit Heater.* Unit heaters can be direct-fired, either with or without a circulating fan behind the heat exchanger surface. Used

Fig. 2. Current types of forced-air furnaces.

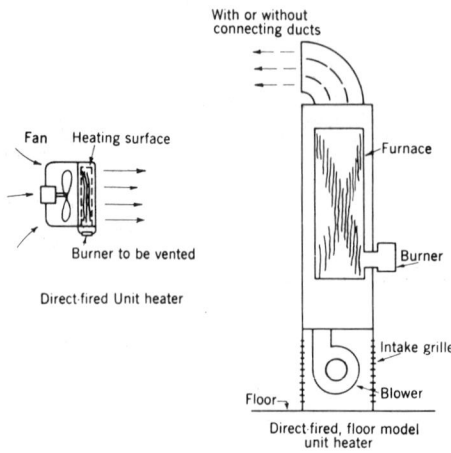

Fig. 3. Allied versions of forced-air furnaces.

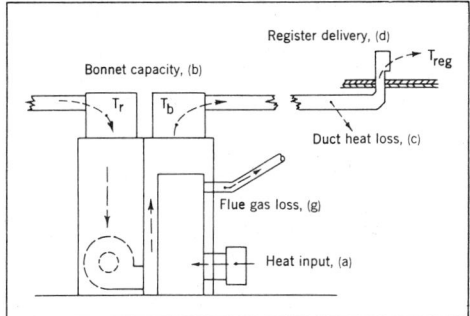

Fig. 4. Terminology related to furnace performance.

mainly for industrial or commercial installations.

(5) *Direct-Fired Floor-Model Unit Heaters.* Extremely large unit heaters are available either with or without duct connections. Used for commercial and industrial buildings. Capacities in excess of 1,000,000 Btu per hr are available.

(6) *Industrial Warm-Air Furnaces.* Industrial warm-air furnaces, or heavy-duty furnaces, are available with capacities in excess of 1,000,000 Btu per hr. Used for schools, churches, commercial buildings, and industrial buildings.

(7) *Aircraft Air Heaters.* Direct-fired air heaters of small size but high capacity are available for aircraft heating. These are usually gasoline fired heaters.

The styles, models, and types of warm-air furnaces are extremely diverse. Models are available for burning wood, coal, coke, oil fuel, kerosene, natural gas, bottled gas, and manufactured gas. The diversity was the result of demand: the demand of special building types and the demands of both the builder and the public for small, efficient, and economical units.

Furnace Performance Terminology

A few terms appear in catalog description of warm-air furnaces which require definition (see also Fig. 4):

(a) Input, or Heat Input, is the rate at which heat is released inside the furnace and is in units of Btu per hr.

For Gas:

$$\left(\begin{array}{c}\text{Input,}\\\text{Btu per hr}\end{array}\right) = \left(\begin{array}{c}\text{Cubic feet of}\\\text{gas per hour}\end{array}\right)\left(\begin{array}{c}\text{Heating Value,}\\\text{Btu per cu ft}\end{array}\right)$$

For Oil:

$$\left(\begin{array}{c}\text{Input,}\\\text{Btu per hr}\end{array}\right) = \left(\begin{array}{c}\text{Gallons of oil}\\\text{per hour}\end{array}\right)\left(\begin{array}{c}\text{Heating Value,}\\\text{Btu per gallon}\end{array}\right)$$

For Coal:

$$\left(\begin{array}{c}\text{Input,}\\\text{Btu per hr}\end{array}\right) = \left(\begin{array}{c}\text{Pounds of coal}\\\text{per hour}\end{array}\right)\left(\begin{array}{c}\text{Heating Value,}\\\text{Btu per pound}\end{array}\right)$$

(b) Capacity, or Bonnet Capacity, is the heat available in the air at the furnace bonnet, in units of Btu per hr.

$$\text{Capacity} = (\text{cfm})_b(60)(0.24)(d_b)(t_b - t_r)$$

in which

(cfm)$_b$ is the air-flow rate in cfm as measured at bonnet temperature,

d_b is the density of air corresponding to bonnet temperature,

t_b is the bonnet-air temperature, deg F, and

t_r is the return-air temperature, deg F.

(c) Duct Heat Loss is the heat loss from the warm-air ducts to the space surrounding the ducts, in units of Btu per hr.

(d) Register Delivery is the heat available in the air that is delivered from the registers into the space to be heated, and is in units of Btu per hr.

$$\text{Delivery} = (\text{cfm})_{\text{reg}}(60)(0.24)(d_{\text{reg}})(t_{\text{reg}} - t_r)$$

in which

(cfm)$_{\text{reg}}$ is the air-flow rate as measured at register-air temperature,

d_{reg} is the density of air corresponding to register-air temperature, in units of lb per cu ft, and

t_{reg} is the register-air temperature, deg F.

(e) Bonnet Efficiency is the ratio of the capacity to the input.

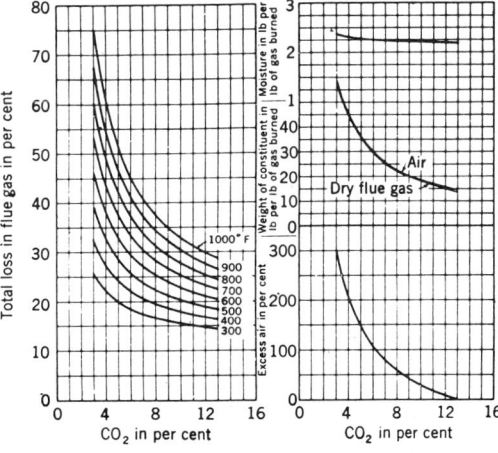

Fig. 5. Flue gas losses with natural gas as fuel. (U. of Ill. Eng. Exp. Sta. Circ. 44)

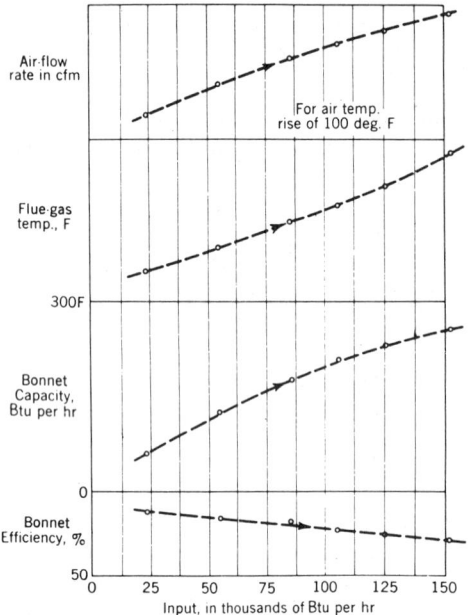

Fig. 6. Typical trends of furnace performance tests.

(f) Duct Transmission Efficiency is the ratio of the register delivery to the capacity.

(g) Flue Gas Loss is the heat loss of the flue gases, expressed as a percentage of the heat input, and includes both sensible and latent heat losses. See Fig. 5 for a typical flue-gas loss curve.

(h) Combustion Efficiency is 100 minus the flue gas loss, expressed as a percentage.

Testing and Rating of Furnaces

A given furnace in combination with a given blower can be operated in a number of different ways and provide almost an indefinitely large number of different capacities and efficiencies. For example, the rate of heat input to the furnace could be successively increased with a fixed speed of the blower and a set of performance data taken. After this, the whole set of tests could be repeated with another speed of the blower. From the standpoint of a commercial rating, it becomes necessary, therefore, to hold some of the variables at a constant value and to vary only one item. This is shown by a typical performance curve for a furnace, as shown by Fig. 6. For this furnace the rate of fuel input was successively increased after each test, but each time the blower speed was so adjusted that the temperature rise of the circulating air over the

furnace was maintained at a constant value of 100 deg. The typical curves in Fig. 6 show some interesting trends, common to all furnace performance tests:

(a) In common with most heat transfer devices, the maximum efficiency occurs at a low value of input. In fact, the efficiency gradually decreases as the input is increased.

(b) The capacity increases with the increase in input, but not in linear relationship. Theoretically, if the input could be made large enough the capacity could be extended indefinitely.

(c) The flue gas temperatures constantly increase with an increase in input since the heat exchanger surface is fixed in amount.

(d) Since the tests were conducted with a constant rise in temperature of the circulating air (100 deg) the required air-flow rate also increased. Note that it would be possible to conduct tests with a constant flow rate and a constantly increasing air temperature.

From the standpoint of the engineer a complete performance curve of the type shown is much more informative and useful than a single-point rating, but a single-point rating is the only kind of rating that is acceptable to the trade in general. Every manufacturer is interested in a rating of the single-point type. The user is also interested in a single-point rating so that he can select a furnace for a given job and have reasonable assurance that it is adequate but not too large.

Acceptable Limits

Theoretically, the furnace whose performance curves are shown in Fig. 6 could be rated at any value between say 10,000 Btu per hr up to 160,000 Btu per hr input. As a matter of fact if no limits of any kind were imposed the furnace could have been tested with higher and higher inputs until finally the furnace disintegrated. It becomes obvious that in order for a manufacturer to specify a single-point rating it is necessary for the testing laboratory to impose some arbitrary limits, or boundaries, as to what is acceptable. The single-point rating that is finally chosen would not necessarily be a point of maximum efficiency, since the maximum efficiency was obtained at an abnormally low input and capacity. On the other hand, the single-point rating would not be one for which the equipment was strained to the last notch. The arbitrary limits that happen to be selected by industry are not

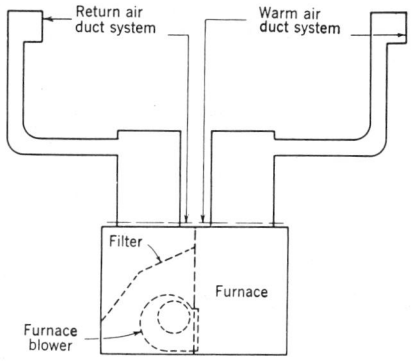

Fig. 7a. Schematic diagram of a forced-air system.

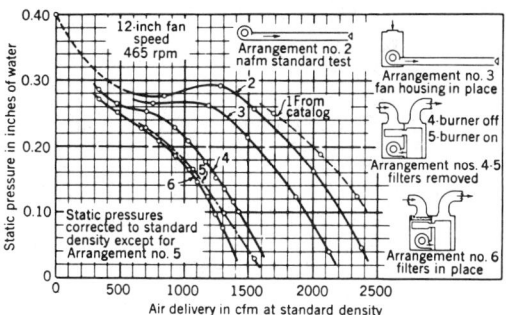

Fig. 7b. Static pressure characteristics of a blower-furnace combination. (U. of Ill. Eng. Exp. Sta. Bul. 300)

important for this discussion. What is important is that industry has sensibly and wisely agreed upon certain limits not only for the safeguard of the consumer, but primarily for the protection of the manufacturer who hopes to stay in the business. Industry could mutually agree upon a value of say 75% as the minimum bonnet efficiency that would be acceptable. Or, a flue-gas temperature of say 800 deg could be considered as the maximum, or the surface temperature of the heat exchanger could be limited to say 900 deg, or the draft could be limited, or all of these limits could be imposed simultaneously. Whatever capacity that corresponded with the most stringent limit would then be the capacity for rating purposes. In practice, an extremely stringent set of limits have been imposed by the manufacturer acting through such laboratories as the American Gas Association Testing Laboratory. In general, the single-point ratings that are given in catalogs are reasonably attainable values for a well-designed installation. A typical single-point rating for a small gas-fired furnace would show:

> Input   90,000 Btu per hr
> Capacity   72,000 Btu per hr

All gas-fired forced-air furnaces that bear the approval label of the American Gas Association show a bonnet efficiency of 80%, no more and no less.

Selection of Furnace for Given Installation

The selection of a furnace for a given installation is simple in principle since it consists of selecting a furnace whose capacity is adequate to take care of the heat losses for the building under design weather conditions. In practice, a number of minor points

arise that require clarification. For example, if the register delivery is to be made equal to the design heat loss, the manufacturer's catalog should either state the delivery or give some means of estimating the duct transmission efficiency. In this connection, the design manuals of the National Environmental Systems Contractors Association (NESCA), arbitrarily assume that the register delivery will be 0.85 times the bonnet capacity or, in other words, that the duct transmission efficiency will be 85%. This assumption may be greatly in error in large installations. Another question that arises in connection with design heat loss is that of determining what portion of the house should be included in the heat loss calculations. For example, should the basement heat loss be considered? If not, what should be done with basement playrooms and hobby rooms?

Rule for Selection

For the sake of uniformity and consistency, one set rule is offered for the selection of furnaces, as follows:

"Calculate the design heat loss for the entire structure, including the basement space, and select a furnace whose bonnet capacity is equal to this design heat loss."

For example, if the design heat loss for the first story rooms is 48,300 Btu per hr and that for the basement is 14,600 Btu per hr, the total heat loss for the entire structure is 62,900 Btu per hr. The furnace for this building should have a capacity equal to or greater than 62,900 Btu per hr. The suggested rule applies to a furnace which is to be located in the space that is to be heated. If, however, the furnace is to be

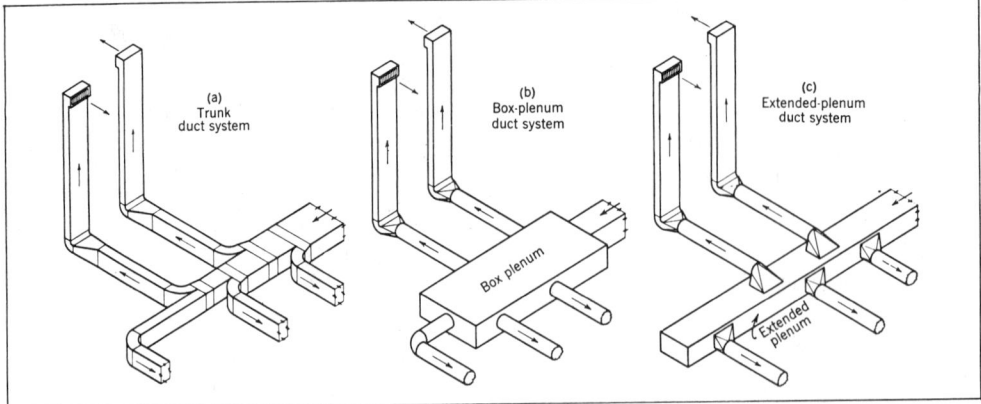

Fig. 8. Three types of air duct systems. (U. of III. Eng. Exp. Sta. Bul. 415)

located in a separate building, such as an attached garage or in the attic space, the heat loss from the ducts will be lost and will not be available for heating the house. In this case, it will be necessary to add a correction, to take into account the heat loss from the ducts and the furnace casing, to the design heat loss from the structure.

Blower Characteristics

The blower is an integral part of the smaller furnace-blower combinations used for residential heating and the air-flow characteristics are not for the blower alone, but for the combination. The

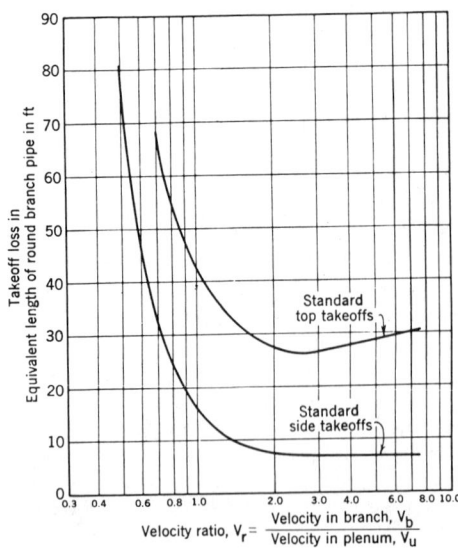

Velocity ratio, $V_r = \dfrac{\text{Velocity in branch, } V_b}{\text{Velocity in plenum, } V_u}$

curves in Fig. 7, for example, show typical performance curves not only for the blower alone but for various arrangements of casing and filter. The performance curve No. 6 is markedly lower than that labelled as No. 2, which is for the blower without any housing. For smaller furnaces, the furnace-blower combination should be capable of providing for an air flow corresponding to a 100 deg temperature rise through the furnace and against a static pressure of 0.20 in. of water, measured external to the furnace-blower combination. In other words, the 0.20 in. of water is for overcoming the frictional losses of the attached duct systems, shown at left in Fig. 7. The internal losses due to frictional effects and expansion losses within the casing and the blower housing have been accounted for by the manufacturer and need not be taken into account by the designer of the duct system.

Blower Sizes

With larger units of the industrial type, the temperature rise through the furnace is usually lower than 100 deg, which corresponds to a larger air-flow rate for a given heat input to the furnace than is characteristic of the small units. For example, temperature rises of the order of 50 to 70 deg are commonly specified for industrial type furnaces. Furthermore, the external static pressure is not limited to 0.20 in. of water and is considerably larger.

Fig. 9. Loss of extended-plenum fittings, in equivalent lengths of branch duct. (U. of III. Eng. Exp. Sta. Bul. 415)

In order to satisfy the rising trend towards summer cooling, some furnace-blower combinations are provided with more than one size of blower for a given size of furnace. In these arrangements a greater range of air-flow rates is possible without over-speeding a small blower or over-loading the blower motor. Hence, if a given furnace-blower combination is to be used in connection with a refrigeration unit, a larger blower can be made available than is normally used for heating purposes alone. This large blower will permit the use of a cooling coil in the air stream without resorting to an auxiliary fan for overcoming the resistance of the cooling coil.

Duct System Characteristics

The ducts which connect the furnace-blower combination with the registers and return intakes are the only part of the entire system that are not factory planned and factory assembled. It is true that certain portions of the duct system can be pre-fabricated, but essentially the duct system for each job is a tailor-made affair, no two of which are exactly alike, except perhaps in a housing project.

The duct system plays an important part in the final cost to the owner since a substantial part of the cost and labor is involved in installing the equipment. In Fig. 8 are shown several alternate types of duct systems, although in an actual installation a combination of these different arrangements might be used. In any case the basic arrangements consist of:

(a) **The trunkduct system** with graduated size of trunk duct. This duct system, which is practical for large installations involves costly fittings and many changes in duct section. This system has proven costly to install in small jobs in spite of the use of prefabricated fittings.

(b) **The box-plenum duct system** has advantages in certain installations where it is possible to construct a box plenum. In this arrangement the air is led to a central box, from which a number of individual ducts are tapped off with butt take-off connections. The most common locations of the box plenum are in a central hallway or below the floor of one wing of a large structure.

(c) **The extended-plenum duct system** is a most practical duct system especially for small installations. In this arrangement the trunk duct is not reduced in cross-sectional area, and a number of branch ducts are connected to the extended plenum with simple branch take-off connections.

(d) **The individual duct system** offers the other extreme in duct arrangements. This system is easy to install, but in many cases proves fairly expensive if a large number of individual ducts is to be run to a far end of the building. Furthermore, the heat loss from the ducts can be large and the duct transmission efficiency can be low. The system is easy to balance.

The air velocity near the furnace can be high, over 1000 fpm, but as one proceeds downstream the branch ducts divert air away from the plenum, and the air velocity in the downstream portions of the plenum become smaller. The reduction in velocity head in the plenum is partially offset by an increase in the static pressure. In fact, the static pressures in an extended-plenum duct system tend to be constant throughout the length of the duct. Furthermore, as shown in Fig. 9, the larger entry losses into the branch ducts occur at the upstream branches rather than at the downstream branches. The net result is, therefore, that a larger flow rate is obtained for a given size of duct at downstream branches than at upstream branches, which is contrary to the usual experience with the reduced trunk system. In small jobs where the additional cost of sheet metal used is of little consequence compared to the saving in installation labor and fabrication, the extended-plenum duct system has proven most practical.

Trends

Some interesting trends had taken place during past years, some of which were conflicting in tendency. In the first place, as a natural outcome of the successful use of small-duct, high-velocity systems for aircraft heating, the idea of using small ducts took hold. In fact, duct sizes ranging from 3 in. in diameter to 6 in. in diameter have been tried. As indicated in Fig. 10, a reduction in duct diameter causes a sharp increase in air friction. Either smaller air-flow rates can be handled through each duct or larger pressure losses can be maintained for the duct system. In fact, as one alternative, some thought has been given to the use of static pressures larger than 0.20 in. of water for the duct system. Such an increase can be handled up to a certain point with commercially available blowers. However, when static pressures much in excess of 0.50 in. of water are contemplated, the usual simple low-cost blower is no longer applicable. Hence, increases in static pressure are limited by types of blowers that are commercially available.

The second alternative is to decrease the air-flow rate handled by the ducts and the entire system.

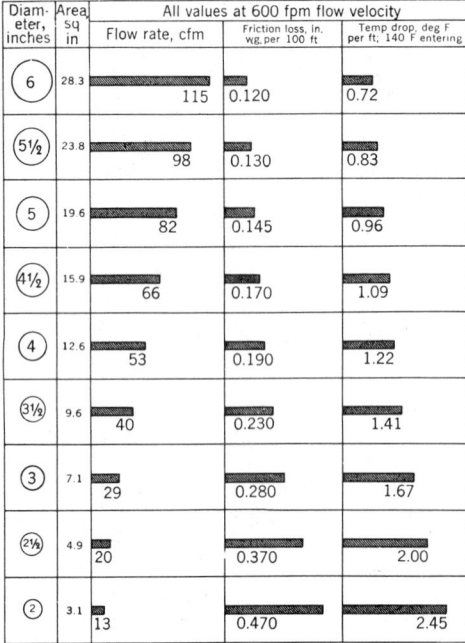

Diameter, inches	Area sq in	All values at 600 fpm flow velocity		
		Flow rate, cfm	Friction loss, in. wg per 100 ft	Temp drop, deg F per ft; 140 F entering
6	28.3	115	0.120	0.72
5½	23.8	98	0.130	0.83
5	19.6	82	0.145	0.96
4½	15.9	66	0.170	1.09
4	12.6	53	0.190	1.22
3½	9.6	40	0.230	1.41
3	7.1	29	0.280	1.67
2½	4.9	20	0.370	2.00
2	3.1	13	0.470	2.45

Fig. 10. Characteristics of small air ducts.

This can be done up to a point. It would be possible to increase the temperature rise through the furnace from the current 100 deg to some value such as 150 deg or even 200 deg. This increase in temperature rise would result in a markedly lower air-flow requirement for a given capacity. Unfortunately, the use of higher bonnet air and register air temperatures brings additional problems. In the first place, the furnaces that are tested for acceptance by laboratories must undergo a 100 deg temperature rise. For any other temperature rise, special tests will have to be conducted and perhaps some changes made in the test provisions. These changes are not easy to make since they must be accepted by a large segment of industry, including the public utilities. If such changes are considered to be a lowering of the standards, rather than an uplifting, the changes are not easy to make. In the second place, the higher air temperatures in the ducts will result in larger temperature drops of the air in the ducts. Small ducts lose a proportionally larger part of the initially available heat than do larger ducts handling the same velocity of air. By increasing the duct air temperature the heat loss is accentuated. In the third place the introduction of high temperature register air creates problems of air stratification in the room. Special diffusers are required in order to

obtain a quick diffusion of the high temperature air as it enters the room so that the heated air does not simply rise and accumulate at the ceiling.

The current trend towards the use of a single system of ducts for both heating and cooling has also affected the matter of duct sizes. In general, the air-flow requirements for summer for a given space are larger than those for winter. If a duct system is to be adapted for year-around air conditioning the small duct system, which may be most suitable for winter heating, may not be adequate for summer cooling. It would be possible, of course, to provide additional small ducts for use in summer cooling alone. In any case, the trend to smaller and smaller ducts has been stopped pending an evaluation of the year-around air conditioning load.

Warm Air Registers

The flexibility of the forced-air heating system is best illustrated by the wide variety of registers that are commercially available and the extremely diverse methods of installation that are possible. In fact, registers can be installed in the floor, the baseboard, above the baseboard, high in the sidewall, in the ceiling, in the window ledge, or any other place to which a duct can be run. The more common types of registers consist of:

(a) Rectangular registers of perforated grille or straight vane type that do not provide for air deflection,

(b) Rectangular registers with movable or adjustable vanes that permit deflection, up, down, sideways, or both vertically and horizontally,

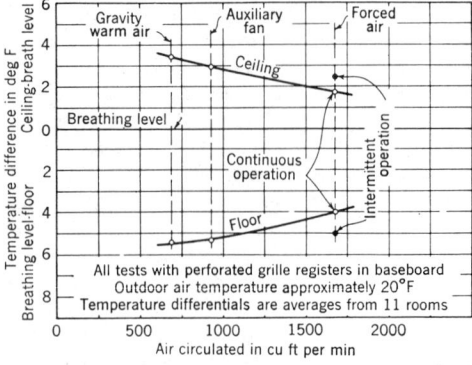

Fig. 11. Average temperature differentials with gravity, auxiliary fan, and forced-air systems. (U. of Ill. Eng. Exp. Sta. Bul. 318)

(c) Baseboard types that have long narrow openings,

(d) Diffuser types that attempt to mix the register air and the room air rapidly before the air stream has travelled far into the room,

(e) Ceiling registers and diffusers of square, round, rectangular, and oval-shaped designs,

(f) Rectangular slots for use in window ledges, in the floors below windows, or in the ceiling above windows,

(g) A multiplicity of special designs for special purposes, such as railroad cars, airplane cabins, etc.

In many respects the registers are the most important aspect of the forced-air system as far as the occupant is concerned, since it is the visible evidence of the system in the living quarters. When improperly located the registers create drafty conditions and provide poor room temperature conditions. When properly located the registers can do much to improve room air temperature conditions. For example, Fig. 11 shows that for a given location of the registers, the indoor temperatures will be much more uniform with a forced-air system than with either a gravity or auxiliary fan system. The reduction in temperature difference with the use of the forced-air system may be attributed not only to the increase in the quantity of air which was circulated, but also to the decrease in the temperature of the air which was introduced into the rooms.

As will be shown in the next section on duct arrangements, the practical location of a register is governed to some extent by the type of structure in which the system is installed. The use of a base-

board register under a window, for example, may not be practical for a system in which the supply ducts are carried overhead. In general, (a) the register should introduce the air into the room in such a manner that the air does not strike an occupant that is seated and at rest, (b) it should tend to overcome the downdrafts that occur below windows and cold walls, and (c) it should not interfere with furniture placement. Room air velocities in the living zone should preferably not exceed 35 fpm and the moving air stream should not deviate by more than 1.0 deg from the room air temperature. The perimeter warm-air system in which the warm air is introduced through floor registers located along the exposed walls of the structure, as shown in Fig. 12, has proven most successful, even for structures built over concrete slab floors or over unheated crawl spaces.

Return-Air Intakes

The function of the return-air intake or grille is to provide a means for the room air to be returned to the furnace to be reheated and recirculated. In smaller residences the use of one or two return intakes has been found to be as satisfactory as the use of a large number of smaller intakes. As a matter of fact, if the warm-air registers are located so that a small difference in temperature exists from floor to ceiling, the location of the return-intake is not too important. Equally satisfactory results have been obtained with the intakes at the ceiling, at the high sidewall, at the low wall, at the baseboard, and at the floor

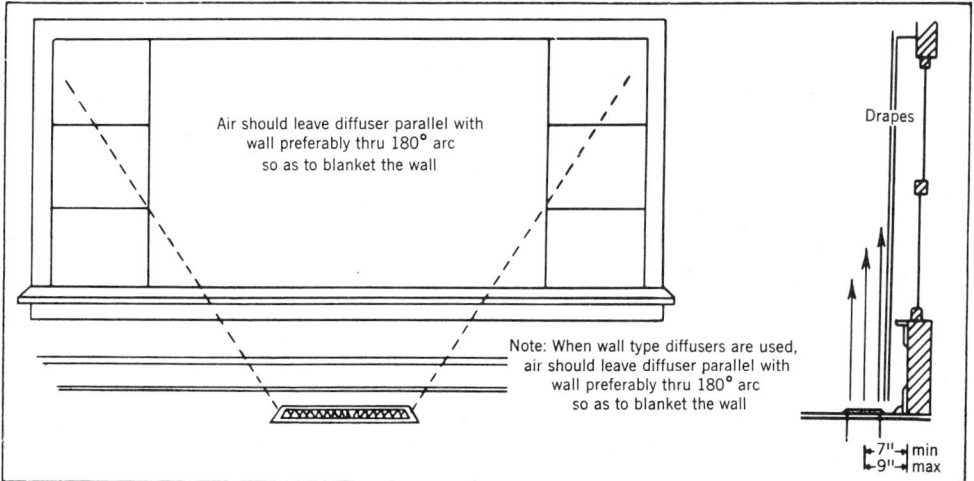

Air should leave diffuser parallel with wall preferably thru 180° arc so as to blanket the wall

Drapes

Note: When wall type diffusers are used, air should leave diffuser parallel with wall preferably thru 180° arc so as to blanket the wall

7" min
9" max

Fig. 12. Location of floor diffuser or wall diffuser below window.

locations. The return air from small bedrooms escapes from the room through the space below the closed door. The only precaution necessary in the location of a relatively large intake is to avoid a place near seated occupants, not only because of air movement but also because of air noise. A hallway location will usually meet the requirements of a central location and of distance from seated occupants.

Contrary to popular belief, room air cannot be "pulled" towards a return intake. In fact, the pulling action is limited to the immediate vicinity of the intake. Fig. 13, for example, shows how the air velocity diminishes rapidly as the distance from the intake increases. Even with 1000 fpm velocity at the opening, at a distance of 12 in. from the opening the velocity was reduced to 90 fpm, and at a distance of 24 in. from the opening the velocity was essentially room-air velocity or 25 fpm.

An outdoor-air intake can be provided in a forced-air system for introducing outdoor ventilation air into the structure. Such provisions are more common in the milder climate regions of the country. This outdoor air should be conveyed through an individual duct leading from the outdoors directly to the return plenum of the furnace and should be provided with a manually-operated damper. The introduction of large quantities of outdoor air into a structure will result in large fuel bills and a lower relative humidity in the structure. In general, even with moderately large amounts of outdoor air introduced into a structure it is not possible to build up indoor pressures sufficiently to stop window infiltration. It should be realized that fireplace openings, chimney openings, door crackages, and window crackages permit air to flow readily from the structure to the outdoors.

Arrangement of Furnace and Ducts

The discussion up to this point has been with the components of the warm-air system, and not with the integration of the parts into a whole. The arrangement of the furnaces, ducts, registers, and intakes will be dependent upon the type of structure. Since each system is a hand-tailored arrangement for that particular job, no two contractors will agree in all respects on the manner in which the job should be installed. A number of arrangements in common use are:

(a) Basement furnace with ductwork at ceiling of basement.

(b) Utility room location of furnace with ductwork at the ceiling of the rooms. This arrangement is common with individual apartment units in which each tenant operates his own heating unit and pays for the fuel bill.

(c) House built over a concrete slab floor. Furnace is located in a utility room and all ducts are embedded in the concrete floor to give a perimeter-loop arrangement. See Figs. 14a and 14c.

(d) Same as (c), but with a radial-feeder arrangement of individual ducts in place of the perimeter-loop arrangement. See Fig. 14b.

(e) Slab floor with numerous air passages provided below the floor through which warm air can be circulated. This is a warm-air floor-panel system, one form of which is shown in Fig. 14d. A warm-air ceiling-panel system is also available.

(f) House built over a crawl space. In one arrangement, individual ducts are carried below the floor to the registers in the floor or in the baseboard. In another arrangement, warm air is pumped into crawl space and permitted to enter the rooms above through floor registers, as indicated in Fig. 14f. With a crawl-space plenum system the foundation wall requires special insulation and the ground must be pro-

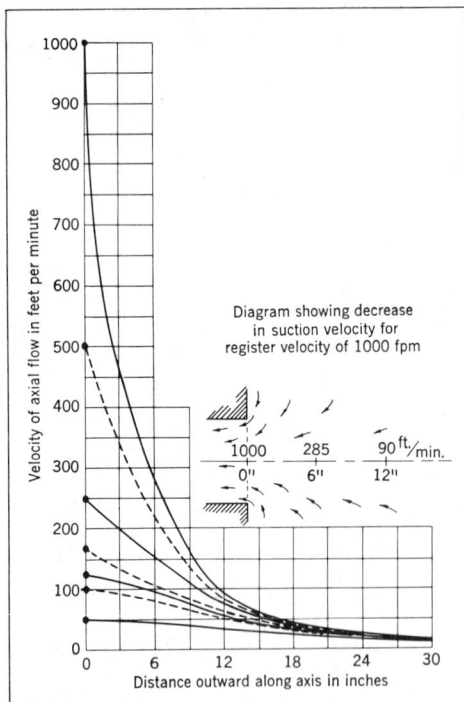

Fig. 13. Air velocity, front of intake. (Data from Dalla Valle, Control of Industrial Dust, Mech. Eng., Oct. 1933)

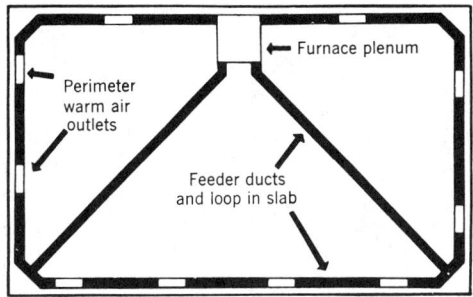

Fig. 14a. Perimeter loop system with furnace over loop.

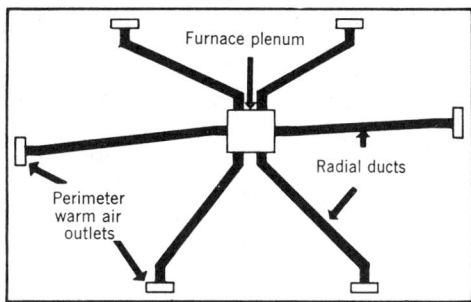

Fig. 14b. Perimeter radial system with feeder ducts in concrete slab or crawl space.

tected against moisture travel from the ground to the crawl space. This arrangement may not be permissible in areas where building codes prohibit ductless systems in combustible areas.

(g) Overhead attic installation of furnace. Furnaces are available for attic space installation. All ducts in the attic space should be heavily insulated since the heat loss from the ducts will not be available for heating the building. See Fig. 14e.

(h) For large buildings the use of multiple furnace units is often possible. Each of several furnaces can be located close to the particular zone to be heated in order to shorten the ductwork. Individual zone temperature control is possible. Multiple furnace systems have been successfully used in the heating of churches, commercial buildings, schools, etc.

The forced-air system is a flexible system and one which has few limitations except that of extreme size. The ducts require no pitching and can be carried upward, downward, or sideways; small air leaks in the ducts are of no consequence; the system needs no purging; the system is extremely responsive to temperature changes outdoors; and all that is required to shut down a plant is to turn off the burner.

A few installation precautions are necessary, as is true of any combustion device. The approval tests for furnaces include specifications for space clearances on the sides and top from combustible construction, and such clearances should be provided in the actual installation. If the furnace is to be located in a small furnace room, or closet, the diagrams in Figs. 15a and 15b show the code requirements as far as combustion air supply and closet ventilation is concerned.

Basic Thermostatic Controls

An almost endless number of control arrangements is possible, but the simplest and the most practical is that which is shown in diagram form in Fig. 16. The components consist of:

(a) **Room thermostat** which should be provided with a differential adjustment setting not to exceed about 1.0 deg.

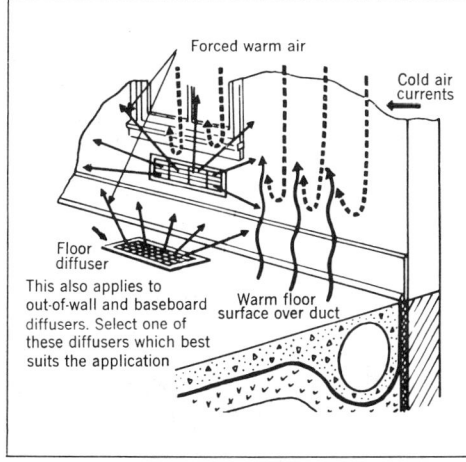

Fig. 14c. Detail of perimeter loop system with floor or baseboard diffuser below window.

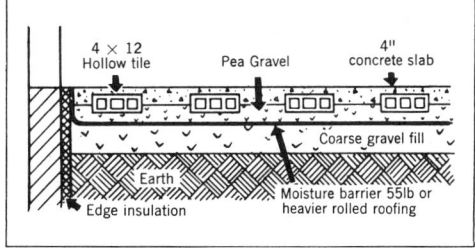

Fig. 14d. One form of warm air floor panel heating.

Horizontal furnace
in attic space
To return air grille
in ceiling or high
on side wall
Down-comer
duct

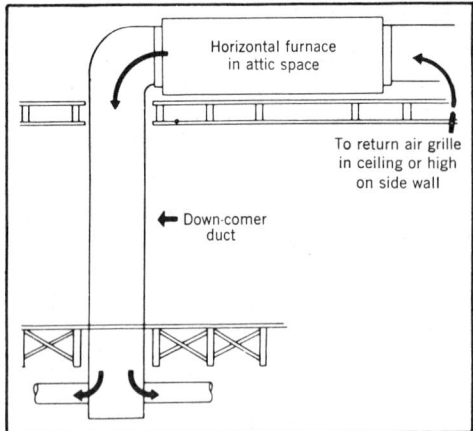

Fig. 14e. Attic furnace installation leading to duct system in crawl space.

Return air
Return
air from
crawl
space
Furnace
Floor diffuser
Plenum
Warm air discharge into crawl space

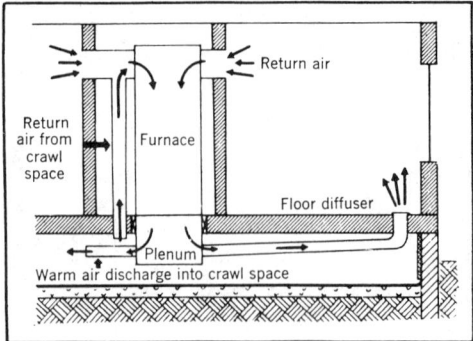

Fig. 14f. Down-flow furnace with crawl space duct system and crawl space plenum system shown.

(b) Limit switch for shutting off the burner whenever the bonnet-air temperature becomes excessively high. The recommended setting is between 175 and 200 deg. In the normal sequence of operation the limit switch is not called upon for action, but is available in case of possible overheating of the furnace.

(c) Burner motor or burner valve is directly connected to the room thermostat circuit. Obviously,

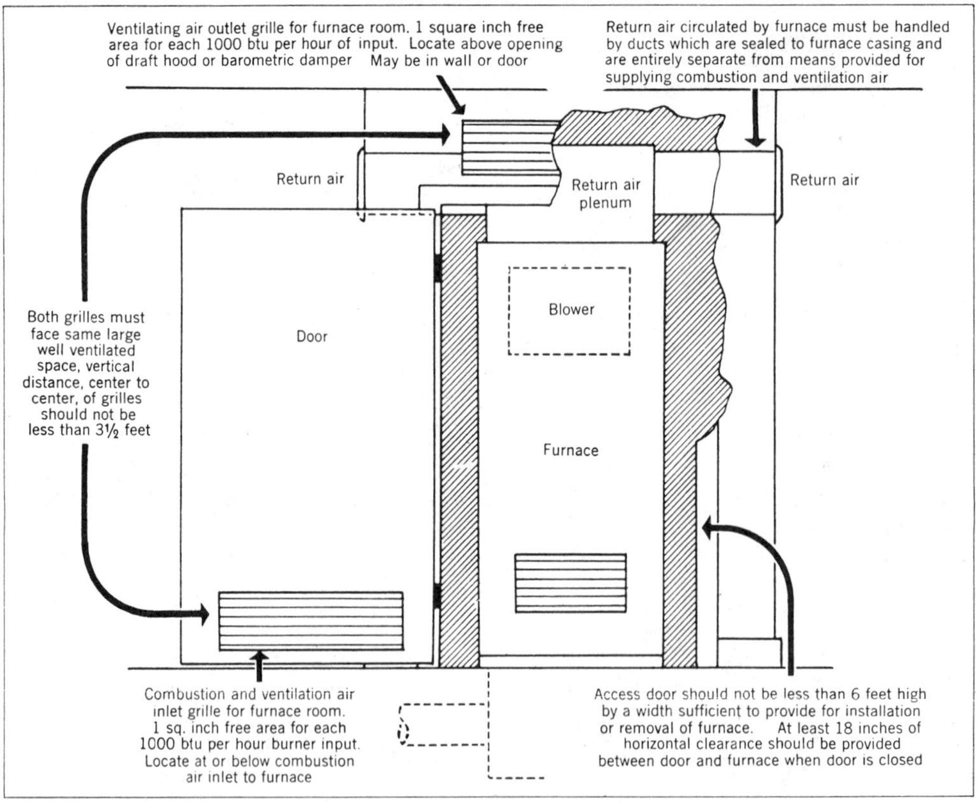
Ventilating air outlet grille for furnace room. 1 square inch free
area for each 1000 btu per hour of input. Locate above opening
of draft hood or barometric damper May be in wall or door
Return air circulated by furnace must be handled
by ducts which are sealed to furnace casing and
are entirely separate from means provided for
supplying combustion and ventilation air
Return air
Return air
plenum
Return air
Both grilles must
face same large
well ventilated
space, vertical
distance, center to
center, of grilles
should not be
less than 3½ feet
Door
Blower
Furnace
Combustion and ventilation air
inlet grille for furnace room.
1 sq. inch free area for each
1000 btu per hour burner input.
Locate at or below combustion
air inlet to furnace
Access door should not be less than 6 feet high
by a width sufficient to provide for installation
or removal of furnace. At least 18 inches of
horizontal clearance should be provided
between door and furnace when door is closed

Fig. 15a. Provisions for combustion and ventilation air for furnace closet.

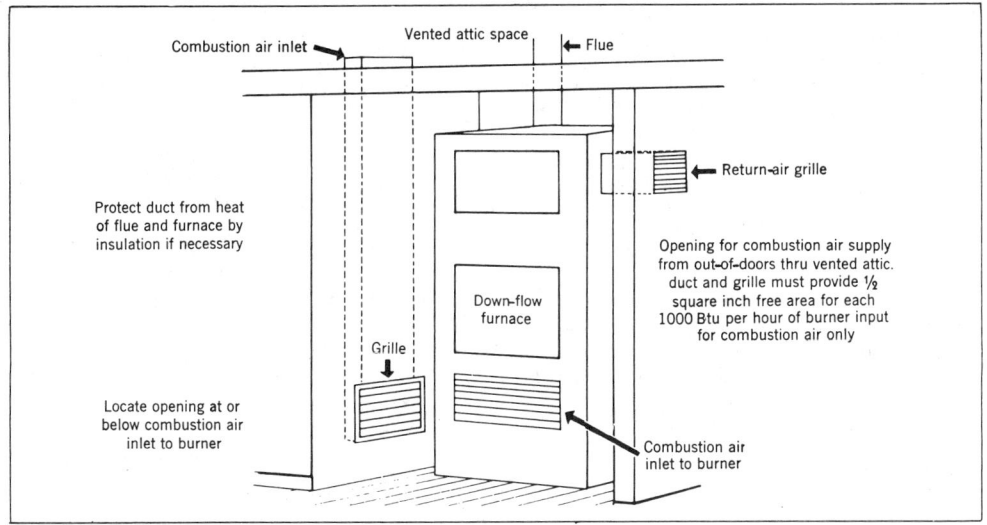

Fig. 15b. Combustion air supply from attic space.

the frequency of operation of the burner is dependent upon the setting and sensitivity of the room thermostat. In a warm-air furnace system, unlike a hot-water or steam system, there is little lag and at the same time no heat storage. In other words, the heat must be generated only as needed and must be distributed as soon as it is generated. For this reason, the sensitivity of the room thermostat must be acute, and the differential settings must be of the order of 0.5 to 1.0 deg. The purpose of the small differential setting is to cause frequent and short operations of the burner. The length of each burner cycle should not, however, be less than about two minutes for a gas burner nor less than about four minutes for an oil burner because shorter operating periods might contribute to large combustion losses due to incomplete burning at the beginning of each burner operation.

(d) Fan switch located in the bonnet is a device which turns on the blower when the air temperature in the bonnet is high and turns off the blower when the bonnet-air temperature is low. The recommended setting of the fan switch is about 110 deg for the cut-in point and about 85 deg for the cut-out point. These settings are of great importance in the successful operation of the entire system.

Continuous Air Circulation Principle (CAC)

In many respects the basic principle of a gravity warm-air system of heating is an ideal one and difficult to surpass even with a mechanical system. As shown in Fig. 17, the gravity warm-air system provides a 24-hour circulation of air, the temperature of which is modulated in accordance with the demand. Furthermore, the quantity of air flow is small in mild weather and large in cold weather, so that the air-flow rate is also modulated. In any case, the system is readily adapted to the changing demands of weather, and such changes occur automatically and without artificial stimulus. If it were possible to operate a forced-air system in exactly the same manner and just as simply, the problem of control would be eased.

With a forced-air system, the flow of air is not dependent upon the weather or the heating demands, but is fixed only by the blower speed, the blower size, and the frictional restrictions in the duct system. It is possible with any given duct system to vary the air-flow rate over a considerable range.

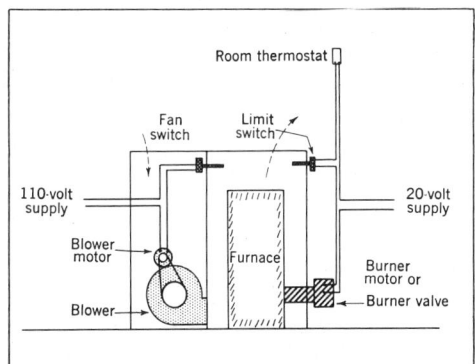

Fig. 16. Basic arrangement of controls.

Furthermore, it is possible to control over a wide range the number of hours that the blower will operate during the day. For example, on a day in which the outdoor air temperature is 30 deg the fan switch can be adjusted so that the blower runs continuously during the 24-hour period, or for 16 hours, or for 8 hours, or for as little as one hour. This wide range of blower operating periods is dependent mainly upon the settings of the fan switch, and to some extent upon the blower speed. Longer operating periods of the blower are desirable since when the blower is operating the control of distribution of heated air is possible; whereas when the blower is not operating, and the system is under the influence of gravity action, no control exists as far as air distribution to the various rooms is concerned.

Continuous Blower Operation

The argument might be raised that, if continuous air circulation is ideal, the system should be operated to give continuous blower operation, as shown in Fig. 18. In this case, the fan switch could be eliminated and the blower operated from a manual switch that is turned on at the start of the heating season and left on all during the season. Or, the fan switch could be left in place and the cut-in point of the fan switch lowered to about 70 deg. The two

possible objections to this arrangement are:

(a) That the blower does not need to be operated in extremely mild weather when little heat demand occurs, and the electrical cost of blower operation could be saved, and

(b) That the temperature of the air issuing from the registers can be as low as room-air temperature and might create objectional drafts in front of the registers.

With high sidewall registers, with floor registers, or with baseboard registers that discharge the air sharply upwards towards the ceiling, this draft possibility does not exist. With such register installations the fan switch cut-in settings can be lowered below 110 deg, and such a lower setting would prove most advantageous from the standpoint of temperature control and uniform distribution of heated air.

Intermittent Blower Operation

For a number of reasons heating contractors have not appreciated the necessity of obtaining almost continuous blower operation in average heating weather. The influence of gravity warm-air heating practice undoubtedly persists. For example, in gravity heating it was common practice in

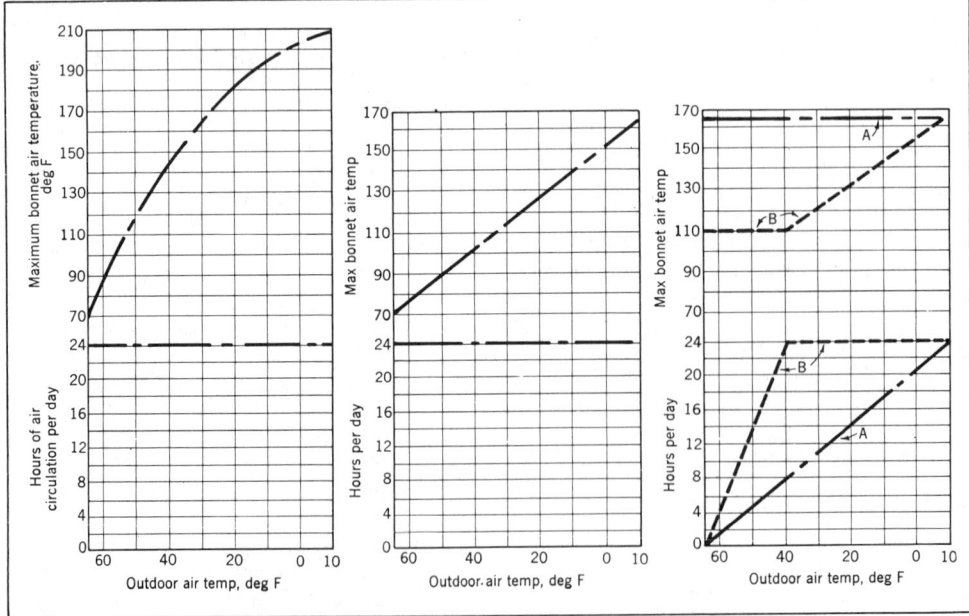

Air flow and air temperature for (left, Fig. 17) gravity warm air system, (center, Fig. 18) continuous blower operation, and (right, Fig. 19) intermittent blower operation.

overcoming the deficiencies of an inadequate plant to raise the bonnet-air temperature, thereby increasing the motive head of circulation, and eventually obtaining a larger air-flow rate.

In the earlier days of forced-air heating a similar idea persisted. For example, since the design temperature for bonnet air was assumed as 165 deg, the contractor assumed that the fan switch cut-in point should be 165 deg. Under this setting the operation obtained was as shown by the curves labelled (A) in Fig. 19. The bonnet-air temperature would average 165 deg in all weather and the blower would operate intermittently in all weather except during design weather. Frequently, in the case of trouble jobs the contractor raised the cut-in temperature even higher than 165 deg and only aggravated troubles, for the system then operated intermittently under all weather conditions. Essentially, a system so controlled is a "booster" system which is partly gravity and partly mechanical in its operation. Since the ducts were designed for blower operation only, it is little wonder that air distribution was poor and the rooms at some distance from the furnace were improperly heated. Many complaints arose that the heating was spasmodic and that when the blower turned on, the register air was cold at first. Unfortunately, such poor control operations still exist in many older plants and the potentialities of a good warm-air system have not been realized.

As shown by the curves labelled as (B) in Fig. 19, a lowering of the fan switch cut-in point causes a reduction in the bonnet-air temperature, especially during average weather and a marked increase in blower operating times. As a matter of fact, if the cut-in point can be lowered below 110 deg, the blower begins to operate continuously at much higher outdoor air temperatures. A desirable method of blower operation is considered to be that in which the blower operates continuously as soon as the outdoor temperature drops to about 40 deg.

The matter of proper control settings is so important for the satisfactory operation of a forced-air system that industry has set up rules of procedure for adjusting the heating system. Essentially, these adjustments[1] consist of:

(a) A differential setting of the room thermostat not to exceed 1.0 deg and preferably only 0.5 to 0.75 deg.
(b) Adjustment of the blower speed to give an air

_____
[1] Complete details on the adjustment procedure are given in a publication entitled, *Manual 6, Service Manual for Continuous Air Circulation Technicians*, by the National Environmental Systems Contractors Assn. (NESCA).

temperature rise through the furnace of 100 deg.
(c) Adjustment of the fan switch to give a cut-in point of about 110 deg and a cut-out point of about 85 deg. If a lower setting can be maintained without danger of objectionable drafts in the rooms, the settings should be lower.

Basic Factors for Designing Warm-Air Systems

A large number of different methods exists for use in designing such diverse types of warm-air systems as: (a) small-duct, high-velocity systems, (b) ceiling warm-air panel systems, (c) floor warm-air panel systems, (d) small-duct perimeter systems, (e) large, commercial systems, etc. Regardless of the system of design most of them are based on a few basic concepts, which will be considered in detail. The assumption will be made that the design heat loss of the spaces to be heated has been established.

Bonnet-Air Temperature

The starting point of any design consists in selecting a value for bonnet-air temperature. As can be seen from the following equation, which was presented earlier, the bonnet-air temperature determines the air-flow rate that will be used to maintain a certain capacity.

Bonnet Capacity = $(60)(0.24)(d_b)(t_b - t_r)$(cfm)

For a given capacity, the air-flow rate increases as the bonnet-air temperature is decreased. For example, for a bonnet capacity of 100,000 Btu per hr the air-flow rate will be 1092 cfm for a bonnet-air temperature of 165 deg, but 1004 cfm for a temperature of 115 deg. As noted before, for small residential units the temperature rise $(t_b - t_r)$ is arbitrarily determined as 100 deg. That is, for a return-air temperature of 65 deg the bonnet-air temperature is assumed to be 165 deg.

For large furnaces used for industrial and commercial applications a temperature rise of less than 100 deg is usually selected. This corresponds to bonnet-air temperatures less than 165 deg. A suggested method for selecting a bonnet-air temperature is by use of Table A-1 for jobs having a heat loss less than 350,000 Btu per hr, and Table A-2 for jobs having a heat loss in excess of 350,000 Btu per hr. In order to use the tables, both the shortest as well as the longest duct lengths, from the bonnet to the register, should be measured from the building plans.

Table A-1 — Register Temperatures and Air Volume for Buildings with Heat Loss Up to 350,000 Btu per Hour

(National Environmental Systems Contractors Association)

Temperature of Bonnet, Deg.	Linear Distance from Bonnet to Register, Feet											
	10	20	30	40	50	60	70	80	90	100	110	120
	Register Temperature, upper figure and Cfm per 1000 Btu per Hour, lower figure											
110	108 / 23.1	107 / 23.6	106 / 24.2	104 / 25.4	103 / 26.0	102 / 26.6	101 / 27.3	100 / 28.0	99 / 28.8	98 / 29.6	97 / 30.5	96 / 31.5
120	118 / 19.0	116 / 19.6	114 / 20.3	112 / 21.1	111 / 21.6	109 / 22.5	108 / 23.1	106 / 24.2	105 / 24.8	103 / 26.0	102 / 26.6	101 / 27.3
130	128 / 16.4	126 / 16.9	123 / 17.6	121 / 18.1	119 / 18.7	117 / 19.3	115 / 20.0	113 / 20.7	111 / 21.6	110 / 22.1	108 / 23.1	107 / 23.6
140	137 / 14.6	134 / 15.1	131 / 15.7	129 / 16.2	126 / 16.9	124 / 17.3	122 / 17.9	120 / 18.4	118 / 19.0	116 / 19.6	114 / 20.3	112 / 21.1
150	147 / 12.9	143 / 13.5	140 / 14.0	137 / 14.6	134 / 15.1	131 / 15.7	129 / 16.2	127 / 16.6	124 / 17.3	122 / 17.9	120 / 18.5	118 / 18.4
160	156 / 11.8	152 / 12.3	149 / 12.7	146 / 13.2	142 / 13.7	139 / 14.2	136 / 14.8	133 / 15.3	130 / 15.9	128 / 16.4	126 / 16.9	123 / 17.6
170	166 / 10.8	161 / 11.3	157 / 11.7	153 / 12.2	150 / 12.6	146 / 13.1	143 / 13.5	140 / 14.0	137 / 14.6	134 / 15.1	131 / 15.7	128 / 16.4
180	175 / 10.0	170 / 10.5	166 / 10.8	161 / 11.3	157 / 11.7	153 / 12.2	150 / 12.6	146 / 13.1	143 / 13.5	140 / 14.0	137 / 14.6	135 / 14.9
190	184 / 9.4	179 / 9.8	174 / 10.1	169 / 10.6	165 / 10.9	160 / 11.4	156 / 11.8	153 / 12.2	149 / 12.7	145 / 13.2	142 / 13.7	139 / 14.2
200	193 / 8.9	187 / 9.2	182 / 9.6	176 / 9.9	171 / 10.4	167 / 10.7	163 / 11.1	158 / 11.6	154 / 12.1	151 / 12.4	147 / 12.9	144 / 13.4

Temperature of Bonnet, Deg.	Linear Distance from Bonnet to Register, Feet											
	130	140	150	160	170	180	190	200	210	220	230	240
	Register Temperature, upper figure and Cfm per 1000 Btu per Hour, lower figure											
110	95 / 32.3	94 / 33.3	93 / 34.6	92 / 35.9	91 / 37.3	90 / 38.6	90 / 38.6	89 / 38.9	89 / 39.9	88 / 41.3	88 / 41.3	87 / 42.7
120	100 / 28.0	99 / 28.8	98 / 29.6	97 / 30.5	96 / 31.5	95 / 32.3	94 / 33.3	93 / 34.6	92 / 35.9	91 / 37.3	90 / 38.6	90 / 38.6
130	105 / 24.8	104 / 25.4	103 / 26.0	102 / 26.6	101 / 27.3	99 / 28.8	98 / 29.6	97 / 30.5	96 / 31.5	95 / 32.3	94 / 33.3	93 / 34.6
140	110 / 22.1	109 / 22.5	107 / 23.6	106 / 24.2	105 / 24.8	103 / 26.0	102 / 26.6	101 / 27.3	100 / 28.0	99 / 28.8	98 / 29.6	97 / 30.5
150	116 / 19.6	114 / 20.3	112 / 21.1	110 / 22.1	109 / 22.5	107 / 23.6	106 / 24.2	105 / 24.8	103 / 26.0	102 / 26.6	101 / 27.3	100 / 28.0
160	121 / 18.1	119 / 18.7	117 / 19.3	115 / 20.0	113 / 20.7	112 / 21.1	110 / 22.1	108 / 23.1	107 / 23.6	106 / 24.2	109 / 25.4	103 / 26.0
170	126 / 16.9	124 / 17.3	122 / 17.9	120 / 18.4	118 / 19.0	116 / 19.6	114 / 20.3	112 / 21.1	110 / 22.1	109 / 22.5	107 / 23.6	106 / 24.2
180	131 / 15.7	128 / 16.4	126 / 16.9	124 / 17.3	122 / 17.9	120 / 18.4	118 / 19.0	116 / 19.6	115 / 20.0	113 / 20.7	111 / 21.6	110 / 22.1
190	136 / 14.7	133 / 15.3	130 / 15.9	128 / 16.4	126 / 16.9	124 / 17.3	121 / 18.1	119 / 18.7	117 / 19.3	115 / 20.0	113 / 20.7	112 / 21.1
200	140 / 14.0	138 / 14.4	135 / 14.9	132 / 15.7	129 / 16.2	127 / 16.6	125 / 17.2	122 / 17.9	120 / 18.4	118 / 19.0	156 / 19.6	114 / 20.3

Table A-2 — Register Temperatures and Air Volume for Building with Heat Loss Greater than 350,000 Btu per Hour

Temperature of Bonnet, Deg.	Linear Distance from Bonnet to Register, Feet											
	10	20	30	40	50	60	70	80	90	100	110	120
	Register Temperature, upper figure and Cfm per 1000 Btu per Hour, lower figure											
140	138 / 14.4	136 / 14.8	134 / 15.1	132 / 15.7	130 / 15.9	128 / 16.4	127 / 16.6	126 / 16.9	125 / 17.2	124 / 17.3	123 / 17.6	121 / 18.1
150	148 / 12.8	146 / 13.1	143 / 13.5	141 / 13.9	139 / 14.2	137 / 14.6	135 / 14.9	133 / 15.3	131 / 15.7	129 / 16.2	128 / 16.4	127 / 16.6
160	158 / 11.6	155 / 12.0	152 / 12.3	150 / 12.6	147 / 12.9	145 / 13.2	143 / 13.5	140 / 14.0	138 / 14.4	136 / 14.8	134 / 15.1	132 / 15.7
170	167 / 10.7	164 / 11.0	161 / 11.3	158 / 11.6	155 / 12.0	152 / 12.3	150 / 12.6	147 / 12.9	145 / 13.2	143 / 13.5	140 / 14.0	138 / 14.4
180	177 / 9.9	173 / 10.2	170 / 10.5	167 / 10.7	164 / 11.0	161 / 11.3	158 / 11.6	155 / 12.0	152 / 12.3	150 / 12.6	147 / 12.9	145 / 13.2
190	186 / 9.3	182 / 9.6	178 / 9.8	175 / 10.0	172 / 10.3	168 / 10.7	165 / 10.9	162 / 11.2	159 / 11.5	156 / 11.8	153 / 12.2	151 / 12.4
200	196 / 8.8	191 / 9.0	187 / 9.2	183 / 9.5	179 / 9.8	176 / 9.9	172 / 10.3	169 / 10.6	166 / 10.8	163 / 11.1	160 / 11.4	157 / 11.7

Temperature of Bonnet, Deg.	Linear Distance from Bonnet to Register, Feet											
	130	140	150	160	170	180	190	200	210	220	230	240
	Register Temperature, upper figure and Cfm per 1000 Btu per Hour, lower figure											
140	120 / 18.4	119 / 18.7	118 / 19.0	117 / 19.3	116 / 19.6	115 / 20.0	114 / 20.3	113 / 20.7	112 / 21.1	111 / 21.6	110 / 22.1	109 / 22.5
150	126 / 16.9	124 / 17.3	123 / 17.6	122 / 17.9	120 / 18.4	119 / 18.7	118 / 19.0	117 / 19.3	116 / 19.6	115 / 20.0	114 / 20.3	113 / 20.7
160	130 / 15.9	128 / 16.4	127 / 16.6	126 / 16.9	125 / 17.2	124 / 17.3	123 / 17.6	121 / 18.1	120 / 18.4	119 / 18.7	118 / 19.0	117 / 19.3
170	136 / 14.8	134 / 15.1	132 / 15.7	130 / 15.9	128 / 16.4	127 / 16.6	126 / 16.9	125 / 17.2	124 / 17.3	123 / 17.6	12 / 18.1	120 / 18.4
180	143 / 13.5	140 / 14.0	138 / 14.4	136 / 14.8	134 / 15.1	132 / 15.7	130 / 15.9	128 / 16.4	127 / 16.6	126 / 16.9	125 / 17.2	124 / 17.3
190	148 / 12.8	146 / 13.1	143 / 13.5	141 / 13.9	139 / 14.2	137 / 14.6	135 / 14.9	133 / 15.3	131 / 15.7	129 / 16.2	128 / 16.4	127 / 16.6
200	154 / 12.1	151 / 12.4	149 / 12.7	147 / 12.9	144 / 13.4	142 / 13.7	140 / 14.0	138 / 14.4	135 / 14.9	133 / 15.3	ʹ131 / 15.7	130 / 15.9

(a) Using the *shortest* actual duct length, read downward in the nearest column in Table A-1 until the lower heavy diagonal line is reached, but do not cross the line. Run horizontally to the left to obtain the value of the maximum bonnet temperature in the first column.

For example, for an actual length of duct of 30 ft, the corresponding maximum bonnet-air temperature is 160 deg.

(b) Now, using the *longest* actual duct length, read downwards in the nearest column in Table A-1 until the heavy diagonal dashed line is just crossed. Run horizontally to the left to obtain the value for the minimum bonnet-air temperature in the first column.

For example, for an actual length of duct of 130 ft, the corresponding minimum bonnet-air temperature is 150 deg.

(c) Select as the *design* bonnet temperature any value between the maximum and the minimum, preferably one of the values shown in the first column of Table A-1. For the example shown, either 160 or 150 deg could be selected.

Table A-2 is used similarly. This method of selecting the bonnet-air temperature merely assures that the final bonnet temperature selected will be between 150 and 110 deg, corresponding to temperature rises through the furnace of 85 and 45 deg. Regardless of whether the design bonnet temperature was selected arbitrarily or by the preceding method it must be used thereafter for determining the air-flow rates.

Temperature Drop of Air in Ducts

The individual boxes in Tables A-1 and A-2 show two values: the upper value is the estimated register-air temperature and the lower value is the cfm air-flow rate for each 1000 Btu per hr heat loss. For example, the box corresponding to a bonnet-air temperature of 110 deg and a duct length of 10 ft gives 108 deg as the register-air temperature and 23.1 cfm for each 1000 Btu per hr room heat loss. It is apparent that the values in Tables A-1 and A-2 take into account the temperature drop of the air as it flows down the duct. The tabular values also indicate that the temperature drop is greater when the bonnet-air temperature is higher. For example, for a bonnet-air temperature of 110 deg the temperature drop is 2 deg in 10 ft of duct, whereas for a bonnet-air temperature of 200 deg, the temperature drop is 7 deg in 10 ft of duct. A further examination of Tables A-1 and A-2 shows that the temperature drop is relatively large for distances close to the bonnet and is relatively small for distances away from the bonnet. In other words, the temperature drop of air in the ducts is not a constant value.

The temperature drop data used in setting up Tables A-1 and A-2 were obtained from the basic data given in Bulletin 351 (Univ. of Ill. Eng. Exp. Sta.). The basic curves, shown in Fig. 20 for various duct velocities, apply to uninsulated, galvanized-iron ducts located in an environment of 70 deg. For the purpose of deriving the values in Tables A-1 and A-2 it was necessary to assume average values for duct velocity (800 fpm) and duct size (10 in. diameter), since neither were known at this point in the design procedure. Theoretically, a correction in temperature drop could be made later for actual duct velocities and actual duct diameters used, but in actual practice the correction is not warranted. The register-air temperatures, shown in the upper part of each box in Tables A-1 and A-2, would be slightly high for velocities less than 800 fpm and for ducts smaller than 10 in. in diameter, and hence would give lower values of cfm requirements for

1000 Btu per hr loss than if corrected values were used. However, calculations made for a 6-in. duct carrying air at 400 fpm velocity indicated that the maximum discrepancy was only of the order of 6%. Conversely, the cfm requirements shown in Tables A-1 and A-2 would be on the safe side for ducts larger than 10 in. in diameter and for air velocities greater than 800 fpm. Hence, in spite of the fact that the values in Tables A-1 and A-2 were based upon a specific diameter and velocity, the results obtained from their use should be well within the accuracy required for ordinary calculations.

Register-Air Temperature and Room-Air Requirements

The register-air temperature is obtained by subtracting the temperature drop from the bonnet-air temperature. This register-air temperature in turn establishes the air-flow rate necessary at the register face to offset the heat loss from the space to be heated. The equation for air-flow rate at the register is:

$$\text{Cfm at register} = \frac{\text{Design heat loss, Btu per hr}}{(60)(0.24)(d_{reg})(t_{reg} - t_r)}$$

A study of the data in Tables A-1 and A-2 shows that a larger air-flow rate is needed for those rooms at some distance from the bonnet. For example, for a bonnet-air temperature of 150 deg, a register that is 10 ft from the bonnet requires 12.9 cfm for each 1000 Btu per hr room heat loss, whereas a register which is 100 ft from the bonnet requires 17.9 cfm for each 1000 Btu per hr heat loss. This is an increase of 39% to take into account the heat loss from the duct.

With the exception of duct systems installed in small homes, where the branch ducts are about the same length, considerable errors will be introduced if a constant register-air temperature is assumed to exist at all registers. Such errors in air-flow requirements will necessitate drastic adjustments in balancing after the duct system is installed. After the air-flow rates have been determined for each register, the total cfm may be considered the same as that handled by the blower even though a slight density correction is theoretically necessary.

Available Static Pressure at Bonnet

In the case of small furnace systems the available static pressure external to the furnace-blower combination is considered to be 0.20 in. w.g. This pres-

Table B — Suggested Bonnet Pressures

Total Cfm Through Any One Duct	Suggested Bonnet Pressure, Inches Water Gage
800 to 1,000	0.10
1,000 to 1,200	0.10
1,200 to 1,800	0.10
1,800 to 2,400	0.13
2,400 to 3,500	0.14
3,500 to 5,000	0.15
5,000 to 7,500	0.25
7,500 to 10,000	0.375
10,000 to 12,000	0.50
12,000 to 14,000	0.75

sure of 0.20 in. w.g. is available for the losses in both the warm-air and return-air sides of the system. For large furnaces, such as are used for commercial and industrial buildings, no fixed value of external static pressure is used. In fact, a wide range of values is permissible, and a choice of several blower sizes is possible. Normally, however, the static pressures do not exceed about 1.0 in. of water. Suggested bonnet pressures are given in Table B.

Before a blower can be selected for the given job, it is necessary to estimate two other items and add them to the bonnet pressure:

(a) The pressure loss for the return duct system, and
(b) The pressure loss for internal losses through the furnace casing and the blower housing. This item should be available from the furnace manufacturer. In any case, the manufacturer should be able to recommend the blower size and speed to handle loss items for the bonnet and the return system.

Before the static pressure at the bonnet is finally fixed, some consideration should be given to the pressure required for the throw of air from registers. This becomes of great importance where the throw is greater than about 30 ft, or where the air must be rapidly dispersed outwards.

Throw of Air from Registers

The data of Tuve and Priester (ASHVE Trans., 1944, p. 153) are the most comprehensive of the references investigated. The basic equation proposed by Tuve and Priester was of the form:

$$V_r = \frac{KQ}{X\sqrt{A_e}} \qquad (1)$$

in which,

V_r is the residual maximum velocity, fpm,
K is a constant to be determined by test,
Q is the volume of air discharged, cfm,
X is the throw of air, ft,
A_e is the effective area of outlet, sq ft.

Furthermore,

$$A_e = A_e/C \qquad (2)$$

and

$$V_c = Q/A_c, \quad \text{or} \quad A_c = Q/V_c \qquad (3)$$

in which,

A_c is the core area of register, sq ft,
C is the coefficient of discharge,
V_c is the core velocity, fpm.

By rearranging equations (1), (2), and (3), the following form was derived:

$$V_c = \frac{X^2}{Q}\left(\frac{CV_r^2}{K^2}\right) \qquad (4)$$

Numerical values for $V_r = 50$ fpm, $K = 2.5$, and $C = 0.61$ were inserted in equation (4) and the relationship was simplified to the form:

$$V_c = \frac{X^2(244)}{Q} \qquad (5)$$

By substituting numerical values for X and Q in equation (5), values of V_c were obtained, and these in turn were substituted in equation (3) to give the required core areas, A_c.

Free Areas

In order to obtain free areas, which are commonly given in manufacturers' catalogs, the assumption was made following consultation with the register manufacturers, that the ratio of free area to core area was 0.78. Hence, by the use of a modified version of the Tuve-Priester equation, relationships between throw, volume, and free areas of registers could be obtained.

A residual velocity of 50 fpm at the wall opposite a register might be considered too large, and an average stream velocity of the order of 15 fpm to 20 fpm might be considered as more reasonable. It is possible to modify equation (5) by inserting a value for the residual velocity of the order of 20 fpm. However, it is questionable whether such an extrapolation to velocities which are practically unmeasurable can be justified. The suggestion has been made that the length of the throw should be considered to be some proportion of the width of

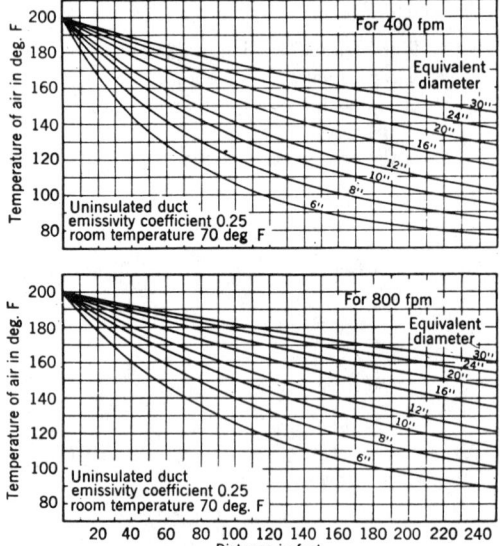

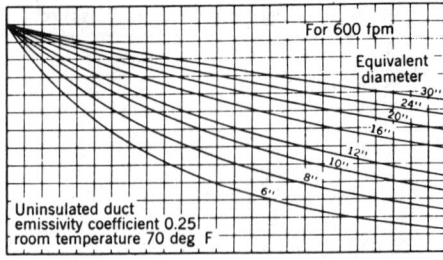

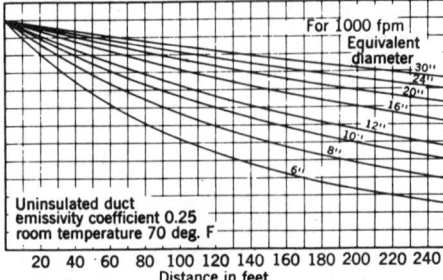

Fig. 20. Temperature drop in ducts for air velocity of (upper left) 400 fpm, (upper right) 600 fpm, (lower left) 800 fpm, and (lower right) 1000 fpm. (U. of Ill. Eng. Exp. Sta. Bul. 351)

the room, say three-quarters of the width of the room, and that the values obtained from equation (5) be applied to this shortened distance. The difficulty is that three-quarters of a room that is 20 ft wide is only 15 ft, leaving 5 ft distance between the far wall and the end of the throw; whereas three-quarters of the distance of a room that is 60 ft wide is 45 ft, leaving a 15 ft distance from the far wall and the end of the throw. In the small room the air may have sufficient momentum to carry on to the far wall, whereas in the larger room the same velocity air may not extend to the far wall. In any case, these are hypothetical objections that have no proof as yet. Hence, unless better evidence exists that some other equation should apply, the modified version of the Tuve-Priester equation is offered for use. The data obtained from equation (7) has been incorporated in a design table, to be shown later.

Pressure Required at Register

For those installations in which relatively long throws of air are required at the register, an available pressure behind the register of 0.10 in. w.g. or more will be encountered. A search of a number of references, including manufacturers' catalogs, indi-

Table C-1 — Capacity and Pressure Loss of Registers

(National Warm Air Heating and Air Conditioning Association)

Capacity, Cfm	Register Size, Inches	Pressure Loss, In. of Water
Up to 59	10 × 4	0.01
	10 × 6	0.01
60–69	10 × 6	0.01
	12 × 4	0.02
70–99	10 × 6	0.02
	12 × 6	0.02
100–119	12 × 6	0.02
	14 × 4	0.02
120–129	12 × 6	0.02
	14 × 6	0.02
130–169	14 × 6	0.02
170–189	14 × 8	0.02

Pressure loss is based on flat face adjustable bar type and does not include stackhead. For other type registers, consult manufacturers' catalogs.

Floor registers: where a velocity of approximately 300 fpm is used, the free area
$$= \frac{\text{cfm} \times 144}{300} \text{ or approximately cfm/2. Assume}$$
a pressure loss of .01 in.

Table C-2 — Register Free Area and Pressure Loss, 22° Deflection of Air
(National Environmental Systems Contractors Association)

Volume, cfm	Distance from Register to Opposite Wall, Ft.							
	Up to 18	19-21	22-24	25-27	28-30	31-34	35-39	40-49
	Free Area, figure at left, and Pressure Loss, in parentheses							
190-209	68 (.02)	49 (.03)	38 (.05)	29 (.08)	24 (.11)	19 (.16)		
210-229	82 (.02)	60 (.03)	45 (.04)	35 (.06)	28 (.10)	23 (.13)		
230-249		71 (.02)	54 (.04)	42 (.06)	34 (.08)	28 (.11)	22 (.17)	
250-269		84 (.02)	63 (.03)	49 (.05)	40 (.07)	32 (.10)	25 (.16)	
270-299		100 (.02)	76 (.03)	60 (.04)	48 (.06)	38 (.09)	30 (.13)	
300-339			96 (.02)	73 (.04)	60 (.05)	48 (.07)	37 (.11)	
340-379			122 (.02)	95 (.03)	78 (.04)	61 (.06)	47 (.09)	33 (.18)
380-419				117 (.02)	94 (.03)	75 (.05)	58 (.08)	39 (.15)
420-459				142 (.02)	113 (.03)	91 (.04)	70 (.06)	47 (.13)
460-499				168 (.02)	135 (.03)	108 (.04)	83 (.06)	56 (.11)
500-539					159 (.02)	126 (.03)	97 (.05)	66 (.10)
540-579					185 (.02)	147 (.03)	113 (.04)	77 (.08)

Volume, cfm	Distance from Register to Opposite Wall, Ft.							
	28-30	31-34	35-39	40-49	50-59	60-69	70-79	80-89
	Free Area, figure at left, and Pressure Loss, in parentheses							
580-619	212 (.02)	169 (.03)	130 (.04)	88 (.08)	51 (.18)			
620-659		192 (.02)	147 (.03)	100 (.07)	59 (.15)			
660-699		217 (.02)	166 (.03)	113 (.06)	75 (.13)			
700-739		243 (.02)	186 (.03)	127 (.06)	85 (.11)			
740-779		271 (.02)	208 (.03)	141 (.05)	94 (.10)			
780-819			230 (.02)	157 (.05)	105 (.09)			
820-859			254 (.02)	173 (.04)	115 (.08)			
860-899			278 (.02)	190 (.04)	126 (.08)			
900-939			304 (.02)	207 (.04)	138 (.07)			
940-979			332 (.02)	226 (.03)	151 (.07)	108 (.12)		
980-1019			360 (.02)	245 (.03)	163 (.06)	118 (.11)		
1020-1059			390 (.02)	265 (.03)	177 (.06)	127 (.11)		
1060-1099				285 (.03)	190 (.96)	137 (.10)		
1100-1139				307 (.03)	204 (.05)	147 (.09)		
1140-1179				330 (.02)	220 (.05)	158 (.09)		
1180-1219				353 (.02)	235 (.04)	169 (.08)		
1220-1259				376 (.02)	251 (.04)	180 (.08)		
1260-1299				401 (.02)	267 (.04)	192 (.07)	136 (.13)	
1300-1339				426 (.02)	284 (.04)	204 (.07)	144 (.13)	
1340-1379				452 (.02)	301 (.04)	217 (.06)	154 (.12)	109 (.22)
1380-1419				479 (.02)	319 (.04)	230 (.06)	163 (.12)	116 (.21)
1420-1459				508 (.02)	338 (.03)	244 (.06)	172 (.11)	122 (.20)
1460-1500				536 (.02)	356 (.03)	256 (.06)	182 (.10)	129 (.19)

If register selected based on distance from register to opposite wall is unsatisfactory on account of size or pressure loss, it is permissable to shift one or more spaces left or right in the table to obtain a more suitable register.

If requirements fall in blank space, select two registers in place of one and divide cfm capacity between the two registers.

Values on the right of line A–A' should not be used in churches, auditoriums, concert halls, and such applications.

Values on right of line B–B' should not be used in residential work, motion picture theaters, court rooms, schools, and such applications.

Pressure loss is based on flat face adjustable bar type and does not include stackhead. For other type registers, see manufacturers' catalogs. For floor registers, where a velocity of approximately 300 fpm is used, the free area $= \dfrac{\text{cfm} \times 144}{300}$ or approximately cfm/2. Assume a pressure loss of .01 in.

Table C-3 — Register Free Area and Pressure Loss, No Deflection of Air
(National Environmental Systems Contractors Association)

Volume, cfm	Up to 21	22–24	25–27	28–30	31–34	35–39	40–49
	Distance from Register to Opposite Wall, Ft.						
	Free Area, figure at left, and Pressure Loss, in parentheses						
190–209	62 (.02)	47 (.03)	37 (.04)	30 (.06)	24 (.09)	18 (.15)	
210–229	76 (.02)	58 (.02)	45 (.04)	36 (.05)	29 (.08)	22 (.12)	
230–249		69 (.02)	53 (.03)	43 (.04)	34 (.07)	26 (.11)	
250–269		81 (.02)	63 (.03)	50 (.04)	40 (.06)	31 (.09)	21 (.18)
270–299		93 (.02)	73 (.02)	59 (.03)	47 (.05)	36 (.08)	24 (.16)
300–339			95 (.02)	77 (.03)	61 (.04)	46 (.06)	32 (.12)
340–379			120 (.02)	97 (.02)	77 (.03)	59 (.05)	40 (.10)
380–419				119 (.02)	95 (.03)	73 (.04)	50 (.08)
420–459				149 (.02)	114 (.03)	88 (.04)	60 (.07)
460–499				171 (.02)	136 (.02)	105 (.03)	71 (.06)
500–539					160 (.02)	123 (.03)	84 (.05)

Volume, cfm	31–34	35–39	40–49	50–59	60–69	70–78	80–89
	Distance from Register to Opposite Wall, Ft.						
	Free Area, figure at left, and Pressure Loss, in parentheses						
540–579	186 (.02)	143 (.02)	97 (.05)	65 (.09)	47 (.17)		
580–619	213 (.02)	164 (.02)	111 (.04)	74 (.08)	54 (.15)		
620–659		180 (.02)	127 (.04)	85 (.07)	61 (.14)		
660–699		210 (.02)	143 (.03)	95 (.07)	69 (.12)	49 (.22)	
700–739		236 (.02)	160 (.03)	107 (.06)	77 (.11)	54 (.21)	
740–779		262 (.02)	179 (.03)	119 (.06)	86 (.10)	61 (.18)	
780–819		291 (.02)	198 (.03)	132 (.05)	95 (.09)	67 (.17)	
820–859		320 (.02)	218 (.03)	145 (.05)	105 (.08)	74 (.15)	
860–899		352 (.02)	240 (.02)	160 (.04)	115 (.08)	81 (.14)	
900–939		385 (.01)	262 (.02)	175 (.04)	126 (.07)	89 (.13)	
940–979		419 (.01)	285 (.02)	190 (.04)	137 (.07)	97 (.12)	
980–1019		455 (.01)	309 (.02)	206 (.04)	148 (.06)	105 (.11)	
1020–1059		493 (.01)	335 (.02)	223 (.03)	161 (.06)	113 (.11)	
1060–1099		530 (.01)	361 (.02)	241 (.03)	173 (.06)	123 (.10)	
1100–1139		571 (.01)	388 (.02)	258 (.03)	186 (.05)	132 (.09)	93 (.17)
1140–1179			416 (.02)	277 (.03)	199 (.05)	141 (.08)	100 (.16)
1180–1219			446 (.02)	297 (.03)	214 (.04)	151 (.08)	107 (.15)
1220–1259			476 (.02)	317 (.03)	228 (.04)	161 (.07)	116 (.14)
1260–1299			507 (.01)	338 (.02)	243 (.04)	172 (.07)	122 (.13)
1300–1339			539 (.01)	359 (.02)	258 (.04)	183 (.07)	130 (.12)
1340–1379				382 (.02)	274 (.04)	195 (.07)	138 (.12)
1380–1419				403 (.02)	290 (.03)	206 (.06)	146 (.12)
1420–1459					308 (.03)	218 (.06)	155 (.10)
1460–1500					324 (.03)	230 (.06)	163 (.10)

If register selected based on distance from register to opposite wall is unsatisfactory on account of size or pressure loss, it is permissable to shift one or more spaces left or right in the table to obtain a more suitable register.

If requirements fall in blank space, select two registers in place of one and divide cfm capacity between the two registers.

Values on the right of line A–A′ should not be used in churches, auditoriums, concert halls, and such applications.

Values on right of line B–B′ should not be used in residential work, motion picture theaters, court rooms, schools, and such applications. Pressure loss is based on flat face adjustable bar type and does not include stackhead. For other type registers, consult manufacturers' catalogs.

For floor registers, where a velocity of approximately 300 fpm is used, the free area $= \dfrac{\text{cfm} \times 144}{300}$ or approximately cfm/2. Assume a pressure loss of .01 in.

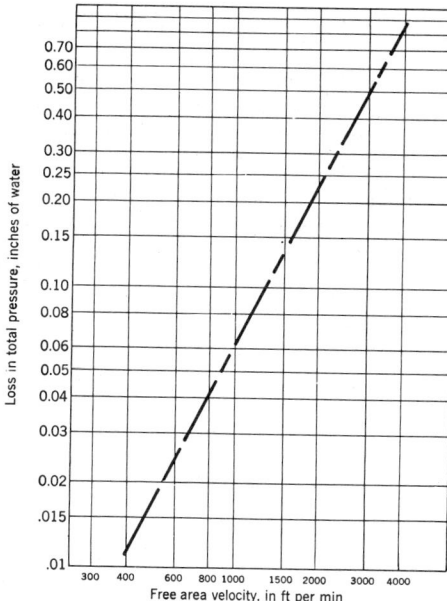

Fig. 21. Total pressure loss of register as affected by free area velocity. (Based on tests with 22 commercial type registers)

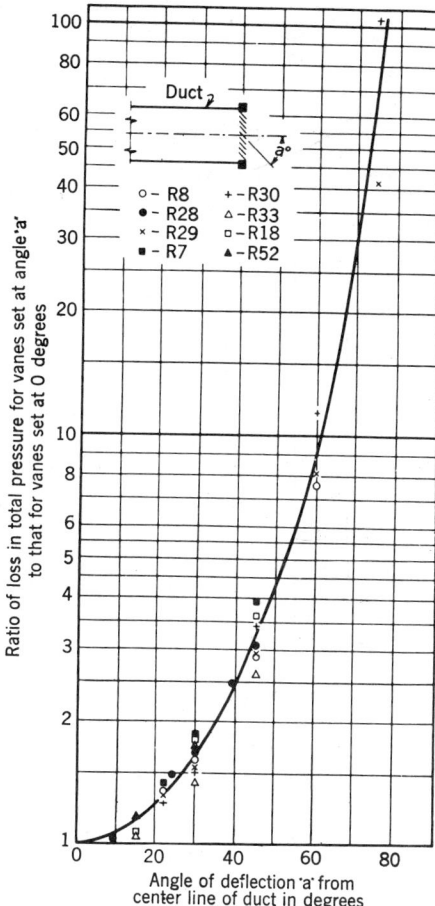

Fig. 22. Increase in pressure loss with different angles of deflection. Symbols represent different experimental runs. (U. of Ill. Eng. Exp. Sta. Bul. 342)

cated the need of a summary table embodying complete performance characteristics of a wide range of register sizes. Wherever such data for a specific model are given in a catalog, they are preferred to any typical results derived for an average register.

In Fig. 21 is shown the relation between the free-area velocity and the total pressure head of the register, based upon tests of 22 commercial types of widely varying design in which *no air deflection* was used. The total pressure ahead of the register included not only the losses due to friction, turbulence, and expansion, but also the velocity head of the air leaving the register.

In Fig. 22 is shown the relation between the deflection angle and pressure loss, based upon tests conducted with eight different commercial types. The pressure loss increased relatively slowly when the deflection angle was increased from 0 deg to about 15 deg, but increased at a greatly accelerated rate for angles larger than 15 deg.

The extensive data given in Bulletin 342 (Univ. of Ill. Eng. Exp. Sta.), part of which are shown in Figs. 21 and 22, as well as the data of Tuve and Priester, were incorporated in Tables C-2 and C-3. The working values in Table C-2 are for 22 deg deflection of air, and are for both register free area

as well as total pressure requirements. These values are shown for various flow rates and for varying distances of the register from the opposite wall. The tabular values do not apply to any specific commercial type of register. Hence, where reliable data can be obtained from a manufacturer for a specific register, such data are preferred to the general information contained in Tables C-2 and C-3. An example of the use of Table C-2 follows:

Given: a room 38 ft wide, one register to handle 300 cfm air flow rate, and a deflection angle of 22 deg to be used.

Solution: From Table C-1, a register having a free area of 37 sq in. and a pressure loss of 0.11 in. will handle the given requirements.

Table D shows average free areas of registers for

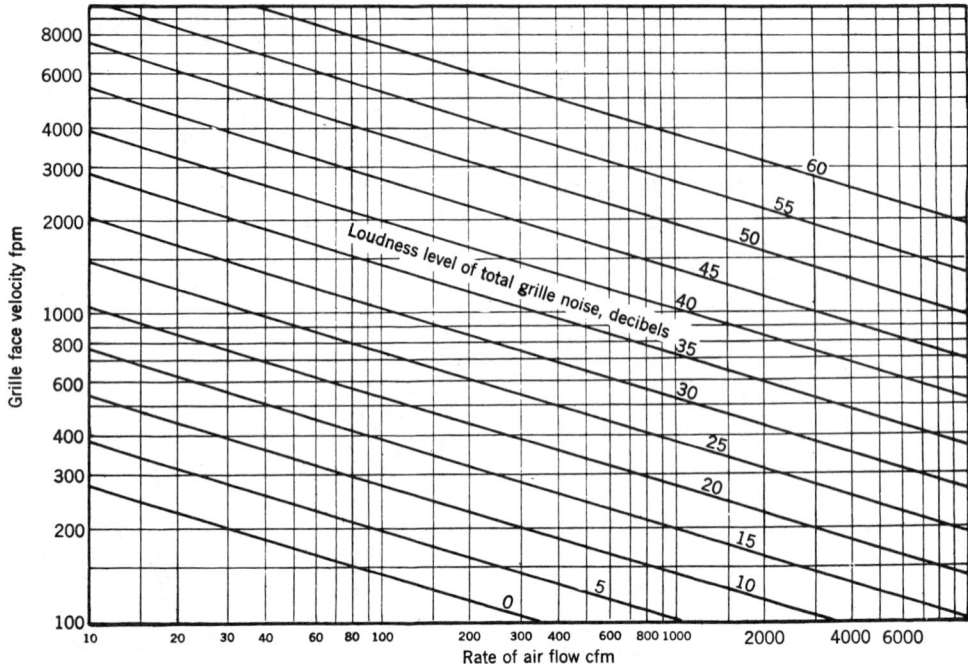

Fig. 23. Average loudness levels for registers with vertical and horizontal fins. (Data from Stewart and Drake)

typical commercial types and should be used only when specific manufacturer's data are not available.

The heavy dividing lines in Tables C-2 and C-3 were included in an attempt to show distinction in sound level. For this purpose the work of Stewart and Drake on sound level was selected from a number of references consulted as the most practical for air conditioning work. The curves of Fig. 23 are based upon the results of tests by Stewart and Drake on registers with horizontal and vertical fins, and supersede those published in the original paper (ASHVE Trans., vol. 43, p. 81).

It is possible to subtract the pressure required at the register from the total available at the bonnet and thereby obtain the pressure available for the duct system alone. For example, (refer to Fig. 24)

Table D — Free Area of Flat Face, Adjustable Bar Type Registers

Nominal Size, In.	Free Area, Sq In., Approximate	Nominal Size, In.	Free Area, Sq In., Approximate
10 × 4	24	14 × 8	75
10 × 5	30	14 × 10	95
10 × 6	38	24 × 4	60
10 × 8	52	24 × 6	91
12 × 4	29	24 × 8	129
12 × 5	37	24 × 10	163
12 × 6	46	24 × 12	200
12 × 8	64	30 × 6	120
12 × 10	82	30 × 8	167
14 × 4	34	30 × 10	207
14 × 5	44	30 × 12	250
14 × 6	54		

This table is not complete and does not list all sizes being manufactured. For specific sizes and free areas, refer to register manufacturers' catalogs.

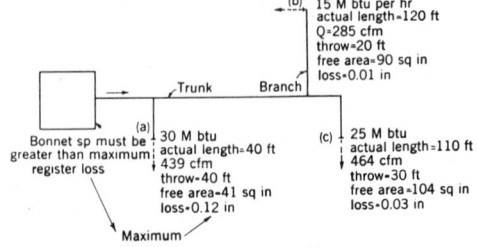

Fig. 24. Maximum register loss and its relation to duct loss and bonnet static pressure.

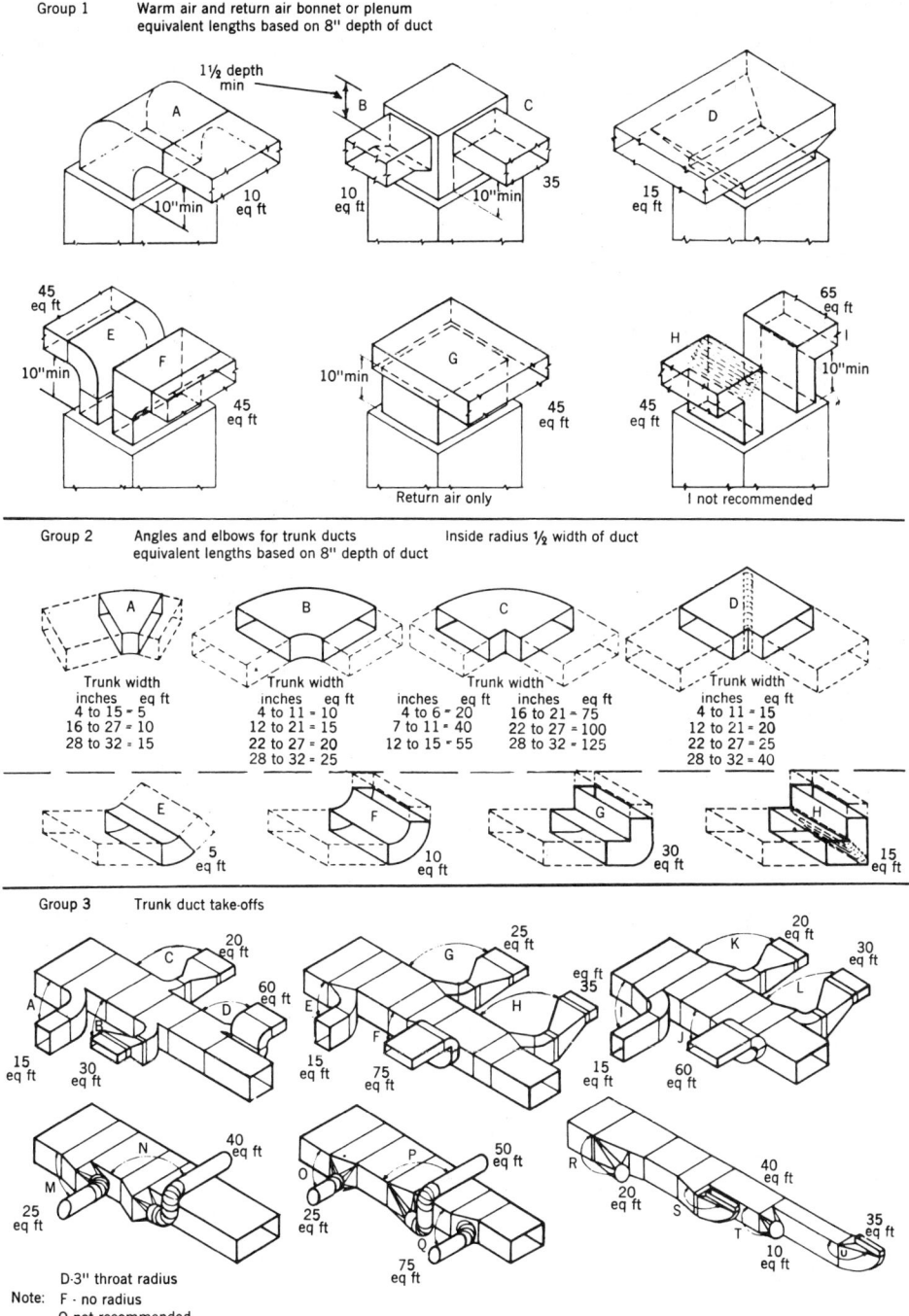

Fig. 25. Equivalent length of duct fittings. (NESCA)

Group 4   Boot fittings · from branch to stack

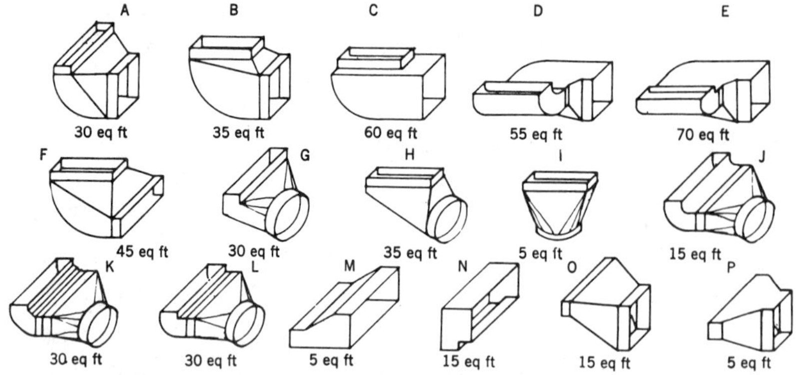

A	B	C	D	E
30 eq ft	35 eq ft	60 eq ft	55 eq ft	70 eq ft

F	G	H	I	J
45 eq ft	30 eq ft	35 eq ft	5 eq ft	15 eq ft

K	L	M	N	O	P
30 eq ft	30 eq ft	5 eq ft	15 eq ft	15 eq ft	5 eq ft

Group 5   Stack angles, elbows, and combinations
A and B have 3" throat radius
D and E have 5" throat radius

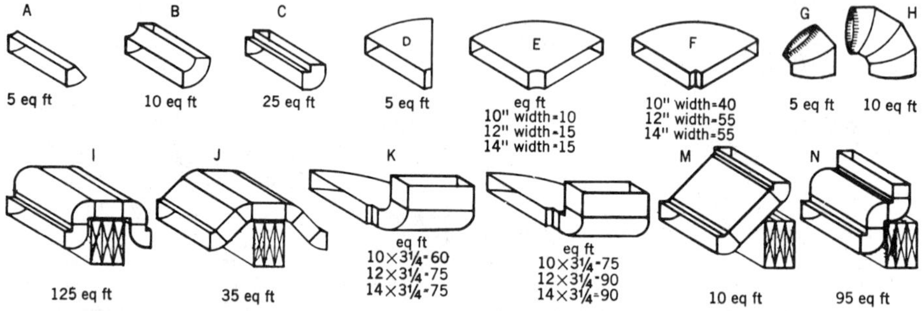

A	B	C	D	E	F	G	H
5 eq ft	10 eq ft	25 eq ft	5 eq ft	eq ft 10" width=10 12" width=15 14" width=15	10" width=40 12" width=55 14" width=55	5 eq ft	10 eq ft

I	J	K	L	M	N
125 eq ft	35 eq ft	eq ft 10×3¼=60 12×3¼=75 14×3¼=75	eq ft 10×3¼=75 12×3¼=90 14×3¼=90	10 eq ft	95 eq ft

Group 6   Return air combinations

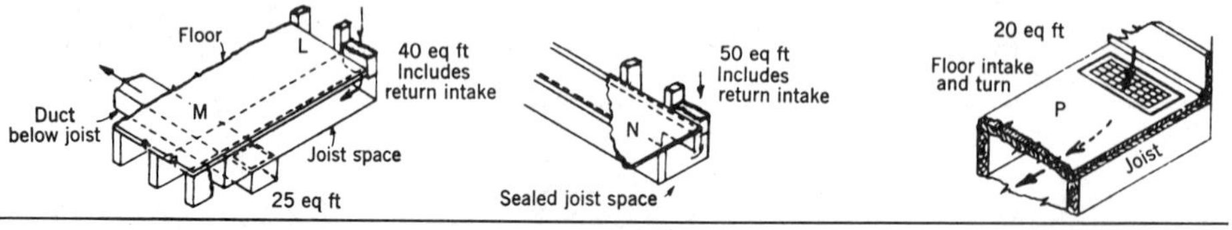

Floor
L
40 eq ft
Includes
return intake
Duct
below joist
M
Joist space
25 eq ft

50 eq ft
Includes
return intake
N
Sealed joist space

20 eq ft
Floor intake
and turn
P
Joist

Group 7   Registers (including losses in stackhead and velocity pressure)

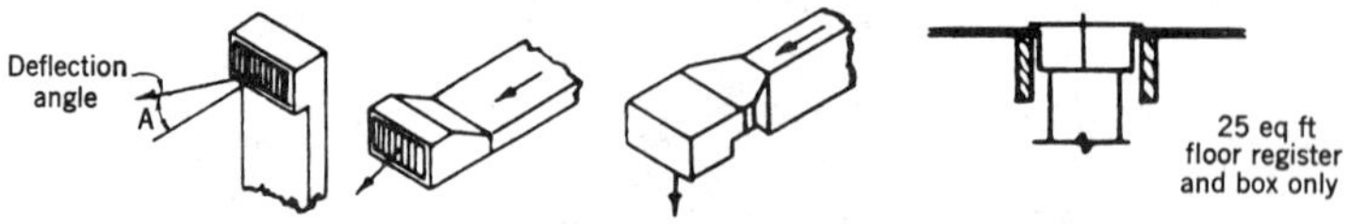

Deflection
angle
A

25 eq ft
floor register
and box only

EQUIVALENT LENGTH FOR REGISTERS

Deflection angle A	0°	15°	22°	30°	45°
Baseboard, high or low sidewall reg	eq ft 35	40	45	60	115
Floor reg with box only	25				

Note:
For equivalent length of
registers, diffusers, and other
types not shown, refer to
mfrs catalog.

For 2-way deflection registers, add the vertical and horizontal deflection angles together
and multiqly by 0.7. Select closest angle 'A' in table 4.

Fig. 26. Equivalent length of duct fittings. (NESCA)

Table E — Sizes and Capacities of Return Intakes

Nominal Size, In.	Approx- imate Capacity, Cfm	Nominal Size, In.	Approx- imate Capacity, Cfm
10 × 4	100	24 × 4	260
10 × 5	125	24 × 5	320
10 × 6	155	24 × 6	390
10 × 8	215	24 × 8	490
12 × 4	120	24 × 10	665
12 × 5	155	24 × 12	795
12 × 6	190	30 × 4	300
12 × 8	260	30 × 5	400
12 × 10	325	30 × 6	485
14 × 4	145	30 × 8	660
14 × 5	180	30 × 10	830
14 × 6	220	30 × 12	1015
14 × 8	305	30 × 14	1160
		30 × 16	1355

This table is not complete and does not list all sizes being manufactured. For other sizes and exact free areas refer to register manufacturers' catalogs.

Free area requirements of baseboard or floor intakes for any capacity can be determined by the following equation:

Required free area, sq. in.

$$= \frac{\text{Required capacity, cfm}}{500} \times 144$$

$$= 0.288 \times \text{cfm}$$

if the pressure loss for register (a) is 0.12 in. w.g., for register (b) is 0.01 in., and for register (c) is 0.03 in., then the maximum pressure loss of the registers is 0.12 in. The required bonnet pressure must be greater than 0.12 in. If, for example, the bonnet pressure selected is 0.25 in., then the available pressure for overcoming resistances in each of the ducts from the bonnet to the register will be for (a) 0.13 in., for (b) 0.24 in., and for (c) 0.22 in.

Return Intake Sizes

As indicated previously, the location of the return intake is not as critical as the locations of the warm-air registers. The air flow from more than one register can be handled through a single return intake. If uniform room-air temperatures are maintained by the proper placement of registers, then the return intakes can be located in the ceiling as well as in the sidewall or in the floor. Usually floor locations are not favored because the intakes will be stepped upon.

Theoretically, the air-flow rate in cfm at the register is greater than that at the return intake because of the less dense air at the supply register. In actual practice the difference of 10 to 15% is not considered material. Furthermore, if the flow rate at the registers is considered to be the same as that at the return intakes, the design of the return duct system will be conservative. Hence, the air flow to be handled by any given return intake is obtained by adding the flow rates for all of the registers in that part of the space. For example, suppose that eight warm-air registers, each handling 150 cfm, are located in a large space, or a total of 1200 cfm of air. Assume that two return intakes are to be located in the space. Then the air-flow requirements of the two return intakes can be any combination that will total 1200 cfm.

A suggested design value for return intakes is 500 fpm through the free area. On this basis the values for Table E were derived.

Sizes of Ducts and Equivalent Lengths of Fittings

Once the total loss in pressure for each run has been determined, the selection of the "friction loss per 100 ft of duct" for each run can be determined by:

$$\frac{\text{Available pressure for each run in inches of water}}{\text{Effective lengths of duct from bonnet to register in units of hundreds of feet}}$$

Now the "effective length of duct" is the summation of the "straight lengths" of duct and the "equivalent lengths" of fittings in each run. For the purpose of determining the equivalent lengths of fittings commonly used in forced-air systems, the values in Figs. 25 and 26 are shown. The values, which are in "equivalent feet of straight duct" are only approximate, and are on the low side for extremely large ducts over 18 in. in diameter, or 18 in. square.

The design procedure from this point forward is no different from that commonly referred to as the "equal friction pressure loss" method. For each branch duct in the system, both the cfm to be handled as well as the friction loss per 100 ft of duct will be known. By means of the common air friction chart for air flow, the size of the ducts can be determined as well as the duct velocity. By means of tables or charts, the round pipe diameters can be converted to rectangular or square pipe sizes that will have the same friction loss characteristics.

STEAM-SUPPLIED UNIT HEATER PERFORMANCE FACTORS

It is accepted practice to rate unit heaters in Btu per hr at an entering-air temperature of 60 F and at a given steady steam pressure within the coil at 2 psig, dry and saturated. Furthermore, heater capacity for such constant entering-air temperature and constant steam pressure machines increases with increase in steam pressure and decreases with increase in entering-air temperature. It is not enough to know capacity change alone to utilize fully and effectively unit heater performance data for space heating. Leaving-air temperature is important, too.

Graphs Accurate Enough

The graph on the page following enables one to correct heater capacity and estimate leaving air temperature, given steam pressure and entering-air temperature. Results are sufficiently accurate for design use and applies to free-delivery conditions with no duct attachments.

Temperature Limits

Tests and actual performance data have proved that leaving-air temperature should be from 50 to 60 F higher than design room temperature. With air entering the steam coil at 60 F, the leaving-air temperature should be from 110 to 120 F; with room air at 70 F, leaving air should be 120 to 130 F. If these criteria are followed, there should be no real concern for air delivery from the units selected.

Too-Buoyant Air

A unit heater with acceptable leaving-air characteristics with steam at 2 psig will deliver excessively heated air if fed high pressure steam. This air, if too high above 120 F, becomes so buoyant that it may never reach the floor to do its heating job.

Rules of Thumb

There are several rules-of-thumb in unit heater selection that may be used to give realistic results: a standard coil unit may be selected for higher steam pressure conditions by the following:

Steam pressure, psig	Leaving-air temperature at standard conditions, F
0 to 25	110 to 120
30 to 125	95 to 105

However, the step-graph may be used to obtain the leaving-air temperature more easily.

Suspended Heaters

When suspended heaters are used with free delivery, higher entering-air temperatures must be used over and above the room temperature at breathing level. Thus, when calculating actual operating capacity when taking air high above the floor, the environmental temperature may vary as much as $\frac{1}{2}$ deg F for each foot of elevation above the breathing line during maximum design conditions, say 0 F outdoors and 70 F indoors.

Example: A heater has catalog performance data as follows: 19,000 Btuh and a leaving-air temperature of 119 F. It is to be supplied with 20 psig steam and will have a new entering-air temperature of 70 F. What is the new capacity and new leaving-air temperature?

Temperature difference from catalog is 119 — 60 = 59 F. Start at 20 psig on "constant steam pressure" scale, extreme right on graph, and move to the left until the entering-air temperature line of 70 F is intersected. Read capacity factor 1.18 below. Locate 59 F sloping line belonging to "leaving minus entering-air" line family. By coincidence it happens to fall on the same point. Thus, the air temperature rise is 70 F. New capacity is 19,000 × 1.18 = 22,420 Btuh and the leaving-air temperature is 70 + 70 = 140 F. Heater is acceptable under these conditions.

Example: Same heater in previous example is to deliver 160 F air for process use. What is the operating steam pressure under new conditions?

Air temperature rise must be 160 — 70 = 90 F. Enter graph at 90 F on "air temperature rise" scale on extreme left of graph. Move horizontally until striking the 59 F line. Then move vertically up the capacity factor line of 1.52 to the entering design line of 70 F. Now read across to steam pressure of 67.5 psig. New capacity is 19,000 × 1.52 = 28,880 Btuh.

Steam-Supplied Unit Heater Performance Factors

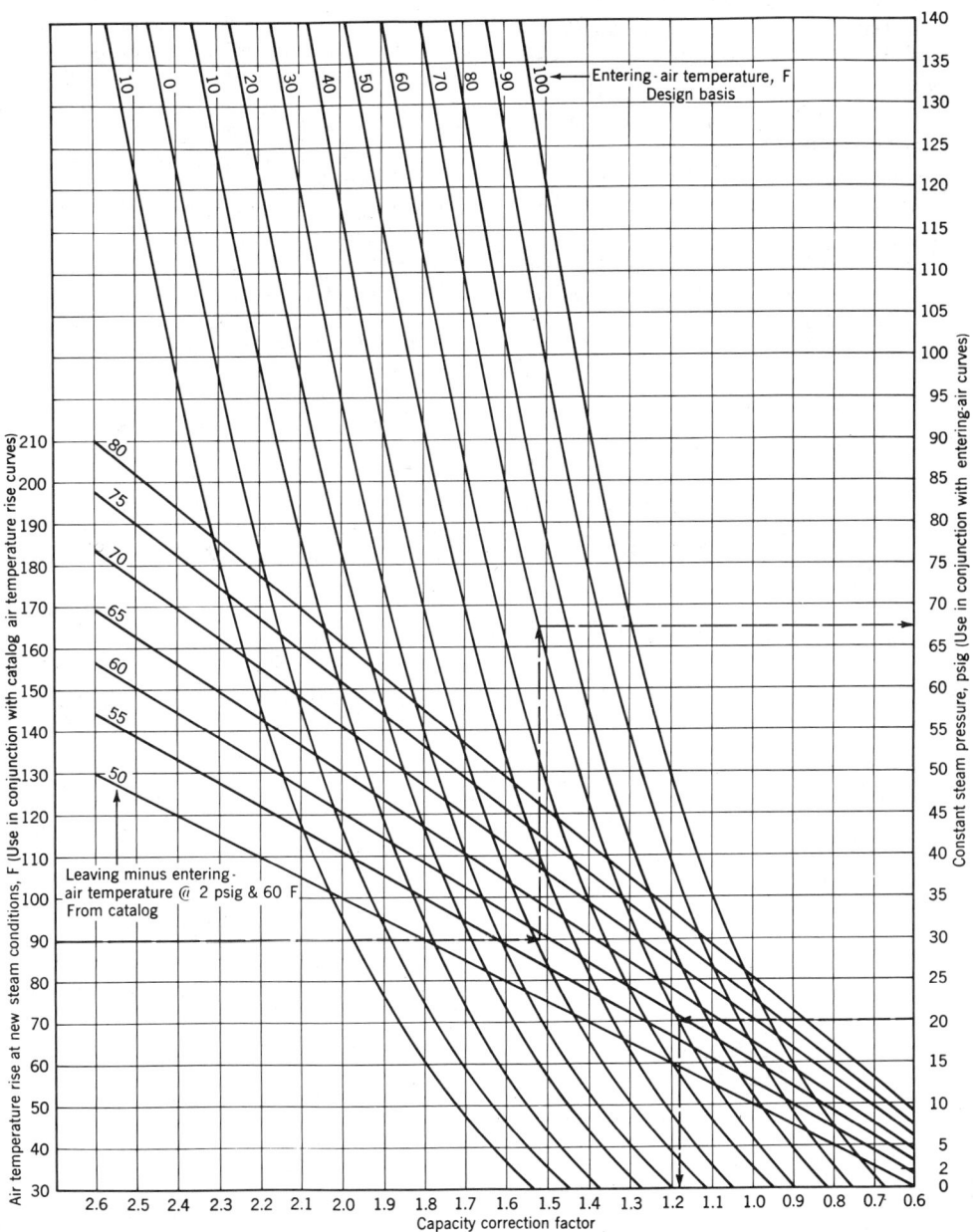

Dashed lines refer to illustrative examples on preceding page.

GAS-FIRED RADIANT (INFRARED) HEATERS

A relatively new method for the space or spot heating of industrial and commercial buildings is by means of overhead gas-burning infrared radiant heaters. These are listed under the *Room Heater* classification of the AGA *Directory of Approved Appliances*.

Overhead radiant heaters have proved successful for heating of warehouses, garage, industrial plants, shopping centers, showrooms, bus stations and repair shops, airport hangars and auditoriums.

Vented Metallic Radiant Overhead Panels. These units are available with inputs from 62,500 to 125,000 Btuh. (Vented heaters tested under ANSI Z21.11 are required to meet a standard thermal efficiency test. Part of this output may be in the form of convected heat.) Metal temperatures are approximately 750 F. Equipment manufacturers claim that heat losses in industrial buildings above the level at which radiant heaters are installed (above 15 ft in high-bay buildings) can be neglected unless there is an overhead wet sprinkler system.

Unvented Ceramic or Metallic Overhead Radiant Heaters. These are available up to 50,000 Btu input. Ceramic surfaces reach a temperature of 1600 to 1650 F. Around the burner assembly is a bright aluminum reflector which may be straight sided or parabolic, and has the purpose of better defining the area receiving radiation. The units may be used singly or in multiple assembly. About 50% of the heat is claimed to be radiated downward, the remainder being transferred by convection and undirected low level radiation from warmed sides of the unit. The latent heat (about 10%) is not recovered.

Since these heaters are usually used in high-bay buildings and are installed well above breathing level, vents are usually provided in the roof to carry off combustion products, the entire roof area serving as a large vent hood.

A minimum 50-sq in. roof vent opening is generally considered desirable for each 100,000 Btuh input. This area may change depending upon local or state codes.

Figure 1 illustrates the principal features of both vented and unvented types of infrared heaters.

Fig. 1. Two types of gas-fired radiant (infrared) heaters. Below, unvented ceramic type; at right, vented metallic type. Mounted overhead, they can be used in systems designed for partial or full area heating.

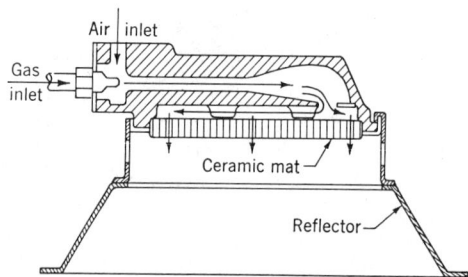

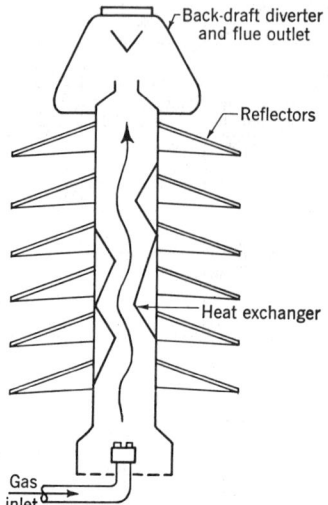

Application

When applying gas-fired infrared units for comfort heating, manufacturer's recommendations as to minimum mounting height and spacing should be followed. These recommendations are based on the pattern of radiant output from the units, and empirical information relating maximum radiation intensity and comfort.

There are basically two ways in which infrared radiant heaters can be employed.

Partial Area Heating or Spot Heating. When only partial areas or spots in a building are heated by overhead radiant panels, the actual ambient temperature in the areas will be lower than when the entire building is heated.

On spot or partial heating jobs, the secondary convection plays a lesser part in the heating and primary radiation becomes the controlling factor. Care must be taken in the placement of the radiant panels to be sure that all persons are "seen" by the panels and, equally important, that they receive radiation on both sides of their bodies.

Area or spot heating should be approached cautiously. There is an indication that approximately 200 to 250 Btuh per sq ft of area are required. As mentioned previously, manufacturers' charts are available to show amount of radiation for various mounting heights. Heat should be provided from two or three directions to the spot being heated. Heaters should be mounted as low as possible to give the spot the greatest infrared intensity.

Full Area Heating Where the entire building area is heated by overhead radiant panels, the actual comfort resulting is due to several factors.

Direct radiation from the panels passes directly through air and is absorbed by the floor and machinery in the building where it is converted to heat and convection is established.

Following are recommendations of a manufacturer of the Schwank infrared heater:

Each burner has a comparatively low output, so many heaters may be required for a large building, each heating a small floor area. The height at which heaters may be mounted depends on the type of building. Heaters may be located high provided reflectors are used to prevent waste of heat on the upper part of outside walls. Heaters near outer walls should be kept as low as possible, and in some cases should be tilted inward.

See Fig. 2.

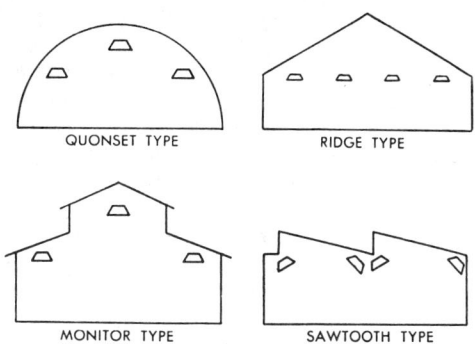

Fig. 2. Building construction determines mounting arrangements.

It is important to avoid high concentration of heat on the heads of occupants, so heaters at low levels should be of smaller size (two or three rayheads), whereas heaters at a 20-40-ft elevation can be larger units (four or more rayheads).

Total installed capacity must be sufficient to compensate for heat losses from the area heated, but since these losses are less than for conventional heating, the input to be installed is correspondingly reduced. In buildings of normal ceiling height, average construction, and for o F design outside temperature, one rayhead for each 100 sq ft of floor space generally does a good job. This gives 120 Btu input per sq ft of floor area. For buildings with high ceilings, leaky walls, no insulation, and similar conditions of high heat loss, the number of heaters to be installed must be increased, while for well insulated buildings and higher design temperatures, the number of units can be decreased.

The highest concentration of heat should be around the periphery of the heated area. For heating spots, or limited areas of large rooms, the capacity installed per square foot must be at least double or triple the amount used when the entire area is heated. In such cases, a number of heaters should be installed around the periphery of the space to be heated. They should have parabolic reflectors and should be inclined toward the center of the area.

While high buildings have higher heat losses per square foot of floor area and require more input than if they had lower ceilings, the input required is not increased to the extent it would be with convection heating, and high bay buildings show the greatest savings over convection systems.

SIZING OF STEAM TRAPS AT AIR HEATING COILS

A heating coil is usually equipped with a throttling control valve. In mild weather, or most of the time, the control valve throttles so that, except for the vacuum breakers installed, the coil would frequently be under vacuum and the condensate would not be discharged.

The vacuum breaker equalizes the pressure between coil and return piping (whether the system be a vacuum or an "open" system, with returns under atmospheric pressure).

When vacuum breakers open and the pressure is equalized, the vertical drop (or static "head of water column") between coil outlet and trap inlet supplies the only force available to discharge condensate through the trap.

Thus, the criterion for sizing traps that drain coils

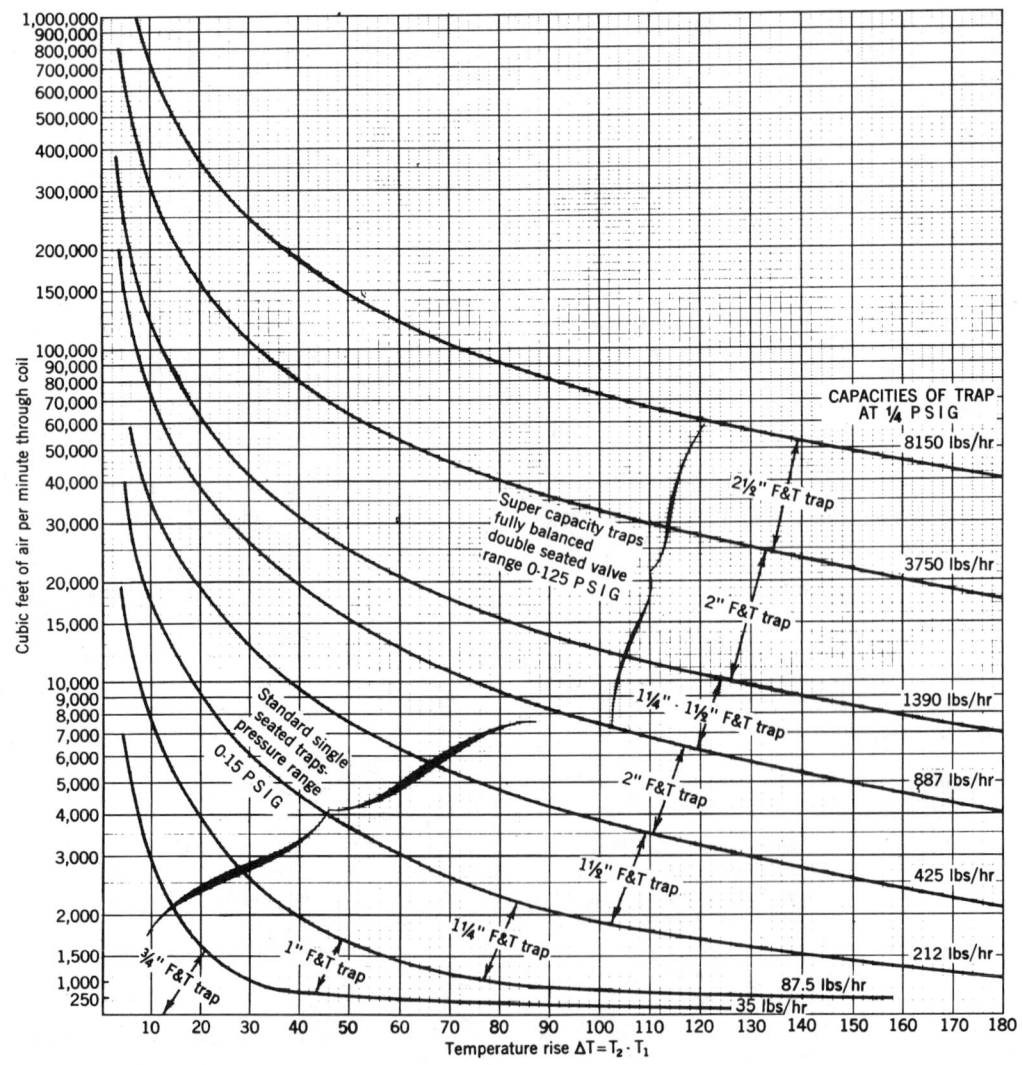

Heavy-duty, double ported float-thermostatic steam traps. Unit at extreme left is suitable for 0-125 psi; unit at right is for 0-175 psi; center illustration is end view. Drawing courtesy of Sarco Co., Inc.

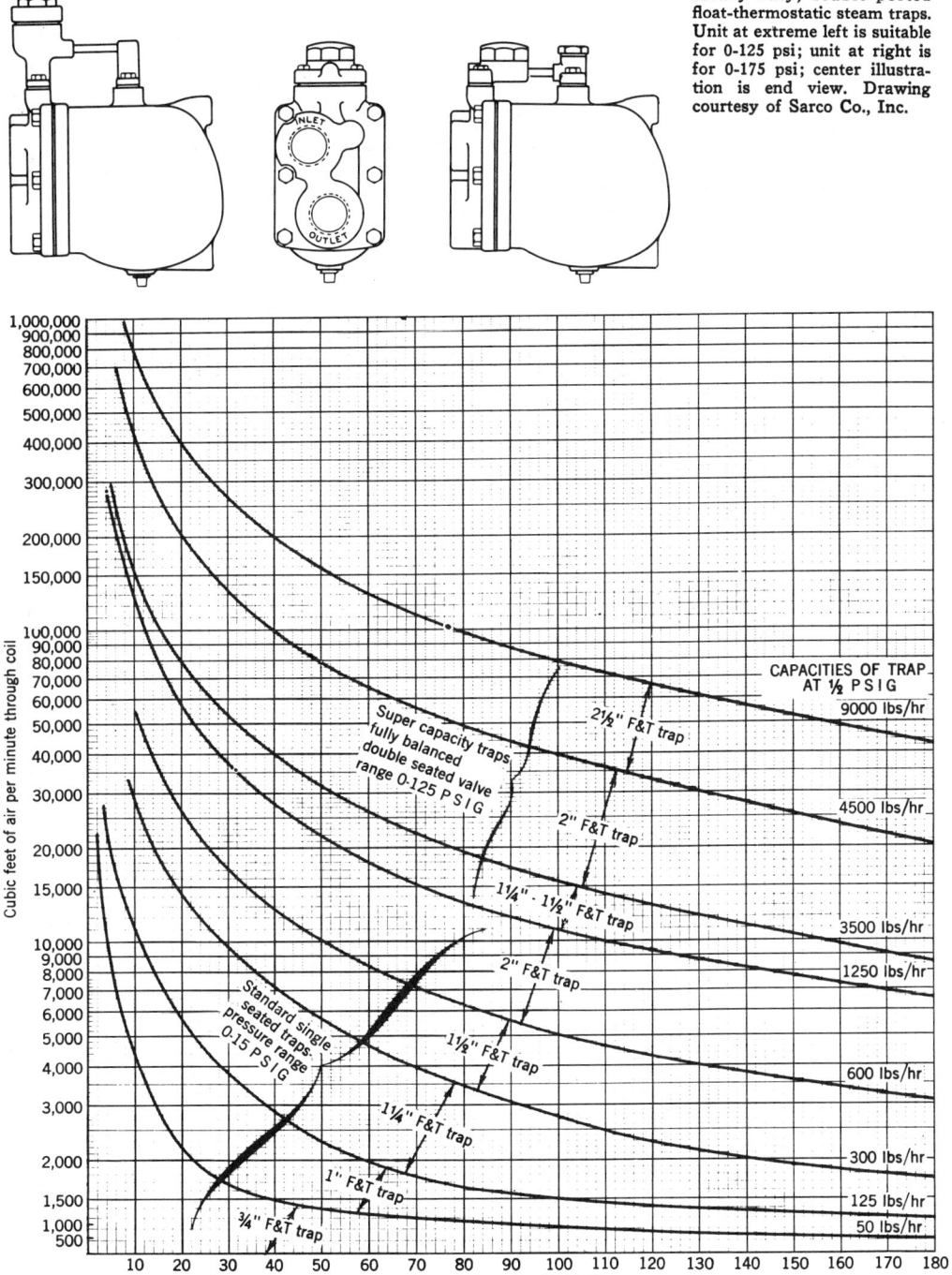

with throttling control is not the steam pressure in the system ahead of the regulator, but this "head of water column."

The four sizing graphs apply to ball float traps with built-in thermostatic air bypass. Each curve shows the maximum quantity of condensate recommended (with a safety factor of 2) for a trap of the type and size shown, according to original standard Steam Heating Equipment Manufacturers Association (SHEMA) ratings. There is one graph for each of four pressure drops through the trap: ¼, ½, 1, and 2 psi. (Note: inches of water column × 0.03613 = psi.) On each graph, air flow rate through coils is plotted against temperature rise of air through the coil.

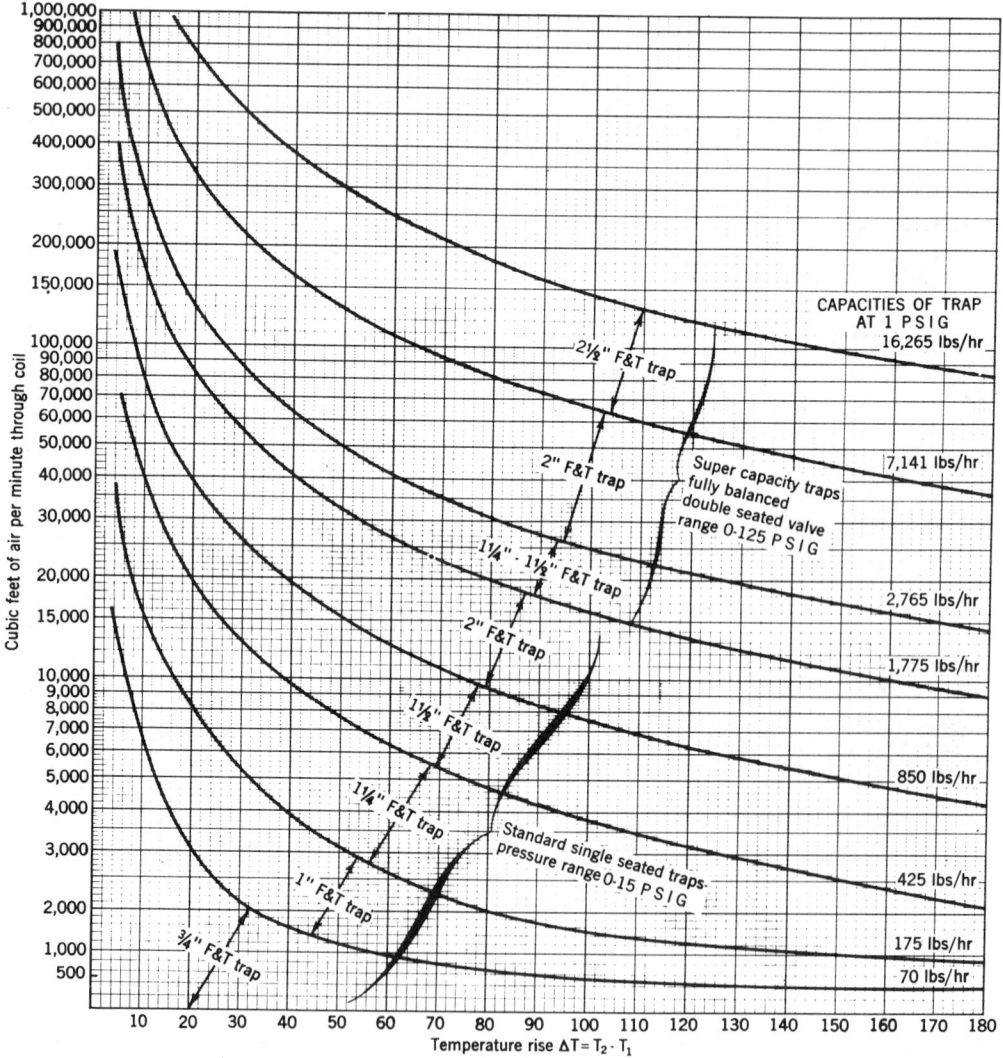

Example: A 12-inch drop exists between coil outlet and trap inlet, corresponding to 0.435 psi. Which chart to use for sizing? If it is decided to size the trap using the chart for ¼ psi with a safety factor of 2, this will give a total safety factor of (0.435/0.250) 2 = 3.5, which is ample, since the volume of steam is considerably less when valve throttles than when control is fully open. Hence, the ¼-psi chart suffices.

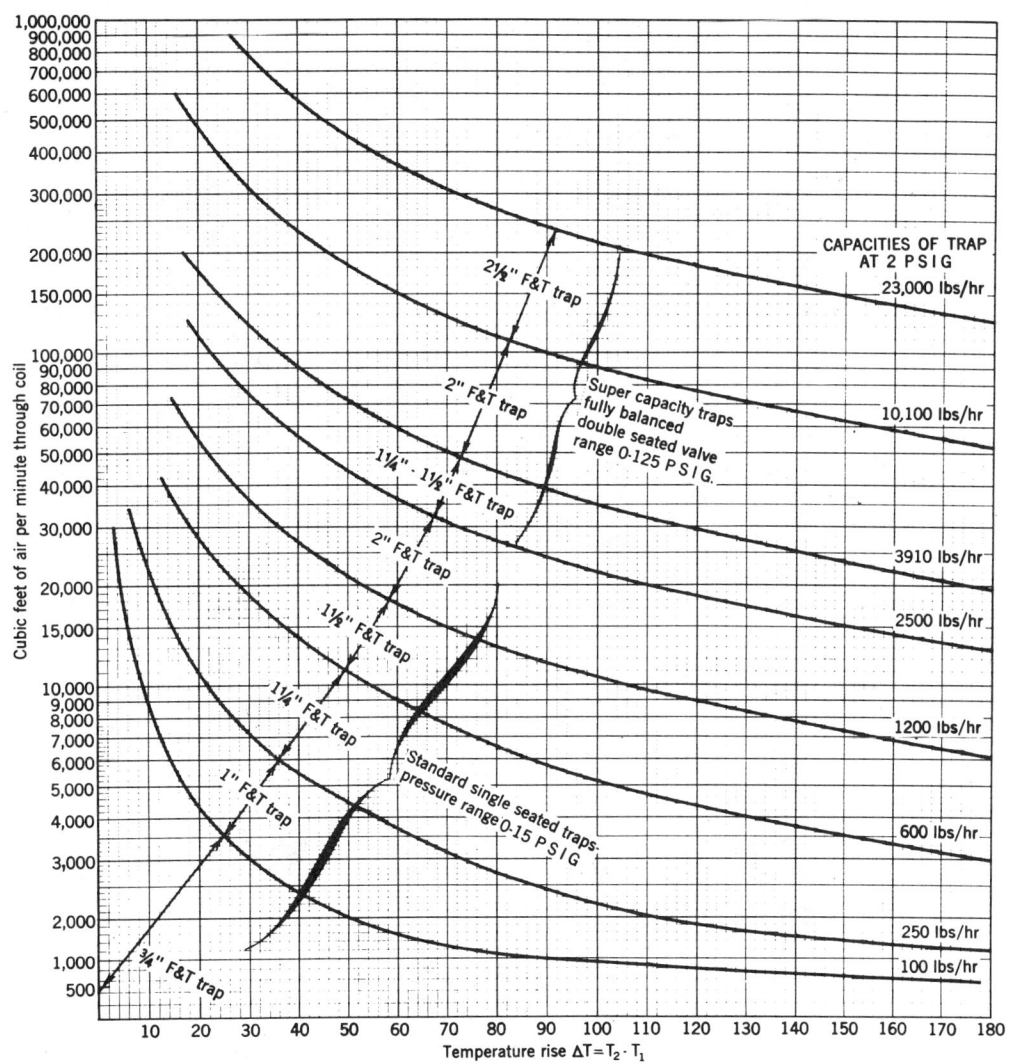

UNIT HEATERS AND OTHER GAS-FIRED AIR HEATERS

Unit heaters described here are self-contained, automatically controlled, vented, gas-burning devices designed for the heating of non-residential space. They may be divided into two groups in two ways; first, according to the static pressure against which they will deliver heated air; and second, according to the method of location.

Types of Unit Heaters

Type	Input Range, MBtu
Suspended Unit Heaters:	
Propeller Fan	25 to 300
Blower type (with or without duct)	25 to 300
Heavy-duty suspended blower	250 to 5,000
Floor or Platform Mounted Unit Heaters:	
Vertical blower type	60 to 750
Heavy-duty floor-mounted blowers	250 to 5,000

Application

Propeller Fan Type. Theoretically, the unit heater with the propeller fan should not operate against any static pressure but should have full opening of the unit heater face. When it becomes necessary to use ductwork with a unit heater in order to direct heat at a desired spot location, it is mandatory that the unit heater be equipped with a blower.

Propeller type unit heaters are, in the main, designed for suspended installation.

Placement of units is governed by the individual heating job. Usually, however, units are spotted in one or two ways:

—placed so they circulate the heated air along the outside walls, or

—located on outside walls so as to pick up colder air from outside walls, heat it and force it toward the center of the area or where heat is most needed.

Figure 1 is a typical suspended gas-fired propeller type unit heater installation.

Suspended-Blower Type. Suspended blower type unit heaters are used much the same as propeller units. Because of the blower type fans, however, they handle larger quantities of air and can be used with or without ductwork. As a result, many special problems in plant heating, such as automatic temperature control of large warehouses, door draft tempering, fog elimination, and many others including some low temperature drying, can be easily accomplished with these units.

Blower units are particularly effective for heating buildings where it is necessary to suspend the units more than 12 ft above the floor. Also, when cold air is to be picked up off the floor to overcome drafty conditions, it is necessary only to install return air supplies to blower type units to accomplish this task. Heat distribution ducts may also be attached to the blower type units and several rooms heated by the one unit. See Fig. 2.

Suspended Heavy-Duty Units. For heavy-duty

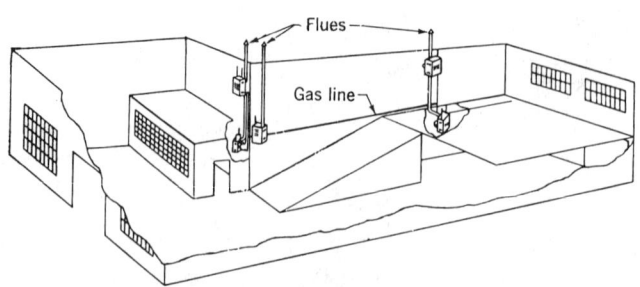

Fig. 1. Typical installation of suspended propeller-type unit heaters.

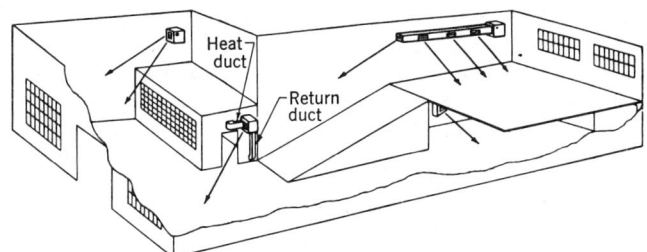

Fig. 2. The same industrial area could be served by suspended blower-type unit heaters in this manner. Gas lines
and flues are not shown.

work, large blower type units ranging in input capacities from 250,000 Btuh to 5,000,000 Btuh and over
are obtainable. For airplane hangars, mills, foundries, factories, warehouses and other large-space installation, where it is necessary to suspend the units
because of codes or to save floor space, these units
effectively blanket large areas at great distances
from the unit with automatically controlled heat.
By use of high-velocity nozzles and specially designed plenums, heat may be directed down and in
any direction so that particular areas may be given
more attention than others.

Floor-Mounted Vertical Blower Units. Vertical
floor type gas heaters have a wide application in the
industrial heating field. Another important usage is
that of supplying heat to office suites and special
enclosures when it is found desirable to have these
separate from the main heating system. This type
equipment is desirable because of the ease with
which the air is filtered. In candy plants, dress
factories, and others of the type where great care
is exercised in the cleanliness of the product, these
units are especially adapted. For larger areas, such
as barracks or large assembling rooms, they are
usually spotted around the outside walls and deliver
the heated air toward the center of the building.
Recirculated air is drawn from the floor at the base
of the unit. In this manner, effective zone control is
obtained.

Floor-Mounted Heavy-Duty Units. Heavy-duty
floor-mounted unit heaters are available:

—without connected ductwork but with high velocity nozzles and

—for application with ductwork.

In the first application, the units are mounted on
the floor and the heated air is delivered around the
8-ft level. Delivery nozzles are rotatable and usually both horizontal and vertical louvers may be
obtained. Placement of the unit is optional, being
governed by the heating requirements of the building as well as by the foundamentals of good heating
practice.

As a general rule, units are either placed around
the outside walls and discharge toward the center,
or wherever possible units are run down the center
of the space and discharge in both directions. In
this latter method, it is many times possible to cut
down on the number of units needed to heat a given
space. Effective delivery of heat with high temperature rise air and high velocity nozzles of this kind
ranges from approximately 110 to 250 ft, depending
upon conditions under which the units are installed.
The blowers usually driven by a single motor, but
some designs are sectionalized so as to provide a
wide variety of output capacities and combinations.

In the second instance, a duct distribution system
is connected directly to the outlet of the unit. Units
of this kind are well suited to extremely large-space
heating, particularly in leaky, draughty industrial
buildings where it is desirable to have reasonably
high leaving temperatures in order to have sufficient
heat at long distances from the duct opening.

Sizing Unit Heaters

Unit heater capacity should be at least 10%
greater than the heat loss of the structure to provide a margin of safety during severe weather and
some pickup capacity. Allowing an equipment efficiency of 80%, Btu input would be 1.4 times the
heat loss. Where temperature is set back at night,

or for intermittent heating of cold loading areas, a greater allowance should be made for pick-up.

When unit heaters are planned for industrial occupancies, there may be included in the heat loss such items as air for process use, heating cold stock and vehicles entering the building and other such heating requirements not normally encountered in residential and commercial building heating.

In the calculation of heat loss where unit heaters are to be used, consideration should be given to the type of occupancy there will be in the building. For example, where a working temperature of 60 F might be required in a shop, there is no point in designing units to maintain a temperature of 80 F. This would result not only in a lot of unnecessary *on-off* operation of the units but would not give as steady a heat or circulation as it would if the units were designed for the specific requirements.

Duct Furnace. Normally installed in distribution ducts of air conditioning systems to supply warm air for heating, duct furnaces are tested to insure that temperatures of combustible materials will not be excessive at points 6 inches from the sides and top, and on the floor at clearances of zero or 2 to 6 inches as specified by the manufacturer. Where floor clearance is greater than zero inches, a non-combustible stand or legs must be provided by the manufacturer.

Duct furnaces may be designed and approved only for suspension, in which case both the installation instructions and a permanent plate attached to the unit must state the required floor clearance. Where draft hoods extend above the top of the casing, a 6-inch clearance must be maintained above the top of the draft hood when designed for horizontal connection and above a standard 90° elbow when designed for a vertical outlet.

Circulating air should not be taken from the same enclosure in which the furnace is located.

Duct furnaces, when used in conjunction with a refrigeration system, should not be located downstream from the refrigeration coil unless specifically approved for such installation.

Enclosed Furnaces. These widely used furnaces are designed primarily for use in such places of public assembly as school rooms, churches and auditoriums. They incorporate an integral total enclosure around the furnace and they use only air from outdoors for combustion.

Typical enclosed furnaces provide automatic room ventilation, blending outside air and recirculation. Such furnaces include insulation for prevention of condensation and to insure quiet operation; they can be used against static pressures up to 0.50 inch w.c.; they normally operate with air temperature rise of 70 to 105 F.

Enclosed furnaces are approved for installation with 0-inch clearance to combustible materials on the floor and one vertical side and, at the option of the manufacturer, may be approved additionally with 0-inch clearance at one or both ends. Surfaces exposed to air in the heated space may not exceed 60 F above room temperature.

Installation of Unit Heaters

Suspended units should be placed as low as permitted by headroom requirements. High mounting increases the difficulty of forcing warm air down to the working plane, and increases the floor-ceiling differential.

An exception to this recommendation concerns unit heaters located in garages for more than 3 cars or in airplane hangars—these should be installed at least 8 ft above the floor.

Suspended type unit heaters should be safely and adequately supported with due consideration given to their weight and vibration characteristics. See Fig. 3.

AGA-approved gas-fired unit heaters should not be located closer to combustible construction than 6 inches above the top when heater has an internal draft hood; not less than 18 inches at the sides, and not less than 12 inches at the bottom, except in cases where the heater is approved for installation with reduced clearances.

Direction of Air Stream

In general, the direction of the air stream leaving a unit heater should be toward the location where the most uniform temperature is desired. This is usually that portion of any space which is most frequently occupied by people at rest or doing work requiring no motion.

Exposed Wall. If unit heaters are directed so that the high temperature air strikes an exposed wall, the temperature on the inside of the wall will be higher than the design temperature along this exposure and will result in greater heat losses to the

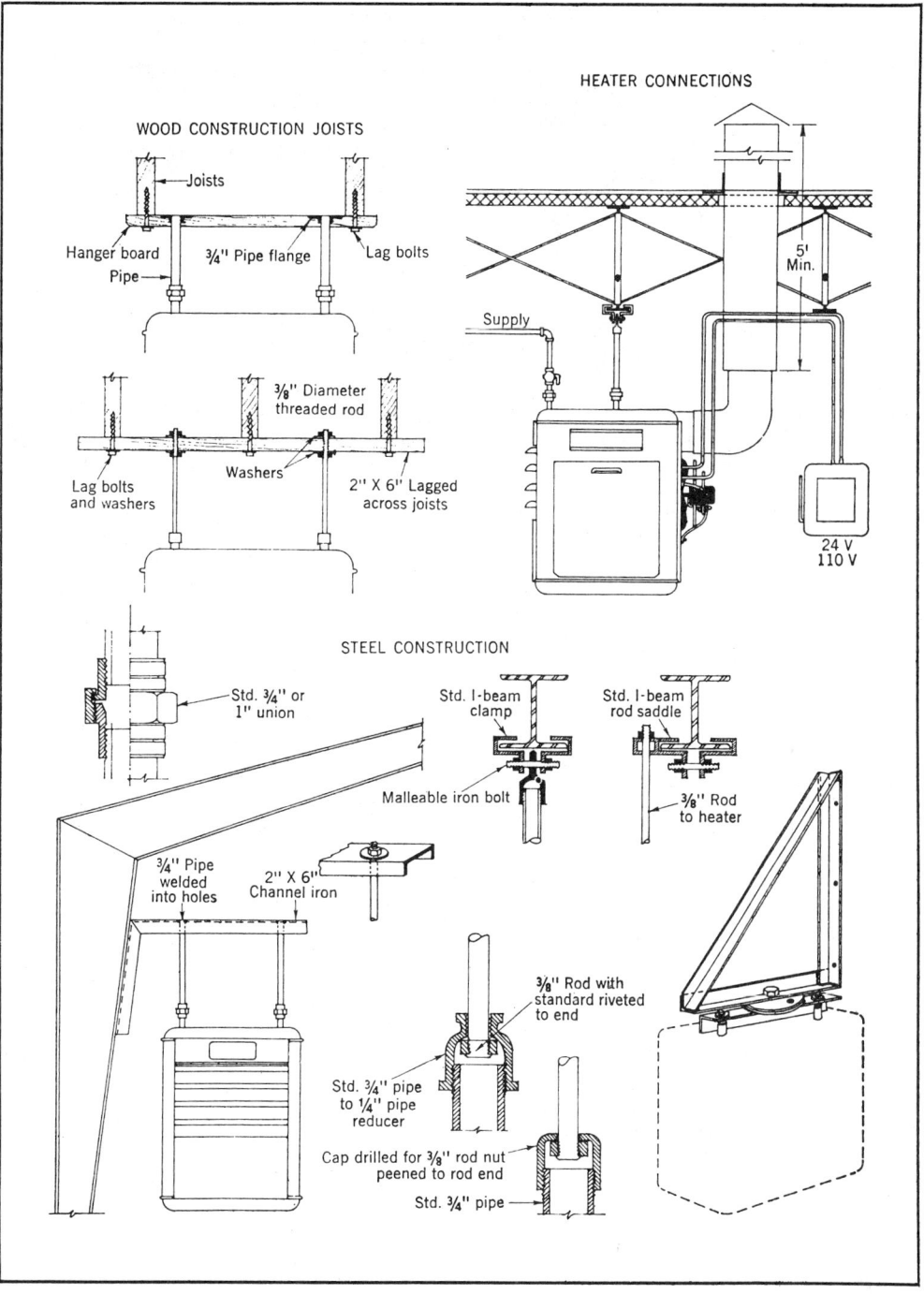

WOOD CONSTRUCTION JOISTS

Joists

Hanger board
Pipe
3/4" Pipe flange
Lag bolts

3/8" Diameter threaded rod

Lag bolts and washers
Washers
2" X 6" Lagged across joists

HEATER CONNECTIONS

5' Min.

Supply

24 V
110 V

STEEL CONSTRUCTION

Std. 3/4" or 1" union

Std. I-beam clamp

Std. I-beam rod saddle

Malleable iron bolt

3/8" Rod to heater

3/4" Pipe welded into holes

2" X 6" Channel iron

3/8" Rod with standard riveted to end

Std. 3/4" pipe to 1/4" pipe reducer

Cap drilled for 3/8" rod nut peened to rod end

Std. 3/4" pipe

Fig. 3. Details of gas-fired unit heater installation, courtesy of ITT Environmental Systems Co.

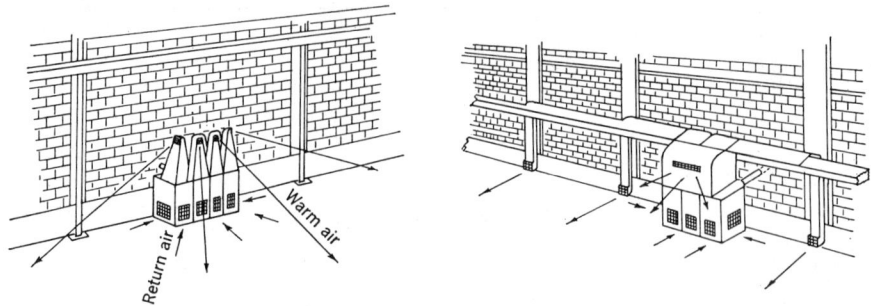

Fig. 4. Heavy-duty unit heater applications show (at left) discharge through
high-velocity nozzles, and (at right) duct-type installation.

outside and consequently, increase fuel consumption. This is especially true if the exposed area consists mainly of glass or some other material with a high transmission factor. The air stream from a unit heater should not be directed against any exposed wall or other cold surface close to the unit.

Circulation of Air. The more complete the circulation of air through the entire space being heated, the more uniform will be the temperature throughout. Where more than one unit is installed in a large space, the units may be placed so that the air will be circulated completely around the inside walls of the building. The air discharged from a unit is projected toward the rear of the next unit, which recieves the air after it is reduced in temperature and projects the reheated air in the same direction.

To maintain good circulation in small rooms, requiring only one unit heater, it is generally good practice to place the unit heater near an unexposed wall with the air stream from the unit directed toward the exposed wall surfaces, but not playing directly on them.

Obstructions. The direction of the air stream from unit heaters will be materially diverted by any obstruction within the building itself. Location of partition walls not extending to the ceiling, displays, piles of material, or other obstructions should be carefully considered. The direction of the air stream leaving the unit should be such that not too great a quantity of the heated air will be diverted by such obstructions.

Occupants. In many places favorable to a unit heater installation, certain portions are set aside for workmen or office workers. Such spaces are usually located near the windows to obtain the best lighting conditions. Most uniform temperatures should be maintained where such occupants work. Careful consideration, however, should be given the direction of the air stream from the unit heaters so that occupants are not subjected to a direct blast of heated air from a unit, as this will result in considerable discomfort.

Location of Thermostats. Thermostats to control the operation of gas-fired unit heaters should be located on an unexposed wall or column at about eye level. If no unexposed walls, columns or other suitable locations for thermostats are available they may be mounted on exposed walls, provided they are mounted on a bracket, or carefully insulated from the wall.

Thermostats should not be located so that a blast of warm air from the unit heater will strike them.

Each unit may be operated by an individual thermostat or, by means of electric relays, any number of units may be operated from one thermostat. It is usually advisable to keep the number of units controlled by one thermostat down to a minimum. Sun effect, wind pressure, lights, shifting of persons from one part of a room to another, can all affect temperature changes within a space, and if all the unit heaters in that space are controlled by one thermostat, it is difficult to keep everybody in the space comfortable.

CHECK LIST FOR CONSERVATION OF FUEL

Part 1.—Physical Features of Buildings

Building Name.......................... Checked by..........................

Address................................ Date................................

Exterior of Building Structure

		Side E. W. N. S.
Exterior walls	Need insulation	
	Have open cracks	
	Unneeded openings should be closed	
	Fan opening needs louvers	
Exterior doorways	Unneeded	
	Frames need caulking...	
	Doors rattle	
	Weatherstripping needed	
	Weatherstripping needs repairs	
	Door check needed....	
	Door check needs repairs	
	Storm door needed.....	
	Storm door needs repairs.	
	Revolving door could be used	
	Revolving door needs re- pairs	
Loading platform	Outer door needed......	
	Door needs repairs......	
	Rear partition needed...	
	Rear partition needs re- pairs	
	Swinging partition door needed	
Roof	Insulation needed	
	Ceiling needed	
	Has open cracks......	
	Unneeded openings	
	Fan opening needs louvers	
Skylight	Glass needed	
	Lower sash needed.....	

Penthouse	Insulation needed ....
	Has open cracks........ .. ..
	Openings unneeded ....
	Frames need caulking... .. ..
	Doors rattle ....
	Door checks needed..... .. .. ..
	Door checks need repairs .. ..
	Windows rattle ....
	Weatherstripping needed. .. ..
	Weatherstripping needs repair ....
	Door locks needed...... .. ..
	Door locks need repairs. .. ..
	Window locks needed... .. ..
	Window locks need repairs .. ..
	Glass needed ....
	Putty needed ....

Interior of Building Structure

		Windows I 2 3
Windows	Window is unneeded......	
	Frame needs caulking.....	
	Sash is loose..............	
	Weatherstripping needed	
	Weatherstripping needs repair	
	Lock needed	
	Lock needs repair.........	
	Glass needed	
	Putty needed	
	Storm sash needed........	
	Storm sash needs repairs....	
	Sash cord or pulleys needed.	
	Weights should be changed..	
	Shade missing	
	Ventilating fan in place....	
Partitions	Door needed	
	Door needs lock	
	Lock needs repairs	
	Transom unneeded	
	Transom needs glass.......	
	Transom lock needs repairs.	

Ceiling	Cracks	Roofs	Add insulation
	Plastering needed		Close openings or cracks
	Insulation needed (top floor)		Install louvers
Floors	Openings not needed.......	Pent-houses	Improvements indicated above under walls, windows, doors and roofs
	Openings required need covers	Skylights	Replace those unneeded with roof
Elevator shafts	Solid walls and doors needed		Install second frame below the first
	Walls and doors need repairs	Interior	Close off elevator shafts and stair wells
Stair wells	Need solid walls and doors..		Install partitions
	Walls and doors need repairs		Install doors in existing partitions
			Close up floor openings
			Close open fireplaces
Basement	(Check in above)..........		

Building Improvements to Save Heat

Evening Inspection

Walls	Add insulation	Windows	Close storm sash ....
	Close openings or cracks		Close shutter* ....
	Install movable louvers on fan outlets and other useful openings		Close ventilator louvers ....
			Remove ventilator* ....
	Shut off "dead air" space openings to basement and roof space		Shut off ventilator motor ....
			Close window ....
			Lock window ....
			Pull shade* ....
			Note difficulties ....
Windows and transoms	Replace those unneeded with wall	Doors	Close storm door tight ....
	Replace missing, broken or cracked lights of glass		Lock storm door* ....
			Close exterior door tight ....
	Install storm sash		Lock door* ....
	Caulk frames		Close transom ....
	Set sash strips to prevent rattle		Set revolving door at 45° ....
	Weatherstrip		Close penthouse door tight ....
	Caulk sash when not opened		Close stair-well doors tight ....
	Install locks if missing		Close elevator partition doors tight ...
	Install new cords if broken		Close partition doors and transoms*...
	Replace window weights if not of proper size		Note difficulties ....
	Install shades	Wall, floor and roof openings	Shut down fan ....
	Remove window ventilating fan		Close louvers or opening ....
			Close roof trap-door tight ....
			Close floor openings ....
			Note difficulties ....
Exterior doors	Replace any unneeded door with wall		
	Weatherstrip		
	Caulk frame	Radiators	Open valve if under control ....
	Reinstall lock-keeper to stop rattle		Shut valve in unoccupied rooms* .....
	Install storm doors—with vestibule, if possible		Note difficulties:
			Valve ....
	Install revolving doors		Air valve ....
	Install door checks		Trap ....
			Steam inlet ....
Loading platforms	Install partition with double swinging door to main part of building		Condensate outlet ....
	Install exterior door if space permits	* If feasible and permissible	

	Suggestions for Operating Economy		Keep daily records of fuel use, degree-days, wind, rain and difficulties Make adjustments when possible during non-business hours
Methods recommended	Prevent the entrance of cold air through: storm sash; windows; doors; ventilating fan openings; transoms; window ventilators; etc. Clean windows on warm days, during non-business hours and with doors from room closed Decrease transmission losses by pulling shades down and closing shutters	Personnel	Instruct and arouse enthusiasm in building employees When in doubt secure advice from: Local steam utility, heating engineer, control contractor, or heating contractor of ability

Part 2.—Records and Data on Past Operations

Building	Heating System
Type	Type ....
Age	Age ....
Height	Kind of Temperature Control ....
Ground Floor Dimensions	....
Major Changes	Major Changes ....
....	....
....	....
Kind of Fuel Used	Degree-days in Normal Year ....
District Heating Service Used	....
Fuel or Service Records furnished by	....

Fuel Record

Period Included Between (Days)	Total Fuel or Service Used	Amount Fuel or Service Non-Heating	Fuel or Service Space Heating	Unit	1000 Btu per Unit	1000 Btu for Space Heating	D.D. in Period	1000 Btu per D.D.	1000 Btu per Normal Year
....	....	....	....	....	....	....	....	....	....
....	....	....	....	....	....	....	....	....	....
....	....	....	....	....	....	....	....	....	....
....	....	....	....	....	....	....	....	....	....
....	....	....	....	....	....	....	....	....	....
....	....	....	....	....	....	....	....	....	....

Conditions During Period

Overheating
    General....... Local...... When...... Where...... Est. Temp...... Cause......

Underheating
    General....... Local...... When...... Where...... Est. Temp...... Cause......
Percentage of building occupied...............Percentage heated..........................
Was there any other considerable source of heat?.....................................
Describe ....

Occupancy Hours
    Weekdays...... to...... Saturdays...... to Sundays...... to...... Holidays...... to......
Temperatures during occupancy...............During non-occupancy......................
Odd hour occupants...
System difficulties such as leaks..
....
Tenants opening windows...
Skill, energy and enthusiasm of operators...
Other tenant peculiarities..

Part 3.—Steam Heating Systems

Basement		Chimney	Repairs needed ....
			Cleaning needed ....
Steam Lines	Unneeded but connected.............		Too small ....
	Uninsulated		
	Flanges uninsulated	Grates	Wrong type ....
	Fittings uninsulated		Defective ....
	Too small		Firing methods ....
	Improperly sloped		
	Sag in pipes	Generating equipment	Added heating surface for boiler ......
	Pocketed.........................		Boiler leaking ....
	Clogged		Flues need cleaning ....
	Broken		Boiler needs insulation ....
	Leaking		Stoker needs repairs ....
	Special service needed		Burner needs repairs ....
	Shut-off valves not accessible		Door and frame settings show leaks ...
	Shut-off valve broken		Air handling equipment needs repairs..
	Shut-off valve clogged		Feed-water regulation faulty ....
	Shut-off valve leaking		Water treating equipment needs repairs
	Relief valves needed		Boiler dirty ....
			Safety valves defective ....
Return Lines	Uninsulated		Vacuum and condensation pumps faulty
	Too small		
	Improperly sloped		Evening Inspection
	Clogged		
	Broken	Wall, floor and roof openings (Incl. Penthouse)	Shut down fan ....
	Leaking		Close louvers or opening ....
			Close roof trap-door tight ....
Traps	Too small		Close floor openings ....
	Faulty		Note difficulties ....
	Cooling leg to short		
		Radiators	Open valve if under control ....
Air vents	Too small		Shut valve in unoccupied rooms* .....
	Faulty		Note difficulties:
	Automatic regulation faulty		Valve ....
			Air valve ....
Radiation	Needed		Trap ....
			Steam inlet ....
Control equipment	Motor faulty		Condensate outlet ....
	Wiring faulty		
	Tubing faulty	* If feasible and permissible	
	Clock faulty		
	Mechanism faulty		Suggestions for Operating Economy
Pumping equipment	Motor faulty		
	Wiring faulty		Run ventilating fans only when needed and no faster than necessary.
	Pump faulty		Check and correct meters.
Pressure regulator	Diaphragm faulty		Heat to no higher temperature than necessary.
	No pressure gauge		
	Improperly installed		
	Governor line faulty		

Turn steam off wherever and whenever not needed; especially in vacant rooms.

Turn steam off at the main or zone valves rather than at the radiators.

Turn valves either entirely off or entirely on.

Check air-fuel proportion.

Carry a low pressure on steam system; a high vacuum on vacuum system.

Carry a protective ash bed on grates.

Promptly move out ash discharged from grate.

Do not heat up the system for a few persons, lend them electric heaters.

Carry on a regular schedule of maintenance.

Study damper control for best operation.

Study methods of heat saving, sources of building heat loss, heating and ventilating system operation, and peculiarities of tenants.

Instruct and arouse enthusiasm in building employees.

Prepare a valve operation schedule for hand control: For use if there is no automatic control, or for use in case of a failure of control to function.

Keep daily records of fuel use, degree-days, wind, rain and difficulties.

When in doubt secure advice from: Local steam utility, heating engineer, control contractor, or heating contractor of ability.

Rooms and Halls

Radiators	Unneeded ....
	Location bad ....
	Oversized ....
	Undersized ....
	Shield obstructs ....

Radiators	Drapes obstruct ....
	Cover obstructs ....
	Water pan obstructs.... .. .. ..
	Metal radiator reflectors .. .. ..
	Painted with metallic paint ....
	Space between fins dirty .. .. ..
	Obstructed by furniture. .. .. ..
	Air valve faulty ....
	Trap faulty ....
	Shut-off valve inaccessible .. .. ..
	Shut-off valve needs packing ....
	Radiator control valve inaccessible ....
	Orifice undersized ....
	Orifice oversized ....
Lines	Steam connection undersized ....
	Steam connection pocketed ....
	Steam connection broken .. .. ..
	Steam connection leaking .. .. ..
	Steam connection uninsulated ....
	Return connection undersized ....
	Return connection pocketed ....
	Return connection broken .. .. ..
	Return connection leaking ....
	Return connection uninsulated ....
	Unneeded ....
	Bare line in room ....
	Thermostat in room ....

Part 4.—Hot Water Heating Systems

Flow and return mains and branches	Improperly graded			Insulation needed ....
	Trapped			Door and frame leaks ....
	Improperly sized			Ash pit dirty ....
	Leaking			Grates need replacing ....
	Uninsulated			Defective parts ....
	Special service needed			How fired ....
				Draft and check dampers bind ....
Boiler	Improperly sized			
	Improper type			
	Leaking	Chimney	Repairs needed ....
	Flues need cleaning		Cleaning needed ....

	Too small in diameter .... Too short .... Obstructed by high trees or surrounding buildings	**Evening Inspection**
		Inspection Checked by .... Date ....
Pressure system	Pressure tank improperly connected ... Pressure tank water-logged .... Pressure relief valve improperly installed Pressure relief valve faulty .... Damper regulator inoperative .... Damper regulator improperly connected or adjusted .... Pressure gauge faulty .... Thermometer faulty .... Pressure reducing valve faulty	Wall, floor and roof openings (Incl. Penthouse) — Shut down fan .... Close louvres or opening .... Close roof trap-door tight .... Close floor openings .... Note difficulties ....
		Radiators — Open valve if under control......... Where feasible, shut valve in unoccupied rooms .... Note difficulties Valve .... Air valve ....
Firing device	Stoker or burner needs repairs....... Proper grade fuel being used? Yes .... No .... Combustion Good .... Bad .... Air Too much .... Too little	
		Suggestions for Operating Economy
Control equipment	Room thermostat faulty .... Improperly located .... Burner control faulty .... Relay faulty .... Thermostat setting Too high .... Too low	Run ventillating fans only when needed and no faster than necessary. Heat to no higher temperature than necessary. Turn heat off wherever and whenever not needed; especially in vacant rooms. Turn radiator valves either entirely off or entirely on.
Water circulating pump	Improperly installed .... Improperly sized .... Motor faulty .... Pump leaking .... Improperly lubricated .... Improperly controlled .... Noisy	Check air-fuel proportion. Carry minimum boiler temperature. Carry a protective ash bed on grates. Promptly move out ash discharged from grate. Carry on a regular schedule of maintenance. Study damper control for best operation.
Radiators	Improperly located .... Improperly sized .... Not needed .... Obstructed by dirty convectors .... Air valve faulty .... Radiator valve faulty .... Circulation through radiators Good .... Bad	Study methods of heat saving, sources of building heat loss, heating and ventilating system operation and peculiarities of tenants. Instruct and arouse enthusiasm in building employees. Keep daily records of fuel use, degree-days, wind, rain and difficulties. When in doubt secure advice from local heating contractor of ability.

CHECK LISTS FOR HEATING SYSTEM SERVICING

The check lists which follow are for locating the cause of troubles with hot water, steam and vapor, gravity warm air and forced circulation warm air systems. It is not the intent to solve all troubles by use of brief tables such as these, but with their help most problems can either be solved or the cause reduced to only a few possibilities.

Such check lists often save time.

Hot Water Systems			
Complaint—Slow Heating or Not Enough Heat Throughout Building			
Possible Cause	Check	Details	Remedy
Insufficient heat input to boiler	Time required to raise temperature of water	Average job—about $1\frac{1}{2}$°F per min. Underrated job—$1\frac{3}{4}$° to 2°F per min. Overrated job—1° to $1\frac{1}{2}$°F per min.	Increase burner input.
Shortage of radiator surface	Amount of radiator surface	Determine heat loss of building and check against actual heat output of surface installed.	Increase burner input and set water temperature control higher. Install a closed system with tank for temperatures over 200F. Add surface.
Slow or sluggish circulation	Piping	Check pipe sizes against piping tables. Check pitch of mains and connections.	Increase burner input. Balance circulation with orifices. Install circulating pump.
	Air vents	Radiators may be airbound.	Open vents to allow air to escape.
	Water level	System may not have enough water.	Fill to proper level.
	Boiler insulation	Circulation is slowed up because high heat loss of boiler reduces water temperature.	Cover with good insulation.
	Piping insulation	Circulation is slowed up because high heat loss of piping reduces water temperature.	Insulate all mains and reduce infiltration to basement.
Radiator covers	Radiator covers	Covers may interfere with air circulation to radiators.	Change design or remove.
Undersized boilers	Boiler rating against total load	Boiler may be large enough for radiator load but because of shortage of radiator surface it is not large enough for building. Boiler may be overloaded.	Increase burner input or install larger burner. Install copper coils in firepot. Install larger boiler.
Oversized boiler	Boiler rating against total load	Greater volume of water will cause greater standby losses.	Increase buner input.
Loss of water from system	Water relief valve and expansion tank	Closed systems without expansion tanks during firing periods will lose hot water which will be replaced with cold water.	Install cushion tank.
	Water level and overflow line	Open systems if too full will overflow on firing periods. If no vent is used in overflow line, system may lose water by siphoning.	Fill to proper level. Install vent on overflow lines.
Complaint—Slow or Insufficient Heat in One or More Rooms			
Possible Cause	Check	Details	Remedy
Insufficient radiator surface	Heat loss and radiator surface	Determine heat loss of room and check against actual output of surface installed.	Paint bronzed radiators. Set water temperature higher and reduce circulation to other radiators. Make more favorable tapping of riser on main and increase pitch. Install more surface.
Slow or sluggish circulation	Piping	Check piping size to radiator, also check pitch and main connection.	Set water temperature higher and reduce circulation to other radiators. Make more favorable tapping of riser on main. Increase pitch. Change pipe size.
	Radiator vents	Air in radiator may interfere with circulation and reduce surface.	Open vents to remove air.
	Water level	System may not be filled with water.	Add water.
	Radiator valve	Dirt in valve opening may reduce circulation. Faulty valve may not open properly.	Clean, repair, or replace valve.
	Radiator connections	Water may be bypassed due to improper connections. Inlet and return may be tapped off same main in two-pipe system.	Make more favorable connections to main and return.
Complaint—Noisy System			
Possible Cause	Check	Details	Remedy
Water boiling	Water temperature	Boiling water will cause a noise like a water hammer. Temperature of boiling in open systems will depend upon height of system.	Provide water temperature limit control and set 215F or lower.
Expansion of piping	Size of pipe holes through floor	Expansion of pipe causes risers to be forced against sides of holes in floors causing a rubbing action.	Enlarge floor holes.

Steam and Vapor Systems			
Complaint—Slow or Insufficient Heat Throughout the Building			
Possible Cause	Check	Details	Remedy
Insufficient heat input to boiler	Time required to raise 1 lb. of steam pressure or to kill vacuum	Atmospheric system with all radiators on—Average Job—20 to 30 min. Underrated job—10 to 20 min. Overrated job—30 to 40 min. Vapor system with all radiators on—Average job—15 min. Underrated job—10 to 20 min. Overrated job—20 min.	Increase burner input to desired steaming rate.
Shortage of radiator surface	Amount of radiator surface	Determine heat loss of building, divide by 240 and check against actual surface installed.	Set pressure control to higher setting and minimum differential and increase burner input. Paint bronzed radiators. Add surface needed.
Slow or sluggish circulation	Piping	Check sizes of all pipes against piping tables for particular system. Check riser connections on mains, pitch of all pipes especially for trapped lines, for possible bypassing.	Correct pipe size, riser connections, pitch and eliminate by-passing if necessary.
	Valves	Make sure nipple and valve openings are not clogged. On one-pipe system see that valves are of type with openings equal to full pipe size. Also check other valvs to see that they are of proper type.	Clean or replace nipples or valves. Replace improper valves.
	Air vents	Look for clogged air openings on vents. Make sure vents do not pass steam. Vents should be located as low as possible on radiators, on ends of mains and returns and on dry returns at top of vertical drop to boiler return. Main and return vents should be of return, quick venting type. On two-pipe vapor systems check operation of air eliminators and traps.	Clean vent openings. Replace corroded valves. Install in correct position if necessary. Clean or repair poorly operating eliminators or traps.
Uncovered or poorly insulated boiler	Boiler insulation	An uncovered or poorly insulated boiler will steam poorly because of increased standby loss and condensation in dome.	Cover with good insulating material.
Uncovered or poorly insulated piping	Pipe insulation and air infiltration to basement	Uncovered or poorly insulated piping will decrease amount of steam especially when basement temperature is low.	Insulate mains and reduce infiltration to basement.
High water line in boiler.	Position of water line	Water line should be in center of gage glass. High water line increases water volume and decreases steam volume and with burner input set for normal water line, input will not be sufficient and slow steaming will result.	Lower water line to normal position. If burner capacity is not sufficient for proper steaming, lower water line a few inches.
Water leaves boiler	Water level and return connections	With a fast steaming boiler, pressure may build up so rapidly that the pressure in boiler will back the water out through returns and low water cut-off will shut off burner before radiator gets hot.	Install a Hartford equalizer loop on return.
Foaming boiler	Action of water line	A jumping water line when boiler is operating indicates foaming caused by oil and dirt in boiler water.	Clean boiler with boiler cleaning compound.
Undersized boiler	Boiler rating against load total	Boiler may be large enough for radiator load but because of shortage of radiator surface it is not large enough for building. Or boiler may be overloaded.	Increase burner input or install larger burner to obtain faster steaming. Install copper coils in firepot. Operate with a lowered water line. Install larger boiler.
Vacuum equipment	Return to condensate	On vacuum jobs faulty action of pumps, traps or air eliminators will cause faulty return of condensate to boiler.	Clean and repair all steam specialties.
Complaint—Slow or Insufficient Heat in One or More Rooms			
Possible Cause	Check	Details	Remedy
Insufficient radiator surface	Heat loss and radiator surface	Determine heat loss of room, divide by 240 and check against actual surface installed.	Paint bronzed radiators. Increase radiator size.
Poor circulation to radiator	Piping	Check branch and riser sizes, pitch of piping from main to radiator, connections to radiators, branches, mains and returns.	Correct pipe sizes, pitch and connections.
	Radiator valve	Check valve and nipple opening for possible clogging by rust or scale. Determine if valve seat has become loose and is only partially open.	Clean valves or nipples. Repair or replace faulty valves.
Poor radiator venting	Radiator vents	Check air vent under normal operating conditions to determine the speed of venting, if radiator is hot all over when venting stops, if vent leaks steam, and if vent is too high.	Clean. If radiator vents slowly replace vent with one with larger opening.
Trapped radiator	Slope of radiator	On one pipe system if radiator slopes down from opening too much condensate will remain in radiator. On two-pipe systems if radiator return is higher than inlet, condensate will flow back into riser on bottom inlet radiators and will collect in radiator on top inlet jobs.	Pitch radiators slightly in proper direction.
Radiator main flooded	Boiler return connections	If end of main is not elevated sufficiently above water line of boiler on systems not having a Hartford loop, condensate may flood end of main, preventing steam from entering end radiator.	Install a Hartford equalizer loop.
Faulty customer operation	Radiator valves	On one-pipe jobs, radiator valves may be kept wide open causing condensate to be trapped while steam is on.	Instruct customer to keep valve wide open.
Faulty radiator traps	Radiator traps	Dirt or scale will interfere with trap operation.	Clean or install new traps.

Forced Circulation Air Systems

Complaint—Slow or Insufficient Heat Throughout the Building

Possible Cause	Check	Details	Remedy
Insufficient input	Bonnet temperature	Bonnet temperature of at least 150°F should be attained.	Increase input of burner.
Undersized furnace	Furnace rating against heat loss	Determine heat loss of building and check against rating of furnace.	Increase burner input. Install larger furnace.
System unbalanced	Room temperature	Too much heat may be supplied to unimportant rooms and too little to important rooms.	Balance system with dampers in supply and returns.
Insufficient air delivery	Air volume	Air circulated may not be adequate to deliver enough heat or provide good heat distribution.	Increase speed of blower. Change blower size.
Dirty filters	Air filters	Dirty filters will decrease the amount of air circulated.	Clean or replace filters.
Poor design	Location of furnace	If furnace is located at one end of a long building, there may be a large temperature drop in the supply duct causing cold air to be delivered to end rooms.	Insulate duct and balance systems. Relocate furnace in center of building.
	Duct sizes	Incorrect size ducts and many bends will cause poor air distribution.	Increase input and air delivery. Balance system.

Complaint—Slow or Insufficient Heat in One or More Rooms

Possible Cause	Check	Details	Remedy
System unbalanced	Temperature in all rooms	If some rooms are overheated and some underheated it is necessary to decrease the flow to overheated rooms and increase it to underheated rooms.	Balance the system with dampers in supply and return ducts.
		If a number of rooms have the correct temperature while the rest are underheated, add more heat to the building.	Increase the burner input and balance the system.
Poor air distribution	Register location	If registers are improperly located the air will tend to stratify and cause cold floors. Registers should be on inside walls while return grilles should be on outside walls.	Try increasing air volume. Relocate registers and return grilles.
	Temperature of air from attic trunkline	Sometimes attic trunklines are used to supply upper floors. If the attic is cold the trunkline will be chilled. Also downward air currents from the attic may cool off riser to trunkline.	Insulate attic trunkline. Seal the shaft in which the riser is installed to prevent cold attic air from reaching the riser. Insulate the riser.

Forced Air Systems—Converted From Gravity

Complaint—Slow or Insufficient Heat Throughout the Building

Possible Cause	Check	Details	Remedy
Insufficient heat input	Bonnet temperature	Bonnet temperatures of at least 150°F should be developed.	Increase input to furnace.
Undersized furnace	Furnace rating against heat loss	Determine heat loss of building and check against rating of furnace.	Increase burner input. Install larger furnace.
System unbalanced	Room temperatures	Too much heat may be supplied to unimportant rooms and little to important rooms.	Balance system by means of dampers in leaders.
Insufficient air volume	Air volume	Blower may not be capable of delivering required amount of air.	Increase speed of blower. Change blower size.
Inadequate return air	Size of grilles and return ducts	Return ducts may be too small or have too much resistance. Grilles may have too small a free area.	Increase fan capacity. Install larger ducts or grilles.
Unbalanced return air	Return air volume	It may be possible that one return is carrying more air than it should.	Balance returns.
Poor return connections	Return connections	Two or more returns should be connected into a manifold to the fan chamber and not enter furnace separately.	Connect all returns to fan chamber.
Poor furnace return connection	Furnace return connection	To even up air distribution in furnace a wide flaring connection with spreaders in it should be used to connect fan chamber to furnace.	Connect fan chamber to furnace with wide flaring connection with spreaders.

Complaint—Slow or Insufficient Heat in One or More Rooms

Possible Cause	Check	Details	Remedy
Unbalanced system	Temperature in all rooms	If some rooms are overheated and some underheated it is necessary to reduce air flow to overheated rooms and decrease it to underheated rooms.	Balance system with leader dampers.
		If some rooms have the proper temperature while the rest are underheated, it will be necessary to add more heat.	Increase the burner input and balance the system.
Poor air distribution	Number of registers and location	Registers located on outside walls or an insufficient number of registers will cause poor air distribution.	Increase fan speed and limit air volumes to other rooms. Provide additional registers. Relocate.
	Cold air return	An undersized or poor return will result in poor distribution.	Install a proper return.
	Leader size	Check leader size against calculated area. Large number of bends will cut down air flow.	Rebalance system. Eliminate unnecessary turns.
	Return grille location	Return grilles should be located on exposed walls. If located on inside walls they will cause cold, drafty floors.	Relocate grilles to exposed walls.

		Gravity Circulation Air Systems	
		Complaint—Slow or Insufficient Heat Throughout the Building	
Possible Cause	Check	Details	Remedy
Insufficient heat input to furnace	Time required to raise register temperature to 140°F	Average job should have living room register air temperature of 140°F within 20 to 30 min. of time burner starts. Poorly circulating jobs should raise air temperature faster.	Increase input of burner.
Slow or sluggish circulation	Leader pipe sizes against heat loss	Determine heat loss of rooms and divide by 111 for first floor, 167 for second, and 200 for third to obtain leader area in inches. Add 2% to area for each foot over 12 ft. of length, 20% for each elbow over one, 10% for every turn over one.	Balance system with dampers. Increase input of burner.
	Wall riser sizes	Wall risers to second and third floors should have area of at least 70% of leader supplying them.	Balance system. Increase burner input. Add risers.
	Cold air returns	Total return duct area should at least equal total leader area. Drops to furnace should be of two 45° ells and round pipe rather than two 90° ells and rectangular pipe. Free area of grilles at least equal to area of duct.	Increasing burner input will overcome defects to some extent. Otherwise changes in returns will be necessary.
		Two returns joining will often cause unbalanced returning and poor air distribution to furnace.	Connect each return separately to furnace casing and balance return air.
	Cold air shoe connection to furnace casing	High point of cold air shoe should not be above the grate level or burner head level.	If shoe is above, protect top of shoe with deflector in upper part or install a radiation shield between firepot and casing extending down to below grate level.
	Number of returns	If only one return is used poor circulation is practically certain even if return area is equal to leader area.	Increase input or burner. Install additional return.
Furnace too small or too large	Furnace rating against heat loss	On undersized furnace an increased burner input is necessary to supply sufficient heat. On an oversized furnace somewhat greater burner input is necessary to heat up more metal and take care of radiation losses.	Increase burner input. Install correct size furnace.
		Complaint—Slow or Insufficient Heat in One Room	
Possible Cause	Check	Details	Remedy
Poor circulation	Leader pipes	Note whether leader has dents, is flattened or has air leaks. Check size. Leaders size in square inches should be the heat loss of room divided by 111, 167, or 200 for first, second, or third floor, respectively. Add 2% to area for each foot over 12 ft. in length, 20% for each elbow over one, 10% for each turn over one.	Remove dents and flattening and seal leaks. If leader is long cover with good insulation. Install small fan in leader and control by room thermostat.
	Leader pitch and turns	Insufficient pitch and sharp turns will slow circulation.	Increase pitch and burner input. Balance system.
	Wall riser	Riser to second or third floor room may have come loose from register boot allowing warm air to escape in wall.	Reconnect wall stack securely.
	Returns to furnace	Unequal distribution of return air in furnace will cut down on air flow to leaders located at points where small amounts of air are returned. This will often occur where only one return is used and where it is of non-flaring type.	Change return shoe to a flaring type with splitters. Change location of leader.
	Leader connection to furnace	On most furnaces there is little circulation at the front and leaders connected here will have poor air supply.	Change location of leader to other part of bonnet.
	Register area	Register free area should at least equal area of leader pipe supplying it.	Replace with register having proper free area.
	Leader pipes, boots and risers	A restriction in either leader, boot or riser will decrease circulation.	Provide full opening throughout run.
	Register location	Register should be on inside wall and direct air towards or along exposed walls.	Change location of register.
	Room location	Rooms on cold side of house may heat poorly even with proper size leaders and returns because of the effect of wind.	Add or increase size of cold air returns.
	Register and return grille location	Registers located too close to return grilles will cause short-cutting of air and poor room circulation.	Relocate registers or return grilles.
	Studding space	Where studding space is used to return air from second floor rooms it is necessary to block off space above grille to stop flue action which tends to carry air up to attic.	Block of studding space above grille.
		Complaint—Cold Floors	
Possible Cause	Check	Details	Remedy
Air stratification	Remove temperatures at various levels	Air stratification may result when the burner is shut off and air circulation has stopped.	Shorten off-periods by using thermostat set for minimum differential.
Cold air traveling across floor	Return grille location	If return grilles are located on inside walls, cold air will travel across floor from the cold walls and windows.	Locate return grilles at outside walls.

Section 8

PIPING

Contents

Authorities

Section Editor: *Marvin Naparstek*, Burns & Roe, Inc.

F. Caplan, Pressure Drop of Air in Pipes
Raymond Durazo, Plastics Pipe
George Jankauskas, Allowable Spaces for Pipes
S. Konzo and *Robert H. Page*, Steam Flow in Pipes
Joachim Nachbar, Natural Gas Flow Tables

Philip J. Olmstead, Characteristics of Centrifugal Pumps; Pump Laws; Centrifugal Pump Selection
Ralph A. Rockwell, Formulas for Sizing Control Valves

FLUID FLOW EQUATIONS

The equation commonly used for calculating pressure drop in pipes, expressed in practical units, is

$$p_1 - p_2 = \frac{.00129\, fl\rho u^2}{d} = 43.75 \frac{fl\rho V^2}{d^5} \quad (1)$$

p_1 = inlet pressure in lb. per sq. in.;
p_2 = outlet pressure in lb. per sq. in.;
f = friction factor;
l = length of pipe in feet;
u = mean velocity of flow in ft. per sec.;
ρ = density in lb. per cu. ft.;
V = volume rate of flow in cu. ft. per sec.;
d = pipe diameter in inches.

The only problem arising in the use of this equation is that of selecting the friction factor. Many empirical relations have been proposed for this friction factor. Such relations, if based on experimental data, will be satisfactory. If, however, the range of conditions is extended beyond the range of the data on which it is based, the empirical formulas will not necessarily apply and may result in serious error. The Reynolds number method for determining the friction factor eliminates this difficulty and makes it possible to use the friction factor determined using one fluid on problems involving some other fluid. Reynolds in his study of the flow problem found that the change of flow from viscous or laminar flow to turbulent flow depended on the dimensionless product $Du\rho/\mu$ which is called the Reynolds number and given the symbol N_R.

$$N_R = \frac{Du\rho}{\mu} \quad (2)$$

where D = diameter in feet, and
μ = viscosity in lb. per ft.-sec.

Expressed in practical units,

$$N_R = 15.3 \frac{V\rho}{d\mu} = 15.3 \frac{W}{d\mu} \quad (3)$$

where W = weight of fluid in lb. per sec.

Regardless of the type of fluid flowing, the friction factor will be the same at the same value of Reynolds number for a given size of pipe. The friction factor cannot be calculated mathematically for any other condition than for the viscous or laminar type flow

where $f = 64/N_R$ for circular pipes. For turbulent flow the friction factor must be determined from experimental data. Based on a summary of data, Figs. 1 and 2 give the value of the friction factor based on the Reynolds number. Table 1 is a key to Figs. 1 and 2 as to which curve to use.

Application of these data requires that the viscosity be known (data on viscosity appears later in this section). An example follows:

Example. Determine the pressure drop for 1000 ft. of 2 inch wrought iron pipe handling 30 gpm of water at 60F with an absolute viscosity of 7.8×10^{-4} lb. per ft. sec.

$$V = \frac{30\ \text{g.p.m.}}{7.48\ \text{gal. per cu. ft.} \times 60\ \text{sec. per min.}}$$
$$= 0.0668\ \text{cu. ft. per sec.}$$

μ (given) $= 7.8 \times 10^{-4}$
$\rho = 62.4$ lb. per cu. ft. at 60F
$d = 2.067$ in. actual diameter of 2 inch pipe.

Then $N_R = 15.3 \dfrac{V\rho}{d\mu} = 3.96 \times 10^4$

Refer to Table 1 and for wrought iron, 2 inch, find that curve 6 is to be used. Enter Fig. 1 for $N_R = 3.96 \times 10^4$, move up to curve 6 and find at the left that f is .025.

Then, from equation (1) $p_1 - p_2 =$

$$43.75 \times \frac{.025 \times 1000 \times 62.4\, (.0668)^2}{2.067^5}$$
$$= 8.06\ \text{lb. per sq. in. pressure drop.}$$

In many flow problems it is desirable to calculate the volume of flow which will result if a given pressure difference is applied. The velocity of flow for this case can be calculated from equation (1) provided f can be determined. The value of f cannot be determined directly by using the Reynolds number since equation (3) involves the rate of flow, which is not known in this case.

However, it can be shown that

$$fN_R{}^2 = 5.4 \frac{(p_1 - p_2)d^3\rho}{l\mu^2} = N_v \quad (4)$$

and this expression is called the velocity number or N_v.

Since f is a function of N_R, then N_v can be

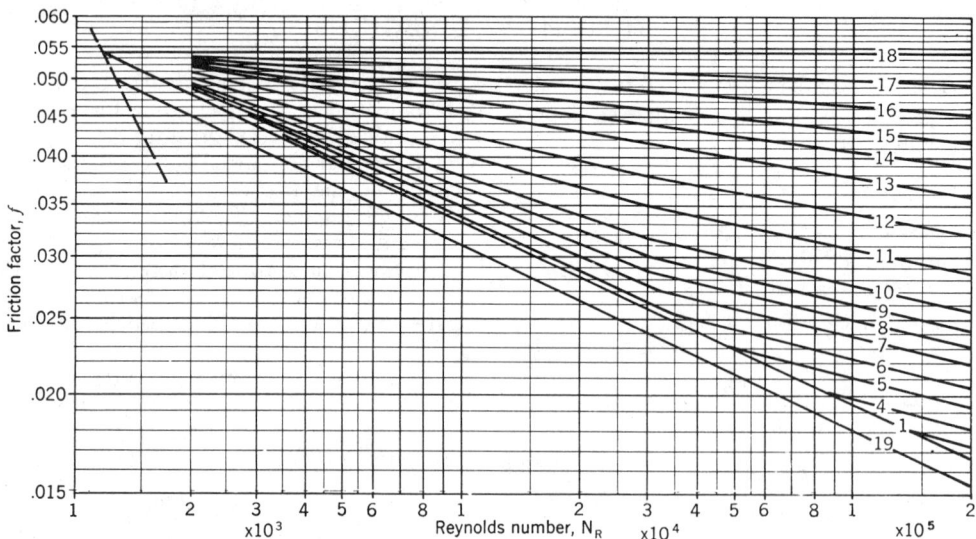

Fig. 1. Reynolds numbers vs. friction factors for values of N_R from 1,000 to 200,000. Refer to Table 1 as to correct curve to use.

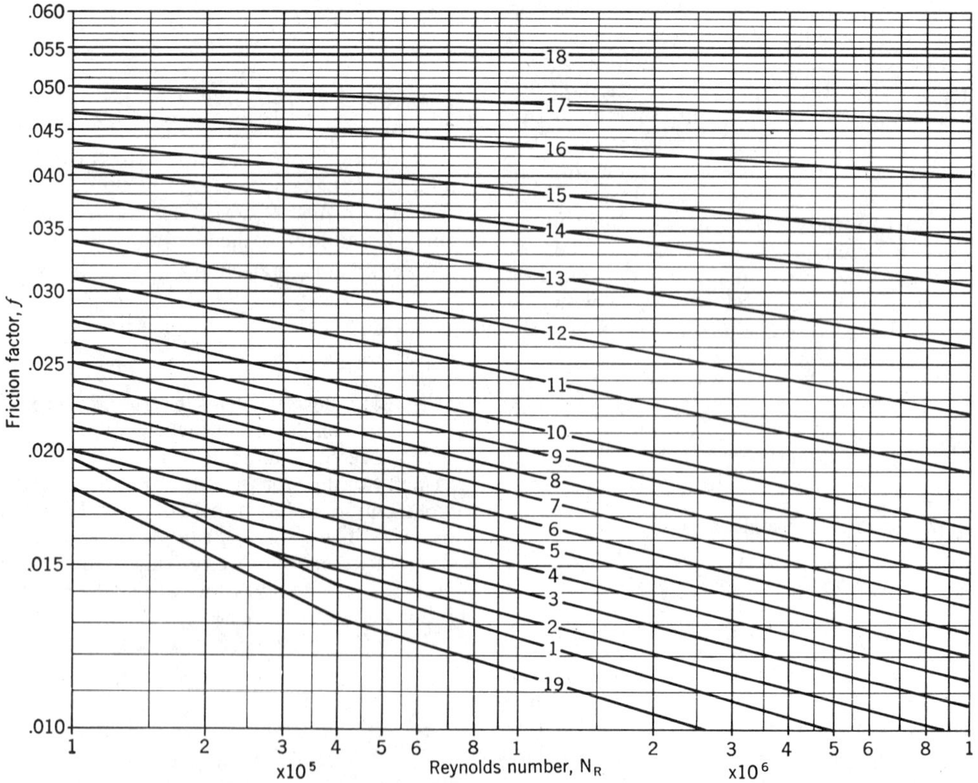

Fig. 2. Reynolds numbers vs. friction factors for values of N_R from 100,000 to 10,000,000. Refer to Table 1 as to correct curve to use.

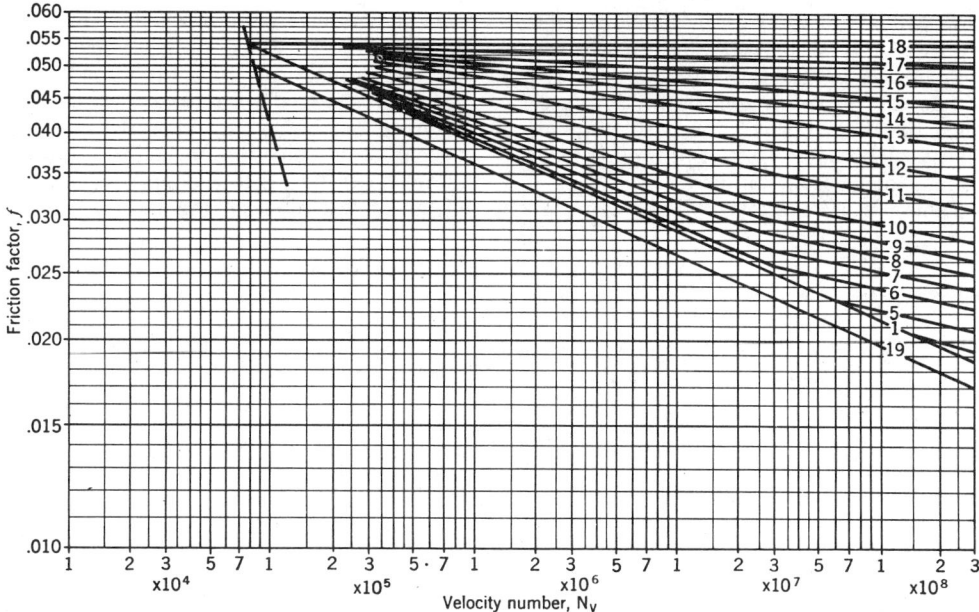

Fig. 3. Velocity numbers vs. friction factors for values of N_V from 10,000 to 300,000,000. Refer to Table 1 as to correct curve to use.

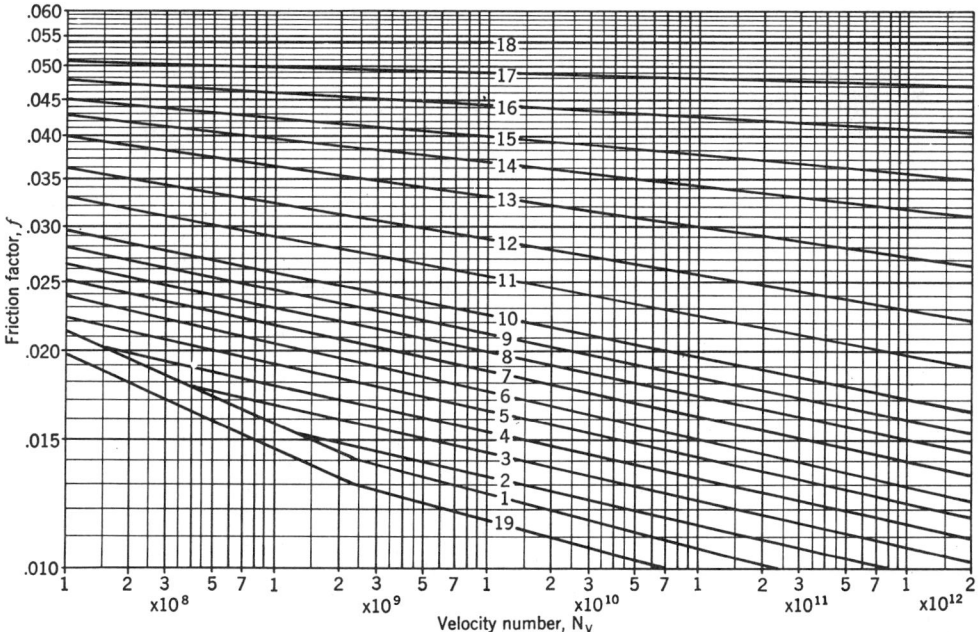

Fig. 4. Velocity numbers vs. friction factors for values of N_V from 100,000,000 to 2,000,000,000,000. Refer to Table 1 as to correct curve to use.

expressed in terms of N_R or if desired, f can be expressed in terms of N_v. Since this velocity number does not contain the rate of flow, it can be used to determine the friction factor when the rate of flow is to be determined. Figures 3 and 4 show a plot of friction factor against the velocity number.

Table 1 is also the key to Figs. 3 and 4.

Designers may achieve a rough sizing of piping before detailed pressure drop calculation is made by using the velocity criteria, as shown in Table 2. Column 4, intended as a guide only, gives the range of normal fluid pipeline velocities. Upper values apply to larger size pipes, where the effects of the pipewall are lessened.

Table 1. — Key to Use of Figs. 1 to 4.

Use Curve No.	Drawn Tubing of Copper, Brass, Tin, Lead, Glass	Clean Steel or Wrought Iron	Clean Galvanized Steel Pipe	Best Cast Iron, Cement, Light Riveted Ducts	Average Cast Iron, Rough Concrete	Good Brick, Heavy Ducts, or Spiral Riveted Pipe
	Pipe Diameter, Inches (Actual Internal Diameter — not Nominal Diameter)					
1	0.35 up	72	....	....	....	....
2	....	48–66	....	....	....	....
3	....	14–42	30	48–60	96	220
4	....	6–12	10–24	20–48	42–96	84–204
5	....	4–5	6–8	12–16	24–36	48–72
6	....	2–3	3–5	5–10	10–20	20–40
7	....	1½	2½	3–4	6–8	16–18
8	....	1–1¼	1½–2	2–2½	4–5	10–14
9	....	¾	1¼	1½	3	8
10	....	½	1	1¼	....	5
11	....	⅜	¾	1	....	4
12	....	¼	½	....	....	3
13	....	⅛	....	....	....	....
14	0.125	....	⅜	....	....	....
15	....	....	....	....	....	....
16	....	....	¼	....	....	....
17	....	....	....	....	....	....
18	0.0625	....	⅛	....	....	....

Table 2.—Velocity Design Criteria

Fluid	Condition	Service	Velocity, Ft per Sec
Water	Cold	Pump Discharge	6 to 9
Water	Cold	Supply Headers	3 to 8
Water	Hot	Boiler Feed Mains	8 to 12
Water	Hot or Cold	Pump Suction and Drains	4 to 7
Acids and Chemicals	All	General	5 to 9
Slurry Mixtures	All	General	3 to 5
Air, Gases, and Steam	Wet (0 to 49 PSIG)	General	65 to 100
Air, Gases, and Steam	Wet (50 PSIG and up)	General	100 to 180
Air, Gases, and Steam	Dry (200 PSIG and up)	General	120 to 350

EQUIVALENT RESISTANCE OF VALVES AND FITTINGS

Nominal Pipe Diameter, Inches	Globe Valve, Open	Gate Valve				Angle Valve, Open	Close Return Bend	Tee		Ordinary Entrance‡
		¾ Closed	½ Closed	¼ Closed	Open			Through Run	Through Side†	
Equivalent Resistance, Feet of Pipe										
½	16	40	10	2	0.3	9	4	1.0	4	0.9
¾	22	55	14	3	0.5	12	5	1.4	5	1.2
1	27	70	17	4	0.6	15	6	1.7	6	1.5
1¼	37	90	22	4	0.8	18	8	2.3	8	2.0
1½	44	110	28	6	0.9	21	10	2.7	9	2.4
2	55	140	35	7	1.2	28	13	3.5	12	3.0
2½	65	160	40	8	1.4	32	15	4.2	14	3.3
3	80	200	50	10	1.6	41	18	5.0	17	4.5
3½	100	240	60	12	2.0	50	21	6.0	19	5.0
4	120	275	69	14	2.2	55	25	7.0	21	6.0
5	140	325	81	16	2.9	70	30	8.5	27	7.5
6	160	400	100	20	3.5	80	36	10.1	34	9.0
8	220	525	131	26	4.5	110	50	14.0	44	12.0
10	—	700	175	35	5.5	140	60	17.0	55	15.0
12	—	800	200	40	6.5	160	72	19.0	65	16.5
14	—	950	238	48	8.0	190	85	23.0	75	20.0
16	—	1100	272	52	9.0	220	100	26.0	88	22.0
18	—	1300	325	65	10.0	250	115	30.0	110	25.0

Nominal Pipe Diameter, Inches	Sudden Contraction§			Borda Entrance	Reducing Tee*		90° Elbow			45° Elbow
	D/d=4	D/d=2	D/d=4/3		D/d=2	D/d=4/3	Standard	Medium Sweep	Long Sweep	
Equivalent Resistance, Feet of Pipe										
½	0.8	0.6	0.3	1.4	1.5	1.3	1.5	1.3	1.0	0.8
¾	1.0	0.8	0.5	1.9	2.0	1.8	2.0	1.8	1.4	1.0
1	1.3	1.0	0.6	2.5	2.6	2.4	2.6	2.4	1.7	1.3
1¼	1.6	1.3	0.8	3.5	3.5	3.1	3.5	3.1	2.3	1.6
1½	2.0	1.5	0.9	4.0	4.5	3.7	4.5	3.7	2.7	2.0
2	2.5	1.9	1.2	5.0	5.3	4.5	5.3	4.5	3.5	2.5
2½	3.0	2.2	1.4	6.0	6.3	5.5	6.3	5.5	4.2	3.0
3	3.7	2.8	1.6	7.5	8.0	6.9	8.0	6.9	5.0	3.7
3½	4.4	3.3	2.0	9.0	9.5	8.0	9.5	8.0	6.0	4.4
4	5.0	3.7	2.2	11.0	11.5	9.5	11.0	9.5	7.0	5.0
5	6.0	4.7	2.9	12.0	12.5	12.0	13.0	12.0	8.5	6.0
6	7.5	5.6	3.5	15.0	16.0	14.0	16.0	14.0	10.1	7.5
8	11.0	7.2	4.5	19.0	20.0	18.0	20.0	18.0	14.0	10.0
10	13.0	9.5	5.5	24.0	25.0	22.0	25.0	22.0	17.0	13.0
12	15.0	11.0	6.5	29.0	31.0	26.0	31.0	26.0	19.0	—
14	17.0	12.5	8.0	34.0	36.0	30.0	36.0	30.0	23.0	—
16	19.0	15.0	9.0	38.0	40.0	35.0	40.0	35.0	26.0	—
18	21.0	16.5	10.0	43.0	45.0	40.0	45.0	40.0	30.0	—

† For flow making 90° turn. ‡ Into pipe of given diameter from tank, etc.
* For run of tee. Use pipe size of small diameter.
§ Use pipe size of small diameter. For Sudden Enlargements, d/D=¼, values are same as for Reducing Tee, D/d=2; d/D=½, values are same as for Tee, Through Run; d/D=¾, values are same as for Sudden Contraction, D/d=4/3.

VOLUME FLOW AT 1 FPM IN PIPE AND TUBE

Nominal Diam. Inches	Schedule 40 Pipe			Schedule 80 Pipe			Type K Copper Tube		
	Cu. Ft. per Minute	Gallons per Minute	Pounds 60 F Water per Min.	Cu. Ft. per Minute	Gallons per Minute	Pounds 60 F Water per Min.	Cu. Ft. per Minute	Gallons per Minute	Pounds 60 F Water per Min.
⅛	0.0004	0.003	0.025	0.0003	0.002	0.016	0.0002	0.0014	0.012
¼	0.0007	0.005	0.044	0.0005	0.004	0.031	0.0005	0.0039	0.033
⅜	0.0013	0.010	0.081	0.0010	0.007	0.061	0.0009	0.0066	0.055
½	0.0021	0.016	0.132	0.0016	0.012	0.102	0.0015	0.0113	0.094
¾	0.0027	0.028	0.232	0.0030	0.025	0.213	0.0030	0.0267	0.189
1	0.0062	0.046	0.387	0.0050	0.037	0.312	0.0054	0.0404	0.338
1¼	0.0104	0.078	0.649	0.0880	0.067	0.555	0.0085	0.0632	0.53
1½	0.0141	0.106	0.882	0.0123	0.092	0.765	0.0196	0.1465	1.22
2	0.0233	0.174	1.454	0.0206	0.154	1.280	0.0209	0.1565	1.31
2½	0.0332	0.248	2.073	0.0294	0.220	1.830	0.0323	0.2418	2.02
3	0.0514	0.383	3.201	0.0460	0.344	2.870	0.0461	0.3446	2.88
3½	0.0682	0.513	4.287	0.0617	0.458	3.720	0.0625	0.4675	3.91
4	0.0884	0.660	5.516	0.0800	0.597	4.970	0.0811	0.6068	5.07
5	0.1390	1.040	8.674	0.1260	0.947	7.940	0.1259	0.9415	7.87
6	0.2010	1.500	12.52	0.1820	1.355	11.300	0.1797	1.3440	11.2
8	0.3480	2.600	21.68	0.3180	2.380	19.800	0.3135	2.3446	19.6
10	0.5470	4.100	34.16	0.5560	4.165	31.130	0.4867	3.4405	30.4
12	0.7850	5.870	48.50	0.7060	5.280	44.040	0.6978	5.2194	43.6
14	1.0690	7.030	58.64	0.8520	6.380	53.180	—	—	—
16	1.3920	9.180	76.58	1.1170	8.360	69.730	—	—	—
18	1.5530	11.120	96.93	1.4180	10.610	88.500	—	—	—
20	1.9250	14.400	120.56	1.7550	13.130	109.510	—	—	—

Nominal Diam. Inches	Type L Copper Tube			Type M Copper Tube			Type DWV Copper Drain Tube		
	Cu. Ft. per Minute	Gallons per Minute	Pounds 60 F Water per Min.	Cu. Ft. per Minute	Gallons per Minute	Pounds 60 F Water per Min.	Cu. Ft. per Minute	Gallons per Minute	Pounds 60 F Water per Min.
⅛	0.0002	0.002	0.014	0.0002	0.002	0.014	—	—	—
¼	0.0005	0.004	0.034	0.0006	0.004	0.036	—	—	—
⅜	0.0010	0.008	0.063	0.0011	0.008	0.069	—	—	—
½	0.0016	0.012	0.101	0.0018	0.013	0.110	—	—	—
¾	0.0034	0.025	0.210	0.0036	0.027	0.224	—	—	—
1	0.0057	0.043	0.358	0.0061	0.045	0.379	—	—	—
1¼	0.0087	0.065	0.545	0.0091	0.068	0.569	0.0091	0.068	0.569
1½	0.0124	0.093	0.770	0.0133	0.100	0.830	0.0129	0.097	0.850
2	0.0215	0.161	1.34	0.0220	0.165	1.35	0.0227	0.170	1.417
2½	0.0331	0.248	2.07	0.0340	0.254	2.12	—	—	—
3	0.0473	0.354	2.96	0.0485	0.363	3.03	0.0502	0.376	3.13
3½	0.0640	0.479	4.00	0.0653	0.488	4.08	—	—	—
4	0.0841	0.622	5.20	0.0834	0.626	5.23	0.0876	0.655	5.47
5	0.1296	0.969	8.10	0.1313	0.982	8.20	0.1352	1.152	8.33
6	0.1862	1.393	11.6	0.1885	1.410	11.78	0.1936	1.448	12.08
8	0.3253	2.434	20.3	0.3304	2.472	20.70	—	—	—
10	0.5050	3.777	21.6	0.5131	3.838	32.10	—	—	—
12	0.7291	5.454	45.6	0.7357	5.503	46.00	—	—	—

FRICTIONAL RESISTANCE OF WATER TO FLOW IN PIPES

In 1944, Prof. Lewis F. Moody embodied the then accepted conclusions of researchers in fluid mechanics on the subject of the friction factors for computing loss of head in pipes. His paper, Friction Factors for Pipe Flow, presented at the semi-annual meeting of The American Society of Mechanical Engineers, contained the Moody Chart which has gained an ever widening stature with working engineers, who find his chart a rapid and convenient tool in sizing piping systems.

Use of the Moody Chart has certain drawbacks, however. The chart curves give friction factor as a function of Reynolds Number plotted on logarithmic coordinates, and curved lines of relative roughness, necessitating two difficult interpolations. In 1947, Prof. Moody presented a formula for friction factor in terms of the same variables which agrees with the most theoretically correct formulation of the friction factor within ± 5% for all values of the variables within the scope of these tables.

The Moody formula is:

$$f = 0.0055 \left[1 + (20{,}000\, e/D + 10^6/N_R)^{1/3} \right]$$

where f = friction factor, dimensionless,
e/D = relative roughness, dimensionless,
e = absolute roughness, ft,
D = inside diameter of pipe, ft,
N_R = Reynolds Number = VD/v
V = mean fluid velocity, fps, and
v = coefficient of kinematic viscosity of the fluid, sq ft per sec; v = 0.00001216 for 60 F water, 0.00000385 for 180 F water, and 0.00000218 for 300 F water.

Friction Loss

In computing these tables, values of f from the above equation were substituted in the familiar Darcy formula for pipe friction loss,

$$H_L = f\,(L/D)\,(V^2/2g)$$

where H_L = pipe friction loss in feet of fluid flowing,
L = length of pipe, ft, and
g = gravitational constant, 32.2 ft per sec²

Pipe length L was taken at 100, so that H_L reads in feet per 100 ft of pipe. In the case of H_L values for 180 and 300 F water, feet of fluid is multiplied by the ratio of the densities of water at the elevated temperature to that of the density of 60 F water, so that all H_L values read in feet of 60 F water, which is common practice.

The resulting tabular values of friction loss are for new, clean steel pipe, with absolute roughness equal to 0.00015 ft, an approximation correct to within ± 5%. Because the absolute roughness of wrought iron pipe is the same, these tables apply equally to wrought iron pipe of the same inside diameter.

As pipe grows old in service, its interior walls may roughen considerably. As can be seen from the Moody equation, an increase of e will increase H_L. Thus, the piping designer must supply a correction factor to the friction loss values in the table, based upon his judgment of what the condition of the pipe will be during its life. *There is no one factor that applies to all "old pipes."* Well-deaerated high temperature water, for instance, has little corrosive effect on steel pipe, even after many years of operation of a high temperature water system. On the other hand, raw cold (or hot) water will, in many instances, materially corrode non-alloy steel pipe, increasing its roughness.

In the case of the smaller pipe, accumulation of corrosion products and scale deposits may significantly reduce its inside diameter. The resultant increase in velocity for the same flow rate increases the friction drop.

Users of the tables will find helpful, also, flow rates in pounds (or thousands of pounds) per hour for each temperature water. Not only does this arrangement present a handy conversion table, but, for heating engineering, the weight flow rates enables rapid calculation of heat transfer, bearing in mind that the specific heat of water is approximately unity for temperatures of 60 and 180 F, and 1.03 Btu per lb of water per deg temperature rise at 300 F.

Example—If 10 gpm of 300 F water is flowing through 200 ft of new, clean, 3/4-inch Schedule 40 pipe, the friction loss, from the tables, is 19.3 feet per 100 ft, or 38.6 ft of 60-deg water for the 200-ft length. Friction loss, in pounds per square inch, is 38.6 times the constant 0.433 psi per ft of 60 F water, or 16.7 psi.

If this pipe serves a heat exchanger, sized for a 100-deg drop, then the heat transfer is the weight flow rate, 4,597 lb per hr (from the table), times 1.03 Btu per lb-deg, times 100 deg, or 473,000 Btuh.

FRICTION LOSS IN ⅛-INCH SCHEDULE 40 STEEL PIPE *

(Note: Weight flow rate is given in pounds per hour.)

Flow Rate		Velocity, Ft per Sec	60-Deg Water		180-Deg Water		300-Deg Water	
Gal per Min	Cu Ft per Sec		Flow, Lb per Hr	Pressure Loss, Ft per 100 Ft	Flow, Lb per Hr	Pressure Loss, Ft per 100 Ft	Flow, Lb per Hr	Pressure Loss, Ft per 100 Ft
.2	0.000446	1.13	100	4.62	97	3.58	92	3.14
.3	0.000668	1.69	150	9.52	146	7.63	138	6.81
.4	0.000891	2.26	200	16.0	194	13.1	184	11.8
.5	0.00111	2.82	250	24.0	243	20.1	230	18.2
.6	0.00134	3.39	300	33.4	291	28.4	276	26.0
.7	0.00156	3.95	350	44.6	340	38.3	322	35.2
.8	0.00178	4.52	400	57.1	389	49.6	368	45.7
.9	0.00201	5.08	450	71.1	437	62.5	414	57.4
1.0	0.00223	5.65	500	86.6	486	76.4	460	70.7
1.2	0.00267	6.77	600	122	583	109	552	101
1.4	0.00312	7.90	700	164	680	148	644	137
1.6	0.00356	9.03	800	213	777	194	735	181
1.8	0.00401	10.2	900	264	874	241	827	225
2.0	0.00446	11.3	1,000	324	972	298	919	278

FRICTION LOSS IN ¼-INCH SCHEDULE 40 STEEL PIPE *

(Note: Weight flow rate is given in pounds per hour.)

Flow Rate		Velocity, Ft per Sec	60-Deg Water		180-Deg Water		300-Deg Water	
Gal per Min	Cu Ft per Sec		Flow, Lb per Hr	Pressure Loss, Ft per 100 Ft	Flow, Lb per Hr	Pressure Loss, Ft per 100 Ft	Flow, Lb per Hr	Pressure Loss, Ft per 100 Ft
0.4	0.000891	1.23	200	3.65	194	2.86	184	2.52
0.6	0.00134	1.85	300	7.54	291	6.11	276	5.47
0.8	0.00178	2.47	400	12.7	389	5.92	368	9.53
1.0	0.00223	3.08	500	19.1	486	16.2	460	14.7
1.2	0.00267	3.70	600	26.8	583	22.9	552	21.0
1.4	0.00312	4.32	700	35.6	680	30.9	644	28.4
1.6	0.00356	4.93	800	45.6	777	39.9	735	36.9
1.8	0.00401	5.55	900	56.9	874	50.2	827	46.5
2.0	0.00446	6.17	1,000	69.4	972	61.8	919	57.2
2.5	0.00557	7.71	1,250	106	1,215	95.4	1,149	89.0
3.0	0.00668	9.25	1,500	150	1,457	137	1,379	128
3.5	0.00780	10.8	1,750	202	1,700	185	1,609	173
4.0	0.00891	12.3	2,000	261	1,943	240	1,839	225
4.5	0.0100	13.9	2,250	328	2,186	304	2,069	285
5.0	0.0111	15.4	2,500	402	2,429	374	2,298	351

* To calculate the velocity (V_x) or the pressure drop (H_x) through pipe other than Schedule 40, use the following relationships:

$$V_x = V_{40}\left(\frac{D_{40}}{D_x}\right)^2; \ H_x = H_{40}\left(\frac{D_{40}}{D_x}\right)^5$$

FRICTION LOSS IN ⅜-INCH SCHEDULE 40 STEEL PIPE*

(Note: Weight flow rate is given in pounds per hour.)

Flow Rate		Velocity, Ft per Sec	60-Deg Water		180-Deg Water		300-Deg Water	
Gal per Min	Cu Ft per Sec		Flow, Lb per Hr	Pressure Loss, Ft Per 100 Ft	Flow, Lb per Hr	Pressure Loss, Ft per 100 Ft	Flow, Lb per Hr	Pressure Loss, Ft Per 100 Ft
0.6	0.00134	1.01	300	1.72	291	1.33	276	1.17
0.8	0.00178	1.34	400	2.88	389	2.28	368	2.02
1.0	0.00223	1.68	500	4.29	486	3.46	460	3.09
1.5	0.00334	2.52	750	8.92	729	7.44	690	6.74
2.0	0.00446	3.36	1,000	15.2	972	12.9	919	11.8
2.5	0.00557	4.20	1,250	22.9	1,215	19.8	1,149	18.2
3.0	0.00668	5.04	1,500	32.2	1,457	28.3	1,379	26.1
3.5	0.00780	5.88	1,750	43.0	1,700	38.2	1,609	35.4
4.0	0.00891	6.72	2,000	55.3	1,943	49.5	1,839	46.0
4.5	0.0100	7.56	2,250	69.2	2,186	62.5	2,069	58.1
5.0	0.0111	8.40	2,500	84.8	2,429	76.9	2,298	71.7
6.0	0.0134	10.1	3,000	120	2,915	110	2,758	102
7.0	0.0156	11.8	3,500	162	3,401	149	3,218	139
8.0	0.0178	13.4	4,000	209	3,887	194	3,677	182
9.0	0.0201	15.1	4,500	263	4,372	245	4,137	230
10.0	0.0223	16.8	5,000	323	4,858	302	4,597	283

FRICTION LOSS IN ½-INCH SCHEDULE 40 STEEL PIPE*

(Note: Weight flow rate is given in pounds per hour.)

Flow Rate		Velocity Ft per Sec	60-Deg Water		180-Deg Water		300-Deg Water	
Gal per Min	Cu Ft per Sec		Flow, Lb per Hr	Pressure Loss, Ft Per 100 Ft	Flow, Lb per Hr	Pressure Loss, Ft per 100 Ft	Flow, Lb per Hr	Pressure Loss, Ft Per 100 Ft
1.0	0.00223	1.06	500	1.38	486	1.08	460	0.946
1.5	0.00334	1.58	750	2.86	729	2.30	690	2.05
2.0	0.00446	2.11	1,000	4.81	972	3.95	919	3.56
2.5	0.00557	2.64	1,250	7.22	1,215	6.05	1,149	5.49
3.0	0.00668	3.17	1,500	10.1	1,457	8.58	1,379	7.83
3.5	0.00780	3.70	1,750	13.4	1,700	11.5	1,609	10.6
4.0	0.00891	4.22	2,000	17.2	1,943	14.9	1,839	13.7
4.5	0.0100	4.75	2,250	21.4	2,186	18.8	2,069	17.3
5.0	0.0111	5.28	2,500	26.1	2,429	23.0	2,298	21.3
6.0	0.0134	6.34	3,000	36.8	2,915	32.9	2,758	30.5
7.0	0.0156	7.39	3,500	49.4	3,401	44.5	3,218	41.4
8.0	0.0178	8.45	4,000	63.2	3,887	57.4	3,677	53.4
9.0	0.0201	9.50	4,500	79.6	4,372	72.7	4,137	67.8
10.0	0.0223	10.6	5,000	97.5	4,858	89.6	4,597	83.7
11.0	0.0245	11.6	5,500	118	5,344	108	5,056	101
12.0	0.0267	12.7	6,000	139	5,830	128	5,516	120
13.0	0.0290	13.7	6,500	162	6,316	151	5,976	141
14.0	0.0312	14.8	7,000	188	6,801	175	6,435	164
15.0	0.0334	15.8	7,500	214	7,287	200	6,895	188

* See footnote, page 8-8, for pipe other than Schedule 40.

FRICTION LOSS IN ¾-INCH SCHEDULE 40 STEEL PIPE*

(Note: Weight flow rate is given in pounds per hour.)

Flow Rate		Velocity, Ft per Sec	60-Deg Water		180-Deg Water		300-Deg Water	
Gal per Min	Cu Ft per Sec		Flow, Lb per Hr	Pressure Loss, Ft Per 100 Ft	Flow, Lb per Hr	Pressure Loss, Ft Per 100 Ft	Flow, Lb per Hr	Pressure Loss, Ft Per 100 Ft
1.0	0.00223	0.602	500	0.358	486	0.269	460	0.231
1.5	0.00334	0.903	750	0.734	729	0.566	690	0.494
2.0	0.00446	1.20	1,000	1.22	972	0.960	919	0.848
2.5	0.00557	1.50	1,250	1.82	1,215	1.46	1,149	1.30
3.0	0.00668	1.81	1,500	2.51	1,457	2.05	1,379	1.84
3.5	0.00780	2.11	1,750	3.32	1,700	2.75	1,609	2.48
4.0	0.00891	2.41	2,000	4.25	1,943	3.54	1,839	3.21
4.5	0.0100	2.71	2,250	5.27	2,186	4.44	2,069	4.04
5.0	0.0111	3.01	2,500	6.40	2,429	5.44	2,298	4.96
6.0	0.0134	3.61	3,000	8.98	2,915	7.73	2,758	7.09
7.0	0.0156	4.21	3,500	12.0	3,401	10.4	3,218	9.57
8.0	0.0178	4.81	4,000	15.3	3,887	13.5	3,677	12.4
9.0	0.0201	5.42	4,500	19.1	4,372	16.9	4,137	15.7
10.0	0.0223	6.02	5,000	23.3	4,858	20.8	4,597	19.3
11.0	0.0245	6.62	5,500	28.0	5,344	25.1	5,056	23.3
12.0	0.0267	7.22	6,000	32.5	5,830	29.7	5,516	27.6
13.0	0.0290	7.82	6,500	38.4	6,316	34.8	5,976	32.4
14.0	0.0312	8.42	7,000	44.3	6,801	40.1	6,435	37.4
15.0	0.0334	9.03	7,500	50.7	7,287	46.2	6,895	43.1
16.0	0.0356	9.63	8,000	57.2	7,773	52.3	7,355	48.8
17.0	0.0379	10.2	8,500	64.4	8,259	59.1	7,814	55.2
18.0	0.0401	10.8	9,000	71.6	8,745	65.9	8,274	61.6
19.0	0.0423	11.4	9,500	79.6	9,231	73.4	8,734	68.6
20.0	0.0446	12.0	10,000	87.9	9,716	81.2	9,193	76.0
22.0	0.0490	13.2	11,000	106	10,688	97.9	10,113	91.8
24.0	0.0535	14.4	12,000	125	11,660	117	11,032	109
26.0	0.0579	15.6	13,000	146	12,631	139	11,951	128
28.0	0.0624	16.8	14,000	169	13,602	158	12,871	148
30.0	0.0668	18.1	15,000	193	14,575	181	13,790	170

* See footnote, page 8-8, for pipe other than Schedule 40.

FRICTION LOSS IN 1-INCH SCHEDULE 40 STEEL PIPE *

(Note: Weight flow rate is given in pounds per hour.)

Flow Rate		Velocity, Ft per Sec	60-Deg Water		180-Deg Water		300-Deg Water	
Gal per Min	Cu Ft per Sec		Flow Rate, Lb per Hr	Pressure Loss, Ft per 100 Ft	Flow Rate, Lb per Hr	Pressure Loss, Ft per 100 Ft	Flow Rate, Lb per Hr	Pressure Loss, Ft per 100 Ft
I	0.00267	0.371	500	0.113	486	0.0830	460	0.0700
2	0.00446	0.742	1,000	0.381	972	0.290	919	0.252
3	0.00668	1.11	1,500	0.778	1,457	0.612	1,379	0.540
4	0.00891	1.48	2,000	1.30	1,943	1.05	1,839	0.936
5	0.0111	1.86	2,500	1.94	2,429	1.59	2,298	1.43
6	0.0134	2.23	3,000	2.71	2,915	2.25	2,758	2.04
7	0.0156	2.60	3,500	3.59	3,401	3.03	3,218	2.75
8	0.0178	2.97	4,000	4.59	3,887	4.08	3,677	3.56
9	0.0201	3.34	4,500	5.70	4,372	4.89	4,137	4.48
10	0.0223	3.71	5,000	6.94	4,858	6.00	4,597	5.51
12	0.0267	4.45	6,000	9.76	5,830	8.54	5,516	7.87
14	0.0312	5.20	7,000	13.0	6,801	11.6	6,435	10.7
16	0.0356	5.94	8,000	16.8	7,773	15.0	7,355	13.9
18	0.0401	6.68	9,000	21.0	8,745	18.9	8,274	17.5
20	0.0446	7.42	10,000	25.7	9,716	23.2	9,193	21.6
22	0.0490	8.17	11,000	30.8	10,688	28.1	10,113	26.1
24	0.0535	8.91	12,000	36.2	11,660	33.1	11,032	30.9
26	0.0579	9.65	13,000	42.5	12,631	38.9	11,951	36.3
28	0.0624	10.4	14,000	48.9	13,603	45.0	12,871	42.0
30	0.0668	11.1	15,000	55.9	14,575	51.6	13,790	48.2
32	0.0713	11.9	16,000	63.2	15,546	58.4	14,709	54.7
34	0.0758	12.6	17,000	71.3	16,518	66.1	15,629	61.9
36	0.0802	13.4	18,000	79.6	17,490	74.0	16,548	69.3
38	0.0847	14.1	19,000	88.2	18,461	82.1	17,467	77.0
40	0.0891	14.8	20,000	97.7	19,433	91.1	18,387	85.4
42	0.0936	15.6	21,000	107	20,404	100	19,306	94.1
44	0.0980	16.3	22,000	118	21,376	110	20,225	103
46	0.102	17.1	23,000	128	22,348	120	21,145	113
48	0.107	17.8	24,000	139	23,319	130	22,064	122
50	0.111	18.6	25,000	151	24,291	141	22,984	133
55	0.123	20.4	27,500	182	26,720	171	25,282	161
60	0.134	22.3	30,000	215	29,149	203	27,580	191
65	0.145	24.1	32,500	252	31,578	241	29,879	224
70	0.156	26.0	35,000	279	34,007	276	32,177	260
75	0.167	27.8	37,500	332	36,437	316	34,475	297
80	0.178	29.7	40,000	379	38,866	360	36,774	339
85	0.189	31.6	42,500	428	41,295	407	39,072	384
90	0.201	33.4	45,000	477	43,724	454	41,370	428
95	0.212	35.3	47,500	531	46,153	506	43,669	477
100	0.223	37.1	50,000	588	48,582	561	45,967	529

* See footnote, page 8-8, for pipe other than Schedule 40.

FRICTION LOSS IN 1¼-INCH SCHEDULE 40 STEEL PIPE*

(Note: Weight flow rate is given in pounds per hour.)

Flow Rate			60-Deg Water		180-Deg Water		300-Deg Water	
Gal per Min	Cu Ft per Sec	Velocity, Ft per Sec	Flow Rate, Lb per Hr	Pressure Loss, Ft per 100 Ft	Flow Rate, Lb per Hr	Pressure Loss, Ft per 100 Ft	Flow Rate, Lb per Hr	Pressure Loss, Ft per 100 Ft
2	0.00446	0.429	1,000	0.102	972	0.0758	919	0.0644
3	0.00668	0.644	1,500	0.208	1,457	0.158	1,379	0.137
4	0.00891	0.858	2,000	0.345	1,943	0.266	1,839	0.233
5	0.0111	1.073	2,500	0.514	2,429	0.404	2,298	0.356
6	0.0134	1.29	3,000	0.708	2,915	0.565	2,758	0.501
7	0.0156	1.50	3,500	0.935	3,401	0.755	3,218	0.676
8	0.0178	1.72	4,000	1.19	3,887	0.972	3,677	0.870
9	0.0201	1.93	4,500	1.47	4,372	1.21	3,137	1.09
10	0.0223	2.15	5,000	1.79	4,858	1.48	4,597	1.34
12	0.0267	2.57	6,000	2.49	5,830	2.10	5,516	1.92
14	0.0312	3.00	7,000	3.31	6,801	2.82	6,435	2.58
16	0.0356	3.43	8,000	4.24	7,773	3.65	7,355	3.35
18	0.0401	3.86	9,000	5.29	8,745	4.59	8,274	4.22
20	0.0446	4.29	10,000	6.43	9,716	5.63	9,193	5.18
22	0.0490	4.72	11,000	7.68	10,688	6.77	10,113	6.26
24	0.0535	5.15	12,000	9.06	11,660	8.02	11,032	7.42
26	0.0579	5.58	13,000	10.5	12,631	9.37	11,951	8.69
28	0.0624	6.01	14,000	12.1	13,603	10.8	12,871	10.0
30	0.0668	6.44	15,000	13.8	14,575	12.4	13,790	11.5
32	0.0713	6.86	16,000	15.6	15,546	14.1	14,709	13.1
34	0.0758	7.29	17,000	17.5	16,518	15.9	15,629	14.7
36	0.0802	7.72	18,000	19.6	17,490	17.7	16,548	16.5
38	0.0847	8.15	19,000	21.7	18,461	19.7	11,467	18.3
40	0.0891	8.58	20,000	23.8	19,433	21.7	18,387	20.3
42	0.0936	9.01	21,000	26.2	20,404	24.0	19,306	22.4
44	0.0980	9.44	22,000	28.6	21,376	26.2	20,225	24.4
46	0.102	9.87	23,000	31.2	22,348	28.7	21,145	26.7
48	0.107	10.3	24,000	34.0	23,319	31.3	22,064	29.2
50	0.111	10.7	25,000	36.8	24,291	33.9	22,984	31.7
55	0.123	11.8	27,500	44.1	26,720	40.8	25,282	38.2
60	0.134	12.9	30,000	52.2	29,149	48.5	27,580	45.5
65	0.145	13.9	32,500	61.1	31,578	56.8	29,879	53.5
70	0.156	15.0	35,000	70.5	34,007	65.8	32,177	61.6
75	0.167	16.1	37,500	80.7	36,437	75.5	34,475	70.8
80	0.178	17.2	40,000	91.5	38,866	85.8	36,774	80.6
85	0.189	18.2	42,500	103	41,295	96.8	39,072	91.0
90	0.201	19.3	45,000	115	43,724	108	41,370	102
95	0.212	20.4	47,500	128	46,153	121	43,669	114
100	0.223	21.5	50,000	142	48,582	134	45,967	126
110	0.245	23.6	55,000	171	53,440	161	50,564	152
120	0.267	25.7	60,000	203	58,298	192	55,160	135
130	0.290	27.9	65,000	237	63,157	225	59,757	159
140	0.312	30.0	70,000	274	68,015	260	64,354	183
150	0.334	32.2	75,000	314	72,873	299	68,951	211

* See footnote, page 8-8, for pipe other than Schedule 40.

FRICTION LOSS IN 1½-INCH SCHEDULE 40 STEEL PIPE *

(Note: Weight flow rate is given in thousands of pounds per hour.)

Flow Rate		Velocity, Ft per Sec	60-Deg Water		180-Deg Water		300-Deg Water	
Gal per Min	Cu Ft per Sec		Flow Rate, 1000 Lb per Hr	Pressure Loss, Ft per 100 Ft	Flow Rate, 1000 Lb per Hr	Pressure Loss, Ft per 100 Ft	Flow Rate, 1000 Lb per Hr	Pressure Loss, Ft per 100 Ft
2	0.00446	0.315	1	.0490	0.972	0.0358	0.919	0.0302
3	0.00668	0.473	1.5	.0993	1.457	0.0741	1.379	0.0632
4	0.00891	0.630	2	.165	1.943	0.125	1.838	0.108
5	0.0111	0.788	2.5	.243	2.429	0.188	2.298	0.164
6	0.0134	0.946	3	.336	2.915	0.263	2.758	0.231
7	0.0156	1.10	3.5	.443	3.401	0.350	3.218	0.308
8	0.0178	1.26	4	.562	3.887	0.449	3.677	0.398
9	0.0201	1.42	4.5	.695	4.372	0.560	4.137	0.498
10	0.0223	1.58	5	.839	4.858	0.682	4.597	0.610
12	0.0267	1.89	6	1.17	5.830	0.962	5.516	0.867
14	0.0312	2.21	7	1.54	6.801	1.29	6.435	1.16
16	0.0356	2.52	8	1.98	7.773	1.66	7.355	1.51
18	0.0401	2.84	9	2.45	8.745	2.09	8.274	1.90
20	0.0446	3.15	10	2.98	9.716	2.55	9.193	2.34
22	0.0490	3.47	11	3.56	10.69	3.08	10.11	2.82
24	0.0535	3.78	12	4.18	11.66	3.63	11.03	3.33
26	0.0579	4.10	13	4.85	12.63	4.25	11.95	3.92
28	0.0624	4.41	14	5.59	13.60	4.91	12.87	4.51
30	0.0668	4.73	15	6.35	14.57	5.60	13.79	5.17
32	0.0713	5.04	16	7.17	15.55	6.35	14.71	5.89
34	0.0758	5.36	17	8.04	16.52	7.15	15.63	6.60
36	0.0802	5.67	18	8.97	17.49	8.00	16.55	7.40
38	0.0847	5.99	19	10.3	18.46	9.21	17.47	
40	0.0891	6,30	20	11.0	19.43	9.84	18.39	9.15
42	0.0936	6.62	21	12.0	20.40	10.8	19.31	10.1
44	0.0980	6.93	22	13.1	21.38	11.8	20.23	11.0
46	0.102	7.25	23	14.3	22.35	12.9	21.14	12.0
48	0.107	7.56	24	15.5	23.32	14.1	22.06	13.1
50	0.111	7.88	25	16.8	24.29	15.3	22.98	14.2
55	0.123	8.67	27.5	20.2	26.72	18.5	25.28	17.2
60	0.134	9.46	30	23.8	29.15	21.8	27.58	20.4
65	0.145	10.2	32.5	27.7	31.58	25.5	29.88	23.8
70	0.156	11.0	35	32.0	34.01	29.5	32.18	27.6
75	0.167	11.8	37.5	36.6	36.44	33.8	34.48	31.7
80	0.178	12.6	40	41.4	38.87	38.5	36.77	36.1
85	0.189	13.4	42.5	46.6	41.29	43.4	39.07	40.7
90	0.201	14.2	45	52.1	43.72	48.6	41.37	45.7
95	0.212	15.0	47.5	57.8	46.15	54.0	43.67	50.8
100	0.223	15.8	50	63.9	48.58	59.8	45.97	56.0
110	0.245	17.3	55	77.0	53.44	72.2	50.56	67.7
120	0.267	18.9	60	91.3	58.30	85.9	55.16	80.6
130	0.290	20.5	65	107	63.16	101	59.76	94.5
140	0.312	22.1	70	123	68.01	117	64.35	110
150	0.334	23.6	75	141	72.87	134	68.95	126
160	0.356	25.2	80	161	77.73	152	73.55	143
170	0.379	26.8	85	181	82.59	172	78.14	162
180	0.401	28.4	90	202	87.45	192	82.74	18.1
190	0.423	29.9	95	225	92.31	214	87.34	20.2
200	0.446	31.5	100	248	97.16	236	91.93	22.3

* See footnote, page 8-8, for pipe other than Schedule 40.

FRICTION LOSS IN 2-INCH SCHEDULE 40 STEEL PIPE *

(Note: Weight flow rate is given in thousands of pounds per hour.)

Flow Rate		Velocity, Ft per Sec	60-Deg Water		180-Deg Water		300-Deg Water	
Gal per Min	Cu Ft per Sec		Flow Rate, 1000 Lb per Hr	Pressure Loss, Ft per 100 Ft	Flow Rate, 1000 Lb per Hr	Pressure Loss, Ft per 100 Ft	Flow Rate, 1000 Lb per Hr	Pressure Loss, Ft per 100 Ft
2	0.00446	0.191	1	0.0150	0.972	0.0108	0.919	0.00897
4	0.00891	0.382	2	0.0498	1.943	0.0368	1.838	0.0313
6	0.0134	0.574	3	0.101	2.915	0.0766	2.758	0.0659
8	0.0178	0.765	4	0.168	3.887	0.130	3.677	0.113
10	0.0223	0.956	5	0.249	4.858	0.196	4.597	0.172
12	0.0267	1.15	6	0.346	5.830	0.275	5.516	0.244
14	0.0312	1.34	7	0.455	6.801	0.366	6.435	0.325
16	0.0356	1.53	8	0.580	7.773	0.471	7.355	0.422
18	0.0401	1.72	9	0.717	8.745	0.588	8.274	0.529
20	0.0446	1.91	10	0.868	9.716	0.718	9.193	0.648
22	0.0490	2.10	11	1.03	10.69	0.861	10.11	0.777
24	0.0535	2.29	12	1.21	11.66	1.02	11.03	0.924
26	0.0579	2.49	13	1.40	12.63	1.18	11.95	1.08
28	0.0624	2.68	14	1.60	13.60	1.36	12.87	1.24
30	0.0668	2.87	15	1.83	14.57	1.56	13.79	1.42
35	0.0780	3.35	17.5	2.43	17.00	2.10	16.09	1.93
40	0.0891	3.82	20	3.11	19.43	2.71	18.39	2.50
45	0.100	4.30	22.5	3.89	21.86	3.42	20.69	3.14
50	0.111	4.78	25	4.74	24.29	4.19	22.98	3.87
55	0.123	5.26	27.5	5.67	26.72	5.05	25.28	4.69
60	0.134	5.74	30	6.68	29.15	5.98	27.58	5.52
65	0.145	6.21	32.5	7.79	31.58	7.00	29.88	6.48
70	0.156	6.69	35	8.96	34.01	8.09	32.18	7.52
75	0.167	7.17	37.5	10.2	36.44	9.26	34.48	8.63
80	0.178	7.65	40	11.6	38.87	10.5	36.77	9.82
85	0.189	8.13	42.5	13.0	41.29	11.9	39.07	11.1
90	0.201	8.60	45	14.5	43.72	13.2	41.37	12.4
95	0.212	9.08	47.5	16.1	46.15	14.7	43.67	13.7
100	0.223	9.56	50	17.8	48.58	16.3	45.97	15.2
110	0.245	10.5	55	21.3	53.44	19.7	50.56	18.4
120	0.267	11.5	60	25.3	58.30	23.4	55.16	21.9
130	0.290	12.4	65	29.5	63.16	27.4	59.76	25.7
140	0.312	13.4	70	34.0	68.01	31.7	64.35	29.7
150	0.334	14.3	75	39.0	72.87	36.4	68.95	34.2
160	0.356	15.3	80	44.2	77.73	41.4	73.55	38.9
170	0.379	16.3	85	49.7	82.59	46.6	78.14	44.0
180	0.401	17.2	90	55.5	87.45	52.1	82.74	48.8
190	0.423	18.2	95	61.7	92.31	58.1	87.34	54.4
200	0.446	19.1	100	68.2	97.16	64.2	91.93	60.2
220	0.490	21.0	110	82.4	106.9	77.7	101.1	72.9
240	0.535	22.9	120	97.6	116.6	92.4	110.3	86.7
260	0.579	24.9	130	114	126.3	108	119.5	102
280	0.624	26.8	140	132	136.0	125	128.7	118
300	0.668	28.7	150	152	145.7	144	137.9	136
320	0.713	30.6	160	171	155.5	163	147.1	154
340	0.758	32.5	170	194	165.2	184	156.3	174
360	0.802	34.4	180	217	174.9	207	165.5	195
380	0.847	36.3	190	242	184.6	230	174.7	217
400	0.891	38.2	200	267	194.3	255	183.9	241

* See footnote, page 8-8, for pipe other than Schedule 40.

FRICTION LOSS IN 2½-INCH SCHEDULE 40 STEEL PIPE *

(Note: Weight flow rate is given in thousands of pounds per hour.)

Flow Rate		Velocity, Ft per Sec	60-Deg Water		180-Deg Water		300-Deg Water	
Gal per Min	Cu Ft per Sec		Flow Rate, 1000 Lb per Hr	Pressure Loss, Ft per 100 Ft	Flow Rate, 1000 Lb per Hr	Pressure Loss, Ft per 100 Ft	Flow Rate, 1000 Lb per Hr	Pressure Loss, Ft per 100 Ft
4	0.00891	0.268	2	0.0215	1.943	0.0157	1.838	0.0131
6	0.0134	0.402	3	0.0432	2.915	0.0322	2.758	0.0274
8	0.0178	0.536	4	0.0715	3.887	0.0542	3.677	0.0465
10	0.0223	0.670	5	0.106	4.858	0.0814	4.597	0.0705
12	0.0267	0.804	6	0.146	5.830	0.113	5.516	0.0990
14	0.0312	0.938	7	0.193	6.801	0.151	6.435	0.133
16	0.0356	1.07	8	0.245	7.773	0.194	7.355	0.172
18	0.0401	1.21	9	0.301	8.745	0.241	8.274	0.214
20	0.0446	1.34	10	0.365	9.716	0.294	9.193	0.262
22	0.0490	1.47	11	0.433	10.69	0.352	10.11	0.315
24	0.0535	1.61	12	0.507	11.66	0.414	11.03	0.372
26	0.0579	1.74	13	0.585	12.63	0.482	11.95	0.434
28	0.0624	1.88	14	0.667	13.60	0.555	12.87	0.501
30	0.0668	2.01	15	0.760	14.57	0.632	13.79	0.572
35	0.0780	2.35	17.5	1.01	17.00	0.849	16.09	0.771
40	0.0891	2.68	20	1.29	19.43	1.10	18.39	1.02
45	0.100	3.02	22.5	1.60	21.86	1.37	20.69	1.26
50	0.111	3.35	25	1.94	24.29	1.68	22.98	1.54
55	0.123	3.69	27.5	2.33	26.72	2.03	25.28	1.86
60	0.134	4.02	30	2.74	29.15	2.40	27.58	2.21
65	0.145	4.36	32.5	3.19	31.58	2.81	29.88	2.59
70	0.156	4.69	35	3.66	34.01	3.24	32.18	3.00
75	0.167	5.03	37.5	4.17	36.44	3.71	34.48	3.43
80	0.178	5.36	40	4.72	38.87	4.21	36.72	3.90
85	0.189	5.70	42.5	5.28	41.29	4.73	39.07	4.38
90	0.201	6.03	45	5.89	43.72	5.29	41.37	4.90
95	0.212	6.37	47.5	6.53	46.15	5.88	43.67	5.46
100	0.223	6.70	50	7.09	48.58	6.51	45.97	6.04
110	0.245	7.37	55	8.63	53.44	7.83	50.56	7.30
120	0.267	8.04	60	10.2	58.30	9.26	55.16	8.63
130	0.290	8.71	65	11.9	63.16	10.9	59.76	10.2
140	0.312	9.38	70	14.3	68.01	12.6	64.36	11.8
150	0.334	10.1	75	15.7	72.87	14.4	68.96	13.5
160	0.356	10.7	80	17.8	77.73	16.4	73.55	15.4
170	0.379	11.4	85	20.1	82.59	18.5	78.14	17.3
180	0.401	12.1	90	22.3	87.45	20.7	82.74	19.4
190	0.423	12.7	95	24.8	92.31	23.1	87.34	21.6
200	0.446	13.4	100	27.3	97.16	25.5	91.93	23.9
220	0.490	14.7	110	33.0	106.9	30.8	101.1	28.9
240	0.535	16.1	120	39.0	116.6	36.6	110.3	34.3
260	0.579	17.4	130	45.7	126.3	43.0	119.5	40.3
280	0.624	18.8	140	52.7	136.0	49.7	128.7	46.7
300	0.668	20.1	150	60.4	145.7	57.0	137.9	53.6
350	0.780	23.5	175	81.7	170.0	77.5	160.9	72.8
400	0.891	26.8	200	107	194.3	101	183.9	95.3
450	1.003	30.2	225	134	218.6	127	206.9	120
500	1.114	33.5	250	164	242.9	157	229.8	148
550	1.225	36.9	275	199	267.2	190	252.8	179
600	1.337	40.2	300	236	291.5	226	275.8	213

* See footnote, page 8-8, for pipe other than Schedule 40.

FRICTION LOSS IN 3-INCH SCHEDULE 40 STEEL PIPE *

(Note: Weight flow rate is given in thousands of pounds per hour.)

Flow Rate			60-Deg Water		180-Deg Water		300-Deg Water	
Gal per Min	Cu Ft per Sec	Velocity, Ft per Sec	Flow Rate, 1000 Lb per Hr	Pressure Loss, Ft per 100 Ft	Flow Rate, 1000 Lb per Hr	Pressure Loss, Ft per 100 Ft	Flow Rate, 1000 Lb per Hr	Pressure Loss, Ft per 100 Ft
5	0.0111	0.217	2.5	0.0113	2.429	0.00732	2.298	0.00681
10	0.0223	0.434	5	0.0375	4.858	0.0281	4.596	0.0241
15	0.0334	0.651	7.5	0.0764	7.287	0.0587	6.895	0.0510
20	0.0446	0.868	10	0.128	9.716	0.0997	9.193	0.0876
25	0.0557	1.09	12.5	0.190	12.15	0.151	11.49	0.134
30	0.0668	1.30	15	0.262	14.57	0.212	13.79	0.189
35	0.0780	1.52	17.5	0.348	17.00	0.284	16.09	0.255
40	0.0891	1.74	20	0.442	19.43	0.365	18.39	0.329
45	0.100	1.95	22.5	0.549	21.86	0.458	20.69	0.413
50	0.111	2.17	25	0.666	24.29	0.559	22.98	0.507
55	0.123	2.39	27.5	0.792	26.72	0.670	25.28	0.609
60	0.134	2.60	30	0.927	29.15	0.789	27.58	0.720
65	0.145	2.82	32.5	1.08	31.58	0.926	29.88	0.847
70	0.156	3.04	35	1.24	34.01	1.06	32.18	0.974
75	0.167	3.25	37.5	1.41	36.44	1.22	34.48	1.12
80	0.178	3.47	40	1.59	38.87	1.38	36.77	1.27
85	0.189	3.69	42.5	1.77	41.29	1.55	39.07	1.42
90	0.201	3.91	45	1.97	43.72	1.73	41.37	1.60
95	0.212	4.12	47.5	2.19	46.15	1.92	43.67	1.77
100	0.223	4.34	50	2.46	48.58	2.16	45.97	1.97
110	0.245	4.77	55	2.89	53.44	2.56	50.56	2.37
120	0.267	5.21	60	3.40	58.30	3.03	55.16	2.81
130	0.290	5.64	65	3.99	63.16	3.55	59.76	3.30
140	0.312	6.08	70	4.56	68.01	4.10	64.35	3.82
150	0.334	6.51	75	5.21	72.87	4.71	68.95	4.38
160	0.356	6.94	80	5.88	77.73	5.33	73.55	4.97
170	0.379	7.38	85	6.61	82.59	6.01	78.14	5.59
180	0.401	7.81	90	7.37	87.45	6.72	82.74	6.27
190	0.423	8.25	95	8.20	92.31	7.50	87.34	7.00
200	0.446	8.68	100	9.01	97.16	8.27	91.93	7.71
220	0.490	9.55	110	10.9	106.9	10.0	101.1	9.34
240	0.535	10.4	120	12.8	116.6	11.9	110.3	11.1
260	0.579	11.3	130	15.0	126.3	13.9	119.5	13.0
280	0.624	12.2	140	17.2	136.0	16.0	128.7	15.0
300	0.668	13.0	150	19.7	145.7	18.4	137.9	17.2
320	0.713	13.9	160	22.4	155.5	20.9	147.1	19.7
340	0.758	14.8	170	25.2	165.2	23.6	156.3	22.1
360	0.802	15.6	180	28.2	174.9	26.4	165.5	24.8
380	0.847	16.5	190	31.3	184.6	29.4	174.7	27.7
400	0.891	17.4	200	34.6	194.3	32.5	183.9	30.6
420	0.936	18.2	210	38.1	204.0	35.9	193.1	33.7
440	0.980	19.1	220	41.7	213.8	39.3	202.2	37.0
460	1.025	20.0	230	45.5	223.5	43.0	211.4	40.4
480	1.069	20.8	240	49.5	233.2	46.8	220.6	43.9
500	1.114	21.7	250	53.6	242.9	50.7	229.8	47.7
600	1.337	26.0	300	76.5	291.5	72.7	275.8	68.4
700	1.560	30.4	350	104	340.1	98.8	321.8	93.1
800	1.782	34.7	400	135	388.7	129	367.7	122
1,000	2.228	43.4	500	210	485.8	202	459.7	190

* See footnote, page 8-8, for pipe other than Schedule 40.

FRICTION LOSS IN 3½-INCH SCHEDULE 40 STEEL PIPE *

(Note: Weight flow rate is given in thousands of pounds per hour.)

Gal per Min	Cu Ft per Sec	Velocity, Ft per Sec	Flow Rate, 1000 Lb per Hr	Pressure Loss, Ft per 100 Ft	Flow Rate, 1000 Lb per Hr	Pressure Loss, Ft per 100 Ft	Flow Rate, 1000 Lb per Hr	Pressure Loss, Ft per 100 Ft
			60-Deg Water		180-Deg Water		300-Deg Water	
5	0.0111	0.162	2.5	0.00564	2.429	0.00406	2.298	0.00337
10	0.0223	0.323	5	0.0188	4.858	0.0139	4.597	0.0118
15	0.0334	0.487	7.5	0.0380	7.287	0.0288	6.895	0.0248
20	0.0446	0.649	10	0.0632	9.716	0.0487	9.193	0.0423
25	0.0557	0.811	12.5	0.0936	12.15	0.117	11.49	0.0642
30	0.0668	0.974	15	0.130	14.57	0.103	13.79	0.0907
35	0.0780	1.14	17.5	0.171	17.00	0.137	16.09	0.121
40	0.0891	1.30	20	0.218	19.43	0.176	18.39	0.157
45	0.100	1.46	22.5	0.269	21.86	0.219	20.69	0.197
50	0.111	1.62	25	0.326	24.29	0.268	22.98	0.241
60	0.134	1.95	30	0.455	29.15	0.379	27.58	0.343
70	0.156	2.27	35	0.604	34.01	0.509	32.18	0.463
80	0.178	2.60	40	0.774	38.87	0.658	36.77	0.602
90	0.201	2.92	45	0.963	43.72	0.827	41.37	0.758
100	0.223	3.25	50	1.17	48.58	1.01	45.97	0.930
110	0.245	3.57	55	1.40	53.44	1.22	50.56	1.12
120	0.267	3.89	60	1.64	58.30	1.44	55.16	1.33
130	0.290	4.22	65	1.91	63.16	1.68	59.76	1.55
140	0.312	4.54	70	2.20	68.01	1.95	64.35	1.80
150	0.334	4.87	75	2.50	72.87	2.22	68.95	2.06
160	0.356	5.19	80	2.82	77.73	2.52	73.55	2.34
170	0.379	5.52	85	3.17	82.59	2.96	78.14	2.63
180	0.401	5.84	90	3.53	87.45	3.17	82.74	2.95
190	0.423	6.17	95	3.91	92.31	3.53	87.34	3.28
200	0.446	6.49	100	4.32	97.16	3.90	91.93	3.63
220	0.490	7.14	110	5.18	106.9	4.70	101.3	4.38
240	0.535	7.79	120	6.12	116.6	5.58	110.3	5.21
260	0.579	8.44	130	7.16	126.3	6.56	119.5	6.13
280	0.624	9.09	140	8.22	136.0	7.55	128.7	7.05
300	0.668	9.74	150	9.39	145.7	8.66	137.9	8.09
320	0.713	10.4	160	10.6	155.5	9.86	147.1	9.22
340	0.758	11.0	170	12.0	165.2	11.1	156.3	10.4
360	0.802	11.7	180	13.4	174.9	12.4	165.5	11.6
380	0.847	12.3	190	14.8	184.6	13.8	174.7	12.9
400	0.891	13.0	200	16.4	194.3	15.3	183.9	14.4
420	0.936	13.6	210	18.1	204.0	16.9	193.1	15.8
440	0.980	14.3	220	21.6	213.8	18.5	202.2	17.3
460	1.025	14.9	230	21.5	223.5	20.2	211.4	18.9
480	1.069	15.6	240	23.4	233.2	21.9	220.6	20.6
500	1.114	16.2	250	30.7	242.9	23.8	229.8	22.4
600	1.337	19.5	300	36.2	291.5	34.2	275.8	32.2
700	1.560	22.7	350	49.1	340.0	46.4	321.8	43.7
800	1.782	26.0	400	63.8	388.7	60.7	367.7	57.1
1,000	2.228	32.5	500	99.2	485.8	94.6	459.7	89.1
1,100	2.451	35.7	550	119	534.4	114	505.6	108
1,200	2.674	38.9	600	142	583.0	136	551.6	128
1,300	2.896	42.2	650	166	631.6	160	597.6	150
1,400	3.119	45.4	700	193	680.1	185	643.5	174
1,500	3.342	48.7	750	221	728.7	212	689.5	200

* See footnote, page 8-8, for pipe other than Schedule 40.

FRICTION LOSS IN 4-INCH SCHEDULE 40 STEEL PIPE*

(Note: Weight flow rate is given in thousands of pounds per hour.)

Flow Rate			60-Deg Water		180-Deg Water		300-Deg Water	
Gal per Min	Cu Ft per Sec	Velocity, Ft per Sec	Flow Rate, 1000 Lb per Hr	Pressure Loss, Ft per 100 Ft	Flow Rate, 1000 Lb per Hr	Pressure Loss, Ft per 100 Ft	Flow Rate, 1000 Lb per Hr	Pressure Loss, Ft per 100 Ft
20	0.0446	0.504	10	0.0345	9.716	0.0262	9.193	0.0226
30	0.0668	0.756	15	0.0706	14.57	0.0550	13.79	0.0481
40	0.0981	1.01	20	0.118	19.43	0.0939	18.39	0.0831
50	0.111	1.26	25	0.176	24.29	0.143	22.98	0.127
60	0.134	1.51	30	0.245	29.15	0.201	27.58	0.180
70	0.156	1.76	35	0.325	34.01	0.269	32.18	0.242
80	0.178	2.02	40	0.414	38.87	0.347	36.77	0.315
90	0.201	2.27	45	0.514	43.72	0.434	41.37	0.395
100	0.223	2.52	50	0.624	48.58	0.532	45.97	0.485
110	0.245	2.77	55	0.741	53.44	0.637	50.56	0.582
120	0.267	3.02	60	0.875	58.30	0.754	55.16	0.692
130	0.290	3.28	65	1.02	63.16	0.882	59.76	0.810
140	0.312	3.53	70	1.16	68.01	1.02	64.35	0.934
150	0.334	3.78	75	1.33	72.87	1.16	68.95	1.07
160	0.356	4.03	80	1.50	77.73	1.32	73.55	1.22
170	0.379	4.28	85	1.68	82.59	1.48	78.14	1.37
180	0.401	4.54	90	1.87	87.45	1.66	82.74	1.54
190	0.423	4.78	95	2.07	92.31	1.84	87.34	1.71
200	0.446	5.04	100	2.28	97.16	2.04	91.93	1.89
220	0.490	5.54	110	2.74	106.9	2.46	101.1	2.28
240	0.535	6.05	120	3.23	116.6	2.91	110.3	2.70
260	0.579	6.55	130	3.76	126.3	3.40	119.5	3.17
280	0.624	7.06	140	4.33	136.0	3.94	128.7	3.67
300	0.668	7.56	150	4.94	145.7	4.51	137.9	4.20
320	0.713	8.06	160	5.59	155.5	5.12	147.1	4.78
340	0.758	8.57	170	6.28	165.2	5.77	156.3	5.38
360	0.802	9.07	180	7.01	174.9	6.46	165.5	6.04
380	0.847	9.58	190	7.81	184.6	7.21	174.7	6.74
400	0.891	10.1	200	8.59	194.3	7.95	183.9	7.44
420	0.936	10.6	210	9.47	204.0	8.75	193.1	8.20
440	0.980	11.1	220	10.3	213.8	9.59	202.2	8.98
460	1.025	11.6	230	11.3	223.5	10.5	211.4	9.82
480	1.069	12.1	240	12.2	233.2	11.4	220.6	10.7
500	1.114	12.6	250	13.3	242.9	12.4	229.8	11.6
600	1.337	15.1	300	18.9	291.5	17.7	275.8	16.6
700	1.560	17.6	350	25.6	340.1	24.1	321.8	22.7
800	1.782	20.2	400	33.2	388.7	31.4	367.7	29.5
900	2.005	22.7	450	41.8	437.2	39.7	413.7	37.4
1,000	2.228	25.2	500	51.5	485.8	48.9	459.7	46.0
1,100	2.451	27.7	550	62.0	534.4	58.9	505.6	55.5
1,200	2.674	30.2	600	73.6	583.0	70.1	551.6	66.2
1,300	2.896	32.8	650	86.5	631.6	82.5	597.6	77.8
1,400	3.119	35.3	700	99.8	680.1	95.3	643.5	89.9
1,500	3.342	37.8	750	115	728.7	109	689.5	103
1,600	3.565	40.3	800	130	777.3	125	735.5	118
1,700	3.788	42.8	850	147	825.9	141	781.4	133
1,800	4.010	45.0	900	165	874.5	158	827.4	149
1,900	4.233	47.9	950	183	923.1	176	873.4	166
2,000	4.456	50.4	1,000	203	971.6	195	919.3	184

* See footnote, page 8-8, for pipe other than Schedule 40.

FRICTION LOSS IN 5-INCH SCHEDULE 40 STEEL PIPE*

(Note: Weight flow rate is given in thousands of pounds per hour.)

Flow Rate			60-Deg Water		180-Deg Water		300-Deg Water	
Gal per Min	Cu Ft per Sec	Velocity, Ft per Sec	Flow Rate, 1000 Lb per Hr	Pressure Loss, Ft per 100 Ft	Flow Rate, 1000 Lb per Hr	Pressure Loss, Ft per 100 Ft	Flow Rate, 1000 Lb per Hr	Pressure Loss, Ft per 100 Ft
20	0.0446	0.321	10	0.0117	9.716	0.00872	9.193	0.00742
40	0.0891	0.641	20	0.0396	19.43	0.0307	18.39	0.0267
60	0.134	0.962	30	0.0816	29.15	0.0649	27.58	0.0575
80	0.178	1.28	40	0.137	38.87	0.111	36.77	0.0996
100	0.223	1.60	50	0.206	48.58	0.170	45.97	0.153
120	0.267	1.92	60	0.286	58.30	0.240	55.16	0.218
140	0.312	2.25	70	0.381	68.01	0.322	64.35	0.293
160	0.356	2.57	80	0.486	77.73	0.415	73.55	0.379
180	0.401	2.89	90	0.604	87.45	0.521	82.74	0.477
200	0.446	3.21	100	0.739	97.16	0.642	91.93	0.589
220	0.490	3.53	110	0.880	106.9	0.768	101.1	0.708
240	0.535	3.85	120	1.04	116.6	0.911	110.3	0.840
260	0.579	4.17	130	1.21	126.3	1.07	119.5	0.983
280	0.624	4.49	140	1.39	136.0	1.23	128.7	1.14
300	0.668	4.81	150	1.58	145.7	1.41	137.9	1.31
320	0.713	5.13	160	1.80	155.5	1.60	147.1	1.48
340	0.758	5.45	170	2.01	165.2	1.80	156.3	1.67
360	0.802	5.77	180	2.24	174.9	2.02	165.5	1.87
380	0.847	6.09	190	2.48	184.6	2.24	174.7	2.08
400	0.891	6.41	200	2.73	194.3	2.47	183.9	2.30
420	0.936	6.74	210	3.00	204.0	2.72	193.1	2.54
440	0.980	7.06	220	3.27	213.8	2.99	202.3	2.78
460	1.02	7.38	230	3.58	223.5	3.26	211.4	3.04
480	1.07	7.70	240	3.88	233.2	3.55	220.6	3.31
500	1.11	8.02	250	4.19	242.9	3.84	229.8	3.58
550	1.23	8.82	275	5.03	267.2	4.63	252.8	4.33
600	1.34	9.62	300	5.96	291.5	5.50	275.8	5.14
650	1.45	10.4	325	6.96	315.8	6.44	298.8	6.03
700	1.56	11.2	350	8.04	340.1	7.47	321.8	7.00
750	1.67	12.0	375	9.18	364.4	8.55	344.8	8.01
800	1.78	12.8	400	10.4	388.7	9.73	367.7	9.11
850	1.89	13.6	425	11.7	412.9	11.0	390.7	10.3
900	2.01	14.4	450	13.1	437.2	12.3	413.7	11.5
950	2.12	15.2	475	14.5	461.5	13.6	436.7	12.8
1,000	2.23	16.0	500	16.1	485.8	15.1	459.7	14.2
1,100	2.45	17.6	550	19.4	534.4	18.2	505.6	17.2
1,200	2.67	19.2	600	23.0	583.0	21.7	551.6	20.4
1,300	2.90	20.8	650	26.9	631.6	25.4	597.6	23.9
1,400	3.12	22.5	700	31.1	680.1	29.4	643.5	27.7
1,500	3.34	24.1	750	35.6	728.7	33.8	689.5	31.8
1,600	3.56	25.7	800	40.3	777.3	38.4	735.5	36.1
1,700	3.79	27.3	850	45.8	825.9	43.6	781.4	41.1
1,800	4.01	28.8	900	50.8	874.5	48.4	827.4	45.7
1,900	4.23	30.5	950	56.6	923.1	54.0	873.4	51.0
2,000	4.46	32.1	1,000	62.9	971.6	60.0	919.3	56.5
2,100	4.68	33.7	1,050	69.0	102.0	66.0	965.3	62.1
2,200	4.90	35.3	1,100	75.7	106.9	72.4	1,011	68.1
2,300	5.12	36.9	1,150	82.7	111.7	79.1	1,057	74.5
2,500	5.57	40.1	1,250	97.8	121.5	93.5	1,149	88.3

* See footnote, page 8-8, for pipe other than Schedule 40.

FRICTION LOSS IN 6-INCH SCHEDULE 40 STEEL PIPE*

(Note: Weight flow rate is given in thousands of pounds per hour.)

Flow Rate			60-Deg Water		180-Deg Water		300-Deg Water	
Gal per Min	Cu Ft per Sec	Velocity, Ft per Sec	Flow Rate, 1000 Lb per Hr	Pressure Loss, Ft per 100 Ft	Flow Rate, 1000 Lb per Hr	Pressure Loss, Ft per 100 Ft	Flow Rate, 1000 Lb per Hr	Pressure Loss, Ft per 100 Ft
20	0.0446	0.222	10	0.00489	9.716	0.00359	9.193	0.00302
40	0.0891	0.444	20	0.0164	19.43	0.0125	18.39	0.0107
60	0.134	0.666	30	0.0336	29.15	0.0262	27.58	0.0228
80	0.178	0.888	40	0.0562	38.87	0.0446	36.77	0.0394
100	0.223	1.11	50	0.0841	48.58	0.0676	45.97	0.0602
120	0.267	1.33	60	0.117	58.30	0.0952	55.16	0.0852
140	0.312	1.55	70	0.155	68.01	0.127	64.35	0.115
160	0.356	1.78	80	0.197	77.73	0.164	73.55	0.148
180	0.401	2.00	90	0.245	87.45	0.206	82.74	0.186
200	0.446	2.22	100	0.297	97.16	0.252	91.93	0.229
220	0.490	2.44	110	0.353	106.9	0.302	101.2	0.275
240	0.535	2.66	120	0.415	116.6	0.355	110.3	0.325
260	0.579	2.89	130	0.484	126.3	0.417	119.5	0.383
280	0.624	3.11	140	0.554	136.0	0.480	128.7	0.440
300	0.668	3.33	150	0.628	145.7	0.547	137.9	0.503
320	0.713	3.55	160	0.710	155.4	0.621	147.1	0.572
340	0.758	3.78	170	0.797	165.2	0.702	156.3	0.647
360	0.802	4.00	180	0.857	174.9	0.756	165.5	0.698
380	0.847	4.22	190	0.983	184.6	0.870	174.7	0.804
400	0.891	4.44	200	1.08	194.3	0.961	183.9	0.889
420	0.936	4.66	210	1.19	204.0	1.06	193.1	0.978
440	0.980	4.89	220	1.29	213.8	1.16	202.2	1.07
460	1.025	5.11	230	1.40	223.5	1.26	211.4	1.17
480	1.07	5.33	240	1.53	233.2	1.37	220.6	1.27
500	1.11	5.55	250	1.64	242.9	1.48	229.8	1.38
600	1.34	6.66	300	2.33	291.5	2.13	275.8	1.98
700	1.56	7.77	350	3.13	340.1	2.87	321.8	2.68
800	1.78	8.88	400	4.06	388.7	3.75	367.7	3.51
900	2.01	9.99	450	5.08	437.2	4.71	413.7	4.41
1,000	2.23	11.1	500	6.24	485.8	5.84	459.7	5.46
1,100	2.45	12.2	550	7.49	534.4	7.03	505.6	6.58
1,200	2.67	13.3	600	8.89	583.0	8.34	551.6	7.82
1,300	2.90	14.4	650	10.4	631.6	9.78	597.6	9.18
1,400	3.12	15.5	700	12.0	680.1	11.4	643.5	10.6
1,500	3.34	16.7	750	13.7	728.7	13.0	689.5	12.2
1,600	3.56	17.8	800	15.6	777.3	14.8	735.5	13.9
1,700	3.79	18.9	850	17.6	825.9	16.7	781.4	15.7
1,800	4.01	20.0	900	19.7	874.5	18.6	827.4	17.5
1,900	4.23	21.1	950	21.9	923.1	20.8	873.4	19.5
2,000	4.46	22.2	1,000	24.2	971.6	23.0	919.3	21.6
2,100	4.68	23.3	1,050	26.7	1,020	25.4	965.3	23.8
2,200	4.90	24.4	1,100	29.3	1,069	27.8	1,011	26.2
2,300	5.12	25.5	1,150	31.8	1,117	30.3	1,057	28.5
2,400	5.35	26.6	1,200	34.7	1,166	33.0	1,103	31.0
2,500	5.57	27.8	1,250	37.8	1,215	35.9	1,149	33.8
2,600	5.79	28.9	1,300	40.8	1,263	38.9	1,195	36.6
2,700	6.02	30.0	1,350	44.0	1,312	41.9	1,241	39.4
2,800	6.24	31.1	1,400	47.0	1,360	44.9	1,287	42.3
3,000	6.68	33.3	1,500	53.7	1,457	51.4	1,379	48.4

* See footnote, page 8-8, for pipe other than Schedule 40.

FRICTION LOSS IN 8-INCH SCHEDULE 40 STEEL PIPE*

(Note: Weight flow rate is given in thousands of pounds per hour.)

Flow Rate		Velocity, Ft per Sec	60-Deg Water		180-Deg Water		300-Deg Water	
Gal per Min	Cu Ft per Sec		Flow Rate, 1000 Lb per Hr	Pressure Loss, Ft per 100 Ft	Flow Rate, 1000 Lb per Hr	Pressure Loss, Ft per 100 Ft	Flow Rate, 1000 Lb per Hr	Pressure Loss, Ft per 100 Ft
20	0.0446	0.128	10	0.00130	9.716	0.000940	9.193	0.000783
40	0.0891	0.257	20	0.00431	19.43	0.00320	18.39	0.00272
60	0.134	0.385	30	0.00881	29.49	0.00667	27.58	0.00573
80	0.178	0.513	40	0.0146	38.87	0.0113	36.77	0.00978
100	0.223	0.641	50	0.0217	48.58	0.0170	45.97	0.0149
120	0.267	0.770	60	0.0301	58.30	0.0238	55.16	0.0210
140	0.312	0.898	70	0.0398	68.01	0.0316	64.35	0.0280
160	0.356	1.03	80	0.0507	77.73	0.0408	73.55	0.0362
180	0.401	1.15	90	0.0625	87.45	0.0507	82.74	0.0453
200	0.446	1.28	100	0.0758	97.16	0.0620	91.93	0.0556
220	0.490	1.41	110	0.0899	106.9	0.0742	101.1	0.0667
240	0.535	1.54	120	0.105	116.6	0.0876	110.3	0.0788
260	0.579	1.67	130	0.122	126.3	0.102	119.5	0.0920
280	0.624	1.80	140	0.140	136.0	0.117	128.7	0.106
300	0.668	1.92	150	0.159	145.7	0.134	137.9	0.121
320	0.713	2.05	160	0.179	155.5	0.151	147.1	0.138
340	0.758	2.18	170	0.200	165.2	0.170	156.2	0.155
360	0.802	2.31	180	0.222	174.9	0.190	165.5	0.173
380	0.847	2.44	190	0.246	184.6	0.210	174.7	0.193
400	0.891	2.57	200	0.269	194.3	0.232	183.9	0.212
450	1.003	2.89	225	0.335	218.6	0.290	206.9	0.267
500	1.11	3.21	250	0.410	242.9	0.358	229.8	0.330
550	1.23	3.53	275	0.488	267.2	0.430	252.8	0.396
600	1.34	3.85	300	0.577	291.5	0.511	275.8	0.472
650	1.45	4.17	325	0.675	315.8	0.599	298.8	0.553
700	1.56	4.49	350	0.773	340.1	0.689	321.8	0.639
750	1.67	4.81	375	0.882	364.4	0.788	344.8	0.731
800	1.78	5.13	400	0.994	388.7	0.896	367.7	0.830
850	1.89	5.45	425	1.12	412.9	1.01	390.7	0.938
900	2.01	5.77	450	1.25	437.2	1.13	413.7	1.05
950	2.12	6.09	475	1.38	461.5	1.25	436.7	1.17
1,000	2.23	6.41	500	1.52	485.8	1.39	459.7	1.29
1,200	2.67	7.70	600	2.16	583.0	1.98	551.6	1.85
1,400	3.12	8.98	700	2.90	680.1	2.69	643.5	2.51
1,600	3.56	10.3	800	3.77	777.3	3.51	735.5	3.28
1,800	4.01	11.5	900	4.64	874.5	4.41	827.4	4.14
2,000	4.46	12.8	1,000	5.84	971.6	5.45	919.3	5.12
2,200	4.90	14.1	1,100	6.98	1,069	6.55	1,011	6.18
2,400	5.35	15.4	1,200	8.32	1,166	7.80	1,103	7.36
2,600	5.79	16.7	1,300	9.72	1,263	9.16	1,195	8.60
2,800	6.24	18.0	1,400	11.2	1,360	10.6	1,287	9.97
3,000	6.68	19.2	1,500	12.9	1,457	12.2	1,379	11.4
3,200	7.13	20.5	1,600	14.6	1,555	13.8	1,471	13.0
3,400	7.58	21.8	1,700	17.4	1,652	15.6	1,563	14.7
3,600	8.02	23.1	1,800	18.4	1,749	17.5	1,655	16.5
3,800	8.47	24.4	1,900	20.5	1,846	19.5	1,747	18.4
4,000	8.91	25.7	2,000	22.6	1,943	21.5	1,839	20.2
4,500	10.03	28.9	2,250	28.5	2,186	27.2	2,069	25.5
5,000	11.1	32.1	2,500	35.2	2,429	33.6	2,298	31.7

* See footnote, page 8-8, for pipe other than Schedule 40.

FRICTION LOSS IN 10-INCH SCHEDULE 40 STEEL PIPE*

(Note: Weight flow rate is given in thousands of pounds per hour.)

Flow Rate		Velocity, Ft per Sec	60-Deg Water		180-Deg Water		300-Deg Water	
Gal per Min	Cu Ft per Sec		Flow Rate, 1000 Lb per Hr	Pressure Loss, Ft per 100 Ft	Flow Rate, 1000 Lb per Hr	Pressure Loss, Ft per 100 Ft	Flow Rate, 1000 Lb per Hr	Pressure Loss, Ft per 100 Ft
100	0.223	0.407	50	0.00725	48.58	0.00555	45.96	0.00478
120	0.267	0.488	60	0.00999	58.30	0.00773	55.16	0.00670
140	0.312	0.570	70	0.0132	68.01	0.0103	64.35	0.00895
160	0.356	0.651	80	0.0167	77.73	0.0131	73.55	0.0115
180	0.401	0.732	90	0.0207	87.45	0.0163	82.74	0.0144
200	0.446	0.814	100	0.0249	97.16	0.0199	91.93	0.0176
220	0.490	0.895	110	0.0296	106.9	0.0239	101.1	0.0211
240	0.535	0.976	120	0.0346	116.6	0.0280	110.3	0.0249
260	0.579	1.06	130	0.0402	126.3	0.0325	119.5	0.0289
280	0.624	1.14	140	0.0459	136.0	0.0374	128.7	0.0333
300	0.668	1.22	150	0.0520	145.7	0.0427	137.9	0.0380
350	0.780	1.42	175	0.0687	170.0	0.0570	160.9	0.0513
400	0.891	1.63	200	0.0878	194.3	0.0733	183.9	0.0663
450	1.003	1.83	225	0.109	218.6	0.0917	206.9	0.0834
500	1.11	2.03	250	0.132	242.9	0.113	229.8	0.102
550	1.23	2.24	275	0.157	267.2	0.135	252.8	0.123
600	1.34	2.44	300	0.185	291.5	0.159	275.8	0.145
650	1.45	2.64	325	0.215	315.8	0.186	298.8	0.171
700	1.56	2.85	350	0.247	340.1	0.214	321.8	0.197
750	1.67	3.05	375	0.281	364.4	0.247	344.8	0.226
800	1.78	3.25	400	0.318	388.7	0.279	367.7	0.257
850	1.89	3.46	425	0.355	412.9	0.312	390.7	0.288
900	2.01	3.66	450	0.395	437.2	0.349	413.7	0.322
950	2.12	3.87	475	0.438	461.5	0.387	436.7	0.360
1,000	2.23	4.07	500	0.483	485.8	0.429	459.7	0.396
1,200	2.67	4.88	600	0.681	583.0	0.614	551.6	0.570
1,400	3.12	5.70	700	0.917	680.1	0.832	643.5	0.771
1,600	3.56	6.51	800	1.18	777.3	1.08	735.5	1.00
1,800	4.01	7.32	900	1.48	874.5	1.36	827.4	1.27
2,000	4.46	8.14	1,000	1.81	971.6	1.68	919.3	1.57
2,200	4.90	8.95	1,100	2.19	1,069	2.03	1,011	1.90
2,400	5.35	9.76	1,200	2.58	1,166	2.40	1,103	2.23
2,600	5.79	10.6	1,300	3.03	1,263	2.82	1,195	2.63
2,800	6.24	11.4	1,400	3.49	1,360	3.25	1,287	3.05
3,000	6.68	12.2	1,500	3.99	1,457	3.74	1,379	3.50
3,200	7.13	13.0	1,600	4.52	1,555	4.23	1,471	3.97
3,400	7.58	13.8	1,700	5.07	1,652	4.78	1,563	4.48
3,600	8.02	14.6	1,800	5.69	1,749	5.36	1,655	5.03
3,800	8.47	15.5	1,900	6.31	1,846	5.97	1,747	5.60
4,000	8.91	16.3	2,000	7.00	1,943	6.63	1,839	6.22
4,500	10.03	18.3	2,250	8.80	2,186	8.34	2,069	7.82
5,000	11.1	20.3	2,500	10.8	2,429	10.3	2,298	9.65
6,000	13.4	24.4	3,000	15.6	2,915	14.8	2,758	13.9
7,000	15.6	28.5	3,500	21.0	3,401	20.0	3,218	18.9
8,000	17.8	32.5	4,000	27.6	3,887	26.2	3,677	24.8
9,000	20.1	36.6	4,500	34.5	4,372	33.1	4,137	31.2
10,000	22.3	40.7	5,000	42.7	4,858	40.9	4,597	38.6

* See footnote, page 8-8, for pipe other than Schedule 40.

FRICTION LOSS IN 12-INCH SCHEDULE 40 STEEL PIPE*

(Note: Weight flow rate is given in thousands of pounds per hour.)

Flow Rate		Velocity, Ft per Sec	60-Deg Water		180-Deg Water		300-Deg Water	
Gal per Min	Cu Ft per Sec		Flow Rate, 1000 Lb per Hr	Pressure Loss, Ft per 100 Ft	Flow Rate, 1000 Lb per Hr	Pressure Loss, Ft per 100 Ft	Flow Rate, 1000 Lb per Hr	Pressure Loss, Ft per 100 Ft
120	0.267	0.344	60	0.00445	58.30	0.00338	55.16	0.00291
140	0.312	0.401	70	0.00583	68.01	0.00448	64.35	0.00385
160	0.356	0.459	80	0.00739	77.73	0.00572	73.55	0.00497
180	0.401	0.516	90	0.00911	87.45	0.00708	82.74	0.00617
200	0.446	0.573	100	0.0110	97.16	0.00864	91.93	0.00751
220	0.490	0.631	110	0.0130	106.9	0.0103	101.1	0.00902
240	0.535	0.688	120	0.0152	116.6	0.0121	110.3	0.0106
260	0.579	0.745	130	0.0176	126.3	0.0140	119.5	0.0123
280	0.624	0.802	140	0.0201	136.0	0.0160	128.7	0.0142
300	0.668	0.860	150	0.0228	145.7	0.0183	137.9	0.0162
400	0.891	1.15	200	0.0384	194.3	0.0312	183.9	0.0279
500	1.11	1.43	250	0.0574	242.9	0.0479	229.8	0.0431
600	1.34	1.72	300	0.0805	291.5	0.0676	275.8	0.0612
700	1.56	2.01	350	0.107	340.1	0.0908	321.8	0.0826
800	1.78	2.29	400	0.136	388.7	0.118	367.7	0.107
900	2.01	2.58	450	0.170	437.2	0.146	413.7	0.135
1,000	2.23	2.87	500	0.207	485.8	0.180	459.7	0.166
1,200	2.67	3.44	600	0.293	583.0	0.258	551.6	0.237
1,400	3.12	4.01	700	0.312	680.1	0.348	643.5	0.320
1,600	3.56	4.59	800	0.504	777.3	0.451	735.5	0.419
1,800	4.01	5.16	900	0.629	874.5	0.567	827.4	0.530
2,000	4.46	5.73	1,000	0.772	971.6	0.700	919.3	0.649
2,200	4.90	6.31	1,100	0.927	1,069	0.840	1,011	0.785
2,400	5.35	6.88	1,200	1.10	1,166	1.00	1,103	0.933
2,600	5.79	7.45	1,300	1.28	1,263	1.17	1,195	1.10
2,800	6.24	8.03	1,400	1.47	1,361	1.35	1,287	1.26
3,000	6.68	8.60	1,500	1.68	1,457	1.55	1,379	1.45
3,200	7.13	9.17	1,600	1.90	1,555	1.77	1,471	1.65
3,400	7.58	9.75	1,700	2.15	1,652	2.00	1,563	1.86
3,600	8.02	10.3	1,800	2.39	1,749	2.21	1,655	2.08
3,800	8.47	10.9	1,900	2.65	1,846	2.47	1,747	2.32
4,000	8.91	11.5	2,000	3.04	1,943	2.73	1,839	2.57
4,500	10.03	12.9	2,250	3.70	2,186	3.47	2,069	3.26
5,000	11.1	14.3	2,500	4.53	2,429	4.27	2,298	3.99
5,500	12.3	15.8	2,750	5.44	2,672	5.13	2,528	4.83
6,000	13.4	17.2	3,000	6.49	2,915	6.12	2,758	5.75
6,500	14.5	18.6	3,250	7.60	3,158	7.17	2,988	6.74
7,000	15.6	20.1	3,500	8.76	3,401	8.33	3,218	7.83
7,500	16.7	21.5	3,750	10.1	3,644	9.55	3,448	8.98
8,000	17.8	22.9	4,000	11.4	3,887	10.9	3,677	10.2
8,500	18.9	24.4	4,250	12.8	4,129	12.3	3,907	11.5
9,000	20.1	25.8	4,500	14.3	4,372	13.7	4,137	12.9
9,500	21.2	27.2	4,750	16.0	4,615	15.2	4,367	14.4
10,000	22.3	28.7	5,000	17.7	4,858	16.9	4,597	16.0
12,000	26.7	34.4	6,000	25.3	5,830	24.2	5,516	22.9
14,000	31.2	40.1	7,000	34.5	6,801	33.0	6,435	31.3
16,000	35.6	45.9	8,000	44.8	7,773	43.2	7,355	40.9
18,000	40.1	51.6	9,000	56.7	8,745	54.6	8,274	51.8
20,000	44.6	57.3	10,000	70.0	9,716	67.5	9,193	63.9

* See footnote, page 8-8, for pipe other than Schedule 40.

FRICTION LOSS IN 14-INCH SCHEDULE 40 STEEL PIPE*

(Note: Weight flow rate is given in thousands of pounds per hour.)

Flow Rate		Velocity, Ft per Sec	60-Deg Water		180-Deg Water		300-Deg Water	
Gal per Min	Cu Ft per Sec		Flow Rate, 1000 Lb per Hr	Pressure Loss, Ft per 100 Ft	Flow Rate, 1000 Lb per Hr	Pressure Loss, Ft per 100 Ft	Flow Rate, 1000 Lb per Hr	Pressure Loss, Ft per 100 Ft
200	0.446	0.474	100	0.00698	97.16	0.00541	91.93	0.00468
250	0.557	0.593	125	0.0104	121.5	0.00814	114.9	0.00751
300	0.668	0.711	150	0.0144	145.7	0.0114	137.9	0.0101
350	0.780	0.830	175	0.0189	170.0	0.0152	160.9	0.0135
400	0.891	0.948	200	0.0242	194.3	0.0196	183.9	0.0174
500	1.114	1.19	250	0.0362	242.9	0.0296	229.8	0.0266
600	1.34	1.42	300	0.0506	291.5	0.0421	275.8	0.0377
700	1.56	1.66	350	0.0672	340.1	0.0565	321.8	0.0509
800	1.78	1.90	400	0.0855	388.7	0.0727	367.7	0.0660
900	2.01	2.13	450	0.107	437.2	0.0913	413.7	0.0835
1,000	2.23	2.37	500	0.129	485.8	0.112	459.7	0.102
1,100	2.45	2.61	550	0.155	534.4	0.135	505.6	0.123
1,200	2.67	2.85	600	0.183	583.0	0.159	551.6	0.146
1,300	2.90	3.08	650	0.212	631.6	0.186	597.6	0.172
1,400	3.12	3.32	700	0.243	680.1	0.214	643.5	0.197
1,500	3.34	3.56	750	0.278	728.7	0.246	689.5	0.227
1,600	3.56	3.79	800	0.314	777.3	0.278	735.5	0.258
1,700	3.79	4.03	850	0.350	825.9	0.312	781.4	0.290
1,800	4.01	4.27	900	0.391	874.5	0.348	827.4	0.323
1,900	4.23	4.50	950	0.432	923.1	0.387	873.4	0.359
2,000	4.46	4.74	1,000	0.478	971.6	0.429	919.3	0.398
2,500	5.57	5.93	1,250	0.732	1,215	0.666	1,149	0.617
3,000	6.68	7.11	1,500	1.04	1,457	0.951	1,379	0.888
3,500	7.80	8.30	1,750	1.40	1,700	1.29	1,609	1.21
4,000	8.91	9.48	2,000	1.81	1,943	1.68	1,839	1.57
4,500	10.02	10.7	2,250	2.28	2,186	2.12	2,069	1.98
5,000	11.1	11.9	2,500	2.79	2,429	2.62	2,298	2.44
6,000	13.4	14.2	3,000	3.96	2,915	3.74	2,758	3.52
7,000	15.6	16.6	3,500	5.35	3,401	5.09	3,218	4.79
8,000	17.8	19.0	4,000	6.99	3,887	6.65	3,677	6.26
9,000	20.1	21.3	4,500	8.78	4,372	8.43	4,137	7.93
10,000	22.3	23.7	5,000	10.8	4,858	10.3	4,597	9.79
11,000	24.5	26.1	5,500	13.1	5,344	12.5	5,056	11.8
12,000	26.7	28.5	6,000	15.6	5,830	14.9	5,516	14.0
13,000	29.0	30.8	6,500	18.2	6,316	17.5	5,976	16.4
14,000	31.2	33.2	7,000	21.0	6,801	20.2	6,435	19.0
15,000	33.4	35.6	7,500	24.2	7,287	23.2	6,895	21.9
16,000	35.6	37.9	8,000	27.6	7,773	26.4	7,355	24.9
17,000	37.9	40.3	8,500	31.0	8,259	29.7	7,814	28.0
18,000	40.1	42.7	9,000	34.8	8,745	33.4	8,274	31.4
19,000	42.3	45.0	9,500	38.7	9,230	37.2	8,734	35.0
20,000	44.6	47.4	10,000	42.5	9,716	41.2	9,193	38.7
22,000	49.0	52.2	11,000	51.6	10,690	49.9	10,110	47.0
24,000	53.5	56.9	12,000	61.4	11,660	59.4	11,030	55.8
26,000	57.9	61.6	13,000	72.1	12,630	69.7	11,950	65.6
28,000	62.4	66.4	14,000	83.6	13,600	80.8	12,870	76.0
30,000	66.8	71.1	15,000	95.9	14,570	92.7	13,790	87.2

* See footnote, page 8-8, for pipe other than Schedule 40.

FRICTION LOSS IN 16-INCH SCHEDULE 40 STEEL PIPE*

(Note: Weight flow rate is given in thousands of pounds per hour.)

Flow Rate		Velocity, Ft per Sec	60-Deg Water		180-Deg Water		300-Deg Water	
Gal per Min	Cu Ft per Sec		Flow Rate, 1000 Lb per Hr	Pressure Loss, Ft per 100 Ft	Flow Rate, 1000 Lb per Hr	Pressure Loss, Ft per 100 Ft	Flow Rate, 1000 Lb per Hr	Pressure Loss, Ft per 100 Ft
300	0.668	0.545	150	0.00757	145.7	0.00590	137.9	0.00516
400	0.891	0.726	200	0.0126	194.3	0.0101	183.9	0.00894
500	1.114	0.908	250	0.0189	242.9	0.0154	229.8	0.0136
600	1.34	1.09	300	0.0263	291.5	0.0215	275.8	0.0193
700	1.56	1.27	350	0.0349	340.1	0.0289	231.8	0.0259
800	1.78	1.45	400	0.0446	388.7	0.0371	367.7	0.0335
900	2.01	1.63	450	0.0552	437.2	0.0465	413.7	0.0423
1,000	2.23	1.82	500	0.0671	485.8	0.0568	459.7	0.0517
1,200	2.67	2.18	600	0.0945	583.0	0.0812	551.6	0.0738
1,400	3.12	2.54	700	0.126	680.1	0.108	643.5	0.0998
1,600	3.56	2.90	800	0.159	777.3	0.140	735.5	0.129
1,800	4.01	3.27	900	0.199	874.5	0.178	827.4	0.163
2,000	4.46	3.63	1,000	0.246	971.6	0.217	919.3	0.201
2,500	5.57	4.54	1,250	0.374	1,215	0.336	1,149	0.311
3,000	6.68	5.45	1,500	0.530	1,457	0.479	1,379	0.447
3,500	7.80	6.35	1,750	0.709	1,700	0.652	1,609	0.605
4,000	8.91	7.26	2,000	0.918	1,943	0.853	1,839	0.790
4,500	10.02	8.17	2,250	1.16	2,186	1.07	2,069	0.999
5,000	11.1	9.08	2,500	1.42	2,429	1.32	2,298	1.23
6,000	13.4	10.9	3,000	2.01	2,915	1.88	2,758	1.76
7,000	15.6	12.7	3,500	2.74	3,401	2.56	3,218	2.40
8,000	17.8	14.5	4,000	3.54	3,887	3.35	3,677	3.13
9,000	20.1	16.3	4,500	4.44	4,372	4.19	4,137	3.96
10,000	22.3	18.2	5,000	5.48	4,858	5.17	4,597	4.88
11,000	24.5	20.0	5,500	6.63	5,344	6.26	5,056	5.90
12,000	26.7	21.8	6,000	7.82	5,830	7.45	5,516	7.03
13,000	29.0	23.6	6,500	9.18	6,316	8.75	5,976	8.24
14,000	31.2	25.4	7,000	10.6	6,801	10.1	6,435	9.55
15,000	33.4	27.2	7,500	12.2	7,287	11.6	6,895	10.9
16,000	35.6	29.0	8,000	13.9	7,773	13.2	7,355	12.4
17,000	37.9	30.9	8,500	15.5	8,259	14.9	7,814	14.1
18,000	40.1	32.7	9,000	17.4	8,745	16.8	8,274	15.8
19,000	42.3	34.5	9,500	19.4	9,231	18.7	8,734	17.6
20,000	44.6	36.3	10,000	21.5	9,716	20.7	9,193	19.5
22,000	49.0	39.9	11,000	26.0	10,690	25.0	10,110	23.5
24,000	53.5	43.6	12,000	31.0	11,660	29.8	11,030	28.0
26,000	57.9	47.2	13,000	36.3	12,630	34.6	11,950	32.8
28,000	62.4	50.8	14,000	41.8	13,600	40.2	12,870	38.1
30,000	66.8	54.5	15,000	47.9	14,570	46.1	13,790	43.7
32,000	71.3	58.1	16,000	54.5	15,550	52.4	14,710	49.7
34,000	75.8	61.7	17,000	61.6	16,520	59.2	15,630	56.1
36,000	80.2	65.4	18,000	69.1	17,490	66.4	16,550	62.9
38,000	84.7	69.0	19,000	77.0	18,460	74.0	17,470	70.2
40,000	89.1	72.6	20,000	85.3	19,430	82.0	18,390	77.7
42,000	93.6	76.2	21,000	94.0	20,400	90.4	19,310	85.7
44,000	98.0	79.9	22,000	103	21,380	99.2	20,230	94.0
46,000	102.5	83.5	23,000	113	22,350	108	21,140	103
48,000	107	87.1	24,000	123	23,320	118	22,060	112
50,000	111	90.8	25,000	133	24,290	128	22,980	121

* See footnote, page 8-8, for pipe other than Schedule 40.

FRICTION LOSS IN 18-INCH SCHEDULE 40 STEEL PIPE *

(Note: Weight flow rate is given in thousands of pounds per hour.)

Flow Rate		Velocity, Ft per Sec	60-Deg Water		180-Deg Water		300-Deg Water	
Gal per Min	Cu Ft per Sec		Flow Rate, 1000 Lb per Hr	Pressure Loss, Ft per 100 Ft	Flow Rate, 1000 Lb per Hr	Pressure Loss, Ft per 100 Ft	Flow Rate, 1000 Lb per Hr	Pressure Loss, Ft per 100 Ft
300	0.668	0.430	150	0.00429	145.7	0.00333	137.9	0.00288
400	0.891	0.574	200	0.00717	194.3	0.00565	183.9	0.00496
500	1.114	0.717	250	0.0106	242.9	0.00854	229.8	0.00754
600	1.34	0.861	300	0.0148	291.5	0.0120	275.8	0.0111
700	1.56	1.00	350	0.0198	340.1	0.0161	321.8	0.0149
800	1.78	1.15	400	0.0250	388.7	0.0207	367.7	0.0186
900	2.01	1.29	450	0.0311	437.2	0.0258	413.7	0.0232
1,000	2.23	1.43	500	0.0378	485.8	0.0315	459.7	0.0285
1,200	2.67	1.72	600	0.0529	583.0	0.0446	551.6	0.0405
1,400	3.12	2.01	700	0.0702	680.1	0.0600	643.5	0.0547
1,600	3.56	2.30	800	0.0901	777.3	0.0776	735.5	0.0710
1,800	4.01	2.58	900	0.112	874.5	0.0972	827.4	0.0893
2,000	4.46	2.87	1,000	0.137	971.6	0.119	919.3	0.110
2,500	5.57	3.59	1,250	0.208	1,215	0.184	1,149	0.170
3,000	6.68	4.30	1,500	0.294	1,457	0.263	1,379	0.244
3,500	7.80	5.02	1,750	0.392	1,700	0.356	1,608	0.331
4,000	8.91	5.74	2,000	0.508	1,943	0.463	1,839	0.431
4,500	10.02	6.45	2,250	0.636	2,186	0.577	2,069	0.538
5,000	11.1	7.17	2,500	0.980	2,429	0.717	2,298	0.670
6,000	13.4	8.61	3,000	1.08	2,915	1.03	2,758	0.961
7,000	15.6	10.0	3,500	1.50	3,401	1.40	3,218	1.31
8,000	17.8	11.5	4,000	1.94	3,887	1.82	3,677	1.71
9,000	20.1	12.9	4,500	2.44	4,372	2.29	4,137	2.15
10,000	22.3	14.3	5,000	3.00	4,858	2.82	4,597	2.66
12,000	26.7	17.2	6,000	4.29	5,830	4.06	5,516	3.81
14,000	31.2	20.1	7,000	5.81	6,801	5.52	6,435	5.20
16,000	35.6	22.9	8,000	7.56	7,773	7.19	7,355	6.78
18,000	40.1	25.8	9,000	9.54	8,745	9.10	8,274	8.60
20,000	44.6	28.7	10,000	11.8	9,716	11.2	9,193	10.6
22,000	49.0	31.6	11,000	14.2	10,680	13.6	10,113	12.8
24,000	53.5	34.4	12,000	16.8	11,660	16.1	11,030	15.2
26,000	57.9	37.3	13,000	19.7	12,630	18.9	11,950	17.8
28,000	62.4	40.2	14,000	22.9	13,600	22.0	12,870	20.7
30,000	66.8	43.0	15,000	26.2	14,570	25.2	13,790	23.8
32,000	71.3	45.9	16,000	29.8	15,550	28.6	14,710	27.0
34,000	75.8	48.8	17,000	33.6	16,520	32.3	15,630	30.5
36,000	80.2	51.6	18,000	37.6	17,490	36.2	16,550	34.2
38,000	84.7	54.5	19,000	41.9	18,460	40.3	17,470	38.1
40,000	89.1	57.4	20,000	46.4	19,430	44.7	18,390	42.2
42,000	93.6	60.2	21,000	51.1	20,410	49.2	19,310	46.5
44,000	98	63.1	22,000	56.1	21,380	54.0	20,230	51.1
46,000	102	66.0	23,000	61.3	22,350	59.0	21,150	55.9
48,000	107	68.9	24,000	66.7	23,320	64.3	22,060	60.7
50,000	111	71.7	25,000	72.2	24,290	69.7	22,980	65.8
55,000	123	78.9	27,500	87.3	26,720	84.3	25,280	79.7
60,000	134	86.1	30,000	104	29,150	100	27,580	94.8
65,000	145	93.2	32,500	122	31,580	118	29,880	111
70,000	156	100.4	35,000	142	34,010	137	32,180	129

* See footnote, page 8-8, for pipe other than Schedule 40.

FRICTION LOSS IN 20-INCH SCHEDULE 40 STEEL PIPE*

(Note: Weight flow rate is given in thousands of pounds per hour.)

| Flow Rate | | | 60-Deg Water | | 180-Deg Water | | 300-Deg Water | |
Gal per Min	Cu Ft per Sec	Velocity. Ft per Sec	Flow Rate, 1000 Lb per Hr	Pressure Loss, Ft per 100 Ft	Flow Rate, 1000 Lb per Hr	Pressure Loss, Ft per 100 Ft	Flow Rate, 1000 Lb per Hr	Pressure Loss, Ft per 100 Ft
300	0.668	0.346	150	0.00255	145.7	0.00195	137.9	0.00169
400	0.891	0.462	200	0.00424	194.3	0.00331	183.9	0.00288
500	1.114	0.577	250	0.00631	242.9	0.00499	229.8	0.00438
600	1.34	0.692	300	0.00879	291.5	0.00701	275.8	0.00619
700	1.56	0.808	350	0.0116	340.1	0.00934	321.8	0.00829
800	1.78	0.923	400	0.0147	388.7	0.0120	367.7	0.0107
900	2.01	1.039	450	0.0183	437.2	0.0150	413.7	0.0134
1,000	2.23	1.15	500	0.0221	485.8	0.0183	459.7	0.0164
1,200	2.67	1.38	600	0.0310	583.0	0.0258	551.6	0.0233
1,400	3.12	1.62	700	0.0410	680.1	0.0347	643.5	0.0315
1,600	3.56	1.85	800	0.0526	777.3	0.0448	735.5	0.0408
1,800	4.01	2.08	900	0.0654	874.5	0.0562	827.4	0.0513
2,000	4.46	2.31	1,000	0.0796	971.6	0.0688	919.3	0.0630
2,500	5.57	2.89	1,250	0.121	1,215	0.106	1,149	0.0973
3,000	6.68	3.46	1,500	0.170	1,457	0.151	1,379	0.139
3,500	7.80	4.04	1,750	0.229	1,700	0.204	1,609	0.189
4,000	8.91	4.62	2,000	0.294	1,943	0.264	1,839	0.246
4,500	10.02	5.19	2,250	0.368	2,186	0.326	2,069	0.324
5,000	11.1	5.77	2,500	0.450	2,429	0.410	2,298	0.381
6,000	13.4	6.92	3,000	0.639	2,915	0.587	2,758	0.548
7,000	15.6	8.08	3,500	0.860	3,401	0.796	3,218	0.744
8,000	17.8	9.23	4,000	1.11	3,887	1.03	3,677	0.968
9,000	20.1	10.39	4,500	1.41	4,372	1.31	4,137	1.23
10,000	22.3	11.5	5,000	1.72	4,858	1.61	4,596	1.51
12,000	26.7	13.8	6,000	2.46	5,830	2.32	5,516	2.17
14,000	31.2	16.2	7,000	4.03	6,801	3.72	6,435	3.51
16,000	35.6	18.5	8,000	4.32	7,773	4.10	7,355	3.86
18,000	40.1	20.8	9,000	5.45	8,745	5.18	8,274	4.88
20,000	44.6	23.1	10,000	6.71	9,716	6.39	9,193	6.02
22,000	49.0	25.4	11,000	8.11	10,690	7.73	10,110	7.27
24,000	53.5	27.7	12,000	9.59	11,660	9.16	11,030	8.64
26,000	57.9	30.0	13,000	11.3	12,630	10.8	11,950	10.2
28,000	62.4	32.3	14,000	13.0	13,600	12.5	12,870	11.8
30,000	66.8	34.6	15,000	14.9	14,570	14.3	13,790	13.5
32,000	71.3	36.9	16,000	17.0	15,550	16.3	14,790	15.4
34,000	75.8	39.2	17,000	19.1	16,520	18.4	15,630	17.3
36,000	80.2	41.5	18,000	21.4	17,490	20.6	16,550	19.4
38,000	84.7	43.9	19,000	23.9	18,460	23.0	17,470	21.7
40,000	89.1	46.2	20,000	26.4	19,430	25.4	18,390	24.0
42,000	93.6	48.5	21,000	29.1	20,400	28.0	19,310	26.5
44,000	98	50.8	22,000	32.0	21,380	30.8	20,230	29.1
46,000	102	53.1	23,000	34.9	22,350	33.6	21,140	31.7
48,000	107	55.4	24,000	38.0	23,320	36.6	22,060	34.5
50,000	111	57.7	25,000	41.2	24,290	39.6	22,980	37.4
60,000	134	69.2	30,000	59.2	29,150	57.1	27,580	53.9
70,000	156	80.8	35,000	80.1	34,010	77.7	32,180	73.4
80,000	178	92.3	40,000	105	38,870	101	36,770	95.4
90,000	201	103.9	45,000	133	43,720	129	41,370	121
100,000	223	115.4	50,000	164	48,580	158	45,970	150

* See footnote, page 8-8, for pipe other than Schedule 40.

FRICTION LOSS IN 24-INCH SCHEDULE 40 STEEL PIPE*

(Note: Weight flow rate is given in thousands of pounds per hour.)

Flow Rate			60-Deg Water		180-Deg Water		300-Deg Water	
Gal per Min	Cu Ft per Sec	Velocity, Ft per Sec	Flow Rate, 1000 Lb per Hr	Pressure Loss, Ft per 100 Ft	Flow Rate, 1000 Lb per Hr	Pressure Loss, Ft per 100 Ft	Flow Rate, 1000 Lb per Hr	Pressure Loss, Ft per 100 Ft
300	0.668	0.239	150	0.00105	145.7	0.000879	137.9	0.000683
400	0.891	0.319	200	0.00175	194.3	0.00134	183.9	0.00116
500	1.114	0.399	250	0.00259	242.9	0.00202	229.8	0.00176
600	1.34	0.479	300	0.00360	291.5	0.00283	275.8	0.00247
700	1.56	0.559	350	0.00474	340.1	0.00376	321.8	0.00330
800	1.78	0.638	400	0.00604	388.7	0.00482	367.7	0.00425
900	2.01	0.718	450	0.00747	437.2	0.00600	413.7	0.00531
1,000	2.23	0.798	500	0.00903	485.8	0.00731	459.7	0.00650
1,200	2.67	0.958	600	0.0125	583.0	0.0103	551.6	0.00917
1,400	3.12	1.12	700	0.0167	680.1	0.0138	643.5	0.0124
1,600	3.56	1.28	800	0.0213	777.3	0.0177	735.5	0.0160
1,800	4.01	1.44	900	0.0265	874.5	0.0222	827.4	0.0201
2,000	4.46	1.60	1,000	0.0321	971.6	0.0272	919.3	0.0247
2,500	5.57	1.99	1,250	0.0485	1,215	0.0417	1,149	0.0380
3,000	6.68	2.39	1,500	0.0683	1,457	0.0593	1,379	0.0544
3,500	7.80	2.79	1,750	0.0909	1,700	0.0797	1,609	0.0733
4,000	8.91	3.19	2,000	0.117	1,943	0.103	1,839	0.0953
4,500	10.02	3.59	2,250	0.146	2,186	0.130	2,069	0.120
5,000	11.1	3.99	2,500	0.178	2,429	0.160	2,298	0.148
6,000	13.4	4.79	3,000	0.253	2,915	0.229	2,758	0.212
7,000	15.6	5.59	3,500	0.340	3,401	0.309	3,218	0.288
8,000	17.8	6.38	4,000	0.439	3,887	0.402	3,677	0.375
9,000	20.1	7.18	4,500	0.550	4,372	0.507	4,137	0.473
10,000	22.3	7.98	5,000	0.674	4,858	0.624	4,597	0.584
12,000	26.7	9.58	6,000	0.957	5,830	0.892	5,516	0.777
14,000	31.2	11.2	7,000	1.30	6,801	1.25	6,435	0.114
16,000	35.6	12.8	8,000	1.68	7,773	1.58	7,355	0.149
18,000	40.1	14.4	9,000	2.13	8,745	2.00	8,274	1.88
20,000	44.6	16.0	10,000	2.61	9,716	2.47	9,193	2.32
22,000	49.0	17.6	11,000	3.15	10,690	2.98	10,110	2.81
24,000	53.5	19.2	12,000	3.73	11,660	3.55	11,030	3.33
26,000	57.9	20.7	13,000	4.38	12,630	4.15	11,950	3.91
28,000	62.4	22.3	14,000	5.06	13,600	4.82	12,870	4.54
30,000	66.8	23.9	15,000	5.80	14,570	5.52	13,790	5.20
34,000	75.8	27.1	17,000	7.39	16,520	7.07	15,630	6.66
38,000	84.7	30.3	19,000	9.24	18,460	8.85	17,470	8.35
42,000	93.6	33.5	21,000	11.3	20,400	10.8	19,310	10.2
46,000	102	36.7	23,000	13.5	22,350	12.9	21,140	12.2
50,000	111	39.9	25,000	15.9	24,290	15.3	22,980	14.4
60,000	134	47.9	30,000	22.8	29,150	22.0	27,580	20.8
70,000	156	55.9	35,000	31.1	34,000	29.9	32,180	28.2
80,000	178	63.8	40,000	40.4	38,870	39.0	36,770	36.8
90,000	201	71.8	45,000	51.2	43,720	49.3	41,370	46.6
100,000	223	79.8	50,000	63.1	48,580	60.9	45,970	57.6
110,000	245	87.8	55,000	76.4	53,440	73.9	50,560	69.8
120,000	267	95.8	60,000	90.5	58,300	87.5	55,160	82.6
130,000	290	103.7	65,000	106	63,160	103	59,760	97.2
140,000	312	112	70,000	123	68,010	120	64,350	113
150,000	334	120	75,000	142	72,870	137	68,950	130

* See footnote, page 8-8, for pipe other than Schedule 40.

STEAM FLOW IN PIPES

Design data on steam flow in pipes have been standardized to such an extent that the same empirical equations and flow tables appear and reappear in reference sources. Since the original data and basic assumptions were presented during the last century, a re-examination of the flow equations seemed overdue to two researchers who, some years ago, made a study of the subject.

These research engineers were Robert H. Page and S. Konzo, assistant in mechanical engineering and professor of mechanical engineering, respectively, at the University of Illinois. A detailed report on their study appeared in HEATING AND VENTILATING, August, 1951. A greatly condensed summary of their findings and 22 working charts resulting from their study follow.

Background of Steam Flow Formulas

The flow relationships that are in common use originated with Darcy and Weisbach, both of whom worked with flow of water. The Darcy-Weisbach equation for drop in pressure head for water flow has been carried forward to this time in its original form which included a friction factor f as follows:

$$\Delta h = \frac{fLv^2}{D2g} \qquad (1)$$

where Δh = head loss in feet of fluid flowing,
 f = friction factor,
 L = length of pipe in feet,
 D = internal diameter of pipe in feet.

In 1857, Darcy proposed that the friction factor should be considered as a function of the diameter alone. In 1876, Unwin modified the Darcy equation but retained the friction factor.

Unwin's equation is the same as that published by Babcock some 14 years later, and which has been commonly referred to as the Babcock equation. Since Babcock gave no derivation nor explanation of the equation, Page and Konzo are of the opinion that the proper reference should be the "Unwin-Babcock" equation.

It is of interest to note that the equation was derived by Unwin for water flow and was adapted by Babcock for steam flow. However, this practice of taking an equation determined for one fluid and applying it to some other fluid was common among early experimenters.

The Unwin-Babcock equation for friction was

$$f' = k(1 + 3.6/d)$$

where f' was the friction factor, k a constant, and d the pipe diameter. In 1894, Unwin re-evaluated the constant and presented a value of 0.0027, based largely on his analysis of experiments made by Riedler and Gutermuth with air flow. This value for k of 0.0027 is still in common use, over 60 years after the time it was first proposed.

One of the significant experiments of this early era was made by Carpenter and Sickles on steam flow. They were able to confirm the k value of 0.0027 only by discarding extreme values. No good reason was given for discarding part of the data. Green and Stratford and McRae also reported experimental results with steam flow and were not successful in confirming the k value of 0.0027. However, since their reports were not widely circulated, any influence they might have had in questioning the Unwin constant was negligible in effect.

By the turn of the century, both the Unwin-Babcock equation and the Unwin constant had become well established and to some extent "frozen." For example, an interesting summary made by Gebhardt in 1909 shows not only that the Unwin-Babcock equation was widely accepted, but that the only other recognized form for friction factor was one in which f was assumed to be a constant and not a function of diameter.

The complete Unwin-Babcock equation was

$$P = \frac{C_1 W^2 L}{\rho d^5} \left(1 + \frac{3.6}{d} \right)$$

where P = pressure drop in pounds per square inch, C_1 = 0.04839k, ρ = density of steam in pounds per cubic foot, L = length of pipe in feet, W = steam flow rate in pounds per minute, and d = internal pipe diameter in feet. This equation is still in common use.

During the early part of the present century, numerous writers extended the application of this equation over a wide range of conditions without regard for the limitations of the empirical constants. For example, Carpenter in 1912 extended the equations to pipe sizes ranging from 1 inch to 60 inches in diameter, and to steam pressures ranging from 0.4 inch of mercury to 500 psi. Such extensions of the equation are highly questionable.

Unwin in 1907 indicated that the friction factor was affected by temperature, and thus the viscosity, of the fluid. For some strange reason, however, in the case of steam flow the restricted Unwin-Babcock concept persisted to such an extent that little work

was done to correlate the existing experimental evidence with the newer concepts of flow.

In 1912, Blasius made a recalculation of the data obtained by two other experimenters with water flow and showed that the factor f in the Darcy-Weisbach equation was a function of $vD\rho/\mu$, where μ is the absolute viscosity of the fluid. This dimensionless expression is referred to as the Reynolds number. Another contribution was that by Stanton and Pannel, who adopted Reynolds' theories of 1883, and established experimentally that the friction factor for the flow of such diverse fluids as oil, water, and air was a function of Reynolds number.

In 1915, Lander presented experimental results with steam flow and water flow on the diagram first suggested by Stanton and Pannel. This research can be considered as the first valid evidence that the friction factor for steam flow was a function of something other than diameter alone.

Since 1914 the developments in fluid flow theory have been made at an accelerated rate. The contributions of Buckingham and Rayleigh in the field of dimensional analysis proved that the friction factor was a function of Reynolds number. In 1939, Keenan obtained experimental results with water and steam, and showed that the friction factor was not only a function of Reynolds number, but was also the same for incompressible and compressible flow. A slight modification in the form of presentation of data was given by Rouse in 1943, but essentially an agreement had been reached that friction factor was a function not of diameter alone, but of Reynolds number and the relative roughness of the pipe.

As a final step in the development of fluid flow up to this time, Moody in 1944 presented a convenient chart showing variations in friction factors for wide ranges of Reynolds number and for various values of the relative roughness of the pipe surface.

Page and Konzo concluded that the best approach to the study of steam flow in pipes consists in utilizing the recent concepts of fluid flow, and in particular the chart prepared by Moody (*Trans. ASME*, 1944). This chart gave the relation between friction factor and Reynolds number for various degrees of roughness of the surface of the pipe. They then proceeded to determine whether the original steam flow data of several investigators could be correlated with the Moody chart.

They showed that the old data of Carpenter and Sickles, originally used to justify the Unwin-Babcock equation, correlated better when analyzed by the Reynolds number concept. They also showed that the Unwin-Babcock equation results in discrepancies

of practical significance. For example, with a 4-inch diameter pipe and a steam flow velocity of 60 fps the results are in substantial agreement with those from the Moody chart. However, with the same velocity of 60 fps and a 1-inch diameter pipe, the Unwin-Babcock values are about 70% greater than those from the Moody chart.

Further, Page and Konzo showed that the extrapolation of the Unwin-Babcock equation to any and all pressure and temperature conditions is not warranted. In some cases, the use of the equation may result in pipe sizes that are too small, and in other cases may result in oversized pipes.

They also concluded that the original data of Carpenter and Sickles, Eberle, and Lander are in good agreement, as judged by the Reynolds number concept, and that the fault lies not in the data, but in the analysis made by others before fluid flow theory had been developed.

Computation Based on Modern Data

By the use of similarity concepts incorporated in the Moody chart for fluid flow, a new method can be derived for the sizing of steam pipes. Twenty-two working charts that summarize the method in graphical form are presented on subsequent pages. A detailed example of the necessary calculations that were made to establish a single point on the charts follows:

If the following variables are known:

    (1) Absolute viscosity, μ
    (2) Density, ρ
    (3) Velocity of steam, V
    (4) Inside diameter of pipe, D
    (5) Absolute roughness of pipe surface, e

then the pressure drop per 100-ft length of pipe may be calculated by the Moody chart and equation (1).

The variables listed are functions of other parameters. For example, the first two variables of absolute viscosity and density depend upon the steam pressure and the steam condition (i.e., wet, dry, or superheated). The fourth variable of inside diameter depends upon the nominal pipe size and the schedule number of the pipe. For example, a two-inch Schedule 40 commercial-steel pipe has an inside diameter of 2.067 inches, while a two-inch Schedule 80 commercial-steel pipe has an inside diameter of 1.939 inch. The fifth variable of absolute surface roughness depends upon the material used in the pipe.

Assume that the pressure drop is desired in 37 ft of three-inch Schedule 40 commercial-steel pipe when 12 lb per min. of saturated steam is flowing at an

initial pressure of 15.7 psia. The first step of the solution is to determine steam properties. This can be done in the ten steps which follow:

(1) Density, ρ, = $1/25.17$ = 0.03973 lb per cu ft (from steam tables by interpolation).

(2) Absolute viscosity, μ, = 9.0 $\times$ 10^{-6} lb per ft sec.

(3) Inside diameter, D, = 0.2557 ft (from piping tables).

(4) Velocity, $V = \dfrac{W}{\rho A}$, (from continuity equation),

$$= \frac{12}{0.03973 \left[\dfrac{\pi}{4} \times (0.2557)^2 \right]}$$

$$= 5,882 \text{ fpm.}$$

where W = pounds of steam flowing per minute,
A = cross-sectional area of inside of pipe.

For the next step involving the Moody chart, the Reynolds number and relative roughness must be computed.

(5) $N_R = \dfrac{\rho V D}{\mu} = \dfrac{0.03973 \times 5,882 \times 0.2557}{9.0 \times 10^{-6} \times 60}$

$= 1.107 \times 10^5$.

(6) Absolute surface roughness, e = 0.00015 ft (from Moody).

(7) Relative roughness, $\dfrac{e}{D} = \dfrac{0.00015}{0.2557} = 0.0005867$.

On the Moody chart, the intersection of the e/D line of 0.0005867 and the N_R line of 1.107×10^5 is located.

(8) Friction factor, f = 0.0206 (from Moody chart).
Appropriate values are substituted in equation (1) to obtain:

(9) Head loss, $\Delta h = \dfrac{fLv^2}{D2g}$

$$= \frac{0.0206 \times 37 \times (5882)^2}{0.2557 \times 2 \times 32.2 \times 3600}$$

$$= 446 \text{ ft.}$$

(10) Pressure drop, $\Delta P = \dfrac{\rho \Delta h}{144}$

$$= \frac{0.03973 \times 446}{144}$$

$$= 0.123 \text{ psi.}$$

The Page-Konzo Working Charts

Tedious calculations of the preceding type can be practically eliminated by the use of 22 working charts as shown on subsequent pages, which were designed by Page and Konzo. The computations were made for Schedule 40 and 80 commercial-steel pipe. Chart scales show the following variables:

(1) Weight rate of flow in pounds per minute (abscissa).

(2) Pressure drop in psi per 100-ft length of pipe (left ordinate).

(3) Pressure drop in ounces per square inch per 100-ft length of pipe, (right ordinate) for the convenience of those who utilize this unit of pressure drop.

(4) Constant velocity lines (sloping downwards to the right).

(5) Constant diameter lines (sloping upwards to the right).

Examples of Use of Charts

Case 1. When the pressure drop is the dependent variable. The sample problem already described can be readily solved by means of the working charts. It was required to determine the pressure drop in 37 ft of 3 inch Schedule 40 commercial-steel pipe when 12 pounds per minute of saturated steam was flowing at an initial pressure of 15.7 psia. The first chart will be used since it applies to Schedule 40 pipe and to a pressure range of 14.7 to 16.6 psia.

Locate the flow rate of 12 pounds per minute on the right ordinate. Proceed to the left on the chart until the 3-inch diameter line is intersected. Then proceed down to obtain a pressure drop of 0.333 psi per 100 ft of length. The pressure drop for 37 ft of pipe is 0.37×0.333, or 0.123 psi. This result is the same as that of the detailed method previously given.

Case 2. When the weight flow rate is the dependent variable. What is the flow rate in pounds per minute of saturated steam at 18.0 psia through a Schedule 40 commercial-steel pipe of 1-inch diameter and 190 ft long, for a pressure drop of 1.1 psi?

The pressure drop per 100 ft length is $1.1 \times 100/190$, or 0.579 psi. Since the initial pressure is 3.3 psig, utilize the second chart. Locate the value of 0.579 psi at the bottom, and proceed vertically until the intersection with the 1-inch diameter line is reached. Then proceed horizontally to the right to obtain a weight flow rate of 0.98 pound per minute.

Case 3. When the pipe diameter is the dependent variable. When given an initial pressure of 19 psia, a pressure drop of 0.5 psi per 100-ft length, and saturated steam to be transported at a rate of 25 pounds per minute, what are the pipe size and steam velocity?

Select the second chart for pressures between 16.7 to 19.7 psia. Locate 0.5 psi on the bottom and proceed vertically. Next locate 25 lb per min. on the ordinate and proceed horizontally. The intersection of these two lines lies between two commercial pipe sizes. A four-inch pipe could be used, but the weight flow is 32.9 lb per min., which is considerably larger than necessary. On the other hand, a 3½-inch pipe gives a weight flow of 23.1 pounds per minute. The 3½-inch pipe will be preferred, since a flow rate of 25 pounds per minute would result in a pressure drop of 0.575 psi, which is only slightly larger than the desired value of 0.5 psi. The steam velocity is 8,000 fpm.

Vertical Pipes

If the pipe in a given piping system rises to a higher level in the direction of flow, an additional pressure drop is involved equal in magnitude to the product of the density and the change in elevation. Similarly, if the pipe drops to a lower level in the direction of flow, a pressure increase is involved equal to the density times the change in elevation. In Table 1 is shown the relationship between the pressure drop (per 10 ft length) and the absolute pressure of saturated steam. For low pressures or small changes in elevation, the correction is negligible, but for high pressures and large changes in elevation the corrections may be significantly large.

Any correction for changes in elevation should be algebraically added to the pressure drop determined for frictional effects alone.

Table 1   Pressure Drop in Saturated Steam for 10-Foot Change in Pipe Elevation

Steam Pressure, psia	Pressure Drop, psi, per 10 Ft. Change
10	.0018
15	.0027
20	.0035
25	.0045
30	.0050
40	.0065
60	.0097
80	.0130
100	.0151
150	.0230
200	.0300
300	.0450
400	.0600
500	.0750

Advantages of Form of Working Charts

Any given form of working charts for steam flow will have inherent advantages as well as limitations. The forms shown in the working charts were finally decided upon after a study of the limitations and advantages of these as well as other forms.

The charts as given have two limitations, which have not been considered as unsurmountable.

(1) For complete coverage over a wide range of steam pressures, a large number of charts are required. The first chart for a range of pressures from 14.7 psia to 16.6 psia was actually based on values for 15.7 psia. Similarly, the second chart for a range of pressures from 16.7 psia to 19.7 psia was based on values for 18.0 psia. For higher steam pressures, successively larger ranges of steam pressures were used to give the same order of accuracy as those embodied in the first two. In this connection, the maximum deviation in all the values shown was limited to seven per cent, which was regarded as reasonable considering the accuracy of the original data involved.

(2) In order to cover the cases of pipes other than Schedule 40 (or, in the case of the last four charts on Schedule 80 pipe) commercial steel or wrought iron, either separate charts or conversion factors would be necessary.

In spite of the obvious limitations of the form used, the advantages were considered to outweigh the disadvantages:

(1) All the data for a particular steam condition are presented on one chart. For most practical applications, the engineer is able to solve a problem without reference to other charts.

(2) The solution to any problem can be started at any place on the chart, providing any two of the four variables are known. As shown by the three sample problems, any variable on the chart can be the dependent variable.

(3) The complete determination of the problem is made. Information is obtained not only about pressure drop, but also weight flow, diameter, and velocity. It is possible to observe how a change in any one of the four variables will affect the other three.

(4) The flow charts are similar in form to those in common use for water and air flow, so that familiarity with such existing charts can be extended to these new charts for steam flow.

Pressure Drop per 100 Ft., Ounces per Sq. In.

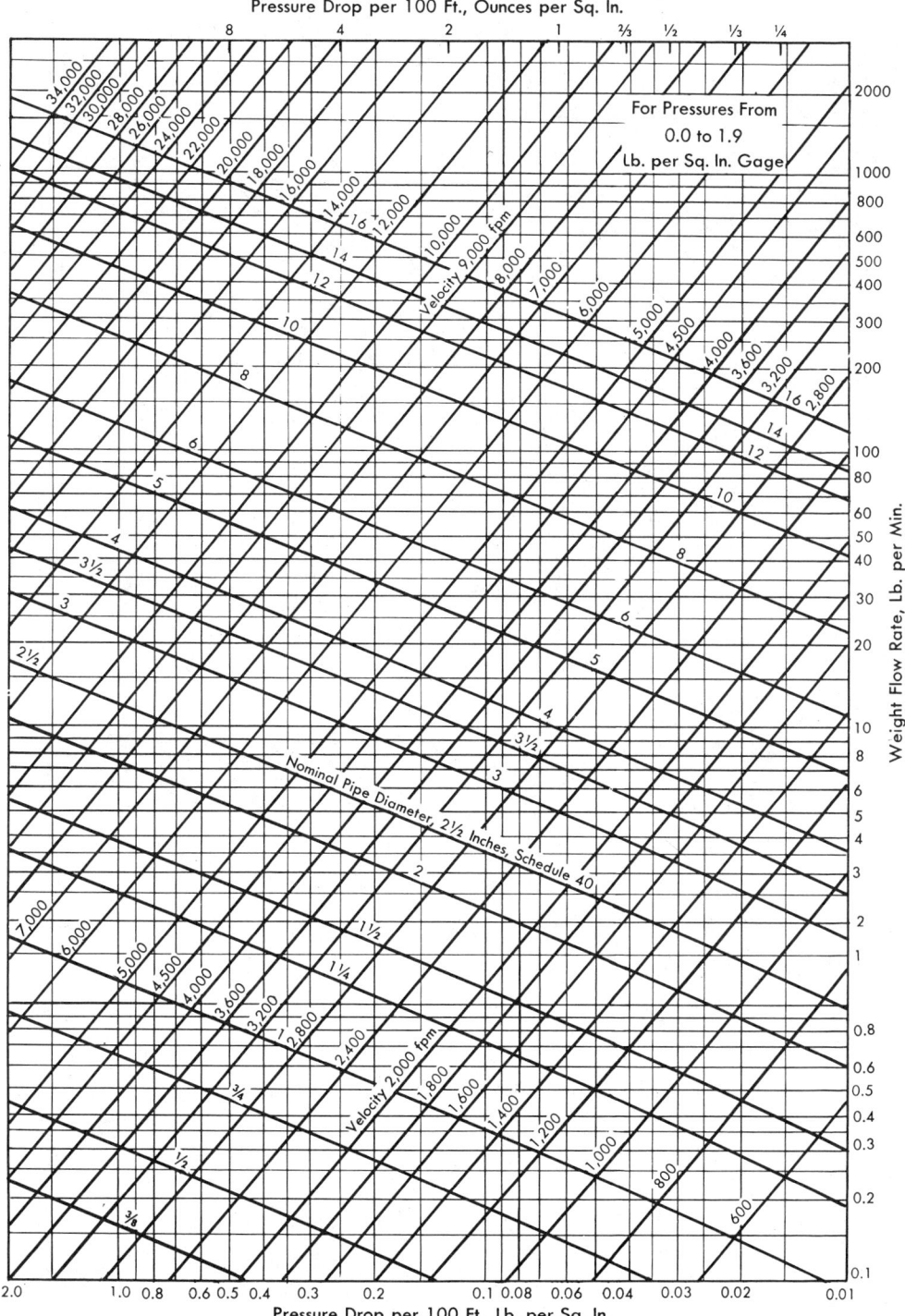

Weight Flow Rate, Lb. per Min.

For Pressures From
0.0 to 1.9
Lb. per Sq. In. Gage

Pressure Drop per 100 Ft., Lb. per Sq. In.

Pressure Drop per 100 Ft., Ounces per Sq. In.

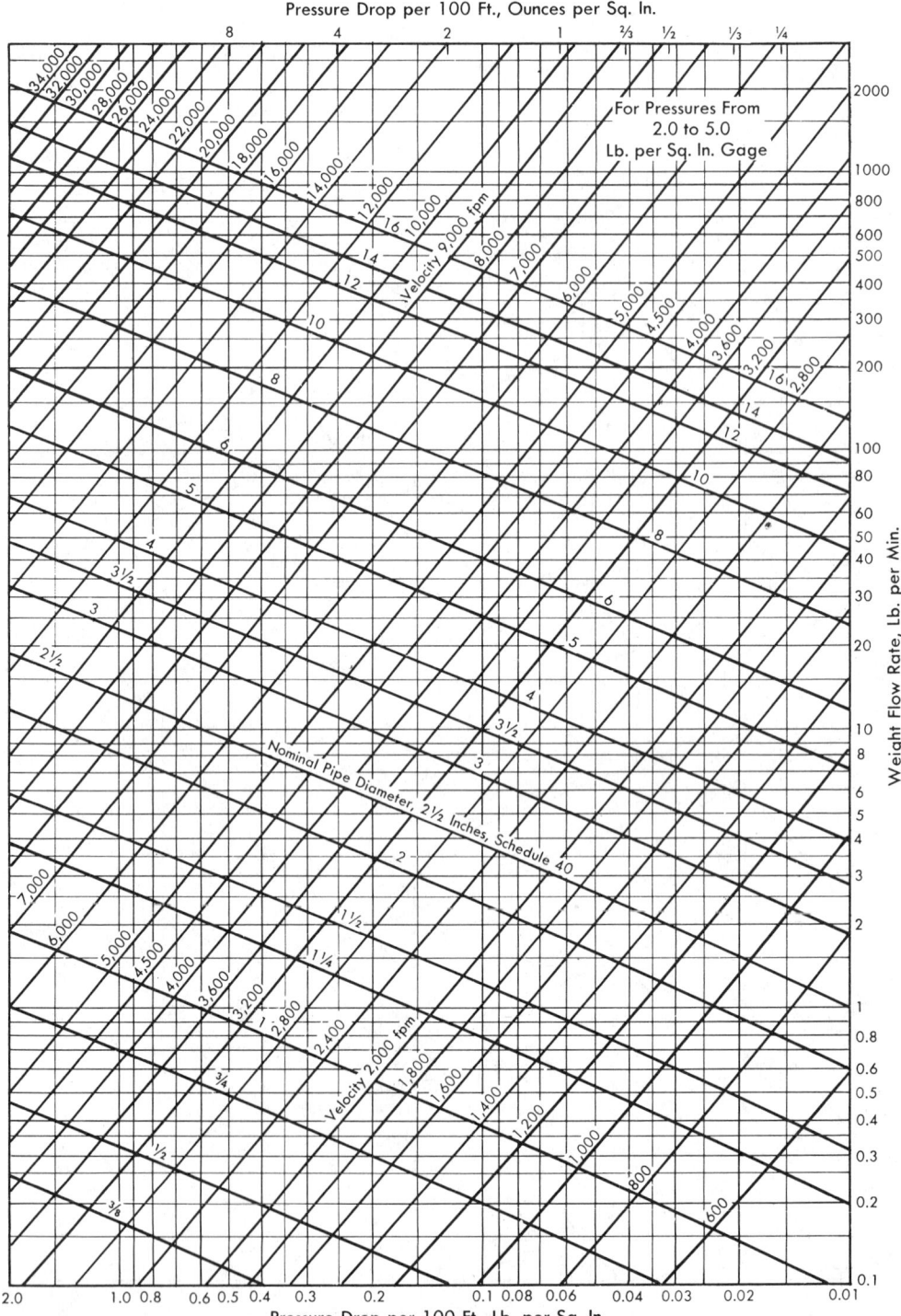

Weight Flow Rate, Lb. per Min.

Pressure Drop per 100 Ft., Lb. per Sq. In.

Pressure Drop per 100 Ft., Ounces per Sq. In.

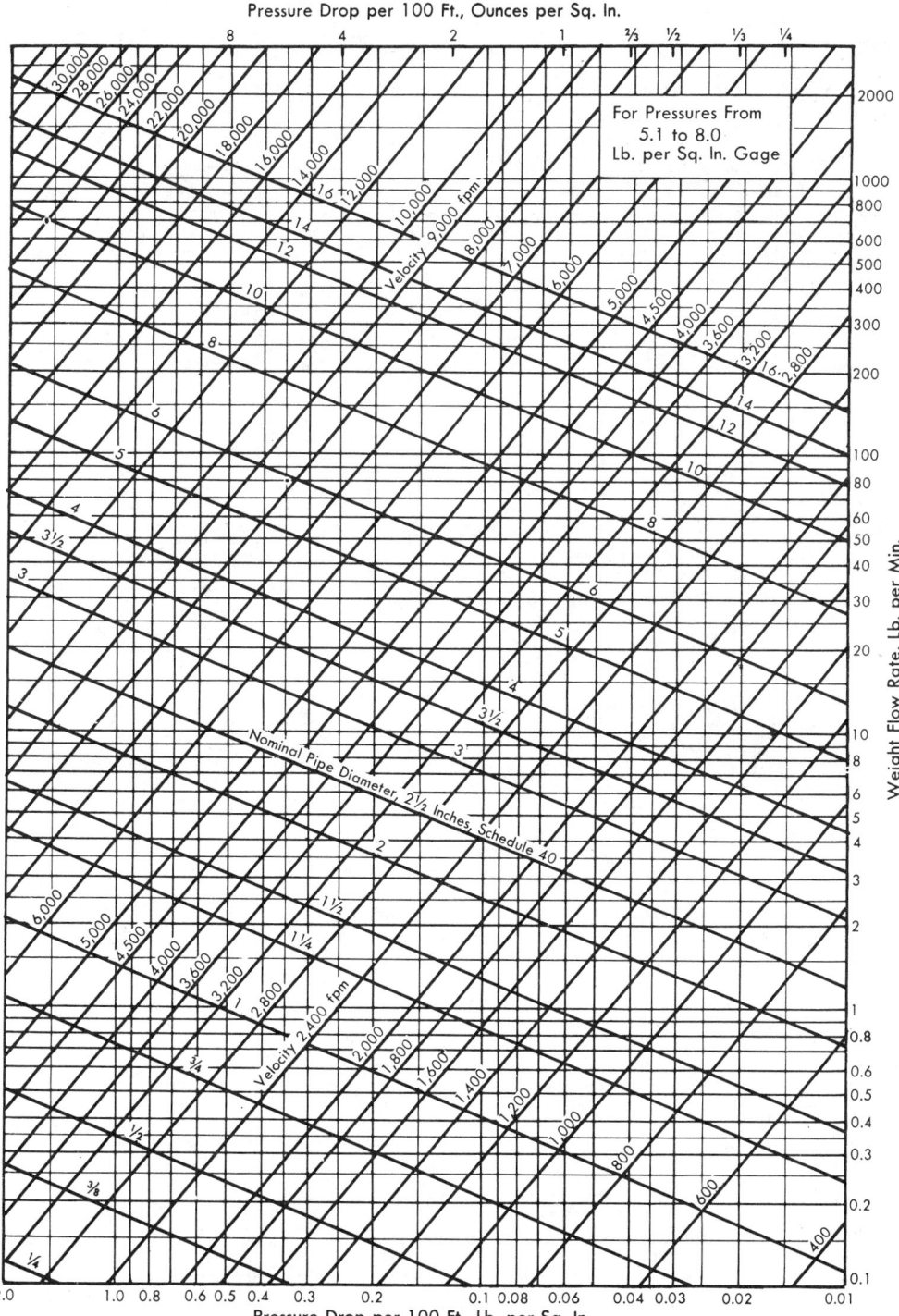

For Pressures From
5.1 to 8.0
Lb. per Sq. In. Gage

Weight Flow Rate, Lb. per Min.

Pressure Drop per 100 Ft., Lb. per Sq. In.

Pressure Drop per 100 Ft., Ounces per Sq. In.

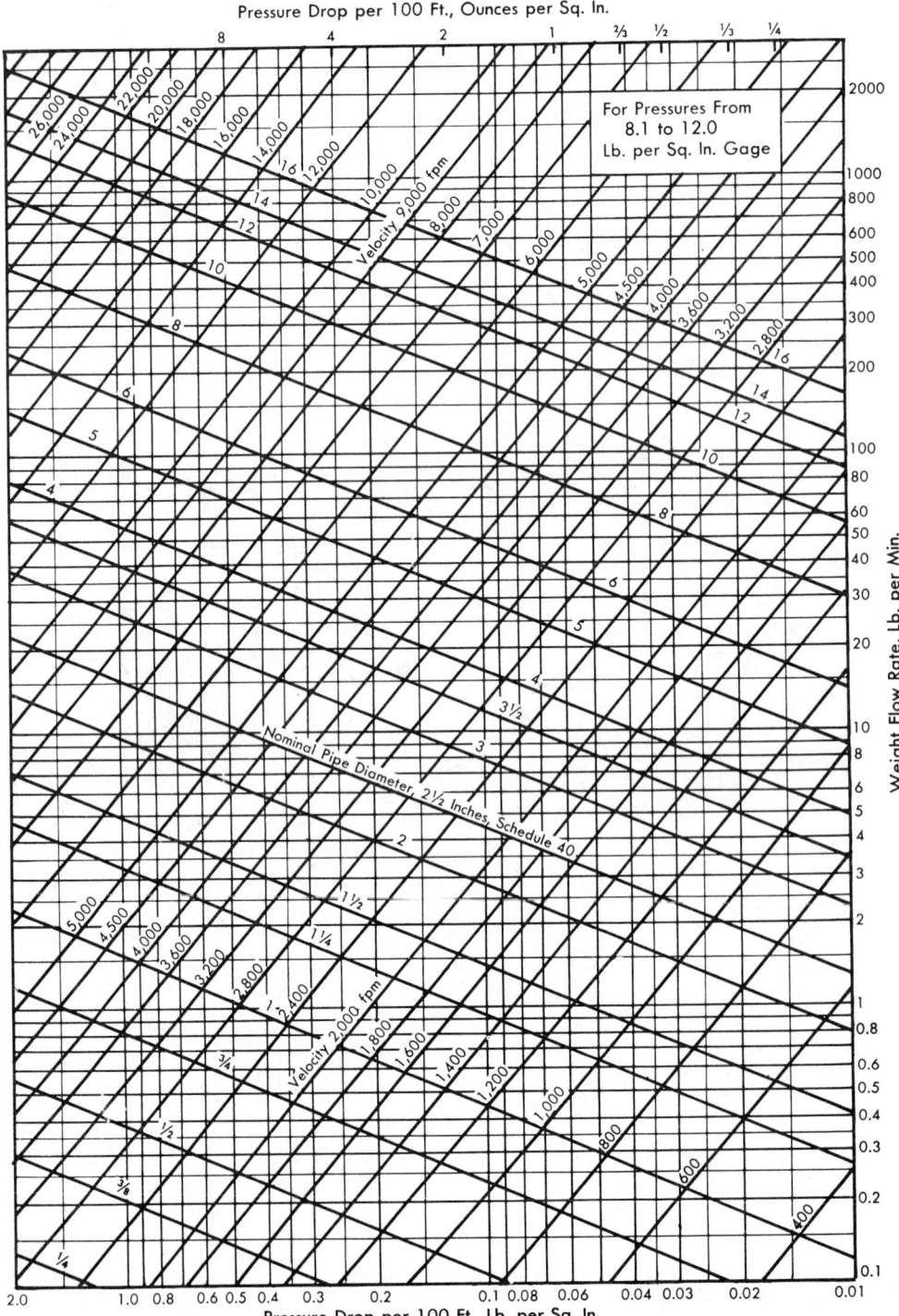

For Pressures From
8.1 to 12.0
Lb. per Sq. In. Gage

Weight Flow Rate, Lb. per Min.

Pressure Drop per 100 Ft., Lb. per Sq. In.

Pressure Drop per 100 Ft., Ounces per Sq. In.

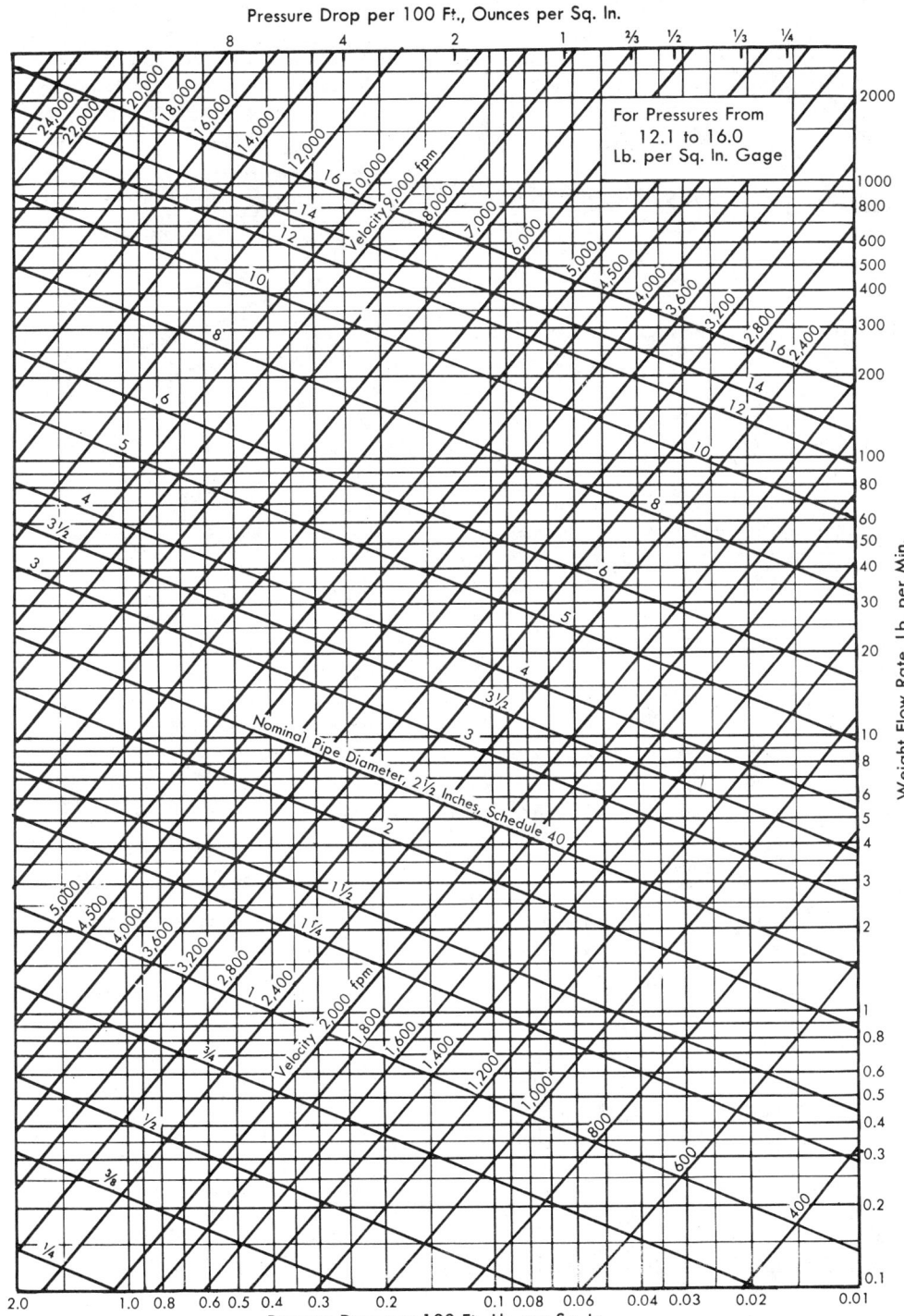

For Pressures From
12.1 to 16.0
Lb. per Sq. In. Gage

Weight Flow Rate, Lb. per Min.

Pressure Drop per 100 Ft., Lb. per Sq. In.

Pressure Drop per 100 Ft., Ounces per Sq. In.

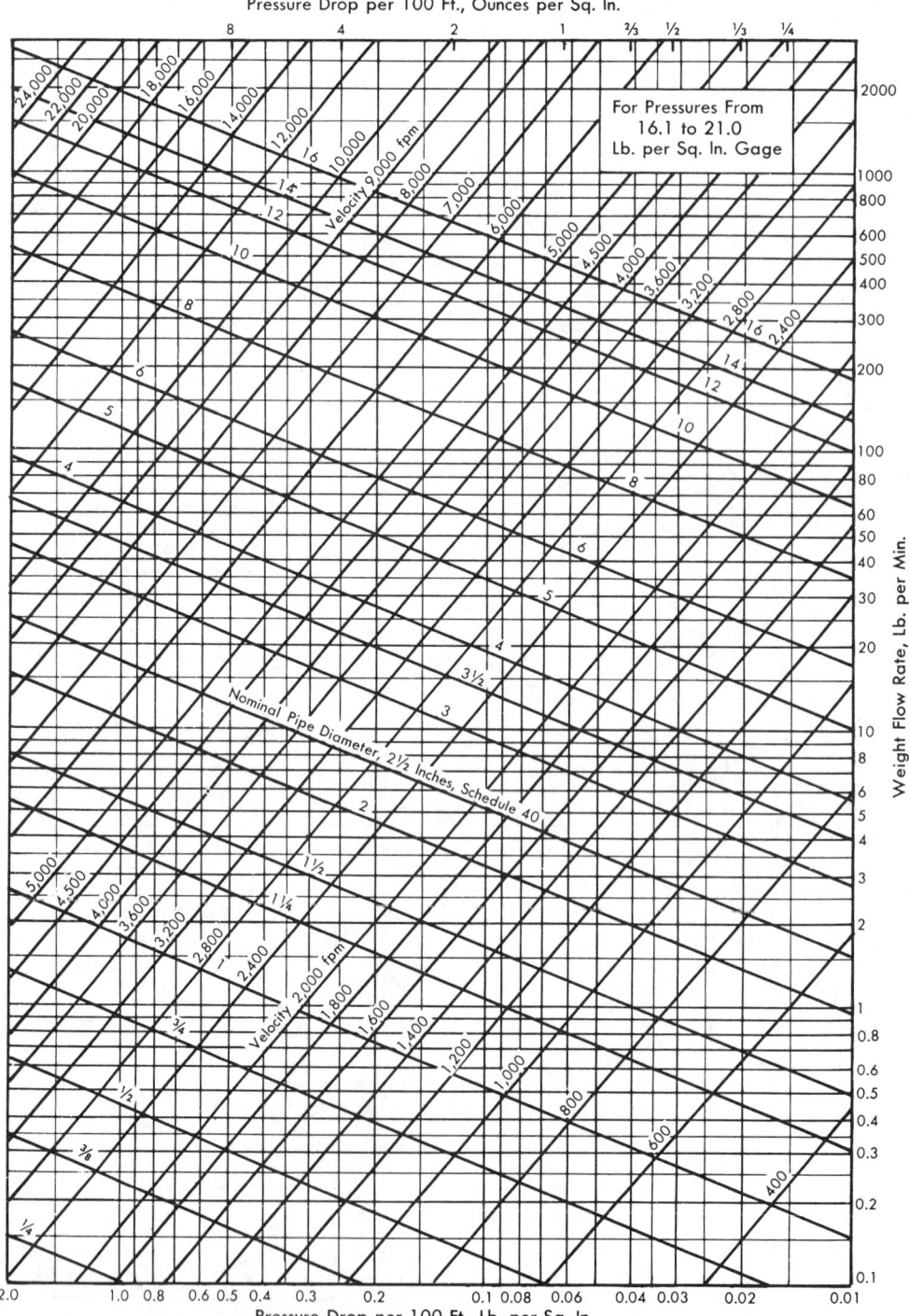

For Pressures From
16.1 to 21.0
Lb. per Sq. In. Gage

Weight Flow Rate, Lb. per Min.

Pressure Drop per 100 Ft., Lb. per Sq. In.

Pressure Drop per 100 Ft., Ounces per Sq. In.

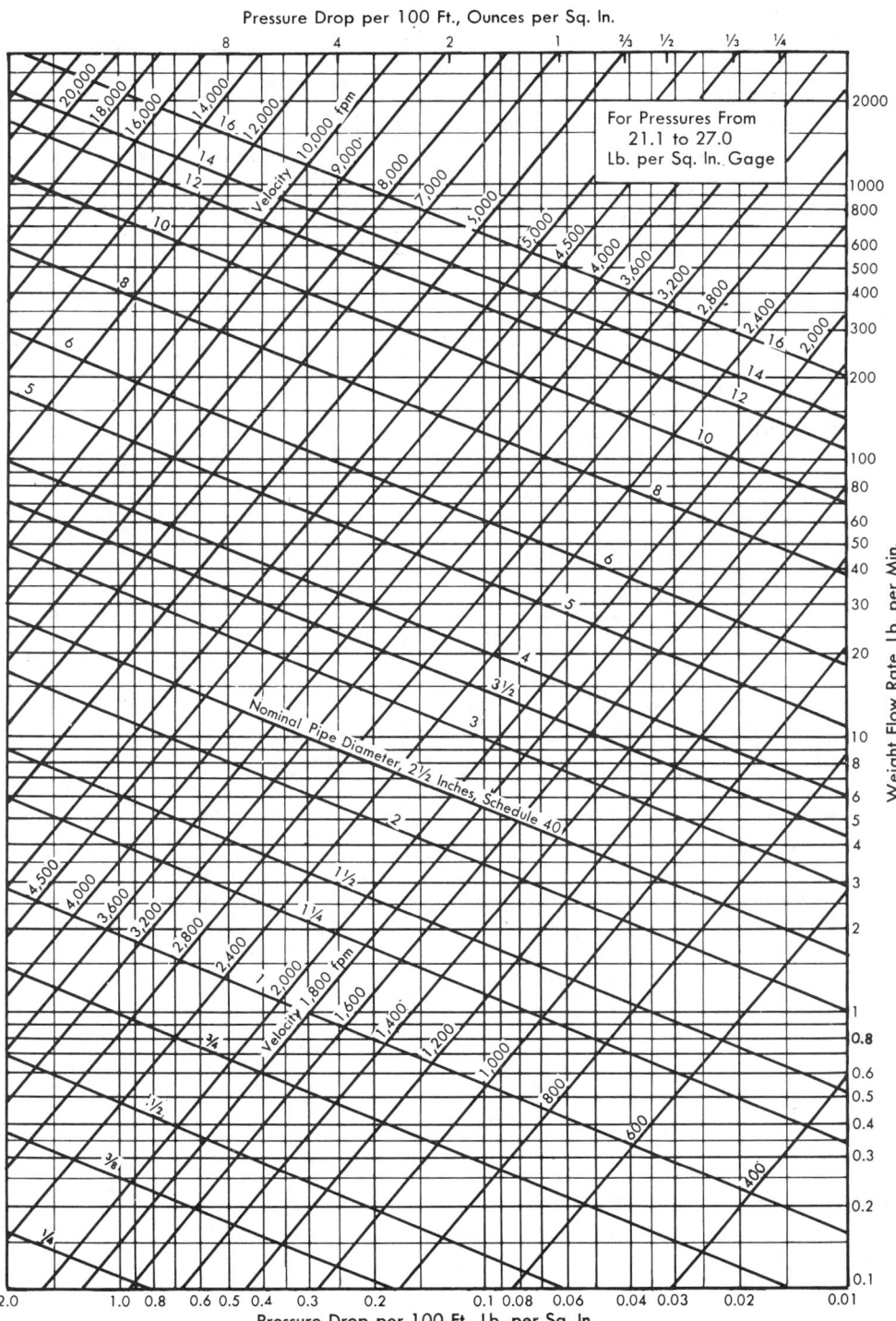

For Pressures From
21.1 to 27.0
Lb. per Sq. In. Gage

Weight Flow Rate, Lb. per Min.

Pressure Drop per 100 Ft., Lb. per Sq. In.

Pressure Drop per 100 Ft., Ounces per Sq. In.

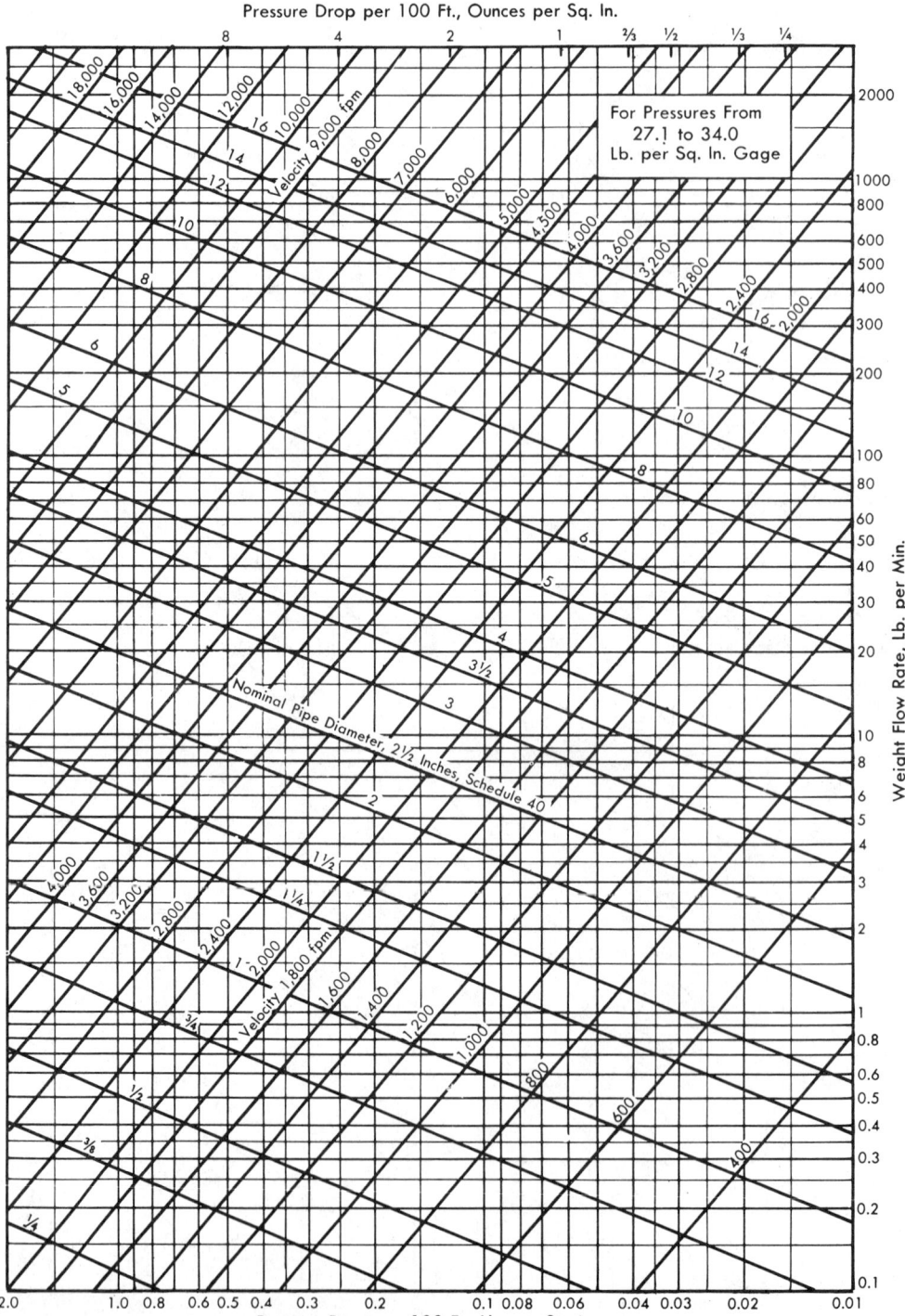

For Pressures From
27.1 to 34.0
Lb. per Sq. In. Gage

Weight Flow Rate, Lb. per Min.

Pressure Drop per 100 Ft., Lb. per Sq. In.

Pressure Drop per 100 Ft., Ounces per Sq. In.

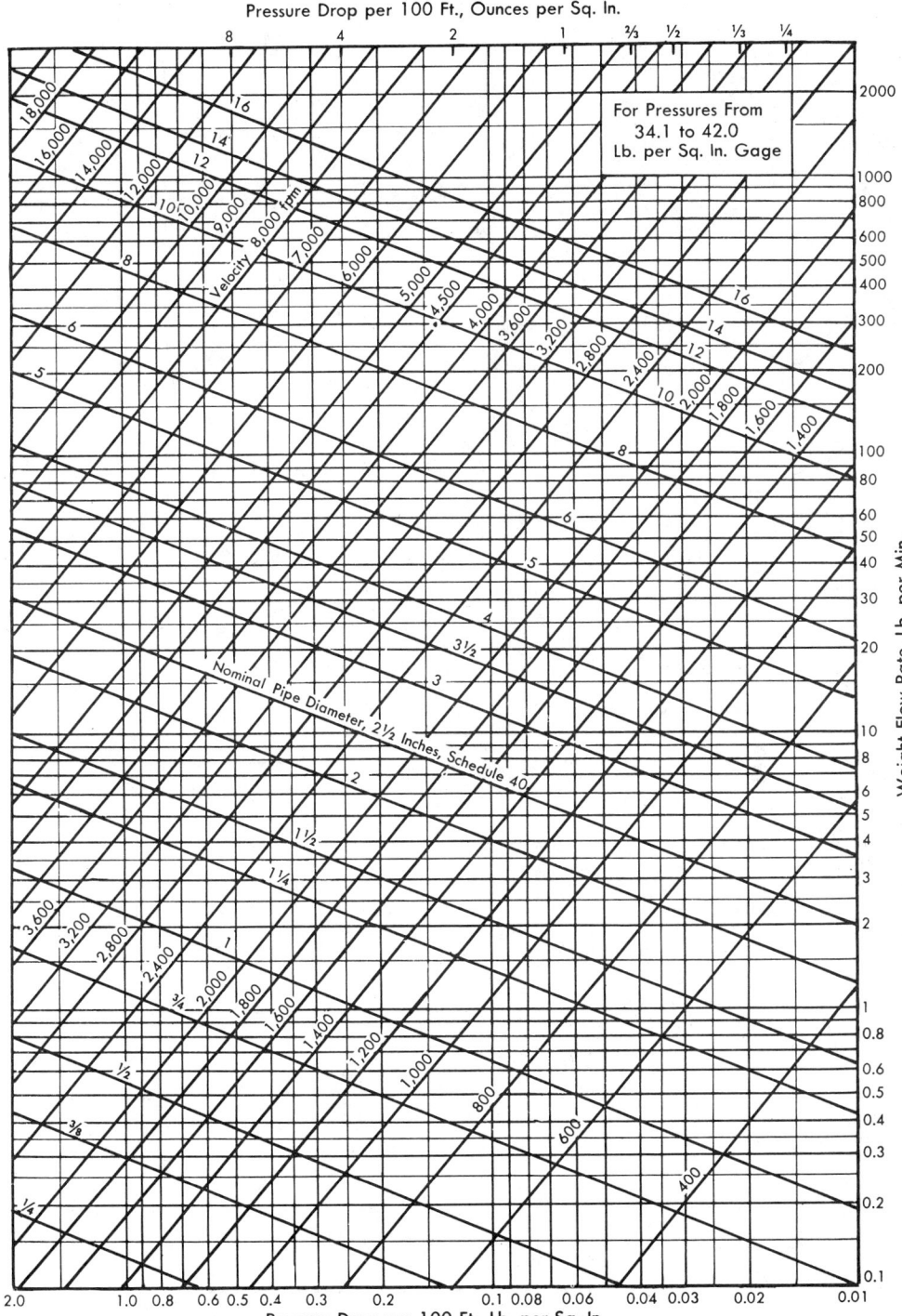

For Pressures From
34.1 to 42.0
Lb. per Sq. In. Gage

Pressure Drop per 100 Ft., Lb. per Sq. In.

Weight Flow Rate, Lb. per Min.

Pressure Drop per 100 Ft., Ounces per Sq. In.

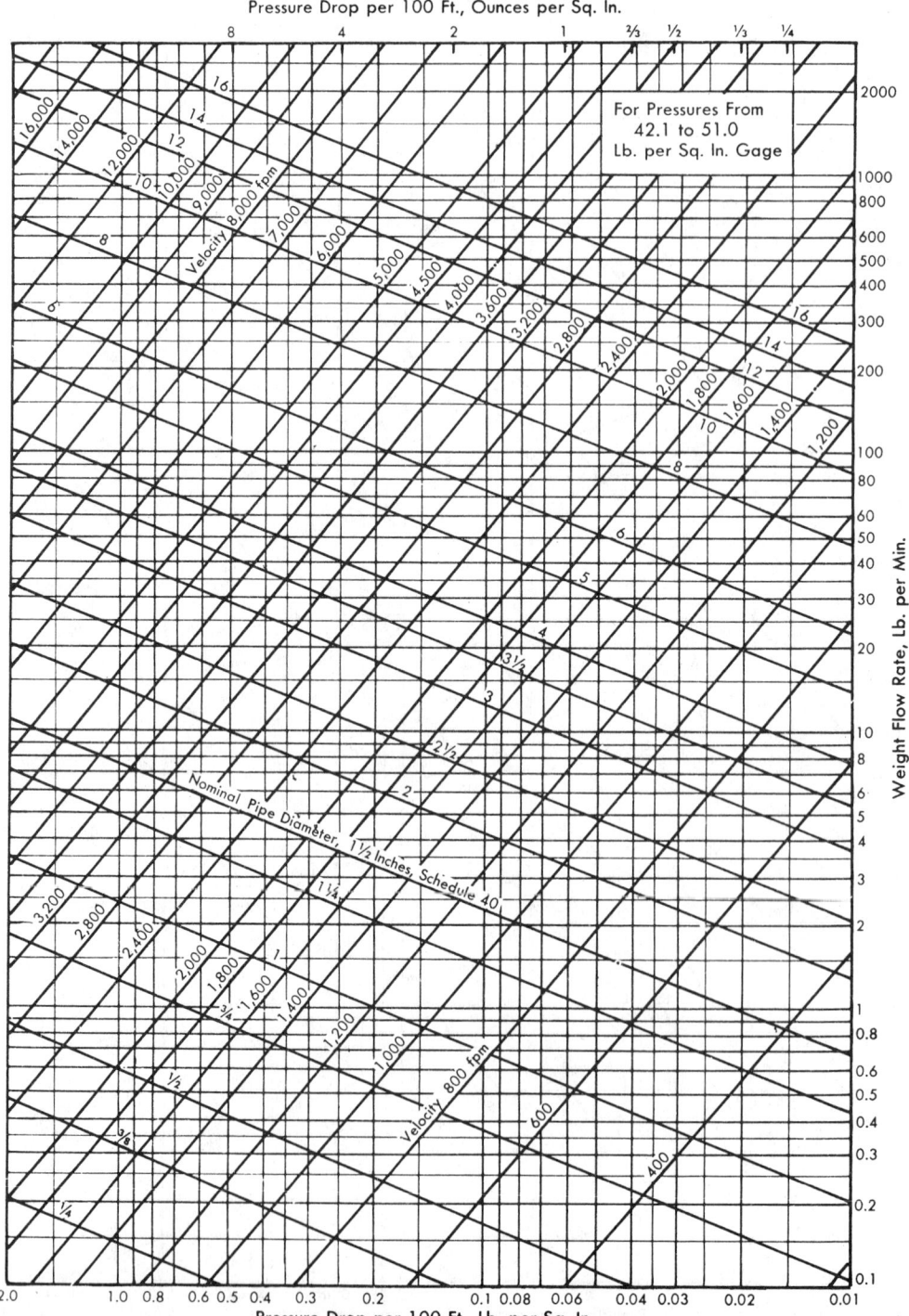

For Pressures From
42.1 to 51.0
Lb. per Sq. In. Gage

Weight Flow Rate, Lb. per Min.

Pressure Drop per 100 Ft., Lb. per Sq. In.

Pressure Drop per 100 Ft., Ounces per Sq. In.

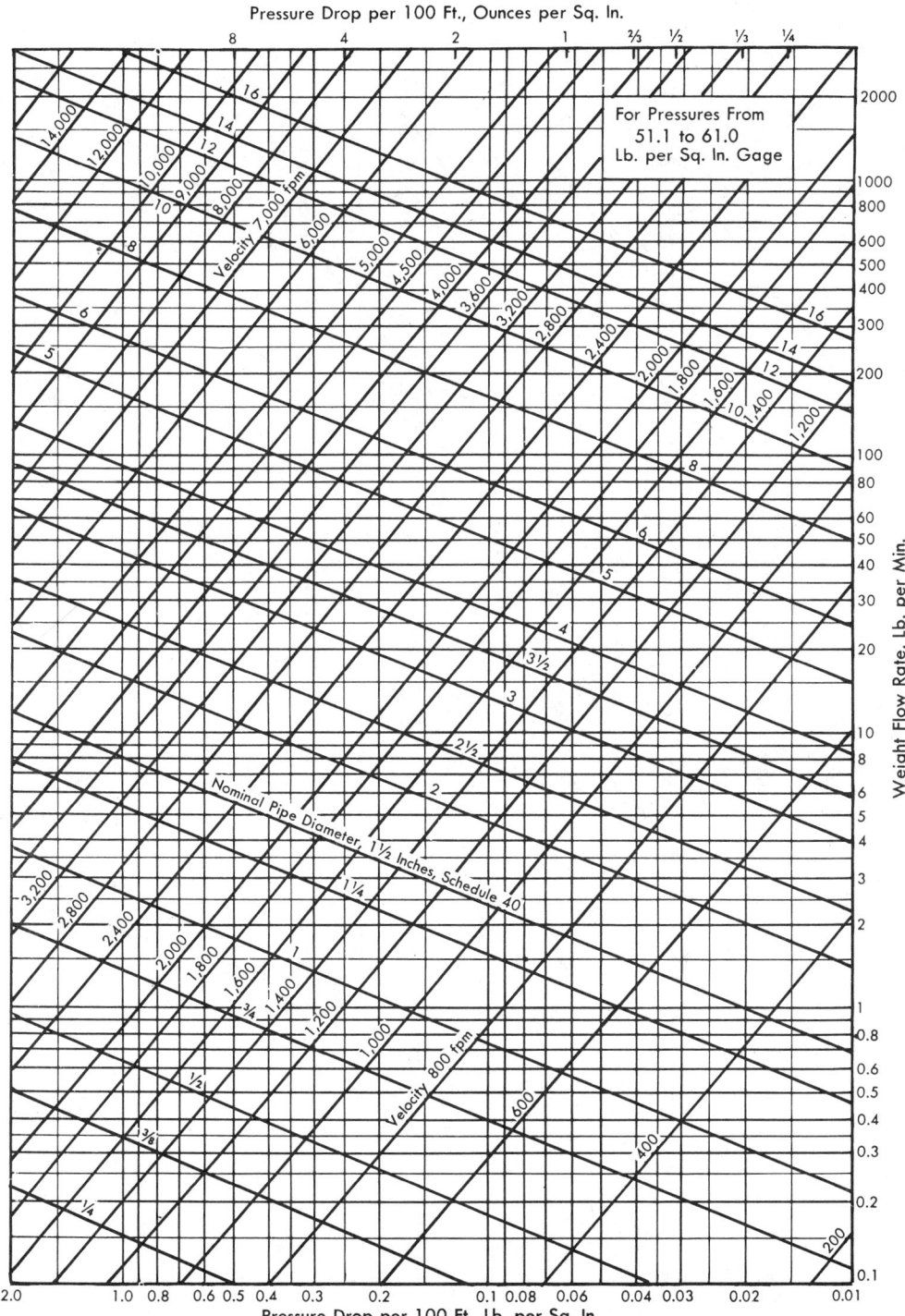

For Pressures From
51.1 to 61.0
Lb. per Sq. In. Gage

Weight Flow Rate, Lb. per Min.

Pressure Drop per 100 Ft., Lb. per Sq. In.

Pressure Drop per 100 Ft., Ounces per Sq. In.

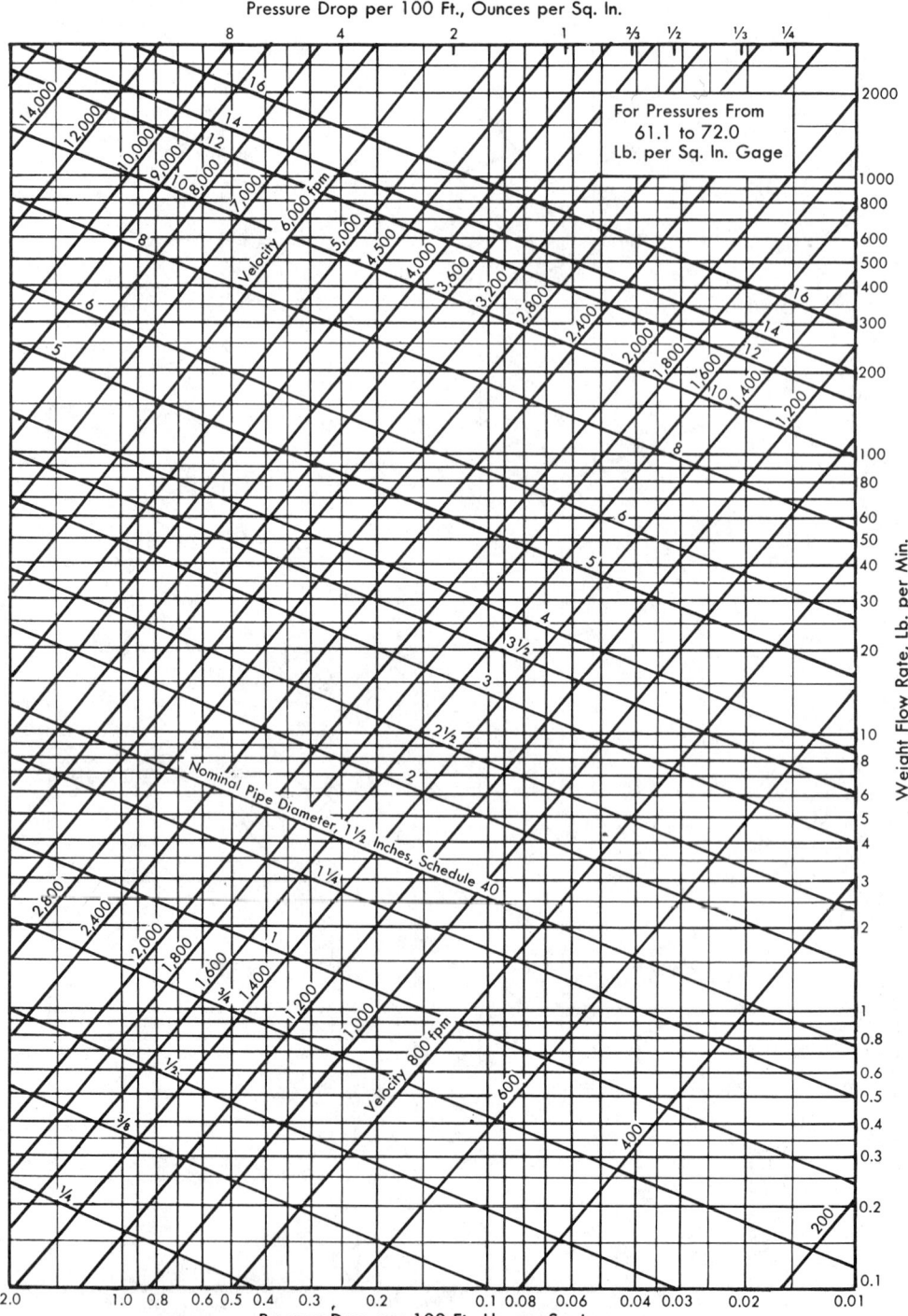

For Pressures From
61.1 to 72.0
Lb. per Sq. In. Gage

Weight Flow Rate, Lb. per Min.

Pressure Drop per 100 Ft., Lb. per Sq. In.

Pressure Drop per 100 Ft., Ounces per Sq. In.

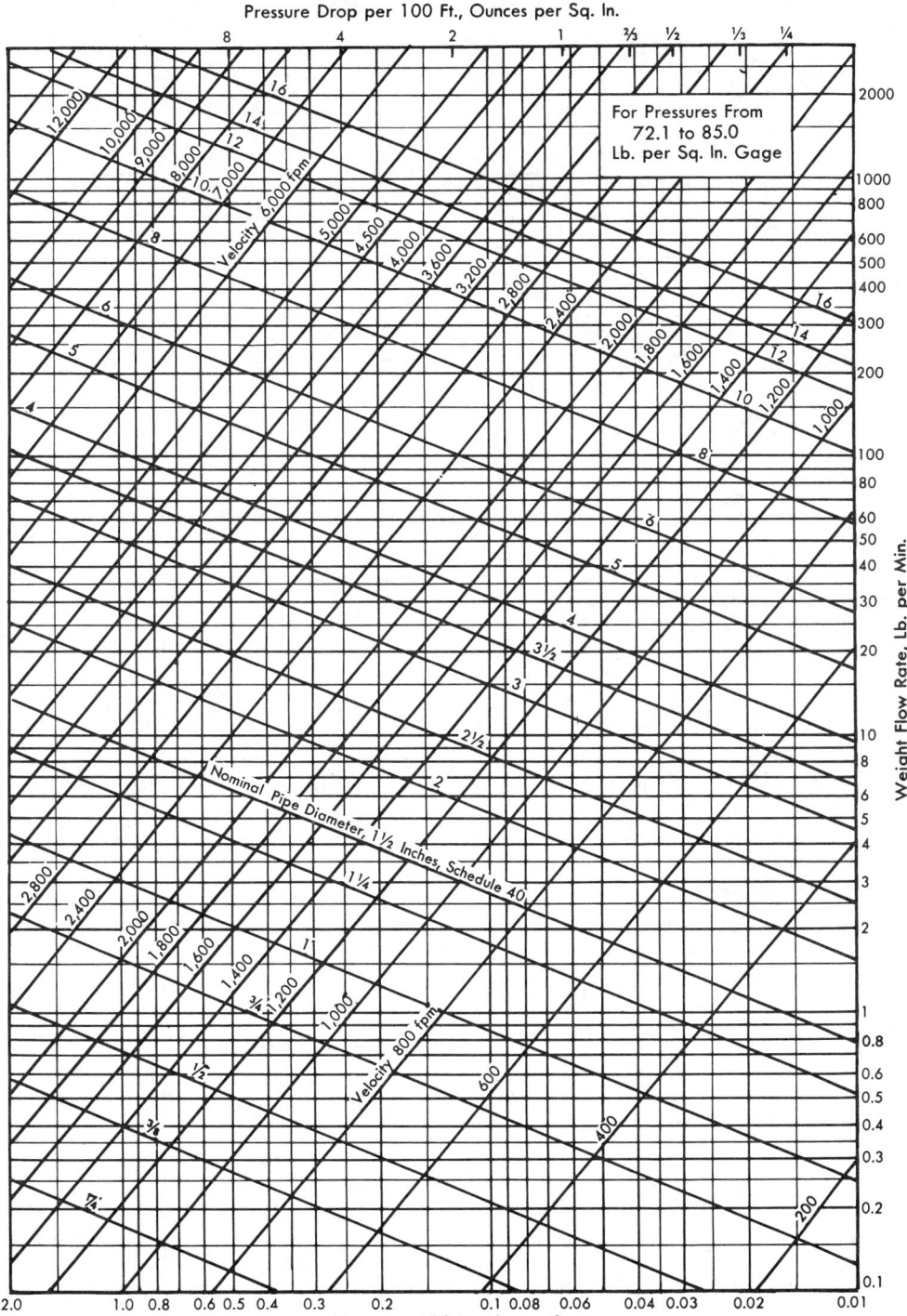

For Pressures From
72.1 to 85.0
Lb. per Sq. In. Gage

Weight Flow Rate, Lb. per Min.

Pressure Drop per 100 Ft., Lb. per Sq. In.

Pressure Drop per 100 Ft., Ounces per Sq. In.

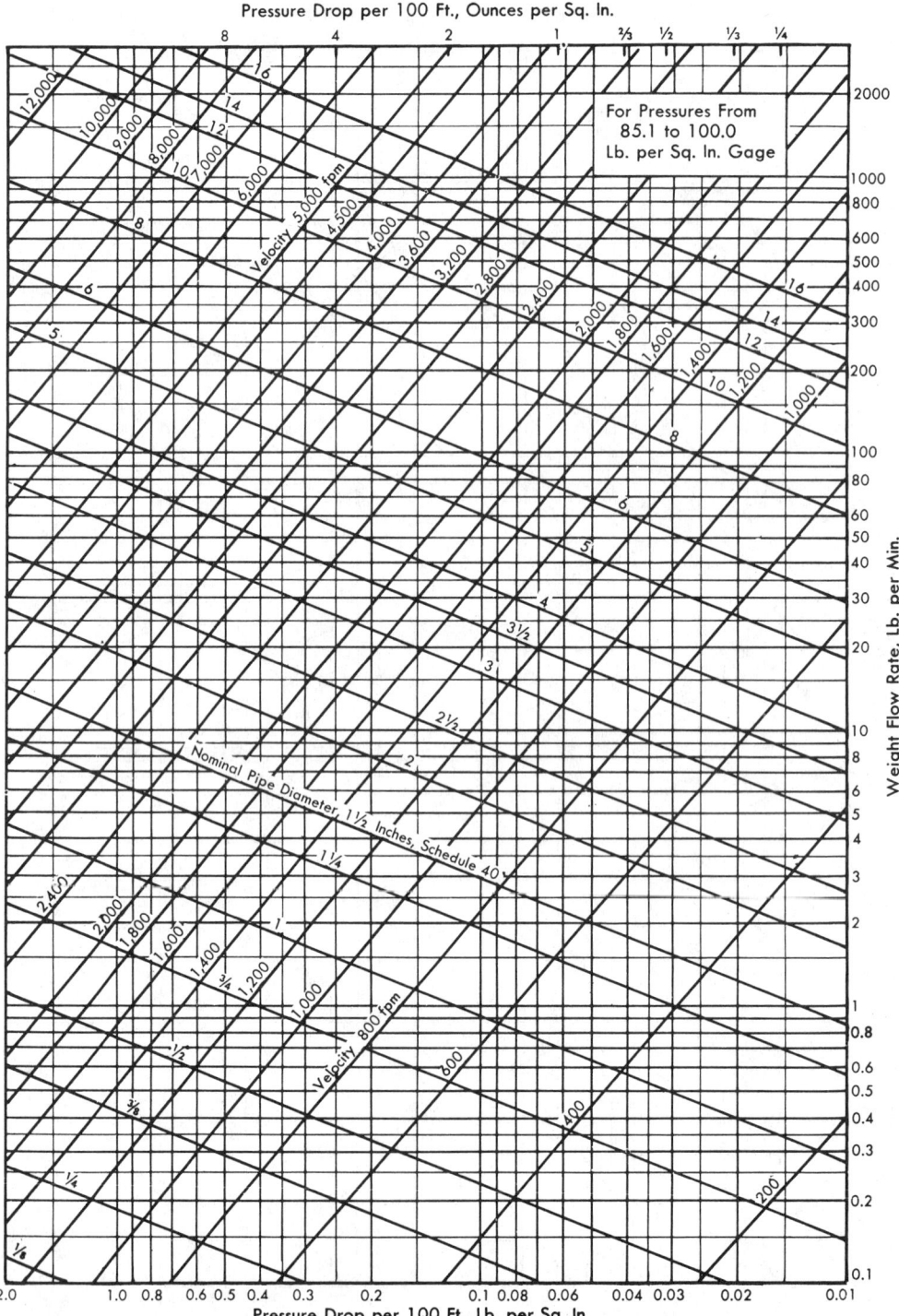

For Pressures From
85.1 to 100.0
Lb. per Sq. In. Gage

Weight Flow Rate, Lb. per Min.

Pressure Drop per 100 Ft., Lb. per Sq. In.

Pressure Drop per 100 Ft., Ounces per Sq. In.

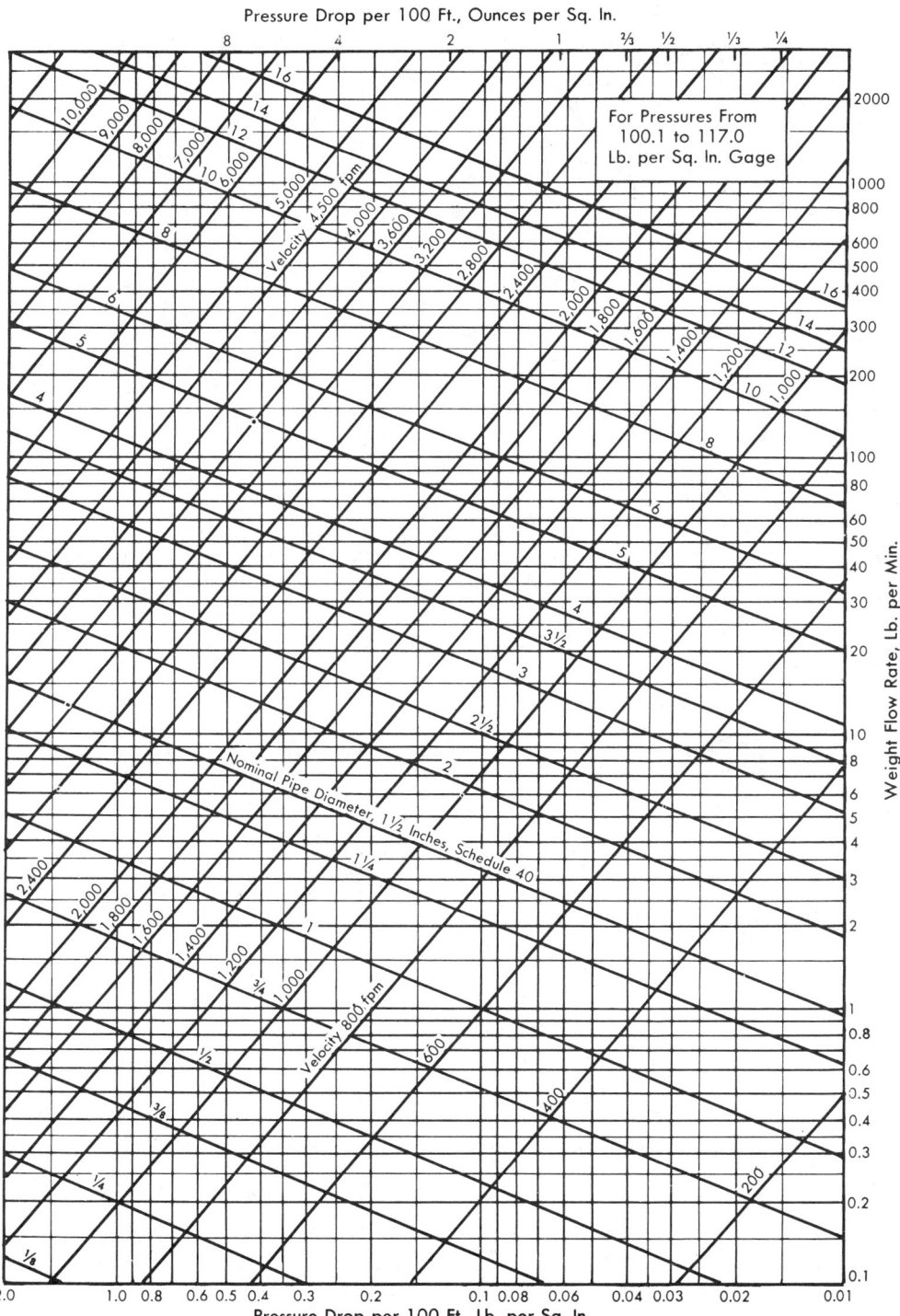

Pressure Drop per 100 Ft., Lb. per Sq. In.

Weight Flow Rate, Lb. per Min.

For Pressures From
100.1 to 117.0
Lb. per Sq. In. Gage

Pressure Drop per 100 Ft., Ounces per Sq. In.

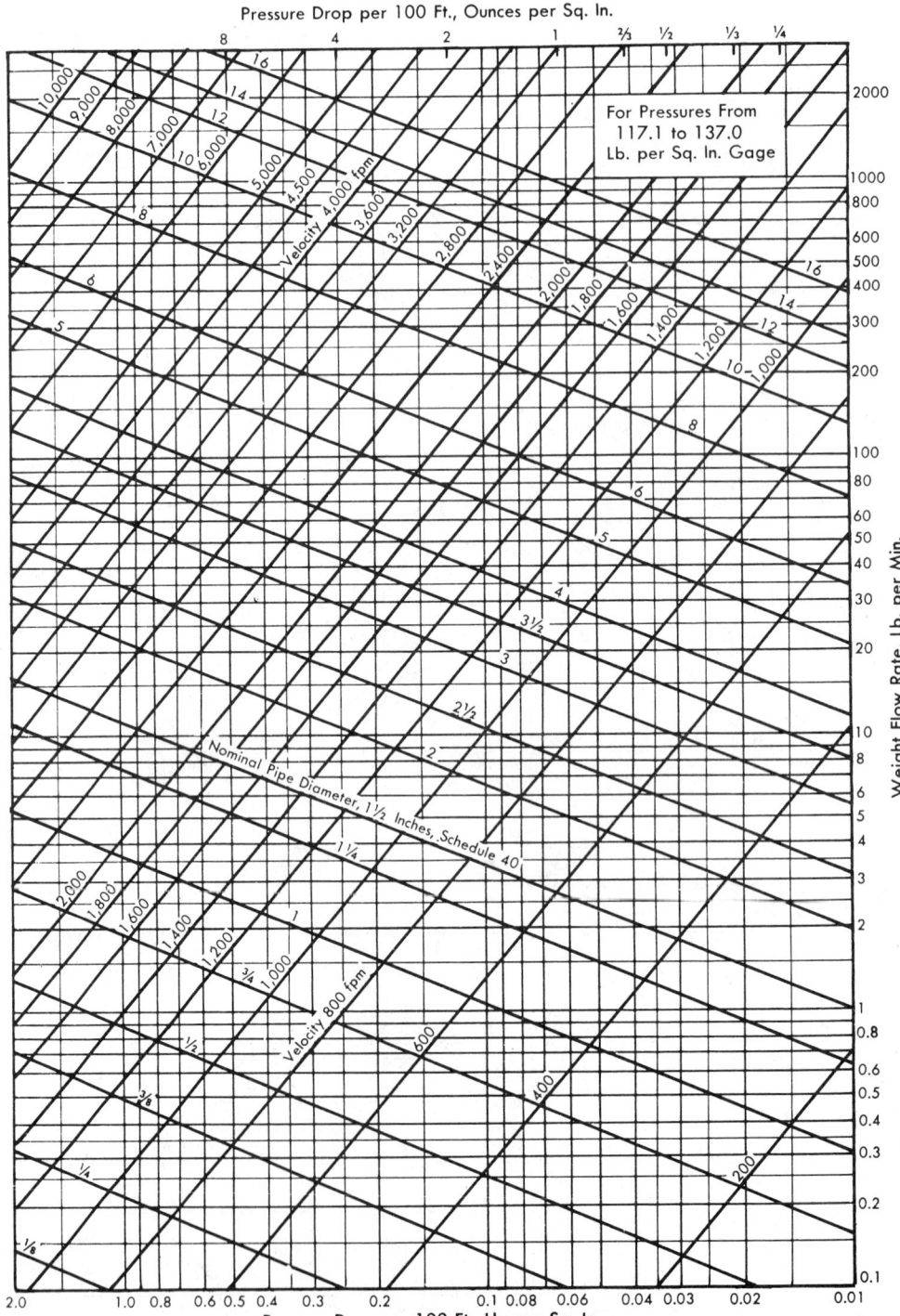

For Pressures From
117.1 to 137.0
Lb. per Sq. In. Gage

Weight Flow Rate, Lb. per Min.

Pressure Drop per 100 Ft., Lb. per Sq. In.

Pressure Drop per 100 Ft., Ounces per Sq. In.

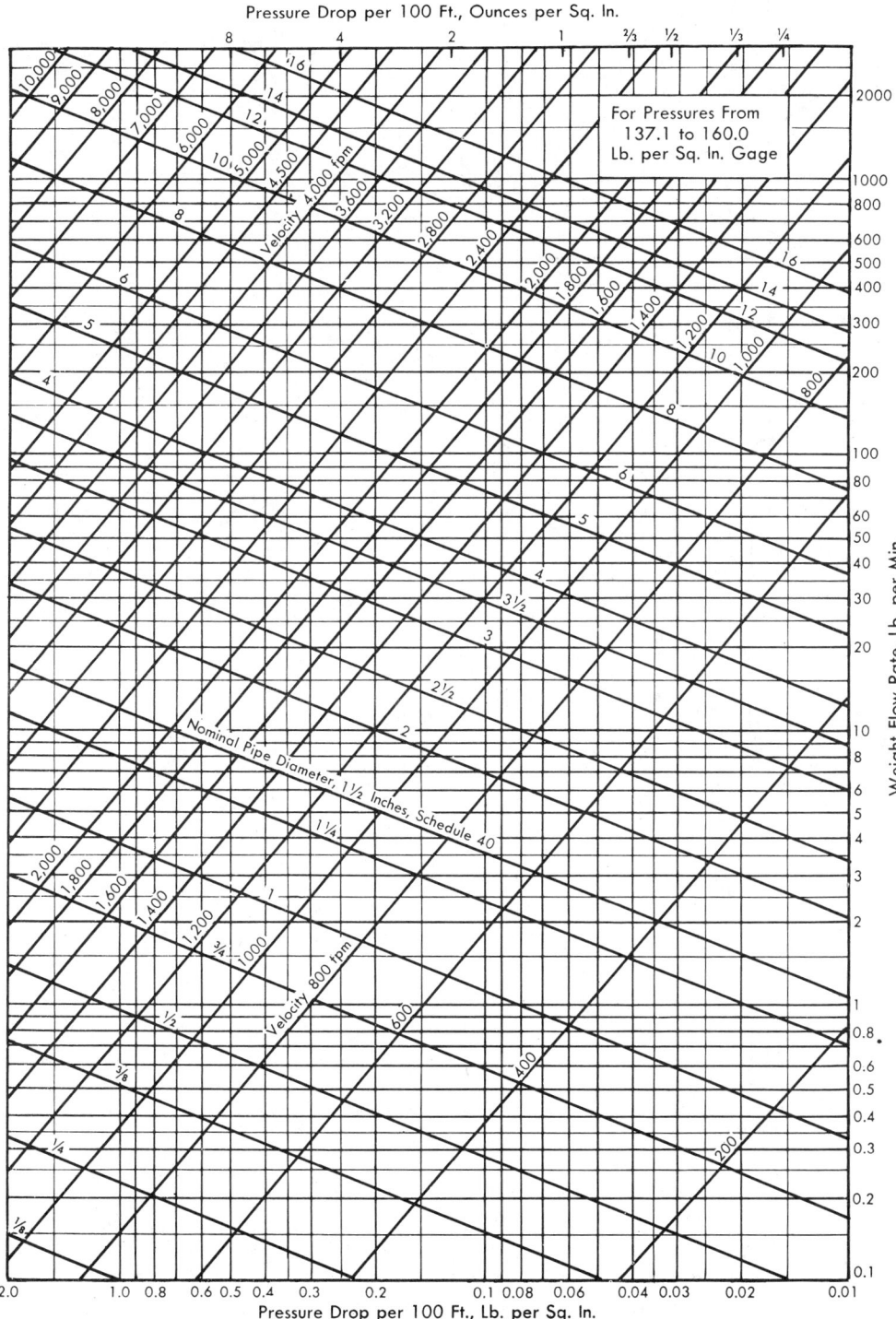

For Pressures From
137.1 to 160.0
Lb. per Sq. In. Gage

Weight Flow Rate, Lb. per Min.

Pressure Drop per 100 Ft., Lb. per Sq. In.

Schedule 40 Pipe

Pressure Drop per 100 Ft., Ounces per Sq. In.

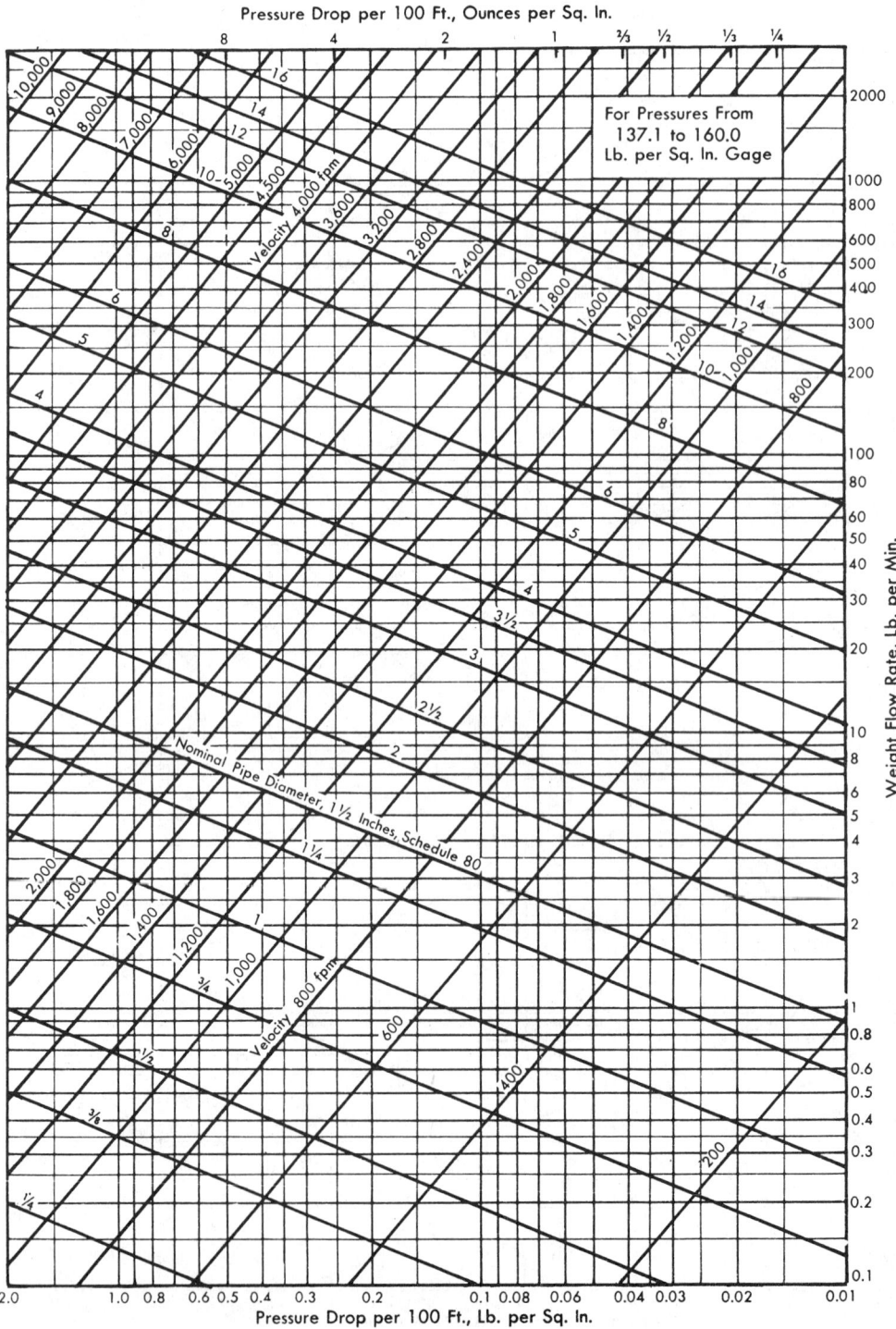

For Pressures From
137.1 to 160.0
Lb. per Sq. In. Gage

Weight Flow Rate, Lb. per Min.

Pressure Drop per 100 Ft., Lb. per Sq. In.

Schedule 80 Pipe

Pressure Drop per 100 Ft., Ounces per Sq. In.

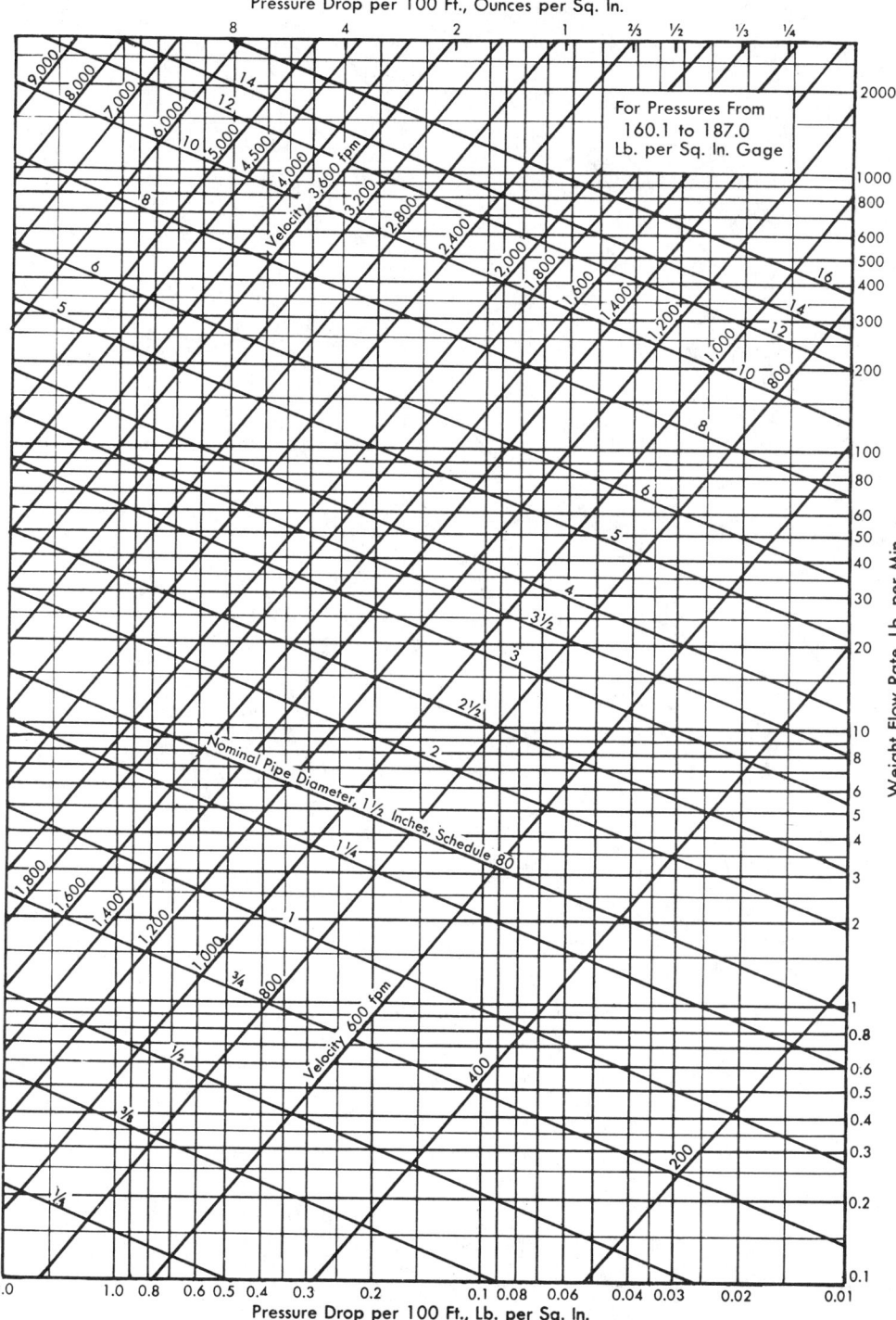

For Pressures From
160.1 to 187.0
Lb. per Sq. In. Gage

Weight Flow Rate, Lb. per Min.

Pressure Drop per 100 Ft., Lb. per Sq. In.

Pressure Drop per 100 Ft., Ounces per Sq. In.

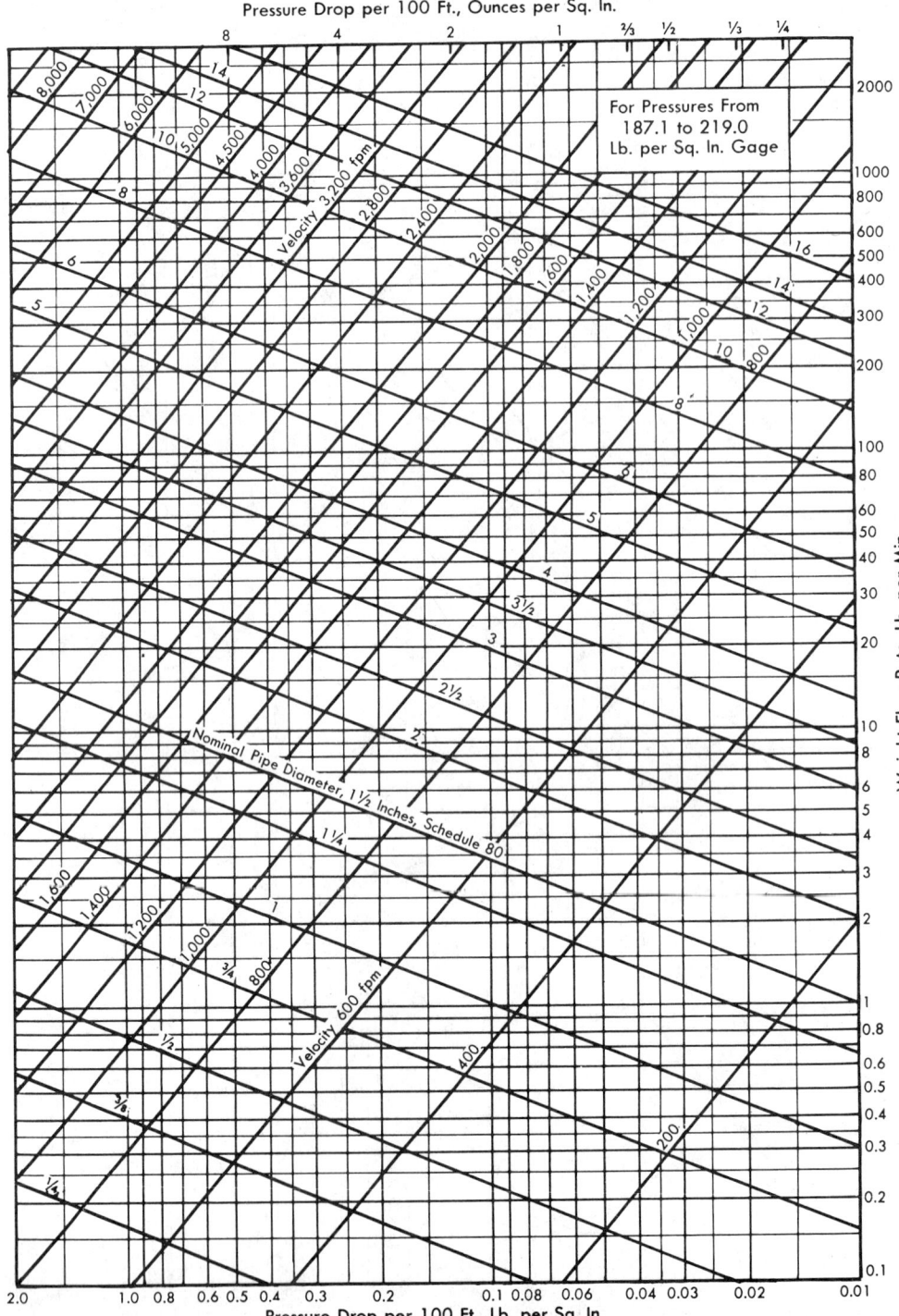

For Pressures From
187.1 to 219.0
Lb. per Sq. In. Gage

Weight Flow Rate, Lb. per Min.

Pressure Drop per 100 Ft., Lb. per Sq. In.

Pressure Drop per 100 Ft., Ounces per Sq. In.

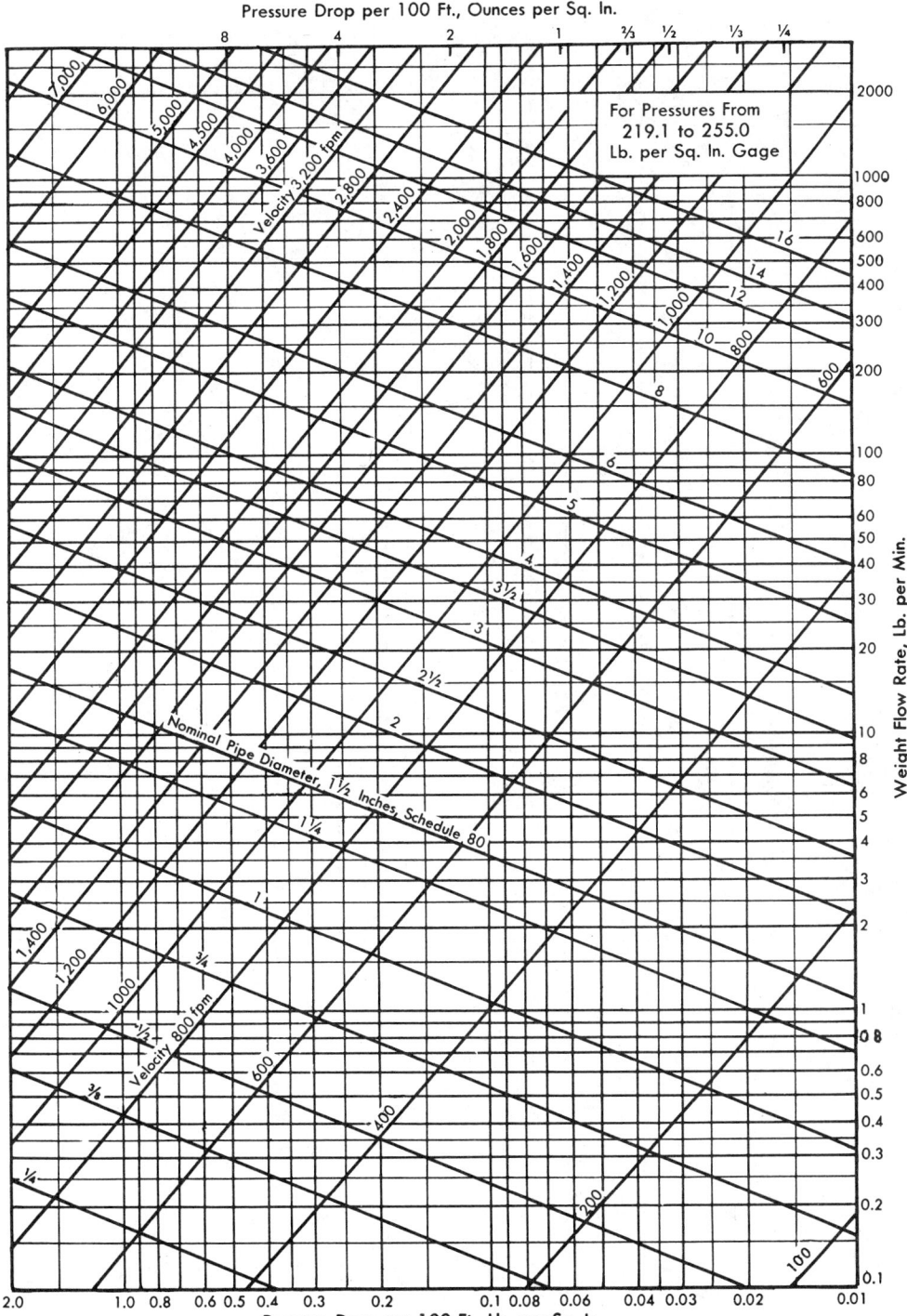

For Pressures From
219.1 to 255.0
Lb. per Sq. In. Gage

Weight Flow Rate, Lb. per Min.

Pressure Drop per 100 Ft., Lb. per Sq. In.

Pressure Drop per 100 Ft., Ounces per Sq. In.

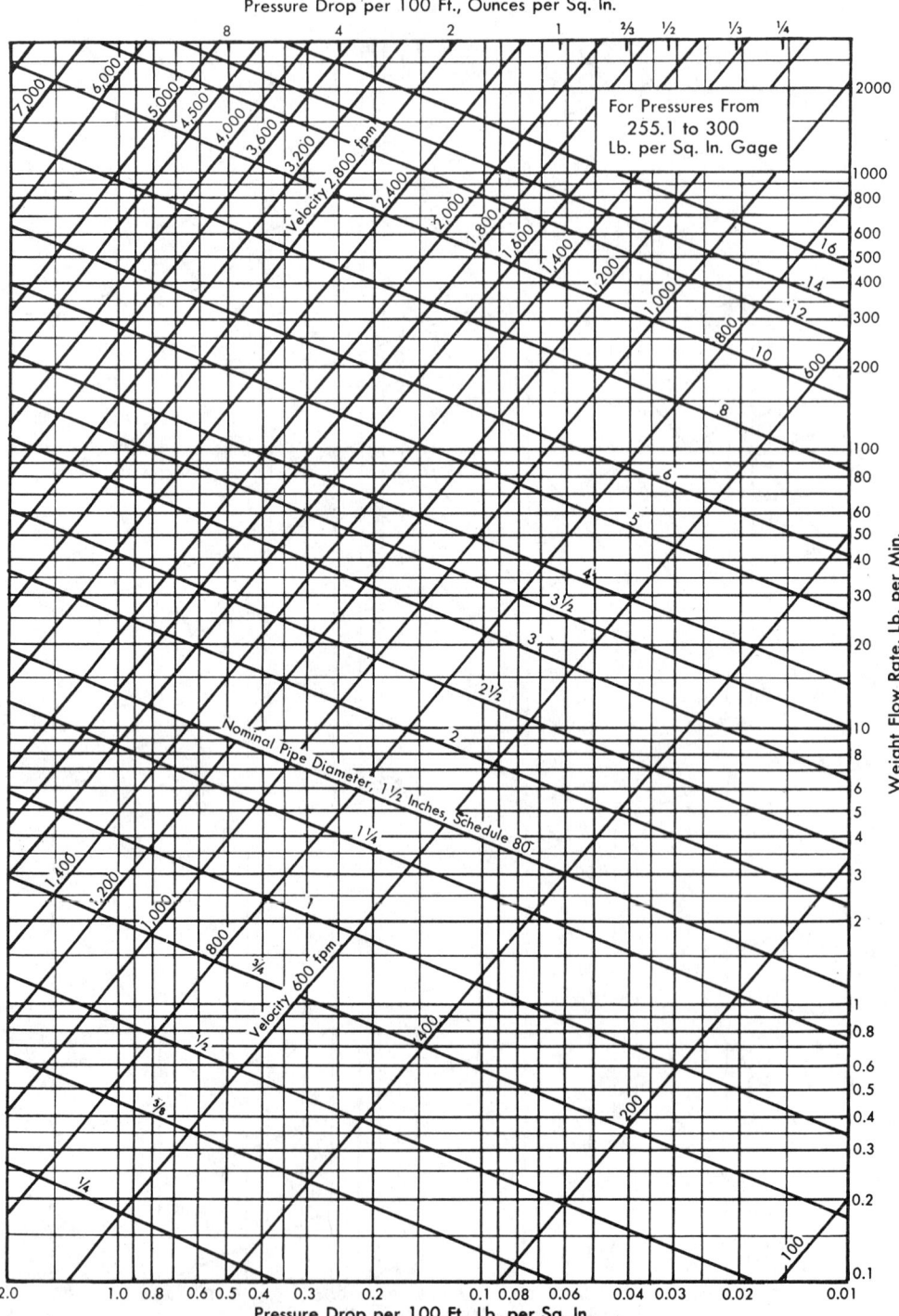

For Pressures From
255.1 to 300
Lb. per Sq. In. Gage

Weight Flow Rate, Lb. per Min.

Pressure Drop per 100 Ft., Lb. per Sq. In.

GAS PIPING FOR BUILDINGS

Residential Piping

Recommendations for the installation of gas piping in buildings other than industrial and commercial are given in ANSI-Z21.30 of the American National Standards Institute, sponsored by American Gas Association. Among the provisions of this standard are the following:

These standards apply only to low pressure (not in excess of ½ pound per square inch) gas piping systems in buildings, extending from the gas meter outlet to the inlet connections of appliances, and the installation and operation of residential and commercial gas appliances supplied through such systems by public utilities. They are intended to cover the design, fabrication, installation, tests and operation of such systems for fuel gases such as natural gas, manufactured gas, liquefied petroleum gas-air or mixtures thereof. They are *not* intended to cover systems distributing undiluted liquefied petroleum gas. They are also *not* intended to cover systems or portions of systems supplying equipment engineered, designed and installed for specific manufacturing, production processing and power generating applications, such as large and high pressure boilers, melting and treating furnaces, and production ovens. For piping in gas distribution systems, in gas manufacturing plants, in gas compressing stations, and in gas processing plants, refer to the latest edition of Code for Fuel Gas Piping, ANSI-B31.2.

It is recommended that before proceeding with the installation of a gas piping system, a piping sketch or plan be prepared showing the proposed location of the piping as well as the size of different branches. Adequate consideration should be given to future demands and provisions made for added gas service.

Piping shall be of such size and so installed as to provide a supply of gas sufficient to meet the maximum demand without undue loss of pressure between the meter and the appliance or appliances.

The size of gas pipe depends upon the following factors:

a. Allowable loss in pressure from meter to appliance.
b. Maximum gas consumption to be provided for.
c. Length of pipe and number of fittings.
d. Specific gravity of the gas.
e. Diversity factor.

It is recommended that the pressure loss in any piping system from the gas meter to any appliance at the maximum probable gas demand not exceed 0.3 inch water column.

The volume of gas to be provided for (in cubic feet per hour) shall be determined, whenever possible, directly from the manufacturer's Btu ratings

Gas Pipe Sizes, Low Pressures, Residential Piping

(For pressures not over ½ psig, pressure drop of 0.3 in. of water and gas of 0.60 specific gravity)

Total Length of Pipe, Feet	Diameter of Pipe, Inches (IPS)									
	½	¾	1	1¼	1½	2	3	4	6	8
	Capacity in Cubic Feet per Hour									
15	76	172	345	750	1220	2480	6500	13880	38700.	79000
30	52	120	241	535	850	1780	4700	9700	27370	55850
45	43	99	199	435	700	1475	3900	7900	23350	45600
60	38	86	173	380	610	1290	3450	6800	19330	39500
75	—	77	155	345	545	1120	3000	6000	17310	35300
90	—	70	141	310	490	1000	2700	5500	15800	32250
105	—	65	131	285	450	920	2450	5100	14620	29850
120	—	—	120	270	420	860	2300	4800	13680	27920
150	—	—	109	242	380	780	2090	4350	12240	25000
180	—	—	100	225	350	720	1950	4000	11160	22800
210	—	—	92	205	320	660	1780	3700	10330	21100
240	—	—	—	190	300	620	1680	3490	9600	19740
270	—	—	—	178	285	580	1580	3250	9000	18610
300	—	—	—	170	270	545	1490	3000	8500	17660
450	—	—	—	140	226	450	1230	2500	7000	14420
600	—	—	—	119	192	390	1030	2130	6000	12480

Multipliers to be Used with Gas Pipe Sizing Tables for Various Specific Gravities

Specific Gravity	Multiplier	Specific Gravity	Multiplier
.35	1.31	1.00	.775
.40	1.23	1.10	.740
.45	1.16	1.20	.707
.50	1.10	1.30	.680
.55	1.04	1.40	.655
.60	1.00	1.50	.633
.65	.962	1.60	.612
.70	.926	1.70	.594
.75	.895	1.80	.577
.80	.867	1.90	.565
.85	.841	2.00	.547
.90	.817	2.10	.535

of the appliances which will be installed and the heating value of the gas to be used.

To obtain the cubic feet per hour of gas required, divide the total Btu input of all appliances by the average Btu heating value per cubic foot of the gas. The average Btu per cubic foot of the gas in the area of installation may be obtained from the local gas company.

Capacities of different sizes and lengths of pipe in cubic feet per hour with a pressure drop of 0.3 inch of water column for gas of 0.60 specific gravity are shown in the accompanying table. In using this table, no allowance for an ordinary number of fittings is necessary.

The accompanying table on residential gas pipe-line capacities is based on the Pole formula:

$$Q = 1350 \left[\frac{d^5 p}{.33 Ls} \right]^{1/2}$$

where Q = flow in cubic feet per hour,
  d = nominal pipe diameter, inches
  p = pressure drop, inches of water, over length L,
  L = length of pipe in feet,
  s = specific gravity of gas

The table is a solution of this formula for various lengths L and pipe sizes when p is $1/2$ in. and s is 0.60. For other values of s, multiply the values of gas flow in the table by the multipliers given in another table.

The diversity factor is the percentage of the total connected load in use at any one time and is an important factor in determining the correct pipe size to be used in multi-family dwellings. It is dependent upon the number and kinds of gas appliances being installed. Consult the local gas company or the authority having jurisdiction for the diversity factor to be used in sizing pipe.

Extensions to existing piping shall conform to the pipe capacity table. Existing piping shall be converted to the proper size of pipe where necessary. In no case shall extensions be made to existing pipe which is smaller than permitted by the pipe capacity table.

Gas piping in buildings shall be wrought iron or steel pipe complying with the American National Standard for Wrought-Steel and Wrought-Iron Pipe, ANSI-B36.10. The connecting of pipe by welding is permissible. Threaded pipe fittings (except stop cocks or valves) shall be malleable iron or steel when used with wrought iron or steel pipe. Where approved by the authority having jurisdiction, cast iron fittings in sizes 4 inches and larger may be used with wrought iron and steel pipe, and copper or brass pipe in iron pipe sizes assembled with threaded fittings of the same materials may be used with gasses not corrosive to such materials.

Gas piping and fittings shall be clear and free rom cutting burrs and defects in structure or threading and shall be thoroughly brushed and scale blown.

Pipe, fittings, valves, etc., removed from any existing installation shall not be used again until they have been thoroughly cleaned, inspected and ascertained to be equivalent to new material.

No pipe smaller than standard $1/2$ inch iron pipe size shall be used in any concealed location.

Concealed piping should be located in hollow rather than in solid partitions.

Piping in solid floors such as concrete shall be laid in channels in the floor suitably covered to permit access to the piping with a minimum of damage to the building. Piping in contact with earth or other material which may corrode the piping shall be protected against corrosion in an approved manner. Piping shall not be laid in cinders.

When installing piping which shall be concealed, unions, running threads, right and left couplings, bushings, and swing joints made by combinations of fittings shall not be used.

When necessary to insert fittings in piping which has been installed in a concealed location, the piping may be reconnected by the use of a ground joint union with the nut "center punched" to prevent loosening by vibration.

The building structure shall not be weakened by the installation of any gas piping. Before any beams or joists are cut or notched, special permission should be obtained from the authority having jurisdiction.

All piping shall be graded not less than $\frac{1}{4}$ inch in 15 feet to prevent traps. All horizontal lines shall grade to risers and from the risers to the meter or to the appliance.

Gas piping shall not be supported by other piping but shall be supported to maintain proper grade with pipe hooks, metal pipe straps, bands, or hangers suitable for the size of pipe, and of proper strength and quality at proper intervals so that the piping cannot be moved accidentally from the installed position.

Spacing of supports in piping installations shall not be greater than the following:

$\frac{1}{2}$ inch pipe	6 feet
$\frac{3}{4}$ inch or 1 inch pipe	8 feet
$1\frac{1}{4}$ inch or longer (horizontal)	10 feet
$1\frac{1}{4}$ inch or longer (vertical)	every floor level

Gas piping shall be protected against freezing temperatures. When piping must be exposed to wide ranges or sudden changes in temperatures, special care shall be taken to prevent stoppages.

Where there are overhanging kitchens or other rooms built beyond foundation walls, in which gas appliances are installed, care shall be taken to avoid placing the piping where it will be exposed to low temperatures (40 deg. F or below for manufactured gas) or to extreme changes of temperatures. In such cases the piping shall be brought up inside the building proper and run around the sides of the room, in the most practical manner.

Pipe shall not be bent. Fittings shall be used when making turns in the gas piping.

A drip shall be provided at any point in the line of pipe where condensate may collect. Where condensation is excessive, a drip should be provided at the outlet of the meter. This drip should be so installed as to constitute a trap wherein an accumulation of condensate will shut off the flow of gas before it will run back into the meter.

All drips shall be installed only in such locations that they will be readily accessible to permit cleaning or emptying. A drip shall not be located where the condensate is likely to freeze. The size of any drip used shall be determined by the capacity and the exposure of the piping which drains to it and in accordance with recommendations of the local gas company.

A tee fitting with the bottom outlet plugged or capped instead of an ell fitting shall be used at the bottom of any riser to catch any dirt or other foreign materials.

Gas pipe inside any building shall not be run in or through an air duct, clothes chute, chimney or flue, ventilating duct, dumbwaiter or elevator shaft.

No device shall be placed inside the gas pipe or fittings that will reduce the cross-sectional area or otherwise obstruct the free flow of gas through the pipe and fittings.

All branch outlet pipes shall be taken from the top or sides of horizontal lines and not from the bottom. Where a branch outlet is placed on a main supply line before it is known what size of pipe will be connected to it, the outlet shall be of the same size as the line which supplies it.

Where piping is run from one building to another, it shall be adequately protected against freezing temperatures and shall be graded toward a suitable drip.

Where manufactured gas is distributed, underground piping shall be one size larger than that specified by the accompanying table, but in no case less than $1\frac{1}{4}$ inch.

Where local conditions require it, underground piping shall be protected against corrosion by coating or by other suitable means.

Lines supplying gas pilots for oil burning appliances shall be accessible, not less than $\frac{1}{2}$ inch standard pipe size and shall not be exposed to extreme temperatures.

In multiple tenant buildings supplied through a master meter, or where meters are not readily accessible from the appliance location, an individual shut-off valve for each apartment or for each separate house line shall be provided at a convenient point of general accessibility.

Before any system of gas piping is finally put in service, it shall be carefully tested to assure that it is gas tight. Where any part of the system is to be enclosed or concealed, this test should precede the work of closing in. To test for tightness the piping may be filled with city gas, air or inert gas but not with any other gas or liquid. *In no case shall oxygen ever be used.* The piping shall stand a pressure of at least 6 inches of mercury measured with a manometer or slope gage, for a period of not less than 10 minutes without showing any drop in pressure.

Commercial-Industrial Piping

The following are among the recommendations of American Gas Association for consumer-owned gas piping *not* covered by American National Standard Requirements for Gas Appliances and Gas Piping Installations, ANSI-Z21.30. They *do not cover* undiluted liquefied petroleum gas, *but do cover* natural gas, manufactured gas, liquefied

petroleum gas-air or mixtures of these gases for in- dustrial installations at any pressure and all other nonresidential installations at operating pressures in excess of ½ psig.

It is recommended that before proceeding with the installation of a gas piping system, a piping sketch or plan be prepared showing the proposed location and size of the piping and the various loads. Adequate consideration should be given to future loads and provisions made for added gas service.

Piping other than entry shall not be buried underground inside of buildings. Avoid running piping through crawl spaces. Where gas piping must be run in crawl spaces, tunnels or unfrequented basements, continuous ventilation should be provided.

Piping in crawl spaces should be treated in the same manner as other above ground piping with respect to painting and insulation if necessary. Entering piping buried inside the building beyond the wall must be encased in a protective pipe of larger diameter. It is recommended that such casing extend at least four inches outside of the building wall, being left open at that point, and sealed at the inside end of the casing and a removable threaded inspection plug be provided in the casing.

When it is practical to do so, outdoor gas piping in industrial plant yards should be installed above ground. Such piping must be securely supported and located where it will be protected from physical damage. Where soil conditions are unstable and settling of piping or foundation walls may occur, adequate measures shall be provided to prevent excessive stressing of the piping. Avoid locating piping in cinder fill.

Where the gas supply pressure is higher than that at which the gas utilization equipment is designed to operate, a gas pressure regulator shall be installed. Where used, gas pressure regulators should maintain the gas pressure to the burner supply line within plus or minus 10% of the operating gage pressure between maximum and minimum firing rates. These limitations do not apply to regulators known as "zero governors" used in connection with gas-air proportioning and mixing systems.

Regulators of the spring loaded, dead weight or pressure balanced type are preferred. Spring, weight loaded or exposed lever type regulators must have springs, levers and weights covered by a suitable housing.

Pressure regulators of the diaphragm type, except zero governors, generally require venting. An adequately sized independent vent to a safe point outside the building should be provided when the regulator is such that a ruptured diaphragm will cause a hazard. Means should likewise be provided to prevent water, insects or foreign materials from entering the vent pipe.

Standard weight or Schedule 40 steel pipe is acceptable and generally used for gas pressures up to 125 psig.

Welded joints should be used wherever practical. Compression or gland type fittings may be used if adequately braced so that neither the gas pressure nor external physical damage will force the joint apart. Pipe shall be fully inserted in gland or compression type of joints before completing the assembly. Screw fittings should be malleable cast iron, steel castings or forgings. In sizes four inches and larger, cast iron fittings may be used.

Cast iron pipe of four inch size and larger equipped with standardized mechanical joints may be used for underground service for pressures up to 50 psig.

For gases containing sulphur compounds, copper and its alloys may prove unsuitable for gas piping. Consult the gas company before using these materials. For copper tubing, brazed (melting point over 1000° F) or flared joints should be used.

For copper and brass pipe, threaded or brazed joints (melting point over 1000° F) should be used. Compression or gland type couplings shall not be used in above ground installations.

Piping shall be of such a size and so installed as to provide a supply of gas sufficient to meet the requirements of demand and pressure at the point of use. The proper size depends upon the following factors:

a. Maximum gas consumption to be accommodated.
b. Diversity of load.
c. Allowable loss in pressure from the start of the pipe to the end.
d. Length of pipe and number and size of fittings, valves and control devices.
e. Specific gravity of the gas.
f. Allowance for any probable future change in consumption, specific gravity or heating value of gas.

The volume of gas to be provided for (in cubic feet per hour) shall be determined directly from the manufacturer's input ratings of the equipment served. When input rating is not indicated, the gas company should be contacted for estimated volume of gas to be supplied.

NATURAL GAS FLOW TABLES

The following tables are designed primarily for use in sizing natural gas systems in industrial plants where, for safety reasons, gas pressures rarely exceed 20 psig. The accuracy of pipe sizing for natural gas in the pressure range 1-20 psig is satisfactory for use in architectural engineering and plant engineering offices.

The Weymouth formula, used for pipes 3 inches and larger in diameter, is

$$Q = 28.0 d^{2.667} \sqrt{\left[\frac{P_1^2 - P_2^2}{GL}\right] \frac{520}{T}}$$

The Cox formula, used for pipes under 3 inches diameter, is

$$Q = 33.3 \sqrt{\frac{(P_1^2 - P_2^2) d^5}{GL}}$$

where

$Q =$ Rate of flow in cubic feet per hour
$d =$ Internal diameter of pipe in inches
$L =$ Length of pipe in miles
$G =$ Specific gravity of gas, here taken as 0.60 (air $= 1.0$)
$T =$ Absolute temperature in deg F of the flowing gas
$P_1 =$ Initial absolute pressure in psi
$P_2 =$ Terminal absolute pressure in psi

Capacities of pipes to carry gas depend, among

other things, on the altitude above sea level of the considered installation. The tables are computed for three specific altitudes, 500, 2500 and 4500 ft above sea level, and are sufficiently accurate within the three selected ranges 0-1500, 1500-3500 and 3500-5500 ft above sea level.

Specific gravity of the gas is selected as 0.60 (air $= 1.0$). For gases having specific gravity other than 0.60, multiply pipe length by correction factor F appearing in Table 4.

For their use in tables, express flow rate in cubic feet per hour (cfh), pressure in pounds per square inch gage (psig), pipe diameter in inches, and length in feet. Equivalent resistance in fittings and valves, expressed in feet of straight pipe, should be added to the actual length of pipe.

EXAMPLE NO. 1

Given: Flow, 50,000 cfh of gas
Pipe length, 1,000 ft
Initial pressure, 10 psig
Pressure drop, 1 psi
Altitude of installation, 500 ft above sea level

Find: Pipe size

(*a*) From Table 2 the *Pressure Drop Factor* corresponding to initial pressure of 10 psig and pressure drop of 1 psi is found to be 48.

(Continued on page 8-65)

Key to Examples

	Follow Example No.						
	1	2	3	4	5	6	7
Flow, cfh	Known	Known	Known	Known	?	Known	Known
Pipe length, feet	Known	Known	Known	?	Known	Known	Known
Initial pressure, psig	Known	Known	?	Known	Known	?	—
Pressure drop, psi	Known	?	Known	Known	Known	—	Known
Pipe size, inches	?	Known	Known	Known	Known	Known	?
Terminal pressure, psig	—	—	—	—	—	Known	Known
Specific gravity	0.60	0.60	0.60	0.60	0.60	other than 0.60	0.60

Table 1—Pressure Drop Factors for Use in Tables 2 and 3

Flow, Cu Ft per Hr	Nominal Pipe Diameter, Inches								
	½ (0.622)	¾ (0.824)	1 (1.049)	1¼ (1.380)	1½ (1.610)	2 (2.067)	2½ (2.469)	3 (3.068)	4 (4.026)
	Pressure Drop Factor per 100 Feet of Pipe								
100	1.10	0.27	0.08	—	—	—	—	—	—
200	4.40	1.08	0.32	0.08	—	—	—	—	—
300	9.9	2.43	0.73	0.18	0.09	—	—	—	—
400	17.6	4.32	1.29	0.33	0.15	—	—	—	—
500	27.5	6.7	2.02	0.51	0.24	0.07	—	—	—
600	39.6	9.7	2.90	0.74	0.34	0.10	—	—	—
700	54.0	13.2	3.95	1.00	0.46	0.13	—	—	—
800	70.4	17.3	5.2	1.31	0.61	0.17	0.07	—	—
900	89.2	21.9	6.5	1.66	0.77	0.22	0.09	—	—
1,000	100	27.0	8.1	2.05	0.95	0.27	0.11	—	—
1,100	133	32.6	9.8	2.48	1.15	0.33	0.13	—	—
1,200	159	38.9	11.6	2.95	1.36	0.39	0.16	—	—
1,300	—	45.6	13.6	3.46	1.60	0.46	0.19	0.06	—
1,400	—	52.9	15.8	4.01	1.86	0.53	0.22	0.07	—
1,500	—	60.7	18.2	4.61	2.13	0.61	0.25	0.08	—
1,600	—	69.1	20.7	5.2	2.43	0.70	0.29	0.09	—
1,700	—	78.8	23.3	5.9	2.74	0.79	0.32	0.11	—
1,800	—	87.4	26.1	6.6	3.07	0.88	0.36	0.12	—
1,900	—	97.4	29.1	7.4	3.42	0.98	0.40	0.13	—
2,000	—	108	32.3	8.2	3.79	1.09	0.45	0.15	—
2,100	—	119	35.6	9.0	4.18	1.20	0.49	0.16	—
2,200	—	131	39.1	9.9	4.58	1.32	0.54	0.18	—
2,300	—	143	42.7	10.8	5.0	1.44	0.59	0.20	—
2,400	—	—	46.5	11.8	5.5	1.57	0.64	0.21	—
2,500	—	—	50.4	12.8	5.9	1.70	0.70	0.23	—
2,600	—	—	54.5	13.8	6.4	1.84	0.76	0.25	—
2,700	—	—	58.8	14.9	6.9	1.98	0.80	0.27	0.06
2,800	—	—	63.2	16.1	7.4	2.13	0.88	0.29	0.07
2,900	—	—	67.8	17.2	8.0	2.28	0.94	0.31	0.07
3,000	—	—	72.6	18.4	8.5	2.44	1.01	0.33	0.08
3,100	—	—	77.5	19.7	9.1	2.61	1.07	0.36	0.08
3,200	—	—	82.6	21.0	9.7	2.78	1.14	0.38	0.09
3,300	—	—	87.8	22.3	10.3	2.96	1.22	0.40	0.09
3,400	—	—	93.3	23.7	11.0	3.14	1.29	0.43	0.10
3,500	—	—	98.8	25.1	11.6	3.33	1.37	0.45	0.11
3,600	—	—	105	26.5	12.3	3.52	1.45	0.48	0.11
3,700	—	—	110	28.0	13.0	3.72	1.53	0.51	0.12
3,800	—	—	116	29.6	13.7	3.92	1.61	0.53	0.13
3,900	—	—	123	31.2	14.4	4.13	1.70	0.56	0.13
4,000	—	—	129	32.8	15.2	4.35	1.79	0.59	0.14
4,100	—	—	136	34.4	15.9	4.57	1.88	0.62	0.15
4,200	—	—	—	36.1	16.7	4.79	1.97	0.65	0.15
4,300	—	—	—	37.9	17.5	5.0	2.07	0.68	0.16
4,400	—	—	—	39.6	18.3	5.3	2.16	0.72	0.17
4,500	—	—	—	41.5	19.2	5.5	2.26	0.75	0.18
4,600	—	—	—	43.3	20.0	5.8	2.36	0.78	0.18
4,700	—	—	—	45.2	20.9	6.0	2.47	0.82	0.19
4,800	—	—	—	47.2	21.8	6.3	2.57	0.85	0.20
4,900	—	—	—	49.2	22.7	6.5	2.68	0.89	0.21
5,000	—	—	—	51.2	23.7	6.8	2.79	0.92	0.22

Table 1 (continued)—Pressure Drop Factors for Use in Tables 2 and 3

Flow, Cu Ft per Hr	Nominal Pipe Diameter, Inches							
	1¼ (1.380)	1½ (1.610)	2 (2.067)	2½ (2.469)	3 (3.068)	4 (4.026)	6 (6.065)	8 (7.981)
	Pressure Drop Factor per 100 Feet of Pipe							
5,200	55.4	25.6	7.3	3.02	1.0	0.24	—	—
5,400	59.7	27.6	7.9	3.26	1.08	0.25	—	—
5,600	64.2	29.7	8.5	3.50	1.16	0.27	—	—
5,800	69.0	31.9	9.1	3.76	1.25	0.29	—	—
6,000	73.8	34.1	9.8	4.02	1.33	0.31	—	—
6,200	78.8	36.4	10.4	4.29	1.42	0.34	—	—
6,400	84.0	38.8	11.1	4.57	1.51	0.36	—	—
6,600	89.3	41.3	11.8	4.87	1.61	0.38	—	—
6,800	94.8	43.8	12.6	5.2	1.71	0.40	—	—
7,000	100	46.4	13.3	5.5	1.81	0.43	—	—
7,200	106	49.1	14.1	5.8	1.92	0.45	—	—
7,400	112	51.9	14.9	6.1	2.02	0.48	—	—
7,600	118	54.7	15.7	6.5	2.13	0.50	—	—
7,800	125	57.6	16.5	6.8	2.24	0.53	—	—
8,000	131	60.6	17.4	7.2	2.37	0.56	—	—
8,200	138	63.7	18.3	7.5	2.49	0.58	—	—
8,400	—	66.8	19.2	7.9	2.61	0.61	—	—
8,600	—	70.0	20.1	8.3	2.73	0.64	0.07	—
8,800	—	73.3	21.0	8.7	2.86	0.67	0.08	—
9,000	—	76.7	22.0	9.1	3.0	0.70	0.08	—
9,200	—	80.2	23.0	9.5	3.13	0.74	0.08	—
9,400	—	83.7	24.0	9.9	3.27	0.77	0.09	—
9,600	—	87.3	25.0	10.3	3.41	0.80	0.09	—
9,800	—	90.9	26.1	10.8	3.55	0.83	0.09	—
10,000	—	94.7	27.2	11.2	3.70	0.87	0.10	—
11,000	—	115	32.9	13.5	4.47	1.05	0.12	—
12,000	—	136	39.1	16.1	5.3	1.25	0.14	—
13,000	—	—	45.9	18.9	6.3	1.47	0.17	—
14,000	—	—	53.2	21.9	7.2	1.70	0.19	—
15,000	—	—	61.1	25.1	8.3	1.95	0.22	—
16,000	—	—	69.6	28.6	9.5	2.22	0.25	—
17,000	—	—	78.5	32.3	10.7	2.51	0.28	—
18,000	—	—	88.0	36.2	12.0	2.81	0.32	0.07
19,000	—	—	98.0	40.3	13.3	3.14	0.35	0.08
20,000	—	—	109	44.7	14.8	3.47	0.39	0.09
21,000	—	—	120	49.3	16.3	3.83	0.43	0.10
22,000	—	—	131	54.1	17.9	4.20	0.47	0.11
23,000	—	—	144	59.1	19.6	4.59	0.52	0.12
24,000	—	—	—	64.3	21.3	5.00	0.56	0.13
25,000	—	—	—	69.8	23.1	5.4	0.61	0.14
26,000	—	—	—	75.5	25.0	5.9	0.66	0.15
27,000	—	—	—	81.4	27.0	6.3	0.71	0.17
28,000	—	—	—	87.6	29.0	6.8	0.77	0.18
29,000	—	—	—	93.9	31.1	7.3	0.82	0.19
30,000	—	—	—	101	33.3	7.8	0.88	0.20
31,000	—	—	—	107	35.5	8.4	0.94	0.22
32,000	—	—	—	114	37.9	8.9	1.0	0.23
33,000	—	—	—	122	40.3	9.5	1.07	0.25
34,000	—	—	—	129	42.7	10.0	1.13	0.26
35,000	—	—	—	137	45.3	10.6	1.20	0.28

Table 1 (continued)—Pressure Drop Factors for Use in Tables 2 and 3

Flow, Cu Ft per Hr	Nominal Pipe Diameter, Inches							
	3 (3.068)	4 (4.026)	6 (6.065)	8 (7.981)	10 (10.02)	12 (12.00)	14 (13.25)	16 (15.25)
	Pressure Drop Factor per 100 Feet of Pipe							
36,000	47.9	11.3	1.27	0.29	—	—	—	—
37,000	50.6	11.9	1.34	0.31	0.09	—	—	—
38,000	53.4	12.5	1.41	0.33	0.10	—	—	—
39,000	56.2	13.2	1.49	0.35	0.10	—	—	—
40,000	59.1	13.9	1.57	0.36	0.11	—	—	—
41,000	62.1	14.6	1.65	0.38	0.11	—	—	—
42,000	65.2	15.3	1.73	0.40	0.12	—	—	—
43,000	68.3	16.1	1.81	0.42	0.13	—	—	—
44,000	71.6	16.8	1.89	0.44	0.13	—	—	—
45,000	74.8	17.6	1.98	0.46	0.14	—	—	—
46,000	78.2	18.4	2.07	0.48	0.14	—	—	—
47,000	81.6	19.2	2.16	0.50	0.15	—	—	—
48,000	85.2	20.0	2.25	0.52	0.16	—	—	—
49,000	88.7	20.9	2.35	0.55	0.16	—	—	—
50,000	92.4	21.7	2.44	0.57	0.17	—	—	—
52,000	99.9	23.5	2.64	0.61	0.18	—	—	—
54,000	108	25.3	2.85	0.66	0.20	—	—	—
56,000	116	27.2	3.07	0.71	0.21	—	—	—
58,000	124	29.2	3.29	0.76	0.23	—	—	—
60,000	133	31.3	3.52	0.82	0.24	—	—	—
62,000	142	33.9	3.76	0.87	0.26	—	—	—
64,000	—	35.6	4.00	0.93	0.28	—	—	—
66,000	—	37.8	4.26	0.99	0.29	—	—	—
68,000	—	40.2	45.2	1.05	0.31	—	—	—
70,000	—	42.6	4.79	1.11	0.33	—	—	—
72,000	—	45.0	5.1	1.18	0.35	0.13	—	—
74,000	—	47.6	5.4	1.24	0.37	0.14	—	—
76,000	—	50.2	5.7	1.31	0.39	0.15	—	—
78,000	—	52.8	6.0	1.38	0.41	0.16	—	—
80,000	—	55.6	6.3	1.45	0.43	0.17	—	—
82,000	—	58.4	6.6	1.52	0.45	0.17	—	—
84,000	—	61.3	6.9	1.60	0.47	0.18	—	—
86,000	—	64.2	7.2	1.68	0.50	0.19	—	—
88,000	—	67.3	7.6	1.76	0.52	0.20	—	—
90,000	—	70.3	7.9	1.84	0.55	0.21	—	—
92,000	—	73.5	8.3	1.92	0.57	0.22	—	—
94,000	—	76.7	8.6	2.00	0.59	0.23	—	—
96,000	—	80.0	9.0	2.09	0.62	0.24	—	—
98,000	—	83.4	9.4	2.18	0.65	0.25	—	—
100,000	—	86.9	9.8	2.27	0.67	0.26	0.15	—
110,000	—	105	11.8	2.74	0.81	0.31	0.18	—
120,000	—	125	14.1	3.27	0.97	0.37	0.22	—
130,000	—	147	16.5	3.83	1.14	0.44	0.26	—
140,000	—	—	19.2	4.45	1.32	0.50	0.30	0.14
150,000	—	—	22.0	5.1	1.51	0.58	0.34	0.16
160,000	—	—	25.0	5.8	1.72	0.66	0.39	0.18
170,000	—	—	28.3	6.6	1.95	0.74	0.44	0.21
180,000	—	—	31.7	7.4	2.18	0.83	0.49	0.23
190,000	—	—	35.3	8.2	2.43	0.93	0.55	0.26
200,000	—	—	39.1	9.1	2.69	1.03	0.61	0.29

Table 1 (continued)—Pressure Drop Factors for Use in Tables 2 and 3

Flow, Cu Ft per Hr	Nominal Pipe Diameter, Inches							
	6 (6.065)	8 (7.981)	10 (10.02)	12 (12.00)	14 (13.25)	16 (15.25)	18 (17.25)	20 (19.25)
	Pressure Drop Factor per 100 Feet of Pipe							
210,000	43.1	10.0	2.97	1.14	0.67	0.32	—	—
220,000	47.3	11.0	3.26	1.25	0.73	0.35	0.18	0.10
230,000	51.7	12.0	3.56	1.36	0.80	0.38	0.20	0.11
240,000	56.3	13.1	3.88	1.48	0.87	0.41	0.21	0.12
250,000	61.1	14.2	4.20	1.61	0.95	0.45	0.23	0.13
260,000	66.1	15.3	4.55	1.74	1.03	0.49	0.25	0.14
270,000	71.3	16.5	4.91	1.88	1.11	0.52	0.27	0.15
280,000	76.7	17.8	5.3	2.02	1.19	0.56	0.29	0.16
290,000	82.2	19.1	5.7	2.17	1.28	0.60	0.31	0.17
300,000	88.0	20.4	6.1	2.32	1.37	0.65	0.33	0.19
310,000	94.0	21.8	6.5	2.47	1.46	0.69	0.36	0.20
320,000	100	23.2	6.9	2.64	1.55	0.73	0.38	0.21
330,000	106	24.7	7.3	2.80	1.65	0.78	0.41	0.23
340,000	113	26.2	7.8	2.98	1.75	0.83	0.43	0.24
350,000	120	27.8	8.2	3.15	1.86	0.88	0.46	0.25
360,000	127	29.4	8.7	3.34	1.97	0.93	0.48	0.27
370,000	134	31.1	9.2	3.52	2.08	0.98	0.51	0.28
380,000	141	32.8	9.7	3.72	2.19	1.04	0.54	0.30
390,000	—	34.5	10.2	3.92	2.31	1.09	0.57	0.32
400,000	—	36.3	10.8	4.12	2.43	1.15	0.60	0.33
420,000	—	40.0	11.9	4.54	2.68	1.27	0.66	0.37
440,000	—	43.9	13.0	4.98	2.94	1.39	0.72	0.40
460,000	—	48.0	14.2	5.5	3.21	1.52	0.79	0.44
480,000	—	52.3	15.5	5.9	3.50	1.65	0.86	0.48
500,000	—	56.7	16.8	6.4	3.79	1.79	0.93	0.52
520,000	—	61.3	18.2	7.0	4.10	1.94	1.00	0.56
540,000	—	66.1	19.7	7.5	4.43	2.10	1.08	0.60
560,000	—	71.1	21.1	8.1	4.76	2.25	1.17	0.65
580,000	—	76.3	22.6	8.7	5.1	2.41	1.25	0.70
600,000	—	81.6	24.2	9.3	5.5	2.58	1.34	0.75
620,000	—	87.2	25.9	9.9	5.8	2.76	1.43	0.80
640,000	—	92.9	27.6	10.5	6.2	2.94	1.52	0.85
660,000	—	98.8	29.3	11.2	6.6	3.13	1.62	0.90
680,000	—	105	31.1	11.9	7.0	3.32	1.72	0.96
700,000	—	111	33.0	12.6	7.4	3.52	1.82	1.02
720,000	—	118	34.9	13.3	7.9	3.72	1.93	1.08
740,000	—	124	36.9	14.1	8.3	3.93	2.04	1.14
760,000	—	131	38.9	14.9	8.8	4.14	2.15	1.20
780,000	—	138	41.0	15.7	9.2	4.37	2.26	1.26
800,000	—	—	43.1	16.5	9.7	4.59	2.38	1.33
820,000	—	—	45.3	17.3	10.2	4.83	2.50	1.39
840,000	—	—	47.5	18.2	10.7	5.1	2.63	1.46
860,000	—	—	49.8	19.0	11.2	5.3	2.75	1.53
880,000	—	—	52.1	20.0	11.8	5.6	2.88	1.61
900,000	—	—	54.5	20.9	12.3	5.8	3.02	1.68
920,000	—	—	57.0	21.8	12.8	6.1	3.15	1.76
940,000	—	—	59.5	22.8	13.4	6.3	3.29	1.83
960,000	—	—	62.0	23.7	14.0	6.6	3.43	1.91
980,000	—	—	64.6	24.7	14.6	6.9	3.57	1.99
1,000,000	—	—	67.3	25.8	15.2	7.2	3.72	2.07

Table 2—Pressure Drop of Gas in Pipe, Initial Pressure Known

Altitude, Feet	Initial Pressure, psig															
3500-5500	1	2	3	4	5	6	7	8	9	10	11	12	13	14	15	16
1500-3500		1	2	3	4	5	6	7	8	9	10	11	12	13	14	15
0-1500			1	2	3	4	5	6	7	8	9	10	11	12	13	14
Pressure Drop, psi	Pressure Drop Factor (From Table 1)															
0.1	3	3	3	3	3	4	4	4	4	4	5	5	5	5	5	6
0.2	5	6	6	7	7	7	8	8	9	9	9	10	10	11	11	11
0.3	8	9	9	10	10	11	12	12	13	13	14	15	15	16	16	17
0.4	11	11	12	13	14	15	15	16	17	18	19	19	20	21	22	23
0.5	13	14	15	16	17	18	19	20	21	22	23	24	25	26	27	28
0.6	—	17**	18*	19	21	22	23	24	25	27	28	29	30	31	33	34
0.7	—	20**	21*	22	24	25	27	28	29	31	32	34	35	36	38	39
0.8	—	22**	24*	26	27	29	30	32	34	35	37	38	40	42	43	45
0.9	—	25**	27*	29	31	32	34	36	38	40	41	43	45	47	49	50
1.0	—	28**	30*	32	34	36	38	40	42	44	46	48	50	52	54	56
1.1	—	30**	33*	35	37	39	41	44	46	48	50	52	55	57	59	61
1.2	—	33**	36*	38	40	43	45	48	50	52	55	57	60	62	64	67
1.3	—	36**	38*	41	44	46	49	51	54	57	59	62	65	67	70	72
1.4	—	38**	41*	44	47	50	52	55	58	61	64	66	69	72	75	78
1.5	—	41**	44*	47	50	53	56	59	62	65	68	71	74	77	80	83
1.6	—	—	47**	50*	53	56	60	63	66	69	72	76	79	82	85	88
1.7	—	—	49**	53*	56	60	63	66	70	73	77	80	83	87	90	94
1.8	—	—	52**	56*	59	63	67	70	74	77	81	85	88	92	95	99
1.9	—	—	55**	59*	63	66	70	74	78	82	85	89	93	97	101	104
2.0	—	—	58**	62*	66	70	74	78	82	86	90	94	98	102	106	110
2.1	—	—	60**	64*	69	73	77	81	85	90	94	98	102	106	111	115
2.2	—	—	63**	67*	72	76	81	85	89	94	98	103	107	111	116	120
2.3	—	—	66**	70*	75	79	84	89	93	98	102	107	112	116	121	125
2.4	—	—	68**	73*	78	83	87	92	97	102	107	111	116	121	126	131
2.5	—	—	71**	76*	81	86	91	96	101	106	111	116	121	126	131	136
2.6	—	—	—	79**	84*	89	94	99	105	110	115	120	125	131	136	141
2.7	—	—	—	81**	87*	92	97	103	108	114	119	124	130	135	141	146
2.8	—	—	—	84**	90*	95	101	106	112	118	123	129	134	140	146	151
2.9	—	—	—	87**	93*	98	104	110	116	122	127	133	139	145	151	155
3.0	—	—	—	89**	95*	101	107	113	119	125	131	137	143	149	155	161
3.1	—	—	—	92**	98*	104	111	117	123	129	135	142	148	154	160	166
3.2	—	—	—	95**	101*	108	114	120	127	133	139	146	152	159	165	171
3.3	—	—	—	97**	104*	111	117	124	130	137	143	150	157	163	170	176
3.4	—	—	—	100**	107*	114	120	127	134	141	147	154	161	168	175	181
3.5	—	—	—	103**	110*	117	124	131	138	145	151	159	166	173	180	186
3.6	—	—	—	—	112**	120*	127	134	141	148	155	163	170	177	184	191
3.7	—	—	—	—	115**	122*	130	137	145	152	159	167	174	182	189	196
3.8	—	—	—	—	118**	125*	133	141	148	156	163	171	179	186	194	201
3.9	—	—	—	—	121**	128*	136	144	152	160	167	175	183	191	199	206
4.0	—	—	—	—	123**	131*	139	147	155	163	171	179	187	195	203	211
4.1	—	—	—	—	126**	134*	142	150	159	167	175	183	191	200	208	216
4.2	—	—	—	—	129**	137*	145	154	162	171	179	187	196	204	213	221
4.3	—	—	—	—	131**	140*	148	157	166	174	183	191	200	209	217	226
4.4	—	—	—	—	134**	143*	151	160	169	178	187	195	204	213	222	231
4.5	—	—	—	—	136**	145*	154	163	172	181	190	199	208	217	226	235
4.6	—	—	—	—	—	148**	157*	167	176	185	194	203	213	222	231	240
4.7	—	—	—	—	—	151**	160*	170	179	188	198	207	217	226	235	245
4.8	—	—	—	—	—	153**	163*	173	182	192	202	211	221	230	240	250
4.9	—	—	—	—	—	156**	166*	176	186	196	205	215	225	235	245	254
5.0	—	—	—	—	—	159**	169*	179	189	199	209	219	229	239	249	259

* These figures valid for altitudes of 1500-3500 and 3500-5500 feet above sea level only.

** These figures valid for altitudes of 3500-5500 feet above sea level only.

Table 2 (continued)

Initial Pressure, psig						Altitude, Feet
17	18	19	20	21	22	3500-5500
16	17	18	19	20	21	1500-3500
15	16	17	18	19	20	0-1500
Pressure Drop Factor (From Table 1)						Pressure Drop, psi
6	6	6	6	7	7	0.1
12	12	13	13	13	13	0.2
18	18	19	19	20	20	0.3
23	24	25	26	27	27	0.4
29	30	31	32	33	34	0.5
35	36	37	39	40	41	0.6
41	42	43	45	46	48	0.7
46	48	50	51	53	54	0.8
52	54	56	58	59	61	0.9
58	60	62	64	66	68	1.0
63	66	68	70	72	75	1.1
69	72	74	76	79	81	1.2
75	77	80	83	85	88	1.3
80	83	86	89	92	94	1.4
86	89	92	98	98	101	1.5
92	95	98	101	104	108	1.6
97	100	104	107	111	114	1.7
103	106	110	113	117	120	1.8
108	112	116	120	123	127	1.9
113	118	122	126	130	134	2.0
119	123	127	132	136	140	2.1
125	129	133	138	142	147	2.2
130	135	139	144	148	153	2.3
135	140	145	150	155	159	2.4
141	146	151	156	161	166	2.5
146	151	157	162	167	172	2.6
151	157	162	168	173	178	2.7
157	162	168	174	179	185	2.8
162	168	174	180	185	191	2.9
167	173	179	185	191	197	3.0
173	179	185	191	197	204	3.1
178	184	191	197	203	210	3.2
183	190	196	203	209	216	3.3
188	195	202	209	215	222	3.4
194	201	208	215	221	228	3.5
199	206	213	220	227	234	3.6
204	211	219	226	233	241	3.7
209	217	224	232	239	247	3.8
214	222	230	238	245	253	3.9
219	227	235	243	251	259	4.0
224	232	241	249	257	265	4.1
229	238	246	255	263	271	4.2
234	243	252	260	269	277	4.3
239	248	257	266	275	283	4.4
244	253	262	271	280	289	4.5
249	259	268	277	286	295	4.6
254	264	273	282	292	301	4.7
259	269	278	288	298	307	4.8
264	274	284	294	303	313	4.9
269	279	289	299	309	319	5.0

(Continued from page 8-59)

(b) *Pressure Drop Factor* per 100 ft is (48 × 100) ÷ 1000 = 4.8

(c) Locate the next smallest *Pressure Drop Factor* per 100 ft from Table 1 in the 50,000 cfh row, here 2.44. The corresponding required pipe size is 6 inches.

EXAMPLE NO. 2

Given: Flow, 300,000 cfh
Pipe length, 2,000 ft
Initial pressure, 15 psig
Pipe size, 10 inches
Altitude of installation, 1,000 ft above sea level

Find: Pressure drop

(a) From Table 1, 6.1 is the *Pressure Drop Factor* per 100 ft corresponding to 300,000 cfh flow and 10-inch pipe size.

(b) For 2,000 ft length the *Pressure Drop Factor* is (6.1 × 2000) ÷ 100 = 122.

(c) From Table 2 under initial pressure of 15 psig, *Pressure Drop Factor* of 122 corresponds to a pressure drop of 2.15 psi.

EXAMPLE NO. 3

Given: Flow, 6,800 cfh
Pipe length, 500 ft
Pressure drop, 0.5 psi
Pipe size, 2½ inches
Altitude of installation, 2,500 ft above sea level

Find: Required initial pressure

(a) From Table 1, 5.2 is the *Pressure Drop Factor* per 100 ft corresponding to 6800 cfh flow and 2½-inch pipe size.

(b) For 500 ft, *Pressure Drop Factor* is (5.2 × 500) ÷ 100 = 26.

(c) From Table 2, initial pressure of 13 psig corresponds to pressure drop of 0.5 psi and *Pressure Drop Factor* of 26.

(Example 4 on page 8-71)

Table 2 (continued)—Pressure Drop of Gas in Pipe, Initial Pressure Known

Altitude, Feet	Initial Pressure, psig															
3500-5500	7	8	9	10	11	12	13	14	15	16	17	18	19	20	21	22
1500-3500	6	7	8	9	10	11	12	13	14	15	16	17	18	19	20	21
0-1500	5	6	7	8	9	10	11	12	13	14	15	16	17	18	19	20
Pressure Drop, psi	Pressure Drop Factor (From Table 1)															
5.1	172**	182*	192	202	213	223	233	243	253	264	274	284	294	304	315	325
5.2	175**	185*	196	206	216	227	237	248	258	268	279	289	300	310	320	331
5.3	178**	188*	199	209	220	231	241	252	262	273	284	294	305	315	326	337
5.4	180**	191*	202	213	224	234	245	256	267	278	288	299	310	321	332	343
5.5	183**	194*	205	216	227	238	249	260	271	282	292	304	315	326	337	348
5.6	186**	197*	208	220	231	242	253	264	276	287	297	309	320	332	343	354
5.7	189**	200*	211	223	234	246	257	268	280	291	302	314	325	337	348	360
5.8	191**	203*	215	226	238	249	261	273	284	296	307	319	331	342	354	365
5.9	194**	206*	218	230	241	253	265	277	289	300	312	324	336	348	359	371
6.0	197**	209*	221	233	245	257	269	281	293	305	317	329	341	353	365	377
6.1	—	212**	224*	236	248	260	273	285	297	309	321	334	346	358	370	383
6.2	—	215**	227*	239	252	264	277	289	301	314	326	339	351	363	376	388
6.3	—	217**	230*	243	255	268	280	293	306	318	331	343	356	369	381	394
6.4	—	220**	233*	246	259	271	284	297	310	323	335	348	361	374	387	399
6.5	—	223**	236*	249	262	275	288	301	314	327	340	353	366	379	392	405
6.6	—	225**	239*	252	265	279	292	305	318	331	345	358	371	384	397	411
6.7	—	228**	242*	255	269	282	295	309	322	336	349	362	376	389	403	416
6.8	—	231**	245*	258	272	286	299	313	326	340	354	367	381	394	408	422
6.9	—	234**	248*	262	275	289	303	317	330	344	358	372	386	400	413	428
7.0	—	237**	251*	265	279	293	307	321	335	349	363	377	391	405	419	433
7.1	—	—	253**	268*	282	296	310	324	339	353	367	381	395	410	424	438
7.2	—	—	256**	271*	285	300	314	328	343	357	372	386	400	415	429	443
7.3	—	—	250**	274*	288	303	318	332	347	361	376	391	404	420	434	449
7.4	—	—	262**	277*	292	306	321	336	351	366	380	395	409	425	440	454
7.5	—	—	265**	280*	295	310	325	340	355	370	385	400	414	430	445	460
7.6	—	—	268**	283*	298	313	328	344	359	374	389	404	419	435	450	465
7.7	—	—	270**	286*	301	316	332	347	363	378	393	409	424	440	455	470
7.8	—	—	273**	289*	304	320	335	351	367	382	398	413	429	445	460	476
7.9	—	—	276**	292*	307	323	339	355	371	386	402	418	434	449	465	481
8.0	—	—	278**	294*	310	326	342	358	374	390	406	422	438	454	470	486
8.1	—	—	—	297**	313*	330	346	362	378	394	411	427	443	459	475	492
8.2	—	—	—	300**	316*	333	349	366	382	398	415	431	448	464	480	497
8.3	—	—	—	303**	319*	336	353	369	386	402	419	436	452	469	485	502
8.4	—	—	—	306**	322*	339	356	373	390	406	423	440	457	474	490	507
8.5	—	—	—	309**	325*	343	360	377	394	410	428	445	462	479	495	512
8.6	—	—	—	311**	328*	346	363	380	397	414	432	449	466	483	500	518
8.7	—	—	—	314**	331*	349	366	384	401	418	436	453	471	488	505	523
8.8	—	—	—	317**	334*	352	370	387	405	422	440	458	475	493	510	528
8.9	—	—	—	320**	337*	355	373	391	409	426	444	462	480	498	515	533
9.0	—	—	—	322**	340*	358	376	394	412	430	448	466	484	502	520	538
9.1	—	—	—	—	343**	361*	379	398	416	434	452	470	489	507	525	543
9.2	—	—	—	—	346**	364*	383	401	420	438	456	475	493	512	530	548
9.3	—	—	—	—	349**	367*	386	405	423	442	460	479	498	516	535	553
9.4	—	—	—	—	352**	370*	389	408	427	446	464	483	502	521	540	558
9.5	—	—	—	—	354**	373*	392	411	430	449	468	487	506	525	544	563
9.6	—	—	—	—	357**	376*	396	415	434	453	472	492	511	530	549	567
9.7	—	—	—	—	360**	379*	399	418	437	457	476	496	515	534	554	573
9.8	—	—	—	—	363**	382*	402	421	441	461	480	500	519	539	559	578
9.9	—	—	—	—	365**	385*	405	425	445	464	484	504	524	544	563	583
10.0	—	—	—	—	368**	388*	408	428	448	468	488	508	528	548	568	588

* These figures valid for altitudes of 1500-3500 and 3500-5500 feet above sea level only.
** These figures valid for altitudes of 3500-5500 feet above sea level only.

Table 3—Pressure Drop of Gas in Pipe, Terminal Pressure Known

Altitude, Feet	Terminal Pressure, psig										
3500-5500	1	2	3	4	5	6	7	8	9	10	11
1500-3500		1	2	3	4	5	6	7	8	9	10
0-1500			1	2	3	4	5	6	7	8	9
Pressure Drop, psi	Pressure Drop Factor (From Table 1)										
0.1	3	3	3	3	3	4	4	4	4	4	5
0.2	5	6	6	7	7	7	8	8	9	9	9
0.3	8	9	9	10	11	11	12	12	13	14	14
0.4	11	12	12	13	14	15	16	16	17	18	19
0.5	14	15	16	17	18	19	20	21	22	23	24
0.6	16	18	19	20	21	22	24	25	26	27	28
0.7	19	21	22	23	25	26	28	29	30	32	33
0.8	22	24	25	27	28	30	32	33	35	36	38
0.9	25	27	29	30	32	34	36	38	39	41	43
1.0	28	30	32	34	36	38	40	42	44	46	48
1.1	31	33	35	37	39	42	44	46	48	50	53
1.2	34	36	38	41	43	46	48	50	53	55	58
1.3	37	39	42	44	47	50	52	55	57	60	63
1.4	39	42	45	48	51	53	56	59	62	65	67
1.5	42	45	48	51	54	57	60	63	66	69	72
1.6	45	49	52	55	58	61	65	68	71	74	77
1.7	48	52	55	59	62	65	69	72	76	79	82
1.8	51	55	59	62	66	69	73	77	80	84	87
1.9	55	58	62	66	70	74	77	81	85	89	93
2.0	58	62	66	70	74	78	82	86	90	94	98
2.1	61	65	69	73	77	82	86	90	94	98	103
2.2	64	68	73	77	81	86	90	95	99	103	108
2.3	67	72	76	81	85	90	95	99	104	108	113
2.4	70	75	80	84	89	94	99	104	108	113	118
2.5	73	78	83	88	93	98	103	108	113	118	123
2.6	76	82	87	92	97	102	108	113	118	123	128
2.7	80	85	90	96	101	107	112	117	123	128	134
2.8	83	88	94	100	105	111	116	122	128	133	139
2.9	86	92	98	104	109	115	121	127	133	138	144
3.0	89	95	101	107	113	119	125	131	137	143	149
3.1	93	99	105	111	117	124	130	136	142	148	155
3.2	96	102	109	115	122	128	134	141	147	154	160
3.3	99	106	113	119	126	132	139	146	152	159	165
3.4	103	109	116	123	130	137	143	150	157	165	171
3.5	106	113	120	127	134	141	148	155	162	169	176
3.6	109	117	124	131	138	145	153	160	167	174	181
3.7	113	120	128	135	142	150	157	165	172	179	187
3.8	116	124	131	139	147	154	162	169	177	185	192
3.9	120	128	135	143	151	159	167	174	182	190	198
4.0	123	131	139	147	155	163	171	179	187	195	203
4.2	130	139	147	155	164	172	181	189	197	206	214
4.4	137	146	155	164	172	181	190	199	208	216	225
4.6	144	154	163	172	181	190	200	209	218	227	236
4.8	152	161	171	180	190	200	209	219	228	238	248
5.0	159	169	179	189	199	209	219	229	239	249	259

Table 3 (continued)—Pressure Drop of Gas in Pipe, Terminal Pressure Known

Altitude, Feet	Terminal Pressure, psig										
3500-5500	12	13	14	15	16	17	18	19	20	21	22
1500-3500	11	12	13	14	15	16	17	18	19	20	21
0-1500	10	11	12	13	14	15	16	17	18	19	20
Pressure Drop, psi	Pressure Drop Factor (from Table 1)										
0.1	5	5	5	5	6	6	6	6	6	7	7
0.2	10	10	11	11	11	12	12	13	13	13	14
0.3	15	15	16	17	17	18	18	19	20	20	21
0.4	20	20	21	22	23	24	24	25	26	27	28
0.5	25	26	27	28	29	30	31	32	33	34	35
0.6	30	31	32	33	34	36	37	38	39	40	42
0.7	35	36	37	39	40	42	43	44	46	47	49
0.8	40	41	43	44	46	48	49	51	52	54	56
0.9	45	47	48	50	52	54	56	57	59	61	63
1.0	50	52	54	56	58	60	62	64	66	68	70
1.1	55	57	59	61	64	66	68	70	72	75	77
1.2	60	62	65	67	70	72	74	77	79	82	84
1.3	65	68	70	73	76	78	81	83	86	89	91
1.4	70	73	76	79	81	84	87	90	93	95	98
1.5	75	78	81	84	87	90	93	96	99	102	105
1.6	81	84	87	90	93	97	100	103	106	109	113
1.7	86	89	93	96	99	103	106	110	113	116	120
1.8	91	95	98	102	105	109	113	116	120	123	127
1.9	96	100	104	108	112	115	119	123	127	131	134
2.0	102	106	110	114	118	122	126	130	134	138	142
2.1	107	111	115	119	124	128	132	136	140	145	149
2.2	112	117	121	125	130	134	139	143	147	152	156
2.3	118	122	127	131	136	141	145	150	154	159	164
2.4	123	128	132	137	142	147	152	156	161	166	171
2.5	128	133	138	143	148	153	158	163	168	173	178
2.6	134	139	144	149	154	160	165	170	175	180	186
2.7	139	144	150	155	161	166	171	177	182	188	193
2.8	144	150	156	161	167	172	178	184	191	196	202
2.9	150	156	162	167	173	179	185	191	196	202	208
3.0	155	161	167	173	179	185	191	197	203	209	215
3.1	161	167	173	179	186	192	198	204	210	217	223
3.2	166	173	179	186	192	198	205	211	218	224	230
3.3	172	179	185	192	198	205	212	218	225	231	238
3.4	177	184	191	198	205	211	218	225	232	239	245
3.5	183	190	197	204	211	218	225	232	239	246	253
3.6	189	196	203	210	217	225	231	239	246	253	261
3.7	194	202	209	216	224	231	239	246	253	261	268
3.8	200	207	215	223	230	238	245	253	261	268	276
3.9	206	213	221	229	237	245	252	260	268	276	284
4.0	211	219	227	235	243	251	259	267	275	283	291
4.2	223	231	239	248	256	265	273	281	290	298	307
4.4	234	243	252	260	269	278	287	296	304	313	322
4.6	246	256	264	273	282	292	301	310	319	328	338
4.8	257	267	276	286	296	305	315	324	334	344	353
5.0	269	279	289	299	309	319	329	339	349	359	369

Table 3 (continued)—Pressure Drop of Gas in Pipe, Terminal Pressure Known

Altitude, Feet	Terminal Pressure, psig										
3500-5500	1	2	3	4	5	6	7	8	9	10	11
1500-3500		1	2	3	4	5	6	7	8	9	10
0-1500			1	2	3	4	5	6	7	8	9
Pressure Drop, psi	Pressure Drop Factor (from Table 1)										
5.1	163	173	183	193	203	214	224	234	244	254	265
5.2	166	177	187	198	208	218	229	239	250	260	270
5.3	170	181	191	202	213	223	234	244	255	266	276
5.4	174	185	195	206	217	228	239	249	260	271	282
5.5	178	189	200	211	222	233	244	255	266	277	288
5.6	181	193	204	215	226	237	249	260	271	282	293
5.7	185	197	208	219	231	242	254	265	276	288	299
5.8	189	201	212	224	235	247	259	270	282	293	305
5.9	193	205	217	228	240	252	264	276	287	299	311
6.0	197	209	221	233	245	257	269	281	293	305	317
6.1	201	213	225	237	249	262	274	286	298	310	323
6.2	205	217	229	242	254	267	279	291	304	316	329
6.3	209	221	234	246	259	272	284	297	309	321	335
6.4	212	225	238	251	264	276	289	302	315	327	340
6.5	216	229	242	255	268	281	294	307	320	333	346
6.6	220	234	247	260	273	286	300	313	326	339	352
6.7	224	238	251	265	278	291	305	318	332	345	358
6.8	228	242	256	269	283	296	310	324	337	351	364
6.9	233	246	260	274	288	302	315	329	343	357	370
7.0	237	251	265	279	293	307	321	335	349	363	377
7.1	241	255	269	283	297	312	326	340	354	368	382
7.2	245	259	274	288	302	317	331	346	360	374	389
7.3	249	264	278	293	307	322	337	351	366	380	395
7.4	253	268	283	297	312	327	342	357	371	386	401
7.5	257	272	287	302	317	332	347	362	377	392	407
7.6	261	277	292	307	322	337	353	368	383	398	413
7.7	266	281	296	312	327	343	358	373	389	404	420
7.8	270	285	301	317	332	348	363	379	395	410	426
7.9	274	290	306	322	337	353	369	385	401	416	432
8.0	278	294	310	326	342	358	374	390	406	422	438
8.1	283	299	315	331	347	364	380	396	412	428	445
8.2	287	303	320	336	353	369	385	402	418	435	451
8.3	291	308	325	341	358	374	391	408	424	441	457
8.4	296	312	329	346	363	380	396	413	430	447	464
8.5	300	317	334	351	368	385	402	419	436	453	470
8.6	304	322	339	356	373	390	408	425	442	459	476
8.7	309	326	344	361	378	396	413	431	448	465	483
8.8	313	331	348	366	384	401	419	436	454	472	489
8.9	318	336	353	371	389	407	425	442	460	478	495
9.0	322	340	358	376	394	412	430	448	466	484	502
9.2	331	350	368	386	405	423	442	460	478	497	515
9.4	340	359	378	397	415	434	453	472	491	509	528
9.6	349	369	388	407	426	445	465	484	503	522	541
9.8	359	378	398	417	437	457	476	496	515	535	555
10.0	368	388	408	428	448	468	488	508	528	548	568

Table 3 (continued)—Pressure Drop of Gas in Pipe, Terminal Pressure Known

Altitude, Feet	Terminal Pressure, psig										
3500-5500	12	13	14	15	16	17	18	19	20	21	22
1500-3500	11	12	13	14	15	16	17	18	19	20	21
0-1500	10	11	12	13	14	15	16	17	18	19	20
Pressure Drop, psi	Pressure Drop Factor (from Table 1)										
5.1	275	285	295	305	316	326	336	346	356	367	377
5.2	281	291	302	312	322	333	343	354	364	374	385
5.3	287	297	308	319	329	340	350	361	372	382	393
5.4	293	303	314	325	336	347	357	368	379	390	401
5.5	299	310	321	332	343	354	365	376	387	398	409
5.6	305	316	327	338	349	361	372	383	394	405	417
5.7	311	322	333	345	356	368	379	390	402	413	425
5.8	317	328	340	351	363	375	386	398	409	421	433
5.9	323	335	346	358	370	382	394	405	417	429	441
6.0	329	341	353	365	377	389	401	413	425	437	449
6.1	335	347	359	371	384	396	408	420	432	445	457
6.2	341	353	366	378	391	403	415	428	440	453	465
6.3	347	360	372	385	398	410	423	435	448	461	473
6.4	353	366	379	392	404	417	430	443	456	468	481
6.5	359	372	385	398	411	424	437	450	463	476	489
6.6	366	379	392	405	418	432	445	458	471	484	498
6.7	372	385	399	412	425	439	452	466	479	492	506
6.8	378	392	405	419	432	446	460	.473	487	500	514
6.9	384	398	412	426	440	453	467	481	495	509	522
7.0	391	405	419	433	447	461	475	489	503	517	531
7.1	397	411	425	439	454	468	482	496	510	525	539
7.2	403	418	432	446	461	475	490	504	518	533	547
7.3	410	424	439	453	468	483	497	512	526	541	556
7.4	416	431	445	460	475	490	505	519	534	549	564
7.5	422	437	452	467	482	497	512	527	542	557	572
7.6	429	444	459	474	489	505	520	535	550	565	581
7.7	435	450	466	481	497	512	527	543	558	574	589
7.8	441	457	473	488	504	519	535	551	566	582	597
7.9	488	464	480	495	511	527	543	559	574	590	606
8.0	454	470	486	502	518	534	550	566	582	598	614
8.1	461	477	493	509	526	542	558	574	590	607	623
8.2	467	484	500	517	533	549	566	582	599	615	631
8.3	474	491	507	524	540	557	574	590	607	623	639
8.4	480	497	514	531	548	564	581	598	615	632	648
8.5	487	504	521	538	555	572	589	606	623	640	657
8.6	494	511	528	545	562	580	597	614	631	648	666
8.7	500	518	535	552	570	587	605	622	639	657	674
8.8	507	524	542	560	577	595	612	630	648	665	683
8.9	514	531	549	567	585	603	620	638	656	673	692
9.0	520	538	556	574	592	610	628	646	664	682	700
9.2	534	552	570	589	607	626	644	662	681	699	718
9.4	547	566	585	603	622	641	660	679	697	716	735
9.6	561	580	599	618	637	657	676	695	714	733	753
9.8	574	594	613	633	653	672	692	711	731	751	770
10.0	588	608	628	648	668	688	708	728	748	768	788

Example No. 4

Given: Flow, 400,000 cfh
Pipe diameter, 12 inches
Initial pressure, 20 psig
Pressure drop, 2 psi
Altitude of installation, 500 ft above sea level

Find: Pipe length causing 2 psi
pressure drop

(a) 134 is the *Pressure Drop Factor* corresponding
to initial pressure of 20 psig and 2 psi pressure
drop as read on Table 2.

(b) From Table 1, 4.12 is the *Pressure Drop Factor*
per 100 ft corresponding to a flow of 400,000
cfh through a 12 inch pipe.

(c) The pipe length is $L = (134 \times 100) \div 4.12$
$= 3,250$ ft.

Example No. 5

Given: Pipe length, 1,500 ft
Initial pressure, 14 psig
Pressure drop, 3 psi
Inside pipe diameter, 12 inches
Altitude of installation, 4,000 ft above sea
level

Find: Flow capacity of pipe

(a) From Table 2, 149 is the *Pressure Drop Factor*
corresponding to 14 psig initial pressure, and 3
psi pressure drop.

(b) The *Pressure Drop Factor* per 100 ft is $(149 \times 100) \div 1500 = 9.9$

(c) From Table 1, 620,000 cfh corresponds to *Pressure Drop Factor* per 100 ft equal to 9.9 under
the *12 inch Diameter* column.

Example No. 6

Given: Specific gravity, 0.65
Flow, 7,400 cfh
Pipe length, 1,850 ft
Terminal pressure, 16 psig
Pipe size, 2½ inches
Altitude of installation, 2,500 ft above sea
level

Find: Necessary initial pressure

(a) From Table 1, 6.1 is the *Pressure Drop Factor*
per 100 ft corresponding to 7400 cfh flow and
2½ inch pipe size.

(b) For specific gravity of 0.65, find factor $F = 1.08$
from Table 4. Multiply length of 1850 × 1.08,
obtaining equivalent length of 2000 ft.

(c) For 2000 ft equivalent length the *Pressure Drop
Factor* is $(6.1 \times 2000) \div 100 = 122$.

(d) From Table 3, under *Terminal Pressure* of 16
psig, the *Pressure Drop Factor* of 122 corresponds to a pressure drop of 2.0 psi. Therefore,
initial pressure is $16 + 2 = 18$ psig.

Example No. 7

Given: Flow, 25,000 cfh
Pipe length, 1,200 ft
Pressure drop, 3 psi
Terminal pressure, 7 psig
Altitude of installation, 4,500 ft above sea
level

Find: Pipe size

(a) From Table 3, the *Pressure Drop Factor* corresponding to terminal pressure of 7 psig and
pressure drop of 3 psi is found to be 125.

(b) *Pressure Drop Factor* per 100 ft is $(125 \times 100) \div 1200 = 10.4$.

(c) Locate the next smallest *Pressure Drop Factor*
per 100 ft from Table 1 in the *25,000 cfh* row,
here 5.4. The corresponding required pipe size
is 4 inches.

**Table 4—Correction Factors
for Specific Gravity**

Specific Gravity	Correction Factor, F	Specific Gravity	Correction Factor, F
0.35	0.58	0.85	1.42
0.40	0.67	0.90	1.50
0.45	0.75	0.95	1.58
0.50	0.83	1.00	1.67
0.55	0.92	1.10	1.83
0.60	1.00	1.20	2.00
0.65	1.08	1.30	2.17
0.70	1.17	1.40	2.33
0.75	1.25	1.50	2.50
0.80	1.33		

PRESSURE DROP OF AIR IN PIPES

The accompanying nomograph permits the rapid calculation of pressure drops in steel pipe and the sizing of air lines.

The Formulas

The Darcy formula for pressure drop of any gas can be written*

$$\Delta P/100 = 0.00705\, f\, (scfm)^2\, S_g^2/d^5\rho \qquad (1)$$

where $\Delta P/100$ = pressure drop per 100 ft of pipe, psi,
f = friction factor,
$scfm$ = cubic feet per minute at standard conditions, (14.7 psia and 60 F),
S_g = specific gravity of gas (air = 1),
d = inside diameter of pipe, inches
ρ = gas density, lb per cu ft.

From the perfect gas law, air density can be written

$$\rho = 2.70\, P/(t + 460) \qquad (2)$$

where P = absolute pressure, psia
t = temperature, F

For fully turbulent flow in steel pipes.

$$f = 0.189\, N_R{}^{-0.187} \qquad (3)$$

where N_R
(Reynolds Number) = $28.92\ (scfm)\ S_g/d\mu$
μ = gas viscosity, centipoises $\qquad (4)$

Substituting (4) into (3),

$$f = 0.1008 d^{0.187}\, \mu^{0.187}/(scfm)^{0.187} \qquad (5)$$

In the temperature range 60 to 150 deg F, $\mu^{0.187}$ for air varies only from 0.472 to 0.482, so that

*See Crane Company's Flow of Fluids.

0.477 can be taken as an average value and Equation (5) becomes

$$f = 0.0481 d^{0.187}/(scfm)^{0.187} \qquad (6)$$

Substituting (6) and (2) into (1),

$$\Delta P/100 = \frac{1.25 \times 10^{-4}\ (t + 460)(scfm)^{1.813}}{(p + 14.7)\ (d^{4.813})} \qquad (7)$$

where p = gage pressure, psig.

The nomograph on the following page is a graphical solution of Equation (7).

Example: What is the pressure drop of 200 scfm of air at 60 deg and 100 psig average pressure flowing through a standard 3-inch steel pipe?

Solution by Formula

Substituting in Equation (7),

$$\frac{\Delta P}{100} = \frac{1.25 \times 10^{-4}(60 + 460)(200)^{1.813}}{(100 + 14.7)(3.068)^{4.813}}$$

$$\frac{\Delta P}{100} = \frac{1.25 \times 10^{-4}(520)(150,000)}{(114.7)(220)}$$

$$\frac{\Delta P}{100} = 0.038\ \text{psi}$$

Solution by Nomograph

The same problem is readily solved by use of the nomograph on the next page. Its solution is drawn on the nomograph to accompany the text immediately underneath.

Solution: Align $t = 60$ with $p = 100$ and read air density, $\rho = 0.60$ lb per cu ft. Align this latter value with $scfm = 200$ and mark intersection on pivot line. Align this intersection with $d = 3$-inch nominal diameter and read $\Delta P = 0.038$ psi per 100 feet of pipe.

NOMOGRAPH FOR PRESSURE DROP OF AIR IN PIPES

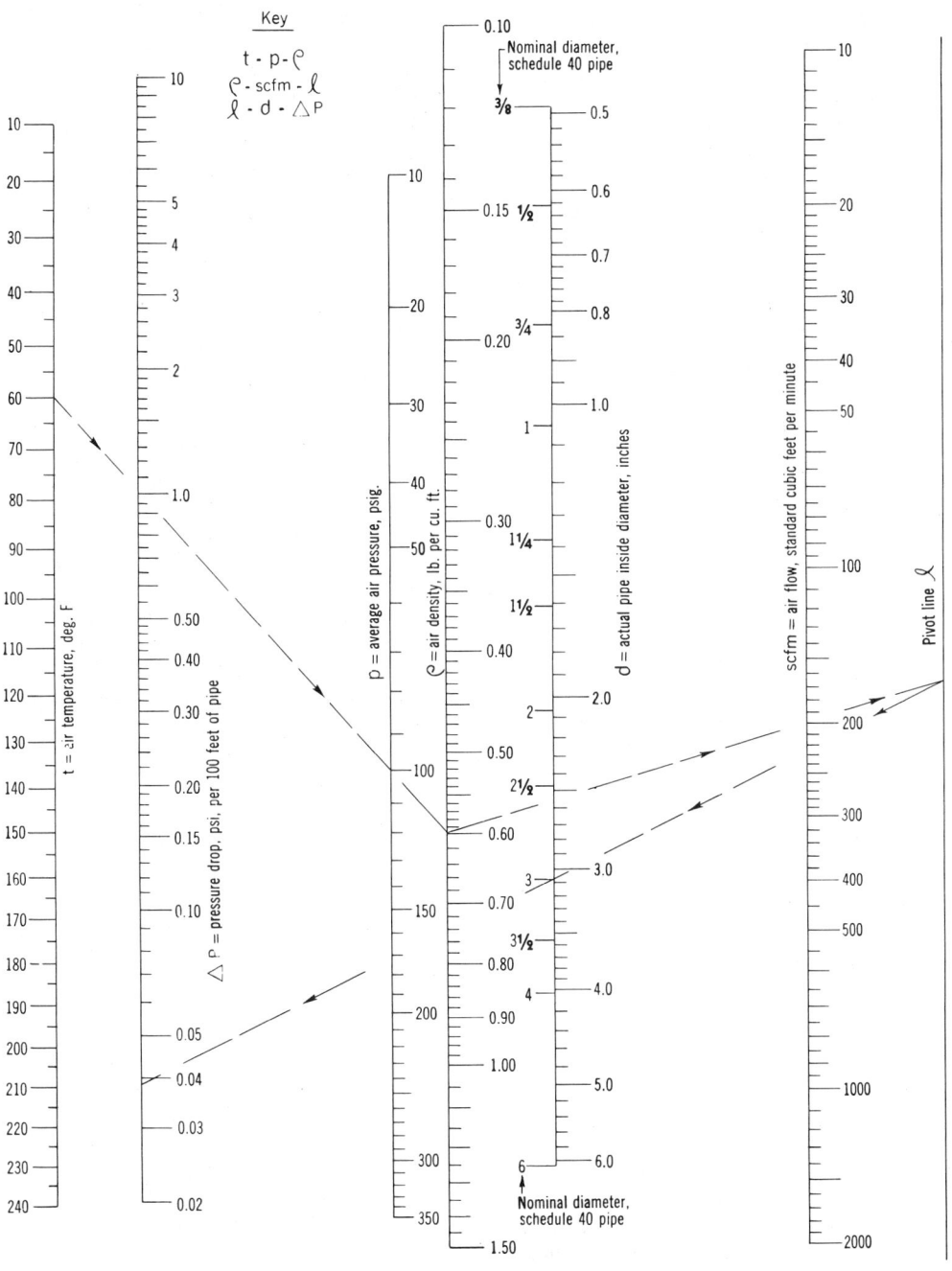

COMPRESSED AIR SYSTEMS

Pressure and quality of compressed air must be determined before design of a system begins.

Instrument air is supplied at pressures of 60-100 psig and generally reduced to 25 psig for most instruments. This air must be filtered and dried.

Cleaning and testing operations utilize full-pressure air up to 100 psig and need not be filtered.

Compressed air quantities for *sewage lift* are determined by the height of the lift and sewage flow in gallons per minute. Compressor capacity may be determined from formula (1):

$$A = S (34 + h)/255 \qquad (1)$$

where A = compressor capacity, cubic feet of free air per minute,

h = lift against which sewage is to be discharged, ft,

S = flow of sewage during discharge, gpm.

Systems with storage tanks are better than tankless compressed air systems for use with pneumatic ejectors. The effect of the tank is to impart 50% more lift, resulting in a more efficient ejector.

Laboratory air requirements are variable. Lab tables should be supplied with 20-psig air, but only 5-10 psig in cases where there is extensive use made of air operated mixers or agitators. Secondary filtering may be advisable. Occasionally, 100-psig air may be required.

Pipe Sizing. Sizes and capacities of compressed air pipe may be computed from formulas (2) or (3) or Table 1.

$$d = 13.5(Q/V)^{1/2} \qquad (2)$$
$$Q = 58(Pd^5/WL)^{1/2} \qquad (3)$$

where d = Pipe ID, inches,

Q = Flow, cfm,

V = Velocity, fpm,

P = Pressure difference, psi

W = Entering air density, lb per cu ft,

L = Length of pipe, ft.

Ascertainment of peak capacities for compressed air systems is largely a matter of judgement. It is necessary to consider the probable use factors, determined by the devices requiring compressed air, the estimated number of people who will be drawing air from the system and the pattern of air use, with especial attention paid to periods of high load.

For example, compressed air demand in a two-man laboratory may reach 20 cfm of free air per laboratory, depending on the number of air outlets (⅛-inch orifices) in use.

Table 2 gives free air discharge through orifices at various pressures.

Where a number of small pipe branches are involved, application of Table 3 is helpful in arriving at relative capacities of pipe headers.

Figure 1, a chart of air quantities *vs.* number of laboratories, was drawn in connection with the design of a specific installation. Similar charts are easily drawn for any installation as an aid to pipe sizing. The chart reflects specific conditions and the designer's judgment in each case. In Fig. 1, two-man laboratories were required with 20-psig supply. A 150% contingency factor is used to cover any peak load.

Air Compressors. Compressors may be single- or two-stage, water- or air-cooled. The single-stage pump is used for general plumbing work with pressures up to 125 psig; the two-stage pump for higher pressures. An air-cooled compressor may be considered for outdoor installation; a water-cooled unit for indoor installation, or where high pressures are required, or where water supply and drainage are conveniently available. Reciprocating or rotary type compressors may be used where oil-free air is required. A compressor oversized 50% of design load is not objectionable.

Receiver tanks are recommended where peak loads are sudden; they also tend to dampen pulsations, maintaining a more uniform pressure. Receivers may be purchased with the compressor as part of a package unit; however, this is recommended only for installations under 75 cfm demand. Capacities of ASME-approved receivers are suggested in Table 4.

Relief valves and gages, along with drains and

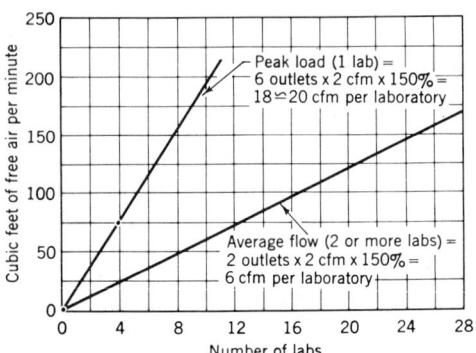

Fig. 1. Chart like this may be drawn for any installation.

Table 1. Loss of Air Pressure in Psi per 1000 Ft of Pipe Due to Friction

Cu Ft Free Air per Min	Equivalent Cu Ft Compressed Air per Min	60-Psig Initial Pressure: Nominal Pipe Dia., Inches								Equivalent Cu Ft Compressed Air per Min	80-Psig Initial Pressure: Nominal Pipe Dia., Inches							
		½	¾	1	1¼	1½	2	2½	3		½	¾	1	1¼	1½	2	2½	3
10	1.96	10.0	1.53	0.43	0.10	—	—	—	—	1.55	7.90	1.21	0.34	—	—	—	—	—
20	3.94	39.7	5.99	1.71	0.39	0.18	—	—	—	3.10	31.4	4.72	1.35	0.31	—	—	—	—
30	5.89	—	13.85	3.86	0.88	0.40	—	—	—	4.65	70.8	10.9	3.04	0.69	0.31	—	—	—
40	7.86	—	24.7	6.85	1.59	0.71	0.19	—	—	6.20	—	19.5	5.40	1.25	0.56	—	—	—
50	9.84	—	38.6	10.7	2.48	1.10	0.30	—	—	7.74	—	30.5	8.45	1.96	0.87	—	—	—
60	11.81	—	55.5	15.4	3.58	1.57	0.43	—	—	9.29	—	43.8	12.16	2.82	1.24	0.34	—	—
70	13.75	—	—	21.0	4.87	2.15	0.57	0.22	—	10.82	—	59.8	16.6	3.84	1.70	0.45	—	—
80	15.72	—	—	27.4	6.37	2.82	0.75	0.29	—	12.40	—	78.2	21.6	5.03	2.22	0.59	—	—
90	17.65	—	—	34.7	8.05	3.57	0.57	0.95	0.37	13.95	—	—	27.4	6.35	2.82	0.75	—	—
100	19.60	—	—	42.8	9.95	4.40	1.18	0.46	—	15.5	—	—	33.8	7.85	3.47	0.93	0.36	—
125	19.4	**3½**	—	46.2	12.4	6.90	1.83	0.71	0.14	19.4	—	—	46.2	12.4	5.45	1.44	0.56	—
150	29.45	0.15	—	—	22.4	9.90	2.64	1.02	0.32	23.2	—	—	76.2	17.7	7.82	2.08	0.81	—
175	34.44	0.20		—	30.8	13.40	3.64	1.40	0.43	27.2		—	—	24.8	10.6	2.87	1.10	—
200	39.40	0.27	**4**	—	39.7	17.60	4.71	1.83	0.57	31.0	**3½**	—	—	31.4	13.9	3.72	1.44	0.45
250	49.20	0.42	0.21		—	27.5	7.37	2.85	0.89	38.7	0.33		—	49.0	21.7	5.82	2.25	0.70
300	58.90	0.60	0.31	**4½**	—	39.6	10.55	4.11	1.30	46.5	0.47	**4**	—	70.6	31.2	8.35	3.24	1.03
350	68.8	0.82	0.42	0.23	—	54.0	14.4	5.60	1.76	54.2	0.65	0.33	—	—	42.5	11.4	4.42	1.39
400	78.8	1.06	0.53	0.30	—	—	18.6	7.30	2.30	62.0	0.84	0.42		—	55.5	14.7	5.76	1.82
450	88.4	1.35	0.70	0.38	—	—	23.7	9.20	2.90	69.7	1.06	0.55	**4½**	—	—	18.7	7.25	2.29
500	98.4	1.67	0.85	0.46	—	—	29.7	11.4	3.60	77.4	1.32	0.67	0.30	—	—	23.3	9.0	2.84
600	118.1	2.40	1.22	0.67	—	—	42.3	16.4	5.17	92.9	1.89	0.96	0.53	—	—	33.4	12.9	4.08
700	137.5	3.27	1.67	0.91	—	—	57.8	22.3	7.00	108.2	2.58	1.32	0.72	—	—	45.7	17.6	5.52
800	157.2	4.26	2.18	1.20	—	—	—	29.2	9.16	124.0	3.36	1.72	0.95	—	—	59.3	23.1	7.15
900	176.5	5.40	2.76	1.51	**5**	**6**	**8**	39.0	11.6	139.5	4.26	2.18	1.19	**5**	**6**		29.2	9.17
1,000	196.0	6.65	3.40	1.87				45.7	14.3	155	5.27	2.68	1.48				36.1	11.3
1,500	294.5	15.0	7.6	4.20	2.32	0.87	0.29	**10**	32.3	232	11.8	6.0	3.32	1.83	0.69	**8**	81.7	25.5
2,000	394.0	26.6	13.6	7.5	4.18	1.53	0.36		57.5	310	21.0	10.7	5.9	3.30	1.21	0.29		45.3
2,500	492	41.7	21.3	11.6	6.4	2.42	0.57	0.17	—	387	32.9	16.8	9.2	5.1	1.91	0.45	**10**	70.9
3,000	589	60.0	30.7	16.7	9.2	3.48	0.81	0.24	—	465	47.4	24.2	13.2	7.3	2.74	0.64	0.19	—
3,500	688	—	41.7	22.6	12.8	4.68	1.07	0.33	—	542	64.5	32.8	17.8	10.1	3.70	0.85	0.26	—
4,000	788	—	54.5	29.7	16.5	6.17	1.44	0.44	**12**	620	—	43.0	23.4	13.0	4.87	1.14	0.34	
4,500	884	—	—	37.9	20.8	7.8	1.83	0.55	0.21	697	—	54.8	29.8	16.4	6.15	1.44	0.43	**12**
5,000	984	—	—	46.4	25.7	9.7	2.26	0.67	0.27	774	—	67.4	36.7	20.3	7.65	1.78	0.53	0.21
6,000	1,181	—	—	—	37.0	13.9	3.25	0.98	0.38	929	—	—	53.0	29.2	11.0	2.57	0.77	0.29
7,000	1,375	—	—	—	50.3	18.7	4.43	1.34	0.51	1,082	—	—	72.1	39.8	14.8	3.40	1.06	0.40
8,000	1,572	—	—	—	—	24.7	5.80	1.73	0.71	1,240	—	—	—	52.1	19.5	4.57	1.36	0.56
9,000	1,765	—	—	—	—	31.3	7.33	2.20	0.87	1,395	—	—	—	65.8	24.7	5.78	1.74	0.69
10,000	1,960	—	—	—	—	38.6	9.05	2.72	1.06	1,550	—	—	—	—	30.5	7.15	2.14	0.84
11,000	2,165	—	—	—	—	46.7	10.9	3.29	1.28	1,710	—	—	—	—	36.8	8.61	2.60	1.01
12,000	2,362	—	—	—	—	55.5	13.0	3.90	1.51	1,860	—	—	—	—	43.8	10.3	3.08	1.19
13,000	2,560	—	—	—	—	—	15.2	4.58	1.77	2,020	—	—	—	—	51.7	12.0	3.62	1.40
14,000	2,750	—	—	—	—	—	17.7	5.32	2.07	2,170	—	—	—	—	60.2	14.0	4.20	1.63
15,000	2,945	—	—	—	—	—	20.3	6.10	2.36	2,320	—	—	—	—	68.5	16.0	4.82	1.86
16,000	3,144	—	—	—	—	—	23.1	6.95	2.70	2,480	—	—	—	—	78.2	18.2	5.48	2.13
18,000	3,530	—	—	—	—	—	29.2	8.80	3.42	2,790	—	—	—	—	—	23.0	6.95	2.70
20,000	3,940	—	—	—	—	—	36.2	10.8	4.22	3,100	—	—	—	—	—	28.6	8.55	3.33
22,000	4,330	—	—	—	—	—	43.7	13.2	5.12	3,410	—	—	—	—	—	34.5	10.4	4.04
24,000	4,724	—	—	—	—	—	51.9	15.6	5.92	3,720	—	—	—	—	—	41.0	12.3	4.69
26,000	5,120	—	—	—	—	—	—	18.3	7.15	4,030	—	—	—	—	—	48.2	14.4	5.6
28,000	5,500	—	—	—	—	—	—	21.3	8.3	4,350	—	—	—	—	—	55.9	16.8	6.5
30,000	5,890	—	—	—	—	—	—	24.4	9.4	4,650	—	—	—	—	—	64.2	19.3	7.5

Reprinted, by permission, from Compressed Air and Gas Handbook, published by Compressed Air and Gas Institute, New York, N. Y.

Table 1. Loss of Air Pressure in Psi per 1000 Ft of Pipe *(Continued)*

Cu Ft Free Air per Min	Equivalent Cu Ft Compressed Air per Min	100-Psig Initial Pressure — Nominal Pipe Dia., Inches								Equivalent Cu Ft Compressed Air per Min	125-Psig Initial Pressure — Nominal Pipe Dia., Inches							
		½	¾	1	1¼	1½	2	2½	3		½	¾	1	1¼	1½	2	2½	3
10	1.28	6.50	.99	0.28	—	—	—	—	—	1.05	5.35	0.82	0.23	—	—	—	—	—
20	2.56	25.9	3.90	1.11	0.25	0.11	—	—	—	2.11	21.3	3.21	0.92	0.21	—	—	—	—
30	3.84	58.5	9.01	2.51	0.57	0.26	—	—	—	3.16	48.0	7.42	2.07	0.47	0.21	—	—	—
40	5.12	—	16.0	4.45	41.03	0.46	—	—	—	4.21	—	13.2	3.67	0.85	0.38	—	—	—
50	6.41	—	25.1	6.96	1.61	0.71	0.19	—	—	5.26	—	20.6	5.72	1.33	0.59	—	—	—
60	7.68	—	36.2	10.0	2.32	1.02	0.28	—	—	6.32	—	29.7	8.25	1.86	0.84	0.23	—	—
70	8.96	—	49.3	13.7	3.16	1.40	0.37	—	—	7.38	—	40.5	11.2	2.61	1.15	0.31	—	—
80	10.24	—	64.5	17.8	4.14	1.83	0.49	0.19	—	8.42	—	53.0	14.7	3.41	1.51	0.40	—	—
90	11.52	—	82.8	22.6	5.23	2.32	0.62	0.24	—	9.47	—	68.0	18.6	4.30	1.91	0.51	0.20	—
100	12.81	—	—	27.9	6.47	2.86	0.77	0.30	—	10.50	—	—	22.9	5.32	2.36	0.63	0.25	—
125	15.82	—	—	48.6	10.2	4.49	1.19	0.46	—	13.15	—	—	39.9	8.4	3.70	0.98	0.38	—
150	19.23	—	—	62.8	14.6	6.43	1.72	0.66	0.21	15.79	—	—	51.6	12.0	5.30	1.41	0.55	0.17
175	22.40	**3½**	—	—	19.8	8.72	2.36	0.91	0.28	18.41	**3½**	—	—	16.3	7.2	1.95	0.75	0.24
200	25.62	0.17	—	—	25.9	11.4	3.06	1.19	0.37	21.05		—	—	21.3	9.4	2.52	0.98	0.31
250	31.64	0.27	**4**	—	40.4	17.9	4.78	1.85	0.58	26.30	0.22	—	—	33.2	14.7	3.94	1.53	0.48
300	38.44	0.39	0.20	**4½**	58.2	25.8	6.85	2.67	0.84	31.60	0.32	**4**	—	47.3	21.2	5.62	2.20	0.70
350	44.80	0.53	0.27		—	35.1	9.36	3.64	1.14	36.80	0.44	0.22	**4½**	—	28.8	7.7	3.00	0.94
400	51.24	0.69	0.35	0.19	—	45.8	12.1	4.75	1.50	42.10	0.57	0.28		—	37.6	10.0	3.91	1.23
450	57.65	0.88	0.46	0.25	—	58.0	15.4	5.98	1.89	47.30	0.72	0.37	0.20	—	47.7	12.7	4.92	1.55
500	63.28	1.09	0.55	0.30	—	71.6	19.2	7.42	2.34	52.60	0.89	0.46	0.25	—	58.8	15.7	6.10	1.93
600	76.88	1.56	0.79	0.44	—	—	27.6	10.7	3.36	63.20	1.28	0.65	0.36	—	—	22.6	8.8	2.76
700	89.60	2.13	1.09	0.59	—	—	37.7	14.5	4.55	73.80	1.75	0.89	0.49	—	—	30.0	11.9	3.74
800	102.5	2.77	1.42	0.78	—	—	49.0	19.0	5.89	84.20	2.28	1.17	0.64	—	—	40.2	15.6	4.85
900	115.3	3.51	1.80	0.99	—	—	62.3	24.1	7.6	94.70	2.89	1.48	0.81	—	—	51.2	19.8	6.2
1,000	128.1	4.35	2.21	1.22	**5**	**6**	76.9	29.8	9.3	105.1	3.57	1.82	1.00	**5**	**6**	63.2	24.5	7.7
1,500	192.3	9.8	4.9	2.73	1.51	0.57	**8**	67.0	21.0	157.9	8.0	4.1	2.25	1.24	0.47	**8**	55.0	17.2
2,000	256.2	17.3	8.8	4.9	2.72	0.99	0.24	—	37.4	210.5	14.2	7.3	4.0	2.24	0.82	0.19	—	30.7
2,500	316.4	27.2	13.8	8.3	4.2	1.57	0.37	**10**	58.4	263.0	22.3	11.4	6.2	3.4	1.30	0.31	**10**	48.0
3,000	384.6	39.1	20.0	10.9	6.0	2.26	0.53		84.1	316	32.1	16.4	9.0	4.9	1.86	0.43		69.2
3,500	447.8	58.2	27.2	14.7	8.2	3.04	0.70	0.22	—	368	47.7	22.3	12.1	6.9	2.51	0.57	0.18	—
4,000	512.4	69.4	35.5	19.4	10.7	4.01	0.94	0.28	**12**	421	57.0	29.2	15.9	8.9	3.30	0.77	0.23	—
4,500	576.5	—	45.0	24.5	13.5	5.10	1.19	0.36		473	—	37.0	20.1	11.1	4.2	0.98	0.29	**12**
5,000	632.8	—	55.6	30.2	16.8	6.3	1.47	0.44	0.17	526	—	45.7	24.8	13.9	5.2	1.21	0.36	
6,000	768.8	—	80.0	43.7	24.1	9.1	2.11	0.64	0.24	632	—	65.7	35.8	19.8	7.5	1.74	0.52	0.20
7,000	896.0	—	—	59.5	32.8	12.2	2.88	0.87	0.33	738	—	—	48.8	26.9	10.0	2.37	0.72	0.27
8,000	1,025	—	—	77.5	42.9	16.1	3.77	1.12	0.46	842	—	—	63.7	35.2	13.2	3.10	0.93	0.38
9,000	1,153	—	—	—	54.3	20.4	4.77	1.43	0.57	947	—	—	—	44.7	16.7	3.93	1.18	0.47
10,000	1,280	—	—	—	67.1	25.1	5.88	1.77	0.69	1,051	—	—	—	55.2	20.6	4.85	1.46	0.57
11,000	1,410	—	—	—	—	30.4	7.10	2.14	0.83	1,156	—	—	—	—	25.0	5.8	1.76	0.68
12,000	1,540	—	—	—	—	36.2	8.5	2.54	0.98	1,262	—	—	—	—	29.7	7.0	2.09	0.81
13,000	1,668	—	—	—	—	42.6	9.8	2.98	1.15	1,368	—	—	—	—	35.0	8.1	2.44	0.95
14,000	1,795	—	—	—	—	49.2	11.5	3.46	1.35	1,473	—	—	—	—	40.3	9.7	2.85	1.11
15,000	1,923	—	—	—	—	56.6	13.2	3.97	1.53	1,579	—	—	—	—	46.5	10.9	3.26	1.26
16,000	2,050	—	—	—	—	64.5	15.0	4.52	1.75	1,683	—	—	—	—	53.0	12.4	3.72	1.45
18,000	2,310	—	—	—	—	81.5	19.0	5.72	2.22	1,893	—	—	—	—	66.9	15.6	4.71	1.83
20,000	2,560	—	—	—	—	—	23.6	7.0	2.74	2,150	—	—	—	—	—	19.4	5.8	2.20
22,000	2,820	—	—	—	—	—	28.5	8.5	3.33	2,315	—	—	—	—	—	23.4	7.1	2.74
24,000	3,080	—	—	—	—	—	33.8	10.0	3.85	2,525	—	—	—	—	—	27.8	8.4	3.17
26,000	3,338	—	—	—	—	—	39.7	11.9	4.65	2,735	—	—	—	—	—	32.6	9.8	3.83
28,000	3,590	—	—	—	—	—	46.2	13.8	5.40	2,946	—	—	—	—	—	37.9	11.4	4.4
30,000	3,850	—	—	—	—	—	53.0	15.9	6.17	3,158	—	—	—	—	—	43.5	13.1	5.1

Table 2. Discharge of Air Through Orifices

Pressure Before Orifice, Psig	Diameter of Orifice, Inches										
	1/64	1/32	1/16	1/8	1/4	3/8	1/2	5/8	3/4	7/8	1
	Discharge, Cubic Feet of Free Air per Minute										
1	0.028	0.112	0.450	1.80	7.18	16.2	28.7	45.0	64.7	88.1	115
2	0.040	0.158	0.633	2.53	10.1	22.8	40.5	63.3	91.2	124	162
3	0.048	0.194	0.775	3.10	12.4	27.8	49.5	77.5	111	152	198
4	0.056	0.223	0.892	3.56	14.3	32.1	57.0	89.2	128	175	228
5	0.062	0.248	0.993	3.97	15.9	35.7	63.5	99.3	143	195	254
6	0.068	0.272	1.09	4.34	17.4	39.1	69.5	109	156	213	278
7	0.073	0.293	1.17	4.68	18.7	42.2	75.0	117	168	230	300
9	0.083	0.331	1.32	5.30	21.1	47.7	84.7	132	191	260	339
12	0.095	0.379	1.52	6.07	24.3	54.6	97.0	152	218	297	388
15	0.105	0.420	1.68	6.72	26.9	60.5	108	168	242	329	430
20	0.123	0.491	1.96	7.86	31.4	70.7	126	196	283	385	503
25	0.140	0.562	2.25	8.98	35.9	80.9	144	225	323	440	575
30	0.158	0.633	2.53	10.1	40.5	91.1	162	253	365	496	648
35	0.176	0.703	2.81	11.3	45.0	101	180	281	405	551	720
40	0.194	0.774	3.10	12.4	49.6	112	198	310	446	607	793
45	0.211	0.845	3.38	13.5	54.1	122	216	338	487	662	865
50	0.229	0.916	3.66	14.7	58.6	132	235	366	528	718	938
60	0.264	1.06	4.23	16.9	67.6	152	271	423	609	828	1,082
70	0.300	1.20	4.79	19.2	76.7	173	307	479	690	939	1,227
80	0.335	1.34	5.36	21.4	85.7	193	343	536	771	1,050	1,371
90	0.370	1.48	5.92	23.7	94.8	213	379	592	853	1,161	1,516
100	0.406	1.62	6.49	26.0	104	234	415	649	934	1,272	1,661
110	0.441	1.76	7.05	28.2	113	254	452	705	1,016	1,383	1,806
120	0.476	1.91	7.62	30.5	122	274	488	762	1,097	1,494	1,951
125	0.494	1.98	7.90	31.6	126	284	506	790	1,138	1,549	2,023

Table 3. Relative Discharge Capacities of Pipe

Relative discharge capacity = (dia of large pipe/dia of small pipe)$^{2.5}$

Nom. Dia of Larger Pipe, Inches	Nominal Diameter of Smaller Pipe, Inches															
	1/8	1/4	3/8	1/2	3/4	1	1 1/4	1 1/2	2	2 1/2	3	3 1/2	4	4 1/2	5	6
	Approximate number of small-pipe flows handled by larger pipe															
1/8	1	—	—	—	—	—	—	—	—	—	—	—	—	—	—	—
1/4	2.1	1	—	—	—	—	—	—	—	—	—	—	—	—	—	—
3/8	4.5	2.1	1	—	—	—	—	—	—	—	—	—	—	—	—	—
1/2	8	3.8	1.8	1	—	—	—	—	—	—	—	—	—	—	—	—
3/4	15	8	3.6	2	1	—	—	—	—	—	—	—	—	—	—	—
1	30	15	6.6	3.7	1.8	1	—	—	—	—	—	—	—	—	—	—
1 1/4	60	25	13	7	3.6	2	1	—	—	—	—	—	—	—	—	—
1 1/2	90	40	20	10	5.5	2.9	1.5	1	—	—	—	—	—	—	—	—
2	165	75	35	20	10	5.5	2.7	1.9	1	—	—	—	—	—	—	—
2 1/2	255	120	55	30	16	8	4.3	2.9	1.6	1	—	—	—	—	—	—
3	440	210	100	55	27	15	7	5	2.7	1.7	1	—	—	—	—	—
3 1/2	630	300	140	80	40	21	11	7	3.9	2.5	1.4	1	—	—	—	—
4	870	400	190	100	55	30	15	10	5.3	3.4	2	1.4	1	—	—	—
4 1/2	1150	540	250	140	70	40	20	13	7	4.5	2.6	1.8	1.3	1	—	—
5	1500	720	330	180	90	50	25	17	9	6	3.5	2.4	1.8	1.3	1	—
6	2400	1130	530	300	150	80	40	28	15	9	5.5	3.8	2.8	2.1	1.6	1

Fig. 2. Pipe Sizing Example. Given: 14 two-man laboratories, 3 air service stations and 6 air-operated ventilation damper controls. Solution: A. Minimum size pipe ⅜ inch for outlets. B. Peak load demand for one laboratory from Fig. 1. C. See Table 2, discharges through orifices, also Fig. 1 for use factors. D, E, F, H, J. Accumulation of air quantities. Header sizes from Table 3. Size J from Table 1. G, I. Hose outlets sized. K. Negligible demand.

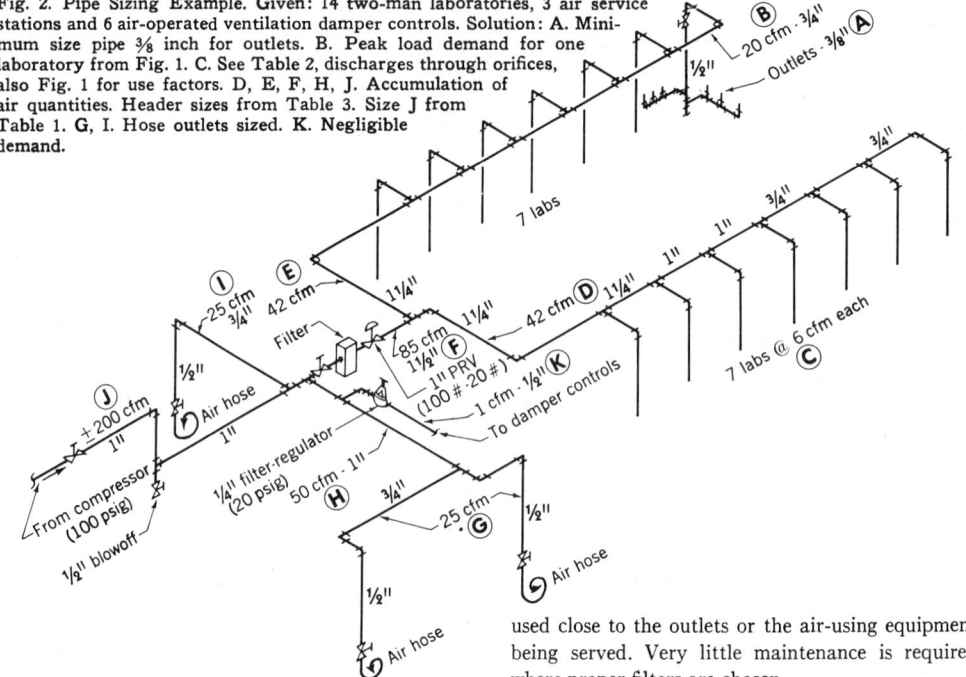

support features, are generally an integral part of the compressor unit.

Filters and Regulators. Filters for the removal of oils, dusts, water and other impurities should be given very careful consideration and selected on the basis of manufacturer's ratings. Regulators are available for reduction of compressor output to any desired pressure. Combination filter-regulators are used close to the outlets or the air-using equipment being served. Very little maintenance is required where proper filters are chosen.

Pipe, Valves, Fittings. Materials for compressed air piping up to 150 psig working pressure, and not over 6 inches diameter, should be Schedule 40 galvanized steel. Type L copper tube and Delrin (Du Pont) plastic pipe with heat fusion joints, are suitable for working pressures to 100 psig. At higher pressures and in the larger pipe sizes, Schedule 40 black steel pipe with welded joints may be considered for an economical installation.

Fittings for steel pipe may be screwed galvanized malleable iron (ANSI-B16.3); or, socket-welding or threaded forged steel (ANSI-B16.11). Valves and checks, ⅜-2 inches, should be brass, 150-lb, with screwed ends (ASTM B62); larger valves (above 2 inches) should be welded or flanged forged steel, 150-lb.

Testing. Piping should be tested hydrostatically at 1½ times the maximum operating pressure, or 50 psig, whichever is greater, then purged dry before the system is placed in service. Testing with air or inert gases may be desirable in some instances, detecting leaks with soap bubbles.

The piping system should be kept dirt-free with filters; however, addition of dirt pickets or blowout valves at the base of piping drops is frequently advisable.

Table 4. Receivers for Compressed Air Systems

*Compressor Capacity, Cfm	ASME Receiver Dimensions		
	Dia, Inches	Length, Ft	Vol., Cu ft
45	14	4	4½
110	13	6	11
190	24	6	19
340	30	7	34
570	36	8	57
960	42	10	96
2115	48	12	151
3120	54	14	223
4400	60	16	314
6000	66	18	428

*Actual compressor capacity, based on constant-speed operation, for which suitable at 40-125 psig, in cubic feet of free air per minute. (From Chicago Pneumatic Tool Co. data.)

VISCOSITY OF LIQUIDS

Viscosity is that property of a fluid which resists any force tending to produce flow or, in other terms, it is the ability to resist shear; it is independent of specific gravity.

The charts below give the absolute and kinematic viscosity to various liquids.

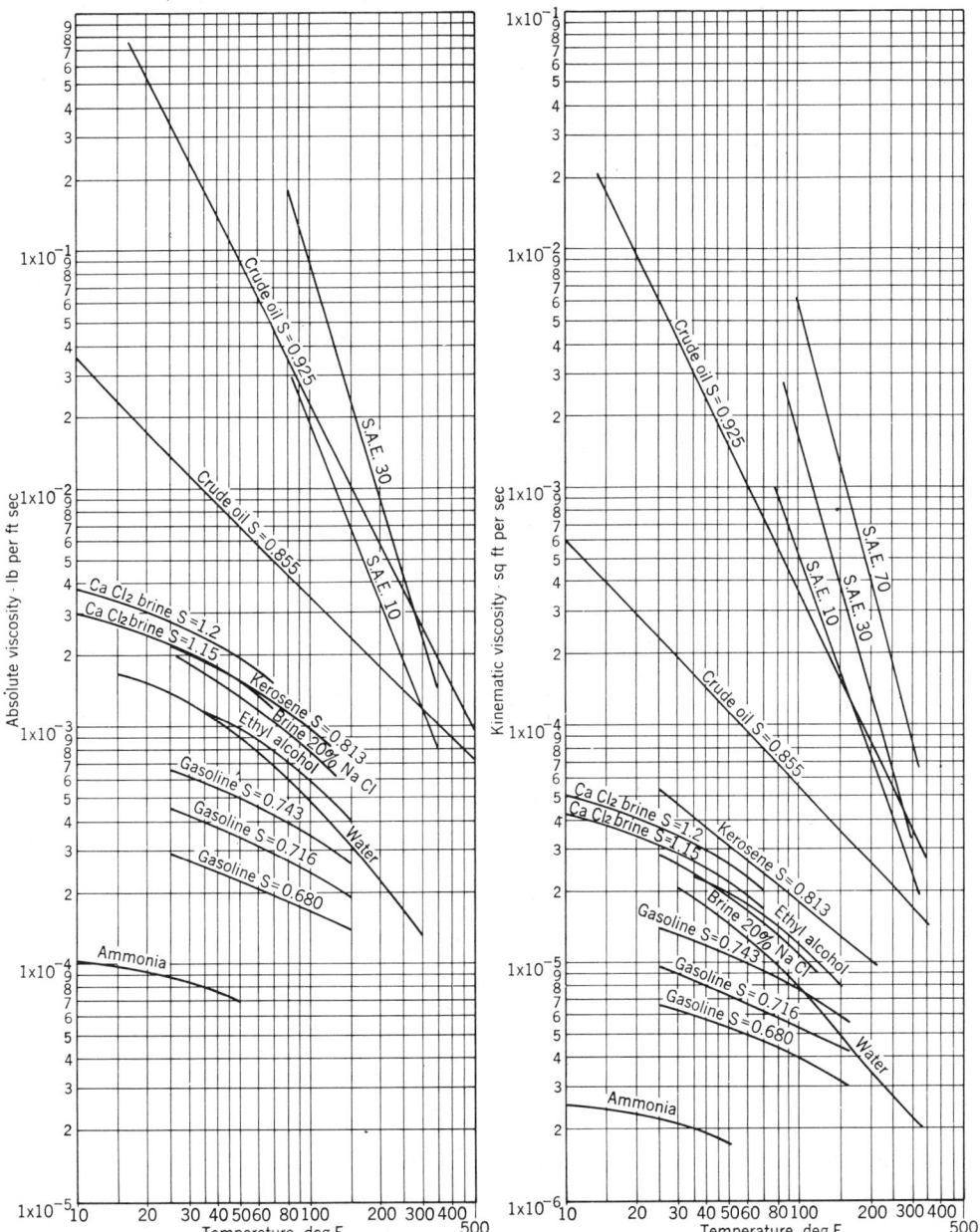

Data from R. L. Daugherty in "Some Physical Properties of Water and Other Fluids," A S M E, 1934.

DIMENSIONAL AND CAPACITY DATA — SCHEDULE 40 STEEL PIPE

Diameter, Inches			Wall Thickness, In.	Cross-Sectional Area, Sq. In.			Weight per Foot, Lb.		
Nominal	Actual Inside	Actual Outside		Outside	Inside	Metal	Of Pipe Alone	Of Water in Pipe	Of Pipe and Water
⅛	0.269	0.405	0.068	0.129	0.057	0.072	0.25	0.028	0.278
¼	0.364	0.540	0.088	0.229	0.104	0.125	0.43	0.045	0.475
⅜	0.493	0.675	0.091	0.358	0.191	0.167	0.57	0.083	0.653
½	0.622	0.840	0.109	0.554	0.304	0.250	0.86	0.132	0.992
¾	0.824	1.050	0.113	0.866	0.533	0.333	1.14	0.232	1.372
1	1.049	1.315	0.133	1.358	0.864	0.494	1.68	0.375	2.055
1¼	1.380	1.660	0.140	2.164	1.495	0.669	2.28	0.649	2.929
1½	1.610	1.900	0.145	2.835	2.036	0.799	2.72	0.882	3.602
2	2.067	2.375	0.154	4.431	3.356	1.075	3.66	1.454	5.114
2½	2.469	2.875	0.203	6.492	4.788	1.704	5.80	2.073	7.873
3	3.068	3.500	0.216	9.621	7.393	2.228	7.58	3.201	10.781
3½	3.548	4.000	0.226	12.568	9.888	2.680	9.11	4.287	13.397
4	4.026	4.500	0.237	15.903	12.730	3.173	10.80	5.516	16.316
5	5.047	5.563	0.258	24.308	20.004	4.304	14.70	8.674	23.374
6	6.065	6.625	0.280	34.474	28.890	5.584	19.00	12.52	31.52
8	7.981	8.625	0.322	58.426	50.030	8.396	28.60	21.68	50.28
10	10.020	10.750	0.365	90.79	78.85	11.90	40.50	34.16	74.66
12	11.938	12.750	0.406	127.67	113.09	15.77	53.60	48.50	102.10
14	13.126	14.000	0.437	153.94	135.33	18.61	63.30	58.64	121.94
16	15.000	16.000	0.500	201.06	176.71	24.35	82.80	76.58	159.38
18	16.876	18.000	0.562	254.47	223.68	30.79	105.00	96.93	201.93
20	18.814	20.000	0.593	314.16	278.01	36.15	123.00	120.46	243.46

Nominal Dia., In.	Circumference, Inches		Sq. Ft. of Surface per Lineal Foot		Contents of Pipe per Lineal Foot		Lineal Feet to Contain		
	Outside	Inside	Outside	Inside	Cu. Ft.	Gal.	1 Cu. Ft.	1 Gal.	1 Lb. of Water
⅛	1.27	0.84	0.106	0.070	0.0004	0.003	2533.775	338.74	35.714
¼	1.69	1.14	0.141	0.095	0.0007	0.005	1383.789	185.00	22.222
⅜	2.12	1.55	0.177	0.129	0.0013	0.010	754.360	100.85	12.048
½	2.65	1.95	0.221	0.167	0.0021	0.016	473.906	63.36	7.576
¾	3.29	2.58	0.275	0.215	0.0037	0.028	270.034	36.10	4.310
1	4.13	3.29	0.344	0.274	0.0062	0.045	166.618	22.28	2.667
1¼	5.21	4.33	0.435	0.361	0.0104	0.077	96.275	12.87	1.541
1½	5.96	5.06	0.497	0.422	0.0141	0.106	70.733	9.46	1.134
2	7.46	6.49	0.622	0.540	0.0233	0.174	42.913	5.74	0.688
2½	9.03	7.75	0.753	0.654	0.0332	0.248	30.077	4.02	0.482
3	10.96	9.63	0.916	0.803	0.0514	0.383	19.479	2.60	0.312
3½	12.56	11.14	1.047	0.928	0.0682	0.513	14.565	1.95	0.233
4	14.13	12.64	1.178	1.052	0.0884	0.660	11.312	1.51	0.181
5	17.47	15.84	1.456	1.319	0.1390	1.040	7.198	0.96	0.115
6	20.81	19.05	1.734	1.585	0.2010	1.500	4.984	0.67	0.080
8	27.09	25.07	2.258	2.090	0.3480	2.600	2.878	0.38	0.046
10	33.77	31.47	2.814	2.622	0.5470	4.100	1.826	0.24	0.029
12	40.05	37.70	3.370	3.140	0.7850	5.870	1.273	0.17	0.021
14	47.12	44.76	3.930	3.722	1.0690	7.030	1.067	0.14	0.017
16	53.41	51.52	4.440	4.310	1.3920	9.180	0.814	0.11	0.013
18	56.55	53.00	4.712	4.420	1.5530	11.120	0.644	0.09	0.010
20	62.83	59.09	5.236	4.920	1.9250	14.400	0.517	0.07	0.008

DIMENSIONAL AND CAPACITY DATA — SCHEDULE 80 STEEL PIPE

Diameter, In.			Wall Thick-ness, In.	Cross-Sectional Area, Sq. In.			Weight per Foot, Lb.		
Nom-inal	Actual Inside	Actual Outside		Out-side	In-side	Metal	Of Pipe	Of Water in Pipe	Of Pipe and Water
⅛	0.215	0.405	0.095	0.129	0.036	0.093	0.314	0.016	0.330
¼	0.302	0.540	0.119	0.229	0.072	0.157	0.535	0.031	0.566
⅜	0.423	0.675	0.126	0.358	0.141	0.217	0.738	0.061	0.799
½	0.546	0.840	0.147	0.554	0.234	0.320	1.087	0.102	1.189
¾	0.742	1.050	0.154	0.866	0.433	0.433	1.473	0.213	1.686
1	0.957	1.315	0.179	1.358	0.719	0.639	2.171	0.312	2.483
1¼	1.278	1.660	0.191	2.164	1.283	0.881	2.996	0.555	3.551
1½	1.500	1.900	0.200	2.835	1.767	1.068	3.631	0.765	4.396
2	1.939	2.375	0.218	4.431	2.954	1.477	5.022	1.280	6.302
2½	2.323	2.875	0.276	6.492	4.238	2.254	7.661	1.830	9.491
3	2.900	3.500	0.300	9.621	6.605	3.016	10.252	2.870	13.122
3½	3.364	4.000	0.318	12.568	8.890	3.678	12.505	3.720	16.225
4	3.826	4.500	0.337	15.903	11.496	4.407	14.983	4.970	19.953
5	4.813	5.563	0.375	24.308	18.196	6.112	20.778	7.940	28.718
6	5.761	6.625	0.432	34.474	26.069	8.405	28.573	11.300	39.873
8	7.625	8.625	0.500	58.426	45.666	12.760	43.388	19.800	63.188
10	9.564	10.750	0.593	90.79	71.87	18.92	64.400	31.130	95.530
12	11.376	12.750	0.687	127.67	101.64	26.03	88.600	44.040	132.640
14	12.500	14.000	0.750	153.94	122.72	31.22	107.000	53.180	160.180
16	14.314	16.000	0.843	201.06	160.92	40.14	137.000	69.730	206.730
18	16.126	18.000	0.937	254.47	204.24	50.23	171.000	88.500	259.500
20	17.938	20.000	1.031	314.16	252.72	61.44	209.000	109.510	318.510

Nom-inal Dia., In.	Circumference, Inches		Sq. Ft. of Surface per Lineal Foot		Contents of Pipe per Lineal Foot		Lineal Feet to Contain		
	Out-side	In-side	Out-side	In-side	Cu. Ft.	Gal.	1 Cu. Ft.	1 Gal.	1 Lb. of Water
⅛	1.27	0.675	0.106	0.056	0.00033	0.0019	3070	527	101.01
¼	1.69	0.943	0.141	0.079	0.00052	0.0037	1920	271	32.26
⅜	2.12	1.328	0.177	0.111	0.00098	0.0073	1370	137	16.39
½	2.65	1.715	0.221	0.143	0.00162	0.0122	616	82	9.80
¾	3.29	2.330	0.275	0.194	0.00300	0.0255	334	39.2	4.69
1	4.13	3.010	0.344	0.251	0.00500	0.0374	200	26.8	3.21
1¼	5.21	4.010	0.435	0.334	0.00880	0.0666	114	15.0	1.80
1½	5.96	4.720	0.497	0.393	0.01230	0.0918	81.50	10.90	1.31
2	7.46	6.090	0.622	0.507	0.02060	0.1535	49.80	6.52	0.78
2½	9.03	7.320	0.753	0.610	0.02940	0.220	34.00	4.55	0.55
3	10.96	9.120	0.916	0.760	0.0460	0.344	21.70	2.91	0.35
3½	12.56	10.580	1.047	0.882	0.0617	0.458	16.25	2.18	0.27
4	14.13	12.020	1.178	1.002	0.0800	0.597	12.50	1.675	0.20
5	17.47	15.150	1.456	1.262	0.1260	0.947	7.95	1.055	0.13
6	20.81	18.100	1.734	1.510	0.1820	1.355	5.50	0.738	0.09
8	27.09	24.000	2.258	2.000	0.3180	2.380	3.14	0.420	0.05
10	33.77	30.050	2.814	2.503	0.5560	4.165	1.80	0.241	0.03
12	40.05	35.720	3.370	2.975	0.7060	5.280	1.42	0.189	0.02
14	47.12	39.270	3.930	3.271	0.8520	6.380	1.18	0.157	0.019
16	53.41	44.970	4.440	3.746	1.1170	8.360	0.895	0.119	0.014
18	56.55	50.660	4.712	4.220	1.4180	10.610	0.705	0.094	0.011
20	62.83	56.350	5.236	4.694	1.7550	13.130	0.570	0.076	0.009

DIMENSIONAL AND CAPACITY DATA — TYPE K COPPER TUBE

Diameter, Inches			Wall Thickness, In.	Cross-Sectional Area, Sq. In.			Weight per Foot, Lb.		
Nominal	Actual Inside	Actual Outside		Outside	Inside	Metal	Of Tube Alone	Of Water in Tube	Of Tube and Water
¼	0.311	0.375	0.035	0.110	0.076	0.034	0.145	0.033	0.167
⅜	0.402	0.500	0.049	0.196	0.127	0.069	0.269	0.055	0.324
½	0.527	0.625	0.049	0.307	0.218	0.089	0.344	0.094	0.438
⅝	0.652	0.750	0.049	0.442	0.334	0.108	0.418	0.145	0.563
¾	0.745	0.875	0.065	0.601	0.436	0.165	0.641	0.189	0.830
1	0.995	1.125	0.065	0.993	0.777	0.216	0.839	0.338	1.177
1¼	1.245	1.375	0.065	1.484	1.217	0.267	1.04	0.53	1.57
1½	1.481	1.625	0.072	2.072	1.722	0.350	1.36	1.22	2.58
2	1.959	2.125	0.083	3.546	3.013	0.533	2.06	1.31	3.37
2½	2.435	2.625	0.095	5.409	4.654	0.755	2.93	2.02	4.95
3	2.907	3.125	0.109	7.669	6.634	1.035	4.00	2.88	6.88
3½	3.385	3.625	0.120	10.321	8.999	1.322	5.12	3.91	9.03
4	3.857	4.125	0.134	13.361	11.682	1.679	6.51	5.07	11.58
5	4.805	5.125	0.160	20.626	18.126	2.500	9.67	7.87	17.54
6	5.741	6.125	0.192	29.453	25.874	3.579	13.9	11.2	25.1
8	7.583	8.125	0.271	51.826	45.138	6.888	25.9	19.6	45.5
10	9.449	10.125	0.338	80.463	70.085	10.378	40.3	30.4	70.7
12	11.315	12.125	0.405	115.395	100.480	14.915	57.8	43.6	101.4

Nominal Dia., In.	Circumference, Inches		Sq. Ft. of Surface per Lineal Foot		Contents of Tube per Lineal Foot		Lineal Feet to Contain		
	Outside	Inside	Outside	Inside	Cu. Ft.	Gal.	1 Cu. Ft.	1 Gal.	1 Lb. of Water
¼	1.178	0.977	0.098	0.081	.00052	.00389	1923	257	30.8
⅜	1.570	1.262	0.131	0.105	.00088	.00658	1136	152	18.2
½	1.963	1.655	0.164	0.138	.00151	.01129	662	88.6	10.6
⅝	2.355	2.047	0.196	0.171	.00232	.01735	431	57.6	6.90
¾	2.748	2.339	0.229	0.195	.00303	.02664	330	37.5	5.28
1	3.533	3.124	0.294	0.260	.00540	.04039	185	24.8	2.96
1¼	4.318	3.909	0.360	0.326	.00845	.06321	118	15.8	1.89
1½	5.103	4.650	0.425	0.388	.01958	.14646	51.1	6.83	0.817
2	6.673	6.151	0.556	0.513	.02092	.15648	47.8	6.39	0.765
2½	8.243	7.646	0.688	0.637	.03232	.24175	30.9	4.14	0.495
3	9.813	9.128	0.818	0.761	.04607	.34460	21.7	2.90	0.347
3½	11.388	10.634	0.949	0.886	.06249	.46745	15.8	2.14	0.257
4	12.953	12.111	1.080	1.009	.08113	.60682	12.3	1.65	0.197
5	16.093	15.088	1.341	1.257	.12587	.94151	7.94	1.06	0.127
6	19.233	18.027	1.603	1.502	.17968	1.3440	5.56	0.744	0.089
8	25.513	23.811	2.126	1.984	.31345	2.3446	3.19	0.426	0.051
10	31.793	29.670	2.649	2.473	.48670	3.4405	2.05	0.291	0.033
12	38.073	35.529	3.173	2.961	.69778	5.2194	1.43	0.192	0.023

DIMENSIONAL AND CAPACITY DATA — TYPE L COPPER TUBE

Diameter, Inches			Wall Thick-ness, In.	Cross-Sectional Area, Sq. In.			Weight per Foot, Lb.		
Nom-inal	Actual Inside	Actual Outside		Out-side	In-side	Metal	Of Tube Alone	Of Water in Tube	Of Tube and Water
¼	0.315	0.375	0.030	0.110	0.078	0.032	0.126	0.034	0.160
⅜	0.430	0.500	0.035	0.196	0.145	0.051	0.198	0.063	0.261
½	0.545	0.625	0.040	0.307	0.233	0.074	0.285	0.101	0.386
⅝	0.666	0.750	0.042	0.442	0.348	0.094	0.362	0.151	0.513
¾	0.785	0.875	0.045	0.601	0.484	0.117	0.455	0.210	0.665
1	1.025	1.125	0.050	0.993	0.825	0.168	0.655	0.358	1.013
1¼	1.265	1.375	0.055	1.484	1.256	0.228	0.884	0.545	1.429
1½	1.505	1.625	0.060	2.072	1.778	0.294	1.14	0.77	1.91
2	1.985	2.125	0.070	3.546	3.093	0.453	1.75	1.34	3.09
2½	2.465	2.625	0.080	5.409	4.770	0.639	2.48	2.07	4.55
3	2.945	3.125	0.090	7.669	6.808	0.861	3.33	2.96	6.29
3½	3.425	3.625	0.100	10.321	9.214	1.107	4.29	4.00	8.29
4	3.905	4.125	0.110	13.361	11.971	1.390	5.38	5.20	10.58
5	4.875	5.125	0.125	20.626	18.659	1.967	7.61	8.10	15.71
6	5.845	6.125	0.140	29.453	26.817	2.636	10.2	11.6	21.8
8	7.725	8.125	0.200	51.826	46.849	4.977	19.3	20.3	39.6
10	9.625	10.125	0.250	80.463	72.722	7.741	30.1	31.6	61.7
12	11.565	12.125	0.280	115.395	104.994	10.401	40.4	45.6	86.0

Nom-inal Dia., In.	Circumference, Inches		Sq. Ft. of Surface per Lineal Foot		Contents of Tube per Lineal Foot		Lineal Feet to Contain		
	Out-side	In-side	Out-side	In-side	Cu. Ft.	Gal.	1 Cu. Ft.	1 Gal.	1 Lb. of Water
¼	1.178	0.989	0.098	0.082	.00054	.0040	1852	250	29.6
⅜	1.570	1.350	0.131	0.113	.00100	.0075	1000	133	16.0
½	1.963	1.711	0.164	0.143	.00162	.0121	617.3	82.6	9.87
⅝	2.355	2.091	0.196	0.174	.00242	.0181	413.2	55.2	6.61
¾	2.748	2.465	0.229	0.205	.00336	.0251	297.6	40.5	4.76
1	3.533	3.219	0.294	0.268	.00573	.0429	174.5	23.3	2.79
1¼	4.318	3.972	0.360	0.331	.00872	.0652	114.7	15.3	1.83
1½	5.103	4.726	0.425	0.394	.01237	.0925	80.84	10.8	1.29
2	6.673	6.233	0.556	0.519	.02147	.1606	46.58	6.23	0.745
2½	8.243	7.740	0.688	0.645	.03312	.2478	30.19	4.04	0.483
3	9.813	9.247	0.818	0.771	.04728	.3537	21.15	2.83	0.338
3½	11.388	10.760	0.949	0.897	.06398	.4786	15.63	2.09	0.251
4	12.953	12.262	1.080	1.022	.08313	.6218	12.03	1.61	0.192
5	16.093	15.308	1.341	1.276	.12958	.9693	7.220	1.03	0.123
6	19.233	18.353	1.603	1.529	.18622	1.393	5.371	0.718	0.0592
8	25.513	24.465	2.126	2.039	.32534	2.434	3.074	0.411	0.0492
10	31.793	30.223	2.649	2.519	.50501	3.777	1.980	0.265	0.0317
12	38.073	36.314	3.173	3.026	.72912	5.454	1.372	0.183	0.0219

DIMENSIONAL AND CAPACITY DATA — TYPE M COPPER TUBE

Diameter, Inches			Wall Thick-ness, In.	Cross-Sectional Area, Sq. In.			Weight per Foot, Lb.		
Nom-inal	Actual Inside	Actual Outside		Out-side	In-side	Metal	Of Tube Alone	Of Water in Tube	Of Tube and Water
3/8	0.450	0.500	0.025	0.196	0.159	0.037	0.145	0.069	0.214
1/2	0.569	0.625	0.028	0.307	0.254	0.053	0.204	0.110	0.314
3/4	0.811	0.875	0.032	0.601	0.516	0.085	0.328	0.224	0.552
1	1.055	1.125	0.035	0.993	0.874	0.119	0.465	0.379	0.844
1 1/4	1.291	1.375	0.042	1.48	1.31	0.17	0.682	0.569	1.251
1 1/2	1.527	1.625	0.049	2.07	1.83	0.24	0.94	0.83	1.77
2	2.009	2.125	0.058	3.55	3.17	0.38	1.46	1.35	2.81
2 1/2	2.495	2.625	0.065	5.41	4.89	0.52	2.03	2.12	4.15
3	2.981	3.125	0.072	7.67	6.98	0.69	2.68	3.03	5.71
3 1/2	3.459	3.625	0.083	10.32	9.40	0.924	3.58	4.08	7.66
4	3.935	4.125	0.095	13.36	12.15	1.21	4.66	5.23	9.89
5	4.907	5.125	0.109	20.63	18.90	1.73	6.66	8.20	14.86
6	5.881	6.125	0.122	29.45	25.15	2.30	8.92	11.78	20.70
8	7.785	8.125	0.170	51.83	47.58	4.25	16.5	20.7	37.2
10	9.701	10.125	0.212	80.46	73.88	6.58	25.6	32.1	57.7
12	11.617	12.125	0.254	115.47	105.99	9.48	36.7	46.0	82.7

Nom-inal Dia., In.	Circumference, Inches		Sq. Ft. of Surface per Lineal Foot		Contents of Tube per Lineal Foot		Lineal Feet to Contain		
	Out-side	In-side	Out-side	In-side	Cu. Ft.	Gal.	1 Cu. Ft.	1 Gal.	1 Lb. of Water
3/8	1.570	1.413	0.131	0.118	.00110	.00823	909	122	14.5
1/2	1.963	1.787	0.164	0.149	.00176	.01316	568	76.0	9.09
3/4	2.748	2.547	0.229	0.212	.00358	.02678	379	37.3	4.47
1	3.533	3.313	0.294	0.276	.00607	.04540	164.7	22.0	2.64
1 1/4	4.318	4.054	0.360	0.338	.00910	.06807	109.9	14.7	1.76
1 1/2	5.103	4.795	0.425	0.400	.01333	.09971	75.02	10.0	1.20
2	6.673	6.308	0.556	0.526	.02201	.16463	45.43	6.08	0.727
2 1/2	8.243	7.834	0.688	0.653	.03396	.25402	29.45	3.94	0.471
3	9.813	9.360	0.818	0.780	.04847	.36256	20.63	2.76	0.330
3 1/2	11.388	10.867	0.949	0.906	.06525	.48813	15.33	2.05	0.246
4	12.953	12.356	1.080	1.030	.08368	.62593	11.95	1.60	0.191
5	16.093	15.408	1.341	1.284	.13125	.98175	7.62	1.02	0.122
6	19.233	18.466	1.603	1.539	.18854	1.410	5.30	0.709	0.849
8	25.513	24.445	2.126	2.037	.33044	2.472	3.03	0.405	0.484
10	31.793	30.461	2.649	2.538	.51306	3.838	1.91	0.261	0.312
12	38.073	36.477	3.173	3.039	.73569	5.503	1.36	0.182	0.217

DIMENSIONAL AND CAPACITY DATA — COPPER DRAINAGE TUBE

Diameter, Inches			Wall Thick-ness, In.	Cross-Sectional Area, Sq. In.			Weight per Foot, Lb.		
Nom-inal	Actual Inside	Actual Outside		Out-side	In-side	Metal	Of Tube Alone	Of Water in Tube	Of Tube and Water
1¼	1.295	1.375	0.040	1.484	1.317	0.167	0.650	0.572	1.222
1½	1.541	1.625	0.042	2.072	1.865	0.207	0.809	0.809	1.618
2	2.041	2.125	0.042	3.546	3.272	0.274	1.07	1.42	2.49
3	3.035	3.125	0.045	7.669	7.234	0.435	1.69	3.14	4.83
4	4.009	4.125	0.058	13.361	12.623	0.738	2.87	5.48	8.35
5	4.981	5.125	0.072	20.626	19.486	1.140	4.43	8.46	12.89
6	5.959	6.125	0.083	29.453	27.890	1.563	6.10	12.10	18.20

Nom-inal Dia., In.	Circumference, Inches		Sq. Ft. of Surface per Lineal Foot		Contents of Tube per Lineal Foot		Lineal Feet to Contain		
	Out-side	In-side	Out-side	In-side	Cu. Ft.	Gal.	1 Cu. Ft.	1 Gal.	1 Lb. of Water
1¼	4.318	4.068	0.360	0.339	.00914	0.0684	109.4	14.6	1.75
1½	5.103	4.841	0.425	0.403	.01294	0.0968	77.3	10.3	1.24
2	6.673	6.412	0.556	0.534	.02271	0.1699	44.0	5.89	0.707
3	9.813	9.535	0.818	0.795	.05020	0.3755	19.92	2.66	0.319
4	12.953	12.595	1.080	1.050	.08760	0.6552	11.42	1.53	0.184
5	16.093	15.648	1.341	1.304	.13523	1.1520	7.39	0.868	0.104
6	19.233	18.721	1.603	1.560	.19356	1.4478	5.17	0.690	0.083

Copper drainage tube has been developed specifically for sanitary drainage service. It can be installed by a simple soldering operation in which the solder is drawn into the joint by capillary action. A complete selection of drainage fittings, featuring long radius elbows, are available for soldered connection to copper tube. These fittings also include fittings with one or more outlets threaded for use with brass or copper pipe or with slip joints. Bends, both adjustable and fixed, with suitable flanges for closet connections, are also available.

Soldered drainage fittings are installed by the same technic that most plumbers have already developed in connection with water lines.

In connection with residences, the actual selection of sizes in most instances is controlled by code requirements, but a 3-inch copper tube stack with not over three branch intervals will accommodate 30 fixture units (not over two water closets). Where not fixed by code, a sanitary engineer or other qualified designer would probably specify a 3-inch stack, which will easily fit between 2 x 4 studs. Copper drainage tube may be used equally well in multi-story buildings. Vent headers and waste lines should be long enough to take up expansion in the stacks.

Type K copper water tube is recommended for use with soldered or compression fittings, for general plumbing involving relatively severe service conditions, and for heating, gas, steam, oil lines and underground service.

Type L copper water tube can also be used with soldered or compression fittings, and is intended for general plumbing work involving relatively less severe conditions than with Type K, the wall thickness being thinner than Type K.

Type M, with thinner walls than either Type K or Type L, is recommended for use with soldered fittings only, for interior application, for heating and for waste, vent, soil, and other non-pressure applications.

DIMENSIONAL DATA—THREADLESS COPPER PIPE

(Of hard-drawn copper; also designated as TP)

Diameter, Inches			Wall Thick-ness. In.	Area of Bore, Sq. In.	Sq. Ft. of Surface per Lineal Foot		Weight per Foot, Lb.	Allowable Internal Pressure, psi		
Nom-inal	Actual Outside	Actual Inside			Out-side	In-side		At 100°	At 200°	At 300°
¼	0.540	0.410	0.065	0.132	0.141	0.107	0.376	1500	1380	1190
⅜	0.675	0.545	0.065	0.233	0.177	0.143	0.483	1170	1070	930
½	0.840	0.710	0.065	0.396	0.220	0.186	0.613	920	850	730
¾	1.050	0.920	0.065	0.665	0.275	0.241	0.780	730	670	580
1	1.315	1.185	0.065	1.10	0.344	0.310	0.989	580	530	460
1¼	1.660	1.530	0.065	1.84	0.435	0.401	1.26	450	420	360
1½	1.900	1.770	0.065	2.46	0.497	0.464	1.45	400	360	310
2	2.375	2.245	0.065	3.96	0.622	0.588	1.83	300	280	240
2½	2.875	2.745	0.065	5.92	0.753	0.719	2.22	250	230	200
3	3.500	3.334	0.083	8.73	0.916	0.874	3.45	260	240	210
3½	4.000	3.810	0.095	11.4	1.05	0.998	4.52	270	250	210
4	4.500	4.286	0.107	14.4	1.18	1.12	5.72	270	240	210
5	5.562	5.298	0.132	22.0	1.46	1.39	8.73	270	250	210
6	6.625	6.309	0.158	31.3	1.73	1.65	12.4	270	250	220
8	8.625	8.215	0.205	53.0	2.26	2.15	21.0	270	250	220
10	10.750	10.238	0.256	82.3	2.81	2.68	32.7	270	250	220
12	12.750	12.124	0.313	115.4	3.34	3.18	47.4	280	260	220

All weights in these tables are calculated values.

Values in these tables for Allowable Internal Pressure are based on the formula

$$P = \frac{2St_m}{D - 0.8t_m}$$

where P = maximum rated internal working pressure, psi

t_m = pipe wall thickness, inches

D = outside diameter of pipe, inches

S = allowable stress in material due to internal pressure, psi

= 6000 at 100°; 5500 at 200°; 4750 at 300°; 3000 at 400°

DIMENSIONAL DATA—REFRIGERATION FIELD SERVICE COPPER TUBE

Diameter, Inches			Wall Thick-ness, In.	Area of Bore, Sq. In.	Sq. Ft. of Surface per Lineal Foot		Weight per Foot, Lb.	Allowable Internal Pressure, psi		
Nom-inal	Actual Outside	Actual Inside			Out-side	In-side		At 100°	At 200°	At 300°
⅛	0.125	0.065	0.030	.0033	.0327	.0170	.0347	3130	2870	2480
³⁄₁₆	0.188	0.128	0.030	.0129	.0492	.0335	.0577	1990	1820	1570
¼	0.250	0.190	0.030	.0284	.0655	.0497	.0804	1450	1330	1150
⁵⁄₁₆	0.312	0.248	0.032	.0483	.0817	.0649	.109	1230	1120	970
⅜	0.375	0.311	0.032	.0760	.0982	.0814	.134	1010	920	800
½	0.500	0.436	0.032	.149	.131	.114	.182	740	680	590
⅝	0.625	0.555	0.035	.242	.164	.145	.251	640	590	510
¾	0.750	0.680	0.035	.363	.196	.178	.305	520	480	410

DIMENSIONAL AND CAPACITY DATA — STANDARD COPPER PIPE

(Not to be confused with copper tube)

Diameter, Inches			Wall Thick-ness, In.	Cross-Sectional Area, Sq. In.			Weight per Foot, Lb.		
Nom-inal	Actual Inside	Actual Outside		Out-side	In-side	Metal	Of Pipe Alone	Of Water in Pipe	Of Pipe and Water
1/8	0.281	0.405	0.062	0.129	0.062	0.067	0.259	0.0269	0.286
1/4	0.376	0.540	0.082	0.229	0.111	0.118	0.457	0.0481	0.505
3/8	0.495	0.675	0.090	0.358	0.192	0.166	0.641	0.0831	0.724
1/2	0.626	0.840	0.107	0.554	0.308	0.246	0.955	0.1338	1.09
3/4	0.822	1.050	0.114	0.866	0.531	0.335	1.30	0.2306	1.53
1	1.063	1.315	0.126	1.358	0.888	0.470	1.82	0.3856	2.21
1 1/4	1.368	1.660	0.146	2.165	1.469	0.696	2.69	0.6375	3.33
1 1/2	1.600	1.900	0.150	2.835	2.011	0.824	3.20	0.8731	4.07
2	2.063	2.375	0.156	4.430	3.343	1.087	4.22	1.451	5.67
2 1/2	2.501	2.875	0.187	6.492	4.913	1.579	6.12	2.133	8.25
3	3.062	3.500	0.219	9.621	7.364	2.257	8.75	3.196	11.95
3 1/2	3.500	4.000	0.250	12.566	9.621	2.945	11.4	4.176	15.6
4	4.000	4.500	0.250	15.904	12.566	3.338	12.9	5.454	18.4
5	5.062	5.562	0.250	24.300	20.122	4.178	16.2	8.734	24.9
6	6.125	6.625	0.250	34.471	29.468	5.003	19.4	12.79	32.2
8	8.001	8.625	0.312	58.426	50.281	8.145	31.6	21.82	53.4
10	10.020	10.750	0.365	90.761	78.854	11.907	46.2	34.23	80.4
12	12.000	12.750	0.375	127.675	113.098	14.577	56.5	49.09	105.6

Nom-inal Dia., In.	Circumference, Inches		Sq. Ft. of Surface per Lineal Foot		Contents of Pipe per Lineal Foot		Lineal Feet to Contain		
	Out-side	In-side	Out-side	In-side	Cu. Ft.	Gal.	1 Cu. Ft.	1 Gal.	1 Lb. of Water
1/8	1.272	0.883	0.106	0.074	.00043	.00322	2325.6	310.6	37.17
1/4	1.696	1.181	0.141	0.098	.00077	.00576	1298.7	173.6	20.79
3/8	2.121	1.555	0.177	0.130	.00133	.00995	751.9	100.5	12.03
1/2	2.639	1.967	0.220	0.164	.00214	.01601	467.3	62.5	7.47
3/4	3.299	2.582	0.275	0.215	.00369	.02760	271.0	36.2	4.34
1	0.131	3.340	0.344	0.278	.00617	.04615	162.1	21.7	2.59
1 1/4	5.215	4.298	0.435	0.358	.01020	.07630	98.0	13.1	1.57
1 1/2	5.969	5.027	0.497	0.419	.01397	.10450	71.6	9.57	1.15
2	7.461	6.481	0.622	0.540	.02322	.17369	43.1	5.76	0.689
2 1/2	9.032	7.857	0.753	0.655	.03412	.25522	29.3	3.92	0.469
3	10.996	9.620	0.916	0.802	.05114	.38253	19.6	2.61	0.313
3 1/2	12.566	10.966	1.047	0.916	.06681	.49974	15.0	2.00	0.239
4	14.137	12.566	1.178	1.047	.08726	.65270	11.5	1.53	0.183
5	17.474	15.903	1.456	1.325	.13974	1.0453	7.2	0.96	0.114
6	20.813	19.242	1.734	1.603	.20464	1.5307	4.9	0.65	0.078
8	27.096	25.136	2.258	2.095	.34917	2.6118	2.9	0.38	0.046
10	33.772	31.479	2.814	2.623	.54760	4.0960	1.8	0.24	0.029
12	40.055	37.699	3.338	3.142	.78540	5.8748	1.3	0.17	0.020

DIMENSIONAL AND CAPACITY DATA — EXTRA STRONG COPPER PIPE

(Not to be confused with copper tube)

Diameter, Inches			Wall Thick-ness, In.	Cross-Sectional Area, Sq. In.			Weight per Foot, Lb.		
Nom-inal	Actual Inside	Actual Outside		Out-side	In-side	Metal	Of Pipe Alone	Of Water Pipe and	Of in Pipe Water
⅛	0.205	0.405	0.100	0.129	0.033	0.096	0.371	0.0144	0.385
¼	0.294	0.540	0.123	0.229	0.067	0.162	0.625	0.0294	0.654
⅜	0.421	0.675	0.127	0.358	0.139	0.219	0.847	0.0606	0.908
½	0.542	0.840	0.149	0.554	0.231	0.323	1.25	0.1000	1.35
¾	0.736	1.050	0.157	0.866	0.425	0.441	1.71	0.1844	1.89
1	0.951	1.315	0.182	1.358	0.710	0.648	2.51	0.3081	2.82
1¼	1.272	1.660	0.194	2.165	1.271	0.894	3.46	0.5519	4.01
1½	1.494	1.900	0.203	2.835	1.753	1.082	4.19	0.7606	4.95
2	1.933	2.375	0.221	4.430	2.934	1.496	5.80	1.273	7.07
2½	2.315	2.875	0.280	6.492	4.209	2.283	.85	1.827	10.7
3	2.892	3.500	0.304	9.621	6.569	3.052	11.8	2.851	14.7
3½	3.358	4.000	0.321	12.566	8.859	3.707	14.4	3.845	18.2
4	3.818	4.500	0.341	15.904	11.451	4.453	17.3	4.970	22.3
5	4.792	5.562	0.375	24.300	18.033	6.267	23.7	7.827	31.5
6	5.751	6.625	0.437	34.471	25.973	8.498	32.9	11.27	44.2
8	7.625	8.625	0.500	58.426	45.663	12.763	49.5	19.82	69.3
10	9.750	10.750	0.500	90.761	74.660	16.101	62.4	32.40	94.8

Nom-inal Dia., In.	Circumference, Inches		Sq. Ft. of Surface per Lineal Foot		Contents of Pipe per Lineal Foot		Lineal Feet to Contain		
	Out-side	In-side	Out-side	In-side	Cu. Ft.	Gal.	1 Cu. Ft.	1 Gal.	1 Lb. of Water
⅛	1.272	0.644	0.106	0.054	.00097	.00726	1030.9	137.7	16.50
¼	1.696	0.924	0.141	0.077	.00023	.00172	4347.8	581.4	69.44
⅜	2.121	1.323	0.177	0.110	.00047	.00352	2127.7	284.1	34.01
½	2.639	1.703	0.220	0.142	.00160	.01197	625.0	83.5	10.00
¾	3.299	2.312	0.275	0.193	.00295	.02207	393.0	45.3	5.42
1	4.131	2.988	0.344	0.249	.00493	.03688	202.8	27.1	3.25
1¼	5.215	3.996	0.435	0.333	.00883	.06605	113.3	15.1	1.81
1½	5.969	4.694	0.497	0.391	.01217	.09103	82.2	11.0	1.31
2	7.461	6.073	0.622	0.506	.02037	.15237	49.1	6.56	0.785
2½	9.032	7.273	0.753	0.606	.02923	.21864	34.2	4.57	0.547
3	10.996	9.086	0.916	0.757	.04562	.34124	21.9	2.93	0.351
3½	12.566	10.549	1.047	0.879	.06152	.46017	16.3	2.17	0.260
4	14.137	11.995	1.178	1.000	.07952	.59481	12.6	1.68	0.201
5	17.474	15.055	1.456	1.255	.12523	.93672	8.0	1.07	0.128
6	20.813	18.067	1.734	1.506	.18037	1.3492	5.5	0.74	0.089
8	27.096	23.955	2.258	1.996	.31710	2.3719	3.2	0.42	0.050
10	33.772	30.631	2.814	2.553	.51847	3.8782	1.9	0.26	0.031

LEAD PIPE†

Water Service Pipe							
Inside Diameter. Inches	Class of Pipe[1]	Commercial Designation[2]		Wall Thickness. Inches	Minimum Outside Circumference, Inches	Weight Per Foot	
		East	West			Pounds	Ounces
3/8	50	A	S	0.143	2- 1/16	1	4
	75	AA	XS	0.175	2- 1/8	1	8
	100	AAA	XXS	0.256	2- 5/8	2	8
1/2	50	A	S	0.149	2- 3/8	1	8
	75	AA	XS	0.188	2- 5/8	2	0
	100	AAA	XXS	0.256	3- 1/16	3	0
5/8	50	A	S	0.197	3- 1/16	2	8
	75	AA	XS	0.228	3- 1/4	3	0
	100	AAA	XXS	0.256	3- 7/16	3	8
3/4	50	A	S	0.203	3- 1/2	3	0
	75	AA	XS	0.231	3-11/16	3	8
	100	AAA	XXS	0.293	4- 1/16	4	12
1	50	A	S	0.214	4- 5/16	4	0
	75	AA	XS	0.246	4- 9/16	4	12
	100	AAA	XXS	0.298	4- 7/8	6	0
1-1/4	50	A	S	0.210	5- 1/8	4	12
	75	AA	XS	0.258	5- 3/8	6	0
	100	AAA	XXS	0.320	5-13/16	7	12
1-1/2	50	A	S	0.242	6- 1/16	6	8
	75	AA	XS	0.288	6- 3/8	8	0
	100	AAA	XXS	0.386	7	11	4
2	50	A	S	0.252	7- 3/4	8	12
	75	AA	XS	0.376	8- 1/2	13	12
	100	AAA	XXS	0.504	9- 5/16	19	8
Soil and Waste Pipe							
1-1/4	—	D	XL	0.118	4- 1/2	2	8
		C	L	0.139	4-11/16	3	0
		B	M	0.171	4- 7/8	3	12
1-1/2	—	D	XL	0.138	5- 7/16	3	8
		C	L	0.165	5- 5/8	4	4
		B	M	0.191	5- 3/4	5	0
2	—	D	XL	0.142	7- 1/16	4	12
		C	L	0.177	7- 1/4	6	0
		B	M	0.205	7- 7/16	7	0
2-1/2	—	D	XL	0.125	8- 1/2	5	0
		B	M	0.250	9- 5/16	10	10
3	—	D	XL	0.125	10- 1/16	6	0
		B	M	0.250	10- 7/8	12	8
4	—	D	XL	0.125	13- 3/16	8	0
		B	M	0.250	14	16	6
5	—	D	XL	0.125	16- 3/8	10	0
		B	M	0.250	17- 1/8	20	4
6	—	D	XL	0.125	19- 1/2	11	12
		B	M	0.250	20- 1/4	24	2

[1] Class of pipe also indicates maximum working pressure in pounds per square inch.
[2] Designations ordinarily used east and west, respectively, of the Illinois-Indiana State line.
† Standard of the Lead Industries Association. Reproduced by permission.

LEAD TRAPS AND BENDS†

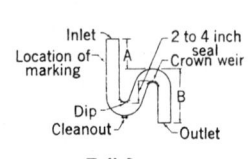

Full S trap

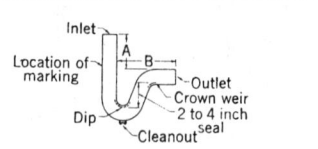

Half S or P trap

Full S Traps / Half S or P Traps / Short and Long Traps

Nominal Inside Diameter, Inches	Weight per Running Foot, Pounds	Short Traps Dimensions, Inches A Inlet	B Outlet	Total Weight Pounds and Ounces		Long Traps Dimensions, Inches A Inlet	B Outlet	Total Weight Pounds and Ounces	
Full S Traps									
1¼	2⅝	4½	7	4	9	4½	19½	7	6
1½	3⅛	4½	7	5	10	4½	19½	8	14
2	4⅛	4½	8	8	5	4½	19½	12	4
Half S or P Traps									
1¼	2⅝	4½	7	4	0	4½	14	5	9
1½	3⅛	4½	7	4	15	4½	14	6	12
2	4⅛	4½	8	7	2	4½	14	9	3

Drum Traps

Trap Sizes, Inches	Thickness of Cap and Flange, Inches	Length of Thread and Ring, Inches	Screw Cap and Ring Thread
4 × 8 4 × 9 4 × 10	5/32	9/16	3-inch straight pipe thread, 8 threads per inch, free-fit (nominal pitch diameter, 3.388 inch).
3 × 8 3 × 9 3 × 10	5/32	7/16	2-inch straight pipe thread, 11½ threads per inch, free-fit (nominal pitch diameter, 2.296 inch).

Bends

1½-inch Nominal Inside Diameter Dimensions, Inches	Total Weight, Pounds and Ounces		3-inch Nominal Inside Diameter Dimensions, Inches	Total Weight, Pounds and Ounces		4-inch Nominal Inside Diameter Dimensions, Inches	Total Weight, Pounds and Ounces	
4 × 7	2	12	5½ × 10	7	3	5½ × 10	9	5
4 × 12	4	1	5½ × 12	8	3	5½ × 12	10	10
4 × 15	4	13	5½ × 15	9	12	5½ × 15	12	10
4 × 18	5	10	5½ × 18	11	4	5½ × 18	14	10
4 × 20	6	2	5½ × 20	12	4	5½ × 20	16	0
7 × 7	3	8	10 × 10	9	8	10 × 10	12	5
7 × 12	4	13	10 × 12	10	8	10 × 12	13	10
7 × 15	5	10	10 × 15	12	0	10 × 15	15	10
7 × 18	6	6	10 × 18	13	8	10 × 18	17	10
7 × 20	6	15	10 × 20	14	8	10 × 20	18	15

Reducing bends shall be considered as meeting this standard only if the reduction in diameter occurs exclusively on the top and sides, leaving the bottom of the bend, when in the installed position, smooth and straight so that it can not form a trap for substances passing through the bend, and if they meet the requirements of this standard in all other respects except total weight.
† Standard of the Lead Industries Association. Reproduced by permission.

PLASTICS PIPE *

The following information covers the characteristics and applications of thermoplastics piping. The word thermoplastics differentiates this group of plastic piping from another composed of different plastics materials which are called *thermosets*. While thermoplastic materials can be softened and hardened repeatedly by heating and cooling them, thermosets cannot be. Thermoset piping materials are not discussed below since they are normally restricted to specialized industrial applications.

Plastics piping technology is still relatively new, hence its technical data tend to become obsolete within a few years. Continued awareness of the latest product information is, therefore, a must for the serious practitioner.

The principal advantages of plastics piping are its light weight, its quick-joining techniques, its high resistance to corrosion, its chemical inertness, and generally lower friction losses.

The following are the major, or most common, thermoplastics piping materials: Acrylonitrile-Butadiene-Styrene (ABS); Polyethylene (PE); Polybutylene (PB); Polypropylene (PP); Polyvinyl Chloride (PVC); and Chlorinated Polyvinyl Chloride (CPVC). Although these plastics piping materials have overlapping characteristics, each does have certain properties and capabilities which are highly suitable for specific applications. Each is also readily available commercially.

Thermoplastic piping can be divided into two categories, *rigid* and *flexible*. A brief description of the more common applications of the major thermoplastics piping materials is as follows:

Acrylonitrile-Butadiene-Styrene (ABS) pipe is a rigid type and is found most commonly in drain, waste, and vent (DWV) systems, building drains, building sewers, well casings, electrical and communications conduit and rainwater conductors. Although ABS can be used as pressure piping, it is generally used in non-pressure systems because of its lower hydrostatic design stress rating.

Polyethylene (PE) is a member of the Polyolefin family of plastics. PE is flexible and is usually found underground in such applications as natural gas distribution lines, gas service, water service,

and turf and agricultural irrigation systems. It also serves as a re-liner for deteriorated gas or sewer mains.

Polybutylene (PB) is a flexible material resembling a high-density Polyethylene but has advantages in pressure and heat resistance performance. In addition to those applications common to PE, PB piping can be used for hot and cold water distribution.

Polypropylene (PP) is another member of the Polyolefin family; it differs from PE and PB in that it is more rigid and may be used at slightly higher temperatures than PE. Also, it is common to chemical waste systems, because of its high degree of chemical resistance.

Polyvinyl Chloride (PVC) is the most widely used thermoplastic pipe. It is a rigid type that is used for DWV systems, water service and water mains, electrical and communications conduit, turf and agricultural irrigation, well casings, building sewers and sewer mains, and chemical process piping.

Chlorinated Polyvinyl Chloride (CPVC) is a variation of PVC sometimes known as Type IV PVC. It is designed to withstand higher temperatures, and therefore has application in both hot and cold water distribution, as well as in hot chemical systems.

Joining Techniques

Common joining techniques for thermoplastics piping include solvent cementing, elastomeric sealing, mechanical coupling, heat fusion, flaring, insert fitting, flanging, and threading.

Solvent cementing is used on ABS, PVC, and CPVC pipe, where it requires socket-type fittings or bell ends. However, PE, PB, and PP pipe, because of their high resistance to chemicals, cannot be joined by this method. The solvent cement is applied to both pipe and fitting, momentarily dissolving the surfaces of each. The pipe and fitting are then joined while the cement is still wet. Once the solvent evaporates (usually in only 2 to 3 minutes), the bond sets permanently. However, it must be remembered that specific cements are designed for use only with their appropriate plastics materials.

---

* Tabular and Graphical Data provided in this Plastics Pipe segment are excerpted from "Plastics Piping Manual," K. C. Ford and F. W. Reinhart; 1976, by permission of the Plastics Piping Institute, N.Y.

Elastomeric seals have applications in joining PVC piping in diameters of 6 inches and above. In these larger pipe sizes solvent cementing becomes difficult. Such piping is manufactured with integral grooved bells on the end of each length of pipe. A specially designed rubber ring is placed into the groove. The straight end of the next pipe is then pushed into the bell, and the elastomeric seal forms a pressure-tight fitting. These types of joints are commonly used in underground applications, such as water, sewer, and irrigation piping.

Mechanical couplings are available in a number of proprietary designs, and can join plastic piping to itself or to other materials. In some cases, stiffeners are inserted into the pipe with a metal clamp or collar which is placed over the pipe and then tightened. Some mechanical couplings afford pull-out resistance while others are designed only to maintain operating pressure.

Heat fusion is used to join PE, PB, or PP piping. Heat fusion with a hot-air torch was a common practice at one time. However, since fittings of nearly every configuration are now easily obtained, it is a technique that is rarely applied today. Various heating tools still are available for fusing pipe ends together with socket fittings or for plain butt-fusion. In butt-fusion the plain ends of pipe are heated, and are pressed or "butted" together while still hot. Cooling brings about a permanently fused joint.

Flaring involves the use of tools designed specifically for that purpose. Flexible plastics such as PE and PB can be cold-flared for joining with flare nuts. This technique is commonly used for fitting transitions from plastic to copper.

Insert fittings consist of metallic or plastic inserts that are forced into the ends of the pipe. A stainless steel hose clamp is then placed on the outside of the pipe and tightened. This type of joint may also be used as a transition fitting when joining plastic pipe to pipe of some other material.

Flanges are often required in industrial process piping systems at various points where frequent assembly and disassembly is necessary. Flanges with bolt-holes can be solvent-cemented to the plain ends of PVC pipe in the make-up of such joints.

Threading can be accomplished on rigid plastic pipe by using special thread cutters. Such pipe must have wall thicknesses conforming to Schedule 80 or 120, since Schedule 40 plastic pipe lacks sufficient wall thickness to support threads. Also, for plastic pipe a *Teflon* tape or an emulsion is used, rather than standard pipe dope. A strap wrench is used to tighten the fitting, with care taken not to overstress the connection.

Standards for Specification and Identification

Although there are standards for plastics piping published by several agencies, the industry has concentrated most of its standards-writing activities with the American Society for Testing and Materials (ASTM).

Volume 34 of the ASTM standards is a complete collection dealing with thermoplastics piping. These standards specify dimensional data and tolerances, pressure and temperature capabilities, as well as other physical characteristics such as impact resistance, crush resistance, and tensile strength. Familiarity with these standards is of great importance when designing and installing plastics piping systems.

Most ASTM standards require that the pipe be clearly marked, with the name of the manufacturer, the plastic material from which the pipe was extruded, the nominal diameter of the pipe, the ASTM standard to which it was manufactured, and with whatever certifying agencies that are required. Pipe markings must be clearly visible throughout the length of the pipe, and should be of sufficient permanence that they will not rub off during normal handling. Pipe or fittings which are not identified in the manner specified should never be accepted. Further information concerning pipe standards and technical data can be obtained from the Society of the Plastics Industry, 355 Lexington Ave., New York, N.Y. 10017.

Design Parameters

The basic design parameters for plastics piping, such as thermal expansion, chemical resistance, flow characteristics, and pressure ratings are essentially the same as for any other piping material. The only difference lies in the numerical values of these parameters.

Coefficients of thermal expansion for a number of piping materials are given in Table 1. The

TABLE 1
COEFFICIENTS OF LINEAR THERMAL EXPANSION OF PIPING MATERIALS

Material[a]	Nominal Coefficients[b] $\times 10^{-5}$, in/in/°F
ABS	4.0 - 6.0
PE — Type I	10.0
Type II	8.0 - 9.0
Type III	7.0
High Molecular Weight (Type IV)	6.0
PVC Type I	3.0 - 3.5
Type II	3.0 - 5.0
CPVC	3.5
SR	3.3 - 5.0
PB	7.2
PP	4.0 - 4.8
CAB	8.0 - 9.5
Acetal	4.5 - 5.0
Steel	0.65
Cast iron	0.56
Copper	0.98

a. The abbreviations for the plastics are in accordance with ASTM D 883, D1600 and F 412, PPI-TR1-NOV 1968, Plastic Pipe Standards, and common usage, and are as follows:

ABS — Acrylonitrile-butadiene-styrene
PVC Type I — Rigid or unplasticized poly(vinyl chloride); covers PVC 1120 and PVC 1220 compounds.
PVC Type II — Rubber modified poly(vinyl chloride); covers PVC 2110, 2112, 2116 and 2120 compounds.
PB — Polybutylene
PE — Polyethylene
CPVC — Chlorinated poly(vinyl chloride). Formerly designated as PVC 4120.
SR — Styrene-rubber
CAB — Cellubose acetate butynate

b. These values are independent of pipe diameter.

data shows that plastic pipe undergoes far greater dimensional change than metal pipe for a given temperature differential. However, the *force* of expansion generated in plastics is so much less than in metal that potentially destructive stresses are also substantially smaller in magnitude. Nevertheless, where temperature differentials of more than 30°F are anticipated, provision for the

control of thermal movement must be made. Some recommendations follow:

Below ground: Small-diameter piping can be "snaked" in the trench; that is, laid in the trench in a sinuous configuration rather than in a straight line. Large-diameter piping must be provided with expansion loops. Prolonged exposure of plastic pipe to sunlight while awaiting burial in the trench will cause heating and therefore expansion. Thus, before the final connection is made, piping should be allowed to cool to operating temperature. Otherwise, after connection, the potential of failure from joint pull-out will exist.

Above ground: Piping systems with many changes of direction and relatively short runs usually do not require expansion control. The DWV system in a single-family dwelling would be one such system. The DWV system in a multi-story building, however, would require some attention to thermal expansion, specifically in vertical stacks which run in a straight line through the entire height of the building. In such non-pressure applications, it is recommended that restraint fittings or O-ring expansion joints be installed at all points of potential strain.

Chemical Resistance: Generally, plastics are impervious to a wider range of chemicals than are metals. However, they are not resistant to all chemicals. The factors which determine the suitability of each particular plastic material when exposed to a certain chemical include:

a. the specific chemical and its concentration;
b. the specific plastic compound within the pipe or fitting;
c. the joining method;
d. the dimensions of the pipe and fittings;
e. the pressure;
f. the temperature;
g. the period of contact with the chemical;
h. other service conditions which may introduce stress concentrations in pipe or fittings.

Water solutions of neutral inorganic salts such as sodium chloride, aluminum potassium sulfate, calcium chloride, copper sulfate, potassium sulfate, and zinc chloride generally have the same effect on thermoplastics piping materials as does water alone. With organic chemicals in a specific series, such as alcohols, ketones, or organic acids, the resistance of a particular plastic increases as the molecular weight of the chemical increases. Thus, one type of PVC is not suitable for transporting ethyl acetate at $73°F$, but it may be used with butyl acetate at that temperature. Complete technical data on chemical resistance can be obtained from the Society of the Plastics Industry. It is strongly recommended, however, that each user satisfy himself by means of tests approximating actual conditions, or from previous experience, before a particular plastic piping system is chosen to transport a specific chemical.

Flow Characteristics: The head losses that occur because of liquid flowing in plastic piping systems are calculated using the William and Hazen formula, in a similar manner as those for other piping materials. In this formula, the constant for inside roughness (C), of plastic, is 150. Because the surface of plastic pipe is smooth and slippery, its frictional resistance is considerably less than that of other piping materials. Further, the C value remains fairly constant over extended periods of use due to the material's inherent resistance to corrosion and to the build-up of scale.

Figure 1 provides a nomograph giving the flow capacity for water in different standard commercial sizes of thermoplastic pipe.

Pressure Ratings: The recommended pressure ratings for plastics pipes that meet current ASTM standards are given in Tables 2 through 12. Threaded PVC and ABS pipe should be pressure rated at 50 percent of the non-threaded rating. The pressure ratings shown make some allowance for surge and water hammer. However, when excessive surges and water hammer are likely to be encountered, extra allowance should be made, or protective devices installed.

The plastics piping industry uses two-dimensional systems for pressure pipe. One is the Schedule system (Sch. 40, 80, and 120) which emulates the Iron Pipe Sizes (IPS) used in metal piping. The other, referred to as the Standard Dimension Ratio (SDR) system, is unique to plastics pipe. This latter system results in uniform pressure ratings regardless of pipe diameter, while in the Schedule system pressure ratings vary according to diameter. Pressure ratings are temperature-dependent. Ratings shown in the Tables are at $73°F$, and a de-rating factor must be employed when considering use at elevated temperatures.

FIGURE 1
FLOW LOSS CHARACTERISTICS OF WATER FLOW THROUGH RIGID PLASTIC PIPE

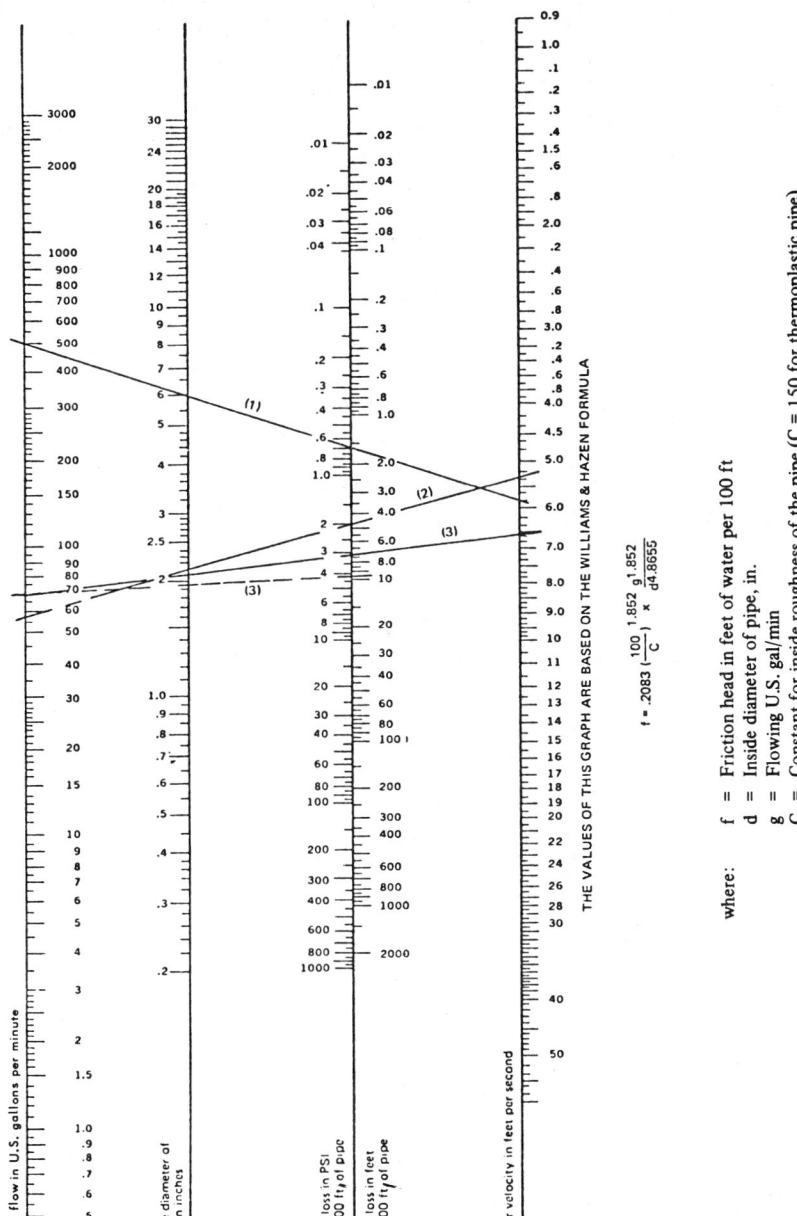

THE VALUES OF THIS GRAPH ARE BASED ON THE WILLIAMS & HAZEN FORMULA

$$f = .2083 \left(\frac{100}{C}\right)^{1.852} \times \frac{g^{1.852}}{d^{4.8655}}$$

where:

f = Friction head in feet of water per 100 ft
d = Inside diameter of pipe, in.
g = Flowing U.S. gal/min
C = Constant for inside roughness of the pipe (C = 150 for thermoplastic pipe).

TABLE 2
STANDARD THERMOPLASTIC PIPE DIMENSION RATIOS (SDR) AND WATER PRESSURE
RATINGS (PR) AT 23°C (73°F) FOR SDR-PR ABS PLASTIC PIPE

Standard Dimension Ratio	ABS Pipe Materials			
	ABS 1316	ABS 2112	ABS 1210	ABS 1208
	Pressure Rating, psi[a]			
13.5	250	200	160	125
17	200	160	125	100
21	160	125	100	—
26	125	100	—	—
Pressure Rating, psi	Standard Dimension Ratio			
250	13.5	—	—	—
200	17	13.5	—	—
160	21	17	13.5	—
125	26	21	17	13.5
100	—	26	21	17

[a] These pressure ratings do not apply for threaded pipe.

TABLE 3
WATER PRESSURE RATINGS AT 23°C (73°F) FOR SCHEDULE 40 ABS PLASTIC PIPE

Nominal Pipe Size In.	Pressure Ratings, psi[a]			
	ABS 1208	ABS 1210	ABS 1316	ABS 2112
½	240	300	480	370
¾	190	240	390	300
1	180	220	360	280
1¼	150	180	290	230
1½	130	170	260	210
2	110	140	220	170
2½	120	150	240	190
3	100	130	210	160
3½	90	120	190	150
4	90	110	180	140
5	80	110	160	120
6	—	90	140	110
8	—	80	120	100
10	—	—	110	90
12	—	—	110	80

[a] These pressure ratings are for unthreaded pipe. The industry does not recommend threading ABS plastic pipe in wall thickness less than Schedule 80.

TABLE 4
WATER PRESSURE RATINGS AT 23°C (73°F) FOR SCHEDULE 80 ABS PLASTIC PIPE

Nominal Pipe Size In.	Pressure Ratings, psi							
	ABS 1208		ABS 1210		ABS 1316		ABS 2112	
	Unthreaded	Threaded	Unthreaded	Threaded	Unthreaded	Threaded	Unthreaded	Threaded
½	340	170	420	210	680	340	530	260
¾	280	140	340	170	550	280	430	210
1	250	130	320	160	500	250	390	200
1¼	210	100	260	130	420	210	330	160
1½	190	90	240	120	380	190	290	150
2	160	80	200	100	320	160	250	130
2½	170	80	210	110	340	170	270	130
3	150	—	190	90	300	150	230	120
3½	140	—	170	90	280	140	220	110
4	130	—	160	—	260	130	200	100
5	120	—	140	—	230	120	180	90
6	110	—	140	—	220	110	170	90
8	100	—	120	—	200	100	150	80
10	90	—	120	—	190	90	150	—
12	90	—	110	—	180	90	140	—

TABLE 5
STANDARD THERMOPLASTIC PIPE DIMENSION RATIOS (SDR) AND WATER PRESSURE RATINGS (PR) AT 23°C (73°F) FOR SDR-PR PE PLASTIC PIPE.
OUTSIDE DIAMETER CONTROL

Standard Dimension Ratio	Pressure Rating, psi[a]		
	PE 3406, PE 3306 PE 2306	PE 2305	PE 1404
11	125	100	80
13.5	100	80	—
17	80	—	—

Pressure Rating psi	Standard Dimension Ratio		
125	11	—	—
100	13.5	11	—
80	17	13.5	11

INSIDE DIAMETER CONTROL

Standard Dimension Ratio	PE 3406, PE 3306, PE 2306	PE 2305	PE 1404
	Pressure Rating, psi[a]		
5.3	200	160	125
7	160	125	100
9	125	100	80
11.5	100	80	—
15	80	—	—

Pressure Rating, psi	Standard Dimension Ratio		
200	5.3	—	—
160	7	5.3	—
125	9	7	5.3
100	11.5	9	7
80	15	11.5	9

[a] The pressure ratings apply only to unthreaded pipe. The industry does not recommend threading PE plastic pipe.

TABLE 6
WATER PRESSURE RATINGS AT 23°C (73°F) FOR SCHEDULE 40 PE PLASTIC PIPE, INSIDE DIAMETER CONTROL

Nominal Pipe Size In.	Pressure Ratings, psi[a]		
	PE 2306, PE 3306, PE 3406	PE 2305	PE 1404
½	190	150	120
¾	150	120	100
1	140	110	90
1¼	120	90	70
1½	100	80	70
2	90	70	60
2½	100	80	60
3	80	70	50
4	70	60	NPR
6	60	NPR	NPR

[a]These pressure ratings apply only to unthreaded pipe. The industry does not recommend threading PE plastic pipe.
NPR = Not pressure rated

TABLE 7
WATER PRESSURE RATINGS AT 23°C (73°F) FOR PE PLASTIC PIPE, SCHEDULES 40 AND 80 OUTSIDE DIAMETER CONTROL

Nominal Pipe Size In.	Pressure Ratings, psi[a]					
	PE 2306, PE 3306, PE 3406		PE 2305		PE 1404	
	SCH 40	SCH 80	SCH 40	SCH 80	SCH 40	SCH 80
½	188	267	149	212	119	170
¾	152	217	120	172	96	137
1	142	199	113	158	90	126
1¼	116	164	92	130	74	104
1½	104	148	83	118	66	94
2	87	127	69	101	55	81
2¼	96	134	76	106	61	85
3	83	118	66	94	53	75
3½	75	109	60	86	50	69
4	70	102	55	81	NPR	65
5	61	91	50	72	NPR	58
6	55	88	NPR	70	NPR	56
8	50	—	NPR	—	NPR	—

[a]These pressure ratings apply only to unthreaded pipe. The industry does not recommend threading PE plastic pipe.
NPR = Not pressure rated

TABLE 8
STANDARD THERMOPLASTIC PIPE DIMENSION RATIOS (SDR) AND WATER PRESSURE RATINGS (PR) AT 23°C (73°F) FOR PVC PLASTIC PIPE

Standard Dimension Ratio	CPVC 4120 PVC 1120 and PVC 1220	PVC 2116	PVC 2112	PVC 2110
	Pressure Ratings, psi[a]			
13.5	315	250	200	160
17	250	200	160	125
21	200	160	125	100
26	160	125	100	—
32.5[b]	125	100	—	—

Pressure Rating, psi	Standard Dimension Ratio			
315	13.5	—	—	—
250	17	13.5	—	—
200	21	16	13.5	—
160	26	21	17	13.5
125	32.5	26	21	17
100	41	32.5	26	21

[a] These pressure ratings do not apply for threaded pipe.
[b] Available only in nominal pipe size diameters of 3 to 4 in.

TABLE 9
WATER PRESSURE RATINGS AT 23°C (73°F) FOR SCHEDULE 40 PVC PLASTIC PIPE

Nominal Pipe Size In.	Pressure Ratings, psi[a]			
	PVC1120 PVC1120 CPVC 4120	PVC 2116	PVC 2110	PVC 2112
½	600	450	300	370
¾	480	390	240	300
1	450	360	220	280
1¼	370	290	180	230
1½	330	260	170	210
2	280	220	140	170
2½	300	240	150	190
3	260	210	130	160
3½	240	190	120	150
4	220	180	110	140
5	190	160	100	120
6	180	140	90	110
8	160	120	80	100
10	140	110	—	90
12	130	110	—	80

[a] These pressure ratings apply only to unthreaded pipe. The industry does not recommend threading PVC plastic pipe in wall thicknesses less than Schedule 80 dimensions in nominal sizes 6 in. and smaller.

TABLE 10
WATER PRESSURE RATINGS AT 23°C (73°F) FOR SCHEDULE 80 PVC PLASTIC PIPE

Nominal Pipe Size In.	Pressure Ratings, psi							
	CPVC 4120 PVC 1120 PVC 1120		PVC 2116		PVC 2112		PVC 2110	
	Unthreaded	Threaded	Unthreaded	Threaded	Unthreaded	Threaded	Unthreaded	Threaded
½	850	420	680	340	530	260	420	210
¾	690	340	550	280	430	210	340	170
1	630	320	500	250	390	200	320	160
1¼	520	260	420	210	320	160	260	130
1½	470	240	380	190	290	150	240	120
2	400	200	320	160	250	130	200	100
2½	420	340	340	170	260	130	210	110
3	370	190	300	150	230	120	190	90
3½	350	170	280	140	220	110	170	90
4	320	160	260	130	200	100	160	80
5	290	140	230	120	180	90	140	–
6	280	140	220	110	170	90	140	–
8	250	120	200	100	150	80	120	–
10	230	120	190	90	150	–	120	–
12	230	110	180	90	140	–	110	–

TABLE 11
WATER PRESSURE RATINGS AT 23°C (73°F) FOR SCHEDULE 120 PVC PLASTIC PIPE

Nominal Pipe Size In.	Pressure Ratings, psi							
	CPVC 4120 PVC 1120 PVC 1220		PVC 2116		PVC 2112		PVC 2110	
	Unthreaded	Threaded	Unthreaded	Threaded	Unthreaded	Threaded	Unthreaded	Threaded
½	1010	510	810	410	630	320	510	250
¾	770	390	620	310	480	240	390	190
1	720	360	570	290	450	220	360	180
1¼	600	300	480	240	370	190	300	150
1½	540	270	430	210	340	170	270	130
2	470	240	380	190	290	150	240	120
2½	470	230	370	190	290	150	230	120
3	440	220	360	180	280	140	220	110
3½	380	190	310	150	240	120	190	100
4	430	220	340	170	270	130	220	110
5	400	200	320	160	250	120	200	100
6	370	190	300	150	230	120	190	90
8	380	180	290	140	230	110	180	90
10	370	180	290	140	230	110	180	90
12	340	170	270	140	210	110	170	80

TABLE 12
STANDARD THERMOPLASTIC PIPE DIMENSION RATIOS (SDR) AND WATER
PRESSURE RATINGS (PR) AT 23°C (73°F) FOR SDR-PR PB PLASTIC PIPE

Standard Dimension Ratio	PB 2110
	Pressure Rating, psi[a]
9	200
11.5	160
15	125

[a] Pressure ratings are for unthreaded PB plastic pipe. The industry does not recomment threading PB plastic pipe.

Installation

Storage Handling: Exposure to direct sunlight for extended periods of time (longer than one year) can have detrimental effects on plastics pipe. Piping which must be stored outdoors should be covered and placed in racks that give continuous support to prevent sagging or draping. Because plastic pipe is softer than metal, it is more susceptible to damage by abrasion or gouging. Therefore, such practices as dragging the pipe over rough ground or over sharp projections, and throwing it from trucks in unloading it, should be avoided. Dropping plastics piping from great heights, particularly in cold weather, should also be avoided since at lower temperatures, plastics tend to lose some of their impact resistance.

Above ground: Rigid plastics piping must be supported above ground at intervals which will prevent sagging, as shown in Table 13. Ordinary pipe hangers and brackets may be used, as long as overtightening is avoided and they do not cut, abrade, or distort the pipe. Valves should be properly anchored so that their use will not apply a torque to the piping. Since the strength of plastics piping is dependent on temperature, such piping should not be installed near hot objects, or in areas of high ambient temperature.

Below ground: The most critical aspects of below-ground installation are the preparation of the trench and the backfilling after installation. The trench bottom must support the pipe evenly throughout its entire length, hence, the use of blocks to attain desired grade is to be avoided. Where a smooth trench bottom cannot be attained in the initial excavation, a bed of sand or fine granular soil should be placed on the bottom of the trench. Care must be exercised to keep rocks and boulders from coming into direct contact with the piping.

Selected backfill materials should be used to surround the pipe. It should be placed in layers; each soil layer should be sufficiently compacted to uniformly develop lateral passive soil forces during the backfill operation. As an added protective measure during installation, it may often be advisable to have the pipe under pressure during backfilling procedures.

Codes and Regulations

The degree to which plastics piping is allowed in mechanical and plumbing systems varies widely from one jurisdiction to another. Therefore, before plastics piping is specified or installed, it is advisable to consult the regulatory agency having jurisdiction. One municipality, for example, may allow the use of plastic DWV piping in any type of building, while another may restrict its use according to building height or occupancy. A clear understanding of any such restrictions well in advance of installation is therefore essential.

Because plastics are combustible materials, penetrations of fire-resistive walls and floors by plastics piping or conduit must be considered to preserve the endurance ratings of rated walls and floors. Industry-sponsored research indicates that after installation such penetrations must be carefully sealed with a non-combustible material. However, a preliminary check should be made to assure that the jurisdictional authority does not prohibit the installation of plastics piping in fire-resistive construction. And if use is permitted, requirements for proper closure should be followed.

TABLE 13
SUPPORT CENTERS FOR HORIZONTAL PIPE (FEET)

Nominal Pipe Size In.	ABS		PE[1]		PVC	
	SDR & SCH. 40	SCH. 80	SDR & SCH. 40	SCH. 80	SDR & SCH. 40	SCH. 80
¼	–	–	1	1½	–	–
⅜	–	–	1	1½	–	–
½	4	5	1¼	2	3½	3½
¾	4	5	1¼	2	3½	4
1	4½	5½	1¼	2	4	4½
1¼	4½	5½	1½	2¼	4½	5
1½	5	6	1½	2¼	4½	5
2	5	6	1½	2¼	4½	5
2	5	6	1¾	2¾	4½	5
2½	–	–	1¾	2¾	–	–
3	6	7	2	3	5½	6
4	6¼	7½	2¼	3½	6¼	7½
6	6¾	8½	2¾	4½	6¾	8½
8	–	–	3½	5¼	7½	9
10	–	–	4½	6½	7¾	9½
12	–	–	5	7½	8	10

[1] These recommended spacings are for PE pipes rated at 80 psi or greater for cold water, at temperatures not exceeding room temperature. For pipes having thinner walls, or when the design temperature will exceed 100°F, continuous support is recommended. Refer to Table 28 for recommended support spacings for polybutylene.

STAINLESS STEEL PIPE†

Dimensions

Nominal Pipe Size	Outside Diameter	Nominal Wall Thickness			
		Schedule 5S‡	Schedule 10S‡	Schedule 40S	Schedule 80S
⅛	0.405	—	0.049	0.068	0.095
¼	0.540	—	0.065	0.088	0.119
⅜	0.675	—	0.065	0.091	0.126
½	0.840	0.065	0.083	0.109	0.147
¾	1.050	0.065	0.083	0.113	0.154
1	1.315	0.065	0.109	0.133	0.179
1¼	1.660	0.065	0.109	0.140	0.191
1½	1.900	0.065	0.109	0.145	0.200
2	2.375	0.065	0.109	0.154	0.218
2½	2.875	0.083	0.120	0.203	0.276
3	3.500	0.083	0.120	0.216	0.300
3½	4.000	0.083	0.120	0.226	0.318
4	4.500	0.083	0.120	0.237	0.337
5	5.563	0.109	0.134	0.258	0.375
6	6.625	0.109	0.134	0.280	0.432
8	8.625	0.109	0.148	0.322	0.500
10	10.750	0.134	0.165	0.365	0.500*
12	12.750	0.156	0.180	0.375*	0.500*

‡ Wall thicknesses of Schedules 5S and 10S do not permit use of ANSI-B2.1 pipe threads.
* These do not conform to ANSI-B36.10.

Weights

Nominal Pipe Size	Weight, Plain Ends, Lb. per Foot*			
	Schedule 5S	Schedule 10S	Schedule 40S	Schedule 80S
⅛	—	0.19	0.24	0.31
¼	—	0.33	0.42	0.54
⅜	—	0.42	0.57	0.74
½	0.54	0.67	0.85	1.09
¾	0.69	0.86	1.13	1.47
1	0.87	1.40	1.68	2.17
1¼	1.11	1.81	2.27	3.00
1½	1.28	2.09	2.72	3.63
2	1.61	2.64	3.65	5.02
2½	2.48	3.53	5.79	7.66
3	3.03	4.33	7.58	10.25
3½	3.48	4.97	9.11	12.51
4	3.92	5.61	10.79	14.98
5	6.36	7.77	14.62	20.78
6	7.60	9.29	18.97	28.57
8	9.93	13.40	28.55	43.39
10	15.23	18.70	40.48	54.74
12	22.22	24.20	49.56	65.42

All dimensions given in inches.
† Extracted from ANSI-B36.19 by permission of the publisher, The American Society of Mechanical Engineers.
* Weights are given in pounds per linear foot and are for carbon steel pipe with plain ends. The different grades of stainless steel permit considerable variations in weight. The ferritic stainless steels may be about 5 per cent less, and the austenitic stainless steels about 2 per cent greater than the values shown in this table which are based on weights for carbon steel.

AMERICAN NATIONAL STANDARD TAPER PIPE THREADS †

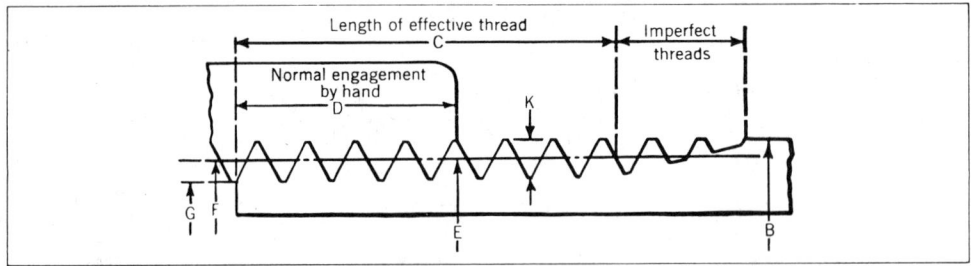

Nominal Pipe Dia.. In.	Outside Dia. In.	No. of Threads per Inch	Pitch Diameter. In.		Length of Effective Thread. In.	Length of Hand-Tight Engagement, In.	Length of Imperfect Threads, In.	Max. Depth of Thread, In.	Root Diameter at Small End, In.
			At End of External Thread	At End of Internal Thread					
	B		F	E	C	D		K	G
1/16	0.3125	27	0.27118	0.28118	0.2611	0.160	0.1285	0.02963	0.2416
1/8	0.405	27	0.36351	0.37476	0.2639	0.180	0.1285	0.02963	0.3339
1/4	0.540	18	0.47739	0.48989	0.4018	0.200	0.1928	0.04444	0.4329
3/8	0.675	18	0.61201	0.62701	0.0478	0.240	0.1928	0.04444	0.5676
1/2	0.840	14	0.75843	0.77843	0.5337	0.320	0.2478	0.05714	0.7013
3/4	1.050	14	0.96768	0.98887	0.5457	0.339	0.2478	0.05714	0.9105
1	1.315	11-1/2	1.21363	1.23863	0.6828	0.400	0.3017	0.06957	1.1441
1-1/4	1.660	11-1/2	1.55713	1.58338	0.7068	0.420	0.3017	0.06957	1.4876
1-1/2	1.900	11-1/2	1.79609	1.82234	0.7235	0.420	0.3017	0.06957	1.7265
2	2.375	11-1/2	2.26902	2.29627	0.7565	0.436	0.3017	0.06957	2.1995
2-1/2	2.875	8	2.71953	2.76216	1.1375	0.682	0.4337	0.10000	2.6195
3	3.500	8	3.34062	3.38850	1.2000	0.766	0.4337	0.10000	3.2406
3-1/2	4.000	8	3.83750	3.88881	1.2500	0.821	0.4337	0.10000	3.7375
4	4.500	8	4.33438	4.38712	1.3000	0.844	0.4337	0.10000	4.2344
5	5.563	8	5.39073	5.44929	1.4063	0.937	0.4337	0.10000	5.2907
6	6.625	8	6.44609	6.50597	1.5125	0.958	0.4337	0.10000	6.3461
8	8.625	8	8.43359	8.50003	1.7125	1.063	0.4337	0.10000	8.3336
10	10.750	8	10.54531	10.62094	1.9250	1.210	0.4337	0.10000	10.4453
12	12.750	8	12.53281	12.61781	2.1250	1.360	0.4337	0.10000	12.4328
14 O.D.	14.000	8	13.77500	13.87262	2.2500	1.562	0.4337	0.10000	13.6750
16 O.D.	16.000	8	15.76250	15.87575	2.4500	1.812	0.4337	0.10000	15.6625
18 O.D.	18.000	8	17.75000	17.87500	2.6500	2.000	0.4337	0.10000	17.6500
20 O.D.	20.000	8	19.73750	19.87031	2.8500	2.125	0.4337	0.10000	19.6375
24 O.D.	24.000	8	23.71250	23.86094	3.2500	2.375	0.4337	0.10000	23.6125

† Extracted from ANSI-B2.1 with the permission of the publisher, The American Society of Mechanical Engineers.

The American Standard for Pipe Threads, originally known as the Briggs Standard was formulated by Robert Briggs about 1862. By 1886, most American manufacturers had adopted this standard, the adoption being formalized by its adoption by the ASME.

The accompanying data are from the ANSI-B2.1 standards adopted by ANSI with ASME and AGA as the sponsor organizations.

For all dimensions see corresponding reference letters in table.

Angle between sides of thread is 60 degrees. Taper of thread, on diameter, is 3/4 inch per foot.

The basic maximum thread depth, K, of the truncated thread is 0.8 × pitch of thread. The crest and root are truncated a minimum of 0.033 × pitch for all pitches. Maximum truncation for crest and root is: 0.096 × pitch for 27 threads per inch, 0.088 × pitch for 18 threads per inch, 0.078 × pitch for 14 threads per inch, 0.073 × pitch for 11½ threads per inch, 0.062 × pitch for 8 threads per inch.

125 AND 250 LB CAST IRON SCREWED FITTINGS†

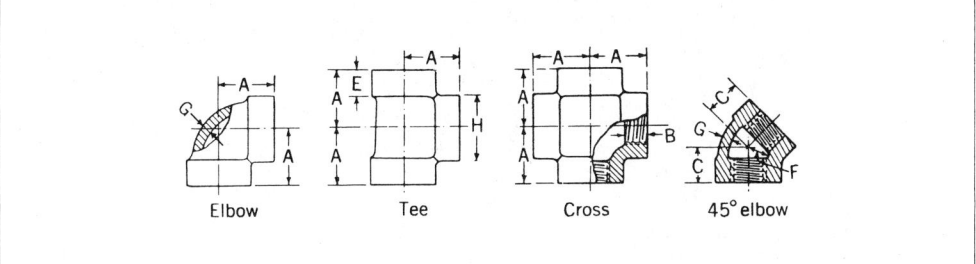

Elbow   Tee   Cross   45° elbow

Nom. Pipe Dia.	A	C	B Min.	E Min.	F Min.	F Max.	G	H Min.
			Fittings for 125 Pounds per Square Inch					
¼	0.81	0.73	0.32	0.38	0.540	0.584	0.110	0.93
⅜	0.95	0.80	0.36	0.44	0.675	0.719	0.120	1.12
½	1.12	0.88	0.43	0.50	0.840	0.897	0.130	1.34
¾	1.31	0.98	0.50	0.56	1.050	1.107	0.155	1.63
1	1.50	1.12	0.58	0.62	1.315	1.385	0.170	1.95
1¼	1.75	1.29	0.67	0.69	1.660	1.730	0.185	2.39
1½	1.94	1.43	0.70	0.75	1.900	1.970	0.200	2.68
2	2.25	1.68	0.75	0.84	2.375	2.445	0.220	3.28
2½	2.70	1.95	0.92	0.94	2.875	2.975	0.240	3.86
3	3.08	2.17	0.98	1.00	3.500	3.600	0.260	4.62
3½	3.42	2.39	1.03	1.06	4.000	4.100	0.280	5.20
4	3.79	2.61	1.08	1.12	4.500	4.600	0.310	5.79
5	4.50	3.05	1.18	1.18	5.563	5.663	0.380	7.05
6	5.13	3.46	1.28	1.28	6.625	6.725	0.430	8.28
8	6.56	4.28	1.47	1.47	8.625	8.725	0.550	10.63
10	*8.08	5.16	1.68	1.68	10.750	10.850	0.690	13.12
12	*9.50	5.97	1.88	1.88	12.750	12.850	0.800	15.47
			Fittings for 250 Pounds per Square Inch					
¼	0.94	0.81	0.43	0.49	0.540	0.584	0.18	1.17
⅜	1.06	0.88	0.47	0.55	0.675	0.719	0.18	1.36
½	1.25	1.00	0.57	0.60	0.840	0.897	0.20	1.59
¾	1.44	1.13	0.64	0.68	1.050	1.107	0.23	1.88
1	1.63	1.31	0.75	0.76	1.315	1.385	0.28	2.24
1¼	1.94	1.50	0.84	0.88	1.660	1.730	0.33	2.73
1½	2.13	1.69	0.87	0.97	1.900	1.970	0.35	3.07
2	2.50	2.00	1.00	1.12	2.375	2.445	0.39	3.74
2½	2.94	2.25	1.17	1.30	2.875	2.975	0.43	4.60
3	3.38	2.50	1.23	1.40	3.500	3.600	0.48	5.36
3½	3.75	2.63	1.28	1.49	4.000	4.100	0.52	5.98
4	4.13	2.81	1.33	1.57	4.500	4.600	0.56	6.61
5	4.88	3.19	1.43	1.74	5.563	5.663	0.66	7.92
6	5.63	3.50	1.53	1.91	6.625	6.725	0.74	9.24
8	7.00	4.31	1.72	2.24	8.625	8.725	0.90	11.73
10	8.63	5.19	1.93	2.58	10.750	10.850	1.08	14.37
12	10.00	6.00	2.13	2.91	12.750	12.850	1.24	16.84

All dimensions given in inches.
† Extracted from ANSI-B16.4 by permission of the publisher, The American Society of Mechanical Engineers.
* This applies to elbows and tees only.

150 LB MALLEABLE IRON SCREWED FITTINGS (1)†

Elbows, Tees and Crosses

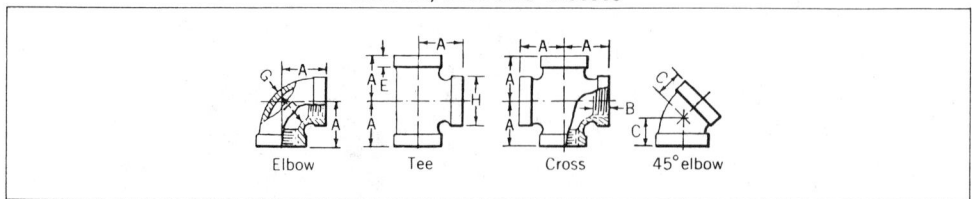

Elbow   Tee   Cross   45° elbow

Pipe Dia.	A	C	B Min.	E Min.	F Min.	F Max.	G Min.	H Min.
⅛	0.69	—	0.25	0.200	0.40	0.43	0.090	0.693
¼	0.81	0.73	0.32	0.215	0.54	0.58	0.095	0.844
⅜	0.95	0.80	0.36	0.230	0.67	0.71	0.100	1.015
½	1.12	0.88	0.43	0.249	0.84	0.89	0.105	1.197
¾	1.31	0.98	0.50	0.273	1.05	1.10	0.120	1.458
1	1.50	1.12	0.58	0.302	1.31	1.38	0.134	1.771
1¼	1.75	1.29	0.67	0.341	1.66	1.73	0.145	2.153
1½	1.94	1.43	0.70	0.368	1.90	1.97	0.155	2.427
.2	2.25	1.68	0.75	0.422	2.37	2.44	0.173	2.963
2½	2.70	1.95	0.92	0.478	2.87	2.97	0.210	3.589
3	3.08	2.17	0.98	0.548	3.50	3.60	0.231	4.285
3½	3.42	2.39	1.03	0.604	4.00	4.10	0.248	4.843
4	3.79	2.61	1.08	0.661	4.50	4.60	0.265	5.401
5	4.50	3.05	1.18	0.780	5.56	5.66	0.300	6.583
6	5.13	3.46	1.28	0.900	6.62	6.72	0.336	7.767

Straight and Reducing Couplings (Cast)

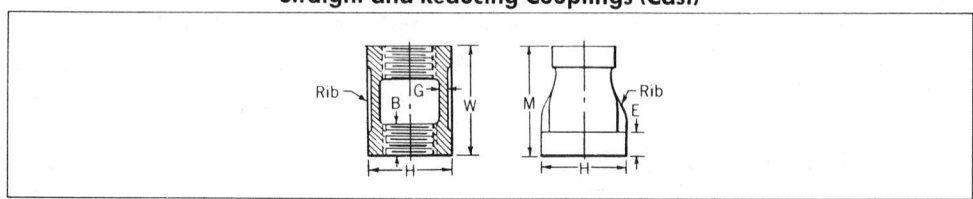

Pipe Dia.	B Min.	E Min.	G Min.	H Min.	Rib Thickness	W	M
⅛	0.25	0.200	0.090	0.693	0.090	0.96	—
¼	0.32	0.215	0.095	0.844	0.095	1.06	1.00
⅜	0.36	0.230	0.100	1.015	0.100	1.16	1.13
½	0.43	0.249	0.105	1.197	0.105	1.34	1.25
¾	0.50	0.273	0.120	1.458	0.120	1.52	1.44
1	0.58	0.302	0.134	1.771	0.134	1.67	1.69
1¼	0.67	0.341	0.145	2.153	0.145	1.93	2.06
1½	0.70	0.368	0.155	2.427	0.155	2.15	2.31
2	0.75	0.422	0.173	2.963	0.173	2.53	2.81
2½	0.92	0.478	0.210	3.589	0.210	2.88	3.25
3	0.98	0.548	0.231	4.285	0.231	3.18	3.69
3½	1.03	0.604	0.248	4.843	0.248	3.43	4.00
4	1.08	0.661	0.265	5.401	0.265	3.69	4.38

All dimensions given in inches.
† Extracted from ANSI-B16.3 by permission of the publisher, The American Society of Mechanical Engineers.

150 LB MALLEABLE IRON SCREWED FITTINGS (2)†

45 deg Y-Branches (Straight Sizes)

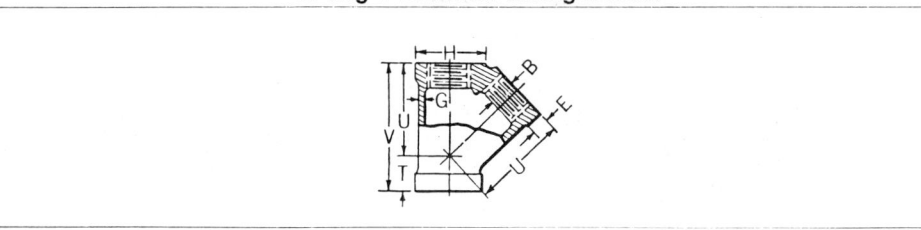

Pipe Dia.	B Min.	E Min.	G Min.	H Min.	T	U	V
3/8	0.36	0.230	0.100	1.015	0.50	1.43	1.93
1/2	0.43	0.249	0.105	1.197	0.61	1.71	2.32
3/4	0.50	0.273	0.120	1.458	0.72	2.05	2.77
1	0.58	0.302	0.134	1.771	0.85	2.43	3.28
1 1/4	0.67	0.341	0.145	2.153	1.02	2.92	3.94
1 1/2	0.70	0.368	0.155	2.427	1.10	3.28	4.38
2	0.75	0.422	0.173	2.963	1.24	3.93	5.17
2 1/2	0.92	0.478	0.210	3.589	1.52	4.73	6.25
3	0.98	0.548	0.231	4.285	1.71	5.55	7.26
3 1/2	1.03	0.604	0.248	4.843	1.85	6.25	8.10
4	1.08	0.661	0.265	5.401	2.01	6.97	8.98

Close, Medium and Open Pattern Return Bends

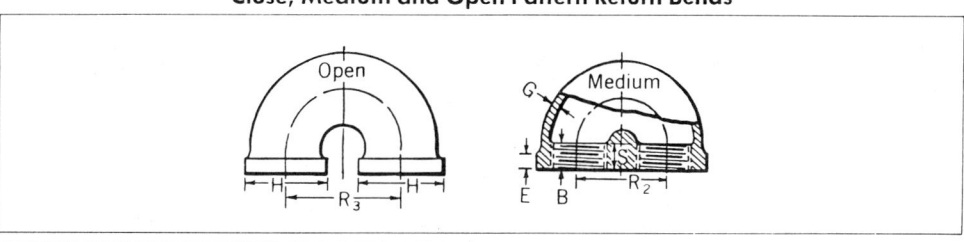

Pipe Dia.	B Min.	E Min.	G Min.	H Min.	R₁ Close Pattern	R₂ Medium Pattern	R₃ Open Pattern
1/2	0.43	0.249	0.116	1.197	1.000	1.25	1.50
3/4	0.50	0.273	0.133	1.458	1.250	1.50	2.00
1	0.58	0.302	0.150	1.771	1.500	1.875	2.50
1 1/4	0.67	0.341	0.165	2.153	1.750	2.25	3.00
1 1/2	0.70	0.368	0.178	2.427	2.188	2.50	3.50
2	0.75	0.422	0.201	2.963	2.625	3.000	4.00
2 1/2	0.92	0.478	0.244	3.589	—	—	4.50
3	0.98	0.548	0.272	4.285	—	—	5.00

All dimensions given in inches.

† Extracted from ANSI-B16.3 by permission of the publisher, The American Society of Mechanical Engineers.

These fittings are rated for maximum saturated steam pressures of 150 psig and for maximum liquid and gas service pressures at 150 deg F of 300 psig.

All fittings are threaded with American Standard Pipe Threads and have taper threads, except 1/8, 1/4 and 3/8 in. wrought couplings and caps which may have straight threads.

300 LB MALLEABLE IRON SCREWED FITTINGS†

Elbows, Tees and Crosses

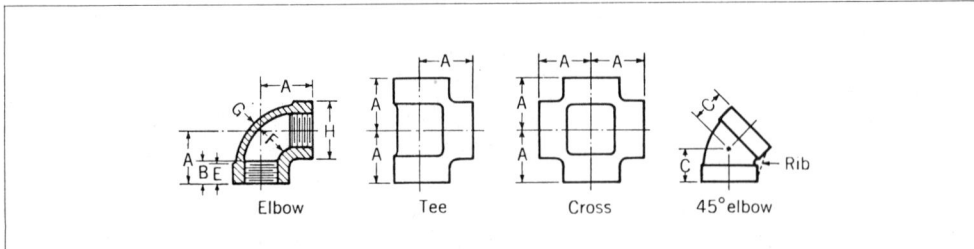

Elbow   Tee   Cross   45° elbow

Pipe Dia.	A	C	B Min.	E Min.	F Min.	F Max.	G	H
¼	0.94	0.81	0.43	0.38	0.540	0.584	0.14	0.93
⅜	1.06	0.88	0.47	0.44	0.675	0.719	0.15	1.12
½	1.25	1.00	0.57	0.50	0.840	0.897	0.16	1.34
¾	1.44	1.13	0.64	0.56	1.050	1.107	0.18	1.63
I	1.63	1.31	0.75	0.62	1.315	1.385	0.20	1.95
1¼	1.94	1.50	0.84	0.69	1.660	1.730	0.22	2.39
1½	2.13	1.69	0.87	0.75	1.900	1.970	0.24	2.68
2	2.50	2.00	1.00	0.84	2.375	2.445	0.26	3.28
2½	2.94	2.25	1.17	0.94	2.875	2.975	0.31	3.86
3	3.38	2.50	1.23	1.00	3.500	3.600	0.35	4.62

Couplings

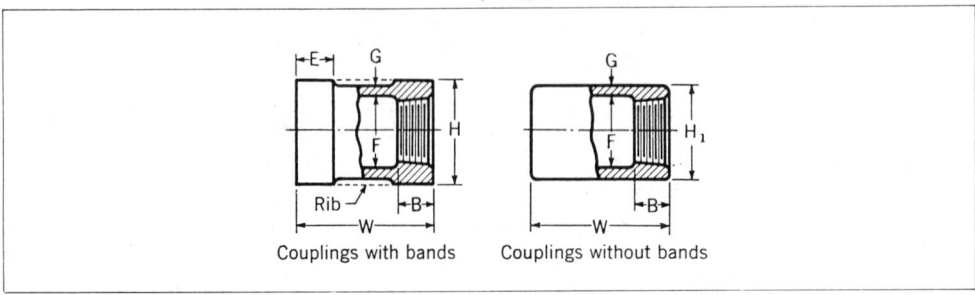

Couplings with bands   Couplings without bands

Pipe Dia.	B Min.	E Min.	F Min.	F Max.	G	H₁ Min.	H Min.	W
¼	0.43	0.38	0.540	0.584	0.14	0.820	0.93	1.375
⅜	0.47	0.44	0.675	0.719	0.15	0.975	1.12	1.625
½	0.57	0.50	0.840	0.897	0.16	1.160	1.34	1.875
¾	0.64	0.56	1.050	1.107	0.18	1.410	1.63	2.125
I	0.75	0.62	1.315	1.385	0.20	1.715	1.95	2.376
1¼	0.84	0.69	1.660	1.730	0.22	2.100	2.39	2.875
1½	0.87	0.75	1.900	1.970	0.24	2.380	2.68	2.875
2	1.00	0.84	2.375	2.445	0.26	2.895	3.28	3.625
2½	1.17	0.94	2.875	2.975	0.31	3.495	3.86	4.125
3	1.23	1.00	3.500	3.600	0.35	4.200	4.62	4.125

All dimensions are given in inches.
† Extracted from ANSI-B16.3 by permission of the publisher, The American Society of Mechanical Engineers.

CLASS 25 CAST IRON FLANGED FITTINGS†

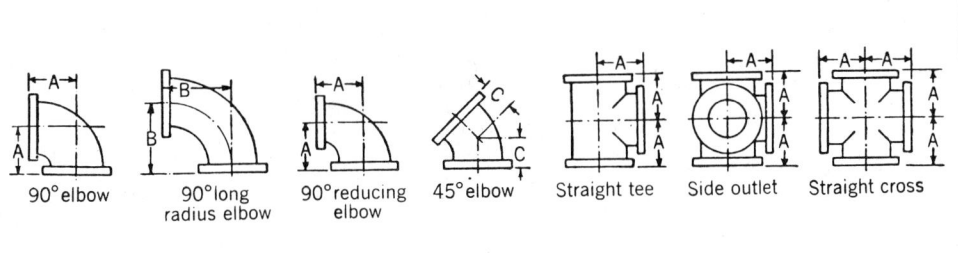

90° elbow   90° long radius elbow   90° reducing elbow   45° elbow   Straight tee   Side outlet   Straight cross

Nominal Pipe Diameter	Center to Face A	Center to Face B	Center to Face C	Face to Face A + A	Dia. of Flange	Thickness of Flange (Min.)	Body Wall Thickness (Min.)
4	6½	9	4	13	9	¾	0.42
5	7½	10¼	4½	15	10	¾	0.44
6	8	11½	5	16	11	¾	0.44
8	9	14	5½	18	13½	¾	0.46
10	11	16½	6½	22	16	⅞	0.50
12	12	19	7½	24	19	1	0.54
14	14	21½	7½	28	21	1⅛	0.57
16	15	24	8	30	23½	1⅛	0.60
18	16½	26½	8½	33	25	1¼	0.64
20	18	29	9½	36	27½	1¼	0.67
24	22	34	11	44	32	1⅜	0.76
30	25	41½	15	50	38¾	1½	0.88
36	28	49	18	56	46	1⅝	0.99
42	31	56½	21	62	53	1¾	1.10
48	34	64	24	68	59½	2	1.26
54	39	71½	27	78	66¼	2¼	1.35
60	44	79	30	88	73	2¼	1.39
72	53	94	36	106	86½	2½	1.62

All dimensions given in inches.
† Extracted from ANSI-B16.1 with the permission of the publisher, The American Society of Mechanical Engineers.

Size of all fittings listed indicates nominal inside diameter of port.

Maximum pressures (gage): Fittings of all sizes, 25 pounds per square inch for saturated steam; sizes 36 inches and smaller, 43 pounds per square inch maximum non-shock working hydraulic pressure or 25 pounds per square inch maximum gas pressure at or near the ordinary range of air temperatures.

The flange diameters, bolt circles, and number of bolts are the same as the 125 lb. American Standard with a reduction in the thickness of flanges and the bolt diameters, thereby maintaining interchangeability with the 125 lb. American Standard flanges. The face to face and center to face dimensions of fittings are the same as the 125 lb. American Standard (Class 125) cast iron flanged fittings.

Special degree elbows, ranging from 1 to 45 degrees inclusive shall have the same center to face dimensions given for 45-degree elbows and those over 45 degrees and up to 90 degrees inclusive shall have the same center to face dimensions as given for 90-degree elbows. The angle designation of an elbow is its deflection from straight line flow and is the angle between the flange faces.

Side outlet elbows and side outlet tees shall have all openings on intersecting center lines.

Tees, side outlet tees, and crosses, 16 inches and smaller, reducing on the outlet or branch, have the same dimensions center to face, and face to face as straight size fittings, corresponding to the size of the larger opening. Sizes 18 inches and larger, reducing on the outlet, are made in two lengths, depending on the size of the outlet.

CLASS 125 CAST IRON PIPE FLANGES†

Nominal Pipe Diameter	Dia. of Flange	Thickness of Flange (Min.)	Dia. of Bolt Circle	Number of Bolts	Dia. of Bolts	Dia. of Drilled Bolt Holes	Length of Bolts	Size of Ring Gasket
1	4-1/4	7/16	3-1/8	4	1/2	5/8	1-3/4	1 × 2-5/8
1-1/4	4-5/8	1/2	3-1/2	4	1/2	5/8	2	1-1/4 × 3
1-1/2	5	9/16	3-7/8	4	1/2	5/8	2	1-1/2 × 3-3/8
2	6	5/8	4-3/4	4	5/8	3/4	2-1/4	2 × 4-1/8
2-1/2	7	11/16	5-1/2	4	5/8	3/4	2-1/2	2-1/2 × 4-7/8
3	7-1/2	3/4	6	4	5/8	3/4	2-1/2	3 × 5-3/8
3-1/2	8-1/2	13/16	7	8	5/8	3/4	2-3/4	3-1/2 × 6-3/8
4	9	15/16	7-1/2	8	5/8	3/4	3	4 × 6-7/8
5	10	15/16	8-1/2	8	3/4	7/8	3	5 × 7-3/4
6	11	1	9-1/2	8	3/4	7/8	3-1/4	6 × 8-3/4
8	13-1/2	1-1/8	11-3/4	8	3/4	7/8	3-1/2	8 × 11
10	16	1-3/16	14-1/4	12	7/8	1	3-3/4	10 × 13-3/8
12	19	1-1/4	17	12	7/8	1	3-3/4	12 × 16-1/8
14 O.D.	21	1-3/8	18-3/4	12	1	1-1/8	4-1/4	14 × 17-3/4
16 O.D.	23-1/2	1-7/16	21-1/4	16	1	1-1/8	4-1/2	16 × 20-1/4
18 O.D.	25	1-9/16	22-3/4	16	1-1/8	1-1/4	4-3/4	18 × 21-5/8
20 O.D.	27-1/2	1-11/16	25	20	1-1/8	1-1/4	5	20 × 23-7/8
24 O.D.	32	1-7/8	29-1/2	20	1-1/4	1-3/8	5-1/2	24 × 28-1/4
30 O.D.	38-3/4	2-1/8	36	28	1-1/4	1-3/8	6-1/4	30 × 34-3/4
36 O.D.	46	2-3/8	42-3/4	32	1-1/2	1-5/8	7	36 × 41-1/4
42 O.D.	53	2-5/8	49-1/2	36	1-1/2	1-5/8	7-1/2	42 × 48
48 O.D.	59-1/2	2-3/4	56	44	1-1/2	1-5/8	7-3/4	48 × 54-1/2
54 O.D.	66-1/4	3	62-3/4	44	1-3/4	2	8-1/2	54 × 61
60 O.D.	73	3-1/8	69-1/4	52	1-3/4	2	8-3/4	60 × 67-1/2
72 O.D.	86-1/2	3-1/2	82-1/2	60	1-3/4	2	9-1/2	72 × 80-3/4
84 O.D.	99-3/4	3-7/8	95-1/2	64	2	2-1/4	10-1/2	84 × 93-1/2
96 O.D.	113-1/4	4-1/4	108-1/2	68	2-1/4	2-1/2	11-1/2	96 × 106-1/4

All dimensions given in inches.

† Extracted from ANSI-B16.1 with the permission of the publisher, The American Society of Mechanical Engineers.

Class 125 cast-iron pipe flanges and fittings are rated as follows for maximum saturated steam service pressures (gage): 125 pounds per square inch for sizes 1 to 12 inches, inclusive; 100 pounds per square inch for sizes 14 to 24 inches, inclusive; and 50 pounds per square inch for sizes 30 to 48 inches, inclusive. For maximum water service pressures (gage) at or near the ordinary range of air temperature: 175 pounds per square inch for sizes 1 to 12 inches, inclusive, and 150 pounds per square inch for sizes 14 to 48 inches, inclusive, for flanges only.

The sizes from 54 to 96 inches are included for convenience where special fittings with larger flanges are required and do not necessarily carry a definite rating.

Bolt holes straddle the center lines and are in multiples of four, so that fittings may be made to face in any quarter. For bolts smaller than 1¾ inches the bolt holes shall be drilled ⅛ inch larger in diameter than the nominal diameter of the bolt. Holes for bolts 1¾ inches and larger shall be drilled ¼ inch larger than nominal diameter of bolts.

The bolt holes of these cast-iron flanges and flanged fittings need not be spot faced for ordinary service except as follows: In sizes 12 inches and smaller when rough flanges, after facing, are oversize more than ⅛ inch in thickness, they shall be spot faced to the specified thickness of flange (minimum) with a plus tolerance of 1/16 inch. In sizes 14 inches to 24 inches, inclusive, when rough flanges, after facing, are oversize more than 3/16 inch in thickness they shall be spot faced to the specified thickness of flange (minimum) with a plus tolerance of 1/16 inch.

CLASS 125 CAST IRON FLANGED FITTINGS (1)†

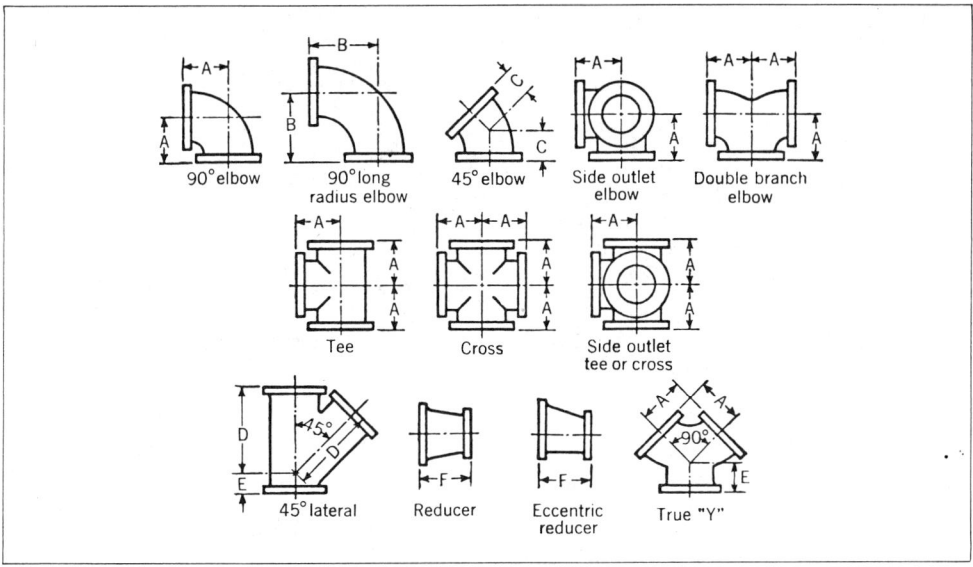

90° elbow   90° long radius elbow   45° elbow   Side outlet elbow   Double branch elbow

Tee   Cross   Side outlet tee or cross

45° lateral   Reducer   Eccentric reducer   True "Y"

Nominal Pipe Diameter	Center to face						Body Wall Thickness	Dia. of Flange	Thickness of Flange
	A	B	C	D	E	F			
1	3-1/2	5	1-3/4	5-3/4	1-3/4	—	5/16	4-1/4	7/16
1-1/4	3-3/4	5-1/2	2	6-1/4	1-3/4	—	5/16	4-5/8	1/2
1-1/2	4	6	2-1/4	7	2	—	5/16	5	9/16
2	4-1/2	6-1/2	2-1/2	8	2-1/2	5	5/16	6	5/8
2-1/2	5	7	3	9-1/2	2-1/2	5-1/2	5/16	7	11/16
3	5-1/2	7-3/4	3	10	3	6	3/8	7-1/2	3/4
3-1/2	6	8-1/2	3-1/2	11-1/2	3	6-1/2	7/16	8-1/2	13/16
4	6-1/2	9	4	12	3	7	1/2	9	15/16
5	7-1/2	10-1/4	4-1/2	13-1/2	3-1/2	8	1/2	10	15/16
6	8	11-1/2	5	14-1/2	3-1/2	9	9/16	11	1
8	9	14	5-1/2	17-1/2	4-1/2	11	5/8	13-1/2	1- 1/8
10	11	16-1/2	6-1/2	20-1/2	5	12	3/4	16	1- 3/16
12	12	19	7-1/2	24-1/2	5-1/2	14	13/16	19	1- 1/4
14 O.D.	14	21-1/2	7-1/2	27	6	16	7/8	21	1-3/8
16 O.D.	15	24	8	30	6-1/2	18	1	23-1/2	1- 7/16
18 O.D.	16-1/2	26-1/2	8-1/2	32	7	19	1- 1/16	25	1- 9/16
20 O.D.	18	29	9-1/2	35	8	20	1- 1/8	27-1/2	1-11/16
24 O.D.	22	34	11	40-1/2	9	24	1- 1/4	32	1- 7/8
30 O.D.	25	41-1/2	15	49	10	30	1- 7/16	38-3/4	2- 1/8
36 O.D.	28*	49	18	—	—	36	1- 5/8	46	2- 3/8
42 O.D.	31*	56-1/2	21	—	—	42	1-13/16	53	2- 5/8
48 O.D.	34*	64	24	—	—	48	2	59-1/2	2- 3/4

All dimensions given in inches.
† Extracted from ANSI-B16.1 with the permission of the publisher, The American Society of Mechanical Engineers.
* Does not apply to true Ys or double branch elbows.

These fittings are for maximum saturated steam service of 125 psig in sizes 1 to 12 in., incl., 100 psig for 14 to 24 in., incl., and 50 psig for sizes 30 to 48 in., incl. The fittings are to be designed to withstand twice these pressures in hydrostatic tests.

CLASS 125 CAST IRON FLANGED FITTINGS (2)†

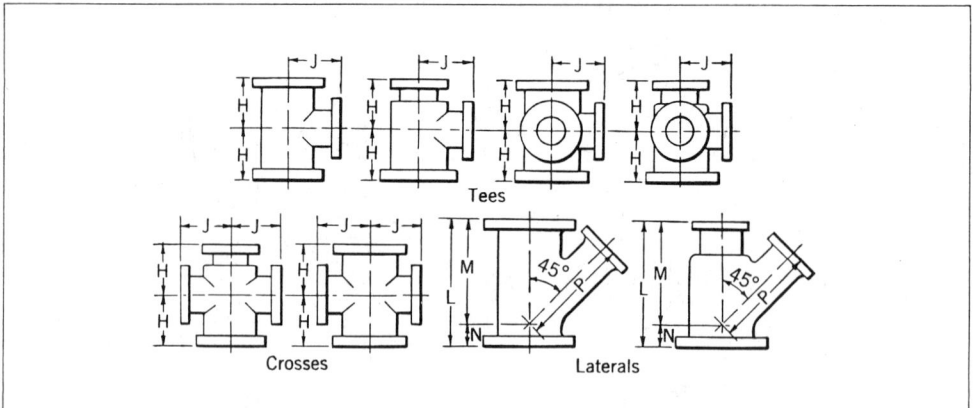

Tees

Crosses   Laterals

Nominal Pipe Diameter	Size of Outlet and Smaller	Center to Face Run H	Center to Face Outlet J	Size of Branch and Smaller	Face to Face Run L	Center to Face Run M	Center to Face Run N	Center to Face Branch P
1	1	3½	3½	1	7½	5¾	1¾	5¾
1¼	1¼	3¾	3¾	1¼	8	6¼	1¾	6¼
1½	1½	4	4	1½	9	7	2	7
2	2	4½	4½	2	10½	8	2½	8
2½	2½	5	5	2½	12	9½	2½	9½
3	3	5½	5½	3	13	10	3	10
3½	3½	6	6	3½	14½	11½	3	11½
4	4	6½	6½	4	15	12	3	12
5	5	7½	7½	5	17	13½	3½	13½
6	6	8	8	6	18	14½	3½	14½
8	8	9	9	8	22	17½	4½	17½
10	10	11	11	10	25½	20½	5	20½
12	12	12	12	12	30	24½	5½	24½
14 O.D.	14	14	14	14	33	27	6	27
16 O.D.	16	15	15	16	36½	30	6½	30
18 O.D.	12	13	15½	8	26	25	1	27½
20 O.D.	14	14	17	10	28	27	1	29½
24 O.D.	16	15	19	12	32	31½	½	34½
30 O.D.	20	18	23	14	39	39	0	42
36 O.D.	24	20	26	—	—	—	—	—

All dimensions given in inches.
† Extracted from ANSI-B16.1 with the permission of the publisher, The American Society of Mechanical Engineers.

The foregoing data apply to reducing tees, crosses and laterals. In the drawing, the tee, extreme left, top row, is reducing on outlet. Top row, next to left, tee is reducing on one run and outlet. Top row, next to right, side outlet tee or cross is reducing on both outlets. Top row, extreme right, side outlet tee or cross is reducing on one run and outlets. Cross, bottom row, extreme left, reducing on both outlets. Next to left, bottom row, cross is reducing on one run and both outlets. Bottom row, next to right, 45° lateral is reducing on branch.

Bottom row, extreme right, 45° lateral is reducing on one run and branch.

Side outlet tees, with outlet at 90 degrees or any other angle, straight or reducing, have the same dimensions center-to-face as regular tees having the same reductions.

In a side outlet tee the larger of the two side outlets governs the center-to-face dimension "J".

Crosses and laterals, both straight and reducing, shall be reinforced where necessary to compensate for the inherent weakness in the casting design.

CLASS 250 CAST IRON FLANGES†

Nominal Pipe Diameter	Dia. of Flange	Thickness of Flange[3] (Min.)	Dia. of Bolt Circle	Dia. of Bolt Holes[1]	Number of Bolts[1]	Size of Bolts	Length of Bolts[2]	Size Ring of Gasket
1	4-7/8	11/16	3-1/2	3/4	4	5/8	2-1/2	1 × 2-7/8
1-1/4	5-1/4	3/4	3-7/8	3/4	4	5/8	2-1/2	1-1/4 × 3-1/4
1-1/2	6-1/8	13/16	4-1/2	7/8	4	3/4	2-3/4	1-1/2 × 3-3/4
2	6-1/2	7/8	5	3/4	8	5/8	2-3/4	2 × 4-3/8
2-1/2	7-1/2	1	5-7/8	7/8	8	3/4	3-1/4	2-1/2 × 5-1/8
3	8-1/4	1- 1/8	6-5/8	7/8	8	3/4	3-1/2	3 × 5-7/8
3-1/2	9	1- 3/16	7-1/4	7/8	8	3/4	3-1/2	3-1/2 × 6-1/2
4	10	1- 1/4	7-7/8	7/8	8	3/4	3-3/4	4 × 7-1/8
5	11	1- 3/8	9-1/4	7/8	8	3/4	4	5 × 8-1/2
6	12-1/2	1- 7/16	10-5/8	7/8	12	3/4	4	6 × 9-7/8
8	15	1- 5/8	13	1	12	7/8	4-1/2	8 × 12-1/8
10	17-1/2	1- 7/8	15-1/4	1- 1/8	16	1	5-1/4	10 × 14-1/4
12	20-1/2	2	17-3/4	1- 1/4	16	1-1/8	5-1/2	12 × 16-5/8
14 O.D.	23	2- 1/8	20-1/4	1- 1/4	20	1-1/8	6	13-1/4 × 19-1/8
16 O.D.	25-1/2	2- 1/4	22-1/2	1- 3/8	20	1-1/4	6-1/4	15-1/4 × 21-1/4
18 O.D.	28	2- 3/8	24-3/4	1- 3/8	24	1-1/4	6-1/2	17 × 23-1/2
20 O.D.	30-1/2	2- 1/2	27	1- 3/8	24	1-1/4	6-3/4	19 × 25-3/4
24 O.D.	36	2- 3/4	32	1-11/16	24	1-1/2	7-3/4	23 × 30-1/2
*30 O.D.	43	3	39-1/4	2	28	1-3/4	8-1/2	29 × 37-1/2
*36 O.D.	50	3- 3/8	46	2- 1/4	32	2	9-1/2	34-1/2 × 44
*42 O.D.	57	3-11/16	52-3/4	2- 1/4	36	2	10-1/4	40-1/4 × 50-3/4
*48 O.D.	65	4	60-3/4	2- 1/4	40	2	10-3/4	46 × 58-3/4

All dimensions given in inches.

† Extracted from ANSI-B16.1 by permission of the publisher, The American Society of Mechanical Engineers.

[1] Drilling templates are in multiples of four, so that fittings may be made to face in any quarter, and bolt holes straddle the center line. For bolts smaller than 1½ in. the bolt holes shall be drilled ⅛ in. larger in diameter than the nominal diameter of the bolt. Holes for bolts 1½ in. shall be drilled ⅛ in. larger in diameter than the nominal diameter of the bolt. Holes for bolts 1¾ in. and larger shall be drilled ¼ in. larger than nominal diameter of bolts.

[2] The bolt holes of these cast-iron flanges and flanged fittings need not be spot faced for ordinary service except as follows: In sizes 12 in. and smaller when rough flanges, after facing, are oversize more than ⅛ in. in thickness, they shall be spot faced to the specified thickness of flange (minimum) with a plus tolerance of ⅛ in. In sizes 14 to 24 in., inclusive, when rough flanges, after facing, are oversize more than ⅛ in. in thickness they shall be spot faced to the specified thickness of flange (minimum) with a plus tolerance of ⅛ in. In sizes 30 in. and larger when rough flanges, after facing, are oversize more than ¼ in. in thickness they shall be spot faced to the specified thickness of flange (minimum) with a plus tolerance of ⅛ in.

[3] All Class 250 cast-iron flanges have a ⅛-in. raised face. This raised face is included in the face-to-face, center-to-face, and the minimum thickness of flange dimensions.

* These sizes are included for convenience where special fittings larger than 24 in. are required.

Class 250 pipe flanges and fittings are rated as follows for maximum saturated steam service pressures (gage): 250 pounds per square inch for sizes 1 to 12 inches, inclusive; 200 pounds per square inch for sizes 14 to 24 inches, inclusive; 100 pounds per square inch for sizes 30 to 48 inches, inclusive. For maximum water service pressures (gage) at or near the ordinary range of air temperature: 400 pounds per square inch for sizes 1 to 12 inches, inclusive; 300 pounds per square inch for sizes 14 to 48 inches, inclusive, for flanges only.

Bolts shall be of carbon steel with American National Standard regular unfinished square heads or American National Standard heavy unfinished hexagonal heads and the nuts shall be of carbon steel with American National Standard heavy hexagonal dimensions, all as specified in American National Standard for Wrench Head Bolts and Nuts and Wrench Openings. For bolts 1¾ in. in diameter and larger, bolt-studs with a nut on each end are recommended.

Hexagonal nuts for pipe sizes 1 in. to 16 in. can be conveniently pulled up with open wrenches of minimum design of heads. Hexagonal nuts for pipe sizes 18 in. to 48 in. can be conveniently pulled up with box wrenches. All bolts, or bolt-studs if used, and all nuts shall be threaded in accordance with American National Standard for Screw Threads Coarse-Thread Series, Class 2 Fit.

Bolt holes of these flanges and fittings need not

CLASS 250 CAST IRON FLANGED FITTINGS†

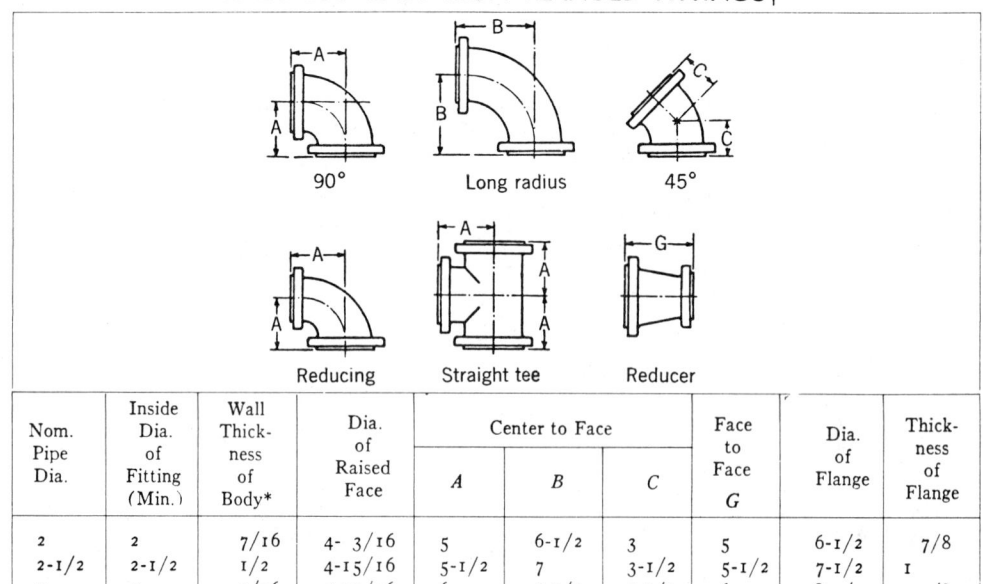

90°   Long radius   45°

Reducing   Straight tee   Reducer

Nom. Pipe Dia.	Inside Dia. of Fitting (Min.)	Wall Thickness of Body*	Dia. of Raised Face	Center to Face			Face to Face G	Dia. of Flange	Thickness of Flange
				A	*B*	*C*			
2	2	7/16	4- 3/16	5	6-1/2	3	5	6-1/2	7/8
2-1/2	2-1/2	1/2	4-15/16	5-1/2	7	3-1/2	5-1/2	7-1/2	1
3	3	9/16	5-11/16	6	7-3/4	3-1/2	6	8-1/4	1-1/8
3-1/2	3-1/2	9/16	6- 5/16	6-1/2	8-1/2	4	6-1/2	9	1-3/16
4	4	5/8	6-15/16	7	9	4-1/2	7	10	1-1/4
5	5	11/16	8- 5/16	8	10-1/4	5	8	11	1-3/8
6	6	3/4	9-11/16	8-1/2	11-1/2	5-1/2	9	12-1/2	1-7/16
8	8	13/16	11-15/16	10	14	6	11	15	1-5/8
10	10	15/16	14- 1/16	11-1/2	16-1/2	7	12	17-1/2	1-7/8
12	12	1	16- 7/16	13	19	8	14	20-1/2	2
14 O.D.	13-1/4	1- 1/8	18-15/16	15	21-1/2	8-1/2	16	23	2-1/8
16 O.D.	15-1/4	1- 1/4	21- 1/16	16-1/2	24	9-1/2	18	25-1/2	2-1/4
18 O.D.	17	1- 3/8	23- 5/16	18	26-1/2	10	19	28	2-3/8
20 O.D.	19	1- 1/2	25- 9/16	19-1/2	29	10-1/2	20	30-1/2	2-1/2
24 O.D.	23	1- 5/8	30- 1/4	22-1/2	34	12	24	36	2-3/4

All dimensions given in inches.
† Extracted from ANSI-B16.1 with the permission of the publisher, The American Society of Mechanical Engineers.
* Wall thickness at no point shall be less than 87½% of dimensions given in table.

be spot-faced except when the rough flanges, after facing, are oversized by certain amounts specified in the standard and depending on the nominal size.

Reducing elbows carry same dimensions, center-to-face, as straight size elbows corresponding to the size of the larger opening.

All Class 250 cast-iron flanges have a 1/16-in. raised face. This raised face is included in the face-to-face, center-to-face, and the minimum thickness of flange dimensions.

Reducing elbows carry the same dimensions center to face as regular straight size elbows corresponding to the size of the larger opening. Tees 16 in. and smaller reducing on the outlet have the same dimensions center to face and face to face as straight size fittings corresponding to the size of

the larger opening. Sizes 18 in. and larger reducing on the outlet are made in two lengths depending on the size of the outlet.

Special degree elbows ranging from 1 to 45 deg, inclusive, have the same center-to-face dimensions given for 45-deg elbows, and those over 45 deg and up to 90-deg, inclusive, shall have the same center-to-face dimensions given for 90-deg elbows. The angle designation of an elbow is its deflection from straight line flow and is the angle between the flange faces.

Reducers, for all reductions, use the same face-to-face dimensions given in the above table of dimensions for the larger opening.

For drilling templates, pressure ratings, and bolt holes, see table headed Class 250 Cast Iron Flanges.

CAST IRON FLANGES FOR REFRIGERANT PIPING†

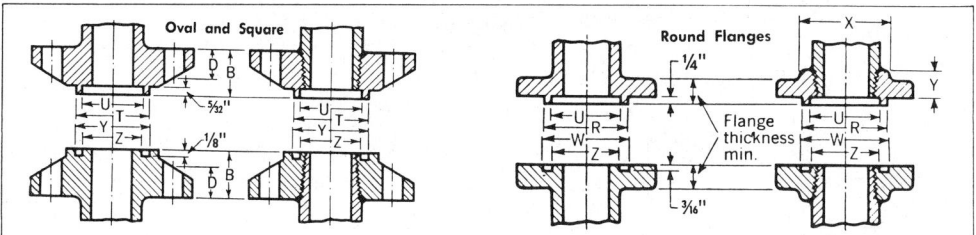

Facing Dimensions, 300 Lb Oval and Square Flanges

Nominal Pipe Size	Tongue			Groove			Flange Thickness, Min		Bolt Spacing
	Height	Outside Diameter T	Inside Diameter U	Depth	Outside Diameter Y	Inside Diameter Z	D	B	
OVAL									
1/4	5/32	1- 3/8	1	1/8	1- 7/16	15/16	1/2	13/16	2- 1/4
3/8	5/32	1- 3/8	1	1/8	1- 7/16	15/16	9/16	15/16	2- 1/2
1/2	5/32	1- 3/8	1	1/8	1- 7/16	15/16	5/8	1	2- 3/4
3/4	5/32	1-11/16	1-5/16	1/8	1- 3/4	1- 1/4	11/16	1- 1/8	3
SQUARE									
1	5/32	1- 7/8	1-1/2	1/8	1-15/16	1- 7/16	—	15/16	2- 5/16
1-1/4	5/32	2- 1/4	1-7/8	1/8	2- 5/16	1-13/16	—	1- 1/16	2- 5/8
1-1/2	5/32	2- 1/2	2-1/8	1/8	2- 9/16	2- 1/16	—	1- 1/8	2-13/16
2	5/32	3- 1/4	2-7/8	1/8	3- 5/16	2-13/16	—	1- 1/4	3- 7/16
2-1/2	5/32	3- 3/4	3-3/8	1/8	3-13/16	3- 5/16	—	1- 3/8	3- 7/8
3	5/32	4- 5/8	4-1/4	1/8	4-11/16	4- 3/16	—	1- 1/2	4- 5/16
3-1/2	5/32	5- 1/8	4-3/4	1/8	5- 3/16	4-11/16	—	1- 5/8	4- 3/4
4	5/32	5-11/16	5-3/16	1/8	5- 3/4	5- 1/8	—	1- 3/4	5- 1/8

Facing Dimensions, 300 Lb Round Flanges

Nominal Pipe Size	Tongue			Groove		
	Height	Outside Diameter R	Inside Diameter U	Depth	Outside Diameter W	Inside Diameter Z
5	1/4	7-5/16	6-5/16	3/16	7- 3/8	6-1/4
6	1/4	8-1/2	7-1/2	3/16	8- 9/16	7-7/16
8	1/4	10-5/8	9-3/4	3/16	10-11/16	9-5/16
10	1/4	12-3/4	11-1/4	3/16	12-13/16	11-3/16
12	1/4	15	13-1/2	3/16	15- 1/16	13-7/16

Nominal Pipe Size	Flange Thickness, Min		Diameter of Bolt Circle	Diameter Hub Min X	Length Through Hub Min Y
	Tongue	Groove			
5	1- 5/8	1- 9/16	9-1/4	7	1- 7/8
6	1-11/16	1- 5/8	10-5/8	8-1/8	1-15/16
8	1- 7/8	1-13/16	13	10-1/4	2- 3/16
10	2- 1/8	2- 1/16	15-1/4	12-5/8	2- 3/8
12	2- 1/4	2- 3/16	17-3/4	14-3/4	2- 9/16

All dimensions are given in inches.

† Extracted from former standard ASA B16.16 by permission of the publisher, The American Society of Mechanical Engineers.

CAST IRON FLANGED FITTINGS FOR REFRIGERANT PIPING†
Oval Flanged Fittings and Screwed Flanges

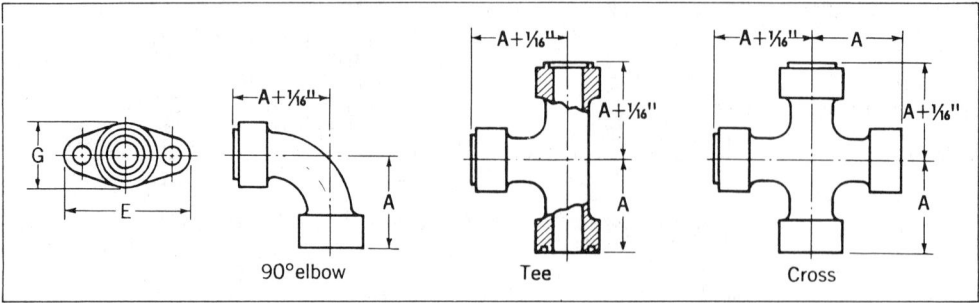

90°elbow   Tee   Cross

Nominal Pipe Size	Inside Diameter of Fitting	Wall Thickness of Fitting	Long Diameter of Flange E	Short Diameter of Flange G	Thickness of Flange, Min B	Thickness of Flange at Bolt Circle D	Center to Groove Flange Edge, Elbow, Tee, and Cross A	Bolt Spacing	Number of Bolts	Size of Bolts
1/4	3/8	1/4	3- 7/16	1-13/16	13/16	1/2	2-3/8	2-1/4	2	1/2
3/8	1/2	1/4	3-11/16	1-13/16	15/16	9/16	2-1/2	2-1/2	2	1/2
1/2	5/8	1/4	3- 7/8	1-13/16	1	5/8	2-3/4	2-3/4	2	1/2
3/4	13/16	1/4	4- 1/8	2- 1/8	1- 1/8	11/16	3-1/4	3	2	1/2

Round Flanged Fittings and Screwed Flanges
(Straight Sizes)

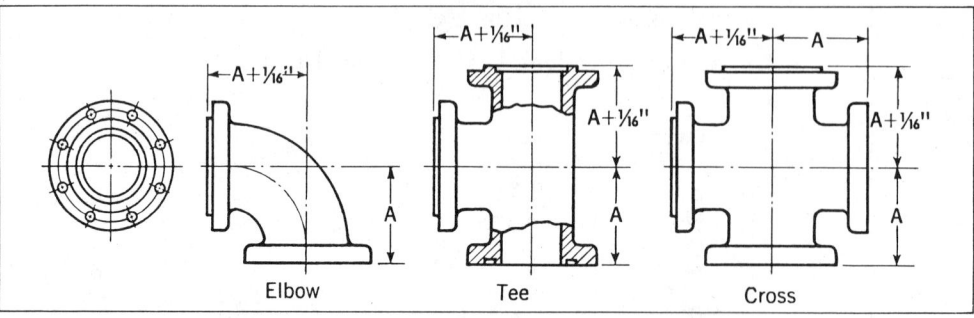

Elbow   Tee   Cross

Nominal Pipe Size	Inside Diameter of Fitting	Wall Thickness of Fitting	Outside Diameter of Flange	Thickness of Flange, Min		Center to Groove, Flange Edge Elbow, Tee and Cross A	Diameter of Bolt Circle	Number of Bolts	Size of Bolts
				Tongue	Groove				
5	5	11/16	11	1- 5/8	1- 9/16	8- 3/16	9-1/4	8	3/4
6	6	3/4	12-1/2	1-11/16	1- 5/8	8-11/16	10-5/8	12	3/4
8	8	13/16	15	1- 7/8	1-13/16	10- 3/16	13	12	7/8
10	10	15/16	17-1/2	2- 1/8	2- 1/16	11-11/16	15-1/4	16	1
12	12	1	20-1/2	2- 1/4	2- 3/16	13- 3/16	17-3/4	16	1-1/8

All dimensions are given in inches.
† Extracted from former standard ASA B16.16 by permission of the publisher, The American Society of Mechanical Engineers.

Square Flanged Fittings and Screwed Flanges

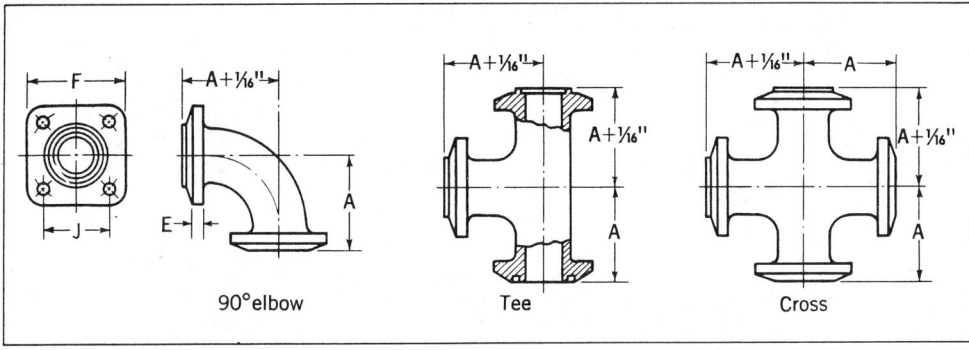

90° elbow   Tee   Cross

Nominal Pipe Size	Inside Diameter of Fitting	Wall Thickness of Fitting	Length and Width of Flange, F	Thickness of Flange, Min B	Thickness of Flange at Edge E	Center to Groove Flange Edge, Elbow, Tee, and Cross A	Bolt Spacing J	Number of Bolts	Size of Bolts
1	1-1/16	1/4	3-9/16	15/16	9/16	3-1/2	2- 5/16	4	1/2
1-1/4	1-3/8	5/16	3-7/8	1- 1/16	5/8	3-3/4	2- 5/8	4	1/2
1-1/2	1-9/16	5/16	4-1/8	1- 1/8	11/16	4-1/4	2-13/16	4	1/2
2	2-1/16	3/8	5	1- 1/4	13/16	4-1/4	3- 7/16	4	5/8
2-1/2	2-1/2	3/8	5-1/2	1- 3/8	7/8	6	3- 7/8	4	5/8
3	3	7/16	6-1/8	1- 1/2	15/16	6	4- 5/16	4	3/4
3-1/2	3-1/2	7/16	6-1/2	1- 5/8	1	6-1/2	4- 3/4	4	3/4
4	4	7/16	7-1/4	1- 3/4	1- 1/8	7	5- 1/8	4	7/8

ANSI-B31.5 Refrigeration Piping Code and ANSI-B9.1 Safety Code for Mechanical Refrigeration require that:

Refrigerant working pressure shall not exceed 250 psi in Schedule-40 ferrous pipe and butt welded pipe shall not exceed 2 in. in size.

Refrigerant liquid ferrous pipe, 1½-in. and smaller, shall be of Schedule-80.

Refrigerant tubing shall not be lighter than Type L for field assembly, or Type M for shop assembly.

Soft annealed copper tubing erected on the premises shall not be used in sizes exceeding ¾ in.

Pipe thinner than Schedule-40 ferrous, or regular brass, shall not be threaded.

For refrigerants, cast iron shall conform to ASTM A126, Class B higher strength gray iron with not less than 31,000 psi tensile strength.

Screwed joints are limited by the Piping Code to not over 3 in. nominal size for refrigerant pressures 250 psi or less, 1¼ in. for refrigerant pressures above 250 psi and not over 4 in. for brine. Screwed-on flanges are not permitted in sizes larger than the above. Bushings are not permitted for reductions of a single pipe size. For screwed brass or copper refrigerant pipe, extra heavy fittings are required.

Flanged joints are used for sizes greater than the above, and in many cases for smaller sizes.

Welding is almost universally accepted as the standard method of assembly of ferrous piping systems today, except that valves are flanged and flanged unions are used where disassembly is frequent.

In soldering for non-ferrous materials the clearance between the two surfaces is maintained within narrow limits. The solder, with suitable flux, is run into this narrow annulus in a molten condition. The joint depends on the large area of contact, and of shear. Hard solder is ordinarily a silver base alloy, melting above 1,000 deg, whereas soft solder is ordinarily a tin base alloy, melting below 500 deg.

Flared compression fittings are permitted by the Piping Code for refrigerant lines not over ¾ in. outside diameter, using annealed copper tubing, providing joints are exposed for visual inspection.

STEEL PIPE FLANGE FACING-ALL PRESSURES†

(Other than Ring Joint Facings)

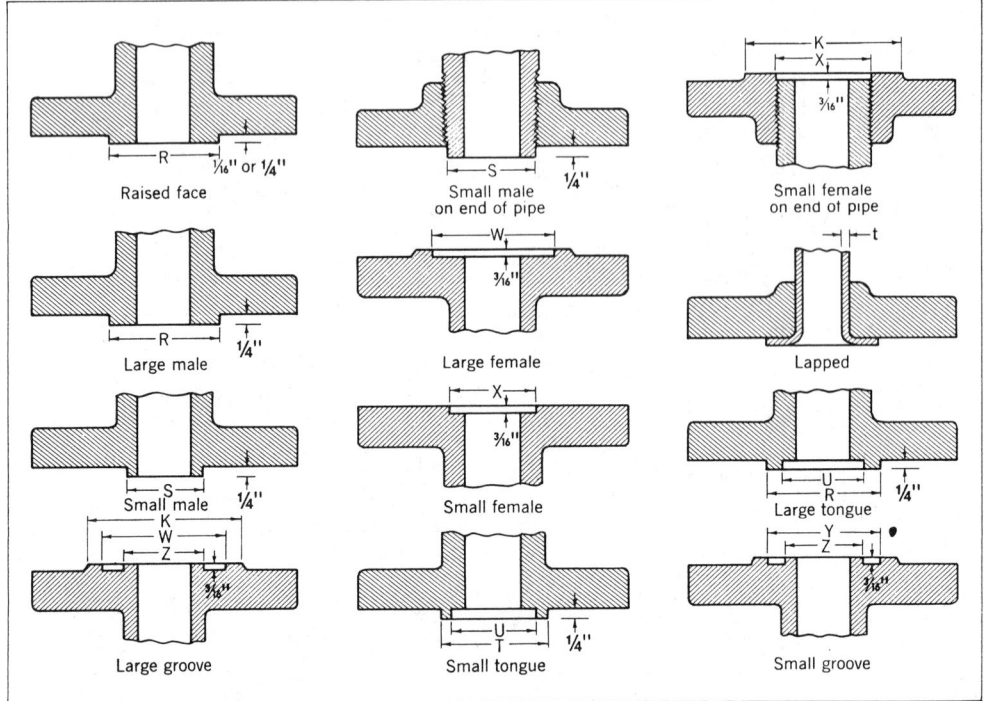

Nom. Pipe Dia.	Dimension								
	R	S	T	U	W	X	Y	Z	K
1/2	1- 3/8	23/32	1- 3/8	1	1- 7/16	25/32	1- 7/16	15/16	1-13/16
3/4	1-11/16	15/16	1-11/16	1-5/16	1- 3/4	1	1- 3/4	1- 1/4	2- 1/8
1	2	1- 3/16	1- 7/8	1-1/2	2- 1/16	1- 1/4	1-15/16	1- 7/16	2- 7/16
1-1/4	2- 1/2	1- 1/2	2- 1/4	1-7/8	2- 9/16	1- 9/16	2- 5/16	1-13/16	2-15/16
1-1/2	2- 7/8	1- 3/4	2- 1/2	2-1/8	2-15/16	1-13/16	2- 9/16	2- 1/16	3- 5/16
2	3- 5/8	2- 1/4	3- 1/4	2-7/8	3-11/16	2- 5/16	3- 5/16	2-13/16	4- 1/16
2-1/2	4- 1/8	2-11/16	3- 3/4	3-3/8	4- 3/16	2- 3/4	3-13/16	3- 5/16	4- 9/16
3	5	3- 5/16	4- 5/8	4-1/4	5- 1/16	3- 3/8	4-11/16	4- 3/16	5- 7/16
3-1/2	5- 1/2	3-13/16	5- 1/8	4-3/4	5- 9/16	3- 7/8	5- 3/16	4-11/16	5-15/16
4	6- 3/16	4- 5/16	5-11/16	5-3/16	6- 1/4	4- 3/8	5- 3/4	5- 1/8	6- 5/8
5	7- 5/16	5- 3/8	6-13/16	6-5/16	7- 3/8	5- 7/16	6- 7/8	6- 1/4	7- 3/4
6	8- 1/2	6- 3/8	8	7-1/2	9- 9/16	6- 7/16	8- 1/16	7- 7/16	8-15/16
8	10- 5/8	8- 3/8	10	9-3/8	10-11/16	8- 7/16	10- 1/16	9- 5/16	11- 1/16
10	12- 3/4	10- 1/2	12	11-1/4	12-13/16	10- 9/16	12- 1/16	11- 3/16	13- 3/16
12	15	12- 1/2	14- 1/4	13-1/2	15- 1/16	12- 9/16	14- 5/16	13- 7/16	15- 7/16
14	16- 1/4	13- 3/4	15- 1/2	14-3/4	16- 5/16	13-13/16	15- 9/16	14-11/16	16-11/16
16	18- 1/2	15- 3/4	17- 5/8	16-3/4	18- 9/16	15-13/16	17-11/16	16-11/16	18-15/16
18	21	17- 3/4	20- 1/8	19-1/4	21- 1/16	17-13/16	20- 3/16	19- 3/16	21- 7/16
20	23	19- 3/4	22	21	23- 1/16	19-13/16	22- 1/16	20-15/16	23- 7/16
24	27- 1/4	23- 3/4	26- 1/4	25-1/4	27- 5/16	23-13/16	26- 5/16	25- 3/16	27-11/16

All dimensions given in inches.

† Extracted from ANSI-B16.5 by permission of the publisher, The American Society of Mechanical Engineers.

All type facings are available in lapped joints. For raised face lapped joints, raised face is equal to t; for large male face, raise is 1/4 inch or t, whichever is greater, where t is the pipe wall thickness; other dimensions as in non-lapped joints.

150 AND 300 LB STEEL FLANGED FITTINGS†
(For other than Ring Joint Facings)

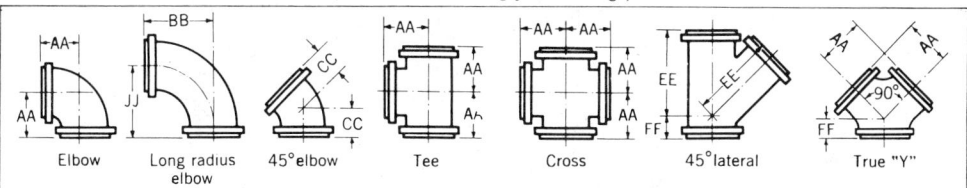

Nom. Pipe Dia.	Inside Dia. of Fitting	Min. Wall Thickness	Flange Dia.	Flange Thickness	Dimensions to Contact Surface of Raised Face				
					AA	BB	CC	EE	FF
150 Pound Fittings									
1	1	1/4	4-1/4	7/16	3-1/2	5	1-3/4	5-3/4	1-3/4
1-1/4	1-1/4	1/4	4-5/8	1/2	3-3/4	5-1/2	2	6-1/4	1-3/4
1-1/2	1-1/2	1/4	5	9/16	4	6	2-1/4	7	2
2	2	1/4	6	5/8	4-1/2	6-1/2	2-1/2	8	2-1/2
2-1/2	2-1/2	1/4	7	11/16	5	7	3	9-1/2	2-1/2
3	3	1/4	7-1/2	3/4	5-1/2	7-3/4	3	10	3
3-1/2	3-1/2	1/4	8-1/2	13/16	6	8-1/2	3-1/2	11-1/2	3
4	4	1/4	9	15/16	6-1/2	9	4	12	3
5	5	9/32	10	15/16	7-1/2	10-1/4	4-1/2	13-1/2	3-1/2
6	6	9/32	11	1	8	11-1/2	5	14-1/2	3-1/2
8	8	5/16	13-1/2	1- 1/8	9	14	5-1/2	17-1/2	4-1/2
10	10	11/32	16	1- 3/16	11	16-1/2	6-1/2	20-1/2	5
12	12	3/8	19	1- 1/4	12	19	7-1/2	24-1/2	5-1/2
14	13-1/4	13/32	21	1- 3/8	14	21-1/2	7-1/2	27	6
16	15-1/4	7/16	23-1/2	1- 7/16	15	24	8	30	6-1/2
18	17-1/4	15/32	25	1- 9/16	16-1/2	26-1/2	8-1/2	32	7
20	19-1/4	1/2	27-1/2	1-11/16	18	29	9-1/2	35	8
24	23-1/4	9/16	32	1- 7/8	22	34	11	40-1/2	9
300 Pound Fittings									
1	1	1/4	4-7/8	11/16	4	5	2-1/4	6-1/2	2
1-1/4	1-1/4	1/4	5-1/4	3/4	4-1/4	5-1/2	2-1/2	7-1/4	2-1/4
1-1/2	1-1/2	1/4	6-1/8	13/16	4-1/2	6	2-3/4	8-1/2	2-1/2
2	2	1/4	6-1/2	7/8	5	6-1/2	3	9	2-1/2
2-1/2	2-1/2	1/4	7-1/2	1	5-1/2	7	3-1/2	10-1/2	2-1/2
3	3	9/32	8-1/4	1- 1/8	6	7-3/4	3-1/2	11	3
3-1/2	3-1/2	9/32	9	1- 3/16	6-1/2	8-1/2	4	12-1/2	3
4	4	5/16	10	1- 1/4	7	9	4-1/2	13-1/2	3
5	5	3/8	11	1- 3/8	8	10-1/4	5	15	3-1/2
6	6	3/8	12-1/2	1- 7/16	8-1/2	11-1/2	5-1/2	17-1/2	4
8	8	7/16	15	1- 5/8	10	14	6	20-1/2	5
10	10	1/2	17-1/2	1- 7/8	11-1/2	16-1/2	7	24	5-1/2
12	12	9/16	20-1/2	2	13	19	8	27-1/2	6
14	13-1/4	5/8	23	2 -1/8	15	21-1/2	8-1/2	31	6-1/2
16	15-1/4	11/16	25-1/2	2- 1/4	16-1/2	24	9-1/2	34-1/2	7-1/2
18	17	3/4	28	2- 3/8	18	26-1/2	10	37-1/2	8
20	19	13/16	30-1/2	2- 1/2	19-1/2	29	10-1/2	40-1/2	8-1/2
24	23	15/16	36	2- 3/4	22-1/2	34	12	47-1/2	10

All dimensions given in inches.
† Extracted from ANSI-B16.5 by permission of the publisher, The American Society of Mechanical Engineers.
A raised face of $\frac{1}{16}$-inch is included in (a) thickness of flanges, (b) "center-to-contact-surface" dimensions; hence the "center-to-contact-surface" dimensions are the same as the "center-to-flange-edge" dimensions for this type of facing. Where facings other than the $\frac{1}{16}$-inch raised face are used, the "center-to-flange-edge" dimensions shall remain unchanged.

400, 900 LB STEEL FLANGED FITTINGS†
(For other than Ring Joint Facings)

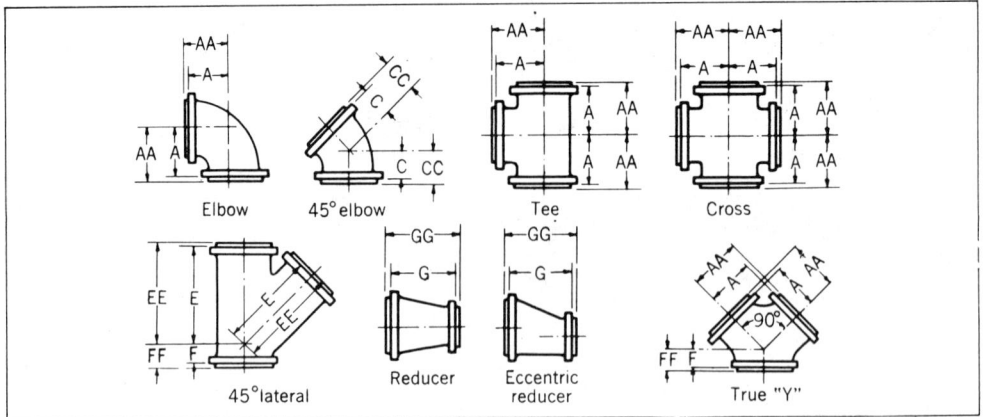

Elbow   45°elbow   Tee   Cross

45°lateral   Reducer   Eccentric reducer   True "Y"

Nominal Pipe Dia.	Flange Dia.	Flange Thickness	Wall Thickness	Inside Diam. of Fitting	Dimensions to Contact Surface of Raised Face				
					AA	CC	EE	FF	GG
400 Pound Fittings									
½ to 3½	Use 600 Pound dimensions in these sizes								
4	10	1-3/8	3/8	4	8	5-1/2	16	4-1/2	8-1/4
5	11	1-1/2	7/16	5	9	6	16-3/4	5	9-1/4
6	12-1/2	1-5/8	7/16	6	9-3/4	6-1/4	18-3/4	5-1/4	10
8	15	1-7/8	9/16	8	11-3/4	6-3/4	22-1/4	5-3/4	12
10	17-1/2	2-1/8	11/16	10	13-1/4	7-3/4	25-3/4	6-1/4	13-1/2
12	20-1/2	2-1/4	3/4	12	15	8-3/4	29-3/4	6-1/2	15-1/4
14	23	2-3/8	13/16	13-1/8	16-1/4	9-1/4	32-3/4	7	16-1/2
16	25-1/2	2-1/2	1/8	15	17-3/4	10-1/4	36-1/4	8	18-1/2
18	28	2-5/8	15/16	17	19-1/4	10-3/4	39-1/4	8-1/2	19-1/2
20	30-1/2	2-3/4	1-1/16	18-7/8	20-3/4	11-1/4	42-3/4	9	21
24	36	3	1-3/16	22-5/8	24-1/4	12-3/4	50-1/4	10-1/2	24-1/2
900 Pound Fittings									
½ to 2½	Use 1500 Pound dimensions in these sizes								
3	9-1/2	1-1/2	13/32	2-7/8	7-1/2	5-1/2	14-1/2	4-1/2	7-3/4
4	11-1/2	1-3/4	1/2	3-7/8	9	6-1/2	17-1/2	5-1/2	9-1/4
5	13-3/4	2	19/32	4-3/4	11	7-1/2	21	6-1/2	11-1/4
6	15	2-3/16	23/32	5-3/4	12	8	22-1/2	6-1/2	12-1/4
8	18-1/2	2-1/2	7/8	7-1/2	14-1/2	9	27-1/2	7-1/2	14-3/4
10	21-1/2	2-3/4	1-1/16	9-3/8	16-1/2	10	31-1/2	8-1/2	16-3/4
12	24	3-1/8	1-1/4	11-1/8	19	11	34-1/2	9	17-3/4
14	25-1/4	3-3/8	1-3/8	12-1/4	20-1/4	11-1/2	36-1/2	9-1/2	19
16	27-3/4	3-1/2	1-9/16	14	22-1/4	12-1/2	40-3/4	10-1/2	21
18	31	4	1-3/4	15-3/4	24	13-1/4	45-1/2	12	24-1/2
20	33-3/4	4-1/4	1-29/32	17-1/2	26	14-1/2	50-1/4	13	26-1/2
24	41	5-1/2	2-9/32	21	30-1/2	18	60	15-1/2	30-1/2

All dimensions given in inches.
† Extracted from ANSI-B16.5 by permission of the publisher, The American Society of Mechanical Engineers.

600 AND 1500 LB STEEL FLANGED FITTINGS†
(For other than Ring Joint Facings)

Nominal Pipe Dia.	Flange Dia.	Flange Thickness	Wall Thickness	Inside Dia. of Fitting	Dimensions to Contact Surface of Raised Edge (As shown for 2500 lb fittings)				
					AA	CC	EE	FF	GG
600 Pound Fittings									
1/2	3-3/4	9/16	1/4	1/2	3-1/4	2	5-3/4	1-3/4	5
3/4	4-5/8	5/8	1/4	3/4	3-3/4	2-1/2	6-3/4	2	5
1	4-7/8	11/16	1/4	1	4-1/4	2-1/2	7-1/4	2-1/4	5
1-1/4	5-1/4	13/16	1/4	1-1/4	4-1/2	2-3/4	8	2-1/2	5
1-1/2	6-1/8	7/8	1/4	1-1/2	4-3/4	3	9	2-3/4	5
2	6-1/2	1	1/4	2	5-3/4	4-1/4	10-1/4	3-1/2	6
2-1/2	7-1/2	1-1/8	9/32	2-1/2	6-1/2	4-1/2	11-1/2	3-1/2	6-3/4
3	8-1/4	1-1/4	5/16	3	7	5	12-3/4	4	7-1/4
3-1/2	9	1-3/8	11/32	3-1/2	7-1/2	5-1/2	14	4-1/2	7-3/4
4	10-3/4	1-1/2	3/8	4	8-1/2	6	16-1/2	4-1/2	8-3/4
5	13	1-3/4	7/16	5	10	7	19-1/2	6	10-1/4
6	14	1-7/8	1/2	6	11	7-1/2	21	6-1/2	11-1/4
8	16-1/2	2-3/16	5/8	7-7/8	13	8-1/2	24-1/2	7	13-1/4
10	20	2-1/2	3/4	9-3/4	15-1/2	9-1/2	29-1/2	8	15-3/4
12	22	2-5/8	29/32	11-3/4	16-1/2	10	31-1/2	8-1/2	16-3/4
14	23-3/4	2-3/4	31/32	12-7/8	17-1/2	10-3/4	34-1/4	9	17-3/4
16	27	3	1-3/32	14-3/4	19-1/2	11-3/4	38-1/2	10	19-3/4
18	29-1/4	3-1/4	1-7/32	16-1/2	21-1/2	12-1/4	42	10-1/2	21-3/4
20	32	3-1/2	1-11/32	18-1/4	23-1/2	13	45-1/2	11	23-3/4
24	37	4	1-19/32	22	27-1/2	14-3/4	53	13	27-3/4
1500 Pound Fittings									
1/2	4-3/4	7/8	1/4	1/2	4-1/4	3	—	—	—
3/4	5-1/8	1	1/4	11/16	4-1/2	3-1/4	—	—	—
1	5-7/8	1-1/8	1/4	7/8	5	3-1/2	9	2-1/2	5
1-1/4	6-1/4	1-1/8	5/16	1-1/8	5-1/2	4	10	3	5-3/4
1-1/2	7	1-1/4	3/8	1-3/8	6	4-1/4	11	3-1/2	6-1/4
2	8-1/2	1-1/2	7/16	1-7/8	7-1/4	4-3/4	13-1/4	4	7-1/4
2-1/2	9-5/8	1-5/8	1/2	2-1/4	8-1/4	5-1/4	15-1/4	4-1/2	8-1/4
3.	10-1/2	1-7/8	5/8	2-3/4	9-1/4	5-3/4	17-1/4	5	9-1/4
4	12-1/4	2-1/8	3/4	3-5/8	10-3/4	7-1/4	19-1/4	6	10-3/4
5	14-3/4	2-7/8	29/32	4-3/8	13-1/4	8-3/4	23-1/4	7-1/2	13-3/4
6	15-1/2	3-1/4	1-3/32	5-3/8	13-7/8	9-3/8	24-7/8	8-1/8	14-1/2
8	19	3-5/8	1-13/32	7	16-3/8	10-7/8	29-7/8	9-1/8	17
10	23	4-1/4	1-23/32	8-3/4	19-1/2	12	36	10-1/4	20-1/4
12	26-1/2	4-7/8	2	10-3/8	22-1/4	13-1/4	40-3/4	12	23
14	29-1/2	5-1/4	2-3/16	11-3/8	24-3/4	14-1/4	44	12-1/2	25-3/4
16	32-1/2	5-3/4	2-1/2	13	27-1/4	16-1/4	48-1/4	14-3/4	28-1/4
18	36	6-3/8	2-13/16	14-5/8	30-1/4	17-3/4	53-1/4	16-1/4	31-1/2
20	38-3/4	7	3-1/8	16-3/8	32-3/4	18-3/4	57-3/4	17-3/4	34
24	46	8	3-23/32	19-5/8	38-1/4	20-3/4	67-1/4	20-1/2	39-3/4

All dimensions given in inches.
† Extracted from ANSI-B16.5 by permission of the publisher, The American Society of Mechanical Engineers.

2500 LB STEEL FLANGED FITTINGS†
(For other than Ring Joint Facings)

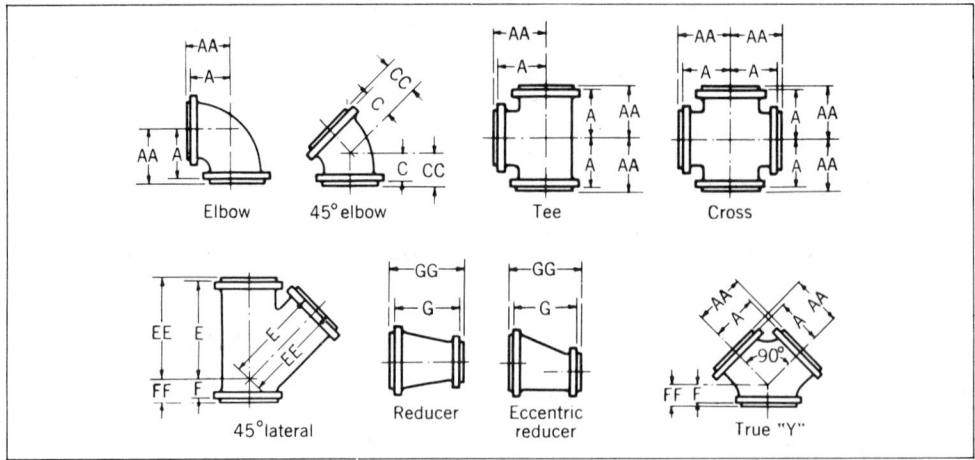

Elbow 45° elbow Tee Cross 45°lateral Reducer Eccentric reducer True "Y"

Nominal Pipe Dia.	Flange Dia.	Flange Thickness	Wall Thickness	Inside Diam. of Fitting	Dimensions to Contact Surface of Raised Face				
					AA	CC	EE	FF	GG
2500 Pound Fittings									
1/2	5-1/4	1-3/16	1/4	7/16	5-3/16	—	—	—	—
3/4	5-1/2	1-1/4	9/32	9/16	5-3/8	—	—	—	—
1	6-1/4	1-3/8	11/32	3/4	6-1/16	4	—	—	—
1-1/4	7-1/4	1-1/2	7/16	1	6-7/8	4-1/4	—	—	—
1-1/2	8	1-3/4	1/2	1-1/8	7-9/16	4-3/4	—	—	—
2	9-1/4	2	5/8	1-1/2	8-7/8	5-3/4	15-1/4	5-1/4	9-1/2
2-1/2	10-1/2	2-1/4	3/4	1-7/8	10	6-1/4	17-1/4	5-3/4	10-1/2
3	12	2-5/8	7/8	2-1/4	11-3/8	7-1/4	19-3/4	6-3/4	11-3/4
4	14	3	1- 3/32	2-7/8	13-1/4	8-1/2	23	7-3/4	13-1/2
5	16-1/2	3-5/8	1-11/32	3-5/8	15-5/8	10	27-1/4	9-1/4	15-3/4
6	19	4-1/4	1-19/32	4-3/8	18	11-1/2	31-1/4	10-1/2	18
8	21-3/4	5	2- 1/16	5-3/4	20-1/8	12-3/4	35-1/4	11-3/4	20-1/2
10	26-1/2	6-1/2	2-19/32	7-1/4	25	16	43-1/4	14-3/4	25-1/2
12	30	7-1/4	3- 1/32	8-5/8	28	17-3/4	49-1/4	16-1/4	29

All dimensions given in inches.

† Extracted from ANSI-B16.5 by permission of the publisher, The American Society of Mechanical Engineers.

Because of the 1/4 inch raised face in these flanges, dimensions A, C, E, and F, are 1/4 inch less than AA, CC, EE, and FF, respectively; since two faces are involved in G, G is 1/2 inch less than GG.

A principle of design in this standard is to maintain a fixed position for the flanged edge with reference to the body of the fitting. The addition of any facing is beyond the outside edge of the flange except for the 1/16-in. raised face in the 150- and 300-lb. standards.

Center to contact surface or center to flange edge dimensions for all openings shall be the same as those of straight size fittings of the largest opening. The contact surface to contact surface or flange edge to flange edge dimensions for all reductions of reducers and eccentric reducers shall be as listed for the larger opening.

Side outlet elbows, side outlet tees, and side outlet crosses, shall have all openings on intersecting center lines and the center to contact surface dimensions of the side outlet shall be the same as for the largest opening. Long radius elbows with side outlet shall have the side outlet on the radial center line of the elbow and the center to contact surface dimension of the side outlet shall be the same as for the regular 90 deg. elbow of the largest opening.

STEEL BUTT-WELDING FITTINGS†

Elbows, Tees and Returns

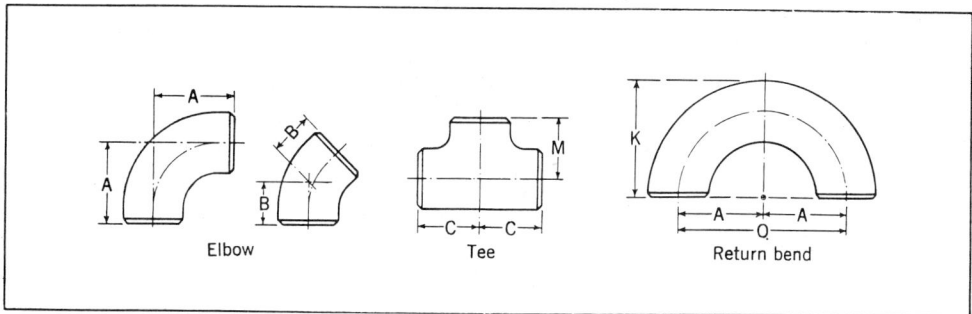

| Elbow | Tee | Return bend |

Nominal Pipe Size	Outside Diameter at Bevel	Center-to-End				180° Returns*	
		90-Deg Elbows A	45-Deg Elbows B	Tees, Run C	Tees, Outlet M	O	K
1	1.315	1½	⅞	1½	1½	3	2³⁄₁₆
1¼	1.660	1⅞	1	1⅞	1⅞	3¾	2¾
1½	1.900	2¼	1⅛	2¼	2¼	4½	3¼
2	2.375	3	1⅜	2½	2½	6	4³⁄₁₆
2½	2.875	3¾	1¾	3	3	7½	5³⁄₁₆
3	3.500	4½	2	3⅜	3⅜	9	6¼
3½	4.000	5¼	2¼	3¾	3¾	10½	7¼
4	4.500	6	2½	4⅛	4⅛	12	8¼
5	5.563	7½	3⅛	4⅞	4⅞	15	10⁵⁄₁₆
6	6.625	9	3¾	5⅝	5⅝	18	12⁵⁄₁₆
8	8.625	12	5	7	7	24	16⁵⁄₁₆
10	10.750	15	6¼	8½	8½	30	20⅜
12	12.750	18	7½	10	10	36	24⅜
14	14.000	21	8¾	11		42	28
16	16.000	24	10	12	Not	48	32
18	18.000	27	11¼	13½	stand-	54	36
20	20.000	30	12½	15	ard	60	40
24	24.000	36	15	17		72	48

All dimensions given in inches.
* Dimension $A = ½$ dimension O.
† Extracted from ANSI-B16.9 by permission of the publisher, The American Society of Mechanical Engineers.

The accompanying data on Steel Butt-Welding fittings is from an American National Standards Institute standard of which the sponsors were Mechanical Contractors Association of America, Manufacturers Standardization Society of the Valve and Fittings Industry, and The American Society of Mechanical Engineers.

The standard covers over-all dimensions, tolerances, and marking for wrought and cast carbon-steel and alloy-steel welding fittings. Welding ends are to be in accordance with ANSI-B16.25.

In this standard "wrought" is used to denote fittings made of pipe, tubing, plate, or forgings.

Fittings shall be designed so that the pressure rating may be calculated as for straight seamless pipe of the same or equivalent material in accordance with the rules established in the various sections of the Code for Pressure Piping (ANSI-B31).

The "size" of the fittings in the accompanying tables is identified by the corresponding "nominal pipe size." For fittings 14 in and larger the OD of the pipe corresponds with the nominal size.

Reducers

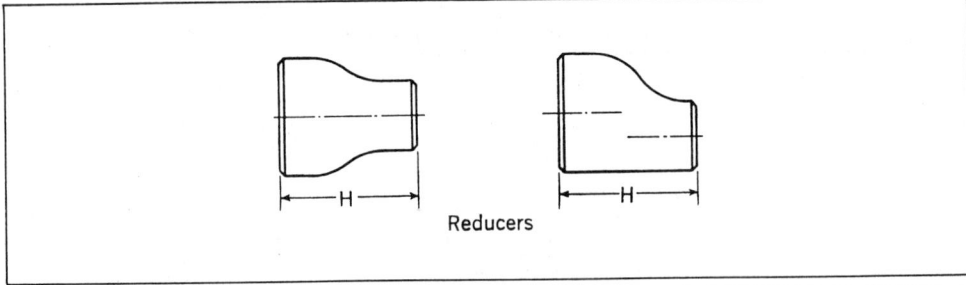

Reducers

Nominal Pipe Size	End-to-End H	Nominal Pipe Size	End-to-End H	Nominal Pipe Size	End-to-End H	Nominal Pipe Size	End-to-End H
1 × ¾	2	3 × 2½	3½	6 × 5	5½	14 × 12	13
1 × ½	2	3 × 2	3½	6 × 4	5½	14 × 10	13
		3 × 1½	3½	6 × 3½	5½	14 × 8	13
1¼ × 1	2	3 × 1¼	3½	6 × 3	5½	14 × 6	13
1¼ × ¾	2			6 × 2½	5½		
1¼ × ½	2	3½ × 3	4			16 × 14	14
		3½ × 2½	4	8 × 6	6	16 × 12	14
1½ × 1¼	2½	3½ × 2	4	8 × 5	6	16 × 10	14
1½ × 1	2½	3½ × 1½	4	8 × 4	6	16 × 8	14
1½ × ¾	2½	3½ × 1¼	4	8 × 3½	6		
1½ × ½	2½					18 × 16	15
		4 × 3½	4	10 × 8	7	18 × 14	15
2 × 1½	3	4 × 3	4	10 × 6	7	18 × 12	15
2 × 1¼	3	4 × 2½	4	10 × 5	7	18 × 10	15
2 × 1	3	4 × 2	4	10 × 4	7		
2 × ¾	3	4 × 1½	4			20 × 18	20
				12 × 10	8	20 × 16	20
2½ × 2	3½	5 × 4	5	12 × 8	8	20 × 14	20
2½ × 1½	3½	5 × 3½	5	12 × 6	8	20 × 12	20
2½ × 1¼	3½	5 × 3	5	12 × 5	8		
2½ × 1	3½	5 × 2½	5			24 × 20	20
		5 × 2	5			24 × 18	20
						24 × 16	20

All dimensions given in inches.

One of the principles of this standard is the maintenance of a fixed position for the welding ends with reference to the center line of the fittings or the over-all dimensions, as the case may be.

The actual bursting pressure of the fittings covered by this standard shall at least equal the computed bursting pressure of seamless pipe of the schedule number (or nominal wall thickness) and material designated by the marking on the fitting. To determine the bursting pressure of the fittings, straight seamless pipe of the designated schedule (or nominal wall thickness) and material shall be welded to each end, each pipe being at least equal in length to twice the outside diameter of the pipe and having proper end closures, applied beyond the minimum length of straight pipe; hydrostatic pressure shall be applied until either the fitting or one of the pipes welded thereto bursts.

Hydrostatic testing of wrought fittings is not required in this standard. Cast welding fittings shall be hydrostatically tested. All fittings shall be capable of withstanding without leakage a test pressure equal to that prescribed in the specification for the pipe with which the fitting is recommended to be used.

Reducing Outlet Tees

Nominal Pipe Size	Center-to-End Run C	Center-to-End Outlet M
1 × 1 × ¾	1½	1½
1 × 1 × ½	1½	1½
1¼ × 1¼ × 1	1⅞	1⅞
1¼ × 1¼ × ¾	1⅞	1⅞
1¼ × 1¼ × ½	1⅞	1⅞
1½ × 1½ × 1¼	2¼	2¼
1½ × 1½ × 1	2¼	2¼
1½ × 1½ × ¾	2¼	2¼
1½ × 1½ × ½	2¼	2¼
2 × 2 × 1½	2½	2⅜
2 × 2 × 1¼	2½	2¼
2 × 2 × 1	2½	2
2 × 2 × ¾	2½	1¾
2½ × 2½ × 2	3	2¾
2½ × 2½ × 1½	3	2⅝
2½ × 2½ × 1¼	3	2½
2½ × 2½ × 1	3	2¼
3 × 3 × 2½	3⅜	3¼
3 × 3 × 2	3⅜	3
3 × 3 × 1½	3⅜	2⅞
3 × 3 × 1¼	3⅜	2¾
3½ × 3½ × 3	3¾	3⅝
3½ × 3½ × 2½	3¾	3½
3½ × 3½ × 2	3¾	3¼
3½ × 3½ × 1½	3¾	3⅛
4 × 4 × 3½	4⅛	4
4 × 4 × 3	4⅛	3⅞
4 × 4 × 2½	4⅛	3¾
4 × 4 × 2	4⅛	3½
4 × 4 × 1½	4⅛	3⅜
5 × 5 × 4	4⅞	4⅝
5 × 5 × 3½	4⅞	4½
5 × 5 × 3	4⅞	4⅜
5 × 5 × 2½	4⅞	4¼
5 × 5 × 2	4⅞	4⅛
6 × 6 × 5	5⅝	5⅜
6 × 6 × 4	5⅝	5⅛
6 × 6 × 3½	5⅝	5
6 × 6 × 3	5⅝	4⅞
6 × 6 × 2½	5⅝	4¾
8 × 8 × 6	7	6⅝

Nominal Pipe Size	Center-to-End Run C	Center-to-End Outlet M
8 × 8 × 5	7	6⅜
8 × 8 × 4	7	6⅛
8 × 8 × 3½	7	6
10 × 10 × 8	8½	8
10 × 10 × 6	8½	7⅝
10 × 10 × 5	8½	7½
10 × 10 × 4	8½	7¼
12 × 12 × 10	10	9½
12 × 12 × 8	10	9
12 × 12 × 6	10	8⅝
12 × 12 × 5	10	8½
14 × 14 × 12	11	
14 × 14 × 10	11	
14 × 14 × 8	11	
14 × 14 × 6	11	
16 × 16 × 14	12	
16 × 16 × 12	12	
16 × 16 × 10	12	
16 × 16 × 8	12	
16 × 16 × 6	12	
18 × 18 × 16	13½	
18 × 18 × 14	13½	
18 × 18 × 12	13½	
18 × 18 × 10	13½	
18 × 18 × 8	13½	
20 × 20 × 18	15	
20 × 20 × 16	15	
20 × 20 × 14	15	
20 × 20 × 12	15	
20 × 20 × 10	15	
20 × 20 × 8	15	
24 × 24 × 20	17	
24 × 24 × 18	17	
24 × 24 × 16	17	
24 × 24 × 14	17	
24 × 24 × 12	17	
14 × 24 × 10	17	

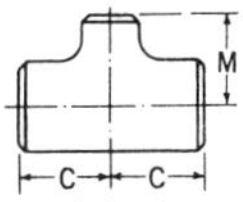

(M column, 14-size and below: — Not Standard —)

Reducing tee

All dimensions given in inches.
Outside diameter at bevel is as given for the diameter concerned under "Elbows, Tees and Returns," for which the dimensions of ½ and ¾ inch was not given but which are 0.840 and 1.050 respectively. The outside bevel diameter for a 1¼ x 1¼ x 1 reducing tee is thus 1.660 (1¼ inch) on the run and 1.315 (1 inch) for the outlet.

STEEL SOCKET WELDED FITTINGS†

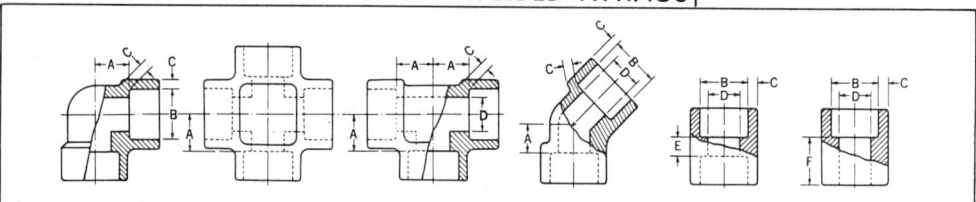

Elbows, Tees, and Crosses

Nominal Pipe Size	Center to Bottom of Socket		Bore Diameter of Socket, Min	Socket Wall Thickness Min			Bore Diameter of Fitting		
	Sched 40 and 80	Sched 160		Sched 40	Sched 80	Sched 160	Sched 40	Sched 80	Sched 160
	A		B	C			D		
1/8	7/16	—	0.420	0.125	0.125	—	0.269	0.215	—
1/4	7/16	—	0.555	0.125	0.149	—	0.364	0.302	—
3/8	17/32	—	0.690	0.125	0.158	—	0.493	0.423	—
1/2	5/8	3/4	0.855	0.136	0.184	0.234	0.622	0.546	0.466
3/4	3/4	7/8	1.065	0.141	0.193	0.273	0.824	0.742	0.614
1	7/8	1-1/16	1.330	0.166	0.224	0.313	1.049	0.957	0.815
1-1/4	1- 1/16	1-1/4	1.675	0.175	0.239	0.313	1.380	1.278	1.160
1-1/2	1- 1/4	1-1/2	1.915	0.181	0.250	0.351	1.610	1.500	1.338
2	1- 1/2	1-5/8	2.406	0.193	0.273	0.429	2.067	1.939	1.689
2-1/2	1- 5/8	2-1/4	2.906	0.254	0.345	0.469	2.469	2.323	2.125
3	2- 1/4	2-1/2	3.535	0.270	0.375	0.546	3.068	2.900	2.626

Minimum depth of socket is 3/8 for sizes up to and including 1/2 in.; 1/2 in. for sizes from 3/4 to 1 1/2, and 5/8 in. for 2, 2 1/2 and 3 in. sizes.

45 Deg Elbows, Couplings, and Half Couplings

Nominal Pipe Size	Center to Bottom of Socket for 45 Deg Ells		Half Couplings, Bottom of Socket to Opposite Face	Bore Diameter of Socket, Min	Socket Wall Thickness Min			Bore Diameter of Fitting		
	Sched 40 and 80	Sched 160			Sched 40	Sched 80	Sched 160	Sched 40	Sched 80	Sched 160
	A		F	B	C			D		
1/8	5/16	—	5/8	0.420	0.125	0.125	—	0.269	0.215	—
1/4	5/16	—	5/8	0.555	0.125	0.149	—	0.364	0.302	—
3/8	5/16	—	11/16	0.690	0.125	0.158	—	0.493	0.423	—
1/2	7/16	1/2	7/8	0.855	0.136	0.184	0.234	0.622	0.546	0.466
3/4	1/2	9/16	15/16	1.065	0.141	0.193	0.273	0.824	0.742	0.614
1	9/16	11/16	1- 1/8	1.330	0.166	0.224	0.313	1.049	0.957	0.815
1-1/4	11/16	13/16	1- 3/16	1.675	0.175	0.239	0.313	1.380	1.278	1.160
1-1/2	13/16	1	1- 1/4	1.915	0.181	0.250	0.351	1.610	1.500	1.338
2	1	1- 1/8	1- 5/8	2.406	0.193	0.273	0.429	2.067	1.939	1.689
2-1/2	1- 1/8	1- 1/4	1-11/16	2.906	0.254	0.345	0.469	2.469	2.323	2.125
3	1- 1/4	1- 3/8	1- 3/4	3.535	0.270	0.375	0.546	3.068	2.900	2.626

All dimensions are given in inches.
Minimum depth of socket same as in footnote of table above.
Couplings distance between bottoms of sockets is 1/4 in. for 1/8 and 1/4 in. sizes, 3/8 for 1/2 and 3/4 sizes, 1/2 for 1, 1 1/4 and 1 1/2 in. sizes, and 3/4 for 2, 2 1/2 and 3 in. sizes.
† Extracted from ASA B16.11 by permission of the publisher, The American Society of Mechanical Engineers.

CAST BRASS SOLDER JOINT FITTINGS†

Dimensions of Solder-Joint Ends

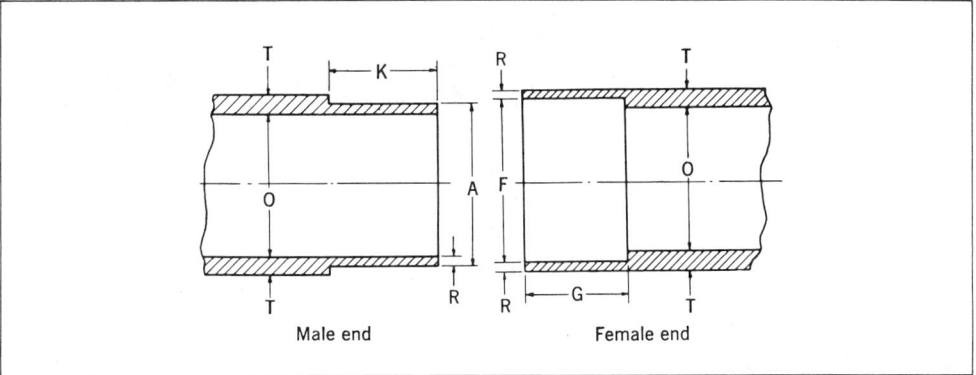

Male end   Female end

Standard Water Tube Size	Male End			Female End			Metal Thickness		Inside Diameter of Fitting
	Outside Diameter A		Length K	Inside Diameter F		Depth G	Body T	Joint R	O
	Min	Max	Min	Min	Max	Min			Min
1/4	0.374	0.376	3/8	0.378	0.380	5/16	0.08	0.048	0.31
3/8	0.499	0.501	7/16	0.503	0.505	3/8	0.09	0.048	0.43
1/2	0.624	0.626	9/16	0.628	0.630	1/2	0.09	0.054	0.54
3/4	0.874	0.876	13/16	0.878	0.880	3/4	0.10	0.060	0.78
1	1.1235	1.1265	31/32	1.1285	1.1305	29/32	0.11	0.066	1.02
1-1/4	1.3735	1.3765	1- 1/32	1.3785	1.3805	31/32	0.12	0.072	1.26
1-1/2	1.623	1.627	1- 5/32	1.629	1.6315	1- 3/32	0.13	0.078	1.50
2	2.123	2.127	1-13/32	2.129	2.1315	1-11/32	0.15	0.090	1.98
2-1/2	2.623	2.627	1-17/32	2.629	2.6315	1-15/32	0.17	0.102	2.46
3	3.123	3.127	1-23/32	3.129	3.1315	1-21/32	0.19	0.114	2.94
3-1/2	3.623	3.627	1-31/32	3.629	3.632	1-29/32	0.20	0.120	3.42
4	4.123	4.127	2- 7/32	4.129	4.132	2- 5/32	0.22	0.132	3.90
5	5.123	5.127	2-23/32	5.129	5.132	2-21/32	0.28	0.168	4.87
6	6.123	6.127	3- 5/32	6.129	6.132	3- 3/32	0.34	0.204	5.84
8	8.123	8.127	4- 1/32	8.129	8.132	3-31/32	0.375	0.310	7.72

All dimensions are given in inches.
† Extracted from ASA B16.18 by permission of the publisher, The American Society of Mechanical Engineers.

Dimensions of Elbows and Tees
(See drawing on next page)

Standard Water Tube Size	Laying Length Tee and Elbow H	Center to External Shoulder 90-Deg Street Elbow I	Laying Length 45 Deg Elbow J	Center to External Shoulder 45-Deg Street Elbow
1/4	1/4	3/8	—	—
3/8	5/16	7/16	3/16	5/16
1/2	7/16	9/16	3/16	5/16
3/4	9/16	11/16	1/4	3/8

Dimensions of Elbows and Tees (Concluded)

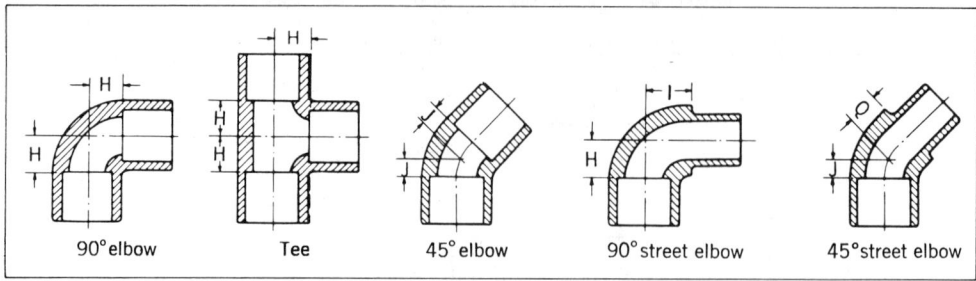

90° elbow Tee 45° elbow 90° street elbow 45° street elbow

Standard Water Tube Size	Laying Length Tee and Elbow H	Center to External Shoulder 90-Deg Street Elbow I	Laying Length 45 Deg Elbow J	Center to External Shoulder 45-Deg Street Elbow
1	3/4	7/8	5/16	7/16
1-1/4	7/8	1	7/16	9/16
1-1/2	1	1- 1/8	1/2	5/8
2	1-1/4	1- 3/8	9/16	3/4
2-1/2	1-1/2	1- 5/8	5/8	—
3	1-3/4	1- 7/8	3/4	—
3-1/2	2	—	7/8	—
4	2-1/4	2- 3/8	15/16	—
5	3-1/8	—	1- 7/16	—
6	3-5/8	—	1- 5/8	—
8	4-7/8	—	2- 1/8	—

BRASS OR BRONZE SCREWED FITTINGS†

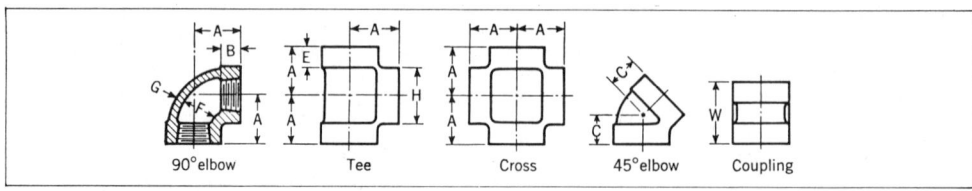

90° elbow Tee Cross 45° elbow Coupling

Nominal Pipe Size	Center to End. Elbows, Tees, and Crosses[1] A	Length of Thread, Min B	Center to End 45-Deg Elbows[1] C	Width of Band. Min E	Inside Diameter of Fitting Min	Inside Diameter of Fitting Max	Metal Thickness, Min G	Outside Diameter of Band, Min H	End to End Coupling W
1/4	0.81	0.32	0.73	0.38	0.54	0.58	0.11	0.93	1.06
3/8	0.95	0.36	0.80	0.44	0.68	0.72	0.12	1.12	1.16
1/2	1.12	0.43	0.88	0.50	0.84	0.90	0.13	1.34	1.34
3/4	1.31	0.50	0.98	0.56	1.05	1.11	0.16	1.63	1.52
1	1.50	0.58	1.12	0.62	1.32	1.38	0.17	1.95	1.67
1¼	1.75	0.67	1.29	0.69	1.66	1.73	0.19	2.39	1.93
1½	1.94	0.70	1.43	0.75	1.90	1.97	0.20	2.68	2.15
2	2.25	0.75	1.68	0.84	2.38	2.45	0.22	3.28	2.53
2½	2.70	0.92	1.95	0.94	2.88	2.98	0.24	3.86	2.88
3	3.08	0.98	2.17	1.00	3.50	3.60	0.26	4.62	3.18
4	3.79	1.08	2.61	1.12	4.50	4.60	0.31	5.79	3.69

Brass or Bronze Screwed Fittings (Concluded)
125 Pound

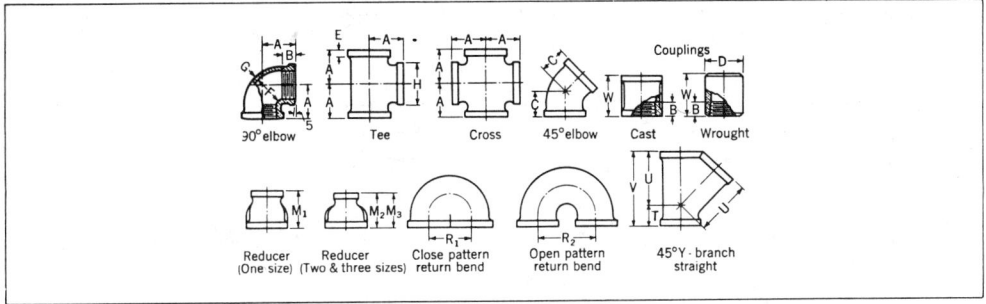

30° elbow   Tee   Cross   45° elbow   Cast   Wrought   Couplings

Reducer (One size)   Reducer (Two & three sizes)   Close pattern return bend   Open pattern return bend   45°Y - branch straight

Nominal Pipe Size	Center to End Elbows. Tees. Crosses A	Length of Thread. Min B	Center to End 45 Deg Elbows C	Band Length. Min E	Inside Diameter of Cast Fitting Min	Inside Diameter of Cast Fitting Max	Metal Thickness. Min G	Band Diameter. Min H	End to End Straight Couplings Cast	End to End Straight Couplings Wrought
					F				W	
1/8	0.54	0.25	0.42	0.14	0.41	0.44	0.08	0.67	0.80	0.83
1/4	0.71	0.32	0.56	0.16	0.54	0.58	0.08	0.81	0.97	1.03
3/8	0.82	0.36	0.63	0.17	0.68	0.72	0.09	1.00	1.05	1.11
1/2	1.01	0.43	0.78	0.19	0.84	0.90	0.09	1.17	1.29	1.36
3/4	1.18	0.50	0.89	0.23	1.05	1.11	0.10	1.42	1.43	1.50
1	1.43	0.58	1.06	0.27	1.32	1.39	0.11	1.72	1.68	—
1 1/4	1.69	0.67	1.22	0.31	1.66	1.73	0.12	2.10	1.86	—
1 1/2	1.84	0.70	1.30	0.34	1.90	1.97	0.13	2.38	1.92	—
2	2.12	0.75	1.45	0.41	2.38	2.45	0.15	2.92	2.20	—
2 1/2	2.70	0.92	1.95	0.48	2.88	2.98	0.17	3.49	2.88	—
3	3.08	0.98	2.17	0.55	3.50	3.60	0.19	4.20	3.18	—
4	3.79	1.08	2.61	0.66	4.50	4.60	0.22	5.31	3.69	—

Nominal Pipe Size	Return Reducers End to End Reducing One Size M₁	Return Reducers End to End Reducing Two Sizes M₂	Return Reducers End to End Reducing Three Sizes M₃	Return Bends Center to Center Close Pattern R₁	Return Bends Center to Center Open Pattern R₂	45 Deg-Y-Branch Center to End Inlet T	45 Deg-Y-Branch Center to End Outlet U	45 Deg-Y-Branch End to End V	Diameter Wrought Coupling D
1/4	0.88	—	—	—	—	—	—	—	9/16
3/8	1.01	0.92	—	—	—	0.50	1.28	1.78	11/16
1/2	1.17	1.13	—	1.00	1.50	0.61	1.58	2.19	27/32
3/4	1.36	1.24	1.24	1.25	2.00	0.72	1.90	2.62	1-1/16
1	1.56	1.49	—	1.50	2.50	0.85	2.33	3.18	1-5/16
1 1/4	1.77	1.65	—	—	3.00	1.02	2.83	3.85	—
1 1/2	1.89	1.80	1.80	—	3.50	1.10	3.14	4.24	—
2	2.06	2.03	2.03	—	4.00	1.24	3.76	5.00	—
2 1/2	3.25	—	—	—	—	—	—	—	—
3	3.69	3.69	—	—	—	—	—	—	—
4	4.38	—	—	—	—	—	—	—	—

All dimensions are given in inches.
† Extracted from ANSI-B16.15 by permission of the publisher, The American Society of Mechanical Engineers.

WROUGHT-COPPER AND WROUGHT BRONZE SOLDER JOINT FITTINGS†

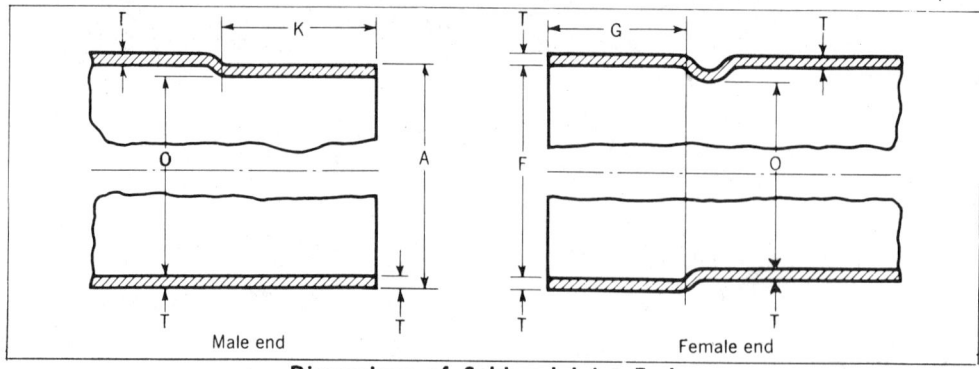

Male end   Female end

Dimensions of Soldered-Joint Ends

Standard Water Tube Size	Male End			Female End			Depth G	Metal Thickness T	Inside Diameter of Fitting O
	Outside Diameter A		Length K	Inside Diameter F					
	Min	Max	Min	Min	Max	Min	Min	Min
1/8	0.249	0.251	3/8	0.253	0.255	5/16	0.022	0.18
1/4	0.374	0.376	3/8	0.378	0.380	5/16	0.026	0.30
3/8	0.499	0.501	7/16	0.503	0.505	3/8	0.031	0.39
1/2	0.624	0.626	9/16	0.628	0.630	1/2	0.036	0.52
5/8	0.749	0.751	11/16	0.753	0.755	5/8	0.038	0.63
3/4	0.874	0.876	13/16	0.878	0.880	3/4	0.041	0.74
1	1.1235	1.1265	31/32	1.1285	1.1305	29/32	0.046	0.98
1-1/4	1.3735	1.3765	1- 1/32	1.3785	1.3805	31/32	0.050	1.23
1-1/2	1.623	1.627	1- 5/32	1.629	1.6315	1- 3/32	0.055	1.47
2	2.123	2.127	1-13/32	2.129	2.1315	1-11/32	0.064	1.94
2-1/2	2.623	2.627	1-17/32	2.629	2.6315	1-15/32	0.074	2.42
3	3.123	3.127	1-23/32	3.129	3.1315	1-21/32	0.083	2.89
3-1/2	3.623	3.627	1-31/32	3.629	3.632	1-29/32	0.093	3.37
4	4.123	4.127	2- 7/32	4.129	4.132	2- 5/32	0.101	3.84

All dimensions are given in inches.
† Extracted from ANSI-B16.22 by permission of the publisher, The American Society of Mechanical Engineers.

Solder Used in Joints

Fluid* and Size	50-50 Tin-Lead				95-5 Tin-Antimony or Lead-Tin				Solders Melting at or Above 1100 Deg
	Service Temperatures, Deg F								
	100	150	200	250	100	150	200	250	350
	Max Service Pressure, psi								
1/8 to 1	200	150	100	85	500	400	300	200	270
1-1/4 to 2	175	125	90	75	400	350	250	175	190
2-1/2 to 4	150	100	75	50	300	275	200	150	155
Steam, all	—	—	—	15	—	—	—	15	120

* Fluid is water except in last line.

BRASS OR BRONZE FLANGES†

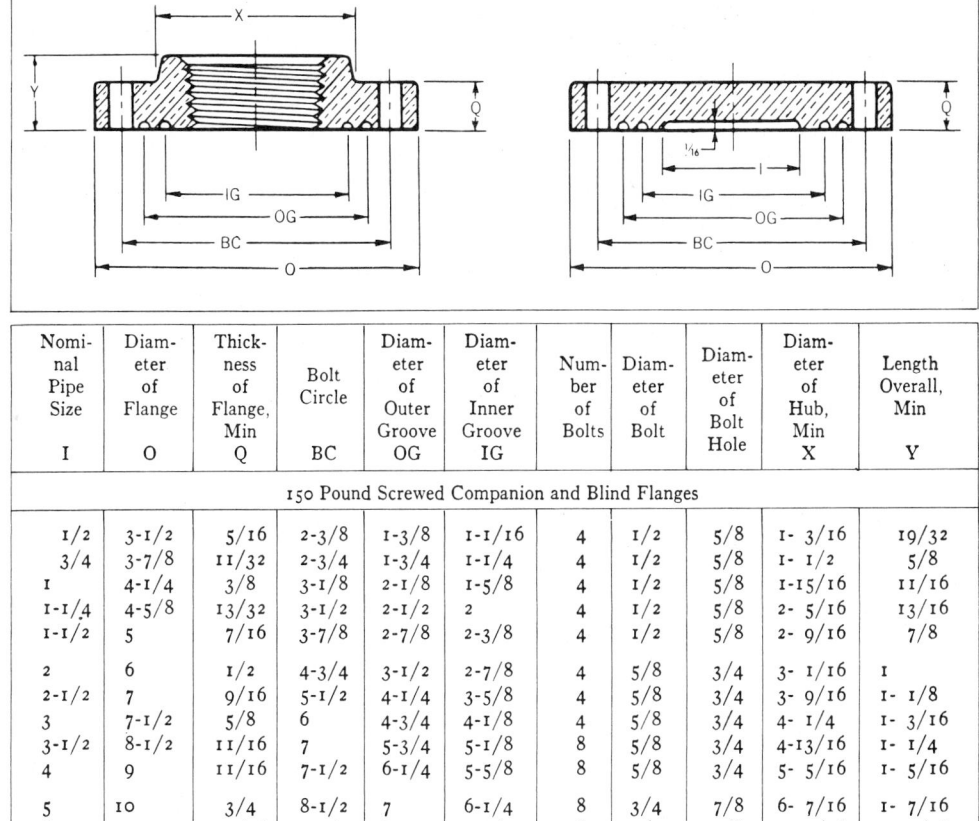

Nominal Pipe Size	Diameter of Flange	Thickness of Flange, Min	Bolt Circle	Diameter of Outer Groove	Diameter of Inner Groove	Number of Bolts	Diameter of Bolt	Diameter of Bolt Hole	Diameter of Hub, Min	Length Overall, Min
I	O	Q	BC	OG	IG				X	Y
150 Pound Screwed Companion and Blind Flanges										
1/2	3-1/2	5/16	2-3/8	1-3/8	1-1/16	4	1/2	5/8	1- 3/16	19/32
3/4	3-7/8	11/32	2-3/4	1-3/4	1-1/4	4	1/2	5/8	1- 1/2	5/8
1	4-1/4	3/8	3-1/8	2-1/8	1-5/8	4	1/2	5/8	1-15/16	11/16
1-1/4	4-5/8	13/32	3-1/2	2-1/2	2	4	1/2	5/8	2- 5/16	13/16
1-1/2	5	7/16	3-7/8	2-7/8	2-3/8	4	1/2	5/8	2- 9/16	7/8
2	6	1/2	4-3/4	3-1/2	2-7/8	4	5/8	3/4	3- 1/16	1
2-1/2	7	9/16	5-1/2	4-1/4	3-5/8	4	5/8	3/4	3- 9/16	1- 1/8
3	7-1/2	5/8	6	4-3/4	4-1/8	4	5/8	3/4	4- 1/4	1- 3/16
3-1/2	8-1/2	11/16	7	5-3/4	5-1/8	8	5/8	3/4	4-13/16	1- 1/4
4	9	11/16	7-1/2	6-1/4	5-5/8	8	5/8	3/4	5- 5/16	1- 5/16
5	10	3/4	8-1/2	7	6-1/4	8	3/4	7/8	6- 7/16	1- 7/16
6	11	13/16	9-1/2	8	7-1/4	8	3/4	7/8	7- 9/16	1- 9/16
8	13-1/2	15/16	11-3/4	10-1/4	9-1/2	8	3/4	7/8	9-11/16	1- 3/4
10	16	1	14-1/4	12-1/2	11-1/2	12	7/8	1	12	1-15/16
12	19	1- 1/16	17	15-1/4	14-1/4	12	7/8	1	14- 3/8	2- 3/16
300 Pound Screwed Companion and Blind Flanges										
1/2	3-3/4	1/2	2-5/8	1-5/8	1-1/8	4	1/2	5/8	1- 3/16	19/32
3/4	4-5/8	17/32	3-1/4	2	1-3/8	4	5/8	3/4	1- 1/2	5/8
1	4-7/8	19/32	3-1/2	2-1/4	1-5/8	4	5/8	3/4	1-15/16	11/16
1-1/4	5-1/4	5/8	3-7/8	2-5/8	2	4	5/8	3/4	2- 5/16	13/16
1-1/2	6-1/8	11/16	4-1/2	3	2-1/4	4	3/4	7/8	2- 9/16	7/8
2	6-1/2	3/4	5	3-3/4	3-1/8	8	5/8	3/4	3- 1/16	1
2-1/2	7-1/2	13/16	5-7/8	4-3/8	3-5/8	8	3/4	7/8	3- 9/16	1- 1/8
3	8-1/4	29/32	6-5/8	5-1/8	4-3/8	8	3/4	7/8	4- 1/4	1- 3/16
3-1/2	9	31/32	7-1/4	5-3/4	5	8	3/4	7/8	4-13/16	1- 1/4
4	10	1- 1/16	7-7/8	6-3/8	5-5/8	8	3/4	7/8	5- 5/16	1- 5/16
5	11	1- 1/8	9-1/4	7-3/4	7	8	3/4	7/8	6- 7/16	1- 7/16
6	12-1/2	1- 3/16	10-5/8	9-1/8	8-3/8	12	3/4	7/8	7- 9/16	1- 9/16
8	15	1- 3/8	13	11-1/4	10-1/4	12	7/8	1	9-11/16	1- 3/4

All dimensions are given in inches, and minimum port diameters of all fittings are the same as nominal pipe size, I.

† Extracted from ANSI-B16.24 by permission of the publisher, The American Society of Mechanical Engineers.

BRASS OR BRONZE FLANGED FITTINGS†

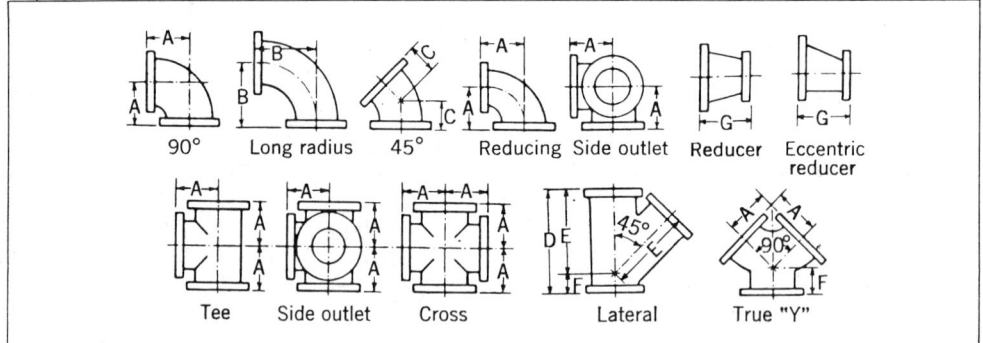

Nominal Pipe Size I	Center to Face A	Face to Face AA	Center to Face B	Center to Face C	Face to Face D	Center to Face E	Center to Face F	Face to Face G	Wall Thickness
150 Pound									
1/2	3	6	—	1-5/8	—	—	—	—	3/32
3/4	3-1/4	6-1/2	—	1-3/4	—	—	—	—	7/64
1	3-1/2	7	5	1-3/4	7-1/2	5-3/4	1-3/4	—	1/8
1-1/4	3-3/4	7-1/2	5-1/2	2	8	6-1/4	1-3/4	—	9/64
1-1/2	4	8	6	2-1/4	9	7	2	—	5/32
2	4-1/2	9	6-1/2	2-1/2	10-1/2	8	2-1/2	5	3/16
2-1/2	5	10	7	3	12	9-1/2	2-1/2	5-1/2	13/64
3	5-1/2	11	7-3/4	3	13	10	3	6	7/32
3-1/2	6	12	8-1/2	3-1/2	14-1/2	11-1/2	3	6-1/2	1/4
4	6-1/2	13	9	4	15	12	3	7	17/64
5	7-1/2	15	10-1/4	4-1/2	17	13-1/2	3-1/2	8	19/64
6	8	16	11-1/2	5	18	14-1/2	3-1/2	9	21/64
8	9	18	14	5-1/2	22	17-1/2	4-1/2	11	13/32
10	11	22	16-1/2	6-1/2	25-1/2	20-1/2	5	12	31/64
12	12	24	19	7-1/2	30	24-1/2	5-1/2	14	9/16
300 Pound									
1/2	3	6	—	1-3/4	—	—	—	—	1/8
3/4	3-1/2	7	—	2-1/4	—	—	—	—	5/32
1	4	8	5	2-1/4	8-1/2	6-1/2	2	—	11/64
1-1/4	4-1/4	8-1/2	5-1/2	2-1/2	9-1/2	7-1/4	2-1/4	—	3/16
1-1/2	4-1/2	9	6	2-3/4	11	8-1/2	2-1/2	—	13/64
2	5	10	6-1/2	3	11-1/2	9	2-1/2	5	1/4
2-1/2	5-1/2	11	7	3-1/2	13	10-1/2	2-1/2	5-1/2	9/32
3	6	12	7-3/4	3-1/2	14	11	3	6	21/64
3-1/2	6-1/2	13	8-1/2	4	15-1/2	12-1/2	3	6-1/2	23/64
4	7	14	9	4-1/2	16-1/2	13-1/2	3	7	13/32
5	8	16	10-1/4	5	18-1/2	15	3-1/2	8	31/64
6	8-1/2	17	11-1/2	5-1/2	21-1/2	17-1/2	4	9	9/16
8	10	20	14	6	25-1/2	20-1/2	5	11	23/32

All dimensions are given in inches, and minimum port diameters of all fittings are the same as nominal pipe size, ◀.
† Extracted from ANSI-B16.24 by permission of the publisher, The American Society of Mechanical Engineers.

CAST IRON SOIL PIPE HUB AND SPIGOT DIMENSIONS†

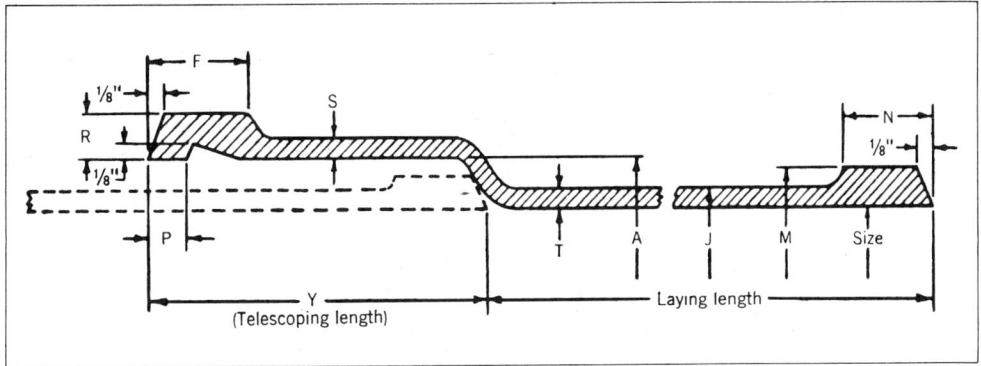

Size	Inside diameter of hub	Outside diameter of spigot	Outside diameter of barrel	Telescoping length	Thickness of barrel		Width of spigot bead
	A	M	J	Y	T (nominal)	T (min)	N
2	2.94	2.62	2.25	2.50	0.18	0.10	0.69
3	3.94	3.62	3.25	2.75	.18	.12	.75
4	4.94	4.62	4.25	3.00	.18	.12	.81
5	5.94	5.62	5.25	3.00	.19	.12	.81
6	6.94	6.62	6.25	3.00	.20	.12	.81
8	9.25	8.75	8.38	3.50	.22	.17	1.12
10	11.38	10.88	10.50	3.50	.26	.21	1.12
12	13.50	12.88	12.50	4.25	.28	.22	1.38
15	16.75	16.00	15.62	4.25	.30	.25	1.38

Size	Thickness of hub		Width of hub bead	Distance from lead groove to end, pipe and fittings		Weight per length of pipe. pounds‡	
	Hub body	Over bead					
	S (min)	R (min)	F	P (min)	P (max)	Single hub	Double hub
2	0.13	0.34	0.75	0.25	0.38	20	21
3	.16	.37	.81	.25	.38	30	31
4	.16	.37	.88	.25	.38	40	42
5	.16	.37	.88	.25	.38	52	54
6	.18	.37	.88	.25	.38	65	68
8	.19	.44	1.19	.25	.38	100	105
10	.27	.53	1.19	.31	.50	145	150
12	.27	.53	1.44	.38	.62	190	200
15	.30	.58	1.44	.38	.62	255	270

All dimensions are given in inches.
† From Commercial Standard 188.53 of the Department of Commerce. Table covers service weight.
‡ Length of single-hub section is such as to lay 5 ft. The laying length is as shown in drawing. Double-hub pipe is of the same overall length as single-hub pipe of the same size. Its laying length shall be 5 ft. minus the telescoping length Y.

CAST IRON SOIL PIPE Y BRANCHES†

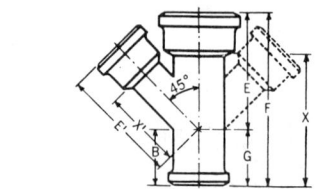

Single and double Y branches

Size	B (min)	E	E'	F	G	X	X'	Weight, single, Pounds	Weight, double, Pounds
				Dimensions					
2	3-1/2	6- 1/2	6- 1/2	10-1/2	4	8	4	6-1/2	9
3	4	8- 1/4	8- 1/4	13-1/4	5	10-1/2	5- 1/2	12-1/2	16
4	4	9- 3/4	9- 3/4	15	5- 1/4	12	6- 3/4	17	23
5	4	11	11	16-1/2	5- 1/2	13-1/2	8	22	28
6	4	12- 1/4	12- 1/4	18	5- 3/4	15	9- 1/4	28	36
8	5-1/2	15- 5/16	15- 5/16	23	7-11/16	19-1/2	11-13/16	55	73
10	5-1/2	18	18	26	8	22-1/2	14- 1/2	94	120
12	7	21- 1/8	21- 1/8	31-1/4	10- 1/8	27	16- 7/8	135	173
15	7	25	25	35-3/4	10- 3/4	31-1/2	20- 3/4	204	262
3 by 2	4	7- 9/16	7- 1/2	11-3/4	4- 3/16	9	5	10	13
4 by 2	4	8- 3/8	8- 1/4	12	3- 5/8	9	5- 3/4	12	15
4 by 3	4	9- 1/16	9	13-1/2	4- 7/16	10-1/2	6- 1/4	14	19
5 by 2	4	8- 7/8	9	12	3- 1/8	9	6- 1/2	14	17
5 by 3	4	9- 5/8	9- 3/4	13-1/2	3- 7/8	10-1/2	7	16-1/2	21
5 by 4	4	10- 5/16	10- 1/2	15	4-11/16	12	7- 1/2	19	24
6 by 2	4	9- 7/16	9- 3/4	12	2- 9/16	9	7- 1/4	16-1/2	20
6 by 3	4	10- 1/8	19- 1/2	13-1/2	3- 3/8	10-1/2	7- 3/4	19	23
6 by 4	4	10-13/16	11- 1/4	15	4- 3/16	12	8- 1/4	22	28
6 by 5	4	11- 9/16	11- 3/4	16-1/2	4-15/16	13-1/2	8- 3/4	24	31
8 by 2	5-1/2	10- 7/8	11	14	3- 1/8	10-1/2	8- 1/2	29	32
8 by 3	5-1/2	11- 9/16	11- 3/4	15-1/2	3-15/16	12	9	32	37
8 by 4	5-1/2	12- 1/4	12- 1/2	17	4- 3/4	13-1/2	9- 1/2	36	42
8 by 5	5-1/2	13	13	18-1/2	5- 1/2	15	10	39	45
8 by 6	5-1/2	13-11/16	13- 1/2	20	6- 5/16	16-1/2	10- 1/2	44	51
10 by 4	5-1/2	13- 7/16	14- 1/8	17	3- 9/16	13-1/2	11- 1/8	53	59
10 by 5	5-1/2	14- 3/16	14- 5/8	18-1/2	4- 5/16	15	11- 5/8	57	63
10 by 6	5-1/2	14- 7/8	15- 1/8	20	5- 1/8	16-1/2	12- 1/8	61	70
10 by 8	5-1/2	16- 1/2	16-15/16	23	6- 1/2	19-1/2	13- 7/16	77	94
12 by 4	7	15- 1/8	15- 7/16	19-1/4	4- 1/8	15	12- 7/16	70	76
12 by 5	7	15- 7/8	15-15/16	20-3/4	4- 7/8	16-1/2	12-15/16	74	81
12 by 6	7	16- 9/16	16- 7/16	22-1/4	5-11/16	18	13- 7/16	80	88
12 by 8	7	18- 3/16	18- 1/4	25-1/4	7- 1/16	21	14- 3/4	96	113
12 by 10	7	19-11/16	19- 5/16	28-1/4	8- 9/16	24	15-13/16	115	142
15 by 6	7	18- 1/4	18- 3/4	22-1/4	4	18	15- 3/4	109	117
15 by 8	7	19- 7/8	20- 9/16	25-1/4	5- 3/8	21	17- 1/16	127	146
15 by 10	7	21- 3/8	21- 5/8	28-1/4	6- 7/8	24	18- 1/8	152	176
15 by 12	7	22-13/16	23- 7/16	31-1/4	8- 7/16	27	19- 3/16	170	210

Y Branches, Single and Double

CAST IRON SOIL PIPE QUARTER BENDS†

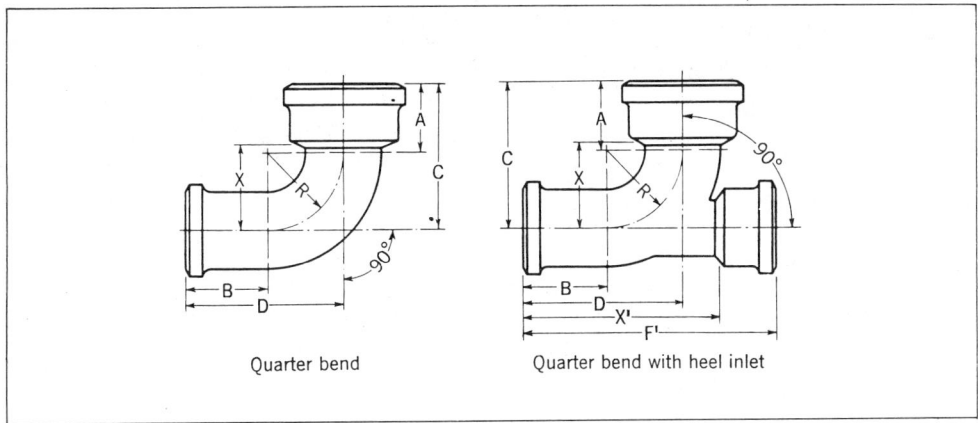

Quarter bend   Quarter bend with heel inlet

Quarter Bends

Size	A	B	C	D	R	X	Weight, Pounds
2	2¾	3	5¾	6	3	3¼	5
3	3¼	3½	6¾	7	3½	4	10
4	3½	4	7½	8	4	4½	15
5	3½	4	8	8½	4½	5	19
6	3½	4	8½	9	5	5½	24
8	4⅛	5½	10⅛	11½	6	6⅝	51
10	4⅛	5½	11⅛	12½	7	7⅝	78
12	5	7	13	15	8	8¾	111
15	5	7	14½	16½	9½	10¼	169

Quarter Bends with Heel Inlets

Size	A	B	C	D	F	R	X	X′	Weight, Pounds
3 × 2	3¼	3½	6¾	7	11½	3½	4	9	13
4 × 2	3½	4	7½	8	13	4	4½	10½	18
4 × 3	3½	4	7½	8	13¼	4	4½	10½	19
5 × 2	3½	4	8	8½	14¼	4½	5	11¾	22
5 × 3	3½	4	8	8½	14½	4½	5	11¾	24
5 × 4	3½	4	8	8½	14¾	4½	5	11¾	25
6 × 2	3½	4	8½	9	15	5	5½	12½	27
6 × 3	3½	4	8½	9	15¼	5	5½	12½	29
6 × 4	3½	4	8½	9	15½	5	5½	12½	30

All dimensions are given in inches.
Dimensions D, X and X¹ are laying lengths.
† From Commercial Standard 188.53 of the Department of Commerce.

Traps in cast iron soil pipe lines shall provide minimum water seal as follows: 2 inch pipe, 2 inch seal; 3 to 6 inches, 2½ inch seal; 8 to 12 inches, 3 inch seal. In the case of 15 inch pipe a 3½ inch seal is required.

Screw plugs and tapped openings in fittings shall have American National Standard taper pipe threads. The threads shall be in accordance with the American National Standard for Pipe Threads, B2.1, or with the National Bureau of Standards Handbook H28, Screw Thread Standards for Federal Services, of the current issue.

CAST IRON SOIL PIPE SWEEPS, 1/8 AND 1/16 BENDS†

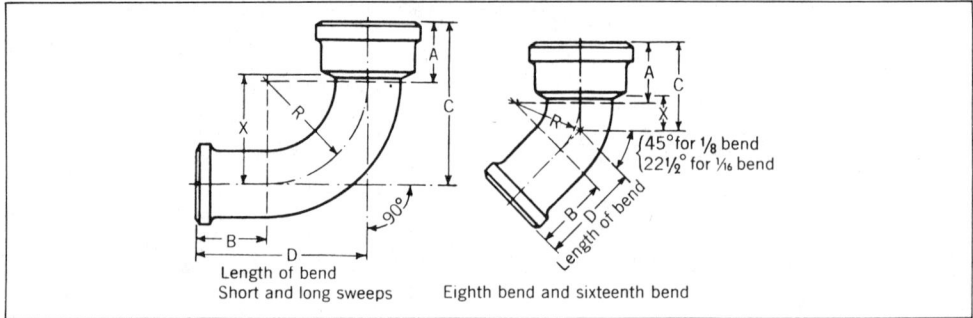

45° for 1/8 bend
22½° for 1/16 bend

Length of bend
Short and long sweeps   Eighth bend and sixteenth bend

Size	Dimensions						Weight, Pounds
	A	B	C	D	R	X	
Short Sweeps, Quarter Bend							
2	2-3/4	3	7- 3/4	8	5	5- 1/4	5
3	3-1/4	3-1/2	8- 3/4	9	5-1/2	6	9
4	3-1/2	4	9- 1/2	10	6	6- 1/2	12-1/2
5	3-1/2	4	10	10- 1/2	6-1/2	7	16
6	3-1/2	4	10- 1/2	11	7	7- 1/2	20
8	4-1/8	5-1/2	12- 1/8	13- 1/2	8	8- 5/8	38
10	4-1/8	5-1/2	13- 1/8	14- 1/2	9	9- 5/8	62
12	5	7	15	17	10	10- 3/4	89
15	5	7	16- 1/2	18- 1/2	11-1/2	12- 1/4	130
Long Sweeps, Quarter Bend							
2	2-3/4	3	10- 3/4	11	8	8- 1/4	6-1/2
3	3-1/4	3-1/2	11- 3/4	12	8-1/2	9	11
4	3-1/2	4	12- 1/2	13	9	9- 1/2	15
5	3-1/2	4	13	13- 1/2	9-1/2	10	19-1/2
6	3-1/2	4	13- 1/2	14	10	10- 1/2	24
8	4-1/8	5-1/2	15- 1/8	16- 1/2	11	11- 5/8	45
10	4-1/8	5-1/2	16- 1/8	17- 1/2	12	12- 5/8	72
12	5	7	18	20	13	13- 3/4	101
15	5	7	19- 1/2	21- 1/2	14-1/2	15- 1/4	147
Eighth Bends							
2	2-3/4	3	4	4- 1/4	3	1- 1/2	3-1/4
3	3-1/4	3-1/2	4-11/16	4-15/16	3-1/2	1-15/16	5-1/2
4	3-1/2	4	5- 3/16	5-11/16	4	2- 3/16	8-1/2
5	3-1/2	4	5- 3/8	5- 7/8	4-1/2	2- 3/8	10-1/2
6	3-1/2	4	5- 9/16	6- 1/16	5	2- 9/16	13
8	4-1/8	5-1/2	6- 5/8	8	6	3- 1/8	28
10	4-1/8	5-1/2	7	8- 3/8	7	3- 1/2	44
12	5	7	8-, 5/16	10- 5/16	8	4- 1/16	64
15	5	7	8-15/16	10-15/16	9-1/2	4-11/16	92

(*Concluded on next page*)

All dimensions are given in inches.
Dimensions D and X are laying lengths.
† From Commercial Standard 188.53 of the Department of Commerce.

Cast Iron Soil Pipe Sweeps (Concluded)

Size	Dimensions A	B	C	D	R	X	Weight Pounds
Sixteenth Bends							
2	2-3/4	3	3- 3/8	3- 5/8	3	7/8	3-1/4
3	3-1/4	3-1/2	3-15/16	4- 3/16	3-1/2	1- 3/16	5-1/2
4	3-1/2	4	4- 5/16	4-13/16	4	1- 5/16	7-1/2
5	3-1/2	4	4- 3/8	4- 7/8	4-1/2	1- 3/8	9
6	3-1/2	4	4- 1/2	5	5	1- 1/2	11-1/2
8	4-1/8	5-1/2	5- 5/16	6-11/16	6	1-13/16	25
10	4-1/8	5-1/2	5- 1/2	6- 7/8	7	2	38
12	5	7	6- 5/8	8- 5/8	8	2- 3/8	56
15	5	7	6- 7/8	8- 7/8	9-1/2	2- 5/8	78

CAST IRON SOIL PIPE CLEANOUT Y BRANCHES†

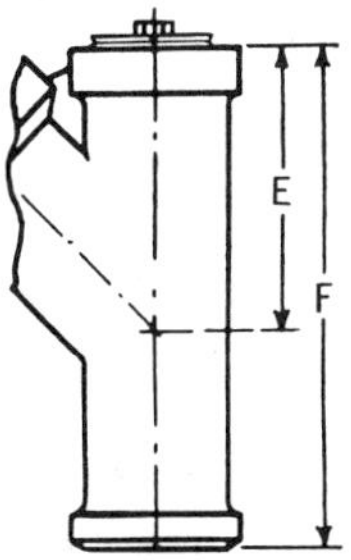

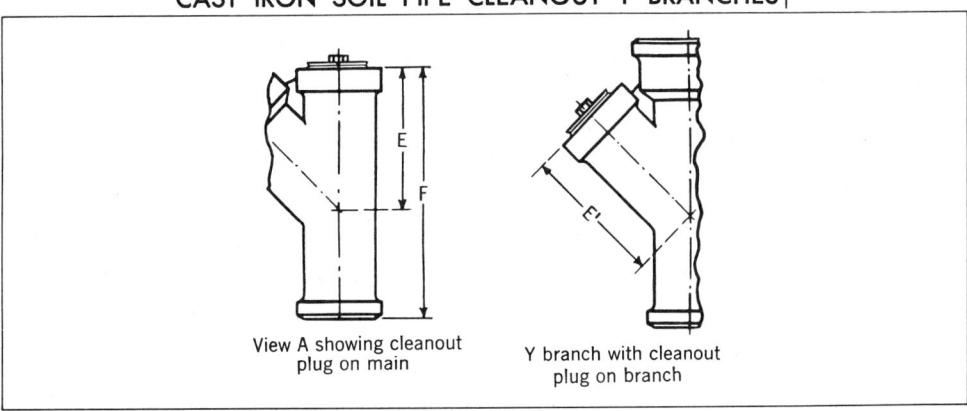

View A showing cleanout plug on main

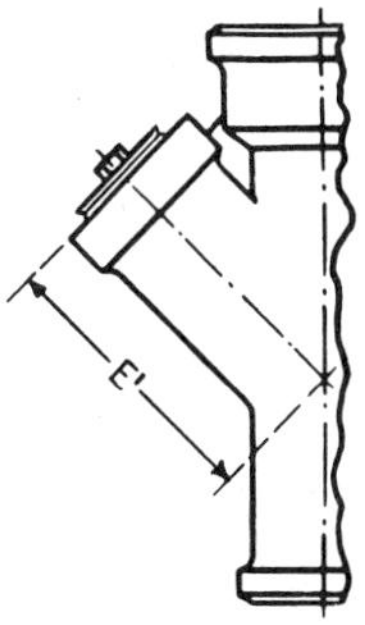

Y branch with cleanout plug on branch

Size (inches)	Dimensions E	E′	F	G	X′	Size of tap-ping	Weight without plug, Pounds
Y Branches, Cleanout, Screw Plug on Main, View A							
2	5¼	6½	9¼	4	4	1½	7¼
3	6⅝	8¼	11⅝	5	5½	2½	10½
4	7⅞	9¾	13⅛	5¼	6¾	3½	14½
5	9⅛	11	14⅝	5½	8	4	19
6	10⅜	12¼	16⅛	5¾	9¼	5	26
Y Branches, Cleanout, Screw Plug on Branch							
2	6½	5¼	10½	4	8	1½	7¼
3	8¼	6⅝	13¼	5	10½	2½	10½
4	9¾	7⅞	15	5¼	12	3½	14½
5	11	9⅛	16½	5½	13½	4	19
6	12¼	10⅜	18	5¾	15	5	26

All dimensions are given in inches.
Dimensions X and X^1 are laying lengths.
† From Commercial Standard 188.53 of the Department of Commerce.

CAST IRON SCREWED DRAINAGE FITTINGS†
Dimensions of Threaded Ends

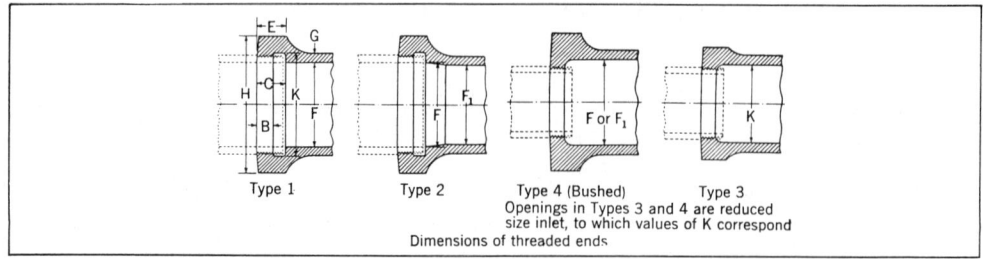

Type 1   Type 2   Type 4 (Bushed)   Type 3
Openings in Types 3 and 4 are reduced
size inlet, to which values of K correspond
Dimensions of threaded ends

Nominal Pipe Size	Length of Threads	Total Length of Thread Chamber to Shoulder	Width of Band	Inside Diameter of Fitting		Metal Thickness	Outside Diameter of Band	Recess Diameter	
								Max	Min
	B	C	E	F	F₁	G	H	K	K
1¼	0.420	0.7068	0.71	1.380	1.25	0.185	2.39	1.730	1.660
1½	0.420	0.7235	0.72	1.610	1.50	0.200	2.68	1.970	1.900
2	0.436	0.7565	0.76	2.067	2.00	0.220	3.28	2.445	2.375
2½	0.682	1.1375	1.14	2.469	2.50	0.240	3.86	2.975	2.875
3	0.766	1.2000	1.20	3.068	3.00	0.260	4.62	3.600	3.500
4	0.844	1.3000	1.30	4.026	4.00	0.310	5.79	4.600	4.500
5	0.937	1.4063	1.41	5.047	5.00	0.380	7.05	5.663	5.563
6	0.958	1.5125	1.51	6.065	6.00	0.430	8.28	6.725	6.625
8	1.063	1.7125	1.71	7.981	8.00	0.550	10.63	8.625	8.725

Dimensions of Tees, Crosses, and Y-Branches

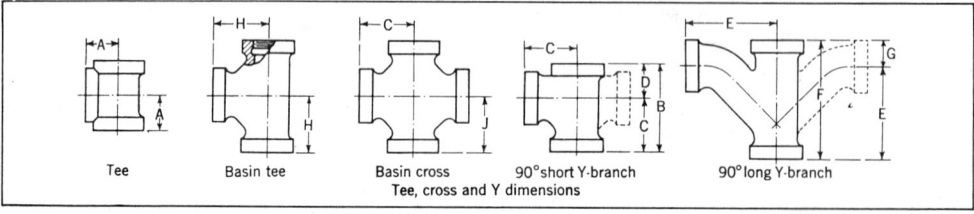

Tee   Basin tee   Basin cross   90°short Y-branch   90° long Y-branch
Tee, cross and Y dimensions

Nominal Pipe Size	Center to End of Tee	90 Deg Short Y-Branch			90 Deg Long Y-Branch			Center to End of Basin Tee	Center to End of Basin Cross
		End to End	Center to End	Center to End	Center to End	End to End	Center to End		
	A	B	C	D	E	F	G	H	J
1-1/4	1- 3/4	3- 3/4	2- 1/4	1- 1/2	3-5/8	4-3/4	1-1/8	2- 5/16	2- 5/16
1-1/2	1-15/16	4- 1/4	2- 1/2	1- 3/4	4-1/8	5-3/8	1-1/4	2-11/16	2- 5/16
2	2- 1/4	5- 3/16	3- 1/16	2- 1/8	5-1/4	7	1-3/4	3- 1/2	3- 1/2
2-1/2	2-11/16	6- 5/16	3-11/16	2- 5/8	6-1/4	8-1/4	2	4- 1/4	—
3	3- 1/16	7- 1/4	4- 1/4	3	7-1/2	9-7/8	2-3/8	—	—
4	3-13/16	8- 3/4	5- 3/16	3- 9/16	9-7/8	13	3-1/8	—	—
5	4- 1/2	10- 5/16	6- 1/8	4- 3/16	12-1/4	15-3/4	3-1/2	—	—
6	5- 1/8	11-15/16	7- 1/8	4-14/16	14-5/8	18-3/4	4-1/8	—	—

All dimensions are given in inches.
† Extracted from ANSI-B16.12 by permission of the publisher, The American Society of Mechanical Engineers.

Center-to-End Dimensions of Elbows

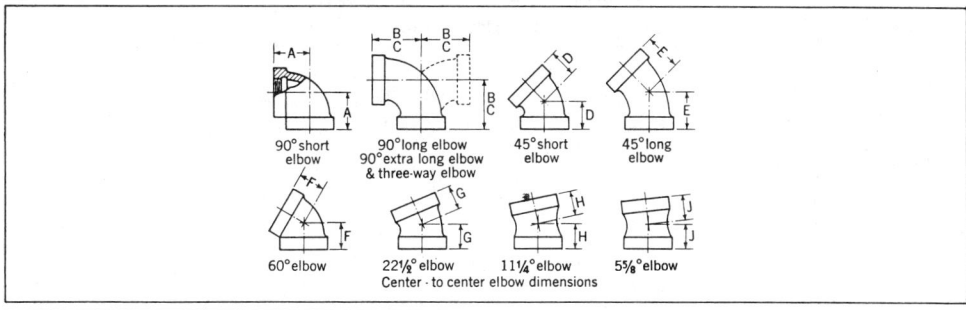

90° short elbow   90° long elbow / 90° extra long elbow & three-way elbow   45° short elbow   45° long elbow

60° elbow   22½° elbow   11¼° elbow   5⅝° elbow
Center - to center elbow dimensions

| Nominal Pipe Size | 90 Deg Elbow | | | 45 Deg Elbow | | 60 Deg Elbow | 22½ Deg Elbow | 11¼ Deg Elbow | 5⅝ Deg Elbow |
| | Short | Long and 3-Way Elbow | Extra Long | Short | Long | | | | |
	A	B	C	D	E	F	G	H	J
1-1/4	1- 3/4	2- 1/4	3	1- 5/16	1- 3/4	1-9/16	1- 1/8	1- 1/16	1-3/16
1-1/2	1-15/16	2- 1/2	3-1/2	1- 7/16	1- 7/8	1-3/4	1- 1/4	1- 1/4	1-5/16
2	2- 1/4	3- 1/16	4	1-11/16	2- 1/4	2-1/16	1- 7/16	1- 3/8	1-1/2
2-1/2	2-11/16	3-11/16	4-1/2	1-15/16	2- 5/8	2-1/2	1- 3/4	1- 5/8	1-5/8
3	3- 1/16	4- 1/4	5-1/4	2- 3/16	2-15/16	2-7/8	2	1-13/16	1-3/4
4	3-13/16	5- 3/16	6-1/4	2- 5/8	3- 1/2	3-3/8	2- 5/16	2	1-7/8
5	4- 1/2	6- 1/8	—	3- 1/16	4- 1/8	3-7/8	2- 5/8	2- 1/4	2
6	5- 1/8	7- 1/8	—	3- 7/16	4- 7/8	4-1/4	2-15/16	2- 3/8	2-1/4
8	6- 9/16	9	—	4- 1/4	—	—	—	—	—

Screw joint drainage fittings were developed about 1880 by the Durham House Drainage Company of New York and are often referred to as Durham fittings. At that time soil pipes and drains had been required by the then recent legislation to be of plumber's cast-iron soil pipe. This was the ruling made in the City of New York, while in Chicago soil pipes were required to be of lead or plumber's cast-iron pipe.

In order to form a continuous passageway with no pockets or obstructions where foreign matter could collect and gradually accumulate, it was necessary to design a special type of screw fitting. These fittings are made with the inside diameter approximately equal to the nominal size of standard weight (Schedule 40, ANSI-B36.10) wrought pipe. The thread chamber is designed so that the end of the pipe, when screwed into the fitting to make a tight joint, will come close to the shoulder in the fitting, thereby making a practically continuous passage. The threading of these fittings required special care and the threads on the pipe were cut to suit the threads in the fitting.

These screwed drainage fittings follow the general proportions of other cast-iron screwed fittings, except that the shapes and center-to-end dimensions are designed to suit the particular service for which they are intended. Manufacturing practice, in the case of the drainage fittings, however, as they developed, differed to a considerable extent in center-to-end dimensions as well as in some other respects, which finally led to an extensive study of the situation by the Manufacturers Standardization Society of the Valve and Fittings Industry. This study resulted in the development of a standard practice of the industry for these fittings as indicated by this present standard.

All fittings shall be threaded with American National Standard Taper Pipe Threads and the variations in threading shall be limited to one turn small and no turn large, except for screwed openings without shoulder which shall be limited to one turn large and one turn small; from the gaging notch on the plug when using working gages.

When gaging internal threads the notch shall be flush with the bottom of the chamfer, which shall be considered as being the intersection of the chamfer cone and the pitch cone of the thread. This depth is approximately equal to one-half thread from the face of the fitting.

CAST BRASS SOLDER JOINT DRAINAGE FITTINGS†

Dimensions of Solder Joint Ends

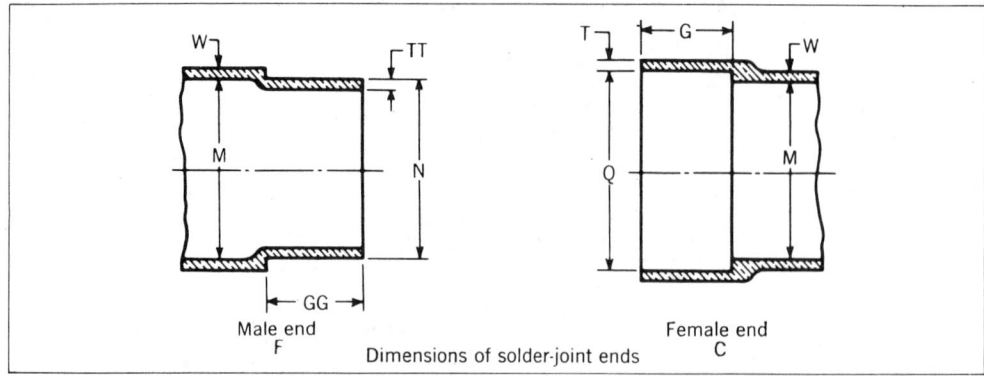

Male end
F

Female end
C

Dimensions of solder-joint ends

Standard Water Tube Sizes	Male End	Female End				Metal Thickness			Inside
	Outside Diameter N. Max	Length GG, Min	Inside Diameter		Depth G, Min	Body W	Joint		Diameter of Fitting M, Min
			Q. Min	Q. Max			T	TT. Max	
1-1/4	1.3765	11/16	1.3785	1.3805	5/8	0.100	0.072	0.090	1.29
1-1/2	1.627	3/4	1.629	1.6315	11/16	0.100	0.078	0.098	1.53
2	2.127	13/16	2.129	2.1315	3/4	0.100	0.090	0.112	2.01
3	3.127	1	3.129	3.1315	15/16	0.120	0.103	0.129	2.98
4	4.127	1-1/4	4.129	4.132	1-3/16	0.120	0.120	0.150	3.93

Dimensions of Screwed Ends

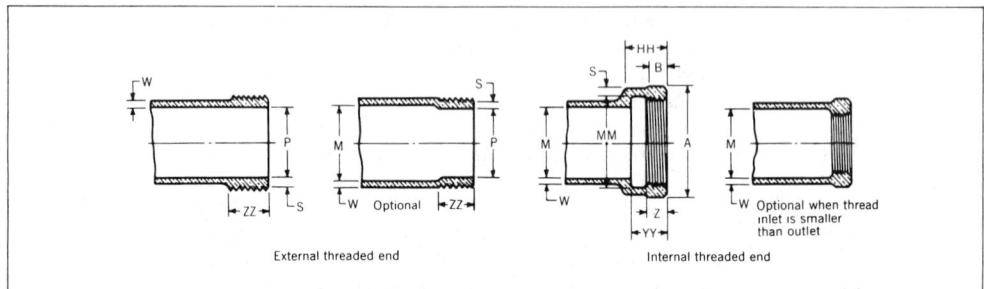

External threaded end

Internal threaded end

Thread Size	Diam of Band A, Min	Band Length B, Min	Internal End						External End		
			Length of Boss HH, Min	Diameter of Recess MM, Min	Thread End Wall S, Min	Body Wall W, Min	End to Shoulder YY	Length of Thread Z	Body Bore M, Max	Thread End Bore P, Max	Length of Thread ZZ, Min
1¼	2.10	0.31	0.83	1.660	0.120	0.100	0.7068	0.420	1.290	1.25	0.71
1½	2.38	0.34	0.89	1.900	0.130	0.100	0.7235	0.420	1.530	1.47	0.73
2	2.92	0.41	0.94	2.375	0.150	0.100	0.7565	0.436	2.010	1.91	0.76
3	4.20	0.55	1.39	3.500	0.190	0.120	1.200	0.766	2.980	2.86	1.20
4	5.31	0.66	1.50	4.500	0.220	0.120	1.300	0.844	3.930	3.89	1.30

All dimensions are given in inches.
† Extracted from ANSI·B16.23 by permission of the publisher, The American Society of Mechanical Engineers.

SURFACE AREA OF FLANGED FITTINGS

STANDARD FLANGED FITTINGS PLUS ACCOMPANYING FLANGES

Nominal Diameter, Inches	90° Elbow Area, Sq. Ft.	Feet of Pipe	Long Radius Ell Area, Sq. Ft.	Feet of Pipe	Tee Area, Sq. Ft.	Feet of Pipe	Flanged Coupling Area, Sq. Ft.	Feet of Pipe	Cross Area, Sq. Ft.	Feet of Pipe
1	.80	2.31	.89	2.59	1.24	3.59	.32	.93	1.62	4.72
1¼	.96	2.20	1.08	2.49	1.48	3.40	.38	.88	1.94	4.47
1½	1.17	2.35	1.34	2.68	1.82	3.64	.48	.95	2.38	4.78
2	1.65	2.65	1.84	2.96	2.54	4.08	.67	1.08	3.32	5.34
2½	2.09	2.78	2.32	3.08	3.21	4.26	.84	1.12	4.19	5.56
3	2.38	2.60	2.68	2.93	3.66	3.99	.95	1.03	4.77	5.70
3½	2.98	2.85	3.28	3.13	4.48	4.28	1.12	1.07	5.83	5.56
4	3.53	2.90	3.96	3.36	5.41	4.59	1.34	1.14	7.03	5.97
4½	3.95	3.01	4.43	3.38	6.07	4.63	1.47	1.13	7.87	6.01
5	4.44	3.05	5.00	3.43	6.81	4.67	1.62	1.11	8.82	6.06
6	5.13	2.95	5.99	3.45	7.84	4.53	1.82	1.05	10.08	5.81
7	6.17	3.09	7.38	3.70	9.37	4.69	2.17	1.10	12.00	6.01
8	6.98	3.09	8.56	3.79	10.55	4.67	2.41	1.07	13.44	5.96
9	8.71	3.46	10.57	4.20	13.18	5.23	3.00	1.19	16.78	6.66
10	10.18	3.61	12.35	4.38	15.41	4.47	3.43	1.22	19.58	6.95
12	13.08	3.92	16.35	4.90	19.67	5.89	4.41	1.32	24.87	7.45
14	16.38	4.47	20.17	5.47	24.81	6.78	5.39	1.47	31.48	8.60
15	18.50	4.72	22.92	5.83	27.91	7.10	6.18	1.57	35.48	9.04
16	20.17	4.82	25.41	6.07	30.32	7.23	6.69	1.60	38.34	9.15

EXTRA-HEAVY FLANGED FITTINGS PLUS ACCOMPANYING FLANGES

Nominal Diameter, Inches	90° Elbow Area, Sq. Ft.	Feet of Pipe	Long Radius Ell Area, Sq. Ft.	Feet of Pipe	Tee Area, Sq. Ft.	Feet of Pipe	Flanged Coupling Area, Sq. Ft.	Feet of Pipe	Cross Area, Sq. Ft.	Feet of Pipe
1	1.02	2.95	1.08	3.15	1.58	4.58	.44	1.27	2.07	6.02
1¼	1.10	2.52	1.34	3.08	1.93	4.43	.51	1.17	2.53	5.82
1½	1.33	2.67	1.87	3.76	2.68	5.38	.73	1.46	3.54	7.11
2	2.01	3.23	2.16	3.47	3.09	4.97	.85	1.36	4.06	6.53
2½	2.57	3.41	2.76	3.67	4.05	5.38	1.11	1.46	5.17	6.87
3	3.49	3.81	3.74	4.08	5.33	5.82	1.48	1.62	6.95	7.58
3½	3.96	3.78	4.28	4.09	6.04	5.77	1.64	1.57	7.89	7.54
4	4.64	3.94	4.99	4.24	7.07	6.00	1.91	1.62	9.24	7.84
4½	5.02	3.83	5.46	4.17	7.72	5.90	2.04	1.56	10.07	7.69
5	5.47	3.76	6.02	4.13	8.52	5.85	2.18	1.50	10.97	7.53
6	6.99	4.03	7.76	4.48	10.64	6.14	2.78	1.60	13.75	7.93
7	8.62	4.32	9.73	4.87	12.33	6.18	3.46	1.73	16.83	8.43
8	9.76	4.32	11.09	4.91	14.74	6.53	3.77	1.67	18.97	8.41
9	11.44	4.54	13.17	5.23	17.23	6.84	4.44	1.76	22.10	8.77
10	13.58	4.82	15.60	5.54	20.41	7.25	5.20	1.85	26.26	9.32
12	17.73	5.31	18.76	5.62	26.65	7.99	6.71	2.01	34.11	10.22
14	22.31	6.08	25.70	7.02	33.63	9.18	8.30	2.26	43.15	11.75
15	25.28	6.43	29.34	7.47	38.04	9.68	9.52	2.43	48.79	12.40
16	27.18	6.48	31.73	7.58	40.94	9.78	10.05	2.40	52.35	12.50

First column is outside surface area in square feet; second column is length of straight pipe in feet with same surface area as in first column

SURFACE AREA OF BUTT WELD FITTINGS

Pipe Diameter, Inches	Long Radius 90° Elbow		Short Radius 90° Elbow		Long Radius 45° Elbow		Cap	
	Area, Sq Ft	Feet of Pipe	Area, Sq Ft	Feet of Pipe	Area, Sq Ft	Feet of Pipe	Area, Sq Ft	Feet of Pipe
½	.04	.18	—	—	.02	.09	—	—
¾	.04	.15	—	—	.02	.07	.01	.04
I	.07	.20	.05	.15	.03	.09	.02	.06
1¼	.11	.25	.07	.16	.05	.12	.02	.05
1½	.15	.30	.10	.20	.07	.14	.02	.04
2	.24	.39	.16	.26	.12	.19	.03	.05
2½	.37	.49	.25	.33	.19	.25	.04	.05
3	.54	.59	.36	.39	.27	.29	.06	.07
3½	.72	.69	.48	.46	.36	.34	.09	.09
4	.93	.79	.62	.53	.46	.39	.10	.08
5	1.43	.98	.95	.65	.72	.49	.16	.11
6	2.04	1.18	1.36	.78	1.02	.59	.22	.13
8	3.55	1.57	2.37	1.05	1.77	.78	.34	.15
10	5.53	1.96	3.69	1.31	2.26	.80	.52	.18
12	7.87	2.35	5.25	1.57	3.93	1.18	.74	.22
14	10.08	2.52	6.72	1.68	5.04	1.26	.89	.22
16	13.16	2.90	8.79	1.93	6.58	1.45	1.21	.26
18	16.66	3.50	11.11	2.33	8.33	1.75	1.42	.30
20	20.56	3.91	13.72	2.61	10.28	1.95	1.77	.34

Pipe Diameter, Inches	Long Radius 180° Elbow		Short Radius 180° Elbow		Cross†		Tee†	
	Area Sq. Ft.	Feet of Pipe	Area, Sq Ft	Feet of Pipe	Area, Sq Ft	Feet of Pipe	Area, Sq Ft	Feet of Pipe
½	.09	.41	—	—	—	—	.04	.18
¾	.08	.29	—	—	—	—	.06	.22
I	.14	.41	.09	.26	—	—	.10	.29
1¼	.21	.48	.14	.32	.18	.41	.16	.37
1½	.29	.47	.20	.32	.26	.42	.22	.35
2	.49	.65	.33	.44	.33	.44	.30	.40
2½	.74	.81	.49	.53	.50	.55	.43	.47
3	1.08	1.03	.72	.69	.63	.60	.57	.54
3½	1.44	1.22	.96	.91	.79	.67	.72	.61
4	1.85	1.57	1.23	1.04	.96	.81	.88	.75
5	2.86	1.96	1.91	1.31	1.36	.93	1.27	.87
6	4.09	2.36	2.72	1.57	1.82	1.05	1.72	.99
8	7.09	3.14	4.73	2.10	2.84	1.26	2.74	1.21
10	11.05	3.92	7.37	2.62	4.20	1.49	4.09	1.45
12	15.73	4.70	10.49	3.14	5.81	1.74	5.69	1.70
14	20.15	5.04	13.42	3.36	7.04	1.76	6.88	1.72
16	26.32	5.79	17.56	3.86	8.39	1.85	8.38	1.84
18	33.31	7.00	22.22	4.67	10.62	2.23	10.61	2.23
20	41.13	7.81	27.43	5.21	13.11	2.49	13.10	2.49

Pipe length is the equivalent length of Schedule 40 pipe, same size, of area given in adjacent column.
† Data for crosses and tees are approximations only.

LAYING LENGTHS OF PIPE WITH SCREWED FITTINGS

The accompanying data for determining the length of pipe between screwed fittings of *malleable iron* eliminate guesswork to a great extent in determining allowances. The data apply to banded as well as plain fittings. Due to variations in manufacture, lubricant used, torque on the wrench, and other factors, the figures are to be considered as close approximations.

Each value in Table 1 represents the distance from the center-line of the fitting to the end of the pipe screwed into the fitting. To determine the pipe length, this distance must be subtracted from the measured distance between the two center-lines for *both ends* of the pipe.

Example: The cen-ter-line distance between a tee and a 90-degree elbow in a 2-inch line is 8 ft 6

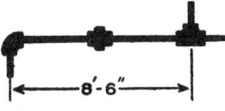

inches. According to dimension A in Table 1 the distance to subtract for the elbow is 1½ inches and the same for the tee. The total distance to subtract is the sum of the two fitting allowances, 3 inches, leaving an actual pipe length of 8 ft 3 inches.

Dimensions G and H in Table 1 show the distance to subtract from pipe length where couplings or unions are inserted in the line.

Example: Suppose it is necessary to use a union in the line in the previous example. The pipe length between the

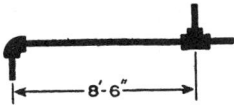

elbow and the tee must still be 8 ft 3 inches with the union added. To take an 8-ft 3-inch length of pipe, cut it, and insert the union would result in a length greater than 8 ft 3 inches. Additional shortening of the length is necessary. According to dimension G in Table 1 the necessary allowance for the 2-inch union is 1-13/16 inches. Subtracting this from 8 ft 3 inches gives us a total length of 8 ft 1-3/16 inches for the two pieces of pipe between the elbow and the tee.

When distances between fittings are short, standard nipples can be used. Nipples are available in lengths up to 12 inches, with graduations at ½-inch intervals for shorter pieces and at inch intervals for longer pieces. (When no nipple is available in the exact length calculated, frequently the distance between fittings can be changed slightly to accommodate the closest available standard nipple size.)

Each pipe diameter has a "close" nipple, lengths of which are given in Table 2. Shortest "long" nipple lengths for each pipe diameter range from ⅜ inch longer than the close nipple for the smaller pipe diameters to 1½ inches longer for the larger pipe diameters.

Dimensions L and M in Table 1 indicate the amount of "offset" to allow between center-lines when using street elbows, 90 degrees or 45 degrees. Center to center distance on a 45° diagonal is 1.412 times the leg distance. If the vertical or horizontal distance between center-lines of fittings is known, the length of the 45-degree measurement between fittings can be calculated. From this must be subtracted fitting allowances taken from Table 1.

Example: A vertical 1½-inch pipe line is to be offset 45 degrees, using two 45-degree elbows. The horizontal distance (a) between the center-lines must be 2 ft 8 inches. The diagonal distance (b) between the two center-lines will be 2 ft 8 in × 1.412 or 3 ft 9¼ inches. From Table 1, dimension B, the fitting allowance to subtract for each 45-degree elbow is ¾ inch, a total of 1½ inches. The actual length of the pipe between the elbows will be 3 ft 9¼ inches less 1½ in., or 3 ft 7¾ in.

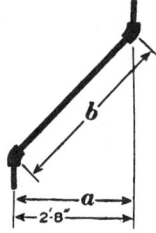

Example: A slight change in the above example makes use of a Y-bend. The fitting allowance for Y-bend, from Table 1, dimension E, is 2⅝ inches, which is greater than the 45-degree elbow allowance. The sum of the two fitting allowances is 3⅜ inches, which makes the actual pipe length 3 ft 5⅞ inches.

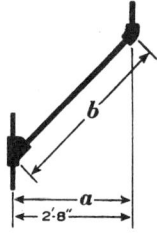

Example: A third example, somewhat more complicated, is shown here. A 45-degree street elbow is screwed into a cross. This shortens a, according to dimension L, Table 1, by 3⅛ inches. Dis-

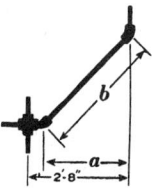

tance *a* is now 2 ft 8 inches less 3⅛ inches, or 2 ft 4⅞ inches. Distance *b* is 2 ft 4⅞ in. × 1.412 or 3 ft 4¾ inches. Fitting allowances from Table 1 are 11/16 inch for the 45-degree street elbow

(dimension *C*) and ¾ inch for the regular 45-degree elbow (dimension *B*), a total of 1-7/16 inches. Pipe length is 3 ft 4¾ inches less 1-7/16 inches, or 3 ft 3-5/16 inches.

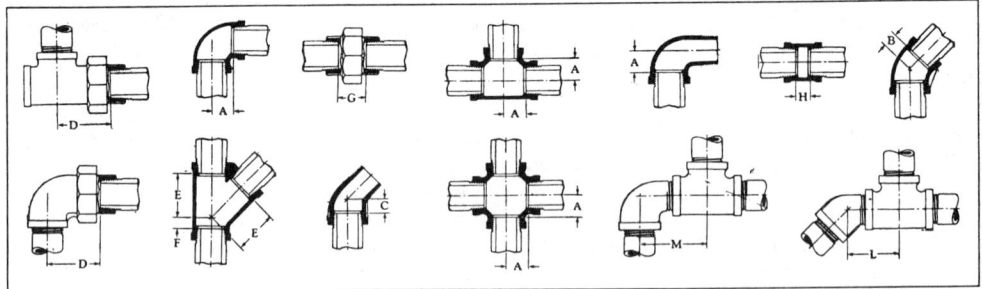

Table 1. Laying Length Data, Malleable Iron Screwed Fittings

Di-men-sion	Nominal Pipe Diameter, Inches										
	1/4	3/8	1/2	3/4	1	1-1/4	1- 1/2	2	2-1/2	3	3-1/2	4
	Dimension, Inches										
A	7/16	9/16	5/8	3/4	13/16	1-1/16	1- 1/4	1- 1/2	1-3/4	2- 1/8	2-3/8	2- 5/8
B	3/8	7/16	3/8	7/16	7/16	5/8	3/4	15/16	1	1- 3/16	1-5/16	1- 1/2
C	1/4	5/16	5/16	3/8	3/8	9/16	11/16	15/16	—	—	—	—
D	1-5/16	1-1/2	1- 5/8	1-7/8	2- 3/16	2-1/2	2-11/16	3- 1/8	—	—	—	—
E	—	1-1/16	1- 3/16	1-1/2	1- 3/4	2-1/4	2- 5/8	3- 1/4	3-3/4	4- 9/16	—	5-13/16
F	—	5/16	1/4	3/16	1/4	7/16	1/2	11/16	5/8	11/16	—	13/16
G	1	1-1/8	1- 3/16	1-5/16	1- 9/16	1-5/8	1-11/16	1-13/16	2-5/16	2- 9/16	—	—
H	5/16	7/16	5/16	3/8	5/16	9/16	3/4	1	1	1- 3/16	—	1- 7/16
L	1-3/8	1-5/8	1-13/16	2-1/16	2- 5/16	2-3/4	3- 1/8	3- 3/4	—	—	—	—
M	1-5/8	2	2- 1/4	2-5/8	2-15/16	3-1/2	3-15/16	4- 3/4	5-9/16	6- 5/8	—	8- 5/16

Table 2. Thread Engagement of Close Nipples

Nominal Pipe Diameter, Inches												
1/4	3/8	1/2	3/4	1	1- 1/4	1- 1/2	2	2- 1/2	3	3-1/2	4	5
Nipple Length, Inches												
7/8	1	1-1/8	1-3/8	1- 1/2	1- 5/8	1- 3/4	2	2- 1/2	2-5/8	2-3/4	2-7/8	3
Thread Engagement, Inches												
3/8	3/8	1/2	9/16	11/16	11/16	11/16	3/4	15/16	1	1-1/16	1-1/8	1-1/4
Laying Length (Nipple Length—2 Thread Engagements)												
1/8	1/4	1/8	1/4	1/8	1/4	3/8	1/2	5/8	5/8	5/8	5/8	1/2

Where valves are concerned, or where fittings not included in the tables are used, the table of standard thread engagements may be of help. The thread engagement distance subtracted from the end-to-center-line dimension of the fitting or valve (obtained by measurement) will give the amount to subtract from the pipe-line measurement in order to determine actual pipe length.

ALLOWABLE SPACES FOR PIPES

The accompanying table gives the allowable minimum distances between (1) two pipes, (2) pipe and wall, and (3) pipe and furring, for two pipes with standard malleable or cast iron screwed fittings in the same elevation.

The table was developed on the basis of the following analysis: Between two pipes of different diameters, there will be two values for the minimum distance which will allow free turning of an elbow or tee. The value used will depend on which pipe is installed first.

In order that there will not be any confusion in the field as to which pipe should be installed first, the condition giving the greater distance between pipes will be considered. It is fairly obvious that turning the fitting on the larger pipe will require the larger space between pipes for clearance. Therefore, less space will be required if the larger pipe is installed first.

It is evident that the minimum distance from the wall will be that distance which will allow turning of a fitting on the pipe. This will be the distance *K* shown in the left column of the table.

The minimum distance from furring (dimension *F* in the table) will not be less than ¼ inch larger than the radius of insulated pipe, because furring is constructed after pipes have been installed.

It should be noted that, for pipes smaller than one inch, the distance allowed for insulation is larger than that required for turning fittings. Therefore, on such small pipes, only the allowance for insulation has been considered in the table.

Example. Find the minimum space required for 1¼- and 4-inch risers if the smaller riser is located near the wall.

Solution. The table shows, under 1¼-inch and opposite 4-inch pipes, that 6 inches is the minimum distance between 4-inch and 1¼-inch pipes.

To the left of 1¼ inches in the second column of the table, find 2⅛ inches as the minimum distance from the wall of the small riser.

The distance from the furring of the larger riser is found (above 4 inches) to be 3¾ inches.

The total space is thus 6 + 2⅛ + 3¾ = 11⅞ inches.

In the same manner it can be shown that the total space would be 6 + 4⅞ + 2 = 12⅞ inches. Therefore it is advisable to locate the smaller pipe near the wall or column.

Minimum Distances Allowable Between Centerlines, Screwed Fittings

Min. Distance from Wall, (*K*), Inches	Nominal Pipe Size, Inches	Minimum Distance from Furring (*F*), Inches											
		1½	1⅝	1⅞	2	2⅛	2½	2¾	3⅛	3¾	4½	5	6
		Nominal Pipe Size, Inches											
		½	¾	1	1¼	1½	2	2½	3	4	5	6	8
		Distance Between Pipes, Inches											
8½	8	9⅛	9⅜	9½	9¾	9⅞	10⅛	10½	10⅞	11⅜	12	12⅝	14
6⅝	6	7⅜	7½	7⅝	7⅞	8	8¼	8⅝	9	9½	10¼	10¾	
5¾	5	6½	6⅝	6¾	7	7⅛	7⅜	7¾	8⅛	8⅝	9¼		
4⅞	4	5½	5⅝	5¾	6	6⅛	6½	6¾	7⅛	7¾			
3⅞	3	4⅝	4¾	4⅞	5⅛	5¼	5½	5⅞	6¼				
3⅜	2½	4	4¼	4⅜	4⅝	4¾	5	5¼					
2⅞	2	3⅝	3¾	3⅞	4	4¼	4½						
2⅜	1½	3⅛	3¼	3⅜	3⅝	3¾							
2⅛	1¼	3	3⅛	3¼	3½								
1⅞	1	2⅞	3	3⅛									
1⅝	¾	2¾	2⅞										
1½	½	2⅝											

EXPANSION OF PIPE

Expansion of pipe in ordinary temperatures (in fact up to at least 1500F) can be calculated according to the formula:

$$L_t = L_o \left[1 + a \left(\frac{t-32}{1000} \right) + b \left(\frac{t-32}{1000} \right)^2 \right]$$

where L_t = the length in feet at F; L_o = the length in feet at 32F; t = the final temperature, F; a and b = constants as follows: steel, a, 0.006212, b, 0.00163; wrought iron, a, 0.006503, b, 0.001622; copper, a, ·0.009278, b, 0.001244; cast iron, a, 0.005441, b, 0.001747.

Example: A 740-foot steel pipeline was installed during 60F weather. What is the computed expansion of the line when 420F steam is turned on in the line?

Solution: Under "Steel," opposite 420F, find 3.423; opposite 60F find 0.449. Subtracting gives 2.974 in. expansion per 100 feet. For 740 feet the expansion is 7.4 × 2.974 = 22.01 in.

Temp. F	Material			Temp., F	Material		
	Steel	Wrought Iron	Copper		Steel	Wrought Iron	Copper
0	0	0	0	390	3.156	3.291	4.532
10	0.075	0.078	0.111	400	3.245	3.383	4.653
20	0.149	0.156	0.222	410	3.334	3.476	4.777
30	0.224	0.235	0.333	420	3.423	3.569	4.899
40	0.299	0.313	0.444	430	3.513	3.662	5.023
50	0.374	0.391	0.556	440	3.603	3.756	5.145
60	0.449	0.470	0.668	450	3.695	3.850	5.269
70	0.525	0.549	0.780	460	3.785	3.945	5.394
80	0.601	0.629	0.893	470	3.874	4.040	5.519
90	0.678	0.709	1.006	480	3.962	4.135	5.643
100	0.755	0.791	1.119	490	4.055	4.231	5.767
110	0.831	0.871	1.233	500	4.148	4.327	5.892
120	0.909	0.952	1.346	520	4.334	4.520	6.144
130	0.987	1.003	1.460	540	4.524	4.715	6.396
140	1.066	1.115	1.575	560	4.714	4.911	6.650
150	1.145	1.198	1.690	580	4.903	5.109	6.905
160	1.224	1.281	1.805	600	5.096	5.309	7.160
170	1.304	1.364	1.919	620	5.291	5.510	7.417
180	1.384	1.447	2.035	640	5.486	5.713	7.677
190	1.464	1.532	2.152	660	5.583	5.917	7.938
200	1.545	1.616	2.268	680	5.882	6.122	8.197
210	1.626	1.701	2.384	700	6.083	6.351	8.460
220	1.708	1.786	2.501	720	6.284	6.539	8.722
230	·.791	1.871	2.618	740	6.488	6.749	8.988
240	1.872	1.957	2.736	760	6.692	6.961	9.252
250	1.955	2.044	2.854	780	6.899	7.175	9.519
260	2.038	2.130	2.971	800	7.102	7.384	9.783
270	2.132	2.218	3.089	820	7.318	7.607	10.056
280	2.207	2.305	3.208	840	7.529	7.825	10.327
290	2.291	2.393	3.327	860	7.741	8.045	10.598
300	2.376	2.481	3.446	880	7.956	8.266	10.872
310	2.460	2.570	3.565	900	8.172	8.489	11.144
320	2.547	2.659	3.685	920	8.389	8.713	11.420
330	2.632	2.748	3.805	940	8.608	8.939	11.696
340	2.718	2.838	3.926	960	8.830	9.167	11.973
350	2.805	2.927	4.050	980	9.052	9.396	12.253
360	2.892	3.017	4.167	1000	9.275	9.627	12.532
370	2.980	3.108	4.289	1100	10.042	10.804	13.950
380	3.069	3.199	4.411	1200	11.598	12.020	15.397

CORROSION RESISTANCE OF METAL PIPES TO VARIOUS FLUIDS*

Data apply to fluid temperatures up to 140 deg F. *S* indicates satisfactory,
L limited, and *U* unsatisfactory

Fluid	Low Carbon Steel	Stainless Steel 302 303 304	316	410 416 430	Cast Iron	Alumi-num	Cop-per	Red Brass	Lead	Monel	Inco-nel
Acetic acid, 10%	U	S	S	L	U	S	U	—	—	S	S
Acetic acid, glacial	U	S	S	U	U	S	U	U	—	—	S
Acetone	S	S	S	U	S	S	S	S	S	S	S
Alcohol, methyl	S	S	S	S	S	L	S	—	S	S	S
Ammonium chloride	L	L	S	L	U	U	L	—	S	S	S
Ammonium sulfate	S	S	S	L	S	L	L	L	—	S	S
Aniline	U	S	S	S	L	U	—	—	—	—	S
Benzene	L	S	S	—	S	S	S	S	—	S	S
Benzoic acid	—	S	S	S	—	S	—	—	—	—	—
Boric acid	U	S	S	L	L	S	S	L	S	S	S
Butyric acid	—	S	S	—	—	S	—	—	—	S	S
Calcium chloride	S	S	L	L	S	U	L	L	—	S	S
Calcium hydroxide	S	S	S	S	S	L	—	—	L	S	S
Carbon tetrachloride	L	S	S	S	L	L	S	S	L	S	S
Chlorine, dry	L	S	S	S	—	U	S	S	—	S	—
Chlorine, wet	U	U	L	U	U	U	L	U	—	L	U
Chloroform	—	S	S	S	—	U	—	—	—	S	S
Chromic acid, 10%	—	L	S	U	U	L	U	U	—	S	U
Chromic acid, 50%	—	U	L	—	U	—	U	U	—	—	—
Citric acid	U	S	S	L	L	S	S	L	—	S	—
Copper chloride	—	S	L	L	—	U	L	—	S	L	L
Copper Sulfate	U	S	S	S	L	U	L	U	—	U	S
Ethyl acetate	L	S	S	S	S	L	—	—	—	S	S
Ethyl chloride	S	S	S	S	U	—	S	—	—	S	S
Fatty acids	—	S	S	S	—	S	—	—	—	S	S
Ferric chloride	U	U	U	U	U	U	U	U	—	U	U
Ferric sulfate	U	S	S	S	U	U	U	U	—	L	S
Formaldehyde	L	S	S	S	—	S	S	S	—	S	S
Formic acid	U	S	S	U	—	U	S	U	—	S	—
Hydrobromic acid	—	U	U	U	—	U	L	—	L	U	U
Hydrochloric acid	U	U	U	U	U	U	—	U	—	—	—
Hydrofluoric acid	U	U	U	U	U	U	L	L	S	S	—
Hydrocyanic acid	L	S	S	S	S	S	L	—	—	S	S
Nickel chloride	U	L	S	U	—	U	L	U	—	S	S
Nickel sulfate	U	S	S	L	—	U	U	U	—	S	S
Nitric acid, 20%	—	—	—	—	U	U	U	U	U	L	U
Nitric acid, 40%	—	—	—	—	U	U	U	U	U	U	L
Nitric acid, 68%	—	—	—	—	U	L	U	U	U	U	L
Oleic acid	L	S	S	S	—	L	S	—	—	S	S
Oxalic acid	U	S	S	L	U	S	S	L	—	S	S
Phosphoric acid, 25%	U	L	S	U	U	U	S	U	S	S	S
Phosphoric acid, 85%	U	U	S	U	U	U	S	U	S	S	S
Picric acid	—	S	S	S	—	—	U	U	—	U	U

* The generalized evaluations given in this table are meant to be an initial guide.

Corrosion is the destruction of a metal by chemical or electrochemical reaction with its environment. In the corrosion process, the reaction products formed may be soluble or insoluble in the attacking environment. Insoluble corrosion products may deposit at or near the attacked area, in which case purely chemical corrosion, although starting rapidly, reduces its rate of attack due to the ob-

| Fluid | Low Carbon Steel | Stainless Steel | | | Cast Iron | Aluminum | Copper | Red Brass | Lead | Monel | Inconel |
		302 303 304	316	410 416 430							
Potassium chloride	—	S	S	S	—	L	S	L	—	S	S
Potassium sulfate ......:....	L	S	S	S	S	S	S	S	S	S	S
Sodium carbonate	S	S	S	S	S	U	S	L	—	S	S
Sodium chloride	L	S	S	L	S	L	S	—	—	S	S
Sodium sulfate	L	S	S	S	S	S	S	S	S	S	S
Stearic acid	L	S	S	S	L	S	S	L	—	S	S
Sulfur dioxide, dry	S	S	S	S	S	—	S	L	—	S	S
Sulfur dioxide, wet	—	S	S	U	—	—	S	U	—	U	U
Sulfuric acid, 10%	U	U	S	U	U	L	U	U	S	L	—
Sulfuric acid, 75%	U	U	U	U	U	U	U	U	S	S	—
Sulfuric acid, 90%	L	L	S	L	L	U	U	U	S	U	—
Trichloroethylene	L	S	S	S	L	S	S	L	—	S	S
Trisodium phosphate	S	S	S	S	—	U	U	U	S	S	S
Water, fresh	L	S	S	S	—	S	S	S	S	S	S
Water, salt	L	S	S	L	S	S	L	L	S	S	S

Source, A. M. Byers Co. and others.

structive layer. Insoluble products of corrosion may also be carried a considerable distance from the point of attack before being deposited. However, rapid destruction of a material may be expected when the corrosion product is soluble in any liquid present.

Underwater Corrosion

Dissolved oxygen, acid gases, and chloride salts are the corrosion accelerators most frequently encountered in water. The corrosion potential of the solution increases in proportion to the amount of the harmful substances present, the temperature, and the rate of flow of the solution over the metal surface. The accompanying table summarizes the relative ability of metal pipes to resist corrosion by harmful substances.

The actual rate at which oxygen reaches the surface is believed to determine the rate of corrosion of ferrous metals. In water, the oxygen penetration to a surface is often restricted by films and scale which form on the surface, and consequently the corrosion rate differs from that of a surface exposed to the atmosphere or to alternately wet and dry conditions. Neutral and slightly alkaline waters saturated with air corrode iron at a rate about triple that for the same water free of air. Hot water containing oxygen will corrode iron three to four times faster than the same water when cold.

Corrosion of iron decreases as the pH of water solutions increases, and practically ceases at a pH of 11.

Steam Condensate Systems

Corrosion in steam condensate systems may be minimized through use of four practical means. These are: (1) Mechanical or chemical elimination of oxygen from boiler feedwater and, therefore, from the steam developed; (2) design of condensing equipment to minimize assimilation in the condensate of non-condensable gases; (3) chemical treatment of the steam or condensate; and (4) use of corrosion-resistant metals.

Cathodic Protection

In addition to the use of special materials and protective coatings for buried pipelines, another form of protection is obtained by rendering the line cathodic to the surrounding soil or water by means of a controlled difference of potential. Protective coatings that insulate a large portion of the metal surface will greatly reduce the total amount of protective current which must be impressed on bare anodic areas to check corrosion.

It is extremely important to know the analysis of the soil before burying pipelines. Soils containing organic matter or high levels of carbon are extremely corrosive. Where high corrosion rates are anticipated, buried piping is usually protected by bitumastic coatings. For extreme conditions, the coating is reinforced by wrappings of impregnated woven fabric. For exceptionally severe corrosion conditions the coated pipes can then be encased in concrete. The concrete must be moisture-proofed with waterproofing membranes.

Where cast iron piping is used to transport corrosive waters such as mine drainage or seawater, a commercially available cement-lined cast iron is commonly utilized.

PIPE SUPPORT SPACING

When a horizontal pipe line is supported at intermediate points, sagging of the pipe occurs between the supports, the amount of sag being dependent upon the weight of the pipe, water, insulation, and valves or fittings which may be included in the line. If the pipe line is installed with no pitch, pockets will be formed in each span in which case condensation may collect if the line is transporting steam. In order to eliminate these pockets, the line must be pitched downward so that the outlet of each span is lower than the maximum sag.

Tests were conducted by Crane Company to determine the deflection of horizontal standard pipe lines filled with water. Results indicate that for pipes larger than 2 inches and with supports more than 10 feet center to center, the resultant deflection is less than that determined by the use of the formula for a uniformly loaded pipe fixed at both ends. For pipe sizes 2 inches and smaller, the test deflection was in excess of that determined by the formula for pipe having fixed ends and approached, for the shorter spans, the deflection for pipe lines having unrestrained ends.

The accompanying chart, by Crane Company, reproduced by permission, gives the deflection of horizontal schedule 40 pipe lines *filled with water,* for varying spans, based upon the results obtained from tests for sizes 2 inches and smaller, and upon the formula for fixed ends for the larger sizes of pipe. The deflection values are twice those obtained from test or calculated, to compensate for any variables including weight of insulation, valves, etc.

The chart can be used (a) to determine the deflection for a given spacing or (b) to determine the slope in the pipe necessary to overcome the deflection. In the former use, for example, the chart shows that the deflection in a 6-inch pipe at a hanger spacing of 20 feet is .061 inches.

A more useful application of the chart is where it is desired to know what hanger spacing is necessary with a given slope and pipe size. For example, if a slope of 1 inch in 30 feet is practicable in a 4-inch pipe line, what hanger spacing is necessary? To find this answer, locate the intersection of the 4-inch pipe size curve with the 1 inch in 30 feet slope curve and find, on the bottom scale, the spacing to be 22 feet.

For slopes not given on the chart, the following formula can be used:

$$y = \frac{L}{4G}$$

where y = deflection in inches,
L = distance between supports, feet
G = gradient of pipe expressed in number of feet in which there is a rise of 1 inch.

As an example, a slope of 1 inch in 25 feet is desired. What hanger spacing is required for 4-inch pipe?

Solve the formula for two points, say $L = 10$ and $L = 60$, and obtain

$$y = \frac{10}{4 \times 25} = .10$$

$$y = \frac{60}{4 \times 25} = .60$$

Plot these points on the graph, and draw a line through them. The intersection of this line with the 4-inch pipe curve is at $y = .24$ and a hanger spacing of 24 feet.

The formula can also be used for exact interpolation of the slope required to overcome deflection of pipe at a given spacing.

For example, the chart shows that a 4-inch pipe supported at 30-foot intervals will have a deflection of 0.62 inch at each span. How much slope must be given the pipe to properly drain the deflection?

The chart shows the slope to be between 10 and 15 feet per inch of fall. Applying the formula:

$$0.62 = \frac{30}{4\,G}$$

$$G = \frac{30}{4 \times .62} = 12.1$$

Therefore, a slope of 1 inch in 12.1 feet will overcome the deflection.

If this slope is too steep for the space available, note that following the 4-inch pipe curve down the chart permits selection of a permissible pitch and corresponding support interval.

If an unusual installation situation should limit the span to, say, 15 ft between supports and at the same time limit the pitch to, say, one inch in 30 feet, the chart shows that a 2-inch pipe is the smallest diameter that can be installed to drain properly. The deflection of any smaller pipe will be too great to be overcome by this pitch. Obviously, any pipe can be pitched at any gradient steeper than the minimum gradients determined from the chart. That is, any pipe larger than 2-inch can be pitched at 1 inch in 30 feet *or steeper* with supports at 15-ft intervals. The values read from the chart are minimum for complete drainage.

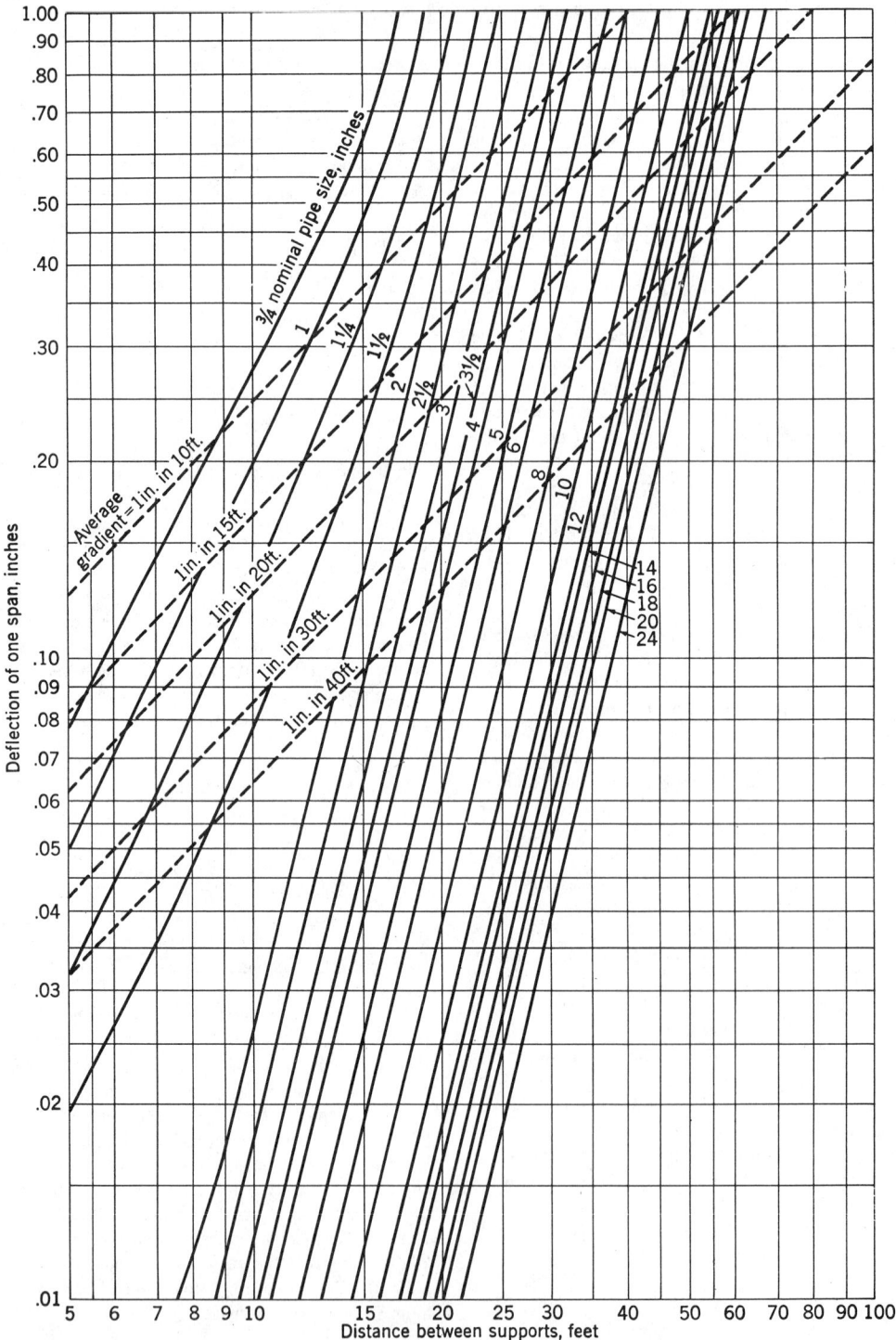

GATE, GLOBE, AND CHECK VALVES

The descriptions and information on gate, globe, and check valves which follow were prepared by B. Krueger and E. N. Morris, both formerly with Guy B. Panero, Consulting Engineers, New York.

Description

Gate valves. The working parts of gate valves include a wedge or gate which fits into the open passageway of the valve between machined seats; a threaded stem or spindle; a handwheel, and packing.

The bonnet may be of one piece construction screwed directly to the valve body, or consist of a union connection screwed to the body. The bonnet also may be bolted to the body, or be constructed as a yoke, exposing the stem or spindle.

Packing fits around the stem in a recess in the top of the bonnet and is held in place by a packing nut which screws to the bonnet, or by a packing gland which is bolted to the bonnet. A packing gland bushing, or follower, installed between the gland or nut and the packing, transmits the force exerted by the packing gland or nut to the packing.

Larger gate valves are the outside screw and yoke (OS&Y) type, with a bolted bonnet, bolted gland, and rising stem. Smaller gate valves are of the inside screw, rising stem, screwed bonnet, screwed packing nut type. Larger valves are flanged on the ends; smaller valves are screwed or solder end valves.

Globe valves. The working parts of globe valves consist of a disc which fits over a circular horizontal opening in the valve passageway into which a "seat" has been fitted or machined; a stem or spindle; a handwheel, and packing.

The same variations that apply to the construction of bonnets, packing glands or nuts and the type of end connections of gate valves also apply to globe valves.

Check valves. The working parts of check valves consist of a hinged disc or clapper which is free to swing upon a hinge pin in only one direction. There are no exterior working parts whatever on a check valve.

Operation

Gate valves. Gate valves are installed in the pipeline, and the passageway of the pipe continues straight through the valve. A wedge (gate) is moved up or down, out of or into the passageway, to permit or prevent flow through the valve. As the handwheel is turned counterclockwise, the threads on the stem or spindle engage the threads in the bonnet and the stem, and the attached wedge is raised out of the passageway. When the handwheel is turned clockwise, the stem screws down through the bonnet and the attached wedge is lowered into the valve passageway, thus preventing flow through the valve. The packing, which is held in place by the packing gland or nut, prevents the fluid within from escaping around the stem, through the bonnet.

Globe valves. Globe valves operate in the same way as gate valves, except that the disc in the globe valves is horizontal and seats against a horizontal opening. When the handwheel is turned counterclockwise, the stem threads through the bonnet, and the disc is lifted off the horizontal valve passageway. When the handwheel is turned clockwise, the stem screws down into the bonnet and the disc is lowered onto the valve passageway, thus preventing flow.

Check valves. Check valves are automatically operated by the fluid flowing through the valve. The clapper is free to swing only in the direction of the flow of the liquid in the pipeline. The pressure exerted by the liquid flowing through the valve lifts the clapper and holds it in an open position. When the flow stops in that direction, the clapper falls back to its original position by gravity, thus preventing back flow.

Recommendations

A valve must never be forced closed. Although strainers are installed in the pipeline to protect valves and other equipment, pipe scale and other foreign material may be present in the liquid in the pipeline, and may become lodged in the valve seat, preventing the valve from closing. To dislodge the

foreign material, the valve is partially opened and closed several times. If the valve does not seat properly after this "washing," the supply to the valve is closed off, the bonnet assembly removed, and the seating surfaces wiped with a clean cloth. The valve is then reassembled and the seating action tested to make sure that the trouble is only dirt.

Note. The "washing" process described is useful only where liquid in the pipe is flowing. It is of no value where the line is flooded and fluid is static or where the line has been drained.

All gate valves are to be either tightly closed or fully opened, as required. When left in a partially opened position, the wedge and seats become grooved and damaged, or "wiredrawn."

The attached piping must be properly supported or hung, so that the valve is not subjected to undue strain resulting from pipe movement.

A gate valve must never be used for throttling or regulating flow; it must be either opened fully or closed completely.

When installing flanged valves, the flanges must be properly aligned. The bolts must not be inserted haphazardly and screwed up on the flange nuts until the flanges are forced into contact. Special care must be taken not to mar gasket faces of flanges by tightening with foreign matter between faces or by accidentally striking with tools.

The piping must not be "sprung" in order to remove or replace a valve. It is better maintenance procedure to cut the pipe about 6 in. from the downstream end of the valve, remove the valve and install a union in the pipe line, before making the valve replacement.

Definitions of Terms

Body. That part of the valve which attaches to the pipe line or equipment either with screwed ends, flanged ends or solder joint ends, and encloses the working parts of the valves.

Bonnet. The top part of the valve housing, through which the stem extends. It provides support and protection to the stem and houses the stem packing. It may be screwed or bolted to the body.

Cap. The top part of the housing of a check valve (equivalent to the bonnet of a gate or globe valve) which may be either screwed or bolted onto the main body.

Check valve. An automatic, self closing valve which permits flow in only one direction. It automatically closes by gravity when liquid ceases to flow in that direction.

Clapper. A common term which is used to describe the disc of a swing type check valve.

Disc. The disc shaped device which is attached to the bottom of the stem and which is brought into contact with, or lifted off of the seating surfaces to close or open a globe valve.

Flanged bonnet. A type of bonnet so constructed that it attaches to the body by means of a flanged, bolted connection. The whole bonnet assembly, including the handwheel, stem and disc, may be quickly removed by unscrewing the nuts from the bonnet stud bolts.

Gate valve. A valve which is used to open or close off the flow of fluid through a pipe. It is so named because of the wedge (gate) which is either raised out of or lowered into a double seated sluice, to permit full flow or completely shut off flow. The passageway through a gate valve is straight through, uninterrupted, and is the full size of the pipe line into which the valve is installed.

Gland bushing. A metal bushing installed between the packing nut and the packing to transmit the force exerted by the packing nut against the packing.

Globe valve. A valve which is used for throttling or regulating the flow through a pipe. It is so named because of the globular shape of the body. The disc is raised off a horizontal seating surface to permit flow, or lowered against the horizontal seating surface to shut off flow. The disc may be lifted completely to permit full flow or may be lifted only slightly in order to throttle or regulate flow. The flow through a globe valve has to make two $90°$ turns.

Handwheel. The wheel-shaped turning device by which the stem is rotated, thus lifting or lowering the disc or wedge.

Hingepin. The pin upon which the disc or clapper of a check valve swings.

Outside screw and yoke. A type of bonnet so constructed that the operating threads of the stem are outside the valve housing, where they may be easily lubricated and do not come into contact with the fluid flowing through the valve.

Packing. A general term describing any yielding material used to effect a tight joint. Valve packing is generally "jam packing"; it is pushed into a stuffing box and adjusted from time to time by tightening down on a packing gland or packing nut.

Packing gland. A device which holds and compresses the packing, and provides for additional compression by manual adjustment of the gland, as wear of the packing occurs. A packing gland may be screwed or bolted in place.

Packing nut. A nut which is screwed into place and presses down upon a gland bushing which transmits the force exerted by the packing nut to the packing. It serves the same purpose as a packing gland.

Rising stem. A threaded stem which is unscrewed or screwed through the bonnet to open or close the valve. The handwheel may rise with the stem, or the stem may rise through the handwheel.

Screwed bonnet. A type of bonnet so constructed that it attaches to the body by means of a screwed joint. It may screw over the body, or inside the body, or may be attached to the body by means of a union-type screwed connection.

Solid wedge. A wedge consisting of one solid piece.

Split wedge. A wedge consisting of two pieces, into which the valve stem is screwed to expand the two pieces against the valve seating surfaces to assure a tight seal when the valve is closed.

Stem. The usually threaded shaft to which is attached the handwheel at the top and the disc or wedge at the lower end. The stem may also be called the spindle.

Stop plug. An adjusting screw, extending through the body of a check valve, which adjusts and controls the extent of movement of the disc or clapper.

Swing check valve. A check valve which uses a hinged disc or clapper to limit the direction of flow. The pressure exerted by the fluid flowing through the valve forces the disc away from the seating surface. When the flow ceases, the clapper falls to its original position, preventing flow in the opposite direction.

Union. A coupling fitting, consisting of three parts (a shoulder piece, a thread piece, and a ring) used for coupling the ends of pipe sections. Adjoining faces of shoulder and thread pieces are lapped together to form a tight joint. Unions permit easy disconnection for repair and replacement of piping and fittings.

Union bonnet. Type of bonnet which is so constructed that the whole bonnet assembly, including the handwheel, stem and disc assembly, may be quickly removed by unscrewing the bonnet union ring from the valve body.

Union ring. The large nut-like ring which secures the union thread piece and the union shoulder piece together. It slips over and against the shoulder piece and screws onto the union thread piece.

Union shoulder piece. That part of the union which is fastened to the pipe and retains the union ring.

Union threaded piece. That part of the union which is fastened to the pipe and has external threads over which the union ring is screwed to effect a coupling.

Wedge. (See also "Disc".) The wedge-shaped device which fits into the seating surfaces of a gate valve and which is drawn out of contact with the seating surfaces to permit flow, or is pushed down into contact with the seating surfaces to close off flow through the valve.

Maintenance Methods

Maintenance personnel in their normal rounds are to remain alert to detect evidence of leakage and/or valve failure. When evidence of leakage and/or valve failure is detected, the inspection procedures described below are to be followed.

Gate valves. (a) Inspect the area around the top of the bonnet and stem for leakage.

(b) Inspect the bolted or screwed bonnet body connection for leakage.

(c) Inspect flanged or screwed connections for leakage.

(d) Check the action of the valve by turning the handwheel counterclockwise to open, and/or clockwise to close. Open and/or close each valve several times to make sure it is not corroded or jammed.

Make sure that the valve is returned to its original operating setting.

Globe valves. (a)Follow the procedure described for the inspection of gate valves.

Check valves. (a) Inspect the gasketed or screwed connections around the cap and both ends of the body for leakage.

Failure of the internal parts of check valves can not be determined by inspection, except by disassembling the valve. Do not disassemble any gasketed connection unless such disassembly is necessary for repair.

Trouble Shooting Methods

Trouble	Cause	Remedy
Gate Valves		
Leakage around stem	Loose packing gland or nut Packing worn out Bent or scored stem	Tighten packing gland or nut Replace packing Replace stem
Leakage around gasketed, screwed or soldered connections	Loose connections Gasket failure	Tighten flange bolts or screwed connections, or resolder joints Replace gasket
Handwheel turns without stopping at extremes	Broken stem	Replace stem
Valve does not close tightly	Wedge damaged Seat damaged Wedge-stem joint damaged Deposits under wedge	Replace wedge Replace valve Replace stem and/or wedge "Wash" or clean
Globe Valves		
Leakage around stem	Loose packing gland or nut Packing worn out Bent or scored stem	Tighten packing gland or nut Replace packing Replace stem
Leakage around gasketed, screwed or soldered connections	Loose connection Gasket failure	Tighten flange bolts Tighten screwed connections or resolder joints Replace gasket
Valve does not close tightly	Disc damaged Seat damaged Disc-stem joint damaged Deposits under disc	Replace disc Replace valve Replace disc or stem "Wash" or clean
Check Valves		
Leakage around gasketed, screwed or soldered connections	Loose connection Gasket failure	Tighten flange nuts or screwed connections or resolder joints Replace gasket

FORMULAS FOR SIZING CONTROL VALVES

The valve flow coefficient, C_v is the standard measure of valve flow capacity, defined as the number of U. S. gallons of water per minute that will flow through the valve in a wide open position with a pressure drop of 1 psi. The C_v rating is determined by flow tests.

While present valve sizing procedures are based on incomplete assumptions and theory, without its use sizing of valves, especially control valves, would be highly complicated. For most engineering purposes, the following procedures are justified:

To determine valve size,

Given: All flowing conditions.

Solve: For C_v and select valve size for the type of valve under consideration from manufacturer's table of C_v rating vs. valve size.

To determine valve capacity,

Given: C_v and flowing conditions.

Solve: For capacity V in U. S. gpm of liquid
Q in cfh gas at 14.7 psia and 60 F.
W in lb steam per hr.

For liquid—

$$C_v = V \sqrt{G/\Delta P}$$

$$V = C_v \sqrt{\Delta P/G}$$

where

$V =$ flow, gpm (U. S.)
$\Delta P =$ pressure drop at maximum flow, psi
$G =$ specific gravity (water = 1.0)
$C_v =$ valve flow coefficient

When flowing temperature is above 200 F, use specific gravity and quantity at flowing condition.

For gas—

$$C_v = Q\sqrt{G T_a}/1360 \sqrt{(\Delta P) P_2}$$

$$Q = 1360 C_v \sqrt{(\Delta P) P_2}/\sqrt{G T_a}$$

where

$Q =$ cu ft per hr at 14.7 psia and 60 F
$\Delta P =$ pressure drop $(P_1 - P_2)$ at maximum flow, psi
$P_1 =$ inlet pressure at maximum flow, psia
$P_2 =$ outlet pressure at maximum flow, psia
$G =$ specific gravity (air = 1.0)

$T_a =$ flowing temperature, absolute (460 + F)
$C_v =$ valve flow coefficient

When P_2 is less than $\frac{1}{2} P_1$, use the value of $P_1/2$ in place of $\sqrt{(\Delta P) P_2}$

For steam—

$$C_v = WK/3 \sqrt{(\Delta P) P_2}$$

$$W = 3 C_v \sqrt{(\Delta P) P_2/K}$$

where

$W =$ lb steam per hr
$\Delta P =$ pressure drop at maximum flow, psi
$P_1 =$ inlet pressure at maximum flow, psia
$P_2 =$ outlet pressure at maximum flow, psia
$K = 1 + (0.0007 \times$ F superheat)
$C_v =$ valve flow coefficient

When P_2 is less than $\frac{1}{2} P_1$, use the value of $P_1/2$ in place of $\sqrt{(\Delta P) P_2}$

The steam formula has been set up using $1/0.00225 P_2$ in place of the specific volume to eliminate the need for steam tables.

Note: Flow of compressible fluids through a restriction reaches a saturation velocity when the differential pressure is increased to approximately 50% of the inlet pressure. This critical pressure ratio varies with the composition of the fluid. The average value of one-half the absolute inlet pressure is well within the tolerance established by the formulas.

For vapors other than steam—general formula-weight basis—

$$C_v = (W/63.4) \sqrt{v_2/\Delta P}$$

$$W = 63.4 C_v \sqrt{\Delta P/v_2}$$

where

$W =$ lb vapor per hr
$\Delta P =$ pressure drop $(P_1 - P_2)$ at maximum flow, psi
$v_2 =$ specific volume (cu ft per lb) at outlet pressure P_2
$P_1 =$ inlet pressure at maximum flow, psia
$P_2 =$ outlet pressure at maximum flow, psia
$C_v =$ valve flow coefficient

When P_2 is less than $\frac{1}{2} P_1$, use the value of $P_1/2$ in place of (ΔP) and use P_2 corresponding to $P_1/2$.

IDENTIFICATION OF PIPING SYSTEMS

Schemes for the identification of piping systems have been developed in the past by a large number of industrial plants and organizations. The schemes arrived at may have given complete satisfaction to those using them but they have suffered from a lack of uniformity. Considerable confusion as well as accidents have occurred to those who change employment from one plant to another and to outside agencies, such as municipal fire departments, when called in to assist.

In order to promote greater safety and lessen the chances of error, confusion, or inaction, especially in times of emergency, a uniform code for identification of piping has been established by the American National Standards Institute ANSI-A13.1, and published by The American Society of Mechanical Engineers. The standard is based on primary identification of the contents of a piping system by stenciled legend and secondary identification through the use of color. It is urged that industry and organizations do not use color as a means of specifying the type of material contained in a piping system unless its use is in conformity with provisions of the standard and supplementary to the use of legends.

Any material transported in a piping system will fall into one of four main classifications:

Fire Protection Materials and Equipment. This classification includes sprinkler systems and other fire fighting or fire protection equipment. The identification for this group may also be used to locate such equipment as alarm boxes, extinguishers, fire doors, hose connections, and hydrants.

Dangerous Materials. This group includes materials which are hazardous to life or property because they are easily ignited, corrosive at high temperatures and pressures, productive of poisonous gases, or are in themselves poisonous.

Safe Materials. This group includes materials involving little or no hazard to life or property in their handling. Classification embraces materials at low pressures and temperatures which are not poisonous and will not produce fire or explosion.

Protective Materials. This group includes materials which are piped through plants for the express purpose of being available to prevent or minimize the hazard of the dangerous materials previously mentioned.

(There formerly was a fifth classification, Extra Valuable Materials, identified by the color purple.)

Method of Identification. Positive identification of contents of a piping system shall be by lettered legend giving the name of the material in full or abbreviated form. Arrows may be used to indicate the direction of flow. Where it is desirable or necessary to give supplementary information such as hazard or use of the piping system contents, this may be done by additional legend or by color applied to the entire piping system or as colored bands. Color identification for the four classifications are shown in Table 1. Recommended width of color bands and height of legend letters for various sizes of pipe are given in Table 2. Table 3 presents a list of typical materials transported in piping systems and their identifying colors.

Table 1. Color Identification of Classifications

Classification	Predominant Color of System	Color of Letters for Legends
F — Fire protection materials and equipment	Red	White
D — Dangerous materials	Yellow (or orange)	Black
S — Safe materials	Green (or the achromatic colors: white, black, gray, or aluminum)	Black
P — Protective materials	Bright blue	White

Table 2. Size of Color Bands and Legend Letters

Outside Diameter of Pipe or Covering, Inches	Width of Color Bands, Inches	Height of Legend Letters, Inches
¾ to 1¼	8	½
1½ to 2	8	¾
2½ to 6	12	1¼
8 to 10	24	2½
Over 10	32	3½

Table 3. Identifying Colors of Typical Materials Transported in Piping Systems

Material	Color	Material	Color	Material	Color
Acetic acid	Yellow	Distilled water	Green	Muriatic acid	Yellow
Acetone	Yellow	Drain oil	Yellow	Naphtha	Yellow
Acetylene gas	Yellow	Drain water	Green	Natural gas	Yellow
Acid	Yellow	Drinking water	Green	Nitric acid	Yellow
Air	Green	Dye	Yellow	Nitrogen	Green
Alcohol	Yellow	Ethane dye	Yellow	Nitrogen oxide	Yellow
Alum	Green	Exhaust air	Green	Oil	Yellow
Ammonia	Yellow	Exhaust gas	Yellow	Oil, lubricating	Yellow
Ammonium nitrate	Yellow	Exhaust system	Yellow	Oil, soluble	Yellow
Amyl acetate	Yellow	Filtered water	Blue	Oxygen	Yellow
Antidote gas	Blue	Fire protection water	Red	Paint	Yellow
Argon	Green	Flue gas	Yellow	Peanut oil	Yellow
Benzol	Yellow	Foamite	Red	Phenol	Yellow
Bisulphite liquor	Yellow	Formalin	Yellow	Process gas	Yellow
Blau gas	Yellow	Freon	Green	Producer gas	Yellow
Bleach liquor	Yellow	Fresh water	Green	Propane gas	Yellow
Blow-off water	Yellow	Fuel gas	Yellow	Raw water	Green
Boiler feed water	Yellow	Fuel oil	Yellow	Refrigerated water	Green
Brine	Green	Gas	Yellow	River water	Green
Burner gas	Yellow	Gasoline	Yellow	Salt water	Green
Butane	Yellow	Glycerine	Green	Sanitary sewer	Green
Butyl alcohol	Yellow	Heating returns	Yellow	Soda ash	Green
Calcium chloride	Blue	Heating steam	Yellow	Solvent	Yellow
Carbon bisulphide	Yellow	Helium	Green	Soybean oil	Yellow
Carbon dioxide	Yellow	Hot water	Yellow	Sprinkler, water	Red
Carbon monoxide	Yellow	Hydrochloric acid	Yellow	Steam	Yellow
Carbonated water	Green	Hydrogen	Yellow	Storm sewer	Green
Caustic soda	Yellow	Hydrogen peroxide	Yellow	Sugar juice	Green
Chlorine	Yellow	Hydrogen sulphide	Yellow	Sulphur' chloride	Yellow
Chlorine gas	Yellow	Industrial water	Green	Sulphur dioxide	Yellow
Chloroform	Yellow	Instrument air	Green	Sulphuric acid	Yellow
Circulating water	Green	Kerosene	Yellow	Tar	Yellow
City gas	Yellow	Lacquer	Yellow	Toluene	Yellow
City water	Green	Lactic acid	Yellow	Toluol	Yellow
Coal gas	Yellow	Linseed oil	Yellow	Trichloroethylene	Yellow
Cold water	Green	Lubricating oil	Yellow	Turpentine	Yellow
Compressed air	Green	Make-up water	Green	Vapor	Yellow
Condensate	Yellow	Mercury	Yellow	Varnish	Yellow
Cooling water	Green	Methyl chloride	Yellow	Vegetable oil	Yellow
Cottonseed oil	Yellow	Mixed acid	Yellow	Waste water	Green
Cutting oil	Green	Mixed gas	Yellow	Water	Green
Diesel oil	Yellow	Monoethanolamine	Green	Water gas	Yellow

CHARACTERISTICS OF CENTRIFUGAL PUMPS

There are two distinct types of centrifugal pumps: (1) the turbine type pump which uses diffusers or guide vanes in the casing for the conversion of the velocity to pressure energy, and (2) the volute type centrifugal pump, most commonly used.

Mechanically, a volute type centrifugal pump consists of an impeller or runner having curved vanes revolving on a shaft and housed in a shell or casing. Liquid enters the impeller axially to the shaft and it has energy imparted to it by rotating vanes of the impeller. The fluid leaves the periphery of the impeller at a relatively high velocity and is collected in the casing or shell. This casing is so designed that the velocity of the liquid is gradually reduced before it is discharged. Here the velocity of the liquid is converted into pressure by reduction of velocity according to Bernoulli's theorem.

The quantity of liquid discharged by the pump is almost always measured in gallons per minute although sometimes the measure is cubic feet per second. In this discussion gallons per minute is used as the unit.

Pressure developed by a centrifugal pump is specified as head in feet of liquid. When h = head in feet and P = pressure in pounds per square inch,

$$h = \frac{2.31P}{s}$$

where s is the specific gravity of the liquid compared to water (water at 60/60F = 1.00).

The head developed by a centrifugal pump is a function of the impeller diameter and the speed of rotation (rpm). Maximum head that can be developed by a centrifugal pump is when the discharge valve is tightly closed and the pump is discharging zero capacity into the system. This is known as the "shut-off" head of the pump. Since there is a predetermined maximum pressure that the pump can develop and this pressure is taken into account by the designer, centrifugal pumps do not require relief valves or other unloading mechanism that are otherwise necessary for the positive displacement type of pumps. The maximum or "shut-off" head h_x of any centrifugal pump can be very closely calculated by the formula:

$$h_x = \left(\frac{D \times N}{1840}\right)^2$$

where D = outside diameter of the impeller in inches and N = rpm.

Operating Characteristics

Hydraulic operating characteristics of a typical centrifugal pump, or performance curve, is shown in Fig. 1. The pressure (or head in feet of liquid) developed by the pump at a specified impeller diameter and at a constant rpm is plotted against the discharge of the pump in gallons per minute (gpm.)

Note that the maximum head developed by the pump is at zero capacity or shut-off as previously mentioned.

The head-capacity curve is a sloping line from shut-off to maximum or wide-open capacity. In other words, as the pump discharges more liquid, its pressure decreases. The slope in the head-capacity curve is due (1) to the curve or shape and the number of vanes in the impeller; (2) to friction or head loss within the pump. As the pump discharges more liquid, there is increased internal friction, and this friction loss is actually a loss in pressure or head at the discharge of the pump — hence, the slope in head capacity curve. The pump designer can control to a certain degree the slope of the head-capacity curve by the shape or warp of the impeller vanes and also the number of vanes. The internal friction, however, is a factor over which the pump designer has very little control.

The efficiency curve rises to a peak within certain capacity limits and then falls off toward the maximum capacity of the pump. The brake horsepower curve is usually as shown; that is, it gradually increases in value as the capacity increases. Maximum efficiency of a centrifugal pump lies within the design range. The pump designer had a definite capacity and head upon which all his calculations were based, and his calculations were such that the maximum efficiency of the pump would be at or very near his design capacity.

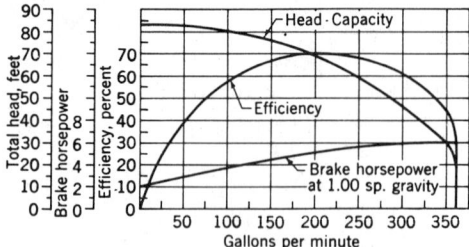

Fig. 1. Performance curves for a typical centrifugal pump — one with 9.5 in. impeller diameter and 1750 rpm constant speed.

PUMP LAWS

The efficiency of a centrifugal pump, as for any machine, is the horsepower output divided by the horsepower input. When the efficiency is known the horsepower requirement of the pump is determined by the formula:

$$HP = \frac{gpm \times DH \times s}{3960 \times E}$$

where DH is the dynamic head in feet, s the specific gravity, and E the efficiency expressed as a decimal.

This formula holds for any liquid since the specific gravity of the liquid as compared with water is inserted in the formula.

Change of Performance

The so-called laws of affinity relating to centrifugal pumps are theoretical rules which apply to the change in performance of a centrifugal pump by a change in the speed of rotation or a change in the impeller diameter of a particular pump. It should always be remembered in using these laws of affinity that they are theoretical and do not always give exact results as compared with tests. However, they are a good guide for predicting the hydraulic performance characteristic of a pump from a known characteristic caused by either altering the speed of rotation or the outside diameter of the impeller. The laws of affinity may be stated as follows:

(A) At a *constant impeller diameter:*

1. The capacity varies directly as the speed;
2. The head varies as the square of the speed;
3. The horsepower varies as the cube of the speed.

In equation form, the foregoing are expressed as:

$$\frac{gpm_y}{gpm_x} = \frac{rpm_y}{rpm_x}$$

$$\frac{head_y}{head_x} = \frac{rpm^2_y}{rpm^2_x}$$

$$\frac{bhp_y}{bhp_x} = \frac{rpm_y^3}{rpm_x^3}$$

(B) At *constant speed:*

1. The capacity varies as the cube of the impeller diameter;
2. The head varies as the square of the impeller diameter; and
3. The horsepower varies as the fifth power of the impeller diameter.

Or, in equation form

$$\frac{gpm_y}{gpm_x} = \frac{d_y}{d_x}$$

$$\frac{head_y}{head_x} = \frac{d_y^3}{d_x^3}$$

$$\frac{bhp_y}{bhp_x} = \frac{d_y^5}{d_x^5}$$

Example. In Table 1 are tabulated actual characteristics of a centrifugal pump at a certain impeller diameter at a constant speed of 1150 rpm. Predict the performance of this particular pump at the increased speed of 1750 rpm by the affinity laws.

Table 1. Performance of a Certain Pump at 1150 RPM

GPM	Total Head, Feet	BHP	Efficiency, %
0	43.2	6.0	—
200	42.9	7.2	30
400	42.0	8.15	52
800	38.0	10.1	76
1200	29.0	10.7	82.3
1400	21.5	11.0	69

Using the subscript x for the unknown performance at 1750 rpm and y for the performance at 1150 rpm, we have, using the first group of formulas, at 400 fpm for 1150 rpm:

$$gpm_x = \frac{gpm_y \times rpm_x}{rpm_y} = \frac{400 \times 1750}{1150} = 610 \text{ gpm}$$

when the pump operates at 1750 rpm, and

$$head_x = \frac{head_y \times rpm_x^2}{rpm_y^2} = \frac{42 \times 1750^2}{1150^2} = 97 \text{ feet}$$

and

$$bhp = \frac{bhp_y \times rpm_x^3}{rpm_y^3} = \frac{8.15 \times 1750^3}{1150^3} = 28.7 \text{ hp.}$$

The 610, 97, and 28.7 are the capacity, head and horsepower of the pump when operating at 1750 rpm. Carrying through the same calculations for other capacities, the data arrived at in Table 2 are found.

Table 2. Performance of the Pump in Table 1 when Speed is 1750 rpm

GPM	Total Head, Feet	BHP	Efficiency, %
0	100	21.1	—
304	99	25.3	30
610	97	28.7	52
1220	88	35.6	76
1825	67	37.7	82
2130	50	38.7	69

CENTRIFUGAL PUMP SELECTION

In the following, the calculations will be on the basis of handling clear, cold water having a specific gravity of 1.00. Assume that the desired capacity or discharge of the pump (in gpm) has already been established.

Total Dynamic Head

In order to select a centrifugal pump for any installation, particular care must be taken in determining the total dynamic head of the system. Careful calculation of the total dynamic head is the most important element of centrifugal pump selection. This fact cannot be stressed too strongly. The total dynamic head (TDH) consists of the following factors all added together to form a total.

1. Static suction lift (H_l) or suction head (H_p): the vertical distance in feet from the free level of the source of supply to the horizontal centerline of the pump.

2. Static discharge head (h_d): the vertical distance in feet from the centerline of the pump to point of free discharge; or, the level of free surface of the discharged liquid, or, in the case of discharging under pressure, the equivalent level of free discharge in feet of liquid.

3. Friction head (h_f): the head in feet of liquid required to overcome the frictional resistance of the piping and fittings of both the suction and discharge lines.

When the pump is above the source of supply and a static suction lift condition prevails,

$$TDH = H_l + h_d + h_f$$

If, however, the source of supply is above the pump the liquid flows to the pump by gravity, a static suction head condition exists and

$$TDH = h_d - h_p + h_f$$

In determining the total dynamic head of a pumping system, the friction head and static head should be calculated at the maximum or extreme conditions of operation. In other words, the static head should be the maximum capacity required by the system. The drawing of a system head or service curve is recommended practice.

Example. Determine the total dynamic head of the following water system: Maximum capacity, 200 gpm; maximum static suction lift, 5 ft; maximum static discharge head, 30 ft; suction piping consisting of 10 ft of 4-in. pipe and one 90 degree 4-in. elbow; discharge piping consisting of 115 ft or 3-in. pipe, five 90-degree 3-in. elbows and one gate valve fully opened.

The total static elevation is 35 ft (composed of 5 ft static suction lift and 30 ft static discharge). The next step is to plot the fricion head at various capacities. From water friction tables secure the following information based on the pipe having a friction coefficient or smoothness factor of 15 year old pipe; one 4-in. 90 degree elbow is equivalent to 11 ft of straight pipe; five 3-in. 90-degree elbows are equivalent to 8 × 5 or 40 ft of straight pipe, one 3 in. gate valve fully open is equivalent to 1.5 ft of straight pipe. Therefore, on the suction of the pump there is an equivalent total of 21 ft of 4-in. pipe and on the discharge an equivalent total of 156.5 ft of 3-in. pipe. Friction for the various capacities in the suction and discharge piping is calculated and tabulated in Table 1.

Table 1. Friction Heads for Illustrative Problem

GPM	Suction Pipe Friction	Discharge Pipe Friction	Total Friction
50	.07	2.16	2.23
100	.26	7.80	8.06
150	.55	16.6	17.15
200	.91	28.2	29.11
250	1.32	42.2	43.52

Plot the friction head data from Table 1 on graph paper and find that at a capacity of 200 gpm a pump is required capable of 64 ft total dynamic head.

To carry the function of the service curve still further, on the service curve can be plotted the hydraulic performance curve of a pump which will develop 200 gpm at 64 ft total dynamic head. This pump cannot pump more than the predetermined maximum capacity of 200 gpm since if 200 gpm is exceeded the total head exceeds the head capabilities of this particular pump at 200 gpm. In other words, the friction head of the system throttles

the pump to a maximum capacity of 200 gpm providing the maximum conditions of static head prevail. If, for example, the static suction lift might at times be zero or the water level at the suction source is level with the pump horizontal centerline, the total static head reduces itself from 35 to 30 ft, but, since the piping arrangement has not been changed the friction head remains unchanged. A second service curve can therefore be plotted based on the minimum static suction condition. Therefore, when this suction condition exists, the pump capacity would be 212 gpm with a varying discharge condition from 200 to 212 gpm depending upon the static suction lift.

Net Positive Suction Head (NPSH)

In addition to selecting a centrifugal pump to perform based on the total dynamic head of the system and the capacity required, it is necessary to check the net positive suction head (NPSH) characteristic of the pump against the NPSH supplied by the system. Pump NPSH is determined by test and computation, and is the energy needed to keep the suction of the pump flooded and solidly liquid. Required NPSH varies with pump type and design, pump size, operating fluid and conditions and its values are normally supplied by the pump manufacturer in the form of a curve covering the entire operating range.

Calculation for NPSH at the suction side of the pump is:

$$\text{NPSH} = Z - \frac{V_s^2}{2g} - h_{fs} + \frac{2.31\,P}{s} - \frac{2.31\,P_{sat}}{s}$$

$Z =$ vertical distance of free level of source of supply to horizontal centerline of pump, ft

$V_s =$ velocity of liquid in suction line

$h_{fs} =$ friction head in suction line only, ft

$P =$ pressure on liquid surface, psia

$P_{sat} =$ saturation pressure at surface of supply source depending on the liquid and its vapor conditions, psia

$s =$ liquid specific gravity

The system NPSH must be higher than the NPSH required by the pump at all operating conditions. When the NPSH in the suction line falls below the value required, vapor bubbles are formed and are carried into the impeller. There the vapor bubbles collapse due to the high pressure level maintained by the pump. This will cause excessive noise and vibration or "cavitation." If this condition persists, excessive impeller erosion, bearing failure, shaft breakage, and other evidences of fatigue and wear failure can appear.

INSULATION TO PREVENT SWEATING ON COLD PIPES

Liquid Temperature, F	Pipe Size, In.	Thermal Cond., k, Btu per Sq. Ft. per Ft. Thickness per F per Hour	Air Temperature, F — 60			70			80			90			100		
			Relative Humidity, % — 50	70	90	50	70	90	50	70	90	50	70	90	50	70	90
			Minimum Thickness of Insulation, Inches														
0	½	.024	.65	1.26	3.55	.73	1.41	3.87	.83	1.51	4.17	.89	1.64	4.50	.95	1.72	4.71
		.029	.74	1.43	4.03	.84	1.59	4.35	.94	1.72	4.75	1.00	1.85	5.08	1.10	1.93	5.13
		.034	.82	1.63	4.62	.95	1.81	4.96	1.06	1.96	5.38	1.15	2.09	5.76	1.22	2.20	6.01
	¾	.024	.69	1.33	3.80	.79	1.44	4.08	.87	1.60	4.43	.94	1.73	4.78	1.01	1.82	4.94
		.029	.78	1.51	4.27	.89	1.68	4.58	.98	1.80	5.00	1.06	1.95	5.37	1.14	2.06	5.57
		.034	.89	1.71	4.84	1.01	1.90	5.21	1.11	2.04	5.68	1.21	2.21	6.10	1.28	2.33	6.3
	1	.024	.71	1.40	3.98	.82	1.56	4.29	.92	1.67	4.68	.98	1.80	4.99	1.05	1.90	5.28
		.029	.81	1.57	4.50	.92	1.74	4.83	1.02	1.88	5.24	1.11	2.04	5.61	1.18	2.16	5.82
		.034	.94	1.79	5.09	1.05	2.01	5.47	1.16	2.14	5.99	1.26	2.32	6.37	1.36	2.44	6.64
	1¼	.024	.73	1.47	4.21	.85	1.64	4.52	.95	1.77	4.89	1.03	1.90	5.26	1.11	2.01	5.47
		.029	.84	1.69	4.74	.97	1.84	5.10	1.10	1.98	5.54	1.16	2.13	5.93	1.25	2.27	7.17
		.034	.97	1.88	5.38	1.11	2.10	5.78	1.21	2.26	6.27	1.32	2.43	6.71	1.42	2.57	6.97
	1½	.024	.75	1.52	4.35	.87	1.68	4.67	.96	1.81	4.86	1.06	1.97	5.43	1.13	2.08	5.67
		.029	.86	1.70	4.90	.99	1.91	5.27	1.09	2.06	5.70	1.19	2.22	6.11	1.28	2.33	6.36
		.034	.98	1.96	5.55	1.12	2.18	5.97	1.27	2.33	6.48	1.38	2.50	6.95	1.50	2.65	7.21
	2	.024	.77	1.59	4.56	.89	1.77	4.90	.99	1.91	5.32	1.09	2.05	5.72	1.17	2.17	5.93
		.029	.89	1.79	5.13	1.00	1.99	5.51	1.13	2.14	5.98	1.22	2.32	6.43	1.33	2.44	6.68
		.034	1.00	2.03	5.82	1.16	2.26	6.26	1.30	2.45	6.78	1.41	2.63	7.29	1.51	2.78	7.57
	3	.024	.82	1.68	4.95	.96	1.89	5.31	1.05	2.05	5.79	1.14	2.21	6.20	1.23	2.36	6.44
		.029	.96	1.85	5.61	1.11	2.17	6.04	1.22	2.36	6.55	1.31	2.53	7.04	1.43	2.67	7.30
		.034	1.08	2.20	6.33	1.24	2.48	6.82	1.37	2.68	7.41	1.49	2.85	7.97	1.61	3.03	8.25
	4	.024	.84	1.73	5.24	.97	1.96	5.60	1.08	2.09	6.12	1.24	2.31	6.57	1.26	2.43	6.76
		.029	.97	1.95	5.91	1.13	2.23	6.31	1.26	2.43	6.90	1.35	2.65	7.45	1.46	2.79	7.69
		.034	1.12	2.27	6.68	1.30	2.56	7.22	1.40	2.81	7.84	1.58	3.03	8.40	1.66	3.22	8.66
	5	.024	.85	1.81	5.51	1.03	2.00	5.91	1.17	2.20	6.43	1.25	2.59	6.90	1.33	2.53	7.15
		.029	1.00	2.06	6.23	1.14	2.31	6.67	1.28	2.48	7.23	1.39	3.06	7.87	1.50	2.87	8.07
		.034	1.17	2.36	7.03	1.33	2.67	7.57	1.48	2.89	8.22	1.61	3.12	8.82	1.73	3.31	9.18
20	½	.024	.38	.86	2.58	.50	1.02	2.94	.61	1.17	3.32	.69	1.29	3.66	.75	1.47	3.91
		.029	.43	.97	2.91	.58	1.12	3.31	.67	1.32	3.73	.77	1.46	4.12	.86	1.60	4.38
		.034	.50	1.09	3.30	.64	1.32	3.75	.77	1.49	4.24	.88	1.66	4.71	.87	1.81	5.00
	¾	.024	.39	.90	2.72	.53	1.09	3.07	.63	1.23	3.51	.73	1.37	3.88	.80	1.49	4.12
		.029	.44	1.03	3.07	.60	1.23	3.59	.72	1.40	4.04	.81	1.54	4.46	.91	1.68	4.62
		.034	.53	1.16	3.49	.67	1.40	3.96	.81	1.57	4.47	.93	1.75	4.94	1.03	1.91	5.26
	1	.024	.41	.95	2.86	.54	1.13	3.24	.65	1.29	3.68	.75	1.44	4.07	.83	1.56	4.33
		.029	.46	1.06	3.23	.60	1.29	3.66	.75	1.45	4.15	.84	1.61	4.60	.95	1.76	4.87
		.034	.55	1.21	3.65	.69	1.45	4.15	.84	1.66	4.71	.97	1.82	5.20	1.06	2.00	5.52
	1¼	.024	.43	.98	3.01	.57	1.19	3.42	.66	1.36	3.89	.77	1.49	4.29	.86	1.65	4.58
		.029	.47	1.12	3.40	.62	1.34	3.87	.76	1.54	4.36	.87	1.71	4.85	.99	1.85	5.16
		.034	.56	1.27	3.85	.72	1.52	4.38	.87	1.74	4.96	1.00	1.92	5.45	1.12	2.10	5.84
	1½	.024	.43	1.00	3.10	.57	1.21	3.53	.67	1.39	4.00	.79	1.53	4.43	.88	1.73	4.71
		.029	.49	1.14	3.50	.66	1.39	3.98	.77	1.57	4.51	.88	1.74	4.99	1.00	1.92	5.31
		.034	.56	1.32	3.97	.73	1.58	4.52	.89	1.79	5.14	1.03	1.99	5.65	1.16	2.18	6.04
	2	.024	.44	1.03	3.26	.58	1.33	3.70	.70	1.46	4.20	.83	1.60	4.65	.91	1.79	4.95
		.029	.49	1.17	3.65	.66	1.43	4.19	.80	1.66	4.74	.92	1.85	5.24	1.03	2.01	5.59
		.034	.57	1.35	4.17	.74	1.66	4.74	.91	1.87	5.37	1.06	2.07	5.94	1.18	2.28	6.32
	3	.024	.44	1.09	3.55	.58	1.35	4.02	.71	1.52	4.55	.84	1.69	5.04	.96	1.89	5.37
		.029	.49	1.24	4.01	.68	1.52	4.55	.84	1.75	5.35	.98	1.96	5.74	1.09	2.17	6.10
		.034	.60	1.45	4.55	.82	1.73	5.15	.98	1.99	5.85	1.10	2.24	6.47	1.26	2.50	6.91
	4	.024	.45	1.13	3.71	.58	1.39	4.25	.71	1.58	4.84	.88	1.78	5.37	.99	1.96	5.67
		.029	.52	1.29	4.23	.68	1.60	4.84	.85	1.85	5.47	1.01	2.05	6.03	1.19	2.23	6.46
		.034	.60	1.51	4.79	.83	1.85	5.45	1.01	2.07	6.18	1.17	2.32	6.84	1.29	2.57	7.31
	5	.024	.47	1.17	3.92	.58	1.44	4.48	.72	1.59	5.08	.88	1.87	5.62	1.03	2.00	6.01
		.029	.56	1.33	4.45	.70	1.64	5.03	.86	1.89	5.73	1.06	2.08	6.33	1.17	2.33	6.72
		.034	.60	1.55	5.03	.84	1.89	5.72	1.03	2.14	6.42	1.20	2.42	7.16	1.36	2.70	7.72

INSULATION TO PREVENT SWEATING ON COLD PIPES

Liquid Temperature, F	Pipe Size, In.	Thermal Cond., k, Btu per Sq. Ft. per Ft. Thickness per F per Hour	60 50	60 70	60 90	70 50	70 70	70 90	80 50	80 70	80 90	90 50	90 70	90 90	100 50	100 70	100 90
35	½	.024	.12	.50	1.72	.28	.71	2.18	.40	.88	2.62	.51	1.04	3.02	.60	1.16	3.30
		.029	.15	.56	1.94	.32	.80	2.46	.46	.99	2.96	.58	1.17	3.40	.67	1.32	3.71
		.034	.18	.65	2.21	.37	.91	2.80	.53	1.08	3.35	.66	1.33	3.87	.77	1.50	4.22
	¾	.024	.12	.52	1.82	.29	.75	2.31	.42	.93	2.77	.52	1.10	3.19	.62	1.23	3.49
		.029	.15	.59	2.06	.34	.85	2.60	.48	1.06	3.12	.60	1.23	3.59	.72	1.39	3.93
		.034	.18	.67	2.32	.38	.96	2.95	.55	1.19	3.54	.69	1.41	4.08	.82	1.58	4.45
	1	.024	.12	.52	1.91	.29	.77	2.41	.42	.95	2.90	.54	1.14	3.34	.63	1.29	3.80
		.029	.15	.59	2.15	.35	.88	2.71	.48	1.09	3.27	.62	1.29	3.91	.73	1.45	4.26
		.034	.18	.69	2.44	.40	1.01	3.10	.56	1.24	3.85	.70	1.46	4.42	.84	1.66	4.82
	1¼	.024	.13	.54	2.01	.31	.81	2.54	.44	1.01	3.05	.57	1.21	3.52	.66	1.35	3.85
		.029	.16	.62	2.27	.35	.92	2.86	.51	1.15	3.45	.63	1.36	3.97	.76	1.53	4.35
		.034	.18	.71	2.56	.41	1.05	3.26	.58	1.30	3.92	.74	1.53	4.49	.87	1.74	4.92
	1½	.024	.13	.56	2.10	.31	.81	2.61	.46	1.05	3.16	.57	1.25	3.65	.68	1.40	3.96
		.029	.16	.64	3.35	.35	.93	2.95	.51	1.18	3.55	.66	1.40	4.09	.78	1.59	4.48
		.034	.19	.73	2.65	.43	1.08	3.36	.59	1.36	4.04	.74	1.60	4.65	.90	1.80	5.11
	2	.024	.14	.57	2.16	.31	.84	2.75	.47	1.06	3.32	.59	1.27	3.82	.70	1.45	4.16
		.029	.16	.66	2.45	.35	.96	3.11	.53	1.22	3.74	.67	1.46	4.29	.79	1.67	4.71
		.034	.19	.74	2.79	.44	1.09	3.51	.61	1.39	4.23	.77	1.67	4.88	.91	1.87	5.35
	3	.024	.14	.58	2.36	.33	.91	2.99	.47	1.08	3.61	.61	1.33	4.15	.70	1.56	4.54
		.029	.18	.70	2.66	.37	1.03	3.38	.53	1.30	3.73	.70	1.56	4.69	.84	1.77	5.13
		.034	.19	.77	3.02	.44	1.17	3.85	.62	1.47	4.61	.82	1.77	5.31	.98	2.00	5.81
	4	.024	.14	.59	3.47	.33	.95	3.12	.47	1.15	3.81	.62	1.39	4.41	.70	1.60	4.84
		.029	.20	.68	2.79	.38	1.06	3.55	.56	1.33	4.30	.71	1.58	4.96	.84	1.82	5.42
		.034	.20	.80	3.21	.45	1.23	4.04	.63	1.55	4.88	.82	1.84	5.53	1.01	2.09	6.12
	5	.024	.14	.59	2.53	.34	.97	3.25	.50	1.23	3.97	.64	1.47	4.58	.72	1.67	5.03
		.029	.20	.70	2.87	.38	1.06	3.76	.56	1.39	4.53	.72	1.64	5.20	.85	1.92	5.71
		.034	.22	.80	3.34	.45	1.25	4.26	.64	1.61	5.08	.84	1.92	5.90	1.02	2.17	6.45
50	½	.024	0	.01	.70	.01	.33	1.33	.16	.56	1.86	.31	.74	2.32	.42	.90	2.66
		.029	0	.01	.80	.02	.37	1.50	.21	.63	2.10	.35	.85	2.60	.48	1.02	3.00
		.034	0	.01	.92	.02	.43	1.70	.24	.72	2.38	.41	.96	2.97	.55	1.16	3.40
	¾	.024	0	.01	.76	.01	.35	1.41	.17	.59	1.96	.33	.79	2.54	.44	.95	2.81
		.029	0	.01	.85	.02	.40	1.59	.21	.65	2.22	.37	.90	2.76	.50	1.08	3.17
		.034	0	.01	.97	.02	.45	1.80	.24	.77	2.52	.43	1.02	3.13	.58	1.18	3.60
	1	.024	0	.01	.79	.01	.34	1.46	.18	.60	2.05	.33	.82	2.57	.46	.99	2.94
		.029	0	.01	.88	.02	.40	1.65	.21	.68	2.31	.38	.94	2.88	.50	1.12	3.31
		.034	0	.01	1.01	.02	.46	1.88	.24	.78	2.63	.44	1.06	3.28	.58	1.25	5.90
	1¼	.024	0	.01	.81	.02	.37	1.55	.19	.62	2.13	.34	.86	2.70	.47	1.04	3.10
		.029	0	.01	.91	.02	.42	1.75	.22	.71	2.43	.39	.98	3.05	.53	1.18	3.50
		.034	0	.01	1.06	.03	.47	1.99	.24	.82	2.77	.46	1.10	3.46	.60	1.35	3.97
	1½	.024	0	.01	.83	.03	.41	1.60	.19	.64	2.23	.35	.87	2.79	.48	1.07	3.20
		.029	0	.01	.93	.03	.43	1.80	.22	.72	2.51	.39	.98	3.14	.53	1.22	3.60
		.034	0	.01	1.07	.03	.50	2.05	.25	.82	2.86	.46	1.14	3.57	.62	1.39	4.10
	2	.024	0	.01	.84	.03	.34	1.67	.19	.66	2.35	.35	.89	2.93	.48	1.13	3.36
		.029	0	.01	.97	.03	.44	1.87	.23	.74	2.63	.40	1.03	3.30	.55	1.24	3.77
		.034	0	.01	1.12	.04	.50	2.13	.25	.85	2.98	.47	1.17	3.75	.63	1.42	4.28
	3	.024	0	.01	.91	.03	.36	1.75	.19	.68	2.56	.35	.96	3.15	.49	1.17	3.66
		.029	0	.01	1.05	.03	.44	2.05	.23	.79	2.87	.40	1.09	3.58	.56	1.31	4.13
		.034	0	.01	1.19	.04	.49	2.32	.26	.91	3.28	.47	1.24	3.94	.67	1.52	4.68
	4	.024	0	.01	.95	.03	.36	1.87	.20	.68	2.65	.36	.97	3.37	.49	1.24	3.87
		.029	0	.01	1.09	.03	.44	2.12	.23	.79	3.04	.43	1.12	3.80	.56	1.40	4.34
		.034	0	.01	1.24	.04	.52	2.43	.27	.95	3.44	.47	1.30	4.30	.68	1.02	4.95
	5	.024	0	.01	.97	.04	.36	1.92	.22	.70	2.73	.36	1.03	3.47	.50	1.25	4.04
		.029	0	.01	1.11	.04	.44	2.17	25	.80	3.14	.45	1.14	3.98	.58	1.45	4.56
		.034	0	.01	1.25	.04	.56	2.50	28	.97	3.64	.50	1.33	4.51	.70	1.67	5.23

HEAT LOSSES FROM BARE PIPE

Heat is lost from pipes by both radiation and convection. The former can be estimated by the Stefan-Boltzmann formula:

$$q_r/A = .174e\left[\left(\frac{T_1}{100}\right)^4 - \left(\frac{T_2}{100}\right)^4\right]$$

where $q_r/A =$ Btu transferred by radiation per hour per square foot of pipe surface,

$e =$ emissivity of the pipe,

$T_1 =$ absolute temperature of the pipe, $F + 460$,

$T_2 =$ absolute temperature of the surroundings, $F + 460$.

Heat transferred by convection can be determined by the Rice-Heilman formula which resulted from work done at Mellon Institute, as follows:

$$q_c/A = C\left(\frac{1}{d}\right)^{0.2}\left(\frac{1}{T_{av}}\right)^{0.181}(t_1 - t_2)^{1.266}$$

where $q_c/A =$ Btu transferred by convection per hour per square foot of pipe surface,

$C =$ a constant with a value of 1.016 for horizontal pipe and 1.394 for vertical plates,

$d =$ outside diameter of pipe, inches,

$T_{av} =$ absolute average temperature of hot body and surrounding air, $F + 460$,

$t_1 =$ temperature of pipe, deg. F,

$t_2 =$ air temperature, deg. F.

The emissivity in the radiation formula is the "effective emissivity," taking into account the absorbtivity of the bodies receiving the radiation. An emissivity of 0.94 was used in the accompanying table on (oxidized) steel pipe; a value of 0.44 is used for tarnished copper tube and 0.08 for new bright copper.

No convection formula is directly available for vertical pipes. However, the emission can be assumed to be closely approximating that of a steel plate. In this connection, note that C becomes 1.394. In the case of vertical pipes, d in the convection formula is not the diameter of the pipe but the height of the plate (pipe) in inches, in which the value of $(1/d)^{0.2}$ becomes constant when $d = 24$ inches. Therefore, the convection formula for vertical pipe or vertical surfaces becomes

$$q_c/A = 1.39 \times .53\left(\frac{1}{T_{av}}\right)^{0.181}(t_1 - t_2)^{1.266}$$

where the terms are as defined previously.

For vertical pipe, the radiation is the same as for horizontal pipe.

The tables are based on the assumption that the outside surface of the pipe is the same as that of the fluid flowing in the pipe; this is not strictly true, but is close enough for all practical estimating purposes.

Also the tables assume an air temperature of 70 deg. and, in the case of radiation, a temperature of 70 deg. for the surrounding walls, machinery, etc. It can be seen, then, that if the surroundings are at temperatures appreciably above or below 70 deg., the heat transferred by both radiation and convection would be appreciably lower than or above, respectively, the values given in the tables. Consequently, the tables should not be used for cases where the air temperatures and surrounding bodies' temperatures are lower than 60 deg. or higher than 80 deg.

The convection formulas are both based on free convection with no appreciable air motion from fans or open doors; in other words, the tables apply to "still" air conditions.

In the formula for radiation, values for e other than those given for steel and iron pipe and copper tube are:

Surface	Emissivity e
Aluminum, polished	0.08
Aluminum paint	0.40
Brass	0.05
Cast iron	0.20
Lead	0.08
Nickel	0.06
Paint	0.94
Tin	0.08
Non-metallic surfaces	0.90

Small iron pipes of ½ inch size, frequently left uninsulated, can be profitably painted with aluminum paint which, with little effort, serves to reduce the emissivity of the pipe and, consequently, the loss by radiation.

Although the emissivity of copper pipe is substantially lower than that for steel pipe, so that the radiation loss is less, the convection loss is the same for copper as for iron where the conditions are the same. Therefore, it is not true that there is no reason to insulate hot lines because they are made of copper.

Many tables are available on heat losses from bare pipe; most of them, however, are on the basis of either "Btu per square foot of pipe surface per hour" or "Btu lineal foot of pipe per degree temperature difference (between pipe surface and air) per hour." Note, then, that the accompanying tables are in Btu per lineal foot per hour, with no additional calculation necessary.

HEAT LOSSES FROM BARE STEEL PIPE

HORIZONTAL PIPES

Diameter of Pipe, Inches	Temperature of Pipe, Deg. F.										
	100	120	150	180	210	240	270	300	330	360	390
	Temperature Difference, Pipe to Air, Deg. F.										
	30	50	80	110	140	170	200	230	260	290	320
	Heat Loss per Lineal Foot of Pipe, Btu per Hour										
½	13	22	40	60	82	106	133	162	193	227	265
¾	15	27	50	74	100	131	163	199	238	280	325
1	19	34	61	90	123	160	199	243	292	343	399
1¼	23	42	75	111	152	198	248	302	362	427	496
1½	27	48	85	126	173	224	280	343	410	483	563
2	33	59	104	154	212	275	344	420	503	594	692
2½	39	70	123	184	252	327	410	502	600	709	827
3	46	84	148	221	303	393	493	601	721	852	994
3½	52	95	168	250	342	444	556	680	816	964	1125
4	59	106	187	278	381	496	621	759	911	1076	1257
5	71	129	227	339	464	603	755	924	1109	1311	1532
6	84	151	267	398	546	709	890	1088	1306	1544	1806
8	107	194	341	509	697	906	1137	1391	1671	1977	2312
10	132	238	420	626	857	1114	1399	1714	2060	2437	2852
12	154	279	491	732	1003	1305	1640	2009	2415	2860	3346
14	181	326	575	856	1173	1527	1918	2350	2826	3347	3918
16	203	366	644	960	1314	1711	2149	2634	3168	3753	4395
18	214	385	678	1011	1385	1802	2266	2777	3339	3958	4635
20	236	426	748	1115	1529	1990	2501	3066	3690	4373	5123

VERTICAL PIPES

Diameter of Pipe, Inches	Temperature of Pipe, Deg. F.										
	100	120	150	180	210	240	270	300	330	360	390
	Temperature Difference, Pipe to Air, Deg. F.										
	30	50	80	110	140	170	200	230	260	290	320
	Heat Loss per Lineal Foot of Pipe, Btu per Hour										
½	11	20	35	52	71	93	116	142	170	201	235
¾	14	25	44	65	89	116	145	177	213	252	294
1	17	31	55	81	111	145	181	222	266	315	368
1¼	22	39	69	103	141	183	230	281	337	398	465
1½	25	45	79	118	161	210	263	321	386	456	532
2	31	56	99	147	201	262	328	401	481	569	665
2½	37	68	120	178	244	317	397	486	583	687	805
3	46	83	146	217	297	386	484	592	710	839	980
3½	52	94	166	248	339	440	552	676	810	958	1119
4	59	106	187	279	382	496	622	760	912	1078	1259
5	72	131	231	344	472	612	768	939	1126	1331	1555
6	86	156	275	410	562	729	915	1119	1342	1587	1853
8	112	203	358	534	731	950	1191	1456	1747	2065	2412
10	140	254	447	667	913	1186	1487	1818	2181	2578	3012
12	166	301	530	790	1081	1404	1761	2154	2584	3054	3567
14	195	354	624	930	1273	1653	2073	2536	3042	3596	4200
16	221	400	705	1051	1438	1868	2343	2865	3437	4063	4745
18	234	425	748	1115	1526	1982	2486	3040	3648	4311	5036
20	260	472	831	1239	1696	2203	2763	3378	4053	4791	5596

HEAT LOSSES FROM BARE BRIGHT COPPER TUBE

HORIZONTAL TUBES											
Nominal Diameter of Tube, Inches	Temperature of Tube, Deg. F.										
	100	120	150	180	210	240	270	300	330	360	390
	Temperature Difference, Tube to Air, Deg. F.										
	30	50	80	110	140	170	200	230	260	290	320
	Heat Loss per Lineal Foot of Tube, Btu per Hr.										
¼	3	6	11	16	22	28	34	40	47	54	61
⅜	4	8	13	20	27	35	43	51	60	69	78
½	5	9	16	24	33	42	51	61	72	82	94
⅝	6	10	19	28	38	49	60	71	83	95	108
¾	7	12	21	32	43	55	68	79	95	109	123
1	8	15	26	39	53	67	83	98	115	133	151
1¼	9	17	31	46	63	80	99	117	137	158	179
1½	10	20	36	53	72	92	113	135	158	181	206
2	13	25	44	66	90	115	141	168	196	226	257
2½	15	29	52	78	107	136	167	200	233	268	305
3	18	34	61	90	123	157	192	229	268	309	352
3½	20	38	68	102	139	178	218	260	305	351	400
4	23	43	77	113	154	198	243	289	339	391	445
5	27	51	91	137	185	237	292	347	407	469	533
6	31	59	106	157	213	270	336	400	464	541	616
8	40	75	134	198	271	347	426	507	594	686	790
10	47	89	159	239	323	413	509	607	710	822	935
12	54	104	184	276	377	482	593	707	827	957	1090

VERTICAL TUBES											
Nominal Diameter of Tube, Inches	Temperature of Tube, Deg. F.										
	100	120	150	180	210	240	270	300	330	360	390
	Temperature Difference, Tube to Air, Deg. F.										
	30	50	80	110	140	170	200	230	260	290	320
	Heat Loss per Lineal Foot of Tube, Btu per Hr.										
¼	2	4	7	10	14	17	21	25	30	34	39
⅜	3	5	9	13	18	23	28	34	40	46	52
½	3	6	11	17	23	29	36	43	50	57	65
⅝	4	7	13	20	27	35	43	51	60	69	78
¾	5	9	16	23	32	40	50	59	70	80	91
1	6	11	20	30	41	52	64	76	89	103	117
1¼	7	14	25	37	50	64	78	93	110	126	143
1½	9	16	29	43	59	75	92	110	129	149	169
2	11	21	38	57	77	98	121	144	169	195	221
2½	14	26	47	70	95	121	150	178	209	241	274
3	17	31	56	84	113	144	178	212	249	286	325
3½	19	36	65	97	131	167	206	246	289	332	377
4	22	41	74	110	149	191	235	280	329	378	429
5	27	51	92	137	185	237	291	348	408	469	533
6	32	61	110	163	221	282	348	415	487	560	636
8	43	81	146	217	294	376	463	553	648	746	847
10	54	101	182	271	366	468	576	687	806	928	1054
12	64	121	217	324	438	560	689	822	965	1110	1260

HEAT LOSSES FROM BARE TARNISHED COPPER TUBE

HORIZONTAL TUBES

Nominal Diameter of Tube, Inches	Temperature of Tube, Deg. F.										
	100	120	150	180	210	240	270	300	330	360	390
	Temperature Difference, Tube to Air, Deg. F.										
	30	50	80	110	140	170	200	230	260	290	320
	Heat Loss per Lineal Foot of Tube, Btu per Hr.										
1/4	4	8	14	21	29	37	46	56	66	77	88
3/8	6	10	18	28	37	48	60	72	85	99	114
1/2	7	13	22	33	45	59	72	88	104	121	139
5/8	8	15	26	39	53	68	85	102	121	141	163
3/4	9	17	30	45	61	79	97	117	139	162	187
1	11	21	37	55	75	97	120	146	173	201	232
1 1/4	14	25	45	66	90	117	145	175	207	242	279
1 1/2	16	29	52	77	105	135	167	203	241	281	324
2	20	37	66	97	132	171	212	257	305	356	411
2 1/2	24	44	78	117	160	206	255	310	367	429	496
3	28	51	92	136	186	240	297	360	428	501	578
3 1/2	32	59	104	156	212	274	340	412	490	573	662
4	36	66	118	174	238	307	381	462	550	644	744
5	43	80	142	212	288	373	464	561	669	783	905
6	51	93	166	246	336	432	541	656	776	915	1059
8	66	120	215	317	435	562	699	848	1010	1184	1372
10	80	146	260	387	527	681	848	1031	1227	1442	1670
12	94	172	304	447	621	802	999	1214	1446	1699	1969

VERTICAL TUBES

Nominal Diameter of Tube, Inches	Temperature of Tube, Deg. F.										
	100	120	150	180	210	240	270	300	330	360	390
	Temperature Difference, Tube to Air, Deg. F.										
	30	50	80	110	140	170	200	230	260	290	320
	Heat Loss per Lineal Foot of Tube, Btu per Hr.										
1/4	3	6	10	15	21	27	34	41	49	57	66
3/8	4	8	14	21	28	36	45	55	65	77	88
1/2	5	10	17	26	35	46	57	69	82	96	111
5/8	6	12	21	31	42	54	68	82	98	114	132
3/4	7	14	24	36	49	64	79	96	114	134	155
1	10	18	31	46	63	82	102	123	147	172	198
1 1/4	12	21	38	57	77	100	125	151	180	210	243
1 1/2	14	25	45	67	91	118	147	178	212	248	287
2	18	33	59	88	120	155	192	233	277	325	375
2 1/2	22	41	73	109	148	191	238	288	343	402	464
3	27	49	87	129	176	227	283	343	408	478	552
3 1/2	31	57	101	150	204	264	328	398	474	554	641
4	35	64	114	171	232	300	374	453	539	631	729
5	43	80	142	212	288	373	464	561	669	783	905
6	52	96	170	253	344	445	554	670	798	934	1080
8	69	127	226	337	458	592	737	892	1063	1244	1438
10	86	158	281	419	570	737	917	1110	1322	1548	1789
12	103	189	336	501	682	881	1097	1328	1582	1851	2140

HEAT LOSSES FROM STEAM PIPING

The accompanying table shows the heat loss (in Btu or in pounds of steam) from bare steam pipes, provided the following are known:

(1) Pipe size and length; (2) steam pressure; (3) whether steam is saturated or superheated and, if the latter, degree of superheat; (4) temperature of air surrounding pipe; and (5) insulation efficiency factor, if loss when pipe is covered is desired.

The tables were computed by Charles A. Pohlig, Assistant to the Superintendent of Steam Distribution, Union Electric Co. of Missouri. Use of the tables is explained in the subsequent examples.

Saturated Steam

Example. One hundred feet of 1 in. bare piping is carrying dry saturated steam at 90 lb pressure, absolute. The surrounding still air temperature is 70 deg. What are the losses per hour expressed in pounds of steam?

Solution. Consult table, column C and proceed downward until 90 lb absolute pressure is reached, thence read across to the right to the 1 in. pipe size column and read 28.40 pounds of steam as the bare piping losses per 100 feet per hour.

If this pipe were insulated with 80% efficiency pipe covering the losses would be

$$(1.00 - .80) \times 28.40 = 5.68 \text{ lb of steam}$$
$$\text{per 100 ft per hr.}$$

Superheated Steam

In case the steam is superheated, proceed as follows:

Example. If 100 feet of 2 in. bare piping is passing steam at 33 lb pressure (absolute) and 25 deg. superheat, surrounding still air temperature is 70F, and insulated pipe covering rated at 85% efficiency is used, what are the losses before and after the covering is applied?

Solution. In column C locate 33 lb pressure absolute and in column B find the corresponding steam temperature at 255 deg. With 25 deg. superheat the steam temperature will be 255 + 25 = 280 deg. Follow down column B to 280 deg. and note that the corresponding absolute steam pressure is 49 lb. Continue to the right and under 2 in. diameter find the uncorrected bare piping losses as 38.61 pounds of steam per hour per 100 ft. In column D the latent heat of vaporization for 49 lb absolute pressure is 924.2 Btu per pound while for

33 absolute pressure the latent heat is 941.2 Btu per hour. The ratio is therefore 924.2 ÷ 941.2 = 0.982. The corrected bare piping loss is therefore 0.982 × 38.61 = 37.92 pounds of steam per hour per 100 ft for the superheated line. The losses after covering with 85% efficiency insulation will be (1.00 − 0.85) × 37.92 = 5.69 pounds of steam per hour

Flat Surfaces

Example. What would be the heat losses from 172 sq ft of virtually flat surface with 90 lb (absolute) steam pressure on one side, and 70 deg. air on the other?

Solution. Continue down column C to 90 lb absolute pressure; follow to the right and under column E find the heat loss to be 737.80 Btu per hour per square foot. For 172 sq ft the heat loss would be 172 × 737.8 = 126,902 Btu per hour.

Steam losses could be found by continuing over to column F, where the loss is found to be 0.8249 lb of steam per square foot; for the whole surface, 172 × 0.8249 = 141.9 lb of steam.

Other Air Temperatures

The table is based on a 70 deg. room air temperature. However, the purpose of column B is to facilitate computations involving other air temperatures.

Example. What would be the steam loss from 100 ft of 1½ inch pipe carrying steam at 42 lb absolute pressure when ambient air is 90 deg?

Solution. The temperature of steam at 42 lb absolute is 270 which gives a 200 deg temperature difference with 70 deg air. The actual temperature difference is 270 − 90 = 180 deg. In column A find 180 deg temperature difference and opposite this, under 1½ inch pipe, find the steam loss to be 24.41 pounds of steam per hour per 100 feet.

Example. What would be the heat loss per hour from 100 feet of 1 inch pipe if the steam pressure is 17 lb absolute and the room temperature 50 deg?

Solution. Under column C find 17 lb and in column B the corresponding steam temperature of 220 deg. The temperature difference is then 220 − 50 = 170 deg. Under column A find 170 deg temperature difference, and opposite this under 1 inch pipe, find 15.52 pounds of steam per hour per 100 feet.

STEAM LOSSES FROM STEEL PIPING — ½ TO 2 INCH PIPE

A	B	C	D	E	F	½	¾	1	1¼	1½	2
	Sat. Steam Temp. (Col. A +70 Deg.)	Abs. Steam Press. for Col. B Temp.	Latent Heat, Btu per Lb.	Heat Loss, Btu per Sq. Ft. per Hr.	Heat Loss, Lb. per Sq. Ft. per Hr.	Nominal Pipe Diameter, Inches					
Temp. Diff., Deg.						Outside Surface Area, Sq. Ft., per 100 Lineal Feet					
						21.99	27.49	34.43	43.46	49.74	62.18
						Pounds of Steam Lost per Hour per 100 Lineal Feet					
50	120	1.692	1025.1	97.50	0.0951	2.09	2.61	3.27	4.13	4.73	5.91
55	125	1.941	1022.2	109.27	0.1069	2.35	2.94	3.68	4.65	5.32	6.65
60	130	2.221	1019.4	121.04	0.1187	2.61	3.26	4.09	5.16	5.90	7.38
65	135	2.536	1016.5	132.81	0.1307	2.87	3.59	4.50	5.68	6.50	8.13
70	140	2.887	1013.6	144.58	0.1426	3.14	3.92	4.91	6.20	7.09	8.87
75	145	3.280	1010.6	156.35	0.1547	3.40	4.25	5.33	6.72	7.69	9.62
80	150	3.716	1007.7	168.12	0.1668	3.67	4.59	5.74	7.25	8.30	10.37
85	155	4.201	1004.7	179.89	0.1790	3.94	4.92	6.16	7.78	8.90	11.13
90	160	4.739	1001.8	191.66	0.1913	4.21	5.26	6.59	8.31	9.52	11.90
95	165	5.334	998.8	203.43	0.2037	4.48	5.60	7.01	8.85	10.13	12.67
100	170	5.990	995.8	215.20	0.2161	4.75	5.94	7.44	9.39	10.75	13.44
105	175	6.716	992.8	229.68	0.2313	5.09	6.36	7.96	10.05	11.50	14.38
110	180	7.510	989.8	244.16	0.2467	5.42	6.78	8.49	10.72	12.27	15.34
115	185	8.382	986.8	258.64	0.2621	5.76	7.21	9.02	11.39	13.04	16.30
120	190	9.336	983.8	273.12	0.2776	6.10	7.63	9.56	12.06	13.81	17.26
125	195	10.385	980.8	287.60	0.2932	6.43	8.06	10.09	12.74	14.58	18.23
130	200	11.525	977.7	302.08	0.3090	6.79	8.49	10.64	13.43	15.37	19.21
135	205	12.772	974.6	316.56	0.3248	7.14	8.93	11.18	14.12	16.16	20.20
140	210	14.123	971.5	331.04	0.3408	7.49	9.37	11.73	14.81	16.95	21.19
145	215	16	968.3	345.52	0.3568	7.85	9.81	12.28	15.51	17.75	22.19
150	220	17	965.1	360.00	0.3730	8.20	10.25	12.84	16.21	18.55	23.19
155	225	19	961.7	377.30	0.3923	8.63	10.78	13.51	17.05	19.51	24.39
160	230	21	958.6	394.60	0.4116	9.05	11.31	14.17	17.89	20.47	25.59
165	235	23	955.0	411.90	0.4313	9.48	11.86	14.85	18.74	21.45	26.82
170	240	25	952.0	429.20	0.4508	9.91	12.39	15.52	19.59	22.42	28.03
175	245	27	949.0	446.50	0.4705	10.35	12.93	16.20	20.45	23.40	29.26
180	250	30	945.2	463.80	0.4907	10.79	13.49	16.89	21.33	24.41	30.51
185	255	33	941.2	481.10	0.5112	11.24	14.05	17.60	22.22	25.43	31.79
190	260	35	938.4	498.40	0.5311	11.68	14.60	18.29	23.08	26.42	33.02
195	265	39	934.4	515.70	0.5519	12.14	15.17	19.00	23.99	27.45	34.32
200	270	42	931.4	533.00	0.5723	12.58	15.73	19.70	24.87	28.47	35.59
205	275	45	928.2	553.48	0.5963	13.11	16.39	20.53	25.92	29.66	37.08
210	280	49	924.2	573.96	0.6210	13.66	17.07	21.38	26.99	30.89	38.61
215	285	53	920.9	594.44	0.6455	14.19	17.74	22.22	28.05	32.11	40.14
220	290	58	917.0	614.92	0.6706	14.75	18.43	23.09	29.14	33.36	41.70
225	295	62	913.4	635.40	0.6956	15.30	19.12	23.95	30.23	34.60	43.25
230	300	67	909.6	655.88	0.7211	15.86	19.82	24.83	31.34	35.87	44.84
235	305	72	906.0	676.36	0.7465	16.42	20.52	25.70	32.44	37.13	46.41
240	310	78	902.1	696.84	0.7725	16.99	21.24	26.60	33.57	38.42	48.03
245	315	84	898.3	717.32	0.7985	17.56	21.95	27.49	34.70	39.72	49.65
250	320	90	894.4	737.80	0.8249	18.14	22.68	28.40	35.85	41.03	51.29
255	325	96	890.6	761.82	0.8554	18.81	23.51	29.45	37.18	42.55	53.19
260	330	103	886.5	785.84	0.8865	19.49	24.37	30.52	38.53	44.09	55.12

.A	B	C	D	E	F	Nominal Pipe Diameter, Inches					
Temp. Diff., Deg.	Sat. Steam Temp. (Col. A +70 Deg.)	Abs. Steam Press. for Col. B Temp.	Latent Heat, Btu per Lb.	Heat Loss, Btu per Sq. Ft. per Hr.	Heat Loss, Lb. per Sq. Ft. per Hr.	½	¾	1	1¼	1½	2
						Outside Surface Area, Sq. Ft., per 100 Lineal Feet					
						21.99	27.49	34.43	43.46	49.74	62.18
						Pounds of Steam Lost per Hour per 100 Lineal Feet					
265	335	110	882.7	809.86	0.9175	20.18	25.22	31.59	39.87	45.64	57.05
270	340	118	878.5	833.88	0.9492	20.87	26.09	32.68	41.25	47.21	59.02
275	345	126	874.4	857.90	0.9811	21.57	26.97	33.78	42.64	48.80	61.00
280	350	135	870.2	881.92	1.0135	22.29	27.86	34.89	44.04	50.41	63.02
285	355	144	865.8	905.94	1.0464	23.01	28.77	36.03	45.48	52.05	65.07
290	360	153	861.7	929.96	1.0792	23.73	29.67	37.16	46.90	53.68	67.10
295	365	163	857.4	953.98	1.1126	24.47	30.59	38.31	48.35	55.34	69.18
300	370	173	853.0	978.00	1.1465	25.21	31.52	39.47	49.83	57.03	71.29
305	375	184	848.6	1007.14	1.1868	26.10	32.63	40.86	51.58	59.03	73.80
310	380	196	844.1	1036.28	1.2277	27.00	33.75	42.27	53.36	61.07	76.34
315	385	208	839.5	1065.42	1.2691	27.91	34.89	43.70	55.16	63.13	78.91
320	390	220	834.9	1094.56	1.3110	28.83	36.04	45.14	56.98	65.21	81.52
325	395	233	830.3	1123.70	1.3534	29.76	37.20	46.60	58.82	67.32	84.15
330	400	248	825.5	1152.84	1.3965	30.71	38.39	48.08	60.69	69.46	86.83
335	405	262	820.6	1181.98	1.4404	31.67	39.60	49.59	62.60	71.65	89.56
340	410	277	815.8	1211.12	1.4846	32.65	40.81	51.11	64.52	73.84	92.31
345	415	292	810.8	1240.26	1.5297	33.64	42.05	52.67	66.48	76.09	95.12
350	420	309	805.8	1269.40	1.5753	34.64	43.30	54.24	68.46	78.36	97.95
355	425	326	800.7	1303.86	1.6284	35.81	44.76	56.07	70.77	81.00	101.25
360	430	344	795.5	1338.32	1.6824	37.00	46.25	57.93	73.12	83.68	104.61
365	435	362	790.2	1372.78	1.7373	38.20	47.76	59.82	75.50	86.41	108.03
370	440	382	784.9	1407.24	1.7929	39.43	49.29	61.73	77.92	89.18	111.48
375	445	402	779.4	1441.70	1.8498	40.68	50.85	63.69	80.39	92.01	115.02
380	450	423	773.8	1476.16	1.9077	41.95	52.44	65.68	82.91	94.89	118.62
385	455	444	768.1	1510.62	1.9667	43.25	54.06	67.71	85.47	97.82	122.29
390	460	467	762.3	1545.08	2.0269	44.57	55.72	69.79	88.09	100.82	126.03
395	465	490	756.4	1579.54	2.0882	45.92	57.40	71.90	90.75	103.87	129.84
400	470	515	750.3	1614.00	2.1511	47.30	59.13	74.06	93.49	107.00	133.76
405	475	540	744.1	1657.66	2.2277	48.99	61.24	76.70	96.82	110.81	138.52
410	480	566	737.8	1701.32	2.3059	50.71	63.39	79.39	100.21	114.70	143.38
415	485	593	731.3	1744.98	2.3861	52.47	65.59	82.15	103.70	118.68	148.37
420	490	622	724.7	1788.64	2.4681	54.27	67.85	84.98	107.26	122.76	153.47
425	495	651	718.0	1832.30	2.5519	56.12	70.15	87.86	110.91	126.93	158.68
430	500	681	711.1	1875.96	2.6381	58.01	72.52	90.83	114.65	131.22	164.04
435	505	712	704.1	1919.62	2.7263	59.95	74.95	93.87	118.48	135.61	169.52
440	510	745	696.9	1963.28	2.8172	61.95	77.44	97.00	122.44	140.13	175.17
445	515	778	689.6	2006.94	2.9103	64.00	80.00	100.20	126.48	144.76	180.96
450	520	813	682.1	2050.60	3.0063	66.11	82.64	103.51	130.65	149.53	186.93
455	525	848	674.6	2104.54	3.1197	68.60	85.76	107.41	135.58	155.17	193.98
460	530	885	666.8	2158.48	3.2371	71.18	88.99	111.45	140.68	161.01	201.28
465	535	923	659.0	2212.42	3.3572	73.82	92.29	115.59	145.90	166.99	208.75
470	540	963	651.0	2266.36	3.4814	76.56	95.70	119.86	151.30	173.16	216.47
475	545	1003	642.8	2320.30	3.6097	79.38	99.23	124.28	156.88	179.55	224.45
480	550	1045	634.5	2374.24	3.7419	82.28	102.86	128.83	162.62	186.12	232.67
485	555	1089	626.1	2428.18	3.8783	85.28	106.61	133.53	168.55	192.91	241.15
490	560	1133	617.5	2482.12	4.0196	88.39	110.50	138.39	174.69	199.93	249.94
495	565	1180	608.7	2536.06	4.1664	91.62	114.53	143.45	181.07	207.24	259.07
500	570	1228	599.7	2590.00	4.3188	94.97	118.72	148.70	187.70	214.82	268.54

STEAM LOSSES FROM STEEL PIPING — 2½ TO 6 INCH PIPE

A	B	C	D	E	F	2½	3	4	5	6
Temp. Diff., Deg.	Sat. Steam Temp. (Col. A +70 Deg.)	Abs. Steam Press. for Col. B Temp.	Latent Heat, Btu per Lb.	Heat Loss, Btu per Sq. Ft. per Hr.	Heat Loss, Lb. per Sq. Ft. per Hr.	: Nominal Pipe Diameter, Inches — Outside Surface Area, Sq. Ft., per 100 Lineal Feet				
						75.27	91.63	117.81	145.64	173.44
						Pounds of Steam Lost per Hour per 100 Lineal Feet				
50	120	1.692	1025.1	97.50	0.0951	7.16	8.71	11.20	13.85	16.49
55	125	1.941	1022.2	109.27	0.1069	8.05	9.80	12.59	15.57	18.54
60	130	2.221	1019.4	121.04	0.1187	8.93	10.88	13.98	17.29	20.59
65	135	2.536	1016.5	132.81	0.1307	9.84	11.98	15.40	19.04	22.67
70	140	2.887	1013.6	144.58	0.1426	10.73	13.07	16.80	20.77	24.73
75	145	3.280	1010.6	156.35	0.1547	11.64	14.18	18.23	22.53	26.83
80	150	3.716	1007.7	168.12	0.1668	12.56	15.28	19.65	24.29	28.93
85	155	4.201	1004.7	179.89	0.1790	13.47	16.40	21.09	26.07	31.05
90	160	4.739	1001.8	191.66	0.1913	14.40	17.53	22.54	27.86	33.18
95	165	5.334	998.8	203.43	0.2037	15.33	18.67	24.00	29.67	35.33
100	170	5.990	995.8	215.20	0.2161	16.27	19.80	25.46	31.47	37.48
105	175	6.716	992.8	229.68	0.2313	17.41	21.19	27.25	33.69	40.12
110	180	7.510	989.8	244.16	0.2467	18.57	22.61	29.06	35.93	42.79
115	185	8.382	986.8	258.64	0.2621	19.73	24.02	30.88	38.17	45.46
120	190	9.336	983.8	273.12	0.2776	20.89	25.44	32.70	40.43	48.15
125	195	10.385	980.8	287.60	0.2932	22.07	26.87	34.54	42.70	50.85
130	200	11.525	977.7	302.08	0.3090	23.26	28.31	36.40	45.00	53.59
135	205	12.772	974.6	316.56	0.3248	24.45	29.76	38.26	47.30	56.33
140	210	14.123	971.5	331.04	0.3408	25.65	31.23	40.15	49.63	59.11
145	215	16	968.3	345.52	0.3568	26.86	32.69	42.03	51.96	61.88
150	220	17	965.1	360.00	0.3730	28.08	34.18	43.94	54.32	64.69
155	225	19	961.7	377.30	0.3923	29.53	35.95	46.22	57.13	68.04
160	230	21	958.6	394.60	0.4116	30.98	37.71	48.49	59.95	71.39
165	235	23	955.0	411.90	0.4313	32.46	39.52	50.81	62.81	74.80
170	240	25	952.0	429.20	0.4508	33.93	41.31	53.11	65.65	78.19
175	245	27	949.0	446.50	0.4705	35.41	43.11	55.43	68.52	81.60
180	250	30	945.2	463.80	0.4907	36.93	44.96	57.81	71.47	85.11
185	255	33	941.2	481.10	0.5112	38.48	46.84	60.22	74.45	88.66
190	260	35	938.4	498.40	0.5311	39.98	48.66	62.57	77.35	92.11
195	265	39	934.4	515.70	0.5519	41.54	50.57	65.02	80.38	95.72
200	270	42	931.4	533.00	0.5723	43.08	52.44	67.42	83.35	99.26
205	275	45	928.2	553.48	0.5963	44.88	54.64	70.25	86.85	103.42
210	280	49	924.2	573.96	0.6210	46.74	56.90	73.16	90.44	107.71
215	285	53	920.9	594.44	0.6455	48.59	59.15	76.05	94.01	111.96
220	290	58	917.0	614.92	0.6706	50.48	61.45	79.00	97.67	116.31
225	295	62	913.4	635.40	0.6956	52.36	63.74	81.95	101.31	120.64
230	300	67	909.6	655.88	0.7211	54.28	66.07	84.95	105.02	125.07
235	305	72	906.0	676.36	0.7465	56.19	68.40	87.95	108.72	129.47
240	310	78	902.1	696.84	0.7725	58.15	70.78	91.01	112.51	133.98
245	315	84	898.3	717.32	0.7985	60.10	73.17	94.07	116.29	138.49
250	320	90	894.4	737.80	0.8249	62.09	75.59	97.18	120.14	143.07
255	325	96	890.6	761.82	0.8554	64.39	78.38	100.77	124.58	148.36
260	330	103	886.5	785.84	0.8865	66.73	81.23	104.44	129.11	153.75

A	B	C	D	E	F	Nominal Pipe Diameter, Inches				
	Sat. Steam Temp.	Abs. Steam Press.	Latent Heat, Btu per Lb.	Heat Loss, Btu per Sq. Ft. per Hr.	Heat Loss, Lb. per Sq. Ft. per Hr.	2½	3	4	5	6
Temp. Diff., Deg.	(Col. A +70 Deg.)	for Col. B Temp.				Outside Surface Area, Sq. Ft., per 100 Lineal Feet				
						75.27	91.63	117.81	145.64	173.44
						Pounds of Steam Lost per Hour per 100 Lineal Feet				
265	335	110	882.7	809.86	0.9175	69.06	84.07	108.09	133.62	159.13
270	340	118	878.5	833.88	0.9492	71.45	96.98	111.83	138.24	164.63
275	345	126	874.4	857.90	0.9811	73.85	89.90	115.58	142.89	170.16
280	350	135	870.2	881.92	1.0135	76.29	92.87	119.40	147.61	175.78
285	355	144	865.8	905.94	1.0464	78.76	95.88	123.28	152.40	181.49
290	360	153	861.7	929.96	1.0792	81.23	98.89	127.14	157.17	187.18
295	365	163	857.4	953.98	1.1126	83.75	101.95	131.08	162.04	192.97
300	370	173	853.0	978.00	1.1465	86.30	105.05	135.07	166.98	198.85
305	375	184	848.6	1007.14	1.1868	89.33	108.75	139.82	172.85	205.84
310	380	196	844.1	1036.28	1.2277	92.41	112.49	144.64	178.80	212.93
315	385	208	839.5	1065.42	1.2691	95.53	116.29	149.51	184.83	220.11
320	390	220	834.9	1094.56	1.3110	98.68	120.13	154.45	190.93	227.38
325	395	233	830.3	1123.70	1.3534	101.87	124.01	159.44	197.11	234.73
330	400	248	825.5	1152.84	1.3965	105.11	127.96	164.52	203.39	242.21
335	405	262	820.6	1181.98	1.4404	108.42	131.98	169.69	209.78	249.82
340	410	277	815.8	1211.12	1.4846	111.75	136.03	174.90	216.22	257.49
345	415	292	810.8	1240.26	1.5297	115.14	140.17	180.21	222.79	265.31
350	420	309	805.8	1269.40	1.5753	118.57	144.34	185.59	229.43	273.22
355	425	326	800.7	1303.86	1.6284	122.57	149.21	191.84	237.16	282.43
360	430	344	795.5	1338.32	1.6824	126.63	154.16	198.20	245.02	291.80
365	435	362	790.2	1372.78	1.7373	130.77	159.19	204.67	253.02	301.32
370	440	382	784.9	1407.24	1.7929	134.95	164.28	211.22	261.12	310.96
375	445	402	779.4	1441.70	1.8498	139.23	169.50	217.92	269.40	320.83
380	450	423	773.8	1476.16	1.9077	143.59	174.80	224.75	277.84	330.87
385	455	444	768.1	1510.62	1.9667	148.03	180.21	231.70	286.43	341.10
390	460	467	762.3	1545.08	2.0269	152.56	185.72	238.79	295.20	351.55
395	465	490	756.4	1579.54	2.0882	157.18	191.34	246.01	304.13	362.18
400	470	515	750.3	1614.00	2.1511	161.91	197.11	253.42	313.29	373.09
405	475	540	744.1	1657.66	2.2277	167.68	204.12	262.45	324.44	386.37
410	480	566	737.8	1701.32	2.3059	173.57	211.29	271.66	335.83	399.94
415	485	593	731.3	1744.98	2.3861	179.60	218.64	281.11	347.51	413.85
420	490	622	724.7	1788.64	2.4681	185.77	226.15	290.77	359.45	428.07
425	495	651	718.0	1832.30	2.5519	192.08	233.83	300.64	371.66	442.60
430	500	681	711.1	1875.96	2.6381	198.57	241.73	310.79	384.21	457.55
435	505	712	704.1	1919.62	2.7263	205.21	249.81	321.19	397.06	472.85
440	510	745	696.9	1963.28	2.8172	212.05	258.14	331.89	410.30	488.62
445	515	778	689.6	2006.94	2.9103	219.06	266.67	342.86	423.86	504.76
450	520	813	682.1	2050.60	3.0063	226.28	275.47	354.17	437.84	521.41
455	525	848	674.6	2104.54	3.1197	234.82	285.86	367.53	454.34	541.08
460	530	885	666.8	2158.48	3.2371	243.66	296.62	381.36	471.45	561.44
465	535	923	659.0	2212.42	3.3572	252.70	307.62	395.51	488.94	582.27
470	540	963	651.0	2266.36	3.4814	262.04	319.00	410.14	507.03	603.81
475	545	1003	642.8	2320.30	3.6097	271.70	330.76	425.26	525.72	626.07
480	550	1045	634.5	2374.24	3.7419	281.65	342.87	440.83	544.97	649.00
485	555	1089	626.1	2428.18	3.8783	291.92	355.37	456.90	564.84	672.65
490	560	1133	617.5	2482.12	4.0196	302.52	368.32	473.55	585.41	697.16
495	565	1180	608.7	2536.06	4.1664	313.60	381.77	490.84	606.79	722.62
500	570	1228	599.7	2590.00	4.3188	325.08	395.73	508.80	628.99	749.05

STEAM LOSSES FROM STEEL PIPING — 7 TO 12 INCH PIPE

A	B	C	D	E	F	Nominal Pipe Diameter, Inches				
						7	8	9	10	12
Temp. Diff., Deg.	Sat. Steam Temp. (Col. A +70 Deg.)	Abs. Steam Press. for Col. B Temp.	Latent Heat, Btu per Lb.	Heat Loss, Btu per Sq. Ft. per Hr.	Heat Loss, Lb. per Sq. Ft. per Hr.	Outside Surface Area, Sq. Ft., per 100 Lineal Feet				
						199.62	225.80	251.98	281.43	333.80
						Pounds of Steam Lost per Hour per 100 Lineal Feet				
50	120	1.692	1025.1	97.50	0.0951	18.98	21.47	23.96	26.76	31.74
55	125	1.941	1022.2	109.27	0.1069	21.34	24.14	26.94	30.08	35.68
60	130	2.221	1019.4	121.04	0.1187	23.69	26.80	29.91	33.41	39.62
65	135	2.536	1016.5	132.81	0.1307	26.09	29.51	32.93	36.78	43.63
70	140	2.887	1013.6	144.58	0.1426	28.47	32.20	35.93	40.13	47.60
75	145	3.280	1010.6	156.35	0.1547	30.88	34.93	38.98	43.54	51.64
80	150	3.716	1007.7	168.12	0.1668	33.30	37.66	42.03	46.94	55.68
85	155	4.201	1004.7	179.89	0.1790	35.73	40.42	45.10	50.38	59.75
90	160	4.739	1001.8	191.66	0.1913	38.19	43.20	48.20	53.84	63.86
95	165	5.334	998.8	203.43	0.2037	40.66	46.00	51.33	57.33	68.00
100	170	5.990	995.8	215.20	0.2161	43.14	48.80	54.45	60.82	72.13
105	175	6.716	992.8	229.68	0.2313	46.17	52.23	58.28	65.09	77.21
110	180	7.510	989.8	244.16	0.2467	49.25	55.70	62.16	69.43	82.35
115	185	8.382	986.8	258.64	0.2621	52.32	59.18	66.04	73.76	87.49
120	190	9.336	983.8	273.12	0.2776	55.41	62.68	69.95	78.12	92.66
125	195	10.385	980.8	287.60	0.2932	58.53	66.20	73.88	82.52	97.87
130	200	11.525	977.7	302.08	0.3090	61.68	69.77	77.86	86.96	103.14
135	205	12.772	974.6	316.56	0.3248	64.84	73.34	81.84	91.41	108.42
140	210	14.123	971.5	331.04	0.3408	68.03	76.95	85.87	95.91	113.76
145	215	16	968.3	345.52	0.3568	71.22	80.57	89.91	100.41	119.10
150	220	17	965.1	360.00	0.3730	74.46	84.22	93.99	104.97	124.51
155	225	19	961.7	377.30	0.3923	78.31	88.58	98.85	110.40	130.95
160	230	21	958.6	394.60	0.4116	82.16	92.94	103.71	115.84	137.39
165	235	23	955.0	411.90	0.4313	86.10	97.39	108.68	121.38	143.97
170	240	25	952.0	429.20	0.4508	89.99	101.79	113.59	126.87	150.48
175	245	27	949.0	446.50	0.4705	93.92	106.24	118.56	132.41	157.05
180	250	30	945.2	463.80	0.4907	97.95	110.80	123.65	138.10	163.80
185	255	33	941.2	481.10	0.5112	102.05	115.43	128.81	143.87	170.64
190	260	35	938.4	498.40	0.5311	106.02	119.92	133.83	149.47	177.28
195	265	39	934.4	515.70	0.5519	110.17	124.62	139.07	155.32	184.22
200	270	42	931.4	533.00	0.5723	114.24	129.23	144.21	161.06	191.03
205	275	45	928.2	553.48	0.5963	119.03	134.64	150.26	167.82	199.04
210	280	49	924.2	573.96	0.6210	123.96	140.22	156.48	174.77	207.29
215	285	53	920.9	594.44	0.6455	128.85	145.75	162.65	181.66	215.47
220	290	58	917.0	614.92	0.6706	133.87	151.42	168.98	188.73	223.85
225	295	62	913.4	635.40	0.6956	138.86	157.07	175.28	195.76	232.19
230	300	67	909.6	655.88	0.7211	143.95	162.82	181.70	202.94	240.70
235	305	72	906.0	676.36	0.7465	149.02	168.56	188.10	210.09	249.18
240	310	78	902.1	696.84	0.7725	154.21	174.43	194.65	217.40	257.86
245	315	84	898.3	717.32	0.7985	159.40	180.30	201.21	224.72	266.54
250	320	90	894.4	737.80	0.8249	164.67	186.26	207.86	232.15	275.35
255	325	96	890.6	761.82	0.8554	170.75	193.15	215.54	240.75	285.53
260	330	103	886.5	785.84	0.8865	176.96	200.17	223.38	249.49	295.91

A	B	C	D	E	F	Nominal Pipe Diameter, Inches				
	Sat. Steam Temp. (Col. A +70 Deg.)	Abs. Steam Press. for Col. B Temp.	Latent Heat, Btu per Lb.	Heat Loss, Btu per Sq. Ft. per Hr.	Heat Loss, Lb. per Sq. Ft. per Hr.	7	8	9	10	12
Temp. Diff., Deg.						Outside Surface Area, Sq. Ft., per 100 Lineal Feet				
						199.62	225.80	251.98	281.43	333.80
						Pounds of Steam Lost per Hour per 100 Lineal Feet				
265	335	110	882.7	809.86	0.9175	183.15	207.17	231.19	258.21	306.26
270	340	118	878.5	833.88	0.9492	189.48	214.33	239.18	267.13	316.84
275	345	126	874.4	857.90	0.9811	195.85	221.53	247.22	276.11	327.49
280	350	135	870.2	881.92	1.0135	202.31	228.85	255.38	285.23	338.31
285	355	144	865.8	905.94	1.0464	208.88	236.28	263.67	294.49	349.29
290	360	153	861.7	929.96	1.0792	215.43	243.68	271.94	303.72	360.24
295	365	163	857.4	953.98	1.1126	222.10	251.23	280.35	313.12	371.39
300	370	173	853.0	978.00	1.1465	228.86	258.88	288.90	322.66	382.70
305	375	184	848.6	1007.14	1.1868	236.91	267.98	299.05	334.00	396.15
310	380	196	844.1	1036.28	1.2277	245.07	277.21	309.36	345.51	409.81
315	385	208	839.5	1065.42	1.2691	253.34	286.56	319.79	357.16	425.63
320	390	220	834.9	1094.56	1.3110	261.70	296.02	330.35	368.95	437.61
325	395	233	830.3	1123.70	1.3534	270.17	305.60	341.03	380.89	451.76
330	400	248	825.5	1152.84	1.3965	278.77	315.33	351.89	393.02	466.15
335	405	262	820.6	1181.98	1.4404	287.53	325.24	362.95	405.37	480.81
340	410	277	815.8	1211.12	1.4846	296.36	335.22	374.09	417.81	495.56
345	415	292	810.8	1240.26	1.5297	305.36	345.41	385.45	430.50	510.61
350	420	309	805.8	1269.40	1.5753	314.46	355.70	396.94	443.34	525.84
355	425	326	800.7	1303.86	1.6284	325.06	367.69	410.32	458.28	543.56
360	430	344	795.5	1338.32	1.6824	335.84	379.89	423.93	473.48	561.59
365	435	362	790.2	1372.78	1.7373	346.80	392.28	437.76	488.93	579.91
370	440	382	784.9	1407.24	1.7929	357.90	404.84	451.77	504.58	598.47
375	445	402	779.4	1441.70	1.8498	369.26	417.68	466.11	520.59	617.46
380	450	423	773.8	1476.16	1.9077	380.82	430.76	480.70	536.88	636.79
385	455	444	768.1	1510.62	1.9667	392.59	444.08	495.57	553.49	656.48
390	460	467	762.3	1545.08	2.0269	404.61	457.67	510.74	570.43	676.58
395	465	490	756.4	1579.54	2.0882	416.85	471.52	526.18	587.68	697.04
400	470	515	750.3	1614.00	2.1511	429.40	485.72	542.03	605.38	718.04
405	475	540	744.1	1657.66	2.2277	444.69	503.01	561.34	626.94	743.61
410	480	566	737.8	1701.32	2.3059	460.30	520.67	581.04	648.95	769.71
415	485	593	731.3	1744.98	2.3861	476.31	538.78	601.25	671.52	796.48
420	490	622	724.7	1788.64	2.4681	492.68	557.30	621.91	694.60	823.85
425	495	651	718.0	1832.30	2.5519	509.41	576.22	643.03	718.18	851.82
430	500	681	711.1	1875.96	2.6381	526.62	595.68	664.75	742.44	880.60
435	505	712	704.1	1919.62	2.7263	544.22	615.60	686.97	767.26	910.04
440	510	745	696.9	1963.28	2.8172	562.37	636.12	709.88	792.84	940.38
445	515	778	689.6	2006.94	2.9103	580.95	657.15	733.34	819.05	971.46
450	520	813	682.1	2050.60	3.0063	600.12	678.82	757.53	846.06	1003.5
455	525	848	674.6	2104.54	3.1197	622.75	704.43	786.10	877.98	1041.4
460	530	885	666.8	2158.48	3.2371	646.19	730.94	815.68	911.02	1080.5
465	535	923	659.0	2212.42	3.3572	670.16	758.06	845.95	944.82	1120.6
470	540	963	651.0	2266.36	3.4814	694.96	786.10	877.24	979.77	1162.1
475	545	1003	642.8	2320.30	3.6097	720.57	815.07	909.57	1015.9	1204.9
480	550	1045	634.5	2374.24	3.7419	746.96	844.92	942.88	1053.1	1249.0
485	555	1089	626.1	2428.18	3.8783	774.19	875.72	977.25	1091.5	1294.6
490	560	1133	617.5	2482.12	4.0196	802.39	907.63	1012.9	1131.2	1341.7
495	565	1180	608.7	2536.06	4.1664	831.70	940.77	1049.8	1172.5	1390.7
500	570	1228	599.7	2590.00	4.3188	862.12	975.19	1088.3	1215.4	1441.6

STEAM LOSSES FROM STEEL PIPING — 14 TO 24 INCH PIPE

A	B	C	D	E	F	Nominal Pipe Diameter, Inches				
	Sat. Steam Temp. (Col. A +70 Deg.)	Abs. Steam Press. for Col. B Temp.	Latent Heat, Btu per Lb.	Heat Loss, Btu per Sq. Ft. per Hr.	Heat Loss, Lb. per Sq. Ft. per Hr.	14	16	18	20	24
Temp. Diff., Deg.						Outside Surface Area, Sq. Ft., per 100 Lineal Feet				
						366.52	418.88	471.24	523.60	628.32
						Pounds of Steam Lost per Hour per 100 Lineal Feet				
50	120	1.692	1025.1	97.50	0.0951	34.86	39.84	44.81	49.79	59.75
55	125	1.941	1022.2	109.27	0.1069	39.18	44.78	50.38	55.97	67.17
60	130	2.221	1019.4	121.04	0.1187	43.51	49.72	55.94	62.15	74.58
65	135	2.536	1016.5	132.81	0.1307	47.90	54.75	61.59	68.43	82.12
70	140	2.887	1013.6	144.58	0.1426	52.27	59.73	67.20	74.67	89.60
75	145	3.280	1010.6	156.35	0.1547	56.70	64.80	72.90	81.00	97.20
80	150	3.716	1007.7	168.12	0.1668	61.14	69.87	78.60	87.34	104.80
85	155	4.201	1004.7	179.89	0.1790	65.61	74.98	84.35	93.72	112.47
90	160	4.739	1001.8	191.66	0.1913	70.12	80.13	90.15	100.16	120.20
95	165	5.334	998.8	203.43	0.2037	74.66	85.33	95.99	106.66	127.99
100	170	5.990	995.8	215.20	0.2161	79.20	90.52	101.83	113.15	135.78
105	175	6.716	992.8	229.68	0.2313	84.78	96.89	109.00	121.11	145.33
110	180	7.510	989.8	244.16	0.2467	90.42	103.34	116.25	129.17	155.01
115	185	8.382	986.8	258.64	0.2621	96.06	109.79	123.51	137.24	164.68
120	190	9.336	983.8	273.12	0.2776	101.75	116.28	130.82	145.35	174.42
125	195	10.385	980.8	287.60	0.2932	107.46	122.82	138.17	153.52	184.22
130	200	11.525	977.7	302.08	0.3090	113.25	129.43	145.61	161.79	194.15
135	205	12.772	974.6	316.56	0.3248	119.05	136.05	153.06	170.07	204.08
140	210	14.123	971.5	331.04	0.3408	124.91	142.75	160.60	178.44	214.13
145	215	16	968.3	345.52	0.3568	130.77	149.46	168.14	186.82	224.18
150	220	17	965.1	360.00	0.3730	136.71	156.24	175.77	195.30	234.36
155	225	19	961.7	377.30	0.3923	143.79	164.33	184.87	205.41	246.49
160	230	21	958.6	394.60	0.4116	150.86	172.41	193.96	215.51	258.62
165	235	23	955.0	411.90	0.4313	158.08	180.66	203.25	225.83	270.99
170	240	25	952.0	429.20	0.4508	165.23	188.83	212.43	236.04	283.25
175	245	27	949.0	446.50	0.4705	172.45	197.08	221.72	246.35	295.62
180	250	30	945.2	463.80	0.4907	179.85	205.54	231.24	256.93	308.32
185	255	33	941.2	481.10	0.5112	187.37	214.13	240.90	267.66	321.20
190	260	35	938.4	498.40	0.5311	194.66	222.47	250.28	278.08	333.70
195	265	39	934.4	515.70	0.5519	202.28	231.18	260.08	288.97	346.77
200	270	42	931.4	533.00	0.5723	209.76	239.73	269.69	299.66	359.59
205	275	45	928.2	553.48	0.5963	218.56	249.78	281.00	312.22	374.67
210	280	49	924.2	573.96	0.6210	227.61	260.12	292.64	325.16	390.19
215	285	53	920.9	594.44	0.6455	236.59	270.39	304.19	337.98	405.58
220	290	58	917.0	614.92	0.6706	245.79	280.90	316.01	351.13	421.35
225	295	62	913.4	635.40	0.6956	254.95	291.37	327.79	364.22	437.06
230	300	67	909.6	655.88	0.7211	264.30	302.05	339.81	377.57	453.08
235	305	72	906.0	676.36	0.7465	273.61	312.69	351.78	390.87	469.04
240	310	78	902.1	696.84	0.7725	283.14	323.58	364.03	404.48	485.38
245	315	84	898.3	717.32	0.7985	292.67	334.48	376.29	418.09	501.71
250	320	90	894.4	737.80	0.8249	302.34	345.53	388.73	431.92	518.30
255	325	96	890.6	761.82	0.8554	313.52	358.31	403.10	447.89	537.46
260	330	103	886.5	785.84	0.8865	324.92	371.34	417.75	464.17	557.01

A	B	C	D	E	F	Nominal Pipe Diameter, Inches				
Temp. Diff., Deg.	Sat. Steam Temp. (Col. A +70 Deg.)	Abs. Steam Press. for Col. B Temp.	Latent Heat, Btu per Lb.	Heat Loss, Btu per Sq. Ft. per Hr.	Heat Loss, Lb. per Sq. Ft. per Hr.	14	16	18	20	24
						Outside Surface Area, Sq. Ft., per 100 Lineal Feet				
						366.52	418.88	471.24	523.60	628.32
						Pounds of Steam Lost per Hour per 100 Lineal Feet				
265	335	110	882.7	809.86	0.9175	336.28	384.32	432.36	480.40	576.48
270	340	118	878.5	833.88	0.9492	347.90	397.60	447.30	497.00	596.40
275	345	126	874.4	857.90	0.9811	359.59	410.96	462.33	513.70	616.44
280	350	135	870.2	881.92	1.0135	371.47	424.53	477.60	530.67	636.80
285	355	144	865.8	905.94	1.0464	383.53	438.32	493.11	547.90	657.47
290	360	153	861.7	929.96	1.0792	395.55	452.06	508.56	565.07	678.08
295	365	163	857.4	953.98	1.1126	407.79	466.05	524.30	582.56	699.07
300	370	173	853.0	978.00	1.1465	420.22	480.25	540.28	600.31	720.37
305	375	184	848.6	1007.14	1.1868	434.99	497.13	559.27	621.41	745.69
310	380	196	844.1	1036.28	1.2277	449.98	514.26	578.54	642.82	771.39
315	385	208	839.5	1065.42	1.2691	465.15	531.60	598.05	664.50	797.40
320	390	220	834.9	1094.56	1.3110	480.51	549.15	617.80	686.44	823.73
325	395	233	830.3	1123.70	1.3534	496.05	566.91	637.78	708.64	850.37
330	400	248	825.5	1152.84	1.3965	511.85	584.97	658.09	731.21	877.45
335	405	262	820.6	1181.98	1.4404	527.94	603.35	678.77	754.19	905.03
340	410	277	815.8	1211.12	1.4846	544.14	621.87	699.60	777.34	932.80
345	415	292	810.8	1240.26	1.5297	560.67	640.76	720.86	800.95	961.14
350	420	309	805.8	1269.40	1.5753	577.38	659.86	742.34	824.83	989.79
355	425	326	800.7	1303.86	1.6284	596.84	682.10	767.37	852.63	1023.2
360	430	344	795.5	1338.32	1.6824	616.63	704.72	792.81	880.90	1057.1
365	435	362	790.2	1372.78	1.7373	636.76	727.72	818.69	909.65	1091.6
370	440	382	784.9	1407.24	1.7929	657.13	751.01	844.89	938.76	1126.5
375	445	402	779.4	1441.70	1.8498	677.99	774.84	871.70	968.56	1162.3
380	450	423	773.8	1476.16	1.9077	699.21	799.10	898.98	998.87	1198.6
385	455	444	768.1	1510.62	1.9667	720.83	823.81	926.79	1029.8	1235.7
390	460	467	762.3	1545.08	2.0269	742.90	849.03	955.16	1061.3	1273.5
395	465	490	756.4	1579.54	2.0882	765.37	874.71	984.04	1093.4	1312.1
400	470	515	750.3	1614.00	2.1511	788.42	901.05	1013.7	1126.3	1351.6
405	475	540	744.1	1657.66	2.2277	816.50	933.14	1049.8	1166.4	1399.7
410	480	566	737.8	1701.32	2.3059	845.16	965.90	1086.6	1207.4	1448.8
415	485	593	731.3	1744.98	2.3861	874.55	999.49	1124.4	1249.4	1499.2
420	490	622	724.7	1788.64	2.4681	904.61	1033.8	1163.1	1292.3	1550.8
425	495	651	718.0	1832.30	2.5519	935.32	1068.9	1202.6	1336.2	1603.4
430	500	681	711.1	1875.96	2.6381	966.92	1105.0	1243.2	1381.3	1657.6
435	505	712	704.1	1919.62	2.7263	999.24	1142.0	1284.7	1427.5	1713.0
440	510	745	696.9	1963.28	2.8172	1032.6	1180.1	1327.6	1475.1	1770.1
445	515	778	689.6	2006.94	2.9103	1066.7	1219.1	1371.4	1523.8	1828.6
450	520	813	682.1	2050.60	3.0063	1101.9	1259.3	1416.7	1574.1	1888.9
455	525	848	674.6	2104.54	3.1197	1143.4	1306.8	1470.1	1633.5	1960.2
460	530	885	666.8	2158.48	3.2371	1186.5	1356.0	1525.5	1694.9	2033.9
465	535	923	659.0	2212.42	3.3572	1230.5	1406.3	1582.0	1757.8	2109.4
470	540	963	651.0	2266.36	3.4814	1276.0	1458.3	1640.6	1822.9	2187.4
475	545	1003	642.8	2320.30	3.6097	1323.0	1512.0	1701.0	1890.0	2268.0
480	550	1045	634.5	2374.24	3.7419	1371.5	1567.4	1763.3	1959.3	2351.1
485	555	1089	626.1	2428.18	3.8783	1421.5	1624.5	1827.6	2030.7	2436.8
490	560	1133	617.5	2482.12	4.0196	1473.3	1683.7	1894.2	2104.7	2525.6
495	565	1180	608.7	2536.06	4.1664	1527.1	1745.2	1963.4	2181.5	2617.8
500	570	1228	599.7	2590.00	4.3188	1582.9	1809.1	2035.2	2261.3	2713.6

HEAT LOSS FROM INSULATED PIPE

For pipe covering in single thicknesses, the heat loss from hot pipes can be expressed as

$$q_1 = \frac{k \ (t_i - t_o)}{r_2 \ log_e \dfrac{r_2}{r_1}} \qquad (1)$$

where q_1 = heat loss is Btu per (hr) (sq ft of outer surface of insulation),

k = conductivity of insulation, Btu per (hr) (sq ft) (deg F) per inch thickness,

t_i = temperature of inner surface of insulation, deg F,

t_o = temperature of outer surface of insulation, deg F,

r_2 = radius of outer surface of insulation, inches,

r_1 = radius of inner surface of insulation or outer surface of pipes, inches.

The objection to the practical application of this formula is that the temperature of the outer surface of the insulation must be known. A more useful formula is

$$q_2 = \frac{t_p - t_a}{\dfrac{r_2 \ log_e \dfrac{r_2}{r_1}}{k} + \dfrac{1}{f_o}} \qquad (2)$$

where q_2 = heat loss in Btu per (hr) (sq ft of outer surface of pipe insulation),

t_p = temperature of pipe surface, deg F,

t_a = temperature of ambient air, deg F,

r_2 = radius of outer surface of insulation, inches,

r_1 = radius of outer surfaces of pipe, inches.

k = conductivity of insulation, Btu per (sq ft) (hr) (deg F) per inch thickness,

f_o = surface conductance of outer surface of insulation, Btu per (sq ft) (hr) (deg F).

Since in most cases the user knows or can assume the temperature of the ambient air, the second equation is more readily applicable than the former.

For very low air movements and low rates of heat transmission the value of $1/f_o$ can be taken at 0.6.

The heat loss figure is usually, in actual problems, desired in Btu per lineal foot of pipe rather than in terms of Btu per square foot of outer surface of radiation. The outer surface of insulation per lineal foot of pipe is $\dfrac{2\pi\ r_2}{12}$, so that the formula becomes

$$q = \frac{.523 \ r_2 \ (t_p - t_a)}{\dfrac{r_2 \ log_e \dfrac{r_2}{r_1}}{k} + 0.6} \qquad (3)$$

where q = Btu loss per (hr) (lineal foot of pipe).

r_2 = outer radius of insulation, inches,

r_1 = outer radius of pipe, inches,

t_p = temperature of pipe, deg F,

Table 1. Heat Conductivity of Pipe Insulating Materials
(Values to be used in the absence of specific data for the exact material and brand name used)

Insulation	Appr. Use Range, Deg F	Appr. Density, Lb per Cu Ft	Mean Temperature, Deg F							
			100	200	300	400	500	600	700	800
			Conductivity, k, Btu per (hr) (sq ft) (deg F) per inch thickness							
85% magnesia	600	11	.35	.38	.42	.46	—	—	—	—
Laminated asbestos	700	30	.40	.45	.50	.55	—	—	—	—
4-ply corrugated asbestos	300	12	.57	.62	—	—	—	—	—	—
Molded asbestos	1000	16	.33	.38	.43	.48	.53	.58	—	—
Mineral fiber,* wire-reinforced	1000	10	.29	.35	.42	.49	.56	.63	—	—
Diatomaceous silica	1600	22	—	—	—	.64	.66	.68	.70	.72
Calcium silicate	1200	11	.32	.37	.42	.46	.51	.56	—	—
Mineral fiber,* molded	350	9	.26	.31	.39	—	—	—	—	—
Mineral fiber,* fine fiber, molded	350	3	.23	.27	.31	—	—	—	—	—
Wool felt	225	20	.33	.37	—	—	—	—	—	—

* Includes glass, mineral wool, rock wool, etc.

t_a = Temperature of ambient air, deg F.
Equation (3) was used to calculate the tables which follow and which show the heat loss for a range of values of k and a range of values for $(t_p - t_a)$.

The problem becomes more complicated when two or more separate layers of pipe insulation are applied. The equation for this condition is

$$q_2 = \frac{t_p - t_a}{\dfrac{r_n \, log_e \dfrac{r_1}{r_p}}{k_1} + \dfrac{r_n \, log_e \dfrac{r_2}{r_1}}{k_2} + \dfrac{r_n \, log_e \dfrac{r_n}{r_{n-1}}}{k_n} + \dfrac{1}{f_o}} \quad (4)$$

where q_2, t_p, t_a and f_o are as given for equation (2) and r_1, $r_2 \ldots r_n$ = outer radius, in inches, of first, second, and nth layer of insulation, respectively; r_p = outer radius of pipe, inches; k_1, $k_2 \ldots k_n$ = conductivity of first, second, and nth layer of insulation, respectively.

The problem of two or more layers of insulation occurs more frequently in high temperature work where, for example, the temperature may be too high for 85% magnesia, so that the first layer may be of calcium silicate or diatomaceous silica.

Values of k

The thermal conductivity k for specific insulations is available from manufacturers and it is suggested that calculations be based on data which apply to the material and brand name of the insulation to be used. In the absence of such data, Table 1 is included as a rough guide. Mean temperatures in Table 1 are arithmetic means between the inside surface and outside surface of insulation.

Effect of Air Velocity

The accompanying tables of heat transmitted through insulated pipe were calculated on the basis of still air with a value of $1/f_o = 0.60$. Actually the surface resistance decreases as the air movement increases and as the rate of heat transfer is increased. The effect of air movement on surface resistance is given in Table 2.

Table 2. Effect of Air Velocity on Surface Resistance

Heat Transmitted, Btu per (Hr) (Sq Ft)	Velocity of Air, Feet per Minute			
	0	100	200	400
	Value of $1/f_o$			
0	—	0.56	0.50	0.41
50	0.60	0.52	0.45	0.39
100	0.55	0.48	0.41	0.36
150	0.50	0.45	0.39	0.34
200	0.48	0.42	0.37	0.32
300	0.43	0.38	0.33	0.29
500	0.36	0.33	0.28	0.25

Cold Surface Temperature

When insulated piping is located where people or animals may come in contact with it, the criterion of economical heat loss in selecting insulation may be outweighed by the need to maintain outside insulation surface temperature at a safe level. The equation

$$t_o = t_i - \left[\frac{\dfrac{x}{k}}{\dfrac{x}{k} + \dfrac{l}{f_o}} \right] (t_i - t_a) \quad (5)$$

gives the temperature of the outside surface of insulation, t_o, or cold surface temperature, in terms of t_i, t_a, k, f_o, as already defined, and x, the thickness of insulation in inches.

Example: Four-inch pipe, carrying water at 300 deg F is to be insulated with 2 inches of 85% magnesia whose density is 12 lb per cu ft. What will be the cold surface temperature of the insulation?

$t_i = 300$

$t_a = 70$ (assumed)

$x = 2$

$k = 0.38$ (from Table 1 and assuming a mean temperature of 200 deg)

$l/f_o = 0.6$

Substituting in Equation 5,

$$t_o = 300 - \left[\frac{\dfrac{2}{0.38}}{\dfrac{2}{0.38} + 0.6} \right] (300 - 70)$$

$$= 300 - 206$$

$$= 94 \text{ deg F.}$$

Since the assumption of a mean temperature of 200 deg was based on the assumption of a cold surface temperature in the order of 100 deg, it can be further assumed that the result of 94 deg for the cold surface temperature is accurate enough for the purpose of gaging whether humans or animals would be endangered by contact. Should the assumption and the result be very different, one would take the first result as a better approximation in assuming a mean temperature for purposes of selecting a value for k in Equation 5.

HEAT LOSS THROUGH PIPE INSULATION — ¾ INCH STEEL PIPE

Insulation Conductivity k	Temperature Difference, Pipe to Air, Deg. F										
	30	50	80	110	140	170	200	230	260	290	320
	Heat Loss per Lineal Foot of Bare Pipe, Btu per Hour										
	15	27	50	74	100	131	163	199	238	280	325
	Heat Loss per Lineal Foot of Insulated Pipe, Btu per Hour										
1 Inch Thick Insulation — $\frac{3}{4}$ Inch Pipe											
0.20	2	6	7	10	13	16	18	21	24	27	29
0.25	3	6	9	12	16	19	23	26	29	33	36
0.30	4	7	11	15	19	23	27	31	35	39	43
0.35	5	8	12	17	21	26	30	35	40	44	49
0.40	5	9	14	19	24	29	34	39	44	50	55
0.45	6	10	15	21	27	32	38	44	49	55	61
0.50	6	10	17	23	29	35	42	48	54	60	67
0.55	7	11	18	25	32	38	45	52	59	65	72
0.60	7	12	19	27	34	41	48	56	63	70	77
$1\frac{1}{2}$ Inch Thick Insulation — $\frac{3}{4}$ Inch Pipe											
0.20	2	4	6	8	10	12	14	16	19	21	23
0.25	3	5	7	10	13	16	18	21	24	27	29
0.30	3	5	9	12	15	19	22	25	28	32	35
0.35	4	6	10	14	18	21	25	29	33	37	40
0.40	4	7	11	16	20	24	29	33	37	41	46
0.45	5	8	13	17	22	27	32	37	41	46	51
0.50	5	9	14	19	25	30	35	40	46	51	56
0.55	6	10	15	21	27	32	38	44	49	55	61
0.60	6	10	16	23	29	35	41	47	53	59	66
2 Inch Thick Insulation — $\frac{3}{4}$ Inch Pipe											
0.20	2	3	5	7	9	11	13	15	17	19	21
0.25	2	4	6	9	11	14	16	18	21	23	26
0.30	3	5	8	11	13	16	19	22	25	28	31
0.35	3	6	9	12	16	19	22	26	29	32	36
0.40	4	6	10	14	18	21	25	29	33	37	40
0.45	4	7	11	16	20	24	28	32	37	41	45
0.50	5	8	12	17	22	26	31	36	40	45	50
0.55	5	8	14	19	24	29	34	39	44	49	54
0.60	5	9	15	20	26	31	37	42	48	53	59
$2\frac{1}{2}$ Inch Thick Insulation — $\frac{3}{4}$ Inch Pipe											
0.20	2	3	5	6	8	10	12	13	15	17	19
0.25	2	3	5	7	9	11	13	15	17	19	21
0.30	3	4	7	10	12	15	17	20	23	25	28
0.35	3	5	8	11	14	17	20	23	26	29	32
0.40	3	6	9	13	16	20	23	26	30	33	37
0.45	3	6	10	14	18	22	26	29	33	37	41
0.50	4	7	11	16	20	24	28	33	37	41	45
0.55	5	8	12	17	22	26	31	36	40	45	50
0.60	5	8	13	18	24	29	34	39	44	49	54

HEAT LOSS THROUGH PIPE INSULATION—1 INCH STEEL PIPE

Insulation Conductivity k	Temperature Difference, Pipe to Air, Deg. F										
	30	50	80	110	140	170	200	230	260	290	320
	Heat Loss per Lineal Foot of Bare Pipe, Btu per Hour										
	19	34	61	90	123	160	199	243	292	343	399
	Heat Loss per Lineal Foot of Insulated Pipe, Btu per Hc										
1 Inch Thick Insulation — 1 Inch Pipe											
0.20	3	5	8	12	15	18	21	24	27	30	34
0.25	4	6	10	14	18	23	26	30	34	37	41
0.30	5	8	12	17	21	26	30	35	40	44	49
0.35	5	9	14	19	24	30	35	40	45	50	56
0.40	6	10	16	22	27	33	39	45	51	57	63
0.45	7	11	17	24	30	37	43	50	56	63	69
0.50	7	12	19	26	33	40	47	55	62	69	76
0.55	8	13	20	28	36	44	51	59	67	74	82
0.60	8	14	22	30	39	47	55	63	72	80	88
1½ Inch Thick Insulation — 1 Inch Pipe											
0.20	2	4	7	9	12	14	17	19	22	24	27
0.25	3	5	8	11	15	18	21	24	27	30	33
0.30	4	6	10	14	17	21	25	29	32	36	40
0.35	4	7	11	16	20	24	29	33	37	41	46
0.40	5	8	13	18	23	27	32	37	42	47	52
0.45	5	9	14	20	25	31	36	41	47	52	58
0.50	6	10	16	22	28	33	39	45	51	57	63
0.55	6	11	17	24	30	37	43	49	56	62	69
0.60	7	12	19	26	32	39	46	53	60	67	74
2 Inch Thick Insulation — 1 Inch Pipe											
0.20	2	4	6	8	10	13	15	17	19	21	24
0.25	3	5	7	10	13	15	18	21	23	26	29
0.30	3	5	9	12	15	18	21	25	28	31	34
0.35	4	6	10	14	17	21	25	29	32	36	40
0.40	4	7	11	15	20	24	28	32	37	41	45
0.45	5	8	13	17	22	27	31	36	41	46	50
0.50	5	9	14	19	24	29	35	40	45	50	55
0.55	6	9	15	21	26	32	38	43	49	55	60
0.60	6	10	16	23	29	35	41	47	53	59	66
2½ Inch Thick Insulation — 1 Inch Pipe											
0.20	2	3	5	7	9	11	13	15	17	19	21
0.25	3	4	7	10	12	15	17	20	23	25	28
0.30	3	5	8	11	14	16	19	22	25	28	31
0.35	3	6	9	12	16	19	22	26	29	32	36
0.40	4	6	10	14	18	22	26	29	33	37	41
0.45	4	7	11	16	20	24	29	33	37	41	46
0.50	5	8	13	17	22	27	32	36	41	46	51
0.55	5	9	14	19	24	29	34	40	45	50	55
0.60	6	9	15	21	26	32	37	43	49	54	60

HEAT LOSS THROUGH PIPE INSULATION—1¼ INCH STEEL PIPE

Insulation Conductivity k	Temperature Difference, Pipe to Air, Deg. F										
	30	50	80	110	140	170	200	230	260	290	320
	Heat Loss per Lineal Foot of Bare Pipe, Btu per Hour										
	23	42	75	111	152	198	248	302	362	427	496
	Heat Loss per Lineal Foot of Insulated Pipe, Btu per Hour										
1 Inch Thick Insulation — 1¼ Inch Pipe											
0.20	4	6	10	14	17	21	25	28	32	36	39
0.25	5	8	12	17	21	26	30	35	39	44	48
0.30	5	9	14	19	25	30	35	41	46	51	57
0.35	6	10	16	22	28	35	41	47	53	59	65
0.40	7	11	18	25	32	39	46	52	59	66	73
0.45	8	13	20	28	35	43	50	58	66	73	81
0.50	8	14	22	30	39	47	55	63	72	80	88
0.55	9	15	24	33	42	50	59	68	77	86	95
0.60	10	16	26	35	45	54	64	73	83	92	102
1½ Inch Thick Insulation — 1¼ Inch Pipe											
0.20	3	5	8	11	14	16	19	22	25	28	31
0.25	4	6	10	13	17	20	24	27	31	35	38
0.30	4	7	11	16	20	24	28	32	37	41	45
0.35	5	8	13	18	23	28	33	37	42	47	52
0.40	6	9	15	20	26	31	37	42	48	53	59
0.45	6	10	16	23	29	35	41	47	53	59	66
0.50	7	11	18	25	32	38	45	52	59	65	72
0.55	7	12	20	27	34	42	49	56	64	71	78
0.60	8	13	21	29	37	45	53	61	69	77	84
2 Inch Thick Insulation — 1¼ Inch Pipe											
0.20	2	4	7	9	12	14	17	19	21	24	26
0.25	3	5	8	11	14	17	20	23	27	30	33
0.30	4	6	10	13	17	21	24	28	32	35	39
0.35	4	7	11	16	19	24	28	32	37	41	45
0.40	5	8	13	18	22	27	32	37	42	47	51
0.45	5	9	14	20	25	30	36	41	46	52	57
0.50	6	10	16	22	27	33	39	45	51	57	63
0.55	6	11	17	24	30	36	43	49	56	62	68
0.60	7	12	19	26	32	39	46	53	60	67	74
2½ Inch Thick Insulation — 1¼ Inch Pipe											
0.20	2	4	6	8	10	12	15	17	19	21	23
0.25	3	5	7	10	13	16	18	21	24	26	29
0.30	3	5	9	12	15	19	22	25	28	32	35
0.35	4	6	10	14	18	21	25	29	33	37	40
0.40	4	7	11	16	20	24	29	33	37	41	46
0.45	5	8	13	18	22	27	32	37	42	46	51
0.50	5	9	14	19	25	30	35	41	46	51	57
0.55	6	10	15	21	27	33	39	44	50	56	62
0.60	6	11	17	23	29	36	42	48	55	61	67

HEAT LOSS THROUGH PIPE INSULATION — 1½ INCH STEEL PIPE

Insulation Conductivity k	Temperature Difference, Pipe to Air, Deg. F										
	30	50	80	110	140	170	200	230	260	290	320
	Heat Loss per Lineal Foot of Bare Pipe, Btu per Hour										
	27	48	85	126	173	224	280	343	410	483	563
	Heat Loss per Lineal Foot of Insulated Pipe, Btu per Hour										
1 Inch Thick Insulation — 1½ Inch Pipe											
0.20	4	7	11	15	19	23	27	31	35	39	43
0.25	5	8	13	18	23	28	32	37	42	47	52
0.30	6	10	16	21	27	33	39	45	50	56	62
0.35	7	11	18	24	31	38	44	51	58	64	71
0.40	7	12	20	27	35	42	50	57	65	72	80
0.45	8	14	22	30	39	47	55	63	72	80	88
0.50	9	15	24	33	42	51	60	69	78	87	96
0.55	10	16	26	36	46	55	65	75	85	94	104
0.60	10	17	28	38	49	59	70	80	90	100	111
1½ Inch Thick Insulation — 1½ Inch Pipe											
0.20	3	5	8	12	15	18	21	24	27	30	34
0.25	4	7	10	14	18	22	26	30	34	38	42
0.30	5	8	12	17	22	26	31	35	40	45	49
0.35	5	9	14	19	25	30	35	41	46	51	57
0.40	6	10	16	22	28	34	40	46	52	58	64
0.45	7	11	18	25	31	38	45	51	58	65	71
0.50	7	12	20	27	34	41	49	56	63	71	78
0.55	8	13	21	29	37	45	53	61	69	77	85
0.60	9	14	23	32	40	49	57	66	75	83	92
2 Inch Thick Insulation — 1½ Inch Pipe											
0.20	3	4	7	10	12	15	18	20	23	25	28
0.25	3	6	9	12	15	19	22	25	29	32	35
0.30	4	7	10	14	18	22	26	30	34	38	42
0.35	5	8	12	17	21	26	30	35	44	49	53
0.40	5	9	14	19	24	29	34	40	45	50	55
0.45	6	10	15	21	27	33	38	44	50	56	61
0.50	6	11	17	23	30	36	43	49	55	62	68
0.55	7	12	18	25	32	39	46	53	60	67	74
0.60	8	13	20	28	35	43	50	58	65	73	80
2½ Inch Thick Insulation — 1½ Inch Pipe											
0.20	2	4	6	9	11	13	16	18	21	23	25
0.25	3	5	8	11	14	17	20	23	25	28	31
0.30	4	6	9	13	16	20	23	27	30	34	37
0.35	4	7	11	15	19	23	27	31	35	39	44
0.40	5	8	12	17	22	26	31	35	40	45	49
0.45	5	9	14	19	24	29	34	40	45	50	55
0.50	6	10	15	21	27	32	38	44	49	55	61
0.55	6	10	17	23	29	35	42	48	54	60	67
0.60	7	11	18	25	32	38	45	52	59	65	72

HEAT LOSS THROUGH PIPE INSULATION — 2 INCH STEEL PIPE

Insulation Conductivity k	Temperature Difference, Pipe to Air, Deg. F										
	30	50	80	110	140	170	200	230	260	290	320
	Heat Loss per Lineal Foot of Bare Pipe, Btu per Hour										
	33	59	104	154	212	275	344	420	503	594	692
	Heat Loss per Lineal Foot of Insulated Pipe, Btu per Hour										
1 Inch Thick Insulation — 2 Inch Pipe											
0.20	5	8	13	17	22	27	32	36	41	46	51
0.25	6	10	15	21	27	32	39	44	50	56	62
0.30	7	11	18	25	32	39	45	52	59	66	73
0.35	8	13	21	29	36	44	52	60	68	75	83
0.40	9	15	23	32	41	49	58	67	76	84	93
0.45	10	16	26	35	45	55	64	74	83	93	103
0.50	10	18	28	39	49	60	70	81	91	102	112
0.55	11	19	30	42	53	64	76	87	99	110	121
0.60	12	20	32	45	57	69	81	93	106	118	130
1½ Inch Thick Insulation — 2 Inch Pipe											
0.20	4	6	10	13	17	21	24	28	32	35	39
0.25	5	8	12	17	21	26	30	35	39	44	48
0.30	5	9	14	20	25	30	36	41	46	52	57
0.35	6	10	16	23	29	35	41	47	53	59	66
0.40	7	12	18	25	32	39	46	53	60	67	74
0.45	8	13	21	28	36	44	51	59	67	75	82
0.50	9	15	23	32	41	49	58	67	75	84	93
0.55	9	15	25	34	43	52	61	71	80	89	98
0.60	10	17	26	36	46	55	66	76	83	96	106
2 Inch Thick Insulation — 2 Inch Pipe											
0.20	3	5	8	11	14	17	20	23	27	30	33
0.25	4	6	10	14	18	22	25	29	33	37	41
0.30	4	8	12	17	21	26	30	35	39	44	48
0.35	5	9	14	19	24	30	35	40	45	50	56
0.40	6	10	16	22	28	33	39	45	51	57	63
0.45	7	11	18	24	31	37	44	51	57	64	70
0.50	7	12	19	27	34	41	48	56	63	70	77
0.55	8	13	21	29	37	45	53	61	69	77	84
0.60	9	14	23	31	40	49	57	66	74	83	92
2½ Inch Thick Insulation — 2 Inch Pipe											
0.20	3	5	7	10	13	15	18	21	23	26	29
0.25	3	6	9	12	16	19	22	26	29	32	36
0.30	4	7	11	15	19	23	27	31	35	39	43
0.35	5	8	12	17	22	26	31	35	40	45	49
0.40	5	9	14	19	25	30	35	40	46	51	56
0.45	6	10	16	22	27	33	39	45	51	57	63
0.50	6	11	17	24	30	37	43	50	56	63	69
0.55	7	12	19	26	33	40	48	54	61	68	76
0.60	8	13	21	29	37	45	53	61	69	77	85

HEAT LOSS THROUGH PIPE INSULATION — 2½ INCH STEEL PIPE

Insulation Conductivity k	Temperature Difference, Pipe to Air, Deg. F										
	30	50	80	110	140	170	200	230	260	290	320
	Heat Loss per Lineal Foot of Bare Pipe, Btu per Hour										
	39	70	123	184	252	327	410	502	600	709	827
	Heat Loss per Lineal Foot of Insulated Pipe, Btu per Hour										
1 Inch Thick Insulation — 2½ Inch Pipe											
0.20	5	9	14	20	25	31	36	42	47	52	58
0.25	7	11	18	24	31	38	44	51	57	64	71
0.30	8	13	21	29	36	44	52	60	68	75	83
0.35	9	15	24	33	42	50	59	68	77	86	95
0.40	10	17	27	37	47	57	67	77	87	97	107
0.45	11	18	29	40	52	63	74	85	96	107	118
0.50	12	21	33	45	57	70	82	94	107	119	131
0.55	13	22	35	48	61	74	87	100	113	126	139
0.60	14	23	37	51	65	79	93	106	120	134	148
1½ Inch Thick Insulation — 2½ Inch Pipe											
0.20	4	7	11	15	19	24	28	32	36	40	44
0.25	5	9	14	19	24	29	34	39	44	50	55
0.30	6	10	16	22	28	35	41	47	53	59	65
0.35	7	12	19	26	33	40	47	54	61	68	75
0.40	8	13	21	29	37	45	53	61	69	77	84
0.45	9	15	23	32	41	50	59	67	76	85	94
0.50	10	16	26	35	45	55	64	74	83	93	103
0.55	10	17	28	38	49	59	70	80	91	101	112
0.60	11	19	30	41	53	64	75	86	98	109	120
2 Inch Thick Insulation — 2½ Inch Pipe											
0.20	3	6	9	13	16	20	23	26	30	33	37
0.25	4	7	11	16	20	24	29	33	37	41	46
0.30	5	9	14	19	24	29	34	39	44	49	54
0.35	6	10	16	22	27	33	39	45	51	57	63
0.40	7	11	18	24	31	38	44	51	58	64	71
0.45	7	12	20	27	35	42	50	57	64	72	79
0.50	8	14	22	30	38	46	55	63	71	79	87
0.55	9	15	24	33	42	50	59	68	77	86	95
0.60	10	16	26	35	45	55	64	74	84	93	103
2½ Inch Thick Insulation — 2½ Inch Pipe											
0.20	3	5	8	11	14	17	20	23	26	29	32
0.25	4	6	10	14	18	21	25	29	33	36	40
0.30	4	7	12	16	21	25	30	34	39	43	48
0.35	5	9	14	19	24	29	35	40	45	50	55
0.40	6	10	16	22	27	33	39	45	51	57	63
0.45	7	11	18	24	31	37	44	50	57	64	70
0.50	7	12	19	27	34	41	48	55	63	70	77
0.55	8	13	21	29	37	45	53	61	69	77	84
0.60	9	14	23	31	40	49	57	66	74	83	92

HEAT LOSS THROUGH PIPE INSULATION — 3 INCH STEEL PIPE

Insulation Conductivity k	Temperature Difference, Pipe to Air, Deg. F										
	30	50	80	110	140	170	200	230	260	290	320
	Heat Loss per Lineal Foot of Bare Pipe, Btu per Hour										
	46	84	148	221	303	393	493	601	721	852	994
	Heat Loss per Lineal Foot of Insulated Pipe, Btu per Hour										
1 Inch Thick Insulation — 3 Inch Pipe											
0.20	6	11	17	23	30	36	42	49	55	61	68
0.25	8	13	21	28	36	44	52	60	67	75	83
0.30	9	15	24	33	43	52	61	70	79	88	97
0.35	10	17	28	38	49	59	69	80	90	101	111
0.40	12	19	31	43	54	66	78	89	101	113	124
0.45	13	21	34	47	60	73	86	99	111	124	137
0.50	14	23	37	51	65	79	93	107	121	135	149
0.55	15	25	40	55	71	86	101	116	131	146	161
0.60	16	27	43	59	76	92	108	124	140	156	173
1½ Inch Thick Insulation — 3 Inch Pipe											
0.20	5	8	13	18	22	27	32	37	41	46	51
0.25	6	10	16	22	27	33	39	45	51	57	63
0.30	7	12	19	26	33	39	46	53	60	67	74
0.35	8	13	22	30	39	46	54	62	70	78	86
0.40	9	15	24	33	42	51	60	69	78	87	96
0.45	10	17	27	37	47	57	67	77	87	97	107
0.50	11	18	29	40	51	62	73	84	95	106	117
0.55	12	20	32	44	56	68	80	92	104	116	128
0.60	13	21	34	47	60	73	86	99	112	124	137
2 Inch Thick Insulation — 3 Inch Pipe											
0.20	4	7	11	15	18	22	26	30	34	38	42
0.25	5	8	13	18	23	28	33	38	43	47	52
0.30	6	10	15	21	27	33	39	45	51	56	62
0.35	7	11	18	25	31	38	45	52	58	65	72
0.40	8	13	20	28	36	43	51	58	66	74	81
0.45	8	14	23	31	40	48	57	65	74	82	91
0.50	9	16	25	34	44	53	62	72	81	90	100
0.55	10	17	27	37	48	58	68	78	88	98	109
0.60	11	18	29	40	51	62	73	84	95	106	117
2½ Inch Thick Insulation — 3 Inch Pipe											
0.20	3	6	9	12	16	19	23	26	29	33	36
0.25	4	7	11	16	20	24	28	33	37	41	45
0.30	5	8	13	19	24	29	34	39	44	49	54
0.35	6	10	16	22	27	33	39	45	51	57	63
0.40	7	11	18	24	31	38	44	51	58	64	71
0.45	7	12	20	27	35	42	50	57	64	72	79
0.50	8	14	22	30	38	46	55	63	71	79	87
0.55	9	15	24	33	42	51	60	69	78	87	96
0.60	10	16	26	36	45	55	65	74	84	94	103

HEAT LOSS THROUGH PIPE INSULATION — 3½ INCH STEEL PIPE

Insulation Conductivity k	Temperature Difference, Pipe to Air, Deg. F										
	30	50	80	110	140	170	200	230	260	290	320
	Heat Loss per Lineal Foot of Bare Pipe, Btu per Hour										
	52	95	168	250	342	444	556	680	816	964	1125
	Heat Loss per Lineal Foot of Insulated Pipe, Btu per Hour										
1 Inch Thick Insulation — 3½ Inch Pipe											
0.20	7	12	19	26	33	40	47	54	61	68	75
0.25	9	14	23	32	40	49	57	66	75	83	92
0.30	10	17	27	37	47	57	67	78	88	98	108
0.35	12	19	31	42	54	65	77	89	100	112	123
0.40	13	22	34	47	60	73	86	99	112	125	138
0.45	14	24	38	52	67	81	95	109	124	138	152
0.50	16	26	41	57	72	88	104	119	135	150	166
0.55	17	28	45	61	78	95	112	128	145	162	179
0.60	18	30	48	66	84	102	119	137	155	173	191
1½ Inch Thick Insulation — 3½ Inch Pipe											
0.20	5	9	14	19	25	30	35	41	46	51	56
0.25	7	11	17	24	30	37	43	50	56	63	70
0.30	8	13	21	28	36	44	51	59	67	75	82
0.35	9	15	24	33	41	50	59	68	77	86	95
0.40	10	17	27	37	47	57	67	77	87	97	107
0.45	11	19	30	41	52	63	74	85	96	107	118
0.50	12	20	32	45	57	69	81	93	105	118	130
0.55	13	22	35	48	62	75	88	101	114	128	141
0.60	14	24	38	52	66	81	95	109	123	138	152
2 Inch Thick Insulation — 3½ Inch Pipe											
0.20	4	7	12	16	20	25	29	33	38	42	46
0.25	5	9	14	20	25	30	36	41	47	52	57
0.30	6	11	17	23	30	36	43	49	55	62	68
0.35	7	12	20	27	34	42	49	56	64	71	79
0.40	8	14	22	31	39	47	56	64	72	81	89
0.45	9	15	25	34	43	53	62	71	80	90	99
0.50	10	17	27	37	48	58	68	78	89	99	109
0.55	11	19	30	41	52	63	74	85	96	108	119
0.60	12	20	32	44	56	68	80	92	104	116	128
2½ Inch Thick Insulation — 3½ Inch Pipe											
0.20	4	6	10	14	18	21	25	29	33	36	40
0.25	5	8	12	17	22	26	31	36	40	45	50
0.30	6	9	15	20	26	31	37	43	48	54	59
0.35	6	11	17	24	30	36	43	49	56	62	68
0.40	7	12	19	27	34	41	49	56	63	70	78
0.45	8	14	22	30	38	46	54	62	70	79	87
0.50	9	15	24	33	42	51	60	69	78	87	96
0.55	10	16	26	36	46	55	65	75	85	95	104
0.60	11	18	28	39	49	60	71	81	92	102	113

HEAT LOSS THROUGH PIPE INSULATION — 4 INCH STEEL PIPE

Insulation Conductivity k	Temperature Difference, Pipe to Air, Deg. F										
	30	50	80	110	140	170	200	230	270	300	320
	Heat Loss per Lineal Foot of Bare Pipe, Btu per Hour										
	59	106	187	278	381	496	621	759	911	1076	1257
	Heat Loss per Lineal Foot of Insulated Pipe, Btu per Hour										
1 Inch Thick Insulation — 4 Inch Pipe											
0.20	8	13	20	28	33	43	51	59	66	74	82
0.25	10	16	25	35	44	54	63	73	82	92	101
0.30	11	19	30	41	52	63	74	86	97	108	119
0.35	13	21	34	47	59	72	85	98	110	123	136
0.40	14	24	38	52	66	81	95	109	123	137	152
0.45	16	26	42	58	73	89	105	121	136	152	168
0.50	17	28	45	62	80	97	114	131	148	165	182
0.55	18	31	49	67	86	104	123	141	159	178	196
0.60	20	33	53	72	92	112	132	151	171	191	211
$1\frac{1}{2}$ Inch Thick Insulation — 4 Inch Pipe											
0.20	6	10	15	21	27	33	38	44	50	56	61
0.25	7	12	19	26	33	40	47	55	62	69	76
0.30	8	14	22	28	39	48	56	64	73	81	90
0.35	10	16	26	36	45	55	65	74	84	94	103
0.40	11	18	29	40	51	62	73	84	95	106	116
0.45	12	20	32	44	56	69	81	93	105	117	129
0.50	13	22	35	49	62	75	88	101	115	128	141
0.55	14	24	38	53	67	81	96	110	124	139	153
0.60	15	26	41	57	72	88	103	118	134	149	165
2 Inch Thick Insulation — 4 Inch Pipe											
0.20	5	8	13	17	22	27	31	36	41	46	50
0.25	6	10	16	21	27	33	39	45	51	57	62
0.30	7	12	19	26	32	39	46	53	60	67	74
0.35	8	13	21	29	37	45	53	61	69	77	85
0.40	9	15	24	33	42	51	60	69	79	88	97
0.45	10	17	27	37	47	57	67	78	88	98	108
0.50	11	19	30	41	52	63	74	85	96	107	118
0.55	12	20	32	44	56	69	81	93	105	117	129
0.60	13	22	35	48	61	74	87	101	114	127	140
$2\frac{1}{2}$ Inch Thick Insulation — 4 Inch Pipe											
0.20	4	7	11	15	19	23	27	31	35	39	44
0.25	5	8	13	18	24	29	34	39	44	49	54
0.30	6	10	16	22	28	34	40	46	52	58	64
0.35	7	12	19	26	32	39	46	53	60	67	74
0.40	8	13	21	29	37	45	52	60	68	76	84
0.45	9	15	24	32	41	50	59	68	76	85	94
0.50	10	16	26	36	45	55	65	75	84	94	104
0.55	11	18	28	39	49	60	71	81	92	102	113
0.60	11	19	31	42	53	65	76	88	99	111	122

HEAT LOSS THROUGH PIPE INSULATION — 5 INCH STEEL PIPE

Insulation Conductivity k	Temperature Difference, Pipe to Air, Deg. F										
	30	50	80	110	140	170	200	230	260	290	320
	Heat Loss per Lineal Foot of Bare Pipe, Btu per Hour										
	71	129	227	339	464	603	755	924	1109	1311	1532
	Heat Loss per Lineal Foot of Insulated Pipe, Btu per Hour										
1 Inch Thick Insulation — 5 Inch Pipe											
0.20	8	14	23	31	40	48	57	65	74	82	90
0.25	10	17	28	38	48	59	69	79	90	100	110
0.30	12	20	32	45	57	69	81	93	105	117	130
0.35	14	23	37	51	65	79	93	106	120	134	148
0.40	15	26	41	57	73	88	104	119	135	150	166
0.45	17	29	46	63	80	97	114	131	148	165	182
0.50	19	31	50	68	87	106	124	143	180	199	217
0.55	20	34	54	74	94	114	134	154	174	194	214
0.60	21	36	57	79	100	122	143	165	186	208	229
1½ Inch Thick Insulation — 5 Inch Pipe											
0.20	6	11	17	23	30	36	42	49	55	61	68
0.25	8	13	21	29	37	44	52	60	68	76	84
0.30	9	15	25	34	43	53	62	71	80	90	99
0.35	11	18	28	39	50	60	71	82	92	103	114
0.40	12	20	32	44	56	68	80	92	104	116	128
0.45	13	22	36	49	62	75	89	102	115	129	142
0.50	15	24	39	53	68	83	97	112	126	141	156
0.55	16	26	42	58	74	90	106	121	137	153	169
0.60	17	28	45	62	80	97	114	131	148	165	182
2 Inch Thick Insulation — 5 Inch Pipe											
0.20	5	9	14	19	24	29	34	40	45	50	55
0.25	6	11	17	23	30	36	43	49	55	62	68
0.30	8	13	20	28	35	43	51	58	66	73	81
0.35	9	15	23	32	41	50	58	67	76	85	93
0.40	10	17	26	36	46	56	66	76	86	96	106
0.45	11	18	29	40	51	62	73	84	95	106	117
0.50	12	20	32	44	56	68	80	92	105	117	129
0.55	13	22	35	48	61	75	88	101	114	128	141
0.60	14	24	38	52	67	81	95	109	124	138	152
2½ Inch Thick Insulation — 5 Inch Pipe											
0.20	4	7	12	16	21	25	29	34	38	43	47
0.25	5	9	15	20	25	31	36	42	47	53	58
0.30	7	11	17	24	30	37	43	50	56	63	69
0.35	8	13	20	28	35	43	50	58	65	73	80
0.40	9	14	23	31	40	48	57	66	74	83	91
0.45	10	16	25	35	45	54	64	73	83	92	102
0.50	11	18	28	39	49	60	70	81	91	102	112
0.55	11	19	31	42	53	65	76	88	99	111	122
0.60	12	21	33	45	58	70	83	95	107	120	132

HEAT LOSS THROUGH PIPE INSULATION — 6 INCH STEEL PIPE

Insulation Conductivity k	Temperature Difference, Pipe to Air, Deg. F										
	30	50	80	110	140	170	200	230	260	290	320
	Heat Loss per Lineal Foot of Bare Pipe, Btu per Hour										
	84	151	267	398	546	709	890	1088	1306	1544	1806
	Heat Loss per Lineal Foot of Insulated Pipe, Btu per Hour										
1 Inch Thick Insulation — 6 Inch Pipe											
0.20	11	18	29	40	51	61	72	83	94	105	115
0.25	13	22	35	48	62	75	88	101	114	128	141
0.30	15	26	41	57	72	88	103	119	134	150	165
0.35	18	29	47	65	82	100	118	135	153	171	188
0.40	20	33	53	72	92	112	132	151	171	191	211
0.45	22	36	58	80	102	123	145	167	189	210	232
0.50	24	39	63	87	110	134	158	181	205	229	252
0.55	25	42	68	93	119	144	170	195	221	246	272
0.60	27	45	73	100	127	154	182	209	236	263	291
1½ Inch Thick Insulation — 6 Inch Pipe											
0.20	8	13	21	29	37	45	53	61	69	77	85
0.25	10	16	26	36	45	55	65	75	84	94	104
0.30	12	19	31	42	54	65	77	88	100	111	123
0.35	13	22	35	49	62	75	88	101	115	128	141
0.40	15	25	40	55	70	84	99	114	129	144	159
0.45	17	28	44	61	77	94	110	127	143	160	176
0.50	18	30	48	66	84	103	121	139	157	175	193
0.55	20	33	52	72	92	111	131	151	170	190	209
0.60	21	35	56	77	99	120	141	162	183	204	225
2 Inch Thick Insulation — 6 Inch Pipe											
0.20	6	11	17	23	30	36	43	49	55	62	68
0.25	8	13	21	29	37	45	53	60	68	76	84
0.30	9	16	25	34	44	53	62	72	81	90	100
0.35	11	18	29	40	50	61	72	83	94	104	115
0.40	12	20	33	45	57	69	81	94	106	118	130
0.45	14	23	36	50	63	77	91	104	118	131	145
0.50	15	25	40	55	70	85	100	114	129	144	159
0.55	16	27	43	60	76	92	108	125	141	157	173
0.60	18	29	47	64	82	99	117	134	152	170	187
2½ Inch Thick Insulation — 6 Inch Pipe											
0.20	5	9	14	20	25	31	36	41	47	52	58
0.25	7	11	18	25	31	38	45	51	58	65	72
0.30	8	13	21	29	37	45	53	61	69	77	85
0.35	9	15	25	34	43	52	62	71	80	89	98
0.40	10	17	28	38	49	59	70	80	91	101	111
0.45	12	19	31	43	54	66	78	89	101	113	124
0.50	13	21	34	47	60	73	86	99	111	124	137
0.55	14	23	37	51	65	79	93	107	121	135	149
0.60	15	25	40	56	71	86	101	116	131	147	162

HEAT LOSS THROUGH PIPE INSULATION — 8 INCH STEEL PIPE

Insulation Conductivity k	Temperature Difference, Pipe to Air, Deg. F										
	30	50	80	110	140	170	200	230	260	290	320
	Heat Loss per Lineal Foot of Bare Pipe, Btu per Hour										
	107	194	341	509	697	906	1137	1391	1671	1977	2312
	Heat Loss per Lineal Foot of Insulated Pipe, Btu per Hour										
1 Inch Thick Insulation — 8 Inch Pipe											
0.20	14	23	36	50	64	78	91	105	119	132	146
0.25	17	28	44	61	78	95	111	128	145	161	178
0.30	20	33	52	72	91	111	130	150	169	189	209
0.35	22	37	59	82	104	126	149	171	193	215	238
0.40	25	42	66	91	116	141	166	191	216	241	266
0.45	27	46	73	101	128	155	183	210	238	265	293
0.50	30	50	79	109	139	169	199	228	258	288	318
0.55	32	53	86	118	150	182	214	246	278	310	342
0.60	34	57	91	126	160	194	229	263	297	331	366
1½ Inch Thick Insulation — 8 Inch Pipe											
0.20	10	16	26	36	46	56	65	75	85	95	105
0.25	12	20	32	44	56	68	80	92	104	117	129
0.30	14	24	38	52	66	81	95	109	123	138	152
0.35	16	27	44	60	76	93	109	125	142	158	174
0.40	18	31	49	68	86	104	123	141	160	178	196
0.45	20	34	54	75	95	116	136	156	177	197	218
0.50	22	37	60	82	104	127	149	171	194	216	238
0.55	24	40	65	89	113	137	161	186	210	234	258
0.60	26	43	69	95	122	148	174	200	226	252	278
2 Inch Thick Insulation — 8 Inch Pipe											
0.20	8	13	21	29	37	45	53	61	68	76	84
0.25	10	16	26	36	46	55	65	75	85	94	104
0.30	12	19	31	42	54	66	77	89	100	112	124
0.35	13	22	36	49	62	76	89	102	116	129	142
0.40	15	25	40	55	70	85	101	116	131	146	161
0.45	17	28	45	62	78	95	112	129	145	162	179
0.50	18	31	49	68	86	104	123	141	160	178	197
0.55	20	33	53	74	94	114	134	154	174	194	214
0.60	22	36	58	79	101	123	144	166	187	209	231
2½ Inch Thick Insulation — 8 Inch Pipe											
0.20	7	11	18	24	31	37	44	51	57	64	71
0.25	8	14	22	30	38	46	55	63	71	79	87
0.30	10	16	26	36	45	55	65	75	84	94	104
0.35	11	19	30	41	53	64	75	86	98	109	120
0.40	13	21	34	47	60	72	85	98	111	123	136
0.45	14	24	38	52	66	81	95	109	123	138	152
0.50	16	26	42	57	73	89	104	120	136	151	167
0.55	17	28	46	63	80	97	114	131	148	165	182
0.60	18	31	49	68	86	105	123	142	160	179	197

HEAT LOSS THROUGH PIPE INSULATION — 10 INCH STEEL PIPE

Insulation Conductivity k	Temperature Difference, Pipe to Air, Deg. F										
	30	50	80	110	140	170	200	230	260	290	320
	Heat Loss per Lineal Foot of Bare Pipe, Btu per Hour										
	132	238	420	626	857	1114	1399	1714	2060	2437	2852
	Heat Loss per Lineal Foot of Insulated Pipe, Btu per Hour										
1 Inch Thick Insulation — 10 Inch Pipe											
0.20	16	27	43	60	76	92	109	125	141	158	174
0.25	20	33	53	73	93	113	133	152	172	192	212
0.30	23	39	62	85	109	132	155	179	202	225	249
0.35	27	44	71	97	124	151	177	204	230	257	283
0.40	30	49	79	109	138	168	198	227	257	287	317
0.45	33	54	87	120	152	185	218	250	283	316	348
0.50	36	59	95	130	166	201	237	272	308	343	379
0.55	38	64	102	140	179	217	255	293	332	370	408
0.60	41	68	109	150	191	232	273	313	354	395	436
1½ Inch Thick Insulation — 10 Inch Pipe											
0.20	12	20	32	44	55	67	79	91	103	115	127
0.25	15	24	39	54	68	83	97	112	127	141	156
0.30	17	29	46	63	81	98	115	132	150	167	184
0.35	20	33	53	73	93	112	132	152	172	192	211
0.40	22	37	59	82	104	126	149	171	193	216	238
0.45	25	41	66	91	115	140	165	189	214	239	264
0.50	27	45	72	99	126	153	180	207	234	261	288
0.55	29	49	78	107	137	166	195	225	254	283	313
0.60	32	53	84	116	147	179	210	242	273	305	336
2 Inch Thick Insulation — 10 Inch Pipe											
0.20	9	16	25	35	44	54	63	73	82	92	101
0.25	12	20	31	43	55	66	78	90	102	113	125
0.30	14	23	37	51	65	79	93	107	120	134	148
0.35	16	27	43	59	75	91	107	123	139	155	171
0.40	18	30	48	66	84	103	121	139	157	175	193
0.45	20	34	54	74	94	114	134	154	174	195	215
0.50	22	37	59	81	103	125	147	169	192	214	236
0.55	24	40	64	88	112	136	160	184	208	232	256
0.60	26	43	69	95	121	147	173	199	225	251	277
2½ Inch Thick Insulation — 10 Inch Pipe											
0.20	8	13	21	29	37	44	52	60	68	76	84
0.25	10	16	26	36	45	55	65	74	84	94	104
0.30	12	19	31	42	54	65	77	89	100	112	123
0.35	13	22	36	49	62	76	89	102	116	129	142
0.40	15	25	40	55	71	86	101	116	131	146	161
0.45	17	28	45	62	79	96	112	129	146	163	180
0.50	19	31	49	68	87	105	124	142	161	179	198
0.55	20	34	54	74	94	115	135	155	175	196	216
0.60	22	36	58	80	102	124	146	168	190	211	233

HEAT LOSS THROUGH PIPE INSULATION — 12 INCH STEEL PIPE

Insulation Conductivity k	Temperature Difference, Pipe to Air, Deg. F										
	30	50	80	110	140	170	200	230	260	290	320
	Heat Loss per Lineal Foot of Bare Pipe, Btu per Hour										
	154	279	491	732	1003	1305	1640	2009	2415	2860	3346
	Heat Loss per Lineal Foot of Insulated Pipe, Btu per Hour										
1 Inch Thick Insulation — 12 Inch Pipe											
0.20	19	32	51	70	89	108	127	146	165	184	204
0.25	23	39	62	85	109	132	155	179	202	225	248
0.30	27	45	73	100	127	155	182	209	237	264	291
0.35	31	52	83	114	145	176	207	239	270	301	332
0.40	35	58	93	127	162	197	232	266	301	336	371
0.45	38	64	102	140	178	217	255	293	331	369	408
0.50	42	69	111	152	194	235	277	319	360	402	443
0.55	45	75	119	164	209	254	298	343	388	433	477
0.60	48	80	127	175	223	271	319	366	414	462	510
1½ Inch Thick Insulation — 12 Inch Pipe											
0.20	14	23	36	50	64	77	91	105	118	132	146
0.25	17	28	45	61	78	95	112	129	145	162	179
0.30	20	33	53	73	92	112	132	152	172	191	211
0.35	23	38	61	83	106	129	152	174	197	220	243
0.40	26	43	68	94	119	145	171	196	222	247	273
0.45	28	47	76	104	132	161	189	217	246	274	302
0.50	31	52	83	114	145	176	207	238	269	300	331
0.55	34	56	90	123	157	191	224	258	291	325	359
0.60	36	60	96	132	169	205	241	277	313	349	385
2 Inch Thick Insulation — 12 Inch Pipe											
0.20	11	18	29	41	52	63	74	85	96	107	118
0.25	14	23	36	50	64	77	91	105	118	132	146
0.30	16	27	43	59	75	92	108	124	140	156	173
0.35	19	31	50	68	87	106	124	143	161	180	199
0.40	21	35	56	77	98	119	140	161	182	203	224
0.45	23	39	62	86	109	133	156	179	203	226	250
0.50	26	43	68	94	120	146	171	197	223	248	274
0.55	28	47	74	102	130	158	186	214	242	270	298
0.60	31	52	83	114	145	176	207	238	269	300	332
2½ Inch Thick Insulation — 12 Inch Pipe											
0.20	9	15	24	34	43	52	61	70	79	89	98
0.25	11	19	30	42	53	64	76	87	98	110	121
0.30	13	22	36	49	63	76	90	103	117	130	144
0.35	16	26	42	57	73	88	104	119	135	151	166
0.40	18	29	47	65	82	100	118	135	153	170	188
0.45	20	33	52	72	92	111	131	151	170	190	210
0.50	22	36	58	79	101	123	144	166	187	209	231
0.55	24	39	63	86	110	134	157	181	204	228	251
0.60	25	42	68	93	119	144	170	195	221	246	272

HEAT LOSS THROUGH PIPE INSULATION — 14 INCH STEEL PIPE

Insulation Conductivity k	Temperature Difference, Pipe to Air, Deg. F										
	30	50	80	110	140	170	200	230	260	290	320
	Heat Loss per Lineal Foot of Bare Pipe, Btu per Hour										
	181	326	575	856	1173	1527	1918	2350	2826	3347	3918

Heat Loss per Lineal Foot of Insulated Pipe, Btu per Hour

1 Inch Thick Insulation — 14 Inch Pipe

k	30	50	80	110	140	170	200	230	260	290	320
0.20	22	36	57	79	100	122	143	165	187	208	230
0.25	26	44	70	96	122	149	175	201	227	253	280
0.30	31	51	82	113	143	174	205	235	266	297	327
0.35	35	58	93	128	163	198	233	268	303	338	373
0.40	39	65	104	143	182	221	260	299	338	377	416
0.45	43	72	114	157	200	243	286	329	372	415	458
0.50	47	78	124	171	218	264	311	357	404	451	497
0.55	50	84	134	184	234	284	335	385	435	485	535
0.60	54	89	143	196	250	303	357	411	464	518	571

1½ Inch Thick Insulation — 14 Inch Pipe

k	30	50	80	110	140	170	200	230	260	290	320
0.20	15	26	41	56	72	87	102	118	133	148	164
0.25	19	31	50	69	88	107	126	145	163	182	201
0.30	22	37	59	82	104	126	148	171	193	215	237
0.35	26	43	68	94	119	145	170	196	221	247	272
0.40	29	48	77	105	134	163	191	220	249	278	306
0.45	32	53	85	117	148	180	212	244	276	307	339
0.50	35	58	93	127	162	197	232	267	301	336	371
0.55	38	63	100	138	176	213	251	289	326	364	402
0.60	40	67	108	148	189	229	270	310	350	391	431

2 Inch Thick Insulation — 14 Inch Pipe

k	30	50	80	110	140	170	200	230	260	290	320
0.20	12	20	31	43	55	66	78	90	102	113	125
0.25	14	24	39	53	68	82	97	111	125	140	154
0.30	17	29	46	63	80	97	114	132	149	166	183
0.35	20	33	53	73	92	112	132	152	171	191	211
0.40	22	37	60	82	104	127	149	171	194	216	238
0.45	25	41	66	91	116	141	166	190	215	240	265
0.50	27	45	73	100	127	155	182	209	236	264	291
0.55	30	49	79	109	138	168	198	227	257	287	317
0.60	32	53	85	117	149	181	213	245	277	309	341

2½ Inch Thick Insulation — 14 Inch Pipe

k	30	50	80	110	140	170	200	230	260	290	320
0.20	10	16	26	36	46	56	65	75	85	95	105
0.25	12	20	32	45	57	69	81	93	105	117	130
0.30	14	24	38	53	67	82	96	111	125	140	154
0.35	17	28	44	61	78	95	111	128	145	161	178
0.40	19	31	50	69	88	107	126	145	164	182	201
0.45	21	35	56	77	98	119	140	161	182	203	225
0.50	23	39	62	85	108	131	154	178	201	224	247
0.55	25	42	67	93	118	143	168	194	219	244	269
0.60	27	45	73	100	127	155	189	209	237	264	291

HEAT LOSS THROUGH PIPE INSULATION -- 16 INCH STEEL PIPE

Insulation Conductivity k	Temperature Difference, Pipe to Air, Deg. F										
	30	50	80	110	140	170	200	230	260	290	320
	Heat Loss per Lineal Foot of Bare Pipe, Btu per Hour										
	203	366	644	960	1314	1711	2149	2634	3168	3753	4395
	Heat Loss per Lineal Foot of Insulated Pipe, Btu per Hour										
1 Inch Thick Insulation — 16 Inch Pipe											
0.20	23	39	62	85	108	131	154	178	201	224	247
0.25	28	47	75	104	132	160	189	217	245	273	302
0.30	33	55	88	122	155	180	188	221	254	287	320
0.35	38	63	101	139	177	215	253	291	329	367	405
0.40	42	70	113	155	197	239	281	324	266	408	450
0.45	46	77	124	170	217	263	310	356	402	449	495
0.50	50	84	135	185	236	286	337	387	438	488	539
0.55	54	91	145	199	254	308	363	417	471	526	580
0.60	58	97	155	213	271	329	387	445	503	561	620
1½ Inch Thick Insulation — 16 Inch Pipe											
0.20	17	28	45	62	79	95	112	129	146	163	180
0.25	21	34	55	76	97	117	138	159	179	200	221
0.30	24	41	65	90	114	138	163	187	212	236	261
0.35	28	47	75	103	131	159	187	215	243	271	329
0.40	32	53	84	116	147	179	210	242	273	305	336
0.45	35	58		128	163	198	333	268	303	338	372
0.50	38	64	102	140	178	217	255	293	331	369	408
0.55	41	69	110	152	193	235	276	317	359	400	442
0.60	44	74	119	163	207	252	296	341	385	430	474
2 Inch Thick Insulation — 16 Inch Pipe											
0.20	13	22	36	49	62	76	89	102	116	129	143
0.25	16	27	44	60	77	93	110	126	143	159	176
0.30	20	33	52	72	91	111	130	150	169	189	208
0.35	23	38	60	83	105	128	150	173	195	218	240
0.40	25	42	68	93	119	144	170	195	220	246	271
0.45	28	47	75	104	132	160	189	217	245	273	302
0.50	31	52	83	114	145	176	207	238	269	300	331
0.55	34	56	90	124	157	191	225	259	292	326	360
0.60	36	61	97	133	170	206	242	278	315	351	388
2½ Inch Thick Insulation — 16 Inch Pipe											
0.20	11	19	30	41	52	63	74	86	97	108	119
0.25	14	23	37	51	64	78	92	106	120	134	147
0.30	16	28	44	60	77	93	109	256	142	159	175
0.35	19	32	51	69	88	107	126	145	164	183	202
0.40	21	36	57	79	100	122	143	165	186	207	229
0.45	24	40	64	88	112	135	159	183	207	231	255
0.50	26	44	70	96	123	149	175	202	228	254	280
0.55	29	48	76	105	134	162	191	220	248	277	306
0.60	31	52	83	114	144	175	206	237	268	299	330

HEAT LOSS THROUGH PIPE INSULATION — 18 INCH STEEL PIPE

Insulation Conductivity k	Temperature Difference, Pipe to Air, Deg. F										
	30	50	80	110	140	170	200	230	260	290	320
	Heat Loss per Lineal Foot of Bare Pipe, Btu per Hour										
	214	385	678	1011	1385	1802	2266	2777	3339	3958	4635
	Heat Loss per Lineal Foot of Insulated Pipe, Btu per Hour										

1 Inch Thick Insulation — 18 Inch Pipe

k	30	50	80	110	140	170	200	230	260	290	320
0.20	27	45	72	99	126	153	180	207	234	261	288
0.25	33	55	88	121	153	186	219	252	285	318	351
0.30	38	64	103	141	180	218	257	295	334	372	411
0.35	44	73	117	161	205	248	292	336	380	424	468
0.40	49	82	130	179	228	277	326	375	424	473	522
0.45	54	90	143	197	251	305	359	412	466	520	574
0.50	58	97	156	214	273	331	390	448	506	565	623
0.55	63	105	168	231	293	356	419	482	545	608	671
0.60	67	112	179	246	313	380	448	515	582	649	716

1½ Inch Thick Insulation — 18 Inch Pipe

k	30	50	80	110	140	170	200	230	260	290	320
0.20	19	31	50	69	88	106	125	144	163	181	200
0.25	23	38	61	84	107	130	153	176	199	222	244
0.30	27	45	72	99	126	153	180	207	234	261	289
0.35	31	52	83	114	145	176	207	238	269	300	331
0.40	35	58	93	128	163	198	233	268	303	338	372
0.45	39	64	103	142	180	219	258	296	335	374	412
0.50	42	71	113	155	197	240	282	324	367	409	451
0.55	46	76	122	168	214	260	306	351	397	443	489
0.60	49	82	131	181	230	279	328	378	427	476	525

2 Inch Thick Insulation — 18 Inch Pipe

k	30	50	80	110	140	170	200	230	260	290	320
0.20	15	25	40	55	70	85	100	115	130	145	160
0.25	18	31	49	68	86	105	123	142	160	178	197
0.30	22	36	58	80	102	124	146	168	190	212	234
0.35	25	42	67	93	118	143	168	193	219	244	269
0.40	28	47	76	104	133	161	190	218	247	275	304
0.45	32	53	84	116	148	179	211	243	274	306	338
0.50	35	58	93	127	162	197	232	266	301	336	371
0.55	38	63	101	138	176	214	252	290	327	365	403
0.60	41	68	109	149	190	231	271	312	553	394	434

2½ Inch Thick Insulation — 18 Inch Pipe

k	30	50	80	110	140	170	200	230	260	290	320
0.20	12	20	33	45	57	69	81	94	106	118	130
0.25	15	25	40	55	70	86	101	116	131	146	161
0.30	18	30	48	66	84	102	120	138	156	174	191
0.35	21	35	55	76	97	117	138	159	180	200	221
0.40	23	39	62	85	108	131	154	177	200	224	247
0.45	26	44	70	96	122	148	174	201	227	253	279
0.50	29	48	77	106	134	161	192	221	250	278	307
0.55	31	52	84	115	146	178	209	240	272	303	334
0.60	34	56	90	124	158	192	226	260	294	327	361

HEAT LOSS THROUGH PIPE INSULATION — 20 INCH STEEL PIPE

Insulation Conductivity k	Temperature Difference, Pipe to Air, Deg. F										
	30	50	80	110	140	170	200	230	260	290	320
	Heat Loss per Lineal Foot of Bare Pipe, Btu per Hour										
	236	426	748	1115	1529	1990	2501	3066	3690	4373	5123
	Heat Loss per Lineal Foot of Insulated Pipe, Btu per Hour										
1 Inch Thick Insulation — 20 Inch Pipe											
0.20	30	49	79	108	138	168	198	227	256	286	316
0.25	36	60	96	132	168	204	240	276	313	349	385
0.30	42	69	111	153	194	236	278	319	361	403	444
0.35	48	80	128	176	224	273	321	369	418	465	513
0.40	54	89	143	197	250	304	358	411	465	519	572
0.45	59	98	157	216	275	334	393	452	511	570	629
0.50	64	107	171	235	299	363	427	491	556	620	683
0.55	69	115	184	253	322	391	460	529	598	667	737
0.60	74	123	196	270	344	417	491	565	638	712	786
1½ Inch Thick Insulation — 20 Inch Pipe											
0.20	21	35	56	77	98	118	139	160	181	202	223
0.25	26	43	69	95	121	146	172	198	224	250	276
0.30	30	51	81	111	141	172	202	232	263	293	323
0.35	35	58	93	127	162	197	232	267	301	336	371
0.40	39	65	104	143	182	222	261	300	339	378	417
0.45	43	72	115	159	202	245	289	332	375	418	462
0.50	47	79	126	174	221	268	316	363	410	458	505
0.55	51	85	137	188	239	291	342	393	444	496	547
0.60	55	92	147	202	257	312	367	422	477	532	587
2 Inch Thick Insulation — 20 Inch Pipe											
0.20	16	27	44	60	76	93	109	125	142	158	174
0.25	20	34	54	74	94	114	134	154	175	195	215
0.30	24	40	64	88	111	135	159	183	207	231	255
0.35	27	46	73	101	128	156	183	211	238	266	293
0.40	31	52	83	114	145	176	207	238	269	300	331
0.45	35	58	92	127	161	196	230	265	299	334	368
0.50	38	63	101	139	177	215	252	290	328	366	404
0.55	41	69	110	151	192	233	274	316	357	398	439
0.60	44	74	118	163	207	251	296	340	385	429	473
2½ Inch Thick Insulation — 20 Inch Pipe											
0.20	14	23	36	50	63	77	90	104	117	131	144
0.25	17	28	45	61	78	95	112	128	145	162	178
0.30	20	33	53	73	93	112	132	152	172	192	212
0.35	23	38	61,	84	107	130	153	176	199	222	244
0.40	26	43	69	95	121	147	173	199	225	251	277
0.45	29	48	77	106	135	164	193	221	250	279	308
0.50	32	53	85	116	148	180	212	244	275	307	339
0.55	35	58	92	127	162	196	231	265	300	335	369
0.60	37	62	100	137	175	212	249	287	324	362	399

Section 9

MOTORS AND STARTERS

Contents

Section Editor: *Howard A. Wolfberg*, Store,
Matakovich & Wolfberg

MOTORS AND STARTERS

NEMA MOTOR CLASSIFICATIONS

NEMA is the National Electrical Manufacturers Association, a nonprofit trade organization supported by manufacturers of electrical apparatus and supplies. NEMA standards are designed to eliminate misunderstandings between the manufacturer and the purchaser. They also assist the purchaser in selecting and obtaining the proper product for his particular need. Some standards included in the *NEMA Standards for Motors and Generators* have frequent application in the air conditioning, heating, ventilating, plumbing, and fire-protection fields. Portions of these standards are described or repeated in the following pages. Reference will be made to these standards in other parts of this section.

The following are NEMA criteria for motor classification:

Classification by Size: A motor may be classified as a fractional-horsepower motor or an integral-horsepower motor.

Classification by Application: Motors normally are classified under one of the following application designations:

General-purpose alternating-current (a-c) motor: an induction motor of open construction, with standard operating characteristics and mechanical construction, for use under usual service conditions without restriction to a particular application or type of application.

Definite-purpose motor: a motor having standard ratings and standard operating characteristics or mechanical construction for use under service conditions other than usual, or for use on a specified type of application.

Special-purpose motor: a motor with special operating characteristics or special mechanical construction, or both.

Industrial Direct-Current (d-c) Motor

Classification by Electrical Type: * The following electrical motor types are recognized:

A-C induction motor

A-C wound-rotor induction motor

---

* A-C or a-c designates alternating current; and D-C or d-c, designates direct current.

A-C synchronous motor

A-C series motor

Single-phase, split-phase, squirrel-cage motor

Single-phase, resistance-start, squirrel-cage motor (a form of split-phase motor)

Single-phase, capacitor, squirrel-cage motor (This may be of a capacitor-start, a permanent-split capacitor, or a two-value capacitor type.)

Single-phase shaded-pole, squirrel-cage motor

Single-phase, repulsion, wound-rotor motor

Single-phase, repulsion-start induction, wound-rotor motor

Single-phase, repulsion-induction, wound-rotor motor

D-C, shunt-wound motor (This may be either a straight shunt-wound motor or a stabilized shunt-wound motor.)

D-C, series-wound motor

D-C, compound-wound motor

Universal, series-wound motor

Universal, compensated series motor

Classification of Polyphase Motors by Design Letter: Polyphase, squirrel-cage, integral-horsepower induction motors may be classified under Design Letters A, B, C, or D. Such motors are designed to withstand full voltage starting and usually have a slip at rated load of less than 5%. However, Design D motors have a slip at rated load of 5% or more, as do Design A and Design B motors with 10 or more poles. Characteristics pertaining to each of the above mentioned Design Letters are given below:

Design A Motor:

Locked-rotor torque in accordance with Table I. Breakdown torque in accordance with Table I.

Design B Motor:

Locked-rotor torque in accordance with Table I. Breakdown torque in accordance with Table I. Locked-rotor current not to exceed Table 2.

Design C Motor:

Locked-rotor torque in accordance with Table I. Breakdown torque in accordance with Table I. Locked-rotor current not to exceed Table 2.

Table 1. Characteristics of Polyphase Squirrel-Cage Motors **

Synchronous Speeds (rpm)	Locked-Rotor Torque, % of Full-Load Torque										Breakdown Torque, % of Full-Load Torque*									
	Designs A and B							Design C			Design B							Design C		
60 cycle (hertz)	3600	1800	1200	900	720	600	514	1800	1200	900	3600	1800	1200	900	720	600	514	1800	1200	900
50 cycle (hertz)	3000	1500	1000	750	...	...	...	1500	1000	750	3000	1500	1000	750	...	...	...	1500	1000	750
hp																				
1/2	...	...	...	140	140	115	110	...	...	...	...	...	...	225	200	200	200	...	...	...
3/4	...	...	175	135	135	115	110	...	...	...	...	...	275	220	200	200	200	...	...	...
1	...	275	170	135	135	115	110	...	...	...	...	300	265	215	200	200	200	...	...	...
1-1/2	175	250	165	130	130	115	110	...	...	...	250	280	250	210	200	200	200	...	...	...
2	170	235	160	130	125	115	110	...	...	...	240	270	240	210	200	200	200	...	...	...
3	160	215	155	130	125	115	110	...	250	225	230	250	230	205	200	200	200	...	225	200
5	150	185	150	130	125	115	110	250	250	225	215	225	215	205	200	200	200	200	200	200
7-1/2	140	175	150	125	120	115	110	250	225	200	200	215	205	200	200	200	200	190	190	190
10	135	165	150	125	120	115	110	250	225	200	200	200	200	200	200	200	200	190	190	190
15	130	160	140	125	120	115	110	225	200	200	200	200	200	200	200	200	200	190	190	190
20	130	150	135	125	120	115	110	200	200	200	200	200	200	200	200	200	200	190	190	190
25	130	150	135	125	120	115	110	200	200	200	200	200	200	200	200	200	200	190	190	190
30	130	150	135	125	120	115	110	200	200	200	200	200	200	200	200	200	200	190	190	190
40	125	140	135	125	120	115	110	200	200	200	200	200	200	200	200	200	200	190	190	190
50	120	140	135	125	120	115	110	200	200	200	200	200	200	200	200	200	200	190	190	190
60	120	140	135	125	120	115	110	200	200	200	200	200	200	200	200	200	200	190	190	190
75	105	140	135	125	120	115	110	200	200	200	200	200	200	200	200	200	200	190	190	190
100	105	125	125	125	120	115	110	200	200	200	200	200	200	200	200	200	200	190	190	190
125	100	110	125	120	115	115	110	200	200	200	200	200	200	200	200	200	200	190	190	190
150	100	110	120	120	115	115	...	200	200	200	200	200	200	200	200	200	200	190	190	190
200	100	100	120	120	115	...	...	200	200	200	200	200	200	200	200	...	...	190	190	190
250	70	80	100	100	...	...	...	...	...	...	175	175	175	175	...	...	...	...	...	...
300	70	80	100	...	...	...	...	...	...	...	175	175	175	...	...	...	...	...	...	...
350	70	80	100	...	...	...	...	...	...	...	175	175	175	...	...	...	...	...	...	...
400	70	80	...	...	...	...	...	...	...	...	175	175	...	...	...	...	...	...	...	...
450	70	80	...	...	...	...	...	...	...	...	175	175	...	...	...	...	...	...	...	...
500	70	80	...	...	...	...	...	...	...	...	175	175	...	...	...	...	...	...	...	...

*Design A values are in excess of those for Design B.

**From NEMA publication MG1-1972. Reproduced by permission.

Design D Motor:
Locked-rotor torque to be 275 percent of full-load torque. Locked-rotor current not to exceed Table 2.

Classification of Single-Phase, Induction Motors by Design Letter: The following single-phase motor types are designed to withstand full-voltage starting and have the characteristics noted:

Design N Motor:
Fractional-horsepower motor with a locked-rotor current not to exceed the values in Table 3.

Design O Motor:
Fractional-horsepower motor with a locked-rotor current not to exceed the values in Table 3.

Design L Motor:
Integral-horsepower motor with a locked-motor current not to exceed the values in Table 4.

Design M Motor:
Integral-horsepower motor with a locked-rotor current not to exceed the values in Table 4.

Torque, Speed, and Horsepower Ratings for Single-Phase Induction Motors: The horsepower rating of a single-phase induction motor is based upon a minimum value of breakdown torque. The range in breakdown torque to be expected for any horsepower and speed is given in Table 5. The breakdown torque of a general-purpose motor is no less than the higher figure in the

Table 2. Characteristics of Three-Phase Squirrel-Cage Induction Motors **

hp	Locked-Rotor Current, Amperes*	Design Letters	hp	Locked-Rotor Current, Amperes*	Design Letters
60 Cycle (hertz), Rated at 230 Volts			**50 Cycle (hertz), Rated at 380 Volts**		
1/2	20	B, D	1 or less	20	B, D
3/4	25	B, D	1-1/2	27	B, D
1	30	B, D	2	34	B, D
1-1/2	40	B, D	3	43	B, C, D
2	50	B, D	5	61	B, C, D
3	64	B, C, D	7-1/2	84	B, C, D
5	92	B, C, D	10	107	B, C, D
7-1/2	127	B, C, D	15	154	B, C, D
10	162	B, C, D	20	194	B, C, D
15	232	B, C, D	25	243	B, C, D
20	290	B, C, D	30	289	B, C, D
25	365	B, C, D	40	387	B, C, D
30	435	B, C, D	50	482	B, C, D
40	580	B, C, D	60	578	B, C, D
50	725	B, C, D	75	722	B, C, D
60	870	B, C, D	100	965	B, C, D
75	1085	B, C, D	125	1207	B, C, D
100	1450	B, C, D	150	1441	B, C, D
125	1815	B, C, D	200	1927	B, C
150	2170	B, C, D			
200	2900	B, C			
250	3650	B			
300	4400	B			
350	5100	B			
400	5800	B			
450	6500	B			
500	7250	B			

*Locked-rotor current of motors designed for voltages other than 230 or 380 volts shall be inversely proportioned to the voltages.

**From NEMA publication MG 1-1972. Reproduced by permission.

Table 3. Characteristics of Single-Phase, Fractional HP 2-, 4-, 6- and 8-Pole, 60-hertz Motors **

hp	Locked-rotor Current, Amperes*			
	115 volts		230 volts	
	Design O	Design N	Design O	Design N
1/6 and smaller	50	20	25	12
1/4	50	26	25	15
1/3	50	31	25	18
1/2	50	45	25	25
3/4	...	61	...	35
1	...	80	...	45

*The locked-rotor currents of single-phase, general-purpose, fractional-hp motors shall not exceed the values for Design N motors.

**From NEMA publication MG 1-1972. Reproduced by permission.

Table 4. Characteristics of Single-Phase, Integral HP Motors *

hp	Locked-Rotor Current, Amperes		
	Design L		Design M
	115 volts	230 volts	230 volts
1/2	45	25	...
3/4	61	35	...
1	80	45	...
1 1/2	...	50	40
2	...	65	50
3	...	90	70
5	...	135	100
7 1/2	...	200	150
10	...	260	200
15	...	390	300
20	...	520	400

*From NEMA publication MG 1-1972. Reproduced by permission.

applicable torque range. The locked-rotor torque of a general-purpose motor is no less than the value given in Table 6 for the applicable horsepower and speed.

Classification by Environmental Protection and Method of Cooling: This classification refers to the amount of protection provided against the environment in which the motor operates and the arrangement of ventilating openings or other means of cooling:

Open Machine: One having ventilating openings which permit passage of external cooling air over and around the windings.

Drip-Proof Machine (Often Referred to as Open Drip-Proof Machine): An open motor in which ventilating openings are so constructed that successful operation is not interfered with when drops of liquid or solid particles strike or enter the enclosure at any angle from 0 to 15 degrees downward from the vertical.

Machine With Encapsulated Windings: An a-c squirrel-cage motor having random windings filled with an insulating resin which also forms a protective coating.

Totally Enclosed Machine: One so enclosed as to prevent free exchange of air between inside and outside of case. The enclosure, however, is not airtight.

Totally Enclosed Fan-Cooled Machine: A totally enclosed motor equipped for exterior cooling by means of a fan, or fans, integral with the motor but external to the enclosing parts.

Explosion-Proof Machine: A totally enclosed or totally enclosed fan-cooled motor designed and built to withstand an explosion of gas or vapor within it and to prevent ignition of gas or vapor surrounding the motor by sparks, flashes, or explosions which occur within the motor casing.

Table 5. Basis of HP and Speed Ratings for Single-Phase Induction Motors [2]

Induction Motors, Except Shaded-Pole and Permanent-Split Capacitor Motors									
60	50	60	50-25	60	50	60	50-25		Frequencies, hertz
3600	3000	1800	1500	1200	1000	900	750		Synchronous Speeds. rpm
3450[1]	2850[1]	1725[1]	1425[1]	1140[1]	950[1]	850[1]	...	hp	Fractional-hp Nominal Speeds, rpm
								Milli hp	
0.35-0.55	0.42-0.66	0.7 -1.1	0.85-1.3	1.1 -1.65	...	...	...	1	
0.55-0.7	0.66-0.85	1.1 -1.45	1.3 -1.75	1.65-2.2	...	...	...	1.5	The figures at
0.7 -1.1	0.85-1.3	1.45-2.2	1.75-2.6	2.2 -3.3	...	...	...	2	left are for
1.1 -1.8	1.3 -2.2	2.2 -3.6	2.6 -4.3	3.3 -5.4	...	...	...	3	motors rated
1.8 -2.7	2.2 -3.2	3.6 -5.4	4.3 -6.6	5.4 -8.1	...	...	...	5	less than 1/20
2.7 -3.6	3.2 -4.3	5.4 -7.2	6.6 -8.6	8.1 -11	...	...	...	7.5	horsepower.
3.6 -5.5	4.3 -6.6	7.2 -11	8.6 -13	11 -17	...	...	...	10	Breakdown
5.5 -9.5	6.6 -11.4	11 -19	13 -23	17 -29	...	...	...	15	torques in oz-
9.5 -15	11.4 -18	19 -30	23 -36	29 -46	...	...	...	25	in.
15 -24	18 -28.8	30 -48	36 -57.6	46 -72	...	...	...	35	
								hp	The figures at
2.0- 3.7	2.4- 4.4	4.0- 7.1	4.8- 8.5	6.0-10.4	7.2-12.4	8.0-13.5	...	1/20	left are for
3.7- 6.0	4.4- 7.2	7.1-11.5	8.5-13.8	10.4-16.5	12.4-19.8	13.5-21.5	...	1/12	fractional-hp
6.0- 8.7	7.2-10.5	11.5-16.5	13.8-19.8	16.5-24.1	19.8-28.9	21.5-31.5	...	1/8	motors.
8.7-11.5	10.5-13.8	16.5-21.5	19.8-25.8	24.1-31.5	28.9-37.8	31.5-40.5	...	1/6	Breakdown
11.5-16.5	13.8-19.8	21.5-31.5	25.8-37.8	31.5-44.0	37.8-53.0	40.5-58.0	...	1/4	torques in oz-
16.5-21.5	19.8-25.8	31.5-40.5	37.8-48.5	44.0-58.0	53.0-69.5	58.0-77.0	...	1/3	ft.
21.5-31.5	25.8-37.8	40.5-58.0	48.5-69.5	58.0-82.5	69.5-99.0	*	*	1/2	
31.5-44.0	37.8-53.0	58.0-82.5	69.5-99.0	5.16- 6.9	*	*	*	3/4	
44.0-58.0	53.0-69.5	5.16- 6.8	6.19- 8.2	6.9- 9.2	*	*	*	1	The figures at
3.6- 4.6	4.3- 5.5	6.8-10.1	8.2-12.1	9.2-13.8	*	*	*	1 1/2	for integral-hp motors.
4.6- 6.0	5.5- 7.2	10.1-13.0	12.1-15.6	13.8-18.0	*	*	*	2	Breakdown
6.0- 8.6	7.2-10.2	13.0-19.0	15.6-22.8	18.0-25.8	*	*	*	3	torques in lb-
8.6-13.5	10.2-16.2	19.0-30.0	22.8-36.0	25.8-40.5	*	*	*	5	ft.
13.5-20.0	16.2-24.0	30.0-45.0	36.0-54.0	40.5-60.0	*	*	*	7 1/2	
20.0-27.0	24.0-32.4	45.0-60.0	54.0-72.0	*	*	*	*	10	

Table 5 (continued). Basis of HP and Speed Ratings

		Shaded-Pole and Permanent-Split Capacitor Motors				
		(Except for Permanent-Split Capacitor Hermetic Motors)				
60	50	60	50	60		Frequencies, hertz
1800	1500	1200	1000	900		Synchronous Speeds, rpm
See MG 1-10.32, Par. A and B					hp	Approximate Full-load Speeds, kpm
					Millihp	
0.89-1.1	1.1-1.3	1.3-1.6	1.6-1.9	1.7-2.1	1	
1.1-1.4	1.3-1.7	1.6-2.1	1.9-2.5	2.1-2.7	1.25	
1.4-1.7	1.7-2.0	2.1-2.5	2.5-3.0	2.7-3.3	1.5	
1.7-2.1	2.0-2.5	2.5-3.1	3.0-3.7	3.3-4.1	2	
2.1-2.6	2.5-3.1	3.1-3.8	3.7-4.6	4.1-5.0	2.5	
2.6-3.2	3.1-3.8	3.8-4.7	4.6-5.7	5.0-6.2	3	
3.2-4.0	3.8-4.8	4.7-5.9	5.7-7.1	6.2-7.8	4	The figures at left are
4.0-4.9	4.8-5.8	5.9-7.2	7.1-8.7	7.8-9.5	5	breakdown torques in oz-in.
4.9-6.2	5.8-7.4	7.2-9.2	8.7-11.0	9.5-12.0	6	
6.2-7.7	7.4-9.2	9.2-11.4	11.0-13.6	12.0-14.9	8	
7.7-9.6	9.2-11.4	11.4-14.2	13.6-17.0	14.9-18.6	10	
9.6-12.3	11.1-14.7	14.2-18.2	17.0-21.8	18.6-23.8	12.5	
12.3-15.3	14.7-18.2	18.2-22.6	21.8-27.1	23.8-29.6	16	
15.3-19.1	18.2-22.8	22.6-28.2	27.1-33.8	29.6-37.0	20	
19.1-23.9	22.8-28.5	28.2-35.3	33.8-42.3	37.0-46.3	25	
23.9-30.4	28.5-36.3	35.3-44.9	42.3-53.9	46.3-58.9	30	
30.4-38.3	36.3-45.6	44.9-56.4	53.9-68.4	58.9-74.4	40	
					hp	
3.20-4.13	3.80-4.92	4.70-6.09	5.70-7.31	6.20-8.00	1/20	
4.13-5.23	4.92-6.23	6.09-7.72	7.31-9.26	8.00-10.1	1/15	
5.23-6.39	6.23-7.61	7.72-9.42	9.26-11.3	10.1-12.4	1/12	
6.39-8.00	7.61-9.54	9.42-11.8	11.3-14.2	12.4-15.5	1/10	
8.00-10.4	9.54-12.4	11.8-15.3	14.2-18.4	15.5-20.1	1/8	The figures at left are
10.4-12.7	12.4-15.1	15.3-18.8	18.4-22.5	20.1-24.6	1/6	breakdown torques in oz-ft.
12.7-16.0	15.1-19.1	18.8-23.6	22.5-28.3	24.6-31.0	1/5	
16.0-21.0	19.1-25.4	23.6-31.5	28.3-37.6	31.0-41.0	1/4	
21.0-31.5	25.4-37.7	31.5-47.0	37.6-56.5	41.0-61.0	1/3	
31.5-47.5	37.7-57.3	47.0-70.8	56.5-84.8	3.81-5.81	1/2	
47.5-63.5	57.3-76.5	4.42-5.88	5.30-7.06	5.81-7.62	3/4	The figures at left are
3.97-5.94	4.78-7.06	5.88-8.88	7.06-10.6	7.62-11.6	1	breakdown torques in lb-ft.
5.94-7.88	7.06-9.56	8.88-11.8	10.6-14.1	11.6-15.2	1 1/2	

NOTE I--The breakdown torque range includes the higher figure down to, but not including, the lower figure.
NOTE II--The horsepower rating of motors designed to operate on two or more frequencies shall be determined by the torque at the highest rated frequency.

*These are ratings for which no torque values have been established.

1 These approximate full-load speeds apply only for fractional-hp motor ratings.
2 From NEMA publication MG 1-1972. Reprinted by permission.

With the exception of the open motor, which is seldom used, the motor enclosure and ventilating types classified above are most frequently utilized in the air conditioning, heating, ventilating, plumbing, and fire-protection fields for other than hermetic or accessible hermetic service.

Other types classified by protection and cooling, but not frequently encountered, are listed below:

Splash-proof machine
Guarded machine

Semiguarded machine
Drip-proof, fully guarded machine
Open, externally ventilated machine
Open pipe-ventilated machine
Weather-protected machine
Machine with sealed windings
Totally enclosed, fan-cooled, guarded machine
Dust-ignition-proof machine
Totally enclosed pipe-ventilated machine
Totally enclosed water-cooled machine
Totally enclosed water-air-cooled machine
Totally enclosed air-to-air cooled machine.

Table 6. Minimum Locked-Rotor Torques of Single-Phase Induction Motors *

| | | Minimum Locked-Rotor Torque | | | | | |
| | | 60-cycle speed, hertz | | | 50-cycle speed, hertz | | |
Synchronous Speed app. f.l. speed		3600 3450	1800 1725	1200 1140	3000 2850	1500 1425	1000 950	Units of Torque
horsepower	1/8	...	24	32	...	29	39	torque in oz.-ft.
	1/6	15	33	43	18	39	51	
	1/4	21	46	59	25	55	70	
	1/3	26	57	73	31	69	88	
	1/2	37	85	100	44	102	120	
	3/4	50	119	8.0	60	143		
	1	61	9.0	9.5	73			
	1 1/2	4.5	12.5	13.0				torque in lb.-ft.
	2	5.5	16.0	16.0				
	3	7.5	22.0	23.0				
	5	11.0	33.0	...				
	7 1/2	16.0	45.0	...				

NOTE: Approximate full-load speeds are for fractional-horsepower motor ratings only. Units for torque (see right-hand column of table), in ounce-feet are above the lines and in pound-feet below the lines.

*From NEMA publication MG 1-1972. Reproduced by permission.

Standard Voltages and Frequencies for Motors

Single-Phase A-C Motors
  60 cycles: 115 and 230 volts
  50 cycles: 110 and 220 volts
A-C Polyphase Motors
  60 cycles: 115, 200, 230, 460, 575, 2300, 4000, 4600 and 6600 volts
  50 cycles: 220 and 380 volts
Universal Motors—115 and 230 volts
D-C Fractional-Horsepower Motors: 115 and 230 volts
Industrial D-C Motors
  1/2 through 10 horsepower:
    120 and 240 volts
  15 through 200 horsepower:
    240 volts
  250 through 500 horsepower:
    250 and 500 volts
  600 through 800 horsepower:
    500 and 700 volts

Permissible Variation from Rated Voltage
  Induction motors: ± 10% (with rated frequency)
Universal motors, except fan motors: ± 6%
D-C motors: ±10%

Permissible Variation from Rated Frequency: ± 5%

Permissible Combined Variation of Voltage and Frequency: ± 10% provided the frequency variation does not exceed 5%.

Locked Rotor kva: The nameplate of every a-c motor, rated 1/20 hp, and larger, is marked with the caption "Code" followed by a letter selected to show locked-rotor kva per hp. Table 7 shows each code letter in relation to its designated range of locked-rotor kva per hp. The majority of

Table 7. Locked-Rotor KVA of A-C Motors **

Designation Letter	Locked-Rotor KVA per hp*	Designation Letter	Locked-Rotor KVA per hp*
A	0 - 3.15	K	8.0 - 9.0
B	3.15 - 3.55	L	9.0 - 10.0
C	3.55 - 4.0	M	10.0 - 11.2
D	4.0 - 4.5	N	11.2 - 12.5
E	4.5 - 5.0	P	12.5 - 14.0
F	5.0 - 5.6	R	14.0 - 16.0
G	5.6 - 6.3	S	16.0 - 18.0
H	6.3 - 7.1	T	18.0 - 20.0
J	7.1 - 8.0	U	20.0 - 22.4
		V	22.4 - and up

*Locked-rotor KVA per horsepower range includes the lower figure up to, but not including the higher figure. For example, everything under 3.15 is in Letter A, and 3.15 and above are in Letter B.

**From NEMA publication MG 1-1972. Reproduced by permission.

polyphase motors used in air conditioning, heating, ventilating, plumbing, and fire-protection work have the code letter "F." This code letter is not related to the design letter classification of an a-c induction motor.

Service Factors: The maximum permissible loading of a motor is obtained by multiplying the rated horsepower by the service factor shown on the nameplate. Open, or open drip-proof,

polyphase, general-purpose motors having a design A, B, or C classification, usually have a service factor of 1.15 for rated horsepowers of 1, or greater. For horsepowers less than 1, the service factors for these motors are generally greater than 1.15 but no greater than 1.4. Totally enclosed, totally enclosed fan-cooled, explosion-proof, and splash-proof motors generally have a service factor of 1.0 and should not be loaded over the nameplate rating.

THE NATIONAL ELECTRICAL CODE

The National Electrical Code (NEC) is sponsored by the National Fire Protection Association (NFPA), under the auspices of the American National Standards Institute (ANSI). The National Electrical Code is purely advisory, as far as the NFPA and ANSI are concerned, but is offered for use in law and for regulatory purposes in the interest of life and property protection.*

The NEC has been adopted in its entirety, in edited versions, or in part, by various state and local agencies throughout the United States having regulatory and enforcement jurisdiction over the installation of electrical work. The Federal Occupational Safety and Health Act (OSHA) has adopted the 1971 National Electrical Code and presumably will adopt subsequent versions of the Code.

Included among the recommendations covered in detail under *Article 430—Motors, Motor Circuits and Controllers* in the 1975 NEC are the following subjects:

Marking on motors and multimotor equipment
Marking on controllers
Sizing of conductors
Motor and branch-circuit running overcurrent
   protection (commonly referred to as overload protection)
Motor branch-circuit short-circuit and ground-fault protection
Motor-feeder short-circuit and ground-fault protection
Motor control circuits
Motor controllers

Disconnecting means
Requirements over 600 volts

Grounding

In addition, the NEC provides for the special considerations necessary for circuits supplying sealed (hermetic-type) motor-compressors. The NEC also gives attention to air conditioning and/or refrigerating equipment incorporating a sealed (hermetic-type) motor-compressor supplied from an individual branch circuit. Included among the recommendations covered in detail under *Article 440—Air Conditioning and Refrigerating Equipment* are the following subjects:

Table 1. Motor HP and Full-Load Currents for Single-Phase Motors *

hp	Full-Load Current, In Amperes	
	115 V	230 V
1/6	4.4	2.2
1/4	5.8	2.9
1/3	7.2	3.6
1/2	9.8	4.9
3/4	13.8	6.9
1	16	8
1 1/2	20	10
2	24	12
3	34	17
5	56	28
7 1/2	80	40
10	100	50

NOTE: To obtain full-load currents of 208- and 200-volt motors, increase corresponding 230-volt, full-load currents by 10 and 15 percent, respectively.

*From Table 430-148, 1978 National Electrical Code, National Fire Protection Association, Boston, Mass. Reproduced with permission.

---

* Editor's Note: For safe practice in design and installation, it is recommended that provisions of either local electrical codes or the NEC be followed—whichever are more strict.

Marking on sealed (hermetic-type) motor com-pressors and equipment

Marking on controllers

Means for disconnecting

Branch-circuit short-circuit and ground-fault protection

Sizing of conductors

Controllers for motor-compressors

Motor-compressor and branch-circuit over-load protection.

NEC tables list full-load currents, versus horse-power for non-hermetic a-c motors and for d-c motors. These tables are used to determine the ampacity of conductors and the rating of switches and branch-circuit overcurrent devices. How-ever, separate motor-running overcurrent (over-load) protection is based on the motor nameplate current rating. The full-load currents listed in

the tables are not necessarily identical with the full-load currents marked on various manufac-turers' motor nameplates for corresponding horse-power ratings. Nevertheless, the table information can be taken as representative for a-c motors that are running at usual speeds, with normal torque characteristics, and for d-c motors run-ning at a base speed. The following identifies the NEC tables repeated in this section:

Table 1: Full Load Current: Single-Phase A-C Motors

Table 2: Full Load Current: Three-Phase A-C Motors

Table 3: Full Load Current: Two-Phase A-C Motors, (4-wire)

Table 4: Full Load Current: D-C Motors.

The NEC also incorporates a locked-rotor cur-rent conversion table relating to the horsepower

Table 2. Motor HP and Full-Load Currents for Three-Phase Motors **

	Full-Load Current In Amperes								
	Induction Type Squirrel-Cage and Wound Rotor					Synchronous Type Unity Power Factor*			
hp	115 V	230 V	460 V	575 V	2300 V	230 V	460 V	575 V	2300 V
1/2	4	2	1	.8					
3/4	5.6	2.8	1.4	1.1					
1	7.2	3.6	1.8	1.4					
1 1/2	10.4	5.2	2.6	2.1					
2	13.6	6.8	3.4	2.7					
3		9.6	4.8	3.9					
5		15.2	7.6	6.1					
7 1/2		22	11	9					
10		28	14	11					
15		42	21	17					
20		54	27	22					
25		68	34	27		53	26	21	
30		80	40	32		63	32	26	
40		104	52	41		83	41	33	
50		130	65	52		104	52	42	
60		154	77	62	16	123	61	49	12
75		192	96	77	20	155	78	62	15
100		248	124	99	26	202	101	81	20
125		312	156	125	31	53	126	101	25
150		360	180	144	37	302	151	121	30
200		480	240	192	49	400	201	161	40

NOTE: For full-load currents of 208- and 200-volt motors, increase the corresponding 230-volt motor full-load current by 10 and 15 percent, respectively.

*For 90 and 80 percent power factor, the above figures shall be multiplied by 1.1 and 1.25, respectively.

**From Table 430-150, 1978 National Electric Code, National Fire Protection Association, Boston, Mass. Reproduced with permission.

Table 3. Motor HP and Full-Load Currents for Two-Phase (4-wire) Motors **

| | Full-Load Current In Amperes | | | | | | | | |
| | Induction Type Squirrel-Cage and Wound Rotor | | | | | Synchronous Type Unity Power Factor* | | | |
hp	115 V	230 V	460 V	575 V	2300 V	220 V	440 V	550 V	2300 V
1/2	4	2	1	0.8					
3/4	4.8	2.4	1.2	1.0					
1	6.4	3.2	1.6	1.3					
1 1/2	9	4.5	2.3	1.8					
2	11.8	5.9	3	2.4					
3		8.3	4.2	3.3					
5		13.2	6.6	5.3					
7 1/2		19	9	8					
10		24	12	10					
15		36	18	14					
20		47	23	19					
25		59	29	24		47	24	19	
30		69	35	28		56	29	23	
40		90	45	36		75	37	31	
50		113	56	45		94	47	38	
60		133	67	53	14	111	56	44	11
75		166	83	66	18	140	70	57	13
100		218	109	87	23	182	93	74	17
125		270	135	108	28	228	114	93	22
150		312	156	125	32		137	110	26
200		416	208	167	43		182	145	35

*For 90 and 80 percent power factor, the above figures should be multiplied by 1.1 and 1.25, respectively.

**From Table 430-149, 1978 National Electrical Code, National Fire Protection Association, Boston, Mass. Reproduced with permission.

and voltage rating of typical motors. The locked rotor values in the table are used in determining ratings of some disconnect devices and controllers for sealed (hermetic-type) motor-compressors. The table values are representative and do not necessarily agree exactly with manufacturers' ac-

Table 4. Full-Load Current for D-C Motors (Amperes)

| | Armature Voltage Rating* | | | | | |
hp	90 V	120 V	180 V	240 V	500 V	550 V
1/4	4.0	3.1	2.0	1.6		
1/3	5.2	4.1	2.6	2.0		
1/2	6.8	5.4	3.4	2.7		
3/4	9.6	7.6	4.8	3.8		
1	12.2	9.5	6.1	4.7		
1 1/2		13.2	8.3	6.6		
2		17	10.8	8.5		
3		25	16	12.2		
5		40	27	20		
7 1/2		58		29	13.6	12.2
10		76		38	18	16
15				55	27	24
20				72	34	31
25				89	43	38
30				106	51	46
40				140	67	61
50				173	83	75
60				206	99	90
75				255	123	111
100				341	164	148
125				425	205	185
150				506	246	222
200				675	330	294

NOTE: The above values of full-load currents are for motors running at base speed.

*These are average direct-current quantities.

**From Table 430-147, 1978 National Electrical Code, National Fire Protection Association, Boston, Mass. Reproduced with permission.

tually locked rotor currents. *Table 5* gives the NEC locked-rotor conversion table. The listed values are applicable to full-voltage (across-the-line) starting, and are useful for relating to various schemes for reduced-voltage and part-winding starting. *Table 2* (in previous section), derived from NEMA, also can be used for this purpose.

Table 5. Locked-Rotor Current Conversions *

Max hp Rating	Single Phase		Two or Three Phase				
	115 V	230 V	115 V	200 V	230 V	460 V	575 V
1/2	58.8	29.4	24	14	12	6	4.8
3/4	82.8	41.4	33.6	19	16.8	8.4	6.6
1	96	48	42	24	21	10.8	8.4
1 1/2	120	60	60	34	30	15	12
2	144	72	78	45	39	19.8	15.6
3	204	102	...	62	54	27	24
5	336	168	...	103	90	45	36
7 1/2	480	240	...	152	132	66	54
10	600	300	...	186	162	84	66
15	...	...	...	276	240	120	96
20	...	...	...	359	312	156	126
25	...	...	...	442	384	192	156
30	...	...	...	538	468	234	186
40	...	...	...	718	624	312	246
50	...	...	...	862	750	378	300
60	...	...	...	1035	900	450	360
75	...	...	...	1276	1110	558	444
100	...	...	...	1697	1476	738	588
125	...	...	...	2139	1860	930	744
150	...	...	...	2484	2160	1080	864
200	...	...	...	3312	2880	1440	1152

Heading: Motor Locked-Rotor Current Amperes

*From Table 430-151, 1978 National Electrical Code, National Fire Protection Association, Boston, Mass. Reproduced with permission.

MOTOR AND LOAD DYNAMICS, AND MOTOR HEATING

Torque-Speed Relationships for a Typical Three-Phase Induction Motor

Figure 1(a) shows the torque-speed curve and Fig. 1(b) the current-speed curve of a 100 hp, 1760 rpm, four-pole, three-phase, NEMA Design B induction motor. It can be seen from Fig. 1(a) that the locked-rotor torque (torque at startup when voltage is applied and the motor has not yet begun to rotate) is 125 percent of the full load torque. As the motor accelerates and picks up speed, the torque increases to a maximum value known as the *breakdown* torque, which for this motor is 200 percent of the full load torque while occurring at about 80 percent of the full load speed.

As the motor continues its acceleration, the torque drops sharply until the motor torque is in balance with the load torque at approximately 98 percent of synchronous speed. Therefore, if the driven load is the same as the motor rating, in this case 100 hp, the motor speed will balance out at the rated speed of 1760 rpm, which is a revolutions slip of 40 from the 1800 rpm synchronous motor speed. If the driven load is less than 100 hp, the revolutions slip will be less than 40; however, even under a no-load condition, the motor speed will not reach the synchronous speed due to friction and windage losses.

In Fig. 1(b) the current at startup (locked-rotor current) is almost 600% of the full load

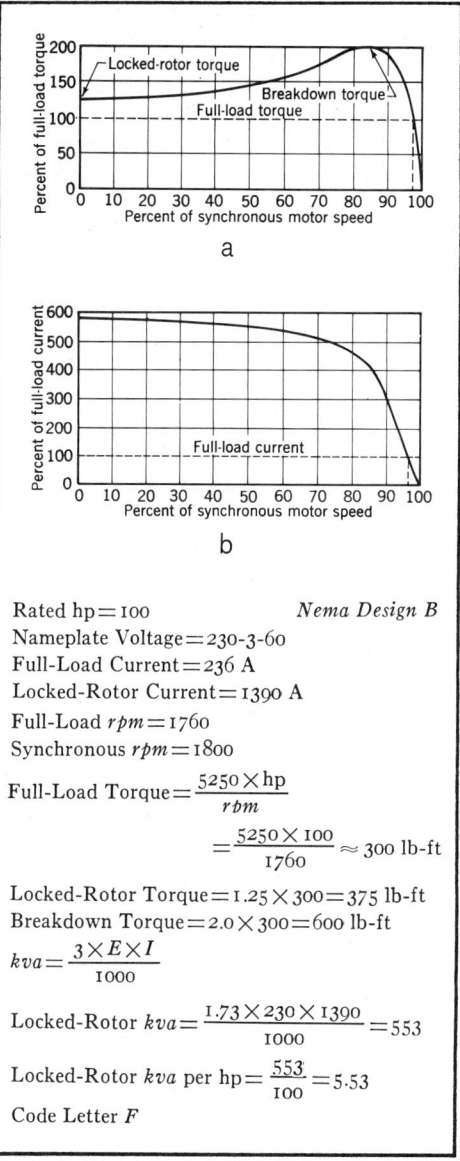

Rated hp = 100   *Nema Design B*

Nameplate Voltage = 230-3-60

Full-Load Current = 236 A

Locked-Rotor Current = 1390 A

Full-Load *rpm* = 1760

Synchronous *rpm* = 1800

$$\text{Full-Load Torque} = \frac{5250 \times \text{hp}}{rpm}$$

$$= \frac{5250 \times 100}{1760} \approx 300 \text{ lb-ft}$$

Locked-Rotor Torque = 1.25 × 300 = 375 lb-ft

Breakdown Torque = 2.0 × 300 = 600 lb-ft

$$kva = \frac{3 \times E \times I}{1000}$$

$$\text{Locked-Rotor } kva = \frac{1.73 \times 230 \times 1390}{1000} = 553$$

$$\text{Locked-Rotor } kva \text{ per hp} = \frac{553}{100} = 5.53$$

Code Letter *F*

Fig. 1. (a) Torque-speed curve and (b) current-speed curve for motor as defined.

point where the motor torque is shown in balance with the load current at approximately 96 percent synchronous speed.

Three-phase induction motors have five NEMA design classes which provide variations in motor characteristics. These variations are obtained by differences in rotor design. The three most commonly used classes are NEMA Design B, NEMA Design C, and NEMA Design D. Figure 2 compares the torque-speed and current-speed curves of a 100 hp motor for the three designs.

Torque-speed relationships of driven loads normally fall within three general categories: constant horsepower, constant torque and variable torque. These are represented by the curves in Fig. 3, the solid lines representing theoretical relationships. For example, Curve A for theoretical

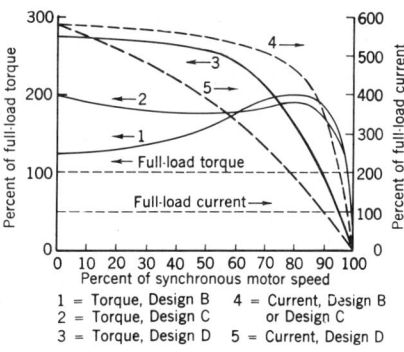

Fig. 2. Torque-current relationships for 100 hp, 1800 rpm motor; locked rotor kva/hp-5.52; code letter F.

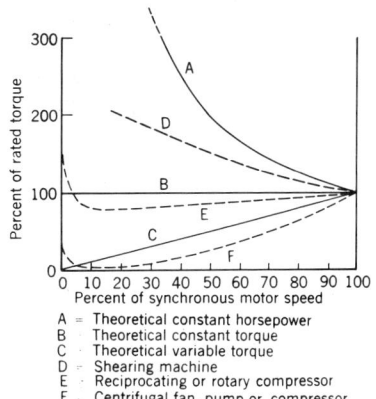

Fig. 3. Load-torque relationships for motor of Fig. 2.

current. As the motor accelerates towards full load speed, there is no significant decrease in the current drawn by the motor until the motor speed is roughly 85 percent of the full load speed. At this point the motor current drops off rapidly as the motor speed continues to increase to the

constant horsepower is derived by making the torque multiplied by the speed equal to a constant. Curve B, by definition, shows a theoretical constant torque at all speeds. Curve C for theoretical variable torque is derived by letting torque be proportional to speed.

This variable-torque relationship is equivalent to the horsepower varying as the square of the speed.

Figure 3 also shows some of the many actually driven loads that approach the theoretical curves. Actual curves, however, are seldom coincident with theoretical curves. For instance, the curve for a shearing machine (similar to that of a lathe) approaches the configuration of the theoretical constant horsepower curve. The curve as shown for a reciprocating or rotary compressor, under constant suction and discharge pressures and without unloaders can be considered close to the configuration of the constant torque curve. In this curve there is a sudden sharp increase in load torque at zero speed to "breakaway" torque, and then a drop-off. The gradual increase in torque thereafter, which is proportional to speed, results from internal friction in the compressor. The curve for a centrifugal fan, pump, or compressor approaches the theoretical variable torque curve. In this actual curve, the torque varies almost as the square of the speed (horsepower varies as the cube of the speed); however, in the theoretical curve the torque is proportional to the speed.

Torque, Inertia, and Acceleration Time

Starting from rest, a motor will accelerate in speed as long as its torque output is greater than the resisting torque of the driven load. Eventually, as the motor continues to pick up speed, its generated torque will decrease. At some point, the motor torque will exactly equal the resisting load torque, and acceleration will cease.

The time to accelerate through the entire starting period is a function of: (1) the difference between the motor output torque and the load torque; and (2) the inertia of the motor-load system. The difference between generated-motor and resisting-load torques through the entire starting period can be analyzed by plotting in graphical form the motor and load torques in relation to motor speed. Manufacturers' data will usually supply the inertia of both a motor and a driven machine.

Torque normally is used in calculations in engineering units of lb-ft. When dealing with rotating bodies, the moment of inertia is frequently referred to as the "flywheel effect." Mathematically it is expressed as Wk^2, in units of pound-feet squared; where W is the weight of the rotating body, and k is its radius of gyration with respect to its rotational axis. Table 1 gives the definition and units of various engineering terms used in motor-load system calculations.

Table 2 illustrates the derivation of the formula for motor acceleration time and its application for a graphical solution. Referring to the graph in Table 2, the torques of the motor and load are plotted against motor speed. The graph then is divided into numerous finite vertical increments in which the average $T_R - T_L$ as well as the $rpm_2 - rpm_1$ of each increment can be measured. The time to accelerate is seconds within each increment is given by the formula: $\Delta t = Wk^2 (rpm_2 - rpm_1)/308(T_R - T_L)$, where, in this instance, Wk^2 is the inertia of the motor plus the load referenced to the motor speed. The total time to accelerate is the sum of all the incremental time periods.

As can be seen from the formula for acceleration time, a large flywheel effect results in a relatively long acceleration period. Centrifugal compressors and large fans are characteristic of loads having large inertias. Acceleration periods approaching one minute for large centrifugal compressors are not uncommon. On the other hand, reciprocating compressors have low inertias. Thus, their acceleration periods are usually of short duration.

Frequently the load speed differs from the motor speed. For example, a V-belt driven fan or compressor usually rotates at a lower speed than the motor which drives it. Also, a centrifugal compressor driven through a gear assembly generally rotates at a higer speed than the motor. In making motor-load system torque and inertia calculations, it is necessary to relate the torque and inertia values to either the motor or the load speed. Industrial practice usually relates these values to the motor speed. By so doing, the load torque vs. speed curve can be shown on the same chart as the motor torque vs. speed curve.

Table 3 lists the mathematical relations for referencing the various torques and inertias to motor and load speeds in situations where the load and motor speeds differ. For example, if a centrifugal compressor is geared to be driven

Table 1. Dynamics of the Motor and the Load

$g=$ Acceleration Due to Gravity, 32.2 ft/sec^2

$k=$ Radius of Gyration, ft

$J_R=$ Polar Moment of Inertia of Motor, lb-ft sec^2

$J_L=$ Polar Moment of Inertia of Load, lb-ft sec^2

$J=$ Polar Moment of Inertia of Motor Plus Load, Referenced to Motor Speed, lb-ft sec^2

$RPM_R=$ Revolutions Per Minute of Motor

$RPM_L=$ Revolutions Per Minute of Load

$RPM_S=$ Revolutions Per Minute of Motor at Synchronous speed

RPM_2-

$RPM_1=$ Revolutions Per Minute Increment of Speed, Referenced to Motor Speed

$RPM=$ Revolutions Per Minute of Motor at Final Speed After Acceleration

$SLIP = (RPM_S - RPM_R)/RPM_S$ Revolution

Slip $= RPM_S - RPM_R$

$T_R=$ Motor Torque, lb-ft

$T_L=$ Load Torque, lb-ft

$T_R - T_L=$ Net Accelerating Torque, Referenced to Motor Speed, lb-ft

$t=$ Accelerating Time, sec

$W_R=$ Weight of Rotating Motor, lb

$W_L=$ Weight of Rotating Load, lb

$W=$ Weight of Rotating Motor and Load, lb

$W_R k_R^2=$ Flywheel Effect of Motor, lb-ft^2

$W_L k_L^2=$ Flywheel Effect of Load, lb-ft^2

$Wk^2=$ Flywheel Effect of Motor Plus Load, Referenced to Motor Speed, lb-ft^2

$Z=$ Distance From the Axis of Rotation to Center of Gravity of an Element, ft

$\omega_R=$ Rotating Speed of Motor, radians per sec

$\omega_L=$ Rotating speed of Load, radians per sec

$\omega=$ Rotating Speed of Motor and Load When Both Rotate at Same Speed, radians per sec

Table 2. Dynamics of the Motor and Load

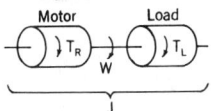

Equivalent Motor-Load System:

$$J = \int Z^2 \frac{dW}{g} = \frac{W}{g} k^2$$

$$T_R = T_L + J \frac{d\omega}{dt}$$

Steady State: $\dfrac{d\omega}{dt} = 0$, $T_R = T_L$

Acceleration Time:

$$dt = J \frac{d\omega}{T_R - T_L} = J \left(\frac{1}{T_R - T_L} \right) d\omega$$

$$t = J \int_{\omega_1}^{\omega_2} \frac{d\omega}{T_R - T_L}$$

Graphical Solution of Acceleration Time:

$$\Delta t = J \left(\frac{1}{T_R - T_L} \right) \Delta \omega$$

$$= \frac{Wk^2}{32.2} \left(\frac{1}{T_R - T_L} \right) \frac{(RPM_2 - RPM_1)\, 2\pi}{60}$$

$$= \frac{Wk^2\, (RPM_2 - RPM_1)}{308\, (T_R - T_L)}$$

$$\boxed{t = \Sigma \frac{Wk^2\, (RPM_2 - RPM_1)}{308\, (T_R - T_L)}}$$

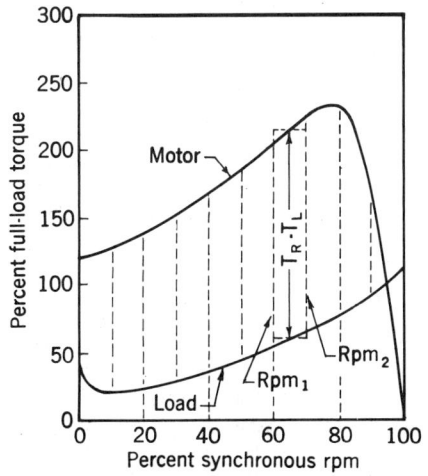

Table 3. Dynamics of the Motor and the Load

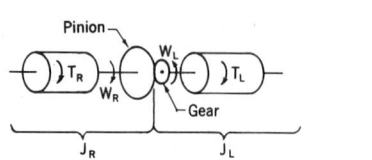

LOAD SPEED DIFFERS FROM MOTOR SPEED

Torque or Inertia	Reference to Motor Speed, ω_R	Reference to Load Speed, ω_L
Motor Torque	T_R	—
Load Torque	$\dfrac{\omega_L}{\omega_R} \times T_L$	T_L
Motor and Pinion Inertia	J_R	—
Load and Gear Inertia	$\left(\dfrac{\omega_L}{\omega_R}\right)^2 \times J_L$	J_L
Motor and Pinion Flywheel Inertia	$W_R k_R^2$	—
Load and Gear Flywheel Inertia	$\left(\dfrac{\omega_L}{\omega_R}\right)^2 \times W_L k_L^2$	$W_L k_L^2$

NOTE: $\dfrac{\omega_L}{\omega_R} = \dfrac{RPM_L}{RPM_R}$

at twice the motor speed, the equivalent inertia (Table 3, line 4) is given by $\left(\dfrac{W_L}{W_R}\right) J_L$. Thus, the equivalent inertia is $(2)^2 \times J_L$, or 4 times the compressor inertia. From this, it can be seen that a small, high-speed centrifugal compressor may have an equivalent inertia which is as great as a large, centrifugal compressor of equal capacity but slower speed. Both may have equally prolonged acceleration periods.

Motor Heating and Motor Life

Table 4 derives mathematically the energy absorbed by the rotor during an increment of acceleration time. It also denotes how the total energy absorbed by the rotor during the acceleration period can be derived graphically with the same motor vs. speed and load vs. speed curves

as used previously for determining the acceleration time.

It is obvious that the energy absorbed by the rotor is greater when the acceleration time increases. Also, a motor will heat up more during starting when it is applied to a load having a greater inertia (Wk^2) and, to some extent, a greater load torque during the starting period.

Dissipation of heat to the surrounding air is limited during starting, thus most of the heat absorbed by the rotor produces a temperature rise in the iron and copper of the motor itself. The newer design motors on the market contain less iron than previous designs and consequently have less thermal storage capacity. When a motor-load system has a prolonged acceleration period, consultation with the motor manufacturer is suggested to determine that motor overheating will not occur.

Frequent starting of a motor-load system also can cause motor overheating, since there may not be sufficient running time between starts to dissipate the heat absorbed during each starting period. And again, this problem becomes more severe when the acceleration periods are prolonged. Good practice is to obtain maximum recommended starts per hour, or starts per day, from the motor manufacturer when frequent starting of a high inertia load is required.

A special, severe case of motor overheating can occur under "locked rotor" conditions. This takes place when the locked-rotor torque of the motor is less than the breakaway torque of the load. In this situation the motor cannot begin to rotate, and the slip remains at 100 percent.

The newer design, open, drip-proof motors have theoretical lives of roughly, say, 20,000 hours when run at rated loads in an ambient temperature of 40°C (104°F). When run at a 1.10 service factor, or a 10 percent overload, the theoretical life is cut to 10,000 hours. When run at 90 percent rated load, the theoretical life increases to 40,000 hours. The decrease in life span results from a high temperature rise within the motor which decreases insulation life.

The most common cause for motor overheating is overloading. Referring to the torque-speed curve of Fig. 1(a), a motor will continue to function without an appreciable increase in slip if the torque or load imposed on the motor is increased almost to the point of double the rated motor load. The current drawn by the motor is almost proportional to the increase in load to

Table 4. Rotor Heating During Starting

Slip is equal to the ratio of the I^2R losses in the rotor to the total non-reactive power delivered to the rotor; therefore I^2R losses at any instant are equal to the non-reactive power multiplied by the slip. Energy absorbed by the rotor during an increment of time is a function of I^2R losses multiplied by the increment of time. Total energy absorbed by the rotor during acceleration is the summation of energy absorption for all increments of time.

For Each Time Increment (Graphical Solution)

Effective power delivered to motor, KW

$$= \frac{T_R \times 2\pi \times RPM_S}{33000} \times .746$$

$$= RPM_S \times \frac{T_R \times 2\pi \times .746}{33000}$$

I^2R rotor loss $=$ power slip loss $=$ slip $\times$ effective power delivered to motor, KW

$$= \frac{RPM_S - \left(\dfrac{RPM_2 + RPM_1}{2}\right)}{RPM_S}$$

$$\times RPM_S \times \frac{T_R \times 2\pi \times .746}{33000}$$

$$= \left[RPM_S - \left(\frac{RPM_2 + RPM_1}{2}\right)\right] \times \frac{T_R \times 2\pi \times .746}{33000}$$

$$\text{where } \left[RPM_S - \left(\frac{RPM_2 + RPM_1}{2}\right)\right]$$

$$= \text{Revolutions slip}$$

Energy absorbed by rotor, KW — seconds

$$= I^2R \text{ Rotor loss} \times \Delta t$$

$$= \left[RPM_S - \left(\frac{RPM_2 + RPM_1}{2}\right)\right]\frac{T_R \times 2\pi \times .746}{33000}$$

$$\times \frac{Wk^2 \,(RPM_2 - RPM_1)}{308 \,(T_R - T_L)}$$

$$= \frac{1}{2.16 \times 10^6}\left[RPM_S - \left(\frac{RPM_2 + RPM_1}{2}\right)\right]$$

$$\times \frac{T_R \times Wk^2 \,(RPM_2 - RPM_1)}{T_R - T_L}$$

Total Energy Absorbed by Rotor During Acceleration, KW — seconds

$$\begin{array}{l}\text{Graphical} \\ \text{Solution}\end{array} = \Sigma \frac{1}{2.16 \times 10^6}\left[RPM_S - \right.$$

$$\left.\left(\frac{RPM_2 + RPM_1}{2}\right)\right]\frac{T_R \times Wk^2 \,(RPM_2 - RPM_1)}{T_R - T_L}$$

Portion of total energy absorbed by rotor for accelerating the rotating mass (load torque neglected)

Kinetic energy, Ft-lb $= \frac{1}{2} JW^2$

Kinetic energy, KW — seconds $= \frac{1}{2} JW^2 \times \dfrac{.746}{550}$

$$= \frac{1}{2} \times \frac{Wk^2}{32.2} \times \frac{(2\pi \times RPM)^2}{3600} \times \frac{.746}{550}$$

$$= \frac{.231 \times Wk^2 \times (RPM)^2}{10^6}$$

---

about 150 percent of rated load where it then increases at a faster rate. The increased current in the stator winding and the increased rotor losses at higher loads raise the motor temperature and reduce motor life.

For a constant load, a reduction in voltage applied to the motor terminals causes an inversely proportional increase in the current drawn by the stator. The increased current, if significant, can cause an excessive temperature rise in the motor. On the other hand, an increase in voltage does not necessarily decrease the total motor current. If the flux density in the motor core has approached saturation, an increase in voltage might require an increase in the magnetizing component of the motor current which is greater than the reduction in the load or effective component of the motor current. The increased magnetizing current not only reduces the power factor but can

cause heating due to additional resistance losses in the stator winding. The increase in flux density causes a greater loss in the motor core contributing to an increase in motor heating. Excessive voltage also can contribute to the breakdown of motor winding insulation. Motor manufacturers generally will not warrant their equipment if the voltage applied to the motor terminals differs by more than 10 percent from the nameplate value.

Other causes of decreased motor life are dust, dirt, chemicals, and oil or grease which can be deposited on motor windings and in motor bearings when the motor is operated in a dirty environment. The erosion of insulation, the decrease in heat dissipation capability, and the contamination of bearings also contribute to motor destruction. Totally enclosed fan cooled (TEFC) or explosion-proof motors generally are recommended for dirty environments.

Either a lack of effective air movement, or high temperatures in spaces surrounding working motors, also can cause excessive temperatures within motors and can reduce motor life.

SINGLE-PHASE MOTORS

If one lead of a three-phase (polyphase) motor is disconnected from the line while the motor is running, it will continue to operate on the phase that remains connected providing the connected load is not too great. However, the speed will be reduced. In this case rotation is maintained by the pulsating field produced by the single phase that is connected to the line and by the field set up by the rotating rotor. A polyphase motor at rest will not start if only two of its leads are connected to the line since the single-phase pulsating field does not rotate and hence is unable to initiate rotation of the rotor.

A single-phase induction motor, then, must have some auxiliary means of providing a rotating field for starting. The most common method of providing this in single-phase induction motors is to add an auxiliary (starting) winding displaced from the main winding in both time and space. This second winding is generally separated in space by 90 electrical degrees from the main winding. Rotation is produced by creating an electrical phase displacement between the currents in the two windings. If the starting winding remained in the circuit under normal load conditions, losses would be excessive, the motor would overheat, and the starting winding would burn out. For this reason the starting winding is disconnected at approximately 75% of the motor's rated speed. This means that some disconnecting means also must be provided in each motor.

Types of Motors

Split-Phase. The split-phase motor is a single-phase induction motor having an auxiliary starting winding that has a high ratio of resistance to reactance as compared to the main winding connected in parallel with it. See Table 1. In this type of motor, the auxiliary or starting winding is disconnected either by a centrifugally-operated starting switch or by a relay when the motor reaches about three quarters of rated speed. These motors have from low to medium starting torque depending on their rating. For ratings above 1/6 hp the locked-rotor or starting torque is kept low because of locked-rotor current limitations, which will be dis-

cussed later. Split-phase motors have a relatively low cost and are simple in construction compared with most other single-phase motors of the same ratings.

Some split-phase motors, for use on infrequently-started devices, are provided with higher starting torques that can be obtained with the normal starting currents for this type of motor. In these cases the starting current is above the NEMA value for single-phase fractional horsepower motors as shown in Tables 3 and 4.

There are various methods of obtaining two-speed operation of split-phase motors. One common means is shown in Table 1.

When the split-phase motor drives through a belt, it is frequently provided with built-in thermal protection to prevent damage from overloading due to a tight belt or incorrect pulley ratio.

Capacitor. The capacitor motor, a modification of the split-phase type, is an induction motor whose auxiliary winding is connected in series with a capacitor. In small sizes, it is generally built in two types—the capacitor-start type and the permanent-split capacitor type. In larger sizes, it may combine the features of both types and is called a capacitor-start and run motor; a capacitor-start-capacitor-run motor, or a two-value capacitor motor.

The principles of operation and the construction of the capacitor-start motor are similar to the split-phase motor except for the addition of the electrolytic capacitor. The capacitor-start fractional horsepower motor is manufactured in ratings of from 1/6 to 3/4 hp. It is a general-purpose motor suitable for most applications.

The permanent-split capacitor motor uses a continuous-duty oil-filled capacitor during both the starting and running periods. This type of motor has low starting current and torque, and the main and auxiliary windings are in the circuit at all times. No centrifugal starting switch or relay is used. The speed control is obtained by impressing different voltages on the main winding by means of a transformer or a reactor (choke coil or autotransformer), or by using a main winding

which is tapped in a manner that reduces the amount of the main winding for higher speed operation.

The capacitor start and run motor generally uses an intermittent-duty electrolytic capacitor during the starting period and a continuous-duty oil-filled capacitor during both the starting and running periods. Advantages of the running capacitor are that it can improve the power factor of the motor and can increase the value of the breakdown torque.

For ratings of ⅓ hp, and above, many capacitor-start and capacitor-start and run motors are manufactured for dual-voltage operation. In the case of a 115/230-volt capacitor-start motor, the main winding is made in two sections. The sections are connected in parallel for 115-volt operation and in series for 230-volt operation. The auxiliary winding and the capacitor connect across one section of the main winding. Wiring connections for operation at the desired voltage can be made at the motor terminal board following instructions supplied by the motor manufacturer.

Shaded-Pole. Another type of single-phase motor is the shaded-pole motor. The starting torque of a shaded-pole motor is provided by an auxiliary, permanently short-circuited winding or windings placed around a portion of the main pole. This winding is frequently a copper strap wrapped around a portion of the pole. The flux through the shaded section of the pole lags behind the flux in the rest of the pole, thus producing the effect of a rotating field and providing a starting force. The starting current and torque of this type motor are very low compared with other types of single-phase motors. The shaded-pole motor is rugged and simple in construction. Its low efficiency, due to losses produced by the shading coils, has limited its application to the smaller ratings.

Repulsion-Start. Although repulsion-start induction motors are in use, they have been supplanted largely by the capacitor-start motor. The repulsion-start induction motor starts on the repulsion principle. Single-phase power is applied to the stator winding. This induces current flow in the rotor winding, which is similar to that of a d-c commutator motor. The brush axis is displaced several degrees from the line of flux flow. This permits an unbalanced current flow in the two halves of the rotor and the resultant torque starts the motor. As it approaches full speed, the armature coils are short circuited by a centrifugal mechanism, and the motor

then operates on the induction principle with the same characteristics as a split-phase or capacitor-start motor. In some motors the brushes are lifted when the armature windings are short circuited in order to reduce brush wear and noise. The advantage of the repulsion-start induction motor is extra high starting torque with low starting current. It has the disadvantages of a wound armature, brushes, armature short-circuiting mechanism, brush-lifting mechanism and their attendant maintenance and increased cost.

This type of motor is characterized by a very high starting torque with relatively low starting current. However, the torque of the motor decays rapidly as it comes up to speed, and even the best designs tend to be deficient in torque at the switching speed. To overcome this, it is common to build the motors with an excess of starting torque, sometimes as high as 500%, so that the torque at switching speed is high enough, say 200%, to insure successful operation.

Repulsion-Induction. The term "repulsion-induction" is applied to a few types of motors in which the rotor carries a wire-wound repulsion winding, with its commutator, and a squirrel-cage winding. The motor gets its starting ability from the repulsion winding and its running torque from the combination of the two rotor windings. The change from the starting to the running condition is made electrically and magnetically so that there is no abrupt change in characteristics.

Commercial motors of this type have a relatively high starting torque, in the 250-325 percent range, and low starting current. Efficiency is competitive with all the other types and the power factor is extremely high. Because of the smooth transfer from the starting to the running condition, this type of motor excels in severe starting and accelerating duty, especially at low voltage.

Large Single-Phase Motors

Large single-phase motors vary in size from ¾ hp through 5 hp, and above. The three general types of large single-phase motors frequently used are the capacitor motor (either capacitor-start or capacitor-start and run), the repulsion-start induction-run motor, and the repulsion-induction motor. Each manufacturer has his own variation of the circuits and the mechanical devices used in the motor.

No practical single-phase motor has yet been

Table 1. Application Data for Fractional

Type of Motor			Wiring Diagram	Horse-power Range	Speed Data		
					Rated Speed	Speed Characteristics	Speed Control
					Alternating Current		
Split Phase	Normal Purpose Low Locked rotor amperes			1/20 to 1/3	1725 1140 860	Constant	None
	High Starting Current			1/4 and 1/3	1725	Constant	None
	Two-speed (Two windings)			1/8 to 1/4	1725/1140 1725/860	Two-speed	1-Pole, double-throw switch
Capac-itor	General Purpose Capacitor-start			1/6 to 3/4	1725 1140 860	Constant	None
	Two-speed Capacitor-start (Two windings)			1/6 to 3/4	1725/1140 1725/860	Two-speed	1-Pole, double-throw switch
	Permanent Split (Single value)			1/20 to 3/4	1625 1075 825	Constant or adjustable varying	Autotrans-former or tapped winding
	Capacitor-Start and Run			1/3 to 3/4	1725 1140 860	Constant	None
Shaded Pole				1/300 to 1/6	1550 1050 800	Constant or adjustable varying	Choke coil, autotrans-former or tapped winding

invented that has the simplicity of the polyphase, squirrel-cage induction motor, and most large single-phase motors have some type of extra device to make them start. This is either integrally mounted or supplied as an external part. Great ingenuity has been shown by designers and manu-facturers in making simple yet rugged devices.

However, the fact remains that single-phase mo-tors are more expensive and more complicated than polyphase induction motors and, therefore, should be used only when polyphase power is not available. A possible exception is in the frac-tional-horsepower field where the driven devices are small and are apt to be of less importance.

Horsepower Single-Phase Motors

Approximate Torque (4-Pole Motors)		Built-in Starting Mechanism	Application Data
Starting	Break-down		
Motors			
Medium to low	Medium	Centrif-ugal switch *	For oil burners, fans, blowers, low locked-rotor current minimizes light flicker making motor suitable for frequent starting. For applications up to 1/3 hp where medium starting and breakdown torques are sufficient.
Medium	High	Centrif-ugal switch *	Ideal for sump pumps. For continuous and intermittent duty where operation is in frequent and locked-rotor current in excess of NEMA values is not objectionable. May cause light flicker on underwired or overloaded lighting circuits.
Medium	Medium	Centrif-ugal switch *	For belted furnace blowers, attic ventilating fans, similar belted medium-torque jobs. Simplicity permits operation with any 1-pole, double-throw switch or relay. Starts well on either speed—thus used with thermostatic or other automatic control. Tight belt or incorrect pulley ratio may overload motor.
High	High	Centrif-ugal switch *	Ideal for all heavy-duty drives, such as compressors, pumps, stokers, refrigerators, air conditioning. All purpose motor for high starting torque, low starting current. Quiet, economical. High efficiency and power factor. Single voltage in 1/6, 1/4 hp. 1725 rpm ratings—dual voltage in others.
Medium	Medium	Centrif-ugal switch *	Supplements line of 2-speed split-phase motors. Used on identical applications requiring horsepower ratings from 1/3 to 3/4 hp.
Very low	Low	None	For direct connected fan drives, particularly unit heaters. Not for belt drives. Same motor adaptable for 115 or 230 volts for 1-speed, 2-speed, or multi-speed service by use of 1-pole, single-throw switch, 2-pole, double-throw switch, or speed controller, respectively. Fan load must be accurately matched to motor output for proper speed control. All rating dual voltage and dual rotation.
Medium to high	High	Centrif-ugal switch *	Same as described above for general-purpose capacitor-start.
Very low	Low	None	Constant speed, switchless motor for low-power applications. Used for fans, small blowers, unit heaters. With fan load accurately matched to motor output, proper speed control can be obtained by means of series choke, autotransformer or tapped winding.

* May use current switching relay 1/6 to 1/4 horsepower and potential switching relay 1/4 to 3/4 horsepower in lieu of centrifugal switch.

Here, the expense of three-phase wiring and switches outweighs the other consideration.

Starting torque in fractional-horsepower single-phase induction motors is proportional to several factors, including the product of the current in the two windings and the size of the angle between these currents.

In a typical split-phase motor, the current (at starting) in the main winding usually leads the current in the auxiliary winding by 20 to 30 degrees. This creates the needed revolving magnetic field during starting. The starting torque of a normal purpose split-phase motor is low, about 140% of full load torque.

In the case of a capacitor-start or capacitor-start and run motor of the same rating, the current (at starting) in the auxiliary winding leads the current in the main winding by 75 to 85 degrees. Because of this, the starting torque of these motors is comparatively high—about 275 to 375% of full load torque. From this it is evident that a capacitor-start motor can be expected to produce from two to three times as much starting torque per ampere as can a split-phase motor.

Since the capacitor-start motor is more expensive, the question might be raised as to why the starting current of the split-phase motor should not be increased to obtain the required starting torque. For example, if 25 oz-ft of starting torque is required, why not obtain it from a split-phase motor with about 40 amperes starting current instead of from a capacitor-start motor with about 15 amperes starting current? One reason is that excessive starting currents on frequent-starting applications may result in objectionable light flicker. In order to avoid this, many utilities restrict the use of single-phase motors with high starting currents.

Frequently used rules are:

(1) Automatically controlled single-phase motors for general use cannot draw starting currents greater than 20 amperes (plus 15% tolerance) at 115 volts. or more than 25 amperes (plus 15% tolerance) at 230 volts.

(2) Manually controlled single-phase motors for general use cannot draw starting currents greater than 40 amperes (plus 15% tolerance) at 115 volts, or more than 50 amperes (plus 15%) at 230 volts.

(3) Motors that draw starting currents greater than these values can be used if approved by the utility.

Tables 3 and 4 give NEMA maximum locked-rotor currents for four single-phase design letter motors; however, the locked-rotor currents of 60 cycle, single-phase general-purpose motors shall not exceed the values shown for the Design N motors. The locked-rotor torques of single-phase general-purpose induction motors, with rated voltage and frequency applied, shall not be less than those given in Section 1, Table 6.

Figures 1(a) through 1(e) show speed-torque curves for various single-phase, fractional horsepower motors. Generally speaking, capacitor-start type motors have the locked-rotor torque and locked-rotor current characteristics required for a single-phase, general-purpose motor classification.

Suggested types of applications, for which each motor is best suited, are summarized in Table 2.

Application

Nameplate Data. When selecting motors for an application, the motor must first be matched to the power supply. The motor nameplate ordinarily gives the name of the manufacturer, rating in horsepower and speed, temperature rise and duty cycle, power-supply frequency and voltage, full-load. amperes, service factor, locked kva code number, and manufacturer's identification number.

Table 2. Suitable Applications

Fig. No. and Type Motor	Suitable Applications
Fig. 1(a) Capacitor start motor	Motor has starting current within the NEMA standard and has high starting torque. It is suitable for general purpose use since it can start heavy loads without causing excessive light flicker on normal circuits.
Fig. 1(b) Permanent-split-capacitor motor.	Motor has very low starting torque. and low starting current. Value of capacitance of the oil-filled capacitor must be selected to provide satisfactory full load operation. With a fan-type load accurately matched to motor characteristics, this motor, with a controller, can provide very good multi-speed and dual-voltage operation.
Fig. 1(c) Split-phase motor.	Motor has a starting current above the NEMA standard and is suitable for frequent starting.
Fig. 1(d) Split-phase motor with high starting current.	Motor has a starting current above the NEMA standard and is suitable for infrequent starting or where the higher starting current is not objectionable.
Fig. 1(e) Shaded-pole motor.	Motor has very low starting torque. Speed when driving a fan-type load can be adjusted by means of a series choke coil or autotransformer or tapped winding. Low efficiency and power factor have limited its use to the smaller ratings.

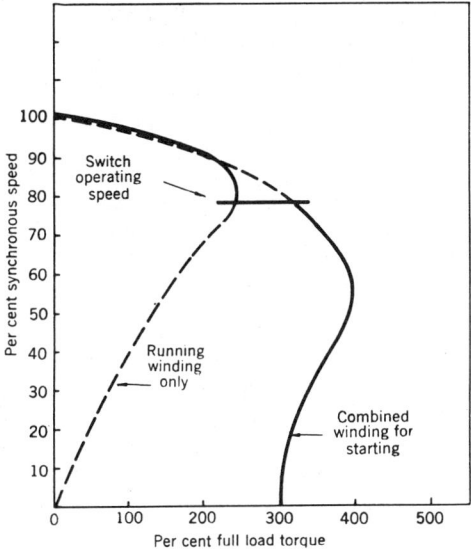

Fig. 1(a). Speed-torque curve for a capacitor-start motor.

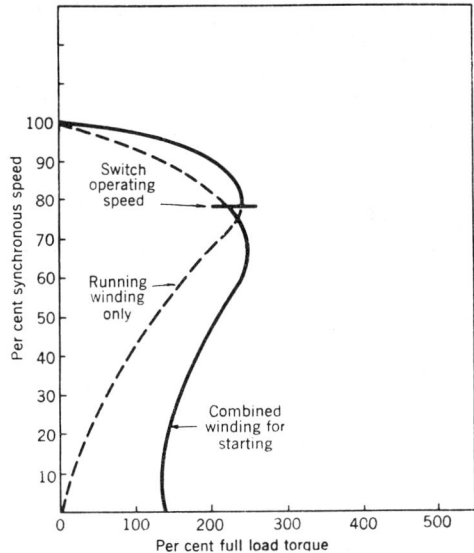

Fig. 1(c). Speed-torque curve for a normal-purpose split-phase motor.

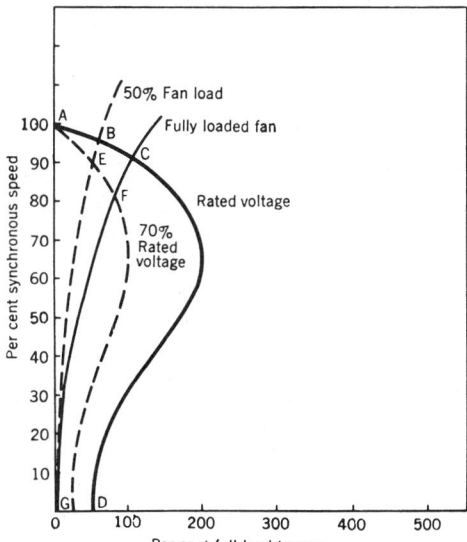

Fig. 1(b). Speed-torque curve for a permanent-split-capacitor motor.

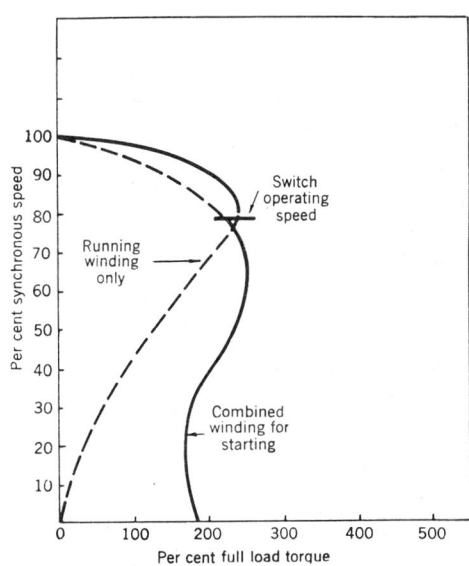

Fig. 1(d). Speed-torque curve for a split-phase motor having a high starting current.

If the motor nameplate is not accessible, the information regarding power supply should be available at the point the apparatus is connected to the line. Power supply information includes voltage,

frequency, and full-load amperes. A motor should be connected only to a power source that matches its rating.

Motors in agreement with NEMA standards can

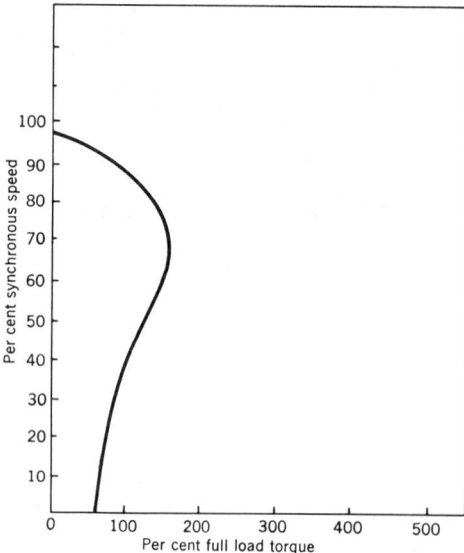

Fig. 1(e). Speed-torque curve for a shaded-pole motor.

be operated successfully at rated load with a variation of plus or minus 10% in voltage or plus or minus 5% in rated frequency, or a combined variation of 10%. However, this operation is not necessarily in accordance with standards established for operation at rated voltage and frequency.

Loading

In general, the most obvious means of checking a load on a motor is by comparing the amperes taken by the motor to that shown on the nameplate. Although this is not the best method of checking load and cannot be relied on entirely, it is usually the one most available and for that reason it is quite often used. Particularly with regard to single-phase motors, the full-load amperes may be only 40 to 50 percent greater than the no-load amperes and in some motors may actually be no greater. Furthermore, variations in applied voltage and in motor construction will have considerable effect on the full-load amperes for single-phase motors.

If it is necessary to calibrate accurately the load on a motor, a much better criterion is to determine the watts taken by the motor. The watts supplied to the motor when driving the required load together with the voltage at which the test is run, with any

duty cycle variations, should be submitted to the manufacturer with the nameplate identification of the motor. With this information the manufacturer can usually give the actual load which the motor is driving. In connection with this, it is usually helpful to submit also the watts input with motor running without load; this permits the manufacturer to estimate the load more accurately. The variation in watt-meter reading between no load and full load normally is from 300 to 700% depending on size of motor.

When applying motors by matching the full-load amperes, it must be remembered that a 15% change in current may mean a 25 to 35% change in load; and when one loads a motor to 115% of its full-load amperes, the motor may be considerably overloaded.

Multi-Speed Operation. An induction motor is used for speed control only in connection with a fan or blower. There are two methods of obtaining this speed control in a single-phase motor. The first method is by changing the number of poles using two different windings—one for the high speed, one for the low. The windings are arranged so that if one connection is used, it will operate at the speed of a four-pole motor, and if another connection is used, it will operate at the speed of six- or eight-pole motor. With this arrangement slight variations in voltage or fan load do not greatly change the operating speeds. Another method of speed control is by varying the voltage to either a shaded-pole or permanent-split capacitor motor or by utilizing a tapped winding. This method should be applied only to direct-connected fans or blowers.

In Fig. 1(b) the solid line $ABCD$ represents the torque produced by the motor at various speeds at rated voltage. The fan or blower curve represents the torque required to drive the fan at each speed. When the output torque of the motor at a given speed equals the torque required to drive the fan at this speed, the motor will operate at the speed indicated by point C. If the voltage to the motor is reduced, the output torque of the motor will be along line $AEFG$. Under these conditions the motor will operate at the speed of point F. This gives speed control by varying the voltage to the motor. This works most successfully only with fan-type loads in which torque required increases with speed and on motors having no starting switch. Any voltage setting can be used to obtain a desired speed. However, it must be realized that with low-voltage settings the starting torque is reduced so that the motor may not start. For this reason, this method

of speed control is applied only to direct connected fans and blowers as the friction in a belt may prevent the motor from starting on low speeds. Also, the fan and motor must be matched to obtain proper speed control. A fan that will not properly load a motor (such as that shown in Fig. 1(b) as 50 percent fan load) will not give proper speed control. On reduced voltage, the speed would only be reduced to point E.

There are many methods for reducing the effective voltage to the motor. Among these are auto-transformers, connecting dual-voltage motors for high voltage and applying the lower voltage, or having an intermediate winding connected into the circuit to reduce effectively the voltage to the main winding.

As can be seen, variation of fan load or applied voltage will result in a variation in motor speed. Multi-speed motors ordinarily operate at a lower full-load speed with the load point nearer the breakdown or maximum torque point since this method facilitates varying the speed.

Motor Protection

Protection of motors is described in appropriate electrical codes and the National Electric Code. For example, most codes require all automatically started motors, regardless of size, to have overload protection. However, attention is directed to a particular type of motor protection called *inherent overheating protection*. This type of protection refers to a temperature or a temperature and current-sensing device that disconnects the motor from the line when the temperature of the winding becomes excessive under any abnormal condition. Since, in many instances, temperature is the reason for motor failure, an inherent device frequently offers improved protection so that short overloads or delayed starting will not cause nuisance outages unless such conditions will severely damage the motor.

In fractional horsepower single-phase motors there are two windings, the starting and the main. The main winding has considerable thermal capacity and operates whenever the motor is turned on. The starting winding has relatively little thermal capacity and is connected only during starting. The inherent protector guards the starting winding by tripping out in a short time when the current is high, such as under locked-rotor conditions. It protects the main winding by being sensitive to temperature: it removes the motor from the line when

the main winding temperature becomes excessive.

Two types of protection are in common use—the automatic-reset and manual-reset types. With the automatic-reset type, the motor cools to a safe temperature and automatically resets and attempts to start. If conditions are changed so that temperatures do not then become excessive, the motor will continue its operation. If conditions are still unsafe, it will again remove the motor from the line. This cycling will continue until the motor is shut off. If for some reason it would be hazardous for a motor to start after once tripping, a manual-reset protector is used in which the operator must manually reset the protector before the motor will again start.

Motor Selection

In selecting the type motor to use for a given application, the characteristics of the motors outlined in Table I should be considered.

Fans and Blowers. For shaft-mounted fans and blowers, little starting torque is required. However, some blowers have relatively high inertia and the motor must come up to speed quickly before the starting winding is damaged. Also, a long acceleration will draw a heavy current and cause lights to dim during acceleration.

A normal-purpose split-phase motor can be used if accelerating time is not excessive. The high-starting current, split-phase motor is the least expensive but should be used only where manual starting is used and starts are infrequent.

A two-speed, split-phase motor can be used if correct speeds can be obtained. Permanent-split capacitor or shaded-pole motor can be used if variable speed is desired. In this case, a shaded-pole motor is a low-performance motor whereas the permanent-split capacitor motor operates at higher efficiencies and power factor and lower noise levels.

Capacitor-start or capacitor-start and run motors are used when starting time must be decreased on high-inertia loads, or on larger motors where the starting torque must be greater than that produced by a normal-purpose split-phase motor.

The same remarks apply to belted fans and blowers, except that shaded-pole motors and permanent-split capacitor motors should be omitted because their inherent low starting torque is not sufficient to start a belted load under various conditions of belt tension.

Compressors. A compressor unit which does not have a sealed (hermetic) motor-compressor

may have a motor which drives a separate compressor by means of a belt. The belt-driven compressor requires a large starting torque as it may have to start under considerable back pressure with considerable friction and low line voltages. The line voltage usually drops when the motor is turned on. For these reasons the capacitor-start, capacitor-start and run, and repulsion a-c motors are used.

Sump Pumps. Sump pumps are usually operated very infrequently and are built as inexpensively as possible. Performance is of little importance and a high starting current split-phase motor will normally suffice. However, for a sump pump that starts regularly and often, a normal-purpose split-phase motor may be required because it has less tendency to produce flickering of lights.

Oil Burners. For oil burners the motor drives a blower and a pump, neither of which requires much starting torque. A normal-purpose split-phase motor is almost always used, except for the large commercial burners.

Hot-Water Pumps Hot water pumps operate at low pressures. The normal-purpose or high-starting-current split-phase motors are satisfactory, depending on frequency of starting and running time.

Analysis of Application

A general guide in selecting a motor to drive a newly-designed machine is provided in Table 1. It is helpful to understand why a certain type of equipment is usually driven by one type of motor.

The belt-connected furnace blower application, for example, can be analyzed as follows:

Power Supply:	Usually single-phase, 60-cycle, alternating current is available.
Horsepower:	1/6, 1/4, 1/3, 1/2 or 3/4 hp depending upon the design.
Torque:	Belted blower load is fairly constant. Starting torque depends upon inertia of blower and tightness of the belt.
Locked-Rotor or Starting Current:	NEMA standards or utility regulations.
Overload Protection:	Desirable to protect the motor against unusual conditions.
Mechanical Features:	Industry standard is open, resilient-mounted motor.

Assume that information on the blower used indicates that it can supply the required amount of air when operated at the desired speed with a motor output of 0.4 hp at 1725 rpm. This would indicate a load of 1.2 times the nameplate rating of a 1/3 hp motor, or 0.8 times the nameplate rating of a 1/2 hp motor. A 1/3 hp, general-purpose, alternating-current motor has a service factor of 1.35 according to NEMA standards, and theoretically this motor could be used. However, the 1/2 hp motor is recommended if the temperature in the area surrounding the motor is consistently hot.

A shaded-pole motor should not be used because of the horsepower required. The starting torque of a permanent-split capacitor motor does not make it suitable for most belt-driven loads. Therefore, there is a choice between a split-phase motor and a capacitor-start motor for this application.

The choice between these two types of motors depends upon the effort required to start the blower, and bring it up to operating speed. This will depend upon the tightness of the belt, the pulley ratio, and the inertia of the blower. If the belt is not too tight and the blower is light weight, it is probable that a split-phase motor will be satisfactory. On this application the motor used should meet the NEMA standards for starting current and thus the high-starting-current type of the split-phase motor is not indicated. If the belt tension is high or the blower is heavy, the capacitor-start motor should be used.

On this application the motor should be provided with inherent overload protection to prevent overheating because of too much belt tension, incorrect pulley size, or incorrect installation.

POLYPHASE MOTORS

Alternating current, three-phase motors are in common use for 1/2 horsepower sizes, and larger, although they are available in smaller sizes. The 60 cycle nominal utility service voltage systems commonly used with three-phase motors are:

A. 120/208-v, 3-phase, 4-wire with grounded neutral,

B. 265/460-v, 3-phase, 4-wire with grounded neutral,

C. 115/230-v, 3-phase, 4-wire with grounded neutral,

D. 230-v, 3-phase, 3-wire,

E. 460-v, 3-phase, 3-wire.

The above voltages are nominal, and actual voltages vary, usually above their nominal values. High-voltage systems (2300-v, 4160-v, and 4800-v) are frequently used in industry for motors above 200 horsepower and for air-conditioning system motors above 1000 horsepower.

Enclosure. The enclosure of a motor refers to the arrangement of the ventilating openings or other means of cooling and the amount of protection provided against the atmosphere in which the motor operates. In the heating and ventilating field, motors of the open drip-proof, totally enclosed, totally enclosed fan-cooled, and explosion-proof enclosures and motors with encapsulated windings might be used.

The *open motor* is the basic standard and its ventilating openings do not protect the motor from dirt, water, or fumes. However, the present trend is to build such motors with considerable protection and in the case of many manufacturers, the open motor meets all the requirements of the drip-proof specification and is called an open drip-proof type.

The *drip-proof type* is essentially an open motor with its ventilating openings arranged so that dripping water will not enter it. An elaboration of this type is the *splash-proof motor*, the openings of which offer further protection against splashing water. The latter usually carries a price premium of about 15% and is sometimes used in locations where splashing may occur followed by a period during which the motor can run in relatively dry air. The present trend is to use either the open drip-proof or totally-enclosed motors, as dictated by the severity of the application, rather than the splash-proof type.

Totally-enclosed, and totally-enclosed fan-cooled motors, as their names imply, have no openings connecting the space around the windings with the outside air, and are generally used where dirt, dust, fumes, dampness, or outdoor conditions prevail. The smaller ratings are non-ventilated, while the larger ones are fan-cooled, by an external fan which blows air over the outside of the motor case. The price is about 150% of the open motor price, but this is often offset by a lower maintenance cost.

Explosion-proof motors are totally-enclosed or totally-enclosed fan-cooled motors so designed that they will not produce an explosion when operated in the flammable atmosphere for which they are approved. They are designed, manufactured, and tested under the strict supervision of Underwriters' Laboratories, Inc. The price is about 180% of that of the open motor of the same rating.

Bearings. Bearings in these types of motors are either of the rolling-contact or sleeve type. Almost all smaller fractional-horsepower motors use wool-packed sleeve bearings while the larger motors usually have ball bearings; but roller or sleeve bearings are sometimes used. The larger sleeve-type bearings are lubricated with oil fed to the journal either by the capillary attraction of the wool waste or by a revolving oil ring, depending on the motor size and speed. If a motor is expected to carry some thrust load, at least one ball bearing or a tapered roller bearing is required unless some very special construction is used.

The pros and cons of ball versus sleeve bearings are legion, but some of the major points follow. Ball bearings will take thrust while the usual sleeve bearing will not. The ball-bearing motor is not so apt to have objectionable end bump of the rotor under pulsating load as a sleeve-bearing motor. A sleeve-bearing motor often has a lower price. A sleeve-bearing motor may be quieter, but this is by no means universally true. A ball bearing generally requires less frequent lubrication, but on the other hand it probably requires a higher degree of skill in lubrication.

Quietness. Certain types of motors are inherently noisier than others, but all commercial motors are reasonably quiet. Sometimes a special quiet motor is specified in the hope of reducing objectionable noise, but in the great majority of such cases the little gain that can be obtained from the special motor is of small consequence. It is better to provide some means of general noise suppression so

that noise of the driven machine is reduced at the same time. On the other hand, poor running balance of the motor may be of considerable importance in producing noise.

In the special case of single-phase motors which have an inherent electrical unbalance, sound-isolating mountings may prove very effective. These are sometimes incorporated in the motor itself by the motor manufacturer, or supplied as an external device. Here, again, it may prove better to isolate a sub-base carrying the motor and the driven machine, rather than the motor alone.

Polyphase, Squirrel-Cage Induction Motors. The induction motor is the most common type of a-c polyphase motor. It is mechanically the simplest of all motors and consists of only a rotor, a stator, and two end-shields with the bearings. It is essentially a constant-speed machine, which means that the speed changes very little with load and cannot be adjusted. Induction motors are classified by starting-torque, locked-rotor current, and full-load speed. The National Electrical Manufacturers Association Standards classify *squirrel-cage induction motors* thus:

Design A. Normal starting torque—normal starting current.
Design B. Normal starting torque—low starting current.
Design C. High starting torque—low starting current.
Design D. High slip.

Design A motors are not commonly found at the present time, and Design B motors are used unless the extra starting torque of the Design C motor is required by the application.

Design B motors usually have 200%, or greater breakdown torque, while the Design C motors have slightly less. Breakdown torque is the greatest torque that the motor will carry at full voltage without a sudden change in speed. However, the maximum peak of the load torque should not exceed two-thirds or three-quarters of the value of the breakdown torque.

Figure 2 in the section on "Motor and Load Dynamics," shows speed-torque curves of 100 horsepower, 1800 rpm, Design B, Design C, and Design D motors. Table 1 of the first section on motor classifications, gives the NEMA standards for horsepowers, speeds, frequencies, and locked-rotor and breakdown torques for Designs A, B, and C polyphase squirrel-cage induction motors.

General-purpose open drip-proof induction motors are permitted by NEMA to have a *service factor* which defines the additional load that the motor will carry without reaching a temperature that will cause undue shortness of insulation life. The service factor multiplied by the nameplate rating of the motor gives the maximum horsepower to which the motor can be overloaded. Service factors are discussed at the beginning of this section. In the case of polyphase squirrel-cage motors, these factors apply to Designs A, B, C, and F motors.

The additional horsepower obtained by applying the service factor can also be used as a safety factor to take care of the potential unknowns that can cause overloading of the motor. Generally, however, it is not advisable to operate a motor above its nameplate rating for air conditioning, heating, or ventilating services. Totally-enclosed types of motors and motors with encapsulated windings normally have a service factor of 1.00. When one of these motors is used, care must be taken that the motor is not overloaded.

NEMA also specifies the *maximum locked-rotor current* for single-speed, three-phase, 60 cycle, and 50 cycle induction motors with rated voltage and frequency impressed. Table 2 again in the first section, gives these data.

The latter part of this section dealing with part winding and reduced voltage starters explains the significance of locked-rotor currents in motor applications and the relationship of locked-rotor currents to locked-rotor torques.

There are occasions when it is necessary to convert locked-rotor current to its equivalent locked-rotor kva (kilovolt-amperes) or to convert full-load running current to its equivalent full-load kva. The formula for a 3-phase motor is given by the following:

$$kva = (1.73\ E\ I)/1000$$

where $E =$ voltage
$I =$ locked-rotor or full-load line current.

NEMA requires that every alternating current motor rated 1/20 hp and larger, except for polyphase wound-rotor motors, shall have a *code letter* marking on its nameplate indicating locked-motor kva per hp. Table 7 of the "Motor Classification section" shows the range in locked-rotor kva per hp for each code letter. In this case, the locked-rotor kva per hp range includes the lower figure, up to, but not including, the higher figure.

The National Electric Code segment, earlier in this section, discusses full-load currents and locked-rotor currents of motors and includes Table 2 which gives the NEC full-load current values for three-phase a-c motors. Table 5, in the same section, repeats the NEC locked-rotor current conversion table. Maximum permissible locked-rotor currents, as given by NEMA, generally differ slightly from the NEC values. The NEMA values for single-speed, three-phase, 60-cycle, constant-speed induction motors rated at 230 volts are given in Table 2 of the first section. Section 2, Table 5, and Section 1, Table 2 are applicable to full-voltage (across-the-line) starting and are useful for relating to various reduced-voltage and part-winding starting schemes.

Speed Control. Two general types of speed control are used for polyphase motors in air conditioning, heating, ventilating, and plumbing work. One type is "two-speed" in which the motor operates at either of two defined constant-speed conditions. The other is variable speed in which speed variation is smooth and stepless.

Two-speed motors frequently are used in conjunction with centrifugal and axial fans applied to variable torque loads (systems resistance varies as the square of the speed or air flow). The centrifugal fan, pump, or compressor curve of Fig. 3 in the section on "Load Dynamics," illustrates such a load. Typical examples would be a central station centrifugal fan operating at half speed for night heating, or an induced-draft, cooling-tower, axial flow fan operating at half speed for capacity reduction and reduced noise. Two-speed motors also are use in conjunction with reciprocating or rotary compressors applied to constant-torque loads when other methods for achieving capacity reduction are not practical. The reciprocating or rotary compressor curve of this Fig. 3, mentioned above, illustrates such a load.

Variable-speed motors and drives frequently are used in conjunction with centrifugal pumps operating against a relatively constant head (constant torque load). A frequent example is a tankless, variable-speed house-pump booster system. In this application, a 20 to 25 percent variation from maximum speed generally is sufficient to reduce the pumping system to minimum capacity. The relatively small required variation in speed is important, since virtually all variable-speed drives generate heat and become inefficient at the lower speeds.

Industrial centrifugal gas compressors operating against relatively constant heads frequently use variable-speed drives. Prior to the acceptance of prerotation vanes, most centrifugal compressors used for refrigeration and water-chilling service used variable-speed drives for capacity control.

Two-speed Polyphase, Squirrel-Cage Induction Motors. Polyphase induction motors are available for multi-speed applications in two, three, or four speeds; however, they seldom can be justified at greater than two speeds for air conditioning, heating and ventilating work.

Most two-speed motors are furnished for full-speed and half-speed operation. The ability to achieve capacity reduction of equipment is the biggest reason for their use, although other benefits of decreased noise and lower operating costs can be very important. For example, a refrigeration compressor without unloaders can operate at 50% capacity when connected to a two-speed motor.

Another example is that of a two-speed motor applied to the axial flow fan of an induced draft cooling tower. The noise level of the cooling tower would markedly decrease at half-speed operation, which can be of tremendous importance in a residential neighborhood during the evening when cooling loads generally are low. The use of a two-speed motor can eliminate or decrease fan cycling with its attendant transient noise effects. Finally, the two-speed motor may decrease operating costs, since its power consumption at half-speed is roughly 1/6 that at full speed.

Two-speed motors come in two types: separate-winding and single-winding, consequent pole. Each type can be furnished with one of the following characteristics:

1. constant horsepower
2. constant torque
3. variable torque

Constant-horsepower motors are used where the torque requirement increases as the speed decreases, such as applications with certain winches and machine tools. At one-half speed, maximum allowable torque is twice the high-speed value. These motors are not used in heating, ventilating or air conditioning work.

Constant-torque motors are suitable for driving machines where the required horsepower varies directly as the speed. This makes the constant-torque motor suitable for reciprocating compressor or rotary compressor applications.

Variable-torque motors are suitable for applications with loads which have torque requirements which are less than, or equal to, a torque requirement which varies directly with the speed. Fans and centrifugal pumps have torque requirements which vary roughly as the square of the speed and, therefore, fall under this category and can be driven by variable-torque motors. A variable-torque motor has a maximum horsepower output at low speed which is one-fourth that at high speed.

A separate, or two-winding, motor is more expensive than a single winding, two-speed motor and usually operates at a lower power factor and efficiency. The two-winding motor does, however, have a simpler starter. A single-winding, consequent pole, two-speed motor can operate only at full speed and one-half of full speed, while the two winding motor can operate at full speed and, if so specified, a second speed which need not be one-half of full speed. Normally, a single-winding motor is used if the desired second speed is one-half full speed.

Figure 1 in this section shows the schematic wiring and starter switching circuits for a two-winding, two-speed motor and for constant-horsepower, constant-torque and variable-torque, single-winding, two-speed motors.

Wound-Rotor Polyphase Induction Motors. The foregoing has pointed out the advantage of variable-speed operation to obtain capacity reduction. The final section of this chapter, dealing with part winding and reduced-voltage starting, points out a frequent requirement for obtaining low starting currents for large motors without jeopardizing starting torques. The wound-rotor motor not only can provide adjustable speed operation but can provide a decrease in starting current with an actual increase in starting torque. Adjustable-speed operation, a decrease in starting current, an increase in starting torque, the acceleration, from start, of a high-inertia load, or a combination of any of these factors, can frequently justify the application of a wound-rotor motor to a load.

A wound rotor induction motor has, as its name implies, a rotor which has polyphase windings or coils instead of the short circuited rotor bars characteristic of a general purpose induction motor. The rotor windings are connected to collector assemblies or slip rings within the motor, which in turn permit the windings to be connected through brushes to external resistors. For an induction motor, at any given value of torque, the slip will vary proportionally with the rotor resistance. Therefore, by adjusting the values of the external resistors, the speed-torque characteristics of the motor can be changed.

Figure 2 shows a family of speed-torque curves and speed-current curves applicable to a single wound-rotor induction motor. Each curve relates to a specific value of rotor resistance. The numbers on the curves indicate the secondary (rotor) resistance in percent of value required to give full-load torque at standstill.

External resistance or reactance may be varied manually or automatically. The most popular type of manual control is a drum switch which cuts in or cuts out banks of external resistors. One method of automatic control uses pilot-controlled contactors which switch the resistor banks in and out. Another method uses a saturable reactor in conjunction with pilot controlled resistor banks.

The external resistors may be designed for intermittent or continuous duty. Intermittent duty resistors are generally utilized when the wound-rotor motor is used to provide a low starting current, extra-heavy starting ability or slow acceleration of a high-inertia load. Continuous-duty resistors are used when speed control is desired and sustained operation at less than maximum speed is anticipated. Wound-rotor motors are seldom operated at less than 50% sustained speed since the power loss through the external resistors becomes severe and the operation is inefficient. Further, the problem of dissipating the heat developed in the resistors is aggravated.

Reference to Fig. 2 shows how ideal a wound-rotor motor can be for satisfying severe power company inrush current requirements, while at the same time providing high starting torques. For example, the "80" secondary circuit resistance curves will provide a starting current which is only 120% of full load current and a starting torque which exceeds 120% of full-load torque. Roughly, this is equivalent to 80% of the full-voltage locked-rotor torque and 21% of the full-voltage locked-rotor current of a NEMA Design B, general purpose, induction motor. In this case, the "torque efficiency" (see the last segment of this section on Part Winding and Reduced Voltage Starting) of the wound-rotor motor would be almost 400%, a performance which cannot be approached by any of the part-winding or reduced-voltage starting methods.

Another advantage of the wound-rotor motor is its ability to accelerate from high-inertia starting loads. In this case, substantially all of the heat energy developed in the rotor circuit during

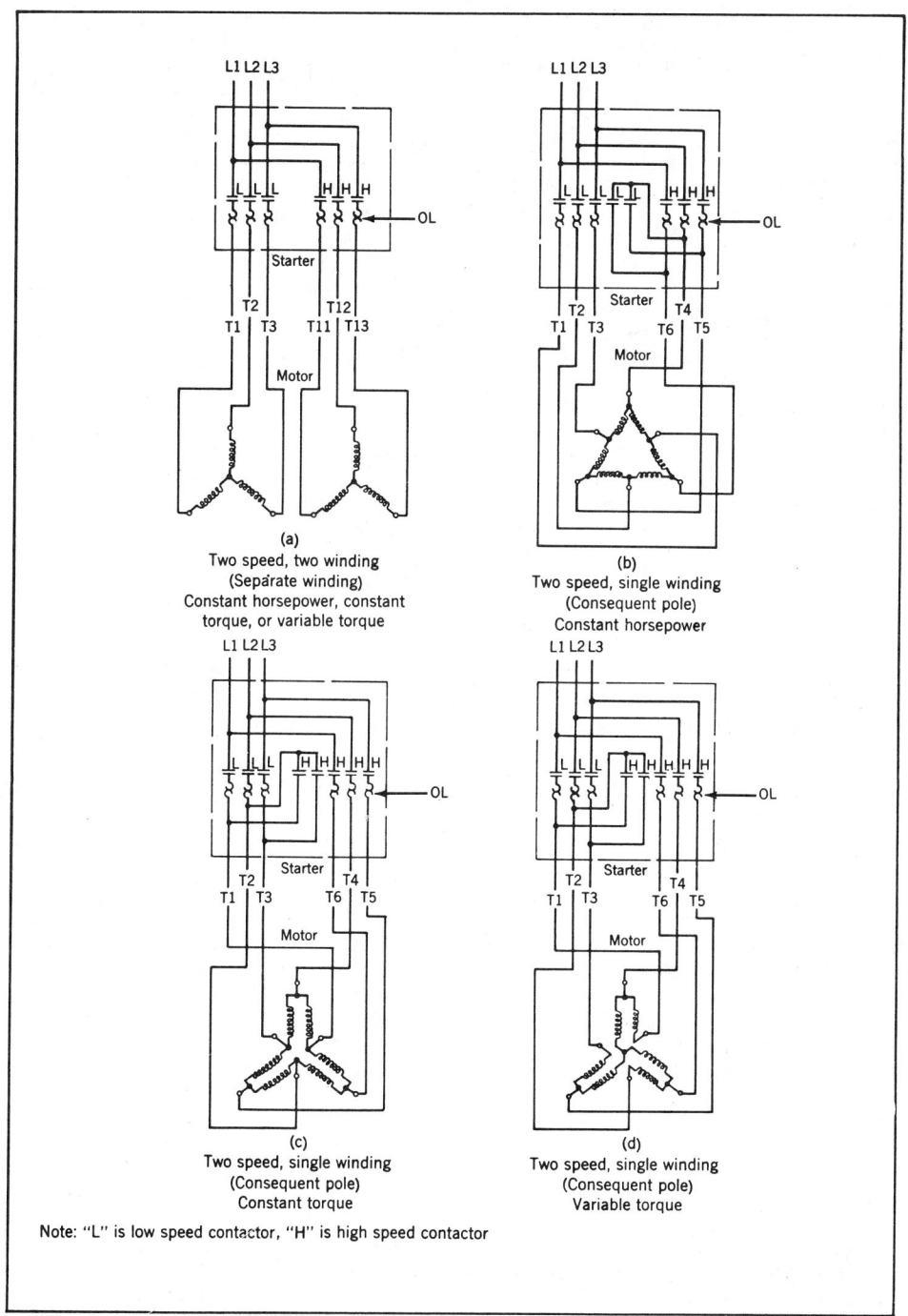

(a)
Two speed, two winding
(Separate winding)
Constant horsepower, constant
torque, or variable torque

(b)
Two speed, single winding
(Consequent pole)
Constant horsepower

(c)
Two speed, single winding
(Consequent pole)
Constant torque

(d)
Two speed, single winding
(Consequent pole)
Variable torque

Note: "L" is low speed contactor, "H" is high speed contactor

Fig. 1 Schematic wiring and starter switching circuits for a two-winding, two-speed motor and for constant horsepower and torque, and variable torque, single winding, two-speed motors.

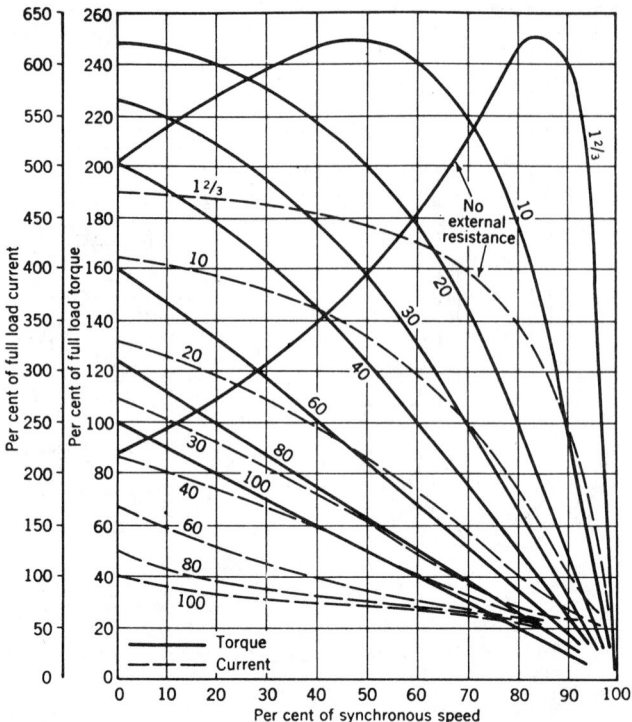

Fig. 2. Family of speed-torque curves and speed-current curves for a single wound-rotor induction motor.

acceleration is dissipated in the external resistors, instead of heating the motor itself.

The wound-rotor motor is particularly well adapted for speed control of centrifugal compressors and centrifugal pumps operating between relatively fixed suction and discharge pressure conditions. In this case, a speed reduction of about 25% brings the equipment to its minimum required speed or to a hunting or surging condition. Prior to the widespread acceptance of pre-rotation vanes, use of wound-rotor motors provided the most acceptable method for varying capacities of centrifugal compressor refrigeration machines. At the present time some manufacturers are offering wound-rotor motors for use with variable-speed, tankless, pump booster systems.

Variable Speed. For polyphase a-c systems, several types of variable-speed drives in addition to wound-rotor motors have been used.

At the present time, a variable speed drive, up to about 100 hp, that frequently sees service uses a modified form of a Design D induction motor in conjunction with a power supply having silicon-controlled rectifiers to vary the effective supply voltage. A typical speed-torque curve for a Design D motor is shown in the section on Motor and Load Dynamics, Fig. 2. The torque varies as the square of the applied voltage, and a relatively small decrease in voltage will cause a significant slip in the presence of a load. Heat is generated when slippage occurs, and the SCR controlled, variable-speed motor must be capable of dissipating this heat.

Other types of variable-speed drives suitable for polyphase alternating current systems are available. These include constant-speed polyphase induction motors which drive their loads through some form of "torque transmitter." One form of torque transmitter is the mechanical fluid drive; another is the electric magnetic drive, sometimes referred to as an eddy-current clutch.

Each of these drives achieves a variation in speed by providing a slip between the driving motor and driven load. The drives are inefficient in that torque

conversion does not take place and the slip constitutes a loss which must be dissipated in the form of heat.

Among the most sophisticated in providing the greatest flexibility in speed and smooth control in conjunction with high efficiency are variations of the Ward Leonard system. In general, these utilize a constant-speed, a-c induction motor which drives a d-c adjustable-voltage generator and a small auxiliary d-c constant voltage generator. The main adjustable voltage generator powers a shunt or compound d-c motor which, in turn, drives the load. The auxiliary d-c generator provides excitation for the main generator and d-c shunt fields.

Speed variation is accomplished by changing the output voltage of the generator through the use of rheostats or amplifiers which vary the current in the shunt field of the generator. Variable-speed drives of this type are used with commercial elevators and with hoists, dredges, machine tools, steel mill drives and other industrial applications. This type of speed control system has not been used in air conditioning, heating, ventilating and plumbing services because applications in these fields do not justify the relative high cost of equipment.

Other types of sophisticated a-c variable-speed drives are available but are too costly for air conditioning, heating, ventilating and plumbing work. These include electronic rectification for use with d-c motors, frequency conversion, and special forms of wound-rotor motors.

Synchronous Motors. For many years, synchronous motors have been direct-connected to large vertical reciprocating ammonia compressors and to the flywheels of large reciprocating horizontal ammonia compressors and air compressors. These machines operate at less than 500 rpm, and the synchronous motors which drive them are known as engine types. In this case, the rotor of the synchronous motor is bolted on to the shaft extension of the compressor or flywheel and there are no rotor bearings or shaft supplied with the motor. Motors of this type are more efficient than induction motors of comparable speed and size, operate at a better power factor or provide power factor correction, and usually are less expensive than the induction motor.

Low inrush currents are characteristic of engine type synchronous motors. Where flywheel effect is lacking in the driven equipment, the rotor can be designed with a large inertia to provide this effect. The amount of flywheel effect required depends on the type of compressor and the method of its unloading during the starting period.

High-speed synchronous motors are designed for operation above 514 rpm, or in other words, have less than 14 poles in conjunction with a 60-cycle a-c power supply. These motors are furnished complete with rotor shaft and bearings. Although these motors have been directly connected to, and have also driven reciprocating compressors via belt drives, their most common application at the present time is to drive open type (non-hermetic) centrifugal refrigeration compressors through speed increasing gears.

Synchronous motor speeds generally are 1200 or 1800 rpm, although 3600 rpm synchronous motors are not uncommon in the extremely large sizes where the speed increaser cost goes up rapidly with an increase in gear ratio. Induction motors of comparable size and speed are less expensive than high speed synchronous motors. In the air conditioning field, synchronous motors can be justified generally only when improved power factor or power factor correction is desired.

Synchronous motors operate at fixed synchronous speed. During starting, they operate as an induction motor, utilizing a rotor amortisseur winding and a separate rotor field winding which is short circuited through a field discharge resistor. At the end of the starting period, a source of direct current excitation is applied to the field winding and the rotor pulls into synchronization. The source of d-c excitation may be a small motor-generator set, a direct or belt drive generator connected to the motor shaft, or a dry-type rectifier.

To provide a synchronous motor design which will pull into synchronization, one must have knowledge of the driven load. Both the required load torque at synchronization speed and the inertia of the load are chief considerations. Maximum slip at time of pull-in runs from 1 to 5 percent, depending on the load characteristics.

Acceleration time from startup to pull-in is directly related to the design of the rotor amortisseur winding. The graphical solution for acceleration time, described earlier under "Motor Heating and Motor Life," is frequently used to determine acceleration time.

Usually, it is necessary for the motor manufacturer and the compressor manufacturer to combine their efforts to provide a satisfactory synchronous motor application.

HERMETIC TYPE MOTOR-COMPRESSORS

Hermetic motors which drive refrigeration compressors are types of definite-purpose motors. There are three main characteristics of hermetic motors which stand out from general-purpose open, or totally enclosed, motors.

1. Unlike a general purpose motor which can be used to drive any number of types of machines, a hermetic motor is specially designed for a particular compressor model of a specific manufacturer.

2. The windings of the hermetic motor are submerged in the refrigerant atmosphere and generally are cooled by suction gas, although some hermetic motors of large size are cooled by discharge gas or refrigerant liquid injection or are water cooled. Thus, motor rating is dependent on the efficiency of refrigerant cooling. Efficiency of refrigerant cooling exceeds that of air cooling, and so the hermetic motor has less copper than an open motor driving a comparable load.

3. The hermetic motor is manufactured without a shaft. The compressor manufacturer attaches the squirrel cage rotor to an extension of the compressor shaft during assembly. The hermetic motor-compressor unit requires no shaft seal.

Refrigerant cooling efficiency is dependent on the design and application of the motor-compressor assembly, and the unit is rated in the amperes it can draw continuously without overheating. Hermetic motors are not horsepower rated, although reference to horsepower is often made to give an indication of approximate capacity.

Because of its decreased copper and its dependence on efficient refrigerant cooling, a hermetic motor is more susceptible to fast overheating and quick burnout in case of a stalled rotor, a malfunction in the refrigerant circuit, an excessive running load, a low voltage condition, or a disruption in the evaporator air flow. As a result, quick acting, sensitive, current and heat-sensing overload devices are desired for motor protection. This is particularly the case for single-phase hermetic motors where there is an increased probability of motor stalling on start-up.

Hermetic Compressors to 5 hp

Hermetic motor-compressor units from fractional horsepower through 5 hp are most unique. For the most part, the few manufacturers who produce the compressor units supply them to a great number of other manufacturers who produce completely assembled air conditioning and refrigeration units, such as window air conditioners and heat pumps, domestic refrigerators and freezers, beverage coolers, vending machines, etc. Motor windings are furnished by a motor manufacturer to the compressor manufacturer who assembles the motor with the compressor. Most units 5 hp and under are hermetically sealed in a steel shell. The shell cannot be opened in the field for inspection or service.

Present trend is for all units through 5 hp to have 2-pole, or 3450-rpm motors.

New units for use with window, residential and small commercial air conditioners through 5 hp (and some heat pumps) utilize *permanent split-capacitor* motors, otherwise known as PSC types. Because of the limited starting torque available with permanent split-capacitor motors, their refrigeration systems must be designed with capillary tube liquid feed to permit equalization on shutdown. Most compressors have a single cylinder in the small sizes and two and three cylinders in the larger sizes.

For single-phase units requiring a higher starting torque, such as those used with refrigeration applications or where equalization on shutdown does not take place, a *resistance start-induction run* (RSIR), *capacitor start-induction run* (CSIR), or a *capacitor start-capacitor run* (CSR) motor must be used. Of the three types, the RSIR provides the lowest torque and its use is generally restricted to refrigerators and freezers in the 1/20 to 1/3-hp range when capillary tube liquid feed is used. The CSIR is generally used from 1/6 to 1/2 hp where high starting torque is required for refrigerators, freezers and vending machines which use thermostatic expansion valves. The CSR is used from 1/3 hp through 1½ hp and larger for such applications as vending machines, beer coolers, reach-in refrigerators and small walk-in refrigerators.

Relays. A centrifugal speed switch is not practical in a hermetic unit; therefore, external hot-wire relays, current sensitive relays or potential type relays are used to switch RSIR, CSIR and CSR units off of their starting windings at the end of the starting period. In addition, kits from the compressor manufacturer often permit the addition of capacitors and potential type relays to PSC units after installation and operation to convert them to CSR units. This sometimes is necessary if the PSC unit is applied to a system which, for some reason, fails to fully equalize on shutdown.

The hot wire relay is used in conjunction with some smaller units and generally has been sup-

planted by the current type relay. Basically, the hot wire relay is a thermal time switch incorporating a straight, stiff wire which quickly heats up when line current passes through it. After a short interval, the elongation of the wire causes a snap switch to open which disconnects the motor starting winding from the line. A second snap switch incorporated in the hot wire relay serves as overload protection and disconnects the entire motor from the line if excessive current should cause the hot wire to further elongate.

The current sensitive relay is more commonly used with RSIR and CSIR motors up through 1/3 hp. A magnetic coil in series with the motor running winding closes a set of movable contacts immediately when the motor is initially energized at the beginning of the starting period. The movable contacts complete a circuit through the starting capacitor to the starting winding in the CSIR motor or to the starting winding in the RSIR motor. As the motor approaches full speed, the current drawn by the running winding decreases. At a predetermined speed, the magnetic coil drops out the movable contacts and disconnects the starting winding from the line.

The potential relay is commonly applied to CSIR and CSR units ½ hp and larger. A potential coil which operates a set of normally closed contacts is connected directly across the start winding. The normally closed contacts connect to the starting capacitor which is in series with the coil. When the motor is first energized, the voltage drop across the capacitor is high, but the voltage drop across the coil is low; therefore, the contacts connect the capacitor to the line. As the motor gains speed, the start winding develops back EMF (a countervoltage) and, at a predetermined speed, the potential coil opens up the set of contacts. This disconnects the capacitor from the line. However, the potential coil keeps the contacts open due to the voltage induced in the open-circuited start winding when the motor is running at high speed.

Overload Protection. Overload protection is carefully selected and coordinated with the motor-compressor unit and furnished with the unit by the manufacturer. Overload protection by use of a hot wire relay already has been described. One prominent manufacturer provides overload protection for units through 5 hp in the following manner:

1. Single-phase units up through 1 hp have an overload device consisting of a snap-acting, thermostatic, bimetal disc and a heater located outside of the compressor shell, but inside of the terminal cover. Line current passes through the heater which heats up the bimetal disc. The disc also is responsive to the heat of the compressor shell. In event of excessive current, excessive shell temperature, or a combination of both, the bimetal disc opens up a pair of line-current switches, stopping the motor.

2. Single-phase units, of 1 hp through 2½ hp, utilize a combination heater and bimetal disc housed within a tiny tin-plated steel can which is inserted in the end turns of the motor windings within the compressor shell. This internal line break overload protector functions in a manner similar to the overload protector described in the preceding paragraph, except that it is more responsive to motor winding temperatures.

3. Single-phase units, 2½ hp through 5 hp, utilize a two-pole contactor with an internal motor winding thermostat and a supplementary exterior overload wired in series to the holding coil of the contactor. The internal thermostat causes the contactor to open and stop the motor in event of excessive motor winding tempertaure. The supplementary overload opens the contactor under excessive locked rotor conditions for which the internal thermostat responds too slowly. The supplementary overload may be a bimetal disc with heater or a special circuit breaker.

4. Three-phase units have overload protection similar to that described in the preceding paragraph, except that two supplementary exterior overloads are used—one in each of two phases. In this case, a single internal thermostat is used.

Wiring Diagrams. The following wiring diagrams show examples of various starting devices and overload protectors used with hermetic motors.

Figure 1 shows a hot wire relay used for starting and overload protection of a RSIR motor.

Figure 2 shows a current sensitive relay used for starting and a line break bimetal disc and heater used for overload and heat protection of a CSIR motor.

Figure 3 shows an internal line break bimetal disc and heater used for overload and heat protection of a PSC motor.

Figure 4 shows a potential relay used for starting and a supplementary interior overload and internal motor winding thermostat and contactor used for overload and heat protection of a CSR motor.

Figure 5 shows two supplementary exterior overloads and an internal motor winding thermo-

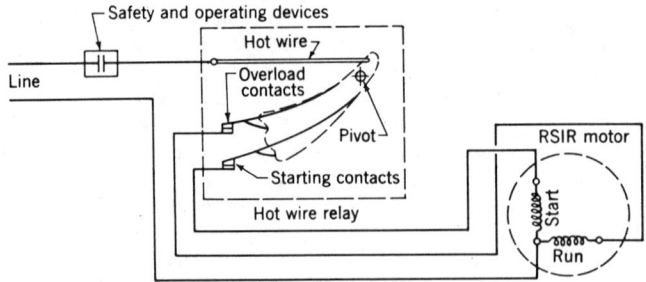

Fig. 1. Hot wire relay with resistance-start, induction-run motor.

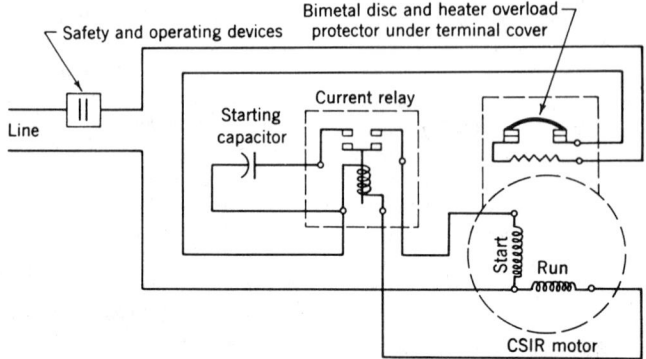

Fig. 2. Current relay and overload with capacitor-start, induction-run motor.

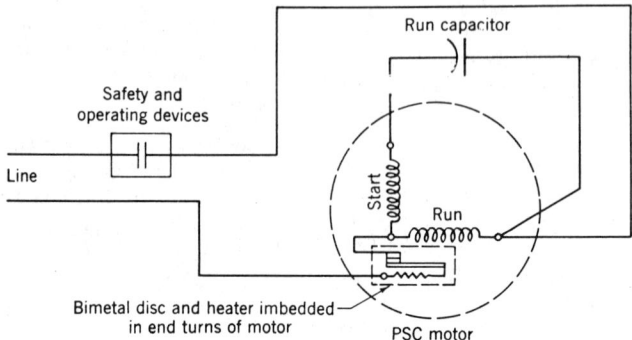

Fig. 3. Permanent split capacitor motor with internal line break overload protection.

stat used for overload and heat protection of a 3-phase motor.

Although most motor-compressor units 5 hp and below are sealed hermetic types, accessible hermetic units can be inspected and serviced in the field, and motors can be replaced or repaired without necessarily removing the compressor.

Hermetic Compressors Above 5 hp

Although sealed hermetic units are available through 10 hp, most units above 7½ hp are of the accessible hermetic type. Motors are 3-phase and generally are furnished by motor manufacturers to compressor manufacturers who market compressors, condensing units and water chilling units to the general air conditioning and refrigeration industry.

Motors generally have internal motor winding thermostats, often one in each phase winding for large

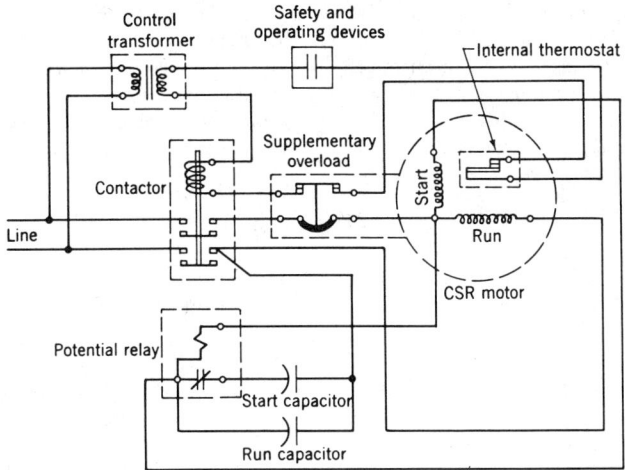

Fig. 4. Capacitor-start, capacitor-run motor with internal protection and potential relay.

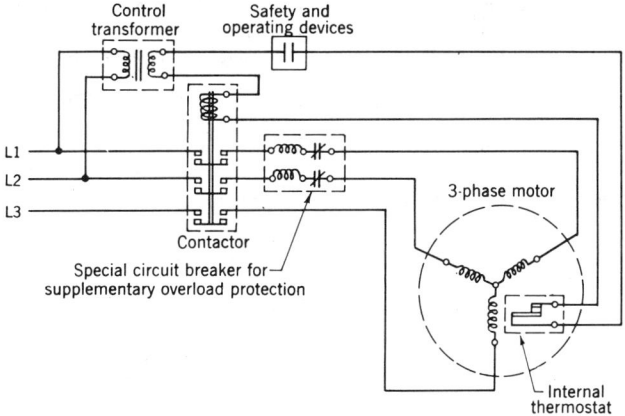

Fig. 5. Three-phase motor with internal protection.

units. Starters are full-voltage part-winding or re-duced-voltage types as required for the particular installation. Starters normally incorporate quick-acting overloads for protection against locked-rotor conditions, excessive loads, and low voltage conditions.

An overload relay utilized in some hermetic motors, makes use of thermistors. The thermistor is a special semiconductor resistor that has a com-paratively low, constant resistance up to a critical temperature point and has a great positive tem-perature coefficient of resistance within a narrow range above the critical temperature. The

thermistor is tiny and can be placed in direct contact with the motor stator winding. It has very good thermal response and reacts much faster to increases in motor winding temperatures than conventional internal motor winding thermo-stats.

The thermistor is used in conjunction with a low-voltage bridge circuit, or a low-voltage, rectified, direct current and a drop-out relay coil wired in series. On a rise in motor winding tem-perature the increase in thermistor resistance above its critical temperature decreases the recti-fied current to the coil of the drop-out relay,

causing the relay contacts to open. The drop-out relay contacts are wired in series with the starter holding coil, causing it to be de-energized and thereby stopping the compressor.

Most hermetic motors above 5 hp are suction gas cooled. Some of the hermetic motors driving the centrifugal compressors of large, water cooling systems are cooled by discharge gas; others are cooled by chilled water circulated through a jacket which forms the shell surrounding the motor. Increased attention is being given to liquid injection for hermetic motor cooling, as large centrifugal compressor water chilling systems tend towards the use of higher pressure refrigerants.

STARTERS

The National Electric Code includes, under the definition of controller, any switch or device normally used to start and stop a motor. The recommendations of the Code cover requirements for motor controllers in addition to requirements for motor running overcurrent protection, motor disconnecting means, motor branch circuit overcurrent protection, motor feeder overcurrent protection and motor conductors. Although the manufacturer of mechanical equipment or the mechanical contractor who installs mechanical equipment frequently furnishes motor controllers and motor running overcurrent protection devices (overloads), requirements for the other devices must be incorporated and coordinated into each installation.

Motor Controllers

National Electric Code requirements for non-hermetic motor controllers can be summarized as follows:

1. Motor controllers are not required for stationary motors, ⅛ hp or smaller, that are so constructed that damage cannot occur from overload or failure to start. An example of this type motor is a clock motor.

2. For portable motors, 1/3 hp or smaller, controller may be an attachment plug and receptacle.

3. For stationary motors, 2 hp or less, and 300 volts or less, the controller may be a general-use switch having an ampere rating at least twice the full-load rating of the motor. For a-c motors, 2 hp or less, and 300 volts or less, an a-c snap switch may be used to control a motor having a full-load current rating not exceeding 80% of the ampere rating of the switch.

4. For motors above 2 hp, or greater than 300 volts, the controller must have a horsepower rating which shall not be lower than the horsepower rating of the motor. Devices which fall under this category include manual, horsepower-rated, pushbutton switches up to about 7½ hp, motor-circuit switches which are rated in horsepower up to 50 hp, and contactors. Also permitted for motors above 2 hp are branch-circuit type circuit breakers rated in amperes only. Each of these devices is capable of interrupting the locked-rotor current of the motor it controls. These devices also are frequently used for motors 2 hp and smaller.

A controller for a hermetic type motor-compressor must have a continuous-duty, full-load, current rating not less than the nameplate rated-load current or branch-circuit selection current, whichever is greater, and locked-rotor current, respectively, of the motor-compressor.

Overcurrent Protection

Except for small, open a-c motors with inherent overload protection and for single-phase and small, polyphase hermetic type motor-compressors, which have separate or inherent overload protection, most controllers incorporate overload protection (motor running overcurrent protection) in their enclosures.

The following summarizes, National Electric Code requirements for motor running overcurrent protection:

1. Continuous duty motors, 1 hp or less, which are manually started and are within sight of the controller location and are not permanently installed, need no motor running overcurrent protection device, providing the branch circuit overcurrent protection device is properly rated for this purpose.

2. Any motor, 1 hp or less, which is permanently installed and any motor of 1 hp or less which is started automatically shall have a motor running overcurrent protection device (overload) which is separate from the motor and responsive to motor current or integral with the motor and approved for use with the motor it protects on the basis that it will prevent dangerous overheating due to overload or failure to start.

An integral device such as just described provides what is known as inherent overload protection. The Code provides exceptions to the requirement for overload protection for high impedance motors, such

as clock motors, and in some other cases where safety controls protect the motor against damage due to stalled rotor current.

3. Each continuous duty motor rated more than 1 hp shall be protected either by a separate motor running overcurrent protection device or by an approved inherent overload protector device.

Starters

Starters used in heating, ventilating, air conditioning and plumbing work most often conform to the following general patterns:

1. Open motors on polyphase a-c systems use magnetic contactors which interrupt all conductors. Separate overload protection is provided within the starter enclosure. Motors about 7½ hp and smaller sometimes use manual pushbutton starters with overloads, although magnetic starters are more popular—even for fractional horsepower motors. Motor-circuit switches using dual element fuses or time delay fuses for overload protection are seldom used.

2. Electrically operated circuit breakers sometimes are used for starters with very large squirrel cage induction motors, large synchronous motors and for the primary controllers of large, wound-rotor induction motors. The circuit breakers usually function also as the motor branch circuit overcurrent protection device. Overcurrent running protection is provided by magnetic overload devices which function to trip the electrically operated circuit breaker in event of overload.

3. Open, single-phase a-c motors up to 1 hp generally use a manual toggle type snap switch motor starter. Overload protection is provided in the starter when the motor does not incorporate inherent overload protection. The speed selector switches of small, multi-speed blower motors often function as the controllers.

4. Hermetic, three-phase motor-compressors over 5 hp, generally use conventional magnetic-motor starters incorporating overloads. Internal motor winding thermostats are wired in series with the overloads for added protection. Hermetic three-phase motor-compressors, 5 hp and smaller, frequently use contactors with the motors having inherent overload protectors or contactors with separate, special overload devices. (Refer to preceding part on hermetic type motor-compressors for more details.)

5. Hermetic, single-phase motor-compressors, larger than 1 hp, frequently use magnetic contactors for motor controllers in conjunction with

inherent overload protection. For motor-compressors, 1 hp and smaller, where a thermostat or operating device can handle the motor locked-rotor current, a manual motor starter can be used. In general, packaged air conditioning units fall into a special category, since all of the electrical components of a specific model are tested and approved by Underwriters' Laboratories as a group installation.

Overload Protection

A shortened insulation life due to excessive temperature rise is the most serious effect of motor overload. Generally, motor overload protection devices have inverse-time characteristics; that is, they respond more rapidly to increased currents. An attempt is made to match the overload protection device to the thermal characteristics of the motor.

Overload protection devices for motors *without* inherent protection fall into the following general categories: thermal and magnetic.

For small motors which are started manually without contactors, the overload device generally consists of a thermal bi-metallic switch or thermal melting-pot switch which directly interrupts the current to the motor. Most frequently, the switch is manually reset in the absence of a motor overload by a snap switch in the same enclosure which functions as the motor controller. Some a-c snap switch controllers used with small, single-phase motors use time delay fuses for overload protection. When motor-circuit switches are used for starting larger motors, time-delay fuses are frequently used for overload protection.

Where contactors are used for motor controllers, the overload devices generally are overload relays. In this case, the thermal or magnetic element of the relay senses line current, but the switch or contact portion of the device interrupts the control circuit to the coil of the magnetic contactor.

Thermal bi-metallic overload relays, thermal melting-pot relays and thermal-induction relays are popular overload protection devices for a-c motors up to around 200 hp in size. Magnetic overload relays are popular for this size motor and larger and are available for motors of smaller size.

The most popular type of magnetic overload relay includes a magnetic coil and plunger which achieves an inverse-time relationship through the use of an oil dashpot which offers resistance to the travel of the plunger. Adjustments to the overload setting can be made by varying the rate of oil flow in the dashpot. Another type of magnetic overload device

uses electro-magnetic induction from a coil to a movable core. No oil is required for this type.

Overload relays are available for manual reset from a pushbutton through the starter cover or enclosure cover or are available for automatic reset. The manual reset type is most commonly used for most air conditioning, heating and ventilating motors; however, the automatic reset overload relay is popular with some types of packaged equipment which use external annunciators, reset relays or high impedance relays to reset safety devices.

Overload protection for non-hermetic motors sometimes use internal thermostats, or internal heaters and bimetal discs, plus supplementary overload protection as described for hermetic type motor-compressors. Thermistors, as described for hermetic type motor-compressors, are also sometimes used.

Starters for Large A-C Motors

Starters for large a-c motors frequently use d-c contactors. For the same static closing force, the frame and armature of a d-c contactor can be smaller than that of an a-c contactor. Furthermore, the impact force on closing of a d-c contactor is smaller than the closing impact force of an a-c contactor of the same size.

This means that for large motors, a d-c magnetic contactor will be smaller and will require less massive support than an a-c contactor. Another advantage is that the coil of a d-c contactor does not have the large inrush current of an a-c coil. Most frequently, the d-c supply for a large starter comes from a full-wave selenium rectifier within the starter enclosure.

For large a-c motors, it becomes impractical to put the full-line motor current through the current-sensing portion of the overload relay. In this case, a current transformer is utilized with each overload relay which causes a small, accurate, percentage of the line motor current to pass through the overload relay. The current transformers normally are mounted within the starter enclosure. Adjustable magnetic overload relays and adjustable thermal-induction relays are commonly used in conjunction with current transformers.

Starters are rated by NEMA in sizes for various horsepower motors at different voltages. Table 1 gives NEMA sizes of full-voltage starters for single-phase and three-phase motors.

Figure 1 shows the *connection diagram* and *elementary diagram* of a full-voltage, Size 1, three-phase magnetic starter. The *connection diagram,*

Table 1. Starter Size vs. Motor Horsepower for Full-Voltage Starting (NEMA)

3-Phase Motors		
Starter Size	200/230 v	460/575 v
	Motor Horsepower	
00	1½ 1½	2
0	3 3	5
I	7½ 7½	10
2	10 15	25
3	25 30	50
4	40 50	100
5	75 100	200
6	150 200	400
7	— 300	600
8	— 450	900

Single-Phase Motors		
Starter Size	115 v	230 v
	Motor Horsepower	
00	1/3	I
0	I	2
I	2	3
IP	3	5
2	3	7½
3	7½	15

sometimes referred to as the wiring diagram, shows all components and devices in their actual physical relationship to each other. The connection diagram gives the information for physically locating components and tracing wires or for wiring up a specific piece of equipment.

On the other hand, the *elementary diagram* shows the system in a simplified manner, with no attempt made to indicate the various components and devices in their actual, relative positions. The elementary diagram gives the information for following easily the operation of the components and devices in the circuit.

Figure 2 shows the elementary diagram of a typical full-voltage, Size 6, three-phase magnetic starter. The contactor has two d-c coils wired in series, and the d-c supply is provided from a full-wave rectifier in the starter. Current transformers are used to reduce the current through the current-sensing elements of the overload relays. A control-circuit transformer is used to decrease the voltage of the control circuit from line voltage to 230 v (shown) or 115 v.

Figure 1 shows an arrangement where a 2-element momentary contact, start-stop pushbutton station

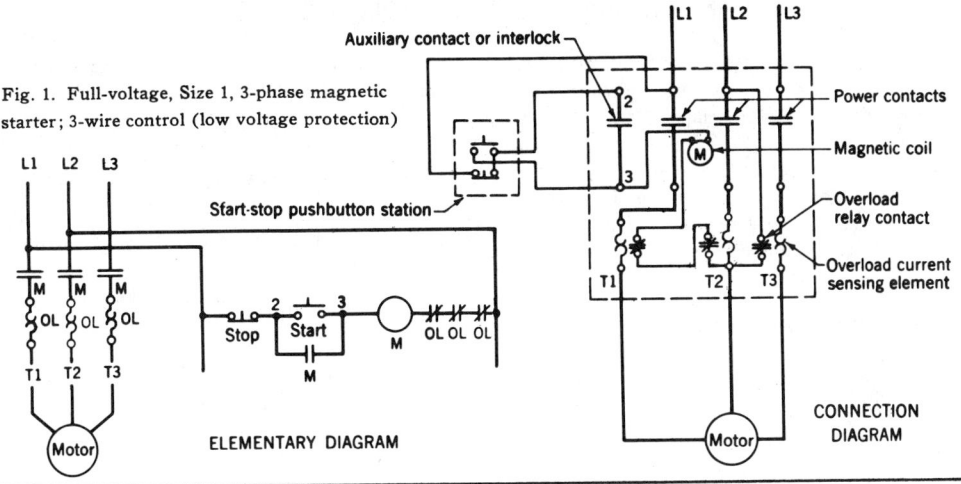

Fig. 1. Full-voltage, Size 1, 3-phase magnetic starter; 3-wire control (low voltage protection)

ELEMENTARY DIAGRAM

CONNECTION DIAGRAM

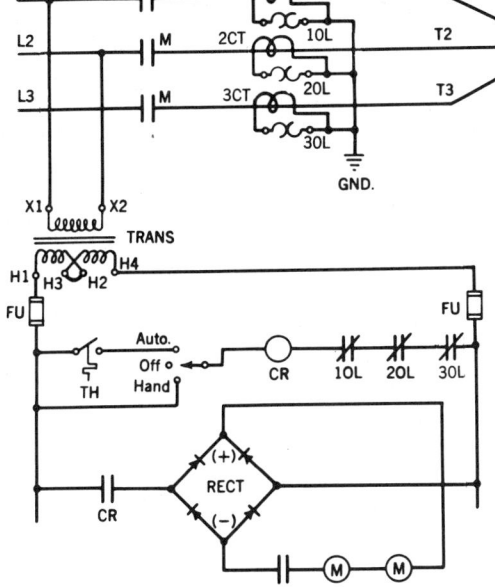

Fig. 2. Full-voltage, Size 6, 3-phase, magnetic starter; 2-wire control (low voltage release).

ELEMENTARY DIAGRAM

M: Contactor
CR: Control relay
10L, 20L, 30L: Overload relay
1CT, 2CT, 3CT: Current transformer
RECT: Full wave rectifier
FU: Fuse
TRANS: Control circuit transformer
TH: Thermostat

control arrangement provides protection to personnel who might be working on a machine after a power outage and has the additional advantage of preventing the simultaneous restarting of many motors in a large installation when voltage is reapplied after a power interruption. Three-wire control from start-stop pushbuttons is common for manually controlled fans and pumps in air conditioning, heating and ventilating applications.

Two-wire control, otherwise known as *low-voltage release* or *no-voltage release*, is used where motors are to be automatically started and stopped from devices such as thermostats, float switches, pressure switches, time clocks, etc. This arrangement permits the motor to restart automatically when voltage returns after a power interruption. Figure 2 shows how a remote thermostat can be used to start and stop, automatically, a large motor using a two-wire control system. The *auto-off-hand* three-position selector switch, which can be remotely located or mounted in the starter cover, is used to bypass the thermostat if automatic operation is not desired.

controls the starter operation. The pushbutton station can be remotely located or mounted in the starter cover and is wired into a normally-open, auxiliary contact of the starter contactor. The control arrangement provides what is known as *three-wire control*, also known as *low-voltage protection* or *no-voltage protection*. In event of a power interruption, the starter will drop out and will not pick up when the power returns. In this case, the start button must be pressed to reclose the starter. This

PART-WINDING AND REDUCED-VOLTAGE STARTING

Part-winding and reduced-voltage starting are used with large squirrel-cage induction motors, large hermetic type motor-compressors, and large synchronous motors to reduce line disturbances to the serving electric utility's distribution system. On some occasions, part-winding and reduced-voltage starting can be very useful in preventing severe shocks to mechanical equipment caused by a motor's high starting torque.

Electric Utility Limitations

Locked-rotor current for a squirrel-cage induction motor will be from $4\frac{1}{2}$ to 7 times full-load current, depending on the design of the particular motor. (See discussion earlier in this section on NEMA code letters). This large starting current, when associated with a big motor, could cause an excessive voltage drop through the electric utility's transformers and distribution system resulting in momentary dimming of lights, an undesired dip in voltage to computers and other electronic machines and, conceivably, opening of magnetic starters serving other motors.

To protect their systems, electric utilities establish limitations on starting of large motors. These limitations generally take one of the following three forms:

1. *A maximum allowable current or power consumption per motor horsepower during any portion of the starting period:* This protects the power company in a general situation. Such limitations are defined in terms of amperes per horsepower and kilovolt-amperes (kva) per horsepower.

2. *A maximum permissible motor horsepower, or a maximum allowable current or power consumption during any portion of the starting period:* Generally, this recognizes the specific size of motor to be used, application of the motor, capacity of the power distribution system serving the motor, and location of the motor in the power distribution system. Such limitations are defined in terms of amperes and kva.

3. *A maximum allowable increase in current or power consumption per unit of time during the starting period:* This type of limitation generally occurs where a power company's automatic voltage regulator is capable of maintaining relatively constant voltage at the distribution point, providing a large change in load is not suddenly applied. Such limitations are defined in terms of amperes, or kva, per unit of time, or amperes, or kva, per horsepower per unit of time, in which case the unit of time is given in terms of seconds or fractions of a second. Generally, there is no definite limit to the ultimate value of the inrush current.

Minimizing Mechanical Shocks

Most motors above 5 hp size have a starting (locked-rotor) torque ranging from 100% to 200% of full-load torque. It is sometimes worthwhile to decrease this starting torque to minimize mechanical shocks when a motor is applied to a high-inertia load. A part-winding or reduced-voltage starter provides this reduction in torque.

A prime example is a large, centrifugal, V-belt driven fan. In many installations, start-up occurs with noticeable belt slippage. Use of an inexpensive part-winding motor and starter provides a "soft start" which decreases or eliminates the slippage.

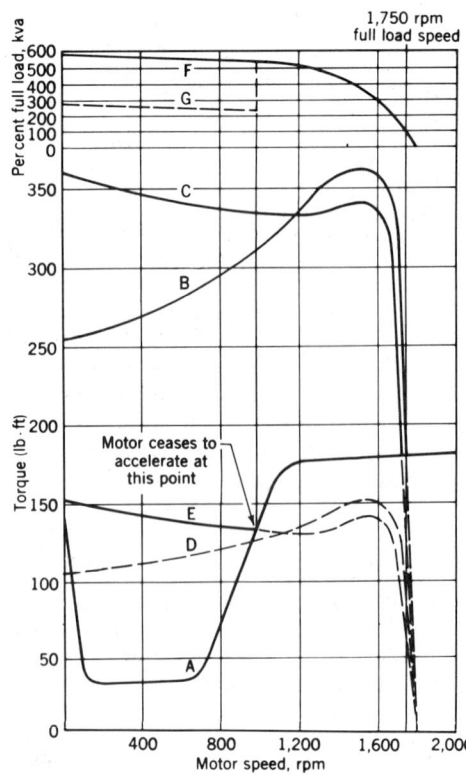

Fig. 1. Torque-speed and kva-speed curves for 60 hp motor and compressor.

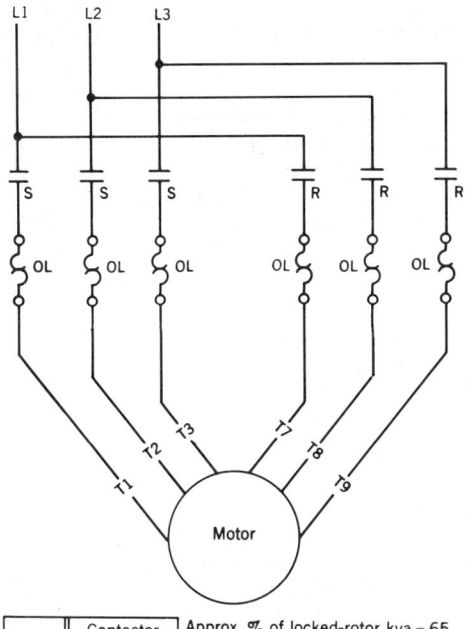

Period	Contactor		Approx. % of locked-rotor kva = 65
	S	R	Approx. % of locked-rotor torque = 45
Start	Close	Open	
Run	Close	Close	Torque efficiency = 69%

Fig. 2. Part winding (one-half winding start).

Application

The power company limitation generally determines the required decrease to motor locked rotor line current. The part-winding or reduced-voltage starting method used to obtain this reduction in starting current will decrease the full-voltage torque-speed characteristics of the motor during the starting period. In most cases, the decreased motor torque must still be enough to start and accelerate the driven load to full-load speed, or close to it, if the power company limitation is to be satisfied. Thus, power company limitation, type of part-winding and reduced-voltage starting used, nature of the driven load and the speed-torque characteristics of the motor are interrelated and must be considered jointly.

An example of these interrelated factors follows:

Curve A, Fig. 1, shows the torque-speed load of an 8-cylinder, reciprocating, refrigeration compressor with automatic unloaders. The compressor is direct-connected to an open 60-hp motor. Starting and stopping is initiated by a medium-differential, suction-pressure switch. The high side of the refrigeration system does not equalize to the low side on shutdown.

Initial "breakaway torque" is required primarily to start two loaded cylinders against the high-side to low-side pressure difference. As the compressor begins to accelerate, the required torque drops to less than 25% of full-load torque, since the remaining six cylinders are unloaded. After a short period, compressor oil pressure builds up sufficiently to load all 8 cylinders, and at this point, the required torque almost equals full-load torque.

Curve B shows the torque-speed curve of a NEMA Design B 60-hp motor, starting at full voltage. Since this curve lies entirely above the compressor torque-speed curve, the motor will start the compressor and accelerate it to full speed without difficulty.

Curve C shows the torque-speed curve of a NEMA Design C motor starting at full voltage. This motor also will start the compressor and bring it up to full load speed without difficulty.

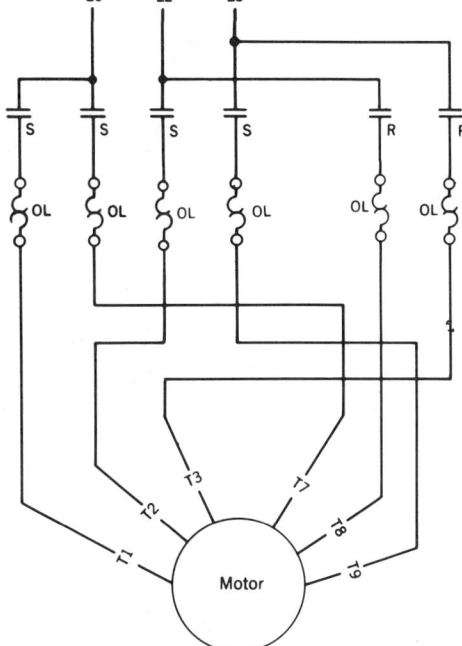

Period	Contactor		Approx. % of locked-rotor kva = 65
	S	R	Approx. % of locked-rotor torque = 45
Start	Close	Open	
Run	Close	Close	Torque efficiency = 69%

Fig. 3. Part winding (two-thirds winding start).

Curve *F* shows the full voltage motor kva, in terms of percent full load kva, in this case applicable to both the NEMA Design B and the NEMA Design C motor. Locked-rotor kva is 5.8 times full-load kva or about 5.5 kva per hp.

Suppose the power company imposes a limitation of 3.0 kva per hp during any portion of the starting period. This would be equivalent to about 3.15 times full-load kva. An autotransformer reduced-voltage starter on the 65% tap would cut the inrush kva down to about 2.7 kva per hp, more or less depending on the magnetizing current of the autotransformer within the starter. However, this reduction in starting kva would be accompanied by a decrease in motor torque.

Curve *D* shows the torque-speed curve of the NEMA Design B motor when energized through the autotransformer starter on the 65% tap. Notice that this motor now *cannot overcome the required breakaway torque* of the compressor. As a result, the motor stalls on the reduced-voltage starting step and, when full voltage is applied on the second step of starting, the inrush is the full 5.5 kva per hp.

Curve *E* shows the torque-speed curve of the NEMA Design C motor when energized through the autotransformer starter on the 65% tap. This motor

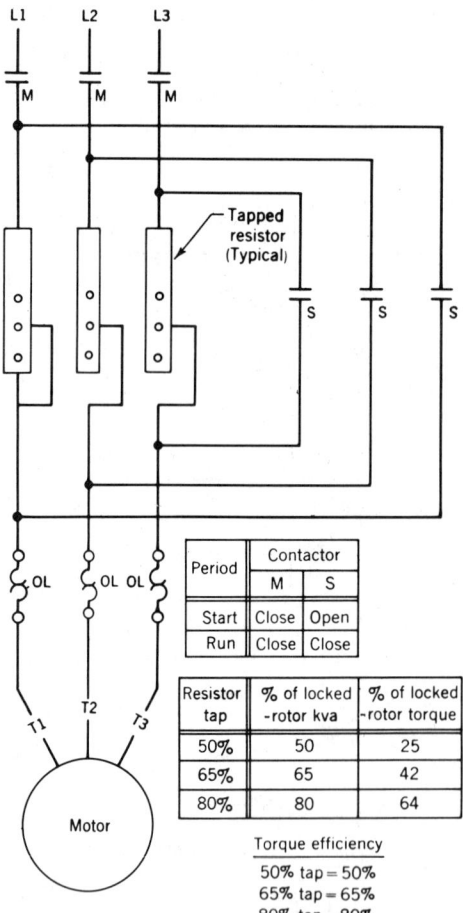

Period	Contactor	
	M	S
Start	Close	Open
Run	Close	Close

Resistor tap	% of locked -rotor kva	% of locked -rotor torque
50%	50	25
65%	65	42
80%	80	64

Torque efficiency
50% tap = 50%
65% tap = 65%
80% tap = 80%

Fig. 4. Primary resistance (two step).

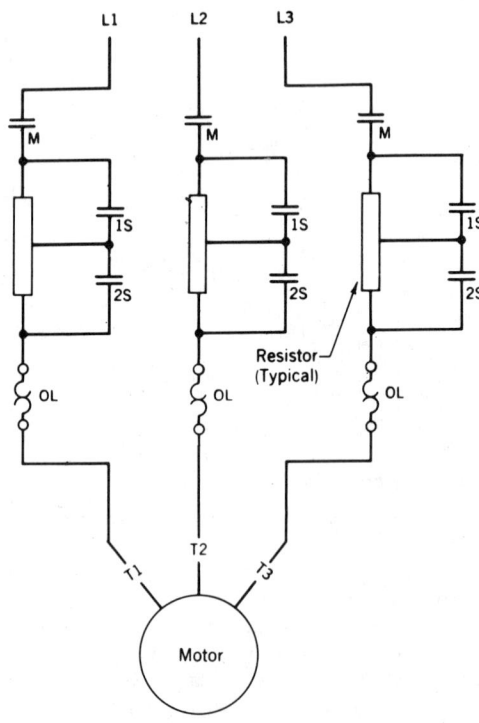

Period	Contactor			Torque efficiency
	M	1S	2S	1st step = 32.5%
1st step	Close	Open	Open	2nd step = 65%
2nd step	Close	Close	Open	
Run	Close	Close	Close	

	% voltage	% of locked -rotor kva	% of locked -rotor torque
1st step	32.5	32.5	10.6
2nd step	65	65	42

Fig. 5. Primary resistance (three step).

can break the compressor away from the dead-stop position, but it cannot maintain full compressor speed after all cylinders load up. When full voltage is applied, the inrush kva will exceed the power company limitation. Curve G shows the percent full-load kva at the point where the motor torque is in balance with the loaded compressor torque and full voltage is applied.

A satisfactory solution to this situation would be to use the autotransformer starter, the NEMA Design C motor and a compressor with externally operable cylinder unloaders. In this case the motor would be capable of starting the compressor and the machine could be kept at 25% capacity during the entire starting period by operating the unloaders through a timing relay or through an interlock on the *run* contactor of the autotransformer starter.

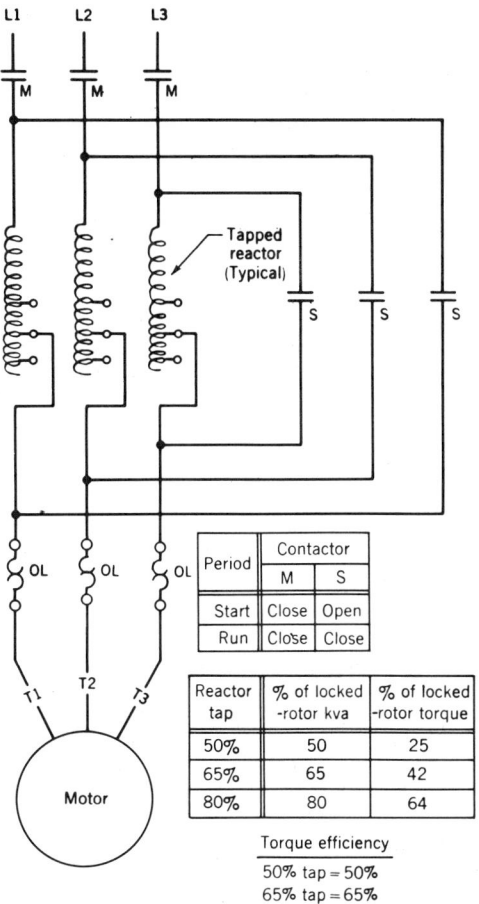

Period	Contactor	
	M	S
Start	Close	Open
Run	Close	Close

Reactor tap	% of locked -rotor kva	% of locked -rotor torque
50%	50	25
65%	65	42
80%	80	64

Torque efficiency

50% tap = 50%
65% tap = 65%
80% tap = 80%

Fig. 6. Primary reactor (two step).

The NEMA Design B motor could be used if the compressor discharge was equalized to its suction during shutdown or at the beginning of start-up. A check valve in the compressor discharge line in conjunction with a solenoid or bleed valve between compressor suction and discharge has been used frequently for this purpose.

Depending upon the type of power company limitation encountered, it is not always necessary to start or break away equipment on the first step of starting. For example, a limitation specifying a maximum allowable increase in current or kva per unit of time or current or kva per horsepower per unit of time does not establish a limit to the ultimate inrush current or kva. Thus, a motor which stalls on the first step of starting would be acceptable if its inrush at this time would not exceed the increment established by the power company.

Generally, centrifugal machines, with their rising torque-speed characteristics, present no problems regarding breakaway torques. However, it is desirable to keep these machines unloaded during the entire starting period to reduce acceleration time, and to approach full-load speed before full voltage is applied. Centrifugal refrigeration compressors almost always have the pre-rotation vanes kept automatically closed during the starting period when used with reduced-voltage starters. Large centrifugal pumps can be started against closed discharge valves to minimize loads during starting.

The effects of the inertia of the driven machine and motor should not be ignored when large equipment is involved. Acceleration time, from dead stop to full load speed, is a function of work load, friction load and inertia of the driven machine as well as the inertia of the motor. Even though the work load is decreased due to unloading, a large inertia (wk^2), in conjunction with a decreased torque due to reduced-voltage starting, may result in an abnormally long acceleration time. An over-extended acceleration time could result in over-heating of the driving motor, overheating of resistors or transformers used in reduced-voltage starters, or lack of lubrication in driven equipment which utilizes shaft-mounted oil pumps.

Types of Starters

Figures 2 through 10 identify and show wiring diagrams of the power switching circuits of the more commonly used part-winding and reduced-voltage starters. Also included, for each starting method, is the sequence in which the various contactors operate during the starting period.

Open Circuit Transition

It will be noticed from Figs. 7 and 8 that both autotransformer and star-delta starters include open-circuit transition types. The open-circuit transition starters function, in effect, to disconnect the motor from the line during the brief instant it takes to switch from the starting step to the running step. This is necessary to insure against

an overlap of contacts which would short-circuit part of the transformer winding in the autotransformer starter or cause a dead short in the star-delta starter.

At the instant full voltage is connected at the end of the open transition, there will be an inrush current of much shorter duration than the inrush which

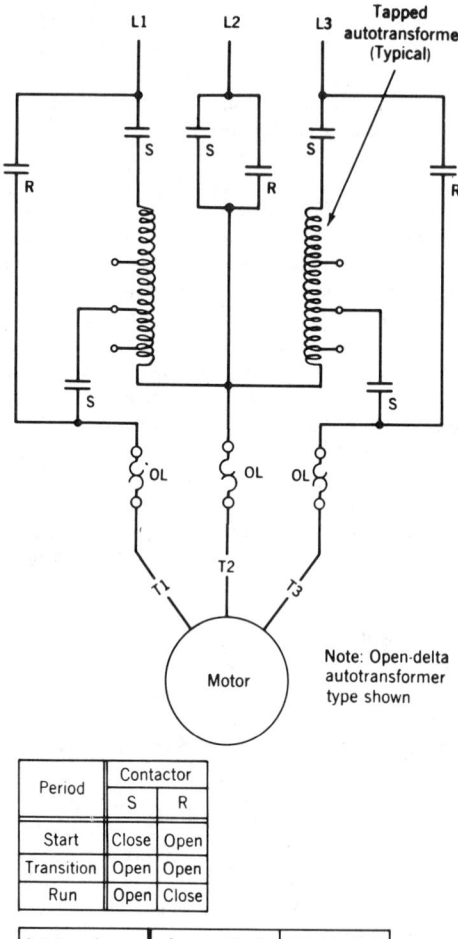

Note: Open-delta autotransformer type shown

Period	Contactor	
	S	R
Start	Close	Open
Transition	Open	Open
Run	Open	Close

Autotransformer tap	Approx. % of locked-rotor kva	% of locked-rotor torque
50%	27	25
65%	45	42
80%	66	64

Torque efficiency = Approx. 95% on all taps

Fig. 7. Autotransformer (open circuit transition).

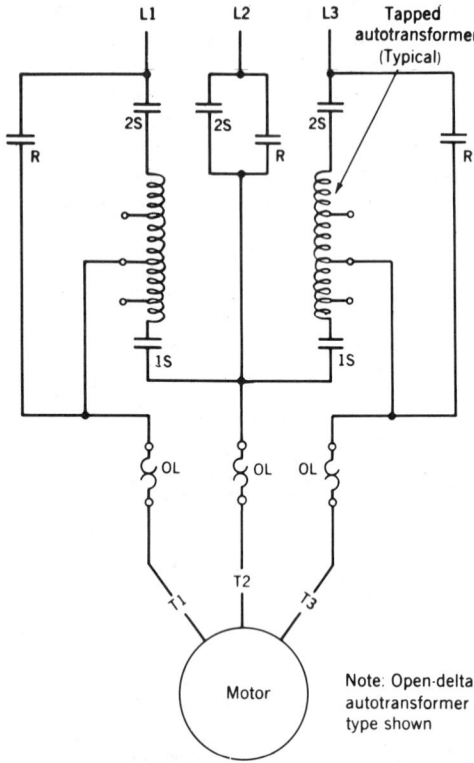

Note: Open-delta autotransformer type shown

Period	Contactor		
	1S	2S	R
Start	Close	Close	Open
Transition	Open	Close	Open
Run	Open	Open	Close

Autotransformer tap	Approx. % of locked-rotor kva	% of locked-rotor torque
50%	27	25
65%	45	42
80%	66	64

Torque efficiency = Approx. 95% on all taps

Fig. 8. Autotransformer (closed circuit transition).

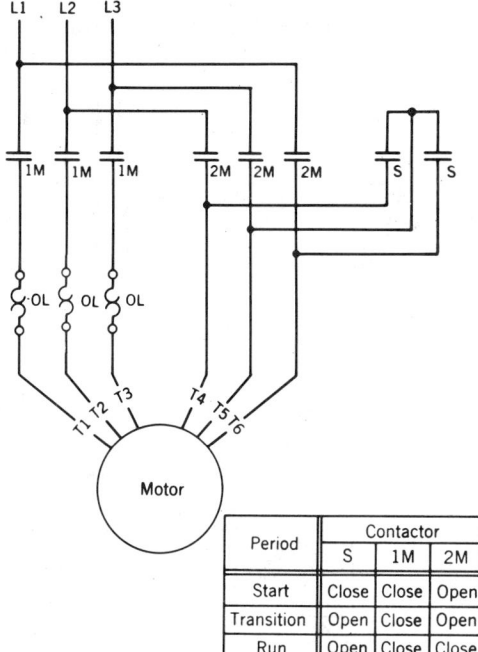

Period	Contactor		
	S	1M	2M
Start	Close	Close	Open
Transition	Open	Close	Open
Run	Open	Close	Close

% of locked-rotor kva = 33 ⅓

% of locked-rotor torque = 33 ⅓

Torque efficiency = 100%

Fig. 9. Star-delta (open circuit transition).

occurs when the motor is first started. This second inrush can be of a magnitude which equals or even exceeds the original inrush. If this second inrush is objectionable to the power company, then a closed-circuit transition starter must be used.

Advantages and Disadvantages

Each type of part-winding and reduced-voltage starter in common use has one or more distinct advantages and usually some disadvantages.

In many situations, the most significant characteristic of the starting method is the ratio of the percent of available motor locked-rotor torque to the percent of motor locked rotor current or kva. This is defined as *torque efficiency* and is given by the formula

$$E_T = \frac{\% \text{ of full-volt. locked-rotor torque}}{\% \text{ of full-volt. locked-rotor kva or line current}}$$

Table 1 data at the beginning of this section includes full-voltage, locked-rotor torques of NEMA Design A, B, and C motors, fours, ranging

sized motors; while Table 2 gives locked-rotor currents for these motors. The data refer to squirrel-cage induction motors.

The tabulated information in Figs. 2 through 10 show the percent of full-voltage, locked-rotor kva and the percent of full-voltage, locked-rotor torque available for each starting method. The particular starting method selected must not only decrease the inrush kva sufficiently to satisfy power company requirements, but must simultaneously permit enough torque to start and accelerate the load if the power company requirement is not an "increment" type of limitation. In other words, a high torque efficiency can often be a significant advantage in the selection of a reduced-voltage starting method.

The following summarizes advantages and disadvantages of the various methods of part winding and reduced voltage starters.

Advantages	Disadvantages

A. *Part Winding (One-Half Winding Start)*

1. Low cost.
2. Closed transition.

1. Frequently will have torque dip at about half-speed during starting.
2. Low torque efficiency.
3. More severe heating and mechanical stresses on motor during starting.

B. *Part Winding (Two-Thirds Winding Start)*

1. Low cost.
2. Closed transition.

1. Low torque efficiency.
2. Unbalanced line currents during starting.
3. More severe heating and mechanical stresses on motor during starting.

C. *Primary Resistance (Two-Step)*

1. Closed transition.
2. Tapped resistors permit field adjustment to inrush kva and starting torque.

1. Relatively expensive.
2. Low torque efficiency.

D. *Primary Resistance (Three-Step)*

1. Closed transition.
2. High power factor during starting step.

1. Expensive.
2. Low torque efficiency.

Note: This starting method is particularly well adapted to power company "increment" type of limitations.

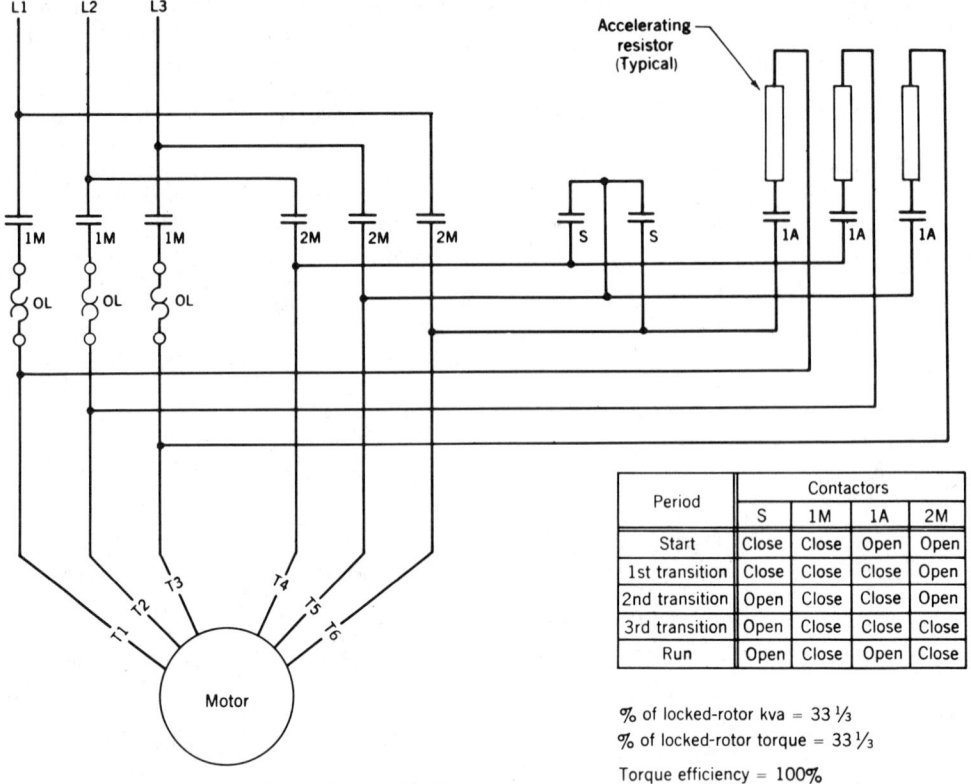

Fig. 10. Star-delta (closed circuit transition).

Period	Contactors			
	S	1M	1A	2M
Start	Close	Close	Open	Open
1st transition	Close	Close	Close	Open
2nd transition	Open	Close	Close	Open
3rd transition	Open	Close	Close	Close
Run	Open	Close	Open	Close

% of locked-rotor kva = 33 ⅓
% of locked-rotor torque = 33 ⅓

Torque efficiency = 100%

Advantages	Disadvantages

E. *Primary Reactor*

1. Closed transition.
2. Tapped reactors permit field adjustment to inrush kva and starting torque.

1. Relatively low cost for high-voltage installations.
2. Low torque efficiency.

Note: This starting method is particularly well adapted to high-voltage systems. Rarely used at low voltage.

F. *Autotransformer (Open-Circuit Transition)*

1. High torque efficiency.
2. Tapped transformer permits field adjustment to inrush kva and starting torque.

1. Relatively expensive.
2. Open-circuit transition.

G. *Autotransformer (Closed-Circuit Transition or Korndorfer Method)*

1. Closed Transition.

1. Expensive.

Advantages	Disadvantages

2. High torque efficiency.
3. Tapped transformer permits field adjustment to inrush kva and starting torque.

H. *Star-Delta (Open-Circuit Transition)*

1. High torque efficiency.
2. Particularly well adapted to long accelerating (starting) periods.

1. Open transition.
2. Low starting torque.

I. *Star-Delta (Closed-Circuit Transition)*

1. Closed transition.
2. High torque efficiency.
3. Particularly well adapted to long accelerating (starting) periods.

1. Expensive.
2. Low starting torque.

USEFUL FORMULAS

Letters and abbreviations used in the following formulas have these meanings:

I Current, amperes

E Potential difference, volts

R Resistance, ohms

W Power, watts

kw Power, kilowatts

va Apparent power, volt-amperes (present only in alternating current circuits with inductance and/or capacitance)

kva Apparent power, kilovolt-amperes

hp Output power, horsepower

eff Efficiency, expressed as a decimal fraction

PF Power Factor, expressed as a decimal fraction (present only in alternating current circuits with inductance and/or capacitance)

T Torque, lb-ft

N Revolutions per minute at rated speed

N_s Revolutions per minute at synchronous speed

Hz Frequency, cycles per second, or Hertz

P Number of poles

s Slip expressed as a fraction of synchronous speed

Direct-Current Circuits

$$I = E/R$$
$$W = IE = I^2R = E^2/R$$

$$kw = W/1000 = IE/1000 = I^2R/1000 = E^2/1000\,R$$
$$hp = (W)(eff)/746 = (kw)(eff)/0.746 = IE\,(eff)/746$$

Single-Phase A-C Circuits

$$va = IE$$
$$kva = va/1000 = IE/1000$$
$$W = (va)(PF) = IE(PF)$$
$$kw = (kva)(PF) = IE(PF)/1000$$
$$hp = (kw)(eff)/0.746 = (kva)(PF)(eff)/0.746 = IE(PF)(eff)/746$$

Three-Phase A-C Circuits

$$va = 1.73IE$$
$$kva = va/1000 = 1.73IE/1000$$
$$W = (va)(PF) = 1.73IE(PF)$$
$$kw = (kva)(PF) = 1.73IE(PF)/1000$$
$$hp = (kw)(eff)/0.746 = (kva)(PF)(eff)/0.746 = IE(PF)(eff)/431$$

Power Transmitted by Shafting

$$hp = TN/5250$$

A-C Motor Speed

$$N_s = \frac{120\ (Hz)}{P}$$

$$N = N_s(1 - s)$$

BUILDING TYPES

Contents

Authorities

Section Editors: *Eugene Stamper* and *Richard L. Koral*

Arthur H. Senner, Greenhouse Heating
Richard S. Arnold, Air Conditioning Printing Plants
Jack F. Schmidt, Church, Theater and Auditorium Air Conditioning
Richard P. Gaulin, Hospital Air Conditioning and Ventilating

C. P. Hedlin, Greenhouse Cooling
Arthur Rose and *S. L. Sloan, Jr.*, Air Conditioning Bowling Centers

SUMMER AIR CONDITIONING FOR RESIDENCES

Year-round air-conditioning systems for residences are designed to keep occupants comfortable during all the hours of occupancy in the year. The calculation of the heating load is quite simple but the calculation of the cooling load is complicated, especially if the latter is determined by the methods used for industrial applications where exact conditions of relative humidity and temperature are required to facilitate manufacturing processes or to obtain proper quality of product.

General Considerations

Heating systems are usually operated at reduced temperatures during sleeping hours for improved sleeping comfort and also for economy. *Cooling systems* are operated continuously during the warm seasons because comfort is just as important at night as during the day.

Interior furnishings and surfaces absorb moisture readily when room relative humidity is high and then release moisture when the system is reducing humidity and temperature of the space.

Reduction of relative humidity occurs only while the cooling system operates. When the compressor is stopped, moisture rise in the space is rapid. In fact, several hours of operation may be required to reduce relative humidity to the desired level but return to the original moisture content may occur within one-half hour after the compressor is stopped. The capacity of the refrigeration equipment should therefore be such that the system will operate continuously during hot, humid weather.

Experience with installations, particularly in the Research Residences at the University of Illinois proved that cooling loads calculated by methods used for commercial and industrial conditions were invariably too large for the residences. One important feature for residential air conditioning is that the selected room temperature for occupant's comfort, usually 75 F, should be maintained throughout the 24-hour daily cycle. This continuous operation makes it possible to take advantage of the ability of the furnishings and the structure to store heat. Some of the stored heat (in the walls, for instance) may never reach the indoors but return to the outdoors during the cool period of the night. The cooling of the structure at night will enable it to store heat early and also through parts of the day. The effect of storage is also to delay the time of peak heat gain for a period which is called the time lag.

In residences it is not necessary to maintain room temperature and humidity as strictly as in industrial installations. Occupants of air-conditioned homes are usually willing to permit some rise in temperature, such as 3 degrees above the setting of the thermostat, at 75 F, when design temperatures prevail. This rise is called "swing." It will usually occur during peak load at design conditions. It should be noted that the rise above 75 F occurs only when the equipment is unable to maintain the desired 75 F temperature.

Residential installations are generally controlled by action of a room thermostat (responsive to dry-bulb temperature) in turning the compressor on and off. If the capacity of the equipment is too large, the compressor will be stopped too much of the time with resultant rise in relative humidity.

By selecting equipment whose capacity is so limited as to permit the 3-degree rise at peak load previously mentioned, the equipment will operate continuously during a large part of the 24-hour period. The desired room temperature—75 F—will be maintained as long as the cooling load does not exceed the equipment capacity and then the temperature will rise 3 degrees for a period.

Factors Affecting Costs

A simple method of calculating cooling load for residences will be presented later. It is important, before applying this method, to consider certain factors that affect the size of the cooling load and, consequently, have an influence on the cost of equipment and also the cost of operation:

1. Awnings and shading devices are particularly important for glass areas exposed to the sun. Since air-conditioned residences are almost always cooled as one zone and it is therefore not possible to shift cooling effect to the different parts of the house to correspond to the load variations in these parts (caused by the travel of the sun), it is important to reduce load variations by use of awnings and shading devices.

2. Insulating materials and constructions that retard flow of heat reduce the heat load.

3. Storm windows should remain in place in summer as well as in winter.

4. Shade from trees and nearby buildings is beneficial, but the question of permanence of the shade from these sources should be raised.

5. Major appliances within the conditioned space

should be vented. This applies to cooking appliances and particularly to domestic clothes dryers.

6. Steps should be taken to reduce the amount of heat entering the attic through the attic roof. These roofs are frequently of lightweight construction and, in addition, have surfaces of dark color that absorb immense quantities of solar heat. The external surface temperature may reach 175 F. Enough heat may be transmitted to produce an attic temperature up to 130 F. Such a high temperature may be reduced by louvers in the end walls but more effectively by enclosing the rafter space by aluminum foil fastened below the rafters so as to form a passage for heated air to flow upward between the rafters to the top of the attic space from where the air may escape from the attic through fans or louvers in the walls.

7. The type of glass used and its arrangement is important in reducing transmitted heat, as will be seen from Table 1.

8. Shading devices are most effective when placed outside of windows, rather than inside, as will be seen in Table 2.

9. Color and finish of building materials have an important effect on the proportion of solar radiation absorbed as shown in Table 3.

Any reduction made in heat gain will reduce not only cost of equipment (because this will be smaller) but also cost of operation.

Simplified Heat Gain Calculation

There has been considerable variation in the methods recommended by associations and individual manufacturers for calculating residence cooling loads. Therefore, at the suggestion of the Federal Housing Administration, a special committee known

Table 1. Effect of Type of Glass on Solar Heat Transmission

Kind of Glass	Relative Heat Transmission
Ordinary single glass	1.000
Double common window	0.851
Single regular plate	0.917
Double regular plate	0.746
Single heat-absorbing plate	0.654
Heat-absorbing plate outside, regular plate inside	0.507

Table 2. Effect of Type of Shading on Heat Gain Through Windows

Type of Shading	Fraction of Gain Through Unshaded Window
Unshaded window	1.00
Inside roller shade	
Half-drawn	0.80
Fully-drawn	0.62
Inside venetian blind	0.65
Canvas awning	0.30
Outside venetian awning	0.15
Outside shading screen	
Solar altitude: 20°	0.38
30°	0.25
40°	0.18

Table 3. Effect of Color and Finish on Absorption of Solar Radiation

Color and Finish	Fraction of Incident Radiation Absorbed
Black	0.92
Red, brown, green	0.73
Yellow	0.60
White, light cream	0.40
Dull metal	0.52
Metallic paint, polished metal	0.40
Highly polished metal	0.25

Surface Color, Type, and Finish	Fraction of Incident Radiation Absorbed*
Slate composition roofing	0.90
Linoleum, red-brown	0.84
Graphite	0.84
Asbestos slate	0.81
Soft rubber, gray	0.65
Concrete	0.60
Red brick	0.55
Porcelain	0.50
Cork	0.45
Vitreous enamel, white	0.39
Polished aluminum	0.30
Aluminum paint	0.20
White Dutch tile	0.18
Anodized aluminum	0.15
Magnesium oxide, evaporated	0.08

*From McAdam's "Heat Transmission."

as the Industry Heat Gain Joint Study Group prepared a recommendation for standardization of as many factors influencing the calculation of residence cooling load as practicable.

As a result of general agreement of industry with the recommendations, methods have been prepared for calculating heat gain that conform to the basic recommendations. The methods for calculating heat gain and the forms and tables adopted are almost identical in the case of Air-Conditioning and Refrigeration Institute (ARI) and National Environmental Systems Contractors Association (NESCA) and, therefore, will be used for illustration in this text. The method of calculating heat gain as developed by Institute of Boiler and Radiator Manufacturers * (I-B-R) is also based upon the recommendations of the committee regarding indoor and outdoor conditions used for design, the type of operation to be desired, the items to be considered, etc., and may be preferred by those who are already familiar with the details of calculation. While there are certain differences in the method of presenting the steps to be taken and also some variation in the details of allowing for effect of shade and solar radiation, it seems probable that all three methods will yield a computed heat gain that is nearer the actual heat gain than the gain obtained by previous methods used for residence calculations.

The method of calculating heat gain that will be described here is based upon the following considerations:

1. Adoption of 75 F dry-bulb as the temperature to be maintained in all parts of the residence, 24 hours a day, during the cooling season.

2. Adoption, for design, of an outdoor dry-bulb temperature of 90, 95, 100, 105 and 110 F (whichever applies for the city in which the installation is to be made). Table 4 shows outdoor design conditions for United States. Designations *L, M,* and *H* for daily temperature range have the following meanings:

Designation	Calculation Value	Application Range
L (Low)	12 deg	less than 15 deg
M (Medium)	20 deg	15 to 25 deg
H (High)	30 deg	more than 25 deg

Wet-bulb design temperatures in Table 4 are used for equipment selection but not for calculating the cooling load.

---

* Currently the Hydronics Institute.

3. A relative humidity of 45 to 55% to be expected only when the compressor is running and has been running long enough to reduce the relative humidity to a stable condition. Relative humidity need not be rigidly controlled.

4. A room temperature rise of 3 F above the thermostat setting is to be permitted and is recommended. This rise (swing) will only occur when the cooling equipment does not have the capacity to maintain temperature at the peak load that occurs occasionally.

Swings of 4½ or 6 F are sometimes used and will result in selection of smaller cooling equipment capactiy. The designer should be aware of the possibility of complaints due to range of indoor temperature encountered.

5. Latent heat load is not calculated but is taken as 30% of the total sensible load.

6. Heat transmission coefficients, by general acceptance, are based on the design heat transmission coefficients for building materials published in the ASHRAE Guide and Data Book and reprinted elsewhere in this handbook.

In order to reduce the space required for publication and to simplify the process of selecting coefficients it is common practice for manufacturers and trade associations to combine several quite similar constructions in a group under one average coefficient when preparing tables of coefficients, such as Table 5. In such tables for further simplification, and in order to eliminate one multiplying operation for the user, the product of the heat transfer coefficient and temperature difference are given as "factors." The user can then obtain the heat gain through an area by multiplying the factor by the area. For example, if the value of the "factor" for a wall is listed as 1.9 and the area of a wall is 20 sq ft, the heat gain is 1.9 x 20 = 38 Btuh.

7. If there is no provision for adding a fixed quantity of outdoor air for ventilation, the natural infiltration may be calculated as follows:

Multiply gross area of outside walls for each room by the factor shown in Table 5, Section 5, to obtain the heat gain in Btuh for the room. The factor should be found in the column of the table headed by the design outdoor temperature assumed for the locality.

If there is mechanical ventilation, the quantity (cfm) of air introduced per hour should be multiplied by the factor for mechanical ventilation found in Table 5, Section 5, in the column headed by the proper design outdoor temperature.

Table 4. Outdoor Design Conditions

Tables 4-9 reprinted, by permission, from ARI Standard 230-62.

State and City	Winter Dry-Bulb Temp. (F)	Summer Dry-Bulb Temp. (F)	Summer Wet-Bulb Temp. (F)	Latitude	Daily Range
ALABAMA					
Birmingham	10	95	78	35	M
Gadsden	10	95	78	35	M
Mobile	20	90	80	30	L
Montgomery	20	95	79	30	M
Tuscaloosa	10	95	80	35	M
ARIZONA					
Bisbee	30	100	72	30	H
Flagstaff	−5	*85	61	35	H
Globe	30	105	76	35	H
Nogales	30	105	72	30	H
Phoenix	35	105	76	35	H
Tucson	30	100	72	30	H
Winslow	−5	95	65	35	H
Yuma	40	110	78	35	H
ARKANSAS					
Bentonville	0	95	76	35	M
Fort Smith	5	95	79	35	M
Hot Springs	10	95	78	35	M
Little Rock	10	95	79	35	M
Pine Bluff	10	95	81	35	M
Texarkana	10	100	79	35	M
CALIFORNIA					
Bakersfield	30	105	71	35	H
El Centro	35	110	80	35	H
Eureka	30	90	65	40	M
Fresno	30	105	72	35	H
Los Angeles	40	90	71	35	M
Montague	15	95	66	40	M
Needles	25	115	75	35	H
Oakland	30	*80	64	40	M
Sacramento	30	95	71	40	H
San Bernardino	30	105	72	35	H
San Diego	45	*80	70	35	L
San Francisco	35	*80	63	40	M
San Jose	40	90	70	35	M
COLORADO					
Boulder	−15	95	64	40	M
Colorado Springs	−10	95	62	40	H
Denver	−10	95	64	40	H
Durango	−5	95	65	35	H
Fort Collins	−15	95	65	40	M
Grand Junction	−5	95	64	40	H
Leadville	−10	95	64	40	M
Pueblo	−15	95	65	40	H
CONNECTICUT					
Bridgeport	0	*85	75	40	L
Hartford	0	90	76	40	M
New Haven	0	*85	76	40	M
New London	5	*85	75	40	L
Torrington	0	90	75	40	M
Waterbury	0	90	75	40	M
DELAWARE					
Dover	10	90	78	40	M
Milford	10	90	78	40	M
Wilmington	5	90	77	40	M
DISTRICT OF COLUMBIA					
Washington	10	90	77	40	M
FLORIDA					
Apalachicola	25	95	80	30	L
Fort Myers	40	95	80	25	M
Gainesville	30	95	78	30	M
Jacksonville	30	95	79	30	M
Key West	55	100	78	25	L
Miami	45	90	79	25	L
Orlando	35	90	79	30	M
Pensacola	25	95	81	30	L
Tallahassee	25	95	79	30	M
Tampa	35	95	79	30	M
GEORGIA					
Athens	10	95	76	35	M
Atlanta	10	95	77	35	M
Augusta	20	100	79	35	M
Brunswick	25	95	78	30	L
Columbus	20	100	76	35	M
Macon	20	95	79	35	M
Rome	10	95	76	35	M
Savannah	25	95	80	30	M
Waycross	25	95	78	30	M
IDAHO					
Boise	−10	95	66	45	H
Idaho Falls	−15	90	64	45	*H
Lewiston	−10	95	65	45	H
Pocatello	−15	90	63	45	*H
Twin Falls	−15	95	65	40	H
ILLINOIS					
Aurora-Joliet	−10	95	76	40	M
Bloomington	−10	95	76	40	M
Cairo	0	100	78	35	M
Champaign-Urbana	−10	95	77	40	M
Chicago	−10	95	76	40	M
Danville	−10	95	77	40	M
Decatur	−10	95	77	40	M
Elgin	−15	95	78	40	M
Moline-Rock Island	−10	95	77	40	M
Peoria	−15	95	77	40	M
Rockford	−15	95	78	40	M
Springfield	−10	95	78	40	M
INDIANA					
Evansville	−5	95	78	40	M
Fort Wayne	−5	95	76	40	M
Indianapolis	−10	90	77	40	M
Lafayette	−10	95	76	40	M
South Bend	−10	95	76	40	M
Terre Haute	−5	95	78	40	M
IOWA					
Burlington	−10	95	78	40	M
Cedar Rapids	−15	95	78	40	M
Charles City	−20	95	75	45	M
Clinton	−15	95	78	40	M
Council Bluffs	−15	100	78	40	M

*Whenever the summer dry-bulb temperatures are less than 90 F, use Factors in Table 5 and 9 for 90 F. When a high daily temperature range is shown for a 90 F design dry-bulb temperature, use factors for 90 F and Medium daily temperature range in Tables 5 and 9.

Table 4. Outdoor Design Conditions (Continued)

State and City	Winter Dry-Bulb Temp. (F)	Summer Dry-Bulb Temp. (F)	Wet-Bulb Temp. (F)	Latitude	Daily Range
IOWA—Continued					
Davenport	−10	95	78	40	M
Des Moines	−15	95	77	40	M
Dubuque	−15	95	78	40	M
Fort Dodge	−15	95	78	40	M
Keokuk	−15	95	78	40	M
Marshalltown	−15	95	78	40	M
Sioux City	−15	95	77	40	M
Waterloo	−15	95	78	40	M
KANSAS					
Atchison-Leavenworth	−10	100	76	40	M
Concordia	−10	95	78	40	M
Dodge City	−10	95	73	40	H
Iola	− 5	100	75	40	M
Salina	−10	100	78	40	M
Topeka	−10	95	78	40	M
Wichita	− 5	100	76	40	M
KENTUCKY					
Bowling Green	0	95	78	35	M
Frankfort	0	95	78	40	M
Hopkinsville	0	95	78	35	M
Lexington	0	95	77	40	M
Louisville	0	95	78	40	M
Owensboro	0	95	78	40	M
LOUISIANA					
Alexandria	20	95	80	30	M
Baton Rouge	20	95	80	30	M
New Orleans	25	95	80	30	L
Shreveport	15	95	80	30	M
MAINE					
Augusta	−15	*85	73	45	L
Bangor	−20	*85	73	45	L
Bar Harbor	−10	*85	73	45	L
Belfast	−10	*85	73	45	L
Eastport	−10	*85	70	45	L
Lewiston	−10	*85	73	45	L
Millinocket	−15	*85	73	45	M
Portland	−10	*85	73	45	M
Rumford	−15	*85	73	45	L
MARYLAND					
Annapolis	10	90	78	40	M
Baltimore	10	90	78	40	M
Cambridge	10	90	78	40	L
Cumberland	0	90	75	40	M
Frederick	5	90	78	40	M
Salisbury	10	90	78	40	M
MASSACHUSETTS					
Amherst	− 5	90	75	40	M
Boston	0	*85	74	40	M
Fall River-New Bedford	0	*85	75	40	L
Fitchburg	− 5	90	75	45	M
Lawrence-Lowell	− 5	*85	74	45	M
Nantucket	0	*85	72	40	L
Pittsfield	−10	90	75	40	M

State and City	Winter Dry-Bulb Temp. (F)	Summer Dry-Bulb Temp. (F)	Wet-Bulb Temp. (F)	Latitude	Daily Range
MASSACHUSETTS—Continued					
Plymouth	0	*85	75	40	L
Springfield	− 5	90	75	40	M
Worcester	− 5	90	73	40	M
MICHIGAN					
Alpena	−10	90	75	45	M
Ann Arbor	− 5	90	75	40	M
Big Rapids	− 5	90	75	45	M
Cadillac	−10	90	75	45	M
Calumet-Houghton	−20	*80	73	45	M
Detroit	− 5	90	75	40	M
Escanaba	−20	*85	74	45	M
Flint	−10	90	75	45	M
Grand Rapids	− 5	90	74	45	M
Kalamazoo	− 5	90	75	40	M
Lansing	−10	90	75	45	M
Ludington	− 5	90	75	45	M
Marquette	−15	*80	73	45	M
Muskegon	− 5	90	74	45	M
Port Huron	−10	90	75	45	M
Saginaw	−10	90	75	45	M
Sault Ste. Marie	−20	*80	71	45	M
MINNESOTA					
Alexandria	−25	*85	74	45	M
Duluth	−25	*80	71	45	M
Minneapolis-St. Paul	−25	90	75	45	M
St. Cloud	−25	90	75	45	M
MISSISSIPPI					
Biloxi	25	90	80	30	L
Columbus	10	95	78	35	M
Corinth	5	95	78	35	M
Hattiesburg	20	95	80	30	M
Jackson	15	95	79	30	M
Meridian	15	95	79	30	M
Natchez	15	95	78	30	M
Vicksburg	15	95	78	30	M
MISSOURI					
Columbia	−10	100	78	40	M
Hannibal	−10	95	77	40	M
Kansas City	−10	100	77	40	M
St. Joseph	−10	100	78	40	M
St. Louis	− 5	95	78	40	M
Springfield	− 5	100	77	40	M
MONTANA					
Billings	−30	90	66	45	*H
Butte	−30	*85	59	45	H
Great Falls	−40	90	63	50	*H
Havre	−40	95	70	50	H
Helena	−40	90	63	45	*H
Kalispell	−30	90	63	50	H
Miles City	−35	95	69	45	H
Missoula	−30	90	63	45	*H
NEBRASKA					
Grand Island	−15	95	75	40	H
Hastings	−15	95	75	40	M

If there is mechanical ventilation, no heat gain by infiltration is considered.

Some designers assume that there should be the equivalent of one half air change per hour in a residence either by infiltration or mechanical means.

8. Heat gain from occupants is assumed to average 300 Btuh per person and to be divided between living and dining rooms.

9. Heat gain from appliances is assumed to be 1200 Btuh and to be added to the kitchen heat gain.

Table 4.   Outdoor Design Conditions (Continued)

State and City	Winter Dry-Bulb Temp. (F)	Summer Dry-Bulb Temp. (F)	Wet-Bulb Temp. (F)	Latitude	Daily Range
NEBRASKA—Continued					
Lincoln	−15	95	77	40	M
Norfolk	−15	95	78	40	M
North Platte	−15	100	73	40	H
Omaha	−15	100	78	40	M
Valentine	−20	95	78	45	M
NEVADA					
Elko	−10	95	63	40	H
Las Vegas	10	110	71	35	H
Reno	5	95	63	40	H
Tonopah	5	90	63	40	M
Winnemucca	−10	95	65	40	H
NEW HAMPSHIRE					
Berlin	−15	*85	73	45	H
Claremont	−15	*85	73	45	M
Franklin	−15	*85	73	45	M
Hanover	−15	*85	73	45	M
Keene	−10	*85	73	45	M
Manchester	−10	*85	74	45	M
Portsmouth	0	*85	74	45	L
NEW JERSEY					
Asbury Park	5	90	78	40	L
Atlantic City	10	90	78	40	L
Bridgeton	5	90	78	40	L
Camden	5	90	78	40	L
Jersey City	0	90	75	40	L
Newark	0	90	76	40	M
New Brunswick	5	90	75	40	L
Paterson	0	90	75	40	L
Trenton	0	90	78	40	L
NEW MEXICO					
Albuquerque	10	95	65	35	M
El Morro	0	*85	65	35	H
Raton	−5	95	65	35	H
Roswell	5	100	71	35	H
Santa Fe	5	90	64	35	M
Tucumcari	5	95	70	35	H
NEW YORK					
Albany	−10	90	74	45	M
Auburn	−10	90	74	45	M
Binghamton	−5	90	72	40	M
Buffalo-Niagara Falls	−5	*85	73	45	M
Elmira	−5	90	73	40	M
Glens Falls	−15	90	73	45	M
Ithaca	−5	90	73	40	M
Jamestown	−5	90	74	40	M
Lake Placid	−15	90	73	45	M
New York	5	90	76	40	M
Ogdensburg	−20	*85	73	45	M
Oneonta	−10	90	73	45	M
Oswego	−5	90	74	45	M
Port Jervis	0	90	75	40	L
Rochester	−5	90	74	45	M
Syracuse	−10	90	74	45	M
Watertown	−15	*85	73	45	M

State and City	Winter Dry-Bulb Temp. (F)	Summer Dry-Bulb Temp. (F)	Wet-Bulb Temp. (F)	Latitude	Daily Range
NORTH CAROLINA					
Asheville	5	90	75	35	M
Charlotte	15	95	77	35	M
Greensboro	10	90	76	35	M
Hatteras	20	90	80	35	L
New Bern	20	95	81	35	L
Raleigh	15	95	78	35	M
Salisbury	10	90	78	35	M
Wilmington	20	90	81	35	M
Winston-Salem	10	90	76	35	M
NORTH DAKOTA					
Bismarck	−30	95	73	45	H
Devils Lake	−30	90	70	50	M
Dickinson	−30	95	70	45	H
Fargo	−30	95	74	45	H
Grand Forks	−30	90	72	50	M
Jamestown	−30	95	73	45	M
Minot	−35	90	71	50	M
Pembina	−35	90	73	50	M
Williston	−35	90	73	50	M
OHIO					
Akron	−5	90	75	40	M
Cincinnati	−5	95	78	40	M
Cleveland	−5	90	75	40	M
Columbus	−5	90	76	40	M
Dayton	−5	90	76	40	M
Lima	−5	90	75	40	M
Marion	−5	90	75	40	M
Sandusky	−5	90	75	40	M
Toledo	−5	90	75	40	M
Warren-Youngstown	−5	90	75	40	M
OKLAHOMA					
Ardmore	5	100	78	35	M
Bartlesville	−5	100	77	35	M
Muskogee	0	95	79	35	M
Oklahoma City	0	100	77	35	M
Tulsa	0	100	78	35	M
Waynoka	−5	105	75	35	M
OREGON					
Arlington	5	95	68	45	M
Baker	−15	90	65	45	M
Eugene	15	90	67	45	*H
Medford	20	95	68	40	H
Pendleton	−10	90	65	45	*H
Portland	10	*85	68	45	M
Roseburg	20	90	66	45	*H
Salem	15	90	68	45	*H
PENNSYLVANIA					
Altoona	−5	90	74	40	M
Bethlehem	0	90	75	40	M
Erie	−5	*85	74	40	M
Harrisburg-York	5	90	75	40	M
New Castle	−5	90	75	40	M
Oil City	−5	90	75	40	M
Philadelphia	5	90	77	40	M
Pittsburgh	−5	90	74	40	M
Reading	5	90	75	40	M

Forms for Calculating Heat Gain

The person making the calculation for cooling load may select a prepared form such as published by ARI, I-B-R,* or NESCA, or may prepare

* Hydronics Institute.

another form. In an illustrative problem that follows will be solved according to the ARI method and will be recorded on ARI Form 230 ALC-1. In proceeding with calculations for heat gain, it is suggested that first a form such as shown in Fig. 1 be made to facilitate listing all window, door, wall,

Table 4. Outdoor Design Conditions (Continued)

State and City	Winter Dry-Bulb Temp. (F)	Summer Dry-Bulb Temp. (F)	Summer Wet-Bulb Temp. (F)	Latitude	Daily Range	State and City	Winter Dry-Bulb Temp. (F)	Summer Dry-Bulb Temp. (F)	Summer Wet-Bulb Temp. (F)	Latitude	Daily Range
PENNSYLVANIA—Continued						**VERMONT**					
Scranton	0	90	75	40	M	Bennington	−10	90	73	45	M
Warren	− 5	90	75	40	M	Burlington	−15	90	73	45	M
Williamsport	− 5	90	75	40	M	Montpelier	−20	90	73	45	M
RHODE ISLAND						Newport	−20	*85	73	45	M
						Rutland	−15	90	73	45	M
Block Island	5	*85	75	40	L						
Kingston	0	*85	75	40	L	**VIRGINIA**					
Providence	0	90	75	40	M	Charlottesville	10	90	78	40	M
SOUTH CAROLINA						Danville	10	90	78	35	M
						Lynchburg	10	90	76	35	M
Charleston	20	90	80	35	L	Norfolk	15	90	78	35	M
Columbia	20	95	78	35	M	Petersburg	10	90	78	35	M
Florence	20	95	79	35	M	Richmond	10	90	78	40	M
Greenville	10	95	76	35	M	Roanoke	5	90	75	35	M
Spartanburg	10	95	77	35	M	Wytheville	5	90	76	35	M
SOUTH DAKOTA						**WASHINGTON**					
Aberdeen	−25	95	75	45	M	Aberdeen	20	*85	64	45	L
Huron	−20	100	75	45	H	Bellingham	10	*80	65	50	L
Pierre	−20	95	74	45	H	Everett	15	*80	65	50	L
Rapid City	−20	95	71	45	H	Olympia	15	*80	65	45	H
Sioux Falls	−20	95	75	45	H	Seattle	15	*80	65	50	M
Watertown	−25	95	74	45	M	Spokane	−15	90	64	50	*H
TENNESSEE						Tacoma	15	*80	64	45	M
						Walla Walla	−10	90	67	45	*H
Chattanooga	10	95	77	35	M	Wenatchee	−10	90	65	50	M
Jackson	5	95	80	35	M	Yakima	− 5	90	67	45	*H
Johnson City	5	95	78	35	M						
Knoxville	5	95	77	35	M	**WEST VIRGINIA**					
Memphis	5	95	79	35	M	Bluefield	0	95	75	35	M
Nashville	5	95	78	35	M	Charleston	0	90	76	40	M
TEXAS						Elkins	− 5	90	73	40	M
						Fairmont	0	90	75	40	M
Abilene	5	95	75	30	M	Huntington	0	90	76	40	M
Amarillo	0	95	71	35	H	Martinsburg	0	90	77	40	M
Austin	15	100	78	30	M	Parkersburg	0	90	77	40	M
Brownsville	30	95	80	25	M	Wheeling	− 5	90	75	40	M
Corpus Christi	25	95	80	30	M						
Dallas-Fort Worth	10	100	78	35	M	**WISCONSIN**					
Del Rio	20	100	79	30	H	Ashland	−25	*80	71	45	M
El Paso	20	100	69	30	M	Beloit	−15	95	78	45	M
Galveston	25	95	81	30	L	Eau Claire	−20	90	74	45	M
Houston	20	95	80	30	M	Green Bay	−20	90	73	45	M
Palestine	10	100	78	30	M	La Crosse	−20	95	76	45	M
Port Arthur	20	95	80	30	M	Madison	−20	90	75	45	M
San Antonio	20	100	77	30	M	Milwaukee	−15	90	75	45	M
Waco	10	100	78	30	M	Oshkosh	−20	90	75	45	M
Wink	10	95	74	30	H	Sheboygan	−20	90	75	45	M
UTAH						**WYOMING**					
						Casper	−25	90	62	45	*H
Logan	−10	95	65	40	H	Cheyenne	−20	90	62	40	*H
Milford	− 5	95	66	40	H	Lander	−30	90	65	45	*H
Ogden	− 5	90	65	40	*H	Sheridan	−30	90	66	45	*H
Salt Lake City	0	95	66	40	H	Yellowstone Park	−35	*85	62	45	H

ceiling, and floor areas in detail showing shaded and sunlit portions of each in each room separately for each wall, in a column showing the direction faced by the wall. In preparing the form for Fig. 1, it is advantageous to make spaces large enough so that the mathematical operation for obtaining each

area can be indicated. This tends to prevent errors and simplifies combining of certain areas when required for use on Fig. 2, the ARI form. Fig. 1 is equally valuable for determining areas or combinations of areas for use on I-B-R (Hydronics Inst.), or NESCA Forms.

Table 5. Residential Cooling Load Estimate Factors

Reprinted, by permission, from ARI Standard 230-62

Item	Heat Gain Factors, Btuh per Sq Ft (for inside design temperatures of 75 F)				
1a. Windows—Solar cooling load	No Shading Devices	Roller Shades Half-Drawn	Draperies or Venetian Blinds	Ventilated Awnings	Outside Shading Screens
(1) Regular Single Glass					
(a) N (or shaded)	19	14	11	16	4
(b) NE and NW	51	37	28	17	12
(c) E and W	77	57	43	18	18
(d) SE and SW	66	48	36	17	9
(e) S	36	26	20	16	5
(2) Regular Double Glass					
(a) N (or shaded)	17	13	10	11	3
(b) NE and NW	45	35	26	12	8
(c) E and W	67	53	40	12	12
(d) SE and SW	57	45	34	12	6
(e) S	31	25	18	11	3
(3) Heat-Absorbing Double Glass					
(a) N (or shaded)	10	8	7	8	2
(b) NE and NW	26	22	18	9	7
(c) E and W	40	33	28	10	10
(d SE and SW	34	28	23	9	5
(e) S	20	16	13	9	3

1b. Window—Air-to-air cooling load	Outdoor Design Temperature, F				
	90	95	100	105	110
(1) Single Glass	8	12	16	20	24
(2) Double Glass	4	7	9	11	13

2. Walls and Doors	Daily Temperature Range									
	Low		Medium				High			
	Outdoor Design Temperature, F									
	90	95	90	95	100	105	95	100	105	110
a. Frame and veneer on frame walls										
(1) No insulation	5.9	7.2	4.8	6.1	7.4	8.7	4.8	6.1	7.4	8.7
(2) Less than 1 in. insulation, or one reflective air space	4.3	5.2	3.5	4.5	5.4	6.4	3.5	4.5	5.4	6.4
(3) 1 in. to 2 in. insulation, or two reflective air spaces	2.9	3.6	2.4	3.1	3.7	4.4	2.4	3.1	3.7	4.4
(4) More than 2 in. insulation, or three reflective air spaces	1.8	2.2	1.5	1.9	2.3	2.7	1.5	1.9	2.3	2.7
b. Masonry walls, 8 in. block or brick										
(1) Plastered or plain	7.3	9.7	5.4	7.8	10.2	12.6	5.4	7.8	10.2	12.6
(2) Furred, no insulation	4.6	6.1	3.4	4.9	6.4	7.9	3.4	4.9	6.4	7.9
(3) Furred, with less than 1 in. insulation, or one reflective air space	3.1	4.1	2.3	3.3	4.3	5.3	2.3	3.3	4.3	5.3
(4) Furred, with 1 in. to 2 in. insulation, or two reflective air spaces	2.1	2.8	1.6	2.3	3.0	3.7	1.6	2.3	3.0	3.7
(5) Furred, with more than 2 in. insulation, or three reflective air spaces	1.4	1.8	1.0	1.5	1.9	2.4	1.0	1.5	1.9	2.4
c. Partitions										
(1) Frame, finished one side only, no insulation	8.4	11.4	6.0	9.0	12.0	15.0	6.0	9.0	12.0	15.0

Table 5. Residential Cooling Load Factors (Concluded)

Item	Low		Medium				High			
	Outdoor Design Temperature									
	90	95	90	95	100	105	95	100	105	110
	Heat Gain Factors, Btuh per Sq Ft									
c. Partitions (continued)										
(2) Frame, finished both sides, no insulation	4.8	6.5	3.4	5.1	6.8	8.5	3.4	5.1	6.8	8.5
(3) Frame, finished both sides, more than 1 in. insulation, or two reflective air spaces.	2.0	2.7	1.4	2.1	2.8	3.5	1.4	2.1	2.8	3.5
(4) Masonry, finished one side, no insulation	2.6	4.4	1.2	3.0	4.7	6.5	1.2	3.0	4.7	6.5
d. Wood doors*	11.3	13.8	9.3	11.8	14.3	16.8	9.3	11.8	14.3	16.8
3. Ceilings and Roofs**										
a. Ceiling under naturally vented attic or vented flat roof										
(1) Uninsulated — dark	9.9	11.0	9.0	10.1	11.3	12.4	9.0	10.1	11.3	12.4
light	8.1	9.2	7.1	8.3	9.4	10.6	7.1	8.3	9.4	10.6
(2) Less than 2 in. insulation, or one reflective air space										
dark	4.3	4.8	3.9	4.4	4.9	5.4	3.9	4.4	4.9	5.4
light	3.5	4.0	3.1	3.6	4.1	4.6	3.1	3.6	4.1	4.6
(3) 2 in. to 4 in. insulation, or two reflective air spaces										
dark	2.6	2.9	2.3	2.6	2.9	3.2	2.3	2.6	2.9	3.2
light	2.1	2.4	1.9	2.2	2.5	2.8	1.9	2.2	2.5	2.8
(4) More than 4 in. insulation, or three or more reflective air spaces										
dark	1.7	1.9	1.6	1.8	2.0	2.2	1.6	1.8	2.0	2.2
light	1.4	1.6	1.2	1.4	1.6	1.8	1.2	1.4	1.6	1.8
b. Built-up roof, no ceiling										
(1) Uninsulated — dark	17.2	19.2	15.6	17.6	19.6	21.6	15.6	17.6	19.6	21.6
light	14.0	16.0	12.4	14.4	16.4	18.4	12.4	14.4	16.4	18.4
(2) 2 in. roof insulation — dark	8.6	9.6	7.8	8.8	9.8	10.8	7.8	8.8	9.8	10.8
light	7.0	8.0	6.2	7.2	8.2	9.2	6.2	7.2	8.2	9.2
(3) 3 in. roof insulation — dark	6.0	6.7	5.5	6.2	6.9	7.6	5.5	6.2	6.9	7.6
light	4.9	5.6	4.3	5.0	5.7	6.4	4.3	5.0	5.7	6.4
c. Ceilings under unconditioned rooms	2.7	3.6	1.9	2.9	3.8	4.8	1.9	2.9	3.8	4.8
4. Floors										
a. Over unconditioned rooms	3.4	4.6	2.4	3.6	4.8	6.0	2.4	3.6	4.8	6.0
b. Over basement, enclosed crawl space, or concrete slab on ground	0	0	0	0	0	0	0	0	0	0
c. Over open crawl space	4.8	6.5	3.4	5.1	6.8	8.5	3.4	5.1	6.8	8.5
5. Outside Air										
a. Infiltration, Btuh per sq ft of gross exposed wall area	1.1	1.5	1.1	1.5	1.9	2.2	1.6	1.9	2.2	2.6
b. Mech. ventilation, Btuh per cfm	16.2	21.6	16.2	21.6	27.0	32.4	21.6	27.0	32.4	37.8
6. People	No. of occupants x 300 Btuh									
7. Kitchen appliance allowance	1200 Btuh									
8. Heat gain to ducts (Btuh)	Subtotal (Item 1 thru 7) x Factor									
9. Total Estimated Sensible Heat Gain (Btuh)	Total (Items 1 thru 8)									
10. Latent Heat Gain Allowance	Item 9 x 0.3									
11. Total Estimated Design Heat Gain (Btuh)	Item 9 + Item 10									

*Consider glass area of doors as windows.
**Factors for walls and roofs include allowance for solar effect and heat lag.

CALCULATION OF HEAT GAIN FOR RESIDENCE — ILLUSTRATIVE PROBLEM

Determine the heat gain for a residence shown in Fig. 3, 4, and 5.

Available Data:

Location: St. Louis, Mo.
Latitude: 40 degrees North.
Temperature range for summer: Medium.
Design outdoor temperature: 95 F db, 78 F wb.
Design indoor temperature: 75 F with 3 F swing.
Occupants: 2

Construction Details:

Windows: Single glass, not weatherstripped, roller shades half-drawn.
Walls: Wood siding, paper, sheathing, 3⅝ inch rockwool, studs and plaster.
Partition: Studs, lath and plaster both sides.
Ceiling: Lath and plaster, 3⅝ inch rock wool, attic space above, no floor in attic, roof dark.
Doors: Wood—13/16 inch thick with 30% glass.
Floor: No heated space underneath.
Roof overhang: 1 ft, 4 inches on south side.

Roof Overhang and Shade Line

If there are roof overhangs on the E, SE, S, SW, or W sides, determine the depth of shade below the overhang and draw the shade line on the elevation of the residence.

Shade line factors given in Table 6 are expressed as the distance the shade line falls beneath the edge of the overhang for each foot of overhang width. The tabulated values are the average of the shade line values for the five hours of maximum solar intensity on each wall orientation shown, for a solar declination of 18° North (August 1). Northeast and northwest-facing windows are not effectively protected by roof overhangs, and in most cases no credit should be taken for shading of them.

The shade line in this problem will be required only for the south wall which has an overhang of 1 ft 4 in. The depth of the shade line below the overhang is the product of the overhang, 1.35 ft, and the factor, 2.6, found in Table 6 in the 40°

latitude column and in the "South" line. This product is 3.5 ft. The ceiling height is 9 inches above the level of the overhang and the tops of the windows are 9 inches below the overhang. The windows are 2 ft, 3 inches high and are therefore shaded. The glass in the doors is also above the shade line.

Table 6. Shade Line Factors
Adapted, by permission, from ARI Standard 230-62

Direction Window Faces	North Latitude, Degrees						
	25	30	35	40	45	50	55
E	0.8	0.8	0.8	0.8	0.8	0.8	0.8
SE	1.9	1.6	1.4	1.3	1.1	1.0	0.9
S	10.1	5.4	3.6	2.6	2.0	1.7	1.4
SW	1.9	1.6	1.4	1.3	1.1	1.0	0.9
W	0.8	0.8	0.8	0.8	0.8	0.8	0.8

Note: Distance shade line falls below the edge of the overhang equals the shade line factor multiplied by width of overhang. Values are averages for five hours of greatest solar intensity on August 1.

Table. 7. Heat Gain to Ducts
Reprinted, by permission, from ARI Standard 230-62

Item	Factor
a. Ducts in attic with 1-inch insulation	0.15
b. Ducts in attic with 2-inch insulation	0.10
c. Ducts in slab floors; insulated ducts in enclosed crawl space, unconditioned basement or furred spaces	0.05

Table 8. Cooling Air Quantity Factors
Reprinted, by permission, from ARI Standard 230-62

*a. Cooling air quantity per ton capacity (cfm)						
350	370	390	400	410	430	450

b. Uncorrected factors						
0.029	0.031	0.032	0.033	0.034	0.036	0.037

c. Capacity multiplier factor (From Table 9)						

d. Corrected factor (Item b × Item c) (Enter in factor column, Item 12)						

*Cfm selection to be that on which the manufacturer's Standard Rating is based.

Determination of Heat Gain

A most important step in determining heat gain in a residence is to calculate the areas of all windows, doors, walls, ceilings and floors in detail showing gross, net, shaded and sunlit parts of each. An orderly procedure is indicated on Fig. 1. These areas should be entered for each room in separate columns showing direction faced by each wall.

Dimensions needed are found in Fig. 3. The shade line height is 3 ft, 9 inches above the floor.

It is advantageous to mark the mathematical

Table 9. Capacity Multiplier Factors
Reprinted, by permission, from ARI Standard 230-62

Outside Design Conditions		Desired Indoor Temperature Swing, F		
Temp., F	Daily Range	6	4½	3
AIR COOLED UNITS				
90	M	0.69	0.83	0.97
	L	0.71	0.84	0.98
95	H	0.74	0.88	1.02
	M	0.75	0.89	1.03
	L	0.77	0.90	1.04
100	H	0.81	0.95	1.08
	M	0.82	0.96	1.09
105	H	0.86	1.01	1.16
	M	0.87	1.02	1.17
110	H	0.92	1.07	1.22

EVAPORATIVELY COOLED UNITS

Outside Design Wet Bulb Temperature, F

		65	70	75	78	80	65	70	75	78	80	65	70	75	78	80
90	M, L	0.74	0.75	0.77	0.78	0.79	0.89	0.90	0.92	0.93	0.94	1.03	1.05	1.07	1.08	1.09
95	H, M, L	0.79	0.80	0.81	0.82	0.83	0.93	0.94	0.96	0.97	0.98	1.07	1.09	1.11	1.12	1.13
100	H, M	—	0.84	0.85	0.87	0.88	—	0.99	1.00	1.02	1.03	—	1.12	1.14	1.15	1.16
105	H, M	—	0.87	0.88	0.89	0.90	—	1.02	1.03	1.05	1.06	—	1.17	1.19	1.20	1.21
110	H	—	—	0.92	0.93	0.94	—	—	1.08	1.09	1.10	—	—	1.23	1.24	1.25

WATER COOLED UNITS

Leaving Water Temperature, F

		90	95	100	105	110	90	95	100	105	110	90	95	100	105	110
90	M, L	0.74	0.77	0.81	0.83	0.86	0.88	0.93	0.97	1.00	1.03	1.03	1.08	1.12	1.16	1.20
95	H, M, L	0.78	0.82	0.85	0.88	0.91	0.93	0.97	1.01	1.05	1.08	1.07	1.12	1.17	1.21	1.25
100	H, M	0.82	0.87	0.91	0.94	0.97	0.97	1.02	1.06	1.09	1.12	1.10	1.16	1.20	1.24	1.28
105	H, M	0.86	0.90	0.94	0.97	1.00	1.00	1.05	1.10	1.14	1.17	1.15	1.20	1.25	1.30	1.34
110	H	0.89	0.94	0.98	1.01	1.04	1.04	1.09	1.14	1.18	1.22	1.18	1.24	1.30	1.34	1.38

This table to be used according to the relationship:
(Calculated Heat Gain) × (Capacity Multiplier Factor) = (Equipment ARI Standard Capacity Rating).

		Living		Dining		Kitchen	Bedroom #1	
Line	Areas	Wall Facing						
		N	W	S	W	S	E	S
1	Window Area — Gross	4×3.83×4.25 65	0	2×3.5×2.25 16	0	3.5×2.25 8	3.5×2.25 8	2×3.5×2.25 16
2	Shaded	65	0	16	0	8	0	16
3	Sunlit	0	0	0	0	0	8	0
4	Door Area — Gross	2.5×6.7 17	0	0	17	0	0	0
5	Glass in Door — Total	30%×(4) 5	0	0	5	0	0	0
6	Shaded	5	0	0	0	0	0	0
7	Sunlit	0	0	0	5	0	0	0
8	Door Area, Net	(4)-(5) 12	0	0	12	0	0	0
9	Shaded	12	0	0	0	0	0	0
10	Sunlit	0	0	0	12	0	0	0
11	Wall Area — Gross	23.5×8 188	5×8 40	11.5×8 92	11×8 88	7.5×8 60	13×8 104	13.5×8 108
12	Net	(11)-(1)-(4) 106	40	(11)-(1) 76	(11)-(4) 71	(11)-(1) 52	(11)-(1) 96	(11)-(1) 92
13	Shaded	106	0	33	0	(12)-(14) 24	0	(12)-(14) 41
14	Sunlit	0	40	3.75×11.5 43	(12)-(13) 71	3.75×7.5 28	(12)-(13) 96	3.75×13.5 51
15	Warm Partition — Area	0	8.5×8 68	0	0	0	0	0
16	Ceiling Area	23.5×13.5 317		11.5×11 126		7.5×11 83	13×13.5 176	
17	Floor Area	317		126		83	176	

All areas are in squre feet. Numbers in parentheses ( ) refer to line numbers on this table.

Bedroom #2		Bedroom #3		Bath	Foyer	Hall	
Wall Facing							Total Area
N	E	N	W	S	S	Inside	
2×3.5×2.25	3.5×225	2×3.5×2.25		3.5×2.25			
16	8	16	0	8	0	0	161
16	0	16	0	8	0	0	145
0	8	0	0	0	0	0	16
0	0	0	0	0	3×7 / 21	0	55
0	0	0	0	0	30% of(4) / 6	0	16
0	0	0	0	0	6	0	11
0	0	0	0	0	0	0	5
0	0	0	0	0	(4)-(5) / 15	0	39
0	0	0	0	0	(8)-(10) / 5	0	17
0	0	0	0	0	3.75×3 / 10	0	22
10.5×8 / 84	13.5×8 / 108	10×8 / 80	1.5×8 / 12	6.83×8 / 55	4×8 / 32	0	1051
(11)-(1) / 68	(11)-(1) / 100	(11)-(1) / 64	12	(11)-(1) / 47	(11)-(4) / 11	0	835
(12)-(14) / 68	0	64	12	21	(12)-(14) / 7	0	376
0	(12)-(13) / 100	0	0	26	3.75×1 / 4	0	459
0	0	0	0	0	0	0	68
10.5×13.5 / 142		10×10.5 / 105		6.83×9.7 / 66	4×11.2 / 45	6×10.3 / 62	1122
142		105		66	45	62	1122

Fig. 1. Area calculation form for heat gain.

ITEM		ENTIRE HOUSE		1 LIVING ROOM		2 DINING ROOM		3 KITCHEN	
Description	Factor 1	Area	Btuh	Area	Btuh	Area	Btuh	Area	Btuh
1a. Windows (Solar) *Shaded*	14	54	756			16	224	8	112
N	14	102	1428	70	980				
Sunlit { E	57	16	912						
S	26	0							
W	57	5	285			5	285		
1b. Windows (Air-to-air)	12	177	2120	70	840	21	252	8	96
2. Walls ~~and Doors~~ *Net*	1.9	835	1585	146	277	147	279	52	99
Doors, *Net*	11.8	39	461	12	142	12	142		
Partition	5.1	68	347	68	347				
3. Ceilings and Roofs	2.6	1120	2916	317	824	126	327	83	216
4. Floors *(Not used)* *(No Gain)*									
●5. Outside Air *Infiltration* } *Gross Wall*	1.5	1051	1577	228	342	180	270	60	90
6. People 2 *4*	300	—	1200	—	600	—	600	—	
7. Appliance Allowance		—	1200	—	—	—	—	—	1200
Sub-Total Sens. Heat Gain 3		—	14787	—	4352	—	2379	—	1813
8. Duct Heat Gain 4	10%	—	1479	—	435	—	238	—	181
9. Sensible Heat Gain (Sub-Total + Item 8)		—	16266	—	4787	—	2617	—	1994
10. Latent Heat Gain Allowance (Item 9 × 0.3)		—	4880	—	1436	—	785	—	598
11. Total Estimated Heat Gain (Item 9 + Item 10)		—	21146	—	6223	—	3402	—	2592
12. Air Quantity (cfm) 5	0.034		719		212		115		88

Notes: 1. Obtain heat gain factors from Table 5.   2. Never use less than three people as a minimum.

Fig. 2. Residential heat gain estimate form, adapted from ARI Standard 230-62.

operation in the space within each area for quick reference to dimensions or later checking of accuracy of computations.

It will be noted that in Fig. 1 the areas (gross, net, sunlit, and shaded) are given in greater detail than required for recording on Form Fig. 2. This has been done to provide a basic area calculation form from which any desired areas can be selected (and combined, if desired) to suit not only the form, Fig. 2, but also I-B-R * Cooling Load Calculation Sheet No. 3001 and the NESCA Work Sheet J.

When entries are completed on Fig. 1, all areas should be transferred to Fig. 2, after which the heat gain factors for each type of surface should be entered in the factor column opposite their corresponding area. These factors have been determined for each type of surface, having been obtained from Table 5.

* Hydronics Institute.

Factors for Heat Gain through Surface Areas for residence in illustrative problem	
Surface	Heat Gain, Btuh per Sq Ft
Windows shaded or north (with roller shades, half-drawn)	14
Windows, east	57
Windows, south (if sunlit)	26
Windows, west	57
Windows, (air-to-air) (This is transfer due to air temp. diff)	12
Walls	1.9
Doors (solid wood part)	11.8
Partitions (next to unconditioned space)	5.1
Ceilings (including roof above)	2.6
Gross wall area (for air infiltration)	1.5

4	BEDROOM #1	5	BEDROOM #2	6	BEDROOM #3	7	BATH #1	8	FOYER	9	HALL
Area	Btuh	Area	Btuh	Area	Btuh	Area	Btuh	Area	Btuh	Area	Btuh
16	224					8	112	6	84		
		16	224	16	224						
8	456	8	456								
24	288	24	288	16	192	8	96	6	72		
188	357	168	319	76	144	47	89	11	21		
176	457	142	369	105	273	66	172	44	114	62	161
212	318	192	288	92	138	55	82	32	48		
—		—		—		—		—		—	
—	—	—	—	—	—	—	—	—	—	—	—
—	2100	—	1944	—	971	—	551	—	516	—	161
—	210	—	194	—	97	—	55	—	52	—	16
—	2310	—	2138	—	1068	—	606	—	568	—	177
—	693	—	642	—	320	—	182	—	171	—	53
—	3003	—	2770	—	1388	—	788	—	739	—	230
	104		94		47		27		25		8

Notes: 3. Total Sections 1a through 7.
4. Multiple Sensible Heat Gain Sub-Total by applicable factor from Table 7 and enter in Btuh column(s).
5. Multiply Section 11 by corrected factor from Table 8.

Each of the Sections of Fig. 2 will be discussed in the order in which it appears on the form, Fig. 2.

It will be noted that Section 1a of Fig. 2 is for solar radiation which reaches glass areas by direct radiation, if the area is sunlit, or by diffuse or sky radiation, if the areas face north or are shaded by roof overhang, other buildings, objects, or trees. Heat gain factors in Table 5, Section 1a, take into account the type of glass, the direction faced, and the effect of shading devices inside and outside.

It is suggested that the two top lines in Section 1a be marked "Shaded" and "N," and that the three lower lines be marked for the compass points E, S, and W, and marked "sunlit." Shaded and N (north) can be combined if desired.

In addition to the heat gain due to solar and diffuse radiation, there is transfer of heat through glass due to the temperature difference between the air on the two sides. The factor for this gain is found in Section 1b of Table 5 for the type of glass (double or single) in the column for 95 F outdoor temperature. The glass area by which to multiply the factor to obtain the heat gain is the gross or total glass area for each room. This gain is entered in Section 1b of Fig. 2.

In Section 2 of Fig. 2, *net wall* and *net door* areas have been placed on separate lines so that the factor for each as obtained from Table 5 can be used with the net wall area and with net door area (gross door area less glass in door). The areas were transferred to Fig. 2 from Fig. 1. Factors for walls include allowance for solar effect and heat lag.

The partition in Section 2 of Fig. 2 is between the living room and unconditioned space. The factor is found in Table 5 and the area was transferred from Fig. 1.

In Section 3 of Fig. 2 the factor for ceiling and roof is that found in Section 3 of Table 5. The area

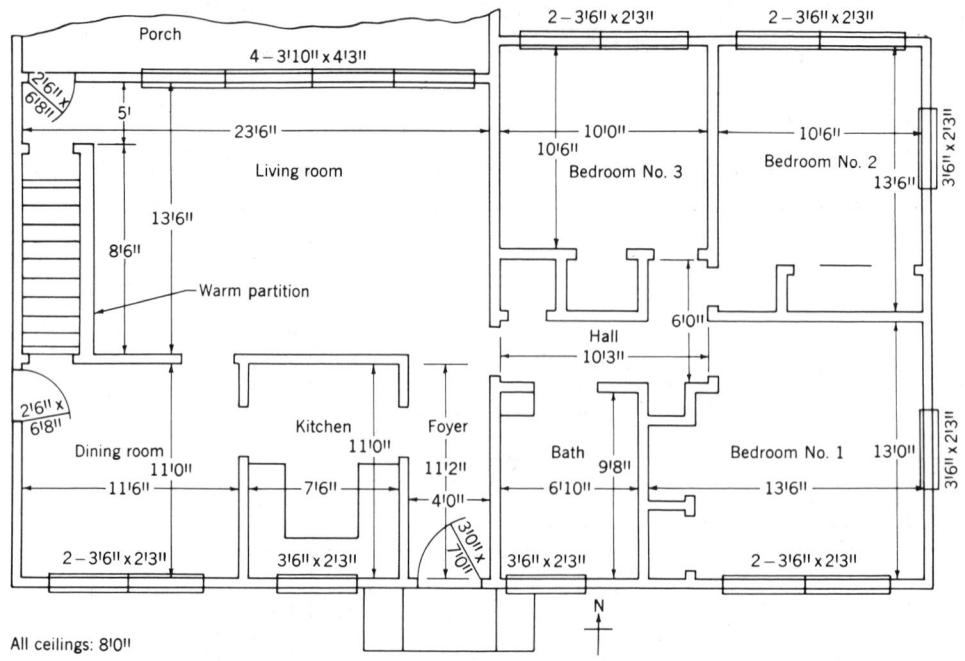

Fig. 3. Residence Floor Plan for Problem.

is transferred from Fig. 1. Factors for ceilings and roof combinations, and for built-up roofs without ceilings include allowances for solar effect and heat lag characteristics.

The floor in this problem has the same area as the ceiling but is considered as having no heat gain, so no entries are made in Section 4 of Fig. 2.

In Section 5, Outdoor Air, on Fig. 2 the heat gain is that due to infiltration. There is no mechanical ventilation air in this problem. The area used is the *gross exposed wall area* and the factor is found in Section 5 of Table 5. The heat gain is the product of the area and the factor, and is expressed in Btuh.

The heat gain from four occupants in Section 6 of Fig. 2 is 4 x 300 = 1200 Btuh and is divided equally between the living and dining rooms.

The appliance heat gain in Section 7 of Fig. 2 is considered to be 1200 Btuh and is included with the kitchen heat gain.

The *duct heat gain* in Section 8 of Fig. 2 is assumed to be equivalent to 10% of the heat gain in Sections 1a to 7 on Fig. 2. It is assumed that the ducts are in the attic and are covered with 2 inches

of insulation. The factor is found in Table 7 (which is also Table B on the back of ARI Form 230 ALC—1).

Total sensible heat gain is the sum of the heat gains in Sections 7 and 8 of Fig. 2 and is entered in Section 9.

Latent heat gain is assumed to be equivalent to 30% of the *sensible heat gain* listed in Section 9, and is entered in Section 10 of Fig. 2.

Total estimated heat gain is the sum of *heat gains* in Sections 9 and 10 of Fig. 2. It is entered in Section 11 of Fig. 2.

The air *quantity* in Section 12 of Fig. 2 is the air (cfm) that must be circulated to remove from the rooms the *total estimated heat gains* shown in Sec. 11 of Fig. 2. It is found by multiplying the *total estimated heat gain* from Section 11 by a corrected *cooling air quantity factor* from Table 8. An uncorrected factor is first found in the column showing the cfm per ton of refrigeration on which the unit being considered is based. The uncorrected factor is then multiplied by a *capacity multiplier factor* found

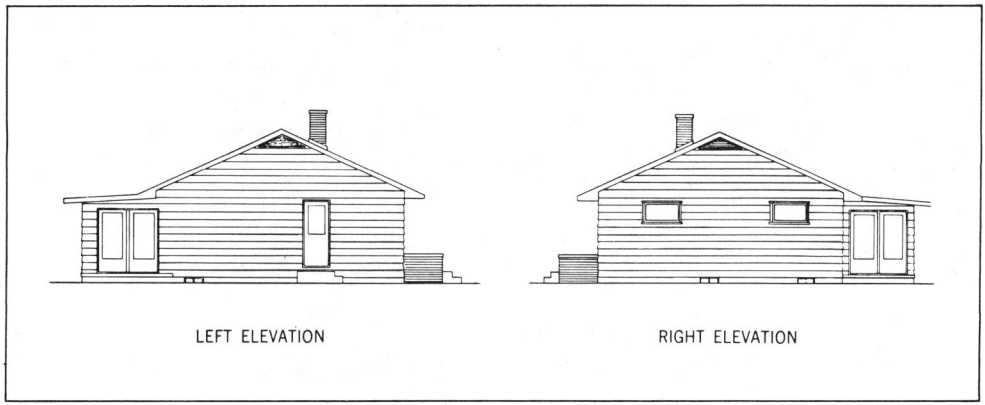

Fig. 4. Left and right elevation for residence.

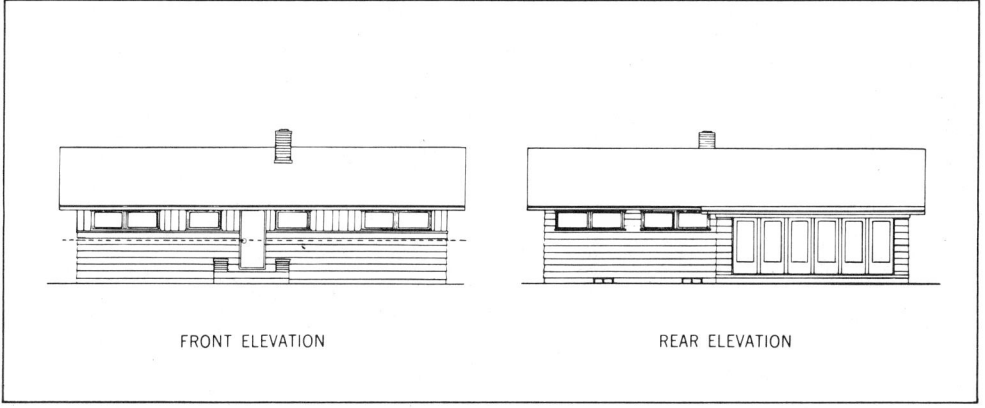

Fig. 5. Front and rear elevation for residence.

in Table 9 for: the type of unit, the desired *indoor temperature swing,* the *outdoor design temperature,* the outdoor wet-bulb design temperature (if unit is evaporatively cooled) and the *leaving water temperature* (if the unit is water cooled). This product is the *corrected cooling air quantity factor* to be entered in line *d* of Table 8 and also entered in the factor column of Fig. 2.

It was understood that the unit selected was air cooled and was operating with a *cooling air quantity* of 400 cfm per ton of refrigeration. Design conditions for the problem were: 95 F outdoor temperature, medium daily temperature range, and a 3 F deg indoor temperature swing.

Equipment Capacity Needed

The equipment capacity needed is obtained by multiplying the Btuh heat gain for the entire residence found in Section 11 on Fig. 2 by the *capacity multiplier factor* from Table 9.

The required capacity is $21,146 \times 1.03 = 21,780$ Btuh.

The quantity of air that must be circulated in a cooling system is that required to minimize frosting of the coil in the unit selected. The system must, therefore, be designed for this required flow and further must not be adjusted in any manner that will restrict the required flow.

GREENHOUSE HEATING

The heating of greenhouses often involves somewhat low air temperatures at which the heat emission of radiators is increased over that at the conventional 70 deg air temperature. In addition, these buildings have an extraordinarily large glass area, so that calculations can frequently be simplified by use of tables based on glass area only, with allowance for infiltration and the relatively small wall surface. Such tables are included here; with the exception of Table 3, they are based on data by Arthur H. Senner of the Department of Agriculture and have been somewhat rearranged.

Table 1. Customary Temperatures for Greenhouses

Type of Plant	Temperature Range, Deg F
Carnation	45-55
Conservatory (general collection, winter, garden, etc.)	60-65
Cool	45-50
Cucumber	65-70
Fern	60-65
Forcing	60-65
General purpose	55-60
Lettuce	40-45
Orchid, warm	65-70
Orchid, cool	50-55
Palm, warm	60-65
Palm, cool	50-55
Propagating	55-60
Rose	55-60
Sweet pea	45-50
Tomato	65-70
Tropical	65-70
Violet	40-45

Table 3. Cast Iron Radiator Surface Required

Greenhouse Temperature, Deg F						
40	45	50	55	60	65	70
Sq Ft of Radiator per 1000 Btu per Hour Low Pressure Steam (220 Deg)						
3.06	3.20	3.38	3.50	3.64	3.82	4.00
Gravity Hot Water (160 Deg)						
5.37	5.68	6.00	6.32	6.70	7.14	7.62
Forced Hot Water (210 Deg)						
3.38	3.50	3.64	3.82	4.00	4.19	4.36

Table 3a. Size of Mains for Gravity Hot Water Pipe Coils

(Based on 160 deg F water, and a total length of supply and return mains of not over 200 ft)

Size of Supply and Return, In	Diameter of Pipe in Coil, Inches	
	3½	2
	Linear Feet of Pipe in Coil	
1¼	50	80
1½	80	140
2	190	320
2½	380	650
3	660	1,100
3½	1,000	1,700
4	1,500	2,600
5	3,000	5,000
6	5,200	8,700

Usual greenhouse inside air temperatures are given in Table 1, and these data are suggested for use where the grower has not specified otherwise.

Table 2. Heat Emission from 1000 Linear Feet of Pipe Radiation

(Based on 220 deg steam, 160 deg gravity hot water, and 210 deg forced hot water)

Pipe Dia., In	Heating Medium	Greenhouse Temperature, Deg F						
		40	45	50	55	60	65	70
		Btu per Hr per 1000 Linear Feet of Pipe						
1¼	Low pressure steam	187,400	180,300	173,300	166,300	160,000	153,300	147,000
2	Gravity hot water	150,200	143,500	135,300	127,700	121,000	113,700	106,800
2	Forced hot water	236,700	227,500	217,500	207,700	199,500	190,800	182,500
3½	Gravity hot water	245,000	232,300	219,700	204,000	193,000	181,800	169,800

Table 4. Heat Required to Maintain Greenhouse Temperature

(Including both heat transmission through glass and allowance for infiltration)

Inside Temperature, Deg F	Outside Temperature, Deg F					
	20	10	0	−10	−20	−30
	Btu Required per Hour per 1000 Sq Ft of Glass					
40	28,600	41,000	53,800	66,800	79,200	92,200
45	34,800	47,500	60,100	72,600	85,300	98,000
50	41,250	53,800	66,450	79,000	91,700	104,500
55	47,500	60,500	72,800	85,300	98,500	111,000
60	53,700	66,600	79,100	91,700	104,500	117,300
65	59,700	72,600	85,400	98,300	111,000	123,000
70	66,250	79,200	91,700	104,800	117,500	129,500

Table 5. Area of Glass Heated per Linear Foot of 1 ¼ Inch Steam Pipe

(Based on 220 deg Steam)

Inside Temp., Deg F	Outside Temperature, Deg F					
	20	10	0	−10	−20	−30
	Area of Glass, Sq Ft					
40	7.57	5.27	4.01	3.23	2.73	2.35
45	5.92	4.34	3.43	2.84	2.42	2.11
50	4.83	3.70	3.00	2.52	2.18	1.91
55	4.08	3.18	2.65	2.27	1.97	1.75
60	3.46	2.79	2.35	2.03	1.78	1.59
65	2.98	2.46	2.09	1.82	1.61	1.45
70	2.54	2.13	1.83	1.61	1.43	1.30

Table 7. Area of Glass Heated per Linear Foot of 3 ½ Inch Gravity Hot Water Pipe

(Based on 160 deg water)

Inside Temp., Deg F	Outside Temperature, Deg F					
	20	10	0	−10	−20	−30
	Area of Glass, Sq Ft					
40	8.60	5.97	4.56	3.68	3.10	2.67
45	6.66	4.88	3.87	3.20	2.73	2.37
50	5.33	4.08	3.31	2.78	2.39	2.11
55	4.32	3.38	2.81	2.41	2.08	1.85
60	3.60	2.92	2.45	2.12	1.85	1.65
65	3.00	2.48	2.10	1.82	1.62	1.46
70	2.56	2.14	1.84	1.62	1.44	1.31

Table 6. Area of Glass Heated per Linear Foot of 2 Inch Gravity Hot Water Pipe

(Based on 160 deg water)

Inside Temp., Deg F	Outside Temperature, Deg F					
	20	10	0	−10	−20	−30
	Area of Glass, Sq Ft					
40	5.28	3.68	2.80	2.26	1.91	1.64
45	4.13	3.03	2.39	1.98	1.69	1.47
50	3.29	2.52	2.04	1.72	1.48	1.30
55	2.72	2.12	1.76	1.51	1.31	1.16
60	2.24	1.82	1.53	1.32	1.16	1.03
65	1.89	1.56	1.33	1.15	1.02	.92
70	1.63	1.36	1.17	1.03	.91	.83

Table 8. Area of Glass Heated per Linear Foot of 2 Inch Forced Hot Water Pipe

(Based on 210 deg water)

Inside Temp., Deg F	Outside Temperature, Deg F					
	20	10	0	−10	−20	−30
	Area of Glass, Sq Ft					
40	8.34	5.80	4.42	3.57	3.00	2.58
45	6.54	4.78	3.79	3.13	2.67	2.33
50	5.28	4.05	3.28	2.77	2.38	.2.09
55	4.40	3.43	2.86	2.45	2.12	1.88
60	3.70	2.98	2.52	2.17	1.91	1.70
65	3.20	2.64	2.24	1.95	1.73	1.55
70	2.77	2.32	1.99	1.75	1.55	1.41

In calculating heat losses from greenhouses, it is possible, due to the low ratio of losses through infiltration and walls to that through glass, simply to add an allowance for air leakage and wall loss and base the whole calculation on glass area. This is done in Table 4. Determine the wall area and assume 2, 3, or 4 sq ft of wall (depending on construction) as equal to 1 sq ft of glass and use Table 4 in which allowance has already been made for infiltration. The figures in the body of Table 4 can also be used to determine pounds of steam required if three digits are crossed off.

Once the load has been calculated, the length of pipe required in the coils, or amount of radiator surface needed, can be found by reference to Table 2 or 3; or directly from Tables 5 to 8.

AIR CONDITIONING PRINTING PLANTS

To make a profit in the printing business, work must be scheduled accurately, production time kept as low as possible, and the last ounce of productivity squeezed from press and printer. It is this last ounce which produces profit. Increased economies do not spring from one source, but rather from the elimination of a host of small delays.

Owners of air conditioned shops have found that where an air conditioning system has been designed to fit a printing plant's needs, handsome dividends will result. These requirements vary according to production volume, physical structure, and type of printing.

There are three basic printing processes:

1. Letterpress, the oldest and most common technique, wherein ink is deposited on raised surfaces,
2. Offset, the process of printing from a flat plane,
3. Gravure, or rotogravure, involves printing from a depressed surface.

The air conditioning engineer should familiarize himself with these aspects, the plant's geographic location, and heat and moisture loads to be handled through the year. There is no one combination of temperature and humidity that can be specified for all printing operations; each is an individual problem. Therefore, close cooperation with the plant owner or manager is valuable and necessary in obtaining the best system at reasonable owning and operating costs. Table 1 lists the air conditioning needs for the various departments of a large printing plant.

Problems peculiar to this application and recommendations for solving them are described below. For simplification, methods for solving climatic hindrances are discussed under product classifications, starting with the smallest unit.

Packaged Units

One of the biggest headaches in any printing plant, regardless of size, is the effect of weather variations on paper. Too much humidity makes paper stock bloat; too little makes it shrink. And when paper is piled in stacks, such as for sheet-fed presses, expansion or contraction can produce wavy edges and make proper feeding difficult, if not impossible.

Moisture content is not the same for all types of paper. Coated stock will absorb less moisture than newsprint, for example. With a variation of 10% relative humidity above or below 50%, paper will absorb or release 1% of its substance weight. But in an air conditioned plant, paper can be left uncovered for weeks without becoming wavy-edged.

Packaged units can provide both heating and cooling either with conventional heat pump operation, or cooling alone. Packages in single or multiple installation permits application in virtually any type office or shop at what is generally the lowest cost. In one installation, a duct from the unit serves a paper storage and supply room to keep paper stock at the same conditions as those in the printing area. Maximum utilization of space, so important in a small shop, is possible because the paper can be used as it comes from the storage area, without a conditioning period in the press room.

This plant owner states that a year-round temperature of 80 F and 60% rh maintains optimum conditions. Paper curling is eliminated and stock holds it size, texture and weight.

He also found that air conditioning helps him save 50% on roller cost. These rollers for the printing presses are made of glue and glycerin; rollers (which pick up and deposit ink) are susceptible to high humidity. It is not uncommon for a seasoned roller to swell ¼ inch in circumference in a relative humidity of 70%, making good press work hopeless.

An Indianapolis, Ind., printer solved his humidity problems with a self-contained unit adapted for the job. A manually-operated bypass was installed in the back of the unit. The design also included a plenum, electric strip heaters for reheat control, and a room thermostat which automatically starts the refrigeration compressor whenever temperature goes above 80 F. This system gives process control plus em-

Table 1—Air Conditioning Needs of a Printing Plant

Plant Area	Air Conditioning Needs
Office	Comfort air conditioning
Offset plate department	Humidification, dehumidification, air cleaning and cooling
Foundry	Ventilation, air cleaning and spot cooling
Composing room	Comfort air conditioning and ventilation
Proof room	Comfort air conditioning
Job press room	Comfort air conditioning and humidification
Offset press room	Humidification, dehumidification, heating, cooling and air cleaning
Web press room	Heating, humidification, and air cleaning
Bindery	Winter humidification
Stock conditioning	Adjustable humidity
Stock room	Same humidity as press room
Shipping room	Heating

ployee comfort. Humidity for process control is held between 47 and 53%. A hygrostat set at 50% relative humidity operates the refrigeration cycle and also the electric strip heaters.

Packaged equipment can be a boon to the printer who has all the problems of a large plant, but whose budget cannot stand investment in a central system, or where future plans may call for add-on equipment.

Accurate Registration

Offset lithography is a printing process using photography as a means of transferring an image to a thin, photo-sensitive metal plate. This plate is then used to print the image. Offset is a rapidly expanding process because of its excellence in color printing.

In color work, from one to four plates may be used. The negatives from which these plates are made are subject to expansion and contraction, and when plates vary in size, colors will overlap and be out of register. Coated plates, used for long press runs, will alter their photographic sensitivity and cause variations in exposure time when atmospheric conditions fluctuate.

National Bureau of Standards and the Lithographic Technical Foundation prescribe that humidity should be held between 45% and 55%, within plus or minus 2%. Temperature, determined by the printer, must also be held constant, within ± 2F.

The Foundation has also labeled static electricity as a frequent, and unsuspected, cause of mis-register. The best answer to this problem, it states, is preconditioned paper and air conditioning with a minimum of 50% relative humidity.

In multi-color printing there may be some time between impressions or runs for respective colors, and unless the paper is kept under controlled conditions during that period, moisture will change the size of the paper so that exact register is impossible. Humidity changes have been known to stretch or shrink large sheets a half inch or more, whereas perfect register requires accuracy within 1/200th of an inch.

Applied Systems

For buildings of moderate size, where load can generally be satisfied by a system having a cooling capacity from 40 to 150 tons, and where space, centralized maintenance, unobtrusiveness and greater durability are required, packaged refrigeration and air handling components are used in what is called an applied or built-up system.

In Scranton, Pa., two ducts from an air treating unit serve two large offset rooms. Room air returns directly to the unit without the use of ducts. Outside air is admitted to the air mixing chamber through minimum and maximum dampers. The minimum outside air damper is opened when the supply fan is started. The maximum damper is closed during summer and is modulated during marginal weather.

A manual summer-winter switch produces this action on the damper, as well as energizing the refrigeration cycle for summer and de-energizing it for winter. A compensated dewpoint controller allows dehumidified air temperature to rise or fall, depending on the room humidity as determined by a humidistat in the master zone.

Zone temperature control is obtained by operation of zone thermostats on their respective reheat coils. This has proven very economical. The compensated feature of the dewpoint control contributes to saving refrigeration in summer and heating in winter.

A temperature of 78 F with relative humidity of 48%, plus or minus 2% is maintained. Close control is secured by a pneumatic system which used the unloading feature of a compressor for its success.

Plate-making rooms present serious problems for the lithographer. The high concentration of fumes from etching acid, which could affect the health and comfort of workers, must be dissipated and removed. Cooling and/or heating coils in the vicinity of this room should have copper tubes and fins to resist acid corrosion. Sprays with proper bleed-off are an additional precaution in preserving coil life.

Central Station Systems

One of the most difficult projects to air condition in the printing field is a newspaper building, a conglomeration of areas with different heat loads and problems.

In winter, a newspaper press room must be heated and ventilated, also humidified. Humidification results in better press work, better looking paper and snappier photo reproductions. One surprising result of humidification is that it may reduce ink consumption, because when paper and rollers are in optimum condition, less ink needs to be run to secure proper color.

When humidity is too low, paper becomes dry and the resultant static electricity makes paper cling to itself and to press rollers. In high-speed presses, static is a major cause of breaks in the web, the continuous roll of paper that is fed through the press. When the web breaks, the press must be stopped and re-threaded.

In a high speed web press, centrifugal force causes particles of ink (mist) and paper (fly) to leave the

press and float about the room, coming to rest and accumulating on walls and equipment. The amount of mist and fly increases with press speed. This is both a nuisance and a fire hazard, since printing ink is composed of kerosene, carbon, linseed oil and other inflammables.

There is an even greater fire hazard with the use of rotogravure (brown) ink. Static sparks have been known to ignite the mixture of mist and fly thrown off by fast gravure presses.

To combat mist, and to solve many other problems caused by weather variations, an air conditioning system should be installed with complete filtration apparatus for the press room.

In one huge press room, 150 ft long, 50 ft wide and 30 ft high, four air cleaning units are spaced along and over both rows of presses. Each unit is equipped with two automatic, dry-type filters with a capacity of 5,000 cfm. Three exhaust fans draw room air through the filter media, a common exhaust duct and a central air shaft. Since only filtered air passes through the duct, fire hazard and maintenance in the duct are reduced.

Ink mist can be handled in two basic ways. One is to use all outside air, with an exhaust system located so that the major portion of the mist is removed and not permitted to diffuse into the room. The second is to use part return air, depending on filters to remove the mist, thus maintaining a low concentration in the space and keeping the cooling coils from being clogged.

Steel wool, glass wool, or other types of non-automatic filters may be used, but they are not recommended for ink mist removal. Steel and glass wool filters clog easily, and the inflammable nature of printing ink adds to the fire hazard.

To benefit press operators, control humidity, and remove ink mist and paper fly, a distribution system was designed introducing air at three different levels. By supplying conditioned air at the lowest level and exhausting it at the highest point immediately above the presses, ink and mist are directed, with the aid of thermal currents of rising heat, from the presses to the point where it is trapped.

Air should be supplied to the press room at a lower rate than is exhausted. The negative air pressure which results is desired to avoid ink leakage through doors and openings into other portions of the newspaper building. Fifteen to twenty air changes per hour is recommended.

Figuring The Cooling Load

Determining the load factors in a web press room is a highly specialized problem. With intermittent operation, such as a newspaper, load factors of fairly large proportion can be applied. But where press operation is almost continuous, such as for magazine or telephone directory printing, the load factor approaches 100% of the operating load. The latter may be considerably lower than the starting load.

To calculate the load factor and room load, it is necessary to carefully analyze (1) ratio of press production to full capacity, (2) starting input to running power requirement, (3) operating time to continuous operation, and (4) frequency of operation.

For example, 31 motors of 40 hp each, plus an additional 142 hp load from small motors, contribute to the heat load at the newspaper plant discussed above. After studying press operation, and considering thermal storage, operating time and fly-wheel effect, a load factor of 0.75 was applied to the nameplate horsepower of the press, and a load factor of 0.10 to the small motors. Each installation must be evaluated individually, but elements of storage, requirements for providing satisfactory press performance at the most reasonable cost, and experience all contribute to establishing load factors.

A reduction in size and cost of air ducts may be effected by a careful scrutiny of the heat load, taking into consideration these factors. Further reduction in duct size may be realized by increasing air velocity, since noise is no problem in a noise-deafening press room.

The load factor varies considerably for each press room, and for different days of the week. This is due to the number of pages printed in each edition. Internal sources of heat will reach their maximum after sundown in a plant printing a morning paper. In an afternoon paper, the maximum will usually coincide with maximum external conditions.

Besides load factor, there is also the important item of heat lag due to the mass of machinery and building, known as "storage." When a press room is operating intermittently, this is of importance in determining the overall load factor to be applied in the air conditioning calculations for reduction of equipment size.

It is advisable to operate the air conditioning after the presses have been shut down until the heat is removed from press frames and the building so that they will be at normal temperature for the next operation. Pre-cooling at somewhat lower temperatures before operation can reduce equipment load.

Taking all outside air increases the load on air conditioning equipment and it is, therefore, necessary to study weather bureau records to determine

wet bulb temperatures that will occur during operating times, be it day or night.

Paper itself picks up large quantities of heat as it passes through the presses. If paper is removed from the press room immediately after it is printed, a sizable reduction in heat load can be effected.

Removing Press Room Heat

An air conditioned Los Angeles newspaper is an excellent example of heat load problems, and may serve as a guide to arriving at appropriate figures for similar installations. Power required to operate the presses is approximately 60% of the motor nameplate rating (load factor). The driving motor is located in the reel room, where rolls of paper are placed during press runs. When the driving motor is outside the press room, motor inefficiency should be deducted from the press room load, but in this case, added to the reel room load. Some motor input is likewise utilized in driving the reels. The deduction under this heading amounts to about 20% of the motor's input power. Or in other words, 80% of the load is delivered to the press room.

When a newsprint roll is kept in a conditioned space equal to or at lower temperatures than the press room, some 15% of the heat dissipated through the press results in from 5 to 10% rise in paper temperature. This heat is carried off as the papers are removed from the press room.

The press itself rises in temperature. Those parts having the greatest mass may increase some 5 F after three hours operation, while some frame parts may rise 15 F. A good average would be slightly above 5 F. This would amount to about 25% of the total heat dissipated in the press. Allowance can be made for this, provided that the presses do not run continuously, and provided that the air conditioning plant is operated to cool the presses while they are not running.

From the above information, we arrive at the following load factor for the press motors:

0.60 × 0.80 × (1 − 0.15 − 0.25) = 0.288

(or 30% of the nameplate rating).

Composing and stereotype rooms are the most vital where human comfort is concerned. Many people in these rooms, brilliant lighting, and heat-producing machinery combine to produce tremendous heat loads. Of primary concern in the composing room is elimination of localized heat gain from linotype and monotype machines. The Mergenthaler Company, leading manufacturer of the linotype machine, states its machines give off 43,700 Btu during an 8-hour day. The pots used for melting lead operate at 550 F. Electric pots do not require special ventilation, but gas pots should be vented to get rid of fumes and heat.

Experience has shown the load factor for linotype machines to be about 65% of total connected horsepower load. The lead melting pots have a load factor of 30% when there is ample positive exhaust for each machine.

Monotype machines are usually rated at 25% of the connected horsepower load. Gas-heated pots radiate about 1,000 Btu per pot. It is necessary to exhaust large quantities of air directly from each machine to remove heat and fumes. Air should be supplied near the floor to permit it to flow up around the operator.

The amount of exhaust and ventilation is often fixed by codes and should be investigated. If it is not covered by codes, a good general practice is to install a ventilating system to handle about 400 cfm per pot to carry off fumes and some of the heat. This figure, however, should be checked with the manufacturer of the machine in question.

Any composing room air conditioning system should be designed so as to eliminate drafts and uneven temperatures. The operators must sit at their machines all day, and are especially sensitive to drafts. Distribution of conditioned air should be designed to blow air over the operator toward the lead pots for removal by the exhaust system for each unit. This will keep fumes and excess heat away from operators.

The stereotype room, where lead plates are cast in a foundry, is the hottest spot in a newspaper plant, but it is also lightly populated. It is not usually practical to cool the entire room to comfort conditions. Spot cooling will give tolerably cool conditions at points where workers stand. Sufficient outside air must be introduced to provide adequate ventilation.

Elimination of Moisture

One problem that occurs, although infrequently, in newspaper and printing plants in northern climes is that of moisture condensation on windows and skylights when outside air is cold and dry. Dirty condensate dripping from the ceiling can cause spoilage and increase building maintenance costs.

Numerous successful applications have been made by vigorously blowing heated air over exposed roof and wall surfaces, thus raising the inside surface temperature and permitting a higher room dewpoint to exist without condensation.

CHURCH, THEATER AND AUDITORIUM AIR CONDITIONING

In air conditioning churches, auditoriums and theaters, the system must be correct not only in the psychrometric sense, but it must also be noiseless.

Load Determination

Factors which influence the cooling load of these applications to a greater extent than in normal air conditioning applications are:

Occupancy Pattern. Because heat gain from people is one of the largest loads imposed, occupancy pattern is very important.

In *churches,* the occupancy pattern will be determined by the number and scheduling of services. Most Protestant churches have one or two services on Sunday, whereas Roman Catholic churches have multiple masses, following each other from early morning until noon. Past summer attendance records are useful in determining the number of people for load purposes, but these must also be compared to seating capacity and attendance records for other seasons. Air conditioning has increased summer attendance in all cases and in many churches the summer attendance has been greater than during other seasons.

Use of the church building during the week is also important and may have an important bearing on the type of system employed.

In *theaters,* the occupancy pattern is determined by the type of house. Some small towns or neighborhood houses put on as few as two shows per week while some city houses run twelve shows a day on a 24-hour per day basis.

Figure 1 shows typical occupancy and cooling load curves for the week-end in a Class *A* movie house enjoying high patronage. These curves are idealized plottings from data taken in several movie houses in various parts of the country.

These curves indicate that the peak load on the air conditioning equipment occurs between 8:00 and 9:30 p.m. This peak occurs at the time of low outside air temperatures and when the solar effects are almost nil, so that it is due to the people load. This is the largest single portion of the air conditioning load. The evening audience consists chiefly of adults, whereas the matinee has a large percentage of children. Test results show the people load in the afternoon is only 75% of that of the evening.

Maximum occupancy is usually determined by the total number of seats. A few theaters may allow standees in the foyer of the theater. When fire ordinances permit this practice, which is rare, an allowance must be made in the system. Usually, the cooling load will be affected only slightly due to the short duration of this condition. However, the air distribution should be designed to provide capacity in the standee area.

The occupancy of *auditoriums* will be determined by the usage. As this usage will vary, good judgment must be used in arriving at the people load. Maximum possible seating capacity is a good starting point. The various possible seating arrangements must be determined; for instance, an auditorium which is used for basketball games or ice shows can also be set up for boxing matches with the number

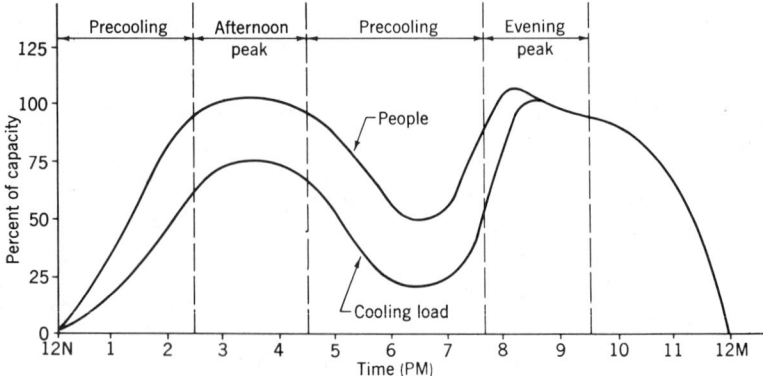

Fig. 1. Occupancy curve for typical Class A theater.

of seats of each being determined by the area of the floor required for the event.

Ventilation. The normally designed quantity of ventilation air for many applications is in excess of true requirements and is reduced in actual operation. Use of more outside air than is necessary leads to excessive first and operating costs, as outside air is a large portion of the load. Ventilation air as low as 5 cfm per person gives very acceptable indoor conditions. However, local codes may require more. This point should be carefully checked before completing the design.

In *churches*, large volume and infrequent use of the building, as well as the absence of smoking provide good conditions with low ventilating rates as compared to commercial applications. In *theaters* and *auditoriums* where smoking is not permitted, the same general condition exists. In theaters the ventilation is further increased due to the interval between the afternoon and evening peaks when the actual ventilation being supplied may be as high as 30 cfm per person.

The outside air intake should be designed so that ventilation air will be provided whenever the fan is operating..A minimum outside air damper or a minimum stop on the outside air damper is usually used. However, outside air ductwork and dampers should be so sized that 100% outside air can be obtained. Then it is possible to flush out the hall, and outside air can be used for cooling during intermediate seasons to save operating costs.

Remote manual control of outside air has proved useful in many cases. Outside air quantity can be set to correspond with attendance or patronage.

Solar and Transmission Heat. Solar heat and tranmission *may* be a factor, depending on the time of day of the peak load.

In *churches*, heat gains from solar and transmission load are substantial parts of the total. This is due to the large window areas and the morning and mid-day occupancy.

In *theaters*, the solar and transmission gain can usually be neglected in the load calculations. As the peak load occurs at 8:00 to 9:30 p.m., the magnitude is small except in theatres with a large exterior wall with a western exposure.

In *auditoriums*, the solar and transmission gains may vary between the two extremes given previously for churches and theaters and thus should be analyzed fully.

Storage Effect. Storage effect is important in these applications, due either to the infrequent operation of the system, as is the case of churches and auditoriums, or due to the peaks and valleys in the total cooling load, as is the case in theaters. The heavy construction, plus the large mass of material in the seats or pews, result in a great amount of heat storage.

At start-up, up to an hour may be required to pull down the temperature of this mass to design temperature. However, this additional load can be quickly compensated for by closing the outside air damper during the pull-down period.

If pre-cooling time is available, some cooling effect can be stored in the structure and furnishings and this cooling used to offset the peak load. Significant storage effect can be accumulated in approximately one hour. By pre-cooling the building and furnishings 2-3F below the temperature allowable at peak load, during peak load the air temperature in the space rises to design conditions and the pre-cooled mass absorbs some portion of the heat load. Thus, equipment can be selected for a capacity less than actual peak load.

Even in Catholic churches, where services follow each other through the morning, since the early masses usually are not as heavily attended as the later ones, and outside temperatures are lower during the early morning, excess capacity can be used for storage which, in turn, will offset the peak load occurring at the late masses. This storage effect can be used only if the peak load is less than three or four hours duration.

In movie *theaters*, the same arrangement can be used. By referring to Fig. 1, it can be seen that there is a period of partial load before each peak load condition during the day.

Stratification. High ceilings and vaulted ceilings in many churches, auditoriums, or theaters make it possible to maintain ideal conditions in the occupied zone without cooling the entire space. The hot air below the ceiling stratifies and most of it never becomes part of the equipment load. Where lighting is at the ceiling level, a large part of the lighting load does not become a part of the cooling load on the equipment.

To obtain stratification, supply outlets must be kept relatively low. When ceiling or sidewall outlets are used close to the ceiling, the entire volume is cooled and no stratification occurs.

When air distribution is from window sill height,

transmission through wall and window areas above the 15-ft level can be neglected. However, when the sun shines through the glass, striking the floor or furnishings, solar heat is absorbed and eventually becomes a part of the equipment load. Note that stained glass windows used in churches and auditoriums are similar to heat absorbing glass. A transmission factor of 0.70 can be used in the absence of precise data.

When side wall outlets are used, transmission through the wall area over 5 ft above the outlets can be disregarded. Again, solar heat gain through the entire glass area must be used.

Due to high temperature of the ceiling surface, radiation to the occupied zone adds to the load. This radiant and convection load is usually taken as one-third of normal roof load.

Inside Design Conditions. Inside design conditions vary with locality and the characteristics of the cooling load.

The inside design dry bulb is usually kept in the range of 75-78F, as this seems to be the most universally accepted. Note that when storage effect due to pre-cooling is used, the thermostat is usually set two to three degrees below design condition, to obtain the pre-cooling necessary.

Design relative humidity for the space can be in the range of 40-60%, without discomfort, when coupled with the range of dry bulb temperatures previously mentioned. Due to the large portion of the load that is made up by people and outside air, sensible heat factor is very low. For that reason, in many cases it is impossible to maintain relative humidities below 55-60% without reheat. As reheating is uneconomical and usually impractical in these applications, it is better to design for a lower dry bulb temperature to obtain an equivalent effective temperature. A design condition of 77F, db and 60% rh is suggested, for maximum economy consistent with satisfactory conditions.

Outdoor Design Conditions. The outdoor design conditions used in the load estimate should be taken at the time of day when the peak load occurs. Note that the peak load comes at the time of peak occupancy. This peak may occur in the morning in churches, or in the evening in theaters and auditoriums. Therefore, the design temperature should be corrected for the time of day. Weather Bureau records can be used to determine the correction factors for various times.

Load Calculation Procedures

Churches. The cooling load calculations is similar to commercial applications, taking into account all of the load sources that are present. Some of the required information has been discussed, such as indoor and outdoor design conditions, population (people load), solar and transmission, ventilation, storage effect and stratification effects.

An extra credit is obtained in churches having one or two services per Sunday where one-half of the outside air supplied is relieved by the vents in the ceiling. This credit can be computed by the following formula:

$$\text{credit (Btuh)} = \tfrac{1}{2} \text{ supply air (cfm)} \times 1.08 \times 10\text{F}$$
$$= 5.4 \text{ supply air (cfm)}$$

The vents usually consist of pressure relief dampers that relieve only when the pressure in the space is higher than outdoors. The stratified air temperature is higher than the conditional space, thus some of the space load is exhausted. In churches having continuous, multiple services, this credit cannot be taken due to the large infiltration that takes place between services when the doors are open for five to ten minutes. In addition, this infiltration load should be estimated. The cooling load of the sanctuary only will vary between 0.045 and 0.060 tons per person, depending on solar and other variable loads. The cooling load on other spaces such as offices, educational units, and fellowship halls will vary according to usage and other factors.

Theaters. Information given in the previous section has made it possible to set up a simplified load estimating method for the basic movie theater. The basic movie theater is defined as the auditorium portion and any other section connected directly to it. This usually does *not* include the manager's office, lobby, rest room areas, and projection booth which are special areas and should be calculated in the normal manner.

Allowing excess ventilation air to spill out from the lobby to the street may be sufficient to cool the lobby and advertise air conditioning to passers-by.

The simplified method requires the following information: Theater location (thus the outdoor design conditions), the number of seats (and standees if any), the minimum outside air quantity (cfm) at peak load conditions, and the inside design condition. With this information, the total tons of refrigeration and dehumidified air quantity can be

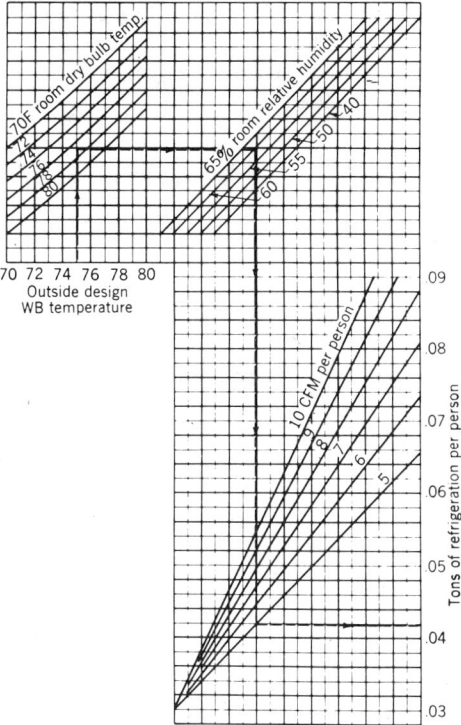

Fig. 2. Cooling estimate chart for movie theaters.

Table 1. Heat Load from People

Room Dry Bulb, F	Sensible Heat, Btu per Hr per Person	Latent Heat, Btu per Hr per Person
70	260	90
71	254	96
72	248	102
73	242	108
74	236	114
75	230	120
76	224	126
77	217	133
78	210	140
79	203	147
80	195	155

Example: Determine the total cooling load for the basic portion of a conventionl theater given

8 p.m. design wet-bulb	= 75F
Room dry-bulb	= 77F
Room humidity	= 60% rh
Ventilation rate	= 5 cfm per person
Number of people	= 1000

Solution: On Fig. 2, project vertically up the 75-F outside design wet-bulb line to 77-F room dry-bulb temperature curve, then horizontally to 60% room relative humidity curve, then vertically downward to 5 cfm per person curve, and, finally, horizontally to read 0.0418 tons of refrigeration per person.

Total basic theater cooling load =
0.0418 tons per person × 1000 people = 41.8 tons

Treated Air Quantity: Total (dehumidified) air quantity is found from the formula:

$$\frac{(\text{room sensible heat per person} \times \text{no. of people})}{(1\text{-}BF)\ (\text{room temp - } ADP)\ (1.08)}$$

where

Room sensible heat (*RSH*) per person is found from Table 3,

BF (bypass factor of cooling coil) must be assumed based on manufacturer's rating of the coil that probably will be selected,

Room temperature is design room dry-bulb, F,

ADP, apparatus dew point, is found from Table 2,

1.08 is a constant.

determined by several simple references to tables and charts and several calculations. This basic information is then used to select the equipment requirements.

Tons per Person. The first step is to find the refrigeration capacity required per person. This is obtained directly from Fig. 2.

This chart is based on the load from people and outside air only. The slight lighting load and the solar and transmission loads are offset by storage. If these loads are abnormally high, such as may be the case with a large western exposure, the chart values should be increased. If any doubt exists as to the time of the peak load, this chart can be used. The chart is based on adult metabolic rates. If children are present, such as at a Saturday matinee, the total load can be corrected for the lower metabolic rate. Table 1 gives the heat load for people.

Tons of Refrigeration. Total refrigeration load is easily calculated as the product of tons per person and number of people.

Table 2. Apparatus Dewpoint (ADP) Temperatures Based on 8:00 pm Design Conditions

3:00 PM DB	3:00 PM WB	8:00 PM DB	8:00 PM WB	Bypass Factor	75 F 55%	75 F 60%	75 F 65%	76 F 55%	76 F 60%	76 F 65%	77 F 55%	77 F 60%	77 F 65%	78 F 55%	78 F 60%	78 F 65%
105	76	98	74	0.10	48	53	58	47	53	58	48	54	59	46	54	59
				0.30	48	54	59	48	55	59	49	56	60	48	56	61
100	78	93	76	0.10	46	52	57	44	53	57	44	53	58	41	52	58
				0.30	39	51	57	40	51	57	41	52	58	—	53	58
95	78	88	76	0.10	44	51	57	43	52	57	38	50	57	40	51	57
				0.30	—	48	55	—	47	56	—	48	57	—	48	57
95	75	88	73	0.10	46	52	57	45	53	57	45	53	58	42	53	58
				0.30	41	52	57	43	52	58	44	53	59	37	54	59
93	75	86	73	0.10	46	52	57	44	53	57	43	54	58	41	52	58
				0.30	39	51	57	40	51	57	—	52	58	—	53	58
90	70	83	68	0.10	48	53	57	47	53	58	47	53	58	45	54	59
				0.30	48	55	59	50	56	60	50	56	60	50	56	61

(Outside Design Temp, F: 3:00 PM and 8:00 PM columns; Bypass Factor; Inside Design Dry-Bulb, F and Relative Humidity; Apparatus Dew Point Temperatures, F)

NOTES:
1. The 8:00 pm design conditions correspond to the 3:00 pm design conditions where the daily range of temperature is 20 F. Table may be used for daily temperature ranges between 15 F and 25 F for the 3:00 pm conditions listed.

2. Table is based on the design conditions shown for 8:00 pm, 0.10 and 0.30 equipment bypass factors, and 7½ cfm per person outside air. Where no ADP values are shown, reheat would be required to maintain these conditions.
   Table is sufficiently accurate for ventilation rates in the range 5 to 10 cfm per person, and for bypass factors in the range 0.05 to 0.35. Use closest bypass factor shown or interpolate between values shown.

3. For conditions not shown, the apparatus dewpoint temperature can be determined as follows:
   (a) $SHF = RSH/(RLH + RSH)$
   where RSH = sensible heat of bypassed portion of outside air plus sensible heat due to people.
   where RLH = latent heat of bypassed portion of outside air plus latent heat due to people.
   (b) ADP temperature read from psychrometic chart for selected room conditions and calculated SHF.

If subsequent equipment selection gives a widely different bypass factor, the calculation can be repeated to obtain a more accurate solution.

If the bypass factor of the coil is not known, the air quantity can be obtained by a trial and error type psychrometric solution using the sensible and latent loads of people as the only room load. The inlet air on the coil is affected by the amount of outside air.

Once the tons and dehumidifier air quantity per person have been calculated for the following set of conditions, the values can be used for any other job:
   Location (outdoor design)
   Inside design
   Outside air quantity

Equipment Determination

Church, theater and auditorium air conditioning systems are usually combined with heating systems to provide year-round air conditioning. There are five main types of cooling systems employed: Chilled water, direct expansion, direct expansion self-contained, ice or water storage, and ice melting. The last two are used primarily for churches. The advantages and disadvantages of church systems are listed in Table 4.

Chilled Water Systems. These systems are widely used today. As refrigerant is restricted to the cooling equipment and not circulated throughout the building being conditioned, there are no problems of

Table 3. Total Room Sensible Heat Per Person Based on 8:00 pm Design Conditions

Outside Design Dry Bulb Temp (F) at 8 PM	Outside Air Per Person, cfm	Inside Design Dry Bulb Temperature, F															
		75				76				77				78			
		Coil Bypass Factor															
		.05	.10	.20	.30	.05	.10	.20	.30	.05	.10	.20	.30	.05	.10	.20	.30
		Room Sensible Heat per Person, Btu per Hr															
98	5	236	242	255	268	230	236	248	260	223	228	240	252	215	221	232	243
	7½	239	249	267	285	233	242	260	278	226	234	251	268	218	226	242	258
	10	242	255	280	305	236	248	272	296	228	240	262	284	221	232	253	274
93	5	235	240	249	258	229	233	242	251	221	226	234	242	214	218	226	234
	7½	237	245	259	273	231	238	252	266	224	230	243	256	216	222	234	246
	10	240	249	269	289	233	242	261	280	226	234	252	270	218	226	242	258
88	5	234	237	244	251	227	231	237	243	220	223	229	235	213	215	221	227
	7½	235	241	251	261	229	234	243	252	222	226	235	244	215	221	232	234
	10	237	244	258	272	231	237	250	263	223	229	241	253	215	221	232	243
86	5	233	236	242	248	227	229	235	241	219	222	227	232	212	214	219	224
	7½	235	239	248	257	228	232	240	248	221	224	232	239	213	217	223	229
	10	236	242	254	266	229	235	246	257	222	227	236	245	214	219	227	235
83	5	232	234	239	244	226	228	232	236	219	220	224	228	212	213	215	217
	7½	233	237	243	250	227	230	235	240	219	222	227	232	212	214	218	222
	10	234	239	247	256	228	232	239	246	220	224	230	236	213	215	221	227

NOTE: Table based on the frequently encountered 8:00 pm outside design dry bulb temperature.

Table 4. Characteristics of Four Church Air Conditioning Systems

Type System	Advantages	Disadvantages
Direct expansion field assembled system	Full capacity at all times; least floor space; simple engineering and installation; high salvage value; lowest operating cost.	Medium first cost.
Direct expansion self-contained system	Full capacity at all times; simplest engineering and installation; highest salvage value; lowest operating cost; low first cost.	
Ice or water storage system	Low operating cost.	Largest floor space requirement; highest first cost; 24 to 48 hours recuperation period needed.
Ice melting system	Lowest first cost.	Higest operating cost; non-automatic operation; poor humidity and temperature control; large space requirements.

refrigerant leakage or necessity for removing structural members to find a tiny leak in the piping system.

Chief advantages provided by water chilling packages are: Pre-engineering to save consultant and contractor time and money; standard control schemes; standard refrigerant piping arrangements, eliminating trouble from improper oil return, faulty operation of expansion valve and flooding of the compressor.

Units using reciprocating compressors of either open or hermetic types are factory assembled and tested, capable of fitting through the average doorway and ready to operate after connection to water and electrical lines. Their capacity ranges between 3 and 100 tons.

Economic analysis together with other factors generally favors centrifugal or absorption machines when the capacity requirements are above 100-200 tons. In the choice between centrifugal and absorption equipment, the initial cost plus installation will often favor centrifugal equipment, slightly. However, this is very likely to be offset by the fact that absorption machines usually do not require a licensed operator and that energy rates should produce lower operating costs. Other factors must also be considered such as availability of steam or high temperature water, and of electric service facilities of the required size. In fact, in many cases, a good analysis may indicate the desirability of a combination of the two machines.

Direct Expansion Systems. The direct expansion systems are basically the same as are used in other large air conditioning installations. These systems have many advantages:

Full facapity is available at all times. This is important because with the addition of air conditioning, church and auditorium usage increases. For instance, air conditioned churches are often used for summer conferences or conventions, which can last for many weeks.

Requires less floor space. Because churches are always trying to find room for another Sunday school class, an extra office, or for storage, this space has tangible value.

This system is simplest from both the engineering and installation viewpoints, as well as maintenance and operation. Broad industry experience with this type of system indicates that it has a high salvage value and for churches that plan to enlarge their facilities at a later date, this may be most important.

The direct expansion system has lowest operating cost. First costs are low to medium, depending on the conditions existing for the particular job.

Package Equipment. These systems are often used when air conditioning is added during remodeling, as it is often easier to find room for self-contained equipment than it is for applied equipment. Self-contained units, particularly the larger sizes, usually prove to be the lowest cost system. Self-contained equipment has also proved advantageous when the life of the present building is limited or future additions are contemplated. These units have high salvage value and can be used in another location at a later date.

Ice Or Water Storage Systems. Ice or water storage systems have been installed in many churches and are popular in many sections of the country. This system combines a chilled water system with a storage tank, for storing cold water, or a commercially available container, for making and storing ice. In this type system a relatively small condensing unit operates continuously 24 to 48 hours prior to the peak load, to store up the necessary capacity.

The advantages of this system are low operating costs, when high demand rates are charged. The disadvantages are large floor space requirements, long recuperation periods (24 to 48 hours for a church) and highest first cost, due to storage tank, extra piping, and controls that are required.

Operating costs of the air conditioning system for churches are low due to the few hours of operation per week. Operating cost, during the cooling season, may run from less than one cent to about two cents per person per week. This cost will vary according to church construction, usage, and system operation. Direct expansion systems are lowest in operating costs in most cases, regardless of the power rate, because the church requires a relatively large amount of cooling for only five to ten hours during the week.

When power rates are based on demand charge for the connected horsepower plus a low rate on the amount of current used, the direct expansion system is still lowest in operating cost, if storage effect and stratification are taken into account. When storage effect is used, the installed horsepower and the demand rate are reduced. The use of the structure to store cooling in an ice storage system is impractical because it would increase the size of plant required and thus increase first and operating costs.

Air Distribution

The main consideration of the air distribution system is to supply the required air quantity without drafts or noise. Types of systems include: ceiling outlets, side wall outlets, floor registers, and under-the-window grilles. Low distribution systems, floor registers, under the window outlets, and side wall outlets, supplying only the occupied zone, reduced the load requirements and therefore are preferred.

In *churches*, special outlets may be required to provide spot cooling at the altar and the pulpit. Without them, the temperature in front of the altar might be 5-8F higher than the temperature in the pews. As the priest or minister wears heavy vestments, he actually requires a lower temperature. The problem is also encountered with choir lofts.

Duct Design. Main duct velocity of 1500 fpm and branch or runout velocities of 1200 fpm are suggested. In all cases, duct turns or wide radius elbows should be used. Velocity is not so important as proper duct design in preventing air-generated noises. Takeoff runs should be carefully designed and installed, especially when supplying air to an outlet. All square elbows or short radius elbows should be equipped with properly designed duct turns.

Return air grille locations are not critical except that they should not be placed near people who are seated. Normal locations are the back of the space, the back of the rear pews, and in steps to the altar or stage. When a theater may be used for "live shows" with orchestral accompaniment, the proscenium, or pit, should not be used for return air. The return air grille should be selected for a free area velocity not exceeding 600 fpm. Return air duct velocity should be 1200 fpm.

Even at the recommended velocities, extreme care must be taken in the manner in which outlets are connected to the supply air system. An outlet should never come from an abrupt take-off. If it is necessary at any time to connect an outlet to a main duct because of space limitations, sound absorbing materials of sufficient density and area should be used to prevent air generated noises.

Control System

Controls should be automatic to the extent that available help can operate the equipment successfully.

Usually, on-off control is adequate for *church* auditoriums. Sunday school rooms or an educational building usually require zoning and other means of room control due to changes in population and orientation effect. Small class rooms can usually be handled by volume control.

In the case of *theaters* and *auditoriums,* the normal control system employed is face and bypass control as this system gives much better control over the space relative humidity than does a simple on-off control system.

On very large jobs, some sort of compressor capacity control is advantageous in order that the system capacity may vary with the air inlet condition, which will change with outdoor temperature fluctuations. Outside air is usually 25 to 33% of the supply air.

To save operating costs, an economizer type of control, which uses outside air for cooling during the intermediate seasons, is often used. Whenever the outside air is cold enough for cooling the space, a change-over thermostat shuts off the refrigeration and allows the room thermostat to control the amount of outside air. At the change-over point, usually 100% outside air is required for cooling. As the outdoor air temperature goes down, the proportion of outside air is reduced until the minimum quantity is reached. At that time the system goes into the heating season cycle.

When Sunday school rooms in the basement below the sanctuary are to be conditioned, and Sunday school is held prior to the service, the cooling loads are not simultaneous. Therefore, the refrigeration equipment can be selected on the basis of the major load, usually the sanctuary. Separate air handling equipment should be selected for each space. This air handling equipment is sized on the individual space load requirement. Capacity control is usually required on the refrigeration machine to prevent coil frosting when only the small air handling unit is operating at reduced loads.

Many times there is sufficient capacity to cool the Sunday school quarters and also pre-cool the sanctuary prior to the service, particularly when the outside air damper on the sanctuary is closed during this time.

In a case such as this, the controls have to be designed so that control can be switched from the Sunday school area to the sanctuary at the proper time.

Tables 1, 2 and 3 and Figures 1 and 2 are reprinted by permission of Carrier Corporation.

HOSPITAL AIR CONDITIONING AND VENTILATING

Hospital air conditioning presents problems that are not usually encountered in ordinary comfort conditioning systems. Other than the hospital staff, the clientele served are often physically debilitated with a lowered resistance to infection and this necessitates certain safeguards for their protection.

Although the outdoor air supply to hospitals may be relatively free of pathogenic contamination, it may become more contaminated after entering the hospital. Therefore, it is desirable that the system be designed to limit the circulation of air from area to area and from patient room to patient room, insofar as is practical.

It is impossible to prescribe a particular system which would be suitable for all hospitals. The design of the hospital air conditioning system will vary with size, type and layout of the structure, space allotments, orientation, climatic conditions and many other factors.

Ventilation Air

Systems discussed are to be considered as basic and not all systems or variation of systems which may be adapted to hospital use are included here. Systems which operate on the principle of induction or aspiration and which provide approximately 20% ventilation air with 80% recirculation of room air lend themselves nicely to minimizing recirculation of air between patient rooms. Using such systems, the corridor serving patient rooms is pressurized with the small amount of air necessary for corridor ventilation. Corridor air is exhausted through all rooms opening on to the corridor. Patient room air is an amount equivalent to primary induction air plus some corridor ventilation air is exhausted through the patient's room toilet or through a similar exhaust system.

Often financial, structural, or other considerations will dictate the use of a design which recirculates the air of a patient area through the central conditioning system and redistributes it to patient rooms. Individual factory-assembled conditioning units with efficient filters in the recirculating system are recommended to confine this recirculation to a limited area such as a nursing unit or floor. Such a system should reduce the hazards of cross infection to a minimum in general medical and surgical nursing units.

The corridors of patient areas should not be used as exhaust air plenums. Such circulation is contrary to most fire safety codes and is not considered desirable for ambulatory patients or personnel who will traverse the corridors.

Special phases of engineering fundamentals and construction peculiar to hospitals and of interest to hospital architects and engineers will now be considered by departments or facilities.

It should be noted that air change rates as specified for specific areas refer to outside air requirements for ventilation and in no way reflect air turnover required for the conditioning process.

For design purposes, the acute general hospital may be divided into its principal service areas. These will vary somewhat from each other in the design treatment which their function will require. These areas are: Administration department, diagnostic and treatment facilities, nursing department, nursery, surgical department, obstetrical department, emergency and service departments.

Administration

This department, consisting of lobby and waiting rooms, offices, library and conference room, record room and toilets, will require no special ventilation treatment. General recirculation of air within the administration area, excepting toilets, would be indicated.

Diagnostic and Treatment

These facilities, some of which require special study, are: laboratory, radiology, physical therapy, occupational therapy, and pharmacy.

Laboratory. The laboratory, depending upon the size of the hospital, may consist of one or several rooms in which are performed functions relating to bacteriology, pathology, serology, chemistry, hematology, and urinalysis.

While odor control in all of the laboratories should not be overlooked, only the bacteriology section and certain areas of pathology will require special attention relative to ventilation. Up to 75% of general laboratory air may be recirculated if properly filtered, except that all air of virus or infectious disease labs should be exhausted to the outside.

Chemical fume hoods with integral ventilation systems designed to void the hood and protect the personnel, will be installed in the laboratory area. Consideration should be given to hood exhaust capacities in computing general air supply and exhaust to the area. Hoods should be connected by

non-corrosive vent lines through the roof and acid resisting fans should be used. It is preferable to place the hood exhaust on the roof to prevent acid fumes being forced into the building.

Bacteriology. As undue air movement is not desirable in the area devoted to bacteriology, care should be exercised to limit air velocities to a minimum. A ventilation rate of 10 air changes per hour is recommended with a negative air pressure.

The Sterile Transfer Room, within or adjoining the bacteriology laboratory, is a room where sterile media is distributed and where some types of specimens may be transferred to culture media. The air supplied to this room must be filtered at the point of entry to the room and velocities at working level should not exceed 25 fpm. A glass fiber filter with 99% efficiency, which is preceded by a dust stop, is recommended.

The Media Room, essentially a kitchen, should be ventilated to remove odors and steam.

The infectious disease (particularly tuberculosis) and virus laboratories of large institutions will require special treatment. To protect the personnel, such laboratories will require ventilation at a rate of 10 air changes per hour with no recirculation. A negative pressure within the area should be established to prevent the spread of the organisms through the movement of air from room to room or to adjoining areas. Area supply and exhaust fans should be electrically interlocked so that the supply will shut off if the exhaust fails, but not the reverse. As protection for personnel, air should be exhausted at points as close to work areas as possible. The entire exhaust from the virus or infectious disease laboratories may be taken from slots or similar openings just above the work counters and from hoods.

Where the nature of the laboratory procedures with infectious microorganisms is such that a ventilated bacteriological hood or safety cabinet is provided, effective sterilization of the outboard exhaust is required. This may be accomplished by the use of electric or gas-fired heaters designed to heat the air to 600 F. Some authorities recommend the installation of glass fiber filters in series with the electric or gas-fired heater to insure against an interruption of work in the event the heater fails for any reason.

In many instances, only glass fiber filters are necessary and, in this case, an integral electric heater element may be provided as a source of 400-F heat for dry sterilization of the filter media prior to changing filters, to protect maintenance personnel.

Pathology. Areas of the pathology section which will require special attention are the autopsy room and, in larger hospitals, the animal quarters.

Autopsy Room. The autopsy room will require a minimum of 10 air changes per hour, with no recirculation and a mechanical exhaust system which will discharge the air above the hospital roof. A negative air pressure is indicated for this room. Where the autopsy room will not be in constant use, it may be desirable to provide local control of the ventilation.

Animal Quarters. The animal room will require 15 air changes per hour with no recirculation and a mechanical exhaust system which will discharge the contaminated air above the hospital roof. Depending upon the type of work performed in the animal quarters, sterilization of outboard exhaust similar to that recommended for infectious disease laboratories may be required. In larger hospitals where the exhaust air from the animal quarters may be carried in a duct system common to the laboratories, throw-away dust stop filters are installed at the animal room exhaust grilles to prevent animal hair from entering the main exhaust system. Air diffusion to avoid drafts is of particular importance in the animal quarters. Temperatures of 75-80 F with 40-50% rh are usually required.

Radiology. In the radiology department, the *fluoroscopic, radiographic, therapy,* and *dark rooms* will require special attention. Among conditions which make positive ventilation mandatory in these areas are the odorous charateristics of certain clinical conditions treated, special construction design to prevent ray leakage, and light-proof window shades which reduce any natural infiltration to the fluoroscopic area.

A positive ventilation rate of 3 air changes per hour with temperatures up to 80 F and 40-50% rh should be provided for *fluoroscopic, radiographic,* and *deep therapy rooms.*

Depending upon the location of air supply and air exhaust louvers, lead lining may be required in supply and return ducts at the points of entry to the deep therapy treatment area to prevent ray leakage to other areas.

The dark room will require 8 air changes per hour. Exhaust from the dark room and the film dryer in it are usually discharged directly to the outside. Because of the intermittent use of this room in small hospitals, local control of the exhaust may be desirable.

Physical Therapy. One or more rooms in the department of physical therapy may be assigned to *electrotherapy, hydrotherapy,* and *exercise.*

The cooling load of the *electrotherapy* section will be affected by the short wave diathermy, infrared, and ultra-violet equipment used in this area.

The *hydrotherapy* section with its various water treatment baths will generally be maintained at temperatures up to 80 F. Potential latent heat build-up of this area should not be overlooked.

As a protection against condensation, ducts exhausting large *hydrotherapy* areas may require soldered joints or be otherwise watertight and should be graded and dipped to an open drain.

The *exercise* section utilizing posture mirrors, bars, steps, ramps, ladders, and other equipment will require no special treatment.

Generally, the recirculation of air of the physical therapy department is permissible and temperatures and humidities within the comfort range are recommended for this area.

Occupational Therapy. In this department the space for activities such as weaving, braiding, art work and sewing will require no special ventilation treatment. Recirculation of air for this department is permissible.

Larger hospitals or those specializing in rehabilitation will have a greater diversification of skills and crafts such as carpentry, metal work, plastics, photography, ceramics, painting and finishing of projects. Such a department will require specific study of the ventilating problem.

Pharmacy. This area will be treated for comfort and will require no special ventilation.

Nursery Area

The nursery area will usually consist of a full-term nursery, a premature nursery, an observation nursery, work space and an examination room.

Full-Term Nursery. A temperature of 75 F and 50% rh are recommended for the full-term nursery, examination room and work space. It is necessary that similar conditions be maintained in all of these areas as care of the full-term infant requires that they be moved periodically from the nursery to the work space and examination room. The maternity nursing section should be treated similarly to protect the infant against temperature changes during visits with the mother.

Premature Nursery. The premature nursery will require 80 F and 55-65% rh. Although it will be equipped with individual incubators which provide close regulation of temperature and humidities, these conditions must be maintained in the area for the infant who is removed from the incubator.

Observation Nursery. The observation nursery (also known as suspect nursery) is a room set aside from the full term nursery for the observation of full term infants who have developed unusual clinical symptoms. This nursery will require 75 F, 50% rh. The air of the observation nursery should *not* be recirculated.

In order to provide maximum safety for the newborn, whose resistance to infection is normally low, the following conditions should be adhered to:

1. The entire nursery should be pressurized to prevent the entry of contaminated air from outside areas.
2. An equivalent of 3 air changes per hour of outdoor air should be supplied.
3. Nursery air, with the exception of the observation nursery, may be recirculated within the nursery when properly filtered.
4. Air from the observation nursery should not be recirculated; it should be discharged to the outside.
5. Air distribution within the area should be carefully planned, and air velocities at floor level should not exceed 25 fpm.
6. Air from other areas of the hospital shall not be circulated through the nurseries under any circumstances.

Nursing Department

This department will include all or most of the following rooms: Patient rooms, isolation unit, psychiatric room, treatment room, nurses' station, utility room, floor pantry, solarium, flower room, toilet, bedpan closet, and bathroom.

Patient Rooms. Because of the ever present possibility of cross-infection particularly of the respiratory tract, and to prevent the spread of odors, it is desirable that air circulation be confined to the individual rooms of this area. However, patient room air may be recirculated through a central conditioning system and be redistributed to the patient rooms with an acceptable degree of safety if properly filtered. A temperature of 75 F and relative humidity of 45-50% are recommended.

A minimum ventilation rate of 1½ air changes per hour is recommended for patient rooms. The ventilation of *toilets, bedpan closets,* and all *interior rooms* and *bathrooms* should conform to local codes.

Isolation Unit. The isolation unit air exhaust system should provide for outboard exhaust of all air when the area is used for contagious disease or ordorous cases and for recirculation of air when not used for this purpose. A minimum ventilation rate of 3 air changes per hour with a negative air pressure is indicated for this area. Room air conditioners which induce a flow of air through the unit are not recommended for this area because of the necessity for the periodic decontamination of coils.

Treatment Rooms: The treatment room may require special ventilation depending upon the nursing area specialty or type of work to be performed in the room. Usual practice will require 4 air changes per hour with temperatures up to 80 F.

Floor Pantry. The ventilation requirements of this area will depend upon the type of food service adopted by the hospital. Where bulk food is dispensed and dishwashing facilities are provided in the pantry, the use of hoods with exhaust to the outside is recommended. Small pantries used for in-between-meal feedings will require no special ventilation. A negative pressure in the area will reduce the spread of odors.

Surgical Department

This department will usually include many or all of the following rooms: Major operating rooms, minor operating rooms, cystoscopic room, anesthesia induction room, recovery room, scrub-up alcove, substerilizing room, central sterilizing and supply room, instrument room, clean-up room, storage closet, stretcher space, janitor closet, surgical supervisor's space, doctors' locker room, nurses' locker room, fracture room darkroom, and anesthesia storage.

The operating rooms, including cystoscopic, fracture, and induction rooms, will require careful design to reduce the hazards of infection and explosion.

Operating Rooms. It is the opinion of some authorities that one of the most dangerous potential sources of surgical infection is the pathogenic bacteria given off from the noses and throats of the occupants of the modern well-run operating rooms. Air disinfection to remove or destroy this source of infection is generally accomplished by ventilation and in some cases by bactericidal irradiation.

Ventilation of the operating rooms as a dilutant or medium for the removal of organisms is the generally accepted installation in modern hospital practice. This practice, as a means of reducing surgical infections, has some substantiation by reports of studies of the subject. Ventilation is accomplished by introduction of 100% filtered outdoor air, preferably taken from above the roof, to the operating room at a rate of 8-12 air changes per hour with 10 air changes recommended. This ventilation rate also reduces the concentration of anesthetic gases thus diminishing the possibilities of explosions and providing a more agreeable atmosphere for the operating team.

Although very dramatic results relative to air disinfection of operating rooms by irradiation have been reported, this method of disinfection has not been generally applied. The reluctance to use irradiation may be accounted for by such factors as special designs required for installation, protective measures necessary for patients and personnel, constant vigilance relative to lamp output, and the maintenance required.

The following conditions are recommended for operating cystoscopic and fracture rooms:

1. Temperature, 72-76 F, as required by the surgeon.
2. Relative humidity, 55%, to reduce the hazard of static spark.
3. Ten air changes per hour.
4. 100% fresh filtered air.
5. No recirculation of operating room air during the time operating rooms are in use.
6. A positive pressure within the operating room to prohibit the entry of air from the corridor or other unsterile areas.
7. Exhaust air taken off from a point near the floor in order to remove dust and the heavy anesthetic gases.
8. Installation to conform to National Fire Protection Association (NFPA) recommendations which pertains to explosion hazards in hospitals.

In order to provide the maximum safety for the patient and to conform to these recommendations, a separate system will be required to serve the operating rooms. The system may be designed to permit recirculation of air during periods when the operating rooms are not in use. Controls governing recirculation of the air should be conveniently located so as to be available to the surgical supervisor at all times. In addition, a visual signal, indicating whether the system is recirculating or not, should be located prominently.

Recovery Rooms. Recovery rooms used in conjunction with operating rooms should provide approximately the same temperature and humidity as the operating rooms so that the patient will not be subjected to any radical change in environment during immediate post-operative recuperation. As anesthesia gas and nausea resulting from anesthesia sometimes create an odor problem in recovery rooms, a ventilation rate of 3 air changes per hour is recommended. The air of the recovery rooms may be recirculated if this air is properly filtered.

Anesthesia Storage. The anesthesia storage room should be ventilated at a rate of two air changes per hour to conform to the regulations set down by NFPA.

Central Sterilizing and Supply

The central sterilizing and supply rooms present no particular problem and comfort requirements of personnel are the prime consideration.

The following items are of importance:
1. Insulation of sterilizers in these areas will materially reduce the heat load.
2. Sterilizer equipment closets should be amply ventilated to remove excess heat.
3. Air exhaust grilles or hoods located immediately above sterilizer doors, will assist in voiding excess heat and steam and will protect wall and ceiling finishes.

Obstetrical Department

This department will include some or all of the following rooms: Delivery rooms, labor rooms, scrub-up alcove, substerilizing room, clean-up room, doctors' lounge, nurses' locker room, nurses' station, non-sterile storage, sterile storage, stretcher storage, and janitors' closet. System design should be patterned after that of the surgical department. The recommendations for operating rooms will apply to delivery rooms.

Emergency Department

This department will usually consist of: emergency operating rooms, toilet, storage and supply closet, and stretcher and wheelchair closet. In addition, larger hospitals will have: office and waiting room, utility room, bath, and observation beds. Emergency operating rooms should be treated the same as the operating rooms in surgery. Observation bed space should conform to the requirements of patient rooms in the hospital Office and waiting rooms will be treated for comfort.

Service Department

This department will include: dietary, housekeeping, mechanical, and employees' facilities.

These areas, with the exception of employees' facilities, are usually not cooled, but adequate ventilation is important to provide sanitation and a wholesome environment. Ventilation of these areas cannot be limited to exhaust systems only; provision for supply air must be incorporated in the design. The best designed exhaust system may prove very ineffective with an inadequate air supply and experience has shown reliance on open windows results only in dissatisfaction during the heating season. The incorporation of air to air heat exchangers in the general ventilation system, other than range hood exhaust, offers possibilities for economical operation in these areas.

Dietary Facilities. In this area the main kitchen, bakery, dietitians' office, dishwashing room, and dining space will require care in design.

Kitchen. The ventilation should conform to local codes, but in the absence of such codes, a minimum of 10 air changes per hour is recommended.

Cooking equipment such as broilers, ovens, ranges, kettles, and steamers are generally grouped for efficient utilization. The area serving this equipment should be vented through a hood to remove heat, odor, and vapors. The ventilation rate through the hood will vary according to design and mounting height, but a rate of 100 cfm per square foot of hood opening is very often used with conventional hoods, mounted 6.5 ft from the floor.

Hoods over cooking equipment should be equipped with grease filters. Safety measures should be incorporated in the design to protect the duct system against fires originating over the stove. Such protection is usually accomplished by one or more of the following: A steam line to the throat of the hood at the duct connection for smothering action, CO_2 outlets at the hood perimeter, or fusible link dampers. The hood exhaust fan should be selected to provide the air volume required over the range of filter resistances.

Location of the system air supply inlets relative to hood exhaust is important to eliminate the possibility of short circuiting between supply and exhaust. Where dietary facilities are located within the hospital structure, exhaust ventilating air should be discharged above the roof.

Dietitians' Office. This office is very often located within the main kitchen or immediately adjacent to it. It is usually completely enclosed to insure privacy and quiet, and will therefore require provision for positive ventilation.

Dishwashing Room. Because of steam and heat liberated, this area will require the same ventilation rate as the kitchen. Depending upon size of the area, all or the greatest part of exhaust air may be taken off through the dishwasher hood which eliminates dissemination of humid air throughout the area. Exhaust ductwork should be noncorrosive and watertight to handle condensation of steam and should be graded and dripped to a convenient drain.

Dining Space. The ventilation of this space will conform to local codes. Where cafeteria service is provided, serving areas and steam tables are usually hooded. Hood air handling capacities will be somewhat less than those required for the cooking area.

Housekeeping Facilities. In this area are central linen room, sewing room, housekeepers' office, soiled linen rooms, and laundry. Of these facilities, the soiled linen rooms and laundry only will require special attention and these items will be enumerated.

Soiled Linen Storage Room. As a storage space prior to laundering, this room is subject to odors from used linens and will require ventilation at a rate of not less than 8 air changes per hour.

Laundry. Ventilation should conform to local codes but in the absence of such codes, a minimum of 10 air changes per hour is recommended.

Design of laundry ventilation systems must receive the study necessary to provide proper safeguards against the recontamination of linen during the cleaning process by airborne organisms.

The activities associated with *sorting of linen* and *loading of washers* are reported to add heavily to the contamination of the laundry air. This contaminated air is then drawn through extractors and tumblers where the linen acts as a filter and collects the organisms. It is reported many of these organisms survive the heat of the ironing process.

To reduce the contamination of laundry air, it is recommended that sorting of linen be done in an enclosed area separated from the main laundry. This area will require 8 air changes per hour and should

be maintained at a negative pressure by the ventilation system.

Of the various laundry equipment, the *flat work ironer* produces the greatest amount of heat and vapor. It is usual practice to install a canopy over the ironer to concentrate and remove the heat immediately at this point. Most laundry layouts will permit the greatest part of the exhaust to be discharged through the canopy although some exhaust should be provided over the washers to remove heat and vapors originating from this equipment. Exhaust from *tumblers* should be carried away independently of the general exhaust system and may require lint filters in addition to those on the tumbler.

All laundry exhaust should be carried to a point above the roof or to an outlet where it will not be obnoxious to other areas of the hospital. This laundry exhaust should not be in close proximity to fresh air ventilation intakes.

Fresh air supply is a requisite to good ventilation of this area and distribution designed to avoid short circuiting is important. The design may lend itself to the use of air-to-air heat exchangers for better economy of operation.

Boiler Room. Air supply to boiler rooms should be computed on the basis of providing both comfortable working conditions and the air quantities required for maximum combustion rates of the particular fuel. Boiler and burner ratings will establish maximum combustion rates, and air quantities can be computed according to the type of fuel. Following are approximate quantities of air required per unit of fuel:

Natural gas	10 cu ft per cubic foot of gas
Oil	1500 cu ft per gallon of oil
Bituminous coal	133 cu ft per pound of coal

Maintenance Shops. Carpenter, machine, electrical, and plumbing shops present no particular difficulties but the ventilation of paint shops and paint storage areas is very important and should conform to local codes. There is the danger of fumes.

Employees' Facilities. This area will include locker rooms, toilets, showers, and rest rooms for nurses, male, and female employees. These rooms will be treated to provide comfort and to conform to local codes.

FACTORY AIR CONDITIONING INVESTMENT ANALYSIS

The investment analysis form which follows was designed by Carrier Corporation to permit an owner to determine whether or not air conditioning of all or part of his factory, for the purpose of providing a comfortable working environment for his employees, will be a profitable investment.

An affirmative answer to two or more of the four questions below is a strong indication that the investment will be a profitable one and that a full analysis is warranted:

(1) Does the plant area have a density of one or more workers per 300 sq ft of floor area?

(2) Is voluntary labor turnover in the plant 15% or more every year?

(3) Are annual wages, including fringe benefits, $25 or more per sq ft of floor area?

(4) Does labor amount to 25% or more of total manufacturing cost?

The cost of air conditioning a man ranges upwards from about $300. The upper limits of investment are influenced by the efficiency of the man, absenteeism, turnover, the difficulty of getting people to work in hot places, industrial hazards and the spoilage that results from fatigue. The potential investment with respect to all these factors is difficult to measure and evaluate, but the potential is believed to warrant investment of not more than 25% of the individuals' gross annual wage. Real interest should exist between $300 and $1,200 per man.

Typically, factory air conditioning represents only about one percent of overall owning and operating costs of new or existing factories, including salaries and wages, machinery and equipment, and the building itself. An increase of 1.5% in plant workers' efficiency is usually sufficient to pay for the installation. Available reports and case histories show that this small improvement should be easy to realize. A study of 19 factories showed that increases in earnings due to air conditioning for worker comfort regained from 20 to 100% of the investment in air conditioning during the first year. Another study reported that increases in worker efficiency averaged 25%, that absenteeism dropped 30% and labor turnover was reduced 40%.

Individual conditioning should be considered under conditions of high concentrated heat load and fixed worker locations, such as crane cabs and pulpits in steel mills. Space isolation, such as a conditioned booth, is one method. The initial cost, including the enclosure, would range from $200 for

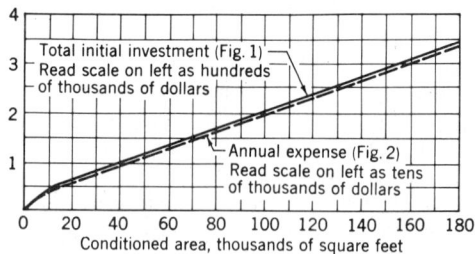

Typical factory comfort air conditioning costs, condensed from Figures 1 and 2 of "How to Quickly Estimate Factory Air Conditioning Owning and Operating Costs" for use with Section IV of Investment Analysis, following.

a small booth up to several thousand for such applications as crane cabs. The most common method of individual conditioning is spot or man cooling, where the worker is blanketed with conditioned air if his movement is limited, or is provided with a station to which he may retire in cases where he has freedom of movement.

Area conditioning is the treatment of one or more segments of a factory because of economic justification or physical need. It avoids the needless conditioning of idle space. The conditioned area does not have to be physically isolated from the rest of the plant, although enclosures or semi-partitions may help reduce operating costs. Such application requires careful arrangement of the air distribution system to limit the extent of the fringe area.

Complete-plant conditioning is usually best for factories with high worker density, such as metal parts fabrication and electronic equipment manufacture. It is worth considering if the space per worker averages less than 500 sq ft. Space assigned to storage should normally be excluded in this evaluation.

The decision to air condition all or part of a plant should be based in part on the investment analysis and in part on a complete plant survey that usually follows an affirmative analysis. Each production area can be evaluated as to the type of system required, initial system cost, operating cost, benefits and return on investment. The survey will provide information on which realistic decisions can be made.

SECTION I. Costs and profits before air conditioning

The purpose of this Section is to establish dollar values for various items of cost, income and profits. The values obtained will be applied in following Sections of the Analysis. Part 1 below is self-explanatory. In order to complete Part 2, it is necessary to estimate from experience the per cent of total cost that should be allocated to each of seven different cost classifications. It is important that the cost classifications be treated within the limits of the definitions supplied. All figures can be round numbers, but they should be accurate enough to reflect true conditions.

Part 1. Basic information

	TOTALS
1. Annual product volume (number of units or pieces).	500,000
2. Annual costs (all costs, including sales, general and administrative as itemized below). . $	2,000,000
3. Average cost per unit (Item 2 divided by Item 1). . . . . $	4.00

4. No. hourly workers 50   5. No. piecework workers 150   6. Total number of all workers 200

Part 2. Cost breakdown, income, profit before Federal taxes

	Per Cent of Total Cost	Dollar Value
7. Hourly labor	8 %	$ 160,000
8. Piecework labor	27 %	$ 540,000
9. Employee benefits	9 %	$ 180,000
10. Materials and parts	20 %	$ 400,000
11. Manufacturing overhead	15 %	$ 300,000
12. Freight	6 %	$ 120,000
13. Sales, General and Administrative	15 %	$ 300,000
14. Other	— %	$ —
15. TOTAL ANNUAL COST	100 %	$ 2,000,000

16. Average markup 20 % | 17. Total selling price $ 2,400,000 | 18. Net profit before taxes $ 400,000

Description of terms

Hourly labor. Includes all factory hourly labor and its supervision directly connected with production.

Piecework labor. All labor paid by the piece or on an incentive basis, plus its supervision.

Employee benefits. All labor expenses not paid to workers, such as insurance, social security, etc.

Materials and parts. All materials and purchased parts used in the manufactured product.

Manufacturing overhead. Normally included are manufacturing engineering, plant and machinery maintenance, depreciation of buildings, jigs, fixtures, tools and machinery, power and steam costs, insurance, property taxes and similar items.

Freight. All transportation charges paid for components or for shipment of finished product.

S.G.&A. Includes such costs as administration, accounting, clerical, product engineering, research, selling, advertising, communications, personnel department.

Other. Any cost items not included elsewhere, except Federal income taxes.

SECTION II. Evaluation of benefits provided by air conditioning

This Section is organized so that plant management can effectively appraise the benefits provided by factory air conditioning, estimate the value of the benefits realistically and express the values in terms of percentages of cost. These steps have been simplified by providing five specific areas for consideration—worker output, worker training, product spoilage, housekeeping and personnel operations. In addition, a means is provided for interpreting economically the effects of air conditioning on piecework production.

Part 1. Increased worker output

The formula below calls for an estimate of the number of minutes saved per hour, per worker, by maintaining a completely comfortable working atmosphere. Minutes saved are then converted to per cent of annual savings in labor. Note that it is necessary to estimate the number of warm weather months during which the air conditioning system will be operated. Normally this is five months (May through September) but local weather and plant conditions may make this number larger or smaller.

In estimating the number of minutes saved per hour, managers should keep in mind these benefits that are assured by constant control of temperature, humidity, ventilation, air cleanliness and circulation:

Less fatigue. Workers are physically and mentally more efficient in a constantly comfortable environment.

Fewer breaks. Number of work stops for trips to water cooler, brow mopping, weather talk, etc. is reduced.

Less absenteeism. On hot days, workers prefer the cool comfort of the plant to hot weather elsewhere.

Less labor turnover. More experienced, efficient labor force becomes stabilized; requires less supervision.

High caliber workers. The air conditioned plant is a preferred place to work; attracts better workers.

Better morale. Improvement of environment helps promote healthy attitudes and better relationships.

19. Minutes saved per hour, per worker, during summer months only: __5__   20. Number of summer months: __5__

21. Calculate annual saving in labor: __6__ (Item 19) $\div$ 60 $\times$ $\dfrac{5 \text{ (Item 20)}}{12}$ = __3.5__ %

Part 2. Reduced training expense

Since air conditioning reduces labor turnover, the cost of training new workers will be reduced. Here it is necessary to determine the amount now spent on training workers, to estimate the probable reduction in labor turnover as a result of air conditioning and to translate this into annual savings in labor.

22. What percentage of total labor cost is now spent on worker training?...................... __5__ %

23. How much, in per cent, do you estimate air conditioning will reduce labor turnover?........ __25__ %

24. Calculate annual saving in labor: __5__ % (Item 22) $\times$ __25__ % (Item 23) = __1.25__ %

Part 3. Reduction of rejects and product spoilage

Many manufacturing operations benefit from humidity control, air cleaning and the ability to keep doors and windows closed in all seasons. Rejects, seconds and product spoilage caused by perspiration, dirt and grime are sharply reduced. These benefits result in both labor savings and conservation of materials. The estimates below reduce these savings to annual percentages of labor and material costs.

25. How much labor is lost thru rejects, etc. __2__ %   26. How much would air conditioning cut loss? __20__ %

27. Calculate annual saving in labor: __2__ % (Item 25) $\times$ __20__ % (Item 26) = __.4__ %

28. How much material is lost thru rejects __2__ %   29. How much would air conditioning cut loss? __20__ %

30. Calculate annual saving in materials: __2__ % (Item 28) $\times$ __20__ % (Item 29) = __.4__ %

Part 4. Reduced housekeeping costs

Sweeping, mopping and similar clean-up expense is a portion of manufacturing overhead that will be cut by air conditioning. Doors and windows are kept closed and indoor air is constantly filtered.

31. What percentage of manufacturing overhead (Item 11) is now spent for cleaning maintenance? ___**4**___ %

32. How much will air conditioning reduce present expenses for cleaning maintenance?........... ___**20**___ %

33. Calculate annual saving in manufacturing overhead: ___**4**___ % (Item 31) × ___**20**___ % (Item 32) = ___**.8**___ %

Part 5. Reduced personnel department expense

When labor turnover is reduced, you not only benefit from more experienced workers and lower training expense but also from a reduction in the cost of recruiting, selecting and processing new workers.

34. What percentage of S. G. and A. (Item 13) is devoted to recruiting and processing workers?.... ___**5**___ %

35. How much do you estimate these costs will be reduced by air conditioning?................. ___**25**___ %

36. Calculate annual saving in S. G. and A.: ___**5**___ % (Item 34) × ___**25**___ % (Item 35) = ___**1.25**___ %

Part 6. Special piecework or incentive pay considerations

In almost all plants that pay for labor on a piecework or production bonus basis, the company as well as the worker benefits from increased worker productivity. The following calculations compute the portion of additional labor output beneficial to the company through a reduction in product labor cost.

37. Present annual cost of piecework labor (from Item 8) .... $ ___**540,000**___

38. Estimate current annual product volume (units) produced by workers paid by piece....... ___**375,000**___ units

39. Calculate average pay per unit: $ ___**540,000**___ (Item 37) ÷ ___**375,000**___ units (Item 38) = $ ___**1.44**___

40. What is your pay per unit for production above current production volume?......... $ ___**1.20**___ per unit

41. Calculate difference in payment per unit that benefits company: (Item 39 minus Item 40) = $ ___**0.24**___

42. Calculate portion of increased production that benefits company: (Item 41 ÷ Item 39) = ___**16.7**___ %

43. Calculate annual company benefit in piecework labor: ___**3.5**___ % (Item 21) × ___**16.7**___ % (Item 42) = ___**.59**___ %

Part 7. Summary of anticipated savings from air conditioning

Expense categories	Cost Reduction	Improvement factor*
44. Hourly labor: ___**3.5**___ % (Item 21) + ___**1.25**___ % (Item 24) + ___**.40**___ % (Item 27) =	___**5.15**___ %	44A. ___**94.85**___ %
45. Piecework labor: ___**.59**___ % (Item 43) + ___**1.25**___ % (Item 24) + ___**.40**___ % (Item 27) =	___**2.24**___ %	45A. ___**97.76**___ %
46. Materials and parts (from Item 30).....................................	___**.40**___ %	46A. ___**99.6**___ %
47. Manufacturing overhead (from Item 33)	___**.80**___ %	47A. ___**99.2**___ %
48. Sales, General and Administrative (from Item 36).....................................	___**1.25**___ %	48A. ___**98.75**___ %

*Improvement factor = 100% minus the percent of Cost Reduction.

SECTION III. Net profit after air conditioning

Before proceeding with this Section of the Analysis, plant management must decide which of two methods will be adopted to capitalize on the increase in worker productivity that will occur after air conditioning. These two methods are: 1. Maintain the same annual rate of production, but employ fewer workers. 2. Increase the rate of production, but employ the same number of workers. Use only the Part below which applies, according to your decision.

Part 1. Maintain current production but employ fewer workers

49. Number of workers before air conditioning   (Item 6) ×   % (Item 44A) =   workers

Calculate cost breakdown after air conditioning:

50. Hourly labor before air conditioning $ (Item 7) ×   % (Item 44A) = $

51. Piecework labor before air conditioning $ (Item 8) ×   % (Item 45A) = $

52. Employee benefits before air conditioning $ (Item 9) ×   % (Item 44A) = $

53. Materials before air conditioning $ (Item 10) ×   % (Item 46A) = $

54. Overhead before air conditioning $ (Item 11) ×   % (Item 47A) = $

55. Freight before air conditioning (Item 12) .... $

56. S. G. & A. before air conditioning $ (Item 13) ×   % (Item 48A) = $

57. Other costs before air conditioning (Item 14) .... $

58. TOTAL ANNUAL COST AFTER AIR CONDITIONING... $

59. TOTAL ANNUAL SELLING PRICE (Same as before air conditioning, from Item 17) ....$

60. NET PROFIT AFTER AIR CONDITIONING, BEFORE FEDERAL TAXES (Item 59 minus Item 58) .....$

61. REVENUE FROM AIR CONDITIONING INVESTMENT (Item 60 minus Item 18) .... $

Part 2. Employ same number of workers but increase production

62. Annual product volume (units) before air conditioning $500,000$ (Item 1) ÷ 94.85% (44A) = $526,000$

Calculate cost breakdown after air conditioning:

63. Hourly labor before air conditioning (Item 7) .... $160,000$

64. Piecework: $ 1.20 (Item 40) × $26,000$ (Item 62 less Item 1) + $$540,000$ Item 8) = $571,200$

65. Employee benefits before air conditioning (Item 9) .... $180,000$

66. Materials before air conditioning $ $400,000$ (10) ÷94.85% (44A) ×99.6(46A) = $420,000$

67. Overhead before air conditioning $ $300,000$ (11) ×99.2% (Item 47A) = $295,000$

68. Freight before air conditioning $ $129,000$ (12) ÷94.85% (Item 44A) = $126,500$

69. S. G. & A. before air conditioning $$300,000$ (13) ÷94.85% (44A) ×98.75% (48A) = $316,000$

70. Other costs before air conditioning (Item 14) .... $ —

71. TOTAL ANNUAL COST AFTER AIR CONDITIONING... $2,068,700$

72. TOTAL ANNUAL SELLING PRICE ($ $2,400,000$ Item 17 ÷ 94.85 % Item 44A) = $2,530,000$

73. NET PROFIT AFTER AIR CONDITIONING, BEFORE FEDERAL TAXES (Item 72 minus Item 71) ..... $461,300$

74. REVENUE FROM AIR CONDITIONING INVESTMENT (Item 73 minus Item 18) .... $ $61,300$

SECTION IV. Total cost of air conditioning

This Section permits the first cost and annual operating expense of the air conditioning system under consideration to be worked into the investment analysis. The total initial investment required for your particular system can be estimated by consulting Figure 1 in the pamphlet entitled "How to Quickly Estimate Factory Air Conditioning Owning and Operating Costs." All you need to know is the number of square feet of factory space the system will serve. Annual expense of the system can be obtained from Figure 2 of the same pamphlet. While these figures admittedly will not be as accurate as those obtained from a detailed professional survey, they will be thoroughly reliable for this purpose. In most cases, the estimated figures will be maximum rather than minimum.

The blue form also permits a quick estimate of the tons of refrigeration and the quantity of conditioned air that will be required by the system. These figures are not essential to the analysis, but they will help you visualize the project realistically.

Please note that the cost figures given below cover a top-quality, long-life system. In some cases the system installed will consist of a central station refrigeration machine which will supply chilled water to air-treating equipment located in various parts of the plant. In other cases, it may be more desirable to install a system utilizing packaged equipment. For certain projects heat pump systems may be indicated. In every case the system will feature well-designed automatic controls to assure good conditions and low operating expense. All auxiliaries and ductwork necessary to the system are covered in the estimates. The air handling apparatus and controls will be connected for winter heating, air cleaning and ventilation even though the benefits evaluated in the Analysis are restricted to summer comfort only.

All expenses to the buyer are included, regardless of classification of work. These are your complete estimated costs for installed air conditioning.

Part 1. Basic information

75. Floor area of factory space under consideration	*30,000* sq ft
76. Total number of workers, including all shifts	*200* workers
77. Floor area per worker	*150* sq ft

Part 2. Initial investment and annual operating cost (Items 78 and 81 from "How to Estimate" folder)

INITIAL INVESTMENT	78. Total initial investment for air conditioning	$ *78,000*
	79. Initial investment per worker (Item 78 ÷ Item 76)	$ *390*
	80. Initial investment per square foot of floor area (78 ÷ 75)	$ *2.60*
ANNUAL EXPENSE	81. Annual expense for air conditioning	$ *7500*
	82. Annual expense per worker (Item 81 ÷ Item 76)	$ *37.50*
	83. Annual expense per square foot of floor area (81 ÷ 75)	$ *0.25*

84. ECONOMIC LIFE OF AIR CONDITIONING SYSTEM	*15* YEARS

Initial investment. Total cost of the system to the buyer includes all installed equipment and related expense, such as construction, wiring and painting.

Annual expense. Includes taxes, insurance, service, parts, operating personnel, water consumed by system, steam or electric power, but not amortization.

SECTION V. Return on air conditioning investment

Here is a summary of the potentialities of the air conditioning system under consideration. Net annual income, cash flow and annual return on investment are the key factors.

85. Annual depreciation

Initial investment $ _78,000_ (Item 78) ÷ _15_ years (Item 84) = $ _5200_

86. Total annual expense

Annual expense $ _7500_ (Item 81) + $ _5200_ (Item 85) = $ _12,700_

87. Taxable annual income

Revenue from air conditioning $ _61,300_ (Item 74) minus $ _12,700_ (Item 86) = $ _48,600_

88. Annual Federal tax

Taxable income $ _48,600_ (Item 87) × _52_ % (Tax rate) = $ _25,272_

89. Net annual income from air conditioning

Taxable income $ _48,600_ (Item 87) minus $ _25,272_ (Item 88) = $ _23,328_

90. Annual cash flow

Net income $ _23,328_ (Item 89) + $ _5200_ (Item 85) = $ _28,528_

91. Pay-back period

Initial investment $ _78,000_ (Item 78) ÷ $ _28,528_ (Item 90) = _2.73_ years

Return on investment

The numerical value of the pay back period (Item 91) is equivalent to a value that indicates the rate of return on the air conditioning investment. To determine rate of return in percent, enter the table below at "years" (Item 84), then read horizontally to figure nearest in value to Item 91, then read up.

92. Annual return on air conditioning investment..................... | **36** %

PRESENT WORTH FACTOR — UNIFORM ANNUAL SERIES

Years	10%	15%	20%	25%	30%	35%	40%	45%	50%	55%	60%	65%	70%	75%	80%	90%	100%	Years
10	6.145	5.019	4.192	3.571	3.092	2.715	2.414	2.168	1.965	1.794	1.650	1.525	1.441	1.363	1.244	1.072	.998	10
11	6.495	5.234	4.327	3.656	3.147	2.752	2.438	2.185	1.977	1.802	1.656	1.530	1.444	1.365	1.245	1.073	.998	11
12	6.814	5.421	4.439	3.725	3.190	2.779	2.456	2.196	1.985	1.807	1.660	1.532	1.446	1.366	1.246	1.073	.998	12
13	7.103	5.583	4.533	3.780	3.223	2.799	2.468	2.204	1.990	1.810	1.662	1.535	1.447	1.367	1.246	1.073	.998	13
14	7.367	5.724	4.611	3.824	3.249	2.814	2.477	2.210	1.993	1.812	1.663	1.536	1.448	1.367	1.246	1.073	.998	14
15	7.606	5.847	4.675	3.859	3.268	2.825	2.484	2.214	1.995	1.813	1.664	1.537	1.448	1.367	1.246	1.073	.998	15
16	7.824	5.954	4.730	3.887	3.283	2.834	2.489	2.216	1.997	1.814	1.665	1.538	1.448	1.367	1.246	1.073	.998	16
17	8.022	6.047	4.775	3.910	3.295	2.840	2.492	2.218	1.998	1.815	1.665	1.538	1.448	1.367	1.246	1.073	.998	17
18	8.201	6.128	4.812	3.928	3.304	2.844	2.494	2.219	1.999	1.815	1.665	1.538	1.448	1.367	1.246	1.073	.998	18
19	8.365	6.198	4.844	3.942	3.311	2.848	2.496	2.220	1.999	1.815	1.665	1.538	1.448	1.367	1.246	1.073	.998	19
20	8.514	6.259	4.870	3.954	3.316	2.850	2.497	2.221	1.999	1.815	1.665	1.538	1.448	1.367	1.246	1.073	.998	20
21	8.649	6.312	4.891	3.963	3.320	2.852	2.498	2.221	2.000	1.815	1.665	1.538	1.448	1.367	1.246	1.073	.998	21
22	8.772	6.359	4.909	3.970	3.323	2.853	2.498	2.222	2.000	1.815	1.665	1.538	1.448	1.367	1.246	1.073	.998	22
23	8.883	6.399	4.925	3.976	3.325	2.854	2.499	2.222	2.000	1.815	1.665	1.538	1.448	1.367	1.246	1.073	.998	23
24	8.985	6.434	4.937	3.981	3.327	2.855	2.499	2.222	2.000	1.815	1.665	1.538	1.448	1.367	1.246	1.073	.998	24
25	9.077	6.464	4.948	3.985	3.329	2.856	2.499	2.222	2.000	1.815	1.665	1.538	1.448	1.367	1.246	1.073	.998	25

GREENHOUSE COOLING

Many greenhouses are equipped with cross-ventilation systems using evaporative cooling of the air. In the fan-pad system, cooling is effected by passing the air through pads of loosely packed material which are kept wet by trickling water through them. In another system, water is sprayed into the incoming air through nozzles operating at 300-600 psi. A bank of nozzles is placed at the air inlet and, if the greenhouse is wide, one or more additional banks may be placed part way across the house. The air is heated as it moves through the greenhouse and the additional spray banks are used to cool it again.

On-off control of spray water is effected by using a humidistat, thermostat, or droplet-sensing device in conjunction with a solenoid valve. Discussion here is limited to fan-pad systems and to spray systems having a single bank of nozzles. Solar radiation, weather conditions, and allowable greenhouse temperatures play important parts in the design of such cooling systems.

In designing an evaporative cooling system, it is first necessary to evaluate air flow and water requirements. The controlling factor in this cooling problem is greenhouse temperature. In the following discussion, evaluation of outdoor design temperatures, and of air flow and water requirements, is based on use of a greenhouse design temperature t_m, which must not be exceeded, even under the most severe conditions. In effect, this means that outdoor temperatures used for system design must be extreme values, available from weather records.

Note that this is not in keeping with the procedure usually followed in other fields, where some leeway is usually allowed by selecting temperatures which are different from the most severe recorded (for instance, the $2\frac{1}{2}\%$ basis, commonly recommended for building heat loss calculations).

During the critical 10 a.m. to 4 p.m. period, heat gain from solar radiation is high. Heat gains from other sources are small in comparison and may reasonably be neglected.

About 50% of solar radiation entering the house becomes sensible heat. Most of the remaining 50% evaporates moisture from wetted surfaces and appears as latent heat. Regardless of which form it takes, it must eventually be absorbed and removed by the ventilating air. Therefore, the cooling load is taken to be the total instantaneous rate of heat gain through a horizontal single sheet of unshaded glass at noon on August 1, with a clear atmosphere.

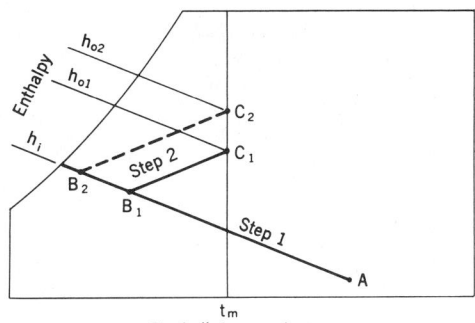

Fig. 1. Skeleton psychrometric charts, showing heating and humidification of ventilation air as it passes through a fan-pad cooler (Step 1) and through greenhouse (Step 2), along two possible paths.

The process of humidification and heating can be divided into two steps, as illustrated in Fig. 1. First, adiabatic humidification by sprays or wetted pads ($A — B_1$, or $A — B_2$) and second, addition of sensible and latent heat, resulting from insolation, in about equal amounts ($B_1 — C_1$ or $B_2 — C_2$).

Design Wet-Bulb and Air Flow

If the cooling load is x Btu per hr per square foot of floor, and $h_o — h_i$ is the enthalpy change of the air passing through the greenhouse, the rate of air flow Q, lb per hr per square foot of floor is

$$Q = x / (h_o — h_i).$$

Since enthalpy of air is closely related to its wet-bulb temperature, the value of h_i, the enthalpy of the incoming air, to be used to find the required air flow rate, depends only on the outside design wet-bulb temperature.

The method of selecting design temperatures was discussed above. Values in Table 1 are the highest noon wet-bulb temperatures found in available records which occurred simultaneously with a sky cover of 0.3 or less. Sky cover factor was included because incident solar radiation, and therefore cooling load, is appreciably reduced by cloud cover. The most severe cooling problems are likely to occur on clear days. The value of 0.3 was rather arbitrarily selected.

As the air moves across the greenhouse, heat is added to it and the highest temperature will be

Table 1. Data for Greenhouse Cooling at Selected Localities

(Based on three years' records)

Location	Cooling Load, Btu per hr per sq ft of floor	Design Dry-Bulb, F	Design Wet-Bulb, F
Columbus, Ohio	275	93	77
Raleigh, N. C.	284	95	79
Sacramento, Calif.	275	97	73
Tallahassee, Fla.	296	94	80
Fort Worth, Tex.	296	103	78
Helena, Mont.	254	89	64

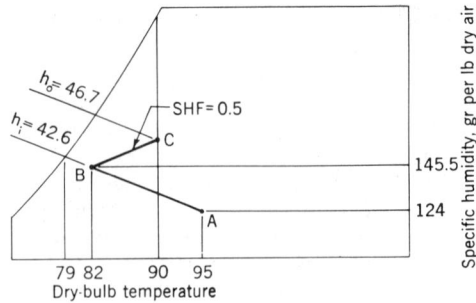

Fig. 2. Skeleton psychrometric chart for Example 1. (A is the initial condition. AB is constant wet-bulb adiabatic saturation through fan-pad cooler which terminates at B, 4/5 of the distance to the saturation curve, because fan-pad is only 80% efficient. From B to C, air through greenhouse picks up both latent and sensible heat, with sensible heat ratio of 0.5, terminating at the design maximum temperature.

reached just as it reaches the outlet. Therefore t_m, which is the maximum temperature to be permitted in the house, limits the temperature and therefore the enthalpy, h_o of the outgoing air. (Fig. 1)

Figure 1 shows that the amount of heat which can be picked up by the air in Step 2 of the process without exceeding t_m depends primarily on how much the air is initially cooled below t_m. If there is an insufficient supply of water to adequately cool the air, either the enthalpy change in Step 2 will have to be reduced, or t_m will be exceeded.

If the air is to be cooled a given number of degrees below t_m, the amount of water required increases as the dry-bulb temperature of the air increases and this increase is more or less independent of the wet-bulb temperature. To be sure of having enough water, a high design dry-bulb temperature should be used. Using a number of years of weather records from several stations, the values given in Table 1 were selected by taking the highest noon dry-bulb temperature which occurred simultaneously with a sky cover of 0.3 or less.

Example 1: *A fan-pad system at Raleigh, N. C. has a humidifying effectiveness of 80%. Air flow and water requirements per square foot of greenhouse floor are to be found. Maximum allowable greenhouse temperature, t_m is 90 F.*

Air flow required is $Q = x / (h_o — h_i)$. From Table 1, cooling load, x is found to be 284 Btu per hr per sq ft. Design wet-bulb from Table 1 is 79 F; therefore enthalpy of the incoming air, h_i is 42.6 Btu per pound of dry air (from the psychrometric chart or tables).

Figure 2 is a skeleton diagram of a psychrometric chart on which h_o, outgoing air enthalpy, is calculated. Design dry-bulb, 95 F, from Table 1 is drawn

in, and its intersection with design wet-bulb, 79 F, is Point A.

Since the humidifying effectiveness is given as 80%, then the output of the fan-pad cooler will be 80% of the way from 95 F to 79 F, or approximately 82 F. Proceeding along the 79-F wet-bulb line ($h_i = 42.6$), its intersection with the 82-F dry-bulb line gives Point B, which represents the air leaving the cooler.

Because it is assumed that 50% of solar heat becomes sensible heat, the air is represented as warming and humidifying along the sensible heat factor line of 0.5 until it reaches the design maximum, 90 F, which yields Point C. Reading directly from the chart, h_o is found to be 46.7.

Therefore, $Q = 284 / (46.7 — 42.6) = 69$ lb air per hr per square foot of greenhouse floor.

Water required to cool the air in the fan-pad cooler from 95 F to 82 F is found by subtracting the specific humidities at Point B from Point A, which quantities are noted on the chart in Fig. 2. The difference is 21.5 grains per pound dry air. Since $Q = 69$ lb per hr, then the water required is 21.5 × 69, or $W = 1480$ gr per hr per sq ft = 0.21 lb per hr per sq ft of floor.

When water is sprayed into the air, the process may be different than illustrated in Fig. 1 and 2. The difference occurs if more water is added to the air than can be evaporated by the sensible heat that is immediately available. If this happens, the air becomes saturated, or nearly so, and water droplets

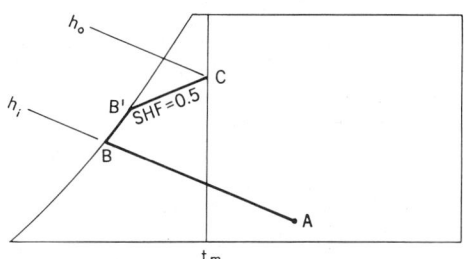

Fig. 3. Skeleton psychrometric chart showing heating and humidification of ventilating air when water droplets (from sprays) are entrained in the air initially. At Point B′ the droplets have finally evaporated and heating and humidification proceeds along the SHF = 0.5 line to C.

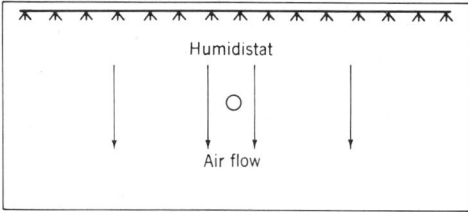

Fig. 4. Plan view of greenhouse in Example 2.

are entrained in it. These droplets may fall out or be evaporated by solar heat as the air moves across the greenhouse.

If it is assumed that the evaporation of water from the wetted surfaces (soil and floor surfaces in the greenhouse) is unaffected by this entrainment, the overall effect will be to increase the fraction of solar heat removed in the latent form. The path of the process is represented in idealized form in Fig. 3. The air is initially saturated (at B) and remains so during the first part of its travel across the house, until the entrained droplets are evaporated (at B′). It then follows the 0.5 SHF line for the remaining distance.

Example 2: *A greenhouse with a spray water evaporative cooling system, Fig. 4, is located at Raleigh, N.C. The sprays are controlled by a humidistat, set at 90% rh, located at the middle of the greenhouse. The air and spray water requirements are to be found.*

Design temperatures are the same as for Example 1. The process is illustrated in Fig. 5.

Air is first adiabatically saturated. Location of

the path is found by assuming that half the enthalpy change in the air has taken place by the time the air reaches the humidistat. Thus, enthalpy of air when it reaches humidistat may be represented by

$$h_{hum} = (h_i + h_o) / 2$$

On the psychrometric chart, of which Fig. 5 is a skeleton diagram, lay out the process. Entering the chart at Point A, the initial dry-bulb and wet-bulb condition, proceed along the path of adiabatic saturation to Point B, complete saturation, and extend the line out past the saturation line an indefinite length to account for entrained moisture. The process proceeds along the saturation curve, but Point B′ is not yet located.

It is assumed that the process proceeds from Point B′ to Point C, somewhere on the $t_m = 90$ F line, along the SHF = 0.5 line. The equation above dictates that this line shall cross the 90% rh line at a point O, such that $DO = OC$, Point D being the intersection of the extension of lines AB and CB′.

Thus, with the slope of the line determined by the given SHF, with its terminus known to be on the 90-deg dry-bulb line, a straight edge can be used, in trial-and-error fashion, to find Points B′ and C, such that $DO = OC$. Then it is found that $h_o = 48.6$ and $Q = 284 \div (48.6 - 42.6) = 47$ lb per hr per sq ft.

The water requirement, based on an initial dry-bulb of 95 F, can be found by extrapolation of CB′ to D on the 79 F wet-bulb line. It is seen that 31 gr per lb dry air are added, so that spray water required is $31 \times 47 = 1460$ gr or 0.21 lb per hr per sq ft of floor.

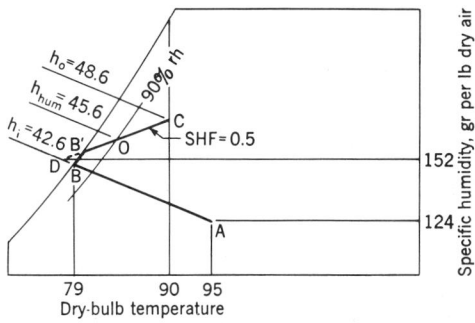

Fig. 5. Skeleton psychrometric chart for Example 2. Line B′C is found by trial and error, given that O, on the 90% rh line, bisects DC, where C is on the line of maximum design temperature.

AIR CONDITIONING VARIETY STORES

The variety type store, with articles from 5 cents to $1 and over, attracts large volumes of traffic-inducing impulsive purchasing.

There is an endless variety of articles on sale: Hard goods, soft goods, sports goods, pets of all description. Among other things, the variety store has become a major food merchandiser, from customary lunch counters to fabulous installations, augmented by stand-up type food centers offering such odorous items as frankfurters, hamburgers, and pizza. Usually, in order to standardize and simplify engineering calculations, inside conditions are 80 F dry bulb with a relative humidity of 50%.

Heat Gain from Lighting. Present-day merchandising demands intense lighting with special lighting for jewelry displays, lighting fixture displays, and other specialties. The wattage can run as high as 6 watts per sq ft. However, in the interest of standardization, 4 watts per sq ft is used in determining heat gain from lighting in all sales areas. Where stock areas are conditioned, a factor of 2 watts per sq ft is used. This applies also to receiving, marking, toilet, and rest room areas.

Heat from Cooking Appliances. The class of store determines the type and variety of appliances used. The restaurant equipment suppliers can be a source of data. (See, also, Commercial Kitchen Ventilation, elsewhere in this Handbook.)

Heat from Special Merchandising Displays. Included are cooking appliances for preparation of popcorn, pizza, frankfurters, hamburgers, doughnuts, roasted chickens and cooked nuts. Heat load from these special merchandisers should be taken from the respective manufacturer's published data.

Heat from Powered Equipment. Check mechanical amusement rides for children.

Population Density. Based on operating experience, it has been found that the following gross sales area figures prevail during periods of steady traffic:

First floor—50 sq ft per person.
Upper floor—100 sq ft per person.
Basement—100 sq ft per person.
Stock room—200 sq ft per person.

Ventilation Requirements. Outside air supply is usually 10 cfm per occupant for sales space and other conditioned areas, 15 cfm per occupant for restaurant spaces separate from sales areas or provided with individual air conditioning.

Total outside air should be figured on the basis of occupants or the exhaust quantities plus one air change, whichever is the greater. A positive pressure should be maintained in the conditioned space to assure exfiltration through the entrance doors and serving doors, or serving area for food.

Selection of the type of equipment to be employed is the prerogative of the landlord. Arrangements are provided in the lease agreement. The most economical system which can be developed will be given first consideration since rentals are based upon total costs for preparing the premises for occupancy.

Among the many things which must be considered are: Floor space that will have to be given up for equipment, cost of operation, cost of maintenance, and simplicity of control. Small stores cannot tolerate space being taken from the sales area to accommodate air conditioning equipment. Therefore, penthouse arrangements, suspended units, or roof-mounted units must be considered. Such equipment is usually self-contained and is factory-assembled; generally, it has proven least expensive, easy to maintain and, in most cases, it has been found highly efficient. (See Unitary Air Conditioning, elsewhere in this Handbook.)

For stores above 16,000 sq ft of sales area, built-up types of systems will be more economical. Due to the greater sales volume of such stores, adequate maintenance to operate this type of system becomes economically feasible. Since high values are placed on possible sales spaces, location of machine room space becomes a major factor. Wherever possible, machine rooms should be located in areas which have the least value. They can be planned for mezzanine, roof, or basement areas. Codes in various cities should be investigated as these regulations will sometimes restrict the choice of design.

Air Distribution

In variety stores, air distribution is not as critical as in a restaurant or a theater, since people are continually moving about and are not affected too much by slight local drafts. However, the full benefits from air conditioning can only be obtained if the temperature and air motion are uniformly maintained throughout the conditioned area. Drafty and stagnant air conditions are particularly objectionable at the cafeteria area where the occupants remain seated in one location while eating.

Good distribution and comfort conditions in the store dictate using approximately six air changes

per hour for the total air quantity. Initial duct velocities can be 2200 to 2400 fpm at the fan outlet. Noise is usually not a factor since the noise level in the store is higher than in the usual office building or department store.

In the smaller stores, ductwork may be unnecessary or held to a minimum if self contained air conditioners are installed and located to give proper air distribution. In larger stores, where a remote or central station system is employed, the volume of air requires ductwork for proper distribution.

Clean air is particularly important because the sale of merchandise depends on a clean, attractive article, and in addition the variety store is usually an eating establishment requiring clean air. Both recirculated and outside air should be filtered. Spun glass, replacement type filters are generally used because they are practical and reasonable in price. However, there are many instances where the cleanable type filter is preferred.

Face velocities at return air grilles should be held at a low velocity not because of noise, but for collection of dust and dirt on the face of the grilles and adjoining walls, ceilings, and display counters located in the path of the returned air. In many cases, several return air openings should be utilized to accomplish best results.

Controls

The store's personnel is selected on a basis of merchandising ability and very seldom are these people even remotely capable of understanding the function of air conditioning. The most practical system is one that is completely automatic and requires nothing more for its operation other than the simple turning of a switch which reads "Day" or "Night."

A typical control sequence is as follows:

The fan system shall be started manually by the "Day-Off-Night" switch and run continuously during occupied periods through a starting device having auxiliary contacts.

A Firestat set at 200 F shall be located in the return air duct and shall stop the fan if the air temperature in the duct exceeds its setting. Provide a freeze protection controller with its 20-ft element fastened to the discharge face of the heating coil and wire in series with the fan starter. Should the discharge temperature leaving the heating coil drop below 35 F, the fan will automatically shut down.

When the system is operating at the day position, the fresh air damper is energized and opens to admit the required amount of fresh air for ventilation.

The space thermostat shall, on a call for heat, cause a modulating control valve at the heating coil to modulate towards an open position so as to maintain space temperature requirements. A low limit discharge controller shall be wired into the heating circuit so that the temperature of the air being discharged to the space shall not be below 60 F, regardles of the thermostat setting. During the cooling cycle, the space thermostat shall, upon a rise in temperature above its set point, operate the cooling system to maintain space temperature requirements by staging of direct expansion valves or modulation of chilled water coils. When the system is on a cooling cycle, the low limit discharge controller shall become inoperative. All elements of the cooling system should be electrically interlocked with the fan starter so that the cooling equipment cannot operate unless the fan is in operation.

When the "Day-Off-Night" switch is moved to the "Night" position during the heating season, a night thermostat shall cycle the fan to maintain a lower night temperature. During this period the fan system shall operate with the fresh air damper closed. With the switch in the "off" position, the fan and cooling system shall be inoperative.

For the larger stores where multiple zoning is required, the same general sequence of control can be employed; it being only necessary to multiplex the controls.

Auxiliary Heating

Entrances handling heavy traffic and large see-through type window areas impose heating loads which the general design of air distribution cannot handle through the normal control system. The sales area thermostat, in many cases, will be satisfied through the lighting load and internal heat gain from people and appliances, and should not be located so as to sense the unusual conditions at the entrances and windows. Therefore, consideration should be given to the use of auxiliary heating sources such as booster coils in the duct serving the frontal area of the store, or individual unit heaters spotted strategically over the affected areas. The control of these sources should be by independent thermostats located in the affected area. Throughout the store, such other areas as receiving entrances, loading docks, stock areas, rest rooms and toilets, may have to have special consideration for comfort of customers and store personnel. These special considerations can involve zoning, additional heat in-put and ventilation employing automatic thermostatic controls.

AIR CONDITIONING BOWLING CENTERS

Air conditioning of bowling alleys presents no unusual problems, but there are several factors which should be kept in mind when designing a system to provide optimum comfort. They are:

1. Bowling alleys are treated much like "spot cooling" applications in that only the space occupied by the bowlers and spectators is conditioned. The *heavy concentration of people* in this portion of the building localizes the load, so that much of the total space within the building need not be cooled. The cooling load in this area is substantial with both high sensible and high latent heat gains.

2. Because of athletic activity and heavy smoking normally associated with the sport, a *high ventilation rate* is necessary to maintain air freshness. Cooling and dehumidification of ventilation air imposes an appreciable load on the system.

3. Most medium-size and larger bowling centers include a restaurant and bar, locker rooms, offices, and sometimes even a nursery. Such spaces deserve special attention when designing anything but a minimum system. Separating such special-purpose rooms from the bowling area by *zoning* techniques is always desirable.

Estimation of Cooling Load

Outside design conditions for all localities are well-established. Design dry bulb and wet bulb temperatures are those which normally occur at 3 p.m. (sun time) on a design day during July. However, the cooling requirements for bowling centers are usually greatest between 7 and 9 p.m. when most of the bowlers are present. The outside air is cooler at this time, but its moisture content usually has not changed appreciably. Unless the proprietor indicates his desire to provide perfect comfort for a full house on a hot July afternoon, the cooling load estimate may be based on outside design temperatures corrected for 8 p.m. The amount of correction depends on the average daily dry-bulb temperature range for the locality, and appropriate Weather Bureau data should be consulted. As an example, in Omaha, Neb., the outside summer design conditions are 95 F db and 78 F wb, with an average daily dry-bulb range of 20 F. Correcting these conditions for 8 p.m. clock time (7 p.m. sun time with daylight saving), the proper design conditions to use in the load estimate would be 90 F db and 76.5 F wb.

Optimum inside design conditions for the bowling and spectator area are somewhat difficult to pin-

point. Commercial practice suggests maximum design conditions of 80 F and 50% relative humidity, but for a deluxe job, design conditions of 76 F and 55% rh could be used. Because bowling centers have high latent heat gains, the choice of a low db and high rh design condition will usually result in an economical equipment selection with little compromise of comfort. Effective Temperature should be between 69 and 73 ET. Generally, an inside design condition of 78 F and 60% rh is suggested as a starting point for the load calculation.

Cooling load due to people present in the space is appreciable, for in addition to the five bowlers on each alley during team play, spectators are usually seated behind the bowlers' benches. The proprietor may be able to estimate the average number of spectators normally present, but in lieu of reliable figures, use eight persons per alley for load calculations. Do not apply the same heat gain factor for each of the eight people. Table 1 gives heat gain due to people in a bowling alley at 78 F, and may be used as a guide. This table assumes that, for each alley, one person will be bowling, one standing and six sitting.

Because of heavy smoking and considerable activity, substantial quantities of outside air must be introduced into the space to combat smoke haze and odors. The absolute minimum of 15 cfm per person is not recommended. A ventilation air quantity of 20 cfm per person is desirable; 25 cfm per person for deluxe jobs.

Under typical conditions adjusted for evening hours, each 1,000 cfm of ventilation air drawn into a bowling center from the outside, requires two tons or more of refrigeration for cooling and dehumidification. It may be definitely advantageous, especially in localities with high outside design temperatures, to place activated carbon or electrostatic filters in the system to reduce the amount of outside air needed. Electrostatic filters will remove tobacco-smoke particles, and activated carbon will remove

Table 1. Heating Gain Due to Bowlers

Activity	Heat Gain per Person, Btu per hr	
	Sensible	Latent
Bowling	485	965
Standing	220	280
Sitting	216	185

odors, but all of the return air need not be treated by such filters. For example, a ventilation rate of 25 cfm per person may be planned for a deluxe job. The actual rate could be reduced to 15 cfm per person by using a by-pass to take 10 cfm of return air per person through an electrostatic filter.

Cool Back of Foul Line. It has become common practice in designing bowling alley systems to cool only the area in back of the foul line. (In back of, or behind the foul line refers to the space occupied by the bowlers and spectators. The alleys themselves are in front of the foul line.) A phantom curtain is considered to exist from ceiling to floor at the foul line, with relatively little heat transfer through it. The load estimate should include this heat gain based on a factor of 6 Btu per hr per sq ft of cross-sectional area (width of the building times ceiling height). In using this factor, it is assumed that the dry bulb temperature difference across the curtain is 5 F. Latent heat gain from this area is not calculated, since in the modern bowling alley equipped with automatic pin-setting machines, there are no appreciable sources of latent heat beyond the foul line. Exhaust air is sometimes taken from the building above the alleys, or through the wall behind the pin-setting machines, and this helps to prevent an excessive build-up of heat in this area.

Some alleys have been designed with plexiglass curtains or frame partitions extending part-way down from the ceiling. These limit heat transmission to some extent, but their primary purpose is to restrict air movement between the conditioned and non-conditioned sections.

Other Considerations. Other heat gains to be estimated include those due to lights, solar radiation, and transmission through walls and roof, but in all cases, only the occupied area behind the foul line is taken into account. Bowling alleys are typically almost windowless structures, so that infiltration through the doors and any windows can usually be offset by exhausting slightly less air than is introduced through the outside air intake. However, if a bowling center has entrance doors at both sides of the row of alleys, simultaneous opening of both doors on a windy night may result in "tons" of cooling literally being swept out the door. The obvious solution is the installation of a vestibule around at least one entrance.

Storage or flywheel effect is usually not of significance in air conditioning bowling alleys because of the nature of the load, and because only a small portion of the total area is conditioned.

Refrigeration Load

An analysis of a number of successful bowling center installations indicates that the cooling load may vary considerably. For jobs on which business competition is keen, package equipment has been used to provide an average 1.1 tons refrigeration and 430 cfm of supply air per alley. For deluxe engineered-equipment jobs, about 1.8 tons of refrigeration and 540 cfm of supply air per alley is usually installed. The average for all types of bowling center jobs is 1.4 tons and 480 cfm per alley.

Eight persons per alley is the average "people load," but for deluxe jobs, as high as 19 persons per alley has been figured. Each of the eight persons is supplied with an average of 20 cfm of fresh air, with the result that ventilation air was between 30 and 37% of the total supply air quantity.

It must be noted that these figures represent averages of data on jobs of many types installed in all parts of the country, and include only the bowling and spectator area of the bowling center. These averages may be used as a rough check for a cooling load estimate, but they should not be used for design in lieu of detailed calculations.

Equipment Selection

The design engineer has the choice of two basic types of cooling systems for bowling center jobs. Related to the type of equipment used, these systems are as follows:

1. Unitary Conditioners. Units are available in relatively fixed increments of capacity. The refrigerant compressor may be included within the unit casing, or it may be part of a package condensing unit located elsewhere. Either water or air could be used as the condensing medium.

2. Engineered-equipped systems. Large fan-coil units, compressors and condensers are separte components matched by the design engineer to suit individual job requirements.

It is possible, of course, to use both unitary and engineered equipment on the same job with some advantage. Packaged units could be used in the restaurant, bar, or other small areas, while an engineered equipment system could be designed for the bowling and spectator area.

Zoning Considerations

Adequate zoning is required for bowling centers if a restaurant or bar is a part of the premises, and if the owner desires acceptable conditions to be

maintained within these spaces at all times. The cooling load in a restaurant or bar differs from the bowling area because these rooms are usually placed at the exterior of the building, and have some glass area. Activity within the spaces is moderate, so a different temperature level is desirable. In addition, the restaurant or bar may be crowded both before and after the evening bowling hours, but more important, it may also be crowded with the bowlers from the first shift while the second shift is bowling. As a general rule, a restaurant or bar should be zoned in some manner when separated from the bowling and spectator area.

One method of zoning special-purpose rooms is to plan an entirely separate system; very often a self-contained unit will be entirely adequate. Another method, possible when designing a custom-engineered system, is to install a separate fan-coil unit to handle areas other than the general bowling space. Both the fan-coil unit for the main space and the unit for the separately-zoned areas are connected to a single compressor or condensing unit with capacity control. Each zone, of course, has its own temperature control system.

A single fan-coil unit for all areas with a reheat coil in the supply branch to the restaurant or bar may also be considered. Because a restaurant normally requires less, and a bar usually requires more ventilation air per person than does the bowling area, this method is not often used. Zoning by volume control is not recommended, nor is the use of a multi-zone fan-coil unit considered necessary in bowling center design.

One aspect of the zoning problem which must not be overlooked is the necessity of providing comfort in the winter as well as the summer. This may be the deciding factor which will make restaurant zoning imperative in northern climates. It is entirely possible that the restaurant, locker rooms and offices at the exposed walls of the building may require *heating* during intermediate seasons, while the bowling and spectator area may require *cooling*. With a properly zoned system, acceptable conditions can be maintained in all areas. If heating performance is a primary consideration in zoning, a split system utilizing baseboard radiation along the outside walls is a method of accomplishing desired results.

In one installation worthy of special mention, sections of the bowling and spectator area are zoned, using separate systems. Fan-coil units connected to remote air-cooled condensing units furnish air to supply ducts furred-in at the ceiling above the foul line. The bowling area is zoned in eight-alley sections, and each section has its own system with individual control. The bowling center manager can regulate the air conditioning in any section at a panel incorporating remote controls for high or low temperature, continuous or automatic fan operation, night set-back for heating, and exhaust air quantity. The obvious advantage of this zoning method is that operating economy is possible through manual adjustment of system operation to air conditioning requirements at all times.

Air Distribution and Ductwork

Because air conditioning of bowling and spectator areas is essentially a spot cooling job, the air-flow pattern within this area must be carefully planned.

Supply Air Distribution with Ductwork. The preferred method of supplying conditioned air to the bowling area is to place the main supply trunk duct at the ceiling above the foul line. See Fig. 1. Extending the width of the alleys, the duct is fed from the center, and usually is furred-in. If the ceiling is particularly high, the duct should be suspended so that it is not more than 12 ft above the floor. Sidewall-type outlets with double-deflection vanes should be

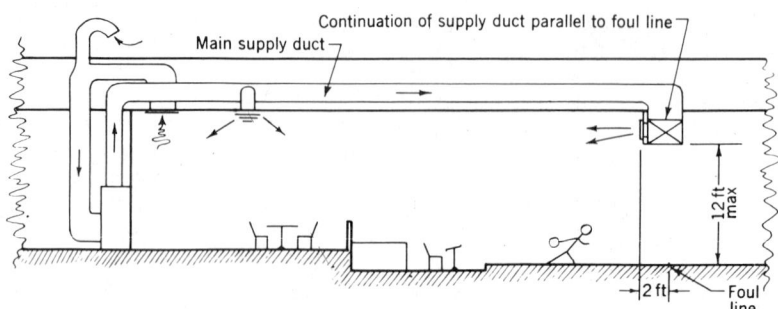

Continuation of supply duct parallel to foul line

Main supply duct

12 ft max

2 ft

Foul line

Fig. 1. Preferred method of supplying cooled air to bowling and spectator area. Double deflection outlets are installed in main duct above foul line.

attached to short collars installed on the side of the duct. Very little air from the unconditioned alley space is induced by the cooled air stream using this type of outlet and duct arrangement. Preferably not more than two alleys should be served by one outlet; for deluxe systems, an outlet above each alley is better. The location of the outlet with respect to the foul line is critical. Usually the outlet face should not be more than 2 ft in front of the foul line, but this is largely dependent on the diffusion characteristics of the outlet and the width of the area it serves. The deflecting vanes of the outlet should be set to throw the conditioned air straight out through the top portion of the grille, and downward toward the bowlers through the bottom portion. Side deflection is adjusted to meet requirements. Ceiling diffusers on branch runs could be used to blanket the areas behind the spectators not reached by the outlets in the main trunk. While this distribution method has been found to give good results, the architect or owner may object to a suspended, furred-in duct. It should be pointed out that the duct enclosure provides an excellent location for mounting advertisement, team records, and the screens used for the projection of score sheets.

With careful design, satisfactory results can be achieved by using round or rectangular ceiling diffusers. Semicircular and two- or three-way rectangular diffusers are particularly desirable. When using these outlets, it is advisable to suspend a curtain or partition from the ceiling at the foul line to restrict movement of cool air to the conditioned area, and to prevent the diffusers from inducing warm air into their jet streams. Heavy transparent plexiglass has been used successfully as a curtain on jobs where the owner desired an unobstructed view of the pin area. A frame partition is a more permanent barrier, and if not extended too low, will still permit both bowlers and spectators to see the action at the pins and the lighted pin indicator at the far end of the alleys. It is not possible to make a general recommendation concerning the height of the partition above the floor. It should extend as far below the ceiling as possible, yet should allow all the specta-

tors behind the players' benches to see the pins, as illustrated by Fig. 2. The partition can also serve as signboard or score-sheet projection screen.

Supply Air Distribution with Packaged Units. When unitary (packaged) air conditioners are chosen for an installation, it is possible to use either of the two duct-work air distribution systems described previously. For the small bowling center, however, it may be less expensive to use self-contained units with discharge plenums for distributing the cooled air. Results are not always as good, but are usually satisfactory for the small job. Inspection of the air conditioner manufacturer's data should be made to determine if the unit or units can adequately cover the area with cooled air.

There are three different arrangements of packaged units with discharge plenums which can be considered. A curtain wall above the foul line is recommended with each of these. In the first arrangement, illustrated by Fig. 3, units of equal capacity are placed at the walls on both sides of the alleys, and the cooled air is directed to the center of the bowling and spectator area. The second arrangement consists of units placed against the wall behind the spectator seating, the air being directed out of the plenum in a spreading pattern. See Fig. 4. Grilles should also be placed on the sides of the plenums to further direct the air to all parts of the conditioned area. The third arrangement consists of a single unit, and is recommended only for bowling centers with less than nine alleys. The best location when only a single unit can be used is at

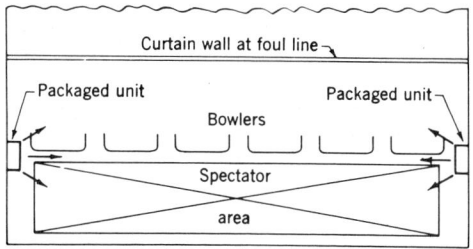

Fig. 3. Self-contained units at both sides of alleys.

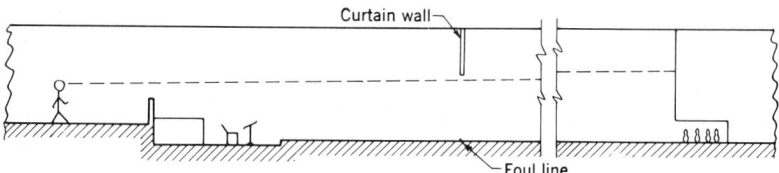

Fig. 2. Suspended curtain should not obstruct spectators line of sight.

one of the side walls, Fig. 5. If space for a unit at the side is not available, the alternative is to center the unit on the wall behind the spectator area. Grilles should be installed on the sides of the discharge plenum, and the air flow must be carefully regulated to provide the best distribution possible with this arrangement.

Return Air

The best location for return-air grilles is the ceiling, or high in the wall behind the spectator area. Multiple grilles should be installed so that air is returned from all parts of the conditioned area. Avoid return air locations near a snack bar because cooking odors may be picked up, even though all appliances may be hooded with connections to an exhaust system. If floor-mounted self-contained units are installed within the conditioned area, return air is usually taken into the unit through grilles in the front panels of the units, or through openings in the rear.

Exhaust Air

The exhaust air location and arrangement assumes substantial importance in the design of a bowling center system. The greater portion of exhaust air should be moved out of the space by fans. Since the purpose of exhausting air is to remove smoke and odors, most of the air should be taken through grilles in the ceiling above the bowling and spectator area, especially if a curtain wall is extended from the ceiling near the foul line. Sometimes a job is designed with fans in the ceiling above the alleys themselves. By thus removing a part of the exhaust air from points above the alleys, there is less possibility of "smoke haze" developing and remaining over the lanes. In addition, the slight air movement from the conditioned to the unconditioned area will help overcome the transfer of convected heat to bowling and spectator area.

Although it may be convenient to exhaust air from the bowling and spectator area out through the locker rooms or kitchen, these spaces should have their own ventilating system. Smoke-laden air tends to collect near the ceiling, and it should be removed at this point.

Ventilation Air

Ventilation (make-up) air should be introduced into the conditioned space through ductwork leading to the return air inlet of the air conditioner. It should be filtered and thoroughly mixed with the

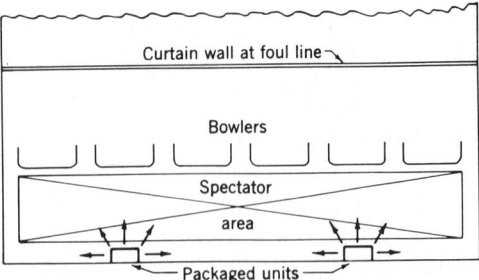

Fig. 4. Units at wall behind spectator seating.

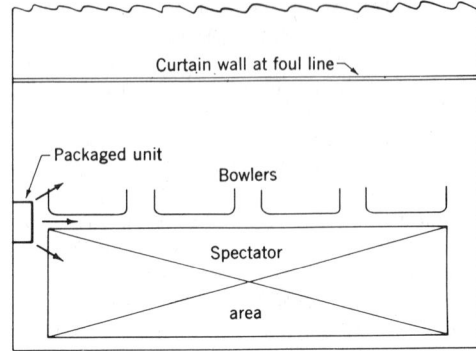

Fig. 5. Unit at one side of small bowling center.

return air before it enters the cooling coil.

On larger jobs it is not uncommon to find a ventilation cycle included in the overall design of the system. By providing for the introduction of fresh air up to the normal air-handling capacity of the unit fan, the system can provide cooling without refrigeration during intermediate seasons, and quick freshening of the interior during the summer and winter. If a ventilation cycle is not provided, consideration could be given to the use of two-speed exhaust fans and two-position outside air dampers to accomplish the quick flushing manually on high speed when required.

Heating

Many bowling center air conditioning units are installed with heating coils, fed from a hot water or steam boiler. Supplementary baseboard fin radiation can also be supplied from the same boiler. Building entrances should receive special attention in cold climate zones, as should the exposed walls at the sides of the bowling and spectator areas.

Gas-fired or electric duct heaters can be installed in the supply air ductwork.

Section 11

NOISE AND RADIATION CONTROL

Contents

Authorities

Section Editors: *L.L. Faulkner, Vito V. Cerami* and *Edwin S. Bishop*

Norman Goldberg, Control of Cooling Tower Noise
Vito V. Cerami, and *N. S. Shataloff*, Acoustical
Problems in High-Velocity Air Distribution

Harrison D. Goodman, and *George Hollands*,
Radiation Protection at Wall Openings for Duct
or Pipe

NOISE AND VIBRATION CONTROL*

Introduction

In addition to fulfilling air-conditioning requirements, an air-handling system must meet acceptable noise and vibration criteria. The purpose of the following is to provide design engineers of air-handling systems with guidelines for selecting the proper criteria for building spaces, and methods for specifying systems to meet design requirements. Common sources of vibrations in buildings are fans, blowers, water chillers, water pumps, and air compressors. Methods for specifying and reducing noise and vibration levels for ducted systems will be given.

Ducted systems treated in this chapter are used to transmit air from a fan or blower to the necessary locations. Air transmission may be for the purpose of central-station air conditioning, an industrial ventilating system, or industrial applications such as process cooling and furnace air supply, among others. Although the fan or blower is usually considered the major source of noise in such systems, the flow of air through duct elements like elbows, turning vanes, and flow control valves can also produce vibration and, therefore, sound.

The design engineer should be aware of noise and vibration principles, since these have become a major concern in the acceptance of air-handling systems. While it is uneconomical to overdesign a system for noise or vibration control, disastrous results can occur if potential problem areas are not considered in the design stage. Optimum design avoids overdesign but results in a system that meets accepted criteria for the use of building space.

Definitions and Terminology

Attenuation—The reduction in sound power as sound propagates through a duct or enclosure. It is expressed in decibels as the ratio of the sound power at a point upstream, to the sound power at a point downstream, in the direction of the sound propagation. For a duct, attenuation is expressed in units of decibels per foot (dB/ft) indicating the average reduction of sound-power level for each foot of duct length.

Band Pressure Level—The band pressure level of a sound for a specified frequency band is the sound-pressure level for the sound contained within the restricted band. The reference pressure must be stated. The band may be specified by its lower and upper cut-off frequencies, or by its geometric center frequency and bandwidth. The width of the band may be indicated by a preceding modifier; e.g., octave-band (sound-pressure) level, half-octave-band level, third-octave-band level, 50-Hz-band level.

Broad-Band Noise—Noise whose energy is distributed over a broad frequency range (generally speaking, more than one octave).

Continuous Noise—On-going noise whose intensity remains at a measurable level, which may vary but is without interruption over an indefinite period or a specified period of time.

Decibel—The decibel is a mathematical scale that is similar to a logarithmic scale used to describe the level of a physical quantity compared to a reference. It compresses the quantity into numbers that are convenient for data presentation. Mathematically, the decibel is ten times the logarithm (base 10) of the ratio of a power or energy quantity with respect to a reference base of the same physical quantity.

Frequency—The frequency of a function periodic in time is the reciprocal of the period. The unit is the cycle per unit time and must be specified. The unit cycle per second is commonly called hertz (Hz).

Insertion Loss—The difference in sound-pressure level at a point after a modification has been made in either the source or the transmission path between the point and the source. It is expressed in decibels as the ratio of sound pressure before the modification to the sound pressure after completing the modification. Since it is quite possible for the sound pressure to increase with the modification, the insertion loss can have a negative value.

Microbar—A unit of pressure commonly used in acoustics. One microbar is equal to 1 dyne per

---

* L. L. Faulkner, ed., *Handbook of Industrial Noise Control*. New York: Industrial Press Inc., 1976.

square centimeter. The SI metric equivalent is one-tenth of a pascal (Pa) or newton per square meter (N/m^2).

Octave Band—A range of frequencies such that the highest frequency in the band is double the lowest frequency. Also, for successive octave bands, the center frequency of the higher band is two times the frequency of the next lower center-frequency band. In acoustics standard center-frequency bands are specified.

Octave-Band Center Frequency—The octave-band center frequency is a specific frequency used to define a particular octave band.

Pitch—A qualitative term used to describe the frequency makeup of a sound. Thus we say that a low-frequency sound has a "low pitch." Although pitch depends primarily upon the frequency, it also depends upon the sound pressure and the waveform of the sound vibration stimulus.

Power Level—Power level, in decibels, is 10 times the logarithm to the base of 10 of the ratio of a given power to a reference power.

Sound Level—A weighted sound-pressure level obtained by the use of metering characteristics and the weightings A, B, or C as specified in the American National Standard Specification for Sound-Level Meters, ANSI S1.4-1971. The weighting employed must be stated.

Sound-Pressure Level—The sound-pressure level, in decibels, of a sound is 20 times the logarithm to the base 10 of the ratio of the pressure of this sound to the reference-pressure. The reference pressure must be stated. *The following reference pressure is in common use for measurements concerned with hearing and with sound in air and liquids:* 2×10^{-5} N/m^2. Unless otherwise explicitly stated, it is to be understood that the sound pressure is the effective (rms) sound pressure.

Noise Criteria

Noise criteria are necessary during the design stage for specifying permissible maximum sound levels in building spaces. Such criteria serve as a guide to designers, manufacturers, customers, and engineers as an indication of the noise performance of a system. Acceptable criteria for various space usage have been developed from the subjective reactions of humans experiencing acceptable noise levels while engaged in specific activities. One of the most widely used sets of design values is the 1957 Noise Criterion Curves (NC Curves), adopted by the American Society of Heating, Refrigerating and Air Conditioning Engineers (ASHRAE). Noise Criterion Curves are plotted in Figure 1. The octave-band sound pressure levels for specific NC values are given in Table 1. A list of recommended NC criteria for a variety of indoor spaces is given in Table 2. Some building codes specify certain of these criteria. There have been objections to the perceived sound levels with octave-band levels that follow exactly the 1957 NC curves. Generally, complaints mention that the sound is both "hissy" and "rumbly." In response to these objections, the NC curves were modified, and the resulting values, known as the "1971 Preferred Noise Criteria Curves" (PNC curves) are shown in Figure 2 and Table 3. While these criteria have yet to be adopted by standardization groups, they are sometimes the recommended criteria. It may be noted from Table 2 that the recommended NC or PNC ratings vary from NC 20 or PNC 20 for suburban homes to NC 60 or PNC 60 for offices with tabulative equipment.

Speech-Interference Criteria

Interference with speech communication is the result of a phenomenon known as *masking*. In the masking process, sounds in the speech frequency range are rendered indistinguishable from non-speech sources by excessive noise levels. The most common speech-interference criterion is called the preferred speech-interference level (PSIL), which is computed from the simple arithmetic average of the background octave-band sound pressure levels having center frequencies of 500, 1000, and 2000 Hz.

Thus,

$$PSIL = \frac{1}{3}[L_{p500} + L_{p1000} + L_{p2000}]$$
$$dB \text{ re } 2 \times 10^{-5} \text{ N/m}^2 \tag{1}$$

where:

L_{p500} = 500 Hz octave-band sound pressure level, re 2×10^{-5} N/m^2

L_{p1000} = 1000 Hz octave-band sound pressure level, re 2×10^{-5} N/m^2

L_{p2000} = 2000 Hz octave-band sound pressure level, re 2×10^{-5} N/m^2

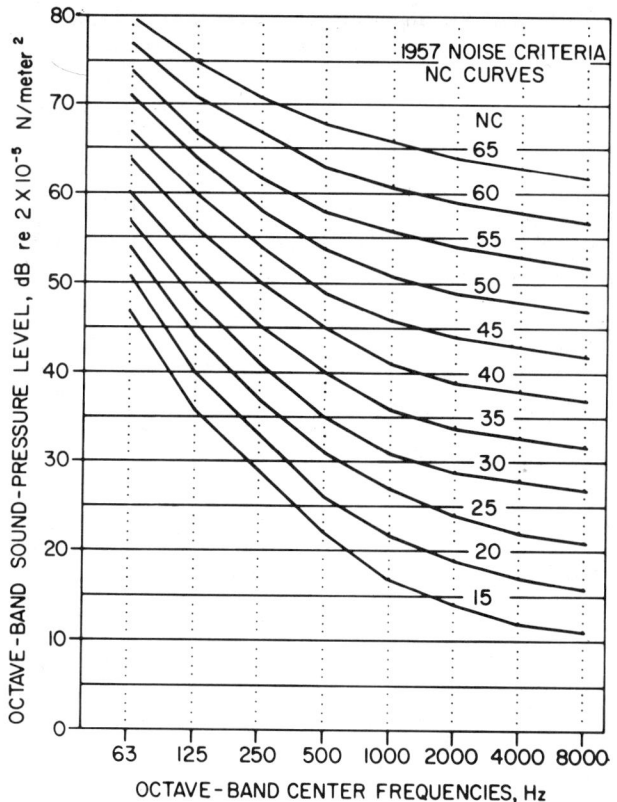

Fig. 1. Noise criteria curves for determination of permissible broad-band background sound levels for indoor spaces. See Table 2 for recommended levels for space usage.

Table 1.   Octave-Band Sound Pressure Levels Defining the 1957 Noise Criterion (NC) Curves as Given in Fig. 1

Noise Criterion	Octave-Band Center Frequency, Hz							
	63	125	250	500	1,000	2,000	4,000	8,000
	Sound-Pressure Levels, dB							
NC-15	47	36	29	22	17	14	12	11
NC-20	51	40	33	26	22	19	17	16
NC-25	54	44	37	31	27	24	22	21
NC-30	57	48	41	35	31	29	28	27
NC-35	60	52	45	40	36	34	33	32
NC-40	64	56	50	45	41	39	38	37
NC-45	67	60	54	49	46	44	43	42
NC-50	71	64	58	54	51	49	48	47
NC-55	74	67	62	58	56	54	53	52
NC-60	77	71	67	63	61	59	58	57
NC-65	80	75	71	68	66	64	63	62

Table 2. Recommended Indoor Noise Criteria

Type of Space	NC or PNC Curve	Approximate dBA Level
RESIDENCES		
Private Homes (rural and suburban)	20 - 30	30 - 38
Private Homes (urban)	25 - 35	34 - 42
Apartment Houses (2 - and 3 - family units)	30 - 40	38 - 47
HOTELS		
Individual Rooms or Suites	30 - 40	38 - 47
Ballrooms, Banquet Rooms	30 - 40	38 - 47
Halls, Corridors, Lobbies	35 - 45	42 - 52
Garages	40 - 50	47 - 56
Kitchens and Laundries	40 - 50	47 - 56
HOSPITALS AND CLINICS		
Private Rooms	25 - 35	34 - 42
Operating Rooms, Wards	30 - 40	38 - 47
Laboratories, Halls Corridors } Lobbies, Waiting Rooms	35 - 45	42 - 52
Washrooms, Toilets	40 - 50	47 - 56
OFFICES		
Board Rooms	20 - 30	30 - 38
Conference Rooms, Meeting Rooms (for 50)	25 - 30	34 - 38
Executive Office, Meeting Rooms (for 20)	30 - 35	38 - 42
Supervisor Office, Reception Room	30 - 45	38 - 52
General Open Offices, Drafting Rooms	35 - 50	42 - 56
Halls, Corridors	35 - 55	42 - 61
Tabulation and Computation Rooms	40 - 60	47 - 66
AUDITORIUMS AND MUSIC HALLS		
Concert and Opera Halls } Studios for Sound Reproduction	20 - 25	30 - 34
Legitimate Theaters, Multi-Purpose Halls	25 - 30	34 - 38
Movie Theaters, TV Audience Studios } Semi-Outdoor Amphitheaters, Lecture Halls, Planetarium	30 - 35	38 - 42
Lobbies	35 - 45	42 - 52
BROADCAST STUDIOS		
(Distant Microphone Pickup)	10 - 20	21 - 30
CHURCHES AND SCHOOLS		
Sanctuaries, Music Rooms	20 - 30	30 - 38
Libraries } Classrooms	25 - 35	34 - 42
Teaching Laboratories	35 - 45	42 - 52
Recreation Halls } Corridors and Halls	35 - 50	42 - 56

Table 2 (Cont.). Recommended Indoor Noise Criteria

Type of Space	NC or PNC Curve	Approximate dBA Level
Kitchens Washrooms and Lavatories	40 - 50	47 - 50
PUBLIC BUILDINGS		
Public Libraries, Museums, Art Galleries Court Rooms	30 - 40	38 - 47
Post Offices, Banking Areas, Lobbiés	35 - 45	38 - 52
Washrooms and Toilets	40 - 50	47 - 56
RESTAURANTS, CAFETERIAS		
Restaurants	35 - 45	42 - 52
Cocktail Lounges	35 - 50	42 - 56
Night Clubs	35 - 45	42 - 52
Cafeterias	40 - 50	47 - 56
STORES		
Clothing Stores Department Stores (upper floors)	35 - 45	42 - 52
Department Stores (main floor) Small Retail Stores Supermarkets	40 - 50	47 - 56
SPORTS FACILITIES		
Coliseums	30 - 40	38 - 47
Bowling Alleys, Gymnasiums	35 - 45	42 - 52
Swimming Pools	40 - 50	47 - 56
TRANSPORTATION (Rail, Bus, Plane)		
Ticket Areas	30 - 40	38 - 47
Lounges, Waiting Rooms	35 - 50	42 - 56
Baggage Areas	40 - 50	47 - 56
MANUFACTURING AREAS		
Supervisors Offices	30 - 45	38 - 52
Laboratories, Engineering Rooms, Drafting Rooms, General Office Areas	40 - 50	47 - 56
Maintenance Shops, Office Equipment Rooms, Computer Rooms, Washrooms and Toilets	45 - 55	52 - 61
Control Rooms, Electrical Equipment Rooms Foreman's Office, Tool Crib, Areas where speech communication is required	50 - 60 Not to exceed 60	56 - 66 Not to exceed 66
Manufacturing Areas, Assembly Lines, Machinery Areas, Packing and Shipping Areas	Use Speech Interference Level or OSHA * Criteria	

* Occupational Safety and Health Act of 1971.

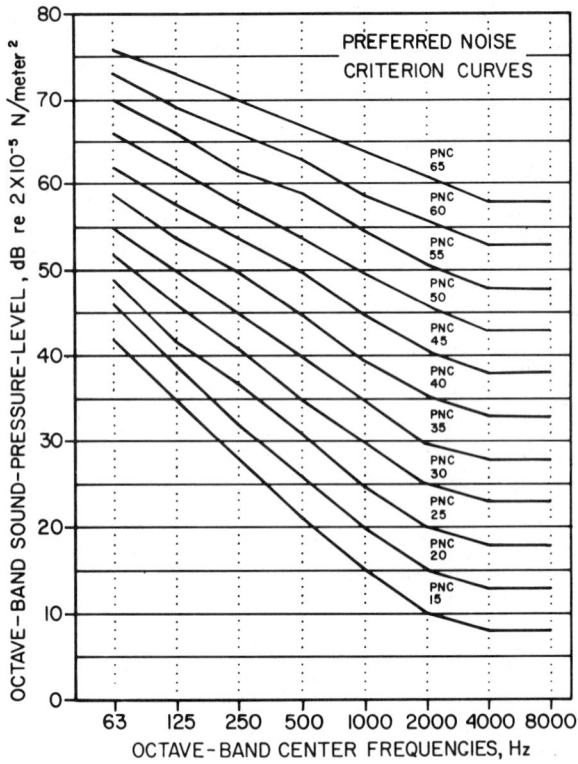

Fig. 2. Preferred noise criteria curves for determination of permissible broad-band background sound levels for indoor spaces. See Table 2 for recommended levels for space usage.

Table 3. Octave-Band Sound Pressure Levels Defining the 1971 Preferred Noise Criterion (PNC) Curves as Given in Fig. 2 *

Preferred Noise Criterion	Octave-band Center Frequency, Hz								
	31.5	63	125	250	500	1,000	2,000	4,000	8,000
	Sound-Pressure Levels, dB								
PNC-15	58	43	35	28	21	15	10	8	8
PNC-20	59	46	39	32	26	20	15	13	13
PNC-25	60	49	43	37	31	25	20	18	18
PNC-30	61	52	46	41	35	30	25	23	23
PNC-35	62	55	50	45	40	35	30	28	28
PNC-40	64	59	54	50	45	40	36	33	33
PNC-45	67	63	58	54	50	45	41	38	38
PNC-50	70	66	62	58	54	50	46	43	43
PNC-55	73	70	66	62	59	55	51	48	48
PNC-60	76	73	69	66	63	59	56	53	53
PNC-65	79	76	73	70	67	64	61	58	58

*These values have not yet been adopted by standardization groups.

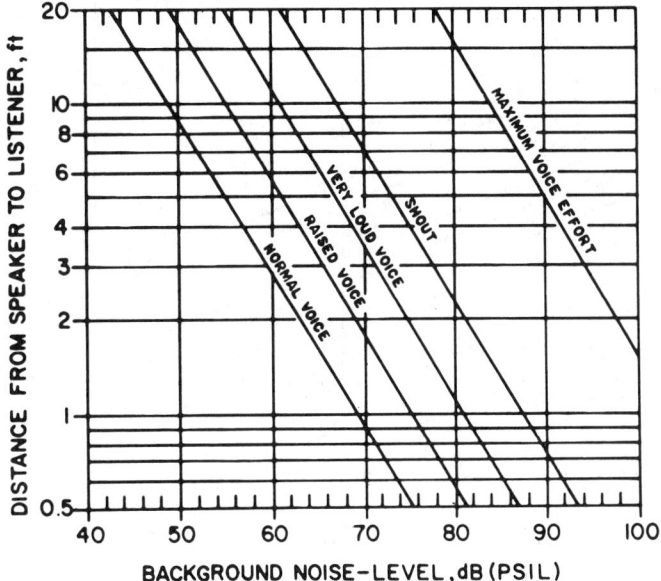

Fig. 3. Relation of required voice effort for communication in terms of PSIL and distance between speaker and listener.

The relationship between PSIL, voice effort required, and distance between speaker and listener, is given in Figure 3. Computation of the preferred speech-interference level requires octave-band sound pressure levels that are measured or calculated. In some instances, only A-weighted sound pressure levels may be available and, in such instances, an approximation to the PSIL value is given by:

$$PSIL = L_p(A) - 7, \text{ dB re } 2 \times 10^{-5} \text{ N/m}^2 \quad (2)$$

where:

$L_p(A) =$ A-weighted sound pressure level, dB re 2×10^{-5} N/m^2

To illustrate the use of the PSIL criteria, assume that the measured octave-band sound pressure levels in a mechanical room were: $L_{p500} = 60$, $L_{p1000} = 62$, $L_{p2000} = 58$, dB re 2×10^{-5} N/m^2; then from Equation (1), PSIL $= \frac{1}{3}$ $(60+62+58) = 60$ dB. From Figure 3 for PSIL equal to 60 dB, the normal voice level for face-to-face communication between two individuals 3 feet (1 m) apart provides satisfactory speech intelligibility. If the two individuals were 6 feet apart, it would be necessary to use raised voices. Note that for good face-to-face communication at 6 feet, using

normal voice levels, the PSIL must be reduced to 53 dB.

Sound Levels of Sources

Noise ratings of air-conditioning machinery are preferred in terms of octave-band sound power levels. Single-figure, overall sound level values were formerly used but are not adequate for proper analysis of building-space sound levels. Sound-power levels in octave bands are necessary to compute octave-band sound pressure levels for comparison to indoor noise criteria. In some instances, a loudness rating in *sones* or *phons,* while not preferred, is used to rate mechanical equipment. The loudness level in sones or phons is calculated from the loudness index I_i for each band as given in Figure 4. The calculation is as follows:

1. Enter the geometric mean frequency, f_i, for each band as the abscissa of Figure 4. From the band pressure level (ordinate of Figure 4) obtain the loudness index I_i for each band.

2. Find the total loudness S_t in sones by means of the formula:

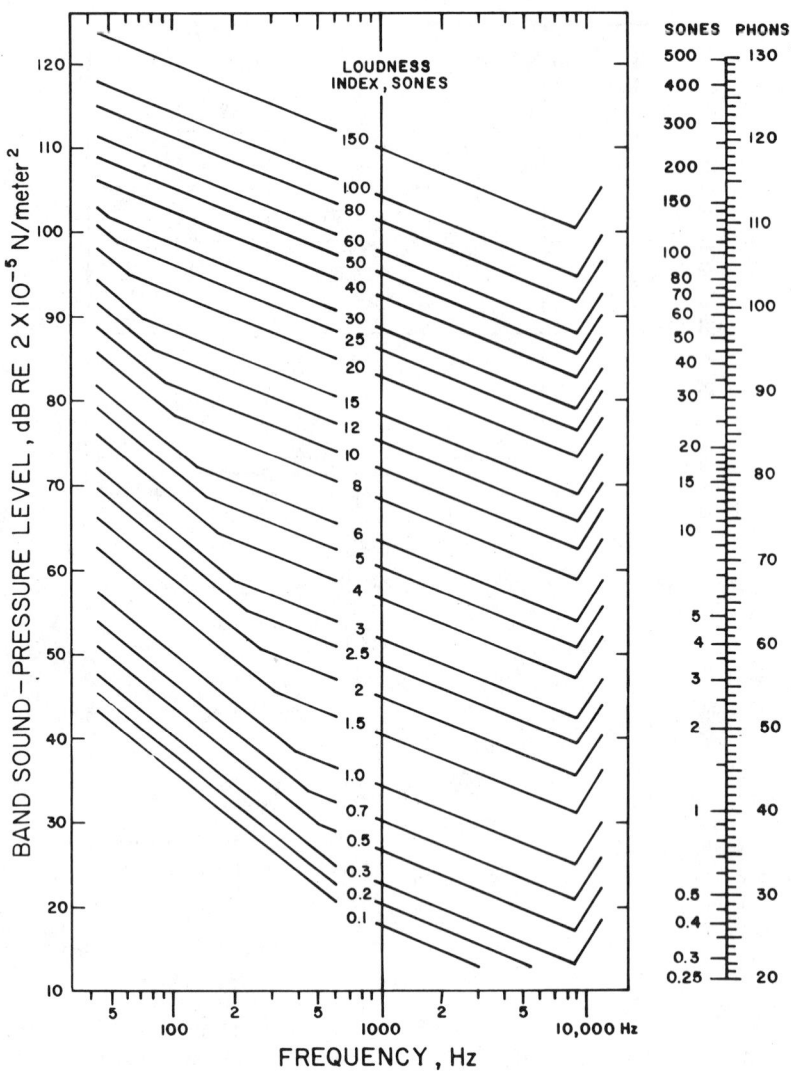

Fig. 4. Contours of equal loudness index to compare noise sources.

$$S_t = I_m + F \left[\left(\sum_{i=l}^{i=n} I_i \right) - I_m \right] \text{sones} \qquad (3)$$

where I_m is the largest of the loudness indexes; ΣI_i is the sum of the indexes for *all* frequencies and F is a factor dependent on the band-width used for the analysis from Table 4.

3. The overall loudness level in phons is ob-

Table 4. Band-width Correction Factor

Band-width	F Value
Third-Octave	0.15
Half-Octave	0.20
Octave	0.30

tained adjacent to the S_t in the nomograph at the right of Figure 4, or by computation from the equation:

$$P = 40 + 10 \log_{10} (S_t) \text{ phons} \qquad (4)$$

Generally, loudness levels are used to compare one product with another with respect to annoyance caused by noise. While this does not provide an absolute scale, it can be helpful in estimating the loudness of one source as compared to another. By rough approximation, a source having a loudness of two sones is twice as loud as one having a rating of one sone. Loudness levels are not recommended for rating air-moving devices and air-conditioning equipment. However, manufacturers are urged to provide octave-band sound power levels in order for the design engineer to make the necessary computations.

Ratings and Standards

The end purpose of noise ratings is to achieve an installed system which, while acceptable, will not disrupt building space usage or be a source of annoyance to occupants. Measurement standards are necessary to insure comparative acoustical quantities, such as sound power levels of equipment, by various manufacturers. Other standards pertain to sound measurements in buildings. A listing of standards related to heating, ventilating and air-conditioning systems is given in Table 5.

Airborne Sound Transmission

When mechanical equipment rooms containing fans, pumps, compressors or other noisy air conditioning equipment are located adjoining critical occupancy areas such as bedrooms, conference rooms and offices, a very severe problem arises. Insofar as the mechanical equipment room may contain equipment with noise output as great as 90 or 100 decibels and the adjoining critical room must have a sound-pressure level not exceeding 25-30 decibels, the construction of the sound barrier separating the two must of necessity take up a great deal of space, be very heavy, be completely airtight from floor to slab above and, in all probability, must be quite expensive. For these reasons and the very important consideration that a small mistake in the design of the partition, construction of the partition or vibration isolation of the equipment may result in something less than the estimated degree of isolation, it is exceedingly advisable to avoid placing mechanical equipment adjacent to very critical areas. Experience indicates that in most cases it is practical to place the equipment room elsewhere than adjoining critical room areas. However, it sometimes becomes uneconomical to move equipment rooms to such parts of the building where adjoining spaces are not extremely critical, such as general offices, public corridors, stenographic offices, or cafeterias.

When the mechanical equipment room is located next to a moderately critical space of the type noted above, airborne sound isolation becomes possible with reasonable reliability and without tremendous expense.

Figure 5 shows typical constructions of floor, wall, and ceiling structures that will provide adequate acoustical isolation between rooms containing either fans or pumps and spaces that will be satisfactory with moderately noisy acoustical environments.

In those critical situations where a noisy mechanical equipment room may be on a floor above a space requiring relatively quiet surroundings, such as executive offices, libraries, and board rooms, the use of floating floor systems has been successful. Shown in Figure 6, A and B, are two

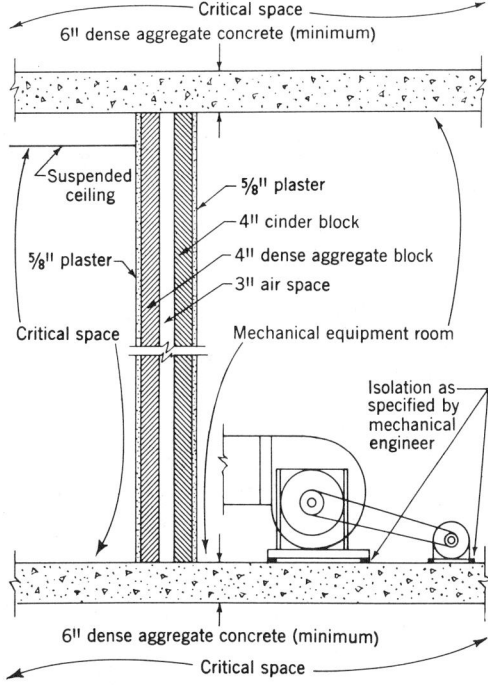

Fig. 5. Suggested partition construction between mechanical equipment room and critical spaces.

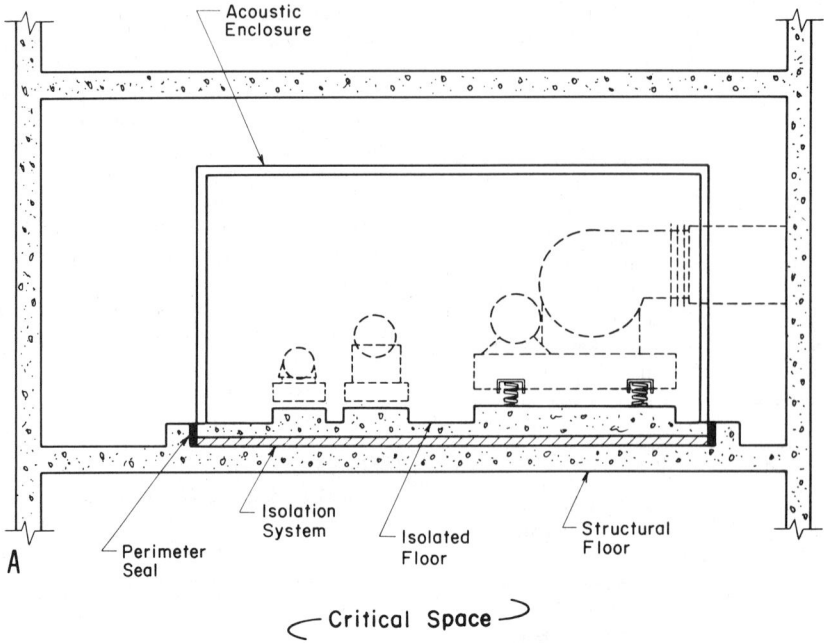

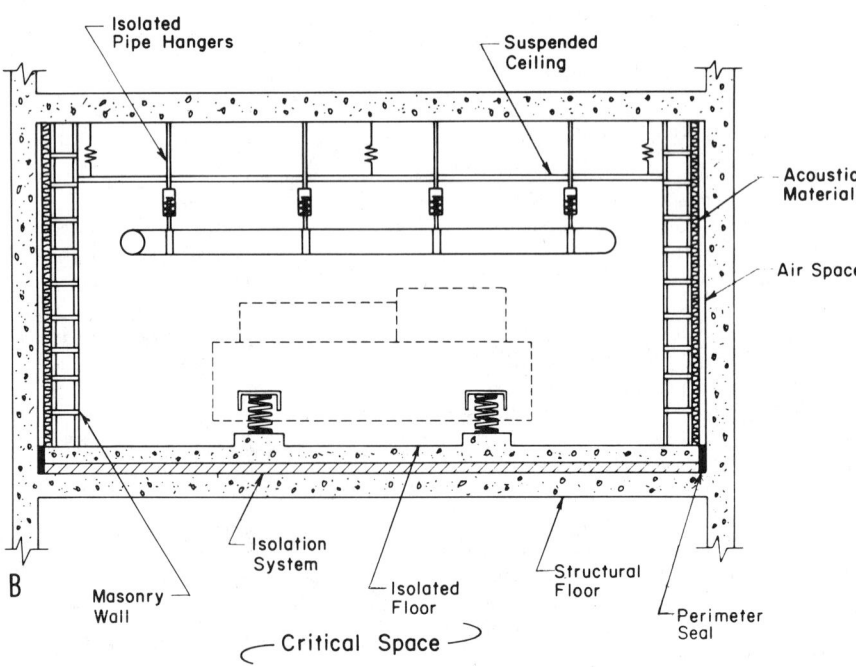

Fig. 6, A and B. Two structural methods for isolating mechanical equipment noise from above or below such critical floor spaces as executive offices, libraries, boardrooms, etc.

Table 5. Standards Pertaining to Heating, Ventilating and Air-Conditioning Systems

Organization *	Standard Designation	Title
ASTM	E413-73	Classification for Determination of Sound Transmission Class
ASTM	E477-73	Testing Duct Liner Materials and Prefabricated Silencers for Acoustical and Airflow Performance
SAE	Standard J952B (1969)	Sound Levels for Engine Powered Equipment
IEEE	85 (1965)	Test Procedure for Airborne Noise Measurements on Rotating Electrical Machinery
ASHRAE	36-62	Measurement of Sound Power Radiated from Heating, Refrigerating, and Air-Conditioning Equipment
ASHRAE	36A-63	Method of Determining Sound Power of Room Air Conditioners
ASHRAE	36B-63	Method of Testing for Rating the Acoustic Performance of Air Control and Terminal Devices and Similar Equipment
ANSI	S1.2 - 1962 (Rev. 1971)	American National Standard Method for the Physical Measurement of Sound
ANSI	S1.6 - 1967 (Rev. 1971)	American National Standard Preferred Frequencies and Band Numbers for Acoustical Measurements
ARI	443 (1970)	Standard for Sound Rating of Room Fan-Coil Air Conditioners
AMCA	300-67	Standard Test Code for Sound Rating Air Moving Devices
ADC	AD-63	Measurement of Room-to-Room Sound Transmission Through Plenum Systems
CAGI	Test Code (1969)	Cagi-Pneurop Test Code for Measurement of Sound from Pneumatic Equipment
ISO	Recommendation R140 (1960)	Field and Laboratory Measurements of Airborne and Impact Sound Transmission
ISO	Recommendation R1680 (1970)	Test Code for the Measurement of Airborne Noise Emitted by Rotating Electrical Machinery

```
*  ANSI --- American National Standards Institute
   ASTM --- American Society for Testing and Materials
   SAE ---- Society of Automotive Engineers
   IEEE --- Institute of Electrical and Electronics Engineers
   ASHRAE - American Society of Heating, Refrigerating and Air Conditioning
            Engineers
   ARI ---- Air Conditioning and Refrigeration Institute
   AMCA --- Air Moving and Conditioning Association
   ADC ---- Air Diffusion Council
   CAGI --- Compressed Air and Gas Institute
   ISO ---- International Organization for Standardization
```

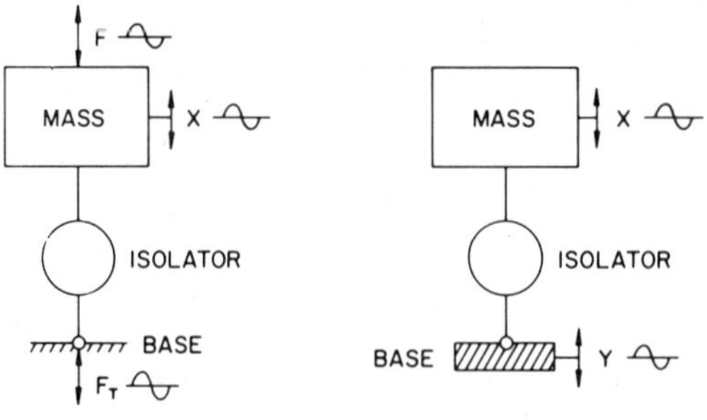

(A) FORCE TRANSMISSION   (B) MOTION TRANSMISSION

Fig. 7. Single-degree-of-freedom vibration isolation systems.

such structural methods for reducing the noise to a floor below mechanical equipment.

Vibration Isolation

By analogy, vibration of mechanical machinery in buildings can be illustrated as a one-degree-of-freedom system (i.e., by motion in the vertical direction only) as shown in Figure 7. The elements of this analogous system are a rigid body or mass, which represents the equipment to be isolated, and an isolator through which it is connected to the structure. The effectiveness of the isolator is determined by the reduction of the transmitted force or motion from the equipment to the structure. The efficiency of isolation devices can be predicted from the formula:

$$E = 100 \left[1 - \frac{1}{(f_d/f_n)^2 - 1} \right] \qquad (5)$$

where:

E = Percentage of vibration isolated
f_d = Disturbing frequency in machinery, Hz or rpm
f_n = Natural frequency of machine on the isolator, Hz or rpm

The disturbing frequency, f_d, should be taken as the rotational speed (Hz or rpm) of the equipment or the driver, whichever is *lower*. For instance, in the case of fan operation the use of rpm times the number of blades in determining the disturbing frequency often will lead to improper isolation, since most often out-of-balance occurs at the rotor speed and not at the blade passage speed. In all cases, isolators must be

selected for the *lowest* rotational speed of motor or equipment.

The natural frequency, f_n, can be computed from the static deflection of the equipment on the isolator system. The formula for natural frequency is

$$f_n = 188 \sqrt{(1/D)}, \text{ Hz} \qquad (6)$$

where D is the static deflection of the machinery on the isolator system, given in inches.

Figures 8 and 9 depict the isolation efficiency as a function of static deflection and disturbing frequency. In most normal situations, the 90% isolation criterion is used, and the appropriate isolator static deflection is obtained from these charts for the lowest disturbing frequency of the equipment to be isolated. Therefore, proper isolators can be selected when the weight of the equipment and the required static deflections are known. Figure 8 shows the isolation efficiency for steel springs which have low internal damping. Since internal damping increases the transmitted force, Figure 9 applies for materials such as cork, rubber, neoprene, or felt, which have greater internal damping than steel springs. These isolation relationships apply for rigid floors which are massive in comparison with the machinery and isolators considered. Table 6 shows appropriate corrections to be made to the isolation efficiency graphs for various floor constructions to account for change in floor mass and flexibility.

Isolation Mount Selection

In selecting isolation systems, it is necessary to obtain all or some of the following information:

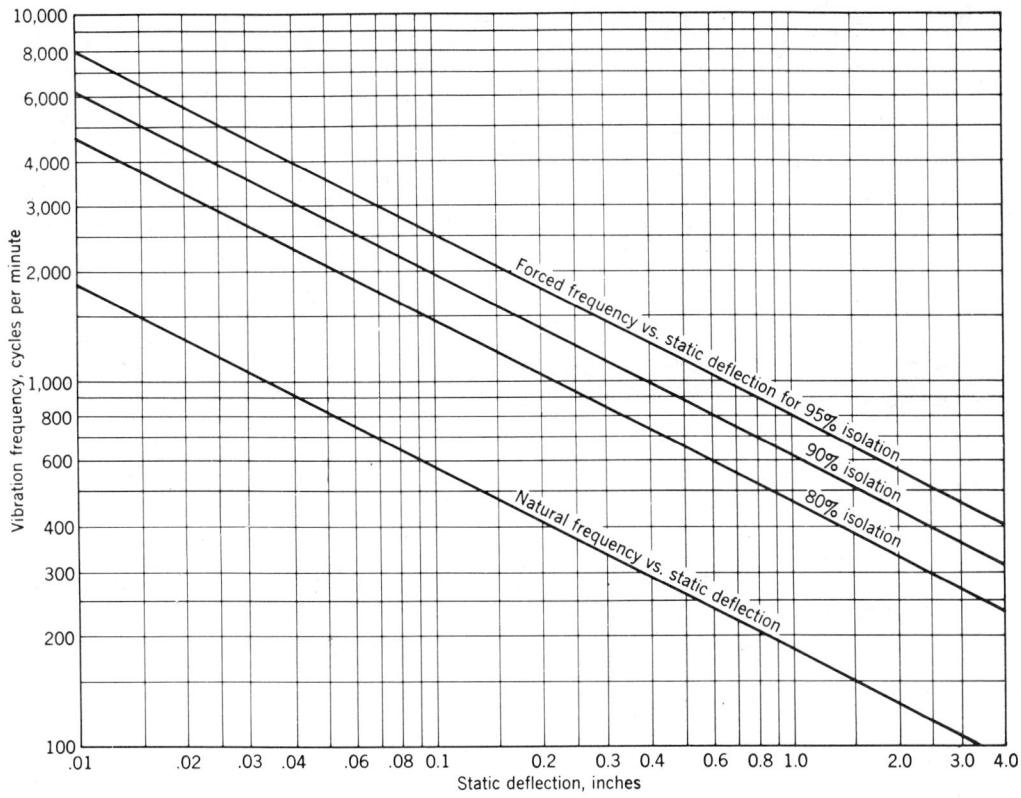

Fig. 8. Isolation efficiency graph for spring mountings.

The extent of needed information depends upon the complexity and seriousness of the resulting vibration or noise problem. For non-critical installations, 80 percent isolation efficiency is usually adequate. In extremely critical situations, 90 to 95 percent isolation efficiency may be necessary. In critical situations, a more detailed analysis should

1. Location where equipment is to be mounted
2. Operating characteristics: power, rpm, strokes per minute, starting torque
3. Direction and magnitude of principal forcing frequencies
4. Equipment weight and weight per mount
5. Number, size, and location of mounting feet
6. Equipment attachments, such as pipe, duct, and electrical conduit
7. Structure characteristics to which equipment is to be mounted
8. Required isolation efficiency.

Table 6. Minimum Mounting Deflections for Various Types of Floors

Operating Speed, Rpm	Type of Floor			
	Basement	Rigid Concrete	Upper Story Light Concrete	Wood
	Deflection, Inches			
300	1.50	3.00	3.50	4.00
600	0.60	1.00	1.50	2.00
900	0.25	0.50	1.00	1.25
1200	0.20	0.45	0.80	1.00
1800	0.10	0.35	0.60	1.00
3600	0.03	0.20	0.50	1.00
7200	0.03	0.20	0.50	1.00

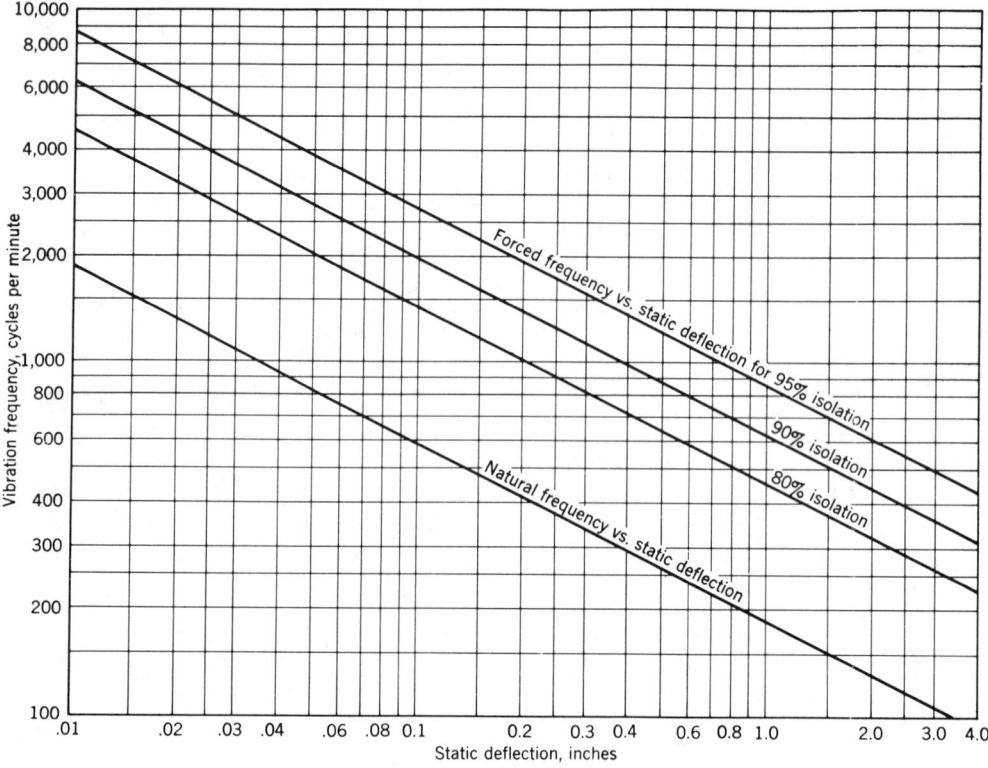

Fig. 9. Isolation efficiency graph for cork, rubber, or felt mountings.

be performed where actual floor vibration amplitudes are calculated and resulting noise levels predicted. These more detailed analysis procedures are necessary when 1. The required isolation efficiency is very high (say, 98 percent or greater); 2. The driving forces are such that they cause rotational as well as translational motions of the machinery; 3. There are combined periodic and shock forces; and 4. Support structures (such as roofs) are flexible.

Recommended mounting systems for a variety of equipment are given in Table 7 with the mount types illustrated in Table 8. Note that there is no standard method to present isolator performance ratings. Most manufacturers will present curves representing load versus deflection, or load versus natural frequency. However, some will only give the natural frequency at the rated load. If the natural frequency at only one load is given

(usually the maximum load), the natural frequency at an intermediate load can be obtained from:

$$f_n' = f_n \sqrt{\frac{W}{W'}}, \text{Hz} \qquad (7)$$

where f_n is the natural frequency at rated load W, and f_n' is the natural frequency at the desired load W'

Airborne Noise Through Ducts

Noise output from a fan is maximum at the fan discharge and almost equally as loud at the fan intake. This, and the fact that acoustically untreated ducts act like speaking tubes or as highly efficient airborne sound transmitters, makes transmission of fan noise through ductwork a major problem. The severity of the problem increases as the fan horsepower increases. In the past, the material most fre-

Table 7. A Guide for Selecting Vibration Isolating Systems for Some Mechanical and Electrical Equipment

Equipment	Power Range	Speed Range, rpm	AREA[6] Mount Type (See Table 8) Min	Normal	Critical	AREA[6] Weight Ratio (See Note 1, below) Min	Normal	Critical	AREA[6] Deflection, in. (See Note 2, below) Min	Normal	Critical
Centrifugal and Axial Flow Fans	<3 hp	All	NN[3]	2	3	...	...	2	...	1.0	1.5
	3-25 hp	<600	2	3	3	...	2	3	1.0	1.5	2.0
		600-1200	2	2	3	...	...	2	0.5	1.0	1.5
		>1200	2	2	3	...	...	2	0.5	1.0	1.0
	>125 hp	<600	2	3	3	...	2	3	1.5	2.0	3.0
		600-1200	2	3	3	...	2	2	1.0	1.5	2.0
		>1200	2	3	3	...	2	2	0.5	1.0	1.0
Motor-Pump Assemblies	<20 hp	450-900	2, 3	3	3	1.5	2-3	3-4	1.5	1.5	3.0
		900-1800	2, 3	3	3	2.5	1.5-2.5	2-3	1.5	1.0	2.0
		>1800	2, 3	3	3	2.5	1.5-2.5	2-3	1.0	0.75	1.5
	20-100 hp	450-900	3	3	3	2-3	2-3	3-4	1.5	1.5	4.0
		900-1800	3	3	3	1.5-2.5	2-3	2-3	1.0	1.0	3.0
		>1800	3	3	3	1.5-2.5	2-3	2-3	0.75	1.0	2.0
	>100 hp	450-900	3	3	3	2-3	3-4	3-4	2.0	2.0	5.0
		900-1800	3	3	3	2-3	2-3	2-3	1.5	1.5	4.0
		>1800	3	3	3	1.5-2.5	2-3	2-3	1.0	1.0	3.0
Internal Combustion Engines and Engine-Driven Equipment	<25 hp	All	2	2	3	...	2-3	3-4	0.35	0.35	2.5
	25-100 hp	All	2	2	3	...	2-3	3-4	0.35	1.75	3.5
	>100 hp	All	2	2	3	...	2-3	3-4	0.35	2.50	5.0
Reciprocating Type Air Compressors, 1 to 2 Cylinders	<20 hp	300-600	4	NR[4]	NR	4-8	...	...	4.0	...	...
		600-1200	4	4	NR	2-4	3-6	...	2.0	4.0	...
		>2400	4	4	4	1-2	2-3	3-6	1.0	2.0	4.0
	20-100 hp	300-600	4	NR	NR	6-10	...	...	5.0	...	...
		600-1200	4	NR	NR	3-6	...	...	3.0	...	...
		>1200	4	4	NR	2-3	3-6	...	1.5	3.0	...
Refrigeration Reciprocating Compressors and Package-Chiller Assemblies	10-50 Tons	600-900	2	3	3	...	2-3	3-4	2.0	2.0	3.0
		900-1200	2	3	3	...	2-3	3-4	1.5	1.5	2.0
		>1200	2	3	3	...	2-3	2-3	1.0	1.5	2.0
	>50 Tons	600-900	3	3	4	2-3	3-4	4-6	2.0	3.0	3.0
		900-1200	2	3	3	...	3-4	3-5	2.0	2.0	2.0
		>1200	2	3	3	...	2-3	3-4	1.5	2.0	2.0
Refrigeration Rotary-Compressor, and Package-Chiller Assemblies	100-500 Tons	>3000	2	2	3	...	...	2-3	0.75	1.0	1.5
	500-1000 Tons	>3000	2	2	3	...	...	3-5	1.0	1.5	1.5
Absorption Refrigeration Assemblies	All Sizes	...	1	2	2	...	...	...	0.25	0.5	1.0
Cooling Towers, Propeller Type (See Note 5, below)	<25 hp	150-300	NN	2	2	...	...	...	...	5.0	5.0
		300-600	NN	2	2	...	...	...	...	3.0	3.0
		>600	NN	2	2	...	...	...	...	3.0	3.0
	25-150 hp	150-300	NN	2	2	...	...	...	...	6.0	6.0
		300-600	NN	2	2	...	...	...	...	4.0	4.0
		>600	NN	2	2	...	...	...	...	3.0	3.0
	>150 hp	150-300	NN	2	2	...	...	...	...	6.0	6.0
		300-600	NN	2	2	...	...	...	...	5.0	5.0
		>600	NN	2	2	...	...	...	...	4.0	4.0
Cooling Towers Centrifugal-Type	<25 hp	450-900	NN	2	2	...	...	...	...	1.0	1.5
		900-1800	NN	2	2	...	...	...	...	0.75	1.0
		>1800	NN	2	2	...	...	...	...	0.75	0.75
	25-150 hp	450-900	NN	2	2	...	...	...	...	1.5	2.0
		900-1800	NN	2	2	...	...	...	...	1.0	1.5
		>1800	NN	2	2	...	...	...	...	0.75	1.0
	>150	450-900	NN	2	2	...	...	...	...	2.0	3.0
		900-1800	NN	2	2	...	...	...	...	1.5	1.5
		>1800	NN	2	2	...	...	...	...	1.0	1.0
Transformers	<10 kva	...	1	1	2	...	...	...	0.13	0.13	0.25
	10-100 kva	...	1	2	2	...	...	...	0.13	0.25	0.5
	>100 kva	...	1	2	2	...	...	...	0.25	0.25	0.5

[1] Weight Ratio = $\dfrac{\text{Weight of Inertia Block or Base}}{\text{Weight of Equipment Mounted on Inertia Block or Base}}$

[2] Deflection of the combined equipment and inertia block on the mount. For normal and critical areas these deflections may have to be increased, depending on the floor flexibility. For normal areas, the deflection should be three to six times the floor deflection and critical areas six to ten times the floor deflection (see text).

[3] NN denotes not needed; mount machine directly on the floor.

[4] NR denotes not recommended, machine cannot be suitably isolated.

[5] Minimum location is on a slab outside of building. To minimize waterfall noise from rooftop location over critical areas, place two or three layers of neoprene pad between tower and springs, or between tower and roof if the propeller and drive assembly are vibration-isolated from the tower.

[6] 1—Minimum Area: Basement or on-grade slab location.

2—Normal Area: Upper floor location, but not above or adjacent to a critical area, private offices, classrooms, conference rooms, operating rooms, etc.

3—Critical Area: Upper floor location above or adjacent to private offices, classrooms, conference rooms, operating rooms, etc.

Table 8. Mount Types

Mount Type	Schematic Illustration	Description
1		Pad mounts of sufficient layers to obtain desired static deflection at load rates of 25–60 psi (45 durometer neoprene), or 145–150 psi (65 durometer neoprene). Floor should be level or self-leveling pads used to insure rated loading.
2	FRAME	Assembly of equipment bolted to stiff steel frame to maintain alignment of components. Frame mounted on springs, general purpose elastomer mounts, or pads, depending on the amount of static deflection required.
3	C.G. INERTIA BLOCK	Assembly of equipment bolted to inertia block which is supported on steel springs with the top of the springs mounted as high as possible on the block. Bottom of springs may rest on pedestals or block can can be recessed into the floor.
4	C.G.	Same as Type 3 except that tops of springs are close to the height of the center of gravity of the combined block and equipment assembly.

quently used to reduce or attenuate fan noise transmission through ductwork was duct lining. However, the introduction of unit sized air conditioning silencers on the market has given the mechanical engineer another tool for the reduction of noise in the ventilating system. In many localities it will be found that the use of a silencer in a duct as small as 12 x 12 inches is less expensive than providing acoustical duct lining to achieve an equal amount of sound attenuation. Recent designs now provide practically negligible pressure drop through air conditioning silencers and therefore do not require increased horsepower and fan sizes.

Detailed methods of calculating duct lining or duct silencer noise reduction required for a specific system are outlined in the following section entitled, "Calculation of Sound Levels from HVAC Systems."

Figure 10 gives an approximate procedure for determining the amount of sound attenuation required in the most critical octave bands; 63, 125, and 250 Hz. This chart is for estimation purposes only and should not replace computations described in the next section. The curves for Figure 10 are based on a fan delivery of 10,000 cfm and a system consisting of 40 feet of duct having two 90 degree elbows and one branch take-off. For other flow rates at the fan, modify the required sound attenuation by the expression: 10 $\log_{10}$ $(Q/10,000)$, where Q is the actual cfm. Note that this expression gives a negative result if Q is less than 10,000 cfm. If less than 40 feet of duct, or fewer take-offs or elbows are used, add 5 dB to the attenuation required. If an axial fan is used instead of a centrifugal fan, add 5 dB to the sound attenuation required.

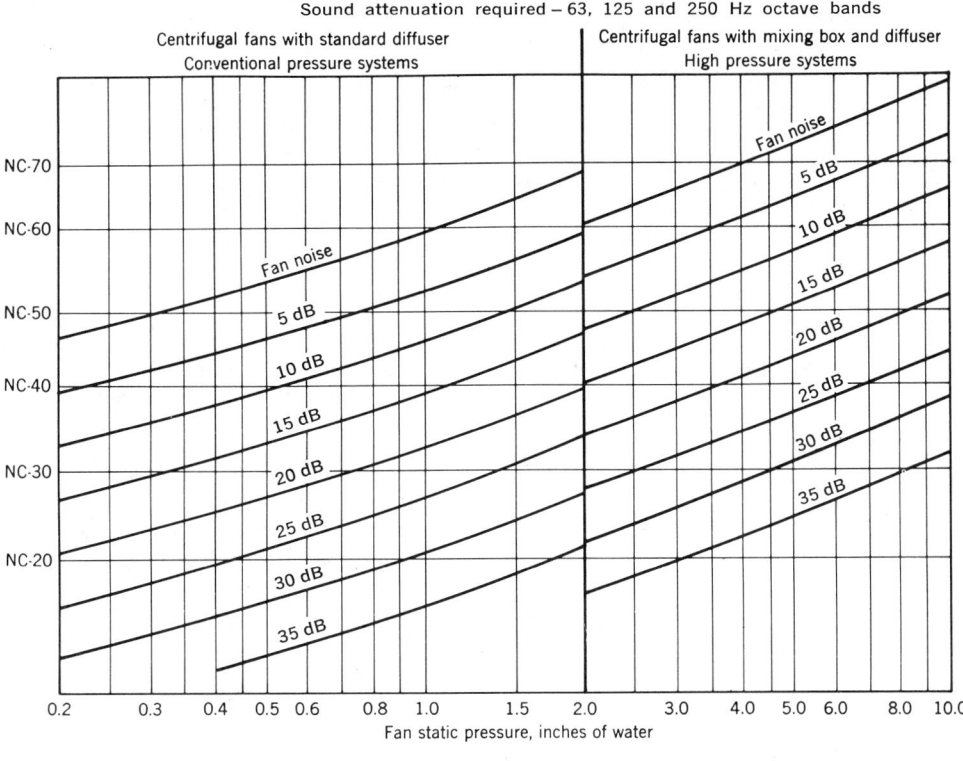

Fig. 10. Sound attenuation required—63, 125, and 250 Hz octave bands.

Regenerated Noise

Airflow, especially at high velocities, is quite capable of creating noise as it passes through the air conditioning system. Middle and high frequency noise can be generated at grilles, diffuser and other orifices; low and middle frequency noise is frequently generated at take-offs, elbows, etc.

Most grille and diffuser manufacturers submit noise output data with their literature giving noise output as related to the velocity of air through the open area of the grille and the pressure drop across the grille.

High velocity systems may be subjected to regenerated noise at various duct fittings. It is strongly recommended that high velocity systems, when used in critical areas, be thoroughly analyzed by an acoustical consultant or other experienced persons.

See "Acoustical Problems in High Velocity Air Distribution" elsewhere in this section of the *Handbook*.

Other Mechanical Noise Sources

Condenser Water and Chilled Water Piping. In addition to the necessity of isolating vibrating piping from the structure to avoid structural transmission of noise from the piping, piping can also be an airborne source of noise. Condenser water or chilled water mains installed horizontally above suspended ceilings will frequently generate and transmit noise in excess of criteria to critical spaces below when the suspended ceiling is transparent to sound as are most acoustical tile ceilings. The solution to this problem is usually to substitute a suspended plaster ceiling for the acoustical tile ceiling and, if necessary, cement the acoustical tile to the underside of the plaster, or to box the piping with wire lath and plaster or equivalent material.

Steam Pressure Reducing Valves. Noise levels as high as 115 dB at a distance of a few feet from a large pressure reducing valve in a high velocity

steam system are possible. Since this noise can be transmitted either structure-borne or airborne, the steam pressure reduction valve and piping for a distance of approximately 30 diameters in each direction should be isolated from the structure.

Adequate sound barriers must also be provided to attenuate the airborne noise transmission. It is strongly recommended that steam pressure reduction valves be located in non-critical areas. Any deviation from this recommendation should be discussed with an experienced acoustical authority.

Transformers. Iron-core transformers also generate noise, usually at the fundamental frequency of 120 Hz. Other harmonics up to 1200 Hz may also be detected. The noise output from transformers is a function of the size of the unit in kva rating. Location of a transformer in excess of 500 kva should be the subject of special analysis.

Cooling Towers. Control of noise of cooling towers is treated separately later in this section.

Specifications

Neglect of acoustical problems in mechanical systems can result in a completely unsatisfactory acoustical environment. Corrective measures taken after a building has been completed are rarely fully satisfactory and always greatly more expensive than if the design had incorporated necessary acoustical precautions.

Mechanical specifications and drawings should clearly define the work to be done. Clauses such as "the contractor shall make the job vibration-less and noiseless" places engineering responsibility upon the contractor in excess of his limited knowledge of the field. Furthermore, such vague wording actually handicaps the more experienced contractor who might, because of his experience, over-engineer the solution at a cost higher than his less experienced competitor.

CALCULATION OF SOUND LEVELS FROM HVAC SYSTEMS

In this section the details necessary for the computation of the sound pressure levels from HVAC systems will be described and illustrated. The approximate method, using Figure 10 of the previous section, is simpler for estimating required noise attenuation. However, for critical situations it is recommended that more accurate, detailed calculations be made, as outlined in this section.

Description of Decibels

The mathematical decibel scale follows a logarithmic pattern. The decibel is used to describe the intensity level or energy level of a physical quantity, and is defined as ten times the logarithm to the base ten of the ratio of a power or energy quantity to a reference value of the same physical quantity:

$$L_E = 10 \log_{10} \left(\frac{E}{E_{ref}} \right) \text{ dB} \qquad (1)$$

where

$L_E =$ Energy or power level in decibels

$E =$ Energy or power quantity of interest, watts

$E_{ref} =$ Reference energy or power quantity, watts

In acoustics, the decibel is used for acoustic power and acoustic pressure expressed by the following:

$$L_W = 10 \log_{10} \left(\frac{W}{W_{ref}} \right) \text{ dB} \qquad (2)$$

$$L_p = 10 \log_{10} \left(\frac{p}{p_{ref}} \right)^2 \text{ dB} \qquad (3)$$

where:

$L_W =$ sound-power level, dB
$L_p =$ sound-pressure level, dB
$W_{ref} = 10^{-12}$ watts
$p_{ref} = 0.0002$ microbar $= 2 \times 10^{-5} \text{ N/m}^2 = 2 \times 10^{-5}$ pascals (Pa).

In Equation 3 it can be seen that the energy is proportional to the square of the pressure; therefore, the pressure term must be squared since the decibel is defined as a ratio of power.

Addition of Decibels

Since decibels are logarithmic quantities, the summation of two decibel values requires the sum of the energy values:

$$L_{W_1} = 10 \log_{10}\left(\frac{W_1}{W_{ref}}\right) \text{ and } L_{W_2} = 10 \log_{10}\left(\frac{W_2}{W_{ref}}\right)$$

$$L_{W_1} + L_{W_2} = 10 \log_{10}\left(\frac{W_1 + W_2}{W_{ref}}\right) \text{ dB} \quad (4)$$

And since power is related to the square of pressure,

$$L_{p_1} = 10 \log_{10}\left(\frac{p_1}{p_{ref}}\right)^2 \text{ and } L_{p_2} = 10 \log_{10}\left(\frac{p_2}{p_{ref}}\right)^2$$

$$L_{p_1} + L_{p_2} = 10 \log_{10}\left(\frac{p_1^2 + p_2^2}{p^2_{ref}}\right) \text{ dB} \quad (5)$$

To add a series of decibel values the antilog first must be taken as well as the logarithm of the sum of the antilogs.

$$L_W = 10 \log_{10}\left(\frac{W}{W_{ref}}\right) \quad (6)$$

$$\frac{W}{W_{ref}} = 10^{\frac{L_W}{10}} \quad (7)$$

Thus,

$$L_{W_1} + L_{W_2} = 10 \log_{10}\left(10^{\frac{L_{W_1}}{10}} + 10^{\frac{L_{W_2}}{10}}\right) \quad (8)$$

The basic equation for the summation of decibels, whether power or pressure level, is given as follows:

$$L_W = 10 \log_{10}\left(10^{\frac{L_{W_1}}{10}} + 10^{\frac{L_{W_2}}{10}} + \ldots 10^{\frac{L_{W_n}}{10}}\right) \quad (9)$$

$$L_p = 10 \log_{10}\left(10^{\frac{L_{p_1}}{10}} + 10^{\frac{L_{p_2}}{10}} + \ldots 10^{\frac{L_{p_n}}{10}}\right) \quad (10)$$

As an example consider the addition of 90 dB and 94 dB sound pressure level:

$$
\begin{aligned}
L_p = 90 \text{ dB} + 94 \text{ dB} &= 10 \log_{10}(10^{90/10} + 10^{94/10}) \\
&= 10 \log_{10}(10^9 + 10^{9.4}) \\
&= 10 \log_{10}(1 \times 10^9 + 2.51 \times 10^9) \\
&= 10 \log_{10}(3.51 \times 10^9) \\
&= 10 (.5453 + 9) \\
90 \text{ dB} + 94 \text{ dB} &= 95.5 \text{ dB}
\end{aligned}
$$

It is obvious that wherever the added dB values in Equation 8 differ by 10 or more, the higher of the values can be chosen as a sufficiently accurate answer. With a modern engineering calculator the addition of logarithmic values can be done quite easily without performing the intermediate steps shown.

Octave Bands and ⅓-Octave Bands

A linear frequency scale (100 Hz, 200 Hz, 300 Hz, etc.) often is inconvenient for assessing acoustical frequencies which range from 20 Hz to 20,000 Hz. The practical method is to separate this audible frequency range into ten unequal segments called *octave bands*. An octave higher means a doubling in frequency; e.g., 63 Hz, 125 Hz, 250 Hz, etc.

For scientific measurement of sound, certain frequency ranges, from low to high, have been established as guides over the range of audible sound. For each octave band the frequency of the upper limit is twice that of the lower band limit, as shown in Table 1. Also, the center frequencies have been standardized for scientific measurements so that each succeeding octave-band center frequency is two times the frequency of the previous band, as shown in Table 1.

Each octave band is sometimes separated into three ranges, referred to as *one-third-octave* bands. These ranges have also been standardized for scientific measurements as given in Table 1. There are portable sound-measuring instruments that will measure the sound-pressure level in decibels for each octave band or one-third-octave band, the octave-band measurements being most commonly used. Octave-band measurements are necessary to determine the NC-criterion of a building space as described previously in this section.

The Sabin

The *sabin* is an important unit of sound absorption expressing the total amount of sound absorption obtained from a given area of material, from a component such as a window or door, or even from occupants in a room. Mathematically the sabin is not a coefficient but is expressed in square feet for a panel of a single material as:

$$A = \alpha S \quad (11)$$

where: A is the absorption of the panel, in sabins (ft^2); α is the absorption coefficient (dimensionless), and S is the surface area of the panel (ft^2).

Table 1. Frequency Limits for Octave Bands and ⅓-Octave Bands

Frequency, Hz					
Octave Bands			⅓-Octave Bands		
Lower Band Limit, Hz	Center Frequency, Hz	Upper Band Limit, Hz	Lower Band Limit, Hz	Center Frequency, Hz	Upper Band Limit, Hz
11	16	22	14.1	16	17.8
			17.8	20	22.4
			22.4	25	28.2
22	31.5	44	28.2	31.5	35.5
			35.5	40	44.7
			44.7	50	56.2
44	63	88	56.2	63	70.8
			70.8	80	89.1
			89.1	100	112
88	125	177	112	125	141
			141	160	178
			178	200	224
177	250	355	224	250	282
			282	315	355
			355	400	447
355	500	710	447	500	562
			562	630	708
			708	800	891
710	1,000	1,420	891	1,000	1,122
			1,122	1,250	1,413
			1,413	1,600	1,778
1,420	2,000	2,840	1,778	2,000	2,239
			2,239	2,500	2,818
			2,818	3,150	3,548
2,840	4,000	5,680	3,548	4,000	4,467
			4,467	5,000	5,623
			5,623	6,300	7,079
5,680	8,000	11,360	7,079	8,000	8,913
			8,913	10,000	11,220
			11,200	12,500	14,130
11,360	16,000	22,720	14,130	16,000	17,780
			17,780	20,000	22,390

The confusion concerning the sabin is that it has the dimensions of square feet but it is *not* related to the sound absorption of a one-square-foot area of material. The amount of absorption in sabins is defined by Equation (11) where the absorption coefficient, α, is determined by a specified method of measurement in an acoustic laboratory.

The absorption of a system composed of a number of materials with different sound absorption coefficients can be determined from the relation:

$$A = \alpha_1 S_1 + \alpha_2 S_2 + \ldots \alpha_n S_n = \sum_{i=1}^{i=n} \alpha_i S_i \quad (12)$$

$A =$ total absorption, in sabins
$\alpha_i =$ absorption coefficient of various materials (dimensionless)
$S_i =$ surface area of various materials (ft^2)

For example, the total absorption of an empty room would be computed from:

$$A = \alpha_{floor} S_{floor} + \alpha_{walls} S_{walls} + \alpha_{ceiling} S_{ceiling}$$

$$+ \alpha_{windows} S_{windows} + \alpha_{door} S_{door} \qquad (13)$$

If there are other materials in the room they must be included. Because α, the absorption coefficient, varies with frequency, the total absorption, A, will also vary with frequency. The usual computational methods use octave-bands with α given for the center frequency of the octave band.

Again, it is important to realize that the absorption coefficient, α, cannot be calculated but must be determined experimentally. Typical values of this coefficient are given in Table 2. Suppliers' catalogs are the usual sources of coefficients for specific materials. A comprehensive source of these data is a bulletin entitled: *Performance Data—Architectural Acoustical Materials,* published by the Acoustical Materials Association.[1] Another source is the *Compendium of Materials for Noise Control.*[2]

---

[1] Acoustical Materials Association, 205 West Touhy Avenue, Park Ridge, Illinois 60068.
[2] For sale by the Superintendent of Documents, U.S. Government Printing Office, Washington, D.C. 20402.

Determination of Sound Pressure Level

The sound pressure level is usually prescribed in octave bands for room usage criteria. This was described previously under "Noise and Vibration Control," with specific values in Table 2 of that section. The sound pressure level can be computed from the sound power level of noise sources, and the room characteristics from the relation:

$$L_p = L_w + 10 \log_{10} \left[\frac{Q}{4\Pi_n} \left(\frac{1}{r_1^2} + \frac{1}{r_2^2} + \ldots \frac{1}{r_{n2}} \right) + \frac{4}{R} \right] + 10 \text{ dB} \qquad (14)$$

where:

L_p = sound pressure level at a location in question in the room, dB

L_w = sound power level, in dB, from sources in the room

Q = directivity factor (dimensionless). (See Figure 1.)

n = number of sound sources of equal sound-power level

r_1, r_2, r_n = distance from sources, feet

R = room constant, in square feet

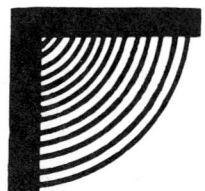

UNIFORM RADIATION OVER 1/4 OF SPHERE, Q=4 TWO REFLECTING SURFACES

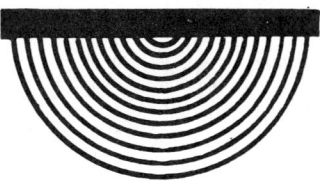

UNIFORM HEMISPHERICAL RADIATION, Q=2 SINGLE REFLECTING SURFACE

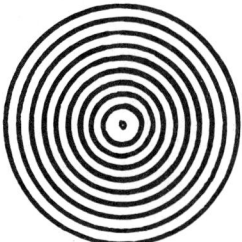

UNIFORM SPHERICAL RADIATION, Q=1 NO REFLECTING SURFACES

NONUNIFORM RADIATION, Q DEPENDS ON DIRECTION

Fig. 1. Directivity factors for various acoustic radiation conditions.

Table 2. Typical Values for Sound Absorption Coefficients of Building Materials

Materials	Frequency, H$_z$						
	125	250	500	1000	2000	4000	8000
	Absorption Coefficient, α						
Cellulose Fiber Ceiling Tile #1 Mounting*	.05 to .10	.20 to .25	.50 to .80	.50 to .99	.45 to .90	.30 to .70	.30 to .75
Same, #7 Mounting*	.30 to .40	.30 to .45	.35 to .65	.45 to .99	.50 to .90	.50 to .75	.50 to .80
Mineral Fiber Ceiling Tile #1 Mounting*	.02 to .10	,15 to .25	.55 to .75	.60 to .99	.60 to .99	.50 to .90	.50 to .95
Same #7 Mounting*	.40 to .70	.40 to .95	.45 to .95	.55 to .99	.65 to .99	.50 to .95	.50 to .95
Brick, Unglazed	.03	.03	.03	.04	.05	.07	.06
Brick, Glazed, Painted	.01	.01	.02	.02	.02	.03	.03
Carpet, Heavy on Concrete	.08	.24	.57	.69	.71	.73	.74
Same with Latex Backing on Hair-Felt or Foam Pad	.08	.27	.39	.34	.48	.63	.72
Light Carpet	.01	.05	.10	.20	.45	.65	.66
Concrete Block, Unpainted	.36	.44	.31	.29	.39	.25	.25
Concrete Block, Painted	.10	.05	.06	.07	.09	.08	.08
Draperies, 10 oz. per sq yd, Hung Straight	.03	.04	.11	.17	.24	.35	.40
14 oz. per sq yd	.07	.31	.49	.75	.70	.60	.60
18 oz per sq yd	.14	.35	.55	.72	.70	.65	.65
Floors, Concrete or Terrazzo	.01	.01	.01	.02	.02	.02	.02
Asphalt, Rubber, or Cork Tile on Concrete	.02	.03	.03	.03	.03	.02	.02
Wood	.15	.11	.10	.07	.06	.07	.06

*A #1 mounting is used to designate ceiling tile supported by 1/8-inch-thick spacers from the ceiling structure. A #7 mounting is the designation for a suspended ceiling system.

Table 2 (Cont.). Typical Values for Sound Absorption Coefficients

	Frequency, H_z						
	125	250	500	1000	2000	4000	8000
Materials	Absorption Coefficient, α						
Glass, Large Panes of Heavy Plate	.18	.06	.04	.03	.02	.02	.02
Common Window Glass	.35	.25	.18	.12	.07	.04	.03
Gypsum Board	.29	.10	.05	.04	.07	.09	.08
Marble or Glazed Tile	.01	.01	.01	.01	.02	.02	.02
Plaster, Smooth Finish	.01	.01	.02	.03	.04	.05	.04
Rough Finish	.14	.10	.06	.05	.04	.03	.03
Sheet Metal Panels	.05	.04	.03	.03	.02	.01	.01
Water, as in a Swimming Pool	.01	.01	.01	.01	.02	.03	.02
Wood Paneling, 3/8" Thick	.28	.22	.17	.09	.10	.11	.09
Air, Sabins per 1000 cubic foot at 50% Relative Humidity	.00	.10	.30	.90	2.30	7.20	28.00
Absorption of Seats and Audience Sabins Per Square Foot of Seating Area Per Unit							
Audience Seated in Upholstered Chairs, per sq ft of Floor Area	.60	.74	.88	.96	.93	.85	.80
Unoccupied Cloth Covered Seats, per sq ft of Floor Area	.49	.66	.80	.88	.82	.70	.65
Wooden Pews, Occupied, per sq ft of Floor Area	.57	.61	.75	.86	.91	.86	.80
Pew Cushions (Without Pews), Sabins Each	.75 to 1.10	1.00 to 1.70	1.45 to 1.90	1.50 to 2.00	1.40 to 1.70	1.40 to 1.60	1.40 to 1.55
Chairs, Unoccupied, Metal or Wood, Sabins Each	.15	.19	.22	.39	.38	.30	.30

The sound power level, L_w, must be provided by the manufacturer of equipment to be installed in the space. For air flow systems L_w must be computed as described later in this section. The directivity factor, Q, is a dimensionless number which accounts for the directional character of the source or the placement of the source in the room. Figure 1 gives directivity factors for various sound radiation conditions. For a sound source near a wall $Q=2$; for a source such as a diffuser suspended from and some distance from a ceiling, a nearly uniform radiation can result with $Q=1$.

The room constant, R, accounts for the amount of acoustic energy absorbed by the materials of the space such as walls, floor, ceiling, and occupants. It is computed from:

$$R = \frac{\bar{\alpha} S}{1 - \bar{\alpha}}$$

where:

$R =$ room constant, ft²

$S =$ total area of sound-absorbing materials, ft²

$\bar{\alpha} =$ average sound absorbing coefficient for all surfaces, dimensionless

The room absorption term $\bar{\alpha} S$ is obtained from the following relation:

$$\bar{\alpha} S = \alpha_1 S_1 + \alpha_2 S_2 + \ldots \alpha_n S_n + A_1 + A_2 + A_n + \alpha_{air} V \quad (16)$$

where

$\bar{\alpha} S =$ room absorption in sabins, ft²

$S_1, S_2, S_n =$ areas of surfaces, ft²

$\alpha_1, \alpha_2, \alpha_n =$ absorption coefficients of room surfaces, dimensionless

$A_1, A_2, A_n =$ absorption of occupants, furnishings, etc., sabins (ft²)

$\alpha_{air} =$ absorption of air in the room per 1000 cubic feet, $\left(\dfrac{1}{\text{ft}}\right)$

$V =$ volume of the room in cubic feet divided by 1000, ft³

The value for $\bar{\alpha}$ for use in Equation (15) can be determined from the right-hand side of Equation (16) by dividing by the total surface area in the room, i.e., $S = S_1 + S_2 + \ldots S_n$.

Example No. 1

Determination of Sound Pressure Level for a Nonducted Source.

As an example of the use of Equation (14) to predict the sound pressure level in a room, consider the executive office, as shown in Figure 2, which has a fan-coil unit. The office has wood paneling on three walls, glass windows on one wall, heavy carpeting, and acoustical ceiling panels. It is desired to compute the noise criteria (NC) value for the office.

First the octave-band sound power levels are to be obtained from the manufacturer of the fan-coil unit. For this particular unit the values shown in Table 3 were provided.

Calculation of the room constant is illustrated in Table 4. For each room surface the acoustic absorption coefficients are obtained from the manufacturer of the material or, in the case of glass, the values are obtained from Table 2. The acoustic absorption of air per 1000 cubic feet is determined from Table 2.

In Table 4, Rows one through four, the absorption (in sabins) for each room surface is computed by multiplying the surface area by the absorption coefficients. The absorption for the air in the room is computed in Row 5 by dividing the room volume—in cubic feet—by 1000 and

Table 3. Manufacturer Provided Sound-Power Levels for Fan-Coil Unit of Figure 2

Octave-Band Center Frequency, Hz	125	250	500	1000	2000	4000	8000
Sound Power Level, L_w, dB re 10^{-12} watts	57	52	49	40	37	31	21

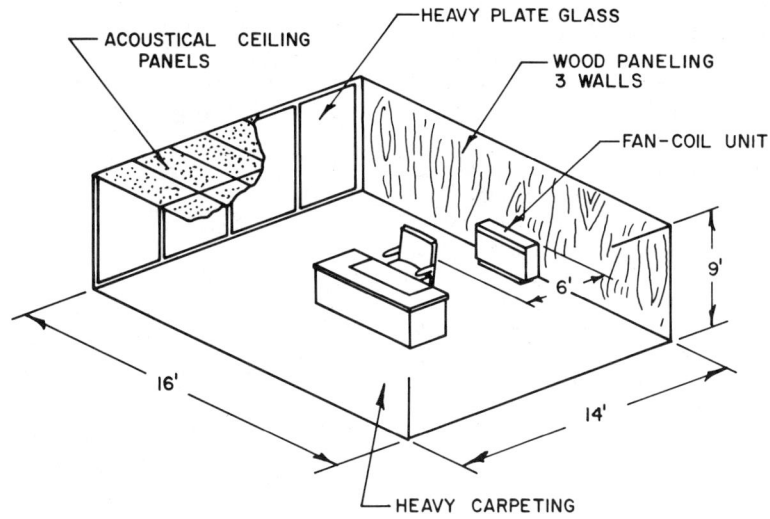

ACOUSTICAL CEILING PANELS

HEAVY PLATE GLASS

WOOD PANELING 3 WALLS

FAN-COIL UNIT

9'

6'

16'

14'

HEAVY CARPETING

Fig. 2. An office for which NC-level is to be computed. See Example No. 1.

Table 4. Calculation of the Room Constant, R, for Example No. 1

Computation	Octave–Band Center Frequency, Hz						
	125	250	500	1000	2000	4000	8000
1. Window Wall: (S_1 = 126 ft^2) α_1 =	.18	.06	.04	.03	.02	.02	.02
$\alpha_1 S_1$ (sabins) =	22.68	7.56	5.04	3.78	2.52	2.52	2.52
2. Paneled Walls: (S_2 = 414 ft^2) α_2 =	.28	.22	.17	.09	.10	.11	.09
$\alpha_2 S_2$ (sabins) =	115.92	91.08	70.38	37.26	41.40	45.54	37.26
3. Floor: (S_3 = 224 ft^2) α_3 =	.08	.27	.39	.34	.48	.63	.74
$\alpha_3 S_3$ (sabins) =	17.92	60.48	87.36	76.16	107.52	141.12	165.76
4. Ceiling: (S_4 = 224 ft^2) α_4 =	.36	.42	.62	.86	.87	.81	.72
$\alpha_4 S_4$ (sabins) =	80.64	94.08	138.88	192.64	194.88	181.44	161.28
5. Air in Room: (V = 2016 ft^3) α_{air} =	.00	.10	.30	.90	2.30	7.20	28.00
α_{air} V/1000 (sabins) =	.00	.20	.60	1.81	4.64	14.52	56.45
7. Calculation of $\bar{\alpha}$ = Σ α_i S_i/S = (S = 988 ft^2). Row 6 divided by S.	.24	.26	.31	.32	.36	.39	.43
8. R = $\dfrac{\bar{\alpha}S}{1-\bar{\alpha}}$ (rounded) =	312	342	438	458	548	631	743

multiplying by the air absorption per 1000 cu. ft. The total absorption of the room is given in Row 6, which is the sum of the absorption (in sabins) of the room surfaces and the air in the room from Rows 1 through 6. The average absorption coefficient, $\bar{\alpha}$, is computed in Row 7 by dividing the total absorption values, in sabins, by the total surface area of the room; i.e., the values in Row 6 are divided by the room surface area to give the values in Row 7. Finally, the room constant, R, is computed in Row 8 from the values in Rows 6 and 7.

Using the values of room constant computed in Table 4, the octave-band sound pressure levels will now be calculated using Equation (14). From Figure 1, the directivity factor, Q, is 2 because the fan-coil unit is against one wall and hemispherical radiation from the wall is assumed. For only one noise source, n is one; and r_1 is the distance from the unit to the receiver location, which is six feet in this example. The sound pressure level is then computed in octave bands from:

$$L_p = L_W + \text{correction}$$

$$L_p = L_W + 10 \log_{10} \left[\frac{Q}{4\Pi n} \left(\frac{1}{r_1^2} \right) + \frac{4}{R} \right] + 10 \text{ dB}$$

For the 125-Hz, octave-band center frequency the computation is as follows:

$$L_W = 57, \text{ from Table 3}$$
$$R = 312, \text{ from Table 4}$$

$Q = 2$, from Figure 1

$$L_p = L_W + 10 \log_{10} \left[\frac{2}{4\Pi 6^2} + \frac{4}{312} \right] + 10$$
$$= L_W + 10 \log_{10} [.0044 + .0128] + 10$$
$$= L_W + 10 \log_{10} [.0172] + 10$$
$$= L_W - 17.6 + 10$$
$$= 57 - 7.6$$
$$L_p = 49 \text{ dB (rounded)}$$

This procedure is repeated for every octave-band center frequency. Results are shown in Table 5. Note that in Row 4 of Table 5 the final values are rounded to the nearest whole decibel.

Determination of the NC-level is accomplished by plotting the resulting sound pressure levels from Table 5 on an NC-criterion chart as shown in Figure 3. Figure 3 indicates that the highest penetration is the NC-35 curve. Thus, the level at the listener location in Figure 2 meets an NC-35 criterion. Referring to Table 2 of the previous section entitled, "Noise and Vibration Control," it is seen that an NC-35 criterion is acceptable for an executive office. If the computed level had exceeded the NC-35 value, alternate systems would have been considered. For instance, a fan-coil unit with lower sound-power levels could have been specified. Or, ceiling panels with larger acoustic absorption values could have been used. Possibly, a combination of both approaches would satisfy the criterion.

Note in this example the necessity of obtaining the octave-band sound-power levels for the unit to be installed in order to compute the sound

Table 5. Calculation of Sound-Pressure Levels for Example No. 1

Computation	Octave-Band Center Frequency, Hz						
	125	250	500	1000	2000	4000	8000
1. $10 \log_{10}[\frac{2}{4\Pi 6^2} + \frac{4}{R}]$, dB (R from Table 4)	-17.6	-17.9	-18.7	-18.8	-19.3	-19.7	-20.1
2. Add ten decibels	+10	+10	+10	+10	+10	+10	+10
3. Correction, dB	-7.6	-7.9	-8.7	-8.8	-9.3	-9.7	-10.1
4. $L_p = L_W$ + Correction, dB (L_W from Table 3) values rounded	49	44	40	31	˙28	21	11

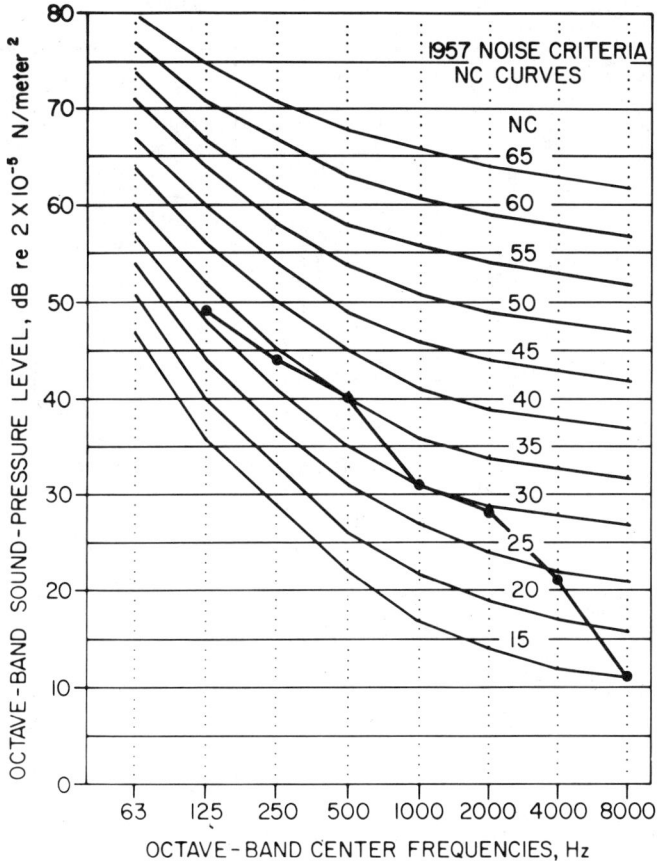

Fig. 3. Determination of noise criteria for Example No. 1.

pressure levels and the NC-criteria. A unit designated as meeting some NC-level is meaningless if the room constant for which the unit is to be installed is not known. Therefore, equipment manufacturers are encouraged to provide octave-band sound power levels, L_w, and *not* NC-levels or sone values.

NOISE IN DUCTED SYSTEMS

In the remainder of this section in-duct noise sources will be treated. An analysis of the sound generation and propagation in a ducted system requires some information in addition to that given in previous sections. The airborne sound inside a ducted system which radiates into an occupied space involves the following factors:

1. Sound power generated by the fan
2. Distribution of sound in the ducted system
3. Attenuation provided by the duct (acoustically treated, or bare duct)
4. Attenuation provided by elbows
5. Attenuation provided by commercial silencers or attenuators
6. Generation of sound due to air flow in fittings, elbows, grilles, etc.
7. Sound transmitted through duct walls.

Fan Noise Generation

Figure 4 gives information on the principal types of fans used in commercial installations. The fan noise is given in terms of the specific sound-power levels in octave bands and the blade

CENTRIFUGAL FANS

TYPE	DESIGN	SPECIFIC SOUND-POWER LEVEL, K_w — CENTER FREQUENCY – Hz, dB re 10^{-12} watt and 1 cfm at 1 inch ftp								BFI	APPLICATIONS
		63	125	250	500	1000	2000	4000	8000		
AIRFOIL		35	35	34	32	31	26	18	10	3	Highest efficiency of all centrifugal fan design contains 10 to 16 blades of airfoil shape. Used for general heating, ventilating, and air-conditioning systems, usually applied to central station units where the horsepower saving will be significant. Can be used on low, medium, and high-pressure systems and will operate satisfactorily in parallel. Is also used in large sizes, for clean-air industrial applications where power savings will be significant. Can be used on industrial exhaust systems, where the air-cleaning system is of high efficiency.
BACKWARD INCLINED BACKWARD CURVED		35	35	34	32	31	26	18	10	3	Efficiency is only slightly less than that of the airfoil fan. Contains 10 to 16 blades. Used for the same general applications as the airfoil fan. Can be used in industrial applications where the gas is essentially clean, but does not meet the standards required for airfoil fan selection.
RADIAL		48	45	43	43	38	33	30	29	5-8	Simplest of all centrifugal fans—relatively low efficiency, usually has 6 to 10 blades; includes both radial blades (R), and modified radial blades (M). Used primarily for industrial exhaust, including dirty gas fans and recirculating gas fans. This design also used for high-pressure industrial applications.
FORWARD CURVED		40	38	38	34	28	24	21	15	2	Efficiency less than the airfoil and backwardly curved fans, this fan is usually fabricated of lightweight and low-cost construction. It may have from 24 to 64 blades. This design will be the smallest of the centrifugal fan types and operates at the lowest speed. Used primarily in low-pressure heating, ventilating, and air-conditioning applications, such as: domestic furnaces, small central station units, and packaged air-conditioning equipment.

AXIAL FANS										
VANEAXIAL		42	39	41	42	40	37	35	25	6-8	High-efficiency axial flow fan with airfoil blades and high pressure capability. Blades may be fixed or adjustable and the hub diameter is usually greater than 50 per cent of the fan tip diameter. There may be from 3 to 16 blades. This fan design has guide vanes downstream from the wheel which permits good air flow pattern on the discharge side of the fan. Used for general heating, ventilating, and air-conditioning applications in low, medium, and high-pressure systems. May also be used in industrial applications such as: drying ovens, paint spray booths, and fume exhaust systems.
TUBEAXIAL	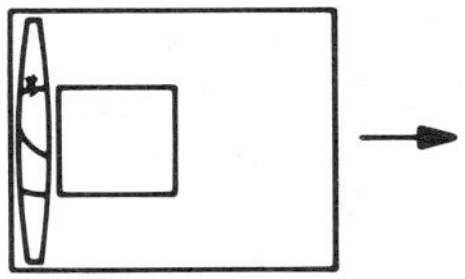	44	42	46	44	42	40	37	30	6-8	This fan is more efficient than the propeller fan design and can develop a more useful pressure capability. The number of blades may vary from 4 to 8 and the hub is usually about 50 per cent of the fan tip diameter. The blades may be of airfoil or single thickness cross-section. The fan is built without downstream guide vanes. Used in low- and medium-pressure ducted heating, ventilating, and air-conditioning applications where the poor air flow pattern downstream from the fan is not detrimental. This fan is also used in some industrial applications such as: drying ovens, paint spray booths and fume exhaust systems.
PROPELLER	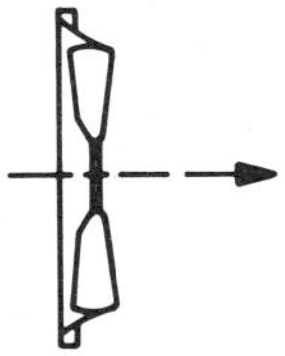	51	48	49	47	45	45	43	31	5-7	Low efficiency wheels are usually of inexpensive construction and are limited to very-low-pressure applications. Usually contains 2 to 8 blades of single thickness construction attached to a relatively small hub. The housing is a simple circular ring or orifice plate. This fan is used for low pressure, high-volume air-moving applications such as air circulation within a space or as exhaust fans in a wall or roof.
TUBULAR CENTRIFUGAL	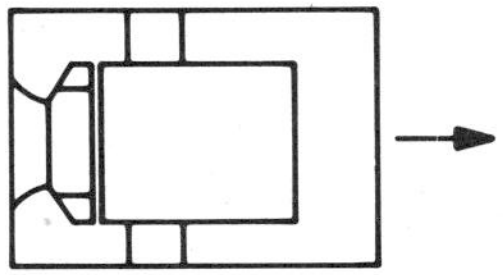	46	43	43	38	37	32	28	25	4-6	This fan usually has a wheel similar to the airfoil or backwardly inclined wheel, described above, which is built into an axial flow type housing. This results in lower efficiencies than the centrifugal fans of similar wheel design. The air is discharged radially from the wheel and must change direction by 90 degrees to flow through the guide vane section. Used primarily for low-pressure return-air systems in heating, ventilating, and air-conditioning applications.

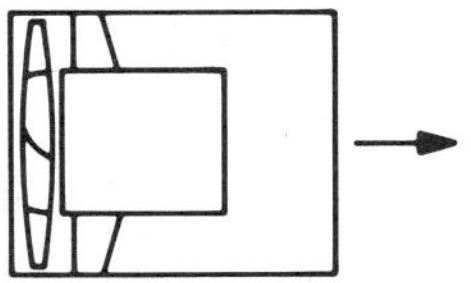

Fig. 4. Acoustic properties of various fan types.

frequency increment (BFI). The *specific sound-power level* (K_W) is defined as the sound power level generated by a fan operating at a flow rate of 1 cfm and a pressure of 1 inch of water. Fans generate a tone at the blade-pass frequency over and above the sound level generated by air flow. In order to account for this blade-pass tone, the sound power level must be increased in the octave band in which the blade frequency occurs. The amount of increase is listed in Figure 4 for each fan under the column labeled *BFI, Blade Frequency Increment*. The *Blade-Pass Frequency, B_f,* is determined from:

$$B_f = \frac{rpm \times No.\ of\ blades}{60} \quad Hz \quad (17)$$

The number of blades and the rpm are to be determined from the catalog used for fan selection to meet design flow requirements.

Estimating Fan Noise

In general, it is desirable to obtain fan sound-power information from the fan manufacturer. There may be times, however, when this is not possible or the fan is to be operated at a condition not covered by manufacturer's data. For such cases the following procedure may be used to estimate the fan-generated sound power.

The specific sound-power levels (K_W) given in Figure 4 provide a basis for estimating the sound power levels of fans with corrections as functions of flow rate (Q), the pressure rise across the fan (P) and the blade frequency increment (BFI). Sound power is estimated from:

$$L_W = K_W + (10\ \log_{10} Q + 20\ \log_{10} P) + BFI\ dB$$
$$(18)$$

where:

$L_W =$ sound power level of the fan, dB reference 10^{-12} watts, in octave bands
$K_W =$ specific sound power level, dB
$Q =$ flow rate, cfm
$P =$ pressure rise across the fan, inches of water
$BFI =$ blade frequency increment, dB

The term $(10\ \log_{10} Q + 20\ \log_{10} P)$ in Equation (18) may be calculated directly or obtained graphically from Figure 5.

Equation (18) predicts the total sound-power level generated by a fan which includes the inlet and outlet combined. If only the inlet sound-power level is desired or only the outlet sound power level, 3 dB must be subtracted from the values computed. This assumes that the inlet and outlet of the fan generate equal sound-power levels. The reason for subtracting 3 dB is a result of how decibel values combine or subtract. As computed from Equation (8), whenever two equal sound-power levels are combined the result is 3 dB greater than the values being added. Conversely, to find the sound-power level which corresponds to one-half the total sound power, we must *subtract* 3 dB from the total sound-power level. This will be illustrated in the following example.

Example No. 2

Estimate the sound-power level from the outlet of a radial-blade fan operating at 10,000 cfm, and 2 inches of H$_2$O. The fan has 16 blades and will operate at 500 rpm.

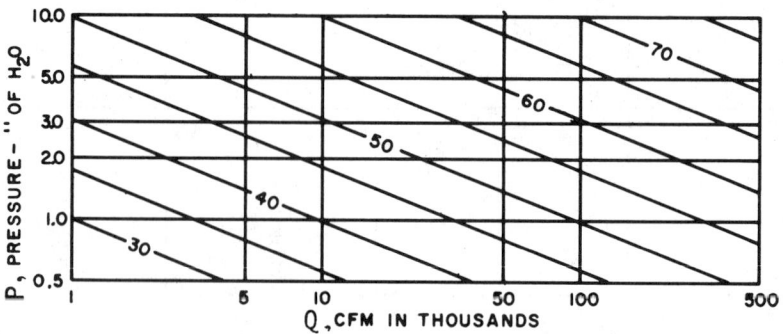

Fig. 5. Correction, in decibels, to be added to sound-power levels listed in Fig. 4 to determine total sound-power level at actual operating conditions.

Step 1. The specific sound-power levels, K_W, for a radial type fan are obtained from Figure 4 and listed in Row 1 of Table 6.

Step 2. Compute the bracketed term in Equation (18) for $Q=10,000$ cfm and $P=2''$ H₂O.

$$(10 \log_{10} Q + 20 \log P)$$
$$= (10 \log_{10} 10,000 + 20 \log_{10} 2)$$
$$= 40 + 6$$
$$= 46 \text{ dB}$$

(or obtain this value from Figure 5).

Thus, 46 dB is listed for every octave band in Row 2 of Table 6.

Step 3. Determine the blade-pass frequency from Equation (17)

$$B_f = \frac{500 \times 16}{60} = 133 \text{ Hz}$$

Note that 133 Hz falls within the 125 Hz octave-band as shown in Table 1 of this section. Figure 4 lists a blade frequency increment, *BFI*, of 5 to 8 dB for a radial fan, therefore the *maximum* value of 8 dB will be listed for the 125 Hz octave-band and zeros are listed for all other octave bands in Row 3 of Table 6.

Step 4. Correct for the sound-power level of the outlet only. Equation (18) predicts the total sound-power level from inlet plus outlet, therefore *subtract* 3 dB, as shown in Row 4, to correct for outlet sound-power level of the fan.

Step 5. Compute the octave-band sound-power level of the outlet by adding Rows 1 through 4 of Table 6. The resulting sound-power level, in decibels, is given in Row 5. These values represent the sound-power levels radiated *into* the duct system by the fan.

Distribution of Sound Power at Branch Take-Offs

At a tee-section, or other branch take-off, the sound power upstream of the take-off propagates into each of the duct elements at the take-off location. The proportion of sound power which transmits from a main air-supply duct into a branch take-off is approximately in the same ratio as the duct area ratios at the branch location. As illustrated by Equation (8), sound levels increase (or decrease) by logarithmic ratios; therefore, if

Table 6. Calculation of Sound-Power Level for Fan Outlet for Example No. 2

Calculations	Octave-Band Center Frequency, Hz							
	63	125	250	500	1000	2000	4000	8000
1. Specific sound level, Kw, dB (Fig. 4)	48	45	43	43	38	33	30	29
2. Correction term, $10 \log_{10} Q + 20 \log_{10} P$, dB	46	46	46	46	46	46	46	46
3. Blade frequency increment, BFI, dB (Fig. 4)	0	8	0	0	0	0	0	0
4. Correction in dB to convert to sound power at fan outlet alone	-3	-3	-3	-3	-3	-3	-3	-3
5. Sound-power level, dB, for fan outlet	91	96	86	86	81	76	73	72

a main duct has airborne sound with a level of 63 dB and this duct divides into two equal area ducts, the sound level will be 60 dB in each. In other words, cutting the sound power in half will result in a 3 dB reduction. The decibel reduction for other area ratios at branch take-offs is given in Table 7.

For multiple outlets, the cfm (cubic feet per minute) at each outlet will be approximately in proportion to the area ratio; therefore, the values in Table 7 can be converted to percentage of cfm of each outlet to the total fan cfm, as given in Table 8.

Attenuation of Untreated Duct

Untreated sheet-metal duct walls will absorb sound energy. For round ducts, the sound attenuation with or without thermal insulation is 0.03 decibels per linear foot for frequencies below 1000 Hz and rises to 0.1 decibel per linear foot at 8000 Hz. Attenuation values for rectangular sheet-metal ducts, round elbows, and square elbows are given in Tables 9, 10, and 11.

Table 7.   Sound-Power Level Division at Branch Take-Offs *

Area of continuing duct, in percent, of the total area of all ducts after branch take-off	5	10	15	20	30	40	50	80
Decibels to be subtracted from power level before take-off in order to get power level in continuing duct	13	10	8	7	5	4	3	1

* Reprinted by permission from *ASHRAE Handbook and Product Directory*, 1976, Table 12.

Table 8.   Allotment of Fan Sound Power to Each Air Outlet *

cfm in Percent of Total Fan cfm	1/5	1/2	1	2	5	10	20	50
Decibels to be subtracted from the up-stream sound-power level to get fan sound-power level per outlet	27	23	20	17	13	10	7	3

*Reprinted by permission from *ASHRAE Guide and Data Book*, 1967, Table 12.

Table 9.   Approximate Natural Attenuation in Bare, Rectangular Sheet-Metal Ducts *†

Duct	Size, in.	Octave-Band Center Frequency, Hz			
		63	125	250	Above 250
		Attenuation, dB/ft			
Small	6 × 6	0.2	0.2	0.15	0.1
Medium	24 × 24	0.2	0.2	0.1	0.05
Large	72 × 72	0.1	0.1	0.05	0.01

*If duct is covered with thermal insulating material, attenuation will be approximately twice the listed values.

† Reprinted by permission from *ASHRAE Handbook and Product Directory*, 1976, Table 13.

Table 10. Approximate Attenuation of Round Elbows *

Diameter or dimensions, in.	Octave-Band Center Frequency, Hz							
	63	125	250	500	1000	2000	4000	8000
	Attenuation, dB							
5 to 10	0	0	0	0	1	2	3	3
11 to 20	0	0	0	1	2	3	3	3
21 to 40	0	0	1	2	3	3	3	3
41 to 80	0	1	2	3	3	3	3	3

* Reprinted by permission from *ASHRAE Handbook and Product Directory*, 1976, Table 14.

Table 11. Attenuation * of Square Elbows without Turning Vanes,** in Decibels (dB) †

	Octave-Band Center Frequency, Hz							
	63	125	250	500	1000	2000	4000	8000
	Attenuation, dB							
(A) No Lining								
5" Duct width (D)	...	...	...	1	5	7	5	3
10" Duct width	...	...	1	5	7	5	3	3
20" Duct width	...	1	5	7	5	3	3	3
40" Duct width	1	5	7	5	3	3	3	3
(B) Lining* Ahead of Elbow								
5" Duct width	...	...	...	1	5	8	6	8
10" Duct width	...	...	1	5	8	6	8	11
20" Duct width	...	1	5	8	6	8	11	11
40" Duct width	1	5	8	6	8	11	11	11
(C) Lining* After Elbow								
5" Duct width	...	...	...	1	6	11	10	10
10" Duct width	...	...	1	6	11	10	10	10
20" Duct width	...	1	6	11	10	10	10	10
40" Duct width	1	6	11	10	10	10	10	10
(D) Lining* Ahead of and After Bend								
5" Duct width	...	...	...	1	6	12	14	16
10" Duct width	...	...	1	6	12	14	16	18
20" Duct width	...	1	6	12	14	16	18	18
40" Duct width	1	6	12	14	16	18	18	18

* Based on lining extending for a distance of at least two duct widths "D" and lining thickness of ten percent of duct width "D." (See Fig. 6). For thinner lining, lined length must be proportionally longer.
** For square elbows with short turning vanes, use average between Tables 9 and 10.
† Reprinted by permission from *ASHRAE Handbook and Product Directory*, 1976, Table 15.

Duct-Lining Attenuation

Direct lining in ducts may be used for both thermal insulation and sound attenuation. The sound attenuation can be estimated from Equation (19).

$$\text{Lined-Duct Attenuation} = 12.6 \, L \, \alpha^{1.4} \left(\frac{P}{S}\right) \text{dB}$$

(19)

where:

L = length of lined duct, in feet.

P = perimeter of duct inside the lining, in inches.

S = cross-sectional area of the duct inside the lining, in square inches.

α = the acoustic absorption coefficient of the lining material (a function of frequency).

The limitations of Equation (19) are:

1. Smallest duct dimension should not exceed 18 in., nor be less than 6 in.
2. Ratio of duct width to height should not exceed 2/1.
3. The absorption coefficient α should be representative for the entire octave band.
4. Air flow velocities should not exceed 4000 fpm.
5. Equation (19) does not allow for line-of-sight propagation of sound which limits high frequency attenuation. In a straight 12 in. duct, for instance, the attenuation in the 8000 Hz octave band will be only about 10 dB for any lining length over 3 ft. The attenuation in the next lower octave band (4000 Hz), will be about midway between 10 dB and the value calculated from Equation (19). The frequency above which the 10 dB limit applies is inversely proportional to the shortest dimension of the duct.

In selecting the duct lining material, the following factors should be considered:

1. For the absorption of frequencies below 500 Hz, the lining material should be 2- to 12-inches thick. Thin materials, particularly when mounted on hard solid surfaces, will absorb only the high frequencies.
2. It has been shown that increased absorption at frequencies below 700 Hz may be realized by using a perforated facing in which the area of the perforations is from 3 to 10

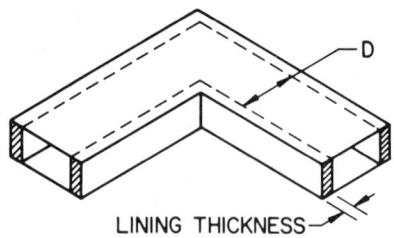

LINING THICKNESS

NOTE: LINING ON SIDES ONLY IS EFFECTIVE FOR ELBOW ATTENUATION

Fig. 6. Illustration of duct width, *D*, for Table 11.

percent of the surface area. Such facings, however, decrease sound absorption at higher frequencies.

3. Any air space behind the lining material will have considerable effect. Absorption coefficients should be based on the particular mounting method intended to be used.

Sound-absorbing materials suitable for use in air ducts are available in the form of blankets, and semirigid boards. Specifications and absorption coefficients for most materials can be obtained from the Acoustical and Insulating Materials Association (AIMA), 205 West Touhy Avenue, Park Ridge, Illinois 60068.

Figure 6 illustrates the correct application of an acoustic lining material in a square or rectangular elbow; the attenuation that can be obtained with this arrangement is given in Table 11.

Example No. 3

Determine the acoustic attenuation of a 24 by 36-in duct, 20 feet long. The lining material is 1-inch glass fiber with absorption coefficients as given in Row 1 of Table 12. These values must be obtained from the supplier of the acoustic material.

Solution: The perimeter of the duct inside the lining is,

$$P = 2 \, (22+34) = 112 \text{ inches}$$

The cross-sectional area inside the lining is,

$$S = 22 \times 34 = 748 \text{ sq in.}$$

For the 20-foot-long duct, the attenuation is computed:

Table 12. Attenuation of a Lined Duct, Example No. 3

Row No.	Calculations	Octave-Band Center Frequency, Hz							
		63	125	250	500	1000	2000	4000	8000
1	Absorption Co-efficient, α, of duct liner material	.08	.11	.34	.70	.81	.86	.85	.89
2	$\alpha^{1.4} =$	.029	.045	.220	.607	.745	.810	.797	.849
3	Attenuation, dB $= 37.7\alpha^{1.4}$	1.1	1.7	8.3	22.9	28.1	30.6	30.1	32.0

Note: The final value for 4000 Hz would be 20 and for 8000 Hz would be 10; per item 5, under "The limitations of Equation (19)," given previously.

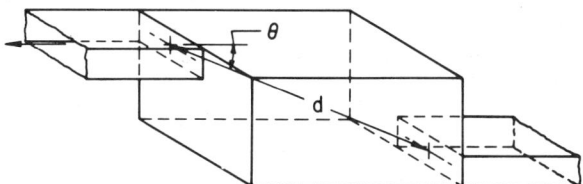

Fig. 7. Diagram of a sound-absorbing plenum.

Attenuation $= 12.6 L \alpha^{1.4}\ (P/S)$ dB/ft

Attenuation $= 12.6 \times 20 \times \alpha^{1.4}\ (112/748) = 37.7\alpha^{1.4}$ dB

The values for $\alpha^{1.4}$ are given in Row 2 of Table 12 and the attenuation values, in decibels, are given in Row 3. In most calculations these values would be rounded to the nearest decibel.

Sound Attenuation of Plenums

When large values of sound attenuation are required, a sound absorption plenum will often be advantageous. The geometry of a sound-absorbing plenum is shown in Fig. 7.

A relation given in Equation (20) has been developed for approximating the acoustic attenuation of a plenum:

Plenum Attenuation $=$

$$10 \log_{10}\left[\cfrac{1}{S_E\left(\cfrac{\cos\theta}{2\pi d^2} + \cfrac{1-\alpha}{\alpha S_w}\right)}\right]\ \text{dB} \quad (20)$$

where:

$\alpha =$ sound absorption coefficient of the lining material (frequency dependent)

$S_E =$ plenum exit area, square feet

$S_w =$ plenum wall surface area, square feet

$d =$ distance between entrance and exit, feet (see Fig. 7)

$\theta =$ the angle d makes with the normal to the entrance opening, degrees (see Fig. 7).

For sound frequencies sufficiently high so that the wavelength of sound given by:

$$\text{wavelength of sound} = \frac{1128}{\text{frequency, Hz}}\text{feet} \quad (21)$$

is less than any of the plenum dimensions (height, width, or length), Equation (20) is accurate within a few decibels. At lower frequencies, when the wavelength of sound becomes greater than the plenum dimensions, Equation (20) is conservative and the actual attenuation exceeds the calculated value by 5 to 10 decibels.

Example No. 4

Determine the attenuation of a plenum as shown in Fig. 7. The exit duct is 10 inches in diameter, the surface area of the plenum is 42 square feet, the distance d is 2.44 feet, and the angle θ is 15 degrees. Compare the attenuation of a bare sheet-metal plenum with the plenum covered with one inch of glass fiber.

Solution: The attenuation, in octave bands, is computed from Equation (20) with the following constants:

$$S_E = \pi \left(\frac{5}{12} \right)^2 = 0.55 \text{ ft}^2$$

$$d^2 = 5.95 \text{ ft}^2$$

$$\cos \theta = 0.966$$

$$S_w = 42 \text{ ft}^2$$

The values for the absorption coefficient, α, of sheet-metal panels are listed in Row 1 of Table 13. These values are taken from Table 2. Note that α for 63 Hz can be considered the same as that for 125 Hz. The calculated values for $1 - \alpha$ are given in Row 2; the values of $\dfrac{1-\alpha}{\alpha S_w}$ are given in Row 4. The value for the argument,

$$\frac{\cos \theta}{2\pi d^2} + \frac{\alpha S_w}{1 - \alpha}$$

from the log quantity of Equation (20) is given in Row 5. The computed values for the attenuation of the plenum as computed from Equation (20) are given in Row 6. These are the values for the attenuation, in decibels, for the various

Table 13. Computation of Sound Attenuation for a Plenum. For Example No. 4

Row No.	Calculations	Octave-Band Center Frequency, Hz							
		63	125	250	500	1000	2000	4000	8000
1	Absorption Coefficient, α, for Sheet-metal Panel.	.05	.05	.04	.03	.03	.02	.01	.01
2	Calculation of $1 - \alpha$	.95	.95	.96	.97	.97	.98	.99	.99
3	Calculation of $\dfrac{1-\alpha}{\alpha}$	19.00	19.00	24.00	32.33	32.33	49.0	99.0	99.0
4	Calculation of $\dfrac{1-\alpha}{\alpha S_w}$	.452	.452	.571	.769	.769	1.167	2.357	2.357
5	$\dfrac{\cos \theta}{2\pi d^2} + \dfrac{1-\alpha}{\alpha S_w}$	.478	.478	.597	.795	.795	1.193	2.383	2.383
6	Attenuation, dB Equation (20)	5.8	5.8	4.8	3.6	3.6	1.8	0 (−1.17)	0 (−1.17)
7	Absorption Coefficient, α for 1-inch Glass Fiber.	.05	.06	.20	.65	.90	.95	.98	1.00
8	Calculation of $1 - \alpha$	.95	.94	.80	.35	.10	.05	.02	.00
9	Calculation of $\dfrac{1-\alpha}{\alpha}$	19.00	15.70	4.00	.54	.11	.053	.020	.00
10	Calculation of $\dfrac{1-\alpha}{\alpha S_w}$	.452	.374	.095	.013	.0026	.0013	.00048	.00
11	$\dfrac{\cos \theta}{2\pi d^2} + \dfrac{1-\alpha}{\alpha S_w}$	.478	.400	.121	.039	.028	.0271	.0263	.0258
12	Attenuation, dB Equation (20)	5.8	6.6	11.8	16.7	18.0	18.3	18.4	18.5

octave bands. For the 4000 and 8000 Hz center-frequency octave bands the resulting attenuation (Row 6) is calculated to be −1.17 dB, which would indicate that sound is generated. Since this violates the definition of attenuation of a plenum, as given by Equation (20), which does not include a term for noise generation, noise cannot be generated and the true attenuation is zero. Thus, a zero value for the attenuation is recorded in Row 6, for the 4000 and 8000 Hz frequencies.

To determine the attenuation of the plenum with the inside surfaces treated with one inch of glass fiber, the calculations are repeated with the absorption coefficients for the glass-fiber treatment as given in Row 7 of Table 13. Those values must be obtained from the supplier of the acoustic absorption material. Values greater than $\alpha = 1.00$ should *not* be used for this calculation because the complete surface area is to be treated and values of α greater than 1.00 are obtained from a sample in a test room. The usual procedure is to use a value of 1.00 for any absorption coefficients for materials which are reported greater than 1.00. This is the case for the value of α, for the 8000-Hz octave band in this example.

The steps shown in Rows 8, 9, 10, and 11 are identical to the computation for the sheet-metal plenum without acoustic treatment. The final values for attenuation are shown in Row 12 which are octave-band values of the plenum attenuation, in decibels. By comparison of Rows 6 and 12, the addition of the sound treatment is most effective in octave bands above 250 Hz. The attenuation of the plenum was not changed in the 63-Hz and 125-Hz octave bands. To increase attenuation at the lower octave bands, a material which has higher values for the absorption coefficient at low frequency must be used, or angle θ must be made larger.

Duct Lining and Elbows

An elbow in a lined duct can improve the sound attenuation. If the wavelength, as calculated by Equation (21), is less than the width of the duct (Fig. 6, dimension D), the sound attenuation is improved as compared to a straight duct of equivalent length. On the other hand, if the wavelength is greater than the duct width, the sound attenuation is not improved. The sound attenuation of an elbow involves the following:

1. Only the lining on the *sides* of the duct is effective in attenuating the sound level, as shown in Fig. 6.
2. The attenuation of the elbow should be added to the attenuation calculated for lengths of lined duct without elbows as calculated by Equation (19).
3. For best results, the sides of the duct should be lined, both before *and* after the elbow, for a length of at least two duct widths. This length is based on a lining thickness of at least 10 percent of the duct width.
4. If the duct is lined only before or after the elbow, there is still some gain in attenuation, rows (B) and (C) of Table 11.
5. The attenuation which can be credited to the duct lining is the difference between the total attenuation in rows (B), (C), or (D), of Table 11, and the attenuation of the unlined elbow Row (A).
6. The listed increases in elbow attenuation are obtained only with duct linings, not with sound traps or lined plenums.

Open-End Reflection Loss

The sudden expansion of the fluid at the open end of a duct into a room or work space will cause some of the sound passing through the duct to reflect back into the system at the opening plane of the duct. Because of this phenomenon, only a portion of the sound energy is radiated from an open duct; the remaining energy is reflected. This effect is most pronounced at low frequencies, but is negligible at high frequencies. In the duct system analysis this result can be treated as an end reflection loss as given as a function of duct size and frequency in Table 14.

Air-Flow Noise

The fan is the major noise source in an air supply system, but as mentioned earlier in this chapter, air-flow interaction with other elements in the system can be of significance. Turbulence in the flow is of major importance and in extreme cases of flow separation, considerable sound power can be produced throughout the system. Estimates can be made for the sound-power generation of typical duct system elements from the following relation:

$$L_W = F + G + H \text{ decibels} \qquad (22)$$

where:

Table 14.   Duct End Reflection Loss *†

Duct Dia, inches	Duct Size, sq in.	Octave-Band Center Frequency, Hz							
		63	125	250	500	1000	2000	4000	8000
		Reflection Loss, dB							
5	25	17	12	8	4	1	0	0	0
10	100	12	8	4	1	0	0	0	0
20	400	8	4	1	0	0	0	0	0
40	1600	4	1	0	0	0	0	0	0
80	6400	1	0	0	0	0	0	0	0

*Applies to ducts terminating flush with wall or ceiling and several duct diameters from other room surfaces. If closer to other surfaces. use entry for next larger duct.

† Reprinted by permission from *ASHRAE Handbook and Product Directory*, 1976, Table 11.

L_w = octave-band sound-power level, in decibels, re 10^{-12} watts.

F = *spectrum function* determined from flow characteristics, in decibels.

G = *velocity function* which accounts for the flow velocity through the duct element, in decibels.

H = *correction function* for the octave band of interest.

The values of the spectrum function, F, are determined from a nondimensional flow parameter called the *Strouhal Number,* computed from,

$$S_t = \frac{5fD}{V} \qquad (23)$$

where:

f = the octave-band center frequency, in Hertz.

D = the duct diameter, in inches, or for rectangular ducts, $D = \sqrt{(4/\pi)(area)}$

V = average air-flow velocity in the duct, in feet per minute.

The value of the *spectrum function, F,* in Equation (22) can be determined for elbows and branch take-offs from Figs. 8, 9, and 10. In all of these figures, the *Strouhal Number, S_t,* must be computed from Equation (23) for entry to the figures.

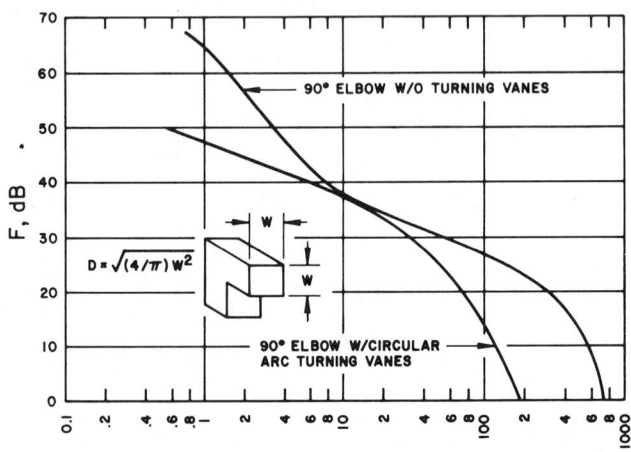

DIMENSIONLESS STROUHAL NUMBER S_T

Fig. 8. Spectrum function, *F*, for square cross-section, 90-degree elbows for use in Equation (22). Reprinted by permission of *ASHRAE Handbook and Product Directory*, 1976, Fig. 11.

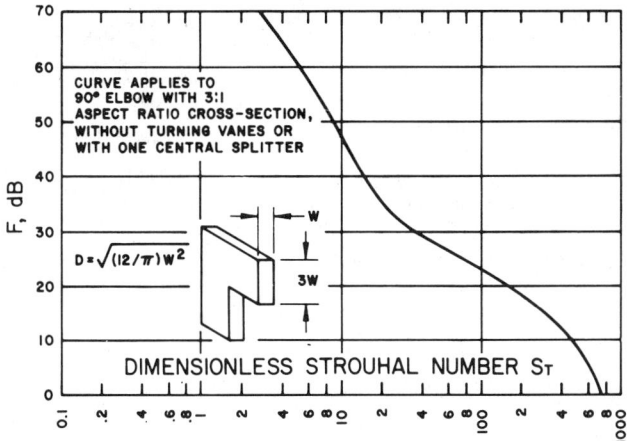

Fig. 9. Spectrum function, F, for rectangular cross-section, 90-degree elbows with 3 to 1 aspect ratio for use in Equation (22). Reprinted by permission of *ASHRAE Handbook and Product Directory*, 1976, Fig. 12.

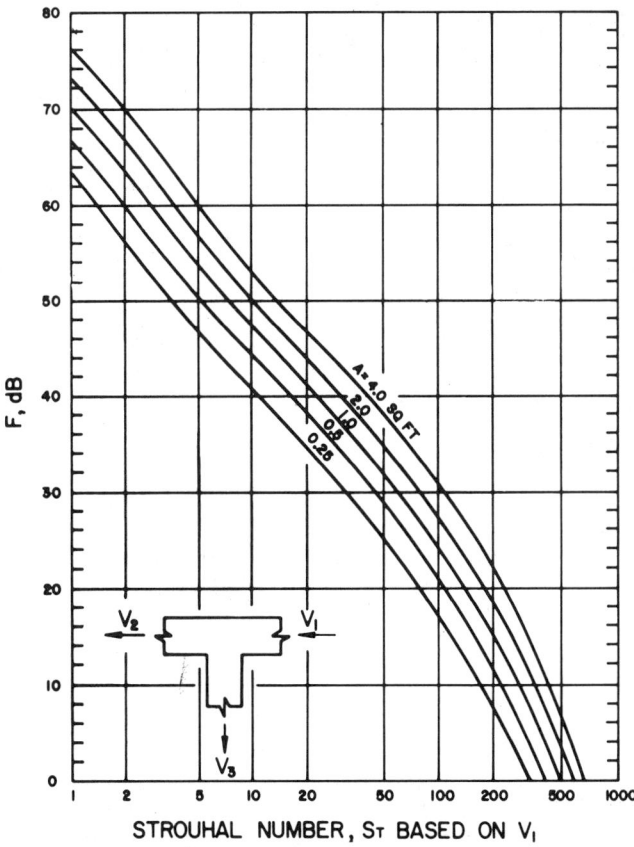

Fig. 10. Spectrum function, F, for 90-degree branch take-offs for use in Equation (22). Reprinted by permission of *ASHRAE Handbook and Product Directory*, 1976, Fig. 15.

Values for the *velocity function, G,* can be determined from Figs. 11 and 12 for elbows and branch take-offs as a function of the average flow velocity in the duct element.

The values for the octave-band width *Correction Function, H,* are obtained from Table 15. The values of H are not a function of the duct element, but rather only depend on the octave band of interest.

The effect of flow-generated sound power for 90-degree elbows is easily seen in Fig. 11. For a doubling of the average flow velocity in the duct element, the sound power increases approximately 12 dB.

Two possibilities exist for controlling noise due to this effect:

1. Size the duct elements to minimize flow velocity and therefore obtain minimum flow noise source generation, or
2. Utilize high-velocity systems with additional silencers to attenuate flow noise. There is a limitation with the second method; the air flow through silencers will also generate noise and thus reduce the attenuation which can be achieved. This will be treated in the next section.

Example No. 5

A 36 by 24-inch rectangular 90-degree elbow without turning vanes handles a flow rate of 8000 cfm air. Determine the sound power generated by the flow in this elbow.

Step 1. The equivalent duct diameter is computed from:

$$D = \left(\frac{4}{\pi} \times 24 \times 36\right)^{1/2} = 33.2 \text{ in.}$$

The average velocity in the duct is:

$$V_{avg} = \frac{Q}{A}$$

$$V_{avg} = \frac{8000}{2 \times 3} = 1333 \text{ fpm.}$$

Using Equation (23), the Strouhal Number is computed for each octave band. Equation (23) becomes:

$$S_t = \frac{5D}{V} f = 0.125 \, f$$

These values are computed and listed in separate rows in Table 16.

Step 2. The spectrum function, F, is obtained from Fig. 8 for square cross-section, 90-degree elbows without turning vanes. This condition is closer to the actual situation than Fig. 9. The values for the spectrum function are given in Row 2 of Table 16.

Step 3. The values for the velocity function, G, are obtained from Fig. 11 for an average velocity of 1333 fpm and a cross-sectional area of 6 square feet. These values are given in Row 3 of Table 16.

Step 4. The values for the octave-band-width function, H, are obtained from Table 15 and listed in Row 4 of Table 16.

Step 5. The sound-power levels for the various octave bands are obtained by summing $F + G + H$, which is the sum of Rows 2,

Table 15. Octave-Band Width Correction Function, H, for Equation (22) *

Octave-Band Center Frequency, hertz	H, dB
63	16
125	19
250	22
500	25
1000	28
2000	31
4000	34
8000	37

* Reprinted by permission from *ASHRAE Handbook and Product Directory,* 1976, Table 21.

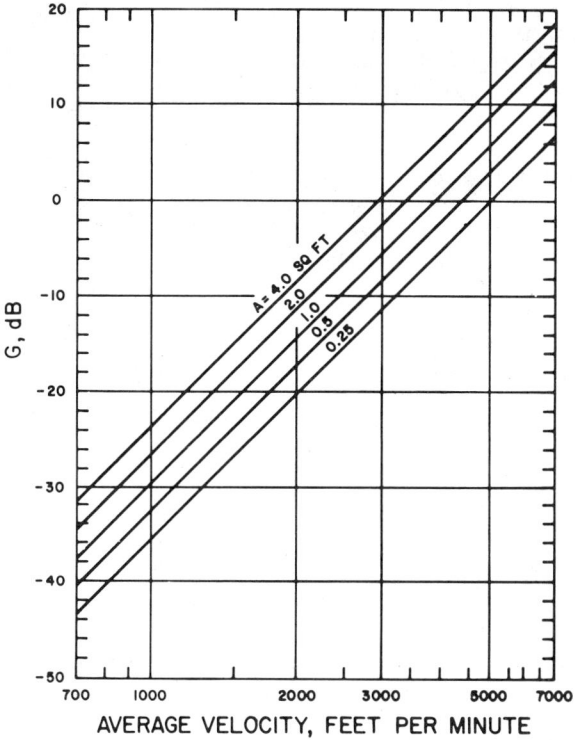

Fig. 11. Velocity function, G, for 90-degree elbows for use in Equation (22). Reprinted by permission of *ASHRAE Handbook and Product Directory*, 1976, Fig. 14.

Table 16. Sound-Power Level Generated by a 90° Elbow, From Example No. 5

Row No.	Calculations	Octave-Band Center Frequency, Hz							
		63	125	250	500	1000	2000	4000	8000
1	S_t [Equation (23)]	7.8	15.5	31	62	125	250	500	1000
2	Spectrum Function, F, from Fig. 8	48	35	33	30	26	20	10	0
3	Velocity Function, G, from Fig. 11	−17	−17	−17	−17	−17	−17	−17	−17
4	Octave-Band Width Function, H. Table 15	16	19	22	25	28	31	34	37
5	Sound-Power Levels, dB, $F + G + H$. Sum of rows 2, 3, and 4	47	37	38	38	37	34	27	20

3, and 4 of Table 16. The result is given in Row 5. The values are the sound-power levels, in decibels, generated by the air flow in the elbow.

Example No. 6

A 36 by 24-inch air duct has a square-edged 90° branch take-off. The branch take-off is 18 by 12 inches. If the upstream main-duct flow rate is 8000 cfm and the branch flow rate is 1000 cfm, determine the octave-band sound-power level downstream of the take-off in the main duct and downstream of the branch take-off.

Step 1. Compute the main air-duct equivalent diameter:

$$D=\left(\frac{4}{\pi}\times36\times24\right)^{1/2}=33.2 \text{ in.}$$

Step 2. Compute the Strouhal Number for each octave band. First, the upstream average velocity is computed from:

$$V_{avg}=\frac{8000}{3\times2}=1333 \text{ fpm}$$

The Strouhal Number is computed for each octave-band center frequency, from Equation (23),

$$S_t=\frac{5fD}{V}=\frac{(5\times33.2)}{1333}=0.125 f$$

These values are recorded in Row 1 of Table 17.

Step 3. Determine the spectrum function, F, from Fig. 10, for the *branch take-off*, using the area $A=12\times18/144=1.5$ square feet. The values are recorded on Row 2 of Table 17.

Step 4. Determine the velocity function, G, from the left-hand plot in Fig. 12. The downstream main-duct velocity is $7000/6=1167$ fpm and the branch take-off velocity is $1000/1.5=667$ fpm. The value for G is found to be -8 dB and is recorded in Row 3 of Table 17.

Table 17. Sound Power Level Generated by a Branch Take-off. For Example No. 6

Row No.	Calculations	Octave-Band Center Frequency, Hz							
		63	125	250	500	1000	2000	4000	8000
1	Strouhal Number, S_t, Equation (9-8).	7.8	15.5	31	62	125	250	500	1000
2	Spectrum Function, F, for Branch, Fig. 9-11.	51	45	38	31	23	13	2	—
3	Velocity Function, G, Fig. 9-13.	−8	−8	−8	−8	−8	−8	−8	−8
4	Octave-Band Width Correction Function, H, Table 9-12.	16	19	22	25	28	31	34	37
5	Sound-Power Level, L_W dB, for the Branch. Sum values in rows 2, 3, and 4.	59	56	52	48	43	36	28	29
6	Spectrum Function, F, for Main Duct, Fig. 9-11.	57	51	44	37	30	20	9	—
7	Sound-Power Level, L_W dB, for the Main Duct. Sum values in rows 3, 4, and 6.	65	62	58	54	50	43	35	29

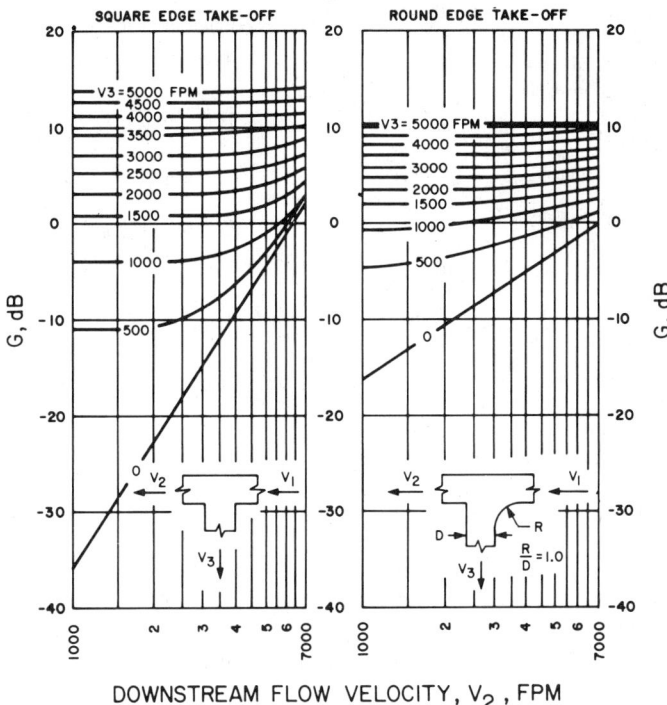

Fig. 12. Velocity function, *G*, for 90-degree branch take-offs for use in Equation (22).
Reprinted by permission of *ASHRAE Handbook and Product Directory*, 1976, Fig. 16.

Step 5. Values for the octave-band-width correction function, *H,* are obtained from Table 15 and are recorded in Row 4 of Table 17.

Step 6. The octave-band sound power levels *Lw* dB are obtained by the sum of *F*+*G*+*H*; Row 2, Row 3, and Row 4 of Table 17. These values are given in Row 5 of the table.

Step 7. The values for the spectrum function, *F,* for the main duct are obtained from Fig. 10 using *A*=6 square feet. The curves in the figure are extrapolated to give values for 6 square feet by drawing a line parallel to the 4-square-feet line and a distance one-half of the spacing of the lines in the figure. The values for *F* are given in Row 6 of Table 17.

Step 8. The values for the velocity function, *G,* and the correction function, *H,* are

the same as given in Rows 3 and 4, respectively, in Table 17.

Step 9. The sound-power levels, *Lw* dB, are obtained by the sum of *F* (main duct)+*G*+*H* which are the sum of Row 3, 4, and 6. The resulting sound-power level generated in the main duct is given in Row 7.

The values in Rows 5 and 7 of Table 17 are the octave-band sound-power levels, in decibels, generated by the air flow in the take-off and main branch, respectively.

Flow Noise Generation of Silencers

Duct silencers are intended to attenuate the sound which has been generated upstream of the silencer unit. In order to obtain maximum sound attenuation in a silencer, the flow is disturbed by baffles, liners, splitters, etc. Air flow over the sound attenuation surfaces can act as a

flow noise generator; therefore a silencer can itself be a generator of sound power. The self-generated sound power of a silencer with a center body can be predicted from Equation (24).

$$L_W = -145 + 55 \log_{10} V + 10 \log_{10} A$$
$$-45 \log_{10} \frac{P}{100} - 20 \log_{10} \frac{460+T}{530} \qquad (24)$$

where:

L_W = Octave-band sound-power level, dB, re 10^{-12} watts.

V = Flow velocity, in feet per minute.

A = Cross-sectional area, in square feet.

P = Percentage of open cross-sectional area in the silencer's open area divided by the total cross-sectional area of the silencer.

T = Temperature of the air, in °F.

Equation (24) predicts the same sound-power level for all octave bands, which agrees with experiments. This equation may be treated as a method of predicting the maximum sound-power level of silencers without regard to flow stream-lining the unit. Streamlining the baffles in the silencer may also reduce the self-generated sound power by as much as 10 dB.

When selecting silencers one should determine if the source attenuation provided by the supplier includes the self noise of the silencer. If the silencer was tested under no-flow conditions, or if the tests were made at a flow velocity different from the application, Equation (24) may be used to estimate the effects of flow on the silencer's acoustic performance.

Example No. 7

Determine the sound power level generated by a triangular duct silencer in a 36- by 36-inch duct with a flow of 8600 cfm of 75 °F air. The silencer has an open area of 80 percent.

Solution: Compute L_W from Equation (24).

Step 1. Determine the flow velocity in the silencer.

$$V = 8600/(3 \times 3) = 956 \text{ fpm}$$

Calculate the term $55 \log_{10} V$ from Equation (24):

$$55 \log_{10} (956) = 55(2.98) = 164 \text{ dB}$$

Step 2. Compute $10 \log_{10} A$, where $A = 9$ sq ft:

$$10 \log_{10} (9) = 10 (.95) = 9.5 \text{ dB}$$

Step 3. Compute the term $-45 \log_{10} \frac{P}{100}$, where $P = 80$ percent:

$$-45 \log_{10} \left(\frac{80}{100} \right) = -45 \log_{10} (.8)$$
$$= -45 (-.097)$$
$$= 4.4 \text{ dB}$$

Step 4. Compute the term $-20 \log_{10} \frac{460+T}{530}$, where $T = 75$ °F:

$$-20 \log_{10} \left(\frac{460+75}{530} \right) = -20 \log_{10} (1.01)$$
$$= -.08 \text{ dB}$$

Step 5. Compute the sound-power level generated by the air flow in the silencer from the above calculated values.

$$L_W = -145 + 164 + 9.5 + 4.4 - .08 \text{ dB}$$
$$L_W = 33 \text{ dB}$$

The value of $L_W = 33$ dB is the sound-power level generated by the air flow through the silencer for each octave band.

Sound Transmission Through Duct Walls

In some instances the sound produced inside a duct system is transmitted through the walls of the duct into a building space causing a noise problem. To correct this condition, the sound transmission loss of the duct walls must be considered. For rectangular-duct cross section the flat surfaces are treated as flat plates and the transmission loss is determined from Equation (25).

$$TL = 20 \log_{10} W + 20 \log_{10} f - 33 \text{ dB} \qquad (25)$$

where:

TL = duct wall transmission loss, dB

W = weight, in pounds per square foot, of the duct wall panel

f = frequency, Hz

The transmission loss, TL, is computed for the octave-band center frequencies and used to determine the reduction in sound power level from the inside to the outside of the duct.

Ducts with round cross-sections behave differently from rectangular cross-sections at low frequencies because of the added stiffness due to curvature. At high frequencies, a round duct has transmission loss much like a rectangular duct

wall. The frequency which separates the low- from the high-frequency region for a circular cross-section duct is called the ring frequency, f_r, given by:

$$f_r = \frac{c_i}{\pi D} \text{ Hz} \qquad (26)$$

where:

$f_r =$ the ring frequency, Hz
$c_i =$ speed of sound in the duct wall material, m/s (5100 m/s for sheet metal)
$D =$ nominal duct diameter, meters

In the frequency region, at the ring frequency or below it, the transmission loss is determined by first finding the nominal transmission loss from:

$$TL_{nominal} = 10 \log_{10}\left(\frac{t}{D}\right) + 50 \text{ dB} \qquad (27)$$

where:

$TL_{nominal} =$ nominal transmission loss at or below the ring frequency for a circular duct, dB
$t =$ wall thickness of the duct
$D =$ nominal duct diameter

t and D must be in the same units.

The nominal transmission loss is corrected for frequencies equal to and below the ring frequency by the values given in Table 18. For frequencies greater than the ring frequency for a circular duct, the transmission loss is computed from Equation (25).

Example No. 8

Compute the transmission loss for an 18-inch diameter sheet metal duct made from 15-gage material which is .0673-inch thick.

Step 1. Compute the ring frequency for the duct. The duct diameter is 18 inches or 0.4572 meter. The speed of sound, c_i, in the duct wall material is 5100 m/s, therefore:

$$f_r = \frac{c_i}{\pi D} = \frac{5100}{\pi \times 0.4572} = 3550 \text{ Hz}$$

From Table 1, this frequency is within the octave band with center frequency of 4000 Hz. This means the circular duct analysis applies for 4000 Hz and below, and the flat plate analysis applies above 4000 Hz.

Step 2. Compute the nominal transmission loss of the duct from Equation (27):

$$\frac{t}{D} = \frac{.0673}{18} = .0037$$

$$TL_{nominal} = 10 \log_{10}\left(\frac{t}{D}\right) + 50$$
$$= 10 \log_{10}(.0037) + 50$$
$$= -24.3 + 50$$
$$= 25.7 \text{ dB}$$

This value is entered in Row 1, Table 19, for every octave band of 4000 Hz or less.

Step 3. Enter the correction function from Table 18 for all octave bands starting with 1f_r equal to 4000 Hz, ½f_r equal to 2000 Hz, etc. These corrections are given in Row 2.

Step 4. Determine the transmission loss for the duct. For 4000 Hz, and below, add the values in Row 1 and Row 2 of Table 19. For the 8000-Hz octave band, which is above the ring frequency, use Equation (25) to compute the transmission loss for standard 15-gage sheet metal;

Table 18. Corrections to the Nominal Transmission Loss for Round Ducts for Frequencies which are Equal To or Less Than the Ring Frequency, f_r

Frequency, f	$\frac{1}{64} f_r$	$\frac{1}{32} f_r$	$\frac{1}{16} f_r$	$\frac{1}{8} f_r$	$\frac{1}{4} f_r$	$\frac{1}{2} f_r$	$1 f_r$
Correction, dB	- 7	- 6	- 5	- 4	- 3	- 2	- 4

Table 19.   Calculation of Duct Wall Transmission-Loss for Example No. 8

Calculations	Octave-Band Center Frequency, Hz							
	63	125	250	500	1000	2000	4000	8000
1. Nominal Transmission Loss, dB. Equation (27).	25.7	25.7	25.7	25.7	25.7	25.7	25.7	- - - -
2. Correction Factors from Table 18 , dB	-7	-6	-5	-4	-3	-2	-4	- - - -
3. Transmission Loss for the Duct Wall, dB.	18.7	19.7	20.7	21.7	22.7	23.7	21.7	54.1

$W = 2.8125$ pounds per square foot and $f = 8000$ Hz.

$$TL = 20 \log_{10} W + 20 \log_{10} f - 33$$
$$= 20 \log_{10} (2.8125) + 20 \log_{10} (8000)$$
$$- 33$$
$$= 9.0 + 78.1 - 33$$
$$= 54.1 \text{ dB}$$

This value is given in Row 3 for the 8000-Hz octave band.

The transmission loss values given by Row 3 of Table 19 can be used to determine the sound-power levels outside a duct if the sound power levels in octave bands inside the duct are known, simply by subtracting the duct transmission loss. The sound pressure levels in a space can thus be computed using Equation (14) with the sound power levels, L_w, outside the duct as the source in the space.

Calculation of Sound Levels in Ducted Systems

Determination of sound levels from ducted systems requires the use of all previous material presented in this section. In the following example the procedure for predicting room space sound levels from a ducted air supply system will be illustrated.

Example No. 9

Compute the sound pressure levels, in octave bands, for the listener location for the system shown in Figure 13. The supply fan is of the radial-blade type with 16 blades operating at 2 inches of water pressure drop and 500 rpm, to supply 10,000 cfm of air.

Step 1. The first step is to obtain octave-band sound-power levels for the fan outlet from the manufacturer. If these values are not available they may be estimated by using the procedure of Example No. 2. The sound power level for the fan is given in Row 1 of Table 20.

Step 2. Determine the attenuation from the fan to the first branch take-off. The attenuation, in dB/ft, is obtained from Table 9, times the length (15 feet) gives the attenuation values in Row 2 of Table 20.

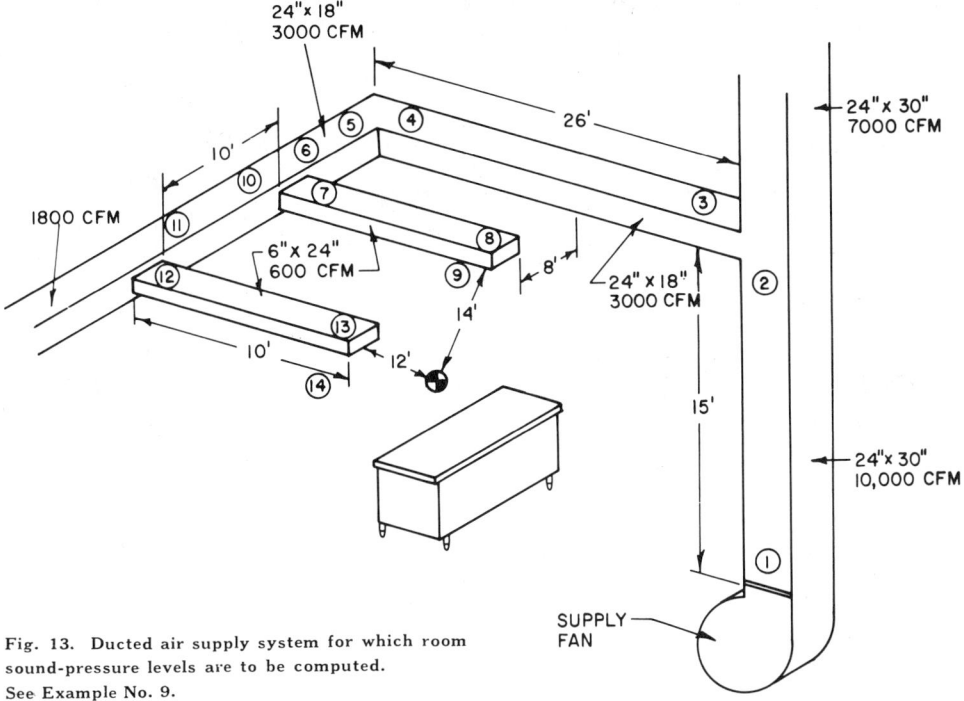

Fig. 13. Ducted air supply system for which room sound-pressure levels are to be computed. See Example No. 9.

Step 3. Determine the attenuation of the branch take-off. The take-off has an area of $24'' \times 18'' = 432$ sq in. and the continuing duct has an area of $24'' \times 30'' = 720$ sq in. The percent area of the take-off to the *total* area of duct after the take-off is $432/1152 = 40\%$ (rounded). From Table 7 the reduction at the take-off is 4 dB as entered in Row 3 of Table 20.

Step 4. Compute the sound power levels at location 3. This is done by adding Rows 1, 2, and 3 with the result given in Row 4.

Step 5. The sound power level of the sound generated by air flow through the first branch take-off must be computed. The procedure is identical to that of Example No. 6. For this example the resulting sound power levels were computed to be as follows: 73, 69, 65, 64, 57, 53, 48, and 37 dB for the octave bands from 63 Hz to

8000 Hz. In general, these values must be combined logarithmically with the values given in Row 4 to obtain the total sound-power levels at location 3 which are due to the fan sound-power transmitted through the duct systems, plus the air-flow generated noise at the branch take-off. In this example, the air-flow generated sound power is 10 dB (or greater) less than the sound power levels given in Row 4, therefore, by use of Equation (8), the sound power level of the combined sources is equal to the *larger* values which are given by Row 4. Thus, the sound power level of the branch take-off can be neglected in this example.* However, this may not always be the case in general.

Step 6. Compute the attenuation from ③ to ④ as shown in Row 5.

* See previous discussion of the addition of decibels at the beginning of this section.

Table 20.   Calculation of Ducted System Sound-Power Levels for Example No. 9

Calculation	Octave-Band Center Frequency, Hz							
	63	125	250	500	1000	2000	4000	8000
1. Fan sound power level from manufacturer, dB	91	96	86	86	81	76	73	72
2. Attenuation from ① to ② dB/ft =	.20	.20	.10	.05	.05	.05	.05	.05
Attenuation, dB =	-3.0	-3.0	-1.5	-.7	-.7	-.7	-.7	-.7
3. Attenuation from ② to ③, dB	-4.0	-4.0	-4.0	-4.0	-4.0	-4.0	-4.0	-4.0
4. Sound power level at ③, dB	84.0	89.0	80.5	81.3	76.3	71.3	68.3	67.3
5. Attenuation from ③ to ④ dB/ft =	.2	.2	.1	.05	.05	.05	.05	.05
Attenuation, dB =	-5.2	-5.2	-2.6	-1.3	-1.3	-1.3	-1.3	-1.3
6. Attenuation of elbow at ④, dB	-0.0	-1.0	-5.0	-7.0	-5.0	-3.0	-3.0	-3.0
7. Sound power level at ⑤, row 4 plus rows 5 and 6	78.8	82.8	72.9	73.0	70.0	67.0	64.0	63.0
8. Attenuation from ⑤ to ⑥ dB/ft =	.2	.2	.1	.05	.05	.05	.05	.05
Attenuation, dB =	-1.6	-1.6	-.8	-.4	-.4	-.4	-.4	-.4
9. Attenuation of take-off at ⑦, dB	-6.0	-6.0	-6.0	-6.0	-6.0	-6.0	-6.0	-6.0
10. Sound power level at ⑦, dB. row 7 plus rows 8 and 9	71.2	75.2	66.1	66.6	63.6	60.6	57.6	56.6
11. Attenuation from ⑦ to ⑧ dB/ft^2 =	.2	.2	.15	.1	.1	.1	.1	.1
Attenuation, dB =	-2.0	-2.0	-1.5	-1.0	-1.0	-1.0	-1.0	-1.0
12. End reflection loss at ⑧, dB	-12.0	-8.0	-4.0	-1.0	0	0	0	0
13. Sound power level at ⑨, dB. row 10 plus rows 11 and 12	57.2	65.2	60.6	64.6	62.6	59.6	56.6	55.6

Step 7. Obtain the attenuation of the square elbow at ④ from Table 11. Use values for a 20″ duct width with no lining and record these values in Row 6.

Step 8. Determine the sound power level at location ⑤ by adding values from Rows 4, 5, and 6. The result is given in Row 7.

Table 20 (Cont.). Calculation of Ducted System Sound-Power Levels for Example No. 9

14. Attenuation from ⑥ to ⑩ , dB	−1.0	−1.0	−1.0	−1.0	−1.0	−1.0	−1.0	−1.0
15. Attenuation from ⑩ to ⑪ , dB/ft =	.2	.2	.1	.05	.05	.05	.05	.05
Attenuation, dB =	−2.0	−2.0	−1.0	−.5	−.5	−.5	.5	−.5
16. Attenuation from ⑪ to ⑫ , dB =	−6.0	−6.0	−6.0	−6.0	−6.0	−6.0	−6.0	−6.0
17. Attenuation from ⑫ to ⑬ , dB	−2.0	−2.0	−1.5	−1.0	−1.0	−1.0	−1.0	−1.0
18. End reflection loss at 13, dB	−12.0	−8.0	−4.0	−1.0	0	0	0	0
19. Sound power level at ⑭ , row 7 plus rows 8, 14, 15, 16, 17, and 18	54.2	62.2	58.6	63.1	61.1	58.1	55.1	54.1

Step 9. Calculate the sound power level of the air-flow noise generated by the square elbow at ④. This is done by a similar calculation to that shown in Example No. 5. For this example the resulting sound power levels are less by 10 dB (or more) than the sound power from the fan as given in Row 7 and will not add to the levels in Row 7.

Step 10. Compute the attenuation from ⑤ to ⑥ as shown in Row 8. The attenuation values in dB/ft are obtained from Table 9.

Step 11. Obtain attenuation of the branch take-off at ⑦ from Table 7. The area of the 6"×24" duct=144 sq in.; the 24"×18" duct=432 sq in. Therefore, the percent area of the take-off to the total area after the branch take-off is 144/576=25%. From Table 7 for 25 percent, interpolate from the table to obtain 6 dB attenuation, as given in Row 9.

Step 12. Calculate the sound power level at location ⑦ by adding the values in Rows 7, 8, and 9.

Step 13. The sound power level for flow generated is due to flow at the branch take-off at location ⑦. For this example the sound power levels will be 10 dB (or more) below the levels given by Row 10 and will not add to the levels in Row 10.

Step 14. Determine the attenuation from ⑦ to ⑧. Use the attenuation values in dB/ft for a small duct from Table 9. The attenuation is given in Row 11.

Step 15. Obtain the open end reflection attenuation for the duct at location 8 from Table 14. Use the 100 sq in. duct size values from Table 14, as listed in Row 12.

Step 16. The sound power level at the diffuser at ⑨ is calculated by adding values in Rows 10, 11, and 12 with the result in Row 13.

At location ⑨ the sound power level generated by air flow through the diffuser must be obtained from manufacturer's data for the required air flow. The sound power levels in decibels must be combined logarithmically with the sound power levels in Row 13 using Equation (8). For this example it will be assumed that the sound power levels due to air flow through the diffuser are

less than 40 decibels for every octave-band, therefore the air-flow noise can be neglected compared to the levels given by Row 13.

The sound power levels for the second diffuser located at ⑭ in Figure 13 will be computed by continuing the calculations as follows:

Step 17. The attenuation from ⑥ to ⑩ is obtained from Table 7 using the same areas as were computed in Step 11. The area of the continuing duct at ⑩ is 576 sq in., the area of duct at ⑦ is 144 sq in., therefore the percent area at ⑩ of the total continuing duct after the take-off (576+144 sq in.) is 576/720=80%. From Table 7 the attenuation is 1 dB, as shown in Row 14.

Step 18. Compute the attenuation from ⑩ to ⑪, as shown in Row 15.

Step 19. Obtain the attenuation of the branch take-off at ⑫ from Table 7 with the percent area of the take-off equal to 25%, resulting in an attenuation of 6 dB by interpolation. The value is given in Row 16.

Step 20. Compute the attenuation from ⑫ to ⑬, as shown in Row 17.

Step 21. Obtain the end reflection loss from Table 14 for the 100-sq-in. duct size. These values are listed in Row 18.

Step 22. The octave-band sound-power levels at location ⑭ are obtained by summation of values in Rows 7, 8, 14, 15, 16, 17, and 18 with the result in Row 19.

The sound power level for the air-flow noise for the diffuser at location ⑭ must be determined and added logarithmically to the levels in Row 19. If this diffuser is identical to the one at location ⑨, the contribution to the sound power levels of Row 19 can be neglected.

Note in the example that the resulting sound power levels at the two diffusers are entirely due to the fan-generated noise which is attenuated through the duct system and air-flow generated noise is negligible. This may not always be true in a ducted system, therefore the flow-generated noise should always be computed and compared to the fan noise at locations of interest.

To compute the sound pressure levels in octave bands at the listener location in Figure 13, Equation (14) must be used for each of the two sources, which are the two diffusers (Locations ⑨ and ⑭), since the sound power level is not the same for each source. These calculations are described below and shown in Table 21.

Step 23. Determine the correction values to convert sound power level, L_w, at the source to sound pressure level, L_p, at the listener location, from Equation (14):

$$L_p = L_w + 10 \log_{10} \left[\frac{Q}{4\pi r_1^2} + \frac{4}{R} \right] + 10$$

For this example the room constant, R, must be determined for the space as illustrated by Example No. 1. For this case the same values for the room constant, R, will be used as in Example No. 1. These values are listed in Row 3 of Table 21. Note that the value for R for the 63-Hz band was computed as in Example No. 1. The value of Q, is determined from Figure 1, to be 2 for a source near the ceiling and $r_1 = 14$ ft, therefore

$$L_p = L_w + 10 \log_{10} \left[\frac{2}{4\pi(14)^2} + \frac{4}{R} \right] + 10$$

$$\text{correction term} = 10 \log_{10} \left[.0008 + \frac{4}{R} \right] + 10 \text{ dB}$$

The correction term for the 63-Hz band becomes

$$= 10 \log \left[.0008 + \frac{4}{243} \right] + 10$$
$$= -17.6 + 10$$
$$= -7.6 \text{ dB}$$

This value is given in Row 4 of Table 21 and subsequent values are computed in Row 4 in a similar manner.

Step 24. Determine the correction values for the source at location ⑭:

$$L_p = L_w + 10 \log_{10} \left[\frac{2}{4\pi(12)^2} + \frac{4}{R} \right] + 10$$

$$\text{correction} = 10 \log_{10} \left[.0011 + \frac{4}{R} \right] + 10 \text{ dB}$$

These values are given in Row 5 of Table 21.

Step 25. Calculate the sound pressure level at the

Table 21. Computation of Sound-Pressure Levels for Example No. 9

Computation	Octave Band Center Frequency, Hz							
	63	125	250	500	1000	2000	4000	8000
1. Sound power level, dB, at ⑨ . From row 13 Table 19	57.2	65.2	60.6	64.6	62.6	59.6	56.6	55.6
2. Sound power level dB, at ⑭. From row 19 Table 19	54.3	62.2	58.6	63.1	61.1	58.1	55.1	54.1
3. Room constant, R, ft² for space	243	312	342	438	458	548	631	743
4. Correction for source at ⑨ , dB	-7.6	-8.6	-9.0	-10.0	-10.2	-10.9	-11.4	-12.1
5. Correction for source at ⑭, dB	-7.5	-8.5	-8.9	-9.9	-10.1	-10.8	-11.3	-11.9
6. Sound pressure level due to source at ⑨, dB, row 1 plus row 4	49.6	56.6	51.6	54.6	52.4	48.7	45.2	43.5
7. Sound pressure level due to source at ⑭, dB. row 2 plus row 5	46.8	53.7	49.7	53.2	51.0	47.3	43.8	42.2
8. Sound pressure level at listener location, dB (rounded)	51	58	54	57	55	51	48	46

listener location due to source at ⑨. This is done by adding Rows 1 and 4 with the results in Row 6.

Step 26. Calculate the listener location sound-pressure level due to source at ⑭, by adding values in Rows 2 and 5. The results are in Row 7.

Step 27. Compute the total sound-pressure level at the listener location by combining logarithmically the values in Rows 6 and 7, using Equation (10). For the 63-Hz band;

$$L_p = 49.6 \text{ dB} + 46.8 \text{ dB}$$
$$= 10 \log_{10} (10^{4.96} + 10^{4.68})$$
$$= 10 \log_{10} (91201 + 47863)$$
$$= 10 \log_{10} (139064)$$
$$= 51 \text{ dB (rounded)}$$

The same procedure is used to obtain the remaining values in Row 8 of Table 21. The usual procedure is to perform calculations rounded to the nearest one-tenth decibel and then round the final sound-pressure levels to the nearest decibel, as in Row 8.

If the sound pressure levels from Row 8, Table 21 are plotted on the Noise Criteria curves, for example as in Figure 3, the resulting value will be NC-54 as determined by the noise criteria which has the highest penetration by the plotted sound-pressure levels. By referring to Table 2 of the Noise and Vibration Control section of this handbook, it is noted under the heading of "OFFICES" that an NC-54 is unacceptable for general open offices. An NC-54 is acceptable for tabulation and computation rooms where office machines are located. In order to reduce the sound pressure levels resulting from the duct system in this example, additional noise attenuation must be provided in the system. To achieve an NC-criteria of below NC-50, a section of lined duct will be sufficient. To calculate lined duct attenuation, see Example No. 2. If either the duct from ① to ② or the duct from ③ to ④ in Figure 13 were lined with a 1″ thickness of glass fiber, the resulting level at the listener location would be below NC-50.

CONTROL OF COOLING TOWER NOISE

Major sources of cooling tower noise are: fan, water, motor, speed-reducing gears or belt, and external sources.

Fan Noise

Sound level produced by a given type of cooling tower is primarily a function of total horsepower of the fans. For induced draft towers with axial flow fans, sound power (See Fig. 1) can be determined from

$$PWL = 95 + 10 \log \text{ fan hp dB}$$
$$\text{re } 10^{-12} \text{ watt} \qquad (1)$$

For forced draft centrifugal towers, sound power (See Fig. 2) is

$$PWL = 84 + 10 \log \text{ fan hp dB}$$
$$\text{re } 10^{-12} \text{ watt} \qquad (2)$$

Sound *power* spectrums from Fig. 1 and 2 and Eq. (1) and (2) can be converted by standard methods to average free field sound *pressure* levels at moderately large distance from the tower.

Sound pressure levels close to induced draft, axial flow towers at the fan discharge are

$$SPL = (105 + 10 \log hp - 3)$$
$$- 10 \log A \text{ dB} \qquad (3)$$

and at the air intake,

$$SPL = (105 + 10 \log hp - 6)$$
$$- 10 \log A \text{ dB} \qquad (4)$$

where A is the area, in square feet, of the discharge and intake, respectively.

Even though the equations and curves are based on field tests performed on a limited number of cooling towers, they serve as valuable design aids in the formative stages of design or when no other data are available. These data can be supplemented, before design is completed, by actual rating data supplied by the cooling tower manufacturer.

Equations (1) and (2) indicate that, for a given tower type, noise generation is a function of fan horsepower; also that, for a given horsepower, the forced draft centrifugal type is quieter than the axial flow induced draft type. Although, for a given duty, the centrifugal requires more horsepower than the axial flow tower, it will be found that, *for a given specific duty, the centrifugal, despite its higher horsepower, is quieter than the axial flow induced draft cooling tower.*

Air-cooled condensers circulate a larger quantity of air per ton than cooling towers. Thus, it can be expected that, for a given cooling capacity, their fans will be noisier than those of cooling towers. Evaporative condensers handle lesser air quantities than cooling towers and, for a given cooling capacity, their fans can be expected to be quieter. This advantage is partially offset because the circulating pump, usually mounted directly adjacent to an evaporative condenser, generates additional noise.

Fan vibrations can be transmitted to the equipment casing or structure. They can be isolated by high deflection springs.

Water Noise

The sound of water flowing through the tower and into the drain pan is of the intensity and general characteristics shown in Fig. 3. Air-borne water noise will ordinarily be a factor only in the high frequency end of the spectrum since its level in the lower frequencies is usually below that of the cooling tower fan. Where the fan is operated at reduced speed in order to achieve lower fan sound levels, the water noise may become audible. The impact of the water will be transmitted as a moderate structure-borne vibration from the tower into the building unless isolating pads are provided beneath the basin.

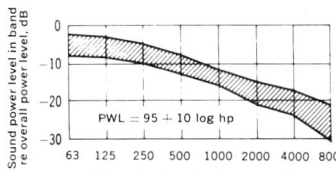

Fig. 1. (left) Sound spectrum for axial flow tower fans.

Fig. 2 (right) Sound spectrum for centrifugal fan towers.

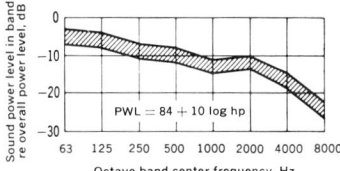

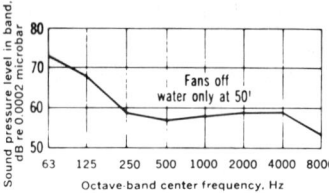

Fig. 3. Typical spectrum, tower water noise.

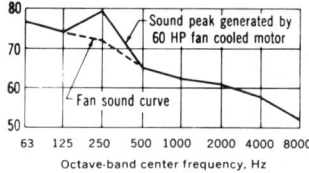

Fig. 4. Noise from tower fan motor.

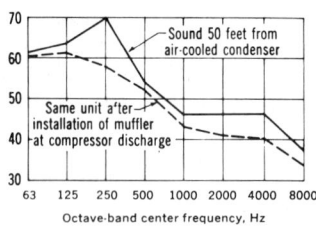

Fig. 5. How muffler reduces sound.

Drive Components

Fan-cooled electric motors can be a source of disturbing sound when large quantities of air are circulated over the motor. Fig. 4 indicates the sound measured at a tower compared with the manufacturer's published data for the tower. The peak in the third octave band was traced to the integral cooling fan for the 60-hp driving motor.

Noise from belt or gear drives is not ordinarily significant unless they are misaligned or damaged. The drives may be vibration generators and should be isolated with the fan and motor.

External Noise Sources

Both air-cooled and evaporative condensers are directly connected to refrigeration compressors. Fig. 5 shows sound levels measured at the side of an air-cooled condenser located approximately 150 ft from the refrigeration compressor it served. The high intensity sound in the third octave band was identified as being produced by pulsations being carried along the refrigeration pipe with the hot gas and radiating from the inlet manifold of the condenser. The lower curve shows measurements taken at the same position after the manufacturer's standard discharge muffler had been installed in the piping near the compressor, resulting in a significant reduction in the critical octave band.

Similar, although usually more moderate, conditions can result from sound generated by condenser water pumps and transmitted through the water or pipe wall.

Configuration Factors

Shape, size, and arrangements of major components all have a bearing on the sound characteristics of a tower. Fig. 6 shows adjustments to sound pressure levels that should be made for different cooling tower configurations of induced and forced draft fan towers.

Induced draft towers are constructed with the fan at the top, discharging vertically upward or, in the case of smaller units, at the sides, discharging horizontally. Air is drawn into louvers near the bottom or along the side of the tower casing. A higher intensity sound, especially at the blade passage frequencies, will be radiated from the fan discharge side of the tower than from either the louvered or encased sides. More sound will be radiated from sides containing louvered openings than from those that are completely encased.

It can be seen that, for an axial flow induced draft type tower, it is desirable to arrange the fan discharge so that it is not overlooked by windows or similar occupied areas and, if possible, to place the encased tower sides in a position where sound criteria are most critical.

Similar adjustments must be made for forced draft towers since these have the fans located along the sides of the casing and have no inlet louvers. As such, they ordinarily produce the highest noise intensity over the top with lesser sound levels on the sides facing the fans and are quietest along the cased sides.

Location

A cooling tower or condenser in an open field radiates sound in all directions. The far field sound levels previously referred to are based on such a condition. Intensity of tower sound will be reduced as it is separated by distance from the observer.

If the equipment is located sufficiently close to a wall or if it is placed within a court so that large, massive surfaces receive the sound waves and reflect them, an observer located opposite the reflecting surfaces wil hear higher sound intensities than he would if both he and the tower were located in a completely open area. For a single-wall surface, the sound levels

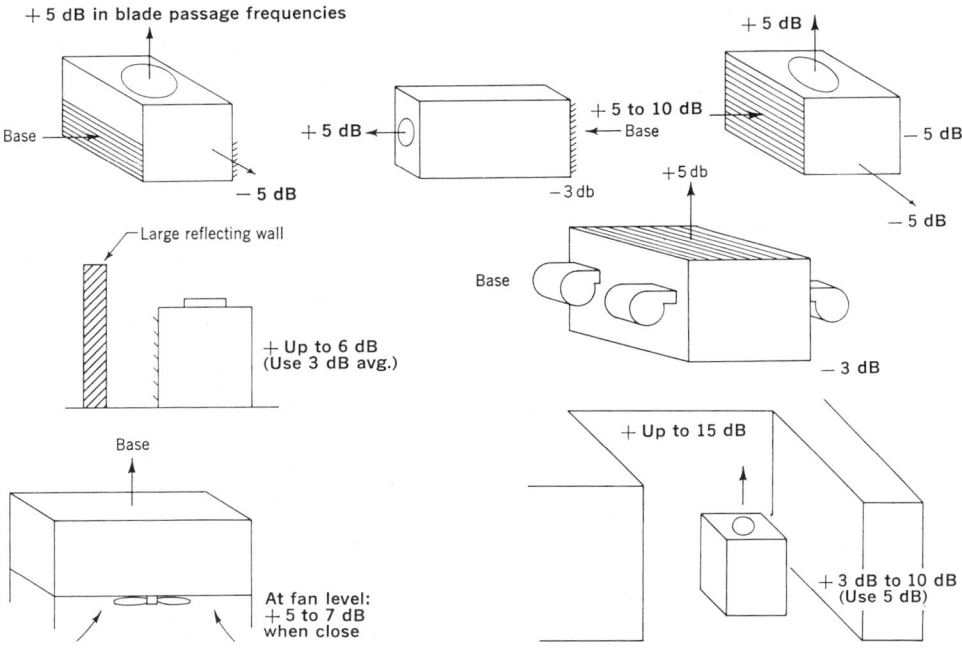

Fig. 6. Effect of tower configuration and location on base sound levels.

received by the observer will be increased as much as 6 dB with an average of approximately 3 dB, a good design value. Where the tower is located in a court, depending on how confined the enclosure, the sound levels can be expected to be anywhere from 3 dB to 10 dB, or more, higher than they would be in an open field; a 5-dB increase may be used for design. Baffle walls will reflect higher frequency sound if they are located close to the equipment, if they are large relative to the equipment dimensions and massive, and have no openings. Otherwise, they are useful primarily for their psychological effect in that they may conceal the sound source and confuse the observer.

Reducing Sound Generated

A given tower operating with lowered fan speed and reduced horsepower will be quieter than when operating at rated speed. The noise reduction is often greater than would be expected due to reduction in horsepower. This indicates that some effects of air velocity on noise generation tend to build up (and reduce) faster than the horsepower. The most significant noise reductions take place in octave bands where sound peaks or pure tones are fan-generated.

Half-speed Operation. When cooling tower or condenser is located where the chief complaints are likely to occur at night and where the nature of the cooling load is such that a reduction in capacity can be tolerated, operation of the cooling tower fans at part-speed during the night or other off-peak hours is an economical and very satisfactory method of achieving sizable reductions in tower sound levels and power consumption. Fig. 7 shows a representative axial flow induced cooling tower operated at both normal and half-speed. Power is reduced by a factor of eight and average noise reduction is almost 12 dB. Significantly, the greatest noise reduction is in the octave band containing the blade passage frequency where sound level peaks often occur. Fan operation should be kept constant at the low level, since changes in sound intensity, due to cycling, are often more disturbing than constant higher sound levels.

Oversizing of the Tower. If conditions are critical and if maximum capacity must be maintained at all hours, the cooling tower may be oversized to permit lower fan horsepower. The degree of oversizing determines sound reduction achieved. Most towers, when operated at half-speed, produce approximately 65% of capacity at full fan speed. Therefore, to

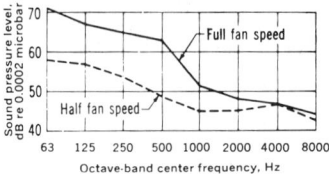

Fig. 7. Effect of half-speed operation.

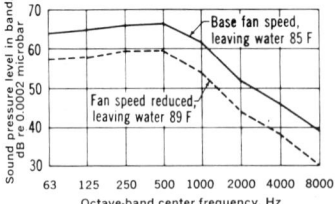

Fig. 8. Effect of changing leaving water conditions on tower sound level.

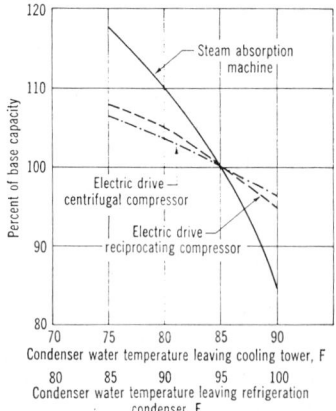

Fig. 9. Change in refrigeration system with changes in condensing temperatures.

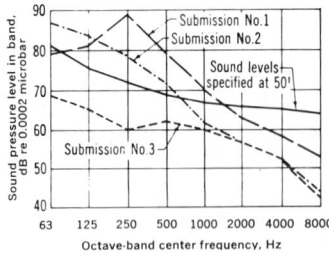

Fig. 10. Manufacturer's cooling tower submissions to meet sound level specification. Severe space and weight limitations were imposed because building was existing one, reflecting wall existed behind only practical tower location, and luxury apartment penthouse was 150 ft from tower.

obtain a noise reduction of as much as 12 dB, a 50% larger tower should be utilized.

Changing Leaving Conditions. The most critical heat transfer condition a cooling tower must satisfy is the approach of the leaving water temperature to the design wet bulb. It is common for designers to require this difference to be as little as 7F. If, all other conditions remaining the same, a standard 7F approach tower selection is made and then the fans are slowed down to permit a wider approach to the wet bulb temperature, reduced sound levels will result. Fig. 8 shows the sound level for the base selection and for that of a different wet bulb approach.

Operating the refrigeration plant with higher temperature condenser water will result in a reduced capacity of the refrigeration equipment. The effect of condenser water temperatures on capacity is dependent on the type of equipment used. Fig. 9 indicates that small capacity reductions result by raising inlet condenser water temperature for reciprocating and centrifugal electric compression units but that sizable reductions in capacity result for absorption equipment.

Sound Absorbers

External sound absorbers applied to cooling tower inlets and discharge reduce the sound radiated from them. It is well to attempt to achieve an installation which will be satisfactory from a sound standpoint without the need to apply this equipment but, if necessary, to make provision for additional attenuation that may be required. In existing installations where modifications to the tower are not practical or where a deficiency between the desired and the actual sound level exists after modifications are made, sound absorbers can be successfully applied. Sound absorber sound ratings and their effect on equipment performance are generally available from the various manufacturers.

Obtaining Desired Sound Levels

In order to insure a satisfactory installation, the sound generated by a cooling tower or remote condenser should be specified as an important performance factor. Required sound levels in each frequency band a given distance from the unit can be set forth after evaluating the characteristics of the area, the location of the tower and the sound characteristics of the available equipment. The sound performance can be confirmed either by manufacturer's certified data or by performing tests at the manufacturer's plant before shipping. See Fig. 10.

ACOUSTICAL PROBLEMS IN HIGH-VELOCITY AIR DISTRIBUTION

With a knowledge only of the noise spectrum generated by various air moving devices and the attenuating factors of duct systems, a low-velocity air distribution system for air conditioning can be designed with reasonable assurance that the end environment for the spaces served will be as predicted.

Not so with high-velocity systems, where excessive system noise is often generated within the distribution system itself and does not originate at the fan. System generated noise is the most costly to rectify.

System noise falls into two categories:

Fan noise, which is either totally or primarily generated by the air moving device (the acoustical designer of low-velocity, conventional fan-duct systems is primarily concerned with the attenuation of fan noise), and

Distribution system generated noise (commonly called regenerated noise) which is other than fan noise and is introduced into the duct system by the turbulent energy of the air stream in either the distributing duct system or terminal devices.

The acoustical designer of high-velocity systems must consider *both* fan noise and distribution system generated noise. (The commonly used term *regenerated noise* is a little misleading in that it implies that it is fan noise which was generated previously and then reintroduced into the system.)

Distribution system generated noise is more commonly found in, though not restricted to, high-velocity systems. It is caused by the turbulent pattern of the air stream, particularly at fittings, generated by air piling up within the duct, which tends to excite the duct panel. Where air pile-up or turbulence at a duct fitting or fire damper occurs, the following conditions have been found:

1. The higher the velocity across a poorly designed duct fitting, the greater the turbulent energy of the air stream acting against the duct panel.

2. The higher the static pressure acting on the duct panel, the greater the resistance of the duct panel to excitation by the turbulent energy of the air stream.

3. The higher the spring rate or stiffness of the duct fitting panel in the area of the air stream turbulence, the greater the resistance to panel excitation. Curved panel faces tend to provide more stable panel conditions and greatly increase the panel's resistance to diaphragming due to turbulence of the air stream.

4. The greater the obstruction to the air stream within the duct, the greater is the air pile-up and turbulent energy of the air stream.

5. The larger an insufficiently braced and supported duct panel, the lower is the frequency, when it occurs, of distribution system generated noise.

6. The larger the duct panel, the more critical is the (a) velocity at the duct fitting, (b) design of the fitting, (c) construction of the duct panel, and the more complex is the aerodynamics at the fitting.

When distribution system generated noise occurs, the resulting transmitted noise is generally characterized by a predominately low-frequency quality. These low-frequency noise levels of high intensity are normally:

Radiated as airborne noise through the duct walls and panels of terminal devices in the area where the generation of noise occurs, and

Transmitted through the distribution system and terminal devices as air- and structure-borne noise.

The curves of Fig. 1 and 2 represent field measured octave band analyses of distribution system generated noise.

Quality of distribution system generated noise is dependent on size and type duct construction, type of fittings and type of terminal devices used. Predominately low-frequency noise, often below 31.5 cycles per second (Hz), is transmitted from large, insufficiently braced, rectangular ducts, where noise generation occurs. As duct size decreases and panel stiffness increases, higher frequency noise (either 1st or 2nd band) is transmitted from noise-generating ducts and/or fittings.

Generally, present day damping compounds, applied either internally, externally or to both surfaces of diaphragming duct which is generating noise due to a turbulent air stream, have been found to have little or no effect on the diaphragming action of the duct panel or the resulting noise transmitted. Unfortunately, acoustical lining and/or factory-built sound traps are also relatively ineffective in attenuating *low-frequency* noise.

Corrective measures aimed at distribution system generated noise usually must involve modification of the duct fittings and/or special sound barrier construction to prevent or reduce transmitted noise through ceiling construction. In cases of noise generated by terminal devices, such as dual-duct mixing

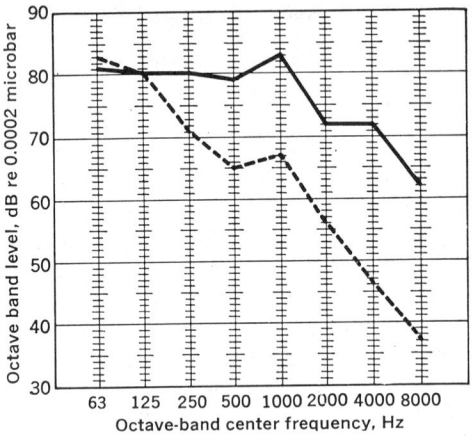

—— Octave band sound pressure levels as measured within mechanical equipment space.

- - - - Octave band sound pressure levels as measured directly under shaft take-off, three floors removed from mechanical equipment space.

Fig. 1. Noise level in first octave band is greater in tenant space, three floors removed from mechanical equipment, than in mechanical equipment space with all equipment operating, despite adequate fan noise attenuation devices before all risers and shaft take-offs.

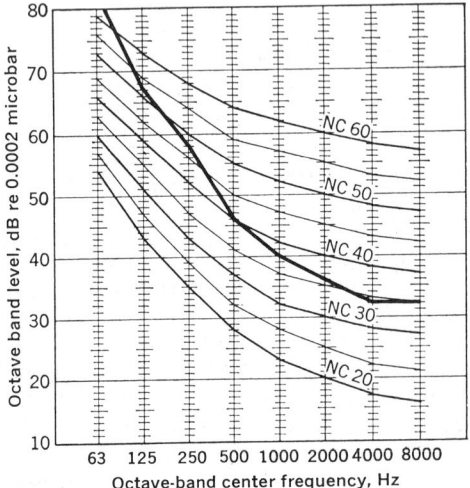

Fig. 2. Typical curve of high-velocity air distribution system generated noise.

units on perimeter walls, reduction of radiated noise from the terminal device and the adjacent high-pressure ductwork may be achieved by special enclosures.

Following are recommendations for acoustical considerations in design and layout of high-velocity systems. These recommendations are based primarily on empirical data and field experience. It is hoped that research in the respective areas will soon minimize reliance on empirical information and reduce all design steps to more rational terms.

Air Handling Apparatus Rooms

Supply Fans. Always select fan equipment for operation at the point of maximum static efficiency. This will prove to be also the optimum point acoustically. For quietest fan operation, it is important that not too large size fan equipment be selected. The design engineer, by selecting a smaller fan operating at higher speed can attain the same air flow rate and static pressure and move the point of operation over to the right on the fan performance curve where there is assurance of staying in the normal operating range despite small changes in system resistance.

Location of fan equipment within casings should provide sufficient clearance between fan inlets and casing walls. Orientation of the fan with respect to fresh air intake air should be such as to provide an approximately balanced air flow to each fan inlet with the best aerodynamic flow conditions to the fan inlets.

In push-through apparatus rooms for single-duct systems, it is important to avoid sudden expansion (greater than 15° included angle) directly at the fan discharge. This condition can set up turbulence, noise and vibration in the sheet metal casings. The method of discharge, with respect to orientation of the connecting supply duct immediately downstream of the fan discharge, should be in accordance with the fan manufacturer's recommendations.

Unfortunately, on blow-through apparatus rooms, the above conditions cannot be met because of the obvious space conditions. The turbulence created, and the high pressures existing, within the casing necessitate special casing construction. This aspect is discussed below.

A compromise of fan quality will always prove to be expensive and will result in a substandard installation.

Selection of Fan Isolation Bases. Fan and driving motor base should be integral and rigid, without flexure caused by belt tension or torque applied.

Base should be sufficiently rigid to maintain belt centers at all times. Separate fan and motor bases or bases inadequately stiffened will result in maintenance problems and vibration/noise transmission.

Mountings and bases should be designed to accept the dynamic thrust introduced by high fan-discharge static pressures.

All isolation elements should be stable, unrestrained (without external snubbers) and must provide adequate spring rate to properly isolate the lowest forcing frequency.

Where large items of fan equipment are supported on lightweight slabs, spring rates of the supporting mountings must be substantially reduced to compensate for the resiliency and the mass of the floor slab.

Apparatus Casings. Inadequately stiffened apparatus casings generate low-frequency noise due to the diaphragming of unstable panel walls. These low frequencies may be airborne through the fan inlets and distribution system or through architectural sound barriers. Where energies are sufficiently great and apparatus casing walls are fixed at floor and roof slabs, casing walls also may transit structurally borne noise and/or vibration. Construction standards developed for low pressure apparatus casings are utterly inadequate for large high-pressure and medium-pressure systems and created serious acoustical problems on a great many installations.

It is recommended, therefore, that apparatus casings subjected to higher pressures be adequately braced and stiffened. On larger apparatus casings handling over, say, 40,000 cfm, one of the following should be considered: (a) double-wall factory insulated panels, (b) cellular casings, or (c) masonry casings.

Masonry casings internally insulated and made airtight will provide the optimum construction to achieve desired acoustical results.

Refer to recommendations contained in SMACNA High Pressure Duct Manual, published by Sheet Metal and Air Conditioning Contractors' National Association, Inc., Arlington, Va.

Shape of High-Pressure Apparatus Casings. On blow-through air handling plants, savings in energy realized by tapered casing designs are rather small. Apparatus casings should be arranged to form plenums of rectangular shape. Rectangular plenums permit better and tighter panel joint constructions, thereby reducing the possibility of air leaks which may cause whistles and hissing noises. These noises

are often transmitted structurally and are also airborne.

For large dual-duct air handling plants, improved acoustical conditions result when the cold and warm plenums are split vertically, or in the direction which produces the smallest span or panel in the direction of the air stream. Less danger of panel flutter vibrations, which generate noise, result. The vertical split of apparatus casings will also provide other functional and maintenance advantages.

Dampers and Air Valves: Avoid hand or automatic dampers in high-velocity systems, since such devices when closed or partially opened produce noise and rumble usually of a low-frequency characteristic. The exception to the above is the off-on damper usually installed in the discharge of multiple supply fans. Such dampers must be of streamlined construction.

Use of air balancing valves in water-air system zone duct mains produce noise. Such equipment noise should be adequately attenuated downstream of the valve.

Flexible Connectors. Install flexible connectors of appropriate construction at all duct connections to high-pressure apparatus casings and to supply and return fans. This will reduce fan casing noise transmitted through the duct wall.

Flexible connectors should be installed with a minimum of one inch greater length than the distance between fan and connecting duct.

Air Distributing Systems

Duct Velocities. In designing high velocity duct systems, every effort should be made to reduce the total fan horsepower and to handle air at moderate velocities. This will assure a quieter system, reduce duct leakages and will approach, in the majority of cases, the point of maximum owning cost economy. The chief justification for high duct velocity is space limitation created by past and present architectural and structural practices. In modern buildings, high duct velocities must be used where space limitations exist. *When the points of restriction are passed, the duct velocities should be reduced.*

Choice of Duct Design Method. For high-velocity duct systems, it is immaterial what duct design method is selected for sizing of high-pressure ducts. None of the commonly used duct design methods will, by themselves, produce an economical and quiet duct system. On large and medium installations, a combination of several design methods may

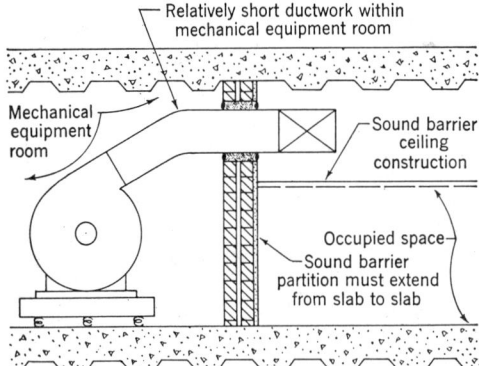

Fig. 3. Where ductwork is short, use special barriers.

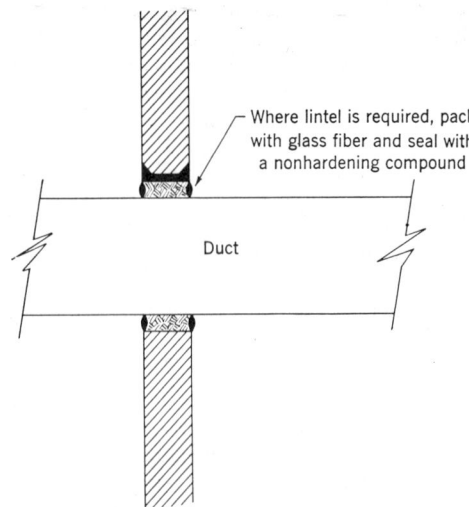

Fig. 4. Duct through partition from mechanical to occupied space.

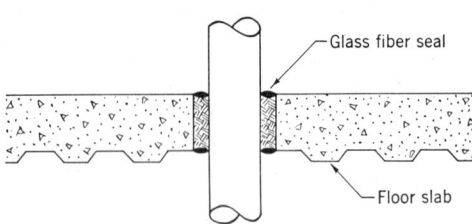

Fig. 5. Penetration of floor slab by duct (outside of shaft).

be used advantageously. The important point to remember is not to use high velocities or friction rates when space conditions will permit a larger duct. If, for example, the equal friction method is selected, it does not mean that the entire (longest) duct run should be sized at the initial high friction rate. Two, three and more friction rates can be selected from consideration of space available for the ductwork.

Ductwork Adjacent to Apparatus Room. Where no space restrictions exist between apparatus casings and duct riser shafts, duct sizes should be increased. An appreciable savings in horsepower requirements can be achieved and the system will be improved acoustically.

When high-velocity mains handling large air quantities penetrate walls containing fire dampers, duct offsets or turns immediately before or after fire dampers should be avoided.

All efforts should be made to reduce turbulence at fire dampers.

Where ductwork within mechanical spaces leading from the fan discharge to the occupied space is relatively short (less than 15 ft) before entering adjoining critical occupied spaces, consideration must be given to the noise radiated from the duct wall. Special sound barrier constructions separating the noise-radiating duct and the occupied space may be required to contain the duct radiated noise. See Fig. 3.

All duct penetrations of partitions separating mechanical spaces from occupied spaces should be linteled if required, sleeved with ½-inch fiberglass

and sealed with a non-hardening compound to reduce structurally borne transmitted duct wall noise into the building structure. All duct penetrations of floor slabs (outside of shafts) shall be similarly treated. See Fig. 4 and 5.

Duct Connections to Apparatus Casings. In order to reduce turbulence, conserve energy, and reduce generation of noise, rectangular and circular high-velocity ducts should be connected to supply plenums with fittings which will reduce abrupt velocity change in the air stream. Fittings shall be prism or shoe-type (for rectangular duct), or bell mouth shaped (for circular duct) connections to apparatus casings.

Type Duct Construction. From the viewpoint of distribution system generated noise, the rank order of design preference of duct construction (assuming the same velocity and static pressure) is (a) circular, (b) flat oval, and (c) rectangular.

Where vertical space is available, installation of circular duct will result in a slightly less expensive and generally quieter distribution system.

Where space is available to permit use of rectangular ducts with large aspect ratios, lower air velocities could be used, which would result in quieter and more economical distribution systems.

The above rank order also holds if one is rating duct construction from the viewpoint of duct wall stiffness to resist duct flutter, where conditions of turbulence occur.

The primary cause of distribution system generated noise, as previously discussed, is due to turbulence within the duct. There have been many high-velocity systems with rectangular ductwork where design acoustical criteria have been maintained primarily because well designed fittings were installed and adequate high-velocity duct construction was used.

Quieter distribution systems using rectangular duct and flat oval construction are provided by using greater care in the *design of duct fittings and duct panel bracing*. The rectangular high pressure duct construction is now standardized and can be found in the SMACNA High Velocity Duct Manual. The newly developed construction will greatly reduce generation of noise in the distributing systems caused by excitation of duct panels due to turbulent air flow.

Fittings for High Velocity Ductwork. It is imperative that, on high-velocity systems, only smoothly designed fittings and take-offs be used. No fittings, which have a sharp or rough edge in the air stream, can be tolerated. Fittings should be designed for gradual velocity changes of the airstream.

Probably the greatest part of acoustical problems on high-velocity systems are created by improperly constructed and installed take-off fittings, especially on rectangular ductwork. The generation of noise is exceptionally severe when holes in the ductwork for mounting fittings are cut smaller than the free area of the fittings, thus exposing raw edges to the air stream. The same problem is encountered with round ductwork when tap-offs have to be mounted in the field in existing ductwork. One type of properly constructed and mounted fittings for rectangular ductwork is shown elsewhere in this

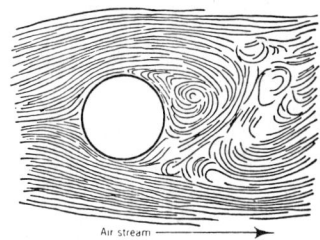

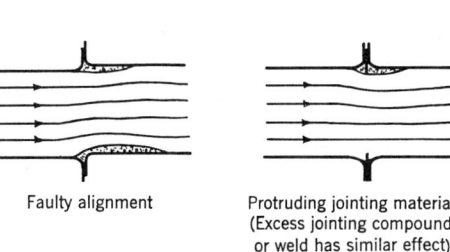

| Faulty alignment | Protruding jointing material (Excess jointing compound or weld has similar effect) |

Fig. 6. Turbulence due to round obstruction, duct misalignment and protruding edge in air stream.

Handbbook in connection with design of high-velocity double-duct systems.

Figure 6 indicates the turbulent pattern of air stream due to faulty duct alignment or obstructions in high-velocity duct systems.

Always allow sufficient length of straight duct between fittings. Successive fittings in high velocity systems generate noise due to the turbulence created in the airstream. Five equivalent diameters should be considered an absolute minimum between successive fittings.

When this is not possible, straightening vanes should be used to improve the condition as shown in Fig. 7.

Square elbows should be vaned and vanes should be of streamlined construction. Vanes should be extended on the downstream side for a minimum of 6 inches. All vanes should be rigidly designed and installed. See Fig. 8, indicating turbulent air pattern in normal, expanded and contracted radiused bends.

Take-Off Fittings (Shafts and Mains). For duct main branch take-offs or shaft take-offs, conical fittings (for circular duct), prism shaped shoe fittings (for rectangular duct), should be used; 90° duct main or shaft take-offs should not be tolerated for larger sizes (say, over 8 inches diameter) and high velocities (say, over 2,000 fpm). All efforts,

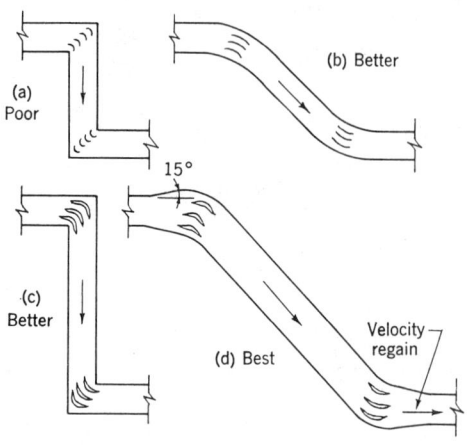

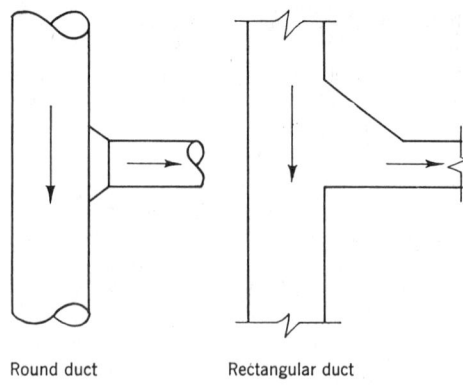

Fig. 9.   Tap-off from main or riser.

Fig. 7.   Air stream turns, good and bad.

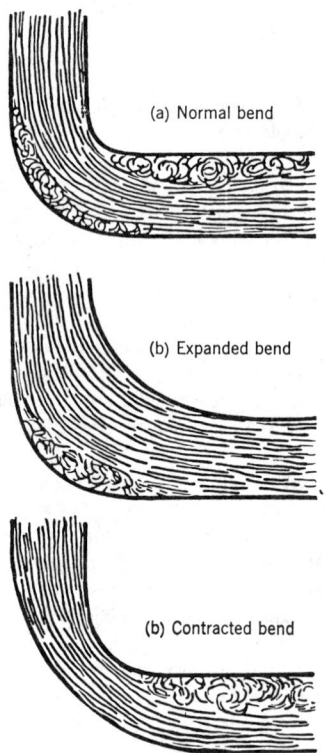

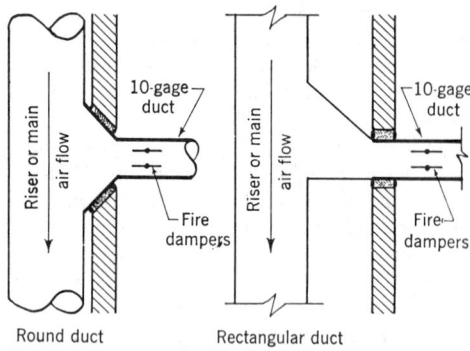

Round duct   Rectangular duct

Fig. 10.   Where shaft take-off space is inadequate, use extra-heavy duct.

Fig. 8. Air pattern in bends of various patterns.

where large duct panels are involved, at shaft take-off and main duct branch take-offs, should be made to reduce the velocity even further. See Fig. 9.

Where shaft take-off conditions do not permit adequate space conditions to provide conical or prism-type taps, it is recommended that 10-gage duct be installed to permit the fire damper to be located further downstream of the tap within the 10-gage duct. The shaft wall thickness may then be utilized to permit the installation of conical or prism-type tap and thereby provide better aerodynamic take-off conditions and approach conditions to the fire damper. See Fig. 10.

When possible, the fire damper should be constructed with blades out of the airstream. Under no

circumstances should the blades be permitted to extended into the shaft airstream.

Dual Duct Area Ratio. The practice of undersizing warm ducts on dual duct systems has resulted in many acoustically unsatisfactory installations. It should be kept in mind that maximum demand on warm duct will be created, not during steady heat flow conditions under maximum winter load, but under a great many partial load operating conditions. The warm shaft ducts and horizonal mains should not be made less than 70-75% of the corresponding cold duct, in area. The warm connections between mixing units and high velocity should always be the same size as cold connections.

Velocities Through Flexible Connections to the Terminal Devices. At no time should excessively high velocities be maintained (maximum of 2200-2500 fpm) in flexible tubing to the terminal devices. Flexible tubing, especially when not straight, generates noise. Flexible tubing must be free of sharp offsets or bends. The inlet collars to the unit must not be distorted in any way or be allowed to project into the air stream.

Limit the Extent of High Velocity Ductwork. In laying out a distributing system for all-air overhead distribution, one must not simply follow the technique used in low-pressure layouts by running high-pressure ductwork over an entire conditioned area. The extent of high-pressure mains on each floor must be limited to well defined areas and air distributed to individual rooms by low-pressure, internally lined, ductwork. One should consider the high-pressure ductwork as a high-velocity plenum which remains intact for the life of the structure. In order to have maximum future flexibility and to avoid the danger of obsolescence of the system, all future changes due to shifting loads and partition changes should be confined to low-pressure ductwork and mixing units.

Dampers as a Noise Generating Source. Dampers are a source of noise generation in all types of fan systems. Volume dampers in conventional systems located at, or in close proximity to, air outlets will generate noise albeit that more than adequate fan noise attenuation has been provided in the distribution system. Adequate natural duct attenuation or acoustical lining should be provided before air outlets.

The use of balancing volume dampers in dual-duct systems should be avoided. They are practically

useless and are a source of noise and an unnecessary expense.

Splitter dampers in high-velocity systems should also be avoided as they generate noise and can cause distribution troubles.

Sound Barrier for High Velocity Ductwork. Adequate sound barrier construction should be installed between all high-velocity ductwork and occupied areas. Particular attention should be given to ceiling constructions in critical areas adjoining shaft take-offs, unavoidably poor duct fittings, and very large terminal devices, and where air valves are located in ceilings of critical spaces.

The relatively new concept of unified ceiling systems, which has been developed to integrate the functions of air distribution within the slab construction, greatly reduces the possibility of problems caused by radiated distribution system noise.

All high-velocity ductwork and terminal devices should be installed free of any contact with ceilings and their supports.

Sound Traps. Factory-built sound traps should be selected for the minimum practical pressure drop, and not greater than 0.4 inch of water. The manufacturer's rating of attenuation with air flow and the sound power generated in the sound trap by the air flow should be balanced to provide useful net effective attenuation. There have been installations where the sound trap was replaced with straight duct and a quieter installation resulted.

Sound traps should be located with good entrance and discharge air-flow conditions. (See Fig. 11.) Poor entrance and discharge conditions will considerably increase the pressure drop and the noise radiation from the duct panels. When space limitations are severe, however, it is less objectionable to reduce the duct length between the trap and the elbow before it, than to reduce the space downstream of the trap.

When mixing unit, induction unit and air valve manufacturers provide data regarding the attenuation properties of their equipment under air flow

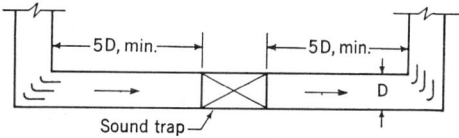

Fig. 11. **Where possible, locate sound traps with long straight inlet and discharge duct sections to provide best approach and leaving-air pattern to trap.**

conditions, it may be possible to reduce acoustical treatment requirements in air distribution systems.

These data may show that, for high-velocity systems, consideration of fan noise in certain duct schemes may be unnecessary and that acoustical lining or sound traps may be necessary only on the low-pressure side of the system. There presently are indications that little or no attenuation is provided by acoustical treatment—subjected to air stream velocities of greater than approximately 3,000 fpm due to the self-generating noise qualities of the acoustical treatment. Maximum attenuation is indicated under no-airflow conditions with the attenuation falling off as the velocity increases across the face of the sound trap or surface area of the acoustical lining.

Cross-Over of Horizontal Dual Duct Mains Near Air Shafts.

Probably the most critical part of a high-velocity, dual-duct layout is the area near the riser shaft. Because of the space limitations, the cross-over of horizontal mains cannot always be made with adequately designed and properly arranged fittings. When this is the case, noise and rumble may be generated in the cross-over ductwork. It may be found desirable on such installations to eliminate the cross-overs entirely by carrying three supply ducts in the shaft instead of two; say, two warm and one cold duct, as shown in Fig. 12.

Testing of High Pressure Ductwork.

Leaks in high-pressure ductwork or apparatus casings create objectionable high-frequency noise in occupied spaces. It is imperative that the high velocity ductwork be pressure tested *before applying insulation and before putting the system in operation.* The ductwork should be tested in convenient sections and the objectionable air leaks eliminated. Many

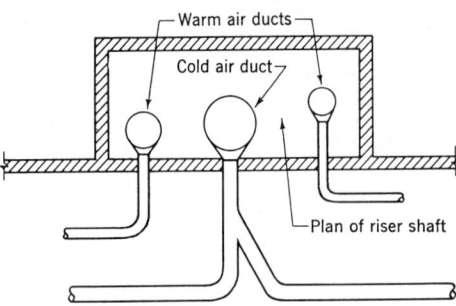

Fig. 12. **Typical duct connections to risers to avoid cross-overs.**

acoustical problems on high-velocity installations have been caused by failure to test the ductwork during installation. All high-velocity duct or apparatus casing joints and seams must be made airtight with a long-life sealant. Cloth plastic or foil adhesive tapes should not be used as the primary sealer but rather should be used in conjunction with mastic sealers.

For a suggested method of sealing and testing, see SMACNA High Velocity Duct Manual.

Terminal Devices

Acoustical Performance. The acoustical performance of high-velocity, all-air or water-air systems depends to a large degree on the acoustical performance of the installed terminal devices. Improper selection or compromise with the acoustical and functional characteristics of terminal devices for these systems, regardless of whether the distribution system and fan noise has been completely attenuated, can create excessive noise and exceed the design criterion.

Similarly, in conventional velocity systems, where considerable balancing adjustment is required to be made at the diffuser or where volume dampers are located in close proximity to outlets, noise generation will occur regardless of the extent of acoustical treatment provided in the distribution duct.

Pressure reducing air valves, single- and dual-duct mixing units, while providing some attenuation of fan noise, also create a self-generating noise which is both radiated from the equipment surface and transmitted through the connecting duct in the form of air- and structure-borne noise. On smaller capacity units, the radiated noise is normally attenuated by most ceiling constructions.

Generally, most air valve or mixing unit equipment manufacturers present acoustical data which reflect only the airborne noise through the equipment outlets. Little data are presently available regarding the *radiated* noise values. Where the *radiated* noise factor of the equipment may be critical, two acoustical performance ratings will be required to fully analyze the condition. Acoustical performance ratings should be conducted:

1. With only the equipment located inside the test chamber and the outlets outside the chamber;

2. With the equipment located outside the test chamber and only the outlet inside the chamber.

Under-window unit induction or air mixing devices, when acoustically tested without enclosures within the chamber, will transmit the total radiated

(through the equipment surfaces) and airborne (through the outlets) noise.

Some protection of radiated mixing unit casing noise may be provided by proper sound attenuation enclosure constructions. These enclosures will result in quieter equipment installations. With window induction units, however, due to the induction air openings, special sound barrier enclosure constructions will provide relatively little radiation noise protection.

Flutter in Dual Duct Mixing Units. To prevent flutter of volume controllers in medium- or high-velocity, dual-duct systems, each mixing unit should be equipped with positive stabilizers. Where bad fittings adjacent to the unit or foreign matter within the duct system occur, flutter of volume controllers in mixing units not equipped with positive stabilizers may take place. The flutter may also be caused by a combination of excessively low pressure at the mixing unit and air pulsations in the air stream. This condition may cause a chain reaction to occur to other units not equipped with stabilizers. Flutter of volume controllers creates objectionable noise transmission to occupied areas.

"Two Motor" Dual Duct Units. Two-motor, dual-duct units may be noisy in operation when the demand on the mixing unit is for warm air only. When this occurs, the constant-volume feature of the unit is suspended and considerably greater volumes of warm air may be handled by the mixing units than the design engineer had intended. These higher volumes of warm air often create velocity conditions within the warm air duct, mixing box and mixing box outlets which causes the equipment to operate with excessive noise.

Inlets to High Velocity Terminal Points. Carelessness on the part of the installer in connecting flexible tubing to high-velocity terminal devices may change the acoustical characteristics of such devices. A sharp bend in the flexible tubing may expose a raw edge of the inlet collar, thus causing a loud hiss or a whistle, changing totally the acoustical characteristics of the units.

Warm Connections to Dual Duct Mixing Units.

Warm run-outs to the mixing units should be made the same size as cold connections, even if the unit is designed to handle less warm air than cold air under normal operating conditions. There will be many sets of operating conditions which will force more warm air through warm connections than designed. This may cause generation of noise in the warm flexible tubing.

Large Terminal Units. Large terminal devices say, handling over 1200-1500 cfm, and operating under high pressures, should be separated from high- and low-velocity ducts by a flexible connection.

Large capacity mixing units may create excessive noise in the immediate area of the equipment radiating noise. Care should be exercised in the location of the large capacity mixing units (1200 cfm and larger) in ceilings of critical spaces. It is preferable that these units be located in non-critical spaces or sound barrier ceiling construction should be installed where they cannot be relocated.

Particular attention should be paid to the radiated noise when the terminal units are exposed in the room with no suspended acoustical ceiling below, where a relatively sound transparent ceiling such as perforated metal pans or very light glass fiber ceiling tiles or boards is used, or where the acoustical design (NC) objective is critical.

Alternatively, a number of smaller units with the same total capacity may be installed to reduce the radiated noise transmission factor. The records of recent installations indicate that there is no real economic justification in using all-air units of large capacities. Only small cost advantages can be gained by increasing units to over 1200-1500 cfm.

Grilles, Registers and Diffusers. Select all terminal outlets at least 5 db below the design criterion required for the space usage.

It must be remembered that, as the static pressure in the branch duct to the terminal air outlets increases, the outlet opening must be decreased to restrict the flow to design volume. At these higher duct pressures, noise generation occurs in the damper restriction. The quality of this noise generation is characterized by high frequency transmission.

RADIATION PROTECTION AT WALL OPENINGS FOR DUCT OR PIPE

The air conditioning engineer should be acquainted with the means for shielding ductwork and other openings that penetrate protective barriers around radiation facilities, particularly X-ray rooms.

Protection against radiation from X-ray tube, cyclotron, radium or other radioactive material is primarily a question of shielding to reduce the level of radiation to a safe or specified amount, of maintaining safe distances from the rays, and/or of limiting the time of exposure.

The prime consideration in preventing penetration of rays is density of the shielding material. Lead is the densest of any commonly available. Where space is at a premium, particularly in modern buildings, and where utmost radiation protection is demanded, lead is invariably used. Lead is useful, especially where neutrons and gamma rays are concerned, in that it does not itself become contaminated and emit harmful rays.

Lead, usually in sheet form, is used to line the walls, floor and often the ceiling of rooms containing radiation facilities. Openings through the barrier for air ductwork, piping, service boxes, conduit, etc., require shielding, usually obtained by a lead barrier around or behind these building utilities of sufficient coverage and thickness to prevent penetraton of these rays.

Shielding of duct and other openings in the protective barrier of radiation facilities depends on energy of radiation, orientation of the beam, dimensions and location of opening in the protective barrier, geometrical relationship between the radiation source and opening, and geometrical relationship between opening and persons, materials or instruments to be protected. The complexity of these factors requires the services of a radiological physicist, who determines extent of shielding, materials for shielding (usually lead or concrete) and the thickness of the shielding material. After the radiological physicist has done the basic design for this shielding, the protective barrier contractor provides the required shielding for the openings.

Design of ductwork, piping, etc., should *anticipate* some of the problems encountered both in the design and installation of shielding. Also, coordination between air conditioning contractor and shielding fabricator can best be achieved by understanding and forethought on the part of the air conditioning designer.

Figures 1-4 give some idea of the *area of shielding* required around ductwork. They show various duct installations that penetrate the protective barrier for walls or partitions of X-ray rooms. Lead shielding is used to cover these openings, the *approximate* extent of which is indicated in terms of simple equations involving the opening dimensions and wall thickness. These are *conservative estimates*, which will aid the air conditioning designer to understand what to expect as to the area of shielded ductwork. The radiological physicist actually determines for *each* case the *lead thickness* and the *exact amount of shielding* required.

Note in Fig. 4 that the protective shielding deals with *primary* radiation, while Figures 1-3 show protection against *scattered* or *secondary* radiation. Primary radiation comes directly from the source; scattered radiation has been deviated in direction; and secondary radiation is emitted by an irradiated material. Primary radiation requires more protection because its energy level is higher.

Because sheet lead is not structurally self-supporting, it must be mounted to prevent sagging by its own weight. For lead thicknesses up to 3.5 mm, sheet lead can be readily shaped around round and small rectangular ducts, say 24-inch maximum diameter or width, with all joints overlapped at least ½ inch. To hold these lead sheets in place, 1-inch wide iron bands should be placed around the periphery of the duct on approximately 12-inch centers, care being taken not to cut into the lead when the bands are bolted up.

When lead thickness is greater than 3.5 mm or duct width exceeds 24 inches, lead shielding should be laminated on a plywood or similar structural core, which is made in sections or panels to conform to the sides of the duct. The laminated sections are mechanically fastened at the seams and corners. These joints are lapped with sheet lead angles or lead strips, the width of which is twice the thickness of the lead, but not less than ½ inch in any case. Nails, bolts, screws or other fasteners used to secure the lead sheet or panel must be covered with lead of thickness equal to the lead sheet. Lead headed nails may be used as shown in Fig. 5.

For lead shielding of 1.0 mm or less, flexible leaded vinyl sheets can be used for easy forming to complex shapes and contours. The flexible leaded vinyl sheets can be applied in layers where heavier than 1.0 mm lead shielding is required. If the duct has a flexible connection or is made of a flexible material, the flexible vinyl sheets could be applied over it more readily than other forms of shielding.

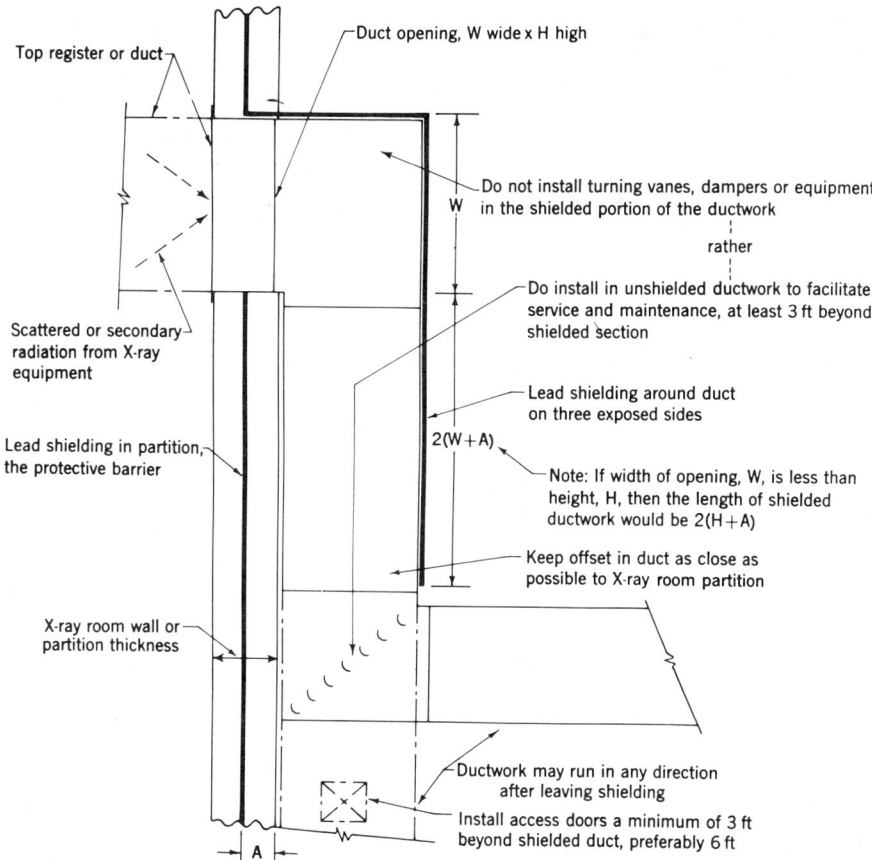

Top register or duct

Duct opening, W wide x H high

Do not install turning vanes, dampers or equipment in the shielded portion of the ductwork

W

rather

Do install in unshielded ductwork to facilitate service and maintenance, at least 3 ft beyond shielded section

Scattered or secondary radiation from X-ray equipment

Lead shielding around duct on three exposed sides

2(W+A)

Lead shielding in partition, the protective barrier

Note: If width of opening, W, is less than height, H, then the length of shielded ductwork would be 2(H+A)

Keep offset in duct as close as possible to X-ray room partition

X-ray room wall or partition thickness

Ductwork may run in any direction after leaving shielding

Install access doors a minimum of 3 ft beyond shielded duct, preferably 6 ft

A

Fig. 1. Plan view of duct running through partition of X-ray room, exposed to secondary or scattered radiation.

Duct hangers are best installed on the outside of the lead shielding so that the hanger rods or straps do not have to pierce the shielding. The lead shielding adds considerably to the weight of the duct and the hangers should be substantial, with such adequate anchoring in the slab above as fish plates. For rectangular ducts, trapeze hangers would be the most practical. For design purposes, estimate each 1/16 inch of lead at 4 lb per sq ft.

Tests for radiation leakage are usually made after the room is finished and the equipment is installed. It is very important to install all shielding properly during the course of construction because of the expense in making corrections to the finished protective barrier. Moreover, equipment such as dampers should never be put in the shielded section of the ductwork, as repairs to this equipment would be very costly if the shielding must be dismantled.

A simple way to avoid penetration of the protective barrier's lead lining by pipes or wires is to offset them as close behind the lead lining as possible so that they can be backed with a lead sheet of sufficient size to prevent passage of the rays at any angle. This lead patch method is also used for electric switch boxes located in the wall.

Medical Installations

The extent of the protective barrier for medical installations is summarized below so that the air conditioning designer can tell whether ducts or pipes running through such spaces are likely to be a problem.

For medical radiographic and fluoroscopic rooms, the lead shielding generally does not extend above a line 7 ft, 0 in. from the finished floor; and if the

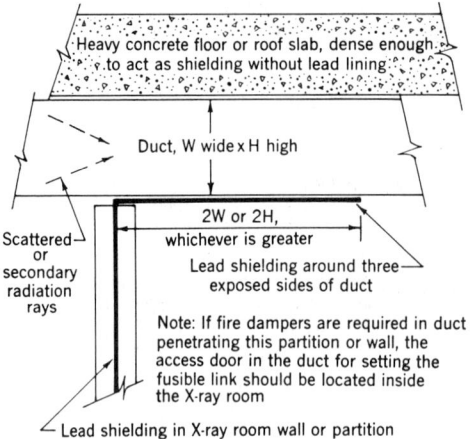

Heavy concrete floor or roof slab, dense enough to act as shielding without lead lining

Duct, W wide x H high

Scattered or secondary radiation rays

2W or 2H, whichever is greater

Lead shielding around three exposed sides of duct

Note: If fire dampers are required in duct penetrating this partition or wall, the access door in the duct for setting the fusible link should be located inside the X-ray room

Lead shielding in X-ray room wall or partition

Fig. 2. Elevation of section through X-ray room partition where duct opening is exposed to secondary radiation.

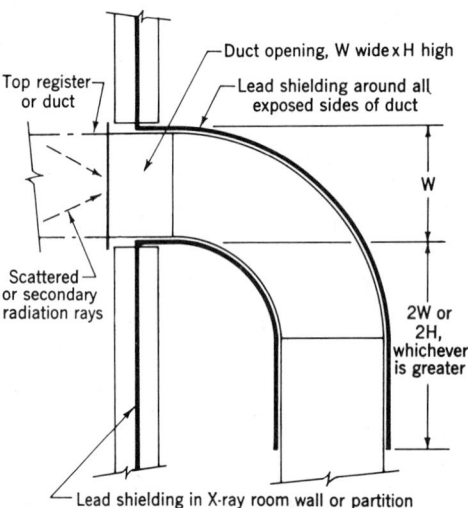

Duct opening, W wide x H high

Top register or duct

Lead shielding around all exposed sides of duct

W

Scattered or secondary radiation rays

2W or 2H, whichever is greater

Lead shielding in X-ray room wall or partition

Fig. 3. Plan view of radius elbow in ductwork running through partition of X-ray room, exposed to secondary or scattered radiation.

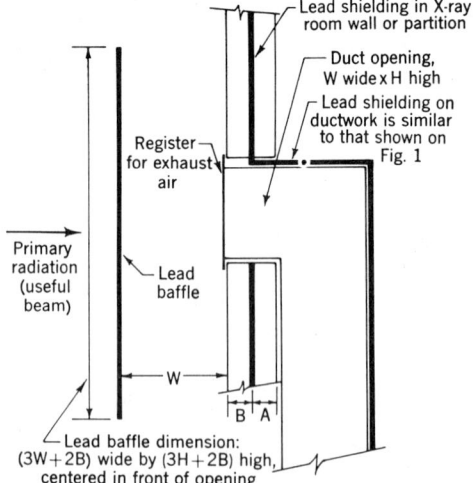

Lead shielding in X-ray room wall or partition

Duct opening, W wide x H high

Lead shielding on ductwork is similar to that shown on Fig. 1

Register for exhaust air

Primary radiation (useful beam)

Lead baffle

W

B A

Lead baffle dimension: (3W + 2B) wide by (3H + 2B) high, centered in front of opening

Note: Where duct runs straight without offset, lead shielding is similar to that shown on Fig. 2

Fig. 4. Plan view of duct running through partition of X-ray room, exposed to primary radiation.

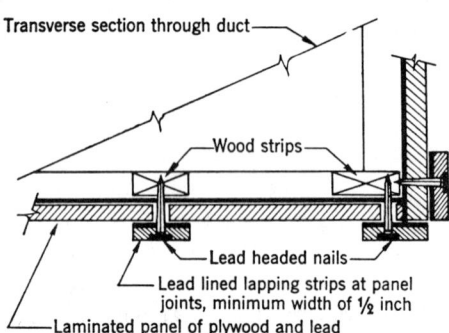

Transverse section through duct

Wood strips

Lead headed nails

Lead lined lapping strips at panel joints, minimum width of ½ inch

Laminated panel of plywood and lead

Note: Let duct rest on wood strips for support. Substantial trapeze hangers should be used for supporting shielding enclosure. Hanger rods must be rigid enough to prevent any sway of the enclosure

Fig. 5. Construction of laminated panel enclosure around shielded ductwork.

service lines and ducts can be located above this line, shielding around them is obviously unnecessary. For X-ray therapy rooms, lead shielding may extend to the ceiling or structural slab. The ceiling or slab above and the floor may also be lead lined, depending upon output of the machine and other conditions. For industrial X-ray work, wall shielding may extend to the ceiling. Both ceiling and floor in some cases will require lead lining.

For shielding in supervoltage rooms, special conditions may apply. In any event, the radiological physicist should be consulted to design the proper protection. Where concrete is considered for the shielding material, it is often more practical to use lead of equivalent value for the shielding of openings. Where recesses occur in concrete barriers for equipment, lead backing, equivalent to the thickness of the concrete removed, should be provided.

MATHEMATICAL DATA, DRAFTING ROOM PRACTICE, AND TERMINOLOGY

Contents

MATHEMATICAL DATA, DRAFTING ROOM PRACTICE, AND TERMINOLOGY

COMMON ABBREVIATIONS

Absolute ....abs
Air horsepower ....air hp
Alternating-current (as adjective) ....a-c
Ampere ....amp
Ampere (SI Metric) ....A
Ampere-hour ....amp-hr
Antilogarithm ....antilog
Atomic weight ....at. wt
Atmosphere ....atm
Average ....avg

Barometer ....bar.
Barrel ....bbl
Baumé ....Bé
Boiler pressure ....bp
Brake horsepower ....bhp
Brake horsepower-hour ....bhp-hr
British thermal unit ....Btu or B

Calorie (or calory) ....cal
Candela ....cd
Candlepower ....cp
Centimeter ....cm
Cubic foot ....cu ft
Cubic feet per minute ....cfm
Cubic feet per second ....cfs
Cubic inch ....cu in.
Cubic meter ....cu m or m³

Decibel ....dB
Degree ....deg or °
Degree Centigrade ....°C or C
Degree Fahrenheit ....°F or F
Degree Kelvin ....K
Degree Rankine ....°R or R
Direct-current (as adjective) ....d-c

Efficiency ....eff
Elevation ....el

Feet per minute ....fpm
Feet per second ....fps

Gallon ....gal
Gallons per minute ....gpm
Gallons per second ....gps
Grain ....spell out
Gram ....g
Gram-calorie ....g-cal

Hertz ....Hz
High-pressure (adjective) ....h-p
Horsepower ....hp
Horsepower-hour ....hp-hr
Hour ....h or hr

Inch ....in.

Joule ....J

Kilocycle ....kc
Kilogram ....kg
Kilograms per second ....kgps
Kilograms per second (SI Metric) ....kg/s
Kilovolt ....kv
Kilovolt (SI Metric) ....kV

Latitude ....lat
Linear foot ....lin ft
Liter ....l or L

Maximum ....max
Mean effective pressure ....mep
Meter (or metre) ....m
Micrometer (micron in SI Metric) ....μm
Micromicron ....μμ or mu mu
Micron ....μ or mu
Miles per hour ....mph
Milliampere ....ma
Milliampere (SI Metric) ....mA
Millimeter ....mm
Millimicron ....mμ or m mu
Minimum ....min
Minute ....min
Minute (angular measure) ....
Molecular weight ....mol. wt

Parts per million ....ppm
Pound ....lb
Pounds per square foot ....lb per sq ft
Pounds per square inch ....lb per sq in.
Pounds per square inch, gage ....psig
Pounds per square inch absolute ....psia

Revolutions per minute ....rpm
Revolutions per second ....rps

Second ....sec
Second (SI Metric) ....s
Second (angular measure) ....″
Specific gravity ....sp gr.
Specific heat ....sp ht
Square foot ....sq ft
Square inch ....sq in.
Square millimeter ....sq mm or mm²

Temperature ....temp

Volt v
Volt (SI Metric) V
Volt-ampere ....va

Watt ....w
Watthour ....whr
Weight ....wt

THE ELEMENTS

Name	Atomic Symbol	Atomic Weight	Name	Atomic Symbol	Atomic Weight
Actinium	Ac	227 *	Mendelevium	Md	258 *
Aluminum	Al	26.98	Mercury	Hg	200.59
Americium	Am	243 *	Molybdenum	Mo	95.94
Antimony	Sb	121.75	Neodymium	Nd	144.24
Argon	A	39.95	Neon	Ne	20.18
Arsenic	As	74.92	Neptunium	Np	237.05
Astatine	At	210 *	Nickel	Ni	58.71
Barium	Ba	137.34	Niobium	Nb	92.91
Berkelium	Bk	247 *	Nitrogen	N	14.01
Beryllium	Be	9.01	Nobelium	No	255 *
Bismuth	Bi	209.98	Osmium	Os	190.2
Boron	B	10.81	Oxygen	O	16.00
Bromine	Br	79.90	Palladium	Pd	106.4
Cadmium	Cd	112.40	Phosphorus	P	30.97
Calcium	Ca	40.08	Platinum	Pt	195.09
Californium	Cf	251 *	Plutonium	Pu	244 *
Carbon	C	12.01	Polonium	Po	209 *
Cerium	Ce	140.12	Potassium	K	39.10
Cesium	Cs	132.91	Praseodymium	Pr	140.91
Chlorine	Cl	35.45	Protactinium	Pa	231.04
Chromium	Cr	52.00	Radium	Ra	226.02
Cobalt	Co	58.93	Radon	Rn	222 *
Copper	Cu	63.55	Rhenium	Re	186.20
Curium	Cm	247 *	Rhodium	Rh	102.91
Dysprosium	Dy	162.50	Rubidium	Rb	85.47
Einsteinium	Es	254 *	Ruthenium	Ru	101.07
Erbium	Er	167.26	Samarium	Sm	150.35
Europium	Eu	151.96	Scandium	Sc	44.96
Fermium	Fm	257 *	Selenium	Se	78.96
Fluorine	F	19.00	Silicon	Si	28.09
Francium	Fr	223 *	Silver	Ag	107.87
Gadolinium	Gd	157.25	Sodium	Na	22.99
Gallium	Ga	69.72	Strontium	Sr	87.62
Germanium	Ge	72.59	Sulfur	S	32.06
Gold	Au	96.97	Tantalum	Ta	180.95
Hafnium	Hf	178.49	Technetium	Tc	97 *
Helium	He	4.003	Tellurium	Te	127.60
Holmium	Ho	164.93	Terbium	Tb	158.92
Hydrogen	H	1.008	Thallium	Tl	204.37
Indium	In	114.82	Thorium	Th	232.04
Iodine	I	126.90	Thulium	Tm	168.93
Iridium	Ir	192.22	Tin	Sn	118.69
Iron	Fe	55.85	Titanium	Ti	47.90
Krypton	Kr	83.8	Tungsten	W	183.85
Lanthanum	La	138.91	Uranium	U	238.03
Lawrencium	Lr	257 *	Vanadium	V	50.94
Lead	Pb	207.19	Xenon	Xe	131.3
Lithium	Li	6.94	Ytterbium	Yb	173.04
Lutetium	Lu	174.97	Yttrium	Y	88.91
Magnesium	Mg	24.31	Zinc	Zn	65.37
Manganese	Mn	54.94	Zirconium	Zr	91.22

* Approximate atomic weight from mass number.

ENGLISH SYSTEM WEIGHTS AND MEASURES

Measures of Length

1 mile = 1760 yards = 5280 feet.

1 yard = 3 feet = 36 inches.   1 foot = 12 inches.

   1 mil = 0.001 inch.   1 fathom = 2 yards = 6 feet.

   1 rod = 5.5. yards = 16.5 feet.   1 hand = 4 inches.   1 span = 9 inches.

   1 micro-inch = one millionth inch or 0.000001 inch.   (1 micron = one millionth meter = 0.00003937 inch.)

Surveyor's Measure

   1 mile = 8 furlongs = 80 chains.

   1 furlong = 10 chains = 220 yards.

   1 chain = 4 rods = 22 yards = 66 feet = 100 links.

   1 link = 7.92 inches.

Nautical Measure

   1 league = 3 nautical miles.

   1 nautical mile = 6076.10 feet = 1.1508 statute miles.   (The *knot*, which is a nautical unit of speed, is equivalent to a speed of 1 nautical mile per hour.)

One degree at the equator = 60 nautical miles = 69.047 statute miles. 360 degrees = 21,600 nautical miles = 24,856.8 statute miles = circumference at equator.

Square Measure

   1 square mile = 640 acres = 6400 square chains.

   1 acre = 10 square chains = 4840 square yards = 43,560 square feet.

   1 square chain = 16 square rods = 484 square yards = 4356 square feet.

   1 square rod = 30.25 square yards = 272.25 square feet = 625 square links.

   1 square yard = 9 square feet.

   1 square foot = 144 square inches.

An acre is equal to a square, the side of which is 208.7 feet.

Measure used for Diameters and Areas of Electric Wires

   1 circular inch = area of circle 1 inch in diameter = 0.7854 square inch.

   1 circular inch = 1,000,000 circular mils.

   1 square inch = 1.2732 circular inch = 1,273,239 circular mils.

A circular mil is the area of a circle 0.001 inch in diameter.

Cubic Measure

   1 cubic yard = 27 cubic feet.

   1 cubic foot = 1728 cubic inches.

The following measures are also used for wood and **masonry:**

   1 cord of wood = 4 × 4 × 8 feet = 128 cubic feet.

   1 perch of masonry = $16\frac{1}{2}$ × $1\frac{1}{2}$ × 1 foot = $24\frac{3}{4}$ cubic feet.

Shipping Measure

For measuring entire internal capacity of a vessel:

   1 register ton = 100 cubic feet.

For measurement of cargo:

   Approximately 40 cubic feet of merchandise is considered a shipping ton, unless that bulk would weigh more than 2000 pounds, in which case the freight charge may be based upon weight.

   40 cubic feet = 32.143 U. S. bushels = 31.16 Imperial bushels.

Dry Measure

1 bushel (U. S. or Winchester struck bushel) = 1.2445 cubic foot = 2150.42 cubic inches.
1 bushel = 4 pecks = 32 quarts = 64 pints.
1 peck = 8 quarts = 16 pints.
1 quart = 2 pints.
1 heaped bushel = $1\frac{1}{4}$ struck bushel.
1 cubic foot = 0.8036 struck bushel.
1 British Imperial bushel = 8 Imperial gallons = 1.2837 cubic foot = 2218.19 cubic inches.

Liquid Measure

1 U. S. gallon = 0.1337 cubic foot = 231 cubic inches = 4 quarts = 8 pints.
1 quart = 2 pints = 8 gills.
1 pint = 4 gills.
1 British Imperial gallon = 1.2009 U. S. gallon = 277.42 cubic inches.
1 cubic foot = 7.48 U. S. gallons.

Old Liquid Measure

1 tun = 2 pipes = 3 puncheons.
1 pipe or butt = 2 hogsheads = 4 barrels = 126 gallons.
1 puncheon = 2 tierces = 84 gallons.
1 hogshead = 2 barrels = 63 gallons.
1 tierce = 42 gallons.
1 barrel = $31\frac{1}{2}$ gallons.

Apothecaries' Fluid Measure

1 U. S. fluid ounce = 8 drachms = 1.805 cubic inch = $\frac{1}{128}$ U. S. gallon.
1 fluid drachm = 60 minims.
1 British fluid ounce = 1.732 cubic inch.

Measures of Weight

Avoirdupois or Commercial Weight

1 gross or long ton = 2240 pounds.
1 net or short ton = 2000 pounds.
1 pound = 16 ounces = 7000 grains.
1 ounce = 16 drachms = 437.5 grains.
The following measures for weight are now seldom used in the United States:
1 hundred-weight = 4 quarters = 112 pounds (1 gross or long ton = 20 hundred-weights); 1 quarter = 28 pounds; 1 stone = 14 pounds; 1 quintal = 100 pounds.

Troy Weight, used for Weighing Gold and Silver

1 pound = 12 ounces = 5760 grains.
1 ounce = 20 pennyweights = 480 grains.
1 pennyweight = 24 grains.
1 carat (used in weighing diamonds) = 3.086 grains.
1 grain Troy = 1 grain avoirdupois = 1 grain apothecaries' weight.

Apothecaries' Weight

1 pound = 12 ounces = 5760 grains.
1 ounce = 8 drachms = 480 grains.
1 drachm = 3 scruples = 60 grains.
1 scruple = 20 grains.

Measures of Pressure

1 pound per square inch = 144 pounds per square foot = 0.068 atmosphere = 2.042 inches of mercury at 62 degrees F. = 27.7 inches of water at 62 degrees F. = 2.31 feet of water at 62 degrees F.

1 atmosphere = 30 inches of mercury at 62 degrees F. = 14.7 pounds per square inch = 2116.3 pounds per square foot = 33.95 feet of water at 62 degrees F.

1 foot of water at 62 degrees F. = 62.355 pounds per square foot = 0.433 pound per square inch.

1 inch of mercury at 62 degrees F. = 1.132 foot of water = 13.58 inches of water = 0.491 pound per square inch.

Miscellaneous

1 great gross = 12 gross = 144 dozen.
1 gross = 12 dozen = 144 units.
1 dozen = 12 units.
1 score = 20 units.

1 quire = 24 sheets.
1 ream = 20 quires = 480 sheets.
1 ream printing paper = 500 sheets.

Circular Mil Gage for Electrical Wires

A.W.G. or B. & S. Gage	Diam. Mils	Circular Mils	A.W.G. or B. & S. Gage	Diam. Mils	Circular Mils	A.W.G. or B. & S. Gage	Diam. Mils	Circular Mils
0000	460	212,000	12	81	6530	27	14.2	202.
000	410	168,000	13	72	5180	28	12.6	160.
00	365	133.000	14	64	4110	29	11.3	127.
0	325	106,000	15	57	3260	30	10.0	101.
1	289	83,700	16	51	2580	31	8.9	79.7
2	258	66,400	17	45	2050	32	8.0	63.2
3	229	52,600	18	40	1620	33	7.1	50.1
4	204	41,700	19	36	1290	34	6.3	39.8
5	182	33,100	20	32	1020	35	5.6	31.5
6	162	26,300	21	28.5	810	36	5.0	25.0
7	144	20,800	22	25.3	642	37	4.5	19.8
8	128	16,500	23	22.6	509	38	4.0	15.7
9	114	13,100	24	20.1	404	39	3.5	12.5
10	102	10,400	25	17.9	320	40	3.1	9.9
11	91	8,230	26	15.9	254	..	....	....

* A circular mil is a unit of area that is applied to electrical wires and cables and is equal to the area of a circle one mil (.001 inch) in diameter. The area of any circle in circular mils is equal to the square of its diameter in mils.

Greek Letters

A	α	Alpha	E	ϵ	Epsilon	I	ι	Iota	N	ν	Nu	P	ρ	Rho	Φ ϕ Phi
B	β	Beta	Z	ζ	Zeta	K	κ	Kappa	Ξ	ξ	Xi	Σ	σ s	Sigma	X χ Chi
Γ	γ	Gamma	H	η	Eta	Λ	λ	Lambda	O	o	Omicron	T	τ	Tau	Ψ ψ Psi
Δ	δ	Delta	$\Theta \vartheta \theta$		Theta	M	μ	Mu	Π	π	Pi	Y	υ	Upsilon	Ω ω Omega

ENGLISH AND NON-SI METRIC CONVERSION FACTORS

Multiply	By	To Obtain
Angstrom units....................	3.939×10^{-9}	Inches
Angstrom units....................	.0001	Microns
Atmospheres........................	760.0	Millimeters of mercury (at 32F)
Atmospheres........................	29.921	Inches of mercury (at 32F)
Atmospheres........................	33.97	Feet of water (at 62F)
Atmospheres........................	14.697	Pounds per square inch
Barrels (U. S. liquid)..............	4.211	Cubic feet
Barrels (U. S. liquid)..............	31.5	Gallons
Barrels (oil).......................	42	Gallons
Boiler horsepower..................	33,475	Btu per hour
Boiler horsepower..................	34.5	Pounds water evaporated from and at 212F
British tons of refrigeration (commerical)	1.1045	Tons of refrigeration (U.S.)
British tons of refrigeration (theoretical).....................	1.1905	Tons of refrigeration (U.S.)
Btu.........................	.252	Calories (large)*
Btu.........................	252	Calories (small)*
Btu.........................	778	Foot pounds
Btu.........................	.000393	Horsepower hour
Btu.........................	.000293	Kilowatt hour
Btu.........................	.0010307	Pounds water evaporated from and at 212F
Btu per 24 hours	$.347 \times 10^{-5}$	Tons of refrigeration
Btu per hour......................	$.2986 \times 10^{-4}$	Boiler horsepower
Btu per hour......................	.252	Calories (large)* per hour
Btu per hour......................	.000393	Horsepower
Btu per hour per square foot........	.2712	Calories (small)* per hour per square centimeter
Btu per hour per square foot per F...	.488	Calories (small)* per hour per square centimeter per C
Btu per pound....................	.5556	Calories (large)* per kilogram
Btu per inch per square foot per hour per F	.0833	Btu per foot per sq. ft. per hour per F
Btu per inch per square foot per hour per F	1.24	Calories (small)* per centimeter per sq. cm. per hr. per C
Calories (large)*....................	3.968	British thermal units
Calories (large)*....................	1.1619	Watt-hours
Calories (large)* per hour...........	3.968	Btu per hour
Calories (large)* per hour...........	.00156	Horsepower

* The large calorie (or kilogram calorie) is equal to 1000 small (or gram) calories.

ENGLISH AND NON-SI METRIC CONVERSION FACTORS

Multiply	By	To Obtain
Calories (small)*per cm. per sq. cm. per hr. per C	.8064	Btu per inch per sq. ft. per hr. per F
Centimeters of mercury.	0.01316	Atmospheres
Centimeters of mercury.	0.4461	Feet of water
Centimeters of mercury.	136.0	Kilograms per square meter
Centimeters of mercury.	27.85	Pounds per square foot
Centimeters of mercury.	0.1934	Pounds per square inch
Cubic feet.	1,728.0	Cubic inches
Cubic feet.	7.48052	Gallons
Cubic feet of water.	62.37	Pounds (at 60F)
Cubic feet per minute.	0.1247	Gallons per second
Feet of water.	0.881	Inches of mercury (at 32F)
Feet of water.	304.8	Kilograms per square meter
Feet of water.	62.37	Pounds per square foot
Feet of water.	0.4335	Pounds per square inch
Feet per minute.	0.01667	Feet per second
Foot-pounds.	0.001286	British thermal units
Gallons (U.S.).	0.1337	Cubic feet
Gallons (U.S.).	231.0	Cubic inches
Gallons of water.	8.3453	Pounds of water (at 60F)
Horsepower.	33,000.0	Foot pounds per minute
Horsepower.	2,546.0	British thermal units per hour
Horsepower.	42.42	British thermal units per minute
Horsepower.	1.014	Horsepower (metric)
Horsepower (boiler)	33,475	British thermal units per hour
Inches of mercury (at 62F).	.03342	Atmospheres
Inches of mercury (at 62F).	13.57	Inches of water (at 62F)
Inches of mercury (at 62F).	1.131	Feet of water (at 62F)
Inches of mercury (at 62F).	345.3	Kilograms per square meter
Inches of mercury (at 62F).	70.73	Pounds per square foot
Inches of mercury (at 62F).	.4912	Pounds per square inch
Inches of water (at 62F)	.002458	Atmospheres
Inches of water (at 62F)	.07355	Inches of mercury
Inches of water (at 62F)	25.40	Kilograms per square meter
Inches of water (at 62F)	.03613	Pounds per square inch
Inches of water (at 62F)	5.202	Pounds per square foot

* The large calorie (or kilogram calorie) is equal to 1000 small (or gram) calories.

ENGLISH AND NON-SI METRIC CONVERSION FACTORS

Multiply	By	To Obtain
Kilowatt-hours	3415.0	British thermal units
Kilowatt-hours	860.5	Kilogram-calories
Latent heat of ice	143.33	British thermal units per pound
Pounds	7,000.0	Grains
Pounds of water (at 60F)	0.01602	Cubic feet
Pounds of water (at 60F)	27.68	Cubic inches
Pounds of water (at 60F)	0.1198	Gallons
Pounds of water evaporated from and at 212F	0.284	Kilowatt-hours
Pounds of water evaporated from and at 212F	0.381	Horsepower hours
Pounds of water evaporated from and at 212F	970.4	British thermal units
Pounds per square foot	4.883	Kilograms per square meter
Pounds per square inch	0.06804	Atmospheres
Pounds per square inch	2.309	Feet of water (at 62F)
Pounds per square inch	2.0416	Inches of mercury (at 62F)
Pounds per square inch	.0703	Kilograms per square centimeter
Tons of refrigeration	12,000.0	British thermal units per hour
Tons of refrigeration	200.0	British thermal units per minute
Tons of refrigeration	50.40323	Calories (large)* per min.
Watts	0.01434	Calories (large)* per min.
Watt-hours	3.415	British thermal units
Watt-hours	0.8605	Calories (large)* per min.

* The large calorie (or kilogram calorie) is equal to 1000 small (or gram) calories.

SI THE INTERNATIONAL SYSTEM OF UNITS—
THE MODERN METRIC SYSTEM

The International System of Units (SI) is a modernized version of the traditional MKSA Metric System (Metre,† Kilogram, Second, Ampere). Like MKSA, the SI system is based on decimal arithmetic. For each physical quantity, units of different sizes are formed by multiplying the value of a base number by powers of 10 or shifting decimal places.

SI Base Units. As a further simplification the SI system is founded on seven base units. Table 12-1 presents: the physical quantity specified by each base unit, the name of the base unit, its symbol and definition.

Table 12-1 SI Base Units

PHYSICAL QUANTITY	NAME OF UNIT	UNIT SYMBOL	DEFINITION
Basic SI Units			
Length	metre	m	1,650,763.73 wavelengths in vacuo of the radiation corresponding to the transition between the energy levels 2p10 and 5d5 of the krypton — 86 atom.
Mass	kilogram	kg	Mass of the international prototype which is in the custody of the Bureau International des Poids et Mesures (BIPM) at Sèvres, near Paris.
Time	second	s	The duration of 9,192,631,770 periods of the radiation corresponding to the transition between the two hyperfine levels of the ground state of the cesium-133 atom.
Electric Current	ampere	A	The constant current which, if maintained in two parallel rectilinear conductors of infinite length, of negligible circular cross section, and placed at a distance of one metre apart in a vacuum, would produce between these conductors a force equal to 2×10^{-7} N/m length.
Thermodynamic Temperature	degree kelvin	K	The fraction 1/273.16 of the thermodynamic temperature of the triple point of water.
Amount of Substance	mole	mol	The amount of substance of a system which contains as many elementary entities as there are atoms in 0.012 kilogram of carbon 12.
Luminous Intensity	candela	cd	The luminous intensity, in the perpendicular direction, of a surface of 1/600,000 square metre of a black body at the temperature of freezing platinum under a pressure of 101,325 newtons per square metre.

† Present usage of metric terminology permits -re or -er suffixes in the spelling of certain metric units: e.g.: metre or meter, and litre or liter are acceptable.

Table 12-2   SI Supplementary Units

Physical Quantity	SI Supplementary Unit Name	Symbol	Definition
Plane angle	radian	rad	The plane angle between two radii of a circle which cut off, on the circumference, an arc equal in length to the radius.
Solid angle	steradian	sr	The steradian is the solid angle which, having its vertex in the center of a sphere, cuts off an area of the surface of the sphere equal to that of a square with sides equal in length to the radius of the sphere.

SI Supplementary Units.   The two SI supplementary units, the radian and steradian, can be considered base units devoted to plane and solid angles, see Table 12-2.

SI Derived Units.   As their name suggests, SI Derived Units are designations of physical quantities obtained from algebraic combinations which can include base or supplementary SI units, or both.

For example, the SI derived unit for acceleration is m/s^2 or metre per second squared.   For angular acceleration, the derived unit is rad/s^2 or radian per second squared.

Certain compound derived units are expressed by or contain special names. In Table 12-3 it will be noted that the name of the SI unit is written in lower case, even though it may be drawn from a proper name.   Symbols are written mostly in lower case letters, except where they derive from a proper name.   In these instances, the first letter of the symbol is capitalized.

USING SI

Convenient Numerical Range.   A user commonly selects a particular decimal multiple or submultiple of an SI unit which is most convenient (see Table 12-4).   For example, multiples or submultiples may be chosen so that numerical values are within a practical range of 0.1 to 1000.   This range tends to eliminate awkward zeros, decimals, and exponential powers of 10.

Examples: Where W=watt, MW=megawatt, m=metre, mm=millimetre, s=second, A=ampere, and nA=nanoampere.

$265,000 \text{ W} = 0.265(10^6)\text{W} = 0.265 \text{ MW}$

$0.0000035 \text{ m}^2 = (3.5 \times 10^{-6})(10^3 \text{ mm})^2 = (3.5 \times 10^{-6})(10^6)\text{ mm}^2 = 3.5 \text{ mm}^2$

$71(10^{-6})\text{ m}^2/\text{s} = (71 \times 10^{-6})(10^3 \text{ mm})^2/\text{s} = (71 \times 10^{-6})(10^6)\text{ mm}^2/\text{s} = 71 \text{ mm}^2/\text{s}$

$9.73 \times 10^{-7}\text{ A} = 973 \times 10^{-9}\text{ A} = 973 \text{ nA}$

Table 12-3 SI Units Having Special Names

SI Units Having Special Names			
Force	newton	$N = kg \cdot m/s^2$	That force which, when applied to a body having a mass of one kilogramme, gives it an acceleration of one metre per second squared.
Work, Energy, Quantity of Heat	joule	$J = N \cdot m$	The work done when the point of application of a force of one newton is displaced through a distance of one metre in the direction of the force.
Power	watt	$W = J/s$	One joule per second.
Electric Charge	coulomb	$C = A \cdot s$	The quantity of electricity transported in one second by a current of one ampere.
Electric Potential	volt	$V = W/A$	The difference of potential between two points of a conducting wire carrying a constant current of one ampere, when the power dissipated between these points is equal to one watt.
Electric Capacitance	farad	$F = C/V$	The capacitance of a capacitor between the plates of which there appears a difference of potential of one volt when it is charged by a quantity of electricity equal to one coulomb.

Table 12-4 Factors and Prefixes for Forming Decimal Multiples and Sub-multiples of the SI Units

Factor by which the unit is multiplied	Prefix	Symbol	Factor by which the unit is multiplied	Prefix	Symbol
10^{12}	tera	T	10^{-2}	centi	c
10^9	giga	G	10^{-3}	milli	m
10^6	mega	M	10^{-6}	micro	μ
10^3	kilo	k	10^{-9}	nano	n
10^2	hecto	h	10^{-12}	pico	p
10	deka	da	10^{-15}	femto	f
10^{-1}	deci	d	10^{-18}	atto	a

Prefixes. Table 12-4 presents the factors, prefixes, and symbols for forming decimal multiples and submultiples of simple and compound SI units.

At present, a preferred method in the selection of multiple and submultiple prefixes for any SI base unit employs factors which differ in orders of 1000 or 10^3.

Example: Applying this preference to the base unit for length, which is the *metre*:

attometre (am), *femto*metre (fm), *pico*metre (pm), *nano*metre (nm), *micro*metre (μm), *milli*metre (mm), *metre* (m), *kilo*metre (km), *mega*metre (Mm), *giga*metre (Gm), *tera*metre (Tm).

Note that double prefixes and double prefix symbols are avoided.

Example: (10^{-12}) m = 1 pm, and *not* 1 $\mu\mu$m.

Also one of the SI Base Units, the kilogram for mass, already contains the prefix "kilo-," as a multiple of the gram. However, all multiples and submultiples for mass units are formed by adding prefixes and their symbols to the "gram."

Example: (10^{-6}) kg = (10^{-3}) g = 1 mg, and *not* 1 μkg.

Units of Mass and Force. Unlike the traditional metric system, which uses the kilogram-force, the unit of force in SI is the *newton*.

In SI, the *kilogram,* and its multiples and submultiples are restricted to mass. Mass indicates the quantity of matter contained within a body. Technically, the term "weight," and therefore the newton, is used when the influence of gravitational force is taken into account. Since the newton is defined as "that force which, when applied to a body having a mass of one kilogram, gives it an acceleration of one metre per second squared," (N = Kg $\cdot$ m/s^2), it is independent of gravitational acceleration "g."

As a result when using SI, the factor g disappears from a wide range of formulas in dynamics. In some formulas in statics, however, in which the weight of a body—rather than mass—is important, g does appear where it was formerly absent. The acceleration of free fall, g, actually varies with both time and location. However, the standard gravitational acceleration in SI is agreed upon as 9.80556 m/s^2.

SI with Non-SI Units. Generally, a combination of SI units with English or other non-SI units should be avoided. An example of this is "kg/gal." However, certain units outside SI are acceptable because of their convenience and importance in various scientific and technical fields.

Some examples of acceptable non-SI units are:

Time (SI unit = s) – minute (min), hour (h), day (d)

Plane angles (SI unit = rad) – degree ($^\circ$), minute ('), second (")

Liquid volume (SI unit = m^3) – litre (l), where 1 l = 1 dm^3

Mass (SI unit = kg) – tonne or metric ton (t), where 1 t = 1000 kg

Fluid pressure (SI unit = Pa) – bar (bar), where 1 bar = 10^5 Pa.

Though the SI thermodynamic temperature is in kelvin (K), where necessary the Celsius (formerly "centigrade") temperature scale can be used:

Kelvin temperature (K) = 273.15 + Celsius Temperature ($^\circ$C).

When using the above equation for different temperature values it will be noted that the interval of change in Kelvin temperature is the same as the interval of change in Celsius.

For an SI temperature interval, either the K or $^\circ$C designation may be used.

SI Metric Conversion Factors

(Symbols of SI units, multiples and submultiples are
given in parentheses in the right-hand column)

Multiply	By	To Obtain
LENGTH		
centimetre	0.03280840	foot
centimetre	0.3937008	inch
fathom	1.8288*	metre (m)
foot	0.3048*	metre (m)
foot	30.48*	centimetre (cm)
foot	304.8*	millimetre (mm)
inch	0.0254*	metre (m)
inch	2.54*	centimetre (cm)
inch	25.4*	millimetre (mm)
kilometre	0.6213712	mile [U. S. statute]
metre	39.37008	inch
metre	0.5468066	fathom
metre	3.280840	foot
metre	0.1988388	rod
metre	1.093613	yard
metre	0.0006213712	mile [U. S. statute]
microinch	0.0254*	micrometre [micron] (μm)
micrometre [micron]	39.37008	microinch
mile [U. S. statute]	1609.344*	metre (m)
mile [U. S. statute]	1.609344*	kilometre (km)
millimetre	0.003280840	foot
millimetre	0.03937008	inch
rod	5.0292*	metre (m)
yard	0.9144*	metre (m)
AREA		
acre	4046.856	metre^2 (m^2)
acre	0.4046856	hectare
centimetre^2	0.1550003	inch^2
centimetre^2	0.001076391	foot^2
foot^2	0.09290304*	metre^2 (m^2)
foot^2	929.0304*	centimetre^2 (cm^2)
foot^2	92,903.04*	millimetre^2 (mm^2)
hectare	2.471054	acre
inch^2	645.16*	millimetre^2 (mm^2)
inch^2	6.4516*	centimetre^2 (cm^2)
inch^2	0.00064516*	metre^2 (m^2)
metre^2	1550.003	inch^2
metre^2	10.763910	foot^2
metre^2	1.195990	yard^2
metre^2	0.0002471054	acre
millimetre^2	0.0001076387	foot^2
millimetre^2	0.001550003	inch^2
yard^2	0.8361274	metre^2 (m^2)

* Where an asterisk is shown, the figure is exact.

SI Metric Conversion Factors *(Continued)*

Multiply	By	To Obtain
VOLUME (including CAPACITY)		
centimetre³	0.06102376	inch³
foot³	0.02831685	metre³ (m³)
foot³	28.31685	litre
gallon [U. K. liquid]	0.004546092	metre³ (m³)
gallon [U. K. liquid]	4.546092	litre
gallon [U. S. liquid]	0.003785412	metre³ (m³)
gallon [U. S. liquid]	3.785412	litre
inch³	16,387.06	millimetre³ (mm³)
inch³	16.38706	centimetre³ (cm³)
inch³	0.00001638706	metre³ (m³)
litre	0.001*	metre³ (m³)
litre	0.2199692	gallon [U. K. liquid]
litre	0.2641720	gallon [U. S. liquid]
litre	0.03531466	foot³
metre³	219.9692	gallon [U. K. liquid]
metre³	264.1720	gallon [U. S. liquid]
metre³	35.31466	foot³
metre³	1.307951	yard³
metre³	1000.*	litre
metre³	61,023.76	inch³
millimetre³	0.00006102376	inch³
yard³	0.7645549	metre³ (m³)
VELOCITY, ACCELERATION, and FLOW		
centimetre/second	1.968504	foot/minute
centimetre/second	0.03280840	foot/second
centimetre/minute	0.3937008	inch/minute
foot/hour	0.00008466667	metre/second (m/s)
foot/hour	0.00508*	metre/minute
foot/hour	0.3048*	metre/hour
foot/minute	0.508*	centimetre/second
foot/minute	18.288*	metre/hour
foot/minute	0.3048*	metre/minute
foot/minute	0.00508*	metre/second (m/s)
foot/second	30.48*	centimetre/second
foot/second	18.288*	metre/minute
foot/second	0.3048*	metre/second (m/s)
foot/second²	0.3048*	metre/second² (m/s²)
foot³/minute	28.31685	litre/minute
foot³/minute	0.0004719474	metre³/second (m³/s)
gallon [U. S. liquid]/min.	0.003785412	metre³/minute
gallon [U. S. liquid]/min.	0.00006309020	metre³/second (m³/s)
gallon [U. S. liquid]/min.	0.06309020	litre/second
gallon [U. S. liquid]/min.	3.785412	litre/minute
gallon [U. K. liquid]/min.	0.004546092	metre³/minute
gallon [U. K. liquid]/min.	0.00007576820	metre³/second (m³/s)
inch/minute	25.4*	millimetre/minute
inch/minute	2.54*	centimetre/minute
inch/minute	0.0254*	metre/minute
inch/second²	0.0254*	metre/second² (m/s²)

* Where an asterisk is shown, the figure is exact.

SI Metric Conversion Factors (*Continued*)

Multiply	By	To Obtain
VELOCITY, ACCELERATION, and FLOW (*Continued*)		
kilometre/hour	0.6213712	mile/hour [U. S. statute]
litre/minute	0.03531466	foot^3/minute
litre/minute	0.2641720	gallon [U. S. liquid]/minute
litre/second	15.85032	gallon [U. S. liquid]/minute
mile/hour	1.609344*	kilometre/hour
millimetre/minute	0.03937008	inch/minute
metre/second	11,811.02	foot/hour
metre/second	196.8504	foot/minute
metre/second	3.280840	foot/second
metre/second^2	3.280840	foot/second^2
metre/second^2	39.37008	inch/second^2
metre/minute	3.280840	foot/minute
metre/minute	0.05468067	foot/second
metre/minute	39.37008	inch/minute
metre/hour	3.280840	foot/hour
metre/hour	0.05468067	foot/minute
metre^3/second	2118.880	foot^3/minute
metre^3/second	13,198.15	gallon [U. K. liquid]/minute
metre^3/second	15,850.32	gallon [U. S. liquid]/minute
metre^3/minute	219.9692	gallon [U. K. liquid]/minute
metre^3/minute	264.1720	gallon [U. S. liquid]/minute
MASS and DENSITY		
grain [1/7000 lb avoirdupois]	0.06479891	gram (g)
gram	15.43236	grain
gram	0.001*	kilogram (kg)
gram	0.03527397	ounce [avoirdupois]
gram	0.03215074	ounce [troy]
gram/centimetre^3	0.03612730	pound/inch^3
hundredweight [long]	50.80235	kilogram (kg)
hundredweight [short]	45.35924	kilogram (kg)
kilogram	1000.*	gram (g)
kilogram	35.27397	ounce [avoirdupois]
kilogram	32.15074	ounce [troy]
kilogram	2.204622	pound [avoirdupois]
kilogram	0.06852178	slug
kilogram	0.0009842064	ton [long]
kilogram	0.001102311	ton [short]
kilogram	0.001*	ton [metric]
kilogram	0.001*	tonne
kilogram	0.01968413	hundredweight [long]
kilogram	0.02204622	hundredweight [short]
kilogram/metre^3	0.06242797	pound/foot^3
kilogram/metre^3	0.01002242	pound/gallon [U. K. liquid]
kilogram/metre^3	0.008345406	pound/gallon [U. S. liquid]
ounce [avoirdupois]	28.34952	gram (g)
ounce [avoirdupois]	0.02834952	kilogram (kg)

* Where an asterisk is shown, the figure is exact.

SI Metric Conversion Factors *(Continued)*

Multiply	By	To Obtain
MASS and DENSITY *(Continued)*		
ounce [troy] ounce [troy]	31.10348 0.03110348	gram (g) kilogram (kg)
pound [avoirdupois]	0.4535924	kilogram (kg)
pound/foot³	16.01846	kilogram/metre³ (kg/m³)
pound/inch³	27.67990	gram/centimetre³ (g/cm³)
pound/gal [U. S. liquid]	119.8264	kilogram/metre³ (kg/m³)
pound/gal [U. K. liquid]	99.77633	kilogram/metre³ (kg/m³)
slug	14.59390	kilogram (kg)
ton [long 2240 lb]	1016.047	kilogram (kg)
ton [short 2000 lb]	907.1847	kilogram (kg)
ton [metric]	1000.*	kilogram (kg)
tonne	1000.*	kilogram (kg)
FORCE and FORCE/LENGTH		
dyne	0.00001*	newton (N)
kilogram-force	9.806650*	newton (N)
kilopond	9.806650*	newton (N)
newton newton newton newton newton newton	0.1019716 0.1019716 0.2248089 100,000.* 7.23301 3.596942	kilogram-force kilopond pound-force dyne poundal ounce-force
newton/metre newton/metre	0.005710148 0.06852178	pound/inch pound/foot
ounce-force	0.2780139	newton (N)
pound-force	4.448222	newton (N)
poundal	0.1382550	newton (N)
pound/inch	175.1268	newton/metre (N/m)
pound/foot	14.59390	newton/metre (N/m)
BENDING MOMENT or TORQUE		
dyne-centimetre	0.0000001*	newton-metre (N · m)
kilogram-metre	9.806650*	newton-metre (N · m)
ounce-inch ounce-inch	7.061552 0.007061552	newton-millimetre newton-metre (N · m)
newton-metre newton-metre newton-metre newton-metre	0.7375621 10,000,000.* 0.1019716 141.6119	pound-foot dyne-centimetre kilogram-metre ounce-inch
newton-millimetre	0.1416119	ounce-inch
pound-foot	1.355818	newton-metre (N · m)

* Where an asterisk is shown, the figure is exact.

SI Metric Conversion Factors (*Continued*)

Multiply	By	To Obtain
MOMENT OF INERTIA and SECTION MODULUS		
moment of inertia [kg · m²]	23.73036	pound-foot²
moment of inertia [kg · m²]	3417.171	pound-inch²
moment of inertia [lb · ft²]	0.04214011	kilogram-metre² (kg · m²)
moment of inertia [lb · inch²]	0.0002926397	kilogram-metre² (kg · m²)
moment of section [foot⁴]	0.008630975	metre⁴ (m⁴)
moment of section [inch⁴]	41.62314	centimetre⁴
moment of section [metre⁴]	115.8618	foot⁴
moment of section [centimetre⁴]	0.02402510	inch⁴
section modulus [foot³]	0.02831685	metre³ (m³)
section modulus [inch³]	0.00001638706	metre³ (m³)
section modulus [metre³]	35.31466	foot³
section modulus [metre³]	61,023.76	inch³
MOMENTUM		
kilogram-metre/second	7.233011	pound-foot/second
kilogram-metre/second	86.79614	pound-inch/second
pound-foot/second	0.1382550	kilogram-metre/second (kg · m/s)
pound-inch/second	0.01152125	kilogram-metre/second (kg · m/s)
PRESSURE and STRESS		
atmosphere [14.6959 lb/inch²]	101,325.	pascal (Pa)
bar	100,000.*	pascal (Pa)
bar	14.50377	pound/inch²
bar	100,000.*	newton/metre² (N/m²)
hectobar	0.6474898	ton [long]/inch²
kilogram/centimetre²	14.22334	pound/inch²
kilogram/metre²	9.806650*	newton/metre² (N/m²)
kilogram/metre²	9.806650*	pascal (Pa)
kilogram/metre²	0.2048161	pound/foot²
kilonewton/metre²	0.1450377	pound/inch²
newton/centimetre²	1.450377	pound/inch²
newton/metre²	0.00001*	bar
newton/metre²	1.0*	pascal (Pa)
newton/metre²	0.0001450377	pound/inch²
newton/metre²	0.1019716	kilogram/metre²
newton/millimetre²	145.0377	pound/inch²
pascal	0.00000986923	atmosphere
pascal	0.00001*	bar
pascal	0.1019716	kilogram/metre²
pascal	1.0*	newton/metre² (N/m²)
pascal	0.02088543	pound/foot²
pascal	0.0001450377	pound/inch²

* Where an asterisk is shown, the figure is exact.

SI Metric Conversion Factors (*Continued*)

Multiply	By	To Obtain
PRESSURE and STRESS (*Continued*)		
pound/foot²	4.882429	kilogram/metre²
pound/foot²	47.88026	pascal (Pa)
pound/inch²	0.06894757	bar
pound/inch²	0.07030697	kilogram/centimetre²
pound/inch²	0.6894757	newton/centimetre²
pound/inch²	6.894757	kilonewton/metre²
pound/inch²	6894.757	newton/metre² (N/m²)
pound/inch²	0.006894757	newton/millimetre² (N/mm²)
pound/inch²	6894.757	pascal (Pa)
ton [long]/inch²	1.544426	hectobar
ENERGY and WORK		
Btu [International Table]	1055.056	joule (J)
Btu [mean]	1055.87	joule (J)
calorie [mean]	4.19002	joule (J)
foot-pound	1.355818	joule (J)
foot-poundal	0.04214011	joule (J)
joule	0.0009478170	Btu [International Table]
joule	0.0009470863	Btu [mean]
joule	0.2386623	calorie [mean]
joule	0.7375621	foot-pound
joule	23.73036	foot-poundal
joule	0.9998180	joule [International U. S.]
joule	0.9999830	joule [U. S. legal, 1948]
joule [International U. S.]	1.000182	joule (J)
joule [U. S. legal, 1948]	1.000017	joule (J)
joule	0002777778	watt-hour
watt-hour	3600. *	joule (J)
POWER		
Btu [International Table]/hour	0.2930711	watt (W)
foot-pound/hour	0.0003766161	watt (W)
foot-pound/minute	0.02259697	watt (W)
horsepower [550 ft-lb/s]	0.7456999	kilowatt (kW)
horsepower [550 ft-lb/s]	745.6999	watt (W)
horsepower [electric]	746. *	watt (W)
horsepower [metric]	735.499	watt (W)
horsepower [U. K.]	745.70	watt (W)
kilowatt	1.341022	horsepower [550 ft-lb/s]
watt	2655.224	foot-pound/hour
watt	44.25372	foot-pound/minute
watt	0.001341022	horsepower [550 ft-lb/s]
watt	0.00134048.3	horsepower [electric]
watt	0.001359621	horsepower [metric]
watt	0.001341022	horsepower [U. K.]
watt	3.412141	Btu [International Table]/hour

* Where an asterisk is shown, the figure is exact.

SI Metric Conversion Factors (*Continued*)

Multiply	By	To Obtain
HEAT and HEAT FLOW†		
Btu†/degree Fahrenheit	1.899101	kilojoule/kelvin (kJ/K)
Btu/degree Rankine	1.899101	kilojoule/kelvin (kJ/K)
Btu/foot^2	11.35653	kilojoule/metre^2 (kJ/m^2)
Btu/foot^3	37.25895	kilojoule/metre^3 (kJ/m^3)
Btu/foot^3-degree Fahrenheit	67.06610	kilojoule/metre^3—kelvin (kJ/m^3•K)
Btu/hour	0.2930711	watt (W)
Btu/hour-foot^2	3.154591	watt/metre^2 (W/m^2)
Btu/hour-foot^2-degree Fahrenheit	5.678263	watt/metre^2-kelvin (W/(m^2•K))
Btu/pound	2.326*	kilojoule/kilogram (kJ/kg)
Btu/pound-degree Fahrenheit	4.1868*	kilojoule/kilogram-kelvin (kJ/kg•K)
Btu/pound-degree Rankine	4.1868*	kilojoule/kilogram-kelvin (kJ/kg•K)
Btu-inch/hour-foot^2-degree Fahrenheit	0.1442279	watt/metre-kelvin {W/(m•K)}
foot^2-hour-degree Fahrenheit/Btu	0.1761102	metre^2-kelvin/watt (m^2•K/W)
foot^2-hour-degree Fahrenheit/Btu-inch	6.933472	metre-kelvin/watt (m•K/W)
foot^2/hour	0.0000258064*	metre^2/second (m^2/s)
kilocalorie†	4.1868*	kilojoule
kilocalorie/degree centigrade	4.1868*	kilojoule/kelvin (kJ/K)
kilocalorie/kilogram-degree centigrade	4.1868*	kilojoule/kilogram-kelvin (kJ/kg•K)
kilocalorie/hour	1.163*	watt (W)
kilocalorie/hour-metre-degree centigrade	1.163*	watt/metre-kelvin {W/(m•K)}
kilocalorie†/hour-metre^2	1.163*	watt/metre^2 (W/m^2)
kilocalorie/hour-metre^2-degree centigrade	1.163*	watt/metre^2-kelvin (W/(m^2•K))
kilocalorie/metre^3-degree centigrade	4.1868*	kilojoule/metre^3-kelvin (kJ/m^3•K)
kilojoule	0.2388459	kilocalorie†
kilojoule/kelvin	0.5265651	Btu†/degree Fahrenheit
kilojoule/kelvin	0.5265651	Btu/degree Rankine
kilojoule/kelvin	0.2388459	kilocalorie/degree centigrade
kilojoule/kilogram	0.4299226	Btu/pound
kilojoule/kilogram-kelvin	0.2388459	Btu/pound-degree Fahrenheit
kilojoule/kilogram-kelvin	0.2388459	Btu/pound-degree Rankine
kilojoule/kilogram-kelvin	0.2388459	kilocalorie/kilogram-degree centigrade

*Where an asterisk is shown, the figure is exact.
†Btu and calorie conversion values are based on the International Table definitions for the Heat and Heat Flow section of this table. Also, the calorie designation used is the "small" or "gram" calorie. Therefore, the kilocalorie, where given, is equal to the "large" or "kilogram" calorie.

SI Metric Conversion Factors (*Continued*)

Multiply	By	To Obtain
	HEAT and HEAT FLOW† (*Continued*)	
kilojoule/metre^2	0.0880551	Btu/foot^2
kilojoule/metre^3	0.02683919	Btu/foot^3
kilojoule/metre^3-kelvin	0.01491066	Btu/foot^3-degree Fahrenheit
kilojoule/metre^3-kelvin	0.2388459	kilocalorie/metre^3-degree centigrade
kilowatt	0.2843451	ton refrigeration
metre-hour-degree centigrade/ kilocalorie	0.8598452	metre-kelvin/watt (m•K/W)
metre-kelvin/watt	0.1442279	foot^2-hour-degree Fahrenheit/ Btu-inch
metre-kelvin/watt	1.163*	metre-hour-degree centigrade/ kilocalorie
metre^2-hour-degree centigrade/ kilocalorie	0.8598452	metre^2-kelvin/watt (m^2•K/W)
metre^2-kelvin/watt	5.678263	foot^2-hour-degree Fahrenheit/ Btu
metre^2-kelvin/watt	1.163*	metre^2-hour-degree centigrade/kilocalorie
metre^2/second	38,750.08	foot^2/hour
ton refrigeration	3.516853	kilowatt (kW)
watt	3.412142	Btu/hour
watt	0.8598452	kilocalorie/hour
watt/foot^2	10.76391	watt/metre^2 (W/m^2)
watt/metre-kelvin	6.933472	Btu-inch/hour-foot^2-degree Fahrenheit
watt/metre-kelvin	0.8598452	kilocalorie/hour-metre-degree centigrade
watt/metre^2	0.3169983	Btu/hour-foot^2
watt/metre^2	0.8598452	kilocalorie-hour-metre^2
watt/metre^2	0.09290304*	watt/foot^2
watt/metre^2-kelvin	0.1761102	Btu/hour-foot^2-degree Fahrenheit
watt/metre^2-kelvin	0.8598452	kilocalorie/hour-metre^2- degree centigrade

TEMPERATURE		
To Convert From	To	Use Formula
temperature Celsius, t_C	temperature Kelvin, t_K	$t_K = t_C + 273.15$
temperature Fahrenheit, t_F	temperature Kelvin, t_K	$t_K = (t_F + 459.67)/1.8$
temperature Celsius, t_C	temperature Fahrenheit, t_F	$t_F = 1.8\,t_C + 32$
temperature Fahrenheit, t_F	temperature Celsius, t_C	$t_C = (t_F - 32)/1.8$
temperature Kelvin, t_K	temperature Celsius, t_C	$t_C = t_K - 273.15$
temperature Kelvin, t_K	temperature Fahrenheit, t_F	$t_F = 1.8\,t_K - 459.67$
temperature Kelvin, t_K	temperature Rankine, t_R	$t_R = 9/5\,t_K$
temperature Rankine, t_R	temperature Kelvin, t_K	$t_K = 5/9\,t_R$

* Where an asterisk is shown, the figure is exact.

SOLUTION OF EQUATIONS

An equation is a statement of equality between two expressions, as $5x = 105$. The unknown quantity in an equation is generally designated by the letter x. If there is more than one unknown quantity, the others are designated by letters also selected at the end of the alphabet, as y, z, u, t, etc.

An equation of the first degree is one which contains the unknown quantity only in the first power, as $3x = 9$. A quadratic equation is one which contains the unknown quantity in the second, but no higher, power, as $x^2 + 3x = 10$.

Solving Equations of the First Degree with One Unknown. — Transpose all the terms containing the unknown x to one side of the equals sign, and all the other terms to the other side. Combine and simplify the expressions as far as possible, and divide both sides by the coefficient of the unknown x. (See the rules given for transposition of formulas.)

Example:
$$22x - 11 = 15x + 10$$
$$22x - 15x = 10 + 11$$
$$7x = 21$$
$$x = 3$$

Solution of Equations of the First Degree with Two Unknowns. The form of the simplified equations is:
$$ax + by = c$$
$$a_1x + b_1y = c_1$$

Then,
$$x = \frac{cb_1 - c_1b}{ab_1 - a_1b} \qquad y = \frac{ac_1 - a_1c}{ab_1 - a_1b}$$

Example:
$$3x + 4y = 17$$
$$5x - 2y = 11$$
$$x = \frac{17 \times (-2) - 11 \times 4}{3 \times (-2) - 5 \times 4} = \frac{-34 - 44}{-6 - 20} = \frac{-78}{-26} = 3.$$

The value of y can now be most easily found by inserting the value of x in one of the equations:
$$5 \times 3 - 2y = 11; \quad 2y = 15 - 11 = 4; \quad y = 2.$$

Solution of Quadratic Equations with One Unknown. If the form of the equation is $ax^2 + bx + c = 0$, then
$$x = \frac{-b \pm \sqrt{b^2 - 4ac}}{2a}$$

Example: Given the equation, $1x^2 + 6x + 5 = 0$, then $a = 1$; $b = 6$ and $c = 5$.
$$x = \frac{-6 \pm \sqrt{6^2 - 4 \times 1 \times 5}}{2 \times 1} = \frac{(-6) + 4}{2} = -1;$$
or
$$\frac{(-6) - 4}{2} = -5$$

Example: A right-angle triangle has a hypotenuse 5 inches long and one side which is one inch longer than the other; find the lengths of the two sides.

Let $x =$ one side and $x + 1 =$ other side; then $x^2 + (x + 1)^2 = 5^2$ or $x^2 + x^2 + 2x + 1 = 25$; or $2x^2 + 2x = 24$; or $x^2 + x = 12$. Now referring to the basic formula, $ax^2 + bx = c$, we find, in this case, that $a = 1$; $b = 1$, and $c = 12$; hence
$$x = \frac{-1 \pm \sqrt{1 + 4 \times 1 \times 12}}{2 \times 1} = \frac{(-1) + 7}{2} = 3$$
or
$$x = \frac{(-1) - 7}{2} = -4$$

Since the positive value (3) would apply in this case, the lengths of the two sides are $x = 3$ inches and $x + 1 = 4$ inches.

Cubic Equations. If the given equation has the form: $x^3 + ax + b = 0$, then
$$x = \left(-\frac{b}{2} + \sqrt{\frac{a^3}{27} + \frac{b^2}{4}}\right)^{\frac{1}{3}} + \left(-\frac{b}{2} - \sqrt{\frac{a^3}{27} + \frac{b^2}{4}}\right)^{\frac{1}{3}}$$

The equation $x^3 + px^2 + qx + r = 0$, may be reduced to the form $x_1^3 + ax_1 + b = 0$ by substituting
$$x_1 - \frac{p}{3} \text{ for } x \text{ in the given equation.}$$

Principal Algebraic Expressions and Formulas

$a \times a = aa = a^2$
$a \times a \times a = aaa = a^3$
$a \times b = ab$
$a^2b^2 = (ab)^2$
$a^2a^3 = a^{2+3} = a^5$
$a^4 \div a^3 = a^{4-3} = a$
$a^0 = 1$
$a^2 - b^2 = (a + b)(a - b)$
$(a + b)^2 = a^2 + 2ab + b^2$
$(a - b)^2 = a^2 - 2ab + b^2$
$\dfrac{a^3}{b^3} = \left(\dfrac{a}{b}\right)^3$
$\dfrac{1}{a^3} = \left(\dfrac{1}{a}\right)^3 = a^{-3}$
$(a^2)^3 = a^{2 \times 3} = (a^3)^2 = a^6$
$a^3 + b^3 = (a + b)(a^2 - ab + b^2)$
$a^3 - b^3 = (a - b)(a^2 + ab + b^2)$
$(a + b)^3 = a^3 + 3a^2b + 3ab^2 + b^3$
$(a - b)^3 = a^3 - 3a^2b + 3ab^2 - b^3$

$\sqrt{a} \times \sqrt{a} = a$
$\sqrt[3]{a} \times \sqrt[3]{a} \times \sqrt[3]{a} = a$
$(\sqrt[3]{a})^3 = a$
$\sqrt[3]{a^2} = (\sqrt[3]{a})^2 = a^{\frac{2}{3}}$
$\sqrt[4]{\sqrt[3]{a}} = \sqrt[4 \times 3]{a} = \sqrt[3]{\sqrt[4]{a}}$
$\sqrt{a} + \sqrt{b} = \sqrt{a + b + 2\sqrt{ab}}$
$\sqrt[3]{ab} = \sqrt[3]{a} \times \sqrt[3]{b}$
$\sqrt[3]{\dfrac{a}{b}} = \dfrac{\sqrt[3]{a}}{\sqrt[3]{b}}$
$\sqrt[3]{\dfrac{1}{a}} = \dfrac{1}{\sqrt[3]{a}} = a^{-\frac{1}{3}}$

Formulas for Arithmetical Progression

To Find	Given			Use Equation
a	d	l	n	$a = l - (n-1)d$
	d	n	S	$a = \dfrac{S}{n} - \dfrac{n-1}{2} \times d$
	d	l	S	$a = \dfrac{d}{2} \pm \dfrac{1}{2}\sqrt{(2l+d)^2 - 8dS}$
	l	n	S	$a = \dfrac{2S}{n} - l$
d	a	l	n	$d = \dfrac{l-a}{n-1}$
	a	n	S	$d = \dfrac{2S - 2an}{n(n-1)}$
	a	l	S	$d = \dfrac{l^2 - a^2}{2S - l - a}$
	l	n	S	$d = \dfrac{2nl - 2S}{n(n-1)}$
l	a	d	n	$l = a + (n-1)d$
	a	d	S	$l = -\dfrac{d}{2} \pm \dfrac{1}{2}\sqrt{8dS + (2a-d)^2}$
	a	n	S	$l = \dfrac{2S}{n} - a$
	d	n	S	$l = \dfrac{S}{n} + \dfrac{n-1}{2} \times d$
n	a	d	l	$n = 1 + \dfrac{l-a}{d}$
	a	d	S	$n = \dfrac{d-2a}{2d} \pm \dfrac{1}{2d}\sqrt{8dS + (2a-d)^2}$
	a	l	S	$n = \dfrac{2S}{a+l}$
	d	l	S	$n = \dfrac{2l+d}{2d} \pm \dfrac{1}{2d}\sqrt{(2l+d)^2 - 8dS}$
S	a	d	n	$S = \dfrac{n}{2}[2a + (n-1)d]$
	a	d	l	$S = \dfrac{a+l}{2} + \dfrac{l^2 - a^2}{2d} = \dfrac{a+l}{2d}(l+d-a)$
	a	l	n	$S = \dfrac{n}{2}(a+l)$
	d	l	n	$S = \dfrac{n}{2}[2l - (n-1)d]$

Arithmetical Progression

An arithmetical progression is a series of numbers in which each consecutive term differs from the preceding one by a fixed amount called the *common difference*, *d*. Thus, 1, 3, 5, 7, etc., is an arithmetical progression where the difference *d* is 2. The difference in this case is *added* to the preceding term, and the progression is called increasing. In the series 13, 10, 7, 4, etc., the difference is (-3), and the progression is called decreasing. In any arithmetical progression (or part of progression) let

a = the first term considered; d = the common difference;
l = the last term considered; S = the sum of n terms.
n = the number of terms;

Then the general formulas are:

$$l = a + (n-1)\,d \quad \text{and} \quad S = \frac{a+l}{2} \times n$$

In these formulas d is positive in an increasing and negative in a decreasing progression. When any three of the five quantities above are given, the other two can be found by the formulas in the accompanying table of arithmetical progression.

Example: — In an arithmetical progression, the first term equals 5, and the last term 40. The difference is 7. Find the sum of the progression.

$$S = \frac{a+l}{2\,d}\,(l+d-a) = \frac{5+40}{2 \times 7}\,(40+7-5) = 135.$$

Geometrical Progression

A geometrical progression or a geometrical series is a series in which each term is derived by multiplying the preceding term by a constant multiplier called the *ratio*. When the ratio is greater than 1, the progression is increasing; when smaller than 1, it is decreasing. Thus, 2, 6, 18, 54, etc., is an increasing geometrical progression with a ratio of 3, while 24, 12, 6, etc., is a decreasing progression with a ratio of ½.

In any geometrical progression (or part of progression) let

a = the first term; r = the ratio of the progression;
l = the last (or nth) term; S = the sum of n terms.
n = the number of terms;

Then the general formulas are:

$$l = ar^{n-1} \quad \text{and} \quad S = \frac{rl-a}{r-1}$$

When any three of the five quantities above are given, the other two can be found by the formulas tabulated in the accompanying table. Geometrical progressions are used for finding the successive speeds in machine tool drives, in interest calculations, etc.

Example: — The lowest speed of a lathe is 20 R.P.M. The highest speed is 225 R.P.M. There are 18 speeds. Find the ratio between successive speeds.

$$\text{Ratio, } r = \sqrt[n-1]{\frac{l}{a}} = \sqrt[17]{\frac{225}{20}} = \sqrt[17]{11.25} = 1.153.$$

Formulas for Geometrical Progression

To Find	Given			Use Equation
a	l	n	r	$a = \dfrac{l}{r^{n-1}}$
	n	r	S	$a = \dfrac{(r-1)S}{r^n - 1}$
	l	r	S	$a = lr - (r-1)S$
	l	n	S	$a(S-a)^{n-1} = l(S-l)^{n-1}$
l	a	n	r	$l = ar^{n-1}$
	a	r	S	$l = \dfrac{1}{r}[a + (r-1)S]$
	a	n	S	$l(S-l)^{n-1} = a(S-a)^{n-1}$
	n	r	S	$l = \dfrac{S(r-1)r^{n-1}}{r^n - 1}$
n	a	l	r	$n = \dfrac{\log l - \log a}{\log r} + 1$
	a	r	S	$n = \dfrac{\log[a+(r-1)S] - \log a}{\log r}$
	a	l	S	$n = \dfrac{\log l - \log a}{\log(S-a) - \log(S-l)} + 1$
	l	r	S	$n = \dfrac{\log l - \log[lr - (r-1)S]}{\log r} + 1$
r	a	l	n	$r = \sqrt[n-1]{\dfrac{l}{a}}$
	a	n	S	$r^n = \dfrac{Sr}{a} + \dfrac{a-S}{a}$
	a	l	S	$r = \dfrac{S-a}{S-l}$
	l	n	S	$r^n = \dfrac{Sr^{n-1}}{S-l} - \dfrac{l}{S-l}$
S	a	n	r	$S = \dfrac{a(r^n - 1)}{r-1}$
	a	l	r	$S = \dfrac{lr - a}{r-1}$
	a	l	n	$S = \dfrac{\sqrt[n-1]{l^n} - \sqrt[n-1]{a^n}}{\sqrt[n-1]{l} - \sqrt[n-1]{a}}$
	l	n	r	$S = \dfrac{l(r^n - 1)}{(r-1)r^{n-1}}$

TRIGONOMETRIC FUNCTIONS

The sides and angles of right-angled triangles are related to each other in a series of ratios such that knowledge of any two sides or of either acute angle and one side allows computation of all other components of the triangle. These ratios, called trigonometric functions, are defined as follows:

The *sine* of an angle equals the opposite side divided by the hypotenuse. Hence, $\sin B = b \div c$ and $\sin A = a \div c$.

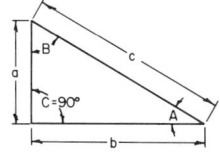

The *cosine* of an angle equals the adjacent side divided by the hypotenuse. Hence, $\cos B = a \div c$, and $\cos A = b \div c$.

The *tangent* of an angle equals the opposite side divided by the adjacent side. Hence, $\tan B = b \div a$, and $\tan A = a \div b$.

The *cotangent* of an angle equals the adjacent side divided by the opposite side. Hence, $\cot B = a \div b$, and $\cot A = b \div a$.

The *secant* of an angle is the reciprocal of the cosine and equals the hypotenuse divided by the adjacent side. *Hence, $\sec B = c \div a$, and $\sec A = c \div b$.*

The *cosecant* of an angle is the reciprocal of the sine and equals the hypotenuse divided by the opposite side. Hence, $\operatorname{cosec} B = c \div b$, and $\operatorname{cosec} A = c \div a$.

Note that the functions of the angles can be found in this manner only when the triangle is right-angled. Note also that the sum of the acute angles in a right-angled triangle is always $90°$, so that $\sin B = \cos(90 - B)$ etc.

Circular Functions

The radius of a circle and projection of that radius on a diameter always form two sides of a right-angled triangle with one of its acute angles at the center of the circle. Trigonometric functions of angles greater than $90°$ can be derived by fixing the position of the diameter and rotating the radius through an arc of $360°$. When conventional coordinate signs are applied with zero origin at the center of the circle, it is found that the functions are periodic in value and change in sign as the radius proceeds from one quadrant to another.

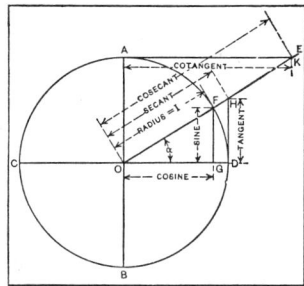

Functions of angles in any quadrant in terms of angles in the first quadrant are as follows:

	$-B$	$90° \pm B$	$180° \pm B$	$270° \pm B$	$360° \pm B$
sin	$-\sin B$	$+\cos B$	$\mp\sin B$	$-\cos B$	$\pm\sin B$
cos	$+\cos B$	$\mp\sin B$	$-\cos B$	$\pm\sin B$	$+\cos B$
tan	$-\tan B$	$\mp\cot B$	$\pm\tan B$	$\mp\cot B$	$\pm\tan B$
cot	$-\cot B$	$\mp\tan B$	$\pm\cot B$	$\mp\tan B$	$\pm\cot B$
sec	$+\sec B$	$\mp\csc B$	$-\sec B$	$\pm\csc B$	$+\sec B$
csc	$-\csc B$	$+\sec B$	$\mp\csc B$	$-\sec B$	$\pm\csc B$

Important Trigonometric Formulas

$$\sin^2 A + \cos^2 A = 1$$

$$\tan A = \frac{\sin A}{\cos A} = \frac{1}{\cot A}$$

$$\cot A = \frac{\cos A}{\sin A} = \frac{1}{\tan A}$$

$$\sec A = \frac{1}{\cos A}$$

$$\operatorname{cosec} A = \frac{1}{\sin A}$$

$$\sin A = \sqrt{1 - \cos^2 A} = \frac{\tan A}{\sqrt{1 + \tan^2 A}} = \frac{1}{\sqrt{1 + \cot^2 A}}$$

$$\cos A = \sqrt{1 - \sin^2 A} = \frac{1}{\sqrt{1 + \tan^2 A}} = \frac{\cot A}{\sqrt{1 + \cot^2 A}}$$

$$\sin(A + B) = \sin A \cos B + \cos A \sin B$$

$$\sin(A - B) = \sin A \cos B - \cos A \sin B$$

$$\cos(A + B) = \cos A \cos B - \sin A \sin B$$

$$\cos(A - B) = \cos A \cos B + \sin A \sin B$$

$$\tan(A + B) = \frac{\tan A + \tan B}{1 - \tan A \tan B}$$

$$\tan(A - B) = \frac{\tan A - \tan B}{1 + \tan A \tan B}$$

$$\cot(A + B) = \frac{\cot A \cot B - 1}{\cot B + \cot A}$$

$$\cot(A - B) = \frac{\cot A \cot B + 1}{\cot B - \cot A}$$

$$\tan A + \tan B = \frac{\sin(A + B)}{\cos A \cos B}$$

$$\tan A - \tan B = \frac{\sin(A - B)}{\cos A \cos B}$$

$$\cot A + \cot B = \frac{\sin(B + A)}{\sin A \sin B}$$

$$\cot A - \cot B = \frac{\sin(B - A)}{\sin A \sin B}$$

$$\sin^2 A - \sin^2 B = \cos^2 B - \cos^2 A = \sin(A + B)\sin(A - B)$$

$$\cos^2 A - \sin^2 B = \cos^2 B - \sin^2 A = \cos(A + B)\cos(A - B).$$

$$\sin A \sin B = \tfrac{1}{2}\cos(A - B) - \tfrac{1}{2}\cos(A + B).$$

$$\cos A \cos B = \tfrac{1}{2}\cos(A - B) + \tfrac{1}{2}\cos(A + B)$$

$$\sin A \cos B = \tfrac{1}{2}\sin(A + B) + \tfrac{1}{2}\sin(A - B)$$

$$\tan A \tan B = \frac{\tan A + \tan B}{\cot A + \cot B}$$

$$\cot A \cot B = \frac{\cot A + \cot B}{\tan A + \tan B}$$

$$\sin A = 2\sin\tfrac{1}{2}A \cos\tfrac{1}{2}A$$

$$\sin 2A = 2\sin A \cos A$$

$$\cos 2A = \cos^2 A - \sin^2 A = 1 - 2\sin^2 A = 2\cos^2 A - 1\,$$

$$\tan 2A = \frac{2\tan A}{1 - \tan^2 A} = \frac{2}{\cot A - \tan A}$$

$$\cot 2A = \frac{\cot^2 A - 1}{2\cot A} = \frac{\cot A - \tan A}{2}$$

$$\sin A = \frac{2\tan\tfrac{1}{2}A}{1 + \tan^2\tfrac{1}{2}A}$$

$$\cos A = \frac{1 - \tan^2\tfrac{1}{2}A}{1 + \tan^2\tfrac{1}{2}A}$$

$$2\sin^2 A = 1 - \cos 2A$$

$$2\cos^2 A = 1 + \cos 2A$$

Segments of Circles

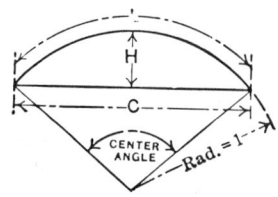

Length of Arc, Height of Segment, Length of Chord and Area of Segment for Angles from 1 to 180 degrees, and Radius = 1. — For other radii, multiply the values of L, H and C in the table by the given radius, and the values for areas, by the square of the radius.

Center Angle, Degrees	L	H	C	Area of Segment	Center Angle, Degrees	L	H	C	Area of Segment
1	0.01745	0.00004	0.01745	0.00000	46	0.803	0.0795	0.781	0.04176
2	0.03491	0.00015	0.03490	0.00000	47	0.820	0.0829	0.797	0.04448
3	0.05236	0.00034	0.05235	0.00001	48	0.838	0.0865	0.813	0.04731
4	0.06981	0.00061	0.06980	0.00003	49	0.855	0.0900	0.829	0.05025
5	0.08727	0.00095	0.08724	0.00006	50	0.873	0.0937	0.845	0.05331
6	0.10472	0.00137	0.10467	0.00010	51	0.890	0.0974	0.861	0.05649
7	0.12217	0.00186	0.12210	0.00015	52	0.908	0.1012	0.877	0.05978
8	0.13963	0.00243	0.13951	0.00023	53	0.925	0.1051	0.892	0.06319
9	0.15708	0.00308	0.15692	0.00032	54	0.942	0.1090	0.908	0.06673
10	0.17453	0.00380	0.17431	0.00044	55	0.960	0.1130	0.923	0.07039
11	0.19199	0.00460	0.19169	0.00059	56	0.977	0.1171	0.939	0.07417
12	0.20944	0.00548	0.20906	0.00076	57	0.995	0.1212	0.954	0.07808
13	0.22689	0.00643	0.22641	0.00097	58	1.012	0.1254	0.970	0.08212
14	0.24435	0.00745	0.24374	0.00121	59	1.030	0.1296	0.985	0.08629
15	0.26180	0.00855	0.26105	0.00149	60	1.047	0.1340	1.000	0.09059
16	0.27925	0.00973	0.27835	0.00181	61	1.065	0.1384	1.015	0.09502
17	0.29671	0.01098	0.29562	0.00217	62	1.082	0.1428	1.030	0.09958
18	0.31416	0.01231	0.31287	0.00257	63	1.100	0.1474	1.045	0.10428
19	0.33161	0.01371	0.33010	0.00302	64	1.117	0.1520	1.060	0.10911
20	0.34907	0.01519	0.34730	0.00352	65	1.134	0.1566	1.075	0.11408
21	0.36652	0.01674	0.36447	0.00408	66	1.152	0.1613	1.089	0.11919
22	0.38397	0.01837	0.38162	0.00468	67	1.169	0.1661	1.104	0.12443
23	0.40143	0.02007	0.39874	0.00535	68	1.187	0.1710	1.118	0.12982
24	0.41888	0.02185	0.41582	0.00607	69	1.204	0.1759	1.133	0.13535
25	0.43633	0.02370	0.43288	0.00686	70	1.222	0.1808	1.147	0.14102
26	0.45379	0.02563	0.44990	0.00771	71	1.239	0.1859	1.161	0.14683
27	0.47124	0.02763	0.46689	0.00862	72	1.257	0.1910	1.176	0.15279
28	0.48869	0.02970	0.48384	0.00961	73	1.274	0.1961	1.190	0.15889
29	0.50615	0.03185	0.50076	0.01067	74	1.291	0.2014	1.204	0.16514
30	0.52360	0.03407	0.51764	0.01180	75	1.309	0.2066	1.217	0.17154
31	0.54105	0.03637	0.53448	0.01301	76	1.326	0.2120	1.231	0.17808
32	0.55851	0.03874	0.55127	0.01429	77	1.344	0.2174	1.245	0.18477
33	0.57596	0.04118	0.56803	0.01566	78	1.361	0.2229	1.259	0.19160
34	0.59341	0.04369	0.58474	0.01711	79	1.379	0.2284	1.272	0.19859
35	0.61087	0.04628	0.60141	0.01864	80	1.396	0.2340	1.286	0.20573
36	0.62832	0.04894	0.61803	0.02027	81	1.414	0.2396	1.299	0.21301
37	0.64577	0.05168	0.63461	0.02198	82	1.431	0.2453	1.312	0.22045
38	0.66323	0.05448	0.65114	0.02378	83	1.449	0.2510	1.325	0.22804
39	0.68068	0.05736	0.66761	0.02568	84	1.466	0.2569	1.338	0.23578
40	0.69813	0.06031	0.68404	0.02767	85	1.483	0.2627	1.351	0.24367
41	0.71559	0.06333	0.70041	0.02976	86	1.501	0.2686	1.364	0.25171
42	0.73304	0.06642	0.71674	0.03195	87	1.518	0.2746	1.377	0.25990
43	0.75049	0.06958	0.73300	0.03425	88	1.536	0.2807	1.389	0.26825
44	0.76795	0.07282	0.74921	0.03664	89	1.553	0.2867	1.402	0.27677
45	0.78540	0.07612	0.76537	0.03915	90	1.571	0.2929	1.414	0.28540

Segments of Circles

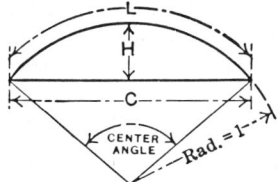

Length of Arc, Height of **Segment**, Length of Chord and Area of Segment for Angles from 1 to 180 degrees, and Radius = 1. — For other radii, multiply the values of L, H and C in the table by the given radius, and the values for areas, by the square of the radius.

Center Angle, Degrees	L	H	C	Area of Segment	Center Angle, Degrees	L	H	C	Area of Segment
91	1.588	0.2991	1.426	0.2942	136	2.374	0.6254	1.854	0.8395
92	1.606	0.3053	1.439	0.3032	137	2.391	0.6335	1.861	0.8545
93	1.623	0.3116	1.451	0.3123	138	2.409	0.6416	1.867	0.8697
94	1.641	0.3180	1.463	0.3215	139	2.426	0.6498	1.873	0.8850
95	1.658	0.3244	1.475	0.3309	140	2.443	0.6580	1.879	0.9003
96	1.675	0.3309	1.486	0.3405	141	2.461	0.6662	1.885	0.9158
97	1.693	0.3374	1.498	0.3502	142	2.478	0.6744	1.891	0.9313
98	1.710	0.3439	1.509	0.3601	143	2.496	0.6327	1.897	0.9470
99	1.728	0.3506	1.521	0.3701	144	2.513	0.6910	1.902	0.9627
100	1.745	0.3572	1.532	0.3803	145	2.531	0.6993	1.907	0.9786
101	1.763	0.3639	1.543	0.3906	146	2.548	0.7076	1.913	0.9945
102	1.780	0.3707	1.554	0.4010	147	2.566	0.7160	1.918	1.0105
103	1.798	0.3775	1.565	0.4117	148	2.583	0.7244	1.922	1.0266
104	1.815	0.3843	1.576	0.4224	149	2.600	0.7328	1.927	1.0427
105	1.833	0.3912	1.587	0.4333	150	2.618	0.7412	1.932	1.0590
106	1.850	0.3982	1.597	0.4444	151	2.635	0.7496	1.936	1.0753
107	1.867	0.4052	1.608	0.4556	152	2.653	0.7581	1.941	1.0917
108	1.885	0.4122	1.618	0.4669	153	2.670	0.7666	1.945	1.1082
109	1.902	0.4193	1.628	0.4784	154	2.688	0.7750	1.949	1.1247
110	1.920	0.4264	1.638	0.4901	155	2.705	0.7836	1.953	1.1413
111	1.937	0.4336	1.648	0.5019	156	2.723	0.7921	1.956	1.1580
112	1.955	0.4408	1.658	0.5138	157	2.740	0.8006	1.960	1.1747
113	1.972	0.4481	1.668	0.5259	158	2.758	0.8092	1.963	1.1915
114	1.990	0.4554	1.677	0.5381	159	2.775	0.8178	1.966	1.2083
115	2.007	0.4627	1.687	0.5504	160	2.792	0.8264	1.970	1.2252
116	2.025	0.4701	1.696	0.5629	161	2.810	0.8350	1.973	1.2422
117	2.042	0.4775	1.705	0.5755	162	2.827	0.8436	1.975	1.2592
118	2.059	0.4850	1.714	0.5883	163	2.845	0.8522	1.978	1.2763
119	2.077	0.4925	1.723	0.6012	164	2.862	0.8608	1.980	1.2933
120	2.094	0.5000	1.732	0.6142	165	2.880	0.8695	1.983	1.3105
121	2.112	0.5076	1.741	0.6273	166	2.897	0.8781	1.985	1.3277
122	2.129	0.5152	1.749	0.6406	167	2.915	0.8868	1.987	1.3449
123	2.147	0.5228	1.758	0.6540	168	2.932	0.8955	1.989	1.3621
124	2.164	0.5305	1.766	0.6676	169	2.950	0.9042	1.991	1.3794
125	2.182	0.5383	1.774	0.6812	170	2.967	0.9128	1.992	1.3967
126	2.199	0.5460	1.782	0.6950	171	2.984	0.9215	1.994	1.4140
127	2.217	0.5538	1.790	0.7090	172	3.002	0.9302	1.995	1.4314
128	2.234	0.5616	1.798	0.7230	173	3.019	0.9390	1.996	1.4488
129	2.251	0.5695	1.805	0.7372	174	3.037	0.9477	1.997	1.4662
130	2.269	0.5774	1.813	0.7514	175	3.054	0.9564	1.998	1.4836
131	2.286	0.5853	1.820	0.7658	176	3.072	0.9651	1.999	1.5010
132	2.304	0.5933	1.827	0.7803	177	3.089	0.9738	1.999	1.5185
133	2.321	0.6013	1.834	0.7950	178	3.107	0.9825	2.000	1.5359
134	2.339	0.6093	1.841	0.8097	179	3.124	0.9913	2.000	1.5533
135	2.356	0.6173	1.848	0.8245	180	3.142	1.0000	2.000	1.5708

Mensuration

In the following tables are given the areas of plane figures, together with other formulas relating to their dimensions and properties; the surfaces of solids; and the volumes of solids. The notation used in the formulas is, as far as possible, given in the illustration accompanying them; where this has not been possible, it is given at the beginning of each set of formulas.

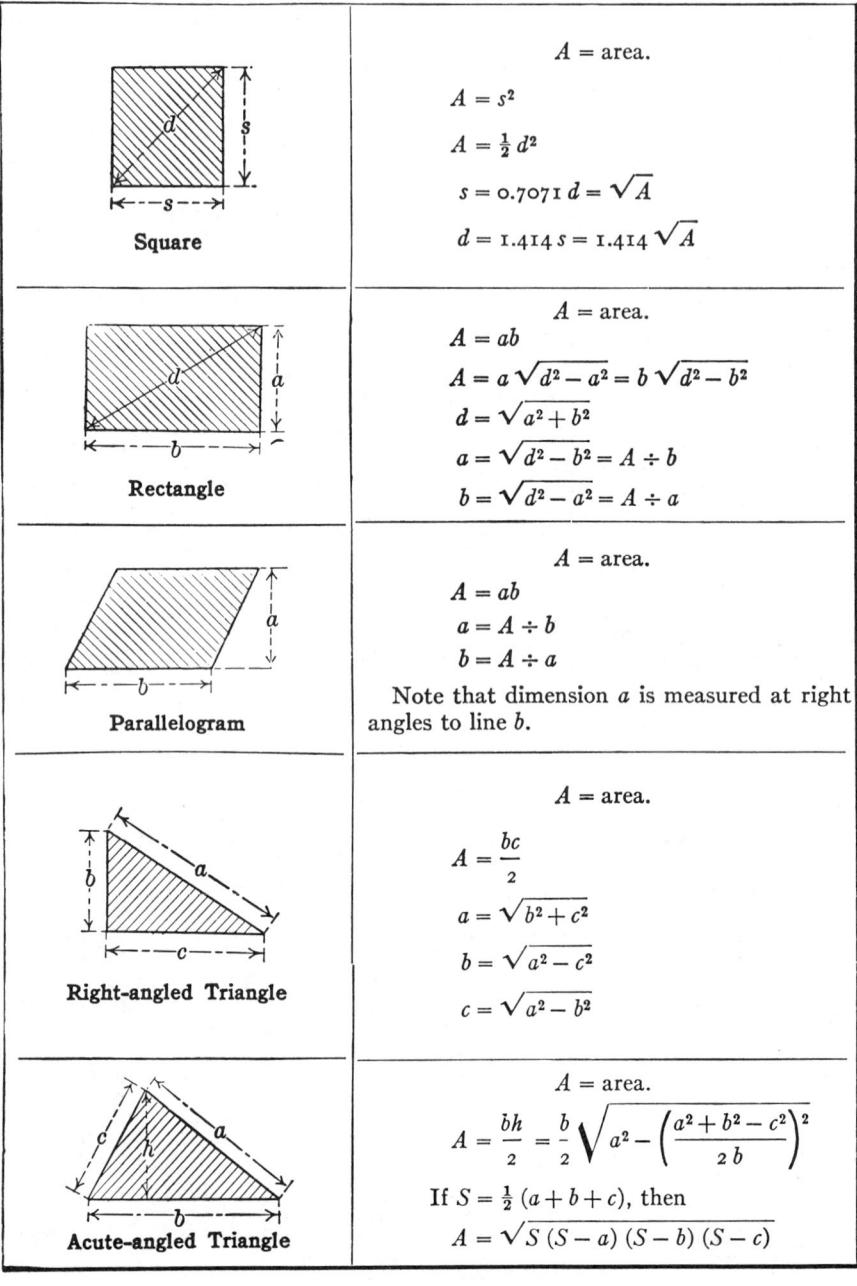

Square

A = area.

$A = s^2$

$A = \tfrac{1}{2} d^2$

$s = 0.7071\, d = \sqrt{A}$

$d = 1.414\, s = 1.414\, \sqrt{A}$

Rectangle

A = area.

$A = ab$

$A = a\sqrt{d^2 - a^2} = b\sqrt{d^2 - b^2}$

$d = \sqrt{a^2 + b^2}$

$a = \sqrt{d^2 - b^2} = A \div b$

$b = \sqrt{d^2 - a^2} = A \div a$

Parallelogram

A = area.

$A = ab$

$a = A \div b$

$b = A \div a$

Note that dimension a is measured at right angles to line b.

Right-angled Triangle

A = area.

$A = \dfrac{bc}{2}$

$a = \sqrt{b^2 + c^2}$

$b = \sqrt{a^2 - c^2}$

$c = \sqrt{a^2 - b^2}$

Acute-angled Triangle

A = area.

$A = \dfrac{bh}{2} = \dfrac{b}{2}\sqrt{a^2 - \left(\dfrac{a^2 + b^2 - c^2}{2b}\right)^2}$

If $S = \tfrac{1}{2}(a + b + c)$, then

$A = \sqrt{S\,(S - a)\,(S - b)\,(S - c)}$

Examples of the Use of the Formulas

Below are given a number of examples showing the use of the formulas on the opposite page. Each section of the page corresponds to the opposite section on the previous page, and the illustration on that page should be referred to. The notation used in the illustrations is also used in the examples given.

Square. — Assume that the side s of a square is 15 inches. Find the area and the length of the diagonal.

$$\text{Area} = A = s^2 = 15^2 = 225 \text{ square inches.}$$
$$\text{Diagonal} = d = 1.414 \ s = 1.414 \times 15 = 21.21 \text{ inches.}$$

The area of a square is 625 square inches. Find the length of the side s and the diagonal d.

$$s = \sqrt{A} = \sqrt{625} = 25 \text{ inches.}$$
$$d = 1.414 \sqrt{A} = 1.414 \times 25 = 35.35 \text{ inches.}$$

Rectangle. — The side a of a rectangle is 12 inches, and the area 70.5 square inches. Find the length of the side b, and the diagonal d.

$$b = A \div a = 70.5 \div 12 = 5.875 \text{ inches.}$$
$$d = \sqrt{a^2 + b^2} = \sqrt{12^2 + 5.875^2} = \sqrt{178.516} = 13.361 \text{ inches.}$$

The sides of a rectangle are 30.5 and 11 inches long. Find the area.

$$\text{Area} = a \times b = 30.5 \times 11 = 335.5 \text{ square inches.}$$

Parallelogram. — The base b of a parallelogram is 16 feet. The height a is 5.5 feet. Find the area.

$$\text{Area} = A = a \times b = 5.5 \times 16 = 88 \text{ square feet.}$$

The area of a parallelogram is 12 square inches. The height is 1.5 inch. Find the length of the base b.

$$b = A \div a = 12 \div 1.5 = 8 \text{ inches.}$$

Right-angled Triangle. — The sides b and c in a right-angled triangle are 6 and 8 inches. Find side a and the area.

$$a = \sqrt{b^2 + c^2} = \sqrt{6^2 + 8^2} = \sqrt{36 + 64} = \sqrt{100} = 10 \text{ inches.}$$
$$A = \frac{b \times c}{2} = \frac{6 \times 8}{2} = \frac{48}{2} = 24 \text{ square inches.}$$

If $a = 10$ and $b = 6$, had been known, but not c, the latter would have been found as follows:

$$c = \sqrt{a^2 - b^2} = \sqrt{10^2 - 6^2} = \sqrt{100 - 36} = \sqrt{64} = 8 \text{ inches.}$$

Acute-angled Triangle. — If $a = 10$, $b = 9$, and $c = 8$ inches, what is the area of the triangle?

$$A = \frac{b}{2} \sqrt{a^2 - \left(\frac{a^2 + b^2 - c^2}{2b}\right)^2} = \frac{9}{2} \sqrt{10^2 - \left(\frac{10^2 + 9^2 - 8^2}{2 \times 9}\right)^2} = 4.5 \sqrt{100 - \left(\frac{117}{18}\right)^2}$$

$$= 4.5 \sqrt{100 - 42.25} = 4.5 \sqrt{57.75} = 4.5 \times 7.60 = 34.20 \text{ square inches.}$$

Mensuration

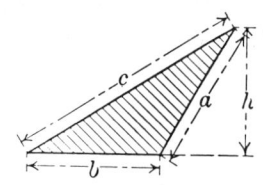

Obtuse-angled Triangle

A = area.

$$A = \frac{bh}{2} = \frac{b}{2}\sqrt{a^2 - \left(\frac{c^2 - a^2 - b^2}{2b}\right)^2}$$

If $S = \frac{1}{2}(a+b+c)$, then

$$A = \sqrt{S(S-a)(S-b)(S-c)}$$

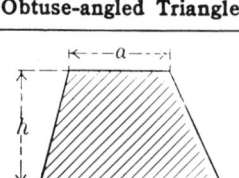

Trapezoid

A = area.

$$A = \frac{(a+b)h}{2}$$

Note: In England, this figure is called a *trapezium* and the one below it is known as a *trapezoid*, the terms being reversed.

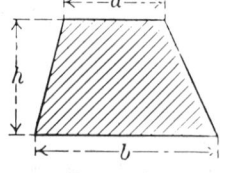

Trapezium

A = area.

$$A = \frac{(H+h)a + bh + cH}{2}$$

A trapezium can also be divided into two triangles as indicated by the dotted line. The area of each of these triangles is computed, and the results added to find the area of the trapezium.

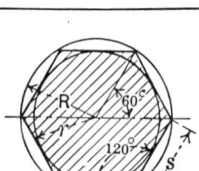

Regular Hexagon

A = area;
R = radius of circumscribed circle;
r = radius of inscribed circle.
$A = 2.598\,s^2 = 2.598\,R^2 = 3.464\,r^2$
$R = s = 1.155\,r$
$r = 0.866\,s = 0.866\,R$
$s = R = 1.155\,r$

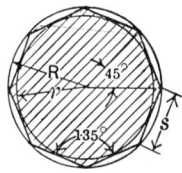

Regular Octagon

A = area;
R = radius of circumscribed circle;
r = radius of inscribed circle.
$A = 4.828\,s^2 = 2.828\,R^2 = 3.314\,r^2$
$R = 1.307\,s = 1.082\,r$
$r = 1.207\,s = 0.924\,R$
$s = 0.765\,R = 0.828\,r$

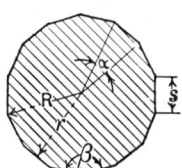

Regular Polygon

A = area; $\qquad n$ = number of sides.
$\alpha = 360° \div n \qquad \beta = 180° - \alpha$

$$A = \frac{nsr}{2} = \frac{ns}{2}\sqrt{R^2 - \frac{s^2}{4}}$$

$$R = \sqrt{r^2 + \frac{s^2}{4}}; \quad r = \sqrt{R^2 - \frac{s^2}{4}}; \quad s = 2\sqrt{R^2 - r^2}$$

Examples of the Use of the Formulas

Obtuse-angled Triangle. — The side $a = 5$, side $b = 4$, and side $c = 8$ inches. Find the area.

$$S = \tfrac{1}{2}(a + b + c) = \tfrac{1}{2}(5 + 4 + 8) = \tfrac{1}{2} \times 17 = 8.5$$

$$A = \sqrt{S(S-a)(S-b)(S-c)} = \sqrt{8.5(8.5-5)(8.5-4)(8.5-8)}$$

$$= \sqrt{8.5 \times 3.5 \times 4.5 \times 0.5} = \sqrt{66.937} = 8.18 \text{ square inches.}$$

Trapezoid. — Side $a = 23$ feet, side $b = 32$ feet, and height $h = 12$ feet. Find the area.

$$A = \frac{(a+b)h}{2} = \frac{(23+32)12}{2} = \frac{55 \times 12}{2} = \frac{660}{2} = 330 \text{ square feet.}$$

Trapezium. — Let $a = 10$, $b = 2$, $c = 3$, $h = 8$, and $H = 12$ inches. **Find** the area.

$$A = \frac{(H+h)a + bh + cH}{2} = \frac{(12+8)10 + 2 \times 8 + 3 \times 12}{2}$$

$$= \frac{20 \times 10 + 16 + 36}{2} = \frac{252}{2} = 126 \text{ square inches.}$$

Regular Hexagon. — The side s of a regular hexagon is 4 inches. Find the area and the radius r of the inscribed circle.

$$A = 2.598\,s^2 = 2.598 \times 4^2 = 2.598 \times 16 = 41.568 \text{ square inches.}$$

$$r = 0.866\,s = 0.866 \times 4 = 3.464 \text{ inches.}$$

What is the length of the side of a hexagon that is described about a circle of 5 inches radius? — Here $r = 5$. Hence,

$$s = 1.155\,r = 1.155 \times 5 = 5.775 \text{ inches.}$$

Regular Octagon. — Find the area and the length of the side of an octagon that is inscribed in a circle of 12 inches diameter.
Diameter of circumscribed circle = 12 inches; hence, $R = 6$ inches.

$$A = 2.828\,R^2 = 2.828 \times 6^2 = 2.828 \times 36 = 101.81 \text{ square inches.}$$

$$s = 0.765\,R = 0.765 \times 6 = 4.590 \text{ inches.}$$

Regular Polygon. — Find the area of a polygon having 12 sides, inscribed in a circle of 8 inches radius. The length of the side s is 4.141 inches.

$$A = \frac{ns}{2}\sqrt{R^2 - \frac{s^2}{4}} = \frac{12 \times 4.141}{2}\sqrt{8^2 - \frac{4.141^2}{4}} = 24.846\sqrt{59.713}$$

$$= 24.846 \times 7.727 = 191.98 \text{ square inches.}$$

Mensuration

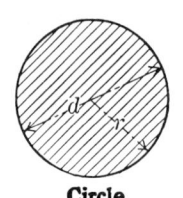

Circle

A = area; C = circumference.

$A = \pi r^2 = 3.1416\, r^2 = 0.7854\, d^2$

$C = 2\,\pi r = 6.2832\, r = 3.1416\, d$

$r = C \div 6.2832 = \sqrt{A \div 3.1416} = 0.564\,\sqrt{A}$

$d = C \div 3.1416 = \sqrt{A \div 0.7854} = 1.128\,\sqrt{A}$

Length of arc for center-angle of $1° = 0.008727\, d$

Length of arc for center-angle of $n° = 0.008727\, nd$

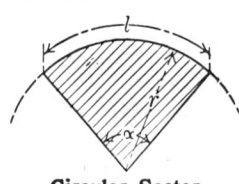

Circular Sector

A = area; l = length of arc; α = angle, in degrees.

$l = \dfrac{r \times \alpha \times 3.1416}{180} = 0.01745\, r\alpha = \dfrac{2\,A}{r}$

$A = \tfrac{1}{2}\,rl = 0.008727\,\alpha r^2$

$\alpha = \dfrac{57.296\, l}{r} \qquad r = \dfrac{2\,A}{l} = \dfrac{57.296\, l}{\alpha}$

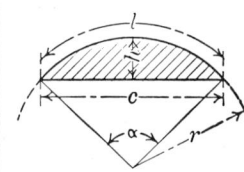

Circular Segment

A = area; l = length of arc; α = angle, in degrees.

$c = 2\,\sqrt{h\,(2\,r - h)} \qquad A = \tfrac{1}{2}\,[rl - c\,(r - h)]$

$r = \dfrac{c^2 + 4\,h^2}{8\,h} \qquad l = 0.01745\, r\alpha$

$h = r - \tfrac{1}{2}\,\sqrt{4\,r^2 - c^2} \qquad \alpha = \dfrac{57.296\, l}{r}$

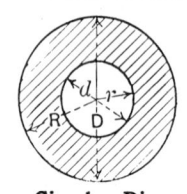

Circular Ring

A = area.

$A = \pi\,(R^2 - r^2) = 3.1416\,(R^2 - r^2)$

$\qquad = 3.1416\,(R + r)\,(R - r)$

$\qquad = 0.7854\,(D^2 - d^2) = 0.7854\,(D + d)\,(D - d)$

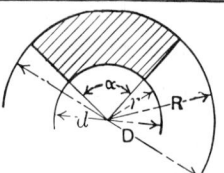

Circular Ring Sector

A = area; α = angle, in degrees.

$A = \dfrac{\alpha\pi}{360}\,(R^2 - r^2) = 0.00873\,\alpha\,(R^2 - r^2)$

$\quad = \dfrac{\alpha\pi}{4 \times 360}\,(D^2 - d^2) = 0.00218\,\alpha\,(D^2 - d^2)$

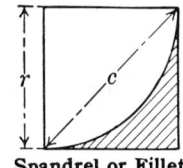

Spandrel or Fillet

A = area.

$A = r^2 - \dfrac{\pi r^2}{4} = 0.215\, r^2$

$\quad = 0.1075\, c^2$

Examples of the Use of the Formulas

Circle. — Find the area A and circumference C of a circle with a diameter of $2\frac{3}{4}$ inches.

$A = 0.7854\, d^2 = 0.7854 \times 2.75^2 = 0.7854 \times 2.75 \times 2.75 = 5.9396$ square inches.

$C = 3.1416\, d = 3.1416 \times 2.75 = 8.6394$ inches.

The area of a circle is 16.8 square inches. Find its diameter.

$d = 1.128\, \sqrt{A} = 1.128\, \sqrt{16.8} = 1.128 \times 4.099 = 4.624$ inches.

Circular Sector. — The radius of a circle is $1\frac{1}{2}$ inch, and angle α of a sector of the circle is 60 degrees. Find the area of the sector and the length of arc l.

$A = 0.008727\, \alpha r^2 = 0.008727 \times 60 \times 1.5^2 = 0.5236 \times 1.5 \times 1.5 = 1.178$ sq. inch.

$l = 0.01745\, r\alpha = 0.01745 \times 1.5 \times 60 = 1.5705$ inch.

Circular Segment. — The radius r of a circular segment is 60 inches and the height h is 8 inches. Find the length of the chord c.

$c = 2\,\sqrt{h\,(2\,r - h)} = 2\,\sqrt{8 \times (2 \times 60 - 8)} = 2\,\sqrt{896} = 2 \times 29.93 = 59.86$ inches.

If $c = 16$, and $h = 6$ inches, what is the radius of the circle of which the segment is a part?

$$r = \frac{c^2 + 4\,h^2}{8\,h} = \frac{16^2 + 4 \times 6^2}{8 \times 6} = \frac{256 + 144}{48} = \frac{400}{48} = 8\tfrac{1}{3} \text{ inches.}$$

Circular Ring. — Let the outside diameter $D = 12$ inches and the inside diameter $d = 8$ inches. Find area of ring.

$A = 0.7854\,(D^2 - d^2) = 0.7854\,(12^2 - 8^2) = 0.7854\,(144 - 64) = 0.7854 \times 80$
$= 62.83$ square inches.

By the alternative formula:

$A = 0.7854\,(D + d)\,(D - d) = 0.7854\,(12 + 8)\,(12 - 8) = 0.7854 \times 20 \times 4$
$= 62.83$ square inches.

Circular Ring Sector. — Find the area, if the outside radius $R = 5$ inches, the inside radius $r = 2$ inches, and $\alpha = 72$ degrees.

$A = 0.00873\,\alpha\,(R^2 - r^2) = 0.00873 \times 72\,(5^2 - 2^2)$
$= 0.6286\,(25 - 4) = 0.6286 \times 21 = 13.2$ square inches.

Spandrel or Fillet. — Find the area of a spandrel, the radius of which is 0.7 inch.

$A = 0.215\,r^2 = 0.215 \times 0.7^2 = 0.215 \times 0.7 \times 0.7 = 0.105$ square inch.

If chord c were given as 2.2 inches, what would be the area?

$A = 0.1075\,c^2 = 0.1075 \times 2.2^2 = 0.1075 \times 4.84 = 0.520$ square inch.

Mensuration

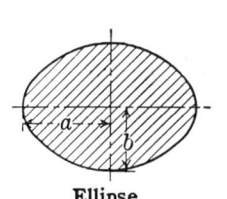

Ellipse

A = area; P = perimeter or circumference.

$A = \pi ab = 3.1416\,ab$.

An approximate formula for the perimeter is:

$$P = 3.1416\sqrt{2\,(a^2+b^2)}$$

A closer approximation is:

$$P = 3.1416\sqrt{2\,(a^2+b^2) - \frac{(a-b)^2}{2.2}}$$

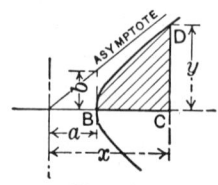

Hyperbola

A = area BCD.

$$A = \frac{xy}{2} - \frac{ab}{2}\,\text{hyp. log}\left(\frac{x}{a}+\frac{y}{b}\right)$$

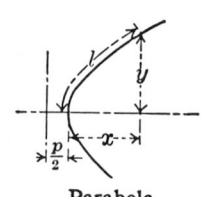

Parabola

l = length of arc.

$$l = \frac{p}{2}\left[\sqrt{\frac{2x}{p}\left(1+\frac{2x}{p}\right)}+\text{hyp. log}\left(\sqrt{\frac{2x}{p}}+\sqrt{1+\frac{2x}{p}}\right)\right]$$

When x is small in proportion to y, the following is a close approximation:

$$l = y\left[1+\frac{2}{3}\left(\frac{x}{y}\right)^2 - \frac{2}{5}\left(\frac{x}{y}\right)^4\right], \text{ or } l = \sqrt{y^2+\frac{4}{3}x^2}$$

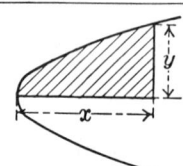

Parabola

A = area.

$$A = \tfrac{2}{3}xy$$

(The area is equal to two-thirds of the rectangle which has x for its base and y for its height.)

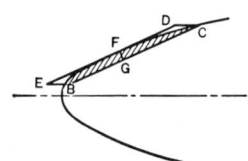

Segment of Parabola

A = area.

Area $BFC = A = \tfrac{2}{3}$ area of parallelogram $BCDE$.

If FG is the height of the segment, measured at right angles to BC, then:

Area of segment $BFC = \tfrac{2}{3}\,BC \times FG$

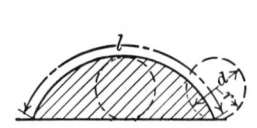

Cycloid

A = area; l = length of cycloid.

$$A = 3\,\pi r^2 = 9.4248\,r^2 = 2.3562\,d^2$$

$$= 3 \times \text{area of generating circle}$$

$$l = 8\,r = 4\,d$$

Examples of the Use of the Formulas

Ellipse. — The larger or major axis is 8 inches. The smaller or minor axis is 6 inches. Find the area and the approximate circumference. Here, then, $a = 4$, and $b = 3$.

$$A = 3.1416\,ab = 3.1416 \times 4 \times 3 = 37.699 \text{ square inches.}$$

$$P = 3.1416\sqrt{2\,(a^2 + b^2)} = 3.1416 \times \sqrt{2\,(4^2 + 3^2)} = 3.1416 \times \sqrt{2 \times 25}$$

$$= 3.1416\sqrt{50} = 3.1416 \times 7.071 = 22.214 \text{ inches.}$$

Hyperbola. — The half-axes a and b are 3 and 2 inches, respectively. Find area shown shaded in illustration for $x = 8$ and $y = 5$.

Inserting the known values in the formula:

$$A = \frac{8 \times 5}{2} - \frac{3 \times 2}{2} \times \text{hyp. log}\left(\frac{8}{3} + \frac{5}{2}\right) = 20 - 3 \times \text{hyp. log } 5.167$$

$$= 20 - 3 \times 1.6423 = 20 - 4.927 = 15.073 \text{ square inches.}$$

Parabola. — If $x = 2$ and $y = 24$ feet, what is the approximate length l of the parabolic curve?

$$l = y\left[1 + \frac{2}{3}\left(\frac{x}{y}\right)^2 - \frac{2}{5}\left(\frac{x}{y}\right)^4\right] = 24\left[1 + \frac{2}{3}\left(\frac{2}{24}\right)^2 - \frac{2}{5}\left(\frac{2}{24}\right)^4\right]$$

$$= 24\left[1 + \frac{2}{3} \times \frac{1}{144} - \frac{2}{5} \times \frac{1}{20,736}\right] = 24 \times 1.0046 = 24.11 \text{ feet.}$$

Parabola. — Let the dimension x in the illustration be 15 inches, and y, 9 inches. Find the area of the shaded portion of the parabola.

$$A = \tfrac{2}{3} \times xy = \tfrac{2}{3} \times 15 \times 9 = 10 \times 9 = 90 \text{ square inches.}$$

Segment of Parabola. — The length of the chord $BC = 19.5$ inches. The distance between lines BC and DE, measured at right angles to BC, is 2.25 inches. This is the height of the segment. Find the area.

$$\text{Area} = A = \tfrac{2}{3}\,BC \times FG = \tfrac{2}{3} \times 19.5 \times 2.25 = 29.25 \text{ square inches.}$$

Cycloid. — The diameter of the generating circle of a cycloid is 6 inches. Find the length l of the cycloidal curve, and the area enclosed between the curve and the base line.

$$l = 4\,d = 4 \times 6 = 24 \text{ inches.}$$

$$A = 2.3562\,d^2 = 2.3562 \times 6^2 = 2.3562 \times 36 = 84.82 \text{ square inches.}$$

Volume of Solids

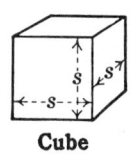

Cube

$V =$ volume.

$V = s^3$

$s = \sqrt[3]{V}$

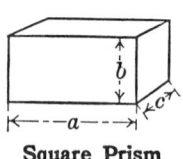

Square Prism

$V =$ volume.

$V = abc$

$$a = \frac{V}{bc} \qquad b = \frac{V}{ac} \qquad c = \frac{V}{ab}$$

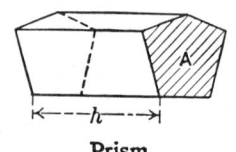

Prism

$V =$ volume; $A =$ area of end surface.

$$V = h \times A$$

The area A of the end surface is found by the formulas for areas of plane figures on the preceding pages. Height h must be measured perpendicular to end surface.

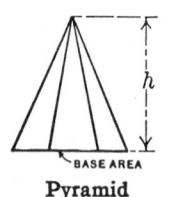

BASE AREA

Pyramid

$V =$ volume.

$$V = \tfrac{1}{3} h \times \text{area of base.}$$

If the base is a regular polygon with n sides, and $s =$ length of side, $r =$ radius of inscribed circle, and $R =$ radius of circumscribed circle, then:

$$V = \frac{nsrh}{6} = \frac{nsh}{6} \sqrt{R^2 - \frac{s^2}{4}}$$

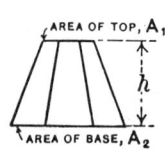

AREA OF TOP, A_1

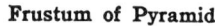

AREA OF BASE, A_2

Frustum of Pyramid

$V =$ volume.

$$V = \frac{h}{3} \left(A_1 + A_2 + \sqrt{A_1 \times A_2} \right)$$

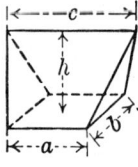

Wedge

$V =$ volume.

$$V = \frac{(2\,a + c)\, bh}{6}$$

Examples of the Use of the Formulas

Cube. — The side of a cube equals 9.5 inches. Find its volume.

Volume $= V = s^3 = 9.5^3 = 9.5 \times 9.5 \times 9.5 = 857.375$ cubic inches.

The volume of a cube is 231 cubic inches. What is the length of the side?

$$s = \sqrt[3]{V} = \sqrt[3]{231} = 6.136 \text{ inches.}$$

Square Prism. — In a square prism, $a = 6$, $b = 5$, $c = 4$. Find the volume.

$$V = a \times b \times c = 6 \times 5 \times 4 = 120 \text{ cubic inches.}$$

How high should a box be made to contain 25 cubic feet, if it is 4 feet long and $2\frac{1}{2}$ feet wide? Here, $a = 4$, $c = 2.5$, and $V = 25$. Then,

$$b = \text{depth} = \frac{V}{ac} = \frac{25}{4 \times 2.5} = \frac{25}{10} = 2.5 \text{ feet.}$$

Prism. — A prism having for its base a regular hexagon with a side s of 3 inches, is 10 inches high. Find the volume.

Area of hexagon $= A = 2.598\, s^2 = 2.598 \times 9 = 23.382$ square inches.

Volume of prism $= h \times A = 10 \times 23.382 = 233.82$ cubic inches.

Pyramid. — A pyramid, having a height of 9 feet, has a base formed by a rectangle, the sides of which are 2 and 3 feet, respectively. Find the volume.

Area of base $= 2 \times 3 = 6$ square feet; $\quad h = 9$ feet.

Volume $= V = \frac{1}{3} h \times$ area of base $= \frac{1}{3} \times 9 \times 6 = 18$ cubic feet.

Frustum of Pyramid. — The pyramid in the previous example is cut off $4\frac{1}{2}$ feet from the base, the upper part being removed. The sides of the rectangle forming the top surface of the frustum are, then, 1 and $1\frac{1}{2}$ foot long, respectively. Find the volume of the frustum.

Area of top $= A_1 = 1 \times 1\frac{1}{2} = 1\frac{1}{2}$ sq. ft. Area of base $= A_2 = 2 \times 3 = 6$ sq. ft.

$$V = \frac{4.5}{3}\left(1.5 + 6 + \sqrt{1.5 \times 6}\right) = 1.5\left(7.5 + \sqrt{9}\right) = 1.5 \times 10.5 = 15.75 \text{ cubic feet.}$$

Wedge. — Let $a = 4$ inches, $b = 3$ inches, and $c = 5$ inches. The height $h = 4.5$ inches. Find the volume.

$$V = \frac{(2a+c)\,bh}{6} = \frac{(2 \times 4 + 5) \times 3 \times 4.5}{6} = \frac{(8+5) \times 13.5}{6} = \frac{13 \times 13.5}{6}$$

$$= \frac{175.5}{6} = 29.25 \text{ cubic inches.}$$

Volume of Solids

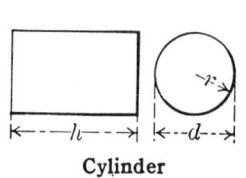

Cylinder

V = volume; S = area of cylindrical surface.

$V = 3.1416\, r^2 h = 0.7854\, d^2 h$

$S = 6.2832\, rh = 3.1416\, dh$

Total area A of cylindrical surface and end surfaces:

$A = 6.2832\, r\,(r + h) = 3.1416\, d\,(\tfrac{1}{2}\, d + h)$

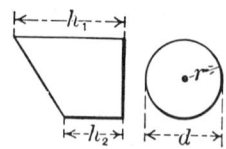

Portion of Cylinder

V = volume; S = area of cylindrical surface.

$V = 1.5708\, r^2\,(h_1 + h_2) = 0.3927\, d^2\,(h_1 + h_2)$

$S = 3.1416\, r\,(h_1 + h_2) = 1.5708\, d\,(h_1 + h_2)$

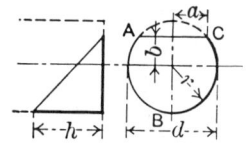

Portion of Cylinder

V = volume; S = area of cylindrical surface.

$$V = \left(\frac{2}{3}\, a^3 \pm b \times \text{area } ABC\right) \frac{h}{r \pm b}$$

$$S = (ad \pm b \times \text{length of arc } ABC)\, \frac{h}{r \pm b}$$

Use $+$ when base area is larger, and $-$ when base area is less than one-half the base circle.

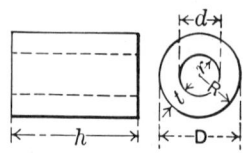

Hollow Cylinder

V = volume.

$V = 3.1416\, h\,(R^2 - r^2) = 0.7854\, h\,(D^2 - d^2)$

$\quad = 3.1416\, ht\,(2R - t) = 3.1416\, ht\,(D - t)$

$\quad = 3.1416\, ht\,(2r + t) = 3.1416\, ht\,(d + t)$

$\quad = 3.1416\, ht\,(R + r) = 1.5708\, ht\,(D + d)$

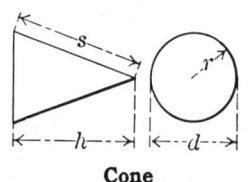

Cone

V = volume; A = area of conical surface.

$$V = \frac{3.1416\, r^2 h}{3} = 1.0472\, r^2 h = 0.2618\, d^2 h$$

$A = 3.1416\, r \sqrt{r^2 + h^2} = 3.1416\, rs = 1.5708\, ds$

$$s = \sqrt{r^2 + h^2} = \sqrt{\frac{d^2}{4} + h^2}$$

Frustum of Cone

V = volume; A = area of conical surface.

$V = 1.0472\, h\,(R^2 + Rr + r^2) = 0.2618\, h\,(D^2 + Dd + d^2)$

$A = 3.1416\, s\,(R + r) = 1.5708\, s\,(D + d)$

$a = R - r \qquad s = \sqrt{a^2 + h^2} = \sqrt{(R - r)^2 + h^2}$

Examples of the Use of the Formulas

Cylinder. — The diameter of a cylinder is $2\frac{1}{2}$ inches. The length or height is 20 inches. Find the volume, and the area of the cylindrical surface S.

$V = 0.7854\, d^2h = 0.7854 \times 2\frac{1}{2}^2 \times 20 = 0.7854 \times 6.25 \times 20 = 98.17$ cubic inches.

$S = 3.1416\, dh = 3.1416 \times 2\frac{1}{2} \times 20 = 157.08$ square inches.

Portion of Cylinder. — A cylinder 5 inches in diameter, is cut off at an angle, as shown in the illustration. Dimension $h_1 = 6$, and $h_2 = 4$ inches. Find the volume and the area S of the cylindrical surface.

$$V = 0.3927\, d^2\, (h_1 + h_2) = 0.3927 \times 5^2 \times (6 + 4) = 0.3927 \times 25 \times 10$$
$$= 98.175 \text{ cubic inches.}$$

$$S = 1.5708\, d\, (h_1 + h_2) = 1.5708 \times 5 \times 10 = 78.54 \text{ square inches.}$$

Portion of Cylinder. — Find the volume of a cylinder so cut off that line AC passes through the center of the base circle — that is, the base area is a half-circle. The diameter of the cylinder = 5 inches, and height $h = 2$ inches. In this case $a = 2.5$; $b = 0$; area $ABC = \frac{1}{2} \times 0.7854 \times 5^2 = 9.82$; $r = 2.5$.

$$V = \left(\frac{2}{3} \times 2.5^3 + 0 \times 9.82 \right) \frac{2}{2.5 + 0} = \frac{2}{3} \times 15.625 \times 0.8 = 8.33 \text{ cubic inches.}$$

Hollow Cylinder. — A cylindrical shell, 28 inches high, is 36 inches in outside diameter, and 4 inches thick. Find its volume.

$$V = 3.1416\, ht\, (D - t) = 3.1416 \times 28 \times 4\, (36 - 4) = 3.1416 \times 28 \times 4 \times 32$$
$$= 11,259.5 \text{ cubic inches.}$$

Cone. — Find the volume and area of conical surface of a cone, the base of which is a circle of 6 inches diameter, and the height of which is 4 inches.

$V = 0.2618\, d^2h = 0.2618 \times 6^2 \times 4 = 0.2618 \times 36 \times 4 = 37.7$ cubic inches.

$A = 3.1416\, r \sqrt{r^2 + h^2} = 3.1416 \times 3 \times \sqrt{3^2 + 4^2} = 9.4248 \times \sqrt{25}$
$$= 47.124 \text{ square inches.}$$

Frustum of Cone. — Find the volume of a frustum of a cone of the following dimensions: $D = 8$ inches; $d = 4$ inches; $h = 5$ inches.

$$V = 0.2618 \times 5\, (8^2 + 8 \times 4 + 4^2) = 0.2618 \times 5\, (64 + 32 + 16)$$
$$= 0.2618 \times 5 \times 112 = 146.61 \text{ cubic inches.}$$

Volume of Solids

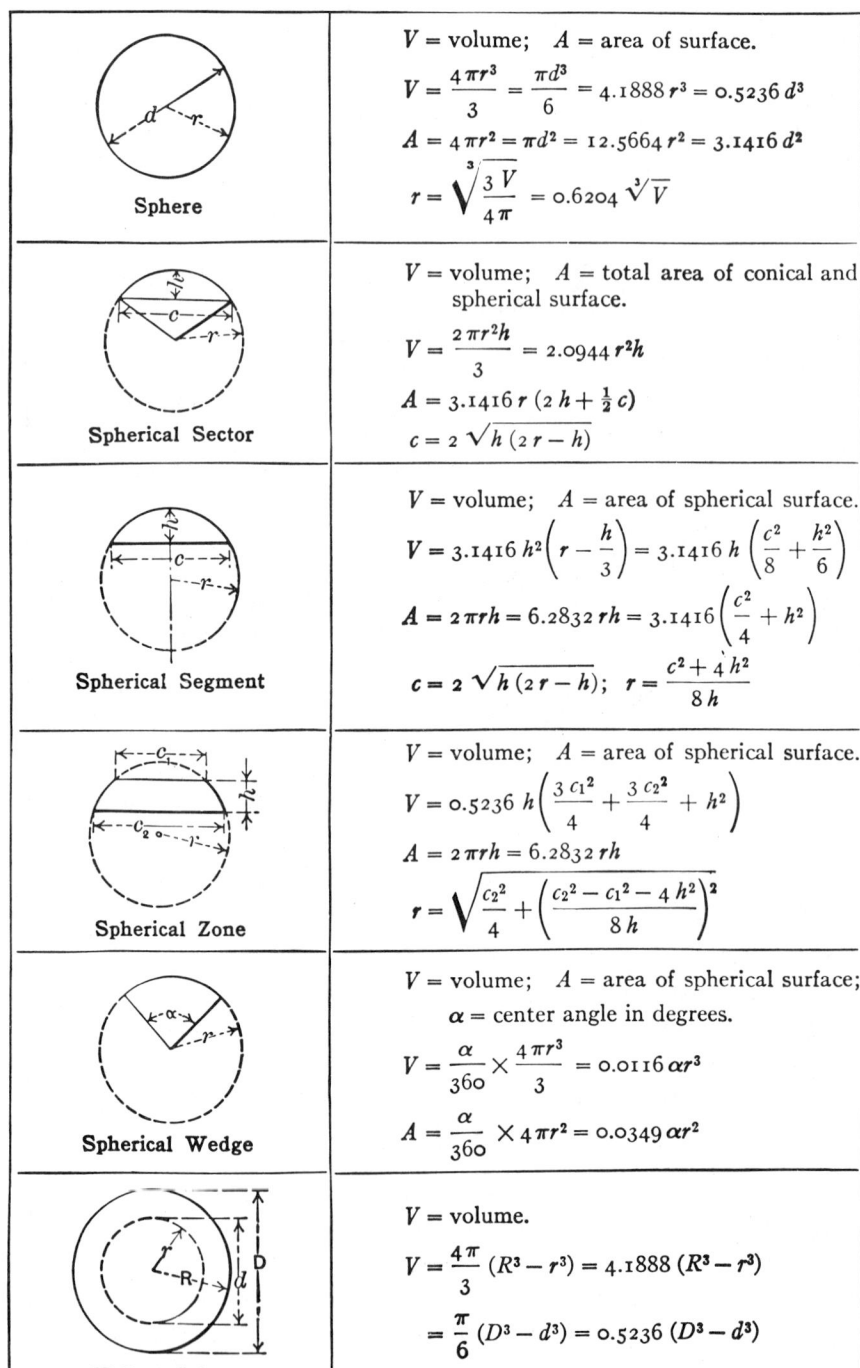

Sphere	V = volume; A = area of surface. $V = \dfrac{4\pi r^3}{3} = \dfrac{\pi d^3}{6} = 4.1888\,r^3 = 0.5236\,d^3$ $A = 4\pi r^2 = \pi d^2 = 12.5664\,r^2 = 3.1416\,d^2$ $r = \sqrt[3]{\dfrac{3\,V}{4\,\pi}} = 0.6204\,\sqrt[3]{V}$
Spherical Sector	V = volume; A = total area of conical and spherical surface. $V = \dfrac{2\pi r^2 h}{3} = 2.0944\,r^2 h$ $A = 3.1416\,r\,(2\,h + \tfrac{1}{2}\,c)$ $c = 2\,\sqrt{h\,(2\,r - h)}$
Spherical Segment	V = volume; A = area of spherical surface. $V = 3.1416\,h^2\left(r - \dfrac{h}{3}\right) = 3.1416\,h\left(\dfrac{c^2}{8} + \dfrac{h^2}{6}\right)$ $A = 2\pi rh = 6.2832\,rh = 3.1416\left(\dfrac{c^2}{4} + h^2\right)$ $c = 2\,\sqrt{h\,(2\,r - h)}$; $r = \dfrac{c^2 + 4\,h^2}{8\,h}$
Spherical Zone	V = volume; A = area of spherical surface. $V = 0.5236\,h\left(\dfrac{3\,c_1^2}{4} + \dfrac{3\,c_2^2}{4} + h^2\right)$ $A = 2\pi rh = 6.2832\,rh$ $r = \sqrt{\dfrac{c_2^2}{4} + \left(\dfrac{c_2^2 - c_1^2 - 4\,h^2}{8\,h}\right)^2}$
Spherical Wedge	V = volume; A = area of spherical surface; α = center angle in degrees. $V = \dfrac{\alpha}{360} \times \dfrac{4\pi r^3}{3} = 0.0116\,\alpha r^3$ $A = \dfrac{\alpha}{360} \times 4\pi r^2 = 0.0349\,\alpha r^2$
Hollow Sphere	V = volume. $V = \dfrac{4\pi}{3}\,(R^3 - r^3) = 4.1888\,(R^3 - r^3)$ $= \dfrac{\pi}{6}\,(D^3 - d^3) = 0.5236\,(D^3 - d^3)$

Examples of the Use of the Formulas

Sphere. — Find the volume and surface of a sphere 6.5 inches in diameter.

$V = 0.5236\,d^3 = 0.5236 \times 6.5^3 = 0.5236 \times 6.5 \times 6.5 \times 6.5 = 143.79$ cubic inches.

$A = 3.1416\,d^2 = 3.1416 \times 6.5^2 = 3.1416 \times 6.5 \times 6.5 = 132.73$ square inches.

The volume of a sphere is 64 cubic inches. Find its radius.

$r = 0.6204\,\sqrt[3]{64} = 0.6204 \times 4 = 2.4816$ inches.

Spherical Sector. — Find the volume of a sector of a sphere 6 inches in diameter, the height h of the sector being 1.5 inch. Also find length of chord c. — Here $r = 3$, and $h = 1.5$.

$V = 2.0944\,r^2h = 2.0944 \times 3^2 \times 1.5 = 2.0944 \times 9 \times 1.5 = 28.27$ cubic inches.

$c = 2\,\sqrt{h\,(2r - h)} = 2\,\sqrt{1.5\,(2 \times 3 - 1.5)} = 2\,\sqrt{6.75} = 2 \times 2.598$
$= 5.196$ inches.

Spherical Segment. — A segment of a sphere has the following dimensions: $h = 2$ inches; $c = 5$ inches. Find the volume V and the radius of the sphere of which the segment is a part.

$$V = 3.1416 \times 2 \times \left(\frac{5^2}{8} + \frac{2^2}{6}\right) = 6.2832 \times \left(\frac{25}{8} + \frac{4}{6}\right) = 6.2832 \times 3.792$$
$= 23.825$ cubic inches.

$$r = \frac{5^2 + 4 \times 2^2}{8 \times 2} = \frac{25 + 16}{16} = \frac{41}{16} = 2\frac{9}{16} \text{ inches.}$$

Spherical Zone. — In a spherical zone, let $c_1 = 3$; $c_2 = 4$; and $h = 1.5$ inch. Find the volume.

$$V = 0.5236 \times 1.5 \times \left(\frac{3 \times 3^2}{4} + \frac{3 \times 4^2}{4} + 1.5^2\right) = 0.5236 \times 1.5 \times \left(\frac{27}{4} + \frac{48}{4} + 2.25\right)$$
$= 0.5236 \times 1.5 \times 21 = 16.493$ cubic inches.

Spherical Wedge. — Find the area of the spherical surface and the volume of a wedge of a sphere. The diameter of the sphere is 4 inches, and the center angle α is 45 degrees.

$V = 0.0116 \times 45 \times 2^3 = 0.0116 \times 45 \times 8 = 4.176$ cubic inches.

$A = 0.0349 \times 45 \times 2^2 = 0.0349 \times 45 \times 4 = 6.282$ square inches.

Hollow Sphere. — Find the volume of a hollow sphere, 8 inches in outside diameter, with a thickness of material of 1.5 inch.
Here $R = 4$; $r = 4 - 1.5 = 2.5$.

$V = 4.1888\,(4^3 - 2.5^3) = 4.1888\,(64 - 15.625) = 4.1888 \times 48.375$
$= 202.63$ cubic inches.

Volume of Solids

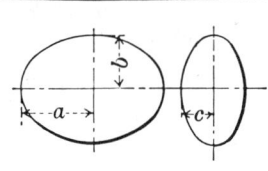

Ellipsoid

$V = $ volume.

$$V = \frac{4\pi}{3} abc = 4.1888\ abc$$

In an ellipsoid of revolution, or spheroid, where $b = c$:

$$V = 4.1888\ ab^2$$

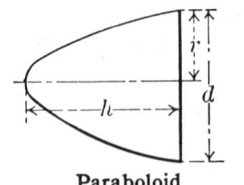

Paraboloid

$V = $ volume; $\quad V = \frac{1}{2}\pi r^2 h = 0.3927\ d^2 h$

$A = $ area; $\quad A = \frac{2\pi}{3p}\left[\sqrt{\left(\frac{d^2}{4} + p^2\right)^3} - p^3\right]$ in which

$$p = \frac{d^2}{8h}$$

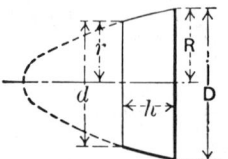

Paraboloidal Segment

$V = $ volume.

$$V = \frac{\pi}{2} h\,(R^2 + r^2) = 1.5708\ h\,(R^2 + r^2)$$

$$= \frac{\pi}{8} h\,(D^2 + d^2) = 0.3927\ h\,(D^2 + d^2)$$

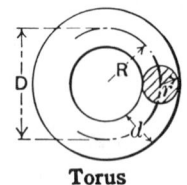

Torus

$V = $ volume; $\quad A = $ area of surface.

$$V = 2\pi^2\ Rr^2 = 19.739\ Rr^2$$

$$= \frac{\pi^2}{4} Dd^2 = 2.4674\ Dd^2$$

$$A = 4\pi^2\ Rr = 39.478\ Rr$$

$$= \pi^2 Dd = 9.8696\ Dd$$

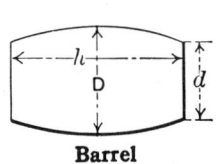

Barrel

$V = $ approximate volume.

If the sides are bent to the arc of a circle:
$$V = \tfrac{1}{12}\pi h\,(2\,D^2 + d^2) = 0.262\ h\,(2\,D^2 + d^2)$$

If the sides are bent to the arc of a parabola:
$$V = 0.209\ h\,(2\,D^2 + Dd + \tfrac{3}{4}\,d^2)$$

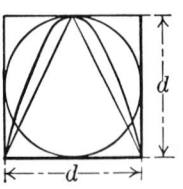

If $d = $ base diameter and height of a cone, a paraboloid and a cylinder, and the diameter of a sphere, then the volumes of these bodies are to each other as below:

Cone: paraboloid: sphere: cylinder $= \tfrac{1}{3} : \tfrac{1}{2} : \tfrac{2}{3} : 1$

Examples of the Use of the Formulas

Ellipsoid or Spheroid. — Find the volume of a spheroid in which $a = 5$, and $b = c = 1.5$ inch.

$$V = 4.1888 \times 5 \times 1.5^2 = 4.1888 \times 5 \times 2.25 = 47.124 \text{ cubic inches.}$$

Paraboloid. — Find the volume of a paraboloid in which $h = 12$ and $d = 5$ inches.

$$V = 0.3927\, d^2 h = 0.3927 \times 5^2 \times 12 = 0.3927 \times 25 \times 12$$
$$= 0.3927 \times 300 = 117.81 \text{ cubic inches.}$$

Segment of Paraboloid. — Find the volume of a segment of a paraboloid in which $D = 5$ inches, $d = 3$ inches, and $h = 6$ inches.

$$V = 0.3927\, h\,(D^2 + d^2) = 0.3927 \times 6 \times (5^2 + 3^2) = 0.3927 \times 6 \times (25 + 9)$$
$$= 0.3927 \times 6 \times 34 = 80.11 \text{ cubic inches.}$$

Torus. — Find the volume and area of surface of a torus in which $d = 1.5$ and $D = 5$ inches.

$$V = 2.4674 \times 5 \times 1.5^2 = 2.4674 \times 5 \times 2.25 = 27.76 \text{ cubic inches.}$$
$$A = 9.8696 \times 5 \times 1.5 = 74.022 \text{ square inches}$$

Barrel. — Find the approximate contents of a barrel, the inside dimensions of which are $D = 24$ inches; $d = 20$ inches; $h = 48$ inches.

$$V = 0.262\, h\,(2\,D^2 + d^2) = 0.262 \times 48 \times (2 \times 24^2 + 20^2) = 0.262 \times 48$$
$$\times (1152 + 400) = 0.262 \times 48 \times 1552 = 19,518 \text{ cubic inches.}$$

Assume, as an example, that the diameter of the base of a cone, paraboloid and cylinder is 2 inches, that the height is 2 inches, and that the diameter of a sphere is 2 inches. Then the volumes, written in formula-form, are as below:

Cone	Paraboloid	Sphere	Cylinder

$$\frac{3.1416 \times 2^2 \times 2}{12} : \frac{3.1416 \times 2^2 \times 2}{8} : \frac{3.1416 \times 2^3}{6} : \frac{3.1416 \times 2^2 \times 2}{4} = \frac{1}{3} : \frac{1}{2} : \frac{2}{3} : 1$$

STANDARD GRAPHICAL SYMBOLS — AIR CONDITIONING

(From American National Standard Z32.2.4, sponsored by ASME and AIEE)

CAPILLARY TUBE		FILTER, LINE	
COMPRESSOR		FILTER AND STRAINER, LINE	
COMPRESSOR, ROTARY (Enclosed crankcase, belted)		FLOAT, HIGH SIDE	
COMPRESSOR, RECIPROCATING (Open crankcase, belted)		FLOAT, LOW SIDE	
COMPRESSOR, RECIPROCATING (Open crankcase, direct-drive)		GAGE	
MOTOR-COMPRESSOR, RECIPROCATING (Direct-connected, enclosed crankcase)		PRESSURESTAT	
MOTOR-COMPRESSOR, ROTARY (Direct-connected, enclosed crankcase)		PRESSURE SWITCH	
MOTOR-COMPRESSOR, RECIPROCATING (Sealed crankcase)		PRESSURE SWITCH (With high pressure cut-out)	
MOTOR-COMPRESSOR, ROTARY (Sealed crankcase)		RECEIVER, HORIZONTAL	
CONDENSING UNIT (Air cooled)		RECEIVER, VERTICAL	
CONDENSING UNIT (Water-cooled)		SCALE TRAP	
CONDENSER, AIR COOLED (Finned, forced Air)		SPRAY POND	
CONDENSER, AIR COOLED (Finned, static)		THERMAL BULB	
CONDENSER, WATER COOLED (Concentric tube in a tube)		THERMOSTAT (Remote bulb)	
CONDENSER, WATER COOLED (Shell and coil)		VALVE, EXPANSION, AUTOMATIC	
CONDENSER, WATER COOLED (Shell and tube)		VALVE, EXPANSION, HAND	
CONDENSER, EVAPORATIVE		VALVE, EXPANSION, THERMOSTATIC	
		VALVE, COMPRESSOR SUCTION PRESSURE LIMITING (Throttling type, compressor side)	
COOLING UNIT, FINNED (Natural convection)		VALVE, CONSTANT PRESSURE, SUCTION	
COOLING UNIT (Forced convection)		VALVE, EVAPORATOR PRESSURE REGULATING (Snap action)	
COOLING UNIT, IMMERSION		VALVE, EVAPORATOR PRESSURE REGULATING (Thermostatic throttling type)	
COOLING TOWER		VALVE, EVAPORATOR PRESSURE REGULATING (Throttling type, evaporator side)	
DRYER		VALVE, MAGNETIC STOP	
EVAPORATOR, CIRCULAR (Ceiling type, finned)		VALVE, SNAP ACTION	
EVAPORATOR, MANIFOLDED (Bare tube, gravity air)		VALVE, SUCTION VAPOR REGULATING	
EVAPORATOR, MANIFOLDED (Finned, forced air)		VALVE, SUCTION	
EVAPORATOR, MANIFOLDED (Finned, gravity air)		VALVE, WATER	
EVAPORATOR, PLATE COILS (Headered or manifolded)		VIBRATION ABSORBER, LINE	

STANDARD GRAPHICAL SYMBOLS — HEATING AND VENTILATING

(From American National Standard Z32.2.4, sponsored by ASME and AIEE)

AIR ELIMINATOR		ACCESS DOOR	AD
ANCHOR	PA	ADJUSTABLE BLANK OFF	TR 20×12
EXPANSION JOINT		ADJUSTABLE PLAQUE	P-20×12-700 cfm / P-20 ∅ -700 cfm
HANGER OR SUPPORT	H		
HEAT EXCHANGER		AUTOMATIC DAMPER	M
HEAT TRANSFER SURFACE (Plan-Indicate type, such as convector)			
PUMP (Indicate type, such as vacuum)	M	CANVAS CONNECTION	
STRAINER		DEFLECTING DAMPER	
TANK (Designate type)	REC		
THERMOMETER		DIRECTION OF FLOW	
THERMOSTAT	T	DUCT (First figure is side shown)	12×20
TRAP, BOILER RETURN		DUCT SECTION (Exhaust or return)	(E or R 20×12)
TRAP, BLAST THERMOSTATIC		DUCT SECTION (SUPPLY)	(S 20×12)
TRAP, FLOAT	F	EXHAUST INLET, CEILING (Indicate type)	CR 20×12 - 700 cfm / CG 20×12 - 700 cfm
TRAP, FLOAT AND THERMOSTATIC		EXHAUST INLET, WALL (Indicate type)	TR - 12×8 700 cfm
TRAP, THERMOSTATIC			
UNIT HEATER (Centrifugal fan type - Plan)		FAN AND MOTOR (With belt guard)	
UNIT HEATER (Propeller fan type - Plan)			
UNIT VENTILATOR, PLAN		INCLINED DROP (In respect to air flow)	D
VALVE, CHECK		INCLINED RISE (In respect to air flow)	R
VALVE, DIAPHRAGM		INTAKE LOUVERS	
VALVE, GATE			
VALVE, GLOBE		LOUVER OPENING	L 20×12 - 700 cfm
VALVE, LOCK AND SHIELD		SUPPLY OUTLET, CEILING (Indicate type)	20" DIAM. 1000 cfm
VALVE, MOTOR-OPERATED		SUPPLY OUTLET, WALL (Indicate type)	TR-12×8 700 cfm
VALVE, PRESSURE-REDUCING		VANES	
VALVE, RELIEF (Either pressure or vacuum)	R		
VENT POINT	VENT	VOLUME DAMPER	

STANDARD GRAPHICAL SYMBOLS — VALVES

(From American National Standard Z32.2.3, sponsored by ASME and AIEE)

NAME OF VALVE	FLANGED	SCREWED	BELL & SPIGOT	WELDED	SOLDERED
ANGLE VALVE, CHECK					
ANGLE VALVE, GATE (ELEVATION)					
ANGLE VALVE, GATE (PLAN)					
ANGLE VALVE, GLOBE (ELEVATION)					
ANGLE VALVE, GLOBE (PLAN)					
AUTOMATIC BY-PASS VALVE					
AUTOMATIC GOVERNOR-OPERATED VALVE					
AUTOMATIC REDUCING VALVE					
CHECK VALVE, STRAIGHT WAY					
COCK					
DIAPHRAGM VALVE					
FLOAT VALVE					
GATE VALVE (Also used for general STOP VALVE symbol when amplified by specification)					
GATE VALVE, MOTOR-OPERATED					
GLOBE VALVE					
GLOBE VALVE, MOTOR-OPERATED					
HOSE VALVE, ANGLE					
HOSE VALVE, GATE					
HOSE VALVE, GLOBE					

STANDARD GRAPHICAL SYMBOLS — VALVES AND PIPING

(From American National Standard Z32.2.3, sponsored by ASME and AIEE)

NAME OF VALVE	FLANGED	SCREWED	BELL & SPIGOT	WELDED	SOLDERED
LOCKSHIELD VALVE					
QUICK-OPENING VALVE		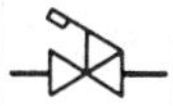		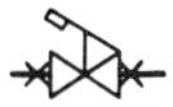	
SAFETY VALVE					

PIPING

AIR CONDITIONING

BRINE RETURN	— — - BR - — —
BRINE SUPPLY	——— B ———
CHILLED OR HOT WATER FLOW (Circulating)	——— CH ———
CHILLED OR HOT WATER RETURN (Circulating)	— — CHR — —
CONDENSER WATER FLOW	——— C ———
CONDENSER WATER RETURN	— — CR — —
DRAIN	——— D ———
HUMIDIFICATION LINE	—·— H —·—
MAKE-UP WATER	—·——·—
REFRIGERANT DISCHARGE	——— RD ———
REFRIGERANT LIQUID	——— RL ———
REFRIGERANT SUCTION	— — RS — —

HEATING

AIR RELIEF LINE	—— - —— - ——
BOILER BLOW-OFF	—— —— ——
COMPRESSED AIR	——— A ———
CONDENSATE DISCHARGE	-o- -o- -o-
FEEDWATER PUMP DISCHARGE	-oo- -oo- -oo-
FUEL-OIL FLOW	——— FOF ———
FUEL-OIL RETURN	- - - FOR - - -
FUEL-OIL TANK VENT	- - - FOV - - -
HIGH PRESSURE RETURN	-//- — -//- —
HIGH PRESSURE STEAM	—//———//—
HOT WATER HEATING RETURN	—— —— —— ——
HOT WATER HEATING SUPPLY	————————
LOW PRESSURE RETURN	—— —— —— ——
LOW PRESSURE STEAM	————————
MAKE-UP WATER	——·——·——
MEDIUM PRESSURE RETURN	-/- — -/- —
MEDIUM PRESSURE STEAM	—/———/—

PLUMBING

ACID WASTE	ACID (above line)
COLD WATER	——·——·——
COMPRESSED AIR	——— A ———
DRINKING WATER FLOW	——·——·——
DRINKING WATER RETURN	——··——··——
FIRE LINE	— F ——— F —
GAS	— G ——— G —
HOT WATER	——··——··——
HOT WATER RETURN	——···——···——
SOIL, WASTE, OR LEADER (Above grade)	————————
SOIL, WASTE, OR LEADER (Below grade)	—— —— —— ——
VACUUM CLEANING	— V ——— V —
VENT	— — — — — —

PNEUMATIC TUBES

TUBE RUNS	════════

SPRINKLERS

BRANCH AND HEAD	—o———o—
DRAIN	— — S — — — S — —
MAIN SUPPLIES	——— S ———

STANDARD GRAPHICAL SYMBOLS — PIPE FITTINGS

(From American National Standard Z32.2.3, sponsored by ASME and AIEE)

NAME OF FITTING	FLANGED	SCREWED	BELL & SPIGOT	WELDED	SOLDERED
BUSHING					
CAP					
CROSS, REDUCING					
CROSS, STRAIGHT SIZE					
CROSSOVER					
ELBOW, 45-DEGREE					
ELBOW, 90-DEGREE					
ELBOW, TURNED DOWN					
ELBOW, TURNED UP					
ELBOW, BASE					
ELBOW, DOUBLE BRANCH					
ELBOW, LONG RADIUS					
ELBOW, REDUCING					
ELBOW, SIDE OUTLET (OUTLET DOWN)					
ELBOW, SIDE OUTLET (OUTLET UP)					
ELBOW, STREET					
JOINT, CONNECTING PIPE					
JOINT, EXPANSION					

CHECK LISTS FOR AIR CONDITIONING DRAWINGS

LEGIBILITY

1 — Is all line work dense, black and of the proper weight?

2 — Are figures and dimensions clear?

3 — Is there any congestion of lines or dimensions?

4 — Will blueprints and reproductions be clear and legible and in keeping with the high quality of the application?

LINE WORK

1 — Are all lines complete so that the drawing cannot possibly be misunderstood?

2 — Are there sufficient views and sections and are they in correct relation to each other?

3 — Are there unnecessary lines and have tabulations been used to the best advantage?

4 — Are center lines, cross section lines, etc., properly shown?

5 — Has the North point been indicated?

DIMENSIONS

1 — Do overall dimensions check with individual dimensions?

2 — Are all dimensions given from a base or center line?

3 — Are dimensions duplicated only where necessary?

4 — Are clearances, fits, weights, etc., correct and given where required?

5 — Are there interferences? Can parts be assembled and fastened together?

6 — Are dimensions and arrow heads correct and neat?

7 — Are "not to scale" dimensions underscored with a "snake" line?

GENERAL

1 — Are all necessary parts and quantities called for and without duplication?

2 — Do drawings and material lists call for commercial parts and standard material in accordance with trade or standard practices?

3 — Are stock parts and materials used wherever possible?

4 — Have standard names been given parts drawings?

5 — Are signatures and dates correctly recorded?

6 — Are reference drawing lists included?

7 — Have correct symbols and abbreviations been used?

8 — Has title, owner, number, scale, drawn by, checked by, and dates been shown on the title block?

9 — If different scales are used on same drawing, is proper scale indicated under each plan, elevation and section?

10 — Are motor characteristics given?

11 — Have consulting engineer's, general contractor's, owner's and/or architect's names and addresses been given if required?

12 — Have overall dimensions of equipment been listed?

13 — Are weights of units necessary?

IDENTIFICATION

1 — Are parts correctly identified?

2 — Are "parts circles" correctly made, located and arranged?

3 — Are notes properly identified and located?

4 — Are all reference parts completely identified by one cross section reference?

5 — Are there any cases where detail information is not complete in any one place?

6 — Are all group and assemblies properly made up?

7 — Should any other parts be so grouped?

ACCESSIBILITY AND CLEARANCE

1 — Have sizes of entrances to rooms where equipment is to be located been checked for clearances?

2 — Is there ample access for servicing equipment?

3 — Do pipes or ducts run through stairs, or in front of windows, doors or electric panels?

4 — Are there pipe joints or fittings in ducts?

5 — Is there space for pulling coils?

ADEQUACY

1 — Are pipes anchored and provision made for expansion loops, joints or offsets?

2 — Has safety relief piping been provided?

3 — Are hangers, stirrups, rollers, etc., adequate?

4 — Machinery foundation adequate?

5 — Does it conform to local and national codes?

6 — Are floor drains provided for?

7 — Are refrigerant, water, steam connection sizes given?

POSSIBLE INTERFERENCES

Electric panel boxes	Overhead track system
Electric conduits	Roof ventilators and skylights
Electric light fixtures	Column caps
Air plenums and ducts	Column drop panels
Air supply and exhaust grilles	Sidewall and overhead radiators
Sprinkler risers and mains	Fire door track and counterweight
Sprinkler branches	Door frames
Vacuum cleaning lines	Pipe trench covers
Special vacuum piping	Pass boxes
Brine lines	Suspended motor, pump, fan and tank platforms
Steam lines — supply and condensate return	
Water lines — hot, cold, and drinking water	Shafting
	Knee braces
	Sash ventilator clearances (pivoted and projected type)
Compressed air lines	
Illuminating gas lines	
Chemical piping	Drying cabinets, fume hoods or any other device with vertical rising doors and counterweight boxes
Roof conductor piping — laterals and stacks	
High and low points of roof	
Drain piping, laterals, and stacks	Desks, benches, tables, etc.

TERMINOLOGY

Absolute filter. A high efficiency mechanical filter built originally for removing radioactive particles and having an efficiency of 99.9% or higher.

Absolute humidity. Not in common use. Weight of water vapor per unit volume of air-steam mixture. Usually expressed in grains per cubic foot.

Absolute pressure. The pressure above an absolute vacuum or, in other words, the sum of the pressure shown by a gage and that indicated by the barometer.

Absolute temperature. Temperature measured from absolute zero; on the Fahrenheit scale, absolute temperature is deg. F + 459.6 and on the Centigrade scale deg. C + 273. See Degrees Rankine and Degrees Kelvin.

Absorption. An assimilation during which the absorbent undergoes a physical or chemical (or both) change. Absorbers such as lithium chloride and calcium chloride are used in dehumidifying. The process is also used in absorption refrigeration.

Absorption refrigeration. A process whereby a secondary fluid absorbs the refrigerant, and in doing so gives up heat, then releases the refrigerant, during which it absorbs heat.

Absorptivity. The ratio of the radiant energy absorbed by a body to that falling upon it. It is equal to the emissivity for radiation of the same wave length.

Acceleration due to gravity. The rate of velocity increase in a freely falling body due to gravity. It varies with latitude and altitude. At sea level and 45° latitude it is 32.174 ft per sec per sec.

Activated carbon or Activated charcoal. An adsorbent of porous structure manufactured by carbonization of carbonaceous material and activation to increase the microporosity, capable of adsorbing molecules of organic substances from gases and liquids. Consists mainly of the element carbon, but exists in a charcoal-type porous structure.

Activation. The processing of charcoal or carbon or other substance to give it a high degree of porosity in pores so small that they will adsorb molecules.

Adiabatic. An action or process during which no heat is added or subtracted.

Adsorption. A phenomenon consisting of the adhesion, in a very thin layer of gas molecules, of dissolved substances or liquids, to the surfaces of solid bodies with which they are in contact. Silica-gel is an adsorber used in dehumidifying while activated carbon is an adsorber used to remove odors from air. During adsorption the adsorber undergoes no permanent physical or chemical change.

Aerosol. A suspension of fine solid or liquid particles in air or a gas.

Air. Normal air in the atmosphere containing approximately 21% oxygen and the remainder, nitrogen and argon. Contains also varying amounts of carbon dioxide, moisture, and traces of contaminants.

Air change. The quantity of infiltration or ventilation air in cubic feet per hour or minute divided by the volume of the room gives the number of so-called air changes during that interval of time, and tables of the recommended number of such air changes for various type rooms are used for estimating purposes.

Air conditioning. The simultaneous production or altering under control of several factors to obtain a desired atmospheric environment. These factors include raising or lowering of temperature, raising or lowering of humidity, distribution of air, and reduction of contaminants, such as dust, fumes, bacteria, odors, etc.

Air filter. A filter for purifying air by removing either particulate matter (this is the ordinary meaning) or gases or vapors which can be removed by chemical action.

Airfoil fan. A fan with an airfoil-shaped blade which moves the air in the general direction of the axis about which it rotates.

Air gap. In plumbing, the distance, vertically, from a faucet opening or pipe opening through the air to the flood-level rim of the fixture.

Air pollution. The contamination of normal atmospheric air with toxic odorous or undesirable substances.

Air washer. A device through which air is passed and washed with water from sprays for the purpose of humidifying, dehumidifying, or cleaning.

Alloy pipe. A steel pipe with one or more elements other than carbon which give it greater resistance to corrosion and more strength than carbon steel pipe.

Ambient. Encompassing on all sides. Thus ambient air is the air surrounding.

Ammonia machine. A mechanical refrigeration machine with ammonia as the refrigerant.

Angle of bend. In a pipe, the angle at the center of the bend between radial lines from the beginning and end of the bend to the center.

Angle valve. A valve, usually of the globe type, in which the axes of the inlet and outlet are at right angles.

Anthracite. A dense hard coal with a very high percentage of fixed carbon, usually above 90%.

Apparatus dew point. A term used interchangeably with average coil surface temperature. It is also the temperature at which the sensible heat ratio line intersects the saturation curve on a psychrometric chart.

Area drain. One installed to collect rain or surface water from an open space.

Aspect ratio. In a duct, the depth of elbow along (parallel) to the axis of bend divided by the width in the plane of the bend.

Atmospheric condenser. One using water at atmospheric pressure.

Atmospheric pressure. Pressure indicated by a barometer. The standard atmospheric pressure is 760 mm of mercury (29.921 inches of mercury) or 14.696 lb per sq in.

Attic fan. A fan so mounted in the attic that it can draw air from the house at night, causing cool outside air to flow into the house, and exhausting the warm air to the outside.

Automatic heating. A central heating system operated without manual attention. Usually means oil fired, gas fired, or stoker fired furnaces and boilers.

Automation. Automatic performance of a controlling function by mechanisms instead of men. The word is applied to many functions in many industries, including automatic handling of materials; control of temperature, pressure, and velocity; automatic processing, and so on.

Auxiliary equipment. Equipment that works in conjunction with a key piece of equipment to use with a resorber for solvent recovery.

Available energy. That part of the total energy which can be usefully employed. In a perfect engine, that part which is converted to work.

Axial flow fan. A fan with a disc or air-foil shaped blade mounted on a shaft and which moves the air in the general direction of the shaft axis.

Backflow. In plumbing, water flow into water-distributing pipes from any source other than that for which the system is designed.

Backflow connection. In plumbing, any connection or arrangement which permits backflow.

Backflow preventer. A device to prevent backflow into a water supply line from the connections at its outlet.

Backing ring. A metal strip used to prevent melted metal from the welding process from entering a pipe when making a butt-welded joint.

Baffle. A partition or wall to divert the flow of a fluid.

Basal metabolism. The metabolism of an organism in the fasting or resting state. The rate at which the organism gives off heat at this state is the basal metabolic rate, and this rate is the lowest for the organism under normal conditions.

Baseboard radiator. An arrangement of heating surface located at or replacing the room baseboard, and applied as an alternative to conventional convectors or radiators.

Baudelot cooler. An arrangement of pipes one above another through which refrigerant flows and is vaporized as it absorbs heat from the water being cooled by trickling over the tubes.

Bell-and-spigot joint. The commonly used joint in cast-iron pipe. Each piece is made with an enlarged diameter or bell at one end into which the plain or spigot end of another piece is inserted when laying. The joint is then made tight by cement, oakum, lead, or rubber calked into the bell around the spigot.

Bellows. An expansible metal device containing a fluid which will volatilize at some desired temperature, expand the device and open or close an opening or a switch, as in controls and steam traps.

Bi-metal. Two metals of different coefficients of expansion welded together so that the piece will bend in one direction when heated, in the other when cooled, and thus can be used in opening and closing electrical circuits, as in thermostats.

Binary cycle. A cycle in which two different media are employed, one superimposed on and augmenting the cycle of the other.

Bituminous coal. Coal covering a wide range of properties, but in general with a fixed carbon content under 80% and volatile matter of over 20%.

Black body. A body that absorbs all the radiant energy falling upon it.

Black pipe. Steel pipe that has not been galvanized.

Blank flange. A flange in which the bolt holes have not been drilled.

Blastcoil. Heat transfer surface, most frequently of an extended surface arrangement, over which air is blown to be heated or cooled, depending on the temperature of the fluid within the pipelike surface.

Blind flange. A flange used to seal off the end of a pipe.

Blow. In connection with air distribution, the distance an air stream travels from an air outlet to the point where the air velocity is reduced to 50 ft per min.

Boiler. A closed vessel in which fuel is burned to generate steam or heat water. The boiler may be made either of steel or, as is frequently the case when low pressure steam is used, of cast iron.

Boiler burner unit. Boiler designed especially for gas or oil and sold integrally with the burner.

Boiler heating surface. That part of the interior surface of a boiler subjected to heat on the one side and transmitting the heat to water (or steam or other fluid) within the boiler. Direct surface is that subjected to direct radiant rays of the burning fuel; other heating surface is termed indirect.

Boiler horsepower. The power required to evaporate 34.5 lb of water from and at 212F per hour, or the equivalent of 33,475 Btu per hr.

Bonnet. Part of a valve used to guide and support the valve stem.

Booster. An accessory to increase capacity or output. In refrigeration, a compressor discharging into the suction line of another unit. In heating, the term is frequently applied to a hot water circulator.

Boot. Sheet metal transformation pieces used in warm air heating and connected to horizontal round leaders on one end and to vertical rectangular stacks on the other.

Branch. The outlet or inlet of a fitting not in line with the run, and taking off at an angle with the run.

Branch interval. In plumbing, a vertical length of waste pipe, usually one-story high, and to which the horizontal branches from that floor are connected.

Branch tee. A tee having many side branches.

Branch vent. A vent connecting one or more individual vents with a vent stack.

Brazed. Joined by hard solder.

Brine. A secondary liquid cooled by refrigeration and circulated to heat transfer surface for absorbing heat from air, water or other fluid.

British thermal unit (Btu). The unit of energy or heat required to raise the temperature of one pound of water from 39F to 40F or from 60F to 61F under standard pressure, or 1/180 of the heat to raise 1 lb from 32F to 212F.

Building drain. The section of the horizontal drain pipe connecting the drainage pipes inside the building with the building sewer outside the inner face of the building wall.

Building sewer. The pipe extending from the outer end of the building drain to the public sewer.

Building storm drain. A drain for carrying rain, surface water, condensate, or similar discharge to the building drain or sewer.

Building subdrain. That part of the drainage system which can not drain by gravity to the building sewer.

Building storm sewer. The pipe from the building storm drain to the public sewer.

Building trap. A trap in the building drain to prevent air circulation between the building drainage system and the building sewer.

Bull head tee. A tee the branch of which is larger than the run.

Bushing. A pipe fitting for connecting a pipe with a female fitting of larger size. It is a hollow plug with internal and external threads.

Butt weld joint. A welded pipe joint made with the ends of the two pipes butting each other, the weld being around the periphery.

Butt weld pipe. Pipe welded along a seam butted edge to edge and not scarfed or lapped.

By-pass. An alternative path, in a duct or pipe, for a fluid to flow from one point to another, with the direction determined by the opening or closing of valves or dampers in the main line as well as in the by-pass.

Calorie. In engineering, the large calorie is usually used, defined as one one-hundredth of the energy or heat required to raise the temperature of one kilogram of water from 0C to 100C; the small calorie is the heat required similarly to raise the temperature of 1 gram of water.

Capacity. In a refrigerating compressor, the refrigerating effect in tons or Btu per hour.

Carbon steel pipe. Steel pipe which owes its properties chiefly to the carbon which it contains.

Central fan system. An indirect system of heating in which the air is heated by steam or hot water at a central location and carried to or from the rooms to be heated by a fan and a system of ducts. Also termed Hot Blast System.

Central heating plant. Either (a) one heating plant in a building serving all or most of the rooms in the building, as distinguished from individual room heaters, or (b) one heating plant serving two or more buildings. The latter is also termed district heating.

Charge. As a noun, the quantity of something introduced into a vessel. In refrigeration, the quantity of refrigerant in a system.

Check valve. A valve designed to allow a fluid to pass through in one direction only. A common type has a plate so suspended that the reverse flow aids gravity in forcing the plate against a seat, shutting off reverse flow.

Chimney effect. See Stack effect.

Circuit vent. A branch vent from two or more traps extending from in front of the last fixture connection of a horizontal branch to the plumbing vent stack.

Class. In refrigeration, a system is classified according to a safety code as Class A if the refrigerant charge is 1000 lb or more, Class B 100 to 1000 lb, Class C 20 to 100 lb, Class D 6 to 20 lb, and Class E 6 lb or less.

Clean room. Refers to a room in which the air is highly purified, particularly with regard to dust and other particulate matter. Clean room techniques are used in the processing of delicate materials and products.

Clearance. In an engine or compressor the space in the cylinder beyond the piston when the latter is at its extreme position in that end.

Clo. A unit of clothing insulation defined as the insulation necessary to keep a sitting man comfortable in a normally ventilated room at 70F and 60% relative humidity. In physical terms, it is equal to 42.7 Btu per sq ft per hour.

Closed cycle. Where the fluid is used over and over without introduction of new fluid, as in a hot water heating or mechanical refrigeration system.

Close nipple, A nipple with a length twice the length of a standard pipe thread.

Coefficient of heat transmission. The quantity of heat (in the U. S. usually Btu) transmitted from fluid to fluid per unit of time (usually 1 hour) per unit of surface (usually 1 sq ft) through a material or arrangement of materials under a unit temperature differential (usually 1F) between fluids. Commonly used for building materials. Unit abbreviation is: Btu/hr-ft²-°F.

Coefficient of performance. In the heat pump, the ratio of effect produced to work input. In refrigeration, the ratio of the refrigerating effect to the work input is a dimensionless quantity.

Cold 70. That condition during the heating season, in spaces where 70F is the desired air temperature, when the air temperature at the thermostat is at 70F but the temperature at the floor or below the thermostat is lower than 70F. This stratification results in relatively long "off" periods of the heating system and a feeling of discomfort even though 70F is maintained at the control.

Combination fixture. A combined sink and laundry tray(s).

Combined sewer. One carrying both storm water and sewage.

Comfort air conditioning. The conditioning of air for the comfort or well-being of human beings, as distinguished from that for the furtherance of a manufacturing or other industrial process.

Comfort zone. That area on the psychrometric chart covered by combinations of temperatures and humidities at which, in subjective tests, more than 50% of the subjects were comfortable.

Companion flange. A pipe flange to connect with another flange or with a flanged valve or fitting. It is attached to the pipe by threads, welding, or other method and differs from a flange which is an integral part of a pipe or fitting.

Complete combustion. Fuel burning where all combustible is burned. Not the same as Perfect Combustion, which see.

Compound compression. Compression by stages in two or more cylinders.

Compression joint. A multi-piece joint with cup shaped threaded nuts which, when tightened, compress tapered sleeves so that they form a tight joint on the periphery of the tubing they connect.

Compression ratio. The ratio of absolute pressure after and before compression.

Concentration. Representing the degree of content of some substance stated in terms such as per cent by weight or volume, parts per million by weight or volume, or mole per cent (molecular per cent). If we have a mixture of 10 parts of alcohol and 90 parts of water, we have a 10% solution of alcohol and water.

Condensate. Liquid formed by condensation of vapor.

Condensers. Condensers are used to liquefy a gas. Two commonly used types of condensers for air conditioning work are (1) Shell and Tube which consists of a large number of tubes inside a steel shell or cylinder. Cooling water circulates inside the tubes while vapor flows into the shell, circulates around outside the tubes, is condensed and drawn off the bottom as a liquid; the other is (2) the Evaporative condenser which consists of a coil over which water is sprayed while air is blown across the tubes of the coil. The refrigeration vapor is cooled, or condensed, inside the tubes due to the extraction of heat caused by the evaporation of the water on the outside.

Condensing heat rejection effect. That portion of the heat rejected at the refrigerating condenser used for condensing the refrigerant vapor to a saturated liquid at the entering refrigerant pressure.

Condensing unit. An assembly attached to one base, and including refrigerating compressor, motor, condenser, receiver, and necessary accessories.

Condition line. The infinite number of combinations of wet and dry bulb temperatures which will satisfy the requirements of an air supply for a given room condition from what is known as the condition line on the psychrometric chart.

Conductance. The quantity of heat (uusally Btu) transmitted per unit of time (usually 1 hour) from a unit (usually a square foot) of surface to an opposite unit of surface of material under a unit temperature (usually 1F) differential between the surfaces. Unit abbreviation is: Btu/hr-ft²-°F.

Conduction. The transmission of heat from one part of a body to another part of the same body, or from one body to another in contact with it, without appreciable displacement of the particles of the body.

Conductivity. The quantity of heat (usually Btu) transmitted per unit of time (usually 1 hour) from a unit surface (usually 1 sq ft) to an opposite unit of surface of one material per unit of thickness (usually 1 inch, but occasionally 1 foot) under a unit temperature differential (usually 1F) between the surfaces.

Consumption charge. That portion of a utility charge based on energy actually consumed, as distinguished from the Demand charge, which see.

Contaminant. Something that is not desired and should be removed. Generally a substance not normally existing in air, water, or the fluid under consideration.

Continuous waste. Waste from more than one fixture connected to a single trap.

Convection. The transfer of heat from one point to another within a fluid by the mixing of one portion of the fluid with another. When the motion is due to differences in density, from temperature differences, the convection is natural; if the motion is imparted mechanically, it is forced convection.

Convector. An arrangement of heat transfer surface contained in an enclosure with openings at top and bottom, thus heating air by convection.

Conversion burners. Fuel-burning devices (usually oil or gas) intended for installation in a wide variety of boilers or furnaces.

Convertor. Usually, a heat exchanger for transferring heat from steam to water.

Cooling tower. A device for cooling water by evaporation in the outside air. A natural draft cooling tower is one where the air flow through the tower is due to the natural chimney effect; a mechanical draft tower employs fans.

Counterflow. In a heat exchanger, where the fluid absorbing heat and the fluid losing heat are so directed that lower and higher temperature of the one is adjacent to the lower and higher temperature of the other, respectively. Ordinarily, the one fluid is flowing in the opposite direction from the other, hence the term.

Coupling. A threaded sleeve used to connect two pipes. They have internal threads at both ends to fit external threads on pipe.

Critical Point. The condition where the liquid and gas of a substance have the same properties. Above the critical point, the phenomena of boiling does not exist.

Critical velocity. The point above which streamline flow becomes turbulent. This usually occurs when the Reynolds number is between 2000 and 3000, in liquids.

Cross. A pipe fitting with four branches in pairs, each pair on one axis, and the axes at right angles.

Cross connection. In plumbing, an arrangement or connection by which the water supply can be contaminated.

Cross-over. A small fitting with a double offset, or shaped like the letter U with the ends turned out. It is only made in small sizes and used to allow pipes to pass each other when the pipes are in the same plane.

Cross valve. A valve fitted on a transverse pipe so as to open communication between two parallel pipes.

Cryogenics. The science of extremely low temperatures.

Cup weld. A pipe weld where one pipe is expanded on the end to allow the entrance of the end of the other pipe. The weld is then circumferential at the end of the expanded pipe.

Cycle. An operation of a number of events which when completed, starts the same series of events over in the same order. Thus a mechanical refrigeration cycle consists of compression, condensing, expansion, evaporation, repeated over and over.

Damper. A valve or plate to regulate the flow of air or some other gas.

Dead end. A branch from a building sewer, vent pipe or soil pipe which is capped or plugged 2 feet or more from the sewer or vent pipe along its developed length.

Decibel. A sound or noise unit, based on the ratio of sound intensity to that at an arbitrary level. One decibel is equal to $10 \log \dfrac{X_2}{X_1}$ where X_2 is the unknown intensity, X_1 the reference intensity. A bel is equal to the $\log \dfrac{X_2}{X_1}$ and is thus one tenth decibel.

Degree-day. The product of one day and the number of degrees F the daily mean temperature is below 65F. Thus on a day when the mean temperature is 40F, there are 25 degree-days. The degree-day unit is used in eliminating the weather variable in determining heating load efficiency and in predicting fuel consumption.

Degree-hour. The product of one hour and (usually) the number of degrees F the hourly mean temperature is above a base point, usually 85F. Thus in an hour when the temperature is 92F, there are 7 degree-hours on an 85F base. The degree-hour is used in measuring, roughly, the cooling load in summer for cases where process heat, heat from human beings, and humidity are relatively unimportant as compared with dry bulb temperature of the outside air.

Degrees Kelvin. The kelvin (K) is the absolute temperature unit in the SI metric system. It is equal to the Celsius temperature (centigrade) plus 273.15.

Degrees Rankine. Absolute temperature on the Fahrenheit scale, or degrees F plus 459.6.

Dehumidification. The process of removing water vapor (steam) from the air.

Dehydrator. A device or material which will remove water from a substance. In refrigeration, a device for removing moisture from refrigerants; also called a drier.

Demand charge. That part of a utility service charged for on the basis of the possible demand as distinguished from the energy actually consumed.

Demand factor. Ratio of the maximum demand to the connected load.

Density. The weight of a unit of volume — usually pounds per cubic foot.

Desiccation. The drying of a substance by evaporation, usually at very low moisture quantities.

Design temperature. The temperature which an apparatus or system is designed to (1) maintain or (2) operate against under most extreme conditions. The former is the inside design temperature, the latter outside design temperature.

Developed length. Length of a section of pipe along the center line through the pipe and fittings.

Dew-point temperature. The temperature at which a given mixture of air and water vapor is saturated.

Diffusion. The movement of individual molecules through narrow spaces by the molecular velocity and bouncing associated with individual gaseous or liquid molecules.

Direct expansion. An arrangement of a refrigerant evaporator where the refrigerant itself expands in an evaporator in the air stream.

Direct-fired heater. A fuel burning device in which the heat from the fuel is transferred through metal to air which is then introduced to the space to be heated.

Direct-fired unit. A heater in which the flame's heat is transferred to metal plates and thence directly to the air to be heated. A warm air furnace is direct fired; a boiler is indirect.

Disc fan. An axial-flow fan with blades formed of flat sheets or plates so bent that when rotated the blades impart to the air a motion along the axis of the fan shaft.

Discomfort Index. Original name of Temperature-Humidity Index (THI). See Temperature-Humidity Index.

District heating. The heating of a multiplicity of buildings from one central boiler plant, from which steam or hot water is piped to the buildings. Can be a private function as in some housing projects, universities, hospitals, etc., or a public utility operation.

Diversity factor. The ratio of the sum of the individual maximum loads during a period to the simultaneous maximum loads of all the same units during the same period. Always unity or more.

Double extra-strong pipe. A schedule of steel or wrought-iron pipe weights in common use, generally Schedule 120.

Double offset. Two offsets in succession in the same pipe line.

Double pipe. In a heat exchanger, one in which two concentric pipes, one within the other, are used with one fluid flowing through the inner pipe, the other through the annular space between.

Double sweep tee. A tee made with easy (long radius) curves between body and branch.

Draft. The pressure difference causing flow of a fluid, usually applied to convection flow, such as chimneys, and usually measured in inches of water.

Draft stabilizer. A device to maintain a constant draft in a fuel burning device.

Drain. A pipe carrying waste water in a building drainage system.

Drainage system. The piping system within the building or premises to the point of disposal to a public sewer or disposal system.

Drip. A pipe connection, usually trapped, between the steam side of a steam piping system and the condensate or return side of the system.

Dry bulb temperature. The temperature of air measured by a conventional thermometer.

Dry return. A condensate line in a steam heating system carrying both water and air, usually located above the boiler water line.

Dry ton. Sensible heat load expressed in tons.

Dual vent. In plumbing, a vent serving two fixtures and connected at the junction of the fixture drains from the two fixtures.

Duct fan. An axial flow fan mounted in, or intended for mounting in, a section of duct. See Tubeaxial fan and Vaneaxial fan.

Dynamic head. The pressure equivalent of the velocity of a fluid. Customary units of "head" are usually taken in feet or inches of H_2O.

Effective opening. The smallest cross-sectional area in the water-supply discharge in terms of the diameter of the circle of a round opening or, if the opening is not round, the diameter of a circle having that cross-sectional area.

Effective temperature. An index developed by the American Society of Heating, Refrigerating and Air-conditioning Engineers and combining, in a single value or arbitrary index number, the effect of temperature, humidity and air motion on the sensation of comfort. The numerical value is that of the temperature of still, saturated air which induces an identical feeling of comfort.

Efficiency. Ratio of output to input. Frequently misused where efficacy, load factor, or coefficient of performance would better apply.

Effluent. A fluid, either gas or liquid, leaving the process, building, or factory.

Ejector. A device in which a high velocity jet acts to entrain mechanically a second fluid to withdraw it from some region of like pressure and to deliver the mixture, with low turbulence, to a region of higher pressure.

Elbow (Ell). A fitting that makes an angle between adjacent pipes. The angle is 90 degrees unless another angle is specified.

Electronic cleaner. An air cleaner in which matter in the air stream is electrically charged, then attracted to surfaces oppositely charged.

Emissivity. The ratio of radiant energy emitted by a body to that emitted by a perfect black body. A perfect black body has an emissivity of 1; a perfect reflector an emissivity of 0.

Enthalpy. It is defined as $u + pv$; where u is the internal energy; p is pressure; v is specific volume. For most applications in HVAC it is a measure of thermal energy.

Entropy. The ratio of heat added to a substance to the absolute temperature at which the heat is added. Specific entropy is the ratio of entropy to weight of substance.

Equivalent direct radiation (EDR). Heat expressed in terms of a square foot of steam radiator surface emitting 240 Btu per hour; in other words, Btu per hour divided by 240.

Equivalent evaporation. The quantity of water which would be evaporated by a given apparatus if the water is received by the apparatus at 212F and vaporized at that temperature under atmospheric pressure. It is expressed in pounds per hour.

Equivalent length. The resistance of a duct or pipe elbow, valve, damper, orifice, bend, fitting, or other obstruction to flow expressed in the number of feet of straight duct or pipe of the same diameter which would have the same resistance.

Equivalent temperature. Also British equivalent temperature (BET). A composite of MRT (see Mean Radiant Temperature) and air temperature, also defined as the mean temperature of the environment effective in controlling the rate of sensible heat loss from a black body in still air when the surface temperature and size of the black body are comparable to those of the human body. Where the enclosure surface (MRT) and air temperatures are equal, this temperature is also the BET; when not equal the BET is that temperature at which a body with an 80F surface temperature will lose sensible heat at the same rate as in the given environment.

Eupatheoscope. A black hollow copper cylinder heated electrically to the surface temperature of a human being. The heat loss from the cylinder in a given environment is equivalent to that lost by a human being; thus, Equivalent temperature, which see.

Evaporative cooling. Cooling by the evaporation of water, heat for which is supplied by the air; feasible where the wet bulb depression is marked, and consequently widely used in dry climates.

Evaporative equilibrium. With a sling psychrometer, when the wet bulb temperature finally reaches a stable point.

Evaporator. That part of a refrigerating system in which the refrigerant is evaporated, absorbing heat from the surrounding fluid.

Exfiltration. Opposite of Infiltration, which see.

Expansion joint. A joint whose primary purpose is not to join pipe but to absorb the longitudinal expansion in the pipeline due to heat.

Expansion loop. A large radius bend in a pipeline to absorb longitudinal expansion in the line due to heat.

Factor of safety. Ratio of a normal working condition to the ultimate condition, such as, in strength of materials, ratio of working stress to ultimate strength. Loosely used as any allowance over and above the estimated required load or condition.

Film coefficient. Also called unit convection conductance or surface coefficient. The heat transferred by convection per unit area per degree temperature difference between the surface and the fluid.

Filter. A device for purifying a fluid, such as air or water, that removes suspended matter and/or gases or dissolved matter.

Finned tube. Extended (also termed gilled) surface in the form of fins on tubes or pipe.

Fixture branch. In plumbing, the branch from the water-distributing pipe in the building to the fixture supply pipe.

Fixture drain. The drain from the trap of a fixture to the point where it connects with another drainpipe.

Fixture supply pipe. The pipe connecting the fixture with the fixture branch at the floor or wall line.

Fixture unit. A base figure or weight in which each type of water-using fixture is assigned a number representing its maximum water demand relative to other fixtures.

Flange. In pipe work, a ring-shaped plate on the end of a pipe at right angles to the end of the pipe and provided with holes for bolts to allow fastening the pipe to a similarly equipped adjoining pipe. The resulting joint is a flanged joint.

Flange faces. Pipe flanges which have the entire face of the flange faced straight across, and use either a full face or ring gasket, are commonly employed for pressures less than 125 pounds on steam and water lines.

Flash point. The temperature of oil where the vapor will ignite but there is not sufficient vapor to allow combustion to continue.

Flooded. A situation when the liquid in a vessel or fixture rises to the flood level.

Flooded system. In refrigeration, a system where the refrigerant passing the evaporator is only partially evaporated.

Flood level. That level at which water begins to overflow the rim of the fixture.

Flood level rim. The top edge of the vessel or fixture above which water overflows.

Flow rate. The flow of fluid such as air per unit of time. Frequently stated as cubic feet per minute.

Flush valves. One for flushing water-closets or similar fixtures.

Foaming. In a boiler, the carry-over of slugs of water into the piping, due to dirty water. Also see Priming.

Fog. Condensed water vapor in very fine particles and lying near the ground, unlike a cloud in this latter respect.

Foot-candle. A point reading of surface illumination. One foot-candle is the illumination at a point one foot from a candle.

Forced convection. See Convection.

Free area. The total area of open space in a grille through which air can pass.

Frost-proof closet. In plumbing, one that has no water in the bowl and with the trap and water supply valve below the frost line.

Fume. Air carrying certain chemical contaminants of very small size ranging from one tenth to one micron in diameter.

Furnace. Either the combustion space in a fuel burning device or a direct-fired air heater. In the latter case, not to be confused with a boiler.

Furnace burner unit. Warm air furnace sold integrally with the burner.

Furnace volume. The cubic contents of the combustion space of a boiler bounded by the grates, direct heating surface, and tube sheet.

Fusible plug. An insert of metal with low melting point placed in boilers, sprinklers and other devices to melt when the temperature becomes dangerously high, so that the melting will relieve pressure, allow water flow, or otherwise tend to alleviate the dangerous condition.

Fusion weld. Joining metals by fusion, using oxyacetylene or electric arc.

Gage pressure. Pressure within a vessel above that shown by the barometer without.

Galvanized pipe. Steel pipe coated with zinc to resist corrosion.

Gas constant. The constant R in the perfect gas equation $PV = RT$. In English units this constant is 1545 divided by the molecular weight of gas.

Gate valve. A valve employing a gate, often wedge-shaped, allowing fluid to flow when the gate is lifted from the seat. Such valves have less resistance to flow than globe valves.

Globe thermometer. A thermometer in a hollow spherical black globe, readings from which show a higher value, due to radiation, than a conventional thermometer, so that the globe device measures the effective radiation temperature.

Globe valve. One with a somewhat globe shaped body with a manually raised or lowered disc which when closed rests on a seat so as to prevent passage of a fluid.

Grade. In piping, the slope of a line compared to horizontal, usually expressed in inches fall per foot of pipe.

Grate area. Area of surface on which fuel is burned, ordinarily measured at the top surface of the grates.

Grease interceptor. A trap designed to collect and retain grease and fats from kitchen waste lines, and installed in the line between the kitchen and building sewer.

Grille. A covering over an air inlet or outlet with openings through which air passes.

Ground joint. One where the parts to be joined are precisely finished and then ground in so that the seal is tight.

Head. A unit of pressure intensity usually given in inches or feet of a column of the fluid under consideration. Thus, 1 foot of water head is the pressure from a column of water 1 ft. high.

Header. A large pipe or drum into which each of a group of boilers is connected. Also used for a large pipe from which a number of smaller ones are connected in line and from the side of the large pipe.

Head pressure. The discharge pressure at the refrigeration compressor; also the condenser pressure.

Heat capacity. A measure of the heat necessary to change the temperature of a quantity of material; weight of material times specific heat.

Heating effect. A term not yet precisely defined but having to do with the ratio of heat output of a radiator or convector in the living zone of a room with the total output measured by steam condensation or water temperature drop.

Heating medium. A fluid used for conveying heat from a heat source to heat dissipating devices; includes air, water and steam.

Heating surface. Broadly, the area intended for transferring heat.

Heat transmission coefficient. See Coefficient of Heat Transmission.

Heat of liquid. The heat necessary to raise the temperature of the liquid, as distinguished from the heat of vaporization.

Heat pump. A mechanical refrigerating system used for air cooling in the summer, and which, when the evaporator and condenser effects are reversed, absorbs heat from the outside air, water, or ground in winter and raises it to a higher potential so that it also can be used for winter heating.

Hermetically sealed space. A tightly closed space, such as a vault, defense shelter, air capsule, or submarine, which either cannot be ventilated using outside air, or need not be.

High capacity. Referring to a product or equipment that will do a great amount of work or has a large or high capacity for work. High performance.

Highside. That portion of a mechanical refrigerating system under relatively high pressure from the compressor to the expansion valve, including the condenser.

Hood. A canopy or collecting structure over a process area or piece of equipment to collect fumes, gases, vapors, or odors which are then exhausted or treated in some manner.

Horizontal branch. In plumbing, the drain line from the fixture drain to the waste stack.

Hot water heating system. One in which water is the heating medium.

Humid heat. Ratio of the increase in total heat per pound of dry air to the rise in temperature, with constant pressure and humidity ratio.

Humidifying effect. The quantity of water evaporated per unit of time (usually 1 hour) times the latent heat of vaporization at the evaporating temperature.

Humidity. Usually, water vapor mixed with dry air. Absolute humidity is the weight of water (or steam) per unit volume of the air-water mixture.

Humidity ratio. See Specific humidity.

Hydrocarbon. A substance containing carbon and hydrogen.

Hydrostatic test. On a boiler, the closing of all openings and pumping water into the boiler at a pressure (such as 50%) greater than the normal operating pressure. The purpose is to locate leaks or prove that there are no leaks.

Hygrometer (also Hygroscope). A device for measuring moisture in the air.

Hygroscopic. Readily absorbing and retaining moisture.

Hygrostat. A device sensitive to humidity changes and arranged to actuate other equipments when a predetermined humidity is attained.

Inch of water. A unit of pressure equivalent to 0.036136 pounds per square inch.

Indirect heating surface. Any type of heating surface which does not come in direct contact with the original source of heat.

Indirect waste pipe. One which drains through a trapped fixture or receptacle to the drainage system.

Individual vent. A pipe venting a trap or waste pipe beneath or back of a fixture served by it, and connected to the main vent system at a point above the fixture.

Industrial air conditioning. Air conditioning in industrial plants where the primary objective is the furtherance of a manufacturing process rather than the comfort of human beings.

Industrial degree-day. A degree-day (which see) unit based on a (usually) 45F or 55F mean daily temperature so as to be applicable to industrial buildings maintained at relatively low temperatures.

Infiltration. Leaking inward. In heat transmission, applies to the air entering the space through cracks around windows, doors, etc. The opposite of Exfiltration.

Influent. A flow into a building or system.

Infra-red. Radiant energy of a wave frequency inferior to visible red light, which, when it falls on a body, causes its temperature to rise. Commonly called radiant heat.

Ingot iron. A commercially pure iron, lower in carbon than carbon steel, used chiefly in sheets.

Initial pressure. In a process, the temperature of the fluid entering the process. Final pressure is that of fluid leaving process.

Insolation. The rate at which direct solar energy is received per unit of horizontal area.

Insulation. Thermal insulation is a material used (a) for the walls, roofs, etc., of buildings or (b) for covering pipes, ducts, vessels, etc., to effect a reduction of transmission of heat to a relatively large degree.

Intercooling. Extraction of heat from a compressed gas between two stages of compression.

Inversion. A reversal of normal atmospheric temperature gradient—an increase of the air temperature with increasing altitude which is contrary to the normal situation.

Isobar. At constant barometric pressure.

Isopsychric. Constant relative humidity.

Isothermal. A change taking place at a constant temperature.

Joule. SI Metric unit for work, energy, and quantity of heat. The joule is the equivalent of the work done when the point of application of one newton is displaced a distance of one meter. Its symbol is J.

Kata thermometer. A thermometer developed by Heberden, Hill, and Griffiths which is heated before being exposed to the environment to be measured and the rate of fall noted. The rate of fall of the dry-bulb is a measure of the radiation and convection; of the wet-bulb, a measure of evaporation plus radiation and convection. If the dry-bulb is cooled in a vacuum bottle the rate is a measure of radiation alone.

Kilogram. Formerly a force unit under the older MKSA Metric System. Under the SI Metric System, force is now designated by the newton, symbol N. In the SI Metric System, the kilogram is used strictly in reference to mass. Its symbol is kg.

Laminar flow. Flow occurring in fluid layers, in a conduit or space, with absence of eddy currents. Each particle of fluid flows directly from one end to the other or from top to bottom, or in some planned manner.

Langley. A unit of intensity of solar radiation, equal to one gram calorie per square centimeter. The abbreviation is ly; 1 ly = 3.69 Btu per sq ft. Weather Bureau sun data are in terms of langleys per minute (or day as the case may be). 1 ly per min = 221.4 Btu per hr.

Lapped joint. A pipe joint made by using loose flanges on lengths of pipe whose ends are turned over or lapped over to produce a bearing surface for a gasket or metal-to-metal joint.

Lap weld pipe. That made by welding along a scarfed longitudinal seam in which one part is overlapped by the other.

Latent heat. The quantity of heat required to produce a change of state at constant temperature; usual cases are latent heat of vaporization of water (970.2 Btu per lb for water to steam at 14.696 lb per sq in. abs pressure) and latent heat of fusion (143.6 Btu per lb for ice to water).

Leader. In warm air heating, the supply duct running horizontally from the furnace to the riser, or stack. In plumbing, a water conductor from the roof to the storm drain.

Lead joint. A joint made by pouring molten lead into the space between a bell and spigot, and making the lead tight by calking.

Lethe. A measure of air purity. The term, coined by Dr. W. F. Wells, University of Pennsylvania, is the equivalent of one complete air change of pure air. The air quantity required for one lethe varies with the size of the room but is independent of time.

Limit switch. A control to limit some function. Examples are pressure limit switches which shut off the fuel burner when the steam pressure reaches a predetermined point; similarly, temperature limit switches for hot water and warm air.

Lip union. A form of union characterized by the lip that prevents the gasket from being squeezed into the pipe so as to obstruct the flow.

Load factor. Ratio of average output during a period to maximum output during the period. Sometimes ratio of output to maximum capacity.

Loop vent. A branch vent serving more than one fixture and looping back to connect with a stack vent instead of the vent stack.

Louvers. Overlapping and sloping slats arranged to prevent entrance or exit of some substances but allow ventilation air to pass.

Lowside. That portion of a mechanical refrigeration system under relatively low pressure, extending from the expansion valve through the evaporator to, but not including, the compressor.

LP gas. Liquified petroleum gas, a gas of high heating content, stored under high pressure in liquid form. Delivered to consumers in containers for use where piped gas is not available and used by gas utilities for peaking (when demand exceeds source of natural gas).

Lumens and Foot-Candles. The basis of light measurement is the lumen, roughly defined as the light emitted by one candle, within a unit solid angle, or steradian. This angle is defined as a cone intercepting a sphere, where the area of intercept is equal to the square of the radius. All this is relevant in that, if the radius is one foot, the area of interception is one square foot, and the light which falls on that area is defined as one foot candle. So, in very approximate terms, a foot candle is the amount of light falling on a one-square-foot surface from a candle one foot away.

Make-up air. Air brought into a building from the outside to replace that exhausted by a ventilating system.

Make-up water. Water supplied to replenish that lost by leaks, evaporation, etc.

Malleable iron. Cast iron heat-treated to reduce its brittleness. The process enables the material to stretch to some extent and to stand greater shock.

Manifold. A fitting with a number of branches in line connecting to smaller pipes. Used largely as an interchangeable term with header.

Mbh. One thousand Btu per hr.

Mean radiant temperature (MRT). That single temperature of all enclosing surfaces which would result in the same heat emission as the same surfaces with various different temperatures.

Mean temperature difference. In a heat exchanger the temperatures of the two fluids change throughout. The mean or average of the temperature differences between the two fluids is the mean temperature difference. In general use, it is taken as the log mean temperature difference.

Medium pressure. When applied to valves and fittings, implies they are suitable for a working pressure of from 125 to 175 pounds per square inch.

Mercury switch. A glass tube employing mercury to establish electrical contact between circuits when the tube is tilted so that the mercury bridges the gap between contacts, and conversely.

Met. A convenient and approximate unit of human heat production equivalent to the average metabolic heat produced by resting man, about 18.50 Btu per sq ft per hr.

Micron. One thousandth of a millimeter.

Milinch. One thousandth of an inch.

Millibar. $\frac{1}{1000}$ of the pressure exerted on a unit area by a column of mercury 29.531 inches high at 32F in latitude 45°. It is 1000 dynes per square centimeter. Millibars on weather maps are (Barometer reading ÷ 29.531) × 1000.

Mill length. Also known as random length. Run-of-mill pipe is 16 to 20 ft in length. Some pipe is made in double lengths of 30 to 35 ft.

Modulating. A control adjusting by increments.

Moisture content. The water content of a substance as measured under specified conditions.

Moisture tons. Latent heat load expressed in tons.

Mollier chart. A form of graph named for its inventor and covering heat properties of a fluid and with entropy as one coordinate and enthalpy as the other.

Multiplier. In refrigeration, the ratio of total cooling load to sensible cooling load.

Natural convection. See Convection.

Needle valve. A valve provided with a long tapering point in place of the ordinary valve disk. The tapering point permits fine graduation of the opening.

Newton. The unit of force in the SI Metric System. It replaces the older metric kilogram-force. The newton is defined as that force which gives to a mass of one kilogram an acceleration of one meter per second per second. Its symbol is N.

Nipple. A tubular pipe fitting usually threaded on both ends and under 12 inches in length. Pipe over 12 inches long is regarded as cut pipe.

Occupied space. Space within a building or structure or other enclosed space which is normally occupied by people engaged in working, other activities, or normal living.

Odor control. The control or removal of odors. The reduction of odors to a tolerable or permissible level.

Offset. An arrangement of fittings or bends changing the line of piping to another line but parallel to the original line.

One-pipe hot water system. One in which the heated and cooled water to and from the radiators are carried in the same "main" but in the same direction, with the hot water on top.

One-pipe steam system. A steam heating system in which steam supplied to radiator travels in same pipe but in opposite direction as the cold water or condensate returning from radiator.

Operative temperature. Operative temperature is that temperature of an imaginary environment in which, with equal wall (enclosing areas) and ambient air temperatures and some standard rate of air motion, the human body would lose the same amount of heat by radiation and convection as it would in some actual environment at unequal wall and air temperatures and for some other rate of air motion. The term originated at the John B. Pierce Laboratory of Hygiene.

Orifice. Opening. Commonly used to apply to discs placed in pipe lines or radiator valves to reduce the fluid flow to desired amount.

Orsat. A device for measuring the percentage of carbon dioxide, oxygen and carbon monoxide in flue gases.

Ozone. A form of oxygen having three atoms (O_3).

Paddle wheel fan. A centrifugal fan with radial blades.

Panel heating. See Radiant heating.

Panel radiator. A radiating surface (see Radiant heating) primarily flat to increase its ability to transmit radiant energy.

Partial pressure. That part of the total pressure of a mixture of gases contributed by one of the constituents.

Pascal. The Pascal, symbol Pa, is a derived SI Metric unit for pressure and stress. IT is equal to one newton per square meter (N/m^2).

Percentage humidity. Ratio of weight of vapor in a mixture of air and water vapor to the weight of vapor when the vapor is saturated at the same temperature. Contrast with Relative humidity.

Perfect combustion. Fuel burning when all combustible is consumed with no excess air so that the theoretical amount of oxygen only is used. Also see Complete Combustion.

Perfect gas. A perfect or ideal gas is one which follows Charles' law, or in other words fits the $PV = RT$ gas equation.

Perimeter heating. A term originally coined by McClatchy & Co., Philadelphia builders, to describe Warren Webster & Company's forced hot water heating system employing series connected baseboard convectors placed the full length of the exposed walls of a building, and currently used in that sense by that manufacturer. The term is also applied by the warm air heating industry to describe a variety of warm air systems all of which introduce the warm air into the rooms by registers or warm air baseboards located along the outer walls, preferably under the windows, close to, or along the floor.

Permeance. The rate of water vapor transmission on a basis of unit time, unit area, and unit vapor pressure difference through a material. Permeability is the rate through a unit thickness of the material. Joy and Queer have coined the term Perm as the unit for 1 grain per (square foot) (hour) (inches of mercury).

pH. Hydrogen ion concentration. The logarithm of the reciprocal of the gram ionic hydrogen equivalents per liter. On the pH scale 7.0 is normal, smaller values to 0 are acid, progressively so, while higher values to 14 are alkaline, progressively so.

Photochemical smog. A type of smog which occurs in many urban areas, and is caused by a combination of sunlight, ozone, nitrogen oxides, and various organic vapors.

Pipe coil. Pipe used for heating, usually in (1) parallel lines connected to heaters at the end, forming a grid; (2) in parallel lines with ends alternately connected to adjacent lines by bends, forming a so-called continuous coil.

Pipe line refrigeration. Piping of refrigeration from one central plant to adjacent buildings.

Plenum. An air space or chamber under pressure.

Plug valve. One with a short section of a cone or tapered plug through which a hole is cut so that fluid can flow through when the hole lines up with the inlet and outlet, but when the plug is rotated 90°, flow is blocked.

Plumbing. The installing of or the installation consisting of piping, fixtures, valves, and traps to carry water to the fixtures and carry liquid and other wastes from these and other fixtures.

Plumbing fixtures. Devices which receive or make available water, or receive liquid or water-borne wastes, and discharge the wastes into a drainage system.

Plumbing system. Arrangements in a building of pipes, fixtures, fittings, valves, and traps to supply water and remove liquid wastes, including storm water.

Pollutant. A substance, either particulate or gaseous or a mixture of substances, that pollutes either air or water because of the fact that the substance is undesirable, toxic, irritating, odorous, or otherwise unwanted or unnatural.

Priming. In a boiler, the excessive carry-over of fine water particles with the steam due to insufficient steam space, faulty boiler design, or faulty operating conditions.

Propeller fan. A fan with airfoil blades and which moves the air in the general direction of the axis of the fan.

Proximate analysis. Fuel analysis on the basis of per cent fixed carbon, volatile matter, moisture and ash.

Psychrometer. A device for measuring the humidity in the air and employing a wet bulb and a dry bulb thermometer.

Psychrometric chart. A graph showing the properties of air-steam mixtures and used in air conditioning.

Quality. The ratio by weight of vapor to liquid plus vapor in a mixture, as in steam.

Radiant heating. Usually a system of heating by surfaces at somewhat higher than body temperatures whereby the rate of heat-loss from the human beings by radiation is controlled.

Radiation. The transfer of energy in wave form from a hot body to a (relatively) cold body independent of matter between the two bodies.

Recessed radiator. One set into the wall but not enclosed.

Recirculated air. Air returned from a space to be heated, conditioned or cleaned, then redistributed to the space.

Reducer. A pipe fitting with a larger size at one end than at the other; the larger size is designated first. Reducers are threaded inside, unless specified flanged, welded, or for some special joint.

Reflectivity. The ratio of radiant energy reflected by a body to that falling upon it.

Refrigerant. A substance which will absorb heat while vaporizing and whose boiling point and other properties make it useful as a medium for refrigeration.

Regain. Moisture regain is the ratio of normal moisture weight to the weight of material holding the moisture, expressed in per cent.

Regenerative heating or cooling. An arrangement whereby heat rejected in one portion of a cycle is usefully employed in another portion of the cycle.

Relative humidity. Ratio of vapor pressure in a mixture of air and water vapor to the vapor pressure when saturated at the same temperature. Contrast with specific humidity.

Relief valve. One designed to open automatically to relieve excess pressure.

Relief vent. A branch from the vent stack connected to the horizontal branch between the first fixture drain and the waste stack to provide air circulation between the vent stack and waste stack.

Reodorant. A masking agent—a type of perfume to cover up or make more pleasant an undesirable odor.

Resistance. In heat transfer, the reciprocal of Conductance, which see.

Resistance to flow. The same as pressure drop. It is the resistance offered by a conduit or device to the flow of fluid.

Resistance weld pipe. Pipe made by bending a plate into circular form and passing electric current through the material to obtain a welding heat.

Resistivity. The reciprocal of Conductivity, which see.

Return air. Air returning to a heater or conditioner from the heated or conditioned space.

Return offset. A double offset is an arrangement which offsets the line, then returns it to the original line.

Reverse cycle refrigeration. See Heat pump.

Reynolds' number. A dimensionless ratio proportional to pipe or duct diameter, velocity, and density of fluid and inversely proportional to the viscosity of the fluid. With English units, a Reynolds' number over 2500 indicates turbulent flow; below that, streamlined flow.

Riser. A water or steam pipe extending vertically one story or more.

Roof drain. One on the roof to collect water on the roof surface and discharge it to leader.

Run. A length of pipe made of more than one piece of pipe; a portion of a fitting having its ends in line or nearly so, in contradistinction to the branch or side opening, as of a tee.

Saddle flange. A flange curved to fit a boiler or tank and to be attached to a threaded pipe. The flange is riveted or welded to the boiler or tank.

Sand interceptor. A trap designed to retain sand.

Sanitary sewer. One carrying sewage but not storm, surface or ground water.

Saturated air. Air containing saturated water vapor with both air and vapor at the same dry bulb temperature.

Saturated steam. Steam at the boiling temperature corresponding to the pressure at which it exists. Dry saturated steam contains no water particles in suspension. Wet saturated steam does.

Saturation pressure. That pressure for a given temperature at which the vapor and the liquid can exist in stable equilibrium.

Schedule number. In piping, an American Standards Association designation of the strength of pipe. It is 1000 (internal service pressure ÷ maximum tensile stress), where the pressure and stress are in pounds per square inch.

Screwed flange. A flange screwed on the pipe which it is connecting to adjoining pipe.

Screwed joint. A pipe joint consisting of threaded male and female parts screwed together.

Seamless pipe. Pipe or tube formed by piercing a billet of steel and then rolling.

Secondary air. Air for combustion introduced over the firebed to oxidize the volatile constituents of the fuel as distinguished from primary air introduced below the firebed.

Self-contained cooling unit. A combination of apparatus for room cooling complete in one package; usually consists of compressor, evaporator, condenser, fan motor and air filter. Requires connection to electric line.

Sensible cooling effect. The difference between the total cooling effect and the dehumidifying effect.

Sensible heat. That heat which when added or subtracted results in a change of temperature, as distinguished from Latent heat, which see.

Service fitting. A street ell or street tee with male threads at one end and female threads at the other.

Service pipe. A pipe connecting water or gas mains with a building.

Sewage. Liquid waste containing animal, vegetable, mineral or chemical matter in suspension or solution.

Short nipple. One whose length is a little greater than that of two threaded lengths or somewhat longer than a close nipple so that it has some unthreaded portion between the two threads.

Sigma function. The enthalpy of an air-steam mixture less the heat of the liquid.

Shoulder nipple. A nipple of any length which has a portion of pipe between two pipe threads. As generally used, however, it is a nipple halfway between the length of a close nipple and a short nipple.

Sleeve weld. A joint made by butting two pipes together and welding a sleeve over the outside.

Slip-on flange. A flange slipped over the end of the pipe and then welded to the pipe.

Socket weld. A pipe joint made by use of a socket weld fitting which has a prepared female end or socket for insertion of the pipe to which it is welded.

Soil pipe. One which carries the discharge from water-closets or fixtures.

Sol-air temperature. That temperature of the outdoor air which, in contact with a completely shaded building surface, would give the same rate of heat entry into the outdoor surface of that material as exists with the actual dry bulb temperature of the outdoor air, the actual intensity of solar and sky radiation incident upon that surface, and the actual wind velocity. This concept and term was originated by Professor C. O. Mackey of Cornell University.

Solar constant. The amount of solar energy received by a surface at right angles to the sun's rays outside the earth's atmosphere. It varies with the month, the mean value being 430 Btu per hour per square foot.

Solder joint. A method of joining tube by use of solder.

Specific enthalpy. See Enthalpy.

Specific heat. The heat absorbed by a unit of weight of substance per unit temperature rise.

Specific humidity. Weight of water vapor, in grains or pounds, per pound of dry air.

Specific speed. The rpm at which a fan, pump, or turbine would run if it were of such diameter as to deliver a unit volume of fluid per minute against a unit pressure.

Specific volume. The volume occupied by a unit of air. Usually given in cubic feet per pound.

Spiral pipe. Pipe made by coiling a plate into a helix and riveting or welding the overlapped edges.

Split system. Historically, a combination of warm air heating and radiator heating; the term is also used for other combinations such as hot water-steam, steam-warm air, etc.

Square foot of radiator. The amount of heating surface in the form of radiators, convectors, unit heaters or other devices which will emit 240 Btu per hour. This term has no exact relationship to the actual area of the surface.

Squirrel cage fan. A centrifugal blower with forward-curved blades.

Stack. In plumbing, a vertical main of soil, waste or vent pipe. In heating work, a chimney.

Stack effect. The impulse of a heated gas to rise in a vertical passage, as in a chimney, a small enclosure, or a building.

Stack height. The height of a convector enclosure measured from the bottom of the enclosure to the top of the outlet.

Stack vent. An extension of a soil or waste stack above the highest horizontal branch connected to the stack.

Stainless steel pipe. An alloy steel pipe with corrosion-resisting properties, usually imparted by nickel and chromium.

Standard air. Air at 70F and 29.921 inches of mercury and weighing 0.07488 lb per cu ft.

Standard conditions. In refrigeration, an evaporating temperature of 5F, condensing temperature of 86F, liquid temperature before the expansion value of 77F, and suction temperature 14F.

Standard pressure. Formerly used to designate cast-iron flanges, fittings, valves, etc., suitable for a maximum working steam pressure of 125 lb. In practice, 14,696 psia is used in determining density at standard pressure and temperature.

Stat. A suffix meaning apparatus to render stationary. For example, thermostat, a device to maintain a constant temperature; pressurestat, device to maintain constant pressure, etc.

Static pressure. The pressure exerted by a fluid in all directions; the pressure which would tend to burst the container; the pressure exerted by the fluid if stationary.

Steam. Water in gaseous or vapor form.

Steam jet refrigeration. A method of cooling involving the use of steam nozzles to reduce the pressure in a water chamber so that the water boils at a low temperature; since heat is drawn from the water, it is thus cooled.

Steam trap. An apparatus for allowing water or air to pass but preventing passage of steam.

Storm sewer. One carrying surface or storm water (but not sewage) from roofs or streets.

Streamline flow. Also termed viscous or laminar flow. When the fluid particles are moving in a straight line parallel to the axis of the pipe or duct. With streamline flow, the friction varies directly as the velocity. Also see Turbulent flow and Critical velocity.

Street elbow. An elbow with male thread on one end, and female thread on the other end.

Subcooling. Cooling of a liquid refrigerant below the condensing temperature at constant pressure.

Subsoil drain. One for collecting sub-surface or seepage water and carrying it to a point of disposal.

Suction line. Refrigerant line leading to the compressor.

Suction pressure. Pressure in the suction line or evaporator of a refrigerating system.

Summer degree hour. The product of one hour in time and the amount the outside dry bulb temperature at that hour is above a base temperature. Some correlation has been found between the number of degree-hours over 85F and summer cooling load in residences.

Sump. A tank or pit to which water is drained and from which it is pumped or ejected to a drainage system.

Superheated steam. Steam at a temperature higher than the boiling temperature corresponding to the pressure at which it exists.

Surface conductance. The heat transmitted from (or to) a surface to (or from) the fluid in contact with the surface in a unit of time per unit of surface area per degree temperature difference between the surface and fluid.

Surface cooling. Air cooling by passing it over cold surfaces.

Surface tension. Due to intermolecular attraction, the interface between two immiscible fluids tends to assume the minimum area consistent with the volume of and pressure on the fluids. To extend this minimum area requires work and the measure of this work (usually in dynes per centimeter) is the surface tension.

Suspended matter. Particles of dust or other types of individual small granules or pieces sufficiently fine that they are suspended in air or carried by air, or other fluid.

Swing joint. An arrangement of screwed fittings and pipe to provide for expansion in pipelines.

Swivel joint. One employing a special fitting designed to be pressure tight under continuous or intermittent movement of the machine or part to which it is connected.

Tee. A pipe fitting, either cast or wrought, that has one side outlet at right angles to the run.

Temperature-Humidity Index. An index that combines sensible temperature in degrees F and humidity in the air to arrive at a number that closely corresponds to the Effective Temperature. When the THI is 70, almost everyone is comfortable. When it is 71, ten percent of the people will be uncomfortable; 72, twenty percent, etc., until, at 80, almost all people are uncomfortable.

Therm. One hundred thousand Btu.

Thermocouple. Two dissimilar metals joined together produce an electromotive force which varies with the temperature. Such an arrangement calibrated for measuring temperature is called a thermocouple.

Throw. Distance from air supply opening measured in the direction of air flow, from the opening to the point where the air velocity is 50 ft per min.

Tons of refrigeration. A common, commercial measure of refrigeration capacity, especially of air conditioning apparatus cooling capacity, the equivalent of the heat required to melt one ton (2,000 lb) of ice (heat of fusion is 144 Btu per lb) in 24 hours, hence, 288,000 Btu per day or 12,000 Btu per hr.

Total cooling effect. The difference between the total heat content of the air-steam mixture entering a conditioner per hour and the total heat of the mixture leaving per hour.

Total heat. The sum of sensible heat and latent heat in a substance or fluid above a base point, usually 32F or 0F.

Total pressure. The sum of the static pressure and velocity pressure of a fluid.

Toxic. Having poisonous effects.

Transmittance. Rate of heat flow per unit area per unit temperature difference.

Trap. In plumbing, a device providing a liquid seal which will allow passage of sewage or waste matter but prevent the back passage of air. Also see Steam trap.

Trap seal. The vertical distance between the bottom of the top bend and the top of the bend of a trap.

Tubeaxial fan. An airfoil (propeller) or disc fan within a cylinder and including driving mechanism supports either for belt drive or direct connection.

Turbulent flow. When the fluid particles are moving in directions other than in a straight line parallel to the axis of the pipe or duct.

Turning vane. Curved strips of short radii placed in a sharp bend or elbow in a rectangular duct to direct the air around the bend in streamlined flow.

Two-pipe steam system. A heating system in which one pipe used for supplying steam to radiator and another pipe is used to return cooled water or condensate to boiler.

Ultimate CO₂. The per cent of CO_2 which would appear in the flue gases if combustion were perfect. Varies with the fuel.

Union. (1) The usual trade term for a device used to connect pipes. It commonly consists of three pieces which are, first, the thread end fitted with exterior and interior threads; second, the bottom end fitted with interior threads and a small exterior shoulder; and third, the ring which has an inside flange at one end while the other end has an inside thread like that on the exterior of the thread end. A gasket is placed between the thread and bottom ends, which are drawn together by the ring. Unions are very extensively used, because they permit of connections with little disturbance of the pipe positions.

Unit heater. A forced convention heating device of two types: (1) an assembly of encased heating surface with fan and motor (or turbine) and for connection to a source of steam or hot water; (2) an assembly of the above plus a fuel burner so that the device is for connection to a source of oil or gas (or supplied with coal) and not to steam or water lines.

Unit ventilator. A unit heater (which see) of type (1) but connected to a source of ventilation air and frequently provided with an air filter. Widely used in school buildings.

Vacuum system. A two-pipe steam heating system equipped with vacuum pumps to permit maintenance of pressures below atmospheric within the radiators.

Vaneaxial fan. An airfoil (propeller) or disc fan within a cylinder and equipped with air guide vanes either before or after the wheel and including driving mechanism supports either for belt drive or direct connection.

Vane ratio. In grilles, the ratio of vane depth to shortest opening width between two adjacent vane bars.

Vapor. The gaseous state of a substance.

Vapor barrier. A material intended to prevent the passage of water vapor through a building wall so as to prevent condensation within the wall.

Vapor pressure. The equilibrium pressure of the vapor of a liquid in contact with the liquid.

Vapor system. A steam heating system operating at pressures very near that of the atmosphere.

Velocity. The rate at which a fluid is flowing. Air velocity is normally stated in feet per second or feet per minute.

Velocity pressure. The pressure exerted by a moving fluid in the direction of its motion. It is the difference between the total pressure and the static pressure.

Vent stack. A vertical vent pipe to provide air circulation to and from the drainage system.

Vent system. Piping to provide a flow of air to or from a drainage system to protect trap seals from siphonage or back pressure.

Ventilation. The art or process of supplying outside (so-called fresh) air to or removing air from an enclosure.

Venturi. A contraction in a pipeline or duct to accelerate the fluid and lower its static pressure. Used for metering and other purposes.

Viscosity. The property of a fluid by which it resists an instantaneous change of shape or arrangement of its molecules. Broadly, resistance to flow.

Viscous flow. See Streamline flow.

Volumetric efficiency. Volume of gas handled by a compressor divided by the displacement.

Warm air furnace. A vessel of steel or cast iron in which fuel is burned and over the surfaces of which air is passed and thus heated. In the heating trade designated as a furnace and not to be confused with a boiler.

Warm air heating system. One whereby air is heated directly in a furnace and circulated to rooms to be heated by gravity or by a fan. The latter is termed Forced warm air heating.

Warming-up allowance. An addition to the capacity of heating system (as calculated for heat loss) to provide for quick pick-up in the morning.

Waste pipe. A line carrying the discharge from any fixture except water-closets and the like, and which conveys the waste to the building drain or waste stack. If it is not directly connected to the building drain or soil pipe it is an Indirect Waste Pipe.

Water distributing pipe. One carrying water from the water service pipe to the fixtures.

Water line. The height of the normal water level in a boiler measured from the floor.

Water main. Street main.

Water service pipe. Pipe from the water main to the building served.

Water supply system. The system of water service pipes, distributing pipes and necessary connections, fittings, and valves.

Water vapor. A term used in psychrometry to denote the water in the air which actually is superheated steam.

Welding fittings. Wrought or forged-steel prefabricated elbows, tees, reducers, saddles, and the like, beveled for welding to pipe.

Welding neck flange. A flange with a relatively long neck beveled for butt welding to pipe.

Wet-bulb temperature. The temperature of the air as measured by a wet-bulb thermometer, and which is lower than the dry bulb temperature (for all cases except when the air is saturated) in inverse proportion to the humidity.

Wet heat. A slang expression for heating systems employing water or steam as the heating medium, as opposed to "dry" heat or warm air systems.

Wet vent. A soil or waste pipe that also serves as a vent.

Wiped joint. A lead joint in which the molten solder is poured upon the joint, after scraping and fitting the parts together, and the joint is wiped up by hand with a cloth pad while the metal is plastic.

Wrought iron. A low carbon iron containing a relatively high proportion of residual slag which gives it ductility and toughness.

Wrought pipe. This term refers to both wrought steel and wrought iron. Wrought in this sense means worked, as in the process of forming furnace-welded pipe from skelp, or seamless pipe from plates or billets. The expression "wrought pipe" is thus used as a distinction from cast pipe. When wrought-iron pipe is referred to, it should be designated by its complete name.

Wye. A pipe fitting with a side outlet at other than 90° to the main axis.

Yoke vent. A pipe extending upward from a waste stack to a vent stack to prevent pressure changes or differences in the stacks.

Zoning. The division of a building or group of buildings into separately-controlled spaces where different conditions can be simultaneously maintained.

GENERAL INDEX

SIXTH EDITION

The Norton

Field Guide
to Writing

SIXTH EDITION

The Norton

Field Guide to Writing

Richard Bullock
WRIGHT STATE UNIVERSITY

Deborah Bertsch
COLUMBUS STATE COMMUNITY COLLEGE

W. W. NORTON & COMPANY
Independent Publishers Since 1923

W. W. NORTON & COMPANY has been independent since its founding in 1923, when William Warder Norton and Mary D. Herter Norton first published lectures delivered at the People's Institute, the adult education division of New York City's Cooper Union. The firm soon expanded its program beyond the Institute, publishing books by celebrated academics from America and abroad. By mid-century, the two major pillars of Norton's publishing program—trade books and college texts—were firmly established. In the 1950s, the Norton family transferred control of the company to its employees, and today—with a staff of four hundred and a comparable number of trade, college, and professional titles published each year—W. W. Norton & Company stands as the largest and oldest publishing house owned wholly by its employees.

Copyright © 2022, 2021, 2019, 2016, 2013, 2009, 2006 by W. W. Norton & Company, Inc.
All rights reserved
Printed in Canada

Editor: Sarah Touborg
Project Editor: Christine D'Antonio
Senior Associate Editor: Claire Wallace
Assistant Editor: Emma Peters
Manuscript Editor: Alice Vigliani
Managing Editor, College: Marian Johnson
Managing Editor, College Digital Media: Kim Yi
Production Managers: Liz Marotta, Stephen Sajdak
Media Editor: Joy Cranshaw
Media Project Editor: Cooper Wilhelm
Media Editorial Assistants: Katie Bolger, Maria Qureshi

Ebook Production Manager: Sophia Purut
Marketing Manager, Composition: Michele Dobbins
Design Director: Rubina Yeh
Book Designers: Anna Palchik, Jen Montgomery
Photo Editor: Ted Szczepanski
Photo Research: Dena Digilio Betz
Director of College Permissions: Megan Schindel
College Permissions Manager: Bethany Salminen
Permissions Clearing: Josh Garvin
Composition: Graphic World
Manufacturing: Transcontinental—Beauceville

Permission to use copyrighted material is included in the Credits section of this book, which begins on page C-1.

Library of Congress Cataloging-in-Publication Data

Names: Bullock, Richard H. (Richard Harvey), author. | Bertsch, Deborah,
   author.
Title: The Norton field guide to writing / Richard Bullock,
   Wright State University ; Deborah Bertsch.
Description: Sixth edition. | New York : W. W. Norton & Company, [2022] |
   Includes bibliographical references and index.
Identifiers: LCCN 2021033654 | **ISBN 9780393533057** (paperback) | ISBN 9780393883794 (epub)
Subjects: LCSH: English language--Rhetoric--Handbooks, manuals, etc. |
   English language--Grammar--Handbooks, manuals, etc. | Report
   writing--Handbooks, manuals, etc.
Classification: LCC PE1408 .B883824 2022 | DDC 808/.042--dc23
LC record available at https://lccn.loc.gov/2021033654

W. W. Norton & Company, Inc., 500 Fifth Avenue, New York, NY 10110
wwnorton.com
W. W. Norton & Company Ltd., 15 Carlisle St., London W1D 3BS

1 2 3 4 5 6 7 8 9 0

Preface

The Norton Field Guide to Writing began as an attempt to offer the kind of writing guides found in the best rhetorics in a format as user-friendly as the best handbooks, and on top of that, to be as flexible as possible. We wanted to create a handy guide to help college students with all their written work. Just as there are field guides for bird watchers, for gardeners, and for accountants, this would be one for writers.

In its first five editions the *Field Guide* has obviously touched a chord with many writing instructors, and it remains the best-selling college rhetoric—a success that leaves us humbled and grateful, but also determined to improve the book with each new edition. To that end, we've added new chapters on writing explorations, creating remixes, organizing your writing and guiding your readers, and reflecting on your writing; new student examples throughout; improved advice on evaluating sources; and (we hope) improved writing in every chapter. New and expanded resources, envisioned and cowritten by Deborah Bertsch, provide practical and inspiring support for instructors and students. But it's not just us: this new edition has also been shaped by our community of adopters and their students. Through formal and informal reviews, emails and letters, and conversations with teachers and students using the *Field Guide*, we've learned from you what you need to teach effectively and what your students need to learn.

The Norton Field Guide still aims to offer both the guidance new teachers and first-year writers need and the flexibility many experienced teachers want. In our own teaching we've seen how well explicit guides to writing work for students and novice teachers. But too often, writing textbooks provide far more information than students need or instructors can assign and as a result are bigger and more expensive than they should be. So we've tried to provide enough structure without too much detail—to give the information college writers need to know while resisting the temptation to tell them everything there is to know.

Most of all, we've tried to make the book easy to use, with menus, directories, a glossary/index, and color-coded links to help students find what they're looking for. The links are also the way we keep the book brief: chapters are short, but the links send students to pages elsewhere in the book if they need more detail.

This new edition also marks an important change in the *Field Guide*'s authorship—we want to welcome Deborah Bertsch as a coauthor. Deb, who teaches at Columbus State Community College, has a background rich in excellent teaching, has directed Columbus State's writing center, and has been an important innovator in dual-credit course design and delivery and in various technologies of teaching writing, including instructional video and online instruction. Deb has contributed new chapters, Remixes (Chapter 22) and Reflecting on Your Writing (Chapter 35), substantially revised the Taking Stock of Your Work questions throughout the book, and led the creation of new videos and interactive activities in InQuizitive for Writers that offer multiple ways for students to understand and apply the book's advice. Along with making numerous other improvements throughout the book and its resources, Deb also re-envisioned the instructor's guide to respond to today's teaching challenges. Thank you, Deb. It's good to have you on the team.

WHAT'S NEW IN THE SIXTH EDITION

Two new Genres chapters help first-year writers adapt to new rhetorical situations:

- A new chapter on Explorations offers advice on using writing to explore approaches to problems (Chapter 21).

- A new chapter on Remixes provides guidance on how to transform writing in one genre to another medium or for another audience (Chapter 22).

New and revised Taking Stock of Your Work questions appear in each genre chapter:

- Every Taking Stock question now connects back to key rhetorical concepts, explicitly encouraging students to reflect on their writing

knowledge and experience so that they can draw on these concepts for future composing tasks.

Three new Processes chapters guide students in using writing as a tool for thinking:

- Writing as a Process offers a succinct and reassuring overview of the different ways writers approach composing tasks (Chapter 26).

- Organizing Your Writing, Guiding Your Readers offers practical advice on creating effective paragraphs, thesis statements, and topic sentences; crafting beginnings and titles and appropriate endings; and tying ideas together with transitions (Chapter 29).

- Reflecting on Your Writing shows students how to reflect on their writing in drafting, in peer review, and in editing what they write; it then walks students through how to write formal reflections (Chapter 35).

New readings throughout. As in previous editions, the *Field Guide* features authentic student writing from a range of two-year and four-year schools across the country, along with high-interest professional writing that demonstrates how the genres and processes learned in first-year writing transfer outside the classroom. These readings reflect the kinds of writing most commonly assigned in first-year composition, and all student writing is documented in the latest MLA or APA style. There are eleven new readings in the rhetoric, including seven by students: new essays in nearly every genre, including two new literacy narratives, a visual-text analysis of how inequality is conveyed through the architecture in the movie *Parasite*, a report on sleep deprivation in college students, an argument advocating for actors with disabilities, and many more (see asterisked readings in the table of contents).

Updated coverage of digital tools

- The Processes chapters have been updated with information on digital tools that support writers' processes, like web-based concept mapping tools that help students generate ideas and text, as well as voice-recording tools, notes apps, and transcription apps for capturing ideas generated on the fly. (See Chapters 26–35.)

- The Processes chapters also include information about ways to use storyboards and mockups as generative tools for multimedia and other types of writing. (See Chapter 28.)

- The Doing Research part now includes more information about online tools for collecting, analyzing, and organizing source material.

- The Media/Design part offers an expanded and thoroughly updated chapter, Writing and Learning Online, with advice on participating in synchronous and asynchronous classes, tips on productivity, and more.

Norton Field Guide ebook.   All new print copies of *The Norton Field Guide* now include access to the interactive ebook, with embedded videos from the new collection (see below). Standalone ebook access is available at digital.wwnorton.com/fieldguide6 or can be offered through Inclusive Access programs on many campuses.

A collection of new videos.   The Sixth Edition offers a new collection of videos on key concepts—from thesis statements to the rhetorical situation—that help students visualize and apply the book's advice. Each video is two to three minutes long, making them ideal for streaming in class or for independent viewing as a quick refresher. Videos are also embedded directly in the ebook, and many are assignable through InQuizitive for Writers activities.

New InQuizitive activities.   New activities on Thesis Statements, Elements of Argument, Rhetorical Situations, Paragraph Development, and Critical Reading Strategies complement the widely used activities on working with sources and editing sentences. Many of the new questions feature embedded videos from the new collection, and direct links to relevant pages in the *Little Seagull Handbook* ebook provide additional support. Access to InQuizitive for Writers and the *Little Seagull Handbook* ebook is now included with all new copies of *The Norton Field Guide to Writing*, and both can be integrated directly into most campus learning management systems.

Norton Teaching Tools. All of the popular resources from A *Guide to Teaching* The Norton Field Guides to Writing can now be downloaded from the searchable and sortable Norton Teaching Tools site in formats that are editable and easier than ever to use. The site also provides guidance for effective use of the new video collection, InQuizitive for Writers assignments, and the *Little Seagull Handbook* ebook, whether in person or online. Highlights include:

- A new collection of common writing assignments comes with transparent prompts that highlight purpose, task, and criteria; templates for peer review; sample rubrics; and sample student responses to assignments.

- New *PowerPoint* slides incorporate suggested class activities; chapter teaching tips have been fully updated as well as reformatted for easier application and assignability, and they now include guidance for dual-enrollment courses.

- New chapters in Norton Teaching Tools and the printed *Guide* support instructors interested in culturally responsive teaching, antiracist pedagogy, transparent assignment design, trauma-informed teaching, and more. Chapters on teaching writing online and contract (labor-based) grading have been extensively updated.

WAYS OF TEACHING WITH
THE NORTON FIELD GUIDE TO WRITING

The Norton Field Guide is designed to give you both support and flexibility. It has clear assignment sequences if you want them, or you can create your own. If, for example, you assign a position paper, there's a full chapter. If you want students to use sources, you can add the appropriate research chapters. If you want them to submit a topic proposal, you can add that chapter.

If you're a new teacher, the Genres chapters offer explicit assignment sequences—and the color-coded links will remind you of detail you may want to bring in. The Norton Teaching Tools site offers advice on creating a syllabus, responding to writing, and more.

If you focus on genres, there are complete chapters on all the genres college students are often assigned. Color-coded links will help you bring in details about research or other writing strategies as you wish.

If you teach a corequisite, IRW, or stretch course, the Academic Literacies chapters offer explicit guidelines to help students write and read in academic contexts, summarize and respond to what they read, and develop academic habits of mind that will help them succeed in college.

If you teach a dual-credit course, the Norton Teaching Tools site offers advice on supporting students in various dual-enrollment contexts, including online, in high schools, and on college campuses.

If you organize your course thematically, Chapter 28 on generating ideas and text can help get students thinking about a theme. You can also assign them to do research on the theme, starting with Chapter 47 on finding sources, or perhaps with Chapter 27 on writing as inquiry. If they then write in a particular genre, there will be a chapter to guide them.

If you want students to do research, there are nine chapters on the research process, including guidelines and sample papers for MLA and APA styles.

If you teach for transfer, three chapters about fields of study help students connect what they're reading, writing, and learning in your class to other college courses. The Considering the Rhetorical Situation questions that appear throughout the book prompt students to reflect on the key concepts of audience, purpose, genre, stance, and media/design, thus helping students develop a framework for analyzing future writing tasks.

If you teach writing about writing, you might use the chapter on literacy narratives as an introduction to your course—and use the Fields section to help students research and analyze the key features of writing in particular disciplines or careers.

If you focus on modes, you'll find chapters on using narration, description, and so on as strategies for many writing purposes, and links that lead students through the process of writing an essay organized around a particular mode.

WHAT'S IN THE BOOK

The Norton Field Guide covers twelve genres often assigned in college. Much of the book is in the form of guidelines, designed to help students consider the choices they have as writers. The book is organized into eight parts:

- ✳ **ACADEMIC LITERACIES.** Chapters 1–4 focus on writing and reading in academic contexts, summarizing and responding, and developing academic habits of mind.

- ◼ **RHETORICAL SITUATIONS.** Chapters 5–9 focus on purpose, audience, genre, stance, and media and design. In addition, almost every chapter includes tips to help students focus on their rhetorical situations.

- ▲ **GENRES.** Chapters 10–21 cover twelve genres, four of them—literacy narrative, textual analysis, report, and argument—treated in greater detail.

- ● **FIELDS.** Chapters 23–25 cover the key features of major fields of study and give guidance on reading and writing in each of those fields.

- ⁘ **PROCESSES.** Chapters 26–35 offer advice for generating ideas and text, drafting, revising and rewriting, editing, proofreading, reflecting, compiling a portfolio, collaborating with others, and writing as inquiry.

- ◆ **STRATEGIES.** Chapters 36–45 cover ways of developing and organizing text—comparing, describing, taking essay exams, and so on.

- **RESEARCH / DOCUMENTATION**. Chapters 46–54 offer advice on how to do academic research; work with sources; quote, paraphrase, and summarize source materials; and document sources using MLA and APA styles. Chapter 53 presents the "official MLA style" introduced in 2021.

- **MEDIA / DESIGN**. Chapters 55–59 give guidance on choosing the appropriate print, digital, or spoken medium; designing text; using images and sound; giving spoken presentations; and writing online.

HIGHLIGHTS

It's easy to use.   Menus, directories, and a glossary/index make it easy for students to find what they're looking for. Color-coded templates and documentation maps even make MLA and APA documentation easy.

It has just enough detail, with short chapters that include color-coded links sending students to more detail if they need more.

It's uniquely flexible for teachers. Short chapters can be assigned in any order—and color-coded links help draw from other chapters as need be.

RESOURCES

Like the book itself, the resources that accompany *The Norton Field Guide to Writing* are designed to provide flexibility, with a wealth of options for student learning and for new and experienced instructors alike.

Ebooks. Included with all new Sixth Edition copies of the print book, *Norton Field Guide* ebook access can also be purchased standalone from the Norton website and can be offered through Inclusive Access programs on many campuses. The ebook's built-in highlighting and note-taking capabilities help students engage with and respond to what they read, and instructors can share notes, videos, or external links with students using the instructor annotation tool. Norton ebooks can be viewed on—and synced among—all computers and mobile devices, and they can be made available for offline reading.

Videos.   New to the Sixth Edition, short videos are now available in the ebook and for streaming online. Informed by feedback from hundreds of composition instructors, topics include the writing process, rhetorical situations, specific kinds of writing, critical reading strategies, and more. Videos are easily assigned from InQuizitive activities or short multiple-choice quizzes—available in Norton Testmaker and for import into an LMS.

InQuizitive for Writers.   Included with all new copies of the *Field Guide*, InQuizitive gives students practice with writing, editing, and research in a low-stakes, feedback-driven environment. The activities are adaptive: students receive additional practice and feedback in the areas where they need more help, with links to relevant pages in the *Little Seagull Handbook* ebook. After practicing with InQuizitive, students will be better prepared to start drafting, find and evaluate sources, and edit their own writing. InQuizitive for the Sixth Edition includes new activities on Rhetorical Situations, Elements of Argument, Thesis Statements, Paragraph Development, Critical Reading Strategies, and Fact-Checking Sources.

Norton Teaching Tools.   Available for the first time with the Sixth Edition, the Norton Teaching Tools site for *The Norton Field Guide* is your first stop when looking for creative, engaging resources to refresh your syllabus or design a new one. All of the revised contents from the popular *A Guide to Teaching with* The Norton Field Guides to Writing can now be found here, including the comprehensive guide to teaching first-year writing, advice for teaching every chapter and reading in the text, and new multimedia like videos and *PowerPoint* slides. Norton Teaching Tools are searchable and can be filtered by chapter or by resource type, making it easy to find exactly what you need, download and customize it, and import it into your LMS course. Contents for the Sixth Edition include:

- A comprehensive guide to teaching first-year writing, with chapters on developing a syllabus, responding to student writing, and more.

- Significantly revised chapters on teaching writing online, contract (labor-based) grading, and more.

- New chapters that address culturally responsive teaching, antiracist pedagogy, trauma-informed teaching, and more.

- An expanded collection of sample writing assignments with new peer-review templates, rubrics, and sample student essays.

- Thoroughly revised tips for teaching every chapter and reading in the book, with new suggestions for teaching dual-enrollment courses.

- *PowerPoint* slides for each chapter with classroom and online activities, as well as genre flowcharts and documentation maps from the book.

- A new collection of videos on topics including thesis statements, rhetorical situations, critical reading strategies, and commonly assigned genres.

- Expanded support for assigning InQuizitive for Writers.

- A collection of sample student essays organized by genre, rhetorical mode, and field, including some that are documented and annotated.

- Worksheets and templates in Word and PDF formats with guidance on editing paragraphs, responding to a draft, and more.

A Guide to Teaching with *The Norton Field Guides*.   Written by Richard Bullock, Deborah Bertsch, Maureen Daly Goggin, and numerous other teachers and scholars, this is a comprehensive guide to teaching first-year writing, from developing a syllabus to facilitating group work, teaching multimodal writing to assessing student writing. It also includes teaching tips for every chapter and reading in the book. The *Guide* is available in both print and PDF formats; the contents are also available in downloadable, editable form on the Norton Teaching Tools site.

Quizzes.   Over 150 quizzes on sentences, language, punctuation/mechanics, paragraph editing, plagiarism, and MLA and APA documentation, including several pre- and post-diagnostics, can be easily imported into your LMS. Additional quizzes for each chapter and reading in the *Field Guide* are also available.

Plagiarism Tutorial.   Now available for integration directly into most campus learning management systems, the Plagiarism Tutorial guides students through why plagiarism matters, what counts as plagiarism, and how to avoid plagiarizing, with activities and short quizzes to assess what they've learned. Students earn 90 percent for completing the tutorial and an additional 10 percent based on their quiz performance. You can see your students' scores, how much time they spent working, and when they finished in your Class Activity Report.

Resources for Your LMS.   Easily add high-quality Norton digital resources to your online, hybrid, or lecture courses. Get started building your course with our easy-to-use integrated resources. All activities can be accessed right within your existing learning management system. Graded activities can be configured to report to the LMS course gradebook. Integration links are available for the *Norton Field Guide* ebook, InQuizitive for Writers, the *Little Seagull Handbook* ebook, the Plagiarism Tutorial, and all videos. Instructors can also add customizable quizzes to their LMS or upload additional resources from the Norton Teaching Tools site, including *PowerPoint* slides, worksheets, sample writing assignments and rubrics, peer-review templates, and model student essays. All resources can be found at digital .wwnorton.com/fieldguide6.

ACKNOWLEDGMENTS

As we've traveled around the country and met many of the students, teachers, and WPAs who are using *The Norton Field Guide*, we've been gratified to hear that so many find it helpful, to the point that some students tell us that they aren't going to sell it back to the bookstore when the term ends—the highest form of praise. As much as we like the positive response, though, we are especially grateful when we receive suggestions for ways the book might be improved. In this sixth edition, as we did in the fifth edition, we have tried to respond to the many good suggestions we've gotten from students, colleagues, reviewers, and editors. Thank you all, both for your kind words and for your good suggestions.

Some people need to be singled out for thanks, especially Marilyn Moller, the guiding editorial spirit of the *Field Guide* through all six editions. When we presented Marilyn with the idea for this book, she encouraged us and helped us conceptualize it—and then taught us how to write a textbook. The quality of the *Field Guide* is due in large part to her knowledge of the field of composition, her formidable editing and writing skills, her sometimes uncanny ability to see the future of the teaching of writing—and her equally formidable, if not uncanny, stamina.

Editor Sarah Touborg guided us through this new edition with good humor and better advice. Just as developmental editor John Elliott did with the third and fourth editions, Sarah shepherded the fifth, and now sixth, editions through revisions and additions with a careful hand and a clear eye for appropriate content and language. Her painstaking editing shows throughout this book, and we're grateful for her ability to make us appear to be better writers than we are.

Many others have contributed, too. Thanks to project editor Christine D'Antonio for her energy, patience, and great skill in guiding the book from manuscript to final pages. Claire Wallace brought her astute eye and keen judgment to all of the readings, while Emma Peters managed the extensive reviewing process and took great care of the manuscript at every stage. *The Norton Field Guide* is more than just a print book, and we thank Joy Cranshaw, Katie Bolger, Maria Qureshi, Sophia Purut, Kim Yi, and Cooper Wilhelm for creating and producing the superb ebook and instructors' site. Anna Palchik designed the award-winning, user-friendly, and attractive interior, Pete Garceau created the beautiful new cover design, and Debra Morton Hoyt and Rubina Yeh further enhanced the design for accessibility in print and online. Liz Marotta and Stephen Sajdak transformed a scribbled-over manuscript into a finished product with extraordinary speed and precision, while Alice Vigliani copyedited. Megan Schindel and Josh Garvin cleared text permissions, coping efficiently with ongoing changes, and Ted Szczepanski cleared permission for the images found by Dena Digilio Betz. Michele Dobbins, Lib Triplett, Elizabeth Pieslor, Annie Stewart, Erin Brown, Kim Bowers, and Emily Rowin helped us all keep our eyes on the market. Thanks to all, and to Mike Wright, Ann Shin, and Julia Reidhead for supporting this proj-

ect in the first place. Last but not least, our profound gratitude goes to the tireless Norton travelers, led so energetically by Erik Fahlgren.

Rich has many, many people at Wright State University to thank for their support and assistance. Jane Blakelock taught Rich most of what he knows about electronic text and writing on and for the web and assembled an impressive list of useful links for the book's website. Adrienne Cassel (now at Sinclair Community College) and Catherine Crowley read and commented on many drafts. Peggy Lindsey shared her students' work and the idea of using charts to show how various genres might be organized. Brady Allen, Deborah Bertsch (now at Columbus State Community College), Vicki Burke, Melissa Carrion, Jimmy Chesire, Carol Cornett, Mary Doyle, Byron Crews, Deborah Crusan, Sally DeThomas, Stephanie Dickey, Scott Geisel, Karen Hayes, Chuck Holmes, Beth Klaisner (now at Colorado State University), Nancy Mack, Marty Maner, Cynthia Marshall, Sarah McGinley, Kristie McKiernan, Michelle Metzner, Kristie Rowe, Bobby Rubin, Cathy Sayer, David Seitz, Caroline Simmons, Tracy Smith, Rick Strader, Mary Van Loveren, and A. J. Williams responded to drafts, submitted good models of student writing, contributed to the instructor's manual, tested the *Field Guide* in their classes, provided support, and shared with Rich some of their best teaching ideas. Henry Limouze and then Carol Loranger, chairs of the English Department, gave him room to work on this project with patience and good humor. Sandy Trimboli, Becky Traxler, and Lynn Morgan, the secretaries to the writing programs, kept him anchored. And he thanks especially the more than 300 graduate teaching assistants and 10,000 first-year students who class-tested various editions of the *Field Guide* and whose experiences helped—and continue to help—to shape it.

Deb is grateful to Columbus State's dean of Arts and Sciences, Allysen Todd, and to current and former English Department chairs Amanda Gradisek and Robyn Lyons-Robinson, who encouraged and supported her work on the *Field Guide*; to her fellow faculty members in the English Department, who show her every day what it means to be a good teacher, writer, collaborator, and colleague; to her high school teaching partners, who help her navigate the world of dual credit, enriching her own teaching in the process; and to her students, whose wit, work ethic, and writing regularly dazzle her. Deb is also incredibly grateful to Rich

Bullock, Joy Cranshaw, Michele Dobbins, Maureen Daly Goggin, Emma Peters, Sarah Touborg, and Lib Triplett for welcoming her to the *Field Guide* team and for teaching her—with patience and kindness—how to do the work.

The Norton Field Guide has also benefited from the good advice and conversations we've had with writing teachers across the country, including (among many others) Anne Champion, Sarah Fish, and Diana Gingo at Collin College; Polina Chemishanova, Cyndi Miecznikowski, and Melissa Schaub at the University of North Carolina at Pembroke; Maureen Mathison, Susan Miller, Tom Huckin, Gae Lyn Henderson, and Sundy Watanabe at the University of Utah; Christa Albrecht-Crane, Doug Downs, and Brian Whaley at Utah Valley State College; Anne Dvorak and Anya Morrissey at Longview Community University; Jeff Andelora at Mesa Community College; Robin Calitri at Merced College; Lori Gallinger, Rose Hawkins, Jennifer Nelson, Georgia Standish, and John Ziebell at the Community College of Southern Nevada; Stuart Blythe at Indiana University–Purdue University Fort Wayne; Janice Kelly at Arizona State University; Jeanne McDonald at Waubonsee Community College; Web Newbold, Mary Clark-Upchurch, Megan Auffart, Matt Balk, Edward James Chambers, Sarah Chavez, Desiree Dighton, Ashley Ellison, Theresa Evans, Keith Heller, Ellie Isenhart, Angela Jackson-Brown, Naoko Kato, Yuanyuan Liao, Claire Lutkewitte, Yeno Matuki, Casey McArdle, Tibor Munkacsi, Dani Nier-Weber, Karen Neubauer, Craig O'Hara, Martha Payne, Sarah Sandman, and Kellie Weiss at Ball State University; Patrick Tompkins at Tyler Community College; George Kanieski and Pamela Hardman at Cuyahoga Community College; Daniela Regusa, Jeff Partridge, and Lydia Vine at Capital Community College; Elizabeth Woodworth, Auburn University–Montgomery; Stephanie Eason at Enterprise Community College; Kate Geiselman at Sinclair Community College; Ronda Leathers Dively at Southern Illinois University; Debra Knutson at Shawnee State University; Guy Shebat and Amy Flick at Youngstown State University; Martha Tolleson, Toni McMillen, and Patricia Gerecci at Collin College; Sylva Miller at Pikes Peak Community College; Dharma Hernandez at Los Angeles Unified School District; Ann Spurlock at Mississippi State University; Luke Niiler at the University of Alabama; and Jeff Tix at Wharton County Junior College.

REVIEWERS

The authors and editors at W. W. Norton are grateful to the many instructors across the country whose astute feedback helped shape this new edition of the *Field Guide* and its digital resources: Carmela Abbruzzese, Regis College; Susan Achziger, Community College of Aurora; Liz Aguilar, Alamo Community College; Shazia Ali, Eastfield College; Kelly Allen, Northampton Community College; Emily Anderson, Columbus State Community College; Susan Austin, Johnston Community College; Stuart Barbier, Delta College; Elizabeth Barnes, Johnston Community College; William Beckham, Horry-Georgetown Technical College; Alyse Bensel, Brevard College; Betty Bettacchi, Collin College; Deb Bloom, Kirkwood Community College; Dean Blumberg, Horry-Georgetown Technical College; Brett Bodily, Dallas College North Lake Campus; Ethel Bowden, Central Maine Community College; Rachel Bragg, West Virginia University Institute of Technology; Laura Brandenburg, Wayland Baptist University; April Brannon, California State University, Fullerton; Crystal Brantley, Louisburg College; Jane Brockman, Rogue Community College; Mark Brumley, Randolph Community College; Keely Byars-Nichols, University of Mount Olive; Robert Canipe, Catawba Valley Community College; Anna Matsen Cantrell, Pellissippi State Community College; Neeta Chandra, Cuyahoga Community College; Catherine Childress, Lees-McRae College; Minda Chuska, Horry-Georgetown Technical College; Heather Clark, Ivy Tech Community College; Billy Clem, Waubonsee Community College; Alison Cope, Lone Star College; Virginia Crisco, California State University, Fresno; Amy Cruickshank, Cuyahoga Community College; Sarah Crump, Southern State Community College; John Daniel, Henry Ford College; Shemika Davis, Johnston Community College; LeeAnn Derdeyn, Dallas College Cedar Valley Campus; Heather Detzner, Spartanburg Community College; Amber Dinquel, Rappahannock Community College; Ivan Dole, Dallas College North Lake Campus; Katie Doughty, Mississippi State University; Holly Dykstra, Laredo College; Sara Edgell, Tarrant County College–Southeast Campus; Ashley Edlin, Wayland Baptist University; Kayleigh Few, Mississippi State University; Maureen Fitzpatrick, Johnson County Community College; Sue Fox, Central New Mexico Community College; Katie Franklin, University of New Orleans; Janet

Frye-Burleigh, San Jacinto College–Generation Park; Jo Gibson, Cuyahoga Community College–Western Campus; Brian Gogan, Western Michigan University; Anival Gonzalez, Northwest Vista College; Marti Grahl, Hagerstown Community College; Barbara Griest-Devora, Northwest Vista College; Kaleena Gross, Columbus State Community College; John Guthrie, Southeastern Community College; Bernabe Gutierrez, Laredo College; Stephanie Hamilton, University of Idaho; Amara Hand, Norfolk State University; Michael Hedges, Horry-Georgetown Technical College; Chelsea Hicken, Dixie State University; Ean High, University of Tennessee at Martin; Lee Ann Hodges, Tri-County Community College; Elizabeth Hogan, University of New Orleans; Monica Hogan, Johnson County Community College; Rachael Holloway, Johnston Community College; Michael Horton, Marshall High School; Deana Hueners, Dakota State University; Kenneth Huffer, Tarrant County College District; Paige Huskey, Clark State College; Patricia Ireland, Pellissippi State Community College; Neshon Jackson, Houston Community College; Kim Jacobs-Beck, University of Cincinnati Clermont; Jo Johnson, Ivy Tech Community College; Lori Johnson, Rappahannock Community College; Margaret Johnson, Langston University; Christine Junker, Wright State University–Lake Campus; Katie Kalisz, Grand Rapids Community College; William Keeth, Mansfield University of Pennsylvania; Jessica Kidd, University of Alabama; Danielle Klafter, Southeast Community College; Stacey Koller, Dixie State University; Christine Kozikowski, Central New Mexico Community College; Jennifer Kraemer, Collin College–Frisco Preston Ridge Campus; Julie Kratt, Cowley College; Alexander Kurian, Dallas College North Lake Campus; Katherine Kysar, University of Alaska Fairbanks; Nina Lambert, Dallas College Eastfield Campus; Kayla Landers, Lehigh University; Linda Lawson, Concord High School; Lindsey Light, University of Dayton; Bronwen Llewellyn, Daytona State College; Kelly Lovelace, Isothermal Community College; Juliette Ludeker, Howard Community College; Crystal Manboard, Northwest Vista College; Terri Mann, El Paso Community College; Diane Marcincin, Northampton Community College; Kelly Ann Martin, Collin College–Plano Campus; Adrienne McClain, Dallas College; Laura McClister, Volunteer State Community College; Jeanne McDonald, Waubonsee Community College; Kerry McGinnis, Columbus State Community College; Megan McIntyre, Sonoma State University; Gena McKinley, Rappahannock

Community College; James McWard, Johnson County Community College; Eileen Medeiros, Johnson & Wales University; Lora Meredith, Colorado Mountain College; Ashley Merrill, Wayne Community College; Dalia El Metwally, Wharton County Junior College; Harriet Millan, Drexel University; Kellie Miller, Santa Fe College; Rosemary Mink, Mohawk Valley Community College; Millie Moncada, Los Angeles Valley College; Michael Morris, Dallas College Eastfield Campus; Rhonda Morris, Santa Fe College; Damon Murrah, Pellissippi State Community College; Mary Newbegin, Lehigh University; Luke Niiler, University of Alabama; Annette Nusbaum, Aiken Technical College; Fabiana Oliveira, Normandale Community College; Tamra Painter, Kanab High School / CE Dixie State University; Diana Paquette, Waukesha County Technical College; Catherine Parisian, University of North Carolina at Pembroke; Teresa Paul, North Arkansas College; Andrea Peck, Cuyahoga Community College; Laurie Phillips, Dallas College El Centro Campus; Dan Portincaso, Waubonsee Community College; Britt Posey, Northwest Vista College; Chris Prentice, Central New Mexico Community College; Sharon Prince, Wharton County Junior College; Jessica Rabin, Anne Arundel Community College; Erin Radcliffe, Central New Mexico Community College; Yasmin Ramirez, El Paso Community College; Kassandra Ramirez-Buck, Eastfield College; Brian Reeves, Lone Star College–University Park; Eric Reyes, Wharton County Junior College; Cynthia Cox Richardson, Central State University; Jim Richey, Tyler Junior College; Jared Riddle, Ivy Tech Community College; Jennifer Riske, Northeast Lakeview College; Kelly Rivers, Pellissippi State Community College; Danielle Roach, Clark State College; Andrew Roback, Illinois Institute of Technology; Michelle Roberti, Rappahannock Community College; Natasha Robinson, Collin College; Brian Roffino, Eastfield College; Andrea Rogers, University of Arkansas; Mark Rooze, Florence-Darlington Technical College; Kathie Russell, Santa Fe College; Mary Rutledge-Davis, Dallas College North Lake Campus; Kelly Savage, Dallas College Eastfield Campus; Molly Scanlon, Nova Southeastern University; Matthew Schmeer, Johnson County Community College; Jessica Schreyer, University of Dubuque; Marcea Seible, Hawkeye Community College; Marilyn Senter, Johnson County Community College; Margaret Shaw, Johnson & Wales University; Chase Sisk, Pellissippi State Community College; Zahir Small, Santa Fe College; Tiffany Smith, Arkansas State University Beebe;

Jason Snart, College of DuPage; Jessi Snider, Wharton County Junior College; Cynthia Soueidan, Gloucester High School / Rappahannock Community College; Pam Stone, North Arkansas College; James (Ty) Stumpf, Central Carolina Community College; Allison Suber, Spartanburg Community College; Eric Sullivan, Metropolitan Community College; Karen Summey, Catawba Valley Community College; Ashley Supinski, Northampton Community College; Gerard Teichman, Bunker Hill Community College; Harun Karim Thomas, Daytona State College; Jessica Thompson, Mississippi State University; Tara Thompson, Johnston Community College; Jeff Tix, Wharton County Junior College; Susan Todd, Jefferson College; Andrew Tolle, Dallas College Eastfield Campus; Tyler Trimm, Mississippi State University; Jennifer Turner, Patrick County High School; Michael Walker, Santa Fe College; Mary Wallen, Big Sandy Community & Technical College; Maureen Walters, Vance-Granville Community College; Angie Weaver, South Beloit High School; Sharon West, Spartanburg Community College; Stephen Whitley, Collin College; Sabine Winter, Dallas College Eastfield Campus; Paul Wise, University of Toledo; Karen Wright, Pellissippi State Community College; Sue Yamin, Pellissippi State Community College; Jason Ziebart, Central Carolina Community College.

We owe a special thanks to the instructors and their students who reviewed this edition of the *Field Guide*, some of whom class-tested it: Jenna Bazzell, Monroe County Community College; Mary Anne Bernal, San Antonio College; Carrie Buttler, Del Mar College; Susan Muaddi Darraj, Harford Community College; Janice Filer, Shelton State Community College; Andrea Green, Motlow State Community College; Yolanda Gonzalez, McLennan Community College; Peaches Henry, McLennan Community College; Lennie Irvin, San Antonio College; Kent Lenz, Del Mar College; Carrie Nartker-Kantz, Monroe Community College; Jim Richey, Tyler Junior College; Ryan Shepherd, Ohio University; Wes Spratlin, Motlow State Community College.

It's customary to conclude by expressing gratitude to one's spouse and family, and for good reason. Writing and revising *The Norton Field Guide* over the past several years, we have enjoyed the loving and unconditional support of our spouses, Barb and Todd, who provide the foundation for all we do. Thank you. We couldn't have done it without you.

How to Use This Book

There's no one way to do anything, and writing is no exception. Some people need to do a lot of planning on paper; others write entire drafts in their heads. Some writers compose quickly and loosely, going back later to revise; others work on one sentence until they're satisfied with it, then move on to the next. And writers' needs vary from task to task, too: sometimes you know what you're going to write about and why, but need to figure out how to do it; other times your first job is to come up with a topic. *The Norton Field Guide* is designed to allow you to chart your own course as a writer, offering guidelines that suit your writing needs. It is organized in eight parts:

1. **ACADEMIC LITERACIES**: The chapters in this part will help you know what's expected in the reading and writing you do for academic purposes, and in summarizing and responding to what you read. One chapter even provides tips for developing habits of mind that will help you succeed in college, whatever your goals.

2. **RHETORICAL SITUATIONS**: No matter what you're writing, it will always have some purpose, audience, genre, stance, and medium and design. This part will help you consider each of these elements, as well as the particular kinds of rhetorical situations created by academic assignments.

3. **GENRES**: Use these chapters for help with specific kinds of writing, from abstracts to explorations to evaluations and more. You'll find more detailed guidance for four especially common assignments: literacy narratives, textual analyses, reports, and arguments.

4. **FIELDS**: The chapters in this part will help you apply what you're learning in this book to your other general education courses or courses in your major.

5. **PROCESSES**: These chapters offer general advice for all writing situations—from generating ideas and text to drafting, organizing, revising and rewriting, compiling a portfolio—and more.

6. **STRATEGIES**: Use the advice in this part to develop and organize your writing—to use comparison, description, dialogue, and other strategies as appropriate.

7. **RESEARCH / DOCUMENTATION**: Use this section for advice on how to do research, work with sources, and compose and document research-based texts using MLA and APA styles.

8. **MEDIA / DESIGN**: This section offers guidance in designing your work and using visuals and sound, and in deciding whether and how to deliver what you write on paper, on screen, or in person.

WAYS INTO THE BOOK

The Norton Field Guide gives you the writing advice you need, along with the flexibility to write in the way that works best for you. Here are some of the ways you can find what you need in the book.

Brief menus.   Inside the front cover you'll find a list of all the chapters; start here if you are looking for a chapter on a certain kind of writing or a general writing issue.

Complete contents.   Pages xxvii–xliv contain a detailed table of contents. Look here if you need to find a reading or a specific section in a chapter.

Guides to writing.   If you know the kind of writing you need to do, you'll find guides to writing twelve common genres in Part 3. These guides are designed to help you through all the decisions you have to make—from coming up with a topic to editing and proofreading your final draft.

Color-coding.   The parts of this book are color-coded for easy reference: light blue for **ACADEMIC LITERACIES**, orange for **RHETORICAL SITUATIONS**, green for **GENRES**, gray for **FIELDS**, lavender for **PROCESSES**, red

for **STRATEGIES**, blue for **RESEARCH/DOCUMENTATION**, and pink for **MEDIA/DESIGN**. You'll find a key to the colors on the front cover flap and also on each left-hand page. When you see a word highlighted in a color, that tells you where you can find additional detail on the topic.

Glossary/index. At the back of the book is a combined glossary and index, where you'll find full definitions of key terms and topics, along with a list of the pages where everything is covered in detail.

Directories to MLA and APA documentation. A brief directory inside the back cover will lead you to guidelines on citing sources and composing a list of references or works cited. The documentation models are color-coded so you can easily see the key details.

WAYS OF GETTING STARTED

If you know your genre, simply turn to the appropriate genre chapter. There you'll find model readings, a description of the genre's Key Features, and a Guide to Writing that will help you come up with a topic, generate text, organize and write a draft, get response, revise, edit, and proofread. The genre chapters also point out places where you might need to do research, use certain writing strategies, design your text a certain way—and direct you to the exact pages in the book where you can find help doing so.

If you know your topic, you might start with some of the activities in Chapter 28, Generating Ideas and Text. From there, you might turn to Chapter 47, for help finding sources on the topic. When it comes time to narrow your topic and come up with a thesis statement, Chapter 29 can help. If you get stuck at any point, you might turn to Chapter 27, Writing as Inquiry; it provides tips that can get you beyond what you already know about your topic. If your assignment or your thesis defines your genre, turn to that chapter; if not, consult Chapter 27 for help determining the appropriate genre, and then turn to that genre chapter.

Contents

New to this edition

▲ **Part 3   Genres** 79

* *New to this edition*

* New to this edition

* New to this edition

New to this edition

*New to this edition

* *New to this edition*

New to this edition

New to this edition

New to this edition

* New to this edition

* New to this edition

SIXTH EDITION

The Norton

Field Guide
to Writing

Part 1
Academic Literacies

Whenever we enter a new community—start a new job, move to a new town, join a new club—there are certain things we need to learn. The same is true upon entering the academic world. We need to be able to **READ** and **WRITE** in certain ways. We're routinely called on to **SUMMARIZE** something we've heard or read and to **RESPOND** in some way. And to succeed, we need to develop certain **HABITS OF MIND**—everyday things such as asking questions and being persistent. The following chapters provide guidelines to help you develop these fundamental academic literacies—and know what's expected of you in academic communities.

Academic Literacies

1 Writing in Academic Contexts

Write an essay arguing whether genes or environment do more to determine people's intelligence. Research and write a report on the environmental effects of electricity-generating windmills. Work with a team to write a proposal and create a multimedia presentation for a sales campaign. Whatever you're studying, you're surely going to be doing a lot of writing, in classes from various disciplines—the above assignments, for example, are from psychology, environmental science, and marketing. Academic writing can serve a number of different purposes—to **ARGUE** for what you think about a topic and why, to **REPORT** on what's known about an issue, to **PROPOSE A SOLUTION** for some problem, and so on. Whatever your topics or purposes, all academic writing follows certain conventions, and learning about them will help you join the conversations going on across campus. This chapter describes what's typically expected of academic writing—and of academic writers.

164–95
140–63
258–68

What's Expected of Academic Writing

Evidence that you've considered the subject thoughtfully. Whether you're composing a report, an argument, or some other kind of writing, you need to demonstrate that you've thought seriously about the topic and done any necessary research. You can use various ways to show that you've considered the subject carefully, from citing authoritative sources to incorporating information you learned in class to pointing out connections among ideas.

An indication of why your topic matters. You need to help your readers understand why your topic is worth exploring and why your writing is worth reading. Even if you are writing in response to an assigned topic, you can better make your point and achieve your purpose by showing your readers why your topic is important and why they should care about it. For example, in the prologue to *Our Declaration*, political philosopher Danielle Allen explains why her topic, the Declaration of Independence, is worth writing about:

> The Declaration of Independence matters because it helps us see that we cannot have freedom *without* equality. It is out of an egalitarian commitment that a people grows—a people that is capable of protecting us all collectively, and each of us individually, from domination. If the Declaration can stake a claim to freedom, it is only because it is so clear-eyed about the fact that the people's strength resides in its equality.
>
> The Declaration also conveys another lesson of paramount importance. It is this: language is one of the most potent resources each of us has for achieving our own political empowerment. The men who wrote the Declaration of Independence grasped the power of words. This reveals itself in the laborious processes by which they brought the Declaration, and their revolution, into being. It shows itself forcefully, of course, in the text's own eloquence.

By explaining that the topic matters because freedom and equality matter—and language gives us the means for empowering ourselves—Allen gives readers reason to read her careful analysis.

A response to what others have said. Whatever your topic, it's unlikely that you'll be the first one to write about it. And if, as this chapter assumes, all academic writing is part of a larger conversation, you are in a way adding your own voice to that conversation. One good way of doing that is to present your ideas as a response to what others have said about your topic—to begin by quoting, paraphrasing, or summarizing what others have said and then to agree, disagree, or both.

❋ academic literacies	● fields	● research
■ rhetorical situations	⁚• processes	● media/design
▲ genres	◆ strategies	

For example, in an essay arguing that the phrase "all lives matter" is manipulative, Diablo Valley College student Kelly Coryell says, "It should be acknowledged that some of those who used the phrase did so naively attempting to unify people in a time of division." But then she responds—and disagrees, arguing that "The phrase 'all lives matter' belies the current racial inequality in America by implying that all lives are at equal risk."

A clear, appropriately qualified thesis. When you write in an academic context, you're usually expected to state your main point explicitly, often in a **THESIS STATEMENT**. Kelly Coryell states her thesis clearly in her essay "All Words Matter: The Manipulation behind 'All Lives Matter'": "Saying 'all lives matter' as a response to "Black lives matter" is, in reality, sending a dangerous message: it steals attention from the systematic oppression of Black Americans and actively distorts the message behind the BLM movement, manipulating the American people into maintaining the oppressive status quo." Often you'll need to **QUALIFY** your thesis statement to acknowledge that the subject is complicated and there may be more than one way of seeing it or exceptions to the generalization you're making about it. Here, for example, is a qualified thesis, from an essay evaluating the movie *The Lovebirds* by Olivia Mazzucato, a student at UCLA: "Though the film's story gets convoluted, Rae and Nanjiani are able to salvage the muddy plot with their razor-sharp comedic wit and hilarious chemistry." Mazzucato makes a claim that *The Lovebirds* achieves success, while acknowledging at the beginning of the sentence that the film's plot is flawed.

347–49

348–49

Good reasons supported by evidence. You need to provide good reasons for your thesis and evidence to support those reasons. For example, Kelly Coryell offers several reasons why saying "all lives matter" as a response to "Black lives matter" sends a dangerous message: systemic racism disproportionately affects the Black community and "Black lives matter" does not mean "only" Black lives matter, but that Black lives need to matter as much as White lives. Evidence to support your reasons sometimes comes

from your own experience but more often from published research and scholarship, research you do yourself, or firsthand accounts by others.

Compared with other kinds of writing, academic writing is generally expected to be more objective and less emotional. You may find *Romeo and Juliet* deeply moving or cry when you watch *A Dog's Journey*—but when you write about the play or the film for a class, you must do so using evidence from the text to support your thesis. You may find someone's ideas deeply offensive, but you should respond to them with reason rather than with emotional appeals or personal attacks.

Acknowledgment of multiple perspectives. Debates and arguments in popular media are often framed in "pro/con" terms, as if there were only two sides to any given issue. Once you begin seriously studying a topic, though, you're likely to find that there are several sides and that each of them deserves serious consideration. In your academic writing, you need to represent fairly the range of perspectives on your topic—to explore three, four, or more positions on it as you research and write. For example, in her report, "Sleepless Nights of a University Student," Drexel University student Renae Tingling examines sleep deprivation in several ways: through academic research, a survey of students, and a personal sleep-deprivation experiment.

A confident, authoritative stance. If one goal of academic writing is to contribute to a larger conversation, your tone should convey confidence and establish your authority to write about your subject. Ways to achieve such a tone include using active verbs ("X claims" rather than "it seems"), avoiding such phrases as "in my opinion" and "I think," and writing in a straightforward, direct style. Your writing should send the message that you've done the research, analysis, and thinking and know what you're talking about. For example, here is the first paragraph of Wilkes University student Brianna Schunk's essay advocating for the casting of actors with disabilities in the roles of characters with disabilities:

* academic literacies
* fields
* research
* rhetorical situations
* processes
* media/design
* genres
* strategies

> When I first saw Gaten Matarazzo in the TV show *Stranger Things* and some of the first words out of his toothless little mouth were, "I have cleidocranial dysplasia," I actually cried with joy. Until then, I had never seen another human not related to my family who had cleidocranial dysplasia on television in a popular TV show. I was so shocked, I made my friend rewind the episode so I could hear him say it again. Having cleidocranial dysplasia myself, it is so much easier to be able to compare myself to Matarazzo when explaining my condition to other people, and I only wish he could have come into my life earlier. Representation—how we portray ourselves and are portrayed by others—is incredibly important to all groups of people, but especially to people with disabilities.

Schunk's use of vivid language ("toothless little mouth," "I was so shocked") and her strong final sentence ("Representation . . . is incredibly important to all groups of people, but especially to people with disabilities") lend her writing a confident tone. Her stance sends the message that she knows what she's talking about.

Carefully documented sources. Clearly acknowledging sources and documenting them carefully and correctly is a basic requirement of academic writing. When you use the words or ideas of others—including visuals, video, or audio—those sources must be documented in the text and in a works-cited or references list at the end. (If you're writing something that will appear online, you may also refer readers to your sources by using hyperlinks in the text; ask your instructor if you need to include a list of references or works cited as well.)

Careful attention to style expectations. In academic contexts, you should almost always write in complete sentences, use capitalization and punctuation as recommended in a handbook or other guide, check your spelling by consulting a dictionary, proofread carefully, and avoid any abbreviations used in texting or other informal writing. Grammar conventions

are important, and it's a good idea to follow them in most kinds of academic writing. If you choose not to, however, you should make clear that you are doing so intentionally, for a particular purpose.

What's Expected of College Writers: The WPA Outcomes

Writing is not a multiple-choice test; it doesn't have right and wrong answers that are easily graded. Instead, your readers, whether they're teachers or anyone else, are likely to read your writing with various questions in mind: does it make sense, does it meet the demands of the assignment, are sources documented appropriately, to name just a few of the things readers may look for. Different readers may notice different things, so sometimes it may seem to you that their response—and your grade—is unpredictable. It should be good to know, then, that writing teachers across the nation have come to some agreement on certain "outcomes," what college students should know and be able to do by the time they finish a first-year writing course. These outcomes have been defined by the National Council of Writing Program Administrators (WPA). Here's a brief summary of these outcomes and how *The Norton Field Guide* can help you meet them:

Knowledge of Rhetoric

- *Understand the rhetorical situation of texts that you read and write.* See Chapters 5–9 and the many prompts for Considering the Rhetorical Situation throughout the book.

- *Read and write texts in a number of different genres, and understand how your purpose may influence your writing.* See Chapters 10–22 for guidelines on writing in twelve genres and remixing what you write in one genre to another.

- *Adjust your voice, tone, level of formality, design, and medium as is necessary and appropriate.* See Chapter 8 on stance and tone and Chapter 9 for help thinking about medium and design.

- *Choose the media that will best suit your audience, purpose, and the rest of your rhetorical situation.* See Chapters 9 and 55.

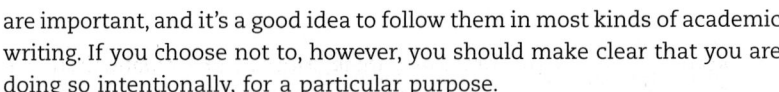

✳ academic literacies   ● fields   ● research

■ rhetorical situations   ⠶ processes   ● media/design

▲ genres   ◆ strategies

Critical Thinking, Reading, and Composing

- *Read and write to inquire, learn, think critically, and communicate.* See Chapters 1 and 2 on academic writing and reading, and Chapter 27 on writing as inquiry. Chapters 10–22 provide genre-specific prompts to help you think critically about a draft.

- *Read for content, argumentative strategies, and rhetorical effectiveness.* Chapter 2 provides guidance on reading texts with a critical eye, Chapter 11 teaches how to analyze a text, and Chapter 48 shows how to evaluate sources.

- *Find and evaluate popular and scholarly sources.* Chapter 47 teaches how to use databases and other methods to find sources, and Chapter 48 shows how to evaluate the sources you find.

- *Use sources in various ways to support your ideas.* Chapter 37 suggests strategies for supporting your ideas, and Chapter 50 shows how to incorporate ideas from sources into your writing to support your ideas.

Processes

- *Use writing processes to compose texts and explore ideas in various media.* Part 5 covers all stages of the processes writers use, from generating ideas and text to drafting, getting response and revising, and editing and proofreading. Each of the thirteen genre chapters (10–22) in Part 3 includes a guide that leads you through the process of writing in that genre or remixing your writing into another genre.

- *Collaborate with others on your own writing and on group tasks.* Chapter 4 offers guidelines for working with others, Chapter 32 provides general prompts for getting and giving response, and Chapters 10–22 provide genre-specific prompts for reading a draft with a critical eye.

- *Reflect on your own writing processes.* Chapters 10–22 provide genre-specific questions to help you take stock of your work, and Chapter 35 offers guidance in reflecting on your writing processes and products.

Knowledge of Conventions

* *Use appropriate grammar, punctuation, and spelling.* Chapter 33 provides tips to help you edit and proofread your writing. Chapters 10–22 offer genre-specific advice for editing and proofreading.

* *Understand and use genre conventions and formats in your writing.* Chapter 7 provides an overview of genres and how to think about them. Part 3 covers twelve genres, describing the key features and conventions of each one, and includes one chapter on remixes.

* *Understand intellectual property and document sources appropriately.* Chapter 51 offers guidance on the ethical use of sources, Chapter 52 provides an overview of documentation styles, and Chapters 53 and 54 provide templates for documenting in MLA and APA styles.

2 Reading in Academic Contexts

We read news sites to know about the events of the day. We read textbooks to learn about history, chemistry, and other academic topics—and other academic sources to do research and develop arguments. We read tweets and blogs to follow (and participate in) conversations about issues that interest us. And as writers, we read our own writing to make sure it says what we mean it to say and proofread our final drafts to make sure they say it the way we intended. In other words, we read many kinds of texts for many different purposes. This chapter offers a number of strategies for various kinds of reading you do in academic contexts.

TAKING STOCK OF YOUR READING

One way to become a better reader is to understand your reading process; if you know what you do when you read, you're in a position to decide what you need to change or improve. Consider the answers to the following questions:

- What do you read for pleasure? for work? for school? Consider all the sorts of reading you do: books, magazines, and newspapers, websites, *Facebook*, texts, blogs, product instructions.

- When you're facing a reading assignment, what do you do? Do you do certain things to get comfortable? Do you play music or seek quiet? Do you plan your reading time or set reading goals for yourself? Do you flip through or skim the text before settling down to read it, or do you start at the beginning and work through it?

- When you begin to read something for an assignment, do you make sure you understand the purpose of the assignment—why you must

read this text? Do you ever ask your instructor (or whoever else assigned the reading) what its purpose is?

- How do you motivate yourself to read material you don't have any interest in? How do you deal with boredom while reading?

- Does your mind wander? If you realize that you haven't been paying attention and don't know what you just read, what do you do?

- Do you ever highlight, underline, or annotate text as you read? Do you take notes? If so, what do you mark or write down? Why?

- When you read text you don't understand, what do you do?

- As you anticipate and read an assigned text, what attitudes or feelings do you typically have? If they differ from reading to reading, why do they?

- What do you do when you've finished reading an assigned text? Write out notes? Think about what you've just read? Move on to the next task? Something else?

- How well do your reading processes work for you, both in school and otherwise? What would you like to change? What can you do to change?

The rest of this chapter offers advice and strategies that you may find helpful as you work to improve your reading skills.

READING STRATEGICALLY

Academic reading is challenging because it makes several demands on you at once. Textbooks present new vocabulary and new concepts, and picking out the main ideas can be difficult. Scholarly articles present content and arguments you need to understand, but they often assume that readers already know key concepts and vocabulary and so don't generally provide background information. As you read more texts in an academic field and begin to participate in its conversations, the reading will become easier, but in the meantime you can develop strategies that will help you read effectively.

Thinking about What You Want to Learn

To learn anything, we need to place new information into the context of what we already know. For example, to understand photosynthesis, we

academic literacies   ● fields   ● research
■ rhetorical situations   ⦂ processes   ● media/design
▲ genres   ◆ strategies

need to already know something about plants, energy, and oxygen, among other things. To learn a new language, we draw on similarities and differences between it and any other languages we know. A method of bringing to conscious attention our current knowledge on a topic and of helping us articulate our purposes for reading is a list-making process called KWL+. To use it, create a table with three columns:

K: What I Know	W: What I Want to Know	L: What I Learned

Before you begin reading a text, list in the "K" column what you already know about the topic. Brainstorm ideas, and list terms or phrases that come to mind. Then group them into categories. Also before reading, or after reading the first few paragraphs, list in the "W" column questions you have that you expect, want, or hope to be answered as you read. Number or reorder the questions by their importance to you.

Then, as you read the text or afterward, list in the "L" column what you learned from the text. Compare your "L" list with your "W" list to see what you still want or need to know (the "+")—and what you learned that you didn't expect.

Previewing the Text

It's usually a good idea to start by skimming a text—read the title and subtitle, any headings, the first and last paragraphs, the first sentences of all the other paragraphs. Study any illustrations and other visuals. Your goal is to get a sense of where the text is heading. At this point, don't stop to look up unfamiliar words; just mark them in some way to look up later.

Adjusting Your Reading Speed to Different Texts

Different texts require different kinds of effort. Some that are simple and straightforward can be skimmed fairly quickly. With academic texts, though, you usually need to read more slowly and carefully, matching the pace of your reading to the difficulty of the text. You'll likely need to skim the text for an overview of the basic ideas and then go back to read

it closely. And then you may need to read it yet again. (But do try always to read quickly enough to focus on the meanings of sentences and paragraphs, not just individual words.) With visual texts, too, you'll often need to look at them several times, moving from gaining an overall impression to closely examining the structure, layout, and other visual features—and exploring how those features relate to any accompanying verbal text.

Looking for Organizational Cues

As you read, look for cues that signal the way the text's ideas are organized and how each part relates to the ones around it.

The introductory paragraph and thesis often offer a preview of the topics to be discussed and the order in which they will be addressed. Here, for example, is a typical thesis statement for a report: *Types of prisons in the United States include minimum and medium security, close security, maximum security, and supermax.* The report that follows should explain each type of prison in the order stated in the thesis.

361-62 ⁂ **Transitions** help GUIDE READERS in following the direction of the writer's thinking from idea to idea. For example, "however" indicates an idea that contradicts or limits what has just been said, while "furthermore" indicates one that adds to or supports it.

Headings identify a text's major and minor sections, by means of both the headings' content and their design.

Thinking about Your Initial Response

Some readers find it helps to make brief notes about their first response to a text, noting their reaction and thinking a little about why they reacted as they did.

What are your initial reactions? Describe both your intellectual reaction and any emotional reaction, and identify places in the text that caused you to react as you did. An intellectual reaction might consist of an evaluation ("I disagree with this position because . . ."), a connection ("This idea

---

❋ academic literacies   ● fields   ● research
■ rhetorical situations   ⁂ processes   ● media/design
▲ genres   ◆ strategies

reminds me of . . ."), or an elaboration ("Another example of this point is . . ."). An emotional reaction could include approval or disapproval ("YES! This is exactly right!" "NO! This is so wrong!"), an expression of feeling ("This passage makes me so sad"), or one of appreciation ("This is said so beautifully"). If you had no particular reaction, note that, too.

What accounts for your reactions? Are they rooted in personal experiences? aspects of your personality? positions you hold on an issue? As much as possible, you want to keep your opinions from interfering with your understanding of what you're reading, so it's important to try to identify those opinions up front.

Dealing with Difficult Texts

Let's face it: some texts are difficult. You may have no interest in the subject matter, or lack background knowledge or vocabulary necessary for understanding the text, or simply not have a clear sense of why you have to read the text at all. Whatever the reason, reading such texts can be a challenge. Here are some tips for dealing with them:

Look for something familiar. Texts often seem difficult or boring because we don't know enough about the topic or about the larger conversation surrounding it to read them effectively. By skimming the headings, the abstract or introduction, and the conclusion, you may find something that relates to something you already know or are at least interested in — and being aware of that prior knowledge can help you see how this new material relates to it.

Look for "landmarks." Reading a challenging academic text the first time through can be like driving to an unfamiliar destination on roads you've never traveled: you don't know where you're headed, you don't recognize anything along the way, and you're not sure how long getting there will take. As you drive the route again, though, you see landmarks along the way that help you know where you're going. The same goes for reading a difficult text: sometimes you need to get through it once just to get some idea of what it's about. On the second reading, now that you have "driven

the route," look for the ways that the parts of the text relate to one another, to other texts or course information, or to other knowledge you have.

Monitor your understanding. You may have had the experience of reading a text and suddenly realizing that you have no idea what you just read. Being able to monitor your reading—to sense when you aren't understanding the text and need to reread, focus your attention, look up unfamiliar terms, take some notes, or take a break—can make you a more efficient and better reader. Keep these questions in mind as you read: What is my purpose for reading this text? Am I understanding it? Does it make sense? Should I slow down, reread, annotate? skim ahead and then come back? pause to reflect?

Be persistent. Research shows that many students respond to difficult texts by assuming they're "too dumb to get it"—and quitting reading. Successful students, on the other hand, report that if they keep at a text, they will come to understand it. Some of them even see difficult texts as challenges: "I'm going to keep working on this until I make sense of it." Remember that reading is an active process, and the more you work at it the more successful you will be.

Annotating

Many readers find it helps to annotate as they read: highlighting keywords, phrases, sentences; connecting ideas with lines or symbols; writing comments or questions in the margin or on sticky notes; circling new words so you can look up the definitions later; noting anything that seems noteworthy or questionable. Annotating forces you to read for more than just the surface meaning. Especially when you are going to be writing about or responding to a text, annotating creates a record of things you may want to refer to.

Annotate as if you're having a conversation with the author, someone you take seriously but whose words you do not accept without question. Put your part of the conversation in the margin, asking questions, talking back: "What's this mean?" "So what?" "Says who?" "Where's evidence?" "Yes!" "Whoa!" or even 😐 or 🙂 or texting shorthand like LOL or INTRSTN. If you're reading a text online, you can use a digital annotation

academic literacies ● fields ● research

■ rhetorical situations ⁙ processes ● media/design

▲ genres ◆ strategies

tool like *Hypothes.is* or *Diigo* to highlight portions of the text and make notes electronically—and even share your annotations with others.

What you annotate depends on your **PURPOSE**, or what you're most interested in. If you're analyzing a text that makes an explicit argument, you would probably underline the **THESIS STATEMENT** and then the **REASONS** and **EVIDENCE** that support that statement. It might help to restate those ideas in your own words in the margins—in order to understand them, you need to put them in your own words! If you're trying to **IDENTIFY PATTERNS**, you might highlight each pattern in a different color or mark it with a sticky note and write any questions or notes about it in that color. You might annotate a visual text by circling and identifying important parts of the image.

There are some texts that you cannot annotate, like library books. Then you will need to use sticky notes or make notes elsewhere, and you might find it useful to keep a reading log for this purpose.

■ 59–60

⁂ 347–49

◆ 413–14

414–22

✳ 23–25

Coding

You may also find it useful to record your thoughts as you read by using a coding system—for example, using "X" to indicate passages that contradict your assumptions, or "?" for ones that puzzle you. You can make up your own coding system, of course, but you could start with this one*:

✔   Confirms what you thought

X   Contradicts what you thought

?   Puzzles you

??   Confuses you

!   Surprises you

☆   Strikes you as important

➔   Is new or interesting to you

You might also circle new words that you'll want to look up later and highlight or underline key phrases.

---

*Adapted from Harvey Daniels and Steven Zemelman, *Subjects Matter: Every Teacher's Guide to Content-Area Reading* (Heinemann, 2004).

A Sample Annotated Text

Here is an excerpt from Justice: What's the Right Thing to Do?, *a book by Harvard professor Michael J. Sandel, annotated by a writer who was doing research for a report on the awarding of military medals:*

What Wounds Deserve the Purple Heart?

✓

On some issues, questions of virtue and honor are too obvious to deny. Consider the recent debate over who should qualify for the Purple Heart. Since 1932, the U.S. military has awarded the medal to soldiers wounded or killed in battle by enemy action. In addition to the honor, the medal entitles recipients to special privileges in veterans' hospitals.

PTSD increasingly common among veterans.

Since the beginning of the current wars in Iraq and Afghanistan, growing numbers of veterans have been diagnosed with post-traumatic stress disorder and treated for the condition. Symptoms include recurring nightmares, severe depression, and suicide. At least three hundred thousand veterans reportedly suffer from traumatic stress or major depression. Advocates for these veterans have proposed that they, too, should qualify for the Purple Heart. Since psychological injuries can be at least as debilitating as physical ones, they argue, soldiers who suffer these wounds should receive the medal.

Argument: Vets with PTSD should be eligible for PH because psych. injuries are as serious as physical.

No PH for PTSD vets? Seems unfair!

After a Pentagon advisory group studied the question, the Pentagon announced, in 2009, that the Purple Heart would be reserved for soldiers with physical injuries. Veterans suffering from mental disorders and psychological trauma would not be eligible, even though they qualify for government-supported medical treatment and disability payments. The Pentagon offered two reasons for its decision: traumatic stress disorders are not intentionally caused by enemy action, and they are difficult to diagnose objectively.

Argument: PTSD is like punctured eardrums, which do get the PH.

Did the Pentagon make the right decision? Taken by themselves, its reasons are unconvincing. In the Iraq War, one of the most common injuries recognized with the Purple Heart has been a punctured eardrum, caused by explosions at close range. But unlike bullets and bombs, such explosions are not a deliberate enemy tactic intended to injure or kill; they are (like traumatic stress) a damaging side effect of

---

✳ academic literacies   ● fields   ● research
■ rhetorical situations   ⋮ processes   ● media/design
▲ genres   ◆ strategies

battlefield action. And while traumatic disorders may be more difficult to diagnose than a broken limb, the injury they inflict can be more severe and long-lasting.

As the wider debate about the Purple Heart revealed, the real issue is about the meaning of the medal and the virtues it honors. What, then, are the relevant virtues? Unlike other military medals, the Purple Heart honors sacrifice, not bravery. It requires no heroic act, only an injury inflicted by the enemy. The question is what kind of injury should count.

PH "honors sacrifice, not bravery." Injury enough. So what kind of injury?

A veteran's group called the Military Order of the Purple Heart opposed awarding the medal for psychological injuries, claiming that doing so would "debase" the honor. A spokesman for the group stated that "shedding blood" should be an essential qualification. He didn't explain why bloodless injuries shouldn't count. But Tyler E. Boudreau, a former Marine captain who favors including psychological injuries, offers a compelling analysis of the dispute. He attributes the opposition to a deep-seated attitude in the military that views post-traumatic stress as a kind of weakness. "The same culture that demands tough-mindedness also encourages skepticism toward the suggestion that the violence of war can hurt the healthiest of minds. . . . Sadly, as long as our military culture bears at least a quiet contempt for the psychological wounds of war, it is unlikely those veterans will ever see a Purple Heart."

Wow: one vet's group insists that for PH, soldier must bleed!

☆

So the debate over the Purple Heart is more than a medical or clinical dispute about how to determine the veracity of injury. At the heart of the disagreement are rival conceptions of moral character and military valor. Those who insist that only bleeding wounds should count believe that post-traumatic stress reflects a weakness of character unworthy of honor. Those who believe that psychological wounds should qualify argue that veterans suffering long-term trauma and severe depression have sacrificed for their country as surely, and as honorably, as those who've lost a limb. The dispute over the Purple Heart illustrates the moral logic of Aristotle's theory of justice. We can't determine who deserves a military medal without asking what virtues the medal properly honors. And to answer that question, we have to assess competing conceptions of character and sacrifice.

Argument based on different ideas about what counts as a military virtue.

—Michael J. Sandel, *Justice: What's the Right Thing to Do?*

Summarizing

Writing a summary, boiling down a text to its main ideas, can help you understand it. To do so, you need to identify which ideas in the text are crucial to its meaning. Then you put those crucial ideas into your own words, creating a brief version that accurately sums up the text. Here, for example, is a summary of Sandel's analysis of the Purple Heart debate:

> In "What Wounds Deserve the Purple Heart?," Harvard professor Michael J. Sandel explores the debate over eligibility for the Purple Heart, the medal given to soldiers who die or are wounded in battle. Some argue that soldiers suffering from post-traumatic stress disorder should qualify for the medal because psychological injuries are as serious as physical ones. However, the military disagrees, since PTSD injuries are not "intentionally caused by enemy action" and are hard to diagnose. Sandel observes that the dispute centers on how "character" and "sacrifice" are defined. Those who insist that soldiers must have had physical wounds to be eligible for the Purple Heart see psychological wounds as reflecting "weakness of character," while others argue that veterans with PTSD and other psychological traumas have sacrificed honorably for their country.

READING CRITICALLY

When we read critically, we apply our analytical skills in order to engage with a text to determine not only what a text says but also what it means and how it works. The following strategies can help you read texts critically.

Believing and Doubting

One way to develop a response to a text is to play the believing and doubting game, sometimes called reading with and against the grain. Your goal is to **LIST** or **FREEWRITE** notes as you read, writing out as many reasons as you can think of for believing what the writer says (reading with the grain) and then as many as you can for doubting it (reading against the grain).

334–35
333–34

First, try to look at the world through the writer's perspective. Try to understand their reasons for arguing as they do, even if you strongly disagree. Then reread the text, trying to doubt everything in it: try to find every flaw in the argument, every possible way it can be refuted—even if you

✳ academic literacies   ● fields   ● research
■ rhetorical situations   ⁖ processes   ● media/design
▲ genres   ◆ strategies

totally agree with it. Developed by writing theorist Peter Elbow, the believing and doubting game helps you consider new ideas and question ideas you already have—and at the same time see where you stand in relation to the ideas in the text you're reading.

Thinking about How the Text Works: What It Says, What It Does

Sometimes you'll need to think about how a text works, how its parts fit together. You may be assigned to analyze a text, or you may just need to make sense of a difficult text, to think about how the ideas all relate to one another. Whatever your purpose, a good way to think about a text structure is by **OUTLINING** it, paragraph by paragraph. If you're interested in analyzing its ideas, look at what each paragraph *says*; if, however, you're concerned with how the ideas are presented, pay attention to what each paragraph *does*.

340–42

What it says.   Write a sentence that identifies what each paragraph says. Once you've done that for the whole text, look for patterns in the topics the writer addresses. Pay attention to the order in which the topics are presented. Also look for gaps—ideas the writer has left unsaid. Such paragraph-by-paragraph outlining of the content can help you see how the writer has arranged ideas and how that arrangement builds an argument or develops a topic. Here, for example, is an outline of Christian B. Miller's proposal, "Just How Dishonest Are Most Students?" (see p. 258). The numbers in the left column refer to the essay's paragraphs.

1   To give exams in online classes is to invite cheating.

2   Papers are good ways to assess learning in philosophy courses but not in the sciences and other fields, so what can be done to deter cheating on exams in online courses?

3   Remote proctoring using tools that monitor the students and their web browsing is one solution, but it assumes that students are dishonest, may infringe on students' privacy, and may be racially biased.

4   A better solution is to extend to online learning honor pledges, which reduce cheating and promote honesty by allowing students to choose not to cheat.

5   Honor pledges and codes can have problems: they may only be instituted for public relations reasons, they may be imposed on faculty and students by administrators, or they may be given lip service but mostly ignored.

6   However, empirical research studies show that cheating is much less common at schools at all levels that take honor codes seriously.

7   Two researchers found that cheating of various types, including plagiarism, crib notes, help with test answers, and other forms of unauthorized help, were all done much less at schools with honor codes than at schools without them.

8   The researchers emphasized that students must be seriously committed to the honor code for it to work.

9   A few schools hold a ceremony at which the students pledge to uphold the code and may also require that the code be affirmed on every assignment.

10   Research as well as our experience in life shows that most people will cheat if they think it will go undetected; but at the same time most people consider themselves honest and know that it's wrong to cheat, even if they sometimes are tempted to do so. A way to create a moral reminder of those basic values is to ask students to sign an honor code.

11   Another research study paid students taking a difficult test money for each correct answer. One group was graded by the researchers, a second group graded themselves, and a third group also graded themselves—but signed an honor pledge before taking the test. Some in the second group cheated by inflating their scores, but no one in the third group did.

12   The research done in face-to-face class environments shows that honor codes work, but their results should be extended to online instruction as well.

13   Honor codes are not a perfect solution, as some "deeply dishonest" students will cheat no matter what; but research and teachers' experience agree that most students, with moral reminders through honor pledges, will behave with honesty.

What it does.   Identify the function of each paragraph. Starting with the first paragraph, ask, What does this paragraph do? Does it introduce

academic literacies ● fields ● research
■ rhetorical situations ⦂ processes ● media/design
▲ genres ◆ strategies

a topic? provide background for a topic to come? describe something? define something? entice me to read further? something else? What does the second paragraph do? the third? As you go through the text, you may identify groups of paragraphs that have a single purpose. Here is a functional outline of Miller's essay (again, the numbers on the left refer to the paragraphs):

1	Introduces the topic by defining a problem; author establishes credibility
2	Describes a flawed solution
3	Describes a second flawed solution
4	States the thesis; introduces a recommended solution
5	Describes potential problems with this solution
6	Offers a reason for adopting the solution
7	Presents evidence: describes a research study's results
8	Outlines what the proposed solution requires to be successful
9	Describes how the proposed solution is implemented
10	Offers another reason: summarizes how the proposed solution achieves its purpose
11	Presents evidence: describes a research study that shows the solution's effectiveness
12	Concludes with a call to action
13	Acknowledges and refutes a counterargument to the proposal

Identifying Patterns

Look for notable patterns in the text—recurring words and their synonyms, as well as repeated phrases, metaphors and other images, and types of sentences. Some readers find it helps to highlight patterns in various colors. Does the author repeatedly rely on any particular writing strategies: **NARRATION**? **COMPARISON**? Something else?

474–82
437–44

Another kind of pattern that might be important to consider is the kind of evidence the text provides. Is it more opinion than facts? nothing but statistics? If many sources are cited, is the information presented in

542–54

any patterns—as **QUOTATIONS**? **PARAPHRASES**? **SUMMARIES**? Are there repeated references to certain experts or sources?

In visual texts, look for patterns of color, shape, and line. What's in the foreground, and what's in the background? What's completely visible, partly visible, or hidden? In both verbal and visual texts, look for omissions and anomalies: What isn't there that you would expect to find? Is there anything that doesn't really fit in?

If you discover patterns, then you need to consider what, if anything, they mean in terms of what the writer is saying. What do they reveal about the writer's underlying premises and beliefs? What do they tell you about the writer's strategies for persuading readers to accept the truth of what they are saying?

See how color-coding an essay by *New York Times* columnist William Safire on the meaning of the Gettysburg Address reveals several patterns in the language Safire uses. In this excerpt from the essay, which was published just before the first anniversary of the September 11, 2001, terrorist attacks, Safire develops his analysis through several patterns. Religious references are colored yellow; references to a "national spirit," green; references to life, death, and rebirth, blue; and places where Safire directly addresses the reader, gray.

> But the selection of this poetic political sermon as the oratorical centerpiece of our observance need not be only an exercise. . . . Now, as then, a national spirit rose from the ashes of destruction.
>
> Here is how to listen to Lincoln's all-too-familiar speech with new ears.
>
> In those 266 words, you will hear the word *dedicate* five times. . . .
>
> Those five pillars of dedication rested on a fundament of religious metaphor. From a president not known for his piety—indeed, often criticized for his supposed lack of faith—came a speech rooted in the theme of national resurrection. The speech is grounded in conception, birth, death, and rebirth.
>
> Consider the barrage of images of birth in the opening sentence. . . .
>
> Finally, the nation's spirit rises from this scene of death: "that this nation, under God, shall have a new birth of freedom." Conception, birth, death, rebirth. The nation, purified in this fiery trial of war, is resurrected. Through the sacrifice of its sons, the sundered nation would be reborn as one. . . .

academic literacies   fields   research
rhetorical situations   processes   media/design
genres   strategies

> Do not listen on Sept. 11 only to Lincoln's famous words and com-
> forting cadences. Think about how Lincoln's message encompasses but
> goes beyond paying "fitting and proper" respect to the dead and the
> bereaved. His sermon at Gettysburg reminds "us the living" of our
> "unfinished work" and "the great task remaining before us"—to
> resolve that this generation's response to the deaths of thousands of
> our people leads to "a new birth of freedom."

The color coding helps us to see patterns in Safire's language, just as Safire
reveals patterns in Lincoln's words. He offers an interpretation of Lincoln's
address as a "poetic political sermon," and the words he uses throughout
support that interpretation. At the end, he repeats the assertion that Lincoln's
address is a sermon, inviting us to consider it differently. Safire's repeated
commands ("Consider," "Do not listen," "Think about") offer additional
insight into how he wishes to position himself in relation to his readers.

READING RHETORICALLY

To read academic texts effectively, you need to look beyond the words on
the page or screen to the **RHETORICAL CONTEXT** of the text and the argu-   ▲ 125–26
ment it makes. Academic texts—both the ones you read and the ones
you write—are parts of ongoing scholarly conversations, in which writers respond to the ideas and assertions of others in order to advance
knowledge. To enter those conversations, you must first read carefully and
critically to understand the rhetorical situation, the larger context within
which a writer wrote, and the argument the text makes.

Considering the Rhetorical Situation

As a reader, you need to think about the message that the writer wants
to articulate, including the intended audience and the writer's attitude
toward that audience and the topic, as well as about the genre, medium,
and design of the text.

PURPOSE   What is the writer's purpose? To entertain? inform? per-   ■ 59–60
                     suade readers to think something or take some action?
                     What is *your* purpose for reading this text?

61–64 ▪	**AUDIENCE**	Who is the intended audience? Are you a member of that group? If not, should you expect that you'll need to look up unfamiliar terms or concepts or that you'll run into assumptions you don't necessarily share? How is the writer addressing the audience—as an expert addressing those less knowledgeable? as an outsider addressing insiders?
65–71 ▪	**GENRE**	What is the genre? Is it a report? an argument? an analysis? something else? Knowing the genre can help you to anticipate certain key features.
72–74 ▪	**STANCE**	Who is the writer, and what is their stance? Critical? Curious? Opinionated? Objective? Passionate? Indifferent? Something else? Knowing the stance affects the way you understand a text, whether you're inclined to agree or disagree with it, to take it seriously, and so on.
75–77 ▪	**MEDIA/DESIGN**	What is the medium, and how does it affect the way you read? If it's a print text, what do you know about the publisher? If it's on the web, who sponsors the site, and when was it last updated? Are there any headings, summaries, or other elements that highlight key parts of the text?

Analyzing the Argument

All texts make some kind of argument, claiming something and then offering reasons and evidence as support for any claim. As a critical reader, you need to look closely at the argument a text makes—to recognize all the claims it makes, consider the support it offers for those claims, and decide how you want to respond. What do you think, and why? Here are some questions to consider when analyzing an argument:

- *What claim is the text making?* What is the writer's main point? Is it stated as a **THESIS** or only implied? Is it limited or **QUALIFIED** somehow? If not, should it have been?

347–49 ⁛
5 ✳

- *How is the claim supported?* What **REASONS** does the writer provide for the claim, and what **EVIDENCE** is given for the reasons? What kind of

413–14 ◆
414–22

✳ academic literacies   ⬣ fields   ⬤ research
▪ rhetorical situations   ⁛ processes   ⬤ media/design
▲ genres   ◆ strategies

evidence is it? Facts? Statistics? Examples? Expert opinions? Images? How convincing do you find the reasons and evidence? Is there enough evidence?

- **What appeals besides logical ones are used?** Does the writer appeal to readers' **EMOTIONS**? try to establish **COMMON GROUND**? demonstrate their **CREDIBILITY** as trustworthy and knowledgeable? How successful are these appeals?

◆ 426–27

424

423–26

- **Are any COUNTERARGUMENTS acknowledged?** If so, are they presented accurately and respectfully? Does the writer concede any value to them or try to refute them? How successfully does the writer deal with them?

◆ 424–25

- **What outside sources of information does the writer cite?** What kinds of sources are they, and how credible do they seem? Are they current and authoritative? How well do they support the argument?

- **Do you detect any FALLACIES?** Fallacies are arguments that involve faulty reasoning. Because they often seem plausible, they can be persuasive. It is important, therefore, that you question the legitimacy of such reasoning when you run across it.

◆ 427–29

Considering the Larger Context

All texts are part of ongoing conversations with other texts that have dealt with the same topic. An essay arguing for an assault-weapons ban is part of an ongoing conversation on gun legislation, which is itself part of a conversation on individual rights and responsibilities. Academic texts document their sources in part to show their relationship to the ongoing scholarly conversation on a particular topic. In fact, anytime you're reading to learn, you're probably reading for some larger context. Whatever your reading goals, being aware of that larger context can help you better understand what you're reading. Here are some specific aspects of the text to pay attention to:

Who else cares about this topic?   Especially when you're reading in order to learn about a topic, the texts you read will often reveal which people or groups are part of the conversation—and might be sources of further reading. For example, an essay describing the formation of Mammoth Cave in Kentucky could be of interest to geologists, cave explorers, travel writers, or tourists. If you're reading such an essay while doing research on the cave, you

should consider how the audience to whom the writer is writing determines the nature of the information provided—and its suitability as a source for your research.

What conversations is this text part of? Does the text refer to any concepts or ideas that give you some sense that it's part of a larger conversation? An argument on airport security measures, for example, is part of larger conversations about government response to terrorism, the limits of freedom in a democracy, and the possibilities of using technology to detect weapons and explosives, among others.

What terms does the writer use? Do any terms or specialized language reflect the writer's allegiance to a particular group or academic discipline? If you run across words like *false consciousness*, *ideology*, and *hegemony*, for example, you might guess that the text was written by a Marxist scholar.

What other writers or sources does the writer cite? Do the other writers have a particular academic specialty, belong to an identifiable intellectual school, share similar political leanings? If an article on politics cites Paul Krugman and Gail Collins in support of its argument, you might assume that the writer holds liberal opinions; if it cites Ross Douthat and Jennifer Rubin, the writer is likely a conservative.

READING VISUAL TEXTS

Photos, drawings, graphs, diagrams, and charts are frequently used to help convey important information and often make powerful arguments themselves. So learning to read and interpret visual texts is just as necessary as it is for written texts.

Taking visuals seriously. Remember that visuals are texts themselves, not just decoration. When they appear as part of a written text, they may introduce information not discussed elsewhere in the text. Or they might illustrate concepts hard to grasp from words alone. In either case, it's important to pay close attention to any visuals in a written text.

✳ academic literacies	● fields	● research
■ rhetorical situations	⁖ processes	● media/design
▲ genres	◆ strategies	

Looking at any title, caption, or other written text that's part of a visual will help you understand its main idea. It might also help to think about its purpose: Why did the writer include it? What information does it add or emphasize? What argument is it making? See, for example, how a psychology textbook uses visuals to help explain two ways that information can be represented:

Analogical and Symbolic Representations

When we think about information, we use two basic types of internal representations: analogical and symbolic.

Analogical representations usually correspond to images. They have some characteristics of actual objects. Therefore, they are analogous to actual objects. For example, maps correspond to geographical layouts. Family trees depict branching relationships between relatives. A clock corresponds directly to the passage of time. **Figure 2.1a** is a drawing of a violin from a particular perspective. This drawing is an analogical representation.

Figure 2.1 Analogical versus Symbolic Representations

(a)   (b)

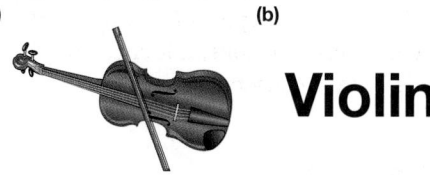

(a) Analogical representations, such as this picture of a violin, have some characteristics of the objects they represent.
(b) Symbolic representations, such as the word "violin," are abstract and do not have relationships to the physical qualities of objects.

By contrast, **symbolic representations** are abstract. These representations usually consist of words or ideas. They do not have relationships to physical qualities of objects in the world. The word "hamburger" is a symbolic representation that usually represents a cooked patty of beef served on a bun. The word "violin" stands for a musical instrument **(Figure 2.1b)**.   —Sarah Grison, Todd Heatherton, and Michael Gazzaniga, *Psychology in Your Life*

The headings tell you the topic: analogical and symbolic representations. The paragraphs define the two types of representation, and the illustrations present a visual example of each type. The visuals make the information in the written text easier to understand by illustrating the differences between the two.

Reading charts and graphs. To read the information in charts and graphs, you need to look for different things depending on what type of chart or graph you're considering. A line graph, for example, usually contains certain elements: title, legend or clearly labeled parts, x axis (horizontal), y axis (vertical), and source information. Figure 2.2 shows one such graph taken from a sociology textbook.

Title: Indicates the topic.

Legend or Labels: Explains the symbols used. Here, colors show the different categories, which are each clearly labeled.

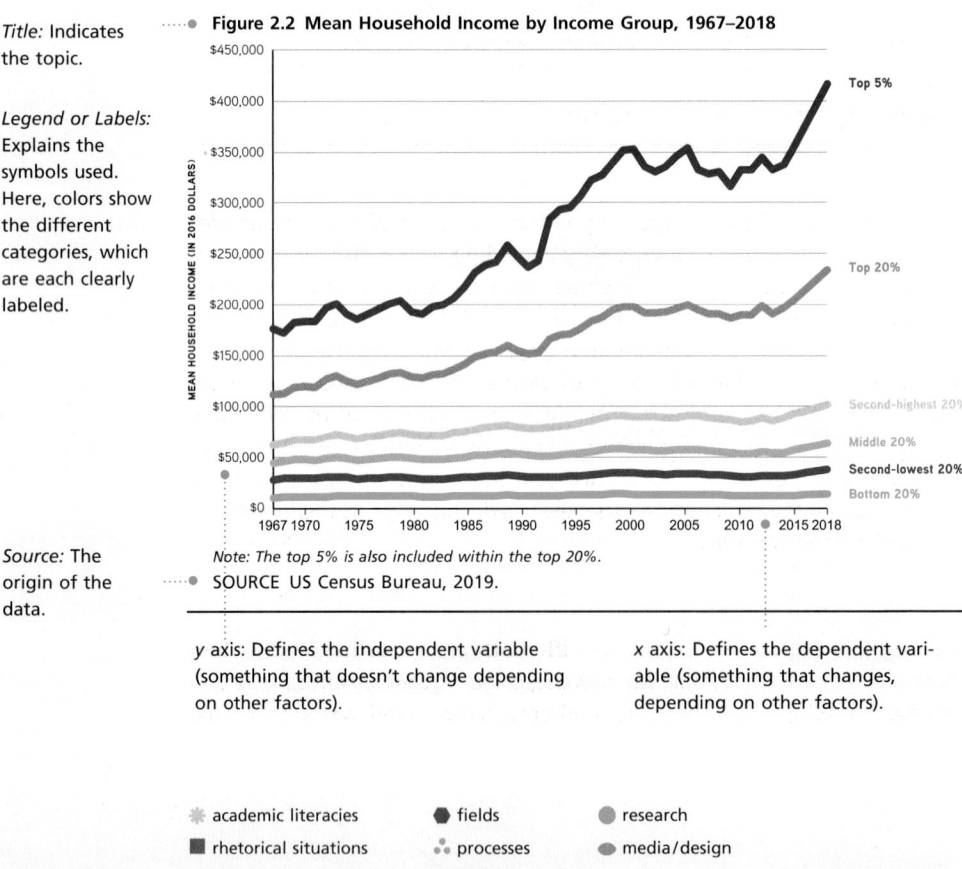

Figure 2.2 Mean Household Income by Income Group, 1967–2018

Note: The top 5% is also included within the top 20%.
SOURCE US Census Bureau, 2019.

Source: The origin of the data.

y axis: Defines the independent variable (something that doesn't change depending on other factors).

x axis: Defines the dependent variable (something that changes, depending on other factors).

※ academic literacies   ● fields   ● research

■ rhetorical situations   ⁝ processes   ● media/design

▲ genres   ◆ strategies

Figure 2.3 Women's Participation in the Labor Force in the United States

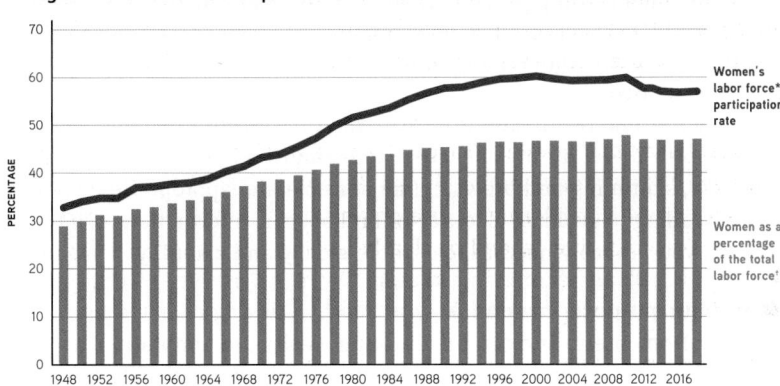

*Women in the labor force as a percentage of all civilian women age 16 and over.
†Women in the labor force as a percentage of the total workforce (both men and women) age 16 and over.
SOURCE US Bureau of Labor Statistics, 2018.

Other types of charts and graphs include some of these same elements. But the specific elements vary according to the different kinds of information being presented, and some charts and graphs can be challenging to read. For example, the chart in Figure 2.3, from the same textbook, includes elements of both bar and line graphs to depict two trends at once: the red line shows the percentage of women in the United States who were in the labor force over a sixty-eight-year period, and the blue bars show the percentage of US workers who were women during that same period. Both trends are shown in two-year increments. To make sense of this chart, you need to read the title, the y-axis legend, and the labels and their definitions carefully.

Reading Onscreen

Research shows that we tend to read differently onscreen than we do when we read print texts: we skim and sample, often reading a sentence or two and then jumping to another site, another text. If we need to scroll the page

to continue, we often don't bother. In general, we don't read as carefully as we do when reading print texts, and we're less likely to reread or take other steps if we find that we don't understand something. Following are some strategies that might help you read effectively onscreen.

Adjust your reading speed and effort to your purpose. Many students use the web to get an overview of a topic and find potential sources. In that case, skimming and browsing are sensible and appropriate tactics. If you're reading to evaluate a source or find specific information on a topic, though, you probably need to read more slowly and carefully.

Keep your purpose in mind as you read. Clicking on hyperlinks and jumping from site to site can be tempting. Resist the temptation! Making a list of specific questions you're seeking to answer can help you stay focused and on task.

Print out longer texts. Some people find reading online to be harder on their eyes than reading pages of print, and many find that they comprehend and remember information in longer texts better if they read them in print. Reading a long text is similar to walking through an unfamiliar neighborhood: we form a mental map of the text as we read and then associate the information with its location in the text, making remembering easier. Since forming such a map is more difficult when reading an electronic text, printing out texts you need to read carefully may be a good strategy.

If you need more help

524–34 ●
333–44 ⦂
367–71
372–79
380–84

See Chapter 48, EVALUATING SOURCES, for questions to help you analyze a text's rhetorical situation. See also Chapter 28 on GENERATING IDEAS AND TEXT; you can adapt those methods as ways of looking at texts, especially clustering and cubing. And see also Chapter 31 on ASSESSING YOUR OWN WRITING, Chapter 32 on GETTING RESPONSE AND REVISING, and Chapter 33 on EDITING AND PROOFREADING if you need advice for reading your own writing.

✳ academic literacies   ● fields   ● research

■ rhetorical situations   ⦂ processes   ● media/design

▲ genres   ◆ strategies

3 Summarizing and Responding
Where Reading Meets Writing

Summarizing a text helps us to see and understand its main points and to think about what it says. Responding to that text then prompts us to think about—*and say*—what we think. Together, summarizing and responding to texts is one way that we engage with the ideas of others. In a history course, you might summarize and respond to an essay arguing that Civil War photographers did not accurately capture the realities of the battle-field. In a philosophy course, you might summarize Plato's "Allegory of the Cave" and respond to its portrayal of knowledge as shadows on a wall.

And in much of the writing that you do, you'll need to cite the ideas of others, both as context for your own thinking and as evidence to support your arguments. In fact, there's probably no topic you'll write about that someone else hasn't already written about—and one way of introducing what you have to say is as a response to something others have said about your topic. A good way of doing that is by summarizing what they've said, using the summary as a launching pad for what you say. This chapter offers advice for summarizing and responding, writing tasks you'll have occasion to do in many of your college classes—and provides a short guide to writing a summary and response essay, a common assignment in composition classes.

SUMMARIZING

In many of your college courses, you'll likely be asked to summarize what someone else has said. Boiling down a text to its basic ideas helps you focus on the text, figure out what the writer has said, and

understand (and remember) what you're reading. In fact, summarizing is an essential academic skill, a way to incorporate the ideas of others into your own writing. Following are some guidelines for summarizing effectively:

Read the text carefully.   To write a good summary, you need to read the original text carefully to capture the writer's intended meaning as clearly and evenhandedly as you can. Start by **SKIMMING** the text to get a general sense of what it's saying. If some parts don't make sense, don't worry; at this point, you're reading just to get the gist. Then reread the text more slowly, **ANNOTATING** it paragraph by paragraph. If there's an explicit **THESIS** stating the main point, highlight it in some way. Then try to capture the main idea of each paragraph in a single sentence.

13

16–17

347–49

State the main points concisely and accurately.   Summaries of a complete text are generally between 100 and 250 words in length, so you need to choose your words carefully and to focus only on the text's main ideas. Leave out supporting evidence, anecdotes, and counterarguments unless they're crucial to understanding the text. For instance, in summarizing "A Fighter's Chance" (see p. 245), Ryan Jones's profile of Rebecca Maine, a student, runner, and boxer, you would omit the lengthy description of her tattoo and her mother's description of her personality.

Describe the text accurately and fairly—and using neutral language. Present the author's ideas evenhandedly and fairly; a summary isn't the place to share your opinion of what the text says. Use neutral verbs such as "states," "asserts," or "concludes," not verbs that imply praise or criticism like "proves" or "complains."

551–54

Use SIGNAL PHRASES to distinguish what the author says from what you say.   Introducing a statement with phrases such as "He says" or "The essay concludes" indicates explicitly that you're summarizing what the author said. When first introducing an author, you may need to say something about their credentials. For example:

* academic literacies ● fields ● research
■ rhetorical situations ∴ processes ● media/design
▲ genres ◆ strategies

In "Our Declaration," political philosopher Danielle Allen analyzes the language of the Declaration of Independence.

Jason Stanley, a Rutgers University philosophy professor, explains the many ways language can be used as a tool for suppression in his *New York Times* opinion piece, "The Ways of Silencing."

Later in the text, you may need to refer to the author again as you summarize specific parts of the text. These signal phrases are typically briefer: "In Stanley's view . . . ," "Allen then argues . . ."

Use quotations sparingly, if at all. You may need to **QUOTE** keywords or memorable phrases, but most or all of a summary should be written in your own words, using your own sentence structures.

● 544–47

DOCUMENT any text you summarize in a works-cited or references list. A summary of a lengthy work should include **IN-TEXT DOCUMENTATION** noting the pages summarized; they aren't needed with a brief text like the one summarized below (see p. 149).

● 560–63
MLA 567–74
APA 618–22

An Example Summary

In "The Reason College Costs More Than You Think," Jon Marcus, a higher-education editor at the *Hechinger Report*, reports that a major reason why college educations are so expensive is the amount of time students stay in college. Although almost all first-year students and their families assume that earning a bachelor's degree will take four years, the reality is that more than half of all students take longer, with many taking six years or more. This delay happens for many reasons, including students changing majors, having to take developmental courses, taking fewer courses per term than they could have, and being unable to register for required courses. As a result, their expenses are much greater—financial aid seldom covers a fifth or sixth year, so students must borrow money to finish—and the additional time they spend in college is time they aren't working, leading to significant losses in wages.

This summary begins with a signal phrase stating the author's name and credentials and the title of the text being summarized. The summary includes only the main ideas, in the summary writer's own words.

RESPONDING

When you summarize a text, you show that you understand its main ideas; responding to a text pushes you to engage with those ideas—and gives you the opportunity to contribute your own ideas to a larger conversation. You can respond in various ways, for instance, by taking a **POSITION** on the text's argument, by **ANALYZING THE TEXT** in some way, or by **REFLECTING** on what it says.

164–95 ▲
104–39
391–401 ⸫

Deciding How to Respond

You may be assigned to write a specific kind of response—an argument or analysis, for instance—but more often than not, the nature of your response is left largely up to you. If so, you'll need to read closely and critically to understand what the text says, to get a sense of how—and how well—it does so, and to think about your own reaction to it. Only then can you decide how to respond. You can respond to what the text says (its ideas), to how it says it (the way it's written), or to where it leads your own thinking (your own personal reaction). Or you might write a response that mixes those ways of responding. You might, for example, combine a personal reaction with an examination of how the writing caused that reaction.

If you're responding to what a text says, you might agree or disagree with the author's argument, supporting your position with good reasons and evidence for your response. You might agree with parts of the argument and disagree with others. You might find that the author has ignored or downplayed some important aspect of the topic that needs to be discussed or at least acknowledged. Here are some questions to consider that can help you think about what a text says:

- What does the writer claim?

413–14 ◆
414–22

- What **REASONS** and **EVIDENCE** does the writer provide to support that claim?

- What parts of the text do you agree with? Is there anything you disagree with—and if so, why?

✳ academic literacies　　● fields　　● research
■ rhetorical situations　　⸫ processes　　● media/design
▲ genres　　◆ strategies

- Does the writer represent any views other than their own? If not, what other perspectives should be considered?
- Are there any aspects of the topic that the writer overlooks or ignores?
- If you're responding to a visual text, how do the design and any images contribute to your understanding of what the text "says"?

Here is a brief response to Jon Marcus's "The Reason College Costs More Than You Think," one that responds to his argument:

> It's true that one reason college costs so much more is that students take longer than four years to finish their degrees, but Jon Marcus's argument in "The Reason College Costs More Than You Think" is flawed in several ways. He ignores the fact that over the past years state governments have reduced their subsidies to state-supported colleges and universities, forcing higher tuition, and that federal scholarship aid has declined as well, forcing students to pay a greater share of the costs. He doesn't mention the increased number of administrators or the costs of fancy athletic facilities and dormitories. Ultimately, his argument places most of the blame for higher college costs on students, who, he asserts, make poor choices by changing majors and "taking fewer courses per term than they could." College is supposed to present opportunities to explore many possible career paths, so changing majors should be considered a form of growth and education. Furthermore, many of us are working full-time to pay the high costs of college, leaving us with little extra time to study for four or five courses at once and sometimes forcing us to take fewer classes per term because that's all we can afford. Marcus is partly right—but he gets much of the problem wrong.

If you're focusing on the way a text is written, you'll consider what elements the writer uses to convey their message—facts, stories, images, and so on. You'll likely pay attention to the writer's word choices and look for any patterns that lead you to understand the text in a particular way. To think about the way a text is written, you might find some of these questions helpful:

- What is the writer's message? Is there an explicit statement of that message?

- How well has the writer communicated the message?

- How does the writer support what they say: by citing facts or statistics? by quoting experts? by noting personal experiences? Are you persuaded?

- Are there any words, phrases, or sentences that you find notable and that contribute to the text's overall effect?

128–34 ▲
- How does the text's design affect your response to it? If it's a **VISUAL TEXT** —a photo or ad, for example—how do the various parts of the text contribute to its message?

Here is a brief response to Marcus's essay that analyzes the various ways it makes its argument:

> In "The Reason College Costs More Than You Think," *Time* magazine writer Jon Marcus argues that although several factors contribute to high college costs, the main one is how long it takes students to graduate. Marcus introduces this topic by briefly profiling a student who is in his fifth year of school and has run out of financial aid because he "changed majors and took courses he ended up not needing." This profile gives a human face to the topic, which Marcus then develops with statistics about college costs and the numbers of students who take more than four years to finish. Marcus's purpose is twofold: to inform readers that the assumption that most students finish college in four years is wrong and to persuade them that poor choices like those this student made are the primary reason college takes so long and costs so much. He acknowledges that the extra costs are "hidden" and "not entirely the student's fault" and suggests that poor high school preparation and unavailable required courses play a role, as do limits on financial aid. However, his final paragraph quotes the student as saying of the extra years, "That's time you're wasting that you could be out making money." As the essay's final statement, this assertion that spending more time in school is time wasted and that the implicit goal of college is career preparation reinforces Marcus's argument that college *should* take only four years and that students who take longer are financially irresponsible.

If you're reflecting on your own reaction to a text, you might focus on how your personal experiences or beliefs influenced the way you

understood the text or on how it reinforced or prompted you to reassess some of those beliefs. You could also focus on how it led you to see the topic in new ways—or note questions that it's led you to wonder about. Some questions that may help you reflect on your own reaction to a text include:

- How did the text affect you personally?
- Is there anything in the text that really got your attention? If so, what?
- Do any parts of the text provoke an emotional reaction—make you laugh or cry, make you uneasy? What prompted that response?
- Does the text bring to mind any memories or past experiences? Can you see anything related to yourself and your life in the text?
- Does the text remind you of any other texts?
- Does the text support (or challenge) any of your beliefs? How?
- Has reading this text given you any new ideas or insight?

Here is a brief response to Jon Marcus's essay that reflects on an important personal issue:

> Jon Marcus's "Why College Costs More Than You Think" made me think hard about my own educational plans. Because I'm working to pay for as much of my education as I can, I'm taking a full load of courses so I can graduate in four years, but truth be told I'm starting to question the major I've chosen. That's one aspect of going to college that Marcus fails to discuss: how your major affects your future career choices and earnings—and whether or not some majors that don't lead immediately to a career are another way of "wasting" your time. After taking several courses in English and philosophy, I find myself fascinated by the study of literature and ideas. If I decide to major in one or both of those subjects, am I being impractical? Or am I "following my heart," as Steve Jobs said in his Stanford commencement speech? Jobs did as he told those graduates to do, and it worked out well for him, so maybe majoring in something "practical" is less practical than it seems. If I graduate in four years and am "out making money" but doing something I don't enjoy, I might be worse off than if I take longer in college but find a path that is satisfying and enriching.

WRITING A SUMMARY AND RESPONSE ESSAY

You may be assigned to write a full essay that summarizes and responds to something you've read. Following is one such essay. It was written by Jacob MacLeod, a student at Wright State University, and responds to a *New York Times* column by Nicholas Kristof, "Our Blind Spot about Guns" (see p. 177).

JACOB MacLEOD

Guns and Cars Are Different

In "Our Blind Spot about Guns," *The New York Times* columnist Nicholas Kristof compares guns to cars in order to argue for sensible gun regulation. Kristof suggests that gun regulations would dramatically decrease the number of deaths caused by gun use. To demonstrate this point, he shows that the regulations governments have instituted for cars have greatly decreased the number of deaths per million miles driven. Kristof then argues that guns should be regulated in the same way that cars are, that car regulation provides a model for gun regulation. I agree with Kristof that there should be more sensible gun regulation, but I have difficulty accepting that all of the regulations imposed on cars have made them safer, and I also believe that not all of the safety regulations he proposes for guns would necessarily have positive effects.

Kristof is right that background checks for those who want to buy guns should be expanded. According to Daniel Webster, director of the Johns Hopkins Center for Gun Policy and Research, state laws prohibiting firearm ownership by members of high-risk groups, such as perpetrators of domestic violence and the mentally ill, have been shown to reduce violence. Therefore, Webster argues, universal background checks would significantly reduce the availability of guns to high-risk groups, as well as reducing the number of guns diverted to the illegal market by making it easier to prosecute gun traffickers.

Kristof also argues that lowering the speed limit made cars safer. However, in 1987, forty states raised their top speed limit from 55 to

65 miles per hour. An analysis of this change by the University of California Transportation Center shows that after the increase, traffic fatality rates on interstate highways in those forty states decreased between 3.4 percent and 5.1 percent. After the higher limits went into effect, the study suggested, some drivers may have switched to safer interstates from other, more dangerous roads, and highway patrols may have focused less on enforcing interstate speed limits and more on activities yielding greater benefits in terms of safety (Lave and Elias 58–61). Although common sense might suggest that lowered speed limits would mean safer driving, research showed otherwise, and the same may be true for gun regulation.

Gun control advocates argue that more guns mean more deaths. However, an article by gun rights advocates Don Kates and Gary Mauser argues that murder rates in many developed nations have no relation to the rate of gun ownership (6). The authors cite data on firearms ownership in the United States and England that suggest that crime rates are lowest where the density of gun ownership is highest and highest where gun density is lowest (8) and that increased gun ownership has often coincided with significant reductions in violence. For example, in the United States in the 1990s, criminal violence decreased, even though gun ownership increased (11–12). However, the authors acknowledge that "the notion that more guns reduce crime is highly controversial" (12).

In fact, a RAND Corporation study of recent research on the relationship between the number of guns and violent crime in the United States concluded that while "most of the new studies provide evidence consistent with the hypothesis that gun prevalence increases violent crime," the difficulty of measuring gun prevalence and other problems with the studies make showing a relationship difficult (Karimov).

All in all, then, Kristof is correct to suggest that sensible gun regulation is a good idea in general, but the available data suggest that some of the particular measures he proposes should not be instituted. I agree that expanding background checks would be a good way to regulate guns and that failure to require them would lead to more guns in the hands of criminals. While background checks are a good form of regulation, however, lower speed limits and trigger locks are not. The problem with this solution is that although it is based on commonsense thinking, the empirical data show that it may not work.

Works Cited

Karimov, Rouslan I. "The Relationship between Firearm Prevalence and Violent Crime." *RAND Corporation: Gun Policy in America,* 2 Mar. 2018, www.rand.org/research/gun-policy/analysis/essays/firearm-prevalence-violent-crime.html.

Kates, Don B., and Gary Mauser. "Would Banning Firearms Reduce Murder and Suicide? A Review of International Evidence." *bepress Legal Repository,* 2006, law.bepress.com/expresso/eps/1564. PDF download, working paper.

Kristof, Nicholas. "Our Blind Spot about Guns." *The New York Times,* 31 July 2014, www.nytimes.com/2014/07/31/opinion/nicholas-kristof-our-blind-spot-about-guns.html.

Lave, Charles, and Patrick Elias. "Did the 65 mph Speed Limit Save Lives?" *Accident Analysis and Prevention,* vol. 26, no. 1, Feb. 1994, pp. 49–62, www.sciencedirect.com/science/article/pii/000145759490068X.

Webster, Daniel. "Why Expanding Background Checks Would, in Fact, Reduce Gun Crime." Interview by Greg Sargent. *The Washington Post,* 3 Apr. 2013, www.washingtonpost.com/blogs/plum-line/wp/2013/04/03/why-expanding-background-checks-would-in-fact-reduce-gun-crime.

In his response, MacLeod both agrees and disagrees with Kristof's argument, using several sources to support his argument that some of Kristof's proposals may not work. MacLeod states his thesis at the end of the first paragraph, after his summary, and ends with a balanced assessment of Kristof's proposals. He cites several sources, both in the text with signal phrases and in-text documentation and at the end in a works-cited section.

Key Features of Summary and Response Essays

A clearly identified author and title. Usually the author and title of the text being summarized are identified in a signal phrase in the first sentence. The author (or sometimes the title) may then be referred to in an abbreviated form if necessary in the rest of the essay: for example, "Kristof argues . . ." or "According to 'Our Blind Spot about Guns' . . ."

✳ academic literacies   ● fields   ● research
■ rhetorical situations   ⦂ processes   ⬬ media/design
▲ genres   ◆ strategies

A concise summary of the text. The summary presents the main and supporting ideas in the text, usually in the order in which they appear. MacLeod, for example, reduces Kristof's argument to four sentences that capture Kristof's main points while leaving out his many examples.

An explicit response. Your essay should usually provide a concise statement (one sentence if possible) of your overall response to the text.

- *If you're responding to the argument,* you'll likely agree or disagree (or both), and so your response itself will constitute an argument, with an explicit thesis statement. For example, MacLeod first agrees with Kristof that "there should be more sensible gun regulation," but then introduces a two-part thesis: that not all automobile regulations have made cars safer and that not all gun regulations would make guns safer.

- *If you're analyzing the text,* you'll likely need to explain what you think the author is saying and how the text goes about conveying that message. An analysis of Kristof's text, for example, might focus on his comparison of automobile regulations with gun regulations.

- *If you're responding with a reflection,* you might explore the ideas, emotions, or memories that the text evokes, the effects of its ideas on your own beliefs, or how your own personal experiences support or contradict the author's position. One response to Kristof's essay might begin by expressing surprise at the comparison of guns to cars and then explore the reasons you find that comparison surprising, leading to a new understanding of the ways regulations can work to save lives.

Support for your response. Whatever your response, you need to offer reasons and evidence to support what you say.

- *If you're responding to what the text says,* you may offer facts, statistics, anecdotal evidence, and textual evidence, as MacLeod does. You'll also need to consider—and acknowledge—any possible counterarguments, positions other than yours.

- *If you're responding to the way the text is written,* you may identify certain patterns in the text that you think mean something, and you'll need to cite evidence from the text itself. For example, Kristof twice invokes a popular slogan among gun rights advocates, "Guns don't kill people. People kill people," changing "guns" to "cars" to advance his argument that regulating guns may make them safer, just as has happened with cars.

- *If you're reflecting on your own reaction to the text,* you may connect its ideas with your own experiences or beliefs or explore how the text reinforced, challenged, or altered your beliefs. A staunch gun-rights advocate, for example, might find in Kristof's essay a reasonable middle ground too often lacking in polarized debates like the one on gun control.

Ways of Organizing a Summary and Response Essay

You can organize a summary and response essay in various ways. You may want to use a simple, straightforward structure that starts out by summarizing the text and then gives the **THESIS** of your response followed by details that develop the thesis.

347–49

[Summary, followed by response]

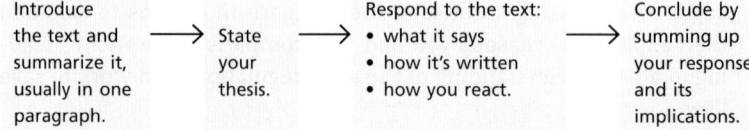

Introduce the text and summarize it, usually in one paragraph. → State your thesis. → Respond to the text: • what it says • how it's written • how you react. → Conclude by summing up your response and its implications.

Or you may want to start out with the thesis and then, depending on whether your response focuses on the text's argument or its rhetorical choices, provide a paired summary of each main point or each aspect of the writing and a response to it.

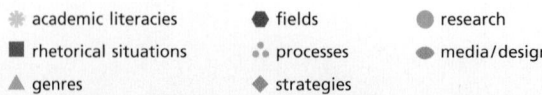

* academic literacies   ● fields   ● research
■ rhetorical situations   ⁘ processes   ● media/design
▲ genres   ◆ strategies

[Introduction and thesis, followed by point-by-point summary and response]

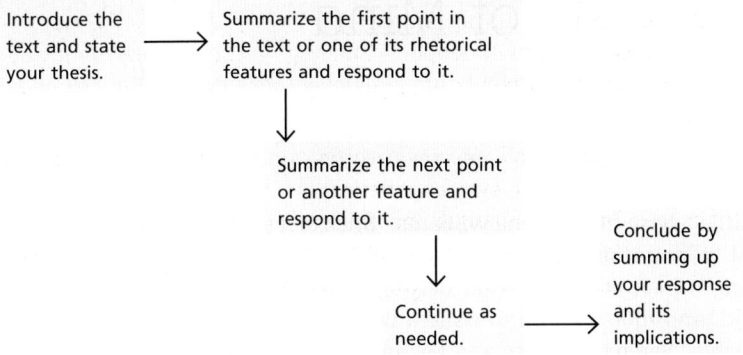

Introduce the text and state your thesis. → Summarize the first point in the text or one of its rhetorical features and respond to it.

↓

Summarize the next point or another feature and respond to it.

↓

Continue as needed. → Conclude by summing up your response and its implications.

If you need more help

See Chapter 30 for guidelines on **DRAFTING** and Chapter 29 for help writing a thesis and coming up with a title. See Chapter 31 on **ASSESSING YOUR OWN WRITING**, Chapter 32 on **GETTING RESPONSE AND REVISING**, and Chapter 33 on **EDITING AND PROOFREADING**. See Chapter 51 on **ACKNOWLEDGING SOURCES, AVOIDING PLAGIARISM**.

364–66
367–71
372–79
380–84
555–59

4 Developing Academic Habits of Mind

A little advice from Serena Williams: "Stick to it and work hard." She wasn't just talking about tennis, and her words resonate for all of us, for everything we set out to do. And here's Michael Jordan, who tells us "Never quit!" and then goes on to issue a warning: "If you quit once, it becomes a habit." Serena Williams and Michael Jordan may be two of the greatest athletes ever, but neither of them was born a champion. They became great by working hard, hanging in there, never giving up.

They succeeded, in other words, by developing certain habits of mind that can serve us all well—and that are especially valuable when it comes to succeeding in school. This chapter is about developing *academic habits of mind*. Just as Serena Williams wasn't born with her powerful serve, none of us was born knowing how to write academic papers or ace exams. But we, too, can learn and can develop the habits we need to succeed. This chapter offers advice for developing habits of mind that writing teachers nationwide have identified as essential for college success.

Engage

We all know people who see school as a series of hoops to jump through, who seem uninvolved—and even bored. We also know people who are passionate about something—a video game, a hobby, a profession—and who invest themselves, their time, and their emotions wholeheartedly in those activities. Successful students make that investment in school. In other words, they engage with what they're studying, what they're doing, and what they're learning.

✳ academic literacies	⬗ fields	⬤ research
◼ rhetorical situations	⁖ processes	⬤ media/design
▲ genres	◆ strategies	

Think about your purpose for being in college. What are your goals? To get a degree? To qualify for a particular job or profession? To find intellectual stimulation? To explore life? Try to define why you are in school, both in larger terms ("to get a degree in accounting") and in terms of the specific courses you're taking ("Learning to write better will help me be a better student in general and communicate effectively at work").

Fight off boredom. Every job, including the job of being a student, involves some tasks that are dull but need to be done. When you encounter such a task, ask yourself how it helps you reach a larger goal. Shooting a hundred free throws for practice may not seem interesting, but it can help you win games. When you're listening to a lecture or reading a textbook, take notes, highlight, and annotate; doing that forces you to pay attention and increases what you remember and learn. Trying to identify the main ideas as you listen to a lecture will help you stay focused. When you're studying, try alternating between different tasks: reading, writing, doing problem sets, drawing, and so on.

If you get distracted, figure out ways to deal with it. It's hard to engage with what you're reading or studying when you're thinking about something else—paying a tuition bill, the last episode of *The Walking Dead*, whatever. Try taking a few moments to write out what's on your mind, in a journal or somewhere else. Sometimes that simple act frees your mind to think about your work, even if it doesn't solve anything.

Raise your hand. When you think you know the answer to a teacher's question (or when you yourself have a question), raise your hand. Most teachers appreciate students who take chances and who participate in class. At the same time, be polite: don't monopolize discussion or interrupt others when they're speaking.

Get involved. Get to know other students; study with them; join a campus organization. People who see themselves as part of something larger, even just a study group with three or four others, engage more in what they're doing than those who try to go it alone.

Be Curious

When we're young, we're curious about everything, and we learn by asking questions (why? why *not?*) and by exploring our surroundings (digging holes, cutting up magazines, investigating attics and basements). As we get older, though, we focus on things that interest us—and may as a result start to ignore other things or even to forget how to explore. In college, you'll be asked to research, study, and write about many topics you know nothing about. Seize the opportunity. Be curious! And take a tip from Dr. Seuss: "The more that you read, the more things you will know. / The more that you learn, the more places you'll go."

Ask questions.   It's tempting to stay within our comfort zones, thinking about things we know and like, listening to those whose views we tend to agree with—to say what we already believe rather than to stop and think. Resist that temptation! Take every opportunity to ask questions, to learn more about things you're already interested in, and especially to learn about things you don't (yet) know anything about. As marine biologist Sylvia Earle says, "The best scientists and explorers have the attributes of kids! They ask questions and have a sense of wonder. They have curiosity. Who, what, where, why, when, and how! They never stop asking questions, and I never stop asking questions, just like a five year old."

Listen!   Pay attention to what others say, including those who you don't necessarily agree with. The words and ideas of others can challenge the way we think, prompt us to rethink what we think—and spark our curiosity: Why does X think that? What do I think—*and why?* Why do my neighbors oppose the Common Core academic standards? What do educators think about it? Paying attention to all sides of an argument, doing research to find out what others have said about a topic, or searching social media to see the latest postings on a trending topic are all ways you can listen in on (and engage in) conversations on important issues of all kinds.

Be Open to New Ideas

No matter where you're in school, you're going to encounter ideas and concepts and even facts that challenge your own beliefs; you're also likely to meet people whose backgrounds and ways of looking at life are very

different from your own. Be open-minded, open to new ideas and to what others think and say. Consider the perspectives and arguments of others. Learning involves accepting new ideas, acknowledging the value of different perspectives, and coming to understand our own beliefs in new ways. Listen to what others say, and think before you respond.

- *Treat the ideas of others with respect,* whether you agree with them or not, and encourage others to do the same. We don't open up if we don't feel safe.
- *Try to withhold judgment.* Be willing to listen to the thoughts of others and to consider ideas that may at first seem alien or weird (or wrong). Remember that your weird is likely someone else's normal—and the reverse.
- *Look for common ground between your perspectives and those of others*—places where you can agree, even in the midst of serious disagreement.

Be Flexible

Being flexible means being adaptable. In college, you'll likely face novel situations and need to find new ways to address problems, such as juggling school, work, and family; adjusting to roommates and making new friends; and figuring out how to do unfamiliar new assignments and how to take tests. You'll even have to do new kinds of writing: lab reports, reflections, literacy narratives, and many, many more; if your school writing up until this point has usually (or always) called for a five-paragraph theme, that's not going to be the case anymore.

Look for new ways to do things. As the saying goes, "If all you have is a hammer, everything looks like a nail." Look for other tools to add to your toolbox: try solving math problems with words or by drawing, or starting a writing project by sketching out the parts.

Try not to see things as right or wrong, good or bad. Be willing to consider alternative points of view and to withhold judgment. Often ideas or actions can only be judged in context; they may be true in some cases or in part, and you often need to understand the larger situation or take into account various perspectives. For example, you may believe that lying is

wrong, but is it excusable if telling the truth will cause someone pain? You find a required reading assignment boring and useless, but why, then, did your instructor assign it?

Approach academic assignments rhetorically.   Analyze each assignment's purpose, intended audience, and the rest of its **RHETORICAL SITUATION**. Think about what's required in terms of content and format—and also about what's appropriate in terms of language and style. And what's expected in the discipline: informal language, sentence fragments, and photographs without captions might be appropriate for a sociology blog, for example, whereas a research project for a history course might have different requirements for how it's organized, formatted, and documented.

57 ■

Be Creative

If you think that creativity is something artists have (and are born with), think again. From the young man selling homemade granola at a local farm market to the woman who puts together an eye-catching outfit from thrift-store bins, many of us are at work expressing ourselves in distinctive ways. Psychologists tell us that acting creatively opens us up to becoming more creative—and it's safe to say that doing so will make your work more productive and very likely more fun.

- *Play with ideas.* Freewrite. Make lists. Try looping, clustering, and the other ways of **GENERATING IDEAS** covered in this book. Take some time to think about the ideas you come up with.

333–44

- *Don't wait until the last minute.* Some students say they do better under the pressure of a deadline. Don't believe it! It's *always* better to give yourself time to think, to explore ideas, to "sleep on" assignments that first stump you.

- *Take risks!* Explore questions and topics that you haven't thought about before. Try out methods you haven't used previously. Challenge yourself to come up with ten ideas. Or twenty.

- *Ask questions.* And remember that there's no such thing as a dumb question.

* academic literacies   ● fields   ● research
■ rhetorical situations   ⁚ processes   ● media/design
▲ genres   ◆ strategies

Persist

Sometimes the key to success is simply sticking to the task at hand: ignoring distractions, hanging in there, forgetting frustration, getting the work done even in the face of setbacks and failures. Here's some advice from actress and singer Julie Andrews: "Perseverance is failing 19 times and succeeding the 20th."

- *Don't quit.* Assume that you can complete the task, and make up your mind to do it. If you're reading a book that seems hopelessly confusing or over your head, for example, keep at it until it starts to make sense. Reread, **OUTLINE CHAPTERS**, **SUMMARIZE**, **ANNOTATE**, and do whatever else you need to do to understand it.

340–42
20
16–17

- *Remember that sometimes you'll encounter setbacks*—and that the goal (a passing grade, a degree, a job) is still reachable if you keep trying. Those who play video games know that failing is an inherent part of playing them; the same is true of many other things as well.

- *Make a plan and establish a schedule,* and stick to them.

- *Break large projects into smaller goals.* Especially with assignments that call for a huge amount of work, approach it in stages: focus on getting through the next chapter, the next draft, the next whatever. It may be good to "keep your eyes on the prize," but it's usually best to take it one step at a time.

- *When you're working on several assignments, tackle the hardest ones first.* Work on them when you're fresh and your mind is clear.

- *If you don't understand something, ask for clarification.* Better to admit to confusion than to act as if you know something when you don't.

- *Ask for help when you need it.* Teachers are usually happy to help students during office hours or by appointment, before or after class, or over email. Get to know your teachers: they're there to help you.

- *Take advantage of whatever help is available* at your school's writing center or learning center and in class. An important part of being persistent is getting help when you need it.

Reflect

Pay attention to the ways you work and make decisions. Reflect on the ways you think and on how you think about the world. This kind of "meta-cognitive" thinking is one of the most important habits of mind, and it's one that will continue to serve you well throughout your life.

- *Figure out when you are most efficient* and do your best work—and try to schedule your work accordingly.

11–32

- *Pay attention to what you're reading and how you're doing it.* Think about why you're reading x, and use **READING STRATEGIES** appropriate to your purpose. If you lose the thread of an explanation or argument, figure out where you got lost and why.

- *After completing an assignment, reflect in writing* on what you did well, what problems you had, and how you solved them (or not). What would you do differently next time—or if you had more time? You'll find prompts for "Taking Stock of Your Work" at the end of Chapters 10–22 that can help.

- *Troubleshoot.* Pay attention to bumps in the road as you encounter them. Try to understand what caused the problem, what you did to solve it, how successful you were and why.

- *Try to focus on your achievements, what you can do*, rather than on what you may not be able to do or to do as well as you'd like.

Collaborate

Whether you're working in a face-to-face group, posting on an online discussion board, or exchanging drafts with a classmate for peer review, you likely spend a lot of time working with others on writing tasks. Even if you do much of your writing sitting alone at a computer, you probably get help from others at various stages in the writing process—and provide help as well. Two (or more) heads can be better than one—and learning to work well with a team is as important as anything else you'll learn in college.

※ academic literacies   ● fields   ● research
■ rhetorical situations   ∴ processes   ● media/design
▲ genres   ◆ strategies

Working with others face-to-face. While informal groups can sometimes accomplish a lot, it's often better to be organized.

- Make sure everyone is facing everyone else and is physically part of the group. Doing that makes a real difference in the quality of the interactions—think how much better conversation works when you're sitting around a table than it does when you're sitting in a row.

- Thoughtfulness, respect, and tact are key, since most writers (as you know) are sensitive and need to be able to trust those commenting on their work. Respond to the contributions of others as you would like others to respond to yours.

- Each meeting needs an agenda—and careful attention paid to time. Appoint one person as timekeeper to make sure all necessary work gets done in the available time.

- Appoint another person to be group leader or facilitator. That person needs to make sure everyone gets a chance to speak, no one dominates the discussion, and the group stays on task.

- Appoint a member of the group to keep a record of the group's discussion, jotting down the major points as they come up and afterward writing a **SUMMARY** of the discussion that the group members then approve.

❋ 33–35

Working with others online. You're likely to work with one or more people online. When sharing writing or collaborating with others online in other ways, consider the following suggestions:

- As with all online communication, remember that you need to choose your words carefully to avoid inadvertently hurting someone's feelings. Without facial expressions, gestures, and other forms of body language and without tone of voice, your words carry all the weight.

- Remember that the **AUDIENCE** for what you write may well extend beyond your group—your work might be forwarded to others, so there's no telling who else might read it.

■ 61–64

- Decide as a group how best to deal with the logistics of exchanging drafts and comments. You can cut and paste text directly into email, send it as an attachment to a message, or post it to your class course management system site or a file-sharing site like *Dropbox* or *Google Docs*. You may need to use a combination of methods, depending on each group member's access to equipment and software. In any case, name your files carefully so that everyone knows which version to use.

- If you're using conferencing software like *Zoom* or *Microsoft Teams*, be aware that only one person at a time can speak and those who chime in quickly can dominate the conversation, sometimes silencing quieter members of the group. Be sure that everyone has a chance to contribute.

Creating group projects.   Creating a document with a team is common in business and professional work and in some academic fields as well. Here are some tips for making collaboration of this kind work well:

- *Define the task as clearly as possible.* Make sure everyone understands and agrees with the stated goals.

- *Divide the task into parts.* Decide which parts can be done by individuals, which can be done by a subgroup, and which need to be done by everyone together.

- *Assign each group member certain tasks.* Try to match tasks to each person's skills and interests and to divide the work equally.

- *Establish a deadline for each task.* Allow time for unforeseen problems before the project deadline.

- *Try to accommodate everyone's style of working.* Some people value discussion; others want to get right down to the writing. There's no best way to get work done; everyone needs to be conscious that their way is not the only way.

- *Work for consensus—not necessarily total agreement.* Everyone needs to agree that the plan to get the writing accomplished is doable and appropriate—if not exactly the way you would do the project if you were working alone.

academic literacies   ● fields   ● research
■ rhetorical situations   ∴ processes   ● media/design
▲ genres   ◆ strategies

- *Make sure everyone performs.* In some situations, your instructor may help, but in others the group itself may have to develop a way to make sure that the work gets done well and fairly. During the course of the project, it's sometimes helpful for each group member to write an assessment both of the group's work and of individual members' contributions.

Making the most of writing conferences.   Conferences with instructors or writing tutors, whether face-to-face or online, can be an especially helpful kind of collaboration. These one-on-one sessions often offer the most strongly focused assistance you can get—and truly valuable instruction. Here are some tips for making the most of conference time:

- *Be prepared.* Have at hand all necessary materials, including the draft you'll be discussing, your notes, any outlines—and, of course, any questions.

- *Be prompt.* Your instructor or tutor has set aside a block of time for you, and once that time is up, there's likely to be another student writer waiting.

- *Listen carefully, discuss your work seriously, and try not to be defensive.* Your instructor or tutor is only trying to help you produce the best piece possible. If you sense that your work is being misunderstood, explain what you're trying to say. Don't get angry! If a sympathetic reader who's trying to help can't understand what you mean, maybe you haven't conveyed your meaning well enough.

- *Take notes.* During the conference, jot down keywords and suggestions. Immediately afterward, flesh out your notes so you'll have a complete record of what was said.

- *Reflect on the conference.* Afterward, think about what you learned. What do you have to do now? Create a plan for revising or doing further work, and write out questions you will ask at your next conference or in a follow-up email or text.

Take Responsibility

In one way or another, all the habits of mind discussed above involve taking responsibility for your own actions. It may be tempting to blame others

or society or bad luck for problems in your academic life, but the more you take ownership of your own learning, the more control you have over the results. Some ways you can enhance your sense of responsibility and demonstrate it include these:

- *Acknowledge that how much you learn and what grades you get depend mostly on you.* Teachers often say that they don't *give* grades, students *earn* grades—an important difference.

- *Treat school as you do a job,* one for which you must show up on time, perform tasks at a certain level of competence, and meet deadlines. In college, where your time is mostly unstructured, you have to become your own boss. So attend class regularly, follow instructions, and turn in assignments on time.

- *Get organized.* Maintain a calendar so you know what's due when. Create a schedule for your day that includes time for class, studying, working, and personal activities. Develop a system for organizing your written work and notes for each course you're taking. Learn where to find the materials you need to do your classwork.

542–54
560–63
- *Use research sources responsibly.* QUOTE, PARAPHRASE, and SUMMARIZE the work of others accurately, and DOCUMENT it correctly. Give appropriate credit to those whose ideas and words you are using.

✳ academic literacies ⬢ fields ⬤ research
■ rhetorical situations ⸫ processes ⬤ media/design
▲ genres ◆ strategies

Part 2
Rhetorical Situations

Whenever we write, whether it's a text to a friend or a toast for a wedding, an English essay or a résumé, we face some kind of rhetorical situation. We have a **PURPOSE**, a certain **AUDIENCE**, a particular **STANCE**, a **GENRE**, and a **MEDIUM** to consider—and, as often as not, a **DESIGN**. All are important elements that we need to think about carefully. The following chapters offer brief discussions of those elements of the rhetorical situation, along with questions that can help you make the choices you need to as you write. See also the **GENRE** chapters for guidelines for considering your rhetorical situation in each of these specific kinds of writing.

Rhetorical Situations

5 Purpose

All writing has a purpose. We write to explore our thoughts and emotions, to express ourselves, to entertain; we write to record words and events, to communicate with others, to try to persuade others to believe as we do or to behave in certain ways. In fact, we often have several purposes at the same time. We may write an essay in which we try to *explain* something to an audience, but at the same time we may be trying to *persuade* that audience of something. Look, for example, at this passage from a *New York Times* opinion essay by Christian B. Miller, a professor of philosophy at Wake Forest University. Miller explores the question of how cheating on exams can be discouraged, especially in online courses and homeschooling, and states, "I suggest that a practice that has been used widely in other educational contexts be extended to the world of online testing: pledging one's honor." He then explores honor codes "from elementary to graduate level":

> Donald McCabe at Rutgers Business School and Linda Treviño at the Smeal College of Business at Penn State found a 23 percent rate of helping someone with answers on a test at colleges without an honor code, versus only 11 percent at schools with an honor code. They reported impressive differences as well for plagiarism (20 percent versus 10 percent), unauthorized crib notes (17 percent versus 11 percent), and unpermitted collaboration (49 percent versus 27 percent), among other forms of cheating.
>
> A serious commitment to the honor code is crucial to its efficacy. As Professors McCabe and Treviño insist, an honor code should be "well implemented and strongly embedded in the student culture."
>
> What does that look like in practice? A few schools start the academic year with an actual commitment ceremony, where each student has to publicly pledge to uphold the school's code. To this can be added a requirement to affirm the honor code on each graded assignment.
>
> —Christian B. Miller, "Just How Dishonest Are Most Students?"

Miller is reporting information here, outlining how honor codes affect students' honesty in schooling. He is also making an argument that honor codes are preferable to other, surveillance- or punishment-focused methods of curbing cheating because "Most students are not deeply dishonest." His main purpose, then, is to persuade readers that schools should adopt honor codes.

Even though our purposes may be many, knowing our primary reason for writing can help us shape that writing and understand how to proceed with it. Our purpose can determine the genre we choose, our audience, even the way we design what we write.

Identify your purpose.   While a piece of writing often has many purposes, a writer usually focuses on one. When you get an assignment or see a need to write, ask yourself what the primary purpose of the writing task is: to entertain? to inform? to persuade? to demonstrate your knowledge or your writing ability? What are your own goals? What are your audience's expectations, and do they affect the way you define your purpose?

Thinking about Purpose

- *What do you want your audience to do, think, or feel?* How will your readers use what you tell them?

- *What does this writing task call on you to do?* Do you need to show that you have mastered certain content or skills? Do you have an assignment that specifies a particular **STRATEGY** or **GENRE** —to compare two things, perhaps, or to argue a position?

- *What are the best ways to achieve your purpose?* What **STANCE** should you take? Should you write in a particular genre? Do you have a choice of **MEDIUM**, and does your text require any special format or **DESIGN** elements?

403 ◆
79 ▲
72–74 ■
657 ●

✳ academic literacies   ● fields   ● research
■ rhetorical situations   ⁞ processes   ● media/design
▲ genres   ◆ strategies

6 Audience

Who will read (or hear) what you are writing? A seemingly obvious but crucially important question. Your audience affects your writing in various ways. Consider a piece of writing as simple as a text from a mother to her son:

Pls. take chicken out to thaw and feed Annye. Remember Dr. Wong at 4.

On the surface, this brief note is a straightforward reminder to do three things. But in fact it is a complex message filled with compressed information for a specific audience. The writer (the mother) counts on the reader (her son) to know a lot that can be left unsaid. She expects that he knows that the chicken is in the freezer and needs to thaw in time to be cooked for dinner; she knows that he knows who Annye is (a pet?), what they are fed, and how much; she assumes that he knows who (and where) Dr. Wong is. She doesn't need to spell out any of that because she knows what her son knows and what he needs to know—and in her text she can be brief. She understands her audience. Think how different such a reminder would be were it written to another audience—a babysitter, perhaps, or a friend helping out while Mom is out of town.

What you write, how much you write, how you phrase it, even your choice of **GENRE** (essay, email, text, social media post, speech)—all are influenced by the audience you envision. And your audience will interpret your writing according to their own expectations and experiences, not yours.

65–71

When you are a student, your audience is most often your teachers, so you need to be aware of their expectations and learn about the conventions (rules, often unstated) for writing in specific academic fields. You may make statements that seem obvious to you, not realizing that your

instructors may consider them assertions that must be proved with evidence of one sort or another. Or you may write more or less formally than teachers expect. Understanding your audience's expectations—by asking outright, by reading materials in your field of study, by trial and error—is important to your success as a college writer.

This point is worth dwelling on. You're probably reading this textbook for a writing course. As a student, you'll be expected to edit and proofread your essays carefully. But if you correspond with family, friends, or coworkers using email and texts, you might not proofread your messages much (or at all)—and your readers are probably fine with that. Whatever the rhetorical situation, your writing must meet the expectations of your audience.

Identify your audience.   Audiences may be defined as *known, multiple,* or *unknown. Known audiences* can include people with whom you're familiar as well as people you don't know personally but whose needs and expectations you do know. You yourself are a known, familiar audience, and you write to and for yourself often. Class notes, to-do lists, reminders, and journals are all written primarily for an audience of one: you. For that reason, they are often in shorthand, full of references and code that you alone understand.

Other known, familiar audiences include anyone you actually know—friends, relatives, teachers, classmates—and whose needs and expectations you understand. You can also know what certain readers want and need, even if you've never met them personally, if you write for them within a specific shared context. Such a known audience might include PC gamers who read cheat codes that you've posted on the internet for beating a game; you don't know those people, but you know roughly what they know about the game and what they need to know, and you know how to write about it in ways they will understand.

You often have to write for *multiple audiences.* Business memos or reports may be written initially for a supervisor, who may pass them along to others. Grant proposals may be reviewed by four to six levels of readers—each, of course, with its own expectations and perspectives. Even writing for a class might involve multiple audiences: your instructor and your classmates.

academic literacies ● fields ● research
■ rhetorical situations ⁙ processes ● media/design
▲ genres ◆ strategies

Unknown audiences can be the most difficult to address since you can't be sure what they know, what they need to know, how they'll react. Such an audience could be your downstairs neighbor, with whom you've chatted occasionally in the laundry room. How will she respond to your letter asking her to sponsor you in an upcoming charity walk? Another unknown audience—perhaps surprisingly—might be many of your instructors, who want—and expect!—you to write in ways that are new to you. While you can benefit from analyzing any audience, you need to think most carefully about those you don't know.

Thinking about Audience

- *Whom do you want to reach?* To whom are you writing (or speaking)?
- *What is your audience's background—their education and life experiences?* It may be important for you to know, for example, whether your readers attended college, fought in a war, or have young children.
- *What are their interests?* What do they like? What motivates them? What do they care about?
- *Is there any demographic information that you should keep in mind?* Consider whether race, gender, sexual orientation, disabilities, occupation, religious beliefs, economic status, and so on should affect what or how you write. For example, writers for *Men's Health*, *InStyle*, and *Out* must consider the particular interests of each magazine's readers.
- *What political circumstances may affect their reading?* What attitudes—opinions, special interests, biases—may affect the way your audience reads your piece? Are your readers conservative, liberal, or middle of the road? Politics may take many other forms as well—retirees on a fixed income may object to increased school taxes, so a letter arguing for such an increase would need to appeal to them differently than would a similar letter sent to parents of young children.
- *What does your audience already know—or believe—about your topic? What do you need to tell them? What is the best way to do so?* Those retirees who oppose school taxes already know that taxes are a burden for them; they may need to know why schools are justified in asking

for more money every few years. A good way to explain this may be with a bar graph showing how property values benefit from good schools with adequate funding. Consider which **STRATEGIES** will be effective—narrative, comparison, something else?

403 ◆

- *What's your relationship with your audience, and how should it affect your language and tone?* Do you know them, or not? Are they friends? colleagues? mentors? adversaries? strangers? Will they likely share your **STANCE**? In general, you need to write more formally when you're addressing readers you don't know, and you may address friends and colleagues more informally than you would a boss.

72–74 ■

- *What does your audience need and expect from you?* Your history professor, for example, may need to know how well you can discuss the economy of the late Middle Ages in order to assess your learning; he may expect you to write a carefully reasoned argument, drawing conclusions from various sources, with a readily identifiable thesis in the first paragraph. Your boss, on the other hand, may need an informal email that briefly lists your sales contacts for the day; she may expect that you list the contacts in the order in which you saw them, that you clearly identify each one, and that you briefly say how well each contact went. What **GENRE** is most appropriate?

79 ▲

- *What kind of response do you want?* Do you want readers to believe or do something? to accept as valid your information on a topic? to understand why an experience you once had still matters to you now?

657 ●

- *How can you best appeal to your audience?* Is there a particular **MEDIUM** that will best reach them? Are there any **DESIGN** requirements? (Elderly readers may need larger type, for instance.)

7 Genre

Genres are kinds of writing. Letters, profiles, reports, position papers, poems, blog posts, instructions, parodies—even jokes—are genres. For example, here is the beginning of a **PROFILE** of a mechanic who repairs a specific kind of automobile:

▲ 245–57

> Her business card reads Shirley Barnes, M.D., and she's a doctor, all right—a Metropolitan Doctor. Her passion is the Nash Metropolitan, the little car produced by Austin of England for American Motors between 1954 and 1962. Barnes is a legend among southern California Met lovers—an icon, a beacon, and a font of useful knowledge and freely offered opinions.

A profile offers a written portrait of someone or something that informs and sometimes entertains, often examining its subject from a particular angle—in this case, as a female mechanic who fixes old cars. While the language in this example is informal and lively ("she's a doctor, all right"), the focus is on the subject, Shirley Barnes, "M.D." If this same excerpt were presented as a poem, however, the new genre would change our reading:

> Her business card reads
> Shirley Barnes, M.D.,
> and she's a doctor, all right
> —a Metropolitan Doctor.
> Her passion is the Nash Metropolitan,
> the little car produced by Austin of England
> for American Motors between 1954 and 1962.
> Barnes is a legend
> among southern California Met lovers
> —an icon,

a beacon,
and a font of useful knowledge and
freely offered opinions.

The content hasn't changed, but the different presentation invites us to read not only to learn about Shirley Barnes but also to explore the significance of the words and phrases on each line, to read for deeper meaning and greater appreciation of language. The genre thus determines how we read and how we interpret what we read.

Genres help us write by establishing features for conveying certain kinds of content. They give readers clues about what sort of information they're likely to find and so help them figure out how to read ("This article begins with an abstract, so it's probably a scholarly source" or "Thank goodness! I found the instructions for editing videos on my phone"). At the same time, genres are flexible; writers often tweak the features or combine elements of different genres to achieve a particular purpose or connect with an audience in a particular way. Genres also change as writers' needs and available technologies change. For example, computers have enabled us to add audio and video content to texts that once could appear only on paper.

Choosing the Appropriate Genre

How do you know which genre you should choose? Often the words and phrases used in writing assignments can give you clues to the best choice. Here are typical terms used in assignments and the genres they usually call for.

81–103 ▲

LITERACY NARRATIVE If you're assigned to explore your development as a writer or reader or to describe how you came to be interested in a particular subject or career, you'll likely need to write a literacy narrative or a variation on one. Some terms that might signal a literacy narrative: "describe a learning experience," "tell how you learned," "trace your development," "write a story."

104–39 ▲
223–35

TEXTUAL ANALYSIS or **LITERARY ANALYSIS** If your assignment calls on you to look at a nonfiction text to see not only what it says but also how it works, you likely need to write a textual analysis. If the text is a short story,

novel, poem, or play, you probably need to write a literary analysis. If you are analyzing a text or texts in multiple media, you might choose either genre or mix the two. Some terms that might signal that a textual or literary analysis is required: "analyze," "examine," "explicate," "read closely," "interpret."

REPORT If your task is to research a topic and then tell your audience in a balanced, neutral way what you know about it, your goal is probably to write a report. Some terms that might signal that a report is required: "define," "describe," "explain," "inform," "observe," "record," "report," "show."

▲ 140–63

POSITION PAPER or **ARGUMENT** Some terms that might signal that your instructor wants you to take a position or argue for or against something: "agree or disagree," "argue," "claim," "criticize," "defend," "justify," "position paper," "prove."

▲ 164–95

SUMMARY If your assignment is to reduce a text into a single paragraph or so, a summary is called for. Some terms that might signal that a summary is expected: "abridge," "boil down," "compress," "condense," "recap," "summarize."

✳ 33–35

EVALUATION If your instructor asks you to say whether or not you like something or whether it's a good or bad example of a category or better or worse than something else, an evaluation is likely required. Some terms that might signal that an evaluation is expected: "assess," "critique," "evaluate," "judge," "recommend," "review."

▲ 214–22

MEMOIR If you're asked to explore an important moment or event in your life, you're probably being asked to write a memoir. Some terms that likely signal that a memoir is desired: "autobiography," "chronicle," "narrate," "a significant personal memory," "a story drawn from your experience."

▲ 234–44

PROFILE If your instructor assigns you the task of portraying a subject in a way that is both informative and entertaining, you're likely being asked to write a profile. Some terms that might indicate that a profile is being asked for: "angle," "describe," "dominant impression," "interview," "observe," "report on."

▲ 245–57

258–68 ▲ **PROPOSAL** If you're asked to offer a solution to a problem, to suggest some action—or to make a case for pursuing a certain project, a proposal is probably in order. Some terms that might indicate a proposal: "argue for [a solution or action]," "propose," "put forward," "recommend."

269–79 ▲ **EXPLORATION** If your assignment calls on you to think in writing about something or to play with ideas, you are likely being asked to write an exploratory essay. Some terms that may mean that an exploration is called for: "consider," "explore," "ponder," "probe," "reflect," "speculate."

280–92 ▲ **REMIX** If you're asked to adapt something you've written to create something new—perhaps for a new purpose, audience, or medium—a remix is probably in order. Some terms that might indicate a remix is asked for: "convert," "modify," "recast," "reimagine," "transform."

Dealing with Ambiguous Assignments

Sometimes even the key term in an assignment doesn't indicate clearly which genre is wanted, so you need to read such an assignment especially carefully. A first step might be to consider whether it's asking for a report or an argument. For example, here are two sample assignments:

> Discuss ways in which the invention of gas and incandescent lighting significantly changed people's daily lives in the nineteenth century.

> Discuss why Willy Loman in *Death of a Salesman* is, or is not, a tragic hero.

Both assignments use the word "discuss," but in very different ways. The first may simply be requesting an informative, researched report: the thesis—new forms of lighting significantly changed people's daily lives in various ways—is already given, and you may just be expected to research and explain what some of these changes were. It's also possible, though, that this assignment is asking you to make an argument about which of these changes were the most significant ones.

In contrast, "discuss" in the second assignment is much more open-ended. It does not lead to a particular thesis but is more clearly asking you to present an argument: to choose a position (Willy Loman is a tragic hero; Willy Loman is *not* a tragic hero; even, possibly, Willy Loman both

✳ academic literacies   ⬢ fields   ⬤ research
◼ rhetorical situations   ⁚⁚ processes   ⬭ media/design
▲ genres   ◆ strategies

is and is not a tragic hero) and to marshal reasons and evidence from the play to support your position. A clue that an argument is being asked for lies in the way the assignment offers a choice of paths.

Other potentially ambiguous words in assignments are "show" and "explore," both of which could lead in many directions. If after a careful reading of the entire assignment you still aren't sure what it's asking for, ask your instructor to clarify the appropriate genre or genres.

Combining Genres

Often your writing will include more than one genre. An **EVALUATION** of mining practices might include a **PROFILE** of a coal company CEO. A **PROPOSAL** to start a neighborhood watch might begin with a **REPORT** on crime in the area. In fact, genres are seldom "pure." When you write, you'll need to consider how you'll choose aspects of various genres to achieve your purpose.

▲ 214–22
245–57
258–68
140–63

Your writing situation will often call for a certain genre that is appropriate for your purpose—an argument, a proposal, a report, a textual analysis, and so forth. Additional genres then play supporting roles. Each genre must contribute to your main point: one genre may serve as the introduction, and others may be woven throughout the text in other ways, but all must address some aspect of the topic and support your central claim. When a text includes several genres, those genres need to fit together clearly and be connected in some way. **TRANSITIONS** do that, and in so doing, they help readers make their way through the text.

⁝ 361–62

It's possible to mix almost any genres together. Following are some of the most commonly mixed genres and how they combine with other genres.

Memoirs. Sometimes a personal anecdote can help support an **ARGU-MENT** or enhance a **REPORT**. Stories from your personal experience can help readers understand your motivations for arguing a certain position and can enhance your credibility as a writer.

▲ 164–95
140–63

Profiles. One way to bring a **REPORT** on an abstract topic to life is to include a profile of a person, place, or event. For example, if you were

▲ 140–63

writing a report for your boss on the need to hire more sales representatives, including a profile of one salesperson's typical day might drive home the point that your sales force is stretched too thin.

Textual analyses. You might need to analyze a speech or other document as part of an **ARGUMENT**, especially on a historical or political topic. For instance, you might analyze speeches by Abraham Lincoln and Jefferson Davis if you're writing about the causes of the Civil War, or an advertisement for e-cigarettes if you're making an argument about teen smoking.

164–95 ▲

Evaluations. You might include an evaluation of something when you write a **PROPOSAL** about it. For example, if you were writing a proposal for additional student parking on your campus, you would need to evaluate the current parking facilities to discuss their inadequacy.

258–68 ▲

Thinking about Genre

- *How does your genre affect what content you can or should include?* Objective information? Researched source material? Your own opinions? Personal experience? A mix?

403 ◆
214–22 ▲
437–44 ◆

- *Does your genre call for any specific* **STRATEGIES**? Profiles, for example, usually include some narration; **EVALUATIONS** often require **COMPARING AND CONTRASTING**.

258–68 ▲

- *Does your genre require a certain organization?* **PROPOSALS**, for instance, usually need to show a problem exists before offering a solution. Some genres leave room for choice. **MEMOIRS** and **LITERACY NARRATIVES** might begin at the beginning, middle, or end of the story.

236–44 ▲
81–103

- *Does your genre affect your tone?* An abstract of a scholarly paper calls for a different **TONE** than a memoir. Should your words sound serious and scholarly? brisk and to the point? objective? opinionated? Sometimes your genre affects the way you communicate your **STANCE**.

73–74 ■
72–74 ■

- *Does the genre require formal (or informal) language?* A letter to the mother of a friend asking for a summer job in her bookstore calls for

✳ academic literacies   ● fields   ● research
■ rhetorical situations   ⁘ processes   ● media/design
▲ genres   ◆ strategies

more formal language than does an email or text to the friend thanking them for the lead.

- **Do you have a choice of medium?** Some genres call for print; others for an electronic medium. Sometimes you have a choice: a résumé, for instance, can be printed to bring to an interview, or it may be downloaded or emailed. Some teachers want reports turned in on paper; others prefer that they be emailed or posted in the class course management system. If you're not sure what MEDIUM you can use, ask.

  657

- **Does your genre have any design requirements?** Some genres call for paragraphs; others require lists. Some require certain kinds of fonts—you wouldn't use **impact** for a personal narrative, nor would you likely use chiller for an invitation to Grandma's sixty-fifth birthday party. Different genres call for different DESIGN elements.

  657

8 Stance

Whenever you write, you have a certain stance, an attitude toward your topic. The way you express that stance affects the way you come across to your audience as a writer and a person. This email from a college student to his father, for example, shows a thoughtful, reasonable stance for a carefully researched argument:

> Hi Dad,
> I'll get right to the point: I'd like to buy a car. I saved over $4,500 from working this summer, and I've found three different cars that I can get for under $3,000. That'll leave me $1,400 to cover the insurance. I can park in Lot J, over behind Monte Hall, for $75 for both semesters. And I can earn gas and repair money by upping my hours at the cafeteria. It won't cost you any more, and if I have a car, you won't have to come and pick me up when I want to come home. May I buy it?
> Love,
> Michael

While such a stance can't guarantee that Dad will give permission, it's more likely to produce results than this version:

> Hi Dad,
> I'm buying a car. A guy in my Western Civ course has a cool Nissan he wants to get rid of. I've got $4,500 saved from working this summer, it's mine, and I'm going to use it to get some wheels. Mom said you'd freak out at me if I did, but I want this car and I'm getting it.
> Michael

The writer of the first email respects his reader and offers reasoned arguments and evidence of research to convince him that buying a car is an action that will benefit them both. The writer of the second, by contrast,

---

☀ academic literacies   ⬤ fields   ⬤ research

■ rhetorical situations   ⦙⦙ processes   ⬤ media/design

▲ genres   ◆ strategies

seems impulsive, ready to buy the first car that comes along, and defiant—he's picking a fight. Each email reflects a certain stance that shows the writer as a certain kind of person dealing with a topic in a certain way and establishing a certain relationship with his audience.

Identify your stance. What is your attitude toward your topic? Objective? Critical? Curious? Opinionated? Passionate? Indifferent? Your stance may be affected by your relationship to your **AUDIENCE**. How do you want them to see you? As a colleague sharing information? a good student showing what you can do? an advocate for a position? Often your stance is affected by your **GENRE**: for example, lab reports require an objective, unemotional stance that emphasizes the content and minimizes the writer's own attitudes. Memoir, by comparison, allows you to reveal your feelings about your topic. Your stance is also affected by your **PURPOSE**, as the two emails about cars show. Your stance in a piece written to entertain will likely differ from the stance you'd adopt to persuade.

■ 61–64

▲ 79

■ 59–60

You communicate (or downplay) your stance through your tone—through the words you use and other ways your text expresses an attitude toward your subject and audience. For example, in an academic essay you would state your position directly—"*The Bachelor* reflects the values of American society today"—using a confident, authoritative tone. In contrast, using qualifiers like "might" or "I think" can give your writing a wishy-washy, uncertain tone: "I think *The Bachelor* might reflect some of America's values." A sarcastic tone might be appropriate for a comment on a blog post but isn't right for an academic essay: "*The Bachelor*'s star has all the personality of a bowling ball."

Like every other element of writing, your tone must be appropriate for your rhetorical situation.

Just as you likely alter what you say depending on whether you're speaking to a boss, an instructor, a parent, or a good friend, so you need to make similar adjustments as a writer. It's a question of appropriateness: we behave in certain ways in various social situations, and writing is a social situation. You might sign an email to a friend with an XO, but in an email to your supervisor you'll likely sign off with a "Many thanks" or "Sincerely." To write well, you need to write with integrity, to say as much

as possible what you wish to say; yet you also must understand that in writing, as in speaking, your stance and tone need to suit your purpose, your relationship to your audience, the way in which you wish your audience to perceive you, and your medium.

In writing as in other aspects of life, the Golden Rule applies: "Do unto audiences as you would have them do unto you." Address readers respectfully if you want them to respond to your words with respect.

Thinking about Stance

- *What is your stance, and how does it relate to your purpose for writing?* If you feel strongly about your topic and are writing an argument that tries to persuade your audience to feel the same way, your stance and your **PURPOSE** fit naturally together. But suppose you're writing about the same topic with a different purpose—to demonstrate the depth of your knowledge about the topic, for example, or your ability to consider it in a detached, objective way. You will need to adjust your stance to meet the demands of this different purpose.

  59–60 ■

- *How should your stance be reflected in your tone?* Can your tone grow directly out of your stance, or do you need to "tone down" your attitude toward the topic or take a different tone altogether? Do you want to be seen as reasonable? angry? thoughtful? gentle? funny? ironic? If you're writing about something you want to be seen as taking very seriously, be sure that your language and even your font reflect that seriousness. Check your writing for words that reflect the tone you want to convey—and for ones that do not (and revise as necessary).

- *How is your stance likely to be received by your audience?* Your tone and especially the attitude it projects toward your **AUDIENCE** will affect how they react to the content of what you say.

  61–64 ■

- *Should you openly discuss your stance?* Do you want or need to announce your own perspective on your topic? Will doing so help you reach your audience, or would it be better not to say directly where you're coming from?

✳ academic literacies   ● fields   ● research

■ rhetorical situations   ⦂ processes   ⬤ media/design

▲ genres   ◆ strategies

9 Media/Design

In its broadest sense, a medium is a go-between: a way for information to be conveyed from one person to another. We communicate through many media, verbal and nonverbal: our bodies (we catch someone's eye, wave, nod); our voices (we whisper, talk, shout, groan); and various technologies, including handwriting, print, phone, radio, video, and digital media.

Each medium has unique characteristics that influence both what and how we communicate. As an example, consider this message: "I haven't told you this before, but I love you." Most of the time, we communicate such a message in person, using the medium of voice (with, presumably, help from eye contact and touch). A phone call will do, though most of us would think it a poor second choice, and a handwritten letter or note would be acceptable, if necessary. Few of us would break such news on a website, with a tweet, or during a radio call-in program.

By contrast, imagine whispering the following sentence in a darkened room: "By the last decades of the nineteenth century, the territorial expansion of the United States had left almost all Native Americans confined to reservations." That sentence starts a chapter in a history textbook, and it would be strange indeed to whisper it into someone's ear. It is appropriate, however, in the textbook, in print or in an e-book, or as a quotation in an oral presentation.

As you can see, we can often choose among various media depending on our purpose and audience. In addition, we can often combine media to create **MULTIMEDIA** texts. And different media enable us to use different ways or modes of expressing meaning, from words to images to sound to hyperlinks, that can be combined in various ways.

661–62

No matter the medium or media, a text's design affects the way it is received and understood. A typed letter on official letterhead sends a different message than the same words handwritten on pastel stationery.

674–80

Classic type sends a different message than *flowery italics*. Some genres and media (and audiences) demand PHOTOS, DIAGRAMS, or color. Some information is easier to explain—and read—in the form of a PIE CHART or a BAR GRAPH than in the form of a paragraph. Some reports and documents are so long and complex that they need to be divided into sections,

670–72

which are then best labeled with HEADINGS. These are some of the elements to consider when thinking about how to design what you write.

Identify your media and design needs. Does your writing situation call for a certain medium and design? A printed essay? An oral report with visual aids? A blog? A podcast? Academic assignments often assume a particular medium and design, but if you're unsure about your options or the degree of flexibility you have, check with your instructor.

Thinking about Media

- *What medium are you using*—print? spoken? electronic? a combination?—and how does it affect the way you will create your text? A printed résumé is usually no more than one page long; an electronic résumé posted on an employer's website has no length limits. An oral presentation should contain detailed information; accompanying slides should provide only an outline.

403

- *How does your medium affect your organization and* STRATEGIES? Long paragraphs are fine on paper but don't work well on the web. On presentation slides, phrases or keywords work better than sentences. In print, you need to define unfamiliar terms; on the web, you can sometimes just add a link to a definition found elsewhere.

- *How does your medium affect your language?* Some print documents require a more formal voice than spoken media; email and texting often invite greater informality.

- *How does your medium affect what modes of expression you use?* Should your text include photos, graphics, audio or video files, or links? Do you need slides, handouts, or other visuals to accompany an oral presentation?

---

* academic literacies   ● fields   ● research
■ rhetorical situations   ⁞ processes   ● media/design
▲ genres   ◆ strategies

Thinking about Design

- *What's the appropriate look for your* RHETORICAL SITUATION? Should your text look serious? whimsical? personal? something else? What design elements will suit your audience, purpose, stance, genre, and medium?

- *What elements need to be designed?* Is there any information you would like to highlight by putting it in a box? Are there any key terms that should be boldfaced? Do you need navigation buttons? How should you indicate links?

- *What font(s) are appropriate* to your audience, purpose, stance, genre, and medium?

- *Are you including any* VISUALS? Should you? Will your AUDIENCE expect or need any? Is there any information in your text that would be easier to understand as a chart or graph? If you need to include video or audio clips, how should the links be presented?

- *Should you include headings?* Would they help you organize your materials and help readers follow the text? Does your GENRE or MEDIUM require them?

- *Should you use a specific format?* MLA? APA?

57

674–80
61–64

65–71
75–77

MLA 564–614
APA 615–55

Part 3

Genres

When we make a shopping list, we automatically write each item we need in a single column. When we email a friend, we begin with a salutation: "Hi, Jordan." Whether we are writing a letter, a résumé, or a proposal, we know generally what it should contain and what it should look like because we are familiar with each of those genres. Genres are kinds of writing, and texts in any given genre share goals and features—a proposal, for instance, generally starts out by identifying a problem and then suggests a certain solution. The chapters in this part provide guidelines for writing in twelve common academic genres. First come detailed chapters on four genres often assigned in writing classes—**LITERACY NARRATIVES**, **TEXTUAL ANALYSES**, **REPORTS**, and **ARGUMENTS**—followed by brief chapters on **EIGHT OTHER GENRES** and one on **REMIXES**.

Genres

10 Writing a Literacy Narrative

Narratives are stories, and we read and tell them for many different purposes. Parents read their children bedtime stories as an evening ritual. College applicants write about significant moments in their lives. In *psychology* courses, you may write a personal narrative to illustrate how individuals' stories inform the study of behavior. In *education* courses, you may share stories of your teaching experiences. And in *computer science* courses, you may write programming narratives to develop programming skills.

This chapter provides detailed guidelines for writing a specific kind of narrative: a literacy narrative. Writers of literacy narratives traditionally explore their experiences with reading or writing, but we'll broaden the definition to include experiences with various literacies, which might include learning an academic skill, a sport, an artistic technique, or something else. For example, the third narrative in this chapter explores one writer's realization that she needs "automotive literacy" to work in her parents' car repair shop. Along with this essay, this chapter includes two additional good examples, the first annotated to point out the key features found in most literacy narratives.

 REA KARIM

Becoming an American Girl

Rea Karim earned an associate's degree in arts and sciences at Bellevue College, where she wrote this literacy narrative. It was nominated for the Norton Writer's Prize. Karim plans to major in political science at a university and then go to law school.

Attention-getting opening.

I stood on the monument's marble steps, surveying my country's capital. Elizabeth eagerly tugged on my denim jacket, "You gotta see this!"

I turned around to see her smiling at the scene before us. We stood, mesmerized by the beauty of the golden hour, as the sun set over Washington, DC, in the summer of 2017. Like all the other girls surrounding me, I belonged here. I was the walking embodiment of my ancestors' hopes and dreams.

In 2005, though, I felt far from belonging. I was four years old, and I could not speak. My life was filled with words in English, Urdu, Hindi, and Bengali. These four languages painted my dining room table and added zest into my young life, but while I could understand, none of the words came out of my mouth. My mom was extremely concerned about my inability to talk but found reassurance from a linguistics professional who told us that my speaking abilities would develop late because of the diversity of voices within our household. But once I could talk, I still felt difference from the kids around me.

Here Karim describes the setting—what hindered her literacy.

I was the only child on my block whose parents came from two different South Asian countries. My dinner plate was a battle between Indian curried vegetables and the juice of Bengali kebabs. Even though I was born in Seattle, I often felt like I was from another planet, struggling to integrate into my Cascadian hometown. At the same time, I didn't feel like I fit in with the other south Asian kids I knew either. I had roots from both India and Bangladesh, and while those countries are close in proximity, their competition could stretch for miles. On top of that, my classmates introduced me to the "Indian" stereotype that all brown kids are naturally smart, good at math, and excellent students. As hard as I tried, I could not fit in to that image. I was late to start speaking, and I couldn't read.

Clearly described details.

Mama would sit with me for hours trying to get me to give up 5 TV and learn to read with her from one of the Dot books, but I didn't like Dot the dog. He was boring, and I was stubborn. Studying always seemed like a chore, and reading didn't spark the same interest in me that it did for my classmates. Kids my age were cruising through Dot and picking up Lemony Snicket, while I was still learning the difference between *p* and *q*.

Because of my reading level, I didn't feel smart enough to be around the students in my class. Even though I made friends, a huge part of me felt like I couldn't fit in. I didn't feel American enough to live in my predominantly white community, and I didn't feel "brown" enough

※ academic literacies   ● fields   ● research

■ rhetorical situations   ⁙ processes   ● media/design

▲ genres   ◆ strategies

to hang out with the kids of my parents' friends. Instead of socializing with my peers, I sought refuge in my elementary school library.

The library's tall walls covered with princess and dragon murals, the dozens of birch bookshelves, and the solitude gave me comfort. Every morning, as soon as the bus parked in the roundabout, I dashed to the library's glass doors. The librarian, Ms. McEldowney, always let me in with a comforting smile. She knew I wasn't the best reader, but she still narrated the blurbs from the newly arrived books and introduced me to some of her old favorites. Usually, I snatched whatever was on display, found a quiet spot next to a shelf, and ran my fingers over the words, trying to sound them out. In that way, I started building a friendship with books. I touched the pictures on the front and back cover, imagining that I was a part of the adventures I held in my hand.

A key person in Karim's narrative.

The library became my safe haven. Days quickly turned into months, and soon I could read a book without pictures. Every day I practiced running my fingers through sentences, absorbing whatever was in front of me and reading the sentences aloud. In a matter of a few months and a dozen McEldowney recommendations, I could read.

The library is a key to her literacy development.

Yet, a huge part of me could not relate to the characters in front of me. I still felt like an outsider. I lacked the courage of the main characters from my books, often feeling like a token side character. I wanted to be fearless like the boys and girls in my stories, but sometimes I felt as if a distant force was holding me back from being fearless outside of the library. I wanted to embody red, white, and blue like the heroes American history books told me about. More than anything, I wanted to belong. Even though I had found a place among these imaginary books, I desired a place in the real world around me.

The conflict in the story.

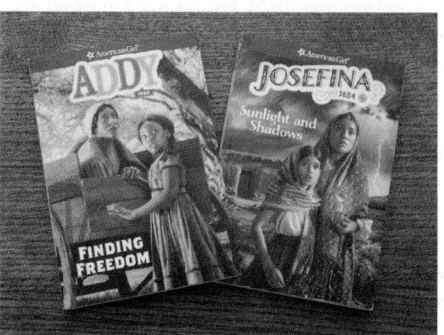

The perceptions I had about being American were all about to ⟨10⟩ change one crisp fall morning during library hour when Ms. McEldowney pulled out a thick set of chapter books. Once a glistening white, their faded covers were now coated in a thick gray dust.

"These" she said with a enthusiastic tone, "are the American Girls."

This brief paragraph emphasizes the moment's significance.

Ms. McEldowney explained to our class that each book was about a different girl in a different period of American History. As she held each one up for us to see, I noticed that not all the girls on the covers were white. Each girl had a unique outfit, her own personality, and her own story.

Her story.

My friends and I rushed to the books. Among the piles, I found *Meet Molly*. On the cover a quirky-looking girl with thick glasses, braided pigtails, and a checkered dress looked back at me. I spent that evening sitting at the dining table reading about Molly's family, her friends, and the new girl from England living with her who was seeking shelter during World War II. I read about how the two girls first felt a cultural clash between them but grew to be fascinated by each other's experiences. Their struggles reminded me of the cultural differences I had with some of my friends. Instead of viewing our lives as polar opposites, I considered sharing with them some of my household's celebrations like Chaand Raat and learning more about holidays like Easter.

More specific details.

What started with just a book became my lifestyle. I went to the ⟨15⟩ library every morning, checked out whatever American Girl book was available, and finished it by the afternoon. From the story of Addy I learned of a family's escape from slavery. From Josephine I learned about living without a mother and having the sole responsibility of raising her Mexican American family in 1824. From Kaya I learned about the impact of the settlers on the Native Americans and the struggle of holding on to what is yours. Each book shaped my perspective of what it means to be an American. I learned that life in this country isn't on one set path. Rather, our country is a mixture of different paths, and each story matters. I started to believe that my story mattered.

The significance of Karim's reading becomes clear to her.

After elementary school, I kept reading. My avid obsession with books grew from the American Girl books to Louisa May Alcott and Ta-Nehisi Coates. Once I entered high school, I pursued the passion for history that the American Girl books instilled in me and tried my hand at AP United States History, World Government, and Comparative Politics.

☀ academic literacies   ● fields   ● research
■ rhetorical situations   ⁙ processes   ● media/design
▲ genres   ◆ strategies

In the summer of 2017, I was accepted to a program that invited sixty girls from all over the country to visit Washington, DC, and learn what it takes for a woman to run for a seat in Congress. As I packed for the trip, I glanced at the old American Girl novels I had purchased, once new and now dusty like the ones I read as a child. I thought about how I spent so much of my childhood wrapped up in the stories of other girls. I realized, then, that I was ready to live my life as the main character. I kept that realization in mind when I met and grew close to the other girls in DC. They came from diverse backgrounds like me, and we shared our passions and stories with each other, each one of us the heroine. By putting myself at the center of my own story, I finally belonged.

As I stood at the steps of the Lincoln Memorial underneath the setting sun, I reflected on my journey to get to that moment. I had fallen in love with the roots my parents planted in Seattle, and I was a smart South Asian American in my own way, well versed not in math, but in history. I was no longer confined to a stereotype. Instead, I embodied the brown color of my skin and belonged to the United States—on my own terms.

The end refers back to the opening anecdote but deepens its significance.

In this literacy narrative, Karim tells the story of how learning to read and reading American Girl novels helped her understand her identity as an American of South Asian heritage. The significance of her story lies in the role that reading played in developing her sense of who she is and where she belongs.

KARLA MARIANA HERRERA GUTIERREZ

Reading, Writing, and Riding
A Literacy Narrative

Karla Mariana Herrera Gutierrez recently graduated from Austin College with a triple major in English, Spanish, and public health. Her literacy narrative was nominated for the Norton Writer's Prize and is documented in MLA style.

Eight faces sleep peacefully as my grandmother surveys the bedroom and closes the door before going outside. Out on the street, the sky is shattered by the soft yellows and pinks that break through the clouds, indicating the start of a new day. My grandmother walks down the sidewalk with her hands grasping her black bag as her mind processes recent events and tries to focus on the task at hand today. *"Toma el autobus a la oficina y pide la pension,"* my grandmother whispers to herself as she approaches the bus stop to go to the police office for her pension.

It's early in the day, but slowly more and more people begin to crowd the streets of Mexico City. The sounds of people talking and the angry honking of horns envelop my grandmother as she looks frantically down the road for the bus. It's only been a few minutes, but she has no time to lose. They told her to be at the office at a certain time, and as a recent widow she does not want to make a bad impression on the people who could make it harder for her to secure the money she needs. Finally, the white and green speck she had fixated on grows closer until she sees it's the bus she has been waiting for. Adjusting her bag, she prepares to step forward but scrunches her brow in concern when the bus doesn't seem to be slowing down. I can picture my grandmother's confused look as the bus passes by without picking her up. My mother said this was my grandmother's breaking point, and she began to cry on the side of the road. "By God's grace," my mother continued, "a lady happened to pass by, and upon seeing my mother's agitated state, she offered her assistance." A simple arm signal stopped the next bus so my grandmother could get on and go to the police department.

During the following two weeks my grandmother learned to no longer fear taking the bus; she got used to traveling downtown to the police station, to the chief's office, sometimes twice in the same day. But her obstacles were only beginning. The process involved in receiving my deceased grandfather's pension was a struggle. The police were reluctant to give it to my grandmother, hoping to wear her out because they didn't care that she had to leave her children to make these trips downtown; they only worried about the money they'd lose.

I can't imagine my grandmother's will to hide her worry as she withstood the harsh glares of the police officers with her back straight and a stone face, determined to fight for what was her family's due. I can't imagine having to trust that the legal documents a stranger presented to her were a written representation of what he was telling

academic literacies   fields   research
rhetorical situations   processes   media/design
genres   strategies

her and not a trap that would make her lose her pension. But I can imagine my grandmother realizing then the importance of knowing how to read and write.

Similar to the collective history of the May family described in literacy scholar Deborah Brandt's article "Accumulating Literacy," my maternal grandmother's experiences provided opportunities for learning that piled up from the moment she hailed that bus. The many obstacles my grandmother faced began with the unfamiliarity of having to navigate new experiences and then intensified when she found herself at the mercy of government officials because she was illiterate.

The importance of literacy became apparent in my mother's household, but obtaining an education was still not a priority. My mother and her siblings learned to read and write, but none finished high school, though one of my uncles later received his GED when he moved to the United States. Literacy and education were two separate things in my mother's youth—but that changed once she moved to the United States with my father and needed to help me when I started school. My own experiences navigating two languages and two cultures expanded my literacy as reading became a key tool not only in learning school subjects but also in learning to navigate a new environment.

I started kindergarten being thrown to the sharks—that is, the English native sharks. Like the *Jaws* theme song, any English word that reached my ears quickened my heartbeat as I turned my head from side to side, hoping to identify the source of such frightening, unfamiliar sounds. I flailed around in the uncharted waters, sobbing and crying for my mother. Every morning my stomach groaned in protest at going to this institution that filled me with dread. However, I wasn't in this alone. In the afternoons I pored over my textbooks and worksheets with my mother and her English-to-Spanish dictionary by my side. Together we conquered my homework, and the sounds that had first put me in fight-or-flight mode gradually became parts of my second language.

It takes more than learning a new language to fully assimilate into a new culture, however. Literacy constitutes more than reading and writing; it also involves the ability to understand and become a part of one's environment. Linguist James Paul Gee calls our "ways of being in the world" Discourses, beginning with the "primary Discourse" we learn at home; he defines literacy as "the mastery or fluent control over a secondary Discourse," the discourse or discourses that include the

social practices and behaviors that allow us to communicate effectively with others and that we learn through becoming members of social institutions, like churches, stores, community groups—and schools (6–9). For the most part, reading books about "typical" non-Hispanic households uncovered the norms of a culture I knew little about, and it also opened a place for conflict to arise when there were conflicting cultural ideas.

Such was the case after I read a book from my favorite series, Junie B. Jones. When Junie B. gets invited to a sleepover, the idea of spending the night at another person's house riveted seven-year-old me. In the book, the sleepover takes place in the giant house of Junie B.'s rich friend, Lucille; however, Junie B.'s crazy antics alter the experience of the sleepover in the way only she knows how to do (Park). I thought my dream would come true when a school friend invited me to come over to her house. I asked my mother for permission even though I was sure she was going to say yes—but I got a lecture instead. My mother refused my request and in fast-spoken Spanish listed all the horrible things that could happen to me if I went to a sleepover. Red-faced and slightly out of breath, she ended her sermon by asking me what was wrong with my bed that I felt the need to sleep in someone else's. My mother, and father, rejected some of the freedoms other kids my age had, such as staying at each other's houses or hanging out after school, because they focused on the potential dangers that could come of such activities. I didn't want to upset my parents, so I only saw my friends at school and read my books to experience what I couldn't in real life.

If reading sometimes caused conflict at home, it was my method 10 for success at school. I was an avid reader and when the school set up a reading program, I took full advantage of it. My school hoped to present reading as a fun activity, so it promoted what was called the Star Program, which offered prizes to students for progressing in the program, with the ultimate prize a ride in a limousine. When the program ended, the twenty kids who had completed the course pressed our young, eager faces against our classroom window as the giant white limo pulled up to the front of the school building. We waited, bouncing in anxious joy, for the teachers to open the school doors and show the wonders of this marvelous car. The driver, in a black suit and sunglasses, walked to the back of the limo and opened the car's door. It could have been the glare of the blazing sun or my young, imaginative mind, but

✳ academic literacies	● fields	● research
■ rhetorical situations	⁘ processes	● media / design
▲ genres	◆ strategies	

the limo seemed to come alive as we caught a glimpse of its sleek black interior. Upon being released to climb in and enjoy our prize, we felt a blast of cold air that cut against the warmth of that spring afternoon. The flash of a camera redirected our attention to our teachers, who positioned us for a picture for the school newspaper.

The Star Program focused on a ride in a limousine, but it failed to mention the bonus prize: feeling like a superstar. My eyes bulged as I took in all the features of this elegant white car as we drove away. The many buttons and compartments for food and drinks (which this ride unfortunately failed to provide) were a novelty to me. Feeling the cold leather of the seats, I traced the shape of the interior until I reached the buttons on the door. With one push the window went down, revealing to the world my awestruck face. The breeze whipped my curly hair around, but I didn't care. I can still see the wondering faces of the people in the cars next to us doing double takes when they saw the stretched car pulling up beside them. As we drove through a roundabout, I felt regal and waved a royal wave to all my subjects because I ruled the world . . . at least for the ride's thirty minutes.

In my elementary years, I thought the Star Program was the epitome of what reading could do, and that limousine ride was only a small taste of what I could achieve by reading and focusing on my academics. Many people would assume that literacy begins like this, by an educational institution, but our literacy actually begins before our time. Brandt observes that literacy "piles up," so that "many generations of literacy . . . now occupy the same social space" (652). This piling up of earlier and more recent literacies produces a "propelling effect"; as Brandt explains, "Schooling typically brings into a family's possession books, manuals, typewriters, and the like that then become the first forms of literacy that the next generation encounters" (659). My literacy began with my grandmother, but it evolved over two generations to include more than simply learning how to read and write—it also meant assimilating into a new culture. Unlike my grandmother's experience with the bus, my ride in the limousine was a marker of my assimilation. It was the first moment that I can remember when I stood out not because of my different background, but because of the reading I had accomplished in a program open to all.

Not only did reading help me assimilate into a new culture, it also helped me with learning a new language. There wasn't someone at home who could teach me how to properly speak and write English, so

it was difficult for me to become fluent. While my mother did her best to help me with her English-to-Spanish dictionary, she couldn't teach me anything that wasn't a part of my homework. It was only through reading that I managed to expand my vocabulary and use the grammar rules I learned in school. By contrast, it was easier for my sister to learn English than it was for me because she had me to converse with at home. My sister's earlier comprehension of English led to her "graduating" from ESL classes in just one year, instead of the four years it took me. My sister had me as her source of English at home, and integrating this new language into our household created a culturally diverse environment that differed from the one my parents had grown up with.

In accumulating new literacies, we improve ourselves for the next generation, and what we improve is our ability to immerse ourselves in new situations because in doing so we elevate the standards of our education to reflect our experiences. But the piling up of literacies isn't an individual effort. My grandmother learned to ride the bus after receiving help from a stranger, and the legal obstacles she faced because she was illiterate pushed her to make sure that my mom and her siblings learned to read. Because my mother knew how to read, she could help me with my homework, and because I learned English, I could converse with my sister in our new language. Each generation is shaped by the collective history of people's literacies because of the opportunities we have to learn from those who came before us and to contribute to those who follow us.

Works Cited

Brandt, Deborah. "Accumulating Literacy: Writing and Learning to Write in the Twentieth Century." *College English*, vol. 57, no. 6, 1995, pp. 649–66.

Gee, James Paul. "Literacy, Discourse, and Linguistics: Introduction." *Journal of Education*, vol. 171, no. 1, 1989, pp. 5–25.

Park, Barbara. *Junie B. Jones Is a Party Animal*. Illustrated by Denise Brunkus, Scholastic, 1997.

Gutierrez traces her literacy development not just to her own life and experiences but to the literacies, or lack of them, of her mother and grandmother as well. Her references to scholarly work on literacy deepen her story and show its significance for many, not just her personally.

* academic literacies　　● fields　　● research
■ rhetorical situations　　⁙ processes　　● media/design
▲ genres　　◆ strategies

ANA-JAMILEH KASSFY

Automotive Literacy

In the following literacy narrative, Ana-Jamileh Kassfy describes an experience that taught her that literacy takes many forms, as well as the importance of knowing what goes on in the family business, auto repair. She wrote this essay in a college writing class at the University of Texas at El Paso and posted it on her class blog.

My father runs a well-known family-owned auto shop here in El Paso, Texas. I come from a family of five, which consists of me, two older brothers, and my parents. My father manages the place, while one of my brothers works as a mechanic in charge of most of the heavy labor and the other spends all day standing by a state inspection machine making sure the cars can run safely on the streets of El Paso. My mother works as the shop's secretary, answering the phone and handling all paperwork. I, on the other hand, was not given the option of being a part of the family business; my job is to graduate from college. And I'll gladly accept going to school and learning in place of spending my days working on cars, even though I spent a lot of time at the shop throughout my childhood.

Since I come from a family whose life revolves around cars, and since I practically lived at the auto shop until I was able to drive, you'd think that I'd understand most of the jargon a mechanic would use, right? Wrong. During my first sixteen years of life, I did manage to learn the difference between a flathead and a Torx screwdriver. I also learned what brake pads do and that a car uses many different colorful fuses. However, rather than paying attention to what was happening and what was being said around me, most of the time I chose to focus on the social aspect of the business. While everyone was running around ordering different pads, filters, and starters or explaining in precise detail why a customer needed a new engine, I preferred to sit and speak with customers and learn their life stories. Being social worked for me—until it didn't.

One day my mother couldn't come to work and decided to have me fill in for her. That was fine with me. I thought to myself, "How hard could it be to answer a phone and say, 'Good afternoon, M & J

Service, how may I help you?' or to greet customers and then turn them over to my dad?"

My morning went by pretty smoothly. I thought I had my duties down to a science. I figured aside from the permanent ringing in my ear from the annoyingly loud air compressors, a few minor paper cuts, and the almost perpetual stench of gasoline and burnt oil, my day was going to fly by.

Then, around lunchtime, my dad left to pick up a part at a car 5 dealership and my brothers went out for lunch. A woman pulled up in a '01 Hyundai Elantra that was desperately in need of a new paint job and walked into the shop. The woman seemed to be in her late forties and was wearing professional clothing with green eye shadow and bright orange lipstick. Her copper-brown hair was feathered out and she wore extremely large gold earrings. And she was angry. She demanded I tell her why she was having a difficult time starting her car. As I began to dust off the file stored in my memory as "Automotive Terms I Will Never Use," I attempted to calm her down, and then I made the mistake of asking her what the problem was.

She said in a harsh tone, "My car doesn't start, your dad just replaced the spark plugs and the motor head, and now my 'blah blah blah' is making noise! I took it to my friend who's a mechanic and he told me your dad fixed the wrong part."

I took a few seconds and just stood there looking at her, trying to add up what I had learned in my life as a "mechanic" and recall what spark plugs were. But it was useless. After a few moments, I gave up. I told her that I couldn't help her, but that she was more than welcome to wait for a mechanic to return.

After giving me an unpleasant look, she proceeded to say, while waving her hands in front of me, "I want my money back! Here is my receipt. I'm taking my car elsewhere. Your dad screwed me over. I told him it was the 'blah blah blah.' Why did he take out the 'blah blah blah'? WHY CAN'T YOU HELP ME? WHY ARE YOU HERE? BLAH BLAH BLAH!"

I stood there, overwhelmed. I began to fidget and push my hair back nervously as I wished some knowledge would kick in. It seemed unbelievable, but despite growing up in a family that lived and breathed automobiles, there was nothing I could do to help her. And my sixteen years of socializing with customers hadn't even paid off because I couldn't calm her down. When I saw my dad walk in, I sighed with

* academic literacies   ● fields   ● research
■ rhetorical situations   ⁙ processes   ● media/design
▲ genres   ◆ strategies

relief, explained the problem to him, and it was resolved. But standing in front of that woman like a deer caught in headlights was so embarrassing. Having somebody shout, "Why can't you help me?" and "Why are you here?" made me feel so ignorant.

Unfortunately, that was only the first of many occasions when 10 people have automatically expected that I know how to take apart and rebuild an engine, or perform some other auto-related task. In reality, I know as much about cars as the next person. After that incident, it became clear to me that I could be literate on very different levels. I'm an expert at running social networks like my *Twitter* feed, and I can zip through and analyze an entire novel written in Spanish, but in other subjects, like automobiles, I am completely illiterate. And when I'm expected to know them, I feel anything but competent.

A confrontation with an irate costumer forces Kassfy to realize that she could be very literate in some situations but almost illiterate in others, and that her lack of knowledge in a workplace context put her at a real disadvantage.

Key Features / Literacy Narratives

A well-told story.   As with most narratives, those about literacy often set up some sort of situation that needs to be resolved. That need for resolution makes readers want to keep on reading. We want to know how Kassfy will deal with an irate customer. Some literacy narratives simply explore the role that developing literacy of some kind played at some time in someone's life, as when Karim discovered the school library and American Girl novels. And some, like Gutierrez's, explore the writer's literacy development in a larger context.

Vivid detail.   Details can bring a narrative to life for readers by giving them vivid mental sensations of the sights, sounds, smells, tastes, and textures of the world in which your story takes place. The details you use when describing something can help readers picture places, people, and events; dialogue can help them hear what is being said. We grasp the importance of books for Karim as she "touched the pictures on the front and back cover." Similarly, we can picture and hear Gutierrez as a little girl who speaks only Spanish in an English-language kindergarten, "sobbing and crying for my mother."

Some indication of the narrative's significance. By definition, a literacy narrative tells something the writer remembers about learning to read, write, or gain competence in a specific area. In addition, the writer needs to make clear why the incident matters to them. You may reveal its significance in various ways. Kassfy comes to understand that to work in an auto repair shop, she needs to understand automotive repair terms. Gutierrez places her literacy in a tradition going back generations—and continuing with her sister's relationship with her. Karim's narrative shows that her literacy was a key to understanding her place in society.

A GUIDE TO WRITING LITERACY NARRATIVES

Choosing a Topic

In general, it's a good idea to focus on a single event that took place during a relatively brief period of time—though sometimes learning to do or understand something may take place over an extended period. In that case, several snapshots or important moments may be needed. Here are some suggestions for topics:

- any early memory about writing, reading, speaking, or another form of literacy that you recall vividly

- someone who taught you to read or write

- someone who helped you understand how to do something

- a book, video game, recording, or other text that has been significant for you in some way

- an event at school that was related to your literacy and that you found interesting, humorous, or embarrassing

- a literacy task that you found (or still find) especially difficult or challenging

- a memento that represents an important moment in your literacy development (perhaps the start of a **LITERACY PORTFOLIO**)

- the origins of your current attitudes about writing, reading, speaking, or doing something

385–90 ⁝

※ academic literacies   ● fields   ● research

■ rhetorical situations   ⁝ processes   ● media/design

▲ genres   ◆ strategies

- learning to text, learning to write an email appropriately, creating and maintaining a *Facebook* page or blog

Make a list of possible topics, and then choose one that you think will be interesting to you and to others—and that you're willing to share with others. If several seem promising, try them out on a friend or classmate. Or just choose one and see where it leads; you can switch to another if need be. If you have trouble coming up with a topic, try **FREEWRITING**, **LISTING**, **CLUSTERING**, or **LOOPING**.

333–36

Considering the Rhetorical Situation

PURPOSE   Why do you want to tell this story? To share a memory with others? To fulfill an assignment? To teach a lesson? To explore your past learning? Think about the reasons for your choice and how they will shape what you write.

59–60

AUDIENCE   Are your readers likely to have had similar experiences? Would they tell similar stories? How much explaining will you have to do to help them understand your narrative? Can you assume that they will share your attitudes toward your story, or will you have to work at making them see your perspective? How much about your life are you willing to share with this audience?

61–64

STANCE   What attitude do you want to project? Affectionate? Neutral? Critical? Do you wish to be sincere? serious? humorously detached? self-critical? self-effacing? something else? How do you want your readers to see you?

72–74

MEDIA / DESIGN   Will your narrative be in print? presented orally? online? Should you use photos, tables, graphs, or video or audio clips? Is there a font that conveys the right tone? Do you need headings?

75–77

Generating Ideas and Text

Good literacy narratives share certain elements that make them interesting and compelling for readers. Remember that your goals are to tell the story as clearly and vividly as you can and to convey the meaning the incident has for you today. Start by thinking about what you already know about writing a literacy narrative. Then write out what you remember about the setting of your narrative and those involved, perhaps trying out some of the methods in the chapter on **GENERATING IDEAS AND TEXT**. You may also want to **INTERVIEW** a teacher or parent or other person who figures in your narrative.

333–44
518–20

Explore what you already know about writing a literacy narrative. Think about recent occasions when you've had to narrate a story, either orally or in writing, in school or out. Take a few moments to think about a couple of those occasions, especially ones involving your reading, writing, speaking, or learning to do something. Why and to whom were you telling these stories? How successful do you think your narratives were? What aspects of telling the story did you feel most confident about or do especially well? What could you have done better? What do you still need to learn about writing a literacy narrative?

Describe the setting. Where does your narrative take place? List the places where your story unfolds. For each place, write informally for a few minutes, **DESCRIBING** what you remember:

456–63

- *What do you see?* If you're inside, what color are the walls? What's hanging on them? What can you see out any windows? What else do you see? Books? Lined paper? Red ink? Are there people? places to sit? a desk or a table?

- *What do you hear?* A radiator hissing? Leaves rustling? The wind howling? Rain? Someone reading aloud? Shouts? Cheers? Children playing? Music? The chime of a text arriving on your phone?

- *What do you smell?* Sweat? Perfume? Incense? Food cooking?

- *How and what do you feel?* Nervous? Happy? Cold? Hot? A scratchy wool sweater? Tight shoes? Rough wood on a bench?

- *What do you taste?* Gum? Mints? Graham crackers? Juice? Coffee?

✳ academic literacies	● fields	● research
■ rhetorical situations	⁚• processes	● media/design
▲ genres	◆ strategies	

Think about the key people. Narratives include people whose actions play an important role in the story. In your literacy narrative, you are probably one of those people. A good way to develop your understanding of the people in your narrative is to write about them:

- *Describe each person in a paragraph or so.* What do the people look like? How do they dress? How do they speak? Quickly? Slowly? With an accent? Do they speak clearly, or do they mumble? Do they use any distinctive words or phrases? You might begin by describing their movements, their posture, their bearing, their facial expressions. Do they have a distinctive scent?

- *Recall (or imagine) some characteristic dialogue.* A good way to bring people to life and move a story along is with **DIALOGUE**, to let readers hear them rather than just hearing about them. Try writing six to ten lines of dialogue between two people in your narrative. If you can't remember an actual conversation, make up one that could have happened. (After all, you are telling the story, and you get to decide how it is to be told.) Try to remember (and write down) some of the characteristic words or phrases that the people in your narrative used.

◆ 464–68

Write about "what happened." At the heart of every good **NARRATIVE** is the answer to the question "What happened?" The action in a literacy narrative may be as dramatic as winning a spelling bee or as subtle as a conversation between two friends; both contain action, movement, or change that the narrative tries to capture for readers. A good story dramatizes the action. Try **SUMMARIZING** the action in your narrative in a paragraph—try to capture what happened. Use active and specific verbs ("pondered," "shouted," "laughed") to describe the action as vividly as possible.

◆ 474–82

● 550–51

Consider the significance of the narrative. You need to make clear the ways in which any event you are writing about is significant for you now. Write a page or so about the meaning it has for you. How did it change or otherwise affect you? What aspects of your life now can you trace to that event? How might your life have been different if this event had not happened or had turned out differently? Why does this story matter to you?

Ways of Organizing a Literacy Narrative

340–42

Start by **OUTLINING** the main events in your narrative. Then think about how you want to tell the story. Don't assume that the only way to tell your story is just as it happened. That's one way—starting at the beginning of the action and continuing to the end. But you could also start in the middle—or even at the end. Karla Gutierrez, for example, could have begun her narrative by discussing her current literacy and then gone back to trace the influences of her family. Several ways of organizing a narrative follow.

[Chronologically, from beginning to end]

Introduce the story. → Describe the setting and people. → Tell what happened. → Say how the story was resolved. → Say something about the significance.

[Beginning in the middle]

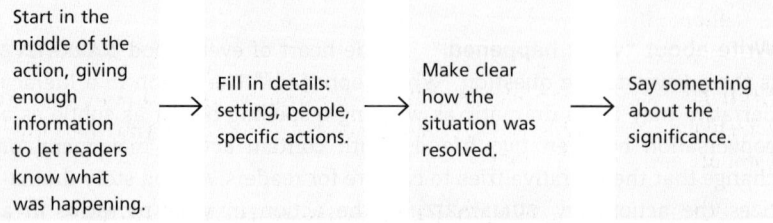

Start in the middle of the action, giving enough information to let readers know what was happening. → Fill in details: setting, people, specific actions. → Make clear how the situation was resolved. → Say something about the significance.

[Beginning at the end]

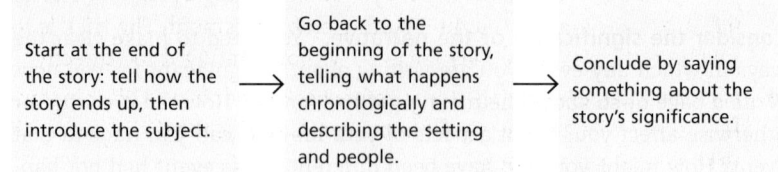

Start at the end of the story: tell how the story ends up, then introduce the subject. → Go back to the beginning of the story, telling what happens chronologically and describing the setting and people. → Conclude by saying something about the story's significance.

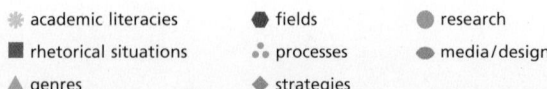

❋ academic literacies ● fields ● research
■ rhetorical situations ⋰ processes ● media/design
▲ genres ◆ strategies

Writing Out a Draft

Once you have generated ideas and thought about how you want to organize your narrative, it's time to begin **DRAFTING**. Do this quickly—try to write a complete draft in one sitting, concentrating on getting the story on paper or screen and on putting in as much detail as you can. Some writers find it helpful to work on the beginning or ending first. Others write out the main event first and then draft the beginning and ending.

364–66

Draft a BEGINNING. A good narrative grabs readers' attention right from the start. Here are some ways of beginning:

346–54

- *Create a question to be answered.* Karim begins her narrative at the Lincoln Memorial in Washington, D.C., leading readers to wonder how this scene connects to her literacy.

- *Describe the context.* You may want to provide background information at the start of your narrative, as Gutierrez does with an anecdote about her illiterate grandmother's struggle to collect a pension.

- *Describe the setting, especially if it's important to the narrative.* Kassfy begins by describing her family's roles working in her father's auto shop.

Draft an ENDING. Think about what you want readers to read last. An effective ending helps them understand the meaning of your narrative. Here are some possibilities:

354–58

- *End where your story ends.* It's up to you to decide where a narrative ends. Karim ends on a field trip in high school.

- *Say something about the significance of your narrative.* Gutierrez explores the meaning of her experience over several paragraphs, and Kassfy discusses her ignorance and resulting embarrassment. The trick is to touch on the narrative's significance without stating it too directly.

- *Refer back to the beginning.* Karim refers back to her trip to Washington, D.C., and Gutierrez to her grandmother and the "piling up" of literacies over several generations.

383–84 **Consider REWRITING.** If you have time and want to explore alternatives, you might try rewriting your draft to see if a different plan or approach might work better.

363 **Come up with a title.** A good **TITLE** indicates something about the subject of your narrative—and makes readers want to take a look. Kassfy's title joins two terms—"automotive" and "literacy"—that aren't usually seen together. Gutierrez uses wordplay—"Reading, Writing, and Riding," to suggest the multiple roles literacy plays in our lives.

Considering Matters of Design

You'll probably write your narrative in paragraph form, but think about the information you're presenting and how you can design it to enhance your story and appeal to your audience.

668–69
- What would be an appropriate **FONT**? Something serious, like Times Roman? Something whimsical, like *Comic Sans*? Something else?

670–72
- Would it help your readers if you added **HEADINGS** in order to divide your narrative into shorter sections?

674–80
- Would photographs or other **VISUALS** show details better than you can describe them with words alone? If you're writing about learning to read, for example, you might scan in an image of one of the first books you read. Or if your topic is learning to write, you could include something you wrote. You could even include a video or audio recording. Would your narrative best be conveyed as a multimedia composition that combines written text, images, and video or audio?

Getting Response and Revising

372–74 The following questions can help you study your draft with a critical eye. **GETTING RESPONSE** from others is always good, and these questions can guide their reading, too. Make sure they know your purpose and audience.

---

* academic literacies   ● fields   ● research

■ rhetorical situations   •• processes   ● media/design

▲ genres   ◆ strategies

- Do the title and first few sentences make readers want to read on? If not, how else might you begin?

- Is the sequence of events in the narrative clear? Does it flow, and are there effective transitions? Does the narrative get sidetracked at any point?

- Is anything confusing?

- Is there enough detail, and is it interesting? Will readers be able to imagine the setting? Can they picture the characters and sense what they're like? Would it help to add some dialogue so that readers can "hear" them?

- Are visuals used effectively and integrated smoothly with the written text? If there are no visuals, would using some strengthen the narrative?

- Have you made the narrative meaningful enough for readers so that they wonder and care about what will happen?

- Do you narrate any actions clearly? vividly? Does the action keep readers engaged?

- Is the significance of the narrative clear?

- Is the ending satisfying? What are readers left thinking?

The preceding questions should identify aspects of your narrative you need to work on. When it's time to **REVISE**, make sure your text appeals to your audience and achieves your purpose as successfully as possible.

375–77

Editing and Proofreading

Once you've revised your draft, follow these guidelines for **EDITING** a narrative:

380–83

- Make sure events are **NARRATED** in a clear order and include appropriate time markers, **TRANSITIONS**, and summary phrases to link the parts and show the passing of time.

474–82
361–62

- Be careful that verb tenses are consistent throughout. If you start your narrative in the past tense ("he *taught* me how to use a computer"), be careful not to switch to the present ("So I *look* at him and *say* . . . ") along the way.

- Check to see that verb tenses correctly indicate when an action took place. If one action took place before another action in the past, for example, you should use the past perfect tense: "I forgot to dot my i's, a mistake I *had made* many times before."

464–68
- Punctuate **DIALOGUE** correctly. Whenever someone speaks, surround the speech with quotation marks ("No way," I said). Periods and commas go inside quotation marks; exclamation points and question marks go inside if they're part of the quotation, outside if they're part of the whole sentence:

INSIDE	Opening the door, Ms. Cordell announced, "Pop quiz!"
OUTSIDE	It wasn't my intention to announce "I hate to read"!

383–84
- **PROOFREAD** your finished narrative carefully before turning it in.

Taking Stock of Your Work

Take stock of what you've written by considering to these questions:

- How did you go about coming up with ideas and generating text? Did you try freewriting, looping, mapping, something else? Which activities were most productive for you?

- How effectively did you use dialogue, description, or other narrative strategies in your writing?

- Did you use photographs or other visual or audio elements? If so, what did they add? If not, why not?

- Did you take any risks with your writing, or did you experiment in some way? Was the experiment a success? A failure? What can you learn from it?

- How did others' responses influence your writing?

☀ academic literacies	● fields	● research
■ rhetorical situations	⁝ processes	● media/design
▲ genres	◆ strategies	

- What did you learn about reading, writing, or other literacies as you worked on your piece? How might you use this learning in the future?

- Overall, what did you do well in this piece? What could still be improved? What would you do differently next time?

If you need more help

See also **MEMOIRS** (Chapter 18), a kind of narrative that focuses more generally on a significant event from your past, and **EXPLORATIONS** (Chapter 21), a kind of essay for thinking about a topic in writing. See Chapter 34 if you are required to submit your literacy narrative as part of a writing **PORTFOLIO**.

236–44
269–79

385–90

11 Analyzing Texts

Both *HuffPost* and *National Review* cover the same events, but each one interprets them differently. All toothpaste ads claim to make teeth "the whitest." The Environmental Protection Agency is a guardian of America's air, water, and soil—or an unconstitutional impediment to economic growth, depending on which politician is speaking. Those are but three examples that demonstrate why we need to be careful, analytical readers of magazines, newspapers, blogs, websites, ads, political documents, even textbooks.

Text is commonly thought of as words, as a piece of writing. In the academic world, however, text can include not only writing but images—photographs, illustrations, videos, films—and even sculptures, buildings, and music and other sounds. And many texts combine words, images, and sounds. We are constantly bombarded with texts: on the web, in print, on signs and billboards, even on our clothing. Not only does text convey information, but it also influences how and what we think. We need to read, then, to understand not only what texts say but also how they say it and how they try to persuade or influence what we think.

Because understanding how texts say what they say and achieve their effects is so crucial, assignments in many disciplines ask you to analyze texts. You may be asked to analyze candidates' speeches in a *political science* course or to analyze the imagery in a poem for a *literature* class. In a *statistics* course, you might analyze a set of data—a numerical text—to find the standard deviation from the mean.

This chapter offers detailed guidelines for writing an essay that closely examines a text both for what it says and for how it does so, with the goal of demonstrating for readers how—and how well—the text achieves its effects. We'll begin with three good examples, the first annotated to point out the key features found in most textual analyses.

* academic literacies   ● fields   ● research
■ rhetorical situations   ⁞ processes   ● media/design
▲ genres   ◆ strategies

PAT FINN

The Architecture of Inequality
On Bong Joon-ho's Parasite

Pat Finn writes for Architizer, an online resource for architects.

In Bong Joon-ho's Academy Award–winning film *Parasite*, architecture isn't silent, it speaks. Again and again, the built environment indicates where people fit into the social hierarchy—or if there is a place for them at all.

The film opens in the apartment of the Kim family, the movie's protagonists. This is a semi-basement apartment, or *banjiha*, a common type of working-class residence in South Korea. "It really reflects the psyche of the Kim family," Bong told Rachel Wallace for *Architectural Digest*. "You're still half overground, so there's this hope and this sense that you still have access to sunlight and you haven't completely fallen to the basement yet. It's this weird mixture of hope and this fear that you can fall even lower."

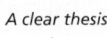

A clear thesis.

The meaning behind the design is explained in a quote by the filmmaker.

An early rendering of the Kim family's semi-basement apartment.

Clear description of the apartment and how the family lives in it.

From the first shot of the film, the viewer can see that this is an improvised living situation, a space where nothing ever feels settled. As the opening credits flash across the screen, the camera lingers over socks hanging from a light fixture that is being used as a makeshift drying rack.

In the next shot, we see the two youngest members of the Kim family, siblings Ki-jung and Ki-woo, frantically scanning the apartment with their phones, searching for an unsecured wi-fi signal they can connect to. They eventually find service in the strange bathroom, in which the toilet is elevated on a tiled platform. Now, it seems, the bathroom will become a new communal space where the family will hang out and browse the internet.

The way the Kim family lives in their space, improvising as necessary, reflects the way they exist economically. Without stable jobs, the Kim family is forced to constantly hustle, make adjustments, and find new ways to make money. If they slow down, even for a second, it might prevent them from making ends meet.

Park So-dam, left, and Choi Woo-shik play the two children of the Kim family.

* academic literacies   ● fields   ● research
■ rhetorical situations   ⁚• processes   ● media/design
▲ genres   ◆ strategies

The Kims' stressful lifestyle is emblematic of the so-called "gig economy" that has emerged in recent years. In his now-classic 2009 book *Capitalist Realism*, the late theorist and blogger Mark Fisher argued that the proliferation of "gig" work in the twenty-first century has taken a serious psychological toll on workers worldwide. "To function effectively as a component of just–in-time production you must develop a capacity to respond to unforeseen events, you must learn to live in conditions of total instability, or 'precarity,' as the ugly neologism has it," Fisher explained. "Periods of work alternate with periods of unemployment. Typically, you find yourself employed in a series of short-term jobs, unable to plan for the future" (38).

This passage, like many others in the book, could have been written about the Kims.

In *Parasite*, the crucial difference between the experience of the working class and the upper class is not opulence but stability. The only other living space that features prominently in the film is a gleaming modern estate enclosed by concrete walls. It belongs to the Park

The family's life is analyzed with support from a scholarly source.

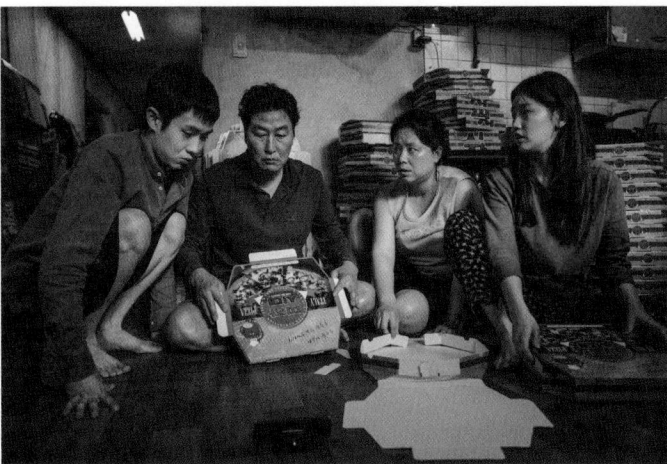

The Kim family's most recent "gig" is folding pizza boxes.

family, for whom the Kims' college-age son, Ki-woo, goes to work as an English tutor. (With fabricated credentials, of course; like his other family members, Ki-woo is adept at hustling, or doing what he needs to survive.)

Much has been written about the Parks' gorgeous house, which is said in the film to have been designed by a famous architect named Namgoong Hyeonja. The space is so convincing that many viewers of the film assumed it was a real house, an architectural marvel. However, it is actually a series of sets designed by Lee Ha Jun, the film's production designer.

"Bong left it all to me in terms of its architectural style," Lee   10
explained in an interview with *Dezeen*. "He showed me a simple floor plan which he sketched whilst writing the script." Instead of sticking to one architectural style, Lee took inspiration from multiple homes that had a minimalist design. The key element he wanted to capture was "great space arrangement."

The fact that the Parks' house was supposedly designed by a famous architect is a significant detail. As soon as Ki-woo steps into the house,

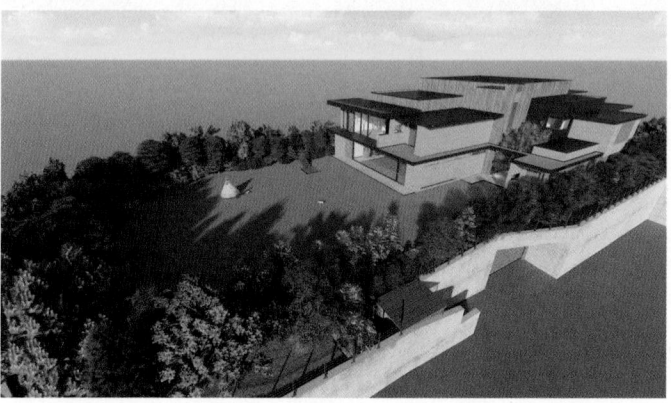

The residence of the wealthy Park family is said to have been built by a famous architect. This is an early rendering developed by production designer Lee Ha Jun.

* academic literacies   ● fields   ● research
■ rhetorical situations   ⋮ processes   ● media/design
▲ genres   ◆ strategies

The illustrations and photos replace written descriptions of the set.

The sets were designed to convey "great space arrangement," according to production designer Lee Ha Jun.

he is leaving the chaotic and improvisatory space of working-class Seoul and entering planned, architectural space. Here, light fixtures do not double as drying racks and bathrooms aren't used for surfing the web. In contrast, every detail of the Parks' house speaks to the logic of its overall design.

Analysis of the significance of the home, contrasted with the apartment.

In *Parasite*, secrets lurk even in spaces that appear meticulously logical in their design.

This is a living space that has carved out its niche in the social order. It asserts its right to exist.

Clear interpretation of the significance of the two dwellings.

However, the seeming stability of the Park household ultimately proves to be an illusion. The shocking ending of the film illustrates that dark secrets lurk in every house, even ones with open floor plans. It also shows that class tensions can only be tamped down for so long. If inequality persists, and workers continue to be unable to take control of their lives, something is eventually going to crack.

Works Cited

Fisher, Mark. *Capitalist Realism: Is There No Alternative?* Zero Books, 2009.

Lee Ha Jun. "*Parasite* House Designed from 'Simple Floor Plan' Sketched by Bong Joon-ho." Interview by Bridget Cogley. *Dezeen*, 16 Apr. 2020, www.dezeen.com/2020/04/16/parasite -film-set-design-interview-lee-ha-jun-bong-joon-ho.

Parasite. Directed by Bong Joon-ho, Barunson E and A, 2019.

Wallace, Rachel. "Inside the House from Bong Joon-ho's *Parasite*." *Architectural Digest,* 31 Oct. 2019, www.architecturaldigest.com /story/bong-joon-ho-parasite-movie-set-design-interview.

✳ academic literacies   ◆ fields   ● research
■ rhetorical situations   ⁚⁚ processes   ● media/design
▲ genres   ◆ strategies

Finn's analysis of the film centers on the two main sets—one family's squalid apartment and another family's lovely home. He compares and contrasts the two dwellings but, interestingly, does not reveal the feature of the house that leads to the "shocking ending of the film."

DANIELLE ALLEN

Our Declaration

Danielle Allen is a political theorist, teaches at Harvard University, and directs Harvard's Edmond J. Safra Center for Ethics. This analysis is a chapter from her book Our Declaration: A Reading of the Declaration of Independence in Defense of Equality.

There's something quite startling about the phrase "We hold these truths to be self-evident." Perhaps it can be made visible most easily with a comparison.

The Catholic Church, too, is committed to a set of truths. At every mass priest and parishioners together recite a list of their beliefs called the *Credo*. One version, called the Apostles' Creed, starts like this: "I believe in God, the Father almighty, creator of heaven and earth. I believe in Jesus Christ, his only son and Lord." Each section begins with the words "I believe," and that's why this recitation is called the *Credo*. Latin, "credo" simply means "I believe."

The Declaration launches its list of truths altogether differently. Jefferson and his colleagues do not say, "I believe," or even "we believe," that all men are created equal. Instead, they say, "We hold these truths to be self-evident," and then they give us a set of either three or five truths, depending on how you count.

What's the difference between "We believe" and "We hold these truths to be self-evident"? In the Catholic *Credo*, when one says, "I believe," the basis for that belief is God's revealed word. In contrast, when Jefferson and his colleagues say, "We hold these truths to be self-evident," they are claiming to know the truths thanks to their own powers of perception and reasoning. These truths are self-evident, and so humans can grasp and hold them without any external or divine assistance.

In order to understand what "We hold these truths to be self-evident" really means, then, it is important to know what "self-evident" means.   5

Sometimes people take it to mean that we can instantly understand an idea, but that's not really right. It's true that sometimes the idea of self-evidence is used for things that we simply perceive. For instance, when I look out my window I immediately perceive that the world includes things like trees and flowers. If outside my window there are many different kinds of tree—hickory and maple and oak, for instance—when I look at them, I nonetheless rapidly perceive that they are all the same kind of thing. That many different kinds of a particular sort of growing thing are all trees is self-evident. We can call this self-evidence from sense perception.

The immediacy of perception, though, is not the same as instantly understanding an idea. And, in fact, to call a proposition self-evident is not at all to say that you will instantly get it. It means instead that if you look into the proposition, if you entertain it, if you reflect upon it, you will inevitably come to affirm it. All the evidence that you need in order to believe the proposition exists within the proposition itself.

This second kind of self-evidence comes not from perception but from logic and how language works.

For instance, we define a chair as an object with a seat and some structure of legs to hold that seat up; and the artifact serves the purpose of having someone sit on it. Then, if I say that a chair is for sitting on, I am expressing a self-evident truth based only on the definition of a chair. Of course a chair is for sitting on! That is how I've defined the word, after all. That's a pretty trivial example of self-evidence. If that were all there were to the idea of self-evidence, it wouldn't be very interesting.

So here is where matters get more interesting: one can string   10 together more than one kind of self-evident proposition—let's call them "premises"—in order to lead to a new piece of knowledge, a conclusion, which will also count as self-evident, since it has been deduced from a few basic self-evident premises.

Aristotle called this method of stringing together valid premises to yield a self-evident conclusion, a syllogism. . . . Here is a basic example:

FIRST PREMISE: *Bill Gates is a human being.*
SECOND PREMISE: *All human beings are mortal.*
CONCLUSION: *Bill Gates is mortal.*

✳ academic literacies   ● fields   ● research
■ rhetorical situations   ⦂ processes   ● media/design
▲ genres   ◆ strategies

This is a bit like math. We can use a Venn diagram to show how the syllogism works. Venn diagrams represent sets of things and how they overlap, and the argument of a syllogism can be thought of as expressing facts about sets and their members. Bill Gates is in the set of human beings. And the set of human beings is entirely contained within the set of mortals. It follows that Bill Gates is in the set of mortals. The validity of this syllogism becomes self-evident when those facts are represented as in this Venn diagram:

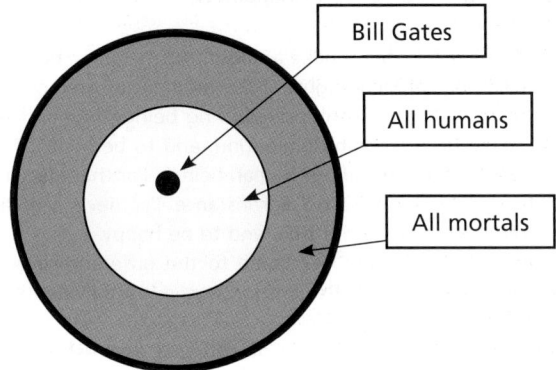

Now, in this syllogism, our two premises are both self-evident truths based on sense perception. We know Bill Gates is a human being by looking at and listening to him. As to the idea that human beings are mortal, we know that human beings die by seeing it happen all around us and never seeing a counterexample. Then we take these two premises, each self-evident through sense perception, and generate a third self-evident proposition, in this case a conclusion, through deduction. From the two premises, we can deduce the certain conclusion that Bill Gates will die.

The Declaration introduces a similar kind of argument when it says, "We hold these truths to be self-evident." At first glance, it looks as if we just have three separate self-evident truths. But if we look closer, we notice that our truths also represent an argument with two premises, which are true from sense perception, and a conclusion that is deduced from them.

Here's how it works.

15

After the Declaration says, "We hold these truths to be self-evident," the text proceeds to identify three truths: one about human beings, one about government, and one about revolution. The truth about human beings, though, is a three-part truth.

It is self-evidently true:

> *that all men are created equal, that they are endowed by their Creator with certain unalienable Rights, that among these are Life, Liberty and the pursuit of Happiness.*

How do these three claims make a single truth? Human beings are equal in all acquiring the same rights at the moment of their creation. From the moment of their emergence as living beings, human beings seek to survive, to be free from domination, and to be happy. This is something we simply observe about human beings. For that matter, we observe it about other animals, too. For instance, I've never seen a cat that didn't want to survive, to be free, and to be happy.

Then, with the next truth, we come to the difference between human beings and animals. The Declaration says, it is self-evidently true

> —*That to secure these rights, Governments are instituted among Men, deriving their just powers from the consent of the governed.*

This is a truly salient point. The signers are saying that, in contrast to the animal kingdom, the world of human beings is indeed full of kingdoms and other kinds of governments. The so-called animal kingdom is a kingdom only metaphorically. There are no governments among animals. Animals have social hierarchies, and they have their own methods for seeking their survival, freedom, and happiness, but human beings use politics. Human beings display self-conscious thought about social organization, and politics is the activity that flows from that self-consciousness about power. Again, this is simply a matter of observation. From the beginning of time to the present day, human beings have formed governments. Human beings have done this just as regularly as birds build nests.

Then the Declaration puts these first two truths together. Since 20 human beings seek their own survival, freedom, and happiness, and since they have a special tool for doing so—namely, the ability to form

governments—it makes sense for them to stick with any particular version of that tool, any particular government, only if it's doing the work it's been built to do.

Compare it to a bird with a nest. What's the point of a bird's staying in a nest if it turns out that the nest has been built out of material inimical or poisonous to the bird? What's the use, in other words, of having a government, if it doesn't serve the purposes of protecting life, liberty, and the pursuit of happiness for which governments are set up in the first place?

The Declaration puts it this way: It is self-evidently true

—That whenever any Form of Government becomes destructive of these ends, it is the Right of the People to alter or to abolish it, and to institute new Government, laying its foundation on such principles and organizing its powers in such form, as to them shall seem most likely to effect their Safety and Happiness.

From the facts, first, that people are simply wired, as are all animals, to seek their survival, freedom, and happiness, and, second, that human beings use governments as their central instrument for protecting their life, liberty, and pursuit of happiness, we can deduce that people have a right to change governments that aren't working for them.

This makes an argument that goes like this:

PREMISE 1: All people have rights to life, liberty, and the pursuit of happiness.

PREMISE 2: Properly constituted government is necessary to their securing these rights.

CONCLUSION: All people have a right to a properly constituted government.

In fact, a philosopher would say that a premise is missing from that argument and that the full formally valid syllogism would look like this:

PREMISE 1: All people have rights to life, liberty, and the pursuit of happiness.

PREMISE 2: Properly constituted government is necessary to their securing these rights.

PREMISE 3:   [All people have a right to whatever is necessary to secure what they have a right to.]

CONCLUSION: All people have a right to a properly constituted government.

Politicians often craft maxims simply by dropping out pieces of their argument. With the missing premise inserted, the Declaration's truths fit together almost like the pieces of a mathematical equation; we intuitively feel the puzzle pieces snap together. That is how self-evidence should feel.

Allen's analysis focuses on the Declaration's second sentence, unpacking its logic through a careful examination of its key term, "self-evident," and explaining how the rest of the sentence forms a syllogism that "snaps together." She looks carefully at every word, restricting her analysis here to a very brief part of the text—but provides insights that illuminate the whole document.

ROY PETER CLARK

Why It Worked
A Rhetorical Analysis of
Obama's Speech on Race

Roy Peter Clark teaches writing at the Poynter Institute. This essay, which Clark describes as "an X-ray reading of the text," appeared online on the Poynter Institute's website, first in 2008 and again in 2017, with a new introduction.

The National Conference of Teachers of English (NCTE) has declared today [October 20, 2017] a National Day on Writing. I celebrate such a day. The introduction of my book *Writing Tools* imagines what America might look like and sound like if we declared ourselves a "nation of writers." After all, what good is freedom of expression if we lack the means to express ourselves?

To mark this day—and to honor language arts teachers every-where—Poynter is republishing an essay I wrote almost a decade ago.

❋ academic literacies   ◆ fields   ● research

■ rhetorical situations   ⁑ processes   ● media/design

▲ genres   ◆ strategies

Remember? It was the spring of 2008 and Barack Obama was running for president. Many of us wondered if America was ready to elect an African-American president (a man with the middle name Hussein).

To dispel the fears of some white Americans and to advance his chances for election, Obama delivered a major address on race in America, a speech that was praised even by some of his adversaries. Obama had / has a gift for language. He is a skilled orator. To neutralize that advantage, his opponents—including Hillary Clinton at one point—would characterize Obama's words as empty "rhetoric"—an elaborate trick of language.

The spring of 2008 seems like such a long time ago. A time just before the Great Recession [that affected the United States from 2008 to 2009]. A time just before the ascendancy of social networks and the trolls who try to poison them. A time before black lives were said to matter in a more assertive way. A time before fake news was anything more dangerous than a piece of satire in the *Onion*. A time before Colin Kaepernick took a knee—except when he was tired. A time before torch-bearing white supremacists marched through the night in Charlottesville, Virginia.

It feels like the perfect time for a restart on a conversation about race. To prepare us, let's take another look at the words of Barack Obama before he was president. Let's review what he said, and, more important, how and why he said it. My X-ray analysis of that speech is meant not as a final word on that historical moment, but as an invitation, a doorway to a room where we can all reflect on American history and the American language.

Have a great National Day on Writing.

More than a century ago, scholar and journalist W. E. B. Du Bois wrote a single paragraph about how race is experienced in America. I have learned more from those 112 words than from most book-length studies of the subject:

After the Egyptian and Indian, the Greek and Roman, the Teuton and Mongolian, the Negro is a sort of seventh son, born with a veil, and gifted with second-sight in this American world, a world which yields him no true self-consciousness, but only lets him see himself through the revelation of the other world. It is a peculiar sensation, this double-consciousness, this sense of always looking at one's self through the eyes

of others, of measuring one's soul by the tape of a world that looks on in amused contempt and pity. One ever feels his two-ness,—an American, a Negro; two souls, two thoughts, two unreconciled strivings; two warring ideals in one dark body, whose dogged strength alone keeps it from being torn asunder.

Much has been said about the power and brilliance of Barack Obama's March 18, 2008, speech on race, even by some of his detractors. The focus has been on the orator's willingness to say things in public about race that are rarely spoken at all, even in private, and his expressed desire to move the country to a new and better place. There has also been attention to the immediate purpose of the speech, which was to reassure white voters that they had nothing to fear from the congregant of a fiery African-American pastor, the Rev. Jeremiah Wright.

Amid all the commentary, I have yet to see an X-ray reading of the text that would make visible the rhetorical strategies that the orator and authors used so effectively. When received in the ear, these effects breeze through us like a harmonious song. When inspected with the eye, these moves become more apparent, like reading a piece of sheet music for a difficult song and finally recognizing the chord changes.

Such analysis, while interesting in itself, might be little more than a 10 scholarly curiosity if we were not so concerned with the language issues of political discourse. The popular opinion is that our current president [George W. Bush], though plain spoken, is clumsy with language. Fair or not, this perception has produced a hope that our next president will be a more powerful communicator, a Kennedy or Reagan, perhaps, who can use language less as a way to signal ideology and more as a means to bring the disparate parts of the nation together. Journalists need to pay closer attention to political language than ever before.

Like most memorable pieces of oratory, Obama's speech sounds better than it reads. We have no way of knowing if that was true of Lincoln's Gettysburg Address, but it is certainly true of Dr. King's "I Have a Dream" speech. If you doubt this assertion, test it out. Read the speech[1] and then experience it in its original setting[2] recited by his soulful voice.

---

1. http://www.americanrhetoric.com/speeches/mlkihaveadream.htm
2. https://archive.org/details/MLKDream

* academic literacies　　● fields　　● research
■ rhetorical situations　　∴ processes　　● media/design
▲ genres　　◆ strategies

The effectiveness of Obama's speech rests upon four related rhetorical strategies:

1. The power of allusion and its patriotic associations.
2. The oratorical resonance of parallel constructions.
3. The "two-ness" of the texture, to use Du Bois's useful term.
4. His ability to include himself as a character in a narrative about race.

Allusion

Part of what made Dr. King's speech resonate, not just for black people, but for some whites, was its framing of racial equality in familiar patriotic terms: "This will be the day when all of God's children will be able to sing with new meaning, 'My country 'tis of thee, sweet land of liberty, of thee I sing. Land where my fathers died, land of the pilgrim's pride, from every mountainside, let freedom ring.'" What follows, of course, is King's great litany of iconic topography that carries listeners across the American landscape: "Let freedom ring from the snowcapped Rockies of Colorado! . . ."

In this tradition, Obama begins with "We the people, in order to form a more perfect union," a quote from the Constitution that becomes a recurring refrain linking the parts of the speech. What comes next is "Two hundred and twenty one years ago," an opening that places him in the tradition of Lincoln at Gettysburg and Dr. King at the Lincoln Memorial: "Five score years ago."

On the first page, Obama mentions the words democracy, Declaration of Independence, Philadelphia convention, 1787, the colonies, the founders, the Constitution, liberty, justice, citizenship under the law, parchment, equal, free, prosperous, and the presidency. It is not as well known as it should be that many black leaders, including Dr. King, use two different modes of discourse when addressing white vs. black audiences, an ignorance that has led to some of the hysteria over some of Rev. Wright's comments. 15

Obama's patriotic lexicon is meant to comfort white ears and soothe white fears. What keeps the speech from falling into a pandering sea of slogans is language that reveals, not the ideals, but the failures of the American experiment: "It was stained by this nation's original sin of slavery, a question that divided the colonies and brought the

convention to a stalemate until the founders chose to allow the slave trade to continue for at least twenty more years, and to leave any final resolution to future generations." And "what would be needed were Americans in successive generations who were willing to do their part . . . to narrow that gap between the promise of our ideals and the reality of their time."

Lest a dark vision of America disillusion potential voters, Obama returns to familiar evocations of national history, ideals, and language:

- "Out of many, we are truly one."
- "survived a Depression."
- "a man who served his country"
- "on a path of a more perfect union"
- "a full measure of justice"
- "the immigrant trying to feed his family"
- "where our union grows stronger"
- "a band of patriots signed that document."

Parallelism

At the risk of calling to mind the worst memories of grammar class, I invoke the wisdom that parallel constructions help authors and orators make meaning memorable. To remember how parallelism works, think of equal terms to express equal ideas. So Dr. King dreamed that one day his four children "will not be judged by the color of their skin but by the content of their character." (*By the content of their character* is parallel to *by the color of their skin*.)

Back to Obama: "This was one of the tasks we set forth at the beginning of this campaign—to continue the long march of those who came before us, a march for a more just, more equal, more free, more caring and more prosperous America." If you are counting, that's five parallel phrases among 43 words.

And there are many more: 20

". . . we may not have come from the same place, but we all want to move in the same direction."

"So when they are told to bus their children to a school across town; when they hear that an African American is getting an advantage in landing a good job or a spot in a good college

※ academic literacies   ● fields   ● research
■ rhetorical situations   ∴ processes   ● media/design
▲ genres   ◆ strategies

because of an injustice that they themselves never committed; when they're told that their fears about crime in urban neighborhoods are somehow prejudiced, resentment builds over time."

". . . embracing the burdens of our past without becoming victims of our past."

Two-ness

I could argue that Obama's speech is a meditation upon Du Bois's theory of a dual experience of race in America. There is no mention of Du Bois or two-ness, but it is all there in the texture. In fact, once you begin the search, it is remarkable how many examples of two-ness shine through:

— "through protests and struggles"
— "on the streets and in the courts"
— "through civil war and civil disobedience"
— "I am the son of a black man from Kenya and a white woman from Kansas."
— "white and black"
— "black and brown"
— "best schools . . . poorest nations"
— "too black or not black enough"
— "the doctor and the welfare mom"
— "the model student and the former gang-banger"
— "raucous laughter and sometimes bawdy humor"
— "political correctness or reverse racism"
— "your dreams do not have to come at the expense of my dreams"

Such language manages to create both tension and balance and, without being excessively messianic, permits Obama to present himself as the bridge builder, the reconciler of America's racial divide.

Autobiography

There is an obnoxious tendency among political candidates to frame their life story as a struggle against poverty or hard circumstances. As satirist Stephen Colbert once noted of presidential candidates, it is not

enough to be an average millionaire. To appeal to populist instincts it becomes de rigueur to be descended from "goat turd farmers" in France.

Without dwelling on it, Obama reminds us that his father was black and his mother white, that he came from Kenya, but she came from Kansas: "I am married to a black American who carries within her the blood of slaves and slave owners—an inheritance we pass on to our two precious daughters. I have brothers, sisters, nieces, nephews, uncles, and cousins, of every race and every hue, scattered across three continents, and for as long as I live, I will never forget that in no other country on Earth is my story even possible."

The word "story" is a revealing one, for it is always the candidate's job (as both responsibility and ploy) to describe himself or herself as a character in a story of his or her own making. In speeches, as in homilies, stories almost always carry the weight of parable, with moral lessons to be drawn.

Most memorable, of course, is the story at the end of the speech— 25 which is why it appears at the end. It is the story of Ashley Baia, a young, white Obama volunteer from South Carolina, whose family was so poor she convinced her mother that her favorite meal was a mustard and relish sandwich.

"Anyway, Ashley finishes her story and then goes around the room and asks everyone else why they're supporting the campaign. They all have different stories and reasons. Many bring up a specific issue. And finally they come to this elderly black man who's been sitting there quietly the entire time. . . . He simply says to everyone in the room, 'I am here because of Ashley.'"

During most of the 20th century, demagogues, especially in the South, gained political traction by pitting working class whites and blacks against each other. How fitting, then, that Obama's story points in the opposite direction through an old black man who feels a young white woman's pain.

Clark traces four patterns of rhetorical strategies used by President Obama in his speech: allusion, parallelism, W. E. B. Du Bois's "two-ness," and auto-biography. His analysis shows how these strategies combine to provide Obama with the opportunity to address and bring together a broad audience through memorable prose.

* academic literacies   ● fields   ● research

■ rhetorical situations   ∴ processes   ● media/design

▲ genres   ◆ strategies

Key Features / Textual Analysis

A summary or description of the text. Your readers may not know the text you are analyzing, so you need to include it or tell them about it before you analyze it. Allen's text, the Declaration of Independence, is well known, so she assumes that her readers already know its first sentences. Texts that are not so well known require a more detailed summary or description. For example, Finn provides just enough information on *Parasite's* plot that readers can see how the sets fit into it.

Attention to the context. Texts don't exist in isolation: they are influenced by and contribute to ongoing conversations, controversies, debates, and cultural trends. To fully engage a particular text, you need to understand its larger context. Clark begins by quoting a 1903 book on race in America, and Finn relates the film's characters and sets to contrasts between social classes.

A clear interpretation or judgment. Your goal in analyzing a text is to lead readers through careful examination of the text to some kind of interpretation or reasoned judgment, sometimes announced clearly in a thesis statement. When you interpret something, you explain what you think it means, as Allen does when she carefully constructs the logical argument underlying the Declaration of Independence. Clark argues that through Obama's use of four rhetorical strategies, he "present[s] himself as the bridge builder, the reconciler of America's racial divide."

Reasonable support for your conclusions. Written analysis of a text is generally supported by evidence from the text itself and sometimes from other sources as well. The writer might support their interpretation by quoting words or passages from a verbal text or referring to images in a visual text. Allen, for example, interprets the term "self-evident" by referring to formal logic, Venn diagrams, and the Catholic *Credo*. Finn offers photos of the sets and quotations from the director and set designer to support his argument. Clark provides lists of phrases from Obama's speech to support his thesis. Note that the support you offer for your interpretation need only be "reasonable"—there is never only one way to interpret something.

A GUIDE TO WRITING TEXTUAL ANALYSES

Choosing a Text to Analyze

Most of the time, you will be assigned a text or a type of text to analyze: a poem in a literature class, the work of a political philosopher in a political science class, a speech in a history or communications course, a painting or sculpture in an art class, a piece of music in a music theory course. If you must choose a text to analyze, look for one that suits the demands of the assignment—one that is neither too large or complex to analyze thoroughly (a Dickens novel or a Beethoven symphony is probably too big) nor too brief or limited to generate sufficient material (a ten-second TV news brief or a paragraph from *Hillbilly Elegy* would probably be too small). You might also choose to analyze three or four texts by examining elements common to all. Be sure you understand what the assignment asks you to do, and ask your instructor for clarification if you're not sure.

Considering the Rhetorical Situation

59–60	**PURPOSE**	Why are you analyzing this text? To demonstrate that you understand it? To show how its argument works—or doesn't? Or are you using the text as a way to make some other point?
61–64	**AUDIENCE**	Are your readers likely to know your text? How much detail will you need to supply?
72–74	**STANCE**	What interests you (or not) about your text? Why? What do you know or believe about it, and how will your own beliefs affect your analysis?
75–77	**MEDIA/DESIGN**	Will your analysis appear in print? on the web? How will your medium affect your analysis? If you are analyzing a visual text, you will probably need to include an image of it.

Generating Ideas and Text

In analyzing a written text, your goal is to understand what it says, how it works, and what it means. To do so, you may find it helpful to follow a

certain sequence: read, respond, summarize, analyze, and draw conclusions from your analysis.

Read to see what the text says. Start by reading carefully, to get a sense of what it says. This means first skimming to **PREVIEW THE TEXT**, rereading for the main ideas, then questioning and **ANNOTATING**.

13
16–17
14–15

Consider your **INITIAL RESPONSE**. Once you have a sense of what the text says, what do you think? What's your reaction to the argument, the tone, the language, the images? Do you find the text difficult? puzzling? Do you agree with what the writer says? disagree? agree *and* disagree? Your reaction to a text can color your analysis, so start by thinking about how you react—and why. Consider both your intellectual and any emotional reactions. Identify places in the text that trigger or account for those reactions. If you think that you have no particular reaction or response, try to articulate why. Whatever your response, think about what accounts for it.

Next, consolidate your understanding of the text by **SUMMARIZING** what it says in your own words. You may find it helpful to **OUTLINE** its main ideas. For instance, Allen carefully maps out the parts of the syllogism at the heart of her analysis.

550–51
340–42

Decide what you want to analyze. Having read the text carefully, think about what you find most interesting or intriguing and why. Does the argument interest you? its logic? its attempt to create an emotional response? its reliance on the writer's credibility or reputation? its use of design to achieve its aims? its context? Does the text's language, imagery, or structure intrigue you? something else? You might begin your analysis by exploring what attracted your notice.

Think about the larger context. All texts are part of larger conversations with other texts that have dealt with the same topic. An essay arguing for handgun trigger locks is part of an ongoing conversation about gun regulation, which is itself part of a conversation on individual rights and responsibilities. Academic texts include documentation in part to weave in voices from the conversation. And, in fact, anytime you're reading to learn, you're probably reading for some larger context. Whatever your reading goals, being aware of that larger context can help you better understand

what you're reading. Here are some specific aspects of the text to pay attention to:

- *Who else cares about this topic?* Especially when you're reading in order to learn about a topic, the texts you read will often reveal which people or groups are part of the conversation—and might be sources of further reading. For example, an essay describing the formation of the Grand Canyon could be of interest to geologists, environmentalists, Native Americans, travel writers, or tourists. If you're reading such an essay while doing research on the canyon, you should consider how the audience addressed determines the nature of the information provided—and its suitability as a source for your research.

- *Ideas.* Does the text refer to any concepts or ideas that give you some sense that it's part of a larger conversation? An argument on airport security measures, for example, is part of larger conversations about government response to terrorism, the limits of freedom in a democracy, and the possibilities of using technology to detect weapons and explosives, among others.

- *Terms.* Is there any terminology or specialized language that reflects the writer's allegiance to a particular group or academic discipline? If you run across words like "false consciousness," "ideology," and "hegemony," for example, you might guess the text was written by a Marxist scholar.

- *Citations.* Whom does the writer cite? Do the other writers have a particular academic specialty, belong to an identifiable intellectual school, share similar political leanings? If an article on politics cites Michael Moore and Maureen Dowd in support of its argument, you might assume the writer holds liberal opinions; if it cites Ben Shapiro and Sean Hannity, the writer is likely a conservative.

Write a brief paragraph describing the larger context surrounding the text and how that context affects your understanding of the text.

Consider what you know about the writer. What you know about the person who created a text can influence your understanding of it. Their **CREDENTIALS**, other work, reputation, stance, and beliefs are all useful windows into understanding a text. You may need to conduct an online search

525

* academic literacies ● fields ● research
■ rhetorical situations ⁚ processes ● media/design
▲ genres ◆ strategies

to find information on the writer. Then write a sentence or two summarizing what you know about the writer and how that information affects your understanding of the text.

Study how the text works. Written texts are made up of various components, including words, sentences, paragraphs, headings, lists, punctuation—and sometimes images as well. Look for patterns in the way these components are used and try to decide what those patterns reveal about the text. How do they affect its message? See the sections on **THINKING ABOUT HOW THE TEXT WORKS** and **IDENTIFYING PATTERNS** for specific guidelines on examining patterns this way. Then write a sentence or two describing the patterns you've discovered and how they contribute to what the text says.

21–23
23–25

Analyze the argument. Every text makes an argument and provides some kind of support for it. An important part of understanding any text is to recognize its argument—what the writer wants the audience to believe, feel, or do. Here are some questions you'll want to consider when you analyze an argument:

- *What is the claim?* What is the main point the writer is trying to make? Is there a clearly stated **THESIS**, or is the thesis merely implied? Is it appropriately qualified?

347–49

- *What support does the writer offer for the claim?* What **REASONS** are given to support the claim? What **EVIDENCE** backs up those reasons? Facts? Statistics? Examples? Testimonials by authorities? Anecdotes or stories? Are the reasons and evidence appropriate, plausible, and sufficient? Are you convinced by them? If not, why not?

413–14
414–22

- *How does the writer appeal to readers?* Do they appeal to your **EMOTIONS**? rely on **LOGIC**? try to establish **COMMON GROUND**? demonstrate **CREDIBILITY**?

426–27
411–22
424
423–426

- *How evenhandedly does the writer present the argument?* Is there any mention of **COUNTERARGUMENTS**? If so, how does the writer deal with them? By refuting them? By acknowledging them and responding to them reasonably? Does the writer treat other arguments respectfully? dismissively?

424–25

427–29 ◆

- *Does the writer use any logical* **FALLACIES**? Are the arguments or beliefs of others distorted or exaggerated? Is the logic faulty?

- *What authorities or other sources of outside information does the writer use?* How are they used? How credible are they? Are they in any way biased or otherwise unreliable? Are they current?

- *How does the writer address you as the reader?* Does the writer assume that readers know something about what is being discussed? Does their language include you or exclude you? (Hint: If you see the word "we," do you feel included?) Do you sense that you and the writer share any beliefs or attitudes? If the writer is not writing to you, what audience is the target? How do you know?

Then write a brief paragraph summarizing the argument the text makes and the main way the writer argues it, along with your reactions to or questions about that argument.

In analyzing a visual text, your goal is to understand its intended effect on viewers as well as its actual effect, the ways it creates that effect, and its relationship to other texts. If the visual text accompanies a written one, you need to understand how the texts work together to convey a message or make an argument.

Describe the text. Your first job is to examine the image carefully. Focus on specific details; given the increasing use of *Photoshop* and other digital image manipulation tools, you can usually assume that every detail in the image is intentional. Ask yourself these questions:

- What kind of image is it? Does it stand alone, or is it part of a group? Are there typical features of this kind of image that it includes—or lacks?

- What does the image show? What stands out? What is in the background? Are some parts of the image grouped together or connected? Are any set apart from one another?

✳ academic literacies   ● fields   ● research

■ rhetorical situations   ⁖ processes   ● media/design

▲ genres   ◆ strategies

- As you look at the image, does the content seem far away, close up, or in between? Are you level with it, looking down from above, or looking up from below? What is the effect of your viewing position?
- Does the image tell or suggest a story about what has happened or is about to happen?
- Does the image allude to or refer to anything else? For example, the Starbucks logo features the image of a Siren, a mythical being who lured sailors to their doom.

Explore your response.   Images, particularly those in advertisements, are often trying to persuade us to buy something or to feel, think, or behave a certain way. News photographs and online videos also try to evoke **EMOTIONAL** responses, from horror over murdered innocents to amusement at cute kittens. Think about your response:

◆ 426–27

- How does the image make you feel? What emotional response, if any, does the image make you feel? Sympathy? Concern? Anger? Happiness? Contentment? Something else?
- What does the image lead you to think about? What connections does it have to things in your life, in the news, in your knowledge of the world?
- Do the image and any accompanying words seem to be trying to persuade you to think or do something? Do they do so directly, such as by pointing out the virtues of a product (Buick Encore: "Sized to fit the way your life moves")? Or indirectly, by setting a tone or establishing a mood (many perfume ads, like this one for Christina Aguilera's Unforgettable: "The feminine essence of timeless fascination, sensual notes like night blooming jasmine make you unforgettable")?
- Does the **GENRE** affect your response? For example, do you expect to laugh at a comic? feel empathy with victims of a tragedy in a photo accompanying a news story? find a satirical editorial cartoon offensive?

▲ 79

Consider the context.   Like written texts, visual texts are part of larger conversations with other texts that have dealt with the same topic or used similar imagery. This editorial cartoon on climate change, for example, is part of an ongoing conversation about how quickly we must act on climate change:

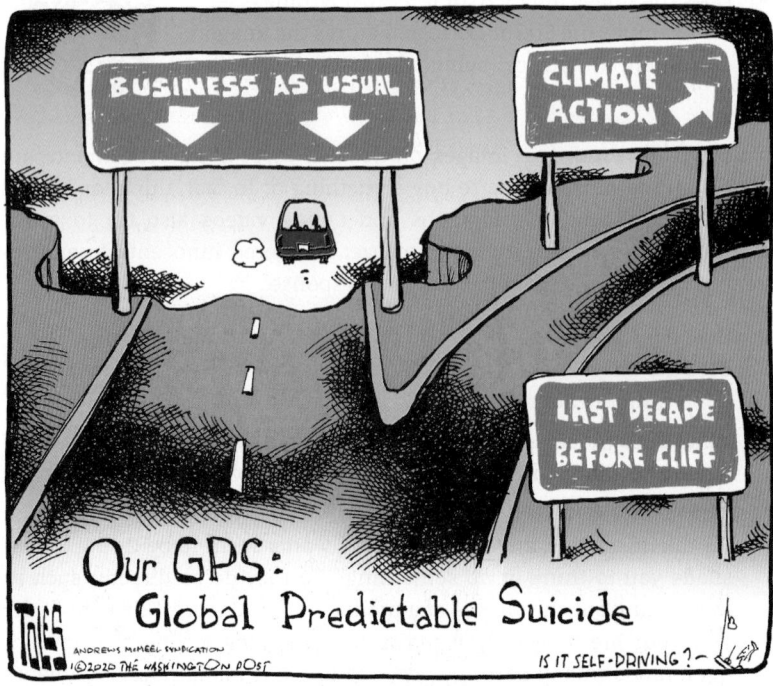

Consider what you know about the artist or sponsor.   Editorial cartoons, like the one above, are usually signed, and information about the artist and their other work is usually readily available on the web. Many commercials and advertisements, however, are created by ad agencies, so the organization or company that sponsored or posted the image should be identified and researched. How does that information affect your understanding of the text?

✳ academic literacies   ● fields   ● research
■ rhetorical situations   ∴ processes   ● media/design
▲ genres   ◆ strategies

Decide on a focus for your analysis. What do you find most interesting about the text, and why? Its details and the way they work together (or not)? The argument it makes? The way it uses images to appeal to its audience? The emotional response it evokes? The way any words and images work together to deliver a message? These are just some ways of thinking about a visual text, ones that can help you find a focus.

However you choose to focus your analysis, it should be limited in scope so that you can zero in on the details of the visual you're analyzing. Here, for example, is an excerpt from an essay from 2014 by an art historian responding to a statement made by former president Barack Obama that manufacturing skills may be worth more than a degree in art history:

> *"I promise you, folks can make a lot more potentially with skilled manufacturing or the trades than they might with an art history degree."* President Barack Obama
>
> Charged with interrogating this quote from the president, I Google "Obama art history." I click on the first result, a video from CNN, in which the quote is introduced by a gray-haired man in a dark and serious suit, standing in front of a bank of monitors in a digitally created nonspace. The camera cuts from this man to President Obama, who stands in shirtsleeves, his tie slightly loosened. His undershirt is visible through his buttondown under the intense light from what I assume is the work-day sun.
>
> Behind him is a crowd of men and women in more casual clothing, some wearing sweatshirts that have the name of a union printed across them. Their presence creates a spectrum of skin tones. Each person was clearly vetted for visual effect, as were the president's and the newscaster's costumes, the size of their flag lapel pins, the shape of the microphones they speak into, and the angle of the light on their faces. The president makes the comment in question, immediately declares his love for art history, and says that he doesn't want to get a bunch of angry emails from art historians. The crowd behind him laughs and the clip cuts off abruptly.
>
> A click away, I find a digitized copy of a handwritten note from President Obama, apologizing to an angry art history professor who emailed him to complain about his comments. The card on which the note is written is plain, undecorated save for two lines of text printed in a conservative, serif font in a shade of blue that is just on the vibrant

side of navy—THE WHITE HOUSE—and under it in smaller letters, WASHINGTON. Its tasteful, minimal aesthetic pulls double duty, meant to convey both populist efficiency (note the absence of gold gilding) and stern superiority (you know where Washington is, right?). It sets up a productive contrast with the friendliness of the president's own handwriting, particularly his looping signature, soft on the outside with a strong slash through the middle.

Like the video of the president's speech, it is a screen-scale tour de force of political imagecraft, certainly produced with the full knowledge that it would be digitized and go viral, at least among a particular demographic.

—Joel Parsons, "Richness in the Eye of the Beholder"

Parsons begins by describing the images—Obama's clothing, the people standing behind him, the letterhead on his note card, his "looping

Barack Obama speaking at a General Electric plant in Waukesha, Wisconsin, 2014.

✳ academic literacies	⬣ fields	⬤ research
■ rhetorical situations	⁘ processes	⬬ media/design
▲ genres	◆ strategies	

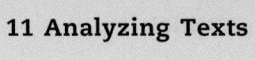

THE WHITE HOUSE

WASHINGTON

Ann —

Let me apologize for my off-the-cuff remarks. I was making a point about the jobs market, not the value of art history. As it so happens, art history was one of my favorite subjects in high school, and it has helped me take in a great deal of joy in my life that

I might otherwise have missed.

So please pass on my apology for the glib remark to the entire department, and understand that I was trying to encourage young people who may not be predisposed to a four year college experience to be open to technical training that can lead them to an honorable career.

Sincerely,

Obama's apology note to Ann Johns, art history professor at the University of Texas at Austin.

signature"—followed by an analysis of how every aspect of the video and the note card was "certainly produced with the full knowledge that it would be digitized and go viral." Notice as well that Parsons's analysis focuses more on the visual aspects of the video and note card than on what was said or written. And in a part of his essay not shown here, he notes that his analysis is grounded in "tools . . . he learned in a first-year art history course"—a not-so-subtle response to what President Obama said.

Coming Up with a Thesis

410–30 ◆
When you analyze a text, you are basically **ARGUING** that the text should be read or seen in a certain way. Once you've studied the text thoroughly, you need to identify your analytical goal: Do you want to show that the text has a certain meaning? uses certain techniques to achieve its purposes? tries to influence its audience in particular ways? relates to some larger context in some significant manner? should be taken seriously—or not?
347–49 ⦂
something else? Come up with a tentative **THESIS** to guide your thinking and analyzing—but be aware that your thesis may change as you continue to work.

Ways of Organizing a Textual Analysis

Examine the information you have to see how it supports or complicates your thesis. Look for clusters of related information that you can use to
340–42 ⦂
structure an **OUTLINE**. Your analysis might be structured in at least two ways. You might, as Clark does, discuss patterns, elements, or themes that run through the text. Alternatively, you might analyze each text or section of text separately, as Allen does. Following are graphic representations of some ways of organizing a textual analysis:

[Thematically]

Introduce your analysis:
- Summarize or describe the text.
- Provide needed context.
- Outline your method or focus.

State your thesis.

→

Analyze the text:
- Identify a theme or pattern.
- Use examples from the text and appropriate contexts as evidence.

Continue as needed.

→

Conclude: Explain what your analysis reveals about the text.

[Part by part, or text by text]

Introduce your analysis:
- Summarize or describe the text.
- Provide needed context.
- Outline your method or focus.

State your thesis.

→ Analyze first section of text.

↓

Analyze next section of text.

↓

Continue as needed.

→

Conclude: Explain what your analysis reveals about the text.

[Spatially, as the text is likely to be experienced by viewers]

Introduce your analysis:
- Summarize or describe the text.
- Provide needed context.
- Outline your method or focus.

State your thesis.

→ Analyze the most prominent part of the text or the part viewers see first.

↓

Continue the analysis in the order the text is experienced:
- top to bottom
- left to right
- center to edges
- foreground to background
- what becomes visible as text is scrolled, video is played, or images are seen in succession

→

Conclude: Explain what your analysis reveals about the text.

Writing Out a Draft

In drafting your analysis, your goal should be to integrate the various parts into a smoothly flowing, logically organized essay. However, it's easy to get bogged down in the details. Consider writing one section of the analysis first, then another and another, until you've drafted the entire middle; then draft your beginning and ending. Alternatively, start by summarizing the text and moving from there to your analysis and then to your ending. However you do it, you need to support your analysis with evidence: from the text itself (as Finn's analysis of the "text" of a film does), or from **RESEARCH** on the larger context of the text (as Allen does), or by comparing the text you are analyzing to another text (as Clark does).

489

346–54

Draft a BEGINNING. The beginning of an essay that analyzes a text generally has several tasks: to introduce or summarize the text for your readers, to offer any necessary information on the larger context, and to present your thesis.

550–51
456–63

- *Summarize or describe the text.* If the text is one your readers don't know, you need to **SUMMARIZE** or **DESCRIBE** it early on to show that you understand it fully. For example, as Finn describes the Kim family apartment and the Parks' house.

- *Show the text.* If you're analyzing a visual text online, consider starting off with an image, a video, or a link to it or something similar, as Finn does by including several illustrations of the sets and photos of scenes from the film.

- *Provide a context for your analysis.* If there is a larger context that is significant for your analysis, you might mention it in your introduction. Allen does this by comparing the Declaration's statement about self-evident truths to the statements of belief in the Apostles' Creed of the Catholic Church.

347–49

- *State your* **THESIS**. Finn begins by asserting that in *Parasite* "the built environment indicates where people fit into the social hierarchy—or if there is a place for them at all." Clark promises to analyze "the rhetorical strategies that the orator and authors used so effectively."

academic literacies   fields   research
rhetorical situations   processes   media/design
genres   strategies

Draft an ENDING. Think about what you want your readers to take away from your analysis, and end by getting them to focus on those thoughts.

354–58

- *Restate your thesis—and say why it matters*. Finn, for example, begins by asserting that *Parasite's* sets show "where people fit into the social hierarchy" and ends by discussing the dangers of "class tensions" and inequality.

- *Explain what your analysis reveals.* Your analysis should tell your readers something about the way the text works or about what it means or says. Allen, for example, concludes by noting that "the Declaration's truths fit together almost like the pieces of a mathematical equation; we intuitively feel the puzzle pieces snap together. That is how self-evidence should feel."

Come up with a TITLE. A good title indicates something about the subject of your analysis—and makes readers want to see what you have to say about it. Finn's title provides a preview of his thesis that the architecture seen in *Parasite* reflects the social inequality at the core of the film, while Clark's title straightforwardly announces his topic.

363

Consider REWRITING.   If you have time and want to explore alternatives, you might try rewriting your draft to see if a different plan or approach might work better.

378–79

Considering Matters of Design

- If you cite written text as evidence, be sure to set long quotations and **DOCUMENTATION** according to the style you're using.

560–63

- If your essay is lengthy, consider whether **HEADINGS** would make your analysis easier for readers to follow.

670–72

- If you're analyzing a visual text, include a copy of the image and a caption identifying it.

- If you're submitting your essay electronically, provide links to whatever text you are analyzing.

- If you're analyzing an image or a screen shot, consider annotating elements of it right on the image.

Getting Response and Revising

372–74

The following questions can help you and others study your draft with a critical eye. Make sure that anyone you ask to read and **RESPOND** to your text knows your purpose and audience.

- Is the beginning effective? Does it make a reader want to continue?
- Does the introduction provide an overview of your analysis? Is your thesis clear?
- Is the text described or summarized clearly and sufficiently?
- Is the analysis well organized and easy to follow? Do the parts fit together coherently? Does it read like an essay rather than a collection of separate bits of analysis?
- Does each part of the analysis relate to and support the thesis?
- Is anything confusing or in need of more explanation?
- Are all quotations accurate and correctly documented?
- Is it clear how the analysis leads to the interpretation? Is there adequate evidence to support the interpretation?
- Does the ending make clear what your analysis shows?

375–77

Then it's time to **REVISE**. Make sure your text appeals to your audience and think hard about whether it will achieve your purpose.

Editing and Proofreading

Once you've revised your draft, edit carefully:

347–49

- Is your **THESIS** clearly stated?

542–54
560–63

- Check all **QUOTATIONS**, **PARAPHRASES**, and **SUMMARIES** for accuracy and form. Be sure that each has the required **DOCUMENTATION**.

361–62

- Make sure that your analysis flows clearly from one point to the next and that you use **TRANSITIONS** to help readers move through your text.

383–84

- **PROOFREAD** your finished analysis carefully before turning it in.

---

* academic literacies   ● fields   ● research
■ rhetorical situations   ⁘ processes   ● media/design
▲ genres   ◆ strategies

Taking Stock of Your Work

Take stock of what you've written and learned by considering these questions:

- What reading strategies did you use as you analyzed your text—annotating? coding? something else? Which strategies were most helpful?
- What writing strategies did you use to analyze your text—defining? comparing? others? Which ones helped you the most?
- Did you provide enough evidence to support your analysis? Can you point to a place in your text where you did an especially good job of providing evidence?
- Did you include any visuals, and if so, what did they add? Could you have shown the same thing with words?
- What did you learn from providing feedback on your peers' work? How did that learning affect your own writing?
- What kinds of analytical writing, if any, have you done in the past? How does your current work compare?
- Was anything new to you about this kind of writing? If so, how did you adapt to the task?
- Are you pleased with your analysis? What did it teach you about the text you analyzed? If you were to write another text analysis, what would you do differently?

If you need more help

See Chapter 30 for guidelines on **DRAFTING**, Chapter 31 on **ASSESSING YOUR OWN WRITING**, Chapter 32 on **GETTING RESPONSE AND REVISING**, and Chapter 33 on **EDITING AND PROOFREADING**. See Chapter 34 if you are required to submit your analysis in a writing **PORTFOLIO**. See Chapter 57 for help using **VISUALS**.

364–66
367–71
372–79
380–84
385–90
673–83

12 Reporting Information

Many kinds of writing report information. Newspapers report on local and world events; textbooks give information about biology, history, writing; websites provide information about products (jcrew.com), people (billieeilish .com), institutions (smithsonian.org). We write out a lot of information ourselves, from a note we post on our door saying we've gone to choir practice to a text we send to tell a friend where to meet us for dinner and how to get there.

College assignments often call for reporting information as well. In a *history* class, you may be assigned to report what you've learned about the state of US relations with Japan just before the bombing of Pearl Harbor. A *biology* course may require you to report the effects of an experiment in which plants are deprived of sunlight for different periods of time. In a *nursing* class, you may have to report the changes in a patient's symptoms after the administration of a particular drug.

This chapter focuses on reports that are written to inform readers about a particular topic. Very often this kind of writing calls for some kind of research: you need to know your subject in order to report on it! When you write to report information, you are the expert. We'll begin with three good examples, the first annotated to show the key features found in most reports.

RENAE TINGLING

Sleepless Nights of a University Student

In the following report, Drexel University student Renae Tingling explains how inadequate sleep can affect college students' performance. Her essay, which is documented in MLA style, originally appeared in The 33rd, *an anthology of student writing published by Drexel Publishing Group.*

☀ academic literacies   ● fields   ● research
■ rhetorical situations   ⁚ processes   ● media/design
▲ genres   ◆ strategies

When can we say we've had "enough" sleep? Does the amount of sleep someone gets in a night make a difference? Is getting enough sleep even worth it? According to Newton's first law of motion, a body in motion will stay in motion unless acted upon by an opposing external force. The human body experiences this opposing force in the form of tiredness, slowly approaching as our day comes to an end. For most college students, every day is jam-packed with an infinite number of responsibilities, ranging from exam preparation and homework assignments to managing student organizations and completing chores. Many students find themselves trapped in a cycle of work and activities with rest on the back burner.

Introduces topic.

As a first-year student, I took on a list of new and exciting ventures during my first week. As the weeks progressed, the announcement of midterms marked the end of the adjustment period. I found myself adding more tasks to my to-do list than I was marking off, while time ran shorter by the day. Surrendering a portion of my sleeping hours to my work was unavoidable. Even though I felt satisfied to have completed my assignments, the lack of sleep did not go unnoticed by my body. My body's demand for compensation resulted in my waking up at some ungodly hour the following Saturday afternoon. This hadn't been common practice for me before college, making me wonder if the same could be said for other students. Is this something that most college students perpetually struggle with? If so, why? How does this kind of schedule affect their lives as students, if at all? It's best if we start by figuring out what the big deal is about sleep.

Offers a personal connection to the topic.

Tingling frames her thesis as questions her report will answer.

Sleep is indeed a big deal. It plays a crucial role in an individual's overall health and wellness. A good and consistent quantity and quality of sleep can help with the regulation of physiological processes (Kabrita and Hajjar-Muça 190), such as cardiovascular activity and glucose regulation (Cheng et al. 270). For students, sleep is in some ways like our very own pot of gold. It is essential for effective processing of information (Kabrita and Hajjar-Muça 190), memorization, and proper immune function (Carter et al. 315). As David Dinges, a professor of psychology and chief of the sleep and chronobiology division in the department of psychiatry at the University of Pennsylvania School of Medicine, notes:

Accurate, well-researched information.

Synthesis of information from several sources.

> We know, for example, that sleep is critical for waking cognition—
> that is, for the ability to think clearly, to be vigilant and alert, and

sustain attention. We also know that memories are consolidated during sleep, and that sleep serves a key role in emotional regulation. (qtd. in Worley 758)

So, what exactly is a good quality of sleep? Some would argue that it's almost entirely subjective. Some students find taking short and frequent naps is an effective way of getting all their hours in, while others couldn't disagree more. According to Dinges, for the average person, the target should be "ideally somewhere between seven to seven and a half hours" of uninterrupted sleep (qtd. in Worley 759). The National Sleep Foundation recommends that young adults get a total of seven to nine hours of sleep with no fewer than six and no more than ten. Anything below six hours and you've reached the danger zone (Suni). Sleep deficiency can have a severe and lasting impact on a student's academic performance. Poor sleep quality can also greatly impact our decision-making process (Pilcher et al.).

Sleep processes explained.

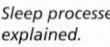

Ask any university student and they would probably agree that sleep is a privilege. Countless factors at play during a student's academic journey can heavily disrupt their sleep schedule. Habitual procrastination is a major cause of increased stress among university students (Sirois), which may result in health issues such as insomnia (difficulty falling and/or staying asleep). Drinking too much coffee can be a problem. A study conducted by two Turkish dieticians finds that though consumption of most caffeinated beverages had no correlation to poor quality of sleep, coffee was the exception: students with poor sleep quality consumed a notably larger amount of coffee than those with good sleep quality (Suna and Ayaz S132). Prolonged internet usage and cigarette smoking are other behavioral factors that can eat away at students' precious hours of sleep. This brings into question those factors that exist beyond the control of the student—and to determine those factors, who better to ask than the students themselves?

Causes of sleep difficulties outlined.

5

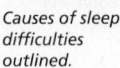

A total of thirty-three students participated in a survey I conducted, ranging from freshman to pre-junior year. The questionnaire collected information pertaining to participants' class level, sleep patterns, and factors that may disrupt their sleep schedule and how those factors affect them academically. Most students (eighty percent) reported sleeping for the recommended number of hours (seven to nine) either a few times or not at all. Behavioral influences, like prolonged internet usage and late-night studying for upcoming exams, were common reasons for

Tingling includes her own research study to give another perspective on the topic.

※ academic literacies   ● fields   ● research

■ rhetorical situations   ∴ processes   ● media/design

▲ genres   ◆ strategies

losing sleep. As for external influences, many students reported that disruptive roommates and outside noise hindered them from sleeping. Some students also pointed out that trying to complete assignments before their fast-approaching deadlines caused heavy stress and prevented them from getting more hours of sleep. Very few placed the blame on social activities. With habitual and external factors such as these, it is understandable that low quality of sleep can lead to poor academic performance (Carter et al. 315). Consequently, many students found it difficult to be attentive and involved during lectures (eighty-seven percent), successfully complete an exam (fifty-six percent), and maintain their involvement in student organizations (sixty-five percent) after a night of little or no sleep. Approximately half of the participants reported that they rely or have relied on medication and/or caffeinated beverages to feel alert during the day.

At this point I decided to try an experiment: for three days I would only allow myself half my normal hours of sleep to see how my mind and body would react. About halfway through my experiment, I didn't mind the thought of having a cup of coffee myself.

The author reports on a third kind of research, a self-study.

I'll admit, getting out of bed at 5:00 a.m. was a bit of a challenge, especially after sleeping for a total of four and a half hours. It wasn't by chance that by the start of the first day, I had already reached my quota of embarrassing absentminded mistakes after having pressed the wrong button on the elevator and wandered into the wrong classroom. This unusual behavior was due to the effect that lack of sleep had on my level of self-control and alertness (Pilcher et al.). This can greatly throw off a person's decision-making process—as it did mine.

By the second day, getting half a night's worth of sleep took a greater toll on me. Most of the day's lectures managed to slip right past me. Too often to count, I found myself opening eyes I hadn't even realized were closed. What little attention span I had left went to counting down the minutes until the end of each class session. Now, we can all agree that conducting an experiment like this during the exam period would not be the wisest thing to do. Knowing this, I was prepared to add this to my list of mistakes. Funnily enough, my lack of sleep did not have its way with the quiz I took that day. In that moment, the effects of my sleep deprivation seemingly vanished. It was as though my brain ensured that I was able to place all my efforts into not failing math. Sadly, those effects came back sometime afterward and with a vengeance. I became so demotivated to do just about anything

She clearly describes the process of sleep deprivation as she experienced it.

afterward that even having dinner with friends seemed like too much. I was completely against the idea of preparing for upcoming lectures and exams and used how tired I felt as an excuse. This response drives home the notion that sleep deprivation can have serious psychological impacts and weaken our ability to concentrate on academic material (Kabrita and Hajjar-Muça 190).

Oddly enough, I felt a lot less tired on the third day. Though I [10] still felt somewhat tired, it was as if my body was getting used to the lack of sleep. At the end of the day my experiment came to a close, allowing me a well-deserved night's rest. What took me by surprise was how exhausted I felt the following morning after a full eight hours of rest. I initially expected to feel rested, having given myself the chance to get the recommended amount of sleep. But I felt as I had on the first morning of the experiment, even though I had double the amount of sleep. Thankfully, a few days of seven to eight hours of sleep was enough to get me back on track. I was able to focus on discussions during lectures and was more motivated to attend club meetings and study with friends for whatever the next exam would be. It was surprising to discover just how greatly each day was affected by the amount of sleep I got several nights before. It wasn't too hard for me to imagine the damaging effects that poor sleep habits would have over time.

Sleep is a beautiful thing; my experiment is a testament to this. Many of us see it as the easiest and most convenient thing to sacrifice on our journey to academic success. How ironic is it that, as we continue to undermine its significance, sleep can potentially determine whether we achieve that success. It should be mentioned that students often find themselves in situations where valuable hours of sleep are stolen from them by circumstances beyond their control. Whether deliberate or unintentional, lost sleep can take a lot more from us than we bargained for. If I had to summarize what I took from my research, it would be that we as students have fallen for the misconception that less sleep is the only way to stay on top of everything. As a result, we tend to make everything a priority except ourselves. We place getting those extra hours of studying above resting the brain that has to take the exam the following day. At the end of the day, the benefits of a good night's rest are limitless, so ask yourself this: Do I get enough sleep?

Conclusion: A summary of the research and its implication for Tingling's audience, college students.

* academic literacies   ● fields   ● research
■ rhetorical situations   ∴ processes   ◖ media/design
▲ genres   ◆ strategies

Works Cited

Carter, Briana, et al. "An Analysis of the Sleep Quality of Under-graduate Students." *College Student Journal*, vol. 50, no. 3, 2017, pp. 315-22. *Gale Academic Onefile*, go.galegroup.com.

Cheng, Shu Hui, et al. "A Study on the Sleep Quality of Incoming University Students." *Psychiatry Research*, vol. 197, Feb. 2012, pp. 270–74.

Kabrita, Colette S., and Theresa A. Hajjar-Muça. "Sex-Specific Sleep Patterns among University Students in Lebanon: Impact on Depression and Academic Performance." *Nature and Science of Sleep*, vol. 8, 2016, pp. 189–96. *DovePress*, https://doi.org/10.2147 /NSS.S104383.

Pilcher, June J., et al. "Interactions between Sleep Habits and Self-Control." *Frontiers in Human Neuroscience*, vol. 9, 11 May 2015, www .frontiersin.org/articles/10.3389/fnhum.2015.00284/full.

Sirois, F. M., et al. "Is Procrastination Related to Sleep Quality? Testing an Application of the Procrastination–Health Model." *Cogent Psychology*, vol. 2, 2015. *Taylor and Francis Online*, https://doi.org/10.1080/23311908.2015.1074776.

Suna, G., and A. Ayaz. "Relationship between Caffeinated Beverages and Sleep Quality in University Students." *Clinical Nutrition*, vol. 37, Sept. 2018, pp. S131–S132. *ScienceDirect*, https://doi .org/10.1016/j.clnu.2018.06.1492. Supplement 1.

Suni, Eric. "How Much Sleep Do We Really Need?" *SleepFoundation .org*, 10 Mar. 2021, www.sleepfoundation.org/how-sleep-works /how-much-sleep-do-we-really-need.

Worley, Susan L. "The Extraordinary Importance of Sleep: The Detrimental Effects of Inadequate Sleep on Health and Public Safety Drive an Explosion of Sleep Research." *Pharmacy and Therapeutics: A Peer-Reviewed Journal for Managed Care and Hospital Formulary Management*, vol. 43, no. 12, 2018, pp. 758–63. *PubMedCentral*, https://pubmed.ncbi.nlm.nih.gov/30559589.

Tingling's report draws on published research, her own survey research, and informal research using personal experience to provide a multifaceted view of the effects of inadequate sleep on college students' performance.

FRANKIE SCHEMBRI
Edible Magic

In the following report, written for a first-year writing class at Massachusetts Institute of Technology, Frankie Schembri explains the science behind how popcorn pops. Originally published in Angles 2016: Selected Essays from Introductory Writing Subjects at MIT, *this report is documented in APA style.*

Life is studded with little pockets of magic. These are the moments with mysterious emergent qualities, when the whole is greater than the sum of the parts, when one plus one somehow equals three. Such magic is even better when it comes in edible form.

When an unassuming little kernel of corn meets hot oil, it is transformed. It is elevated with an unmistakable "pop!" into a fluffy cloud of goodness ready to be dressed with butter, caramel, or whatever your heart desires.

This magic is called popcorn.

Popcorn exploded in popularity in the early 20th century but surprisingly only made it into movie theaters with the advent of sound films, or "talkies," in 1927. Silent films catered to a smaller, more exclusive clientele and owners worried that the sound of the snack being munched would detract from the experience (Geiling, 2013).

By the 1940s, popcorn had become an inextricable part of going   5
to the movies, and from then on, over half of the popcorn consumed yearly in America was eaten

at movie theaters (Geiling, 2013). Fast-forward some 70 years and the relationship remains unbreakable. Step into any movie theater lobby across America and your senses are bombarded with the unmistakable scent of salt and butter.

Popcorn snuck into the American household in the

* academic literacies   ● fields   ● research

■ rhetorical situations   ⁘ processes   ● media/design

▲ genres   ◆ strategies

1960s with Jiffy Pop, a self-contained stovetop popper including kernels, oil, and even the pan, and it flourished with the popularity of microwave ovens in the 1970s (Smith, 1999, p. 124). Popcorn cultivated an important relationship with microwaves in the latter half of the 20th century, important enough that popcorn was eventually given the rare honor of its own designated microwave button (Smith, 1999, p. 127).

"But how?" you might ask. How do these ordinary kernels magically spring to life with a little heat and oil, enchanting kids and grown-ups alike?

Like most acts of magic, popcorn's "pop" can be understood with a little science.

Botanically speaking, popcorn is a type of maize, the only domesticated subgroup in the genus *Zea*, a group of plants in the grass family. The different types of maize are classified based on their kernel's size, shape, and composition (Smith, 1999, p. 6).

Corn kernels have three main structural components. First, there is the germ (from the Latin *germen*, meaning seed or sprout), a small pocket of genetic material that is essentially a baby corn plant waiting to grow. The germ is surrounded by the endosperm (the Greek *endon*, within, and *sperma*, seed), a larger parcel of water mixed with soft and hard starch granules that make up most of the kernel's weight and would provide food for the corn plant if it were to sprout. Finally, the germ-endosperm complex is surrounded by the pericarp (from the Greek *peri*, around, and *karpos*, fruit), the hard shell that winds up stuck between your molars after you enjoy a bag of popcorn (Ghose, 2015).

Popcorn kernels are unique in that they are relatively small, they have endosperms containing a larger number of hard starch granules, and their pericarps are hard and impermeable, essentially sealing off the contents of the kernel from the outside environment. These characteristics have endowed popcorn kernels with the ability to pop (Ghose, 2015).

In an attempt to better understand the physics of popcorn, aeronautical engineer Emmanuel Virot and physicist Alexandre Ponomarenko, who seem to have also fallen under popcorn's spell (judging by the language of their published report), experimented with the temperature at which a kernel pops. As Virot and Ponomarenko (2015) explained, when heat is applied to the kernels via hot oil on a stovetop or in a microwave, the temperature of the kernels begins to rise accordingly. Most

affected by this increase in heat is the water stored between the starch granules in the endosperm. Much like a bubbling pot of water brought to a boil, the water in the kernel begins to change from liquid into gas.

While liquid water is content to stay put, gaseous water in the form of steam craves space to move, but the hard shell of the pericarp effectively keeps the water trapped inside the kernel. As a result, the kernel acts like a tiny steamer and the starch granules inside are cooked into a gooey mass. As the mass gets hotter and hotter, the steam presses harder and harder on the inside surface of the kernel's shell like the hands of a million tiny creatures trapped inside a bubble (Virot & Ponomarenko, 2015).

The tension is palpable. The kernel begins to shake with anticipation. It rocks back and forth, back and forth, faster and faster, and faster still. Then finally . . . pop! The bubble bursts, the lid flies off the pot, and fireworks explode as the hard shell of the pericarp cracks and the steam breaks free from its kernel prison. The starchy goop also bubbles out into the world, where it meets cold, fresh air and rapidly hardens into spongy cloudlike shapes. Just like that, in just one-fifteenth of a second, a new piece of popcorn is born (Virot & Ponomarenko, 2015).

Virot and Ponomarenko (2015) determined that the temperature at which kernels typically pop is 180 degrees Celsius. The pair also determined that the resulting popped kernel can be up to 40 times its unpopped volume, although usually the kernel's radius merely doubles.

But what propels the kernel, with what Virot and Ponomarenko (2015) called "all the grace of a seasoned gymnast," into the air as it pops? When the pericarp fractures, it does so in only one place first, giving some of the steam and starchy goop a head start on escaping. The starch released first extends to create a leg of sorts, off of which the rest of the kernel springboards, launching it somersaulting into the air like an Olympic gymnast. Popcorn jumps typically only reach a height of a few centimeters, but still manage to create endless entertainment for the hungry viewer.

Where does the "pop" sound come from? Arguably the most important part of the whole experience, popcorn's characteristic noise is not, contrary to popular belief, the sound of the kernel's shell breaking open. The popping sound is created by the release of trapped water vapor resonating in the kernel, similar to how, when removed, a champagne cork makes a popping sound that resonates in the glass bottle (Virot & Ponomarenko, 2015).

※ academic literacies   ● fields   ● research
■ rhetorical situations   ⦂ processes   ◖ media/design
▲ genres   ◆ strategies

Making popcorn hardly seems like an opportunity to learn about physics, but the kernels' unique transformation illustrates some principles of thermodynamics, biomechanics, and acoustics, as the properties of different materials dictate how they respond to pressure and heat.

Life is studded with little pockets of magic. From the enticing smell during a night at the movies to the rising staccato sound of a bag coming to life in your microwave, popcorn is magic-meets-science in its most delicious form. And it always leaves you hungry for more.

References

Geiling, N. (2013, October 3). Why do we eat popcorn at the movies? *Smithsonianmag.com.* https://www.smithsonianmag.com/arts-culture/why-do-we-eat-popcorn-at-the-movies-475063/

Ghose, T. (2015, February 10). *The secret acrobatics of popcorn revealed.* Live Science. https://www.livescience.com/49768-mechanics-of-popcorn.html

Smith, A. (1999). *Popped culture: A social history of popcorn in America.* University of South Carolina Press.

Virot, E., & Ponomarenko, A. (2015). Popcorn: Critical temperature, jump and sound. *Journal of the Royal Society Interface, 12*(104). https://doi.org/10.1098/rsif.2014.1247

Schembri introduces her subject by placing it in a cultural context of the movies, Jiffy Pop, and microwave popcorn. She then distinguishes popcorn from other varieties of corn or maize and goes on to describe what happens when popcorn kernels are heated, pop, and fly into the air. To explore her subject from these various perspectives, she draws on scientific, historical, and popular sources.

JON MARCUS
The Reason College Costs More Than You Think

Writing online for Time in 2014, Hechinger Report editor Jon Marcus examines the length of time students take to graduate and how that affects the cost of getting a degree.

When Alex Nichols started as a freshman at the University of Mississippi, he felt sure he'd earn his bachelor's degree in four years. Five years later, and Nichols is back on the Oxford, Mississippi, campus for what he hopes is truly his final semester.

"There are a lot more students staying another semester or another year than I thought there would be when I got here," Nichols says. "I meet people once a week who say, 'Yes, I'm a second-year senior,' or, 'I've been here for five years.'"

They're likely as surprised as Nichols still to be toiling away in school.

Nearly nine out of 10 freshmen think they'll earn their bachelor's degrees within the traditional four years, according to a nationwide survey conducted by the Higher Education Research Institute at UCLA. But the U.S. Department of Education reports that fewer than half that many actually will. And about 45 percent won't have finished even after six years.

That means the annual cost of college, a source of so much anxi-   5
ety for families and students, often overlooks the enormous additional expense of the extra time it will actually take to graduate.

"It's a huge inconvenience," says Nichols, whose college career has been prolonged for the common reason that he changed majors and

The Lyceum, the oldest building at the University of Mississippi.

* academic literacies   ● fields   ● research
■ rhetorical situations   ∴ processes   ● media/design
▲ genres   ◆ strategies

took courses he ended up not needing. His athletic scholarship—Nichols was a middle-distance runner on the cross-country team—ran out after four years. "I had to get some financial help from my parents."

The average added cost of just one extra year at a four-year public university is $63,718 in tuition, fees, books, and living expenses, plus lost wages each of those many students could have been earning had they finished on time, according to the advocacy group Complete College America.

A separate report by the Los Angeles–based Campaign for College Opportunity finds that the average student at a California State University campus who takes six years instead of four to earn a bachelor's degree will spend an additional $58,000 and earn $52,900 less over their lifetime than a student who graduates on time, for a total loss of $110,900.

"The cost of college isn't just what students and their families pay in tuition or fees," says Michele Siqueiros, the organization's executive director. "It's also about time. That's the hidden cost of a college education."

So hidden that most families still unknowingly plan on four years 10 for a bachelor's degree, says Sylvia Hurtado, director of the Higher Education Research Institute at UCLA.

Although the institute does not poll parents in its annual survey, "that high percentage of freshmen [who are confident they'll finish in four years] is probably reflecting their parents' expectation—'This is costing me a lot, so you're going to be out in four years.' So the students think, 'Sure, why not?' I don't think the parents even initially entertain or plan for six years or some possible outcome like that."

Yet many students almost immediately doom themselves to taking longer, since they register for fewer courses than they need to stay on track. Surveys of incoming freshmen in California and Indiana who said they expected to graduate in four years found that half signed up for fewer courses than they'd needed to meet that goal, according to a new report by the higher-education consulting firm HCM Strategists.

It's not entirely the students' fault.

More than half of community-college students are slowed down by having to retake subjects such as math and reading that they should have learned in high school, says Complete College America. And at some schools, budget cuts have made it difficult to register for the

courses students do need to take. Two-thirds of students at one California State University campus weren't able to get into their required courses, according to a 2010 study by the University of California's Civil Rights Project.

Most state financial-aid programs, meanwhile, cover only four years. "They do not fund a fifth or sixth year," says Stan Jones, president of Complete College America and a former Indiana commissioner of higher education. "And by that time the parents' resources and the students' resources have run out. So that fifth year is where you borrow." 15

Students at the most elite colleges and universities tend not to have this problem, which means that schools with some of the highest annual tuition can turn out to be relative bargains. These schools "would have a revolt if their students had to go a fifth year," Jones says. "But that recognition has really not hit the public sector yet, about the hidden cost of that extra year."

Policymakers urge speeding students through remedial classes more quickly, adding more sections of required courses so students can get in when they need them, and encouraging students to take 15 credits per semester instead of the typical 12.

Change won't come soon enough for Nichols, who is determined that it won't take more than one extra semester to finish his degree in integrated marketing communications.

"That's time you're wasting," he says, "that you could be out making money."

Marcus combines information from various research institutes, advocacy groups, surveys, and academic sources to support his argument. His statistics are given a human face by quotations from a student who is taking longer to graduate than he expected.

Key Features / Reports

A tightly focused topic.   The goal of this kind of writing is to inform readers about something without digressing—and without, in general, bringing in the writer's own opinions. All three examples focus on a particular topic—sleep deprivation, popping popcorn, and the cost of college—and present information about the topics evenhandedly.

* academic literacies
◼ rhetorical situations
▲ genres

⬣ fields
⁙ processes
◆ strategies

⬤ research
⬤ media/design

Accurate, well-researched information. Reports usually require research. The kind of research depends on the topic. Sometimes internet research will suffice, though reports done for college courses may require library research to locate scholarly sources—Tingling, for example, uses various sources available through her university library's databases, and Schembri uses a mix of scholarly and popular sources. Other topics may require or benefit from field research—interviews, observations, surveys, and so on. In addition to doing library and online research, for example, Marcus interviewed a student, and Tingling conducted a survey of students.

A synthesis of ideas. Reports seldom rely on a single source of information. Rather, they draw on several sources, making connections among the facts and ideas found in them. For example, Schembri combines information from a magazine article and a book to provide a brief history of the growth of popcorn's popularity in the United States. Marcus compares undergraduate students' expectations of finishing college in four years with statistics showing that more than half will take longer, using information from a university study and a government report as well as an interview with a student.

Various writing strategies. Presenting information usually requires various organizing patterns—defining, comparing, classifying, explaining processes, analyzing causes and effects, and so on. Schembri explains the process governing popcorn popping and classifies different kinds of maize. Marcus analyzes the financial effects of delaying graduation, and Tingling traces the effects of inadequate sleep on college students' performance.

Clear definitions. Reports need to provide clear definitions of any key terms that their audience may not know. Schembri defines several components of corn kernels, including the germ, endosperm, and pericarp. Tingling includes a definition of "waking cognition."

Appropriate design. Reports often combine paragraphs with information presented in lists, tables, diagrams, and other illustrations. When you're

presenting information, you need to think carefully about how to design it —numerical data, for instance, can be easier to understand and remember in a table than in a paragraph. Often a photograph can bring a subject to life, as does the photo on page 146, which accompanies "Edible Magic." Online reports offer the possibility of video and audio clips as well as links to source materials and more detailed information.

A GUIDE TO WRITING REPORTS

Choosing a Topic

Whether you get to choose your topic or are working with an assigned one, see if you can approach the topic from an angle that interests you.

If you get to choose.   What interests you? What do you wish you knew more about? The possible topics for informational reports are limitless, but the topics that you're most likely to write well on are those that engage you. They may be academic in nature or reflect your personal interests or both. If you're not sure where to begin, here are some places to start:

- an intriguing technology: driverless cars, touchscreens, tooth whiteners
- sports: soccer, snowboarding, ultimate Frisbee, basketball
- an important world event: the Arab Spring, the fall of Rome, the COVID-19 pandemic
- a historical period: the African diaspora, the Middle Ages, the Ming dynasty, the Great Depression
- a common object: hoodies, gel pens, mascara, Post-it notes
- a significant environmental issue: melting Arctic ice, deer overpopulation, mercury and the fish supply
- the arts: rap, outsider art, the Crystal Bridges Museum of American Art, Savion Glover, Mary Cassatt

334–35 ⁙   **LIST** a few possibilities, and then choose one that you'd like to know more about—and that your audience might find interesting, too. You might start

---

⁙ academic literacies   ● fields   ● research
■ rhetorical situations   ⁙ processes   ● media/design
▲ genres   ◆ strategies

out by phrasing your topic as a question that your research will attempt to answer. For example:

How is *Google* different from *Yahoo!*?

How was the Great Pyramid constructed?

What kind of training do football referees receive?

If your topic is assigned. If your assignment is broad—"Explain some aspect of the US government"—try focusing on a more limited topic within the larger topic: federalism, majority rule, political parties, states' rights. Even if an assignment seems to offer little flexibility—"Explain the physics of roller coasters"—your task is to decide how to research the topic, and sometimes even narrow topics can be shaped to fit your own interests and those of your audience.

Considering the Rhetorical Situation

PURPOSE	Why are you presenting this information? To teach readers about the subject? To demonstrate your research and writing skills? For some other reason?	■ 59–60
AUDIENCE	Who will read this report? What do they already know about the topic? What background information do they need in order to understand it? Will you need to define any terms? What do they want or need to know about the topic? Why should they care about it? How can you attract their interest?	■ 61–64
STANCE	What is your own attitude toward your subject? What interests you most about it? What about it seems important?	■ 72–74
MEDIA/DESIGN	What medium are you using? What is the best way to present the information? Will it all be in paragraph form, or is there information that is best presented as a chart, table, or infographic? Do you need headings? Would diagrams, photographs, or other illustrations help you explain the information?	■ 75–77

Generating Ideas and Text

Good reports share certain features that make them useful and interesting to readers. Remember that your goal is to present information clearly and accurately. Start by exploring your topic.

333–36

Explore what you already know about your topic. Write out whatever you know or want to know about your topic, perhaps by **FREEWRITING**, **LISTING**, or **CLUSTERING**. Why are you interested in this topic? What questions do you have about it? Such questions can help you decide what you'd like to focus on and how you need to direct your research efforts.

Narrow your topic. To write a good report, you need to narrow your focus—and to narrow your focus, you need to know a fair amount about your subject. If you are assigned to write on a subject like biodiversity, for example, you need to know what it is, what the key issues are, and so on. If you do, you can simply list or brainstorm possibilities, choose one, and start your research. If you don't know much about the subject, though, you need to do some research to discover focused, workable topics. This research may shape your thinking and change your focus. Start

501–23

with **SOURCES** that can give you a general sense of the subject, such as a *Wikipedia* entry, a magazine article, a website, perhaps an interview with an expert. Your goal at this point is simply to find out what issues your topic might include and then to focus your efforts on an aspect of the topic you will be able to cover.

347–49

Come up with a tentative thesis. Once you narrow your topic, write out a statement that explains what you plan to report or explain. A good **THESIS** is potentially interesting (to you and your readers) and limits your topic enough to make it manageable. Schembri phrases her thesis as a question: "How do these ordinary kernels magically spring to life with a little heat and oil?" Marcus presents detailed information on the effects of taking six years to earn a bachelor's degree and then offers a simple thesis: "It's not entirely the students' fault." At this point, however, you need only a tentative thesis that will help focus any research you do.

* academic literacies   ● fields   ● research
■ rhetorical situations   ⋮ processes   ● media/design
▲ genres   ◆ strategies

Do any necessary research, and revise your thesis. To focus your research efforts, **OUTLINE** the aspects of your topic that you expect to discuss. Identify any aspects that require additional research, and develop a research plan. Expect to revise your outline as you do your research, since more information will be available for some aspects of your topic than others, some may prove irrelevant to your topic, and some may turn out to be more than you need. You'll need to revisit your tentative thesis once you've done any research, to finalize your statement.

340–42

Ways of Organizing a Report

Reports can be organized in various ways. Here are three common organizational structures:

[Reports on topics that are unfamiliar to readers]

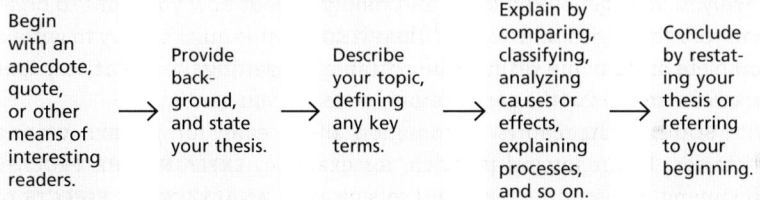

Begin with an anecdote, quote, or other means of interesting readers. → Provide background, and state your thesis. → Describe your topic, defining any key terms. → Explain by comparing, classifying, analyzing causes or effects, explaining processes, and so on. → Conclude by restating your thesis or referring to your beginning.

[Reports on events or procedures]

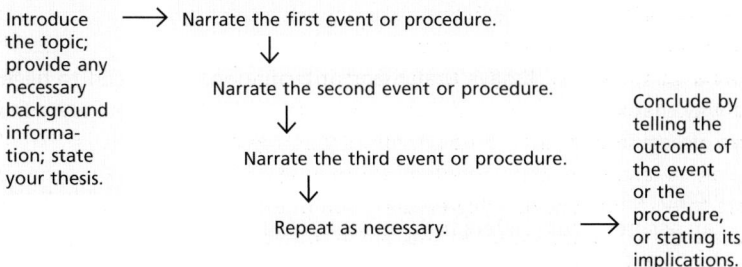

Introduce the topic; provide any necessary background information; state your thesis. → Narrate the first event or procedure. ↓ Narrate the second event or procedure. ↓ Narrate the third event or procedure. ↓ Repeat as necessary. → Conclude by telling the outcome of the event or the procedure, or stating its implications.

[Reports that compare and contrast]

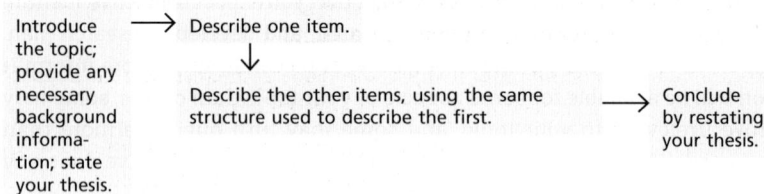

Introduce the topic; provide any necessary background information; state your thesis. → Describe one item. ↓ Describe the other items, using the same structure used to describe the first. → Conclude by restating your thesis.

Many reports use a combination of organizational structures; don't be afraid to use whatever method of organization best suits your material and your purpose.

Writing Out a Draft

364–66

Once you have generated ideas and thought about how you want to organize your report, it's time to start **DRAFTING**. Do this quickly—try to write a complete draft in one sitting, concentrating on getting the report on paper or screen and on putting in as much detail as you can.

469–73
405–9

437–44

Writing that reports information often calls for certain writing strategies. The report on popcorn, for example, **EXPLAINS THE PROCESS** of popping, whereas the report on college costs **ANALYZES THE EFFECTS** of delaying college graduation. When you're reporting on a topic your readers aren't familiar with, you may wish to **COMPARE** it with something more familiar; you can find useful advice on these and other writing strategies in Part 6 of this book.

346–54

Draft a BEGINNING. Essays that report information often need to begin in a way that will get your audience interested in the topic. Here are a few ways of beginning:

- *Simply state your thesis.* Opening with a thesis works well when you can assume your readers have enough familiarity with your topic that you don't need to give much detailed background information.

* academic literacies   ● fields   ● research
■ rhetorical situations   •• processes   ● media/design
▲ genres   ◆ strategies

- *Begin by providing a context for your report.* Tingling starts her report on sleep deprivation by describing college students' busy lives, which often cause them to get inadequate sleep.

- *Start with something that will provoke readers' interest.* Marcus's report begins with an anecdote about a college student.

- *Begin with an illustrative example.* Schembri evokes the childhood wonder of popping corn before exploring its history and physics.

Draft an ENDING.   Think about what you want your readers to read last. An effective ending leaves them thinking about your topic.   354–58

- *Summarize your main points.* This is a good way to end when you've presented several key points you want readers to remember. Tingling ends this way, summarizing the results of her research.

- *Point out the implications of your report.* Tingling ends by noting that lost sleep can have greater effects than simply sleepiness: it "can take a lot more than bargained for."

- *Frame your report by referring to its introduction.* Marcus begins and ends his report by quoting the same student, and Schembri returns to her introductory evocation of popcorn as "little pockets of magic."

- *Tell what happened.* If you are reporting on an event, you could conclude by telling how it turns out.

Come up with a title.   You'll want a title that tells readers something about your subject—and makes them want to know more. Tingling's title, "Sleepless Nights of a University Student," clearly indicates what's to come in her report. Marcus suggests that his essay will disclose the reason college costs more than you think—but doesn't tell us in the title. See the chapter on **ORGANIZING YOUR WRITING AND GUIDING YOUR READERS** for tips on coming up with titles that are informative and enticing enough to make readers wish to read on.   345–63

Considering Matters of Design

You'll probably write the main text of your report in paragraph form, but think about what kind of information you're presenting and how you can design and format it to make it as easy as possible for your readers to understand. You might ask yourself these questions:

668–69
- What is an appropriate **FONT**? A font like Times New Roman that is easy to read in print? A font like Arial or **Verdana** that looks good onscreen? Something else?

670–72
- Would it help your readers if you divided your report into shorter sections and added **HEADINGS**?

669–70
- Is there any information that would be easier to follow in a **LIST**?

676
- Could any of your information be summarized in a **TABLE** or **FIGURE**?

676
- Do you have any data that readers would more easily understand in the form of a bar **GRAPH**, line graph, or pie chart?

674–80
- Would **ILLUSTRATIONS** (diagrams, photos, drawings, and so on), video or audio clips, or links help you explain anything in your report?

Getting Response and Revising

372–74
The following questions can help you study your draft with a critical eye. **GETTING RESPONSE** from others is always good, and these questions can guide their reading, too. Make sure they know your purpose and audience.

- Do the title and opening sentences get readers' interest? If not, how might they do so?
- What information does this text provide, and for what purpose?
- Does the introduction explain why this information is being presented? Does it place the topic in a larger context?
- Are all key terms defined that need to be?
- Do you have any questions? Where might more explanation or an example help you understand something better?

🔆 academic literacies　　● fields　　● research
■ rhetorical situations　　⁝ processes　　● media/design
▲ genres　　◆ strategies

- Is any information presented visually, with a chart, graph, table, drawing, or photograph? If so, is it clear how the illustration relates to the written text? Is there any text that would be more easily understood if it were presented visually?

- Is any information presented through digital media, such as hyperlinks, video clips, or audio files? If so, is the relation of these elements to the written text made clear? Would any aspect of the report be clearer if presented using such elements?

- Does the organization help make sense of the information? Does the text include description, comparison, or any other writing strategies? Does the topic or rhetorical situation call for any particular strategies that should be added?

- If the report cites any sources, are they quoted, paraphrased, or summarized effectively (and with appropriate documentation)? Is information from sources introduced with **SIGNAL PHRASES**? 551–54

- Does the report end in a satisfying way? What are readers left thinking?

These questions should identify aspects of your report you need to work on. When it's time to **REVISE**, make sure your report appeals to your audience and achieves your purpose as successfully as possible. If you have time and want to explore alternatives, you might try **REWRITING** your draft to see if a different plan or approach might work better. 375–77

378–79

Editing and Proofreading

Once you've revised your draft, follow these guidelines for **EDITING** a report: 380–83

- Check your use of key terms. Repeating key words is acceptable in reports; using synonyms for unfamiliar words may confuse readers, while the repetition of key words or the use of clearly identified **PRONOUNS** for them can be genuinely helpful. Glossary

- Check to be sure you have **TRANSITIONS** where you need them. 361–62

670–72 • If you have included **HEADINGS**, make sure they're parallel in structure and consistent in design.

674–80 • Make sure that any photos or other **ILLUSTRATIONS** have captions, that charts and graphs have headings—and that all are referred to in the main text. Use white space as necessary to separate sections of your report and to highlight graphic elements.

560–63 • Check any **DOCUMENTATION** to see that it follows the appropriate style.

383–84 • **PROOFREAD** and spell-check your report carefully.

Taking Stock of Your Work

Take stock of what you've written by considering these questions:

• How well did you convey the information? Is it complete enough for your audience's needs?

• What writing strategies (for example, defining, comparing, classifying) did you rely on, and how did they help you achieve your purpose?

• What element of your rhetorical situation (purpose, audience, genre, stance, media/design) was the easiest to figure out for this writing project? What element was the hardest to figure out? Why?

• How did you go about researching the information for this piece?

• How did you go about drafting this piece? For example, did you write out an outline or plan? Do something else? How could you make the drafting process more productive in the future?

• How well did you synthesize information from your sources? Can you point to a specific place in your text where you did an especially good job of making connections among the facts and ideas in your sources?

• Did you use tables, graphs, diagrams, photographs, illustrations, or other graphics? If so, what did they add to the piece? If not, can you think of such elements you might have used?

❋ academic literacies   ● fields   ● research
■ rhetorical situations   ⁚⁚ processes   ● media/design
▲ genres   ◆ strategies

- What did you do to keep yourself engaged with your writing task? If you weren't engaged, how could you improve in the future?

- What did you learn about yourself as a writer through your work on this project? How can you apply that learning to future writing tasks?

If you need more help

See Chapter 34 if you are required to submit your report in a writing **PORTFOLIO**. See also Chapter 14 on **ABSTRACTS** if your report requires one; and Chapter 19 on **PROFILES**, a report based on firsthand research.

385–90
196–200
245–57

13 Arguing a Position

Everything we say or do presents some kind of argument, takes some kind of position. Often we take overt positions: "Everyone in the United States is entitled to affordable health care." "The university needs to offer more language courses." "Photoshopped images should carry disclosure notices." But arguments can be less direct and specific as well, from empty chairs that honor fallen veterans and COVID-19 victims to a yellow smiley face, which might be said to argue for a good day.

In college course work, you are constantly called on to argue positions: in an *English* class, you may argue for a certain interpretation of a poem; in a *business* course, you may argue for the merits of a flat tax; in a *linguistics* class, you may argue that English is now a global language. All of those positions are arguable—people of goodwill can agree or disagree with them and present reasons and evidence to support their positions.

This chapter provides guidelines for writing an essay that argues a position. We'll begin with three good examples, the first one annotated to point out key features of this kind of writing.

KELLY CORYELL

All Words Matter
The Manipulation behind "All Lives Matter"

Kelly Coryell wrote this essay in her first-year writing course at Diablo Valley College in Pleasant Hill, California. An English major, Coryell works as a tutor in the college's learning center and serves as a supplemental instruction leader for multilingual students. This essay was nominated for the Norton Writer's Prize and is documented in MLA style.

* academic literacies   ● fields   ● research
■ rhetorical situations   ⁚ processes   ● media/design
▲ genres   ◆ strategies

I've never understood the popular saying "Sticks and stones may break my bones, but words will never hurt me." I grew up as a tomboy; I've had more than my fair share of scrapes, bruises, and stitches. But I've found that words inflict the most painful injuries. On sleepless nights when I toss and turn, I'm not replaying the time I broke my foot over and over in my head—I'm thinking about some embarrassing thing I said that still makes me physically cringe or a time someone said something hurtful to me. Broken bones heal—words stay with us.

This is because words have power. A skilled wordsmith can influence us by using evocative words that elicit an emotional response. The meanings of these "loaded" words aren't located in a dictionary. There is a context surrounding them that implies a meaning beyond the basic information they convey. Loaded words and phrases appeal to our emotions, not our logic—they enter our hearts, not our minds. They can manipulate, so sometimes people use loaded language to distract us from a flawed argument.

Such is the case when people use the phrase "all lives matter" to oppose the phrase "Black lives matter." In 2012, neighborhood watch coordinator George Zimmerman shot and killed Trayvon Martin, an unarmed African American teenager returning home from a late-night snack run. In 2013, Zimmerman was acquitted of all charges. This sparked the birth of the Black Lives Matter (BLM) movement, when activists Alicia Garza, Opal Tometi, and Patrisse Cullors, frustrated with the systemic inequality and oppression exemplified by Zimmerman's trial, started using the hashtag #BlackLivesMatter on *Twitter*. As more and more Black people died during police confrontations or in police custody—Eric Garner, Michael Brown, Tamir Rice, Walter Scott, Freddie Gray—the movement, and the phrase "Black lives matter," gained momentum, eventually becoming a major talking point in the 2016 presidential election. And in the spring of 2020, the deaths of Ahmaud Arbery, Breonna Taylor, George Floyd, Rayshard Brooks, Dominique "Rem'mie" Fells, and Riah Milton reignited the BLM movement, leading to an overdue reckoning with racism in the United States.

As a direct response, Americans who either disagreed with the BLM movement as a whole or supported the movement but were uncomfortable with its slogan began to chant their own phrase: "all lives matter." It should be acknowledged that some of those who have used the phrase "all lives matter" did so naïvely aiming to unify people in a time

Coryell defines an underlying premise: words have power.

Background: Events leading to Black Lives Matter movement.

Background: "All lives matter" as response to BLM.

A banner in Yellow Springs, Ohio.

Thesis: "All lives matter" sends a dangerous message.

of division, thinking of the words as a positive affirmation to which no one could object. However, saying "all lives matter" as a response to "Black lives matter" is, in reality, sending a dangerous message: it steals attention from the systematic oppression of Black Americans and actively distorts the message behind the BLM movement, manipulating the American people into maintaining the oppressive status quo.

First reason supporting thesis.

The phrase "all lives matter" belies racial inequality in America by   5
implying that all lives are at equal risk. The racism and prejudice endured by African Americans didn't end when slavery was abolished in 1865, it didn't end when Congress passed the Civil Rights Act of 1964, and it didn't end when a Black man was elected president of the United States in 2008. Racial inequality in America persists to this day, ingrained and interwoven so deeply in American society that our prison systems and manufacturing industries depend on it. It's so expertly hidden under layers of celebrity gossip in our media and blatant and bold lies from our politicians, and so "normalized" in our culture, that many people may not even be aware of the plight of Black Americans.

Evidence to support first reason.

But the statistics speak for themselves. According to the NAACP, African Americans are incarcerated at six times the rate of White Americans for drug crimes ("Criminal Justice Fact Sheet"). The Center for American Progress reported that in 2016, the median wealth of Black families was only ten percent of the median wealth of White families (Hanks et al.). Police shoot and kill Black Americans at two and a half times the rate of White Americans, according to a 2016 report (Lowery). All these statistics point to the fact that Black lives

※ academic literacies   ● fields   ● research

■ rhetorical situations   ∴ processes   ● media/design

▲ genres   ◆ strategies

in America are valued less than White lives. The BLM movement aims
to change this depressing reality by insisting that "Black lives matter."

Responding to "Black lives matter" with "all lives matter" ignores
the unique prejudices and discrimination Black people experience
in America. In an essay explaining why she calls herself a feminist,
Chimamanda Ngozi Adichie observes that "[f]eminism is, of course,
part of human rights in general—but to choose to use the vague
expression *human rights* is to deny the specific and particular problem
of gender." Adichie's point, that not every human being suffers the
oppression that women suffer, parallels the way that emphasizing
all lives takes focus away from the oppression specifically felt only
by Black Americans.

"All lives matter" implies that all lives endure an equal amount of
hardship; therefore, the struggles of Black Americans deserve no more
attention than the struggles of White Americans—which is statistically
false. Daniel Victor illustrates the harmful way the phrase "all lives matter" removes focus from the hardships of Black Americans in his article
"Why 'All Lives Matter' Is Such a Perilous Phrase." He writes:

> Those in the Black Lives Matter movement say black people are in
> immediate danger and need immediate attention. . . .
>
> Saying "All Lives Matter" in response would suggest to them
> that all people are in equal danger, invalidating the specific con-
> cerns of Black people.
>
> "You're watering the house that's not burning but you're
> choosing to leave the house that's burning unattended," said Allen
> Kwabena Frimpong, an organizer for the New York chapter of
> Black Lives Matter. "It's irresponsible."

To put it simply, focusing too much on the whole can divert much-
needed attention from crises affecting a specific group.

Using "all lives matter" as a response to "Black lives matter" perpetuates the misconception that BLM activists do not believe both
sentiments. The core of the BLM movement is the belief that *all* lives
matter—*including* Black lives, which are treated differently than White
lives, not just by law enforcement but by America as a whole. As
J. Clara Chan explains in her article "What Is 'All Lives Matter'? A Short
Explainer":

Second reason.

*Here she makes
the analogy clear
and extends her
argument with
another analogy.*

*Summary of this
argument.*

*Third reason:
"All lives matter"
misinterprets the
meaning of "Black
lives matter."*

BLM supporters stress that the movement isn't about believing no other races matter. Instead, the movement seeks to highlight and change how racism disproportionately affects the black community, in terms of police brutality, job security, socioeconomic status, educational opportunities, and more.

Here and below, evidence supporting the third reason. Also, an appeal to readers' shared values—in this case, fairness.

The BLM movement's focus on particular inequities doesn't mean other races don't matter, but the phrase "all lives matter" is loaded language that implies they don't. As Dave Bry explains, "Understanding 'black lives matter' to mean 'only' black lives matter has been a misinterpretation from the beginning." "All lives matter" is the point of the BLM movement—that all lives do matter; therefore, Black lives need to matter as much as White lives. It isn't *"only* Black lives matter"; it's "Black lives matter, *too."*

"All lives matter" manipulates people into believing those who say "Black lives matter" are against other lives. Jason Stanley explains the many ways language can be used as a tool for suppression: "Words are misappropriated and meanings twisted. I believe that these tactics are not really about making substantive claims, but rather play the role of silencing. They are, if you will, linguistic strategies for stealing the voices of others." In other words, a group can be silenced if the language vital to their ability to express themselves and their beliefs is co-opted by another group. Supporters of "all lives matter" co-opted the language of the BLM movement, forcing a separation between the two phrases. As a result, BLM activists cannot say "all lives matter" without sounding like they oppose the views of the BLM movement, since "all lives matter" is no longer just a phrase but a rebuttal—a counterargument.

The essay refers back to its underlying premise, that words have power—but focuses on the power of "all lives matter" to shape perception.

10

Saying "all lives matter" is premature, since it presupposes that equality has already been achieved. If Black lives really did matter as much as White lives in American society, then saying so would be as uncontroversial as stating any other fact. In fact, America is not a postracial society: poverty, education, health, incarceration rates—all are unequal between different races in America. We must acknowledge this ugly truth if we are ever going to change it. "All lives matter" invokes a naïve reality in which all races are holding hands in a great big circle, singing "This Land Is Your Land" as the sun smiles down on

the world. "All lives matter," like the statement "I don't see race, I just see people," may, at first, seem like a beautiful sentiment. Those who say such things may truly do so with the intent of creating a world where everyone is treated equally—regardless of race, gender, creed, orientation, and so on.

But upon close examination, these phrases—however well-meaning—are harmful because, aside from robbing people of their racial identities, they ignore race-based discrimination. How can one notice racism when one does not "see race"? How can one point out discrimination against African Americans if the only response one gets is "everyone matters"? If one insists that a goal has already been met when it hasn't, why would anyone put in more effort toward reaching that goal? This is why "all lives matter" is so manipulative and damaging. It's an attempt to convince us we don't have to keep reaching for equality and justice, even though every fact around us tells us this is far from the case. Only by consciously acknowledging racial inequality will we ever be able to put an end to it.

A counterargument is presented and refuted.

The root of "all lives matter" boils down to one thing: privilege, and the reluctance to give it up in the name of equality. As Chris Boeskool puts it, "Equality can *feel* like oppression. But it's not. What you're feeling is just the discomfort of losing a little bit of your privilege." Evening the playing field can feel like a bad thing to the person benefiting from the imbalance. In the 1960s, a White bus rider may have felt oppressed when she started having to compete with Black bus riders for the front seat—she lost the privilege of getting a guaranteed front seat because of her race. In the 1970s, a man might have complained that he could no longer "compliment" his female coworkers without being accused of sexual harassment—he lost the privilege of being able to say offensive things because of his gender. If American society starts to accept that Black lives matter as much as White lives, White Americans will lose some of the privileges afforded to them as a result of the oppression of Black Americans. Therefore, people who either do not realize they have this privilege, or simply don't want to give it up, chant "all lives matter" in an effort to hold on to that privilege. If the BLM movement is silenced, the status quo of racial inequality will be maintained. The privileged must resist this unethical temptation and not mistake oppression for equality.

Strong conclusion.

Works Cited

Adichie, Chimamanda Ngozi. "I Decided to Call Myself a Happy Feminist." *The Guardian*, 17 Oct. 2014, www.theguardian.com /books/2014/oct/17/chimamanda-ngozi-adichie-extract-we-should -all-be-feminists.

Boeskool, Chris. "When You're Accustomed to Privilege, Equality Feels Like Oppression." *HuffPost*, 14 Mar. 2016, www.huffpost .com/entry/when-youre-accustomed-to-privilege_b_9460662.

Bry, Dave. "'All Lives Matter' Is and Always Was Racist—and This Weekend's Trump Rally Proved It." *The Guardian*, 23 Nov. 2015, www.theguardian.com/commentisfree/2015/nov/23/all-lives -matter-racist-trump-weekend-campaign-rally-proved-it.

Chan, J. Clara. "What Is 'All Lives Matter'? A Short Explainer." *The Wrap*, 13 July 2016, www.thewrap.com/what-is-all-lives -matter-a-short-explainer.

"Criminal Justice Fact Sheet." *NAACP*, www.naacp.org/criminal -justice-fact-sheet. Accessed 25 Feb. 2019.

Hanks, Angela, et al. *Systematic Inequality: How America's Structural Racism Helped Create the Black–White Wealth Gap.* Center for American Progress, Feb. 2018, cdn.americanprogress.org /content/uploads/2018/02/20131806/RacialWealthGap-report.pdf.

Lowery, Wesley. "Aren't More White People Than Black People Killed by Police? Yes, but No." *The Washington Post*, 11 July 2016, www.washingtonpost.com/news/post-nation/ wp/2016/07/11/arent-more-white-people-than-black-people -killed-by-police-yes-but-no.

Stanley, Jason. "The Ways of Silencing." *The New York Times*, 25 June 2011, opinionater.blogsnytimes.com/2011/06/25 /the-ways-of-silencing. Opinionator.

Victor, Daniel. "Why 'All Lives Matter' Is Such a Perilous Phrase." *The New York Times*, 15 July 2016, www.nytimes.com/2016/07 /16/us/all-lives-matter-black-lives-matter.html.

Coryell bases her argument on a clear premise, that "words have power" and that the phrase "all lives matter" reflects a misunderstanding and negative manipulation of the Black Lives Matter movement.

✳ academic literacies   ● fields   ● research

■ rhetorical situations   ⁘ processes   ● media/design

▲ genres   ◆ strategies

BRIANNA SCHUNK

Come Look at the Freaks
Examining and Advocating for Disability Theatre

In this essay, Wilkes University student Brianna Schunk argues that our culture should support disability theatre in various ways. Schunk's essay was nominated for the Norton Writer's Prize, and it is documented in MLA style.

When I first saw Gaten Matarazzo in the TV show *Stranger Things* and some of the first words out of his toothless little mouth were, "I have cleidocranial dysplasia" ("Chapter One" 00:35:21), I actually cried with joy. Until then, I had never seen another human not related to my family who had cleidocranial dysplasia on television in a popular TV show. I was so shocked, I made my friend rewind the episode so I could hear him say it again. Having cleidocranial dysplasia myself, it is so much easier to be able to compare myself to Matarazzo when explaining my condition to other people, and I only wish he could have come into my life earlier. Representation—how we portray ourselves and are portrayed by others—is incredibly important to all groups of people, but especially to people with disabilities.

Disability, until recently, has rarely been studied through the lens of culture. Up until the last fifty years, disability was studied in the fields of medicine and science by doctors and surgeons. However, the world has recently begun to publish works not related to science and medicine, and this study of the true experience of disability is called *disability culture*. Disability culture has taken root in many fields, such as fiction, nonfiction, the analysis of stories, sociological studies of the disability experience, and visual, musical, and theatre arts. Champions like Rosemarie Garland-Thomson have brought disability culture to a wider audience. In her book *Extraordinary Bodies*, Garland-Thomson uses the lens of feminist critique as well as comparison to the study of race and gender to explain disability culture in America. Her publications are regarded as some of the earliest works on the emerging field of disability culture. *Disability* refers to an incredibly wide spectrum of

things, ranging from physical conditions to unseeable mental illnesses and cognitive impairment, but all are part of disability culture. Using studies and reports like Garland-Thomson's, we can share our experiences of being disabled with both able-bodied and disabled people.

Disability theatre is about making theatre for and with people with disabilities, and it is an important topic to me because I am an actor with a disability. Cleidocranial dysplasia affects me physically but is not immediately noticeable and manifests in less visible ways, such as hearing loss in my left ear and years of corrective orthodontic surgery. Throughout most of my life I didn't even consider myself disabled because my disability does not affect my life as much as some other people's disabilities affect theirs. In addition, people around me discouraged me from considering myself disabled, likely because the idea of disability is so stigmatizing. This stems from the power difference found between people with disabilities and people without. Garland-Thomson asserts that people who have "ordinary" (non-disabled) bodies are considered superior to those who do not, and these "normates" have social power and authority (8). However, by considering myself an actress with a disability—an identity important to me—I assert my power. Disability is not bad, it's just different. In the same way able-bodied people are different from one another, so people with disabilities are different from one another and from people who are able-bodied. However, people advocate less for people with disabilities than for any other minority group, and especially so in the theatre and entertainment world. As Lennard Davis notes in a *HuffPost* article, many people turn a blind eye when an able-bodied actor portrays a disabled person, but if anyone does blackface today it is widely acknowledged to be racist.

Many people in the field of disability theatre think it is acceptable for able-bodied actors to portray people with disabilities. This is a strange gray area for two main reasons. One, there are often not enough actors with disabilities available to play the roles; and two, sometimes the character needs to go through a period of time when they are not disabled, such as the story of Stephen Hawking (a role played by able-bodied actor Eddie Redmayne), whose disability developed as he grew older—a situation that has no definite solution. Unfortunately, "the current reality is that non-disabled actors get to

play whatever roles they want whereas disabled actors don't" (Davis). That there aren't enough disabled actors stems from a vicious cycle identified by disability scholar Carrie Sandahl, who observes that most theatre departments are inaccessible to people with disabilities. They are often denied admission to acting programs because educators are not willing to work with them or the programs do not have the means to accommodate them (255). Also, many training programs can be a challenge to people with disabilities. Sandahl notes that the origin of the "neutral" body in many acting techniques stems from the study of body normalcy (261). Bodies were considered "damaged physically and emotionally from the process of living, and those bodies capable of cure are suitable actors"—disabled bodies could not be "cured" and therefore could not be actors (262). These limitations then lead to fewer actors with disabilities in the professional world. And directors are not willing to work with or adapt stories for actors with disabilities, thinking "their impairments would detract from the playwright's or director's intent for a nondisabled character" (255).

We can get to the root of the problem by recognizing and eliminating disability stereotypes. Marilyn Dahl's essay "The Role of the Media in Promoting Images of Disability" explores common metaphors surrounding people with disabilities, such as villains who are marked as evil because of a missing body part or victims who are completely helpless and dependent on others. However, Dahl's essay also highlights how news reports do the same thing in creating fantastic stories out of the lives of people with disabilities, creating what Dahl calls "heroes by hype." Jack Nelson has also done research into stereotyped characters, even going so far as to categorize them: the "supercrip," the disabled villain, the burden, the misfit, and the victim. However, Nelson gives examples of a few positive representations as well, such as Nemo and his "lucky fin" in the children's animated movie *Finding Nemo*. Nelson stresses that "what most people with disabilities ask is that they [should] be portrayed as real people with real problems that often are not connected to their disability" (289).

Some actors and directors worry that in casting people with disabilities in disabled roles they face "typecasting" them, an idea that stems from years of stereotyping characters with disabilities. When people

deny actors with disabilities opportunities to expand their repertoire and play parts traditionally played by able-bodied actors, they close their minds to creating a whole new level of theatre. Having someone with a disability play a typically able-bodied role can add much depth to the character and to the story and can bring to life the character's journey and problems through the medium of the human body. For example, when I was in a local production of the musical *Side Show*, a story about a pair of conjoined twins on the freak show circuit who later achieved vaudeville and film performance fame, I played one of the "World's Tiniest Cossacks," among other ensemble roles. I resonated deeply with my character, because even though I do not have dwarfism, my short stature relates to my disability, and I took to playing a "freak" easily because my condition leads me to entertain people. Playing a character who makes a living by entertaining people with their different body structure was easy, because I drew on my real-life experience. It wasn't the best thing that we didn't have actors with dwarfism to play that part, but at least there was a disabled actor playing the part—a part that honored disability. In the same way that our having actual twin sisters play the main characters added a whole new level of connection between the sisters in our production, so I felt that my playing the part of a disabled character created a whole new connection to the show for myself and for the audience.

People with disabilities originally had a connection to disability theatre through the world of freak shows. Freak shows "challenged the boundaries of the individual" (Garland-Thomson 59) to audiences by presenting bodies different from their own. Those shows were among the early origins of the celebration of disability culture. They also provided a place for people to "identify with nonconformity" (Garland-Thomson 68)—specifically, people wounded and missing limbs from war. When the shows gradually shut down in the mid-twentieth century, people with disabilities couldn't find jobs to sustain themselves because the people who took away their celebration of bodily difference refused to hire them to work anywhere else. They found some work in the developing film industry, in films such as Tod Browning's 1932 movie *Freaks*. It features a cast of actors with disabilities, including people with dwarfism, a pair of conjoined twins, and "pinheads," or actors with microcephaly.

* academic literacies
■ rhetorical situations
▲ genres
● fields
∴ processes
◆ strategies
● research
● media/design

This movie, in addition to providing jobs and representation to actors with disabilities, also challenged traditional stereotypes about people with disabilities. As the movie progresses and one sees the freaks going about their daily lives, one comes to root for them and value their lives and ambitions as humans, not as freaks or disabled characters. The script even recognizes this, with the characters calling out the ableism of one able-bodied woman in the movie by saying, "Cleopatra ain't one of us. Why, we're just filthy things to her" (*Freaks* 00:25:21–23). Nowadays, we have countless organizations and forms of media that celebrate disability. Derrick Vanmeter, a theatre professor at Clayton State University, lists several organizations, such as Full Radius Dance, which puts on dance shows with people in wheelchairs, and That Uppity Theatre Company, which introduces artists with disabilities in their work *The DisAbility Project*. The Star Wars movies portray people losing limbs—becoming disabled—without any accompanying storylines or stereotypes, instead tying their disabilities to the overarching theme of keeping the universe in balance, which video blogger Hank Green of the vlogbrothers channel summarizes as "when someone cuts someone's hand off in *Star Wars* . . . they're gonna get their hand chopped off too" ("13 Severed Hands" 00:01:21–26). Even musicals such as *Sideshow*, *Next to Normal* (a story about a woman with bipolar disorder), and Deaf West's production of *Spring Awakening* create stories using characters with disabilities and support them as humans with desires and goals. In 2019, actor Ali Stroker, who played Ado Annie in the Broadway revival of the classic musical *Oklahoma!*, became the first person who uses a wheelchair to win a Tony award, proving actors with disabilities can excel in starring roles created for able-bodied characters.

I encourage other people with disabilities to do theatre and see theatre and seek out opportunities to participate in theatre, and I encourage everyone to support us in our endeavors and to support disability theatre on a whole. Directors: cast more disabled actors; open up your shows to a completely different take by using an actor with a disability. Choose to perform more shows about people with disabilities, and honor those roles with actors who truly know the disabled experience, not just able-bodied actors who can fake a limp. Theatre is the place for everyone to express themselves and tell their stories, and we as a

Ali Stroker as Ado Annie in the 2019 Broadway revival of *Oklahoma!*

theatre community need to honor all people. As a performer with a skeletal condition, my whole life is technically disability theatre. I am evidence that these two worlds can work together, and I want to show that to the world. No one can stop me from doing what I love to do, and I hope I can extend this confidence and belief to every other human that I meet. We can all be brilliant.

Works Cited

"Chapter One." *Stranger Things*, season 1, episode 1, 21 Laps Entertainment / Monkey Massacre, 2016. *Netflix*, www .netflix.com.

Dahl, Marilyn. "The Role of the Media in Promoting Images of Disability: Disability as Metaphor: The Evil Crip." *Canadian Journal of Communication*, vol. 18, no. 1, 1993. *ProQuest Central*, https://doi.org/10.22230/cjc.1993v18n1a718.

* academic literacies   ● fields   ● research
■ rhetorical situations   ⋰ processes   ● media/design
▲ genres   ◆ strategies

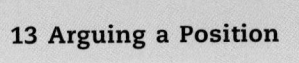

Davis, Lennard J. "Let Actors with Disabilities Play Characters with Disabilities." *HuffPost*, 6 Dec. 2017, www.huffpost.com/entry /let-actors-with-disabilit_b_380266.

Freaks. Directed by Tod Browning, Metro-Goldwyn Mayer, 1932.

Garland-Thomson, Rosemarie. *Extraordinary Bodies: Figuring Physical Disability in American Culture and Literature*. Columbia UP, 1997.

Nelson, Jack A. "Invisible No Longer: Images of Disability in the Media." *Images That Injure: Pictorial Stereotypes in the Media*, edited by Susan Dente Ross and Paul Martin Lester, 3rd ed., Praeger, 2011, pp. 274–91.

Sandahl, Carrie. "The Tyranny of Neutral: Disability and Actor Training." *Bodies in Commotion: Disability and Performance*, edited by Sandahl and Philip Auslander, U of Michigan P, 2005, pp. 255–67.

"13 Severed Hands!" *YouTube*, uploaded by vlogbrothers, 30 Sept. 2009, youtube.com/watch?v=lDlqwbMCBg4.

Vanmeter, Derrick. "Disability and Theatre: Access for All." Southeastern Theatre Conference, 3 Mar. 2017, Hilton Hotel, Lexington, Kentucky.

Schunk argues that actors with disabilities should be cast in the roles of characters with disabilities—and in roles of nondisabled characters as well. To make her argument, she draws on recent scholarship on disability culture and her own experiences.

NICHOLAS KRISTOF

Our Blind Spot about Guns

In this essay, which first appeared in the New York Times in 2014, columnist Nicholas Kristof argues that if guns and their owners were regulated in the same way that cars and their drivers are, thousands of lives could be saved each year.

If we had the same auto fatality rate today that we had in 1921, by my calculations we would have 715,000 Americans dying annually in vehicle accidents.

Instead, we've reduced the fatality rate by more than 95 percent—not by confiscating cars, but by regulating them and their drivers sensibly.

We could have said, "Cars don't kill people. People kill people," and there would have been an element of truth to that. Many accidents are a result of alcohol consumption, speeding, road rage or driver distraction. Or we could have said, "It's pointless because even if you regulate cars, then people will just run each other down with bicycles," and that, too, would have been partly true.

Yet, instead, we built a system that protects us from ourselves. This saves hundreds of thousands of lives a year and is a model of what we should do with guns in America.

Whenever I write about the need for sensible regulation of guns, some readers jeer: *Cars kill people, too, so why not ban cars? Why are you so hypocritical as to try to take away guns from law-abiding people when you don't seize cars?*   5

That question is a reflection of our national blind spot about guns. The truth is that we regulate cars quite intelligently, instituting evidence-based measures to reduce fatalities. Yet the gun lobby is too strong, or our politicians too craven, to do the same for guns. So guns and cars now each kill more than 30,000 in America every year.

One constraint, the argument goes, is the Second Amendment. Yet the paradox is that a bit more than a century ago, there was no universally recognized individual right to bear arms in the United States, but there was widely believed to be a "right to travel" that allowed people to drive cars without regulation.

A court struck down an early attempt to require driver's licenses, and initial attempts to set speed limits or register vehicles were met with resistance and ridicule. When authorities in New York City sought in 1899 to ban horseless carriages in the parks, the idea was lambasted in the *New York Times* as "devoid of merit" and "impossible to maintain."

Yet, over time, it became increasingly obvious that cars were killing and maiming people, as well as scaring horses and causing accidents. As a distinguished former congressman, Robert Cousins, put it in 1910: "Pedestrians are menaced every minute of the days and nights by a

wanton recklessness of speed, crippling and killing people at a rate that is appalling."

Courts and editorial writers alike saw the carnage and agreed that something must be done. By the 1920s, courts routinely accepted driver's license requirements, car registration and other safety measures. 10

That continued in recent decades with requirements of seatbelts and air bags, padded dashboards and better bumpers. We cracked down on drunken drivers and instituted graduated licensing for young people, while also improving road engineering to reduce accidents. The upshot is that there is now just over 1 car fatality per 100 million miles driven.

Yet as we've learned to treat cars intelligently, we've gone in the opposite direction with guns. In his terrific new book, *The Second Amendment: A Biography,* Michael Waldman, the president of the Brennan Center for Justice at the New York University School of Law, notes that "gun control laws were ubiquitous" in the nineteenth century. Visitors to Wichita, Kansas, for example, were required to check their revolvers at police headquarters.

And Dodge City, symbol of the Wild West? A photo shows a sign on the main street in 1879 warning: "The Carrying of Fire Arms Strictly Prohibited."

The National Rifle Association supported reasonable gun control for most of its history and didn't even oppose the landmark Gun Control Act of 1968. But, since then, most attempts at safety regulation have stalled or gone backward, and that makes the example of cars instructive.

"We didn't ban cars, or send black helicopters to confiscate them," notes Waldman. "We made cars safer: air bags, seatbelts, increasing the drinking age, lowering the speed limit. There are similar technological and behavioral fixes that can ease the toll of gun violence, from expanded background checks to trigger locks to smart guns that recognize a thumbprint, just like my iPhone does." 15

Some of these should be doable. A Quinnipiac poll this month found 92 percent support for background checks for all gun buyers.

These steps won't eliminate gun deaths any more than seatbelts eliminate auto deaths. But if a combination of measures could reduce the toll by one-third, that would be 10,000 lives saved every year.

Dodge City, Kansas, 1879. The sign reads, "The Carrying of Fire Arms strictly prohibited."

A century ago, we reacted to deaths and injuries from unregulated vehicles by imposing sensible safety measures that have saved hundreds of thousands of lives a year. Why can't we ask politicians to be just as rational about guns?

Kristof argues that because regulating cars has made them much safer, guns should be regulated similarly. He supports his argument with data on fatality rates and the history of automobile and gun regulation in the United States.

Key Features / Arguments

A clear and arguable position. At the heart of every argument is a claim with which people may reasonably disagree. Some claims are not arguable because they're completely subjective, matters of taste or opinion ("I hate

academic literacies　　● fields　　● research
■ rhetorical situations　　● processes　　● media / design
▲ genres　　◆ strategies

sauerkraut"), because they are a matter of fact ("The first *Star Wars* movie came out in 1977"), or because they are based on belief or faith ("There is life after death"). To be arguable, a position must reflect one of at least two points of view, making reasoned argument necessary: guns should (or should not) be regulated; selling human organs should be legal (or illegal). In college writing, you will often argue not that a position is correct but that it is plausible—that it is reasonable, supportable, and worthy of being taken seriously.

Necessary background information.   Sometimes we need to provide some background on a topic we are arguing so that readers can understand what is being argued. Coryell recounts the origin of the phrase "Black lives matter" before launching her argument that "all lives matter" misrepresents the BLM phrase and movement; Kristof describes the history of automobile regulation.

Good reasons.   By itself, a position does not make an argument; the argument comes when a writer offers reasons to back up the position. There are many kinds of good reasons. Kristof makes his argument by comparing cars to guns. Schunk offers several reasons why actors with disabilities should get more parts: using non-disabled actors in roles of characters with disabilities is discriminatory; if there is a shortage of actors with disabilities, it derives from barriers to participation of those actors; and stereotypes limit the roles available to actors with disabilities.

Convincing evidence.   Once you've given reasons for your position, you then need to offer evidence for your reasons: facts, statistics, expert testimony, anecdotal evidence, case studies, textual evidence. All three arguments use a mix of these types of evidence. Coryell cites statistics on the differences between Black Americans and White Americans in terms of their incarceration, wealth, and incidence of being shot and killed by police. Kristof shows how regulating cars led to dramatic decreases in driving deaths and injuries. Both Schunk and Coryell cite anecdotes from their own experience to support their arguments.

Appeals to readers' values. Effective arguers try to appeal to readers' values and emotions. Kristof appeals to the idea that reducing traffic deaths—and, by extension, shooting deaths—is a worthy goal. Coryell appeals to her readers' sense of fairness, that "Black lives matter, *too*"—as much as the lives of White people. These values are deeply held and may be seen as common ground we share with one another.

A trustworthy tone. Arguments can stand or fall on the way readers perceive the writer. Very simply, readers need to trust the person who's making the argument. One way of winning this trust is by demonstrating that you know what you're talking about. Kristof offers plenty of facts to show his knowledge of the history of automotive regulation—and he does so in a self-assured tone. There are many other ways of establishing yourself (and your argument) as trustworthy—by showing that you have some experience with your subject, that you're fair, and of course that you're honest.

Careful consideration of other positions. No matter how reasonable and careful we are in arguing our positions, others may disagree or offer counterarguments. We need to consider those other views and to acknowledge and, if possible, refute them in our written arguments. Coryell, for example, acknowledges some people's discomfort with the phrase "Black lives matter," but she counters that to say instead that "all lives matter" is to make a mistake—"Black lives matter" asserts that racism against Black people needs to be pointed out and that Black lives are as worthy as everyone else's lives.

A GUIDE TO WRITING ARGUMENTS

Choosing a Topic

A fully developed argument requires significant work and time, so choosing a topic in which you're interested is very important. Students often find that widely debated topics such as "animal rights" or "abortion" can

※ academic literacies   ● fields   ● research

■ rhetorical situations   ⸪ processes   ⬤ media/design

▲ genres   ◆ strategies

be difficult to write on because they don't feel any personal connection to them. Better topics include those that

- interest you right now
- are focused, but not too narrowly
- have some personal connection to your life

One good way to **GENERATE IDEAS** for a topic that meets those three criteria is to explore your own roles in life.

333–44

Start with your roles in life. Make four columns with the headings "Personal," "Family," "Public," and "School." Then **LIST** the roles you play that relate to each heading. Here is a list one student wrote:

334–35

Personal	Family	Public	School
gamer	son	voter	college student
dog owner	younger	homeless-shelter	work-study employee
old-car owner	brother	volunteer	dorm resident
male	grandson	American	primary-education major
White		resident of Texas	
middle class			

Identify issues that interest you. Think, then, about issues or controversies that may concern you as a member of one or more of those groups. For instance, as a primary-education major, this student cares about the controversy over whether teachers' jobs should be focused on preparing kids for high-stakes standardized tests. As a college student, he cares about the costs of a college education. Issues that stem from these subjects could include the following: Should student progress be measured by standardized tests? Should college cost less than it does?

Pick four or five of the roles you list. In five or ten minutes, identify issues that concern or affect you as a person who takes on of each of those roles. It might help to word each issue as a question starting with "Should."

Frame your topic as a problem.   Most position papers address issues that are subjects of ongoing debate—their solutions aren't easy, and people disagree on which ones are best. Posing your topic as a problem can help you think about the topic, find an issue that's suitable to write about, and find a clear focus for your essay.

For example, if you wanted to write an argument on the lack of student parking at your school, you could frame your topic as one of several problems: What causes the parking shortage? Why are the university's parking garages and lots limited in their capacity? What might alleviate the shortage?

Choose one issue to write about.   Remember that the issue should be interesting to you and have some connection to your life. It is a tentative choice; if you find later that you have trouble writing about it, simply go back to your list of roles or issues and choose another.

Considering the Rhetorical Situation

59–60 ■   **PURPOSE**   Do you want to persuade your audience to do something? Change their minds? Consider alternative views? Accept your position as plausible—see that you have thought carefully about an issue and researched it appropriately?

61–64 ■   **AUDIENCE**   Who is your intended audience? What do they likely know and believe about this issue? How personal is it for them? To what extent are they likely to agree or disagree with you—and with one another? Why? What common ground can you find with them?

72–74 ■   **STANCE**   What's your attitude toward your topic, and why? How do you want your audience to perceive your attitude? How do you want your audience to perceive you? As an authority on your topic? As someone much like them? As calm? reasonable? impassioned or angry? something else?

※ academic literacies   ● fields   ● research
■ rhetorical situations   ⁘ processes   ● media/design
▲ genres   ◆ strategies

MEDIA / DESIGN   What media will you use, and how do your media affect your argument? Does your print or online argument call for photos or charts? If you're giving an oral presentation, should you put your reasons and support on slides? If you're writing electronically, should you include audio or video evidence or links to counter-arguments or your sources?

75–77

Generating Ideas and Text

Most essays that successfully argue a position share certain features that make them interesting and persuasive. Remember that your goal is to stake out a position and convince your readers that it is plausible.

Explore what you already know about the issue.   Write out whatever you know about the issue by **FREEWRITING** or as a **LIST** or **OUTLINE**. Why are you interested in this topic? What is your position on it at this point, and why? What aspect do you think you'd like to focus on? Where do you need to focus your research efforts? This activity can help you discover what more you need to learn. Chances are you'll need to learn a lot more about the issue before you even decide what position to take.

333–34
334–35
340–42

Do some research.   At this point, try to get an overview. Start with one **GENERAL SOURCE** of information that will give you a sense of the ins and outs of your issue, one that isn't overtly biased. *The Atlantic, Time, Slate,* and other online newspapers and magazines can be good starting points on current issues. For some issues, you may need to **INTERVIEW** an expert. For example, one student who wanted to write about chemical abuse of animals at 4-H competitions interviewed an experienced show competitor. Use your overview source to find out the main questions raised about your issue and to get some idea about the various ways in which you might argue it.

513

518–20

Explore the issue strategically.   Most issues may be argued from many different perspectives. You'll probably have some sense of the different views that exist on your issue, but you should explore multiple perspectives

before deciding on your position. The following methods are good ways of exploring issues:

445–55 ◆

- As a matter of **DEFINITION**. What is it? How should it be defined? How can "organic" or "genetically modified food" be defined? How do proponents of organic food define it—and how do they define "genetically modified food"?How do advocates of "genetically modified food" define it—and how do they define "organic food"? Considering such definitions is one way to identify different perspectives on the topic.

431–36 ◆

- As a matter of **CLASSIFICATION**. Can the issue be divided into categories? Are there different kinds of, or different ways of, producing organic foods and genetically modified foods? Do different categories suggest particular positions or perhaps a way of supporting a certain position? Are there other ways of categorizing foods?

437–44 ◆

- As a matter of **COMPARISON**. Is one subject being considered better than another? Is organic food healthier or safer than genetically modified food? Is genetically modified food healthier or safer than organic? Is the answer somewhere in the middle?

469–73 ◆

- As a matter of **PROCESS**. Should somebody do something? What? Should people buy and eat more organic food? More genetically modified food? Should they buy and eat some of each?

Reconsider whether the issue can be argued. Is this issue worth discussing? Why is it important to you and to others? What difference will it 410–30 ◆ make if one position or another prevails? Is it **ARGUABLE**? At this point, you want to be sure that your topic is worth arguing about.

Draft a thesis. Having explored the possibilities, decide your position, and write it out as a complete sentence. For example:

Parents should be required to have their children vaccinated.

Pod-based coffeemakers should be banned.

Genetically modified foods should not be permitted in the United States.

✳ academic literacies   ● fields   ● research
■ rhetorical situations   ⁞ processes   ● media/design
▲ genres   ◆ strategies

Qualify your thesis. Rather than taking a strict pro or con position, in most cases you'll want to **QUALIFY YOUR POSITION** —in certain circumstances, with certain conditions, with these limitations, and so on. This is not to say that we should settle, give in, sell out; rather, it is to say that our position may not be the only "correct" one and that other positions may be valid as well. **QUALIFYING YOUR THESIS** also makes your topic manageable by limiting it. For example:

◆ 412

∴ 348–49

> Parents should be required to have their children vaccinated, with only medical exemptions allowed.

> Pod-based coffeemakers should be banned unless the pods are recyclable.

> Genetically modified foods should not be permitted in the United States if a link between GMOs and resistance to antibiotics is proven.

Come up with good reasons. Once you have a thesis, you need to come up with good **REASONS** to convince your readers that it's plausible. Write out your position, and then list several reasons. For instance, if your thesis is that pod-based coffeemakers should be banned, two of your reasons might be:

◆ 413–14

> The pods cannot be recycled.

> Other methods of making coffee are more environmentally sound.

Think about which reasons are best for your purposes. Which seem the most persuasive? Which are most likely to be accepted by your audience? Which seem to matter the most now? If your list of reasons is short or you think you'll have trouble developing them enough to write an appropriate essay, this is a good time to rethink your topic—before you've invested too much time in it.

Develop support for your reasons. Next you have to come up with **EVIDENCE** to support your reasons: facts, statistics, examples, testimony by authorities and experts, anecdotal evidence, scenarios, case studies and observation, and textual evidence. For some topics, you may want or need to use evidence in visual form like photos, graphs, and charts; online,

◆ 414–22

you could also use video or audio evidence and links to evidence in other websites.

What counts as evidence varies across audiences. Statistical evidence may be required in certain disciplines but not in others; anecdotes may be accepted as evidence in some courses but not in engineering. Some audiences will be persuaded by emotional appeals while others will not. For example, if you argue that foods produced from genetically modified organisms (GMOs) should be allowed to be sold because they're safe, you could support that reason with *facts*: GMOs are tested thoroughly by three separate US government agencies. Or you could support it with *statistics*: a study of 29 years of data on livestock fed GMO feed found that GMO-fed cattle had no adverse health effects on people who ate them. *Expert testimony* might include R. E. Goodman of the Department of Food Science and Technology at the University of Nebraska–Lincoln, who writes that "there is an absence of proof of harm to consumers from commercially available GMOs."

Identify other positions.   Now think about positions other than yours and the reasons people are likely to give for those positions. Be careful to represent their points of view as accurately and fairly as you can. Then decide whether you need to acknowledge or to refute each position.

424–25 ◆

Acknowledging other positions.   Some positions can't be refuted but are too important to ignore, so you need to **ACKNOWLEDGE** concerns and objections they raise to show that you've considered other perspectives. For example, in an essay arguing that vacations are necessary to maintain good health, medical writer Alina Tugend acknowledges that "in some cases, these trips—particularly with entire families in tow—can be stressful in their own way. The joys of a holiday can also include lugging around a ridiculous amount of paraphernalia, jet-lagged children sobbing on airplanes, hotels that looked wonderful on the Web but are in reality next to a construction site." Tugend's acknowledgment moderates her position and makes her argument appear more reasonable.

☀ academic literacies	● fields	● research
■ rhetorical situations	⣿ processes	● media/design
▲ genres	◆ strategies	

Refuting other positions. State the position as clearly and as fairly as you can, and then **REFUTE** it by showing why you believe it is wrong. Perhaps the reasoning is faulty or the supporting evidence inadequate. Acknowledge the merits of the position, if any, but emphasize its shortcomings. Avoid logical **FALLACIES** such as attacking the person holding the position or bringing up a competing position that no one seriously entertains.

◆ 426

◆ 427–29

Ways of Organizing an Argument

Readers need to be able to follow the reasoning of your argument from beginning to end; your task is to lead them from point to point as you build your case. Sometimes you'll want to give all the reasons for your argument first, followed by discussion of any other positions. Alternatively, you might discuss each reason and any opposing arguments together.

[Reasons to support your argument, followed by opposing arguments]

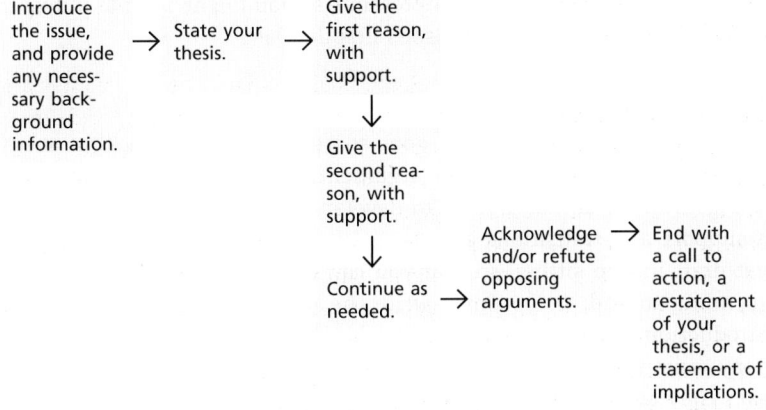

Introduce the issue, and provide any necessary background information. → State your thesis. → Give the first reason, with support.

↓

Give the second reason, with support.

↓

Continue as needed. → Acknowledge and/or refute opposing arguments. → End with a call to action, a restatement of your thesis, or a statement of implications.

[Reason/opposing argument, reason/opposing argument]

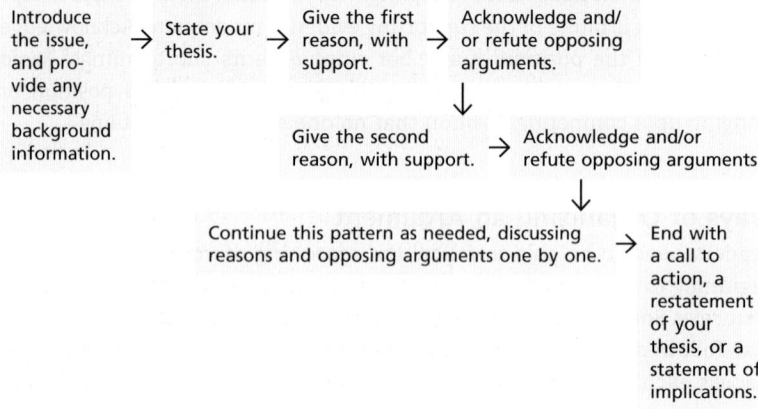

Introduce the issue, and provide any necessary background information. → State your thesis. → Give the first reason, with support. → Acknowledge and/or refute opposing arguments.

↓

Give the second reason, with support. → Acknowledge and/or refute opposing arguments.

↓

Continue this pattern as needed, discussing reasons and opposing arguments one by one. → End with a call to action, a restatement of your thesis, or a statement of implications.

Consider carefully the order in which you discuss your reasons. Usually what comes last makes the strongest impression on readers and what comes in the middle the least impression, so you might want to put your most important or strongest reasons first and last.

Writing Out a Draft

364–66

Once you have generated ideas, done some research, and thought about how you want to organize your argument, it's time to start **DRAFTING**. Your goal in the initial draft is to develop your argument—you can fill in support and transitions as you revise. You may want to write your first draft in one sitting, so that you can develop your reasoning from beginning to end. Or you may write the main argument first and the introduction and conclusion after you've drafted the body of the essay; many writers find that beginning and ending an essay are the hardest

346–58

tasks they face. Here is some advice on how you might **BEGIN AND END** your argument:

* academic literacies   ● fields   ● research
■ rhetorical situations ∴ processes   ● media/design
▲ genres   ◆ strategies

Draft a beginning. There are various ways to begin an argument essay, depending on your audience and purpose. Here are a few suggestions:

- *Offer background information.* You may need to give your readers information to help them understand your position. Coryell first establishes the truth of the assertion that "Words have power" and then provides a brief history of the events leading to the formation of the Black Lives Matter movement.

- *Begin with an anecdote.* Schunk starts by describing the first time she saw on television an actor who had the same disability that she has.

- *Define a key term.* You may need to show how you're using certain keywords. Coryell carefully defines "loaded" words as "evocative words that elicit an emotional response. . . . They can manipulate, so sometimes people use loaded language to distract us from a flawed argument."

- *Explain the context for your position.* All arguments are part of a larger, ongoing conversation, so you might begin by showing how your position fits into the arguments others have made. Kristof places his argument about guns in the **CONTEXT** of government regulation of other dangerous technologies.

▲ 125–26

- *Begin with something that will get readers' attention.* Kristof begins with an eye-popping statistic: "If we had the same auto fatality rate today that we had in 1921, by my calculations we would have 715,000 Americans dying annually in vehicle accidents."

Draft an ending. Your conclusion is the chance to wrap up your argument in such a way that readers will remember what you've said. Here are a few ways of concluding an argument essay:

- *Summarize your main points.* Especially when you've presented a complex argument, it can help readers to **SUMMARIZE** your main point. Schunk sums up her argument by stating, "Theatre is the place for

● 550–51

everyone to express themselves and tell their stories, and we as a theatre community need to honor all people."

- *Call for action.* Kristof does this by asking politicians to consider "sensible safety measures." Schunk calls for theater and film directors to cast more actors with disabilities.

- *Frame your argument by referring to the introduction.* Kristof ends by returning to the country's response "to death and injuries from unregulated vehicles," suggesting that "sensible safety measures" similar to those adopted 100 years ago for cars could be devised for guns.

Come up with a title.   Most often you'll want your title to tell readers something about your topic—and to make them want to read on. Kristof's title, "Our Blind Spot about Guns," entices us to find out what that blind spot is. Schunk's "Come Look at the Freaks: Examining and Advocating for Disability Theatre" offers a shocking suggestion—and then a subtitle phrased in scholarly language. Coryell also offers a two-part title, the first part a play on the two slogans she discusses in the essay and the second a clear description of the contents. See the chapter on 345–63 **ORGANIZING YOUR WRITING AND GUIDING YOUR READER** for more advice on composing a good title.

Considering Matters of Design

You'll probably write the main text of your argument in paragraph form, but think about what kind of information you're presenting and how you can design it to make your argument as easy as possible for your readers to understand. Think also about whether any visual or audio elements would be more persuasive than written words.

- 668–69 What would be an appropriate **FONT**? Something serious like Times Roman? Something traditional like Courier? Something else?

- 670–72 Would it help your readers if you divided your argument into shorter sections and added **HEADINGS**?

☀ academic literacies   ● fields   ● research
■ rhetorical situations   ⁚⁚ processes   ● media/design
▲ genres   ◆ strategies

- If you're making several points, would they be easier to follow if you set them off in a **LIST**?

● 669–70

- Do you have any supporting evidence that would be easier to understand in the form of a bar **GRAPH**, line graph, or pie chart?

● 676

- Would **ILLUSTRATIONS**—photos, diagrams, or drawings—add support for your argument? Online, would video, audio, or links help?

● 674–80

Getting Response and Revising

At this point you need to look at your draft closely, and if possible **GET RESPONSE** from others as well. Following are some questions for looking at an argument with a critical eye.

⁂ 372–74

- Is there sufficient background or context?
- Have you defined terms to avoid misunderstandings?
- Is the thesis clear and appropriately qualified?
- Are the reasons plausible?
- Is there enough evidence to support these reasons? Will readers accept the evidence as valid and sufficient?
- Can readers follow the steps in your reasoning?
- Have you considered potential objections or other positions? Are there any others that should be addressed?
- Have you cited enough sources, and are these sources credible?
- Are source materials documented carefully and completely, with in-text citations and a works-cited or references section?
- Are any visuals or links that are included used effectively and integrated smoothly with the rest of the text? If there are no visuals or links, would using some strengthen the argument?

Next it's time to **REVISE**, to make sure your argument offers convincing evidence, appeals to readers' values, and achieves your purpose. If you have

⁂ 375–77

380–83

time and want to explore alternatives, you might try **REWRITING** your draft to see if a different plan or approach might work better.

Editing and Proofreading

Once you've revised your draft, follow these guidelines for **EDITING** an argument:

- Check to see that your tone is appropriate and consistent throughout, reflects your **STANCE** accurately, and enhances the argument you're making.

- Be sure readers will be able to follow the argument; check to see you've provided **TRANSITIONS** and summary statements where necessary.

- Make sure you've smoothly integrated **QUOTATIONS**, **PARAPHRASES**, and **SUMMARIES** from source material into your writing and **DOCU-MENTED** them accurately.

- Look for phrases such as "I think" or "I feel" and delete them; your essay itself expresses your opinion.

- Make sure that **ILLUSTRATIONS** have captions and that charts and graphs have headings—and that all are referred to in the main text.

- If you're writing online, make sure all your links work.

- **PROOFREAD** and spell-check your essay carefully.

Taking Stock of Your Work

Take stock of what you've written by considering these questions:

- How did you go about researching your topic? What would you do differently next time?

- Did your stance change as you researched and wrote your argument? Why or why not?

- What types of evidence did you use in your argument—facts? statistics? expert testimony? something else? Why did you choose those types of evidence?

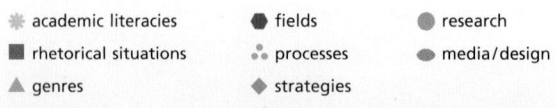

☀ academic literacies　　● fields　　● research
■ rhetorical situations　　⁝ processes　　◗ media/design
▲ genres　　◆ strategies

margin references:
378–79
380–83
72–74 ■
361–62
542–54 ●
560–63
673–83 ◗
383–84

- How did feedback from peers, teachers, tutors, or others influence your writing?

- In what ways were you open to other perspectives as you worked on your argument?

- If you were going to revise this argument for a different audience (say, a letter to an online forum or a newspaper), what changes would you need to make?

- What was most meaningful to you about this piece or about your experience of writing it?

If you need more help

See Chapter 34 if you are required to submit your argument as part of a writing **PORTFOLIO**. See also Chapter 11 on **ANALYZING TEXTS**, Chapter 16 on **EVALUATIONS**, and Chapter 20 on **PROPOSALS** for advice on writing those specific types of arguments.

385–90
104–39
214–22
258–68

14 Abstracts

140–63 ▲

Abstracts are summaries written to give readers the gist of a **REPORT** or presentation. Sometimes they are published in conference proceedings or databases. In courses in the *sciences*, *social sciences*, and *engineering*, you may be asked to create abstracts of your proposed projects and completed reports and essays. Abstracts are brief, typically 100–200 words, sometimes even shorter. Two common kinds are *informative abstracts* and *proposal abstracts*.

INFORMATIVE ABSTRACTS

Informative abstracts state in one paragraph the essence of a whole paper about a study or a research project. That one paragraph must mention all the main points or parts of the paper: a description of the study or project, its methods, the results, and the conclusions. Here is an example of the abstract accompanying a seven-page article that appeared in the *Journal of Clinical Psychology*:

> The relationship between boredom proneness and health-symptom reporting was examined. Undergraduate students ($N = 200$) completed the Boredom Proneness Scale and the Hopkins Symptom Checklist. A multiple analysis of covariance indicated that individuals with high boredom-proneness total scores reported significantly higher ratings on all five subscales of the Hopkins Symptom Checklist (Obsessive–Compulsive, Somatization, Anxiety, Interpersonal Sensitivity, and Depression). The results suggest that boredom proneness may be an important element to consider when assessing symptom reporting. Implications for

* academic literacies   ● fields   ● research
■ rhetorical situations   ⁘ processes   ● media/design
▲ genres   ◆ strategies

determining the effects of boredom proneness on psychological- and physical-health symptoms, as well as the application in clinical settings, are discussed.

—Jennifer Sommers and Stephen J. Vodanovich,
"Boredom Proneness"

The first sentence states the nature of the study being reported. The next summarizes the method used to investigate the problem, and the following one gives the results: students who, according to specific tests, are more likely to be bored are also more likely to have certain medical or psychological symptoms. The last two sentences indicate that the paper discusses those results and examines the conclusion and its implications.

PROPOSAL ABSTRACTS

Proposal abstracts contain the same basic information as informative abstracts, but their purpose is very different. You prepare proposal abstracts to persuade someone to let you write on a topic, pursue a project, conduct an experiment, or present a paper at a scholarly conference. This kind of abstract is not written to introduce a longer piece but rather to stand alone, and often the abstract is written before the paper itself. Titles and other aspects of the proposal deliberately reflect the theme of the proposed work, and you may use the future tense, rather than the past, to describe work not yet completed. Here is a possible proposal for doing research on boredom:

Undergraduate students will complete the Boredom Proneness Scale and the Hopkins Symptom Checklist. A multiple analysis of covariance will be performed to determine the relationship between boredom-proneness total scores and ratings on the five subscales of the Hopkins Symptom Checklist (Obsessive–Compulsive, Somatization, Anxiety, Interpersonal Sensitivity, and Depression).

Key Features / Abstracts

A summary of basic information. An informative abstract includes enough information to substitute for the report itself, and a proposal abstract gives an overview of the planned work.

Objective description. Abstracts present information on the contents of a report or a proposed study; they do not present arguments about or personal perspectives on those contents. The informative abstract on boredom proneness, for example, offers only a tentative conclusion: "The results *suggest* that boredom proneness *may* be an important element to consider."

Brevity. Although the length of abstracts may vary, journals and organizations often restrict them to 120–200 words—meaning you must carefully select and edit your words.

A BRIEF GUIDE TO WRITING ABSTRACTS

Considering the Rhetorical Situation

<table>
<tr><td>59–60 ■</td><td>**PURPOSE**</td><td>Are you giving a brief but thorough overview of a completed study? only enough information to create interest? a proposal for a planned study or presentation?</td></tr>
<tr><td>61–64 ■</td><td>**AUDIENCE**</td><td>For whom are you writing this abstract? What information about your project will your readers need?</td></tr>
<tr><td>72–74 ■</td><td>**STANCE**</td><td>Regardless of stance in the longer work, your abstract must be objective.</td></tr>
<tr><td>75–77 ■</td><td>**MEDIA / DESIGN**</td><td>How will you set off your abstract from the rest of the text? If you are publishing it online, should it be on a separate page? What format do your readers expect?</td></tr>
</table>

* academic literacies
■ rhetorical situations
▲ genres

● fields
∴ processes
◆ strategies

● research
● media/design

Generating Ideas and Text

Write the paper first, the abstract last. You can then use the finished work as the guide for the abstract, which should follow the same basic structure. *Exception:* you may need to write a proposal abstract months before the work it describes will be complete.

Copy and paste key statements. If you've already written the work, highlight your **THESIS**, objective, or purpose; basic information on your methods; your results; and your conclusion. Copy and paste those sentences into a new document to create a rough version of your abstract.

347–49

Pare down the rough abstract. **SUMMARIZE** the key ideas in the document, editing out any nonessential words and details. In your first sentence, introduce the overall scope of your study. Also include any other information that seems crucial to understanding your paper. Avoid phrases that add unnecessary words, such as "It is concluded that." In general, you probably won't want to use "I"; an abstract should cover ideas, not say what you think or will do.

550–51

Conform to any requirements. In general, an informative abstract should be at most 10 percent of the length of the entire work and no longer than the maximum length allowed. Proposal abstracts should conform to the requirements of the organization calling for the proposal.

Ways of Organizing an Abstract

Organizing abstracts is straightforward: in a single paragraph, briefly state the nature of the report or presentation, followed by an overview of the paper or the proposal.

[An informative abstract]

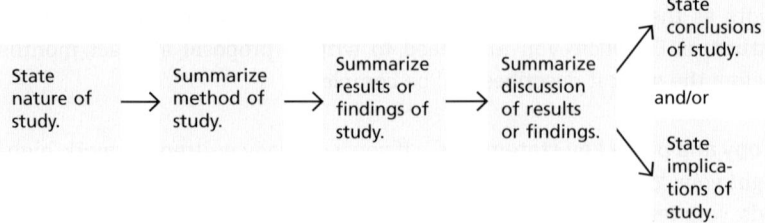

State nature of study. → Summarize method of study. → Summarize results or findings of study. → Summarize discussion of results or findings. → State conclusions of study. and/or State implications of study.

[A proposal abstract]

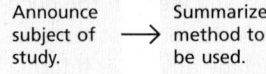

Announce subject of study. → Summarize method to be used.

Taking Stock of Your Work

Take stock of what you've written by considering these questions:

- How did you decide on the type of abstract—informative or proposal—to write?

- Describe your process for writing your abstract. What was easy, and what was challenging?

- What writing skills did you develop or refine by working on this piece? How might you use those skills in other writing situations?

- What did you do well in this piece? What could still be improved? What would you do differently next time?

If you need more help

364–66
367–71
372–79
380–84

See Chapter 30 for guidelines on **DRAFTING**, Chapter 31 on **ASSESSING YOUR OWN WRITING**, Chapter 32 on **GETTING RESPONSE AND REVISING**, and Chapter 33 on **EDITING AND PROOFREADING**.

* academic literacies   ● fields   ● research
■ rhetorical situations   •• processes   ● media/design
▲ genres   ◆ strategies

15 Annotated Bibliographies and Reviews of Scholarly Literature

When we do research, we may consult annotated bibliographies to evaluate potential sources and literature reviews when we need an overview of the important research ("literature") on a topic. In some courses, you may be asked to create annotated bibliographies or literature reviews to demonstrate that you have researched your topic thoroughly. This chapter offers advice on writing both.

ANNOTATED BIBLIOGRAPHIES

Annotated bibliographies describe, give publication information for, and sometimes evaluate each work on a list of sources. There are two kinds of annotations, *descriptive* and *evaluative*; both may be brief, consisting only of phrases, or more formal, consisting of sentences and paragraphs. Sometimes an annotated bibliography is introduced by a short statement explaining its scope.

Descriptive annotations simply summarize the contents of each work, without comment or evaluation. They may be very short, just long enough to capture the flavor of the work, like the examples in the following excerpt from a bibliography of books and articles on teen films, published in the *Journal of Popular Film and Television*.

MICHAEL BENTON, MARK DOLAN,
AND REBECCA ZISCH

Teen Film$

In the introduction to his book *The Road to Romance and Ruin*, Jon Lewis points out that over half of the world's population is currently under the age of twenty. This rather startling fact should be enough to make most Hollywood producers drool when they think of the potential profits from a target movie audience. Attracting the largest demographic group is, after all, the quickest way to box-office success. In fact, almost from its beginning, the film industry has recognized the importance of the teenaged audience, with characters such as Andy Hardy and locales such as Ridgemont High and the 'hood.

Beyond the assumption that teen films are geared exclusively toward teenagers, however, film researchers should keep in mind that people of all ages have attended and still attend teen films. Popular films about adolescents are also expressions of larger cultural currents. Studying the films is important for understanding an era's common beliefs about its teenaged population within a broader pattern of general cultural preoccupations.

This selected bibliography is intended both to serve and to stimulate interest in the teen film genre. It provides a research tool for those who are studying teen films and their cultural implications. Unfortunately, however, in the process of compiling this list we quickly realized that it was impossible to be genuinely comprehensive or to satisfy every interest.

Doherty, Thomas. *Teenagers and Teenpics: The Juvenilization of American Movies in the 1950s*. Unwin Hyman, 1988.
> Historical discussion of teenagers as a film market.

Foster, Harold M. "Film in the Classroom: Coping with 'Teenpics.'" *English Journal*, vol. 76, no. 3, Mar. 1987, pp. 86–88.
> Evaluation of the potential of using teen films such as *Sixteen Candles*, *The Karate Kid*, and *The Breakfast Club* to instruct adolescents on the difference between film as communication and film as exploitation.

Washington, Michael, and Marvin J. Berlowitz. "Blaxploitation Films and
High School Youth: Swat Superfly." *Jump Cut*, vol. 9, Oct.–Dec.
1975, pp. 23–24.
Marxist reaction to youth-oriented Black action films. Article
seeks to illuminate the negative influences the films have on
students by pointing out the false ideas about education, morality, and the Black family espoused by the heroes in the films.

*These annotations are purely descriptive; the authors express none of their own
opinions. They describe works as "historical" or "Marxist" but do not indicate
whether they're "good." The bibliography entries are documented in MLA style.*

Evaluative annotations offer opinions on a source as well as describe it.
They are often helpful in assessing how useful a source will be for your
own writing. The following evaluative annotations are from a bibliography by Kelly Green, a student at Arizona State University. Following her
instructor's directions, she labeled each required part of her annotation—
summary, degree of advocacy, credibility, and reliability.

KELLY GREEN

Researching Hunger and Poverty

Abramsky, Sasha. "The Other America, 2012: Confronting the Poverty Epidemic." *The Nation*, vol. 294, no. 20, 25 Apr. 2012, www.thenation
.com/article/other-america-2012-confronting-poverty-epidemic/.
The author presents the image of American poverty in 2012
with examples from various families living in poverty. The
author explores the conditions that make up the new recession and suggests that people in America notice the scale of
the issue and take action to solve it [Summary]. The author
advocates poverty reform and shows bias toward the interests
of low-income families. He acknowledges other perspectives
on the issue respectfully [Degree of Advocacy]. Abramsky is
a freelance journalist with experience in several magazines
and newspapers. He has written several books on the topic of

poverty [Credibility]. *The Nation* is one of the oldest-running magazines in the United States and contains opinions on politics and culture [Reliability].

Ambler, Marjane. "Sustaining Our Home, Determining Our Destiny." *Tribal College Journal*, vol. 13, no. 3, spring 2002, www.tribalcollegejournal .org/sustaining-home-determining-destiny/.

> The author examines the causes of poverty on Native American reservations, the factors that lead to solutions to Native American poverty, and the ways in which tribal colleges have helped improve life on reservations [Summary]. The author is strongly biased toward Native American interests and advocates that effective solutions to poverty originate within the reservations, especially in tribal colleges and universities [Degree of Advocacy]. Marjane Ambler was an editor for the *Tribal College Journal* for nine years and worked in national park service for nearly a decade [Credibility]. This article was published in 2002 in the *Tribal College Journal*, a national magazine published by the American Indian Higher Education Consortium [Reliability].

These annotations not only summarize the sources in detail but also evaluate their bias, or "degree of advocacy"; credibility; and reliability.

Key Features / Annotated Bibliographies

A statement of scope. Sometimes you need or are asked to provide a brief introductory statement to explain what you're covering. The authors of the bibliography on teen films introduce their bibliography with three paragraphs establishing a context for the bibliography and announcing their purpose for compiling it.

Complete bibliographic information. Annotations should provide all the information about each source using one documentation system (MLA, APA, or another one) so that you, your readers, or other researchers will be able to find the source easily. It's a good idea to include sources' URLs or **PERMALINKS** to make accessing online sources easier.

499 ●

A concise description of the work. A good annotation describes each item as carefully and objectively as possible, giving accurate information

---

※ academic literacies   ● fields   ● research
■ rhetorical situations   ∴ processes   ● media / design
▲ genres   ◆ strategies

and showing that you understand the source. These qualities will help to build authority—for you as a writer and for your annotations.

Relevant commentary.   If you write an evaluative bibliography, your comments should be relevant to your purpose and audience. The best way to achieve relevance is to consider what questions a potential reader might have about each source: What are the main points of the source? What is its argument? How even-handed or biased is it? How current and reliable is it? Will the source be helpful for your project?

Consistent presentation.   All annotations should follow a consistent pattern: if one is written in complete sentences, they should all be. Each annotation in the teen films bibliography, for example, begins with a phrase (not a complete sentence) characterizing the work.

A BRIEF GUIDE TO WRITING ANNOTATED BIBLIOGRAPHIES

Considering the Rhetorical Situation

PURPOSE   Will your bibliography need to demonstrate the depth or breadth of your research? Will your readers actually track down and use your sources? Do you need or want to convince readers that your sources are good?   ■ 59–60

AUDIENCE   For whom are you compiling this bibliography? What does your audience need to know about each source?   ■ 61–64

STANCE   Are you presenting yourself as an objective describer or evaluator? Or are you expressing a particular point of view toward the sources you evaluate?   ■ 72–74

MEDIA/DESIGN   If you are publishing the bibliography electronically, will you provide links from each annotation to the source itself? Online or offline, should you distinguish the bibliographic information from the annotation by using a different font?   ■ 75–77

Generating Ideas and Text

Decide what sources to include. You may be tempted to include in a bibliography every source you find or look at. A better strategy is to include only those sources that you or your readers may find potentially useful in researching your topic. For an academic bibliography, consider the qualities in the list below. Some of these qualities should not rule a source in or out; they simply raise issues you need to think about.

- *Appropriateness.* Is this source relevant to your topic? Is it a primary source or a secondary source? Is it aimed at an appropriate audience? General or specialized? Elementary, advanced, or somewhere in between?

- *Credibility.* Is the author reputable? Is the publication, publishing company, or sponsor of the site reputable? Do the ideas more or less agree with those in other sources you've read?

- *Balance.* Does the source present enough evidence for its assertions? Does it show any particular bias? Does it present opposing arguments fairly?

- *Timeliness.* Is the source recent enough? Does it reflect current thinking or research about the subject?

501–23 ●

If you need help **FINDING SOURCES**, see Chapter 47.

MLA 564–614 ●
APA 615–55

Compile a list of works to annotate. Give the sources themselves in whatever documentation style is required; see the guidelines for **MLA** and **APA** styles in Chapters 53 and 54.

Determine what kind of bibliography you need to write. Will your bibliography be descriptive or evaluative? Will your annotations be in the form of phrases? complete sentences? paragraphs? The form will shape your reading and note taking. If you're writing a descriptive bibliography, your reading goal will be just to understand and capture the writer's message as clearly as possible. If you're writing an evaluative bibliography, you will

※ academic literacies　　● fields　　● research
■ rhetorical situations　　⋮ processes　　● media / design
▲ genres　　◆ strategies

also need to assess the source as you read in order to include your own opinions of it.

Read carefully. To write an annotation, you must understand the source's argument; but when writing an annotated bibliography as part of a **PROPOSAL**, you may have neither the time nor the need to read the whole text. Here's a way of quickly determining whether a source is likely to serve your needs:

▲ 258–68

- Check the publisher or sponsor (university press? scholarly journal? popular magazine? website sponsored by a reputable organization?).
- Read the preface (of a book), abstract (of a scholarly article), introduction (of an article in a nonscholarly magazine or a website).
- Skim the table of contents or the headings.
- Read the parts that relate specifically to your topic.

Research the writer, if necessary. If you are required to indicate the writer's credentials, you may need to do additional research. You may find information by typing the writer's name into a search engine or looking up the writer in *Gale Literature: Contemporary Authors*. In any case, information about the writer should take up no more than one sentence in your annotation.

Summarize the work in a sentence or two. **DESCRIBE** it as objectively as possible: even if you're writing an evaluative annotation, you can evaluate the central point of a work better by stating it clearly first. *If you're writing a descriptive annotation, you're done.*

◆ 456–63

Establish criteria for evaluating sources. If you're **EVALUATING SOURCES** for a project, you'll need to evaluate them in terms of their usefulness for your project, their **STANCE**, and their overall credibility.

● 524–34

■ 72–74

Write a brief evaluation of the source. If you can generalize about the worth of the entire work, fine. You may find, however, that some parts are useful while others are not, and what you write should reflect that mix.

Be consistent—in content, sentence structure, and format.

- *Content.* Try to provide about the same amount of information for each entry. If you're evaluating, don't evaluate some sources and just describe others.

- *Sentence structure.* Use the same style throughout—complete sentences, brief phrases, or a mix.

668–69
- *Format.* Use one documentation style throughout; use a consistent **FONT** for each element in each entry—for example, italicize or underline all book titles.

Ways of Organizing an Annotated Bibliography

Depending on their purpose, annotated bibliographies may or may not include an introduction. Most annotated bibliographies cover a single topic and so are organized alphabetically by author's or editor's last name. When a work lacks a named author, alphabetize it by the first important word in its title. Consult the documentation system you're using for additional details about alphabetizing works appropriately.

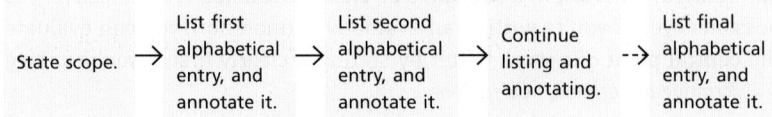

State scope. → List first alphabetical entry, and annotate it. → List second alphabetical entry, and annotate it. → Continue listing and annotating. ⇢ List final alphabetical entry, and annotate it.

Sometimes an annotated bibliography needs to be organized into several subject areas (or genres, periods, or some other category); if so, the entries are listed alphabetically within each category. For example, a bibliography about terrorism breaks down into subject categories such as "Global Terrorism" and "Weapons of Mass Destruction."

- ☀ academic literacies
- ■ rhetorical situations
- ▲ genres
- ⬟ fields
- ⁙ processes
- ◆ strategies
- ● research
- ⬤ media/design

[Multicategory bibliography]

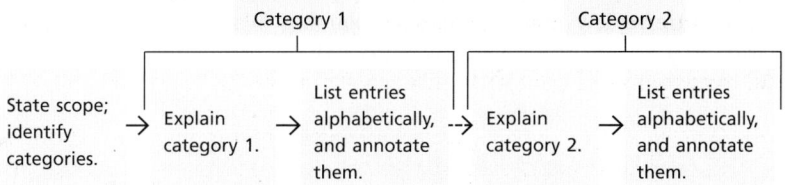

Taking Stock of Your Work

Take stock of what you've written by considering these questions:

- Did curiosity about your topic play a role in your research and writing? If so, how?

- What previous experience did you have with research? In what ways, if any, did you draw on that experience for your annotated bibliography?

- How did you go about researching the entries in this bibliography? What databases or other tools did you use? How did you decide on keywords for your search?

- How did you approach sources that were difficult to understand? What reading strategies did you use to make sense of them?

- How did you decide which sources to include in your bibliography and which ones to leave out?

- Did you cite sources accurately?

- Overall, what was challenging about writing an annotated bibliography? What was rewarding?

REVIEWS OF SCHOLARLY LITERATURE

Reviews of scholarly literature describe and evaluate important research ("literature") available on a topic. In writing a literature review, your goal is to give an overview of the literature on a topic. You do that by discussing the literature that is most relevant to your topic and your purposes, providing

clear and accurate summaries of appropriate source material, and describing relationships among facts and concepts. Here is a brief excerpt from a literature review that describes scholarship in zombie movies; it was written by a student at the University of Mary Washington.

CAMERON CARROLL

Zombie Film Scholarship
A Review of the Literature

Zombies are the shambling undead creatures that attack in hordes and strike terror into the hearts of moviegoers with their mindless aggression. Monsters on the big screen fascinate American audiences and scholars alike as representations of cultural fears and reflections of public perceptions, and they have done so since they first shuffled into theaters. The walking dead first haunted cinemas in 1932 with the release of *White Zombie* and have gone through cycles of popularity both at the box office and in scholarly debate. Without a doubt, the height of zombie scholarship began in the mid-2000s with only shallow mentions made in horror analyses before then. Only the English discipline gave the zombie credit as a subject worthy of study. As author and English professor Kyle Bishop wrote in 2010, "The zombie phenomenon has yet to be plumbed to its depths by the academic and literary markets."[1] Although significant zombie research has only surfaced in the 2000s, a valiant effort is currently being made to explore zombie symbolism and its historical and cultural context across nontraditional scholarly fields, such as anthropology, psychology, sociology, and philosophy.

. . .

Zombies, as creatures that waver between living and dead, necessarily bring forth questions on the nature of being and thus also fall into the realm of philosophy as seen in *Zombies, Vampires, and*

academic literacies   fields   research

rhetorical situations   processes   media/design

genres   strategies

Philosophy: New Life for the Undead published in 2010 and edited by Richard Greene and K. Silem Mohammed, professors of philosophy and English, respectively. Philosophers in the study of zombies seek to answer such questions as, is it better to be undead or dead? Richard Greene argues that since death is a state of nonexistence and Undeath is at least a primal existence, the question comes down to the classic "to be, or not to be?" For him, Undeath is the only option for a state of being, since death is nonexistence. Existence in any form is inherently better than utterly ceasing to be.[16] Greene's philosophy colleague, William Larkin, counters that zombies are driven to actions seen as evil and mindless, both states that humans do not usually choose. He cites how zombie films repeatedly depict survivors who ask their friends and family to kill them should it appear that they will turn into zombies, so desperate are they to avoid Undeath.[17] Simon Clark, a writer with a master's degree in Fine Arts, discusses morality in relation to zombies: if zombies are nonmoral and whether or not they are the freest creatures because of their lack of morality. Clark writes that they become ultimately free in their modern incarnations from turn-of-the-millennium cinema.[18] Zombies break through barriers that survivors put up to keep them confined and are so primal that they cannot be held accountable for their actions, making them truly liberated. A fictional creature encourages real discussions in philosophical debate, demonstrating yet another area that inspires scholars to reexamine their own field because of zombies.

. . .

The message to take away from zombie scholarship is that what it represents is evolutionary and infinitely debatable, but the victory for zombie studies is that zombies are being studied at all. After decades of being underrepresented in horror scholarship, zombies are finally getting their due as a cultural icon, complete with varying opinions and interpretations. Scholarly debate proves that the zombie is a valid resource for understanding American culture and worthy of the enthusiastic pursuit of interdepartmental scholarship. As shown within the scholarly works above, the debate over the meaning of the living dead is quite lively, indeed.

Notes

1. Kyle William Bishop, *American Zombie Gothic: The Rise and Fall (and Rise) of the Walking Dead in Popular Culture* (Jefferson, NC: McFarland, 2010), 7.

. . .

16. Richard Greene and K. Silem Mohammed, eds., *Zombies, Vampires, and Philosophy: New Life for the Undead* (Chicago: Open Court, 2010), 13.

17. Greene and Mohammed, 20.

18. Greene and Mohammed, 208.

Carroll begins by establishing a context for her discussion and then focuses on her topic, the scholarship of zombies in film. In her review, she discusses the history of zombie scholarship in general and then in psychology, social trends, fiction, and, in this excerpt, philosophy. A history major, she follows Chicago style; in addition to the notes, she included a bibliography listing all her sources, including the two noted here.

Key Features / Reviews of Scholarly Literature

Careful, thorough research.   A review of scholarly literature demands that you research all the major literature on the topic—or at least the major literature available to you, given the time you have.

Accurate, objective summaries of the relevant literature.   Readers expect a literature review to objectively summarize the main ideas or conclusions of the texts reviewed.

Critical evaluation for the literature.   A literature review offers considered selection of the most important, relevant, and useful sources of information on its topic, so you must evaluate each source to decide whether it should be included and then to determine how it advances understanding of the topic.

535–41

SYNTHESIS of the scholarship.   A literature review differs from an annotated bibliography in that the review identifies key concepts, similarities, and differences within the body of literature, showing how the sources relate to one another by method, study findings, themes, main ideas, or something else.

* academic literacies   ● fields   ● research
■ rhetorical situations   ⁚ processes   ● media / design
▲ genres   ◆ strategies

A clear focus. Because a literature review provides an overview of your topic's main issues and explains the main concepts underlying your research, it must be carefully organized and clearly focused on your specific topic.

Taking Stock of Your Work

Take stock of what you've written by considering these questions:

- Did curiosity about your topic play a role in your research and writing? If so, how?

- What previous experience did you have with research? In what ways, if any, did you draw on that experience for your literature review?

- How did you approach sources that were difficult to understand? What reading strategies did you use to make sense of them?

- What helped you the most in finding, keeping track of, and evaluating the scholarship? What can you do next time to make the process more efficient and productive?

- How did you decide which sources to include in your literature review and which ones to leave out?

- What led you to group related sources together as you did?

- How did you go about synthesizing similar sources?

- Overall, what was challenging about writing a literature review? What was rewarding?

If you need more help

See Chapter 30 for guidelines on **DRAFTING**, Chapter 31 on **ASSESSING YOUR OWN WRITING**, Chapter 32 on **GETTING RESPONSE AND REVISING**, and Chapter 33 on **EDITING AND PROOFREADING**. See Chapter 34 if you are required to submit your bibliography in a writing **PORTFOLIO**.

364–66
367–71
372–79
380–84
385–90

16 Evaluations

Techspot reviews computer gaming equipment. The *Princeton Review* and *U.S. News & World Report* evaluate colleges and universities. You probably consult such sources to make decisions, and you probably evaluate things all the time—when you recommend a film (or not) or a teacher (ditto). An evaluation is basically a judgment; you judge something according to certain criteria, supporting your judgment with reasons and evidence. You need to give your reasons for evaluating it as you do because often your evaluation will affect your audience's actions: they must see this movie,

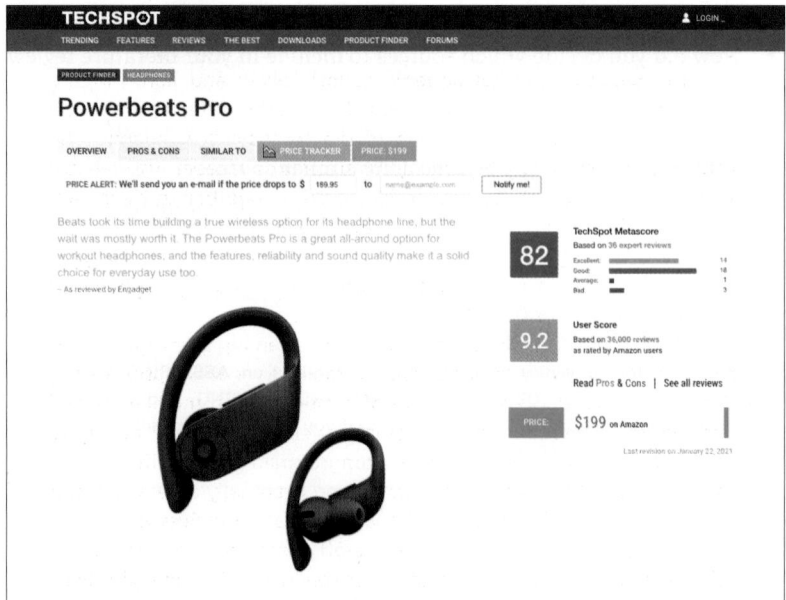

Techspot aggregates reviews of electronics from across the web and *Amazon*.

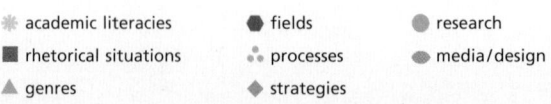

- ☀ academic literacies
- ■ rhetorical situations
- ▲ genres
- ⬢ fields
- ⦂ processes
- ◆ strategies
- ⬤ research
- ⬬ media/design

needn't bother with this book, should be sure to have the Caesar salad at this restaurant, and so on.

In college courses, students in *literature, film, drama,* and *art* classes may be assigned to evaluate poems, fiction, movies, plays, and artwork, and those in *business* and *political science* classes may be asked to evaluate advertising or political campaigns or plans for business or public-policy initiatives. In a review that follows, written for *The Daily Bruin,* UCLA's student newspaper, Olivia Mazzucato offers her evaluation of the film *The Lovebirds.*

OLIVIA MAZZUCATO

Difficult-to-Follow Narrative Redeemed by Well-Executed Comedy in The Lovebirds

Issa Rae and Kumail Nanjiani are two of the brightest stars in modern comedy—pairing them together is an obvious winning combo.

The two play the beleaguered couple Leilani and Jibran in *The Lovebirds,* the first film to be released on Netflix after its theatrical premiere was canceled as a result of the coronavirus pandemic. The central pair unwittingly become accomplices to a murder just minutes after breaking up and quickly decide to work together to clear their names, leading them down a rabbit hole of blackmail and black-tie cult gatherings. Though the film's story gets convoluted, Rae and Nanjiani are able to salvage the muddy plot with their razor-sharp comedic wit and hilarious chemistry.

The plot of *The Lovebirds* starts out with an air of intrigue but is soon bogged down with unnecessary complications and loose threads, becoming less a mystery and more of a mess. Within minutes, Leilani and Jibran are wrapped up in a complex web of conspiracy, having to confront an absurd lineup of foes—a congressman and his wife, a blackmail ring operating out of a fraternity, and eventually a nefarious cult that's equal parts Illuminati, *The Handmaid's Tale,* and *Eyes Wide Shut.*

All the individual capers with one-off characters make for entertaining vignettes but fail to tie together cohesively. The final plot twist fails to make sense of the jumble, barely registering as an anticlimactic reveal. In a film full of inventive quips and conversations, the finale is

Rae and Nanjiani in a scene from *The Lovebirds*.

far from the most memorable scene in the film, which deflates a lot of
the momentum the story generates.

Luckily, Rae and Nanjiani's easy chemistry and awkward humor   5
transcend the narrative jumble. Whether they're arguing over the odds
of orgy schedulers using a calendar app or intimidating answers out of
a terrified fraternity brother, their back and forth feels effortless as they
operate on the same wavelength of frenetic, weird energy.

Their palpable onscreen connection makes it easy to both believe
the love between the two as well as understand why they've started to
grate on each other after four years together. They might argue with
comical conviction about whether or not they'd do well on *The Amazing
Race*, but the conversation hints at the deeper emotional insecurities
about incompatibility that plague their relationship.

The two comedians also bring versatility and depth to their characters. Within the first fifteen minutes of the movie, Leilani and Jibran are
falling in love, breaking up, and witnessing their first murder. The actors
take the fast pacing in stride, matching it with equally fast-paced jokes.

✳ academic literacies	⬡ fields	● research
◼ rhetorical situations	⁝ processes	⬭ media/design
▲ genres	◆ strategies	

Though *The Lovebirds* isn't necessarily a traditional romantic comedy, the movie makes Leilani and Jibran's breakup more than just a plot point—it's a scene with heartbreaking gravity. The quiet tension that fills the car after another blowout argument adds a sense of realism to their relationship, making it impossible not to root for their reconciliation as they go on their adventure. The narrative arc of the film may fail, but the emotional arc is where the film finds its footing.

Not to mention, *The Lovebirds* is flat-out funny.

From the ridiculous scenarios Leilani and Jibran find themselves   10
in to the one-liners Rae and Nanjiani scatter throughout the film, *The Lovebirds* never fails to earn a laugh, even in its most serious scenes. The easily enjoyable film is able to span the range of comedy by blending physical slapstick comedy with witty dialogue to produce a compact eighty-six-minutes full of levity and heart. The film is in no way groundbreaking, but it puts Rae and Nanjiani firmly into the spotlight, demonstrating their potential for larger silver-screen roles.

The two transform an easily mediocre film into a fun, enjoyable comedy that offers a brief escape from reality.

Mazzucato provides a brief summary of the plot of The Lovebirds *and the context of its release and then evaluates the film using clear criteria: its plot, the chemistry between the leading actors, their comedic and acting skill, and its ability to generate laughs.*

Key Features / Evaluations

A concise description of the subject. You should include just enough information to let readers who may not be familiar with your subject understand what it is; the goal is to evaluate, not summarize. Depending on your topic and medium, some of this information may be in visual or audio form. In her introduction, Mazzucato briefly describes *The Lovebirds'* main plot to provide what readers need to understand the context of her evaluation, and then she goes into more detail when evaluating the plot and its shortcomings.

Clearly defined criteria. You need to determine clear criteria as the basis for your judgment. In reviews or other evaluations written for a broad audience, you can integrate the criteria into the discussion as reasons for your assessment, as Mazzucato does in her evaluation of *The Lovebirds*. In more formal evaluations, you may need to announce your criteria explicitly. Mazzucato evaluates the film based on the stars' performances, the complexity and coherence (or lack thereof) of the plot, the acting ability of the main characters and their chemistry together, and the film's ability to generate laughs.

A knowledgeable discussion of the subject. To evaluate something credibly, you need to show that you know it yourself and that you understand its context. Mazzucato cites many examples from *The Lovebirds*, demonstrating her knowledge of the film. Some evaluations require that you research what other authoritative sources have said about your subject. Mazzucato might have referred to other reviews of the film to show that she'd researched others' views.

A balanced and fair assessment. An evaluation is centered on a judgment. Mazzucato concedes that "the film's story gets convoluted" and does not "tie together cohesively," and that "the finale . . . deflates a lot of the momentum the story generates." Nevertheless, the overall humorousness and the chemistry between the leads "transform an easily mediocre film into a fun, enjoyable comedy." It is important that any judgment be balanced and fair. Seldom is something all good or all bad. A fair evaluation need not be all positive or all negative; it may acknowledge both strengths and weaknesses. For example, a movie's soundtrack may be wonderful while the plot is not.

Well-supported reasons. You need to argue for your judgment, providing reasons and evidence that might include visual and audio as well as verbal material. Mazzucato gives several reasons for her overall positive assessment of *The Lovebirds*—the problems with the plot, the strong performances of Issa Rae and Kumail Nanjiani, the film's emotional arc, that the film "is flat-out funny"—and she supports these reasons with several examples from the film.

* academic literacies
■ rhetorical situations
▲ genres

● fields
•• processes
◆ strategies

● research
● media/design

A BRIEF GUIDE TO WRITING EVALUATIONS

Choosing Something to Evaluate

You can more effectively evaluate a limited subject than a broad one: review certain dishes at a local restaurant rather than the entire menu; review one film or scene rather than all the films by Alfred Hitchcock or all seventy-three *Game of Thrones* episodes. The more specific and focused your subject, the better you can write about it.

Considering the Rhetorical Situation

PURPOSE	Are you writing to affect your audience's opinion of a subject? to help others decide what to see, do, or buy? to demonstrate your expertise in a field?	59–60
AUDIENCE	To whom are you writing? What will your audience already know about the subject? What will they expect to learn from your evaluation of it? Are they likely to agree with you or not?	61–64
STANCE	What is your attitude toward the subject, and how will you show that you have evaluated it fairly and appropriately? Think about the tone you want to use: Should it be reasonable? passionate? critical?	72–74
MEDIA/DESIGN	How will you deliver your evaluation? In print? Electronically? As a speech? Can you show images or audio or video clips? If you're submitting your text for publication, are there any format requirements?	75–77

Generating Ideas and Text

Explore what you already know. FREEWRITE to answer the following questions: What do you know about this subject or subjects like it? What are your initial or gut feelings, and why do you feel as you do? How does this subject reflect or affect your basic values or beliefs? How have others evaluated subjects like this?

333–34

Identify criteria. Make a list of criteria you think should be used to evaluate your subject. Think about which criteria will likely be important to your **AUDIENCE**. You might find **CUBING** and **QUESTIONING** to be useful processes for thinking about your criteria.

61–64
336
337

Evaluate your subject. Study your subject closely to determine to what extent it meets each of your criteria. You may want to list your criteria and take notes related to each one, or you may develop a rating scale for each criterion to help stay focused on it. Come up with a tentative judgment.

Compare your subject with others. Often, evaluating something involves **COMPARING AND CONTRASTING** it with similar things. We judge movies in comparison with the other movies we've seen and french fries with the other fries we've tasted. Sometimes those comparisons can be made informally. For other evaluations, you may have to do research—to try on several pairs of jeans before buying any, for example—to see how your subject compares.

437–44

State your judgment as a tentative thesis statement. Your **THESIS STATEMENT** should be one that addresses both pros and cons. "*Hawaii Five-O* is fun to watch despite its stilted dialogue." "Of the five sport-utility vehicles tested, the Toyota 4Runner emerged as the best in comfort, power, and durability, though not in styling or cargo capacity." Both of these examples offer a judgment but qualify it according to the writer's criteria.

347–49

Anticipate other opinions. I think Anthony Anderson is a comic genius whose TV show *black·ish* is first-rate. You think Anderson is a terrible actor whose show is unwatchable. How can I write a review of his latest series that you will at least consider? One way is by **ACKNOWLEDGING** other opinions—and **REFUTING** those opinions as best I can. I may not persuade you to watch Anderson's show, but I can at least demonstrate that by certain criteria he should be appreciated. You may need to **RESEARCH** how others have evaluated your subject.

424–25
426
489

☀ academic literacies   ● fields   ● research
■ rhetorical situations   ⁖ processes   ● media/design
▲ genres   ◆ strategies

Identify and support your reasons. Write out all the **REASONS** you can think of that will convince your audience to accept your judgment. Review your list to identify the most convincing or important reasons. Then review how well your subject meets your criteria, and decide how best to **SUPPORT** your reasons: through examples, authoritative opinions, statistics, visual or audio evidence, or something else.

◆ 413–14

◆ 414–22

Ways of Organizing an Evaluation

Evaluations are usually organized in one of two ways. One way is to introduce what's being evaluated, followed by your judgment, discussing your criteria along the way. This is a useful strategy if your audience may not be familiar with your subject.

[Start with your subject]

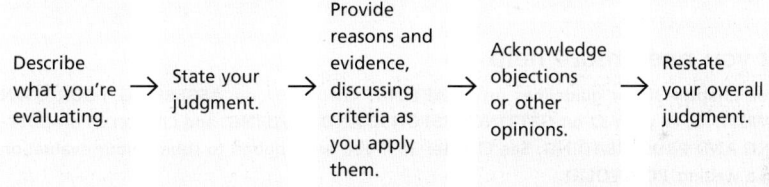

Describe what you're evaluating. → State your judgment. → Provide reasons and evidence, discussing criteria as you apply them. → Acknowledge objections or other opinions. → Restate your overall judgment.

You might also start by identifying your criteria and then follow with a discussion of how well your subject meets those criteria. This strategy foregrounds the process by which you reached your conclusions.

[Start with your criteria]

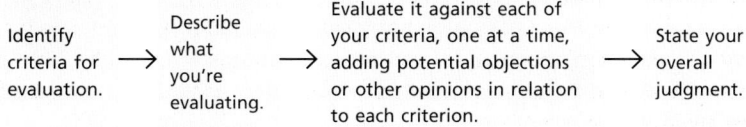

Identify criteria for evaluation. → Describe what you're evaluating. → Evaluate it against each of your criteria, one at a time, adding potential objections or other opinions in relation to each criterion. → State your overall judgment.

Taking Stock of Your Work

Take stock of what you've written by considering these questions:

- How did you decide on your stance for your evaluation? How well does your evaluation communicate that stance?

- How did you develop criteria for your evaluation?

- In working on this piece, did you use a process or strategy you've used before? If so, what was it? How well did it work? What new strategies or processes do you want to try in the future?

- What academic habits of mind—like creativity, flexibility, persistence, or another habit—did you need in order to write this evaluation? What specific examples can you give?

- How satisfied are you with your writing? What would you do differently next time?

If you need more help

364–66
367–71
372–79
380–84
385–90

See Chapter 30 for guidelines on **DRAFTING**, Chapter 31 on **ASSESSING YOUR OWN WRITING**, Chapter 32 on **GETTING RESPONSE AND REVISING**, and Chapter 33 on **EDITING AND PROOFREADING**. See Chapter 34 if you are required to submit your evaluation in a writing **PORTFOLIO**.

☀ academic literacies ● fields ● research
■ rhetorical situations ⁘ processes ● media/design
▲ genres ◆ strategies

17 Literary Analyses

Literary analyses are essays that examine literary texts closely to understand their messages, interpret their meanings, appreciate their techniques, or understand their historical or social contexts. Such texts traditionally include novels, short stories, poems, and plays but may also include films, TV shows, videogames, music, and comics. You might read Gwendolyn Brooks's poem "We Real Cool" and notice how she uses slang to get across her message. You could explore the distinctive point of view in Ambrose Bierce's story "An Occurrence at Owl Creek Bridge." Or you could point out the differences between Stephen King's *The Shining* and Stanley Kubrick's screenplay based on that novel. In all these cases, you use specific analytical techniques to go below the surface of the work to deepen your understanding of how it works and what it means.

You may be assigned to analyze works of literature in courses in *English, film, drama,* and many other subjects. Here is a poem by the twentieth-century American poet Robert Frost, followed by a student's analysis of it written for a literature course at the University of South Dakota and chosen as a winner of the 2017 Norton Writer's Prize.

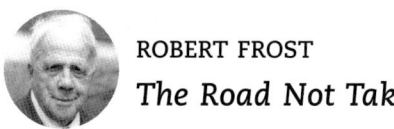

ROBERT FROST
The Road Not Taken

Two roads diverged in a yellow wood,
And sorry I could not travel both
And be one traveler, long I stood
And looked down one as far as I could
To where it bent in the undergrowth;   5

Then took the other, as just as fair,
And having perhaps the better claim,
Because it was grassy and wanted wear;
Though as for that the passing there
Had worn them really about the same,   10

And both that morning equally lay
In leaves no step had trodden black.
Oh, I kept the first for another day!
Yet knowing how way leads on to way,
I doubted if I should ever come back.   15

I shall be telling this with a sigh
Somewhere ages and ages hence:
Two roads diverged in a wood, and I—
I took the one less traveled by,
And that has made all the difference.   20

MATTHEW MILLER

Frost's Broken Roads

"The Road Not Taken" by Robert Frost is arguably one of the most popular poems ever written. Read at graduations, eulogies, even in movies and car commercials, it is often interpreted as an ode to individualism. Frost's image of the road "less traveled" has become synonymous with daring life choices that make "all the difference" in living a fulfilling life (lines 19–20). Some may latch on to this as the poem's deeper meaning. However, this convenient conclusion ignores several conflicting, yet beautiful, details that lead the poem down a path of broken metaphor and temporal inconsistency. To truly recognize what Frost is building in this poem, a few nagging inconsistencies must be considered.

In the first line of the poem, the traveler is depicted hiking in a "yellow wood" wherein he finds a fork in his path (1). This setting is the foundation of a common metaphor, but transposing familiar notions of a figurative fork in the road onto Frost's poem requires acceptance of the traveler's metaphoric natural world and its "temporal scheme,"

※ academic literacies   ● fields   ● research
■ rhetorical situations   ⁞ processes   ● media/design
▲ genres   ◆ strategies

or some form of unified movement through time, as the critic Cleanth Brooks describes it in his classic book on poetic structure, *The Well Wrought Urn* (203). If the path that is covered "[i]n leaves no step had trodden black" (Frost, line 12) makes "all the difference" (20) in a life, common sense indicates the decision to choose that path has to occur early enough during a lifetime to properly affect its outcome. The speaker/traveler affirms this, telling the reader that he "shall be telling" them the story of his decision "with a sigh / Somewhere ages and ages hence" (16–17). In other words, the traveler must be fairly young, otherwise he wouldn't have "ages and ages" (17) left to tell his story. It may not be apparent at a surface level, but this complicates our understanding of the poem's implicit story.

It is common knowledge that morning, the beginning of the day, is often paralleled figuratively with the beginning of other things (e.g., "the dawn of civilization"). In this sense, the setting helps to solidify the poem's established temporal sense; the earliness of the day—"morning" (16)—parallels the early point in life when the traveler makes his life-altering decision. However, the description of the scene, the "yellow wood" (1) where the ground "lay / In leaves" (11–12), indicates that late autumn has set in around the forest. Parallels between seasonal progression and the human life cycle saturate literature and art with such metaphors as the "springtime of youth" and the "hoary winter of old age" (Kammen 23). With that in mind, the end of autumn represents the bitter end of productive years and the first step into the cold and death of winter. So, embedded in the poem is a temporal inconsistency; the traveler is young, with "ages and ages" yet to live, but the autumn setting implies his world is quickly coming to an end.

Another question that complicates a traditional understanding of the poem is that of identity. The narrator claims he is "sorry" he "could not travel both" paths "and be one traveler" (Frost, lines 2–3). For him to stray off his path would equal becoming a different person; he "could not travel both / *And be one traveler*" (2–3; emphasis added). However, one person can easily be two different travelers in a lifetime (i.e., one person can take both a cruise vacation and a backpacking trip: two very different traveling styles). The traveler/speaker even admits this is possible by saying he *could* keep "the first for another day" (13). His excuse for not traveling both paths was not that it was impossible but rather that "knowing how way leads on to way, / I doubted if I should

ever come back" (14–15). Although he believes one of the paths has "the better claim, / Because it was grassy and wanted wear" (7–8), in practically the same breath he casts doubt on the claim that it is "less traveled" (19), admitting other travelers have worn down the two trails "really about the same" (10). The speaker in this poem is not Ralph Waldo Emerson's self-reliant transcendentalist, who can "speak what [he] think[s] to-day in words as hard as cannon balls, and to-morrow speak what to-morrow thinks in hard words again, though it contradict every thing [he] said to-day" (214).

Frost's syntax and punctuation add even more nuance. The poem consists of one sentence that takes up the first two stanzas and part of the third, in which our traveler deliberates and, eventually, makes his decision; and the three sentences of the third and fourth stanzas, in which our traveler lives with his choice. As the poem progresses to the point where our traveler makes his decision, we see punctuation that is irregular when compared with the rest of the poem: an exclamation point when the traveler finally makes a decision (possibly showing excess emotion); an em dash (—) followed by a repeated word when retelling his story that functions almost like a stutter, possibly showing regret/ lack of confidence in his choice. With this in mind, it seems as though our traveler had a tough time making his decision, and afterward, there is no obvious approval or happiness with it, only that it "made all the difference" (Frost, line 20), which could be positive or negative.

Placing the poem in its historical context further complicates these questions. According to an article by the poet Katherine Robinson, "The Road Not Taken" was actually written "as a joke for a friend, the poet Edward Thomas." Robinson writes:

> Indeed, when Frost and Thomas went walking together, Thomas would often choose one fork in the road because he was convinced it would lead them to something. . . . In a letter, Frost goaded Thomas, saying, "No matter which road you take, *you'll always sigh*, and wish you'd taken another." (emphasis added)

Introducing the poet's biography might be considered by many a sin against the work, especially for those espousing Cleanth Brooks's celebration of poetic unity and universal meaning. However, knowing

this information fills in several gaps about the poem; instead of confusing inconsistencies and paradoxical meanings, the poem can now be viewed—at least partly—as teasing from a friend. It is easy to imagine an indecisive Thomas, standing and staring down the fork in the path, afraid of what he'll miss if he picks the wrong trail, and then on the way home "telling" Frost, "with a sigh" (Frost, line 16), about all he swore he missed on *the road he didn't take*. This being said, Frost published this work knowing full well of its depth and epistemological possibilities, saying in a letter, "My poems . . . are all set to trip the reader head foremost into the boundless" (qtd. in Robinson).

Frost's timeless poem is an elegant narrative, filled with serene imagery and laced with layers of mystery. Readers of this canonical work can easily find themselves slipping into the easy, traditional reading of an ode to individualism. Upon closer inspection, there is only one thing clear about "The Road Not Taken," which is said best in words Frost loved to tell his readers regarding his classic poem: "[Y]ou have to be careful of that one; it's a tricky poem—very tricky" (qtd. in Robinson).

Works Cited

Brooks, Cleanth. *The Well Wrought Urn: Studies in the Structure of Poetry*. Harcourt, 1975.

Emerson, Ralph Waldo. "Self-Reliance." *The Norton Anthology of American Literature*, Robert S. Levine et al., general editors, 9th ed., vol. B, W. W. Norton, 2017, pp. 236–53.

Frost, Robert. "The Road Not Taken." *Poetry Foundation,* www.poetryfoundation.org/poems/44272/the-road-not-taken.

Kammen, Michael G. *A Time to Every Purpose: The Four Seasons in American Culture*. U of North Carolina P, 2004.

Robinson, Katherine. "Robert Frost: 'The Road Not Taken.'" *Poetry Foundation,* 27 May 2016, www.poetryfoundation.org/articles/89511/robert-frost-the-road-not-taken.

Miller focuses his analysis on "tricky" aspects of Frost's poem. In addition, he uses aspects of Frost's biography and letters to resolve some of the seeming contradictions and tensions in the poem.

Key Features / Literary Analyses

An arguable thesis.   A literary analysis is a form of argument; you are arguing that your analysis of a literary work is valid. Your thesis, then, should be
arguable, as Miller's is: "To truly recognize what Frost is building in this poem,
a few nagging inconsistencies must be considered." A mere summary—
"Frost writes about someone trying to decide which road to take"—would
not be arguable and therefore would not be a good thesis.

Careful attention to the language of the text.   The key to analyzing
a text is looking carefully at the language, which is the foundation of its
meaning. Specific words, images, metaphors—these are where analysis
begins. You may also bring in contextual information, such as cultural,
historical, or biographical facts, or you may refer to similar texts. But the
words, phrases, and sentences that make up the text you are analyzing
are your primary source when dealing with texts. That's what literature
teachers mean by "close reading": reading with the assumption that every
word of a text is meaningful.

Attention to patterns or themes.   Literary analyses are usually built on
evidence of meaningful patterns or themes within a text or among several
texts. These patterns and themes reveal meaning. In Frost's poem, images
of diverging roads and yellow leaves create patterns of meaning, while
the regular rhyme scheme ("wood"/"stood"/"could," "both"/"undergrowth")
creates patterns of sound and structure that may contribute to the overall
meaning.

A clear interpretation.   A literary analysis demonstrates the plausibility
of its thesis by using evidence from the text and, sometimes, relevant contextual evidence to explain how the language and patterns found there
support a particular interpretation. When you write a literary analysis, you
show readers one way the text may be read and understood; that is your
interpretation.

MLA style.   Literary analyses usually follow MLA style. Miller's essay
includes a works-cited list and refers to line numbers using MLA style.

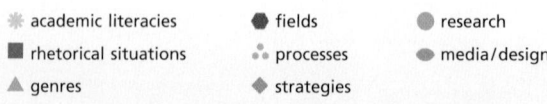

※ academic literacies   ● fields   ● research
■ rhetorical situations   ∴ processes   ● media/design
▲ genres   ◆ strategies

A BRIEF GUIDE TO WRITING LITERARY ANALYSES

Considering the Rhetorical Situation

PURPOSE   What do you need to do? Show that you have examined the text carefully? Offer your own interpretation? Demonstrate a particular analytical technique? Or some combination? If you're responding to an assignment, does it specify what you need to do?   59–60

AUDIENCE   What do you need to do to convince your readers that your interpretation is plausible and based on sound analysis? Can you assume that readers are already familiar with the text you are analyzing, or do you need to tell them about it?   61–64

STANCE   How can you see your subject through interested, curious eyes—and then step back in order to see what your observations might mean?   72–74

MEDIA/DESIGN   Will your analysis focus on an essentially verbal text or one that has significant visual content, such as a graphic novel? Will you need to show visual elements in your analysis? Will it be delivered in a print, spoken, or electronic medium? Are you required to follow MLA or some other style?   75–77

Generating Ideas and Text

Look at your assignment.   Does it specify a particular kind of analysis? Does it ask you to consider a particular theme? To use any specific critical approaches? Look for any terms that tell you what to do, words like "analyze," "compare," "interpret," and so on.

Study the text with a critical eye.   When we read a literary work, we often come away with a reaction to it: we like it, we hate it, it made us cry or laugh, it perplexed us. That may be a good starting point for a literary

21–23 ※

analysis, but to write about literature you need to go beyond initial reactions, to think about **HOW THE TEXT WORKS**: What does it *say*, and what does it *do*? What elements make up this text? How do those elements work together or fail to work together? Does this text lead you to think or feel a certain way? How does it fit into a particular context (of history, culture, technology, genre, and so on)?

Choose a method for analyzing the text. There are various ways to analyze your subject. Three common focuses are on the text itself, on your own experience reading it, and on other cultural, historical, or literary contexts.

445–55 ◆
456–63
474–82

- *The text itself.* Trace the development and expression of themes, characters, and language through the work. How do they help to create the overall meaning, tone, or effect for which you're arguing? To do this, you might look at the text as a whole, something you can understand from all angles at once. You could also pick out parts from the beginning, middle, and end as needed to make your case, **DEFINING** key terms, **DESCRIBING** characters and settings, and **NARRATING** key scenes. Miller's essay about "The Road Not Taken" offers a text-based analysis that looks at Frost's treatment of time in the poem. You might also examine the same theme in several different works.

- *Your own response as a reader.* Explore the way the text affects you or develops meanings as you read through it from beginning to end. By doing such a close reading, you're slowing down the process to notice how one element of the text leads you to expect something, confirming earlier suspicions or surprises. You build your analysis on your experience of reading the text—as if you were pretending to drive somewhere for the first time, though in reality you know the way intimately. By closely examining the language of the text as you experience it, you explore how it leads you to a set of responses, both intellectual and emotional. If you were responding in this way to the Frost poem, you might discuss how the narrator keeps trying to assert that one road is preferable to another but admits that both are the same, so that his willful assertion that one is "less traveled by" and that his choice "made all the difference" is no difference at all.

- *Context.* Analyze the text as part of some larger **CONTEXT** —as part of a certain time or place in history or as an expression of a certain culture (how does this text relate to the time and place of its creation?), as one of many other texts like it, a representative of a genre (how is this text like or unlike others of its kind? how does it use, play with, or flout the conventions of the genre?). A context-based approach to the Frost poem might look at Frost's friendship with another poet, Edward Thomas, for whom Frost wrote the poem, and its influence on Thomas's decision to enlist in the army at the start of World War I.

▲ 114

Read the work more than once. Reading literature, watching films, or listening to speeches is like driving to a new destination: the first time you go, you need to concentrate on getting there; on subsequent trips, you can see other aspects—the scenery, the curve of the road, other possible routes—that you couldn't pay attention to earlier. When you experience a piece of literature for the first time, you usually focus on the story, the plot, the overall meaning. By experiencing it repeatedly, you can see how its effects are achieved, what the pieces are and how they fit together, where different patterns emerge, how the author crafted the work.

To analyze a literary work, then, plan to read it more than once, with the assumption that every part of the text is there for a reason. Focus on details, even on a single detail that shows up more than once: Why is it there? What can it mean? How does it affect our experience of reading or studying the text? Also, look for anomalies, details that *don't* fit the patterns: Why are they part of the text? What can they mean? How do they affect the experience of the text? See the **READING IN ACADEMIC CONTEXTS** chapter for several different methods for reading a text.

✳ 11–32

Compose a strong thesis. The **THESIS** of a literary analysis should be specific, limited, and open to potential disagreement. In addition, it should be analytical, not evaluative: avoid thesis statements that make overall judgments, such as a reviewer might do: "Virginia Woolf's *The Waves* is a failed experiment in narrative" or "No one has equaled the achievement of *The Lego Movie*." Rather, offer a way of seeing the text: "The choice presented

⁖ 347–49

in Robert Frost's 'The Road Not Taken' ultimately makes no difference"; "The plot of *The Lego Movie* reflects contemporary American media culture."

Read the text carefully. When you analyze a text, you need to find specific, brief passages that support your interpretation. Then you should interpret those passages in terms of their language, their context, or your reaction to them as a reader. To find such passages, you must read the text closely, questioning it as you go, asking, for example:

- What language provides evidence to support your thesis?
- What does each word (phrase, passage) mean exactly?
- Why does the writer choose *this* language, *these* words? What are the implications or connotations of the language? If the language is dense or difficult, why might the writer have written it that way?
- What images, similes, or metaphors are used? What is their effect on the meaning?
- What patterns of language, imagery, or plot do you see? If something is repeated, what significance does the repetition have?
- How does each word, phrase, or passage relate to what precedes and follows it?
- How does the experience of reading the text affect its meaning?
- What words, phrases, or passages connect to a larger **CONTEXT**? What language demonstrates that this work reflects or is affected by that context?
- How do these various elements of language, image, and pattern support your interpretation?

11–32 ✷

Your analysis should focus on analyzing and interpreting your subject, not simply summarizing or paraphrasing it. Many literary analyses also use the strategy of **COMPARING** two or more works.

437–44 ◆

✷ academic literacies   ● fields   ● research
■ rhetorical situations   ∴ processes   ● media/design
▲ genres   ◆ strategies

Find evidence to support your interpretation.   The parts of the text you examine in your close reading become the evidence you use to support your interpretation. Some think that we're all entitled to our own opinions about literature. And indeed we are. But when writing a literary analysis, we're entitled only to our own *well-supported* and *well-argued* opinions. When you analyze a text, you must treat it like any other **ARGUMENT**: you need to discuss how the text creates an effect or expresses a theme, and then you have to show **EVIDENCE** from the text—significant plot or structural elements; important characters; patterns of language, imagery, or action—to back up your argument.

◆ 410–30

◆ 414–22

Pay attention to matters of style.   Literary analyses have certain conventions for using pronouns and verbs.

- In informal papers, it's OK to use the first person: "I believe Frost's narrator has little basis for claiming that one road is 'less traveled.'" In more formal essays, make assertions directly; claim authority to make statements about the text: "Frost's narrator has no basis for claiming that one road is 'less traveled.'"

- Discuss textual features in the present tense even if quotations from the text are in another tense: "When Nick finds Gatsby's body floating in the pool, he says very little about it: 'the laden mattress moved irregularly down the pool.'" Describe the historical context of the setting in the past tense: "In the 1920s, such estates as Gatsby's were rare."

Cite and document sources appropriately.   Use **MLA** citation and documentation style unless told otherwise. Format **QUOTATIONS** properly, and use **SIGNAL PHRASES** to introduce quoted material.

● MLA 564–614

544–47

551–54

Think about format and design.   Brief essays do not require **HEADINGS**; text divisions are usually marked by **TRANSITIONS** between paragraphs. In longer papers, though, headings can be helpful.

● 670–72

⁙ 361–62

Organizing a Literary Analysis

[Of a single text]

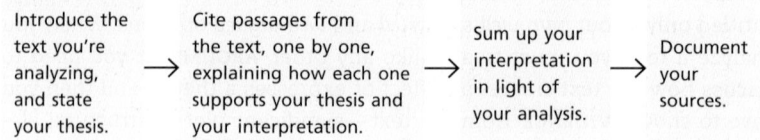

Introduce the text you're analyzing, and state your thesis. → Cite passages from the text, one by one, explaining how each one supports your thesis and your interpretation. → Sum up your interpretation in light of your analysis. → Document your sources.

[Comparing two texts]

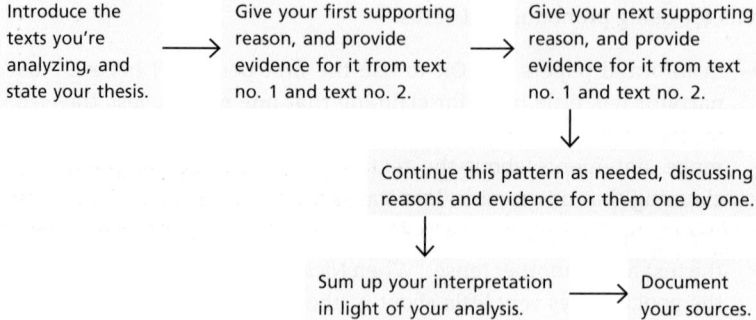

Introduce the texts you're analyzing, and state your thesis. → Give your first supporting reason, and provide evidence for it from text no. 1 and text no. 2. → Give your next supporting reason, and provide evidence for it from text no. 1 and text no. 2.

↓

Continue this pattern as needed, discussing reasons and evidence for them one by one.

↓

Sum up your interpretation in light of your analysis. → Document your sources.

Taking Stock of Your Work

Take stock of what you've written by considering these questions:

- How did you go about analyzing the text? Did you focus on the text itself? your own response? the larger context? something else? Why?

- How did you come up with your thesis? Did you need to revise, expand, or narrow it?

- Did you provide enough evidence from the text or its context to support your analysis?

☀ academic literacies   ⬣ fields   ● research
■ rhetorical situations   ⦂ processes   ● media/design
▲ genres   ◆ strategies

- How did your audience affect the language and tone you used in your piece?

- Did you try anything new in your writing process for this piece? A new way of generating ideas and text? A new way of drafting or revising? Something else? If so, what was the result?

- How did you plan and manage your time for this project? How well did you stick to your plan? If changes were needed, how did you adapt?

- What did you learn about yourself as a reader through your work on this piece? How can you apply that learning to future reading tasks?

- Are you pleased with your analysis? What did it teach you about the text you analyzed? Did it make you want to read more works by the same author?

If you need more help

See Chapter 30 for guidelines on **DRAFTING**, Chapter 31 on **ASSESSING YOUR OWN WRITING**, Chapter 32 on **GETTING RESPONSE AND REVISING**, and Chapter 33 on **EDITING AND PROOFREADING**. See Chapter 34 if you are required to submit your analysis in a writing **PORTFOLIO**.

364–66
367–71
372–79
380–84
385–90

18 Memoirs

We write memoirs to explore our past—about shopping for a party dress with Grandma, or driving a car for the first time, or breaking up with our first love. Memoirs focus on events and people and places that are important to us. We usually have two goals when we write a memoir: to capture an important moment and to convey something about its significance for us. You may be asked to write memoirs or personal reflections that include memoirs in *psychology*, *education*, and *English* courses. The following example is from *All Over but the Shoutin'* by Pulitzer Prize–winning journalist Rick Bragg. Bragg grew up in Alabama, and in this memoir he recalls when, as a teenager, he paid a final visit to his dying father.

RICK BRAGG
All Over but the Shoutin'

He was living in a little house in Jacksonville, Alabama, a college and mill town that was the closest urban center—with its stoplights and a high school and two supermarkets—to the country roads we roamed in our raggedy cars. He lived in the mill village, in one of those houses the mills subsidized for their workers, back when companies still did things like that. It was not much of a place, but better than anything we had ever lived in as a family. I knocked and a voice like an old woman's, punctuated with a cough that sounded like it came from deep in the guts, told me to come on in, it ain't locked.

It was dark inside, but light enough to see what looked like a bundle of quilts on the corner of a sofa. Deep inside them was a ghost of a man, his hair and beard long and going dirty gray, his face pale and cut with deep grooves. I knew I was in the right house because

---

🔆 academic literacies   ⬤ fields   ⬤ research

■ rhetorical situations   ⋮ processes   ⬤ media/design

▲ genres   ◆ strategies

my daddy's only real possessions, a velvet-covered board pinned with medals, sat inside a glass cabinet on a table. But this couldn't be him.

He coughed again, spit into a can and struggled to his feet, but stopped somewhere short of standing straight up, as if a stoop was all he could manage. "Hey, Cotton Top," he said, and then I knew. My daddy, who was supposed to be a still-young man, looked like the walking dead, not just old but damaged, poisoned, used up, crumpled up and thrown in a corner to die. I thought that the man I would see would be the trim, swaggering, high-toned little rooster of a man who stared back at me from the pages of my mother's photo album, the young soldier clowning around in Korea, the arrow-straight, good-looking boy who posed beside my mother back before the fields and mop handle and the rest of it took her looks. The man I remembered had always dressed nice even when there was no cornmeal left, whose black hair always shone with oil, whose chin, even when it wobbled from the beer, was always angled up, high.

I thought he would greet me with that strong voice that sounded so fine when he laughed and so evil when, slurred by a quart of corn likker, he whirled through the house and cried and shrieked, tormented by things we could not see or even imagine. I thought he would be the man and monster of my childhood. But that man was as dead as a man could be, and this was what remained, like when a snake sheds its skin and leaves a dry and brittle husk of itself hanging in the Johnson grass.

"It's all over but the shoutin' now, ain't it, boy," he said, and when ⁵ he let the quilt slide from his shoulders I saw how he had wasted away, how the bones seemed to poke out of his clothes, and I could see how it killed his pride to look this way, unclean, and he looked away from me for a moment, ashamed.

He made a halfhearted try to shake my hand but had a coughing fit again that lasted a minute, coughing up his life, his lungs, and after that I did not want to touch him. I stared at the tops of my sneakers, ashamed to look at his face. He had a dark streak in his beard below his lip, and I wondered why, because he had never liked snuff. Now I know it was blood.

I remember much of what he had to say that day. When you don't see someone for eight, nine years, when you see that person's life red on their lips and know that you will never see them beyond this day, you listen close, even if what you want most of all is to run away.

"Your momma, she alright?" he said.

I said I reckon so.

"The other boys? They alright?"   10

I said I reckon so.

Then he was quiet for a minute, as if trying to find the words to a question to which he did not really want an answer.

"They ain't never come to see me. How come?"

I remember thinking, fool, why do you think? But I just choked down my words, and in doing so I gave up the only real chance I would ever have to accuse him, to attack him with the facts of his own sorry nature and the price it had cost us all. The opportunity hung perfectly still in the air in front of my face and fists, and I held my temper and let it float on by. I could have no more challenged him, berated him, hurt him, than I could have kicked some three-legged dog. Life had kicked his ass pretty good.

"How come?"   15

I just shrugged.

For the next few hours—unless I was mistaken, having never had one before—he tried to be my father. Between coughing and long pauses when he fought for air to generate his words, he asked me if I liked school, if I had ever gotten any better at math, the one thing that just flat evaded me. He asked me if I ever got even with the boy who blacked my eye ten years ago, and nodded his head, approvingly, as I described how I followed him into the boys' bathroom and knocked his dick string up to his watch pocket, and would have dunked his head in the urinal if the aging principal, Mr. Hand, had not had to pee and caught me dragging him across the concrete floor.

He asked me about basketball and baseball, said he had heard I had a good game against Cedar Springs, and I said pretty good, but it was two years ago, anyway. He asked if I had a girlfriend and I said, "One," and he said, "Just one?" For the slimmest of seconds he almost grinned and the young, swaggering man peeked through, but disappeared again in the disease that cloaked him. He talked and talked and never said a word, at least not the words I wanted.

He never said he was sorry.

He never said he wished things had turned out different.   20

He never acted like he did anything wrong.

Part of it, I know, was culture. Men did not talk about their feelings in his hard world. I did not expect, even for a second, that he would bare his soul. All I wanted was a simple acknowledgment that he was wrong, or at least too drunk to notice that he left his pretty wife and sons alone again and again, with no food, no money, no way to get any, short of begging, because when she tried to find work he yelled, screamed, refused. No, I didn't expect much.

After a while he motioned for me to follow him into a back room where he had my present, and I planned to take it and run. He handed me a long, thin box, and inside was a brand-new, well-oiled Remington .22 rifle. He said he had bought it some time back, just kept forgetting to give it to me. It was a fine gun, and for a moment we were just like anybody else in the culture of that place, where a father's gift of a gun to his son is a rite. He said, with absolute seriousness, not to shoot my brothers.

I thanked him and made to leave, but he stopped me with a hand on my arm and said wait, that ain't all, that he had some other things for me. He motioned to three big cardboard egg cartons stacked against one wall.

Inside was the only treasure I truly have ever known.   25

I had grown up in a house in which there were only two books, the King James Bible and the spring seed catalog. But here, in these boxes, were dozens of hardback copies of everything from Mark Twain to Sir Arthur Conan Doyle. There was a water-damaged Faulkner, and the nearly complete set of Edgar Rice Burroughs's *Tarzan*. There was poetry and trash, Zane Grey's *Riders of the Purple Sage*, and a paperback with two naked women on the cover. There was a tiny, old copy of *Arabian Nights*, threadbare Hardy Boys, and one Hemingway. He had bought most of them at a yard sale, by the box or pound, and some at a flea market. He did not even know what he was giving me, did not recognize most of the writers. "Your momma said you still liked to read," he said.

There was Shakespeare. My father did not know who he was, exactly, but he had heard the name. He wanted them because they were pretty, because they were wrapped in fake leather, because they looked like rich folks' books. I do not love Shakespeare, but I still have those books. I would not trade them for a gold monkey.

"They's maybe some dirty books in there, by mistake, but I know you ain't interested in them, so just throw 'em away," he said. "Or at least, throw 'em away before your momma sees 'em." And then I swear to God he winked.

I guess my heart should have broken then, and maybe it did, a little. I guess I should have done something, anything, besides mumble "Thank you, Daddy." I guess that would have been fine, would not have betrayed in some way my mother, my brothers, myself. But I just stood there, trapped somewhere between my long-standing, comfortable hatred, and what might have been forgiveness. I am trapped there still.

Bragg's memoir illustrates all the features that make a memoir good: how the son and father react to each other creates the kind of suspense that keeps us reading; vivid details and rich dialogue bring the scene to life. His later reflections make the significance of that final meeting very clear.

Key Features / Memoirs

A good story.   Your memoir should be interesting, to yourself and others. It need not be about a world-shaking event, but your topic—and how you write about it—should interest your readers. At the center of most good stories stands a conflict or question to be resolved. The most compelling memoirs feature some sort of situation or problem that needs resolution. That need for resolution is another name for suspense. It's what makes us want to keep reading.

Vivid details.   Details bring a memoir to life by giving readers mental images of the sights, sounds, smells, tastes, and textures of the world in which your story takes place. The goal is to show as well as tell, to take readers there. When Bragg describes a "voice like an old woman's, punctuated with a cough that sounded like it came from deep in the guts," we can hear his dying father ourselves. A memoir is more than simply a report of what happened; it uses vivid details and dialogue to bring the events of the past to life, much as good fiction brings to life events that the writer makes up or embellishes. Depending on your topic and medium, you may want to provide some of the details in audio or visual form.

academic literacies   fields   research

rhetorical situations   processes   media/design

genres   strategies

Clear significance. Memories of the past are filtered through our view from the present: we pick out some moments in our lives as significant, some as more important or vivid than others. Over time, our interpretations change, and our memories themselves change.

A good memoir conveys something about the significance of its subject. As a writer, you need to reveal something about what the incident means to you. You don't, however, want to simply announce the significance as if you're tacking on the moral of the story. Bragg tells us that he's "trapped between [his] long-standing, comfortable hatred, and what might have been forgiveness," but he doesn't come right out and say that's why the incident is so important to him.

A BRIEF GUIDE TO WRITING MEMOIRS

Choosing an Event to Write About

LIST several events or incidents from your past that you consider significant in some way. They do not have to be earthshaking; indeed, they may involve a quiet moment that only you see as important—a brief encounter with a remarkable person, a visit to a special place, a memorable achievement (or failure), something that makes you laugh whenever you think about it. Writing about events that happened at least a few years ago is often easier than writing about recent events because you can more easily step back and see those events with a clear perspective. To choose the event that you will write about, consider how well you can recall what happened, how interesting it will be to readers, and whether you want to share it with an audience.

334–35

Considering the Rhetorical Situation

PURPOSE	What is the importance of the memory you are trying to convey? How will this story help you understand yourself and your readers understand you, as you were then and as you are now?

59–60

61–64 ■ **AUDIENCE**   Who are your readers? Why will they care about your memoir? What do you want them to think of you after reading it? How can you help them understand your experience?

72–74 ■ **STANCE**   What impression do you want to give, and how can your words contribute to that impression? What tone do you want to project? Sincere? Serious? Humorous? Detached? Self-critical?

75–77 ■ **MEDIA/DESIGN**   Will your memoir be a print document? a speech? Will it be posted on a website? Can you include photographs, audio or video clips, or other visual texts?

Generating Ideas and Text

Think about what happened.   Take a few minutes to write out an account of the incident: **WHAT** happened, **WHERE** it took place, **WHO** else was involved, what was said, how you feel about it, and so on. Can you identify any tension or conflict that will make for a compelling story? If not, you might want to rethink your topic.

337 ⁚

Consider its significance.   Why do you still remember this event? What effect has it had on your life? What makes you want to tell someone else about it? Does it say anything about you? What about it might interest someone else? If you have trouble answering these questions, you should probably find another topic. But in general, once you have defined the significance of the incident, you can be sure you have a story to tell—and a reason for telling it.

Think about the details.   The best memoirs connect with readers by giving them a sense of what it was like to be there, leading them to experience in words and images what the writer experienced in life. Spend some time **DESCRIBING** the incident, writing what you see, hear, smell, touch, and taste when you envision it. Do you have any photos or memorabilia or other **VISUAL** materials you might include in your memoir? Try writing out **DIALOGUE**, things that were said (or, if you can't recall exactly, things

456–63 ◆

674–80 ●
464–68 ◆

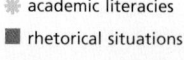

☀ academic literacies   ● fields   ● research
■ rhetorical situations   ⁚ processes   ● media/design
▲ genres   ◆ strategies

that might have been said). Look at what you come up with—is there detail enough to bring the scene to life? anything that might be called vivid? If you don't have enough detail, you might reconsider whether you recall enough about the incident to write about it. If you have trouble coming up with plenty of detail, try **FREEWRITING**, **LISTING**, or **LOOPING**.

333–35

Ways of Organizing Memoirs

[Tell about the event from beginning to end]

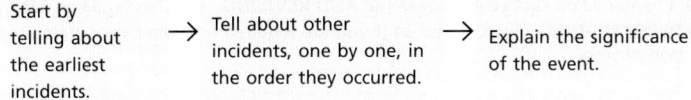

Start by telling about the earliest incidents. → Tell about other incidents, one by one, in the order they occurred. → Explain the significance of the event.

[Start at the end and tell how the event came about]

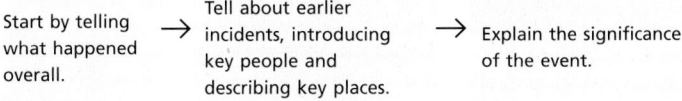

Start by telling what happened overall. → Tell about earlier incidents, introducing key people and describing key places. → Explain the significance of the event.

Taking Stock of Your Work

Take stock of what you've written by considering to these questions:

- How did you decide on your topic and determine the conflict or question to be resolved in your memoir?

- How effectively did you use dialogue, description, or other narrative strategies in your memoir?

- Did you use photographs or other visual or audio elements? What did they add? Can you think of such elements you might have used?

- How did you go about revising your work? How effective do you think your revisions were?

- What previous writing experiences, if any, did you draw on in writing your memoir? Did those experiences help you or hinder you?

- How much did you reflect on your story and what it means to you—and might mean to others?

- What was your most important achievement in this piece? What's the best part of this memoir? Why?

If you need more help

364–66
367–71
372–79
380–84
385–90

See Chapter 30 for guidelines on **DRAFTING**, Chapter 31 on **ASSESSING YOUR OWN WRITING**, Chapter 32 on **GETTING RESPONSE AND REVISING**, and Chapter 33 on **EDITING AND PROOFREADING**. See Chapter 34 if you are required to submit your memoir in a writing **PORTFOLIO**.

☀ academic literacies   ● fields   ● research

■ rhetorical situations   ⁙ processes   ● media/design

▲ genres   ◆ strategies

19 Profiles

Profiles are written portraits—of people, places, events, or other things. We find profiles of celebrities, travel destinations, and offbeat festivals in magazines and newspapers, on radio and TV, on the web and through various social media. A profile presents a subject in an entertaining way that conveys its significance, showing us something or someone that we may not have known existed or that we see every day but don't know much about. In college *journalism* classes, students learn to create profiles using words and, in many cases, photos and video as well. Here is a profile of a young woman whose path to college included hopping freight trains, battling addiction, and becoming a boxer. This profile originally appeared in a longer version in *The PennStater*, the alumni magazine of Penn State University, where writer Ryan Jones is the senior editor.

RYAN JONES

A Fighter's Chance

The end of the world begins on Rebecca Maine's right shoulder, an apocalypse of skulls and trees and snarling beasts that wraps around her arm until it reaches the back of her hand, where a serpent clenches an eyeball in its jaws.

The scene is a tattoo depiction of Ragnarok, the violent denouement in ancient Norse mythology. Her mother has traced the family's genealogy back to the Vikings, and Maine appreciates the connection to her lineage. But mostly, she says, "I just think it's badass."

Maine began work on the tattoo nearly a decade ago, but only got it finished last year. She's got a number of other pieces—the comic-book

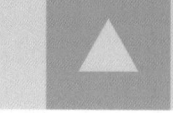

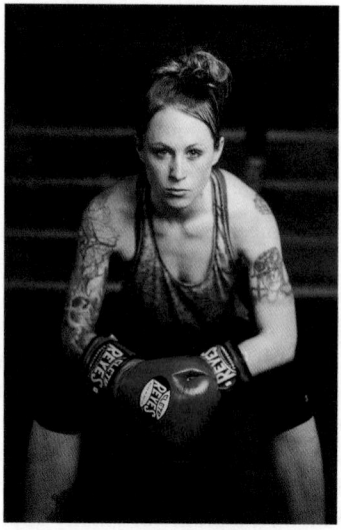

heroine Tank Girl on her left arm, others of varying complexity on her neck, legs, and back—but she's particularly fond of that full sleeve. "It's silly," she says, "but finishing it felt like an accomplishment." The greater achievement is that she lived long enough to see it through.

It's difficult to reconcile the 26-year-old dean's-list student-athlete Maine is today with the young woman—still a girl, really—she was eight or nine years ago, the one who left a difficult home life looking for adventure, acceptance, and escape. She carried little more than the clothes on her back and a burgeoning taste for substances that could numb her pain. She ended up hopping freight trains, traveling the country that way for the better part of three years. And she fell deep into the throes of heroin addiction that nearly killed her.

In March, near the end of her first year in the physical therapist assistant program at Penn State DuBois, she celebrated five years sober. 5

A few months earlier, Maine had completed an award-winning season on the campus cross-country team; in May, she fought in the national Golden Gloves tournament. Still, the darkest parts of her past are never far out of mind.

"I feel like I've had to keep accomplishing more and more, trying to kind of bury all that," she says. "But I can't get too removed from it. I need it, as a reminder."

So she has embraced the pain and loss and anger and guilt, because the future she's set her mind to requires nothing less. Talking about it took longer, because it meant exposing herself to judgment, but she has embraced that, too. She sees the value of sharing her experience, both for herself and, she hopes, for others, kids who are where she was: in a small town, without many options, in the midst of a national

☀ academic literacies   ● fields   ● research
■ rhetorical situations   ⁘ processes   ● media/design
▲ genres   ◆ strategies

epidemic of opioid abuse, bearing emotional burdens that few around them are willing or able to understand. "I don't think she recognized how big of a deal her story is—it hits home for a lot of people, especially in our community," says DuBois cross-country coach Kyle Gordon. "It's a story of hope."

Rebecca started drinking in middle school in Punxsutawney, PA, both to fit in and to cope with the stress of family life: Though close with her mother, Rebecca describes her father as a functional alcoholic who was unfaithful, abusive, and often absent, and it was for the best when her parents' marriage ended when she was in high school. But it was an unrelated incident during her freshman year that helped send Rebecca spiraling: She describes herself as a reluctant member of the popular crowd, hanging with kids who "just drank all the time." Her mother, Louise—a teacher at the local high school—found photos of a booze-fueled underage party on Rebecca's phone, and decided her best move—the right move—would be calling the other kids' parents. "I thought I should talk to them about what was happening," Louise says. "But they didn't welcome that."

Instead, their kids targeted Rebecca for retribution. "They tormented me for it," she says. Louise remembers her daughter coming to her classroom to eat lunch, moody and miserable but desperate to avoid the relentless bullying of the classmates she had once considered friends. Ostracized from her peer group, Rebecca began hanging out with older kids, many of whom had already moved beyond drinking to more dangerous substances. She followed suit: "By my sophomore year," Rebecca says, "I did anything I could get my hands on." 10

Around this time, Louise found drug paraphernalia in Rebecca's room, but she says she still didn't fully grasp the turmoil that was driving her daughter to self-medicate. Rebecca got a reprieve of sorts when her mother allowed her to finish high school via online courses. Louise was skeptical, thinking her daughter was just looking for a way to quit school altogether, but Rebecca stuck it out and finished ahead of schedule—a sign of the persistence she'd show years later.

But Rebecca was also entrenched in a social circle that enabled her worst tendencies, including substance abuse. Not long after she turned 18, she and her boyfriend moved 80 miles southwest to Pittsburgh. Once there, she toyed with the idea of culinary school or a veterinary tech

program, but nothing took. And then, she says, "I pretty much snapped. I had suppressed so much, and I just felt like I had to do . . . something."

She sold everything she owned, minus a backpack full of clothes, and bought a plane ticket to New Orleans. She felt guilty leaving her mom and brother, and she hated herself for feeling like she deserved the turmoil that defined her life. She had no idea what came next. She only knew that running away felt like something she could control.

They were on the road to Scranton for a cross-country meet last fall, the coach and his three runners, one of them a 26-year-old freshman. Kyle Gordon was struck by the fact that the young woman beside him seemed so excited to visit a city famous for its train museum. "I thought it was a little bizarre," he says with a laugh. "That's when she told us about her train-hopping."

She had left Pittsburgh and landed in New Orleans and spent a 15 couple of months there with friends, thinking she'd hitchhike across the country. And then she linked up with the amorphous community of train-hoppers — "travelers," as she calls them, a group she describes as an underground society, hidden in plain sight. "I was a little punk rocker, and it was the most punk-rock thing I'd ever seen," she says. "I wanted in."

All these years later, there are aspects of her nearly three years on the rails that provide fond memories. She learned all the tricks, like which trains ran faster or offered better hiding places. How to wait until the conductors' shift change, when trains would generally be left unattended on the edge of a yard, and she and fellow hoppers would wait, hidden, until they were sure it was clear. And of course, she got to see so much of the country for free. But it was often dangerous, and always illegal, and it's impossible to fully separate the good parts from the bad. About a month into traveling, she says, she was "eating, sleeping, drinking heroin."

Money was always tight: Maine quickly burned through the money she'd made selling her things, and the spare change she made playing music on the mandolin she traveled with never lasted long. She remembers one bleak stretch near the high desert town of Twentynine Palms, a few hours east of Los Angeles, when it occurred to her that she'd barely eaten in two weeks. "Del Taco had 49-cent burritos, and

✹ academic literacies   ● fields   ● research
■ rhetorical situations   ⁚• processes   ● media/design
▲ genres   ◆ strategies

I'd get one of those, once or twice a week," she says. "All my money was going elsewhere."

There were other wake-up calls: Maine says she was "Narcanned" several times, referring to emergency doses of the medication carried by police and EMTs to block the effects of opioid overdose. There was the traveling companion, someone she thought she could trust, who she learned had offered to "sell" her for sex. There were friends whose addictions got the best of them, who weren't as tough, or as lucky, and who never made it home.

She never thought of herself as homeless, not as long as her mom was back in Punxsutawney, wishing for her safe return. They would talk by phone, and Louise held out hope that the travel might be good for Rebecca, that at least she'd be far from the roots of her problems. But Rebecca could only maintain the facade for so long.

More than once, Rebecca tried to get sober. She returned to Punx- 20 sutawney, moving back in with her mom until she started using again, then moving in with a friend. She tried Pittsburgh again, even tried Brooklyn, where she played in a folk-punk band that hoped to record an album. Each time, she backslid. "She would be OK for a while," Louise says, "but you always know when they relapse. It would be immediate." She tried therapy but hated it. She got a part-time job as a hospital housekeeper near her home. She made regular trips to and from Pittsburgh, scoring drugs for herself and occasionally re-selling the surplus to support her habit. She'd ruined the veins in her arms and started shooting into her neck.

Finally, Rebecca says, something—some combination of guilt, desperation, and exhaustion—clicked. She deleted her dealer's number from her phone. She took all the time off she could afford from her job at the hospital, and she went cold turkey. It was February in Punxsutawney, in a house with no heat, and she burrowed into her couch, chain-smoking cigarettes, flinging off the blankets when the sweats came, piling them back on when they passed. Her roommate had box sets of the first five seasons of *Sons of Anarchy*, and she watched every episode. It took two weeks before she was able to function, to keep any food down, and to tell her mom—and for the first time, mean it—that she needed help.

And here Rebecca knows her experience might not be fully transferable: She did this largely on her own. Her mom was a rock, family and

close friends helped hold her accountable, but she never felt comfortable in 12-step programs, and so she focused instead on staying busy. She was about a month sober, still wracked with cold sweats, when she took seasonal work as a merchandiser for a garden wholesaler. Around the same time, she and her mom went to watch a local boxing showcase; Louise was dating the physician who sat ringside for the bouts, and she thought the discipline and physical activity would be good for her daughter. They were introduced to a trainer, who saw before him a young woman who was, in her own words, "un-athletic and a wreck."

She told him she wanted to learn how to fight.

Rebecca dove right in, going through an intensive workout her first day at the gym. The next day, she says with a laugh, "I couldn't walk." But she stuck with it, never missing practice, and along the way she found that the sport had a way of helping other issues. She had struggled for years with body image, and she'd battled both bulimia and anorexia; training for fights forced her to eat well and made her proud of the athlete she was becoming.

Through endless hours at the gym, Maine transformed her body 25 and her life, and before long she'd decided she wanted to try to make a career as a professional fighter. But that hope was nearly derailed with a pair of diagnoses, first hepatitis C—a relic from her days of sharing needles—and then endometriosis. She's since undergone treatment for both conditions and says the silver lining was that she developed a curiosity about her body—how it worked and how to take care of it. "Boxing was everything—boxing may have saved my life—but I didn't realize maybe it was a stepping stone to the next thing," Rebecca says. It turned out the next thing was Penn State.

The letter arrived not long before the start of the school year, informing Rebecca Maine that she could pick up two additional kinesiology credits if she participated in an intercollegiate sport at Penn State DuBois. *I want to go for a doctorate someday*, Maine thought to herself, *so I might as well get as many credits as I can.*

She had yet to complete a single credit in anything, of course, but Maine was thinking ahead. She was already running as part of her boxing training; joining the DuBois cross-country team didn't seem like too much of a stretch. The team featured just three runners, and Maine was

* academic literacies   ● fields   ● research
■ rhetorical situations   ⦂ processes   ● media/design
▲ genres   ◆ strategies

the only woman, a 26-year-old freshman who'd never raced competing against 18- and 19-year-olds fresh off their high school teams. She'd seen and endured so much more of life than most of her opponents ever would, and yet when she lined up for her first race, she was terrified.

"I'm on the starting line, and there's all these young people—all these young *athletes*," she says. "I was scared. I didn't know what to expect."

And yet she thrived, finishing sixth in her first race, and first among runners from Penn State campuses. She went on to finish fourth in the PSUAC championship meet, earning a spot on the all-conference team and a trip to nationals. And it was there that her coach finally convinced her to tell her story. Gordon asked Rebecca if she'd mind him nominating her for a conference award. As they'd gotten to know each other over the course of the semester, she had grudgingly offered Gordon bits of her past. For this, she agreed to open up a bit more.

It was in Virginia Beach in November, at the U.S. Collegiate Athletic Association banquet, that strangers first heard pieces of Maine's story. In the write-up honoring her as the USCAA's Student-Athlete of the Year, the key details stood out: *addicted to heroin . . . hitchhiking and train-hopping . . . boxing as a means to turn her life around.*

And there were more achievements to come. Maine would go on to compete in the 2018 Golden Gloves boxing championship, winning two fights in her weight class before falling in the semifinals. Back at DuBois, she earned the highest GPA among first-semester freshmen, started a campus PT assistant's club, and led the campus LGBTQA support group.

When we first spoke last winter, Maine was reeling, her eyes welling a half dozen times as she laid out her story to a stranger for the first time. "It was things surfacing that I'd repressed, things I don't want to remember," she said then. "All the positive things people are saying now, it's new to me. It's very alien."

But by the spring, having had time to process her feelings and appreciate the value of her story, she had come around. She knew the stigma around issues like mental health and addiction, knew how it felt to believe there was something "wrong" with her, and not that she simply needed help. She understood that her story could help others and that telling it could help her, too.

Jones's profile traces Rebecca Maine's unusual life from her teen years to the present. Because it's a profile, the piece focuses on a part of her life, rather than trying to describe her entire life story. Jones provides details, from Maine's tattoos to her first boxing workout, that give us snapshots of Maine's life and character.

Key Features / Profiles

An interesting subject.   The subject may be something unusual, or it may be something ordinary shown in an intriguing way. You might profile an interesting person (like a local celebrity), a place (like a company that makes paper straws), an event (like the rescue of a sea turtle), or an object (like a rare automobile).

Any necessary background.   A profile usually includes just enough information to let readers know something about the subject's larger context. Jones provides some details about why Maine left home (an alcoholic and abusive father, ostracism by her peers), but he leaves out other details about her life that don't matter for this profile.

An interesting angle.   A good profile captures its subject from a particular angle. Sometimes finding an angle will be fairly easy because your topic—like the history of a local coffee shop—is offbeat enough to be interesting in and of itself. Other topics, though, may require you to find a particular aspect that you can focus on. For example, a profile of a person might focus on the important work the person does or a challenging hobby they pursue; it would likely ignore aspects of the person's life that don't relate to that angle.

A firsthand account.   Whether you're writing about a person, place, object, or event, you need to spend time observing and interacting with your subject. With a person, interacting means watching and conversing. Journalists tell us that "following the guy around," getting your subject to do something and talk about it at the same time, yields excellent material for a profile. When one writer met Theodor Geisel (Dr. Seuss) before profiling him, she asked him not only to talk about his characters but also to

---

☀ academic literacies   ● fields   ● research
■ rhetorical situations   ∴ processes   ● media / design
▲ genres   ◆ strategies

draw one—resulting in an illustration for her profile. With a place, object, or event, interacting may mean visiting and participating, although sometimes you may gather even more information by playing the role of the silent observer.

Engaging details. You need to include details that bring your subject to life. These may include *specific information* ("She learned all the tricks, like which trains ran faster or offered better hiding places. How to wait until the conductors' shift change, when trains would generally be left unattended on the edge of a yard, and she and fellow hoppers would wait, hidden, until they were sure it was clear"); *sensory images* ("she burrowed into her couch, chain-smoking cigarettes, flinging off the blankets when the sweats came, piling them back on when they passed"); *figurative language* ("The end of the world begins on Rebecca Maine's right shoulder, an apocalypse of skulls and trees and snarling beasts"); *dialogue* ("'Boxing was everything—boxing may have saved my life—but I didn't realize maybe it was a stepping stone to the next thing,' Rebecca says"); and *anecdotes* (Rebecca started training for boxing, "going through an intensive workout her first day at the gym. The next day, she says with a laugh, 'I couldn't walk.'"). Choose details that show rather than tell—that let your audience see and hear your subject rather than merely read an abstract description of it. Sometimes you may let them see and hear it literally, by including *photographs* or *video and audio clips*. And be sure all the details create some *dominant impression* of your subject: the impression that we get out of this profile, for example, is of a strong-willed, ambitious young woman who has persevered despite many challenges and setbacks.

A BRIEF GUIDE TO WRITING PROFILES

Choosing a Suitable Subject

A person, a place, an object, an event—whatever you choose, make sure it's something that arouses your curiosity and that you're not too familiar with. Knowing your subject too well can blind you to interesting details. **LIST** five to ten interesting subjects that you can experience firsthand. Obviously,

334–35

you can't profile a person who won't be interviewed or a place or activity that can't be observed. So before you commit to a topic, make sure you'll be able to carry out firsthand research.

Considering the Rhetorical Situation

59–60 **PURPOSE**
Why are you writing the profile? What angle will best achieve your purpose? How can you inform and engage your audience?

61–64 **AUDIENCE**
Who is your audience? How familiar are they with your subject? What expectations of your profile might they have? What background information or definitions do you need to provide? How interested will they be—and how can you capture and hold their interest?

72–74 **STANCE**
What view of your subject do you expect to present? Sympathetic? Critical? Sarcastic? Will you strive for a carefully balanced perspective?

75–77 **MEDIA/DESIGN**
Will your profile be a print document? electronic? an oral presentation? Can (and should) you include images or any other visuals? Will it be recorded as an audio file or a multimodal text?

Generating Ideas and Text

Explore what you already know about your subject. Why do you find this subject interesting? What do you know about it now? What do you expect to find out about it from your research? What preconceived ideas about or emotional reactions to this subject do you have? Why do you have them? It may be helpful to try some of the activities in the chapter 333–44 on **GENERATING IDEAS AND TEXT**.

Visit your subject. If you're writing about an amusement park, go there; if you're profiling the man who runs the carousel, make an appointment to meet and interview him. Get to know your subject—if you profile the owner of a local BBQ place, sample the ribs! Take photos or videos if there's

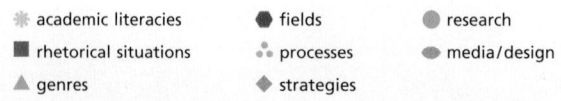

* academic literacies   ● fields   ● research
■ rhetorical situations   ⁚ processes   ● media/design
▲ genres   ◆ strategies

anything you might want to show visually in your profile. Find helpful hints for **OBSERVING** and **INTERVIEWING** in the chapter on finding sources.

520–21
518–20

If you're planning to interview someone, prepare questions. Jones likely asked Rebecca Maine such questions as, "What triggered your interest in train-hopping? You did it for three years; what are some of your good memories of that time? What are some of the bad ones?" See the **INTERVIEWING** guidelines in Chapter 47 for help with planning good questions and conducting interviews.

518–20

Do additional research. You may be able to write a profile based entirely on your field research. You may, though, need to do some library or web **RESEARCH** as well, to deepen your understanding, get a different perspective, or fill in gaps. Often the people you interview can help you find sources of additional information; so can the sponsors of events and those in charge of places. To learn more about a city park, for instance, contact the government office that maintains it. Download any good photos of your subject that you find online, both to refer to as you write and to illustrate your profile.

489

Analyze your findings. Look for patterns, images, recurring ideas or phrases, and engaging details. Look for contrasts or discrepancies: between a subject's words and actions, between the appearance of a place and what goes on there, between your expectations and your research findings. Jones may have expected to meet an athlete with an unusual background—but he may not have expected Maine's life to be as difficult as it turned out to be. You may find the advice in the **READING IN ACADEMIC CONTEXTS** chapter helpful here.

11–32

Come up with an angle. What's most memorable about your subject? What most interests you? What will interest your audience? Jones focuses on the aspects of Maine's life that set her apart from most young women. Sometimes you'll know your angle from the start; other times you'll need to look further into your topic. You might try **CLUSTERING**, **CUBING**, **FREE-WRITING**, and **LOOPING**, to help you see your topic from many different angles.

333–36

456–63
437–44
464–68

Note details that support your angle. Use your angle to focus your research and generate text. Try **DESCRIBING** your subject as clearly as you can, **COMPARING** your subject with other subjects of its sort, writing **DIALOGUE** that captures your subject. Jones, for instance, describes Maine's tattoos, compares her to the much younger runners on her team, and includes several quotes from Maine that capture her impressions—and her spirit. Engaging details will bring your subject to life for your audience. Together, these details should create a dominant impression of your subject.

Ways of Organizing a Profile

474–82

One common way to organize a profile is by **NARRATING**. For example, if you're profiling a chess championship, you may write about it chronologically, creating suspense as you move from start to finish.

[As a narrative]

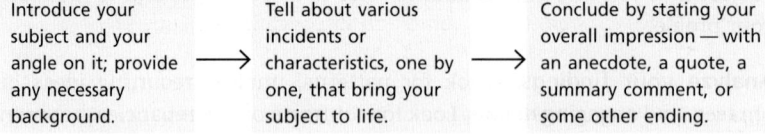

| Introduce your subject and your angle on it; provide any necessary background. | → | Tell about various incidents or characteristics, one by one, that bring your subject to life. | → | Conclude by stating your overall impression — with an anecdote, a quote, a summary comment, or some other ending. |

456–63

Sometimes you may organize a profile by **DESCRIBING**—a person or a place, for instance.

[As a description]

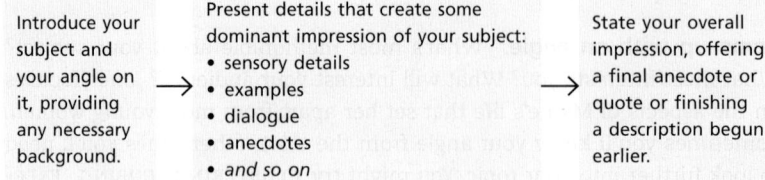

| Introduce your subject and your angle on it, providing any necessary background. | → | Present details that create some dominant impression of your subject:
• sensory details
• examples
• dialogue
• anecdotes
• *and so on* | → | State your overall impression, offering a final anecdote or quote or finishing a description begun earlier. |

* academic literacies   ● fields   ● research
■ rhetorical situations   ⁜ processes   ● media/design
▲ genres   ◆ strategies

Taking Stock of Your Work

Take stock of what you've written by considering these questions:

- How did you go about choosing a subject—and an angle on that subject?

- What strategies did you use for the beginning and ending of your profile? How did you decide on those strategies?

- What design or media choices did you make? Why did you make them, and how effective were they?

- In what ways did you reflect on your writing as you worked on it? How did the act of reflecting help you in your writing?

- What setbacks did you face as you worked on your writing? How did you overcome them and persist as a writer?

- How did you use concepts or methods you learned elsewhere (in other classes, for example) as you worked on your profile?

- What were your writing goals for yourself in this piece? How well did you achieve them? How could you improve in the future?

If you need more help

See Chapter 30 for guidelines on **DRAFTING**, Chapter 31 on **ASSESSING YOUR OWN WRITING**, Chapter 32 on **GETTING RESPONSE AND REVISING**, and Chapter 33 on **EDITING AND PROOFREADING**. See Chapter 34 if you are required to submit your profile in a writing **PORTFOLIO**.

364–66
367–71
372–79
380–84
385–90

20 Proposals

Proposals are part of our personal lives: lovers propose marriage, friends propose sharing dinner and a movie, you offer to pay half the cost of a car and insurance if your parents will pay the other half. They are also part of our academic and professional lives: student leaders lobby for lights on bike paths. Musicians, artists, writers, and educators apply for grants. Researchers in all fields of the humanities, social sciences, sciences, and technology seek funding for their projects. In business, contractors bid on building projects, and companies and freelancers solicit work from potential clients. These are all examples of proposals, ideas put forward for consideration that say, "Here is a solution to a problem" or "This is what ought to be done." For example, here is a proposal for reducing cheating among college students, written by a philosophy professor at Wake Forest University who directs a team of scholars examining the role of honesty in our lives. The team's undertaking is called The Honesty Project.

CHRISTIAN B. MILLER

Just How Dishonest Are Most Students?

I teach philosophy to college students, and there was no way I was going to give them exams this semester, with our classes being held online. Why not? Simple—cheating. It is nothing personal with these particular students, but I have read enough psychological research to know that it would be very hard for them to resist looking for help in places where they are not supposed to, such as their notes, their friends, and the internet.

<table>
<tr><td>☀ academic literacies</td><td>● fields</td><td>● research</td></tr>
<tr><td>■ rhetorical situations</td><td>⁚ processes</td><td>● media/design</td></tr>
<tr><td>▲ genres</td><td>◆ strategies</td><td></td></tr>
</table>

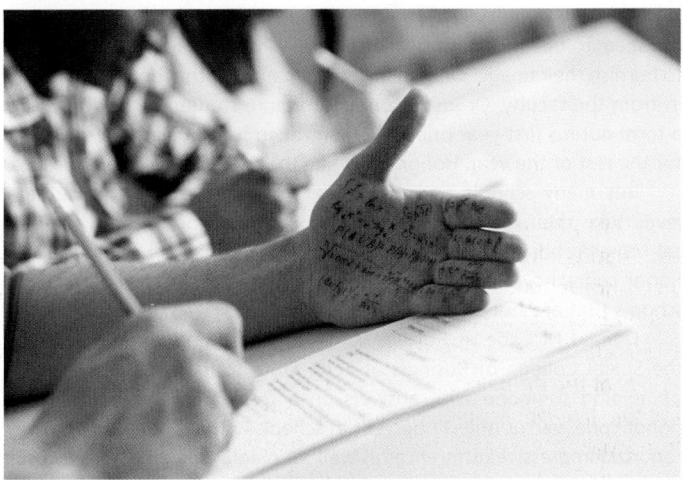

I am fortunate that papers are a great alternative means of assessment in philosophy courses. But they do not work so well in certain other fields, like the sciences. In this time of widespread online learning and home-schooling, what can be done to curb cheating on exams?

One solution is remote proctoring, where the student is video-recorded during the exam, with any suspicious web browsing reported. That might be effective, but it strikes me as a crude approach, relying as it does on active surveillance, which creates an overt atmosphere of distrust. Naturally enough there are also privacy concerns (Hubler; Patil and Bromwich), as well as some anecdotal evidence (Patil and Bromwich; Swauger) that remote proctoring technology encodes racial biases.

Instead I suggest that a practice that has been used widely in other educational contexts be extended to the world of online testing: pledging one's honor. Honor pledges not only are surprisingly effective in curbing cheating; they also promote honesty. Students who abide by them refrain from cheating not because they can't, but because they choose not to.

It is easy to be cynical about honor pledges and honor codes. They can seem to be—and sadly too often are—PR stunts for schools looking to burnish their image. Or administrative mandates that do not have buy-in from the faculty. Or just a formality, where students check a box on a form during first-year orientation and then never give it any thought for the rest of the year. Honor codes like these are indeed mere facades.

But many schools and programs, from elementary to graduate level, take their honor codes seriously. And for good reason. Empirical research has repeatedly found that schools that are committed to honor codes have significantly reduced cheating rates compared with schools that are not.

Donald McCabe at Rutgers Business School and Linda Treviño at the Smeal College of Business at Penn State found a 23 percent rate of helping someone with answers on a test at colleges without an honor code, versus only 11 percent at schools with an honor code. They reported impressive differences as well for plagiarism (20 percent versus 10 percent), unauthorized crib notes (17 percent versus 11 percent), and unpermitted collaboration (49 percent versus 27 percent), among other forms of cheating (224).

A serious commitment to the honor code is crucial to its efficacy. As Professors McCabe and Treviño insist, an honor code should be "well implemented and strongly embedded in the student culture" (224).

What does that look like in practice? A few schools start the academic year with an actual commitment ceremony, where each student has to publicly pledge to uphold the school's code. To this can be added a requirement to affirm the honor code on each graded assignment.

When I was an undergraduate at Princeton, every paper we turned in had to have the honor code written out and then signed. Now as a professor at Wake Forest, I make my class recite aloud with me before each exam our entire honor code and then sign it.

Signing an honor code can, among other things, serve as a moral reminder. As we know from both ordinary life and recent experimental findings, most of us are willing to cheat to some extent if we think it would be rewarding and we can get away with it. At the same time, we also want to think of ourselves as honest people and genuinely believe that cheating is wrong. But our more honorable intentions can be pushed to one side in our minds when tempting opportunities arise to come out ahead, even if by cheating. What a moral reminder does, then, is help to place our values front and center in our minds.

* academic literacies   ● fields   ● research
■ rhetorical situations   ⦙ processes   ◖ media/design
▲ genres   ◆ strategies

This is borne out by recent findings in the lab. In a widely cited study, Nina Mazar at the Questrom School of Business at Boston University and her colleagues had one group of students take a 20-problem test where they would be paid 50 cents per correct answer. It was a hard test—students averaged only 3.4 correct answers. A second group of students took the same test, but they graded their own work and reported their "scores" with no questions asked. The average in this group was 6.1 correct answers, suggesting some cheating. The third and most interesting group, though, began by signing an honor code and then took the test, followed by grading their own work. The result? An honorable 3.1 correct answers. Cheating was eliminated at the group level. Signing the honor code did the job (636–37).

Studies of honor codes and cheating have typically been conducted in face-to-face environments. But as we settle into the routine of online instruction, we should consider trying to extend the impact of an honor code virtually as well.

Honor codes won't eliminate cheating. Deeply dishonest students will not be deterred. But fortunately, the research confirms what experience suggests: most students are not deeply dishonest.

Works Cited

Hubler, Shawn. "Keeping Online Testing Honest? Or an Orwellian Overreach?" *The New York Times*, 10 May 2020, www.nytimes .com/2020/05/10/us/online-testing-cheating-universities -coronavirus.html.

Mazar, Nina, et al. "The Dishonesty of Honest People: A Theory of Self-Concept Maintenance." *Journal of Marketing Research*, vol. 45, 2008, pp. 633–44.

McCabe, Donald L., et al. "Cheating in Academic Institutions: A Decade of Research." *Ethics & Behavior*, vol. 11, no. 3, 2001, pp. 219–32.

Patil, Anushka, and Jonah Engel Bromwich. "How It Feels When Software Watches You Take Tests." *The New York Times*, 29 Sept. 2020, www.nytimes.com/2020/09/29/style/testing-schools -proctorio.html.

Swauger, Shea. "Software That Monitors Students during Tests Perpetuates Inequality and Violates Their Privacy." *MIT Technology Review*, 7 Aug. 2020, www.technologyreview .com/2020/08/07/1006132/software-algorithms-proctoring -onlinetests-ai-ethics.

This proposal clearly defines the problem — students cheat on exams — and offers a solution: honor pledges and honor codes. The author provides arguments and evidence from research to show why this is a good solution.

Key Features / Proposals

A well-defined problem.   Some problems are self-evident and relatively simple, and you would not need much persuasive power to make people act — as with the problem "This university discards too much paper." While some people might see nothing wrong with throwing paper away, most are likely to agree that recycling is a good thing. Other issues are controversial: some people see them as problems while others do not, such as this one: "Motorcycle riders who do not wear helmets risk serious injury and raise health-care costs for everyone." Some motorcyclists believe that wearing or not wearing a helmet should be a personal choice; you would have to present arguments to convince your readers that not wearing a helmet is indeed a problem needing a solution. Any written proposal must establish at the outset that there is a problem — and that it's serious enough to require a solution. For some topics, visual or audio evidence of the problem may be helpful.

A recommended solution.   Once you have defined the problem, you need to describe the solution you're suggesting and to explain it in enough detail for readers to understand what you are proposing. Again, photographs, diagrams, or other visuals may help. Sometimes you might suggest several solutions, weigh their merits, and choose the best one.

A convincing argument for your proposed solution.   You need to convince readers that your solution is feasible — and that it is the best way to solve the problem. Sometimes you'll want to explain in detail how your proposed solution would work: "The honor code proposal offers several reasons why honor codes work, drawing on both empirical research and understanding of human nature." Visuals may strengthen this part of your argument as well.

✳ academic literacies	● fields	● research
■ rhetorical situations	⁘ processes	● media/design
▲ genres	◆ strategies	

A response to anticipated questions. You may need to consider any questions readers may have about your proposal—and to show how its advantages outweigh any disadvantages. Had the honor code proposal been written for college judicial officers, it would have needed to anticipate and answer questions about how such a code would be implemented.

A call to action. The goal of a proposal is to persuade readers to accept your proposed solution. This solution may include asking readers to take action.

An appropriate tone. Since you're trying to persuade readers to act, your tone is important—readers will always react better to a reasonable, respectful presentation than to anger or self-righteousness.

A BRIEF GUIDE TO WRITING PROPOSALS

Deciding on a Topic

Choose a problem that can be solved. Complex, large problems, such as poverty, hunger, or terrorism, usually require complex, large solutions. Most of the time, focusing on a smaller problem or a limited aspect of a large problem will yield a more manageable proposal. Rather than tackling the problem of world poverty, for example, think about the problem faced by people in your community who have lost jobs and need help until they find employment.

Considering the Rhetorical Situation

PURPOSE	Do you have a stake in a particular solution, or do you simply want to eliminate the problem by whatever solution might be adopted?	▇ 59–60
AUDIENCE	Do your readers share your view of the problem as a serious one needing a solution? Are they likely to be open to possible solutions or resistant? Do they have the authority to carry out a proposed solution?	▇ 61–64

72–74 ■
STANCE　How can you show your audience that your proposal is reasonable and should be taken seriously? How can you demonstrate your own authority and credibility?

75–77 ■
MEDIA/DESIGN　How will you deliver your proposal? In print? Electronically? As a speech? Would visuals, or video or audio clips, help support your proposal?

Generating Ideas and Text

Explore potential solutions to the problem. Many problems can be solved in more than one way, and you need to show your readers that you've examined several potential solutions. You may develop solutions on

489 ●
your own; more often, though, you'll need to do **RESEARCH** to see how others have solved—or tried to solve—similar problems. Don't settle on a

437–44 ◆
single solution too quickly—you'll need to **COMPARE** the advantages and disadvantages of several solutions in order to argue convincingly for one.

Decide on the most desirable solution(s). One solution may be head and shoulders above others—but it's important to be open to rejecting all the possible solutions on your list and starting over if you need to, or to combining two or more potential solutions in order to come up with an acceptable fix.

Think about why your solution is the best one. Why did you choose your solution? Why will it work better than others? What has to be done to enact it? What will it cost? What makes you think it can be done? Writing out answers to these questions will help you argue for your solution: to show that you have carefully and objectively outlined a problem, analyzed the potential solutions, weighed their merits, and determined the reasons the solution you propose is the best.

Ways of Organizing a Proposal

You can organize a proposal in various ways, but always you will begin by establishing that there is a problem. You may then identify several possible

※ academic literacies　● fields　● research
■ rhetorical situations　∴ processes　● media/design
▲ genres　◆ strategies

solutions before recommending one of them or a combination of several. Sometimes, however, you might discuss only a single solution.

[Several possible solutions]

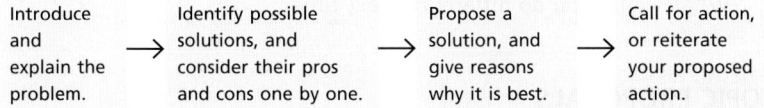

| Introduce and explain the problem. | → | Identify possible solutions, and consider their pros and cons one by one. | → | Propose a solution, and give reasons why it is best. | → | Call for action, or reiterate your proposed action. |

[A single solution]

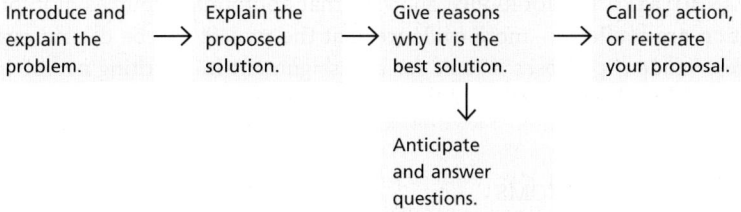

Introduce and explain the problem. → Explain the proposed solution. → Give reasons why it is the best solution. → Call for action, or reiterate your proposal.

↓

Anticipate and answer questions.

Taking Stock of Your Work

Take stock of what you've written by considering these questions:

- How did you determine your audience for this piece?
- How did you decide on and limit your topic?
- How did you determine the best solution to the problem you defined?
- What writing strategies did you use to guide your reader through your text? How effectively did you use them?
- What editing and proofreading processes did you use, and how effective were they?
- Based on your experiences with this piece, what academic habit of mind—for example, being flexible, creative, and so on—do you want to work on as you tackle your next writing project?

- What did you learn about writing as you wrote your proposal? How might you use that learning as you write in other contexts now or in the future?

- What did you do well in this piece? What could still be improved? What would you do differently next time?

TOPIC PROPOSALS

201–9

Instructors often ask students to write topic proposals to ensure that their topics are appropriate or manageable. Some instructors may also ask for an **ANNOTATED BIBLIOGRAPHY** showing that appropriate sources of information are available—more evidence that the project can be carried out. Here a student proposes a topic for an assignment in a writing course in which she has been asked to take a position on a global issue.

CATHERINE THOMS

Social Media and Data Privacy

The relationship between social media and data privacy is an issue that has recently risen to the forefront of many social media users' concerns. While we have been posting and sharing online, major companies like Facebook, Twitter, and Google have been silently collecting information about our personal lives and leaving that data vulnerable to potentially harmful exposure. As someone who has had their private information compromised multiple times by large-scale data breaches, I feel compelled to share my research on the dangers of such security breaches and how they can happen.

In this paper, I will argue that it is crucial for consumers to be aware of the ways in which social media platforms can gain access to consumers' private data and what those platforms can do with that data. In March 2018, it was revealed that 87 million Facebook profiles were exposed to manipulation by the political consulting firm Cambridge

---

☀ academic literacies	⬢ fields	⬤ research
■ rhetorical situations	⁖ processes	⬤ media/design
▲ genres	◆ strategies	

Analytica during the 2016 presidential election. I will use this scandal as an example of how the unlawful exposure of personal data can have far-reaching effects on external world events, and I plan to analyze how this scandal has functioned as a global catalyst for further scrutiny into the privacy mechanisms of social media.

I will concentrate on the specific ways personal data may be misused. Although sharing thoughts and pictures on social media may not seem like anything to think twice about, many people don't realize just how much is at stake should enough of that personal data fall into the wrong hands. Much of my research will be done on specific social media applications such as Facebook and Google Analytics, as that is where the heart of the issue resides and where consumers can learn the most about their personal data exposure and protection.

Thoms defines and narrows her topic (from data breaches to the specific ways consumers' data may be misused), discusses her interest, outlines her argument, and discusses her research strategy. Her goal is to convince her instructor that she has a realistic writing project and a clear plan.

Key Features / Topic Proposals

You'll need to explain what you want to write about, why you want to explore it, and what you'll do with your topic. Unless your instructor has additional requirements, here are the features to include:

A concise discussion of the subject. Topic proposals generally open with a brief discussion of the subject, outlining any important areas of controversy or debate associated with it and clarifying the extent of the writer's current knowledge of it. In its first two paragraphs, Thoms's proposal includes a concise statement of the topic she wishes to address.

A clear statement of your intended focus. State what aspect of the topic you intend to write on as clearly as you can, narrowing your focus appropriately. Thoms does so by stating her intended topic—data breaches— and then showing how she will focus on the specific ways consumers' data may be misused.

A rationale for choosing the topic. Tell your instructor why this topic interests you and why you want to write about it. Thoms both states what made her interested in her topic and hints at a practical reason for choosing it: plenty of information is available.

Mention of resources. To show your instructor that you can achieve your goal, you need to identify the available research materials.

If you need more help

364–66
367–71
371–79
380–84
385–90

See Chapter 30 for guidelines on **DRAFTING**, Chapter 31 on **ASSESSING YOUR OWN WRITING**, Chapter 32 on **GETTING RESPONSE AND REVISING**, and Chapter 33 on **EDITING AND PROOFREADING**. See Chapter 34 if you are required to submit your proposal in a writing **PORTFOLIO**.

* academic literacies ● fields ● research
■ rhetorical situations ∴ processes ● media/design
▲ genres ◆ strategies

21 Explorations

Sometimes we write essays just to think about something—to play with an idea, develop a thought, explore possible solutions to a problem, or even to figure out what the problem is. Exploratory essays are our attempt to think something through by writing about it and to share our thinking—or the progression of our understanding—with others. In college, you might be asked in courses across the curriculum to write formal or informal exploratory essays, journals, design reports, or learning log entries. Have a look at one example of an exploratory essay written for a first-year composition class at Columbus State Community College. The author, Brian Diehl, was a high school student taking college courses when he wrote his exploration.

BRIAN DIEHL

Talking with Granddad

An overcast sky chokes the usual radiant light of summer on this Sunday afternoon—an ironic backdrop to the sign in front of me, which reads "Sunrise on the Scioto." Upon entering the nursing home's lobby, I'm met with a clash of two scents: the buttery aroma of a nearby popcorn machine and that distinct "old people" smell. The latter always seems to outlast its opponent and seeps into my nostrils around every corner of the building. I make sure to grab a few hard candies before taking an elevator up to the third floor.

Knock, knock, knock.

No answer. Peeking into the doorway, I see a man slouched in his bed absorbing the sounds of a televised college football game.

"Hey granddad, it's me!" A slow turn allows him to see the boy now in front of him. "You know who I am, right?"

He stares blankly.   5

"I'm Brian, your grandson!"

The scruff around his face rises, revealing a smile of recognition. ". . . Yes," he responds.

"How've you been holding up?" I ask. The drone of the TV fills the silence. "Hey, I have a girlfriend now! Would you like to see her?"

I bring the image on my phone closer to him. After some scrutiny, he exclaims, "Well, I'll be damned!"

"Has grandma stopped by recently?"   10

". . . Well, I'll be damned!"

He repeats this phrase after every question I ask. After a while, I lightly wrap my arms around his motionless body, and we part ways.

For the longest time, there has been no medical phenomenon more intriguing to me than dementia. One may be familiar with Alzheimer's disease, the most common type of dementia characterized by plaques and tangles that spread throughout the brain, destroying tissue and eventually affecting memory, language, and other neurological functions ("Dementia"). When my grandfather showed signs of Alzheimer's in its early stages, I found myself struggling in my attempts to talk with him; my words sometimes seemed to bounce off a brick wall. Knowing that my other grandparents could be at risk of dementia as well, I wondered: How could I communicate more effectively with family members who may exhibit memory loss in the future?

Of course, communication is a two-way street, so before I could explore verbal tactics on my end, I decided to first gain an understanding of how people with dementia might express their mental state—how *they* try to communicate their experiences and perceptions. This is when I came across the story of artist William Utermohlen, who was diagnosed with Alzheimer's in 1995. As a way of understanding the disease, the artist chose to paint himself. Below is a series of Utermohlen's self-portraits, revealing changes in his perception of himself over time.

Comparing the 1996 portrait to the one drawn pre-diagnosis, we   15
can notice an immediate disappearance of details such as hair texture and facial shadows. Utermohlen's unicolor portrayal of himself inside a green framework reflects neurologist Bruce Miller's comments on artistic

❋ academic literacies   ● fields   ● research
■ rhetorical situations   ⁝ processes   ● media/design
▲ genres   ◆ strategies

A collage showing details from William Utermohlen's self-portraits. The first (top left) was made in 1967, and the rest were made following his diagnosis of Alzheimer's disease: the top middle in 1996, the top right in 1997, and the bottom row, from left to right, in 1998, 1999, and 2000.

creativity among people with brain diseases: "Alzheimer's affects the right parietal lobe in particular, which is important for visualizing something internally and then putting it onto a canvas. The art becomes more abstract, the images are blurrier and vague, more surrealistic" (qtd. in Grady). This surrealism becomes more apparent in the next portrait, where the eyes are positioned far too low, the forehead and ear are enlarged, and the frown of the artist becomes more solemn. The true erasure of identity begins in the 1998 portrait, revealing a fragmented Utermohlen with his most melancholic facial expression yet. Perhaps the most devastating work is the fifth portrait; other than an oval serving as the bare structure of a human head, there is hardly a distinguishable facial feature contained in the canvas. The final portrait—with its pencil scratches, eraser smudges, and incompleteness—suggests a dramatic alteration of Utermohlen's sense of self. What can perhaps be represented by this artist's works are the combined voices of those just like him—those who cannot verbalize their reality. As heartbreaking as Utermohlen's portraits are, they offer caregivers and others a bridge toward understanding dementia from the perspective of those who experience it.

After gaining insight into the possible mental state of someone with dementia, I sought to find concrete methods of verbal communication

that could best meet the needs of individuals with Alzheimer's. I came across a study published in the *Journal of Speech, Language, and Hearing Research* conducted by Jeff Small and colleagues, who set out to evaluate the effectiveness of ten common strategies for communicating with people with Alzheimer's. As part of the study, the researchers audio-recorded twenty-two couples—individuals with Alzheimer's and their spouses—as the couples engaged in everyday tasks like setting the table, getting an item upon request, and using the telephone. The success of each communication strategy was measured by the number of communication breakdowns that occurred after its usage. Small and colleagues report that "eliminating distractions, speaking in simple sentences, and employing yes/no questions" resulted in fewer breakdowns (364). On the other hand, using a slower rate of speech caused an increase in the number of breakdowns. This study suggests the viability of some verbal techniques over others, but I was still left with a burning question: Doesn't communication effectiveness ultimately depend on what personally works for the individual with Alzheimer's?

At this point, I decided to reach out to somebody who could shed light on my question: my grandmother, Andrea Diehl. Being the embodiment of positivity that she is, her continual presence by my grandfather's side at the nursing home was one of the greatest factors that helped keep him stable. Below is a portion of our interview that highlights their communication dynamic in the midst of his dementia.

> **Brian:** Did you make any changes to your communication with granddad during his state of dementia? If so, what methods worked for you?
>
> **Andrea:** Anything that worked for me was something that was familiar already to him, something that *he* wanted to say. I always made sure he was smiling. Anything that I knew was going to please him was the object of my time with him. I would give him a hot chocolate and give him a little hug, and then when I put a marshmallow in that hot chocolate, he would smile! There were many things that I tried to do to keep him above board and happy. There were many little tricks like that. I wanted him to have the memories that were still living in both of our minds.
>
> **Brian:** In what ways did grandpa communicate with you?

※ academic literacies   ● fields   ● research

■ rhetorical situations   ∴ processes   ● media/design

▲ genres   ◆ strategies

Andrea: He would reminisce about old times. For example, we had taken seven cruises on one cruise line, and we would relive those because he remembered those! He would ask, "Do you remember that meal we had? Do you remember when I found just the right things to make the drink I wanted to make?" He could remember those things and then ask me if I remembered them! It was a joy, every minute.

Reflecting on my grandmother's experiences, it becomes clear that one does not need to be a master of psychological techniques in order to properly approach communication with a person affected by dementia. What's needed is a repeated effort to be present with the individual, expressing joy over what *they* say instead of sweating over anything *we* could ever say. Even if one were to find a golden strategy that makes conversation a breeze, its use alone will never mitigate the frustration of identity loss portrayed so vividly by Utermohlen. Instead, the most important action one can take is a demonstration of love. Ranging from hugs to marshmallow drops, small acts of kindness honor the humanity of the person with Alzheimer's. It is the enjoyment of their company that fosters communication. It is relationship, not repetition, that gives way to a signal of understanding, even if that signal is as simple as a smile.

Looking back on my initial visit with my grandfather, I knew that he processed every word I said. Those shining blue eyes and unwavering smile told me that he fully knew of my presence, but he was unable to express that awareness in my world. I like to think that my grandfather's mind found an entire universe overflowing with worlds unique to him, each planet representing a blissful memory of a life full of love.

Did dementia ever take his [20] universe away? No.

Well, I'll be damned!

Works Cited

"Dementia." *The Gale Encyclopedia of Psychology*, edited by Jacqueline L. Longe, 3rd ed., Gale, 2016. *Credo Reference*, search.credoreference.com/content/entry/galegp/dementia/0.

Diehl, Andrea. Interview with author. 11 Nov. 2020.

Grady, Denise. "Self-Portraits Chronicle a Descent into Alzheimer's." *The New York Times*, 24 Oct. 2006, www.nytimes.com/2006/10/24/health/24alzh.html.

Small, Jeff A., et al. "Effectiveness of Communication Strategies Used by Caregivers of Persons with Alzheimer's Disease during Activities of Daily Living." *Journal of Speech, Language, and Hearing Research*, vol. 46, no. 2, Apr. 2003, pp. 353–67. EBSCOhost, https://doi.org/10.1044/1092-4388(2003/028).

Diehl's challenges in conversing with a grandparent with Alzheimer's disease spur him to explore ways to communicate effectively with people with dementia. Through his exploration, Diehl ponders the merits of several communication strategies, including those recommended by experts and those used by his own grandmother.

Key Features / Explorations

A topic that intrigues you. An exploratory essay has a dual purpose: to ponder something you find interesting or puzzling, and to share your thoughts with an audience. Your topic may be anything that interests you. You might write about someone you've never met and are curious about, an object or occurrence that makes you think, a problem you need to solve—or one you need to define. Your goal is to explore the meaning that the person, object, event, or problem has for you in a way that will interest others. One way to do that is by making connections between your personal experience and more general ones that readers may share. Diehl writes about his interactions with his own grandfather, but in so doing he raises questions and offers insights about how others communicate with people affected by dementia.

* academic literacies   ● fields   ● research
■ rhetorical situations   ∴ processes   ● media/design
▲ genres   ◆ strategies

Some kind of structure. An exploratory essay can be structured in many ways, but it needs to *be* structured. It may seem to wander, but all its paths and ideas should relate, one way or another. The challenge is to keep your readers' interest as you explore your topic and to leave readers satisfied that the journey was pleasurable, interesting, and profitable. Often doing this requires COMBINING GENRES. For example, Diehl begins with a narrative that includes snippets of dialogue, moves into a visual analysis of a series of self-portraits, summarizes a study of communication strategies, and then recounts an interview with his grandmother. He ends by proposing that demonstrating love for a person with dementia is more important than using any particular strategy for communication.

■ 69–70

Specific details. You'll need to provide specific details to help readers understand and connect with your subject, especially if it's an abstract or unfamiliar one. Diehl notes how his grandfather is "slouched in his bed, absorbing the sounds of a televised college football game" and how the "scruff around his face rises, revealing a smile of recognition." Later, Diehl captures an anecdote from his grandmother to help bring her relationship with her husband to life: "'I always made sure he was smiling. Anything that I knew was going to please him was the object of my time with him. I would give him a hot chocolate and give him a little hug, and then when I put a marshmallow in that hot chocolate, he would smile!'" In addition to details and anecdotes, exploratory essays may include researched information, as Diehl's does when it mentions a study of communication strategies among couples affected by Alzheimer's. In some cases, photographs or other visuals—like the self-portraits Diehl analyzes—help provide examples as well as set a certain tone for an exploration.

A questioning, speculative tone. In an exploratory essay, you are working toward answers, not providing them neatly organized and ready for consumption. So your tone is usually tentative and open, demonstrating a willingness to entertain, accept, and reject various ideas as your essay

progresses from beginning to end. Often this tone is achieved by asking questions, like Diehl does: "Knowing that my other grandparents could be at risk of dementia as well, I wondered: How could I communicate more effectively with family members who may exhibit memory loss in the future?" "This study suggests the viability of some verbal techniques over others, but I was still left with a burning question: Doesn't communication effectiveness ultimately depend on what personally works for the individual with Alzheimer's?"

A BRIEF GUIDE TO WRITING EXPLORATIONS

Deciding on a Topic

Choose a subject you want or need to explore. Write a list of things that you think about, wonder about, find puzzling or annoying. They may be big things—life, relationships—or little things—quirks of certain people's behavior, curious objects, everyday events. Try clustering one or more of those things, or begin by freewriting to see what comes to mind as you write. If you need to define and solve a problem, try googling the topic to see what others have said about it and how you might think about it—and to see what potential solutions have already been proposed.

Considering the Rhetorical Situation

PURPOSE — What's your goal in writing this essay? To explore a topic that interests you? To work through an issue or problem that concerns you? To provoke readers to think about something? What aspects of your subject do you want to ponder and reflect on?

59–60

AUDIENCE — Who is the audience? How familiar are they with your subject? How will you introduce it in a way that will interest them? Even if your primary audience is yourself, how will you help readers understand your thinking?

61–64

* academic literacies
* fields
* research
* rhetorical situations
* processes
* media/design
* genres
* strategies

STANCE	What is your attitude toward the topic you plan to explore—questioning? playful? critical? curious? something else?	■ 72–74
MEDIA/DESIGN	Will your essay be a print document? an oral presentation? Will it be posted on a website or a blog? Would it help to include any visuals or video or audio files?	■ 75–77

Generating Ideas and Text

Explore your subject in detail.   Exploratory essays often include descriptive details. Diehl, for example, describes Utermohlen's self-portraits as he speculates about their meaning and effect: "This surrealism becomes more apparent in the next portrait, where the eyes are positioned far too low, the forehead and ear are enlarged, and the frown of the artist becomes more solemn." Diehl also compares and contrasts the portraits, pointing out the changes between the pre-diagnosis portrait and those that follow. You may also make your point by defining, narrating, even classifying. Virtually any organizing pattern will help you explore your subject.

Back away.   Ask yourself why your subject matters: Why is it important or intriguing or significant? You may try listing or outlining possibilities, or you may want to start drafting to see where the first round of writing takes your thinking. Your goal is to think onscreen (or on paper) about your subject, to play with its possibilities.

Think about how to keep readers with you.   Explorations may seem loose or unstructured, but they must be carefully crafted so that readers can follow your train of thought. It's a good idea to sketch out a rough thesis to help focus your thoughts. You might not include the thesis in the essay itself, but every part of the essay should in some way relate to it. If Diehl's essay has a thesis, it's this: "One does not need to be a master of psychological techniques in order to properly approach communication with a person affected by dementia. What's needed is a repeated effort to be present with the individual, expressing joy over what *they* say instead of sweating over anything *we* could ever say."

Ways of Organizing an Exploratory Essay

Exploratory essays may be organized in many ways because they mimic the way we think, associating one idea with another in ways that make sense but do not necessarily form a "logical" progression. In general, you might consider organizing an exploration using this overall strategy:

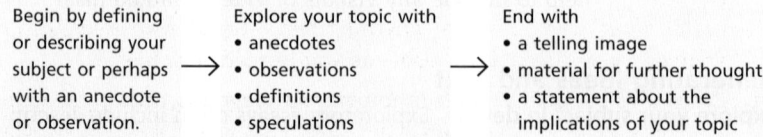

Begin by defining or describing your subject or perhaps with an anecdote or observation. → Explore your topic with
• anecdotes
• observations
• definitions
• speculations → End with
• a telling image
• material for further thought
• a statement about the implications of your topic

Another way to organize this type of essay is as a series of brief anecdotes and ideas that together create an overall impression:

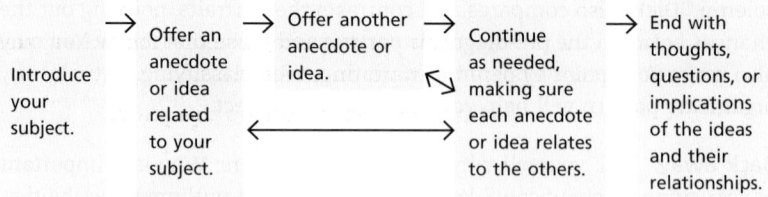

Introduce your subject. → Offer an anecdote or idea related to your subject. → Offer another anecdote or idea. → Continue as needed, making sure each anecdote or idea relates to the others. → End with thoughts, questions, or implications of the ideas and their relationships.

If you're pondering solutions to a problem, you might organize your essay like this:

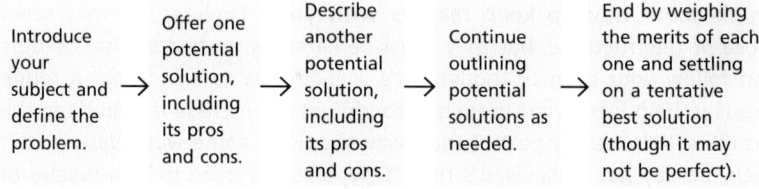

Introduce your subject and define the problem. → Offer one potential solution, including its pros and cons. → Describe another potential solution, including its pros and cons. → Continue outlining potential solutions as needed. → End by weighing the merits of each one and settling on a tentative best solution (though it may not be perfect).

* academic literacies   ● fields   ● research
■ rhetorical situations   ⁝ processes   ● media/design
▲ genres   ◆ strategies

Taking Stock of Your Work

Take stock of what you've written by considering these questions:

- How did you decide on your topic?

- Why did you organize your piece as you did?

- Describe your revision process for this piece. What worked, and what didn't work? Did you learn anything new—about your topic, about yourself, about something else—as you revised your piece?

- Did you use visual elements (photographs, diagrams, drawings), audio elements, or links? If so, what did they add? If not, what might you have used?

- How did curiosity play a role in the writing of your exploration? Can you point to a specific part of your text that illustrates your curiosity about your topic?

- What have you learned about your writing ability from writing this piece? What do you need to work on in the future?

- What was most rewarding about your work on this piece?

If you need more help

See Chapter 30 for guidelines on **DRAFTING**, Chapter 31 on **ASSESSING YOUR OWN WRITING**, Chapter 32 on **GETTING RESPONSE AND REVISING**, and Chapter 33 on **EDITING AND PROOFREADING**.

364–66
367–71
372–79
380–84

▲

22 Remixes

Just as musicians alter existing tracks to create new ones, writers sometimes "remix" or refashion their writing for new contexts. A mechanical engineer uses elements of her technical report about a new engine part to write instructions for the machinists who assemble the part. A historian turns his scholarly article on slavery and disability into a podcast episode for the general public. A sociology student transforms her essay on food insecurity into an infographic to persuade her college administration to start a food pantry. In each case, the writer modifies all or part of their earlier work for a new purpose, audience, genre, stance, or medium.

In college courses across the curriculum, you might be asked to do the same—for example, by writing a researched report for a class assignment and then sharing your work through a multimedia presentation for your classmates or others. For a capstone project in your major, you might be asked to return to some earlier writing you've done, reimagining or extending it for a new purpose.

This chapter offers suggestions for creating remixes—texts that grow out of and "play on" earlier texts you've composed. Let's start by looking at an example: a profile originally published in a university alumni magazine that gets remixed into two different videos for the university's website.

Original Text: A Print Profile

"Rare Earth," a profile that inspired two remixes, traces Dr. Rattan Lal's path from refugee to premier soil scientist. Following is an excerpt from the piece, which appeared in the *Ohio State Alumni Magazine*. The author, Todd Jones, is a senior writer for the magazine.

✳ academic literacies   ● fields   ● research

■ rhetorical situations   ∴ processes   ● media/design

▲ genres   ◆ strategies

TODD JONES
Rare Earth

The thin man in the Ohio State ballcap hunches down near his home garden, in the backyard with the two buckeye trees. He's holding a clump of dirt. "No, no. Not dirt," he says. "This is soil. Soil is not the same as dirt. Soil is a living entity. That's the part we must understand."

His hands are cupped, as if holding a baby bird. He gazes in admiration, points to roots and a worm. He tells us a tablespoon of soil contains billions of microorganisms. "Soil is one. Indivisible. Interconnected," he says. "Soil is the basis of all terrestrial life. Every living thing on the planet depends on it, and yet this material is underappreciated and unrecognized."

His face is wrinkled and weathered, a testament to years spent researching in the fields, forging wisdom about our place in the world. "We belong to nature. We belong to soil," he says. "Therefore, we have a duty and obligation to soil." He has researched soil for five decades on five continents, traveling to 106 countries to spread a message: that we must "make an investment to restore [soil]. You always want to leave something behind," the concept underlying regenerative agriculture. But more work must be done to spread understanding about the critical role of soil in the health of the planet. "That challenge," he says, "is my motivator, my driving force."

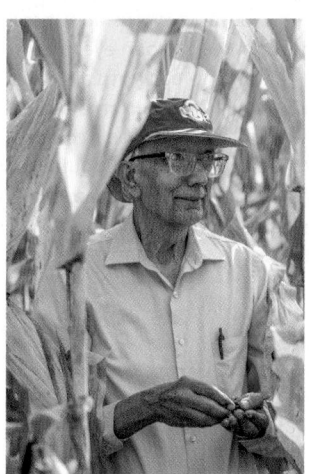

Lal surveys an Ohio cornfield.

Thomson Reuters has long listed Dr. Rattan Lal, Ohio State Distinguished University Professor of Soil Science, among the top 1 percent of the most frequently cited agricultural

researchers in the world. In 2007, Lal was recognized for his contributions to the Intergovernmental Panel on Climate Change, which shared
the 2007 Nobel Peace Prize with former U.S. vice president Al Gore. In
2019, Lal was awarded the Japan Prize, one of the most prestigious
honors in science and technology. That same year, he won the Alumni
Medalist Award, the highest honor bestowed by The Ohio State University Alumni Association. Such status suggests that Lal deserves a gilded
lair, but he sits behind an ordinary metal desk in a painted cinder-block
room measuring 9 feet by 12 feet. "My office is fine," Lal says. "The
office is in proportion to your contributions."

His modesty also masks a fierce inner drive. Three heart attacks   5
have not slowed him, and praise from international peers has not
inflated him. He has more speeches to give, papers to write, conference calls to join, classes to teach, research to conduct, and students
to mentor. "This is who he is," says Sukhvarsha Sharma, his wife of 48
years. "He has very high expectations of himself. He wants to make a
difference in the environment, and through that, make a difference
in peoples' lives."

There are times when Lal looks like a symphony conductor as he speaks,
his voice accompanied by slowly moving hands, as if words about soil
were notes in a musical sonnet. In those moments, your eyes might
fall on an ancient symbol in faded blue ink on the back of Lal's right
hand. He thinks he was about 8 years old when a stranger on a bicycle
came to his school in India offering cheap tattoos. The boy couldn't
resist. Nearly seven decades later, he regrets his choice. "I would not
get it done now," Lal says. "But a small decision about a tattoo, that's
very little in importance. Decisions about soil are very, very important.
Those are decisions that influence other people's lives and have very
long-lasting impact."

The impact is intensifying because the world's population is projected to increase nearly 50 percent in the next 80 years—an ominous
forecast that Lal tells Ohio State students on a summer afternoon while
lecturing in an Introduction to Soil Science class. Humanity, Lal warns,
is facing a dire necessity to advance food security by protecting finite,
fragile soil resources. "Agriculture and soil are, can be, and must be a
solution," he says, clasping and shaking his hands for emphasis.

☀ academic literacies   ● fields   ● research
■ rhetorical situations   ∴ processes   ◗ media/design
▲ genres   ◆ strategies

In that moment, you could see Lal's tattoo, an ink version of the symbol for om, considered the most sacred of mantras in his Hindu religion. Through the word's three distinct sounds — "a-u-m" — the speaker is meant to understand the essence of the universe: Everything is connected. Om is the sound of Lal's life.

"There's just kind of an equanimity, a sense of mental peace, about him," says Ellen Maas, one of Lal's current doctoral students. "His manner is not one that is focused on himself. He just cares so deeply about humanity and the planet. He finds an inner strength in caring about things bigger than himself."

Lal's family experience shaped his worldview. The youngest of three, he was raised by his father after his mother died of an illness when Lal was 3 months old. When India was partitioned to form Pakistan in 1947, his family fled as refugees to the Indian village of Rajaund, about 120 miles northwest of New Delhi. They left a 9-acre farm and settled on one totaling only 1.5 acres, a change that marked Lal with a different tattoo, just as permanent but invisible.

10

Lal holds his undergraduate diploma from Punjab Agricultural University.

"It taught me that the land is very precious," Lal says. "You cannot let it waste. Your life depends on it. Our land was so small, but it was not a unique situation. There are right now in the world about 600 million farm families that have less than 2 acres. In a way, what I do and what I think relates back to those people and my experience."

Lal helped grow rice, wheat, cotton, and sugarcane on his family farm. But unlike his siblings, he also was sent to school by his father. The child felt chosen, obligated to make good on a privileged opportunity. In 1959, his diligence and excellent grades met

Ohio State in serendipity: he enrolled at the Government Agricultural College and Research Institute (which became Punjab Agricultural University his senior year). Ohio State had helped establish the institute only four years earlier. "One of the professors, Dr. [Dev Raj] Bhumbla, had just returned with a PhD from Ohio State, and he started teaching a soil class," Lal says. "I was in that class. My best subject was botany, but he told me to go into soil science. In fact, he recommended my name to faculty at Ohio State to be a student there."

In 1965, Lal boarded a plane for the first time and traveled alone 7,300 miles to pursue his doctorate. He brought along an insatiable hunger to succeed. At Punjab Agricultural University, he had run 4 miles— yes, run them—to school from home every day, studied by the light of a kerosene lamp, sometimes held up by his grandmother. He found an agricultural book in the library so essential that he spent two months hand-copying it to have his own edition. "Nothing can be achieved without the strongest possible desire to achieve it," Lal says.

His desire and degree mixed to fuel a career trek around the globe. Ohio State connections landed him research jobs, first in Sydney, Australia, for a year, and then in Nigeria, where the Rockefeller Foundation asked him to set up a soil science laboratory at the International Institute of Tropical Agriculture. Lal and his wife—only a year into their arranged marriage—went in spite of a bloody civil war. They stayed 18 years, even as three Nigerian governments were toppled by coups. "Once, I was at the house of a general," Lal says. "He had invited me to come talk about suggestions to control soil erosion. Two days later, he was hanged and his family shot. People have said if I was there when they came, I would not be here today."

One moment can make a difference. So can one person. This is 15 why he accepts invitations to speak about soil, whether the audience is filled with heads of power or young burgeoning minds. "The students are ambassadors of Ohio State," Lal says. "They represent us. They carry on the work that we do."

There are twenty-two students in this day's class. Lal speaks to them without notes, standing next to a podium, hand in pocket. His tone is

measured, but his message linking soil management to the nourishment of humankind is urgent. "The story has to be told in a language that farmers and policymakers understand and can relate to. That's critical," Lal says. "That's going to take education. We have to change opinions in a humble way. That falls on your generation."

Remix 1: A Video Profile

Eventually, the writer and photographer of "Rare Earth" collaborated with others to remix the article into a video profile. The remix uses recorded footage to showcase Dr. Lal in action as he and his students assess the results of no-till farming. Below are a few frames from the video; you can view the complete video at digital.wwnorton.com/fieldguide6.

(gentle music)

and what practices can be shown are not sustainable

And we're gonna quantify the total fresh weight

If we can do no-till and we can have better yields

Remix 2: An Explainer Video

The original profile piece was remixed into yet another video, this time a motion-graphics video that explains the science behind regenerative agriculture. Below are some frames from the full video, available at digital.wwnorton.com/fieldguide6.

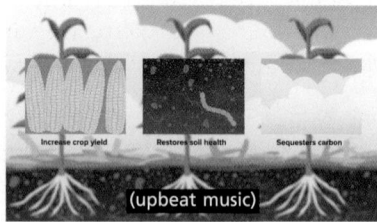

All three texts remix parts of the others, but each has a different purpose, audience, and medium/design. The print article is designed for magazine subscribers and aims to profile Dr. Lal's personal and professional achievements as an alumnus of the university. The video profile is produced for the web and provides a firsthand look at Dr. Lal's work in agriculture, while the explainer video helps viewers understand a crucial concept in Dr. Lal's research. Together the remixes offer a fuller picture of the scientist and his contributions to his field.

Key Features / Remixes

A connection to the original subject. A remix always hearkens back to the original subject in some way. Sometimes the connection is obvious— Dr. Lal takes center stage in both the video profile and the print profile, though each profile shares a different aspect of his story. Other times, just a fragment of the original subject inspires the remix. For example, the science behind Dr. Lal's work is mentioned only briefly in the print article but becomes the sole focus of the explainer video.

A modification of the earlier work. By definition, a remix involves change. Something in the writing situation changes—like the writer's purpose or audience—and that change requires other modifications as well. The writer might need to work within a new medium, or adopt a different stance, or write in a new genre to get their message across most effectively. When the creators of the "Rare Earth" article needed to explain the complex concept of regenerative agriculture, they determined that a visual medium would better suit their purpose than a print-based one.

Attention to genre and media / design expectations. Often remixes require a change in genre or a change in medium (or both), and effective remixes meet audience expectations for the new genre or medium. The profile video of Dr. Lal, for example, uses background music, on-screen titles, and clips of recorded footage—all of which are typical features of a profile (genre) delivered through a video (medium).

SOME TYPICAL WAYS OF REMIXING

Your writing situation—especially your purpose and audience—will determine the type of remix you might create. Following, though, are some common ways of remixing texts for new genres or media.

Memoirs. With their vivid details and scenes, print memoirs lend themselves to visual remixes as photo montages, graphic memoirs, comic strips, and digital stories. You might also repurpose parts of a memoir into a

new genre—for example, by sharing an anecdote to support a point in an argument.

Profiles. A print profile that showcases an interesting person, place, or event might get remixed into a mini documentary or a podcast to bring the subject to life even more fully for an audience. You could also incorporate a profile into another genre—like a report—to make an abstract concept more concrete.

Arguments. With their focus on making and supporting claims, arguments are often remixed as newspaper op-eds, public service announcements (PSAs) for radio or television, or print or online petitions. With instructor permission, you might remix an argument you've written on gun control for your first-year writing class into a proposal for your public health class.

Proposals. A written proposal might morph into a slide show for a presentation and then into a fund-raising letter to secure financial support for a proposed project. In your college classes, you might repurpose parts of a topic proposal for a report, an argument, or some other genre you're asked to write—as long as your instructors approve.

Reports. Research findings from reports often get remixed into infographics, poster presentations, timelines, and listicles (articles presented in list form). In college courses, you might remix a report you've written into an oral presentation to your class.

In creating a remix as part of your college coursework, be sure to follow your school's academic integrity policies. Most policies forbid students from reusing or resubmitting their previous work without instructor permission. If you plan to remix something you've created for one class into a project for another class, get approval from *both* instructors beforehand.

A BRIEF GUIDE TO CREATING REMIXES

Determining What Will Change in Your Remix

When you compose a remix, you're not starting from scratch—you already have a general topic and a text to work from. Still, you need to determine how your remix will be different from your earlier work.

A good place to start is to think about what has changed in your writing situation. Has your instructor asked you to prepare a *Prezi* version of an academic essay you've written for your course? If so, the media/design elements of your rhetorical situation have changed. Or maybe you've written a report for a class assignment, and now you want to inspire a particular group of people to take action on your topic. In this case, your purpose and audience have changed. Usually, a change in one element of the rhetorical situation affects the others, so it's important to carefully consider all those elements as you plan your remix.

Considering the Rhetorical Situation

PURPOSE	Why are you creating the remix? To share the same message from your earlier work through a different genre or medium? To make a new point about your topic? To explain a concept from your earlier work? To reach a different audience? To reach the same audience in a new way? To persuade or inform or entertain?	■ 59–60
AUDIENCE	Who will read, view, or listen to your remix? Are they the same audience as before? a different audience? Why should they care about your message? What will they already know? How can you engage their interest?	■ 61–64
GENRE	Will you work within a new genre or the same one as earlier? Will you mix multiple genres? Which genres would work best for your purpose and audience? What expectations will your audience have for genre?	■ 65–71

72–74 ■	**STANCE**	Will your stance in the remix be the same as or different from your earlier stance? What is your relationship with your audience? Are you an equal, a student, an interested or concerned outsider, something else? What tone is appropriate—objective, impassioned, respectful, informal, something else?
75–77 ■	**MEDIA/DESIGN**	Will you work within a new medium or the same one as earlier? What medium would be most appropriate for your audience and purpose? Will you use multiple media, combining sounds, visuals, words, movement? What expectations will your audience have for design and format? What steps can you take to make your
666–68 ●		composition **ACCESSIBLE** to diverse audiences?

Generating Ideas and Text

Return to your earlier work. Remember, you're not starting from a blank slate. You can go back to your earlier work to decide what to keep, delete, change, or add as you create your remix. Are there key ideas you want to be sure to include in your remix? Will you need to condense some information from your earlier work? explore a new angle on your topic? reframe an argument? conduct more research? find or create images or sounds? Try **FREEWRITING**, **LISTING**, or **CLUSTERING** to figure out what you can use from your earlier work and what you'll need to reimagine or create anew.

If you're remixing for a new genre, examine sample texts in your chosen genre. Whether you're creating a brochure, a business plan, an infographic, or something else, find samples that you think are effective. Then study the genre features: What do you notice about the use of headings, bullets, photos, and other **DESIGN** elements? What do you notice about the formality of the language, the lengths of sentences and paragraphs, the **TONE**? How is source information documented? How are **VISUALS** or sounds acknowledged or cited? Consider how you'll make use of the genre features in your own remixed composition.

333–36 ⁘

75–77 ■
73–74
674–80 ●

⁘ academic literacies	⬢ fields	● research
■ rhetorical situations	⁘ processes	● media/design
▲ genres	◆ strategies	

If you're remixing for a new medium, consider your capabilities and your options. Maybe you want to create a website, a video, a podcast, or some other digital text, but you don't have access to the tools you need or you're not skilled in using them. Your school might have a multimedia support center or other resources to help you get started. Another option: with your instructor's permission, you could perhaps create a "pre-production" document instead of a full-fledged multimedia piece. For example, you might make a mock-up of a website in a *Microsoft Word* document. Or you might write a script for a podcast or create a detailed **STORYBOARD** for a video.

338–39

Remember, too, that online tools and templates can make composing with sound and visuals easier. Some tools you might consider using (with a free trial) include:

For brochures, advertisements, posters
Adobe Spark, Canva

For infographics
Venngage, Piktochart, Canva

For interactive line charts
Knight Lab Storyline

For interactive timelines
Tiki Toki, Knight Lab Timeline

For photo montages
Fotor

For picture books
Storybird, Storyjumper

For podcasts
Podbean

For slide presentations
Prezi, Canva

For comics, storyboards, graphic novels
Pixton

For videos
Animoto, Powtoon

For virtual reality stories
Knight Lab Scene

For websites
Adobe Spark, Google Sites, Weebly, Wix

For free music and sound effects
Free Music Archive, Orange Free Sounds

For free images
Pexels, Pixabay

Ways of Organizing a Remix

Your organizational strategy will depend on your genre and medium, and you may be able to consult the relevant genre chapter as a guide. If appropriate, you might also consider creating a simple **STORYBOARD** to plan out a sequence of text and images for your remix.

338–39

Taking Stock of Your Work

Take stock of what you've created by considering these questions:

- What elements of your rhetorical situation changed as you moved from your original work to your remix—your purpose? audience? genre? stance? medium? In what specific ways does your remixed composition respond to those changes?

- What part of the composing process was most enjoyable as you worked on your remix? Why?

- What technological, logistical, design, or other challenges did you face in creating your remix, and how did you address them?

- What previous composing experiences (print-based and multimedia ones), if any, did you draw on in creating your remix? Did those experiences help you or hinder you?

- What resources did you use to complete your remix? How did those resources help you, and how might you use them in future composing situations?

- What new things did you learn about your topic in the process of remixing your earlier work?

- What new things did you learn about composing? about genres, media, and other elements of rhetorical situations?

- What are you most proud of in this piece?

If you need more help

659–63
664–72
673–83
684–93
391–401

See Chapter 55 for guidelines on **CHOOSING MEDIA**, Chapter 56 on **DESIGNING TEXT**, Chapter 57 on **USING VISUALS, INCORPORATING SOUND**, and Chapter 58 on **WRITING AND LEARNING ONLINE**. If you're required to write a formal **REFLECTION** to accompany your remix, see Chapter 35.

☀ academic literacies	● fields	● research
■ rhetorical situations	⠿ processes	● media/design
▲ genres	◆ strategies	

Part 4

Fields

When we study at a college or university, we take courses in many academic fields of study, or disciplines, in the humanities, social sciences, the sciences, and various career-oriented fields. Each field of study has its own methods of doing research and communicating, requiring us to learn how to **READ** and **WRITE** in the ways appropriate to that field. The chapters that follow offer advice on how to adapt your reading and writing to the demands of various fields of study.

Fields

23 Fields of Study

Most colleges and universities are organized around fields of study, often referred to as *disciplines*, that share ways of seeing the universe, doing research, and presenting information—each through a different lens. For example, historians and economists differ in the ways they examine human behavior; biologists and physicists differ in the ways they study the natural world. The various majors that colleges and universities offer invite you to learn to see the world in their distinctive ways and join professions that use distinctive skills. Marketing majors, for example, learn ways of influencing human behavior, while education majors learn ways of teaching students and music majors learn the intricacies of playing an instrument or singing. As you immerse yourself in a major, you come to see the world with a particular perspective.

Academic Fields and General Education

Your general education courses have a different goal: to prepare you for the demands of living in a complex and changing world—a world in which you'll need to use your ability to understand a wide range of facts, theories, and concepts to interact with a wide variety of people, work in various jobs, and make important decisions. In other words, these courses are designed to teach you how to learn and think. They are also helpful if you're not sure what you'd like to major in, because you can sample courses in various disciplines to help you decide. In your first couple of years, you will likely take general education courses in the following subject areas:

- the humanities, which include the arts, communications, history, literature, music, philosophy, and religion
- the social sciences, which may include anthropology, economics, political science, geography, psychology, and sociology

- the sciences, which usually include such disciplines as astronomy, biology, chemistry, geology, mathematics, and physics

It's important to note that these lists are far from comprehensive; schools differ in the courses and majors they offer, so some schools may offer courses in fields that aren't on any of these lists—and some of these courses might not be available at your school. You may also take courses in career-oriented fields such as education, business, engineering, and nursing.

Studying, Reading, and Writing in Academic Fields

Each field focuses on the study of particular subjects and issues. In *psychology*, you study the human mind—what it is and why we behave as we do. In *sociology*, you study the way society shapes our actions. In *biology*, you study life itself, from bacteria and fungi to organisms interacting with their environment. In *history*, you study the past to understand the present. In *nursing*, you study best ways of providing patient care within the health-care system. In *engineering*, you learn how to solve technical problems and design engineering systems.

Each field also examines the world through a distinctive lens, using its own methods to study and analyze those subjects. For example, scientists test hypotheses with experiments designed to prove or disprove their accuracy; sociologists study groups by using statistical evidence; and historians examine diaries, speeches, or photographs from the past. In addition, disciplines develop technical terms and ways of using language that allow scholars to understand one another—but that can be hard to understand. For example, consider the word "significant." When people say that something is *significant* in day-to-day conversation, they usually mean that it's important, a big deal; in statistics, though, a *significant* result is one that is probably true—but it may or may not be important.

Disciplines also present information using methods standard in that discipline. In *business* courses, you're likely to read case studies of specific companies and write business cases and other communications. In *education* courses, you'll read and write lesson plans; in *science* courses, you'll read and write laboratory reports; in *English* courses, you'll read fiction,

---

* academic literacies   ● fields   ● research
■ rhetorical situations   :● processes   ● media/design
▲ genres   ◆ strategies

poetry, and plays and write literary analyses. In each, you'll present information in ways used in conversations among scholars in the discipline. And that means that you may need to adapt your reading and writing to the genres, concepts, and methods of the various academic fields you encounter.

THINKING ABOUT READING AND WRITING IN THE FIELDS

- What reading assignments do you have in your other courses?
- What makes them easy or hard to read? What helps you read them?
- Do you enjoy reading some genres or subjects more than others? Why?
- What writing assignments do you have in your other courses?
- Do you alter or adjust your writing processes in different courses? Why or why not?
- What makes writing in some courses hard or easy?
- Do you enjoy writing in certain genres or on certain topics? Why?

24 Reading across Fields of Study

We read shopping lists differently from graphic novels and operating instructions differently from poems. For that matter, we read novels and poems differently—and go about reading textbooks differently still. Just as we write using various processes—generating ideas and text, drafting, getting responses and revising, editing, and proofreading—we vary our reading processes from task to task. We will likely read a mathematics textbook differently from a case study in nursing, and read a speech by Abraham Lincoln in a history course differently from the way we'd read it in a rhetoric course. This chapter offers advice on how to engage with texts in a variety of fields.

Considering the Rhetorical Situation

A good way to approach reading in academic fields is to treat reading as a process that you adapt to the demands of each new situation. Instead of diving in and focusing your reading on the details of the content, it's often useful to take some time to consider the text's purpose, your purpose, the audience, the author, the author's stance, the genre, and the medium and design. This information can help you decide how to get the most out of the text.

59–60 ■

- **The text's PURPOSE**. When was the text written, and why? Is it a textbook? a classic work that lays out concepts that have become fundamental to the field? a work proposing a new way of looking at an issue?

- **Your purpose as a reader.** Why are you reading this text? To study for a test? To find sources for a research project? To learn about something on your own?

※ academic literacies   ● fields   ● research

■ rhetorical situations   ∴ processes   ● media/design

▲ genres   ◆ strategies

- *The* **AUDIENCE**. For whom was the text written? Students like you? Scholars in the field? Interested nonspecialists? Readers who lived at some point in the past or in a particular location? What facts and concepts does the text assume its readers already know and understand?   61–64

- *The author.* Who wrote what you are reading? Does knowing the author's identity matter? Is the author reliable and credible? For example, is the author a respected scholar? Is the author known for a particular point of view?

- *The author's* **STANCE**. What is the author's stance toward the subject? toward the reader? Approving? Hostile? Critical? Passionate? What elements of the text reveal the author's stance?   72–74

- *The* **GENRE**. What is the text's genre? How was it meant to be read? Is that how you're reading it? What are the genre's key features? Knowing the key features of the genre can help you understand the text's content as you read and predict the text's organization.   65–71

- *The* **MEDIUM** *and* **DESIGN**. The medium and design of a text affect how you must read it. You may need to read a textbook differently from an article found on the internet, for example. An online textbook may require different commenting and note-taking strategies from a print text. Understanding the limitations and the advantages of a particular genre, medium, and design will help you get the most out of your reading.   75–77

Advice for Reading across Fields of Study

Becoming knowledgeable in a field of study requires learning its specialized ways of thinking, writing, and reading, including the terminology specific to the field and the kinds of texts the field is likely to produce and study. However, there is some general advice that can help you as you begin to read the kinds of texts you're likely to encounter in college.

Pay attention to vocabulary.   All disciplines require that you learn their vocabulary, which includes not only concepts and facts but also the names of important figures in the field and what they represent. Scholars typically write for other scholars in the same field, so they may assume that readers understand these terms and references and refer to them without

explanation—and even "popular" writing often makes the same assumption. You'll find that much of the work in all your courses consists of learning terms and concepts. So it's important to take good notes in lectures, writing down terms your instructor emphasizes as they are defined, or, if they aren't, writing them down to look up later. Your textbooks introduce and define key terms and often include a glossary in the back that provides definitions.

In addition, be aware that disciplinary vocabularies use different kinds of words. In the sciences, the language is likely very technical, with many terms based on Greek and Latin roots (for example, the Greek prefix *gastr-* forms "gastropod" [snail], "gastric" [stomach], and "gastroscopy" [examination of the abdomen and stomach]). In the humanities, the vocabulary often alludes to complex concepts. Consider, for example, these sentences from *The Swerve*, a book by literary historian Stephen Greenblatt:

> The household, the kinship network, the guild, the corporation—these were the building blocks of personhood. Independence and self-reliance had no cultural purchase.

Almost every term, from "household" to "guild" to "personhood" to "cultural purchase," carries a wealth of information that must be understood. While understanding the scientific terms requires a good dictionary, these particular terms require an encyclopedia—or the knowledge gained by much reading.

Consider the author—or not. In some disciplines, it's important to know who wrote a text, while in others the identity or **STANCE** of the author is 72–74 ■ considered irrelevant. Historians, for example, need to know who the author of a text is, and the perspective that author brings to the text is central to reading history. Literary scholars may try not to consider the author at all, focusing solely on a close reading of the text itself. Scientists may consider the author's identity or the school or lab where the author works to decide whether or not to read a text. Mathematicians resolutely ignore the author, focusing solely on the information in the text itself. So to make sense of a text in a discipline, you need to find out how that discipline sees the role of the author.

✳ academic literacies   ⬟ fields   ⬤ research

◾ rhetorical situations   ⦙ processes   ⬤ media/design

▲ genres   ◆ strategies

Identify key ideas and make comparisons. Whatever the discipline, look for the main ideas its texts present. Rather than highlight or underline them, write out the key ideas in your own words. As you read texts on the same subject, you may want to develop a matrix to help you **COMPARE AND CONTRAST** as well as **SYNTHESIZE** the main ideas. The following chart, adapted from one developed by two librarians, presents the main ideas in two sources and then synthesizes the two versions. This example synthesizes one main idea found in each of two articles on how social media affect political participation among youths and then, in the third column, synthesizes those ideas:

◆ 437–44
● 535–41

Source #1: Skorik, M. M. "Youth Engagement in Singapore"	**Source #2:** Ahmad, K. "Social Media and Youth Participatory Politics"	**Synthesis: Main Ideas**
Summary of 1st main idea: Singapore's government has consistently applied controls on traditional media outlets but has left social media outlets untouched and unregulated.	**Summary of 1st main idea:** Online participation in political campaigns and issues was almost five times greater than traditional participation. Researchers concluded that this "could provide the participant with anonymity, in turn less vulnerability to political vengeance."	**Synthesis of 1st main ideas:** Government control and censorship of mainstream media has caused protesters to look for alternative communication tools.
Summary of 2nd main idea:	**Summary of 2nd main idea:**	**Synthesis of 2nd main ideas:**
Summary of 3rd main idea:	**Summary of 3rd main idea:**	**Synthesis of 3rd main ideas:**

Adapted from Michelle Chiles and Emily Brown, "Literature Review." Bristol Community College, Fall River, MA, 2015. Unpublished *PowerPoint*.

More main ideas can be summarized and synthesized in the rows below; add as many as needed.

You might also get into the habit of skimming the list of works cited or references at the end of articles and books. As you do so, you'll begin to see who the most important people are in the discipline and what counts as evidence.

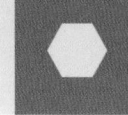

Build a map of the discipline. To make sense of a new idea, you need to have some way of fitting it into what you already know. Reading in an unfamiliar discipline can be hard because you likely won't have a sense of where the information in the text fits into the conversations of the discipline, its history, or your goals—what you'd like to know or do in this field. It's useful to visualize the discipline so that you can place readings into the appropriate context. Possible ways of organizing a discipline follow.

Draw a word map. Using your textbook and perhaps some online sources as guides, draw an overview of the field. Here's one for psychology:

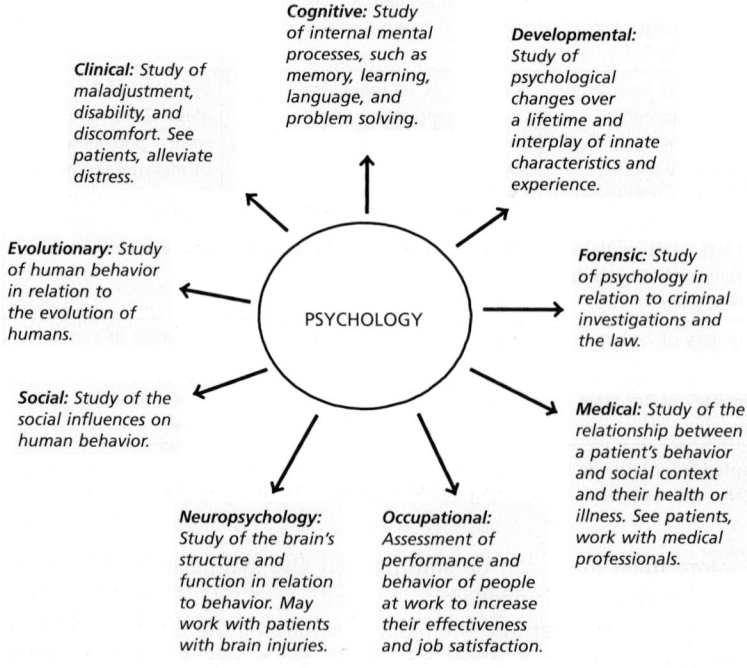

Cognitive: *Study of internal mental processes, such as memory, learning, language, and problem solving.*

Developmental: *Study of psychological changes over a lifetime and interplay of innate characteristics and experience.*

Clinical: *Study of maladjustment, disability, and discomfort. See patients, alleviate distress.*

Evolutionary: *Study of human behavior in relation to the evolution of humans.*

Forensic: *Study of psychology in relation to criminal investigations and the law.*

PSYCHOLOGY

Social: *Study of the social influences on human behavior.*

Medical: *Study of the relationship between a patient's behavior and social context and their health or illness. See patients, work with medical professionals.*

Neuropsychology: *Study of the brain's structure and function in relation to behavior. May work with patients with brain injuries.*

Occupational: *Assessment of performance and behavior of people at work to increase their effectiveness and job satisfaction.*

※ academic literacies   ◆ fields   ● research

■ rhetorical situations   ⋮ processes   ● media/design

▲ genres   ◆ strategies

Create a timeline. Sometimes it's helpful to understand how a discipline developed over time, how its research methods or emphases came to be, so you can place its texts and key developments within that history. Here, for example, is a timeline of the behaviorist approach in psychology:

Timeline of the behaviorist approach, 1900–2000

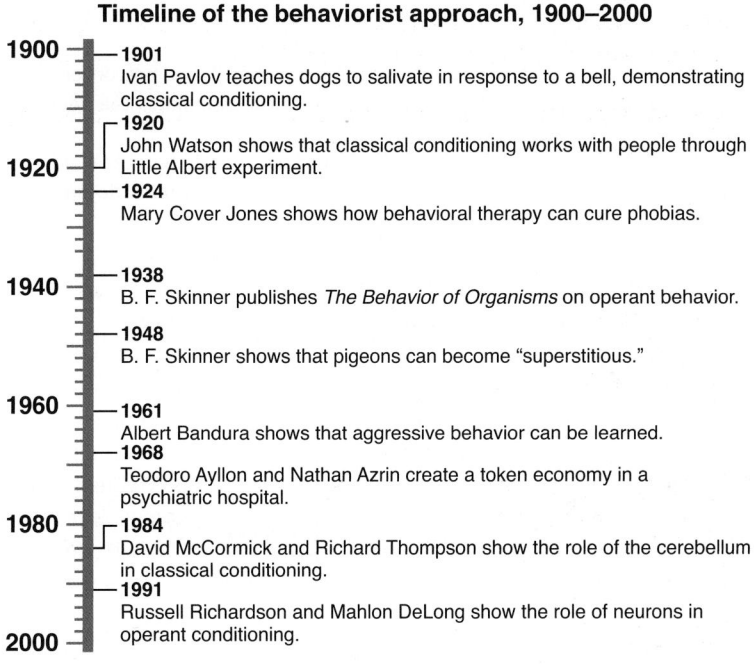

1901
Ivan Pavlov teaches dogs to salivate in response to a bell, demonstrating classical conditioning.

1920
John Watson shows that classical conditioning works with people through Little Albert experiment.

1924
Mary Cover Jones shows how behavioral therapy can cure phobias.

1938
B. F. Skinner publishes *The Behavior of Organisms* on operant behavior.

1948
B. F. Skinner shows that pigeons can become "superstitious."

1961
Albert Bandura shows that aggressive behavior can be learned.

1968
Teodoro Ayllon and Nathan Azrin create a token economy in a psychiatric hospital.

1984
David McCormick and Richard Thompson show the role of the cerebellum in classical conditioning.

1991
Russell Richardson and Mahlon DeLong show the role of neurons in operant conditioning.

Write a conversation among researchers. To better understand the similarities and differences among various schools of thought within a discipline, choose some of the most prominent proponents of each school and write a fictional conversation among them, outlining where they agree and disagree and where they published or disseminated their theories and findings. Here, for instance, is a snippet of a dialogue that might have

taken place between developmental psychologist Jean Piaget and behavioral psychologist B. F. Skinner:

> **Piaget:** Children learn by adapting to conflicts between what they already know and challenges to that knowledge.
> **Skinner:** Bosh! We can't know what's in someone's mind; we have to focus on their behavior.
> **Piaget:** If a child sees something that doesn't match her preconceptions, she learns from it and through her behavior assimilates the new information into her thinking.
> **Skinner:** But if we reinforce a certain behavior by rewarding it, the child will do it more, and if we ignore it, the behavior will stop. It doesn't matter what, if anything, the child is thinking.

TIPS FOR READING IN VARIOUS FIELDS OF STUDY

Though groups of disciplines (the sciences, the arts and humanities, and so on) within each field share many similarities, each discipline has features unique to it. While you should expect to read difficult texts in any field several times, taking notes in your own words rather than highlighting or underlining, you need to be aware that reading, say, a biology text likely requires different reading techniques than a history text would. Here are some tips that can help you read effectively in various disciplines.

The Humanities

- Much of the scholarship in the humanities is based on texts of one sort or another—the Constitution, the Bible, Toni Morrison's novels, Mozart's piano concertos, Georgia O'Keeffe's paintings. When reading in the humanities, then, be aware of the differences between reading **PRIMARY SOURCES** like those and **SECONDARY SOURCES** (commentaries, analyses, evaluations, and interpretations of primary sources).

501–2

- When reading a primary source, you may need to ask questions like these: What kind of document is this? Who created it? When and

✺ academic literacies   ● fields   ● research
■ rhetorical situations   ⁚● processes   ● media/design
▲ genres   ◆ strategies

where was the source produced? Who was it intended for? What is its historical context? Why did the creator of the source create it? How was it received at the time of its creation? Has that reception changed?

- Read primary sources carefully and expect to reread them, perhaps several times. The way a text is organized, its sentence structure, its wording, and its imagery can all contribute to its overall message, and that's true of documents like the Declaration of Independence, the Torah, and Descartes's *Principles of Philosophy*.

- Similarly, plan to reread secondary sources. The argument of interpretations, analyses, or critical works requires close attention to their logic and use of evidence.

- Reading secondary sources requires that you examine them with a critical eye. The questions on pages 532–33 can help you understand the source and evaluate its accuracy, perspective or angle, and usefulness for your own work.

- Remember that in the humanities, "criticism" doesn't necessarily mean looking for flaws and errors. Rather, it's another term for analyzing or evaluating.

- Some key terms you need to know in order to read texts in the humanities include: "analogy," "allusion," "argument," "deductive reasoning," "inductive reasoning," "irony," "metaphor," "natural law," "natural rights," "political rights," "premise." If you don't understand the meaning of any of these terms, be sure to look up their definitions.

The Social Sciences

- As you read in the social sciences, you need to pay attention to the hypotheses, claims, reasons, and evidence presented in the texts. Are they persuasive? How do they compare or contrast with other arguments?

- Be sure you understand what a theory in the social sciences is: a way of organizing information to help enhance our understanding of behavior, events, phenomena, or issues. For example, in sociology, social exchange theory assumes that we behave according to our sense

of whether we'll be rewarded or punished by those with whom we interact. Remember that a theory is not an unsupported opinion but rather a coherent, logical frame for understanding and describing.

209–13

- Many articles include a **LITERATURE REVIEW** that summarizes and critiques previous work on the topic. Since scholarly work always grows out of the work of others, authors need to connect their work to articles and books previously published on the topic. Many literature reviews discuss limitations and problems in previous studies, ultimately identifying missing elements or gaps in them (often identified with words like "but," "however," and "although"), leading to a rationale for why this new work is needed.

- The "Discussion" and "Conclusion" sections (which are sometimes combined into a single section) explain and interpret the results of a study in the context of the previous literature, note any limitations in the current study, and recommend future research to address those limitations. In other words, these sections discuss what the author thinks the study means. As you read, consider possible flaws or omissions in the author's thinking as well as in the research design; for example, the study is too broad, the research question is poorly defined, or the study doesn't address its significance—the "so what?" question. Your insights could lead you to a better understanding of the topic—or to a research project of your own.

- Some key terms you need to know in order to read texts in the social sciences include "adaptation," "aggregate," "alienation," "capital," "class system," "deviance," "interest," "markets," "motivation," "norm," "power," "schema," "supply and demand," and "value." If you don't understand the meaning of any of these terms, be sure to look up their definitions.

The Sciences

- Remember that scientific texts are not collections of established facts. They make claims and argue for them, using reasons and evidence to support those arguments. Scientific texts report the results of studies

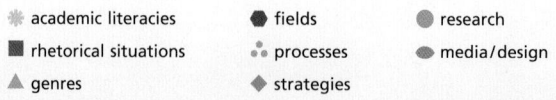

✳ academic literacies   ⬡ fields   ⬤ research
◼ rhetorical situations   ⋰ processes   ⬤ media/design
▲ genres   ◆ strategies

and experiments and are written by scientists for an audience of other scientists in the same field. These texts are different from science writing, which is often a form of nonfiction sometimes written by scientists, journalists, or other writers and aimed at general audiences to inform about scientific topics.

- If you're examining potential sources for a research project, read selectively: use articles' **ABSTRACTS** to decide whether an article is likely to be useful for your project, and skim the text to find the information relating to your needs. Many scientific articles follow a structure nicknamed IMRaD (Introduction, Methods, Results, and Discussion), so you may need to read only the sections of the article that discuss what you're looking for.
▲ 196–200

- Pay attention to the sample size—the number of units of whatever is being studied that are included in the study (a larger sample size is generally better than a smaller one)—and error bars in graphs, which show the uncertainty in the findings of the study being described.

- Scientists often make their arguments through visuals as much as through words, so read images, graphs, and charts as carefully as you would the words in the text.

- For math problems or exercises, read the entire problem. Draw a picture or diagram to help visualize the problem. Read it again, identifying the most important information; be sure you understand what the problem is asking for. Then decide on a method for solving it, and come up with an answer. If you get stuck, think about how you'd deal with this information if it weren't a math problem. In that case, how would you go about solving it? Finally, reread the problem, and ask yourself if you've answered the question that was asked.

- Some key terms you need to know in order to read texts in the sciences include "skepticism," "data," "evidence/observation," "hypothesis," "variables," "biodiversity," "falsifiability," "theory," "evolution," "experimentation," "population," "qualitative," "quantitative," "repeatability," "empirical," "paradigm," "rational/agnostic," "cosmos." If you don't understand the meaning of any of these terms, be sure to look up their definitions.

A Note on Career-Focused Fields

In general, the advice here for reading texts in general-education courses holds for reading in every major, including career-oriented disciplines, such as business administration, nursing, teaching, and education. At the same time, it's worth paying attention to the ways of thinking and the priorities of every field as you read in them. Here are a few examples:

Both business administration and education focus on processes. In business, much reading discusses how businesspeople do things, manage workers, set up systems, follow best accounting practices, and the like; in education, your reading will likely focus on how to write lesson plans, how to present information appropriately, and so on. Drawing flowcharts and diagrams of processes can help you understand the steps involved and help you remember them.

Engineers solve problems. So when reading engineering texts, look for relationships among concepts and ideas, and think about ways those concepts can be used to solve complex problems. Pay close attention to charts, graphs, diagrams, and visuals, as they often pack considerable information into a single image. When reading a graph, for example, consider not only the data as presented but also relationships in the data, and look for inferences and predictions you can make from the data presented.

Engineering and nursing texts contain a lot of information, much of which you won't need right away, so don't try to master every fact or procedure. Instead, read the chapter introductions, summaries, and questions, and skim, looking for the answers to the questions. You might also review a study guide for the NCLEX nursing licensing exam or the NCEES PE exam in engineering and focus your reading on the subject areas the exam focuses on. In general, work to understand the concepts you need rather than trying to memorize an avalanche of information; look for patterns in the information and how concepts are interrelated.

✳ academic literacies	⬣ fields	⬤ research
◼ rhetorical situations	⁖ processes	⬤ media/design
▲ genres	◆ strategies	

25 Writing in Academic Fields of Study

In a *literature* course, you're asked to write an analysis of a short story. In a *biology* course, you must complete several lab reports. In a *management* course, you may create a detailed business plan. In fact, just about every course you take in college will require writing, and to write successfully, you'll need to understand the rhetorical situation of your writing in the course—and in the larger field of study. That understanding will come in time, especially as you begin writing in your major. For now, this chapter offers help in determining the general expectations of writing done in various academic fields of study.

Considering the Rhetorical Situation

To write in academic fields, you need to use the same processes and strategies you're asked to use in your writing classes, including analyzing the **RHETORICAL SITUATION** in which you're writing. These questions can help:

■ 57

■ 59–60

PURPOSE   Why do people in this discipline write? To share scholarship and research findings? persuade? teach or provide guidance? show learning or mastery? track progress? propose solutions or plans of action? explore ideas or the self? earn grants or other rewards? something else?

AUDIENCE To whom do people in this discipline write? To colleagues and other scholars? students? managers? employees? customers? clients? granting agencies? the public? others? What do they already know about the discipline and the topic? What specialized terms or concepts do they understand, and which ones need to be defined or explained? How much evidence or support is required, and what kinds (empirical data, research findings, logical analysis, personal testimony, something else) will they accept?

GENRE What genres—reports, analyses, arguments, instructions, case studies, résumés, to name only a few—are typically used in this discipline? Are they organized in a certain way, and do they contain specific kinds of information? How much flexibility or room for innovation and creativity is allowed? What counts as evidence or support for assertions, and how is it cited (in citations in the text, in footnotes, in a works-cited page, informally in the text, or in some other way)?

STANCE What attitude is considered appropriate in this discipline? Objective? Unemotional? Critical? Passionate? Should you write as a good student showing what you can do? an instructor of others? an advocate for a position? someone exploring an idea? something else? Does the discipline require a certain tone? formal or informal language? Can you include your personal perspective and write using "I"? Should you write only in the third person and use passive voice?

MEDIA/DESIGN What media are typically used in this discipline? Print? Spoken? Electronic? A combination? Are certain design elements expected? to be avoided? Are visuals commonly used? What kinds—charts, graphs, photos, drawings, video or audio clips, or something else? In which genres? How much design freedom do you have?

WRITING IN ACADEMIC FIELDS OF STUDY

Generalizing about the requirements of writing in academic disciplines is tricky; what constitutes a discipline is sometimes unclear, and universities group academic fields together in various ways. For example, in some universities psychology is considered a science, while in others it's a social science. Economics is sometimes part of a college of business administration, sometimes in a college of arts and sciences. In addition, the writing required in, say, history differs from that required in English literature, though both are considered parts of the humanities.

Furthermore, certain genres of writing, like *case studies* and *research reports*, can share the same name but have very different organizational structures and content, depending on the discipline in which they're used. For example, research reports in psychology and the natural sciences include a review of relevant scholarly literature in the introduction; in reports in sociology and other social sciences, the literature review is a separate section. A case study in business identifies a problem or issue in an organization; provides background information; includes a section, "Alternatives," that discusses possible solutions to the problem and why they were rejected; outlines and argues for a proposed solution; and proposes specific strategies for achieving the proposed solution. A case study in nursing, on the other hand, includes three sections: patient status, which gives an overview of the patient's condition and treatment; the nurse's assessment of the patient's symptoms and their possible causes; and a plan for helping the patient improve. The guide below offers general advice on how to write in broad academic disciplines, but as the differences between two disciplines' expectations for case studies show, it's always a good idea to ask all your professors for guidance on writing for their particular fields.

WRITING IN THE ARTS AND HUMANITIES

The arts and humanities focus on human culture and the expressions of the human mind, and the purpose of writing in these fields is to explore and analyze aspects of the human experience across time and sometimes

to create original works of literature, music, and art. The methods used in these disciplines include careful reading, critical analysis, historical research, interpretation, questioning, synthesis, and imitation. Courses in the arts and humanities typically include fine arts, architecture, music, dance, theater, film, photography, literature, history, classical and modern languages, linguistics, and philosophy.

Writing in the arts and humanities generally serves a broad audience that includes professors and scholars, other students, the general public, and oneself. Genres may include **ANNOTATED BIBLIOGRAPHIES**, **ANALYSES**, **ARGUMENTS**, essays, **EVALUATIONS**, **JOURNALS**, personal narratives, **REPORTS**, **PRESENTATIONS**, **PROPOSALS**, **REFLECTIONS**, and **LITERATURE REVIEWS**, as well as fiction and poetry. Support is often based on textual and observational evidence and personal insight, though in some fields empirical evidence and data are also valued. Writers in the arts and humanities tend to use modifiers (such as "arguably" or "perhaps") to acknowledge that their insights and conclusions are interpretive, not definitive. Documentation is usually done in **MLA** or Chicago style. Elements of style favored in writing in the arts and humanities may include the use of "I"; the active voice; an informal vocabulary, if appropriate; and vivid language.

A Sample of Writing in History: A Researched Essay

Identifies a problem in current understanding of a historical event.

The Pueblo Revolt of 1680 was one of the most significant yet misrepresented events in the history of American Indians. After three generations of being oppressed by Spanish rule, the Pueblo Indians throughout the southwest region of North America banded together,

Offers a narrative of a past event.

organizing a widespread rebellion in the blistering summer heat of 1680 and successfully liberating themselves from their oppressors by springtime. When examining the causes of the revolt, the lack of

Demonstrates familiarity with relevant sources.

authentic Pueblo voices within the written records challenges the validity of the available sources and makes one wonder if we will ever

---

✳ academic literacies   ● fields   ● research

■ rhetorical situations   ⸫ processes   ● media/design

▲ genres   ◆ strategies

know what went on through the eyes of the Pueblo. Although in the traditional narrative, the Spaniards are regarded as missionaries sent by God to "save" the "barbaric" Pueblos, the event, if seen from the Pueblo perspective, can be understood as a violent retaliation by the Pueblo against the Spanish oppression. The Pueblo uprisings, from burning down churches to the violent deaths of Catholic friars, reveal spiritual abuse as the major cause of the revolt. Moreover, without texts written by the Pueblo, their architecture and spatial organization provide valuable insight into the causes of the revolt era and help to overcome the veneer of Spanish colonialism.

Offers a strong thesis.

Clear, engaging writing style.

Adapted from "Letting the Unspoken Speak: A Reexamination of the Pueblo Revolt of 1680," by E. McHugh, April 2015, Armstrong Undergraduate Journal of History 5, no. 1, digitalcommons.georgiasouthern.edu/aujh/vol5/iss1/5.

Typical Organization of Arts and Humanities Essays

Typical essays in the arts and humanities include these elements:

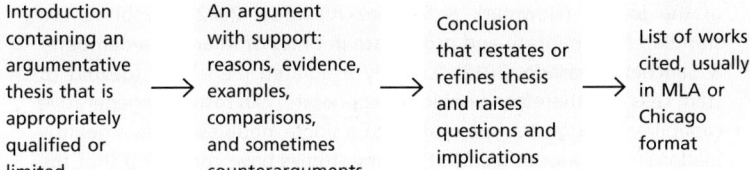

Introduction containing an argumentative thesis that is appropriately qualified or limited → An argument with support: reasons, evidence, examples, comparisons, and sometimes counterarguments → Conclusion that restates or refines thesis and raises questions and implications → List of works cited, usually in MLA or Chicago format

WRITING IN SCIENCE AND MATHEMATICS

The sciences include biology, chemistry, geology, earth sciences, and physics. Mathematics may include statistics and logic as well. All these fields aim to increase our knowledge of the physical and natural world and its phenomena through observation, experiment, logic, and computation.

314 FIELDS

196–200
209–13
140–63
164–95
694–705
258–68
322

APA 615–55

Scientists and mathematicians typically write **ABSTRACTS**, **LITERATURE REVIEWS**, **REPORTS**, **ARGUMENTS**, poster **PRESENTATIONS**, **PROPOSALS**, and **LAB REPORTS** for audiences that may include other researchers, granting agencies, teachers, students, and the general public. Support in the sciences most often consists of repeatable empirical evidence; in mathematics, careful reasoning and the posing and solving of problems; in both, careful attention to the work of previous researchers. The writing in these fields focuses on the subject of the study, not the researcher, so most often the passive voice is used. Source material is paraphrased and summarized and cited in CSE or **APA** style.

A Sample of Scientific Writing: A Scientific Proposal in Biology

Careful reference to previous sentence.

Specialized disciplinary vocabulary.

Sources paraphrased and summarized.

Planarians, flatworms widely known for their incredible regenerative capabilities, are able to restore an entire organism from even a small fragment of tissue. This ability to regenerate is attributed solely to neoblasts, pluripotent adult stem cells located throughout the parenchyma of the animal (Newmark & Sanchez Alvarado, 2002). Neoblasts are stimulated to migrate and proliferate in times of injury (Guedelhoefer & Sanchez Alvarado, 2012). Lethally irradiated planarians (devoid of stem cells and therefore unable to regenerate) can restore regenerative capability through transplantation of a single neoblast from a healthy planarian (Wagner et al., 2011). Many studies have concluded that the population of neoblasts is not homogenous (Scimone et al., 2014), and there are different responses to different injury types. Wenemoser and Reddien (2012) observed a body-wide increase in mitotic activity, such as cell division and migration, with any injury.

Passive voice, third person.

Adapted from "Identifying Genes Involved in Suppression of Tumor Formation in the Planarian Schmidtea mediterranea," by E. Dorsten, 2015, Best Integrated Writing, 2, https://corescholar.libraries.wright.edu/biw/vol2/iss1/6/.

* academic literacies
■ rhetorical situations
▲ genres
● fields
•• processes
◆ strategies
● research
● media/design

Typical Organization of Research Reports in the Sciences

Typical reports in the sciences include elements that follow the IMRaD structure: Introduction, Methods, Results, and Discussion. They also include an abstract and a list of references.

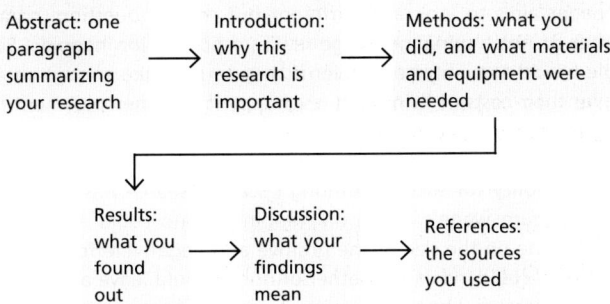

WRITING IN THE SOCIAL SCIENCES

Anthropology, archaeology, criminal justice, cultural studies, gender studies, geography, psychology, political science, and sociology are considered social sciences because they all explore human behavior and society using observation, experimentation, questionnaires, and interviews.

Social scientists typically write for fellow scholars, teachers, students, and the general public. They may write in several genres: **ABSTRACTS**, **ANNOTATED BIBLIOGRAPHIES**, **ANALYSES**, **ARGUMENTS**, case studies, ethnographies, **LITERATURE OR RESEARCH REVIEWS**, **REPORTS**, **SUMMARIES**, and **PRESENTATIONS**. Claims are typically supported by empirical evidence, fieldwork done in natural settings, observation, and interviews. Writers in these fields strive for an objective tone, often using the passive voice. Sources may be cited in **APA** or Chicago style.

▲ 196–200
201–9
104–39
164–95
209–13
140–63
✳ 33–35
● 694–705

● APA 615–55

A Sample of Writing in the Social Sciences: A Research Report

Objective tone.

Traditional economic theory states that a minimum wage above the marginal product of labor will lead to increased unemployment. . . . This paper aims to look at a different but related question, namely, whether or not a minimum wage makes a population happier. Since people would arguably be happier if they could make enough money to cover their costs of living and less happy if the unemployment rate

Research question.

rose, the answer to such a question could help determine which effect is the dominant force and if an overall increase in the minimum wage is a good policy for society. Although scant research has been done

Specialized language.

on a minimum wage's effect on happiness, one could assume that research done on the size of the positive and negative effects of minimum wages could indicate whether or not it would leave a population

Literature review.

happier. Therefore, I begin by reviewing relevant economic theory and research on the effects of a minimum wage increase to provide background information and describe what related questions have been

Empirical method.

approached and answered. I then describe the data and method used to answer this question, followed by the interpretation of such results

Analysis and evaluation of results.

as well as the implications.

Adapted from "The Effect of Minimum Wages on Happiness," by J. Nizamoff, *Beyond Politics* 2014, pp. 85–95, *beyondpolitics.nd.edu/wp-content/uploads /2015/03/2013-14-Full-Journal.pdf.*

Typical Organization of a Research Report in the Social Sciences

Typical research reports in social science courses might include the following elements, though the order and names of the elements may differ from discipline to discipline. For example, in psychology, the literature review is part of the introduction, not a separate section as shown here.

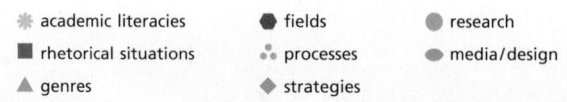

* academic literacies
* rhetorical situations
* genres
* fields
* processes
* strategies
* research
* media/design

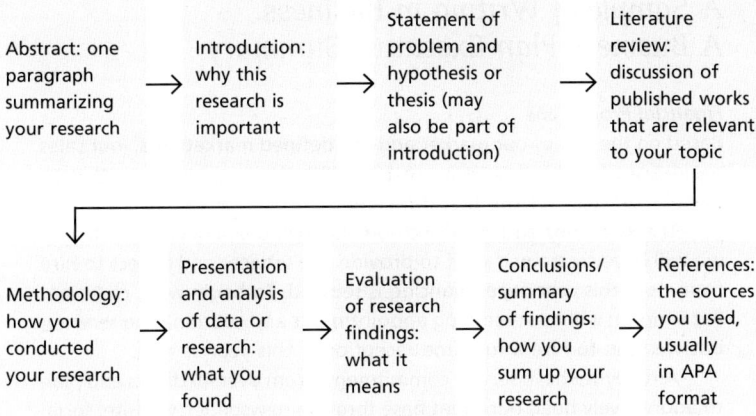

WRITING IN BUSINESS

The focus of the academic discipline of business is business-management principles and their application, and the purpose of writing in business is often to motivate readers to make a decision and then act on it. The primary methods used include problem solving, planning, and experiential learning, or learning by doing. Courses typically taught include finance, economics, human resources, marketing, operations management, and accounting.

The audiences for writing in business usually include colleagues, employees in other departments, supervisors, managers, clients, customers, and other stakeholders—often several at the same time as a text moves through an organization. Genres may include memos, emails, letters, case studies, executive summaries, RÉSUMÉS, business plans, REPORTS, and ANALYSES. Support usually takes the form of facts and figures, examples, narratives, and expert testimony, and documentation is usually done in APA or Chicago style. Elements of style favored in business writing include these features: the main point is presented early; the language is simple, direct, and positive; and the active voice is used in most cases.

Glossary

▲ 140–63

104–39

● APA 615–55

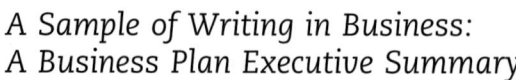

A Sample of Writing in Business:
A Business Plan Executive Summary

Financial Projections

Precise numbers, confidently stated.

Based on the size of our market and our defined market area, our sales projections for the first year are $340,000. We project a growth rate of 10% per year for the first three years.

Clear, direct writing, free of jargon and hedging.

The salary for each of the co-owners will be $40,000. On start up we will have six trained staff to provide pet services and expect to hire four more this year once financing is secured. To begin with, co-owner Pat Simpson will be scheduling appointments and coordinating services, but we plan to hire a full-time receptionist this year as well.

Positive tone.

Already we have service commitments from over 40 clients and plan to aggressively build our client base through newspaper, website, social media, and direct mail advertising. The loving on-site professional care that Pet Grandma Inc. will provide is sure to appeal to cat and dog owners throughout the West Vancouver area.

Adapted from "Business Plan Executive Summary Sample," by S. Ward, March 29, 2017, The Balance, *www.thebalancesmb.com/business-plan -executive-summary-example-2948007.*

Typical Organization of Business Plans

A common assignment in business courses is a business plan. Business plans typically include these sections:

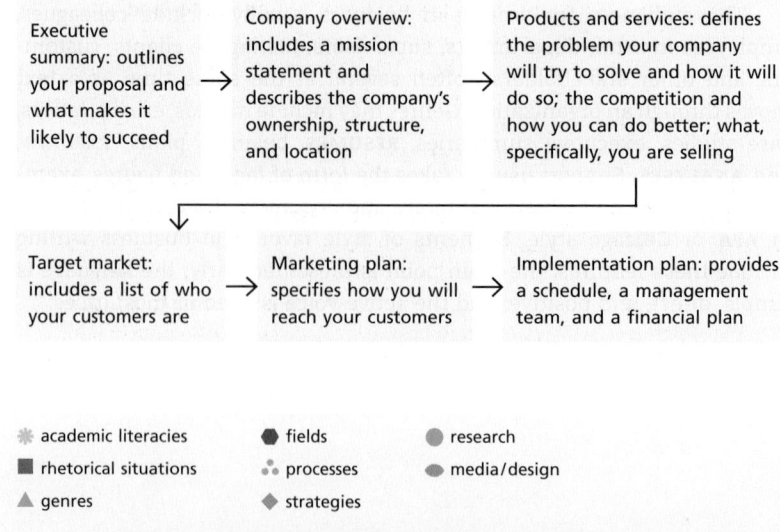

Executive summary: outlines your proposal and what makes it likely to succeed → Company overview: includes a mission statement and describes the company's ownership, structure, and location → Products and services: defines the problem your company will try to solve and how it will do so; the competition and how you can do better; what, specifically, you are selling

Target market: includes a list of who your customers are → Marketing plan: specifies how you will reach your customers → Implementation plan: provides a schedule, a management team, and a financial plan

* academic literacies
● fields
● research
■ rhetorical situations
⁝ processes
● media/design
▲ genres
◆ strategies

WRITING IN EDUCATION

The focus of study in education is how people learn and how to teach effectively. Its primary methods include observation, problem solving, and practice teaching. Courses typically center on teaching methods, the philosophy of education, educational measurement and assessment, educational psychology, and instructional technology, among others.

Educators typically write for audiences that include their students, parents, other teachers, administrators, and the public. Genres may include lesson plans, **SUMMARIES**, **REPORTS**, **ANNOTATED BIBLIOGRAPHIES**, **PORTFOLIOS**, and **REFLECTIONS**. Support for claims may include facts, statistics, test scores, personal narratives, observations, and case studies. Sources are documented in **APA** style. Clarity and correctness are important in writing in education; "I" may be used in reflective writing and informal communication, while in formal writing the third person is preferred.

☀ 33–35

▲ 140–63

   201–9

⁂ 385–90

   391–401

● APA 615–55

A Sample of Writing in Education: A Teaching Philosophy Statement

My Image of the Child: •⋯⋯⋯⋯⋯⋯⋯⋯⋯⋯⋯⋯⋯⋯⋯⋯⋯⋯⋯⋯⋯

Sections labeled with headings.

I believe that the student should be at the center of the instructional •⋯⋯ process. I have an image of children as strong and capable beings. The classroom is a place where the teacher serves as a facilitator and guide as the students construct their own understanding of the world around them. Although it is the teacher's role to plan lessons and evaluate students' progress, it is of the utmost importance to always take the children and their own unique needs into consideration. For my second field experience, I was placed at Margaret Manson Elementary. Their school motto is that "the children come first." When children are the priority in teaching, an amazing amount of learning can take place. I believe in creating opportunities for students to develop to their fullest potential while developing and expanding their horizons and worldviews. In order to accomplish this, there must be a welcoming, positive environment that is open and honest. When students feel comfortable

Argument is constructed to show teaching priorities.

Writing carefully crafted and proofread.

As a reflective piece, "I" is appropriate.

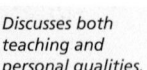
Discusses both teaching and personal qualities. ·········•

at school they will surely be more engaged and responsive to class activities. I also consider it essential to be passionate and enthusiastic about learning so that the students can have a most relevant and meaningful experience.

Adapted from "Statement of Teaching Philosophy," by K. Tams (n.d.), Kelly Tams' Teaching Portfolio, tams.yolasite.com/my-philosophy-of-education .php.

Typical Organization of Lesson Plans in Education

Frequent assignments in education courses are lesson plans, which typically include these elements:

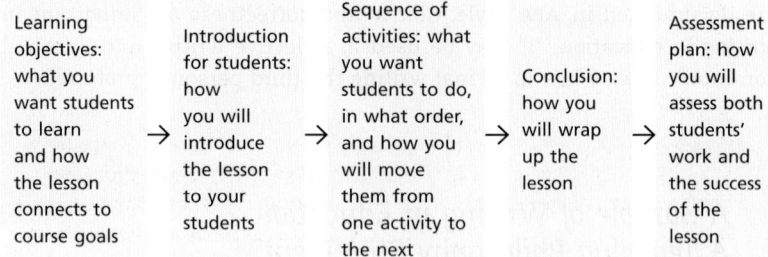

| Learning objectives: what you want students to learn and how the lesson connects to course goals | → | Introduction for students: how you will introduce the lesson to your students | → | Sequence of activities: what you want students to do, in what order, and how you will move them from one activity to the next | → | Conclusion: how you will wrap up the lesson | → | Assessment plan: how you will assess both students' work and the success of the lesson |

WRITING IN ENGINEERING AND TECHNOLOGY

In the fields of engineering and technology, the focus is how to create and maintain useful structures, systems, processes, and machines. Engineers and technicians define problems as well as solve them, weigh various alternatives, and test possible solutions before presenting them to clients. This is a broad set of disciplines that may include civil, computer, electrical, mechanical, and structural engineering; computer science; and various technology specialties such as HVAC and automotive technology.

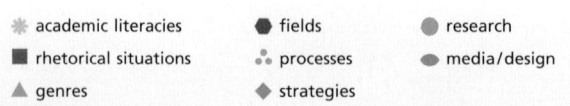

✴ academic literacies	● fields	● research
■ rhetorical situations	⁖ processes	● media/design
▲ genres	◆ strategies	

Engineers and technicians typically write for their peers and team members, their clients, and the public. Writing tasks may include **ABSTRACTS**, **EVALUATIONS**, instructions, **LITERATURE REVIEWS**, memos, **PROPOSALS**, **REPORTS**, and **SUMMARIES**. Support usually includes data, examples, mathematical and logical reasoning, and experimental results, and sources are usually cited in **APA** format. Engineers and technicians value writing that includes logical ordering of ideas and precise language. Tables, charts, figures, illustrations, and headings and subheadings within the writing—all ways of quickly and efficiently conveying information— are also valued.

▲ 196–200

214–22

209–13

258–68

140–63

✳ 33–35

● APA 615–55

A Sample of Writing in Engineering: A Research Report

2. MATERIALS AND METHODS
To begin testing, an ATV test bed was designed (see Figure 1). To secure the machine, a loose rope was attached to the front of the machine and then to the testing platform. An additional rope was then attached at a 90° angle to the front of the machine to act as the lifting force. The test bed platform could be raised to a maximum of 60°, which simulated hills or steep terrain. Each test was started at 0° and then increased by increments of 10 (angles were determined by a digital level attached to platform). Once the machine was at the appropriate angle, a lift force was applied to observe turnover weight. Once the machine's tires lifted off of the platform, the scale was read to determine the amount of weight. Each machine was tested from front to rear and side to side.

Headings used.

Charts, graphs, and photos included.

Precise description of procedure.

Technical language used for precision.

Adapted from *"Analysis of All Terrain Vehicle Crash Mechanisms,"* by *S. Tanner, M. Aitken, and J. N. Warnock, 2008–10, Journal of Undergraduate Research in Bioengineering, docplayer.net/32747021-Analysis-of-all-terrain-vehicle-crash-mechanisms.html.*

Typical Organization of Lab Reports in Engineering

Lab reports, a typical assignment in engineering classes, usually include the IMRaD elements, along with an abstract and a list of references. This format may vary depending on the engineering field and the requirements of the experiment or task.

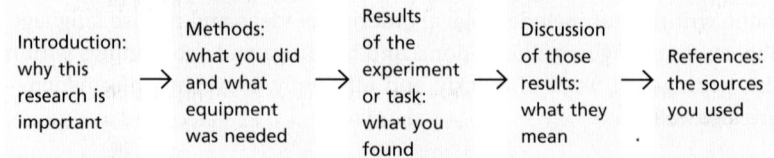

Introduction: why this research is important → Methods: what you did and what equipment was needed → Results of the experiment or task: what you found → Discussion of those results: what they mean → References: the sources you used

WRITING IN HEALTH SCIENCES AND NURSING

Health sciences and nursing is a broad set of fields that may include nursing, anatomy, physiology, nutrition, and pharmacology as well as athletic training, exercise science, physical or occupational therapy, and speech pathology. Consequently, the methods used are also broad and varied, and they may include study of theories and techniques, observation, role-playing, and experiential learning.

196–200 ▲
201–9
164–95
140–63
391–401 ⫶•
209–13 ▲
33–35 ✳
456–63 ◆

APA 615–55 ●

Writing in these fields may include **ABSTRACTS**, **ANNOTATED BIBLIOGRAPHIES**, **ARGUMENTS**, case studies, instructions, personal narratives, **REPORTS**, **REFLECTIONS**, **LITERATURE OR RESEARCH REVIEWS**, **SUMMARIES**, and charts **DESCRIBING** patients' conditions and care. The audiences for this writing may include other patient care providers, clinic and hospital administrators and staff, insurance companies, and patients or clients. Support for assertions typically includes scholarly research, observation, and description, and high value is placed on accurate information and detail. Other aspects of this writing include a preference for writing in the third person, paraphrased source information, and the use of headings and subheadings. Sources are usually cited in **APA** format.

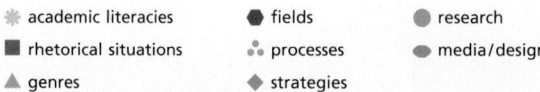

✳ academic literacies   ● fields   ● research
■ rhetorical situations   ⫶• processes   ● media/design
▲ genres   ◆ strategies

A Sample of Writing in Nursing: A Case Study

Patient status.

Ms. D is a morbidly obese 67 year old female, 240 lbs, 5'2" with type II diabetes mellitus. She was transferred from a nursing home to the hospital for pneumonia, but also suffers from congestive heart disease, sleep apnea, psoriasis, and osteoarthritis. She has a weak but productive cough with tonsil suction, and she was on breathing treatments with albuterol. Her skin is very dry and thin with several lesions and yeast infections, and the deep folds of her lower abdomen bled during the bed bath. She did not want to wear her breathing mask at night and refused to get out of bed. She cried when encouraged to use the bathroom or to move her legs. She expressed great fear of returning to the nursing home.

Careful description of patient's condition.

Detailed observation using precise terms.

Behavior as well as medical conditions taken into account.

From the outset, we realized that Ms. D needed care beyond physical therapy and treatment for pneumonia; we realized that her obesity and refusal to participate in her health care expressed important patterns of her life. Morbid obesity does not happen overnight; it is a progressive pattern associated with activity levels, diet, and self-care practices, as well as other possible physiological and psychosocial dimensions. Johnson's (1980) Behavioral System Model, which outlines seven behavioral subsystems, was helpful in providing a perspective of the complexity of Ms. D's health needs. We also assessed that Ms. D lacked confidence in taking care of herself (reflecting the *achievement subsystem*) and lacked a sense of family support from her two sons (*affiliative subsystem*). Her fear of returning to the nursing home coupled with her need for ongoing care challenged her sense of interdependency as addressed in Johnson's *dependency subsystem*.

Patient assessment.

Use of paraphrased scholarly source to aid assessment.

Adapted from "Esthetic Knowing with a Hospitalized Morbidly Obese Patient," by R. Brinkley, K. Ricker, and K. Tuomey, Fall 2007, Journal of Undergraduate Nursing Scholarship, 9, no. 1, http://www.juns.nursing.arizona.edu/articles/Fall%202007/Esthetic%20knowing.htm. Accessed July 21, 2010.

Typical Organization of Case Studies in Health Sciences and Nursing

Case studies, typical assignments in these fields, usually include the following elements:

| Patient status: symptoms, lab findings, history, doctor's orders, how much help the patient needs | → | Assessment: why the patient is being cared for; origins of current situation; when symptoms started, how the patient has been treated, the expected outcome | → | Care plan: what has been done, how current care is working, how care might be improved |

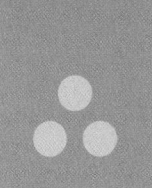

Part 5
Processes

To create anything, we generally break the work down into a series of steps. We follow a recipe (or the directions on a box) to bake a cake; we break a song down into different parts and the music into various chords to arrange a piece of music. So it is when we write. We rely on various processes to get from a blank screen or page to a finished product. The chapters that follow offer advice on some of these processes—from **WRITING AS INQUIRY** and **GENERATING IDEAS** to **DRAFTING** to **GETTING RESPONSE** to **EDITING** to **COMPILING A PORTFOLIO**, and more.

Processes

26 Processes of Writing

Many people think that good writers can just sit down in front of a screen or piece of paper and create a finished piece of writing. After all, in the movies, newspaper reporters pound out stories that then appear on the front page; novelists crank out hundreds of pages without breaking a sweat. That kind of talent for getting it right the first time happens on occasion, or for formulaic writing like business memos. Most of the time, though, most of us write the way we learn to draw, play piano, drive a car, or bake a cake: by working, practicing, making mistakes, and changing what we do as a result. In a way, because we usually need to generate ideas, draft, revise, and edit what we write, we learn to write every time we write.

You may have learned in school about a writing process that looks something like this: prewrite, outline, draft, revise, edit—five steps, to be done in order. That's the way the chapters in this book look, too; but that ordering is somewhat misleading. Sometimes we have to do considerable research before we can begin to figure out what we need to say. Sometimes we have to revise several times, getting responses from others between drafts. Sometimes we reach a certain point in our writing and realize that we need to stop, abandon the work in progress, and start over. And sometimes we can knock out a draft quickly, revising and editing as we go.

In other words, the processes writers use can vary from text to text, genre to genre, task to task. What is easy to describe as a set of lockstep activities:

research → generate ideas → write a draft → get response → revise → edit

often looks more like this:

And it will differ from text to text, assignment to assignment. As writer and writing teacher Donald M. Murray once wrote, "Any process is legitimate if it produces an effective text."

Still, it's good to know the various processes you can use to create an effective text. The chapters that follow can help you understand writing as a way of learning; give you ways to generate ideas and text; offer advice on organizing, drafting, revising, and editing your writing; help you assess your own writing and get response from others; compile your writing into a portfolio; and reflect on your writing and learning.

27 Writing as Inquiry

Sometimes we write to say what we think. Other times, however, we write in order to figure out what we think. Much of the writing you do in college will be the latter. Even as you learn to write, you will be writing to learn. This chapter is about writing with a spirit of inquiry—approaching writing projects with curiosity, moving beyond the familiar, keeping your eyes open, tackling issues that don't have easy answers. It's about starting with questions and going from there—and taking risks. As Mark Twain once said, "Sail away from the safe harbor. . . . Explore. Dream. Discover." This chapter offers strategies for doing just that with your writing.

Starting with Questions

The most important thing is to start with questions—with what you *don't* know rather than with what you *do* know. Your goal is to learn about your subject and then to learn more. If you're writing about a topic you know well, you want to expand on what you already know. In academic writing, good topics arise from important questions, issues, and problems that are already being discussed. As a writer, you need to find out what's being said about your topic and then see your writing as a way of entering that larger conversation.

So start with questions, and don't expect to find easy answers. If there were easy answers, there would be no reason for discussion—or for you to write. For purposes of inquiry, the best questions can't be answered by looking in a reference book or searching online. Instead, they are ones that help you explore what you think—and why. As it happens, many of the strategies in this book can help you ask questions of this kind. Following are some questions to get you started:

445-55 ◆ **How can it be DEFINED?**   What is it, and what does it do? Look it up in a dictionary; check *Wikipedia*. Remember, though, that these are only starting points. How *else* can it be defined? What more is there to know about it? If your topic is being debated, chances are that its very definition is subject to debate. If, for instance, you're writing about gay marriage, how you define marriage will affect how you approach the topic.

456-63 ◆ **How can it be DESCRIBED?**   What details should you include? From what vantage point should you describe your topic? If, for example, your topic is the physiological effects of running a marathon, what are those effects—on the lungs, heart muscles, nerves, brain, and so on? How will you describe the physical experience of running over twenty-six miles from the runner's point of view?

469-73 ◆ **How can it be EXPLAINED?**   What does it do? How does it work? If you're investigating the use of performance-enhancing drugs by athletes, for example, what exactly are the effects of these drugs? What makes them dangerous—and are they always dangerous or only in certain conditions? Why are they illegal—and should they be illegal?

437-44 ◆ **What can it be COMPARED with?**   Again using performance-enhancing drugs by athletes as an example, how does taking such supplements compare with wearing high-tech footwear or uniforms? Does such a comparison make you see taking steroids or other performance-enhancing drugs in a new light?

405-9 ◆ **What may have CAUSED it? What might be its EFFECTS?**   Who or what does it affect? What causes cerebral palsy in children, for example? What are its symptoms? If children with cerebral palsy are not treated, what might be the consequences?

431-36 ◆ **How can it be CLASSIFIED?**   Is it a topic or issue that can be placed into categories of similar topics or issues? What categories can it be placed into? Are there legal and illegal performance-enhancing supplements (human growth hormone and steroids, for instance), and what's the difference? Are some safe and others less safe? Classifying your topic in this way can help you consider its complexities.

❋ academic literacies   ● fields   ● research

■ rhetorical situations   ⁂ processes   ● media/design

▲ genres   ◆ strategies

How can it be ANALYZED?   What parts can the topic be divided into? For ▲ 104–39 example, if you're exploring the health effects of cell phone use, you might ask these questions: What evidence, if any, suggests that cell phone radiation causes cancer? What cancers are associated with cell phone use? What do medical experts and phone manufacturers say? How can cell phone users reduce their risk?

How can it be interpreted?   What does it really mean? How do you interpret it, and how does your interpretation differ from those of other writers? What evidence supports your interpretation, and what argues against it? Imagine you're exploring the topic of sports injuries among young women. Do these injuries reflect a larger cultural preoccupation with competition? anatomical differences between male and female bodies? something else?

What expectations does it raise?   What will happen next? What makes you think so? If this happens, how will it affect those involved? For instance, will the governing bodies of professional sports require more testing of athletes' blood, urine, and hair than they do now? Will such tests be unfair to athletes taking drugs for legitimate medical needs?

What are the different POSITIONS on it?   What controversies or disagree- ▲ 164–95 ments exist, and what evidence is offered for the various positions? What else might be said? Are there any groups or individuals who seem especially authoritative? If so, you might want to explore what they have said.

What are your own feelings about it?   What interests you about the topic? How much do you already know about it? For example, if you're an athlete, how do you feel about competing against others who may have taken performance-enhancing supplements? If a friend has problems with drugs, do those problems affect your thinking about drugs in sports? How do you react to what others say about the topic? What else do you want to find out?

Are there other ways to think about it?   Is what seems true in this case also true in others? How can you apply this subject in another situation? Will what works in another situation also work here? What do you have

to do to adapt it? Imagine you're writing about traffic fatalities. If replacing stop signs with roundabouts or traffic circles reduced traffic fatalities in England, could doing so also reduce accidents in the United States?

337

You can also start with the journalist's **QUESTIONS**: *Who? What? When? Where? Why? How?* Asking questions from these various perspectives can help you deepen your understanding of your topic by leading you to see it from many angles.

Keeping a Journal

343–44

One way to get into the habit of using writing as a tool for inquiry is to keep a **JOURNAL**. You can use a journal to record your observations, reactions, whatever you wish. Some writers find journals especially useful places to articulate questions or speculations. You may be assigned by teachers to do certain work in a journal, but in general you can use a journal to write for yourself. Note your ideas, speculate, digress—go wherever your thoughts lead you.

Keeping a Blog

687

You may also wish to explore issues or other ideas online in the form of a **BLOG**. Most blogs have a comments section that allows others to read and respond to what you write, leading to potentially fruitful discussions. You can also include links to other websites, helping you connect various strands of thought and research. The blogs of others, along with online discussion forums and groups, may also be useful sources of opinion on your topic, but keep in mind that they probably aren't authoritative research sources. There are a number of search engines that can help you find blog posts related to specific topics, including blogsearchengine.org and *Blogarama*. You can create your own blog on sites such as *Blogger*, *Tumblr*, *Svbtle*, or *WordPress*.

---

* academic literacies ● fields ● research
■ rhetorical situations ⁘ processes ● media/design
▲ genres ◆ strategies

28 Generating Ideas and Text

All good writing revolves around ideas. Whether you're writing a job-application letter, a sonnet, or an essay, you'll always spend time and effort generating ideas. Some writers can come up with a topic, put their thoughts in order, and flesh out their arguments in their heads; but most of us need to write out our ideas, play with them, tease them out, and examine them from some distance and from multiple perspectives. This chapter offers activities that can help you do just that. *Freewriting*, *looping*, *listing*, *clustering*, and *talking* can help you explore what you already know about a subject; *cubing*, *questioning*, and *using visuals* nudge you to consider a subject in new ways; and *outlining*, *letter writing*, *journal keeping*, and *discovery drafting* offer ways to generate a text.

Freewriting

An informal method of exploring a subject by writing about it, freewriting ("writing freely") can help you generate ideas and come up with materials for your draft. Here's how to do it:

1. Write as quickly as you can without stopping for 5 to 10 minutes (or until you fill a screen or page).

2. If you have a subject to explore, write it at the top and then start writing about it; but if you stray, don't worry—just keep writing. If you don't have a subject yet, just start writing and don't stop until the time is up. If you can't think of anything to say, write that ("I can't think of anything to say") again and again until you do—and you will!

3. Once the time is up, read over what you've written, and underline or highlight passages that interest you.

4. Write some more, starting with one of those underlined or highlighted passages as your new topic. Repeat the process until you've come up with a usable topic.

Looping

Looping is a more focused version of freewriting; it can help you explore what you already know about a subject. You stop, reflect on what you've written, and then write again, developing your understanding in the process. It's good for clarifying your knowledge and understanding of a subject and finding a focus. Here's what to do:

1. Write for 5 to 10 minutes on whatever you know about your subject. This is your first loop.

2. Read over what you wrote, and then write a single sentence summarizing the most important or interesting idea. You might try completing one of these sentences: "I guess what I was trying to say was . . . " or "What surprises me most in reading what I wrote is . . ." This will be the start of another loop.

3. Write again for 5 to 10 minutes, using your summary sentence as your beginning and your focus. Again, read what you've written, and then write a sentence capturing the most important idea—in a third loop.

Keep going until you have enough understanding of your topic to be able to decide on a tentative focus—something you can write about.

Listing

Some writers find it useful to keep lists of ideas that occur to them while they're thinking about a topic. Follow these steps:

1. Write a list of potential ideas about a topic. Don't try to limit your list—include anything that interests you.

2. Look for relationships among the items on your list: What patterns do you see? If other ideas occur to you, add them to the list.

❋ academic literacies   ● fields   ● research

■ rhetorical situations   ⁖ processes   ● media/design

▲ genres   ◆ strategies

3. Arrange the items in an order that makes sense for your purpose and can serve as the beginning of an outline for your writing.

Clustering or Mapping Ideas

Clustering (also called "idea mapping," "mind mapping," or "concept mapping") is a way of generating and connecting ideas visually. It's useful for seeing how various ideas relate to one another and for developing subtopics. You can create a cluster with pencil and paper or use an online mapping tool like *Coggle*, *Bubble.us*, or *MindMup*. Either way, the technique is simple:

1. Write your topic in the middle of a page or computer screen.

2. Write ideas relating to that topic around it, and connect them to the central idea with lines or shapes.

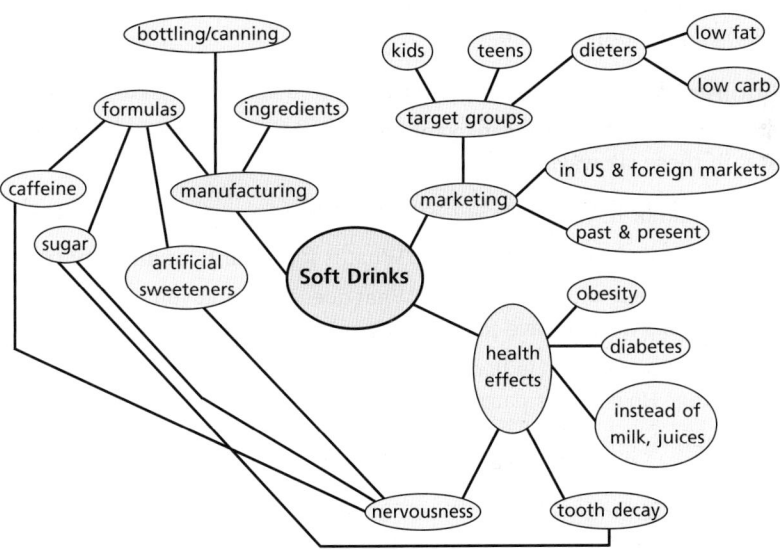

3. Write down examples, facts, or other details relating to each idea, and join them to connecting topics.

4. Keep going until you can't think of anything else relating to your topic.

You should end up with various ideas about your topic, and the clusters will enable you to see how they relate to one another. In the example cluster on the topic of "soft drinks" on the previous page, note how some ideas link not only to the main topic or related topics but also to other ideas.

Talking

Sometimes it's helpful to generate ideas by talking rather than writing. You might talk about your topic with a classmate, a teacher, a tutor, a friend—or even just talk out loud to yourself. If you do, be sure to take notes during or after talking, or use a voice-recording app to capture the conversation. You can also use free transcription apps, like *Otter* or *Google Voice*, to generate a print version of the conversation. Some apps, like *Otter*, even tag keywords from the transcript, helping you see themes and patterns in your conversation.

Cubing

A cube has six sides. You can examine a topic as you might a cube, looking at it in these six ways:

- **DESCRIBE** it. What's its color? shape? age? size? What's it made of? 456–63 ◆

- **COMPARE** it to something else. What is it similar to or different from? 437–44 ◆

- Associate it with other things. What does it remind you of? What connections does it have to other things? How would you **CLASSIFY** it? 431–36 ◆

- **ANALYZE** it. How is it made? Where did it come from? Where is it going? How are its parts related? 104–39 ▲

- Apply it. What is it used for? What can be done with it?

- **ARGUE** for or against it. Choose a position relating to your subject, and defend it. 164–95 ▲

☀ academic literacies   ● fields   ● research
■ rhetorical situations   ⁝ processes   ● media/design
▲ genres   ◆ strategies

Questioning

It's always useful to ask questions. One way is to start with *What? Who? When? Where? How?* and *Why?* A particular method of exploring a topic is to ask questions as if the topic were a play. This method is especially useful for exploring literature, history, the arts, and the social sciences. Start with these questions:

- *What?* What happens? How is it similar to or different from other actions?

- *Who?* Who are the actors? Who are the participants, and who are the spectators? How do the actors affect the action, and how are they affected by it?

- *When?* When does the action take place? How often does it happen? What happens before, after, or at the same time? Would it be different at another time? Does the time have historical significance?

- *Where?* What is the setting? What is the situation, and what makes it significant?

- *How?* How does the action occur? What are the steps in the process? What techniques are required? What equipment is needed?

- *Why?* Why did this happen? What are the actors' motives? What end does the action serve?

Using Visuals

Some writers use visuals—photographs, presentation slides, drawings—as tools for sparking or capturing ideas or for thinking about a topic in a new way. Here are some suggestions for doing so:

- Search for images related to your topic on a stock photo site like *Pixabay* or *Pexels*. Then freewrite about the images you find. What do the images make you think of? Do they help you see your topic in a different light? Do they suggest details you hadn't considered?

Here's an example of how a stock image helped one student generate ideas for a memoir about memories of her grandmother raising chickens and selling eggs.

 Freewrite: This picture is beautiful, the way the eggs are almost glowing in the basket. It makes me think of bounty—of having more than enough. Maybe that's part of the significance of my story? I didn't realize it when I was young, but Grandma sold the eggs to make her own money—to afford "extras" and to have some financial freedom apart from Grandpa. And then there was all the bounty she left us in stories and memories and recipes. Maybe I can play on this theme somehow in my writing?

- Make a *PowerPoint* or *Keynote* slide about your topic. Choose keywords for headings and subheadings, find images or icons for each, and experiment with different arrangements of the material. Doing so might help you make new connections or associations among your ideas.

- Do what filmmakers and animators do when developing a new project: create a storyboard—a scene-by-scene plan of what the final product might look like. You could draw scenes by hand (stick figures are fine!) or use stock images to represent each scene or section of your draft. Search for storyboard templates online to get started.

Here's an example storyboard for the student memoir mentioned earlier. Created with a template and stock images from the graphic design app *Canva*, the storyboard helped the student generate memories and details and then organize them before drafting.

⁕ academic literacies   ● fields   ● research
■ rhetorical situations   ∴ processes   ● media/design
▲ genres   ◆ strategies

Storyboard: "Egg Money" Memoir

SCENE #1
I feed the chickens while Grandma gathers the eggs, placing them gently in the pockets of her tattered apron.

SCENE #2
Grandma shows me how to separate an egg, whisking out the last drop of egg white with her finger. "No use wasting," she says. Later, after we make our egg deliveries, we'll come home to Grandma's angel food cake.

SCENE #3
Grandma and I place the eggs in cartons and wrap each in newspaper and twine.

SCENE #4
We load up the old wheelbarrow with egg cartons and set out to deliver our goods to Grandma's customers.

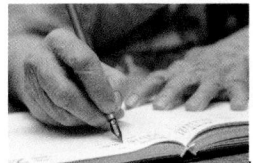

SCENE #5
After the egg deliveries, Grandma records every transaction in her cash ledger. I sit next to her, eating my angel food cake.

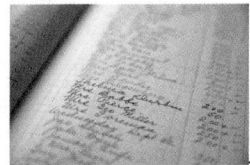

SCENE #6
Going through Grandma's attic as an adult, I find her recipe books, her county-fair ribbons, her cash ledgers documenting forty years' worth of chicken raising and egg selling.

Using Genre Features

Genres typically include particular kinds of information and organize it in particular ways. One way to generate ideas and text, then, is to identify the key features of the genre in which you're writing and use them to guide you as you write. Of course, you may alter the genre's features or combine two or more genres in order to achieve your purpose, but the overall shape and content of the genre can give you a way to develop and organize your ideas and research.

Outlining

You may create an *informal outline* by simply listing your ideas and numbering them in the order in which you want to write about them. You might prefer to make a *working outline*, to show the hierarchy of relationships among your ideas. While still informal, a working outline distinguishes your main ideas and your support, often through simple indentation:

First main idea
  Supporting evidence or detail
  Supporting evidence or detail

Second main idea
  Supporting evidence or detail
  Supporting evidence or detail

A *formal outline* shows the hierarchy of your ideas through a system of indenting, numbering, and lettering. Remember that when you divide a point into more specific subpoints, you should have at least two of them—you can't divide something into only one part. Also, try to keep items at each level parallel in structure. Formal outlines work this way:

Thesis statement

  I. First reason
    A. Supporting evidence
      1. Detail of evidence
      2. Detail of evidence
    B. Supporting evidence
  II. Another reason

※ academic literacies   ● fields   ● research
■ rhetorical situations   ⁖ processes   ● media/design
▲ genres   ◆ strategies

Here is a formal outline of the first part of the **MLA** research report by Dylan Borchers, "Against the Odds: Harry S. Truman and the Election of 1948," that shows how he organized it: MLA 608–14

    I. Introduction: Outcome of 1948 election
   II. Bad predictions by pollsters
       A. Pollsters stopped polling.
       B. Dewey supporters became overconfident.
       C. Truman supporters were either energized or stayed home.
  III. Dewey's campaign overly cautious
       A. He was overconfident.
       B. His message was vague—he avoided taking stands.
  IV. Dewey's public appearances poor
       A. He was seen as aloof, uncomfortable with crowds.
       B. He made blunders.
       C. His speeches were dull.

Writing out a formal outline can be helpful when you're dealing with a complex subject; as you revise your drafts, though, be flexible and ready to change your outline as your understanding of your topic develops.

Another way to outline is to use an online organizing tool like *Trello*, *Jira*, or *Wrike*. These tools let you create a digital bulletin board to which you post "lists" and "cards." Below is an example *Trello* board that outlines Dylan Borchers's research report. Each color-coded "card" represents a supporting detail for the main point listed at the top of a column of cards. You can drag cards to different lists, depending on where the cards fit

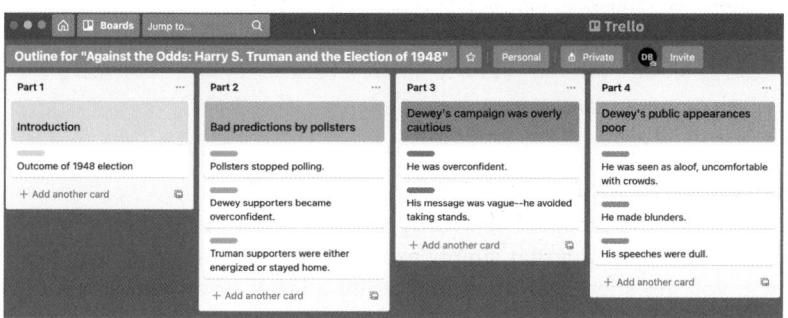

best. You can expand a card to jot down notes, save website links, and add attachments, as in the following example, which includes notes and links to sources that support one of Borchers's points.

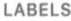 **He was seen as aloof, uncomfortable with crowds.**

in list <u>Part 4</u>

LABELS

   +

≡ **Description**   Edit

Comparison to plastic groom on top of a wedding cake -- Hamby.

Stiff and cold demeanor -- McCullough

@ **Attachments**

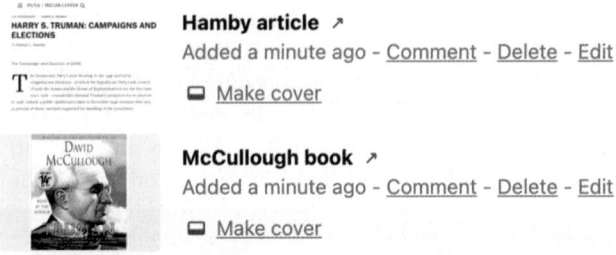

Hamby article ↗

Added a minute ago - <u>Comment</u> - <u>Delete</u> - <u>Edit</u>

▭ <u>Make cover</u>

McCullough book ↗

Added a minute ago - <u>Comment</u> - <u>Delete</u> - <u>Edit</u>

▭ <u>Make cover</u>

Add an attachment

Letter Writing

Sometimes the prospect of writing a report or an essay can be intimidating. You may find that simply explaining your topic to someone will help you get started. In that case, write a letter or an email or a text

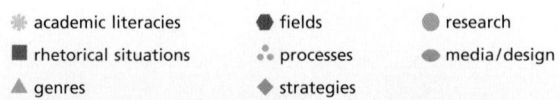

to someone you know—your best friend, a parent or grandparent, a sibling—in which you discuss your subject. Explain it in terms that your reader can understand. Use the unsent letter to rehearse your topic; make it a kind of rough draft that you can then revise and develop to suit your actual audience.

Keeping a Journal

Some writers find that writing in a journal helps them generate ideas. Making note of your ideas, thoughts, feelings, or the events of your day can provide a wealth of topics, and a journal can also be a good place to explore what you think and why you think as you do.

Journals are private—you are the only audience—so you can feel free to write whatever comes to mind. And you can do more than write. If you choose a paper journal, doodle or draw in it, and keep clippings or scraps of paper between the pages; if you keep your journal on a computer, copy and paste interesting images or text you find online. Whatever form your journal takes, feel free to play with its contents, and don't worry about errors or grammar. The goal is to generate ideas; let yourself wander without censoring yourself or fretting that your writing is incorrect or incomplete or incoherent. That's OK.

One measure of the success of journaling and other personal writing is length: journal entries, **FREEWRITING**, **LISTING**, **CUBING**, and other types of informal writing are like warm-up exercises to limber you up and get you thinking. If you don't give them enough time and space, they may not do what you want them to. Often, students' best insights appear at the end of their journal entries. Had they stopped before that point, they would never have had those good ideas.

333–34
334–35
336

After you've written several journal entries, one way to study the ideas in them is to highlight useful patterns in different colors. For example, journal entries usually include some questioning and speculating, as well as summarizing and paraphrasing. Try color-coding each of these, sentence by sentence, phrase by phrase: yellow for summaries or paraphrases, green for questions, blue for speculations. Do any colors dominate? If, for example, your entries are mostly yellow, you may be restating the course

content or quoting from the textbook too much and perhaps need to ask more questions. If you're generating ideas for an essay, you might assign colors to ideas or themes to see which ones are the most promising.

Discovery Drafting

364–66

Some writers do best by jumping in and writing. Here are the steps to take if you're ready to write a preliminary **DRAFT**:

1. Write your draft quickly, in one sitting if possible.

2. Assume that you are writing to discover what you want to say and how you need to say it—and that you will make substantial revisions in a later part of the process.

3. Don't worry about grammatical or factual correctness—if you can't think of a word, leave a blank space to fill in later. If you're unsure of a date or spelling, put a question mark in parentheses as a reminder to check it later. Just write.

If you need more help

79

See each of the **GENRE** chapters for specific strategies for generating text in each genre.

---

* academic literacies
■ rhetorical situations
▲ genres

● fields
⁘ processes
◆ strategies

● research
● media/design

29 Organizing Your Writing, Guiding Your Readers

Traffic lights, street signs, and lines on the road help drivers find their way. Readers need similar guidance—to know, for example, whether they're reading a report or an argument, evaluation, or proposal. They also need to know what to expect: What will the report be about? What perspective will it offer? What will this paragraph cover? What about the next one? How do the two paragraphs relate to each other—and to the rest of the essay?

When you write, then, you need to organize your ideas and provide cues to help your readers navigate your text and understand the points you're trying to make. This chapter offers advice on organizing your writing and guiding your readers through it by organizing the text and its paragraphs; creating strong thesis statements, clear topic sentences, and helpful beginnings and titles; ending appropriately; and tying ideas together with transitions. See also the **GENRE** chapters for advice on organizing texts in specific genres.

▲ 79

Outlining

At some point as writers develop their ideas and collect their information, they often create an **OUTLINE** as a plan for drafting a text. You may create an *informal outline* by simply listing your ideas and numbering them in the order in which you want to write about them. You might prefer to make a *working outline* to show the hierarchy of relationships among your ideas. Or you might need to create a *formal outline*, which shows the hierarchy of your ideas through a system of indenting, numbering, and lettering. And you might use an online organizing tool.

⁝ 340–42

Beginning

Whenever we pick up something to read, we generally start by looking at the first few words or sentences to see if they grab our attention, and based on them, we decide whether to keep reading. Beginnings, then, are important—both for attracting readers and for giving them some information about what's to come. How you begin depends on your **RHETORICAL SITUATION**, especially your purpose and audience. Academic audiences generally expect a text's introduction to establish context, explaining how the text fits into some larger conversation, addresses certain questions, or explores an aspect of the subject. Most introductions also offer a brief description of the text's content, often in the form of a thesis statement. The following opening paragraph of a report on cyberloafing—that is, using an employer's internet access to do non-work-related tasks—does all of this:

> Increased technological advances and widespread internet access have revolutionized the workplace. But all this innovation has come with at least one questionable stowaway: cyberloafing—the act of browsing the internet for personal use while at work. It's becoming essential for organizations to understand cyberloafing: what it is, which factors lead employees to do it, and how it affects productivity. At its worst, cyberloafing can jeopardize an organization's information security, so awareness about cyberloafing is essential to both employees and employers in order to prevent its most negative consequences.
>
> —Rocia Celeste Mejia Avila,
> "Cyberloafing: Distraction or Motivation?"

If you're writing for a nonacademic audience or genre—for a newspaper or a website, for example—your introduction may need to entice your readers to read on by connecting your text to their interests through shared experiences, anecdotes, or some other attention-getting device. Cynthia Bass, writing a newspaper article about the Gettysburg Address on its 135th anniversary, connects that date—the day her audience would read it—to Lincoln's address. She then develops the rationale for thinking about the

57 ■

* academic literacies   ● fields   ● research
■ rhetorical situations   ⁙ processes   ◗ media/design
▲ genres   ◆ strategies

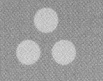
speech and introduces her specific topic: debates about the writing and delivery of the Gettysburg Address:

> November 19 is the 135th anniversary of the Gettysburg Address. On that day in 1863, with the Civil War only half over and the worst yet to come, Abraham Lincoln delivered a speech now universally regarded as both the most important oration in U.S. history and the best explanation—"government of the people, by the people, for the people"—of why this nation exists.
>
> We would expect the history of an event so monumental as the Gettysburg Address to be well established. The truth is just the opposite. The only thing scholars agree on is that the speech is short—only ten sentences—and that it took Lincoln under five minutes to stand up, deliver it, and sit back down.
>
> Everything else—when Lincoln wrote it, where he wrote it, how quickly he wrote it, how he was invited, how the audience reacted—has been open to debate since the moment the words left his mouth.
>
> —Cynthia Bass, "Gettysburg Address: Two Versions"

Thesis Statements

A thesis identifies the topic of a text along with the claim being made about it. A good thesis helps readers understand an essay. Working to create a sharp thesis can help you focus both your thinking and your writing. Here are four steps for moving from a topic to a thesis statement:

1. **State your topic as a question.** You may have an idea for a topic, such as "gasoline prices," "analysis of 'real women' ad campaigns," or "famine." Those may be good topics, but they're not thesis statements, primarily because none of them actually makes a statement. A good way to begin moving from topic to thesis statement is to turn your topic into a question:

   What causes fluctuations in gasoline prices?

   Are ads picturing "real women" who aren't models effective?

   What can be done to prevent famine in Somalia?

2. **Then turn your question into a position.** A thesis statement is an assertion—it takes a stand or makes a claim. Whether you're writing a report or an argument, you are saying, "This is the way I see . . ." or "My research shows . . ." or "This is what I believe about . . ." Your thesis statement announces your position on the question you're raising about the topic, so a relatively easy way of establishing a thesis is to answer your own question:

Gasoline prices fluctuate for several reasons.

Ads picturing "real women" instead of models are effective because most women can easily identify with them.

The threat of famine in Somalia could be avoided if certain measures are taken.

3. **Narrow your thesis.** A good thesis is specific, guiding you as you write and showing readers exactly what your essay will cover. The preceding thesis statements need to be qualified and focused—they should be made more specific. For example:

Gasoline prices fluctuate because of production procedures, consumer demand, international politics, and oil companies' policies.

Dove's "Campaign for Self-Esteem" and Aerie's ads featuring Iskra Lawrence work because consumers can identify with the women's bodies and admire their confidence in displaying them.

The current threat of famine in Somalia could be avoided if the government and humanitarian organizations increase the availability of water, provide shipments of food, and supply medical and nutrition assistance.

337 A good way to narrow a thesis is to ask **QUESTIONS** about it: *Why* do gasoline prices fluctuate? *How* could the Somalia famine have been avoided? The answers will help you craft a narrow, focused thesis.

4. **Qualify your thesis.** Sometimes you want to make a strong argument and state your thesis bluntly. Often, however, you need to acknowledge

---

* academic literacies   ● fields   ● research

■ rhetorical situations   ⁝ processes   ● media/design

▲ genres   ◆ strategies

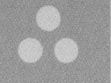

that your assertions may be challenged or may not be uncondition-
ally true. In those cases, consider limiting the scope of your thesis by
adding to it such terms as "may," "probably," "apparently," "very likely,"
"sometimes," and "often."

Gasoline prices *very likely* fluctuate because of production procedures,
consumer demand, international politics, and oil companies' policies.

Dove's and Aerie's ad campaigns featuring "real women" *may* work
because consumers can identify with the women's bodies and admire
their confidence in displaying them.

The current threat of famine in Somalia could *probably* be avoided if
the government and humanitarian organizations increase the availability of water, provide shipments of food, and supply medical and
nutrition assistance.

Thesis statements are typically positioned at or near the end of a text's
introduction, to let readers know at the outset what is being claimed and
what the text will be aiming to prove. A thesis doesn't necessarily forecast
your organization, which may be more complex than the thesis itself. For
example, Notre Dame University student Sarah Dzubay's essay, "An Out-
break of the Irrational," contains this thesis statement:

The movement to opt out of vaccinations is irrational and dangerous
because individuals advocating for their right to exercise their personal
freedom are looking in the wrong places for justification and ignoring
the threat they present to society as a whole.

The essay that follows this thesis statement includes discussions of herd
immunity; a socioeconomic profile of parents who choose not to have their
children vaccinated; outlines of the rationales those parents use to justify
their choice, which include fear of autism, fear of causing other health
problems, and political and ethical values; and a conclusion that parents
who refuse to have their children vaccinated are being unreasonable and
selfish. The paper delivers what the thesis promises but includes important
information not mentioned in the thesis itself.

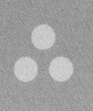
More Ways of Beginning

Explain the larger context of your topic. Most essays are part of an ongoing conversation, so you might begin by outlining the context of the subject to which your writing responds. An essay exploring the "emotional climate" of the United States after Barack Obama became president begins by describing the national moods during some of his predecessors' administrations:

> Every president plays a symbolic, almost mythological role that's hard to talk about, much less quantify — it's like trying to grab a ball of mercury. I'm not referring to using the bully pulpit to shape the national agenda but to the way that the president, as America's most inescapably powerful figure, colors the emotional climate of the country. John Kennedy and Ronald Reagan did this affirmatively, expressing ideals that shaped the whole culture. Setting a buoyant tone, they didn't just change movies, music, and television; they changed attitudes. Other presidents did the same, only unpleasantly. Richard Nixon created a mood of angry paranoia, Jimmy Carter one of dreary defeatism, and George W. Bush, especially in that seemingly endless second term, managed to do both at once.
>
> —John Powers, "Dreams from My President"

Forecast your organization. You might begin by briefly summarizing the way in which you will organize your text. The following example from a scholarly paper on the role of immigrants in the US labor market offers background on the subject and describes the points that the writer's analysis will discuss:

> Debates about illegal immigration, border security, skill levels of workers, unemployment, job growth and competition, and entrepreneurship all rely, to some extent, on perceptions of immigrants' role in the U.S. labor market. These views are often shaped as much by politics and emotion as by facts.
>
> To better frame these debates, this short analysis provides data on immigrants in the labor force at the current time of slowed immigration,

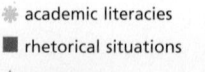 academic literacies   fields 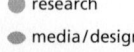 research
rhetorical situations processes media/design
genres strategies

high unemployment, and low job growth and highlights eight industries where immigrants are especially vital. How large a share of the labor force are they and how does that vary by particular industry? How do immigrants compare to native-born workers in their educational attainment and occupational profiles?

The answers matter because our economy is dependent on immigrant labor now and for the future. The U.S. population is aging rapidly as the baby boom cohort enters old age and retirement. As a result, the labor force will increasingly depend upon immigrants and their children to replace current workers and fill new jobs. This analysis puts a spotlight on immigrant workers to examine their basic trends in the labor force and how these workers fit into specific industries and occupations of interest.

—Audrey Singer,
"Immigrant Workers in the U.S. Labor Force"

Offer background information. If your readers may not know as much as you do about your topic, giving them information to help them understand your position can be important, as journalist David Guterson does in an essay on the Mall of America:

Last April, on a visit to the new Mall of America near Minneapolis, I carried with me the public-relations press kit provided for the benefit of reporters. It included an assortment of "fun facts" about the mall: 140,000 hot dogs sold each week, 10,000 permanent jobs, 44 escalators and 17 elevators, 12,750 parking places, 13,300 short tons of steel, $1 million in cash disbursed weekly from 8 automatic-teller machines. Opened in the summer of 1992, the mall was built on the 78-acre site of the former Metropolitan Stadium, a five-minute drive from the Minneapolis–St. Paul International Airport. With 4.2 million square feet of floor space—including twenty-two times the retail footage of the average American shopping center—the Mall of America was "the largest fully enclosed combination retail and family entertainment-complex in the United States."

—David Guterson,
"Enclosed. Encyclopedic. Endured. One Week at the Mall of America"

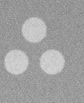

A roller coaster in the Mall of America.

Visuals can also help provide context. For example, Guterson's essay on the Mall of America might have included a photo like the one above to convey the size of the structure.

445–55 ◆

Define key terms or concepts. The success of an argument often hinges on how key terms are **DEFINED**. You may wish to provide definitions up front—as an advocacy website, *Health Care without Harm*, does in a report on the hazards of fragrances in health-care facilities:

> To many people, the word "fragrance" means something that smells nice, such as perfume. We don't often stop to think that scents are chemicals. Fragrance chemicals are organic compounds that volatilize, or vaporize into the air—that's why we can smell them. They are added to products to give them a scent or to mask the odor of other ingredients. The volatile organic chemicals (VOCs) emitted by fragrance products can contribute to poor indoor air quality (IAQ) and are associated with a variety of adverse health effects.
>
> —"Fragrances," *Health Care without Harm*

* academic literacies   ● fields   ● research
■ rhetorical situations   ∴ processes   ● media/design
▲ genres   ◆ strategies

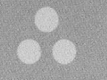
Connect your subject to your readers' interests or values. You'll always want to establish **COMMON GROUND** with your readers, and sometimes you may wish to do so immediately, in your introduction, as in this example:

◆ 424

> We all want to feel safe. Most Americans lock their doors at night, lock their cars in parking lots, try to park near buildings or under lights, and wear seat belts. Many invest in expensive security systems, carry pepper spray or a stun gun, keep guns in their homes, or take self-defense classes. Obviously, safety and security are important issues in American life.
>
> —Andy McDonie, "Airport Security: What Price Safety?"

Start with something that will provoke readers' interest. Writer and columnist Anna Quindlen opens an essay on feminism with the following eye-opening assertion:

> Let's use the F word here. People say it's inappropriate, offensive, that it puts people off. But it seems to me it's the best way to begin, when it's simultaneously devalued and invaluable.
>   Feminist. Feminist, feminist, feminist.
>
> —Anna Quindlen, "Still Needing the F Word"

Start with an anecdote. Sometimes a brief **NARRATIVE** helps bring a topic to life for readers. See, for example, how an essay on student loans and their effects on individuals and society begins:

◆ 474–82

> Rodney Spangler first enrolled at the University of North Texas in 2001. There, he pursued a degree in what the school now calls "integrative studies," focusing on history, philosophy, and criminal justice. Rodney also worked full time, and so attended UNT off and on until 2007. For every semester of classes, Spangler took out student loans. When he left—without a degree—he estimates that he had about $30,000 in outstanding student debt.
>
> —Charles Fain Lehman,
> "The Student Loan Trap: When Debt Delays Life"

Ask a question. Instead of opening with a thesis statement, you might open with a question about the topic your text will explore, as this study of the status of women in science does:

> Are women's minds different from men's minds? In spite of the women's movement, the age-old debate centering around this question continues. We are surrounded by evidence of de facto differences between men's and women's intellects—in the problems that interest them, in the ways they try to solve those problems, and in the professions they choose. Even though it has become fashionable to view such differences as environmental in origin, the temptation to seek an explanation in terms of innate differences remains a powerful one.
>
> —Evelyn Fox Keller, "Women in Science: A Social Analysis"

Jump right in. Occasionally you may wish to start as close to the key action as possible. See how one writer jumps right in when opening his profile of a blues concert:

> Long Tongue, the Blues Merchant, strolls onstage. His guitar rides side-saddle against his hip. The drummer slides onto the tripod seat behind the drums, adjusts the high-hat cymbal, and runs a quick, off-beat tattoo on the tom-tom, then relaxes. The bass player plugs into the amplifier, checks the settings on the control panel, and nods his okay. Three horn players stand off to one side, clustered, lurking like brilliant sorcerer-wizards waiting to do magic with their musical instruments.
>
> —Jerome Washington, "The Blues Merchant"

Ending

Endings are important because they're the last words readers read. How you end a text will depend in part on your **RHETORICAL SITUATION**. You may end by wrapping up loose ends, or you may wish to give readers something to think about. Some endings do both, as Nicholas Kristof's does in arguing that guns should be regulated in the same way cars are. In the first of two paragraphs, Kristof acknowledges that his proposals aren't a cure-all for gun violence but notes that thousands of lives would be saved. In the second, he sums up his entire argument:

57 ▪

* academic literacies   ● fields   ● research
▪ rhetorical situations   ∴ processes   ● media/design
▲ genres   ◆ strategies

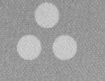

These steps won't eliminate gun deaths any more than seatbelts eliminate auto deaths. But if a combination of measures could reduce the toll by one-third, that would be 10,000 lives saved every year.

A century ago, we reacted to deaths and injuries from unregulated vehicles by imposing sensible safety measures that have saved hundreds of thousands of lives a year. Why can't we ask politicians to be just as rational about guns?

—Nicholas Kristof, "Our Blind Spot about Guns"

Kristof's final question provides readers with food for thought and challenges legislators either to come up with "sensible safety measures" or to provide reasons why they can't or won't.

Ways of Ending

Restate your main point. Sometimes you'll simply summarize your central idea, as in this example from an essay arguing that people have no "inner" self and that they should be judged by their actions alone:

The inner man is a fantasy. If it helps you to identify with one, by all means, do so; preserve it, cherish it, embrace it, but do not present it to others for evaluation or consideration, for excuse or exculpation, or, for that matter, for punishment or disapproval.

Like any fantasy, it serves your purposes alone. It has no standing in the real world which we share with each other. Those character traits, those attitudes, that behavior—that strange and alien stuff sticking out all over you—*that's the real you!*

—Willard Gaylin, "What You See Is the Real You"

Discuss the implications of your argument. The following conclusion of an essay on the development of Post-it notes leads readers to consider how failure sometimes leads to innovation:

Post-it notes provide but one example of a technological artifact that has evolved from a perceived failure of existing artifacts to function without frustrating. Again, it is not that form follows function but, rather, that the form of one thing follows from the failure of another thing to function as we would like. Whether it be bookmarks that fail

to stay in place or taped-on notes that fail to leave a once-nice surface clean and intact, their failure and perceived failure is what leads to the true evolution of artifacts. That the perception of failure may take centuries to develop, as in the case of loose bookmarks, does not reduce the importance of the principle in shaping our world.

—Henry Petroski, "Little Things Can Mean a Lot"

474–82 ◆

End with an anecdote. If you take this approach, you might finish a NARRATIVE that was begun earlier in your text or add one that illustrates the point you're making. See how Sarah Vowell uses a story to end an essay on students' need to examine news reporting critically:

I looked at Joanne McGlynn's syllabus for her media studies course, the one she handed out at the beginning of the year, stating the goals of the class. By the end of the year, she hoped her students would be better able to challenge everything from novels to newscasts, that they would come to identify just who is telling a story and how that person's point of view affects the story being told. I'm going to go out on a limb here and say that this lesson has been learned. In fact, just recently, a student came up to McGlynn and told her something all teachers dream of hearing. The girl told the teacher that she was listening to the radio, singing along with her favorite song, and halfway through the sing-along she stopped and asked herself, "What am I singing? What do these words mean? What are they trying to tell me?" And then, this young citizen of the republic jokingly complained, "I can't even turn on the radio without thinking anymore."

—Sarah Vowell, "Democracy and Things Like That"

Refer to the beginning. One way to bring closure to a text is to bring up something discussed in the beginning; often the reference adds to or even changes the original meaning. For example, Amy Tan opens an essay on her Chinese mother's English by establishing herself as a writer and lover of language who uses many versions of English in her writing:

I am not a scholar of English or literature. I cannot give you much more than personal opinions on the English language and its variations in this country or others.

I am a writer. And by that definition, I am someone who has always loved language. I am fascinated by language in daily life. I spend a

---

☀ academic literacies   ● fields   ● research

■ rhetorical situations   ∴ processes   ● media/design

▲ genres   ◆ strategies

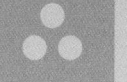

great deal of my time thinking about the power of language—the way it can evoke an emotion, a visual image, a complex idea, or a simple truth. Language is the tool of my trade. And I use them all—all the Englishes I grew up with.

At the end of her essay, Tan repeats this phrase, but now she describes language not in terms of its power to evoke emotions, images, and ideas but in its power to evoke "the essence" of her mother. When she began to write fiction, she says,

> [I] decided I should envision a reader for the stories I would write. And the reader I decided upon was my mother, because these were stories about mothers. So with this reader in mind—and in fact she did read my early drafts—I began to write stories using all the Englishes I grew up with: the English I spoke to my mother, which for lack of a better term might be described as "simple"; the English she used with me, which for lack of a better term might be described as "broken"; my translation of her Chinese, which could certainly be described as "watered down"; and what I imagined to be her translation of her Chinese if she could speak in perfect English, her internal language, and for that I sought to preserve the essence, but neither an English nor a Chinese structure. I wanted to capture what language ability tests can never reveal: her intent, her passion, her imagery, the rhythms of her speech and the nature of her thoughts.
>
> —Amy Tan, "Mother Tongue"

Note how Tan not only repeats "all the Englishes I grew up with" but also provides parallel lists of what those Englishes can do for her: "evoke an emotion, a visual image, a complex idea, or a simple truth" on the one hand, and on the other, capture her mother's "intent, her passion, her imagery, the rhythms of her speech and the nature of her thoughts."

Propose some action. A good example of this strategy is the following conclusion of a report on the consequences of binge drinking among college students:

> The scope of the problem makes immediate results of any interventions highly unlikely. Colleges need to be committed to large-scale and

long-term behavior-change strategies, including referral of alcohol abusers to appropriate treatment. Frequent binge drinkers on college campuses are similar to other alcohol abusers elsewhere in their tendency to deny that they have a problem. Indeed, their youth, the visibility of others who drink the same way, and the shelter of the college community may make them less likely to recognize the problem. In addition to addressing the health problems of alcohol abusers, a major effort should address the large group of students who are not binge drinkers on campus who are adversely affected by the alcohol-related behavior of binge drinkers.

—Henry Wechsler et al.,
"Health and Behavioral Consequences of Binge Drinking in
College: A National Survey of Students at 140 Campuses"

Paragraphs

Paragraphs are groups of sentences (and sometimes a single sentence) that relate to one main idea. Paragraphs in some of your textbooks may contain several hundred words, while paragraphs in news articles often contain just a single sentence or even, occasionally, a phrase. In English, we show paragraphs by indenting the first word; this is one way that the white space of a text conveys meaning. In fact, we sometimes use white space to create emphasis, especially with very brief paragraphs.

This one, for instance.

In academic writing, paragraphs usually contain a topic sentence that defines the main idea. Then the paragraph is developed by presenting examples or using defining, narrating, classifying, and other writing **STRATEGIES**. Following is advice on writing topic sentences and organizing paragraphs.

403 ◆

Topic Sentences

Just as a thesis statement announces the topic and position of an essay, a topic sentence states the subject and focus of a paragraph. Good paragraphs focus on a single point, which is summarized in a topic sentence. Usually, but not always, the topic sentence begins the paragraph:

☀ academic literacies	● fields	● research
■ rhetorical situations	⁘ processes	● media/design
▲ genres	◆ strategies	

Graduating from high school or college is an exciting, occasionally even traumatic event. Your identity changes as you move from being a high school teenager to a university student or a worker; your connection to home loosens as you attend school elsewhere, move to a place of your own, or simply exercise your right to stay out later. You suddenly find yourself doing different things, thinking different thoughts, fretting about different matters. As recent high school graduate T. J. Devoe puts it, "I wasn't really scared, but having this vast range of opportunity made me uneasy. I didn't know *what* was gonna happen." Jenny Petrow, in describing her first year out of college, observes, "It's a tough year. It was for all my friends."

> —Sydney Lewis, *Help Wanted: Tales from the First Job Front*

Sometimes the topic sentence may come at the end of the paragraph or even at the end of the preceding paragraph, depending on the way the paragraphs relate to one another. At other times, a topic sentence will summarize or restate a point made in the previous paragraph, helping readers understand what they've just read as they move on to the next point. See how linguist Deborah Tannen does this in the first paragraphs of an article on differences in men's and women's conversational styles:

I was addressing a small gathering in a suburban Virginia living room—a women's group that had invited men to join them. Throughout the evening, one man had been particularly talkative, frequently offering ideas and anecdotes, while his wife sat silently beside him on the couch. Toward the end of the evening, I commented that women frequently complain that their husbands don't talk to them. This man quickly concurred. He gestured toward his wife and said, "She's the talker in our family." The room burst into laughter; the man looked puzzled and hurt. "It's true," he explained. "When I come home from work I have nothing to say. If she didn't keep the conversation going, we'd spend the whole evening in silence."

This episode crystallizes the irony that although American men tend to talk more than women in public situations, they often talk less at home. And this pattern is wreaking havoc with marriage.

> —Deborah Tannen, "Sex, Lies, and Conversation:
> Why Is It So Hard for Men and Women to Talk to Each Other?"

Of course, simply stating the topic of a paragraph isn't enough; you need to support your main point with evidence to prove that your idea or claim is worth considering. Such **EVIDENCE** might include reasons, facts, statistics, quotations or citations from authorities, anecdotes, textual evidence, observations, and examples. For example, in the paragraph below, linguist Dennis Baron uses quotations and examples to make his point that **"THEY"** as a singular pronoun ("Everyone forgets *their* passwords," rather than "*his or her*" or merely "*his*") has been in use for a long time:

414–22 ◆

Glossary

> Faced with the onslaught of coined pronouns in the 1880s, singular *they* began to seem less and less objectionable, and more observers recognized the old but much-maligned singular *they* as the people's choice. For example, a writer in 1884 recommended *they* as an interim solution to be used while experts pondered *thon* and *lin.* According to another writer, C. K. Maddox, there was no need for invented pronouns because *they* has been singular "for ages" in the writing of standard authors like Dickens. Maddox criticized experts who withheld their approval: "Our grammarians and dictionary makers are very conservative and often positively stupid" for rejecting a term "so natural to the genius of our language that hardly one in a hundred has noticed it as an intrusion."
>
> —Dennis Baron, *What's Your Pronoun? Beyond He and She*

The topic sentence begins the paragraph, and then information from two different sources—a paraphrase of one writer's solution and a quotation from another—provides support for the assertion made in the topic sentence.

Paragraph Length and Number

The simplest answer to the questions "How long should my paragraphs be?" and "How many paragraphs should my essays contain?" is this: "as long as you need" and "as many as you need." In other words, the length and number of paragraphs in a text depend on your content, on what you need to say. Generally, your paragraphs should signal that groups of sentences are related in meaning and should help readers make those connections. For those reasons, too many brief paragraphs may make a

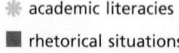

text hard to read because each sentence stands alone, unrelated to the surrounding sentences, while long paragraphs or an essay of a single, very long paragraph forces readers to try to group related ideas together themselves. In other words, the number of paragraphs and the number of sentences within a paragraph depend on the message you're trying to convey—not on some formula or arbitrary number. Here are some reasons to begin a new paragraph:

- to introduce a new subject or idea
- to emphasize an idea
- to give readers a needed pause
- to signal a new speaker when writing dialogue

Transitions

Transitions help readers move from thought to thought—from sentence to sentence, paragraph to paragraph. You're likely to use a number of transitions as you draft; when **EDITING**, you should make a point of checking transitions. Here are some common ones:

380–83

Causes and effects	Changes in direction or expectations	Comparisons
accordingly	although	also
as a result	but	in the same way
because	even though	like
consequently	however	likewise
hence	in contrast	similarly
so	instead	
then	nevertheless	**Examples**
therefore	nonetheless	for example
thus	on the contrary	for instance
	on the one hand . . . on the other hand	indeed
	still	in fact
	yet	such as

Sequences or similarities	Summary or conclusion	Time relations
again	as a result	after
also	as we have seen	as soon as
and	finally	at first
and then	in a word	at the same time
besides	in any event	before
finally	in brief	eventually
first, . . . second, . . . third, etc.	in conclusion	finally
furthermore	in other words	immediately
last	in short	later
moreover	in the end	meanwhile
next	in the final analysis	next
too	in the whole	simultaneously
	therefore	so far
	thus	soon
	to summarize	then
		thereafter

Transitions can also help readers move from paragraph to paragraph and, by summing up the previous paragraph's main point, show how the paragraphs are connected. A common way to summarize is to use phrases like "this _____" and "such _____." Here's an example from an anthropologist's study of American college students:

> When I asked students in interviews whether they felt they had a "community" at AnyU, most said yes. But what they meant by community were these personal networks of friends that some referred to as "my homeys." It was these small, ego-centered groups that were the backbone of most students' social experience in the university.
>
> On a daily basis these personal networks were easily recognizable within the dorm and on campus. "Where are you now?" says the cell phone caller walking back to the dorm from class. "I'm on my way home, so ask Jeffrey and Mark to come, and I'll meet you at my room at 8." Such conversations are everywhere.
>
> —Rebekah Nathan, *My Freshman Year*

"these personal networks" ties the second paragraph to the preceding one.

"Such conversations" sums up the example as a single concept.

* academic literacies   ● fields   ● research

■ rhetorical situations   ⁞ processes   ● media/design

▲ genres   ◆ strategies

Titles

A title serves various purposes, naming a text and providing clues to the content. It also helps readers decide whether they want to read further, so it's worth your while to come up with a title that attracts interest. Some titles include subtitles. You generally have considerable freedom in choosing a title, but you'll always want to consider the **RHETORICAL SITUATION** to be sure your title serves your purpose and appeals to the audience you want to reach.

■ 57

Some titles simply announce the subject of the text:

"Black Men and Public Space"

The Pencil

"Why Colleges Shower Their Students with A's"

"Does Texting Affect Writing?"

Some titles provoke readers or otherwise entice them to read:

"Kill 'Em! Crush 'Em! Eat 'Em Raw!"

"Thank God for the Atom Bomb"

"Just How Dishonest Are Most Students?"

Sometimes writers add a subtitle to explain or illuminate the title:

Aria: Memoir of a Bilingual Childhood

"It's in Our Genes: The Biological Basis of Human Mating Behavior"

"From Realism to Virtual Reality: Images of America's Wars"

Sometimes when you're starting to write, you'll think of a title that helps you generate ideas and write. More often, though, a title is one of the last things you'll write, when you know what you've written and can craft a suitable name for your text.

30 Drafting

At some point, you need to write out a draft. By the time you begin drafting, you've probably written quite a bit—in the form of notes, lists, outlines, and other kinds of informal writing. This chapter offers some hints on how to write a draft—and reminds you that as you draft, you may well need to get more information, rethink some aspect of your work, or follow new ideas that occur to you as you write.

Establishing a Schedule with Deadlines

Don't wait until the last minute to write. Computers crash, printers jam. Life intervenes in unpredictable ways. You increase your chances of success immensely by setting and meeting deadlines: research done by ____; rough draft done by ____; revisions done by ____; final draft edited, proofread, and submitted by ____. How much time you need varies with each writing task; but trying to compress everything into twenty-four or forty-eight hours before the deadline is asking for trouble.

Getting Comfortable

When are you at your best? When do you have your best ideas? For major writing projects, consider establishing a schedule that lets you write when you stand the best chance of doing good work. Schedule breaks for exercise and snacks. Find a good place to write—a place where there's a good surface on which to spread out your materials, good lighting, a comfortable chair, and the right tools (computer, pen, paper) for the job. Often, however, we must make do: you may have to do your drafting in a busy computer lab or classroom. The trick is to make yourself as comfortable as you can manage. Sort out what you *need* from what you *prefer*.

---

* academic literacies   ● fields   ● research
■ rhetorical situations   ⁙ processes   ● media/design
▲ genres   ◆ strategies

Starting to Write

All of the above advice notwithstanding, don't worry so much about the trappings of your writing situation that you don't get around to writing. Write. Start by **FREEWRITING**, start with a first sentence, start with awful writing that you know you'll discard later—but write. That's what gets you warmed up and going.

333–34

Write quickly in spurts.   Write quickly with the goal of generating a complete draft, or a complete section of a longer draft, in one sitting. If you need to stop in the middle, make some notes about where you were headed when you stopped so that you can easily pick up your train of thought when you begin again.

Break down your writing task into small segments.   Big projects can be intimidating. But you can always write one section or, if need be, one paragraph or even a single sentence—and then another and another. It's a little like dieting. If I think I need to lose twenty pounds, I get discouraged and head for the doughnuts; but if I decide that I'll lose one pound and I lose it, well, I'll lose another—*that* I can do.

Expect surprises.   Writing is a form of thinking; the words you write lead you down certain roads and away from others. You may end up somewhere you didn't anticipate. Sometimes that can be a good thing—but sometimes you can write yourself into a dead end or out onto a tangent. Just know that this is natural, part of every writer's experience, and it's OK to double back or follow a new path that opens up before you.

Expect to write more than one draft.   A first sentence, first page, or first draft represents your attempt to organize into words your thoughts, ideas, feelings, research findings, and more. It's likely that some of that first try will not achieve your goals. That's OK—having writing onscreen or on paper that you can change, add to, and cut means you're part of the way there. As you revise, you can fill in gaps and improve your writing and thinking.

Dealing with Writer's Block

You may sit down to write but find that you can't—nothing occurs to you; your mind is blank. Don't panic; here are some ways to get started writing again:

- Think of the assignment as a problem to be solved. Try to capture that problem in a single sentence: "How do I explain the context for my topic?" "What is the best way to organize my argument?" "What am I trying to do in the conclusion?"

- Start early and break the writing task into small segments drafted over several days. Waiting until the night before an assignment is due can create panic—and writer's block.

- Stop trying: take a walk, take a shower, do something else. Come back in a half hour, refreshed.

- Open a new document on your computer or get a fresh piece of paper and **FREEWRITE**, or try **LOOPING** or **LISTING**. What are you trying to say? Just let whatever comes come—you may write yourself out of your box.

  333–35

- If you usually write on your computer, turn it off, get out paper and pencil, and write by hand.

- Try a graphic approach: try **CLUSTERING**, or draw a chart of what you want to say; draw a picture or a comic strip; doodle.

  335–36

- Do some **RESEARCH** on your topic to see what others have said about it.

  489

- **TALK** to someone about what you are trying to do. If there's a writing center at your school, talk to a tutor: **GET RESPONSE**. If there's no one to talk to, then talk to yourself. It's the act of talking—using your mouth instead of your hands—that can free you up.

  336
372–74

If you need more help

333–44
367–71
372–79

See Chapter 28 on **GENERATING IDEAS AND TEXT** if you find you need more material. And once you have a draft, see Chapter 31 on **ASSESSING YOUR OWN WRITING** and Chapter 32 on **GETTING RESPONSE AND REVISING** for help evaluating your draft.

☀ academic literacies   ● fields   ● research

■ rhetorical situations   ⁝ processes   ● media/design

▲ genres   ◆ strategies

31 Assessing Your Own Writing

In school and out, our work is continually assessed by others. Teachers judge whether our writing meets their expectations; supervisors decide whether we merit raises or promotions; even friends and relatives size up in various ways the things we do. As writers, we need to assess our own work—to step back and see it with a critical eye. By developing standards of our own and being conscious of the standards others use, we can assess—and shape—our writing, making sure it does what we want it to do. This chapter will help you assess your own written work.

What we write for others must stand on its own because we usually aren't present when it is read—we rarely get to explain to readers why we did what we did and what it means. So we need to make our writing as clear as we can before we submit, post, display, or publish it. It's a good idea to assess your writing in two stages, first considering how well it meets the needs of your particular rhetorical situation, then studying the text itself to check its focus, argument, organization, and clarity. Sometimes some simple questions can get you started:

What works?
What still needs work?
Where do I need to say more (or less)?

Considering the Rhetorical Situation

PURPOSE   What is your purpose for writing? If you have multiple purposes, list them, and then note which ones are the most important. How well does your draft achieve your

■ 59–60

purpose(s)? If you're writing for an assignment, what are its requirements, and does your draft meet those requirements?

61–64 **AUDIENCE** To whom are you writing? What do those readers need and expect, as far as you can tell? Does your draft answer their needs? Do you define any terms and explain any concepts they won't know?

65–71 **GENRE** What is the genre, and what are the key features of that genre? Does your draft include each of those features? If not, is there a good reason?

72–74 **STANCE** Is your attitude toward your topic and your audience clear? Does your language project the personality and tone that you want?

75–77 **MEDIA/DESIGN** What medium (print? spoken? electronic?) or combination of media is your text intended for, and how well does your writing suit it? How well does the design of the text suit your purpose and audience? Does it meet any requirements of the genre or of the assignment, if you're writing for one?

Examining the Text Itself

Look carefully at your text to see how well it says what you want it to say. Start with its focus, and then examine its support, organization, and clarity, in that order. If your writing lacks focus, the revising you'll do to sharpen the focus is likely to change everything else; if it needs more support, the organization may well change.

Consider your focus. Your writing should have a clear point, and every part of the writing should support that point. Here are some questions that can help you see if your draft is adequately focused:

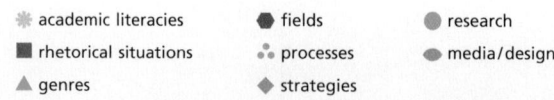

✳ academic literacies   ◗ fields   ⬤ research
■ rhetorical situations   ⠶ processes   ⬤ media/design
▲ genres   ◆ strategies

- What is your main point? Even if it's not stated directly, you should be able to summarize it for yourself in a single sentence.

- If your genre calls for a thesis, is it narrow or broad enough to suit the needs and expectations of your audience?

- How does the **BEGINNING** focus attention on your thesis or main point?

  346–54

- Does each paragraph support or develop that point? Do any paragraphs or sentences stray from your focus?

- Does the **ENDING** leave readers thinking about your main point? Is there another way of concluding the essay that would sharpen your focus?

  354–58

Consider the support you provide. Your writing needs to give readers enough information to understand your points and see the logic of your thinking. How much information is enough will vary according to your audience. If they already know a lot about your subject or are likely to agree with your point of view, you may need to give less detail. If, however, they are unfamiliar with your topic or are skeptical about your views, you will probably need to provide much more.

- If your text makes an argument, what **REASONS** and **EVIDENCE** do you give to support your thesis? Where might more information be helpful? If you're writing online, could you provide links to it?

  413–14
  414–22

- What key terms and concepts do you **DEFINE**? Are there any other terms your readers might need to have explained? Could you do so by providing links?

  445–55

- Where might you include more **DESCRIPTION** or other detail?

  456–63

- Do you make any **COMPARISONS**? Especially if your readers won't be familiar with your topic, it can help to compare it with something more familiar.

  437–44

- If you include **NARRATIVE**, how is it relevant to your point?

  474–82

- See Part 6 for other useful **STRATEGIES**.

  403

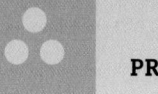
Consider the organization. As a writer, you need to lead readers through your text, carefully structuring your material so that they will be able to follow your ideas.

340–43
- Analyze the structure by **OUTLINING** it. An informal outline will do since you mainly need to see the parts, not the details.

196–200
MLA 574–605
- Is your text complete? Does your genre require an **ABSTRACT**, a **WORKS-CITED LIST**, or any other elements?

361–62
- What **TRANSITIONS** help readers move from idea to idea and paragraph to paragraph? Do you need to add more?

670–72
- If there are no **HEADINGS**, would adding them help orient readers?

Check for clarity. Nothing else matters if readers can't understand what you write. Following are some questions that can help you see whether your meaning is clear and your text is easy to read:

363
- Does your **TITLE** announce the subject of your text and give some sense of what you have to say? If not, would a more direct title give readers a better sense of your topic?

- If your writing requires a thesis, do you state it directly? If not, will readers easily understand your main point? Try stating your thesis outright, and see if it makes your ideas easier to follow.

346–54
354–58
- Does your **BEGINNING** tell readers what they need to understand your text, and does your **ENDING** help them make sense of what they've just read?

361–62
- How does each paragraph relate to the ones before and after? Are those relationships clear—or do you need to add **TRANSITIONS**?

- Do you vary your sentences? If all the sentences are roughly the same length or follow the same subject-verb-object pattern, your text probably lacks any clear emphasis and might even be difficult to read.

674–80
- Are **VISUALS** clearly labeled, positioned near the text they relate to, and referred to clearly in the text?

❋ academic literacies   ● fields   ● research

■ rhetorical situations   ⁞ processes   ● media/design

▲ genres   ◆ strategies

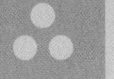

- If you introduce materials from other sources, have you clearly distinguished **QUOTED**, **PARAPHRASED**, or **SUMMARIZED** ideas from your own? 542–54

- Do you **DEFINE** all the words that your readers may not know? 445–55

- Does your punctuation make your writing more clear or less? Incorrect punctuation can make writing difficult to follow or, worse, change the meaning from what you intended. As a best-selling punctuation manual reminds us, there's a considerable difference between "eats, shoots, and leaves" and "eats shoots and leaves."

You may be asked to go into greater detail in an essay-length **REFLECTION** on your writing. If so, these questions may give you a good starting point. 391–401

Assessing a Body of Your Work

If you're required to submit a portfolio of your writing as part of a class, you'll likely need to write a reflective letter or essay that introduces the portfolio's contents, describes the processes you used to create them, and **ASSESSES THE WRITING IN YOUR PORTFOLIO**. See Chapter 35 for detailed advice. 385–90

32 Getting Response and Revising

If we want to learn to play a song on the guitar, we play it over and over again until we get it right. If we play basketball or baseball, we likely spend hours shooting foul shots or practicing a swing. Writing works the same way. Making meaning clear can be tricky, and you should plan on revising and, if need be, rewriting in order to get it right. When we speak with someone face-to-face or on the phone or text a friend, we can get immediate response and restate or adjust our message if we've been misunderstood. In most other situations when we write, that immediate response is missing, so we need to seek out responses from readers to help us revise. This chapter includes a list of guidelines for those readers to consider, along with various strategies for subsequent revising and rewriting.

Giving and Getting Peer Response

When you meet with other students in pairs or small groups to respond to one another's work, in class or online, you have the opportunity to get feedback on your work from several readers who can help you plan revisions. At the same time, you learn from reading others' work how they approached the writing task—you're not writing in a vacuum. Some students wonder why class time is being taken up by peer response, assuming that their instructor's opinion is the only one that counts, but seeing the work of others and learning how others see your work can help you improve the clarity and depth of your writing. The key to responding effectively is to be as specific in your response as possible and avoid being either too harsh or too complimentary. These guidelines can help:

※ academic literacies   ● fields   ● research
■ rhetorical situations   ⁘ processes   ● media/design
▲ genres   ◆ strategies

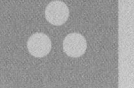
- Read your peer review partner's draft first from beginning to end as an interested reader, trying to understand the information and ideas. Don't look for problems. In fact, a good rule of thumb is this: read your partner's drafts in the same spirit that you want yours to be read.

- Before starting a second reading, ask your partner what questions they have about the draft or if you should focus on a particular aspect or part of the draft.

- As you read the draft again, take notes on a separate sheet of paper or save your comments to a different file. Your notes might include positive comments ("I like the way you . . . "), negative comments ("This sentence seems out of place"; "_____ doesn't seem like the best word to use"), and questions ("I'm not sure what you mean by _____"; "Would this paragraph work better on p. 2?").

- When you can, do more than identify issues. Offer suggestions or possible alternatives.

- When it's your draft's turn to be discussed, read or listen carefully to your partner's responses, take notes, and ask for clarification if necessary. Do not take issue with your partner's responses or argue over them; even if you're sure that what you wrote is perfectly clear, it's worth taking a second look if your partner has had trouble understanding it.

Getting Effective Response

Ask your readers to consider some of the specific elements in the list below, but don't restrict them to those elements. Caution: if a reader says nothing about any of these elements, don't be too quick to assume that you needn't think about them yourself.

- What did you think when you first saw the **TITLE**? Is it interesting? informative? appropriate? Will it attract other readers' attention?

363

- Does the **BEGINNING** grab your attention? If so, how does it do so? Does it give enough information about the topic? offer necessary background information? How else might the piece begin?

346–54

347–49
- Is there a clear **THESIS**, if the genre or assignment calls for one? What is it?

411–22
- Is there sufficient **SUPPORT** for the thesis? Is there anywhere you'd like to have more detail? Is the supporting material sufficiently 560–63 **DOCUMENTED**?

345–63
- Does the text have a clear pattern of **ORGANIZATION**? Does each part relate to the thesis or main idea? Does each part follow from the one preceding it? Was the text easy to follow? How might the organization be improved?

354–58
- Is the **ENDING** satisfying? What did it leave you thinking? How else might the piece end?

72–74
- Can you tell the writer's **STANCE** or attitude toward the subject and audience? What words convey that attitude? Is it consistent throughout?

61–64
59–60
- How well does the text meet the needs and expectations of its **AUDIENCE**? Where might readers need more information, guidance, or clarification? How well does it achieve its **PURPOSE**? Does every part of the text help achieve the purpose? Could anything be cut? Should anything be added? Does the text meet the requirements of 65–71 its **GENRE**? Should anything be added, deleted, or changed to meet those requirements?

445–55
464–68
- Do any terms need **DEFINING**? Would examples, additional detail, explanations, **DIALOGUE**, or some other strategies help you understand the draft?

676
673–83
- Are **CHARTS**, **GRAPHS**, or **TABLES** clear and readable? If there are no **VISUALS**, should there be?

361–62
- Are sentences complete and grammatical? Are **TRANSITIONS** helpful or needed? Is the punctuation correct?

- Can any words or phrases be sharpened? Are verbs mostly active? Is language that refers to others appropriate? Are all words spelled correctly?

* academic literacies   ● fields   ● research
■ rhetorical situations   ⁑ processes   ● media/design
▲ genres   ◆ strategies

Revising

Once you have studied your draft with a critical eye and, if possible, gotten responses from other readers, it's time to revise. Major changes may be necessary, and you may need to generate new material or do some rewriting. But assume that your draft is good raw material that you can revise to achieve your purposes. Revision should take place on several levels, from global (whole-text issues) to particular (the details). Work on your draft in that order, starting with the elements that are global in nature and gradually moving to smaller, more particular aspects. This allows you to use your time most efficiently and take care of bigger issues first. In fact, as you deal with the larger aspects of your writing, many of the smaller ones will be taken care of along the way.

Give yourself time to revise. When you have a due date, set **DEADLINES** for yourself that will give you time—preferably several days, but as much as your schedule permits—to work on the text before it has to be delivered. Also, get some distance. Often when you're immersed in a project, you can't see the big picture because you're so busy creating it. If you can, get away from your writing for a while and think about something else. When you return to it, you're more likely to see it with fresh eyes. If there's not enough time to put a draft away for several days or more, even letting it sit overnight or for a few hours can help.

364

As you revise, assume that nothing is sacred. Bring a critical eye to all parts of a draft, not only to those parts your reviewers point out. Content, organization, sentence patterns, individual words—all are subject to improvement. Be aware that a change in one part of the text may require changes in other parts.

At the same time, don't waste energy struggling with writing that simply doesn't work; you can always discard it. Look for the parts of your draft that *do* work—the parts that match your purpose and say what you want to say. Focus your efforts on those bright spots, expanding and developing them.

368–69
59–60
340–42

Revise to sharpen your FOCUS.   If your draft includes a thesis, make sure it matches your **PURPOSE** as you now understand it. Read each paragraph to ensure that it contributes to your main point; you may find it helpful to **OUTLINE** your draft to help you see all the parts. One way to do this is to highlight one sentence in each paragraph that expresses the paragraph's main idea. Then copy and paste the highlighted sentences into a new document. Does one state the thesis of the entire essay? Do the rest relate to the thesis? Are they in the best order? If not, you need to either modify the parts of the draft that don't advance your main idea or revise your thesis to reflect your draft's focus and to rearrange your points so they advance your discussion more effectively.

346–58

Read your **BEGINNING AND ENDING** carefully; make sure that the first paragraphs introduce your topic and provide any needed contextual information and that the final paragraphs provide a satisfying conclusion.

369

Revise to strengthen the argument.   If readers find some of your claims unconvincing, you need to provide more information or more **SUPPORT**. You may have to define terms you've assumed they will understand, offer additional examples, or provide more detail by describing, explaining pro-

403

cesses, adding dialogue, or using some other **STRATEGIES**. Make sure you show as well as tell—and don't forget that you might need to do so literally, with visuals like photos, graphs, or charts. You might try freewriting, clus-

333–44
489

tering, or other ways of **GENERATING IDEAS AND TEXT**. If you have to provide additional evidence, you might need to do additional **RESEARCH**.

345–63

Revise to improve the ORGANIZATION.   If you've outlined your draft, it helps to number each paragraph and make sure each one follows from the one before. If anything seems out of place, move it—or if necessary,

361–62
670–72

cut it completely. Check to see if you've included appropriate **TRANSITIONS** or **HEADINGS** to help readers move through the text, and add them as needed. Check to make sure your text meets readers' expectations of

65–71

the **GENRE** you're writing in.

370–71
363

Revise for CLARITY.   Be sure readers will be able to understand what you're saying. Look closely at your **TITLE** to be sure it gives a sense of

---

❋ academic literacies   ● fields   ● research
■ rhetorical situations   ⁝ processes   ● media/design
▲ genres   ◆ strategies

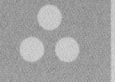
what the text is about and at your **THESIS**, if you state one directly: Will readers recognize your main point? If you don't state a thesis directly, consider whether you should. Provide any necessary background information, and **DEFINE** any key terms. Make sure you've integrated any **QUOTATIONS**, **PARAPHRASES**, or **SUMMARIES** into your text smoothly. Are all **PARAGRAPHS** focused around one main point? Do the sentences in each paragraph contribute to that point? Finally, consider whether there are any data that would be more clearly presented in a **CHART**, **TABLE**, or **GRAPH**.

347–49

445–55
542–54
368–71

676

One way to test whether your text is clear is to switch audiences: write what you're trying to express as if you were talking to an eight-year-old. Your final draft probably won't be written that way, but the act of explaining your ideas to a young audience or readers who know nothing about your topic can help you discover any points that may be unclear.

Revise VISUALS.   Make sure images are located as close as possible to the discussion to which they relate and that the information in each visual is explained in your text. Each image should be numbered and have a title or caption that identifies it and explains its significance. Each part of a **CHART**, **GRAPH**, or **TABLE** should be clearly labeled to show what it represents. If you didn't create the image yourself, make sure to cite its source, and if you're posting your work online, obtain permission from the copyright owner.

673–83

676

Read and reread—and reread.   Take some advice from writing theorist Donald Murray:

> Nonwriters confront a writing problem and look away from the text to rules and principles and textbooks and handbooks and models. Writers look at the text, knowing that the text itself will reveal what needs to be done and what should not yet be done or may never be done. The writer reads and rereads and rereads, standing far back and reading quickly from a distance, moving in close and reading slowly line by line, reading again and again, knowing that the answers to all writing problems lie within the evolving text.
>
> —Donald Murray, *A Writer Teaches Writing*

Rewriting

Some writers find it useful to try rewriting a draft in various ways or from various perspectives just to explore possibilities. Try it! If you find that your original plan works best for your purpose, fine. But you may find that another way will work better. Especially if you're not completely satisfied with your draft, consider the following ways of rewriting. Experiment with your rhetorical situation:

- Rewrite your draft from different points of view—through the eyes of different people, perhaps, or through the eyes of an animal or even from the perspective of an object. See how the text changes (in the information it presents, its perspective, its voice).

61–64 ■
- Rewrite for a different **AUDIENCE**. How might an email detailing a recent car accident be written to a friend, an insurance agent, a parent?

73–74 ■
- Rewrite in a different **TONE**. If the first draft was written in formal academic prose, rewrite it more informally.

65–71 ■
75–77
- Rewrite the draft in a different **GENRE** or **MEDIUM**. Rewrite an essay as a letter, story, poem, speech, comic strip, *PowerPoint* presentation. Which genre and medium work best to reach your intended audience and achieve your purpose?

Ways of rewriting a narrative

464–68 ◆
- Rewrite one scene completely in **DIALOGUE**.

- Start at the end of the story and work back to the beginning, or start in the middle and fill in the beginning as you work toward the end.

Ways of rewriting a textual analysis

437–44 ◆
- **COMPARE** the text you're analyzing with another text (which may be in a completely different genre—film, TV, song lyrics, computer game, poetry, fiction, whatever).

- Write a parody of the text you're analyzing. Be as silly and as funny as you can while maintaining the structure of the original text. Alternatively, write a parody of your analysis, using evidence from the text to support an outrageous analysis.

✳ academic literacies	● fields	● research
■ rhetorical situations	⁚ processes	● media/design
▲ genres	◆ strategies	

Ways of rewriting a report

- Rewrite for a different **AUDIENCE**. For example, explain a concept to your grandparents; describe the subject of a profile to a visitor from another planet, a different time, or a different culture.

  ■ 61–64

- Be silly. Rewrite the draft as if for *The Onion* or a Dave Chappelle skit, or rewrite it as if it were written by Bart Simpson.

Ways of rewriting an argument

- Rewrite taking another **POSITION**. Argue as forcefully for that position as you did for your actual one, acknowledging and refuting your original position. Alternatively, write a rebuttal to your first draft from the perspective of someone with different beliefs.

  ▲ 164–95

- Rewrite your draft as a **STORY** —make it real in the lives of specific individuals. (For example, if you were writing about abortion rights, you could write a story about a young pregnant woman trying to decide what she believes and what to do.) Or rewrite the argument as a fable or parable.

  ◆ 474–82

- Rewrite the draft as a letter responding to a hostile reader, trying at least to make them understand what you have to say.

- Rewrite the draft as an angry letter to someone or as a table-thumping dinner-with-the-relatives discussion. Write from the most extreme position possible.

- Write an **ANALYSIS** of the topic of your argument in which you identify, as carefully and as neutrally as possible, the various positions people hold on the issue.

  ▲ 104–33

Once you've rewritten a draft in any of these ways, see whether there's anything you can use. Read each draft, considering how it might help you achieve your purpose, reach your audience, and convey your stance. Revise your actual draft to incorporate anything you think will make the text more effective, whether it's other genres or a different perspective.

33 Editing and Proofreading

No matter what you write, you need to communicate in ways that are appropriate to your rhetorical situation. When you text your friends, for example, you might use minimal punctuation, incomplete sentences, and mainly lowercase letters—and your friends won't bat an eye. If you write a job application letter and résumé, on the other hand, your readers may expect clear, error-free writing in the belief that careful attention to editing—to grammar, punctuation, language, and formality—may indicate similar care in the work you do on the job. Most college instructors expect similar attention to editing and proofreading in formal assignments like research papers. This chapter offers strategies for editing and proofreading your writing carefully.

Editing

Editing is the stage where you work on the details of your paragraphs, sentences, words, and punctuation to make your writing as clear, precise, and appropriate as possible. Your goal is not to achieve "perfection" (whatever that may be) so much as to make your writing as effective as possible for your particular purpose and audience. Consult a good writing handbook for detailed advice, but use the following guidelines to help you check the paragraphs, sentences, and words in your draft.

Editing paragraphs

359–60

- Does each paragraph focus on one point? Does it have a **TOPIC SENTENCE** that announces that point, and if so, where is it located? If it's not the first sentence, should it be? If there's no clear topic sentence, should there be one?

- Does every sentence relate to the main point of the paragraph? If any sentences do not, should they be deleted, moved, or revised?

* academic literacies   ● fields   ● research
■ rhetorical situations   ⁘ processes   ● media/design
▲ genres   ◆ strategies

- Is there enough detail to develop the paragraph's main point? How is the point developed—with narrative? definition? some other **STRATEGY**?   ◆ 403

- Where have you placed the most important information—at the beginning? at the end? in the middle? The most emphatic spot is at the end, so in general that's where to put information you want readers to remember. The second most emphatic spot is at the beginning.

- Are any paragraphs especially long or short? Consider breaking long paragraphs if there's a logical place to do so—maybe an extended example should be in its own paragraph, for instance. If you have paragraphs of only a sentence or two, see if you can add to them or combine them with another paragraph, unless you're using a brief paragraph to provide emphasis.

- Check the way your paragraphs fit together. Does each one follow smoothly from the one before? Do you need to add any **TRANSITIONS**?   361–62

- Do the **BEGINNING** paragraphs catch readers' attention? In what other ways might you begin your text?   346–54

- Do the final paragraphs provide a satisfactory **ENDING**? How else might you conclude your text?   354–58

Editing sentences

- Is each sentence complete? Does it have someone or something (the subject) performing some sort of action or expressing a state of being (the verb)? Does each sentence begin with a capital letter and end with a period, question mark, or exclamation point?

- Check your use of the **PASSIVE VOICE**. Although there are some rhetorical situations in which the passive voice ("The prince was killed by a rival") is more appropriate than the active voice ("A rival killed the prince") because you want to emphasize an action rather than who performed it, you'll do well to edit it out unless you have a good reason for using it.   Glossary

- Check for **PARALLELISM**. Items in a list or series should be parallel in form—all nouns ("lions," "tigers," "bears"), all verbs ("hopped,"   Glossary

"skipped," "jumped"), all clauses ("he came, he saw, he conquered"), and so on.

- Do many of your sentences begin with "it" or "there"? Too often these words make writing wordy and vague or even conceal needed information. Why write "There are reasons we voted for him" when you can say "We had reasons to vote for him"?

361–62

- Are your sentences varied? If they all start with the subject or are the same length, your writing might be dull and maybe even hard to read. Try varying your sentence openings by adding **TRANSITIONS**, introductory phrases, or clauses. Vary sentence lengths by adding detail to some or by combining some sentences.

- Make sure you've used **COMMAS** correctly. Is there a comma after each introductory element? ("After the lead singer quit, the group nearly disbanded. However, they then produced a string of hits.") Do commas set off nonrestrictive elements—parts that aren't needed to understand the sentence? ("The books I read in middle school, like the Harry Potter series, became longer and more challenging.") Are compound sentences connected with a comma? ("I'll eat broccoli steamed, but I prefer it roasted.")

Editing words

- Are you sure of the meaning of every word? Use a dictionary; be sure to look up words whose meanings you're not sure about. And remember your audience—do you use any terms they'll need to have defined?

- Is any of your language too general or vague? Why write that you competed in a race, for example, if you could say you ran the 4×200 relay?

73–74

- What about the **TONE**? If your stance is serious (or humorous or critical or something else), make sure that your words all convey that attitude.

Glossary

- Do any pronouns have vague or unclear **ANTECEDENTS**? If you use "he" or "they" or "it" or "these," will readers know whom or what the words refer to?

Glossary

- Have you used any **CLICHÉS**—expressions that are used so frequently that they're no longer fresh? "Live and let live," avoiding something

---

- academic literacies
- rhetorical situations
- genres
- fields
- processes
- strategies
- research
- media/design

"like the plague," and similar expressions are so predictable that your writing will almost always be better off without them.

- Be careful with language that refers to others. Make sure that your words do not stereotype any individual or group. Mention age, gender, race, religion, sexual orientation, and so on only if they are relevant to your subject. When referring to an ethnic group, make every effort to use the terms members of the group prefer. Also, whenever possible, refer to people by the **PRONOUNS** they use for themselves. If you don't know their pronouns and you need to refer to them with pronouns, it's best to ask. If you can't ask or don't think you should, your best alternative is to use the **SINGULAR "THEY."**

  Glossary

  Glossary

- Edit out language that might be considered sexist. Have you used words like "manpower" or "policemen" to refer to people who may be female? If so, substitute less gendered words such as "personnel" or "police officers." Do your words reflect any gender stereotypes—for example, that all engineers are male, or all nurses female? If you mention someone's gender, is it even necessary? If not, eliminate the unneeded words.

- How many of your verbs are forms of "be" and "do"? If you rely too much on these words, try replacing them with more specific verbs. Why write "She did a proposal for" when you could say "She proposed"?

- Do you ever confuse "its" and "it's"? Use "it's" when you mean "it is" or "it has." Use "its" when you mean "belonging to it."

Proofreading

Proofreading is the final stage of the writing process, when you check for misspelled words, mixed-up fonts, missing pages, and so on. It's the time to pay extra attention to detail. Most readers will excuse an occasional error, but by and large, academic readers and many employers are an intolerant bunch: too many errors will lead them to declare your writing—and maybe your thinking—flawed. There goes your credibility. So careful proofreading helps ensure that your message is taken as seriously as you want it to be.

Up to this point, you've been told *not* to read individual words on the page and instead to read for meaning. Proofreading demands the opposite: you must slow down your reading so that you can see every word, every punctuation mark.

- Use your computer's grammar checker and spelling checker, but only as a first step, and know that they're not very reliable. Computer programs don't read writing; instead, they rely on formulas and banks of words, so what they flag (or don't flag) as mistakes may or may not be accurate. If you were to write, "My brother was diagnosed with a leaning disorder," "leaning" wouldn't be flagged as misspelled because it's a word (and might even be a disorder), even though it's the wrong word in that sentence.

- To keep your eyes from jumping ahead, place a ruler or piece of paper under each line as you read. Use your finger or a pencil as a pointer.

- Some writers find it helpful to read the text one sentence at a time, beginning with the last sentence and working backward.

- Temporarily change the font or size of your text as you proofread; doing so can help you notice problems you may have overlooked.

- Read the text out loud to yourself—or better, to others, who may *hear* problems you can't see. Alternatively, have someone else read your text aloud while you follow along on the screen or page.

- Ask someone else to read your text. The more important the writing is, the more important this step is.

- If you find a mistake after you've printed out your text and are unable to print out a corrected version, make the change as neatly as possible in pencil or pen.

34 Compiling a Portfolio

Artists maintain portfolios of their work to show gallery owners, collectors, and other potential buyers. Money managers work with investment portfolios of stocks, bonds, mutual funds, and other products. And often as part of a writing class, student writers compile portfolios of their work. As with a portfolio of paintings or drawings, a portfolio of writing includes a writer's best work and, sometimes, preliminary and revised drafts of that work, along with a statement by the writer articulating why they consider it good. The *why* is as important as the work, for it provides you with an occasion for assessing your overall strengths and weaknesses as a writer. This chapter offers guidelines to help you compile both a *writing portfolio* and a *literacy portfolio*, a project that writing students are sometimes asked to complete as part of a literacy narrative.

Considering the Rhetorical Situation

As with the writing you put in a portfolio, the portfolio itself is generally intended for a particular audience but could serve a number of different purposes. It's a good idea, then, to consider these and the other elements of your rhetorical situation when you begin to compile a portfolio.

PURPOSE Why are you creating this portfolio? To show your learning? To create a record of your writing? As the basis for a grade in a course? To organize your research? To explore your literacy? For something else?

59–60

AUDIENCE Who will read your portfolio? What will your readers expect it to contain? How can you help them understand the context or occasion for each piece of writing you include?

61–64

65–71 **GENRE** What genres of writing should the portfolio contain? Do you want to demonstrate your ability to write one particular type of writing or in a variety of genres? Will your introduction to or assessment of the portfolio be in the form of a letter or an essay?

72–74 **STANCE** How do you want to portray yourself in this portfolio? What items should you include to create this impression? What stance do you want to take in your written assessment of its contents? Thoughtful? Enthusiastic? Something else?

75–77 **MEDIA/DESIGN** Will your portfolio be in print? Or will it be electronic? Will it include multiple media? Whichever medium you use, how can you help readers navigate its contents? What design elements will be most appropriate to your purpose and medium?

A WRITING PORTFOLIO

What to Include

A portfolio developed for a writing course typically contains examples of your best work in that course, including any notes, outlines, preliminary drafts, and so on, along with your own assessment of your performance in the course. You might include any of the following items:

- freewriting, outlines, and other work you did to generate ideas
- drafts, rough and revised
- in-class writing assignments
- source material—copies of articles and online sources, observation notes, interview transcripts, and other evidence of your research
- tests and quizzes
- responses to your drafts

※ academic literacies   ● fields   ● research
■ rhetorical situations   ⁘ processes   ● media/design
▲ genres   ◆ strategies

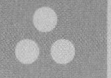

- conference notes, error logs, lecture notes, and other course materials
- electronic material, including visuals, blogs, and multimedia texts
- reflections on your work

What you include will vary depending on what your instructor asks for. You may be asked to include three or four of your best papers or everything you've written. You may also be asked to show work in several different genres. In any case, you'll usually need to choose, and to do that you'll need to have criteria for making your choices. Don't base your decision solely on grades (unless grades are one criterion); your portfolio should reflect *your* assessment of your work, not your instructor's. What do you think is your best work? your most interesting work? your most ambitious work? Whatever criteria you use, *you* are the judge.

Organizing a Portfolio

If you set up a way to organize your writing at the start of the course, you'll be able to keep track of it throughout the course, making your job at term's end much easier. Remember that your portfolio presents you as a writer, presumably at your best. It should be neat, well organized, and easy to navigate. Your instructor may provide explicit guidelines for organizing your portfolio. If not, here are some guidelines:

Paper portfolios.   Choose something in which to gather your work. You might use a two-pocket folder, a three-ring binder, or a file folder, or you may need a box, basket, or some other container to accommodate bulky or odd-shaped items.

Label everything. Label each piece at the top of the first page, specifying the assignment, the draft, and the date: "Proposal, Draft 1, 9/12/21"; "Text Analysis, Final Draft, 10/10/21"; "Portfolio Self-Assessment, Final Draft, 11/11/21"; and so on. Write this information neatly on the page, or put it on a Post-it note. For each assignment, arrange your materials chronologically, with your earliest material (freewriting, for example) on the bottom, and each successive item (source materials, say, then your outline, then your first draft, and so on) on top of the last, ending with your final draft on top. That way, readers can see how your writing changed from draft to draft.

Electronic portfolios. You might also create an electronic portfolio, or e-portfolio. E-portfolios typically consist of a network of **LINKED** documents that might include not only your writing and reflections on that writing, but also sources, writing, and art you did for other courses or for your own enjoyment; audio and video clips; and other resources. Tools that can help you create an e-portfolio include:

681–82

- *Courseware.* Your school may use a learning platform, such as *Blackboard*, *Canvas*, or *Moodle*, that allows you to create an e-portfolio of your work.

- *Online tools.* Several websites, including *Weebly* and *Wix*, offer free tools to help you create a preformatted e-portfolio. For example, *Google-Sites* provides templates you can use to build an e-portfolio, uploading documents, images, and videos from your computer.

- *Blogging tools.* You can create an e-portfolio using a blogging platform, like *Tumblr* or *WordPress*, which allows you to upload files and create a network of linked pages. Readers can then comment on your e-portfolio, just as they might on your blog entries.

It's also possible to create an electronic portfolio using word processing, spreadsheet, or presentation software. The programs available for your use and the requirements for publishing your portfolio vary from school to school and instructor to instructor; ask your instructor or your school's help desk for assistance (and see Chapter 58 on **WRITING AND LEARNING ONLINE** for general guidance).

684–93

Assessing Your Writing Portfolio

An important part of your portfolio is your written self-assessment of your work. This is an opportunity to assess your work with a critical eye and to think about what you're most proud of, what you most enjoyed doing, what you want to improve. It's your chance to think about and say what you've learned during the class. Some instructors may ask you to write out your **ASSESSMENT** in essay form, as an additional sample of your writing; others will want you to put it in letter form, which usually allows for a more relaxed and personal tone. See Chapter 31 for detailed

367–71

☀ academic literacies   ● fields   ● research
■ rhetorical situations   ⁙ processes   ● media/design
▲ genres   ◆ strategies

advice on assessing your writing and Chapter 35 for advice on composing a formal reflection.

A LITERACY PORTFOLIO

As a writing student, you may be asked to think back to the time when you first learned to read and write or to remember significant books or other texts you've read, and perhaps to put together a portfolio that chronicles your development as a reader and writer. You may also be asked to put together a literacy portfolio to accompany a **LITERACY NARRATIVE**.

▲ 81–103

What you include in such a portfolio will vary depending on what you've kept over the years and what your family has kept. You may have all of your favorite books, stories you dictated to a preschool teacher, notebooks in which you practiced writing the alphabet. Or you may have almost nothing. What you have or don't have is unimportant in the end: what's important is that you gather what you can and arrange it in a way that shows how you think about your development and growth as a literate person. What have your experiences been with reading and writing? What's your earliest memory of learning to write? If you love to read, what led you to love it? Who was most responsible for shaping your writing ability? Those are some of the questions you'll ask if you write a literacy narrative. You might also compile a literacy portfolio as a good way to generate ideas and text for that assignment.

What to Include in a Literacy Portfolio

* school papers
* drawings and doodles from preschool
* favorite books
* photographs you've taken
* drawings
* poems
* letters

- journals and diaries
- lists
- reading records or logs
- electronic texts you've created
- marriage vows
- speeches you've given
- awards you've received
- workplace writing

Organizing a Literacy Portfolio

You may wish to organize your material chronologically, but there are other methods of organization to consider as well. For example, you might group items according to where they were written (at home, at school, at work), by genre (stories, poems, essays, letters, notes), or even by purpose (pleasure, school, work, church, and so on). Arrange your portfolio in the way that best conveys who you are as a literate person. Label each item you include, perhaps with a Post-it note, to identify what it is, when it was written or read, and why you've included it in your portfolio. Or you might create an e-portfolio, scanning print items to include in it along with electronic items.

Reflecting on Your Literacy Portfolio

- Why did you choose each item?
- Is anything missing? Are there any other important materials that should be here?
- Why is the portfolio organized as it is?
- What does the portfolio show about your development as a reader and writer?
- What patterns do you see? Are there any common themes you've read or written about? any techniques you rely on? any notable changes over time?
- What are the most significant items, and why?

✳ academic literacies	⬢ fields	● research
■ rhetorical situations	⁝ processes	⬤ media/design
▲ genres	◆ strategies	

35 Reflecting on Your Writing

Reflecting on experiences—looking back on them to understand and learn from them—is an important part of our work, school, and personal lives. Employees reflect on their job performance as they develop goals for the future. Students reflect on their coursework and extracurricular activities as they prepare applications for scholarships or internships. Lots of people reflect on their daily lives in diaries or journals. And writers reflect on their writing—they look back on their work and consider what they've learned, what they've accomplished, and how they can use their new knowledge and experience in future writing situations. But reflection isn't just an activity that writers engage in *after* they've finished writing. It's also a powerful tool for thinking and learning throughout the writing process. This chapter gives advice for making the most of that tool.

REFLECTING WHILE WRITING

When teachers ask you to reflect during your writing process, they're asking you to pause, step back, and pay attention to what you're doing and thinking as you write. This activity—sometimes called "metacognition"—can help you solve problems in your writing, adjust or adapt your strategies, and make decisions about how to move forward.

Following are questions you can ask yourself in order to prompt reflection at several points in your writing process. You might record your reflections in a process log, or talk through them with someone else, or just mull them over in your mind.

- *Reflecting as you approach a new writing assignment or task.* Have you encountered a similar assignment or task in the past? If so, what processes or strategies worked for you then, and how might you adapt them for your new writing situation now? What goals can you set for yourself? What problems might you encounter, and what resources can help you solve them?

- *Reflecting as you generate ideas and compose a draft.* What's the most intriguing idea you've generated? the most puzzling or surprising? What connections do you see among ideas? What do you like about your draft so far?

- *Reflecting as you participate in peer review.* Which comments from your peers are most helpful? Which comments do you think you'll ignore, and why? What did you learn from reviewing your classmates' drafts, and how can you apply that learning to your own draft?

- *Reflecting as you revise and edit*: What feedback did your teacher provide on your previous work, and can you apply that feedback to your current writing situation? What revision strategies have worked for you in the past, and can you use or adapt them now? What resources can you draw on for help with editing and proofreading?

REFLECTING AFTER WRITING

At times, you may be asked to write a formal reflection on your work—perhaps after completing a writing project, after compiling a portfolio, or at the end of a course. Formal reflection assignments go by many names: portfolio cover letters, reflective self-assessments, companion pieces, statements of goals and choices, and others. Regardless of the name, reflection-after-writing assignments typically ask you to share insights about your work and the process you used to create it—and to support your insights with evidence. Considering your rhetorical situation will help you make decisions about how to approach your particular reflection task.

❋ academic literacies	● fields	● research
■ rhetorical situations	⁖ processes	● media/design
▲ genres	◆ strategies	

Considering the Rhetorical Situation

PURPOSE
Why are you writing your reflection? To examine your growth and development as a writer over time? To assess your achievement of course learning goals? To consider how you might use your new knowledge in future writing contexts? Are you reflecting on just one piece of writing or on a body of work?

59–60

AUDIENCE
Who will read your reflection, and what's your relationship to your reader? Is your course instructor your primary reader? Does your audience include other instructors or external evaluators whom you may not know? Will you need to introduce or contextualize the writing you're reflecting on? What examples or evidence of your learning will your readers most want to see?

61–64

GENRE
Has a genre been assigned to you? If so, what are the key features of the genre, and how can you use them in your reflection? Can you choose a genre? If so, what will you choose—a letter, a literacy narrative, a textual analysis, something else? What expectations will your audience have for your chosen genre?

65–71

STANCE
How will you portray yourself in your reflection—as introspective? curious? self-critical? What stance will you take toward the work you're reflecting on—objective? evaluative? analytical? What tone will you use—relaxed and conversational? serious and formal?

72–74

MEDIA/DESIGN
What medium will best suit your purpose and audience? A print document? An electronic text like a blog post, a video diary, or a podcast? Will visuals be helpful? Does your audience expect, or your instructor require, a certain format or type of documentation?

75–77

Generating Insights for Your Reflection

If you've responded to the "Taking Stock of Your Work" prompts at the end of Chapters 10 through 22, you can return to those responses as a starting point to generate insights for an essay-length reflection. The following activities might also help you develop ideas and text:

Take an inventory of your writing. List everything you wrote for your course or for a particular project—freewrites, peer reviews, notes, drafts, revisions, and so on. Then gather as much of the material as you can, and read it carefully. Pretend you're looking at your writing as an outsider, and try **ANNOTATING** or **CODING** it: notice the strengths and weaknesses of individual pieces of writing; pay attention to the language, organization, and other choices you made; note whether your writing changed over time. In other words, *study* your own writing—be curious about it! Take notes so you can harness insights from your inventory.

16–19

Create a map or flowchart of your current writing process. Think about a writing project you recently completed, and make a visual representation of the process you engaged in to create the work. Begin by listing the writing tools you used, the spaces where you worked, the types of tasks you completed—anything you remember about your writing process. Then represent those items with stick figures, symbols, or something else. Use arrows to show how you moved from one point in your writing process to another, and perhaps mark the times when you struggled and the times when you felt productive. When you finish your map, ask yourself what it reveals: Which parts of your writing process were most and least enjoyable? What would you change about your process if you could? What insights can you gain about what you do when you write—and what you might do better in the future? The sample map on the next page represents one writer's process for a research assignment.*

---

* Adapted from Tim Lockridge and Derek Van Ittersum, *Writing Workflows: Beyond Word Processing.*

* academic literacies　　● fields　　● research
■ rhetorical situations　　⁘ processes　　● media/design
▲ genres　　◆ strategies

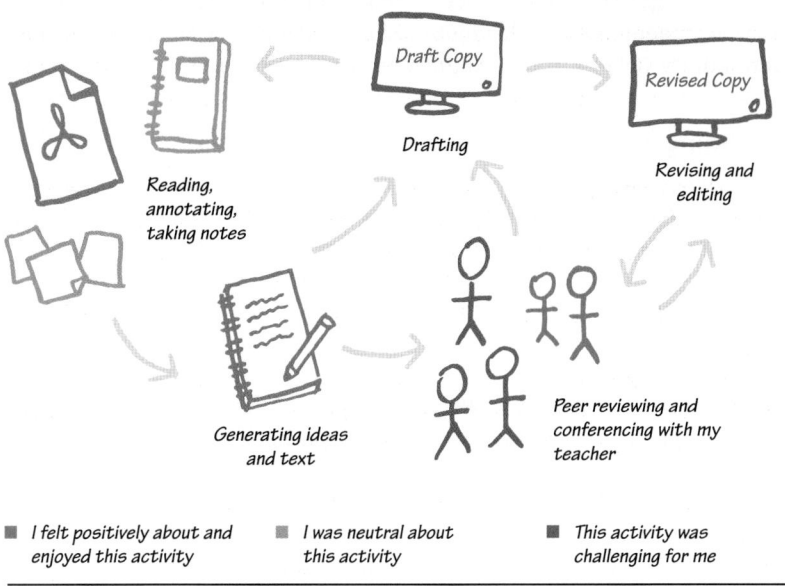

Draft Copy

Revised Copy

Drafting

Revising and editing

Reading, annotating, taking notes

Generating ideas and text

Peer reviewing and conferencing with my teacher

■ I felt positively about and enjoyed this activity

■ I was neutral about this activity

■ This activity was challenging for me

Think of words, phrases, or images that capture your experiences as a writer. Start by listing key activities or concepts that guided your work: maybe **REVISING** was especially important to your writing process, or **AUDIENCE AWARENESS** was something you worked hard to develop. Maybe a particular **ACADEMIC HABIT OF MIND** —like openness or curiosity—was a hallmark of your work. Think, too, about other words or phrases that represent your lived experiences as a writer—like the name of the café where you did your best writing, or the tool (blue notebook, *Google Docs*) that you relied on to get your work done, or the mantra that got you through your course. Freewrite about what those key words or phrases mean to you in the context of your writing and learning.

375–77

61–64

46–56

You might even use images to help brainstorm ideas for your reflection, as student Jennifer Martinez did for a writing class at Columbus State Community College in the example on the next page.

Providing Support for Your Reflection

Once you've generated insights about your writing and learning, the next step is to find evidence to support and illustrate those insights. Fortunately, you'll have a wealth of material to draw on—your own writing, your interactions with classmates, and more. In fact, you can mine your entire course experience for evidence to support your reflective insights. Some types of evidence you might use in your reflection include:

Examples and anecdotes. Recall memories from your course experience—a time when you exhibited a particular habit of mind, for example, or when you solved a problem in your writing, or when you helped a peer re-see their work. Describe the specific instance, or narrate the story behind it.

Textual evidence. Consider all the texts that make up your learning experience: drafts and revisions, peer-review comments, discussion board posts, written feedback from your teacher, and so on. All those texts provide fodder to support the insights you share in your reflection. For example, if you want to illustrate your revising skills, you might quote a passage from an early draft of a writing project and compare it to a revised passage from a later draft. If you want to support a claim about your rhetorical knowledge, you might quote from your writing to highlight particular language choices you made to meet readers' needs. To illustrate your openness to others' ideas, you could quote a comment a peer reviewer made, and then explain how you used that comment to rethink something in your own writing. To show your evolving self-awareness as a writer, you could quote from the "Taking Stock" responses you wrote throughout the course.

Jennifer's Reflection Brainstorming

This whole semester I struggled with **SELF-DOUBT** and overwhelming anxiety at times. Slowly I was able to overcome some of these issues by trusting the writing process and just knowing that I will get through it. I have learned to give myself more time by starting earlier so I am able to work through those moments and to also take breaks when I start to feel the anxiety creep in.

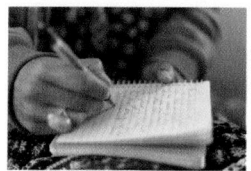

Before this class I would have dreaded the thought of writing a **JOURNAL** entry. Once writing became a normal routine, I started looking forward to our journal entries. I even decided I'm going to buy my mom a daily writing prompt book. You can't be wrong with a journal; it's just your thoughts and your feelings.

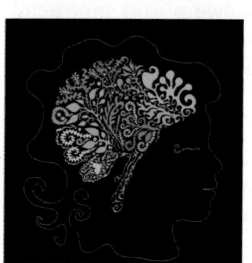

One of the most helpful lessons I received from this class was a more positive attitude toward the things I struggle with. After we watched the video on **GROWTH MIND-SET** at the beginning of this semester, something clicked and I started looking at all of my struggles a little differently; I understood that the struggles were helping to make me smarter. Once I accepted that my brain was like a muscle and that I should treat it that way, it was easier for me to work through rough times in my writing.

Statistics. Some students enjoy collecting and interpreting data about their writing processes and products. You might use the "Compare" feature in *Word* to tabulate the number of changes you made from one draft to the next. You might do a search through your process work to see how many times the word "genre" or "audience" or some other key term appeared. You might look for themes in your peer reviewers' comments and tabulate, for example, the percentage of comments related to focus or organization or editing. You might then use the numerical data as support for a particular insight you want to share or a claim you want to make in your reflection.

Visuals. Sometimes visual evidence can help support your reflective insights and claims. You might use photographs of notes, drafts, sketches, and storyboards to support an insight about your writing process. You might create a graph or chart to illustrate numerical data you've generated about your own writing, if you're using such data. You might develop a **WORD CLOUD** — a visual representation of words in a text — to highlight key themes in your writing. Remember that visuals almost never speak for themselves, so you'll need to explain them in your reflection. For example, if you include a graph, you might say something like, "The graph in figure 1 depicts how many times my peer reviewers and instructor mentioned key elements — like thesis, organization, or editing — in response to each of my essays."

506–7

Ways of Organizing a Reflection

The organization of your reflection will depend on your genre, and you may be able to consult the organizational diagrams in the genre chapters of this book for help. Some common genres used for reflections include letters, **LITERACY NARRATIVES**, **TEXTUAL ANALYSES**, **EXPLORATIONS**, and **REPORTS**.

81–103
104–39
269–79
140–63

✳ academic literacies   ● fields   ● research
■ rhetorical situations   ⁘ processes   ● media/design
▲ genres   ◆ strategies

A Sample Reflection

In the letter below, Nathaniel Cooney reflects on the strengths and weaknesses of the writing he produced in his first-year composition class at Wright State University.

2 June 2021

Dear Reader,

It is my hope that in reading this letter you will gain an understanding of the projects contained in this portfolio. I enclose three works that I have submitted for an introductory writing class at Wright State University, English 102, Writing in Academic Discourse: an informative report, an argument paper, and a remix project based largely on the content of the argument paper. I selected the topics of these works for two reasons: First, they address issues that I believe to be relevant in terms of both the intended audience (peers and instructors of the course) and the times when they were published. Second, they speak to issues that are important to me personally. Below I present general descriptions of the works, along with my review of their strengths and weaknesses.

My purpose in writing the informative report "Higher Standards in Education Are Taking Their Toll on Students" was to present a subject in a factual manner and to support it with well-documented research. My intent was not to argue a point. However, because I chose a narrowly focused topic and chose information to support a thesis, the report tends to favor one side of the issue over the other. Because as a student I have a personal stake in the changing standards in the formal education system, I chose to research recent changes in higher education and their effects on students. Specifically, I examine students' struggles to reach a standard that seems to be moving further and further beyond their grasp.

I believe that this paper could be improved in two areas. The first is a bias that I think exists because I am a student presenting

information from the point of view of a student. It is my hope, however, that my inclusion of unbiased sources lessens this problem somewhat and, furthermore, that it presents the reader with a fair and accurate collection of facts and examples that supports the thesis. My second area of concern is the overall balance in the paper between outside sources supporting my own thoughts and outside sources supporting opposing points of view. Rereading the paper, I notice many places where I may have worked too hard to include sources that support my ideas.

The second paper, "Protecting Animals That Serve," is an argument intended not only to take a clear position on an issue but also to argue for that position and convince the reader that it is a valid one. That issue is the need for legislation guaranteeing that certain rights of service animals be protected. I am blind and use a guide dog. Thus, this issue is especially important to me. During the few months that I have had him, my guide dog has already encountered a number of situations where intentional or negligent treatment by others has put him in danger. At the time I was writing the paper, a bill was being written in the Ohio House of Representatives that, if passed, would protect service animals and establish consequences for those who violated the law. The purpose of the paper, therefore, was to present the reader with information about service animals, establish the need for the legislation in Ohio and nationwide, and argue for passage of such legislation.

I think that the best parts of my argument are the introduction and the conclusion. In particular, I think that the conclusion does a good job of not only bringing together the various points but also conveying the significance of the issue for me and for others. In contrast, I think that the area most in need of further attention is the body of the paper. While I think the content is strong, I believe the overall organization could be improved. The connections between ideas are unclear in places, particularly in the section that acknowledges opposing viewpoints. This may be due in part to the fact that I had difficulty understanding the reasoning behind the opposing argument.

The argument paper served as a starting point for the remix project, for which the assignment was to revise one paper written for this class in a different genre. My remix project consists of a poster and a brochure. As it was for the argument paper, my primary goal was to convince my audience of the importance of a particular issue and viewpoint—specifically, to convince my audience to support House Bill 369, the bill introduced in the Ohio legislature that would create laws to protect the rights of service animals in the state.

Perhaps both the greatest strength and the greatest weakness of the remix project is my use of graphics. Because of my blindness, I was limited in my use of some graphics. Nevertheless, the pictures were carefully selected to capture the attention of readers and, in part, to appeal to their emotions as they viewed and reflected on the material.

I put a great deal of time, effort, and personal reflection into each project. While I am hesitant to say that they are finished and while I am dissatisfied with some of the finer points, I am satisfied with the overall outcome of this collection of works. Viewing it as a collection, I am also reminded that writing is an evolving process and that even if these works never become exactly what I envisioned them to be, they stand as reflections of my thoughts at a particular time in my life. In that respect, they need not be anything but what they already are, because what they are is a product of who I was when I wrote them. I hope that you find the papers interesting and informative and that as you read them, you, too, may realize their significance.

Respectfully,

Nathaniel J. Cooney

Nathaniel J. Cooney

Enclosures (3)

Cooney describes each of the works he includes in his portfolio and considers their strengths and weaknesses, citing examples from his texts to support his assessment and reflection.

Part 6
Strategies

Whenever we write, we draw on many different strategies to articulate what we have to say. We may **DEFINE** key terms, **DESCRIBE** people or places, and **EXPLAIN** how something is done. We may **COMPARE** one thing to another. Sometimes we may choose a pertinent story to **NARRATE**, and we may even want to include some **DIALOGUE**. The chapters that follow offer advice on how to use these and **OTHER BASIC STRATEGIES** for developing and organizing the texts you write.

Strategies

36 Analyzing Causes and Effects

Analyzing causes helps us think about why something happened, whereas thinking about effects helps us consider what might happen. When we hear a noise in the night, we want to know what caused it. Children poke sticks into holes to see what will happen. Researchers try to understand the causes of diseases. Writers often have occasion to consider causes or effects as part of a larger topic or sometimes as a main focus: in a **PROPOSAL**, we might consider the effects of reducing tuition or the causes of recent tuition increases; in a **MEMOIR**, we might explore why the person we had a date with failed to show up.

258–68
236–44

Usually we can only speculate about *probable* causes or *likely* effects. In writing about causes and effects, then, we are generally **ARGUING** for those we consider plausible, not proven. This chapter will help you analyze causes and effects in writing—and to do so in a way that suits your rhetorical situation.

410–30

Determining Plausible Causes and Effects

What causes ozone depletion? sleeplessness? obesity? And what are their effects? Those are of course large, complex topics, but whenever you have reason to ask why something happened or what could happen, there will likely be several possible causes and just as many predictable effects. There may be obvious causes, though often they will be less important than others that are harder to recognize. (Eating too much may be an obvious cause of being overweight, but *why* people eat too much has several less obvious causes: portion size, advertising, lifestyle, and psychological disorders are only a few possibilities.) Similarly, short-term effects are often less important than long-term ones. (A stomachache may be an effect of

eating too much candy, but the chemical imbalance that can result from consuming too much sugar is a much more serious effect.)

334–35
335–36
340–43
489
59–60
61–64

LISTING, CLUSTERING, and OUTLINING are useful processes for analyzing causes. And at times you might need to do some RESEARCH to identify possible causes or effects and to find evidence to support your analysis. When you've identified potential causes and effects, you need to analyze them. Which causes and effects are primary? Which seem to be secondary? Which are most relevant to your PURPOSE and are likely to convince your AUDIENCE? You will probably have to choose from several possible causes and effects for your analysis because you won't want or need to include all of them.

Arguing for Causes or Effects

410–30

Once you've identified several possible causes or predictable effects, you need to ARGUE that some are more plausible than others. You must provide convincing support for your argument because you usually cannot *prove* that x causes y or that y will be caused by z; you can only show, with good reasons and appropriate evidence, that x is *likely* to cause y or that y will *likely* follow from z. See, for example, how an essay on the psychological basis for risk taking speculates about two potential causes for the popularity of extreme sports:

> Studies now indicate that the inclination to take high risks may be hardwired into the brain, intimately linked to arousal and pleasure mechanisms, and may offer such a thrill that it functions like an addiction. The tendency probably affects one in five people, mostly young males, and declines with age. It may ensure our survival, even spur our evolution as individuals and as a species. Risk taking probably bestowed a crucial evolutionary advantage, inciting the fighting and foraging of the hunter-gatherer. . . .
>
> As psychologist Salvadore Maddi, PhD, of the University of California at Davis warns, "High-risk takers may have a hard time deriving meaning and purpose from everyday life." Indeed, this peculiar form of dissatisfaction could help explain the explosion of high-risk sports in America and other postindustrial Western nations. In unstable cultures, such as those at war or suffering poverty, people rarely seek

✳ academic literacies   ● fields   ● research
■ rhetorical situations   ⋰ processes   ● media/design
▲ genres   ◆ strategies

out additional thrills. But in a rich and safety-obsessed country like America, land of guardrails, seat belts, and personal-injury lawsuits, everyday life may have become too safe, predictable, and boring for those programmed for risk taking.

—Paul Roberts, "Risk"

Roberts suggests that genetics is one likely cause of extreme sports and that an American obsession with safety is perhaps a cause of their growing popularity. Notice, however, that he presents these as likely or possible, not certain, by choosing his words carefully: "studies now *indicate*"; "the inclination to take high risks *may* be hardwired"; "[r]isk taking *probably* bestowed a crucial evolutionary advantage"; "this . . . dissatisfaction *could help* explain." Like Roberts, you'll almost always need to qualify what you say about causes and effects—to say that something *could explain* (rather than saying it "explains") or that it *suggests* (rather than "shows"). Causes and effects can seldom be proved definitively, so it's important to acknowledge that your argument is not the last word on the subject.

Ways of Organizing an Analysis of Causes and Effects

Your analysis of causes and effects may be part of a proposal or some other genre of writing, or you may write a text whose central purpose is to analyze causes or speculate about effects. While there are many ways to organize an analysis of causes and effects, three common ways are to state a cause and then discuss its effects, to state an effect and then discuss its causes, and to identify a chain of causes and effects.

Identify a cause and then discuss its effects.   If you were writing about climate change, you might first show that many scientists fear it will have several effects, including more violent storms, the extinction of various kinds of plants, and elevated sea levels.

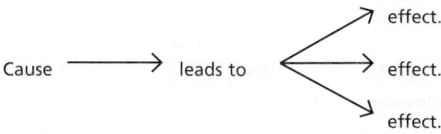

Identify an effect and then trace its causes. If you were writing about school violence, for example, you might argue that it is a result of sloppy dress, informal teacher-student relationships, low academic standards, and disregard for rules.

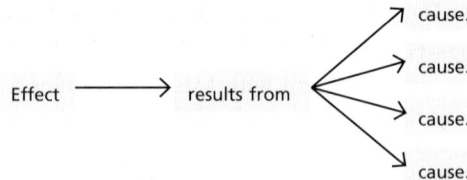

Identify a chain of causes and effects. You may sometimes discuss a chain of causes and effects. If you were writing about the right to privacy, for example, you might consider the case of Megan's law. A convicted child molester raped and murdered a girl named Megan; the crime caused New Jersey legislators to pass the so-called Megan's law (an effect), which requires that convicted sex offenders be publicly identified. As more states enacted versions of Megan's law, concern developed for the rights of those who are identified—the effect became a cause of further effects.

First cause → leads to → first effect, → which leads to → next effect. --> Continue the chain as need be.

Considering the Rhetorical Situation

As a writer or speaker, it's important to think about the message you want to articulate, the audience you want to reach, and the larger context you are writing in.

59–60 ■

PURPOSE   Your main purpose may be to analyze the causes and effects of something. But sometimes you'll have another goal that calls for such analysis—a business report, for example, might need to explain what caused a decline in sales.

✳ academic literacies   ● fields   ● research

■ rhetorical situations   ∴ processes   ● media/design

▲ genres   ◆ strategies

AUDIENCE   Who is your intended audience, and how will analyz-   61–64
                 ing causes help you reach them? Do you need to tell
                 them why some event happened or what effects
                 resulted?

GENRE   Does your genre require you to analyze causes? Pro-   65–71
                 posals, for example, often need to consider the effects
                 of a proposed solution.

STANCE   What is your stance, and could analyzing causes or   72–74
                 effects show that stance? Could it help demonstrate
                 your seriousness or show that your conclusions are
                 reasonable?

MEDIA/DESIGN You can rely on words to analyze causes, but some-   75–77
                 times a drawing will help readers *see* how causes lead
                 to effects.

If you need more help

See also the **PROCESSES** chapters for help with generating ideas, drafting, and so on if you   325
need to write an entire text whose purpose is to analyze causes or speculate about effects.

37 Arguing

Basketball fans argue about who's better, LeBron James or Steph Curry. Political candidates argue that they have the most experience or best judgment. A toilet paper ad argues that "you deserve a little luxury in your life, and so does your bottom." As you likely realize, we are surrounded by arguments, and much of the work you do as a college student requires you to read and write arguments. When you write a **LITERARY ANALYSIS**, for instance, you argue for a particular interpretation. In a **PROPOSAL**, you argue for a particular solution to a problem. Even a **PROFILE** argues that a subject should be seen in a certain way. This chapter offers advice on some of the key elements of making an argument, from developing an arguable thesis and identifying good reasons and evidence that supports those reasons to building common ground and dealing with viewpoints other than your own.

223–35 ▲
258–68
245–57

Reasons for Arguing

We argue for many reasons, and they often overlap: to convince others that our position on a subject is reasonable, to influence the way they think about a subject, to persuade them to change their point of view or take some sort of action. In fact, many composition scholars and teachers believe that all writing makes an argument.

As a student, you'll be called on to make arguments continually: when you participate in class discussions, when you take an essay exam, when you post a comment to an online discussion or a blog. In all these instances, you're adding your opinions to some larger conversation, arguing for what you believe — and why.

* academic literacies
■ rhetorical situations
▲ genres

● fields
∴ processes
◆ strategies

● research
● media/design

Arguing Logically: Claims, Reasons, and Evidence

The basic building blocks of argument are claims, reasons, and evidence that supports those reasons. Using these building blocks, we can construct a strong logical argument, also known as *logos*.

Claims. Good arguments are based on arguable claims—statements that reasonable people may disagree about. Certain kinds of statements cannot be argued:

- *Verifiable statements of fact.* Most of the time, there's no point in arguing about facts like "the earth is round" or "George H. W. Bush was America's forty-first president." Such statements contain no controversy, no potential opposition—and so no interest for an audience. However, you might argue about the basis of a fact. For example, until recently it was a fact that our solar system had nine planets; but when further discoveries led to a change in the definition of "planet," Pluto no longer qualified.

- *Issues of faith or belief.* By definition, matters of faith cannot be proven or refuted. If you believe in reincarnation or don't believe there is an afterlife, there's no way I can convince you otherwise. However, in a philosophy or religion course you may be asked to argue, for example, whether or not the universe must have a cause.

- *Matters of simple opinion or personal taste.* If you think cargo pants are ugly, no amount of arguing will convince you to think otherwise. If you've downloaded every Taylor Swift album and think she's the greatest singer ever, you won't convince your Nirvana-loving parents to like her, too. If matters of taste are based on identifiable criteria, though, they may be argued in an **EVALUATION**, where "Tom Cruise is a terrible actor" is more than just your opinion—it's an assertion you can support with evidence.

▲ 214–22

You may begin with an opinion: "I think wearing a helmet makes riding a bike more dangerous, not less." As it stands, that statement can't be

considered a claim—it needs to be made more reasonable and informed. To do that, you might reframe it as a question—"Do bike riders who wear helmets get injured more often than those who don't?"—that may be answered as you do research and start to write. Your opinion or question should lead you to an arguable claim, however, one that could be challenged by another thoughtful person. In this case, for example, your research might lead you to a focused, qualified claim: "Contrary to common sense, wearing a helmet while riding a bicycle increases the chances of injury, at least to adult riders."

Qualifying a claim. According to an old saying, there are two sides to every story. Much of the time, though, arguments don't sort themselves neatly into two sides, pro and con. No matter what your topic, your argument will rarely be a simple matter of being for or against; in most cases, you'll want to qualify your claim—that it is true in certain circumstances, with certain conditions, with these limitations, and so on. Qualifying your claim shows that you're reasonable and also makes your topic more manageable by limiting it. The following questions can help you qualify your claim:

- *Can it be true in some circumstances or at some times but not others?* For example: "Freedom of speech should generally be unrestricted, but individuals can sue for slander or libel."

- *Can it be true only with certain conditions?* For instance: "Cell phones and computer monitors should be recycled, but only by licensed, domestic recyclers."

- *Can it be true for some groups or individuals but not others?* For example: "The Keto and Atkins diets can lead to weight loss for many people, but people with kidney or liver conditions should avoid them."

SOME WORDS FOR QUALIFYING A CLAIM

sometimes	nearly	it seems/seemingly
rarely	usually	some
in some cases	more or less	perhaps
often	for the most part	possibly
routinely	in many cases	in most cases

Drafting a thesis statement. Once your claim is focused and appropriately qualified, it can form the core of your essay's **THESIS STATEMENT**, which announces your position and forecasts the path your argument will follow. For example, here is the opening paragraph of an essay by the executive director of the National Congress of American Indians arguing that the remains of Native Americans should be treated with the same respect given to others. The author outlines the context of her argument and then presents her thesis (here, in italics):

347–49

> What if museums, universities and government agencies could put your dead relatives on display or keep them in boxes to be cut up and otherwise studied? What if you believed that the spirits of the dead could not rest until their human remains were placed in a sacred area? The ordinary American would say there ought to be a law—and there is, for ordinary Americans. *The problem for American Indians is that there are too many laws of the kind that make us the archeological property of the United States and too few of the kind that protect us from such insults.*
>
> —Suzan Shown Harjo, "Last Rites for Indian Dead:
> Treating Remains Like Artifacts Is Intolerable"

Reasons. Your claim must be supported by reasons that your audience will accept. A reason can usually be linked to a claim with the word "because":

CLAIM	+	BECAUSE	+	REASON
College students should strive to graduate		because		they will earn far more over their lifetimes than those who do not.

Keep in mind that you likely have a further reason, a rule or principle that underlies the reason you link directly to your claim. In this argument, the underlying reason is that graduating from college leads to a boost in lifetime income because employers value college graduates. If your audience doesn't accept that principle, you may have to back it up with further reasons or evidence.

To come up with good reasons, start by stating your position and then answering the question "Why?"

CLAIM: College students should strive to graduate. *Why?*

REASON: (Because) They will earn far more over their lifetimes than those who do not. *Why?*

UNDERLYING REASON: The economy values college graduates and pays them more.

As you can see, this exercise can continue indefinitely as the underlying reasons grow more and more general and abstract. You can do the same with other positions:

CLAIM: Smoking should be banned. *Why?*

REASON: (Because) It is harmful to smokers and also to nonsmokers.

UNDERLYING REASON: People should be protected from harmful substances.

Evidence. Evidence to support your reasons can come from various sources. In fact, you may need to use several kinds of evidence to persuade your audience that your claim is true. Some of the most common types of evidence include facts, statistics, examples, authorities, anecdotes, scenarios, case studies, textual evidence, and visuals.

Facts are ideas that are proven to be true. Facts can include observations or scholarly research (your own or someone else's), but they need to be accepted as true. If your audience accepts the facts you present, they can be powerful means of persuasion. For example, an essay on the problems faced by people who lose their senses of smell and taste offers these facts to demonstrate the seriousness of this loss:

Smell is intimately tied to both taste and appetite, and anosmia often robs people of the pleasure of eating. But the sudden absence also may have a profound impact on mood and quality of life.

Studies have linked anosmia to social isolation and anhedonia, an inability to feel pleasure, as well as a strange sense of detachment and

isolation. Memories and emotions are intricately tied to smell, and the olfactory system plays an important though largely unrecognized role in emotional well-being, said Dr. Sandeep Robert Datta, an associate professor of neurobiology at Harvard Medical School.

—Roni Caryn Rabin,
"Some Covid Survivors Haunted by Loss of Smell and Taste"

Rabin quotes a specialist, which gives her facts credibility; citing the studies she refers to would provide even greater credibility.

Statistics are numerical data, usually produced through research, surveys, or polls. Statistics should be relevant to your argument, as current as possible, accurate, and obtained from a reliable source. An argument advocating that Americans should eat less meat presents these data to support the writer's contention that we eat far too much of it:

> Americans are downing close to 200 pounds of meat, poultry, and fish per capita per year (dairy and eggs are separate, and hardly insignificant), an increase of 50 pounds per person from 50 years ago. We each consume something like 110 grams of protein a day, about twice the federal government's recommended allowance; of that, about 75 grams come from animal protein. (The recommended level is itself considered by many dietary experts to be higher than it needs to be.) It's likely that most of us would do just fine on around 30 grams of protein a day, virtually all of it from plant sources.
>
> —Mark Bittman, "Rethinking the Meat-Guzzler"

Bittman's statistics demonstrate the extent to which Americans have increased their meat consumption over the last half century, the proportion of our diets that comes from meat, and, by comparison, how much protein our bodies require—and summarize the heart of his argument in stark numeric terms.

Examples are specific instances that illustrate general statements. In a book on life after dark in Europe, a historian offers several examples to demonstrate his point that three hundred years ago, night—without artificial lighting—was treacherous:

Even sure-footed natives on a dark night could misjudge the lay of the land, stumbling into a ditch or off a precipice. In Aberdeenshire, a fifteen-year-old girl died in 1739 after straying from her customary path through a churchyard and tumbling into a newly dug grave. The Yorkshireman Arthur Jessop, returning from a neighbor's home on a cold December evening, fell into a stone pit after losing his bearings.

—A. Roger Ekirch, *At Day's Close: Night in Times Past*

Ekirch illustrates his point and makes it come alive for readers by citing two specific individuals' fates.

Authorities are experts on your subject. To be useful, authorities must be reputable, trustworthy, and qualified to address the subject. You should carefully **EVALUATE** any authorities you consult to be sure they have the credentials necessary for readers to take them seriously. When citing experts, you should clearly identify them and the origins of their authority in a **SIGNAL PHRASE**, as does the author of an argument that deforested land can be reclaimed:

524–34

551–54

Reed Funk, professor of plant biology at Rutgers University, believes that the vast areas of deforested land can be used to grow millions of genetically improved trees for food, mostly nuts, and for fuel. Funk sees nuts used to supplement meat as a source of high-quality protein in developing-country diets.

—Lester R. Brown, *Plan B 2.0: Rescuing a Planet under Stress and a Civilization in Trouble*

Brown cites Funk, an expert on plant biology, to support his argument that humans need to rethink the global economy in order to create a sustainable world. Without the information on Funk's credentials, though, readers would have no reason to take his proposal seriously.

474–82

Anecdotes are brief **NARRATIVES** that your audience will find believable and that contribute directly to your argument. Anecdotes may come from your personal experience or the experiences of others. In an essay arguing that it's understandable when athletes give in to the temptation to

✳ academic literacies   ● fields   ● research
■ rhetorical situations   ⁘ processes   ● media/design
▲ genres   ◆ strategies

use performance-enhancing drugs, sports blogger William Moller uses an anecdote to show that the need to perform can outweigh the potential negative consequences of using drugs:

> I spent my high school years at a boarding school hidden among the apple orchards of Massachusetts. Known for a spartan philosophy regarding the adolescent need for sleep, the school worked us to the bone, regularly slamming us with six hours of homework. I pulled a lot more all-nighters (of the scholastic sort) in my years there than I ever did in college. When we weren't in class, the library, study hall, or formal sit-down meals, we were likely found on a sports field. We also had school on Saturday, beginning at 8 a.m. just like every other non-Sunday morning.
>
> Adding kindling to the fire, the students were not your laid-back types; everyone wanted that spot at the top of the class, and social life was rife with competition. The type A's that fill the investment banking, legal, and political worlds—those are the kids I spent my high school years with.
>
> And so it was that midway through my sophomore year, I found myself on my third all-nighter in a row, attempting to memorize historically significant pieces of art out of E. H. Gombrich's *The Story of Art*. I had finished a calculus exam the day before, and the day before that had been devoted to world history. And on that one cold night in February, I had had enough. I had hit that point where you've had so little sleep over such a long time that you start seeing spots, as if you'd been staring at a bright light for too long. The grade I would compete for the next day suddenly slipped in importance, and I began daydreaming about how easy the real world would be compared to the hell I was going through.
>
> But there was hope. A friend who I was taking occasional study breaks with read the story in the bags beneath my eyes, in the slump of my shoulders, the nervous drumming of my fingers on the chair as we sipped flat, warm Coke in the common room. My personal *deus ex machina*,* he handed me a small white pill.
>
> I was very innocent. I matured way after most of my peers, and was probably best known for being the kid who took all the soprano solos

---

* *Deus ex machina:* In ancient Greek and Roman drama, a god introduced into the plot to resolve complications.

away from the girls in the choir as a first-year student. I don't think I had ever been buzzed, much less drunk. I'd certainly never smoked a cigarette. And knowing full well that what I was doing could be nothing better than against the rules (and less importantly, illegal) I did what I felt I needed to do, to accomplish what was demanded of me. And it worked. I woke up and regained focus like nothing I'd ever experienced. Unfortunately, it also came with serious side effects: I was a hypersensitized, stuffed-up, sweaty, wide-eyed mess, but I studied until the birds started chirping. And I aced my test.

Later I found out the pill was Ritalin, and it was classified as a class 3 drug.* I did it again, too—only a handful of times, as the side effects were so awful. But every time it was still illegal, still against the rules. And as emphasized above, I was much more worried about the scholastic consequences if I were discovered abusing a prescription drug than the fact that I was breaking the law. Though I was using it in a far different manner than the baseball players who would later get caught with it in their systems, it was still very clearly a "performance-enhancing drug."

Just like every other person on this planet, I was giving in to the incentive scheme that was presented to me. The negative of doing poorly on the test was far greater than the negative of getting caught, discounted by the anesthetic of low probability.

—William Moller, "We, the Public,
Place the Best Athletes on Pedestals"

Moller uses this anecdote to demonstrate the truth of his argument—that given the choice between "breaking the rules and breaking my grades" or "getting an edge" in professional sports, just about everyone will choose to break the rules.

Scenarios are hypothetical situations. Like anecdotes, "what if?" scenarios can help you describe the possible effects of particular actions or offer new ways of looking at a particular state of affairs. For example, a mathematician presents this lighthearted scenario about Santa Claus in a tongue-in-cheek argument that Christmas is (almost) pure magic:

---

* *Class 3 drug:* Drug that is illegal to possess without a prescription.

Let's assume that Santa only visits those who are children in the eyes of the law, that is, those under the age of 18. There are roughly 2 billion such individuals in the world. However, Santa started his annual activities long before diversity and equal opportunity became issues, and as a result he doesn't handle Muslim, Hindu, Jewish and Buddhist children. That reduces his workload significantly to a mere 15% of the total, namely 378 million. However, the crucial figure is not the number of children but the number of homes Santa has to visit. According to the most recent census data, the average size of a family in the world is 3.5 children per household. Thus, Santa has to visit 108,000,000 individual homes. (Of course, as everyone knows, Santa only visits good children, but we can surely assume that, on an average, at least one child of the 3.5 in each home meets that criterion.)

— Keith Devlin, "The Mathematics of Christmas"

Devlin uses this scenario, as part of his mathematical analysis of Santa's yearly task, to help demonstrate that Christmas is indeed magical—because if you do the math, it's clear that Santa's task is physically impossible.

Case studies and observations feature detailed reporting about a subject. Case studies are in-depth, systematic examinations of an occasion, a person, or a group. For example, in arguing that class differences exist in the United States, sociologist Gregory Mantsios presents studies of three "typical" Americans to show "enormous class differences" in their lifestyles.

Observations offer detailed descriptions of a subject. Here's an observation of the emergence of a desert stream that flows only at night:

At about 5:30 water came out of the ground. It did not spew up, but slowly escaped into the surrounding sand and small rocks. The wet circle grew until water became visible. Then it bubbled out like a small fountain and the creek began.

— Craig Childs, *The Secret Knowledge of Water*

Childs presents this and other observations in a book that argues (among other things) that even in harsh, arid deserts, water exists, and knowing where to find it can mean the difference between life and death.

542–54

Textual evidence includes QUOTATIONS, PARAPHRASES, and SUMMARIES. Usually, the relevance of textual evidence must be stated directly, as excerpts from a text may carry several potential meanings. For example, here is an excerpt from a student essay analyzing the function of the raft in *Huckleberry Finn* as "a platform on which the resolution of conflicts is made possible":

> [T]he scenes where Jim and Huck are in consensus on the raft contain the moments in which they are most relaxed. For instance, in chapter 12 of the novel, Huck, after escaping capture from Jackson's Island, calls the rafting life "solemn" and articulates their experience as living "pretty high" (Twain 75–76). Likewise, subsequent to escaping the unresolved feud between the Grangerfords and Shepherdsons in chapter 18, Huck is unquestionably at ease on the raft: "I was powerful glad to get away from the feuds. . . . We said there warn't no home like a raft, after all. Other places do seem so cramped up and smothery, but a raft don't. You feel mighty free and easy and comfortable on a raft" (134).
>
> —Dave Nichols, "'Less All Be Friends': Rafts as Negotiating Platforms in Twain's *Huckleberry Finn*"

Huck's own words support Nichols's claim that he can relax on a raft. Nichols strengthens his claim by quoting evidence from two separate pages, suggesting that Huck's opinion of rafts pervades the novel.

674–80

Visuals can be a useful way of presenting evidence. Remember, though, that charts, graphs, photos, drawings, and other VISUAL TEXTS seldom speak for themselves and thus must be explained in your text. For example, at the top of the facing page, is a photograph of a poster carried by demonstrators at the 2008 Beijing Summer Olympics, protesting China's treatment of Tibetans. If you were to use this photo in an essay, you would need to explain that the poster combines the image of a protester standing before a tank during the 1989 Tiananmen Square uprising with the Olympic logo, making clear to your readers that the protesters are likening China's

---

treatment of Tibetans to its brutal actions in the past. Similarly, you could use the image below of an American flag made from license plates in an argument about America's dependence on the auto industry.

57 ■

Choosing appropriate evidence. The kinds of evidence you provide to support your argument depend on your **RHETORICAL SITUATION**. If your purpose is, for example, to convince readers to accept the need for a proposed solution, you'd be likely to include facts, statistics, and anecdotes. If you're writing for an academic audience, you'd be less likely to rely on anecdotes, preferring authorities, textual evidence, statistics, and case studies instead. And even within academic communities, different disciplines and genres may focus primarily on different kinds of evidence. If you're not sure what counts as appropriate evidence, ask your instructor for guidance.

Arguing with a Hostile Audience

Academic arguments are often presented to an audience that is presumed to be open-minded and fair, and the goal of such arguments is to demonstrate that your position is plausible and reasonable. Sometimes, though, your goal is to change people's minds, to try to get them to see some issue differently. That can be more of a challenge, because your audience likely has good reasons for thinking as they do, or their views reflect their basic values, and they may well feel defensive, angry, or threatened when you challenge them. These situations call for different argumentative strategies, such as Rogerian argument.

Rogerian argument.   This method of presenting an argument is based on the work of psychologist Carl Rogers. The method assumes that common ground—areas of shared values or beliefs—exists between people who disagree. If they can find that common ground, they're more likely to come to some agreement or compromise position.

Since the goal of a Rogerian argument is to find compromise, it is organized differently than a traditional argument. First you must show that you understand your opponents' position, and then you offer your own position. Here's how a typical Rogerian argument is organized; be aware that each section may require several paragraphs:

☀ academic literacies ● fields ● research
■ rhetorical situations ⁝ processes ● media/design
▲ genres ◆ strategies

Introduction. Here you introduce the issue, acknowledging the various sides of the controversy and being as fair as you can.

Describe the opposing view. In this section, you try to capture the opposing view as accurately and as neutrally as you can. There may be circumstances when that view might be valid, and you should include them, too. Your goal here is to show that you understand the opposing view, and why your readers hold that view, convincingly enough that they will agree that it's accurate. By doing so, you establish your credibility and honest desire to understand those whose views differ from yours.

State your THESIS and support it. Now it's time to outline your position and defend it with reasons and evidence. As with a traditional argument, your goal here is to show that you have thought carefully about your position and researched it thoroughly. As with all arguing, your TONE should show your openness to ideas and avoid sounding as if you and you alone know the truth about the issue or that you are intellectually or morally superior to your opponent.

347–49

73–74

Conclude. In your concluding section, you bring together the two perspectives you've just outlined to show how, while your opponent's views have merit, your position deals with the issue or solves the problem in a better way—or how a compromise position, somewhere in the middle, allows both sides to benefit.

You may find value in using Rogerian techniques in traditional arguments, including seeking common ground, describing issues and positions in neutral terms, and addressing those with opposing views respectfully and with goodwill.

Convincing Readers You're Trustworthy

For your argument to be convincing, you have to establish your own credibility with readers (also known as *ethos*)—to demonstrate your knowledge

about the topic, to show that you and your readers share some common ground, and to show yourself to be evenhanded in the way you present your argument.

Building common ground.   One important element of gaining readers' trust involves identifying some common ground, some values you and your audience share. For example, to introduce a book arguing for the compatibility of science and religion, author Chet Raymo offers some common memories:

> Like most children, I was raised on miracles. Cows that jump over the moon; a jolly fat man that visits every house in the world in a single night; mice and ducks that talk; little engines that huff and puff and say, "I think I can"; geese that lay golden eggs. This lively exercise of credulity on the part of children is good practice for what follows—for believing in the miracle stories of traditional religion, yes, but also for the practice of poetry or science.
>
> —Chet Raymo, *Skeptics and True Believers: The Exhilarating Connection between Science and Religion*

Raymo presents childhood stories and myths that are part of many people's shared experiences to help his readers find a connection between two realms that are often seen as opposed.

To show that you have carefully considered the viewpoints of others, including those who may agree or disagree with you, it's important to incorporate those viewpoints into your argument by acknowledging, accommodating, or refuting them.

Acknowledging other viewpoints.   One essential part of establishing your credibility is to acknowledge that there are viewpoints different from yours and to represent them fairly and accurately. Rather than weakening your argument, acknowledging possible objections to your position shows that you've thought about and researched your topic thoroughly.

❈ academic literacies   ● fields   ● research
■ rhetorical situations   ⁘ processes   ● media/design
▲ genres   ◆ strategies

For example, in an essay about his experience growing up homosexual, writer Andrew Sullivan admits that not every young gay man or woman has the same experience:

> I should add that many young lesbians and homosexuals seem to have had a much easier time of it. For many, the question of sexual identity was not a critical factor in their life choices or vocation, or even a factor at all.
>
> —Andrew Sullivan, "What Is a Homosexual?"

In response to a reasonable objection, Sullivan qualifies his assertions, making his own stance appear to be reasonable.

Accommodating other viewpoints. You may be tempted to ignore views you don't agree with, but in fact it's important to demonstrate that you're aware of them and have considered them carefully. You may find yourself conceding that opposing views have some merit and qualifying your claim or even making them part of your own argument. See, for example, how a philosopher arguing that torture is sometimes "not merely permissible but morally mandatory" addresses a major objection to his position:

> The most powerful argument against using torture as a punishment or to secure confessions is that such practices disregard the rights of the individual. Well, if the individual is all that important—and he is— it is correspondingly important to protect the rights of individuals threatened by terrorists. If life is so valuable that it must never be taken, the lives of the innocents must be saved even at the price of hurting the one who endangers them.
>
> —Michael Levin, "The Case for Torture"

Levin acknowledges his critics' argument that the individual is indeed important; but he then asserts that if the life of one person is important, the lives of many people must be even more important. In effect, he uses an opposing argument to advance his own.

Refuting other viewpoints.   Often you may need to refute other arguments and make a case for why you believe they are wrong. Are the values underlying the argument questionable? Is the reasoning flawed? Is the evidence inadequate or faulty? For example, an essay arguing for the elimination of college athletics scholarships includes this refutation:

> Some argue that eliminating athletics scholarships would deny opportunity and limit access for many students, most notably black athletes. The question is, access to what? The fields of competition or an opportunity to earn a meaningful degree? With the six-year graduation rates of black basketball players hovering in the high 30% range, and black football players in the high 40% range, despite years of "academic reform," earning an athletics scholarship under the current system is little more than a chance to play sports.
>
> —John R. Gerdy, "For True Reform,
> Athletics Scholarships Must Go"

Gerdy bases his refutation on statistics showing that for more than half of African American college athletes, the opportunity to earn a degree by playing a sport is an illusion.

427–29 ◆ When you incorporate differing viewpoints, be careful to avoid the **FALLACIES** of attacking the person making the argument or refuting a competing position that no one seriously entertains. It's also important that you not distort or exaggerate opposing viewpoints. If *your* argument is to be persuasive, other arguments should be represented fairly.

Appealing to Readers' Emotions

Logic and facts, even when presented by someone who seems reasonable and trustworthy, may not be enough to persuade readers. Many successful arguments include an emotional component that appeals to readers' hearts as well as to their minds. Advertising often works by appealing to its audience's emotions, as in this paragraph from a Volvo ad:

> Choosing a car is about the comfort and safety of your passengers, most especially your children. That's why we ensure Volvo's safety research

examines how we can make our cars safer for everyone who travels in them—from adults to teenagers, children to babies. Even those who aren't even born yet.

—*Volvo.com*

This ad plays on the fear that children—or a pregnant mother—may be injured or killed in an automobile accident.

Keep in mind that emotional appeals, also known as *pathos*, can make readers feel as though they're being manipulated and, consequently, less likely to accept an argument. For most kinds of academic writing, use emotional appeals sparingly.

Checking for Fallacies

Fallacies are arguments that involve faulty reasoning. It's important to avoid fallacies in your writing because they often seem plausible but are usually unfair or inaccurate and make reasonable discussion difficult. Here are some of the most common fallacies:

- *Ad hominem* arguments attack someone's character rather than address the issues. ("Ad hominem" is Latin for "to the man.") It's an especially common fallacy in political discourse and elsewhere: "Jack Turner has no business talking about the way we run things in this city. He's just another liberal snowflake." Whether or not Turner is a "liberal snowflake" has no bearing on the worth of his argument about "the way we run things in this city"; insulting one's opponents isn't an argument against their positions. A variation, the "you too" or "whataboutism" fallacy, results when someone responds to criticism by accusing the attacker of hypocrisy but not answering the criticism itself: "You say I'm lying, but what about your use of a company car to take a family vacation?"

- *Bandwagon appeals* argue that because others think or do something, we should, too. For example, an advertisement for a breakfast cereal claims that it is "America's favorite cereal." It assumes that readers want to be part of the group and implies that an opinion that's popular must be correct.

- *Begging the question* is a circular argument. It assumes as a given what is trying to be proved, essentially asserting that A is true because A is true. Consider this statement: "Affirmative action can never be fair or just because you cannot remedy one injustice by committing another." This statement begs the question because in trying to prove that affirmative action is unjust, it assumes that it is an injustice.

- *Either-or* arguments, also called *false dilemmas*, are oversimplifications that assert there can be only two possible positions on a complex issue. For example, "Those who oppose our actions in this war are enemies of freedom" inaccurately assumes that if someone opposes the war in question, they oppose freedom. In fact, people might have many other reasons for opposing the war.

- *False analogies* compare things that resemble each other in some ways but not in the most important respects—for example, "Trees pollute the air just as much as cars and trucks do." Although it's true that plants emit hydrocarbons, and hydrocarbons are a component of smog, plants also produce oxygen, whereas motor vehicles emit gases that combine with hydrocarbons to form smog. Vehicles pollute the air; trees provide the air that vehicles' emissions pollute.

- *Faulty causality,* also known as *post hoc, ergo propter hoc* (Latin for "after this, therefore because of this"), assumes that because one event followed another, the first event caused the second—for example, "Legalizing same-sex marriage in Sweden led to a decline in the marriage rate of opposite-sex couples." The statement contains no evidence to show that the first event caused the second.

- *Straw man* arguments misrepresent an opposing position to make it sound ridiculous or extreme and thus easy to refute, rather than dealing with the actual position. For example, if someone argues that funding for supplemental nutrition assistance should be cut, a straw man response would be "You want the poor to starve," transforming a proposal to cut a specific program into an exaggerated argument that the proposer hasn't made.

- *Hasty generalizations* are conclusions based on insufficient or inappropriately qualified evidence. This summary of a research study is a

good example: "Twenty randomly chosen residents of Brooklyn, New York, were asked whether they found graffiti tags offensive; fourteen said yes, five said no, and one had no opinion. Therefore, 70 percent of Brooklyn residents find tagging offensive." In Brooklyn, a part of New York City with a population of over two million, twenty residents is far too small a group from which to draw meaningful conclusions. To be able to generalize, the researcher would have had to survey a much greater percentage of Brooklyn's population.

- *Slippery slope* arguments assert that one event will inevitably lead to another, often cataclysmic event without presenting evidence that such a chain of causes and effects will in fact take place. Here's an example: "If the state legislature passes this 2 percent tax increase, it won't be long before all the corporations in the state move to other states and leave thousands unemployed." According to this argument, if taxes are raised, the state's economy will be ruined—not a likely scenario, given the size of the proposed increase.

Considering the Rhetorical Situation

To argue effectively, it's important to think about the message you want to articulate, the audience you want to persuade, the effect of your stance, and the larger context you are writing in.

PURPOSE   What do you want your audience to do? To think a certain way? To take a certain action? To change their minds? To consider alternative views to their current ones? To accept your position as plausible? To see that you have thought carefully about an issue and researched it appropriately?

◼ 59–60

AUDIENCE   Who is your intended audience? What do they likely know and believe about your topic? How personal is it for them? To what extent are they likely to agree or disagree with you? Why? What common ground can you find with them? How should you incorporate other viewpoints they might have? What kinds of evidence are they likely to accept?

◼ 61–64

65–71 ■ **GENRE** What genre will help you achieve your purpose? A position paper? An evaluation? A review? A proposal? An analysis?

72–74 ■ **STANCE** What's your attitude toward your topic, and why? What strategies will help you to convey that stance? How do you want your audience to perceive you? As an authority on your topic? As someone much like them? As calm? reasonable? impassioned or angry? something else?

75–77 ■ **MEDIA/DESIGN** What media will you use, and how will your choice of media affect your argument? If you're writing on paper, does your argument call for photos or charts? If you're giving an oral presentation, should you put your reasons and support on slides? If you're writing online, should you add links to sites representing other positions or containing evidence that supports your position?

If you need more help

164–95 ▲
79
325 ⋮⋮

See also Chapter 13 for advice on writing an **ARGUMENT** essay. And see other appropriate **GENRE** chapters and all the **PROCESSES** chapters for guidelines on drafting, revising, and so on for help writing an essay that includes arguments.

☀ academic literacies ● fields ● research
■ rhetorical situations ⋮⋮ processes ● media/design
▲ genres ◆ strategies

38 Classifying and Dividing

Classification and division are ways of organizing information: various items may be classified according to their similarities, or a single topic may be divided into parts. We might classify different kinds of flowers as annuals or perennials, for example, and classify the perennials further as daylilies, irises, roses, and peonies. We might also divide a garden into distinct areas: for herbs, flowers, and vegetables.

Writers often use classification and division as ways of developing and organizing material. This book, for instance, classifies comparison, definition, description, and several other common ways of thinking and writing as strategies. It divides the information it provides about writing into eight parts: "Rhetorical Situations," "Genres," and so on. Each part further divides its material into various chapters. Even if you never write a book, you will have occasion to classify and divide material in **ANNOTATED BIBLIOGRAPHIES**, essays **ANALYZING TEXTS**, and other kinds of writing. This chapter offers advice for classifying and dividing information for various purposes—and in a way that suits your own rhetorical situation.

▲ 201–13
104–39

Classifying

When we classify something, we group it with similar things. A linguist would classify French, Spanish, and Italian as Romance languages, for example—and Russian, Polish, and Bulgarian as Slavic languages. In a phony news story from the *Onion* about a church bake sale, the writer classifies the activities observed there as examples of the seven deadly sins:

GADSDEN, AL—The seven deadly sins—avarice, sloth, envy, lust, gluttony, pride, and wrath—were all committed Sunday during the twice-annual bake sale at St. Mary's of the Immaculate Conception Church.

— "All Seven Deadly Sins Committed at Church Bake Sale,"
Onion

The article goes on to categorize the participants' behavior in terms of the sins, describing one parishioner who commits the sin of pride by bragging about her cookies and others who commit the sin of envy by resenting the popularity of the prideful parishioner's baked goods (the eating of which leads to the sin of gluttony). In all, the article humorously notes, "347 individual acts of sin were committed at the bake sale," and every one of them can be classified as one of the seven deadly sins.

Dividing

As a writing strategy, division is a way of breaking something into parts—and a way of making the information easy for readers to follow and understand. See how this example about children's ways of nagging divides their tactics into seven categories:

James U. McNeal, a professor of marketing at Texas A&M University, is considered America's leading authority on marketing to children. In his book *Kids as Customers* (1992), McNeal provides marketers with a thorough analysis of "children's requesting styles and appeals." He [divides] juvenile nagging tactics into seven major categories. A *pleading* nag is one accompanied by repetitions of words like "please" or "mom, mom, mom." A *persistent* nag involves constant requests for the coveted product and may include the phrase "I'm gonna ask just one more time." *Forceful* nags are extremely pushy and may include subtle threats, like "Well, then, I'll go and ask Dad." *Demonstrative* nags are the most high risk, often characterized by full-blown tantrums in public places, breath holding, tears, a refusal to leave the store. *Sugar-coated* nags promise

☀ academic literacies   ● fields   ● research
■ rhetorical situations   ⁛ processes   ● media/design
▲ genres   ◆ strategies

affection in return for a purchase and may rely on seemingly heartfelt declarations, like "You're the best dad in the world." *Threatening* nags are youthful forms of blackmail, vows of eternal hatred and of running away if something isn't bought. *Pity* nags claim the child will be heartbroken, teased, or socially stunted if the parent refuses to buy a certain item. "All of these appeals and styles may be used in combination," McNeal's research has discovered, "but kids tend to stick to one or two of each that prove most effective . . . for their own parents."

—Eric Schlosser, *Fast Food Nation:
The Dark Side of the All-American Meal*

Here the writer announces the division scheme of "seven major categories." Then he names each tactic and describes how it works. Notice the italics: each tactic is italicized, making it easy to recognize and follow. Take away the italics, and the divisions would be less visible.

Creating Clear and Distinct Categories

When you classify or divide, it's important to create clear and distinct categories. If you're writing about music, you might divide it on the basis of the genre (rap, rock, classical, gospel), artist (male or female, group or solo), or instruments (violins, trumpets, bongos, guitars). These categories must be distinct so that no information overlaps or fits into more than one category, and they must include every member of the group you're discussing. The simpler the criteria for selecting the categories, the better. The nagging categories in the example from *Fast Food Nation* are based on only one criterion: a child's verbal behavior.

Sometimes you may want to highlight your categories visually to make them easier to follow. Eric Schlosser does that by italicizing each category: the *pleading* nag, the *persistent* nag, the *forceful* nag, and so on. Other **DESIGN** elements—bulleted lists, pie charts, tables, images—might also prove useful.

664–72

Sometimes you might show categories visually, as in the following chart. The differences among the six varieties pictured are visible at a glance, and the chart next to the photos shows the best uses for each variety—and its level of tartness.

	EATING	BAKING	COOKING	
 Red Delicious	✔	✔		LESS TART
 Honeycrisp	✔	✔	✔	
 Red Rome		✔	✔	
 Braeburn	✔		✔	
 McIntosh	✔	✔	✔	MORE TART

All photos © New York Apple Association

The photographs allow us to see the differences among the varieties at a glance. Although the varieties shown here are arranged according to the tartness of their flavor, they could have been arranged in other ways, too—for example, alphabetically or by shape or size.

* academic literacies
■ rhetorical situations
▲ genres
● fields
∴ processes
◆ strategies
● research
● media/design

For another example, see how *The World of Caffeine* authors Bennett Alan Weinberg and Bonnie K. Bealer use a two-column list to show what they say are the differing cultural connotations of coffee and tea:

Coffee Aspect	Tea Aspect
Male	Female
Boisterous	Decorous
Indulgence	Temperance
Hardheaded	Romantic
Topology	Geometry
Heidegger	Carnap
Beethoven	Mozart
Libertarian	Statist
Promiscuous	Pure

—Bennett Alan Weinberg and Bonnie K. Bealer,
The World of Caffeine

Considering the Rhetorical Situation

As a writer or speaker, it's important to think about the message you want to articulate, the audience you want to reach, and the larger context you are writing in.

PURPOSE Your purpose for writing will affect how you classify or divide information. Weinberg and Bealer classify coffee as "boisterous" and tea as "decorous" to help readers understand the cultural styles the two beverages represent, whereas J. Crew might divide sweaters into cashmere, wool, and cotton to help shoppers find and buy clothing from its website.

■ 59–60

61–64 **AUDIENCE**   Who do you want to reach, and will classifying or dividing your material help them follow your discussion?

65–71 **GENRE**   Does your genre call for you to categorize or divide information? A long report might need to be divided into sections, for instance.

72–74 **STANCE**   Your stance may affect the way you classify information. Weinberg and Bealer's classification of coffee as "Beethoven" and tea as "Mozart" reflects a stance that focuses on cultural analysis (and assumes an audience familiar with the difference between the two composers). If the authors were botanists, they might categorize the two beverages in terms of their biological origins ("seed based" and "leaf based").

75–77 **MEDIA/DESIGN**   You can classify or divide in paragraph form, but sometimes a pie chart or list will show the categories better.

If you need more help

334–36
325
See also **CLUSTERING**, **CUBING**, and **LOOPING**, three methods of generating ideas discussed in Chapter 28 that can be especially helpful for classifying material. And see all the **PROCESSES** chapters for guidelines on drafting, revising, and so on if you need to write a classification essay.

* academic literacies   ● fields   ● research
■ rhetorical situations   ∴ processes   ● media/design
▲ genres   ◆ strategies

39 Comparing and Contrasting

Comparing things looks at their similarities; contrasting them focuses on their differences. It's a kind of thinking that comes naturally and that we do constantly—for example, comparing Houston with Dallas, iPhones with Androids, or three paintings by Renoir. And once we start comparing, we generally find ourselves contrasting—Houston and Dallas have differences as well as similarities.

As a student, you'll often be asked to compare and contrast paintings or poems or other things. As a writer, you'll have cause to compare and contrast in most kinds of writing. In a **PROPOSAL**, for instance, you'll need to compare your solution with other possible solutions; or in an **EVALUATION**, such as a movie review, you might contrast the film you're reviewing with some other film. This chapter offers advice on ways of comparing and contrasting things for various writing purposes and for your own rhetorical situations.

258–68
214–22

Most of the time, we compare obviously similar things: laptops we might purchase, three competing political candidates, two versions of a film. Occasionally, however, we might compare things that are less obviously similar. See how John McMurtry, a former football player, compares football with war in an essay arguing that the attraction football holds for spectators is based in part on its potential for violence and injury:

> The family resemblance between football and war is, indeed, striking. Their languages are similar: "field general," "long bomb," "blitz," "take a shot," "front line," "pursuit," "good hit," "the draft," and so on. Their principles and practices are alike: mass hysteria, the art of intimidation, absolute command and total obedience, territorial aggression, censorship, inflated insignia and propaganda, blackboard maneuvers and strategies, drills, uniforms, marching bands, and training camps.

And the virtues they celebrate are almost identical: hyperaggressiveness, coolness under fire, and suicidal bravery.

—John McMurtry, "Kill 'Em! Crush 'Em! Eat 'Em Raw!"

McMurtry's comparison helps focus readers' attention on what he's arguing about football in part because it's somewhat unexpected. But the more unlikely the comparison, the more you might be accused of comparing apples and oranges. It's important, therefore, that the things we compare be *legitimately* compared—as is the case in the following comparison of the health of the world's richest and poorest people:

World Health Organization (WHO) data indicate that roughly 1.2 billion people are undernourished, underweight, and often hungry. At the same time, roughly 1.2 billion people are overnourished and overweight, most of them suffering from excessive caloric intake and exercise deprivation. So while 1 billion people worry whether they will eat, another billion should worry about eating too much.

Disease patterns also reflect the widening gap. The billion poorest suffer mostly from infectious diseases—malaria, tuberculosis, dysentery, and AIDS. Malnutrition leaves infants and small children even more vulnerable to such infectious diseases. Unsafe drinking water takes a heavier toll on those with hunger-weakened immune systems, resulting in millions of fatalities each year. In contrast, among the billion at the top of the global economic scale, it is diseases related to aging and lifestyle excesses, including obesity, smoking, diets rich in fat and sugar, and exercise deprivation, that cause most deaths.

—Lester R. Brown, *Plan B 2.0: Rescuing a Planet
under Stress and a Civilization in Trouble*

While the two groups of roughly a billion people each undoubtedly have similarities, this selection from a book arguing for global action on the environment focuses on the stark contrasts.

Two Ways of Comparing and Contrasting

Comparisons and contrasts may be organized in two basic ways: block and point by point.

* academic literacies   ● fields   ● research
■ rhetorical situations   ∴ processes   ● media/design
▲ genres   ◆ strategies

The block method. One way is to discuss separately each item you're comparing, giving all the information about one item and then all the information about the next item. A report on Seattle and Vancouver, for example, compares the firearm regulations in each city by using a paragraph about Seattle and then a paragraph about Vancouver:

> Although similar in many ways, Seattle and Vancouver differ markedly in their approaches to the regulation of firearms. In Seattle, handguns may be purchased legally for self-defense in the street or at home. After a thirty-day waiting period, a permit can be obtained to carry a handgun as a concealed weapon. The recreational use of handguns is minimally restricted.
>
> In Vancouver, self-defense is not considered a valid or legal reason to purchase a handgun. Concealed weapons are not permitted. Recreational uses of handguns (such as target shooting and collecting) are regulated by the province, and the purchase of a handgun requires a restricted-weapons permit. A permit to carry a weapon must also be obtained in order to transport a handgun, and these weapons can be discharged only at a licensed shooting club. Handguns can be transported by car, but only if they are stored in the trunk in a locked box.
>
> —John Henry Sloan et al., "Handgun Regulations, Crime, Assaults, and Homicide: A Tale of Two Cities"

The point-by-point method. The other way to compare things is to focus on specific points of comparison. In this paragraph, humorist David Sedaris compares his childhood with his partner's, discussing corresponding aspects of the childhoods one at a time:

> Certain events are parallel, but compared with Hugh's, my childhood was unspeakably dull. When I was seven years old, my family moved to North Carolina. When he was seven years old, Hugh's family moved to the Congo. We had a collie and a house cat. They had a monkey and two horses named Charlie Brown and Satan. I threw stones at stop signs. Hugh threw stones at crocodiles. The verbs are the same, but he definitely wins the prize when it comes to nouns and objects. An eventful day for my mother might have involved a trip to the dry cleaner or a conversation with the potato-chip deliveryman. Asked one ordinary Congo afternoon what she'd done with her day, Hugh's mother

answered that she and a fellow member of the Ladies' Club had visited
a leper colony on the outskirts of Kinshasa.

—David Sedaris, "Remembering My Childhood
on the Continent of Africa"

Using Graphs and Images to Present Comparisons

674–80 ●

Some comparisons can be easier to understand if they're presented visually, as a **CHART**, **GRAPH**, or **ILLUSTRATION**. For example, the following line
graph, from *Business Insider*, compares the incomes of Black and White
Americans between 2002 and 2018.

It would be possible to write out this information in a paragraph—but
it's much easier to understand it this way:

Per capita income of Black and White Americans

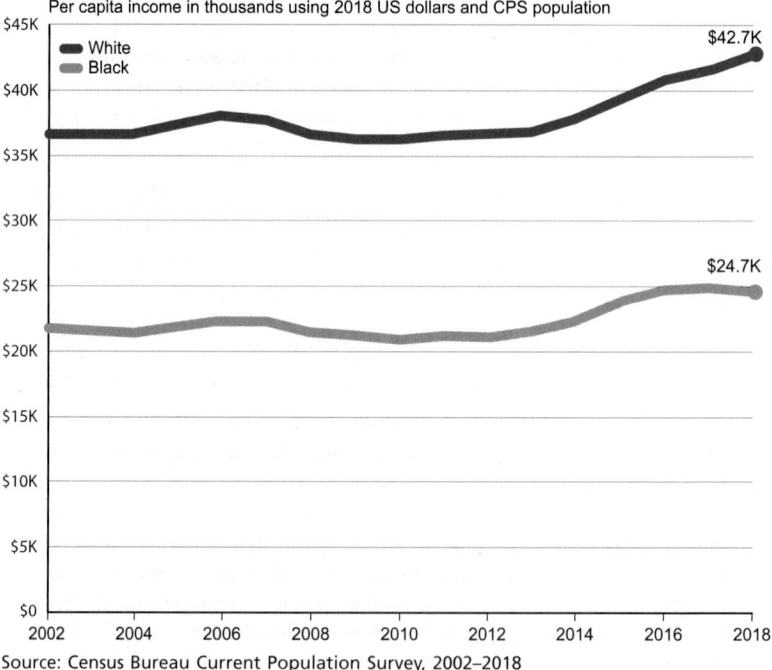

Per capita income in thousands using 2018 US dollars and CPS population

Source: Census Bureau Current Population Survey, 2002–2018

✳ academic literacies   ● fields   ● research
■ rhetorical situations   ⁙ processes   ● media/design
▲ genres   ◆ strategies

Roadway in Houston before the 2017 hurricane (left) and after (right).

Sometimes photographs can make a comparison. The two photos above show a street in Houston before and after Hurricane Harvey in 2017.

Using Figurative Language to Make Comparisons

Another way to make comparisons is with figurative language: words and phrases used in a nonliteral way to help readers see a point. Three kinds of figurative language that make comparisons are similes, metaphors, and analogies. When Robert Burns wrote that his love was "like a red, red rose," he was comparing his love with a rose and evoking an image—in this case, a simile—that helps us understand his feelings for her. A simile makes a comparison using "like" or "as." In the following example, from an article in the food section of the *New York Times*, a restaurant critic uses several similes (underlined) to help us visualize an unusual food dish:

> Once upon a time, possibly at a lodge in Wyoming, possibly at a butcher shop in Maurice, Louisiana, or maybe even at a plantation in South Carolina, an enterprising cook decided to take a boned chicken, a boned duck, and a boned turkey, stuff them one inside the other <u>like Russian dolls</u>, and roast them. He called his masterpiece turducken. . . .

A well-prepared turducken is a marvelous treat, a free-form poultry terrine layered with flavorful stuffing and moistened with duck fat. When it's assembled, it looks like a turkey and it roasts like a turkey, but when you go to carve it, you can slice through it like a loaf of bread. In each slice you get a little bit of everything: white meat from the breast, dark meat from the legs, duck, carrots, bits of sausage, bread, herbs, juices, and chicken, too.

—Amanda Hesser,
"Turkey Finds Its Inner Duck (and Chicken)"

Metaphors make comparisons without such connecting words as "like" or "as." See how desert ecologist Craig Childs uses a metaphor to help us understand the nature of water during a flood in the Grand Canyon:

Water splashed off the desert and ran all over the surface, looking for the quickest way down. It was too swift for the ground to absorb. When water flows like this, it will not be clean tap water. It will be a gravy of debris, snatching everything it finds.

—Craig Childs, *The Secret Knowledge of Water*

Calling the water "a gravy of debris" allows us to see the murky liquid as it streams through the canyon.

Analogies are extended similes or metaphors that compare something unfamiliar with something more familiar. Arguing that corporations should not patent parts of DNA whose function isn't yet clear, a genetics professor uses the familiar image of a library to explain an unfamiliar concept:

It's like having a library of books and randomly tearing pages out. You may know which books the pages came from but that doesn't tell you much about them.

—Peter Goodfellow, quoted in John Vidal and
John Carvel, "Lambs to the Gene Market"

Sometimes analogies are used for humorous effect as well as to make a point, as in this passage from a critique of history textbooks:

※ academic literacies   ● fields   ● research
■ rhetorical situations   ⁛ processes   ● media/design
▲ genres   ◆ strategies

Another history text—this one for fifth grade—begins with the story of how Henry B. Gonzalez, who is a member of Congress from Texas, learned about his own nationality. When he was ten years old, his teacher told him he was an American because he was born in the United States. His grandmother, however, said, "The cat was born in the oven. Does that make him bread?"

—Frances FitzGerald, *America Revised: History Schoolbooks in the Twentieth Century*

The grandmother's question shows how an intentionally ridiculous analogy can be a source of humor—and can make a point memorably.

Considering the Rhetorical Situation

As a writer or speaker, it's important to think about the message you want to articulate, the audience you want to reach, and the larger context you are writing in.

PURPOSE   Sometimes your main purpose for writing will be to compare two or more things. Other times, you may want to compare several things for some other purpose—to compare your views with those of others in an argument essay or to compare one text with another as you analyze them.   ◼ 59–60

AUDIENCE   Who is your audience, and will comparing your topic with a more familiar one help them to follow your discussion?   ◼ 61–64

GENRE   Does your genre require you to compare something? Evaluations often include comparisons—one book to another in a review, or ten different cell phones in *Consumer Reports*.   ◼ 65–71

STANCE   Your stance may affect any comparisons you make. How you compare two things—evenhandedly, or clearly favoring one over the other, for example—will reflect your stance.   ◼ 72–74

75–77 ■

MEDIA/DESIGN　Some things you'll want to compare with words alone (lines from two poems, for instance), but sometimes you may wish to make comparisons visually (two images juxtaposed on a page, or several numbers plotted on a line graph).

If you need more help

334
336
325

See **LOOPING** and **CUBING**, two methods of generating ideas discussed in Chapter 28 that can be especially helpful for comparing and contrasting. If you're writing an essay whose purpose is to compare two or more things, see also the **PROCESSES** chapters for help with drafting, revising, and so on.

❋ academic literacies　　● fields　　● research
■ rhetorical situations　　⁙ processes　　● media/design
▲ genres　　◆ strategies

40 Defining

Defining something says what it is—and what it is not. A terrier, for example, is a kind of dog. A fox terrier is a small dog now generally kept as a pet but once used by hunters to dig for foxes. Happiness is a pink frosted doughnut, at least according to Homer Simpson. All of those are definitions. As writers, we need to define any terms our readers may not know. And sometimes you'll want to stipulate your own definition of a word in order to set the terms of an **ARGUMENT**—as Homer Simpson does with a definition that's not found in any dictionary. This chapter details strategies for using definitions in your writing to suit your own rhetorical situations.

◆ 410–30

Formal Definitions

Sometimes to make sure readers understand you, you'll need to provide a formal definition. If you are using a technical term that readers are unlikely to know or if you are using a term in a specific way, you should say then and there what the word means. The word *mutual*, for example, has several meanings: it can mean *shared* or *reciprocal*, as when two people have mutual interests or have mutual respect. It can mean having the *same relationship toward one another*, as when two people are mutual enemies. And it can refer to a type of organization in which profits, losses, and expenses are shared by the organization's members—a common form of insurance in which policyholders pay into a common fund that is used to pay insurance claims. (Examples might include companies like Mutual of Omaha and Liberty Mutual Insurance.)

The first two meanings are commonly understood and probably require no definition. But if you were to use *mutual* in the third sense, it might need to be defined, depending on your audience. A general audience would probably need the definition; an audience from the insurance industry would not. A website that gives basic financial advice to an audience of non-

specialists, for instance, offers a specific definition of the term "mutual fund":

> *Mutual funds* are financial intermediaries. They are companies set up to receive your money and then, having received it, to make investments with the money.
>
> —Bill Barker, "A Grand, Comprehensive
> Overview to Mutual Funds Investing"

But even writers in specialized fields routinely provide formal definitions to make sure their readers understand the way they're using certain words. See how two writers define the word "stock" as it pertains to their respective (and very different) fields:

> Stocks are the basis for sauces and soups and important flavoring agents for braises. Admittedly, stock making is time consuming, but the extra effort yields great dividends.
>
> —Tom Colicchio, *Think Like a Chef*

> Want to own part of a business without having to show up at its office every day? Or ever? Stock is the vehicle of choice for those who do. Dating back to the Dutch mutual stock corporations of the sixteenth century, the modern stock market exists as a way for entrepreneurs to finance businesses using money collected from investors. In return for ponying up the dough to finance the company, the investor becomes a part owner of the company. That ownership is represented by stock—specialized financial "securities," or financial instruments, that are "secured" by a claim on the assets and profits of a company.
>
> —"Investing Basics: Stocks," *Motley Fool*

To write a formal definition

- Use words that readers are likely to be familiar with.
- Don't use the word being defined in the definition.
- Begin with the word being defined; include the general category to which the term belongs and the attributes that make it different from the others in that category.

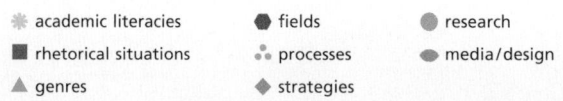

For example:

Term	General Category	Distinguishing Attributes
"Stock" is	a specialized financial "security"	that is "secured" by a claim.
"Photosynthesis" is	a process	by which plants use sunlight to create energy.
"Astronomers" are	scientists	who study celestial objects and phenomena.
"Zach Galifianakis,"	an actor,	has been featured in several films, including *The Hangover* and *Birdman*.

Note that the category and distinguishing attributes cannot be stated too broadly; if they were, the definition would be too vague to be useful. It wouldn't be helpful in most circumstances, for example, to say, "Zach Galifianakis is a man who has acted" or "Photosynthesis is something having to do with plants."

Extended Definitions

Sometimes it's useful to provide a more detailed definition. Extended definitions may be several sentences long or several paragraphs long and may include pictures or diagrams. Sometimes an entire essay is devoted to defining a difficult or important concept. Here is one writer's extended definition of "soul food":

> I've been eating soul food all my life and cooking it my whole career. I don't just know soul food. Soul food is *in* my soul. . . .
>
> By definition, soul food refers to the dishes of the Cotton Belt of Georgia, Mississippi, and Alabama that traveled out to the rest of the country during the Great Migration. (The term itself came around the middle of the twentieth century.) You know what travels well? Fried chicken. Mac and cheese. Delicious, but not what anyone's meant to eat every day. I'm here to redefine soul food, to reclaim it.
>
> Soul food is the true food of African-Americans.

The roots of our cooking are in West Africa. And from there, the American South, from the slave ports along the eastern coast to the southern border. We relied on seasonal vegetables, beans, and grains, with meat on rare occasions. Let's be clear: those were horrible times of suffering under the most unspeakable evil. I don't want to romanticize any of it. Not even the food. Remember, we didn't get to choose what we ate. But we made the most delicious dishes from what little we had. And what we cooked for the slave owners effectively became what we know as "American" food today.

—Carla Hall, *Carla Hall's Soul Food*

That definition includes a description of the basic features of soul food, examples of it, and the origin of the term. We can assume that it's written for a general audience, one that doesn't know much about soul food.

Abstract concepts often require extended definitions because by nature they are more complicated to define. There are many ways of writing an extended definition, depending in part on the term being defined and on the writer's audience and purpose. The following examples show some of the methods that can be used for composing extended definitions of "democracy."

Explore the word's origins.   Where did the word come from? When did it first come into use? In the following example, from an essay considering what democracy means in the twenty-first century, the writer started by looking at the word's first known use in English. Though it's from an essay written for a first-year writing course and thus for a fairly general audience, it's a definition that might pique any audience's interest:

According to the *Oxford English Dictionary*, the term *democracy* first appeared in English in a thirteenth-century translation of Aristotle's works—specifically, in his *Politics*, where he stated that the "underlying principle of democracy is freedom" and that "it is customary to say that only in democracies do men have a share in freedom, for that is what every democracy makes its aim." By the sixteenth century, the word was used much as it is now. One writer in 1586, for instance, defined it in this way: "where free and poore men being the greater number, are lords of the estate."

—Susanna Mejía, "What Does Democracy Mean Now?"

* academic literacies   ● fields   ● research
■ rhetorical situations   ∴ processes   ● media/design
▲ genres   ◆ strategies

Norman Rockwell's 1943 painting *Freedom of Speech* presents a visual definition of democracy: a citizen stands to speak at a public meeting while his fellow citizens listen attentively.

Here's another example, this one written for a scholarly audience, from an essay about women, participation, democracy, and the information age:

> The very word *citizenship* carries with it a connotation of place, a "citizen" being, literally, the inhabitant of a city. Over the years the word has, of course, accumulated a number of associated meanings . . . and the word has come to stand in for such concepts as participation, equality, and democracy. The fact that the concept of locality is deeply embedded in the word *citizen* suggests that it is also fundamental to our current understanding of these other, more apparently abstract words.
>
> In Western thought, the concepts of citizenship, equality, and democracy are closely interlinked and can be traced back to a common source, in Athens in the fifth century BCE. Perhaps it is no accident that it was the same culture which also gave us, in its theater, the concept of the unity of time and space. The Greek city-state has been represented for centuries as the ideal model of democracy, with free and equal access for all citizens to decision making. Leaving aside, for the moment, the question of who was included, and who excluded from this notion of citizenship, we can see that the sense of place is fundamental to this model. Entitlement to participate in the democratic process is circumscribed by geography; it is the inhabitants of the geographical entity of the city-state, precisely defined and bounded, who have the rights to citizenship. Those who are not defined as inhabitants of that specific city-state are explicitly excluded, although, of course, they may have the right to citizenship elsewhere.
>
> —Ursula Huws, "Women, Participation, and Democracy in the Information Society"

Provide details. What are its characteristics? What is it made of? See how a historian explores the basic characteristics of democracy in a book written for an audience of historians:

> As a historian I am naturally disposed to be satisfied with the meaning which, in the history of politics, men have commonly attributed to the word—a meaning, needless to say, which derives partly from the experience and partly from the aspirations of mankind. So regarded, the term *democracy* refers primarily to a form of government, and it

has always meant government by the many as opposed to government by the one—government by the people as opposed to government by a tyrant, a dictator, or an absolute monarch. . . . Since the Greeks first used the term, the essential test of democratic government has always been this: the source of political authority must be and remain in the people and not in the ruler. A democratic government has always meant one in which the citizens, or a sufficient number of them to represent more or less effectively the common will, freely act from time to time, and according to established forms, to appoint or recall the magistrates and to enact or revoke the laws by which the community is governed.

—Carl Becker, *Modern Democracy*

Compare it with other words. How is this concept like other similar things? How does it differ? What is it *not* like? **COMPARE AND CONTRAST** it. See how a political science textbook defines a *majoritarian democracy* by comparing its characteristics with those of a *consensual democracy*:

◆ 437–44

A majoritarian democracy is one

1. having only two major political parties, not many

2. having an electoral system that requires a bare majority to elect one clear winner in an election, as opposed to a proportional electoral system that distributes seats to political parties according to the rough share of votes received in the election

3. a strong executive (president or prime minister) and cabinet that together are largely independent of the legislature when it comes to exercising the executive's constitutional duties, in contrast to an executive and cabinet that are politically controlled by the parties in the legislature and therefore unable to exercise much influence when proposing policy initiatives.

—Benjamin Ginsberg, Theodore J. Lowi, and Margaret Weir,
We the People: An Introduction to American Politics

And here's an example in which democracy is contrasted with various other forms of governments of the past:

Caesar's power derived from a popular mandate, conveyed through established republican forms, but that did not make his government any

the less a dictatorship. Napoleon called his government a democratic republic, but no one, least of all Napoleon himself, doubted that he had destroyed the last vestiges of the democratic republic.

—Carl Becker, *Modern Democracy*

Give examples. See how the essayist E. B. White defines democracy by giving some everyday examples of considerate behavior, humility, and civic participation—all things he suggests constitute democracy:

> It is the line that forms on the right. It is the don't in "don't shove." It is the hole in the stuffed shirt through which the sawdust slowly trickles; it is the dent in the high hat. Democracy is the recurrent suspicion that more than half of the people are right more than half of the time. . . . Democracy is a letter to the editor.
>
> —E. B. White, "Democracy"

White's definition is elegant because he uses examples that his readers will know. His characteristics—metaphors, really—define democracy not as a conceptual way of governing but as an everyday part of American life.

431–36 ◆ **Classify it.** Often it's useful to divide or **CLASSIFY** a term. The ways in which democracy unfolds are complex enough to warrant entire textbooks, of course; but the following definition, from a political science textbook, divides democracy into two kinds, representative and direct:

> A system of government that gives citizens a regular opportunity to elect the top government officials is usually called a representative democracy or republic. A system that permits citizens to vote directly on laws and policies is often called a direct democracy. At the national level, America is a representative democracy in which citizens select government officials but do not vote on legislation. Some states, however, have provisions for direct legislation through popular referendum. For example, California voters in 1995 decided to bar undocumented immigrants from receiving some state services.
>
> —Benjamin Ginsberg, Theodore J. Lowi, and Margaret Weir,
> *We the People: An Introduction to American Politics*

Stipulative Definitions

Sometimes a writer will stipulate a certain definition, essentially saying, "This is how I'm defining x." Such definitions are not usually found in a dictionary—and at the same time are central to the argument the writer is making. Here is one example, from an essay by Toni Morrison. Describing a scene from a film in which a newly arrived Greek immigrant, working as a shoe shiner in Grand Central Terminal, chases away an African American competitor, Morrison calls the scene an example of "race talk," a concept she then goes on to define:

> This is race talk, the explicit insertion into everyday life of racial signs and symbols that have no meaning other than pressing African Americans to the lowest level of the racial hierarchy. Popular culture, shaped by film, theater, advertising, the press, television, and literature, is heavily engaged in race talk. It participates freely in this most enduring and efficient rite of passage into American culture: negative appraisals of the native-born black population. Only when the lesson of racial estrangement is learned is assimilation complete. Whatever the lived experience of immigrants with African Americans—pleasant, beneficial, or bruising—the rhetorical experience renders blacks as noncitizens, already discredited outlaws.
>
> All immigrants fight for jobs and space, and who is there to fight but those who have both? As in the fishing ground struggle between Texas and Vietnamese shrimpers, they displace what and whom they can. Although U.S. history is awash in labor battles, political fights and property wars among all religious and ethnic groups, their struggles are persistently framed as struggles between recent arrivals and blacks. In race talk the move into mainstream America always means buying into the notion of American blacks as the real aliens. Whatever the ethnicity or nationality of the immigrant, his nemesis is understood to be African American.
>
> —Toni Morrison, "On the Backs of Blacks"

The following example is from a book review of Nancy L. Rosenblum's *Membership and Morals: The Personal Uses of Pluralism in America*, published in the *American Prospect*, a magazine for readers interested in political analysis. In it a Stanford law professor outlines a definition of "the democracy of everyday life":

Democracy, in this understanding of it, means simply treating people as equals, disregarding social standing, avoiding attitudes of either deference or superiority, making allowances for others' weaknesses, and resisting the temptation to respond to perceived slights. It also means protesting everyday instances of arbitrariness and unfairness—from the rudeness of the bakery clerk to the sexism of the car dealer or the racism of those who vandalize the home of the first black neighbors on the block.

—Kathleen M. Sullivan, "Defining Democracy Down"

Considering the Rhetorical Situation

As a writer or speaker, it's important to think about the message you want to articulate, the audience you want to reach, and the larger context you are writing in.

59–60 ■ **PURPOSE** Your purpose for writing will affect any definitions you include. Would writing an extended definition help you explain something? Would stipulating definitions of key terms help you shape an argument? Could an offbeat definition help you entertain your readers?

61–64 ■ **AUDIENCE** What audience do you want to reach, and are there any terms your readers are unlikely to know (and therefore need to be defined)? Are there terms they might understand differently from the way you're defining them?

65–71 ■ **GENRE** Does your genre require you to define terms? Chances are that if you're reporting information you'll need to define some terms, and some arguments rest on the way you define key terms.

72–74 ■ **STANCE** What is your stance, and do you need to define key terms to show that stance clearly? How you define "fetus," for example, is likely to reveal your stance on abortion.

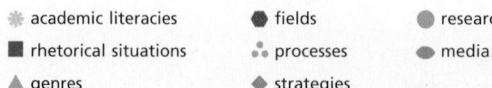

✳ academic literacies   ● fields   ● research
■ rhetorical situations   ⦂ processes   ● media/design
▲ genres   ◆ strategies

MEDIA/DESIGN   Your medium will affect the form your definitions
take. In a print text, you'll need to define terms in your
text; if you're giving a speech or presentation, you
might also provide images of important terms and
their definitions. In an electronic text, you may be able
to define terms by linking to an online dictionary
definition.

■ 75–77

If you need more help

See also the **PROCESSES** chapters for help with generating ideas, drafting, revising, and
so on if you are writing a whole essay dedicated to defining a term or concept.

•• 325

41 Describing

When we describe something, we indicate what it looks like—and sometimes how it sounds, feels, smells, and tastes. Descriptive details are a way of showing rather than telling, of helping readers see (or hear, smell, and so on) what we're writing about—that the sky is blue, that Miss Havisham is wearing an old yellowed wedding gown, that the chemicals in the beaker have reacted and smell like rotten eggs. You'll have occasion to describe things in most of the writing you do—from describing a favorite hat in a **MEMOIR** or **REFLECTION** to detailing a chemical reaction in a lab report. This chapter will help you work with description—and, in particular, help you think about the use of *detail*, about *objectivity and subjectivity*, about *vantage point*, about creating a clear *dominant impression*, and about using description to fit your rhetorical situation.

236–44 ▲
391–401 ⸫

Detail

The goal of using details is to be as specific as possible, providing information that will help your audience imagine the subject or make sense of it. See, for example, how Nancy Mairs, an author with multiple sclerosis, describes the disease in clear, specific terms:

> During its course, which is unpredictable and uncontrollable, one may lose vision, hearing, speech, the ability to walk, control of bladder and/or bowels, strength in any or all extremities, sensitivity to touch, vibration, and/or pain, potency, coordination of movements—the list of possibilities is lengthy and, yes, horrifying. One may also lose one's sense of humor. That's the easiest to lose and the hardest to survive without.
>
> In the past ten years, I have sustained some of these losses. Characteristic of MS are sudden attacks, called exacerbations, followed by remissions, and these I have not had. Instead, my disease has been slowly progressive. My left leg is now so weak that I walk with the aid of a brace and a cane, and for distances I use an Amigo, a variation on the

---

✳ academic literacies   ● fields   ● research

■ rhetorical situations   ⸫ processes   ● media/design

▲ genres   ◆ strategies

electric wheelchair that looks rather like an electrified kiddie car. I no longer have much use of my left hand. Now my right side is weakening as well. I still have the blurred spot in my right eye. Overall, though, I've been lucky so far.

—Nancy Mairs, "On Being a Cripple"

Mairs's gruesome list demonstrates, through *specific details*, how the disease affects sufferers generally and her in particular. We know far more after reading this text than we do from the following more general description, from a National Multiple Sclerosis Society brochure:

Multiple sclerosis is a chronic, unpredictable disease of the central nervous system (the brain, optic nerves, and spinal cord). It is thought to be an autoimmune disorder. This means the immune system incorrectly attacks the person's healthy tissue.

MS can cause blurred vision, loss of balance, poor coordination, slurred speech, tremors, numbness, extreme fatigue, problems with memory and concentration, paralysis, and blindness. These problems may be permanent, or they may come and go.

—National Multiple Sclerosis Society, *Just the Facts*

Specific details are also more effective than labels, which give little meaningful information. Instead of saying that someone is a "loser" or "really smart," it's better to give details so that readers can understand the reasons behind the label: What does this person *do* or *say* that makes them deserve this label? See, for example, how the writer of a memoir focused on his pet snake describes the way his snake, Woohoo, and he interacted:

He wound around my wrist, hung in the air, pulled himself up and into the arm of my shirt, through which he slithered until he decided on another opening, the neck, from which his head emerged, and then his forked tongue, darting out, shaking, tasting and retreating into his mouth. I gave his cold blood my heat and flexed hello to his constrictions. My fingers followed the grain of his scales down his undulating length, which grew to four feet from ten inches as the years passed, its colors freshly radiant after every shedding.

—Paul McAdory, "Snakes"

The writer might simply have said, "The snake moved around under my shirt." Instead, he shows the snake as it moves from place to place ("He wound around my wrist, hung in the air, pulled himself up and into the arm of my shirt") and describes his own physical reaction ("I . . . flexed hello to his constrictions," "My fingers followed the grain of his scales")—all details that create a far more vivid description.

Sensory details help readers imagine sounds, odors, tastes, and physical sensations in addition to sights. In the following example, writer Scott Russell Sanders recalls sawing wood as a child. Note how visual details, odors, and even the physical sense of being picked up by his father mingle to form a vivid scene:

> As the saw teeth bit down, the wood released its smell, each kind with its own fragrance, oak or walnut or cherry or pine—usually pine because it was the softest, easiest for a child to work. No matter how weathered and gray the board, no matter how warped and cracked, inside there was this smell waiting, as of something freshly baked. I gathered every smidgen of sawdust and stored it away in coffee cans, which I kept in a drawer of the workbench. When I did not feel like hammering nails I would dump my sawdust on the concrete floor of the garage and landscape it into highways and farms and towns, running miniature cars and trucks along miniature roads. Looming as huge as a colossus, my father worked over and around me, now and again bending down to inspect my work, careful not to trample my creations. It was a landscape that smelled dizzyingly of wood. Even after a bath my skin would carry the smell, and so would my father's hair, when he lifted me for a bedtime hug.
>
> —Scott Russell Sanders, *The Paradise of Bombs*

Whenever you describe something, you'll select from many possible details you might use. Simply put, to exhaust all the details available to describe something is impossible—and would exhaust your readers as well. To focus your description, you'll need to determine the kinds of details appropriate for your subject. They will vary, depending on your **PURPOSE**. See, for example, how the details might differ in three different genres:

59–60 ■

236–44 ▲

- For a **MEMOIR** about an event, you might choose details that are significant for you, that evoke the sights, sounds, and other sensations that give personal meaning to the event.

※ academic literacies   ⬢ fields   ● research

■ rhetorical situations   ⁝ processes   ● media/design

▲ genres   ◆ strategies

- For a **PROFILE**, you're likely to select details that will reinforce the dominant impression you want to give, that portray the event from the perspective you want readers to see.

▲ 245–57

- For a lab report, you should give certain specifics—what equipment was used, what procedures were followed, what exactly were the results.

Deciding on a focus for your description can help you see it better, as you'll look for details that contribute to that focus.

Objectivity and Subjectivity

Descriptions can be written with objectivity, with subjectivity, or with a mixture of both. Objective descriptions attempt to be uncolored by personal opinion or emotion. Police reports and much news writing aim to describe events objectively; scientific writing strives for objectivity in describing laboratory procedures and results. See, for example, the following objective account of what happened at the World Trade Center on September 11, 2001:

> **World Trade Center Disaster—Tuesday, September 11, 2001**
>
> On Tuesday, September 11, 2001, at 8:45 a.m. New York local time, One World Trade Center, the north tower, was hit by a hijacked 767 commercial jet airplane loaded with fuel for a transcontinental flight. Two World Trade Center, the south tower, was hit by a similar hijacked jet eighteen minutes later, at 9:03 a.m. (In separate but related attacks, the Pentagon building near Washington, D.C., was hit by a hijacked 757 at 9:43 a.m., and at 10:10 a.m. a fourth hijacked jetliner crashed in Pennsylvania.) The south tower, WTC 2, which had been hit second, was the first to suffer a complete structural collapse, at 10:05 a.m., 62 minutes after being hit itself, 80 minutes after the first impact. The north tower, WTC 1, then also collapsed, at 10:29 a.m., 104 minutes after being hit. WTC 7, a substantial forty-seven-story office building in its own right, built in 1987, was damaged by the collapsing towers, caught fire, and later in the afternoon also totally collapsed.
>
> —"World Trade Center," *GreatBuildings.com*

Subjective descriptions, in contrast, allow the writer's opinions and emotions to come through. A house can be described as comfortable, with a lived-in look, or as run-down and in need of a paint job and a new roof.

Here's a subjective description of the planes striking the World Trade Center, as told by a woman watching from a nearby building:

> Incredulously, while looking out [the] window at the damage and carnage the first plane had inflicted, I saw the second plane abruptly come into my right field of vision and deliberately, with shimmering intention, thunder full-force into the south tower. It was so close, so low, so huge and fast, so intent on its target that I swear to you, I swear to you, I felt the vengeance and rage emanating from the plane.
>
> —Debra Fontaine, "Witnessing"

Vantage Point

Sometimes you'll want or need to describe something from a certain vantage point. Where you locate yourself in relation to what you're describing will determine what you can perceive (and so describe) and what you can't. You may describe your subject from a *stationary vantage point*, from which you (and your readers) see your subject from one angle only, as if you were a camera. This description of one of three photographs that captured a woman's death records only what the camera saw from one angle at one particular moment:

> The first showed some people on a fire escape—a fireman, a woman and a child. The fireman had a nice strong jaw and looked very brave. The woman was holding the child. Smoke was pouring from the building behind them. A rescue ladder was approaching, just a few feet away, and the fireman had one arm around the woman and one arm reaching out toward the ladder.
>
> —Nora Ephron, "The Boston Photographs"

By contrast, this description of a drive to an Italian villa uses a *moving vantage point*; the writer recounts what he saw as he passed through a gate in a city wall, moving from city to country:

> La Pietra—"the stone"—is situated one mile from the Porta San Gallo, an entry to the Old City of Florence. You drive there along the Via Bolognese, twisting past modern apartment blocks, until you come to a gate, which swings open—and there you are, at the upper end of a long lane of cypresses facing a great ocher palazzo; with olive groves

✳ academic literacies   ● fields   ● research
■ rhetorical situations   ∴ processes   ● media/design
▲ genres   ◆ strategies

spreading out on both sides over an expanse of fifty-seven acres. There's something almost comically wonderful about the effect: here, the city, with its winding avenue; there, on the other side of a wall, the country, fertile and gray green.

—James Traub, "Italian Hours"

The description of quarries in the following section uses *multiple vantage points* to capture the quarries from many perspectives.

Dominant Impression

With any description, your aim is to create some dominant impression—the overall feeling that the individual details add up to. The dominant impression may be implied, growing out of the details themselves. For example, Scott Russell Sanders's memory of the smell of sawdust creates a dominant impression of warmth and comfort: the "fragrance . . . as of something freshly baked," sawdust "stored . . . away in coffee cans," a young boy "lifted . . . for a bedtime hug." Sometimes, though, a writer will state the dominant impression directly, in addition to creating it with details. In an essay about Indiana limestone quarries, Sanders makes the dominant impression clear from the start: "They are battlefields."

The quarries will not be domesticated. They are not backyard pools; they are battlefields. Each quarry is an arena where violent struggles have taken place between machines and planet, between human ingenuity and brute resisting stone, between mind and matter. Waste rock litters the floor and brim like rubble in a bombed city. The ragged pits might have been the basements of vanished skyscrapers. Stones weighing tens of tons lean against one another at precarious angles, as if they have been thrown there by some gigantic strength and have not yet finished falling. Wrecked machinery hulks in the weeds, grimly rusting, the cogs and wheels, twisted rails, battered engine housings, trackless bulldozers and burst boilers like junk from an armored regiment. Everywhere the ledges are scarred from drills, as if from an artillery barrage or machine-gun strafing. Stumbling onto one of these abandoned quarries and gazing at the ruins, you might be left wondering who had won the battle, men or stone.

—Scott Russell Sanders, *The Paradise of Bombs*

The rest of his description, full of more figurative language ("like rubble in a bombed city," "like junk from an armored regiment," "as if from an artillery barrage or machine-gun strafing") reinforces the direct "they are battlefields" statement.

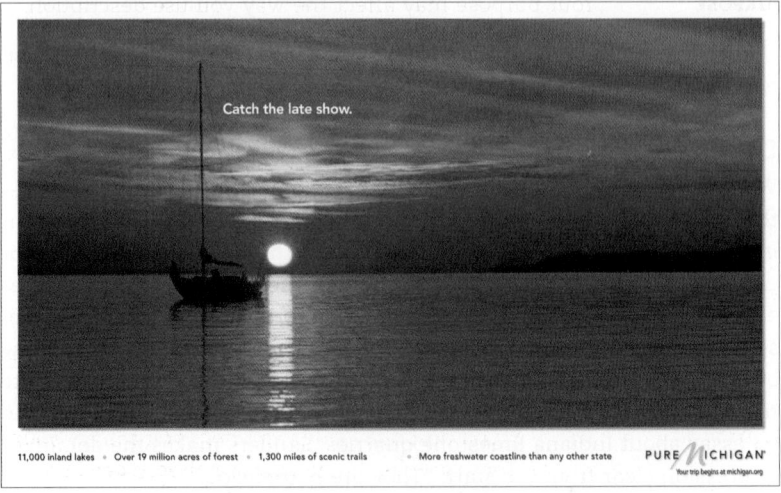

The orange sunset and expanse of sky and water in this Michigan tourism ad create a dominant impression of spaciousness and warmth, while the text invites readers to visit a Michigan beach and enjoy watching the sun set, rather than watching television.

Organizing Descriptions

You can organize descriptions in many ways. When your description is primarily visual, you'll probably organize it spatially: from left to right, top to bottom, outside to inside. One variation on this approach is to begin with the most significant or noteworthy feature and move outward from that center, as Ephron does in describing a photo. Or you may create a chronological description of objects as you move past or through them in space, as Traub does in his description of his drive. You might even pile up details to create a dominant impression, as Sanders and Mairs do, especially if your description draws on senses besides vision.

academic literacies   fields   research

rhetorical situations   processes   media/design

genres   strategies

Considering the Rhetorical Situation

As a writer or speaker, it's important to think about the message you want to articulate, the audience you want to reach, and the larger context you are writing in.

PURPOSE	Your purpose may affect the way you use description. If you're arguing that a government should intervene in another country's civil war, for example, describing the anguish of refugees from that war could make your argument more persuasive. If you're analyzing a painting, you will likely need to describe it.

■ 59–60

AUDIENCE	Who is your audience, and will they need detailed description to understand the points you wish to make?

■ 61–64

GENRE	Does your genre require description? A lab report generally calls for you to describe materials and results; a memoir about Grandma should probably describe her— her smile, her favorite outfit, her apple pie.

■ 65–71

STANCE	The way you describe things can help you convey your stance. For example, the details you choose can show you to be objective (or not), careful or casual.

■ 72–74

MEDIA/DESIGN	Your medium will affect the form your description can take. In a print or spoken text, you'll likely rely on words, though you may also include visuals. In an electronic text, you can easily provide links to visuals as well as audio clips and so may need fewer words of your own.

■ 75–77

If you need more help

See also **FREEWRITING**, **CUBING**, and **LISTING**, three methods of generating ideas that can be especially helpful for developing detailed descriptions. Sometimes you may be assigned to write a whole essay describing something: see the **PROCESSES** chapters for help with drafting, revising, and so on.

333–34
336
334–35
325

42 Using Dialogue

236–44 ▲
245–57
223–35
164–95

Dialogue is a way of including people's own words in a text, letting readers hear those people's voices—not just what you say about them. **MEMOIRS** and **PROFILES** often include dialogue, and many other genres do as well: **LITERARY ANALYSES** often quote dialogue from the texts they analyze, and essays **ARGUING A POSITION** might quote an authoritative source as support for a claim. This chapter provides brief guidelines for the conventions of paragraphing and punctuating dialogue and offers some good examples of how you can use dialogue most effectively to suit your own rhetorical situations.

Why Add Dialogue?

Dialogue is a way of bringing in voices other than your own, of showing people and scenes rather than just telling about them. It can add color and texture to your writing, making it memorable. Most important, however, dialogue should be more than just colorful or interesting. It needs to contribute to your rhetorical purpose, to support the point you're making. See how dialogue is used in the following excerpt from a magazine profile of the Mall of America, how it gives us a sense of the place that the journalist's own words could not provide:

> Two pubescent girls in retainers and braces sat beside me sipping coffees topped with whipped cream and chocolate sprinkles, their shopping bags gathered tightly around their legs, their eyes fixed on the passing crowds. They came, they said, from Shakopee—"It's nowhere," one of them explained. The megamall, she added, was "a buzz at first, but now it seems pretty normal. 'Cept my parents are like Twenty Questions every time I want to come here. 'Specially since the shooting."
>
> On a Sunday night, she elaborated, three people had been wounded when shots were fired in a dispute over a San Jose Sharks

* academic literacies   ● fields   ● research
■ rhetorical situations   •• processes   ● media/design
▲ genres   ◆ strategies

jacket. "In the *mall*," her friend reminded me. "Right here at megamall. A shooting."

"It's like nowhere's safe," the first added.

—David Guterson, "Enclosed. Encyclopedic. Endured. One Week at the Mall of America"

Of course, it was the writer who decided whom and what to quote, and Guterson deliberately chose words that capture the young shoppers' speech patterns, quoting fragments ("In the *mall*. . . . Right here at megamall. A shooting"), slang ("a buzz at first," "my parents are like Twenty Questions"), even contractions ("'Cept," "'Specially").

Integrating Dialogue into Your Writing

There are certain conventions for punctuating and paragraphing dialogue:

- *Punctuating*. Enclose each speaker's words in quotation marks, and put any end punctuation—periods, question marks, and exclamation marks—inside the closing quotation mark. Whether you're transcribing words you heard or making them up, you'll sometimes need to add punctuation to reflect the rhythm and sound of the speech. In the last sentence of the example below, see how Chang-Rae Lee adds a comma after "Well" and italicizes "practice" to show intonation—and attitude.

- *Paragraphing*. When you're writing dialogue that includes more than one speaker, start a new paragraph each time the speaker changes.

- *Signal phrases*. Sometimes you'll need to introduce dialogue with SIGNAL PHRASES —"I said," "she asked," and so on—to make clear who is speaking. At times, however, the speaker will be clear enough, and you won't need any signal phrases.

551–54

Here is a conversation between a mother and her son that illustrates each of the conventions for punctuating and paragraphing dialogue:

"Whom do I talk to?" she said. She would mostly speak to me in Korean, and I would answer back in English.

"The bank manager, who else?"

"What do I say?"

"Whatever you want to say."

"Don't speak to me like that!" she cried.

"It's just that you should be able to do it yourself," I said.

"You know how I feel about this!"

"Well, maybe then you should consider it *practice*," I answered lightly, using the Korean word to make sure she understood.

—Chang-Rae Lee, "Coming Home Again"

Interviews

Interviews are a kind of dialogue, with different conventions for punctuation. When you're transcribing an interview, give each speaker's name each time they speak, starting a new line but not indenting, and do not use quotation marks. Here's an excerpt from an interview that graduate student Emma Peters conducted with Brenda Ouattara, a recent two-year college graduate, while researching how the COVID-19 pandemic affected college education and online learning.

Emma Peters: What school did you go to, and what year were you in when the pandemic started?

Brenda Ouattara: I attended LaGuardia Community College in Queens, and it was during my first year there. I was only on campus for a few months before COVID, so I actually finished almost my whole two-year college experience online instead of in person.

Peters: Did you take online classes before the pandemic? How were they different?

Ouattara: I was in hybrid classes before, where we'd meet once a week in person and the rest was online. The online classes during the pandemic were more challenging—it felt like they had more work. Because, in my opinion, the professors felt like we were going to be missing out on the full experience of their class, so they would give us more work. And we also had to learn how to manage doing online courses, and how to figure out and navigate the websites.

Peters: What do you think is lost with an online learning experience that you might have had if you were in person?

✳ academic literacies	● fields	● research
■ rhetorical situations	⁂ processes	● media/design
▲ genres	◆ strategies	

Ouattara: The experience of meeting new students, meeting your actual professors, just being face to face to get the help that you need . . . and having the opportunities to have student fairs or any other events. All of those things were just not happening for us. I feel like it was hard for some students to really be motivated at home. I know a few of my friends actually just decided to take the semesters off. I kept going, but I wasn't as motivated to after a while because it's online, so it's not as exciting as walking in the buildings and meeting people and going straight to your professors for help.

Peters: What do you think are the benefits of online classes?

Ouattara: It's a little bit easier to manage with your personal life. You can have a regular work schedule and then come home and do assignments. But when you're in person you have to be physically in a classroom. It was definitely a benefit being able to set my own schedule.

Peters: How did being online affect your relationship with your professors?

Ouattara: The communication was different—you had to speak through email and online calls, and sometimes it was harder to get a fast response. Some professors were also struggling to use online resources, and sometimes I would ask them a question and they would ask me a question—and we were helping each other out. [Laughs]

Peters: Do you think anything will change permanently going forward after the pandemic?

Ouattara: I don't think it will go back to exactly how it was before, even though many colleges are trying. I feel like I've spent two years doing online learning and now it would almost be out of my comfort zone to go back to in-person classes—this is normal for me now, you know? But everyone is different.

In preparing the interview transcript for publication, Peters had to add punctuation, which of course was not part of the oral conversation, and probably deleted pauses and verbal expressions such as "um" and "uh." At the same time, she kept informal interjections ("you know?") and included expressions of emotion ("Laughs") to maintain the oral flavor of the interview and reflect Ouattara's voice. More tips on interviewing may be found in Chapter 47.

Considering the Rhetorical Situation

As a writer or speaker, it's important to think about the message you want to articulate, the audience you want to reach, and the larger context of your writing.

59–60 ■ **PURPOSE** Your purpose will affect any use of dialogue. Dialogue can help bring a profile to life and make it memorable. Interviews with experts or firsthand witnesses can add credibility to a report or argument.

61–64 ■ **AUDIENCE** Whom do you want to reach, and will dialogue help? Sometimes actual dialogue can help readers hear human voices behind facts or reason.

65–71 ■ **GENRE** Does your genre require dialogue? If you're evaluating or analyzing a literary work, for instance, you may wish to include dialogue from that work. If you're writing a profile of a person or event, dialogue can help you bring your subject to life. Similarly, an interview with an expert can add credibility to a report or argument.

72–74 ■ **STANCE** What is your stance, and can dialogue help you communicate that stance? For example, excerpts of an interview may enable you to challenge someone's views and make your own views clear.

75–77 ■ **MEDIA/DESIGN** Your medium will affect the way you present dialogue. In a print text, you'll present dialogue through written words. In an oral or electronic text, you might include actual recorded dialogue.

If you need more help

518–20 ●
542–54. See also the guidelines on **INTERVIEWING EXPERTS** for advice on setting up and recording interviews and those on **QUOTING, PARAPHRASING,** and **SUMMARIZING** for help with deciding how to integrate dialogue into your text.

☀ academic literacies   ● fields   ● research
■ rhetorical situations   ∴ processes ● media/design
▲ genres   ◆ strategies

43 Explaining Processes

When you explain a process, you tell how something is (or was) done—how a bill becomes a law, how an embryo develops; or you tell someone how to do something—how to throw a curve ball, how to write a memoir. This chapter focuses on those two kinds of explanations, offering examples and guidelines for explaining a process in a way that works for your rhetorical situation.

Explaining a Process Clearly

Whether the process is simple or complex, you'll need to identify its key stages or steps and explain them one by one, in order. The sequence matters because it enables readers to follow your explanation; it is especially important when you're explaining a process that others are going to follow. Most often you'll explain a process chronologically, from start to finish. **TRANSITIONS** —words like "first," "next," "then," and so on—are often necessary, therefore, to show readers how the stages of a process relate to one another and to indicate time sequences. Finally, you'll find that verbs matter; they indicate the actions that take place at each stage of the process.

361–62

Explaining How Something Is Done

All processes consist of steps, and when you explain how something is done, you describe each step, generally in order, from first to last. Here, for

example, is an explanation of how french fries are made, from an essay published in the *New Yorker*:

> Fast-food French fries are made from a baking potato like an Idaho russet, or any other variety that is mealy, or starchy, rather than waxy. The potatoes are harvested, cured, washed, peeled, sliced, and then blanched—cooked enough so that the insides have a fluffy texture but not so much that the fry gets soft and breaks. Blanching is followed by drying, and drying by a thirty-second deep fry, to give the potatoes a crisp shell. Then the fries are frozen until the moment of service, when they are deep-fried again, this time for somewhere around three minutes. Depending on the fast-food chain involved, there are other steps interspersed in this process. McDonald's fries, for example, are briefly dipped in a sugar solution, which gives them their golden-brown color; Burger King fries are dipped in a starch batter, which is what gives those fries their distinctive hard shell and audible crunch. But the result is similar. The potato that is first harvested in the field is roughly 80 percent water. The process of creating a French fry consists, essentially, of removing as much of that water as possible—through blanching, drying, and deep-frying—and replacing it with fat.
>
> —Malcolm Gladwell, "The Trouble with Fries"

Gladwell clearly explains the process of making french fries, showing us the specific steps—how the potatoes "are harvested, cured, washed, peeled, sliced," and so on—and using clear transitions—"followed by," "then," "until," "when"—and action verbs to show the sequence. His last sentence makes his stance clear, pointing out that the process of creating a french fry consists of removing as much of a potato's water as possible "and replacing it with fat."

Explaining How to Do Something

In explaining how to do something, you are giving instruction so that others can follow the process themselves. See how cookbook author Amanda Hesser explains the process of making french fries. She starts by listing the ingredients and then describes the steps:

4 large long Idaho potatoes
Peanut oil
Kosher or coarse sea salt

1. Peel the potatoes. Place in a bowl, cover with water, and refrigerate for 8 hours.

2. Slice the potatoes lengthwise into ¼-inch-thick sticks. Place in a
   bowl, cover with water, and refrigerate for 8 hours more.

3. Drain the potato sticks and lay out on dish towels to dry. Be sure
   they are completely dry before frying.

4. Heat 2 inches of oil to 300 degrees in a large deep pot with a frying basket and a deep-frying thermometer clipped to the side. Add
   just enough potatoes to cover the bottom of the frying basket and
   cook until slightly limp, 1½ to 2 minutes; do not brown. Lift the
   basket and drain the fries, then transfer the potatoes to a wire
   rack set over a baking sheet and separate the sticks. Repeat with
   the remaining potatoes.

5. Increase the heat to 375 degrees. Again add the potatoes in batches
   to the oil and fry until chestnut brown on the edges and crisp. Drain
   and transfer to a bowl lined with paper towels. Immediately season
   with salt, tossing to coat.

—Amanda Hesser, "French Fries"

Hesser's explanation leaves out no details, giving a clear sequence of
steps and descriptive verbs that tell us exactly what to do: "Peel," "Slice,"
"Heat," and so on. After she gives the recipe, she offers variations, including adding a slice of bacon to the oil "for a light pork flavor" or adding
some herb leaves into the oil and then serving the fries "with the crisped
herbs."

Explaining a Process Visually

Some processes are best explained **VISUALLY**, with diagrams or photographs. See, for example, how a blogger explains one process of shaping
dough into a bagel—giving the details in words and then showing us in
photos how to do it:

674–80

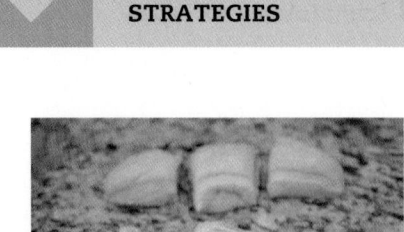

Gently press dough to deflate it a bit and divide into 6 equal portions.

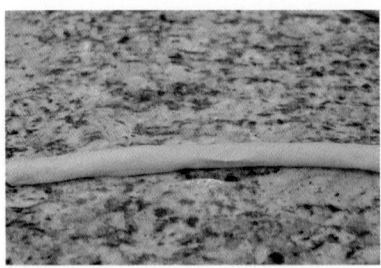

Roll each portion into a rope about 1/2-inch in diameter.

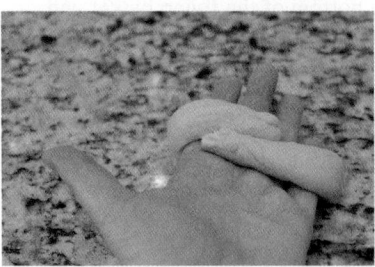

Wrap the dough around your hand like this.

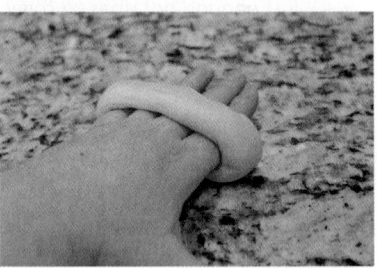

Seal the ends together by rolling back and forth on the counter a few times.

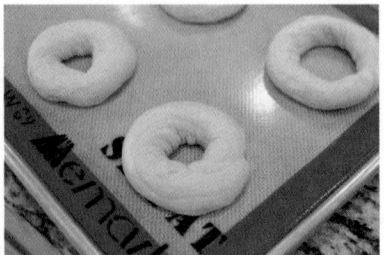

Place bagels on a lined sheet pan. Allow to rise, uncovered.

—Patricia Reitz, *ButterYum*

Photos by Patricia Reitz (butteryum.org).

* academic literacies ● fields ● research
■ rhetorical situations ⁞ processes ● media/design
▲ genres ◆ strategies

Considering the Rhetorical Situation

As always, it's important to think about the message you want to articulate, the audience you want to reach, and the larger context you are writing in.

PURPOSE   Your purpose for writing will affect the way you explain a process. If you're arguing that we should avoid eating fast food, you might explain the process by which chicken nuggets are made. But to give information about how to fry chicken, you would explain the process quite differently.   59–60

AUDIENCE   Whom are you trying to reach, and will you need to provide any special background information or to interest them in the process before you explain it?   61–64

GENRE   Does your genre require you to explain a process? In a lab report, for example, you'll need to explain processes used in the experiment. You might want to explain a process in a profile of an activity or a proposal for a solution.   65–71

STANCE   If you're giving practical directions for doing something, you'll want to take a straightforward "do this, and then do that" perspective. If you're writing to entertain, try taking a clever or amusing stance.   72–74

MEDIA/DESIGN   Your medium will affect the way you explain a process. In a print text, you can use both words and images. On the web, you may have the option of showing a video of the process as well.   75–77

If you need more help

See also **PROFILES** if you are writing about an activity that needs to be explained. See **NARRATING** for more advice on organizing an explanation chronologically. Sometimes you may be assigned to write a whole essay or report that explains a process; see **PROCESSES** for help with drafting, revising, and so on.   245–57
          474–82
          325

44 Narrating

164–95 ▲

245–57 ▲

Narratives are stories. As a writing strategy, a good narrative can lend support to most kinds of writing—in a **POSITION PAPER** arguing for Title IX compliance, for example, you might include a brief narrative about an Olympic sprinter who might never have had an opportunity to compete on a track-and-field team without Title IX. Or you can bring a **PROFILE** of a favorite coach to life with an anecdote about a pep talk they once gave before a championship track meet. Whatever your larger writing purpose, it's important to make sure that any narratives you add support that purpose—they should not be inserted simply to tell an interesting story. You'll also need to compose them carefully—to put them in a clear *sequence*, include *pertinent detail*, and make sure they are appropriate to your particular rhetorical situation.

Sequencing

When we write a narrative, we arrange events in a particular sequence. Writers typically sequence narratives in chronological order, reverse chronological order, or as a flashback.

Use chronological order. Often you may tell the story chronologically, starting at the beginning of an event and working through to the end, as Maya Angelou does in this brief narrative from an essay about her high school graduation:

> The school band struck up a march and all classes filed in as had been rehearsed. We stood in front of our seats, as assigned, and on a signal from the choir director, we sat. No sooner had this been accomplished than the band started to play the national anthem. We rose again and sang the song, after which we recited the pledge of allegiance. We

✳ academic literacies   ⬣ fields   ⬤ research
◼ rhetorical situations   ⁘ processes   ⬤ media/design
▲ genres   ◆ strategies

remained standing for a brief minute before the choir director and the principal signaled to us, rather desperately I thought, to take our seats.

—Maya Angelou, "Graduation"

Use reverse chronological order. You may also begin with the final action and work back to the first, as Aldo Leopold does in this narrative about cutting down a tree:

> Now our saw bites into the 1890s, called gay by those whose eyes turn cityward rather than landward. We cut 1899, when the last passenger pigeon collided with a charge of shot near Babcock, two counties to the north; we cut 1898, when a dry fall, followed by a snowless winter, froze the soil seven feet deep and killed the apple trees; 1897, another drouth year, when another forestry commission came into being; 1896, when 25,000 prairie chickens were shipped to market from the village of Spooner alone; 1895, another year of fires; 1894, another drouth year; and 1893, the year of "the Bluebird Storm," when a March blizzard reduced the migrating bluebirds to near zero.
>
> —Aldo Leopold, *A Sand County Almanac*

RÉSUMÉS* are one genre where we generally use reverse chronological order, listing the most recent jobs or degrees first and then working backward. Notice, too, that we usually write these as narratives—telling what we have done rather than just naming positions we have held:

Glossary

Sept. 2020–present	*Student worker*, Department of Information Management, Central State University, Wilberforce, OH. Compile data and format reports using Excel, Word, and university database.
June–Sept. 2020	*Intern*, QuestPro Corporation, West Louisville, KY. Assisted in development of software.
Sept. 2019–June 2020	*Bagger*, Ace Groceries, Elba, KY. Bagged customers' purchases.

---

*See the ebook at digital.wwnorton.com/fieldguide6 for a chapter on Résumés and Job Letters.

Use a flashback. You can sometimes put a flashback in the middle of a narrative, to tell about an incident that illuminates the larger narrative. Terry Tempest Williams does this in an essay about the startling incidence of breast cancer in her family: she recalls a dinnertime conversation with her father right after her mother's death from cancer, when she learned for the first time what caused all of the cancer in her family:

> Over dessert, I shared a recurring dream of mine. I told my father that for years, as long as I could remember, I saw this flash of light in the night in the desert. That this image had so permeated my being, I could not venture south without seeing it again, on the horizon, illuminating buttes and mesas.
>
> "You did see it," he said.
>
> "Saw what?" I asked, a bit tentative.
>
> "The bomb. The cloud. We were driving home from Riverside, California. You were sitting on your mother's lap. She was pregnant. In fact, I remember the date, September 7, 1957. We had just gotten out of the Service. We were driving north, past Las Vegas. It was an hour or so before dawn, when this explosion went off. We not only heard it, but felt it. I thought the oil tanker in front of us had blown up. We pulled over and suddenly, rising from the desert floor, we saw it, clearly, this golden-stemmed cloud, the mushroom. The sky seemed to vibrate with an eerie pink glow. Within a few minutes, a light ash was raining on the car."
>
> I stared at my father. This was new information to me.
>
> —Terry Tempest Williams,
> "The Clan of the One-Breasted Women"

Williams could have simply announced this information as a fact—but see how much more powerful it is when told in narrative form.

Use time markers. Time markers help readers follow a sequence of events. The most obvious time markers are those that simply label the time, as the narrative entries in a diary, journal, or log might. For example, here is the final part of the narrative kept in a diary by a doomed Antarctic explorer:

✴ academic literacies   ● fields   ● research

■ rhetorical situations   ⁚ processes   ● media/design

▲ genres   ◆ strategies

WEDNESDAY, MARCH 21: Got within eleven miles of depot. Monday night; had to lay up all yesterday in severe blizzard. Today forlorn hope, Wilson and Bowers going to depot for fuel.

MARCH 22 and 23: Blizzard bad as ever—Wilson and Bowers unable to start—tomorrow last chance—no fuel and only one or two [days] of food left—must be near the end. Have decided it shall be natural—we shall march for the depot with or without our effects and die in our tracks.

THURSDAY, MARCH 29: Since the 21st we have had a continuous gale from W.S.W. and S.W. We had fuel to make two cups of tea apiece and bare food for two days on the 20th. Every day we have been ready to start for our depot eleven miles away, but outside the door of the tent it remains a scene of whirling drift. I do not think we can hope for any better things now. We shall stick it out to the end, but we are getting weaker, of course, and the end cannot be far. It seems a pity, but I do not think I can write more. . . .

Last Entry: For God's sake look after our people.

—Robert F. Scott, *Scott's Last Expedition: The Journals*

More often you will integrate time markers into the prose itself, as is done in this narrative about a woman preparing and delivering meals to workers at a cotton gin:

She made her plans meticulously and in secret. <u>One early evening</u> to see if she was ready, she placed stones in two five-gallon pails and carried them three miles to the cotton gin. She rested a little, and then, discarding some rocks, she walked in the darkness to the sawmill five miles farther along the dirt road. <u>On her way back</u> to her little house and her babies, she dumped the remaining rocks along the path.

<u>That same night</u> she worked into the early hours boiling chicken and frying ham. She made dough and filled the rolled-out pastry with meat. <u>At last</u> she went to sleep.

<u>The next morning</u> she left her house carrying the meat pies, lard, an iron brazier, and coals for a fire. <u>Just before lunch</u> she appeared in an empty lot behind the cotton gin. <u>As the dinner noon bell rang</u>, she dropped the savors into boiling fat, and the aroma rose and floated

over to the workers who spilled out of the gin, covered with white lint, looking like specters.

—Maya Angelou,
Wouldn't Take Nothing for My Journey Now

Use transitions. Another way to help readers follow a narrative is with
361–62
TRANSITIONS, words like "first," "then," "meanwhile," "at last," and so on. See how the following paragraphs from Langston Hughes's classic essay about meeting Jesus use transitions (and time markers) to advance the action:

> <u>Suddenly</u> the whole room broke into a sea of shouting, <u>as</u> they saw me rise. Waves of rejoicing swept the place. Women leaped in the air. My aunt threw her arms around me. The minister took me by the hand and led me to the platform.
>
> <u>When</u> things quieted down, in a hushed silence, punctuated by a few ecstatic "Amens," all the new young lambs were blessed in the name of God. <u>Then</u> joyous singing filled the room. <u>That night</u>, for the last time in my life but one—for I was a big boy twelve years old—I cried.

—Langston Hughes, "Salvation"

Including Pertinent Detail

When you include a narrative in your writing, you must decide which details you need—and which ones you don't. For example, you don't want to include so much detail that the narrative distracts the reader from the larger text. You must also decide whether you should include any background, to set the stage for the narrative. The amount of detail you include depends on your audience and purpose: How much detail does your audience need? How much detail do you need to make your meaning clear? In an essay on the suspicion African American men often face when walking at night, a journalist deliberately presents a story without setting the stage at all:

> My first victim was a woman—white, well dressed, probably in her late twenties. I came upon her late one evening on a deserted street

---

☀ academic literacies   ● fields   ● research

■ rhetorical situations   ⁘ processes   ● media/design

▲ genres   ◆ strategies

in Hyde Park, a relatively affluent neighborhood in an otherwise mean, impoverished section of Chicago. As I swung onto the avenue behind her, there seemed to be a discreet, uninflammatory distance between us. Not so. She cast back a worried glance. To her, the youngish black man—a broad six feet two inches with a beard and billowing hair, both hands shoved into the pockets of a bulky military jacket—seemed menacingly close. After a few more quick glimpses, she picked up her pace and was soon running in earnest. Within seconds she disappeared into a cross street.

—Brent Staples, "Black Men and Public Space"

Words like "victim" and phrases like "came upon her" lead us to assume the narrator is scary and perhaps dangerous. We don't know why he is walking on the deserted street because he hasn't told us: he simply begins with the moment he and the woman encounter each other. For his purposes, that's all the audience needs to know at first, and details of his physical appearance that explain the woman's response come later, after he tells us about the encounter. Had he given us those details at the outset, the narrative wouldn't have been nearly so effective. In a way, Staples lets the story sneak up on us, as the woman apparently felt he had on her.

Other times you'll need to provide more background information, as an MIT professor does when she uses an anecdote to introduce an essay about young children's experiences with electronic toys. First the writer tells us a little about *Merlin*, the computer tic-tac-toe game that the children in her anecdote play with. As you'll see, the anecdote would be hard to follow without the introduction:

Among the first generation of computational objects was Merlin, which challenged children to games of tic-tac-toe. For children who had only played games with human opponents, reaction to this object was intense. For example, while Merlin followed an optimal strategy for winning tic-tac-toe most of the time, it was programmed to make a slip every once in a while. So when children discovered strategies that allowed them to win and then tried these strategies a second time, they usually would not work. The machine gave the impression of not being "dumb enough" to let down its defenses twice. Robert, seven, playing

with his friends on the beach, watched his friend Craig perform the "winning trick," but when he tried it, Merlin did not slip up and the game ended in a draw. Robert, confused and frustrated, threw Merlin into the sand and said, "Cheater. I hope your brains break." He was overheard by Craig and Greg, aged six and eight, who salvaged the by-now very sandy toy and took it upon themselves to set Robert straight. "Merlin doesn't know if it cheats," says Craig. "It doesn't know if you break it, Robert. It's not alive." Greg adds, "It's smart enough to make the right kinds of noises. But it doesn't really know if it loses. And when it cheats, it don't even know it's cheating." Jenny, six, interrupts with disdain: "Greg, to cheat you have to know you are cheating. Knowing is part of cheating."

—Sherry Turkle, "Cuddling Up to Cyborg Babies"

Opening and Closing with Narratives

346–54 ⁘

Narratives are often useful as **BEGINNINGS** to essays and other kinds of writing. Everyone likes a good story, so an interesting or pithy narrative can be a good way to get your audience's attention. In the following introductory paragraph, a historian tells a gruesome but gripping story to attract our attention to a subject that might not otherwise merit our interest—bubonic plague:

In October 1347, two months after the fall of Calais, Genoese trading ships put into the harbor of Messina in Sicily with dead and dying men at the oars. The ships had come from the Black Sea port of Caffa (now Feodosiya) in the Crimea, where the Genoese maintained a trading post. The diseased sailors showed strange black swellings about the size of an egg or an apple in the armpits and groin. The swellings oozed blood and pus and were followed by spreading boils and black blotches on the skin from internal bleeding. The sick suffered severe pain and died quickly, within five days of the first symptoms. As the disease spread, other symptoms of continuous fever and spitting of blood appeared instead of the swellings or buboes. These victims coughed and sweated heavily and died even more quickly, within three days or less, sometimes in twenty-four hours. In both types everything that issued from the body—breath, sweat, blood from the buboes and lungs, bloody urine,

✳ academic literacies   ● fields   ● research
■ rhetorical situations   ⁘ processes   ● media/design
▲ genres   ◆ strategies

and blood-blackened excrement—smelled foul. Depression and despair accompanied the physical symptoms, and before the end "death is seen seated on the face."

—Barbara Tuchman,
"This Is the End of the World: The Black Death"

Imagine how different the preceding paragraph would be if it weren't in the form of a narrative. Imagine, for example, that Tuchman had begun by defining bubonic plague. Would that have gotten your interest? The piece was written for a general audience; how might it have been different if it had been written for scientists? Would they need (or appreciate) the story told here?

Narrative can be a good way of **ENDING** a text, too, by winding up a discussion with an illustration of the main point. Here, for instance, is a concluding paragraph from an essay on American values and Las Vegas weddings:

354–58

> I sat next to one . . . wedding party in a Strip restaurant the last time I was in Las Vegas. The marriage had just taken place; the bride still wore her dress, the mother her corsage. A bored waiter poured out a few swallows of pink champagne ("on the house") for everyone but the bride, who was too young to be served. "You'll need something with more kick than that," the bride's father said with heavy jocularity to his new son-in-law; the ritual jokes about the wedding night had a certain Panglossian character, since the bride was clearly several months pregnant. Another round of pink champagne, this time not on the house, and the bride began to cry. "It was just as nice," she sobbed, "as I hoped and dreamed it would be."
>
> —Joan Didion, "Marrying Absurd"

No doubt Didion makes her points about American values clearly and cogently in the essay. But concluding with this story lets us *see* (and hear) what she is saying about Las Vegas wedding chapels, which sell "'niceness,' the facsimile of proper ritual, to children who do not know how else to find it, how to make the arrangements, how to do it 'right.'"

Considering the Rhetorical Situation

As a writer or speaker, it's important to think about the message you want to articulate, the audience you want to reach, and the larger context you are writing in.

59–60 ■ **PURPOSE**

Your purpose will affect the way you use narrative. For example, in an essay about seat belt laws, you might tell about the painful rehabilitation of a teenager who wasn't wearing a seat belt and was injured in an accident in order to show readers why seat belt use is mandatory in all but one state.

61–64 ■ **AUDIENCE**

Whom do you want to reach, and do you have an anecdote or other narrative that will help them understand your topic or persuade them that your argument has merit?

65–71 ■ **GENRE**

Does your genre require you to include narrative? A memoir about an important event might be primarily narrative, whereas a reflection about an event might focus more on the significance of the event than on what happened.

72–74 ■ **STANCE**

What is your stance, and do you have any stories that would help you convey that stance? A funny story, for example, can help create a humorous stance.

75–77 ■ **MEDIA / DESIGN**

In a print or spoken text, you will likely be limited to brief narratives, perhaps illustrated with photos or other images. In an electronic text, you might have the option of linking to full-length narratives or videos available on the web.

If you need more help

325 ⁘
322 ⬣
Glossary

See also the **PROCESSES** chapters if you are assigned to write a narrative essay and need help with drafting, revising, and so on. Two special kinds of narratives are **LAB REPORTS** (which use narrative to describe the steps in an experiment from beginning to end) and **RÉSUMÉS** (which essentially tell the story of the work we've done, at school and on the job).

⁕ academic literacies	⬡ fields	⬤ research
■ rhetorical situations	⁘ processes	⬤ media/design
▲ genres	◆ strategies	

45 Taking Essay Exams

Essay exams present writers with special challenges. You must write quickly, on a topic presented to you on the spot, to show your instructor what you know about a specific body of information. This chapter offers advice on how to take essay exams.

Considering the Rhetorical Situation

PURPOSE	In an essay exam, your purpose is to show that you have mastered certain material and that you can analyze and apply it in an essay. You may need to make an argument or simply convey information on a topic.	59–60
AUDIENCE	Will your course instructor be reading your exam, or will a teaching assistant? Sometimes standardized tests are read by groups of trained readers. What specific criteria will your audience use to evaluate your writing?	61–64
GENRE	Does the essay question specify or suggest a certain genre? In a literature course, you may need to write a compelling literary analysis of a passage. In a history course, you may need to write an argument for the significance of a key historical event. In an economics course, you may need to contrast the economies of the North and South before the Civil War. If the essay question doesn't specify a genre, look for keywords such as "argue," "evaluate," or "explain," which point to a certain genre.	65–71

72–74 **STANCE** In an essay exam, the most appropriate stance is usually unemotional, thoughtful, and critical.

75–77 **MEDIA/DESIGN** If you're taking a test online, write your answers in a word processor, edit there, and then paste them into the exam. If you're handwriting on lined paper or in an exam booklet, write as legibly as you can.

Analyzing Essay Questions

Essay questions usually include key verbs that specify the kind of writing you'll need to do—*argue* a position, *compare* two texts, and so on. Here are some of the most common kinds of writing you'll be asked to do on an essay exam:

104–39
- *Analyze.* Break an idea, theory, text, or event into its parts and examine them. For example, a world history exam might ask you to **ANALYZE** European imperialism's effect on Africa in the late nineteenth century and discuss how Africans responded.

- *Apply.* Consider how an idea or concept might work out in practice. For instance, a film studies exam might ask you to apply the concept of auteurism—a theory of film that sees the director as the primary creator, whose body of work reflects a distinct personal style—to two films by Clint Eastwood. An economics exam might ask you to apply the concept of opportunity costs to a certain supplied scenario.

410–30
- *Argue/prove/justify.* Offer reasons and evidence to support a position. A philosophy exam, for example, might ask you to **ARGUE** whether or not all stereotypes contain a "kernel of truth" and whether believing a stereotype is ever justified.

431–36
- *Classify.* Group something into categories. For example, a marketing exam might ask you to **CLASSIFY** shoppers in categories based on their purchasing behavior, motives, attitudes, or lifestyle patterns.

✳ academic literacies   ⬡ fields   ● research
■ rhetorical situations   ⁑ processes   ● media/design
▲ genres   ◆ strategies

- *Compare/contrast.* Explore the similarities and/or differences between two or more things. An economics exam, for example, might ask you to **COMPARE** the effectiveness of patents and tax incentives in encouraging technological advances.

  ◆ 437–44

- *Critique.* **ANALYZE** and **EVALUATE** a text or argument, considering its strengths and weaknesses. For instance, an evolutionary biology exam might ask you to critique John Maynard Smith's assertion that "scientific theories say nothing about what is right but only about what is possible" in the context of the theory of evolution.

  ▲ 104–39
  214–22

- *Define.* Explain what a word or phrase means. An art history exam, for example, might ask you to **DEFINE** negative space and discuss the way various artists use it in their work.

  ◆ 445–55

- *Describe.* Tell about the important characteristics or features of something. For example, a sociology exam might ask you to **DESCRIBE** Erving Goffman's theory of the presentation of self in ordinary life, focusing on roles, props, and setting.

  ◆ 456–63

- *Evaluate.* Determine something's significance or value. A drama exam, for example, might ask you to **EVALUATE** the setting, lighting, and costumes in a filmed production of *Macbeth*.

  ▲ 214–22

- *Explain.* Provide reasons and examples to clarify an idea, argument, or event. For instance, a rhetoric exam might ask you to explain the structure of the African American sermon and discuss its use in writings of Frederick Douglass and Martin Luther King Jr.

- *Summarize/review.* Give the major points of a text or idea. A political science exam, for example, might ask you to **SUMMARIZE** John Stuart Mill's concept of utilitarianism and its relation to freedom of speech.

  ● 550–51

- *Trace.* Explain a sequence of ideas or order of events. For instance, a geography exam might ask you to trace the patterns of international migration since 1970 and discuss how these patterns differ from those of the period between 1870 and World War I.

Some Guidelines for Taking Essay Exams

Before the exam

16–19

- *Read over* your class notes and course texts strategically, **ANNOTATING** them to keep track of details you'll want to remember.

- *Collaborate* by forming a study group to help one another master the course content.

535–39

- *Review* key ideas, events, terms, and themes. Look for common themes and **CONNECTIONS** in lecture notes, class discussions, and any readings—they'll lead you to important ideas.

- *Ask* your instructor about the form the exam will take: how long it will be, what kind of questions will be on it, how it will be evaluated, and so on. Working with a study group, write questions you think your instructor might ask, and then answer the questions together.

333–34

- *Warm up* just before the exam by **FREEWRITING** for 10 minutes or so to gather your thoughts.

During the exam

- *Scan the questions* to determine how much each part of the test counts and how much time you should spend on it. For example, if one essay is worth 50 points and two others are worth 25 points each, you'll want to spend half your time on the 50-point question.

- *Read over* the entire test before answering any questions. Start with the question you feel most confident answering, which may or may not be the first question on the test.

- *Don't panic.* Sometimes when students first read an essay question, their minds go blank, but after a few moments they start to recall the information they need.

- *Plan.* Although you won't have much time for revising or editing, you still need to plan and allow yourself time to make some last-minute changes before turning in the exam. So apportion your time. For a three-question essay test in a two-hour test period, you might divide your time like this:

☀ academic literacies   ● fields   ● research

■ rhetorical situations   ⁘ processes   ● media/design

▲ genres   ◆ strategies

Total Exam Time—120 minutes
Generating ideas—20 minutes (6–7 minutes per question)
Drafting—85 minutes (45 for the 50-point question,
                         20 for each 25-point question)
Revising, editing, proofreading—15 minutes

Knowing that you have built in time at the end of the exam period can help you remain calm as you write, because you can use that time to fill in gaps or reconsider answers you feel unsure about.

- *Jot down the main ideas* you need to cover in answering the question on scratch paper or on the cover of your exam book, numbering those ideas in the order you think makes sense—and you have an outline for your essay. If you're worried about time, plan to write the most important parts of your answers early on. If you don't complete your answer, refer your instructor to your outline to show where you were headed.

- *Turn the essay question into your introduction,* like this:

    Question: How did the outcomes of World War II differ from those of World War I?

    Introduction: The outcomes of World War II differed from those of World War I in three major ways: World War II affected more of the world and its people than World War I did, distinctions between citizens and soldiers were eroded, and the war's brutality made it impossible for Europe to continue to claim cultural superiority over other cultures.

- *State your thesis explicitly,* provide **REASONS** and **EVIDENCE** to support your thesis, and use **TRANSITIONS** to move logically from one idea to the next. Restate your main point in your conclusion. You don't want to give what one professor calls a "garbage truck answer," dumping everything you know into your answer and expecting the instructor to sort it all out.

    ◆ 413–14
    414–22
    ⁝ 361–62

- *Write on every other line* and only on one side of each page so that you'll have room to make additions or corrections. If you're typing on a computer, be sure to double-space.

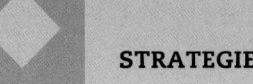

- *If you have time left, go over your exam,* looking for ideas that need elaboration as well as for grammatical and punctuation errors.

After the exam. Your instructor may not return your exam or may not provide comments on it. So consider asking for a conference to go over your work so you can learn what you did well and where you need to improve—important knowledge to take with you into your next exam.

Part 7
Doing Research

We do research all the time, for many different reasons. We search the web for information about a new computer, ask friends about the best place to get coffee, try on several pairs of jeans before deciding which ones to buy. You have no doubt done your share of library research before now, and you probably have visited a number of schools' websites before deciding which college you wanted to attend. Research, in other words, is something you do every day. The following chapters offer advice on the kind of research you'll need to do for your academic work and, in particular, for research projects.

Doing Research

46 Getting a Start on Research

When you need to do research, it's sometimes tempting to jump in and start looking for information right away. However, doing research is complex and time-consuming. Research-based writing projects usually require you to follow several steps. You need to come up with a topic (or to analyze the requirements of an assigned topic) and come up with a research question to guide your research efforts. Once you do some serious, focused research to find the information you need, you'll be ready to turn your research question into a tentative thesis and sketch out a rough outline. After doing whatever additional research you need to fill in your outline, you'll write a draft—and get some response to that draft. You may then need to do even more research before revising. Once you revise, you'll need to edit and proofread. In other words, there's a lot to do. You need a schedule.

Establishing a Schedule and Getting Started

A good way to start a research project is by creating a timeline for getting all this work done, perhaps using the form on the next page. Once you have a schedule, you can get started. The sections that follow offer advice on considering your rhetorical situation, coming up with a topic, and doing preliminary research; developing a research question, a tentative thesis, and a rough outline; and creating a working bibliography and keeping track of your sources. The chapters that follow offer guidelines for **FINDING SOURCES**, **EVALUATING SOURCES**, and **SYNTHESIZING IDEAS**.

501–23
524–34
535–41

Scheduling a Research Project

Complete by:

Analyze your rhetorical situation.	_____
Choose a possible topic or analyze the assignment.	_____
Plan a research strategy, and do preliminary research.	_____
Come up with a research question.	_____
Schedule interviews and other field research.	_____
Find and evaluate sources.	_____
Read sources, and take notes.	_____
Do any field research.	_____
Come up with a tentative thesis and outline.	_____
Write a draft.	_____
Get response.	_____
Do any additional research.	_____
Revise.	_____
Prepare a list of works cited.	_____
Edit.	_____
Proofread the final draft.	_____
Submit the final draft: your deadline.	_____

You may find it useful to start with your deadline—when the final project is due—and work backward from there, so you'll know how much time you have and can estimate how much time you'll need to do each task.

Considering the Rhetorical Situation

As with any writing task, you need to start by considering your purpose, your audience, and the rest of your rhetorical situation:

59–60

PURPOSE   Is this project part of an assignment—and if so, does it specify any one purpose? If not, what is your broad purpose? To inform? argue? analyze? A combination?

※ academic literacies   ● fields   ● research
■ rhetorical situations   ⁂ processes   ● media/design
▲ genres   ◆ strategies

AUDIENCE	To whom are you writing? What does your audience likely know about your topic, and is there any background information you'll need to provide? What opinions or attitudes do your readers likely hold? What kinds of evidence will they find persuasive? How do you want them to respond to your writing?	■ 61–64
GENRE	Are you writing to report on something? to compose a profile? to make a proposal? an argument? What are the requirements of your genre in terms of the number and kind of sources you must use?	■ 65–71
STANCE	What is your attitude toward your topic? What accounts for your attitude? How do you want to come across? As curious? Critical? Positive? Something else?	■ 72–74
MEDIA/DESIGN	What medium or media will you use? Print? Spoken? Electronic? Will you need to create any charts, photographs, video, presentation software slides, or other visuals?	■ 75–77

Coming Up with a Topic

If you need to choose a topic, consider your interests as they relate to the course for which you're writing. What do you want to learn about? What do you have questions about? What topics from the course have you found intriguing? What community, national, or global issues do you care about? Once you've thought of a potential topic, use the questions in Chapter 27, **WRITING AS INQUIRY**, to explore it and find an angle on it that you can write about—and want to.

⁘ 329–32

If your topic is assigned, make sure you understand exactly what it asks you to do. Read the assignment carefully, looking for keywords: Does it ask you to **ANALYZE**, **COMPARE**, **EVALUATE**, **SUMMARIZE**, or **ARGUE**? If the assignment offers broad guidelines but allows you to choose within them, identify the requirements and the range of possible topics, and define your topic within those constraints.

▲ 104–39
◆ 437–44
▲ 214–22
● 550–51
◆ 410–30

For example, in an American history course, your instructor might ask you to "discuss social effects of the Civil War." Potential but broad topics

might include poverty among Confederate soldiers or former slaveholders, the migration of members of those groups to Mexico or Northern cities, the establishment of independent African American churches, or the spread of the Ku Klux Klan—to name only a few of the possibilities.

Think about what you know about your topic.   Chances are you already know something about your topic, and articulating that knowledge can help you see possible ways to focus the topic or come up with potential sources of information. FREEWRITING, LISTING, CLUSTERING, and LOOPING are all good ways of tapping your knowledge of your topic. Consider where you might find information about it: Have you read about it in a textbook? heard stories about it on the news? visited websites focused on it? Do you know anyone who knows about this topic?

333–36

Narrow the topic.   As you consider possible topics, look for ways to narrow your topic's focus to make it specific enough to discuss in depth. For example:

> **Too general:** fracking

> **Still too general:** fracking and the environment

> **Better:** the potential environmental effects of extracting natural gas through the process of hydraulic fracturing, or fracking

If you limit your topic, you can address it with specific information that you'll be more easily able to find and manage. In addition, a limited topic will be more likely to interest your audience than a broad topic that forces you to use abstract, general statements. For example, it's much harder to write well about "the environment" than it is to address a topic that explores a single environmental issue.

Consulting with Librarians and Doing Preliminary Research

Consulting with a reference librarian at your school and doing some preliminary research in the library can save you time in the long run. Reference librarians can direct you to the best scholarly sources for your topic

and help you focus the topic by determining appropriate search terms and **KEYWORDS** —significant words that appear in the title, abstract, or text of potential sources and that you can use to search for information on your topic in library catalogs, in databases, and on the web. These librarians can also help you choose the most appropriate reference works, sources that provide general overviews of the scholarship in a field. General internet searches can be time-consuming, as they often result in thousands of possible sites—too many to weed out efficiently, either by revising your search terms or by going through the sites themselves, many of which are unreliable. Library databases, in contrast, include only sources that already have been selected by experts, and searches in them usually present manageable numbers of results.

● 506–9

Reading a *Wikipedia* entry on your topic can help you understand background information, come up with new keywords to search, and discover related topics. A list of references—often with hyperlinks—at the bottom of each entry can lead to potential sources. *Google*, *Google Scholar*, and other free search tools can also be helpful as you begin to explore your topic. See pages 509–11 for advice on using these and other online search tools.

Coming Up with a Research Question

Once you've surveyed the territory of your topic, you'll likely find that your understanding of it has become broader and deeper. You may find that your interests have changed and your research has led to surprises and additional research. That's okay: as a result of exploring avenues you hadn't anticipated, you may well come up with a better topic than the one you'd started with. At some point, though, you need to develop a research question—a specific question that you will then work to answer through your research.

To write a research question, review your analysis of the **RHETORICAL SITUATION** to remind yourself of any time constraints or length considerations. Generate a list of questions beginning with "what," "when," "where," "who," "how," "why," "would," "could," and "should." Here, for example, are some questions about the tentative topic "the potential environmental

■ 57

effects of extracting natural gas through the process of hydraulic fracturing, or fracking":

> *What* are the environmental effects of fracking?
>
> *When* was fracking introduced as a way to produce natural gas?
>
> *Where* is fracking done, and *how* does this affect the surrounding people and environment?
>
> *Who* will benefit from increased fracking?
>
> *How* is fracking done, and *how much* energy does the process use?
>
> *Why* do some environmental groups oppose fracking?
>
> *Would* other methods of extracting natural gas be safer?
>
> *Could* fracking cause earthquakes?
>
> *Should* fracking be expanded, regulated, or banned?

Select one question from your list that you find interesting and that suits your rhetorical situation. Use the question to guide your research.

Drafting a Tentative Thesis

347–49 Once your research has led you to a possible answer to your research question, try formulating that answer as a tentative **THESIS**. You need not be committed to the thesis; in fact, you shouldn't be. The object of your research should be to learn about your topic, not to find information that simply supports what you already think you believe. Your tentative thesis may (and probably will) change as you learn more about your subject, consider the many points of view on it, and reconsider your topic and, perhaps, your goal: what you originally planned to be an argument for considering other points of view may become a call to action. However tentative, a thesis allows you to move forward by clarifying your purpose for doing research. Here are some tentative thesis statements on the topic of fracking:

> Fracking is a likely cause of earthquakes in otherwise seismically stable regions of the country.

* academic literacies   ● fields   ● research
■ rhetorical situations   ● processes   ● media/design
▲ genres   ◆ strategies

The federal government should strictly regulate the production of natural gas by fracking.

Fracking can greatly increase our supplies of natural gas, but other methods of producing energy should still be pursued.

As with a research question, a tentative thesis should guide your research efforts—but you should be ready to revise it as you learn still more about your topic. Research should be a process of **INQUIRY** in which you approach your topic with an open mind, ready to learn and possibly change. If you hold too tightly to a tentative thesis, you risk focusing only on evidence that supports your own view—a tendency called **CONFIRMATION BIAS** that can make your writing biased and unconvincing.

329–32

Glossary

Creating a Rough Outline

After you've created a tentative thesis, write out a rough **OUTLINE** for your research project. This outline can be a simple list of topics you want to explore, something that will help you structure your research efforts and organize your notes and other materials. As you read your sources, you can use the outline to keep track of what you need to find and where the information you do find fits into your argument. Then you'll be able to see if you've covered all the ideas you intended to explore—or whether you need to rethink the categories on your outline.

340–42

Keeping a Working Bibliography

A working bibliography is a record of all the sources you consult. You should keep such a record so that you can find sources easily when you need them and then cite any that you actually use. Your library likely offers tools to store source information you find in its databases and catalog, and software such as *Zotero* or *EasyBib* can also help you save, manage, and cite your sources. You may find it helpful to print out bibliographical data you find useful or to keep your working bibliography on index cards or in a notebook. However you decide to compile your working bibliography, include all the information you'll need later to document any sources you use; follow the **DOCUMENTATION** style you'll use when you write so that you won't need to go back to your sources to find the information. Some

560–63

Information for a Working Bibliography

FOR A BOOK

Library call number
Author(s) or editor(s)
Title and subtitle
Publication information: city, publisher, year of publication
Other information: edition, volume number, translator, and so on
If your source is an essay in a collection, include its author, title, and page numbers.

FOR A SOURCE FROM A DATABASE

Publication information for the source, as listed above
Name of database
DOI (digital object identifier) or URL of original source, such as the periodical in which an article was published
Stable URL or permalink for database
Date you accessed source

FOR AN ARTICLE IN A PRINT PERIODICAL

Author(s)
Title and subtitle
Name of periodical
Volume number, issue number, date
Page numbers

FOR A WEB SOURCE

URL
Author(s) or editor(s) if available
Name of site
Sponsor of site
Date site was first posted or last updated
Date you accessed site
If the source is an article or book reprinted on the web, include its title, the title and publication information of the periodical or book, where it was first published, and any page numbers.

academic literacies   fields   research
rhetorical situations   processes   media/design
genres   strategies

databases make this step easy by preparing rough-draft citations in several styles that you can copy, paste, and edit.

On the previous page is most of the basic information you'll want to include for each source in your working bibliography. Go to digital.wwnorton.com/fieldguide6 for templates you can use to keep track of this information.

Keeping Track of Your Sources

- *Staple together photocopies and printouts.* It's easy for individual pages to get shuffled or lost on a desk or in a backpack. Keep a stapler handy, and fasten pages together as soon as you copy them or print them out.

- *Bookmark web sources,* or save them using a free bookmark management tool available through several library databases. For database sources, use the *DOI* or *stable URL, permalink,* or *document URL* (the terms used by databases vary)—not the URL in the "Address" or "Location" box in your browser, which will expire after you end your online session.

- *Label everything.* Label your copies with the source's author and title.

- *Highlight sections you plan to use.* When you sit down to draft, your goal will be to find what you need quickly, so as soon as you decide you might use a source, highlight the paragraphs or sentences that you think you'll use. If your instructor wants copies of your sources to see how you used them, you've got them ready. If you're using PDF copies or websites, you can highlight or add notes using an online annotation tool like *Hypothesis* or *Diigo.*

- *Use your rough outline to keep track of what you've got.* In the margin of each highlighted section, write the number or letter of the outline division to which the section corresponds. (It's a good idea to write it in the same place consistently so you can flip through a stack of copies and easily see what you've got.) Alternatively, attach sticky notes to each copy, using a different color for each main heading in your outline.

- *Store all your research material in one place.* If you prefer to work digitally, create computer folders and subfolders that correspond to your rough outline. Or use web apps like *PowerNotes* or *Scrible* to collect, annotate, and organize your sources. If you prefer to work with paper, use an accordion file folder or a box. Having your material in one place will make writing a draft easier—and will serve you well if you're required to create a portfolio that includes your research notes, copies of sources, and drafts.

- *Use a reference manager.* Web-based reference or citation management software allows you to create and organize a personal database of resources. You can import references from databases to a personal account, organize them, and draft citations in various formats. *RefWorks, EndNote, Mendeley,* and *Zotero* are four such systems; check with your librarian to see what system your library supports, or search online, as several of them are available for free. Be aware, though, that the citations generated are often inaccurate and need to be checked carefully for content and format. So treat them as rough drafts, and plan to edit them.

If you need more help

501–23
524–34

See the guidelines on **FINDING SOURCES** once you're ready to move on to in-depth research and those on **EVALUATING SOURCES** for help thinking critically about the sources you find.

✳ academic literacies   ⬤ fields   ⬤ research

◼ rhetorical situations   ⁖ processes   ⬤ media/design

▲ genres   ◆ strategies

47 Finding Sources

To analyze media coverage of the 2020 Democratic National Convention, you examine news stories and blogs published at the time. To write an essay interpreting a poem by Maya Angelou, you study the poem and read several critical interpretations in literary journals. To write a report on career opportunities in psychology, you interview a graduate of your university who is working in a counseling center. In each of these cases, you go beyond your own knowledge to consult additional sources of information. Depending on your topic, then, you'll need to choose from many sources for your research—from reference works, books, periodicals, and websites to surveys, interviews, and other kinds of field research that you yourself conduct.

This chapter offers guidelines for locating a range of sources—print and online, general and specialized, published and firsthand. Keep in mind that as you do research, finding and **EVALUATING SOURCES** are two activities that usually take place simultaneously. So this chapter and the next one go hand in hand.

● 524–34

Kinds of Sources

Primary and secondary sources. Your research will likely lead you to both primary and secondary sources. *Primary sources* include historical documents, literary works, eyewitness accounts, field reports, diaries, letters, and lab studies, as well as any original research you do through interviews, observation, experiments, or surveys. *Secondary sources* include scholarly books and articles, reviews, biographies, textbooks, and other works that interpret or discuss primary sources. Novels and films are primary sources; articles interpreting them are secondary sources. The Declaration of Independence is a primary historical document; a historian's

description of the events surrounding the Declaration's writing is secondary. A published report of scientific findings is primary; a critique of that report is secondary.

Whether a work is considered primary or secondary sometimes depends on your topic and purpose. If you're analyzing a poem, a critic's article interpreting the poem is a secondary source—but if you're investigating that critic's work, the article would be a primary source for your own study and interpretation.

Secondary sources are often useful because they can help you understand and evaluate primary source material. Whenever possible, however, you should find and use primary sources, because secondary sources can distort or misrepresent the information in primary sources. For example, a seemingly reputable secondary source describing the 1948 presidential election asserted that the *New York Times* ran a headline reading, "Thomas E. Dewey's Election as President Is a Foregone Conclusion." But the actual article was titled "Talk Is Now Turning to the Dewey Cabinet," and it began by noting "[the] *popular view that* Gov. Thomas E. Dewey's election as President is a foregone conclusion." Here the secondary source not only got the headline wrong but also distorted the source's intended meaning by leaving out an important phrase. Your research should be as accurate and reliable as possible; using primary sources whenever you can helps ensure that it is.

Scholarly and popular sources. Scholarly sources are written by academic experts or scholars in a particular discipline and are *peer-reviewed*— evaluated by other experts in the same discipline for their factual accuracy and lack of bias. They are also written largely *for* experts in a discipline, as a means of sharing research, insights, and in-depth analysis with one another; that's why they must meet high standards of accuracy and objectivity and adhere to the discipline's accepted research methods, including its style for documenting sources. Scholarly articles are usually published in academic journals; scholarly books may be published by university presses or by other academically focused publishers.

Popular sources include just about all other online and print publications, from websites to magazines to books written for nonspecialists. These sources generally explain or provide opinion on current events or topics of general interest; when they discuss scholarly research, they tend

to simplify the concepts and facts, providing definitions, narratives, and examples to make them understandable to nonspecialist audiences. They are often written by journalists or other professional writers who may specialize in a particular area but who report or comment on the scholarship of others rather than doing any themselves. Their most important difference from scholarly sources is that popular sources are not reviewed by other experts in the field being discussed, although editors or fact-checkers review the writing before it's published.

That said, the distinction can be blurry: many scholars write books for a general readership that are informed by those authors' own scholarship, and many writers of popular sources have extensive expertise in the subject. Even if it's not a requirement, citing scholarly sources often contributes to your own authority as a writer, demonstrating that you are familiar with important research and scholarship and that your own writing is informed by it.

In most of your college courses, you'll be expected to rely primarily on scholarly sources rather than popular ones. However, if you're writing about a very current topic or need to provide background information on a topic, a mix of scholarly and popular sources may be appropriate. To see how scholarly and popular sources differ in appearance, look at the Documentation Map for scholarly journals (p. 583) and at the illustrations on pages 504–5. Here's a guide to determining whether or not a potential source is scholarly:

Identifying scholarly sources: what to look for

- *The author:* Look for the author's credentials, including affiliations with academic or other research-oriented institutions. Make sure the author has expertise on the topic being discussed; an English professor may have academic credentials but not be an expert on the environment, while a journalist may have studied environmental issues for years and know a great deal.

- *The publisher:* Look for academic journals, university presses, and professional organizations such as the Modern Language Association or the American Psychological Association; popular sources are published in general-interest publications such as the *Atlantic* or *Slate* or trade publishers such as Penguin Random House or HarperCollins.

Scholarly Source

Published in an academic journal.

Includes an abstract.

Cites academic research with consistent documentation style.

Describes research methods; includes numerical data.

Lists multiple authors who are academics.

Includes complete references list.

Journal List ▸ NIHPA Author Manuscripts ▸ PMC2918968

NIH Public Access
Author Manuscript
Accepted for publication in a peer reviewed journal

About Author manuscripts | Submit a manuscript

J Res Pers. Author manuscript; available in PMC 2011 August 1.
Published in final edited form as:
J Res Pers. 2010 August 1; 44(4): 478–484.
doi: 10.1016/j.jrp.2010.06.001

PMCID: PMC2918968
NIHMSID: NIHMS216233

Sounds like a Narcissist: Behavioral Manifestations of Narcissism in Everyday Life

Nicholas S. Holtzman, Simine Vazire, and Matthias R. Mehl

Author information ▸ Copyright and License information ▸

See other articles in PMC that cite the published article.

Abstract   Go to: ⊡

Little is known about narcissists' everyday behavior. The goal of this study was to describe how narcissism is manifested in everyday life. Using the Electronically Activated Recorder (EAR), we obtained naturalistic behavior from participants' everyday lives. The results suggest that the defining characteristics of narcissism that have been established from questionnaire and laboratory-based studies are borne out in narcissists' day-to-day behaviors. Narcissists do indeed behave in more extraverted and less agreeable ways than non-narcissists, skip class more (among narcissists high in exploitativeness/entitlement only), and use more sexual language. Furthermore, we found that the link between narcissism and disagreeable behavior is strengthened when controlling for self-esteem, thus extending prior questionnaire-based findings (Paulhus, Robins, Trzesniewski, & Tracy, 2004) to observed, real-world behavior.

Keywords: narcissism, behavior, personality traits, sexual behavior, language use

Narcissists love attention. Lucky for them, they have recently received a considerable amount of it from academic psychologists, especially in laboratory settings (e.g., Back, Schmukle, & Egloff, 2010; Bushman & Baumeister, 1998; Campbell, Foster, & Finkel, 2002; Miller et al., 2009). This laboratory research has led to several wide-reaching theories about why narcissists do what they do (Holtzman & Strube, 2010a; Morf & Rhodewalt, 2001; Twenge & Campbell, 2009; Vazire & Funder, 2006). Despite all this attention from researchers, however, we still know little about what narcissists actually do in their everyday lives. The aim of this paper is to help create an empirical basis for a more complete understanding of narcissism by exploring behavioral manifestations of narcissism in everyday life. Thus, we intend to answer a simple, yet largely unanswered question: What do narcissists do on a day-to-day basis?

Method   Go to: ⊡

Participants

Participants were 80 undergraduate students at the University of Texas at Austin (79 provided valid EAR data), recruited mainly from introductory psychology courses and by flyers in the psychology department. The sample was 54% female, and the ethnic composition of the sample was 65% White, 21% Asian, 11% Latino, and 3% of another ethnicity. Participants ranged from 18 to 24 years old (M = 18.7, SD = 1.4). Participants were compensated $50. Data from this sample were also reported in Vazire and Mehl (2008), where further information can be found about the study.¹

Narcissistic Personality Inventory (NPI)

The NPI is a 40-item test of narcissism that is reliable and well-validated (Raskin & Terry, 1988). The items on this forced-choice test contain pairs of statements such as "Sometimes I tell good stories" (non-narcissistic) versus "Everybody to hear my stories" (narcissistic). In our study, the NPI exhibited good reliability (α = .83). As seen in Table 1, we also calculated means and reliabilities for four facets (Emmons, 1987).

Table 1
Means, Standard Deviations, Gender-Differences, and Reliabilities
for the NPI and NPI Facets

Contributor Information   Go to: ⊡

Nicholas S. Holtzman, Washington University in St. Louis.

Simine Vazire, Washington University in St. Louis.

Matthias R. Mehl, University of Arizona.

References   Go to: ⊡

1. Back MD, Schmukle SC, Egloff B. Why are narcissists so charming at first sight? Decoding the narcissism-popularity link at zero acquaintance. Journal of Personality and Social Psychology. 2010;98:132–145. [PubMed]
2. Baumeister RF, Vohs KD, Funder DC. Psychology as the science of self-reports and finger movements: Whatever happened to actual behavior? Perspectives on Psychological Science. 2007;2:396–403.

PubReader format: click here to try

Formats:
Article | PubReader | sPub (beta) | PDF (338K)

Related citations in PubMed
Impulsivity and the self-defeating behavior of narcissists. [Pers Soc Psychol Rev. 2006]
Why are narcissists so charming at first sight? Decoding the narcissism-popularity link at zero acqu[J Pers Soc Psychol. 2010]
The performance of narcissists rises and falls with perceived opportunity for glory. [J Pers Soc Psychol. 2002]
An empirical typology of narcissism and mental health in late adolescence. [J Adolesc. 2006]
Animal models of obsessive-compulsive disorder: rationale to understanding psychobiology and ; [Psychiatr Clin North Am. 2006]
See reviews...
See all...

Cited by other articles in PMC
Evidence for the criterion validity and clinical utility of the Pathological Narcissism Inventory [Assessment. 2012]
See all...

Links
MedGen
PubMed

Recent activity   Turn Off Clear
Sounds like a Narcissist: Behavioral Manifestations of Narcissism in Everyday Li... PMC
See more...

Does self-love lead to love for others? A story of narcissistic game playing. [J Pers Soc Psychol. 2002]
Interpersonal and intrapsychic adaptiveness of trait self-enhancement: a mixed blessing? [J Pers Soc Psychol. 1998]

Knowing me, knowing you: the accuracy and unique predictive validity of self-ratings and other-ratin[J Pers Soc Psychol. 2008]

A principal-components analysis of the Narcissistic Personality Inventory and further evidence of its [J Pers Soc Psychol. 1988]
Narcissism theory and measurement. [J Pers Soc Psychol. 1987]

✳ academic literacies   ⬟ fields   ● research

■ rhetorical situations   ⁚• processes   ● media/design

▲ genres   ◆ strategies

Popular Source

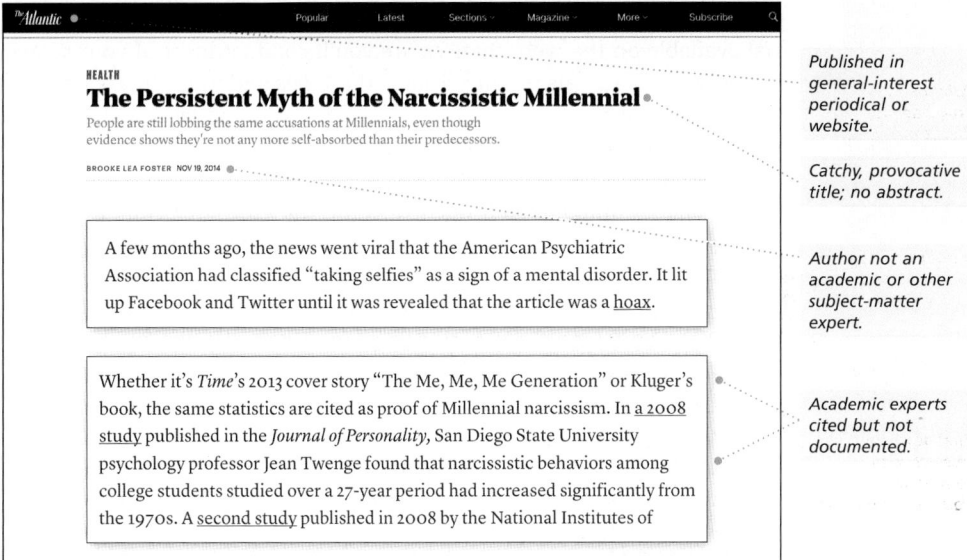

The Atlantic   Popular   Latest   Sections ˅   Magazine ˅   More ˅   Subscribe   🔍

HEALTH
The Persistent Myth of the Narcissistic Millennial
People are still lobbing the same accusations at Millennials, even though
evidence shows they're not any more self-absorbed than their predecessors.

BROOKE LEA FOSTER NOV 19, 2014

> A few months ago, the news went viral that the American Psychiatric
> Association had classified "taking selfies" as a sign of a mental disorder. It lit
> up Facebook and Twitter until it was revealed that the article was a hoax.

> Whether it's *Time*'s 2013 cover story "The Me, Me, Me Generation" or Kluger's
> book, the same statistics are cited as proof of Millennial narcissism. In a 2008
> study published in the *Journal of Personality,* San Diego State University
> psychology professor Jean Twenge found that narcissistic behaviors among
> college students studied over a 27-year period had increased significantly from
> the 1970s. A second study published in 2008 by the National Institutes of

*Published in
general-interest
periodical or
website.*

*Catchy, provocative
title; no abstract.*

*Author not an
academic or other
subject-matter
expert.*

*Academic experts
cited but not
documented.*

- *Peer review.* Look for a list of reviewers at the front of the journal or on the journal's or publisher's website. If you don't find one, the source is probably not peer-reviewed.

- *Source citations.* Look for a detailed list of works cited or references at the end of the source and citations either parenthetically within the text or in footnotes or endnotes. (Popular sources may include a reference list but seldom cite sources within the text, except in signal phrases.)

- *Language and content.* Look for abstracts (one-paragraph summaries of the contents) at the beginning of articles and for technical or specialized language and concepts that readers are assumed to be familiar with.

- *Other clues.* Look for little or no advertising on websites or within the journal; for a plain design with few or no illustrations, especially in print sources; and for listing in academic databases when you limit your search to *academic*, *peer-reviewed*, or *scholarly sources*.

Searching Effectively Using Keywords

Whether you're searching for books, articles in periodicals, or other material available on the web, chances are you'll conduct most of your search online. Most materials produced since the 1980s and most library catalogs are online, and most periodical articles can be found by searching electronic indexes and databases. In each case, you can search for authors, titles, or subjects.

To search online, you'll need to come up with keywords. Keywords are significant words that stand for an idea or concept. The key to searching efficiently is to use keywords and combinations of them that will focus your searches on the information you need—but not too much of it. Often you'll start out with one general keyword that will yield far too many results; then you'll need to switch to more specific terms or combinations (*homeopathy* instead of *medicine*, or *secondary education Japan* instead of *education Japan*).

Other times your keyword search won't yield enough sources; then you'll need to use broader terms or combinations (*education Japan* instead of *secondary education Japan*) or substitute synonyms (*home remedy* instead of *folk medicine*). Sometimes you'll need to learn terms used in academic disciplines or earlier in history for things you know by other names, such as *myocardial infarction* rather than *heart attack* or *the Great War* instead of *World War I*. Or look through the sources that turn up in response to other terms to see what keywords you might use in subsequent searches. Searching requires flexibility, in the words you use and the methods you try.

Finding keywords using word clouds.   One way to find keywords to help you narrow and focus your topic is to create a word cloud, a visual representation of words used in a text; the more often a word is used, the larger it looks in the word cloud. Several websites, including *Tagxedo*, *Wordle*, and *TagCrowd*, let you create word clouds. Examining a word cloud created from an article in a reference work may help you see what terms are used to discuss your topic—and may help you see new possible ways

✳ academic literacies   ● fields   ● research
■ rhetorical situations   ⁚ processes   ● media/design
▲ genres   ◆ strategies

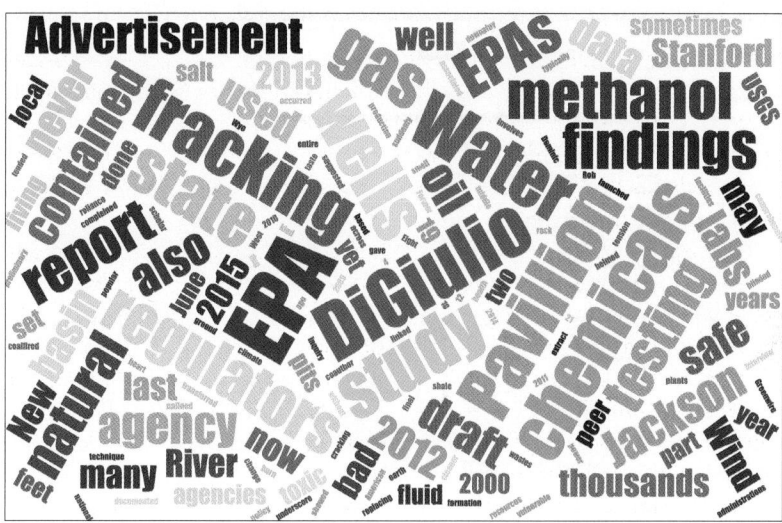

to narrow it. Above, for example, is a word cloud derived from an article in *Scientific American* discussing fracking. Many of the terms—"fracking," "water," "gas," "wells," "drilling"—are just what you'd expect. However, some terms—"USGS," "DiGiulio," "coalfired"—may be unfamiliar and lead to additional possibilities for research. For instance, "DiGiulio" is the last name of an expert on fracking whose publications might be worth examining, while "USGS" is an acronym for the United States Geological Survey, a scientific government agency.

Finding keywords using databases.   Once you've begun searching for and finding possible sources, you can expand your list of possible keywords by skimming the "detailed record" or "metadata" page for any scholarly articles you find, where full bibliographic information on the source may be found. A search for *fracking* resulted in this source:

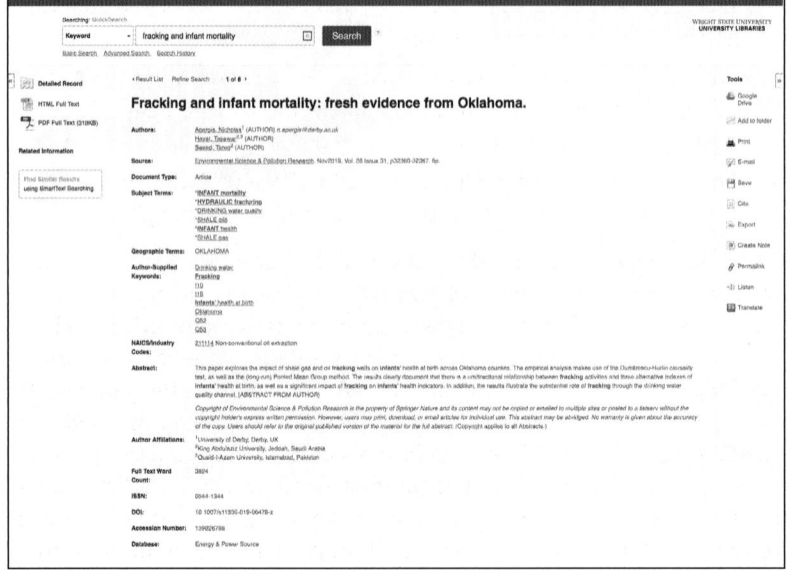

Note the list of author-supplied keywords, which offers options for narrowing and focusing your topic. Each keyword is a link, so simply clicking on it will produce a new list of sources. Also look for "permalink," which will allow you to find the source again quickly.

Advanced keyword searching. Most search sites have "advanced search" options that will help you focus your research. Some allow you to ask questions in conversational language: *What did Thomas Jefferson write about slavery?* Others allow you to focus your search by using specific words or symbols. Here are some of the most common ones:

- Type quotation marks around words to search for an exact phrase— "Thomas Jefferson."

- Type AND to specify that more than one keyword must appear in sources: Jefferson AND Adams. Some search engines require a plus sign instead: +Jefferson +Adams.

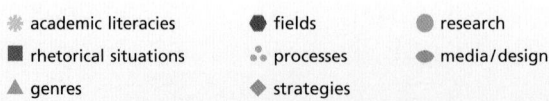

- Type OR if you're looking for sources that include any of several terms: Jefferson OR Adams OR Madison.

- Type NOT to find sources *without* a certain word: Jefferson NOT Adams. Some search engines call for a minus sign (actually, a hyphen) instead: +Jefferson –Adams.

- Type an asterisk to search for words in different form. For example, teach* will yield sources containing "teacher"/"teaching."

In addition, look for features that allow you to filter in order to limit results to a specific date range; only full-text articles (articles that are available in full online); and, in library databases, only scholarly, peer-reviewed sites.

If a search turns up too many sources, be more specific (*homeopathy* instead of *medicine*). If your original keywords don't generate good results, try synonyms (*home remedy* instead of *folk medicine*). Keep in mind that searching requires flexibility, both in the words you use and in the methods you try. And remember that the first items in a results list are not necessarily the best; scan what appears beyond page 1 of your results. Exercise "click restraint" by scanning through many items in the results and choosing only the ones that look most promising for your needs. Open each potential source in its own separate tab for evaluation.

Searching Using Popular Sites and Search Engines

When we need to find something out, most of us begin with *Google* or *Wikipedia*. If you aren't familiar with your research topic, these sites provide general information and a variety of perspectives. However, using general search engines can present challenges: paywalls limit access, it's easy to be overwhelmed by the number of sources, and determining what's trustworthy can be difficult. So you'll want to rely primarily on your library's resources to find accurate, up-to-date, scholarly sources; but the following sites can still serve as helpful starting points.

Wikipedia   An open, online reference, *Wikipedia* often serves as a starting point for preliminary research. Reading a *Wikipedia* entry on a topic can help you understand background information, come up with new **KEYWORDS** to search, and discover related topics. A list of references—

506–9

often hyperlinked—at the bottom of each entry can lead to potential sources. Be sure to check information you find on *Wikipedia* for accuracy, since entries can be written or edited by anyone.

Google, Google Scholar, and other free search tools. To search *Google*, use simple and specific KEYWORDS. Googling "fracking environmental effects" will turn up more focused results than googling "what is fracking." Scan the list of results and open sources that look promising in a new tab instead of stopping to read the first result that catches your eye. Be sure to look beyond the first page of results; higher placement doesn't always mean most relevant to your topic.

506–9

Google Scholar limits your search results to scholarly literature such as researched articles, theses, books, abstracts, and court opinions. Do a keyword search in *Google Scholar*, and scan the results to get an initial idea of the scholarly authors, publications, and databases publishing on your topic. You can sort results by publication date—a good way to check if there's recent research. Some articles and books in *Google Scholar* are full-text, but many are behind paywalls so must be accessed through your campus library.

Here are several more tools for searching online, outside of the library, to get started:

- *Keyword search engines*. Search engines use algorithms to tailor results to you, so search engines—and the results they show you—are not all the same. Compare results from *Google*, *Microsoft Edge*, *DuckDuckGo*, and other search sites to get more varied results.

- *Nonlibrary, academic searches*. In addition to Google Scholar, for peer-reviewed academic writing in many disciplines, try *Microsoft Academic* or *BASE*. *Science.gov* offers results from more than 15 US federal agencies, and *CORE* collects open-access research papers.

- *Social media searches*. *Google Social Search*, *Social Searcher*, and *Talkwalker* let you search various social media platforms. Use HASHTAGS (#) or *Tagboard*, a hashtag-based search engine, to search the contents of posts.

Glossary

- academic literacies
- rhetorical situations
- genres
- fields
- processes
- strategies
- research
- media/design

Image, video, and audio platforms.　*Google Images* and *TinEye* allow you to search for images by keyword or by uploading an image. *Google Video Search* lets you search for videos by keyword, and *YouTube's* search function allows you to find videos by relevance, upload date, view count, or rating. *Audioburst Search* provides results from podcast and radio segments.

News sites.　Sites for newspapers, magazines, and radio and TV stations provide both up-to-the-minute information and also archives of older stories online. Through news-aggregating sites like *Google News* and *Apple News* you can sort through worldwide news from a wide range of sources. *Google News Archive* has news archives extending back to the 1700s. Beware that **MISINFORMATION** and lies are spread most often as news, so **FACT-CHECKING** sources that *appear* to be news is essential.

Glossary
● 526–32

Government sites.　Many government agencies and departments maintain websites where you can find government reports, statistics, legislative information, and other resources. *USA.gov* offers resources from the US government.

Digital archives.　You can find primary sources from the past, including drawings, maps, recordings, speeches, and historic documents, at sites maintained by the National Archives, the Library of Congress, the New York Public Library, and others.

Although many websites provide authoritative information, content found online varies greatly in stability and reliability: what you see on a site today may be different (or gone) tomorrow, so save copies of all pages you use. In addition, many reference and news sites are behind paywalls, their content unavailable unless you pay a fee or subscribe. If you find that a source you need is behind a paywall, check to see if it's available through your library or through interlibrary loan.

Searching in Academic Libraries

College and university libraries typically offer several ways to search their holdings. Their websites may look very different from one another, but

most include search boxes like this one, from the Wright State University Libraries:

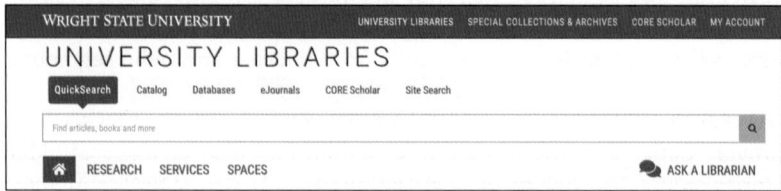

This box allows you to search through all the library's holdings at once, an option that may be a good way to get started. You may already know, though, that you need to focus your search on one type of source, such as scholarly articles, or search using a particular method. In that case, it's best to use a more advanced search box like this one:

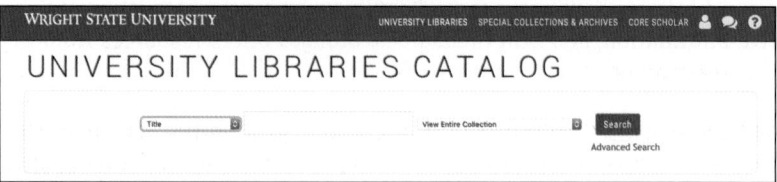

This box lets you shape and limit your search in several ways: the drop-down menu on the left lets you choose to search by title, keyword, author, or several other identifiers; the drop-down menu on the right lets you choose to search the library's entire collection, journals, books, media, or even materials in a specific location, such as course reserves or archives; and "Advanced Search" gives even more options. When in doubt, take advantage of "Ask a Librarian" services, which may include texting, chat, email, and phone conversations with reference librarians, when you need help but aren't working in the library.

Reference Works

The reference section of your school's library is the place to find encyclopedias, dictionaries, atlases, almanacs, bibliographies, and other reference

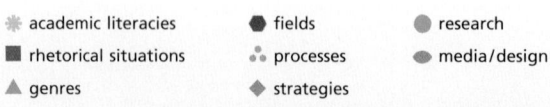

works in print. Many of these sources are also online and can be accessed from any computer that is connected to the internet. Others are available only in the library. Remember, though, that whether in print or online, reference works are only a starting point, a place where you can get an overview of your topic.

General reference works. Consult encyclopedias for general background information on a subject, dictionaries for definitions of words, atlases for maps and geographic data, and almanacs for statistics and other data on current events. These are some works you might consult:

The New Encyclopaedia Britannica
The Columbia Encyclopedia
Webster's Third New International Dictionary
Oxford English Dictionary
National Geographic Atlas of the World
Statistical Abstract of the United States
The World Almanac and Book of Facts

Specialized reference works. You can also go to specialized reference works, which provide in-depth information on a single field or topic. These may also include authoritative bibliographies, leading you to more specific works. A reference librarian can refer you to specialized encyclopedias in particular fields, but good places to start are online collections of many topic-specific reference works that offer overviews of a topic, place it in a larger context, and sometimes provide links to potential academic sources. Collections that are available through libraries include the following:

CQ Researcher offers in-depth reports on topics in education, health, the environment, criminal justice, international affairs, technology, the economy, and social trends. Each report gives an overview of a particular topic, outlines of the differing positions on it, and a bibliography of resources on it.

Gale eBooks offers thousands of full-text specialized encyclopedias, almanacs, articles, and ebooks.

Oxford Reference contains hundreds of dictionaries, encyclopedias, and other reference works on a wide variety of subjects, as well as timelines with links to each item mentioned on each timeline.

SAGE Knowledge includes many encyclopedias and handbooks on topics in the social sciences.

Bibliographies.   Bibliographies provide an overview of what has been published on a topic, listing published works along with the information you'll need to find each work. Some are annotated with brief summaries of each work's contents. You'll find bibliographies at the end of scholarly articles and books, and you can also find book-length bibliographies, both in the reference section of your library and online. Check with a reference librarian for bibliographies on your research topic.

Books / Searching the Library Catalog

The library catalog is your primary source for finding books. Almost all library catalogs are computerized and can be accessed through the library's website. You can search by author, title, subject, or keyword. The image below shows the result of a keyword search for material on looted art in Nazi Germany. This search of the library's catalog revealed six items—print books

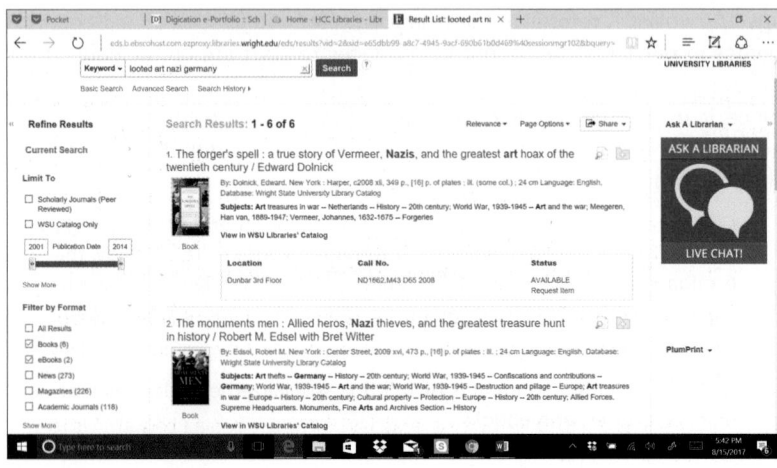

* academic literacies
* rhetorical situations
* genres
* fields
* processes
* strategies
* research
* media/design

and ebooks—on the topic; to access information on each one, the researcher must simply click on the title or thumbnail image. The image below shows detailed information for one source: bibliographic data about author, title, and publication; related subject headings (which may lead to other useful materials in the library)—and more. Library catalogs also supply a call number, which identifies the book's location on the library's shelves.

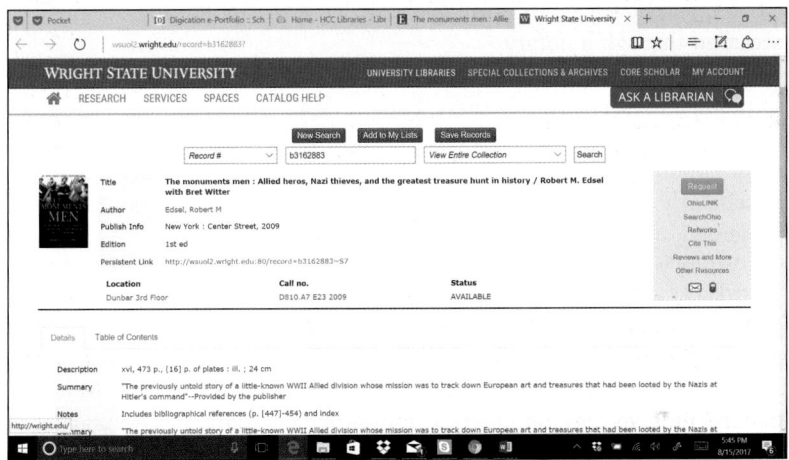

Ebooks / Finding Books Online

Many books in the library catalog are available online. Some may be downloaded to a tablet or mobile device. In addition, thousands of classic works that are in the public domain—no longer protected by copyright—may be read online. *Bartleby*, *Google Books*, *Open Library*, and *Project Gutenberg* are four collections of public-domain works. Here are some other sources of ebooks:

HathiTrust Digital Library offers access to millions of ebooks, about a third of them in the public domain, contributed by university libraries.

Internet Archive includes millions of ebooks as well as audio, video, music, software, images, and the Wayback Machine, which archives historical webpages.

Gale eBooks, *Oxford Reference*, and *SAGE Knowledge* all contain large ebook collections.

Periodicals/Searching Indexes and Databases

To find journal, magazine, and newspaper articles, you will need to search periodical indexes and databases. Indexes provide listings of articles organized by topics; many databases provide the full texts. Some indexes are in print and can be found in the reference section of the library; most are online. Some databases are available for free; most of the more authoritative ones, however, are available only by subscription and so must be accessed through a library.

Many databases now include not only scholarly articles but also dissertations, theses, book chapters, book reviews, and conference proceedings. Dissertations and theses are formal works of scholarship done as requirements for graduate degrees; book reviews offer critical evaluations of scholarly and popular books; and conference proceedings are papers presented, usually orally, at scholarly meetings.

When you access a source through a database, the URL or link address is different each time you log in, so if you want to return to a source, look for a *stable URL, permalink,* or *DOI* option and choose it to copy and paste into your list of sources.

General indexes and databases.   A reference librarian can help you determine which databases will be most helpful for your needs, but here are some useful ones:

> *Academic Search Complete* and *Academic Search Premier* are multi-disciplinary indexes and databases containing the full text of articles in thousands of journals and indexing of even more, with abstracts of their articles.

> *InfoTrac* offers millions of full-text articles in a broad spectrum of disciplines and on a wide variety of topics from thousands of scholarly and popular periodicals, including the *New York Times*.

> *JSTOR* archives scanned copies of entire publication runs of scholarly journals in many disciplines, but it may not include current issues of the journals.

> *ProQuest Central* provides access to full-text articles from thousands of books, scholarly journals, conference papers, magazines, newspapers,

---

* academic literacies   ● fields   ● research
■ rhetorical situations   ∴ processes   ● media/design
▲ genres   ◆ strategies

blogs, podcasts, and websites and a large collection of dissertations and theses.

Single-subject indexes and databases.   The following are just a sample of what's available; check with a reference librarian for indexes and databases in the subject you're researching.

America: History and Life indexes scholarly literature on the history and culture of the United States and Canada.

BIOSIS Previews provides abstracts and indexes for thousands of sources on a wide variety of biological and medical topics.

ERIC is the US Department of Education's Educational Resource Information Center database. It includes hundreds of journal titles as well as conference papers, technical reports, and other resources on education.

Historical Abstracts includes abstracts of articles on the history of the world, excluding the United States and Canada, since 1450.

Humanities International Index contains bibliographic references to thousands of journals dealing with the humanities.

MLA International Bibliography indexes scholarly articles on modern languages, literature, folklore, and linguistics.

PsycINFO indexes scholarly literature in a number of disciplines relating to the behavioral and social sciences.

PubMed includes millions of citations for biomedical literature, many with links to full-text content.

Print indexes.   You may need to consult print indexes to find articles published before the 1980s. Here are six useful ones:

The Readers' Guide to Periodical Literature (print, 1900–; online, 1983–)

InfoTrac Magazine Index (print, 1988–; online, 1973–)

The New York Times Index (print and online, 1851–)

Humanities Index (print, 1974–; online, 1984–)

Social Sciences Index (print, 1974–; online, 1983–)

General Science Index (print, 1978–; online, 1984–)

Images, Sound, and More

Your library likely subscribes to various databases that allow you to find and download video, audio, and image files. Here is a sampling:

AP Images provides access to photographs taken for the Associated Press, the cooperative agency of thousands of newspapers and radio and television stations worldwide.

ArtStor provides images in the arts, architecture, humanities, and sciences.

Dance Online: Dance in Video offers hundreds of videos of dance productions and documentaries on dance.

Bird Sounds from the Borror Laboratory of Bioacoustics includes digitized recordings of the sounds of over 1,000 species of animals.

Music Online: Smithsonian Global Sound for Libraries is a "virtual encyclopedia of the world's musical and aural traditions."

Naxos Music Library contains more than 130,000 classical, jazz, and world music recordings, as well as libretti and synopses of hundreds of operas and other background information.

Theatre in Video provides videos of hundreds of performances of plays and film documentaries.

Doing Field Research

Sometimes you'll need to do your own research, to go beyond the information you find in published sources and gather data by doing field research. Three kinds of field research you might want to consider are interviews, observations, and questionnaires.

Interviewing experts.   Some kinds of writing—a profile of a living person, for instance—almost require that you conduct an interview. And sometimes you may just need to find information that you haven't been able to find in published sources. To get firsthand information on the experience of serving as a soldier in Afghanistan, you might interview your cousin who served a tour of duty there; to find current research on

academic literacies   fields   research
rhetorical situations   processes   media/design
genres   strategies

pesticide residues in food, you might need to interview a toxicologist. Whatever your goal, you can conduct interviews in person, using video-calling software such as *Skype* or *FaceTime*, by telephone, through email, or by mail. In general, you will want to use interviews to find information you can't find elsewhere. Below is some advice on planning and conducting an interview.

BEFORE THE INTERVIEW

1. Once you identify someone you want to interview, email or phone to ask the person, stating your **PURPOSE** for the interview and what you hope to learn.

■ 59–60

2. Once you've set up an appointment, send a note or email confirming the time and place. If you wish to record the interview, be sure to ask for permission to do so. If you plan to conduct the interview by mail or email, state when you will send your questions.

3. Write out questions. Plan questions that invite extended response and supporting details: "What accounts for the recent spike in gasoline prices?" forces an explanation, whereas "Is the recent spike in gas prices a direct result of global politics?" is likely to elicit only a yes or a no.

AT THE INTERVIEW

4. Record the full name of the person you interview, along with the date, time, and place of the interview; you'll need this information to cite and document the interview accurately.

5. Take notes, even if you are recording the interview.

6. Keep track of time: don't take more than you agreed to beforehand unless both of you agree to keep talking. End by thanking your subject and offering to provide a copy of your final product.

AFTER THE INTERVIEW

7. Flesh out your notes with details as soon as possible after the interview, while you still remember them. What did you learn? What surprised you? Summarize both the interviewee's words and your impressions.

8. Make sure you've reproduced quotations from the interview accurately and fairly. Avoid editing quotations in ways that distort the speaker's intended meaning.

9. Be sure to send a thank-you note or email.

Observation.   Some writing projects are based on information you get by observing something. For a sociology report, you may observe how students behave in large lectures. For an education course, you may observe one child's progress as a writer over a period of time. The following advice can help you conduct observations.

BEFORE OBSERVING

59–60 ■

1. Think about your research PURPOSE: What are you looking for? What do you expect to find? How will your presence as an observer affect what you observe? What do you plan to do with what you find?

2. If necessary, set up an appointment. You may need to ask permission of the people you wish to observe and of your school as well. (Check with your instructor about your school's policy in this area.) Be honest and open about your goals and intentions; college students doing research assignments are often welcomed where others may not be.

WHILE OBSERVING

3. If you're taking notes on paper, you may want to divide each page down the middle vertically and write only on the left side of the page, reserving the right side for information you will fill in later. If you're using a laptop, you can set up two columns or a split screen.

4. Note descriptive details about the setting. What do you see? What do you hear? Do you smell anything? Get down details about color, shape, size, sound, and so on. Consider photographing or making a sketch of what you see.

456–63 ◆

5. Who is there, and what are they doing? DESCRIBE what they look like, and make notes about what they say. Note any significant demographic details—about gender, race, occupation, age, dress, and so on.

✳ academic literacies	● fields	● research
■ rhetorical situations	⁘ processes	● media/design
▲ genres	◆ strategies	

6. What is happening? Who's doing what? What's being said? Make note of these kinds of **NARRATIVE** details.

◆ 474–82

AFTER OBSERVING

7. As soon as possible after you complete your observations, use the right side of your notes to fill in gaps and include additional details.

8. **ANALYZE** your notes, looking for patterns. Did some things appear or happen more than once? Did anything stand out? surprise or puzzle you? What did you learn?

▲ 104–39

Questionnaires and surveys.   Various kinds of questionnaires and surveys can provide information or opinions from a large number of people. For a political science course, you might conduct a survey to ask students who they plan to vote for. Or, for a marketing course, you might distribute a questionnaire asking what they think about an advertising campaign. The advice in this section will help you create useful questionnaires and surveys.

Define your goal.   The goal of a questionnaire or survey should be limited and focused, so that every question will contribute to your research question. Also, people are more likely to respond to a brief, focused survey.

Define your sample.   A survey gets responses from a representative sample of the whole group. The answers to these questions will help you define that sample:

1. Who should answer the questions? The people you contact should represent the whole population. For example, if you want to survey undergraduate students at your school, your sample should reflect your school's enrollment in terms of gender, year, major, age, ethnicity, and so forth as closely as possible.

2. How many people make up a representative sample? In general, the larger your sample, the more the answers will reflect those of the whole group. But if your population is small—200 students in a history course, for example—your sample must include a large percentage of that group.

Decide on a medium. Will you ask the questions face-to-face? over the phone? on a website such as *SurveyMonkey*? by mail? by email? Face-to-face questions work best for simple surveys or for gathering impersonal information. You're more likely to get responses to more personal questions with printed or online questionnaires, which should be neat and easy to read. Phone interviews may require well-thought-out scripts that anticipate possible answers and make it easy to record these answers.

Design good questions. The way you ask questions will determine the usefulness of the answers you get, so take care to write questions that are clear and unambiguous. Here are some typical question types:

MULTIPLE-CHOICE
What is your current age?

_____ 15–20   _____ 21–25   _____ 26–30   _____ 31–35   _____ Other

RATING SCALE
How would you rate the service at the campus bookstore?

_____ Excellent   _____ Good   _____ Fair   _____ Poor

AGREEMENT SCALE
How much do you agree with the following statements?

	Strongly Agree	Agree	Disagree	Strongly Disagree
The bookstore has sufficient numbers of textbooks available.	❐	❐	❐	❐

	Strongly Agree	Agree	Disagree	Strongly Disagree
Staff at the bookstore are knowledgeable.	❐	❐	❐	❐
Staff at the bookstore are courteous.	❐	❐	❐	❐

OPEN-ENDED
How often do you visit the campus bookstore?

How can the campus bookstore improve its service?

❋ academic literacies   ● fields   ● research
■ rhetorical situations   ⁙ processes   ● media/design
▲ genres   ◆ strategies

Include all potential alternatives when phrasing questions to avoid biasing the answers. And make sure each question addresses only one issue—for example, "Bookstore staff are knowledgeable and courteous" could lead to the response "knowledgeable, agree; courteous, disagree."

When arranging questions, place easier ones at the beginning and harder ones near the end (but if the questions seem to fall into a different natural order, follow it). Make sure each question asks for information you will need—if a question isn't absolutely necessary, omit it.

Include an introduction.   Start by stating your survey's purpose and how the results will be used. It's also a good idea to offer an estimate of the time needed to complete the questions. Remind participants of your deadline.

Test the survey or questionnaire.   Make sure your questions elicit the kinds of answers you need by asking three or four people who are part of your target population to answer them. They can help you find unclear instructions, questions that aren't clear or that lack sufficient alternatives, or other problems that you should correct to make sure your results are useful. But if you change the questionnaire as a result of their responses, don't include their answers in your total.

If you need more help

See **EVALUATING SOURCES** for help determining their usefulness. See also Chapter 50 for help **TAKING NOTES** on your sources.

524–34

542–44

48 Evaluating Sources

Searching the *Health Source* database for information on the incidence of meningitis among college students, you find thirty-three articles. A *Google* search on the same topic produces over 13 million hits. How do you decide which sources to read? This chapter presents advice on evaluating sources—first to determine whether a source might be useful for your purposes and is worth looking at more closely, and then to read with a critical eye the ones you choose.

Considering Whether a Source Might Be Useful

59–60 ■

61–64 ■

414–22 ◆

As you consider potential sources—sources you've found online or from library resources—keep your **PURPOSE** in mind. Are you trying to persuade readers to believe or do something? If so, look for sources representing various positions. Are you reporting on a topic? If so, you'll likely need sources that are factual or informative. Consider your **AUDIENCE**. What kinds of sources will they find persuasive? If you're writing for readers in a particular field, what counts as **EVIDENCE** in that field? Following are some questions that can help you judge whether a possible source you've found deserves your time and attention—but remember to consider most or all of the questions, rather than relying on a single criterion:

- *Is it relevant?* How well does the source relate to your purpose? What would it add to your work? To see what it covers, look at the title and at any introductory material (such as a preface or an abstract).

- *Is it reliable?* Is it scholarly? peer-reviewed? published in a reputable journal or magazine, or by a reputable publisher? Did you find it in a library database? on the web? Evaluating web-based texts may require more work than using results from library databases. Whatever kind of search you do, skim the results quickly to evaluate their reliability— and use the techniques described below to fact-check them. You might

※ academic literacies   ● fields   ● research
■ rhetorical situations   ∴ processes   ● media/design
▲ genres   ◆ strategies

also check the *Media Bias/Fact Check* website, which ranks news organizations on their political slant and factual reporting.

- *What are the author's credentials?* How is the author qualified to write on the subject? Has the author written other works on this subject? Are they known for a particular position on it? As an expert? A scholar? A journalist? Will your audience find the author credible? If the credentials aren't stated, do a search to see what reliable sources say about the author.

- *What is the* STANCE*?* Does the source explain various points of view or advocate only one perspective? Does its title suggest a certain slant? If it's online, check to see whether it includes links to other sites—and if so, what their perspectives are. You'll want to consult sources with various viewpoints and understand the bias of each source you use.

72–74

- *Who is the publisher or sponsor?* If it's a book, what kind of company published it; if an article, what kind of periodical did it appear in? Books published by university presses and articles in scholarly journals are reviewed by experts before they're published. Many books and articles written for the general public also undergo careful fact-checking, but some do not—and they may lack the kind of in-depth discussion that is useful for research.

- *At what level is it written?* Can you understand the material? Texts written for a general audience might be easier to understand but not authoritative enough for academic work. Scholarly texts will be more authoritative but may be hard to comprehend.

- *How current is it?* Check to see when books and articles were published and when websites were last updated. (If a site lists no date, see if links to other sites still work; if not, the site is probably too dated to use.) A recent publication date or update, however, doesn't necessarily mean that a potential source is good—some topics require current information; others call for older sources.

- *Is it cited in other works?* If so, you can probably assume that some other writers regard it as trustworthy.

- *Does it include other useful information?* Is there a bibliography that might lead you to other sources? How current or authoritative are the sources it cites?

- *Is it available?* If it's a book and your school's library doesn't have it, can you get it through interlibrary loan? If it's online and there's a paywall, can your library get you access?

Fact-Checking Popular Sources Online

Scholarly books are edited; scholarly articles are reviewed by experts; established news sources employ fact-checkers to verify information. But many websites and social media sites aren't checked by anyone. As a result, when you're using *Google* and other nonlibrary resources online to get started, you may find misleading information, data, images, and videos.

Sometimes such misleading information takes the form of satires, such as Jonathan Swift's "A Modest Proposal," stories in the *Onion*, or episodes of *The Daily Show*. Such satires offer humorous exaggerations to expose and criticize people and governments. Other misleading or false news stories are malicious, created usually for political ends. World War II Nazi propaganda and recent fraudulent stories such as "Pope Francis Endorses Donald Trump," a false story that went viral on *Facebook*, are intended to mislead readers. Further complicating things, some politicians and other public figures have taken to calling news reports with which they disagree "fake news"—even though the stories reported are verifiably true and the news outlets are considered trustworthy.

Although false news has been around at least since the invention of the printing press, the internet and social media have led to a huge increase in false news stories, seriously challenging and muddying "real" news. Each false story can rapidly multiply over sites such as *Facebook* and *Twitter* and through email. As a reader and researcher, you need to be able to determine whether a potential news source is false. As a writer, you risk harming your credibility if you cite a false news story as evidence. To conduct sound research—and to use good information to form our own ideas—we need to be able to determine what is true, what is false, and what is trying to manipulate us when reading popular online sources. Professional fact-checkers, people who investigate the truthfulness of assertions made in published work, use the following methods to evaluate their accuracy.

Fact-checkers begin not by immediately reading something they find online that looks relevant, but by first moving *outside* the source—

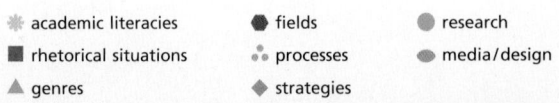

opening new tabs in the same browser—to search and see what other reliable sites say about it. Only after they've done some research to confirm a source's accuracy do they read the source itself. When you find something online that looks relevant, first stop and ask yourself if you know the website publishing the source and have reason to trust it. If not, take the following steps to assess the source before reading any further:

- *Look up the main* **CLAIMS**. Copy and paste the title or a few key words of the article, essay, or post into *Google* or another browser to see if a reliable fact-checking site—*Snopes, FactCheck, Politifact, Allsides,* or *mediabiasfactcheck*—has already evaluated it. If several reputable sources report the same information, it's probably true. If the information appears in only one source—even if many other people are quoting it—it's worth taking more steps to check whether it's reliable and even-handed enough for your academic writing.

  ◆ 411–13

- *Investigate the author's expertise and* **STANCE**. Use *Wikipedia, Google News,* or *Google Scholar* to see if the author is an expert on the topic they're weighing in on. Some signs that the author has expertise include: an academic position in a relevant field, other pieces published in reputable sources on the same topic, or considerable experience in the area being written about. If you can't find any evidence of an author's expertise on the topic, the source may still be useful to you, but don't depend on the information as authoritative. Also, the author's affiliations—the groups they're members of or the places they've published—can help you determine the author's stance.

  ■ 72–74

- *Check out who runs the site.* Don't rely on the site's About page alone. Instead, google the organization's name in quotation marks and add keywords like "who sponsors" or "funding." What do other reputable sites say about the publisher or sponsoring organization? Take note of information on controversy or a specific agenda attached to the organization. Do reliable sources say the organization has a good reputation? a lot of members? a long history? Those are all good signs the publication is trustworthy.

- *Check the date of the story and dates within the story.* Sometimes old stories resurface as "news." They aren't any truer now than they were then.

59–60
- *Take stock, and use your judgment.* Consider your **PURPOSE** and what you've found about the source's main claims, author, and publisher. Does the source seem accurate and trustworthy, given what you know about the world and how it works? Do the claims seem designed to generate a strong emotional response, especially anger or fear, rather than appealing to logic or sound evidence? If taking these steps shows 72–74 you that a source has a strong **STANCE** or bias, that doesn't automatically mean you should discard it, but it does mean you should keep that stance in mind as you work with the source.

If even just one of these steps reveals troubling information, it's best to move on to find a different, more solidly reliable source. As you use these fact-checking moves, don't just click on the first item in a list of search results. Instead, scan the results to get a broad sense of what others are saying, and choose the most trustworthy ones to read.

The graphics on the following pages show a good example of how fact-checking a popular source found on the web works.

Fact-checking photos and video. Visual media—especially images and videos posted online—can also be manipulated to present misleading information. Hoax photos are images taken out of context or doctored in order to deceive using *Photoshop* or another photo manipulation program. For example, this misleading image of presidential candidate John Kerry and antiwar activist and actor Jane Fonda was widely distributed during the 2004 election campaign:

The image was created by combining two photos taken more than a year apart on two different occasions and was used to discredit Kerry's patriotism despite his service in Vietnam.

* academic literacies   ● fields   ● research
■ rhetorical situations ⁖ processes   ● media/design
▲ genres   ◆ strategies

However, doctoring sometimes isn't even needed: the camera lenses used or the camera's position can lead to very different views. For example, during the coronavirus pandemic, the photo on the left purported to show people ignoring social distancing guidance to stay at least six feet apart on a beach in Florida:

The photo on the right, though, shows that the beachgoers were in fact maintaining a safe distance from one another.

To identify the source of most photographs, you can do a reverse image search using these tools:

- *Google Image.* Click on the small camera icon to the right of the search box, and paste in the image's web address. The results will show other uses of the image as well as similar images.

- *TinEye.* Upload the image or enter its URL in the search box to get a list of other places where the image has appeared online.

Hoax videos are altered or presented in misleading ways. The *Washington Post* identifies three ways hoax videos are made:

- *Missing context.* The video itself hasn't been altered, but it is misrepresented in order to mislead the viewer—using an incorrect date, location, or summary. Or it's a brief clip from a longer video taking an event out of context.

- *Deceptive editing.* Portions of the video have been edited out and the remaining parts spliced together to create a misleading narrative. Or videos shot on two different occasions have been spliced together to create a single false narrative.

Fact-Checking Popular Sources

Once you've got a source to fact-check, begin by looking outside the site itself following the steps below.

Look Up the Claim

Other reliable news organizations report similar claims from various sources.

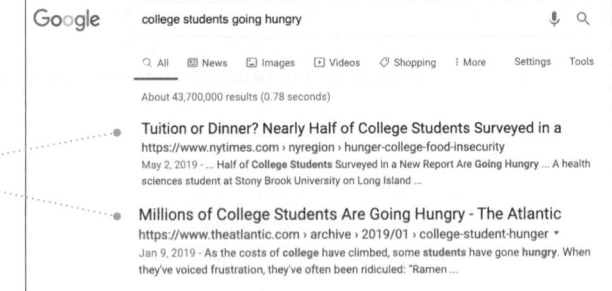

Check Out the Website

Wikipedia says the site began in 2008 and closed in 2019. Be skeptical of a new site or one that no longer exists.

A New York Times article confirms the site was funded by a well-established academic publisher and won two awards for quality, so it appears trustworthy.

Investigate the Author

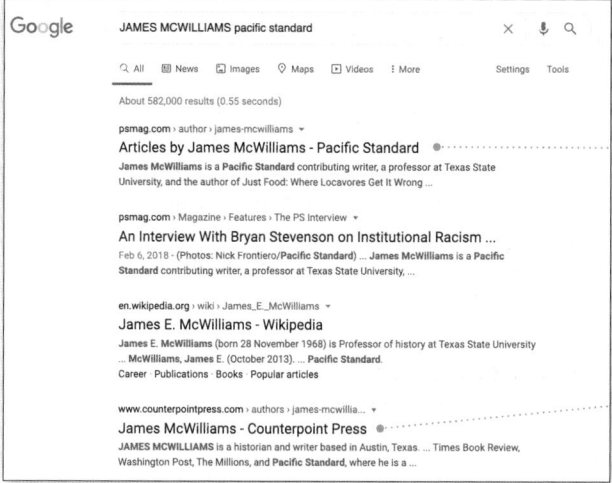

The author has published several more pieces on the same site.

The author's bio from his book publisher appears at the bottom of the first search results page.

Opening the author's academic webpage shows he's a professor at Texas State University, teaching courses in American history and writing about the history of eating habits.

An Amazon *page shows the author's published a book on food and eating. So he appears credible on this topic.*

- *Malicious transformation.* The video itself has been altered to deceive the viewer, or artificial intelligence has been used to create fake images and audio.

In addition, videos may simply present lies as factual information. For example, during the coronavirus pandemic, online videos included false statements including "The pharmaceutical companies have a cure but won't sell it," "Coronavirus is caused by 5G mobile phone network signals," and "The coronavirus is being used for worldwide population control." Some of these videos are viewed by millions of people, so the popularity of a video may not reflect its accuracy or truthfulness, either.

To evaluate the accuracy of videos, type a brief summary of the video into *Google* to see if fact-checking sites like *Snopes* have reported on it. Also, do your own analysis: Does anything in the video seem doctored? Are the events depicted believable or unlikely? Is the language used inflammatory or demeaning? If the video appears on several websites, do its details change from site to site? Use *Youtube DataViewer*, available at citizenevidence.amnestyusa.org, to see where else a *YouTube* video has been posted, and do a reverse image search on stills from a video.

Reading Sources with a Critical Eye

Approach your sources with an open mind, but consider their arguments with a critical eye. Pay attention to what they say, to the reasons and evidence they offer to support what they say, and to whether they address viewpoints other than their own. Assume that each author is responding to some other argument. Ask some of these questions of each of your sources:

164–95 ▲
- *What **ARGUMENTS** does the author make?* Does the author present several different positions or argue for a particular position? What arguments is the author responding to?

413–14 ◆
414–22
526–28 ●
- *How persuasive do you find the argument to be?* What **REASONS** and **EVIDENCE** does the author provide to support their positions? Are there citations or links—and if so, are they **CREDIBLE**? Is any evidence presented without citations? Are any of the author's assumptions questionable? How thoroughly does the author consider alternative arguments?

※ academic literacies   ● fields   ● research
■ rhetorical situations   ⁘ processes   ● media/design
▲ genres   ◆ strategies

- *What is the author's* **STANCE**? Does it seem objective, or does the content or language reveal a particular bias? Are opposing views considered and treated fairly? Do the headlines and text try to elicit an emotional reaction?   72–74

- *Do you recognize ideas you've run across in other sources?* Does the source leave out any information or perspective that other sources include—or does it include any that they leave out?

- *What can you tell about the intended* **AUDIENCE** *and* **PURPOSE**? Are you a member of the audience for which the source was written—and if not, does that affect the way you interpret what you read? Is the main purpose to inform readers about a topic or to argue a certain point?   61–64
  59–60

- *Does this source support or challenge your own position—or does it do both?* Does it support your thesis? Does it support a different argument altogether? Does it represent a position you need to address? Don't reject a source just because it challenges your views; work to understand the various perspectives or positions taken on your topic to avoid **CONFIRMATION BIAS**, which is our tendency to believe things that match what we already believe and to discount information that doesn't. Be wary of assuming a story or article is trustworthy just because you agree with the author—you need to step back and make sure your own beliefs aren't clouding your judgment.   Glossary

Comparing Sources

You may find that two or more of your sources present similar information or arguments. How do you decide which one to use? Compare them, using these questions as a guide:

- *Which source is most current?* Generally, a more recent source is better than an older one, because the newer source includes information or data that is more up to date—and may include (or refute) the information in the earlier source. Be aware, though, that in some fields, such as literary criticism, decades-old sources may still be important.

- *Which argument is more persuasive?* Examine the **CLAIMS**, **REASONS**, and **EVIDENCE** presented in each source. Which source's argument is most logical? Which one has the best supporting reasons and   411–22

424–26 ◆

evidence? Which one best **ACKNOWLEDGES**, **ACCOMMODATES**, or **REFUTES** opposing arguments?

- *Which author or authors are most authoritative?* An expert in the subject is more authoritative than, say, a journalist writing on the subject. An article published in a scholarly journal is more authoritative than one published in a general-circulation magazine or website. The journalist's article may be easier to read and more interesting, but you're best off looking for the best information—not the best read.

72–74 ■

- *Which source has the most appropriate* **STANCE**? In general, look for sources that strive for objectivity, rather than a particular bias. Also, be aware that we all tend to favor information that agrees with our views—and that may lead us to choose sources we agree with rather than sources that present all sides of an issue.

- *Which source best fits your needs?* All other things being equal, the best source to choose is the one that will give you the information you need, support the argument you're making, and perhaps provide useful quotations to add to your writing. The best source will show that you've done appropriate research and will enhance your own credibility as a thinker, reader, researcher, and writer.

If you need more help

542–54 ●
555–59

See Chapter 50, **QUOTING, PARAPHRASING, AND SUMMARIZING**, for help in taking notes on your sources and deciding how to use them in your writing. See also Chapter 51, **ACKNOWLEDGING SOURCES, AVOIDING PLAGIARISM**, for advice on giving credit to the sources you use.

☀ academic literacies   ⬣ fields   ● research

■ rhetorical situations   ⁝ processes   ⬤ media/design

▲ genres   ◆ strategies

49 Synthesizing Ideas

To **ANALYZE** a collection of poetry, you show how the poet uses similar images in three different poems to explore a recurring concept. To solve a crime, a detective studies several eyewitness accounts to figure out who did it. To trace the history of photojournalism, a professor **COMPARES** the uses of photography during the Civil War and during the Vietnam War. These are all cases where someone synthesizes—brings together material from two or more sources in order to generate new information or to support a new perspective. When you do research, you need to go beyond what your sources say; you need to use what they say to inspire and support what you want to say. This chapter focuses on how to synthesize ideas you find in other sources as the basis for your own ideas.

104–39

437–44

Reading for Patterns and Connections

Your task as a writer is to find as much information as you can on your topic—and then to sift through all that you have found to determine and support what you yourself will write. In other words, you'll need to synthesize ideas and information from the sources you've consulted to figure out first what arguments *you* want to make and then to provide support for those arguments.

When you synthesize, you group similar bits of information together, looking for patterns or themes or trends and trying to identify the key points. In the brief report on the following page, writer Jude Stewart synthesizes several pieces of research on boredom. Stewart's report originally appeared in the *Atlantic*, which uses an abbreviated documentation style.

Boredom Is Good for You

Boredom has, paradoxically, become quite interesting to academics lately. The International Interdisciplinary Boredom Conference gathered humanities scholars in Warsaw for the fifth time in April. In early May, its less scholarly forerunner, London's Boring Conference, celebrated seven years of delighting in tedium. At this event, people flock to talks about toast, double yellow lines, sneezing, and vending-machine sounds, among other snooze-inducing topics.

This question allows Stewart to bring together four ways boredom can be measured.

What, exactly, is everybody studying? One widely accepted psychological definition of boredom is "the aversive experience of wanting, but being unable, to engage in satisfying activity." [1] But how can you quantify a person's boredom level and compare it with someone else's? In 1986, psychologists introduced the Boredom Proneness Scale, [2] designed to measure an individual's overall propensity to feel bored (what's known as "trait boredom"). By contrast, the Multidimensional State Boredom Scale, [3] developed in 2008, measures a person's feelings of boredom in a given situation ("state boredom"). A German-led team has since identified five types of state boredom: indifferent, calibrating, searching, reactant, and apathetic (indifferent boredom—characterized by low arousal—was the mellowest, least unpleasant kind; reactant—high arousal—was the most aggressive and unpleasant). [4] Boredom may be miserable, but let no one call it simple.

Stewart synthesizes seven different studies under one category, "behavior issues."

Boredom has been linked to behavior issues, including bad driving, [5] mindless snacking, [6] binge-drinking, [7] risky sex, [8] and problem gambling. [9] In fact, many of us would take pain over boredom. One team of psychologists discovered that two-thirds of men and a quarter of women would rather self-administer electric shocks than sit alone with their thoughts for 15 minutes. [10] Probing this

One study builds on a previous study.

phenomenon, another team asked volunteers to watch boring, sad, or neutral films, during which they could self-administer electric shocks. The bored volunteers shocked themselves more and harder than the sad or neutral ones did. [11]

Again, Stewart creates a category that includes two different studies.

But boredom isn't all bad. By encouraging contemplation and daydreaming, it can spur creativity. An early, much-cited study gave participants abundant time to complete problem-solving and word-association exercises. Once all the obvious answers were exhausted, participants gave more and more inventive answers to fend off boredom. [12] A British study took these findings one step further, asking subjects

* academic literacies   ● fields   ● research
■ rhetorical situations   ∴ processes   ● media/design
▲ genres   ◆ strategies

to complete a creative challenge (coming up with a list of alternative uses for a household item). One group of subjects did a boring activity first, while the others went straight to the creative task. Those whose boredom pumps had been primed were more prolific. [13]

    In our always-connected world, boredom may be an elusive state, but it is a fertile one. Watch paint dry or water boil, or at least put away your smartphone for a while. You might unlock your next big idea.

Stewart shows how one study relates to an earlier one's findings.

Conclusion brings together all the research, creating a synthesis of the findings of all 13 sources.

The Studies

[1] Eastwood et al., "The Unengaged Mind" (*Perspectives on Psychological Science*, Sept. 2012)

[2] Farmer and Sundberg, "Boredom Proneness" (*Journal of Personality Assessment*, Spring 1986)

[3] Fahlman et al., "Development and Validation of the Multidimensional State Boredom Scale" (*Assessment*, Feb. 2013)

[4] Goetz et al., "Types of Boredom" (*Motivation and Emotion*, June 2014)

[5] Steinberger et al., "The Antecedents, Experience, and Coping Strategies of Driver Boredom in Young Adult Males" (*Journal of Safety Research*, Dec. 2016)

[6] Havermans et al., "Eating and Inflicting Pain Out of Boredom" (*Appetite*, Feb. 2015)

[7] Biolcati et al., " 'I Cannot Stand the Boredom' " (*Addictive Behaviors Reports*, June 2016)

[8] Miller et al., "Was Bob Seger Right?" (*Leisure Sciences*, Jan. 2014)

[9] Mercer and Eastwood, "Is Boredom Associated with Problem Gambling Behaviour?" (*International Gambling Studies*, April 2010)

[10] Wilson et al., "Just Think: The Challenges of the Disengaged Mind" (*Science*, July 2014)

[11] Nederkoorn et al., "Self-Inflicted Pain Out of Boredom" (*Psychiatry Research*, March 2016)

[12] Schubert, "Boredom as an Antagonist of Creativity" (*Journal of Creative Behavior*, Dec. 1977)

[13] Mann and Cadman, "Does Being Bored Make Us More Creative?" (*Creativity Research Journal*, May 2014)

—Jude Stewart, "Boredom Is Good for You: The Surprising Benefits of Stultification"

Here are some tips for reading to identify patterns and connections:

550–51

- Read all your sources with an open mind. Withhold judgment, even of sources that seem wrong-headed or implausible. Don't jump to conclusions.

- Take notes, and write a brief **SUMMARY** of each source to help you see relationships, patterns, and connections among your sources. Take notes on your own thoughts, too.

333–36

340–42

- Pay attention to your first reactions. You'll likely have many ideas to work with, but your first thoughts can often lead somewhere that you'll find interesting. Try **FREEWRITING**, **CLUSTERING**, or **LISTING** to see where they lead. How do these thoughts and ideas relate to your topic? Where might they fit into your rough **OUTLINE**?

- Try to think creatively, and pay attention to thoughts that flicker at the edge of your consciousness, as they may well be productive.

- Be playful. Good ideas sometimes come when we let our guard down or take ideas to extremes just to see where they lead.

Ask yourself these questions about your sources:

- What sources make the strongest arguments? What makes them so strong?

- Do some arguments crop up in several sources?

424–25

- Which arguments do you agree with? disagree with? Of those you disagree with, which ones seem strong enough that you need to **ACKNOWLEDGE** them in your text?

- Are there any disagreements among your sources?

- Are there any themes you see in more than one source?

- Are any data—facts, statistics, examples—or experts cited in more than one source?

- Do several of your sources use the same terms? Do they use the terms similarly, or do they use them in different ways?

---

* academic literacies   ● fields   ● research
■ rhetorical situations   ⁝ processes   ● media/design
▲ genres   ◆ strategies

- What have you learned about the topic? How have your sources affected your thinking on the topic? Do you need to adjust your **THESIS**? If so, how?

347–49

- Have you discovered new questions you need to investigate?

- Keep in mind your **RHETORICAL SITUATION** — have you found the information you need that will achieve your purpose, appeal to your audience, and suit your genre and medium?

57

What is likely to emerge from this questioning is a combination of big ideas, including new ways of understanding your topic and insights into recent scholarship about it, and smaller ones, such as how two sources agree with each other but not completely and how the information in one source supports or undercuts the argument of another. These ideas and insights will become the basis for your own ideas and for what *you* have to say about the topic.

Synthesizing Ideas Using Notes

You may find that identifying connections among your sources is easier if you examine them together rather than reading them one by one. For example, taking notes on note cards and then laying the cards out on a desk or table (or on the floor) lets you see passages that seem related. Doing the same with photocopies or printouts of your sources can help you identify similarities as well.

In doing research for an essay arguing that the sale of assault weapons should be banned, you might find several sources that address the scope of US citizens' right to bear arms. On the next page are notes taken on three such sources: Joe Klein, a journalist writing in *Time.com*; Antonin Scalia, a former US Supreme Court justice, quoted in an online news article; and Drew Westen, a professor of psychology writing in a blog sponsored by the *New York Times*. Though the writers hold very different views, juxtaposing these notes and highlighting certain passages show a common thread running through the sources. In this example, all three sources might be used to support the thesis that restrictions on the owning of weapons—but not an outright ban—are both constitutional and necessary.

Source 1

<u>Limits of gun ownership</u>

Although the U.S. Constitution includes the right to bear arms, that right is not absolute. "No American has the right to own a stealth bomber or a nuclear weapon. Armor-piercing bullets are forbidden. The question is where you draw a reasonable bright line."
— Klein, "How the Gun Won" — quote

Source 4

<u>Limits of gun ownership</u>

Supreme Court justice Antonin M. Scalia has noted that when the Constitution was written and ratified, some weapons were barred. So limitations could be put on owning some weapons, as long as the limits are consistent with those in force in 1789.
— Scalia, quoted in Woods — paraphrase

Source 3

<u>Limits of gun ownership</u>

Westen's "message consulting" research has shown that Americans are ambivalent about guns but react very positively to a statement of principle that includes both the right to own guns and restrictions on their ownership, such as prohibiting large ammunition clips and requiring all gun purchasers to undergo background checks for criminal behavior or mental illness.
— Westen — paraphrase

Synthesizing Information to Support Your Own Ideas

If you're doing research to write a **REPORT**, your own ideas will be communicated primarily through which information you decide to include from the sources you cite and how you organize that information. If you're writing a **TEXTUAL ANALYSIS**, your synthesis may focus on the themes, techniques, or other patterns you find. If you're writing a research-based **ARGUMENT**, in contrast, your synthesis of sources must support the position you take in that argument. No matter what your genre, the challenge is to synthesize information from your research to develop ideas about your topic and then to support those ideas.

▲ 140–63

▲ 104–39

▲ 164–95

Entering the Conversation

As you read and think about your topic, you will come to an understanding of the concepts, interpretations, and controversies relating to the topic—and you'll become aware that there's a larger conversation going on. When you begin to find connections among your sources, you'll begin to see your own place in that conversation, to discover your own ideas and your own stance on the topic. This is the exciting part of a research project, for when you write out your own ideas on the topic, you will find yourself entering that conversation. Remember that your **STANCE** as an author needs to be clear: simply stringing together the words and ideas of others isn't enough. You need to show readers how your source materials relate to one another and to your thesis.

■ 72–74

If you need more help

See Chapter 50, **QUOTING, PARAPHRASING, AND SUMMARIZING**, for help in integrating source materials into your own text. See also Chapter 51 on **ACKNOWLEDGING SOURCES, AVOIDING PLAGIARISM** for advice on giving credit to the sources you cite.

● 542–54

555–59

50 Quoting, Paraphrasing, and Summarizing

In an oral presentation about the rhetoric of Abraham Lincoln, you quote a memorable line from the Gettysburg Address. For an essay on the Tet Offensive in the Vietnam War, you paraphrase arguments made by several commentators and summarize some key debates about that war. When you work with the ideas and words of others, you need to clearly distinguish those ideas and words from your own and give credit to their authors. This chapter will help you with the specifics of quoting, paraphrasing, and summarizing source materials that you use in your writing.

Taking Notes

When you find material you think will be useful, take careful notes. How do you determine how much to record? You need to write down enough information so that when you refer to it later, you will be reminded of its main points and have a precise record of where it comes from.

- *Use a computer file, note cards, or a notebook,* labeling each entry with the information that will enable you to keep track of where it comes from—author, title, and the pages or the URL (or DOI, the digital object identifier). You needn't write down full bibliographic information (you can abbreviate the author's name and title) since you'll include that information in your **WORKING BIBLIOGRAPHY**.

497–99

- *Take notes in your own words, and use your own sentence patterns.* If you make a note that is a detailed **PARAPHRASE**, label it as such so that you'll know to provide appropriate **DOCUMENTATION** if you use it.

547–50
560–63

- *If you find wording that you'd like to quote,* be sure to enclose it in quotation marks to distinguish your source's words from your own.

❋ academic literacies   ⬢ fields   ⬤ research

◼ rhetorical situations   ⁚ processes   ⬤ media/design

▲ genres   ◆ strategies

Double-check your notes to be sure any quoted material is accurately quoted—and that you haven't accidentally **PLAGIARIZED** your sources.

● 555–59

- *Label each note with a number to identify the source and a subject heading* to relate the note to a subject, supporting point, or other element in your essay. Doing this will help you to sort your notes easily and match them up with your rough outline. Restrict each note to a single subject.

Here are a few examples of one writer's notes on a source discussing synthetic dyes, bladder cancer, and the use of animals to determine what causes cancers. Each note includes a subject heading and brief source information and identifies whether the source is quoted or paraphrased.

Source 3

<u>Synthetic dyes</u>

The first synthetic dye was mauve, invented in 1854 and derived from coal. Like other coal-derived dyes, it contained aromatic amines.
Steingraber, "Pesticides," 976 — paraphrase

Source 3

<u>Synthetic dyes & cancer</u>

Bladder cancer was common among textile workers who used dyes. Steingraber: "By the beginning of the twentieth century, bladder cancer rates among this group of workers had skyrocketed."
Steingraber, "Pesticides," 976 — paraphrase and quote

> **Source 3**
>
> Synthetic dyes & cancer
>
> In 1938, Wilhelm Hueper exposed dogs to aromatic amines and showed that the chemical caused bladder cancer. Steingraber, "Pesticides," 976 — paraphrase

Deciding Whether to Quote, Paraphrase, or Summarize

364–66

When it comes time to **DRAFT**, you'll need to decide *how* to use any source you want to include—in other words, whether to quote, paraphrase, or

544–47

summarize it. You might follow this rule of thumb: **QUOTE** texts when the wording is worth repeating or makes a point so well that no rewording will do it justice, when you want to cite the exact words of a known authority on your topic, when an authority's opinions challenge or disagree with those of others, or when the source is one you want to emphasize.

547–50
550–51

PARAPHRASE sources that are not worth quoting but contain details you need to include. **SUMMARIZE** longer passages whose main points are important but whose details are not. Whatever your choice, you'll need to introduce

551–54

any words or ideas that aren't your own with a **SIGNAL PHRASE** in order to clearly distinguish what your sources say from what *you* have to say.

Quoting

Quoting a source is a way of weaving someone else's exact words into your text. You need to reproduce the source exactly, though you can modify it to omit unnecessary details (with ellipses) or to make it fit smoothly into your text (with brackets). You also need to distinguish quoted material from your own by enclosing short quotations in quotation marks, setting

551–54

off longer quotes as a block, and using appropriate **SIGNAL PHRASES**.

Incorporate short quotations into your text, enclosed in quotation marks.

MLA 606–7

If you are following **MLA STYLE**, short quotations are defined as four typed

lines or fewer; if using **APA STYLE**, as below, short means fewer than forty words.

APA 646

> Gerald Graff (2003) has argued that colleges make the intellectual life seem more opaque than it needs to be, leaving many students with "the misconception that the life of the mind is a secret society for which only an elite few qualify" (p. 1).

If you are quoting three lines or fewer of poetry, run them in with your text, enclosed in quotation marks. Separate lines with slashes, leaving one space on each side of the slashes.

> Emma Lazarus almost speaks for the Statue of Liberty with the words inscribed on its pedestal: "Give me your tired, your poor,/Your huddled masses yearning to breathe free,/The wretched refuse of your teeming shore" (lines 10–12).

Set off long quotations block style.   If you are using MLA style, set off quotations of five or more typed lines by indenting the quote one-half inch from the left margin. If you are using APA style, indent quotations of forty or more words one-half inch (or five to seven spaces) from the left margin. In either case, do not use quotation marks, and put any parenthetical documentation *after* any end punctuation.

> Nonprofit organizations such as Oxfam and Habitat for Humanity rely on visual representations of the poor. What better way to get our attention, asks rhetorician Diana George:
>
> > In a culture saturated by the image, how else do we convince Americans that—despite the prosperity they see all around them—there is real need out there? The solution for most nonprofits has been to show the despair. To do that they must represent poverty as something that can be seen and easily recognized: fallen-down shacks and trashed-out public housing, broken windows, dilapidated porches, barefoot kids with stringy hair, emaciated old women and men staring out at the camera with empty eyes. (210)

If you are quoting four or more lines of poetry, they need to be set off block style in the same way.

Indicate any omissions with ellipses.   You may sometimes delete words from a quotation that are unnecessary for your point. Insert three ellipsis marks (leaving a space before the first and after the last one) to indicate the deletion. If you omit a sentence or more in the middle of a quotation, put a period before the three ellipsis dots. Be careful not to distort the source's meaning, however.

> Faigley points out that Gore's "Information Superhighway" metaphor "associated the economic prosperity of the 1950s and . . . 1960s facilitated by new highways with the potential for vast . . . commerce to be conducted over the Internet" (253).

> According to Welch, "Television is more acoustic than visual. . . . One can turn one's gaze away from the television, but one cannot turn one's ears from it without leaving the area where the monitor leaks its aural signals into every corner" (102).

Indicate additions or changes with brackets.   Sometimes you'll need to change or add words in a quotation—to make the quotation fit grammatically within your sentence, for example, or to add a comment. In the following example, the writer changes the passage "one of our goals" to clarify the meaning of "our."

> Writing about the dwindling attention among some composition scholars to the actual teaching of writing, Susan Miller notes that "few discussions of writing pedagogy take it for granted that one of [writing teachers'] goals is to teach how to write" (480).

Here's an example of brackets used to add explanatory words to a quotation:

> Barbosa observes that Buarque's lyrics have long included "many a metaphor of *saudades* [yearning] so characteristic of *fado* music" (207).

Use punctuation correctly with quotations.   When incorporating a quotation into your text, be sure to think about the end punctuation in the quoted material and also about any punctuation you need to add when inserting the quote into your own sentence.

※ academic literacies   ● fields   ● research
■ rhetorical situations   ⦂ processes   ● media/design
▲ genres   ◆ strategies

Periods and commas. Put periods or commas *inside* closing quotation marks, except when you have parenthetical documentation at the end, in which case you put the period or comma after the parentheses.

> "Country music," Tichi says, "is a crucial and vital part of the American identity" (23).

After long quotations set off block style with no quotation marks, however, the period goes *before* the documentation, as in the example on page 545.

Question marks and exclamation points. These go *inside* closing quotation marks if they are part of the quoted material but *outside* when they are not. If there's parenthetical documentation at the end of the quotation, any punctuation that's part of your sentence comes after it.

> Speaking at a Fourth of July celebration in 1852, Frederick Douglass asked, "What have I, or those I represent, to do with your national independence?" (35).

> Who can argue with W. Charisse Goodman's observation that media images persuade women that "thinness equals happiness and fulfillment" (53)?

Colons and semicolons. These always go *outside* closing quotation marks.

> It's hard to argue with W. Charisse Goodman's observation that media images persuade women that "thinness equals happiness and fulfillment"; nevertheless, American women today are more overweight than ever (53).

Paraphrasing

When you paraphrase, you restate information from a source in your own words, using your own sentence structures. Paraphrase when the source material is important but the original wording is not. Because it includes all the main points of the source, a paraphrase is usually about the same length as the original.

Here is a paragraph about synthetic dyes and cancer, followed by two paraphrases of it that demonstrate some of the challenges of paraphrasing:

ORIGINAL SOURCE

In 1938, in a series of now-classic experiments, exposure to synthetic dyes derived from coal and belonging to a class of chemicals called aromatic amines was shown to cause bladder cancer in dogs. These results helped explain why bladder cancers had become so prevalent among dyestuffs workers. With the invention of mauve in 1854, synthetic dyes began replacing natural plant-based dyes in the coloring of cloth and leather. By the beginning of the twentieth century, bladder cancer rates among this group of workers had skyrocketed, and the dog experiments helped unravel this mystery. The International Labor Organization did not wait for the results of these animal tests, however, and in 1921 declared certain aromatic amines to be human carcinogens. Decades later, these dogs provided a lead in understanding why tire-industry workers, as well as machinists and metalworkers, also began falling victim to bladder cancer: aromatic amines had been added to rubbers and cutting oils to serve as accelerants and antirust agents.

—Sandra Steingraber, "Pesticides, Animals, and Humans"

The following paraphrase borrows too much of the language of the original or changes it only slightly, as the highlighted words and phrases show:

UNACCEPTABLE PARAPHRASE: WORDING TOO CLOSE

Now-classic experiments in 1938 showed that when dogs were exposed to aromatic amines, chemicals used in synthetic dyes derived from coal, they developed bladder cancer. Similar cancers were prevalent among dyestuffs workers, and these experiments helped to explain why. Mauve, a synthetic dye, was invented in 1854, after which cloth and leather manufacturers replaced most of the natural plant-based dyes with synthetic dyes. By the early twentieth century, this group of workers had skyrocketing rates of bladder cancer, a mystery the dog experiments helped to unravel. As early as 1921, though, before the test results proved the connection, the International Labor Organization had labeled certain aromatic amines carcinogenic. Even so, decades later many metalworkers, machinists, and tire-industry workers began developing bladder cancer. The animal tests helped researchers understand that rubbers and cutting oils contained aromatic amines as accelerants and antirust agents (Steingraber 976).

* academic literacies   ● fields   ● research
■ rhetorical situations   ⁘ processes   ● media/design
▲ genres   ◆ strategies

The next paraphrase uses original language but follows the sentence structure of Steingraber's text too closely:

UNACCEPTABLE PARAPHRASE: SENTENCE STRUCTURE TOO CLOSE

In 1938, several pathbreaking experiments showed that being exposed to synthetic dyes that are made from coal and belong to a type of chemicals called aromatic amines caused dogs to get bladder cancer. These results helped researchers identify why cancers of the bladder had become so common among textile workers who worked with dyes. With the development of mauve in 1854, synthetic dyes began to be used instead of dyes based on plants in the dyeing of leather and cloth. By the end of the nineteenth century, rates of bladder cancer among these workers had increased dramatically, and the experiments using dogs helped clear up this oddity. The International Labor Organization anticipated the results of these tests on animals, though, and in 1921 labeled some aromatic amines carcinogenic. Years later these experiments with dogs helped researchers explain why workers in the tire industry, as well as metalworkers and machinists, also started dying of bladder cancer: aromatic amines had been put into rubbers and cutting oils as rust inhibitors and accelerants (Steingraber 976).

Patchwriting, a third form of unacceptable paraphrase, combines the other two. Composition researcher Rebecca Moore Howard defines it as "copying from a source text and then deleting some words, altering grammatical structures, or plugging in one-for-one synonym-substitutes." Here is a patchwrite of the first two sentences of the original source: (The source's exact words are shaded in yellow; paraphrases are in blue.)

PATCHWRITE

Scientists have known for a long time that chemicals in the environment can cause cancer. For example, in 1938, in a series of important experiments, being exposed to synthetic dyes made out of coal and belonging to a kind of chemicals called aromatic amines was shown to cause dogs to develop bladder cancer. These experiments explain why this type of cancer had become so common among workers who handled dyes.

Here is an acceptable paraphrase of the entire passage:

ACCEPTABLE PARAPHRASE

Biologist Sandra Steingraber explains that pathbreaking experiments in 1938 demonstrated that dogs exposed to aromatic amines (chemicals used in coal-based synthetic dyes) developed cancers of the bladder that were similar to cancers common among dyers in the textile industry. After mauve, the first synthetic dye, was invented in 1854, leather and cloth manufacturers replaced most natural dyes made from plants with synthetic dyes, and by the early 1900s textile workers had very high rates of bladder cancer. The experiments with dogs proved the connection, but years before, in 1921, the International Labor Organization had labeled some aromatic amines carcinogenic. Even so, years later many metalworkers, machinists, and workers in the tire industry started to develop unusually high rates of bladder cancer. The experiments with dogs helped researchers understand that the cancers were caused by aromatic amines used in cutting oils to inhibit rust and in rubbers to speed up the manufacturing process (976).

SOME GUIDELINES FOR PARAPHRASING

- *Use your own words and sentence structure.* It's acceptable to use some words from the original; but as much as possible, the phrasing and sentence structures should be your own.

551–54
- *Introduce paraphrased text with* SIGNAL PHRASES.

- *Put in quotation marks any of the source's original phrasing that you use.*

- *Indicate the source.* Although the wording may be yours, the ideas and information come from another source; be sure to name the author, and include **DOCUMENTATION** to avoid the possibility of **PLAGIARISM**.

MLA 567–74
APA 618–22
551–54

Summarizing

A summary states the main ideas in a source concisely and in your own words. Unlike a paraphrase, a summary does *not* present all the details, and it is generally as brief as possible. Summaries may boil down an entire

* academic literacies   ● fields   ● research
■ rhetorical situations   ∴ processes   ● media/design
▲ genres   ◆ strategies

book or essay into a single sentence, or they may take a paragraph or more to present the main ideas. Here, for example, is a one-sentence summary of the Steingraber paragraph:

> Steingraber explains that experiments with dogs demonstrated that aromatic amines, chemicals used in synthetic dyes, cutting oils, and rubber, cause bladder cancer (976).

In the context of an essay, the summary might take this form:

> Medical researchers have long relied on experiments using animals to expand understanding of the causes of disease. For example, biologist and ecologist Sandra Steingraber notes that in the second half of the nineteenth century, the rate of bladder cancer soared among textile workers. According to Steingraber, experiments with dogs demonstrated that synthetic chemicals in dyes used to color the textiles caused the cancer (976).

SOME GUIDELINES FOR SUMMARIZING

- *Include only the main ideas; leave out the details.* A summary should include just enough information to give the reader the gist of the original. It is always much shorter than the original, sometimes even as brief as one sentence.

- *Use your own words.* If you quote phrasing from the original, enclose the phrase in quotation marks.

- *Indicate the source.* Although the wording may be yours, the ideas and information come from another source. Name the author, either in a signal phrase or in parentheses, and include an appropriate **IN-TEXT CITATION** to avoid the possibility of **PLAGIARISM**.

MLA 567–74

APA 618–22

551–54

Introducing Source Materials Using Signal Phrases

You need to introduce quotations, paraphrases, and summaries clearly, usually letting readers know who the author is—and, if need be, something about their credentials. Consider this sentence:

> Professor and textbook author Elaine Tyler May argues that many high
> school history books are too bland to interest young readers (531).

The beginning ("Professor and textbook author Elaine Tyler May argues")
functions as a *signal phrase*, telling readers who is making the assertion
and why she has the authority to speak on the topic—and making clear
that everything between the signal phrase and the parenthetical citation
comes from that source. Since the signal phrase names the author, the
parenthetical citation includes only the page number; had the author not
been identified in the signal phrase, she would have been named in the
parentheses:

> Even some textbook authors believe that many high school history
> books are too bland to interest young readers (May 531).

MLA and APA have different conventions for constructing signal
phrases. In MLA, the language you use in a signal phrase can be neutral—
like "X says"/"Y thinks"/"according to Z." Or it can suggest something
about the **STANCE** —the source's or your own. The example above referring to the textbook author uses the verb "argues," suggesting that what
she says is open to dispute (or that the writer believes it is). How would it
change your understanding if the signal verb were "observes"/"suggests"?

72–74 ■

In addition to the names of sources' authors, signal phrases often give
readers information about institutional affiliations and positions authors
have, their academic or professional specialties, and any other information that lets readers judge the credibility of the sources. You should craft
each signal phrase you use so as to highlight the credentials of the author.
Here are some examples:

> A study done by Keenan Johnson, professor of psychology at Duke
> University, showed that . . .

The signal phrase identifies the source's author, his professional position,
and his university affiliation, emphasizing his title.

> Science writer Isaac McDougal argues that . . .

✳ academic literacies   ● fields   ● research
■ rhetorical situations   ⁖ processes   ● media/design
▲ genres   ◆ strategies

This phrase acknowledges that the source's author may not have scholarly credentials but is a published writer; it's a useful construction if the source doesn't provide much information about the writer.

> Writing in *Psychology Today,* Amanda Chao-Fitz notes that . . .

This is the sort of signal phrase to use if you have no information on the author; you establish credibility on the basis of the publication in which the source appears.

If you're writing using APA style, signal phrases are typically briefer, giving only the author's last name and the date of publication:

> According to Benzinger (2010), . . .
> Quartucci (2011) observed that . . .

Signal verbs.   The verbs used in signal phrases do more than simply introduce a quote, paraphrase, or reference; they also provide information on the author's intended meaning and give your reader a clue as to how the source should be read. They also can give your writing variety and interest.

SOME COMMON SIGNAL VERBS

acknowledge	comment	dispute	point out
admit	conclude	emphasize	reason
advise	concur	grant	reject
agree	confirm	illustrate	report
argue	contend	imply	respond
assert	declare	insist	state
believe	deny	note	suggest
charge	disagree	observe	think
claim			

Verb tenses.   MLA and APA also have different conventions regarding the tenses of verbs in signal phrases. MLA requires present-tense verbs ("writes," "asserts," " notes") in signal phrases to introduce a work you are quoting, paraphrasing, or summarizing.

In *Poor Richard's Almanack*, Benjamin Franklin <u>notes</u>, "He that cannot obey, cannot command" (739).

If, however, you are referring to the act of writing or saying something rather than simply quoting someone's words, you might not use the present tense. The writer of the following sentence focuses on the year in which the source was written — therefore, the verb is necessarily in the past tense:

Back in 1941, Kenneth Burke <u>wrote</u> that "the ethical values of work are in its application of the competitive equipment to cooperative ends" (316).

If you are following APA style, use the past tense to introduce sources composed in the past. If you are referring to a past action that didn't occur at a specific time or that continues into the present, you should use the present perfect.

Dowdall et al. (2020) <u>observed</u> that women attending women's colleges are less likely to engage in binge drinking than are women who attend coeducational colleges (p. 713).

Many researchers <u>have studied</u> drinking habits on college campuses.

APA requires the present tense, however, to discuss the results of an experiment or to explain conclusions that are generally agreed on.

The findings of this study <u>suggest</u> that excessive drinking has serious consequences for college students and their institutions.

The authors of numerous studies <u>agree</u> that smoking and drinking among adolescents are associated with lower academic achievement.

If you need more help

See Chapter 51 for help **ACKNOWLEDGING SOURCES** and giving credit to the sources you use. See also the **SAMPLE RESEARCH PAPERS** to see how sources are cited in MLA and APA styles. And see Chapter 3 if you're writing a **SUMMARY/RESPONSE** essay.

555–59
MLA 608–14
APA 648–55
33–45

* academic literacies   ● fields   ● research
■ rhetorical situations   ⁝ processes   ● media/design
▲ genres   ◆ strategies

51 Acknowledging Sources, Avoiding Plagiarism

Whenever you do research-based writing, you find yourself entering a conversation—reading what many others have had to say about your topic, figuring out what you yourself think, and then putting what you think in writing—"putting in your oar," as the rhetorician Kenneth Burke once wrote. As a writer, you need to *acknowledge* any words and ideas that come from others—to give credit where credit is due, to recognize the various authorities and many perspectives you have considered, to show readers where they can find your sources, and to situate your own arguments in the ongoing conversation. Using other people's words and ideas without acknowledgment is *plagiarism*, a serious academic and ethical offense. This chapter will show you how to acknowledge the materials you use and avoid plagiarism.

Acknowledging Sources

When you insert in your text information that you've obtained from others, your readers need to know where your source's words or ideas begin and end. Therefore, you should usually introduce a source by naming the author in a **SIGNAL PHRASE** and then provide brief **DOCUMENTATION** of the specific material from the source in a parenthetical reference following the material. (Sometimes you can put the author's name in the parenthetical reference as well.) You need only brief documentation of the source here, since your readers will find full bibliographic information about it in your list of **WORKS CITED** or **REFERENCES**.

● 551–54
MLA 567–74
APA 618–22

● MLA 574–605
APA 623–44

Sources that need acknowledgment.   You almost always need to acknowledge any information that you get from a specific source. Material you should acknowledge includes the following:

- *Direct quotations.* Unless they are well known (see p. 558 for some examples), any quotations from another source must be enclosed in quotation marks, cited with brief bibliographic information in parentheses, and usually introduced with a signal phrase that tells who wrote or said it and provides necessary contextual information, as in the following sentence:

  In a dissenting opinion on the issue of racial preferences in college admissions, Supreme Court justice Ruth Bader Ginsburg argues, "The stain of generations of racial oppression is still visible in our society, and the determination to hasten its removal remains vital" (*Gratz v. Bollinger*).

- *Arguable statements and information that may not be common knowledge.* If you state something about which there is disagreement or for which arguments can be made, cite the source of your statement. If in doubt about whether you need to give the source of an assertion, provide it. As part of an essay on "fake news" programs, for example, you might make the following assertion:

  The satire of *The Daily Show* complements the conservative bias of Fox News, since both have abandoned the stance of objectivity maintained by mainstream news sources, contends Michael Hoyt, executive editor of the *Columbia Journalism Review* (43).

  Others might argue with the contention that the Fox News Channel offers biased reports of the news, so the source of this assertion needs to be acknowledged. In the same essay, you might present information that should be cited because it's not widely known, as in this example:

  According to a report by the Pew Research Center, 25 percent of Americans got information about the 2016 presidential campaign from comedy shows like *The Late Show with Stephen Colbert* and *Saturday Night Live* (2).

- *The opinions and assertions of others.* When you present the ideas, opinions, and assertions of others, cite the source. You may have rewritten

academic literacies   fields   research

rhetorical situations   processes   media/design

genres   strategies

the concept in your own words, but the ideas were generated by someone else and must be acknowledged, as they are here:

David Boonin, writing in the *Journal of Social Philosophy*, asserts that, logically, laws banning marriage between people of different races are not discriminatory since everyone of each race is affected equally by them. Laws banning same-sex unions are discriminatory, however, since they apply only to people with a certain sexual orientation (256).

- *Photographs, visual images, or video.* If you didn't create a photo, image, or video yourself, cite the source. If possible, use the original source, not a version that appears in another source, to ensure that the version you're using hasn't been altered.

- *Any information that you didn't generate yourself.* If you didn't do the research or compile the data yourself, cite your source. This goes for interviews, statistics, graphs, charts, visuals, photographs—anything you use that you didn't create. If you create a chart using data from another source, you need to cite that source.

- *Collaboration with and help from others.* In many of your courses and in work situations, you'll be called on to work with others. You may get help with your writing at your school's writing center or from fellow students in your writing courses. Acknowledging such collaboration or assistance, in a brief informational note, is a way of giving credit—and saying thank you. See guidelines for writing notes in the **MLA** and **APA** sections of this book.

MLA 574
APA 622–23

Sources that don't need acknowledgment.   Widely available information and common knowledge do not require acknowledgment. What constitutes common knowledge may not be clear, however. When in doubt, provide a citation, or ask your instructor whether the information needs to be cited. You generally do not need to cite the following:

- *Information that most readers are likely to know.* You don't need to acknowledge information that is widely known or commonly accepted as fact. For example, in a literary analysis, you wouldn't cite a source saying that Harriet Beecher Stowe wrote *Uncle Tom's Cabin*; you can assume your readers already know that. However, you should cite the

source from which you got the information that the book was first published in installments in a magazine and then, with revisions, in book form, because that information isn't common knowledge. As you do research in areas you're not familiar with, be aware that what constitutes common knowledge isn't always clear; the history of the novel's publication would be known to Stowe scholars and would likely need no acknowledgment in an essay written for them. In this case, too, if you aren't sure whether to acknowledge information, it's best to do so.

- *Information and documents that are widely available.* If a piece of information appears in several sources or reference works or if a document has been published widely, you needn't cite a source for it. For example, the date when astronauts Neil Armstrong and Buzz Aldrin landed a spacecraft on the moon can be found in any number of reference works. Similarly, the Declaration of Independence and the Gettysburg Address are reprinted in thousands of sources, so the ones where you found them need no citation.

- *Well-known quotations.* These include such famous quotations as Lady Macbeth's "Out, damned spot!" and John F. Kennedy's "Ask not what your country can do for you; ask what you can do for your country." Be sure, however, that the quotation is correct. Winston Churchill is said to have told a class of schoolchildren, "Never, ever, ever, ever, ever, ever, ever give up. Never give up. Never give up. Never give up." His actual words, however, are much different and begin "Never give in."

- *Material that you created or gathered yourself.* You need not cite photographs that you took, graphs that you composed based on your own findings, or data from an experiment or survey that you conducted—though you should make sure readers know that the work is yours.

A good rule of thumb: *when in doubt, cite your source.* You're unlikely to be criticized for citing too much—but you may invite charges of plagiarism by citing too little.

Avoiding Plagiarism

In North America, authors' words and ideas are considered to belong to them, so using others' words or ideas without acknowledging the source

is considered to be a serious offense called plagiarism. In fact, plagiarism is often committed unintentionally—as when a writer paraphrases someone else's ideas in language that is too close to the original. It is essential, therefore, to know what constitutes plagiarism: (1) using another writer's words or ideas without acknowledging the source, (2) using another writer's exact words without quotation marks, and (3) paraphrasing or summarizing someone else's ideas using language or sentence structures that are too close to theirs.

To avoid plagiarizing, take careful **NOTES** as you do your research, clearly labeling as quotations any words you quote directly and being careful to use your own phrasing and sentence structures in paraphrases and summaries. Be sure you know what source material you must **DOCUMENT**, and give credit to your sources, both in the text and in a list of **REFERENCES** or **WORKS CITED**.

542–44

560–63

APA 623–44

MLA 574–605

Be aware that it's easy to plagiarize inadvertently when working with online sources, such as full-text articles, that you've downloaded or cut and pasted into your notes. Keep careful track of these materials, since saving copies of your sources is so easy. Later, be sure to check your draft against the original sources to make sure your quotations are accurately worded—and take care, too, to include quotation marks and document the source correctly. Copying online material right into a document you are writing and forgetting to put quotation marks around it or to document it (or both) is all too easy to do. You must acknowledge information you find on the web just as you must acknowledge all other source materials.

And you must recognize that plagiarism has consequences. Scholars' work will be discredited if it too closely resembles another's. Journalists who are found to have plagiarized lose their jobs, and students routinely fail courses or are dismissed from their school when they are caught cheating—all too often by submitting as their own essays that they have purchased from online "research" sites. If you're having trouble completing an assignment, seek assistance. Talk with your instructor; or if your school has a writing center, go there for advice on all aspects of your writing, including acknowledging sources and avoiding plagiarism.

52 Documentation

In everyday life, we are generally aware of our sources: "I read it on Megan McArdle's blog." "Amber told me it's your birthday." "If you don't believe me, ask Mom." Saying how we know what we know and where we got our information is part of establishing our credibility and persuading others to take what we say seriously.

The goal of a research project is to study a topic, combining what we learn from sources with our own thinking and then composing a written text. When we write up the results of a research project, we cite the sources we use, usually by quoting, paraphrasing, or summarizing, and we acknowledge those sources, telling readers where the ideas came from. The information we give about sources is called documentation, and we provide it not only to establish our credibility as researchers and writers but also so that our readers, if they wish to, can find the sources themselves.

Understanding Documentation Styles

The Norton Field Guide covers the documentation styles of the Modern Language Association (MLA) and the American Psychological Association (APA). MLA style is used chiefly in the humanities; APA is used in the social sciences, sciences, education, and nursing. Both are two-part systems, consisting of (1) brief in-text parenthetical documentation for quotations, paraphrases, or summaries and (2) more-detailed documentation in a list of sources at the end of the text. MLA and APA require that the end-of-text documentation provide the following basic information about each source you cite:

- author, editor, or creator of the source
- title of source (and of publication or site where it appears)
- version or edition of source
- name of publisher
- date of publication
- retrieval information (for online sources)

MLA and APA are by no means the only documentation styles. Many other publishers and organizations have their own style, among them the University of Chicago Press and the Council of Science Editors. We focus on MLA and APA here because those are styles that college students are often required to use. On the following page are examples of how the two parts—the brief parenthetical documentation in your text and the more detailed information at the end—correspond in each of these systems.

The examples here and throughout this book are color-coded to help you see the crucial parts of each citation: tan for author and editor, yellow for titles, and gray for publication information: name of publisher, date of publication, page number(s), DOI or URL, and so on.

As the examples of in-text documentation show, in either MLA or APA style you should name the author either in a signal phrase or in parentheses following the source information. But there are several differences between the two styles in the details of the documentation. In MLA, the author's first and last names are used in a signal phrase at first mention; in APA, only the last name is used. In APA, the abbreviation "p." is used with the page number; in MLA, there is no abbreviation before a page number. Finally, in APA the date of publication always appears just after the author's name.

Comparing the MLA and APA styles of listing works cited or references also reveals some differences: MLA includes an author's first name while APA gives only initials; MLA puts the date near the end while APA places it right after the author's name; MLA capitalizes most of the words in a book's title and subtitle while APA capitalizes only the first words and proper nouns and proper adjectives in each.

MLA Style

IN-TEXT DOCUMENTATION

As Lester Faigley puts it, "The world has become a bazaar from which to shop for an individual 'lifestyle'" (12).

As one observer suggests, "The world has become a bazaar from which to shop for an individual 'lifestyle'" (Faigley 12).

WORKS-CITED DOCUMENTATION

Faigley, Lester. *Fragments of Rationality: Postmodernity and the Subject of Composition.* U of Pittsburgh P, 1992.

APA Style

IN-TEXT DOCUMENTATION

As Faigley (1992) suggested, "The world has become a bazaar from which to shop for an individual 'lifestyle'" (p. 12).

As one observer has noted, "The world has become a bazaar from which to shop for an individual 'lifestyle'" (Faigley, 1992, p. 12).

REFERENCE-LIST DOCUMENTATION

Faigley, L. (1992). *Fragments of rationality: Postmodernity and the subject of composition.* University of Pittsburgh Press.

TITLE   AUTHOR   PUBLICATION

Some of these differences are related to the nature of the academic fields in which the two styles are used. In humanities disciplines, the authorship of a text is emphasized, so both first and last names are included in MLA documentation. Scholarship in those fields may be several years old but still current, so the publication date doesn't appear in the in-text citation. In APA style, as in many documentation styles used in the sciences, education, and engineering, emphasis is placed on the date of publication because in these fields, more recent research is usually preferred over older studies. However, although the elements are arranged differently, both MLA and APA—and other documentation styles as well—require similar information about author, title, and publication.

53 MLA Style

MLA style calls for (1) brief in-text documentation and (2) complete bibliographic information in a list of works cited at the end of your text. The models and examples in this guide draw on the ninth edition of the *MLA Handbook*, published by the Modern Language Association of America in 2021. For additional information, or if you're citing a source that isn't covered in this guide, visit style.mla.org.

A DIRECTORY TO MLA STYLE

Formatting a Research Paper   605

Sample Research Paper   607

Throughout this chapter, you'll find color-coded templates and examples to help you see how writers include source information in their texts and in their lists of works cited: tan for author, editor, translator, and other contributors; yellow for titles; blue for publication information—date of publication, page number(s), DOIs, and other location information.

IN-TEXT DOCUMENTATION

Whenever you **QUOTE**, **PARAPHRASE**, or **SUMMARIZE** a source in your writing, you need to provide brief documentation that tells readers what you took from the source and where in the source you found that information. This brief documentation also refers readers to the full entry in your works-cited list, so begin with whatever comes first there: the author, the title, or a description of the source.

544–47

You can mention the author or title either in a signal phrase—"Toni Morrison writes," "In *Beowulf*," "According to the article 'Every Patient's Nightmare'"—or in parentheses—(Morrison). If relevant, include pages or other details about where you found the information in the parenthetical reference: (Morrison 67).

Shorten any lengthy titles or descriptions in parentheses by including the first noun with any preceding adjectives and omitting any initial articles (*Norton Field Guide* for *The Norton Field Guide to Writing*). If the title

doesn't start with a noun, use the first phrase or clause (*How to Be* for *How to Be an Antiracist*). Use the full title if it's short.

The first two examples below show basic in-text documentation of a work by one author. Variations on those examples follow. The examples illustrate the MLA style of using quotation marks around titles of short works and italicizing titles of long works.

1. AUTHOR NAMED IN A SIGNAL PHRASE

551–54

If you mention the author in a **SIGNAL PHRASE**, put only the page number(s) in parentheses. Do not write "page" or "p." The first time you mention the author, use their first and last names. You can usually omit any middle initials.

> David McCullough describes John Adams's hands as those of someone used to manual labor (18).

2. AUTHOR NAMED IN PARENTHESES

If you do not mention the author in a signal phrase, put the author's last name in parentheses along with any page number(s). Do not use punctuation between the name and the page number(s).

> Adams is said to have had "the hands of a man accustomed to pruning his own trees, cutting his own hay, and splitting his own firewood" (McCullough 18).

Whether you use a signal phrase and parentheses or parentheses only, try to put the parenthetical documentation at the end of the sentence or as close as possible to the material you've cited—without awkwardly interrupting the sentence. When the parenthetical reference comes at the end of the sentence, the period follows it.

3. TWO OR MORE WORKS BY THE SAME AUTHOR

If you cite multiple works by one author, include the title of the work you are citing either in the signal phrase or in parentheses.

Robert Kaplan insists that understanding power in the Near East requires "Western leaders who know when to intervene, and do so without illusions" (*Eastward to Tartary* 330).

Put a comma between author and title if both are in the parentheses.

Understanding power in the Near East requires "Western leaders who know when to intervene, and do so without illusions" (Kaplan, *Eastward to Tartary* 330).

4. AUTHORS WITH THE SAME LAST NAME

Give each author's first and last names in any signal phrase, or add the author's first initial in the parenthetical reference.

"Imaginative" applies not only to modern literature but also to writing of all periods, whereas "magical" is often used in writing about Arthurian romances (A. Wilson 25).

5. TWO OR MORE AUTHORS

For a work with two authors, name both. If you first mention them in a signal phrase, give their first and last names.

Lori Carlson and Cynthia Ventura's stated goal is to introduce Julio Cortázar, Marjorie Agosín, and other Latin American writers to an audience of English-speaking adolescents (v).

For a work by three or more authors that you mention in a signal phrase, you can either name them all or name the first author followed by "and others" or "and colleagues." If you mention them in a parenthetical reference, name the first author followed by "et al."

Phyllis Anderson and colleagues describe British literature thematically (A54-A67).

One survey of British literature breaks the contents into thematic groupings (Anderson et al. A54-A67).

6. ORGANIZATION OR GOVERNMENT AS AUTHOR

In a signal phrase, use the full name of the organization: American Academy of Arts and Sciences. In parentheses, use the shortest noun phrase, omitting any initial articles: American Academy.

> The US government can be direct when it wants to be. For example, it sternly warns, "If you are overpaid, we will recover any payments not due you" (Social Security Administration 12).

7. AUTHOR UNKNOWN

If you don't know the author, use the work's title in a signal phrase or in a parenthetical reference.

> A powerful editorial in *The New York Times* asserts that healthy liver donor Mike Hurewitz died because of "frightening" faulty postoperative care ("Every Patient's Nightmare").

8. LITERARY WORKS

When referring to common literary works that are available in many different editions, give the page numbers from the edition you are using, followed by information that will let readers of any edition locate the text you are citing.

NOVELS AND PROSE PLAYS. Give the page number followed by a semicolon and any chapter, section, or act numbers, separated by commas.

> In *Pride and Prejudice,* Mrs. Bennet shows no warmth toward Jane when she returns from Netherfield (Austen 105; ch. 12).

VERSE PLAYS. Give act, scene, and line numbers, separated with periods.

> Shakespeare continues the vision theme when Macbeth says, "Thou hast no speculation in those eyes / Which thou dost glare with" (*Macbeth* 3.3.96-97).

POEMS. Give the part and the line numbers, separated by periods. If a poem has only line numbers, use the word "line" or "lines" in the first reference; after that, give only numbers.

> Walt Whitman sets up not only opposing adjectives but also opposing nouns in "Song of Myself" when he says, "I am of old and young, of the foolish as much as the wise, / . . . a child as well as a man" (16.330-32).

> One description of the mere in *Beowulf* is "not a pleasant place" (line 1372). Later, it is labeled "the awful place" (1378).

9. WORK IN AN ANTHOLOGY

Name the author(s) of the work, not the editor of the anthology.

> "It is the teapots that truly shock," according to Cynthia Ozick in her essay on teapots as metaphor (70).

> In *In Short: A Collection of Creative Nonfiction*, readers will find both an essay on Scottish tea (Hiestand) and a piece on teapots as metaphors (Ozick).

10. ENCYCLOPEDIA OR DICTIONARY

Acknowledge an entry in an encyclopedia or dictionary by giving the author's name, if available. For an entry without an author, give the entry's title.

> According to *Funk and Wagnall's New World Encyclopedia*, early in his career, most of Kubrick's income came from "hustling chess games in Washington Square Park" ("Kubrick, Stanley").

11. LEGAL DOCUMENTS

For legal cases, give whatever comes first in the works-cited entry. If you're citing a government document in parentheses and multiple entries in your

works-cited list start with the same government author, give as much of the name as you need to differentiate the sources.

> In 2015, for the first time, all states were required to license and
> recognize the marriages of same-sex couples (United States, Supreme
> Court).

12. SACRED TEXT

When citing a sacred text such as the Bible or the Qur'an for the first time, give the title of the edition as well as the book, chapter, and verse (or their equivalent), separated by periods. MLA recommends abbreviating the names of the books of the Bible in parenthetical references. Later citations from the same edition do not have to repeat its title.

> The wording from *The New English Bible* follows: "In the beginning of
> creation, when God made heaven and earth, the earth was without form
> and void . . ." (Gen. 1.1-2).

13. MULTIVOLUME WORK

If you cite more than one volume of a multivolume work, each time you cite one of the volumes, give the volume *and* the page number(s) in parentheses, separated by a colon and a space.

> Carl Sandburg concludes with the following sentence about those
> paying last respects to Lincoln: "All day long and through the night the
> unbroken line moved, the home town having its farewell" (4: 413).

If you cite an entire volume of a multivolume work in parentheses, give the author's last name followed by a comma and "vol." before the volume number: (Sandburg, vol. 4). If your works-cited list includes only a single volume of a multivolume work, give just the page number in parentheses: (413).

14. TWO OR MORE WORKS CITED TOGETHER

If you're citing two or more works closely together, you will sometimes need to provide a parenthetical reference for each one.

> Dennis Baron describes singular "they" as "the missing word that's
> been hiding in plain sight" (182), while Benjamin Dreyer believes
> that "singular 'they' is not the wave of the future; it's the wave of the
> present" (93).

If you are citing multiple sources for the same idea in parentheses, separate the references with a semicolon.

> Many critics have examined great works of literature from a cultural
> perspective (Tanner 7; Smith viii).

15. SOURCE QUOTED IN ANOTHER SOURCE

When you are quoting text that you found quoted in another source, use the abbreviation "qtd. in" in the parenthetical reference.

> Charlotte Brontë wrote to G. H. Lewes, "Why do you like Miss Austen so
> very much? I am puzzled on that point" (qtd. in Tanner 7).

16. WORK WITHOUT PAGE NUMBERS

For works without page or part numbers, including many online sources, no number is needed in a parenthetical reference.

> Studies show that music training helps children to be better at
> multitasking later in life ("Hearing the Music").

If you mention the author in a signal phrase, or if you mention the title of a work with no author, no parenthetical reference is needed.

> Arthur Brooks argues that a switch to fully remote work would have a
> negative effect on mental and physical health.

If the source has chapter, paragraph, or section numbers, use them with the abbreviations "ch.," "par.," or "sec.": ("Hearing the Music," par. 2). Don't count lines or paragraphs on your own if they aren't numbered in the source. For an ebook, use chapter numbers. For an audio or video recording, give the hours, minutes, and seconds (separated by colons) as shown on the player: (00:05:21-31).

17. AN ENTIRE WORK OR A ONE-PAGE ARTICLE

If you cite an entire work rather than a part of it, or if you cite a single-page article, there's no need to include page numbers.

> Throughout life, John Adams strove to succeed (McCullough).

NOTES

Sometimes you may need to give information that doesn't fit into the text itself—to thank people who helped you, to provide additional details, to refer readers to other sources, or to add comments about sources. Such information can be given in a footnote (at the bottom of the page) or an endnote (on a separate page with the heading "Notes" or "Endnotes" just before your works-cited list). Put a superscript number at the appropriate point in your text, signaling to readers to look for the note with the corresponding number. If you have multiple notes, number them consecutively throughout your paper.

TEXT

This essay will argue that giving student athletes preferential treatment undermines educational goals.[1]

NOTE

[1] I want to thank those who contributed to my thinking on this topic, especially my teacher Vincent Yu.

LIST OF WORKS CITED

A works-cited list provides full bibliographic information for every source cited in your text. See page 607 for guidelines on formatting this list and page 613 for a sample works-cited list.

Core Elements

MLA style provides a list of core elements for documenting sources in a works-cited list. Not all sources will include each of these elements; include as much information as is available for any title you cite. For

guidance about specific sources you need to document, see the templates and examples on pages 580–605, but here are some general guidelines for how to treat each of the core elements.

CORE ELEMENTS FOR ENTRIES IN A WORKS-CITED LIST

- Author
- Title of the source
- Title of any "container," a larger work in which the source is found—an anthology, a website, a journal or magazine, a database, a streaming service like *Netflix*, or a learning management system, among others
- Editor, translator, director, or other contributors
- Version
- Number of volume and issue, episode and season
- Publisher
- Date of publication
- Location of the source: page numbers, DOI, permalink, URL, etc.

The above order is the general order MLA recommends, but there will be exceptions. To document a translated essay that you found in an anthology, for instance, you'd identify the translator after the title of the essay rather than after that of the anthology. You may sometimes need additional elements as well, either at the end of an entry or in the middle—for instance, a label to indicate that your source is a map, or an original year of publication. Remember that your goal is to tell readers what sources you've consulted and where they can find them. Providing this information is one way you can engage with readers—and enable them to join in the conversation with you and your sources.

AUTHORS AND CONTRIBUTORS

- An author can be any kind of creator—a writer, a musician, an artist, and so on.
- If there is one author, put the last name first, followed by a comma and the first name: Morrison, Toni.

- If there are two authors, list the first author last name first and the second one first name first: Lunsford, Andrea, and Lisa Ede. Put their names in the order given in the work. For three or more authors, give the first author's name followed by "et al.": Greenblatt, Stephen, et al.

- Include any middle names or initials: Toklas, Alice B.

- If the author is a group or organization, use the full name, omitting any initial article: United Nations.

- If an author uses a handle that is significantly different from their name, include the handle in square brackets after the name: Ocasio-Cortez, Alexandria [@AOC].

- If there's no known author, start the entry with the title.

- If there's an editor but no author, put the editor's name in the author position and specify their role: Lunsford, Andrea, editor.

- If you're citing someone in addition to an author—an editor, translator, director, or other contributors—specify their role. If there are multiple contributors, put the one whose work you wish to highlight before the title, and list any others you want to mention after the title. If you don't want to highlight one particular contributor, start with the title and include any contributors after the title. For contributors named before the title, specify their role after the name: Fincher, David, director. For those named after the title, specify their role first: Directed by David Fincher.

TITLES

- Include any subtitles, and capitalize all the words except for articles ("a," "an," "the"), prepositions ("to," "at," "from," and so on), and coordinating conjunctions ("and," "but," "for," "or," "nor," "so," "yet")—unless they are the first or last word of a title or subtitle.

- Italicize the titles of books, periodicals, websites, and other long works: *Pride and Prejudice, Wired.*

- Put quotation marks around the titles of articles and other short works: "Letter from Birmingham Jail."

- To document a source that has no title, describe it without italics or quotation marks: Letter to the author, Photograph of a tree. For a short, untitled email, text message, tweet, or poem, you may want to include the text itself instead: Dickinson, Emily. "Immortal is an ample word." *American Poems*, www.americanpoems.com/poets/emilydickinson /immortal-is-an-ample-word.

VERSIONS

- If you cite a source that's available in more than one version, specify the one you consulted in your works-cited entry. Write ordinal numbers with numerals, and abbreviate "edition": 2nd ed. Write out names of specific versions, and capitalize following a period or if the name is a proper noun: King James Version, unabridged version, director's cut.

NUMBERS

- If you cite a book that's published in multiple volumes, indicate the volume number. Abbreviate "volume," and write the number as a numeral: vol. 2.
- Indicate volume and issue numbers (if any) of journals, abbreviating both "volume" and "number": vol. 123, no. 4.
- If you cite a TV show or podcast episode, indicate the season and episode numbers: season 1, episode 4.

PUBLISHERS

- Write publishers', studios', and networks' names in full, but omit initial articles and business words like "Inc." or "Company."
- For academic presses, use "U" for "University" and "P" for "Press": Princeton UP, U of California P. Spell out "Press" if the name doesn't include "University": MIT Press.
- Many publishers use "&" in their name: Simon & Schuster. MLA says to use "and" instead: Simon and Schuster.

- If the publisher is a division of an organization, list the organization and any divisions from largest to smallest: Stanford U, Center for the Study of Language and Information, Metaphysics Research Lab.

DATES

- Whether to give just the year or to include the month and day depends on the source. In general, give the full date that you find there. If the date is unknown, simply omit it.

- Abbreviate the months except for May, June, and July: Jan., Feb., Mar., Apr., Aug., Sept., Oct., Nov., Dec.

- For books, give the publication date on the copyright page: 1948. If a book lists more than one date, use the most recent one.

- Periodicals may be published annually, monthly, seasonally, weekly, or daily. Give the full date that you find there: 2019, Apr. 2019, 16 Apr. 2019. Do not capitalize the names of seasons: spring 2021.

- For online sources, use the copyright date or the full publication date that you find there, or a date of revision. If the source does not give a date, use the date of access: Accessed 6 June 2020. Give a date of access as well for online sources you think are likely to change, or for websites that have disappeared.

LOCATION

- For most print articles and other short works, give a page number or range of pages: p. 24, pp. 24-35. For articles that are not on consecutive pages, give the first page number with a plus sign: pp. 24+.

- If it's necessary to specify a section of a source, give the section name before the page numbers: Sunday Review sec., p. 3.

- Indicate the location of an online source by giving a DOI if one is available; if not, give a URL—and use a permalink if one is available. MLA notes that URLs are not always reliable, so ask your instructor if you should include them. DOIs should start with "https://doi.org/"—but no need to include "https://" for a URL, unless you want the URL to be a hyperlink.

- For a geographical location, give enough information to identify it: a city (Houston), a city and state (Portland, Maine), or a city and country (Manaus, Brazil).

- For something seen in a museum, archive, or elsewhere, name the institution and its location: Maine Jewish Museum, Portland, Maine.

- For performances or other live presentations, name the venue and its location: Mark Taper Forum, Los Angeles.

PUNCTUATION

- Use a period after the author name(s) that start an entry (Morrison, Toni.) and the title of the source you're documenting (*Beloved*.).

- Use a comma between the author's last and first names: Ede, Lisa.

- Some URLs will not fit on one line. MLA does not specify where to break a URL, but we recommend breaking it before a punctuation mark. Do *not* add a hyphen or a space.

- Sometimes you'll need to provide information about more than one work for a single source—for instance, when you cite an article from a periodical that you access through a database. MLA refers to the periodical and database (or any other entity that holds a source) as "containers" and specifies certain punctuation. Use commas between elements within each container, and put a period at the end of each container. For example:

  Semuels, Alana. "The Future Will Be Quiet." *The Atlantic*, Apr. 2016,
  pp. 19-20. *ProQuest*, search.proquest.com/docview
  /1777443553?accountid+42654.

The guidelines that follow will help you document the kinds of sources you're likely to use. The first section shows how to acknowledge authors and other contributors and applies to all kinds of sources—print, online, or others. Later sections show how to treat titles, publication information, location, and access information for many specific kinds of sources. In general, provide as much information as possible for each source—enough to tell readers how to find a source if they wish to access it themselves.

SOURCES NOT COVERED

These guidelines will help you cite a variety of sources, but there may be sources you want to use that aren't mentioned here. If you're citing a source that isn't covered, consult the MLA style blog at style.mla.org, or ask them a question at style.mla.org/ask-a-question.

Authors and Contributors

When you name authors and other contributors in your citations, you are crediting them for their work and letting readers know who's in on the conversation. The following guidelines for citing authors and contributors apply to all sources you cite: in print, online, or in some other media.

1. ONE AUTHOR

> Author's Last Name, First Name. *Title.* Publisher, Date.

> Anderson, Chris. *The Long Tail: Why the Future of Business Is Selling Less of More.* Hyperion, 2006.

2. TWO AUTHORS

> 1st Author's Last Name, First Name, and 2nd Author's First and Last Names. *Title.* Publisher, Date.

> Lunsford, Andrea, and Lisa Ede. *Singular Texts/Plural Authors: Perspectives on Collaborative Writing.* Southern Illinois UP, 1990.

3. THREE OR MORE AUTHORS

> 1st Author's Last Name, First Name, et al. *Title.* Publisher, Date.

> Sebranek, Patrick, et al. *Writers INC: A Guide to Writing, Thinking, and Learning.* Write Source, 1990.

4. TWO OR MORE WORKS BY THE SAME AUTHOR

Give the author's name in the first entry, and then use three hyphens in the author slot for each of the subsequent works, listing them alphabetically by the first word of each title and ignoring any initial articles.

Author's Last Name, First Name. *Title That Comes First
Alphabetically.* Publisher, Date.

---. *Title That Comes Next Alphabetically.* Publisher, Date.

Kaplan, Robert D. *The Coming Anarchy: Shattering the Dreams of the
Post Cold War.* Random House, 2000.

---. *Eastward to Tartary: Travels in the Balkans, the Middle East, and the
Caucasus.* Random House, 2000.

5. AUTHOR AND EDITOR OR TRANSLATOR

Author's Last Name, First Name. *Title.* Role by First and Last
Names, Publisher, Date.

Austen, Jane. *Emma.* Edited by Stephen M. Parrish, W. W. Norton, 2000.

Dostoevsky, Fyodor. *Crime and Punishment.* Translated by Richard Pevear
and Larissa Volokhonsky, Vintage Books, 1993.

Start with the editor or translator, followed by their role, if you are focusing on that contribution rather than the author's. If there is a translator
but no author, start with the title.

Pevear, Richard, and Larissa Volokhonsky, translators. *Crime and
Punishment.* By Fyodor Dostoevsky, Vintage Books, 1993.

Beowulf. Translated by Kevin Crossley-Holland, Macmillan, 1968.

6. NO AUTHOR OR EDITOR

When there's no known author or editor, start with the title.

The Turner Collection in the Clore Gallery. Tate Publications, 1987.

"Being Invisible Closer to Reality." *The Atlanta Journal-Constitution*, 11
Aug. 2008, p. A3.

7. ORGANIZATION OR GOVERNMENT AS AUTHOR

> Organization Name. *Title*. Publisher, Date.

> Diagram Group. *The Macmillan Visual Desk Reference*. Macmillan, 1993.

For a government publication, give the name that is shown in the source.

> United States, Department of Health and Human Services, National
> Institute of Mental Health. *Autism Spectrum Disorders*. Government
> Printing Office, 2004.

When a nongovernment organization is both author and publisher, start with the title and list the organization only as the publisher.

> *Stylebook on Religion 2000: A Reference Guide and Usage Manual*.
> Catholic News Service, 2002.

If a division of an organization is listed as the author, give the division as the author and the organization as the publisher.

> Center for Workforce Studies. *2005-13: Demographics of the U.S.*
> *Psychology Workforce*. American Psychological Association, July 2015.

Articles and Other Short Works

Articles, essays, reviews, and other shorts works are found in journals, magazines, newspapers, other periodicals, and also in books—all of which you may find in print, online, or in a database. For most short works, you'll need to provide information about the author, the titles of both the short work and the longer work where it's found, any page numbers, and various kinds of publication information, all explained below.

8. ARTICLE IN A JOURNAL

PRINT

> Author's Last Name, First Name. "Title of Article." *Name*
> *of Journal*, Volume, Issue, Date, Pages.

> Cooney, Brian C. "Considering *Robinson Crusoe*'s 'Liberty of Conscience' in an
> Age of Terror." *College English*, vol. 69, no. 3, Jan. 2007, pp. 197-215.

TITLE   AUTHOR   PUBLICATION

Documentation Map (MLA)
Article in a Print Journal

Marge Simpson, Blue-Haired Housewife: •················· TITLE OF ARTICLE
Defining Domesticity on *The Simpsons*

JESSAMYN NEUHAUS   •·············· AUTHOR

MORE THAN TWENTY SEASONS AFTER ITS DEBUT AS A SHORT ON *THE Tracy Ullman Show* in 1989, pundits, politicians, scholars, journalists, and critics continue to discuss and debate the meaning and relevance of *The Simpsons* to American society. For academics and educators, the show offers an especially dense pop culture text, inspiring articles and anthologies examining *The Simpsons* in light of American religious life, the representation of homosexuality in cartoons, and the use of pop culture in the classroom, among many other topics (Dennis; Frank; Henry "The Whole World's Gone Gay"; Hobbs; Kristiansen). Philosophers and literary theorists in particular are intrigued by the quintessentially postmodern self-aware form and content of *The Simpsons* and the questions about identity, spectatorship, and consumer culture it raises (Alberti; Bybee and Overbeck; Glynn; Henry "The Triumph of Popular Culture"; Herron; Hull; Irwin et al.; Ott; Parisi).

Simpsons observers frequently note that this TV show begs one of the fundamental questions in cultural studies: can pop culture ever provide a site of individual or collective resistance or must it always ultimately function in the interests of the capitalist dominant ideology? Is *The Simpsons* a brilliant satire of virtually every cherished American myth about public and private life, offering dissatisfied Americans the opportunity to critically reflect on contemporary issues (Turner 435)? Or is it simply another TV show making money for the Fox Network? Is *The Simpsons* an empty, cynical, even nihilistic view of the world, lulling its viewers into laughing hopelessly at the pointless futility of

right-side labels
NAME OF JOURNAL
VOLUME
ISSUE
PAGES
YEAR

The Journal of Popular Culture, Vol. 43, No. 4, 2010, pp.761-81 •··············
© 2010, Wiley Periodicals, Inc.

Neuhaus, Jessamyn. "Marge Simpson, Blue-Haired Housewife: Defining Domesticity on *The Simpsons*." *The Journal of Popular Culture*, vol. 43, no. 4, 2010, pp. 761-81.

● 582–89
for more
on citing
articles
MLA style

Documentation Map (MLA)
Article in an Online Magazine

URL

NAME OF MAGAZINE

TITLE OF
ARTICLE

AUTHOR

DATE

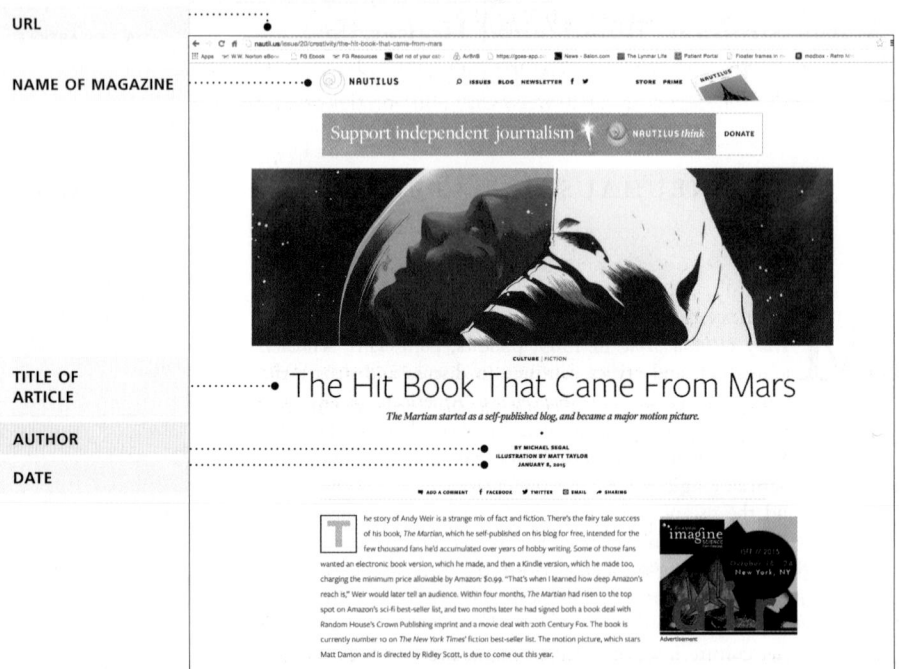

595–98 for more on citing websites MLA style

Segal, Michael. "The Hit Book That Came from Mars." *Nautilus*, 8 Jan. 2015, nautil.us/issue/20/creativity/the-hit-book-that -came-from-mars.

TITLE   AUTHOR   PUBLICATION

ONLINE

> Author's Last Name, First Name. "Title of Article." *Name of*
> *Journal*, Volume, Issue, Date, DOI *or* URL.

> Schmidt, Desmond. "A Model of Versions and Layers." Digital Humanities
> Quarterly, vol. 13, no. 3, 2019, www.digitalhumanities.org/dhq
> /vol/13/3/000430/000430.html.

9. ARTICLE IN A MAGAZINE

PRINT

> Author's Last Name, First Name. "Title of Article." *Name*
> *of Magazine*, Volume (if any), Issue (if any), Date, Pages.

> Burt, Tequia. "Legacy of Activism: Concerned Black Students' 50-Year
> History at Grinnell College." *Grinnell Magazine*, vol. 48, no. 4,
> summer 2016, pp. 32-38.

ONLINE

> Author's Last Name, First Name. "Title of Article." *Name*
> *of Magazine*, Volume (if any), Issue (if any), Date, DOI *or* URL.

> Brooks, Arthur C. "The Hidden Toll of Remote Work." *The Atlantic*,
> 1 Apr. 2021, www.theatlantic.com/family/archive/2021/04/zoom
> -remote-work-loneliness-happiness/618473.

10. ARTICLE IN A NEWS PUBLICATION

PRINT

> Author's Last Name, First Name. "Title of Article." *Name*
> *of Publication*, Date, Pages.

> Saulny, Susan, and Jacques Steinberg. "On College Forms, a Question of
> Race Can Perplex." *The New York Times*, 14 June 2011, p. A1.

To document a particular edition of a newspaper, list the edition before
the date. If a section name or number is needed to locate the article,
put that detail after the date.

Burns, John F., and Miguel Helft. "Under Pressure, YouTube Withdraws Muslim Cleric's Videos." *The New York Times*, late ed., 4 Nov. 2010, sec. 1, p. 13.

ONLINE

Author's Last Name, First Name. "Title of Article." *Name of Publication*, Date, URL.

Banerjee, Neela. "Proposed Religion-Based Program for Federal Inmates Is Canceled." *The New York Times,* 28 Oct. 2006, www.nytimes .com/2006/10/28/us/28prison.html.

11. ARTICLE ACCESSED THROUGH A DATABASE

Author's Last Name, First Name. "Title of Article." *Name of Periodical*, Volume, Issue, Date, Pages. *Name of Database*, DOI or URL.

Stalter, Sunny. "Subway Ride and Subway System in Hart Crane's 'The Tunnel.'" *Journal of Modern Literature*, vol. 33, no. 2, Jan. 2010, pp. 70-91. *JSTOR*, https://doi.org/10.2979/jml.2010.33.2.70.

12. ENTRY IN A REFERENCE WORK

PRINT

Author's Last Name, First Name (if any). "Title of Entry." *Title of Reference Book*, edited by First and Last Names (if any), Edition number, Volume (if any), Publisher, Date, Pages.

Fritz, Jan Marie. "Clinical Sociology." *Encyclopedia of Sociology*, edited by Edgar F. Borgatta and Rhonda J. V. Montgomery, 2nd ed., vol. 1, Macmillan Reference USA, 2000, pp. 323-29.

"California." *The New Columbia Encyclopedia*, edited by William H. Harris and Judith S. Levey, 4th ed., Columbia UP, 1975, pp. 423-24.

Documentation Map (MLA)
Journal Article Accessed through a Database

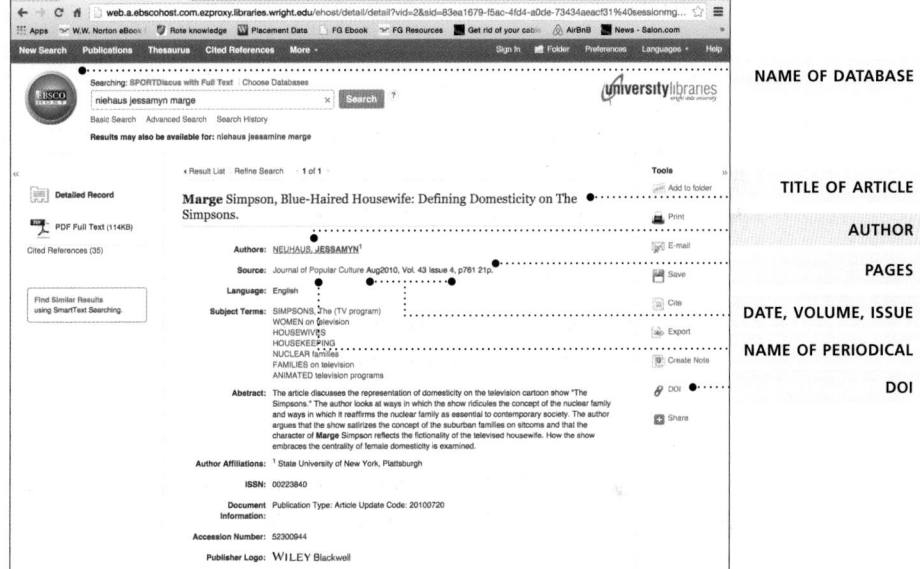

NAME OF DATABASE

TITLE OF ARTICLE

AUTHOR

PAGES

DATE, VOLUME, ISSUE

NAME OF PERIODICAL

DOI

Neuhaus, Jessamyn. "Marge Simpson, Blue-Haired Housewife:
Defining Domesticity on *The Simpsons*." *Journal of Popular
Culture*, vol. 43, no. 4, Aug. 2010, pp. 761-81. *EBSCOhost*,
https://doi.org/10.1111/j.1540-5931.2010.00769.x.

ONLINE

Document online reference works the same as print ones, adding the URL after the date of publication.

> "Baseball." *The Columbia Electronic Encyclopedia*, edited by Paul Lagassé, 6th ed., Columbia UP, 2012, www.infoplease.com/encyclopedia.

13. EDITORIAL OR OP-ED

EDITORIAL

> Editorial Board. "Title." *Name of Periodical*, Date, Page *or* URL.

> Editorial Board. "A New Look for Local News Coverage." *The Lakeville Journal*, 13 Feb. 2020, p. A8.

> Editorial Board. "Editorial: Protect Reporters at Protest Scenes." *Los Angeles Times*, 11 Mar. 2021, www.latimes.com/opinion/story/2021 -03-11/reporters-protest-scenes.

OP-ED

> Author's Last Name, First Name. "Title." *Name of Periodical*, Date, Page *or* URL.

> Okafor, Kingsley. "Opinion: The First Step to COVID Vaccine Equity Is Overall Health Equity." *The Denver Post*, 15 Apr. 2021, www.denverpost.com/2021/04/15/covid-vaccine-equity-kaiser.

If it's not clear that it's an op-ed, add a label at the end.

> Balf, Todd. "Falling in Love with Swimming." *The New York Times*, 17 Apr. 2021, p. A21. Op-ed.

14. LETTER TO THE EDITOR

> Author's Last Name, First Name. "Title of Letter (if any)." *Name of Periodical*, Date, Page *or* URL.

> Pinker, Steven. "Language Arts." *The New Yorker*, 4 June 2012, p. 10.

If the letter has no title, include "Letter" after the author's name.

> Fleischmann, W. B. Letter. *The New York Review of Books*, 1 June 1963,
> www.nybooks.com/articles/1963/06/01/letter-21.

15. REVIEW

PRINT

> Reviewer's Last Name, First Name. "Title of Review." *Name
> of Periodical*, Date, Pages.

> Frank, Jeffrey. "Body Count." *The New Yorker*, 30 July 2007, pp. 86-87.

ONLINE

> Reviewer's Last Name, First Name. "Title of Review." *Name of
> Periodical*, Date, URL.

> Donadio, Rachel. "Italy's Great, Mysterious Storyteller." *The New York
> Review of Books*, 18 Dec. 2014, www.nybooks.com/articles/2014/12
> /18/italys-great-mysterious-storyteller.

If a review has no title, include the title and author of the work being reviewed after the reviewer's name.

> Lohier, Patrick. Review of *Exhalation*, by Ted Chiang. *Harvard Review
> Online*, 4 Oct. 2019, www.harvardreview.org/book-review
> /exhalation.

16. COMMENT ON AN ONLINE ARTICLE

> Commenter's Last Name, First Name *or* Username. Comment on "Title of
> Article." *Name of Periodical*, Date posted, Time posted, URL.

> ZeikJT. Comment on "The Post-Disaster Artist." *Polygon*, 6 May 2020,
> 4:33 a.m., www.polygon.com/2020/5/5/21246679/josh-trank-capone
> -interview-fantastic-four-chronicle.

Books and Parts of Books

For most books, you'll need to provide information about the author, the title, the publisher, and the year of publication. If you found the book inside a larger volume, a database, or some other work, be sure to specify that as well.

17. BASIC ENTRIES FOR A BOOK

PRINT

Author's Last Name, First Name. *Title*. Publisher, Year of publication.

Watson, Brad. *Miss Jane*. W. W. Norton, 2016.

EBOOK

Author's Last Name, First Name. *Title*. Ebook ed., Publisher, Year of Publication.

Watson, Brad. *Miss Jane*. Ebook ed., W. W. Norton, 2016.

ON A WEBSITE

Author's Last Name, First Name. *Title*. Publisher, Year of publication, DOI *or* URL.

Ball, Cheryl E., and Drew M. Loewe, editors. *Bad Ideas about Writing*. West Virginia U Libraries, 2017, textbooks.lib.wvu.edu/badideas /badideasaboutwriting-book.pdf.

WHEN THE PUBLISHER IS THE AUTHOR

Title. Edition number (if any), Publisher, Year of publication.

MLA Handbook. 9th ed., Modern Language Association of America, 2021.

18. ANTHOLOGY OR EDITED COLLECTION

Last Name, First Name, editor. *Title*. Publisher, Year of publication.

Kitchen, Judith, and Mary Paumier Jones, editors. *In Short: A Collection of Brief Nonfiction*. W. W. Norton, 1996.

TITLE   AUTHOR   PUBLICATION

Documentation Map (MLA)
Print Book

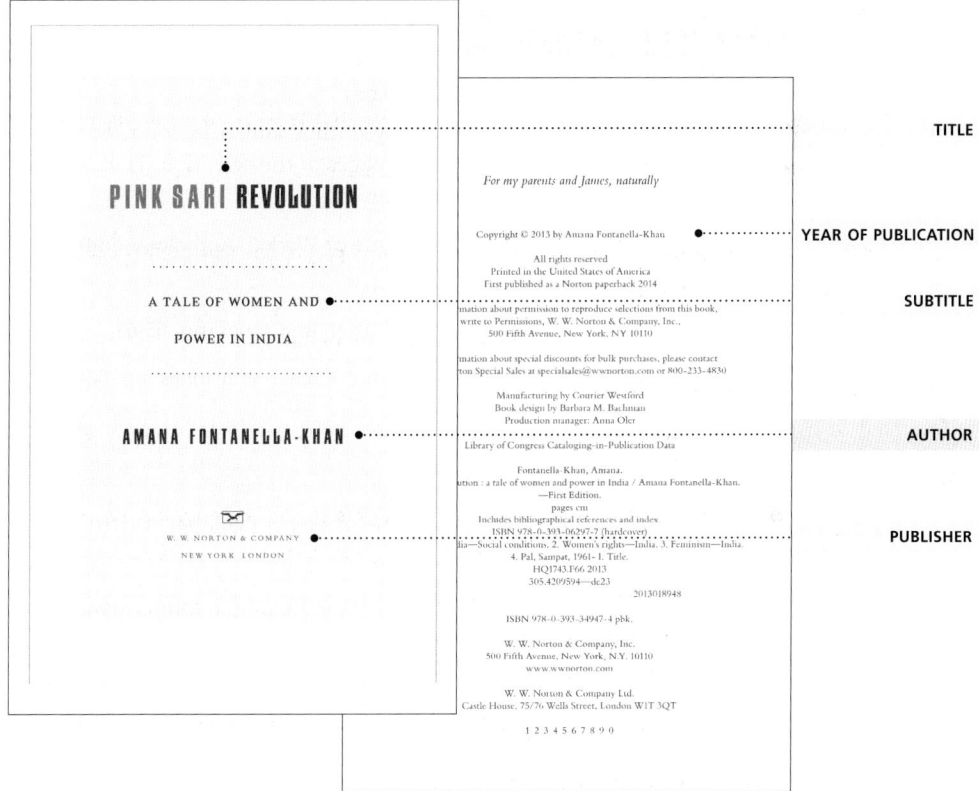

TITLE

YEAR OF PUBLICATION

SUBTITLE

AUTHOR

PUBLISHER

Fontanella-Khan, Amana. *Pink Sari Revolution: A Tale of Women and Power in India.* W. W. Norton, 2013.

590–95 for more on citing books MLA style

19. WORK IN AN ANTHOLOGY

> Author's Last Name, First Name. "Title of Work." *Title*
>> *of Anthology*, edited by First and Last Names, Publisher, Year
>> of publication, Pages.

> Achebe, Chinua. "Uncle Ben's Choice." *The Seagull Reader: Literature*,
>> edited by Joseph Kelly, W. W. Norton, 2005, pp. 23-27.

TWO OR MORE WORKS FROM ONE ANTHOLOGY

Prepare an entry for each selection by author and title, followed by the anthology editors' last names and the pages of the selection. Then include an entry for the anthology itself (see no. 18).

> Author's Last Name, First Name. "Title of Work." Anthology Editors'
>> Last Names, Pages.

> Hiestand, Emily. "Afternoon Tea." Kitchen and Jones, pp. 65-67.

> Ozick, Cynthia. "The Shock of Teapots." Kitchen and Jones, pp. 68-71.

20. MULTIVOLUME WORK

ALL VOLUMES

> Author's Last Name, First Name. *Title of Work.* Publisher, Year(s)
>> of publication. Number of vols.

> Churchill, Winston. *The Second World War.* Houghton Mifflin, 1948-53.
>> 6 vols.

SINGLE VOLUME

> Author's Last Name, First Name. *Title of Work.* Vol. number, Publisher,
>> Year of publication.

> Sandburg, Carl. *Abraham Lincoln: The War Years.* Vol. 2, Harcourt, Brace
>> and World, 1939.

If the volume has its own title, include it after the author's name, and indicate the volume number and series title after the year.

Caro, Robert A. *Means of Ascent*. Vintage Books, 1990. Vol. 2 of *The Years of Lyndon Johnson*.

21. BOOK IN A SERIES

Author's Last Name, First Name. *Title of Book*. Edited by First and Last Names, Publisher, Year of publication. Series Title.

Walker, Alice. *Everyday Use*. Edited by Barbara T. Christian, Rutgers UP, 1994. Women Writers: Texts and Contexts.

22. GRAPHIC NARRATIVE OR COMIC BOOK

Author's Last Name, First Name. *Title*. Publisher, Year of publication.

Barry, Lynda. *One! Hundred! Demons!* Drawn and Quarterly, 2005.

If the work has both an author and an illustrator, start with the one you want to highlight, and label the role of anyone who's not an author.

Pekar, Harvey. *Bob and Harv's Comics*. Illustrated by R. Crumb, Running Press, 1996.

Crumb, R., illustrator. *Bob and Harv's Comics*. By Harvey Pekar, Running Press, 1996.

To cite several contributors, you can also start with the title.

Secret Invasion. By Brian Michael Bendis, illustrated by Leinil Yu, inked by Mark Morales, Marvel, 2009.

23. SACRED TEXT

If you cite a specific edition of a religious text, you need to include it in your works-cited list.

The New English Bible with the Apocrypha. Oxford UP, 1971.

The Torah: A Modern Commentary. W. Gunther Plaut, general editor, Union of American Hebrew Congregations, 1981.

24. EDITION OTHER THAN THE FIRST

> Author's Last Name, First Name. *Title.* Edition name *or*
> number, Publisher, Year of publication.

Smart, Ninian. *The World's Religions.* 2nd ed., Cambridge UP, 1998.

25. REPUBLISHED WORK

> Author's Last Name, First Name. *Title.* Year of original
> publication. Current publisher, Year of republication.

Bierce, Ambrose. *Civil War Stories.* 1909. Dover, 1994.

26. FOREWORD, INTRODUCTION, PREFACE, OR AFTERWORD

> Part Author's Last Name, First Name. Name of Part. *Title of Book*, by
> Author's First and Last Names, Publisher, Year of publication, Pages.

Tanner, Tony. Introduction. *Pride and Prejudice*, by Jane Austen, Penguin,
1972, pp. 7-46.

27. PUBLISHED LETTER

> Letter Writer's Last Name, First Name. "Title of letter." Day Month
> Year. *Title of Book*, edited by First and Last Names, Publisher,
> Year of publication, Pages.

White, E. B. "To Carol Angell." 28 May 1970. *Letters of E. B. White*,
edited by Dorothy Lobrano Guth, Harper and Row, 1976, p. 600.

28. PAPER HEARD AT A CONFERENCE

> Author's Last Name, First Name. "Title of Paper." Conference, Day
> Month Year, Location.

Hern, Katie. "Inside an Accelerated Reading and Writing Classroom."
Conference on Acceleration in Developmental Education, 15 June
2016, Sheraton Inner Harbor Hotel, Baltimore.

TITLE   AUTHOR   PUBLICATION

29. DISSERTATION

> Author's Last Name, First Name. *Title.* Year. Institution, PhD
> dissertation. *Name of Database*, URL.

> Simington, Maire Orav. *Chasing the American Dream Post World War II:*
> *Perspectives from Literature and Advertising.* 2003. Arizona State U,
> PhD dissertation. *ProQuest*, search.proquest.com/docview/305340098.

For an unpublished dissertation, end with the institution and a description of the work.

> Kim, Loel. *Students Respond to Teacher Comments: A Comparison of*
> *Online Written and Voice Modalities.* 1998. Carnegie Mellon U, PhD
> dissertation.

Websites

Many sources are available in multiple media—for example, a print periodical that is also on the web and contained in digital databases—but some are published only on websites. A website can have an author, an editor, or neither. Some sites have a publisher, and some do not. Include whatever information is available. If the publisher and title of the site are essentially the same, omit the name of the publisher.

30. ENTIRE WEBSITE

> Editor's Last Name, First Name, role. *Title of Site.* Publisher, Date, URL.

> Proffitt, Michael, chief editor. *The Oxford English Dictionary.* Oxford UP,
> 2021, www.oed.com.

PERSONAL WEBSITE

> Author's Last Name, First Name. *Title of Site.* Date, URL.

> Park, Linda Sue. *Linda Sue Park: Author and Educator.* 2021,
> lindasuepark.com.

Documentation Map (MLA)

Work on a Website

URL

TITLE OF SITE

TITLE OF WORK

DATE

AUTHOR

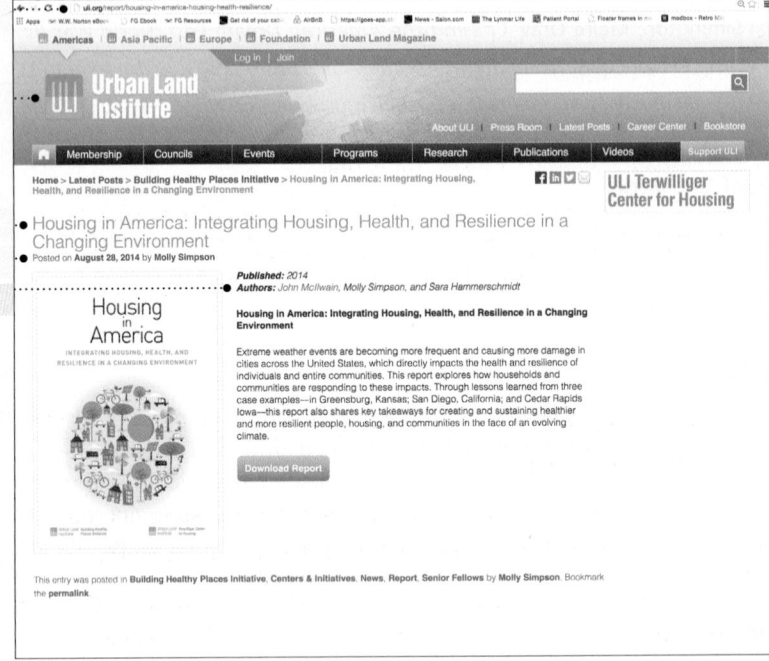

595–97
for more
on citing
websites
MLA style

McIlwain, John, et al. "Housing in America: Integrating Housing,

   Health, and Resilience in a Changing Environment." *Urban*

   Land Institute, 28 Aug. 2014, uli.org/report/housing-in

   -america-housing-health-resilience.

TITLE   AUTHOR   PUBLICATION

If the site is likely to change, if it has no date, or if it no longer exists, include a date of access.

> *Archive of Our Own.* Organization for Transformative Works,
> archiveofourown.org. Accessed 23 Apr. 2021.

31. WORK ON A WEBSITE

> Author's Last Name, First Name (if any). "Title of Work." *Title*
> *of Site*, Publisher (if any), Date, URL.

> Cesareo, Kerry. "Moving Closer to Tackling Deforestation at Scale."
> *World Wildlife Fund*, 20 Oct. 2020, www.worldwildlife.org/blogs
> /sustainability-works/posts/moving-closer-to-tackling-deforestation
> -at-scale.

32. BLOG ENTRY

> Author's Last Name, First Name. "Title of Blog Entry." *Title*
> *of Blog*, Date, URL.

> Hollmichel, Stefanie. "Bring Up the Bodies." *So Many Books*, 10 Feb.
> 2014, somanybooksblog.com/2014/02/10/bring-up-the-bodies.

Document a whole blog as you would an entire website (no. 30) and a comment on a blog as you would a comment on an online article (no. 16).

33. WIKI

> "Title of Entry." *Title of Wiki*, Publisher, Date, URL.

> "Pi." *Wikipedia*, Wikimedia Foundation, 28 Aug. 2013, en.wikipedia.org
> /wiki/Pi.

Personal Communication and Social Media

34. PERSONAL LETTER

> Sender's Last Name, First Name. Letter to the author. Day Month Year.

> Quindlen, Anna. Letter to the author. 11 Apr. 2013.

35. EMAIL OR TEXT MESSAGE

> Sender's Last Name, First Name. Email *or* Text message to First Name
> Last Name *or* to the author. Day Month Year.

> Smith, William. Email to Richard Bullock. 19 Nov. 2013.

> Rombes, Maddy. Text message to Isaac Cohen. 4 May 2021.

> O'Malley, Kit. Text message to the author. 2 June 2020.

You can also include the text of a short email or text message, with a label at the end.

> Rust, Max. "Trip to see the cows tomorrow?" 27 Apr. 2021. Email.

36. POST TO *TWITTER*, *INSTAGRAM*, OR OTHER SOCIAL MEDIA

> Author. "Title." *Title of Site*, Day Month Year, URL.

> Oregon Zoo. "Winter Wildlife Wonderland." *Facebook*, 8 Feb. 2019,
> www.facebook.com/80229441108/videos/2399570506799549.

If there's no title, you can use a concise description or the text of a short post.

> Millman, Debbie. Photos of Roxane Gay. *Instagram*, 18 Feb. 2021, www
> .instagram.com/p/CLcT_EnhnWT.

> Obama, Barack [@POTUS44]. "It's been the honor of my life to serve you.
> You made me a better leader and a better man." *Twitter*, 20 Jan.
> 2017, twitter.com/POTUS44/status/822445882247413761.

Audio, Visual, and Other Sources

37. ADVERTISEMENT

PRINT

Description of ad. *Title of Periodical*, Date, Page.

Advertisement for Grey Goose. *Wine Spectator*, 18 Dec. 2020, p. 22.

VIDEO

"Title." *Title of Site*, uploaded by Company, Date, URL.

"First Visitors." *YouTube*, uploaded by Snickers, 20 Aug. 2020,
www.youtube.com/watch?v=negeco0b1L0.

38. ART

ORIGINAL

Artist's Last Name, First Name. *Title of Art.* Year created, Location.

Van Gogh, Vincent. *The Potato Eaters.* 1885, Van Gogh Museum,
Amsterdam.

IN A BOOK

Artist's Last Name, First Name. *Title of Art.* Year created, Location. *Title of
Book*, by First and Last Names, Publisher, Year of publication, Page.

Van Gogh, Vincent. *The Potato Eaters.* 1885, Scottish National Gallery.
*History of Art: A Survey of the Major Visual Arts from the Dawn of
History to the Present Day*, by H. W. Janson, Prentice Hall / Harry N.
Abrams, 1969, p. 508.

ONLINE

Artist's Last Name, First Name. *Title of Art.* Year created. *Title
of Site*, URL.

Warhol, Andy. *Self-portrait.* 1979. *J. Paul Getty Museum*, www.getty
.edu/art/collection/objects/106971/andy-warhol-self-portrait
-american-1979.

39. CARTOON

PRINT

> Author's Last Name, First Name. Cartoon *or* "Title of Cartoon." *Name of Periodical*, Date, Page.

> Mankoff, Robert. Cartoon. *The New Yorker*, 3 May 1993, p. 50.

ONLINE

> Author's Last Name, First Name. Cartoon *or* "Title of Cartoon." *Title of Site*, Date, URL.

> Munroe, Randall. "Up Goer Five." *xkcd*, 12 Nov. 2012, xkcd.com/1133.

40. SUPREME COURT CASE

> United States, Supreme Court. *First Defendant v. Second Defendant*. Date of decision. *Title of Source Site*, Publisher, URL.

> United States, Supreme Court. *District of Columbia v. Heller*. 26 June 2008. *Legal Information Institute*, Cornell Law School, www.law.cornell.edu/supremecourt/text/07-290.

41. FILM

Name individuals based on the focus of your project—the director, the screenwriter, or someone else.

> *Title of Film*. Role by First and Last Names, Production Company, Date.

> *Breakfast at Tiffany's*. Directed by Blake Edwards, Paramount, 1961.

ONLINE

> *Title of Film*. Role by First and Last Names, Production Company, Date. *Title of Site*, URL.

> *Interstellar*. Directed by Christopher Nolan, Paramount, 2014. *Amazon Prime Video*, www.amazon.com/Interstellar-Matthew-McConaughey/dp/B00TU9UFTS.

If your essay focuses on one contributor, you may put their name before the title.

> Edwards, Blake, director. *Breakfast at Tiffany's*. Paramount, 1961.

42. TV SHOW EPISODE

Name contributors based on the focus of your project—director, creator, actors, or others. If you don't want to highlight anyone in particular, don't include any contributors.

BROADCAST

> "Title of Episode." *Title of Program*, role by First and Last Names (if any), season, episode, Production Company, Date.

> "The Storm." *Avatar: The Last Airbender*, created by Michael Dante DiMartino and Bryan Konietzko, season 1, episode 12, Nickelodeon Animation Studios, 3 June 2005.

DVD

> "Title of Episode." Broadcast Date. *Title of DVD*, role by First and Last Names (if any), season, episode, Production Company, Release Date, disc number. DVD.

> "The Storm." 2005. *Avatar: The Last Airbender: The Complete Book 1 Collection*, created by Michael Dante DiMartino and Bryan Konietzko, episode 12, Nickelodeon Animation Studios, 2006, disc 3. DVD.

STREAMING ONLINE

> "Title of Episode." *Title of Program*, role by First and Last Names (if any), season, episode, Production Company, Date. *Title of Site*, URL.

> "The Storm." *Avatar: The Last Airbender*, season 1, episode 12, Nickelodeon Animation Studios, 2005. *Netflix*, www.netflix.com.

STREAMING ON AN APP

"Title of Episode." *Title of Program,* role by First and Last Names (if
any), season, episode, Production Company, Date. *Name of* app.

"The Storm." *Avatar: The Last Airbender,* season 1, episode 12,
Nickelodeon Animation Studios, 2005. *Netflix* app.

If you're focusing on a contributor who's relevant specifically to the
episode you're citing, include their name after the episode title.

"The Storm." Directed by Lauren MacMullan. *Avatar: The Last Airbender,*
season 1, episode 12, Nickelodeon Animation Studios, 3 June 2005.

43. ONLINE VIDEO

"Title of Video." *Title of Site,* uploaded by Uploader's Name, Day
Month Year, URL.

"Everything Wrong with *National Treasure* in 13 Minutes or Less."
YouTube, uploaded by CinemaSins, 21 Aug. 2014, www.youtube
.com/watch?v=1ul-_ZWvXTs.

44. PRESENTATION ON *ZOOM* OR OTHER VIRTUAL PLATFORM

MLA doesn't give specific guidance on how to cite a virtual presentation,
but this is what we recommend. See style.mla.org for more information.

Author's Last Name, First Name. "Title." Sponsoring Institution, Day
Month Year. *Name of Platform.*

Budhathoki, Thir. "Cross-Cultural Perceptions of Literacies in Student
Writing." Conference on College Composition and Communication,
9 Apr. 2021. *Zoom.*

45. INTERVIEW

If it's not clear that it's an interview, add a label at the end. If you are citing
a transcript of an interview, indicate that at the end as well.

PUBLISHED

Subject's Last Name, First Name. *"Title of Interview."* Interview by First Name Last Name (if given). *Name of Publication*, Date, Pages *or* URL.

Whitehead, Colson. "Colson Whitehead: By the Book." *The New York Times*, 15 May 2014, www.nytimes.com/2014/05/18/books/review /colson-whitehead-by-the-book.html. Interview.

PERSONAL

Subject's Last Name, First Name. Concise description. Day Month Year.

Bazelon, L. S. Telephone interview with the author. 4 Oct. 2020.

46. MAP

If the title doesn't make clear it's a map, add a label at the end.

Title of Map. Publisher, Date.

Brooklyn. J. B. Beers, 1874. Map.

47. MUSICAL SCORE

Composer's Last Name, First Name. *Title of Composition.* Publisher, Year of publication.

Frank, Gabriela Lena. *Compadrazgo.* G. Schirmer, 2007.

48. ORAL PRESENTATION

Presenter's Last Name, First Name. "Title of Presentation." Sponsoring Institution, Date, Location.

Cassin, Michael. "Nature in the Raw—The Art of Landscape Painting." Berkshire Institute for Lifelong Learning, 24 Mar. 2005, Clark Art Institute, Williamstown, Massachusetts.

49. PODCAST

If you accessed a podcast on the web, give the URL.

"Title of Episode." *Title of Podcast*, hosted by First Name Last
Name, season, episode, Production Company, Date, URL.

"DUSTWUN." *Serial*, hosted by Sarah Koenig, season 2, episode 1, WBEZ /
Serial Productions, 10 Dec. 2015, serialpodcast.org/season-two/1
/dustwun.

THROUGH AN APP

"DUSTWUN." *Serial*, hosted by Sarah Koenig, season 2, episode 1, WBEZ /
Serial Productions, 10 Dec. 2015. *Spotify* app.

50. RADIO PROGRAM

"Title of Episode." *Title of Program*, hosted by First Name Last
Name, Station, Day Month Year.

"In Defense of Ignorance." *This American Life*, hosted by Ira Glass, WBEZ,
22 Apr. 2016.

51. SOUND RECORDING

If you accessed a recording on the web, give the URL.

Artist's Last Name, First Name. "Title of Work." *Title of Album*, Label,
Date, URL.

Beyoncé. "Pray You Catch Me." *Lemonade*, Parkwood Entertainment /
Columbia Records, 2016, www.beyonce.com/album/lemonade
-visual-album/songs.

THROUGH AN APP

Simone, Nina. "To Be Young, Gifted and Black." *Black Gold*, RCA Records,
1969. *Spotify* app.

ON A CD

Artist's Last Name, First Name. "Title of Work." *Title of Album*, Label,
     Date. CD.

Brown, Greg. "Canned Goods." *The Live One*, Red House, 1995. CD.

52. VIDEO GAME

Title of Game. Version, Distributor, Date of release.

Animal Crossing: New Horizons. Version 1.1.4, Nintendo, 6 Apr. 2020.

FORMATTING A RESEARCH PAPER

Name, course, title. MLA does not require a separate title page, unless
your paper is a group project. In the upper left-hand corner of your first
page, include your name, your instructor's name, the course name and
number, and the date. Center the title of your paper on the line after the
date; capitalize it as you would a book title. If your paper is a group project,
include all of that information on a title page instead, listing all the authors.

Page numbers. In the upper right-hand corner of each page, one-half inch
below the top of the page, include your last name and the page number.
If it's a group project and all the names don't fit, include only the page
number. Number pages consecutively throughout your paper.

Font, spacing, margins, and indents. Choose a font that is easy to read
(such as Times New Roman) and that provides a clear contrast between regular text and italic text. Set the font size between 11 and 13 points. Double-
space the entire paper, including your works-cited list and any notes. Set
one-inch margins at the top, bottom, and sides of your text; do not justify
your text. The first line of each paragraph should be indented one-half inch
from the left margin. End punctuation should be followed by one space.

Headings. Short essays do not generally need headings, but they can be useful in longer works. Use a large, bold font for the first level of heading, and smaller fonts and italics to signal lower-level headings. MLA requires that headings all be flush with the left margin.

First-Level Heading

Second-Level Heading

Third-Level Heading

Long quotations. When quoting more than three lines of poetry, more than four lines of prose, or dialogue between characters in a drama, set off the quotation from the rest of your text, indenting it one-half inch (or five spaces) from the left margin. Do not use quotation marks, and put any parenthetical documentation *after* the final punctuation.

> In *Eastward to Tartary*, Robert Kaplan captures ancient and contemporary Antioch for us:
>
> > At the height of its glory in the Roman-Byzantine age, when it had an amphitheater, public baths, aqueducts, and sewage pipes, half a million people lived in Antioch. Today the population is only 125,000. With sour relations between Turkey and Syria, and unstable politics throughout the Middle East, Antioch is now a backwater—seedy and tumbledown, with relatively few tourists. I found it altogether charming. (123)

In the first stanza of Matthew Arnold's "Dover Beach," the exclamations make clear that the speaker is addressing someone who is also present in the scene:

> Come to the window, sweet is the night air!
> Only, from the long line of spray
> Where the sea meets the moon-blanched land,
> Listen! You hear the grating roar
> Of pebbles which the waves draw back, and fling. (lines 6-10)

Be careful to maintain the poet's line breaks. If a line does not fit on one line of your paper, put the extra words on the next line. Indent that line an additional quarter inch (or two spaces). If a citation doesn't fit, put it on the next line, flush with the right margin.

Tables and illustrations. Insert illustrations and tables close to the text that discusses them, and be sure to make clear how they relate to your point. For tables, provide a number (Table 1) and a title on separate lines above the table. Below the table, provide a caption with source information and any notes. Notes should be indicated with lowercase letters. For graphs, photos, and other figures, provide a figure number (Fig. 1) and caption with source information below the figure. Use a lowercase "f" when referring to a figure in your text: (fig. 1). If you give only brief source information, use commas between elements—Zhu Wei, *New Pictures of the Strikingly Bizarre #9*, print, 2004—and include full source information in your list of works cited. If you give full source information in the caption, don't include the source in your list of works cited. Punctuate as you would in the works-cited list, but don't invert the author's name: Berenice Sydney. *Fast Rhythm*. 1972, Tate Britain, London.

List of works cited. Start your list on a new page, following any notes (see p. 574). Center the title, Works Cited, and double-space the entire list. Begin each entry at the left margin, and indent subsequent lines one-half inch (or five spaces). Alphabetize the list by authors' last names (or by editors' or translators' names, if appropriate). Alphabetize works with no author or editor by title, disregarding "A," "An," and "The." To cite more than one work by a single author, list them as in number 4 on page 580.

SAMPLE RESEARCH PAPER

The following report was written by Dylan Borchers for a first-year writing course. It's formatted according to the guidelines of the MLA (style.mla.org).

Sample Research Paper, MLA Style

Last name and page number.

$\frac{1}{2}$"

1"

Dylan Borchers

Professor Bullock

English 102, Section 4

4 May 2019

Title centered.

Against the Odds:

Harry S. Truman and the Election of 1948

Double-spaced throughout.

Just over a week before Election Day in 1948, a *New York Times* article noted "[t]he popular view that Gov. Thomas E. Dewey's election as President is a foregone conclusion" (Egan). This assessment of the race

No page number for one-page article.

between incumbent Democrat Harry S. Truman and Dewey, his Republican challenger, was echoed a week later when *Life* magazine published a photograph whose caption labeled Dewey "The Next President" (Photograph). In a *Newsweek* survey of fifty prominent political writers, each predicted Truman's defeat, and *Time* correspondents declared Dewey would carry 39 of the 48 states (Donaldson 210). Nearly every major US

1"

1"

Author named in signal phrase, page numbers in parentheses.

media outlet endorsed Dewey. As historian Robert Ferrell observes, even Truman's wife, Bess, thought he would lose (270).

The results of an election are not easily predicted, as is shown in the famous photograph in which Truman holds up a newspaper proclaiming Dewey the victor (fig. 1). Not only did Truman win, but he won by a significant margin: 303 electoral votes and 24,179,259 popular votes to Dewey's 189 electoral votes and 21,991,291 popular votes (Donaldson 204-07). In fact, many historians and political analysts argue that Truman would have won by an even greater margin had third-party candidates Henry A. Wallace and Strom Thurmond not won votes (McCullough 711). Although Truman's defeat was predicted, those predictions themselves, Dewey's passiveness as a campaigner, and Truman's zeal turned the tide for a Truman victory.

1"

Borchers   2

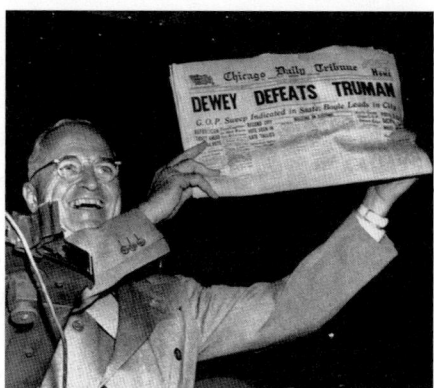

Illustration is positioned close to the text to which it relates, with figure number, caption, and parenthetical documentation.

Fig. 1. President Harry S. Truman holds up an edition of the *Chicago Daily Tribune* that mistakenly announced "Dewey Defeats Truman." Byron Rollins, *Dewey Defeats Truman*, photograph, 1948.

In the months preceding the election, public opinion polls predicted that Dewey would win by a large margin. Pollster Elmo Roper stopped polling in September, believing there was no reason to continue, given a seemingly inevitable Dewey landslide. Although the margin narrowed as the election drew near, other pollsters predicted a Dewey win by at least 5 percent (Donaldson 209). Many historians believe that these predictions aided the president. First, surveys showing Dewey in the lead may have prompted some Dewey supporters to feel overconfident and therefore to stay home from the polls. Second, the same surveys may have energized Democrats to mount late get-out-the-vote efforts (Lester). Other analysts believe that the overwhelming predictions of a Truman loss kept home some Democrats who saw a Truman loss as inevitable. According to

No signal phrase; author and page number in parentheses.

Short title given for author with multiple works cited.

political analyst Samuel Lubell, those Democrats may have saved Dewey from an even greater defeat (Hamby, *Man* 465). Whatever the impact on the voters, the polls had a decided effect on Dewey.

Historians and political analysts alike cite Dewey's overly cautious campaign as a main reason Truman was able to win. Dewey firmly believed in public opinion polls. With all signs pointing to an easy victory, Dewey and his staff believed that all he had to do was bide his time and make no foolish mistakes. Dewey himself said, "When you're leading, don't talk" (Smith 30). As the leader in the race, Dewey kept his remarks faultlessly positive, with the result that he failed to deliver a solid message or even mention Truman. Eventually, Dewey began to be perceived as aloof, stuffy, and out of touch with the public. One observer compared him to the plastic groom on top of a wedding cake

Two works cited within the same sentence.

(Hamby, "Harry S. Truman"), and others noted his stiff, cold demeanor (McCullough 671-74).

Through the autumn of 1948, Dewey's speeches failed to address the issues, with the candidate declaring that he did not want to "get down in the gutter" (Smith 515). When fellow Republicans said he was losing ground, Dewey insisted that his campaign stay the course. Even *Time* magazine, though it endorsed and praised him, conceded that his speeches were dull (McCullough 696). According to historian Zachary Karabell, they were "notable only for taking place, not for any specific message" (244). Dewey's poll numbers slipped before the election, but he still held a comfortable lead. Truman's famous whistle-stop campaign would make the difference.

Paragraphs indented $\frac{1}{2}$ inch or 5–7 spaces.

Few candidates in US history have campaigned for the presidency with more passion and faith than Harry Truman. In the autumn of 1948, Truman wrote to his sister, "It will be the greatest campaign any President

Borchers 4

ever made. Win, lose, or draw, people will know where I stand." For thirty-three days, he traveled the nation, giving hundreds of speeches from the back of the *Ferdinand Magellan* railroad car. In the same letter, Truman described the pace: "We made about 140 stops and I spoke over 147 times, shook hands with at least 30,000 and am in good condition to start out again tomorrow." David McCullough writes of Truman's campaign:

> No President in history had ever gone so far in quest of support from the people, or with less cause for the effort, to judge by informed opinion. . . . As a test of his skills and judgment as a professional politician, not to say his stamina and disposition at age sixty-four, it would be like no other experience in his long, often difficult career, as he himself understood perfectly. (655)

He spoke in large cities and small towns, defending his policies and attacking Republicans. As a former farmer, Truman was able to connect with the public. He developed an energetic style, usually speaking from notes rather than from a prepared speech, and often mingled with the crowds, which grew larger as the campaign progressed. In Chicago, over half a million people lined the streets as he passed, and in St. Paul the crowd numbered over 25,000. When Dewey entered St. Paul two days later, only 7,000 supporters greeted him (McCullough 842). Reporters brushed off the large crowds as mere curiosity seekers wanting to see a president (682). Yet Truman persisted, even if he often seemed to be the only one who thought he could win. By connecting directly with the American people, Truman built the momentum to surpass Dewey and win the election.

The legacy and lessons of Truman's whistle-stop campaign continue to be studied, and politicians still mimic his campaign methods by scheduling multiple visits to key states, as Truman did. He visited California, Illinois, and Ohio 48 times, compared with 6 visits to those states

Quotations of more than 4 lines indented $\frac{1}{2}$ inch (5 spaces) and double-spaced.

Parenthetical reference after final punctuation.

by Dewey. Political scientist Thomas Holbrook concludes that his strategic campaigning in those states and others gave Truman the electoral votes he needed to win (61, 65).

The 1948 election also had an effect on pollsters, who, as Roper admitted, "couldn't have been more wrong." *Life* magazine's editors concluded that pollsters as well as reporters and commentators were too convinced of a Dewey victory to analyze the polls seriously, especially the opinions of undecided voters (Karabell 256). Pollsters assumed that undecided voters would vote in the same proportion as decided voters—and that turned out to be a false assumption (257). In fact, the lopsidedness of the polls might have led voters who supported Truman to call themselves undecided out of an unwillingness to associate themselves with the losing side, further skewing the polls' results (McDonald et al. 152). Such errors led pollsters to change their methods significantly after the 1948 election.

Many political analysts, journalists, and historians concluded that the Truman upset was in fact a victory for the American people. And Truman biographer Alonzo Hamby notes that "polls of scholars consistently rank Truman among the top eight presidents in American history" (*Man* 641). But despite Truman's high standing, and despite the fact that the whistle-stop campaign remains in our political landscape, politicians have increasingly imitated the style of the Dewey campaign, with its "packaged candidate who ran so as not to lose, who steered clear of controversy, and who made a good show of appearing presidential" (Karabell 266). The election of 1948 shows that voters are not necessarily swayed by polls, but it may have presaged the packaging of candidates by public relations experts, to the detriment of public debate on the issues in future presidential elections.

Work by 3 or more authors is shortened using "et al."

1″   Borchers   6

Works Cited

Donaldson, Gary A. *Truman Defeats Dewey.* UP of Kentucky, 1999.

Egan, Leo. "Talk Is Now Turning to the Dewey Cabinet." *The New York Times,* 20 Oct. 1948, p. E8.

Ferrell, Robert H. *Harry S. Truman: A Life.* U of Missouri P, 1994.

Hamby, Alonzo L. "Harry S. Truman: Campaigns and Elections." *Miller Center,* U of Virginia, millercenter.org/president/biography /truman-campaigns-and-elections. Accessed 17 Mar. 2019.

---. *Man of the People: A Life of Harry S. Truman.* Oxford UP, 1995.

Holbrook, Thomas M. "Did the Whistle-Stop Campaign Matter?" *PS: Political Science and Politics,* vol. 35, no. 1, Mar. 2002, pp. 59-66.

Karabell, Zachary. *The Last Campaign: How Harry Truman Won the 1948 Election.* Alfred A. Knopf, 2000.

Lester, Will. "'Dewey Defeats Truman' Disaster Haunts Pollsters." *Los Angeles Times,* 1 Nov. 1998, www.latimes.com/archives/la-xpm-1998 -nov-01-mn-38174-story.html.

McCullough, David. *Truman.* Simon and Schuster, 1992.

McDonald, Daniel G., et al. "The Spiral of Silence in the 1948 Presidential Election." *Communication Research,* vol. 28, no. 2, Apr. 2001, pp. 139-55.

Photograph of Truman. *Life,* 1 Nov. 1948, p. 37. *Google Books,* books.google .com/books?id=ekoEAAAAMBAJ&printsec=frontcover#v=onepage &q&f=false.

Rollins, Byron. *Dewey Defeats Truman.* 1948. *Harry S. Truman Library and Museum,* National Archives and Records Administration, www .trumanlibrary.gov/photograph-records/95-187.

Heading centered.

Double-spaced.

Alphabetized by authors' last names.

Each entry begins at the left margin; subsequent lines are indented.

Multiple works by a single author listed alphabetically by title. For subsequent works, replace author's name with three hyphens.

Borchers　7

Roper, Elmo. "Roper Eats Crow; Seeks Reason for Vote Upset." *Evening*
　　Independent, 6 Nov. 1948, p. 10. *Google News*, news.google.com
　　/newspapers?nid=PZE8UkGerEcC&dat=19481106&printsec
　　=frontpage&hl=en.

Smith, Richard Norton. *Thomas E. Dewey and His Times.* Simon and
　　Schuster, 1982.

Truman, Harry S. "Campaigning,　Letter, October 5, 1948." *Harry S. Truman,*
　　edited by Robert H. Ferrell, CQ Press, 2003, p. 91. American Presidents
　　Reference Series.

*Every source
used is in the list
of works cited.*

54 APA Style

American Psychological Association (APA) style calls for (1) brief documentation in parentheses near each in-text citation and (2) complete documentation in a list of references at the end of your text. The models in this chapter draw on the *Publication Manual of the American Psychological Association, 7th edition* (2020). Additional information is available at www.apastyle.org.

A DIRECTORY TO APA STYLE

Throughout this chapter, you'll find models and examples that are color-coded to help you see how writers include source information in their texts and reference lists: tan for author or editor, yellow for title, blue for publication information—publisher, date of publication, page number(s), DOI or URL, and so on.

IN-TEXT DOCUMENTATION

Brief documentation in your text makes clear to your readers precisely what you took from a source. If you are quoting, provide the page number(s) or other information that will help readers find the quotation in the source. You're not required to give the page number(s) with a paraphrase or summary, but you may want to do so if you are citing a long or complex work.

542–54
PARAPHRASES and **SUMMARIES** are more common than **QUOTATIONS** in APA-style projects. As you cite each source, you will need to decide whether to name the author in a signal phrase—"as McCullough (2020) wrote"—or in parentheses—"(McCullough, 2020)." Note that APA requires

551–54
you to use the past tense for verbs in **SIGNAL PHRASES**, or the present perfect if you are referring to a past action that didn't occur at a specific time or that continues into the present: "Moss (2019) argued," "Many authors have argued."

1. AUTHOR NAMED IN A SIGNAL PHRASE

Put the date in parentheses after the author's last name, unless the year is mentioned in the sentence. Put any page number(s) you're including in parentheses after the quotation, paraphrase, or summary. Parenthetical documentation should come *before* the period at the end of the sentence and *after* any quotation marks.

> McCullough (2001) described John Adams as having "the hands of a man accustomed to pruning his own trees, cutting his own hay, and splitting his own firewood" (p. 18).

> In 2001, McCullough noted that John Adams's hands were those of a laborer (p. 18).

> John Adams had "the hands of a man accustomed to pruning his own trees," according to McCullough (2001, p. 18).

If the author is named after a quotation, as in this last example, put the page number(s) after the date within the parentheses.

2. AUTHOR NAMED IN PARENTHESES

If you do not mention an author in a signal phrase, put the name, the year of publication, and any page number(s) in parentheses at the end of the sentence or right after the quotation, paraphrase, or summary.

> John Adams had "the hands of a man accustomed to pruning his own trees, cutting his own hay, and splitting his own firewood" (McCullough, 2001, p. 18).

3. AUTHORS WITH THE SAME LAST NAME

If your reference list includes more than one first author with the same last name, include initials in all documentation to distinguish the authors from one another.

> Eclecticism is common in modern criticism (J. M. Smith, 1992, p. vii).

4. TWO AUTHORS

Always mention both authors. Use "and" in a signal phrase, but use an ampersand (&) in parentheses.

> Carlson and Ventura (1990) wanted to introduce Julio Cortázar, Marjorie Agosín, and other Latin American writers to an audience of English-speaking adolescents (p. v).

> According to the Peter Principle, "In a hierarchy, every employee tends to rise to his level of incompetence" (Peter & Hull, 1969, p. 26).

5. THREE OR MORE AUTHORS

When you refer to a work by three or more contributors, name only the first author followed by "et al.," Latin for "and others."

> Peilen et al. (1990) supported their claims about corporate corruption with startling anecdotal evidence (p. 75).

6. ORGANIZATION OR GOVERNMENT AS AUTHOR

If an organization name has a familiar abbreviation, give the full name and the abbreviation the first time you cite the source. In subsequent references, use only the abbreviation. If the organization does not have a familiar abbreviation, always use its full name.

FIRST REFERENCE
> The American Psychological Association (APA, 2020)

> (American Psychological Association [APA], 2020)

SUBSEQUENT REFERENCES
> The APA (2020)

> (APA, 2020)

7. AUTHOR UNKNOWN

Use the complete title if it's short; if it's long, use the first few words of the title under which the work appears in the reference list. Italicize the title if it's italicized in the reference list; if it isn't italicized there, enclose the title in quotation marks.

> According to *Feeding Habits of Rams* (2000), a ram's diet often changes from one season to the next (p. 29).

> The article noted that one donor died because of "frightening" postoperative care ("Every Patient's Nightmare," 2007).

8. TWO OR MORE WORKS TOGETHER

If you document multiple works in the same parentheses, place the source information in alphabetical order, separated by semicolons.

> Many researchers have argued that what counts as "literacy" is not necessarily learned at school (Heath, 1983; Moss, 2003).

Multiple authors in a signal phrase can be named in any order.

9. TWO OR MORE WORKS BY ONE AUTHOR IN THE SAME YEAR

If your list of references includes more than one work by the same author published in the same year, order them alphabetically by title, adding lowercase letters ("a," "b," and so on) to the year.

> Kaplan (2000a) described orderly shantytowns in Turkey that did not resemble the other slums he visited.

10. SOURCE QUOTED IN ANOTHER SOURCE

When you cite a source that was quoted in another source, add the words "as cited in." If possible, cite the original source instead.

> Thus, Modern Standard Arabic was expected to serve as the "moral glue" holding the Arab world together (Choueri, 2000, as cited in Walters, 2019, p. 475).

11. WORK WITHOUT PAGE NUMBERS

Instead of page numbers, some works have paragraph numbers, which you should include (preceded by the abbreviation "para.") if you are referring to a specific part of such a source.

> Russell's dismissals from Trinity College at Cambridge and from City College in New York City have been seen as examples of the controversy that marked his life (Irvine, 2006, para. 2).

In sources with neither page nor paragraph numbers, point to a particular part of the source if possible: (Brody, 2020, Introduction, para. 2).

12. AN ENTIRE WORK

You do not need to give a page number if you are directing readers' attention to an entire work.

> Kaplan (2000) considered Turkey and Central Asia explosive.

When you're citing an entire website, give the URL in the text. You do not need to include the website in your reference list. To document a webpage, see number 18 on page 632.

> Beyond providing diagnostic information, the website for the Alzheimer's Association (http://www.alz.org) includes a variety of resources for the families of patients.

13. PERSONAL COMMUNICATIONS

Document emails, telephone conversations, personal interviews, personal letters, messages from nonarchived online discussion sources, and other personal texts as "personal communication," along with the person's initial(s), last name, and the date. You do not need to include such personal communications in your reference list.

> L. Strauss (personal communication, December 6, 2013) told about visiting Yogi Berra when they both lived in Montclair, New Jersey.

NOTES

You may need to use footnotes to give an explanation or information that doesn't fit into your text. To signal a content footnote, place a superscript numeral at the appropriate point in your text. Include this information

in a footnote, either at the bottom of that page or on a separate page with the heading "Footnotes" centered and in bold, after your reference list. If you have multiple notes, number them consecutively throughout your text. Here is an example from *In Search of Solutions: A New Direction in Psychotherapy* (2003).

TEXT WITH SUPERSCRIPT

> An important part of working with teams and one-way mirrors is taking the consultation break, as at Milan, BFTC, and MRI.[1]

FOOTNOTE

> [1]It is crucial to note here that while working within a team is fun, stimulating, and revitalizing, it is not necessary for successful outcomes. Solution-oriented therapy works equally well when working solo.

REFERENCE LIST

A reference list provides full bibliographic information for every source cited in your text with the exception of entire websites, common computer software and mobile apps, and personal communications. See page 646 for guidelines on preparing such a list; for a sample reference list, see page 654.

Key Elements for Documenting Sources

To document a source in APA style, you need to provide information about the author, the date, the title of the work you're citing, and the source itself (who published it; volume, issue, and page numbers; any DOI or URL). The following guidelines explain how to handle each of these elements generally, but there will be exceptions. For that reason, you'll want to consult the entries for the specific kinds of sources you're documenting; these entries provide templates showing which details you need to include. Be aware, though, that sometimes the templates will show elements that your source doesn't have; if that's the case, just omit those elements.

AUTHORS

Most entries begin with the author's last name, followed by the first and any middle initials: Smith, Z. for Zadie Smith; Kinder, D. R. for Donald R. Kinder.

- If the author is a group or organization, use its full name: Black Lives Matter, American Historical Association.
- If there is no author, put the title of the work first, followed by the date.
- If the author uses a screen name, first give their real name, followed by the screen name in brackets: Scott, B. [@BostonScott2]. If only the screen name is known, leave off the brackets: AvalonGirl1990.

DATES

Include the date of publication, in parentheses right after the author. Some sources require only the year; others require the year, month, and day; and still others require something else. Consult the entry in this chapter for the specific source you're documenting.

- For a book, use the copyright year, which you'll find on the copyright page. If more than one year is given, use the most recent one.
- For most magazine or newspaper articles, use the full date that appears on the work, usually the year followed by the month and day.
- For a journal article, use the year of the volume.
- For a work on a website, use the date when the work was last updated. If that information is not available, use the date when the work was published.
- If a work has no date, use "n.d." for "no date."
- For online content that is likely to change, include the month, day, and year when you retrieved it. No need to include a retrieval date for materials that are unlikely to change.

TITLES

Capitalize only the first word and any proper nouns and adjectives in the title and subtitle of a work. But sometimes you'll also need to provide the title of a periodical or website where a source was found, and those are

treated differently: capitalize all the principal words (excluding articles and prepositions).

- For books, reports, webpages, podcasts, and any other works that stand on their own, italicize the title: *White fragility*, *Radiolab*, *The 9/11 report*. Do not italicize the titles of the sources where you found them, however: NPR, ProQuest.

- For journal articles, book chapters, TV series episodes, and other works that are part of a larger work, do not italicize the title: The snowball effect, Not your average Joe. But do italicize the title of the larger work: *The Atlantic*, *Game of thrones*.

- If a work has no title, include a description in square brackets after the date: [Painting of sheep on a hill].

- If the title of a work you're documenting includes another title, italicize it: *Frog and Toad* and the self. If the title you're documenting is itself in italics, do not italicize the title within it: *Stay, illusion!: The* Hamlet *doctrine*.

- For untitled social media posts or comments, include the first twenty words as the title, in italics, followed by a bracketed description: *TIL pigeons can fly up to 700 miles in one day* [Tweet].

SOURCE INFORMATION

This indicates where the work can be found (in a database or on a website, for example, or in a magazine or on a podcast) and includes information about the publisher; any volume, issue, and page numbers; and, for some sources, a DOI or URL. DOIs and URLs are included in all the templates; if the work you are documenting doesn't have one, just leave it off.

- For a work that stands on its own (a book, a report, a webpage), the source might be the publisher, a database, or a website.

- For a work that's part of a larger work (an article, an episode in a TV series, an essay in a collection), the source might be a magazine, a TV series, or an anthology.

- Give the volume and issue for journals and magazines that include that information. No need to give them for newspapers.

- Include a DOI for any work that has one, whether you accessed the source in print or online. For an online work with no DOI, include a working URL unless the work is from an academic database. You can use a shortDOI (https://shortdoi.org/) or a URL shortened using an online URL shortener, as long as the shorter DOI or URL leads to the correct work. No need to include a URL for a print work with no DOI.

Authors and Other Contributors

Most entries begin with authors—one author, two authors, or twenty-five. And some include editors, translators, or others who've contributed. The following templates show you how to document the various kinds of authors and other contributors.

1. ONE AUTHOR

Author's Last Name, Initials. (Year of publication). *Title.* Publisher. DOI *or* URL

Lewis, M. (2003). *Moneyball: The art of winning an unfair game.* W. W. Norton.

2. TWO AUTHORS

First Author's Last Name, Initials, & Second Author's Last Name, Initials. (Year of publication). *Title.* Publisher. DOI *or* URL

Montefiore, S., & Montefiore, S. S. (2016). *The royal rabbits of London.* Aladdin.

3. THREE OR MORE AUTHORS

For three to twenty authors, include all names.

First Author's Last Name, Initials, Next Author's Last Name, Initials, & Final Author's Last Name, Initials. (Year of publication). *Title.* Publisher. DOI *or* URL

Greig, A., Taylor, J., & MacKay, T. (2013). *Doing research with children: A practical guide* (3rd ed.). Sage.

For a work by twenty-one or more authors, name the first nineteen authors, followed by three ellipsis points, and end with the final author.

Gao, R., Asano, S. M., Upadhyayula, S., Pisarev, I., Milkie, D. E., Liu, T.-L., Singh, V., Graves, A., Huynh, G. H., Zhao, Y., Bogovic, J., Colonell, J., Ott, C. M., Zugates, C., Tappan, S., Rodriguez, A., Mosaliganti, K. R., Sheu, S.-H., Pasolli, H. A., . . . Betzig, E. (2019, January 18). Cortical column and whole-brain imaging with molecular contrast and nanoscale resolution. *Science, 363*(6424). https://doi.org/10.1126/science.aau8302

4. TWO OR MORE WORKS BY THE SAME AUTHOR

If the works were published in different years, list them chronologically.

Lewis, B. (1995). *The Middle East: A brief history of the last 2,000 years.* Scribner.

Lewis, B. (2003). *The crisis of Islam: Holy war and unholy terror.* Modern Library.

If the works were published in the same year, list them alphabetically by title (ignoring "A," "An," and "The"), adding "a," "b," and so on to the year.

Kaplan, R. D. (2000a). *The coming anarchy: Shattering the dreams of the post Cold War.* Random House.

Kaplan, R. D. (2000b). *Eastward to Tartary: Travels in the Balkans, the Middle East, and the Caucasus.* Random House.

5. AUTHOR AND EDITOR

If a book has an author and an editor who is credited on the cover, include the editor in parentheses after the title.

> Author's Last Name, Initials. (Year of publication). *Title* (Editor's Initials Last Name, Ed.). Publisher. DOI *or* URL (Original work published Year)

> Dick, P. F. (2008). *Five novels of the 1960s and 70s* (J. Lethem, Ed.). Library of America. (Original works published 1964–1977)

6. AUTHOR AND TRANSLATOR

> Author's Last Name, Initials. (Year of publication). *Title* (Translator's Initials Last Name, Trans.). Publisher. DOI *or* URL (Original work published Year)

> Hugo, V. (2008). *Les misérables* (J. Rose, Trans.). Modern Library. (Original work published 1862)

7. EDITOR

> Editor's Last Name, Initials (Ed.). (Year of publication). *Title*. Publisher. DOI *or* URL

> Jones, D. (Ed.). (2007). *Modern love: 50 true and extraordinary tales of desire, deceit, and devotion*. Three Rivers Press.

8. UNKNOWN OR NO AUTHOR OR EDITOR

> *Title.* (Year of Publication). Publisher. DOI *or* URL

> *Feeding habits of rams.* (2000). Land's Point Press.

> Clues in salmonella outbreak. (2008, June 21). *The New York Times*, A13.

If the author is listed as "Anonymous," use that as the author's name in the reference list.

9. ORGANIZATION OR GOVERNMENT AS AUTHOR

Sometimes an organization or a government agency is both author and publisher. If so, omit the publisher.

> Organization Name *or* Government Agency. (Year of publication).
>> *Title.* DOI *or* URL

> Catholic News Service. (2002). *Stylebook on religion 2000: A reference guide.*

Articles and Other Short Works

Articles, essays, reviews, and other short works are found in periodicals and books—in print, online, or in a database. For most short works, provide information about the author, the date, the titles of both the short work and the longer work, plus any volume and issue numbers, page numbers, and a DOI or URL if there is one.

10. ARTICLE IN A JOURNAL

> Author's Last Name, Initials. (Year). Title of article. *Title of Journal, volume*(issue), page(s). DOI *or* URL

> Gremer, J. R., Sala, A., & Crone, E. E. (2010). Disappearing plants: Why they hide and how they return. *Ecology, 91*(11), 3407–3413. https://doi.org/10.1890/09-1864.1

11. ARTICLE IN A MAGAZINE

If a magazine is published weekly, include the year, month, and day. Put any volume number and issue number after the title.

> Author's Last Name, Initials. (Year, Month Day). Title of article. *Title of Magazine, volume*(issue), page(s). DOI *or* URL

> Klump, B. (2019, November 22). Of crows and tools. *Science, 366*(6468), 965. https://doi.org/10.1126/science.aaz7775

12. ARTICLE IN A NEWSPAPER

If page numbers are consecutive, separate them with an en dash. If not, separate them with a comma.

> Author's Last Name, Initials. (Year, Month Day). Title of article.
> *Title of Newspaper,* page(s). URL

> Schneider, G. (2005, March 13). Fashion sense on wheels. *The Washington Post,* F1, F6.

13. ARTICLE ON A NEWS WEBSITE

Italicize the titles of articles on CNN and other news websites. Do not italicize the name of the website.

> Author's Last Name, Initials. (Year, Month Day). *Title of article.*
> Name of Site. URL

> Travers, C. (2019, December 3). *Here's why you keep waking up at the same time every night.* HuffPost. https://bit.ly/3drSwAR

14. JOURNAL ARTICLE FROM A DATABASE

> Author's Last Name, Initials. (Year). Title of article. *Title of Journal,*
> *volume*(issue), pages. DOI

> Simpson, M. (1972). Authoritarianism and education: A comparative approach. *Sociometry, 35*(2), 223–234. https://doi.org/10.2307/2786619

15. EDITORIAL

Editorials can appear in journals, magazines, and newspapers. If the editorial is unsigned, put the title in the author position.

> Author's Last Name, Initials. (Year, Month Day). Title of editorial
> [Editorial]. *Title of Periodical.* DOI *or* URL

> *The Guardian* view on local theatres: The shows must go on [Editorial].
> (2019, December 6). *The Guardian.* https://bit.ly/2VZHIUg

Documentation Map (APA)

Article in a Journal with a DOI

ETHICS & BEHAVIOR, 23(4), 324–337
Copyright © 2013 Taylor & Francis Group, LLC
ISSN: 1050-8422 print / 1532-7019 online
DOI: 10.1080/10508422.2013.787359

Routledge
Taylor & Francis Group

TITLE OF JOURNAL

VOLUME AND ISSUE

PAGES

PUBLISHER

DOI

TITLE OF ARTICLE

AUTHOR

Smart Technology and the Moral Life

Clifton F. Guthrie

Department of Science and Humanities
Husson University

Smart technology is recording and nudging our intuitive and behavioral reactions in ways that are not fully shaped by our conscious ethical reasoning and so are altering our social and moral worlds. Beyond reasons to worry, there are also reasons to embrace this technology for nudging human behavior toward prosocial activity. This article inquires about four ways that smart technology is shaping the individual moral life: the persuasive effect of promptware, our newly evolving experiences of embodiment, our negotiations with privacy, and our experiences of risk and serendipity.

Keywords: persuasive technology, morality, ethics, virtue

PERSUASIVE TECHNOLOGY

For some time, cars have worked to shape our behaviors, beeping to warn us when a door is unlocked or a seat belt unfastened, or giving us fuel efficiency feedback. These straightforward but persuasive sensor systems nudge us toward a repertoire of safe driving behaviors, and we often cannot override them even if we want to. Newer cars include an increasing number of smart technologies that interact with us more intelligently. Some detect the presence of electronic keys and make it impossible for drivers to lock themselves out. Others use sensors to monitor approaching obstacles or lane boundaries and give warnings or even apply the brakes. We are seeing the emergence of street intersections that communicate directly with cars and cars that can communicate with one another (Dean, Fletcher, Porges, & Ulrich, 2012). These are so-called smart technologies because they draw data from the environment and from us, and often make decisions on our behalf. A leading researcher in automated driving noted, "The driver is still in control. But if the driver is not doing the right thing, the technology takes over" (Markoff & Sengupta, 2013).

As cars become smarter they are helping to lead us into what technologists describe as a pervasive, ambient, or calm computing environment. In 1991, Mark Weiser of the Palo Alto Research Center presciently called it "ubiquitous computing" or "ubicomp" in a much-quoted article from *Scientific American*, in which he outlined what has come to be accepted as a standard interpretation of the history of human interaction with computers. This is the age in which computers are increasingly liberated from manual input devices like laptops and cell phones to become an invisible, interactive, computational sensorium. Early examples include motion sensors, smart

Correspondence should be addressed to Clifton F. Guthrie, Department of Science and Humanities, Husson University, 1 College Circle, Bangor, ME 04401. E-mail: cfguthrie@gmail.com

Guthrie, C. F. (2013). Smart technology and the moral life. *Ethics & Behavior*, 23(4), 324–337. https://doi.org/10.1080 /10508422.2013.787359

629-34

for more
on citing
articles
APA style

16. REVIEW

Use this general format to document a review that appears in a periodical or on a blog.

> Reviewer's Last Name, Initials. (Year, Month Day). Title of review [Review of the work *Title*, by Author's Initials Last Name]. *Title of Periodical or Name of Blog.* DOI *or* URL

> Joinson, S. (2017, December 15). Mysteries unfold in a land of minarets and magic carpets [Review of the book *The city of brass,* by S. A. Chakraborty]. *The New York Times.* https://nyti.ms/2kvwHFP

For a review published on a website that is not associated with a periodical or a blog, italicize the title of the review and do not italicize the website name.

17. COMMENT ON AN ONLINE PERIODICAL ARTICLE OR BLOG POST

> Writer's Last Name, Initials [username]. (Year, Month Day). Text of comment up to 20 words [Comment on the article "Title of work"]. Title of Publication. DOI *or* URL

> PhyllisSpecial. (2020, May 10). How about we go all the way again? It's about time . . . [Comment on the article "2020 Eagles schedule: Picking wins and losses for all 16 games"]. *The Philadelphia Inquirer.* https://rb.gy/iduabz

Link to the comment if possible; if not, include the URL of the article.

18. WEBPAGE

> Author's Last Name, Initials. (Year, Month Day). *Title of work.* Title of Site. URL

> Pleasant, B. (n.d.). *Annual bluegrass.* The National Gardening Association. https://garden.org/learn/articles/view/2936/

Documentation Map (APA)
Webpage

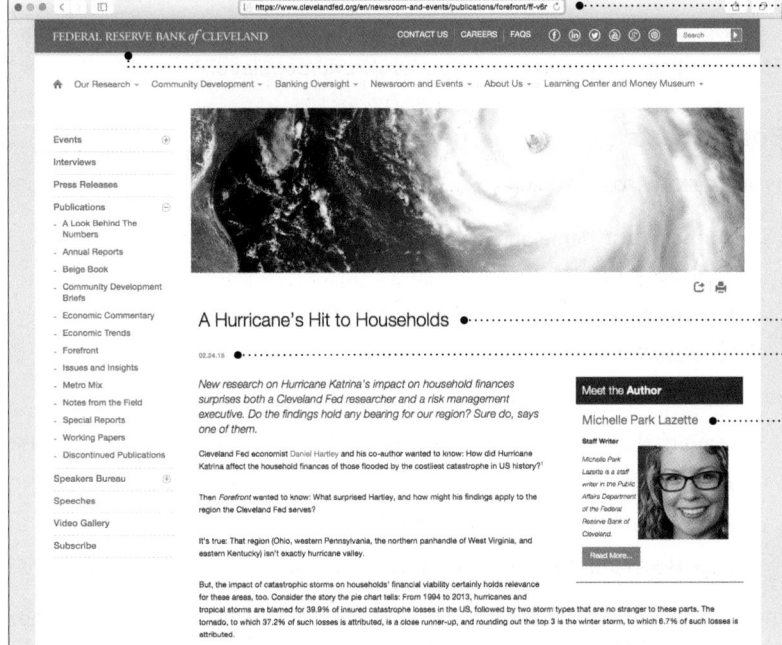

URL

TITLE OF SITE

TITLE OF WORK

DATE OF PUBLICATION

AUTHOR

Lazette, M. P. (2015, February 24). *A hurricane's hit to households.*
Federal Reserve Bank of Cleveland. https://www.clevelandfed.org
/en/newsroom-and-events/publications/forefront/ff-v6n01
/ff-20150224-v6n0107-a-hurricanes-hit-to-households.aspx

632–34 for more on citing webpages APA style

If the author and the website name are the same, use the website name as the author. If the content of the webpage is likely to change and no archived version exists, use "n.d." as the date and include a retrieval date.

> Worldometer. (n.d.). *World population*. Retrieved February 2, 2020, from https://www.worldometers.info/world-population/

Books, Parts of Books, and Reports

19. BASIC ENTRY FOR A BOOK

> Author's Last Name, Initials. (Year of publication). *Title*.
> Publisher. DOI *or* URL

PRINT BOOK

> Schwab, V. E. (2018). *Vengeful*. Tor.

EBOOK

> Jemisin, N. K. (2017). *The stone sky*. Orbit. https://amzn.com/B01N7EQOFA

AUDIOBOOK

> Obama, M. (2018). *Becoming* (M. Obama, Narr.) [Audiobook]. Random
> House Audio. http://amzn.com/B07B3JQZCL

Include the word "Audiobook" in brackets and the name of the narrator only if the format and the narrator are something you've mentioned in what you've written.

20. EDITION OTHER THAN THE FIRST

> Author's Last Name, Initials. (Year). *Title* (Name *or* number ed.).
> Publisher. DOI *or* URL

> Burch, D. (2008). *Emergency navigation: Find your position and shape
> your course at sea even if your instruments fail* (2nd ed.).
> International Marine/McGraw-Hill.

Documentation Map (APA)
Book

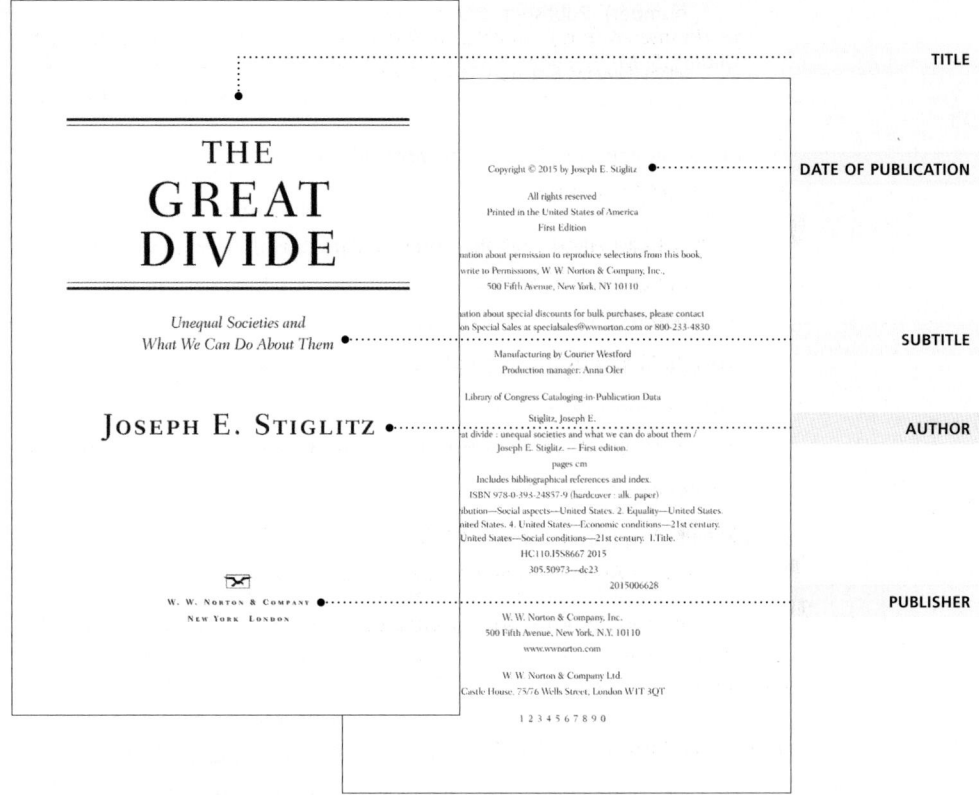

Stiglitz, J. E. (2015). *The great divide: Unequal societies and what we can do about them.* W. W. Norton.

634–37
for more
on citing
books APA
style

21. EDITED COLLECTION OR ANTHOLOGY

> Editor's Last Name, Initials (Ed.). (Year). *Title* (Name *or* number ed., Vol. number). Publisher. DOI *or* URL

> Gilbert, S. M., & Gubar, S. (Eds.). (2003). *The Norton anthology of literature by women: The traditions in English* (3rd ed., Vol. 2). W. W. Norton.

22. WORK IN AN EDITED COLLECTION OR ANTHOLOGY

> Author's Last Name, Initials. (Year of edited edition). Title of work. In Editor's Initials Last Name (Ed.), *Title of collection* (Name *or* number ed., Vol. number, pp. pages). Publisher. DOI *or* URL (Original work published Year)

> Baldwin, J. (2018). Notes of a native son. In M. Puchner, S. Akbari, W. Denecke, B. Fuchs, C. Levine, P. Lewis, & E. Wilson (Eds.), *The Norton anthology of world literature* (4th ed., Vol. F, pp. 728–743). W. W. Norton. (Original work published 1955)

23. CHAPTER IN AN EDITED BOOK

> Author's Last Name, Initials. (Year). Title of chapter. In Editor's Initials Last Name (Ed.), *Title of book* (pp. pages). Publisher. DOI *or* URL

> Amarnick, S. (2009). Trollope at fuller length: Lord Silverbridge and the manuscript of *The duke's children*. In M. Markwick, D. Denenholz Morse, & R. Gagnier (Eds.), *The politics of gender in Anthony Trollope's novels: New readings for the twenty-first century* (pp. 193–206). Routledge.

24. ENTRY IN A REFERENCE WORK (DICTIONARY, THESAURUS, OR ENCYCLOPEDIA)

If the entry has no author, use the name of the publisher as the author. If the reference work has an editor, include their name after the title of the entry. If the entry is archived or is not likely to change, use the publication date and do not include a retrieval date.

Author's Last Name, Initials. (Year). Title of entry. In Editor's Initials
   Last Name (Ed.), *Title of reference work* (Name *or* number ed., Vol.
   number, pp. pages). Publisher. URL

Merriam-Webster. (n.d.). Epoxy. In *Merriam-Webster.com dictionary*.
   Retrieved January 29, 2020, from https://www.merriam-webster
   .com/dictionary/epoxy

25. BOOK IN A LANGUAGE OTHER THAN ENGLISH

Author's Last Name, Initials. (Year). *Title of book* [English translation of
   title]. Publisher. DOI *or* URL

Ferrante, E. (2011). *L'amica geniale* [My brilliant friend]. Edizione E/O.

26. ONE VOLUME OF A MULTIVOLUME WORK

Author's Last Name, Initials. (Year). *Title of entire work* (Vol. number).
   Publisher. DOI *or* URL

Spiegelman, A. (1986). *Maus* (Vol. 1). Random House.

If the volume has a separate title, include the volume number and title
in italics after the main title.

Ramazani, J., Ellmann, R., & O'Clair, R. (Eds.). (2003). *The Norton
   anthology of modern and contemporary poetry: Vol. 1. Modern
   poetry* (3rd ed.). W. W. Norton.

27. RELIGIOUS WORK

Do not include an author for most religious works. If the date of original
publication is known, include it at the end.

Title. (Year of publication). Publisher. URL (Original work published Year)

New American Bible. (2002). United States Conference of Catholic
   Bishops. http://www.vatican.va/archive/ENG0839/_INDEX.HTM
   (Original work published 1970)

28. REPORT BY A GOVERNMENT AGENCY OR OTHER ORGANIZATION

> Author's Last Name, Initials. (Year). Title (Report No. number).
> Publisher. DOI *or* URL

> Centers for Disease Control and Prevention. (2009). *Fourth national report on human exposure to environmental chemicals.* US Department of Health and Human Services. https://www.cdc.gov/exposurereport/pdf/fourthreport.pdf

Include the year, month, and day if the report you're documenting includes that information. Omit the report number if one is not given. If more than one government department is listed as the publisher, list the most specific department as the author and the larger department as the publisher.

29. DISSERTATION

> Author's Last Name, Initials. (Year). Title (Publication No. number)
> [Doctoral dissertation, Name of School]. Database *or* Archive
> Name. URL

> Solomon, M. (2016). *Social media and self-examination: The examination of social media use on identity, social comparison, and self-esteem in young female adults* (Publication No. 10188962) [Doctoral dissertation, William James College]. ProQuest Dissertations and Theses Global.

If the dissertation is in a database, do not include a URL. Include a URL if it is published elsewhere online. If it is unpublished, write "Unpublished doctoral dissertation" in brackets, and use the name of the school in place of the database.

30. PAPER OR POSTER PRESENTED AT A CONFERENCE

> Presenter's Last Name, Initials. (Year, Month First Day–Last Day).
> *Title* [Paper or Poster presentation]. Name of Conference, City,
> State, Country. URL

Dolatian, H., & Heinz, J. (2018, May 25–27). *Reduplication and finite-state technology* [Paper presentation]. The 53rd Annual Meeting of the Chicago Linguistic Society, Chicago, IL, United States. http://shorturl.at/msuB2

Audio, Visual, and Other Sources

If you are referring to an entire website, do not include it in your reference list; simply mention the website's name in the body of your paper and include the URL in parentheses. Do not include email, personal communication, or other unarchived discussions in your list of references.

31. *WIKIPEDIA* ENTRY

Wikipedia has archived versions of its pages, so give the date when you accessed the page and the permanent URL of the archived page, which is found by clicking "View history."

Title of entry. (Year, Month Day). In *Wikipedia.* URL

List of sheep breeds. (2019, September 9). In *Wikipedia.* https://en.wikipedia .org/w/index.php?title=List_of_sheep_breeds&oldid=914884262

32. ONLINE FORUM POST

Author's Last Name, Initials [username]. (Year, Month Day). *Content of the post up to 20 words* [Online forum post]. Name of Site. URL

Hanzus, D. [DanHanzus]. (2019, October 23). *GETCHA DAN HANZUS. ASK ME ANYTHING!* [Online forum post]. Reddit. https://bit.ly/38WgmSF

33. BLOG POST

Author's Last Name, Initials [username]. (Year, Month Day). Title of post. Name of Blog. URL

gcrepps. (2017, March 28). Shania Sanders. *Women@NASA.* https://blogs .nasa.gov/womenatnasa/2017/03/28/shania-sanders/

If only the username is known, use it without brackets.

34. ONLINE STREAMING VIDEO

Uploader's Last Name, Initials [username]. (Year, Month Day). *Title* [Video]. Name of Video Platform. URL

CinemaSins. (2014, August 21). *Everything wrong with* National treasure *in 13 minutes or less* [Video]. YouTube. https://www.youtube.com /watch?v=1uI-_ZWvXTs

Whoever uploaded the video is considered the author, even if someone else created the content. If only the username is known, use it without brackets.

35. PODCAST

Host's Last Name, Initials (Host). (First Year–Last Year). *Podcast name* [Audio podcast]. Production Company. URL

Poor, N., Woods, E., & Thomas, R. (Hosts). (2017–present). *Ear hustle* [Audio podcast]. PRX. https://www.earhustlesq.com/

36. PODCAST EPISODE

Host's Last Name, Initials (Host). (Year, Month Day). Episode title (No. number) [Audio podcast episode]. *In Podcast name.* Production Company. URL

Tamposi, E., & Samocki, E. (Hosts). (2020, January 8). The year of the broads [Audio podcast episode]. In *The broadcast podcast*. Podcast One. https://podcastone.com/episode/the-year-of-the-broads

Omit the episode number if one is not given.

37. FILM

Director's Last Name, Initials (Director). (Year). *Title* [Film]. Production Company. URL

Jenkins, B. (Director). (2016). *Moonlight* [Film]. A24; Plan B; PASTEL.

Cuarón, A. (Director). (2016). *Harry Potter and the prisoner of Azkaban* [Film; two-disc special ed. on DVD]. Warner Brothers.

List the director as the author of the film. Indicate how you watched the film only if the format is relevant to what you've written.

38. TELEVISION SERIES

Executive Producer's Last Name, Initials (Executive Producer). (First Year– Last Year). *Title of series* [TV series]. Production Company. URL

Iungerich, L., Gonzalez, E., & Haft, J. (Executive Producers). (2018–present). *On my block* [TV series]. Crazy Cat Lady Productions.

Indicate how you watched the TV series (two-disc DVD set, for example) only if the format is relevant to your essay.

39. TELEVISION SERIES EPISODE

Writer's Last Name, Initials (Writer), & Director's Last Name, Initials (Director). (Year, Month Day). Title of episode (Season number, Episode number) [TV series episode]. In Executive Producer's Initials Last Name (Executive Producer), *Title of series*. Production Company. URL

Siegal, J. (Writer), Morgan, D. (Writer), & Sackett, M. (Director). (2018, December 6). Janet(s) (Season 3, Episode 10) [TV series episode]. In M. Schur, D. Miner, M. Sackett, & D. Goddard (Executive Producers), *The good place*. Fremulon; 3 Arts Entertainment; Universal Television.

40. MUSIC ALBUM

Artist's Last Name, Initials. (Year). *Title of album* [Album]. Label. URL

Jonas Brothers. (2019). *Happiness begins* [Album]. Republic.

41. SONG

> Artist's Last Name, Initials. (Year). Name of song [Song]. *On Title of album.* Label. URL
>
> Giddens, R. (2015). Waterboy [Song]. On *Tomorrow is my turn*. Nonesuch.

42. *POWERPOINT* SLIDES

> Author's Last Name, Initials. (Year, Month Day). *Title of presentation* [PowerPoint slides]. Publisher. URL
>
> Pavliscak, P. (2016, February 21). *Finding our happy place in the internet of things* [PowerPoint slides]. Slideshare. https://bit.ly/3aOcfs7

43. RECORDING OF A SPEECH OR WEBINAR

> Author's Last Name, Initials. (Year, Month Day *or* Year). *Title* [Speech audio recording or Webinar]. Publisher. URL
>
> Kennedy, J. F. (1961, January 20). *Inaugural address* [Speech audio recording]. American Rhetoric. https://bit.ly/339Gc3e

For a speech, include the year, month, and day. For a webinar, include only the year.

44. PHOTOGRAPH

> Photographer's Last Name, Initials. (Year). *Title of photograph* [Photograph]. Name of Site. URL
>
> Kudacki, P. (2013). [Photograph of Benedict Cumberbatch]. Time. http://content.time.com/time/covers/asia/0,16641,20131028,00.html

Use this format to document a photograph that is not in a museum or on a museum website. For a photograph with no title, include a description of the photograph in brackets after the date.

45. MAP

> Mapmaker's Last Name, Initials. (Year). *Title of map* [Map].
>> Publisher. URL

> Daniels, M. (2018). *Human terrain: Visualizing the world's population, in 3D* [Map]. The Pudding. https://pudding.cool/2018/10/city_3d/

46. SOCIAL MEDIA POSTS

If only the username is known, do not use brackets. List any audiovisual content (e.g., videos, images, or links) in brackets. Replicate emoji or include a bracketed description. Follow the spelling and capitalization of the post.

> Author's Last Name, Initials [@username]. (Year, Month Day).
>> *Content of post up to 20 words* [Description of any audiovisual content] [Type of post]. Platform. URL

TWEET

> Baron, D. [@DrGrammar]. (2021, March 12). *Got vaxxed and now I'm coming to you in 5g.* [Image attached] [Tweet]. Twitter. https://bit .ly/3sxEELx

***INSTAGRAM* PHOTOGRAPH OR VIDEO**

> Jamil, J. [@jameelajamilofficial]. (2018, July 18). *Happy Birthday to our leader. I steal all my acting faces from you. @kristenanniebell* [Face with smile and sunglasses emoji] [Photograph]. Instagram. https://www.instagram.com/p/BlYX5F9FuGL/

***FACEBOOK* POST**

> Raptor Resource Project. (2020, May 8). *Happy Fri-yay, everyone! We'll keep the news short and sweet: today Decorah eaglets D34 and D35 turn 33 days* [Images attached]. Facebook. https://bit.ly/3icwFzN

47. DATA SET

> Author's Last Name, Initials. (Year). *Title of data set* (Version number)
> [Data set]. Publisher. DOI *or* URL

> Pew Research Center. (2019). *Core trends survey* [Data set]. https://www
> .pewresearch.org/internet/dataset/core-trends-survey/

Omit the version number if one is not given. If the publisher is the author, no need to list it twice; omit the publisher.

48. SUPREME COURT CASE

> Name of Case, volume US pages (Year). URL

> Plessy v. Ferguson, 163 US 537 (1896). https://www.oyez.org/cases
> /1850-1900/163us537

> Obergefell v. Hodges, 576 US ___ (2015). https://www.oyez.org
> /cases/2014/14-556

The source for most Supreme Court cases is the *United States Reports*, which is abbreviated "US" in the reference list entry. If the case does not yet have a page number, use three underscores instead.

Sources Not Covered by APA

To document a source for which APA does not provide guidelines, look at models similar to the source you have cited. Give any information readers will need in order to find the source themselves—author; date of publication; title; and information about the source itself (including who published it; volume, issue, and page numbers; and a DOI or URL). You might want to check your reference note to be sure it will lead others to your source.

FORMATTING A PAPER

Title page. APA generally requires a title page. The page number should go in the upper right-hand corner. Center the full title of the paper in bold in the top half of the page. Center your name, the name of your department and school, the course number and name, the instructor's name, and the due date on separate lines below the title. Leave one line between the title and your name.

Page numbers. Place the page number in the upper right-hand corner. Number pages consecutively throughout.

Fonts, spacing, margins, and indents. Use a legible font that will be accessible to everyone, either a serif font (such as Times New Roman or Bookman) or a sans serif font (such as Calibri or Verdana). Use a sans serif font within figure images. Double-space the entire paper, including any notes and your list of references; the only exception is footnotes at the bottom of a page, which should be single-spaced, and text within tables and images, the spacing of which will vary. Leave one-inch margins at the top, bottom, and sides of your text; do not justify the text. The first line of each paragraph should be indented one-half inch (or five to seven spaces) from the left margin. APA recommends using one space after end-of-sentence punctuation.

Headings. Though they are not required in APA style, headings can help readers follow your text. The first level of heading should be bold and centered; the second level of heading should be bold and flush with the left margin; the third level should be bold, italicized, and flush left. Capitalize all headings as you would any other title within the text.

First-Level Heading

Second-Level Heading

Third-Level Heading

Abstract. An abstract is a concise summary of your paper that introduces readers to your topic and main points. Most scholarly journals require an abstract; an abstract is not typically required for student papers, so check your instructor's preference. Put your abstract on the second page, with the word "Abstract" centered and in bold at the top. Unless your instructor specifies a length, limit your abstract to 250 words or fewer.

Long quotations. Indent quotations of forty or more words one-half inch (or five to seven spaces) from the left margin. Do not use quotation marks, and place the page number(s) or documentation information in parentheses *after* the end punctuation. If there are paragraphs in the quotation, indent the first line of each paragraph another one-half inch.

> Kaplan (2000) captured ancient and contemporary Antioch:
>> At the height of its glory in the Roman-Byzantine age, when it had an amphitheater, public baths, aqueducts, and sewage pipes, half a million people lived in Antioch. Today the population is only 125,000. With sour relations between Turkey and Syria, and unstable politics throughout the Middle East, Antioch is now a backwater—seedy and tumbledown, with relatively few tourists. (p. 123)
>
> Antioch's decline serves as a reminder that the fortunes of cities can change drastically over time.

List of references. Start your list on a new page after the text but before any notes. Title the page "References," centered and in bold, and double-space the entire list. Each entry should begin at the left margin, and subsequent lines should be indented one-half inch (or five to seven spaces). Alphabetize the list by authors' last names (or by editors' names, if appropriate). Alphabetize works that have no author or editor by title, disregarding "A," "An," and "The." Be sure every source listed is cited in the text; do not include sources that you consulted but did not cite.

Tables and figures. Above each table or figure (charts, diagrams, graphs, photos, and so on), provide the word "Table" or "Figure" and a number, flush

left and in bold (e.g., **Table 1**). On the following line, give a descriptive title, flush left and italicized. Below the table or figure, include a note with any necessary explanation and source information. Number tables and figures separately, and be sure to discuss them in your text so that readers know how they relate.

TABLE 1
Hours of Instruction Delivered per Week

	American classrooms	Japanese classrooms	Chinese classrooms
First grade			
Language arts	10.5	8.7	10.4
Mathematics	2.7	5.8	4.0
Fifth grade			
Language arts	7.9	8.0	11.1
Mathematics	3.4	7.8	11.7

Note. Adapted from *Peeking Out from under the Blinders: Some Factors We Shouldn't Forget in Studying Writing* (Occasional Paper No. 25), by J. R. Hayes, 1991, National Center for the Study of Writing and Literacy (https://archive.nwp.org/cs/public/print/resource/720). Copyright 1991 by the Office of Educational Research and Improvement.

SAMPLE RESEARCH PAPER

The following sample pages are from "The Benefits of Prison Nursery Programs," a paper written by Analisa Johnson for a first-year writing course. It is formatted according to the guidelines of the *Publication Manual of the American Psychological Association*, 7th edition (2020). While APA guidelines are used widely in linguistics and the social sciences, exact requirements may vary from discipline to discipline and course to course. If you're unsure about what your instructor wants, ask for clarification.

1

Page number.

Title bold and centered.

The Benefits of Prison Nursery Programs

Author's name, school name and department, course number and name, instructor's name, and due date.

Analisa Johnson

Writing Program, Boston University

WR 150: Burning Questions: Human Expression

Samantha Myers

May 1, 2017

2

Abstract

The rising population of women in prisons has resulted in the births of some 2,000 babies per year to women behind bars. Female prisoners suffer from a number of inadequacies in their health care, but changes in birthing practices and the provision of nursery programs in prisons could yield important benefits. Currently, nine states offer such programs, and research conducted in these states has shown a number of positive effects. Fully 86.3% of women who have come through these programs remain in their communities after 3 years. Likewise, preschool performance of their children shows better emotional / behavioral adjustment than that of children who have been sent to foster care. Finally, estimates show that the annual costs of such programs are approximately 40% less than those of foster care.

Keywords: birthing practices, correctional, foster care, health care, incarceration, nursery program, prenatal care, preschool, prison, recidivism, sentencing project, shackles

Heading bold and centered.

Limited to 250 words or fewer.

"Keywords:" in italics, indented.

Title bold and centered.

Double-spaced throughout.

1"

The Benefits of Prison Nursery Programs

Over the past 40 years or so, the United States has seen a steady increase in incarcerated individuals, with 2.2 million people currently in prisons and jails nationwide, according to statistics on prisons and the criminal justice system provided by the Sentencing Project (2017, p. 2). In particular, the number of incarcerated females has risen dramatically, at a rate 50% higher than that of men since the early 1980s. As recently as 2015, there were nearly 112,000 incarcerated women across the nation (Sentencing Project, 2017, p. 4). While there is a plethora of health care issues that women face when locked up, one of the most concerning is that of reproductive health, specifically pregnancy and birth in prison. Roughly 1 in 25 women entering prison or jail is pregnant (Yager, 2015). As a result, the number of babies born behind bars has also grown at an alarming rate.

1"   It is estimated that up to 2,000 infants are born to incarcerated mothers   1"
each year, only to be taken from them a scant 24 hours after birth and placed either with a family member or, more often, in the foster care system (Sufrin, 2012). Current scholarly sources have proven prison nursery programs—which allow mothers to keep their infants with them while they serve out their sentences—to be a very effective method in dealing with the issue of incarcerated mothers. Despite this fact, there are only nine nursery programs currently operational in America. In order to make prison nursery programs more prevalent, better education is needed for

First-level headings bold and centered.
correctional administrators about the effectiveness of nursery programs.

Inadequacies of Prison Health Care for Pregnant Women

Paragraphs indented $\frac{1}{2}$ inch or 5–7 spaces.
When it comes to women's health care in the prison system, there are many inadequate areas, but two that are of great importance are prenatal care and birthing practices.

1"

4

Prenatal Care

Author named in a signal phrase, publication date in parentheses after the name.

Hotelling (2008) discussed the lack of quality health care provided to expectant mothers behind bars. Despite adequate health care being mandated to all inmates through the Eighth Amendment to the Constitution, women still make up a lesser percentage of total incarcerated individuals than men—a fact that is used by correctional staff to justify providing scarcer health care and rehabilitative programs for incarcerated women. In addition, prisons are not subject to any sort of external review of their standards of inmate care, so they are not encouraged to improve health care services. As a result, many incarcerated women face unnecessarily high-risk pregnancies.

Birthing Practices

Second-level headings bold and flush with left margin.

Another practice that increases the risk of complications in pregnancy in prison is the custom of shackling female inmates during labor, delivery, and postpartum. This practice is both degrading and inhumane and can pose a problem for health care providers in case of an emergency. In an official position statement, the Association of Women's Health, Obstetric and Neonatal Nurses (2011) noted that the unnecessary practice can interfere with the ability of nurses and health care providers to deliver the proper care and treatment (p. 817). Only 18 states in the United States currently ban the shackling of expectant mothers in prison while they give birth (American Civil Liberties Union Foundation, 2012). The remaining 32 states are left to their own devices, in some cases shackling mothers with no regard to the recommendations of nurses and other health care providers.

Prison Nursery Programs

In order to effectively spread awareness of prison nursery programs, it must be clear exactly what they are and how they serve incarcerated

5

mothers. Prison nursery programs offer women who become incarcerated while pregnant the option to keep and parent their child while they serve their sentences. Getting into these programs can be a rigorous process; with limited spots available, prospective mothers must have a nonviolent conviction, no record of child abuse, and be roughly within 18 months of completing their sentence—which is the maximum amount of time a child can stay with their mother behind bars (Stein, 2010, p. 11).

Author name in parentheses when no signal phrase is used.

Benefits

When mothers do get into these programs, there are many benefits to be had for both themselves and their children. For starters, mothers are provided with parenting classes, support groups, substance abuse counseling, and complementary day-care services to attend these classes. Many prisons also provide high school and college courses for those mothers who have not yet completed their education (Wertheimer, 2005). Last, vocational programs are also offered, which aid in the job search once the women are released from prison.

Recidivism

As a result of the programs, many of the mothers have been shown to have a reduced recidivism rate. An astounding 86.3% of women exiting a prison nursery program remained in the community 3 years following their release (Goshin et al., 2013, Results section, para. 3). This is a proven positive for both the mothers and their children.

In-text citation of journal article without pagination.

Cost of Prison Nursery Programs

While it may seem as if implementing prison nursery programs across the country would be an expensive endeavor, it is important to consider by comparison the costs of putting children in foster care. On average, the total financial cost for one child to remain in foster care per

6

year in Oregon is about $26,600 (Fixsen, 2011, p. 3). Conversely, based on an evaluation of Nebraska's prison nursery program, expenses for the nursery program would be roughly 40% less per year than foster care expenses would be for the same babies, and even more money would be saved if the nursery program reduced recidivism (Carlson, 1998, as cited in Yager, 2015, para. 34). Rapidly increasing the number of prison nursery programs across the country may result in some extra money spent in the short run, but it will provide benefits in the long run, both financially and socially.

Source quoted in another source identified with "as cited in."

7

References

American Civil Liberties Union Foundation. (2012). *The shackling of pregnant women & girls in U.S. prisons, jails & youth detention centers* [Briefing paper]. https://www.aclu.org/files/assets/anti-shackling _briefing_paper_stand_alone.pdf

Association of Women's Health, Obstetric and Neonatal Nurses. (2011). *Shackling incarcerated pregnant women* [Position statement]. http://www.jognn.org/article/S0884-2175(15)30763-2/pdf

Fixsen, A. (2011). *Children in foster care.* A Family for Every Child. http://www.afamilyforeverychild.org/wp=content/uploads/2018/04 /children_in_foster_care.pdf

Goshin, L. S., Byrne, M. W., & Henninger, A. M. (2013). Recidivism after release from a prison nursery program. *Public Health Nursing, 33*(2), 109–117. https://doi.org/10.1111/phn.12072

Hotelling, B. A. (2008). Perinatal needs of pregnant incarcerated women. *The Journal of Perinatal Education, 17*(2), 37–44. https://doi .org/10.1624/105812408X298372

Sentencing Project. (2017). *Trends in U.S. corrections* [Fact sheet]. http://sentencingproject.org/wp-content/uploads/2016/01 /Trends-in-US-Corrections.pdf

Stein, D. J. (2010, July/August). Babies behind bars: Nurseries for incarcerated mothers and their children. *Children's Voice, 19*(4), 10–13.

Sufrin, C. (2012, July 1). *Incarcerated women and reproductive health care* [Video]. YouTube. https://www.youtube.com /watch?v=WNx1ntLyI2Q

8

Wertheimer, L. (2005, November 5). *Prenatal care behind bars* [Radio
  broadcast]. NPR. http://www.npr.org/templates/story/story
  .php?storyId=4990886

Yager, S. (2015, July/August). Prison born. *The Atlantic.* https://www
  .theatlantic.com/magazine/archive/2015/07/prison-born/395297/

*All sources cited in
the text are listed.*

Part 8

Media / Design

Consciously or not, we design all the texts we write, choosing typefaces, setting up text as lists or charts, deciding whether to add headings—and then whether to center them or align them on the left. Sometimes our genre calls for certain design elements—essays begin with titles, letters begin with salutations ("Dear Auntie Em"). Other times we design texts to meet the demands of particular audiences, formatting documentation in MLA or APA or some other style, setting type larger for young children, and so on. And our designs always depend on our medium. A memoir might take the form of an essay in a book, be turned into a bulleted list for a slide presentation, or include links to images or other pages if presented on a website. The chapters in this part offer advice for **CHOOSING MEDIA**; working with **DESIGN**, **IMAGES**, and **SOUND**; **WRITING AND LEARNING ONLINE**; and **GIVING PRESENTATIONS**.

Media/Design

55 Choosing Media

USA Today reports on contract negotiations between automakers and auto-workers with an article that includes a large photo and a colorful graph; the article on the same story on the *New York Times* website includes a video of striking workers. In your economics class, you give a presentation about the issue that includes *Prezi* slides.

These examples show how information about the same events can be delivered using three different media: print (*USA Today*), digital (nytimes.com), and spoken (the main medium for your class presentation). They also show how different media offer writers different modes of expressing meaning, ranging from words to images to sounds and hyperlinks. A print text can include written words and still visuals; online, the same text can also incorporate links to moving images and sound as well as to other written materials. A presentation with slides can include both spoken and written words, can incorporate video and audio elements — and can also include print handouts.

In college writing, the choice of medium often isn't up to you: your instructor may require a printed essay or a classroom talk, a website, or some combination of media. Sometimes, though, you'll be the one deciding. Because your medium will play a big part in the way your audience receives and reacts to your message, you'll need to think hard about what media best suit your audience, purpose, and message. This chapter will help you choose media when the choice is yours.

Print

When you have a choice of medium, print has certain advantages over spoken and digital text in that it's more permanent and doesn't depend on audience access to technology. Depending on your own access to technology, you can usually insert photos or other visuals and can present data and other information as graphs or charts. Obviously, though, print documents are more work than digital ones to update or change, and they don't allow for sound, moving images, or links to other materials.

Digital

Online writing is everywhere: on course learning management systems and class websites; in virtual discussion groups and wikis; in emails, text messages, tweets, and other social media. And when you're taking an online course, you are, by definition, always using a digital medium. Remember that this medium has advantages as well as limitations and potential pitfalls. You can add audio, video, and links—but your audience may not have the same access to technology that you do. These are just some of the things you'll need to keep in mind when deciding, say, whether to include or link to videos or a site that has restricted access. Also, digital texts that circulate online through blogs, websites, email, or social media can take on a life of their own; other viewers may forward, like, retweet, or repost your text to much larger audiences than you originally considered.

Spoken

If you deliver your text orally, as a speech or presentation, you have the opportunity to use your tone of voice, gestures, and physical bearing to establish credibility. But you must write your text so that it's easy to understand when it is heard rather than read. Speaking from memory or from notecards, rather than reading a script or essay, often makes it easier for the audience to follow your talk. The spoken medium can be used alone with a live, face-to-face audience, but it's often combined with print in the form of handouts, or with digital media in the form of presentation

academic literacies   ● fields   ● research
■ rhetorical situations   ∴ processes   ● media/design
▲ genres   ◆ strategies

software like *PowerPoint* or *Prezi*, or designed for remote audiences in formats like webcasts, webinars, podcasts, or videoconferencing platforms such as *Zoom*.

Multimedia

It's increasingly likely that you'll be assigned to create a multimedia text, one that includes some combination of print, oral, and digital elements. It's also possible that you'll have occasion to write a multimodal text, one that uses more than one mode of expression: words, images, audio, video, links, and so on. The words "multimedia" and "multimodal" are often used interchangeably, but "multimodal" is the term that's used most often in composition classes, whereas "multimedia" is the one used in other disciplines and in industry. In composition classes, the word generally refers to writing that includes more than just words.

For example, let's say that in a US history class you're assigned to do a project about the effects of the Vietnam War on American society. You might write an essay using words alone to discuss such effects as increased hostility toward the military and government, generational conflict within families and society at large, and increased use of recreational drugs. But you could also weave such a text together with many other materials to create a multimodal composition.

If you're using print, for example, you could include photographs from the Vietnam era, such as of antiwar protests or military funerals. Another possibility might be a timeline that puts developments in the war in the context of events going on simultaneously elsewhere in American life, such as in fashion and entertainment or in the feminist and civil rights movements. If you're posting your project online, you might also incorporate video clips of TV news coverage of the war and clips from films focusing on it or its social effects, such as *Apocalypse Now* or *Easy Rider*. Audio elements could include recorded interviews with veterans who fought in the war, people who protested against it, or government officials who were involved in planning or overseeing it. Many of these elements could be inserted into your document as links.

If your assignment specifies that you give an oral presentation, you could play some of the music of the Vietnam era, show videos of government officials defending the war and demonstrators protesting it, maybe hang some psychedelic posters from the era.

You might do something similar with your own work by creating an electronic portfolio, or e-portfolio. Tips for compiling an **E-PORTFOLIO** may be found in Chapter 34.

388 ⁘

Considering the Rhetorical Situation

59–60 ■

PURPOSE   What's your purpose, and what media will best suit that purpose? A text message or email may be appropriate for inviting a friend to lunch, but neither would be ideal for demonstrating to a professor that you understand a complex historical event; for that, you'd likely write a report, either in print or online—and you might include photos, maps, or other such texts to provide visual context.

61–64 ■

AUDIENCE   What media are your audience likely to expect—and be able to access? A blog may be a good way to reach people who share your interest in basketball or cupcakes, but to reach your grandparents you may want to send them an email message—or a handwritten note in the mail. Some employers and graduate school admissions officers require applicants to submit résumés and applications online, while others prefer to receive them in print form.

65–71 ■

GENRE   Does your genre require a particular medium? If you're giving an oral presentation, you'll often be expected to include slides. Academic essays are usually formatted to be printed out, even if they are submitted electronically. An online essay based on field research might include audio files of those you've interviewed, but if your essay

☀ academic literacies   ● fields   ● research
■ rhetorical situations   ⁘ processes   ● media/design
▲ genres   ◆ strategies

were in print, you'd need to quote (or paraphrase or summarize) what they said.

STANCE   If you have a choice of media, think about whether a particular medium will help you convey your stance. A print document in MLA format, for instance, will make you seem scholarly and serious. Tweeting or blogging, however, might work better for a more informal stance. Presenting data in charts will sometimes help you establish your credibility as a knowledgeable researcher.

72–74

Once you decide on the media and modes of expression you're using, you'll need to design your text to take advantage of their possibilities and to deal with their limitations. The next chapters will help you do that.

56 Designing Text

You're trying to figure out why a magazine ad you're looking at is so funny, and you realize that the text's font is deliberately intended to make you laugh. An assignment for a research paper in psychology specifies that you are to follow APA format. Your classmates complain that the *PowerPoint* slides you use for a presentation are hard to read because there's not enough contrast between the words and the background. Another student says you include too many words on each slide. Whether you're putting together your résumé, creating a website for your intramural soccer league, or writing an essay for a class, you need to think about how you design what you write.

Sometimes you can rely on established conventions: in MLA and APA styles, for example, there are specific guidelines for margins, headings, and the use of single-, double-, or triple-spaced lines of text. But often you'll have to make design decisions on your own—and not just about words and spacing. If what you're writing includes photos, charts, tables, graphs, or other visuals, you'll need to integrate these with your written text in the most attractive and effective way; online, you may also need to decide where and how to include video clips and links. You might even use scissors, glue, and staples to attach objects to a poster or create pop-ups in a brochure.

No matter what your text includes, its design will influence how your audience responds to it and therefore how well it achieves your purpose. This chapter offers general advice on designing print and online texts.

Considering the Rhetorical Situation

As with all writing tasks, your rhetorical situation should affect the way you design a text. Here are some points to consider:

✳ academic literacies   ◆ fields   ● research

■ rhetorical situations   ∴ processes   ● media/design

▲ genres   ◆ strategies

PURPOSE	How can you design your text to help achieve your purpose? If you're reporting information, for instance, you may want to present statistical data in a chart or table rather than in the main text to help readers grasp it more quickly. If you're trying to get readers to care about an issue, a photo or **PULL QUOTE**—a brief selection of text "pulled out" and featured in a larger font—might help you do so.	59–60 Glossary
AUDIENCE	How can you make your design appeal to your intended audience? By using a certain font style or size to make your text look stylish, serious, or easy to read? What kind of headings—big and bold, simple and restrained?—would your readers expect or find helpful? What colors would appeal to them?	61–64
GENRE	Are you writing in a genre that has design conventions, such as an abstract, an annotated bibliography, or a résumé? Do you need to follow a format such as those prescribed in MLA or APA style?	65–71
STANCE	How can your design reflect your attitude toward your audience and subject? Do you need a businesslike font or a playful one? Would tables and graphs help you establish your credibility? How can illustrations help you convey a certain tone?	72–74

Some Basic Principles of Design

Be consistent. To keep readers oriented while reading documents or browsing multiple webpages, any design elements should be used consistently. In a print academic essay, that task may be as simple as using the same font throughout for your main text and using boldface or italics for headings. If you're writing for the web, navigation buttons and other major elements should be in the same place on every page. In a presentation, each slide should use the same background and the same font unless there's a good reason to introduce differences.

Keep it simple. One of your main design goals should be to help readers see quickly—even intuitively—what's in your text and how to find specific information. Adding headings to help readers see the parts, using consistent colors and fonts to help them recognize key elements, setting off steps in lists, using white space to set off blocks of text or to call attention to certain elements, and (especially) resisting the temptation to fill pages with fancy graphics or unnecessary animations—these are all ways of making your text simple to read.

Look, for example, at a furniture store's simple, easy-to-understand webpage design on the next page. This webpage contains considerable information: a row of links across the top, directing readers to various products; a search option; a column down the right side that provides details about the chair shown in the wide left-hand column; thumbnail photos below the chair and ordering information; and more details across the bottom. Despite the wealth of content, the site's design is both easy to figure out and, with the generous amount of white space, easy on the eyes.

Aim for balance. On the webpage on the next page, the photo takes up about a quarter of the screen and is balanced by a narrower column of text, and the product information tabs and text across the page bottom balance the company logo and search box across the top. For a page without images, balance can be created through the use of margins, headings, and spacing. In the journal page shown on page 583, notice how using white space around the article title and the author's name, as well as setting both in larger type and the author's name in all capital letters, helps to balance them vertically against the large block of text below. The large initial letter of the text also helps to balance the mass of smaller type that follows. MLA and APA styles have specific design guidelines for academic research papers that cover these elements. A magazine page might create a sense of balance by using PULL QUOTES and illustrations to break up dense vertical columns of text.

Glossary

Use color and contrast carefully. Academic readers usually expect black text on a white background, with perhaps one other color for headings. Presentation slides and webpages are most readable with a plain, light-colored

✳ academic literacies   ● fields   ● research
■ rhetorical situations   ⦂ processes   ● media/design
▲ genres   ◆ strategies

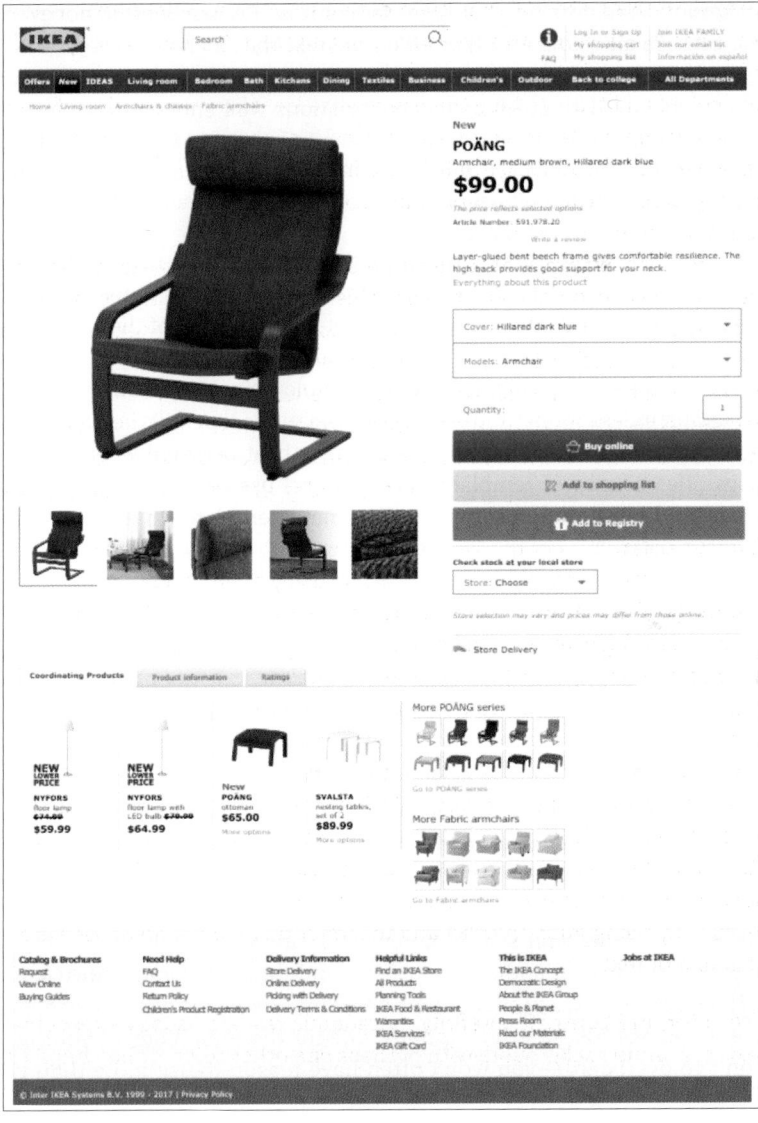

background and dark text that provides contrast. Remember that not everyone can see all colors and that an online text that includes several colors might be printed out and read in black and white; make sure your audience will be able to distinguish any color variations well enough to grasp your meaning. Colored lines on a graph, for example, should be distinguishable even if readers cannot see the colors. Red-green contrasts are especially hard to see and should be avoided.

Use available templates. Good design takes time, and most of us don't have training as designers. If you're pressed for time or don't feel up to the challenge of designing your own text, take advantage of the many templates available. In *Microsoft Word*, for example, you can customize "styles" to specify the font, including its size and color; single- or double-spacing; paragraph indentations; and several other features that will then automatically apply to your document. Websites that host personal webpages and blogs offer dozens of templates that you can use or modify to suit your needs. And presentation software offers many templates that can simplify creating slides.

Some Elements of Design

Fonts. You can usually choose from among many fonts, and the one you choose will affect how well the audience can read your text and how they will perceive your TONE. Times Roman will make a text look businesslike or academic; *Comic Sans* will make it look playful. For most academic writing, you'll want to use a font size between 10 and 12 points and a serif font (such as Times Roman or Bookman) rather than a sans serif font (such as Arial, Verdana, or Calibri) because serif fonts are generally easier to read. Reserve sans serif for headings and parts of the text that you want to highlight. Decorative fonts (such as *Magneto*, *Amaze*, Chiller, and Jokerman) should be used sparingly and only when they're appropriate for your audience, purpose, and the rest of your RHETORICAL SITUATION. If you use more than one font, use each one consistently: one for HEADINGS, one for captions, one for the main body of your text. Don't go overboard—you won't often have reason to use more than two or, at most, three fonts in any one text.

73–74 ■

57 ■

670–72 ●

✳ academic literacies   ● fields   ● research
■ rhetorical situations   ⁘ processes   ● media/design
▲ genres   ◆ strategies

Every font has regular, **bold**, and *italic* forms. In general, choose regular for the main text and lower-level headings, bold for major headings, and italic within the main text to indicate titles of books and other long works and, occasionally, to emphasize words or brief phrases. Avoid italicizing or boldfacing entire sentences or paragraphs, especially in academic writing. If you are following **MLA**, **APA**, or some other style format, be sure your use of fonts conforms to its requirements.

MLA 564–614
APA 614–55

Finally, consider the line spacing of your text. Generally, academic writing is double-spaced, whereas job application letters and **RÉSUMÉS*** are usually single-spaced. Some kinds of **REPORTS** may call for single-spacing; check with your instructor if you're not sure. You'll often need to add extra space to set off parts of a text—items in a list, for instance, or headings.

140–63

Layout.   Layout is the way text is arranged on a page. An academic essay, for example, will usually have a title centered at the top, one-inch margins all around, and double-spacing. A text can be presented in paragraphs— or in the form of **LISTS**, **TABLES**, **CHARTS**, **GRAPHS**, and so on. Sometimes you'll need to include other elements as well: headings, images and other graphics, captions, lists of works cited.

669–70
676

Paragraphs.   Dividing text into paragraphs focuses information for readers and helps them process it by dividing it into manageable chunks. If you're writing a story for a print newspaper with narrow columns, for example, you'll divide your text into shorter paragraphs than you would if you were writing an academic essay. In general, indent paragraphs five to seven spaces (one-half inch) when your text is double-spaced; either indent or skip a line between single-spaced paragraphs.

Lists.   Put into list form information that you want to set off and make easily accessible. Number the items in a list when the sequence matters (in instructions, for example); use bullets when the order is not important. Set off lists with an extra line of space above and below, and add extra space between the items if necessary for legibility. Here's an example:

---

*See the ebook at digital.wwnorton.com/fieldguide6 for a chapter on Résumés and Job Letters.

Darwin's theory of how species change through time derives from three postulates, each of which builds on the previous one:

1. The ability of a population to expand is infinite, but the ability of any environment to support populations is always finite.
2. Organisms within populations vary, and this variation affects the ability of individuals to survive and reproduce.
3. The variations are transmitted from parents to offspring.

—Robert Boyd and Joan B. Silk, *How Humans Evolved*

Do not set off text as a list unless there's a good reason to do so, however. Some lists are more appropriately presented in paragraph form, especially when they give information that isn't meant to be referred to more than once. In the following example, there's no reason to highlight the information by setting it off in a list—and bad news is softened by putting it in paragraph form:

I regret to inform you that the Scholarship Review Committee did not approve your application for a Board of Rectors scholarship for the following reasons: your grade-point average did not meet the minimum requirements; your major is not among those eligible for consideration; and the required letter of recommendation was not received before the deadline.

Presented as a list, that information would be needlessly emphatic.

Headings.   Headings make the structure of a text easier to follow and help readers find specific information. Some genres require standard headings—announcing an **ABSTRACT**, for example, or a list of **WORKS CITED**. Other times you'll want to use headings to provide an overview of a section of text. You may not need any headings in brief texts, but when you do, you'll probably want to use one level at most, just to announce major topics. Longer texts, information-rich genres such as brochures or detailed **REPORTS**, and websites may require several levels of headings. If you decide to include headings, you will need to decide how to phrase them, what fonts to use, and where to position them.

Phrase headings concisely.   Make your headings succinct and parallel in

196–200 ▲
607 ●
140–63 ▲

academic literacies   ● fields   ● research
■ rhetorical situations   °● processes   ● media/design
▲ genres   ◆ strategies

structure. You might make all the headings nouns **(Mushrooms)**, noun phrases **(Kinds of Mushrooms)**, gerund phrases **(Recognizing Kinds of Mushrooms)**, or questions **(How Do I Identify Mushrooms?)**. Whatever form you decide on, use it consistently for each heading. Sometimes your phrasing will depend on your purpose. If you're simply helping readers find information, use brief phrases:

Head	**Forms of Social Groups among Primates**
Subhead	*Solitary Social Groups*
Subhead	*Monogamous Social Groups*

If you want to address readers directly with the information in your text, consider writing your headings as questions:

How can you identify edible mushrooms?
Where can you find edible mushrooms?
How can you cook edible mushrooms?

Make headings visible.   Headings need to be visible, so if you aren't following an academic style like MLA or APA, consider making them larger than the regular text, putting them in **bold** or *italics*, or using underlining— or a different font. For example, you could use a serif font like Times Roman for your main text and a sans serif font like Arial for your headings. On the web, consider making headings a different color from the body text. When you have several levels of headings, use capitalization, bold, and italics to distinguish among the various levels:

First-Level Head
Second-Level Head
Third-level head

APA format requires that each level of heading appear in a specific style: centered bold uppercase and lowercase for the first level, flush-left bold uppercase and lowercase for the second level, and so on.

● APA 645

Position headings appropriately.   If you're following APA format, center first-level headings. If you aren't following a prescribed format, you get to decide where to position your headings: centered, flush with the left margin, or even alongside the text in a wide left-hand margin. Position each

level of head consistently throughout your text. Generally, online headings are positioned flush left.

White space.   Use white space to separate the various parts of a text. In general, use one-inch margins for the text of an essay or report. Unless you're following MLA or APA format, include space above headings, above and below lists, and around photos, graphs, and other visuals. See the two sample research papers in this book for examples of the formats required by **MLA** and **APA**.

MLA 606–14
APA 648–55

Evaluating a Design

59–60

Does the design suit your PURPOSE?   Does the overall look of the design help convey the text's message, support its argument, or present information?

61–64

How well does the design meet the needs of your AUDIENCE?   Will the overall appearance of the text appeal to the intended readers? Is the font large enough for them to read? Are there headings to help them find their way through the text? Does the design help readers find the information they need?

65–71

How well does the text meet any GENRE REQUIREMENTS?   Can you tell by looking at the text that it is an academic essay, a lab report, a résumé, a blog? Do its fonts, margins, headings, and page layout meet the requirements of **MLA**, **APA**, or whatever style you're using?

MLA 605–7
APA 645–47
72–74

How well does the design reflect your STANCE?   Do the page layout and fonts convey the appropriate tone—serious, playful, adventuresome, conservative, or whatever other tone you intended?

✳ academic literacies   ◆ fields   ● research
■ rhetorical situations   ⁂ processes   ● media/design
▲ genres   ◆ strategies

57 Using Visuals, Incorporating Sound

For an art history class, you write an essay comparing two paintings by Willem de Kooning. For a business class, you create a proposal to improve department communication in a small local firm and incorporate diagrams to illustrate the new procedure. For a visual rhetoric class, you take a photograph of yourself and include a two-page analysis of how the picture distills something essential about you. For an engineering class project, you design a model of a bridge and give an in-class presentation explaining the structures and forces involved, which you illustrate with slides. For a psychology assignment, you interview several people who've suffered foreclosures on their homes in recent years about how the experience affected them—and then create an online text weaving together a slideshow of photos of the people outside their former homes, a graph of foreclosure rates, video and audio clips from the interviews, and your own insights.

All of these writing tasks require you to incorporate and sometimes to create visuals and sound. Many kinds of visuals can be included in print documents: photos, drawings, diagrams, graphs, charts, and more. And with writing that's delivered online or as a spoken presentation, your choices expand to include audio and video, voice-over narration, and links to other materials.

Visuals and sound aren't always appropriate, however, or even possible—so think carefully before you set out to include them. But they can help you make a point in ways that words alone cannot. Election polling results are easier to see in a bar graph than in a paragraph; photos of an event may convey its impact more powerfully than words alone; an audio clip can make a written analysis of an opera easier to understand. This

chapter provides some tips for using visuals and incorporating sound in your writing.

Considering the Rhetorical Situation

Use visuals and sounds that are appropriate for your audience, purpose, and the rest of your **RHETORICAL SITUATION**. If you're trying to persuade voters in your town to back a proposal on an issue they don't know much about, for example, you might use dramatic pictures just to get their attention. But when it's important to come across as thoughtful and objective, maybe you need a more subdued look—or to make your points with written words alone. A newspaper article on housing prices might include a bar or line graph and also some photos. A report on that topic for an economics class would probably have graphs with no photos; a spoken presentation for a business class might use a dynamic graph that shows prices changing over time and an audio voice-over for pictures of a neighborhood; a community website might have graphs, links to related sites, and a video interview with a home owner.

In your academic writing, especially, be careful that any visuals you use support your main point—and don't just serve to decorate the text. (Therefore, avoid clip art, which is primarily intended as decoration and comes off as unsophisticated and childish.) Images should validate or exemplify what you say elsewhere with written words and add information that words alone can't provide as clearly or easily.

Using Visuals

Photos, drawings, diagrams, videos, tables, pie charts, bar graphs—these are many kinds of visuals you could use. Visuals can offer support, illustration, evidence, and comparison and contrast in your document.

Photographs. Photos can support an **ARGUMENT**, illustrate **NARRATIVES** and **PROCESSES**, present other points of view, and help readers "place" your information in time and space. You may use photos you take yourself, or you can download photos and other images from the internet—within limits. Most downloadable photos are copyrighted, meaning that you can use

57

410–30
474–82
469–73

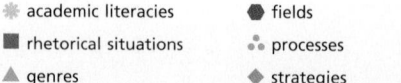

* academic literacies   ● fields   ● research
■ rhetorical situations   ⁖ processes   ● media/design
▲ genres   ◆ strategies

An essay discussing the theme of mother and child might compare this painting from the Italian Renaissance (left) with a modern photograph such as Dorothea Lange's *Migrant Mother* (right).

them without obtaining permission from the copyright owner only if you're doing so for academic purposes, to fulfill an assignment. If you're going to publish your text, either in print or on the web, you must have permission. You can usually gain permission by emailing the copyright holder, but that often entails a fee, so think carefully about whether you need the image. Consider, too, the file size of digital images; large files can clog readers' email in-boxes, take a long time to display on their screens, or be hard for you to upload in the first place, so you may have to compress an image in a zip file or reduce its resolution (which can diminish its sharpness).

Videos.   If you're writing online, you can include video clips for readers to play. If you're using a video already available online, such as on *YouTube*, you can show the opening image with an arrow for readers to click on to start the video, or you can simply copy the video's URL and paste it into your text as a **LINK**. In either case, be sure to introduce the video in your text with a **SIGNAL PHRASE**. As with any other source, be sure to provide an in-text citation and full documentation.

681–82
551–54

If you want to include a video you made yourself, you can edit it using such programs as *iMovie* or *Shotcut*. Once you're ready to insert it into your online document, the easiest way is to first upload it to *YouTube*, choosing the Private setting so only those you authorize may view it, and then create a link in your document.

Graphs, charts, and tables. Statistical and other numerical information is often best presented in graphs, charts, and tables. If you can't find the right one for your purpose, you can create your own, as long as it's based on sound data from reliable sources. To do so, you can use various spreadsheet programs such as *Excel* or online chart and graph generators such as *Plotly* or *Venngage*.

In any case, remember to follow basic design principles: be **CONSISTENT**, label all parts clearly, and **KEEP THE DESIGN SIMPLE** so readers can focus on the information and not be distracted by a needlessly complex design. In particular, use color and contrast wisely to emphasize what's most significant. Choose **COLORS** that are easy to distinguish from one another — and that remain so if the graph or chart is printed out in black and white. (Using distinct gradations of color from light to dark will show well in black and white.) Some common kinds of graphs, charts, and tables are shown on the next pages.

Diagrams, maps, flowcharts, and timelines. Information about place and time is often presented in diagrams, maps, flowcharts, and timelines. If you're using one of these visuals from the web or elsewhere, be sure to **DOCUMENT** it. Otherwise you can create one of these yourself using online tools like *Canva* or the Shape and Chart tabs in your word processing program. Make diagrams and maps as simple as possible for the point you're emphasizing. Unnecessarily complex maps can be more of a distraction than a help. For timelines, make sure the scale accurately depicts the passage of time you're writing about and avoids gaps and bunches.

665

666

666–68

560–63

⁑ academic literacies   ● fields   ● research

■ rhetorical situations   ⁑ processes   ● media/design

▲ genres   ◆ strategies

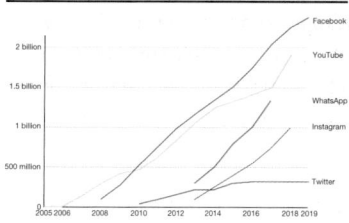

Line graphs are a good way of showing changes in data over time. Each line here represents the number of users logging on to a different social media platform. Plotting the lines together allows readers to compare the data at different points in time. Be sure to label the *x* and *y* axes, and limit the number of lines to avoid confusion.

China's expanding auto market

Number of vehicles sold in China and the U.S., in millions:

■ Sold in U.S.   ■ Sold in China

20   16.6   18.5
15   12.8
10   7.2
5
     '06   '07   '08   '09   '10   '11

Source: China Association of Automobile Manufacturers, Automotive News
Graphic: Los Angeles Times   © 2012 MCT

Bar graphs are useful for comparing quantitative data, measurements of how much or how many. The bars can be horizontal or vertical. This graph compares the number of cars sold in China and the United States over a period of five years. Some software offers 3-D and other special effects, but simple graphs are often easier to read.

Costly walk down the aisle

According to The Wedding Report, the average cost of a wedding in the U.S. is about $25,300. Here's a breakdown of spending:

Venue, catering and rentals: $11,230

Beauty and spa $124

Gifts and favors $641

Invitations $817

Entertainment $1,230

Flowers and decorations $1,510

Planner/consultant $1,540

Jewelry $3,920

Photography and video $2,670

Attire and accessories $1,610

Source: The Wedding Report
Graphic: Dallas Morning News   © 2014 MCT

Pie charts can be used to show how a whole is divided into parts or how parts of a whole relate to one another. This pie chart shows the average cost of a traditional wedding in the United States. The segments in a pie should always add up to 100 percent, and each segment should be clearly labeled.

The wage gap in numbers

Year	Percent of men's income earned by women	Wage gap in dollars	Year	Percent of men's income earned by women	Wage gap in dollars
2015	80%	$10,470	1985	65%	$18,011
2010	77%	$12,027	1980	60%	$20,314
2005	77%	$11,564	1975	59%	$21,052
2000	74%	$13,474	1970	59%	$19,851
1995	71%	$13,908	1965	60%	$16,851
1990	72%	$13,822	1960	61%	$14,790

Source: U.S. Census Bureau   Graphic: Tribune News Service

Tables are useful for displaying numerical information concisely, especially when several items are being compared. This table shows the gap between men's and women's incomes over a fifty-five-year period. Presenting information in columns and rows permits readers to find data and identify relationships among them.

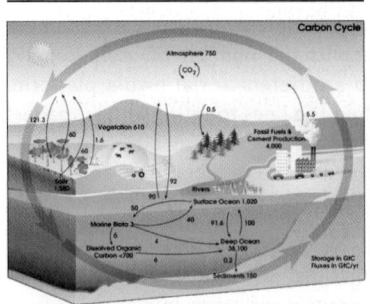

Diagrams and flowcharts are ways of showing relationships and processes. This diagram shows how carbon moves between the Earth and its atmosphere. Flowcharts can be made using widely available templates; diagrams, in contrast, can range from simple drawings to works of art. Some simple flowcharts may be found in the genres chapters.

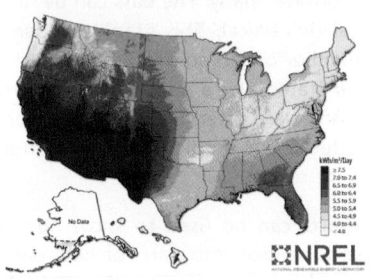

Maps show physical locations. This map shows the annual average direct solar resources for each of the US states. Maps can be drawn to scale or purposefully out of scale to emphasize a point.

405–9 ◆

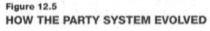

Figure 12.5
HOW THE PARTY SYSTEM EVOLVED

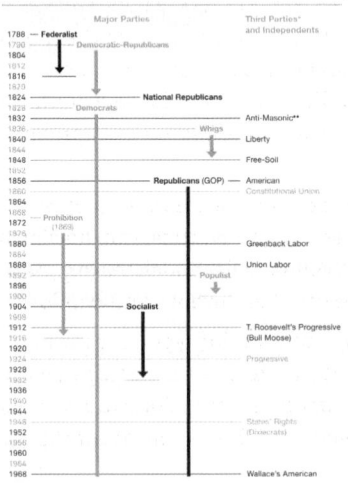

Timelines show change over time. These are useful to demonstrate **CAUSE** and **EFFECT** relationships or evolution. This timeline depicts how the American party system has evolved from 1788 to 2016. Timelines can be drawn horizontally or vertically.

☀ academic literacies	■ fields
■ rhetorical situations	⁖ processes
▲ genres	◆ strategies

● research
● media/design

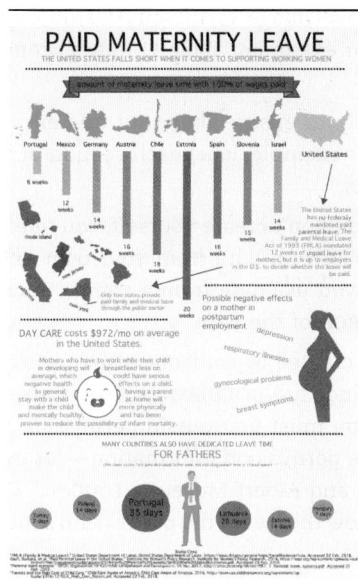

Infographics are eye-catching visual representations of information that help simplify a complicated subject. Infographics typically include an engaging title; a design that reflects the content being displayed; and a balance of charts, graphs, and other elements; readable fonts; and the sources of your information. Several websites offer templates to help you design and execute infographics, including *Piktochart*, *Infogram*, and several others.

SOME TIPS FOR USING VISUALS

- Position images as close as possible to the discussion to which they relate. In *Microsoft Word*, simply position your cursor where you want to insert an image; click the appropriate type of visual from the menu on the Insert tab; choose the appropriate image from your files; and click Insert. You may then need to adjust the way the text flows or wraps around the image: in the Page Layout tab, choose the appropriate option in Wrap Text.

- In academic writing, number all images, using separate sequences of numbers for figures (photos, graphs, diagrams, video clips, and drawings) and tables: Fig. 1, Fig. 2; Table 1, Table 2. Follow style guidelines (e.g., **MLA** or **APA**) for where to place the figure or table number and caption.

MLA 607

APA 646–47

- Explain in your written text whatever information you present in an image—don't expect it to speak for itself. Refer to the image before it appears, identifying it and summarizing its point. For example: "As Table 1 shows, Italy's economic growth rate has been declining for thirty years."

- Provide a title or caption for each image to identify it and explain its significance for your text. For example: "Table 1: Italy's Economic Growth Rate, 1990–2020."

- Label the parts of visuals clearly to ensure that your audience will understand what they show. For example, label each section of a pie chart to show what it represents.

- Cite the source of any images you don't create yourself. You need not document visuals you create, based on data from your own experimental or field research; but if you use data from a source to create a graph or chart, **CITE THE SOURCE** of the data.

555–58

- In general, you may use visuals created by someone else in your academic writing as long as you include full **DOCUMENTATION**. If you post your writing online, however, you must first obtain permission from the copyright owner and include permission information—for example: "Photo courtesy of Victoria and Albert Museum, London." Copyright holders will often tell you how they want the permission sentence to read.

560–63

Incorporating Sound

Audio clips, podcasts, and other sound files can serve various useful purposes in online writing and spoken presentation. Music, for example, can create a mood for your text, giving your audience hints about how to interpret the meaning of your words and images or what emotional response you're evoking. Other types of sound effects—such as background conversations, passing traffic, birdsongs, crowd noise at sports events—can provide a sense of immediacy, of being part of the scene or event you're describing. Spoken words can serve as the primary way you present an online text or as an enhancement of or even a counterpoint to a written text. (And if your audience includes visually impaired people, an audio track can allow or help them to follow the text.)

You can download or link to various spoken texts online, or you can record voice and music as podcasts using programs such as *GarageBand* and *Audacity*. Remember to provide an **IN-TEXT CITATION** and full **DOCUMENTATION** of any sound material you obtain from another source.

MLA 567–74
APA 618–22
560–63

- academic literacies
- rhetorical situations
- genres
- fields
- processes
- strategies
- research
- media/design

Adding Links

If you're writing an online text in which you want to include images, video, or sound material available on the web, it's often easier and more effective to create links to them within the text than to embed them by copying and pasting. Rather than provide the URL for the link, use relevant words to make it easier for a reader to decide to click on the link. (See the example below where the links are marked by blue color and words such as "Francis Davis Millet and Millet family papers.") Such links enable readers to see the materials' original context and to explore it if they wish. Be selective in the number of links you include: too many links can dilute a text.

The example below shows a blog post from the Archives of American Art with links to additional detail and documentation.

John Singer Sargent

This lively caricature from the Francis Davis Millet and Millet family papers features an artist fervently painting his subject, just in the background. Most likely it is John Singer Sargent at work on his painting *Carnation, Lily, Lily, Rose*. His posture and the expression on his face suggest an exuberance that matches the action of the paint dripping and splashing as it prepares to meet the canvas with energetic strokes.

Caricature of an artist painting vigorously, ca. 1885-1886. Francis Davis Millet and Millet family papers. Archives of American Art, Smithsonian Institution.

SOME TIPS FOR CREATING LINKS

551–54

- Indicate links with underlining and color (most often blue), and introduce them with a **SIGNAL PHRASE**.

- Don't include your own punctuation in a link. In the example on page 681, the period is not part of the link.

- Try to avoid having a link open in a new browser window. Readers expect links to open in the same window.

Editing Carefully—and Ethically

You may want to edit a photograph, cropping to show only part of it or using *Photoshop* or similar programs to enhance the colors or otherwise alter it. Similarly, you may want to edit a video, podcast, or other audio file to shorten it or remove irrelevant parts. If you are considering making a change of this kind, however, be sure not to do so in a way that misrepresents the content. If you alter a photo, be sure the image still represents the subject accurately; if you alter a recording of a speech or interview, be sure the edited version maintains the speaker's intent. Whenever you alter an image, a video, or a sound recording, tell your readers how you have changed it.

The same goes for editing charts and graphs. Changing the scale on a bar graph, for example, can change the effect of the comparison, making the quantities being compared seem very similar or very different, as shown in the two bar graphs of identical data in Figures 1 and 2.

Both charts show the increase in average housing costs in the United States between 2000 and 2020. However, by making the baseline in Figure 1 $200,000 instead of zero, the increase appears to be far greater than it was in reality. Just as you shouldn't edit a quotation or a photograph in a way that might misrepresent its meaning, you should not present statistical data in a way that could mislead readers.

academic literacies   fields   research

rhetorical situations   processes   media/design

genres   strategies

Figure 1. Average housing prices in the United States, 2000–2020 (exaggerated).

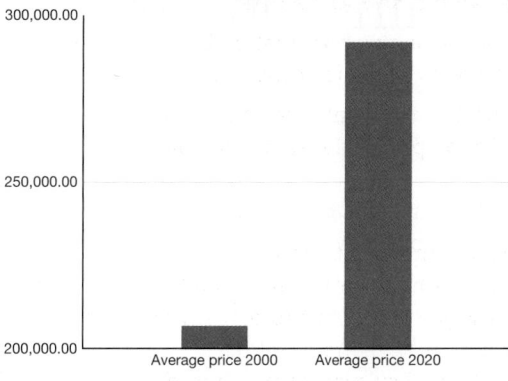

Figure 2. Average housing prices in the United States, 2000–2020 (presented accurately).

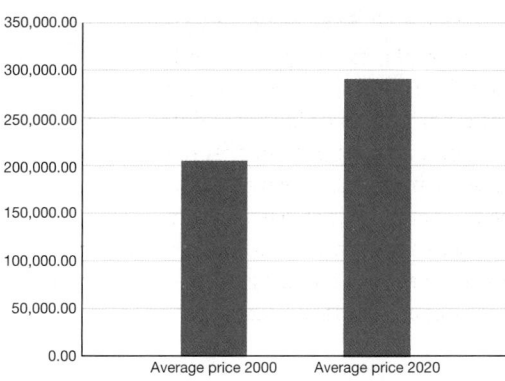

58 Writing and Learning Online

Email. *Facebook. Snapchat, Twitter, Instagram, Pinterest, YouTube.* Texts. Tweets. It may seem as if almost all writing is done online now. In college courses, you may still be required to turn in some writing assignments on paper, but more and more writing is not only done online but submitted that way too through learning management systems, email, or some other online system. You may rarely use email, but your professor may require you to do so. And many classes are being taught online, with little or no face-to-face communication between instructors and students.

Online, your instructor and classmates often cannot see or hear you—and that matters more than you might think. A puzzled look or a smile of recognition can start an important conversation in a face-to-face class, but in an online environment your audience may have only your written words to respond to. Therefore, you need to express your thoughts and feelings as clearly as you can *in writing.*

So it's useful to think about how the digital medium affects the way we write and learn—and how we can do both effectively online. This chapter provides some advice.

WRITING ONLINE

For most of us, email, texting, and social networking sites like *Facebook* and *Twitter* are already parts of everyday life. But using them for academic purposes may require some careful attention. Following are some guidelines.

* academic literacies   ● fields   ● research
■ rhetorical situations   ∴ processes   ● media/design
▲ genres   ◆ strategies

Email and Texts

When you're sending emails or texts to friends, you can be informal, using abbreviations, acronyms, emojis, and other shorthand ways of communicating. When you're texting or emailing your instructors, bosses, or colleagues, though, you need to write more formally: in emails, use an appropriate salutation ("Dear Professor McWilliams") and craft a specific subject line; instead of writing "Question about paper," be specific: "Profile organization question." In both email messages and texts, write clearly and concisely in complete sentences; use standard capitalization and punctuation; and proofread. If you're writing about group work or a specific course, identify the course or group explicitly. If you're writing to request something—for example, to read a draft or to write a letter of recommendation—be sure to give your readers adequate time to do what you ask before any deadlines. And be careful before you hit Send—you want to be good and sure that your email neither says something you'll regret later (don't send an email when you're angry—cool down and reread it!) nor includes anything you don't want the whole world reading (don't put confidential or sensitive information in an email).

Social Media

You may take courses that use social media or a **LEARNING MANAGEMENT SYSTEM** as a way for class members to communicate with the instructors or one another. When you do, be sure to consider your online rhetorical situation to make sure your course postings represent you as a respectful (and respectable) member of the class. Also, remember that many employers and graduate school administrators routinely check job applicants' social media pages, so don't post writing or photos on social media that you wouldn't want a potential employer to see.

689–91

Websites

You may need or want to create a website for school, a business you've started, or a hobby. While it's possible to create your own websites from

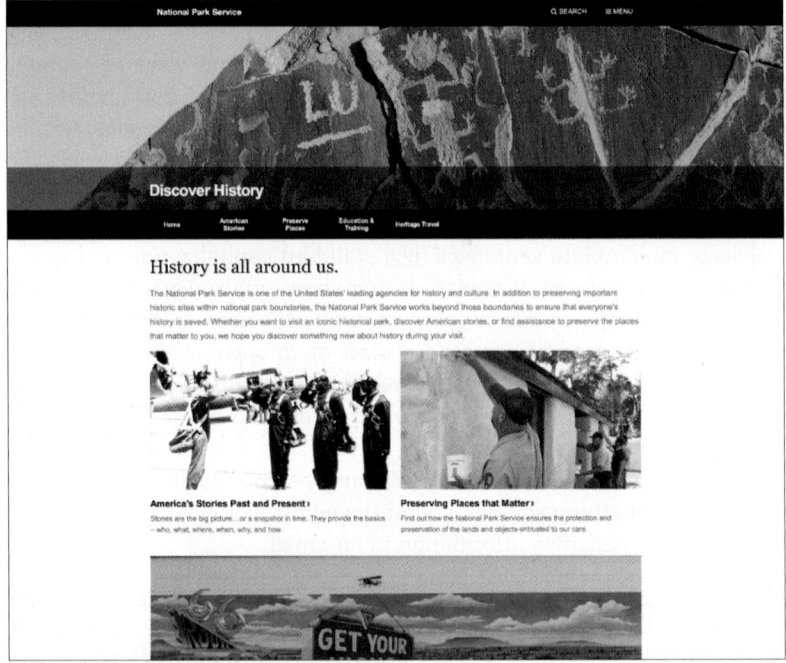

The homepage of *Discover History*, the National Park Service cultural resource program, provides a navigation menu that leads to various sections of the site. Links connect to pages describing Park Service programs.

scratch, free website builders such as *Wix*, *Weebly*, or *Duda* make it easy to create a site by providing templates for homepages, page designs, and navigation systems.

One key element in a website is the use of links to bring material from other sources into your text. You can link to the definition of a key term, for instance, rather than defining it yourself, or you can summarize a source and link to the full text rather than quoting or paraphrasing it. Providing links lets readers decide whether they need or want to see more detailed information—or not.

* academic literacies
* rhetorical situations
* genres
* fields
* processes
* strategies
* research
* media/design

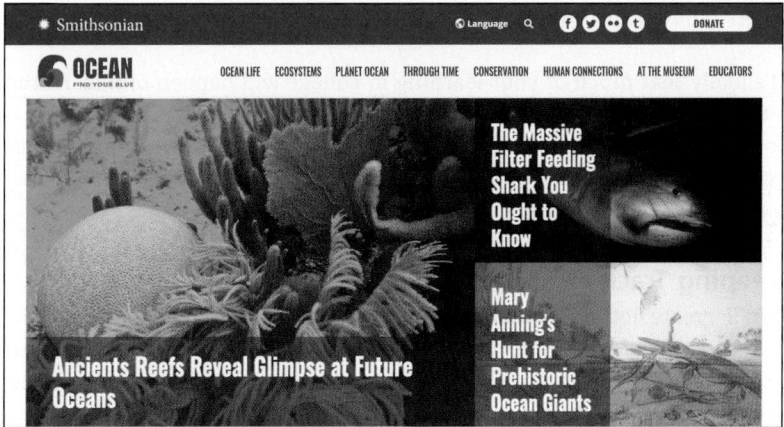

This blog, hosted by the Smithsonian Institution, focuses on marine biology and includes video, audio, slideshows, and written narratives.

Blogs

Blogs are websites that are maintained and updated regularly by individuals or groups who post opinions, reflections, information, and more—with writing, photos, video and audio files, and links to other sites. Blogs are an easy way to share your writing with others—and to invite response. Free blog hosting sites such as *WordPress*, *Wix*, or *Blogger* offer templates that let you create a blog and post to it easily, and some learning management systems include blogging capability as well.

If your blog is public, anyone can read it, including potential employers; so just as with *Facebook* and other social media, you'll want to be careful about how you present yourself and avoid posting anything that others could see as offensive. (Think twice before posting when you're angry or upset.) You may want to activate privacy settings that let you restrict access to some of the content or that make your blog unsearchable by *Google* and other search tools. Also, assume that what you post in a blog is permanent: your friends, family, employer—anyone—may read a posting years in the future, even if the blog is no longer active.

LEARNING ONLINE

It's likely that much of your learning in college will happen online. At the very least, you'll access course material through an online learning platform. But you might also take fully online courses—or hybrid courses that combine in-person and online learning. Following is advice for managing such learning effectively.

Keeping Track of Files

You'll create lots of documents as a college student. In a single writing course, for example, you may write three or four drafts of four essays—that's twelve to sixteen documents. To keep track of your files, you'll need to create folders, establish consistent file names, and back up your work.

Creating folders.　Create a folder for each course, and name it with the course title or number and the academic term: ENG 101 Fall 2021. Within each course folder, create a folder for each major assignment. Save your work in the appropriate folder, so you can always find it.

Naming and saving files.　If you're expected to submit files electronically, your instructor may ask you to name them in a certain way. If not, devise a system that will let you easily find the files, including multiple drafts of your writing. For example, you might name your files using *Your last name + Assignment + Draft number + Date:* Jones Evaluation Draft 2 10-5-2021.docx. You'll then be able to find a particular file by looking for the assignment, the draft number, or the date. Saving all your drafts as separate files will make it easy to include them in a portfolio; also, if you lose a draft, you'll be able to use the previous one to reconstruct it.

Backing up your work.　Hard drives fail, laptops and tablets get dropped, flash drives are left in public computers. Because files stored in computers can be damaged or lost, you should save your work in several places: on your computer, on a flash drive or portable hard drive, in space supplied by your school, or online. You can also ensure that an extra copy of your work exists by emailing a copy to yourself.

* academic literacies
■ rhetorical situations
▲ genres
● fields
⠂ processes
◆ strategies
● research
● media/design

Finding Basic Course Information

You'll need to learn some essential information about any online courses you take:

- *The phone number for the campus help desk* or technology center. Check the hours of operation, and keep the number handy.

- *The syllabus,* list of assignments, and calendar with deadlines.

- *Where to find tutorials* for your school's learning management system and other programs you may need help with.

- *Whether the course is synchronous or asynchronous.* In a *synchronous* class, you'll spend at least some of your course time in live virtual meetings with your classmates and teacher. In an *asynchronous* class, you'll interact with your peers and instructor through discussion boards, blogs, or other tools that don't require you to be logged in at a certain time. In both formats, course materials like videos, presentation slides, lecture notes, and assignments are usually available at any time through a learning management system.

- *How and when you can contact your instructor*—in person during office hours? by phone or email? through a videoconferencing platform like *Zoom?*—and how soon you can expect a response.

- *What file format you should use* to submit assignments—.doc, .docx, .rtf, .pdf, something else?—and how to submit them.

- *How to use the spell-check function* on your word processor or learning management system.

- *How to participate in online discussions*—will you use a discussion board? a chat function in a learning management system? a video-conferencing platform? something else?

Using Learning Management Systems

Whether you're in a face-to-face, hybrid, or online class, you may be asked to do some or all of your classwork online using a learning management system (LMS) such as *Blackboard, D2L, Moodle,* or *Canvas.* An LMS is a web-based educational tool that brings together all the course information your instructor wants you to have, along with features that allow you to

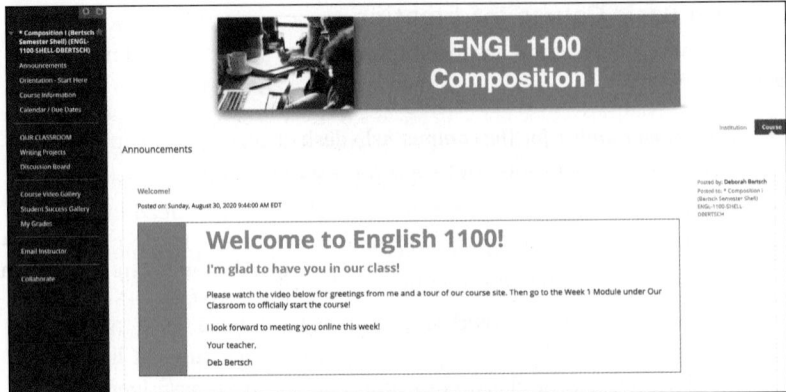

A course homepage from Columbus State Community College's LMS, *Blackboard*.

participate in the class in various ways. Your school's LMS likely includes the following features that you'll be expected to use:

A course homepage contains posts from your instructor; a calendar with due dates for assignments; and links to the course syllabus, other course content, and additional features available on the site.

A discussion board enables you to communicate with classmates even if everyone isn't logged in to the board at the same time. These conversations may be organized in "threads" so that posts on a particular topic appear together and may be read in order. When you contribute to a threaded discussion, treat it as an ongoing conversation: you need not introduce the topic but can simply add your comments.

A chat tool allows you to engage in written conversations in real time, with all participants logged in simultaneously. In a classroom, doing this may be like texting with many others at once, so the rules for class discussion apply: be patient while waiting for a response; focus on the topic being discussed; avoid sarcasm or personal attacks.

☀ academic literacies   ● fields   ● research

■ rhetorical situations   ⁚ processes   ◗ media/design

▲ genres   ◆ strategies

A dropbox is a place where you submit assignments online. If your course dropbox has folders for each assignment, be sure to upload your assignment into the correct folder. Keep in mind that systems go down, so don't wait until the last minute to submit a file. It's a good idea to double-check that the file you've submitted has been uploaded; often you can simply exit the dropbox and then return to it to see that your file is where it should be.

A web conferencing platform. If your course meets partly or completely online, you may attend class through a conferencing tool built into your LMS.

Online portfolios. Many LMSs allow you to create an online portfolio where you may post your coursework as well as photos, personal information, and links to other websites.

Additional features. An LMS may also include email; a space to keep a journal; a whiteboard for posting images, graphics, and presentations; a web conferencing tool; video and audio uploading capabilities; a gradebook; a social network (sometimes called a Ning) for class members only; and other features that can help you keep track of your work in a class.

Managing Challenges of Online Learning

Online classes offer a great deal of freedom and flexibility. You can work on your class almost anywhere—and often at your own pace (while still meeting deadlines, of course). Even so, online classes can present challenges, like managing your time effectively and keeping yourself motivated, disciplined, and engaged. Here are some tips for addressing those challenges and succeeding as an online learner.

- *Schedule your own "do dates."* Online classes may have only one or two due dates per week, but successful students create their own "do dates"—dates when they'll *do* the work. Try breaking the work into chunks each week and assigning yourself a specific day and time for each chunk: one day for reading the assigned texts, another day for generating ideas for a writing project, and so on. Put reminders in your

phone, or ask friends or family to hold you accountable. The sense of accomplishment you get in completing one task can motivate you to move on to the next.

- *Find a good place to work, and avoid multitasking.* Having a dedicated work space can help you stay focused. Find a place that suits you best—a quiet corner in your home, a favorite booth in a cafe, a desk at your school library. Turn off notifications on your phone, and consider using a website blocker like *Freedom* or *Cold Turkey* to minimize distractions further.

- *Take breaks to increase productivity.* Working long stretches online can be exhausting. Be sure to take breaks from the work—and from the screen time. If you can, take a full day off from your course each week. And when you're in the midst of a work session, take micro breaks to help you recharge. Consider setting a timer on your phone—say twenty-five minutes of focused work, followed by a five-minute break. Then repeat the cycle, with a longer break after several work sessions. Or use a productivity app like *TomatoTimer* to schedule your focus times and your break times.

- *Connect with your classmates—and your teacher.* You might not see your online classmates in person, but you can still get to know them and draw on them for support. In online writing courses, you'll typically have lots of opportunities to interact via discussion boards, peer review, and maybe web-based meetings or break-out rooms. As you get acquainted with your peers, try to identify at least one who can be your go-to person for reminders, tips, and general help. Then check in with each other via email, text, *FaceTime*, or other tools. Take advantage of opportunities to connect with your teacher, too. Use the email, chat, or videoconferencing tools available in your learning management system to reach out to your instructor whenever you need help.

- *Participate actively in online class meetings.* If your class requires real-time meetings via *Zoom*, *Google Meet*, or another platform, try to approach the online session as you would an in-person one: do your homework, write down questions for your teacher or classmates, and show up on time. During the session, keep yourself engaged

※ academic literacies   ● fields   ● research
■ rhetorical situations   ⁝ processes   ● media/design
▲ genres   ◆ strategies

and focused by taking notes, speaking during discussions, or making comments in the chat box. You might also volunteer to moderate the chat conversation and alert your teacher to important questions or comments.

Using Web Conferencing Tools

Zoom, *Skype*, *Microsoft Teams*, and *Google Hangouts* are all online platforms that enable groups of people to see and talk with one another. You may be in a class that meets partly or solely in a web conference session, or you may use one of these programs to meet with friends or coworkers. Since you're on camera, consider both your own appearance and the background. Here are some additional hints for participating in an online video conference:

- Consider adding a virtual background instead of the room behind you. *Zoom*, for example, offers several; or you can download photos of your own.

- Keep your microphone muted when you aren't speaking. Background noise, coughing, even setting drink glasses on the desk during a session can distract everyone.

- Consider recording online class sessions and meetings so you can review them later.

- Remember that online video discussions are different from face-to-face meetings or even conference calls. In *Zoom* and other programs, only one person can speak at a time, and people who are quick to respond can sometimes drown out and silence those who need more time before responding. So it's important to make sure that everyone gets the opportunity to speak.

- Also, it's harder to "read the room" by gauging others' reactions (and in large online classes, figuring out who's talking may be difficult). So you need to monitor yourself, making sure you're not holding the floor too long, being repetitive, or going off topic.

59 Giving Presentations

In a marketing class, you give a formal presentation that includes slides and handouts as part of a research project on developing brand loyalty to clothing labels among college students. As a candidate for student government, you deliver several speeches to various campus groups that are simultaneously broadcast over the web. At a good friend's wedding, after you make a toast to the married couple, another friend who couldn't attend in person toasts them remotely using *Zoom*; a third guest records both toasts on his cell phone and uploads them to *Facebook*.

Whether or not you include digital and print media, whenever you are called on to give a spoken presentation, you need to make your points clear and memorable. This chapter offers guidelines to help you prepare and deliver effective presentations. We'll start with two good examples.

ABRAHAM LINCOLN

Gettysburg Address

Given by the sixteenth president of the United States, at the dedication of the Gettysburg battlefield as a memorial to those who died in the Civil War, this is one of the most famous speeches ever delivered in the United States.

> Four score and seven years ago our fathers brought forth on this continent, a new nation, conceived in Liberty, and dedicated to the proposition that all men are created equal.
>
> Now we are engaged in a great civil war, testing whether that nation, or any nation so conceived and so dedicated, can long endure. We are met on a great battle-field of that war. We have come to

dedicate a portion of that field, as a final resting place for those who here gave their lives that that nation might live. It is altogether fitting and proper that we should do this.

But, in a larger sense, we can not dedicate—we can not consecrate—we can not hallow—this ground. The brave men, living and dead, who struggled here, have consecrated it, far above our poor power to add or detract. The world will little note, nor long remember what we say here, but it can never forget what they did here. It is for us the living, rather, to be dedicated here to the unfinished work which they who fought here have thus far so nobly advanced. It is rather for us to be here dedicated to the great task remaining before us—that from these honored dead we take increased devotion to that cause for which they gave the last full measure of devotion—that we here highly resolve that these dead shall not have died in vain—that this nation, under God, shall have a new birth of freedom—and that government of the people, by the people, for the people, shall not perish from the earth.

You won't likely be called on to deliver such an address, but the techniques Lincoln used—brevity, rhythm, recurring themes—are ones you can use in your own spoken texts. The next example represents the type of spoken text we are sometimes called on to deliver at important occasions in the lives of our families.

JUDY DAVIS

Ours Was a Dad . . .

This short eulogy was given at the funeral of the writer's father, Walter Boock. Judy Davis lives in Davis, California, where she was for many years the principal of North Davis Elementary School.

Elsa, Peggy, David, and I were lucky to have such a dad. Ours was a dad who created the childhood for us that he did not have for himself. The dad who sent us airborne on the soles of his feet, squealing with delight. The dad who built a platform in the peach tree so we could

eat ourselves comfortably into peachy oblivion. The dad who assigned us chores and then did them with us. The dad who felt our pain when we skinned our knees.

Ours was the dad who took us camping, all over the U.S. and Canada, but most of all in our beloved Yosemite. The one who awed us with his ability to swing around a full pail of water without spilling a drop and let us hold sticks in the fire and draw designs in the night air with hot orange coals.

Our dad wanted us to feel safe and secure. On Elsa's eighth birthday, we acquired a small camping trailer. One very blustery night in Minnesota, Mom and Dad asleep in the main bed, David suspended in the hammock over them, Peggy and Elsa snuggled in the little dinette bed, and me on an air mattress on the floor, I remember the most incredible sense of well-being: our family all together, so snug, in that little trailer as the storm rocked us back and forth. It was only in the morning that I learned about the tornado warnings. Mom and Dad weren't sleeping: they were praying that when morning came we wouldn't find ourselves in the next state.

Ours was the dad who helped us with homework at the round oak table. He listened to our oral reports, taught us to add by looking for combinations of 10, quizzed us on spelling words, and when our written reports sounded a little too much like the *World Book* encyclopedia, he told us so.

Ours was a dad who believed our round oak table that seated twelve when fully extended should be full at Thanksgiving. Dad called the chaplain at the airbase, asked about homesick boys, and invited them to join our family. Or he'd call International House in Berkeley to see if someone from another country would like to experience an American Thanksgiving. We're still friends with the Swedish couple who came for turkey forty-five years ago. Many people became a part of our extended family around that table. And if twelve around the table was good, then certainly fourteen would be better. Just last fall, Dad commissioned our neighbor Randy to make yet another leaf for the table. There were fourteen around the table for Dad's last Thanksgiving.

Ours was a dad who had a lifelong desire to serve. He delivered Meals on Wheels until he was eighty-three. He delighted in picking up the day-old doughnuts from Mr. Rollen's shop to give those on his route an extra treat. We teased him that he should be receiving those meals

5

※ academic literacies   ● fields   ● research
■ rhetorical situations   ⁞ processes   ● media/design
▲ genres   ◆ strategies

himself! Even after walking became difficult for him, he continued to drive and took along an able friend to carry the meals to the door.

Our family, like most, had its ups and downs. But ours was a dad who forgave us our human failings as we forgave him his. He died in peace, surrounded by love. Elsa, Peggy, David, and I were so lucky to have such a dad.

This eulogy, in honor of the writer's father, provides concrete and memorable details that give the audience a clear image of the kind of man he was. The repetition of the phrase "ours was a dad" provides a rhythm and unity that moves the text forward, and the use of short, conventional sentences makes the text easy to understand—and deliver.

Key Features / Spoken Presentations

A clear structure. Spoken texts need to be clearly organized so that your audience can follow what you're saying. The **BEGINNING** needs to engage their interest, make clear what you will be talking about, and perhaps forecast the central parts of your talk. The main part of the text should focus on a few main points—only as many as your listeners can be expected to absorb and retain. (Remember, they can't go back to reread!) The **ENDING** is especially important: it should leave your audience with something to remember, think about, or do. Davis ends as she begins, saying that she and her sisters and brother "were so lucky to have such a dad." Lincoln ends with a dramatic resolution: "that government of the people, by the people, for the people, shall not perish from the earth."

346–54

354–358

Signpost language to keep your audience on track. You may need to provide cues—signposts—to help listeners follow your text, especially **TRANSITIONS** that lead them from one point to the next. Sometimes you'll also want to stop and **SUMMARIZE** a complex point to help your audience keep track of your ideas and follow your development of them.

361–62
550–51

A tone to suit the occasion. Lincoln spoke at a serious, formal event, the dedication of a national cemetery, and his address is formal and even solemn. Davis's eulogy is more informal in **TONE**, as befits a speech given

73–74

for friends and loved ones. In a presentation to a panel of professors, you probably would want to take an academic tone, avoiding too much slang and speaking in complete sentences. If you had occasion to speak on the very same topic to a neighborhood group, however, you would likely want to speak more casually.

Repetition and parallel structure. Even if you're never called on to deliver a Gettysburg Address, you will find that repetition and parallel structure can lend power to a presentation, making it easier to follow— and more likely to be remembered. "We can not dedicate—we can not consecrate—we can not hallow": the repetition of "we can not" and the parallel forms of the three verbs are one reason these words stay with us more than 150 years after they were written and delivered. These are structures any writer can use. See how the repetition of "ours was a dad" in Davis's eulogy creates a rhythm that engages listeners and at the same time unifies the text.

Slides and other media. Depending on the way you deliver your presentation, you will often want or need to use other media—*PowerPoint*, *Prezi*, or other presentation slides, video and audio clips, handouts, flip charts, whiteboards, and so on—to present certain information and to highlight key points.

Considering the Rhetorical Situation

As with any writing, you need to consider your rhetorical situation when preparing a presentation:

58–60 ■

PURPOSE   Consider what your primary purpose is. To inform? persuade? entertain? evoke another kind of emotional response?

61–64 ■

AUDIENCE   Think about whom you'll be addressing and how well you know them. Will they be interested, or will you need to get them interested? Are they likely to be friendly? How can you get and maintain their attention, and how can you establish common ground with them?

✳ academic literacies ● fields ● research
■ rhetorical situations ⁂ processes ● media/design
▲ genres ◆ strategies

How much will they know about your subject—will you need to provide background or define any terms?

GENRE The genre of your text will affect the way you structure and present it. If you're making an argument, for instance, you'll need to consider counterarguments—and, depending on the way you're giving the presentation, perhaps to allow for questions and comments from members of the audience who hold other opinions. If you're giving a report, you may have reasons to prepare handouts with detailed information you don't have time to cover in your spoken text, or links to online documents or websites.

■ 65–71

STANCE Consider the attitude you want to express. Is it serious? thoughtful? passionate? well informed? humorous? something else? Choose your words and any other elements of your presentation accordingly. Whatever your attitude, your presentation will be received better by your listeners if they perceive you as comfortable and sincere.

■ 72–74

A Brief Guide to Writing Presentations

Whether you're giving a poster presentation at a conference or an oral report in class, what you say will differ in important ways from what you might write for others to read. Here are some tips for composing an effective presentation.

Budget your time. A five-minute presentation calls for about two and a half double-spaced pages of writing, and ten minutes means only four or five pages. Your introduction and conclusion should each take about one-tenth of the total time available; time for questions (if the format allows for them) should take about one-fifth; and the body of the talk, the rest. In a ten-minute presentation, then, allot one minute for your introduction, one minute for your conclusion, and two minutes for questions, leaving six minutes for the body of your talk.

Organize and draft your presentation. Readers can go back and reread if they don't understand or remember something the first time through a text. Listeners can't. Therefore, it's important that you structure your presentation so that your audience can follow your text—and remember what you say.

- *Craft an introduction* that engages your audience's interest and tells them what to expect. Depending on your rhetorical situation, you may

346–54 ◆

  want to **BEGIN** with humor, with an anecdote, or with something that reminds them of the occasion for your talk or helps them see the reason for it. In any case, you always need to summarize your main points, provide any needed background information, and outline how you'll proceed.

- *In the body of your presentation,* present your main points in more

413–14 ◆
414–22

  detail and support them with **REASONS** and **EVIDENCE**. As you draft, you may well find that you have more material than you can present in the time available, so you'll need to choose the most important points to focus on and leave out the rest.

- *Let your readers know you're concluding* (but try to avoid saying "in conclusion"), and then use your remaining time to restate your main points and to explain why they're important. End by saying "thank you" and offering to answer questions or take comments if the format allows for them.

Consider whether to use visuals. You may want or need to include some visuals to help listeners follow what you're saying. Especially when you're presenting complex information, it helps to let them see it as well as hear it. Remember, though, that visuals should be a means of conveying information, not mere decoration.

DECIDING ON THE APPROPRIATE VISUALS

- *Slides* are useful for listing main points and for projecting illustrations, tables, and graphs.

- *Videos, animations, and sounds* can add additional information to your presentations.

❋ academic literacies   ● fields   ● research
■ rhetorical situations   ∴ processes   ● media/design
▲ genres   ◆ strategies

- *Flip charts, whiteboards, or chalkboards* allow you to create visuals as you speak or to keep track of comments from your audience.

- *Posters* sometimes serve as the main part of a presentation, providing a summary of your points. You then offer only a brief introduction and answer any questions. You should be prepared to answer questions from any portion of your poster.

- *Handouts* can provide additional information, lists of works cited, or copies of any slides you show.

What visual tools (if any) you decide to use is partly determined by how your presentation will be delivered. Will you be speaking to a crowd or a class, delivering your presentation through a podcast, or creating an interactive presentation for a web conference? Make sure that any necessary equipment and programs are available—and that they work. If at all possible, check out any equipment in the place where you'll deliver your presentation before you go live. If you bring your own equipment for a live presentation, make sure you can connect to the internet if you need to and that electrical outlets are in reach of your power cords. Also, make sure that your visuals can be seen. You may have to rearrange the furniture or the screen to make sure everyone can see.

And finally, have a backup plan. Computers fail; projector bulbs burn out; marking pens run dry. Whatever your plan is, have an alternative in case any problems occur.

Presentation software.   *PowerPoint*, *Keynote*, and other presentation software can include images, video, and sound in addition to displaying written text. They are most useful for linear presentations that move audiences along one slide at a time. Cloud-based programs like *Prezi* also allow you to arrange words or slides in various designs, group related content together, and zoom in and out. Here are some tips for writing and designing slides:

- *Use LISTS or images, not paragraphs.* Use slides to emphasize your main points, not to reproduce your talk onscreen: keep your audience's attention focused on what you're saying. A list of brief points,

669–70

presented one by one, reinforces your words. An image can provide additional information that your audience can take in quickly.

668–69

- *Make your text easy for your audience to read.* FONTS should be at least 18 points, and larger than that for headings. Projected slides are easier to read in sans serif fonts like Arial, Helvetica, and Tahoma than in serif fonts like Times New Roman. And avoid using all capital letters, which can be hard to read.

- *Choose colors carefully.* Your text and any illustrations must contrast with the background. Dark content on a light background is easier to read than the reverse. And remember that not everyone sees all colors; be sure your audience doesn't need to be able to see particular colors or contrasts in order to get your meaning. Red-green and blue-yellow contrasts are especially hard for some people to see and should be avoided.

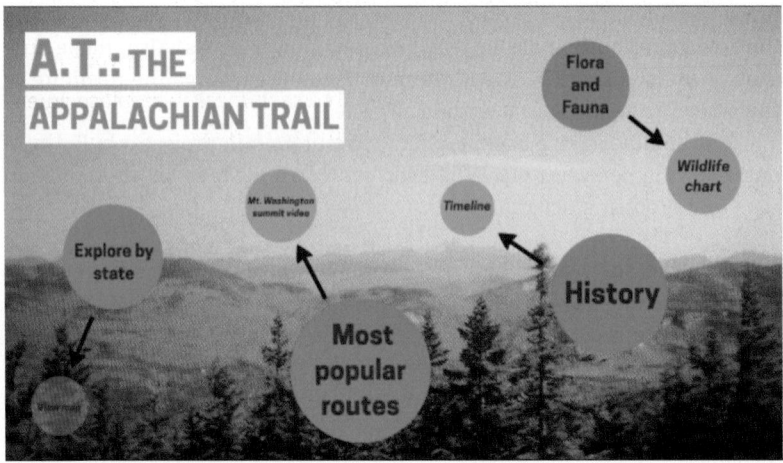

This *Prezi* presentation rotates, includes audio and video, and zooms in and out to let viewers take a closer look.

- ✴ academic literacies
- ■ rhetorical situations
- ▲ genres
- ◗ fields
- ⸫ processes
- ◆ strategies
- ● research
- ● media/design

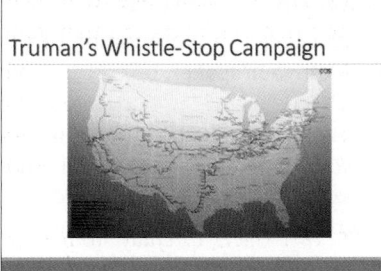

Two *PowerPoint* slides on the US presidential election of 1948. The slide on the left outlines the main points; the one on the right shows a map of Truman's whistle-stop campaign, providing a graphic illustration of the miles he traveled as he campaigned to be president.

- *Use bells and whistles sparingly, if at all.* Presentation software offers lots of decorative backgrounds, letters that fade in and out or dance across the screen, and sound effects. These features can be more distracting than helpful; use them only if they help to make your point.

- *Mark your text.* In your notes or prepared text, mark each place where you need to click a mouse to call up the next slide.

Handouts. When you want to give your audience information they can refer to later—reproductions of your visuals, bibliographic information about your sources, printouts of your slides—do so in the form of handouts. Refer to the handouts in your presentation, but unless they include material your audience needs to consult before or as you talk, wait until you are finished to distribute them so as not to distract listeners. Clearly label everything you give out, including your name and the date and title of the presentation. You might even upload your handouts to a blog or website and give its URL, share them using a program such as *Canva*, or provide your email address and invite interested audience members to request your handouts.

Delivering a Presentation

The success of a presentation often hinges on how it's delivered. As you work on your spoken texts, bear in mind the following points:

Practice. Practice, practice, and then practice some more. The better you know your talk, the more confident you will be, and your audience will respond positively to that confidence. If you're reading a prepared text, try to write it as if you were talking. Then practice by recording it as you read it; listen for spots that sound as if you're reading, and work on your delivery to sound more relaxed. As you practice, pay attention to keeping within your time limit. If possible, rehearse your talk with a small group of friends to test their response and to get used to speaking in front of an audience.

Speak clearly. When you're giving a spoken presentation, your first goal is to be understood by your audience. If listeners miss important words or phrases because you don't pronounce them distinctly, your talk will not succeed. Make sure, too, that your pace matches your audience's needs. Often you'll need to speak more slowly than usual to explain complex material (or to compensate for nerves); sometimes you may need to speed up to keep your audience's attention. In general, though, strive for a consistent pace throughout, one that ensures you don't have to rush at the end.

Pause for emphasis. In writing, you have white space and punctuation to show readers where an idea or discussion ends. When speaking, you need to pause to signal the end of a thought, to give listeners a moment to consider something you've said, or to get them ready for a surprising or amusing statement.

Stand up (or sit up) straight, and look at your audience. If you're in the same physical space as your audience, try to maintain some eye contact with them. If that's uncomfortable, fake it: pick a spot on the wall just above the head of a person in the back of the room, and focus on it. You'll appear as if you're looking at your audience even if you're not looking them in the eye. And if you stand or sit up straight, you'll project the sense that you have confidence in what you're saying. If you appear to believe in your words,

---

※ academic literacies   ● fields   ● research

■ rhetorical situations   ⁘ processes   ● media/design

▲ genres   ◆ strategies

others will, too. If you're speaking via an online forum like *Zoom*, look at the computer's camera—not at the screen. Also, make sure the camera is positioned at your eye level, so you aren't looking down at it (and showing your viewers the ceiling behind you!).

Use gestures for emphasis. If you're not used to speaking in front of a group, you may let your nervousness show by holding yourself stiffly, elbows tucked in. To overcome some of that nervousness, take some deep breaths, try to relax, and move your arms and the rest of your body as you would if you were talking to a friend. Use your hands for emphasis: most public speakers use one hand to emphasize specific points and both hands to make larger gestures. Watch politicians on C-SPAN to see how people who speak on a regular basis use gestures as part of their overall delivery.

Credits

PHOTOGRAPHS

Chapter 2: p. 16: deisgorelkin/Depositphotos. **Chapter 3: p. 40**: Jacob MacLeod. **Chapter 10: p. 81**: Kinsey Glanz; **p. 83**: W. W. Norton; **p. 85**: Leslie Erwin; **p. 91**: Ana-Jamileh Kassfy. **Chapter 11: p. 105** (top): Erica Shires, (bottom): © 2019 CJ ENM CORPORATION, BARUNSON E&A ALL RIGHTS RESERVED; **pp. 106–10**: © 2019 CJ ENM CORPORATION, BARUNSON E&A ALL RIGHTS RESERVED; **p. 111**: Courtesy of Danielle Allen. Laura Rose Photography; **p. 116**: Photo by Kenneth Irby. Courtesy of The Poynter Institute; **p. 129**: urbanbuzz/Shuttershock.com; **p. 130**: TOLES © 2020 The Washington Post. Reprinted with permission of ANDREWS MCMEEL SYNDICATION. All rights reserved; **p. 132**: AP Photo/Jeffrey Phelps; **p. 133**: Courtesy of Ann C. Johns Ph.D./University of Texas at Austin. **Chapter 12: p. 140**: Renae Tingling; **p. 146** (top): Courtesy of Frankie Schembri, (bottom): Coki10/Shutterstock; **p. 149**; Courtesy of Jon Marcus; **p. 150**: Wesley Hitt/Getty Images. **Chapter 13: p. 164**: Kelly Coryell; **p. 166**: Richard Bullock; **p. 171**: Karen McQuait; **p. 176**: Sara Krulwich/The New York Times/Redux; **p. 177**: Evan Agostini/Invision/AP Photo; **p. 180**: Bettmann/Corbis via Getty Images. **Chapter 15: p. 203**: Photo courtesy of www.donlaurencephotography.com. **Chapter 16: p. 214**: Techspot; **p. 215**: Olivia Mazzucato; **p. 216**: Skip Bolen/© Netflix/Courtesy Everett Collection. **Chapter 17: p. 223**: Keystone Pictures USA/Alamy Stock Photo; **p. 224**: Matthew M. Miller. **Chapter 18: p. 236**: John Amis/AP Photo.

Chapter 19: p. 245: Ryan Jones; **p. 246**: Ryan Smith. **Chapter 20: p. 258**: Christian Miller; **p. 259**: mediaphotosiStock/Getty Images. **Chapter 21: p. 269**: Keira Hubschman; **p. 271**: Images courtesy of the Utermohlen Estate and GV Art London; **p. 273**: Brial Diehl. **Chapter 22: p. 281** (top): Todd Jones, (bottom): Photos provided by Ohio State Alumni Magazine, published by The Ohio State University Alumni Association (OSUAA). They appear here with the permission of OSUAA, which retains full ownership of the photos; **p. 283**: Courtesy of Rattan Lal. **Chapter 28: p. 338**: Roman Sigaev/Alamy Stock Photo; **p. 339** (clockwise from top left): jjspring/depositphotos, ffolas/Depositphotos, John D. Ivanko/Alamy Stock Photo, Charlotte Lake/Shutterstock, lifefoto/Shutterstock, celiafoto/Depositphotos; **pp. 341–42**: Trello. **Chapter 29: p. 352**: Alvis Upitis/Alamy Stock Photo. **Chapter 35: p. 397** (top to bottom): Anna Elesina/Shutterstock, Charlie Blacker/Alamy Stock Photo, Space Oleandr/Alamy Stock Vector. **Chapter 37: p. 421** (top): © Reagan Louie, (bottom): AaronFoster.com. **Chapter 38: p. 434**: All photos © New York Apple Association. **Chapter 39: p. 441** (left): F. Carter Smith/Bloomberg via Getty Images, (right): Richard Carson/Reuters/Alamy. **Chapter 40: p. 449**: Photo by Swim Ink 2 LLC/Corbis via Getty Images. **Chapter 41: p. 462**: Michigan Economic Development Corporation. **Chapter 43: p. 472**: Patricia Reitz (butteryum.org). **Chapter 47: p. 504**: Nicholas S. Holtzman, Simine Vazire, Matthias R. Mehl. "Sounds Like a Narcissist: Behavioral Manifestations of Narcissism in

Glossary/Index

A

abstract, 196–200 A writing GENRE that summarizes a book, an article, or a paper, usually in 100–200 words. Authors in some academic fields must provide, at the top of a report submitted for publication, an abstract of its content. The abstract may then appear in a journal of abstracts, such as *Psychological Abstracts*. An *informative abstract* summarizes a complete report; a briefer *descriptive abstract* provides only a brief overview; a *proposal abstract* (also called a TOPIC PROPOSAL) requests permission to conduct research, write on a topic, or present a report at a scholarly conference. Key Features: SUMMARY of basic information • objective description • brevity
    key features, 198
    reading: "Boredom Proneness," 196–97
    types and examples
        informative, 196–97
        proposal, 197
        science fields, 307
    writing guide, 198–99

academic fields. *See* field of study
academic habits of mind, 46–56
Academic Search Complete, 516
Academic Search Premier, 516
academic searches, nonlibrary, 505

academic writing, 3–10 Writing done in an academic or scholarly context, such as for course assignments. Key Features: evidence that you've carefully considered the subject • clear, appropriately qualified THESIS • response to what others have said • good reasons supported by evidence • acknowledgment of multiple perspectives • carefully documented sources • confident, authoritative STANCE • indication of why your topic matters • careful attention to correctness.
    in arts and humanities, 311–13
    in business, 317–18
    in education, 319–20
    in engineering and technology, 320–22
    in health sciences and nursing, 322–24
    rhetorical situation, 309–10
    in science and mathematics, 313–15
    in social sciences, 315–17

accommodating other viewpoints, 54, 425, 554
acknowledging other viewpoints. *See* multiple perspectives, incorporating
acknowledging sources, 555–58

action verb, 471 A VERB that expresses a physical or mental action (jump, consider).

research and scholarship on a topic. Each entry includes complete publication information and a **SUMMARY** or an **ABSTRACT**. A *descriptive annotation* summarizes the content of a source without commenting on its value; an *evaluative annotation* gives an opinion about the source along with a description of it. Key Features: statement of the scope • complete bibliographic information • relevant commentary • consistent presentation • brevity
  key features, 204–5
  types and examples
      descriptive: "Teen Film$," 202–3
      evaluative: "Researching Hunger
        and Poverty," 203–4
  writing guide, 205–8

annotating The process of marking up a text in order to better understand it by underlining or highlighting key words and phrases, connecting ideas with lines and symbols, and writing comments and questions in the margins.
  as reading strategy, 16–19
  textual analyses and, 125

antecedent, 382 The **NOUN** or **PRONOUN** to which a pronoun refers. In "<u>Maya</u> lost <u>her</u> wallet," "Maya" is the antecedent of "her." *See also* pronoun

AP Images, 518

APA style, 615–55 A system of **DOCUMENTATION** used in the social sciences. APA stands for the American Psychological Association.
  comparison with MLA style, 561–63
  directory, 615–17
  fields of study
      business, 317
      education, 319–20
      engineering and technology, 321

      health sciences and nursing, 322
      science and mathematics, 314
      social sciences, 314
  formatting
      headings, 645
      notes, 622–23
      papers, 645–47
      quotations, 545
  in-text documentation, 618–22
  reference list
      articles and other short works,
        629–34
      audio, visual, and other sources,
        639–44
      authors and other contributors,
        626–29
      books, parts of books, and reports,
        634–39
      sources not covered by APA, 644
  sample research paper, 647–55
  verb tense, 554

Apple News, 511
appropriate sources, 206
"The Architecture of Inequality: On Bong
    Joon-ho's *Parasite*" (Finn), 105–11
arguable positions. *See also* claim
    in literary analyses, 228
    in position papers, 180–81, 186

arguing, 410–30 A **STRATEGY** that can be used in any kind of writing to support a claim with **REASONS** and **EVIDENCE**.
    building blocks
        claims, 411–13
        evidence, 414–22
        reasons, 413–14
    establishing credibility
        accommodating other viewpoints,
            425

arguing a position, 164–95 A writing **GENRE** that uses **REASONS** and **EVIDENCE** to support a **CLAIM** or **POSITION** and, sometimes, to persuade an **AUDIENCE** to accept that position. Key Features: clear and arguable position • necessary background • good reasons • convincing support for each reason • appeal to readers' values • trustworthy **TONE** • careful consideration of other positions

article The word "a," "an," or "the," used to indicate that a **NOUN** is indefinite (a̲ writer, a̲n author) or definite (the̲ author).

chronological order A way of ORGANIZING text that proceeds from the beginning of an event to the end. Reverse chronological order proceeds in the other direction, from the end to the beginning.

citation, 126, 233, 555–58 In a text, the act of giving information from a source. A citation and its corresponding parenthetical DOCUMENTATION, footnote, or endnote provide minimal information about the source; complete bibliographic information appears in a list of WORKS CITED or REFERENCES at the end of the text.

claim, 180–82, 185–89, 411–13 A statement that asserts a belief or position. In an ARGUMENT, a claim needs to be stated in a THESIS or clearly implied, and requires support by REASONS and EVIDENCE.

classifying and dividing, 330, 431–36 A STRATEGY that either groups (classifies) numerous individual items into categories by their similarities (for example, classifying cereal, bread, butter, chicken, cheese, cream, eggs, and oil as carbohydrates, proteins, and fats) or breaks (divides) one large category into small categories (for example, dividing food into carbohydrates, proteins, and fats). Classification and/or division can serve as the ORGANIZING principle for a whole text.

clause A group of words that consists of at least a SUBJECT and a PREDICATE; a clause may be either INDEPENDENT or SUBORDINATE. *See also* independent clause; subordinate clause

cliché, 382–83 An expression used so frequently that it is no longer fresh: busy as a bee.

clustering, 335–36 A PROCESS for GENERATING IDEAS AND TEXT, in which a writer visually connects thoughts by jotting them down and drawing lines between related items.

coding, 17 The process of developing and using a system of codes—like stars or other symbols—to track one's thoughts about a text.

collaboration, 52–55, 557 The PROCESS of working with others.

collective noun A NOUN—such as "committee," "crowd," "family," "herd," or "team"—that refers to a group.

colons, 547

color, in design, 666–68

The Columbia Encyclopedia, 513

"Come Look at the Freaks: Examining and Advocating for Disability Theatre" (Schunk), 171–77

comfortable, getting, 364

"Coming Home Again" (Lee), 465–66

comma splice Two or more INDEPENDENT CLAUSES joined with only a comma: I came, I saw, I conquered.

commas, 382, 465, 547

commentary, 205

common ground, 424 Shared values. Writers build common ground with AUDIENCES by acknowledging others' POINTS OF VIEW, seeking areas of compromise, and using language that includes, rather than excludes, those they aim to reach.

comparing and contrasting, 330, 437–44 A STRATEGY that highlights the similarities and differences between items. Using the *block method* of comparison-contrast, a writer discusses all the points about one item and then all the same points about the other item; using the *point-by-point method*, a writer discusses one point for both items before going on to discuss the next point for both items, and so on. Sometimes comparison and/or contrast serves as the ORGANIZING principle for a whole text.

    block method, 439

    cubing and, 336

    in evaluations, 220

figurative language, 441–43

    methods and organization, 438–40

    point-by-point method, 439–40

    reading across fields of study, 301

    in reports, 158

    research sources, 533–34

    transitions, 361

    visuals, 440–41

complement A NOUN, NOUN PHRASE, PRONOUN, or ADJECTIVE that modifies either the SUBJECT or the direct OBJECT of a sentence. A subject complement follows a LINKING VERB and tells more about the subject: She is a <u>good speaker</u>. She is <u>eloquent</u>. An object complement describes or renames the direct object: Critics called the movie a <u>masterpiece</u>. We found the movie <u>enjoyable</u>.

complete sentences, 381

compound sentences, 382

conclusions. *See* endings

conference proceedings, 516

conferencing software, 52, 693

confirmation bias, 533 The tendency to look for information that supports what you already believe and to ignore information that doesn't.

connections, 287

consensus, 54 A position or opinion generally accepted by a group of people.

consistency, 206, 665

context Part of any RHETORICAL SITUATION, conditions affecting the text such as what else has been said about a topic; social, economic, and other factors; and constraints such as due date and length.

    beginning with, 351–52

    in literacy narratives, 99

contrast, in design, 666–68

contrasting. *See* comparing and contrasting

conversation among researchers approach, 303–4

Cooney, Nathaniel J., 399–401

coordinating conjunction One of these words—"and," "but," "or," "nor," "so," "for," or "yet"—used to join two elements in a way that gives equal weight to each one (bacon <u>and</u> eggs; pay up <u>or</u> get out).

correlative conjunction A pair of words used to connect two equal elements: "either . . . or," "neither . . . nor," "not only . . . but also," "just as . . . so," and "whether . . . or."

Coryell, Kelly, 164–70

Council of Science Editors documentation style, 561

count noun A word that names something that can be counted (one book, two books). *See also* noncount noun

counterargument, 182, 189, 424–26 In ARGU-MENT, an alternative POSITION or objection to the writer's position. The writer of an argument should not only acknowledge counterarguments but also, if at all possible, accept, accommodate, or refute each one.

course management system (CMS). *See* learning management system.

CQ Researcher, 513

credentials

researching, for annotated bibliographies, 207

textual analyses and, 126–27

credibility The sense of trustworthiness that a writer conveys in their text.

accommodating other viewpoints and, 425

acknowledging other viewpoints and, 424–25

common ground and, 424

emotional appeals and, 426–27

refuting other viewpoints, 426

of sources, 206, 526–32

criteria, 207, 218, 220, 221 In EVALUATION, the standards against which something is judged.

cubing, 336 A PROCESS for GENERATING IDEAS AND TEXT in which a writer looks at a topic in six ways—to DESCRIBE it, to COMPARE it to something else, to associate it with other things or CLASSIFY it, to ANALYZE it, to apply it, and to ARGUE for or against it.

"Cuddling Up to Cyborg Babies" (Turkle), 479–80

cueing devices

beginning, 346–54

ending, 354–58

in presentations, 697–98

thesis statements, 347–49

titles, 363

topic sentences, 358–60

transitions, 361–62

"Cyberloafing: Distraction or Motivation?" (Mejia Avila), 346

D

definition, 330, 445–55 A STRATEGY that says what something is. *Formal definitions* identify the category that something belongs to and tell what distinguishes it from other things in that category: a worm as an invertebrate (a category) with a long, rounded body and no appendages (distinguishing features). *Extended definitions* go into more detail: a paragraph or even an essay explaining why a character in a story is tragic. *Stipulative definitions* explain a writer's distinctive use of a term, one not found in a dictionary. Definition can serve as the ORGANIZING principle for a whole text.

description, 330, 456–63 A STRATEGY that tells how something looks, sounds, smells, feels, or tastes. Effective description creates a clear DOMINANT IMPRESSION built from specific details. Description can be *objective, subjective,* or both. Description can serve as the ORGANIZING principle for a whole text.

design, 664–73 The way a text is arranged and presented visually. Elements of design include font, color, illustration, layout, and white space. One component of a RHETORICAL SITUATION, design plays an important

to emphasize logical reasons and evidence more than emotion.

e-portfolio, 388 An electronic collection of writing and other evidence selected by a writer to show their work, including a statement assessing the work and explaining what it demonstrates.

essential element, A word, PHRASE, or CLAUSE with information that is necessary for understanding the meaning of a sentence: French is the only language that I can speak.

ethical appeal, 423–26 In ARGUMENT, a way a writer establishes credibility with readers, such as by demonstrating knowledge of the topic; pointing out common ground between the writer's values and those of readers; or incorporating the views of others, including opposing views, into the argument.

ethos. *See* ethical appeal

evaluation, 214–22 A writing GENRE that makes a judgment about something (for example, a source, poem, film, restaurant) based on certain CRITERIA. Key Features: description of the subject • clearly defined criteria • knowledgeable discussion of the subject • balanced and fair assessment

evidence, 414–22 In ARGUMENT, the data you present to support your REASONS. Such data may include statistics, calculations, examples, ANECDOTES, QUOTATIONS, case studies, or anything else that will convince your reader that your reasons are compelling. Evidence should be sufficient (enough to show that the reasons have merit) and relevant (appropriate to the argument you're making).

    in academic writing, 5

    assessing own writing, 369

    choosing appropriate, 422

    genres and

        evaluations, 218, 220

        literary analyses, 233

        position papers, 181, 187–89

    in paragraphs, 360

    reflection on, 396

    types of

        anecdotes, 416–18

        authorities, 416

        case studies and observations,

           419–20

        examples, 415–16

        facts, 414–15

        scenarios, 418–19

        statistics, 415

        textual evidence, 420–21

        visuals, 420–21

examples

    beginning with, 159

    in extended definitions, 452

    reflection on, 396

    as type of evidence, 415–16

exclamation points, 547

executive summary, business plan, 318

expectations, 331

explainer videos, 286

explaining a process, 469–73 A STRATEGY for telling how something is done or how to do something. An explanation of a process can serve as the ORGANIZING principle for a whole text.

    rhetorical situation, 473

    types of

        explaining visually, 471–72

        how something is done,

           469–70

        how to do something, 470–71

explanation, 330

expletive A word such as "it" or "there" used to introduce information provided later in a sentence: <u>It</u> was difficult to drive on the icy road. <u>There</u> is plenty of food in the refrigerator.

exploration, 269–79 A writing genre that presents a writer's thoughtful, personal consideration of a subject. Key Features: topic intriguing to the writer • some kind of structure • specific details • speculative tone.

    choosing, 68

    genres

        evaluations, 219

        literacy narratives, 96

        position papers, 185–86

        profiles, 254

        textual analyses, 129

    key features, 274–76

    reading: "Talking with Granddad,"

        269–74

    writing guide, 276–79

exploratory essay

    choosing, 68

    key features, 274–76

fused sentence Two or more INDEPENDENT CLAUSES with no punctuation between them: I came I saw I conquered.

G

Gale eBooks, 513, 515
Gaylin, Willard, 355
general education field, 295–96
general indexes and databases, 516–17
general language, editing, 382
general reference works, 513
General Science Index, 517

generating ideas and text, 333–44 A set of PROCESSES that help writers develop a topic, examples, REASONS, EVIDENCE, and other parts of a text.

    clustering or mapping ideas, 335–36
    cubing, 336
    discovery drafting, 344
    freewriting, 333–34
    journaling, 343–44
    letter writing, 342–43
    listing, 334–35
    looping, 334
    outlining, 340–42
    questioning, 337
    talking, 336
    using genre features, 340
    using visuals, 337–39

genre, 65–71, 79–292 A kind of writing marked by and expected to have certain key features and to follow certain conventions of style and presentation. In literary texts, readers recognize such genres as short stories and novels and poems; in academic and workplace settings, readers and writers focus on such genres as ABSTRACTS, ANNOTATED BIBLIOGRAPHIES, ARGUMENTS, EVALUATIONS, LAB REPORTS, LITERACY NARRATIVES, LITERARY ANALYSES, PROFILES, PROPOSALS, REFLECTIONS, RÉSUMÉS, REPORTS, and TEXTUAL ANALYSES.

    for ambiguous assignments, 68–69
    combining, 69–70
    definition, 65–66
    list of genres, 66–68
    in remixes, 287, 290

Gerdy, John R., 426

gerund A VERB form ending in "-ing" that functions as a NOUN: Swimming improves muscle tone and circulation.

"Gettysburg Address" (Lincoln), 694–95
"Gettysburg Address: Two Versions" (Bass), 347
Ginsberg, Benjamin, 451, 452
Gladwell, Malcolm, 470
Goodfellow, Peter, 442
Google, research using, 510
Google Books, 515
Google Hangouts, 693
Google Images, 511, 529
Google News, 511
Google News Archive, 511
Google Scholar, 510
Google Video Search, 511
government sites, research with, 511
"Graduation" (Angelou), 474–75
"A Grand, Comprehensive Overview to Mutual Funds Investing" (Barker), 445–46
graphs
    for comparing, 440–41
    design of, 676, 677
    reading, 30–31

GreatBuildings.com, 459
Green, Kelly, 203–4
Greenblatt, Stephen, 300
Grison, Sarah, 29
guiding your reader. *See* cueing devices
"Guns and Cars Are Different" (MacLeod),
40–42
Guterson, David, 351, 464–65

H

habits of mind, academic, 46–56
Hall, Carla, 447–48
"Handgun Regulations, Crime, Assaults, and
Homicide: A Tale of Two Cities" (Sloan),
439
handouts for presentations, 703
Harjo, Suzan Shown, 413

hashtag, 510 A number sign (#) in front of a
word or unspaced phrase (#BlackLivesMat-
ter), used in social media to mark posts by
KEYWORD or theme and make them searchable by these tags. Also used to add commentary to texts.

hasty generalization, 428–29 A **FALLACY** that
reaches a conclusion based on insufficient
or inappropriately qualified **EVIDENCE.**

HathiTrust Digital Library, 515
headings
as design element, 670–72
in literary analyses, 233
reading, 14
"Health and Behavioral Consequences of
Binge Drinking in College" (Wechsler et
al.), 357–58
Health Care without Harm (website), 352
health sciences, writing for, 322–24

Help Wanted: Tales from the First Job Front
(Lewis), 359

helping verb A **VERB** that works with a main
verb to express a **TENSE** or **MOOD.** Helping
verbs include "do," "have," "be," and **MODALS:**
Elvis has left the building. Pigs can fly.

Herrera Gutierrez, Karla Mariana, 85–90
Hesser, Amanda, 441–42, 470–71
Historical Abstracts, 517
history, researched essay for, 312–13
hoax videos, 529, 532
Hughes, Langston, 478
humanities
academic writing, 311–13
reading texts in, 304–5
Humanities Index, 517
Humanities International Index, 517
Huws, Ursula, 450
Hypothesis, 499

I

idea mapping. *See* clustering
"Identifying Genes Involved in Suppression
of Tumor Formation in the Planarian
Schmidtea mediterranea" (Dorsten), 314

identity-first language A way of referring to
individuals (or groups) with specific characteristics by stating the characteristic, either
exclusively or as the first word in a description. "Blind musicians" rather than "musicians who are blind." Some people may
prefer to be described using identity-first
language; if you don't know or can't confirm
this preference, use **PERSON-FIRST LANGUAGE.**

illustration. *See* visual
image platforms, research using, 511

literature Literary works—including fiction, poetry, drama, and some nonfiction; also, the body of written work produced in a given field.

literature review, 209–13, 306 A GENRE of writing that surveys and synthesizes the prior research on a TOPIC. In the sciences, a literature review is a required part of the introduction to an IMRAD report; in all disciplines, scholars write article-length literature reviews devoted to specific topics. Key Features: careful, thorough research • accurate, objective SUMMARIES of the relevant LITERATURE • synthesis of the scholarship • critical EVALUATION of the literature • a clear focus

logical appeal, 411–22 In ARGUMENT, an appeal to readers based on the use of logical reasoning and of evidence such as facts, statistics, statements from authorities on the subject, and so on.

logical fallacy. *See* fallacy, logical
logos. *See* logical appeal

looping, 334 A PROCESS for GENERATING IDEAS AND TEXT in which a writer writes about a topic quickly for several minutes and then reads the results and writes a one-sentence summary of the most important or interesting idea. The summary becomes the beginning of another round of writing and summarizing, and so on, until the writer finds a tentative focus for writing.

M

medium, 659–63 A way that a text is delivered. Some examples include in print, with speech, or online.

modal A HELPING VERB—such as "can," "could," "may," "might," "must," "ought to," "should," "will," or "would"—that does not change form for person or number and indicates probability or necessity.

modifier A word, PHRASE, or CLAUSE that describes or specifies something about another word, phrase, or clause (a <u>long, informative</u> speech; the actors spoke <u>in unison</u>; the <u>man who would be king</u>).

mood A characteristic of VERBS that indicates a writer's attitude toward the action or condition the verb expresses. The *indicative mood* is used to state fact or opinion: I'm waiting to buy tickets. The *imperative mood* is used to give commands or directions: Sit down, and take off your shoes. The *subjunctive mood* is used to express wishes, requests, or requirements or to indicate unlikely conditions: I wish the ticket line were shorter. I suggest that you be ready for a long wait.

N

outlining, 345 A PROCESS for GENERATING IDEAS AND TEXT or for organizing or examining a text. An *informal outline* simply lists ideas and then numbers them in the order that they will appear; a *working outline* distinguishes supporting from main ideas by indenting the former; a *formal outline* is arranged as a series of headings and indented subheadings, each on a separate line, with letters and numerals indicating relative levels of importance.

P

parallelism A writing technique that puts similar items into the same grammatical structure. For example, every item on a to-do list might begin with a command: "clean," "wash," "buy"; or a discussion of favorite hobbies might name each as a GERUND: running, playing basketball, writing poetry.

paraphrasing Rewording someone else's text using about the same number of words but not the phrasing or sentence structure of the original. Paraphrasing is generally called for when a writer wants to include the details of a passage but does not need to quote it word for word. Like a QUOTATION or SUMMARY, a paraphrase requires DOCUMENTATION.

predicate In a sentence or **CLAUSE**, the **VERB** and the words that tell more about the verb—**MODIFIERS, COMPLEMENTS**, and **OBJECTS**. In the sentence "Mario forcefully stated his opinion," the predicate is "forcefully stated his opinion."

premise A statement that assumes something is true and serves as the basis for an argument.

preposition A word or group of words that tells about the relationship of a **NOUN** or **PRONOUN** to another word in the sentence. Some common prepositions are "after," "at," "before," "behind," "between," "by," "for," "from," "in," "of," "on," "to," "under," "until," "with," and "without."

present participle A **VERB** form used with a **HELPING VERB** to create progressive **TENSES** (is <u>writing</u>) or used alone as an **ADJECTIVE** (a <u>living</u> organism). The present participle of a verb always ends in "-ing."

present perfect A **TENSE** used to indicate actions that took place at no specific time in the past or that began in the past and continue into the present: I <u>have</u> often <u>wondered</u> how I can make my love of language into a career. He <u>has cried</u> every day since his companion of fifty years died.

primary source, 304–5, 501–2 A source such as a literary work, historical document, work of art, or performance that a researcher examines firsthand. Primary sources also include experiments and FIELD RESEARCH. A researcher would likely consider the Declaration of Independence a primary source and a textbook's description of how the document was written a SECONDARY SOURCE.

process, 325–401 In writing, a series of actions that may include GENERATING IDEAS AND TEXT, ORGANIZING, DRAFTING, REVISING, EDITING, and PROOFREADING. *See also* explaining a process *and specific processes*

profile, 245–57 A GENRE that presents an engaging portrait of a person, place, or event based on firsthand FIELD RESEARCH. Key Features: interesting subject • necessary background • interesting angle • firsthand account • engaging details

pronoun A word that takes the place of a NOUN, such as "she," "anyone," "whoever."

proofreading, 383–84 The final PROCESS of writing, when a writer checks for correct spelling and punctuation as well as for page order, missing text, and consistent use of FONTS. *See also* editing; revision; rewriting

proposal, 258–68 A GENRE that argues for a solution to a problem or suggests some action. Key Features: well-defined problem • recommended solution • answers to anticipated questions • call to action • appropriate TONE. *See also* topic proposal

pull quote, 665 A brief excerpt set off within
a text in order to highlight certain information. Pull quotes are often set in a different
FONT or color.

purpose, 59–60 A writer's goal: to explore
ideas, to express oneself, to entertain, to
demonstrate learning, to inform, to persuade, and so on. Purpose is one element of
the RHETORICAL SITUATION.

Q

qualifying word, 412, A word such as "frequently," "often," "generally," "sometimes,"
or "rarely" that limits a CLAIM in some way.

questioning A PROCESS of GENERATING IDEAS
AND TEXT about a topic—asking, for exam-
ple, "What?," "Who?," "When?," "Where?,"
"How?," and "Why?" or other questions.

quotation The use of someone else's words
exactly as they were spoken or written. Quoting is most effective when wording makes a
point so well that no re-wording will do it
justice. Quotations need to be acknowledged
with DOCUMENTATION.

Rogerian argument, 422–23 A technique for ARGUING that aims to solve conflicts by seeking common ground among people who disagree.

secondary source, 304–5, 501–2 An ANALYSIS or INTERPRETATION of a PRIMARY SOURCE. In writing about the Revolutionary War, a researcher would likely consider the Declaration of Independence a primary source and a textbook's description of how the document was written a secondary source.

stance, 72–74 A writer's or speaker's attitude toward their subject—for example, reasonable, neutral, angry, curious. Stance is conveyed through TONE. *See also* rhetorical situation

storyboard, 291, 338, 339 A visual representation of images and text for a planned multimedia project.

strategy, 403–88 A pattern for ORGANIZING text to ANALYZE CAUSE AND EFFECT, ARGUE, CLASSIFY AND DIVIDE, COMPARE AND CONTRAST, DEFINE, DESCRIBE, EXPLAIN A PROCESS, or NARRATE.

straw man fallacy, 189, 428
structure
 explorations, 275
 presentations, 697
"The Student Loan Trap: When Debt Delays
 Life" (Lehman), 353
study group, 486
style, literary analyses, 233

subject A word or word group, usually including at least one NOUN or PRONOUN plus its MODIFIERS, that tells who or what a sentence or CLAUSE is about. In the sentence "A frustrated group of commuters waited for the late bus," the subject is "A frustrated group of commuters."
 in explorations, 277
 in profiles, 252–55

subjective descriptions, 459–60

subordinate clause A clause that begins with a SUBORDINATING WORD and therefore cannot stand alone as a sentence: She feels good when she exercises. My roommate, who was a physics major, tutors students in science.

subordinating word A word, such as a RELATIVE PRONOUN or a subordinating conjunction, that introduces a SUBORDINATE CLAUSE: The ice sculpture melted because the ballroom was too hot. Common subordinating words include "although," "as," "because," "if," "since," "that," "which," and "why."

Sullivan, Andrew, 424–25
Sullivan, Kathleen M., 453–54

summary, 33–35, 550–51 The use of one's own words and sentence structure to condense someone else's text into a briefer version that gives the main ideas of the original. As with PARAPHRASING and QUOTATION, summarizing requires DOCUMENTATION.
 choosing, 67
 essay questions, 485
 genres
  abstracts, 198, 199
  annotated bibliographies, 207
  literacy narratives, 97
  literature reviews, 212
  position papers, 191–92
  presentation, 697
  reports, 159
  textual analyses, 123, 125
 integrating sources, 544, 550–51
 as reading strategy, 20
 signal phrases, 551–53
 textual evidence, 420
 transitions, 362

summary and response essay, 33–45 A GENRE of writing that demonstrates one's ability to convey a text's main ideas in condensed form and to engage with those ideas by ARGUING a position, ANALYZING the text, or writing a REFLECTION on its content. Key Features: clearly identified author and title • concise summary of the text • an explicit response • support for that response
 key features, 42–44
 sample essay: "Guns and Cars Are
  Different" (MacLeod), 40–42
 writing guide, 44–45

whataboutism, 427 A **FALLACY** that results
when someone responds to criticism by
accusing the attacker of hypocrisy, but not
answering the criticism itself. Also known
as "You too."

working bibliography, 497–99 A record of all
sources consulted in research providing all
the bibliographic information necessary for
DOCUMENTATION including author, title, and
publication information.

The Norton Writer's Prize

"I have something to say to the world, and I have taken English 12 in order to say it well." —**W. E. B. Du Bois**

The Norton Writer's Prize recognizes outstanding original nonfiction by undergraduates. All entries are considered for possible publication in Norton texts—in fact, many of the essays that appear in this book were nominated for the prize.

The contest is open to students age 17 and above who are enrolled in an accredited 2- or 4-year college or university. Three cash prizes of $1,000 apiece are awarded annually for coursework submitted during the academic year, one in each of the following three categories:

- Writing by a first-year student in a 2- or 4-year college or university
- Writing by a student in a 2-year college or university
- Writing by a student in a 4-year college or university

Submissions must be between 1,000 and 3,000 words in length. Literacy narratives, literary and other textual analyses, reports, profiles, evaluations, arguments, memoirs, proposals, multimodal pieces, and other forms of original nonfiction will be considered if written by a student age 17 or above in fulfillment of an undergraduate course requirement at an eligible institution. Entries submitted in accordance with the Official Contest Rules will be considered for all applicable prizes, but no more than one prize will be awarded to any single entry.

For full contest rules, eligibility, and instructions on how to enter or nominate students, please visit **wwnorton.com/norton-writers-prize**. For questions, please email us at **composition@wwnorton.com**

A Directory to MLA Style